Sixty Years of Double Beta Decay

From Nuclear Physics to Beyond Standard Model Particle Physics

Sixty Years of Double Beta Decay

From Nuclear Physics to Beyond Standard Model Particle Physics

H. V. Klapdor-Kleingrothaus

Max-Planck-Institut für Kernphysik, Germany

World Scientific
Singapore • New Jersey • London • Hong Kong

Published by

World Scientific Publishing Co. Pte. Ltd.
P O Box 128, Farrer Road, Singapore 912805
USA office: Suite 1B, 1060 Main Street, River Edge, NJ 07661
UK office: 57 Shelton Street, Covent Garden, London WC2H 9HE

British Library Cataloguing-in-Publication Data
A catalogue record for this book is available from the British Library.

SIXTY YEARS OF DOUBLE BETA DECAY: FROM NUCLEAR PHYSICS TO BEYOND STANDARD MODEL PARTICLE PHYSICS

ISBN 981-02-3779-0

Printed in Singapore by Uto-Print

publisher.

to World Scientific Publishing Company

Preface

This book describes the history of double beta decay and its potential as research tool in particle physics. Double beta decay was investigated theoretically for the first time shortly after the introduction of the neutrino. Its history is related with fundamental problems and discoveries of particle physics such as lepton-number conservation, parity non-conservation, and gauge theories. Being considered in the beginning as a purely nuclear physics problem, it was understood early as the most sensitive means of probing the Majorana mass of the neutrino. Only rather late, in the early eighties, far after the experimental discovery of the neutrino, and after the introduction of gauge theories, its potential to determine quantitatively the (Majorana) mass of the neutrino was understood. The sensitivity of double beta decay experiments has reached now a level, at which it starts to 'dramatically constrain the form of the neutrino mass matrix' [Geo2000]. The at present most sensitive experiment *already now* excludes, as various theoretical papers since 1997 show, the small angle *MSW* solution of the solar neutrino problem, if neutrinos are degenerate, and also according to some papers, with some reservation, the large angle *MSW* solution, in this case, 'forcing us into the vacuum oscillation solution' [Ell99a].

A nice confirmation could be that recent Superkamiokanda solar neutrino results also claim, that the small angle MSW solution, (as well as the vacuum oscillation solution, and oscillation into sterile neutrinos) are ruled out with 95% c.l. [Suz2000] (although some authors also make less definite conclusions [Gon2000]). Favoured remain the large angle solution and the *LOW* solution. On the other hand it becomes visible that the various solar and other neutrino oscillation experiments may not finally be able to alone solve the problem of the neutrino mass matrix which makes double beta decay indispensable in helping to solve this problem (and in particular the problem of fixing an absolute mass scale).

For this purpose future double beta experiments with strongly increased sensitivity are required such as *GENIUS* (and other experiments prompted by this proposal). The proposal of *GENIUS* in 1997 has given a great impact to the field of double beta decay, particularly also to deeper theoretical investigations of the potential of the field for neutrino physics and other beyond standard model physics.

Future experiments at the sensitivity level of *GENIUS* would help to discriminate various neutrino mass scenarios, among them the large mixing angle MSW solution. By using it as a real time solar neutrino detector for pp neutrinos, *GENIUS* also could probe the *LOW* solution. As a by-product it will give the chance to search for cold dark matter with unprecedented sensitivity.

In recent years **double beta decay** has become ONE OF THE MOST IMPORTANT FIELDS OF NON-ACCELERATOR PARTICLE PHYSICS, by contributing to probing also other physics beyond the standard model in various directions already now on the **TeV scale**, where new physics should manifest itself.

Future experiments such as *GENIUS* will have the potential, to probe the multi-TeV range for other beyond standard model topics such as supersymmetry, compositeness, leptoquarks, violation of Lorentz invariance, equivalence principle, etc. They could yield important complementary information to the physics to be expected from future high-energy colliders and satellite experiments.

The book does not intend to give all historical details. If in some cases the contributions of some researchers to the progress of the field may not have been always properly weighted, this may be not only the fault of the editor, first because of space restrictions, but also - and more important - since some publishers did not allow reprinting some important articles from their journals. The editor regrets this deeply. Another natural problem of such a type of book is, that the usually most valuable review articles can be reprinted only to a limited extent because of space reasons. We gave in these cases, as also for books and for some other important articles which could not be fully included, as a hint and help for the reader at least the first pages. In the Introductory Chapter I, which gives an overview of the history, status and perspectives of the field, for the convenience of the reader, articles which are fully reprinted in this book, are marked with an asterisk (*), followed by a Roman number indicating the subchapter of Chapter 2, where it can be found. In case that only the first page(s) are reprinted, the article is marked with two asterisks (**).

It is hoped, that in any case the lines of development of the field are clearly shown, and also, in particular, the great future potential of this kind of non-accelerator particle physics becomes transparent. This is the main aim of this book.

We hope that the book will be useful and stimulating for students and researchers in the field of particle physics.

It is the pleasure of the editor to thank his colleagues from the Heidelberg Double Beta and Dark Matter Group, and the many scientific colleagues and guests from many countries, for many years of fruitful and pleasant collaboration on the subject of this book.

The editor is most indebted to Dr. Irina Krivosheina, for giving the inspiration for the edition of this book and for her invaluable help in its realization.

H.V. Klapdor-Kleingrothaus
Heidelberg, 5 July, 2000

Acknowledgments

The author and publisher gratefully acknowledge permission to reproduce previously published material, as granted by authors and publishers, and as indicated by citations. They have attempted to trace the copyright holders of all material reproduced from all sources and apologize to any copyright holders whose prior permission might not have been obtained.

We are grateful in particular to
American Association for the Advancement of Science: Science
Associate Publishers, American Physical Society: Phys. Rev, Phys. Rev. Lett.
Elsevier Science Publishers B.V.: Phys. Lett., Nucl. Phys., for the permission to reprint the authors own articles
European Physical Society: Europhysics Letters, Europhysics News
Gordon and Breach Science Publishers, Inc.: Comments in Nuclear and Particle Physics
High Energy Physics Group, University of Hawaii
Institute of Physics Publishing Co., Bristol
Nature Publishing Group, London: Nature
Nuovo Cimento, Italy
Permissions Department, Annual Review Inc.: Annual Reviews in Nucl. and Part. Science
Publication Office: Progress of Theoretical Physics, Kyoto, Japan
Rossiskaja Academija Nauk, Isdatelstwo Nauka, Yadernaja Fizika (Physics of Atomic Nuclei)
Spektrum der Wissenschaft Verlagsgesellschaft mbH, Heidelberg
Springer Verlag GmbH & Co, KG: Zeitschrift für Physik, Europ. Phys. J.
Wiley-VCH Verlag GmbH, Weinheim: Physik in unserer Zeit, Physik. Blätter
World Scientific Publishing Co Pte Ltd., Singapore: Int. Journal of Modern Physics

Contents

Chapter 1

Double Beta Decay — Historical Retrospective and Perspectives

The history of double beta decay using the nucleus as a complicated laboratory for a wide range of particle physics starts more than 60 years ago. This history is closely connected with fundamental discoveries of particle physics, such as the discovery of parity non-conservation and of the gauge theories, and double beta decay research has become one of the most important fields of non–accelerator particle physics. Table 1.1 sketches this historical development. Starting from purely nuclear physics questions, double beta decay yields at present a broad access to many topics of physics beyond the Standard Model of particle physics at the TeV range, on which new physics should manifest itself (see Fig. 1.1 and Table 1.2.). Future experiments should give access to the multi-TeV range, could complement in many ways the search for new physics at the Large Hadron Collider (*LHC*) and at Next Linear Colliders (*NLC*), and could serve as an important bridge between the physics that will be gleaned from high energy accelerators such as *LHC* and *NLC* on the one hand, and satellite experiments such as *MAP* and *PLANCK* on the other.

Concerning neutrino physics, without double beta decay there will be no solution of the nature of the neutrino (Dirac or Majorana particle) and of the structure of the neutrino mass matrix, since neutrino oscillation experiments measure only differences of neutrino mass eigenstates. Only investigation of neutrino oscillations *a n d* double beta decay together can lead to an absolute mass scale.

It is the aim of this book to describe the way double beta research went and to give an insight into the physical potential of double beta decay for present and future particle physics and cosmology.

1.1 From the Early Days until the Gauge Theory Era

1.1.1 *The First Steps in Double Beta Research*

The long and close association between the phenomenon of nuclear double beta decay, the question of lepton number conservation and the nature and mass of the neutrino began in *1935* shortly after the 'discovery' of the neutrino by W. Pauli in

DATE	DISCOVERY	NAMES
1935	First Calculation of $2\nu\beta\beta$ Decay	M. Goeppert–Mayer [Goe35*-I]
1939	First consideration of $0\nu\beta\beta$ Decay	W.H. Furry [Fur39*-I]
$\sim$ **1950**	First Experiments	E.L. Fireman, D. Schwarzer [Fir48, 49*-I, 52*-I]; M.G. Inghram, J.H. Reynolds [Ing49*-I, 50*-I]; C.A. Levine, A. Ghiorso, G.T. Seaborg [Lev50*-I]
1956	Discovery of Parity Violation $\Rightarrow$ Interest Decreased in $\beta\beta$	T.D. Lee, C.N. Yang [Lee56]; S. Wu et al. [Wu57]
since late **1960's**	Verification of existence of $2\nu\beta\beta$ decay	N. Takaoka [Tak66**-I]; T. Kirsten [Kir68*-I]; O.K. Manuel [Hen75], [Man86]; T. Bernatowicz [Ber92*-I]
later **1970's**	GUTs Predict Non–Vanishing Neutrino Mass $\Rightarrow$ Renaissance of $\beta\beta$ Research	H. Fritzsch, P. Minkowski [Fri75]; H. Georgi [Geo75] et al.
1979-80	See–Saw Mechanism	M. Gell–Mann, P. Ramond, R. Slanskij [Gel79]; T. Yanagida [Yan79*-I]; R. Mohapatra, G. Senjanovic [Moh80*-I]
Early 1980's	$0\nu\beta\beta$ in gauge theories $\Longleftrightarrow m_\nu \neq 0$	T. Kotani [Doi80, 85**-I]; J. Schechter, J.W.F. Valle [Sch80**-I, 82]; W. Haxton [Hax81*-I, 84**-I]; L. Wolfenstein [Wol81]; B. Kayser [Kay89**-I]; S.P. Rosen [Ros88**-I]
since **1984**	Solution of nuclear matrix element problem	H.V. Klapdor–Kleingrothaus, K. Grotz, K. Muto [Kla84*-II], [Gro85a*,b*-II], [Gro86a**-II]; [Mut89a*-II]; P. Vogel [Vog86*-II]; T. Tomoda, A. Faessler [Tom87]; S. Stoica [Sto93a*-II]; F. Simkovich, G. Pantis [Sim96]; [Sto2000**-II]
1987	First discovery of $2\nu\beta\beta$ decay by direct detection	M. Moe [Ell87a*-V]

Table 1.1 **60 years of Double Beta Decay**

1930 (see [Pau30*-I], [Bet34*-I]). The motivation of M. Goeppert-Mayer (Fig. 1.2) [Goe35*-I], however, when performing the first calculations of the half–life of double beta decay, was not the nature of the neutrino, not conservation of leptons, but the stability of even–even nuclei over geological time.

Applying the new Fermi theory of beta decay [Fer34**-I], to the two–neutrino

DATE	DISCOVERY	NAMES
1997	Extension to SUSY: $0\nu\beta\beta \Longleftrightarrow \tilde{m}_\nu \neq 0$	H.V. Klapdor–Kleingrothaus, M. Hirsch, S. Kovalenko [Hir97a**-IV,b**-IV], [Hir98a*-IV]
1986–2000	EXPLOSION OF POTENTIAL FOR PARTICLE PHYSICS $0\nu\beta\beta$ Probe of SUSYs, Leptoquarks, Compositeness... $0\nu\beta\beta$ – **One of Most Powerful Non-Accelerator Probes of Particle Physics Beyond Standard Model**	J. Schechter, J.W.F. Valle [Sch82a*-I]; R.N. Mohapatra [Moh86*-IV, 86a*-IV]; J. D. Vergados [Ver87]; H.V. Klapdor–Kleingrothaus, M. Hirsch, S. Kovalenko, H. Päs, U. Sarkar, [Hir95**-IV, 96*-IV], [Hir96d*-IV, 98a*-IV], [Kla99c*-IV]; E. Takasugi [Tak98]; O. Panella [Pan95, 2000*-IV]; G. Bélanger [Bel98*-IV]
since $\sim$ **1990**	SECOND GENERATION EXPERIMENTS Using LARGE SOURCE STRENGTH OF ENRICHED $\beta\beta$ Emitter Material	HEIDELBERG-MOSCOW [Kla87**-A, 92*-V, 99*-VI], [HM2000**-III] NEMO [Bar97*-V]; IGEX [All99a]; Te–cryo [Ale94, 95, 98]; ...
since **1991**	THIRD GENERATION EXPERIMENTS PROPOSED	1991 BOREXINO [Bor91]; 1995 KAMLAND [Rag94*-V], [Suz95**-V]; **1997 GENIUS [Kla97**-VI, 99*-VI]**; 1998 CUORE [Fio98, Giu99a*-V]; 1999 MOON [Eji99b*-V]; 2000 EXO [Dan2000a]

Continued Table 1.1. 60 years of Double Beta Decay

double beta decay – a second order effect of the weak interaction –

$$(A, Z) \longrightarrow (A, Z+2) + 2e^- + 2\bar{\nu} \tag{1.1}$$

she obtained a lifetime in excess of 10^{17} years even if the daughter nucleus (A, Z+2) "were more stable by 20 times the electron mass".

In *1939* W.H. Furry [Fur39*-I] showed that the "symmetrical" theory of neutrino and antineutrino proposed by E. Majorana [Maj37**-I] (Fig. 1.3) could give rise to another process, not observed until now, namely no–neutrino double beta decay (Fig. 1.3)

$$(A, Z) \longrightarrow (A, Z+2) + 2e^- \tag{1.2}$$

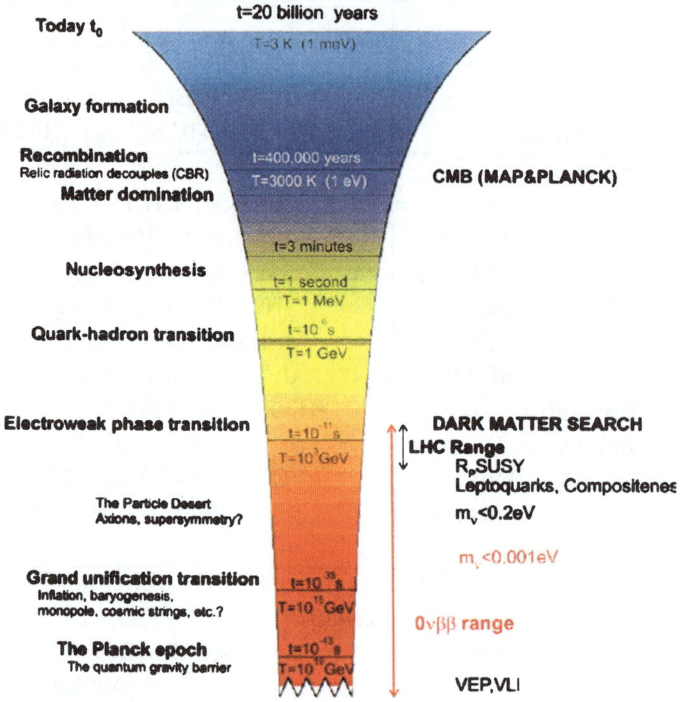

Fig. 1.1 The potential contribution of double β decay to particle physics and cosmology. The energy scales correspond to the masses of the right-handed neutrinos in the see-saw mechanism.

Fig. 1.2 H. Bethe in 1935 (left), Enrico Fermi together with Maria Goeppert-Mayer in 1935 (middle) and Wolfgang Pauli in 1958 (right).

which would have a lifetime shorter than that for the two–neutrino decay "by a factor which ranges from 10^5 to 10^{15} or more". The neutrino would be a virtual particle in this case, the signal in the $2e^-$ spectrum a sharp line, in contrast to the continuous spectrum of process (1.1).

At the time when E. Fermi and Mayer wrote their papers, little distinction was made between neutrino and antineutrino. While beta minus emitters were

Observ.	Restrictions	Topics investigated
0ν:	**via ν exchange:** Neutrino mass Light Neutrino Heavy Neutrino Test of Lorentz invariance and equivalence principle Right handed weak currents **via photino, gluino, zino (gaugino) or sneutrino exchange:** R-parity breaking, sneutrino mass **via leptoquark exchange** leptoquark-Higgs interaction	Beyond the standard model and SU(5) model; early universe, matter–anti-matter asymmetry, Dark matter, L–R –symmetric models (e.g. SO(10)), compositeness. Lorentz invariance in the neutrino sector, new gravitational interactions. $V + A$ interaction, $W_R^{\pm}$ masses SUSY models: Bounds for parameter space beyond the range of accelerators leptoquark masses and models
$0\nu\chi$:	existence of the Majoron	Mechanism of (B-L) breaking -explicit -spontaneous breaking of the local/global B-L symmetry new Majoron models

Table 1.2 *The potential contribution of double β decay to particle physics and cosmology*

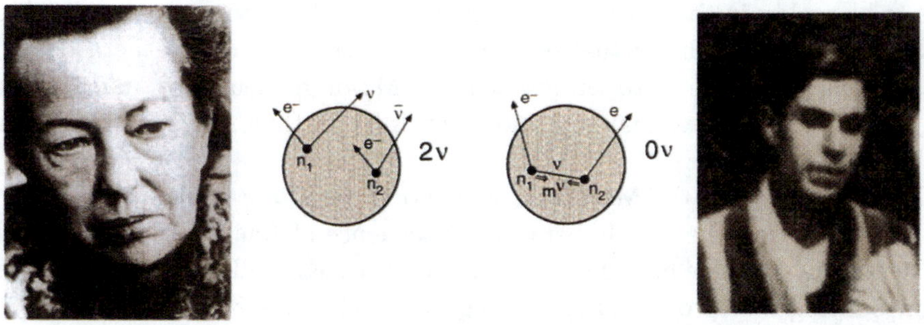

Fig. 1.3 Maria Goeppert–Mayer (left), who first investigated (theoretically) in 1935, 2ν double β decay, and Ettore Majorana, 1932.

known to occur naturally, beta plus emitters had only just been observed by F. Joliot and I. Curie [Jol34]. De Broglie and C.C. Wick recognized in *1934* that the neutral particles associated with the two processes could be different, and de Broglie introduced the term antineutrino (see [Jol34], [Wic34]), but it was not until the work of Majorana [Maj37**-I], and its elaboration by G. Racah [Rac37*-I], that the possibility of a clear physical distinction, or alternatively, of a complete identity, between neutrinos and antineutrinos was better understood.

Racah observed that if the neutrino is a Majorana particle, it must have no magnetic moment and the same neutral particle is emitted in both beta minus and beta plus decay. To test the latter property he proposed to take the neutral particle from one beta minus decay and to see whether it could induce another beta minus decay.

Twenty years later R. Davis [Dav55] carried out this test using a reactor neutrino source and the famous reaction

$$neutrino + \ ^{37}Cl \longrightarrow e^- \ +^{37} Ar \tag{1.3}$$

as the stimulated emission.

Fig. 1.4 Left: Ray Davis, 1997 at the Solar neutrino Conference, Heidelberg, (foto author). Right: Ray Davis with his experiment in the Homestake underground laboratory.

Furry realized that the neutrino in the two–stage process did not necessarily have to be real as in the reactor experiment, but could be virtual – in neutrinoless double beta decay. The virtual exchange in neutrinoless double beta decay has finally proved to be the more sensitive test for Majorana neutrinos, mainly because the phase space of the virtual neutrino is much larger than for the real neutrino in the Davis experiment.

The alternative to the Majorana neutrino is the Dirac neutrino. If neutrinos are Dirac particles, the double beta minus sequence of Racah would be forbidden and neutrinoless double beta decay could not occur. Formally this can be described by introducing a "leptonic charge" being additively conserved in beta decay and having different sign for particle and antiparticle. This charge we call lepton number. Electron and neutrino are assigned $L = 1$, positron and antineutrino $L =$ -1. Lepton number L appears, as baryon number B, conserved empirically to the experimental precision reached today, although there is no theoretical 'reason' for it, i.e. no underlying symmetry is known. On the other hand in most grand unified theories (*GUTs*) B and L are not conserved. While in minimal *SU(5)* $B - L$ is still conserved, which forbids neutrinoless double beta decay, also this condition is relaxed in more general theories, e.g. already in the simplest left–right symmetric model, *SO(10)*.

Fig. 1.5 Fred Reines, at the Neutrino'92 Conference, Granada, Spain, (from left to right, first row: A. Wagner (DESY), ..., F. Reines, the author, T. Kotani, second row: G.T. Zatsepin, K. Winter, R. Mössbauer, P. Vogel).

The first experiments on double beta decay were undertaken, even before the existence of the neutrino was proved *directly*, in the famous reactor experiment by G.A. Cowan and F. Reines [Cow56], [Rei59] in *1955*. The first experiment on double beta decay seemed to give a hint on neutrinoless double beta decay. E.L. Fireman [Fir48], [Fir49*-I] searched for electrons from the transition $^{124}Sn \longrightarrow {}^{124}Te$, using coincidence counters and observed a signal which corresponded to a half life between *4* and $9{\times}10^{15}$ *years*. Other experiments stimulated in the period *1949–1952* by this result, were, however, not able to confirm this result. E.L. Fireman and D. Schwarzer concluded [Fir52*-I], that it was "probably caused by a small trace of an impurity having a coincidence activity" in the ^{124}Sn source. With this statement these authors stressed one of the major problems confronting all efforts to detect double beta decay: traces of radioactive impurities as low as one part in 10^{13} or less, become significant sources of background, when half lives of the order of 10^{20} years or more are to be measured.

While all experiments at this time looked for the decay electrons, a remarkable exception was the one performed by M.G. Inghram and J.H. Reynolds [Ing49*-I], [Ing50*-I]. They looked for the daughter nucleus and exploited the fact that measurable amounts of the daughter might accumulate over geological time in ores which are rich in the corresponding parent nucleus. They analyzed a tellurium ore from Boliden, Sweden, which was about *1.5 billion years* old and reported evidence for the transition $^{130}Te \longrightarrow {}^{130}Xe$ with a half-life of 1.4×10^{21} *years*. They attributed their result to the two–neutrino decay of ^{130}Te.

Fig. 1.6 Till Kirsten, in Gran Sasso, 30 November, 1990, at the official opening of the GALLEX experiment.

Fig. 1.7 Oliver K. Manuel, University of Arkansas.

Fig. 1.8 Enrico Bellotti, first director of LNGS, in Gran Sasso, at the official opening of the GALLEX experiment, 30 November, 1990.

Another early approach was to look for radioactive daughter nuclei which in principle are detectable in much smaller quantities than stable rare gases. One of the first experiments of this type investigated the double beta decay of ^{238}U [Lev50*-I].

The experiment of M.G. Inghram and J.H. Reynolds was the forerunner of a series of geochemical experiments which definitely proved the occurrence of the double beta decay process and confirmed their value within a factor of about two [Tak66**-I], [Kir67**-I], [Kir68*-I], [Hen75], [Man86], [Man91*-I], [Ber92*-I], [Ber93**-I]. N. Takaoka and G. Ogata used tellurium ores from the *Oya* gold mine in Japan, T. Kirsten et al. investigated the decay of ^{130}Te in an ore from the *Good Hope* mine in Colorado, E.W. Hennecke et al. an ore from the *Kalgoorlie* mine in Australia, while T. Bernatowicz et al. investigated probes from various mines. A recent experiment by [Tak96**-I] determined the half–life of ^{128}Te to be 2.2×10^{24} y, essentially confirming earlier results [Ber92*-I], [Ber93**-I], [Man91*-I]. These experiments, looking only for the daughter nucleus, naturally cannot differentiate between the neutrinoless and the two–neutrino decay processes.

An early 'direct' (non–geochemical or radiochemical) experiment like the one by the Kiev group [Zde80a**-I] obtained a limit for the neutrinoless decay of ^{130}Te of 1.2×10^{21} *years.*

Also the radiochemical investigation of ^{238}U has been improved more recently [Tur91*-I].

The first active source experiment (see section V) was the one by E. der Mateosian and M. Goldhaber using CaF_2 [Mat66*-I].

The first observation of two–neutrino double beta decay in direct experiments was claimed in *1980* for ^{82}Se [Moe80], and finally achieved by Mike Moe (Fig. 1.9) and collaborators in *1987* [Ell87a*-V]. Further early experiments are listed up in some detail in [Pri59**-I], [Zde80**-I], [Hax84**-I].

Our understanding of double beta decay underwent a profound change when parity non–conservation was discovered in *1957* [Lee56], [Wu57], and the helicity of neutrinos was observed [Gol58]. The helicity mismatch between the antineutrino

Fig. 1.9 Mike Moe (left) at the *Hoover Dam*, the site of his double beta laboratory, in 1994, with the author (foto author).

Fig. 1.10 M. Goldhaber (middle) at the WEIN'86 Conference, Heidelberg, with the author (left) (foto author).

emitted by one neutron and the neutrino to be absorbed by another neutron now forbids this process also for a Majorana neutrino as long as the latter is massless, and as long as there are no right–handed currents.

Fig. 1.11 Tsung-Dao Lee in 1956 (left), Chien-Shung Wu in 1957 (middle) and Chen-Ning Yang in 1956 (right).

This point of view dominated for about *20 years* and only in the late seventies the interest in the existence of Majorana neutrinos – and that is in double beta decay – began to revive. One of the main driving forces was the occurrence of grand unification theories *(GUTs)*, according to which the electroweak and strong interactions are separate manifestations of a single force, and which allow in contrast to the usual standard model for non–vanishing neutrino masses.

1.1.2 *Double Beta Decay, Gauge Theories and Neutrino Mass*

1.1.2.1 *Origins of Neutrino Masses*

Already the simplest of *GUTs*, *SU(5)*, allows for breaking of baryon number and lepton number conservation, and the simplest left–right symmetric *GUT* model, *SO(10)* – introduced in *1975* [Fri75], [Geo75] – also for breaking of *B–L*, thus allowing for double beta decay. In *SO(10)*, requiring the existence of a right–handed neutrino, a nonvanishing Majorana neutrino mass seems to be inevitable. The see-saw mechanism then explains the lightness of the left–handed (Majorana) neutrino by the heavy or superheavy right–handed neutrino at the other end of the balance [Gel79], [Yan79*-I], [Moh80*-I] (see also [Lan88*-I], [Moh97*-I]). Most other beyond *SU(5)* models also predict in a natural way non–vanishing Majorana neutrino masses. (However, also small Dirac masses as 'Superstring Exotica' have been discussed [Mas86], [Mas87], [Cam87].)

Fig. 1.12 Rabindra N. Mohapatra, at the Dark'98 Conference, Heidelberg, Germany, 1998, (foto author).

Fig. 1.13 Tsutomu Yanagida (foto kindly presented by T. Yanagida).

Fig. 1.14 Murray Gell–Mann.

According to the seesaw mechanism the neutrino acquires a small Majorana mass of size m_l^2/M, where the light mass m_l is typically the Dirac mass of a quark or a charged lepton, and $M >> m_l$ is the mass of the heavy right–handed neutrino, associated with the large grand unified mass scale governing the breaking of lepton–number conservation. Thus double beta experiments may probe mass scales M of the order of $10^3 - 10^9$ *GeV*, i.e. ranges far beyond the highest energies directly accessible in present and future colliders (see Fig. 1.1).

Another attractive and most natural way to produce small neutrino Majorana masses has been shown by R. Mohapatra and F. Senjanovic [Moh80*-I], [Moh81a]. They obtain in a left-right symmetric model with spontaneous parity non–conservation, based on the gauge group $SU_L(2) \times SU_R(2) \times U(1)$, a relation between the neutrino mass and the mass m_{W_R} of the right-handed W boson $m_{\nu_e} \approx m_e{}^2/gm_{W_R}$. A similar formula holds for leptons in each generation. This formula is suggestive in the sense that, in the limit of m_{W_R} going to infinity, the neutrino mass goes to zero and we have at the same time a pure *V–A* theory of weak interactions.

In superstring models the smallness of the Majorana neutrino masses must be traced back to much more complicated mixing mechanisms than the seesaw mechanism, since there are no Higgs fields to generate a sufficiently large Majorana mass for the right–handed neutrino. For this, an intermediate mass scale of order of magnitude *TeV* is required. This may be identified with the scale of supergravity breaking [Moh86b*-I], [Moh88a*-I], [Val87], [Ber87]. The essential role of string-derived symmetries for ensuring light neutrino masses has been stressed by J.C. Pati [Pat96].

Neutrino masses in superstring-derived standard–like models have been discussed in [Far93]. Ultralight neutrinos and R-parity in superstring models were considered by [Val87a].

Superstring *GUTs* lead – since they do not contain high–dimensional Higgs representations – naturally to the flipped *SU(5)* theory [Ant87], [Ant88], based on a $SU(5) \times U(1)$ group [Ant92], [Lop94a]. The discussion of neutrino mass patterns in a flipped *SU(5)* model derived from a 'realistic' string theory has revived in the light of recent ***Superkamiokande*** data [Ell98**-I], [Ell99]. A superstring-neutrino dominated universe has been discussed by [Nan87].

Fig. 1.15 Left: John Ellis (CERN). In middle: Miriam Cvetic, at the BEYOND'99 Conference, Castle Ringberg, Germany, 1999, (with the author)(foto author). Right: Zurab Berezhiani, at the LEPTON-BARYON'98 Conference, Trento, Italy, 1998, (with the author) (foto author).

Neutrino masses in SUSY models have been widely discussed, e.g. by [Leo95**-I], [Vem2000], [Che2000**-I], [Der2000**-I], [Bla99**-I], [Cho97a], [Ach95], [Asa96].

A stringy origin of neutrino masses and a 'gravity–induced' seesaw mechanism within the minimal supersymmetric standard model has been discussed by [Cve92], [Cve92a*-I]. It is based on an interplay of nonrenormalizable and renormalisable terms in the superpotential and may be realized in a class of string vacua.

Light neutrino masses are naturally explainable also in theories with large *TEV*-scale extra dimensions [Die2000**-I], [Die2000a]. Here the heavy right-handed neutrino will reside in the bulk of Kaluza-Klein states which leads to a power-law suppression of the neutrino mass relative to the masses of all other fermions. Oscillations could become possible without ν masses in $D > 5$ theories. Double beta decay in such scenario has been considered by [Moh99].

Another kind of string origin of neutrino masses, and of sterile neutrinos was discussed by Z. Berezhiani and R.N. Mohapatra, and R. Foot and R.R. Volkas [Ber95*-I], [Foo95], [Ber99a**-I]. The request for a light sterile neutrino naturally leads to the concept of a shadow world, assuming exact duplication of the standard model in both the gauge field and the fermion content. The only bridge connecting known and shadow sector is here gravitation, and the mirror neutrinos mix with the ordinary ones through Planck scale induced higher–order operators.

One may exclude the possibility of neutrinos being of Dirac type, if one postulates "a strong naturality" in which no global conservation laws are assumed a priori [Sch80**-I]. More generally, while within the framework of gauge theories the conservation of electric charge forces the charged fermions to be Dirac particles, no such strong symmetry is required for neutrinos, and consequently their description as Majorana particles might be more fundamental [Doi85**-I].

Together with the increasing presumption that neutrinos were actually massive Majorana neutrinos, it was realized by J. Schechter and J.W.F. Valle [Sch81] that the magnetic ***transition*** moments, Majorana neutrinos can have, would allow an inhomogeneous magnetic field to rotate both spin *and* flavour of a neutrino. In this case the spin rotation changes particle to antiparticle – an effect discussed as one of the possible solutions of the solar neutrino puzzle (e.g. [Akh97]). The results from solar neutrino observation were claimed already to disfavour a (diagonal) magnetic moment of the neutrino thus favouring its Majorana nature [Ans92].

Since the idea of the see–saw mechanism to explain the smallness of the electron neutrino mass requires left–right symmetric models, the existence of right–handed $(V + A)$ currents is a natural precondition.

Fig. 1.16 Peter Rosen, at the NEUTRINO'96 Conference, Helsinki, Finland, 1996, (foto author).

Fig. 1.17 Sergei Petcov, at the NEUTRINO'98 Conference, Takayama, Japan, 1998, (foto author).

Fig. 1.18 T. Kotani, at the WEIN'95 Conference, Osaka, Japan, 1995, (foto author).

Thus the ban for double beta decay due to the left–handedness of the weak interaction can be circumvented by a right–handed component of the weak leptonic current, or by a change in the handedness of the neutrino between emission and absorption, induced by a finite Majorana mass. In *GUT* models, both solutions are not independent. In these models a right–handed component is only effective in simultaneous association with a Majorana mass [Doi85**-I], [Moh86c**-I], [Ros88**-I], [Kay89**-I].

It has been shown already in *1982* by J. Schechter and J.W.F. Valle [Sch82a*-I], that under the general assumption that the weak interactions are described by a local gauge theory, the existence of neutrinoless double beta decay implies a non–zero neutrino mass and vice versa. This Schechter–Valle theorem is valid *independently* of the mechanism of neutrinoless double beta decay (see below, and Fig. 1.19). It has been realized by T. Kotani and coworkers [Doi80], that since the electron–neutrino can be an admixture of two or more mass eigenstates, the mass eigenstates will contribute coherently to the amplitude of neutrinoless double beta decay. Since the relative phases of these contributions depend on the relative *CP* eigenvalues of the states, there can be, as pointed out by L. Wolfenstein [Wol81], large cancellations in the transition amplitude produced by the 'effective mass' $\langle m_\nu \rangle$ triggering the transition.

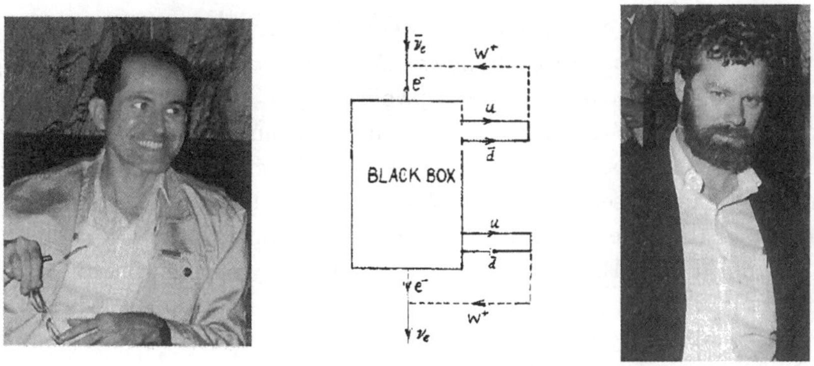

Fig. 1.19 Jose W.F. Valle at the BEYOND'97 Conference, Castle Ringberg, Germany, 1997, (left) (foto author). Diagram showing how any neutrinoless double beta decay, irrespective of which mechanism induces the process, induces a $\bar{\nu}_e$ to ν_e transition, that is an effective Majorana mass term, [Sch82a*-I] (middle). Paul Langacker at the WEIN'86 Conference, Heidelberg, Germany, (foto author).

While as mentioned double beta decay is not possible by a Dirac neutrino, which can be thought of as formed from two degenerate Majorana neutrinos with opposite *CP* parity, a contribution, though strongly suppressed, could be made by socalled pseudo–Dirac neutrinos, a pair of Majorana neutrinos with approximately, but not exactly degenerate masses [Wol81a], [Pet82], [Val83*-I], [Doi83a], see also [Mut88*-I].

1.1.2.2 *The Double Beta Decay Half Life and the Neutrino Mass*

The half–life for neutrinoless double beta decay induced by the exchange of a light Majorana neutrino is given by (see [Doi85**-I], [Mut88*-I])

$$[T^{0\nu}_{1/2}(0^+_i \to 0^+_f)]^{-1} = C_{mm}\frac{\langle m_\nu\rangle^2}{m_e^2} + C_{\eta\eta}\langle\eta\rangle^2 + C_{\lambda\lambda}\langle\lambda\rangle^2 + C_{m\eta}\langle\eta\rangle\frac{m_\nu}{m_e}$$

$$+C_{m\lambda}\langle\lambda\rangle\frac{\langle m_\nu\rangle}{m_e} + C_{\eta\lambda}\langle\eta\rangle\langle\lambda\rangle \qquad (1.4)$$

where $C_{mm}, C_{\lambda\lambda}$, ... are products of a nuclear matrix element squared and a phase space integral, η and λ are right–handed weak current parameters, and $\langle m_\nu\rangle$ denotes the effective neutrino mass

$$\langle m_\nu\rangle = \sum_i m_i U^2_{ei}, \qquad (1.5)$$

U_{ei} being terms of the neutrino mixing matrix U, which is given for Majorana neutrinos by

$$\begin{pmatrix} c_{12}c_{13} & s_{12}c_{13}e^{-i\delta_{12}} & s_{13}e^{-i\delta_{13}} \\ -s_{12}c_{23}e^{i\delta_{12}} - c_{12}s_{23}s_{13}e^{i(\delta_{13}+\delta_{23})} & c_{12}c_{23} - s_{12}s_{23}s_{13}e^{i(\delta_{23}+\delta_{13}-\delta_{12})} & s_{23}c_{13}e^{i\delta_{23}} \\ s_{12}s_{23}e^{i(\delta_{23}+\delta_{13})} - c_{12}c_{23}s_{13}e^{i(\delta_{23}+\delta_{13})} & -c_{12}s_{23}e^{i\delta_{23}} - s_{12}c_{23}s_{13}e^{i(\delta_{13}-\delta_{12})} & c_{23}c_{13} \end{pmatrix} \qquad (1.6)$$

Here $s_{ij} = \sin\theta_{ij}$, $c_{ij} = \cos\theta_{ij}$, and δ is a *CP* violating phase.

The effective neutrino mass could be *smaller* than m_i *for all* i for appropriate *CP* phases of the mixing coefficients U_{ei} [Wol81].

On the other hand it has been shown that there must be at least one neutrino mass eigenstate with mass larger than or equal to the effective mass [Kay89**-I].

It has been shown (see e.g. [Lan88*-I], [Lan92]) that the effective mass may not be dramatically different from the electron neutrino mass in not "pathological" grand unified models. On the other hand this is experimentally still an open question – it would indeed be true if we assume the small angle solution of the solar neutrino problem to be valid. In any case the dependence of the double beta observable $\langle m_\nu\rangle$ from the neutrino mixing matrix produces an exciting connection between double beta decay and the running neutrino oscillation experiments (see below).

A detailed outline of the theory and the formalism of double beta decay in the context of modern gauge theories has been given, starting from the early eighties, in the fundamental papers of T. Kotani et al. ([Doi80], [Doi81a*-I], [Doi81b**-I], [Doi83**-I], [Doi83b**-I], [Doi85**-I], [Doi88**-I], [Doi93**-I]), see also [Ros81], [Ver81**-I], [Hax84**-I], [Ver86], [Kot86**-I], [Mut88*-I], [Gro89/90**-I], [Tom91**-I], [Ros 92].

Great impact on the understanding of double beta decay in the preceeding earlier post–parity–non–conservation era had the work of H. Primakoff and S.R. Rosen and E. Greuling and R.C. Whitten, [Pri59**-I], [Gre60**-I], see also [Pri81**-I]. Some

Fig. 1.20 Boris Kayser (left), with the author and Pran Nath (right) at the BEYOND'97 Conference, Castle Ringberg, Germany, 1997, (foto author).

of the approximations of pregauge theorie have been investigated by J. Vergados [Ver81**-I].
For the half-life of the two-neutrino mode we refer to [Gro83a*-II], [Kla84*-II], [Doi85**-I], for a derivation see [Gro89/90**-I]. The experimental study of the two-neutrino mode which is the rarest process observed so far in nature, is important for understanding nuclear structure problems in their own right, and in calculating the nuclear matrix elements for the neutrinoless mode in eq. (1.4).
Already by [Doi81a*-I] a systematic study was made of the $0^+ \longrightarrow J^+$ $\beta\beta$ transitions for the two-neutrino and the neutrinoless modes. It was shown that for the neutrinoless mode, only the $0^+ \longrightarrow 0^+$ transition is allowed in the two-nucleon mechanism, if there is no right-handed interaction, while the $0^+ \longrightarrow 2^+$ transition can become similarly important for sizable right-handed interactions. The $0^+ \longrightarrow 2^+$ neutrinoless double beta transitions thus are relatively more sensitive to the right-handed weak current parameters η and λ and less sensitive to m_ν, compared with $0^+ \longrightarrow 0^+$ transitions (although the absolute $0^+ \longrightarrow 2^+$ decay rate is much smaller than the $0^+ \longrightarrow 0^+$ rate) [Tom91**-I]. These two transition types are the important ones for learning about particle physics from double beta decay. Since in this book the weight will be put on the particle physics aspects, only little attention will be given to transitions to other excited states, which have recently found some attention (see, e.g. [Suh98]), and also to the two-neutrino decay mode.

A general Lorentz-invariant parameterization for the long range and the short range parts of the interactions contributing to the neutrinoless double beta decay rate (see Fig. 1.21) has been derived recently by H. Päs et al. [Pae99*-I], [Pae2000**-I].

In summary it became clear only in the early eighties [Doi80], [Doi81a*-I], [Hax82a*-I], [Hax84**-I], [Doi85**-I], [Kot86**-I], that and how a neutrino mass can be deduced from a neutrinoless double beta amplitude. This opened the way to use double beta decay as the most practical and most sensitive means of searching for Majorana neutrinos and their masses.

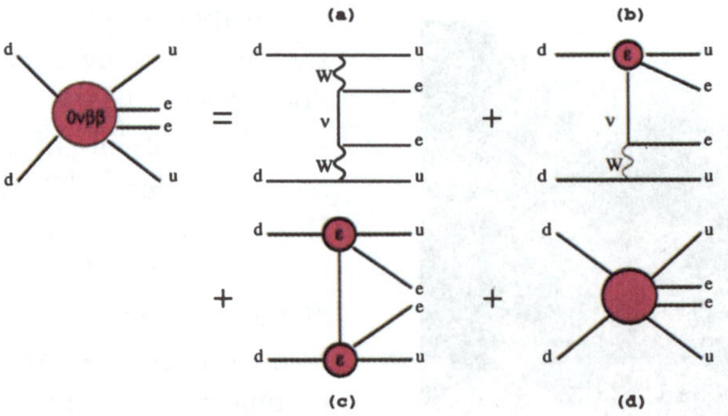

Fig. 1.21 Feynman graphs of the general double beta decay rate, with long range (a-c) and short range interactions (d) (from [Pae99*-I]).

Fig. 1.22 Eiichi Takasugi (right) and R.N. Mohapatra (left), at the Double- Beta Decay and Related Topics Conference, Trento, Italy, 1995. In the second row (from left to right): S. Jullian, S. Kovalenko, M. Hirsch.

Fig. 1.23 Wick C. Haxton (right), with the author and John Vergados (left) at the International School of Nuclear Physics (Neutrinos in Astro and Nuclear Physics), Erice, Sicily, Italy, 1997.

T. Kotani and coworkers at Osaka were the first to deduce – in *1980* – a neutrino mass from double beta decay [Doi80], [Doi81a*-I], analyzing the ratio of double beta decay half lives for ^{128}Te and ^{130}Te measured by E.W. Hennecke, O.K. Manuel and D.D. Suba in *1975* [Hen75], and deduced a value of *34 eV*, using nuclear matrix elements of J. Vergados [Ver76]. This opened a short era of finite neutrino masses. Another analysis by W.C. Haxton et al. [Hax82a*-I] basing on the B. Pontecorvo assumption [Pon68], that the matrix elements of similar nuclei (as ^{128}Te and ^{130}Te) should be equal, also assuming a vanishing right–handed contribution, yielded a value of *10 eV*.

Already earlier H. Primakoff and S.P. Rosen [Pri69*-I] had analyzed this ratio taken from earlier measurements by N. Takaoka and G. Ogata [Tak66**-I] and T. Kirsten et al. [Kir67**-I], [Kir68*-I] and had deduced an upper limit of 10^{-3} on the admixture η of a right–handed leptonic current into the left–handed *(V–A)* Hamiltonian. Later limits were deduced by [Tom86], and in [Suh93].

Also at that time a mass *limit* of *15 eV* was deduced [Hax81*-I] from experiments on ^{76}Ge and Se, and an upper limit of *40* eV for the light neutrino and a lower limit of 10^4 GeV for a heavy neutrino from ^{48}Ca experiments by J. Vergados [Ver82a].

The value deduced by T. Kotani et al. [Doi80], [Doi81a*-I] showed a coincidence with the neutrino mass of *20–45* eV claimed that time by the V.A. Lubimov group at ITEP, Moscow from tritium decay [Lub80], [Bor87]. But already in *1983* a new measurement of the ratio of the ^{128}Te and ^{130}Te isotopes [Kir83*-I] (also relying on the B. Pontecorvo argument) yielded only an upper limit for the neutrino mass of *5.6* eV, and also the tritium decay now claimed only upper limits, the Zurich group of W. Kuendig a limit of *18* eV [Fri86].

Among the mass limits from some double beta emitters discussed in the review by [Hax84**-I] the sharpest limit from direct (non–geochemical) experiments was that obtained from ^{76}Ge, $\langle m_\nu \rangle < 5.8\ eV$ deduced from the best half–life limit at that time of $> 5 \times 10^{22}$ *years* [Bel83], [Bel83a].

All of this was still consistent with the upper mass limit of *25* to *30* eV deduced [Sat87], [Kol87], [Arn87] from the observed time– and energy spread of the neutrinos from *SN87a*, the first naked–eye supernova since the invention of the telescope, discovered on *February 24, 1987* in the Large Magellanic Cloud.

However, these early limits from double beta decay suffered from a serious drawback. The Heidelberg group [Gro83a*-II] among others were the first to stress, that the calculation of the nuclear matrix elements entering into the analysis was by far not solved to a degree of reliability (the 'best' calculations of two–neutrino double beta decay rates at that time – which were shell model calculations [Hax82a*-I], [Hax84**-I] – disagreed with experiment by up to two orders magnitude) that reliable conclusions could be drawn on the neutrino mass at this stage. This was the beginning of a decade of hard work devoted to solve the problem of the matrix elements.

1.2 The Nuclear Physics Side – Nuclear Matrix Elements

The at that time non–understood discrepancy between calculation and experiment for the two–neutrino double beta decay lead into the dead–end of introducing a scaling procedure for 2ν *and* 0ν matrix elements [Hax84**-I], which as understood later, transferred the error in the calculations of the two–neutrino matrix elements to the 0ν matrix elements. Another serious drawback critisized by [Gro83a*-II], was that often the socalled closure approximation was used in the calculations (see [Hax82a*-I], [Hax84**-I]). Also the B. Pontecorvo argument underwent some criticism [Hax82a*-I].

A discussion of some other earlier approaches to calculate double beta decay matrix elements can be found in [Huf70], [Hax84**-I], [Ver86], [Gro86a**-II], [Boe87], [Hax93].

The clou to solve these problems was found by K. Grotz and H.V. Klapdor in *1984* [Kla84*-II], [Gro86a**-II] who realized that the spin–isospin and quadrupole–

Fig. 1.24 The author (standing in left) at the International Conference on Nuclear Structure, in Dubna, USSR, June 1976.

quadrupole ground state correlations very sensitively influence the nuclear matrix elements. This work based on earlier extensive work on the shape of the beta strength distribution in nuclei, starting essentially in *1976* [Kla76], which led already to important consequences in ***astrophysics*** (element synthesis, age of the universe, cosmological constant, solar neutrino detection), ***nuclear physics*** (single beta decay half lives far from stability, beta–delayed particle emission and fission [Sta92a], reactor antineutrino spectra, nuclear charge distribution), and ***nuclear technology*** (decay heat of nuclear reactors) [Kla81], [Kla82a], [Kla82b], [Kla83], [Gro83b], [Kla86], [Kla86a], [Gro86], [Mad89], [Sta90b], [Kla91a], [Hir92], [Hir93a].

Fig. 1.25 Klaus Grotz in Heidelberg, 1986, (foto author).

Fig. 1.26 From left to right: Sabin Stoica, the author, Kazuo Muto, Martin Hirsch, Takeshi Oda, Tokyo, Japan, 1995, (foto author).

1.2.1 *A Breakthrough to the Understanding of the Matrix Elements*

Grotz and Klapdor calculated for the first time half–lives for two–neutrino and zero–neutrino decay for all possible double beta emitters with mass number A > 70, using two different models [Kla84*-II], [Gro85a*-II], [Gro85b*-II], [Gro86a**-II]. One was a *QRPA* calculation with a schematic Gamow–Teller force, which is the part of the proton–neutron interaction most important for Gamow–Teller transitions (the particle–particle force is neglected), the other is an extended nucleon–number projected *BCS* calculation with a residual Gamow–Teller force, in a model space consisting of zero– and four–quasiparticle states, as well as an additional quadrupole–quadrupole force, which is another important part of the proton–neutron interaction. These calculations included further virtual excitations of the delta resonance, the Gamow–Teller force being extended to a generalized spin–isospin force in a simple quark model. Earlier calculations already yielded the important result that it is not possible to describe the Δ-quenching mechanism by a constant renormalization of the axial vector current [Gro83c*-II] (investigations of the contributions through the Δ resonance have been made later also, e.g., by [Ver88], [Ver89]).

This research showed the way to the solution of the above-mentioned long-standing problem in the calculation of two-neutrino double beta decay rates. The quadrupole–quadrupole force which was found to strongly suppress the two-neutrino matrix element [Kla84*-II], [Gro86a**-II], leading to a remarkable improvement between calculation and experiment, is contained as a constituent in the particle–particle interaction of the $J^\pi = 1^+$ mode, which enhances further the amplitude of the spin–isospin correlations.

1.2.2 *QRPA Calculations Including the pp–Force*

It was therefore natural to investigate the particle–hole and particle-particle forces separately which has been done in the *QRPA* calculations by the Caltech, Tübingen and Heidelberg groups [Vog86*-II], [Civ87], [Tom87], [Tom88], [Mut88*-I], [Eng88**-II], [Mut89**-II], [Mut89a*-II], [Sta90a*-II], [Tom91**-I] in which a particle–particle force was included in the *QRPA* calculation. Reviews of this new development were given by K. Muto and H.V. Klapdor [Mut88*-I] and T. Tomoda [Tom91**-I]. From these calculations only the Caltech group [Vog86*-II], [Eng88**-II], [Vog96] did not use a realistic nucleon–nucleon force. An analysis of the dependence of the calculation on the nucleon–nucleon force and various kinds of renormalization has been performed by [Sta90*-II], [Sta92**-II].

The calculations using a realistic nucleon–nucleon force can reliably calculate neutrinoless double beta decay, which more softly depends on ground state correlations, while the two–neutrino mode is up to now somewhat less predictive (we refer to the discussion in [Kla95/98**-II]). The great advantage of the *QRPA* method compared to shell model calculations is that it can be used in large model spaces.

This general picture also does not change essentially with some refinements that have been discussed in the last years such as particle number projection [Gro86a**-II], [Suh92], higher–order *QRPA* (boson expansion for the transition operators) [Rad91], [Rad91a], renormalization of the *pnQRPA* (analog to the known renormalization of usual *QRPA* (see [Rin80]) to avoid the early collapse of the *QRPA* solutions resulting from overestimation of ground state correlations [Toi95*-II], or inclusion of proton–neutron pairing (see [Sim96], correcting an earlier paper [Pan96]). Momentum-dependent induced nucleon currents such as weak magnetism and pseu-

Fig. 1.27 At the International Double Beta Decay and Related Topics Conference, Trento, Italy, April 1995. From left to right: E. Takasugi, Petr Vogel, Ettore Fiorini, Irina Krivosheina, Ben Mottelson, F. Piquemal, Amand Faessler, Jouni Suhonen, the author, Sabin Stoica and Heinrich Päs.

Ref.	[Hax84 **-I] SM	[Sta90 *-II] QRPA	[Tom91 **-I] QRPA	[Cau96] *-II] LSSM	[Sim96] QRPA	[Sim99] QRPA	[Sto2000 **-II] QRPA
$M^{0\nu}$	3.41	4.25	4.26	1.57	1.92	2.80	2.36/3.62
$\langle m_\nu \rangle$eV							
90 %	0.43	0.35	0.33	0.94	0.77	0.53	0.63/0.41
68 %	0.34	0.27	0.26	0.74	0.60	0.42	0.49/0.32

Table 1.3 Nuclear matrix element $M^{0\nu}$ of neutrinoless double beta decay of ^{76}Ge, according to different theroretical approaches and calculations, and deduced effective neutrino masses $\langle m_\nu \rangle$ for a half-life limit of $T_{1/2}{}^{0\nu} = 1.9 \cdot 10^{25}$ y (90% c.l.) and $T_{1/2}{}^{0\nu} = 3.1 \cdot 10^{25}$ y (68% c.l.), as measured by [HM2000**-III]. SM and LSSM stand for Shell Model and Large Scale Shell Model. The non-dimensional values of the matrix elements are reconverted, when necessary, to allow a comparison.

doscalar couplings change the matrix elements also only by the order of 30% [Sim99]. The second and third approaches violate the Ikeda sum rule. The *QRPA* calculations including pn–pairing and renormalization, do not fulfill the Ikeda sum rule by about 30% (see e.g. [Sim96], [Toi97]). Moreover the violation of the sum rule is

Fig. 1.28 Fodor Simkovich (upper left), Peter Herczeg (center), Vadim Bednyakov (upper right), A.V. Derbin (down left), Darrel Smith (down middle) at NANP'97, Dubna, Russia.

Fig. 1.29 Left: T.T.S. Kuo, in Heidelberg, 1990. Right: Cheng-rui Ching (left) and Tso-Hsiu Ho (right) in Heidelberg, 1993, with the author and his son (fotos author).

found to be mass-dependent. The calculated matrix elements thus are correspondingly smaller than from calculations fulfilling the sum rule [Mut89a*-II], [Sta90a*-II]. It has been tried to cure this violation of the Ikeda sum rule in the recent models in a Second *pnQRPA* approach improving the approach of [Rad91], [Rad91a] by avoiding states produced by three–boson operators [Sto93a*-II], [Sto93b], [Sto96**-II], [Sto2000**-II]. The second *QRPA* approach has also been applied to calculate for the first time double beta decay transitions to first excited 2^+ daughter states [Sto94]. A comparison of the results yielded by the various *QRPA* methods is given in [Sim96], [Pan98a**-II], [Sto2000**-II]. Recent general reviews can be found in [Fae98**-II], [Suh98]. A new field theory approach trying to avoid the problems in the various *QRPA* approaches has been presented by F. Šimkovic and G. Pantis [Sim99a**-II]. For a comparison of matrix elements from different calculations and approaches see also [HM99*-III], [Sto2000**-II]. Table 1.3 gives a useful compilation for the example of ^{76}Ge decay. The matrix elements show a impressive agreement. If one considers the approximations made in the various approaches (violation of Ikeda sum rule in [Cau96*-II], [Sim96], [Sim99], [Sto2000**-II], etc.), the tendency goes to a variation of within only a factor of *1.5*.

1.2.3 *Shell Model Calculations*

An effort to calculate double beta matrix elements in heavy deformed nuclei has been made in a *pseudo–SU(3)* model by [Hir94a].

Although the possibilities of recent shell model calculations improved considerably with the upcome of new shell model codes [Cau96*-II], [Zha90**-II], [Suh97], [Nak96] and with larger computer facilities, it remains still difficult to perform realistic shell model calculations without severe truncations and reliable calculations are impossible *beyond* and partly *in* the *fp* shell. The imposed truncations on the model spaces exclude configurations responsible for the reduction by destructive interference, such as spin–orbit partners (see [Fae98**-II]). So, still recent shell model

Fig. 1.30 Left: Takeshi Oda (middle), Hisashi Horie and the author, in Tokyo 1983. Middle: Cheng-rui Ching (left) and Tso-Hsiu Ho in Heidelberg, 1993. Right: X.R. Wu, in Heidelberg, 1991, (fotos author).

Monte Carlo [Rad96*-II], and the socalled large shell–model calculations [Cau96*-II] suffer from too small configuration space and the latter again do not fulfill the Ikeda sum rule by about 50%, thus leading to too large half–life predictions. A careful analysis [Mut91*-II] shows, however, that *pnQRPA* results using realistic nucleon–nucleon interactions are consistent with shell model approaches, where the latter are possible.

1.2.4 *The Operator Expansion Method*

Another approach to calculate the matrix elements has used the operator expansion method *(OEM)*, which could decrease the dependence of the matrix elements on the strength of the particle–particle force and which has the further advantage to avoid the explicit sum over the intermediate nuclear states [Chi88*-II], [Chi89*-II], [Chi89a], [Sim89**-II], [Gmi90*-II], [Wu91*-II], [Wu92*-II], [Wu93*-II], [Hir93**-II], [Hir94b**-II], [Hir95a*-II], [Sim98**-II] (see also [Fae98**-II]). A calculation of the matrix elements for all potential double beta minus emitters has been performed for the two–neutrino mode by [Wu93*-II] (see also [Hir94b**-II]), while an extension of the method to calculation of neutrinoless double beta decay has been made in [Hir93**-II], [Hir95a*-II]. Efforts continue to derive a fully 'consistent' formulation of *OEM* (see, e.g. [Fae98**-II], [Sim98**-II]).

1.2.5 $\beta^+\beta^+, EC/EC, \beta^+/EC$ Decay

A calculation of matrix elements for double positron emission ($\beta^+\beta^+$) and double electron capture *(EC/EC)* and the mixed mode (β^+/EC) has been performed, in the *pnQRPA* model, by the Heidelberg group [Sta91*-II], [Hir94*-II] for the most promising isotopes for these modes.

The latter investigation led to the discovery of a beautiful method to decide to what extent the double beta decay is dominated by the mass mechanism or by the right–handed current interaction. In principle a way to decide this question would be to measure the angular distribution of the outgoing leptons in addition to the half-life, since the angular distributions for the $\langle\lambda\rangle$ – terms differ from the ones for the mass mechanism. However, such a measurement would require large statistics and moreover can not be done within an experiment using semiconductors (like ^{76}Ge), that is within the at present most sensitive experiments. However, measurement of the $\beta^-\beta^-$ decay of one double beta emitter (e.g. ^{76}Ge) *and* of the 0ν β^+/EC decay of another one (e.g. ^{124}Xe) offers a striking possibility to fix the decay mechanism, once neutrinoless double beta decay would be observed.

1.2.6 *Matrix Elements for Exchange of Heavy Particles*

The calculation of nuclear matrix elements for double beta decay remained an important task until the most recent time – namely for other neutrinoless decay modes than the simplest one discussed above, for example for Majoron–accompanied modes, for the decay modes where a heavy or composite neutrino is exchanged, and for those ones, where no neutrino is exchanged but supersymmetric particles, lepto-quarks or others instead (see below). Such calculations have been performed only in the late nineties by the Heidelberg group (e.g. [Hir96*-IV], [Hir96e*-IV], [Hir96d*-IV], [Pae99*-I]). Here double beta decay also sheds light on interesting dynamical aspects of nuclear couplings at very short distances.

The matrix elements for heavy particle exchange deserved special attention because of short–range repulsive correlations due to the nucleon hard core (see, e.g. [Ver82a], [Hir96*-IV]) or, in principle more precisely, matrix elements of (approximately) four–fermion (four–quark) operators [Heu94], [Min96**-II], [Gre96**-II]. The effect of the latter is severely overestimated by these authors. On the other hand the influence of the matrix elements is often limited, since e.g. in *SUSY*-induced double beta decay, the dependence of the deduced results e.g. about the Yukawa coupling λ'_{111} depends only from the square root of the matrix element, and the result about sfermion masses only on the fourth root. All this has been discussed recently in [Hir96*-IV] and [Kla2000g*-II]. Some of the new matrix elements have been calculated only recently [Pae99*-I] in connection with the development of a general Lorentz–invariant parameterization of the long and short range parts of the neutrinoless double beta decay rate in terms of effective B–L couplings, which allows to deduce limits on arbitrary lepton number violating theories from double beta decay [Pae99*-I], [Pae2000**-I].

1.3 Double Beta Decay, Neutrino Mass Models and Cosmological Parameters – Status and Prospects

With the understanding of the reliable calculation of nuclear matrix elements for neutrinoless double beta decay the way was free to deduce effective neutrino masses $\langle m_\nu \rangle$ and right-handed coupling parameters $\langle \eta \rangle, \langle \lambda \rangle$ from double beta decay experiments.

Fig. 1.31 Left: Alexey Yu. Smirnov, at the BEYOND'97 Conference, Castle Ringberg, Germany, 1997. Right: Yoji Totsuka, director of Superkamiokande, near his detector, in 1995, with the author (fotos author).

In the last years, with the increased sensitivity of experiments, double beta decay started to take striking influence on presently discussed neutrino mass scenarios which arose in connection with the recent solar and atmospheric neutrino oscillation results from *Superkamiokande* and other detectors, and in connection with the *COBE* results on the cosmic microwave background radiation and the related discussion of neutrinos as hot dark matter - and took also striking impact on related cosmological models (see, e.g. [Cal93*-III], [Pet94], [Lee94], [Ioa94], [Smi96a**-III], [Kla98a**-A], [Kla99e**-III], [HM99*-III], [HM2000**-III], [Adh98*-III], [Min97], [Min97a**-III], [Min2000**-III], [Giu99*-III], [Ell99a], [Giu99b], [Bil99], [Bar99a], [Ma99*-III], [Vis99*-III], [Kla99*-VI], [Kla99a**-B], [Smi99], [Geo2000*-III], [Kla-2000*-III]). Statistically not relevant hints on a positive neutrinoless double beta signal [Kla93**-III] prompted some speculative theoretical papers (e.g. [Cal94**-III]). We give a few examples describing the *present* situation.

The first sub–eV values for the neutrino mass limit as given at present by the Heidelberg–Moscow experiment (Fig. 1.32) [Kla92*-V], [Kla96a*-V], [HM97a], [HM99*-III], [Kla99b**-A], [Kla99e**-III], [HM2000**-III], (the most sensitive experiment at present and for some future), according to [Adh98*-III] exclude already

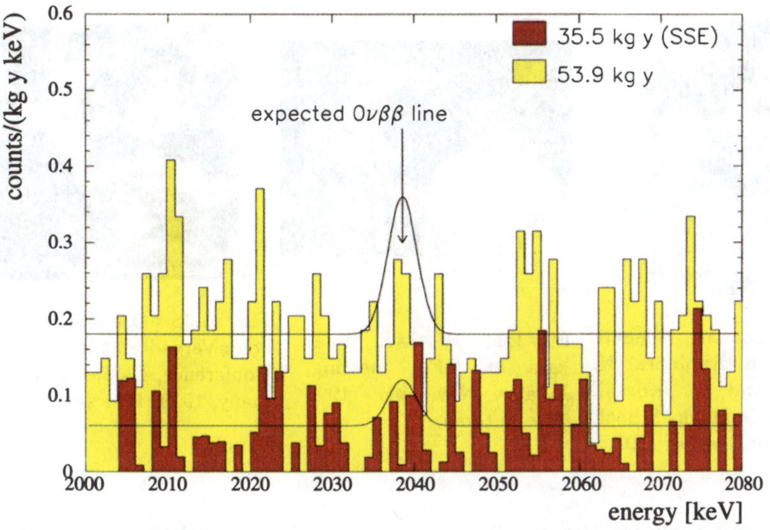

Fig. 1.32 HEIDELBERG-MOSCOW experiment, June 2000: Sum spectrum of all five detectors with 53.9 kg y and SSE spectrum (with pulse shape analysis) with 35.5 kg y in the region of interest for the $0\nu\beta\beta$ - decay. The experiment yields $T_{1/2}^{0\nu} > 1.3 \times 10^{25}$ y (90% C.L.), $T_{1/2}^{0\nu} > 2.2 \times 10^{25}$ y (68% C.L.), for the full spectrum and $T_{1/2}^{0\nu} > 1.9 \times 10^{25}$ y (90% C.L.), $T_{1/2}^{0\nu} > 3.1 \times 10^{25}$ y (68% C.L.) for the SSE spectrum. The curves correspond to the excluded signals at 90% C.L. The corresponding Majorana neutrino mass limits are (for the SSE spectrum) 0.35 eV (90% C.L.) and 0.27 eV (68% C.L.) (from [HM2000]).

now simultaneous 3ν solutions for hot dark matter, the atmospheric neutrino problem and the small mixing angle *MSW* (Mikheyev–Smirnov–Wolfenstein) solution of the solar neutrino problem. This means that *if* we insist on neutrinos as hot dark matter candidates, Majorana neutrinos would be ruled out, if the small mixing angle solution of the solar neutrino problem should be borne out.

According to [Min97], [Min97a**-III], [Yas99*-III] degenerate neutrino mass schemes for hot dark matter, solar and atmospheric anomalies and for the *CHOOZ* result are already now almost excluded for the small *and* the large mixing angle *MSW* solutions (without unnatural finetuning). If starting from recent dark matter models [Pri98], including in addition to cold and hot dark matter also a cosmological constant $\Lambda \neq 0$, these conclusions remain also valid, except for the large angle solution, which would not yet be excluded by double beta decay (see [Kla2000*-III]). Also a study by J. Ellis and S. Lola [Ell99a] finds that the limit

Fig. 1.33 Samoil Bilenkii, (middle) with Rabindra N. Mohapatra and T. Kotani, at the Erice Neutrino School, Sicily, Italy, September 1997.

Fig. 1.34 Hisakazu Minakata at KEK meeting, Japan, November 1997 (foto author).

Fig. 1.35 Osamu Yasuda and John Vergados at the BEYOND'99 Conference, Castle Ringberg, Germany, 1999 (foto author).

by $0\nu\beta\beta$ decay excludes the small angle *and* large-angle *MSW* solution if neutrinos are near-degenerate, "forcing us into the vacuum-oscillation solution, in which the neutrino mass degeneracy must be at the level of one part in 10^{10}". They warn, however, that degenerate neutrino mass textures may have many problems when renormalization effects are taken into account. Also H.M. Georgi and S.L. Glashow [Geo2000*-III] stress a dramatical constraint of the form of the neutrino mass matrix from the recent results on neutrinoless double beta decay from the HEIDELBERG-MOSCOW experiment. They also claim that the small-angle *MSW* solution of the solar neutrino oscillations is ruled out, if relic neutrinos comprise at least three percent of the critical mass density of the universe. These results have been dramatically confirmed - in a more general way - by the recent exclusion of the small mixing angle *MSW* solution for the solar neutrino problem as reported by Superkamiokande [Suz2000] (see, however, also, e.g. [Gon2000]).

A model producing the neutrino masses based on a heavy scalar triplet instead of the seesaw mechanism derives from the solar small angle *MSW* allowed range of mixing, and accomodating the atmospheric neutrino problem, $\langle m_\nu \rangle = 0.17 - 0.31\ eV$ [Ma99*-III]. Also this model is already close to be disfavoured. Looking into *4-neutrino scenarios*, including in addition to the three known neutrino flavours a fourth 'sterile' neutrino, there are according to [Giu99*-III], [Bil99], [Kla2000*-III] only two schemes, that can accomodate the results of *all* neutrino oscillation experiments (including the Los Alamos experiment *LSND*). In the first of these schemes, where $m_1 < m_2 << m_3 < m_4$, with solar (atmospheric) neutrinos oscillating between m_3 and m_4 (m_1 and m_2), and $\Delta m^2{}_{LSND} = \Delta m^2{}_{41}$, the Heidelberg–Moscow $0\nu\beta\beta$ bound excludes [Giu99*-III] the small mixing angle *MSW* solution of the solar neutrino problem, for both $\nu_e - \nu_\tau$, and $\nu_e - \nu_s$ transitions. Including recent astrophysical data yielding in big bang nucleosynthesis for the number of neutrino flavours a value of $N_{\nu_{BBN}} \leq 3.2$ (95% c.l.) [Bur99], the oscillations of solar neutrinos occur mainly in the $\nu_e - \nu_s$ channel, and *only* the small angle solution is allowed by the fit of the solar neutrino data [Bah98a], [Fuk99]. This means that double beta decay excludes the first whole scheme.

Figure 1.76 (ChapterVI) illustrates not only that in degenerate neutrino mass models cosmological models with cold and hot dark matter (*CHDM models*) and also those including a non–vanishing cosmological constant Λ ($\Lambda CHDM$ *models*) are excluded for the case of the small angle mixing *MSW* solution of the solar neutrino problem, but that already now, if assuming the small angle solution to be valid, the Heidelberg–Moscow experiment would rule out the whole range of sensitivity of the future satellite experiments *MAP* and *PLANCK* for cosmological models of the above types involving degenerate neutrinos as dark matter [Kla2000*-III].

In the case of inverse hierarchical neutrino mass scenarios, with only two neutrinos contributing to dark matter, only ΛCDM models with a small Hubble constant ($h = 0.5$) are marginally not yet excluded.

Figure 1.76 (Chapter VI) also demonstrates, that combined with the neutrino oscillation results from solar and atmospheric experiments and with precise determinations of cosmological parameters, double beta decay is obviously the only way to obtain precise informations about the neutrino mixing and the absolute mass scale in partially degenerate and degenerate neutrino mass scenarios. A general discussion of the potential of double beta decay and its decisive role for further investigation of the neutrino mass matrix is given in [Kla2000*-III], [Vis99*-III], [Giu99*-III], [Bil99], [Yas99*-III], [Min2000**-III], [Kal2000**-III], [Kla2000i**-III] (see also Fig. 1.72 in section VI.)

It might be mentioned in this place that some double beta experiments can in addition to the information on hot dark matter (neutrinos), because of their extreme low background in the low-energy part of the spectrum, also give important information on *c o l d* dark matter, competing with dedicated dark matter experiments. If *WIMPs* (Weakly Interacting Massive Particles) populate the halo of our galaxy, they could be looked for by elastic scattering off the *Ge* nuclei in a ^{76}Ge double beta experiment and the following ionization by the recoiling nucleus (for an early discussion see [Cal87a]). The deposited energy for neutralinos with masses between *10 GeV* and *1 TeV* is below *100 keV*. The best current limits on *WIMP* nucleon cross sections come from the *DAMA* experiment [Ber97], [Bel99], from *CDMS* [Gai2000] and from the Heidelberg–Moscow experiment [HM98**-III], the latter experiment yielding the most stringent limits for using raw data without pulse shape analysis. All of these experiments at present just marginally touch only the upper part of the parameter space predicted for neutralinos as cold dark matter (see [Kla99a**-B], [Bed96], [Bed97a], [Kla98c], [Bed98], [Bed99a], [Bed99b] and Fig. 1.71 in Chapter VI).

The low background achieved in the $\beta\beta$–experiments allowed further to deduce the sharpest laboratory bounds for electron decay of $\tau_e > 1.63 \times 10^{25}\ y$ [Bal93].

1.4 Other Beyond Standard Model Physics: From SUSY and Leptoquarks to Compositeness and Quantum Foam

1.4.1 *General*

Another breathtaking development opening – in addition to neutrino physics – a wide field of further beyond standard model physics to become accessible by double beta decay started on the side of theory in the early eighties and culminated in the last five years. This development was pushed by researchers like J.W.F. Valle, R.N. Mohapatra, J. Vergados, and in the last five years mainly by the Heidelberg group. Recent reviews have been given by [Cal89**-V], [Cal91a], [Kla91b**-V], [Moh96*-IV], [Kla96c**-V], [Val96], [Kla96b**-IV], [Kla96d], [Moh98], [Kla98a**-A], [Kla98b*-VI], [Kla99*-VI], [Kla99b**-A], [Kla99e**-III], [Kla2000e*-IV] (see also [Kla95/98**-II], [Moh91**-IV], [Kla97/2000**-IV]). It was a lucky coincidence that in the last decade then experiments developped to such a sensitivity that this broad new theoretical potential of double beta decay for research into beyond standard model physics *really could be explored* – in several fields *competitive to the largest present colliders and beyond.*

We first outline the theoretical development and then look, in the next section, into the various lines of experimental development with some emphasis on the last *10–15 years*, to understand the essential points which made the progress possible, and with the aim to understand the most promising lines to be followed for the future.

It was recognized already in the early *80's*, that gauge theories can yield diagrams contributing to neutrinoless double beta decay, but not containing any neutrino [Sch82a*-I]. Instead, other particles like supersymmetric particles etc. are exchanged.

It has been shown on the other hand, as mentioned earlier, that *a n y* mechanism what so ever generating $0\nu\beta\beta$ decay, as long as the weak interactions are described by a local gauge theory, implies a non–vanishing Majorana mass of the electron neutrino [Sch82a*-I], [Tak84], [Nie84]. A generalisation of this fundamental Schechter–Valle theorem to supersymmetry has been given by [Hir97a**-IV], [Hir97b**-IV], and [Hir98a*-IV]. The latter theorem claims for the neutrino Majorana mass, the B–L violating mass of the sneutrino and the neutrinoless double beta decay amplitude: If one of them is non–zero, also the others are non–zero and vice versa, independent of the mechanism of $0\nu\beta\beta$ decay and (s–)neutrino mass generation. This theorem connects double beta decay research with new processes potentially observable at future colliders like *NLC* (Next Linear Collider) [Hir97b**-IV], [Hir98a*-IV], [Hir98c*-IV]. For example, finding a finite "Majorana" mass of the sneutrino in future collider experiments would be equivalent to the proof that the neutrino is a Majorana particle (see also [Kol98], [Hir98d*-IV]).

Fig. 1.36 John Vergados, Rabindra Mohapatra and the author, at the DARK96 Conference in Heidelberg, Germany, 1996, (foto author).

1.4.2 *Doubly Charged Higgs and Pion Double Charge Exchange, and Double Beta Decay*

It was discussed by R.N. Mohapatra and J. Vergados [Moh81*-IV] (Fig. 1.36), that in a general class of gauge models the decay of doubly charged Higgs particles may induce double beta decay. This process does not involve a Majorana neutrino, but proceeds by decay of a doubly charged Higgs boson to electrons. This process is, however, negligible because of the weak coupling of the physical Higgs particle to quarks [Sch82a*-I], and also since the matrix element for this process is suppressed [Hax82b].

The decay of two pions in flight between two nucleons has been described by J. Vergados [Ver82*-IV], [Ver86]. The latter showed that, if the mixing between light and heavy neutrinos is not negligible, the double charge exchange of pions in flight between the two nucleons gives an important contribution to the neutrinoless double beta decay mediated by a heavy neutrino.

1.4.3 *Superstring-Inspired Models, R-Parity Breaking Supersymmetry and Double Beta Decay*

Lepton phenomenology and neutrinoless double beta decay has already early been investigated in superstring–inspired theories by G.K. Leontaris and J.D. Vergados [Leo89**-IV]. Neutrinoless double beta decay in the presence of extra dimensions has been discussed by R. Mohapatra and A. Pérez-Lorenzana [Moh99]. They show that the higher Kaluza-Klein modes of the right-handed W-boson provide new contributions to this process. In this way correlated limits on m_{W_R} and the inverse size of the extra dimensions can be obtained from double beta decay.

New contributions to neutrinoless double beta decay in R–parity violating supersymmetric theories [Hal84], [Aul83], [Ros85], [Ell85], were discussed by R.N. Mohapatra [Moh86a*-IV] and by J. Vergados [Ver87]. Now the exchanged particles are gluinos or neutralinos, etc. This work has been extended – within the minimal supersymmetric standard model – by including diagrams previously not considered and calculating the relevant matrix elements by the Heidelberg group [Hir95**-IV], [Hir95b**-IV], [Hir95c*-IV], [Hir96*-IV], [Hir96f], [Hir97]. This decay mode yields with the limit from the most sensitive experiment – the Heidelberg–Moscow experiment – the sharpest constraint on *R–parity* violation, more precisely on the Yukawa coupling λ'_{111} in the *R-parity* violating part of the superpotential, more stringent then the values from the largest accelerators like *TEVATRON* or *HERA* (see Fig. 1.79 in Chapter VI). This λ'_{111} value from double beta decay immediately excludes the formation of squarks of first generation (of *R*–parity violating *SUSY*) in the high Q^2 events seen at *HERA* [Alt97], [Bab97], [Kal97**-IV], [Cho97], [Hir98c*-IV], [Dre98**-IV]. Using, instead of a MSSM with supergravity–mediated *SUSY* breaking by means of additional assumptions relating sfermion and gaugino masses, as in [Hir96*-IV], or an MSSM using *GUT* constraints [Wod99a**-IV], an MSSM with gauge–mediated *SUSY* breaking, leads to qualitatively new constraints and even much stronger limits on λ'_{111} [Wod99*-IV]. It has been shown by S.K. Kang [Kan98], that stronger limits on λ'_{111} than in the first case are also obtained from neutrinoless double beta decay in the non–minimal supersymmetric model (NMSSM), as a result of the extended neutralino sector.

A *SUSY*–accompanied neutrino exchange mechanism has been considered by K.S. Babu and R.N. Mohapatra [Bab95*-IV], [Moh96*-IV], and by the Heidelberg group [Hir96a*-IV], [Pae99a*-IV]. Taking into account that the *SUSY* partners of the left– and right–handed quark states can mix with each other, one can derive stringent limits on different combinations of λ' such as $\lambda'_{113} \times \lambda'_{131}$ and $\lambda'_{112} \times \lambda'_{121}$.

1.4.4 *R–Parity Conserving Supersymmetry and Double Beta Decay*

Also *R–parity* conserving softly broken supersymmetry has been found (by the Heidelberg group, [Hir97a**-IV], [Hir97b**-IV], [Hir97c], [Hir98a*-IV]) to give contributions to neutrinoless double beta decay, via the *(B–L)* violating sneutrino mass term being a generic ingredient of any weak–scale *SUSY* model with a Majorana neutrino mass. These contributions are realized on the level of sneutrino box diagrams (see Fig. 1.37). This allows to obtain upper limits for the sneutrino 'Majorana' mass which are – for first generation sneutrinos – sharper than obtainable from Next Linear Colliders *(NLCs)* [Hir98a*-IV], [Hir98c*-IV], [Kol98], [Hir98d*-IV], [Kol2000]. This rolls up again the question of whether heavy sneutrinos could be candidates for cold dark matter [Kla2000h**-IV].

1.4.5 *Leptoquarks and Double Beta Decay*

It has been found out, again by the Heidelberg group [Hir96d*-IV], [Hir97], that also scalar or vector leptoquarks can contribute to double beta decay (see Fig. 1.37). A condition is that leptoquarks couple to the standard model Higgs doublet

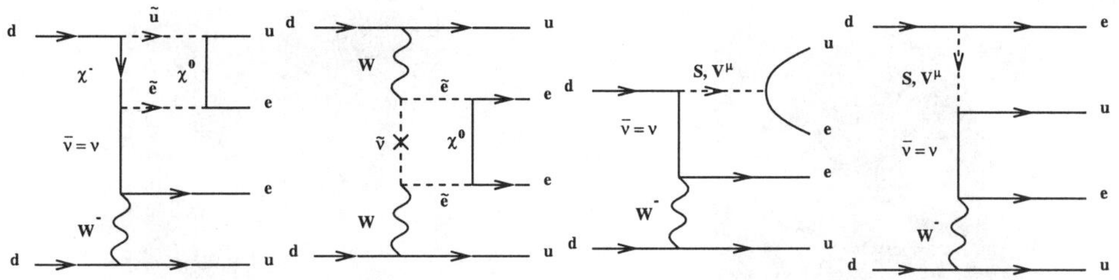

Fig. 1.37 Examples of R_P conserving SUSY contributions to $0\nu\beta\beta$ decay (two diagrams on the left) and for $0\nu\beta\beta$ decay within leptoquark models. S and V^{μ} stand symbolically for scalar and vector LQs, respectively (two diagrams on the right).

field on the same footing as to the quark- lepton field [Hir96c*-IV]. This allows to deduce constraints on the effective leptoquark parameters, which are of interest in connection with the recently discussed evidence for new physics from the high Q^2 events from *HERA* (see e.g. [Alt97], [Hew97**-IV], [Bab97], [Kal97**-IV], [Cho97]). Assuming that actually leptoquarks have been produced at *HERA*, double beta decay (the Heidelberg–Moscow experiment) would allow to fix the leptoquark–Higgs coupling to a few 10^{-6} [Hir98c*-IV].

1.4.6 *Superheavy Neutrinos and Double Beta Decay*

Important information came also from the consideration of processes induced by heavy (left– and right–handed), composite and sterile neutrinos. It was discussed already since *1980* by Mohapatra [Moh80*-I], [Moh86*-IV] that not only exchange of a left–handed neutrino, but also of a heavy *right–handed neutrino*, which naturally occurs in left–right symmetric models, could induce neutrinoless double beta decay. This process which was discussed in more detail later by the Heidelberg group [Hir96b*-IV], [Hir96g], yields at present the most restrictive lower bound on the mass of a right–handed *W* boson [Hir96b*-IV], [Kla99e**-III], [Kla2000e*-IV].

In the case of the exchange of heavy or superheavy *left–handed* neutrinos one can exploit the mass dependence of the matrix element (see e.g. [Mut88*-I]) to obtain *l o w e r* limits on the latter [HM95].

The deduced lower limit [Kla99*-VI], [Kla99e**-III], [Kla2000e*-IV] is $\langle m_H \rangle \geq 9 \times 10^7\ GeV$. Assuming the bound on the mixing matrix $U^2_{ei} = 5 \times 10^{-3}$ [Nar95], [Bel96], [Bel98*-IV] and assuming no cancellation between the involved states, the limit implies a bound on the mass eigenstate $M_i > 5 \times 10^5\ GeV$. To obtain the same information by inverse double beta decay (first considered in this context by [Riz82], (see [Riz93]) $e^-e^- \longrightarrow W^-W^-$ at a Next Linear Collider, the latter should have a center of mass energy of two *TeV* [Bel96], [Bel98*-IV] (Fig. 1.78 in Chapter VI).

Fig. 1.38 Orlando Panella, at the BEYOND'97 Conference, Castle Ringberg, Germany, June 1997, (foto author).

Fig. 1.39 Geneviève Bélanger, at the First International LEPTON–BARYON Conference, Trento, Italy, April 1998, (foto author).

1.4.7 *Compositeness and Double Beta Decay*

Also the exchange of excited composite Majorana neutrinos has been considered. O. Panella and Y.N. Srivastava [Pan95] were the first to derive bounds on the compositeness parameters from the non-observation of neutrinoless double beta decay. More recently this topic has been discussed by O. Panella et al. [Pan97**-IV], [Pan98] and E. Takasugi [Tak98]. The potential of double beta decay for compositeness has been compared to the corresponding potential of present and future colliders by [Pan99], [Pan2000*-IV]. It is shown that the Heidelberg–Moscow experiment [Kla99*-VI], [Kla99e**-III] exceeds already now the sensitivity of *LEPII* in probing compositeness, while the future project *GENIUS* (see below) will reach the sensitivity of *LHC* (Fig. 1.78 in Chapter VI). The fact that neutrinoless double beta decay might constrain composite models, was also discussed already much earlier by R. Barbieri, R. Mohapatra and A. Masiero [Bar81], but within the framework of a particular model.

1.4.8 *Sterile Neutrinos, Majorons and Double Beta Decay*

Double beta decay proceeding by exchange of a heavy sterile Majorana neutrino (mass scale in the *GeV* range or higher) has been considered by [Bam94], [Moh96*-IV].

Double beta decay has also been used to test the existence of new bosons, called Majorons, which could play a significant role in beyond standard model physics. The possibility of a 0ν mode associated with Majoron emission was pointed out by [Geo81], [Gel82], the Majoron being a light pseudo–scalar Goldstone boson arising from spontaneous breaking of the global *B–L* symmetry [Chi80*-IV], [Chi81], [Gel81]. Depending on the model there are singlet [Chi80*-IV], doublet [Aul83] and triplet Majorons [Gel81]. After the measurement of the Z^0 width at *LEP* ruled out triplet and doublet Majorons (see, e.g. the discussion in [Kla95/98**-II]) and on the other hand pushed by claims of experimental Majoron observation in double beta decay [Avi87], [Wal87] (which later have been disproved [Fis87], [Cal87*-V], [Ell87*-IV], [Als88], [Cal89**-V], [Bec93*-IV], [Tan93]), various new Majoron models were introduced by [Moh88], [Car93], [Bur93**-IV], [Bur95**-IV], [Bam94a] whose main novel features are, that they can carry leptonic charge, need not be Goldstone bosons, and that emission of two Majorons can occur (scalar– or fermion-mediated). All of them, however, have been ruled out by double beta decay by the Heidelberg group [Hir96e*-IV], [Gun96**-IV], [HM2000**-III]. The essential point here was the calculation of the corresponding new types of matrix elements, which have been calculated for the first time – as also for the cases of leptoquark–induced, *SUSY*- induced and right–handed neutrino–induced double beta decay – by the Heidelberg group [Hir96e*-IV, Hir96d*-IV, Hir96*-IV, Hir96b*-IV] (see also Chapter II).

1.4.9 *Lepton Number Violating Interactions, Non–Exponential Decay and Time Dependence of the Weak Interaction*

Also lepton number violating interactions, in the context of their possible effects on neutrino oscillation experiments, have been discussed with respect to the constraints of the effective couplings of the relevant operators by double beta decay [Ber2000**-IV].

It had been discussed already early, to what extent non-exponential decay or the socalled *Zenon* effect may lead to a measurable slowing down of such extremely rare decay process as double beta decay [Gro84*-IV]. Recently, possibilities to obtain information on a possible time dependence of the weak interaction constant from comparison of direct (counter) and geochemical experiments on two–neutrino double beta decay have been discussed [Bar98], [Bar99**-IV].

1.4.10 *Test of Lorentz Invariance, Equivalence Principle and Quantum Foam*

In the last two years exciting connections of double beta decay to special relativity and equivalence principle and the effects of a possible contribution of new gravitational interactions to neutrinoless double beta decay, and also effects of the quantum space time foam in the neutrino sector have been investigated in [Kla99c*-IV], [Kla2000a*-IV], [Kla2000f*-IV], [Kla2000b*-IV].

Violations of Lorentz invariance, equivalence principle and *CPT* could occur as consequences of superstring theories. In contrast to relativistic point–particle field theories these conservation laws are no more valid a priori in superstring theories. String theories could lead to violations of the equivalence principle by contributions of the dilaton field (excitations in compactified space–time dimensions) by flavour–characteristic coupling to gravitation [Dam94]. This effect can lead to oscillations in the neutrino sector [Hal98]. Since theories of quantum gravity allow for quantum fluctuations of the metric, the vacuum gets a non–trivial space–time structure.

Fig. 1.40 Utpal Sarkar (left) and Rabindra N. Mohapatra (right) with the author, at the DARK'98 Conference, Heidelberg, Germany, July 1998, (foto author).

Fig. 1.41 Sheldon Glashow, at the NEUTRINO'98 Conference, Takayama, Japan, June 1998, (foto author).

Thus particles propagating through the quantum foam show energy–dependent recoil and vacuum polarization effects, which lead to violation of Lorentz invariance, and the interaction with microscopic black holes could transfer pure into mixed quantum states by a *CPT* violating effect [Ell96a], [Ell96b], [Ell96c]. The underlying formalisms have been applied to neutrino oscillations by [Col97], [Liu97**-IV].

Precision tests of local Lorentz invariance have been performed in various sectors of the standard model through vacuum Cerenkov radiation [Gas89b], photon decay [Col97], neutrino oscillations [Gla97*-IV], [Gas89a], [Hal91], [Hal96], [But93] and K physics [Ham98], [Goo61]. It has been shown recently [Kla99c*-IV], [Kla2000f*-IV] that neutrinoless double beta decay yields unique constraints on Lorentz invariance in the neutrino sector particularly in the range of small mixing, which is not accessible by neutrino oscillation experiments.

The equivalence principle implies that spacetime is described by unique operational geometry and hence universality of the gravitational coupling for all species of matter. In recent years there have been attempts to constrain a possible amount of violation of the equivalence principle *(VEP)* in the neutrino sector from neutrino oscillation experiments [Gas89], [Hal91], [Hal96], [But93]. However, these bounds do not apply, when the gravitational and the weak eigenstates have small mixing. It has been observed, however, that neutrinoless double beta decay can constrain the amount of *VEP* for small mixing angles, for which there does not exist any bound at present [Kla99c*-IV].

A general formalism for the study of the effects of new gravitational interactions in neutrinoless double beta decay, which allows to constrain the amount of violation of the gravitational laws is given in [Kla2000a*-IV].

An investigation of quantum foam effects in the neutrino sector for the cases of neutrinoless double beta decay and oscillations of neutrinos from astrophysical sources (supernovae and active galactic nuclei) [Kla2000b*-IV] shows, that the latter has an impressive potential for the search for quantum foam, while the potential of double beta decay is negligible. The reason is that the propagation distance is found to be the crucial parameter entering any bounds on the *CPT* violating parameters (*EHNS* parameters).

1.5 The Experimental Race: From the Late Eighties to the Future

1.5.1 *General*

Parallel to the rapid development on the theoretical side much fantasy was involved in the improvement of the sensitivity of experiments. In fact the attack to approach all of the exciting particle physics beyond the standard model sketched above was possible only with the upcoming of a new, second generation of experiments using enriched material and large source strengths, since the early nineties. Only the large increase in sensitivity reached in this way allowed to exploit the theoretical potential of double beta decay to an extent that beyond standard model physics information could be obtained which is out of reach of the present (and partly also of future) accelerators. Therefore it is useful to have a brief look into detector development and particularly into the development during the last ten years, before we consider the future of the field.

A chronology of experiments from *1948* to the early eighties is given by Yu. Zdesenko [Zde80**-I] and W.C. Haxton and G.J. Stephenson [Hax84**-I]. More recent reviews of experimental approaches are given by D.O. Caldwell [Cal86a*-V], [Cal89**-V], [Cal91**-V], [Cal91a], T. Kirsten [Kir88], H.V. Klapdor-Kleingrothaus [Kla90**-V], [Kla91b**-V], [Kla94*-V], M. Moe [Moe95], M. Moe and P. Vogel [Moe94**-V], [Kla96b**-IV], [Kla96c**-V], [Kla98a**-A], [Fio99a], [Kla99e**-III], [Kla99*-VI] and in [Kla86b**-V], [Boe87**-II], [Kla88a**-V], [Kla88**-V], [Tre95**-V], [Kla95/98**-II], [Kla96**-V].

The sensitivity for neutrinoless double beta decay has increased over four decades from 3×10^{15} *years* in *1948* to 5×10^{23} *years* in *1987*. The improvement from *1974* to *1987* was about a factor of *100*. Since *1987* another factor of *100* to *a few* 10^{25} *years* has been achieved [HM99*-III], [HM2000**-III].

Since the background of the cosmic radiation and radioactive impurities represent serious problems, the experiments are usually carried out in underground laboratories. In addition there are limited possibilities for studying large amounts of source material since one is restricted to the about *35* double β^- and a few double β^+ emitters [Gro85a*-II], [Gro86a**-II], [Sta90a*-II] existing in nature. While when studying proton decay it is possible to select favourable materials such as water with several thousand tonnes of source material, double beta experiments have until now been restricted to a maximum of several kilograms.

The various experiments can be classified into four groups: active source, passive source, geochemical and radiochemical experiments.

The geochemical experiments starting with the early experiments of [Ing49*-I], [Ing50*-I] yielded the first experimental evidence for double beta decay and have been discussed already. They yielded with the half life limit of 7.7×10^{24} *years* [Ber 92*-I], [Ber93**-I], or the half life of 2.2×10^{24} *years* [Tak96**-I], for the decay of ^{128}Te for long time the most restrictive limits on the electron neutrino mass.

Analogously to geochemical experiments, radiochemical experiments involve a search for an enrichment of the daughter nuclei in a sample, but exploiting the advantage of radioactive daughter nuclei which are detectable in much smaller quantities than stable rare gases. So one needs no longer geological integration periods and the measurements may be carried out over much shorter time scales of a few years. Two typical candidates for radiochemical experiments on double beta decay are ^{232}Th and ^{238}U. One of the first experiments of this type was the one of [Lev50*-I] investigating the decay of ^{238}U. The same decay has been investigated later by [Tur91*-I].

Nowadays the potential of geochemical experiments is limited compared to other types of experiments since they cannot differentiate between two-neutrino and zero–neutrino decay, and thus the former will form an impenetrable "background" when searching for no–neutrino decay. The same is true for radiochemical experiments. Geochemical experiments might, however, still find an exciting application in searching for a time dependence of the weak coupling constant, as mentioned already, by comparing results for two–neutrino double beta decay of one and the same isotope from a geochemical and a 'counter' experiment. Since the geochemical experiment measures the decay over geological times, whereas the counter experiment measures for the present time, such comparison allows in principle to search for a time dependence of the weak interaction [Bar98], [Bar99**-IV]. Suitable isotopes are ^{82}Se, ^{96}Zr, ^{130}Te and some others.

Fig. 1.42 Mike Moe in front of his TPC.

Fig. 1.43 Hiro Ejiri (director of RCNP, Osaka), Tsvetan Vylov (vice–director of JINR, Dubna, UdSSR), the author, and Yurij Gaponov (Kurchatov Institute) at the III. International *WEIN* Conference, Dubna, 1992.

The different decay modes in double beta can be distinguished in counter experiments, which dominate the present and future experimental efforts.

In the *active source* counter experiments source and detector are identical. Examples are CaF_2 as used in the experiment by [Mat66*-I], and later in the *ELEGANT VI* experiment [Eji98], in *Ge* experiments looking for the decay of ^{76}Ge in *Ge* detectors, started by E. Fiorini [Fio67], [Bel83], [Bel83a], (see also [Ant60]) and in later experiments using *enriched* ^{76}Ge starting (after first attempts to built up such experiment were made by the Kiev group of Yu. Zdesenko and by L. Popeko et. al. from LNPI, Gatchina, Russia) in the late eighties by a Moscow–Armenian group under I.V. Kirpichnikov [Vas90*-V] and – as the most sensitive one – by the Heidelberg–Moscow collaboration [Kla87**-A], [Kla91*-A], [Kla92*-V], [HM99*-III], [Kla99e**-III], [Kla99*-VI], [HM2000**-III]. This type of active source experiments proves to the be most sensitive method up to now. Other examples are the Kiev experiment using $GdWO_4$ crystals [Dan89], [Dan95], [Bur96*-V], [Dan99], [Dan2000**-V], and time projection chambers using beta–beta active counting gas, for example ^{136}Xe (see e.g. [Won91*-V], [Vui93**-V]). In contrast to such type of experiments, in the *passive source* experiments usually only relatively small amounts of the source can be investigated without deterioration of the energy resolution of the detector. A direct detection of double beta decay and the differentiation between different modes is possible only in active source or passive source, counter experiments.

1.5.2 *Ionisation and Time Projection Chambers, and Combinations with Plastic Scintillators*

The first direct observation of two–neutrino double beta decay in a direct experiment was claimed in 1980 for ^{82}Se [Moe80], and finally achieved by Mike Moe and his colleagues in *1986* [Ell86**-I], [Ell86a], [Ell87a*-V]. This experiment used a *TPC* (time projection chamber). A sample foil of *14 g* of isotope–enriched selenium was placed in the middle of the chamber, the trajectories of the beta–beta electrons emitted from the foil in the gas–filled chamber being observed.

Other groups used combinations of drift chambers, plastic scintillators and *NaI* counters, so the ones installed in the Kamioka mine by the group of H. Ejiri (*ELEGANTS* detectors) [Eji91], [Eji92], [Kud92], [Eji96], [Eji97*-V], [Eji98] which used a foil of enriched ^{100}Mo or ^{116}Cd as source. Recently the *ELEGANTS V* and *VI* detectors have been built up in the Oto tunnel [Eji97*-V], [Eji98]. The half life limits for the neutrinoless mode lie in the range of a few times 10^{22} *years*.

An interesting variant of the gas detectors involves the use of *Xenon* as a counting and simultaneously $\beta\beta$ source in proportional counters, time projection chambers and drift chambers. The energy resolution is approximately an order of magnitude worse than in germanium semiconductors (see below); however, ^{136}Xe has the advantage of somewhat larger *Q*–value and can be relatively easily enriched in ul-

Fig. 1.44 Takayuki Watanabe (with the author) at the *ELEGANTS* installation in the KAMIOKA underground laboratory, 1988, (foto author).

Fig. 1.45 *ELEGANTS-V* in the Oto tunnel, 1997 (with K. Takahisa and the author), (foto author).

Fig. 1.46 *ELEGANTS-VI* in the Oto tunnel (left), and its head T. Kishimoto (right) with the author, 1997 (fotos author).

tracentrifuges. The double beta decay of ^{136}Xe was investigated using scintillation counters [Bar86] and ionization chambers [Bar89]. A group in Milan installed a multicell proportional counter in the Gran Sasso underground laboratory (*3500 m* water equivalent shielding) [Bel91]. A high-pressure multi-wire wall-less proportional counter was used recently at Baksan (Russia) [Gav2000**-V], to look for two-neutrino decay of ^{136}Xe.

In the Swiss Gotthard underground laboratory (*3000 m* water equivalent shielding) a time projection chamber with an active volume of *180 l* was built to study the double beta decay of ^{136}Xe (see Fig. 1.47). This experiment yielded the sharpest limits of the ^{136}Xe experiments of a few 10^{23} *years* for the half life of the neutrinoless decay mode [Won91*-V], [Tre91], [Vui93**-V], [Far97**-V], [Lue98].

A *TPC* for use of up to ten *10 kg* of ^{136}Xe (see Fig. 1.48) was proposed by the group of O. Zeldovich at *ITEP*, Moscow [Art89]. A small prototype in operation at atmospheric pressure [Art91], [Art95] has measured among others the two–neutrino double beta decay of ^{136}Xe [Art91], [Art92*-V], and ^{150}Nd [Art95], [Art96]. The *10 kg* xenon version is expected to reach a half–life limit for 0ν decay of $> 10^{24}$ *years* [Art92*-V]. The full detector is planned to go into operation by end of the year 2000. Important research on ion transport in xenon gas in context with double beta decay

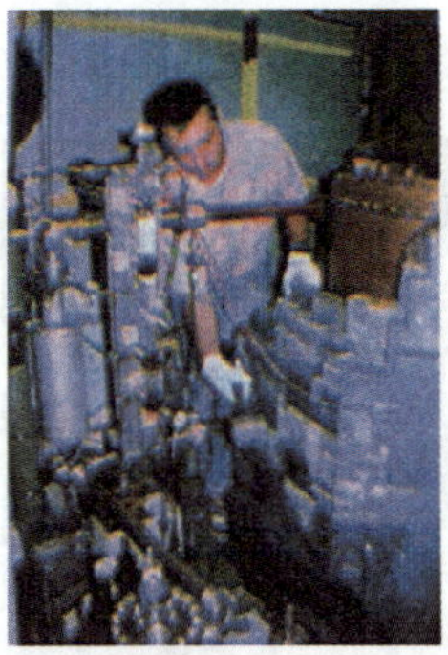

Fig. 1.47 The ^{136}Xe *TPC* at the *Gotthard* tunnel.

Fig. 1.48 Construction of the large *TPC* for use with *10 kg* of ^{136}Xe at *ITEP*, Moscow, Russia

Fig. 1.49 Olga Zeldovich, head of the ITEP TPC project, and the author, at the NEUTRINO 2000 Conference, Sudbury, Canada, June 2000 (foto author).

Fig. 1.50 Alexandr S. Barabash (left), John Simpson (right) and the author, at the NEUTRINO 2000 Conference, Sudbury, Canada, June 2000 (foto author).

research, which may be important for future applications, is being performed by [Mad95].

A not realized proposal to detect double beta decay of ^{136}Xe suggested to exploit the occurrence of the daughter nucleus as an additional signature. The idea was to detect the ionized ^{136}Ba, which is formed in the decay of ^{136}Xe, via its laser fluorescence, in a *liquid Xe TPC*. The coincident detection of the daughter nuclide and the beta particle would practically eliminate the background completely [Moe91*-V]. The aim of the M. Moe proposal was to reach a limit of the effective electron neutrino mass of *0.01 eV* with one ton of enriched ^{136}Xe.

The line of using a *liquid Xenon TPC* had also been investigated by [Gir92] who estimated the sensitivity of a large volume Xenon double beta decay detector and tested a two–dimensional *liquid Xenon TPC* [Apr92].

Research on the decay products of double beta decay of ^{136}Xe in liquid *Xe* has been performed by [Miy94].

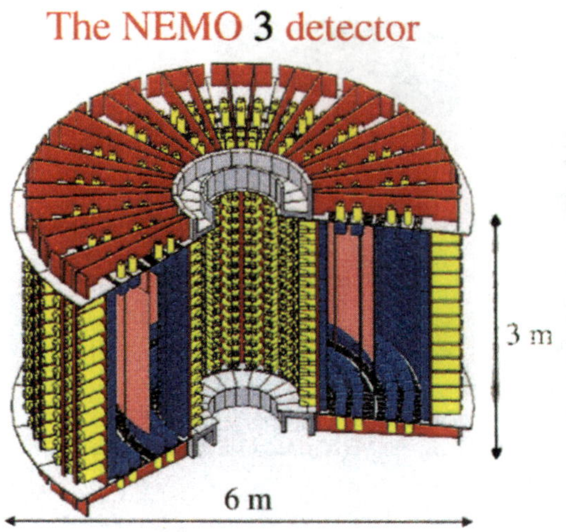

Fig. 1.51 Left: View of the *NEMO3* detector at present under construction. Right: Serge Julian, head of the *NEMO* experiment, at the NANP'97 Conference, Dubna, Russia, 1997 (first line, on the left).

A liquid argon ionization chamber has been used by the group of A. Barabash in the Gran Sasso to investigate the double beta decay of ^{100}Mo, yielding a limit for neutrinoless double beta decay of 3.5×10^{21} *years* [Ash99**-V].

The *NEMO* (Neutrino Experiments in Molybdenum) detector (see Fig.1.51) in the *Frejus* underground laboratory is a three–dimensional electron tracking detector where tracking is accomplished with long open Geiger cells working in Helium gas and ethyl alcohol. A calorimeter covering the tracking volume is made of plastic scintillators using low–radioactivity PMTs. The *NEMO-I* detector is described in [Das91]. The *NEMO-2* detector [Arn95] was operated from *1992* to *1997*, and is now dismounted. The three isotopes ^{100}Mo, ^{116}Cd, and ^{82}Se have been investigated [Das95**-V], [Arn96], [Bar97*-V], [Sar99]. The detector has in principle the advantage of measuring in addition to the summed energy spectra, electron angular distributions and single electron spectra. The results for the neutrinoless decay modes are of the order of a few times 10^{21} to 10^{22} *years*. *NEMO-3* is at present under construction with 6180 Geiger cells, the calorimeter being equipped with *1960* low–radioactivity PMTs. The detector will be able to accomodate up to *10 kg* of various double beta emitter candidates (^{100}Mo, ^{116}Cd, ^{82}Se, ^{130}Te, ^{96}Zr, ^{150}Nd, etc). Since the efficiency of the detector is about 20%, the effective mass is, however, considerably smaller. This means that this brilliant detector unfortunately cannot really be used to exploit large source strengths. The scientific goal aimed at for the year *2005* is an upper limit for the effective neutrino mass of $<$ $0.3 - 0.7\ eV$ [Bar97*-V], [NEM2000], i.e. of the order of the *present* limit of the Heidelberg–Moscow experiment (see below). To what extent this will be reached depends critically on the energy resolution to be reached (see Fig. 1 in [Tre95**-

V]). As a consequence the discussion given in [Bed97c] concerning a future *SUSY* potential of *NEMO* is also premature.

A magnetic tracking detector using a drift chamber, *DCBAI and II*, with the goal to reach a neutrino mass limit of *0.1 - 0.5 eV* in the decay of ^{150}Nd has been proposed at KEK, Japan by [Ish96].

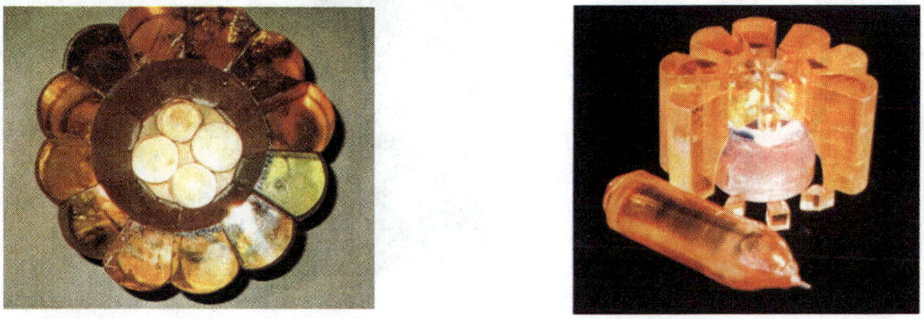

Fig. 1.52 The four enriched $CdWO_4$ crystals surrounded by *14* ***natural*** $CdWO_4$ crystals used at present in the ***Solotvina*** salt mine in the Ukraine (foto Yu. Zdesenko).

Fig. 1.53 Solotvina mine, Ukraina (left). Yurij Zdesenko in Heidelberg, with the author and his son, 1993 (right) (fotos Yu. Zdesenko).

1.5.3 *Scintillation Detectors*

Scintillation counters have an energy resolution lower than semiconductors and in addition, the photomultipliers are often a source of background. In Brookhaven as the first active source experiment the decay of ^{48}Ca was studied exploiting the fact that calcium in the form of CaF_2 may be used as a scintillation crystal [Mat66*-I]. ^{48}Ca benefits from a very large *Q*-value of *4.271 MeV*, however this isotope is very rare (0.187% natural abundance). The original experiment of *1966* had a source strength of *11.4 g* ^{48}Ca. New measurements with *37.4 kg* CaF_2 scintillation crystals (*43 g* of ^{48}Ca), were carried out in a *512 m* deep colliery shaft near Beijing

(China) [You91], and an experiment (*ELEGANTS VI*) using *31 g* of ^{48}Ca is planned at Osaka (Japan) [Kum96], [Eji97*-V], [Eji98] in the Oto tunnel. *ELEGANTS VI* uses CaF_2 scintillators surrounded by CsI scintillators (see Fig. 1.46).

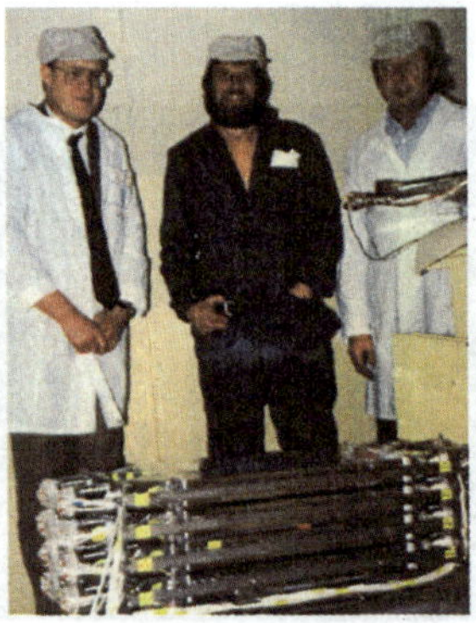

Fig. 1.54 Anatoly Smolnikov (middle), at his SISTEMA I/II installation at the Baksan Underground laboratory, Kaukasus, USSR, in 1987 with Igor Kondurov from Gatchina (right) and the author (foto author).

Fig. 1.55 Anatoly Smolnikov (left) and his colleagues at Baksan Valley (Kaukasus): Boris Prituchenko, Vladimir Novikov, the author, Victor Skljarov and Ludvik Popeko, in 1987 (foto author).

In an analogous way, ^{116}Cd can be built into scintillator material ($CdWO_4$). This line of research has been followed by Yu. Zdesenko and coworkers in Kiev [Zde91], [Dan89], [Dan95], [Bur96*-V], [Dan96**-V]. At present this experiment runs with four enriched $CdWO_4$ crystals (total mass *339 g*) surrounded by an active shield of *14* natural $CdWO_4$ crystals (see Fig. 1.52, 1.53) in the *Solotvina* salt mine [Dan98], [Dan99], [Dan99a], [Dan2000**-V]. The present half life limit is 7×10^{22} years which is the best at present for this nucleus, and a neutrino mass limit of $\langle m_\nu \rangle < 2.6\ eV$.

Fig. 1.56 Igor Kirpichnikov, with the author, at the NEUTRINO'92 Conference, Granada, Spain, 1992.

Fig. 1.57 Frank Avignone, Irina Krivosheina (middle), Masato Morita (left), and Yurij Lyutostanskii (right) on the Volga during LEWI Conference, Dubna, Russia, September 1990 (foto author).

The nuclei ^{100}Mo and ^{150}Nd have been investigated also by using plastic scintillators sandwiching the isotopically enriched samples (*SISTEMA I/II*) by A. Smolnikov et al. [Kli86], [Kli89], [Vas93] (see Fig. 1.54).

1.5.4 *Semiconductor Detectors*

Semiconductor detectors because of their outstanding energy resolution are suited excellently for the study of double beta decay, and in particular for neutrinoless double beta decay which is manifest by a sharp line in the total energy spectrum. Semiconductor devices may be used to study external probes, or as active detectors where the detector material at the same time is the double beta emitter.

An example of the first type is a setup consisting of a sandwich detector from silicon counters separated by molybdenum sheets [Oka88], [Als89], [Als93*-V]. This setup yielded a half–life limit of $> 4.4 \times 10^{22}$ *years* for $0\nu\beta\beta$ decay of ^{100}Mo.

A particularly favourable case of the second type is represented by the double beta candidate ^{76}Ge. This germanium isotope occurs with an abundance of 7.8% in natural germanium, from which large high–resolution detectors can be manufactured. Thus, germanium can be used simultaneously as source and detector allowing for large source strength without spoiling the high energy resolution, which is about *3 keV* in the region of the decay energy of *2.04 MeV*.

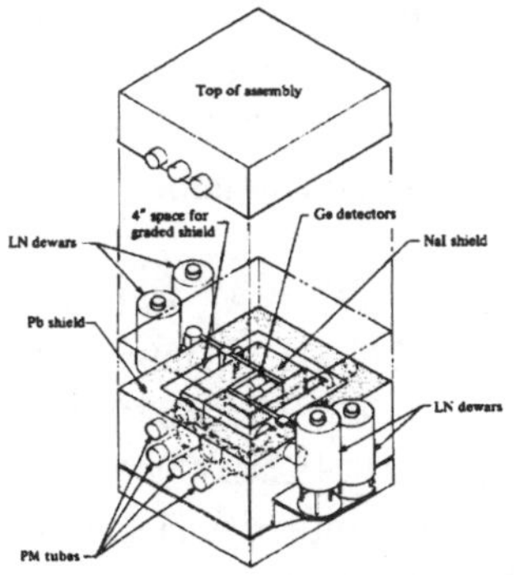

Fig. 1.58 Set-up of the *USCB–LBL Ge* experiment of David Caldwell and collaborators (left) and David Caldwell, at BEYOND'97 Conference, Castle Ringberg, Germany, June 1997 (right) (foto author).

The most sensitive experiment using detectors from *n a t u r a l* Germanium was that of D. Caldwell (see Fig. 1.58) et al. [Cal86], [Cal86a*-V], [Cal87*-V], [Cal91**-V] in California located in the *Oroville* dam (*600 m.w.e.* underground) leading to a halflife limit of 1.2×10^{24} *years*, corresponding to a neutrino mass limit $\langle m_\nu \rangle < 1 - 3\ eV$. This was at the same time the most sensitive first generation experiment through many years, until in 1990 the Russian-Armenian, and in 1992/93 the HEIDELBERG-MOSCOW experiment took over (see below).

Fig. 1.59 Left foto: Ludvik Popeko (left) with the author and Igor Kondurov at LNPI, Gatchina, Russia, February, 1987. Center foto: Ludvik Popeko with part of his active silicon shielding construction at the Leningrad Institute of Nuclear Physics, in February, 1987. Right foto: The author with Ludvik Popeko (right) and Igor Kondurov near Elbrus, Baksan, Kaukasus, investigating the optimal site for the later HEIDELBERG–MOSCOW experiment, in August, 1987 (fotos author).

An in principle very powerful detector setup suggesting use of enriched *Ge* detectors in an active shielding of silicon was discussed and constructed by L. Popeko et al. [Pop86*-V], [Pop89] (see Fig. 1.59), but finally used only with a natural *Ge* detector supplied by the Heidelberg group. This shielding was one of the early, however not realized, options for the later Heidelberg-Moscow experiment. The Russian/Armenian group [Kir87] under I. Kirpichnikov used two *0.5 kg* enriched *Ge* detectors (enriched in ^{76}Ge to 85%) in a salt mine in *Yerevan* at a depth of *645 m*. They observed for the first time two–neutrino decay of ^{76}Ge [Vas90*-V], confirmed by a Russian–American collaboration [Avi91] and later by the Heidelberg–Moscow collaboration [Bal94], [HM97*-V].

The **Heidelberg-Moscow** experiment [Leg86*-A], [Kla87**-A], [Mem88*-A], [Add98*-A], [Kla90**-V], [Kla91b**-V], [Kla92*-V], [Kla93**-III], [Kla94*-V], [Bal 95], [Kla 96a*-V], [Kla96b**-IV], [HM97*-V], [HM97a], [Kla98], [Kla98a**-A], [Kla-98g], [HM99*- III], [Kla99*-VI], [Kla99b**-A], [Kla99e**-III], [Kla2000c*-A], [Kla-2000e*-IV], [HM2000**-III] using for the first time large amounts of enriched double beta emitter material, was the first 'second generation' experiment, starting a new era of double beta experiments. It found large (sometimes also curious, see [Sci91*-A]), public attention [Rad86*-A], [Rad87*-A], [Kla87a*-A], [Bun88*-A], [Kla89*-A], [Bun90*-A], [NEU90*-A], [Kla91*-A], [Kla91c*-A], [CER91*-A], [MPG94*-A], [Sci97*-A], [Kla98b*-IV], [CER2000*-A]. After some exploration of the most suitable location for the experiment, it finally started operation with the first detector in *1990* in the Gran Sasso underground laboratory (*3500 meter water equivalent*) (see Fig. 1.62 – 1.63) where since *1995* a setup of *5* detectors with in total *11.5 kg* of enriched ^{76}Ge are operated [HM97*-V]. This amount of enriched material makes the experiment at least as sensitive as an experiment using *1.2 tonnes* of natural germanium. The experiment has the lowest background of all double beta experiments of this type, which since 1996 still has been reduced by a new method of pulse shape discrimination distinguishing between single site (double beta) and multiple site events [HM97a], [Hel99], to about *0.06 counts/kg year keV*. A new method of background reduction exploiting the potential of neuronal networks has been developed recently [Maj99**-V]. The present half–life limit for the neutrinoless decay mode is 1.9×10^{25} *years* (90% c.l.), 3.1×10^{25} *years* (68% c.l.) yielding the most restrictive limit on the effective Majorana neutrino mass of *0.35 (0.27) eV* (90% c.l. and 68% c.l., respectively) [HM99*-III], [HM2000**-III], (see also [Kla99e**-III], [Kla2000c*-A]).

Fig. 1.60 Left foto: from left to right – S.T. Belyaev, the author and V. Lebedev, at the Gran–Sasso, 1989. Foto in center: Signing the agreement for the transfer of the world's largest sample of high purity ^{76}Ge (*86%*) from Moscow's Kurchatov institute for the **HEIDELBERG–MOSCOW** experiment are (seated) the two collaboration ***spokesman*** **Hans V. Klapdor–Kleingrothaus** of Heidelberg (Germany) and **Spartak T. Belyaev** of Moscow, USSR (March 1989) in Heidelberg. Foto right: L. Popeko with author in Baksan, near the Elbrus, Kaukasus, (fotos author).

Fig. 1.61 Left foto: Signing the agreement for the **Heidelberg-Moscow** experiment in Kurchatov Institute in Moscow, in December 1988, (from left to right): H.V. Klapdor-Kleingrothaus, S.T. Belyaev and A. Balysh. Foto in center: Arrival of the first part of the enriched ^{76}Ge from Moscow in Heidelberg by train in December 1988. Right foto: at Max Planck Institute, Heidelberg, March 1989, S.T. Belyaev (left) and H.V. Klapdor-Kleingrothaus (fotos author) with the second, major portion of ^{76}Ge (white package on the right).

Another enriched ^{76}Ge experiment has been set up partially at the *Baksan* underground laboratory in Russia and at the *Canfranc* laboratory in Spain (*2450 m* water equivalent) using *6 kg* of ^{76}Ge. This experiment meanwhile claims a limit of 0.7 to 0.8×10^{25} *years* for the neutrinoless decay mode [Aal99], [Aal99a**-V]. Although this experiment had already *2.1 kg* of enriched detectors at disposal in

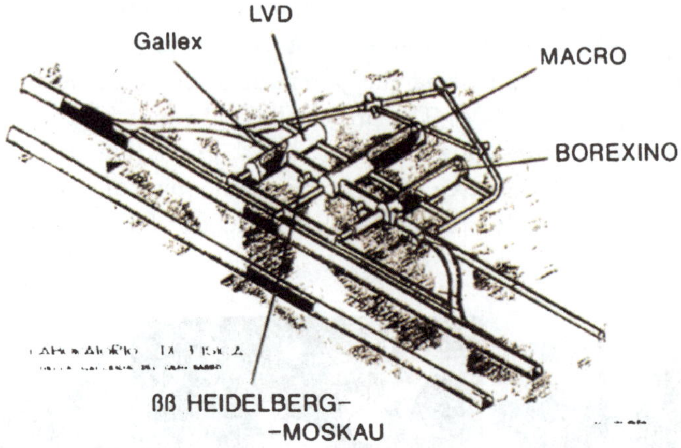

Fig. 1.62 Gran-Sasso Underground Laboratory, Schematical view, Gran-Sasso, Italy

Fig. 1.63 The laboratory of the **HEIDELBERG-MOSCOW** experiment at *Gran–Sasso* – on top the author with the visiting Prof. Masato Morita.

1992 [Bro92], *4.1 kg* in 1993 [Avi93], and *8 kg* in 1995 [Aal96], it collected until 1999 only a statistics of *5.7 kg y*, corresponding to not much more than *0.5 years* of data taking (for some discussion see [Kla2000j]).

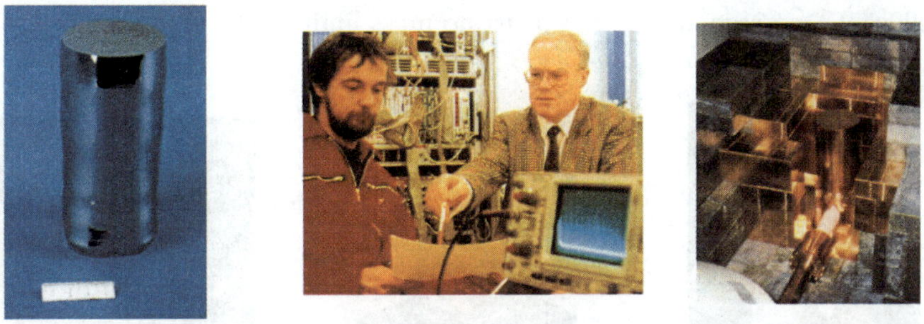

Fig. 1.64 Left: The **first high-purity enriched ^{76}Ge crystal worldwide.** Center: A. Müller with the author in the Gran-Sasso laboratory, in November 1990. Right: One of the enriched ^{76}Ge detectors in its shielding of electrolytic copper (fotos author.)

Fig. 1.65 The **Heidelberg-Moscow** experiment – at the time of installation of the last of the five *Ge* detectors, end 1994, beginning 1995. In middle the second enriched detector, at the time of its production the largest high-purity Germanium detector worldwide.

1.5.5 *Cryogenic Detectors*

Another recent approach is to use cryogenic detectors. The socalled bolometers were explored for the use in the search for double beta decay by E. Fiorini and T.O. Niinikoski [Fio84], [Fio91a**-V], [Fio91b], [Giu91**-V], [Pre99**-V]. They involve the use of a pure diamagnetic or dielectric crystal, whose thermal capacity at low temperatures is so small, that even the energy released in a single double beta decay event causes a measurable increase in the temperature. Four *334 g* heavy TeO_2 crystals at *10 mK* have been operated as bolometers in the Gran Sasso laboratory in the search for double beta decay of ^{130}Te [Ale94]. The half–life limit obtained for the neutrinoless mode was 1.8×10^{22} *years*.

The Milano group is at present operating a *20 detector* array [Ale95], [Ale98], [Ale99], [Ale99a] whose single modules have a mass of *340g* of TeO_2. At low energies the energy resolution is *1.5 to 2.5 keV*, at *two MeV* it is a factor of *2–3* worse. The background is *0.5 counts/keV kg year* in the energy region of the hypothetical neutrinoless double beta decay line, i.e. about a factor of ten higher than in the Heidelberg–Moscow ^{76}Ge experiment. This setup has yielded, after a measurement time of *0.35 kg y*, a half–life limit of 7.7×10^{22} *years* for the neutrinoless mode

[Giu99a*-V], corresponding to a neutrino mass limit of *2.5 to 5.2 eV*.

Fig. 1.66 Left foto: Raju S. Raghavan, at the NEUTRINO'98 Conference, Takayama, Japan, June 1998. Center: Atsuto Suzuki (leader of the KAMLAND project), at the LEPTON-BARYON Conference, Trento, Italy, June 1998. Right: Andrea Giuliani, at the LEPTON-BARYON Conference at Trento, Italy, April 1998 (fotos author).

For the future an extension of this *20 detector* array is proposed. The proposed *CUORE* experiment [Fio98], [Giu99a*-V] suggests use of 1000 TeO_2 bolometers with a mass of *750 g* each, i.e. a total mass of *0.75 tons* of natural TeO_2 (with the natural abundance of ^{130}Te of 34% it corresponds to a source strength of *250 kg* of ^{130}Te). Up to now an energy resolution of *9 keV* has been reached. In a measuring time of *t e n y e a r s* this would lead, with the background of the present *20 detector* array, to a half life limit of 1×10^{25} *years*, corresponding in sensitivity for the neutrino mass to the *p r e s e n t* result of the *Heidelberg–Moscow* experiment.

An improvement of the background even by a factor of *100* would *n o t* increase the sensitivity of *CUORE* such that a neutrino mass considerably *b e l o w 0.1 eV* could be probed. However, such an improvement of the background will be a formidable task because of noise arising from the thermometer coupled thermally to the detector, the degree of radiopurity of the cryostat components, etc. So this heroic approach to use large source strengths may *n o t* allow a major step into really new regions of sensitivity.

An interesting candidate was for some time a NdF_3 bolometer exploiting the relatively large phase space of ^{150}Nd (see [Moe95]), but the project never surpassed a *100 g* crystal.

For a possible far–future application there is may be in principle another kind of cryodetectors exploiting the phenomenon of superconductivity, exhibited by some potential double beta emitters. A possible detector would consist, for example, of a large number of small granules of such a substance (e.g. zirconium, molybdenum, cadmium,...) with a typical diameter in the micron area in an overheated superconducting state. The energy deposited in these granules as a result of the decay is sufficient to induce a transition from the superconducting state to a state of normal conductivity. This phase transition can be detected via the Josephson effect or

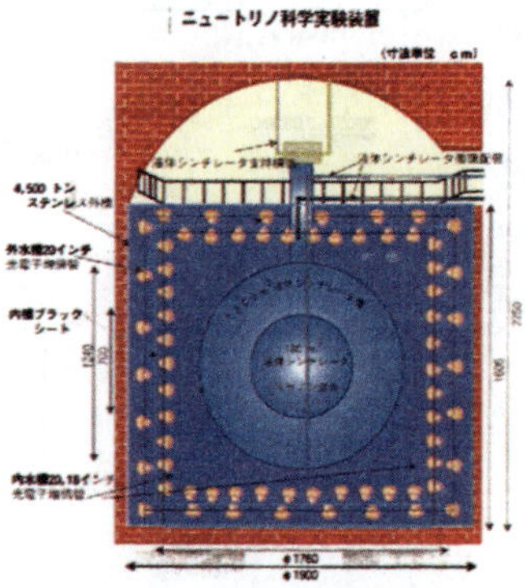

Fig. 1.67 Left: Atsuto Suzuki and his coworkers at Sendai, November 1997, with the author (foto author). Right: The *KAMLAND* project, artists view (from [Suz95**-V]).

via the change in the magnetic flux. Reviews for cryogenic detectors are given by [Pre90*-V], [Fio91a**-V], [Pre93**-V], [Pre93a**-V], [Pre99**-V], [Fio99b].

1.5.6 *Other Large Source Strength Detectors*

Another approach entering into the use of large source strength has been proposed by R.S. Raghavan [Rag94*-V]. He proposed use of *200 kg* of enriched or *2 tons* of *natural Xe* added to the scintillator of the *KAMIOKANDE* detector or similar amounts added to *BOREXINO* (both primarily devoted to solar neutrino investigation). The sensitivity of such application of *BOREXINO* would allow to set a half life limit for ^{136}Xe of about 3.8×10^{25} *years*, probing the neutrino mass down to about *0.3 eV* in two years of measurement (see [Bor91], [Cac2000]). The idea for *KAMIOKA* is going to be realized in the *KAMLAND* (Kamioka Liquid Scintillator AntiNeutrino Detector) detector [Suz95**-V], [Suz99a**-V], [Suz99b], [Suz99c]. The sensitivity would allow to test the Majorana neutrino mass down to *0.15 eV* according to [Suz95**-V].

A comparison of the sensitivities of the most sensitive present double beta experiments is given in Fig. 1.68, including the future approaches *CUORE, KAMLAND*. It is obvious that none of them really yields greater future perspectives in the sense that none of them has a chance to surpass the border of *0.1 eV* of the neutrino mass to lower values – except for the *GENIUS* project, also shown in the figure and to be discussed below, and perhaps the recently proposed far- future projects *MOON* [Eji99a], [Eji99b*-V] and *EXO* [Dan2000a], [Bre2000-V].

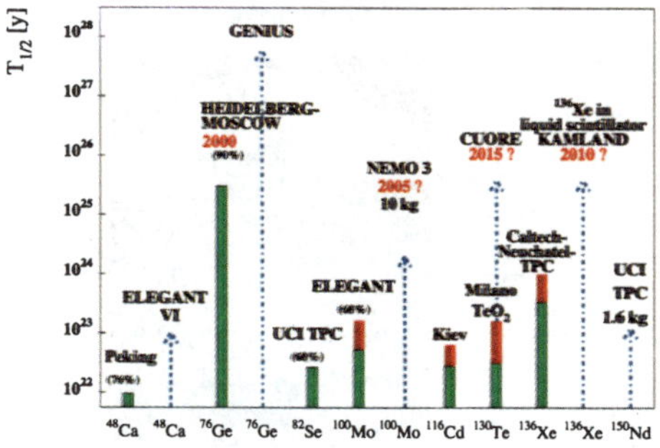

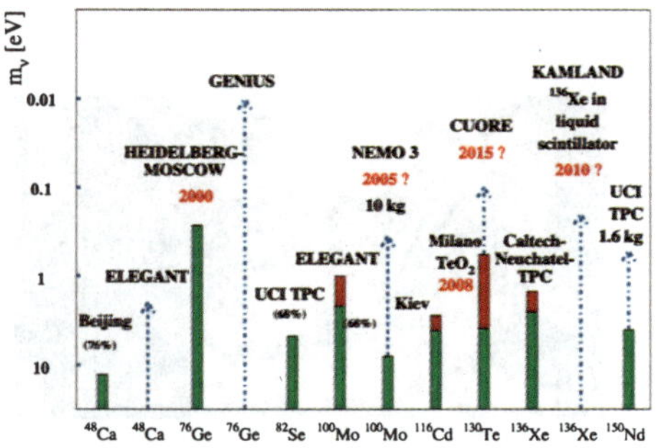

Fig. 1.68 **Present situation, 2000, and expectation** for the near future and beyond, of the most promising $\beta\beta$-experiments concerning accessible ***half life (left)*** and ***(effective) neutrino mass limits (right).*** The green parts of the bars correspond to the present status, the red parts of the bars to expectations for running experiments, dashed lines to experiments under construction and dash-dotted lines to proposed experiments (from [Kla99*-VI]).

1.6 The Future of Double Beta Decay

1.6.1 *General*

The time of the small smart experiments is over. The big step in sensitivity increase achievable by replacing natural by enriched material with the effect, that with a 'few kg experiment' the sensitivity of a 'order of ton of natural material experiment' can be reached, has been made. Example is the Heidelberg–Moscow experiment using 11 kg of *enriched* ^{76}Ge, corresponding to more than **1.2 tonnes** of *natural Ge.* While the improvement in the neutrino mass sensitivity went with the square root of the degree of enrichment of the detector material, after this step having been done, the improvement in the neutrino mass bound now goes with the fourth root of the detector mass, of the background and of the measuring time. The half–life bound reachable in experiment is $T \approx a(mt/\Delta B)^{1/2}$, with percentage of enrichment of the double beta emitter a in the detector, detector mass m, background B, energy resolution Δ and measuring time t. The neutrino mass limit $m^{-1} \approx T^{1/2}$. Therefore, the experimental future of double beta decay now becomes 'large' and leads unavoidably into the tonne and many tonnes region.

Since there are natural limits for the exposure time (of say not more than 10 years), an *e s s e n t i a l* improvement requires work on mass *a n d* background simultaneously. E.g., improving the mass limits alone, would require a dramatic increase of the source strength to multi–ton devices without a large increase in sensitivity – such as the *CUORE* and possibly also the *MOON* approach. Similarly the reduction of background alone without providing the possibility to use

Fig. 1.69 Conference foto of the *BEYOND'97* Conference, Castle Ringberg, Germany, where the *GENIUS* proposal has first been presented.

large source strength as in the *NEMO* project is not sufficient. *CUORE* can reach in perhaps *10 years* from now with a mass of one tonne at most a limit of 0.1 eV, *MOON* plans to reach with 100 tonnes of natural or 10 tons of enriched of *Mo* detector material, a limit of 0.03 eV. *NEMO* goes the other way, improving the background, but being seriously restricted by the emitter mass which can be used (not more than 10–20 kg). Its best limit would be 0.3–0.7 eV in five years from now [NEM2000].

Improving only *o n e* of these parameters thus *c a n n o t* lead to any decisive progress – and also would run quickly into other limitations. For example the detector mass of enriched (and also natural) material can certainly not exceed by much the annual world production, not to talk about the economical problems.

Consequently, there has to be made simultaneously with an increase of the source strength some revolutionary step in reduction of the background to obtain a major step of improvement in sensitivity. Exactly this is the idea underlying the *GENIUS* (GErmanium in Liquid NItrogen Underground Setup) project which has been proposed in 1997 [Kla97**-VI], [Kla97c], [Kla98], [Kla98a**-A], [Kla99*-VI], [Kla99b**-A] see also [CER97*-B], [CER97a*-B] and which has prompted the proposals of several other large scale experiments, and also gave a large push to further theoretical investigations of the potential of double beta decay. It is also the idea of the new ^{136}Xe time projection chamber proposal [Dan2000a] basing on the old idea of Mike Moe [Moe91*-V], mentioned earlier, who proposed to exploit individual detection of the daughter nucleus by laser tagging in a liquid Xenon *TPC* for background reduction, and on the experience of the *ITEP* and Gotthard groups with TPCs. The new proposal is, to use this technique, but using a *5* to *10 atm*, *40* m^3 *TPC* which would contain *1* to *2 tons* of enriched (*65%*) ^{136}Xe. Going to the limit of the world market potential of 10 tonnes of enriched (*65%*) ^{136}Xe, and considering installation of *5* to *10 TPC's* of the above size, M. Danilov and coworkers

$\beta\beta$ – Isotope	Name	Status	Mass (tonnes)	Assumed backgr. † events/ kg y keV, ‡ events/kg y FWHM, * events /yFWHM	Running Time (tonn. years)	Results limit for $0\nu\beta\beta$ half-life (years)	$<m_\nu>$ (eV)
^{76}Ge	**HEIDELBERG MOSCOW [Kla99e**] [HM2000*] [-III]**	**running**	0.011 (enriched)	† 0.06 ‡ 0.24 * 2	**35.5 kg y**	**$1.9 \cdot 10^{25}$ 90% c.l. $3.1 \cdot 10^{25}$ 68% c.l. NOW !!**	**< 0.35 ** 90% c.l. < 0.27 ** 68% c.l. NOW !!**
^{100}Mo	NEMO III [NEM2000]	*under constr.*	~0.01 (enri--ched)	† **0.0005** ‡ 0.2 * 2	50 kg y	10^{24-25}	0.3-0.7
^{130}Te	CUORE$^{\nabla}$ [Gui99a* -VI]	Proposal	0.75 (natural)	† 0.5 ‡ 4.5 * 1000	5	$9 \cdot 10^{24}$	0.2-0.5
^{130}Te	CUORE	Proposal	0.75 (natural)	† 0.005 ‡ 0.045 * 45	5	$9 \cdot 10^{25}$	0.07-0.2
^{100}Mo	MOON [Eji99b* -VI]	idea	10 (enrich.) 100 (nat.)	?	30 300	?	0.03
^{136}Xe	EXO	Pro-	1	* 0.4	5	$8.3 \cdot 10^{26}$	0.05-0.14
^{136}Xe	[Dan2000a]	posal	10	* 0.6	10	$1.3 \cdot 10^{28}$	0.01-0.04

Table 1.4 Some key numbers of future double beta decay experiments (and of the HEIDELBERG-MOSCOW experiment). Explanations: ∇ - assuming the background of the present pilot project. ** - with matrix element from [Sta90*-II], [Tom91**-I], [Hax84**-I], [Wu91*-II], [Wu92*-II] (see Table II in [HM99*-III]), and Table 1.3 in Chapter II.

[Dan2000a] claim that a limit of $m_\nu \approx$ 0.01–0.04 eV could be reached in a measuring time of ten years. Reaching of this goal, however, still requires development of this novel technique of background suppression from scratch. *GENIUS*, proposed by [Kla97**-VI], has been investigated in detail concerning its experimental realization and physics potential in [Kla97a*-VI], [Kla97b], [Hel97], [Kla98], [Kla99b**-A], [Kla99d], [Kla99e**-III], [Kla98b*-IV], [Kla98d**-VI], [Kla98e**-VI], [Bau99], [Kla99*-VI], [Bau99a], [Kla99a**-B] and [Kla2000*-III]. It has the advantage of re-

$\beta\beta$ – Isotope	Name	Status	Mass (tonnes)	Assumed backgr. † events/ kg y keV, ‡ events/kg y FWHM, * events /y FWHM	Running Time (tonn. years)	Results limit for $0\nu\beta\beta$ half-life (years)	$<m_\nu>$ (eV)
^{76}**Ge**	**GENIUS** [Kla97** -VI]	Proposal	1 (enrich.)	† $\mathbf{0.04 \cdot 10^{-3}}$ ‡ $\mathbf{0.15 \cdot 10^{-3}}$ * **0.15**	1	$\mathbf{5.8 \cdot 10^{27}}$	**0.02-0.05**
			1	* **1.5**	10	$\mathbf{2 \cdot 10^{28}}$	**0.01-0.028**
^{76}**Ge**	**GENIUS** [Kla97**-VI]	Proposal	10 (enrich.)	‡ $\mathbf{0.15 \cdot 10^{-3}}$	10	$\mathbf{6 \cdot 10^{28}}$	**0.006 - 0.016**
				$\mathbf{0^{\triangle}}$	10	$\mathbf{5.7 \cdot 10^{29}}$	**0.002 - 0.0056**

Continued table 1.4. Some key numbers of future double beta decay experiments. Explanations: $\triangle$ - this case shown to demonstrate **the ultimate limit** of such experiments.

lying on conservative techniques and to be in sensitivity more than competitive, and it may be realized in a foreseeable future. Table 1.4 gives some key numbers of future projects under discussion.

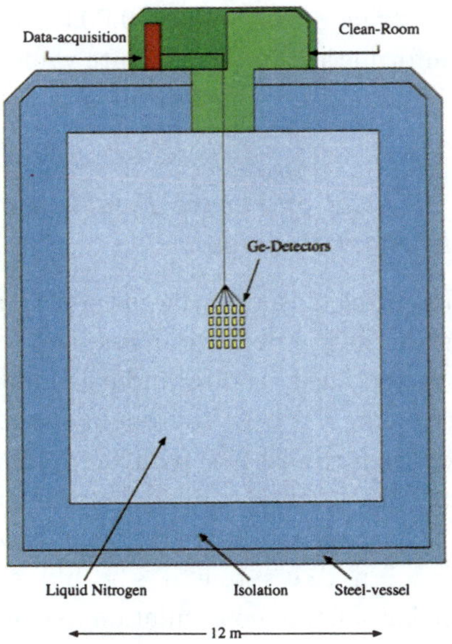

Fig. 1.70 Schematic view of the proposed **GENIUS** experiment.

1.6.2 *GENIUS*

The idea of *GENIUS* (see Fig. 1.70) is to use a large source strength of about *1 ton* of enriched ^{76}Ge detectors and to reduce at the same time the background by about a factor of *1000* by using *'naked'* detectors in liquid nitrogen. That *Ge* detectors operate in liquid nitrogen has been demonstrated in the Heidelberg low–level laboratory [Hel97], [Kla98d**-VI], [Bau99] and the overall feasibility of the project has been shown by [Kla98d**-VI], [Bau99], [Kla99a**-B], [Kla99b**-B]. Potential sites for the *GENIUS* project would be the *Gran Sasso* underground laboratory in Italy, the *WIPP Underground Facility* in Carlsbad, New Mexico, or others.

GENIUS in a first step using *100 kg* of natural *Ge* detectors, would be an extremely sensitive cold dark matter detector, covering a large part of the parameter space of the Minimal Supersymmetric Standard Model *(MSSM)* for the prediction of neutralinos as dark matter [Bed97a], [Kla97b], [Kla98c], [Bed98], [Bed99a], [Bed99b] and allowing easily to check the recently claimed positive evidence for dark matter by the *DAMA* group [Ber97], [Bel99], [Ber99], by direct detection of a signal *as well as by* proving it by the modulation method. (Fig. 1.71). This will be possible for no other present or proposed dedicated dark matter detector, including *CDMS, CRESST.*

In a second step using *1 ton* of enriched ^{76}Ge detectors it would extend the half life limit for neutrinoless double beta decay in one year of measurement to 6×10^{27} *years*, corresponding to an upper limit on the neutrino mass of *0.02 eV*. An ultimate experiment, using ten tons of enriched ^{76}Ge might in ten years of measurement probe the neutrino mass down to 0.003 eV. This would probably be the ultimate information on the absolute value of the effective Majorana neutrino mass which can be obtained from double beta decay.

1.6.3 *The Physics Potential of Future Double Beta Decay for Beyond Standard Model Physics*

Naturally, the physics potential of such a large step in sensitivity as opened by the new projects as *GENIUS* for the neutrino mass, and for other beyond standard model physics research is huge. The beyond standard model physics to be covered by the third generation experiments has been thoroughly investigated and discussed by the Heidelberg group and other groups (see [Kla2000*-III], [Kla99e**-III], [Kla99*-VI], [Kla99a**-B], [Giu99*-III], [Cza99], [Bil99], [Vis99*-III], [Kal2000**-III]).

It has been pointed out by many theorists (see Chapter III and [Kla99*-VI]) that such experiments are indispensable to solve, together with and complementary to the large neutrino oscillation experiments *Superkamiokande, MINOS, Miniboone, SNO, KAMLAND*, and *ICARUS* the problem of the neutrino mass matrix in the cases of degenerate and hierarchical neutrino mass models. This is underlined by the present results from Superkamiokande on solar neutrinos,

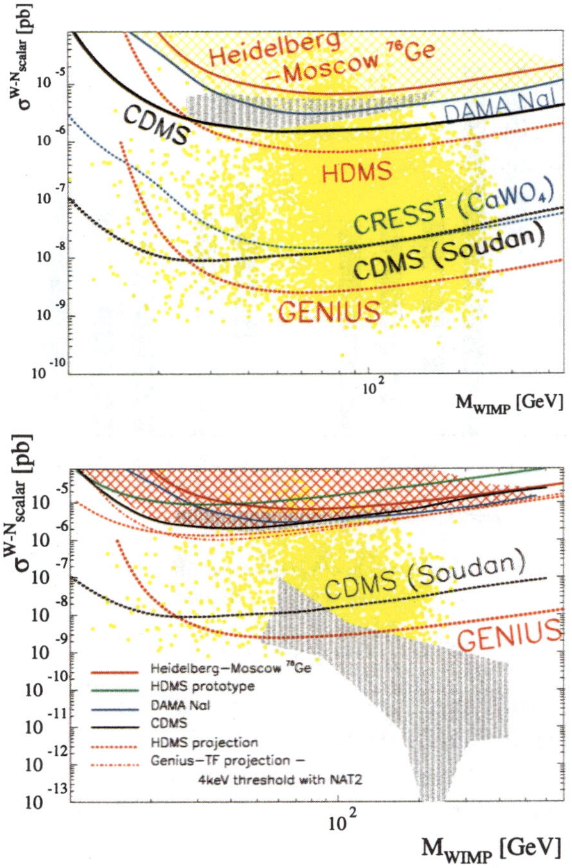

Fig. 1.71 *WIMP*–nucleon cross section limits as a function of the *WIMP* mass. The red region (upper yellow in the upper figure) is excluded by present experiments (HEIDELBERG-MOSCOW, DAMA, HDMS-prototype, CDMS). The dashed lines are expectations for recently started or future experiments, like *GENIUS*, *HDMS*, *CDMS* and *CRESST*. The upper shaded contour area represents the 2σ evidence region of the *DAMA* experiment. The expectation for the *GENIUS* project corresponds to a background level of *0.01 counts/(keV kg y)*, an energy threshold of *11 keV* and an exposure of *300 kg yr*. The experimental limits are compared to expectations (scatter plot) for *WIMP*–neutralinos calculated in the *MSSM* framework with non-universal scalar mass unification [Bed99b] (yellow points in the upper and lower figures) and to predictions in a minimal supergravity-inspired *MSSM*, with universal scalar and gaugino masses, respecting *LEP* and other accelerator constraints [Ell2000] in the lower figure (black-white shaded area). In the case of the second model, only *GENIUS* has a chance to probe (part of) the relevant parameter space (see also [Kla2000k]).

which after *1100 days* of data taking, are reported to rule out the small angle *MSW* solution with 95% c.l. [Suz2000], but somehow also indicate some limitations of such experiments, beyond which the results cannot be improved statistically much further. Fig. 1.72 shows the expected values for the effective neutrino mass in various neutrino mass schemes and the potential of various future double beta

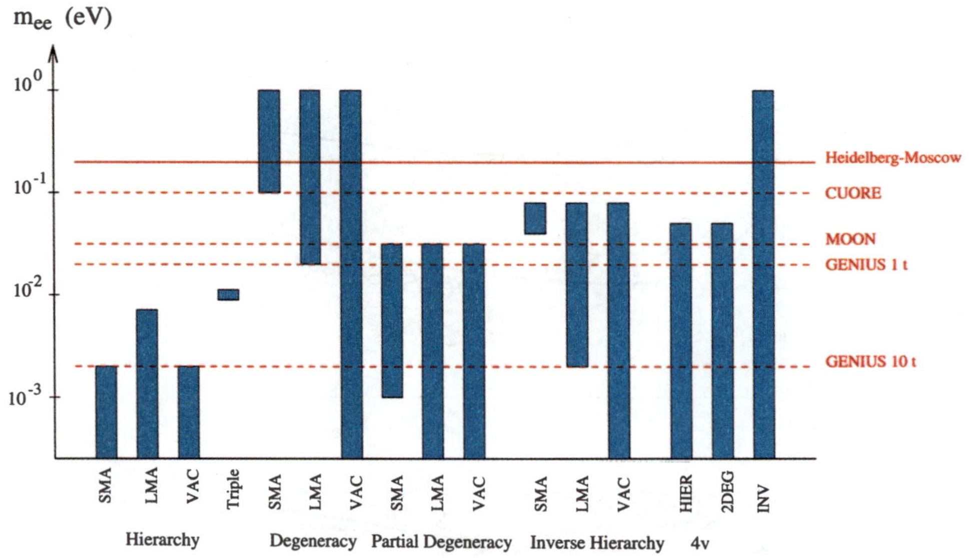

Fig. 1.72 Summary of expected values for the *effective neutrino mass* m_{ee} in different neutrino mass schemes and the potential of future double beta experiments to probe them (from [Kla2000*-III]).

decay experiments to probe these scenarios. Fig 1.73 shows for comparison the discriminative power for various neutrino mass scenarios of the *SNO* experiment by the neutral current to charged current double ratio according to an investigation by J. Bahcall et al. [Bah2000]. Figs. 1.75 show examples of the interplay of the above experiments with double beta decay.

The sensitivity of *GENIUS* for neutrino oscillation parameters is complementary and competitive to present and future terrestrial neutrino oscillation experiments [Kla97**-VI], [Kla97a*-VI]. *GENIUS* could yield a test of *LSND* in case of quasi–degenerate mass scenarios (see Fig. 1.74). In its larger version *GENIUS* could probe directly the large angle solution of the solar neutrino problem, and for rather degenerate but also some hierarchical scenarios, together with long base-line experiments also the small angle solution [Kla99*-VI], [Kla2000*-III] (see Figs. 1.74, 1.75, 1.76). The former is now of particular interest after the recent Superkamiokande results favour the *LMA* and the *LOW* solution [Suz2000].

In strongly hierarchical neutrino mass schemes the magnitude of the double beta decay observable depends crucially on the assumed solution for the solar neutrino problem. While with assuming the *MSW* small angle solution of the solar neutrino problem the observation of double beta decay with $\langle m \rangle > 6 \times 10^{-3}\ eV$ would rule out this scheme, in scenarios with *MSW* large angle solutions *GENIUS (10 tonnes)* should observe a positive signal, an information being complementary to the search for day–night effects in present and future solar neutrino experiments such as *Superkamiokande, SNO or ICARUS*. Even more stringent restrictions may

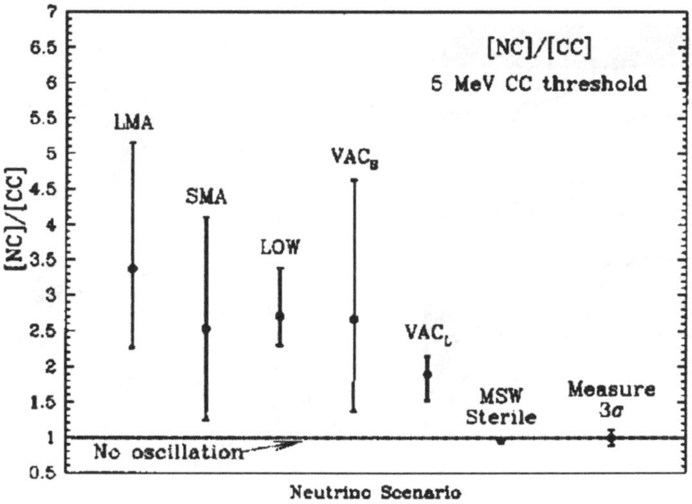

Fig. 1.73 Expected discriminative power of the *SNO* solar neutrino experiment by the neutral current to charged current double ratio for various neutrino mass scenarios according to [Bah2000]. Shown is the predicted neutral current (NC), to charged current (CC) double ratio for various neutrino mass schemes, which are globally consistent at the 99% C.L. with all of the available neutrino data.

be obtained in partially or completely degenerate schemes (Fig. 1.76), motivated by giving sizable contributions to the *hot* dark matter in the universe. *GENIUS* could help to determine the mixing in such schemes with extreme accuracy, providing informations being complementary to precision tests of cosmological parameters by the satellite experiments *MAP* and *PLANCK*. In four–neutrino scenarios *GENIUS* has good perspectives for testing the possible values of the effective Majorana mass in the range of the *LSND* result for the case of the *MSW LMA* solution of the solar neutrino problem (Fig. 1.77).

GENIUS might also marginally allow to check the prediction of shadow world models [Ber95* -I], which naturally lead to light sterile neutrinos and would predict an effective electron neutrino mass of *0.002 eV* [Moh97*-I], [Moh97a], [Ber99a**-I].

For further recent discussions of the potential of *GENIUS* for neutrino physics we refer to [Kla99*-VI], [Kla99a**-B], [Kla2000*-III], [Giu99*-III], [Cza99], [Vis99*-III], [Bil99], [Kal2000**-III].

The experiments would further allow a breakthrough into the *multi–TeV* range for many (other) beyond standard models (see [Kla97**-VI], [Kla97a*-VI], [Kla99*-VI], [Kla99a**-B], [Kla99e**-III]). The sensitivity for composite models of quarks and leptons would be as large as that of *LHC* [Pan99], [Pan2000*-IV] (see Fig. 1.78 (left)). The sensitivity of *GENIUS* for left–handed heavy or superheavy neutrinos corresponds to that of a future *5 to 8 TeV* Linear Collider [Bel98*-IV] (Fig. 1.78 (right)). The sensitivity for right–handed *W* bosons occurring in left–right

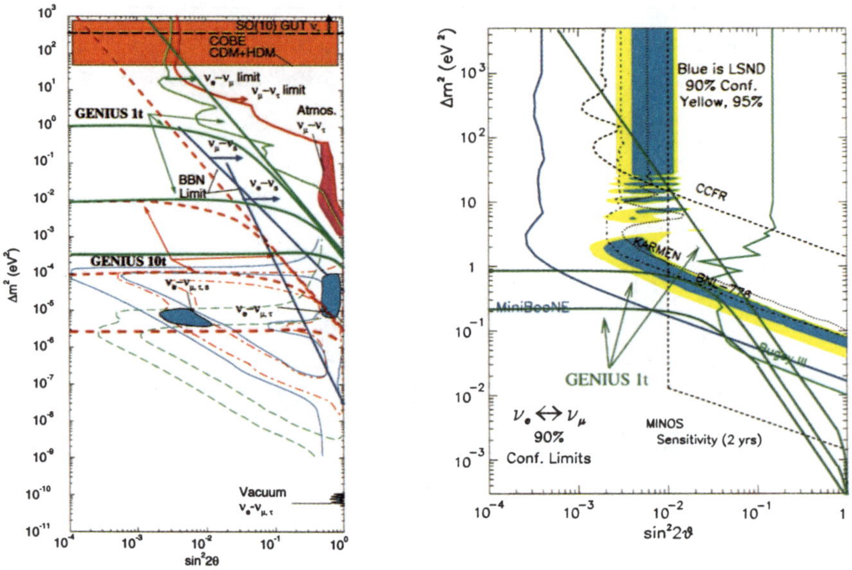

Fig. 1.74 Left: Summary of currently known constraints on neutrino oscillation parameters (for two generations of neutrinos) for the solar neutrino problem, the atmospheric neutrino problem and various reactor and accelerator limits on neutrino oscillations. The thick lines indicate the sensitivity of *GENIUS* (full lines *1 tonne*, broken lines *10 tonnes*) to neutrino oscillation parameters for three values of neutrino mass ratios $R = 0,\ 0.1$ and 0.01 (from top to bottom). For *GENIUS 10 tonnes* also the contour line for $R{=}0.5$ is shown. The region beyond the lines would be excluded (see [Kla97a*-VI], [Kla98a**-A], [Kla99*-VI], [Kla99e**-III]). Right: *LSND* compared to the sensitivity of *GENIUS 1 tonne* for three ratios R_{12}, from top to bottom $R_{12} = 0, 0.01, 0.02$. *GENIUS* should see a signal if *LSND* is right, in quasi-degenerate models (see [Kla97a*-VI], [Kla99*-B]).

symmetric models reaches up to $m_{W_R} \geq 5.3\ TeV$, or even $m_{W_R} \geq 18\ TeV$ for the *10 ton* version. This has to be compared to the final sensitivity of *LHC* of $m_{W_R} \geq 5 - 6\,TeV$.

Also in the determination of the R–parity violating Yukawa coupling λ'_{111} in the R–parity violating term of the superpotential in supersymmetric theories the sensitivity of *GENIUS* corresponds to that of *LHC* (see also [Wod99*-IV], [Wod99a*-IV]) (Fig. 1.79). The potential of *GENIUS* for R–parity breaking coupling products is discussed in [Bha99**-VI] and exceeds those from the *HEIDELBERG–MOSCOW* experiment by *1 to 2 orders* of magnitude. Concerning R–parity conserving *SUSY* models, *GENIUS* would allow to extract limits on the *'Majorana'* mass of sneutrinos lower by factors of *7* to *20* compared with present constraints [Hir97a**-IV], [Hir97b**-IV], [Hir98a*-IV]. For leptoquarks in the range of *200 GeV* the leptoquark–Higgs coupling could be probed down to a level of (a few) 10^{-8}. In

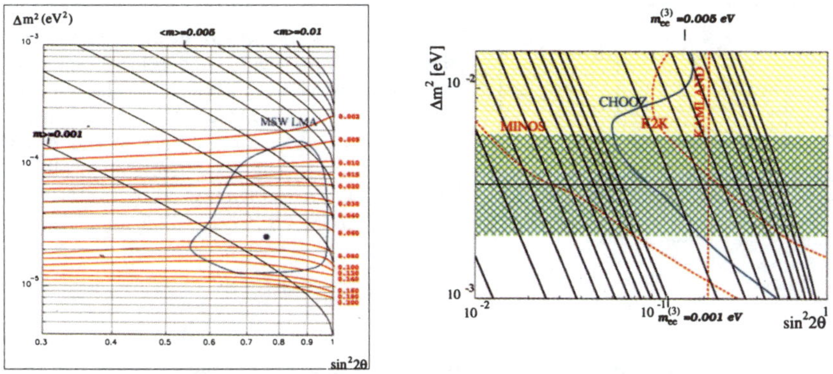

Fig. 1.75 The interplay of double beta decay search with neutrino oscillation experiments for hierarchical neutrino mass patterns: Left: *GENIUS 10 tonnes* can test the Large Mixing Angle (*LMA*) solution of the solar neutrino problem. *GENIUS* should see a double beta signal, and could together with a complementary day–night effect in present and future solar neutrino experiments, fix the mass scale of this hierarchical model. Right: *GENIUS* could together with the long baseline experiment *MINOS* test a hierarchical model assuming the Small Mixing Angle (*SMA*) *MSW* solution for the solar neutrino problem (see [Kla2000*-III].)

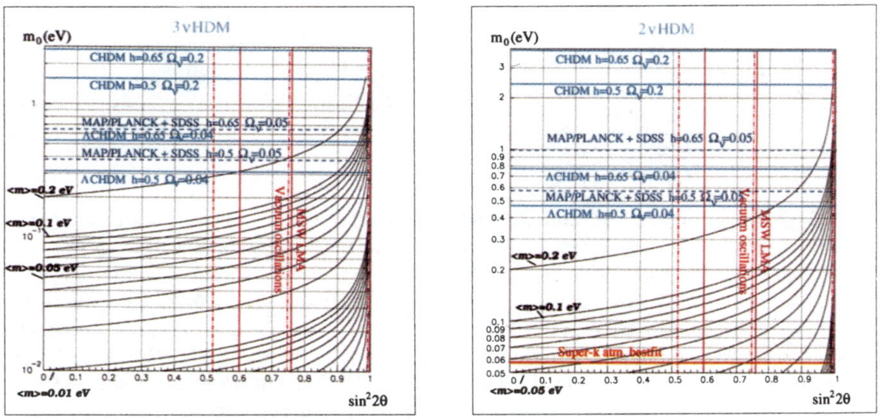

Fig. 1.76 The potential of double beta decay for degenerate neutrino mass scenarios (ranges above the effective mass contour lines will be excluded). Left: Almost totally degenerate scenario. For the *SMA MSW* solution of the solar neutrino problem already the *HEIDELBERG–MOSCOW* experiment excludes the whole range of sensitivity of the future satellite experiments *MAP* and *PLANCK* for cosmological models involving neutrinos as hot dark matter. Right: Partially degenerate scenario with two neutrinos contributing to hot dark matter (see [Kla2000*-III]).

other words, if leptoquarks interact with the standard model Higgs boson with a coupling of the order *O(1)*, either neutrinoless double beta decay should be found, or leptoquarks must be heavier than (several) *10 TeV*. The sharp limits set by double beta decay already now on violation of Lorentz invariance and equivalence principle in the neutrino sector would be improved by one to two orders of magnitude, and probe an otherwise totally unconstrainable region in the parameter space of mixing

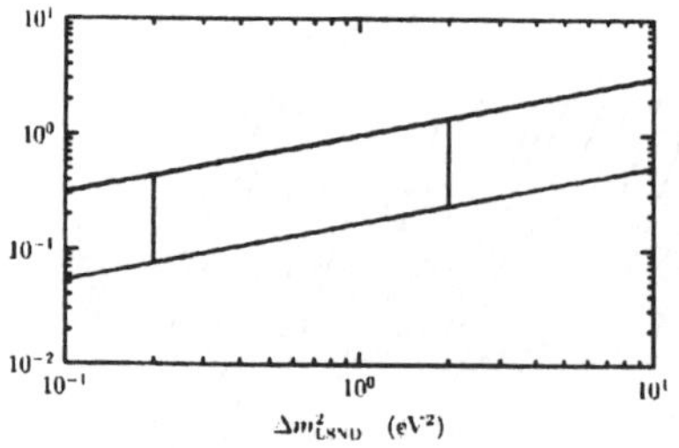

Fig. 1.77 ***Four neutrinos in the scheme with direct mass hierarchy:*** The central area between the two vertical lines shows the possible value of the effective Majorana mass $\langle m \rangle$ in the range of Δm^2_{LSND} for the case of the *MSW LMA* solution of the solar neutrino problem. This case can be easily checked by the *1 tonne* version of *GENIUS* (from [Bil99]).

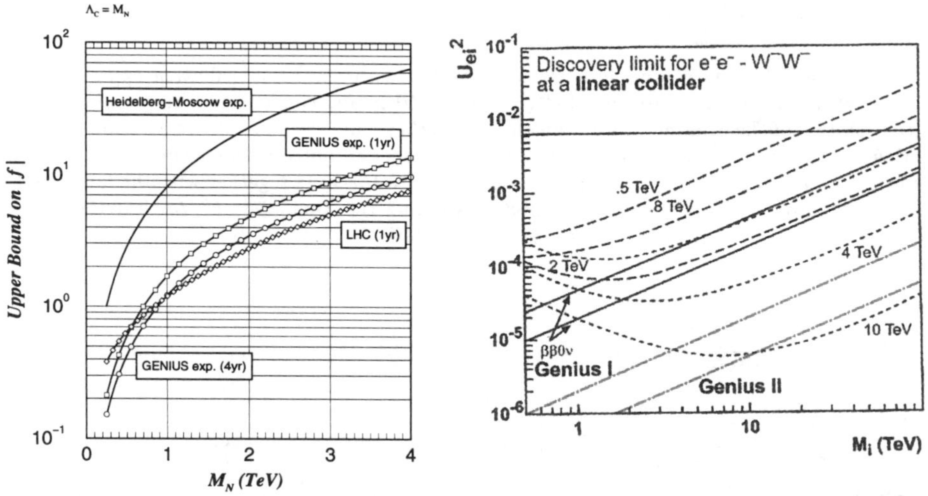

Fig. 1.78 Left: Sensitivity of *LHC* and *GENIUS* (*1 tonne*) to compositeness parameters (assuming $\Lambda_C = M_N$). Regions above the curves are excluded. The *LHC* bound is weaker than the *GENIUS* bound for $M_N < 550(1000)$ GeV. (from [Pan2000*-IV]). Right: Discovery limit for $e^-e^- \rightarrow W^-W^-$ at a linear collider as function of the mass M_i of a heavy left–handed neutrino, and of U^2_{ei}, for $\sqrt{s}$ between *500 GeV* and *10 TeV*. In all cases the parameter space above the line corresponds to observable events. Also shown are the limits set by the *HEIDELBERG–MOSCOW* $0\nu\beta\beta$ experiment as well as the prospective limits from *GENIUS*. The areas above the $0\nu\beta\beta$ contour lines are excluded. The horizontal line denotes the limit on neutrino mixing, U^2_{ei}, from *LEP*. Here the parameter space above the line is excluded. (from [Bel98*-IV]).

of weak and gravitational eigenstates [Kla99c*-IV]. The potential for probing new gravitational interactions and effects of quantum space–time foam in the neutrino sector have been discussed in [Kla2000a*-IV], [Kla2000b*-IV].

Finally the potential of *GENIUS* for solar neutrinos might be mentioned. A large target mass of at least one ton of natural or enriched Germanium in combination with the low background in the energy range below *100 keV* opens the possibility to measure the solar *pp* and ^7Be neutrino flux with a very low energy threshold *i n r e a l t i m e* [Bau99a]. Thus *GENIUS* could serve as the first real–time detector of solar *pp* neutrinos. The detection reaction would be the elastic

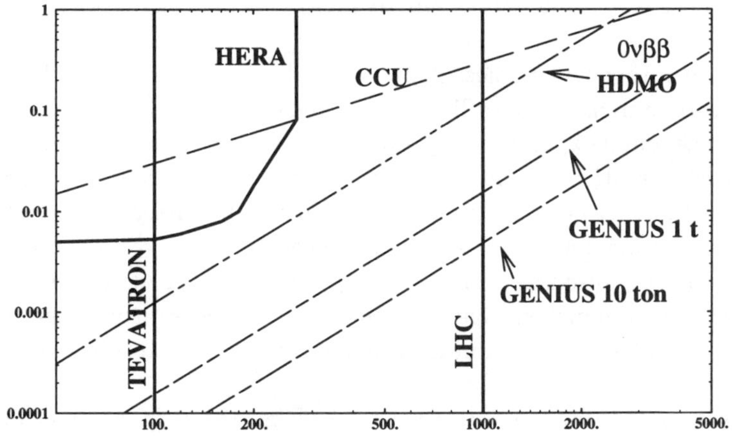

Fig. 1.79 Comparison of sensitivities of existing and future experiments on $R\!\!\!/_p$ *SUSY* models in the plane $\lambda'_{111} - m_{\tilde{q}}$. ***Note the double logarithmic scale!*** Shown are the areas currently excluded by the experiments at the *TEVATRON* and *HERA*, the limit from charged–current universality, denoted by *CCU*, and the limit from absence of $0\nu\beta\beta$ decay from the *HEIDELBERG–MOSCOW* collaboration ($0\nu\beta\beta$ *HDMO*). In addition, the estimated sensitivity of *LHC* is compared to the one expected for *GENIUS* in the *1 tonne* and the *10 tonne* version (from [Kla97a*-VI]). Excluded are the areas beyond and left from the contour lines.

scattering process $\nu + e^- \longrightarrow \nu + e^-$, the threshold is *11 keV*, the signal rates would be (in the standard solar model) *2 events per day (40 SNU) for 1 ton*, and about *20 events* per day for *10 tons*. These rates are larger than those of previous and present (non–real time) detectors such as *GALLEX, SAGE or GNO*. In the case of a day–night variation of the solar neutrino flux, *GENIUS* would be particularly sensitive to the socalled *LOW MSW* solution of the solar neutrino problem, which may be of particular interest after the recent Superkamiokande results [Suz2000]. It would also allow for the first time to measure the *1.3 keV* predicted shift of the average energy of the beryllium neutrino line. This shift is a direct measure of the central temperature of the Sun.

1.7 Conclusion

Concluding, this book describes the history and the broad present and future potential of double beta decay for particle physics and cosmology, as a probe yielding important information on beyond standard model physics complementary to future colliders such as *LHC* and *NLC*, and to satellite experiments such as *MAP* and *PLANCK*.

We have seen that neutrinoless double beta decay, which has not been observed up to now, sets extremely sharp limits for a wealth of beyond standard model parameters. Once found experimentally, double beta decay would have proven lepton–number nonconservation, a nonvanishing neutrino mass and the neutrino to be a Majorana particle. No other experiment could do the latter. It would, however, require further experiments to disentangle the contributions to the process from the various mechanisms discussed. The information on the neutrino mass from double beta decay is an indispensable requirement, complementary to the information obtainable by neutrino oscillation experiments, to fix the absolute neutrino mass scale.

The question put from time to time (so in [Cal91**-IV]), whether double beta decay has a future, has to be answered nowadays clearly positively. The ultimate sensitivity of the method is probably a limit for the effective Majorana neutrino mass of some 10^{-3} eV, two orders of magnitude more than the best present limit.

The cost of such an experiment (*GENIUS* with one ton of ^{76}Ge) would be of the order of other large scale non-accelerator experiments, such as Superkamiokande, or of the order of the neutrino beam from CERN to Gran Sasso.

The scientific community is fully aware of the great impact of such a huge step for modern neutrino physics and more general modern particle physics, in particular at a time when elementary particle physics is imposing very extreme requirements on new generations of accelerators, with the natural consequence of increasing importance of non–accelerator experiments (see, e.g. [Kla95/98**-II], [Kla97/2000**-IV], [Kla98f**-VI], [Kla99f**-VI], [Her99**-VI], [Kla2000d**-VI]). When this crucial step forward in particle physics will be done, will depend as usual, on whether we will have or not have the competent, far–seeing and courageous leaders and personalities in the large Scientific Societies and in the decision making committees, which can speed up the progress of science or can slow it down for decades and more.

H.V. Klapdor-Kleingrothaus
Max-Planck Institut für Kernphysik
Heidelberg, Germany
5 July, 2000

Bibliography

[Aal96] [Aal96] C. E. Aalseth et al. (IGEX Collaboration) *"Recent Results from the IGEX Double-Beta Decay Experiment"*, in Proc. of 5th International Workshop on Topics in Astroparticle and Underground Physics (TAUP 95), Toledo, Spain, 17 - 21 September 1995, eds. *A. Morales J. Morales and J. A. Villar*, *Nucl. Phys. Proc. Suppl.* **48** (1996) 223 - 225

[Aal99] [Aal99] C. E. Aalseth, F. T. Avignone III, R. L. Brodzinski, J. I. Collar D. Gonzáles, E. Garcia, F. Hasenbalg, W. K. Hensley, I. V. Kirpichnikov, A. A. Klimenko, H. S. Miley, A. Morales, J. Morales, A. Ortiz de Solorzano, S. B. Osetrov, V. S. Pogosov J. Puimedon, J. H. Reeves, A. Salinas, M. L. Sarsa, A. A. Smolnikov, A. S. Starostin, A. G. Tamanyan, A. A. Vasenko, S. I. Vasilev and J. A. Villar *"The Status of the IGEX ^{76}Ge Double-Beta Decay Experiment in 1997"*, in Proc. of 5th International Workshop on Topics in Astroparticle and Underground Physics (TAUP 97), Gran Sasso, Italy, 7 - 11 September 1997, eds. *A. Bottino, A. di Credico and P. Monacelli*, *Nucl. Phys. Proc. Suppl.* **70** (1999) 236 - 238

[Aal99a] C. E. Aalseth (IGEX Collaboration) *"Neutrinoless Double-β Decay of ^{76}Ge: First Results from the International Germanium Experiment (IGEX) with Six Isotopically Enriched Detectors"*, *Phys. Rev.* **C 59** (1999) 2108 - 2113

[Add98] Addendum to Protokoll about HEIDELBERG-MOSCOW Collaboration (1998)

[Adh98] R. Adhikari and G. Rajasekaran *"Constraints on Mixing Angles of Majorana Neutrinos"*, *Phys. Rev.* **D 61** (2000) 031301-1–031301-5 and *Preprint hep-ph/* **9812361** (1998)

[Ach95] Y. Achiman and T. Greiner *"Predicting the Neutrino Spectrum in SUSY SO(10)"*, *Nucl. Phys.* **B 443** (1995) 3 - 19 and *Preprint hep-ph/* **9502304** (1995)

[Akh97] E. Kh. Akhmedov *"The Neutrino Magnetic Moment and Time Variations of the Solar Neutrino Flux"*, Proc. 4th International Solar Neutrino Conference, Heidelberg, Germany, 8–11 Apr., 1997 388 - 405 and *Preprint hep-ph/* **9705451** (1997)

[Ale94] A. Alessandrello, C. Brofferio, D. V. Camin, O. Cremonesi, E. Fiorini, E. Garcia, A. Giuliani, P. de Marcillac, A. Nucciotti, M. Pavan, G. Pessina, E. Previtali and L. Zanotti *"A New Search for Neutrinoless $\beta\beta$ Decay with a Thermal Detector"*, *Phys. Lett.* **B 335** (1994) 519 - 525

[Ale95] A. Alessandrello, C. Brofferio, O. Cremonesi, E. Garcia, A. Giuliani, A. Nucciotti, S. Parmeggiano, M. Pavan, G. Pessina E. Previtali and P. de Marcillac *"First Tests on a Large Mass, Low Temperature Array Detector"*, *Nucl. Instrum. & Meth. Phys. Res.* **A 360** (1995) 363 - 366

[Ale98] A. Alessandrello, C. Brofferio, D. V. Camin, O. Cremonesi, E. Fiorini, A. Giuliani, A. Nucciotti, M. Pavan, G. Pessina, S. Pirro, E. Previtali, L. Zanotti and C. Bucci *"Preliminary Resultats on Double Beta Decay of* ^{130}Te *With an Array of Twenty Cryogenic Detectors"*, *Phys. Lett.* **B 433** (1998) 156 – 162

[Ale99] A. Alessandrello, C. Brofferio, C. Bucci, D. V. Camin, P. Caspani, P. Colling, O. Cremonesi, E. Fiorini, A. Giuliani, A. Nucciotti, M. Pavan, G. Pessina, S. Pirro, E. Previtali and L. Zanotti *"Measure of Low Radioactive Contamination and Nuclear Recoil Quenching Factor Using Cryogenic Particle Detectors"*, in Proc. of 5th International Workshop on Topics in Astroparticle and Underground Physics (TAUP 97), Gran Sasso, Italy, 7 – 11 Sep. 1997, eds. *A. Bottino, A. Di Credico and P. Monacelli*, *Nucl. Phys. Proc. Suppl.* **70** (1999) 96 – 100

[Ale99a] A. Alessandrello, C. Brofferio, D. V. Camin, O. Cremonesi, E. Fiorini, A. Giuliani, A. Nucciotti, M. Pavan, G. Pessina, S. Pirro, E. Previtali, M. Vanzini and L. Zanotti *"Preliminary results of the 20* TeO_2 *Crystals Detector Arrary, the first step toward the CUORE Experiment"*, in Proc. of 2nd International Conference on Dark Matter in Astrophysics and Particle Physics (DARK98), Heidelberg, Germany, 20 – 25 July, 1998, ed by H. V. Klapdor–Kleingrothaus and L. Baudis, *IOP* (1999) 785 – 789

[Als88] M. Alston-Garnjost, B. Dougherty, R. W. Kenney, J. Krivicich, R. Tripp, H. Nicholson, S. Sutton, B. Dieterle, J. Kang and C. Leavitt *"Limit on Majoron Emission in Beta Beta Decay of* ^{100}Mo*"*, *Phys. Rev. Lett.* **60** (1988) 1928 – 1931

[Als89] M. Alston–Garnjost, B. L. Dougherty, R. W. Kenney, R. D. Tripp, J. M. Krivicich, H. W. Nicholson, C. S. Sutton, B. D. Dieterle, S. D. Foltz, J. Kang, C. P. Leavitt, R. A. Reeder, J. D. Baker and A. J. Caffrey *"Search for Neutrinoless Double-*β *Decay of* ^{100}Mo*"*, *Phys. Rev. Lett.* **63** (1989) 1671 – 1673

[Als93] M. Alston–Garnjost, B. L. Dougherty, R. W. Kenney, R. D. Tripp, J. M. Krivicich, H. W. Nicholson, C. S. Sutton, B. D. Dieterle, S. D. Foltz, C. P. Leavitt, R. A. Reeder, J. D. Baker and A. J. Caffrey *"Experimental Search for Neutrinoless Double-*β *Decay of* ^{100}Mo*"*, *Phys. Rev. Lett.* **71** (1993) 831 – 834

[Alt97] G. Altarelli, J. Ellis, G. F. Giudice, S. Lola and M. L. Managano *"Pursuing Interpretations of the HERA Large* Q^2 *Data"*, *Preprint hep–ph/* **9703276** (1997) *CERN–TH/* **97–040** and *Nucl. Phys.* **B 506** (1997) 3 – 28

[Ans92] P. Anselmann et al. (GALLEX Collaboration) *"Implications of the GALLEX Determination of the Solar Neutrino Flux"*, *Phys. Lett.* **B 285** (1992) 390 – 397 *"Solar Neutrinos Observed by GALLEX at Gran Sasso"*, *Phys. Lett.* **B 285** (1992) 376 – 389

[Ant87] I. Antoniadis, J. Ellis, J. S. Hagelin and D. V. Nanopoulos *"Supersymmetric Flipped SU(5) Revitalized"*, *Phys. Lett.* **B 194** (1987) 231 – 235

[Ant88] I. Antoniadis, J. Ellis, J. S. Hagelin and D. V. Nanopoulos *"An Improved SU(5)*$\times$ *U(1) Model from Four–Dimensional String"*, *Phys. Lett.* **B 208** (1988) 209 – 215

[Ant92] I. Antoniadis, J. Rizos and K. Tamvakis *"Naturally Light Neutrinos in the Flipped SU(5)*$\times$ *U(1) Superstring Model"*, *Phys. Lett.* **B 279** (1992) 281 – 284

[Ant60] G. T. Dell'Antonio and E. Fiorini *"Experimental and Theoretical Remarks on the Double* β*–Decay"*, *Nuovo. Cim. Suppl.* **17** (1960) 132 – 193

[Apr92] E. Aprile, D. Chen, M. Moulson, R. Mukherjee and M. Suzuki *"Test of a Two–Dimensional Liquid Xenon TPC"*, *Nucl. Instrum. Meth.* **A 316** (1992) 29 – 37

[Arn87] W. D. Arnett and J. L. Rosner *"Neutrino Mass Limits from SN1987A"*, *Phys. Rev. Lett.* **58** (1987) 1906 – 1909

[Arn95] R. Arnold, A. Barabash, D. Blum, V. Brudanin, J. E. Campagne, F. Danevich, D. Dassié, V. Egorov, R. Eschbach, J. L. Guyonnet, F. Hubert, Ph. Hubert, M. C. Isaac, C. Izac, S. Jullian, O. Kochetov, V. N. Kornoukhov, V. Kouts, V. Kovalenko,

D. Lalanne, T. Lamhamdi, F. Laplanche, F. Leccia, Yu. Lepikhin, I. Linck, C. Longuemare, F. Mauger, P. Mennrath, F. Natchez, H. W. Nicholson, A. Nozdrin, G. Pichenot, F. Piquemal, F. Scheibling, C. S. Sutton, G. Szklarz, V. I. Tretyak, V. Umatov, I. Vanyushin, Yu. Vassilyev, A. Vareille, Ts. Vylov, Sh. Zaparov and Yu. Zdesenko *"Performance of a Prototype Tracking Detector for Double Beta Decay Measurements"*, *Nucl. Instrum. & Meth. Phys. Res.* **A 354** (1995) 338 - 351

[Arn96] R. Arnold et. al. (NEMO Collaboration) *"Double-β Decay of ^{116}Cd"*, *Z. Phys.* **C 72** (1996) 239 - 247

[Arn96] R. Arnold et. al. (NEMO Collaboration) *"Limits on Different Majoron Decay Modes of ^{100}Mo, ^{116}Cd, ^{82}Se and ^{96}Zr for Neutrinoless Double Beta Decays in the NEMO-2 experiment"*, *Preprint-ITEP, Moscow* **15 – 00** (2000) 1 -14

[Art89] V. A. Artemev, E. V. Brakchman, M. A. Ivanovsky, A. K. Karelin, V. V. Kirichenko, V. M. Knyazev, O. M. Kozodaeva, V. A. Lubimov, A. I. Mitin, V. P. Nikolaev, T. N. Tsvetkova and O. Ya. Zeldovich *"The ^{136}Xe Double-Beta-Decay Track Experiment"*, *Preprint* **ITEP– 89–186** (1989) 34pp and *Nucl. Instrum. & Meth. Phys. Res.* **A 303** (1991) 309 - 322

[Art91] V. A. Artemev, E. V. Brakchman, M. A. Ivanovsky, A. K. Karelin, V. V. Kirichenko, V. M. Knyazev, O. M. Kozodaeva, V. A. Lubimov, A. I. Mitin, V. P. Nikolaev, T. N. Tsvetkova and O. Ya. Zeldovich *"The Search for ^{136}Xe $2\beta 2\nu$-Decay at the Proportional-Drift Detector"*, *Preprint* **ITEP–91–34** (1991) 16pp. and *Phys. Lett.* **B 280** (1992) 159 - 162 and *Yad. Fiz.* **54** (1991) 1485- 1486

[Art92] V. Artemev, E. Brakhman, M. Ivanovsky, A. Karelin, V. Kirichenko, V. Knyazev, O. Kozodaeva, V. Lubimov, A. Mitin, V. Nikolaev, T. Tsvetkova and O. Zeldovich *"^{136}Xe Double Beta Decay from ITEP"*, in Proc. of 3rd International Symposium on Weak and Electromagnetic Interactions in Nuclei (WEIN 92), Dubna, USSR, 16 - 22 June, 1992, ed *Ts. D. Vylov*, *World Scientific* (1993) 570 - 574

[Art95] V. A. Artemev, E. V. Brakhman, A. K. Karelin, V. V. Kirichenko, A. A. Klimenko, O. M. Kozodaeva, V. A. Lyubimov, A. I. Mitin, S. B. Osetrov, V. V. Paramokhin, A. A. Pomansky, A. A. Smolnikov, T. N. Tsvetkova, S. I. Vasiliev and O. Ya. Zeldovich *"Half-Life measurement for ^{150}Nd $2\beta 2\nu$ decay in the Time Projection Chamber experiment"*, *Phys. Lett.* **B 345** (1995) 564 - 568

[Art96] V. A. Artemev, E. V. Brakhman, A. K. Karelin, V. V. Kirichenko, A. A. Klimenko, O. M. Kozodaeva, V. A. Lyubimov, A. I. Mitin, V. P. Nikolaev, S. B. Osetrov, V. V. Paramokhin, A. A. Pomansky, A. A. Smolnikov, T. N. Tsvetkova, S. I. Vasiliev and O. Ya. Zeldovich *"Measurement of the ^{150}Nd Half-Life $T_{1/2}(2\beta 2\nu)$ with a Time Projection Chamber in a Magnetic Field"*, *Phys. of Atom. Nucl.* **59** *No* **1** (1996) 6 - 13 and *Yad. Fiz.* **59** *No* **1** (1996) 10 - 17

[Asa96] G. M. Asatrian and A. N. Ioannisian *"The Neutrino Masses in SUSY GUT"*, *Phys. Atom. Nucl.* **59** (1996) 1768 - 1773 *Preprint hep-ph/* **9508273** (1995)

[Ash99] V. D. Ashitkov, A. S. Barabash, S. G. Belogurov, G. Carugno, S. I. Konovalov, F. Massera, I. O. Pilugin, G. Puglierin, R. R. Saakian, V. N. Stekhanov, V. I. Umatov and I. A. Vanyushin *"Investigation of Double-Beta Decay of ^{100}Mo with the Liquid-Argon Ionization Chamber"*, *Phys. of Atom. Nucl.* **62** *No* **12** (1999) 2044 - 2047 and *Yad. Fiz.* **62** *No* **12** (1999) 2217 - 2220

[Aul83] C. S. Aulakh and R. N. Mohapatra *"The Neutrino as the Supersymmetric Partner of the Majoron"*, *Phys. Lett.* **B 119** (1983) 136 - 140

[Avi87] F. T. Avignone III, H. S. Miley, R. L. Brodzinski and J. H. Reeves *"Experimental Evidence Suggestive of Neutrinoless Beta Beta Decay With a Near Massless Particle"*, Talk in American Physical Society, Salt Lake City, January (1987), AIP Conference Procedings *AIP, New York* 359 - 369

[Avi91] F. T. Avignone III, R. L. Brodzinski, C. K. Guerard, I. V. Kirpichnikov, H. S. Miley, V. S. Pogosov, J. H. Reeves, A. S. Starostin and A. G. Tamanyan *"Confirmation of the Observation of $2\nu\beta\beta$ decay of ^{76}Ge"*, *Phys. Lett.* **B 256** (1991) 559 – 561

[Avi93] F. T. Avignone III, et al. (IGEX Collaboration) *"The International Germanium Experiment (IGEX) in 1993"*, in Proc. of 3th International Workshop on Teoretical and Phenomenological Aspects of Underground Physics (TAUP 97), Gran Sasso, Italy, 19 – 23 September 1993, eds. *C. Arpesella E. Bellotti and A. Bottino, Nucl. Phys. Proc. Suppl.* **35** (1994) 354 – 357

[Bab95] K. S. Babu and R. N. Mohapatra *"New Vector–Scalar Contributions to Neutrinoless Double Beta Decay and Constraints on R–Parity Violation"*, *Phys. Rev. Lett.* **75** (1995) 2276 – 2279 and *Preprint hep-ph/* **9506354** (1995)

[Bab97] K. S. Babu, Ch. Kolda, J. March–Russell and F. Wilczek *"Comments on the high Q^2 HERA anomaly"*, *Phys. Lett.* **B 402** (1997) 367 – 373 and *Preprint hep-ph/* **9703299** (1997)

[Bah98a] J. N. Bahcall, P. I. Krastev and A. Yu. Smirnov *"Where do we Stand with Solar Neutrino Oscillations?"*, *Phys. Rev.* **D 58** (1998) 096016-1–096016-22 and *Preprint hep-ph/* **9807216** (1998)

[Bah2000] J. N. Bahcall, P. I. Krastev and A. Yu. Smirnov *"SNO: Predictions for Ten Measurable Quantities"*, *Preprint hep-ph/* **0002293** (2000)

[Bal93] A. Balysh, M. Beck, S. Belyaev, F. Bensch, J. Bockholt, A. Demehin, A. Gurov, G. Heusser, H. V. Klapdor–Kleingrothaus, I. Kondratenko, V. Lebedev, B. Maier, A. Müller, F. Petry, A. Piepke, H. Strecker and K. Zuber *"New Experimental Limits for Electron Decay and Charge Conservation"*, *Phys. Lett.* **B 298** (1993) 278 – 282

[Bal94] A. Balysh, M. Beck, S. Belyaev, F. Bensch, J. Bockholt, A. Demehin, A. Gurov, G. Heusser, H. V. Klapdor–Kleingrothaus, I. Kondratenko, D. Kotel'nikov, V. I. Lebedev, B. Maier, A. Müller, F. Petry, A. Piepke, A. Pronsky, H. Strecker, M. Völlinger and K. Zuber *"Measurement of the $\beta\beta2\nu$ Decay of ^{76}Ge"*, *Phys. Lett.* **B 322** (1994) 176 – 181

[Bal95] A. Balysh, M. Beck, S. T. Belyaev, J. Bockhold, A. Demehin, A. Gurov, J. Hellmig, G. Heusser, M. Hirsch, Ch. Hoffmann, H. V. Klapdor–Kleingrothaus, I. Kondratenko, D. Kotel'nikov, V. I. Lebedev, B. Maier, A. Müller, H. Päs, F. Petry, E. Scheer, H. Strecker and M. Völlinger *"Sub–eV Limit for the Neutrino Mass From ^{76}Ge Double Beta Decay by the HEIDELBERG–MOSCOW Experiment"*, *Phys. Lett.* **B 356** (1995) 450 – 455

[Bam94] P. Bamert, C. P. Burgess and R. N. Mohapatra, *"Heavy Sterile Neutrinos and Neutrinoless Double Beta Decay"*, *Preprint hep-ph/* **9408367** (1994) and *Nucl. Phys.* **B 438** (1995) 3 – 16

[Bam94a] P. Bamert, C. P. Burgess and R. N. Mohapatra *"Multi–Majoron Modes for Neutrinoless Double Beta Decay"*, *Preprint hep-ph/* **9412365** (1994) and *Nucl. Phys.* **B 449** (1995) 25 – 48

[Bar86] I. R. Barabanov, V. N. Gavrin, S. V. Girin, V. N. Kornoukhov and A. M. Pshukov *"Double Beta Decay of ^{136}Xe"*, *JETP Lett.* **43** (1986) 210 – 211

[Bar89] A. S. Barabash V. M. Lobashev, V. V. Kuzminov, V. M. Novikov, B. M. Ovchinnikov and A. A. Pomansky *"Results of the Experiment on the Search for Double Beta Decay of ^{136}Xe, ^{134}Xe and ^{124}Xe"*, *Phys. Lett.* **B 223** (1989) 273 – 276

[Bar97] A. S. Barabash for the NEMO Collaboration *"NEMO Collaboration: Latest Results and Perspectives for the Future"*, in Proc. 17th International Conference on Neutrino Physics and Astrophysics, Helsinki, Finland, June 13 – 19, 1996, eds. *K. Enqvist, K. Huitu and J. Maalampi, World Scientific, Singapore,* (1997) 374 – 380

[Bar98] A. S. Barabash *"Does the Weak Interaction Constant Depend on Time?"*, *JETP Letters* **68** (1998) 1 – 6 and *Pisma Zh. Eksp. Teor. Fiz.* **68** (1998) 3 – 8

[Bar99] A. S. Barabash *"Is the Weak Interaction Constant Really Constant?"*, *Preprint ITEP-Moscow* **39-99** (1999) 0 – 11 and *Eur. Phys. J.* (2000)

[Bar81] R. Barbieri, R. N. Mohapatra and A. Masiero *"Compositeness and a Left–Right Symmetric Electroweak Model without Broken Gauge Interactions"*, *Phys. Lett.* **B 105** (1981) 369 – 374

[Bar99a] V. Barger and K. Whisnant *"Majorana Neutrino Masses from Neutrinoless Double Beta Decay and Cosmology"*, *Phys. Lett.* **B 456** (1999) 194 – 200

[Bar94] S. D. Barthelmy, L. M. Bartlett, N. Gehrels, M. Leventhal, B. J. Teegarden, J. Tueller, S. Belyaev, V. Lebedev and H. V. Klapdor–Kleingrothaus *"GRIS Background Reduction Results Using Isotopically Enriched Ge"*, *Astr. Journ.* **427** (1994) 519 – 521

[Bau99] L. Baudis, G. Heusser, B. Majorovits, Y. Ramachers, H. Strecker and H. V. Klapdor–Kleingrothaus *"Background and Technical Studies for GENIUS as a Dark Matter Experiment"*, *Nucl. Instrum. Meth.* **A 426** (1999) 425 – 435 and *Preprint hep–ex/* **9811040** (1998)

[Bau99a] L. Baudis and H. V. Klapdor–Kleingrothaus *"Detection of Low-Energy Solar Neutrinos with HP Germanium"*, *Eur. Phys. J.* **A 5** (1999) 441 – 443

[Bec93] M. Beck, F. Bensch, J. Bockholt, G. Heusser, M. Hirsch, H. V. Klapdor–Kleingrothaus, B. Maier, F. Petry, A. Piepke, H. Strecker, M. Völlinger, K. Zuber, A. Balysh, S. T. Belyaev, A. Demekhin, A. Gurov, I. Kondratenko, V. I. Lebedev and A. Mül ler *"Investigation of the Majoron Accompanied Double Beta Decay Mode of ^{76}Ge"*, *Phys. Rev. Lett.* **70** (1993) 2853 – 2855

[Bed96] V. A. Bednyakov, H. V. Klapdor–Kleingrothaus and S. G. Kovalenko *"Perspectives of Research of Cosmical Dark Matter in Underground Experiments"*, *Yad. Fiz.* **59, No 10** (1996) 1777 – 1787

[Bed97a] V. A. Bednyakov, H. V. Klapdor–Kleingrothaus, S. G. Kovalenko and Y. Ramachers *"Is SUSY Accessible by Direct Dark Matter Detection?"*, *Eur. Phys. J.* **A 357** (1997) 339 – 347

[Bed97c] V. A. Bednyakov, V. B. Brudanin, S. G. Kovalenko and Ts. D. Vylov *"On Prospects for Exploration of Supersymmetry in Double Beta Decay Experiments"*, *Mod. Phys. Lett.* **A 12** (1997) 233 – 241 and *Preprint hep-ph/* **9812528** (1998)

[Bed98] V. A. Bednyakov and H. V. Klapdor–Kleingrothaus *"Direct Dark Matter Detection in the Next–to Minimal Supersymmetric Standard Model"*, *Phys. Rev.* **D 59** (1998) 023514-1 – 023514-8

[Bed99a] V. A. Bednyakov and H. V. Klapdor–Kleingrothaus *"About Possibility of Direct Dark Matter Detection in Next–to–Minimal Supersymmetric Standard Model"*, *Yad. Fiz.* **62, No 6** (1999) 1 – 9

[Bed99b] V. A. Bednyakov and H. V. Klapdor–Kleingrothaus *"SUSY Spectrum Constraints on Direct Dark Matter Detection"*, *Preprint hep-ph/* **9908427** (1999) 1 - 12

[Bel96] G. Bélanger, F. Boudjema, D. London and H. Nadeau *"Inverse Neutrinoless Double β Decay Reexamined"*, *Phys. Rev.* **D 53** (1996) 6292 – 6301 and *Preprint hep-ph/* **9508317** (1996)

[Bel98] G. Bélanger *"Inverse Neutrinoless Double Beta Decay and $\Delta L = 2$ Processes at Linear Colliders"*, in Proc. "Lepton and Baryon Number Violation in Particle Physics, Astrophysics and Cosmology", eds. H.V. Klapdor–Kleingrothaus and I.V. Krivosheina, *International Workshop at ECT,* Trento, Italy, April 20 – April 25, 1998, *IOP, Bristol* (1998) 605 – 620

[Bel99] P. Belli, R. Bernabei, F. Montecchia, C. J. Dai, L. K. Ding, H. H. Kuang, J. M. Ma, W. Di Nicolantonio, G. Ignesti, A. Incicchitti, D. Prosperi, M. Angelone, P. Batistoni and M. Pillon *"The DAMA Experiments: Status Report"*, in Proc. of 2nd International Conference on Dark Matter in Astrophysics and Particle Physics (DARK98), Heidelberg, Germany, 20 – 25 July, 1998, ed by H. V. Klapdor–Kleingrothaus and L. Baudis, *IOP* (1999) 711 – 718

[Bel83] E. Bellotti, E. Fiorini, C. Liguori, A. Pullia, A. Sarracino and L. Zanotti *"Recent Results Obtained in the Mt. Blanc Experiment on Beta Beta Decay of ^{76}Ge"*, in Proc. "Science Underground", Los Alamos (1982), *AIP*, eds. *M. M. Nieto et.al.*, (1983) 430 – 433

[Bel83a] E. Bellotti, E. Fiorini, C. Liguori, A. Pullia, A. Sarracino and L. Zanotti *"A New Experimental Limit on Lepton Non–conservation in Double Beta Decay"*, *Phys. Lett.* **B 121** (1983) 72 – 76

[Bel91] E. Bellotti, O. Cremonesi, E. Fiorini, G. Gervasio, S. Ragazzi, L. Rossi, P. P. Sverzellati, J. Szarka, T. Tabarelli de Fatis and L. Zanotti *"A Search for Two Neutrino and Neutrinoless Double Beta Decay of ^{136}Xe in the Gran Sasso Underground Laboratory"*, *Phys. Lett.* **B 266** (1991) 193 – 200

[Bet34] H. Bethe and R. Peierls *"The "Neutrino""*, *Nature* **133** (1934) 532 and 689 – 690

[Ber95] Z. G. Berezhiani and R. N. Mohapatra *"Reconciling Present Neutrino Puzzles: Sterile Neutrinos as Mirror Neutrinos"*, *Phys. Rev.* **D 52** (1995) 6607 – 6611 and *Preprint hep-ph/* **9505385** (1995)

[Ber99a] Z. Berezhiani *"Exotic Mechanisms for Neutrino Masses"*, in Proc. "Lepton and Baryon Number Violation in Particle Physics, Astrophysics and Cosmology", eds. *H. V. Klapdor–Kleingrothaus and I. V. Krivosheina, International Workshop at ECT,* Trento, Italy, April 20 – April 25, 1998, *World Scientific* (1999) 147 – 172

[Ber2000] S. Bergmann, H. V. Klapdor–Kleingrothaus and H. Päs *"Lepton Number Violation Interactions and their Effects on Neutrino Oscillation Experiments"*, *Preprint, hep-ph/* **0004048** (2000) 28pp.

[Ber87] J. Bernabeu, A. Santamaria, J. Vidal, A. Mendez and J. W. F. Valle *"Lepton Flavor Non–conservation at High Energies in a Superstring Inspired Standard Model"*, *Phys. Lett.* **B 187** (1987) 303 – 308

[Ber97] J. Bernabei, P. Belli, F. Montecchia, W. Di Nicolantonio, A. Incicchitti, D. Prosperi, C. Bacci, C. J. Dai, L. K. Ding, H. H. Kuang and J. M. Ma *"WIMPs Search by Scintillators: Possible Strategy for Annual Modulation Search with Large Mass Highly Radiopure NaI(Tl)"*, in Proc. of 5th International Workshop on Topics in Astroparticle and Underground Physics (TAUP 97), Gran Sasso, Italy, 7 – 11 Sep. 1997, eds. *A. Bottino, A. Di Credico and P. Monacelli, Nucl. Phys. Proc. Suppl.* **70** (1999) 79 – 84

[Ber99] J. Bernabei (DAMA Collaboration) *"On a Further Search for a Yearly Modulation of the Rate in Particle Dark Matter Direct Search"*, *Phys. Lett.* **B 450** (1999) 448 – 455

[Ber92] T. Bernatowicz, J. Brannon, R. Brazzle, R. Cowsik, C. Hohenberg and F. Podosek *"Neutrino Mass Limits from a Precise Determination of $\beta\beta$ – Decay Rates of ^{128}Te and ^{130}Te"*, *Phys. Rev. Lett.* **69** (1992) 2341 – 2344

[Ber93] T. Bernatowicz, J. Brannon, R. Brazzle, R. Cowsik, C. Hohenberg and F. Podosek *"Precise Determination of Relative and Absolute $\beta\beta$– Decay Rates of ^{128}Te and ^{130}Te"*, *Phys. Rev.* **C 47** (1993) 806 – 825

[Bha99] G. Bhattacharyya, H. V. Klapdor-Kleingrothaus and H. Päs *"Neutrino Mass and Magnetic Moment in Supersymmetry without R–Parity in the Light of Recent Data"*, *Phys. Lett.* **B 463** (1999) 77 – 82 and *Preprint hep-ph/* **9907432** (1999)

[Bil99] S. M. Bilenky, C. Giunti, W. Grimus, B. Kayser and S. T. Petcov *"Constraints from Neutrino Oscillation Experiments on the Effective Majorana Mass in Neutrinoless Double β–Decay"*, *Phys. Lett.* **B 465** (1999) 193 – 202 and *Preprint hep-ph/* **9907234** (1999)

[Bla99] T. Blazek, S. Raby and K. Tobe *"Neutrino Oscillations in a predictive SUSY GUT"*, *Phys. Rev.* **D 60** (1999) 113001-1 – 9 and *Preprint hep-ph/* **9903340** (1999)

[Boe87] F. Boehm and P. Vogel "Physics and Massive Neutrinos", *Cambridge: Combridge University Press*, 211pp.

[Bor87] S. D. Boris, A. I. Golutvin, L. P. Laptin, V. A. Lyubimov, V. V. Nagovitsyn, E. G. Novikov, V. Z. Nozik, V. A. Soloshchenko, I. N. Tikhomirov, E. F. Tretyakov and N. F. Myasoedov *"The Neutrino Mass Determination from the Beta Spectrum of Tritium in Valine (ITEP-84)"*, *Phys. Rev. Lett.* **58** (1987) 2019 – 2022

[Bor91] BOREXINO Collaboration *"BOREXINO at Gran Sasso"*, eds. *G. Bellini, M. Campanella, D. Giugni and R. Raghavan*, Proposal **Vol. 1** (August, 1991) p. 97

[Bre2000] M. Breidenbach et all. *"Proposal: EXO – an advanced Enriched Xenon Double–Beta Decay Observatory,*

[Bro92] R. L. Brodzinski, et al. (IGEX Collaboration) *Status Report on the International Germanium Experiment*, in Proc. of 15th International Conference on Neutrino Physics and Astrophysics (NEUTRINO 92), Granada, Spain, 7 – 12 June, 1992, ed. *A. Morales*, *Nucl. Phys. Proc. Suppl.* **31** (1993) 76 – 79

[Bun88] *"Deutsch–Sowjetisches Projekt zur Untersuchung der Neutrinomasse Gestartet"*, *Pressereferat, Der Bundesminister für Forschung und Technologie* **Nr. 111/88**, Bonn, den 09. Dezember 1988

[Bun90] *"Hauptphase eines Deutsch–Sowjetischen Projektes zur Untersuchung der Neutrinomasse beginnt"*, *Pressereferat, Der Bundesminister für Forschung und Technologie* **Nr. 69/90**

[Bur96] S. Ph. Burachas, F. A. Danevich, A. Sh. Georgadze, H. V. Klapdor–Kleingrothaus, V. V. Kobychev, B. N. Kropivyansky, V. N. Kuts, A. Müller, V. V. Muzalevsky, A. S. Nikolaiko, O. A. Ponkratenko, V. D. Ryzhikov, A. S. Sai, I. M. Solsky, V. I. Tretyak and Yu. Zdesenko *"Large Volume $CDWO_4$ Crystal Scintillators"*, *Nucl. Instrum. Meth.* **A 369** (1996) 164 – 168

[Bur93] C. P. Burgess and J. M. Cline *"A New Class of Majoron–Emitting Double–Beta Decays"*, *Preprint hep-ph/* **9307316** (1993) and *Phys. Rev.* **D 49** (1994) 5925 – 5944

[Bur95] C. P. Burgess *"Scalar – Emitting Models in Double – Beta Decay"*, *Preprint hep-ph/* **9505349** (1995) and in Proc. "Double Beta Decay and Related Topics", eds. *H. V. Klapdor–Kleingrothaus and S. Stoica, Intern. Workshop at ECT,* Trento, Italy, April 24 – May 5, 1995, *World Scientific* (1996) 110 – 129

[Bur99] S. Burles, K. M. Nollett, J. N. Truran and M. S. Turner *"Sharpening the Predictions of Big–Bang Nucleosynthesis"*, *Phys. Rev. Lett.* **82** (1999) 4176 – 4179 and *Preprint astro-ph/* **9901157** (1999)

[But93] M. N. Butler, S. Nozawa, R. A. Malaney and A. I. Boothroyd *"Gravitationally Induced Neutrino Oscillations"*, *Phys. Rev.* **D 47** (1993) 2615 – 2618
J. Pantaleone, A. Halprin and C. N. Leung *"Neutrino Mixing due to a Violation of the Equivalence Principle"*, *Phys. Rev.* **D 47** (1993) 4199 – 4202
K. Iida, H. Minakata and O. Yasuda *"A Test of the Equivalence Principle by Long Baseline Neutrino Oscillation Experiments"*, *Mod. Phys. Lett.* **A 8** (1993) 1037 – 1047 and *Preprint hep-ph/* **9211328** (1992)

[Cac2000] B. Caccianiga and M. G. Giammarchi *Neutrinoless Double Beta Decay with ^{136}Xe in BOREXINO and the Borexino Counting Test Facility (CTF)"*, *INFN-BE-*

00-01 (Jan 2000) 32pp

[Cal86] D. O. Caldwell, R. M. Eisberg, D. M. Grumm, D. L. Hale, M. S. Witherell, F. S. Goulding, D. A. Landis, N. W. Madden, D. F. Malone, R. H. Pehl and A. R. Smith *"Half-Life Limits on the Zero-Neutrino and Two-Neutrino Double-β Decay of ^{76}Ge"*, *Phys. Rev.* **D 33** (1986) 2737 - 2739

[Cal86a] D. O. Caldwell, *"Review of Double Beta Decay Results from ^{76}Ge"*, in Proc. 12th International Conference on "Neutrino Physics and Astrophysics", Sendai, Japan, June 3 - 8, 1986, eds *T. Kitagaki and H. Yuta, World Scientific, Singapore* (1986) 77 - 92

[Cal87] D. O. Caldwell, R. M. Eisberg, D. M. Grumm, M. S. Witherell, F. S. Goulding and A. R. Smith *"Limits on Neutrinoless $\beta\beta$ Decay Including That with Majoron Emission"*, *Phys. Rev. Lett.* **59** (1987) 419 - 422

[Cal87a] D. O. Caldwell *"Restrictions on Dark Matter from Germanium Detectors"*, in Proc. "New and Exotic Phenomena", Les Arcs (1987) 229 - 232

[Cal89] D. O. Caldwell *"Review of Double Beta Decay Experiments"*, *Int. J. Mod. Phys.* **A 4** (1989) 1851 - 1869

[Cal91] D. O. Caldwell *"Double Beta Decay - Present and Future"*, in Proc. of 14th Europhysics Conference on Nuclear Physics: Rare Nuclear Decays and Fundamental Physics, Bratislava, Czechoslovakia, 22 - 26 October, 1990, ed *P. Povinec, J. Phys.* **G 17** (1991) *Suppl.* S137 - S144

[Cal91a] D. O. Caldwell *"Searching for Double Beta Decay"*, in Neutrino Physics, ed *K. Winter, Cambridge University Press* (1991) 125 - 144 *(Cambridge Monographs on Particle Physics, Nuclear Physics, and Cosmology, 1)*

[Cal93] D. O. Caldwell and R. N. Mohapatra *"Neutrino Mass Explanations of Solar and Atmospheric Neutrino Deficits and Hot Dark Matter"*, *Phys. Rev.* **D 48** (1993) 3259 - 3263

[Cal94] D. O. Caldwell and R. N. Mohapatra *"Accommodating Solar and Atmospheric Neutrino Deficits, Hot Dark Matter, and a double β Decay Signal"*, *Phys. Rev.* **D 50** (1994) 3477 - 3482

[Cam87] B. A. Campbell, J. Ellis, K. Enqvist, M. K. Gaillard and D. V. Nanopoulos *"Superstring Models Challenged by Rare Processes"*, *Int. J. Mod. Phys.* **A 2** (1987) 831 - 890

[Car93] C. D. Carone *"Double Beta Decay with Vector Majorons"*, *Preprint hep-ph/* **9310 365** (1993) and *Phys. Lett.* **B 308** (1993) 85 - 88

[Cau96] E. Caurier, F. Nowacki, A. Poves and J. Retamosa *"Shell Model Studies of the Double Beta Decays of ^{76}Ge, ^{82}Se and ^{136}Xe"*, *Preprint nucl-th /* **9601017** (1996) and *Phys. Rev. Lett.* **77** (1996) 1954 - 1957

[CER91] *"GRAN SASSO - Enriched Germanium in Action"*, *CERN-Courier* **31, No 10** (1991) 14 - 17

[CER97] H. V. Klapdor-Kleingrothaus *"Physics Monitor"*, *CERN-Courier* **37, No 9** (1997) 16 - 18

[CER97a] *"HEIDELBERG Debut of GENIUS"*, *CERN-Courier* **37, No 10** (1997) 19

[CER2000] *"Majorana Mass Limit Reaches an All-Time Low"*, *CERN-Courier* **40, No 1** (2000) 8

[Che2000] K. Cheung and O. C. W. Kong *"Zee Neutrino Mass Model in a SUSY Framework"*, *Phys. Rev.* **D 61** (2000) 113012-1 - 10 *Preprint hep-ph/* **9912238** (1999)

[Chi80] Y. Chikashige, R. N. Mohapatra and R. D. Peccei *"Spontaneously Broken Lepton Number and Cosmological Constraints on the Neutrino Mass Spectrum"*, *Phys. Rev. Lett.* **45** (1980) 1926 - 1929

[Chi81] Y. Chikashige, R. N. Mohapatra and R. D. Peccei *"Are there Real Goldstone Bosons Associated with Broken Lepton Number?"*, *Phys. Lett.* **B 98** (1981) 265 -

268

[Chi88] C. R. Ching and T. H. Ho *"An Alternative Approach in Nuclear Double Beta Decay Theory"*, *Commun. Theor. Phys.* **10** (1988) 45 - 51

[Chi89] C. R. Ching and T. H. Ho *"A Discussion of the Question in Connecxion with the Expansion of a Divergent Series Encountered in Calculating Nuclear* $2\nu - 2\beta$ *Decay Matrix Element"*, *Commun. Theor. Phys.* **11** (1989) 495 - 497

[Chi89a] C. R. Ching, T. H. Ho and X. - R. Wu *"Operator Expansion Method and the Double Beta Decay of* ^{48}Ca*"*, *Phys. Rev.* **C 40** (1989) 304 - 313

[Civ87] O. Civitarese, A. Fässler and T. Tomoda *"Suppression of the Two Neutrino Double* β *Decay"*, *Phys. Lett.* **B 194** (1987) 11 - 14

[Cho97a] K. C. Chou and Yue-Liang Wu *"A Solution to the Puzzles of CP Violation, Neutrino Oscillation, Fermion Masses and Mixing in an SUSY GUT Model Small* $\tan\beta$*"*, *Nucl. Phys. Proc. Suppl.* **52 A** (1997) 159 - 162 and *Preprint hep-ph* **9610300** (1996)

[Cho97] D. Choudhury and S. Raychaudhuri *"R- Parity Violation at HERA?"*, *Preprint hep-ph* **9702392** (1997) and *Phys. Lett.* **B 401** (1997) 54 - 61

[Col97] S. Coleman and S. L. Glashow *"Cosmic Ray and Neutrino Tests of Special Relativity"*, *hep-ph/* **9703240** (1997) and *Phys. Lett.* **B 405** (1997) 249 - 252

[Cow56] C. L. Cowan, F. Reines, F. B. Harrison, H. W. Kruse and A. D. McGuire *"Detection of the Free Neutrino: a Confirmation"*, *Science* **124** (1956) 103 - 104

[Cve92] M. Cvetic *"Stringly Origin of Neutrino Masses Within the Minimal Supersymmetric Standard Model"*, in Proc. 26th International Conference on High Energy Physics (ICHEP 92), Dallas, TX, 6 - 12 August 1992, *Preprint hep-ph* **9210201** (1992) and *Preprint Dallas HEP* **QCD161:H51:1992** (1992) 1178 - 1186

[Cve92a] M. Cvetic and P. Langacker *"Neutrino Masses Within the Minimal Supersymmetric Standard Model"*, *Preprint hep-ph* **9205029** (1992) and *Phys. Rev.* **D 46** (1992) 2759 - 2763

[Cza99] M. Czakon, J. Gluza and M. Zralek *"Nature of Neutrinos in the Light of Present and Future Experiments"*, *Phys. Lett.* **B 465** (1999) 211 - 218 and *Preprint hep-ph* **9906381** (1999)

[Dam94] T. Damour and A. M. Polyakov *"The String Dilation and a Least Coupling Principle"*, *Nucl. Phys.* **B 423** (1994) 532 - 558 and *Preprint/hep-th* **9401069** (1994)

[Dan89] F. A. Danevich, Yu. Zdesenko A. S. Nikolaiko and V. I. Tretyak *"Search for* 2β *Decay of* ^{116}Cd *With the Help of a* $^{116}CdWO_4$ *Scintillator"*, *JETP Lett.* **49** (1989) 476 - 479

[Dan95] F. A. Danevich, A. Sh. Georgadze, V. V. Kobychev, B. N. Kropivyansky, V. N. Kuts, A. S. Nikolaiko, V. I. Tretyak and Yu. Zdesenko *"The Research of* 2β *Decay of* ^{116}Cd *with Enriched* $^{116}CdWO_4$ *Crystal Scintillators"*, *Phys. Lett.* **B 344** (1995) 72 - 78

[Dan96] F. A. Danevich, A. Sh. Georgadze, J. Hellmig, M. Hirsch, H. V. Klapdor-Kleingrothaus, V. V. Kobychev, B. N. Kropivyansky, V. N. Kuts, A. Müller, A. S. Nikolaiko, F. Petry, O. A. Ponkratenko, H. Strecker, V. I. Tretyak, M. Völlinger and Yu. Zdesenko *"Investigation of* $\beta^+\beta^+$ *and* β^+/EC *Decay of* ^{106}Cd*"*, *Z. Phys.* **A 355** (1996) 433 - 437

[Dan98] F. A. Danevich, A. Sh. Georgadze, V. V. Kobychev, B. N. Kropivyansky, A. S. Nikolaiko, O. A. Ponkratenko, V. I. Tretyak and Yu. Zdesenko *"Limits on Majoron Models of* ^{116}Cd *Neutrinoless* 2β *Decay"*, *Nucl. Phys.* **A 643** (1998) 317 - 328

[Dan99] F. A. Danevich, A. Sh. Georgadze, V. V. Kobychev, B. N. Kropivyansky, A. S. Nikolaiko, O. A. Ponkratenko, V. I. Tretyak and Yu. Zdesenko *"Status of the INR*

Experiment on 2β Decay of ^{116}Cd", in Proc. of 5th International Workshop on Topics in Astroparticle and Underground Physics (TAUP 97), Gran Sasso, Italy, 7 – 11 Sep. 1997, eds. *A. Bottino, A. di Credico and P. Monacelli, Nucl. Phys. Proc. Suppl.* **70** (1999) 246 – 248

[Dan99a] F. A. Danevich, V. V. Kobychev, B. N. Kropivyansky, A. S. Nikolaiko, O. A. Ponkratenko, V. I. Tretyak and Yu. Zdesenko *"New Phase of the* ^{116}Cd *2β Decay Experiment With* $^{116}CdWO_4$ *Scintillators"*, in the Proceedings of the NANP'99 (Dubna, 1999), eds. V. Bednyakov et al. (2000)

[Dan2000] F. A. Danevich, A. Sh. Georgadze, V. V. Kobychev, B. N. Kropivyansky, A. S. Nikolaiko, O. A. Ponkratenko, V. I. Tretyak, S. Yu. Zdesenko, Yu. G. Zdesenko, P. G. Bizzeti, T. F. Fazzini and P. R. Maurenzig *"New Results of* ^{116}Cd *Double β Decay Study with* $^{116}CdWO_4$ *Scintillators"*, Subm. to *Phys. Rev.* **C** (2000) and *Preprint nucl-ex/* **0003001** (2000)

[Dan2000a] M. Danilov, R. DeVoe, A. Dolgolenko, G. Giannini, G. Gratta, P. Picchi, A. Piepke, F. Pietropaolo, P. Vogel, J.Ł. Vuilleumier, Y. F. Wang and O. Zeldovich *"Detection of Very Small Neutrino Masses in Double–Beta Decay Using laser Tagging"*, *Preprint hep-ex/* **0002003** (2000) and *Phys. Lett.* **B 480** (2000) 12 – 18

[Das91] D. Dassié, P. Hubert, M. C. P. Isaac, C. Izac, F. Leccia, P. Mennrath, C. Longuemare, D. Blum, J. Busto, J. E. Campagne, S. Jullian, D. Lalanne, F. Laplanche, F. Natchez, G. Szklarz, T. Lamhamdi, I. Linck and F. Scheibling *"Double Beta Decay Prototype Detector with Multiwire Drift Tubes in the Geiger Mode"*, *Nucl. Instrum. Meth.* **A 309** (1991) 465 – 475

[Das95] D. Dassié, R. Eshbach, F. Hubert, Ph. Hubert, M. C. Isaac, C Izac, F. Leccia, P. Mennrath, A. Vareille, C. Longuemare, F. Mauger, F. Danevich, V. Kouts, V. I. Tretyak, Yu. Vassilyev, Yu. Zdesenko, A. Barabash, V. N. Kornoukhov, Yu. B. Lepikhin, V. I. Umatov, I. A. Vanushin, C. Augier, D. Blum, J. E. Campagne, S. Jullian, D. Lalanne, F. Laplanche, F. Natchez, G. Pichenot, G. Szklarz, R. Arnold, J. L. Guyonnet, T. Lamhamdi, I. Linck, F. Piquemal, F. Scheibling, V. Brudanin, V. Egorov, O. Kochetov, A. Nozdrin, Ts. Vylov, Sh. Zaparov, H. W. Nicholson and C. S. Sutton *"Two- Neutrino Double β Decay Measurement of* ^{100}Mo*" Phys. Rev.* **D 51** (1995) 2090 – 2100

[Dav55] R. Davis Jr. *"Attempt to Detect the Antineutrinos from a Nuclear Reactor by the* ^{37}Cl $(\bar{\nu}, e^-)Ar^{37}$ *Reaction"*, *Phys. Rev.* **97** (1955) 766 – 769

[Der2000] R. Dermíšek and S. Raby *"Fermion Masses and Neutrino Oscillations in SO(10) SUSY GUT With* $D_3 \times U(1)$ *Family Symmetry"*, *Phys. Rev.* **D 62** (2000) 015007-1 – 9 *Preprint hep-ph/* **9911275** (1999)

[Die2000] K. R. Dienes *"New Directions for New Dimensions: from Strings to Neutrinos to Axions to ..."*, *Preprint hep-ph/* **0004129** (2000)

[Die2000a] K. R. Dienes *"Neutrinos and Extra Dimensions"*, in Proc. of International Conference "Neutrino Physics and Astrophysics",NEUTRINO'2000, Sudbury, Canada, 16 –21 June, 2000, ed *A. McDonald, Nucl. Phys. Proc. Suppl.* **B** (2000)

[Doi80] M. Doi, T. Kotani, H. Nishiura, K. Okuda and E. Takasugi *"Neutrino Masses and the Double β Decay"*, Osaka Preprint **OS–GE–27** *University of Osaka, Japan, Phys. Lett.* **B 103** (1981) 219 – 224

[Doi81a] M. Doi, T. Kotani, H. Nishiura, K. Okuda and E. Takasugi *"Neutrino Mass, the Right – Handed Interaction and the Double Beta Decay. I."*, *Progr. of Theor. Phys.* **66** (1981) 1739 – 1764

[Doi81b] M. Doi, T. Kotani, H. Nishiura, K. Okuda and E. Takasugi *"Neutrino Mass, the Right – Handed Interaction and the Double Beta Decay. II."*, *Progr. of Theor. Phys.* **66** (1981) 1765 – 1788

[Doi83] M. Doi, T. Kotani, H. Nishiura and E. Takasugi *"Double Beta Decay"*, *Progr. of Theor. Phys.* **69** (1983) 602 – 635

[Doi83a] M. Doi, T. Kotani, H. Nishiura and E. Takasugi *"Pseudo Dirac Neutrino"*, *Progr. of Theor. Phys.* **70** (1983) 1331 – 1374

[Doi83b] M. Doi, M. Kenmoku, T. Kotani, H. Nishiura and E. Takasugi *"The Energy Spectrum and the Angular Correlation in the $\beta\beta$ Decay"*, *Progr. of Theor. Phys.* **70** (1983) 1353 – 1374

[Doi85] M. Doi, T. Kotani and E. Takasugi *"Double Beta Decay and Majorana Neutrino"*, *Prog. of Theor. Phys. Suppl.* **83** (1985) 1 – 175

[Doi88] M. Doi, T. Kotani and E. Takasugi *"Neutrinoless Double Beta-Decay with Majoron Emission"*, *Phys. Rev.* **D 37** (1988) 2575 – 2589

[Doi93] M. Doi and T. Kotani *"Neutrinoless Modes of Double Beta Decay"*, *Prog. of Theor. Phys.* **89** (1993) 139 – 159

[Dre98] H. Dreiner, M. Krämer and P. Morawitz *"High Q^2 DIS at HERA and Squark Production"*, in Proc. "Beyond the Desert'97: Accelerator and Nonaccelerator Approaches, eds. *H. V. Klapdor–Kleingrothaus and H. Päs,* "Conference on Physics Beyond the Standard Model, Tegernsee, Germany, 8 – 14 June, 1997, *IOP* (1998) 159 – 172

[Eji91] H. Ejiri, K. Fushimi, T. Kamada, H. Kinoshita, H. Kobiki, H. Ohsumi, K. Okada, H. Sano, T. Shibata, T. Shima, N. Tanabe, J. Tanaka, T. Taniguchi, T. Watanabe and N. Yamamoto *"Double Beta Decays of ^{100}Mo"*, *Phys. Lett.* **B 258** (1991) 17 – 23

[Eji92] H. Ejiri, K. Hayashi, K. Fushimi, R. Hazama, N. Kudomi, K. Kume, K. Nagata, T. Nitta, H. Ohsumi, K. Okada, T. Shima, K. Shinmyo and J. Tanaka *"Double Beta Decay of ^{100}Mo and Charge Nonconserving Transportations in ^{127}I Studied by Means of ELEGANT-V"*, in Proc. 3th International Workshop on Theoretical and Phenomenological Aspects of Underground Physics (TAUP 91), *Nucl. Phys. Proc. Suppl.* **A 28** (1992) 219 – 222

[Eji96] H. Ejiri, K. Fushimi, K. Hayashi, R. Hazama, N. Kudomi, K. Kume, K. Nagata, T. Nitta, H. Ohsumi, K. Okada, T. Shima, K. Shinmyo and J. Tanaka *"$0\nu\beta\beta$ Decays of ^{100}Mo by Means of ELEGANT V"*, in Proc. of 4th International Workshop on Theoretical and Phenomenological Aspects of Underground Physics (TAUP 95), Toledo, Spain, 17 – 21 Sep. 1995, eds. *A. Morales, J. Morales and J.A. Villar*, *Nucl. Phys. Proc. Suppl.* **48** (1996) 229 – 231

[Eji97] H. Ejiri *"Double Beta Decays and Dark Matters Studied by ELEGANT Detectors"*, in Proc. of 17th International Conference on Neutrino Physics and Astrophysics, Helsinki, Finland, June 13 – 19, 1996, eds. *K. Enqvist, K. Huitu and J. Maalampi*, *World Scientific, Singapore*, (1997) 342 – 346

[Eji98] H. Ejiri *"Double Beta Decay and Neutrino Nuclear Responses"*, in Proc. of International Conference "Neutrino Physics and Astrophysics",NEUTRINO'98, Takayama, Japan, 4 – 9 June, 1998, eds. *Y. Suzuki and Y. Totsuka*, *Nucl. Phys.* **B 77** *Proc. Suppl.* (1999) 346 – 351

[Eji99a] H. Ejiri inProc. of 6th International Workshop on Theoretical and Phenomenological Aspects of Underground Physics (TAUP 99), Paris (1999)

[Eji99b] H. Ejiri, J. Engel, R. Hazama, P. Krastev, N. Kudomi, Res. and R. G. H. Robertson *"Spectroscopy of Double-Beta and Inverse-Beta Decays from ^{100}Mo for Neutrinos "*, *Preprint nucl-ex/* **9911008** (1999)

[Ell86] S. R. Elliot, A. A. Hahn, M. K. Moe *"Experimental Investigation of Double–Beta Decay in ^{82}Se"*, *Phys. Rev. Lett.* **56** (1986) 2582 – 2585

[Ell86a] S. R. Elliot, A. A. Hahn, M. K. Moe *"An Experimental Search for Double Beta Decay in ^{82}Se"*, in Proc. WEIN'86 International Symposium "Weak and Electromagnetic

Interactions in Nuclei", Heidelberg 1986, Germany, *ed. H. V. Klapdor, Springer–Verlag Berlin, Heidelberg, New York, London, Paris, Tokyo,* (1986) 692 – 695

[Ell87] S. R. Elliot, A. A. Hahn and M. K. Moe *"Limit on Neutrinoless Double–Beta Decay with Majoron Emission in ^{82}Se"*, *Phys. Rev. Lett.* **59** (1987) 1649 – 1651

[Ell87a] S. R. Elliott, A. A. Hahn and M. K. Moe *"Direct Evidence for Two–Neutrino Double–Beta Decay in ^{82}Se"*, *Phys. Rev. Lett.* **59** (1987) 2020 – 2023

[Ell85] J. Ellis, G. Gelmini, C. Jarlskog, G. G. Ross and J. W. F. Valle *"Phenomenology of Supersymmetry With Broken R–Parity"*, *Phys. Lett.* **B 150** (1985) 142 – 148

[Ell96a] J. Ellis, N. E. Mavromatos and D. V. Nanopoulos *"CPT and Superstring"*, *Preprint hep-ph/* **9607434** (1996)

[Ell96c] J. Ellis, N. E. Mavromatos and D. V. Nanopoulos *"CPT Violation in String Modified Quantum Mechanics and the Neutral Kaon System"*, *Int. J. Mod. Phys.* **A 11** (1996) 1489 – 1508 and *Preprint hep-ph/* **9212057** (1992)

[Ell98] J. Ellis, G. K. Leontaris, S. Lola and D. V. Nanopoulos *"Neutrino Textures in the Light of SUPERKAMIOKANDE Data and a Realistic String Model"*, *Eur. Phys. J.* **C 9** (1999) 389 – 408 and *Preprint hep-ph/* **9808251** (1998)

[Ell99] J. Ellis, S. Lola and D. V. Nanopoulos *"Leptogenesis in the Light of Super-Kamiokande Data and a Realistic String Model"*, *Phys. Lett.* **B 452** (1999) 87 – 97 and *Preprint hep-ph/* **9902364** (1999)

[Ell99a] J. Ellis and S. Lola *"Can Neutrinos be Degenerate in Mass?"*, *Phys. Lett.* **B 458** (1999) 310 – 321 and *Preprint hep-ph/* **9904279** (1999)

[Ell2000] J. Ellis A. Ferstl and A. Olive *Re-Evaluation of the Elastic Scattering of Supersymmetric Dark Matter"*, *Phys. Lett.* **B 481** (2000) 304 – 314

[Eng88] J. Engel, P. Vogel, and M. R. Zirnbauer *"Nuclear Structure Effects in Double–Beta Decay"*, *Phys. Rev.* **C 37** (1988) 731 – 746

[Fae98] A. Faessler and F. Simkovic *"Double Beta Decay"*, *J. Phys.* **G 24** (1998) 2139 – 2178 and *Preprint hep-ph/* **9901215** (1999) 1 –32

[Far93] A. E. Faraggi and E. Halyo *"Neutrino Masses in Superstring Derived Standard–Like Models"*, *Phys. Lett.* **B 307** (1993) 311 – 317 and *Preprint hep-ph/* **9303060** (1993)

[Far97] J. Farine *"Search for $\beta\beta$ Decay in ^{136}Xe Gotthard Results"*, in Proc. of 17th International Conference on Neutrino Physics and Astrophysics, Helsinki, Finland, June 13 – 19, 1996, eds. *K. Enqvist, K. Huitu and J. Maalampi, World Scientific, Singapore,* (1997) 347 – 351

[Fer34] E. Fermi *"Versuch einer Theorie der β-Strahlen. I"*, *Z. Phys.* **88** (1934) (21. Februar) 161 – 177

[Fio67] E. Fiorini, A. Pullia, G. Bertolini, F. Cappellani and G. Restelli *"A Search for Lepton Non–Conservation in Double Beta Decay With a Germanium Detector"*, *Phys. Lett.* **B 45** (1967) 602 – 603

[Fio84] E. Fiorini and T. O. Niinikoski *"Low Temperature Calorimetry for rare Decays"*, *Nucl. Instrum. Meth.* **A 224** (1984) 83 – 96

[Fio91a] E. Fiorini *"Cryogenic Thermal Detectors in Subnuclear Physics and Astrophysics"*, *Physica* **B 169** (1991) 388 – 399

[Fio91b] E. Fiorini *"Double Beta Decay: Past, Present and Future"*, *Nucl. Phys. News* **1, No. 5** (1991) 17 – 21

[Fio98] E. Fiorini *"CUORE: A Cryogenic Underground Observatory for Rare Events"*, 3rd International Symposium on Sources and Detection of Dark Matter in the Universe (DM 98), Santa Monica, CA, 18 – 20 Feb. (1998), *Phys. Rept.* **307** (1998) 309 – 317

[Fio99a] E. Fiorini *"Double Beta Decay"*, in Proc. of 8th International Workshop On Neutrino Telescopes, 23 – 26 February, 1999, Venezia, Italy, ed *M. Baldo Ceolin*, Padova,

Papergraf (1999) 35 – 51

[Fio99b] E. Fiorini *"Weak Interaction Searches with Cryogenic Detectors"*, 8th International Workshop On Low Temperature Detectors (LTD 8), 15 – 20 August, 1999, Dalfsen, The Netherlands

[Fir48] E. L. Fireman *"Double Beta Decay"*, *Phys. Rev.* **74** (1948) 1238

[Fir49] E. L. Fireman *"A Measurement of the Half-Life of Double Beta-Decay from- $_{50}Sn^{124}$"*, *Phys. Rev.* **75** (1949) 323 – 324

[Fir52] E. L. Fireman and D. Schwarzer *"A Re-Investigation of the Double Beta-Decay from Sn^{124}"*, *Phys. Rev.* **86** (1952) 451 – 453

[Fis87] P. Fisher, F. Boehm, H. Henrikson, E. Bovet, J. P. Egger, J. L. Vuilleumier and K. Gabathuler *"Limit on Double Beta Decay with Majoron Emission"*, *Phys. Lett.* **B 192** (1987) 460 – 462

[Foo95] R. Foot and R. R. Volkas *"Neutrino Physics and the Mirror World: How Exact Parity Symmetry Explains the Solar Neutrino Deficit, the Atmospheric Neutrino Anomaly and the LSND Experiment"*, *Phys. Rev.* **D 52** (1995) 6595 – 6606 and *Preprint hep-ph/* **9505359** (1995)

[Fri75] H. Fritzsch and P. Minkowski *"Unified Interactions of Leptons and Hadrons"*, *Annals Phys.* **93** (1975) 193 – 266 and in Gauge Theories of Fundamental Interactions, eds. *R. N. Mohapatra and C. H. Lai* 432 – 505

[Fri86] M. Fritschi, E. Holzschuh, W. Kündig, J. W. Petersen, R. E. Pixley, H. Stussi *"An Upper Limit for the Mass of Anti-Electron-Neutron from Tritium Beta Decay"*, *Phys. Lett.* **B 173** (1986) 485 – 489

[Fuk99] The Super-Kamiokande Collaboration: Y. Fukuda, T. Hayakawa, E. Ichihara and K. Inoue *"Constraints on Neutrino Oscillation Parameters from the Measurement of Day-Night Solar Neutrino Fluxes at Super-Kamiokande"*, *Phys. Rev. Lett.* **82** (1999) 1810 – 1814 and *Preprint hep-ex/* **9812009** (1998)

[Fur39] W. H. Furry *"On Transition Probabilities in Double Beta-Disintegration"*, *Phys. Rev.* **56** (1939) 1184 – 1193

[Gai2000] R. Gaitskell *"Latest Results from the CDMS Collaboration"*, in Proc. "Dark Matter 2000", Los Angeles, February 2000, ed D. B. Cline, *Springer, Heidelberg* (2000)

[Gas89a] M. Gasperini, *"Testing the Principle of Equivalence with Neutrino Oscillations"*, *Phys. Rev.* **D 38** (1988) 2635 – 2637 and *"Experimental Constraints on a Minimal and Nonminimal Violation of the Equivalence Principle in the Oscillations of Massive Neutrinos"*, *Phys. Rev.* **D 39** (1989) 3606 – 3611

[Gas89b] M. Gasperini *"Constraint on Deviations from Universality in the Coupling to Gravity of Photons and High-Energy Cosmic Rays"*, *Phys. Rev. Lett.* **62** (1989) 1945 – 1947 *Erratum-ibid.* **63** (1989) 1658

[Gav2000] Ju. M. Gavriljuk, V. V. Kuzminov, N. Ya. Osetrova and S. S. Ratkevich *"Results of a search for the two neutrino double β decay of ^{136}Xe with proportional counters"*, *Phys. Rev.* **C 61** (2000) 035501-1 – 6

[Gel79] M. Gell-Mann, P. Ramond and R. Slansky *"Complex Spinor and Unified Theories"*, in Proc. of the Supergravity Workshop at Stony Brook, 27–29 September, (1979) ed. by P. Van Nieuwenhuizen D. Z. Freedman, *North - Holland Publ. Comp.* (1979) 315 – 321

[Gel81] G. B. Gelmini and M. Roncadelli *"Left-Handed Neutrino Mass Scale and Spontaneously Broken Lepton Number"*, *Phys. Lett.* **B 99** (1981) 411 – 415

[Gel82] G. B. Gelmini S. Nussinov and M. Roncadelli *"Bounds and Prospects for the Majoron model of Left-Handed Neutrino Masses"*, *Nucl. Phys.* **B 209** (1982) 157 – 173

[Geo75] H. M. Georgi *"The State of the Art–Gauge Theories"*, in "Particles and Fields", ed *C. E. Carlson, A.I.P. New York*, (1975) 575 – 582

[Geo81] H. M. Georgi, S. L. Glashow and S. Nussinov *"Unconventional model of Neutrino Masses"*, *Nucl. Phys.* **B 193** (1981) 297 – 316

[Geo2000] H. Georgi and S. L. Glashow *"Neutrinos on Earth and in the Heavens"*, *Phys. Rev.* **D 61** (2000) 097301-1 – 4

[Gir92] T. A. Girard, E. Aprile and R. C. Fernholz *"Sensitivity of a Large Volume LXE TPC to Neutrinoless Double Beta Decay"*, *Nucl. Instrum. Meth.* **A 316** (1992) 44 – 50

[Giu91] A. Giuliani *"Low Temperature Detectors in Non–Accelerator Particle Physics"*, in Proc. of 14th Europhysics Conference on Nuclear Physics: Rare Nuclear Decays and Fundamental Physics, Bratislava, Czechoslovakia, 22 – 26 October, 1990, ed *P. Povinec, J. Phys.*. **G 17** (1991) S309 – S323

[Giu99a] A. Giuliani (CUORE Collaboration) *"The CUORE Project: a Large Observatory for Neutrinoless Double Beta Decay and Other Rare Events"*, in Proc. "Lepton and Baryon Number Violation in Particle Physics, Astrophysics and Cosmology", eds. *H. V. Klapdor–Kleingrothaus and I. V. Krivosheina, International Workshop at ECT,* Trento, Italy, April 20 – April 25, 1998, *IOP, Bristol* (1999) 302 – 308

[Giu99] C. Giunti *"Neutrinoless Double–β Decay with Three or Four Neutrino Mixing"*, *Phys. Rev.* **D 61** (2000) 036002-1–036002-8 and *Preprint hep–ph/* **9906275** (1999)

[Giu99b] C. Giunti *"Neutrino Oscillations and Neutrinoless Double–β Decay"*, *Preprint hep–ph/* **9912427** (1999) in Proc. "TAUP'99", 6 – 10 September 1999, College de France, Paris, France

[Gla97] S. L. Glashow, A. Halprin, P. I. Krastev, C. N. Leung and J. Panteleone *"Remarks on Neutrino Tests of Special Relativity"*, **Phys. Rev. D 56** (1997) 2433 – 2434 *Preprint hep–ph/* **9703454** (1997)

[Gmi90] M. Gmitro and F. Šimkovic *"New Developments in the Theory of Double β Decay"*, *Bull. Acad. Sci. USSR, Phys. Ser.* **54** (1990) **No. 9** 110 – 117, *Izv. Akad. Nauk SSSR, Fiz.* **54** (1990) 1780 – 1786

[Goe35] M. Goeppert–Mayer *"Double Beta-Disintegration"*, *Phys. Rev.* **48** (1935) 512 – 516

[Gon2000] M. Gonzales-Garcia *"Global Analysis of Solar Data"*, invited talk on NEUTRINO'2000, in Proc. of International Conference "Neutrino Physics and Astrophysics", NEUTRINO'2000, Sudbury, Canada, 16 –21 June, 2000, ed *A. McDonald, Nucl. Phys. Proc. Suppl.* **B** (2000)

[Goo61] M. L. Good *"$K_2{}^0$ and the Equivalence Principle"*, *Phys. Rev.* **121** (1961) 311 – 313

O. Nachtmann *"CP Violation and Cosmological Fields"*, *Acta Physica Austriaca, Supp.* **VI** *Particle Physics* ed. *P. Urban* (1969) p. 485 – 500

S. H. Aronson, G. J. Bock, H-Y Cheng and E. Fishbach *"Determination of the Fundamental Parametres of the K^0 $\bar{K}^0$ System in the Energy Range 30 GeV – 110 GeV"*, *Phys. Rev. Lett.* **48** (1982) 1306 – 1309

"Energy Dependence of the Fundamental Parametres of the K^0 $\bar{K}^0$ System. 2. Theoretical Formalism", *Phys. Rev.* **D 28** (1983) 495 – 706

I. R. Kenyon *"A Recalculation of the Gravitational Mass Difference Between the K^0 and $\bar{K}^0$ Mesons"*, *Phys. Lett.* **B 237** (1990) 274 – 277

R. J. Hughes *"New Test of the Equivalence Principle for the Anti–Proton"*, *Phys. Rev.* **D 46** (1992) R2283 – R2286

[Gol58] M. Goldhaber, L. Grodzins and A. W. Sunyar *"Helicity of Neutrinos"*, *Phys. Rev.* **109** (1958) 1015 – 1017

[Gre60] E. Greuling and R. C. Whitten *"Lepton Conservation and Double Beta–Decay"*, *Ann. Phys. (N.Y.)* **11** (1960) 510 – 533

[Gre96] C. Greub and P. Minkowski *"Heavy Majorana Neutrinos in e^-e^- Collisions"*, *Int. J. Mod. Phys.* **A 13** (1998) 2363 – 2381 and *Preprint hep–ph/* **9612340** (1996)

[Gro83a] K. Grotz, H. V. Klapdor and J. Metzinger *"Double Beta Decay of $^{128}Te/^{130}Te$ and the Question of a Neutrino Majorana Mass"*, *J. Phys.* **G 9** (1983) L169 – L173

[Gro83b] K. Grotz, H. V. Klapdor, J. Metzinger, R. Madey, W. Pairsuwan, B. D. Anderson, A. R. Baldwin, B. Flanders, C. Lebo, J. W. Watson and C. C. Foster *"The Gamow – Teller Strength Distribution in ^{208}Bi – A Probe of the Charge Distribution in ^{208}Pb?"*, *Phys. Lett.* **B 126** (1983) 417 – 420

[Gro83c] K. Grotz, H. V. Klapdor and J. Metzinger *"The Quenching of low–Energetic Nuclear Gamow-Teller Strength by Δ-Excitations"*, *Phys. Lett.* **B 132** (1983) 22 – 26

[Gro84] K. Grotz and H. V. Klapdor *"Time Scale of Short Time Deviations from Exponential Decay"*, *Phys. Rev.* **C 30** (1984) 2098 – 2100

[Gro85a] K. Grotz and H. V. Klapdor *"Predictions of 2ν and 0ν Double Beta Decay Rates for Nuclei with $A \geq 70$"*, *Phys. Lett.* **B 157** (1985) 242 – 246

[Gro85b] K. Grotz and H. V. Klapdor *" The Neutrino Mass from Double Beta Decay"*, *Phys. Lett.* **B 153** (1985) 1 – 4

[Gro86] K. Grotz, H. V. Klapdor and J. Metzinger *"Microscopic Calculation of Neutrino Capture Rates in $^{69,71}Ga$ and the Detection of Solar and Galactic Neutrinos"*, *Phys. Rev.* **C 33** (1986) 1263 - 1269

[Gro86a] K. Grotz and H. V. Klapdor *"Calculation of Double Beta Decay Rates and the Neutrino Mass"*, *Nucl. Phys.* **A 460** (1986) 395 – 436

[Gro89/90] K. Grotz and H. V. Klapdor "Die Schwache Wechselwirkung in Kern-, Teilchen- und Astrophysik", *B.G. Teubner, Stuttgard* (1989), "The Weak Interaction in Nuclear, Particle and Astrophysics", *IOP Bristol* (1990), *Moscow, MIR* (1992) and *China* (1998)

[Gun96] M. Günther, J. Hellmig, G. Heusser, M. Hirsch, H. V. Klapdor–Kleingrothaus, B. Maier, H. Päs, F. Petry, Y. Ramachers, H. Strecker, M. Völlinger, A. Balysh, S. T. Belyaev, A. Demehin, A. Gurov, I. Kodratenko, D. Kotel'nikov, V. I. Lebedev and A. Müller *"Bounds on New Majoron Models from the HEIDELBERG–MOSCOW Experiment"*, *Phys. Rev.* **D 54** (1996) 3641 – 3644

[Hal84] L. J. Hall and M. Suzuki *"Explicit R–Parity Breaking in Supersymmetric Models"*, *Nucl. Phys.* **B 231** (1984) 419 – 444

[Hal91] A. Halprin and C. N. Leung *"Can the Sun Shed Light on Neutrino Gravitational Interactions?"*, *Phys. Rev. Lett.* **67** (1991) 1833 – 1835
"Solar Neutrino Oscillations and Neutrino–Gravity Couplings", *Nucl. Phys. Proc. Supp.* **B 28A** (Proc. Supp.)(1992) 139 – 141
J. N. Bahcall, P. I. Krastev, and C. N. Leung, *"Solar Neutrinos and the Principle of Equivalence"*, *Phys. Rev.* **D 52** (1995) 1770 – 1779
R. B. Mann and U. Sarkar *"Test of the Equivalence Principle from Neutrino Oscillation Experiments"*, *Phys. Rev. Lett.* **76** (1996) 865 – 868 *Preprint hep-ph/* **9505353** (1995)

[Hal96] A. Halprin, C. N. Leung and J. Pantalone *"A Possible Violation of the Equivalence Principle by Neutrinos"*, *Phys. Rev.* **D 53** (1996) 5365 – 5376

[Hal98] A. Halprin and C. N. Leung *"Neutrino Oscillations from String Theory"*, *Phys. Lett.* **B 416** (1998) 361 – 364 and *Preprint hep-ph/* **9707407** (1997)

[Ham98] T. Hambye, R. B. Mann and U. Sarkar *"Test of Special Relativity from K Physics"*, *Phys. Lett.* **B 421** 105 (1998) 105 – 108 and *Preprint hep–ph/* **9709350** (1997)
"Test of Special Relativity and Equivalence Principle from K Physics", *Phys. Rev.*

D 58 (1998) 025003 – 025000-8 and *Preprint hep-ph/* **9802300** (1998)

[Hax81] W. C. Haxton, G. J. Stephenson and D. Strottman *"Double Beta Decay and the Majorana Mass of the Electron Neutrino"*, *Phys. Rev. Lett.* **47** (1981) 153 – 156

[Hax82a] W. C. Haxton, G. J. Stephenson and D. Strottman *"Lepton–Number Conservation and the Double-β Decay of ^{128}Te and ^{130}Te"*, *Phys. Rev.* **D 25** (1982) 2360 – 2369

[Hax82b] W. C. Haxton, S. P. Rosen and G. J. Stephenson, Jr. *"Higgs–Boson–Exchange Contributions to Neutrinoless Double–β Decay"*, *Phys. Rev.* **D 26** (1982) 1805 – 1808

[Hax84] W. C. Haxton and G. J. Stephenson *"Double Beta Decay"*, *Prog. Part. Nucl. Phys.* **12** (1984) 409 – 479

[Hel97] J. Hellmig and H. V. Klapdor–Kleingrothaus *"A Large Scale Double Beta and Dark Matter Experiment: GENIUS"*, *Z. Phys.* **A 359** (1997) 351 – 359 and *Preprint nucl-ex/* **9801004** (1998)

[Hel99] J. Hellmig and H. V. Klapdor–Kleingrothaus *"Identification of Single Site Events in Germanium Detectors by Digital Pulse Shape Analysis"*, *Subm. to Nucl. Instr. Meth.* (2000)

[Hen75] E. W. Hennecke, O. K. Manuel and D. D. Sabu *"Double Beta Decay of ^{128}Te"*, *Phys. Rev.* **C 11** (1975) 1378 – 1384

[Heu94] C. A. Heusch and P. Minkowski *"Lepton Flavor Violation Induced by Heavy Majorana Neutrinos"*, *Nucl. Phys.* **B 416** (1994) 3 – 45

[Her99] Proc. "Physics Beyond the Standard Model", Santa Fe, New Mexico, USA, June 14 – 19, 1998, *Singapore: World Scientific* (1999) eds. *P. Herczeg, C. M. Hoffman and H. V. Klapdor–Kleingrothaus*, 784pp.

[Hew97] J. L. Hewett and T. G. Rizzo *"Much Ado About Leptoquarks: A Comprehensive Analysis"*, *Phys. Rev.* **D 56** (1997) 5709 – 5724 and *Preprint hep-ph/* **9703337** (1997)

[Hir92] M. Hirsch, A. Staudt and H. V. Klapdor–Kleingrothaus *"Prediction of Average β and γ Energies and Probabilities of β–Delayed Neutron Emission in the Region of Fission Products"*, *At. Nucl. Data Tables* **51** (1992) 243 – 271

[Hir93] M. Hirsch, X. R. Wu, H. V. Klapdor–Kleingrothaus, Ching Cheng–Rui and Ho Tso–Hsiu *"Matrix Elements for $0\nu\beta\beta$ Decay Calculated with the Operator Expansion Method and QRPA Wave Functions"*, *Z. Phys.* **A 345** (1993) 163 – 169

[Hir93a] M. Hirsch, A. Staudt, K. Muto and H. V. Klapdor–Kleingrothaus *"Microscopic predictions of β^+/EC -Decay half–lives"*, *At. Data Nucl. Data Tables* **53** (1993) 165 – 193

[Hir94] M. Hirsch, K. Muto, T. Oda and H. V. Klapdor–Kleingrothaus *"Nuclear Structure Calculation of $\beta^+\beta^+$, β^+/EC and EC/EC Decay Matrix Elements"*, *Z. Phys.* **A 347** (1994) 151 – 160

[Hir94a] J. G. Hirsch, O. Castaños and P. O. Hess *"Neutrinoless Double Beta Decay in Heavy Deformed Nuclei"*, *Preprint nucl-th/* **9407022** (1994) and *Nucl. Phys.* **A 582** (1995) 124 – 140

[Hir94b] M. Hirsch, X. R. Wu, H. V. Klapdor–Kleingrothaus, Ching Cheng–Rui and Ho Tso–Hsiu *"Operator Expansion Method and Nuclear $\beta\beta$ Decay"*, *Phys. Rep.* **242** (1994) 403 – 422

[Hir95] M. Hirsch, H. V. Klapdor–Kleingrothaus and S. G. Kovalenko *"New Constraints on R–Parity–Broken Supersymmetry from Neutrinoless Double Beta Decay"*, *Phys. Rev. Lett.* **75** (1995) 17 – 20

[Hir95a] M. Hirsch, O. Kadowaki, H. V. Klapdor–Kleingrothaus, K. Muto and T. Oda *"Extended Operator Expansion Method for Neutrinoless Double Beta Decay"*, *Z. Phys.*

A 352 (1995) 33 - 45

[Hir95b] M. Hirsch, H. V. Klapdor–Kleingrothaus and S. G. Kovalenko *"New Contributions to Supersymmetric Mechanism of Neutrinoless Double Beta Decay", Phys. Lett.* **B 352** (1995) 1 - 7 and *Preprint hep-ph/* **9502315** (1995)

[Hir95c] M. Hirsch and H. V. Klapdor–Kleingrothaus *"Test Von Supersymmetrie: Doppelbetazerfall und Teilchenbeschleuniger", Phys. Bl.* **51** (1995) 418 - 419

[Hir96] M. Hirsch, H. V. Klapdor–Kleingrothaus and S. G. Kovalenko *"Supersymmetry and Neutrinoless Double Beta Decay", Phys. Rev.* **D 53** (1996) 1329 - 1348

[Hir96a] M. Hirsch, H. V. Klapdor–Kleingrothaus and S. G. Kovalenko *"On the SUSY Accompanied Neutrino Exchange Mechanism of Neutrinoless Double Beta Decay", Phys. Lett.* **B 372** (1996) 181 - 186 and *Erratum - ibid* **B 381** (1996) 488

[Hir96b] M. Hirsch, H. V. Klapdor–Kleingrothaus and O. Panella *"Double Beta Decay in Left–Right Symmetric Models", Phys. Lett.* **B 374** (1996) 7 - 12

[Hir96c] M. Hirsch, H. V. Klapdor–Kleingrothaus and S. G. Kovalenko *"New Low–Energy Leptoquark Interactions", Phys. Lett.* **B 378** (1996) 17 - 22

[Hir96d] M. Hirsch, H. V. Klapdor–Kleingrothaus and S. G. Kovalenko *"New Leptoquark Mechanism of Neutrinoless Double β Decay", Phys. Rev.* **D 54** (1996) R4207 - R4210

[Hir96e] M. Hirsch, H. V. Klapdor–Kleingrothaus, S. G. Kovalenko and H. Päs *"On the Observability of Majoron Emitting Double Beta Decays", Phys. Lett.* **B 372** (1996) 8 - 14

[Hir96f] M. Hirsch, H. V. Klapdor–Kleingrothaus and S. G. Kovalenko *"The R-Parity Violating Supersymmetric Mechanism of Neutrinoless Double Beta Decay",* in Proc. of International Workshop "Double Beta Decay and Related Topics", Trento, Italy, April 24 – May 5, 1995", eds. *H. V. Klapdor–Kleingrothaus and S. Stoica: Singapore: World Scientific* (1996) 91 - 109

[Hir96g] M. Hirsch and H. V. Klapdor–Kleingrothaus *"Constraints on Left–Right Symmetric Models from Neutrinoless Double Beta Decay",* in Proc. International Workshop "Double Beta Decay and Related Topics", Trento, Italy, April 24 – May 5, 1995, eds. *H. V. Klapdor–Kleingrothaus and S. Stoica, Singapore: World Scientific* (1996) 175 - 191

[Hir97] M. Hirsch and H. V. Klapdor–Kleingrothaus *"Probing Physics Beyond the Standard Model with Neutrinoless Double Beta Decay", Nucl. Phys. B (Proc. Suppl.)* **A 52** (1997) 257 - 262

[Hir97a] M. Hirsch, H. V. Klapdor–Kleingrothaus and S. G. Kovalenko *"B–L–Violating Masses in Softly Broken Supersymmetry", Preprint hep-ph/* **9701253** (1997) and *Phys. Lett.* **B 398** (1997) 311 - 314

[Hir97b] M. Hirsch, H. V. Klapdor–Kleingrothaus and S. G. Kovalenko *"Sneutrino Oscillations and Neutrinoless Double Beta Decay", Phys. Lett.* **B 403** (1997) 291 - 296

[Hir97c] M. Hirsch, H. V. Klapdor–Kleingrothaus and S. G. Kovalenko *"B–L–Violating Masses in Supersymmetry and Neutrinoless Double Beta Decay",* in Proc. "Beyond the Desert'97": Accelerator and Nonaccelerator Approaches, eds. *H. V. Klapdor–Kleingrothaus and H. Päs,* "Conference on Physics Beyond the Standard Model, Tegernsee, Germany, 8 - 14 June, 1997, *IOP* (1998) 237 - 244

[Hir98a] M. Hirsch, H. V. Klapdor–Kleingrothaus and S. G. Kovalenko *"R–Parity-Conserving Supersymmetry, Neutrino Mass, and Neutrinoless Double Beta Decay", Phys. Rev.* **D 57** (1998) 1947 - 1961 and *Preprint hep - ph/* **9707207** (1997)

[Hir98c] M. Hirsch, H. V. Klapdor–Kleingrothaus and S. G. Kovalenko *"Neutrinoless Double Beta Decay and the HERA 'Anomaly' ",* in Proc. "Beyond the Desert'97": Accelerator and Nonaccelerator Approaches, eds. *H. V. Klapdor–Kleingrothaus and H. Päs,* "Conference on Physics Beyond the Standard Model, Tegernsee, Germany, 8 – 14 June, 1997, *IOP* (1998) 322 – 330

[Hir98d] M. Hirsch, H. V. Klapdor–Kleingrothaus, St. Kolb and S. G. Kovalenko *"Phenomenological Implications of "Majorana" Sneutrinos at Future Accelerators", Phys. Rev.* **D 57** (1998) 2020 – 2023

[HM95] HEIDELBERG–MOSCOW Collaboration, *"Sub-eV Limit for the Neutrino Mass from ^{76}Ge Double Beta Decay by the HEIDELBERG–MOSCOW Experiment", Phys. Lett.* **B 356** (1995) 450 – 455

[HM97] HEIDELBERG–MOSCOW Collaboration, M. Günther, J. Hellmig, G. Heusser, M. Hirsch, H. V. Klapdor–Kleingrothaus, B. Maier, H. Päs, F. Petry, Y. Ramachers, H. Strecker, M. Völlinger, A. Balysh, S. T. Belyaev, A. Demehin, A. Gurov, I. Kondratenko, D. Kotel'nikov, V. I. Lebedev and A. Müller *"HEIDELBERG - MOSCOW $\beta\beta$ Experiment With ^{76}Ge: Full Setup With Five Detectors", Phys. Rev.* **D 55** (1997) 54 – 67

[HM97a] HEIDELBERG–MOSCOW Collaboration, L. Baudis, M. Gunther, J. Hellmig, G. Heusser, M. Hirsch, H. V. Klapdor–Kleingrothaus, H. Päs, Y. Ramachers, H. Strecker, M. Völlinger, A. Bakalyarov, A. Balysh, S. T. Belyaev, V. I. Lebedev, S. Zhukov and S. Kolb *"The HEIDELBERG - MOSCOW Experiment: Improved Sensitivity for ^{76}Ge Neutrinoless Double Beta Decay", Phys. Lett.* **B 407** (1997) 219 – 224

[HM98] HEIDELBERG–MOSCOW Collaboration, L. Baudis, J. Hellmig, G. Heusser, H. V. Klapdor–Kleingrothaus, S. Kolb, B. Majorovits, H. Päs, Y. Ramachers, H. Strecker, V. Alexeev, A. Balysh, A. Bakalyarov, S. T. Belyaev, V. I. Lebedev and S. Zhukov *"New Limits on Dark–Matter Wearkly Interacting Particles from the Heidelberg-Moscow Experiment", Phys. Rev.* **D 59** (1998) 022001-1 – 022001-5

[HM99] HEIDELBERG–MOSCOW Collaboration, L. Baudis, A. Dietz, G. Heusser, H. V. Klapdor–Kleingrothaus, I. V. Krivosheina, St. Kolb, B. Majorovits, V. F. Melnikov, H. Päs, F. Schwamm, H. Strecker, V. Alexeev, A. Balysh, A. Bakalyarov, S. T. Belyaev, V. I. Lebedev and S. Zhukov *"Limits on the Majorana Neutrino Mass in the 0.1 eV Range", Phys. Rev. Lett.* **83** (1999) 41 – 44

[HM2000] HEIDELBERG–MOSCOW Collaboration, H. V. Klapdor–Kleingrothaus, L. Baudis, A. Dietz, I. V. Krivosheina, G. Heusser, St. Kolb, B. Majorovits, H. Päs, H. Strecker, H. Tu, V. Alexeev, A. Balysh, A. Bakalyarov, S. T. Belyaev, V. I. Lebedev and S. Zhukov *"Double Beta Decay of ^{76}Ge: New Results from the HEIDELBERG-MOSCOW Experiment", Subm. for publ.* (2000)

[Huf70] A. H. Huffman *"Nuclear Matrix Elements in the Double Beta Decay $Te^{130} \longrightarrow Xe^{130}$", Phys. Rev.* **C 2** (1970) 742 – 747

[Ing49] M. G. Inghram and J. H. Reynolds *"On the Double Beta – Process", Phys. Rev.* **76** (1949) 1265 – 1266

[Ing50] M. G. Inghram and J. H. Reynolds *"Double Beta–Decay of Te^{130}", Phys. Rev.* **78** (1950) 822 – 823

[Ioa94] A. Ioannisyan and J. W. F. Valle *"SO(10) Grand Unification Model for Degenerate Neutrino Masses", Phys. Lett.* **B 322** (1994) 93 – 99 and *Preprint hep-ph/* **9402333** (1994)

[Ish96] N. Ishihara, T. Ohama and Y. Yamada *"A Proposed Detector DCBA for Double Beta Decay Experiments", Nucl. Instrum. & Meth. in Phys. Res.* **A 373** (1996) 325 – 332

[Jol34] F. Joliot and I. Curie *"Artificial Production of a New Kind of Radio–Element", Nature* 133 (Febr. 10. 1934) 201 – 202

[Kal97] J. Kalinowski, R. Rückl, H. Spiesberger and P. M. Zerwas *"Leptoquark/Squark Interpretation of HERA Events: Virtual Effects in e^+e^- Annihilation to Hadrons"*, *Preprint hep-ph/* **9703288** (1997) *DESY* **97 – 038** and *Z. Phys.* **C 74** (1997) 595 – 603

[Kan98] S. K. Kang *"Neutrinoless Double Beta Decay in the Nonminimal Supersymmetric Standard Model"*, *Phys. Lett.* **B 435** (1998) 264 – 271

[Kal2000] A. Kalliomaki and J. Maalampi *"Neutrinoless Double Beta Decay in Four-Neutrino Models"*, *Preprint hep-ph/* **0003281** (2000)

[Kay89] B. Kayser *"The Physics of Massive Neutrinos"*, *World Scientific, World Scientific Lecture Notes in Physics* **25** (1989)

[Kir87] I. Kirpichnikov, in Proc. of International Conference "Underground Physics", Baksan, Russia, August 1987, eds. E.N. Alekseev et.al.

[Kir67] T. Kirsten, W. Gentner and O. A. Schaeffer *"Massenspektrometrischer Nachweis von $\beta\beta$- Zerfallsprodukten"*, *Z. Phys.* **202** (1967) 273 – 292

[Kir68] T. Kirsten, O. A. Schaeffer, E. Norton and R. W. Stoenner *"Experimental Evidence for the Double-Beta Decay of Te^{130}"*, *Phys. Rev. Lett.* **20** (1968) 1300 – 1303

[Kir83] T. Kirsten, H. Richter and E. Jessberger *"Rejection of Evidence for Nonzero Neutrino Rest Mass from Double Beta Decay"*, *Phys. Rev. Lett.* **50** (1983) 474 – 477

[Kir88] T. Kirsten *"Upcoming Experiments and Plans in Low Energy Neutrino Physics"*, in Proceedings of 13th International Conference "Neutrino Physics and Astrophysics", Boston, June 5 – 11, 1988, eds. *J. Schneps, T. Kafka, W. A. Mann and P. Nath, World Scientific* (1989) 742 – 764

[Kla76] H. V. Klapdor *"The Structure of the Gamow-Teller Giant Resonance and Consequences for Beta-Delayed Neutron Spectra and Element Synthesis"*, *Phys. Lett.* **B 65** (1976) 35 – 38

[Kla81] H. V. Klapdor *"Reactor Neutrino Oscillation Experiments and the Shape of the Beta Strength Function"*, *Phys. Rev.* **C 23** (1981) 1269 – 1271

[Kla82a] H. V. Klapdor and J. Metzinger *"Calculation of the Anti-Neutrinos Spectrum from Thermal Fission of ^{235}U"*, *Phys. Lett.* **B 112** (1982) 22 – 26

[Kla82b] H. V. Klapdor and J. Metzinger *"Antineutrino Spectrum from the Fission Products of ^{239}Pu"*, *Phys. Rev. Lett.* **48** (1982) 127 – 131

[Kla83] H. V. Klapdor *"The Shape of the Beta Strength Function and Consequences for Nuclear Physics and Astrophysics"*, *Prog. Part. Nucl. Phys.* **10** (1983) 131 – 225

[Kla84] H. V. Klapdor and K. Grotz *"Calculation of Double Beta Decay of ^{76}Ge,^{82}Se, $^{128,130}Te$"*, *Phys. Lett.* **B 142** (1984) 323 – 328

[Kla86] H. V. Klapdor *"Nuclear Beta Strength, Neutrino Mass and Cosmology"*, *Prog. Part. Nucl. Phys.* **17** (1986) 419 – 455

[Kla86a] H. V. Klapdor and K. Grotz *"Evidence for a Nonvanishing Energy Density of the Vacuum (or Cosmological Constant)"*, *Astr. J.* **301** (1986) L39 – L43

[Kla86b] ed. H. V. Klapdor Weak and Electromagnetic Interactions in Nuclei", *Proc. of the International Symposium*, Heidelberg, July 1 – 5, 1986, *Springer-Verlag, Berlin, Heidelberg, New York, London, Paris, Tokyo*, 1110pp.

[Kla87] H. V. Klapdor *"Untersuchung des Doppelbetazerfalls von angereichertem ^{76}Ge zur Bestimmung der Neutrinomasse"*, Vorschlag eines Experiments, *MPI H* **V 17** (1987) 18pp

[Kla87a] H. V. Klapdor *"Weak and Electromagnetic Interactions in Nuclei – (Evolution of Nuclear Physics into the Particle Domain)"*, Europh. News **18, N 2** (1987) 25 – 27

[Kla88] H. V. Klapdor "Neutrinos", *Springer, Verlag, Heidelberg* (1988) 338p.

[Kla88a] H. V. Klapdor and B. Povh "Neutrino Physics", *Proceedings of an International Workshop Held in Heidelberg*, October 20 – 22, 1987, *Springer, Verlag, Heidelberg*

(1988) 333p.

[Kla89] H. V. Klapdor *"Deutsch-Sowjetisches Projekt zur Untersuchung der Neutrinomasse Gestartet"*, *Phys. Bl.* **45 Nr. 3** (1989) 86 – 87

[Kla90] H. V. Klapdor–Kleingrothaus *"Spectroscopy with Enriched Detectors: Double Beta Decay and Perspectives in Astrophysical γ -Ray Spectroscopy and in Dark Matter detection"*, in Proc. International Conference on "Gamma–Ray Line Astrophysics", Paris, France, *AIP* **232** (1990) 464 – 476

[Kla91] H. V. Klapdor–Kleingrothaus *"Neuer Vorstoß zur Bestimmung der Neutrinomasse"*, *Phys. Bl.* **47 Nr. 3** (1991) 206 – 207

[Kla91a] H. V. Klapdor–Kleingrothaus *"Beta Decay Far from Stability and Double Beta Decay, and Consequences for Astrophysics"*, in "Nuclei in the Cosmos", ed G. Oberhummer *Heidelberg: Springer* (1991) 199 – 232

[Kla91b] H. V. Klapdor–Kleingrothaus *"Concluding Remarks"*, in Proc. of 14th Europhysics Conference on Nuclear Physics: "Rare Nuclear Decays and Fundamental Physics", Bratislava, Czechoslovakia, 22 – 26 October, 1990, ed *P. Povinec*, *J. Phys.* **G 17** (1991) S537 – S543

[Kla91c] H. V. Klapdor–Kleingrothaus *"Weitere Eingrenzung der Neutrinomasse über den Doppelten Beta–Zerfall"*, *Spektrum der Wissenschaft* (October 1991)

[Kla92] H. V. Klapdor–Kleingrothaus *"The Heidelberg–Moscow Double Beta Decay Experiment with Enriched ^{76}Ge : First Results"*, *Nucl. Phys.* **A 28** *Proc. Suppl.* (1992) 207 – 209

[Kla93] H. V. Klapdor–Kleingrothaus *"The HEIDELBERG - MOSCOW Double Beta Experiment with Enriched ^{76}Ge: Status and Perspectives"*, in Proc. of 13th International Conference on "Particles and Nuclei", Perugia, Italy, 28 June – 2 July, 1993, *World Scientific, Singapore* (1994) 283 – 289

[Kla94] H. V. Klapdor–Kleingrothaus *"Introductory Remarks – Workshop Session on Double Beta Decay"*, in Proc. "TAUP'93" Conference, Gran–Sasso, Italy, *Nucl. Phys.* **B 38** *Proc. Suppl.* (1994) 351 – 353

[Kla95/98] H. V. Klapdor–Kleingrothaus and A. Staudt "Teilchenphysik ohne Beschleuniger", *B.G. Tübner, Stuttgart* (1995), "Non–Accelerator Particle Physics", *IOP Publishing, Bristol and Philadelphia* (1995) and 2. ed. (1998) translated by *S. S. Wilson* and *Moscow, Nauka, Fizmalit* (1998) translated by *V. A. Bednyakov*

[Kla96] H. V. Klapdor–Kleingrothaus and S. Stoica *eds.* "Double Beta Decay and Related Topics", Proc. of International Workshop, Trento, Italy, April 24 – May 5, 1995, *Singapore: World Scientific* (1996) 509pp.

[Kla96a] H. V. Klapdor–Kleingrothaus *"Double Beta Decay – Physics at Beyond Accelerator Energies"*, in to Proc. of 4th International Workshop on Theoretical and Phenomenological Aspects of Underground Physics (TAUP 95), Toledo, Spain, 17 – 21 Sep. 1995, *Nucl. Phys. Proc. Suppl.* **48** (1996) 216 – 222

[Kla96b] H. V. Klapdor–Kleingrothaus *"Double Beta Decay – Physics Beyond the Standard Model"*, in Proc. of the 17th International Conference "Neutrino Physics and Astrophysics", NEUTRINO'96, Helsinki, Finland, June 13 – 19, 1996, eds. *K. Enqvist, K. Huitu and J. Maalampi, World Scientific, Singapore*, (1997) 317 – 341

[Kla96c] H. V. Klapdor–Kleingrothaus *"Double Beta Decay – Physics at Beyond Accelerator Energies"*, in Proc."Double Beta Decay and Related Topics", International Workshop, Trento, Italy, April 24 – May 5, 1995, eds. *H. V. Klapdor–Kleingrothaus and S. Stoica: Singapore: World Scientific* (1996) 3 – 43

[Kla96d] H. V. Klapdor–Kleingrothaus *"Double Beta Decay: Physics at Beyond Accelerator Energies"*, in Proc. International Workshop on Future Prospects of Baryon Instability Search, Oak Ridge, Tennessee, USA, March 1996, eds. *S.J. Ball and Y.A.*

Kamishkov, OPNL (1996) 125 – 165

[Kla97] H. V. Klapdor–Kleingrothaus *"Double Beta Decay – Physics Beyond the Standard Model Now, and in Future (GENIUS)"*, in Proc. "Beyond the Desert'97": Accelerator and Nonaccelerator Approaches, eds. *H. V. Klapdor–Kleingrothaus and H. Päs, "Conference on Physics Beyond the Standard Model,* Tegernsee, Germany, 8 – 14 June, 1997, *IOP* (1998) 485 – 531

[Kla97a] H. V. Klapdor–Kleingrothaus and M. Hirsch *"A Large Scale Double Beta and Dark Matter Experiment: on the Physics Potential of GENIUS"*, *Z. Phys.* **A 359** (1997) 361 – 372

[Kla97b] H. V. Klapdor–Kleingrothaus and Y. Ramachers *"Future of Dark Matter Direct Detection Experiments"*, in Proc. "Beyond the Desert'97": Accelerator and Nonaccelerator Approaches, eds. *H. V. Klapdor–Kleingrothaus and H. Päs, "Conference on Physics Beyond the Standard Model,* Tegernsee, Germany, 8 – 14 June, 1997, *IOP* (1998) 786 – 801

[Kla97c] H. V. Klapdor–Kleingrothaus, J. Hellmig and M. Hirsch *"GENIUS - a New Experiment with Large Discovery Potential for Particle and Astrophysics,* Proposal **November 20, 1997, first draft**, (1997)

[Kla97/2000] H. V. Klapdor–Kleingrothaus and K. Zuber Teilchenastrophysik, *Teubner Studienbücher, Stuttgard* (1997), Particle Astrophysics, *Institute of Physics Publishing (IOP)*, Bristol and Philadelphia (1997) 507pp., translated by *S.M. Foster and B. Foster* and Astrophysical Chastizi, *Nauka, Fizmatlit, Moscow* (2000), translated by *V.A. Bednyakov*

[Kla98] H. V. Klapdor–Kleingrothaus *"Double Beta Decay–Physics Beyond the Standard Model Now, and in Future (GENIUS)"*, *Progr. in Part. and Nucl. Phys.* **40** (1998) 265 – 282

[Kla98a] H. V. Klapdor–Kleingrothaus *"Status and Perspectives of Double Beta Decay – Window to New Physics Beyond the Standard Model of Particle Physics"*, *Intern. Journ of Modern Phys.* **A 13, No. 23** (1998) 3953 – 3992

[Kla98b] H. V. Klapdor–Kleingrothaus *"Doppelbetazerfall – Physik jenseits des Standardmodells"*, *Phys. in unserer Zeit* **3** (1998) 123 – 130

[Kla98c] H. V. Klapdor–Kleingrothaus and Y. Ramachers *"Experiments Aiming at Direct Detection of Dark Matter"*, *Eur. Phys. J.* **A 3** (1998) 85 – 92

[Kla98d] H. V. Klapdor–Kleingrothaus, J. Hellmig and M. Hirsch *"Future Perspectives of Double Beta Decay and Dark Matter Search – GENIUS"*, *J. Phys.* **G 24** (1998) 483 – 516

[Kla98e] H. V. Klapdor–Kleingrothaus and Yu. G. Zdesenko *"Ice Shielding in the Large Scale GENIUS Experiment for Double Beta Decay and Dark Matter Search"*, *Eur. Phys. J.* **A 3** (1998) 107 – 108

[Kla98f] H. V. Klapdor–Kleingrothaus and H. Päs in Proc. "Beyond the Desert'97": Accelerator and Nonaccelerator Approaches, eds. *H. V. Klapdor–Kleingrothaus and H. Päs, "Conference on Physics Beyond the Standard Model"*, Tegernsee, Germany, 8 – 14 June, 1997, *IOP* (1998) 1 – 992

Kla98g [Kla98g] H. V. Klapdor–Kleingrothaus *"Status and Perspectives of Double Beta Decay and Dark Matter Search – Windows to New Physics,* in Proc. "Physics Beyond the Standard Model", Santa Fe, New Mexico, USA, June 14 – 19, 1998, eds. *P. Herczeg, C. M. Hoffman and H. V. Klapdor–Kleingrothaus, World Scientific* (1999) 275 – 311

[Kla99] H. V. Klapdor–Kleingrothaus *"Double Beta and Dark Matter Search – Window to New Physics Beyond the Standard Model of Particle Physics"*, in Proc. "Lepton and Baryon Number Violation in Particle Physics, Astrophysics and Cosmology", eds. *H.*

V. Klapdor–Kleingrothaus and I. V. Krivosheina, International Workshop at ECT, Trento, Italy, April 20 – April 25, 1998, *IOP, Bristol* (1999) 251 – 301

[Kla99a] H. V. Klapdor–Kleingrothaus L. Baudis, G. Heusser, B. Majorovits and H. Päs "GENIUS - a Supersensitive Germanium Detector System for Rare Events", *Proposal* **August 1999 second draft**, *Preprint hep-ph/* **9910205** (1999) and in Proc. "Beyond the Desert'99: Accelerator, Non–accelerator and Space Approaches into the Next Millenium, BEYOND 2000", International Conference, Castle Ringberg, Tegernsee, Germany, June 6 – 12, 1999, eds. *H. V. Klapdor–Kleingrothaus and I. V. Krivosheina, IOP* (2000) 915 – 1015

[Kla99b] H. V. Klapdor–Kleingrothaus *"Double Beta Decay with Ge–detectors – and the Future of Double Beta and Dark Matter Search (GENIUS)"*, in Proc. of International Conference "Neutrino Physics and Astrophysics", NEUTRINO'98, Takayama, Japan, 4 – 9 June, 1998, eds. *Y. Suzuki and Y. Totsuka, Nucl. Phys.* **B 77** *Proc. Suppl.* (1999) 357 – 368

[Kla99c] H. V. Klapdor–Kleingrothaus, H. Päs and U. Sarkar *"Test of Special Relativity and Equivalence Principle from Neutrinoless Double Beta Decay", Eur. Phys. J.* **A 5** (1999) 3 – 6

[Kla99d] H. V. Klapdor–Kleingrothaus and S. Kolb *"Neutrinoless Double Beta Decay and Dark Matter Search with GENIUS"*, in the Proceedings of the NANP'99 (Dubna, 1999), ed by V. Bednyakov et al. *Phys. Atom. Nucl.* **63, N 7** (2000)

[Kla99e] H. V. Klapdor–Kleingrothaus *"Perspectives of Double Beta and Dark Matter Search as Windows to New Physics"*, in the Proceedings "Symmetries in Intermediate High Energy Physics", eds. *A. Faessler, T. Kosmas and G. K. Leontaris, Springer–Verlag, Berlin, Heidelberg Springer Tracts in Modern Physics* **63** (2000) 69 – 104

[Kla99f] Proc. "Lepton and Baryon Number Violation in Particle Physics, Astrophysics and Cosmology", eds. *H. V. Klapdor–Kleingrothaus and I. V. Krivosheina, International Workshop at ECT,* Trento, Italy, April 20 – April 25, 1998, *IOP, Bristol* (1999) 760pp.

[Kla2000] H. V. Klapdor–Kleingrothaus, H. Päs and A. Yu. Smirnov *"Neutrino Mass Spectrum and the Future of Neutrinoless Double Beta Decay", Preprint hep–ph/* **0003219** (2000) 48pp.

[Kla2000a] H. V. Klapdor–Kleingrothaus, H. Päs and U. Sarkar *"Effects of New Gravitational Interactions on Neutrinoless Double Beta Decay", Preprint hep–ph/* **0002215** (2000) 10pp.

[Kla2000b] H. V. Klapdor–Kleingrothaus, H. Päs and U. Sarkar *"Effects of Quantum Space Time Foam in the Neutrino Sector", Preprint hep–ph/* **0004123** (2000) 5pp.

[Kla2000c] H. V. Klapdor–Kleingrothaus *"Ten Years of Heidelberg–Moscow Experiment – a Fresh Look"*, in Proc. International Symposium "Advanced in Nuclear Physics", Bucharest, Romania, 9 – 10 December 1999, *World Scientific* (2000), ed. by *D. N. Poenaru, S. Stoica et al.* 69 – 104

[Kla2000d] H. V. Klapdor–Kleingrothaus and I. V. Krivosheina (eds.) Proc. "Beyond the Desert99": Accelerator, Non–accelerator and Space Approaches, *"Conference on Physics Beyond the Standard Model,* Tegernsee, Germany, 6 – 12 June, 1999, *IOP* (2000) 1 – 1200

[Kla2000e] H. V. Klapdor–Kleingrothaus and H. Päs *"Neutrinoless Double Beta Decay and New Physics in the Neutrino Sector"*, in Proc. "COSMO 99": 3rd International Conference on Particle Physics and the Early Universe, Trieste, Italy, 27 September - 3 October 1999 and *Preprint hep–ph/* **0002109** (2000)

[Kla2000f] H. V. Klapdor–Kleingrothaus, H. Päs and U. Sarkar *"Comment on "Closing the Neutrinoless Double Beta Decay Window into VEP and/or VLI", Preprint hep–ph/*

0002105 (2000)

[Kla2000g] H. V. Klapdor–Kleingrothaus and H. Päs *"On Nuclear Matrix Element Uncertainties in Short Range $0\nu\beta\beta$ Decay"*, *Preprint hep–ph/* **0005045** (2000) 3p.

[Kla2000h] H. V. Klapdor–Kleingrothaus, St. Kolb and V. A. Kuzmin *"Light Lepton Number Violating Sneutrinos and the Baryon Number of the Universe"*, *Preprint hep–ph/* **9909546** (1999) and *in Press in Phys. Rev.* **D** (2000)

[Kla2000i] H. V. Klapdor–Kleingrothaus and H. Päs *"The Quest for the Neutrino Mass Spectrum"*, *Subm to Comments in Nucl. and Part. Phys.* and *Preprint physics/* **0006024** (2000) 18pp.

[Kla2000j] H. V. Klapdor–Kleingrothaus *"Comment" Subm. to Phys. Rev.* **D** (2000)

[Kla2000k] H. V. Klapdor–Kleingrothaus et al. *"Results on Cold Dark Matter from the HDMS Prototype Detector"*, *Subm. for Publ.* (2000)

[Kli86] A. A. Klimenko, S. B. Osetrov, A. A. Pomansky, A. A. Smolnikov and S. I. Vasilev *"Search for Beta Beta Decay of ^{150}Nd: Next Step"*, in Proc. "Weak and electromagnetic interactions in nuclei (WEIN'86)", Heidelberg, Germany, (1986), *ed. H. V. Klapdor, Springer–Verlag Berlin, Heidelberg, New York, London, Paris, Tokyo,* (1986) 701 – 702

[Kli89] A. A. Klimenko, S. B. Osetrov, A. A. Pomansky, A. A. Smolnikov and S. I. Vasilev *"The Baksan Experiment on ^{100}Mo Double Beta Decay (Systema-II)"*, in Proc. International Workshop on the Theoretical and Phenomenological Aspects of Underground Physics,, L'Aquila, Italy, 25 – 28 Sep. 1989, *in L'Aquila 1989, TAUP'89* (1989) 241 – 246, eds. *A. Bottino and P. Monacelli* and in Proc. "Weak and electromagnetic interactions in nuclei (WEIN'89)", Montreal, Canada, (1989) 697 – 699

[Kol87] E. W. Kolb, A. J. Stebbins and M. S. Turner *"How Reliable are Neutrino Mass Limits Derived from SN1987A?"*, *Phys. Rev.* **D 35** (1987) 3598 – 3606

[Kol98] S. Kolb, H. V. Klapdor–Kleingrothaus, M. Hirsch and S. G. Kovalenko *"Observing Majorana–Like Sneutrinos at Future Colliders"*, in Proc. "Beyond the Desert'97": Accelerator and Nonaccelerator Approaches, eds. *H. V. Klapdor–Kleingrothaus and H. Päs,* "Conference on Physics Beyond the Standard Model, Tegernsee, Germany, 8 – 14 June, 1997, *IOP* (1998) 245 – 253

[Kol2000] S. Kolb, M. Hirsch, H. V. Klapdor–Kleingrothaus and O. Panella *"Collider Signatures of Sneutrino Cold Dark Matter"*, *Phys. Lett.* **B 478** (2000) 262 – 268 and *Preprint hep–ph/* **9910542** (1999)

[Kol98] S. Kolb, M. Hirsch, H. V. Klapdor–Kleingrothaus and O. Panella *"Collider Signatures of Sneutrino Cold Dark Matter"*, *Phys. Lett.* **B 478** (2000) 262 – 268 and *Preprint hep-ph/* **9910542** (1999)

[Kot86] T. Kotani *"Are Massive Neutrinos Majorana Particles?"*, in Proc. of 12th International Conference on "Neutrino Physics and Astrophysics", Sendai, Japan, June 3 - 8, 1986, eds *T. Kitagaki and H. Yuta, World Scientific, Singapore* (1986) 114 – 134

[Kud92] N. Kudomi, H. Ejiri, K. Nagata, K. Okada, T. Shibata, T. Shima and J. Tanaka *"Double Beta Decay of ^{100}Mo to Excited States in ^{100}Ru"*, *Phys. Rev.* **C 46** (1992) R2132 – R2135

[Kum96] K. Kume, K. Fushimi, R. Hazama, M. Kawasaki, N. Kudomi, K. Nagata, T. Nitta, H. Ohsumi, K. Okada H. Sano, T. Senoo, T. Shibata, T. Shima, K. Shinmyo, N. Suzuki, J. Tanaka, H. Ejiri, K. Hayashi, V. Kuts and Yu. Zdesenko *"Double Beta Decays Studied by ELEGANT"*, in Proc. International Workshop "Double Beta Decay and Related Topics", Trento, Italy, April 24 – May 5, 1995", eds. H. V. Klapdor–Kleingrothaus and S. Stoica: *Singapore: World Scientific* (1996) 490 – 495

[Lan88] P. Langacker *"Massive Neutrinos in Gauge Theories"*, in "Neutrinos", ed. *H. V. Klapdor, Springer–Verlag Berlin, Heidelberg, New York, London, Paris, Tokyo*

(1988) 71 – 115

[Lan92] P. Langacker *"Scenarios for Neutrino Mass"*, in Proc. of 4th International Symposium on "Neutrino Telescopes", Venice, Italy, March 10 – 13, ed M. Baldo-Ceolin, (1992) 73 – 111

[Lee56] T. D. Lee and C. N. Yang *"Question of Parity Conservation in Weak Interactions"*, *Phys. Rev.* **104** (1956) 254 – 258

[Lee94] D.-G. Lee and R. N. Mohapatra *"Natural Doublet – Triplet Splitting in Supersymmetric SO(10) Models"*, *Phys. Lett.* **B 324** (1994) 376 – 379 *Preprint hep-ph/* **9310371** (1993)

[Leg86] V. A. Legasov Letter from Academik Academii Nauk USSR, V. A. Legasov *Kurchatov Institute*, (July 8, 1986)

[Leo89] G. K. Leontaris and J. D. Vergados *"Lepton Phenomenology and Neutrinoless Double β – Decay in Superstring Inspired Theories"*, *Z. Phys.* **C 41** (1989) 623 – 629

[Leo95] G. K. Leontaris and S. Lola *"Neutrino Masses in Supersymmetric Theories"*, *Preprint hep-ph/* **9509351** (1995), *Preprint HD–THEP–95–31A* (Sep. 1995) and 5th Hellenic School and Workshops on Elementary Particle Physics, Corfu, Greece, 3 – 24 September, 1995

[Let88] *Letters from the German Ministry of Research and Technology about funding the HEIDELBERG-MOSCOW experiment, September 1988 and October 1988* (1988)

[Lev50] C. A. Levine, A. Ghiorso and G. T. Seaborg *"Half–Life for Double Beta–Decay"*, *Phys. Rev.* **77** (1950) 296

[Liu97] Y. Liu, L.-Z. Hu and M.-L. Ge *"Effect of Violation of Quantum Mechanics on Neutrino Oscillation"*, *Phys. Rev.* **D 56** (1997) 6648 – 6652

[Lop94a] J. L. Lopez, D. V. Nanopoulos, G. T. Park, X. Wang and A. Zichichi *"Experimental aspects of SU(5) $\times$ U(1) supergravity"*, *Phys. Rev.* **D 50** (1994) 2164 – 2190

[Lub80] V. A. Lubimov, E. G. Novikov, V. Z. Nozik, E. F. Tretyakov and V. S. Kosik *"An Estimate of the ν_e mass from the β-spectrum of tritium in the Valine Molecule"*, *Phys. Lett.* **B 94** (1980) 266 – 268

[Lue98] R. Luescher, J. Farine, F. Boehm, J. Busto, K. Gabathuler, G. Gervasio, H. E. Henrikson, V. Jorgens, K. Lou, A. Paic, D. Schenker, A. Tadsen, M. Treichel, J. L. Vuilleumier, J. M. Vuilleumier and H. Wong *"Search for $\beta\beta$ Decay in ^{136}Xe: New results from the Gotthard Experiment"*, *Phys. Lett.* **B 434** (1998) 407 – 414

[Ma99] E. Ma *"Simple Connection Between Atmospheric and Solar Neutrino Vacuum Oscillations"*, *Phys. Rev. Lett.* **83** (1999) 2514 – 2517 and *Preprint hep-ph/* **9902392**

[Mad95] T. M. Maddern and L. W. Mitchell *"Ion Transport in Xenon Gas: Applications to Double Beta Decay"*, *Nucl. Instrum. Meth.* **A 359** (1995) 506 – 510

[Mad89] R. Madey, B. S. Flanders, B. D. Anderson, A. R. Baldwin, J. W. Watson, S. M. Austin, C. C. Foster, H. V. Klapdor and K. Grotz, *"Low – Lying Structures in the Gamow – Teller Strength Functions for the Double – Beta – Decaying Nuclei ^{76}Ge, ^{82}Se, ^{128}Te, and ^{130}Te"*, *Phys. Rev.* **C 40** (1989) 540 – 552

[Maj37] E. Majorana, *"Teoria Simmetrica dell'Elettrone e del Positrone"*, *Nuovo Cimento* **14** (1937) 171 – 184

[Maj99] B. Majorovits and H. V. Klapdor-Kleingrothaus *"Digital Pulse Shape Analysis by Neural Networks for the Heidelberg–Moscow-Double-Beta–Decay-Experiment"*, *Eur. Phys. J.* **A 6** (1999) 463 – 466 and *Preprint hep-ex/* **9911001** (1999)

[Man86] O. K. Manuel *"Geochemical Measurements of Double–Beta Decay"*, in Proc. International Symposium "Nuclear beta decays and neutrino", eds. *T. Kotani, H. Ejiri and E. Takasugi,* Osaka, Japan, 1986, *World Scientific Singapore* (1986), 71 – 80

[Man91] O. K. Manuel *"Geochemical Measurements of Double–Beta Decay"*, in Proc. of 14th Europhysics Conference on "Nuclear Physics: Rare Nuclear Decays and Fundamental Physics", Bratislava, Czechoslovakia, 22 – 26 October, 1990, ed *P. Povinec*, *J. Phys.* **G 17** (1991) S221 – S229

[Mas86] A. Masiero, D. V. Nanopoulos and A. I. Sanda *"Observable Physics from Superstring Exotica: Small Dirac Neutrino Mass"*, in Proc. *23rd* Int. Conf. on High Energy Physics, Berkeley, Calif., July 16-23, (1986)

[Mas87] A. Masiero *"Small Dirac Neutrino Masses in Superstrings"*, in Proc. "Neutrino Masses and Neutrino Astrophysics", Ashland 1987, 90 – 98

[Mat66] E. der Mateosian and M. Goldhaber *"Limits for Lepton–Conserving and Lepton–Nonconserving Double Beta Decay in Ca^{48}"*, *Phys. Rev.* **146** (1966) 810 – 815

[Mem88] Memorandum between MPI, Heidelberg and Kurchatov Institute, Moscow (1988)

[Min97] H. Minakata and O. Yasuda *"Dark Matter Neutrinos Must Come with Degenerate Masses"*, *Nucl. Phys.* **B 523** (1998) 597 – 610 and *Preprint hep-ph/* **9712291**

[Min97a] H. Minakata and O. Yasuda *"Constraining Almost Degenerate Three-Flavor Neutrinos"*, *Phys. Rev.* **D 56** (1997) 1692 – 1697

[Min2000] H. Minakata *"Answering the Sphinx's Questions on Neutrinos"*, *Preprint hep-ph/* **0004249** (2000)

[Min96] P. Minkowski *" $e^-e^- \longrightarrow W^-W^-$ – Mediated by Heavy Neutrino Flavors"*, *Int. J. Mod. Phys.* **A 11** (1996) 1591 – 1605

[Miy94] M. Miyajima, S. Sasaki and H. Tawara *"Search for Double Beta – Decay Products of ^{136}Xe in Liquid Xenon"*, *IEEE Trans. Nucl. Scien.* **41** (1994) 835 – 839

[Moe80] M. Moe and D. D. Lowenthal *"Double Beta Decay of ^{82}Se"*, *Phys. Rev.* **C 22** (1980) 2186 – 2203

[Moe91] M. Moe *"New Approach to the Detection of Neutrinoless Double Beta Decay"*, *Phys. Rev.* **C 44** (1991) R931 – R934

[Moe94] M. Moe and P. Vogel *"Double Beta Decay"*, *Annu. Rev. Nucl. Part. Sci.* **44** (1994) 247 – 283

[Moe95] M. Moe *"Double Beta Decay"*, in Proc. of the 16th International Conference on Neutrino Physics and Astrophysics, "NEUTRINO'94", Eilat, Israel, 29 May – 3 June, 1994, eds. *A. Dar, G. Eilam and M. Gronau*, *Nucl. Phys.* **B 38** *Proc. Suppl.* (1995) 36 – 44

[Moh80] R. N. Mohapatra and G. Senjanović *"Neutrino Mass and Spontaneous Parity Nonconservation"*, *Phys. Rev. Lett.* **44** (1980) 912 – 915

[Moh81] R. N. Mohapatra and J. D. Vergados *"New Contribution to Neutrinoless Double Beta Decay in Gauge Models"*, *Phys. Rev. Lett.* **47** (1981) 1713 – 1716

[Moh81a] R. N. Mohapatra and G. Senjanović *"Neutrino Masses and Mixings in Gauge Models with Spontaneous Parity Violation"*, *Phys. Rev.* **D 23** (1981) 165 – 180

[Moh86] R. N. Mohapatra *"Limits on the Mass of the Right–Handed Majorana Neutrino"*, *Phys. Rev.* **D 34** (1986) 909 – 910

[Moh86a] R. N. Mohapatra *"New Contributions to Neutrinoless Double–Beta Decay in Supersymmetric Theories"*, *Phys. Rev.* **D 34** (1986) 3457 – 3461

[Moh86b] R. N. Mohapatra and J. W. F. Valle *"Neutrino Mass and Baryon–Number Nonconservation in Superstring Models"*, *Phys. Rev.* **D 34** (1986) 1642 – 1645

[Moh86c] R. N. Mohapatra "Unification and Supersymmetry", the Frontiers of Quark Lepton Physics, (Contemporary Physics), *Springer – Verlag* (1986)

[Moh88] R. N. Mohapatra and E. Takasugi, *"Neutrinoless Double Beta Decay with Double Majoron Emission"*, *Phys. Lett.* **B 211** (1988) 192 – 196

[Moh88a] R. N. Mohapatra *"Neutrinos in Left–Right Symmetric, SO(10) and Superstring Inspired Models"*, "Neutrinos", Graduate texts in contemporary physics", ed. *H. V.*

Klapdor, Berlin, Germany: Springer (1988) 117 – 146

[Moh91] R. N. Mohapatra and P. B. Pal "Massive Neutrinos in Physics and Astrophysics", *Singapore, Singapore: World Scientific, World Scientific lecture notes in physics*, **41** (1991) 318 pp.

[Moh96] R. N. Mohapatra *"Neutrinoless Double Beta Decay and Physics Beyond the Standard Model"*, in Proc. International Workshop "Double Beta Decay and Related Topics", Trento, Italy, April 24 – May 5, 1995, eds. *H. V. Klapdor–Kleingrothaus and S. Stoica: Singapore: World Scientific* (1996) 44 – 68

[Moh97] R. N. Mohapatra *"Neutrino Mass Textures and the Nature of New Physics Implied by Present Neutrino Data"*, in Proc. of 17th International Conference on Neutrino Physics and Astrophysics, Helsinki, Finland, June 13 – 19, 1996, eds. *K. Enqvist, K. Huitu and J. Maalampi, World Scientific, Singapore*, (1997) 290 – 302

[Moh97a] R. N. Mohapatra *"Neutrino Puzzles and their Implications for the Nature of New Physics"*, in Proc of First International Workshop on Particle Physics and the Early Universe", Ambleside, England, September 15 – 19, 1997, ed. L. Roszkowski, *World Scientific, Singapore* (1998) 71 – 84 and in Proc of International School of Nuclear Physics: 19th Course: "Neutrinos in Astro, Particle, and Nuclear Physics", Erice, Italy, 16 – 24 September, 1997, ed. A. Faessler *Prog. Part. Nucl. Phys.* **40** (1998) 55 – 69

[Moh98] R. N. Mohapatra *"Particle Physics Implications of Neutrinoless Double Beta Decay"*, in Proc. of International Conference "Neutrino Physics and Astrophysics",- NEUTRINO'98, Takayama, Japan, 4 – 9 June, 1998, eds. *Y. Suzuki and Y. Totsuka, Nucl. Phys.* **B 77** *Proc. Suppl.* (1999) 376 – 383

[Moh99] R. N. Mohapatra and A. Pérez-Lorenzana *"Breaking Parity Symmetry Using Extra Dimentions", Preprint hep-ph/* **9909389** (1999) 1 – 10 and *Phys. Lett.* **B 468** (1999) 195 – 200

[MPG94] H. V. Klapdor–Kleingrothaus *"Suchaktion im Tunnel", MPG–Spiegel* **6** (1994) 6 – 10

[Mut88] K. Muto and H. V. Klapdor *"Double Beta Decay, Neutrino Mass and Nuclear Structure"*, in "Neutrinos", Graduate texts in contemporary physics", ed. *H. V. Klapdor Berlin, Germany: Springer* (1988) 183 – 238

[Mut89] K. Muto, E. Bender and H. V. Klapdor *"Effects of Ground – State Correlations on $2\nu\beta\beta$ Decay Rates and Limitations of the QRPA Approach", Z. Phys.* **A 334** (1989) 177 – 186

[Mut89a] K. Muto, E. Bender and H. V. Klapdor *"Nuclear Structure Effects on the Neutrinoless Double Beta Decay", Z. Phys.* **A 334** (1989) 187 – 194

[Mut91] K. Muto, E. Bender and H. V. Klapdor-Kleingrothaus *"A Comparative Study of Double Beta Decay by Shell Model and Quasiparticle RPA", Z. Phys.* **A 339** (1991) 435 – 444

[Mut93] K. Muto *"Coulomb Effect in Operator Expansion Method for Two–Neutrino Double Beta Decay", Phys. Rev.* **C 48** (1993) 402 – 412

[Nak96] H. Nakada, T. Sebe and K. Muto *"Ealistic Shell Model Calculation of Two Neutrino Double Beta Nuclear Matrix Elements and Role of Shell Structure in Intermediate States", Nucl. Phys.* **A 607** (1996) 235 – 249

[Nan87] D. V. Nanopoulos and K. A. Olive *"A Superstring Neutrino–Dominated Universe?", Nature* **327** (1987) 487 – 489

[Nar95] E. Nardi, E. Roulet and D. Tommasini *"New Neutral Gauge Bosons and New Heavy Fermions in the Light of the New LEP Data", Phys. Lett.* **B 344** (1995) 225 – 232

[NEM2000] NEMO Collaboration *" Status Report on the Double-β Decay Experiment NEMO-3"*, Contributed paper for XIX International Conference of Neutrino Physics and Astrophysics, NEUTRINO 2000, Sudbury, Canada, June 16 - 21, 2000 *LAL* **00-31** (2000) 1 – 10

[Neu78] W. Neuhauser, M. Hohenstatt, P. Toschek and H. Dehmett *"Optical–Sideband Cooling of Visible Atom Cloud Confined in Parabolic Well"*, *Phys. Rev. Lett.* **41** (1978) 233 – 236

[NEU90] *"Das Letzte Rätsel der Physik"*, *Neuheiten* **6, ISSN 0938–4480** (Dez./Jan. 1990/1991) p.58

[Nie84] J. F. Nieves *"Dirac and Pseudo-Dirac Neutrinos and Neutrinoless Double Beta Decay"*, *Phys. Lett.* **B 147** (1984) 375 – 379

[Oka88] K. Okada, H. Ejiri, T. Shibata, Y. Nagai, N. Takahashi, N. Kamikubota, T. Watanabe, T. Shima, J. Tanaka, T. Taniguchi, T. Kamada, M. Kobiki *"Double Beta Decay Measurements for ^{76}Ge and ^{100}Mo"*, *Nucl. Phys.* **A 478** (1988) 447 – 452

[Pae99] H. Päs, M. Hirsch, H. V. Klapdor–Kleingrothaus and S. G. Kovalenko *"Towards a Superformula for Neutrinoless Double Beta Decay"*, *Phys. Lett.* **B 453** (1999) 194 – 198

[Pae99a] H. Päs, M. Hirsch, H. V. Klapdor-Kleingrothaus *"Improved Bounds on SUSY Accompanied Neutrinoless Double Beta Decay"*, *Phys. Lett.* **B 459** (1999) 450 – 454 and *hep-ph/* **9810382** (1998)

[Pae2000] H. Päs, M. Hirsch, H. V. Klapdor–Kleingrothaus and S. G. Kovalenko *"A Superformula for Neutrinoless Double Beta Decay II: The Short Range Part"*, *to be Publ.* (2000)

[Pan95] O. Panella and Y. N. Srivastava *"Bounds on Compositeness from Neutrinoless Double β Decay"*, *Phys. Rev.* **D 52** (1995) 5308 – 5313 and *Preprint hep-ph/* **9411224** (1994)

[Pan96] G. Pantis, F. Šimkovic, J. D. Vergados and A. Fässler *"Neutrinoless Double Beta Decay within QRPA with Proton–Neutron Pairing"*, *Phys. Rev.* **C 53** (1996) 695 – 707

[Pan98a] G. Pantis, F. Šimkovic and A. Fässler *"QRPA and RQRPA Calculations of Neutrinoless Double–Beta Decay beyond the Point of Collapse"*, *Yad. Fizika* **61** (1998) 1311 – 1317 and *Phys. Atom. Nucl.* **61** (1998) 1211 – 1217

[Pan97] O. Panella, C. Carimalo, Y. N. Srivastava and A. Widom *"Neutrinoless Double Beta Decay with Composite Neutrinos"*, *Preprint hep–ph/* **9701251** (1997) and *Phys. Rev.* **D 56** (1997) 5766 – 5775

[Pan98] O. Panella, C. Carimalo, Y. N. Srivastava, and A. Widom *"Bounds on Compositeness from Neutrino–Less Double Beta Decay"*, in Proc. "Beyond the Desert'97": Accelerator and Nonaccelerator Approaches, eds. *H. V. Klapdor–Kleingrothaus and H. Päs,* "Conference on Physics Beyond the Standard Model, Tegernsee, Germany, 8 – 14 June, 1997, *IOP*, (1998) 339 – 367

[Pan99] O. Panella, C. Carimalo and Y. N. Srivastava *"Low– and High–Energy Approaches to Lepton Number Violation"*, in Proc. "Beyond the Desert'99": Accelerator, Non–accelerator and Space Approaches, eds. *H. V. Klapdor–Kleingrothaus and I. V. Krivosheina, "Conference on Physics Beyond the Standard Model",* Tegernsee, Germany, 6 – 12 June, 1999, *IOP* (2000)

[Pan2000] O. Panella, C. Carimalo and Y. N. Srivastava *"Production of Like Sign Dileptons in p–p Collisions through Composite Majorana Neutrinos"*, *Preprint hep–ph/* **9903253** (1999) and *Phys. Rev.* **D 62** (2000) 015013-1 – 15

[Pat96] J. C. Pati *"The Essential Role of String–Derived Symmetries in Ensuring Proton–Stability and Light Neutrino Masses"*, *Phys. Lett.* **B 388** (1996) 532 – 542 and *Preprint hep–ph/* **9607446** (1996)

[Pau30] W. Pauli *"Brief an die Gruppe der Radioaktiven ..."*, Zürich, 4 Dezember, 1930 (1930)

[Pet82] S. T. Petcov *"On Pseudodirac Neutrinos, Neutrino Oscillations and Neutrinoless Double Beta Decay"*, *Phys. Lett.* **B 110** (1982) 245 – 249

[Pet94] S. T. Petcov and A. Yu. Smirnov *"Neutrinoless Double Beta Decay and the Solar Neutrino Problem"*, *Phys. Lett.* **B 322** (1994) 109 – 118

[Pon68] B. Pontecorvo *"Superweak Interactions and Double Beta Decay"*, *Phys. Lett.* **B 26** (1968) 630 – 632

[Pop86] L. A. Popeco, A. V. Derbin, I. A. Kondurov, V. V. Martyanov, H. V. Klapdor and J. Metzinger *"New Possibilities in a Double Beta Decay Experiment Using Enriched ^{76}Ge Inside of an Active Si(Li) Shielding"*, in Proc. WEIN'86 International Symposium "Weak and Electromagnetic Interactions in Nuclei", Heidelberg, Germany, July 1 – 5, 1986 *ed. H. V. Klapdor, Springer–Verlag Berlin, Heidelberg, New York, London, Paris, Tokyo,* (1986) 703 – 704

[Pop89] L. A. Popeco, A. V. Derbin, A. V. Chernij, S. B. Bachlanov, V. N. Muratov, G. A. Shishkina and H. V. Klapdor *"Super Low Background Installation with Germanium Shielding"*, *Pisma v JETF* **50** (1989) 222 – 225

[Pre90] K. P. Pretzl *"Superconducting Granule Detectors"*, *Part. World* **1** (1990) 153 – 162

[Pre93] K. P. Pretzl *"Bringing Dark Matter in from the Dark"*, *Europhys. News* **24** (1993) 167 – 171

[Pre93a] K. Pretzl *"Superconducting Granule Detectors"*, *Journal of Low–Temperature Physics* **93, No3/4** (1993) 439 – 448

[Pre99] K. Pretzl *"Cryogenic Calorimeters in Astro and Particle Physics"*, *Preprint hep–ex/* **9911028** (1999) and in Proc. of 8th International Conference on "Calorimetry in High–Energy Physics (CALOR 99)", Lisbon, Portugal, 13 – 19 June, 1999

[Pri98] J. R. Primack *"Cold + Hot Dark Matter after Superkamiokande"*, in Proc. of 10th Rencontres de Blois: "The Birth of Galaxies" Blois, France, 28 Jun - 4 July, 1998 and *Preprint astro–ph/* **9810204** (1998)

[Pri59] H. Primakoff and S. P. Rosen *"Double Beta Decay"*, *Rep. Progr. Phys.* **22** (1959) 121 – 166

[Pri69] H. Primakoff and S. P. Rosen *"Nuclear Double–Beta Decay and a New Limit on Lepton Nonconservation"*, *Phys. Rev.* **184** (1969) 1925 – 1933

[Pri81] H. Primakoff and S. P. Rosen *"Baryon Number and Lepton Number Conservation Laws"*, *Ann. Rev. Nucl. Part. Sci.* **31** (1981) 145 – 192

[Pro88] MPI, Heidelberg and Kurchatov Institute, Moscow *"Protokoll: Über die Durchführung eines Gemeinsamen Experimentes zur Suche nach dem Doppelbetazerfall von ^{76}Ge"*, *MPI, Heidelberg and Kurchatov Institute, Moskau, für die Jahre 1988 – 1992* (1988) 1

[Rac37] G. Racah *"Sulla Simmetria tra Particelle e Antiparticelle"*, *Nuovo Cimento* **14** (1937) 322 – 328

[Rad96] P. B. Radha, D. J. Dean, S. E. Koonin, T. T. S. Kuo, K. Langanke, A. Poves, J. Retamosa and P. Vogel *"Shell Model Monte Carlo Method for Two-Neutrino Double Beta Decay"*, *Phys. Rev. Lett.* **76** (1996) 2642 – 2645 and *Preprint nucl-th/* **9510052** (1995)

[Rad86] H. V. Klapdor *"Die Rätsel um die kleinsten und größten Dimensionen im Kosmos,"*, Gespräch zwischen Dr. P. Zaun und Prof. Dr. H. V. Klapdor, *Gesendet vom*

Norddeutschen Rundfunk (NDR) in der Reihe "Wissenschaft aktuell", des 3. Programms, am 15. 11. 1986, (1986) 1 – 11

[Rad87] H. V. Klapdor *"Keine Neutrinos beim Doppel-Betazerfall?"*, Prof. Dr H. V. Klapdor Interviewed von Dr. P. Zaun, *Gesendet vom RIAS Berlin in der Reihe "Wissenschaft aktuell" des 3. Programms*, am 15. 7. 1987, (1987) 1 – 4

[Rad91] A. A. Raduta, A. Faessler, S. Stoica and W. A. Kaminski *"The 2ν $\beta\beta$ Decay Rate within a Higher RPA Approach"*, *Phys. Lett.* **B 254** (1991) 7 – 12

[Rad91a] A. A. Raduta, A. Faessler and S. Stoica *"The Two Neutrino Beta–Beta Decay Rate Within a Boson Expansion Formalism"*, *Nucl. Phys.* **A 534** (1991) 149 – 172

[Rag94] R. S. Raghavan *"New Approach to the Search for Neutrinoless Double Beta Decay"*, *Phys. Rev. Lett.* **72** (1994) 1411 – 1414

[Rei59] F. Reines and C. L. Cowan *"Free Antineutrino Absorption Crosssection. I: Measurement of the Free Antineutrino Absorption Crosssection by Protons"*, *Phys. Rev.* **113** (1959) 273 – 279

[Rin80] P. Ring and P. Schuck "The Nuclear Many Body Problem", *Heidelberg, Berlin, New York: Springer* (1980)

[Riz82] Th. G. Rizzo *"Inverse Neutrinoless Double Beta Decay"*, *Phys. Lett.* **B 116** (1982) 23 – 28

[Riz93] Th. G. Rizzo *"Inverse Neutrinoless Double Beta Decay ay the NLC?"*, *Preprint hep–ph/* **9306276** (1993) and in Proc. of 2nd International Workshop on "Physics and Experiments with Linear e^+e^- Colliders", Waikoloa, HI, 26 – 30 April, 1993, **0520-523**

[Ros81] S. P. Rosen *"Lepton Nonconservation and Double Beta Decay: Constraints on the Masses and Couplings of Majorana Neutrinos"* in Proc. International Conference "NEUTRINO'81", Hawaii **v.2** (1981) 76 – 92, eds. *R. J. Cence, E. Ma. and A. Roberts,* Publ. *by High Energy Physics Group Department of Physics and Astrophysics University of Hawaii*

[Ros88] S. P. Rosen *"On Seeing the First Double Beta Decay"*, *Comments Nucl. Part. Phys.* **18** (1988) 31 – 54

[Ros92] S. P. Rosen *"Double Beta Decay"*, *Preprint hep–ph/* **9210202** (1992)

[Ros85] G. G. Ross and J. W. F. Valle *"Sypersymmetric Models Without R Parity"*, *Phys. Lett.* **B 151** (1985) 375 – 381

[Sat87] K. Sato and H. Suzuki *"Analysis of Neutrino Burst from the Supernova in LMC"*, *Phys. Lett.* **58** (1987) 2722 – 2735

[Sar99] X. Sarazin for the NEMO Collaboration *"Double-β Decays With the NEMO Experiment: Final Results of NEMO–2 with Various Nuclei and Status of NEMO–3"*, in Proc. of 5th International Workshop on Topics in Astroparticle and Underground Physics (TAUP 97), Gran Sasso, Italy, 7 – 11 Sep. 1997, eds. *A. Bottino, A. Di Credico and P. Monacelli, Nucl. Phys. Proc. Suppl.* **70** (1999) 239 – 241

[Sch80] J. Schechter and J. W. F. Valle *"Neutrino Masses in SU(2) $\otimes$ U(1) Theories"*, *Phys. Rev.* **D 22** (1980) 2227 – 2235

[Sch81] J. Schechter and J. W. F. Valle *"Majorana Neutrinos and Magnetic Fields"*, *Phys. Rev.* **D 24** (1981) 1883 – 1889

[Sch82a] J. Schechter and J. W. F. Valle *"Neutrinoless Double-β Decay in SU(2) $\times$ U(1) Theories"*, *Phys. Rev.* **D 25** (1982) 2951 – 2954

[Sci91] J. A. Emison *Unique crystal 'grown' in OR could play role in basic physics"*, *Science-Business, the Oak Ridge*, (February 10, 1991) page 1D

[Sci97] D. Normile *New Experiments Step Up Hunt for Neutrino Mass"*, *Science* **276** (1997) 1795

[Sim89] F. Šimkovic *"A New Method of the Calculation of the Two-Neutrino Double Beta Decay Amplitudes"*, *Dubna, JINR Communications* **39** (1989) 21 – 29

[Sim96] F. Šimkovic, J. Schwinger, M. Veselský, G. Pantis and A. Fässler *"Non-Collapsing Renormalized QPRA with Proton-Neutron Pairing for Neutrinoless Double Beta Decay", Preprint nucl-th/* **9612037** (1996) and in *Phys. Lett.* **B 393** (1997) 267 – 273

[Sim98] F. Šimkovic, G. Pantis and A. Faessler *"Two-Neutrino Double Beta Decay: Critical Analysis"*, in Proc. of 1st International Workshop on Nonaccelerator New Physics (NANP 97), *Dubna, Moscow Region, Russia, 7 – 11 July, 1997, Yad. Fiz.* **61** (1998) 1318 – 1328 and *Phys. Atom. Nucl.* **61** (1998) 1218 – 1228, and *Preprint nucl-th/* **9711060** (1997)

[Sim99] F. Šimkovic, G. Pantis J. D. Vergados and A. Faessler *"Additional Nucleon Current Contributions to Neutrinoless Double Beta Decay", Phys. Rev.* **C 60** (1999) 055502 and *Preprint hep-ph/* **9905509** (1999)

[Sim99a] F. Šimkovic, G. Pantis *"A Field Theory Approach to Two-Neutrino Double Beta Decay", Yad. Fiz.* **62, N 2** (1999) 632 – 638 and *Phys. Atom. Nucl.* **62** (1999) 585 – 591

[Smi96a] A. Yu. Smirnov *"Neutrino Masses and Oscillations"*, in Proc. 28th International Conference on "High-energy Physics (ICHEP 96)", Warsaw, Poland, 25 – 31 July, 1996, *ICHEP '96*, **vol. 1** 288 – 306

[Smi99] A. Yu. Smirnov *Lepton Mixing: Small. Large, Maximal?", Preprint hep-ph/* **9907296** (1999)

[Sta90] A. Staudt, T. T. S. Kuo and H. V. Klapdor–Kleingrothaus *"$\beta\beta$ Decay of ^{76}Ge with Renormalized Effective Interaction Derived from Paris, Bonn and Reid Potentials", Phys. Lett.* **B 242** (1990) 17 – 23

[Sta90a] A. Staudt, K. Muto and H. V. Klapdor-Kleingrothaus *"Calculation of 2ν and 0ν Double-Beta Decay Rates", Europhys. Lett.* **13 (1)** (1990) 31 – 36

[Sta90b] A. Staudt, E. Bender, K. Muto and H. V. Klapdor-Kleingrothaus *"Second-Generation Microscopic Predictions of Beta-Decay Half-Lives of Neutron-Rich Nuclei", At. Nucl. Data Tables* **44** (1990) 79 – 132

[Sta91] A. Staudt, K. Muto and H. V. Klapdor–Kleingrothaus *"Nuclear Matrix Elements for Double Positron Emission", Phys. Lett.* **B 268** (1991) 312 – 316

[Sta92] A. Staudt, T. T. S. Kuo and H. V. Klapdor–Kleingrothaus *"$\beta\beta$ Decay of ^{128}Te, ^{130}Te, and ^{76}Ge with Renormalized Effective Interactions Derived from Paris and Bonn Potentials", Phys. Rev.* **C 46** (1992) 871 – 883

[Sta92a] A. Staudt and H. V. Klapdor–Kleingrothaus *"Calculation of β-Delayed Fission Rates of Neutron-Rich Nuclei far off Stability", Nucl. Phys.* **A 549** 254 – 264

[Sto93a] S. Stoica and W. A. Kaminski *" Gamow-Teller Matrix Elements for Two-Neutrino Double Beta Decay within a Second Quasi-Random-Phase Approximation", Phys. Rev.* **C 47** (1993) 867 – 869

[Sto93b] S. Stoica and W. A. Kaminski *"Double-β-Decay Rates within a Second Quasi-Random-Phase Approximation", Nuovo Cimento* **A 106** (1993) 723 – 733

[Sto94] S. Stoica *"Half-Lives for Two Neutrino Double-Beta-Decay Transitions to First 2^+ Excited States", Phys. Rev.* **C 49** (1994) 2240 – 2243

[Sto96] S. Stoica *"Second-QRPA Calculations for Two Neutrino Double-Beta Decay"*, in Proc. "Double Beta Decay and Related Topics", *International Workshop*, Trento, Italy, April 24 – May 5, 1995, eds. *H. V. Klapdor-Kleingrothaus and S. Stoica, Singapore, Singapore: World Scientific* (1996) 382 – 398

[Sto2000] S. Stoica and H. V. Klapdor–Kleingrothaus *"Double-Beta Decay Matrix Elements for ^{76}Ge", to be Published* (2000)

[Suh92] J. Suhonen, O. Civitarese and A. Faessler *"Description of the $0^+ \to 0^+$ Neutrinoless Double-Beta Decay Transition in ^{76}Ge: Particle-Number-Projected Quasiparti-*

cle Random Phase Approximation", *Nucl. Phys.* **A 543** (1992) 645 – 660

[Suh93] J. Suhonen and O. Civitarese *"Estimation of Bounds for Left–Right Mixing from Nuclear Double Beta Decay Processes"*, *Phys. Lett.* **B 312** (1993) 367 – 371

[Suh97] J. Suhonen, P. C. Divari, L. D. Skouras and I. P. Johnstone *"Double Beta Decay of* ^{92}Mo*: Comparison of the Shell Model and the Quasiparticle Random–Phase Approximation"*, *Phys. Rev.* **C 55** (1997) 714 – 719

[Suh98] J. Suhonen and O. Civitarese *"Weak–Interaction and Nuclear–Structure Aspects of Nuclear Double Beta Decay"*, *Phys. Rept.* **300** (1998) 123 – 214

[Suz95] A. Suzuki (KAMLAND Collaboration) *KAMLAND Proposal* (1995) (in Japanese)

[Suz99a] A. Suzuki (KAMLAND Collaboration) *"Solar Neutrinos, Atmospheric Neutrinos and Proton Decays in Super–Kamiokande and KamLAND Project"*, in Proc. "Lepton and Baryon Number Violation in Particle Physics, Astrophysics and Cosmology", eds. *H. V. Klapdor–Kleingrothaus and I. V. Krivosheina, International Workshop at ECT,* Trento, Italy, April 20 – April 25, 1998, *World Scientific* (1999), 189 – 209.

[Suz99b] A. Suzuki (KAMLAND Collaboration) *"Present Status of KAMLAND"*, in Proc. of International Conference "Neutrino Physics and Astrophysics",NEUTRINO'98, Takayama, Japan, 4 – 9 June, 1998, eds. *Y. Suzuki and Y. Totsuka, Nucl. Phys. Proc. Suppl.* **77** (1999) 171 – 176

[Suz99c] A. Suzuki (KAMLAND Collaboration) *"The Present Status of KAMLAND"*, in Proc. of 8th International Workshop on "Neutrino Telescopes", Venezia, Italy, February 23 – 26, 1999, ed. *M. Baldo–Ceolin,* (1999) 325 – 336

[Suz2000] Y. Suzuki (for the SUPERKAMIOKANDE collaboration) *"Super-Kamiokande",* in Proc. of International Conference "Neutrino Physics and Astrophysics",NEUTRINO'-2000, Sudbury, Canada, 16 –21 June, 2000, ed *A. McDonald, Nucl. Phys. Proc. Suppl.* **B** (2000)

[Tak66] N. Takaoka and K.. Ogata *"The Half–life of* ^{130}Te *Double* β*–Decay"*, *Z. Naturforsch* **A 21** (1966) 84 – 90

[Tak96] N. Takaoka, Y. Motomura and K. Nagao *"Half–Life of* ^{130}Te *Double–*β *Decay Measured with Geologically Qualified Samples"*, *Phys. Rev.* **C 53** (1996) 1557 – 1561

[Tak84] E. Takasugi *"Can the Neutrinoless Double Beta Decay Take Place in the Case of Dirac Neutrinos?"*, *Phys. Lett.* **B 149** (1984) 372 – 376

[Tak98] E. Takasugi *"Double Beta Decay Constraint on Composite Neutrinos"*, in Proc. "Beyond the Desert'97": Accelerator and Nonaccelerator Approaches, eds. *H. V. Klapdor–Kleingrothaus and H. Päs*, "Conference on Physics Beyond the Standard Model, Tegernsee, Germany, 8 – 14 June, 1997, *IOP* (1998) 360 – 366

[Tan93] J. Tanaka and H. Ejiri *"Limits on Single and Double Majoron Emission Processes in Neutrinoless Double* β *Decay of* ^{100}Mo*"*, *Phys. Rev.* **D 48** (1993) 5412 – 5415

[Toi95] J. Toivanen and J. Suhonen *"Renormalized Proton–Neutron Quasiparticle Random–Phase Approximation and Its Application to Double Beta Decay"*, *Phys. Rev. Lett.* **75** (1995) 410 – 413,

[Toi97] J. Toivanen and J. Suhonen *"Study of Several Double–Beta–Decaying Nuclei Using the Renormalized Proton Neutron Quasiparticle Random–Phase Approximation"*, *Phys. Rev.* **C 55** (1997) 2314 – 2323

[Tom86] T. Tomoda, A. Fässler, K. W. Schmid and F. Grümmer, *"Neutrinoless* $\beta\beta$ *Decay and a New Limit on the Right–Handed Current"*, *Nucl. Phys.* **A 452** (1986) 591 – 620

[Tom87] T. Tomoda and A. Faessler *"Suppression of the Neutrinoless* $\beta\beta$ *Decay ?"*, *Phys. Lett.* **B 191** (1987) 475 – 481

[Tom88] T. Tomoda *" $0^+ \to 2^+$ Neutrinoless $\beta\beta$ Decay of ^{76}Ge"*, *Nucl. Phys.* **A 484** (1988) 635 - 646

[Tom91] T. Tomoda *"Double Beta Decay"*, *Rept. Prog. Phys.* **54** (1991) 53 – 126

[Tre95] V. I. Tretyak and Yu. G. Zdesenko *"Tables of Double Beta Decay Data"*, *At. Data Nucl. Data Tables* **61** (1995) 43 – 62

[Tre91] M. Treichel, C. Broggini, D. Reusser, L. Fluri, V. Joergens, L. W. Mitchell, C. Nussbaum, J. L. Vuilleumier, F. Boehm, P. Fisher, H. Henrikson and K. Gabathuler *"Double Beta Decay and Dark Matter in the Gotthard Germanium Experiment"*, in Proc. of 14th Europhysics Conference on "Nuclear Physics: Rare Nuclear Decays and Fundamental Physics", Bratislava, Czechoslovakia, 22 – 26 October, 1990, ed *P. Povinec*, *J. Phys.* **G 17** (1991) *Suppl.* S193 – S201

[Tur91] A. L. Turkevich, T. E. Economou and G. A. Cowan *"Double Beta Decay of ^{238}U"*, *Phys. Rev. Lett.* **67** (1991) 3211 – 3214

[Val83] J. W. F. Valle *"Neutrinoless Double β Decay With Quasi-Dirac Neutrinos"*, *Phys. Rev.* **D 27** (1983) 1672 – 1674

[Val87] J. W. F. Valle *"Neutrino Mass and New Light Gauge Boson in Superstring Models"*, *Phys. Lett.* **B 196** (1987) 157 – 162

[Val87a] J. W. F. Valle *"Ultralight Neutrinos and R Parity in Superstring Models"*, *Phys. Lett.* **B 196** (1987) 73 – 84

[Val96] J. W. F. Valle *"Neutrino–less Double Beta Decay and Beyond the Standard Model Physics"*, in Proc. "Double Beta Decay and Related Topics", International Workshop, Trento, Italy, April 24 – May 5, 1995, eds. *H. V. Klapdor-Kleingrothaus and S. Stoica, Singapore, Singapore: World Scientific* (1996) 69 – 90

[Vas90] A. A. Vasenko, I. V. Kirpichnikov, V. A. Kuznetsov, A. S. Starostin, A. G. Dzhanian, G. E. Markosian, V. M. Oganessian, V. S. Pogosov, A. G. Tamanian and S. R. Shakhazizian *"New Results in the ITEP/YePI Double Beta–Decay Experiment with Enriched Germanium Detectors"*, *Mod. Phys. Lett.* **A 5** (1990) 1299 – 1306

[Vas93] S. I. Vasilev, A. A. Klimenko, S. B. Osetrov, A. A. Pomansky, A. A. Smolnikov *"Search for a two Neutrino Mode of Double Beta decay of the isotope ^{150}Nd"*, *JETP Lett.* **58** (1993) 178 – 180 and *Pisma Zh. Eksp. Teor. Fiz.* **58** (1993) 177 – 179

[Vem2000] S. K. Vempati *"Neutrino Masses from SUSY: Different Contributions and their Implications"*, *Pramana* **54** (2000) 133 - 146

[Ver76] J. D. Vergados *"Double β–Decay Nuclear Matrix Elements and Lepton Coservation"*, *Phys. Rev.* **C 13** (1976) 865 – 871

[Ver81] J. D. Vergados *"Lepton Violating Double β Decay in Modern Gauge Theories"*, *Phys. Rev.* **C 24** (1981) 640 – 653 and *Preprint IOANNINA* **-149** (Oct. 1980)

[Ver82] J. D. Vergados *"Pion–Double–Charge–Exchange Contribution to Neutrinoless - Double–β Decay"*, *Phys. Rev.* **D 25** (1982) 914 – 917

[Ver82a] J. D. Vergados *"A Limit on the Neutrino Masses from Double β Decay"*, *Phys. Lett.* **B 109** (1982) 96 – 100 and *Erratum-ibid.* **B 113** (1982) 513

[Ver86] J. D. Vergados *"The Neutrino Mass and Family, Lepton and Baryon Non Conservation in Gauge Theories"*, *Phys. Reports* **133** (1986) 1 – 216

[Ver87] J. D. Vergados *"Neutrinoless Double β–Decay Without Majorana Neutrinos in Supersymmetric Theories"*, *Phys. Lett.* **B 184** (1987) 55 – 62

[Ver88] J. D. Vergados A. Fässler and T. Tomoda *"The Δ (3/2, 3/2) Contribution to the $0^+ \longrightarrow 0^+$ $\beta\beta$ Decay Transitions"*, *Nucl. Phys.* **A 490** (1988) 556 –570

[Ver89] J. D. Vergados *"Does the Δ(3/2, 3/2) Resonance Contribute to $0^+ \longrightarrow 0^+$ $\beta\beta$ Decay?"*, *Phys. Lett.* **B 218** (1989) 119 – 123

[Vis99] F. Vissani *"Signal of Neutrinoless Double Beta Decay, Neutrino Spectrum and Oscillation Scenarios"*, *JHEP* **9906** (1999) 022 and *Preprint hep-ph/* **9906525** (1999)

1 – 14

[Vog86] P. Vogel and M. R. Zirnbauer *"Suppression of the Two–Neutrino Double-Beta Decay by Nuclear–Structure Effects"*, *Phys. Rev. Lett.* **57** (1986) 3148 – 3151

[Vog96] P. Vogel *"QRPA Methods in Double Beta Decay and Related Processes:*, in Proc. "Double Beta Decay and Related Topics", International Workshop, Trento, Italy, April 24 – May 5, 1995, eds. *H. V. Klapdor-Kleingrothaus and S. Stoica, Singapore, Singapore: World Scientific* (1996) 323 – 338

[Vui93] J. -C. Vuilleumier, J. Busto, J. Farine, V. Jörgens, L. W. Mitchell, M. Treichel, J. -L. Vuilleumier, H. T. Wong, F. Böhm, P. Fisher, H. E. Henrikson, D. A. Imel, M. Z. Iqbal, B. M. O'Callaghan-Hay, J. Thomas and K. Gabathuler *"Search for Neutrinoless Double–β Decay in ^{136}Xe with a Time Projection Chamber"*, *Phys. Rev.* **D 48** (1993) 1009 – 1020

[Wal87] M. M. Waldrop *"Possible First Hints of Double Beta Decay"*, *Science* **235** (1987) 534

[Wic34] G. C. Wick *Rendiconti di Academia Nazionale dei Lincei*, **19** (1934) 319

[Wod99] A. Wodecki and W. A. Kamiński *Limits on R-Parity Nonconservation from the Neutrinoless Double Beta Decay in the MSSM with Gauge Mediated Breaking"*, *Phys. Rev.* **C 59** (1999) R1232 – R1236 *Preprint hep–ph/* **9806288** (1998)

[Wod99a] A. Wodecki W. A. Kamiński and F. Šimkovic *Grand Unified Theory Constrained Supersymmetry and Neutrinoless Double β decay*, *Phys. Rev.* **D 60** (1999) 115007-1 – 115007-18

[Wol81] L. Wolfenstein *"CP Properties of Majorana Neutrinos and Double Beta Decay"*, *Phys. Lett.* **B 107** (1981) 77 – 79

[Wol81a] L. Wolfenstein *"Different Varieties of Massive Dirac Neutrinos"*, *Nucl. Phys.* **B 186** (1981) 147 – 1527

[Won91] H. T. Wong, F. Böhm, P. Fisher, K. Gabathuler, H. E. Henrikson, D. A. Imel, M. Z. Iqbal, V. Jörgens, L. W. Mitchell, B. M. O'Callaghan-Hay, J. Thomas, M. Treichel, J. -C. Vuilleumier and J. -L. Vuilleumier *"New Limit on Neutrinoless Double β Decay in ^{136}Xe with a Time Projection Chamber"*, *Phys. Rev. Lett.* **67** (1991) 1218 – 1221

[Wu57] C. S. Wu, E. Ambler, R. W. Hayward, D. D. Hoppes and R. P. Hudson *"Experimental Test of Parity Conservation in Beta Decay"*, *Phys. Rev.* **105** (1957) 1413 – 1414

[Wu91] X. R. Wu, A. Staudt, H. V. Klapdor–Kleingrothaus, Cheng-Rui Ching and Tso–Hsiu Ho *"Two Neutrino Double Beta Decay with Operator Expansion Method"*, *Phys. Lett.* **B 272** (1991) 169 – 172

[Wu92] X. R. Wu, A. Staudt, T. T. S. Kuo and H. V. Klapdor–Kleingrothaus *"Tensor Force and Operator Expansion Method for Nuclear Double Beta Decay"*, *Phys. Lett.* **B 276** (1992) 274 – 278

[Wu93] X. R. Wu, M. Hirsch, A. Staudt, H. V. Klapdor–Kleingrothaus, Cheng–Rui Ching and Tso–Hsiu Ho *"New Theoretical Results of $2\nu\beta\beta$ Decay with the Operator Expansion Method"*, *Commun. Theor. Phys.* **20** (1993) 453 – 460

[Yan79] T. Yanagida *"Horizontal Gauge Symmetry and Masses of Neutrinos"*, in Proceedings of the Workshop on "Unified Theory and the Baryon Number in the Universe, 13 – 14 February, 1979, eds. *O. Sawada and A. Sugamoto*, KEK, Tsukuba, Japan (1979) 95 - 98

[Yas99] O. Yasuda *"Constraining Degenerate Neutrino Mass Model and Implications"*, in Proc. "Beyond the Desert'99": Accelerator, Non–accelerator and Space Approaches, *"Conference on Physics Beyond the Standard Model*, eds. *H. V. Klapdor–Kleingrothaus and I. V. Krivosheina*, Tegernsee, Germany, 6 – 12 June, 1999, *IOP* (2000)

[You91] K. You, Y.-c. Zhu, J.-g. Lu, H.-s. Sun, W.-h. Tian, W.-h. Zhao, Z.-p. Zheng, M.-h. Ye C.-r. Ching, T.-h. Ho, F.-z. Cui, C.-j. Yu and G.-j. Jiang *"A Search for Neutrinoless Double Beta Decay of ^{48}Ca"*, *Phys. Lett.* **B 265** (1991) 53 – 56

[Zde80] Yu. G. Zdesenko *"Double β Decay and Conservation of Lepton Charge"*, *Sov. J. Part. Nucl.* **11** (1980) 542 – 563

[Zde80a] Yu. G. Zdesenko, I. A. Mytsyk, A. S. Nikolaiko and V. N. Kuts *"Study of the Double Beta Decay of ^{130}Te"*, *Yad. Fiz.* **32** (1980) 607 – 616 and *Sov. J. Nucl. Phys.* **32** (1980) 312 – 317

[Zde91] Yu. G. Zdesenko *"Double Beta Decay Experiments at Kiev"*, in Proc. of 14th Europhysics Conference on "Nuclear Physics: Rare Nuclear Decays and Fundamental Physics", Bratislava, Czechoslovakia, 22 – 26 October, 1990, ed. *P. Povinec*, *J. Phys.* **G 17** (1991) *Suppl.* S243 – S249

[Zha90] L. Zhao, B. A. Brown and W. A. Richter *"Shell Model Calculation for Two-Neutrino Double Beta Decay of ^{48}Ca"*, *Phys. Rev.* **C 42** (1990) 1120 – 1125

CHAPTER 2

Original Articles

2.1 From the Early Days until the Gauge Theory Era

2.1.1 The First Steps in Double Beta Research

Physikalisches Institut
der Eidg. Technischen Hochschule
Zürich

Zürich, 4. Dez. 1930
Gloriastr.

Liebe Radioaktive Damen und Herren,

wie der Überbringer dieser Zeilen, den ich huldvollst anzuhören bitte, Ihnen des näheren auseinandersetzen wird, bin ich angesichts der „falschen" Statistik der N- und Li 6-Kerne, sowie des kontinuierlichen β-Spektrums auf einen verzweifelten Ausweg verfallen, um den „Wechselsatz" der Statistik und den Energiesatz zu retten. Nämlich die Möglichkeit, es könnten elektrisch neutrale Teilchen, die ich Neutronen nennen will, in den Kernen existieren, welche den Spin 1/2 haben und das Ausschließungsprinzip befolgen und sich von Lichtquanten außerdem noch dadurch unterscheiden, daß sie nicht mit Lichtgeschwindigkeit laufen. Die Masse der Neutronen müßte von derselben Größenordnung wie die Elektronenmasse sein und jedenfalls nicht größer als 0,01 Protonenmasse. — Das kontinuierliche β-Spektrum wäre dann verständlich unter der Annahme, daß beim β-Zerfall mit dem Elektron jeweils noch ein Neutron emittiert wird, derart, daß die Summe der Energien von Neutron und Elektron konstant ist.

Nun handelt es sich weiter darum, welche Kräfte auf die Neutronen wirken. Das wahrscheinlichste Modell für das Neutron scheint mir aus wellenmechanischen Gründen (näheres weiß der Überbringer dieser Zeilen) dieses zu sein, daß das ruhende Neutron ein magnetischer Dipol von einem gewissen Moment μ ist. Die Experimente verlangen wohl, daß die ionisierende Wirkung eines solchen Neutrons nicht größer sein kann als die eines γ-Strahls, und dann darf μ wohl nicht größer sein als $e \cdot 10^{-13}$ cm.

Ich traue mich vorläufig aber nicht, etwas über diese Idee zu publizieren, und wende mich erst vertrauensvoll an Euch, liebe Radioaktive, mit der Frage, wie es um den experimentellen Nachweis eines solchen Neutrons stände, wenn dieses ein ebensolches oder etwa 10mal größeres Durchdringungsvermögen besitzen würde wie ein γ-Strahl.

Ich gebe zu, daß mein Ausweg vielleicht von vornherein wenig wahrscheinlich erscheinen mag, weil man die Neutronen, wenn sie existieren, wohl längst gesehen hätte. Aber nur wer wagt, gewinnt, und der Ernst der Situation beim kontinuierlichen β-Spektrum wird durch einen Ausspruch meines verehrten Vorgängers im Amte, Herrn Debye, beleuchtet, der mir kürzlich in Brüssel gesagt hat: „O, daran soll man am besten gar nicht denken, so wie an die neuen Steuern." Darum soll man jeden Weg zur Rettung ernstlich diskutieren. — Also, liebe Radioaktive, prüfet, und richtet. — Leider kann ich nicht persönlich in Tübingen erscheinen, da ich infolge eines in der Nacht vom 6. zum 7. Dez. in Zürich stattfindenden Balles hier unabkömmlich bin. — Mit vielen Grüßen an Euch, sowie auch an Herrn Back, Euer untertänigster Diener

W. Pauli

532 NATURE APRIL 7, 1934

The "Neutrino"

THE view has recently been put forward[1] that a neutral particle of about electronic mass, and spin $\frac{1}{2}\hbar$ (where $\hbar = h/2\pi$) exists, and that this 'neutrino' is emitted together with an electron in β-decay. This assumption allows the conservation laws for energy and angular momentum to hold in nuclear physics[2]. Both the emitted electron and neutrino could be described either (*a*) as having existed before in the nucleus or (*b*) as being created at the time of emission. In a recent paper[3] Fermi has proposed a model of β-disintegration using (*b*) which seems to be confirmed by experiment.

According to (*a*), one should picture the neutron as being built up of a proton, an electron and a neutrino, while if one accepts (*b*), the rôles of neutron and proton would be symmetrical[4] and one would expect that positive electrons could also sometimes be created together with a neutrino in nuclear transformations. Therefore the experiments of Curie and Joliot[5] on an artificial positive β-decay give strong support to method (*b*), as one can scarcely assume the existence of positive electrons in the nucleus.

Why, then, have positive electrons never been found in the natural β-decay? This can be explained by the fact that radioactivity usually starts with α-emission and therefore leads to nuclei the charge of which is too small compared with their weight. The artificial β-emission was found for two unstable nuclei (most probably N^{13} and P^{30}) formed by capture of an α-particle and emission of a neutron, and therefore having too high a charge for their mass.

A consequence of assumption (*b*) is that two isobares differing by 1 in atomic number can only be stable if the difference of their masses is less than the mass of electron and neutrino together. For otherwise the heavier of the two elements would disintegrate with emission of a neutrino and either a positive or negative electron. There will be only a limited region on the mass defect curve, probably at medium atomic weight, where such small differences are possible. In fact, neighbouring isobares have only been found with the mass numbers 87, 115, 121, 123, (187), (203), while isobares with atomic numbers differing by 2 are very frequent. In the first case, one of the two nuclei (Rb) is known to emit β-rays. In each of the last two cases one of the two isobares is stated to be exceedingly rare and its identification might be due to experimental error. The other three cases actually lie close together and have medium weight. A particular case of isobares are proton and neutron. Since all experimentally deduced values of the neutron mass lie between 1·0068 and 1·0078, they are certainly both stable even if the mass of the neutrino should be zero.

The possibility of creating neutrinos necessarily implies the existence of annihilation processes. The most interesting amongst them would be the following: a neutrino hits a nucleus and a positive or negative electron is created while the neutrino disappears and the charge of the nucleus changes by 1.

The cross section σ for such processes for a neutrino of given energy may be estimated from the lifetime *t* of β-radiating nuclei giving neutrinos of the same energy. (This estimate is in accord with Fermi's model but is more general.) Dimensionally, the connexion will be

$$\sigma = A/t$$

where A has the dimension cm.^{2} sec. The longest length and time which can possibly be involved are $\hbar/mc$ and $\hbar/mc^2$. Therefore

$$\sigma < \frac{\hbar^3}{m^3 c^4 t}$$

For an energy of $2 \cdot 3 \times 10^6$ volts, t is 3 minutes and therefore $\sigma < 10^{-44}$ cm.^{2} (corresponding to a penetrating power of 10^{16} km. in solid matter). It is therefore absolutely impossible to observe processes of this kind with the neutrinos created in nuclear transformations.

With increasing energy, σ increases (in Fermi's model[3] for large energies as $(E/mc^2)^2$) but even if one assumes a very steep increase, it seems highly improbable that, even for cosmic ray energies, σ becomes large enough to allow the process to be observed.

If, therefore, the neutrino has no interaction with other particles besides the processes of creation and annihilation mentioned—and it is not necessary to assume interaction in order to explain the function of the neutrino in nuclear transformations—one can conclude that there is no practically possible way of observing the neutrino.

H. BETHE.
R. PEIERLS.

Physical Laboratory,
University,
Manchester.
Feb. 20.

[1] W. Pauli, quoted repeatedly since 1931, to be published shortly in "Rapports du Septième Conseil Solvay, Brussels", 1933.
[2] C. D. Ellis and N. F. Mott, *Proc. Roy. Soc.*, A, **141**, 502; 1933.
[3] E. Fermi, *La Ricerca Scientifica*, **2**, No. 12; 1933.
[4] This point of view was first put forward by I. Curie and F. Joliot at the Conseil Solvay, 1933.
[5] I. Curie and F. Joliot, NATURE, **133**, 201, Feb. 10, 1934.

Z. Phys. **88** (1934) (21. Februar) 161 – 177 161

Versuch einer Theorie der β-Strahlen. I[1]).

Von **E. Fermi** in Rom.

Mit 3 Abbildungen. (Eingegangen am 16. Januar 1934.)

Eine quantitative Theorie des β-Zerfalls wird vorgeschlagen, in welcher man die Existenz des Neutrinos annimmt, und die Emission der Elektronen und Neutrinos aus einem Kern beim β-Zerfall mit einer ähnlichen Methode behandelt, wie die Emission eines Lichtquants aus einem angeregten Atom in der Strahlungstheorie. Formeln für die Lebensdauer und für die Form des emittierten kontinuierlichen β-Strahlenspektrums werden abgeleitet und mit der Erfahrung verglichen.

1. Grundannahmen der Theorie.

Bei dem Versuch, eine Theorie der Kernelektronen sowie der β-Emission aufzubauen, begegnet man bekanntlich zwei Schwierigkeiten. Die erste ist durch das kontinuierliche β-Strahlenspektrum bedingt. Falls der Erhaltungssatz der Energie gültig bleiben soll, muß man annehmen, daß ein Bruchteil der beim β-Zerfall frei werdenden Energie unseren bisherigen Beobachtungsmöglichkeiten entgeht. Nach dem Vorschlag von W. Pauli kann man z. B. annehmen, daß beim β-Zerfall nicht nur ein Elektron, sondern auch ein neues Teilchen, das sogenannte „Neutrino" (Masse von der Größenordnung oder kleiner als die Elektronenmasse; keine elektrische Ladung) emittiert wird. In der vorliegenden Theorie werden wir die Hypothese des Neutrinos zugrunde legen.

Eine weitere Schwierigkeit für die Theorie der Kernelektronen besteht darin, daß die jetzigen relativistischen Theorien der leichten Teilchen (Elektronen oder Neutrinos) nicht imstande sind, in einwandfreier Weise zu erklären, wie solche Teilchen in Bahnen von Kerndimensionen gebunden werden können.

Es scheint deswegen zweckmäßiger, mit Heisenberg[2]) anzunehmen, daß ein Kern nur aus schweren Teilchen, Protonen und Neutronen, besteht. Um trotzdem die Möglichkeit der β-Emission zu verstehen, wollen wir versuchen, eine Theorie der Emission leichter Teilchen aus einem Kern in Analogie zur Theorie der Emission eines Lichtquants aus einem angeregten Atom beim gewöhnlichen Strahlungsprozeß aufzubauen. In der Strahlungstheorie ist die totale Anzahl der Lichtquanten keine Konstante: Lichtquanten entstehen, wenn sie von einem Atom emittiert werden, und verschwinden, wenn sie absorbiert werden. In Analogie hierzu wollen wir der β-Strahlentheorie folgende Annahmen zugrunde legen:

[1]) Vgl. die vorläufige Mitteilung: La Ricerca Scientifica **2**, Heft 12, 1933. — [2]) W. Heisenberg, ZS. f. Phys. **77**, 1, 1932.

SEPTEMBER 15, 1935 PHYSICAL REVIEW VOLUME 48

Double Beta-Disintegration

M. GOEPPERT-MAYER, *The Johns Hopkins University*

(Received May 20, 1935)

From the Fermi theory of β-disintegration the probability of simultaneous emission of two electrons (and two neutrinos) has been calculated. The result is that this process occurs sufficiently rarely to allow a half-life of over 10^{17} years for a nucleus, even if its isobar of atomic number different by 2 were more stable by 20 times the electron mass.

1. INTRODUCTION

IN a table showing the existing atomic nuclei it is observed that many groups of isobars occur, the term isobar referring to nuclei of the same atomic weight but different atomic number. It is unreasonable to assume that all isobars have exactly the same energy; one of them therefore will have the lowest energy, the others are unstable. The question arises why the unstable nuclei are in reality metastable, that is, why, in geologic time, they have not all been transformed into the most stable isobar by consecutive β-disintegrations.

The explanation has been given by Heisenberg[1] and lies in the fact that the energies of nuclei of fixed atomic weight, plotted against atomic number, do not lie on one smooth curve, but, because of the peculiar stability of the α-particle are distributed alternately on two smooth curves, displaced by an approximately constant amount against each other (the minimum of each curve is therefore at, roughly, the same atomic number). For even atomic weight the nuclei of even atomic number lie on the lower curve, those with odd atomic number on the higher one. One β-disintegration then brings a nucleus from a point on the lower curve into one of the upper curve, or *vice versa*. The nuclei on the upper curve are all of them unstable. But it may happen that a nucleus on the lower curve, in the neighborhood of the minimum, even though it is not the most stable one, cannot emit a single β-particle, since the resultant isobar, whose energy lies on the upper curve, has higher energy. This nucleus would then be metastable, since it cannot go over into a more stable one by consecutive emission of two electrons. This explanation is borne out by the fact that almost only isobars of even difference in atomic number occur.

A metastable isobar can, however, change into a more stable one by simultaneous emission of two electrons. It is generally assumed that the frequency of such a process is very small. In this paper the propability of a disintegration of that kind has been calculated.

The only method to attack processes involving the emission of electrons from nuclei is that of Fermi[2] which associates with the emission of an electron that of a neutrino, a chargeless particle of negligible mass. Thereby it is possible to explain the continuous β-spectrum and yet to have the energy conserved in each individual process by adjusting the momentum of the neutrino. In this theory the treatment of a β-disintegration is very similar to that of the emission of light by an excited atom.

A disintegration with the simultaneous emission of two electrons and two neutrinos will then be in strong analogy to the Raman effect, or, even more closely, to the simultaneous emission of two light quanta,[3] and can be calculated in essentially the same manner, namely, from the second-order terms in the perturbation theory. The process will appear as the simultaneous occurrence of two transitions, each of which does not fulfill the law of conservation of energy separately.

The following investigation is a calculation of the second-order perturbation, due to the interaction potential introduced by Fermi between neutrons, protons, electrons and neutrinos. As far as possible the notation used is that of Fermi. For a more detailed discussion and justification of this mathematical form and the assumptions involved reference must be made to Fermi's paper.

[1] W. Heisenberg, Zeits. f. Physik **78**, 156 (1932).

[2] E. Fermi, Zeits. f. Physik **88**, 161 (1934).

[3] M. Goeppert-Mayer, Ann. d. Physik (V) **9**, 273 (1931).

2. The Mathematical Apparatus

The nucleus is assumed not to contain any electrons and neutrinos but to be built up out of neutrons and protons only. Neutron and proton are regarded as not essentially different from one another, but to represent two different quantum states of the heavy particle. The two kinds of light particles outside of the nucleus, the electrons and neutrinos are treated according to the method of superquantization. The stationary states of the electrons are taken to be those of positive energy H_s in the Coulomb field of the nucleus, described by four Dirac functions $\psi_s=(\psi_s^1, \psi_s^2, \psi_s^3, \psi_s^4)$. Since the neutrinos are not affected by the field of the nucleus their eigenfunctions are represented by plane Dirac waves $\varphi_\sigma=(\varphi_\sigma^1, \varphi_\sigma^2, \varphi_\sigma^3, \varphi_\sigma^4)$ with energies K_σ. The Pauli principle is assumed to hold for both electrons and neutrinos, so that the number N_s of electrons in a stationary state s and the number M_σ of neutrinos in a stationary state σ can be 0 or 1 only.

A quantum state N of the total system is then determined by these numbers N_s, M_σ and by the quantum state n, with energy W_n of the nucleus; n means in this case not only the quantum states in the ordinary sense but includes the quantum states of all possible isobars. The energy of a quantum state N is given by

$$E_N=\sum_s N_s H_s+\sum_\sigma M_\sigma K_\sigma+W_n. \qquad (1)$$

The interaction energy between the heavy particles within the nucleus and the light particles without is constructed in such a way that the transition from neutron to proton is necessarily accompanied by the emission of an electron and a neutrino and *vice versa*. A matrix element of H corresponding to the transition of a neutron with eigenfunction u_n to a proton with eigenfunction v_m is different from 0 only if at the same time two numbers N_s, M_σ change from 0 to 1, and is then given by

$$H_M^N=H_{m\cdots 1_s\cdots 1_\sigma\cdots}^{n\cdots 0_s\cdots 0_\sigma\cdots}$$

$$=(-1)^{N_1+\cdots+N_{s-1}+M_1+\cdots+M_{\sigma-1}}\tilde{\psi}_s\delta\varphi_\sigma^* H_{nm}, \qquad (2)$$

with the abreviation

$$\tilde{\psi}\delta\varphi^*=-\psi^{*1}\varphi^{*2}+\psi^{*2}\varphi^{*1}+\psi^{*3}\varphi^{*4}-\psi^{*4}\varphi^{*3}, \qquad (3)^*$$

$$H_{nm}=g\int v_m^* u_n d\tau. \qquad (4)$$

The functions ψ and φ are taken at the place of the nucleus, the assumption being made that they do not vary considerably over its range. For an "allowed" transition $\int v_m^* u_n d\tau$, taken over the volume of the nucleus, has the order of magnitude 1. The value of the proportionality constant g can be roughly obtained by equating calculated and observed intensity of β-disintegration. Fermi determines it to be

$$g=4\times 10^{-50}\ \text{cm}^3\ \text{ergs}. \qquad (5)$$

3. Calculations

The probability of simultaneous emission of two electrons and two neutrinos is obtained from the second order terms of the perturbation theory. In this approximation the amplitude of the transition probability is given by

$$a_M{}^N=\sum_K \frac{H_K{}^N H_M{}^K}{E_K-E_N}\left\{\frac{e^{(2\pi i/h)(E_M-E_N)t}-1}{E_M-E_N}-\frac{e^{(2\pi i/h)(E_M-E_K)t}-1}{E_M-E_K}\right\}. \qquad (6)$$

We want to consider transitions of such a type that in the beginning state, N, no electrons or neutrinos are present; in the end state, M, the charge of the nucleus has been increased by 2, that is two neutrons have been transformed into protons, and two electrons have been emitted into states s and t, two neutrinos into states σ and τ. The intermediate state K then must be such that one neutron has changed into a proton, one electron and one neutrino have been emitted. This means that the intermediate state k of the nucleus is a quantum state of the isobar with atomic number larger by 1 than the original nucleus. The assumption that ordinary β-disintegration is energetically impossible means that $W_{kn}=W_k-W_n>-mc^2$; the first denominator E_K-E_N in (6) will therefore always be positive. The process of double β-disintegration is, however, energetically permissible if $W_{mn}=W_m-W_n<-2mc^2$. In this case the second

* A star denotes the conjugate complex of a quantity.

denominator, $E_M - E_N$, the difference of energy between beginning and end state, will, for those processes for which the energy is conserved, approach zero, and only then has $a_M{}^N$ appreciable values. As in the theory of radiation the second term in the bracket in (6) can then be neglected with respect to the first, since $E_M - E_K$ has always a finite value.

The summation over the intermediate states of the light particles can be made explicitly; there are only four intermediate states possible, namely, electron s may have been emitted before or after electron t, and neutrino σ before or after neutrino τ. The power of -1 in the operators $a_s{}^*$, $b_\sigma{}^*$ introduces a difference in sign in the different cases. The result is:

$$a_M{}^N = \pm \sum_k H_{nk} H_{km} \left\{ \left[\frac{1}{W_{kn}+H_s+K_\sigma} + \frac{1}{W_{kn}+H_t+K_\tau} \right] \bar{\psi}_s \delta \varphi_\sigma{}^* \bar{\psi}_t \delta \varphi_\tau{}^* - \left[\frac{1}{W_{kn}+H_s+K_\tau} + \frac{1}{W_{kn}+H_t+K_\sigma} \right] \bar{\psi}_s \delta \varphi_\tau{}^* \bar{\psi}_t \delta \varphi_\sigma{}^* \right\} \frac{\exp\left[(2\pi i/h)(W_{mn}+H_s+H_t+K_\sigma+K_\tau)t\right]-1}{W_{mn}+H_s+H_t+K_\sigma+K_\tau}. \quad (7)$$

To obtain the transition probability $|a_M{}^N|^2$ has to be summed over all possible states s, σ, t and τ. The first step to this is to average the absolute value squared of (7) over all possible directions of the plane waves of the neutrino. This is done most easily by assuming that the neutrinos are confined to a large space Ω. In that case, neglecting the neutrino mass, $\overline{\varphi_\sigma{}^\nu \varphi_\sigma{}^\mu} = (1/4\Omega)\delta_{\nu\mu}$. We obtain terms of the type

$$\overline{|\bar{\psi}_s \delta \varphi_\sigma{}^*|^2} = (1/4\Omega)(\bar{\psi}_s \psi_s), \quad \overline{\bar{\psi}_s \delta \varphi_\sigma{}^* \cdot \bar{\psi}_t{}^* \delta \varphi_\sigma} = (1/4\Omega)(\bar{\psi}_s \psi_t),$$

where the abbreviation $(\bar{\psi}_s \psi_t) = \sum_\nu \psi_s{}^{\nu *} \psi_t{}^\nu$ has been used.

Furthermore, from the properties of the ψ's, the relativistic eigenfunctions of the Coulomb field, it follows that, if the spin of the state s is parallel to that of t,

$$|(\bar{\psi}_s \psi_t)|^2 = (\bar{\psi}_s \psi_s)(\bar{\psi}_t \psi_t);$$

if the spins of states s and t are antiparallel

$$|\bar{\psi}_s \psi_t|^2 = 0.\dagger$$

We therefore obtain

$$|a_M{}^N|^2 = \frac{1}{16\Omega^2} A (\bar{\psi}_s \psi_s)(\bar{\psi}_t \psi_t) \cdot \frac{4 \sin^2\left[(\pi/h)(W_{mn}+H_s+H_t+K_\sigma+K_\tau)t\right]}{(W_{mn}+H_s+H_t+K_\sigma+K_\tau)^2},$$

where A is the following abbreviation:

For parallel spin of states s and t,

$$A = A_1 = \left| \sum_k H_{nk} H_{km} \left[\frac{1}{W_{kn}+H_s+K_\sigma} + \frac{1}{W_{kn}+H_t+K_\tau} - \frac{1}{W_{kn}+H_s+K_\tau} - \frac{1}{W_{kn}+H_t+K_\sigma} \right] \right|^2. \quad (8a)$$

For antiparallel spin

$$A = A_2 = \left| \sum_k H_{nk} H_{km} \left[\frac{1}{W_{kn}+H_s+K_\sigma} + \frac{1}{W_{kn}+H_t+K_\tau} \right] \right|^2 + \left| \sum_k H_{nk} H_{km} \left[\frac{1}{W_{kn}+H_s+K_\tau} + \frac{1}{W_{kn}+H_t+K_\sigma} \right] \right|^2. \quad (8b)$$

Here the sum over k runs over all possible quantum states of the intermediate isobar.

† This is true only for ψ functions with the angular quantum number $j = 1/2$ in the neighborhood of $r = 0$, with which we are concerned. But all other ψ's are negligible at these distances from the origin.

The summation over all possible states of the neutrinos is now replaced by an integration over the energies K_σ, K_τ. The number of neutrino states whose energy lies between K_σ and $K_\sigma+dK_\sigma$ is given by

$$(8\pi\Omega/h^3c^3)K_\sigma^2dK_\sigma.$$

In the first integration, i.e., over K_τ, only those energies contribute which lie in the neighborhood of the place $E_M-E_N=0$, which corresponds to conservation of energy. The integral becomes proportional to the time t in the usual manner. By neglecting the dependence of A on the energies, the probability of emission of one electron into state s, one into the state t becomes

$$P^{st}=(1/t)\sum_{\sigma,\tau}|a_M{}^N|^2=(8\pi^4/h^7c^6)A(\bar{\psi}_s\psi_s)(\bar{\psi}_t\psi_t)\int K_\sigma^2(W_{mn}+H_s+H_t+K_\sigma)^2dK_\sigma$$
$$=(4\pi^4/15)(A/h^7c^6)(\bar{\psi}_s\psi_s)(\bar{\psi}_t\psi_t)(W_{nm}-H_s-H_t)^5. \quad (9)$$

Finally the summation over s and t is replaced by a double integration over the energy, that is over the continuous spectrum of the electron in the Coulomb field of the nucleus. It is somewhat more convenient to express the energies in units of mc^2, namely, to introduce the dimensionless quantities $mc^2\epsilon=W_{nm}>0$, $mc^2h_s=H_s$, $mc^2h_t=H_t$.

According to Fermi the value of ψ at the outer edge of the nucleus, that is at a distance $\rho=9\times10^{-13}$ cm from the center has to be used; the sum of $(\bar{\psi}_s\psi_s)$ over all states with a fixed direction of spin and energy in the range dh_s is given by:

$$\sum_{dh_s}\bar{\psi}_s(\rho)\psi_s(\rho)=\frac{16\pi}{|\Gamma(3+2S)|^2}\frac{m^3c^3}{h^3}\left(\frac{4\pi mc\rho}{h}\right)^{2S}\cdot h_s(h_s^2-1)^{\frac{1}{2}+S}$$
$$\exp\left[\pi\gamma\frac{h_s}{(h_s^2-1)^{\frac{1}{2}}}\right]\left|\Gamma\left(1+S+i\gamma\frac{h_s}{(h_s^2-1)^{\frac{1}{2}}}\right)\right|^2dh_s, \quad (10)$$

where Z is atomic number, $\gamma=Z/137$, $S=(1-\gamma^2)^{\frac{1}{2}}-1$. S is for any occurring value of nuclear charge negative and small compared to 1; $\gamma=0.3$, $S=-0.05$ correspond to $Z\simeq41$. (10) can therefore be approximated by a much simpler form by neglecting S against 1/2 in the exponent of (h_s^2-1) and S against 1 in the Γ-function, since

$$\exp\left(\pi\gamma\frac{h_s}{(h_s^2-1)^{\frac{1}{2}}}\right)\left|\Gamma\left(1+i\gamma\frac{h_s}{(h_s^2-1)^{\frac{1}{2}}}\right)\right|^2\simeq2\pi\gamma\frac{h_s}{(h_s^2-1)^{\frac{1}{2}}}.$$

The transition probability of the nucleus obtained by summation of (9) over all states and the two cases of parallel and antiparallel spin of the electrons, then turns out to be

$$P=\frac{4^6\pi^8\gamma^2}{15|\Gamma(3+2S)|^4}\frac{m^{11}c^{10}}{h^{13}}\left(\frac{4\pi mc\rho}{h}\right)^{4S}(A_1+A_2)\int_1^{\epsilon-1}h_s^2\int_1^{\epsilon-h_s}h_t^2(\epsilon-h_s-h_t)^5dh_tdh_s \quad (11)$$
$$=\frac{4^6\pi^8\gamma^2}{6\cdot7\cdot15|\Gamma(3+2S)|^4}\frac{m^{11}c^{10}}{h^{13}}\left(\frac{4\pi mc\rho}{h}\right)^{4S}(A_1+A_2)F(\epsilon-2),$$

where

$$F(x)=x^7\left(1+\frac{1}{2}x+\frac{1}{9}x^2+\frac{1}{9.10}x^3+\frac{1}{2\cdot9\cdot10\cdot11}x^4\right).$$

ϵ is the difference of energy between the beginning and end states of the nucleus in units mc^2. A_1 and A_2 are given by (8); they contain the dependence of the process on the energy levels of the intermediate nucleus. It turns out, however, that the values of those energies do not greatly influence the probability. Roughly one can put

$$A_1+A_2 \backsimeq \left|\frac{H_{nk}H_{km}}{W_k-W_n}\right|^2 \sim \frac{g^4}{m^2c^4},$$

with g given by (5). Numerical evaluation for $Z=31$ leads to

$$P=1.15\times10^{-35}F(\epsilon-2)\ \text{sec.}^{-1}$$
$$=3.6\times10^{-28}F(\epsilon-2)\ \text{year}^{-1}. \qquad (12)$$

As seen from the general formula (11) P is almost independent of Z.

The value of F for some arguments is given in the following table:

$\epsilon=$	4	6	8	10
$F(\epsilon+2)=$	0.37×10^2	9.2×10^4	3.4×10^6	4.2×10^7

$\epsilon=$	12	20
$P(\epsilon-2)=$	3.3×10^8	1×10^{11}

The author wishes to express her gratitude to Professor E. Wigner for suggesting this problem, and for the interest taken in it.

SEPTEMBER 15, 1935 PHYSICAL REVIEW VOLUME 48

Nuovo Cimento 14 (1937) 171 – 184

TEORIA SIMMETRICA DELL'ELETTRONE E DEL POSITRONE

Nota di ETTORE MAJORANA

Sunto. - *Si dimostra la possibilità di pervenire a una piena simmetrizzazione formale della teoria quantistica dell'elettrone e del positrone facendo uso di un nuovo processo di quantizzazione. Il significato delle equazioni di* DIRAC *ne risulta alquanto modificato e non vi è più luogo a parlare di stati di energia negativa; nè a presumere per ogni altro tipo di particelle, particolarmente neutre, l'esistenza di « antiparticelle » corrispondenti ai « vuoti » di energia negativa.*

L'interpretazione dei cosidetti « stati di energia negativa » proposta da DIRAC (1) conduce, come è ben noto, a una descrizione sostanzialmente simmetrica degli elettroni e dei positroni. La sostanziale simmetria del formalismo consiste precisamente in questo, che fin dove è possibile applicare la teoria girando le difficoltà di convergenza, essa fornisce realmente risultati del tutto simmetrici. Tuttavia gli artifici suggeriti per dare alla teoria una forma simmetrica che si accordi con il suo contenuto, non sono del tutto soddisfacenti; sia perchè si parte sempre da una impostazione asimmetrica, sia perchè la simmetrizzazione viene in seguito ottenuta mediante tali procedimenti (come la cancellazione di costanti infinite) che possibilmente dovrebbero evitarsi. Perciò abbiamo tentato una nuova via che conduce più direttamente alla meta.

Per quanto riguarda gli elettroni e i positroni, da essa si può veramente attendere soltanto un progresso formale; ma ci sembra importante, per le possibili estensioni analogiche, che venga a cadere la nozione stessa di stato di energia negativa. Vedremo infatti che è perfettamente possibile costruire, nella maniera più naturale, una teoria delle particelle neutre elementari senza stati negativi.

(1) P. A. M. DIRAC, « Proc. Camb. Phil. Soc. », **30**, 150, 1924. V. anche W. HEISENBERG, « ZS. f. Phys. », **90**, 209, 1934.

SULLA SIMMETRIA TRA PARTICELLE E ANTIPARTICELLE

Nota di GIULIO RACAH

Sunto. - *Si mostra che la simmetria tra particelle e antiparticelle porta alcune modificazioni formali nella teoria di* FERMI *sulla radioattività* β, *e che l'identità fisica tra neutrini ed antineutrini porta direttamente alla teoria di* E. MAJORANA.

Nella prima parte del presente lavoro si pone in rilievo una certa arbitrarietà che ancora sussiste nella trasformazione delle autofunzioni di DIRAC associata a un cambiamento di assi nello spazio-tempo, e si mostra come si possa eliminare questa arbitrarietà aggiungendo al postulato dell'invarianza relativistica quello della simmetria tra particelle e antiparticelle. Si perviene così ad una legge di trasformazione che differisce in alcuni casi da quella generalmente ammessa (1), e ad una conseguente modificazione dell'interazione proposta da FERMI nella sua teoria dei raggi β (2). Gli effetti di tale modificazione non sono verificabili sperimentalmente, perchè tendono a zero con la massa del neutrino, ma hanno una certa importanza teorica, perchè eliminano una dissimmetria che era stata rilevata da KONOPINSKI e UHLENBECK (3).

Nella seconda parte si considera l'ipotesi (che dovrà essere un giorno verificata sperimentalmente) che nel caso particolare dei neutrini non si abbia una semplice simmetria, ma addirittura una identità fisica tra neutrini ed antineutrini, e si mostra come questa ipotesi porti automaticamente al formalismo di E. MAJORANA (4). Si rende così evidente il contenuto fisico assolutamente nuovo della teoria di E. MAJORANA, e si indica come l'esperienza potrà decidere della sua validità.

(1) W. PAULI, « Handbuch der Physik », vol. XXIV/1, pp. 220-224.
(2) E. FERMI, « Nuovo Cimento », **11**, 1, 1934.
(3) E. J. KONOPINSKI e G. E. UHLENBECK, « Phys. Rev. », **48**, 7, 1935.
(4) E. MAJORANA, « Nuovo Cimento », **14**, 171, 1937.

Nella terza parte si discutono alcune difficoltà che si oppongono all'estensione della teoria di E. MAJORANA ai neutroni.

* * *

1. È noto che si ottiene l'invarianza relativistica dell'equazione di DIRAC associando ad ogni trasformazione di LORENTZ una sostituzione lineare sulle quattro componenti delle autofunzioni; lo studio più completo di tali sostituzioni è stato fatto da PAULI ([1]), e a tale studio converrà perciò riferirsi, per vedere quale arbitrarietà presentino i resultati che comunemente si utilizzano, e come di tale arbitrarietà si possa convenientemente disporre.

PAULI ha dimostrato che la sostituzione lineare

(P-23) $$\psi_\rho' = \sum_\sigma \Lambda_{\rho\sigma}\psi_\sigma \qquad (\text{o più brevemente } \psi' = \Lambda\psi)$$

che bisogna associare alla trasformazione di LORENTZ

(P-20) $$x_\mu' = \sum_\nu a_{\mu\nu} x_\nu ,$$

per avere l'invarianza formale dell'equazione di DIRAC

(P-22) $$\sum_\mu \gamma^\mu \frac{\partial\psi}{\partial x_\mu} + \frac{2\pi mc}{h}\psi = 0,$$

è sempre univocamente determinata a meno di un fattore numerico. Tale fattore numerico può essere normalizzato con la condizione

(P-25) $$\text{Det. } \Lambda = 1 ;$$

ma, essendo Λ una sostituzione del quarto ordine, resta ancora indeterminato un fattore $\sqrt[4]{1}$, cioè una potenza di i ([2]).

Questa potenza di i può esser determinata con considerazioni di continuità per le trasformazioni di LORENTZ in senso ristretto, ma non è per ora determinata se si hanno riflessioni degli assi.

Per quest'ultime trasformazioni PAULI mostra che si può porre

(P-28a) per $x_k' = -x_k, \quad x_4' = +x_4, \quad \Lambda = \gamma^4,$

(P-28b) per $x_k' = +x_k, \quad x_4' = -x_4, \quad \Lambda = \gamma^5\gamma^4 = \gamma^1\gamma^2\gamma^3 ;$

([1]) W. PAULI, «Ann. Inst. Poincaré», 6, 109, 1936; i numeri preceduti da P- si riferiscono alle formule di tale lavoro.

([2]) Un fattore di modulo 1 nelle autofunzioni è considerato generalmente unessenziale, ed è anzi noto che l'indeterminazione della fase è legata alla conservazione delle particelle; ma appunto perciò il fattore di fase assume importanza in quei problemi, come l'emissione β, in cui si ha creazione di particelle.

ma nulla vieta a priori di sostituire alla prima

(1a) per $x_k' = -x_k$, $x_4' = +x_4$, $\Lambda = i\gamma^4$,

e alla seconda

(1b) per $x_k' = +x_k$, $x_4' = -x_4$, $\Lambda = i\gamma^5\gamma^4$.

Si vede immediatamente che un fattore i nelle espressioni di Λ non influisce sulle equazioni (P-26), (P-26') e (P-28), mentre fa sì che si debba sostituire alla (P-27) la relazione

(2) $$\bar{\Lambda} B \Lambda = -B.$$

L'arbitrarietà tra le trasformazioni (P-28a) e (P-28b) e le trasformazioni (1a) e (1b) può essere elimanata, se al postulato dell'invarianza relativistica si aggiunge quello della simmetria tra particelle e antiparticelle. Questa simmetria è dovuta alla possibilità di stabilire una corrispondenza biunivoca e relativisticamente invariante tra le soluzioni dell'equazione di Dirac a energia positiva e quelle a energia negativa; e questa possibilità è legata, come ha mostrato Pauli, all'esistenza di una matrice di C tale che una funzione φ definita dalla relazione

(P-32) $$\varphi^* = C\psi,$$

sia soggetta alla stessa equazione (P-22) e alla stessa legge di trasformazione (P-23) di ψ.

Per soddisfare la prima condizione dev'essere (a meno di un fattore numerico inessenziale)

(P-33) $$C = \bar{A}^{-1} B \gamma^4 \gamma^5 ;$$

questa relazione, applicando successivamente le (P-15) e (P-3), diventa

$$C = \bar{A}^{-1} \bar{\gamma}^4 B \gamma^5 = A^{-1} (\bar{\gamma}^4)^{-1} B \gamma^5 = (\overline{A\gamma^4})^{-1} B \gamma^5,$$

e tenendo conto della definizione di β,

(3) $$C = \bar{\beta}^{-1} B \gamma^5.$$

Per soddisfare la seconda condizione dev'essere

(4) $$\Lambda^* C = C \Lambda.$$

Questa condizione è certamente soddisfatta dalla (3) per le trasformazioni di Lorentz in senso ristretto, perchè si ha

$$\Lambda^* C = \Lambda^* \bar{\beta}^{-1} B \gamma^5 = \bar{\Lambda}^{+} \bar{\beta}^{-1} B \gamma^5 = (\overline{\Lambda^{+-1}\beta})^{-1} B \gamma^5,$$

ed utilizzando successivamente le (P-26), (P-27) e (P-28),

(5) $$\Lambda^* C = (\overline{\beta \Lambda})^{-1} B \gamma^5 = \bar{\beta}^{-1} \bar{\Lambda}^{-1} B \gamma^5 = \bar{\beta}^{-1} B \Lambda \gamma^5 = \bar{\beta}^{-1} B \gamma^5 \Lambda = C \Lambda.$$

Il richiedere che la (4) valga anche per le riflessioni determina la scelta tra (P-28a) e (1a), tra (P-28b) e (1b).

Infatti osserviamo che nel caso della riflessione degli assi spaziali si ha nelle trasformazioni (5) un cambiamento di segno dovuto alla (P-28): per compensarlo bisognerà poter sostituire la (2) alla (P-27), e si dovrà perciò scegliere la (1a) anzichè la (P-28a).

Nel caso della riflessione dell'asse temporale si ha nelle trasformazioni (5) un secondo cambiamento di segno dovuto alla (P-26'): si dovrà dunque lasciare ferma la (P-27), e perciò la (P-28b).

Per avere una completa simmetria tra particelle e antiparticelle sarà dunque necessario porre in definitiva

$$(6) \qquad \begin{cases} \text{per } x_k' = -x_k, \quad x_4' = +x_4, \quad \Lambda = i\gamma^4, \\ \text{per } x_k' = +x_k, \quad x_4' = -x_4, \quad \Lambda = \gamma^5\gamma^4. \end{cases}$$

Cambiando le leggi di trasformazione delle autofunzioni, cambieranno anche quelle delle combinazioni bilineari di due autofunzioni, e non saranno più valide le leggi di coviaranza date dalle (P-31). Per ottenere le nuove leggi di covarianza si puo procedere per diverse vie: la più semplice è forse l'osservare che, poichè la (P-32a) è ora invariante anche per le riflessioni, si può introdurre quel valore di ψ^* nell'espressione (P-30) di ψ^+; tenendo conto che la (P-32a) si può anche scrivere (¹)

$$\psi^* = \varphi\overline{C},$$

abbiamo, con la successiva applicazione delle (3) e (P-16),

$$(7) \qquad \psi^+ = i\varphi\overline{C}\beta = i\varphi\overline{B\gamma^5} = -i\varphi B\gamma^5.$$

Confrontando con le (P-29) si vede che si hanno i coviarianti seguenti: uno scalare

$$(8_1) \qquad \Omega_1' = \varphi\overline{C}\beta\psi = -\varphi B\gamma^5\psi,$$

un vettore

$$(8_2) \qquad S_\mu' = i\varphi\overline{C}\beta\gamma^\mu\psi = -i\varphi B\gamma^5\gamma^\mu\psi,$$

un tensore doppio antisimmetrico

$$(8_3) \qquad M'_{[\mu\nu]} = -\varphi\overline{C}\beta\gamma^{[\mu\nu]}\psi = \varphi B\gamma^5\gamma^{[\mu\nu]}\psi,$$

un tensore di volume o pseudovettore

$$(8_4) \qquad S'_{[\lambda\mu\nu]} = i\varphi\overline{C}\beta\gamma^{[\lambda\mu\nu]}\psi = -i\varphi B\gamma^5\gamma^{[\lambda\mu\nu]}\psi,$$

e uno pseudoscalare

$$(8_5) \qquad \Omega_2' = i\varphi\overline{C}\beta\gamma^5\psi = -i\varphi B\psi.$$

(¹) Cfr. la dimostrazione di (P-$\overline{23}$).

Nella teoria dei raggi β si deve dunque sostituire al quadrivettore (11) di FERMI, corrispondente a (P-21_2), il quadrivettore

$$(9) \qquad \begin{cases} A_0' = -\psi_1\varphi_4 + \psi_2\varphi_3 + \psi_3\varphi_2 - \psi_4\varphi_1, \\ A_1' = \psi_1\varphi_1 - \psi_2\varphi_2 - \psi_3\varphi_3 + \psi_4\varphi_4, \\ A_2' = i\psi_1\varphi_1 + i\psi_2\varphi_2 - i\psi_3\varphi_3 - i\psi_4\varphi_4, \\ A_3' = -\psi_1\varphi_2 - \psi_2\varphi_1 + \psi_3\varphi_4 + \psi_4\varphi_3, \end{cases}$$

corrispondente a (8_2). Questo fa sì che nelle espressioni (28), (31) e (32) di FERMI il segno — vada sostituito da un segno +. Al medesimo resultato erano giunti KONOPINSKI e UHLENBECK, ammettendo che nell'emissione β ad un elettrone sia accompagnato un antineutrino, e considerando perciò nell'interazione il quadrivettore

$$i\varphi^*\beta\gamma^\mu\psi.$$

Si può porre in maggior evidenza la relazione tra le interazioni corrispondenti ai punti di vista di FERMI e di KONOPINSKI e UHLENBECK, scrivendole rispettivamente nella forma

$$(10a) \qquad H_F = g[Q\psi \cdot C\varphi + Q^*\psi^* \cdot C^*\varphi^*],$$

e

$$(10b) \qquad H_{KU} = g[Q\psi \cdot \varphi^* + Q^*\psi^* \cdot \varphi],$$

e si può facilmente dimostrare che le due interazioni portano sempre esattamente ai medesimi resultati, anche se la massa del neutrino è diversa da zero. Questa è la riprova della necessità di sostituire la (1a) alla (P-28a) per avere completa simmetria tra particelle e antiparticelle.

* * *

2. Simmetria tra particelle e antiparticelle significa che una particella può essere arbitrariamente considerata come vera particella o come antiparticella, ma non significa che i due tipi di particelle siano fisicamente indistinguibili. Questa affermazione vale non soltanto per gli elettroni, che la differenza di carica permette di distinguere dai positroni, ma anche per i neutrini, per cui la mancanza di carica sembra a prima vista (1) rendere impossibile tale distinzione.

Le interazioni (10a) e (10b) danno infatti luogo, oltre che ai processi di radioattività β spontanea, anche a dei processi di radioattività indotta dall'urto dei neutrini contro i nuclei; in questi processi avviene l'assorbimento del neutrino contemporaneamente all'emis-

(1) J. DESTOUCHES, « Comptes Rendus », **198**, 467, 1934.

sione dell'elettrone o del positrone. Ci si rende però immediatamente conto che, quando si consideri l'interazione tra particelle pesanti e particelle leggere sia nella forma (10a) che (10b), un neutrino emesso in un processo β^- può indurre per assorbimento soltanto un processo β^+, e viceversa. Le interazioni (10a) e (10b) portano perciò all'esistenza di due tipi di neutrini perfettamente distinti.

Se l'esperienza dovesse un giorno dimostrare che tale distinzione in natura non esiste, cioè che qualsiasi neutrino può produrre indifferentemente emissione di elettroni e di positroni, bisognerebbe cambiare la forma di interazione tra particelle pesanti e particelle leggere, e precisamente assumere la somma di H_F e H_{KU}:

$$H = H_F + H_{KU} = g[Q\psi \cdot (C\varphi + \varphi^*) + Q^*\psi^* \cdot (C^*\varphi^* + \varphi)]. \tag{11}$$

Questa posizione equivale ad applicare ai neutrini il formalismo di Majorana, perchè quando si assuma per le matrici di Dirac appunto quel particolare schema in cui è $C = I$, l'interazione diventa

$$H = g[Q\psi \cdot (\varphi + \varphi^*) + Q^*\psi^*(\varphi + \varphi^*)], \tag{12}$$

ed in essa entra soltanto un'onda neutrinica reale.

L'imposizione della realtà delle onde neutriniche non è dunque soltanto un metodo matematicamente più semplice per rappresentare teoricamente i neutrini, ma è una logica conseguenza dell'ipotesi dell'identità fisica dei neutrini e degli antineutrini; si vede così che la teoria di E. Majorana non ha soltanto un interesse formale, ma porta a delle conseguenze fisiche essenzialmente diverse da quelle derivanti dalla teoria di Fermi.

***

3. Assai differente è la posizione della teoria di E. Majorana di fronte ai neutroni, perchè alla sua estensione a tali particelle si oppongono due difficoltà.

La prima è costituita dal momento magnetico del neutrone: è noto che l'equazione di Dirac può esser completata, senza disturbarne l'invarianza relativistica e elettromagnetica, in modo da tener conto di un momento magnetico anche nel caso di particelle elettricamente neutre; si verifica però facilmente che nell'equazione così completata è impossibile separare la parte reale dalla parte immaginario di ψ, come è richiesto dalla teoria di E. Majorana.

La seconda difficoltà è costituita dal fatto che se nella teoria dei raggi β anche l'autofunzione neutronica fosse reale, i neutroni avrebbero uguale probabilità di trasformarsi in protoni per emissione β^- o in antiprotoni per emissione β^+; e questo, almeno fino ad oggi, deve considerarsi in contraddizione con l'esperienza.

Da un punto di vista più fisico possiamo riassumere queste considerazioni dicendo che la teoria di E. MAJORANA equivale a identificare le particelle con le antiparticelle, e che se tale identificazione può farsi per i neutrini, essa non sembra possibile per i neutroni, perchè l'antineutrone dovrebbe differire dal neutrone e per il segno del momento magnetico e per la capacità di trasformarsi per processo β in antiprotone anzichè in protone. Ricordando l'ipotesi di WICK (1) sull'origine del momento magnetico del neutrone, si vede che le due difficoltà non sono indipendenti.

Desidero ringraziare il prof. W. PAULI per interessanti e proficue discussioni sugli argomenti di questa nota.

Firenze, Istituto Fisico di Arcetri, Luglio 1937-XV.

(1) G. C. WICK, « Rend. Lincei », **21**, 170, 1935.

DECEMBER 15, 1939 PHYSICAL REVIEW VOLUME 56

On Transition Probabilities in Double Beta-Disintegration

W. H. FURRY
Physics Research Laboratory, Harvard University, Cambridge, Massachusetts

(Received October 16, 1939)

The phenomenon of double β-disintegration is one for which there is a marked difference between the results of Majorana's symmetrical theory of the neutrino and those of the original Dirac-Fermi theory. In the older theory double β-disintegration involves the emission of four particles, two electrons (or positrons) and two antineutrinos (or neutrinos), and the probability of disintegration is extremely small. In the Majorana theory only two particles—the electrons or positrons—have to be emitted, and the transition probability is much larger. Approximate values of this probability are calculated on the Majorana theory for the various Fermi and Konopinski-Uhlenbeck expressions for the interaction energy. The selection rules are derived, and are found in all cases to allow transitions with $\Delta i = \pm 1, 0$. The results obtained with the Majorana theory indicate that it is not at all certain that double β-disintegration can never be observed. Indeed, if in this theory the interaction expression were of Konopinski-Uhlenbeck type this process would be quite likely to have a bearing on the abundances of isotopes and on the occurrence of observed long-lived radioactivities. If it is of Fermi type this could be so only if the mass difference were fairly large ($\epsilon \gtrsim 20$, $\Delta M \gtrsim 0.01$ unit).

I. INTRODUCTION

THE probability of double β-disintegration was calculated some years ago by Goeppert-Mayer[1] on the basis of the Fermi theory.[2, 3] The result obtained was extremely small, corresponding to a lifetime of the order of 10^{25} years in the case of two isobars whose masses differ by 0.002 mass unit and whose atomic numbers differ by two units. Thus one can account for the large number of isobaric pairs with $\Delta Z = 2$, as compared to the scarcity of isobars with $\Delta Z = 1$. Although not strictly stable, the heavier isobar of a pair with $\Delta Z = 2$ may be supposed to be metastable, having a lifetime large compared to geologic time.

An inspection of the calculations shows that the results would not be changed by any significant factor by the use of an expression for the interaction energy involving derivatives of the neutrino wave function, as suggested by Konopinski and Uhlenbeck.[4] The same is true as regards the generalizations in structure of this expression,[5] which make it possible to obtain selection rules[6] for ordinary β-decay decidedly different from those originally given by Fermi. The original Fermi picture of the fundamental interaction processes concerned in β-decay has now been generally supplanted by a picture in which mesons play the role of mediaries between the heavy particles and the electrons and neutrinos. In this case also, as is evident from a consideration of the arguments of Yukawa,[7] the results of Goeppert-Mayer will remain unchanged.

The situation is, however, decidedly altered if one admits a change in the theory of the neutrino itself as an elementary particle. Such a change was suggested by Majorana[8] in a paper on the symmetry properties of the Dirac theory. Majorana's suggestions have been more generally developed in the case of the positron theory by Kramers,[9] and for the neutral particle by the writer.[10] Racah[11] has also discussed the application to the neutrino theory of β-decay.

The essential difference between the Majorana theory of the neutrino and the usual form of the Dirac theory is that in the former there are only

[1] M. Goeppert-Mayer, Phys. Rev. **48**, 512 (1935).
[2] E. Fermi, Zeits. f. Physik **88**, 161 (1934).
[3] For a review of the theory and its various modifications and applications up to the beginning of 1936, see pages 186–206 of the article by Bethe and Bacher, Rev. Mod. Phys. **8** (1936).
[4] E. J. Konopinski and G. E. Uhlenbeck, Phys. Rev. **48**, 7 (1935).
[5] Cf. reference 3, pp. 190–192.
[6] G. Gamow and E. Teller, Phys. Rev. **49**, 895 (1936).
[7] H. Yukawa, Proc. Phys.-Math. Soc. Japan, **17**, 55–56 (1935); H. Yukawa, S. Sakata, M. Kobayasa and M. Taketani, Proc. Phys.-Math. Soc. Japan **20**, 731–734 (1938).
[8] E. Majorana, Nuovo Cimento **14**, 171 (1937).
[9] H. A. Kramers, Proc. Amsterdam Akad. **40**, 814 (1937).
[10] W. H. Furry, Phys. Rev. **54**, 56 (1938), referred to as *N*.
[11] G. Racah, Nuovo Cimento **14**, 322 (1937).

two states for a given momentum, corresponding to the two possibilities for the spin; there are no negative energy neutrinos, and no "holes," or antineutrinos. If in applying the usual Dirac-Fermi theory to β-decay one assumes that the emission of a neutrino accompanies that of a positron, then the emission of an electron must be accompanied by the emission of an *antineutrino*, or else the *absorption* of a neutrino. In the Majorana theory the emission of *either* an electron *or* a positron is accompanied by *either* the emission *or* the absorption of a neutrino.

It can be shown that the use of the Majorana form of neutrino theory instead of the usual theory makes no difference in the case of ordinary β-decay.[12] For the double β-disintegration, however, there is a marked qualitative difference between the results of the two theories. In the ordinary form of the theory four particles must be emitted in such a process: two neutrinos (or antineutrinos) must accompany the emission of two positrons (or electrons). In the Majorana theory there can occur not only these four-particle disintegrations, but also disintegrations in which only the two charged particles—electrons or positrons—are emitted, unaccompanied by neutrinos. In these two-particle disintegrations the neutrino plays only a transitory or virtual part, such as is played by electron-positron pairs in certain hypothetical radiative processes.[13] Subject to the usual limitations on the meaning of such language, one can say that a (virtual) neutrino is emitted together with one of the electrons (or positrons), and reabsorbed when the other electron (or positron) is emitted.

In calculating the probability amplitude for such a process, one must integrate over the momentum space of the neutrino. This introduces a quantity which is a sort of "form factor" of the nucleus. In most cases the integral converges, giving a finite "form factor" without any arbitrary "cutting off" procedure; one has to resort to "cutting off" only for certain types of Konopinski-Uhlenbeck expressions for the interaction energy. The results obtained for the disintegration probability are, nevertheless, greater than those for the four-particle process by a factor which ranges from 10^5 to 10^{15} or more, depending on the particular interaction expression used. The fact that the neutrino momentum is not limited by the amount of energy available in the transition also makes the probability less strongly dependent on the energy.

In the following section we carry through the calculation for the simplest case, using the "scalar-type" interaction energy expression[14] and assuming that the nuclear spin does not change and that the transition is of even-even or odd-odd type. In Section III we consider the effects of using various other expressions for the interaction energy, and determine the selection rules governing changes in the nuclear spin i and the even-odd or even-even, odd-odd character of the transitions.

Before we continue beyond the introduction it should be remarked that the Majorana form of the theory is not the only one which permits this new type of disintegration. The same sort of result can be obtained with the more usual theory if one introduces an interaction expression involving a linear combination of the neutrino wave function and its conjugate. Such an introduction has never been suggested, however, in the neutrino-antineutrino type of theory. The Majorana theory is the natural one to use for our purpose, both because, in the elimination of the antineutrino, it provides an independent motive for the use of such expressions, and because it provides, so to speak, a canonical form for them.

II. Scalar-Type Interaction without Derivatives; $\Delta i=0$, Even-Even, Odd-Odd Transitions

We consider the transition of two neutrons in the nucleus into two protons, with the emission of electrons into states having wave functions ψ_s and ψ_t and energies H_s and H_t. The formula for the probability amplitude must contain two terms, since either electron may be emitted first. These terms are of opposite signs because of the antisymmetry of the electronic wave functions; whether or not they actually cancel strongly depends on the interaction expression used.

[12] Reference 10, pp. 66–67; but see limitation stated in footnote 22 of reference 10.

[13] Cf. for example M. Delbrück, Zeits. f. Physik **84**, 144 (1933); N. Kemmer, Helv. Phys. Acta, **10**, 112 (1937); H. Euler, Ann. d. Physik **26**, 398 (1936).

[14] Cf. Eq. (208a) of reference 3.

The neutrino wave functions are taken to be expanded in terms of plane waves subject to a periodic boundary condition in a box of volume V, as expressed in Eqs. (30) and (31) of N. Only the neutrino amplitudes are regarded as quantized, and all other wave functions are treated as c-numbers.

Apart from a resonance factor, the square of whose absolute value may be set equal to $(2\pi t/\hbar)\delta(E_N+H_s+H_t-E_M)$, the probability amplitude is

$$a_{N\leftarrow M}=g^2\sum(\mathbf{k})\sum_L\int\Psi_N{}^\dagger\{\sum_i\beta_iQ_i{}^\dagger(\mathbf{r}_i|\mathbf{k})\}\Psi_Ld\tau$$

$$\cdot\int\Psi_L{}'^\dagger\{\sum_j\beta_jQ_j{}^\dagger(\mathbf{r}_j{}'|-\mathbf{k})\}\Psi_M{}'d\tau'$$

$$\cdot\left[\frac{(\psi_t{}^\dagger\beta a(\mathbf{k}))(\psi_s{}^\dagger\beta a(-\mathbf{k}))}{E_M-E_L-H_s-K_k}-\frac{(\psi_s{}^\dagger\beta a(\mathbf{k}))(\psi_t{}^\dagger\beta a(-\mathbf{k}))}{E_M-E_L-H_t-K_k}\right]. \quad (1)$$

Here g is Fermi's constant; K_k is the neutrino energy in the virtual intermediate state; and ψ_M, ψ_N, and ψ_L are the wave functions of the nucleus in the initial, final, and intermediate states. They are functions not only of the space coordinates of all the heavy particles, but also of the four-valued Dirac spin coordinates on which matrices such as β operate, and of a two-valued fifth coordinate for each particle which indicates whether it is a neutron or a proton. The indicated integrations must be understood as including summation over the fourth and fifth coordinates. It is on the fifth coordinates that the Fermi operators $Q_i^\dagger$ act in such a way as to change a neutron into a proton; if the ith particle is already a proton, the operator $Q_i^\dagger$ gives the result zero. The operators $\beta_iQ_i^\dagger$ are symmetrized by summing over all the particles in the nucleus, and the factor $(\mathbf{r}_i|\mathbf{k})$ appears because the neutrino wave function must be given its value at the position of the heavy particle. The electron wave functions ψ_s and ψ_t are to be evaluated at the "outer edge of the nucleus," as is usual in calculations on β-decay.[1, 2]

According to the Majorana theory, the four-component neutrino amplitudes $a(\mathbf{k})$ are given by Eqs. (36), (50), and (53) of N. We have to substitute from these equations into (1), and then put

$$B_iB_j=0,\quad B_i{}^*B_j=0 \quad (2)$$

and

$$B_i(\mathbf{k})B_j{}^*(\mathbf{k})=\delta_{ij},\quad B_i{}^*B_j{}^*=0. \quad (3)$$

Eq. (2) corresponds to the initial condition that no neutrinos are present, and (3) indicates our intention to calculate the process in which no neutrinos are emitted and not that in which two neutrinos are emitted. As indicated in N, the calculation can be simplified by using Eq. (51) of N instead of (50), and then discarding all terms in b_3, b_4, $b_3{}^*$, and $b_4{}^*$. Our conditions (2) and (3) can be satisfied by using instead of Eq. (51) of N just

$$b(\mathbf{k})\rightarrow b(\mathbf{k}) \quad (4)$$

for the left-hand factor and

$$b(\mathbf{k})\rightarrow T^*(\mathbf{k})b^*(-\mathbf{k}) \quad (4')$$

for the right-hand factor. Then, remembering that in the Majorana form of β-decay theory it is consistent to regard the factors $(2)^{-\frac{1}{2}}$ given by Eq. (53) of N as absorbed in the disposable constant g, we obtain

$$(\psi_t{}^\dagger\beta a(\mathbf{k}))(\psi_s{}^\dagger\beta a(-\mathbf{k}))$$
$$=\sum_{i=1,2}(\psi_t{}^\dagger\beta S(\mathbf{k}))_i(\psi_s{}^\dagger\beta S(-\mathbf{k})T^*(-\mathbf{k}))_i$$
$$=\psi_t{}^\dagger\beta S(\mathbf{k})\cdot\tfrac{1}{2}(1+\rho_3)\cdot T^\dagger(-\mathbf{k})S'(-\mathbf{k})\beta'\psi_s{}^*. \quad (5)$$

Here ρ_3 is a matrix originally defined by Dirac,[15] and quoted in N, Eq. (39). Then we can use the value of $T(\mathbf{k})$ as given in N, Eq. (43), and obtain

$$(\psi_t{}^\dagger\beta a(\mathbf{k}))(\psi_s{}^\dagger\beta a(-\mathbf{k}))=\psi_t{}^\dagger\beta S(\mathbf{k})$$
$$\cdot\tfrac{1}{2}(1+\rho_3)S^\dagger(\mathbf{k})A^\dagger\beta'\psi_s{}^*=\psi_t{}^\dagger M\psi_s{}^*. \quad (6)$$

It is readily verified that (6) has the same invariance properties as Eq. (56) of N with respect to changes in the form of the matrices α_i, β, and $S(\mathbf{k})$, and that if the neutrino mass is equal to zero M is a symmetrical matrix, independently of such changes. We shall use the representation of α_i, β which was originally given by Dirac, and has usually been employed in explicit calculations; it is quoted in N, Eqs. (59), (62). We can accordingly substitute Eqs.

[15] P. A. M. Dirac, Proc. Roy. Soc. **A117**, 614 (1928).

(60)–(62) of N into (6); we shall, however, replace both ϵ and $\epsilon+1$ by k, which corresponds to setting the neutrino mass equal to zero. The result is

$$M=-\tfrac{1}{2}k^{-1}\{k+(\boldsymbol{\alpha}\cdot\mathbf{k})\}\cdot i\alpha_y\beta=M'. \qquad (7)$$

It is certainly correct to set the neutrino mass equal to zero, since important contributions to the result will come only from large values of the neutrino momentum. This step leads to a formal difficulty, however, because in N the vector $\mathbf{k}$ represents the propagation vector in units $mc/\hbar$, where m is the neutrino mass. We must, accordingly, change the units of k; this makes no difficulty, because (7) is homogeneous of degree zero in k. We shall from now on take k to be expressed in c.g.s. units, and replace Eq. (31) of N by

$$(\mathbf{r}|\mathbf{k})=V^{-\frac{1}{2}}\exp(i\mathbf{k}\cdot\mathbf{r}). \qquad (8)$$

When (7) is substituted into (1), we see that on account of the symmetry of the matrix M the two numerators are equal. The minus sign due to the antisymmetry of the electronic wave function thus causes a strong cancellation, and we get

$$a_{N\leftarrow M}=g^2V^{-1}\sum(\mathbf{k})\sum_L\int\Psi_N^\dagger\{\sum_i\beta_iQ_i^\dagger\exp(i\mathbf{k}\cdot\mathbf{r}_i)\}\Psi_Ld\tau\cdot\int\Psi_L'^\dagger\{\sum_j\beta_jQ_j^\dagger\exp(-i\mathbf{k}\cdot\mathbf{r}_j')\}\Psi_N'd\tau$$
$$\cdot(H_s-H_t)(E_M-E_L-H_s-K_k)^{-1}(E_M-E_L-H_t-K_k)^{-1}\psi_t^\dagger M\psi_s^*. \qquad (9)$$

The main difficulty in the calculation occurs in the evaluation of the factor

$$\Gamma=g^2\sum_L\int\Psi_M^\dagger\{\sum_i\beta_iQ_i^\dagger\exp(i\mathbf{k}\cdot\mathbf{r}_i)\}\Psi_Ld\tau\cdot\int\Psi_L'^\dagger\{\sum_j\beta_jQ_j^\dagger\exp(-i\mathbf{k}\cdot\mathbf{r}_j')\}\Psi_M'd\tau'\cdot f(E_L) \qquad (10)$$

where

$$f(E_L)=(E_M-E_L-H_s-K_k)^{-1}(E_M-E_L-H_t-K_k)^{-1}.$$

For values of the neutrino energy small enough to allow us to put $\exp(i\mathbf{k}\cdot\mathbf{r}_i)=1$, the first factor of the summand is

$$\int\Psi_M^\dagger\{\sum_i\beta_iQ_i\}\Psi_Ld\tau. \qquad (11)$$

This is just the sort of factor which occurs in calculations on single β-disintegration. It is usually assumed to have the value 1 if the angular momentum selection rule is obeyed, and, of course, the value 0 otherwise. Actually this is a decidedly crude assumption, as has been pointed out by Nordheim and Yost;[16] for moderately heavy nuclei it seems likely that the expression (11) cannot have a value larger than perhaps 0.1. Since, however, the value of g given by Fermi was assigned to agree with experiment on the basis of setting an expression such as (11) equal to 1, we shall not make any very large net error if we use this value of g and assume that there is at least one value of L for which both of the first two factors of the summand in (10) can be set equal to 1. There still remains, however, the question as to how many terms in the summation will yield similarly appreciable contributions to the value of Γ.

At first one might suppose that there would be a great many such values of L, because of the large number of energy levels possesed by any moderately heavy nucleus and because $\mathbf{k}$ is not restricted to small values, so that limitations on the angular momentum of the state L do not appear in the usual way. Such a conclusion, however, would in all probability be erroneous. The nature of nuclear states is such that any appreciable excitation energy is practically certain to be distributed effectively among a considerable number of particles, so that a factor $\exp(i\mathbf{k}\cdot\mathbf{r}_i)$, which involves the coordinates of one particle only, is unlikely to be very effective in removing the orthogonality of two wave functions. The assumption that there is only one intermediate state L which makes an important contribution to the sum is accordingly not an unreasonable one; we shall proceed in this way, with the reservation that our result is perhaps best regarded as a lower limit, from which the correct result should not differ by any very large factor.

[16] L. W. Nordheim and F. L. Yost, Phys. Rev. **51**, 942 (1937). Very recently (Phys. Rev. **56**, 519 (1939)) the matrix elements for a number of light radioactive elements have been calculated explicitly by Wigner.

In accordance with the above considerations, we shall use in our further calculations the simple formula

$$\Gamma=g^2f(E_L)(4\pi\rho^3/3)^{-2}\iiint_{r<\rho}\exp(i\mathbf{k}\cdot\mathbf{r})d\tau\cdot\iiint_{r'<\rho}\exp(-i\mathbf{k}\cdot\mathbf{r}')d\tau'. \tag{12}$$

Here the integral is over the interior of a three-dimensional sphere of radius ρ, the nuclear radius. The factor $(4\pi\rho^3/3)^{-2}$ comes from normalization. The formula has been written on the assumption that the angular momentum of the state L is the same as that of M and N, but this assumption has actually rather little to do with our final result. If, for example, $i_L-i_M=i_L-i_N=\pm1$, we should have to use some such expression as

$$\Gamma'=g^2f(E_L)(4\pi\rho^3/3)^{-2}\cdot3\cdot\iiint_{r<\rho}\cos\theta\cdot\exp(i\mathbf{k}\cdot\mathbf{r})d\tau\cdot\iiint_{r'<\rho}\cos\theta'\exp(-i\mathbf{k}\cdot\mathbf{r}')d\tau', \tag{12'}$$

and we shall verify later that this makes no decided difference.

Further simplification in Eq. (9) can easily be made. Since k is in c.g.s. units, the energy of the neutrino is $K_k=\hbar ck$. The main contributions to (9) come from values of k so large that the other energies can be neglected compared to K_k; the occurrence of a vanishing denominator is of course ruled out by the postulated impossibility of single β-disintegration for the nucleus in question. Accordingly we set

$$f(E_L)=K_k^{-2}=(\hbar c)^{-2}k^{-2}. \tag{13}$$

The sum over the allowed values of $\mathbf{k}$ in the box of volume V can be replaced by an integral in the usual way:

$$\sum(\mathbf{k})\cdots=(2\pi)^{-3}V\int_0^\infty k^2dk\iint d\Omega\cdots. \tag{14}$$

Substituting (12), (13) and (14) in (9), we obtain

$$a_{N\leftarrow M}=-(2\pi)^{-3}(g/\hbar c)^2C^{0,0}\cdot(H_s-H_t)\psi_t^\dagger\cdot\tfrac{1}{2}i\alpha_y\beta\psi_s^*, \tag{15}$$

where

$$C^{0,0}=\int_0^\infty dk\iint d\Omega\left|(3/4\pi\rho^3)\iiint_{r<\rho}\exp(i\mathbf{k}\cdot\mathbf{r})d\tau\right|^2=(9/4\pi\rho^6)\int_0^\infty dk\left|\iiint_{r<\rho}e^{ikz}d\tau\right|^2; \tag{16}$$

the superscripts 0,0 refer to the values of i_L-i_M and i_L-i_N. The second term in (7) is omitted from (15), since it makes no contribution for even-even or odd-odd transitions.

Following Goeppert-Mayer, we express the energies of the electrons and the total available energy in terms of dimensionless quantities: $H_s=h_smc^2$, $H_t=h_tmc^2$, $E_M-E_N=\epsilon mc^2$, where m is the mass of an electron. Let us also put $p_s=(h_s^2-1)^{\frac{1}{2}}$, $p_t=(h_t^2-1)^{\frac{1}{2}}$, and make the abbreviation

$$R_s=(16\pi/|\Gamma(3+2S)|^2)(mc/h)^3(2mc\rho p_s/\hbar)^{2S}\cdot\exp(\pi\gamma h_s/p_s)|\Gamma(1+S+i\gamma h_s/p_s)|^2, \tag{17}$$

where $\gamma=Z/137$, $S=(1-\gamma^2)^{\frac{1}{2}}-1$. We follow Goeppert-Mayer in replacing R_s by the following approximation, valid because $|S|\ll1$:

$$R_s\cong(16\pi/|\Gamma(3+2S)|^2)(mc/h)^3(2mc\rho/\hbar)^{2S}\cdot2\pi\gamma h_s/p_s. \tag{17'}$$

A similar definition and approximation hold for R_t.

When the resonance factor is included, the squared amplitude for the transition to two particular states s and t is $(2\pi t/\hbar)|a_{N\leftarrow M}|^2\delta(\epsilon mc^2-h_smc^2-h_tmc^2)$. The total probability per unit time for transitions in which one electron gets an energy between h_smc^2 and $(h_s+dh_s)mc^2$ and the other receives the rest of the energy ϵmc^2 is then

$$P(h_s)dh_s=(2\pi/\hbar mc^2)(dh_t)^{-1}\sum_{dh_t}\sum_{dh_s}|a_{N\leftarrow M}|^2. \tag{18}$$

Here the symbol $\sum_{dh_t}$ means the sum over all states having energy between $h_t mc^2$ and $(h_t+dh_t)mc^2$; the result is of course proportional to dh_t. Using the Dirac wave functions for an electron in the Coulomb field, evaluated at $r=\rho$, we obtain, to the lowest order in γ^2:

$$\sum_{dh_s}\sum_{dh_t}|\psi_t^\dagger\cdot\tfrac{1}{2}i\alpha_y\beta\psi_s^*|^2 = \tfrac{1}{4}R_sR_tp_sp_t(h_sh_t-1)dh_sdh_t. \quad (19)$$

Then

$$P(h_s)dh_s=2^{-7}\pi^{-5}g^4\hbar^{-5}mc^{-2}|C^{0,0}|^2 \times R_sR_tp_sp_t(h_sh_t-1)(h_s-h_t)^2dh_s. \quad (20)$$

The energy spectrum of the electrons is, of course, symmetrical around the value $\frac{1}{2}\epsilon mc^2$. The factor $(h_s-h_t)^2$ in (20), which makes the intensity vanish for the energy $\frac{1}{2}\epsilon mc^2$, comes directly from the cancellation due to the antisymmetry of the electronic wave function.

To get the total probability of disintegration per unit time, we must integrate (20) from $h_s=1$ to $h_s=\epsilon-1$. According to the approximation (17′) the quantity R_sp_s/h_s is essentially independent of h_s. This enables us to write for the total probability of disintegration per unit time:

$$P_1=\int_1^{\epsilon-1}P(h_s)dh_s=2^{-7}\pi^{-5}g^4\hbar^{-5}mc^{-2}|C^{0,0}|^2(R_sp_s/h_s)^2\cdot\int_1^{\epsilon-1}h_s(\epsilon-h_s) \times\{h_s(\epsilon-h_s)-1\}(\epsilon-2h_s)^2dh_s. \quad (21)$$

Then

$$P_1=2^{-7}\pi^{-5}g^4\hbar^{-5}mc^{-2}|C^{0,0}|^2(R_sp_s/h_s)^2\cdot\varphi(\epsilon-2), \quad (22)$$

where

$$\varphi(x)=(x^4/3)\{1+(11x/10)+(x^2/5)+(x^3/70)\}. \quad (23)$$

We shall now evaluate P_1 numerically for $Z=31$, the value used as an example by Goeppert-Mayer. Taking $\rho=8\times10^{-13}$ cm, we get from (17′) the value $R_sp_s/h_s=1.52\times10^{30}$ cm^{-3}. The integral (16) can be evaluated by a rather tedious elementary calculation, or more elegantly by means of Fourier's integral theorem:

$$\int_0^\infty dk\left|\iiint_{r<\rho}e^{ikz}d\tau\right|^2=\tfrac{1}{2}\int_{-\infty}^\infty dk\cdot\iiint_{r<\rho}e^{ikz}dzdxdy\cdot\iiint_{r'<\rho}e^{-ikz'}dz'dx'dy'=\pi\int\!\!\int\!\!\int\!\!\int\!\!\int dxdydx'dy'dz,$$

where in the last expression the limits for x' and y' depend on the value of z in the same way as those for x and y. Then

$$\int_0^\infty dk\left|\iiint_{r<\rho}e^{ikz}d\tau\right|^2=\pi\int_{-\rho}^{\rho}\pi^2(\rho^2-z^2)^2dz=16\pi^3\rho^5/15.$$

Thus

$$C^{0,0}=12\pi^2/5\rho. \quad (24)$$

A similar calculation shows that if (12′) were used instead of (12) we should have simply to replace $C^{0,0}$ by a quantity $C^{1,1}=9\pi^2/5\rho=(3/4)C^{0,0}$. Using $g=4\times10^{-50}$ erg cm^3, we obtain finally

$$P_1=1.07\times10^{-28}\varphi(\epsilon-2)\ \text{sec}^{-1}. =3.4\times10^{-21}\varphi(\epsilon-2)\ \text{year}^{-1}. \quad (25)$$

Some values of φ are as follows:

$\epsilon=$	3	4	6	8	10	12	20
$\varphi(\epsilon-2)=$	0.77	22	8.1×10^2	7.7×10^3	4.1×10^4	1.54×10^5	5.9×10^6

(26)

For small values of ϵ the P found here is about 10^6 times that given by Goeppert-Mayer. For larger values of ϵ the ratio is not so large. This is natural, since for large ϵ there is plenty of energy available for the emission of neutrinos along with the electrons.

III. Other Interaction Expressions

In the preceding section we calculated the value of P by using the scalar-type expression for the interaction energy. We now want to consider what would be the effect of using any of the various other energy expressions which have been proposed.

Since we are interested only in the order of magnitude of P, we shall neglect the effect of using a different matrix instead of $\frac{1}{2}i\alpha_y\beta$ in calculating the expression corresponding to (19). This would at most replace the factor $(h_s h_t-1)$ by some factor such as $h_s h_t$ or $(h_s h_t+1)$, and for $\epsilon-2 \gtrsim 1$ such a change amounts to only a factor of perhaps two in the answer. Hence we shall simply keep the factor $(h_s h_t-1)$ throughout.

To begin with we shall consider the two main differences in the size of P which can arise from changes in the interaction energy expression. We shall then list the various expressions and indicate the main types of terms which they give in the integrand of the equation corresponding to Eq. (1) in the general case. This classification makes it possible to state the selection rules for the different types of interaction expression.

Differences in order of magnitude of P

The first important difference which may come from the use of a different interaction expression is the absence of the strong cancellation due to the antisymmetry of the electronic wave function which occurred in Section II. If the matrix[17] M is not purely symmetric, the main contribution to $a_{N\leftarrow M}$ will come from the antisymmetric part of M, $M_A=\frac{1}{2}(M-M')$. Eq. (15) is then replaced by

$$a_{N\leftarrow M}=(2\pi)^{-3}(g^2/\hbar c)D^{0,0}\psi_t{}^\dagger M_A\psi_s{}^* \qquad (27)$$

and Eq. (20) by

$$P(h_s)dh_s=2^{-7}\pi^{-5}g^4\hbar^{-3}m^{-1}c^{-4}|D^{0,0}|^2 \times R_sR_tp_sp_t(h_sh_t-1)dh_s, \qquad (28)$$

where

$$D^{0,0}=\int_0^\infty 2kdk\int\int d\Omega \times\left|(3/4\pi\rho^3)\int\int\int_{r<\rho}\exp(i\mathbf{k}\cdot\mathbf{r})d\tau\right|^2. \qquad (29)$$

The spectrum now has a maximum at $h_s=\frac{1}{2}\epsilon$, instead of vanishing there.

The total probability of disintegration per unit time is now

$$P_2=2^{-7}\pi^{-5}g^4\hbar^{-3}m^{-1}c^{-4}|D^{0,0}|^2 \times(R_sp_s/h_s)^2\cdot\chi(\epsilon-2), \qquad (30)$$

where

$$\chi(\epsilon-2)=\int_1^{\epsilon-1}h_s(\epsilon-h_s)\{h_s(\epsilon-h_s)-1\}dh_s$$

or

$$\chi(x)=x^2\{1+(7x/6)+(x^2/3)+(x^3/30)\}. \qquad (31)$$

Integration of (29) by elementary methods gives

$$D^{0,0}=18\pi/\rho^2. \qquad (32)$$

Then

$$P_2=1.43\times10^{-24}\chi(\epsilon-2)\text{ sec.}^{-1} =4.6\times10^{-17}\chi(\epsilon-2)\text{ year}^{-1}. \qquad (33)$$

$\epsilon=$	3	4	6	8	10	12	20	
$\chi(\epsilon-2)=$	2.5	19.7	2.1×10^2	9.8×10^2	3.1×10^3	7.9×10^3	1.05×10^5	(34)

Another important effect on the disintegration probability is obtained if we use a Konopinski-Uhlenbeck form of interaction expression. The use of such an expression is subject to certain theoretical difficulties,[18] and it has been suggested[19] that the experimental results can perhaps be fully explained without using an interaction expression involving derivatives; nevertheless it is not without interest to see what such expressions give for the double β-disintegration.

The use of a Konopinski-Uhlenbeck expression means essentially the insertion of a factor k^2 into the integrand in (16) or (29), and the replacement of the constant g by a constant which differs from it by the dimensions of a length. The length by which g is multiplied to give the new constant should be $(2\pi)^{-1}$ times a typical de Broglie wave-length of the neutrino in an ordinary β-disintegration, in order that the probabilities for single β-disintegration may be kept the same. Hence we shall replace g by $(\hbar/2mc)g$. The total disintegration probability per unit time is then

$$P_3=2^{-11}\pi^{-5}g^4\hbar^{-1}m^{-3}c^{-6}|F^{0,0}|^2 \times(R_sp_s/h_s)^2\cdot\varphi(\epsilon-2), \qquad (35)$$

[17] In general it is actually a matrix *function* which has to be used, as is indicated below in Eq. (45). In the case treated in Section II this function could conveniently be split into factors, only one of which is a matrix (Cf. Eq. (46)). In any case it is only the symmetry or antisymmetry under the interchange of two four-valued indices which comes into question in determining the presence or absence of the cancellation.

[18] Cf. M. Fierz, Helv. Phys. Acta **10**, 123, 284 (1937).

[19] H. A. Bethe, F. Hoyle and R. Peierls, Nature **143**, 200 (1939).

where

$$F^{0,0}=\int_0^\infty k^2dk\int\int d\Omega \times\left|(3/4\pi\rho^3)\int\int\int_{r<\rho}\exp(i\mathbf{k}\cdot\mathbf{r})d\tau\right|^2 \quad (36)$$

if M is a pure symmetric matrix. If M is asymmetric we get

$$P_4=2^{-11}\pi^{-5}g^4\hbar m^{-5}c^{-8}|G^{0,0}|^2 \times(R_sp_s/h_s)^2\chi(\epsilon-2), \quad (37)$$

where

$$G^{0,0}=\int_0^\infty 2k^3dk\int\int d\Omega \times\left|(3/4\pi\rho^3)\int\int\int_{r<\rho}\exp(i\mathbf{k}\cdot\mathbf{r})d\tau\right|^2. \quad (38)$$

The integral (36) can be evaluated elegantly by using Fourier's integral theorem:

$$F^{0,0}=(3/4\pi\rho^3)^2\int_{-\infty}^\infty dk_x\int_{-\infty}^\infty dk_y\int_{-\infty}^\infty dk_z \cdot\int\int\int_{r<\rho}\exp(i\mathbf{k}\cdot\mathbf{r})dxdydz \cdot\int\int\int_{r'<\rho}\exp(-i\mathbf{k}\cdot\mathbf{r}')dx'dy'dz'$$

$$=(3/4\pi\rho^3)^2\cdot(2\pi)^3\cdot\int\int\int_{r<\rho}dxdydz=6\pi^2/\rho^3. \quad (39)$$

Numerical computation then gives:

$$P_3=2.28\times10^{-22}\varphi(\epsilon-2)\ \text{sec.}^{-1} =7.17\times10^{-15}\varphi(\epsilon-2)\ \text{year}^{-1}. \quad (40)$$

The integral (38) diverges logarithmically, so that in order to obtain a finite value it is necessary to "cut off" at some value of k. It is generally assumed that the value at which the theory becomes unreliable and the cutting off should be done is about mc^2/e^2. In evaluating $G^{0,0}$ it is particularly convenient to cut off at a value which makes $k\rho=\pi$. One then readily finds by elementary methods that

$$G^{0,0}=(72\pi/\rho^4)\left\{\int_0^\pi \sin^2 y\cdot dy/y-\tfrac{1}{2}\right\} \cong 162/\rho^4. \quad (41)$$

The cut-off occurs rather beyond the first maximum of the integrand. The range $0<k<\pi/\rho$ includes the essential contributions to the integrals $C^{0,0}$, $D^{0,0}$, and $F^{0,0}$, so that their values are genuinely independent of whether one cuts off or not.

Using the value (41), one obtains

$$P_4=3.99\times10^{-18}\chi(\epsilon-2)\ \text{sec.}^{-1} =1.26\times10^{-10}\chi(\epsilon-2)\ \text{year}^{-1}. \quad (42)$$

We have now listed four typical formulas for the disintegration constant P. Next we must classify the various possible expressions for the interaction energy in such a way as to show what value of P each would give.

Classification of interaction energy expressions

By methods analogous to those used in deriving Eq. (7) one can readily calculate the value of $a_{N\leftarrow M}$ corresponding to any particular expression for the interaction energy. The formulas obtained are, however, for the most part very long and cumbersome. Accordingly we shall indicate only the typical characteristics of the dominant terms, which suffice to determine the order of magnitude of the disintegration constant and to determine the selection rules.

Usually in calculations using the neutrino theory of β-decay one neglects terms involving the small components of the wave functions of the heavy particles compared to those which involve only the large components. This is done because the small components are of order $\bar{v}/c$ compared to the large, $\bar{v}$ being a typical velocity of a heavy particle, and $\bar{v}/c$ is only about 0.2. For our present purposes we can neglect such terms usually, but not always; if the terms involving only the large components give purely symmetric matrices M and other terms give asymmetric matrices, then the latter will be more important because of the absence of any strong cancellation. This situation arises with Fermi's original "vector-type" interaction expression.

Usually only the product of the time components is used in calculations, because the product of the space components brings in small components of the heavy particle wave functions. In our present case, however, such a procedure gives a symmetric matrix M, and the space components appear in the actual dominant terms.

To avoid useless multiplication of the number of types listed, we shall write the typical expressions in terms of just the matrices 1 and $\boldsymbol{\sigma}$ for the heavy particles, whose wave functions may then be thought of as two-component Pauli wave functions Any occurrence of the small components in the dominant terms will then be indicated by a suitable power of $(\bar{v}/c)$. Explicitly, so far as the heavy particles are concerned we set

$$\beta\rightarrow 1,\quad \gamma_5=\rho_1\rightarrow(\bar{v}/c)$$
$$\beta\boldsymbol{\sigma}\rightarrow\boldsymbol{\sigma},\quad \boldsymbol{\alpha}\rightarrow(\bar{v}/c)\boldsymbol{\sigma},\quad \beta\boldsymbol{\alpha}\rightarrow(v/c)\boldsymbol{\sigma}. \tag{43}$$

We use the abbreviations:

$$\begin{aligned}\Psi_L{}^\dagger\{\sum_i Q_i{}^\dagger(\mathbf{r}_i|-\mathbf{k})\}\Psi_M&=a_1,\\ \Psi_L{}^\dagger\{\sum_i \sigma_i Q_i{}^\dagger(\mathbf{r}_i|-\mathbf{k})\}\Psi_M&=\mathbf{b}_1,\\ \Psi_N{}^\dagger\{\sum_i Q_i{}^\dagger(\mathbf{r}_i|\mathbf{k})\}\Psi_L&=a_2,\\ \Psi_N\{\sum_i \sigma_i Q_i{}^\dagger(\mathbf{r}_i|\mathbf{k})\Psi_L&=\mathbf{b}_2.\end{aligned} \tag{44}$$

The quantity $a_{N\leftarrow M}$ can always be expressed in the form

$$a_{N\leftarrow M}=g^2\sum(\mathbf{k})\int d\tau\int d\tau'\sum_L\sum_{lm}\psi_{tl}{}^*\times\left\{\frac{\Lambda_{lm}{}^L(\mathbf{r},\mathbf{r}';\mathbf{k})}{E_M-E_L-H_s-K_k}-\frac{\Lambda_{ml}{}^L(\mathbf{r},\mathbf{r}';\mathbf{k})}{E_M-E_L-H_t-K_k}\right\}\psi_{sm}{}^*. \tag{45}$$

In the case treated in Section II the matrix function Λ was simply

$$\Lambda(\mathbf{r},\mathbf{r}';\mathbf{k})=a_2(\mathbf{r};\mathbf{k})M(\mathbf{k})a_1(\mathbf{r}';\mathbf{k}), \tag{46}$$

with M given by Eq. (7). Eq. (9) gives the result of substituting (46) and (44) in (45), provided (43) is taken into account.

We now list the types of expression which can appear as the dominant terms of the matrix function Λ:

Symmetric Terms:

$$S_1=a_2\cdot i\alpha_y\beta a_1, \tag{47a}$$
$$S_2=a_2(\boldsymbol{\alpha}\cdot\mathbf{n})i\alpha_y\beta a_1\quad(\mathbf{n}=\mathbf{k}/k). \tag{47b}$$

Many other types of symmetric term occur, but only in combination with antisymmetric terms which give the dominant contributions to the result.

Antisymmetric Terms:

$$A_1=i([\mathbf{b}_1\times\mathbf{b}_2]\cdot\boldsymbol{\sigma})i\alpha_y\beta, \tag{48a}$$
$$A_2=i([\mathbf{b}_1\times\mathbf{b}_2]\cdot\mathbf{n})\rho_1 i\alpha_y\beta, \tag{48b}$$
$$A_3=i([\mathbf{b}_1\times\mathbf{b}_2]\cdot\mathbf{n})(\mathbf{n}\cdot\boldsymbol{\sigma})i\alpha_y\beta, \tag{48c}$$
$$A_4=\beta([\mathbf{b}_1\times\mathbf{n}]\cdot[\mathbf{b}_2\times\mathbf{n}])i\alpha_y\beta, \tag{48d}$$
$$A_5=i([(a_1\mathbf{b}_2-a_2\mathbf{b}_1)\times\mathbf{n}]\cdot\boldsymbol{\sigma})i\alpha_y\beta. \tag{48e}$$

We give the different expressions for the interaction energy the same designations as used by Bethe and Bacher.[3] The types of the dominant terms in Λ and the order of magnitude of the disintegration constant P are given in the following list:

Fermi Expressions:

Scalar: $\Lambda\sim S_1, S_2;\quad P\approx P_1,$ (49a)

Vector: $\Lambda\sim(\bar{v}/c)A_5;\quad P\approx(\bar{v}/c)^2P_2,$ (49b)

Tensor: $\Lambda\sim A_1, A_2;\quad P\approx P_2,$ (49c)

Pseudovector: $\Lambda\sim A_1, A_2;\quad P\approx P_2,$ (49d)

Pseudoscalar:

$$\Lambda\sim(\bar{v}/c)^2S_1,\ (\bar{v}/c)^2S_2;\quad P\approx(\bar{v}/c)^4P_1. \tag{49e}$$

Konopinski-Uhlenbeck Expressions:

Vector: $\Lambda\sim qS_1, qS_2;\quad P\approx P_3,$ (50a)

Tensor: $\Lambda\sim qA_2, qA_3;\quad P\approx P_4,$ (50b)

Pseudovector:

$$\Lambda\sim qA_1, qA_2, qA_3, qA_4;\quad P\approx P_4. \tag{50c}$$

Here

$$q=k^2(\hbar/2mc)^2 \tag{51}$$

is the essential factor which distinguishes the Konopinski-Uhlenbeck from the Fermi formulas, and which was taken into account in calculating P_3 and P_4.

Selection rules

The expressions

$$\sum_L \psi_t^\dagger S_j \psi_s^* \quad \text{and} \quad \sum_L \psi_t^\dagger A_j \psi_s^*$$

are all scalars with respect to the three-dimensional rotation-reflection group: so far as transformation properties are concerned we can think of $i\alpha_y\beta\psi_s^*$ as replaced by a function ψ_σ.[20] To find the selection rules corresponding to the expressions (47) and (48) we may inspect the transformation properties of either the factors involving a_n, $\mathbf{b}_n$, and $\mathbf{n}$ or the matrix factors, apart from the factors $i\alpha_y\beta$. We obtain:

S_1 and A_4: Scalars: $\Delta i=0$; odd-odd and even-even transitions.

A_2: Scalar, but with reflection character -1: $\Delta i=0$; odd-even and even-odd transitions.

S_2: Polar vector: $\Delta i=0$ or ± 1, but not $0\to 0$; odd-even and even-odd transitions.

A_1, A_3, and A_5: Axial vectors: $\Delta i=0$ or ± 1, but not $0\to 0$; odd-odd and even-even transitions.

The selection rules for each interaction expression can then be read off from (49) or (50). The selection rules in general bear more resemblance to those of Gamow and Teller than to those of Fermi, since some $\Delta i=\pm 1$ transitions are always allowed. Transitions forbidden by these rules can occur only if at least one electron is emitted in a state with $j\neq\frac{1}{2}$, and their probabilities will be decidedly smaller than those given by (49) and (50). For $\epsilon-2\gtrsim 2$ the probabilities will be decreased by a factor of roughly $\{(\epsilon-2)\rho mc/2\hbar\}^2$ for a change in parity type of the transition or for each extra unit of Δi. For smaller values of $\epsilon-2$ the decrease will be more drastic.

[20] Cf. discussion in *N*, p. 66; also W. H. Furry, Phys. Rev. **51**, 125 (1937), Eq. (8) with $A=i\alpha_y\beta$. The free electron functions form a complete set.

Dependence of transition probability on atomic number

The numerical results which we have given were all computed for $Z=31$, $\rho=8\times 10^{-13}$ cm. The quantity $(R_s p_s/h_s)$ is roughly proportional to Z, and ρ may be taken to be proportional to $Z^{\frac{1}{3}}$. Approximate relations for changing the results to different values of Z or a different estimate of the nuclear radius are, accordingly:

$$P_1\propto Z^2\rho^{-2}\propto Z^{4/3}, \qquad P_2\propto Z^2\rho^{-4}\propto Z^{\frac{2}{3}},$$
$$P_3\propto Z^2\rho^{-6}\propto \text{cons.}, \qquad P_4\propto Z^2\rho^{-8}\propto Z^{-\frac{2}{3}}. \quad (52)$$

The results as calculated are for emission of two negative electrons. In the case of positron emission the probabilities will be smaller, because R_s contains an additional factor $\exp(-2\pi\gamma h_s/p_s)$, and R_t a similar factor. These factors decidedly complicate the evaluation of the integrals from which the functions φ and χ have to be determined. It is evident, however, that for $\epsilon-2\gtrsim 2$ the decrease will be by a factor not much smaller than $e^{-4\pi\gamma}=e^{-0.092Z}$. As is the case for the forbidden transitions, the decrease is more decided for smaller values of $\epsilon-2$.

IV. Conclusion

We have seen that the phenomenon of double beta-disintegration is one for which there is a decided difference between the results of the Majorana theory and those of the older theory of the neutrino. According to the older theory it seemed certain that double beta-disintegration could never be capable of observation because of its extremely minute probability, but the Majorana theory indicates that this is by no means necessarily the case. Indeed, if the interaction expression were of Konopinski-Uhlenbeck type this process would be quite likely to have a bearing on the abundances of isotopes and on the occurrence of observed long-lived radioactivities. If it is of Fermi type this could be so only if the mass difference were fairly large ($\epsilon\gtrsim 20$, $\Delta M\gtrsim 0.01$ unit).

A Measurement of the Half-Life of Double Beta-Decay from $_{50}Sn^{124}$ *

E. L. FIREMAN
Department of Physics, Princeton University, Princeton, New Jersey
November 29, 1948

If two isobars differ by two units in atomic number, the heavier may decay into the lighter by double beta-decay.[1,2] This is the simultaneous emission of two negatrons if the heavier has lower atomic number or the simultaneous emission of two positons, 1 positon+1K capture, or 2K captures if the heavier has higher atomic number. The half-life depends markedly upon whether or not two neutrinos are emitted in the process. If no neutrinos are emitted, the half-life is of the order of 10^{10} times shorter than if two neutrinos are emitted (see Table I). The reason for this large difference in half-life arises from the number of cells in phase space available for the transition.

TABLE I. Theoretical half-life for allowed double negaton emission.

Atomic mass difference	0	0.52 Mev	1.04 Mev	1.56 Mev	.2.08 Mev	2.60 Mev
2 neutrinos	∞	$2.6\cdot10^{27}$ yr.	$2.4\cdot10^{25}$ yr.	$1.3\cdot10^{24}$ yr.	$2.1\cdot10^{23}$ yr.	$4.3\cdot10^{22}$ yr.
No neutrinos	∞	$2.1\cdot10^{16}$ yr.	$2.7\cdot10^{15}$ yr.	$6.5\cdot10^{14}$ yr.	$2.2\cdot10^{14}$ yr.	$8.3\cdot10^{13}$ yr.

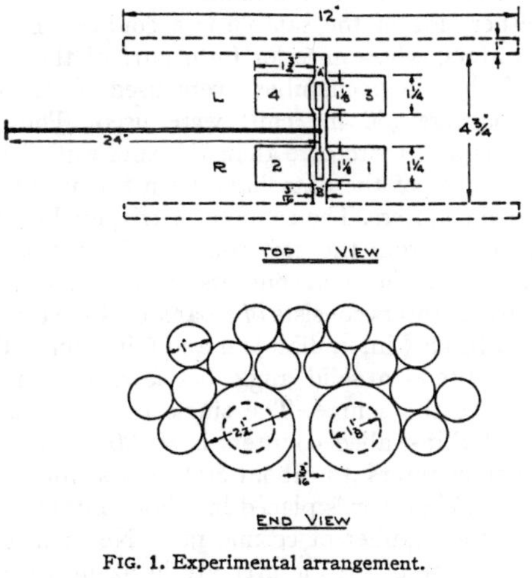

FIG. 1. Experimental arrangement.

In the present work an experimental investigation of $_{50}Sn^{124}$ which belongs to isobaric triplet $_{50}Sn^{124}-{}_{52}Te^{124}-{}_{54}Xe^{124}$ has been carried out. The experimental arrangement consists of four thin window counters (3 mg/cm^2 mica) connected in pairs to coincidence circuits L and R. These counters are shielded by anticoincidence counters and by an Fe and Pb box. Two specimens** of Sn (25 g) identical in all respects except for isotopic constitution are placed between the thin window counters. These specimens are called A and B. Specimen A contains 54 percent of Sn^{124}; specimen B contains 0.4 percent of Sn^{124}. The position of the specimens between the counters is interchanged by rotating the specimen holder through 180° (see Fig. 1).

Coincidences and single counts from both specimens are recorded simultaneously. The specimen holder is rotated through 180° every other hour and the positions of the specimens in the holder are interchanged every 20 hours. These data are summarized in Table II.

In all situations specimen A gives 2 coincidence counts/hr. more than specimen B. By repeating this type of measurement with Al absorbers over one side of each specimen an absorption curve is obtained. This absorption curve is similar to that of electrons from a spectrum with an energy end point between 1.0 Mev and 1.5 Mev. The single counts from specimens A and B both give 6.5±0.3 counts/min. If one interprets this effect as double beta-decay from Sn^{124}, one obtains a half-life between $0.4\cdot10^{16}$ yr. and $0.9\cdot10^{16}$ yr. Other alternative explanations for these observations have been considered but none have been found to be plausible. This result would indicate that double beta-decay is unaccompanied by neutrinos. A further consequence of these results pointed out to the author by Professor J. R. Oppenheimer is that the neutron-proton charge difference is exactly equal to the electron charge.

TABLE II. $L(A)$ gives the coincidences from specimen A between counters L, and $R(B)$ gives the coincidences from B between counters R. Holder 0° and 180° are the two holder positions. Positions 1 and 2 are positions for the specimens in the holder.

Pos. 1	Holder 0°	$L(A)$	$R(B)$
	Coin. counts/hr.	16.4±0.3	14.3±0.3
	Holder 180°	$L(B)$	$R(A)$
	Coin. counts/hr.	14.4±0.3	15.9±0.3
Pos. 2	Holder 0°	$L(B)$	$R(A)$
	Coin. counts/hr.	14.6±0.3	16.4±0.3
	Holder 180°	$L(A)$	$R(B)$
	Coin. counts/hr.	16.4±0.3	13.9±0.3

A detailed report of this work is being prepared for publication in the Physical Review.

The author is grateful to Professor R. Sherr for accepting the supervision of this research and to Professor E. P. Wigner for many profitable discussions.

* This work is assisted by the Office of Naval Research.
** These isotopes were obtained from Oak Ridge.

[1] Maria Goeppert-Mayer, Phys. Rev. 48, 512 (1935).
[2] W. H. Furry, Phys. Rev. 56, 1184 (1939).

PHYSICAL REVIEW VOLUME 86, NUMBER 4 MAY 15, 1952

A Re-Investigation of the Double Beta-Decay from Sn^{124}

E. L. FIREMAN AND D. SCHWARZER
Brookhaven National Laboratory, Upton, New York
(Received February 5, 1952)

Radiations from natural tin and tin enriched with the 124 isotope are examined in a magnetic field with a helium filled cloud chamber that is triggered by internal counters. Only three pictures out of more than four thousand photographs are pictures of two electrons coming out of the same point in the tin and entering the counters, and even these may be pictures of multiply scattered electrons passing through the tin. However, one may set a lower limit to the half-life of double beta-decay from Sn^{124} as 10^{17} years. This is a decay rate less than one-tenth of a value previously reported by one of the authors.

INTRODUCTION

THEORIES of the neutrino predict different properties for the double beta-decay process. Some[1] predict that two neutrinos accompany the two electrons. Others[2–5] require no neutrinos. Therefore, the deciding point between the two types of theories is the sum of the energies of the two electrons. This sum is either constant and equal to the mass difference or has a distribution of values. The two types of theories also predict different half-lives for the process; however, the fourth power of the nuclear matrix element enters into the half-life calculation. This matrix element must be less than one; otherwise its magnitude is quite uncertain. Therefore, a measurement of a minimum half-life for the process is not a conclusive way of distinguishing between the theories.

Attempts[6–9] have been made to detect double beta-decay from various isotopes. In the main these have given negative results, i.e., minimum half-lives for the process. In a previous letter one of the authors[6] had interpreted a slight coincidence activity from a Sn^{124} sample as a possible double beta-activity. It was therefore decided to re-investigate Sn^{124} with the present apparatus, which is more sensitive for detecting rare coincidence electrons than the previous apparatus. The present apparatus also measures the energies of the electrons.

DESCRIPTION OF APPARATUS[10]

A cylindrical cloud chamber of $12\frac{1}{8}$-in. inside diameter and $4\frac{5}{8}$-in. depth is filled with helium and ethyl alcohol at 108-cm pressure at 20°C temperature. The tin sample is placed across the center of the chamber. This sample acts both as a source and as the chamber clearing field. Three different samples were used in the experiment: a sheet of tin of normal isotopic constitution $10\frac{1}{2}$ in. long, 4 in. high, and 0.003 in. thick; another sheet of the same tin rolled to 0.0008-in. thickness; and a tin sample enriched in the 124 isotopes of 2.2205-g weight, and 0.0015-in. thickness. The enriched sample was obtained from Oak Ridge[11] and contained the 124 isotope in 95.04 percent abundance.

On either side of the sample is a counter. Initially open counters, whose interior form part of the visible region of the cloud chamber, were used. Later thin-walled counters (30 mg/cm²) were used. The open counters require more care than the thin-walled ones; therefore, most of the data have been taken with the thin-walled counters. The chamber is triggered by coincidence pulses from the two counters (1 microsecond resolving time). The open counters are $10\frac{1}{2}$ in. long. The thin-walled counters consists of a bank of three counters in parallel (each $\frac{3}{4}$ in. in diameter and 6 in. long). These counters subtend a solid angle between one and two steradians at the source. The single counting rate in each set of thin-walled counters is 80–90 counts/min; in the open counters it is about 500 counts/min.

The cloud chamber is placed in a horizontal position to reduce the number of cosmic rays. Nevertheless, a large percentage of the pictures are horizontal cosmic-ray mesons and showers. Rather than eliminate these events by anticoincidence counters, it was decided to keep these as an interesting sideline.

Helmholtz coils about the chamber give a magnetic field uniform to 1 percent over the chamber volume. The magnetic field is on continuously. Magnetic fields up to 900 gauss were used. However, most pictures were taken with the field in the neighborhood of 200 gauss. A larger magnetic field cannot be profitably used since low energy electrons would then curl up before reaching the counters. The coincidence rate is dependent upon the magnetic field strength. At zero field strength, there is one coincidence per minute; at 185 gauss there is one

[1] M. Goeppert Mayer, Phys. Rev. **48**, 512 (1935).
[2] W. H. Furry, Phys. Rev. **56**, 1184 (1939).
[3] B. Touschek, Z. Physik. **125**, 108 (1948).
[4] A careful analysis of neutrino theories and their effects on the double beta-process is given by Jayme Tiomono in a Princeton thesis (1950).
[5] L. A. Sliv, Zhur. Eksp. Teoret. Fiz. S.S.S.R. **20**, 11, 1039 (1950).
[6] E. L. Fireman, Phys. Rev. **74**, 1238 (1948); **75**, 323 (1949).
[7] M. G. Inghram and J. H. Reynolds, Phys. Rev. **76**, 1265 (1949); **78**, 822 (1950).
[8] Levine, Ghiorso, and Seaborg, Phys. Rev. **77**, 296 (1950).
[9] Marvin I. Kalkstein, University of Chicago thesis (1951).
[10] The apparatus is similar in some respects to one described by E. L. Fireman and G. M. McHaney, Rev. Sci. Instr. **21**, 813 (1950).

[11] The authors wish to thank Y-12 Branch of the Oak Ridge Stable Isotope Division for a four-week loan of this specimen. Presumably, this is the same specimen used by Kalkstein in his experiment.

TABLE I. Tin of normal isotopic constitution (0.003-in. thickness).

Magnetic field (gauss)	Mesons	Electrons	Scattered electrons	Showers	Double electrons	Coincidence pictures taken
350	3	3	6	3	none	26
390	27	18	17	25	none	272
450	24	21	14	23	none	267

coincidence every two and one-half minutes; and at 900 gauss there is one coincidence every three minutes. This dependence is caused mainly by the bending of low energy electrons from the chamber walls.

The sensitive time of the chamber is less than 0.1 second. Therefore at 185-gauss field strength the triggered picture is equivalent to more than 1500 random ones.

For accurate energy measurements, the magnetic curvature should be large compared to the multiple scattering curvature. According to Bethe's formula,[12] the multiple scattering from our chamber gas with our average track length corresponds to a field of $H_s=31/\beta$ gauss, where β is the ratio of the velocity of the electron to the velocity of light. For 340-kev electrons β is 0.8 and H_s is 39 gauss. It is interesting to note that the ethyl alcohol vapor makes a larger contribution to the multiple scattering than the helium.

A mirror gives a stereoscopic view in addition to the direct view of the chamber. The photographs are reprojected through an optical system identical to the one used in taking the pictures.

RESULTS

Coincidence tracks are classified in the following manner; if they are straight, they are called mesons. If they are bent in the form of a "C," they are called electrons. If they are scattered in the source giving the form "3," they are called scattered electrons; these could also be electron-positron pairs arising in the source. If the picture contains more than four tracks with at least one passing through both counters, it is called a shower. And finally, if the track has an "S" shape with the sign of curvature correct for electrons leaving the source, it is called a double electron track.

Tracks of the "S" type could be due to the double beta-process or to a beta-internal conversion electron of some contaminant.[13] Multiple scattering from the gas could also give a double electron track by changing the sign of the curvature of an electron passing through the source on the side which it approaches the source.

If sufficient "S" type tracks are observed, the double beta-process could be distinguished from a beta-internal conversion process by the energies of the electrons. In the beta-internal conversion process, one of the electrons has a constant energy. This is not true of the double beta-process either with or without neutrinos.

The data are summarized according to the above classification in Tables I, II, and III.

Only three pictures are those of double electrons coming out of the same point in the source and entering the counters. The curvatures and energies of these tracks are given in Table IV.

In addition to these three, there are two photographs of the "S" form whose curvatures (18.8 cm, 16.8 cm) and (12.8 cm and 14.2 cm) have the wrong sign for electrons leaving the source. These pictures can only be interpreted as an electron passing through the source and having the sign of its curvature on the side where it leaves the source changed by gas scattering. This leads us to the belief that some of the three double electron pictures are probably multiply scattered electrons with the wrong sign of curvature rather than double beta-electrons from tin 124. It should also be noted that the sum of the energies of the electrons given in Table IV is not constant. All of the coincidence tracks except a few of the meson tracks have close to minimum ionization. In several of the shower pictures there are a number of alpha-tracks.

If the three photographs are interpreted as double

TABLE II. Tin of normal isotopic constitution (0.0008-in. thickness).

Magnetic field (gauss)	Mesons	Electrons	Scattered electrons	Showers	Double electrons	Coincidence pictures taken
0	35	3	28	23	none	424
420	45	43	42	84	none	633

TABLE III. Tin enriched with 124 isotope to 95.04 percent abundance (0.0016-in. thickness).

Magnetic field (gauss)	Mesons	Electrons	Scattered electrons	Showers	Double electrons	Coincidence pictures taken
0	1	3	4	none	none	16
80	2	none	1	none	none	6
160	6	7	9	5	none	38
185	156	257	183	171	3	1904
310	39	59	29	58	none	520
400	5	15	4	14	none	100
860	3	2	1	4	none	70

TABLE IV. Curvatures and energies of two electrons out of tin.

Magnetic curvature (cm)	Electron energies (kilovolt)	Sum of energies (kilovolts)
10.2	255	750
15.8	495	
11.0	290	785
15.8	495	
8.8	195	330
7.0	135	

[12] H. A. Bethe, Phys. Rev. 70, 822 (1946).

[13] The tin samples were examined spectroscopically for impurities; no impurities of significance were found.

beta-events, a half-life of $2-5\times10^{17}$ years is obtained for the process. A result of three pictures in a field of 175 gauss when the multiple scattering corresponds to a field of 39 gauss is certainly a negative result, and the experiment should be interpreted as giving a half-life greater than 10^{17} years for the double beta-decay process. It was thought worthwhile to report these results, since they differ significantly from those reported in reference 6 and agree with those in reference 9.

The previous result[6] may have been caused by a small trace of an impurity having a coincidence activity in the enriched sample. In fact, some of the three double electron pictures in this experiment may have the same origin.

The authors wish to thank E. Bolze and B. R. Gibbs for aid in designing the apparatus and Anthony Del Duca who designed and constructed the electronics equipment.

On the Double Beta-Process

MARK G. INGHRAM AND JOHN H. REYNOLDS
University of Chicago and Argonne National Laboratory, Chicago, Illinois

August 31, 1949

A STUDY has been made of the isotopic constitution of xenon extracted from a pre-Cambrian tellurium ore, in order to determine the half-lives of the double beta-transitions $Te^{128} \rightarrow Xe^{128}$ and $Te^{130} \rightarrow Xe^{130}$. In this ore, measurable amounts of these two isotopes of xenon should have accumulated during geological time if the transitions take place at a rate comparable to that reported by Fireman[1] for the similar transition $Sn^{124} \rightarrow Te^{124}$. The importance of double beta-studies lies in the fact that the half-lives calculated from the Majorana theory of the neutrino[2] and the Dirac theory of the neutrino[3] differ by a large factor. Table I gives half-lives so calculated; the energies available for the transitions have been estimated from the decay schemes of the intermediate nuclei and from considerations of the nuclear energy surface[4] and are conservative.

TABLE I. Theoretical half-lives for allowed double beta-disintegrations.

Transition	$Te^{128} \rightarrow Xe^{128}$	$Te^{130} \rightarrow Xe^{130}$
Atomic mass difference in Mev	0.5	1.6
Majorana half-life in years	2×10^{16}	6.0×10^{14}
Dirac half-life in years	3×10^{27}	1.1×10^{24}

The ore used for this study was the mineral tellurobismuthite (Bi_2Te_3) in andalusite and sericite rock blasted from an outcrop in Mångfallberget, near Boliden, Sweden. This telluride deposit has been described in detail by Grip and Ödman.[5] A rough estimate by Dr. K. Rankama places the age of the original sulfide mineralization at 1500±500 million years.

TABLE III. Measured minimum half-lives based on crystal age of 1.5×10^9 years.

Transition	$Te^{128} \rightarrow Xe^{128}$	$Te^{130} \rightarrow Xe^{130}$
Max. no. radiogenic xenon atoms per atom of Xe^{134}	0.07	0.014
Max. no. radiogenic xenon atoms in sample	3.0×10^{12}	5.0×10^{11}
No. of parent atoms in sample	3.8×10^{22}	4.2×10^{22}
Minimum half-life in years	1.3×10^{19}	8.0×10^{19}

The sample (430 grams containing 6 percent tellurium) was crushed in an iron mortar and transferred to a quartz bottle which was then evacuated and heated to well above the melting point of Bi_2Te_3. The gas evolved was collected and purified in an apparatus patterned closely after that of Epstein and co-workers[6] in their experiments with fission xenon. The final rare gas mixture was analyzed in a conventional 60° single-focusing mass spectrometer. By calibrating with artificially prepared mixtures of argon and xenon, it was possible to measure both the chemical composition and the isotopic constitution of the gas.

The mixture was found to consist of $1.3 \pm 0.3 \times 10^{-5}$ cc S.T.P. xenon plus $1.3 \pm 0.3 \times 10^{-2}$ cc S.T.P. argon. The results of the isotopic analysis are represented in Table II together with Nier's values[7] for normal xenon. All abundances are referred to mass 134. It will be noted that the only evidence for the presence of radiogenic xenon occurs at mass 130. For this minute sample of xenon, however, the accuracy of the measurements is such that the apparent excess xenon at this mass can only be viewed as intriguing (even though a measurement of normal xenon immediately thereafter checked Nier's value of 0.386 exactly). At masses 124, 126 and 128 the recorder peaks were so close to background level that only upper limits for these abundances can be given.

These data can be used, however, to place lower limits on the half-lives of the double beta-transitions. The results are presented in Table III. A corresponding calculation for the energetically questionable transition $Te^{126} \rightarrow Xe^{126}$ gives a minimum half-life of 6×10^{19} years.

On comparison with the half-lives presented in Table I, it appears that these results support the Dirac antineutrino theory. However, the values are subject to qualification on two counts. First, the possibility exists that radiogenic xenon may have escaped from the crystals of ore since the time of the original mineralization, either by simple diffusion or during later alteration of the mineral. Considering the retentivity of helium by various minerals,[8,9] it appears that the former possibility is remote. A hydrothermal alteration of the telluride, on the other hand, would certainly result in loss of xenon, and this possibility cannot be dismissed for the present ore. If this has occurred, the accumulation of daughter xenon dates only from the time of alteration and the half-lives in Table III must be reduced in proportion. Secondly, the theoretical half-lives presented in Table I are essentially minimum values; favorable forms of the interaction terms have been assumed and the calculations have been made for completely allowed transitions. As a result, the Majorana half-lives may be longer in the particular transitions we have investigated, due, for example, to a possible difference in parity between the ground states of Te^{130} and Xe^{130}. That these two effects, acting in combination, could overpower the factor of 10^5 by which the measured minimum half-life exceeds the calculated Majorana half-life in the transition at mass 130 is improbable, but it must be admitted as a possibility.

These results are preliminary to more extensive investigations of the matter.

The writers take pleasure in acknowledging helpful discussions with Professors W. H. Newhouse and K. Rankama of the Geology Department and A. J. Dempster and E. Fermi of the Physics Department of the University of Chicago.

1 E. L. Fireman, Phys. Rev. 75, 323 (1949).
2 W. H. Furry, Phys. Rev. 56, 1184 (1939).
3 M. G. Mayer, Phys. Rev. 48, 512 (1935).
4 E. Feenberg, Rev. Mod. Phys. 19, 239 (1947).
5 E. Grip and O. Ödman, Sveriges Geologiska Undersökning 36, No. 4 (1942).
6 S. Epstein, Proc. Conf. on Nuclear Chem., McMaster University, pp. 108–116 (May 1947).
7 A. O. Nier, Phys. Rev. 52, 933 (1937).
8 P. M. Hurley and C. Goodman, Bull. Geol. Soc. Am. 54, 305 (1943).
9 N. B. Keevil, Proc. Am. Acad. 73, 311 (1940).

TABLE II. Relative abundances of the xenon isotopes.

Mass	124	126	128	129	130	131	132	134	136
Present study	<0.015	<0.018	<0.25	2.47 ±0.03	0.394 ±0.004	2.00 ±0.02	2.55 ±0.02	1.000	0.847 ±0.008
Normal xenon*	0.0089	0.0083	0.180	2.49	0.386	2.009	2.558	1.000	0.849

* See reference 7.

Double Beta-Decay of Te^{130}

MARK G. INGHRAM AND JOHN H. REYNOLDS
University of Chicago and Argonne National Laboratory, Chicago, Illinois
April 27, 1950

IN a previous letter[1] the authors described the preliminary results of a study of the double beta-transition $Te^{130} \rightarrow Xe^{130}$ by the method of isotopic analysis of xenon extracted from geologically old tellurium ores. The ore on which that work was based was blasted from an outcrop at Mångfallberget, Sweden, and it contained 12 percent by weight of the mineral Bi_2Te_3. Although the age of the telluride minerals in Mångfallberget and in Boliden, Sweden is known to be 1500±500 million years, there exists an uncertainty in the "xenon age" of the Mångfallberget material owing to the possibility of comparatively recent crystal alteration by percolating surface waters.

As a result of the generous cooperation of Dr. Erland Grip of the Boliden Mining Company, we have recently obtained some excellent samples of 70 percent rich Bi_2Te_3 from the 240-meter level of the Boliden mine. According to geologists, it is improbable that the crystals of this Bi_2Te_3 have been affected by a recent alteration. A 371-g sample of this material, containing 124 g of tellurium, was broken up to pea size and vacuum roasted at temperatures high enough to decompose the mineral and produce vigorous boiling of the molten bismuth and tellurium. The rare gases evolved were purified as before,[1] and were found to consist of 2.4×10^{-2} cc S.T.P. argon plus 2.6×10^{-7} cc S.T.P. xenon. The results of an isotopic analysis of this xenon are presented in Table I. The percentage composition of the excess or radiogenic xenon has been given in the last row of the table. These percentages were calculated by assuming that all of the Xe^{132} present was normal (atmospheric), and by subtracting corresponding amounts at the other mass positions. It is evident from the table that, aside from the somewhat questionable excess xenon at mass 134 and 136 (possibly due to uranium fission), the excess xenon is distributed among the isotopes Xe^{129}, Xe^{130}, and Xe^{131}.

TABLE I. Isotopic composition of xenon extracted from 371 g of a 1.5×10^9 year old mineral containing 70 percent Bi_2Te_3. The total xenon content was 2.6×10^{-7} cc S.T.P.

Mass	Bi_2Te No. 4	Normal	Diff.	Difference normalized
124	<0.004	0.0005	<0.0035	<0.29 —
126	<0.005	0.0005	<0.0045	<0.38 —
128	<0.014	0.011	<0.003	<0.25 —
129	≡1.000	0.1525	0.8475	71.66±0.7
130	0.0652	0.0236	0.0416	3.52±0.17
131	0.4088	0.1235	0.2853	24.12±0.3
132	0.1566	≡0.1566	≡0.0000	≡0.00 —
134	0.0651	0.0611	0.0040	0.34±0.25
136	0.0562	0.0519	0.0043	0.36±0.16

The excess Xe^{129} and Xe^{131} is probably caused by (n,γ) reactions on Te^{128} and Te^{130}. To account for the neutron "flux" required to produce this much xenon in 1.5×10^9 years, it is necessary to assume that there was considerable uranium in the immediate neighborhood of the tellurium mineral. Dr. Grip informs us that unusually large amounts of the uranium mineral thucholite have been found in the stope from which the Bi_2Te_3 was taken. Thus we ascribe the excess Xe^{129} and Xe^{131} to the proximity of such a deposit.

An interesting discrepancy appears in the ratio of Xe^{129} to Xe^{131} found in the sample. The present measurements show this to be 3.0, whereas one would expect a ratio of 0.6 from Seren's values[2] for the tellurium cross sections. One possible explanation is that Xe^{129} was also produced by the decay of small amounts of I^{129}($\sim3\times10^7$ yr.) present in the mineral. This nuclide, as yet undetected in nature, may have been formed originally in amounts comparable to that of I^{127}.

The excess Xe^{130} is attributed to double beta-decay of Te^{130}. There appears to be no other explanation for its formation. Assuming an age of 1.5×10^9 years for the Bi_2Te_3, the excess Xe^{130} present corresponds to double beta-decay of Te^{130} with a half-life of 1.4×10^{21} years. This result is to be compared with theoretical half-lives of 6×10^{14} years and 10^{24} years. the former computed from the Majorana theory of the neutrino, and the latter from the Dirac theory. Both calculations are for allowed transitions with 1.6 Mev of available energy.

[1] M. G. Inghram and J. H. Reynolds, Phys. Rev. **76**, 1265 (1949).
[2] Seren, Friedlander, and Turkel, Phys. Rev. **72**, 888 (1947).

Half-Life for Double Beta-Decay*

C. A. Levine, A. Ghiorso, and G. T. Seaborg
Department of Chemistry and Radiation Laboratory, University of California, Berkeley, California
November 28, 1949

FIREMAN[1] has reported the results of a rather difficult beta-particle coincidence counting experiment in which the decay of ${}_{50}Sn^{124}$ to ${}_{52}Te^{124}$ by the simultaneous emission of two negative beta-particles, with a half-life between 4×10^{16} and 9×10^{16} years, seems to have been observed. This note reports the results obtained from a different and somewhat simpler method of looking for the phenomenon of simultaneous emission of two beta-particles. These results are negative so far and show that this process is considerably less probable in the case chosen by us than in that reported by Fireman.

Our method consists of looking in uranium samples for 90-year Pu^{238} which would come from U^{238} by the double beta-particle mechanism since Np^{238} is heavier than U^{238}, which in turn is substantially heavier than Pu^{238}, in the isobaric triplet ${}_{92}U^{238}-{}_{93}Np^{238}-{}_{94}Pu^{238}$. This chemical method of investigation is particularly applicable to this isobaric triplet because there appears to be no other mechanism to account for the Pu^{238} should it be found. The energetics of the situation are summarized in the following diagram, where the disintegration energies (in Mev) are derived from sources which may be traced through a recent compilation.[2]

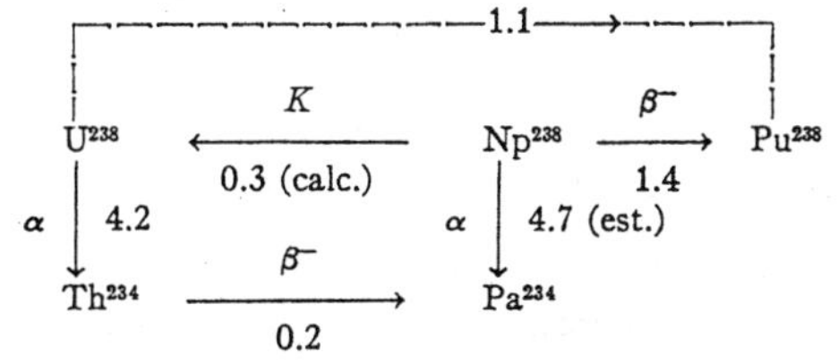

The alpha-disintegration energy of Np^{238} is estimated from alpha-decay systematics.[3]

Our experiment consisted of taking 14 kg of very pure, 6-year old, UO_3 and extracting and separating the plutonium fraction by chemical means. The method consisted essentially of dissolving the oxide in nitric acid and precipitating Pu(IV) with lanthanum fluoride, followed by solution of the lanthanum fluoride and oxidation of the plutonium to Pu(VI), which was extracted into diethyl ether and then re-extracted into water. Similar cycles were repeated five times in order to separate completely from UX_1 and to reduce the amount of lanthanum carrier, after which the final sample was plated out on flat platinum with total carrier weight probably less than 50 micrograms. The use of tracer Pu^{239} in this separation established that the chemical yield amounted to 10 percent.

This final sample was measured for the presence of the 5.51-Mev alpha-particles of Pu^{238} on the alpha-pulse analyzer apparatus in this laboratory.[4] This analysis showed that the counting rate of the Pu^{238} alpha-particles at essentially 50 percent counting yield amounted to 0.00 ± 0.01 counts per minute above background. This indicates that the "half-life" of U^{238} for simultaneous emission of two beta-particles, for which a total energy of 1.1 Mev is available, is greater than 6×10^{18} years.

This experiment could be extended to reach longer half-lives through the use of larger and older sources of uranium such as pitchblende ore. In this case, of course, the plutonium fraction so isolated will contain a certain amount of Pu^{239} as has already been demonstrated.[5] The extraction of plutonium from a ton of pitchblende (50 percent uranium) with 10 percent yield could detect a half-life as long as some 10^{22} years for the same limits of counting accuracy.

This result appears to disagree with that of Fireman although it may not be possible to be positive about this in view of the difference in energies and atomic numbers and possible difference in degree of prohibition involved. The theory for the double beta-decay process sets widely differing ranges of half-life depending upon whether the process can take place without neutrine emission.[6] Fireman's results are in the range predicted for double beta-decay without neutrino emission while our half-life limit seems to be above this predicted range and perhaps points toward the emission of two neutrinos in this process.

A recent investigation[7] of the double beta-transition ${}_{52}Te^{130}\rightarrow{}_{54}Xe^{130}$, by a similar method in which the xenon present with a tellurium ore was analyzed, also points toward a two-neutrino process, but is subject to experimental uncertainty with respect to the age of and the degree of xenon retention by the ore.

It is a pleasure to acknowledge the assistance of Dr. L. B. Magnusson in the chemical procedure.

* This work was performed under the auspices of the AEC.
1 E. L. Fireman, Phys. Rev. 75, 323 (1949).
2 G. T. Seaborg and I. Perlman, Revs. Mod. Phys. 20, 585 (1948).
3 Perlman, Ghiorso, and Seaborg, Phys. Rev. 74, 1730 (1948); Phys. Rev. 77, 26 (1950).
4 Ghiorso, Jaffey, Robinson, and Weissbourd, National Nuclear Energy Series, Plutonium Project Record Vol. 14B "The Transuranium Elements: Research Papers," Paper No. 16.8 (McGraw-Hill Book Company, Inc., New York, 1949).
5 G. T. Seaborg and M. L. Perlman, J. Am. Chem. Soc. 70, 1571 (1948); National Nuclear Energy Series, Plutonium Project Record Vol. 14B "The Transuranium Elements: Research Papers," Paper No. 1.3 (McGraw-Hill Book Company, Inc., New York, 1949).
6 W. H. Furry, Phys. Rev. 56, 1184 (1939).
7 M. G. Inghram and J. H. Reynolds, Phys. Rev. 76, 1265 (1949).

The Half-life of ^{130}Te Double β-decay

N. Takaoka and K. Ogata

Department of Physics, Faculty of Science, Osaka University.
Kita-toneyama, Toyonaka-shi, Osaka-fu, Japan

(Z. Naturforschg. **21 a**, 84—90 [1966]; received 15 September 1965)

Dedicated to Prof. J. Mattauch *on his 70th birthday*

In order to determine the half-life of the ^{130}Te double β-decay, the amounts and isotopic composition of xenon extracted from tellurium ores, from the Oya gold mine in Japan, have been measured with a high-sensitivity mass spectrometer. Compared with atmospheric xenon an excess was definitely found at mass numbers 129, 130 and 131 in the extracted xenon. The excess of ^{130}Xe is predominant, the average amount in three samples being $(1.32 \pm 0.09) \times 10^{-11}$ ccSTP/g ^{130}Te. Attributing the excess ^{130}Xe to the double β-decay of ^{130}Te, the half-life is estimated to be $(8.20 \pm 0.64) \times 10^{20}$ years, assuming an age of $(9.06 \pm 0.29) \times 10^{7}$ years for the Te ores. The latter value is the K-Ar age of porphyrite, which is in close geological connection with the Te ores.

In order to investigate the other excesses than that of ^{130}Xe, isotopic analyses were also carried out on Xe from three other Te ores from the same mine. The ratios $(^{129}\text{Xe}/^{131}\text{Xe})_{\text{excess}} = 1.58$ and $(^{129}\text{Xe}/^{130}\text{Xe})_{\text{excess}} = 2.1$ were found to be the same for all samples. The origin of these excesses is discussed.

In addition a small excess of ^{128}Xe was found. If this is attributed to ^{128}Te double β-decay, the half-life of ^{128}Te is estimated to be 3×10^{22} years, a value shorter by about three orders of magnitude than the theoretically expected half-life. The above estimated half-life may be a lower limit of the ^{128}Te half-life.

The general tendency of the isotopic abundances (except for the above excesses), of the xenon extracted from Te ores seems to be to slightly increase in excess as one moves toward the lighter isotopes (as compared with atmospheric xenon).

In the mass Table there are about fifty nuclei for which double β-decay is energetically possible. ^{130}Te is one of them. The energy level scheme for its double β-decay [1] is shown in Fig. 1. Double β-decay has been studied theoretically by many physicists [2–8]. Accordding to their theoretical analyses, its probability is extremely small, with strong dependence on the decay energy. According to Primakoff and Rosen [8], the half-life of the ^{130}Te double β-decay is $2 \times 10^{21 \pm 2}$ years or $8 \times 10^{15 \pm 2}$ years for two-neutrino or no neutrino emission respectively. If it could be proven from a half-life determination that the double β-decay is accompanied by two-neutrino emission, the neutrino may be represented by the Dirac theory or the two component theory of the neutrino, and the total lepton number is conserved. On the other hand, if there is no neutrino emission, the neutrino may be represented by the Majorana theory, and the total lepton number is not conserved. The double β-decay may also yield information on the type of nucleon-lepton interaction, as well as the nature of the neutrino and the conservation of the total lepton number.

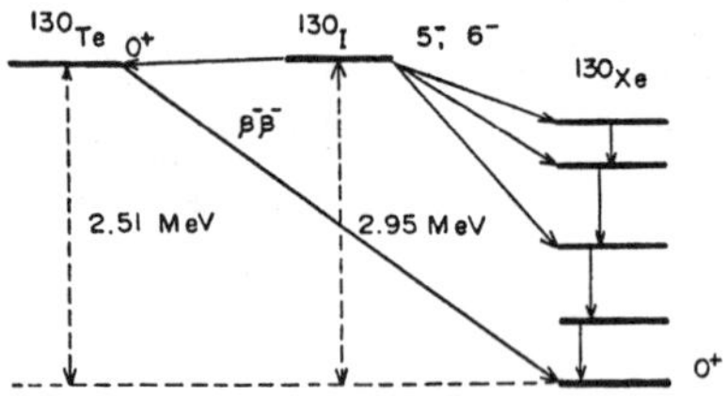

Fig. 1. Energy level scheme of ^{130}Te double β-decay [1].

Many experimental studies [9–19] of double β-decay have been undertaken using mass spectrometric methods, cloud chambers, photographic emulsions,

1 N. Zeldes, M. Gronau, and A. Lev, Nucl. Phys. **63**, 1 [1965].
2 M. G. Mayer, Phys. Rev. **48**, 512 [1935].
3 G. Racah, Nuovo Cim. **14**, 322 [1937].
4 W. H. Furry, Phys. Rev. **56**, 1184 [1939].
5 H. Primakoff, Phys. Rev. **85**, 888 [1953].
6 J. H. Maclenan, Phys. Rev. **106**, 821 [1957].
7 K. M. Case, Phys. Rev. **107**, 307 [1957].
8 H. Primakoff and S. P. Rosen, Rept. Prog. Phys. **22**, 121 [1959].
9 E. L. Fireman, Phys. Rev. **74**, 1238 [1948].
10 E. L. Fireman, Phys. Rev. **75**, 323 [1949].
11 J. A. McCarthy, Phys. Rev. **97**, 1234 [1955].
12 J. A. McCarthy, Phys. Rev. **90**, 853 [1953].
13 M. G. Inghram and J. H. Reynolds, Phys. Rev. **78**, 822 [1950].
14 R. J. Hayden and M. G. Inghram, Mass Spectr. Phys. Res., Nat. Bur. Stand. Circ. **522**, 189 [1953].
15 M. I. Kalkstein and W. F. Libby, Phys. Rev. **85**, 368 [1952].
16 E. L. Fireman and D. Schwarzer, Phys. Rev. **86**, 451 [1952].
17 M. Awschalom, Phys. Rev. **101**, 1041 [1956].
18 E. I. Bobrokotov et al., Soviet Phys.-Doklady **1**, 600 [1958].
19 E. I. Dobrokotov et al., Soviet Phys.-JETP **9**, 54 [1959].

Zeitschrift für Physik 202, 273–292 (1967)

Massenspektrometrischer Nachweis von ββ-Zerfallsprodukten

T. Kirsten
Max-Planck-Institut für Kernphysik, Heidelberg, und State University of New York at Stony Brook
W. Gentner
Max-Planck-Institut für Kernphysik, Heidelberg
O. A. Schaeffer
State University of New York at Stony Brook

Eingegangen am 7. März 1967

The principles and methods of mass-spectrometric detection of $\beta\beta$-decay products and the theoretical purposes for such studies are discussed in detail.
We report also the results of our new measurements on two tellurides, one native tellurium sample and one selenide. Concerning the decay $Te^{130} \xrightarrow{\beta^-\beta^-} Xe^{130}$, we have found that Xe^{130} in a native tellurium sample amounts to 67% of the total xenon with an absence of any other anomaly, which is an important improvement over previous measurements. — For the first time the decay $Se^{82} \xrightarrow{\beta^-\beta^-} Kr^{82}$ was studied by mass-spectrometry. A Kr^{82}-excess was positively found in an copper selenide. A detailed discussion of all processes other than double β-decay which might contribute to the Xe^{130} and Kr^{82}-anomalies is undertaken. It can be shown that these effects are small compared to the $\beta\beta$-decay process. It can be said now positively that $\beta\beta$-decay occurs in nature. Discussing all available data we obtained the effective half-life of Te^{130} to be $6 \times 10^{20 \pm 0.3}$ y and that of Se^{82} equals to $6 \times 10^{19 \pm 0.3}$ y. The results are in agreement with the theoretical predictions for the 2ν-$\beta\beta$-decay. Nevertheless, no-neutrino $\beta\beta$-decay is still possible. However, from the data we calculated that the lepton-nonconservation is at most 0.7%.

Diese Arbeit verfolgt zwei Ziele: Erstens soll ein Überblick über die Methode und den gegenwärtigen Stand des massenspektrometrischen Nachweises von $\beta\beta$-Zerfallsprodukten gegeben werden.

Zweitens soll über unsere neuen Meßergebnisse berichtet werden. Dabei wird außer dem Zerfall $Te^{130} \xrightarrow{\beta^-\beta^-} Xe^{130}$ erstmals auch der Zerfall $Se^{82} \xrightarrow{\beta^-\beta^-} Kr^{82}$ behandelt.

1. Theoretische Grundlagen

Die Theorie des doppelten β-Zerfalls wurde entwickelt von Goeppert-Mayer[1], Furry[2], Konopinski[3] und Primakoff und Rosen[4]. Zusam-

[1] Goeppert-Mayer, M.: Phys. Rev. **48**, 512 (1935).
[2] Furry, W. H.: Phys. Rev. **56**, 1184 (1939).
[3] Konopinski, E. J.: USAEC Report Los Alamos (1949).
[4] Primakoff, H., and S. P. Rosen: Repts. Progr. in Phys. **22**, 121 (1959).

VOLUME 20, NUMBER 23 PHYSICAL REVIEW LETTERS 3 JUNE 1968

EXPERIMENTAL EVIDENCE FOR THE DOUBLE-BETA DECAY OF Te^{130}†

T. Kirsten and O. A. Schaeffer
Department of Earth and Space Sciences, State University of New York, Stony Brook, New York

and

E. Norton and R. W. Stoenner
Department of Chemistry, Brookhaven National Laboratory, Upton, New York
(Received 1 April 1968)

It is shown that double-beta decay occurs in nature. The half life of Te^{130} is $10^{21.34\pm0.12}$ yr.

Attempts to detect double-beta decay directly by means of coincidence techniques or nuclear emulsions have not led to an actual observation of double-beta decay.[1] Lower limits for the half lives of different isotopes were established. The most studies were made on Ca^{48}.[1]

Another approach is to detect the accumulation of the decay product during geological time periods (~10^9 yr) in those minerals which are rich in a suspected $\beta\beta$-active isotope. Two very suitable cases for this method are the decays of $Te^{130} \xrightarrow{\beta\beta} Xe^{130}$ and $Se^{82} \xrightarrow{\beta\beta} Kr^{82}$.[2]

An anomalous isotopic composition of xenon extracted from tellurium minerals has been reported repeatedly.[3,4] In all cases, surplus amounts of Xe^{129}, Xe^{130}, and Xe^{131} were superimposed on the general pattern of xenon of atmospheric composition. These findings have not been accepted[1] as proof for the double-beta decay of Te^{130} since the Xe^{130} anomaly was always accompanied by Xe^{129} and Xe^{131} excesses not yet completely understood. Rather, it was suspected that all three anomalies might result from the same unknown mechanism. In some cases, the xenon spectrum was even more confused by an additional fission xenon component which complicated the computation of the atmospheric Xe^{130} correction. The Xe^{130} excess was arrived at only by a rather involved calculation which included large uncertainties. The magnitude of the Xe^{130} excess has been less than 6% of the total xenon amount in all samples studied by other authors. In addition, the calculated half lives of Te^{130} were based on assumed geological ages of the minerals rather than on age determinations of the tellurium ores themselves.

To prove that Te^{130} is $\beta\beta$ active one has to show that the Xe^{130} excess is unambiguously due to double-beta decay. In addition, a calculated half-life can only be accepted if it is based on radioactive dating of the mineral.

We have analyzed the isotopic composition of xenon extracted from a native tellurium ore from the Good Hope mine (Colorado).[5] We found a large excess of Xe^{130} not accompanied by any other anomalies.[6] In uncrushed samples not exposed to air contamination, the excess Xe^{130} amounts to about 70% of the total xenon and the ratio ${Xe_{excess}}^{130}/{Xe_{atmospheric}}^{130}$ is higher than 50. The Xe^{130} excess does not constitute a small anomaly on the border of detectability but determines the pattern of the xenon spectrum (Fig. 1).

The absence of other xenon anomalies rules out processes other than double-beta decay which might possibly result in a Xe^{130} excess.[7,2] The production of Xe^{130} via two successive single-beta decays is impossible since the mass excess of I^{130} is 477 ± 35 keV larger than the mass excess of Te^{130}.[8] We must then conclude that the Xe^{130} excess is due to double-beta decay of Te^{130}.

From the amount of ${Xe_{excess}}^{130}$, the tellurium concentration, and the gas-retention age of the mineral, the half-life of Te^{130} can be calculated. We determined the K-Ar age of the ore and obtained an age of 1.31 ± 0.14 Gyr. The age is considered to be reliable since it is consistent with the geological situation of the ore deposit.[9]

VOLUME 20, NUMBER 23 PHYSICAL REVIEW LETTERS 3 JUNE 1968

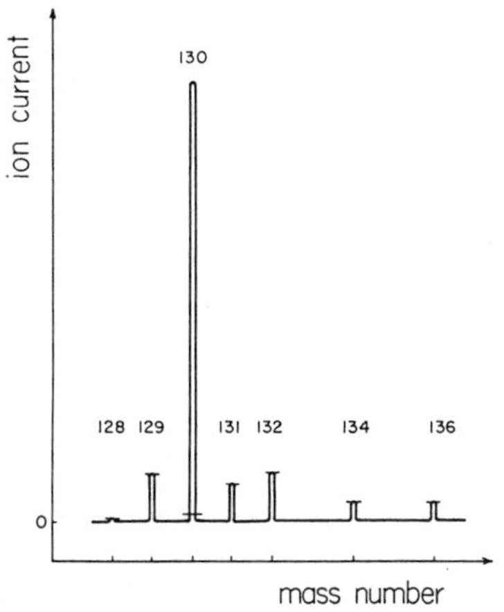

FIG. 1. Isotopic composition of xenon extracted from native tellurium ore (run No.2). The horizontal lines indicate the maximum contribution of atmospheric xenon.

The resulting half-life of Te^{130} is $10^{21.34\pm0.12}$ yr and is in agreement with the theoretically predicted[10] half-life for the lepton-conserving two-neutrino mode of decay ($T_{\frac{1}{2}, 2\nu, \text{theor}} = 10^{22.5\pm2.5}$ yr). However, mainly because of the uncertainty in the theoretical prediction, a slight contribution of the lepton-nonconserving neutrinoless decay mode (predicted half-life $T_{\frac{1}{2}, \text{no}\ \nu, \text{theor}} = 10^{16.3\pm2}$yr)[10] cannot be ruled out. The maximum interaction amplitude which does not conserve leptons is limited by $\alpha \leqslant (T_{\frac{1}{2}, \text{no}\ \nu, \text{theor}}/T_{\frac{1}{2}, \text{exp}})^{1/2} \approx 10^{0.5(16.3-21.3)} \approx 3\times10^{-3}$.

Experimental details.—The available sample was a compact piece of fresh-appearing native tellurium, 17 g in weight. It was supplied to us by Professor C. Frondel from the Harvard mineral collection (sample No. 98589). The surface was cleaned with a dental drill. The sample was then broken and two solid inner pieces (about 1 g each) were used for the mass spectrometer runs Nos. 1 and 2. The remaining 15 g were crushed and screened to yield 12 g of ore having grain sizes between 62 and 405 μ. This sample was split in two parts. One part was used for the chemical analysis. From the other part, three samples were prepared for the mass spectrometer runs Nos. 3, 4, and 5. The chemical bulk analysis gave 99.4± 0.6% Te with a trace of Fe.[11] K and U[12] as well as Th were determined by neutron activation. The results were 5.6± 0.2 ppm K, 21.0± 2 parts per billion (ppb) U, and Th ≦ 10 ppb.

The techniques involved in handling and analyzing minute amounts of rare gases by mass spec-

Table I. Results of the rare-gas analysis.

Run	Sample	Extraction Time (min)	Extraction Temp. (°C)	Rare Gas Amounts (10^{-8}cc STP/gram) Radiogenic Gases He^4	Ar^{40}	Xe^{130}	Gases of Atm. Composition[a] Ar	Kr	Xe	$Xe^{130}_{rad}/Xe^{130}_{atm}$
1	0.880 g whole piece	10	1100	10.3	1.03	0.00283	4.0	0.00049	0.00141	48.1
2	0.866 g whole piece	10	1100	40.4	1.56	0.00272	23.9	0.00095	0.00118	56.5
3 a	0.894 g 62μ-405μ	~1	~600[b]	42	4.27	0.00233	36.1	0.01340	0.01690	3.38
b		10	1100	2.7	0.37	0	2.1	0.00123	0.00248	-
total				44.7	4.64	0.00233	38.2	0.01463	0.01938	2.95
4 a	1.745 g 62μ-405μ	800	200	2.1	0	0.00011	15.9	0.00554	0.00298	0.90
b		800	200	2.5	0.15	0.00002	4.1	0.00168	0.00136	0.36
c		270	310	12.0	1.02	0.00082	8.7	0.00503	0.00593	3.39
d		10	1100	34.1	2.89	0.00118	10.7	0.00368	0.00255	11.4
total				50.7	4.06	0.00213	39.4	0.01593	0.01282	4.07
5 a	0.779 g 62μ-405μ	1400	200	3.0	0	0.00015	16.4	0.00575	0.00422	0.87
b		10	1100	52.0	4.12	0.00184	22.1	0.01025	0.00737	6.11
total				55.0	4.12	0.00199	38.5	0.01600	0.01159	4.24
3-5	weighted averages			50.3 ± 6	4.22 ±0.4					
1-5	weighted average					0.00240 ±0.0004				

[a]Except for He^4, Ar^{40} and Xe^{130} no other isotopic anomalies have been found.
[b]Native tellurium melts at 452°C.

VOLUME 20, NUMBER 23 PHYSICAL REVIEW LETTERS 3 JUNE 1968

trometry are described, except for one improvement,[13] elsewhere.[14,2,4] In runs Nos. 1 and 2, we extracted the rare gases in only one step by melting the ore. In runs Nos. 3, 4, and 5 the gases were extracted stepwise as described in Table I which contains the results. The mean value for Xe^{130} excess is $(2.4 \pm 0.4)\times 10^{-11}$ cc STP/g; the radiogenic Ar^{40} content of the crushed sample is $(4.22 \pm 0.4)\times 10^{-8}$ cc STP/g, and the He^4 content is $(50.3 \pm 6)\times 10^{-8}$ cc STP/g.

Runs Nos. 1 and 2 had the lowest yield of atmospheric xenon since no inner surfaces were exposed to air contamination. In these cases, the Xe^{130} anomaly was about 5000%. For dating purposes it is absolutely necessary to prepare a finely-crushed and well-mixed average sample in order to exclude the effects of K and U inhomogeneities. This necessity is underlined by the different radiogenic Ar^{40} contents of the crushed and the uncrushed samples. The amount of the atmospheric xenon due to surface adsorption is higher in the crushed samples but becomes relatively lower after outgassing at moderate temperatures (Table I). The absolute amounts of excess Xe^{130} are in reasonable agreement for all five runs. From the data we obtained the following gas-retention ages: K-Ar age = 1.31 ± 0.14 Gyr; U-He^4 age = 205 ± 55 Myr. The results indicate that the thermal history of the ore was such that helium is partially lost by diffusion but not Ar. If this is true, one may then conclude that probably no Xe^{130} loss occurred. A high Xe^{130} retentivity is also supported by the stepwise heating experiments. The half-life for the double-beta decay of Te^{130} is then $10^{21.34 \pm 0.12}$ yr.

Despite the fact that $\beta\beta$ decay was not observed with certainty, half-lives for Te^{130} have been calculated previously.[3] Based on geological assumptions for the gas-retention ages of the tellurium ores, half-lives between $10^{20.48}$ yr and $10^{21.15}$ yr have been reported.[3] The true ages of these minerals can now be calculated by introducing the Te^{130}-Xe^{130} age-determination method, based on the half-life obtained in this work.

We wish to thank Professor C. Frondel for providing us with the rare mineral, Dr. Z. Peterman, Dr. J. Olson, Dr. D. Hedlund, and Professor S. Goldich for valuable information concerning the geology of the ore deposit, and Dr. R. Davis for helpful suggestions.

---

†Work supported by the U. S. Atomic Energy Commission, partially under Contract No. At-(30)-3629.

[1]Earlier work reviewed by S. P. Rosen and H. Primakoff, in Alpha-, Beta-, and Gamma-Ray Spectroscopy, edited by K. Siegbahn (North-Holland Publishing Company, Amsterdam, The Netherlands, 1965), Vol. 2, p. 1499; and by V. R. Lazarenko, Usp. Fiz. Nauk 90, 601 (1966) [translation: Soviet Phys. Usp. 9, 860 (1967)]. For recent references see R. K. Bardin, P. J. Gollon, J. D. Ullman, and C. S. Wu, Phys. Letters 26B, 112 (1967); C. Y. Chang, G. B. Yodh, R. Ehrlich, R. Plano, and A. Zinchenko, Phys. Rev. Letters 20, 510 (1968).

[2]T. Kirsten, W. Gentner, and O. A. Schaeffer, Z. Physik 202, 273 (1967).

[3]M. G. Inghram and J. H. Reynolds, Phys. Rev. 76, 1265 (1949), and 78, 822 (1950); N. Takaoka and K. Ogata, Z. Naturforsch. 21a, 84 (1966); E. K. Gerling, J. A. Shukoljukov, and G. S. Askinadze, Yadern. Fiz. 6, 311 (1967) [translation: Soviet J. Nucl. Phys. 6, 226 (1968)], and Geokhimiya 9, 1038 (1967).

[4]T. Kirsten, W. Gentner, and O. Müller, Z. Naturforsch. 22a, 1783 (1967).

[5]A part of the data of this investigation has been published in a preliminary state (Ref. 2). Reference 2 contains a detailed discussion of the principles, methods, and techniques of mass spectrometric detection of $\beta\beta$-decay products and of their theoretical interpretation.

[6]The results show that the Xe^{129} and Xe^{131} excesses observed in other tellurium ores are due to the contamination of the samples by U. Measured U concentrations in other tellurides are 196 and 225 ppm (Refs. 3 and 4). This is four orders of magnitude above the U concentration in the sample discussed here. A further discussion of the Xe^{129} and Xe^{131} anomalies in U-rich tellurides may be found elsewhere.[2,4]

[7]If Xe^{130} were produced by spontaneous fission of uranium, one should see about 10^5 times more Xe^{136} than Xe^{130}. Cosmic-ray–induced spallation reactions would produce much more Xe^{124} and Xe^{126} than Xe^{130}, but no Xe^{124} and Xe^{126} anomalies are observed. In addition, the ore was shielded by the atmosphere and rocks above the mine. Meson-induced reactions do not lead to Xe^{130}. Neutron-induced reactions would primarily form Xe^{129} and Xe^{131}, but no such anomalies exist in our sample. The absence of Xe^{129} and Xe^{131} anomalies sets limits on the maximum internal neutron flux. From this limit and the highest possible estimate of the cesium concentration, it follows that the maximum contribution of the reaction

$$Cs^{133}(n,\alpha)I^{130}\xrightarrow{\beta^-}Xe^{130}$$

is far below the detection limit. Natural α radiation would produce Xe^{128}, Xe^{129}, and Xe^{131} but not Xe^{130}, since natural α energies are insufficient for the reaction $Te^{128}(\alpha,2n)Xe^{130}$.

[8]$\Delta M(I^{130}) = -868\,90 \pm 19$ keV, H. Daniel, M. Kuntze, B. Martin, P. Schmidlin, and H. Schmitt, Nucl. Phys. 63, 145 (1965). $\Delta M(Te^{130}) = -873\,67 \pm 16$ keV, N. Zeldes, M. Gronau, and A. Lev, Nucl. Phys. 63, 1 (1965).

[9]The Good Hope Mine (Vulcan mining district, N38°20′ 45″-107°00′15″W) belongs geologically to the Gunnison River (Colorado) Series. Veins rich in pyrite, gold and copper tellurides, and native tellurium

VOLUME 20, NUMBER 23 PHYSICAL REVIEW LETTERS 3 JUNE 1968

occur in Precambrian amphibole schists (Dubois greenstone); E. B. Eckel, U. S. Geol. Surv. Bull. 1114, 1 (1961); J. F. Hunter, U. S. Geol. Surv. Bull. 777, 1 (1925). No radioactive dating on the Precambrian basement near Vulcan has been done. However, for the basement in the nearby Monarch Pass–Gunnison area, Rb-Sr ages between 1.3 and 1.6 Gyr have been reported by G. W. Wetherill and M. E. Bickford, Trans. Am. Geophys. Union 45, 116 (1964). The stage of mineralization in the Vulcan district is probably also Precambrian [W. S. Burbank and C. T. Pierson, U. S. Geol. Surv. Circ. 236, 1 (1953); Z. Peterman, J. Olson, and D. Hedlund, U. S. Geological Survey, Denver, Colorado, private communication].

[10]Calculated from the formula given by Rosen and Primakoff (Ref. 1) and new mass excess data by N. Zeldes, M. Gronau, and A. Lev, Nucl. Phys. 63, 1 (1965). The mass difference between Te^{130} and Xe^{130} is 2509 ± 19 keV.

[11]A similar result on a native tellurium sample from the Good Hope Mine is reported by W. P. Headden, Colo. Sci. Soc. Proc. 7, 141 (1903).

[12]R. W. Stoenner and J. Zähringer, Geochim. Cosmochim. Acta 15, 40 (1958); J. W. Morgan and J. F. Lovering, Anal. Chim. Acta 28, 405 (1963).

[13]The xenon calibration was improved by the use of xenon-gas standards.

[14]T. Kirsten, in Potassium-Argon Dating, edited by O. A. Schaeffer and J. Zähringer (Springer-Verlag, Berlin, Germany, 1966), p. 7.

J. Phys. G: Nucl. Part. Phys. **17** (1991) S221–S229. Printed in the UK

Geochemical Measurements of Double-Beta Decay

Oliver K. Manuel

Chemistry Department, University of Missouri, Rolla, MO 65401, USA

ABSTRACT. A review of recent geochemical measurements on the double-beta decay of ^{82}Se, ^{128}Te and ^{130}Te suggests that the current "best" value for the decay rate of ^{128}Te relative to that of ^{130}Te is 4×10^{-4} and the current "best" values for individual half-lives are as follows: 1×10^{20} y for ^{82}Se, 2×10^{24} y for ^{128}Te, and 8×10^{20} y for ^{130}Te.

1. INTRODUCTION

The "geochemical" method of measuring double-beta decay depends on the accumulation of decay products from this rare nuclear process in a natural mineral of the parent element. The method is most likely to succeed if little or none of the daughter element was incorporated when the mineral of the parent element formed. The likelihood of this separation occurring in nature is enhanced if the chemical nature of the double-beta decay product is radically different from that of the parent element. Success also depends on the mineral being geologically old and having survived intact over a long period of time so that detectable quantities of the daughter have accumulated in the mineral. Hard, high temperature minerals have a favored chance of surviving variations in the pressure and temperature to which minerals are exposed in nature.

To date, all geochemical measurements of double-beta decay are based on the detection of noble gas isotopes that have accumulated in minerals of selenium and tellurium. Generally, these minerals formed under conditions that efficiently separated the parent chalcogen elements from noble gases --- the chemical form of their double-beta decay products. This facilitates detection of minute amounts of decay products that have accumulated in these minerals. However, most minerals of these two chalcogen elements are soft, low-temperature formations that may readily undergo recrystallization in a deformation episode subsequent to mineralization. This may be accompanied by loss of radiogenic noble gas isotopes and thus reset their Se-Kr and Te-Xe age dating systems.

0954-3899/91/0S0221 + 09 $03.50

As noted by Richardson *et al* (1986), uncertainties in the gas retention age of Te-minerals produce a spread of about a factor of 3 in values obtained by the geochemical method for the double-beta decay of ^{130}Te. In spite of these obvious difficulties, the geochemical method was the only means available until recently to detect the occurrence of such rare nuclear processes. Further, the initial order-of-magnitude discrepancy between the results obtained with the UCI cloud chamber (Moe and Lowenthal, 1980) and geochemical measurements of the decay of ^{82}Se was ultimately resolved in favor of the geochemical data (Elliott *et al* 1987). Kirsten *et al* (1986) note that the geochemical method has now gained wide acceptance. The technique is particularly well suited for measuring the ratio of double-beta decay rates, e.g., the ratio of the half-lives of ^{82}Se, ^{128}Te and ^{130}Te (Takaoka and Sagawa 1988, Lin *et al* 1988a,b). Such measurements are much less subject to error from gas loss than are experiments that use the geochemical method to determine the absolute decay rate.

Interest in double-beta decay concerns information that the decay mode provides on basic laws of physics. In this regard, direct detection methods have a decided advantage over geochemical ones. The geochemical method shows the amount of decay product that has accumulated but does not indicate the energy carried away by the beta particles or the number of neutrinos that accompany these. In spite of this, the geochemical method continues to provide important information on rare processes that cannot be detected by other means. This review will focus on measurements since the Osaka and Tokyo meetings (Kirsten *et al* 1986, Manuel 1986, Takaoka and Sagawa 1988, Chiou and Manuel 1988).

2. THE RATIO OF ^{130}Te AND ^{128}Te HALF-LIVES

Table 1 shows the range of values that has been reported for the ratio of the decay rate of ^{128}Te relative to that of ^{130}Te from geochemical measurements. The differences shown there arise, not from uncertainties in the gas retention age of these minerals, but from instrumental problems in detecting the extremely small amounts of radiogenic ^{128}Xe that have accumulated in these samples over geologic time. Some of the problems that have been identified are background peaks in the mass spectrometers at m/q = 128, excess ^{128}Xe from sources other than the decay of ^{128}Te, and large shifts in the apparent abundance of excess ^{128}Xe induced by relatively minor corrections for fission products.

Takaoka and Ogata (1966) cautioned that the unexpectedly high excess of ^{128}Xe observed in their experiment might be from instrument "memory" of a monoisotopic ^{128}Xe spike used earlier in their laboratory. Subsequent measurements suggest that this was the case. Hennecke *et al* (1975) reported the first convincing evidence of radiogenic ^{128}Xe from the decay of ^{128}Te, but results of later studies suggest that about one-third of the excess ^{128}Xe observed by Hennecke *et al* was from other than the decay of ^{128}Te.

Table 1. Values reported for the ratio of the double-beta decay rate of ^{128}Te relative to that of ^{130}Te

Reference	$^{130}T_{1/2}/^{128}T_{1/2}$
Takaoka and Ogata 1966	2.7×10^{-2}
Srinivasan *et al* 1972	$<9 \times 10^{-3}$
Hennecke *et al* 1975	6.3×10^{-4}
Kirsten 1983	1.0×10^{-4}
Kirsten *et al* 1986	1.0×10^{-4}
Manuel 1986	5.0×10^{-4}
Takaoka and Sagawa 1988	$<6 \times 10^{-4}$
Lin *et al* 1988b	3.9×10^{-4}
Takaoka and Sagawa 1990	3.2×10^{-4}
Lee *et al* 1990a	4.3×10^{-4}
Current best estimate	$\mathbf{4 \times 10^{-4}}$

The measurement reported by Kirsten (1983) on excess ^{128}Xe and ^{130}Xe in native Te from the Good Hope Mine is difficult to interpret. Manuel (1986) notes that the enrichment of radiogenic ^{130}Xe in gas extracted from the sample was small, the sample was too young to have accumulated readily measurable amounts of radiogenic ^{128}Xe, and the experimental data appear to be compatible with the measurements by Hennecke *et al* (1975).

Geochemical measurements made at Yamagata and Rolla since the earlier reviews are tabulated in the bottom part of table 1. From these the current best value for the half-life of ^{130}Te relative to that of ^{128}Te was estimated. The value shown at the bottom of table 1 for this ratio, 4×10^{-4}, is about 20% less than the value concluded by Manuel (1986) and essentially identical to the value concluded by Chiou and Manuel (1988). This is in good agreement with the values calculated by Vogel and Zirnbauer (1986) and Engel *et al* (1988) with QRPA (quasi-particle random phase approximation) for 2ν $\beta\beta$-decay.

3. THE HALF-LIFE OF ^{130}Te

As noted earlier, absolute values of individual half-lives measured by the geochemical method are subject to uncertainties because the gas retention age of minerals is usually not well known. This was suggested by the work of Richardson *et al* (1986), and it has been confirmed by recent measurements here comparing the gas release patterns of radiogenic ^{40}Ar and ^{130}Xe (Lee *et al* 1990b). In the latter study, it was shown that the release of radiogenic ^{40}Ar from several previously studied Te-rich minerals does not parallel that of radiogenic ^{130}Xe. Differences in the release patterns of these two radiogenic gases indicate that the potassium resides in micro-minerals that are imbedded in, but distinct from, the mineral containing the chalcogen element. From this study, it was concluded that K-Ar ages may contain valuable information on the time span during which radiogenic gases have accumulated in the K-rich micro-minerals, but these K-Ar ages do not necessarily correspond to the period of time available for the accumulation of gaseous products of double-beta decay within the bulk, Te-rich mineral.

As shown in table 2, there have been no recent reports of geochemical measurements for the half-life of ^{130}Te. Srinivasan *et al* (1972) and Kirsten (1983) reported half-life values that now seem to be systematically high. This probably indicates that they have overestimated the gas-retention ages of the soft, low-temperature Te-minerals. Manuel (1986) attempted to extract a meaningful K-Ar age for tellurium from the Good Hope Mine by correcting for a trapped Ar component having $^{40}Ar/^{36}Ar = 320$. However, from the results of the recent study by Lee *et al* (1990b) on the release of radiogenic ^{40}Ar and ^{130}Xe from tellurium-rich minerals, it appears doubtful that the K-Ar age calculated by Manuel (1986) for Good Hope tellurium contains reliable information on its retention time for radiogenic ^{130}Xe. For this reason, the 1986 K-Ar age estimate was discarded in estimating the current best value for the half-life of ^{130}Te. The resulting estimate is only about half of the value recommended by Kirsten *et al* (1986), 14% higher than the value estimated in our earlier reviews (Manuel 1986, Chiou and Manuel 1988), and indistinguishable from the value reported much earlier by Takaoka and Ogata (1966).

Table 2 Values reported for the half-life of ^{130}Te

Reference	Half-life
Inghram and Reynolds 1950	1.4×10^{21} y
Takaoka and Ogata 1966	8.2×10^{20} y
Srinivasan *et al* 1972	2.5×10^{21} y
Hennecke *et al* 1975	9.7×10^{20} y
Kirsten 1983	2.6×10^{21} y
Richardson *et al* 1986	$<1 \times 10^{21}$ y
Kirsten *et al* 1986	1.6×10^{21} y
Manuel 1986	7×10^{20} y
Chiou and Manuel 1988	7×10^{20} y
Current best estimate	**8×10^{20} y**

4. THE HALF-LIFE OF ^{128}Te

Since the earlier reviews on geochemical measurements of double-beta decay (Kirsten *et al* 1986, Manuel 1986, Takaoka and Sagawa 1988, Chiou and Manuel 1988), there have been two additional measurements (Takaoka and Sagawa 1990, Lee *et al* 1990a) that provide information on the half-life of ^{128}Te. These are shown in table 3, together with the results of earlier measurements. In some cases the value reported is the ratio of the decay rate of ^{128}Te relative to that of ^{130}Te. For these, the estimate from that laboratory for the half-life of ^{130}Te was used to compute the half-life values of ^{128}Te shown in table 3.

It can be seen from the data tabulated there that most geochemical measurements, except for the very early ones by Takaoka and Ogata (1966) and those made by the Heidelberg group during the 1980s (Kirsten 1983, Kirsten *et al* 1986), are consistent with the value shown at the bottom of table 3 as the current best estimate of the half-life of ^{128}Te. This is significantly smaller than the value recommended by Kirsten but essentially the same value concluded in other recent reviews.

Table 3 Values reported for the half-life of ^{128}Te

Reference	Half-life
Takaoka and Ogata 1966	3×10^{22} y
Srinivasan *et al* 1972	$>3 \times 10^{23}$ y
Hennecke *et al* 1975	1.5×10^{24} y
Kirsten 1983	$>8 \times 10^{24}$ y
Kirsten *et al* 1986	$>5 \times 10^{24}$ y
Manuel 1986	1.4×10^{24} y
Chiou and Manuel 1988	1.8×10^{24} y
Lin *et al* 1988b	1.8×10^{24} y
Takaoka and Sagawa 1990	2.5×10^{24} y
Lee *et al* 1990a	1.6×10^{24} y
Current best estimate	**2×10^{24} y**

5. THE HALF-LIFE OF ^{82}Se

Selenium-82 has a shorter half-life than any other double-beta nuclide that has been studied by the geochemical method. This is the only nuclide for which the results obtained by the geochemical method can be compared with counting measurements made in the laboratory. The value shown in table 4 from Elliott *et al* (1987) was obtained from measurements made on ^{82}Se in UCI's time projection chamber. All other values are from geochemical measurements. Except for the high value reported by Srinivasan *et al* (1973), which probably reflects gas loss from the selenide minerals, it can be seen that all values reported for the half-life of ^{82}Se are close to that shown at the bottom of table 4 as the current best estimate.

The half-lives of both ^{82}Se (table 4) and ^{130}Te (table 2) display the factor of 3 spread in values that Richardson *et al* (1986) attribute to uncertainties in the minerals' gas retention age. In spite of this inherent uncertainty in the geochemical method of determining double beta-decay, the results of recent geochemical measurements are in remarkably agreement with the values concluded earlier by Chiou and Manuel (1988) and with the results obtained by direct detection of ββ-decay in the laboratory (Elliott *et al* 1987).

Table 4. Values reported for the half-life of ^{82}Se

Reference	Half-life
Kirsten and Mueller 1969	1.4×10^{20} y
Srinivasan *et al* 1973	2.8×10^{20} y
Kirsten 1983	1.5×10^{20} y
Marti and Murty 1985	1.2×10^{20} y
Kirsten *et al* 1986	1.3×10^{20} y
Manuel 1986	1.0×10^{20} y
Murty and Marti 1987	1.0×10^{20} y
Elliott *et al* 1987	1.1×10^{20} y
Chiou and Manuel 1988	1.0×10^{20} y
Lin *et al* 1988a	1.2×10^{20} y
Current best estimate	**1×10^{20} y**

6. CONCLUSIONS

Recent geochemical measurements of double-beta decay generally confirm the ratios of decay rates and the individual half-life values concluded in our earlier review (Chiou and Manuel 1988). The only exceptions are 10% increases in the values concluded for the half-lives of ^{130}Te and ^{128}Te. From a comparison of geochemical measurements of ββ-decay with the theoretical values calculated with QRPA (Vogel and Zirnbauer 1986, Engel *et al* 1988), it appears that the 2υ of ββ-decay is dominant for ^{82}Se, ^{130}Te and ^{128}Te. The comparison also indicates that ^{128}Te will have the largest contribution from 0υ ββ-decay, if this decay mode occurs.

The importance of independent confirmation in the laboratory of the geochemical measurement of the double-beta decay rate of ^{82}Se (Elliott *et al* 1987) can hardly be exaggerated. Geochemical measurements of the ratio of decay rates are not plagued by problems of gas loss over geologic time. Thus, this laboratory confirmation of the decay rate of ^{82}Se provides just reason for confidence also in values obtained by the geochemical method for the half-lives of ^{130}Te and ^{128}Te, especially if these are measured relative to the half-life of ^{82}Se.

ACKNOWLEDGMENTS

The author would like to acknowledge the assistance of Professors T. Kirsten, K. Marti, M. K. Moe and N. Takaoka for responding to a request to share any recent developments in their laboratories and the assistance of my student, Mr. J. T. Lee, for preparing graphs to illustrate differences in the release patterns of radiogenic ^{40}Ar and radiogenic ^{130}Xe from telluride minerals.

REFERENCES

Chiou K Y and Manuel O K 1988 in Proceedings of the XVI INS International Symposium, *Neutrino Mass and Related Topics* ed S Kato and T Ohshima (World Scientific: Singapore) p 178
Elliott S R Hahn A A and Moe M K 1987 *Phys.Rev. Lett.* **59** 2020
Engel J Vogel P and Zirnbauer M R 1988 *Phys. Rev.* **C37** 731
Hennecke E W Manuel O K and Sabu D D 1975 *Phys. Rev.* **C11** 1378
Inghram M G and Reynolds J H 1950 *Phys. Rev.* **78** 822
Kirsten T 1983 *AIP Conference* **96** 396
Kirsten T Heuser E Kaether D Oehm J Pernicka E and Richter M 1986 in *Nuclear beta decays and neutrino* ed T Kotani, H Ejiri and E Takasugi (World Scientific: Singapore) p 81
Kirsten T and Mueller H W 1969 *Earth Planet. Sci. Lett.* **6** 271
Lee J T Manuel O K and Thorpe R I 1990a *Nucl. Phys.* (submitted)
Lee J T and Manuel O K 1990b To be published
Lin W J Manuel O K Cumming G L Krstic D and Thorpe R I 1988a *Nucl. Phys.* **A481** 477
Lin W J Manuel O K Muangnoicharoen S and Thorpe R I 1988b *Nucl. Phys.* **A481** 484
Manuel O K 1986 in *Nuclear beta decays and the neutrino* ed T Kotani, H Ejiri and E Takasugi (World Scientific: Singapore) p 71
Marti K and Murty S V S 1985 *Phys. Lett.* **163B** 71
Moe M K and Lowenthal D D 1980 *Phys. Rev.* **C22** 2186
Murty S V S and Marti K 1987 *Geochim. Cosmochim Acta* **51** 163
Richardson J F Manuel O K Sinha B and Thorpe R I 1986 *Nucl. Phys.* **A453** 26
Srinivasan B Alexander E C and Manuel O K 1972 *J. Inorg. Nucl. Chem.* **34** 2381
Srinivasan B Alexander E C Beaty R D Sinclair D E and Manuel O K 1973 *Econ. Geol.* **68** 252
Takaoka N and Ogata K 1966 *Z. Naturforschg.* **21A** 84

Takaoka N and Sagawa H 1988 in Proceedings of the XVI INS International Symposium, *Neutrino Mass and Related Topics* ed S Kato and T Ohshima (World Scientific: Singapore) p 183
Takaoka N and Sagawa H 1990 *45th Annual Mtg. Phys. Soc. Japan* Abstract 31a-S-11 p 48
Vogel P and Zirnbauer M R 1986 *Phys. Rev. Lett.* **57** 3148

VOLUME 69, NUMBER 16 PHYSICAL REVIEW LETTERS 19 OCTOBER 1992

Neutrino Mass Limits from a Precise Determination of $\beta\beta$-Decay Rates of ^{128}Te and ^{130}Te

T. Bernatowicz,[(1),(2),(3)] J. Brannon,[(1),(3)] R. Brazzle,[(1),(2)] R. Cowsik,[(1),(2),(4)] C. Hohenberg,[(1),(2)] and F. Podosek[(1),(3)]

[(1)] *McDonnell Center for the Space Sciences, Washington University, St. Louis, Missouri 63130*
[(2)] *Department of Physics, Washington University, St. Louis, Missouri 63130*
[(3)] *Department of Earth and Planetary Sciences, Washington University, St. Louis, Missouri 63130*
[(4)] *Tata Institute of Fundamental Research, Bombay-400005, India*
(Received 29 June 1992; revised manuscript received 27 July 1992)

Double beta decay of ^{128}Te has been confirmed and the ratio of half-lives for $\beta\beta$ decay of ^{130}Te and ^{128}Te has been determined as $^{130}\mathrm{Te}_{1/2}/^{128}\mathrm{Te}_{1/2}=(3.52\pm0.11)\times10^{-4}$ by ion-counting mass spectrometry of Xe in ancient Te ores. The Xe measurements, combined with common Pb dating, yield a ^{130}Te half-life of $(2.7\pm0.1)\times10^{21}$ yr and thus a ^{128}Te half-life of $(7.7\pm0.4)\times10^{24}$ yr. These results give limits on the effective Majorana mass of the neutrino (<1.1–1.5 eV) and right-handed currents ($|\langle\eta\rangle|<5.3\times10^{-8}$) comparable to the best obtained from direct neutrinoless $\beta\beta$-decay searches. They also imply new limits on nonstandard Majorons not constrained by measurements of the Z^0 decay width.

PACS numbers: 23.40.Bw, 14.60.Gh, 14.80.Gt, 27.60.+j

In recent years a considerable amount of effort has been devoted to the study of $\beta\beta$ decay of various nuclei. The motivation has been provided by the implication of such studies for the Majorana mass of the neutrino, the possibility of right-handed currents in weak interactions and violation of lepton number conservation, and more generally for the physics beyond the standard model of electroweak interactions. We present here the main results of our study of $\beta\beta$ decay of tellurium, which provides some of the most stringent constraints available on these issues.

In this paper we are concerned with the following possible decays,

$$^{128}\mathrm{Te}\rightarrow{}^{128}\mathrm{Xe}+2e^-+2\bar{\nu}\ (0.87\ \mathrm{MeV})\,, \tag{1a}$$

$$^{128}\mathrm{Te}\rightarrow{}^{128}\mathrm{Xe}+2e^-+0\bar{\nu}\,, \tag{1b}$$

$$^{128}\mathrm{Te}\rightarrow{}^{128}\mathrm{Xe}+2e^-+0\bar{\nu}+\phi\,, \tag{1c}$$

where ϕ is the Majoron, and the analogous decay modes for ^{130}Te (2.53 MeV). The advantages that the system of Te isotopes provides in the geochemical determination of the lifetimes and for the interpretation of the data in terms of neutrino mass, etc., have been well recognized [1,2]. The decay product Xe is a noble gas with an extremely low abundance in most terrestrial rocks, typically with a mass fraction $<10^{-14}$. Ores of native Te or with high Te content (such as $\mathrm{AuTe_2}$) are available with ages in excess of 10^9 yr, so that a relatively high abundance of the daughter Xe is accumulated. Contributions to ^{128}Xe and ^{130}Xe through nuclear interactions are far less than those from $\beta\beta$ decay in Te ores shielded from cosmic rays by more than 10 m overburden of rock [3]. The only significant interference for these isotopes is from Xe initially trapped in the sample at the time of ore formation (generally having the composition of atmospheric Xe). All of these points permit an accurate determination of the amounts of ^{128}Xe and ^{130}Xe generated in Te $\beta\beta$ decay. One disadvantage of the geochemical method, however, is that it cannot directly determine the $\beta\beta$-decay mode but only the sum of the contributing decay channels.

From the point of view of interpreting the observed lifetimes of ^{128}Te and ^{130}Te, the rather large difference in their decay energies simplifies the analysis. The phase-space dependence of the 2ν decay rate of a given nucleus can be represented as a polynomial in T_0 (where T_0 is the total kinetic energy carried off by the leptons) that varies sharply as T_0^7 to T_0^{11}. The 0ν $\beta\beta$ process varies somewhat less rapidly as $T_0\langle m\rangle^2$ to $T_0^5\langle m\rangle^2$, where $\langle m\rangle$ is the effective Majorana mass of the neutrino [4]. This implies that the relative contribution of 0ν $\beta\beta$ decay to the ^{128}Te decay is much larger than that for ^{130}Te for reasonable values of $\langle m\rangle$. These advantages provided by the Te system have prompted many groups to study it, and although the decay of ^{130}Te is well established, there has existed a long-standing controversy over whether $\beta\beta$ decay of ^{128}Te has actually been observed. Several studies by the University of Missouri at Rolla group [5] have led to positive claims, but work on Te from the Colorado Good Hope Mine by the Heidelberg group [1] does not support those observations.

We have measured the Xe isotopic composition in several samples, some also taken from the same ore deposits studied by these two groups, in order to resolve this question. Attention was focused on two crucial problems, namely, reduction of the level of trapped initial Xe in the Te ores and establishment of reliable sample ages. First, we note that a significant fraction of the initially trapped Xe is probably present in fluid inclusions and other defects in the Te ores. We found that a sizable fraction of the trapped Xe could be released by crushing the sample into a fine powder *in vacuo,* and the mass spectrum of this Xe could thus be checked for the presence of mass fractionation, etc. On the other hand, the radiogenic and

TABLE I. Ages and relative and absolute $\beta\beta$-decay half-lives for Te ores.

Sample	Radiogenic ^{130}Xe (10^{-12} cm^3/g Te)	Common Pb age (10^9 yr)^a	$T_{1/2}^{130}/T_{1/2}^{128}$ (10^{-4})	$T_{1/2}^{130}$ (10^{21} yr)	$T_{1/2}^{128}$ (10^{24} yr)
Native Te (American Mine, CO)	21.20	1.66	3.59 ± 0.24	3.2	9.2
Native Te (Good Hope Mine, CO)	23.37	1.60	3.41 ± 0.29	2.8	8.0
Native Te (Vulcan Mine, CO)	23.77	1.61	3.44 ± 0.18	2.8	7.9
Altaite; $PbTe_2$ (Mattagami Lake, Quebec)	42.62	2.67	4.70 ± 0.58	2.6	7.3
Krennerite; $(Au,Ag)Te_2$ (Kalgoorlie, Australia)	93.73		3.48 ± 0.23		
Mean value			3.52 ± 0.11^b	2.7 ± 0.1^c	$7.7 \pm 0.4^{c,d}$

^{a}Computed using the two-stage Stacey-Kramers [22] common Pb evolution model.
^{b}Mean ratio of half-lives and 1σ error in mean. Data are corrected ($\leq 15\%$) for cosmic-ray-muon contributions to Te-derived ^{128}Xe [9].
^{c}American Mine data excluded from mean due to apparent 10% Xe loss from this ore [9]. Stated uncertainty is for 1σ dispersion about the mean.
^{d}Computed from $T_{1/2}(130)$ and mean half-life ratio given in (b).

fissiogenic Xe are relatively tightly bound in the crystal lattice and are released at elevated temperatures and completely only upon sublimation of the sample. Thus, apart from indicating the composition of the initially trapped Xe, the crushing procedure makes the signals of the less abundant Xe components in the sample stand out more clearly against the reduced background of this gas. In our experiments we devised the means of crushing the samples into fine powders, and then transferring them into a quartz oven without any break in the vacuum. The complete release of Xe was effected by the standard procedure of stepwise heating. The Xe isotopic composition for each extraction step was measured using a high-sensitivity ion-counting mass spectrometer [6] which had not been used for any previous analyses except of atmospheric Xe, thus ensuring that the analyses of the Te ores were not affected by spectrometer "memory" effects caused by release of previously implanted Xe. Second, we have attempted to improve the specification of sample ages by determining the common Pb ages by thermal ionization mass spectrometry [7], instead of relying only on the age deduced from geological context, or on techniques involving lighter noble gases such as Ar or He that are likely to be unreliable for such samples [8]. The list of samples used in the present study, their ages, Te $\beta\beta$-decay half-lives, and Xe concentrations are given in Table I. Additional experimental details are discussed elsewhere [9].

The ratios of isotopes ^{128}Xe/^{132}Xe and ^{130}Xe/^{132}Xe as obtained for the various stages of Xe extraction for all the samples are shown in Fig. 1. The ^{132}Xe data and ^{128}Xe data have been corrected for minor contributions from ^{238}U spontaneous fission and from reactions induced by cosmic-ray muons and their secondaries on Te, respectively [9]. The amounts of ^{128}Xe and the ^{130}Xe from $\beta\beta$ decay of ^{130}Te are linearly correlated and yield a consistent radiogenic ^{128}Xe/^{130}Xe ratio for all samples studied. The least-squares line ($\chi^2 = 1.04$) passes through the point corresponding to the composition of Xe in air and the slope of the line gives the ^{128}Xe/^{130}Xe ratio generated from Te $\beta\beta$ decay as $(3.30 \pm 0.10) \times 10^{-4}$, corresponding to a ratio of ^{130}Te to ^{128}Te half-lives of $(3.52 \pm 0.11) \times 10^{-4}$. The fact that ^{128}Xe/^{132}Xe ratios in the stepwise-heating data are often substantially in excess of the atmospheric value assures us that radiogenic contributions to ^{128}Xe are definitely observed. More detailed analysis shows that there are no other nuclear reactions contributing significantly to the production of ^{128}Xe [9];

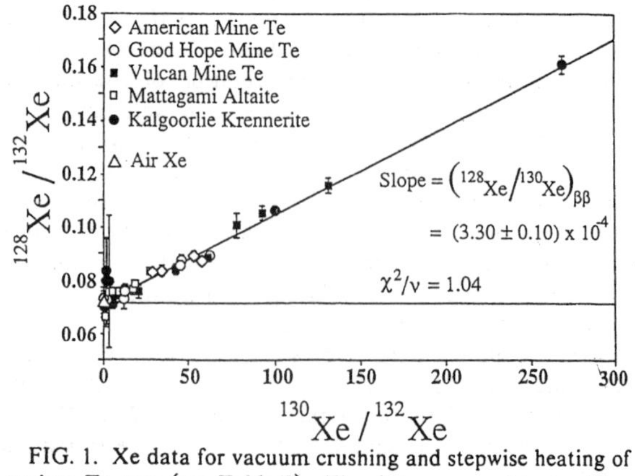

FIG. 1. Xe data for vacuum crushing and stepwise heating of ancient Te ores (see Table I). The data are consistent with a mixture of Xe from air and from $\beta\beta$ decay of ^{128}Te and ^{130}Te. See text for discussion.

VOLUME 69, NUMBER 16 PHYSICAL REVIEW LETTERS 19 OCTOBER 1992

in particular, the most prolific likely interference from $^{127}I(n,\gamma)$ can be ruled out because the neutron fluences for our samples vary by an order of magnitude (on the basis of differences in $^{129}Xe/^{131}Xe$ from neutron capture) but the corrected $^{128}Xe/^{130}Xe$ ratios have a dispersion of only a few percent about their mean value. We note that our mean half-life ratio of $(4.02\pm0.09)\times10^{-4}$, exclusive of corrections for cosmic-ray-muon reactions, is in good agreement with the values $(4.2\pm0.8)\times10^{-4}$ and $(4.4\pm0.8)\times10^{-4}$ recently reported by Lee, Manuel, and Thorpe [5] for krennerite and altaite. In that work, however, no measurements of ^{126}Xe were made, precluding assessment of muon-induced contributions to ^{128}Xe. In comparing our Good Hope Te results with those of Kirsten, Richter, and Jessberger [1], we attribute the failure of the Heidelberg group to detect ^{128}Te $\beta\beta$ decay to a much greater concentration of atmospheric Xe in their sample.

By summing over all of the Xe emitted at various temperatures and subtracting out the initial trapped Xe and a small correction to the ^{128}Xe from cosmic-ray-muon-induced reactions, the absolute amounts of ^{128}Xe and ^{130}Xe generated by $\beta\beta$ decay of Te are established. From these it is straightforward to calculate the half-lives for these decays (given in Table I). The weighted averages correspond to decay widths for these two nuclides of

$$^{128}\Gamma_{tot}=(9.0\pm0.5)\times10^{-26}\ \mathrm{yr}^{-1},\qquad ^{130}\Gamma_{tot}=(2.6\pm0.1)\times10^{-22}\ \mathrm{yr}^{-1}. \tag{2}$$

The implications of these results are briefly discussed below.

(a) The $\beta\beta$ decay of ^{128}Te has been firmly established and its half-life has been determined to be $(7.7\pm0.4)\times10^{24}$ yr without any ambiguity due to trapped Xe interferences. This is the longest radioactive lifetime ever to have been measured.

(b) Theoretical calculations [10–15] fail to correctly reproduce the long half-lives determined for ^{128}Te and ^{130}Te and underestimate them by 1 to 2 orders of magnitude, pointing to a real suppression in the 2ν decay rate of these isotopes.

(c) Despite the inaccuracy of the absolute 2ν-decay-rate calculations implied by comparison of theory with the present experimental results, most $\beta\beta$-decay models predict a *ratio* of 2ν decay widths, $\rho_{2\nu}=\Gamma_{2\nu}^{128}/\Gamma_{2\nu}^{130}$, which is in fair agreement with observation. In particular, these models give $\rho_{2\nu}\geq2\times10^{-4}$, compared to our result $\rho_{total}=(\Gamma^{128}/\Gamma^{130})_{total}=(3.52\pm0.11)\times10^{-4}$. While it is tempting to ascribe any difference between the total decay ratio and the predicted 2ν decay ratio to the presence of neutrinoless decay channels in ^{128}Te decay, we resist doing so because the theoretical calculations generally overestimate the 2ν matrix elements for both ^{128}Te and ^{130}Te by large factors, with widely varying results. However, most calculations of the 0ν matrix elements are in fair agreement, at least on a factor-of-2 level, and we can have better confidence in using the measured ^{128}Te decay rate *alone* to set limits on the neutrinoless decay rate. We thus take the total decay width $\Gamma_{total}^{128}=(9.0\pm0.5)\times10^{-26}\ \mathrm{yr}^{-1}$ as a conservative upper bound on the 0ν decay width,

$$\Gamma_{0\nu}^{128}\leq\Gamma_{total}^{128}, \tag{3a}$$

whence a lower limit on the half-life for 0ν decay of ^{128}Te is given by the experimentally determined ^{128}Te half-life,

$$(T_{1/2}^{128})_{0\nu}\geq(7.7\pm0.4)\times10^{24}\ \mathrm{yr}. \tag{3b}$$

In theoretical calculations of 0ν $\beta\beta$ decay [4,10–14], the relationship between the decay half-life and neutrino mass is usually presented in the form

$$[(T_{1/2}^{128})_{0\nu}]^{-1}=C_{mm}\langle m\rangle^2+C_{m\eta}\langle m\rangle\langle\eta\rangle+C_{\eta\eta}\langle\eta\rangle^2+\cdots, \tag{4}$$

where $\langle m\rangle$ is the effective Majorana mass of the neutrino and $\langle\eta\rangle$ the parameter which scales with the assumed strength of the right-handed weak leptonic currents. The coefficients C are tabulated by various authors. Using the lower limit on the neutrinoless decay half-life [Eq. (3b)] and neglecting right-handed currents, we obtain (Table II) a range of upper limits on the Majorana neutrino mass with Eq. (4), from <1.1 eV to <1.5 eV for various estimates of C_{mm}. We also derive the limit $|\langle\eta\rangle|<5.3\times10^{-8}$ on the basis of coefficients calculated by Suhonen, Khadkikar, and Faessler [14], which give the least restrictive of the estimates. These limits are comparable to the best currently obtained from direct neutrinoless $\beta\beta$-decay searches.

(d) The standard Majoron is now ruled out by the LEP measurements of the Z^0 decay width [16]. However, nonstandard Majorons have been postulated that would evade the LEP limits but nonetheless contribute to $\beta\beta$ decay. To illustrate the sensitivity of our ^{128}Te measurements to three-body decays, we give the standard Majoron coupling $|\langle g_B\rangle|$ corresponding to our total ^{128}Te half-life of 7.7×10^{24} yr for the Majoron process [17] with 0ν $\beta\beta\phi$ in the final state [Eq. (1c)],

$$|\langle g_B\rangle|<3\times10^{-5}. \tag{5}$$

TABLE II. Upper limits on the effective Majorana neutrino mass (eV). Derived from observational limit $T_{1/2}^{0\nu}>7.7\times10^{24}$ yr for ^{128}Te $\beta\beta$ decay, with $|\langle\eta\rangle|=0$.

Theory (Ref.)	Upper Limit on $\langle m_\nu\rangle$
Haxton [4] [a]	1.1
Tomoda [13]	1.1
Suhonen [14]	1.5

[a]Calculated with $g_A/g_V=1.0$

A coupling at this limit would then generate the following rates in other $\beta\beta$-decay nuclei:

$$\begin{aligned} T_{1/2}(^{48}\text{Ca}) &> 3.8\times 10^{23}\,\text{yr}\,, \\ T_{1/2}(^{76}\text{Ge}) &> 4.9\times 10^{23}\,\text{yr}\,, \\ T_{1/2}(^{82}\text{Se}) &> 7.1\times 10^{22}\,\text{yr}\,, \\ T_{1/2}(^{100}\text{Mo}) &> 3.7\times 10^{21}\,\text{yr}\,, \\ T_{1/2}(^{150}\text{Nd}) &> 2.1\times 10^{21}\,\text{yr}\,. \end{aligned} \tag{6}$$

The $\beta\beta$-decay Majoron searches [18–20] that have been performed using these nuclei have yielded bounds far less stringent than those of Eq. (6). Although nonstandard Majoran models may involve somewhat different nuclear physics, qualitatively it is clear that the total ^{128}Te decay rate is presently the best constraint on three-body decays.

(e) In our experiments we have observed excess ^{126}Xe generated by cosmic-ray muons and their secondaries [3,9]; we note that any Te ore that is to be used in a test of the standard solar model as proposed by Haxton [21] should be shielded by ~4400 meters of water equivalent of rock to reduce the cosmic-ray contributions to below 10% of the neutrino-induced production of ^{126}Xe.

We thank O. K. Manuel of the University of Missouri at Rolla for samples of altaite and krennerite and Peter Dunn of the National Museum of Natural History for the native Te samples. We are happy to acknowledge the guidance and support of R. M. Walker of the McDonnell Center for the Space Sciences and are especially grateful to W. C. Haxton of the University of Washington at Seattle for many discussions of the theoretical aspects of $\beta\beta$ decay.

[1] T. Kirsten, H. Richter, and E. Jessberger, Phys. Rev. Lett. **50**, 474 (1983).

[2] See reviews by F. T. Avignone, III, and R. L. Brodzinski, p. 147, and by K. Muto and H. V. Klapdor, p. 183, in *Neutrinos,* edited by H. V. Klapdor (Springer-Verlag, New York, 1988), and M. Doi, T. Kotani, and E. Takasugi, Prog. Theor. Phys. Suppl. **83**, 1 (1985). See also Ref. [4].

[3] R. Cowsik, Yash Pal, and S. N. Tandon, Proc. Ind. Acad. Sci. **63**, 217 (1967); S. Miyake, Proc. Int. Cosmic Ray Conf. **5**, 3638 (1973); L. B. Bezrukov and E. V. Bugaev, Proc. Int. Cosmic Ray Conf. **4**, 90 (1981); G. N. Fowler and A. W. Wolfendale, Prog. Elem. Part. Cosmic Ray Phys. **IV**, 107 (1958); J. K. Rowley *et al.*, in *The Ancient Sun,* edited by R. O. Pepin, J. A. Eddy, and R. B. Merrill (Pergamon, New York, 1980), p. 45.

[4] W. C. Haxton and G. J. Stephenson, Prog. Part. Nucl. Phys. **12**, 409 (1984).

[5] E.g., E. W. Hennecke, O. K. Manuel, and D. D. Sabu, Phys. Rev. C **11**, 1378 (1975); J. T. Lee, O. K. Manuel, and R. I. Thorpe, Nucl. Phys. **A529**, 29 (1991).

[6] C. M. Hohenberg, Rev. Sci. Instrum. **51**, 1075 (1980).

[7] J. C. Brannon *et al.*, Geochim. Cosmochim. Acta **55**, 1407 (1991).

[8] J. T. Lee and O. K. Manuel, Earth Planet. Sci. Lett. (to be published).

[9] T. J. Bernatowicz *et al.* (to be published).

[10] P. Vogel and M. R. Zirnbauer, Phys. Rev. Lett. **57**, 3148 (1986).

[11] K. Grotz and H. V. Klapdor, Nucl. Phys. **A460**, 395 (1986).

[12] K. Muto, in *Proceedings of the International Symposium on Nuclear Beta Decays and Neutrino,* edited by T. Kotani, H. Ejiri, and E. Takasugi (World Scientific, Singapore, 1986), p. 177.

[13] T. Tomoda and A. Faessler, Phys. Lett. B **199**, 475 (1987).

[14] J. Suhonen, S. B. Khadkikar, and A. Faessler, Nucl. Phys. **A535**, 509 (1991).

[15] A. G. Williams and W. C. Haxton, in *Contribution of the Axial Charge Operator to 2ν ββ Decay,* Proceedings of the Third International Conference on Intersections between Particle and Nuclear Physics, edited by G. Bunce, AIP Conf. Proc. No. 176 (AIP, New York, 1988), p. 924.

[16] LEP Collaboration, D. Decamp *et al.*, Phys. Lett. B **276**, 247 (1992); for theoretical motivation, see Georgi, Glashow, and Nussinov in Ref. [17].

[17] H. M. Georgi, S. L. Glashow, and S. Nussinov, Nucl. Phys. **B193**, 297 (1981); J. D. Vergados, Phys. Lett. **109B**, 96 (1982); M. Doi, T. Kotani, and E. Takasugi, Phys. Rev. D **37**, 2575 (1988); also Ref. [13]; see Muto and Klapdor (Ref. [2]).

[18] D. O. Caldwell *et al.*, Nucl. Phys. B (Proc. Suppl.) **13**, 547 (1990); A. S. Starostin, in Proceedings of the Twenty-Fifth International Conference on High Energy Physics, Singapore, 1990 (to be published); H. Ejiri *et al.*, J. Phys. G **13**, 839 (1987); F. T. Avignone, III, *et al.*, Phys. Lett. B **198**, 253 (1987); P. Fisher *et al.*, Phys. Lett. B **192**, 460 (1987).

[19] M. K. Moe, Bull. Am. Phys. Soc. **37**, 925 (1992); S. R. Elliot, A. A. Hahn, and M. K. Moe, Phys. Rev. Lett. **59**, 2020 (1987).

[20] F. Avignone, Bull. Am. Phys. Soc. **37**, 1024 (1992).

[21] W. C. Haxton, Phys. Rev. Lett. **65**, 809 (1990).

[22] J. S. Stacey and J. D. Kramers, Earth Planet. Sci. Lett. **26**, 207 (1975).

PHYSICAL REVIEW C VOLUME 47, NUMBER 2 FEBRUARY 1993

Precise determination of relative and absolute $\beta\beta$-decay rates of ^{128}Te and ^{130}Te

T. Bernatowicz,(1,2,3) J. Brannon,(1,3) R. Brazzle,(1,2) R. Cowsik,(1,2,4) C. Hohenberg,(1,2) and F. Podosek(1,3)
(1)*McDonnell Center For the Space Sciences, Washington University, St. Louis, Missouri 63130*
(2)*Department of Physics, Washington University, St. Louis, Missouri 63130*
(3)*Department of Earth and Planetary Sciences, Washington University, St. Louis, Missouri 63130*
(4)*Tata Institute of Fundamental Research, Bombay-400005, India*

(Received 18 September 1992)

Double beta decay of ^{128}Te has been confirmed and the ratio of half-lives for $\beta\beta$ decay of ^{130}Te and ^{128}Te has been precisely determined as $T^{130}_{1/2}/T^{128}_{1/2}=(3.52\pm0.11)\times10^{-4}$ by ion-counting mass spectrometry of Xe in ancient Te ores, using techniques that reduce interferences due to trapped Xe. We have also detected excesses of ^{126}Xe originating in high energy reactions of cosmic ray muons and their secondaries on Te; such reactions make minor contributions to the measured ^{128}Xe excesses in the Te ores. The Xe measurements, combined with common Pb dating of the ores, yield a ^{130}Te half-life of $(2.7\pm0.1)\times10^{21}$ yr and thus a ^{128}Te half-life of $(7.7\pm0.4)\times10^{24}$ yr, the longest radioactive decay lifetime measured to date. These results give limits on the effective Majorana mass of the neutrino ($<1.1-1.5$ eV) and right-handed currents ($|\langle\eta\rangle|<5.3\times10^{-8}$) comparable to the best obtained from direct neutrinoless $\beta\beta$-decay searches. They also imply new limits on unconventional Majorons not constrained by measurements of the Z^0 decay width.

PACS number(s): 23.40.Bw, 14.60.Gh, 14.80.Gt, 27.60.+j

I. INTRODUCTION

Double beta ($\beta\beta$) decay has long been recognized as providing important constraints on the nature of extensions to the standard model of electroweak interactions, in particular on the issues of neutrino mass and conservation of lepton number [1]. Nuclear $\beta\beta$ decay occurs in even-even nuclei for which the pairing force between like nucleons energetically forbids ordinary single β decay to an adjacent odd-odd isobar. In principle, there are at least three modes in which such decay can take place: a 2ν mode in which two antineutrinos are emitted along with two electrons,

$$(A,Z)\rightarrow(A,Z+2)+2e^-+2\bar{\nu}\ , \quad (1)$$

a neutrinoless (0ν) mode in which no neutrinos are emitted (resulting from the emission of a virtual neutrino from one neutron and its absorption by another),

$$(A,Z)\rightarrow(A,Z+2)+2e^-+0\bar{\nu}\ , \quad (2)$$

and a 0ν mode accompanied by the emission of a Goldstone boson (Majoron) ϕ,

$$(A,Z)\rightarrow(A,Z+2)+2e^-+0\bar{\nu}+\phi\ . \quad (3)$$

In $\beta\beta$ decay the neutrino may either be a Dirac particle ($\nu\neq\bar{\nu}$) or a Majorana particle ($\nu=\bar{\nu}$), where $\bar{\nu}$ denotes the charge-conjugate state or antiparticle. Observation of the decay mode in Eq. (2) would imply that the neutrino is a Majorana particle and that at least one neutrino eigenstate has a nonzero Majorana mass [2].

$\beta\beta$ decay is a second-order weak interaction, and, consequently, it is one of the slowest processes in nature, with half-lives normally in excess of 10^{20} yr. Although the 2ν decay mode of Eq. (1) has been observed in direct-counting experiments for several isotopes, the low decay rates make laboratory observation challenging and in some cases essentially impossible. The geochemical method of observing decay through daughter isotope excesses in natural samples makes use of geological times over which decays can be integrated. However, in order to be observable the daughter product must make a measurable change in the isotopic composition of the daughter element. In practice, the noble gases, which are normally present in extremely low concentrations in terrestrial materials ($<10^{-8}$ ppm for Xe), are the only group of elements sufficiently scarce to make the geochemical detection of $\beta\beta$ decay experimentally feasible. For example, as early as 1949 Inghram and Reynolds [3] were able to successfully use the noble gas Xe to study the $\beta\beta$ decay of Te. At present, the occurrence of $\beta\beta$ decay for the reactions ^{130}Te$\rightarrow$^{130}Xe and ^{82}Se$\rightarrow$^{82}Kr have been established beyond reasonable doubt, and half-lives for these decays have been determined to be 10–30 and 1–2, respectively, in units of 10^{20} yr [4]. The disadvantage of the geochemical method is that it does not directly determine the mode by which the $\beta\beta$ decay occurs, but only gives the sum of all decay channels.

In this study, we are concerned with determination of both absolute and relative $\beta\beta$-decay rates for ^{128}Te and ^{130}Te. This particular isotopic system represents a fortuitous combination of two circumstances favorable to the evaluation of neutrino mass limits, one theoretical and the other experimental. The phase-space dependence of the 2ν decay rate of a given nucleus can be represented as a polynomial in T_0 (where T_0 is the total kinetic energy carried off by the leptons) that varies sharply as T_0^7 to T_0^{11}. The 0ν $\beta\beta$ process varies somewhat less rapidly as

PHYSICAL REVIEW C VOLUME 53, NUMBER 4 APRIL 1996

Half-life of ^{130}Te double-β decay measured with geologically qualified samples

Nobuo Takaoka,[1,*] Yoshinobu Motomura,[1] and Keisuke Nagao[2]
[1]*Department of Earth and Planetary Sciences, Kyushu University, Hakozaki, Fukuoka 812-81, Japan*
[2]*Institute for Study of the Earth's Interior, Okayama University, Misasa, Tottori 682-01, Japan*
(Received 20 September 1995)

Tellurobismuthite (Bi_2Te_3) has been analyzed for Xe isotopes to determine the half-life for double-β decay of ^{130}Te. Excess ^{130}Xe amounts to $(6.18\pm0.18)\times10^7$ atom/g with 47.8 wt. % Te, or $(1.24\pm0.04)\times10^{13}$ for the parent/daughter ratio ^{130}Te/^{130}Xe. With $(9.3\pm1.1)\times10^7$ yr for the Xe retention age of Te mineral, this provides $(7.9\pm1.0)\times10^{20}$ yr for the absolute half-life of ^{130}Te double-β decay. With this and a literature ratio $T_{1/2}(130)/T_{1/2}(128)$ of $(3.52\pm0.11)\times10^{-4}$, we have $(2.2\pm0.3)\times10^{24}$ yr for the absolute half-life of ^{128}Te double-β decay.

PACS number(s): 21.10.Tg, 23.40.−s, 27.60.+j

I. INTRODUCTION

Isobars ^{130}Te, ^{130}I, and ^{130}Xe belong to a system of nuclides for which successive single-β decays are energetically forbidden but disintegration through double-β decay (DBD) is possible from ^{130}Te to ^{130}Xe. Since the first report on the possibility of DBD by Goeppert-Mayer [1], many experimental as well as theoretical studies have been done. The first experimental result on the ^{130}Te DBD half-life has been reported by Inghram and Reynolds [2]. Since then, many data have been reported on the ^{130}Te half-life determined by a geochemical method. However, there are long-standing problems of discrepancies in the half-life: Experimentally some workers have reported 2 to 3×10^{21} yr [3–7] and others 7 to 9×10^{20} yr [8–10]. These experimental results are significantly larger by one to two orders of magnitude than theoretical estimates [11]. However, this does not imply that the geochemical method is unreliable because it has given the double-β decay rate for ^{82}Se ([10] and references cited therein) in excellent agreement with that determined by a counting method [12]. In order to solve the discrepancies both in experimental results and between experimental and theoretical half-lives, further works are awaited with samples that are well documented on their occurrences and geological relationships. By the geochemical method, geologically old Te minerals are analyzed for the decay product ^{130}Xe by high sensitivity and high precision mass spectrometry. The amount of parent ^{130}Te is determined from the chemical composition of the sample mineral and the isotopic abundance for ^{130}Te. That the parent to daughter ratio ^{130}Te/^{130}Xe reported by different workers are in good agreement with one another for samples collected at the same locality (i.e., the same mine) indicates that problems are not in analytical techniques for Xe isotopes and Te contents, but in determination of the time interval in which Te minerals accumulated and retained radiogenic ^{130}Xe produced by *in situ* decay of ^{130}Te [7,13].

In this paper, we will report the half-life of ^{130}Te determined from Xe data for Te and associated mineral separates, along with the Xe retention age deduced from K-Ar age data for rocks in genetically close relation to the Te samples. We will also give a brief description of field relationships between Te ores and country rocks at the Oya mine, because they are of great importance to decipher the chronological sequence in the formation of mineral deposits and to estimate the Xe retention age for the Te mineral.

II. EXPERIMENT

A. Samples

Samples used in this work are pure tellurobismuthite (Bi_2Te_3) grains of typically 1–2 mm across, separated from a specimen of Te-bearing ores collected at the Oya mine. The ore specimen was carefully crushed to isolate the mineral grains by hand picking. The Te minerals are embedded in quartz veins, in close relationships with native gold, arsenopyrite (FeAsS) and other minerals. The quartz veins, filling fissures, are hosted in the Triassic formation composed mainly of sandstone and slate. The deposits at Oya are characterized by abundant arsenopyrite and the zonal distribution of other minerals. Tellurobismuthite is richer in close association with native gold in the deeper portions of veins. From viewpoints of ore genesis [14–16], it seems that silica-Au-Te-As-bearing hydrothermal fluids were introduced along fissures to deposit the quartz veins therein. These elements were extracted in part from magma and/or in part from country rocks through interactions with water heated by quartzdiorite magma. The fissures were formed both in circumference hoods of a quartzdiorite cupola and in roof rocks surrounding it (i.e., the Triassic sandstone and slate). A quartzdiorite mass crops out near the northern part of the mine and is believed to underlie the ore deposits, because bosses regarded as apophyses of the quartzdiorite outcrop at the mine and some veins cut them. The fissures were initiated by doming up of quartzdiorite magma and formed after at least part of outer rims of the quartzdiorite body had solidified. Field observations give no definite evidence that the Te-bearing veins have experienced thermal (igneous) events postdating the quartzdiorite intrusion, implying that radiogenic ^{130}Xe was retained quantitatively after the mineralization.

*To whom correspondence should be addressed.

Study of the double β decay of ^{130}Te

Yu. G. Zdesenko, I. A. Mytsyk, A. S. Nikolaĭko, and V. N. Kuts

Institute of Geochemistry and Mineral Physics, Ukrainian Academy of Sciences
(Submitted 21 January 1980)
Yad. Fiz. **32**, 607–616 (September 1980)

A direct experiment searching for the neutrinoless double β decay of ^{130}Te is described. A half-life limit equal to 1.2×10^{21} years at the 68% confidence level is established. The leptonic charge nonconservation parameter corresponding to this limit is calculated on the basis of a two-nucleon model and found to be $(3–5) \times 10^{-4}$.

PACS numbers: 23.40.Bw, 27.60. + j

INTRODUCTION

The phenomenon of double β decay, which as long ago as 1935 was predicted to exist,[1] consists of a nucleus simultaneously emitting two electrons and in so doing changing its charge by two units. For 2β decay to occur it is necessary that the mass of the daughter nucleus be less than the mass of the initial nucleus, which in addition must be stable against ordinary β decay.

Theoretically two decay channels are possible:

a) two-neutrino decay consistent with all conservation laws,

$$(A, Z) \to (A, Z+2) + 2e^- + 2\bar{\nu}_e; \qquad (1)$$

b) neutrinoless decay violating lepton conservation,

$$(A, Z) \to (A, Z+2) + 2e^-. \qquad (2)$$

Usually double β decay is considered as a second-order effect in the weak interaction, in the framework of which two mechanisms for this process have been proposed: the resonance mechanism[2] and the two-nucleon mechanism.[3]

The resonance mechanism is based on the existence in certain nuclei of baryon resonances.[4,5] Rosen and Primakoff[2] first pointed out the role of the $\Delta(1232)$ resonance ($J = \frac{3}{2}^+, I = \frac{3}{2}, m = 1232$ MeV) in the problem of 2β decay. The $\Delta(1232)$ resonance, if it is present in the parent nucleus (A, Z) and the daughter nucleus $(A, Z+2)$, could emit and absorb a virtual neutrino with simultaneous emission of two electrons, i.e., it could lead eventually to neutrinoless 2β decay of the nucleus (A, Z):

$$\Delta^- \to p + e^- + e^-,$$
$$n \to \Delta^{++} + e^- + e^-.$$

Somewhat later Smith, Picciotto and Bryman showed[6,7] that two-neutrino double β decay can also occur with participation of the $\Delta(1232)$ resonance (on the basis of the transition $\pi \to \pi + e^- + \bar{\nu}_e$):

$$\Delta^- \to p + e^- + e^- + \bar{\nu}_e + \bar{\nu}_e,$$
$$n \to \Delta^{++} + e^- + e^- + \bar{\nu}_e + \bar{\nu}_e.$$

The two-nucleon mechanism of 2β decay was proposed by Furry in 1939.[3] According to this mechanism double β decay with participation of neutrinos is accomplished as follows:

$$n_1 + n_2 \to n_1 + p_2 + e_2^- + \bar{\nu}_e \to p_1 + p_2 + e_1^- + e_2^- + \bar{\nu}_e + \bar{\nu}_e.$$

Neutrinoless decay occurs with exchange of a virtual neutrino, which is emitted by one neutron and absorbed by another, so that in the final state only two electrons are present:

$$\begin{array}{l} n_1 \to p_1 + e_1^- + \bar{\nu}_e \\ \qquad\qquad\quad \downarrow \\ \qquad\quad n_2 + \bar{\nu}_e \to p_2 + e_2^-. \end{array} \qquad (3)$$

The process (3) is possible only for the condition of identity of the neutrino and antineutrino, as a result of which, double β decay was first discussed as a means of distinguishing the Dirac neutrino ($\nu_e \neq \bar{\nu}_e$) from the Majorana neutrino ($\nu_e \equiv \bar{\nu}_e$). After the discovery of parity nonconservation and the creation of the two-component theory of neutrinos, in accordance with which the neutrino and antineutrino are massless particles which are completely and oppositely polarized, the situation was greatly complicated.

While decay by the two-neutrino channel is possible also in the framework of the contemporary theory of weak interactions, for realization of neutrinoless 2β decay there is required, in addition to violation of leptonic charge conservation, either depolarization of the neutrino in the intermediate virtual state or a nonzero mass of the stationary neutrino. The experimental data do not refute the possibility of incomplete polarization of neutrinos ($\leqslant 10\%$; see Ref. 8) and admit the existence in it of a finite rest mass ($m_{\bar{\nu}_e} \leqslant 35$ eV (Ref. 9) and $m_{\nu_e} \leqslant 6$ keV (Ref. 10)).

A fundamentally different treatment of neutrinoless double β decay was proposed by Pontecorvo.[11] According to his idea this process is considered to be an effect of first order in a hypothetical superweak interaction which does not preserve leptonic charge. In this

 0038-5506/80/090312-06$02.20

VOLUME 67, NUMBER 23 PHYSICAL REVIEW LETTERS 2 DECEMBER 1991

Double Beta Decay of ^{238}U

Anthony L. Turkevich,(1),(2),(4) Thanasis E. Economou,(1) and George A. Cowan(3),(4)

(1)*Enrico Fermi Institute, University of Chicago, Chicago, Illinois 60637*
(2)*Department of Chemistry, University of Chicago, Chicago, Illinois 60637*
(3)*Santa Fe Institute, Santa Fe, New Mexico 87501*
(4)*Los Alamos National Laboratory, Sante Fe, New Mexico 87501*

(Received 5 November 1990; revised manuscript received 26 August 1991)

The half-life for the decay of ^{238}U to ^{238}Pu has been measured to be $(2.0 \pm 0.6) \times 10^{21}$ yr by chemically isolating and measuring, from the resultant alpha particles, the amount of plutonium that had accumulated in 33 yr from 8.47 kg of purified uranyl nitrate. Other sources of ^{238}Pu have been studied and found negligible.

PACS numbers: 23.40.Bw, 14.60.Gh, 27.90.+b

There has recently been much interest in the rate at which certain nuclei decay by changing their nuclear charge by two units, a process usually considered to be the simultaneous emission of two beta particles (double beta decay). The rate of this process is of interest in considerations of the numbers and properties of neutrinos, of fundamental interactions and conservation laws, and of the existence of other than presently known particles. Such decays also test our understanding of the properties of excited states of nuclei since these are involved in calculating the rates of double beta decay. At present, there are about thirty experiments, worldwide, which study various aspects of this process. However, it is one of the slowest spontaneous decays in nature ($\lambda \lesssim 10^{-20}$ yr^{-1}) and there have been, to date, only five systems for which experimental evidence has been obtained for its actual existence. In addition to the geochemical work on ^{82}Se, ^{130}Te, and ^{128}Te, some of which goes back about forty years, recently there have been published counter experiments on ^{76}Ge [1,2], ^{82}Se [3], and ^{100}Mo [4]. Recent reviews of the experimental situation are to be found in Refs. [4] and [5]. Theoretical reviews of double beta decay have been made by Haxton and Stephenson [6], Doi *et al.* [7], and Muto, Bender, and Klapdor and Tomada [8]. Compared to the other systems that have been studied, the decay of ^{238}U to ^{238}Pu by double beta decay has several special features. The Q value of the decay (1.1 MeV) is the lowest of the systems presently being studied. This leads to theoretical predictions of half-life in the 10^{23}-yr range. The low-Q value, however, allows competition by other than standard 2ν processes (such as 0ν or Majorana emission) to benefit from phase-space considerations. Thus, Staudt, Muto, and Klapdor-Kleingrothaus [9] predict that a neutrino with a value of the Majorana mass parameter $\langle m_\nu \rangle$ equal to 3 eV would lead to a zero neutrino double beta decay rate ($t_{1/2} = 3 \times 10^{22}$ yr) faster than that calculated (5.2×10^{22} yr) for the conventional two-neutrino mode using a favorable value of the nuclear pairing parameter (g_{pp}).

A special feature of geochemical and milking experiments, such as the present one, is that they measure the total transformation rate of a nucleus ^{A}Z to $^{A}(Z+2)$ irrespective of the particular mechanism or particles involved. This consideration may be pertinent in view of recent [10] indications of the participation of 17-keV neutrinos in single beta decay.

The only previous experimental study of such a decay of ^{238}U is early work by Levine, Ghiorso, and Seaborg [11]. They used a radiochemical technique that was similar to the one used in the present work and set a lower limit of 6×10^{18} yr for the half-life. Theoretical estimates on the half-life based on conventional two-neutrino emission have ranged from 2.2×10^{19} [12] to greater than 5×10^{22} yr by Staudt, Muto, and Klapdor-Kleingrothaus [9]. This last estimate reflects the consideration [13] that cancellations by different contributions to the nuclear matrix elements can strongly suppress the rate of the conventional two-neutrino double beta decay process.

A favorable feature of the ^{238}U system is the high Z of the nuclei involved. This and the strong fission competition minimize the possibility of competing nuclear reactions forming ^{238}Pu. The most important such reaction is $^{238}U(p,n)^{238}Np$. The ^{238}Np then decays with a 2.117-d half-life to ^{238}Pu. As part of the studies of competing reactions, the cross section for this process was determined at Los Alamos.

A serious practical consideration in dealing with the $^{238}U(\beta\beta)^{238}Pu$ system is the worldwide fallout of ^{238}Pu from atmospheric tests and satellites that have had ^{238}Pu power sources and then, on reentry, burned out in the atmosphere. A typical fallout concentration is 10^6 atoms of ^{238}Pu per cm^2 of surface all over the world [14]. Since the present experiments produce only $\sim 10^5$ atoms of ^{238}Pu, such fallout introduction must be avoided. The present work used uranium nitrate that had been purified and isolated before much ^{238}Pu had been introduced into the atmosphere.

Haxton, Cowan, and Goldhaber [15] revived interest in the ^{238}U system and the present work was started as a result of their publication. The experiment involves the extraction of the accumulated plutonium from uranium salt that had been purified and isolated from fallout for 33 yr. This amount of uranyl nitrate (1.02×10^{25} atoms of ^{238}U) produces 2.3×10^5 atoms of ^{238}Pu in 33 yr if the half-life for double beta decay is 10^{21} yr. The chemically

VOLUME 67, NUMBER 23 PHYSICAL REVIEW LETTERS 2 DECEMBER 1991

isolated and purified plutonium is examined in low-background alpha counters. In the region of 5.5 MeV only ^{241}Am and ^{222}Rn can interfere with the ^{238}Pu alpha particles. The former was not prevalent in 1956 and is excluded by the chemistry. The latter, and its parents, should also be eliminated by the chemistry. In addition, its presence can be identified by the simultaneous presence of the daughter alpha radioactivities.

Chemical operations.— The uranyl nitrate used in this experiment had been purified by Shattuck Chemical Co. of Denver on about 1 July 1956. At the time there had not been much ^{238}Pu introduced into the atmosphere by large explosions and the large-scale use of ^{238}Pu for power sources on space vehicles had not started. Moreover, the chemical purification of the uranyl nitrate at the time probably used ether extractions which should have removed any fallout plutonium. The uranyl nitrate had been kept in a plastic bag inside a sealed cardboard container in Chicago until the summer of 1989. Several smaller samples of this material had previously been examined for ^{238}Pu content with negative results [16,17]. In August 1989, 8.47 kg of the salt were dissolved in deionized water and a known amount of ^{239}Pu tracer solution was added to this acid solution. This tracer had been prepared by neutron irradiation of uranium in a low enough flux so that the ratio of ^{238}Pu alpha particles to those of ^{239}Pu was less than 10^{-5}.

This solution was allowed to stand for a few days and then solid $NaHCO_3$ was added. A precipitate formed as neutrality was approached and then completely dissolved as the pH neared 7.0. The final approach to the desired pH of 7.1 was made by the addition of a saturated Na_2CO_3 solution. A total of 4.3 mole of $CO_3{}^{2-}$ was used per mole of U and the final solution was about 40 L.

The isolation and purification of the plutonium involved primarily relatively standard chemical procedures. The most novel was the first step that concentrated the plutonium from the large mass of uranium by extracting the cupferron complex [18] into chloroform from the slightly alkaline carbonate solution. There followed chromatographic column removals of ^{234}Th and Fe^{3+}, column purifications of the plutonium according to the procedure of Hoffman [19], and then evaporation of the nearly-mass-free plutonium-containing solution on a platinum disc.

The final sample was measured on three different silicon alpha detectors. The first two were rather large-area detectors and showed 5 and 2 times the counting rate of ^{239}Pu of the third small detector. This was due to the rather wide dispersal of the sample over the platinum disk. The smaller detector (MCA1) had more than 10 times lower background and gave the most definitive results. Three different counters were used to minimize the uncertainty in the stability of the background rates when dealing with measurement times (many months) as long as those involved in this experiment.

Results.— Over the last three years, successively larger samples (experiments A-3, A-4, and A-5) of the 1956 uranyl nitrate have been worked up. Preliminary results [16] from A-3, giving a double beta decay half-life limit of 5×10^{19} yr, were presented in 1988. The value from our work on sample A-4, presented by Moe [17], was a limit of $>10^{20}$ yr.

The data obtained from the present experiment (A-5), using the University of Chicago counter MCA1, are presented in Fig. 1. Shown are the number of events as a function of the energy of the alpha particles. The large peak on the left of Fig. 1 is from the ^{239}Pu tracer added at the start of the experiment. This served as a measure of the chemical recovery of the plutonium and of the particular counting efficiency.

Just to the right of this main group are the small number of alpha particles that can be identified by their energy as ^{210}Po ($T_\alpha=5.3$ MeV), always present as contamination. Adequately separated from the ^{210}Po alpha particles, in the energy region between 5.35 and 5.55 MeV, are fourteen events (the correct region to be alpha particles of ^{238}Pu). Their distribution with energy is consistent with that expected from ^{238}Pu.

At somewhat higher energy, up to 6 MeV, are a small number of events whose origin is not completely understood. They all appeared in the first 70 d of measurement on this counter (months after sample preparation) and do not appear to be bunched up enough at 6 MeV for all to be ^{218}Po. In contrast, six of the fourteen observed 5.5-MeV events appeared in the last 78 d. There were no events registered between 6 and 7 MeV, thus setting adequate limits on other possible contaminants.

Table I summarizes the experimental data on the sample from experiment A-5 as obtained on three different alpha counters. All indicate excesses at the 5.5-MeV region. The excesses, corrected for the counter efficiencies, are consistent in indicating about 2.6 dis/d in the 8.47-kg uranyl nitrate sample. Interpreted as ^{238}Pu, they corre-

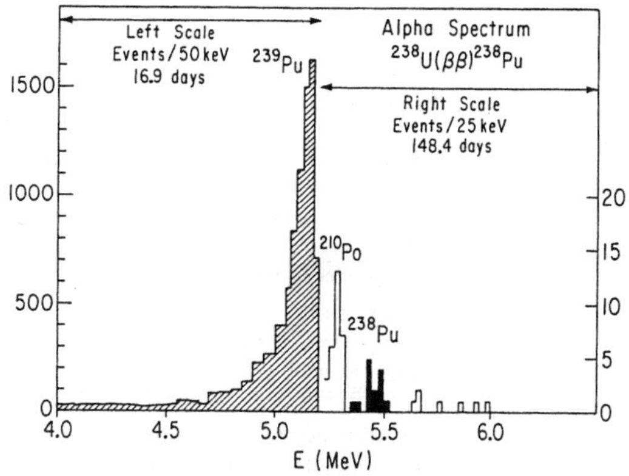

FIG. 1. A sample of the results from experiment A-5 on the University of Chicago counter MCA1 on the production of ^{238}Pu from ^{238}U. Shown are the numbers of alpha particles as a function of energy.

spond to the production of $(1.2\times0.4)\times10^5$ atoms of ^{238}Pu, or a half-life of 2.0×10^{21} yr for the decay of ^{238}U to ^{238}Pu. This rate is a factor of 40 longer than our previously published limit.

The possible contamination by ^{238}Pu fallout of the procedure or by chemicals has been mentioned. Evidence against this are the results from two previous experiments (A-3 and A-4) that, although starting with smaller amounts of uranyl nitrate, had, after the first steps, used mostly the same chemicals and procedures as experiment A-5. The net counting rates were 0.03 ± 0.04 and 0 ± 0.04 event/d on counter MCA1. After the A-5 experiment a blank run involving $NaHCO_3$ from the same large batch of salt and comparable in amount was put through almost the same chemistry. The final sample, corrected to the A-5 yield, measured 0.006 ± 0.014 event/d. Finally, a sample of the La^{3+} solution and HF (the only chemicals not used in previous experiments or the blank run) was tested and found to have no ^{238}Pu. These experiments make it very unlikely that the ^{238}Pu measured in A-5 is the result of fallout contamination of the chemicals or procedure.

As a final test, the measured sample of A-5 was removed from the platinum and subjected to an additional column purification that is specific for plutonium [19]. Within statistics, the resulting sample had the same ratio of ^{238}Pu to ^{239}Pu alpha particles.

The production rate of ^{238}Pu indicated by the present experiment is so low (~10 atoms/d) that reactions other than double beta decay for producing them must be considered. Since the sample analyzed contained no atoms heavier than ^{238}U, this nucleus would have to be the target of competing processes. The most serious competing reaction is the (p,n) process on ^{238}U leading to ^{238}Np. The protons could originate either from the intrinsic radioactivity of the sample or from external radiation (cosmic rays or other external particles). Because the (p,n) cross section on ^{238}U was not known, it was determined at the Los Alamos Van de Graaf accelerator. This cross section rises from negligible values at 6 MeV to ~5 mb at 12 MeV. A literature [20] value of 4 ± 2 mb at 14 MeV is consistent with this behavior. Using these data, the conversion of protons born inside a large $UO_2(NO_3)_2\cdot6H_2O$ sample into ^{238}Np was calculated to be only $\sim10^{-6}$ at 12 MeV. This is in contrast to the 50–100 times larger (p,n) probability in large samples of C_2Cl_4 or $8M$ $GaCl_3$ solution [21].

In spite of the high radioactivity of the sample (several mCi of alpha and beta activity and $\sim10^3$ spontaneous fissions per minute) the charged particles from this radioactivity are either too low in energy or too few at high energy to convert the ^{238}U to significant amounts of ^{238}Pu directly or indirectly.

The neutrons emitted in spontaneous fission are somewhat more serious because of their larger numbers. However, less than 1% are above 8 MeV, only a few percent of these will give rise to protons above this energy, and the low (p,n) yield per proton in $UO_2(NO_3)_2\cdot6H_2O$ leads to negligible numbers of ^{238}Pu. Thus no internal source for producing anywhere near 10^5 atoms of ^{238}Pu from 10^{25} atoms of U in 33 yr has been identified.

The possible production of ^{238}Np by cosmic rays during the long sea-level storage of the sample was estimated using data furnished by Davis [21]. In a large tank of C_2Cl_4, Davis found a conversion into ^{37}Ar of 6.6×10^{-23} per atom/yr due to cosmic rays and a similar number for the conversion of ^{71}Ga to ^{71}Ge in $8M$ $GeCl_3$. Interpreting these numbers as due to protons born inside the sample due to the different components of cosmic rays, and taking into account the different cross sections and stop-

TABLE I. Alpha measurements on sample A-5.

Counter	Chicago MCA2A	Los Alamos TA-48, No. 84	Chicago MCA1
Efficiency [a]	0.123	0.045	0.022
Background rate [b]			
Events observed (d)	12 (43.2)	14 (86)	4 (370)
Events/d	0.28 ± 0.08	0.16 ± 0.05	0.011 ± 0.006
Sample A-5			
Events observed (d)	16 (37.7)	14 (36.8)	14 (148)
Net events/d	0.14 ± 0.14	0.22 ± 0.11	0.084 ± 0.027
^{238}Pu produced (dis/d)	1.1 ± 1.1	4.9 ± 2.5	3.8 ± 1.2
Average dis/d [c]		2.6 ± 0.8	
Average atoms $(\times10^{-5})$		1.2 ± 0.4	
$t_{1/2}(\beta\beta)\times10^{-21}$ yr [d]		2.0 ± 0.6	

[a]The efficiency is the product of the chemical recovery of the added ^{239}Pu tracer (5.76 dis/min) and the counter efficiency.
[b]All event rates are applicable to the 5.5-MeV region of ^{238}Pu. The tabulated background value is the result of 4 events in 298 d before experiment A-5 and 0 event in 72 d after A-5.
[c]This is the weighted average; the errors are the statistical (1σ) errors.
[d]Calculated from the number of atoms (1.02×10^{25}) and the decay time (33 yr).

VOLUME 67, NUMBER 23 PHYSICAL REVIEW LETTERS 2 DECEMBER 1991

ping powers involved, we calculate a production of only 200 atoms of ^{238}Np during the 33-yr storage of our sample.

An additional possible contaminating nuclear reaction must be considered in our case because, for about 14 of the 33 yr, the sample was stored near the control room of the 450-MeV University of Chicago cyclotron. A check of the radiation exposure records of operating personnel suggests a very conservative upper limit of 1 neutron/cm^{2}sec of 15–30-MeV neutrons during this period, which would lead to a production of less than 10^4 atoms of ^{238}Pu in our sample.

Thus no source of contaminating nuclear reactions has been identified that would produce even 10% of the 5.5-MeV alpha particles that have been observed.

Discussion.—The observation of a half-life of 2×10^{21} yr for the transformation of ^{238}U to ^{238}Pu is only the sixth nuclear system for which definitive evidence has been obtained for the occurrence of a double beta decay process. This half-life can be considered in comparison with the half-lives of other cases, with theoretical predictions, and in relation to current uncertainties about the numbers and types of neutrinos.

The comparison with other cases of double beta decay is best made via the implied matrix elements for the relevant nuclear transformation on the assumption that the process responsible is the most conventional type of two-neutrino, two-electron decay. The removal of the phase-space factor takes care of the approximately seventh-power dependence of the rate on the Q value of the transition. Using the phase-space factors of Boehm and Vogel [22] leads to M^2 values for the ^{238}U transition in the same range as deduced for ^{130}Te and ^{76}Ge. On the other hand, the latest theoretical estimates [9] give an upper limit that is 10 times lower. This large discrepancy implies either a defect in the calculations or the presence of a faster path than the standard two-neutrino mode in this case.

In considering possible alternate scenarios, the rate of transformation of uranium to plutonium as determined by this experiment has special features. The rate determined is the sum of all possible paths to the final state. The low-Q value enhances the relative importance of processes with small numbers of emitted particles. The large-Z environment also favors the concentration of the available energy into smaller numbers of particles. It is thus important to confirm or resolve the present discrepancy between the observed and calculated rates of transformation of ^{238}U to ^{238}Pu by the standard two-neutrino process.

We are pleased to acknowledge the help of W. Haxton in early discussions, of F. Lawrence and S. Knight of Los Alamos in the development of the plutonium chemistry, of G. Butler and the counting facilities at Los Alamos, and of E. Blume and K. Wielgoz in the chemistry and counting operation at Chicago. J. Wilhelmy of Los Alamos kindly arranged the measurements of the ^{238}U(p,n) cross section at the Los Alamos Van de Graaf accelerator. A. Staudt and H. V. Klapdor kindly furnished the results of their calculation before publication. This work was supported by the U.S. Department of Energy, Grant No. DE-FG02-88ER40450.M003.

[1] A. A. Vasenko *et al.*, Mod. Phys. Lett. **A5**, 1299 (1990).

[2] H. S. Miley *et al.*, Phys. Rev. Lett. **65**, 3092 (1990).

[3] S. R. Elliott, A. A. Hahn, and M. K. Moe, Phys. Rev. Lett. **59**, 2020 (1987).

[4] M. K. Moe, in *Neutrino '90*, Proceedings of the Fourteenth International Conference on Neutrino Physics and Astrophysics, CERN, Geneva, Switzerland, 10–15 June 1990, edited by J. Parman and K. Winter [Nucl. Phys. B (Proc. Suppl.) **19** (1991)], pp. 158–176.

[5] F. T. Avignone, III, and R. L. Brodzinski, J. Prog. Part. Nucl. Phys. **21**, 99 (1988).

[6] W. C. Haxton and G. J. Stephenson, Prog. Part. Nucl. Phys. **12**, 409 (1984).

[7] M. Doi *et al.*, Prog. Theor. Phys. Suppl. **83**, 1 (1985).

[8] K. Muto, E. Bender, and H. V. Klapdor, Z. Phys. A **334**, 187 (1989); T. Tomada, Rep. Prog. Phys. **54**, 53 (1991).

[9] A. Staudt, K. Muto, and H. V. Klapdor-Kleingrothaus, Europhys. Lett. **13**, 31 (1990).

[10] J. J. Simpson, Phys. Rev. Lett. **54**, 1891 (1985); J. J. Simpson and A. Hime, Phys. Rev. D **39**, 1825 (1989); A. Hime and J. J. Simpson, Phys. Rev. D **39**, 1837 (1989).

[11] C. A. Levine, A. Ghiorso, and G. T. Seaborg, Phys. Rev. **77**, 296 (1950).

[12] K. Grotz and H. V. Klapdor, Phys. Lett. **157B**, 242 (1985).

[13] P. Vogel and M. R. Zirnbauer, Phys. Rev. Lett. **57**, 3148 (1986).

[14] B. G. Bennet, IAEA Report No. SM 199/40, 1971 (unpublished).

[15] W. C. Haxton, G. A. Cowan, and M. Goldhaber, Phys. Rev. C **28**, 467 (1983).

[16] A. Turkevich, G. A. Cowan, T. Economou, F. Lawrence, and S. Knight, in *Proceedings of the Workshops on Fundamental Symmetries and Nuclear Structure, Santa Fe, New Mexico, 12 October 1988*, edited by J. N. Ginocchio and S. P. Rosen (World Scientific, Singapore, 1989), pp. 86–99.

[17] A. Turkevich, reported by M. K. Moe, in *Neutrino '90* (Ref. [4]).

[18] Suggested by G. A. Cowan (private communication).

[19] D. C. Hoffmann, Los Alamos Report No. LA 1721, 1990, 5th ed., p. I-184.

[20] G. H. McCormick and B. L. Cohen, Phys. Rev. **96**, 722 (1954).

[21] R. A. Davis (private communication).

[22] F. Boehm and P. Vogel, *Massive Neutrinos* (Cambridge Univ. Press, Cambridge, 1987), pp. 143 and 144.

PHYSICAL REVIEW VOLUME 146, NUMBER 3 17 JUNE 1966

Limits for Lepton-Conserving and Lepton-Nonconserving Double Beta Decay in Ca^{48}†

E. DER MATEOSIAN AND M. GOLDHABER
Brookhaven National Laboratory, Upton, New York
(Received 10 February 1966)

A search has been made for double beta decay in Ca^{48} using a new technique which involves the use of CaF_2 crystals as scintillation detectors. These crystals, enriched in either Ca^{48} or Ca^{40}, were used to detect ionizing particles originating within the crystals. The advantages of this technique are (1) a 4π geometry for a source without self-absorption and (2) good resolution which is independent of the relative angular and energy distribution of the two electrons. A lower limit of 2×10^{20} yr is placed on the lepton-nonconserving double beta decay without neutrinos. The determination of a lower limit for the Dirac-type double beta decay (with emission of two neutrinos) is complicated by impurities in the Ca^{48} used. By resorting to the usual coincidence type of search for double beta decay, which makes the effect of the impurities negligible, we established a lower limit of 5×10^{18} yr for this process. This is comparable with the lower limit of the theoretical estimates.

INTRODUCTION

PHENOMENOLOGICALLY, double beta decay[1] is concerned with a search for two possible reactions:

$$(A,Z)\to(A,Z+2)+2e^-, \qquad (1)$$

$$(A,Z)\to(A,Z+2)+2e^-+2\bar{\nu}. \qquad (2)$$

It has been emphasized, especially by Pauli[2] and Greuling and Whitten,[3] that the existence or nonexistence of reaction (1) is probably the most sensitive test for lepton conservation. Both reactions are expected to take place as the result of a second-order effect due to the same nucleon-lepton (weak) interaction which gives rise to the usual single beta decay.[4] It has been pointed out,[5] however, that reaction (1) could also conceivably take place in the first order if there existed certain lepton-nonconserving interactions. Prior to 1957, when only parity-conserving interactions were considered in beta decay theory, the double beta decay process was regarded as a possible means of determining whether the Majorana or Dirac description of the neutrino was the correct one. Both of the above reactions are allowed by the Majorana theory, but only reaction (2) can take place in the Dirac theory. In the particular case of Ca^{48}, which appears to be one of the most favorable nuclei to study, these theories predicted[4] a half-life of $3\times10^{15\pm2}$ yr for reaction (1) and $1\times10^{21\pm2}$ yr for reaction (2). Further, the two reactions are distinguished by the fact that the sum of the two electron energies in reaction (1) is constant and equal to the transition energy while it varies continuously in reaction (2) up to the same energy as its limit. Many searches were made in this period for double beta decay and several early apparently positive results were later disproved except one obtained by an indirect chemical method.

† Work performed under the auspices of the U. S. Atomic Energy Commission.

[1] G. F. Dell' Antonio and E. Fiorini, Nuovo Cimento **17**, Suppl. I, 132 (1960).

[2] W. Pauli, Nuovo Cimento **6**, 204 (1957).

[3] E. Greuling and R. C. Whitten, Ann. Phys. (N. Y.) **11**, 510 (1960).

[4] S. P. Rosen and H. Primakoff, *Alpha-, Beta- and Gamma-ray Spectroscopy* (North-Holland Publishing Company, Amsterdam, 1965), pp. 1499–1516.

[5] G. Feinberg and M. Goldhaber, Proc. Natl. Acad. Sci. **45**, 1301 (1959).

The direct, counting measurements were made with sufficient sensitivity so that lifetimes shorter than 10^{18} yr could be excluded and this led to the conclusion[6] that decay (1) did not take place. With the discovery of parity nonconservation, the Majorana theory in its simplest form had to be given up, but recent reformulations[3,7] of the theory again allow under certain assumptions for the possibility of the occurrence of reaction (1) as a second-order effect, adding further interest to the search for both reactions (1) and (2) with improved techniques.

About a dozen years ago we realized that increased sensitivity could be gained in the investigation of double beta decay by giving up coincidence counting techniques for a method which has the advantage of a 4π geometry.[8] Specifically, we looked for the double beta decay of Ca^{48} by using a large single crystal of CaF_2 as a scintillator and recording events whose origins were within the crystal. Since our source, Ca^{48}, was an integral constituent of the crystal the method offered the further advantages of the equivalent of a "thin source" and fairly good resolution independent of the relative angular or energy distribution of the two electrons.

The total available energy for double beta decay in Ca^{48} has been calculated to be 4.27 MeV from atomic mass measurements made by Giese and Benson.[9] Energies based on (pn) reaction data[10] did not agree with this value until Chasman, Jones, and Ristinen[11] showed that the $Ca^{48}(p,n)$ reaction leads to an excited state of Sc^{48}, 131 keV above the ground state. The addition of this transition energy to the Q value of the reaction made the reaction data consistent with the mass spectroscopic data.[12] Single beta decay of Ca^{48} is energetically possible to the ground state and to two known excited states of Sc^{48} but has not been observed.[13] These transitions are highly forbidden because of the large difference of spin between the ground state of Ca^{48} and the levels in Sc^{48}, as shown in Fig. 1. In neutrino-less double beta decay the two electrons would share the total energy difference between Ca^{48} and Ti^{48}.

Our earliest attempt to detect double beta decay in Ca^{48} was made with a large single crystal of pure unenriched CaF_2 borrowed from the Harshaw Chemical Company. This crystal was a right cylinder 4.0 in. high, 4.5 in. in diameter and 2951 g in weight. Figure 2 shows the pulse-height spectrum obtained with this crystal and gives details of the geometry of the experiment in the insert. The total counting rate in the 4.27-MeV energy region as delimited by the resolution of the crystal for electrons leads to a lower limit of 2.5×10^{16} yr for the half-life for neutrino-less double beta decay. A conservative statistical treatment of our data, estimating the background by extrapolation, enabled us to set a lower limit of 10^{17} yr, similar to the values obtained at that time by other authors.[14,15]

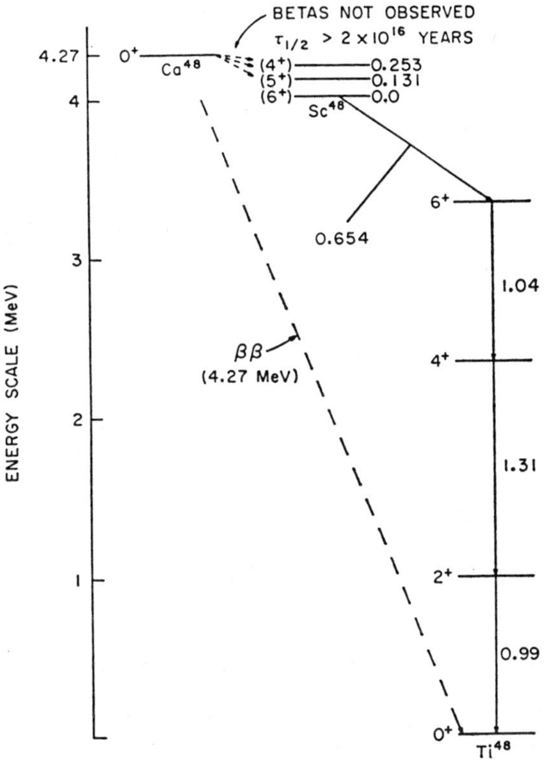

FIG. 1. Energy diagram of $A=48$ mass chain.

NEUTRINO-LESS DOUBLE BETA DECAY

At our request the Isotope Separation Department at Oak Ridge National Laboratory produced sufficient enriched Ca^{48}, assaying 96.59%, to enable us to have a crystal of CaF_2 grown[16] containing about 10.6 g of Ca^{48}. The performance of this crystal as a scintillator was improved over that of pure CaF_2 by the addition of ~0.5% Eu as an activator. Crystals of both normal

[6] J. S. Allen, *The Neutrino* (Princeton University Press, Princeton, New Jersey, 1958), Chap. 6.
[7] K. M. Case, Phys. Rev. **107**, 307 (1957).
[8] Preliminary reports of these results have been given as follows: M. Goldhaber, Bull. Am. Phys. Soc. 8, 46 (1963); E. der Mateosian and M. Goldhaber, in *Proceedings of the 12th Annual International Conference on High-Energy Physics, Dubna, 1964* (Atomizdat, Moscow, 1965); Bull. Am. Phys. Soc. **9**, 717 (1964); and Proceedings of the Neutrino Conference Cern, Geneva, 1965 (unpublished).
[9] C. F. Giese and J. L. Benson, Phys. Rev. **110**, 712 (1958).
[10] L. A. König, J. H. E. Mattauch, and A. H. Wapstra, Nucl. Phys. **31**, 18 (1962) [C. M. Johnson (private communication)].
[11] C. Chasman, K. W. Jones, and R. A. Ristinen, Phys. Rev. **140**, B212 (1965).
[12] The latest results are included in the 1964 Atomic Mass Table given by J. H. E. Mattauch, W. Thiele, and A. H. Wapstra, in Nucl. Phys. **67**, 1 (1965), from which one obtains an energy of 4.267 MeV for the double beta process in Ca^{48}.
[13] J. W. Jones and T. P. Kohman, Phys. Rev. **85**, 941 (1952).
[14] J. A. McCarthy, Phys. Rev. **97**, 1234 (1955).
[15] M. Awschalom, Phys. Rev. **101**, 1041 (1956).
[16] All crystals were grown by the Harshaw Chemical Company, Cleveland, Ohio.

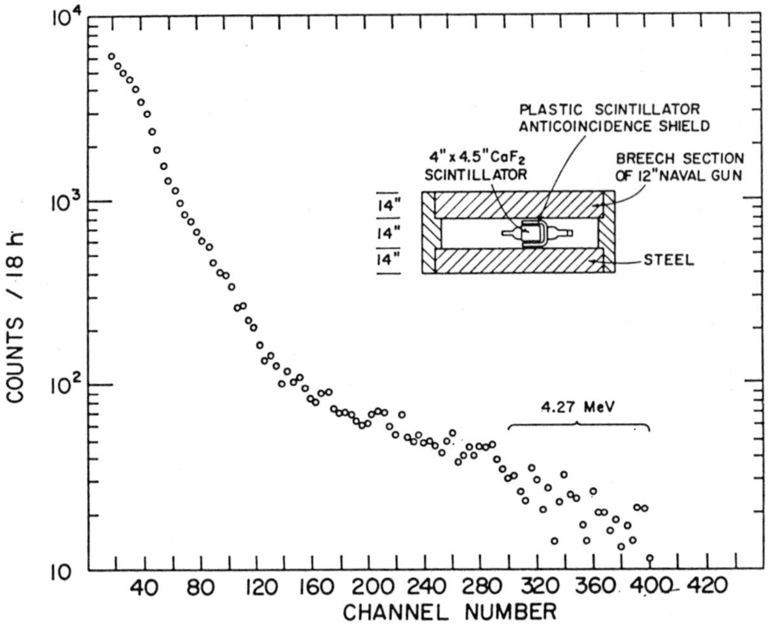

FIG. 2. Search for double beta decay of Ca^{48} in single crystal of normal CaF_2. Spectrum of counts seen with a 4-in. by 4.5-in. diam. normal CaF_2 crystal used as a scintillator to detect pulses originating within the crystal. Neutrino-less double beta decay would yield a peak at 4.27 MeV.

CaF_2 and $Ca^{40}F_2$ were also grown under identical conditions and these were used as controls. Figure 3 shows the spectrum observed with the crystal enriched in Ca^{48}. The very definite peaks in the region from channels 40 to 100 were due to either electron- or alpha-particle-emitting impurities. The light output of this crystal for electrons and alpha particles was measured and was found to be four times larger for an electron of a given energy (in the MeV region) than for an alpha particle of the same energy. Hence, the peak in channel 100 represents either a 2-MeV electron or an 8-MeV alpha particle. Strong sources of homogeneous 2-MeV electrons are exceedingly rare, but a homogeneous alpha-particle group of 7.68 MeV is emitted by Po^{214}, which is a member of the uranium series of radioactive elements. An impurity of 1 part per million of uranium would be sufficient to give the observed peak in channel 100, and a suitable mixture of thorium and uranium impurities

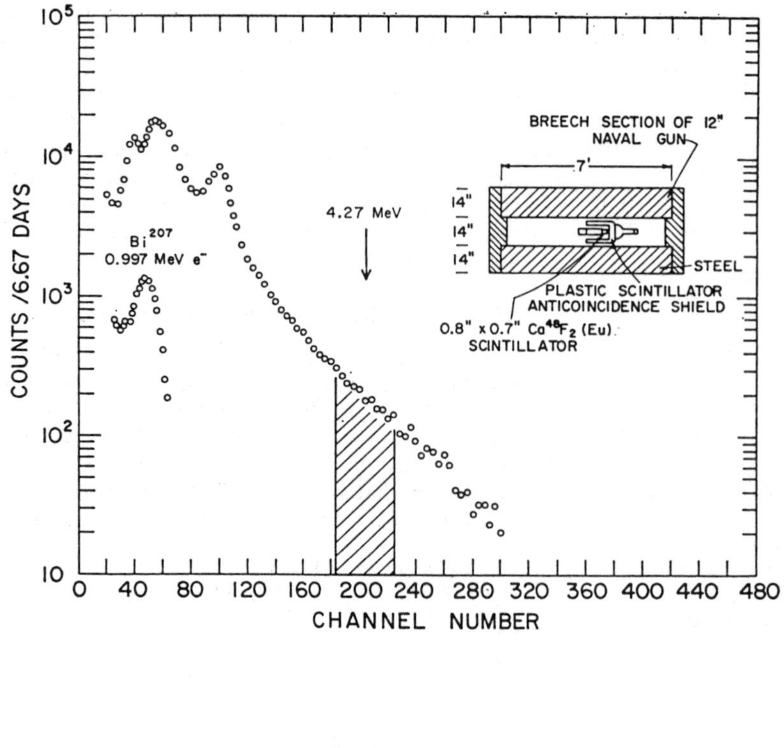

FIG. 3. Spectrum of counts seen with a crystal of $Ca^{48}F_2$ (Eu) enriched to 96.59% Ca^{48}. An external source of Bi^{207} conversion electrons was used for calibration.

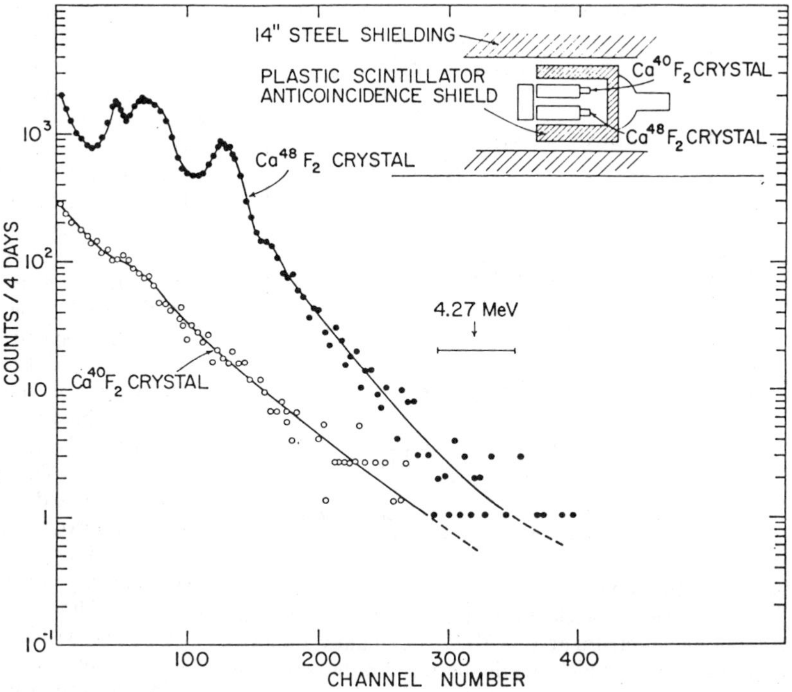

FIG. 4. Spectrum of counts seen with a crystal of $Ca^{48}F_2$ (Eu) enriched to 96.59% Ca^{48}. This run differs from that of Fig. 3 in that the Ca^{48} was repurified and a new crystal was grown. The reduced counting rate can be judged by referring to the scales of each figure. A run with a control crystal of $Ca^{40}F_2$ (Eu) is also shown for comparison.

could account for the main features of the spectrum observed.

The Ca^{48} was sent back to Oak Ridge for purification and a second crystal was grown which contained 11.4 g of Ca^{48}. Again control crystals of normal CaF_2 and $Ca^{40}F_2$ were grown. Figure 4 shows that a lower counting rate was obtained with this crystal, as a comparison of the scales of Figs. 3 and 4 will verify. In the region where neutrino-less double beta decay would be observed (4.27 MeV) the counting rate is low and close to a background count obtained with a similar crystal of $Ca^{40}F_2$. After subtracting this background, we find that the total remaining counts represent a lower limit of 1.2×10^{19} yr for the double-beta-decay half-life. Experimental details of the run are again shown in an insert in Fig. 4. In this run data were accumulated simultaneously with both the $Ca^{48}F_2$ and the $Ca^{40}F_2$ detectors, which were placed side by side within a plastic scintillator operated in anticoincidence with both detectors. The entire assembly was housed within a section of a naval gun with 14-in.-thick walls of steel.

Five such runs were made totaling 28.7 days of running time. The difference of counts of the combined runs in the Ca^{48} and Ca^{40} crystals is plotted against channel number in Fig. 5.

In neutrino-less double beta decay one would expect to see a peak at 4.27 MeV (channel 140 in Fig. 5) with a resolution which may be deduced from the resolutions of either the impurity peaks in the Ca^{48} spectra or externally impressed internal conversion electron peaks. A conservative estimate of the lower limit for neutrino-less double beta decay was made by adding all the counts in a 0.6 MeV wide region centered around 4.27 MeV. A lower limit of 2.4×10^{19} yr is found either by adding the experimentally obtained counts or by taking the total counts under the fitted smooth curve. A somewhat better lower limit may be established by considering the fluctuation of the experimental points about the smooth curve, representing the extrapolated background. The average deviation of the points about the line may be calculated and twice this deviation may be considered to be the minimum deviation a point would

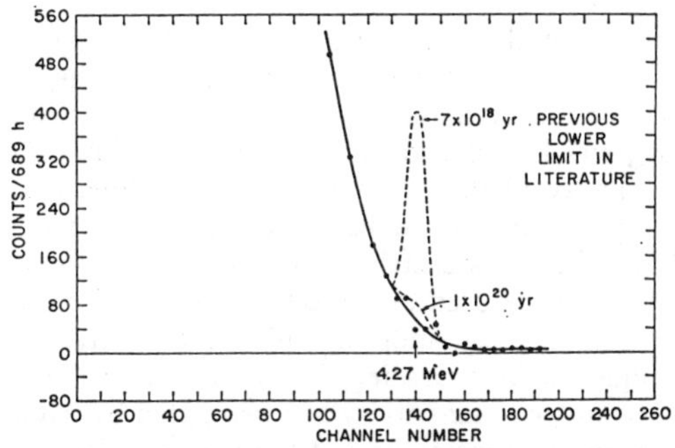

FIG. 5. Difference of counts in Ca^{48} and Ca^{40} crystals. The dashed curve labeled 1×10^{20} yr indicates the deviation from the smooth curve that would have been noticed if neutrino-less double beta decay took place in Ca^{48} with $\tau_{1/2}=1\times10^{20}$ yr. The larger dashed peak labeled 7×10^{18} yr represents neutrino-less double beta decay with a lifetime corresponding to a previously published lower limit. (See Ref. 17.) Actually, this peak should be about three times larger. (See text and Ref. 17.)

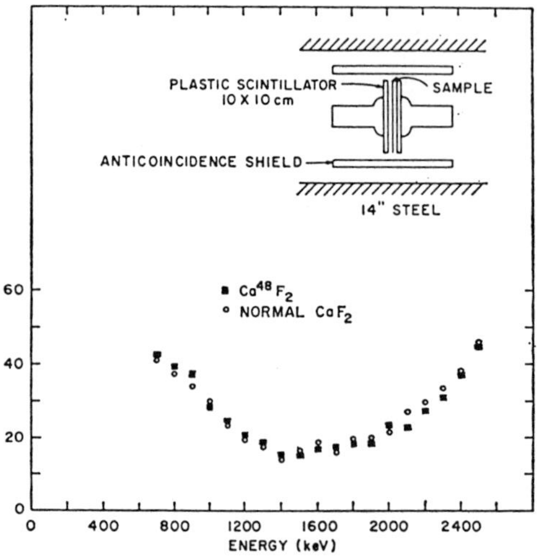

FIG. 6. Spectra of the combined energies of coincidence pulses for Ca^{48} and for normal Ca.

the α particles no longer constituted a measurable portion of the radiations that were able to emerge from the sample and be counted in the detectors, and the single beta rays did not contribute directly to the coincidence counting rate. The detectors were two plastic scintillators with 100 cm^2 of counting area which were mounted within a plastic scintillator anticoincidence shield in the 14-in.-thick steel shield. A sample of normal CaF_2 was used as a control. Similar spectra were observed with a Ca^{48} source and the normal Ca source. Two runs were made, one for 4 days, the other for 6 days, and in one run the Ca^{48} "spectrum" had more counts than the normal Ca "spectrum" in the region of concern while the reverse was true in the second run. These fluctuations were not statistically significant. The two runs were combined and are plotted in Fig. 6. Again experimental details of the counting arrangements are given in an insert. The spectra shown were obtained by demanding coincidences between the two detectors which were not accompanied by counts in the anticoincidence shield, adding the outputs of the two detectors and accumulating these in a multichannel pulse-height analyzer. The spectrum observed with the Ca^{48} sample in place is, within statistics, identical to the spectrum with the normal Ca sample in place. The difference between the spectrum found for Ca^{48} and normal Ca is plotted in Fig. 7 where it is compared with the theo-

need in order to be significantly different from the line. Then, barring the accidental presence of a dip in the background spectrum at 4.27 MeV, the smallest double-beta-decay peak which would be detectable could be considered to be one that is at least equivalent to twice the average deviation. This criterion gives a lower limit of 2.3×10^{20} yr. Still a third criterion, which has been used by previous investigators, is to take the statistical accuracy with which the total counts in the region under investigation are known. If twice the standard deviation is taken one obtains $\tau_{1/2}>1.8\times10^{20}$ yr. For comparison, in Fig. 5 a peak corresponding to 1×10^{20} yr is superimposed on the background (dashed line). Therefore, 2×10^{20} yr appears to be a reasonable lower limit. A larger peak corresponding to the previous lower limit of Dobrokhotov *et al.*[17] is also shown for comparison.

DOUBLE BETA DECAY WITH TWO NEUTRINOS

Because of the impurities in the Ca^{48} we were not able to obtain a significant lower limit for the Dirac type of double beta decay (two electrons plus two neutrinos), which would yield a continuum ending at 4.27 MeV. To avoid the interference of the alpha-particle emitting impurities we had the Ca^{48} powdered and packaged in a moderately thin layer (0.162 g/cm^2) so that it could be used as an external source for coincidence counting. By adopting this more generally used technique for investigating double beta decay we lost the advantage of sensitivity but benefitted by avoiding the background from the radioactive impurities, since

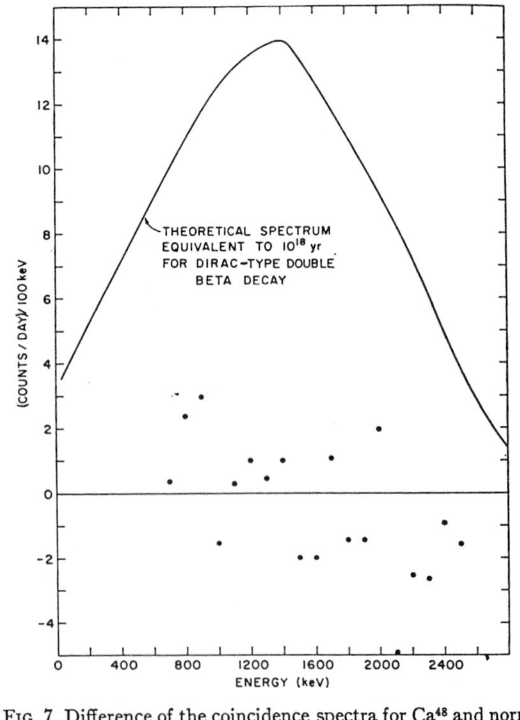

FIG. 7. Difference of the coincidence spectra for Ca^{48} and normal Ca of Fig. 6. A theoretical shape for the spectrum of double beta decay with neutrinos is shown which is equivalent to a half-life of 10^{18} yr. This curve is shifted 400 keV to the left to account for self-absorption of the electrons in the sample.

[17] E. I. Dobrokhotov, V. R. Lazarenko, and S. Yu Luk'yanov, Zh. Eksperim. i Teor. Fiz. **36**, 76 (1959) [English transl.: Soviet Phys.—JETP **9**, 54 (1959)]. {Since the preparation of this paper, new results have been reported by the last two authors of this reference. See V. R. Lazarenko and S. Yu. Luk'yanov, Zh. Eksperim. i Teor. Fiz. **49**, 751 (1965) [English transl.: Soviet Phys.—JETP **22**, 521 (1966)]}.

TABLE I. Theoretical predictions and experimental limits for double beta decay of Ca^{48}.

Nature of lifetime determination	Half-life (years) Neutrino-less	Two neutrinos (Dirac type)	Authors
Theory	$3\times10^{15\pm2}$	$1\times10^{21\pm2}$	Rosen and Primakoff[a]
		$>10^{19}$	Meichsner[b]
		10^{19}	Beliaev and Zakharev[c]
		$>2\times10^{19\pm1}$	Greuling and Whitten[d]
Experiment	$>2\times10^{18}$ e		Awschalom[f]
	$>7\times10^{18}$ e		Dobrokhotov, Lazarenko, and Luk'yanov[g]
	$>5\times10^{19}$ e	$>3\times10^{18}$	Lazarenko and Luk'yanov[g]
	$>4\times10^{18}$	$>4\times10^{18}$	Shapiro, Frankel, Koicki, Wales, and Wood[h] (improved values by private communication)
	$>2>10^{20}$	$>5\times10^{18}$ yr	Present work

a See Ref. 4.
b See Ref. 19.
c See Ref. 20.
d See Ref. 3.
e These values are a factor 3 too high due to the expected angular correlation between the electrons. See Refs. 17 and 18.
f See Ref. 15.
g See Ref. 17.
h See Ref. 21.

retically expected curve[3] equivalent to a half-life of 10^{18} yr. A quantitative estimate for the lower limit was made by taking the total number of counts under the Ca^{48} curve, correcting for the fact that the instrument has a low-energy cutoff at 700 keV, taking twice the statistical fluctuation of this number as the measure of the lower limit and correcting for the coincidence geometry. This yields $\tau_{1/2}>5\times10^{18}$ yr.

CONCLUSION

In Table I[18–21] we have summarized some of the significant theoretical estimates[3,4,19,20] and experimental observations[15,17,21] concerning the double beta decay lifetime of Ca^{48}. The "e" following some of the experimental values listed for the lower limit to the half-life calls attention to a fact pointed out by Dobrokhotov *et al.*[17] that those lifetime values which were obtained by coincidence measurements ought to be reduced by a factor of 3 because of the angular correlation[18] expected between the two electrons in the neutrino-less double beta emission.

The search for double beta decay in Ca^{48} has still not led to the observation of this phenomenon. One can see from Table I that the experimentally determined lower limit for the half-life for the Dirac-type (two-neutrino) double beta decay is just approaching the more optimistic theoretical estimates. This higher order effect must be expected to exist, but it is difficult to predict its half-life since the matrix elements involved are hard to estimate. It has been reported[22] that the possible existence of double beta decay has been detected in Te^{130} by an indirect chemical method.

Our results on the neutrino-less double beta decay can be considered an improved test of lepton nonconservation. If lepton-nonconserving interactions exist[5] which allow double beta decay as a first-order process, our results indicate that the coupling strength of such interactions must be more than 10^{14} times weaker than the Fermi coupling. Thus, the continued search for neutrino-less double beta decay has helped to establish the law of lepton conservation on essentially as firm a basis as charge conservation and baryon conservation.

ACKNOWLEDGMENTS

The authors wish to thank Dr. George L. Rogosa of the U. S. Atomic Energy Commission for his help in obtaining separated isotopes of Ca. L. O. Love, Supervisor of the Electromagnetic Isotope Separation Department and members of the Chemistry Group at Oak Ridge National Laboratory, especially H. R. Gwinn and R. L. Bailey, have been very cooperative and helpful in the preparation and purification of the Ca^{48} isotope. Finally, the authors wish to express their appreciation of the very extensive cooperation of the Harshaw Chemical Company, especially, E. C. Stewart and Dr. C. F. Swinehart, who tackled the difficult task of growing the scintillation crystals of enriched Ca isotopes without loss of material.

[18] H. Primakoff, Phys. Rev. **85**, 888 (1952).
[19] L. Meichsner, Phys. Rev. **117**, 489 (1960); **120**, 552 (1960).
[20] V. B. Beliaev and B. N. Zakharev, Zh. Eksperim. i Teor. Fiz. **34**, 505 (1958) [English transl.: Soviet Phys.—JETP **7**, 347 (1958)].
[21] M. H. Shapiro, S. Frankel, S. Koicki, W. Wales, and G. T. Wood, Bull. Am. Phys. Soc. **10**, 424 (1965). The values in the table were given in an invited paper presented by M. H. Shapiro at the Summer Meeting of the American Physical Society, New York, 1965, and in private communication.
[22] R. J. Hayden and M. G. Inghram, Natl. Bur. Std. Circ. No. **522**, 189 (1953); N. Takaoka and J. Okano, Shitsuryo Bunseki **12**, 195 (1965).

Double β decay and conservation of lepton charge

Yu. G. Zdesenko

Institute of Nuclear Research, Ukrainian Academy of Sciences, Kiev
Fiz. Elem. Chastits At. Yadra **11**, 1369–1420 (November–December 1980)

The present state, tendencies, and prospects for the investigation of double β decay are described. The basic properties and most widely accepted schemes for classifying leptons are considered briefly, and also the methods for investigating possible nonconservation of the lepton charge. The fundamentals of the theory of double β decay are presented, and estimates of the lepton nonconservation parameter based on investigations of 2β decay are given. The experimental data are discussed extensively. The main attention is devoted to recent experiments. The possibility and prospects of further investigations of double β decay in direct experiments are considered.

PACS numbers: 23.40.Bw

INTRODUCTION

The part played by the law of conservation of the lepton charge in elementary-particle physics,[1] cosmology, astrophysics, and nuclear physics[2-3] is well known. So too is the fact that the results of investigation of 2β decay depend sensitively on even a very weak possible violation of this law.[2-5]

The possibility of realizing 2β decay was pointed out for the first time by Goeppert-Mayer[6] as a process in which "... a metastable isobar can change into a more stable one by simultaneous emission of two electrons." This work was published in 1935, i.e., only a year after the appearance of Fermi's theory of β decay.[7] Thus, the problem of 2β decay is as "ancient" as the problem of weak interactions. Nevertheless, despite an immense number of sometimes extremely subtle and complex experiments so far made, 2β decay has not been observed directly in any of them. It is only recently that indirect "geological" methods have shown that ^{130}Te and ^{82}Se may undergo 2β decay.[7-18]

The problem of 2β decay has been the subject of several excellent reviews[19-24] of varied length and differing in the particular emphasis. Since the study of 2β decay involves an exceptionally large number of questions, the present review is an attempt at a comparatively complete consideration of all the most important theoretical and experimental results bearing on this problem.

In addition, recent successes in the development of experimental techniques make it possible to plan experiments with a sensitivity that appeared unattainable even a few years ago. It has therefore become very necessary to review the achievements and tendencies in the development of facilities for studying 2β decay in order to estimate the possibilities and prospects of further progress in this field.

1. CONSERVATION OF LEPTON CHARGE

Properties and classification of leptons. It is well known that in nature there are four charged ($e^{\pm}, \mu^{\pm}$) and four neutral ($\nu_e, \tilde{\nu}_e, \nu_\mu, \tilde{\nu}_\mu$) leptons[1)] and that these particles are subject to only the weak and electromagnetic interactions. The main properties of the leptons are given in Table I.

The limit on the mass of the electron antineutrino is established in experiments which measure the profile of the β-decay spectrum of tritium: $m_{\nu_e} \leqslant 35$ eV.[26] For the electron neutrino, $m_{\nu_e} \leqslant 6$ keV.[27] An upper limit for the mass of the muon neutrino is given in Ref. 28: $m_{\nu_\mu} \leqslant 0.65$ MeV.

All decays and processes with the participation of leptons are conveniently systematized by the introduction of a lepton charge, which is assumed to be conserved universally.[29]

There exist several different schemes by means of which leptons are classified. We shall consider the ones that are most widely used: the additive scheme, the multiplicative scheme, and the Konopinski–Mahmoud–Zel'dovich scheme.

The additive scheme[30] is usually employed in the theory of weak interactions and appears to be in the best agreement with the totality of the experimental data. In this scheme, one introduces two lepton charges, the electron L_e and muon L_μ, which are conserved separately and have the values

$$L_e = \begin{cases} +1 - e^-, \ \nu_e; \\ -1 - e^+, \ \tilde{\nu}_e; \end{cases} \quad L_\mu = \begin{cases} +1 - \mu^-, \ \nu_\mu; \\ -1 - \mu^+, \ \tilde{\nu}_\mu. \end{cases}$$

The particles $e^-, \nu_e, \mu^-, \nu_\mu$ are leptons, while $e^+, \nu_e, \mu^+, \nu_\mu$ are antileptons. The neutrinos ν_e and ν_μ have helicity -1, and the antineutrinos $\bar{\nu}_e$ and $\bar{\nu}_\mu$ have helicity $+1$.

In the multiplicative scheme,[31,32] it is not the charges L_e and L_μ separately that are conserved but their sum $\sum(L_e + L_\mu)$ and the sign of $(-1)^{L_\mu}$. Leptons have the same values of the charges as in the additive scheme.

TABLE I. Basic properties of leptons.

Lepton	Mass	Electric Charge	Spin	Mean lifetime	S, B
$e^{\mp}$	0.511 MeV	∓ 1	$1/2\pm$	Stable	0
$\tilde{\nu}_e$, ν_e	35 eV, 6 keV	0	1/2	Stable	0
$\mu^{\pm}$	105.650 MeV	± 1	$1/2\mp$	2.2×10^{-6} sec	0
ν_μ, $\tilde{\nu}_\mu$	0.65 MeV	0	1/2	Stable	0

[1)] In this review, we shall not consider the properties of the heavy lepton τ^+ with mass of about 1.9 GeV.[25]

 0090-4759/80/060542-22$02.80

2.1.2 Double Beta Decay, Gauge Theories and Neutrino Mass:

Origins of Neutrino Masses

Horizontal Gauge Symmetry and Masses of Neutrinos

Tsutomu YANAGIDA

Department of Physics, Tohoku University, Sendai 980

PROCEEDINGS OF THE WORKSHOP
ON
"THE UNIFIED THEORY AND THE BARYON NUMBER IN THE UNIVERSE"

Held at National Laboratory for High Energy Physics (KEK),
February 13 - 14, 1979
Edited by

Osamu SAWADA and Akio SUGAMOTO

Recently several authors have studied a possible unification of electronic and muonic matter by adding the horizontal local-symmetry, $SU_F(2)$,[1)∼3)] to the weak and electromagnetic $SU(2)\times U(1)$.[4)] As a consequence of gauging the symmetry,the conservation of muon number is violated. The exchange of horizontal gauge bosons, S^a_μ, also induce the superweak type of CP-nonconservation.[1)] From the data on CP-violation in $K^0_L \to 2\pi$ decay, the effective coupling constant, G_S, of S^a_μ with leptons and quarks is determined as $G_S \sim 10^{-15}$ GeV^{-2} unless the accidental cancellation occurs.[1)] The strength is enough weak to avoid unwanted flavour-changing transitions.

NATIONAL LABORATORY FOR
HIGH ENERGY PHYSICS
OHO-MACHI, TSUKUBA-GUN
IBARAKI, JAPAN

If there exist six leptons and six quarks, we extend the horizontal $SU_F(2)$ to $SU_F(3)$. The weak-SU(2) doublet-and singlet-fermions transform as triplets under the horizontal $SU_F(3)$. The triangle anomalies[5)] appearing in the lepton sector can be removed by assuming right-handed neutrinos.[*)]

The purpose of this short note is to point out the possibility that the spontaneous breakdown of the symmetry generates the masses of right-handed and left-handed neutrinos and each neutrino becomes a massive Majorana particle. The mass of each particle may be $m_\xi \sim 10^5$ GeV and $m_\zeta \sim 10 \sim 10^4$ eV, respectively.

The assignment for leptons is the following:

$$\updownarrow SU(2) \quad \psi_L = \begin{pmatrix} \nu_e & \nu_\mu & \nu_\tau \\ e & \mu & \tau \end{pmatrix}_L , \qquad \begin{matrix} \nu_R = [\nu_e \;\; \nu_\mu \;\; \nu_\tau]_R , \\ (3, \;\; 1, \;\; 0) \\ \ell_R = [e \;\; \mu \;\; \tau]_R . \\ (3, \;\; 1, \;\; -2) \end{matrix} \tag{1}$$
$$(3, \;\; 2, \;\; -1)$$
$$\leftarrow SU_F(3) \rightarrow \qquad\qquad \leftarrow SU_F(3) \rightarrow$$

Here the first two values in each parenthesis denote the representation dimensions of $SU_F(3) \times SU(2)$ and the last one the U(1) hypercharge.

In order to make the horizontal gauge bosons, S_μ^a $(a = 1 \sim 8)$, heavy sufficiently we introduce a Higgs scalar, $\chi_{ij} = (6, 1, 0)$. Another Higgs scalar, $\phi^a = (8, 2, -1)$ is also assumed to break the symmetry $SU(2) \times U(1)$ surviving down to the electromagnetic one.

---

*) It is possible to assign fermions as triplets of $SU_F(2)$. In this case the anomalies are not generated without right-handed neutrinos (see Ref.2)).

The general form of the neutrino's mass term is given by

$$\mathcal{L}^{\nu}_{mass} = \frac{1}{2}G^{\nu}_{\chi}\ \overline{(\nu^{i}_{R})^{c}}\ \langle x_{ij}\rangle\ \nu^{i}_{R} + G^{\nu}_{\phi}\ \bar{\psi}_{L}\langle\phi^{a}\rangle\lambda^{a}\nu_{R} + h.c, \qquad (2)$$

where $(\nu^{i}_{R})^{c}$ denotes the charge-conjugated field of ν^{i}_{R} and $(\nu^{1}\ \nu^{2}\ \nu^{3})_{R}$ corresponds $(\nu_{e}\ \nu_{\mu}\ \nu_{\tau})_{R}$. Eq(1) represents a 6×6 mass matrix among six Majorana particles, $\xi^{i} = \nu^{i}_{R} + (\nu^{i}_{R})^{c}$ and $\zeta^{i} = \nu^{i}_{L} + (\nu^{i}_{L})^{c}$ $(i = 1\sim 3)$. The masses of these neutrinos are roughly obtained as

$$m_{\xi} \sim G^{\nu}_{\chi}\langle\chi\rangle\ , \qquad (3)$$

$$m_{\zeta} \sim \frac{G^{\nu}_{\phi}\langle\phi\rangle}{G^{\nu}_{\chi}\langle\chi\rangle}\cdot G^{\nu}_{\phi}\langle\phi\rangle\ , \qquad (4)$$

where $\langle\chi\rangle$ and $\langle\phi\rangle$ are vacuum-expectation values averaged and $\langle\phi\rangle/\langle\chi\rangle\sim 10^{-5}$.[1)] To estimate magnitudes of these masses we tentatively assume that all Yukawa coupling constants are same order, $G^{\nu}_{\chi} \sim G^{\nu}_{\phi} \sim G^{\ell}_{\phi}$, where G^{ℓ}_{ϕ} is the coupling constant for leptons, $\ell = (e\ \mu\ \tau)$. Then we find $m_{\xi} \sim 10^{5}$ GeV and $m_{\zeta} \sim 10 \sim 10^{4}$ eV. Neutrino oscillations are also expected, but the oscillation length depends on the details of the mass matrix.

Finally we stress that the present scheme of the symmetry breaking is a realistic one in the sense that the Higgs scalars ϕ^{a} and χ_{ij} can be considered as bound states of fermion-antifermion $(\bar{\nu}_{R}\psi_{L}+\bar{\ell}_{R}\psi_{L}+\cdots)$ and fermion-fermion $(\nu_{R}\nu_{R})$, respectively.[6)] It is, therefore, due to the large violation of the horizontal symmetry that right-handed neutrinos disappear at low energy regions.

References

1) T. Maehara and T. Yanagida, Prog. Theor. Phys. 60 (1978), 822; 61 (1979), No.5.

2) F. Wilzek and A. Zee, Phys. Rev. Letters 42 (1979), 421.

3) K. Akama, Y. Chikashige and T. Matsuki, INS-Report-288 (1977). H. Terazawa, Y. Chikashige and K. Akama, Phys. Rev. D15 (1977), 480.

4) S. Weinberg, Phys. Rev. Letters 19 (1967), 1264. A. Salam, in Elementary Particle Physics, edited by N. Svartholm (Stockholm, 1968), P367.

5) C. Bouchiat, J. Iliopoulos and Ph. Meyer, Phys. Letter 38B (1972) 519. D. Gross and R. Jackiew, Phys. Rev. D6 (1972), 477.

6) Y. Nambu and G. Jona-Lasinio, Phys. Rev. 122 (1961) 345; 124 (1961) 246.

Massive Neutrinos in Gauge Theories

P. Langacker

Deutsches Elektronen Synchrotron, DESY,
D-2000 Hamburg 52, Fed. Rep. of Germany
Permanent address: Department of Physics,
University of Pennsylvania, Philadelphia, PA 19104, USA

in "Neutrinos", ed.
H. V. Klapdor, Springer-Verlag Berlin, Heidelberg, New York, London, Paris, Tokyo (1988) 71 – 115

The present status of several aspects of neutrino physics are summarized, including the weak interactions of neutrinos, neutrino counting, and the theoretical expectations for and experimental constraints on neutrino mass.

1 Introduction

Neutrinos have long been amongst the most important probes of the fundamental interactions. In the last fifteen years, in particular, neutrinos have helped establish the standard $SU_2 \times U_1$ electroweak model as correct to first approximation, have been important probes of the structure of the nucleon and of the strong interactions, and have set stringent limits on new physics beyond the standard model. Furthermore, the question of whether the neutrino has a nonzero mass is one of the most important issues in both particle physics and astrophysics: most extensions of the standard model predict a nonzero mass at some level. Masses in the 10 eV range could account for the dark matter of the universe, while masses $\leq 10^{-2} eV$ could resolve the Solar neutrino problem.

In this talk I will describe several aspects of neutrino physics starting with the weak interactions of neutrinos. It will be seen that both the charged and neutral current processes are very well described by the standard model. I will then turn to the question of neutrino counting: indirect evidence leaves little doubt as to the existence of the τ-neutrino, while a number of laboratory and cosmological constraints strongly suggest that the number of neutrinos (with mass $<< \frac{M_Z}{2}$ is less than 0(3-5)). Finally, I will consider the complicated subject neutrino mass: the principle theoretical models and their implications will be described, and the experimental situation will be briefly summarized.

2 The Weak Interactions of Neutrinos

The Glashow-Weinberg-Salam standard electroweak model [1] is based on the gauge group $SU_2 \times U_1$, with gauge couplings g and g' for the two factors and gauge bosons $(W^{\pm}, W^0, B)$. It incorporates the Fermi theory of the charged current weak interactions [2] and quantum electrodynamics (QED), and successfully predicted the existence and properties of a new neutral current interaction (Fig. 1). The charged and neutral current interactions are mediated by the massive gauge bosons $W^{\pm}$ and Z, respectively, while QED is mediated by the massless photon A, where

$$\begin{aligned} A &\equiv \cos\theta_W B + \sin\theta_W W^0 \\ Z &\equiv -\sin\theta_W B + \cos\theta_W W^0 \end{aligned} \tag{1}$$

*Permanent address: Department of Physics, University of Pennsylvania, Philadelphia, PA 19104.

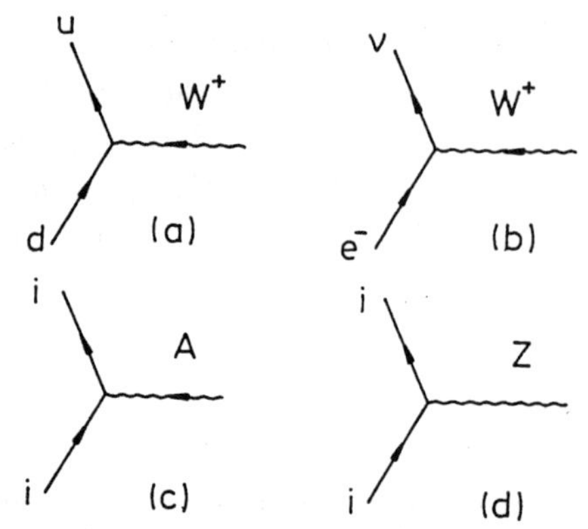

Figure 1: Charged current, QED, and neutral current interactions. The vertex factors are (a) $-\frac{ig}{2\sqrt{2}}\gamma_\mu(1+\gamma^5)V_{ud}$, (b) $-\frac{ig}{2\sqrt{2}}\gamma_\mu(1+\gamma^5)$, (c) $-ieq_i\gamma_\mu$, and (d) $-\frac{ig}{2\cos\theta_W}\gamma_\mu\left[t_{3L}(i)(1+\gamma^5)-2\sin^2\theta_W q_i\right]$. V_{ud} is an element of the quark mixing matrix, q_i is the electric charge of fermion i (in units of e), and $t_{3L}(i)$, the eigenvalue of the third generator of SU_2, is $+\frac{1}{2}$ for (u,ν) and $-\frac{1}{2}$ for (d,e^-).

In (1), $\theta_W \equiv \tan^{-1}(g'/g)$ is the weak angle. e, the positron electron charge, is related by

$$e = g\sin\theta_W \tag{2}$$

The W and Z masses are predicted in terms of $\sin^2\theta_W$, which can be determined independently from deep inelastic neutrino scattering. One has

$$\begin{aligned} M_W &= \frac{A_o}{\sin\theta_W(1-\Delta r)^{\frac{1}{2}}} \\ M_Z &= \frac{M_W}{\cos\theta_W} \end{aligned} \tag{3}$$

where $A_o = (\pi\alpha/\sqrt{2}G_F)^{1/2} = 37.281\ GeV$. Δr is a higher order correction, mainly due to A, W, and Z self-energy diagrams. It is predicted to be 0.0713 ± 0.0013 for top quark and Higgs boson masses of 45 GeV and 100 GeV, respectively, while $\Delta r \to 0$ for $m_t \sim 245\ GeV$. The predictions of (3) are in striking agreement with the data from the UA1 [3] and UA2 [4] groups at CERN, and even provide a rough confirmation of the radiative corrections [5] (Table 1). The production cross sections, couplings, and angular distributions ($\Rightarrow$ spin) are also in agreement with expectations.

Table 1: The measured W and Z masses (in GeV), compared with the theoretical expectations [5] from deep inelastic scattering with and without radiative corrections. (The radiative corrections include Δr from (3) as well as to the value of $\sin^2\theta_W$ extracted from experiment).

	M_W	M_Z
UA1 + UA2	80.9 ± 1.4	91.9 ± 1.8
prediction (with radiative corrections)	80.2 ± 1.1	91.6 ± 0.9
prediction (without radiative corrections)	75.9 ± 1.0	87.1 ± 0.7

2.1 The Charged Current

The weak charged current interaction is described by the coupling

$$L = -\frac{g}{2\sqrt{2}}(J_W^\mu W_\mu^- + J_W^{\mu\dagger} W_\mu^+) \tag{4}$$

between the massive $W^\pm$ bosons and the charged fermion current J_W^μ, given (for massless neutrinos), by

$$J_W^{\mu\dagger} = (\bar{u}\ \bar{c}\ \bar{t})\ \gamma^\mu(1+\gamma^5)V \begin{pmatrix} d \\ s \\ b \end{pmatrix} + (\bar{\nu}_e\ \bar{\nu}_\mu\ \bar{\nu}_\tau)\ \gamma^\mu(1+\gamma^5) \begin{pmatrix} e^- \\ \mu^- \\ \tau^- \end{pmatrix}. \tag{5}$$

The weak charged current is purely $V-A$, which means that it involves only the left-chiral $(1+\gamma^5)$ projections of the quark and lepton fields [6]. In (5),

$$V \equiv \begin{pmatrix} V_{ud} & V_{us} & V_{ub} \\ V_{cd} & V_{cs} & V_{cb} \\ V_{td} & V_{ts} & V_{tb} \end{pmatrix} \tag{6}$$

is the unitary Cabibbo-Kobayashi-Maskawa [7] (CKM) quark mixing matrix, which is due to the mismatch between the weak interactions and the quark mass matrix. V_{ij} describes the relative amplitude for the transition $d_j \to u_i$. Experimentally,

$$V \simeq \begin{pmatrix} \cos\theta_c & \sin\theta_c & 0 \\ -\sin\theta_c & \cos\theta_c & 0 \\ 0 & 0 & 1 \end{pmatrix} + O(\theta_c^2), \tag{7}$$

where $\sin\theta_c \simeq 0.23$ is the sine of the Cabibbo angle. For massless neutrinos there is no analogue of V in the leptonic current. Since there are no mass terms to define the neutrino flavours one can simply define ν_e as the state produced in weak transitions involving the electron, etc. For momenta small compared to M_W, weak charged current processes can be described by an effective four-fermion interaction (Fig. 2) $-L_{eff}^{CC} = \frac{G_F}{\sqrt{2}} J_W^{\mu\dagger} J_{W\mu}$, where the fermi constant G_F is given (to lowest order) by

$$G_F = \frac{\sqrt{2}g^2}{8M_W^2} = 1.16637\times10^{-5} GeV^{-2}. \tag{8}$$

The numerical value is determined from muon decay.

The standard model predictions for the weak charged current have been extensively tested in a variety of processes [8]. In particular, there have been many precise tests in the purely leptonic

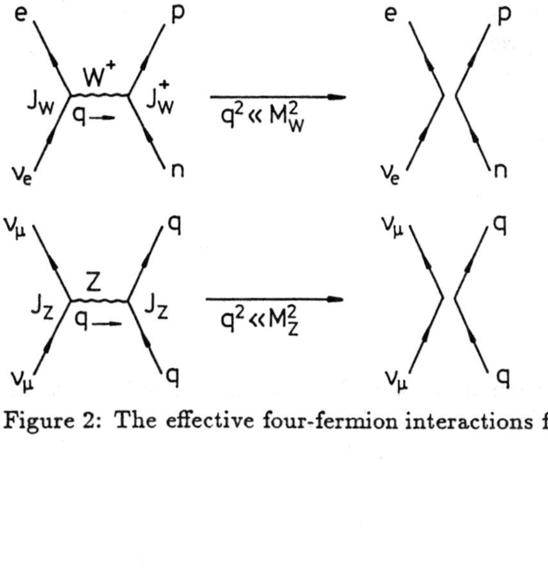

Figure 2: The effective four-fermion interactions for $\nu_e n \to e^- p$ and $\nu_\mu q \to \nu_\mu q$.

sector (which is free from any uncertainties from the strong interactions), including μ and τ decay and $\nu_\mu e \to \mu^- \nu_e$ scattering. In a recent model-independent analysis of muon decay and inverse decay data, Fetscher, Gerber, and Johnson[9] have considered the most general local derivative-free four-fermion interaction for muon decay, assuming only Lorentz invariance, separately conserved electron and muon lepton numbers [10], and massless neutrinos. They found that the data uniquely require $V - A$ couplings for the leptonic interactions (Fig. 3 and Table 2). The other invariants, involving $V + A$ as well as scalar, pseudoscalar, and tensor operators are all required to be small, with stringent limits on the coefficients of all operators except the scalar interaction involving left-chiral e and μ.

One way of seeing to what extent pure $V - A$ is required is provided by a series of measurements of polarized μ^+ decay asymmetries at TRIUMF [11]. They find that the mass of W_R, a hypothetical gauge boson coupling to right-chiral ($V + A$) current [12] in μ decay, must exceed 400 GeV, in contrast to the ordinary W (coupling to $V - A$) mass of 80.9 ± 1.4 GeV. The same results can be used [9] to infer that $1 - |h_{\nu_\mu}| < 0.0032$, where h_{ν_μ} is the helicity of ν_μ produced in $\pi_{\mu 2}$ decays. This is in striking agreement with the $V - A$ prediction [13] of $h_{\nu_\mu} = -1$.

Similarly, L^{CC}_{eff} has been extensively tested in a variety of semi-leptonic decay processes, such as β, hyperon, π, K, c, and b decays. The results are in impressive agreement with the predictions of the standard model. In particular, the $V - A$ nature of the charged current interaction and the relative strength of the various weak processes, as predicted in (5), are quantitatively confirmed. For example, from μ, β, K, hyperon, and b decays one can extract the CKM matrix elements $|V_{ud}|$, $|V_{us}|$, and $|V_{ub}|$. One finds [14]

Table 2: Limits on the branching ratios $Q^\gamma_{\epsilon\mu}$ for muon decay via scalar ($\gamma = S$), vector ($\gamma = V$), and tensor ($\gamma = T$) interactions, from Fetscher et al.[9] ϵ and μ are the chiralities of the e and μ, respectively. The $Q^\gamma_{\epsilon\mu}$ are related to the couplings in Fig. 3 by $Q^\gamma_{\epsilon\mu} = \lambda_\gamma |g^\gamma_{\epsilon\mu}|^2$, where $\lambda_S = \frac{1}{4}$, $\lambda_V = 1$, and $\lambda_T = 3$.

Quantity	Limit (90% $c.l.$)
$Q^S_{RR} + Q^V_{RR}$	< 0.002
$Q^S_{LR} + Q^V_{LR} + Q^T_{LR}$	< 0.008
$Q^S_{RL} + Q^V_{RL} + Q^T_{RL}$	< 0.04
$Q^S_{LL} + Q^V_{LL}$	> 0.95
Q^S_{LL}	< 0.21
Q^V_{LL}	> 0.79

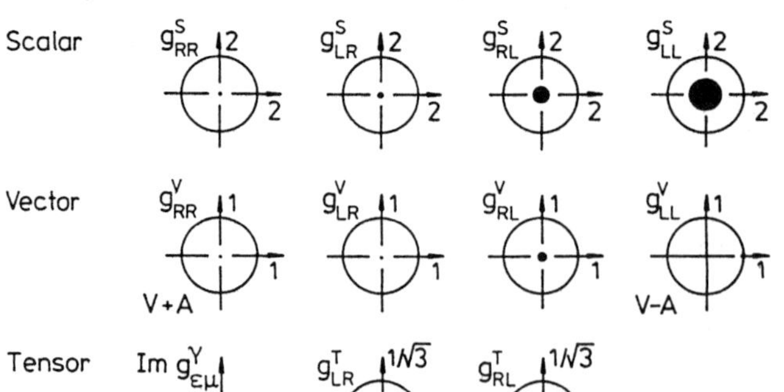

Figure 3: Values of scalar, vector, and tensor interactions in muon decay, as determined by Fetscher et al [9]. The subscripts refer to the chiralities of the e and μ, respectively.

$$|V_{ud}^2| + |V_{us}^2| + |V_{ub}^2| = 0.9979 \pm 0.0021, \tag{9}$$

in remarkable agreement with the expectation of unity from universality [15] (i.e. from the unitarity of V).

The leptonic and semi-leptonic data combined leave little room for any deviation from the standard model in charged current processes. In particular, we can be certain that neutrinos are produced by (almost) canonical $V - A$ interactions in weak decays.

The semi-leptonic charged current interaction has also been extensively tested [16] in neutrino scattering processes such as quasi-elastic $\nu_\mu n \to e^- p$ and deep inelastic $\nu_\mu N \to \mu^- X$. These processes are more useful as probes of the hadron than of the neutrino. They have been very useful in testing the QCD-improved proton model and in measuring the relative amount of u and d quarks, antiquarks, and strange quarks in the proton, as well as in determining the CKM elements V_{cd} and V_{cs}. L_{eff}^{CC} has also been qualitatively tested in $|\Delta S| = 1$ nonleptonic decays and $\Delta S = 0$ parity violating interference effects, but in these cases hadronic uncertainties obscure the interpretation of the experiments. Higher order weak effects have been semi-quantitatively tested in the $K_L - K_S$ mass difference, the CP-violating parameters ϵ and ϵ' observed in K decays [17,18], and, recently, in the $B_d^0 \leftrightarrow \bar{B}_d^0$ oscillations observed by the ARGUS collaboration [19] at DESY.

2.2 The Neutral Current

The weak neutral current interaction is

$$L = -\frac{g}{2\cos\theta_W} J_Z^\mu Z_\mu, \tag{10}$$

where

$$\begin{aligned} J_Z^\mu &= \tfrac{1}{2}\bar{u}\gamma^\mu(1+\gamma^5)u - \tfrac{1}{2}\bar{d}\gamma^\mu(1+\gamma^5)d \\ &+ \tfrac{1}{2}\bar{\nu}\gamma^\mu(1+\gamma^5)\nu - \tfrac{1}{2}\bar{e}\gamma^\mu(1+\gamma^5)e \\ &- 2\sin^2\theta_W J_{EM}^\mu \end{aligned} \tag{11}$$

(+ heavy fermion terms), and

$$J_{EM}^\mu = \sum_i q_i\ \bar{\psi}_i\gamma^\mu\psi_i = \frac{2}{3}\bar{u}\gamma^\mu u - \frac{1}{3}\bar{d}\gamma^\mu d - \bar{e}\gamma^\mu e + \cdots \tag{12}$$

is the (purely vector) electromagnetic current. Z couples to both left and right-chiral fermions, but with different strength. For low momenta compared to M_Z, (10) implies the effective four-fermion interaction

$$-L_{eff}^{NC} = \frac{G_F}{\sqrt{2}} J_Z^\mu J_{Z\mu}, \tag{13}$$

The neutral current interaction has been observed and quantitatively tested in a wide variety of weak processes, including deep inelastic $\overset{(-)}{\nu}_\mu N$ scattering from isoscalar and proton targets, elastic $\overset{(-)}{\nu}_\mu p$ scattering, coherent $\nu N \to \nu\pi^0 N$ scattering, elastic $\overset{(-)}{\nu}_i e$ ($i = e, \mu$) scattering, and $e^+e^- \to$ hadrons. In addition, weak-electromagnetic interference has been studied in polarized eD and μC scattering, atomic parity violation, and forward-backward asymmetries in $e^+e^- \to e^+e^-$, $\mu^+\mu^-$, $\tau^+\tau^-$, $c\bar{c}$, and $b\bar{b}$. All processes are in excellent agreement with the standard model predictions, as can be seen in Fig. 4 and Table 3. Combined with the W and Z masses the standard model is quantitatively confirmed over an enormous momentum range, $10^{-6}\ GeV^2 < |Q^2| < 10^4\ GeV^2$. It is almost certainly correct to first approximation.

Let us now examine the neutral current interactions of neutrinos in more detail. It is convenient to write the terms in $-L_{eff}^{NC}$ relevant to ν-hadron processes in a form that is valid in an arbitrary gauge theory (assuming massless left-handed neutrinos). One has

$$-L^{\nu H} = \frac{G_F}{\sqrt{2}}\bar{\nu}\gamma^\mu(1+\gamma^5)\nu \left\{\sum_i [\epsilon_L(i)\,\bar{q}_i\gamma_\mu(1+\gamma^5)q_i + \epsilon_R(i)\,\bar{q}_i\gamma_\mu(1-\gamma^5)q_i]\right\}, \quad (14)$$

where in the standard model [20]

$$\begin{aligned} \epsilon_L(u) &= \frac{1}{2} - \frac{2}{3}\sin^2\theta_W \\ \epsilon_L(d) &= -\frac{1}{2} + \frac{1}{3}\sin^2\theta_W \\ \epsilon_R(u) &= -\frac{2}{3}\sin^2\theta_W \\ \epsilon_R(d) &= +\frac{1}{3}\sin^2\theta_W \end{aligned} \quad (15)$$

It is also convenient to define the variables

$$\begin{aligned} g_L^2 &\equiv \epsilon_L(u)^2 + \epsilon_L(d)^2 \simeq \frac{1}{2} - \sin^2\theta_W + \frac{5}{9}\sin^4\theta_W \\ g_R^2 &\equiv \epsilon_R(u)^2 + \epsilon_R(d)^2 \simeq \frac{5}{9}\sin^4\theta_W, \end{aligned} \quad (16)$$

and

$$\theta_i \equiv \tan^{-1}(\epsilon_i(u)/\epsilon_i(d)),\ i = L \text{ or } R \quad (17)$$

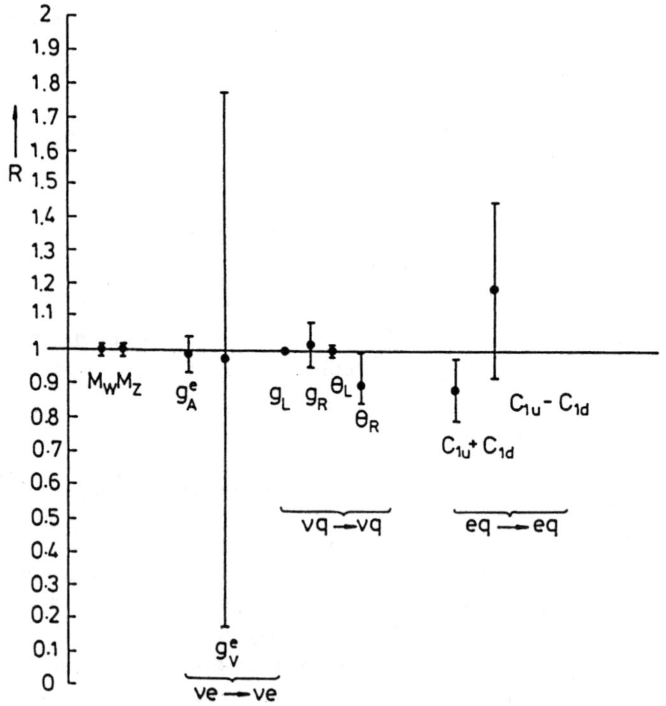

Figure 4: Experimental values of the W and Z masses and the neutral current couplings, relative to the standard model predictions for the global best fit value $\sin^2\theta_W = 0.230$ (the value of g_L should be regarded as the major determinant of $\sin^2\theta_W$ rather than a prediction). $C_{1i}, i = u, d$ are the coefficients in $-L_{eff}^{NC}$ of the parity-violating eq interaction $\frac{G_F}{\sqrt{2}}\bar{e}\gamma_\mu\gamma^5 e\bar{q}_i\gamma^\mu q_i$. The other quantities are defined in the text. The error bars on g_V^e are large only because the predicted value (-0.045) is so small.

Table 3: Values of the model independent neutral current parameters, compared with the standard model prediction for $\sin^2\theta_W = 0.230$. Correlations are not given for the neutrino-hadron couplings because of the non-Gaussian χ^2 distributions. However, the neutrino-hadron constraints are accurately represented by the ranges of the variable g_i^2 and θ_i, $i = L, R$, which are very weakly correlated.

Quantity	Experimental Value	Standard Model Prediction	Correlation	
$\epsilon_L(u)$	$0.339 \pm .017$	0.345		
$\epsilon_L(d)$	$-0.429 \pm .014$	-0.427		
$\epsilon_R(u)$	$-0.172 \pm .014$	-0.152		
$\epsilon_R(d)$	$-0.011^{+.081}_{-.057}$	0.076		
g_L^2	0.2996 ± 0.0044	0.301		
g_R^2	0.0298 ± 0.0038	0.029		
θ_L	2.47 ± 0.04	2.46		
θ_R	$4.65^{+0.48}_{-0.32}$	5.18		
g_A^e	$-0.498 \pm .027$	-0.503		-0.08
g_V^e	$-0.044 \pm .036$	-0.045		
C_{1u}	-0.249 ± 0.071	-0.191	-0.98	-0.88
C_{1d}	0.381 ± 0.064	0.340		0.88
$C_{2u} - \frac{1}{2}C_{2d}$	0.19 ± 0.37	-0.039		

At present the most precise determinations of $\sin^2\theta_W$ are from deep inelastic neutrino scattering from (approximately) isoscalar targets. The ratio $R_\nu \equiv \sigma_{\nu N}^{NC}/\sigma_{\nu N}^{CC}$ of neutral to charged current cross sections has been measured to 1% accuracy by the CDHS [21] and CHARM [22] collaborations, so it is important to obtain theoretical expressions for R_ν and $R_{\bar\nu} \equiv \sigma_{\bar\nu N}^{NC}/\sigma_{\bar\nu N}^{CC}$ (as functions of $\sin^2\theta_W$) to comparable accuracy. Fortunately, most of the uncertainties concerning the strong interactions (as well as neutrino spectra) cancel in the ratio. For neutral current parameters in the vicinity of the standard model $\simeq 90\%$ of R_ν can be predicted from isospin alone [23]. The remaining 10% (from such effects as quark mixing and the s sea) is strongly constrained by independent measurements involving deep inelastic e, μ, and charged-current ν scattering, including dimuon production, and can be estimated to the necessary (10%) accuracy.

A simple zeroth order approximation (ignoring quark mixing, the s and c sea, and certain tiny higher twist effects) is

$$\begin{aligned} R_\nu &= g_L^2 + g_R^2 r \\ R_{\bar\nu} &= g_L^2 + \frac{g_R^2}{r}, \end{aligned} \tag{18}$$

where $r \equiv \sigma_{\bar\nu N}^{CC}/\sigma_{\nu N}^{CC}$ is the ratio of $\bar\nu$ and ν charged current cross sections, which can be measured directly. (In the simple parton model, ignoring hadron energy cuts, $r \simeq (\frac{1}{3}+\epsilon)/(1+\frac{1}{3}\epsilon)$, where $\epsilon \sim 0.125$ is the ratio of the fraction of the nucleon's s momentum carried by antiquarks to that carried by quarks. i.e. $\epsilon \equiv (\bar U + \bar D)/(U+D)$, where $U \equiv \int_0^1 xu(x)dx$ is the first moment of the u quark distribution.) In practice, (18) must be corrected for quark mixing, the s and c seas, c quark threshold effects (which mainly affect σ^{CC} - these turn out to be the largest theoretical uncertainty), non-isoscalar target effects, $W - Z$ propagator differences, and radiative corrections (which lower the extracted value of $\sin^2\theta_W$ by ~ 0.009.). Details of the neutrino spectra, experimental cuts, x

and Q^2 dependence of structure functions, and longitudinal structure functions enter only at the level of these corrections and therefore lead to very small uncertainties. Altogether, the theoretical uncertainty is $\Delta\sin^2\theta_W \sim \pm 0.005$, which would be very hard to improve in the future.

There are also a number of measurements [24] of deep inelastic $\overset{(-)}{\nu}_\mu$ scattering from non-isoscalar targets, which are useful for determining the isospin structure of the neutral current interaction. [25] The most recent result (from BEBC [26]) determines the ratio of neutral to charged current cross sections to around 7% accuracy for both ν_μ and $\bar{\nu}_\mu$.

The differential cross sections for elastic $\overset{(-)}{\nu}_\mu p \rightarrow \overset{(-)}{\nu}_\mu p$ scattering have been precisely measured in the BNL E734 experiment [27]. Four groups [24] have measured the cross section for coherent $\nu N \rightarrow \nu \pi^0 N$, for which the hadronic matrix elements can be estimated fairly reliably [28] using PCAC.

From these results [29] the neutrino-hadron couplings can be determined uniquely and (for the left-handed couplings) precisely. The extracted couplings, shown in Fig. 5 and Table 3, are in impressive agreement with the standard model predictions.

Similarly, for an arbitrary gauge theory with massless left-handed neutrinos, the four-fermion interaction for $\overset{(-)}{\nu}_\mu e$ scattering is

$$-L^{\nu e} = \frac{G_F}{\sqrt{2}} \bar{\nu}_\mu \gamma^\mu (1+\gamma^5) \nu_\mu \, \bar{e} \gamma_\mu (g_V^e + g_A^e \gamma^5) e \tag{19}$$

(for $\overset{(-)}{\nu}_e e$ the charged current contribution must be included). In the standard model

$$\begin{aligned} g_V^e &= -\tfrac{1}{2} + 2\sin^2\theta_W \\ g_A^e &= -\tfrac{1}{2}, \end{aligned} \tag{20}$$

up to radiative corrections.

The laboratory cross section for $\overset{(-)}{\nu}_\mu e \rightarrow \overset{(-)}{\nu}_\mu e$ elastic scattering is

$$\frac{d\sigma_{\nu_\mu,\bar{\nu}_\mu}}{dy} = \frac{G_F^2 m_e E_\nu}{2\pi} \left[(g_V^e \pm g_A^e)^2 + (g_V^e \mp g_A^e)^2 (1-y)^2 - (g_V^{e2} - g_A^{e2}) \frac{y m_e}{E_\nu} \right], \tag{21}$$

where the upper (lower) sign refers to $\nu_\mu(\bar{\nu}_\mu)$, $y \equiv T_e/E_\nu$ (which runs from 0 to $(1+\frac{m_e}{2E_\nu})^{-1}$) is the ratio of the kinetic energy of the recoil electron to the incident $\overset{(-)}{\nu}$ energy, and $G_F^2 m_e/2\pi = 4.31 \times 10^{-42}\ cm^2/GeV$. For $E_\nu \gg m_e$ this yields a total cross section

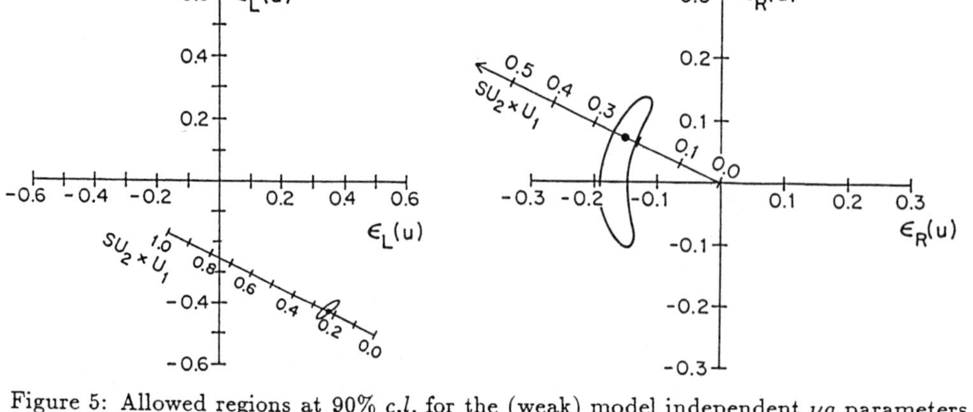

Figure 5: Allowed regions at 90% *c.l.* for the (weak) model independent νq parameters $\epsilon_i(u)$ and $\epsilon_i(d)$, $i = L$ or R and the predictions of the standard model as a function of $\sin^2\theta_W$.

$$\sigma = \frac{G_F^2 m_e E_\nu}{2\pi}\left[(g_V^e \pm g_A^e)^2 + \frac{1}{3}(g_V^e \mp g_A^e)^2\right]$$
$$\sim \frac{G_F^2 m_e E_\nu}{2\pi}\begin{cases} 1 - 4\sin^2\theta_W + \frac{16}{3}\sin^4\theta_W, & \nu_\mu e \\ \frac{1}{3} - \frac{4}{3}\sin^2\theta_W + \frac{16}{3}\sin^4\theta_W, & \bar{\nu}_\mu e \end{cases} \quad (22)$$

The most accurate leptonic measurements [30,31] of $\sin^2\theta_W$ are from the ratio $R \equiv \sigma_{\nu_\mu e}/\sigma_{\bar{\nu}_\mu e}$, in which many of the systematic uncertainties cancel. Radiative corrections, which are small compared to the precision of present experiments, increase the extracted $\sin^2\theta_W$ by $\simeq 0.002$.

The $\bar{\nu}_e e$ cross section was measured a decade ago at the Savannah River reactor [32], while $\nu_e e \to \nu_e e$ has been measured recently at Los Alamos [33]. These are not nearly so precise as the $\overset{(-)}{\nu}_\mu e$ measurements, but are interesting because they involve both neutral and charged current contributions. (The cross sections for $\overset{(-)}{\nu}_e e$ may be obtained from (21) by replacing $g_{V,A}^e$ by $g_{V,A}^{e\prime} \equiv g_{V,A}^e + 1$, where the 1 is due to the charged current.)

In fact, the Los Alamos result strongly supports destructive interference ($g_A^e < 0$) between the two amplitudes and rules out constructive interference ($g_A^e > 0$).

The results of the various reactions [5] for the νe couplings are shown in Fig. 6. The $\bar{\nu}_\mu e$ data alone allow four solutions (which differ by $g_i^e \leftrightarrow -g_i^e$ and $g_V^e \leftrightarrow g_A^e$). The reactor $\bar{\nu}_e e$ results eliminate C, while the Los Alamos $\nu_e e$ experiment eliminates solutions C and D. The remaining two solutions (axial dominant (A) and vector dominant (B)) are consistent with all νe data. However, solution (B) is eliminated by the $e^+e^- \to \mu^+\mu^-$ forward-backward asymmetry under the (now very reasonable) assumption that the neutral current is dominated by the exchange of a single Z. The remaining solution (A) is in excellent agreement with the standard model prediction, as can be seen in Table 3.

The ν - hadron and νe interactions are therefore uniquely determined and are consistent with the standard model within uncertainties. Similar statements hold for the e-hadron and e^+e^- couplings [5]. Having established the standard model couplings as correct to first approximation, the neutral current and boson mass results can be used to test the standard model more stringently and to set limits on possible new physics.

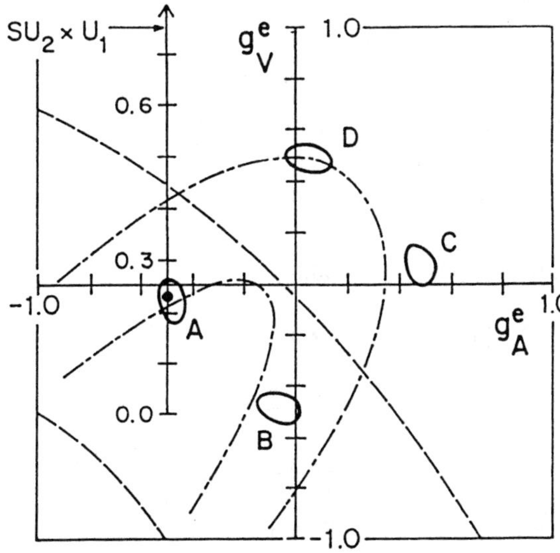

Figure 6: Allowed regions (90% c.l.) for the νe parameters g_V^e and g_A^e, for $\bar{\nu}_\mu e$ (solid lines), reactor $\bar{\nu}_e e$ (dot-dash), and $\nu_e e$ (dash).

The values of $\sin^2\theta_W$ and, equivalently, M_Z (using (3)) determined from various processes are shown in Table 4 and Fig. 7. They are in impressive agreement with each other, reconfirming the quantitative success of the standard model. The best fit to all data yields [34] $\sin^2\theta_W = 0.230 \pm 0.0048$ and $M_Z = 92.0 \pm 0.7\ GeV$, where the errors include full statistical, systematic, and theoretical uncertainties.

As can be seen in Fig. 7 consistency of the various $\sin^2\theta_W$ values (especially those obtained from deep inelastic νN and the W, Z masses) depends sensitively on the top quark mass, which enters the radiative corrections. In fact, one can use these results to set an upper limit [5] $m_t < 200\ GeV$ (90% *c.l.*), with similar limits applying to the splitting between the masses of possible fourth generation fermions. Similarly, the deep inelastic neutrino data can be combined with the W and Z masses to determine Δr in (3). One finds [5] $\Delta r = 0.077 \pm 0.037$, in excellent agreement with the value 0.0713 ± 0.0013 predicted for $m_t = 45\ GeV$ and $M_H = 100\ GeV$, and providing a rough test of the theory at the level of radiative corrections (see also Table 1).

The best fit value of $\sin^2\theta_W \equiv 1 - \frac{M_W^2}{M_Z^2}$ corresponds to the modified minimal subtraction value [35]

$$\sin^2\hat{\theta}_W(M_W) = 0.228 \pm 0.0044 \tag{23}$$

This is larger by $\simeq 2.5\ \sigma$ than the prediction $0.214^{+0.003}_{-0.004}$ of minimal SU_5 (for $\Lambda^{(4)}_{\overline{MS}} = 150^{+150}_{-75}\ MeV$) and other "great desert" models. Similar conclusions hold for all values of m_t and M_H, as can be seen in Fig. 8. Of course, the simplest grand unified theories (GUTs) have been excluded for some time by the nonobservation of proton decay [36], but the additional evidence is welcome, especially since variations on the simplest GUTs can yield much longer lifetimes.

The fact that the $\sin^2\hat{\theta}_W(M_W)$ value in (23) is close to but not identical with the SU_5 prediction can be taken as a hint that the basic ideas of GUTs may be roughly correct, but that there is additional structure in the desert. For example, (23) is closer to (but still somewhat below)

Table 4: Determination of $\sin^2\theta_W$ and M_Z (in GeV) from various reactions. The central values of all fits assume $m_t = 45\ GeV$ and $M_H = 100\ GeV$ in the radiative corrections. Where two errors are shown the first is experimental and the second (in square brackets) is theoretical, computed assuming 3 fermion families, $m_t < 100\ GeV$, and $M_H < 1\ TeV$. In the other cases the theoretical and experimental uncertainties are combined.

Reaction	$\sin^2\theta_W$	M_Z
Deep inelastic (isoscalar)	$0.233 \pm .003 \pm [.005]$	$91.6 \pm 0.4 \pm [0.8]$
$\overset{(-)}{\nu}_\mu p \to \overset{(-)}{\nu}_\mu p$	$0.210 \pm .033$	95.0 ± 5.2
$\overset{(-)}{\nu}_\mu e \to \overset{(-)}{\nu}_\mu e$	$0.223 \pm .018 \pm [.002]$	93.0 ± 2.7
W, Z	$0.228 \pm .007 \pm [.002]$	92.3 ± 1.1
Atomic parity violation	$0.209 \pm .018 \pm [.014]$	95.1 ± 3.9
SLAC eD	$0.221 \pm .015 \pm [.013]$	93.3 ± 2.7
μC	$0.25 \pm .08$	89.6 ± 9.7
All data	0.230 ± 0.0048	92.0 ± 0.7

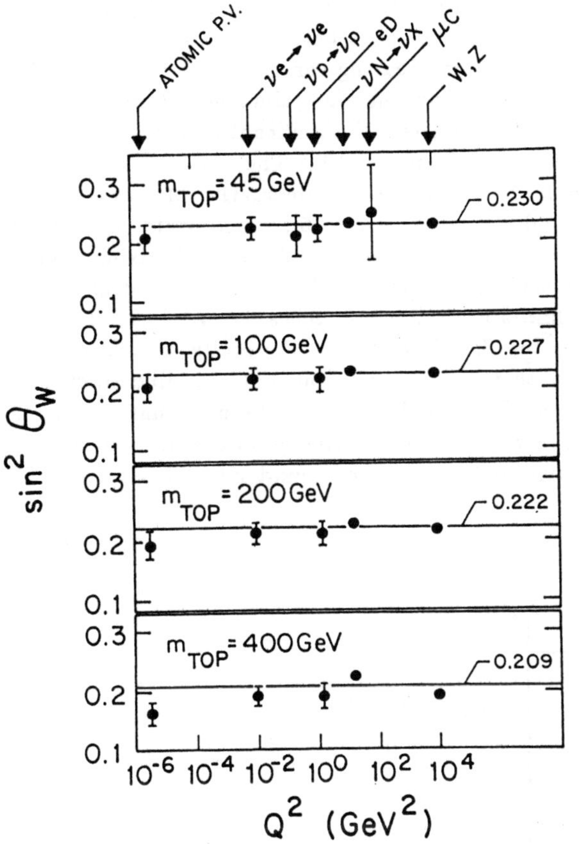

Figure 7: (a) $\sin^2\theta_W$ for various reactions as a function of the typical Q^2, determined for $m_t = 45\ GeV$. The best fit line $\sin^2\theta_W = 0.230$ is also shown. (b-d) $\sin^2\theta_W$ values determined for $m_t = 100,\ 200$, and $400\ GeV$.

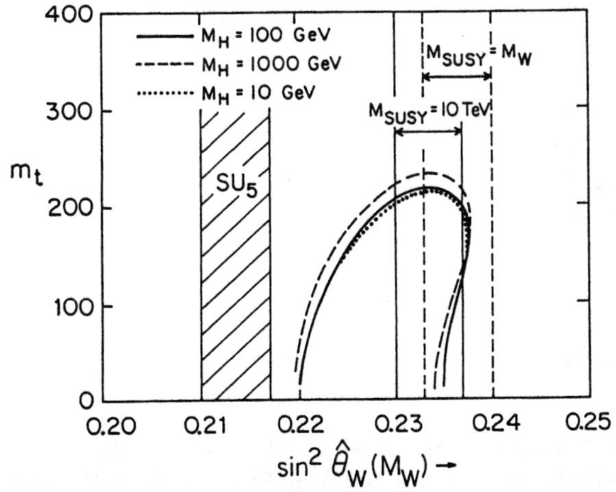

Figure 8: Allowed regions (90% *c.l.*) in $\sin^2\hat{\theta}_W(M_W)$and m_t for fixed values of M_H. Also shown are the predictions of ordinary and supersymmetric GUTs, assuming no new thresholds between M_W or M_{SUSY} and the unification scale.

the prediction of the simplest supersymmetric GUTs. (Typically $0.237^{+0.003}_{-0.004}$ for $M_{SUSY} \sim M_W$, decreasing by ~ 0.003 for $M_{SUSY} \sim 10\ TeV$). The agreement is better for larger m_t (Fig. 8). Similarly, SO_{10} models [36] with three stages of symmetry breaking can be compatible with (23).

The neutral current data can be used to place rather stringent constraints on certain deviations from the standard model, such as the existence of Higgs triplets with significant vacuum expectation values [5], or the mixing between ordinary and exotic fermions [10]. The $e^+e^- \to b\bar{b}$ forward-backward asymmetry [37] excludes all topless models not involving exotic quarks. Many extensions of the standard model predict the existence of additional Z bosons [5], which could conceivably be light enough to be experimentally relevant. Some limits on the masses M_2 and mixing angle θ between the new and ordinary Z are shown for a class of E_6 models in Fig. 9. These neutral current limits are somewhat more stringent [38] than limits from direct searches $\bar{p}p \to Z_2 + X$, $Z_2 \to l^+l^-$ at the $Sp\bar{p}S$ except for a small region in β near the Z_η. Nevertheless, the limits (typically $120 - 300\ GeV$) are still relatively weak. In constrast, there is a non-rigorous but plausible lower limit [39] from the $K_L - K_S$ mass difference of several TeV on the mass of the new charged bosons in many $SU_{2L} \times SU_{2R} \times U_1'$ models. This situation will presumably change in the near future: for example, the FNAL $\bar{p}p$ collider should be sensitive to bosons up to around 400 GeV and the SSC would be sensitive up to several TeV.

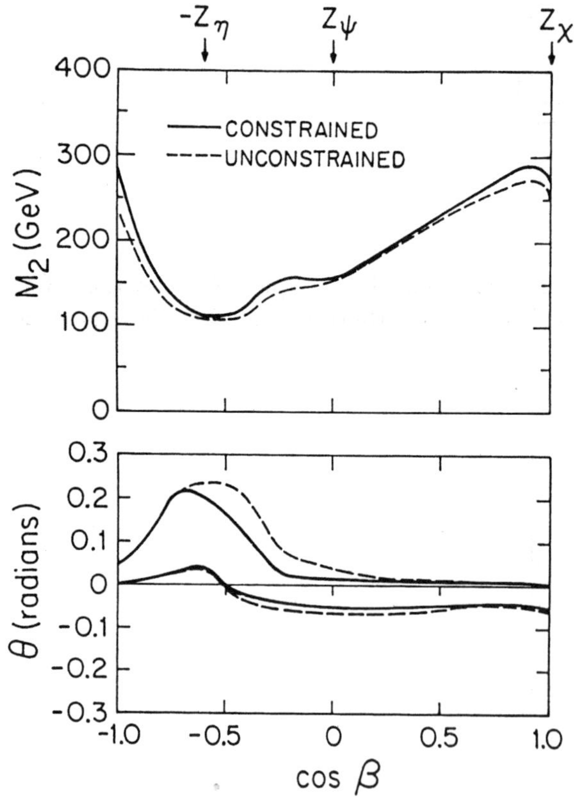

Figure 9: Lower limits on M_2 and allowed θ range (both at 90% $c.l.$) for an E_6 boson $Z(\beta) = \cos\beta\ Z_\chi + \sin\beta\ Z_\psi$, where Z_χ and Z_ψ refer to the breaking patterns $SO_{10} \to SU_5 \times U_{1\chi}$ and $E_6 \to SO_{10} \times U_{1\psi}$, respectively, and $Z_\eta = -Z(\pi - \tan^{-1}\sqrt{\frac{5}{3}})$ occurs in many superstring models. Constrained and unconstrained refer to whether or not it is assumed that SU_2 breaking is due to Higgs doublets only.

3 Neutrino Counting

Table 5: Limits on the number N_ν of neutrino flavors and the mass ranges to which they apply. The laboratory limits are at 90% *c.l.*

N_ν	mass range	source	reference
$N_\nu \geq 2$	–	direct	
$N_\nu \geq 3$	–	τ properties	
$N_\nu \leq 4$	$m_\nu < 1\ MeV$	nucleosynthesis	[40]
$N_\nu \leq 6$	$m_\nu < O(MeV)$	SN1987A energetics	[41]
$N_\nu \leq \begin{cases} 7.5 \text{ (ASP)} \\ 4.9 \text{ (combined)} \end{cases}$	$m_\nu < O(5\ GeV)$	$e^+e^- \not\to \gamma\nu\bar{\nu}$	[42]
$N_\nu \leq \begin{cases} 5,\ m_t < 40\ GeV \\ 3,\ m_t > 50\ GeV \end{cases}$	$m_\nu < O(40\ GeV)$	R	[43]

Constraints on the number of neutrino flavors are listed in Table 5.

There is direct laboratory proof for the existence of only two neutrinos, ν_e and ν_μ. However, indirect evidence leaves little doubt as to the separate existence of the ν_τ. If there were no ν_τ then, up to mixing effects, the τ_L^- would have to be a singlet under SU_2 transformations. Including mixing, the two left-handed lepton doublets and one charged singlet would be

$$\begin{pmatrix} \nu_1 \\ U_{1i}e_i^- \end{pmatrix}_L \begin{pmatrix} \nu_2 \\ U_{2i}e_i^- \end{pmatrix}_L U_{3i}e_{iL}^- \tag{24}$$

where $(e_1, e_2, e_3)_L \equiv (e, \mu, \tau)_L$ and U is a unitary matrix. However, one knows that the μ and e weak interactions are canonical - there is little room for mixing with an SU_2 singlet. From μ, β, K, and hyperon decays and the W mass one can show [10]

$$|U_{13}|, |U_{23}| < 0.05 \tag{25}$$

(this is confirmed by the absence of $\tau^- \to \mu^-\mu^-\mu^+$ decays). On the other hand, the τ lifetime [44] $\tau_\tau = (3.07 \pm 0.09) \times 10^{-13}$ sec, which agrees at least roughly with the value $(2.87 \pm 0.06) \times 10^{-13}$sec expected if ν_τ exists, implies

$$|U_{13}|^2 + |U_{23}|^2 = 0.94 \pm 0.04, \tag{26}$$

in clear conflict with (25). An independent argument is that A^τ, the axial vector coupling of τ in the weak neutral current, is determined from the e^+e^- forward-backward asymmetry to be $A^\tau = -0.46 \pm 0.05$. This is in agreement with the value $-\frac{1}{2}$ expected if the τ_L^- is in an SU_2 doublet with its own partner (ν_τ), and disagrees with the value (zero) expected if τ_L^- and τ_R^- are in SU_2 singlets. [45] Hence, the ν_τ almost certainly exists, but it would nevertheless be desirable to observe it directly.

There are several upper limits on the number of neutrinos with normal weak interactions. An upper limit of $N_\nu \leq 4$ neutrino flavors with masses $\leq 1\ MeV$ is determined by nucleosynthesis [40] (the abundance primordial ^4He). Extra neutrino flavors [46] would cause the universe to expand faster, causing the $\nu_e n \leftrightarrow e^- p$ reactions to freeze out earlier (when there are more neutrons), leading to too much ^4He.

Limits can also be set from the cross section for $e^+e^- \to \gamma\nu\bar{\nu}$ (with only the photon observed), which effectively sums the number of neutrinos. The ASP experiment at PEP obtained [42] $N_\nu < 7.5$. Combined with cross section limits from MAC and CELLO this implies $N_\nu < 4.9$ (90% *c.l.*) sensitive to masses less than several GeV. Finally, the Z width increases by 170 MeV for each new neutrino with mass $\leq 40\ GeV$. Indirect limits on Γ_Z already exist from the ratio

$$R = \frac{\sigma_{\bar{p}p \to W} B_{W \to l\nu}}{\sigma_{\bar{p}p \to Z} B_{Z \to l^+l^-}} = \frac{\sigma_{\bar{p}p \to W}}{\sigma_{\bar{p}p \to Z}} \frac{\Gamma_{W \to l\nu}}{\Gamma_{Z \to l^+l^-}} \frac{\Gamma_Z}{\Gamma_W}. \tag{27}$$

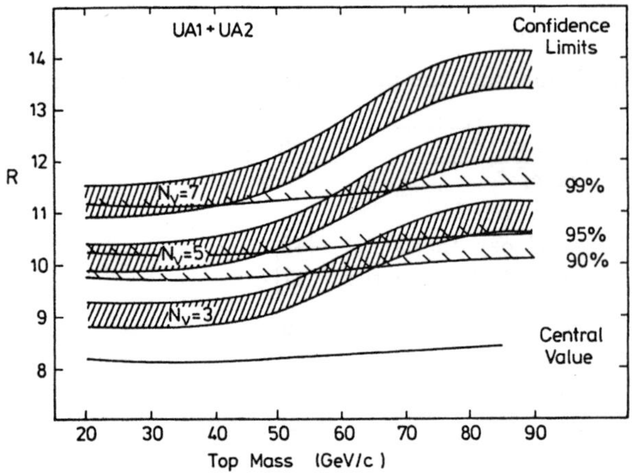

Figure 10: The value of R (27) as a function of N_ν and m_t, and the experimental results form UA1 and UA2.

Using the measured R and theoretical values for the cross section ratio and leptonic widths one determines Γ_Z/Γ_W, which is sensitive to both N_ν and the t quark mass. Recent estimates [43] typically yield $N_\nu \leq 5$ for $m_t \leq 40\ GeV$ and $N_\nu \leq 3$ for $m_t \geq 50\ GeV$, (the larger m_t range is favored by $B-\bar{B}$ oscillations [19] and the non-observation [47] of the t by UA1), and incidentally suggest the upper limit $m_t \leq 65\ GeV$. These limits are suggestive but should be viewed with caution. As can be seen in Fig. 10 the bounds essentially disappear if one increases the uncertainty in either R itself or the cross section ratio.

Future direct measurements of Γ_Z at SLC and LEP should ultimately yield a precision of $\Delta\Gamma_Z \simeq 35\ MeV$, which is equivalent to an uncertainty [48] of $\Delta N_\nu \sim 0.2$. It should be possible to obtain an independent measurement of $\Gamma_{Z\to\nu\bar{\nu}}$ accurate to $\simeq 50\ MeV$ by measuring $e^+e^-\to\gamma Z\to\gamma\nu\bar{\nu}$ above the Z pole.

4 Neutrino Mass

In the minimal $SU_2 \times U_1$ model the neutrinos are predicted to be massless. However, extensions of the standard model involving new SU_2-singlet neutral fermions (the right-handed neutrino partners needed for Dirac mass terms) or new Higgs representations (to generate Majorana masses) allow non-zero masses. [49] In fact, most extensions of the standard model (e.g. most grand unified theories other than SU_5) involve one or both of these mechanisms. Furthermore, non-zero masses could have important implications for the missing Solar neutrinos and/or the missing (dark) matter of the universe.

4.1 Weyl, Majorana, and Dirac Neutrinos.

For the weak interactions it is convenient to deal with Weyl two-component spinors ψ_L or ψ_R, each of which represents two physical degrees of freedom. The field ψ_L can annihilate a left-handed (L) particle or create a right-handed (R) antiparticle, while $\psi_L^\dagger$ annihilates a L-particle or creates an R-antiparticle. For a ψ_R field the roles of L and R are reversed. An ordinary four-component Dirac field ψ can be written as the sum $\psi = \psi_L + \psi_R$ of two Weyl fields, where ψ_L and ψ_R are just the chiral projections

$$\psi_{L,R} = P_{L,R}\psi, \tag{28}$$

with $P_{L,R} = (1 \pm \gamma_5)/2$.

Alternatively, one can consider Weyl fermions that do not have distinct partners of the opposite chirality. We will see below that such spinors correspond to particles that are either massless or carry no conserved quantum numbers.

In the free field limit a Weyl field ψ_L can be written as

$$\psi_L(x) = \sum_{\vec{p}} \left[b_L(\vec{p}) u_L(\vec{p}) e^{-ip\cdot x} + d_R^\dagger(\vec{p}) v_R(\vec{p}) e^{+ip\cdot x} \right], \tag{29}$$

where $\sum_{\vec{p}}$ represents $\int d^3\vec{p}/\sqrt{(2\pi)^3 2E}$. In(29), b_L and d_R are annihilation operators for L particles and R-antiparticles, respectively, and u_L and v_R are the corresponding (4-component) spinors satisfying $P_L u_L = u_L$, $P_L v_R = v_R$, $P_R u_L = P_R v_R = 0$. For a ψ_R spinor one simply interchanges L and R. Equation (29) differs from an ordinary (Dirac) free field in that there is no sum over spin.

It is apparent from (29) that each left-handed (right-handed) particle is necessarily associated with a right-handed (left-handed) antiparticle. The right-handed antiparticle [50] field ψ_R^c is not independent of ψ_L, but is closely related to $\psi_L^\dagger$. One has

$$\psi_R^c = C\bar{\psi}_L^T, \tag{30}$$

where C is the charge conjugation matrix, defined by $C\gamma_\mu C^{-1} = -\gamma_\mu^T$. Similarly, for a R-Weyl spinor, $\psi_L^c = C\bar{\psi}_R^T$. In the special case that ψ_L is the chiral projection $P_L\psi$ of a Dirac field ψ, ψ_R^c is just the R-projection $P_R\psi^c$ of the antiparticle field $\psi^c = C\bar{\psi}^T$.

If ψ_R and ψ_R^c both exist, they have the opposite values for all additive quantum numbers. Since the quarks and charged leptons carry conserved quantum numbers (e.g. color and electric charge), they must be Dirac fields - i.e. ψ_R and ψ_R^c must be distinct. The only quantum number associated with the neutrinos is lepton number, however, and it is conceivable that that is violated in nature. As we will see, that will allow for two very different possibilities for neutrino mass.

The known neutrinos of the first family are the left-handed electron neutrino ν_{eL} and its CP partner, the right-handed "antineutrino" $\nu_{eR}^c = C\bar{\nu}_{eL}^T$. These are associated with the e_L^- and e_R^+, respectively, in ordinary charged current weak interactions.

Mass terms always take left- and right-handed fields into each other. If one introduces a new field N_R (distinct from ν_R^c) and its CP conjugate $N_L^c = C\bar{N}_R^T$ into the theory, then one can write a Dirac (lepton number conserving) mass term

$$-L_{Dirac} = m_D \bar{\nu}_L N_R + h.c., \tag{31}$$

which connects N_R and ν_L. In this case ν_L, N_R, N_L^c and ν_R^c form a four component Dirac particle - i.e. one can define $\nu \equiv \nu_L + N_R$, $\nu^c = N_L^c + \nu_R^c = C\bar{\nu}^T$, so that

$$-L_{Dirac} = m_D \bar{\nu}\nu. \tag{32}$$

Clearly lepton number is conserved in this case, because there is no transition between ν and ν^c. In the free field limit the Dirac neutrino field ν has the canonical expression

$$\nu_{Dirac}(x) = \sum_{\vec{p}} \sum_{S=L,R} \left[b_S(\vec{p}) u_S(\vec{p}) e^{-ip\cdot x} + d_S^\dagger(\vec{p}) v_S(\vec{p}) e^{+ip\cdot x} \right], \tag{33}$$

Usually, the N_R is an $SU_2 \times U_1$ singlet, with m_D generated by an ordinary Higgs doublet, and $L = L_e + L_\mu + L_\tau$ is conserved in the three family generalization. This possibility is most similar to the way in which masses are generated for the other fermions (e^-, u, d, etc.) in the standard model, but it is difficult to understand why m_{ν_e} is so small in this case.

Another possibility [51] is that N_R is a known doublet neutrino, such as $\nu_{\tau R}^c$. This is a variation on the Konopinski-Mahmoud model. [52] Then ν_{eL}, $\nu_{\tau R}^c$, $\nu_{\tau L}$ and ν_{eR}^c can be combined to form a Dirac neutrino with $L_e - L_\tau$ conserved.

For the generalization of (31) to F fermion families one has

$$- L_{Dirac} = \bar{n}_L^0 m_D N_R^0 + h.c., \tag{34}$$

where m_D is an arbitrary [53] $F \times F$ mass matrix, and n_L^0 and N_R^0 are F-component vectors; thus $n_L^0 = (n_{1L}^0 \, n_{2L}^0 \ldots n_{FL}^0)^T$, where n_{iL}^0 are the "weak eigenstate" neutrinos - i.e. n_{iL}^0 is associated with e_{iL}^- in weak transitions. The weak eigenstate neutrinos are related to the neutrinos n_{iL}, N_{iR} of definite mass by unitary transformations

$$\begin{aligned} n_L^0 &= V_L n_L \\ N_R^0 &= V_R N_R. \end{aligned} \tag{35}$$

V_L and V_R are $F \times F$ unitary matrices, determined by

$$V_L^\dagger m_D V_R = m_d = \mathrm{diag}(m_1 \, m_2 \cdots m_F) \tag{36}$$

where m_d is the diagonal matrix of physical neutrino masses. V_L and V_R can be determined by

$$V_L^\dagger m_D m_D^\dagger V_L = V_R^\dagger m_D^\dagger m_D V_R = m_d^2 \tag{37}$$

($m_D m_D^\dagger$ and $m_D^\dagger m_D$ are Hermitian). In general V_L and V_R are unrelated. If there are no degeneracies then V_L and V_R are determined uniquely by (37) up to diagonal phase matrices; i.e. if $V_{L,R}$ satisfy (37) then so do $V_{L,R}K_{L,R}$, where $K_{L,R}$ are diagonal phase matrices associated with the unobservable phases of the n_{iL} and N_{jR} fields. Usually one chooses K_L to put V_L into a simple conventional form. Then K_R is determined by the requirement that m_d be real.

V_L modifies the leptonic weak charged current in (5) to [54]

$$J_W^{\mu\dagger} = (\bar{\nu}_e \, \bar{\nu}_\mu \, \bar{\nu}_\tau) \; V_L^\dagger \gamma^\mu (1 + \gamma^5) \begin{pmatrix} e^- \\ \mu^- \\ \tau^- \end{pmatrix} \tag{38}$$

so that $V_L^\dagger$ is just the analogue of the CKM quark mixing matrix. It describes the relative strengths [55] of the weak transition between the various charged leptons and neutrinos of definite mass.

In a Majorana (lepton number violating) mass term one avoids the need for a new fermion field by coupling the ν_L to its CP conjugate ν_R^c:

$$\begin{aligned} -L_M &= \tfrac{1}{2} m \bar{\nu}_L \nu_R^c + h.c. \\ &= \tfrac{1}{2} m \bar{\nu}_L C \bar{\nu}_L^T + h.c. \end{aligned} \tag{39}$$

L_M can be thought of as creating or annihilating two neutrinos, and violates lepton number by $\Delta L = \pm 2$. ν_L and ν_R^c can be combined to form a two component Majorana neutrino $\nu = \nu_L + \nu_R^c$, so that $-L_M = \frac{1}{2} m \bar{\nu} \nu$. From (30) we see that $\nu = C\bar{\nu}^T$, i.e. a Majorana neutrino is its own antiparticle. In the free field limit ν is just

$$\nu(x) = \sum_{\vec{p}} \sum_{S=L,R} \left[b_S(\vec{p}) u_S(\vec{p}) e^{-ip\cdot x} + b_S^\dagger(\vec{p}) v_S(\vec{p}) e^{+ip\cdot x} \right], \tag{40}$$

i.e. it has the same form as for a free Dirac field (cf (33)) except that there is no distinction between b and d annihilation operators.

The Majorana mass m in (39) can be generated by the vacuum expectation value (VEV) of a new Higgs triplet [56] or as a higher order effective operator. Majorana masses are popular amongst theorists because they are so different from quark and lepton masses, and there is therefore the possibility of explaining why m_{ν_e} is so small (if it is non-zero).

For F fermion families, the Majorana mass term is

$$-L_M = \tfrac{1}{2}\bar{n}_L^0 M n_R^{0c} + h.c. \tag{41}$$

where M is an $F \times F$ Majorana mass matrix and n_L^0 and n_R^{0c} are F component vectors: i.e. $n_L^0 = (n_{1L}^0 \cdots n_{FL}^0)^T$, $n_R^{0c} = (n_{1R}^{0c} \cdots n_{FR}^{0c})^T$, where n_{iL}^0 and n_{iR}^{0c} are weak eigenstate neutrinos and "antineutrinos", related by

$$n_{iR}^{0c} = C\bar{n}_{iL}^{0T} \tag{42}$$

From (42) one can prove the identity

$$\bar{n}_{iL}^0 n_{jR}^{0c} = \bar{n}_{jL}^0 n_{iR}^{0c}, \tag{43}$$

from which it follows that the Majorana mass matrix M must be symmetric: $M = M^T$. Proceeding in analogy to the Dirac case, one can relate the n_{iL}^0 and n_{jR}^{0c} to mass eigenstate neutrino fields by

$$\begin{aligned} n_L^0 &= U_L n_L \\ n_R^{oc} &= U_R n_R^c, \end{aligned} \tag{44}$$

where U_L and U_R are $F \times F$ unitary matrices chosen so that

$$U_L^\dagger M U_R = M_d = \text{diag}(m_1\ m_2\ \ldots m_F), \tag{45}$$

where M_d is a diagonal matrix of Majorana mass eigenvalues. Unlike the Dirac case (for which m_D was an arbitrary matrix and V_L and V_R unrelated), the symmetry of M implies a relation between U_L and U_R, viz

$$U_L = U_R^* K^\dagger, \tag{46}$$

where K is unitary and symmetric. That is, just as in the Dirac case, U_L is determined from

$$U_L^\dagger M M^\dagger U_L = M_d^2 \tag{47}$$

to be of the form $U_L = \hat{U}_L K_L$, where K_L is a matrix of phases that can be chosen for convenience. U_R is then determined from (46), where K is chosen so that M_d is real and positive. If there are no degeneracies then K is just a matrix of phases. [57] One can always pick K_L such that $K = I$, but it is not always convenient to do so.

In terms of the mass eigenstates, (41) reduces to

$$\begin{aligned} -L_M &= \tfrac{1}{2}\sum_{i=1}^{F} m_i \bar{n}_{iL} n_{iR}^c + h.c. \\ &= \tfrac{1}{2}\sum_{i=1}^{F} m_i \bar{n}_i n_i, \end{aligned} \tag{48}$$

where $n_i = n_{iL} + n_{iR}^c$ is the ith Majorana mass eigenstate. [58] Written in terms of the n_{iL}, the weak charged current assumes a form analogous to (38), with $U_L^\dagger$ replacing $V_L^\dagger$ to describe the leptonic mixing. [59]

There are several physical distinctions between Dirac and Majorana neutrinos. If the ν_e is Majorana, for example, one could have the sequence $\pi^+ \to e^+\nu_e$ followed by $\nu_e p \to e^+ n$. The combined process violates lepton number by two units and is allowed for Majorana but not Dirac neutrinos. Similarly, a hypothetical heavy neutrino N would undergo the decays $N \to e^+ q_1 \bar{q}_2$ and $N \to e^- \bar{q}_1 q_2$ with equal rates if it is Majorana, while for a Dirac particle one would have $N \to e^- \bar{q}_1 q_2$, $N^c \to e^+ q_1 \bar{q}_2$ only [60]. There are differences due to Fermi statistics in the production of $\nu\nu$ (Majorana) or $\nu\nu^c$ (Dirac) pairs near threshold [61], and finally Majorana neutrinos cannot have electromagnetic form factors, such as magnetic moments [62].

It is important to keep in mind, however, that these distinctions must all disappear in the limit that the neutrino mass can be neglected. For $m_\nu \to 0$ the ν_R component of a Dirac neutrino decouples, and both Majorana and Dirac neutrinos reduce to Weyl two-component neutrinos - there is no difference between them. [63] In particular, lepton number conservation is reestablished smoothly as $m_\nu \to 0$ for a Majorana neutrino, because in that limit helicity - which is conserved up to corrections of order m_ν/E_ν - plays the role of an approximate lepton number. For example, the ν_e produced in $\pi^+ \to e^+\nu_e$ has $h_\nu = -1$ up to corrections of order $(m_\nu/E_\nu)^2$ (in rate), while the reaction $\nu_e p \to e^+ n$ has a cross section that is suppressed by $(m_\nu/E_\nu)^2$ for the wrong (negative) helicity.

In many models Dirac and Majorana mass terms are both present. For one doublet neutrino ν_L^0 (with $\nu_R^{0c} = C\bar{\nu}_L^{0T}$) and one new singlet N_R^0 (with $N_L^{0c} = C\bar{N}_R^{0T}$), for example, one could have the general mass term

$$-L = \tfrac{1}{2}\begin{pmatrix}\bar{\nu}_L^0 & \bar{N}_L^{0c}\end{pmatrix}\begin{pmatrix} m_t & m_D \\ m_D^T & m_S \end{pmatrix}\begin{pmatrix}\nu_R^{0c} \\ N_R^0\end{pmatrix} + h.c., \tag{49}$$

where $m_D = m_D^T$ is a Dirac mass generated by a Higgs doublet (analogous to (31)), m_t is a Majorana mass for ν_L^0 generated by a Higgs triplet or effective interaction (cf. (39)), and m_S is a Majorana mass for N_R^0, generated by a Higgs singlet or bare mass. Similarly, for F families (49) still holds provided one interprets ν_L^0 , N_L^{0c} , ν_R^{0c} , and N_R^0 as F component vectors, and m_t, m_D, and m_S as $F \times F$ matrices (with $m_t = m_t^T$, $m_S = m_S^T$). Then, (49) becomes simply

$$-L = \tfrac{1}{2}\bar{n}_L^0 M n_R^{oc} + h.c., \tag{50}$$

where $n_L^0 \equiv (\nu_L^0,\ N_L^{0c})^T$ and $n_R^{0c} \equiv (\nu_R^{0c},\ N_R^0)^T$ are $2F$ component vectors and

$$M = \begin{pmatrix} m_t & m_D \\ m_D^T & m_S \end{pmatrix} \tag{51}$$

is a symmetric $2F \times 2F$ Majorana mass matrix. Equation (50) can be diagonalized in exact analogy with (41-48), yielding finally

$$-L = \tfrac{1}{2}\sum_{i=1}^{2F} m_i \bar{n}_{iL} n_{iR}^c + h.c. \tag{52}$$

i.e. there are in general $2F$ Majorana neutrinos, related to n_L^0, n_R^{0c} by unitary transformations similar to (44). Unlike the pure Majorana case, however, there is now mixing between particles with different weak interaction properties (e.g. $n_{iL} = (U_L^\dagger)_{ij} n_{jL}^0$ is a mixture of SU_2 doublets and singlets), which can have important consequences for neutrino oscillations [64] and decays.

It is instructive to see how the Dirac case ($m_t = m_S = 0$) emerges as a limiting case of (49). For a single family one has

$$M = m_D \begin{pmatrix} 0 & 1 \\ 1 & 0 \end{pmatrix}. \tag{53}$$

Since M is Hermitian (for m_D real) one can diagonalize it by a unitary transformation U_L. One finds

$$U_L^\dagger M U_L = m_D \begin{pmatrix} 1 & 0 \\ 0 & -1 \end{pmatrix}, \tag{54}$$

with $U_L = \frac{1}{\sqrt{2}}\begin{pmatrix} 1 & 0 \\ 0 & -1 \end{pmatrix}$; i.e. the mass eigenstates are

$$\begin{aligned} n_{1L} &= \frac{1}{\sqrt{2}}(\nu_L^0 + N_L^{0c}) \\ n_{2L} &= \frac{1}{\sqrt{2}}(\nu_L^0 - N_L^{0c}) \\ n'_{1R} &= \frac{1}{\sqrt{2}}(\nu_R^{0c} + N_R) \\ n'_{2R} &= \frac{1}{\sqrt{2}}(\nu_R^{0c} - N_R). \end{aligned} \tag{55}$$

The negative mass eigenvalue in (54) can be removed by redefining [65] the right-handed fields $n_{1R} = n'_{1R}$, $n_{2R} = -n'_{2R}$. This is nothing more than taking

$$U_L^\dagger M U_R = m_d = m_D \begin{pmatrix} 1 & 0 \\ 0 & 1 \end{pmatrix}, \tag{56}$$

where U_R is given by (46) with $K = \begin{pmatrix} 1 & 0 \\ 0 & -1 \end{pmatrix}$. Finally, the two Majorana states $n_1 = n_{1L} + n_{1R}^c$ and $n_2 = n_{2L} + n_{2R}^c$ are degenerate. We can therefore reexpress L in the new basis

$$\begin{aligned} \nu &\equiv \frac{1}{\sqrt{2}}(n_1 + n_2) = \nu_L^0 + N_R^0 \\ \nu^c &\equiv \frac{1}{\sqrt{2}}(n_1 - n_2) = N_L^{0c} + \nu_R^{0c}, \end{aligned} \tag{57}$$

yielding

$$\begin{aligned} -L &= \tfrac{1}{2} m_D(\bar{n}_{1L} n_{1R}^c + \bar{n}_{2L} n_{2R}^c) + h.c. \\ &= m_D \bar{\nu}_L^0 N_R^0 + h.c. \\ &= m_D \bar{\nu}\nu. \end{aligned} \tag{58}$$

This is just a standard Dirac mass term, with a conserved lepton number (i.e. no transition between ν and ν^c). A Dirac neutrino is therefore nothing but a pair of degenerate two-component Majorana neutrinos (n_1 and n_2), combined to form a 4-component neutrino with a conserved lepton number.

Similarly, the Dirac limit for F families ($m_t = m_S = 0$ in (51)), can be obtained by choosing

$$U_L = \frac{1}{\sqrt{2}}\begin{pmatrix} V_L & V_L \\ V_R^* & -V_R^* \end{pmatrix} \tag{59}$$

and $K = \begin{pmatrix} I & 0 \\ 0 & -I \end{pmatrix}$, where V_L and V_R are the $F \times F$ unitary matrices that diagonalize m_D (in (36). One then obtains

$$U_L^\dagger \begin{pmatrix} 0 & m_D \\ m_D^T & 0 \end{pmatrix} U_R = \begin{pmatrix} m_d & 0 \\ 0 & m_d \end{pmatrix}, \tag{60}$$

so that one obtains F pairs of degenerate Majorana neutrinos, which can be combined into F Dirac neutrinos.

One sometimes refers to a pseudo-Dirac neutrino, which is just a Dirac neutrino to which is added as small lepton number-violating perturbation. For example, for $F = 1$ one could modify the Dirac mass in (53) to

$$M = \begin{pmatrix} \epsilon & m_D \\ m_D & 0 \end{pmatrix}, \tag{61}$$

with $\epsilon << m_D$. One then finds two Majorana mass eigenstates $n_\pm$, with

$$\begin{aligned} n_{+L} &= n_{1L} + \frac{\epsilon}{4} n_{2L} \\ n_{-L} &= -\frac{\epsilon}{4} n_{1L} + n_{2L}, \end{aligned} \tag{62}$$

(n_{1L} and N_{2L} are defined in (55)), with masses $m_D \pm \frac{\epsilon}{2}$.

Other important special cases of (51) are considered below.

4.2 Models of Neutrino Mass

There are many models for neutrino mass [49], all of which have good and bad features. The major classes of models are listed in Table 6, along with the most natural scales for the neutrino masses and for $\langle m_{\nu_e} \rangle$, an effective mass relevant to neutrinoless double β decay.

Table 6: Models of neutrino mass, along with their most natural scales for the light neutrino masses.

Model	m_{ν_e}	$\langle m_{\nu_e} \rangle$	m_{ν_μ}	m_{ν_τ}
Dirac	$1 - 10\ MeV$	0	$100\ MeV - 1\ GeV$	$1 - 100\ GeV$
pure Majorana [56] (Higgs triplet)	arbitrary	m_{ν_e}	arbitrary	arbitrary
GUT seesaw [66,67] ($M \sim 10^{14}\ GeV$)	$10^{-11}\ eV$	m_{ν_e}	$10^{-6}\ eV$	$10^{-3}\ eV$
intermediate seesaw [68] ($M \sim 10^9\ GeV$)	$10^{-7}\ eV$	m_{ν_e}	$10^{-2}\ eV$	$10\ eV$
$SU_{2L} \times SU_{2R} \times U_1$ seesaw [69] ($M \sim 1\ TeV$)	$10^{-1}\ eV$	m_{ν_e}	$10\ KeV$	$1\ MeV$
light seesaw [70] ($M \ll 1\ GeV$)	$1 - 10\ MeV$	$\ll m_{\nu_e}$	-	-
charged Higgs [71]	$< 1\ eV$	$\ll m_{\nu_e}$	-	-

Dirac neutrinos are exactly like other fermions. They involve a conserved total lepton number (though the individual L_e, L_μ, and L_τ lepton numbers are violated by mixing in general) and therefore do not lead to neutrinoless double beta decay. The problem with Dirac neutrinos is that it is hard to understand why the neutrinos are so much lighter than the other fermions. In the standard model Dirac mass are generated by the vacuum expectation value (VEV) $v = \sqrt{2}\langle \varphi^0 \rangle \simeq (\sqrt{2} G_F)^{-1/2} \simeq 246\ GeV$ of the neutral component of a doublet [72] of Higgs scalar fields. One has

$$m_D = h_\nu v, \tag{63}$$

where h_ν is the Yukawa coupling

$$L = -\sqrt{2} h_\nu (\bar{\nu}_L\ \bar{e}_L) \begin{pmatrix} \varphi^0 \\ \varphi^- \end{pmatrix} N_R + h.c. \tag{64}$$

of the neutrino to φ^0.

A ν_e mass in the 20 eV range would require an anomalously small Yukawa coupling $h_{\nu_e} \leq 10^{-10}$. Moreover, h_{ν_e} would have to be smaller by $m_{\nu_e}/m_e \leq 10^{-4}$ than the analogous Yukawa coupling for the electron. Of course, we do not understand the masses of the other fermions either (or why they range over at least five orders of magnitude), so it is hard to totally exclude the possibility that h_{ν_e} is simply small. Nevertheless, the possibility seems sufficiently ugly that it is hard to take seriously unless some mechanism (other than fine-tuning) for the smallness is proposed.

One possibility is that h_ν is actually zero to lowest order (tree level) due to some new symmetry, and that h_ν is only generated as a higher order correction (i.e. so that m_ν/m_e is some power of α.) This is a very attractive possibility, but no particularly compelling models to implement it have emerged. The idea has recently been resurrected in some superstring inspired models [73], which have difficulty incorporating the seesaw type ideas described below.

Majorana mass terms for the ordinary SU_2-doublet neutrinos involve a transition from ν_R^c ($t_3 = -\frac{1}{2}$) into ν_L ($t_3 = +\frac{1}{2}$), and therefore must be generated by an operator transforming as a triplet under weak SU_2.

The simplest possibility is the Gelmini-Roncadelli model [56], in which one introduces a triplet of Higgs fields $\vec{\varphi}_t = (\varphi_t^0, \varphi_t^-, \varphi_t^{--})$ into the theory. The Yukawa coupling

$$\begin{aligned} L &= \tfrac{1}{2}h_t(\bar{\nu}_L\ \bar{e}_L)\vec{\tau}\cdot\vec{\varphi}_t\begin{pmatrix} e_R^c \\ -\nu_R^c \end{pmatrix} \\ &= \tfrac{1}{2}h_t(\bar{\nu}_L\ \bar{e}_L)\begin{pmatrix} \varphi_t^- & \sqrt{2}\varphi_t^0 \\ \sqrt{2}\varphi_t^{--} & -\varphi_t^- \end{pmatrix}\begin{pmatrix} e_R^c \\ -\nu_R^c \end{pmatrix} \end{aligned} \tag{65}$$

then generates a Majorana mass

$$m_t = h_t v_t \tag{66}$$

for the ν, where $v_t = \sqrt{2}\langle\varphi_t^0\rangle$ is the VEV of the Higgs triplet. Since both h_t and v_t are unknown the neutrino mass is unrelated to the other fermions and can in principle be arbitrarily small, at least at tree level.

However, small m_{ν_e} is not explained in such models - it is merely parametrized and in fact is almost as problematic as a Dirac mass. The weak neutral current (and W and Z masses) require [5] $v_t \le 0.08v \sim 20\ GeV$. For v_t close to this limit one requires $h_t \le 10^{-9}$, i.e. almost as bad a fine-tuning as the Dirac case. For $v_t \ll v$ one can tolerate more reasonable values for h_t, but then it is difficult to understand the large hierarchy in vacuum expectation values. One generally expects all non-zero VEV's to be comparable in magnitude unless fine-tunings are performed on the parameters in the Higgs potential. Even if one does this, higher order corrections are likely to upset the hierarchy. [74]

The VEV $\langle\varphi_t^0\rangle \neq 0$ necessarily violates lepton number conservation by two units (the Yukawa coupling in (65) does not by itself violate L because φ_t can be regarded as carrying two units of L). If the rest of the Lagrangian conserves L then lepton-number is spontaneously broken, and there will be an associated massless Goldstone boson, the triplet-Majoron. (This is the version of the model that is usually considered [56].) In this case limits based on stellar energy loss (carried off by Majorons) require [75] $v_t \le 2 - 10\ KeV$. Implications of the Majoron for neutrino decay and annihilation, cosmology, and neutrinoless double beta decay will be mentioned below.

It is also possible introduce other couplings into the Higgs triplet model which explicitly break lepton number conservation, such as a cubic interaction between $\vec{\varphi}_t$ and two Higgs doublets. (This violates L since $\vec{\varphi}_t$ was assigned $L = 2$ to make (65) invariant). In that case all of the new scalar particles associated with φ_t become massive - i.e. there is no Majoron.

Another mechanism for introducing a Majorana mass is to consider the induced interaction (Fig. 11).

$$L_{eff} = \tfrac{1}{2}\frac{C}{M}(\bar{\nu}_L\ \bar{e}_L)\vec{\tau}\begin{pmatrix} e_R^c \\ -\nu_R^c \end{pmatrix}\cdot(\varphi^-\ \ -\varphi^0)\vec{\tau}\begin{pmatrix} \varphi^0 \\ \varphi^- \end{pmatrix} \tag{67}$$

between two leptons and two Higgs doublets. The Higgs fields in (67) are arranged to transform as an SU_2 triplet, so L_{eff} is $SU_2 \times U_1$ invariant; however, L_{eff} is non-renormalizable, as is evidenced by the dimensional coupling C/M, where M is a mass. L_{eff} cannot therefore be an elementary

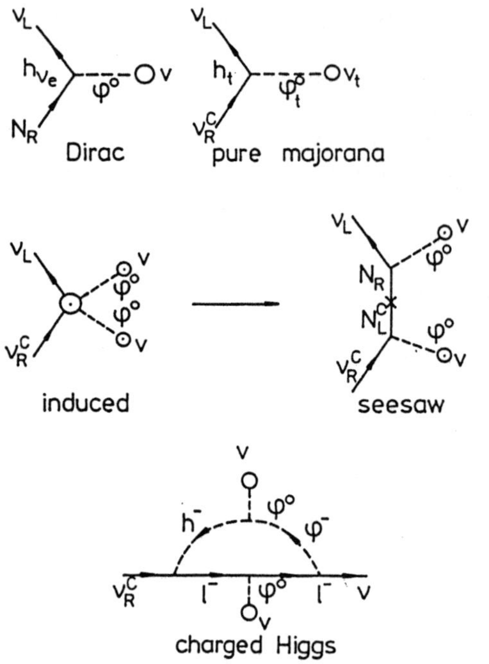

Figure 11: Dirac, pure Majorana, induced, and charged Higgs generated neutrino masses.

coupling, but it could be an effective four-particle interaction induced [76] by new physics at some large mass scale M (just as the four-fermion weak interaction is a nonrenormalizable effective interaction that is really generated by W and Z exchange). When φ^0 is replaced by its vacuum expectation value, (67) yields an effective Majorana mass $m \sim Cv^2/M$, which is naturally small for $M \gg v$. For example, if (67) were somehow induced by quantum gravity one would expect $M \sim 10^{19}\ GeV$ (the Planck scale). Then for $C \sim 1$ one would have $m_\nu \sim 10^{-5}\ eV$.

The most popular realisation of this idea is the seesaw model, [66] in which the underlying physics is the exchange of a very heavy SU_2-singlet Majorana neutrino N_R^0, as indicated in Fig. 11. The seesaw model for one family is a special case of the general mass matrix in (49), in which m_D is a typical Dirac mass (typically assumed to be comparable to m_u or m_e for the first family) connecting ν_L^0 to a new SU_2-singlet N_R^0 and $m_S \gg m_D$ is a Majorana mass for N_R^0, presumably comparable to some new (large) physics scale. One typically assumes that $m_t = 0$ in the seesaw model, i.e. that there is not a Higgs triplet as well. [77] In that case, (49) yields two Majorana mass eigenstates n_1 and n_2 with

$$\begin{aligned}
\nu_L^0 &= n_{1L}\cos\theta + n_{2L}\sin\theta \\
N_L^{0c} &= -n_{1L}\sin\theta + n_{2L}\cos\theta \\
\nu_R^{0c} &= -(n_{1R}^c\cos\theta + n_{2R}^c\sin\theta) \\
N_R^0 &= -n_{1R}^c\sin\theta + n_{2R}^c\cos\theta.
\end{aligned} \tag{68}$$

The physical masses [78] are

$$\begin{aligned}
m_1 &\simeq \frac{m_D^2}{m_S} \ll m_D \\
m_2 &\simeq m_S
\end{aligned} \tag{69}$$

and the mixing angle is

$$\tan\theta = \left(\frac{m_1}{m_2}\right)^{1/2} \simeq \frac{m_D}{m_S} \ll 1. \tag{70}$$

Hence, one naturally obtains one very light neutrino, which is mainly the ordinary SU_2 doublet (ν_L^0, ν_R^{0c}), and one very heavy neutrino, which is mainly the singlet (N_L^{0c}, N_R^0).

If one does allow $m_t \neq 0$ (but $\ll m_S$) then there are still two Majorana neutrinos with masses $|a - \frac{m_D^2}{m_S}|$ and m_S, respectively, while $\theta \sim m_D/m_S \ll 1$ still holds. (The minus sign in ν_R^{0c} is removed if $a - \frac{m_D^2}{m_S}$ is positive). In this case, however, one loses the natural explanation of why m_1 is so small, unless m_t is itself induced by the underlying physics and is of the same order as m_D^2/m_S.

The seesaw model is easily generalized to F families. One then has the general $2F \times 2F$ Majorana mass matrix in (51). Assuming that the eigenvalues of m_S are all much larger than any of the components of m_D or m_t (if it is non-zero) one can calculate the eigenvalues and mixing matrices to leading order in m_S^{-1}. One finds that there are F light Majorana neutrinos (consisting of the F doublets (ν_L^0, ν_R^{0c}), up to corrections of order $m_D m_S^{-1}$ and F heavy Majorana neutrinos (consisting of the singlets (N_L^{0c}, N_R^0), to $O(m_D m_S^{-1})$). That is, one can write

$$\begin{aligned} \begin{pmatrix} \nu_L^0 \\ N_L^{0c} \end{pmatrix} &= U_L \begin{pmatrix} n_{lL} \\ n_{hL} \end{pmatrix} \\ \begin{pmatrix} \nu_R^{0c} \\ N_R^0 \end{pmatrix} &= U_R \begin{pmatrix} n_{lR}^c \\ n_{hR}^c \end{pmatrix}, \end{aligned} \tag{71}$$

where n_{lL} and n_{hL} are F component vectors of light and heavy Majorana mass eigenstates, respectively, and similarly for n_{lR}^c, n_{hR}^c. As usual, U_L and U_R are $2F \times 2F$ unitary matrices which diagonalize M in (51), viz

$$U_L^\dagger \begin{pmatrix} m_t & m_D \\ m_D^T & m_S \end{pmatrix} U_R = m_d = \begin{pmatrix} m_l & 0 \\ 0 & m_h \end{pmatrix}, \tag{72}$$

where m_l and m_h are diagonal $F \times F$ matrices of the F light and F heavy eigenvalues, respectively. To leading order in m_S^{-1} one can write $U_L^\dagger$ and U_R in block diagonal form

$$U_L^\dagger = K U_R^T = \begin{pmatrix} K_1 & 0 \\ 0 & K_2 \end{pmatrix} \begin{pmatrix} A^T & -A^T m_D m_S^{-1} \\ D^T m_S^{-1\dagger} m_D^\dagger & D^T \end{pmatrix}, \tag{73}$$

where A^T and D^T are unitary (to leading order) $F \times F$ matrices defined by

$$\begin{aligned} m_l &= K_1 A^T (m_t - m_D m_S^{-1} m_D^T) A \\ m_h &= K_2 D^T m_S D \end{aligned} \tag{74}$$

i.e. the mass matrix for the light neutrinos is $m_t - m_D m_S^{-1} m_D^T$, which is diagonalized by A, while that for the heavy neutrinos is m_S, diagonalized by D. K_1 and K_2 are diagonal phase matrices which ensure that m_l and m_h are real and positive. We see from (71-74) that indeed there are F heavy states with masses of $O(m_S)$, and in the simplest case $m_t = 0$ there are F states which are naturally very light $(O(m_D^2 m_S^{-1}))$. (For $m_t \neq 0$ one must separately assume m_t is small). Furthermore, the mixing between the light and heavy sectors is very small (of $O(m_D m_S^{-1})$), while the matrices A and D, which describe mixings within the two sectors, are in general arbitrary.

There are several classes of seesaw models [66], depending on the scale of m_S. In simple grand unified models one assumes that the scale is a typical GUT unification scale of around 10^{14} GeV. In many such models (e.g. SO_{10}) one has that the neutrino Dirac mass matrix m_D is the same as m_u/k where m_u is the u-quark mass matrix and $k \simeq 4.7$ represents the running of the Yukawa couplings between the GUT scale and low energies. If one makes the somewhat ad-hoc assumption that the matrix m_S is just $M_X I$, where $M_X \sim 10^{14}\ GeV$ is the unification scale and I is the identity matrix, one has (for $m_t = 0$) the light eigenvalues

$$m_{\nu_i} \sim \frac{m_{u_i}^2}{M_X k^2} \tag{75}$$

$\sim 10^{-11}\ eV$, $10^{-6}\ eV$, $10^{-3}\ eV$, i.e. the neutrino masses are naturally expected to be extremely tiny, and to scale like the squares of the u, c, and t quark masses. (Equation (75) was computed for $m_{top} \sim 50\ GeV$). Several caveats are in order: the assumption of $m_S \sim M_X I$ was quite arbitrary. One could easily imagine that the eigenvalues of m_S are smaller than M_X due to small Yukawa coupling couplings (increasing m_{ν_i}). Also, they need not be the same. For example, if the m_S eigenvalues followed the same family hierarchy as the ordinary fermions (i.e. $m_{S_i} \propto m_{u_i}$) then one would have m_{ν_i} scaling as m_{u_i} rather than $m_{u_i}^2$. (A similar linear hierarchy ensues in some variant GUTs in which m_S is zero at tree level but is generated at higher orders [79,36]. Of course, more complicated patterns for m_S and m_D (in (74)) are also possible. Furthermore, in many cases loop corrections to the (GUT) Higgs potential may induce [77] VEV's for Higgs representations that can yield a non-zero triplet terms m_t in (72). These are most likely to affect the smallest masses (e.g. m_{ν_e}). Equation (75) should therefore be regarded only as a typical order of magnitude.

If one does assume that $m_S = M_X I$, however, then m_u^2/M_X is diagonalized by the same transformations that diagonalize m_u. Since one also has equal electron and d-quark mass matrices (i.e. $m_e = m_d/k$) in most simple GUTs the final result is that flavor mixing in the lepton sector (analogous to (38)) is described by the same mixing matrix as the CKM quark mixing matrix. This result continues to hold [67] approximately for a far wider class of m_S than does the simple mass prediction in (75).

Lower mass scales for m_S imply larger values for the light neutrino masses (and generally less predictive power for m_D). Several authors [68] have suggested that the heavy Majorana scale could be the intermediate range $10^8 - 10^{12}\ GeV$ associated with invisible axions. For $m_{D_i} \sim m_{e_i}$ and $m_S \sim 10^9\ GeV I$, for example, one obtains the values $\sim 10^{-7}\ eV$, $10^{-2}\ eV$, $10\ eV$ for m_{ν_e}, m_{ν_μ}, m_{ν_τ}, respectively.

If m_S is in the several TeV range (as expected in some left-right symmetric [80] $SU_{2L} \times SU_{2R} \times U_1$, models [69], for example) one typically expects (for $m_{D_i} \sim m_{e_i}, m_S \propto I$) $m_{\nu_e}, m_{\nu_\mu}, m_{\nu_\tau}$ to have relatively large values $10^{-1}\ eV$, $10\ KeV$, and $1\ MeV$, respectively. As we will see, such models run into severe cosmological difficulties unless the mass hierarchy is somehow modified or a fast decay mechanism is found for the ν_μ and ν_τ. Of course, one could also have m_S much smaller than the $SU_{2L} \times SU_{2R} \times U_1$ scale (e.g. in the $10\ GeV - 100\ GeV$ range), with corresponding larger masses for the light neutrinos. Similar statements apply to models with extra Z bosons in the $100\ GeV - 10\ TeV$ range, which usually also have heavy Majorana neutrinos.

Finally, one can consider light seesaw models, in which typically $m_S \ll 1\ GeV$. Such models are very artificial and abandon the principal advantages of the seesaw, because both m_D and m_S must be taken unnaturally small to obtain an acceptable ν_e mass. Their only virtue is that they yield strongly suppressed neutrinoless double beta decay rates, even though the neutrinos are Majorana.

Seesaw models were first introduced in GUT type models in which lepton number is explicitly violated by the gauge interactions. One can also consider non-gauge seesaw models [81] in which lepton number is spontaneously broken by the VEV of the Higgs field which generates m_S. Such models imply the existence of a massless Goldstone boson, the singlet-Majoron. [82] Unlike the triplet-Majoron in the Gelmini-Roncadelli model,[56] which can couple strongly to the ordinary neutrinos (coupling $\sim h_t$), the singlet-Majoron effectively decouples from ordinary particles. That is, it couples strongly to the heavy neutrino, with a coupling of order $m_D m_S^{-1}$ to off-diagonal $n_l n_h$ vertices, and with strength $(m_d m_S^{-1})^2$ to light neutrinos.

It is difficult to implement the seesaw model in most superstring inspired models, because there is no Higgs field available to generate a large m_S. It has been suggested [83] that m_S could be generated by a higher order effective operator, but such model may run into serious cosmological problems [84].

There have also been variant seesaw models constructed [85] in which the light neutrinos occur in degenerate pairs which can be combined from Dirac neutrinos with a conserved L.

Finally, I mention the charged Higgs models [71], in which small Majorana masses are generated by loop diagrams involving new charged Higgs bosons with explicit L-violating couplings (Fig. 11). Viable versions often lead to pseudo-Dirac neutrinos. The approximately conserved lepton number is typically $L_e - L_\mu + L_\tau$, for example, rather than L. The actual mass scale depends on unknown Yukawa couplings and masses.

4.3 Experimental Constraints

There are a number of excellent reviews [49] of the experimental status of neutrino mass. My major purpose in this section is to comment on the implications of the various theoretical models for the different types of experiments.

4.4 Kinematic Tests

Direct kinematic limits on the masses of the ν_e, ν_μ, and ν_τ are given in Table 7. The ITEP group [86] has long claimed evidence for a non-zero ν_e mass in the 20 eV range from tritium β decay, but this has not been confirmed by other groups, and in fact the Zurich-SIN measurement is on the verge of conflicting with the ITEP result. In addition the neutrinos from supernova 1987A observed by the Kamiokande [93] and IMB [94] experiments place upper limits in the 20 eV range on the ν_e mass (otherwise the arrival times of the detected neutrinos would be spread out more than is observed), but it is hard to make this limit precise because it depends on the details of the neutrino emission [90].

A 20 eV neutrino mass is just in the range that would be most interesting cosmologically, so clearly it is essentially to resolve the situation. Hopefully, the current and next generation of tritium β decay experiments will be sensitive down to a few eV, but it is doubtful whether experiments of this type will ever be able to probe to much lower scales. As can be seen in Table 6, none of the models really predict m_{ν_e} in the 20 eV range (the $SU_{2L} \times SU_{2R} \times U_1$ models come closest), but most can accomodate masses in this range by fine-tuning parameters.

As can be seen in Table 7, the direct kinematic limits on m_{ν_μ} (from $\pi_{\mu 2}$ decay) and on m_{ν_τ} (from $\tau \rightarrow \nu_\tau + 5\pi$) are relatively weak. The experiments are extremely difficult (the mass scales being probed are very much smaller than the energies released in the decays), so it is unlikely that these measurements will improve by much more than a factor of two.

Table 7: Kinematic limits/values on neutrino masses.

$17\ eV < m_{\nu_e} < 40\ eV$	ITEP [86]
$m_{\nu_e} < 18\ eV$	Zurich [87]
$m_{\nu_e} < 27\ eV$	LANL [88]
$m_{\nu_e} < 32\ eV$	INS-Tokyo [89]
$m_{\nu_e} < O(20\ eV)$	SN1987A [90]
$m_{\nu_\mu} < 0.25\ MeV$	SIN [91]
$m_{\nu_\tau} < 50\ MeV$	ARGUS [92]

4.5 Heavy Neutrinos

There are many limits [49,95] on possible small admixtures of heavy neutrino states in the ν_e o ν_μ, including universality tests in nuclear β decay, searches for secondary peaks or distortions o the lepton spectra in β, π, and K decay, searches for the decay products of heavy neutrinos (e.g $\nu_h \to \nu_e e^+ e^-$) produced in beam dumps, e^+e^- annihilation, or neutrino scattering [96]. The limits on the mass m_i versus mixing angle U_{ai}, $a = e$ or μ, where

$$\nu_a^0 = \sum_i U_{ai} \nu_i \tag{76}$$

are shown in Fig. 12.

It is seen that the constraints on $|U_{ai}|^2$ are quite impressive, especially for m_i in the range $10\ MeV - 10\ GeV$, where they are comparable to the expectations in (70) of a seesaw model with $m_1 \sim 10\ eV$ and $m_2 = m_i$. The lower part of this range corresponds to the masses expected in the "light-seesaw" model (Table 6), while the $1\ GeV - 1\ TeV$ range is consistent with $SU_{2L} \times SU_{2R} \times U_1$ models. [69]

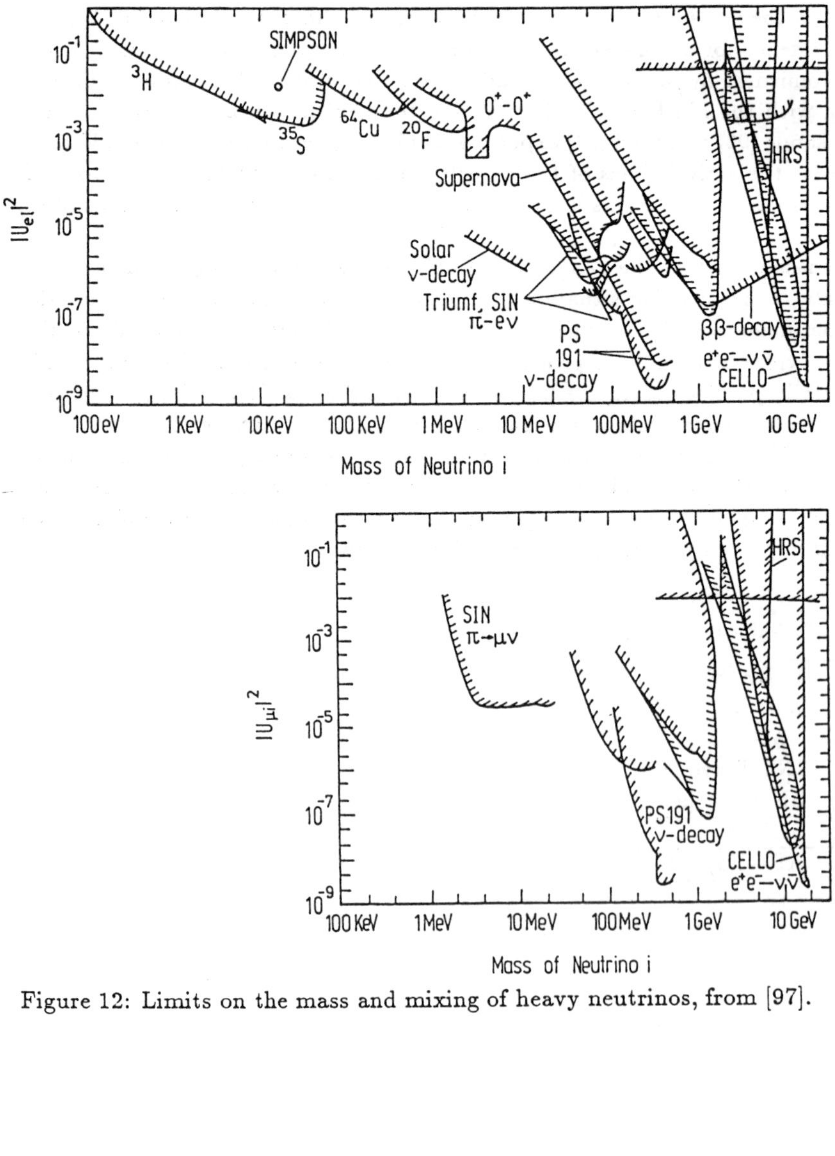

Figure 12: Limits on the mass and mixing of heavy neutrinos, from [97].

Also, most models with extra Z bosons in the 100 $GeV - 1\ TeV$ range predict [98] the existence of heavy SU_2-singlet Majorana or Dirac neutrinos [95,99,100]. The extra Z's typically couple to these new neutrinos and other exotic fermions much more strongly than to the ordinary fermions. Future hadron colliders should therefore be able to extend the search for heavy neutrinos via

$$\begin{aligned} \overset{(-)}{p}\, p &\rightarrow Z' \rightarrow NN^c \\ \overset{(-)}{p}\, p &\rightarrow W_R \rightarrow Nl \end{aligned} \tag{77}$$

into the several hundred GeV range. The subsequent decays of the N's should be a superb probe of underlying physics. In models with just an extra Z, for example, the N is expected to decay due to mixing with the light neutrinos. The N can then decay [95,99] via virtual W or Z exchange [101] into such modes as $3\nu_1$, $\nu_1 l^+ l^-$, $\nu_1 q\bar{q}$, and $\nu_1 e^+\mu^-$. On the other hand, in $SU_{2L} \times SU_{2R} \times U_1$ models the N will generally decay via virtual W_R exchange [100], and for the lightest N the decay should usually be into $l^\pm q\bar{q}$. Moreover, the decay modes should easily establish whether the heavy neutrino is Majorana or Dirac, because in the former case the decays $N \rightarrow l^+\bar{q}q$ and $N \rightarrow l^- q\bar{q}$ would be equally likely [60] (though with different angular distributions).

It is of course also possible that a heavy neutrino could simply be a massive 4th generation neutrino.

As has already been mentioned, heavy neutrinos in the $GeV - TeV$ range are likely to give too large m_{ν_μ} and m_{ν_τ} unless the typical seesaw hierarchy $m_{\nu_i} \propto m_{l_i}^n$ or $m_{u_i}^n$, $n = 1$ or 2, for the light neutrinos is avoided or new physics is invoked to ensure fast decays or annihilations for the ν_μ and ν_τ. On the other hand, if such new physics is present some of the limits in Fig. 12 (those based on decays) may no longer be valid, because in many cases the heavy neutrinos will decay rapidly into unobservable channels (e.g. $\nu_h \rightarrow \nu_l +$ Majoron) before reaching the detector.

4.6 Neutrino Oscillations

Neutrino oscillations are a beautiful example of a common quantum phenomenon: viz that if one starts at time $t = 0$ in a state that is not an energy eigenstate [102] then at later times it can oscillate into another (orthogonal) state. For example, suppose that the ν_e^0 and a second neutrino ν_a^0 (e.g. $\nu_a^0 = \nu_\mu^0$ or ν_τ^0) are mixtures of two mass eigenstates ν_1 and ν_2 with mixing angle θ:

$$\begin{aligned} \nu_e^0 &= \cos\theta\ \nu_1 + \sin\theta\ \nu_2 \\ \nu_a^0 &= -\sin\theta\ \nu_1 + \cos\theta\ \nu_2 \end{aligned} \tag{78}$$

If at time $t = 0$ the weak eigenstate ν_e^0 is produced (e.g. in the process $\pi^+ \rightarrow \pi^0 e^+ \nu_e^0$) then at time t it will have evolved into the state

$$\begin{aligned} \nu_e^0(t) &= \cos\theta\ \nu_1 e^{-iE_1 t} + \sin\theta\ \nu_2 e^{-iE_2 t} \\ &\simeq \cos\theta\ \nu_1 e^{\frac{-im_1^2 t}{2p}} + \sin\theta\ \nu_2 e^{\frac{-im_2^2 t}{2p}}. \end{aligned} \tag{79}$$

In the second form I have assumed relativistic neutrinos $E_i = \sqrt{p^2 + m_i^2} \sim p + m_i^2/2p$ with definite momentum [103] $p >> m_i$, and have neglected an irrelevant overall phase $\exp(-ipt)$. The state $\nu_e^0(t)$ has a non-trivial overlap with ν_a^0. After traveling a distance $L \sim t$, there will be a probability

$$\begin{aligned} P(\nu_e \rightarrow \nu_a) &= |\langle \nu_a^0 \mid \nu_e^0(t)\rangle|^2 \\ &= \sin^2 2\theta \sin^2\left(\frac{\Delta m^2 L}{4p}\right) \\ &= \sin^2 2\theta \sin^2\left(\frac{1.27\Delta m^2(eV^2)L(m)}{p(MeV)}\right) \end{aligned} \tag{80}$$

that the state will have evolved into ν_a^0 (as can be observed in the process $\nu_a N \to e_a N'$, for example), and a probability

$$P(\nu_e \to \nu_e) = 1 - P(\nu_e \to \nu_a) \tag{81}$$

that the state will remain a ν_e^0. In (80), $\Delta m^2 = m_1^2 - m_2^2$, and the last form is valid for Δm^2 in eV^2, L in m, and p in MeV. It is seen that the $\nu_e \to \nu_a$ probability depends on both the mixing angle θ and on $\Delta m^2 L/p$. For moderate values of the latter quantity the probability oscillates as a function of L and p, while for very large values the oscillations are averaged by a finite-sized detector or non-monochromatic source, (the second factor in (80) averages to 1/2). It is easy to generalize [49] (80) to the case that the initial neutrino is a mixture of more than two mass eigenstates, as in (76). One obtains

$$P(\nu_e \to \nu_a) = \sum_i |U_{ei} U_{ai}^*|^2 + Re \sum_{i \neq j} U_{ei} U_{ai}^* U_{ej}^* U_{aj} e^{\frac{-i(m_i^2 - m_j^2)L}{2p}} \tag{82}$$

Neutrino oscillations can be searched for in (a) appearance experiments, in which one looks for the interactions of ν_a in a detector, and (b) disappearance experiments, in which one looks for a reduced ν_e flux. In both cases one can compare the observed counting rate with the expectation from known backgrounds (appearance) or from the expected flux (disappearance) as determined, for example, by measuring the electron spectrum from $n \to p e^- \bar{\nu}_e$ in reactor $\bar{\nu}_e$ oscillation experiments. A much cleaner technique is to search for actual oscillations in the appearance or disappearance probabilities as a function of L or p, such as by using two detectors at different distances form the source.

There are many limits on neutrino oscillations from accelerator experiments [49] (e.g. counter and emulsion experiments and beam dumps, searching for $\nu_\mu \to \nu_e$, $\nu_\mu \to \nu_\tau$ and $\nu_e \to \nu_\tau$, as well as ν_μ disappearance), and reactors [49] ($\bar{\nu}_e$ disappearance), as well as on the oscillations of ν_μ produced in cosmic ray interactions in the atmosphere [104]. (Implications for the Solar neutrino problem are discussed below). The results of these searches [105] are summarized in Fig. 13. The Bugey reactor experiment [106] reports a positive signal for $\bar{\nu}_e$ disappearance, but their results are contradicted by the Gösgen experiment [107]. Similarly, the CERN PS-191 counter experiment [108] reports an excess of ν_e events in a ν_μ beam, but their signal is in conflict with several other $\nu_\mu \to \nu_e$ experiments [97]. Clearly, a clarification of the situation is essential .

From Fig. 13 it is clear that there are stringent limits on neutrino mixings for $|\Delta m^2|$ above $\simeq 1\ eV^2$. This should be contrasted with the suggested value $m_{\nu_e} \sim 17 - 40\ eV$ by the ITEP experiment [86]. If the ITEP result is correct then most likely the ν_e could not have any significant mixing with other neutrinos (the alternative possibility, that the ν_e is almost degenerate with another neutrino flavor so that $|\Delta m^2| << m_{\nu_e}^2$, seems rather contrived but cannot be excluded). A comparison of Fig. 13 with the expectations of various models (Table 6) suggests that $\nu_\mu \to \nu_\tau$ oscillations may be the most optimistic possibility for the future. Many of the seesaw-type models predict that the lepton mixing angles are roughly correlated with the corresponding quark mixing angles. This would suggest $\sin^2 2\theta \sim 10^{-4}$, 10^{-2}, 10^{-1} for $\nu_e \leftrightarrow \nu_\tau$, $\nu_\mu \leftrightarrow \nu_\tau$, and $\nu_e \leftrightarrow \nu_\mu$, respectively.

Oscillations between ordinary SU_2 doublet neutrinos ($\nu_e^0, \nu_\mu^0, \nu_\tau^0$, and possible fourth family ν's), known as first class or flavor oscillations, occur for pure Dirac and pure Majorana neutrinos, as well as in the multi-family seesaw models. In models involving both Dirac and Majorana mass terms of comparable magnitude, however, there can be additional light neutrinos, and the mass eigenstates can have significant admixtures of both SU_2 doublets and singlets. In this case second class oscillations [64] can occur, in which the ordinary neutrinos oscillate into SU_2 singlets with negligible interactions. These "sterile" neutrinos are essentially undetectable, so second class oscillations can be observed [109] only in disappearance experiments. Of course, first and second class oscillations can occur simultaneously. For three families, for example, there could be oscillations between six Majorana neutrinos (3 doublets and 3 singlets).

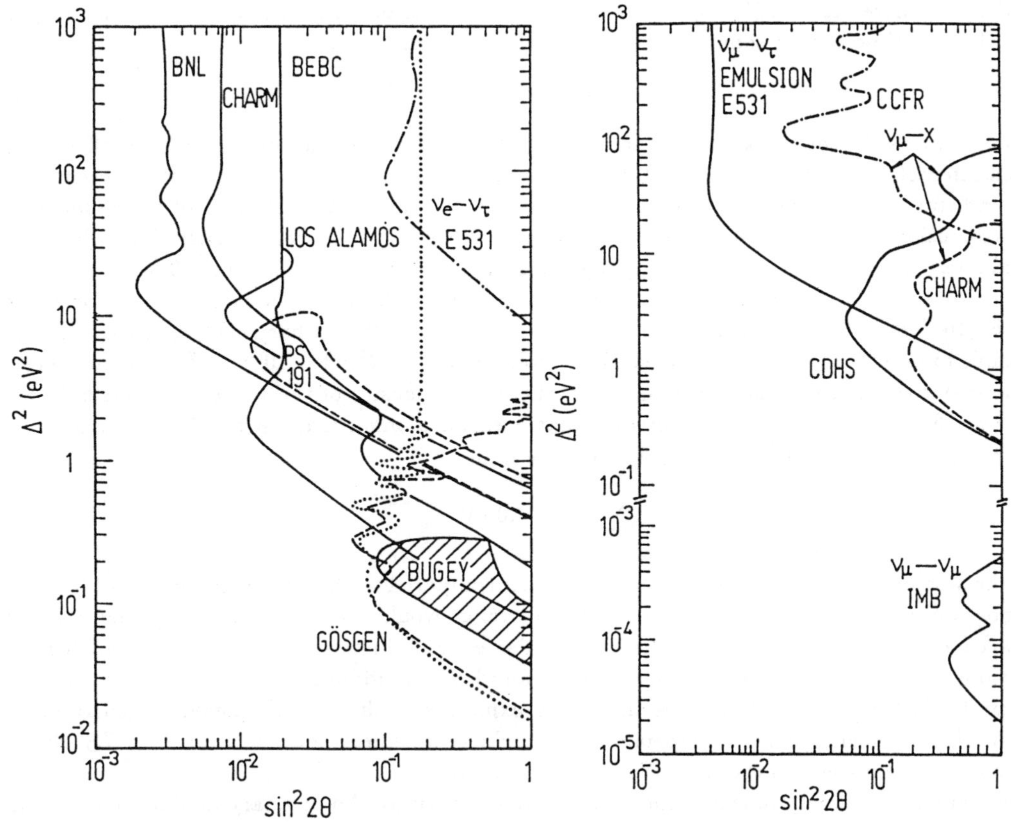

Figure 13: 90% *c.l.* limits on neutrino oscillations, from [97]. (a) $\nu_\mu \to \nu_e$ (BNL, CHARM, BEBC, Los Alamos, PS-191), $\nu_e \to \nu_\tau$ (E531), and $\bar{\nu}_e \to \nu_X$ (Bugey, Gösgen). (b) $\nu_\mu \to \nu_\tau$, ν_X, ν_μ. The Bugey [106] and PS-191 [108] regions are allowed by positive results. The other contours are exclusion plots (the regions to the right are excluded).

Yet another possibility [110] are models in which the ordinary neutrinos have small mixings with heavy neutrinos. In that case the neutrinos actually produced in weak processes are the projections of the weak eigenstates onto the subspace of light or massless neutrinos. It can easily occur that the projections of the ν_e^0 and ν_μ^0, for example, are not orthogonal. The result is that a ν_μ^0 could produce an e^- in a subsequent reaction. Such a non-orthogonality would mimic the effects of oscillation appearance experiments, even if the masses of the light neutrinos are zero or negligible.

4.7 Cosmology

There are many limits on neutrino mass and decays from cosmology [111]. Ordinary light or massless neutrinos would have been produced by such weak processes as $e^+e^- \leftrightarrow \nu\nu^c$ in the early universe. As long as the weak reaction rate [112]

$$\Gamma_{weak} \sim \langle \sigma v \rangle n_T \sim G_F^2 T^5 \tag{83}$$

($\langle \sigma v \rangle \sim G_F^2 T^2$ is the thermally averaged cross section times relative velocity, and $n_T \sim T^3$ is the density of target particles, where T is the temperature) was large compared to the expansion rate

$H \sim T^2/m_p$ (where $m_p = G_N^{-1/2} \sim 10^{19}\ GeV$ is the Planck scale) the number of neutrinos stayed in equilibrium. However, as soon as T dropped below the temperature

$$T_D \sim (G_F^2 m_P)^{-1/3} \simeq 3\ MeV \tag{84}$$

for which $\Gamma_{weak} \sim H$, the weak rate became negligible and the neutrinos decoupled, i.e. effectively stopped interacting. According to most models these neutrinos should remain in the present universe, undisturbed from the first second of the big bang except for a redshifting of their momenta by the expansion of the universe. They are analogous to the $2.7°K$ microwave radiation (which decoupled later). If the neutrino masses are much less than $1\ eV$ there should be $\simeq 50$ neutrinos/cm^3 of each type (ν_{eL}, ν_{eR}^c, etc) with momenta characterized by a thermal spectrum with temperature $\simeq 1.9°K$ ($10^{-4}\ eV$). Despite the large number of neutrinos ($\simeq 10^{10}$ per baryon) they are essentially impossible to detect [113] - [115] because their cross section $\sim G_F^2 E_\nu^2 \sim 10^{-62} cm^2$ is so low. [116]

The major cosmological bound is based on the energy density of the present universe. There are predicted to be so many relic neutrinos that even for a small mass in the $10\ eV$ range they would be important. Limits on the energy density imply

$$\sum_i m_{\nu_i} < 40\ eV \tag{85}$$

where the sum extends over the light, stable (at least compared to the age of the universe) doublet neutrinos. Conversely, a neutrino with mass in this range would dominate the energy density and could account for the dark (missing) matter in galaxies and clusters [117]. In particular, for the ITEP value $m_{\nu_e} \sim (17-40)\ eV$, the ν_e would be an ideal candidate for the dark matter, but one would probably then have to find a mechanism to explain why the ν_e is the heaviest neutrino.

Similarly, the energy density associated with light or massless neutrinos for $T \sim T_D$ affects nucleosynthesis and leads to the limit $N_\nu \leq 4$ (section III).

There are also a variety of constraints on unstable neutrinos. An ordinary doublet mass eigenstate neutrino ν_2 (with $m_{\nu_2} > m_{\nu_1}$) is expected to decay into

$$\begin{aligned} \nu_2 &\to \nu_1\gamma, \quad (m_{\nu_2} < 2m_e) \\ \nu_2 &\to \nu_1 e^+e^-, \quad (2m_e < m_{\nu_2} < m_\mu + m_e). \end{aligned} \tag{86}$$

The first decay occurs at one loop, while the second occurs at tree level. Both decays are very slow for small m_{ν_2} and the decay products are detectable. There are a large variety of cosmological and astrophysical constraints [118] on m_{ν_2} and τ_{ν_2} from the present energy density, the growth of galaxies, the distortion of the $2.7°K$ background radiation, the non-observation of the decay photons, supernovae, and nucleosynthesis and breakup. For reasonable mixing angles these limits exclude the range $40\ eV - (20-40)\ MeV$ for ordinary neutrinos [119] decaying according to (86). Combined with laboratory limits this implies [118,120] that the ν_μ and ν_τ (i.e. their dominant mass eigenstate components) should be lighter than $40\ eV$. In particular, this poses serious problems for the TeV scale seesaw model.

Most of the cosmological limits can be evaded if new physics is invoked to allow fast and invisible (except for the relativistic energy of the decay products) decays or annihilation for the heavy neutrinos. One possibility is the decay $\nu_2 \to 3\nu_1$. However, the rate for this mode from off-diagonal Z couplings [121] is too slow, while models in which the couplings of a Higgs triplet [122] (present in $SU_{2L} \times SU_{2R} \times U_1$) are arranged to allow a fast decay generally run into problems [123] with $\mu \to 3e$.

More promising are models in which $\nu_2 \to \nu_1 G$, where G is a Goldstone boson [124]-[127] associated with a spontaneously broken global symmetry. Likely examples are the case that G is a familon [124] (a Goldstone boson associated with a broken family symmetry) or a triplet-Majoron [125]. In fact, for triplet-Majorons one expects the annihilation process $\nu\nu \to MM$ (which begins when T drops below v_t) to have removed any relic neutrinos from the present universe [56]. In familon models some care must be taken to avoid unacceptably large flavor changing neutral current effects. The decay $\nu_2 \to \nu_1 M$ is too slow in the simpler versions of the singlet-Majoron model [126] to avoid cosmological problems.

The role of spontaneous L violation in Majoron models in reducing possible initial large lepton asymmetries to cosmologically interesting values at the time of nucleosynthesis is discussed in [128].

4.8 Double Beta Decay

Another important source of information on the ν_e mass (if it is Majorana) is neutrinoless double beta decay ($\beta\beta_{0\nu}$).

First consider the lepton-number conserving two-neutrino ($\beta\beta_{2\nu}$) process $(Z,N) \leftrightarrow (Z+2, N-2)e^-e^-\nu_e^c\nu_e^c$, which can be thought of as two ordinary beta decays occurring in the same nucleus (Fig. 14). In the context of neutrino mass this process is mainly of interest as a calibration of the calculated nuclear matrix elements that are needed for the neutrinoless case. There has long been a two order of magnitude discrepancy between the predicted rates [129], e.g. for $^{130}Te \to {}^{130}Xe$, and indirect measurements by geochemical techniques [130]. Within the last year, however, this discrepancy has gone away. The geochemical measurements were confirmed by the first laboratory observation of double beta decay (at Irvine [131].) In addition, several groups [129] have found that previously neglected ground state correlation effects could suppress the matrix element by the required order of magnitude. Furthermore, there is no analogous uncertainty in the $\beta\beta_{0\nu}$ case.

The neutrinoless double beta decay process $(Z,N) \to (Z+2, N-2)e^-e^-$, which violates lepton number by two units, can proceed through the second diagram [132] in Fig. 14. In the absence of mixing the quantity $\langle m_{\nu_e}\rangle$, the effective Majorana neutrino mass, is

$$\langle m_{\nu_e}\rangle = \begin{cases} 0, & \text{Dirac neutrino} \\ m_{\nu_e}, & \text{unmixed Majorana neutrino} \end{cases} \tag{87}$$

Although the matrix element is proportional to $\langle m_{\nu_e}\rangle$, which is necessarily very small, $\beta\beta_{0\nu}$ has an enormous advantage in phase space over $\beta\beta_{2\nu}$ and could be observable for $\langle m_{\nu_e}\rangle$ in the eV range. Of course, the sum of the electron energies should be a sharp peak in $\beta\beta_{0\nu}$ (and a continuum for $\beta\beta_{2\nu}$), so the principal difficulty is controlling the background. [133,49] Currently, the most sensitive experiments are for $^{76}Ge \to {}^{76}Se\ e^-e^-$. No evidence for $\beta\beta_{0\nu}$ has been observed, [134] and the lower limit on the lifetime is [97] $\tau_{1/2} > 9 \times 10^{23}\ yr$ (68% $c.l.$). According to several calculations of the nuclear matrix elements [135] this implies $\langle m_{\nu_e}\rangle \le 1\ eV$. However, a recent estimate by Engel et al. [136] yielded a much weaker limit $\langle m_{\nu_e}\rangle \le 11\ eV$, so caution is advisable.

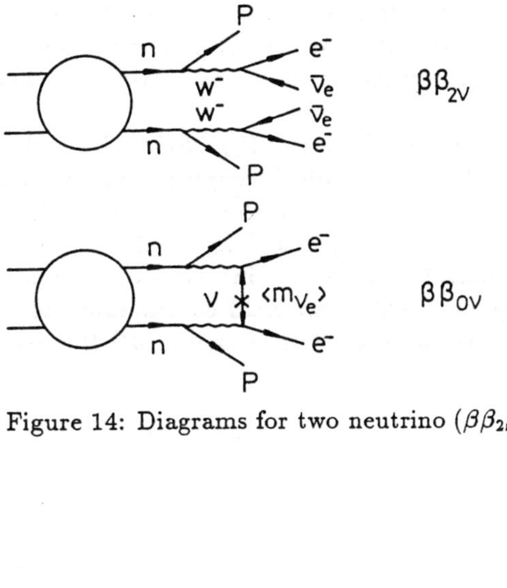

Figure 14: Diagrams for two neutrino ($\beta\beta_{2\nu}$) and neutrinoless ($\beta\beta_{0\nu}$) double beta decay.

Even the largest value $\langle m_{\nu_e}\rangle \leq 11\ eV$ is smaller than the range $m_{\nu_e} \sim (17-40)\ eV$ suggested by the ITEP experiment. If the latter is correct the simplest possibility is that the ν_e is Dirac. Another possibility [137] is that the ν_e^0 is a mixture of Majorana mass eigenstate neutrinos, as in (44). Then, $\langle m_{\nu_e}\rangle$ becomes

$$\langle m_{\nu_e}\rangle = \sum_i m_i U_{Lei}^2 \xi_i F(m_i, A), \tag{88}$$

where $m_i \geq 0$ is the physical mass of the ith mass eigenstate, U_{Lei} is the mixing matrix element ($\nu_{eL}^0 = \sum U_{Lei}\nu_{iL}$) and $\xi_i = \pm 1$ is the CP parity of ν_{iL}. ξ_i is just K_{ii} in (46), and a negative value $\xi_i = -1$ means simply that the eigenvalue of M in (41) was negative before choosing K to redefine ν_R^c. In (88), $F(m_i, A)$ is a nucleus dependent propagator correction, [138] defined by

$$F(m_i, A) = \frac{\langle e^{-m_i r}/r\rangle}{\langle 1/r\rangle}. \tag{89}$$

It is ~ 1 for $m_i \ll 10\ MeV$. For $m_i \gg 10\ MeV$, $F(m_i, A) \ll 1$ (it falls as m_i^{-2}) and allows the possibility [139] of A dependence of $\langle m_{\nu_e}\rangle$.

Because of the possibility of negative contributions to $\langle m_{\nu_e}\rangle$ it is conceivable that there are cancellations so that $\langle m_{\nu_e}\rangle$ is much smaller than the mass of the dominant Majorana component of ν_e (e.g. $m_1 \sim (17-40)\ eV$). Such a cancellation is actually not so contrived as it might first appear. If all of the m_i are small enough that $F(m_i, A) = 1$ then from (45) $\langle m_{\nu_e}\rangle$ is just the M_{ee} component of the original Majorana mass matrix in (41). As we have seen, M_{ee} must be generated by a Higgs triplet and vanishes in many models. In fact, the light seesaw model of Table 6 automatically leads to $\langle m_{\nu_e}\rangle = 0$ for sufficiently small m_i. For two neutrinos, for example, $\langle m_{\nu_e}\rangle = m_1\cos^2\theta - m_2\sin^2\theta$, which vanishes by (69) and (70). However, the light seesaw model was devised just in order to give $\langle m_{\nu_e}\rangle = 0$. For seesaw models with more natural scales $m_2 \gg 10\ MeV$ one has that $F(m_i, A) \ll 1$ and $U_{e1} \sim 1$, so that $\langle m_{\nu_e}\rangle \sim m_{\nu_e}$. In most Majorana models, therefore, one expects $\langle m_{\nu_e}\rangle \sim m_{\nu_e}$ unless fine-tuned deviations from the seesaw formula are invoked.

Whether or not the cancellation of the terms in (88) is natural, one can consider whether it is phenomenologically viable. For two neutrinos, for example, the conditions $m_1 \sim 20\ eV$, and $\langle m_{\nu_e}\rangle \ll m_1$ imply

$$\tan^2\theta = \frac{m_1}{m_2 F(m_2, A)}, \tag{90}$$

where $m_1 \leq m_2$, $\tan^2\theta \leq 1$ since the ITEP experiment presumably measures the dominant component of ν_e. However, the reactor oscillation limits in Fig. 13 allow only two possibilities. One is that $m_1 \simeq m_2$; $0 \sim 45^0$. In that case ν_1 and ν_2 can be combined to form a Dirac neutrino (or pseudo-Dirac if the degeneracy is not exact), possibly with a non-canonical lepton number (such as $L_e - L_\mu + L_\tau$) conserved. Alternatively, one can have $m_2 \geq 450\ eV$. However, the various laboratory and cosmological limits exclude [70] almost all values of m_2 except for small windows around 40 MeV and 2 GeV. Hence, if the ITEP results turn out to be correct they would almost certainly imply either (a) the ν_e is Dirac, or (b) there is new physics (such as a Majoron) that evades the cosmological bounds.

There are additional contributions to neutrinoless double beta decay in $SU_{2L}\times SU_{2R}\times U_1$ models [140]. Typically, such models contain additional charged $W_R^\pm$ bosons which couple to right-handed currents $\bar{e}_R\gamma^\mu N_R$, where N_R is a heavy Majorana neutrino. The exchange of a N_R (rather than a ν_L in Fig. 14) yields a new contribution $M_N F(M_N, A)(M_{W_L}/M_{W_R})^4$ to $\langle m_{\nu_e}\rangle$, which sets non-trivial constraints [69] on M_N and M_{W_R}. Furthermore, mixed contributions involving one ordinary left-handed current $\bar{e}_L\gamma^\mu\nu_L$ and one right-handed current $\bar{e}_R\gamma^\mu N_R$ can yield contributions to to $0^+ \to 2^+$ decay amplitudes that are not directly proportional to a neutrino mass [141] However, the relevant amplitudes are of order [49,140]

$$\left(\frac{M_{W_L}}{M_{W_R}}\right)^2 \theta, \ \ \zeta\theta, \tag{91}$$

where θ is a light-heavy neutrino mixing angle and ζ is the $W_L - W_R$ mixing angle. One typically expects $(M_{W_L}/M_{W_R})^2$ and ζ to be less than 10^{-3}. Since we expect $\theta \sim m_D/M_{W_R} \leq 10^{-4} - 10^{-5}$ in a typical TeV-seesaw, the expected values for the quantities in (91) are smaller than the experimental limits (of $\sim 10^{-6}$).

One typically has $\langle m_{\nu_e}\rangle \ll m_{\nu_e}$ for the charged Higgs models [71] because the antisymmetry of the relevant Yukawa coupling forces M_{ee} to vanish.

4.9 The Solar Neutrino Problem

For some years the event rate in the $^{37}Cl \rightarrow {}^{37}Ar$ Solar neutrino experiment [142] (2.0 ± 0.3 SNU [143]) has been considerably below the prediction [144] 5.8 ± 2.2 SNU of the standard Solar model. The discrepancy has recently been confirmed by the Kamiokande group which reports [145] an upper limit on the ν_e flux (from $\nu_e e$ elastic scattering) that is less than half the expected event rate. One explanation for the discrepancy is the existence of vacuum oscillations of the ν_e into other neutrinos. These could be important for neutrino mass-squared differences [146] $\Delta m^2 \equiv m_1^2 - m_2^2$ as small as $\Delta m^2 \sim (10^{-11} - 10^{-10})$ eV^2, but only if the mixing angles are large.

Another possibility [147] is that the ν_e is a Dirac particle with a magnetic moment in the range $\mu_{\nu_e} \sim (0.6 - 10) \times 10^{-10}\mu_B$. The ν_e spin could then process in the Solar magnetic field into a sterile right-handed ν_e, thus reducing the observed flux by a factor $\simeq 2$. The necessary value of μ_{ν_e} is barely consistent with laboratory limits [148] but is probably excluded by astrophysical constraints from nucleosynthesis and stellar cooling [149] (Table 8). The worst objection, however, is that the necessary μ_{ν_e} is unnaturally high. In the standard model with a Dirac mass one expects [150]

$$\mu_\nu = \frac{3G_F m_\nu m_e}{4\pi^2\sqrt{2}}\mu_B \sim 3\times 10^{-19}\left(\frac{m_\nu}{1eV}\right)\mu_B \tag{92}$$

which is many orders of magnitude too small. Non-standard models [151] can yield larger μ_ν, but to obtain a sufficiently large value appears highly contrived.

Other canonical explanations involve non-standard Solar models. The existing experiments are mainly sensitive to the relatively high energy (from 0.81 MeV up to 14 MeV) neutrinos from ^8B decay. The flux of these ^8B neutrinos depends very sensitively on the temperature of the Solar core and could be changed significantly by modifications of the standard Solar model. Recently, there has been much attention to the possibility that weakly interacting massive particles (WIMPs),

Table 8: Limits on the neutrino magnetic moments. A value $\mu_{\nu_e} \sim (0.6 - 10) \times 10^{-10}\mu_B$ would be needed to resolve the Solar ν problem.

laboratory [148]	$\mu_{\bar{\nu}_e} < 1.5\times 10^{-10}\mu_B$
	$\mu_{\nu_\mu} < 9.5\times 10^{-10}\mu_B$
Stellar cooling [149] ($\gamma \rightarrow \nu\bar{\nu}$)	$\mu_\nu < 0.8\times 10^{-11}\mu_B$
Nucleosynthesis [149] (ν_R produced by spin precession)	$\mu_\nu < 0.5\times 10^{-10}\mu_B$
Standard model [150] (Dirac mass)	$\mu_\nu \sim 3\times 10^{-19}\left(\frac{m_\nu}{1\ eV}\right)\mu_B$

which could form the dark matter, could carry energy out of the Solar core and lower the central temperature slightly. [111] Less exotic modifications of the standard model are also possible.

A $^{71}Ga \rightarrow {}^{71}Ge$ experiment could distinguish the nonstandard Solar model from the first two possibilities. Most of the expected ^{71}Ga event rate is from the low energy pp neutrinos, the flux of which can be inferred from the over-all Solar luminosity and is relatively insensitive to the temperature of the Solar core. The predicted ^{71}Ga event rate of $\simeq 107$ SNU can be reduced at most to around 78 SNU in most non-standard Solar models [144,152]. The traditional view has been that a flux lower than this would imply large vacuum oscillations, which would reduce the ^{71}Ga rate by a factor comparable to the ^{37}Cl event rate reduction for most oscillation parameters (e.g. to around 40 SNU).

Yet another possibility, i.e. that neutrinos decay between the Sun and the Earth, is all but excluded by the survival of neutrinos from supernova 1987A, except in some two-component models with large mixing angles. [153]

Recently, Mikheyev and Smirnov [154] have proposed an elegant new solution to the Solar neutrino problem, in which even tiny vacuum mixing angles can be amplified by the coherent interactions of ν_e with matter.

Considering $\nu_e \leftrightarrow \nu_\mu$ oscillations for definiteness, the vacuum oscillation equation in (79) can be described in terms of the weak basis states $|\nu_e\rangle$ and $|\nu_\mu\rangle$ by

$$|\nu(t)\rangle = \nu_e(t)|\nu_e\rangle + \nu_\mu(t)|\nu_\mu\rangle, \tag{93}$$

where the coefficients satisfy the Schrödinger-like equation

$$i\frac{d}{dt}\begin{pmatrix} \nu_e(t) \\ \nu_\mu(t) \end{pmatrix} = M_0 \begin{pmatrix} \nu_e(t) \\ \nu_\mu(t) \end{pmatrix}, \tag{94}$$

with

$$M_0 = \begin{pmatrix} \frac{\Delta m^2}{2p}\cos 2\theta & \frac{-\Delta m^2}{4p}\sin 2\theta \\ \frac{-\Delta m^2}{4p}\sin 2\theta & 0 \end{pmatrix}, \tag{95}$$

where an irrelevant term proportional to the identity (which only affects the overall phase) has been dropped. Wolfenstein pointed out [155] that in the presence of matter, M_0 is replaced by the M', where

$$M' = M_0 + \begin{pmatrix} \sqrt{2}G_F n_e & 0 \\ 0 & 0 \end{pmatrix} \tag{96}$$

and n_e is the density of electrons. The new term [155] - [157] is the effect of the coherent forward scattering amplitude for $\nu_e e^- \rightarrow \nu_e e^-$ via the charged current. The effects of neutral current scattering from e^-, p, and n have been neglected because they are the same for ν_e and ν_μ and only contribute to the overall phase. For $\Delta m^2 < 0$ (i.e. $m_{\nu_e} < m_{\nu_\mu}$ [157]) there is a critical density [158] $n_e^{crit} = -\Delta m^2 \cos 2\theta/(2\sqrt{2}G_F p)$ for which the diagonal elements of M' are equal (i.e. zero). At that density a resonance occurs, i.e. even a tiny off diagonal mixing term leads to large mixing effects.

In particular, if n_e in the Sun varies slowly an adiabatic approximation applies [154,159]. ν_e's produced in the core of the Sun (where $n_e > n_e^{crit}$) correspond to the larger mass eigenstate ν_2 of M' (Fig. 15). Outside the Sun, on the other hand, the higher energy state ν_2 corresponds to ν_μ for $\Delta m^2 < 0$. Hence, if the variation of n_e with the distance from the center of the Sun is sufficiently slow, the initial ν_e will be adiabatically converted to ν_μ as they pass through the resonance.

A number of authors [152],[154],[159] - [161] have analyzed the implications of the Mikheyev-Smirnov-Wolfenstein (MSW) effect for the Solar neutrinos quantitatively. It is found that there are three classes of parameters which can explain the reduction of ^8B neutrinos observed in the ^{37}Cl experiment. These roughly form the sides of a triangle, as is illustrated schematically in Fig. 16. For solution (a) corresponding to $|\Delta m^2| \sim 5 \times 10^{-5}$ eV^2, $\sin^2 2\theta \geq 4 \times 10^{-4}$, the adiabatic

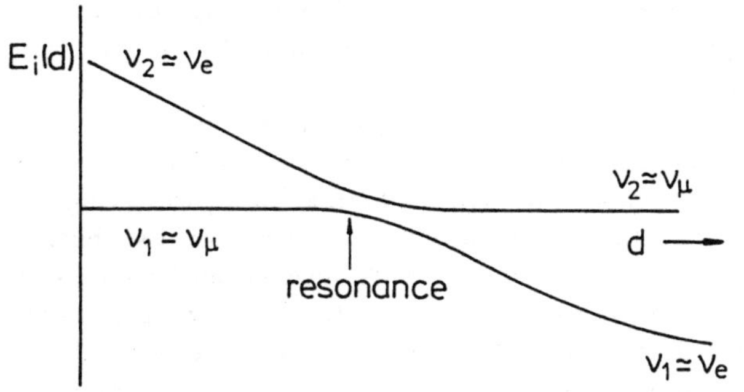

Figure 15: The energy eigenvalues of M' as a function of d, the distance from the center of the Sun.

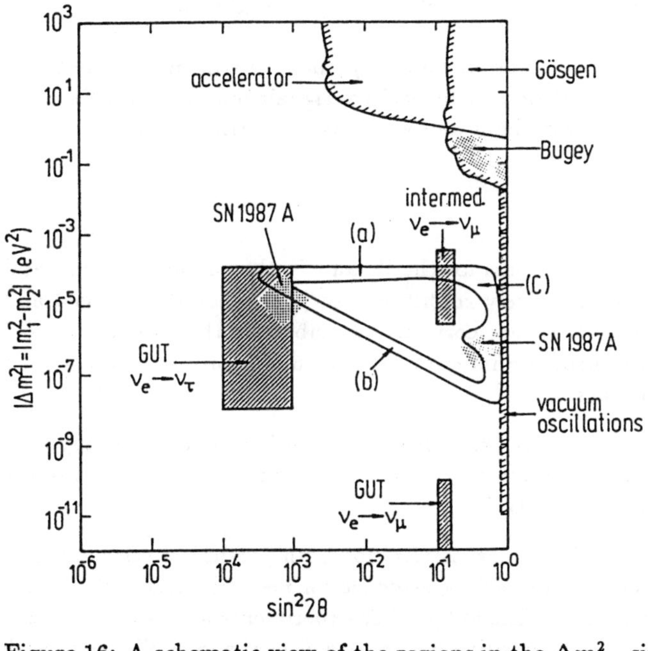

Figure 16: A schematic view of the regions in the Δm^2 - $\sin^2 2\theta$ plane which can explain the Solar neutrino problem via the Mikheyev-Smirnov-Wolfenstein (MSW) effect.

hypothesis is valid and $\simeq$ 100% conversion occurs. However, only the high energy ^8B neutrinos actually encounter a resonance layer (the central density is too low for the low energy neutrinos) and are converted. For this parameter range one expects little reduction in the counting rate for the gallium experiment (i.e. the effect is similar to non-standard Solar models in that respect).

For solution (b), extending down to $|\Delta m^2| \sim 10^{-8}\ eV^2$ one has [160] $|\Delta m^2| \sin^2 2\theta \sim 10^{-7.5}\ eV^2$. Here the adiabatic approximation starts to break down. All neutrino energies are affected, but the conversion probability is less than unity. For these solutions one expects a significant reduction in the gallium counting rate, similar to large vacuum oscillations or magnetic moments. Solution (c), corresponding to large vacuum mixing angles, is an extension of the vacuum oscillation solution. In the middle of region (c) one expects a large day-night asymmetry in the ν_e counting rate due to MSW regeneration in the Earth at night [152].

The MSW solution is very elegant, but it severely complicates the task of sorting out which if any of the proposed solutions to the Solar ν problem is correct. It will take an ambitious program of experiments [152,162] to clarify the matter.

The first two events from SN1987A observed by the Kamiokande experiment [93] point back towards the supernova. They are consistent with ν_e from the initial neutronization burst, scattering via $\nu_e e^- \rightarrow \nu_e e^-$. However, they could also be $\bar{\nu}_e$ from the subsequent thermal burst, scattering via $\bar{\nu}_e p \rightarrow e^+ n$, which has a much larger cross section (and which produces an isotropic distribution of positrons). If they are indeed ν_e they are problematic for the MSW mechanism because one expects $\nu_e \rightarrow \nu_\mu$ conversions on the way out of the supernova. However, there are two parameter regions (shown in Fig. 16), which would still be consistent [163], corresponding respectively to incomplete conversion in the supernova and reconversion in the Earth. Unfortunately there is no way to determine whether the two events are really ν_e's.

The MSW mechanism is consistent with the expectation of GUT [67] and intermediate scale [68] seesaws. As is illustrated in Fig. 16 the predictions of the GUT seesaw are consistent with $\nu_e \leftrightarrow \nu_\tau$ conversions in the Sun. In this case, the mass ranges are too small to ever see any direct laboratory effects of neutrino mass. The intermediate scale seesaw could account for the Solar ν problem via $\nu_e \leftrightarrow \nu_\mu$ conversions. In that case, $\nu_\mu \leftrightarrow \nu_\tau$ oscillations could well be observable in the laboratory.

It is also possible that the Solar ν problem could be explained by small neutrino masses, but that neutrino appearance experiments might nevertheless yield positive signals due to non-orthogonal neutrino states (induced by mixings between very light and very heavy neutrinos. [110])

5 Summary

- The predictions of the standard $SU_2 \times U_1$ model for the W and Z, the charged current, and the neutral current interaction are qualitatively confirmed. In particular, the charged and neutral current interactions of the neutrino are correctly described by the standard model to excellent precision. Furthermore, neutrino interactions are superb probes of the strong interactions and of possible new physics.

- Indirect arguments indicate that the ν_τ must exist. Nucleosynthesis constraints imply that there are no more than 4 neutrinos with masses $\leq 1\ MeV$. e^+e^- and $\bar{p}p$ constraints imply $\leq (3-5)$ neutrinos with masses up to $\sim 40\ GeV$.

- The question of whether the neutrinos have mass is vital for both particle physics and cosmology. However, there is at present no compelling evidence for neutrino mass. The ITEP result $m_{\nu_e} \sim (17-40)\ eV$ has not been confirmed by other experiments and is on the verge of being excluded. Although there are two positive indications of neutrino oscillations (with different parameters), these are contradicted by other experiments. The negative results suggest $m_{\nu_i} \simeq O(1\ eV)$ unless there are very small mixings or degeneracies. There are also stringent limits on incoherent mixing with heavy neutrinos.

- The MSW solution to the Solar neutrino problem would work for ν_μ or μ_τ in the $10^{-2}\ eV$ range, consistent with intermediate mass or GUT seesaws.

- The nonobservation of neutrinoless double beta decay implies $\langle m_{\nu_e} \rangle < 1 - 11\ eV$. If the ITEP result is correct this would most likely imply a Dirac neutrino or new physics to evade cosmological bounds.

- A ν mass in the 5 - 40 eV range would dominate the energy density of the universe and would be an excellent candidate for the dark matter, though other mechanisms would have to be invoked to explain the initial formation of galaxies. Conversely, the light stable neutrinos must be lighter than $\simeq$ 40 eV. A variety of astrophysical, laboratory, and cosmological bounds exclude unstable neutrinos up to $\sim$ 40 MeV (unless new physics is invoked for fast, invisible decays or annihilations), implying that m_{ν_μ}, $m_{\nu_\tau} < 40$ eV.

Acknowledgement

It is a pleasure to thank the Alexander von Humboldt-Stiftung, the DESY theoretical group, the Max Planck Inst. für Kernphysik, and the U.S. Department of Energy grant DE-AC02-ER0-3071 for support during the preparation of this paper.

References

[1] S. Weinberg, Phys. Rev. Lett. 19, 1264 (1967); A. Salam in *Elementary Particle Theory*, ed. N. Svartholm (Almquist and Wiksells, Stockholm, 1969) p. 367; S.L. Glashow, J. Iliopoulos, and L. Maiani, Phys. Rev. D2, 1285 (1970).

[2] As modified to include parity violation, quarks, flavor mixing, and the intermediate vector boson.

[3] UA1: G. Arnison et al, Phys. Lett. 166B, 484 (1986).

[4] UA2: R. Ansari et al., Phys. Lett. 186B, 440 (1987).

[5] The results given here are from U. Amaldi, A. Böhm, L.S. Durkin, P. Langacker, A.K. Mann, W.J. Marciano, A. Sirlin, and H.H. Williams, Phys. Rev. D36, 1385 (1987). Very similar conclusions are reached in an analysis by G. Costa, J. Ellis, G.L. Fogli, D.V. Nanopoulos, and F. Zwirner, CERN preprint CERN-TH.4675/87.

[6] Left (L) and right (R) chiral $(1 \pm \gamma^5)$ are equivalent to negative and positive helicity, respectively, for relativistic fermions, up to corrections of order m/E.

[7] For a recent review, see A. Sirlin, *1987 Int. Symp. on Lepton and Photon Interactions at High Energies*, Hamburg, July 1987.

[8] For reviews, see G. Barbiellini and C. Santoni, Riv. Nuo. Cim. 9(2), 1 (1986); E.D. Commins and P.H. Bucksbaum, *Weak interactions of leptons and quarks*, (Cambridge Univ. Press, Cambridge, 1983).

[9] W. Fetscher, H.-J. Gerber, and K.F. Johnson, Phys. Lett. 177B, 102 (1986); H.-J. Gerber, *Int. Europhysics Conference on High Energy Physics*, Uppsala (Sweden), June, 1987; W. Fetscher, 12th *Int. Conf. of Neutrino Physics and Astrophysics*, Sendai, Japan, June 1986.

[10] For a generalization, see P. Langacker and D. London, to be published.

[11] A. Jodidio et al., Phys. Rev. D34, 1967 (1986); see also I. Beltrami et al., Phys. Lett. B194, 326 (1987).

[12] The limit on W_R assumes that the neutrinos coupled to e_R^- and μ_R^- in the $V + A$ current are sufficiently light to be produced in μ decay, and does not apply if the $V + A$ current involves heavy majorana neutrinos.

[13] It is known independently that $h_{\nu_\mu} < 0$.

[14] A. Sirlin, Phys. Rev. D35, 3423 (1987), and [7].

[15] The result (9) also tests the radiative corrections to the standard model, which can only be consistently calculated within a gauge theory (they would be infinite in a 4-Fermi or intermediate vector boson theory). Without the radiative corrections, the right hand side of (9) would be 1.036, in apparent violation of unitarity [14].

[16] See, for example, F. Sciulli, *1985 Int. Symposium on Lepton and Photon Interactions at High Energies*, eds. M. Konuma and K. Takahashi (Nissha, Kyoto, 1986) p.8.

[17] I. Manelli et al., to be published.

[18] CP violation is incorporated in the standard model by a phase in the (small) components of V.

[19] H. Albrecht et al., Phys. Lett. 192B, 245 (1987).

[20] The expressions in (15) are modified slightly by radiative corrections. See [5].

[21] CDHS: H. Abramowicz et al., Phys. Rev. Lett. 57, 298 (1986).

[22] CHARM: J.V. Allaby et al., Phys. Lett. 177B, 446 (1986).

[23] C.H. Llewellyn Smith, Nucl. Phys. B228, 205 (1983).

[24] For a review, see [5].

[25] It is crucial to use the ratio of valence quark moments $D_v/U_v = 0.39 \pm 0.06$ determined from charged current scattering, [5,16] rather than the naive ratio 0.5.

[26] BBCIMOU: G.F. Jones et al., Phys. Lett. 178B, 329 (1986).

[27] E734: L.A. Ahrens et al., Phys. Rev. D35, 785 (1987).

[28] D. Rein and L.M. Sehgal, Nucl. Phys. B223, 29 (1983).

[29] Other reactions, such as inclusive and exclusive incoherent pion production and $\bar{\nu}_e D \to \bar{\nu}_e np$ are in qualitative agreement with the standard model, but are not used in the analysis because of large uncertainties in the hadronic matrix elements.

[30] CHARM: F. Bergsma et al., Phys. Lett. 147B, 481 (1984).

[31] BNL E734: L.A. Ahrens et al., Phys. Rev. Lett. 54, 18 (1985).

[32] Savannah River: F. Reines et al., Phys. Rev. Lett. 37, 315 (1976).

[33] LANL ILM: R.C. Allen et al., Phys. Rev. Lett. 55, 2401 (1985).

[34] The radiative corrections are computed using the Sirlin definition [24] $\sin^2\theta_W \equiv 1 - \frac{M_W^2}{M_Z^2}$ of the renormalized weak angle.

[35] Here the value is for fixed $m_t = 45\ GeV$, $M_H = 100\ GeV$, to facilitate comparison with SU_5.

[36] For a recent review, see P. Langacker, *Weak and Electromagnetic Interactions in Nuclei*, ed. H.V. Klapdor (Springer-Verlag, Berlin, 1986) p. 879; Phys. Rep. 72, 185 (1981).

[37] R. Marshall, Rutherford preprint RAL-87-031.

[38] V. Barger et al., Phys. Rev. D35, 2893 (1987).

[39] G. Beall, M. Bander, and A. Soni, Phys. Rev. Lett. 48, 848 (1982).

[40] G. Steigman et al., Phys. Lett.176B, 35 (1986).

[41] J. Ellis and K. Olive, Phys. Lett. 193B, 525 (1987).

[42] For reviews, see D.L. Burke, *Proc. of the Theoretical Advanced Study Institute in Particle Physics*, Santa Cruz, July 1986, SLAC-PUB-4284; N.G. Deshpande, *Neutrino Physics and Neutrino Astrophysics*, ed. V. Barger et al., (World, Singapore, 1987) p. 78.

[43] UA1: C. Albajar et al., Phys. Lett. 198B, 271 (1987); A.D. Martin, R.G. Roberts, and W.J. Stirling, Phys. Lett. 189B, 220 (1987); F. Halzen, C.S. Kim, and S. Willenbrock, MAD/PH/342 (1987).

[44] H.-J. Gerber, [9]; K. Mursula and F. Scheck, Nucl. Phys. B253, 189 (1985).

[45] A^τ also rules out the possibility that τ_R^- is assigned to an SU_2 doublet with $\nu_{\tau R}$ ($A^\tau = 0$ or $\frac{1}{2}$, depending on whether τ_L is in a doublet or singlet, respectively). These models are also excluded by the energy spectrum in leptonic τ decays [44].

[46] The nucleosynthesis constraint applies to all light neutrino degrees of freedom that were present in significant numbers at $T \sim 1\ MeV$. It does not apply to the right handed (SU_2 singlet) partners of the ordinary neutrinos that must be introduced if they are to have small Dirac masses, unless new interactions (e.g. new charged or neutral gauge bosons) are introduced as well. Otherwise, the singlet neutrinos have essentially no interactions and would not have been produced in significant numbers in the early universe. See G. Steigman, K.A. Olive, and D. Schramm, Phys. Rev. Lett. 43, 239 (1979); Nucl. Phys. B180, 497 (1981); J. Ellis et al., Phys. Lett. 167B, 457 (1986).

[47] UA1: C. Albajar et al., to be published.

[48] It is important to obtain an accuracy on ΔN_ν of much less than one, because other neutral weakly interacting particles can contribute fractional values to the apparent value of N_ν. For example, supersymmetric scalar neutrinos contribute $\Gamma_Z \sim 85\ MeV$, equivalent to $\frac{1}{2}$ unit of N_ν.

[49] There are several recent reviews of neutrino mass. See, for example, F. Boehm and P. Vogel, *Physics of Massive Neutrinos*, (Cambridge Univ. Press, Cambridge, 1987); J.D. Vergados, Phys. Rev. 133, 1 (1986); S.M. Bilenky and S.T. Petcov, Rev. Mod. Phys. 59, 671 (1987); R. Eichler, *1987 Int. Symp. on Lepton and Photon Interactions at High Energies.*

[50] Under a CP transformation $\psi_L(\vec{x},t) \to \gamma^0 \psi_R^c(-\vec{x},t)$, so ψ_L and ψ_R^c are essentially CP conjugates.

[51] L. Wolfenstein, Nucl. Phys. B186, 147 (1981).

[52] E.S. Konopinski and M. Mahmoud, Phys. Rev. 92, 1045 (1953).

[53] Fermion mass matrices need not be Hermitian. One could generalize still further and introduce $K (\neq F)$ N_R^0 fields, in which case m_D would be $F \times K$ dimensional.

[54] We have chosen a basis for the charged leptons so that the analogue of V_L for the electrons is the identity matrix.

[55] $V_L^\dagger$ contains F^2 real parameters, of which $F(F-1)/2$ are angles, analogous to the Cabibbo angle, which describe transitions between families and lead to such effects as neutrino oscillations. $2F-1$ parameters are unobservable phases, which can be removed by appropriate choices of phases for the n_{iL} and e_{iL}^- fields (corresponding phases must be chosen for the n_{iR} and e_{iR} to ensure real masses). The remaining $(F-1)(F-2)/2$ parameters are phases which could lead to CP violation in the leptonic sector.

[56] G.B. Gelmini and M. Roncadelli, Phys. Lett. 99B, 411 (1981); H. Georgi et al., Nucl. Phys. B193, 297 (1983).

[57] More generally, K is block diagonal, the blocks being unitary symmetric matrices in the subspaces of degenerate eigenvalues.

[58] One has $n_{iR}^c = K_{ji} C \bar{n}_{jL}^T$, so that n_i is self-conjugate under the generalized C transformation $n_i \rightarrow K_{ji} C \bar{n}_j^T$.

[59] The counting of angles in $U_L^\dagger$ is the same as for the Dirac case. However, there are an additional $F-1$ CP violating phases associated with the fact that U_L and U_R are related by (46). For example, if one chooses K_L so that $K = I$, then the phases of the n_{iL} are fixed and cannot be redefined to remove phases from U_L. See S.M. Bilenky, J. Hosek, and S.T. Petcov, Phys. Lett. 94B, 495 (1980); M. Doi et al., Phys. lett. 102B, 323 (1981).

[60] W.Y. Keung and G. Senjanovic, Phys. Rev. Lett. 50, 1427 (1983); J.F. Gunion and B. Kayser, *Proc. 1984 Summer Study on the SSC*, eds R. Donaldson and J.G. Morfin (APS, 1984) p. 153.

[61] H. Goldberg, Phys. Rev. Lett. 50, 1419 (1983); L. M. Krauss, Phys. Lett. 128B, 37 (1983), Nucl. Phys. B227, 556 (1983).

[62] For a review, see B. Kayser, Comm. Nucl. Part. Phys. 14, 69 (1985).

[63] New interactions could in principle distinguish the two cases even for $m_\nu = 0$, depending on whether they conserved lepton number.

[64] S.M. Bilenky and B. Pontecorvo, Lett. Nuovo Cimento 17, 569 (1976). The implications of the existence of Dirac and Majorana neutrino mass terms were also studied in: V. Barger et al., Phys. Rev. Lett. 45, 692 (1976); S.M. Bilenky, J. Hosek and S.T. Petcov, Phys. Lett. 94B, 495 (1980); J. Schechter and J.W. F. Valle, Phys. Rev. D22, 2227 (1980); T.P. Cheng and L.F. Li, Phys. Rev. D22, 2860 (1980); T. Yanagida and M. Yoshimura, Progr. Th. Phys. 64, 1870 (1980).

[65] The relative sign of n_{1R} and n_{2R} will reemerge, however, in neutrinoless double beta decay, where it will be seen to be observable.

[66] M. Gell-Mann, P. Ramond, and R. Slansky, in *Supergravity*, eds. F. van Nieuwenhuizen and D. Freedman, (North Holland, Amsterdam, 1979) p. 315; T. Yanagida, Prog. Th. Physics B135, 66 (1978).

[67] P. Langacker, S. T. Petcov, G. Steigman, and S. Toshev, Nucl. Phys. B282, 589 (1987).

[68] R.N. Mohapatra and G. Senjanovic, Z. Phys. C17, 53 (1983); Q. Shafi and F.W. Stecker, Phys. Rev. Lett. 53, 1292 (1984); K. Kang and M. Shin, Nucl. Phys. B287, 687 (1987); K. Kang and A. Pantziris, Phys. Lett. 193B, 467 (1987); P. Langacker, R.D. Peccei, and T. Yanagida, Mod. Phys. Lett. A1, 541 (1986).

[69] R.N. Mohapatra and G. Senjanovic, Phys. Rev. D23, 165 (1981). For constraints on the heavy neutrino mass, see R.N. Mohapatra; Phys. Rev. D34, 909 (1986).

[70] P. Langacker, B. Sathiapalan, and G. Steigman, Nucl.Phys. B266, 669 (1986).

[71] A. Zee, Phys. Lett. 93B, 389 (1980); 161B, 141 (1985); Nucl. Phys. B264, 99 (1986); K. Tamvakis and J. Vergados, Phys. Lett. 155B, 373 (1985); G.K. Leontaris and K. Tamvakis, Phys. Lett. 191B, 421 (1987); M. Fukugita and T. Yanagida, Phys. Rev. Lett. 58, 1807 (1987).

[72] Since ν_L transforms according to a doublet representation of SU_2 and N_R is a singlet, the Higgs field coupling ν_L to N_R must be an SU_2 doublet.

[73] A. Masiero, D.V. Nanopoulos, and A.Sanda, Phys. Rev. Lett. 57, 663 (1986).

[74] Unlike similar hierarchy problems in comparing the weak and unification scale, we know that there is no supersymmetry to control the radiative corrections.

[75] D.S.P. Dearborn, D.N. Schramm, and G. Steigman, Phys. Rev. Lett. 56, 26 (1986); H.Y. Cheng, Phys. Rev. D36, 1649 (1987); and references therein.

[76] R. Barbieri, J. Ellis, and M.K. Gaillard, Phys. Lett. 90B, 249 (1980).

[77] In many models which implement the seesaw mechanism Higgs triplet fields are also present. Even if these have zero VEV at tree level they may obtain VEV's due to higher order corrections to the Higgs potential. See M. Magg and C. Wetterich, Phys. Lett. 94B, 61 (1980); D. Chang and R.N. Mohapatra, Phys. Rev. D32, 1248 (1985); J. Bijnens and C. Wetterich, DESY 87-042 (1987).

[78] A simple diagonalization of $\begin{pmatrix} 0 & m_D \\ m_D & 0 \end{pmatrix}$ yields eigenvalues $-m_D^2/m_S$ and m_S. The sign of the smaller eigenvalue is reversed by choosing $K = \begin{pmatrix} -1 & 0 \\ 0 & 1 \end{pmatrix}$, which accounts for the minus sign in ν_R^{0c}.

[79] E. Witten, Phys. Lett. 91B, 81 (1980).

[80] Some versions of these models [77] are expected to have induced values for m_t in (72).

[81] Y. Chikashige, R.N. Mohapatra, and R.D. Peccei, Phys. Lett. 98B, 265 (1981).

[82] In many of the intermediate mass seesaw models [68] the singlet-Majoron and the invisible axion are the same particle.

[83] S. Nandi and U. Sarkar, Phys. Rev. Lett. 56, 564 (1986); J. Derendinger, L. Ibanez, and H.P. Nilles, Nucl. Phys. B267, 365 (1986).

[84] J. Ellis et al., Phys. Lett. 188B, 415 (1987).

[85] M. Roncadelli and D. Wyler, Phys. Lett. 133B, 325 (1983); A.S. Joshipura et al., Phys. Lett. 156B, 353 (1985); P. Roy et al, Phys. Rev. D30, 2385 (1984); O. Shanker, Nucl. Phys. B250, 351 (1985); J. Oliensis and C.H. Albright, Phys. Lett. 160B, 121 (1985).

[86] S. Boris et al., Phys. Rev. Lett. 58, 2019 (1987).

[87] M. Fritschi et al., Phys. Lett. 173B, 485 (1986).

[88] J.F. Wilkerson et al, Phys. Rev. Lett. 58, 2023 (1987).

[89] H. Kawakami et al., *'86 Massive Neutrinos in Physics and Astrophysics*, ed. O. Fackler and J. Tran Than Van.

[90] See, for example, D. Schramm, *1987 Int. Symp. on Lepton and Photon Interactions at High Energies*; H. Meyer, *X^{th} Workshop on Particles and Nuclei*, Heidelberg, Oct. 1987; E.W. Kolb et al., Phys. Rev. D35, 3598 (1987).

[91] R. Abela et al., Phys. Lett. B146, 431 (1984).

[92] ARGUS collaboration, quoted by R. Eichler, [49].

[93] K. Hirakata et al., Phys. Rev. Lett. 58, 1490 (1987).

[94] R.M. Bionta et al., Phys. Rev. Lett. 58, 1494 (1987).

[95] For reviews and a general discussion, see M. Gronau, C.N. Leung, and J.L. Rosner, Phys. Rev. D29, 2539 (1984); F.J. Gilman and S.H. Rhie, Phys. Rev. D32, 324 (1985). F.J. Gilman, Comm. Nucl. Part. Phys. 16, 231 (1986); R.E. Shrock, Phys. Rev. D24, 1232, 1275 (1981).

[96] CCFRR: S.R. Mishra et al., Phys. Rev. Lett. 59, 1397 (1987).

[97] R. Eichler, [49].

[98] For references, see [5].

[99] M.J. Duncan and P. Langacker, Nucl. Phys. B277, 285 (1986).

[100] B. Kayser and P. Langacker, to be published.

[101] There will be off-diagonal $\bar{\nu}_1\nu_2 Z$ vertices because of the mixing between SU_2 doublets and singlets.

[102] A similar phenomenon occurs for a system of coupled classical oscillators.

[103] For a discussion of quantum mechanical subtleties, such as the uncertainty in p, see F. Boehm and P. Vogel, [49].

[104] IMB, J.M. Lo Secco et al., Phys. Lett. 184B, 305 (1987).

[105] The limits in Fig. 13 are obtained by analyzing the data in terms of the formula in (80) for oscillations between two types of neutrino only. Slightly weaker limits are obtained if one allows mixings between 3 neutrino flavors (ν_e, ν_μ, ν_τ). See K. Kleinknecht, Comm. Nucl. Part. Phys. 16, 267 (1986).

[106] J.F. Cavaignac et al., Phys. Lett. 148B, 387 (1984).

[107] G. Zacek et al., Phys. Rev. D34, 2621 (1986).

[108] G. Bernardi et al., Phys. Lett. 181B, 173 (1986).

[109] Another manifestation of the underlying physics would be off-diagonal neutral currents, so that a heavy neutrino could decay into three light neutrinos.

[110] D. Wyler and L. Wolfenstein, Nucl. Phys. B218, 205 (1983); B.W. Lee and R.E. Shrock, Phys. Rev. D16, 1444 (1977); P. Langacker and D. London, to be published.

[111] For a review, see G. Gelmini, this volume.

[112] Both Γ_{weak} and H have additional factors involving the number of types of relativistic particles at temperature T.

[113] P. Langacker, J.P. Leveille, and J. Sheiman, Phys. Rev. D27, 1228 (1983).

[114] N. Cabibbo and L. Maiani, Phys. lett. 114B, 115 (1982).

[115] For a recent review, see B. Müller, *Proc. X^{th} Workshop on Particles and Nuclei: Neutrino Physics*, Heidelberg, Oct. 1987.

[116] For masses in the 10 eV range the $\nu's$ would be clustered in the galaxy and their cross sections would be increased by 10^8, but direct detection still appears virtually impossible.

[117] Massive neutrinos could easily account for the scales of galactic clusters. However, it is hard to see how such structures could have fragmented into galaxies in the time available, unless another ingredient, such as cosmic strings, provided the seed.

[118] For a recent review, see M. Roos, *X^{th} Workshop on Particles and Nuclei*; M. Turner, AIP Conf. Proc. 72, 266 (1980).

[119] Most of these limits apply to light SU_2-singlet neutrinos as well, because in many cases their small mixing with ordinary doublet neutrinos is sufficient to produce them prolifically in the early universe. See [70].

[120] S. Sarker and A.M. Cooper, Phys. Lett. B148, 347 (1984); L. Krauss, Phys. Rev. Lett. 53, 1976 (1984); R. Cowsik, Phys. Lett. 151B, 52 (1985); H. Harari and Y. Nir, Phys. Lett. 188B, 163 (1987), Nucl. Phys. B292, 251 (1987).

[121] Y. Hosotani, Nucl. Phys. B191, 411 (1981); M. Davis et al., AP.J. 250, 423 (1981); P. Hut and S. White, Nature 310, 637 (1984); P. Binetruy et al., Phys. Lett. 134B, 147 (1984); A.A. Natale, Phys. Lett. 141B, 323 (1984).

[122] M. Roncadelli and G. Senjanovic, Phys. Lett. 107B, 59 (1983).

[123] P.B. Pal, Nucl. Phys. B227, 237 (1983); H. Harari and Y. Nir, [120].

[124] D.B. Reiss, Phys. Lett. 115B, 217 (1982); F. Wilczek, Phys. Rev. Lett. 49, 1549 (1982); G. Gelmini, S. Nussinov, and T. Yanagida, Nucl. Phys. B219, 31 (1983).

[125] See [56].

[126] [81] and J. Schechter and J.W.F. Valle, Phys. Rev. D25, 774 (1982); G.B. Gelmini and J.W.F. Valle, Phys. Lett. 142B, 181 (1984); G.B. Gelmini et al., Phys. Lett. 146B, 311 (1984); A. Kumar and R. N. Mohapatra, Phys. Lett. 150B, 191 (1985).

[127] For recent reviews, see [75].

[128] P. Langacker, G. Segre, and S. Soni, Phys. Rev. D26, 3425 (1982).

[129] For a recent review, see A. Faessler, *X^{th} Workshop on Particles and Nuclei.*

[130] T. Kirsten et al., Phys. Rev. Lett. 50, 474 (1983), Z. Phys. C16, 189 (1983).

[131] M.K. Moe, *X^{th} Workshop on Particles and Nuclei.*

[132] Neutrinoless double beta decay could conceivably occur through other mechanisms, such as Higgs exchange, or could be accompanied by the emission of a triplet-Majoron. [134] See G.K. Leontaris and J.D. Vergados, CERN-TH.4835/1987. J.D. Vergados, [49].

[133] F.T. Avignone, this volume.

[134] The PNL/USC group [133] has reported evidence for $^{76}Ge \rightarrow\ ^{76}Se\ e^-e^-+$ Majoron, but their results are in contradiction with other experiments.

[135] W.C. Haxton and G.J. Stephenson, Prog. Part. Nucl. Phys. 12, 409 (1984); T. Tomoda et al., Nucl. Phys. A452, 591 (1986); K. Grotz and H.V. Klapdor, Nucl. Phys. A460, 395 (1986); T. Tomoda and A. Faessler, to be published.

[136] J. Engel, P. Vogel, and M.R. Zirnbauer, to be published.

[137] L. Wolfenstein, Phys. Lett. 107B, 77 (1981); M. Doi et al., Phys. Lett. 102B, 323 (1981).

[138] See W.C. Haxton and G.J. Stephenson, [135], J.D. Vergados, [49].

[139] A. Halprin, S.T. Petcov, and S.P. Rosen, Phys. Lett. 125B, 335 (1983); C.N. Leung and S.T. Petcov, Phys. Lett. 145B, 416 (1984).

[140] M. Doi, T. Kotani, and E. Takasugi, Prog. Th. Phys. (Supp) 83, 1 (1985).

[141] However, in a gauge theory some neutrino must be massive. Otherwise the diagrams cancel. See B. Kayser, *7th Moriond Workshop on Searches for New and Exotic Phenomena*, Les Arcs, Jan. 1987.

[142] J.K. Rowley, B.T. Cleveland and R. Davis, AIP Conference Proc. 126, 1 (1985).

[143] A Solar neutrino unit (SNU) is defined as 10^{-36} interactions per target nucleus per second.

[144] J.N. Bahcall et al., Rev. Mod. Phys. 54, 767 (1982); J.N. Bahcall, AIP Conference Proc. 126, 60 (1985).

[145] H. Hirata et al, to be published.

[146] For recent discussions, see V. Barger et al., Phys. Rev. D24, 538, (1981); S.L. Glashow and L.M. Krauss, Phys. Lett. 190B, 199 (1987); L. Krauss and F. Wilczek, Phys.Rev. Lett. 55, 122 (1985).

[147] M.B. Voloshin and M.I Vysotsky, ITEP 86-1(1986); L.B. Okun et al., Zh. Eksp. Teor. Fiz. 91, 754 (1986) (JETP 64, 446 (1986)). For earlier references, see [67].

[148] A.V. Kyuldjiev, Nucl. Phys. B243, 387 (1987); K. Abe et al., Phys. Rev. Lett. 58, 636 (1987).

[149] For a recent discussion, see M. Fukujita and S. Yazaki, Kyoto preprint. RIFP-709.

[150] B.W. Lee and R. Shrock, Phys. Rev. D16, 1444 (1977); K. Fujikawa and R. Shrock, Phys. Rev. Lett. 45, 963 (1980).

[151] M.J. Duncan et al., Phys. Rev. Lett. 191B, 304 (1987); J. Liu, Phys. Rev. D35, 3447 (1987); M. Fukugita and T. Yanagida, [71].

[152] For a review, see W. Hampel, this volume.

[153] F.A. Frieman, H.E. Haber, and K. Freese, Phys. Lett. 200B, 115 (1988).

[154] S.P. Mikheyev and A. Yu Smirnov, Yad Fiz. 42, 1441 (1985) (Sov. Jour. Nucl. Phys. 42, 1441 (1985)); Nuo. Cim. 9C, 17 (1986).

[155] L. Wolfenstein, Phys. Rev. D17, 2369 (1968); D20, 2634 (1979).

[156] V. Barger et al., Phys. Rev. D22, 2718 (1980).

[157] A crucial sign error was corrected in [113].

[158] The scattering amplitude in M' changes sign for antineutrinos. Hence, resonance can occur for either ν_e or $\bar{\nu}_e$, depending on the sign of Δm^2.

[159] H.A. Bethe, Phys. Rev. Lett. 56, 1305 (1986).

[160] S.P. Rosen and S.M. Gelb, Phys. Rev. D34, 969 (1986).

[161] P. Langacker, S.T. Petcov, G. Steigman, and S. Toshev, Nucl. Phys. B282, 589 (1987).

[162] E.W. Beier et al., 1986 Snowmass Workshop, Pennsylvania UPR-0140E.

[163] J. Arafune et al., Phys. Rev. Lett. 59, 1864 (1987); P.O. Lagage et al., Phys. Lett. 193B, 127 (1987); L. Wolfenstein, Phys. Lett. 194B, 197 (1987); D. Nötzgold, Phys. Lett. 196B, 315 (1987); S. P. Rosen, Los Alamos LA-UR-87-1296.

290

NEUTRINO MASS TEXTURES AND THE NATURE OF NEW PHYSICS IMPLIED BY PRESENT NEUTRINO DATA

R. N. MOHAPATRA
Department of Physics, University of Maryland, College Park Maryland 20742

If all the indications for neutrino oscillations observed in the solar, atmospheric neutrino data as well as in the LSND experiment are borne out by the ongoing and future experiments, then they severely constrain the neutrino mass texture. In particular, the need for an extra ultra-light sterile neutrino species is hard to avoid. Such an extra neutrino has profound implication not only for physics beyond the standard model but even perhaps for physics beyond conventional grand unification. We discuss a scenario involving a parallel (or shadow) universe that interacts with the familiar universe only via the gravitational interactions where the ultra-lightness of the sterile neutrino follows from the same physics that explains the near masslessness of the familiar neutrinos.

1 Introduction

There are several different observations involving neutrinos which receive a plausible and satisfactory explanation if the neutrinos are massive. First is the well-known solar neutrino deficit[1], observed by four different experiments[2]. Second is the deficit of muon neutrinos relative to electron neutrinos produced in the atmosphere, as measured by three experiments[3]. Third is the reported evidence for $\bar{\nu}_\mu$ to $\bar{\nu}_e$ oscillation from the Los Alamos Liquid Scintillation Neutrino Detector (LSND) experiment[5]. Finally, there is the likely need for a neutrino component of the dark matter of the universe to understand the structure and density on all distance scales[6]. Since the highly successful standard model of particle physics predicts zero mass for all the neutrinos, confirmation of any one of the above observations by ongoing and future experiments will already be a major step towards decoding the nature of new physics beyond the standard model. If however all the above findings are substantiated in future, one can reasonably expect the nature of this new physics to fall into only very few categories. In this talk, I will assume the validity of all the above findings (although it is clear that they must be considered tentative until further confirmation by ongoing and future experiments) and argue first that it severely restricts the neutrino mass texture and in particular requires the existence of a new ultra-light sterile neutrino. I will then outline the two see-saw formulae for understanding the small neutrino masses, and discuss their implementation in $SO(10)$ models. I will then argue that the simplest scenario which explains the lightness of the sterile neutrino in a natural manner is one that involves a

in Proc. NEUTRINO'96, Helsinki, Finland, June 1996, eds. *K. Enqvist, K. Huitu and J. Maalampi, W.S., Singapore,* (1997) 290 - 302 shadow universe which has identical particle and force content as the familiar universe but with the weak scale being somewhat higher.

1.1 Solar Neutrino Deficit

For massive neutrinos which can oscillate from one species to another, the solar electron neutrino observations[2] can be understood if the neutrino mass differences and mixing angles fall into one of the following ranges[1], where the Mikheyev-Smirnov-Wolfenstein (MSW) mechanism is included:

$$\begin{array}{ll} \text{a)} & \text{Small} - \text{angle MSW}, \; \Delta m^2_{ei} \sim 5\times 10^{-6} - 10^{-5}\text{eV}^2, \; \sin^2 2\theta_{ei} \sim 7\times 10^{-3}, \\ \text{b)} & \text{Large} - \text{angle MSW}, \; \Delta m^2_{ei} \sim 9\times 10^{-6}\text{eV}^2, \; \sin^2 2\theta_{ei} \sim 0.6, \end{array} \quad (1)$$

(1)

If the solar neutrinos oscillate into sterile neutrinos, the MSW effect is different from the ν_e to ν_μ case and the large angle solution is no more allowed. The above results are based on the approximation that only two of the neutrino species are involved in the oscillation.

1.2 Atmospheric Neutrino Deficit

The second set of experiments indicating non-zero neutrino masses and mixings has to do with atmospheric ν_μ's and ν_e's arising from the decays of π's and K's and the subsequent decays of secondary muons produced in the final states of the π and K decays. In the underground experiments the ν_μ and $\bar{\nu}_\mu$ produce muons and the ν_e and $\bar{\nu}_e$ lead to $e^\pm$. Observations of $\mu^\pm$ and $e^\pm$ indicate a far lower value for ν_μ and $\bar{\nu}_\mu$ than suggested by naïve counting arguments which imply that $N(\nu_\mu + \bar{\nu}_\mu) = 2N(\nu_e + \bar{\nu}_e)$ [3]. If one assumes that the oscillation of ν_μ to ν_τ provides an explanation of these results, then to fits to both the sub-GeV and multi-GeV data require that[4]

$$\Delta m^2_{\mu\tau} \approx 0.025 \text{ to } 0.005 \text{ eV}^2, \; sin^2 2\theta_{\mu\tau} \approx .6 \;\; to \;\; 1. \quad (2)$$

A recent reanalysis of the data[7] seems to imply that the data allows an upper limit on the Δm^2 upto .25 eV^2 at 90% confidence level.

1.3 Results from the LSND experiment

Recently, the LSND collaboration has published the results of their search for $\bar{\nu}_\mu$ to $\bar{\nu}_e$ oscillation using the liquid scintillation detector at Los Alamos. Combining their results which indicate a positive result with the negative results by the E776 group and the Bugey reactor data, one can conclude that a mass difference squared between the ν_e and the ν_μ lies between

$$0.27\ eV^2 \leq \Delta m^2 \leq 2.3\ eV^2 \qquad (3)$$

with points at 6 and 10 eV^2 also perhaps allowed.

1.4 Hot Dark Matter

There is increasing evidence that more than 90% of the mass in the universe must be detectable so far only by its gravitational effects. This dark matter is likely to be a mix of $\sim$ 20% of particles which were relativistic at the time of freeze-out from equilibrium in the early universe (hot dark matter) and $\sim$ 70% of particles which were non-relativistic (cold dark matter). Such a mixture gives the best fit of any available model to the structure and density of the universe on all distance scales, such as the anisotropy of the microwave background, galaxy-galaxy angular correlations, velocity fields on large and small scales, correlations of galaxy clusters, etc. A very plausible candidate for hot dark matter is one or more species of neutrinos with total mass of $m_{\nu_H} = 93h^2F_H\Omega = 4.8$ eV, if $h = 0.5$ (the Hubble constant in units of 100 km·s^{-1}·Mpc^{-1}), $F_H = 0.3$ (the fraction of dark matter which is hot), and $\Omega = 1$ (the ratio of density of the universe to closure density).

It is usually assumed that the ν_τ would supply the hot dark matter. However, if the atmospheric ν_μ deficit is due to $\nu_\mu \to \nu_\tau$, the ν_τ alone cannot be the hot dark matter, since the ν_μ and ν_τ need to be closer to each other in mass. It is interesting that instead of a single $\sim$ 4.8 eV neutrino, sharing that $\sim$ 4.8 eV between two or among three neutrino species provides a better fit to the universe structure and particularly a better understanding of the variation of matter density with distance scale[6].

It is worth noting that an equally popular picture adopts the hypothesis that there is a large cosmological constant ($\Omega_\Lambda = .8$ or so) in a low density baryon plus CDM universe to make up $\Omega = 1$[8]. This has been inspired by reported large values ($h_0 = .7 - .8$ or so) of the Hubble parameter from several observations[9] which have hard time fitting the age of the universe (e.g. from globular clusters) with $\Omega = 1$ without a cosmological constant. There are however other observations that give a lower value for h_0 ($h_0 \simeq .5$). The final verdict on the dark matter picture of the universe will therefore have to wait. It is nevertheless heartening that there is a compelling case for a neutrino mass in the eV range from structure formation in the universe.

In understanding the detailed implications of these data for physics beyond the standard model, one must also take into account other constraints on neutrinos, from nucleosynthesis, the Heidelberg-Moscow[11] $\beta\beta_{0\nu}$ experiment searching for the Majorana mass of the neutrino using enriched ^{76}Ge and the synthesis of heavy elements supposedly by the rapid neutron capture process (the so-called *r-process*) around supernovae[12].

1.5 Other costraints:

(i) While the Z^0 width limits the number of weakly interacting neutrino species to three, the nucleosynthesis limit[10] of about 3.3 on the number of light neutrinos is more useful here, since it is independent of the neutrino interactions. Invoking a fourth neutrino, ν_s, which is sterile, meaning it does not have the usual weak interaction, must be done with parameters such that it will not lead to overproduction of light elements in the early universe. For example, the atmospheric ν_μ problem cannot be explained by $\nu_\mu \to \nu_s$, since $\sin^2 2\theta_{\mu s} \approx 0.5$ is too large for the $\Delta m^2_{\mu s}$ involved, and that ν_s would have been brought into equilibrium in the early universe . On the other hand, the solar ν_e problem can be explained by $\nu_e \to \nu_s$ for either the small-angle MSW or the vacuum oscillation solutions, but not for the less favored large-angle MSW solution.

(ii) The Heidelberg-Moscow ^{76}Ge experiment[11] has been conducting a high precision search for neutrinoless double beta decay for the past several years and have at present set the most stringent upper limits on the effective Majorana mass of the neutrino: $< m_\nu > \leq .56$ eV.

(iii) It has been pointed out that in minimal model with three massive neutrinos, supernova r-processes provide a very stringent constraint on the neutrino mixings for eV mass range or higher. The origin of this constraint can be understood as follows. Inside the supernova, the MSW phenomenon enhances the conversion of the muon neutrinos (which have higher energy) to electron neutrinos if the mass difference square $\Delta m^2 \geq 4\ (eV)^2$ while leaving the $\bar{\nu}_\mu$'s unaffected. The newly born high energy ν_e's deplete the neutron content of the supernova environment via the reaction $\nu_e + n \to e^- + p$. This reduction of the neutron content slows down the r-process making it difficult to understand the heavy element abundance of the present universe. This result crucially hinges on the assumption that $m_{\nu_\mu} \geq m_{\nu_e}$ and that there are neutrinos that ν_μ mixes with. In fact in the presence of sterile neutrinos, its mixing with ν_μ can lead to MSW enhancement of ν_μ to ν_s conversion deeper in the supernova providing a way out of this constraint[13].

2 Neutrino mass textures consistent with data

In discussing the neutrino mass textures in this section, we will assume that all the neutrinos are Majorana particles, since it is easier to understand the smallness of Majorana masses of neutrinos within the framework of grand unified theories. Before going to a detailed discussion of the allowed mass matrices, let us note two generic requirements for the allowed mass matrices dictated by the data: *(i) at least two neutrinos must be degenerate in mass; and (ii) there is a very compelling case for the existence of a sterile neutrino in the present data.*

2.1 Are neutrinos degenerate?

If only two of the above hints (either solar and atmospheric data or solar and HDM) are taken seriously, then one can maintain a hierarchical picture for neutrino masses i.e. $m_{\nu_e} \ll m_{\nu_\mu} \ll m_{\nu_\tau}$. Such a pattern emerges very naturally in one class of the see-saw models (see below). However, if we take any three of the above four hints for neutrino masses, then we must have at least two neutrinos degenerate[15]. To see the case for a sterile neutrino, let us first note that it is not easy to write down a neutrino mass matrix within the three generation picture that can accomodate all the above observations as well as constraints. The main obstacle comes from the conflict between the LSND data and the MSW resolution of the solar neutrinodata. The first requires that the $\Delta m^2_{\nu_e - \nu_\mu}$ is in the eV range which is much larger than the mass difference required to explain the solar neutrino data.. If we ignore the LSND data, the solar, atmospheric data and the HDM neutrino can be accomodated in a three neutrino scenario by assuming that all three neutrinos are degenerate in mass.[15]. One very marginal possibility[14] has been advocated recently using a variant of this, to accomodate the LSND results in this picture provided the LSND Δm^2 is chosen to be around .3 eV^2. First point to note is that since solar neutrino puzzle requires that $\Delta m^2_{e-\mu} \simeq 10^{-5}$ eV^2, to understand the LSND results in this scenario, one must use the complete three neutrino oscillation keeping all mixing angles[17]. This requires first that $\nu_e - \nu_\tau$ mixing angle is not too small. Secondly, we must have $\Delta^2_{\mu-\tau}$ be $\approx .3\ eV^2$. Thus the oscillation frequency is determined by ν_e-ν_τ mass difference. The main problem for this scenario comes from the atmospheric neutrino data, since the original analysis of the Kamiokande sub-GeV and the multi-GeV data by the Kamiokande group excludes $\Delta m^2 \geq .1\ eV^2$ at 90% confidence level (c.l.). As mentioned earlier, a subsequent analysis [7] extends this range further excuding only the $\Delta m^2 \geq 0.25\ eV^2$ at 90% confidence level while allowing it at 95% c.l. level. While at its face a value of $m_{\nu_e} \simeq 1.6$ eV may appear to be in conflict with the neutrinoless double beta decay limit [11], one can hide under the uncertainties of nuclear matrix element calculations which typically could be as much as a factor of 2-3. As the precision in $\beta\beta_{0\nu}$ search improves further (say to the level of 0.1 eV), nuclear matrix element uncertainties cannot come to the rescue and this mass texture will then be ruled out. One can write the neutrino Majorana mass matrix for this case as follows:

$$M = \begin{pmatrix} m + \delta_1 s_1^2 & -\delta_1 c_1 c_2 s_1 & -\delta_1 c_1 s_1 s_2 \\ -\delta_1 c_1 c_2 s_1 & m + \delta_1 c_1^2 c_2^2 + \delta_2 s_2^2 & \delta_1 c_1^2 c_2 s_2 - \delta_2 c_2 s_2 \\ -\delta_1 c_1 s_1 s_2 & \delta_1 c_1^2 s_2 c_2 - \delta_2 c_2 s_2 & m + \delta_1 c_1^2 s_2^2 + \delta_2 c_2^2 \end{pmatrix}, \tag{4}$$

where $c_i = \cos\theta_i$ and $s_i = \sin\theta_i$, $m = 1.6$ eV; $\delta_1 \simeq 1.5 \times 10^{-5}$ eV; $\delta_2 \simeq .1$ eV; $s_1 \simeq 0.05$; and $s_2 \simeq 0.4$ for the small-angle MSW solution.

2.2 The need for a sterile neutrino

We thus see that if the above scenario is ruled out, for instance by the tightening of the double beta decay limit on the Majorana mass of ν_e or by the atmospheric neutrino data, then the only way to understand all neutrino results will be to assume the existence of an additional neutrino species which in view of the LEP data must not couple (or couple extremely weakly) to the Z-boson.We will call this the sterile neutrino. The picture then would be as follows[15,16]: the solar neutrino puzzle is explained by the $\nu_e - \nu_s$ oscillation; atmospheric neutrino data would be explained by the $\nu_\mu - \nu_\tau$ oscillation. The LSND data would set the overall scale for the masses of ν_μ and ν_τ (which are nearly degenerate) and if this scale is around 2 to 3 eV as is allowed by the data[5], then the $\nu_{\mu,\tau}$ would constitute the hot dark matter of the universe. The mass matrix in this case would be in the basis (ν_s, ν_e, ν_μ, ν_τ),

$$M = \begin{pmatrix} \mu_1 & \mu_3 & \epsilon_{11} & \epsilon_{12} \\ \mu_3 & \mu_2 & \epsilon_{21} & \epsilon_{22} \\ \epsilon_{11} & \epsilon_{21} & m & \delta/2 \\ \epsilon_{12} & \epsilon_{22} & \delta/2 & m+\delta \end{pmatrix}. \tag{5}$$

For simplicity, we set the $\epsilon_{12} = \epsilon_{22} = 0$ and $\mu_2 \ll \mu_1 \simeq 10^{-3}\ eV$. The ϵ_{11} term is responsible for the $\nu_e - \nu_\mu$ oscillation that can explain the LSND data. The apparent problem for such a scenario comes from the supernova r-process nucleosynthesis. But it has been argued[13] that in such a scenario, the ν_μ can oscillate into the ν_s at a smaller protoneutron star radius before it reaches the radius where ν_μ to ν_e MSW transition occurs. This may enable one to evade the r-process bound for $\Delta m^2_{e-\mu} \geq 4\ eV^2$. Clearly the crucial test of the sterile neutrino scenario will come when SNO collaboration obtains their results for neutral current scattering of solar neutrinos. One would expect that $\Phi_{CC} = \Phi_{NC}$ if the ν_e oscillation to ν_s is responsible for the solar neutrino deficit. There should be no signal in $\beta\beta_{0\nu}$ search. Precision measurement of the energy distribution in charged current scattering of solar neutrinos at Super-Kamiokande can also shed light on this issue.

Before proceeding to the discussion of the theoretical implications of the mass textures outlined above, we want to note that if the atmospheric neutrino data is excluded but LSND, HDM and solar neutrino constraints are kept, a theoretical explanation for them can be found also with an inverted mass texture[18] for neutrinos where the $m_{\nu_e} \simeq m_{\nu_\tau} \simeq 2.4\ eV \gg m_{\nu_\mu}$ and which does not invoke the sterile neutrino. This texture is consistent with the supernova r-process constraints and uses the $\nu_e \to \nu_\tau$ large angle MSW solution to explain the solar neutrino data. This could therefore be tested once Super Kamiokande results for the neutrino energy spectrum as well as the data on day-night variation is in.

3 Implications for higher unification and two types of see-saw mechanism

In this section we address the question of what implications the small nonzero neutrino masses and in particular any of the scenarios discussed above have for the nature for the nature of new physics beyond the standard model. To start with let us remind the reader that in the standard model the presence of an exact global B-L symmetry combined with the absence of the right handed neutrino leads to zero mass for all neutrinos. The simplest way to generate a nonzero neutrino mass is therefore to add three right handed neutrinos N_i, one per generation. It is easy to see that as soon as the N_i are included, the maximal anomaly-free gaugeable symmetry becomes $SU(2)_L \times SU(2)_R \times U(1)_{B-L} \times SU(3)_c$ which can eventually lead to an $SO(10)$ grand unification of fermions. As has been shown during the past decade and half, this class of models provide the most natural framework for describing the neutrino masses[19,20]. What happens in these models is that as the $SU(2)_R \times U(1)_{B-L}$ gauge group breaks down to $U(1)_Y$, not only the right-handed gauge bosons but also the right handed neutrinos of all three generations acquire a mass fv_R proportional to the $B-L$ breaking scale $v_R \gg v_W$, where v_W is the electroweak symmetry breaking scale of the standard model. At this stage, the left handed neutrinos are massless. At the scale v_W, the Dirac mass for the neutrino that connects the left and the right handed neutrino is generated with a value given by hv_W which is expected to be of the order of the masses of the charged fermions m_f which also arise at that scale. This leads to the see-saw matrix for the neutrinos[21,19,20],

$$M = \begin{pmatrix} 0 & m_f \\ m_f & M_N \end{pmatrix} \tag{6}$$

The diagonalization matrix leads to the generic formula for neutrino masses

$$m_{\nu_i} \simeq \frac{(m_f)^2}{M_{N_i}} \tag{7}$$

This formula has two interesting implications: (i) the first is that the neutrinos, which are now necessarily Majorana fermions have masses which are suppressed compared to the masses of the charged fermions of the corresponding genearation and (ii) the neutrino masses show a generationwise hierarchical pattern linked to the square of the masses of the charged fermions of the corresponding generation (i.e. $m_{\nu_e} \ll m_{\nu_\mu} \ll m_{\nu_\tau}$) We will call this the type I see-saw formula.

It was pointed out in[20] that when the spontaneous symmetry breaking of the left-right model (or its $SO(10)$ grand unified version) is carefully analyzed, one actually gets a modified neutrino mass matrix given by

$$\begin{pmatrix} \frac{\lambda v_W^2}{v_R} & m_f \\ m_f & M_N \end{pmatrix} \tag{8}$$

Diagonalization of the above mass matrix leads to what I call the type II see-saw formula for the neutrino masses:

$$m_{\nu_i} \simeq \frac{\lambda v_W^2}{v_R} - \frac{m_f^2}{M_{N_i}} \tag{9}$$

Note that the first term in the above formula is practically generation independent. Therefore, the neutrino mass pattern in this case is not hierarchical and could lead to a nearly degenerate spectrum as has been advocated in the previous section.

There are conditions under which the type II see-saw formula reduces to a type I see-saw formula[22]: for instance when the discrete parity symmetry of the left-right or $SO(10)$ models is broken at a scale higher than the $SU(2)_R \times U(1)_{B-L}$ gauge symmetry, then the first term in the type II see-saw formula is replaced by $\lambda v_W v_R^2/M_P^2$ (M_P being the scale of discrete parity breaking), which can clearly be arranged to yield the hierarchical mass pattern. Another class of models where the type I see-saw can emerge are some supersymmetric models with restricted Higgs representations.

A very interesting point worth emphasizing here is that if we look at the typical masses needed to solve the solar as well as the atmospheric neutrino puzzles and use the see-saw formula to find the scale of $B-L$ symmetry breaking, we find that $v_R \approx 10^{12} - 10^{13}$ GeV. It may be more than a mere coincidence that the B-L breaking scale of $v_R \sim 10^{13}$ GeV emerges naturally from constraints of $\sin^2 2\theta_W$ and α_s in non-supersymmetric SO(10) grandunified theories[25], as well as supersymmetric SO(10)[23] theories. In the least it enhances the reason for an SO(10) scenario. It is however possible to construct TeV scale right handed neutrino scenarios[24] where the suppression of the neutrino mass originates from the fact that the Dirac masses are radiatively induced. To summarize this section, it is reasonable to conclude that evidence for a small neutrino mass would indicate the existence of a local $B-L$ symmetry in nature and perhaps even a left-right symmetry, which will be a major new dimension to our understanding of particle physics. Secondly, the generic class of grand unified models where the see-saw mechanism (both type I and type II) is naturally implemented are based on the $SO(10)$ GUT group with the type I see-saw leading to a hierarchical pattern for neutrino masses whereas the type II leads to a near degenerate pattern. In the next section, we explore whether definite predictions can be made for the neutrino masses and mixing angles in this class of models.

4 Predictions from minimal SO(10) grand unification models

If there are only three light neutrinos, it is both economical and elegant to work within simple $SO(10)$ grand unified models. While simple electroweak gauge theories without additional symmetries do not have the capability to predict fermion masses, the assumption of grand unification improves this record somewhat (e.g. the prediction of b-quark mass in $SU(5)$). In the minimal $SO(10)$ models, the neutrino Dirac mass and the up-quark mass matrices become equal since they both arise from the Yukawa couplings of the fermions (which belong to the **16** dim representations) to the Higgs boson in the **10**-dim representation, thereby reducing the number of free parameters. This raises the possibility for a prediction of the neutrino masses in these models. The problem however is that the Majorana mass of the N_i arises from the couplings of the fermions to the $\mathbf{\bar{126}}$ dimensional Higgs bosons. Since these couplings are arbitrary, in general no specific predictions can be made. It was however pointed out by Babu and this author[26] that in the minimal $SO(10)$ models, the standard model doublets arise from an admixture of the $SU(2)_L$ doublets in **10** and $\mathbf{\bar{126}}$ dimensional Higgs bosons. Therefore, the $\mathbf{\bar{126}}$ Yukawa couplings (as well as those to **10**) get predicted in terms of the quark, lepton masses and their mixings. This model (which is a realization of the type I see-saw mechanism) therefore leads to numerical predictions for the neutrino masses and mixings. The reader is refered to the original papers for the detailed predictions for the non-supersymmetric[26,27] as well as supersymmetric versions[28] of the model. There are actually six solutions depending on the relative signs of the various quark masses. Here we simply want to note that there are predictions in both versions that can accomodate the small angle MSW solution to the solar neutrino puzzle but not the atmospheric nor the LSND nor the HDM neutrino.

There is another class of $SO(10)$ models[29] where additional symmetries are imposed to fix the heavy Majorana mass matrix for the right handed neutrinos and different popular quark mass textures are used for the neutrino Dirac masses. They also implement the type I see-saw formula and give generic predictions that can accomodate only the small angle MSW solutions to the solar neutrino puzzle.

Finally a different class $SO(10)$ models were studied[30] where the type II see-saw mechanism was implemented. Using an additional S_4-permutation symmetry on the fermions and the Higgs bosons, it was possible to obtain a realization of the degenerate neutrino mass mixing angle predictions that can solve both the solar as well as the atmospheric neutrino problem.

5 Beyond grand unification: Into the shadow universe

Once we admit the possibility of light sterile neutrinos, one needs to go beyond simple grand unified models to understand why the sterile neutrino is so light. The reason for this is that the sterile neutrino by definition is an $SU(2)_L \times U(1)_Y$ singlet and therefore is allowed to have an arbitrary mass unless there are some compelling new symmetries that keep it massless. Attempts have been made using additional $U(1)$'s and supersymmetry[31] etc to achieve this goal. But it is perhaps fair to say that there are no compelling motivations for such symmetries. To circumvent such arguments, it was proposed in [32] to make the conjecture that there is an exact duplication of the standard model in both the gauge as well as the fermion content i.e. an extra $G'_{standard}$ with $Q', u^{c\prime}, d^{c\prime}, L', e^{c\prime}$ etc. (this adds a new sector to the world of elementary particles, which will be called the shadow sector). It is then clear that we have three extra neutrinos which do not interact with the Z-boson. We further assume that the only interactions that connect the known and the shadow sector is the gravitational interaction.

Within this framework, it is easy to understand that the shadow neutrinos (which will be the sterile neutrinos) are massless in the renormalizable theory for exactly the same reason that the ordinary neutrinos are (i.e. the existence of a $B' - L'$ symmetry in the shadow standard model sector). We may assume that there is a "shadow" see-saw mechanism which operates exactly the same way to give tiny masses to the shadow neutrinos. The next question is how do the shadow (or sterile) neutrinos acquire small masses and mix with the known neutrinos ? Here we use the existing lore that all global symmetries are broken by Planck scale effects. It was already pointed out[33] that one can write Planck scale induced operators such as LH_uLH_u/M_P, $LH_uL'H'_u/M_P$ and $L'H'_uL'H'_u/M_P$ which violate both $B - L$ as well as $B' - L'$ symmetries and after electrweak symmetry breaking in both the sectors lead to $\nu - \nu'$ mixing. If we now make the additional assumption that $v'_W \simeq 30v_W$, the resulting $\nu_e - \nu'_e$ mass matrix gives a solution to the solar neutrino puzzle with small mixing angles. When this idea is combined with the postulate that there exists an $L_e + L_\mu - L_\tau$ symmetry (instead of the overall $B - L$ symmetry) that is broken by Planck scale effects, we come up with a neutrino mass matrix that explains all neutrino puzzles using the four neutrino mass texture noted in Ref.[15].

The next interesting feature of these models is that if the $m_{\nu_\mu} \simeq m_{\nu_\tau} \simeq 2$ eV or so, then the $m_{\nu'_\mu} \simeq m_{\nu'_{tau}} \simeq 2$ keV. Thus $\nu'_{\mu,\tau}$ can qualify as warm (or cool) dark matter of the universe, a possibility which does not appear to have been ruled out present cosmological observations. Such models have many interesting implications for cosmology[34], which we will not go into here.

6 Conclusions

The solar, atmospheric and LSND neutrino data, along with a need for some hot dark matter, if all are due to neutrino mass have two very profound implications: (i) at least two neutrinos must be degenerate in mass, a feature nor shared by charged fermions and not expected in the minimal SO(10) type models; (ii) there is a very good possibility that there is need for a sterile neutrino. In this talk, I have considered the various implications of these conclusions for physics beyond the standard model, such as the simple $SO(10)$ scenarios and conclude that one needs to go beyond such simple models if all the present indications neutrino mass are correct. I then outline the recent suggestion of Z. Berezhiani and this author that a scenario with a shadow universe with identical gauge and fermion structure (but with an asymmetric weak scale) can explain all the neutrino puzzles without the need for any other ingredients.

Acknowledgement

This work is supported by a grant from the National Science Foundation.

I also want to thank the organizers of the Neutrino'96 conference for financial support.

1. J. Bahcall and A. Smirnov, these proceedings.
2. R. Davis et al., Proceedings of the 21st International Cosmic Ray Conference, Vol. 12, edited by R.J. Protheroe (Univ. of Adelaide Press, Adelaide, 1990), p. 143; K.S. Hirata et al., Phys. Rev. **44**, 2241 (1991); A.I. Abrazov et al., Phys. Rev. Lett.**67**, 3332 (1991) and V.N. Gavrin in TAUP 93 Workshop, Gran Sasso, Italy, 1993 (unpublished); P. Anselmann et al., Phys. Lett. **B285**, 376 (1992) and **B314**, 445 (1993); for recent reviews, see Y. Suzuki, T. Kirsten, V. Gavrin these proceedings.
3. K.S. Hirata et al., Phys. Lett. **B280**, 146 (1992); R. Becker-Szendy et al., Phys. Rev. **D 46**, 3720 (1992); P.J. Litchfield in International Europhysics Conference on High Energy Physics, Marseille, France, 1993 (unpublished); E. J. Peterson, these proceedings.
4. Y. Fukuda *et al*, Phys. Lett. **B 335**, 237 (1994).
5. C. Athanassopoulos et al. hep-ex/9605001; D. Caldwell, these proceedings.
6. J. Primack, these proceedings.
7. O. Yasuda and H. Minakata, hep-ph/9602386.
8. J. Ostriker and P. Steinhardt, Nature, **377**, 600 (1995).
9. W. L. Freedman et al, Nature, **371**, 757 (1994).
10. For a recent review of nucleosynthesis and its implications for new physics, see S. Sarkar, Rep. on Prog. in Phys. (to appear).
11. H. Klapdor-Kleingrothaus, these proceedings and *Double Beta Decay and Related Topics*, ed. H. Klapdor-Kleingrothaus and S. Stoica, World Scientific, (1995) p. 3; A. Balysh et al., Phys. Lett. **B283**, 32 (1992).
12. Y-Z Qian and G. M. Fuller, Phys. Rev. **D52**, 656 (1995).
13. J. Peltoniemi, Helsinki preprint (1995); D. Caldwell, private communication.
14. C. Cardall and G. M. Fuller, astro-ph/9606024; H. Minakata and O. Yasuda, hep-ph/9609276.
15. D.O. Caldwell and R.N. Mohapatra, Phys. Rev. **D 48**, 3259 (1993).
16. J. Peltoniemi and J. W. F. Valle, Nucl. Phys. **B 406**, 409 (1993).
17. G. Fogli, E. Lisi and D. Montanino, Phys. Rev. **D 49**, 3626 (1994); C. Giunti, C. W. Kim and J. D. Kim, Phys. Lett. **B352**, 357 (1995); S. Goswami, K. Kar and A. Raychoudhuri, hep-ph/9505395; K. S. Babu, J. C. Pati and F. Wilczek, Phys. Lett. **B359**, 351 (1995).
18. G. G. raffelt and J. Silk, Phys. Lett. **B366**, 429 (1996); D. Caldwell and R. N. Mohapatra, Phys. Lett. **B354**, 371 (1995).
19. M. Gell-Mann, P. Ramond, and R. Slansky in Supergravity, edited by

D. Freedman et al. (1979).
20. R. Mohapatra and G. Senjanović, Phys. Rev. Lett. **44**, 912 (1980); Phys. Rev. **D 23**, 165 (1981).
21. T. Yanagida, in KEK Lectures, edited by O. Sawada et al. (1979).
22. D. Chang, R. N. Mohapatra, J. Gipson, R. E. Marshak and M. K. Parida, Phys. Rev. **D 31**, 1718 (1985).
23. D. G. Lee and R. N. Mohapatra, Phys. Rev. **D 52**, 4215 (1995).
24. K. S. Babu and R. N. Mohapatra, Phys. Lett. **267 B**, 400 (1991); R. N. Mohapatra, *Particle Physics and Cosmology at the interface* ed. J. C. pati et al. World scientific, (1994), p.273.
25. D. Chang and R. N. Mohapatra, Phys. Rev. **D 32**, 1248 (1985).
26. K. S. Babu and R. N. Mohapatra, Phys. Rev. Lett. **70**, 2845 (1993).
27. L. Lavoura Phys. Rev. **D 48**, 5440 (1993).
28. D. G. Lee and R. N. Mohapatra, Phys. Rev. **D 51**, 1353 (1995).
29. Y. Achiman and T. Greiner, Nucl. Phys. **B443**, 3 (1995).
30. D. G. Lee and R. N. Mohapatra, Phys. Lett. **B329**, 463 (1994).
31. E. J. Chun, A. Joshipura and A. Smirnov, Phys. Lett. **B357**, 608 (1995).
32. Z. Berezhiani and R. N. Mohapatra, Phys. Rev. **D 52**, 6607 (1995).
33. R. Barbieri, J. Ellis and M. K. Gaillard, Phys. Lett. **90 B**, 249 (1980); E. Akhmedov, Z. Berezhiani and G. Senjanović, Phys. Rev. Lett. **69**, 3013 (1992).
34. R. N. Mohapatra and V. L. Teplitz, Ap. J. (to appear).

VOLUME 44, NUMBER 14 PHYSICAL REVIEW LETTERS 7 APRIL 1980

Neutrino Mass and Spontaneous Parity Nonconservation

Rabindra N. Mohapatra
Department of Physics, City College of New York, New York, New York 10031

and

Goran Senjanović
Department of Physics and Astronomy, University of Maryland, College Park, Maryland 20742
(Received 10 December 1979)

In weak-interaction models with spontaneous parity nonconservation, based on the gauge group $SU(2)_L \otimes SU(2)_R \otimes U(1)$, we obtain the following formula for the neutrino mass: $m_{\nu_e} \simeq m_e^2/gm_{W_R}$, where W_R is the gauge boson which mediates right-handed weak interactions. This formula, valid for each lepton generation, relates the maximality of observed parity nonconservation at low energies to the smallness of neutrino masses.

PACS numbers: 11.30.Er, 11.30.Ly, 12.30.Ez, 14.60.Gh.

It is attractive to suppose that observed parity nonconservation in weak interactions is only a low-energy phenomenon, which ought to disappear at high energies. This idea has been implemented in unified gauge theories of electroweak interactions based on the gauge group $SU(2)_L \otimes SU(2)_R$

VOLUME 44, NUMBER 14 PHYSICAL REVIEW LETTERS 7 APRIL 1980

$\otimes$ U(1),[1] where parity nonconservation arises from the spontaneous breaking of the gauge symmetry. The suppression of the right-handed weak currents in this model owes its origin to the large mass of the right-handed gauge bosons. As far as the structure of the neutrino neutral-current interactions[2] and the parity-nonconserving electron-hadron weak interactions[3] are concerned, this model is indistinguishable from the standard $SU(2)_L \otimes U(1)$ model[4] at the present level of experimental accuracy. There exists, however, a fundamental distinction between the left-right-symmetric models and the pure left-handed $SU(2)_L \otimes U(1)$ models: In the former the neutrino has an arbitrary but finite mass, whereas in the latter it is massless. It is therefore important to understand the smallness of the neutrino mass in left-right symmetric models. Furthermore, it is very suggestive in the context of these models that there may be a connection between the smallness of the neutrino mass and the suppression of the right-handed weak interactions. In this Letter, we propose a model of spontaneous parity nonconservation based on the $SU(2)_L \otimes SU(2)_R \otimes U(1)$ gauge group, where this connection is brought out explicitly. We obtain the following estimate that relates the neutrino mass to the mass of the right-handed gauge bosons (see below for the detailed nature of the approximations):

$$m_{\nu_e} \simeq m_e^2/g m_{W_R}. \tag{1}$$

A similar formula holds for leptons in each generation. This formula is very illuminating in the sense that, in the limit of $m_{W_R} \to \infty$, the neutrino mass goes to zero and we have at the same time a pure $V-A$ theory of weak interactions.

We now proceed to derive Eq. (1) for one generation of leptons and repeat the same procedure for each generation. The main new ingredient of our proposal is that we start with two Majorana[5,6] neutrinos ν and N and choose the left- and right-handed lepton multiplets prior to spontaneous breakdown to be

$$\psi_L = \begin{pmatrix} \nu_L \\ e_L \end{pmatrix}, \quad \psi_R = \begin{pmatrix} N_R \\ e_R \end{pmatrix}, \tag{2}$$

with $SU(2)_L \otimes SU(2)_R \otimes U(1)$ representation numbers $(\frac{1}{2}, 0, -1)$ and $(0, \frac{1}{2}, -1)$, respectively. The quarks are assigned to left-right doublets as before.[1] We impose the left-right symmetry on the Lagrangian; under this symmetry $\psi_L \leftrightarrow \psi_R$ and this demands that at the tree level, $g_L = g_R$. We now introduce the Higgs multiplets[7] to break the gauge symmetry down to $U(1)_{em}$: φ transforms as the $(\frac{1}{2}, \frac{1}{2}, 0)$ representation of the gauge group; $\Delta_L \equiv (1, 0, 2)$ and $\Delta_R \equiv (0, 1, 2)$. Under left-right discrete symmetry $\varphi \leftrightarrow \varphi^\dagger$ and $\Delta_L \leftrightarrow \Delta_R$. We have already shown[8] earlier that, starting with a left-right-symmetric potential, it is possible to find a domain of coupling parameters in the theory for which we have

$$\langle\varphi\rangle = \begin{pmatrix} \kappa & 0 \\ 0 & \kappa' \end{pmatrix}, \quad \langle\Delta_L\rangle = 0, \quad \langle\Delta_R\rangle = \begin{pmatrix} 0 \\ 0 \\ v \end{pmatrix}. \tag{3}$$

It is easy to see that for $v \gg \kappa', \kappa$, after the first stage of the symmetry breakdown, the local symmetry group is reduced to $SU(2)_L \otimes U(1)$, where U(1) corresponds to $T_{3R} + Y$, which is finally broken down to $U(1)_{em}$ by $\langle\varphi\rangle \neq 0$.

We now proceed to discuss the main result of our paper, i.e., calculation of the neutrino masses. For simplicity, we also assume that $\kappa' \ll \kappa$. The gauge-invariant Yukawa couplings can be written as

$$\mathcal{L}_Y = h_1 \bar{\psi}_L \varphi \psi_R + h_2 \bar{\psi}_L \tilde{\varphi} \psi_R + h_3(\psi_L{}^T C i \tau_2 \Delta_L \psi_L + \psi_R{}^T C i \tau_2 \Delta_R \psi_R) + \text{H.c.}, \tag{4}$$

where $\tilde{\varphi} \equiv \tau_2 \varphi^* \tau_2$ and

$$\Delta_{L,R} = \begin{pmatrix} \frac{1}{\sqrt{2}}\delta^+ & \delta^{++} \\ \delta^0 & -\frac{1}{\sqrt{2}}\delta^+ \end{pmatrix}_{L,R},$$

and C is the Dirac charge-conjugation matrix. From (3) and (4) we find for the electron mass

$$m_e \simeq h_2 \kappa \tag{5}$$

and the mass matrix[9] for the ν-N sector is

$$\begin{matrix} & \bar{\nu} & \bar{N} \\ \begin{matrix} \nu \\ N \end{matrix} & \multicolumn{2}{c}{\begin{pmatrix} 0 & h_1\kappa \\ h_1\kappa & h_3 v \end{pmatrix}} \end{matrix}, \tag{6}$$

where we have used the property of Majorana particles $\nu^c = \nu$, $N^c = N$ in showing that $N_R{}^T C N_R$ and $\nu_L{}^T C \nu_L$ are mass terms. We further assume that Yukawa couplings h_1 and h_2 are of the same order of magnitude, i.e., $h_1 \simeq h_2$. It then follows from (6) that $m_N \simeq h_3 v$ and

$$m_\nu = (h_1\kappa)^2/m_N = g m_e^2/h_3 m_{W_R}. \tag{7}$$

This is the main result of our paper. Choosing a reasonable value for h_3, e.g., $h_3 \simeq g^2$, we obtain (1). For the second and third generations of leptons, the corresponding formulas are

$$m_{\nu_\mu} \simeq m_\mu^2/g m_{W_R} \text{ and } m_{\nu_\tau} = m_\tau^2/g m_{W_R}, \tag{8}$$

VOLUME 44, NUMBER 14 PHYSICAL REVIEW LETTERS 7 APRIL 1980

where we assume the heavy Majorana mass to be generation independent and we ignore generation mixings. This admittedly arbitrary assumption is taken only for illustrative purposes; therefore the values for m_{ν_μ} and m_{ν_τ} should be taken with this in mind.

To point out that these formulas lead to quite reasonable upper limits on the neutrino masses, we note that the analysis of charged-[10] and neutral-current[2] phenomena puts a lower bound on the right-handed W-boson mass, i.e., $m_{W_R} \gtrsim 3m_{W_L}$ (or $m_{W_R} \gtrsim 250-300$ GeV). If we choose $m_{W_R} \gtrsim 300$ GeV, Eqs. (1) and (8) yield $m_{\nu_e} \lesssim 1.5$ eV, $m_{\nu_\mu} < 56$ keV, and $m_{\nu_\tau} \lesssim 18$ MeV, in accord with the present laboratory experiments.[11]

We now comment on the quark sector of the model. As noted earlier, if we restrict ourselves to one generation, the left and right doublets are $(u_L, d_L) \equiv (\frac{1}{2}, 0, \frac{1}{3})$ and $(u_R, d_R) \equiv (0, \frac{1}{2}, \frac{1}{3})$. They obtain their mass through coupling to the Higgs multiplet φ. However, because the U(1) quantum number for quarks is $\frac{1}{3}$, $\Delta_{L,R}$ do not couple to quarks.

We now wish to remark on the following aspects of the model:

(a) At low energies, our model is indistinguishable from the standard model, since $\langle\Delta_R\rangle \neq 0$ still keeps $Y' \equiv T_{3R} + Y$ and $\vec{T}_L$ unbroken; that is, $SU(2)_L \otimes U(1)$ is unbroken after the first stage of symmetry breaking. We obtain the following mass relations between the neutral and charged gauge mesons (in the limit $v^2 \gg \kappa^2 + \kappa'^2$):

$$m_{W_L}{}^2 \simeq m_{Z_L}{}^2 \cos^2\theta_W,$$
$$m_{W_R}{}^2 \simeq m_{Z_R}{}^2 (2\cos 2\theta_W / \cos^2\theta_W). \tag{9}$$

Note the factor 2 in (9), which is characteristic of the triplet Higgs boson Δ_R which breaks $SU(2)_R$.

(b) In order to suppress lepton-number-changing processes such as $\mu \to e\gamma$ and $\mu \to 3e$, it turns out to be desirable to make the mixings of electron generation with μ and τ generation as small as possible. We make it zero naturally by imposing the following discrete symmetry on the Lagrangian

$$\begin{aligned} &\psi_{1L} \to -\psi_{1L}, \quad \psi_{1R} \to -\psi_{1R}, \\ &\psi_{2L} \to \psi_{2L}, \quad \psi_{2R} \to \psi_{2R}, \\ &\psi_{3L} \to \psi_{3L}, \quad \psi_{3R} \to \psi_{3R}, \\ &\varphi \to \varphi, \quad \Delta_L \to \Delta_L, \quad \Delta_R \to \Delta_R, \end{aligned} \tag{10}$$

where $i = 1, 2, 3$ counts the electron, the muon, and the τ generation, respectively, i.e.,

$$\psi_{1L} = \begin{pmatrix} \nu_e \\ e \end{pmatrix}_L, \quad \psi_{2L} = \begin{pmatrix} \nu_\mu \\ \mu \end{pmatrix}_L,$$

etc. This symmetry obviously forbids e-μ and e-τ mixings and, since it is unbroken, that will be true to all orders in perturbation theory. Therefore, in this model $\mu \to e\gamma$ and $\mu \to 3e$ are forbidden processes. By the same token, neutrino oscillations $\nu_e \leftrightarrow \nu_\mu$, $\nu_e \leftrightarrow \nu_\tau$ are absent in this model (but not $\nu_\mu \leftrightarrow \nu_\tau$). The interesting physical consequences when the above symmetry is relaxed will be dealt with elsewhere.

(c) A further characteristic of our model is the existence of doubly charged Higgs bosons $\delta_L{}^{++}$ and $\delta_R{}^{++}$. However, it is easily seen that their masses are of order m_{W_R} and therefore they are not expected to play an important role at low energies.

(d) The presence of Majorana neutrinos will allow for neutrinoless double β decay.[12,13] The contribution coming from the exchanges of W_L involves the light Majorana neutrino ν_e and is known[13] to be a few orders of magnitude below the experimentally allowed value. However, it is possible to exchange W_R's in which case the process goes through the heavy Majorana particle N as an internal state. We just mention that in this case one obtains the limit on N using the analysis of Ref. 13: $m_N \gtrsim (m_{W_L}/m_{W_R})^4 \times 10^4$ GeV $\gtrsim 10^2$ GeV, which is definitely satisfied, since $m_N \simeq m_{W_R} \simeq 300$ GeV. It should be emphasized, though, that more precise measurements of double β decay could in principle provide a more stringent lower bound on m_N, or in turn on m_{W_R}.

In summary, we have constructed a realistic and simple model with spontaneous parity nonconservation, where the suppression of $V+A$ currents is proportional to neutrino mass. The model provides, therefore, an understanding[14] of a tiny neutrino mass. We believe that it makes the search for the effects due to finite m_{W_R} even more warranted than before.

We thank N. P. Chang, R. E. Marshak, and E. Witten for useful discussions. This work was supported in part by the National Science Foundation, Grants No. PHY-78-24888 and PHY-76-16562 A02, and in part by CUNY-PSC-BHE, research award No. RF-13096.

---

[1]J. C. Pati and A. Salam, Phys. Rev. D 10, 275 (1974);

R. N. Mohapatra and J. C. Pati, Phys. Rev. D 11, 566, 2558 (1975); G. Senjanović and R. N. Mohapatra, Phys. Rev. D 12, 1502 (1975). For a detailed discussion of these models, see G. Senjanović, Nucl. Phys. B153, 334 (1979). For a review and extensive list of references see, R. N. Mohapatra, in *New Frontiers in High Energy Physics*, edited by A. Perlmutter and L. F. Scott (Plenum, New York, 1978), p. 337.

[2]J. C. Pati, S. Rajpoot, and A. Salam, Phys. Rev. D 17, 131 (1978). See also, Senjanović, Ref. 1.

[3]A. Janah, to be published; A. Costa, M. D'Anna, and P. Marcolungo, to be published; J. C. Pati and S. Rajpoot, to be published; Senjanović, Ref. 1.

[4]S. Weinberg, Phys. Rev. Lett. 19, 1264 (1967); A. Salam, in *Elementary Particle Theory: Relativistic Groups and Analyticity*, edited by N. Svartholm (Wiley, New York, 1969). See also S. L. Glashow, Nucl. Phys. 22, 579 (1961).

[5]For a thorough discussion of the properties of Majorana theory of massive fermions, see K. M. Case, Phys. Rev. 107, 307 (1957). See also R. E. Marshak, Riazuddin, and C. P. Ryan: *Theory of Weak Interactions in Particle Physics* (Wiley, New York, 1969).

[6]For earlier suggestions that neutrinos may be Majorana particles in the context of gauge theories, see H. Fritzsch, M. Gell-Mann, and P. Minkowski, Phys. Lett. 59B, 256 (1975); T. P. Cheng, Phys. Rev. D 14, 1367 (1976). Recently, this idea has been put forward in the context of O(10) grand unified theory by M. Gell-Mann, P. Ramond, and R. Slansky, unpublished; E. Witten, unpublished.

[7]The Higgs multiplets we have introduced can be bound states of the fundamental fermion fields of our model, e.g., $\phi=\bar{\psi}_L\psi_R$ and $\Delta_L=\psi_L{}^TC^{-1}\psi_L$ and $\Delta_R=\psi_R{}^TC^{-1}\psi_R$. The symmetry breakdown in our model could therefore be entirely dynamical in origin.

[8]Senjanović and Mohapatra, Ref. 1.

[9]Similar mass matrices for Majorana neutrinos have been considered before in the context of other gauge models by the authors of Ref. 6 [for a review see F. Wilczek, in Proceedings of the Lepton-Photon Conference, Fermilab, 1979 (unpublished)]. After completing this work we became aware of the fact that if the $SU(2)_L \otimes SU(2)_R \otimes U(1)$ electroweak model is embedded in O(10) grand unified theory, the corresponding Higgs system would involve only superheavy 126 and light 10 (126 is used by Gell-Mann *et al.*, Ref. 6). We point out that 10 leads to unsatisfactory mass relations, namely, $m_e=m_d$ and $m_\mu=m_s$. We emphasize again that the philosophy of our work is to illustrate the connection between neutrino mass and the maximality of parity nonconservation in weak interactions, by paying the minimal price in enlarging the standard $SU(2)_L \otimes U(1)$ gauge model.

[10]M. A. Bég, R. V. Budny, R. N. Mohapatra, and A. Sirlin, Phys. Rev. Lett. 38, 1252 (1977), and 39, 54(E) (1977).

[11]If we accept the astrophysical upper limit of $m_\nu < 10$ eV for all species of neutrinos [see K. Cowsik and J. McClelland, Phys. Rev. Lett. 29, 669 (1972)], this imposes a much more stringent lower bound on m_{W_R}: $m_{W_R} \geq 10^7-10^8$ GeV. We note that such large values of m_{W_R} arise in the context of SO(10) grand unified theory; see Q. Shafi and C. Wetterich, to be published; T. Goldman and D. A. Ross, to be published; R. N. Mohapatra and G. Senjanović, to be published.

[12]See, for example, H. Primakoff and S. P. Rosen, Phys. Rev. D 184, 1925 (1969).

[13]A. Halprin, P. Minkowski, H. Primakoff, and S. P. Rosen, Phys. Rev. D 13, 2567 (1976).

[14]An attempt to understand small or vanishing neutrino mass in left-right-symmetric gauge models with neutrinos as Dirac particles was made by G. C. Branco and G. Senjanović, Phys. Rev. D 18, 1621 (1978); P. Ramond and D. Reiss, to be published; T. P. Cheng and L.-F. Li, Phys. Rev. D 17, 2375 (1978).

PHYSICAL REVIEW D VOLUME 34, NUMBER 5 1 SEPTEMBER 1986

Neutrino mass and baryon-number nonconservation in superstring models

R. N. Mohapatra
Department of Physics and Astronomy, University of Maryland, College Park, Maryland 20742

J. W. F. Valle
Departamento de Fisica Teorica, Universitat Autonoma de Barcelona, Bellaterra, Barcelona, Spain
(Received 3 March 1986)

We propose new mechanisms for understanding neutrino masses in superstring models that contain E_6-singlet zero-mass fields after compactification. We show that the low-energy gauge group of these models can be phenomenologically acceptable. We then comment on $\Delta B=1$ and $\Delta B=2$ baryon-number-violating processes in these models.

Recently, there has been a great deal of activity in superstring theories with the gauge group $E_8\times E_8'$.[1-5] The zero-slope limit of these theories leads to an anomaly-free ten-dimensional $E_8\times E_8'$ Yang-Mills theory coupled to $N=1$ supergravity. When six extra dimensions are compactified[3] to a Calabi-Yau manifold with SU(3) holonomy, an $N=1$ locally supersymmetric four-dimensional grand unified theory based on gauge group $E_8'\times E_6$ emerges with N_g copies of massless {27}-dimensional (under E_6) and $b_{1,1}$ pairs of $\{27\}+\{\overline{27}\}$ chiral superfields (where $b_{1,1}$ is the Betti-Hodge number). By an appropriate choice of[3,4] the Calabi-Yau space, one can have $N_g=3$ or 4. One can then assume the observed matter fields (quarks and leptons) to belong to the {27}-dimensional representations of E_6. This model, therefore, has all the right ingredients for being a candidate theory that unifies all matter and all interactions in nature.[5]

Even though this program of unification appears very attractive, several potential difficulties exist: too fast proton decay, potentially large neutrino masses, the problem of light Higgs multiplets, and the lack of a proper mechanism for supersymmetry breaking. In this paper, we will concern ourselves with only the first three problems. We will exhibit mechanisms for understanding small neutrino masses using light E_6-singlet fields. These models require the existence of light Higgs fields with specific quantum numbers. We show that this requirement can be satisfied for phenomenologically acceptable low-energy gauge groups. In some of these models, the $SU(2)_L$ Higgs doublet responsible for symmetry breaking must arise from matter superfields. This leads us to discuss the question of baryon nonconservation such as proton decay and neutron-antineutron oscillation in these models. We comment on the possibility that neutron-antineutron oscillation may be observable in this class of models under certain circumstances while avoiding catastrophic proton decay.

The first difficulty in obtaining an acceptable pattern of neutrino masses in superstring models arises from the presence of new exotic neutral fermions beyond the usual left- and right-handed neutrinos. This can be seen from the decomposition of the {27}-dimensional representation of E_6 under the [SO(10), SU(5)] subgroups:

$$\begin{aligned}\{27\}= &\ [16,10] + [16,\bar{5}] + [16,1] + [10,5] + [10,5] + [1,1]\\ &(u,d;u^c,e^c) + (d^c;\nu,e) + \nu^c + (D^c,N,E^-) + [D,N^c,E^+] + n_0\,.\end{aligned} \tag{1}$$

The various particles are identified below each group representation. In what follows, we represent a matter multiplet by ψ and a Higgs multiplet by H. The five neutral leptons are (ν,ν^c,N,N^c,n_0). The new neutral leptons (N,N^c) must be massive enough so that their contribution to the present energy density of the Universe is below the critical density. As far as n_0 is concerned, it can be massless or superheavy depending on whether it couples to superheavy or light gauge bosons. Assuming that N, N^c, and n_0 decouple from low energies, we are still left with a Dirac neutrino obtained by combining ν_L and ν_L^c with a mass $m_D\simeq O(m_e)\approx 1$ MeV. To solve this problem, we seek ways by which ν^c acquires a large mass. A Majorana mass for ν^c would require breaking $(B-L)$ by two units and is therefore not possible to have at the tree level since the conventional "seesaw"[6] mechanism is not available in this case. They can, however, be induced in either of the following ways.

(i) A higher-dimensional term[7] of the form $(1/M)\times\{27\}_\psi\times\{\overline{27}\}_H\times\{27\}_\psi\times\{\overline{27}\}_H$ leads to an effective Majorana mass for the right-handed neutrino ν^c: $M_{\nu^c}=V_{BL}^2/M$ where $\langle\tilde{\nu}_H^c\rangle=V_{BL}$. This leads to a light neutrino mass $m_\nu\approx m_D^2M/V_{BL}^2$. Choosing $M\simeq M_{\rm Planck}\simeq 10^{18}$ GeV, $m_D\simeq 1$ MeV requires that $V_{BL}\simeq 10^{11}$ GeV leads to $m_\nu\simeq 0.1$ eV.

(ii) The second possibility is to use another neutral fermion which is $B-L$ neutral to form a $\Delta(B-L)=1$ Dirac mass term. This mechanism was used in Ref. 8, where the extra neutral fermion chosen was the gaugino corresponding to $B-L$ symmetry. In this paper, we replace the gaugino by an E_6-neutral fermion[9] (denoted by S) that may be present in superstring models. The scenario outlined below realizes this mechanism.

Now we present the two models for neutrino masses. We first outline the details of the models that are relevant only to the discussion of neutrino masses and in a subsequent section discuss the associated low-energy gauge group. Both the models we present will require $b_{1,1}=2$, i.e., two pairs of $\{27\}+\{\overline{27}\}$ representations that act as Higgs fields denoted by H and J, respectively. We will then assume that the SO(10)-singlet components (denoted by n_0) and [SO(10),SU(5)] representation [16,1] (denoted by ν^c) remain light. The reason for this is that we would like to give them intermediate scale vacuum expectation values (VEV's) (without breaking supersymmetry, i.e., maintaining a D-flat direction).

Model I. To write down the most general low-energy superpotential, we use the [SO(10),SU(5)] notation of Eq. (1) and denote $Q=(u,d)$, $l=(\nu,e^-)$, $E^c=(N,E^-)$, and $E=(E^+,N^c)$. Denoting the components of the Higgs field by a subscript J and H and suppressing all generation indices, we can write the superpotential as $P_{\rm I}=P_0+P_{\rm I}'$, where

$$P_0=\lambda_1 QQD+\lambda_2 Qu^cE+\lambda_3 Qd^cE^c+\lambda_4 QD^cl+\lambda_5 u^cd^cD^c+\lambda_6 u^cDe^c+\sum_{a=\text{matter},H}\lambda_7^a d^cD\nu_a^c+\sum_{a=\text{matter},H}\lambda_8^a D^cDn_{0,a}$$
$$+\sum_{a=\text{matter},H}\lambda_9^a lE\nu_a^c+\sum_{a=\text{matter},H}\lambda_{10}^a EE^cn_{0,a}+\lambda_{11}lE^ce^c+\beta_1\nu_H^c\bar{\nu}_J^cS+\beta_2 n_{0,H}\bar{n}_{0,J}S\ , \tag{2a}$$

and

$$P_{\rm I}'=\lambda_{12}n_0\bar{n}_{0,J}S+\lambda_{13}\nu^c\bar{\nu}_J^cS\ . \tag{2b}$$

In $P_{\rm I}'$, Higgs field H contributions are eliminated by imposing a discrete symmetry under which the E_6 singlet S and $J,\bar{J}$ are odd and all other fields are even. We first assume that the SO(10)-singlet component of H acquires a VEV V_6 along a D-flat direction; subsequently, two components of $\bar{J}$, one along the SO(10) singlet and another along the [16,1] acquire the effective VEV $\langle\bar{J}(n_0)\rangle=\mu$ and $\langle\bar{J}(\nu^c)\rangle=V_{BL}$ with $\mu\ll V_{BL}\ll V_6$. The neutral fermions N and N^c as well as the D and D^c quarks pick up large mass V_6 and decouple from low energies. We further assume $\beta_1\ll\lambda_{13}$. Defining, $\nu^{c'}=\nu^c+(\beta_1/\lambda_{13})\nu_H^c$ and $n_0'=(\beta_2n_{0H}+\lambda_{12}n_0)/(\beta_2{}^2+\lambda_{12}{}^2)^{1/2}$, we find (the combinations orthogonal to $\nu c'$ and n_0' remain massless and invisible), the 4×4 mass matrix for the remaining neutral fermions ν, $\nu^{c'}$, n_0', and S to be of the form

$$\begin{array}{c|cccc} & \nu & n_0' & \nu^{c'} & S\\ \nu & 0 & 0 & m_D & 0\\ n_0' & 0 & 0 & 0 & \mu\\ \nu^{c'} & m_D & 0 & 0 & V_{BL}\\ S & 0 & \mu & V_{BL} & 0\end{array}\ . \tag{3}$$

It is easy to see that on diagonalizing this matrix we obtain two Dirac particles with masses: $m_1=m_D\mu/V_{BL}$ and $M_2=V_{BL}$. If we choose $V_{BL}\simeq10^{12}$ GeV, this implies that for $\mu<10^6$ GeV, $m_1<1$ eV. Thus, in this picture the neutrino is a Dirac particle with a naturally small mass. We point out that this is a completely new mechanism for generating light Dirac neutrinos and could be useful in general supersymmetric models. In this model, neutrinoless double-beta decay will be forbidden. Furthermore, we envisage this large intermediate scale having its origin in dimension-four terms in the superpotential.

Model II. This model differs from model I in two respects. First, we add two E_6-singlet fields S_1 and S_2 and impose the discrete symmetry under which J, $\bar{J}$, and S_2 fields are odd and all other fields are even. The superpotential for this model has the form $P_{\rm II}=P_0+P_{\rm II}'$ where

$$P_{\rm II}'=\lambda_{14}n_0\bar{n}_{0,J}S_2+\lambda_{15}\nu^c\bar{\nu}_J^cS_2+\lambda_{16}n_0\bar{n}_{0,H}S_1+\lambda_{17}\nu^c\bar{\nu}_H^cS_1\ . \tag{4}$$

The second difference from model I is in the pattern of symmetry breaking which we assume to be $\langle n_{0,H}\rangle=\langle\bar{n}_{0,H}\rangle=V_6$ and $\langle\nu_J^c\rangle=\langle\nu_{\bar{J}}^c\rangle=V_{BL}$. As a result, N, N^c, and n_0 disappear from the low-energy spectrum, leaving a 3×3 mass matrix of the type (assuming $\beta_1\ll\lambda_{15}$)

$$\begin{array}{c|ccc} & \nu & \nu^c & S_2\\ \nu & 0 & m_D & 0\\ \nu^c & m_D & 0 & V_{BL}\\ S_2 & 0 & V_{BL} & 0\end{array}\ . \tag{5}$$

This gives a massless Majorana neutrino: $\nu_{\rm phys}\approx\nu-(m_D/V_{BL})S_2$ and a heavy Dirac neutrino with mass $\approx V_{BL}$. In this approximation, the amplitude for neutrinoless double-beta decay vanishes identically. However, supersymmetry breaking could induce a Majorana mass m_S for S_2, leading to a small Majorana mass for the neutrino, $m_\nu\simeq m_S(m_D/V_{BL})^2$. For $m_D\approx1$ MeV for the first generation, $V_{BL}\gtrsim1$ TeV, a value of $m_S\approx10^2$–10^3 GeV would lead to $m_{\nu_e}\lesssim0.1$ to 1 eV, which is consistent with all observations. Again, this mechanism can be trivially extended to higher generations by adding two E_6 singlets per generation.

We also note that one can construct a variant of this model without S_1, with the same result for neutrino masses. In this case, n_0 remains massless and invisible since it does not couple to any light gauge bosons. This will also not contribute to the expansion of the Universe at the nucleosynthesis epoch.

In the aforementioned discussion, we have assumed certain light Higgs multiplets to obtain realistic neutrino masses. We have to show that this can happen without enlarging the gauge group to an unacceptable level. The procedure for deciding this has been outlined in Ref. 10. One has to make sure that, under the discrete group $\mathcal{H}\subset E_6$ which is a subgroup of the discrete group in the Calabi-Yau manifold K, the light fields must remain invariant, i.e.,

$$\psi(gx)=U_g\psi(x)\equiv\psi(x)\ , \tag{6}$$

where $x\subset K$ and $U_g\subset\mathcal{H}$. It is known that, in order to leave $SU(3)_c\times SU(2)_L$ as an unbroken subgroup following the "flux-loop"-breaking mechanism,[11] the U_g is parametrized by six numbers $(x_1,\ldots,x_6)$ (i.e., $U_g=e^{ix_iH_i}$ where H_i belong to the Cartan subalgebra of E_6) with

TABLE I. Low-energy gauge groups below the Planck scale for different choice of discrete symmetries.

Discrete symmetry	Gauge groups below Planck scale (prior to intermediate scale breaking)
Z_2	$SU(6)\times SU(2)_L$
Z_3	$SU(3)_c\times SU(3)_L\times SU(3)_R$
Z_4	$SU(5)\times SU(2)_L\times U(1)$
Z_5	$SU(5)\times SU(2)_N\times U(1)$
Z_6	$SU(3)_c\times U(1)_L\times SU(2)_L\times SU(3)_R$
$Z_n, n\geq 7$	$SU(3)_c\times SU(2)_L\times SU(2)_N\times U(1)\times U(1)$

$(x_1,\ldots,x_6)=(-c,c,a,b,c,0)$. It is then straightforward to check that for ν^c and n_0 components to remain light, we must have $a=-c$ and $b=3c$. The unbroken low-energy group after flux breaking is then given by $SU(3)_c\times SU(2)_L\times SU(2)\times U(1)\times U(1)$, if the discrete group $\mathcal{H}$ is Z_n, $n\geq 7$, and $SU(3)_c\times SU(3)_L\times SU(3)_R$, if $\mathcal{H}=Z_3$. In Table I, we list the low-energy groups for other discrete symmetries. An important point to note is that the low-energy electroweak gauge group after the intermediate scale is phenomenologically acceptable in all cases except the ones corresponding to Z_2, Z_4, and Z_5 symmetry, where the SU(5) or SU(6) gauge group survives below the Planck scale.

The next question to ask is where do the light Higgs doublets that break $SU(2)_L\times U(1)_Y$ symmetry and give mass to fermions come from? This depends on the discrete symmetry in question. If the discrete symmetry is Z_3, we find that two $SU(2)_L$ doublets (N,E^-) and (E^+,N^c) from the $\{27\}+\{\overline{27}\}$ Higgs multiplets remain light and can, therefore, serve as light doublets that break $SU(2)_L\times U(1)_Y$ symmetry. In this case, the constraints of $\sin^2\theta_W$ require that $V_{BL}\approx V_6\approx 10^{14}$ GeV.

On the other hand, in other cases, where no light $SU(2)_L$ doublet survives from the $\{27\}+\{\overline{27}\}$ pair we propose that they come from matter multiplets; more specifically we have in mind the two doublets (per generation) (N,E^-) and (N^c,E^+). One can assign VEV's to $\tilde{N}$ and $\tilde{N}^c$ to break $SU(2)_L\times U(1)$. From Eq. (2a), we see that the λ_2, λ_3, λ_9, and λ_{11} terms can then lead to fermion masses. In this class of models, where light Higgs doublets (E,E^c) arise as part of the matter multiplets, we assume as before that both axial-vector $B-L$ and vector $B-L$ symmetry are broken at an intermediate scale. We then have to tune $\lambda_{10}\simeq 0$ to keep the E and E^c light. But since they do not couple to ν and ν^c, it does not affect our discussion of the neutrino mass matrix.

We now discuss baryon-number violation in these models. A typical diagram that makes a dominant contribution to proton decay is shown in Fig. 1 and we estimate the $\Delta B=1$ amplitude

$$A_{\Delta B=1}\simeq\frac{\lambda_1\lambda_4 m_{\tilde{g}}}{M_D M_{sq}^2}\left[\frac{\alpha_s}{4\pi}\right]. \qquad (7)$$

where $m_{\tilde{g}}$, M_{sq}, and M_D represent the gluino, scalar-quark, and D-quark masses respectively and λ_i are coupling constants in Eq. (4). Choosing $M_D\simeq 10^{12}$ GeV, $m_{\tilde{g}}\approx 10$ GeV, we find $A_{\Delta B=1}\approx 10^{-19}\lambda_1\lambda_4\ \text{GeV}^{-2}$. Since the couplings λ_i in our superpotential Eq. (4a) are related at the Planck scale, it is reasonable to expect them to be $\approx 10^{-5}$–10^{-6}. Choosing $\lambda_1\approx\lambda_4\simeq 10^{-6}$, we find $A_{\Delta B=1}\simeq 10^{-31}\ \text{GeV}^{-2}$, which is consistent with present experiments. We expect the photino to be heavier than the proton so that proton decay via photino emission is avoided.

Turning now to $\Delta B=2$ transitions, we first note that it requires $\langle\nu_H^c\rangle$, or $\langle\nu_\psi^c\rangle\neq 0$. Therefore, in the two models for neutrino mass that we have presented here, since $\langle\nu_H^c\rangle=0=\langle\nu_\psi^c\rangle$, the $\Delta B=2$ transition is forbidden. On the other hand, in our model II (as well as in other models discussed in the literature[7]), one might expect $\langle\nu_H^c\rangle=V_{BL}^H\neq 0$ or $\langle\nu_\psi^c\rangle\neq 0$. In such a case, a nonzero $\Delta B=2$ amplitude arises from the diagram[12] in Fig. 2. Its magnitude is

$$A_{\Delta B=2}\simeq\left[\frac{\lambda_5\lambda_7^H V_{BL}^a}{M_D}\right]^2\frac{4\pi\alpha_s}{M_{sq}^4 m_{\tilde{g}}},\quad a=H\text{ or }\psi. \qquad (8)$$

The corresponding n-$\bar{n}$ mixing strength is given by $\delta m_{n\text{-}\bar{n}}\simeq A_{\Delta B=2}|\psi(0)|^4$. In our model II, we prefer $V_{BL}^H\ll M_D$ in order not to spoil the neutrino mass results. For instance, if we choose, $V_{BL}^H\leq 10^6$ GeV, using $|\psi(0)|^4\simeq 10^{-3}\ \text{GeV}^6$, we find, $\delta m_{n\text{-}\bar{n}}\simeq 10^{-25}\lambda_5^2\lambda_7^2$ GeV, which for $\lambda_5\approx\lambda_7\approx 10^{-2}$ can lead to mixing times $\tau_{n\text{-}\bar{n}}\simeq 6\times 10^9$ sec. This may be barely accessible in future experiments with intense cold-neutron beams.[13] In the type-II models of neutrino masses,[7] $V_{BL}^H\approx 10^{12}$ GeV. In such models, smaller values of λ ($\sim 10^{-5}$) can also lead to an observable n-$\bar{n}$ oscillation. Finally, it is possible that all three fields in the matter field self-coupling in Eq. (4) do not belong to the same generation, thereby weakening the constraints on λ_1 and λ_4. This may improve the situation with respect to n-$\bar{n}$ oscillations.

Note added in proof. It has been brought to our atten-

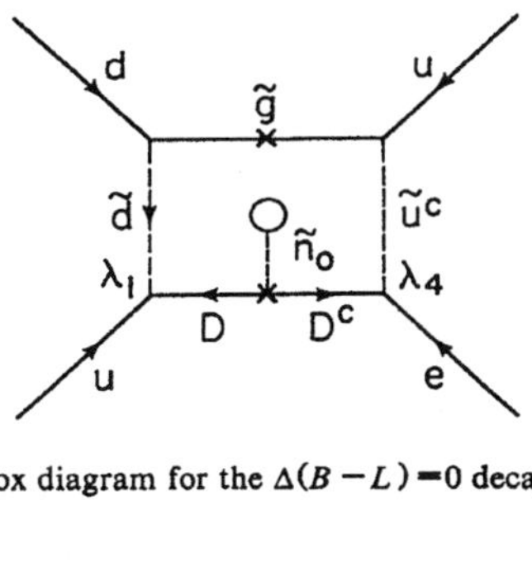

FIG. 1. Box diagram for the $\Delta(B-L)=0$ decay mode of the proton.

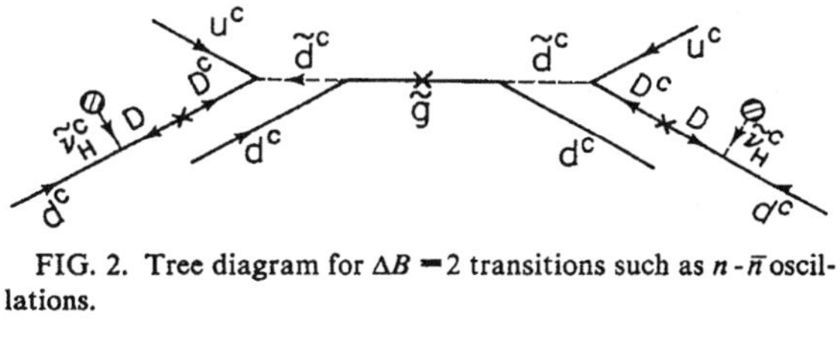

FIG. 2. Tree diagram for $\Delta B=2$ transitions such as n-$\bar{n}$ oscillations.

tion by L. Wolfenstein that the mass matrix in Eq. (3) was considered by M. Roncadelli and D. Wyler [Phys. Lett. **133B**, 325 (1983)] and P. Roy and O. Shankar [Phys. Rev. Lett. **52**, 713 (1984)], and the one in Eq. (5) was considered by L. Wolfenstein and D. Wyler [Nucl. Phys. **B218**, 205 (1983)]. We thank Professor Wolfenstein for bringing these works to our attention.

One of us (R.N.M.) would like to thank P. K. Mohapatra for useful discussions, and we also acknowledge useful discussions with G. Ross and L. E. Ibanez. The work of one of the authors (R.N.M.) was supported by a grant from the National Science Foundation. The other (J.W.F.V.) was supported by the Spanish Ministry of Science.

---

[1]M. Green and J. Schwarz, Phys. Lett. **149B**, 117 (1984); **151B**, 21 (1985).

[2]D. Gross, J. Harvey, E. Martinec, and R. Rohm, Phys. Rev. Lett. **54**, 502 (1985); Nucl. Phys. **B256**, 251 (1985).

[3]P. Candelas, G. Horowitz, A. Strominger, and E. Witten, Nucl. Phys. **B158**, 46 (1985).

[4]E. Witten, Nucl. Phys. **B258**, 75 (1985).

[5]For phenomenological studies of these models, see Witten, Ref. 4; M. Dine, V. Kaplunovsky, M. Mangano, C. Nappi, and N. Seiberg, Nucl. Phys. **B259**, 549 (1985); J. Breit, B. Ovrut, and G. Segré, Phys. Lett. **158B**, 33 (1985); J. P. Derendinger, L. Ibanez, and H. P. Nilles, CERN Report No. TH-4228, 1985 (unpublished); F. del Aguila, G. Blair, M. Daniel, and G. G. Ross, CERN report, 1985 (unpublished); S. Cecotti, J. P. Derendinger, S. Ferrara, L. Girardello, and M. Roncadelli, Phys. Lett. **156B**, 318 (1985); C. Nappi and V. Kaplunovsky, Comments Nucl. Part. Phys. (to be published); P. Binetruy, S. Dawson, I. Hinchliffe, and M. Sher, Lawrence Berkeley Laboratory Report No. LBL-20317, 1985 (unpublished). For earlier work on E_6 grand unification, see F. Gursey, P. Sikivie, and P. Ramond, Phys. Lett. **60B**, 177 (1976); F. Gursey and M. Serdaroglu, Lett. Nuovo Cimento **21**, 28 (1978); Y. Achiman and B. Stech, Phys. Lett. **77B**, 389 (1978); Q. Shafi, *ibid.* **79B**, 301 (1979); J. Rosner, Comments Nucl. Part. Phys. **15**, 195 (1986).

[6]M. Gell-Mann, P. Ramond, and R. Slansky, in *Supergravity*, edited by P. Van Nieuwenhuizen and D. Freedman (North-Holland, Amsterdam, 1980); T. Yanagida, in *Proceedings of Workshop on Unified Theory and Baryon Number of the Universe*, edited by O. Sawada *et al.* (KEK, Tsukuba, Japan, 1979); R. N. Mohapatra and G. Senjanović, Phys. Rev. Lett. **44**, 912 (1980).

[7]Derendinger *et al.*, Ref. 5; S. Nandi and U. Sarkar, Phys. Rev. Lett. **56**, 564 (1986).

[8]R. N. Mohapatra, Phys. Rev. Lett. **56**, 561 (1986).

[9]E. Witten, Princeton report, 1985 (unpublished).

[10]Breit *et al.*, Ref. 5; Witten, Ref. 4.

[11]Y. Hosotani, Phys. Lett. **129B**, 193 (1983).

[12]F. Zwirner, Phys. Lett. **132B**, 103 (1983); R. Barbieri and A. Masiero, LPTENS Report No. 85/20, 1985 (unpublished).

[13]R. N. Mohapatra, in Proceedings of the Harvard Workshop on Neutron-Antineutron Oscillation, 1982, edited by M. Goodman (unpublished).

Neutrinos in Left-Right Symmetric, SO(10) and Superstring Inspired Models

R.N. Mohapatra *

Department of Physics and Astronomy, University of Maryland,
College Park, MD 20742, USA

"Neutrinos", Graduate texts in contemporary physics", ed.
H. V. Klapdor, Berlin, Germany: Springer (1988) 117 - 146

We review the properties of neutrinos such as masses, decays and magnetic moments in the left-right symmetric as well as superstring inspired models.

1. Introduction

Since Pauli postulated the neutrino more than fifty years ago, it has played a key role in the development of weak interaction theory as well as our understanding of the evolution of the universe to its present form. We know that, neutrinos are unique in that they are nearly massless, (or perhaps massless?) neutral spin half particles that participate only in weak and gravitational interactions. Furthermore, leaving aside the proton, neutron, electron and the photon, the only other kind of matter that fills the universe as abundantly as radiation is the neutrino: the neutrino to photon ratio in the universe today is about one-third. Besides, it also plays a major role in the energy loss from stellar objects starting from the main sequence stars (such as the sun) to the exploding supernovae (such as the SN 1987a). Our own sun bombards us (on earth) with nearly sixty billion neutrinos per cm^2 per sec. We practically "swim" in the ocean of neutrinos. Yet other than its spin, some of its interactions and the fact that there are 3 kinds of neutrinos we really know very little else about is properties. In the absence of a full picture (as, say, we have for the electron or the proton), the discussion of the properties of the neutrinos involve clever combinations of results from null experiments, theoretical model building and often, sheer intuition and guess work. In this review, I discuss the properties of neutrinos that can be inferred from some popular models of elementary particle physics available today. The two classes of model we will discuss are the left-right symmetric models[1,2] and superstring models.[3] Both of these models are extensions of physics beyond that predicted by the standard model of Glashow, Weinberg and Salam[4] and have been subjects of extensive investigation in the recent past.

This paper is organized as follows: in sec. II, we outline the phenomenological (and cosmological) constraints on massive neutrinos and their life times; in sec. III, we briefly introduce the various possibilities for neutrino masses such as Dirac, Majorana and pseudo-Dirac masses and their connection to $B-L$ symmetry; in sec. IV, we begin the discussion of left-right symmetric models and phenomenological constraints on various parameters of the model; sec. V is devoted to the discussion of Dirac vrs. Majorana masses in left-right

*Work supported by a grant from the National Science Foundation.

symmetric models as well as the ν-decays in both cases; in sec. VI we discuss the embedding of these models in SO(10) grandunified theories; in sec. VII, we discuss neutrino masses in superstring models and the various possibilities, that exist for neutrino masses in these models.

2. Observational Constraints on Massive Neutrinos:

There are several kinds of observational constraints on the properties of massive neutrinos: those implied by (a) cosmological observations; (b) solar neutrino puzzle; (c) supernova, SN1987a and (d) those implied by accelerator and reactor experiments attempting to detect neutrino stability. Below, we briefly summarize the first three cases.

(a) Cosmological Constraints: The cosmological constraints date back to the pioneering works of Cowsik and Mclelland[5] and Gershtein and Zeldovich[6] who pointed out that, there exists a neutrino sea all around us, with a temperature of $1.9^o\,K \simeq 1.6 \times 10^{-4}$ ev. Usual considerations of Fermi-Dirac statistics then leads to the conclusion that the present number density of each neutrino species is given by $n_\nu \simeq \frac{3}{11} n_\gamma \simeq 109/cm^3$. If neutrinos have mass m_{ν_i} and are stable, then their contribution to the mass density of the universe must be smaller than the critical mass density of the universe leading to an upper limit on the sum of the masses of the stable neutrinos:

$$\left(m_{\nu_e} + m_{\nu_\mu} + m_{\nu_\tau} + \cdots\right) n_\nu \le 10^4\, h_o^2\, ev$$

where h_o denotes the uncertainty in the Hubble constant (i.e. $H_o^{-1} = 10^{10}/h_o$ years) (or the age of the universe) and we have $.4 \le h_o \le 1$, with smaller h_o corresponding to an older universe. For $h_o \approx .5$ to 1, we get,

$$\Sigma m_{\nu_i} \le 25 \text{ to } 100\ ev\ . \tag{2.1}$$

Later on it was pointed out[7] that for stable heavier neutrinos, there is a lower bound of about 2 GeV. The situation is, however, very different for unstable neutrinos as was first pointed out by Dicus, Kolb and Teplitz.[8] The reason of course is that, if a neutrino species decays in the course of the evolution of the universe, it no longer contributes to mass density at present; the decay products, however, carry the information about the mass in their energy but energy, unlike the mass gets redshifted with the cosmological evolution and if the decay happens early enough, the energies could be sufficiently redshifted to be harmless. In this way, one gets a constrained joint bound on mass and lifetime as follows:

$$m_{\nu_H}\left(\frac{\tau_{\nu_H}}{\tau_U}\right)^{1/2} \le 25\, ev \text{ to } 100\ ev\ . \tag{2.2}$$

There are three principal decay modes of heavier neutrinos that one may consider to avoid the bound in eqn. (2.1):

a) $\nu_H \rightarrow \nu_e\gamma$: Cosmology implies the following bound on this decay mode. Unless the half-life is less than[8a] 10^4 sec., the decay photons will disintegrate the deuterium and affect the isotropy of the observed black body radiation. In gauge models, these decay times are too long to be useful in avoiding the cosmological bounds. (See later). Therefore, much attention has been focussed in the following two decay modes:

b) $\nu_H \rightarrow \nu_e + \chi$, where χ is the Majoron,[9] which represents the Goldstone boson corresponding to spontaneous breaking of global $B - L$ quantum number. These decay modes are known[10] to lead to life-times much shorter than the photonic decay mode.

c) $\nu_H \rightarrow 3\nu_e$: This mode, in certain models such as the left-right symmetric ones, is important for consistency of neutrino masses with cosmology but often requires fine tuning of parameters in gauge models[11] since by an $SU(2)_L$ rotation, $\nu_\mu \rightarrow 3\nu_e$ gets related to $\mu \rightarrow 3e$ decay, which is known to have a tiny upper bound on its branching ratio ($\lesssim 10^{-11}$). In some models with Majorana neutrinos $\nu_H \rightarrow 3\nu_e$ can be induced via flavor changing neutral current coupling Z, but these effects are small.

It is important to point out that arbitrary neutrino mass can be generated by including the right-handed neutrino in the standard model; but the decay modes in cases (b) and (c) involve new physics beyond this and are therefore extremely important. For instance, evidence for neutrino mass in the forbidden range (10^{-7} to 2 GeV) combined with cosmological constraints will imply the existence of the massless Majoron or an ultralight neutral Higgs boson (in the Gev range).[12]

(b) Constraints Implied by Solar Neutrino Puzzle: Another class of constraints on the properties of the neutrino can be inferred from recent discussions of the solar neutrino puzzle. To present those constraints, we briefly remind the reader about the essential aspects of solar neutrino puzzle. It is well-known that sun shines because of hydrogen burning to helium in the hot core of the sun ($T_{core} \sim 1.5 \times 10^7 \,^\circ K$, $E \sim 1.3$ kev). In this process, two electron neutrinos are emitted for every 26.5 Mev of radiation emitted. From the total luminosity of the sun, one can conclude that there is a flux of roughly 10^{10} ν_e/cm^2 sec. on earth. These neutrinos cover the entire energy range from zero to 14 Mev, with most of them concentrated in the lower energy range.

The experiment of Davis[13] exposes C_2Cl_4 to this neutrino beam, and is sensitive to the higher energy Boron ν_e's, which cause the reaction $\nu_e + {}^{37}Cl \rightarrow {}^{37}Ar + e^-$. According to extensive theoretical calculation of Bahcall et al[14] using the standard solar model, the Davis experiment should see 7.9 ± 2.4 SNU's (1 SNU = 10^{-36} captures/atom sec.) On the other hand, only 2.1 ± 0.4 SNU's are detected in the Davis experiment. This discrepancy between theory and observation is known as the solar neutrino puzzle.

It may be that the solar core where the high energy Boron neutrinos are produced via the reaction ${}^8B \rightarrow {}^8B_e^* + e^+ + \nu_e$ has a temperature (T_c) lower than is assumed in the solar model due to some yet unknown mechanism. A 15% reduction in T_c would be enough to reconcile data with theoretical predictions.

A more interesting possibility from the point of view of particle physics is that the discrepancy has to do with the properties of the neutrino. If we accept this point of view, an immediate conclusion is that the neutrinos must be massive. Furthermore, they must have one of the following properties such that after their production in the central core, most of them oscillate into another paticle which has no observable interaction with chlorine. The possible candidates for such particles in known particle physics models are: ν_μ, ν_τ, right-handed neutrino ν_R or singlet neutrino ν_s. Depending on which particle we assume, we obtain a certain constraint on the observed left-handed neutrino properties, summarized below:

(i) vacuum oscillation:

If ν_e, ν_μ and ν_τ mix among themselves, a fraction of ν_e will change their flavor in transit from sun to the earth. This probability is given by (for the case of two flavors)

$$|\langle\nu_e(t) \mid \nu_e(o)\rangle|^2 = 1 - \frac{1}{2} sin^2 2\theta \left(1 - cos \frac{\Delta m^2 t}{2k}\right) \tag{2.3}$$

Using $t = L/c$, (L is the Earth-Sun distance) one can immediately conclude that a sixty percent reduction in the neutrino signal corresponds to $\Delta m^2 \simeq 10^{-10}\ ev^2$. If we do not assume any finetuning of parameters, this implies that, $m_{\nu_e} << m_{\nu_\mu} \simeq 10^{-5}$ eV, for maximal mixing ($\theta = \frac{\pi}{4}$). This kind of neutrino spectrum is not accessible to laboratory experiments.

(ii) Vacuum oscillation into sterile neutrinos:

An interesting variation of the above idea is that neutrino is a pseudo-Dirac particle (see sec. 3), with a Majorana mass mixing $\delta m \simeq 10^{-11}$ eV. In such a case, the ν_{eL} will oscillate into its Dirac partner in transit from sun to the earth if $m_{\nu_e}\delta m \simeq 10^{-10}\ eV^2$ or for $m_{\nu_e} \simeq 10$ ev.[16] If the Dirac partner happens to be sterile with respect to weak interaction, it will escape detection. This scheme has two advantages over case (i); first, here the neutrino mass being in the eV range is accessible to in the laboratory experiments; secondly, the maximal mixing (i.e. $\theta = 45°$) arises in a natural manner.[16]

(iii) Neutrino oscillation in matter:

This solution was proposed by Mikheyev and Smirnov[17] using a mechanism proposed by Wolfenstein.[17] According to this proposal, neutrino scattering in a medium changes the form of the neutrino mass matrix. As a result, neutrino flavor oscillation can occur for a much wider range of mass differences for smaller values of laboratory mixing angles between ν_e and $\nu_{\mu,\tau}$. Detailed analysis has yielded[18] the range of masses and mixings that can solve the solar neutrino puzzles to be the following:

$$\Delta m^2_{ei} \simeq 3 \times 10^{-5} - 10^{-4}\ ev^2$$

and

$$sin^2 2\theta_{ei} \geq 10^{-3}\ . \tag{2.4}$$

Again, we assume no finetuning, this would imply that, $m_{\nu_e} << m_{\nu_\mu} \simeq 10^{-2}$ ev or $m_{\nu_e} << m_{\nu_e}$ $\simeq 10^{-2}$ ev. As is clear, this spectrum (to be called MSW spectrum in this article) is different from the above two cases.

(iv) ν_e-Decay: If the electron neutrino decays on its way to the earth with an appropriate life time that would also provide a resolution of the solar neutrino puzzle. This possibility was originally proposed by Bahcall et al;[19] but at that time no plausible particle physics model was known for such decay. With the introduction of the idea of the Majoron,[9] this possibility was revived again, since a model where the decay of type $\nu_e \rightarrow \nu_{\mu,\tau}$ + Majoron occurs could now be constructed.[20] A solution of the solar neutrino puzzle requires that

$$\gamma \tau_{\nu_e} \simeq 500 \ sec. \tag{2.5}$$

where γ is the relativistic time dilation factor.

(v) Magnetic Moment of the Neutrino:

A quite fascinating resolution of the solar neutrino puzzle with several interesting and testable predictions is the suggestion[21,22] that the neutrino is a Dirac particle with a non-vanishing magnetic moment μ_ν. In such a case in the magnetic field of about 10^4 Gauss in the convective zone of the sun, the left-handed neutrino emitted from the solar core undergoes precession thereby becoming a right-handed neutrino which has no visible weak interactions. A value of $\mu_\nu \approx (3-1) \times 10^{-10}\ \mu_B$ (where $\mu_B = e\hbar/2m_e c$) is required to flip the helicity of 60% of the emitted neutrinos. This value of the magnetic moment is close to the upper bound, obtained from considerations of energy loss from stars[23] as well as nucleosynthesis,[24] and was therefore acceptable at the time it was proposed. This model predicts that the intensity of solar neutrinos observed on earth should be correlated with the 11 year sun-spot cycle and that, there should be a half yearly variation. However, as we will see below, such a large value of μ_ν may not be consistent with observations from supernova.

(c) SN1987a Constraints:

Let us now turn to the information gained from the observations of neutrino signals in the underground detectors at Kamioka mine in Japan[25] and by the IMB group in Ohio[26] from the supernova 1987a. As is well known, neutrino emission is one of the primary mechanisms for energy loss in stars. In the case of a dying star, whose iron core collapses leading to a neutron star or a black hole, the neutrino emission accounts practically for 99% of the emitted energy, Q. Since the core contracts mainly through gravitational attraction, this energy Q represents the binding energy of the core. Using the information that core mass $M_c \simeq (1.4-2)M_O$ and core radius $R_c \simeq 10$ Km., the binding energy $B \cdot E \equiv Q$ is

$$Q = \frac{G_N M_C^2}{R_C} \simeq (2-4) \times 10^{53}\ .ergs \tag{2.6}$$

In the standard model of quark-lepton interactions, the neutrinos carry this energy to the earth in the form of ν_e, $\bar{\nu}_e$, ν_μ, $\bar{\nu}_\mu$, ν_τ and $\bar{\nu}_\tau$. These neutrinos interact with water Cherenkov detector in the earth to produce electrons (via $\nu_e e^- \rightarrow \nu_e e^-$ reaction) or positrons via $\bar{\nu}_e p \rightarrow e^+ n$ reactions. From the observed $e^\pm$ energy and angular distribution one can infer the energy of the incident ν_e and $\bar{\nu}_e$. From the references 25 and 26, we conclude that perhaps all the events are due to $\bar{\nu}_e$ interactions with incident $\bar{\nu}_e$ energy $E_{\bar{\nu}_e} \simeq$ 10-20 Mev. From this, one can infer that almost all the energy released in gravitational collapse of the core is emitted in the form of neutrinos. It is, therefore, to be expected that any new property of the neutrino not implied by the standard model will be constrained by these observations. We summarize some of these constraints below.

(i) Constraints on m_{ν_e}: For an anti-neutrino, $\bar{\nu}_e$ of mass $m_{\bar{\nu}_e}$ and energy $E_{\bar{\nu}_e}$, its transit time is given by

$$t \simeq \frac{D}{c}\left[1+\frac{1}{2}\frac{m^2_{\bar{\nu}_e}}{E^2_{\bar{\nu}_e}}\right] . \tag{2.7}$$

Since most of the neutrino arrive (at Kamioka) in a bunch of about 8 within a time space of 2 sec., and typical core collapse time is of the order of a second or two, the $\bar{\nu}_e$-mass must be constrained[27] to be less than 10-30 eV.

(ii) τ_{ν_e}: From the mere fact that the electron neutrinos arrive on earth after transversing 50 Kpc (which is about 1.7×10^5 light years), one easily concludes that,

$$\gamma\tau_{\bar{\nu}_e} \geq 1.7 \times 10^5 \; yrs. \tag{2.8}$$

By CPT theorem, ν_e must have the same bound on it lifetime, thereby ruling out neutrino decay as a solution of the solar neutrino puzzle.

(iii) Magnetic moment of the neutrino μ_{ν_e}:

It has recently been shown[28] that SN1987a observations severely constrain μ_{ν_e}. The reason for this is that for large values of μ_{ν_e}, the degenerate ν_{eL}'s in the supernova core can flip helicity to ν_{eR} via the photon exchange reaction $\nu_L e^- \to \nu_R e^-$. But since ν_R interctions are very weak, the meanfree path of ν_R's produced exceeds the core radius (unless there is a neutral right-handed gauge boson Z_R of mass $M_{Z_R} \leq 300$ Gev). The emitted ν_R's, therefore, have high energy (i.e. $E_{\nu_R} \approx$ 100-200 Mev) and they escape the core without further interaction, thus providing an additional mechanism of energy loss from the supernova. Demanding that $Q_{\nu_R} < Q$, we obtain[28]

$$\mu_{\nu_e} \leq 10^{-12}\,\mu_B \tag{2.9}$$

It further turns out that in the Galactic magnetic fields of 10^{-6} Gauss, the emitted high energy ν_R's can flip helicity to become ν_L before they reach the earth, providing possibilities of high energy neutrino signals ($E_{\nu_e} \geq 100$ Mev) in the underground detectors. Again absence of any such signals implies a stronger bound

$$\mu_{\nu_e} \leq 10^{-13}\,\mu_B . \tag{2.10}$$

This rules out neutrino magnetic moment as an explanation of the solar neutrino puzzle, (unless, of course, there is a Z_R of mass less than 300 Gev in which case, the emitted ν_R's get trapped in the supernova core).

(iv) Constraints on neutrino radiative decays and neutrino mixings:: Finally, we turn to the constraints on radiative neutrino decays and neutrino flavor mixing from supernova observations. If the heavier neutrinos ν_μ and ν_τ decayed into ν_e + photon, photons would be observable; however, lack of any signal in the kev and Mev range[29,30] has been shown to imply rather stringent constraints on allowed values of neutrino mixings.

Laboratory experiments also have led to constraints on neutrino masses as follows:[31] $m_{\nu_e} \leq 18$ ev, $m_{\nu_\mu} \leq .25$ Mev and $m_{\nu_\tau} \leq 70$ Mev. There exist constraints on the mixing angles from various accelerator experiments, which we do not include in this article, since we will not be using it in the subsequent sections.

3. Majorana, Dirac or Pseudo-Dirac Neutrino Masses:

Unlike a boson or an electrically charged fermion, a neutral spin half object such as the neutrino can have several kinds of masses. To understand this, we consider 2-component Weyl spinor ν to denote the chiral spin 1/2 neutrino emitted in the process of β-decay. It transforms as a $\left(\frac{1}{2}, 0\right)$ representation of the Lorentz group and admits a Lorentz invariant mass term such as:

$$\mathcal{L}^{(1)}_{mass} = i\, m\, \nu^T\, \sigma_2 \nu + \text{hermitean conjugate}\ . \tag{3.1}$$

If we assume that the neutrino carries a global quantum number such as lepton number (under which $\nu \rightarrow e^{i\alpha}\,\nu$), then $\mathcal{L}^{(1)}_{mass}$ violates lepton number by two units. In this case, the neutrino is said to have a Majorana mass and is called a Majorana particle. If the only spin half neutral fermion (with properties similar to the neutrino) around is ν, then no other mass term is possible for the neutrino. So, the neutrino will either remain massless or become a Majorana particle. On the other hand, if there is another spin 1/2 neutral lepton N present, then, one can envisage a generalized form for the mass term:

$$\mathcal{L}^{(2)}_{mass} = i\left(\nu^T\, N^T\right)\begin{pmatrix} m_{LL} & m_{LR} \\ m_{LR} & m_{RR} \end{pmatrix}\sigma_2 \begin{pmatrix} \nu \\ N \end{pmatrix} \tag{3.2}$$

(We will denote this 2×2 mass matrix in eqn. (3.2) by M). If all elements of M are not-vanishing, again no global symmetry will be respected by $\mathcal{L}^{(2)}_{mass}$; therefore, in this case, the neutrino will also be a Majorana particle and lepton number will be violated by two units. One can, however, have special situations:

a) $m_{LL} = m_{RR} = 0$: In this case, one can define the lepton number $U(1)$ symmetry as follows: $\nu \rightarrow e^{i\alpha}\,\nu$ and $N \rightarrow e^{-i\alpha}\,N$. $\mathcal{L}^{(2)}_{mass}$ will then respect lepton number. The ν and N combine to form a four-component Dirac neutrino. In the two component notation, one has the following eigenstates:

$$\nu_{\pm} = \frac{\nu \pm N}{\sqrt{2}} \tag{3.3}$$

with mass eigenvalues $\pm m_{LR}$; making the mass positive changes ν_- to $i\nu_-$. In the four component language, we write,

$$\psi = \begin{pmatrix} \nu \\ i\sigma_2\, N^* \end{pmatrix}\ . \tag{3.4}$$

Then,

$$\psi^c = C\bar{\psi}^T = -\gamma_2\psi^* = \begin{pmatrix} N \\ i\sigma_2\nu^* \end{pmatrix} . \qquad (3.5)$$

We then define:

$$\psi_\pm = \frac{1}{\sqrt{2}}\left(\psi \pm \psi^c\right) = \begin{pmatrix} \nu_\pm \\ i\sigma_2\nu_\pm^* \end{pmatrix} . \qquad (3.5)$$

From (3.5), it is obvious that, ψ_+ and ψ_- are even and odd under charge conjugation respectively.

b) $m_{LR} >> m_{LL} = m_{RR}$: In this case, the neutrino is called a pseudo-Dirac particle.[32] In this case there is a slight mixing between the CP-odd and CP-even eigenstates $\psi_\pm$; although the neutrino is a predominantly Dirac particle, it has a very slight admixture of Majorana character. This situation is of particular interest in the study of evolution of a chiral left-handed neutrino state, such as that emitted in beta decay. When m_{LL} or m_{RR} are nonzero (and small, m_{LL} , $m_{RR} << m_{LR}$), the eigenstates of M are still $\nu_\pm$; but their masses are split by an amount δm , i.e.

$$m_\pm = \mp m_{LR} + m_{LL} . \qquad (3.7)$$

The left-handed neutrino oscillates into the extra sterile lepton N , with a probability given by

$$P_{\nu N}(t) \simeq sin^2 \frac{1}{2}\left(\frac{m_{LL}}{R}\right)\left(\frac{m_{LR}}{E}\right) t . \qquad (3.8)$$

As discussed earlier if $m_{LL} \times m_{LR} \simeq 10^{-10}$ eV^2 , this can provide an alternative solution to the solar neutrino problem.[16]

c) $m_{LL} = 0$; $m_{LR} << m_{RR}$: In this case, the neutrino is a Majorana particle. But unlike cases (a) and (b) where the two neutrino states are nearly degenerate, in this case, there are two neutrinos, one heavy ($m_H \simeq m_{RR}$) and one light ($m_L \simeq m_{LR}^2/m_{RR}$) . This is well known as the "see-saw" mechanism[33] and provides a "naturally" small value for the neutrino mass without any finetuning of parameters. The eigenstates are also nearly pure, i.e.

$$\nu_{Light} \simeq \nu + \left(\frac{m_{LR}}{m_{RR}}\right) \cdot N$$

$$\nu_{Heavy} \simeq N - \left(\frac{m_{LR}}{m_{RR}}\right) \nu . \qquad (3.9)$$

Before closing this section, we wish to reemphasize the point that Majorana mass terms require the breaking of Lepton number, L. Since in general gauge theories the only gauge-anomaly free combination of quantum numbers is $B-L$, we will talk about breaking of $B-L$ symmetry. In the context of gauge theories, there are three possible ways to achieve this breaking:

(i) Explicit breaking of $B-L$ symmetry, where Lagrangian contains terms that break $B-L$ symmetry.

(ii) Spontaneous breaking of local $B-L$ symmetry[2]

(iii) Spontaneous breaking of global $B - L$ symmetry.[9]

In sec. 4-7 of the article, we will focus on the local $B - L$ symmetry models (case ii) and its connection to neutrino mass. In sec. 8, we return to case (iii), where a Goldstone boson (the Majoron[9]) appears in the theory coupled to the neutrino. This particle leads to many interesting phenomenological and cosmological implications for neutrino decay and neutrino scattering. For instance, heavier neutrinos can now decay via the emission of Majoron such as $\nu_H \to \nu_e + \chi$; there can also be neutrino annihilation such as $\nu\nu \to \chi\chi$. These new processes effect cosmological constraints on the neutrino masses. In the laboratories, one can have new modes for neutrinoless double β-decay such as $(N, Z) \to (N-2, Z+2) + 2e^- + \chi$. We discuss these questions in sec. 8.[17]

4. Brief Description of Left-Right Symmetric Models:

In the standard model of Glashow, Weinberg and Salam, only one helicity component is introduced into the theory to fit observations in muon and beta decay. Since the model has exact $B - L$ symmetry, it follows from the discussions of sec. 3 that neutrinos must be massless. In this model, the left-handed nature of weak interactions is also built in by hand, because of which no fundamental understanding of the origin of parity violation emerges. In order to provide an understanding of parity violation, the left-right symmetric theories of weak interactions were developed in a series of papers[1] in 1974-75. In these models, the basic interactions are invariant under parity transformations: the observed near maximal parity violation at low energies is then understood to be a consequence of spontaneous symmetry breaking. Before proceeding to the discussion of neutrino masses in this model, we present the basic ideas of the model and review the constraints on the new parameters of the model such as the masses of the right-handed gauge bosons, right-handed neutrino masses, etc.

Motivation and description of the Model

There is a rather compelling reason why left-right symmetry ought to be a fundamental and natural symmetry of the quark-lepton world. It is well-known that weak interactions are symmetric between quarks and leptons. We can use this symmetry to write a unified formula for electric charge in terms of weak isospin and $B - L$ quantum number, in a manner analogous to the Gell-Mann-Nishijima formula for the hadronic world, i.e.

$$Q = I_{3W} + \frac{B-L}{2}\,. \tag{4.1}$$

Since at low energies electroweak processes involve only the left-handed weak isospin, we must rewrite the eqn. (4.1) as

$$Q = I_{3L} + I_{3R} + \frac{B-L}{2}\,. \tag{4.2}$$

It is then natural to assume that Nature realizes the full right-handed weak isospin in the same way as it realizes the left-handed one. This then leads to the left-right symmetric model of weak interactions, based on the gauge group $SU(2)_L \times SU(2)_R \times U(1)_{B-L}$.[1] The quarks and leptons are assigned to the gauge group as follows: (denote $Q \equiv (u, d)$ and $\psi \equiv (\nu, e^-)$)

$$Q_L : (2,1,1/3)\,;\quad Q_R : (1,2,1/3)\,;\quad \psi_L : (2,1,-1)\,;\quad \psi_R : (1,2,-1)\,. \tag{4.3}$$

(where the numbers within the bracket represent the transformation property under the gauge group). It is clear that gauge interactions are symmetric between left and right-handed fermions; therefore, prior to spontaneous breaking of gauge symmetry, weak interactions like strong, electromagnetic and gravitational interactions conserve parity. This can be seen explicitly, by writing the gauge interactions

$$\mathcal{L}_{WK} = \frac{g}{2}\left[\vec{J}_{\mu,L}\cdot\vec{W}_L^{\mu} + \vec{J}_{\mu,R}\cdot\vec{W}_R^{\mu}\right] + g'V_{\mu}^{B-L}\cdot B^{\mu}\,. \tag{4.4}$$

The $\vec{W}_L$, $\vec{W}_R$ and B are the gauge bosons corresponding to $SU(2)_L$, $SU(2)_L$ and $U(1)_{B-L}$ gauge symmetries. As long as the W_L and W_R have the same mass, the weak interactions conserve parity. The observed parity violation at low energies is then attributed to vacuum being parity asymmetric, which manifests itself in $m_{W_R} >> m_{W_L}$. The symmetry breaking pattern responsible for the parity violation observed at low energies can be obtained using the following generic pattern for the Higgs multiplets: (we drop the reference to color gauge group from now on) ϕ: $(2,2,0)$, $H_L\,(a,1,b) + H_R\,(1,a,b)$ where a and b will be fixed later. The first stage of the symmetry breaking is implemented by the neutral component of the right-handed multiplet H_R acquiring a vacuum expectation value (v.e.v.) i.e., $\langle H_R^{\circ}\rangle = v_R$. This reduces the electroweak symmetry to $SU(2)_L \times U(1)_Y$, which is subsequently broken by ϕ acquiring a v.e.v.

$$\langle\phi\rangle = \begin{pmatrix} \kappa & 0 \\ 0 & \kappa' e^{i\alpha} \end{pmatrix}. \tag{4.5}$$

The second stage of the symmetry breaking can induce[18] a v.e.v. for $\langle H_L^{\circ}\rangle = v_L \simeq \gamma\,\kappa^2/v_R$ assuming $\kappa' >> \kappa$; the quarks and leptons acquire a mass at the second stage, as do the W_L and Z-bosons. The form of the low energy charged current interaction in this model is as follows:

$$H_{WK}^{c.c} = \frac{G_F}{\sqrt{2}}\left[\left(\cos^2\zeta + \eta\sin^2\zeta\right) J_{\mu L}^{+} J_L^{\mu^-} + \left(\eta\cos^2\zeta + \sin^2\zeta\right) J_{\mu R}^{+} J_R^{\mu^-}\right.$$
$$\left. + e^{i\alpha}\cos\zeta\sin\zeta\,(1-\eta)\,J_{\mu L}^{+} J_R^{\mu^-} + \text{h.c.}\right] \tag{4.6}$$

where $\eta = \left(\frac{m_{W_L}}{m_{W_R}}\right)^2$ and ζ is the mixing angle between the charged W-bosons and is given in terms of symmetry breaing parameters as $\zeta \simeq \left(\kappa\kappa'/m_{W_R}^2\right)$. The currents $J_{\mu L,R}$ are given as follows:

$$J_{\mu L}^{+} = \overline{P}\gamma_{\mu}\,(1+\gamma_5)\,U_L N + \overline{E}^{\circ}\gamma_{\mu}\,(1+\gamma_5)\,V_L E^{-} \tag{4.7}$$

where P, N, E° and E^{-} denote the up quark, down quark, neutrino and electron column vector involving all generations and U_L and V_L denote the charged current mixing matrices for the quark and lepton sector. The corresponding right-handed currents are obtained by replacing L by R and γ_5 by $-\gamma_5$. We would now like to confront the Hamiltonian in eqn.

(4.6) with experiment; and since weak processes involving leptons are less complicated by strong interactions, we would like to study purely leptonic and semileptonic processes to gain information and η and ζ. For this purpose, we need to know the nature of the neutrino and the structure of the weak right-handed current, specifically the quark and lepton mixing matrix in the right handed sector. As we saw in sec. 3, in a theory with two spin half neutral leptons, many possibilities exist for the neutrino masses. We must therefore study the experimental limits on W_R, Z_R and ν_R for various cases separately.

Limits on the Masses of the Z_2, W_R and the Right-Handed Neutrino:

To search for the signatures of new physics associated with the right-handed symmetry of weak interactions, we must determine to what extent known low energy physics restricts the mass of Z_2, W_R and ν_R. The most model independent limit arises on the mass of the Z_2-boson. The form of the neutral current interaction of Z_1 and Z_2 can be written in these models in the form[35,36]

$$\mathcal{L}_{WK}^{N.C.} = \frac{g}{\cos\theta_W}\left[Z_{1\mu}\left\{J_L^\mu + \eta_Z J^\mu\right\} + \frac{1}{(\cos 2\theta_W)^{1/2}}\, Z_{2\mu}J^\mu\right] \tag{4.8}$$

where $\eta_Z = (M_{Z_1}/M_{Z_2})^2$ and $J = \sin^2\theta_W\, J_L + \cos^2\theta_W\, J_R$; $J_{L,R} = \left(I_{3L,R} - Q\sin^2\theta_W\right)$. We, then, see that the effects of right-handed bosons vanish as $M_{Z_2} \to \infty$ and we obtain the neutral current interaction of the standard model. Typical precision of neutral current experiments is about 10%, which allows one to have a rather light Z_2; a recent analysis[36] yields $M_{Z_2} \geq 275$ Gev for these models.

As far as the charged W_R-boson is concerned, we consider both the light Dirac and Majorana cases. And, we will further assume that the left and right handed mixing matrices are equal. This is the case of manifest left-right symmetry[37] which emerges for the simplest choice of Higgs multiplets ϕ. In this case, if the neutrino is a Dirac particle with ν_L and ν_R forming this particle or is a Majorana particle such that $m_{\nu_R} \leq 10$ Mev, the most model independent bounds on m_{W_R} and ζ come from the analysis of a recent TRIUMF muon decay experiment.[38] In this experiment the endpoint spectrum of the positrons moving along a direction opposite to the spin of a stopped 100% polarized μ^+ is measured. In pure $V-A$ theory, this spectrum vanishes at the end point, thus any non-vanishing would be an indication of the presence of $V+A$ currents. The quantity measured this way, to be called R is given in terms of the muon decay parameters δ, ξ and ρ as

$$R = 1 - \frac{\delta\xi}{\rho}\, P_\mu \tag{4.9}$$

where in terms of the η and ζ defined after eqn. (4.6) we get

$$R = 4\eta^2 + 2\zeta^2 + 4\eta\zeta\,. \tag{4.10}$$

The experimental result for $1 - R = .99863 \pm .00046$ (stat) $\pm$ .00075 (syst.). This implies that $m_{W_R} \geq 432$ Gev for arbitrary ζ and $m_{W_R} \geq 514$ Gev for $\zeta = 0$; for arbitrary m_{W_R}, -0.050 $\leq$

$\zeta \leq .035$ and $|\zeta| \leq .035$ for $m_{W_R} \to \infty$. These analyses have recently been extended[39] to the case of non-manifest left-right symmetry,[40] where bounds on m_{W_R} become weaker.

Turning now to the case of the heavy right-handed neutrino (i.e. m_{ν_R} in the Gev range), the only bound comes from non-leptonic decays such as K-decays[44] and $K_L - K_S$ mass difference[42] (ΔM_K), the latter case leading to the most stringent bound. The reason for this bound is that calculation of the one loop $\Delta S = 2$ effective Hamiltonian arising from the $W_L - W_R$ exchange leads to an enhanced contribution to ΔM_K as follows

$$\Delta m_K = \Delta m_{K_{LL}} [1 - 430\,\eta] \ . \tag{4.11}$$

Thus we see that unless $\eta \leq 2.3 \times 10^{-3}$ or $m_{W_R} \geq 1.6$ Tev, eqn. (4.11) will give the wrong sign and contradict observations. This is the most stringent bound on m_{W_R}.

Coming now to the bound on the mixing parameter ζ, we have already seen that for Dirac neutrinos muon decay experiments provide a bound $\zeta < 3.5\%$. For heavy right-handed neutrinos, the most model independent bound comes from the study[43] of y-distribution in deep inelastic scattering of antineutrinos off nuclei. In the absence of $V + A$ current, the valence quarks lead to $(1 - y)^2$ type distribution, whereas the left-right mixing leads to flat y-distribution proportional to ζ. Analysis of existing experiments[44] lead to $\zeta < .1$. Other more model dependent bounds can be obtained by considering deviations from current algebra relations in non-leptonic decays, $(\zeta \leq 4 \times 10^{-3})$ as well as from beta decays[45] of nucleus combined with unitarity of the KM-matrix $(\zeta < .005)$. Finally, we note that with the simplest Higgs structure (one ϕ and $\Delta_{L,R}$), one has the following theoretical upperbound[46] on the mixing parameter $\zeta \leq \left(\frac{m_{W_L}}{m_{W_R}}\right)^2$. If we use the bound from the $K_L - K_S$ mass difference, it implies $\zeta \leq 2 \times 10^{-3}$, which is the most stringent, though model dependent bound on this parameter.

Bound on the mass of the Majorana Neutrino ν_R:

As mentioned earlier, the most natural way to understand small neutrino mass is to assume that the right-handed neutrino is a heavy Majorana particle. While the smallness of m_ν implies that m_{ν_R} is in the tens of Gev range, we would like to discuss the limits on m_{ν_R} from phenomenological constraints. An interesting limit that correlates $(m_{\nu_R})_{min}$ with m_{W_R} arises from present experimental limits[47] on neutrinoless double β-decay.[48] The point is that a heavy right-handed neutrino contributes to $(\beta\beta)_{ov}$ decay via the exchange of W_R boson with an amplitude proportional to $G_F^2\, \eta^2\, m_{\nu_R}\, \langle e^{-m_{\nu_R}^r/r} \rangle_{nuc.}$ and this contribution adds incoherently with the usual light neutrino contribution. Using the available nuclear matrix element calculations,[49] we obtain the bound shown in fig. 1. An upper limit on $m_{\nu_R} \leq m_{W_R}$ arises from considerations of vacuum stability, which is the right-hand line in fig. 1. Also, we infer from Fig. 1 that, combination of vacuum stability and $(\beta\beta)_{ou}$ decay lead to a lower bound on M_{W_R} of 800 Gev.

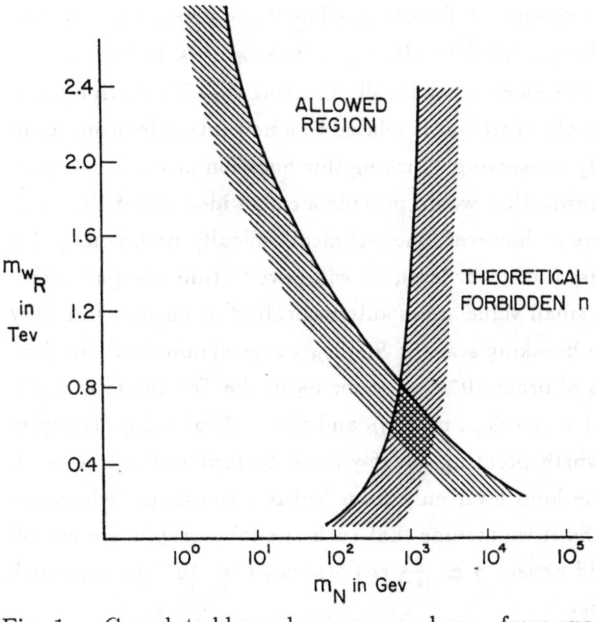

Fig. 1. Correlated bounds on m_{N_R} and m_{W_R} from experiments on neutrinoless double beta decay (left arm) and considerations of vacuum stability. The shaded region is forbidden.

5. Dirac vs. Majorana Neutrino in Left-Right Models:

Due to left-right symmetry, ν_L is accompanied by ν_R leading to a rich set of possibilities for the neutrino masses. The detailed nature, however, depends on the pattern of symmetry breaking which must lead to the standard model at low energies. We see that a crucial role is played by the Higgs boson that is responsible for breaking the $SU(2)_R$ symmetry, i.e. the nature of the Higgs multiplets H_L and H_R. Two possible choices of H_L and H_R can be considered: the first one which we will call the canonical choice was introduced in ref. 35 (case i) and it leads to Majorana neutrinos and provides an understanding of the smallness of the neutrino mass being related to the suppression of $V + A$ currents. The second one we mention here leads to light Dirac neutrinos (case ii) at the price of expanding the model to include singlet leptons.[50]

Case (i): Here, H are chosen to be triplets under the weak gauge group and are denoted by $\vec{\Delta}_L(3,1,2) + \vec{\Delta}_R(1,3,2)$. They couple to leptons only as follows:

$$\mathcal{L}'_Y = h\left\{\psi_L^T c^{-1}\tau_2\vec{\tau}\cdot\vec{\Delta}_L\psi_L + \psi_R^T C^{-1}\tau_2\vec{\tau}\cdot\vec{\Delta}_R\psi_R\right\} + \text{h.c.} \tag{5.1}$$

Choosing $\langle\Delta_R^\circ\rangle = v_R$ and assuming that the neutrino Dirac mass coming from the $SU(2)_L$ breaking is m_e, we obtain the following 2×2 mass matrix for $(\nu, N)'$

$$\begin{array}{cc} & \begin{array}{cc}\nu & N\end{array} \\ \begin{array}{c}\nu \\ N\end{array} & \begin{pmatrix} h\gamma\kappa^2/v_R & m_e \\ m_e & h\,V_R \end{pmatrix} \end{array}. \tag{5.2}$$

This mass matrix provides a qualitative explanation for the small neutrino mass, which comes out to be $m_\nu \simeq (h\gamma\kappa^2 - m_e^2)/v_R$; according to this formula, $m_\nu \to 0$ as $v_R \to \infty$ in the absence of right-handed weak interactions; it is, therefore, aesthetically pleasing that if neutrino has a small mass, it is not just a freak accident of nature but is related to a new intrinsic property of weak interactions being ultimately parity conserving. Turning this question around, we could say that once mass of the neutrino is known, that would provide a rough idea about the mass scale for right-handed interactions. There is, however, one technical difficulty with eqn. (5.2); to get acceptable values for neutrino mass (i.e. ev or less), we will have to tune the parameter $\gamma \simeq 10^{-6}$ or so. It turns out that such a small value arises automatically,[51] if parity symmetry is broken at a scale M_P higher than the breaking scale of $SU(2)_R$ gauge symmetry.[52] In fact, in this case, we find, $v_L \simeq \frac{\kappa^2 v_R}{M_P^2}$ which is of order 10^{-10} or so for v_R in the Tev range and $M_P \simeq 10^9$ Gev. This has the impact that at $\mu \simeq m_{W_R}$, $g_L \neq g_R$ and the multiplet Δ_L decouples from low energies. Alternatively, it is worth pointing out if γ is set to zero at the tree level, then it receives contributions at the one loop level only from Yukawa couplings, which are much smaller than one ($\sim 10^{-2}$ or so). So if we assume that such one loop graphs are cut off in momentum at $\mu \simeq M_{Planck}$, we would expect $\gamma \simeq \frac{h^4}{16\pi^2}\,\ell n\,(M_P/m_W) \lesssim 10^{-8}$ or so, which may justify the finetuning adopted earlier.

Turning to eqn. (4.4), we note that the right-handed leptonic current involves the right-handed Majorana neutrino. As a result, whether the right-handed leptonic currents manifest themselves in low energy decay processes depends on the mass, $m_{N_R} \simeq h v_R$, which for natural values of coupling parameters is expected to be in the tens to hundreds of Gev range. We will derive constraints on this mass from neutrinoless double beta decay.

A further point worth discussing in connection with the mass matrix in eqn. (5.2) is that if the scale v_R is high, then for arbitrary γ, we get $m_{\nu_i} \simeq h\gamma\cdot\kappa^2/v_R$. We find $m_{\nu_i} \approx 10$ eV requires for $h \simeq 10^{-3} - 10^{-4}$, $\gamma \simeq 10^{-1}$, $v_R \simeq 10^8 - 10^9$ Gev. An important characteristic of this point of view is that all light neutrinos can be nearly degenerate in mass, which could manifest themselves in substantial neutrino oscillations.

The mass eigenstates for neutrinos in this case (ignoring mixings between generations) is given by (setting $\gamma \simeq 0$)

$$\begin{aligned} \nu_e &\simeq \nu + \xi N \\ N_e &\simeq N - \xi \nu \end{aligned} \tag{5.3}$$

where

$$\xi \simeq (m_{\nu_e}/m_{N_e})^{1/2} .$$

Due to this admixture of the right-handed neutrinos, the GIM cancellation in this sector does not occur; as a result, one obtains flavor changing neutral current decays of Z such as $Z \to \nu_e\bar{\nu}_\mu$. This, then, opens up a channel for heavier neutrinos to decay via Z-exchange, e.g. $\nu_\mu \to \nu_e\bar{\nu}_e\nu_e$. This decay amplitude is, however, small in practice (see below).

Turning to the neutrino masses, we find (in the simplest approximation), the following mass formula for the neutrinos, i.e.

$$m_{\nu_i} \simeq m_{e_i}^2/m_{N_i} \tag{5.4}$$

where i denotes the generation and ℓ_i denotes the charged lepton. Assuming m_{N_i} to be generation independent, we find that, $m_{\nu_e} : m_{\nu_\mu} : m_{\nu_\tau} = m_e^2 : m_\mu^2 : m_\tau^2$. This relation was first proposed in ref. 18. If we assume, $m_{\nu_e} \simeq 1$ to 5 ev, we obtain from the above scaling law, $m_{\nu_\mu} \simeq 40$ - 200 kev and $m_{\nu_\tau} \simeq 10$ - 50 mev. Now, recalling our discussion in sec. 2, we see that, for stable neutrinos, these values are in the forbidden range. Therefore, for the model to be acceptable, ν_μ and ν_e must decay with typical life times of about $\tau_{\nu_\mu} \simeq 10^{12}$ sec. or less. The photonic decay mode, which occurs via a one loop graph[54] is bigger than 10^{16} sec. for $m_{\nu_\mu} \simeq 200$ kev. This is clearly much too long for our purpose. Let us, therefore, turn to the $.3\nu_e$ decay mode of ν_μ. This decay mode can arise in two ways in left-right symmetric models: one, via the flavor changing Z-couplings to neutrinos and two, via the exchange of neutral Higgs boson Δ_L°. The first mechanism leads to an effective $\nu_\mu \to 3\nu_e$ interaction with effective couplings strength $\sim G_F \left(m_{\nu_\mu}/m_{N_e}\right)^{1/2}$; since $G_F \left(m_{\nu_\mu}/m_\mu\right)^{1/2} \simeq 10^{-8}$ Gev^{-2}. This leads to a lifetime of order

$$\tau_{\nu_\mu} \geq 10^{17} \ sec.. \tag{5.5}$$

In the second mechanism, suggested in ref. 11 the strength of the four neutrino interaction is of order $\sim \left(h_{e\mu} h_{ee}/m^2_{\Delta_L^\circ}\right)$, where $m_{\Delta_L^\circ}$ is the mass of the neutral Higgs boson, $h_{\mu e}$ and h_{ee} are the Yukawa couplings of Δ_L with leptons. Since $m_{\Delta_L^\circ}$ is unknown, it can be arbitrary. To see how big it is, we note that, to satisfy eqn. (2.2), we find the following constraint:[12]

$$h_{e\mu} h_{ee}/M^2_{\Delta_L^\circ} \geq 6 \times 10^{-6} \, Gev^{-2} \left(\frac{m_{\nu_\mu}}{40 \, kev}\right)^{-3/2} . \tag{5.6}$$

By an $SU(2)_L$-rotation, the Δ_L^{++} gives rise to the highly suppressed decay mode $\mu \to 3e$, leading to the constraint that,

$$h_{\mu e} h_{ee}/m^2_{\Delta_L^{++}} \leq 10^{-10} \, Gev^{-2} . \tag{5.7}$$

Since Δ_L^{++} and Δ_L° are members of the same $SU(2)_L$ multiplet, their mass difference can at most be 250 GeV. This, together with eqn. (5.6) and (5.7) implies that,[12] $m_{\Delta_L^\circ} \simeq$ few Gev. This can therefore be tested once the Z-width is measured since $B\left(Z \to \Delta_L^\circ \Delta_L^{\circ *}\right) / B\left(Z \to \nu\bar{\nu}\right) \simeq 2$.

Finally, we turn to the Majoron decay mode of ν_μ and ν_τ in left-right symmetric models.[55] If the left-right symmetric model is extended by the inclusion of the left-right symmetric doublets $\chi_L\left(\frac{1}{2}, 0, 1\right)$ and $\chi_R\left(0, \frac{1}{2}, 1\right)$, then, the model can admit an extra global symmetry, (similar to the $B-L$ quantum number). Spontaneous breaking of this symmetry leads to the existence of a mass less Goldstone boson χ. The Goldstone boson couples to neutrinos and allows for the decay mode $\nu_\mu \to \nu_e\chi$. The typical strength of this decay mode can be big enough[54] to avoid the constraints from cosmological mass density.

Case (ii) Now, we turn to the Dirac neutrinos in left-right symmetric models. As mentioned earlier, this requires the introduction of new singlet fermions θ_1 and θ_2 and new set of Higgs bosons $\chi_L\left(\frac{1}{2},0,1\right) \oplus \chi_R\left(0,\frac{1}{2},1\right)$. The existence of these multiplets enables one to write the following Yukawa coupling:

$$\mathcal{L}_Y = f\left(\bar{\psi}_L\,\chi_L\,\theta_1 + \bar{\psi}_R\,\chi_R\,\theta_2\right) + \mu\bar{\theta}_1\theta_2 + h.c.\,. \tag{5.8}$$

The first point we wish to note about eqn. (5.8) is that the parameter μ does not receive infinite contributions from higher loops and is, therefore radiatively stable. The significance of this observation is that if we fix μ to be $\simeq 1 - 10$ Mev it will not change in higher orders. Minimization of the potential leads to $\langle\chi_R\rangle \simeq v_R$, $\langle\chi_L\rangle \simeq \frac{\mu'\kappa}{V_R} \approx 10^{-2}\,\mu'$ where we choose $\mu' \approx \mu$. (This is because as $\mu \to 0$, $\mu' \to 0$). Then, $\langle\chi_L\rangle \simeq 10^{-4} - 10^{-5}$ Gev. In the subsequent discussion, we will ignore this. We then obtain the following 4×4 mass matrix:

$$\begin{array}{c} \\ \nu \\ \theta_1 \\ N \\ \theta_2 \end{array}\begin{array}{c} \begin{array}{cccc} \nu & \theta_1 & N & \theta_2 \end{array} \\ \left(\begin{array}{cccc} 0 & 0 & m_D & 0 \\ 0 & 0 & 0 & \mu \\ m_D & 0 & 0 & f\,V_R \\ 0 & \mu & f\,V_R & 0 \end{array}\right) \end{array}. \tag{5.9}$$

This matrix leads to two Dirac neutrinos, one heavy with mass $\sim f\,v_R$ and another light, with mass, $m_\nu \sim m_D\mu/f\,v_R$. This light four component spinor has the correct weak interaction properties to be identified as the neutrino.

Coming to the mass spectrum, we find that for $f \simeq 1$ and $v_R \sim$ Tev, $m_\nu \simeq 1$ eV. To extend this model to include th higher generations, we include 3 pairs of $(\theta_1\,,\,\theta_2)$. If we assume f and μ to be generation independent, then neutrino masses scale linearly with charged lepton masses. Thus, if $m_{\nu_e} \sim 10$ eV, we would have $m_{\nu_\mu} \sim 2$ kev and $m_{\nu_\tau} \sim 40$ kev. We again see the need for ν_μ and ν_τ to decay to avoid cosmological mass constraints.

Let us discuss the magnetic moment of neutrino. It is well-known that a Dirac neutrino can have a magnetic moment, μ_ν. In fact in the case conventional left-right models with Dirac neutrinos, μ_ν has been calculated.[55] The interest in this parameter is due to recent observations by Okun, Voloshin and Vysotski[56] that the observed deficit in the number of solar neutrinos reaching the earth can be explained if $\mu_\nu \simeq 10^{-11}\,\mu_e$. In conventional left-right[1] symmetric models with no extra singlet neutrinos, the principal diagram contributing to μ_ν is shown in fig. 2 and has the magnitude

$$\mu_\nu \simeq \frac{\alpha\,m_e}{8\pi\,sin^2\theta_W\,m_{W_L}^2}\,\xi \simeq 8\times 10^{-11}\,\xi\cdot\mu_e\,. \tag{5.10}$$

If we take the phenomenological upperbounds of 10^{-2} on ξ, $\mu_\nu \simeq 10^{-12}\,\mu_e$, which is smaller than that required for solar neutrino puzzle. Note that in a model like this, the smallness of the neutrino mass is put in by hand. If, however, we consider a model like the one proposed in this section, (where small m_ν arises without finetuning of parameters), we estimate

$$\mu_\nu \simeq \frac{\alpha}{8\pi\,sin^2\theta_W}\cdot\left(\frac{m_e}{m_W^2}\right)\left(\frac{m_D}{m_{W_R}}\right)\cdot\xi\,. \tag{5.11}$$

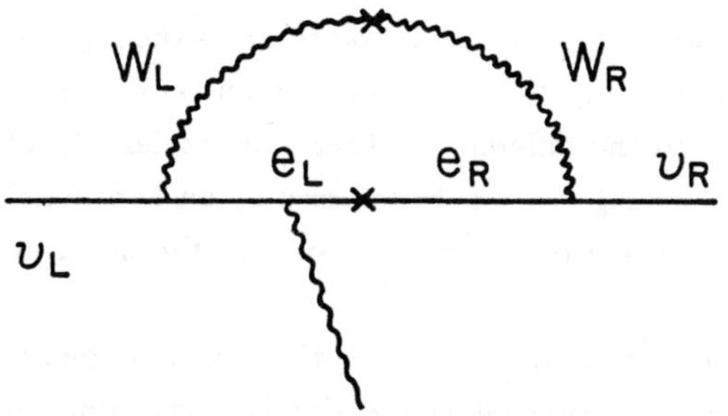

Fig. 2. The diagram contributing to $\nu_H \to \nu_e \gamma$ decay as well as the magnetic moment of the neutrino in left-right symmetric models.

Due to the additional suppression factor (m_D/m_{W_R}), which is of order 10^{-5}, implying that, $\mu_\nu \simeq 10^{-17}\,\mu_e$, making it completely irrelevant for any discussion of solar neutrino puzzle.

Finally, we wish to note that one can construct a left-right symmetric model by addition of weak iso-singlet quarks and leptons such that the Dirac mass of the neutrino vanishes at the tree level and arises at the one-loop level via $W_L - W_R$ mixing.[40] In this case, the Dirac neutrino mass is given by:

$$m_{\nu_i} = \left(\frac{\alpha}{4\pi\, sin^2\,\theta_W}\right)^2 \left(\frac{m_b m_t}{m_{W_R}^2}\right) \cdot m_{\ell_i}\,. \tag{5.12}$$

For $m_{W_R} \simeq 10$ Tev (say), this will lead to $m_{\nu_\mu} \simeq 5\times 10^{-4}$ ev, (and $m_{\nu_e} \simeq 5\times 10^{-6}$ eV) which is in the range of interest for the MSW solution to the solar neutrino puzzle.

6. $SO(10)$ Embedding:

In this section, we discuss the $SO(10)$ embedding of the left-right symmetric models discussed in the previous section and any new possibilities for neutrino masses, that may arise in these models. It is of course well known that, the simplest grandunification model that leads to left-right symmetric models at low energies is the $SO(10)$ model. The fermions (both quarks and leptons) belong to the 16-dimensional spinor representation of $SO(10)$. The symmetry breaking pattern of interest to us is the following one:

$$SO(10) \longrightarrow G \longrightarrow SU(3)_C \times SU(2)_L \times SU(2)_R \times U(1)_{B-L}$$
$$\downarrow$$
$$G_{std.}$$

We will first consider the scenario that leads to Majorana neutrinos: this is achieved by including the Higgs multiplets, which transform as $\{54\}$ (S), $\{45\}$ (A), $\{126\}$ (Σ), and $\{10\}$ (H) dimensional representations. (The symbols within the parenthesis denote those representations in what follows). The $\{54\}$-dimensional representation breaks $SO(10)$ down to $SU(4)_C \times SU(2)_L \times SU(2)_R$, whereas $\{45\}$ breaks this group further down to $SU(3)_C \times SU(2)_L \times SU(2)_R \times U(1)_{B-L}$ without the discrete D-parity symmetry that relates the left and right-handed gauge couplings. The {126}-dimensional representation contains the Δ_L and Δ_R defined earlier and therefore serves to break not only $SU(2)_R$ symmetry but also gives a heavy mass to the right-handed neutrinos. Due to the difference between the scales M_P of D-parity breaking and M_R of $SU(2)_R$-breaking, the Δ_L° v.e.v. is suppressed by an additional factor of $(v_R/M_P)^2$ with respect to κ^2/v_R. This gives the real see-saw form to the neutrino mass matrix.[59]

Having outlined the overall scenario, we mention two interesting specific cases that emerge in this model: in one case, the D-parity breaking scenario enables the $sin^2\theta_W$ constraints to be satisfied even with a low M_{W_R} (in the Tev range) leading neutrino mass spectrum of ev-kev-Mev type as discussed in sec. 5; in the second case, if the $B-L$ breaking scale (which coincides with $SU(2)_R$ scale in the cases we consider) is identified with the Peccei-Quinn scale,[58] then a unique $SO(10)$ model emerges, that simultaneously provides a solution to solar neutrino problem via the MSW mechanism.

In order to comment on these two possibilities, we first explain the meaning of D-parity. When $SO(10)$ breaks down to one of its maximal subgroups, $SO(4)\times SO(6)$, there is an additional discrete symmetry which can also remain unbroken. We call this discrete symmetry as D-parity. Under D-parity, $q\rightarrow q^C$ and $\vec{W}_L\rightarrow\vec{W}_R$; so as far as the quark sector is concerned it acts as charge conjugation or parity; but its effect on Higgs bosons is more complicated; therefore we prefer to call it D-parity rather than C or P. Its practical impact is to maintain the equality of gauge couplings constant of the $SU(2)_L$ and $SU(2)_R$ groups. Until the work of ref. 52, it was not realized that $SU(2)_R$ and D-parity could be broken separately. If $SU(2)_R$ and D-parity were broken together, the observed value of $sin^2\theta_W$ could only be consistent with $SO(10)$ model if the value of m_{W_R} is rather high ($\sim 10^{10}-10^{12}$ Gev). On the other hand, if D-parity was broken "strongly" (i.e. if g_L/g_R differed very much from 1 at the scale m_{W_R}), one could have $m_{W_R}\simeq$ from TeV's without spoiling the agreement with $sin^2\theta_W$. Since in the neutrino mass matrix, m_{W_R} determines the mass of the right-handed neutrino, m_{RR}, we would get an ev-Kev-Mev type neutrino spectrum. The question, then, is: do we have a built-in mechanism to achieve sufficiently fast neutrino decay? This is now more subtle due to the existence of D-parity breaking; since the mass of the entire $\vec{\Delta}_L$ triplet is now lifted to the D-parity breaking scale M_P. Thus, $\nu_\mu\rightarrow 3\nu_e$ decay modes are suppressed. On the other hand, one can have the Majoron decay modes present by including Higgs multiplets {16}, if we forbid the $\{16\}\cdot\{16\}\cdot\{126\}$ Higgs coupling a là ref 54. In such a case, a consistent $SO(10)$ model with m_{ν_e} in the ev range could be constructed.

Let us now turn to the second scenario for the neutrino mass suggested in ref. 58. The theoretical motivation for the model is the observation that while the vectorial $B-L$ symmetry is anomaly-free and is, in fact, a part of $SO(10)$ symmetry, the axial $B-L$ is plagued with anomalies arising from the QCD gluon contributions and is the source of the well-known strong CP-problem. The suggestion of Peccei and Quinn was of course to make axial $B-L$ into a continuous global symmetry of the gauge model (operating both on quarks and Higgs bosons) and use it to convert θ into symmetry parameter, in which case $\theta=0$ is as physical as $\theta\neq 0$. The breaking of this symmetry leads to axion and the presently consistent axion models[59] seem to require that 10^9 Gev $\leq M_{PQ}\leq 10^{12}$ GeV. Since PQ symmetry and vectorial gauge

$B - L$ symmetry as so closely related, it was suggested in ref. 58 that perhaps they may be broken at the same scale. In fact, $SO(10)$ models, as mentioned before, are required (without or with a mild D-parity breaking) to have $M_{B-L} \simeq 10^{12}$ Gev.

This has profound implications as far as neutrino mass is concerned. The neutrino masses are now given by:

$$m_{\nu_i} \simeq m_{u_i}^2 / 9\, M_{BL} \tag{6.1}$$

where m_{u_i} is the up-quark mass of the corresponding generation and with $M_{BL} \simeq 10^{13}$ Gev, we get, $m_{\nu_e} \simeq 10^{-8}$ ev; $m_{\nu_\mu} \simeq 10^{-3}$ ev, $m_{\nu_\tau} \simeq 4$ ev. We then see that m_{ν_μ} has mass in the right range to provide a solution to the solar neutrino problem a là MSW. It is worth pointing out that this model is rather unique and leads to a proton lifetime of 2.8×10^{34} years.[58,60] This could therefore be tested in the next generation of proton decay experiments.

Finally, it should also be noted that the left-right symmetric model for the Dirac neutrino also has a simple extension to $SO(10)$, where we include a single extra singlet fermion θ and replace the $\{126\}$ Higgs by a $\{16\}$-dimensional Higgs boson, which, however, does not have a coupling of type $\{16\}\{16\}\{10\}$. This leads to the following mass matrix for (ν, N, θ):

$$\begin{array}{c} \\ \nu \\ N \\ \theta \end{array} \begin{array}{c} \begin{array}{ccc} \nu & N & \theta \end{array} \\ \begin{pmatrix} 0 & m_D & 0 \\ m_D & 0 & V_R \\ 0 & V_R & 0 \end{pmatrix} \end{array} . \tag{6.2}$$

This leads to a massless neutrino. On the other hand, if a mass term for θ $\left(\mu \theta^T C^{-1} \theta\right)$ is added, it leads to a light Majorana neutrino ν_e, with $m_{\nu_e} \simeq \frac{m_D^2 \mu}{V_R^2}$, which is ultralight. This last case is similar to solution proposed by the author[61] to solve the neutrino mass problem in superstring models.

7. Neutrinos in Superstring Inspired Models:

In this section, we consider the properties of neutrinos in the superstring model. To make this discussion useful, we first give a miniature review of the gross features of superstring inspired models. These models imply a supersymmetric E_6-grandunified model with matter fields belonging to $\{27\}$-dimensional representations of the E_6-group and with Higgs fields also transforming as $\{27\} + \{\overline{27}\}$ fields under E_6. There are additional E_6-singlet fields. In particular note the absence of any E_6 representation such as $\{351\}$, which contains the $\{126\}$-dimensional representation of $SO(10)$ that was essential to the see-saw mechanism for small neutrino masses. Therefore, the nature of neutrino masses are likely to be more complicated in superstring models. To facilitate this discussion, we first write down the decomposition of the $\{27\}$-dim. representation of E_6 under $SO(10)$ as well as $SU(3)_c \times SU(3)_L \times SU(3)_R$.

$$E_6 \longrightarrow SO(10) \times U(1)_X \ :$$

$$\{27\} = \{16\}_1 \oplus \{10\}_{-2} \oplus \{1\}_4$$

$$\text{Particle Content}: \quad \{16\}\,; \quad \begin{pmatrix} u_1 & u_2 & u_3 & \nu \\ d_1 & d_2 & d_3 & e^- \end{pmatrix}_{L+R}$$

$$\begin{aligned} \{10\} \ &: \ g\,,\, g^c \ \text{singlet quark} \\ & \quad \begin{pmatrix} H_u^o & H_u^+ \\ H_d^- & H_d^o \end{pmatrix} \\ \{1\} \ & \quad n_o \end{aligned} \tag{7.1}$$

$$E_6 \supset SU(3)_c \times SU(3)_L \times SU(3)_R \ :$$

$$\{27\} \supset (3,3,1) \oplus (\bar{3},1,\bar{3}) \oplus (1,\bar{3},3)$$

$$(3,3,1) \ : \ \begin{pmatrix} u \\ d \\ g \end{pmatrix} \ ; \ (\bar{3},1,\bar{3}) \ : \ \begin{pmatrix} u^c \\ d^c \\ g^c \end{pmatrix}$$

$$(1,\bar{3},3) \ : \ \begin{pmatrix} H_u^o & H_u^+ & e^+ \\ H_d^- & H_d^o & \nu^c \\ e^- & \nu & n_o \end{pmatrix} \equiv L \tag{7.2}$$

We see from eqn. (7.1) and (7.2) that aside from the known fields, there are eleven extra fermions out of which three are neutral. In total, each generation have 5 neutral leptons. Add to this the neutral gauginos (4 colorless ones) and many Higgsinos plus E_6 singlet fermions, which makes a total of more than 20 neutral fermions. All these in general can mix making any understanding of the nature of the neutrino next to impossible in general. We will therefore simplify the picture by first ignoring generation mixing and secondly by assuming that there exist intermediate mass scales, which decouple the neutral leptons $H_u^o - H_d^o$. We also further assume that, in these models, $B - L$ symmetry scale is either an intermediate scale such as $10^{11} - 10^{12}$ Gev or a Tev. In this simplified approximation, four mechanisms have been proposed to understand neutrino masses, which we briefly summarize below.

(i) R-parity violation Mechanism:[61]

In this mechanism, it is assumed that the low energy gauge group is given by $SU(3)_c \times SU(2)_L \times U(1)_{I_{3R}} \times U(1)_{B-L}$ and that the superpartner of ν^c (the anti right-handed neutrino) acquires a v.e.v. of order V_R to break I_{3R} and $B-L$. Because of this, the gaugino corresponding to the broken generator (denoted by $\lambda_{Y'}$) couples to the ν^c and leads to the following mass matrix gaugino-neutrino mass matrix:

$$\begin{array}{c} \\ \nu \\ \nu^c \\ \lambda_{Y'} \end{array} \begin{array}{c} \begin{array}{ccc} \nu & \nu^c & \lambda_{Y'} \end{array} \\ \begin{pmatrix} 0 & m_D & 0 \\ m_D & 0 & g\,v_R \\ 0 & g\,v_R & \mu \end{pmatrix} \end{array} . \tag{7.3}$$

Here μ is the Majorana mass for λ_Y-gaugino that can arise in the two-loop order. This matrix can be diagonalized leading to a pseudo-Dirac 4-component heavy lepton ($m \simeq g\,v_R$) and a light 2-component Majorana neutrino

$$\nu_e \simeq \nu + \frac{m_D}{g\,v_R}\,\nu^c \tag{7.4}$$

with mass

$$m_{\nu_e} \simeq \frac{m_D^2 \cdot \mu}{(g\,v_R)^2}\,. \tag{7.5}$$

Usually, in supergravity models, $\mu \simeq 100$ Gev or so; assuming $(g\,v_R) \simeq 1$ Tev, we get $m_{\nu_e} \simeq 10^{-1}$ ev. If, on the other hand, we choose $v_R \simeq 10^{12}$ Gev, $m_{\nu_e} \simeq 10^{-9}$ ev. Thus, smallness of neutrino mass is understood without any unnatural fine tuning of parameters. It has been pointed out[45] that, in these models, the H_u^o-Higgsino mixes with the ν with mixing proportional to $h_\nu^D\,v_R$; this modified eqn. (7.5) and m_ν has a component which grows with v_R. Thus, if we want to understand the small m_ν, we must have an upperbound on v_R of a few Tev. A limitation of this method so that it works only for one generation.

(ii) E_6-Singlet Mechanism:

Use of E_6-singlets (S) for studying the neutrino mass problem was hinted at by Witten[63] and was analyzed in detail in ref. 64. here, one has to rely on possible symmetries that may arise in superstring models on specific Calabi-Yau spaces to restrict the Yukawa couplings in such a way that we have couplings of the following type in the superpotential: using the notation of (7.2),

$$W = W_o + \lambda_1\,LLL_H + \lambda_2\,L\bar{L}_H\,S \tag{7.6}$$

where subscript H stands for the Higgs field. If we then give v.e.v. to $\tilde{\nu}_H^c$, and $(H_u^o)_{Higgs}$, then, for each generation we get the following type of 3×3 mass matrix:

$$\begin{array}{c} \\ \nu \\ \nu^c \\ S \end{array}\begin{array}{c} \begin{array}{ccc} \nu & \nu^c & S \end{array} \\ \begin{pmatrix} 0 & m_D & 0 \\ m_D & 0 & \lambda_2\,v_R \\ 0 & \lambda_2\,v_R & 0 \end{pmatrix} \end{array}\,. \tag{7.7}$$

This, as discussed in sec. 6, leads to a massless neutrino and a massive Dirac neutrino. If we include a possible Majorana mass term for the S-fermion of order of 100 Gev, the mass matrix looks like that in eqn. (7.3) with similar values for m_ν. Note that these values fall outside the range needed to solve the solar neutrino puzzle for $v_R \simeq 1$ Tev or 10^{11} Gev.

(iii) Higher Dimensional Operators:

It was proposed in ref. 65 that one may solve the neutrino mass problem in superstring models with two intermediate mass scales corresponding to $\langle \tilde{\nu}_H^c \rangle \approx \langle \tilde{n}_{o_H} \rangle \approx 10^{11} - 10^{12}$ Gev. In this case, a higher dimension term in the superpotential of the form $L\bar{L}_H \cdot \bar{L}_H / M_{Pl}$ leads to an effective see-saw-like mechanism with right-handed neutrino mass $m_{N_R} \simeq \langle \tilde{\nu}^c \rangle^2 / M_{Pl} \simeq 10^4 - 10^6$ Gev. Using the standard see-saw formula, we get $m_{\nu_e} \simeq m_e^2/m_{N_R} \simeq 10^{-2}$ ev $- 10^{-4}$ ev $m_{\nu_\mu} \simeq 400$ ev $- 4$ ev and $m_{\nu_\tau} \simeq 160$ Kev $- 1.6$ Kev. Since in the Superstring model, there is no known way for ν_μ and ν_τ to decay, this scenario will be ruled out by cosmology. However, there exist arguments from proton decay that $\langle \nu_H^c \rangle \simeq \langle n_{oH} \rangle$ may be of order 10^{14} Gev, leading to $m_{N_R} \simeq 10^{10}$ Gev in which case, $m_{\nu_e} \simeq 10^{-8}$ ev, $m_{\nu_\mu} \simeq 4 \times 10^{-4}$ ev and $m_{\nu_\tau} \simeq .16$ ev. This kind of a spectrum is cosmologically acceptable and with slight variation of coupling parameters could be useful for the solution of solar neutrino puzzle via the MSW mechanism.

(iv) One Loop Induced Neutrino Mass:[66]

It has also been proposed that existence of some accidental symmetries may force the tree level Dirac mass of the neutrino to vanish. As a result, the dominant contribution to neutrino masses can arise at the one loop level from diagrams of the type shown in fig. 3. Here also the neutrino masses are expected to be small and could be of interest for discussion of the MSW effect.

Whereas in the above discussion, we have kept the different mechanisms (i), (ii) and (iii) separate, one may keep all these mechanisms together as has been recently attempted.[67] The above conclusions remain unaffected in this general analysis.

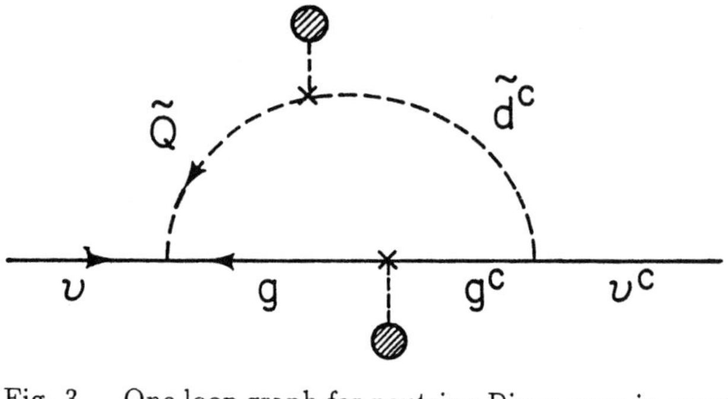

Fig. 3. One loop graph for neutrino Dirac mass in some versions of superstring inspired models.

8. Spontaneous Breaking of Global B-L Symmetry and the Majoron:

As we saw earlier in this article, if the neutrino is a Majorana particle, the low enegy theory must break $B-L$ symmetry. A novel possibility first suggested by Chikashige, Mohapatra and Peccei[9] is that the Majorana character of the neutrino may arise from spontaneous breaking of a global $B-L$ symmetry. This leads, by Nambu-Goldstone theorem to the existence of a massless scalar boson coupled to leptons, called the Majoron[68] in ref. 9. While apriori existence of a massless boson would appear to be disastrous since, it would lead to $1/r$ type long range forces, it was pointed out in ref. 9 that as a consequence of shift invariance associated with Goldstone bosons, their diagonal couplings in the non-relativistic limit are always spin dependent, i.e. $[(\vec{\sigma}_1 \cdot \vec{\nabla}_1)(\vec{\sigma}_2 \cdot \vec{\nabla}_2)] \frac{1}{R}$ type and are not therefore easily observable. A concrete realization of this idea was proposed in reference 9 by a minimal extension of the standard model that includes the right handed neutrino and a lepton number carrying neutral singlet Higgs boson. Let us start by describing this model, which is based on the standard model gauge group $SU(2)_L \times U(1)_Y$ with usual particle assignments: Quarks: $Q_L \equiv (\frac{1}{2}, \frac{1}{3})$; $u_R \equiv (0, \frac{4}{3})$, $d_R \equiv (0, -\frac{2}{3})$; leptons: $\psi_L \equiv (\frac{1}{2}, -1)$, $e_R \equiv (0, -2)$; the Higgs field: $\varphi \equiv (\frac{1}{2}, +1)$. To this we add the right handed neutirno $\nu_R = (0,0)$ (one for each generation) and a neutral singlet Higgs boson $\Delta^\circ \equiv (0,0)$. We choose the Yukawa couplings and the Higgs potential so that the model has a global lepton-number symmetry with Δ° having $L = 2$, i.e. in our lagrangian we include the couplings $\nu_R^T c^{-1} \nu_R \Delta^\circ$, $\bar{\psi}_L \varphi \nu_R$ and forbid the scalar self-coupling $\varphi^+ \varphi \Delta$, (which is otherwise allowed by the gauge symmetry). The Higgs potential is then so chosen as to lead to a minimum where the following Higgs fields have vacuum expectation values:

$$\langle \varphi \rangle = \begin{pmatrix} 0 \\ v/\sqrt{2} \end{pmatrix} \quad \text{and} \quad \langle \Delta^\circ \rangle = v_{BL} \,. \tag{8.1}$$

The global lepton number (or more precisely $B - L$) symmetry is then broken spontaneously. The associated Goldstone boson (denoted by χ) is given by $\chi = I_m \Delta^\circ$ and couples in the non-diagonal basis to ν_R's. Furthermore, the neutrino mass matrix (per generation) takes the following see-saw form,[9]

$$M = \begin{pmatrix} 0 & h_\ell v/\sqrt{2} \\ h_\ell v/\sqrt{2} & h^M v_{BL} \end{pmatrix} \tag{8.2}$$

leading to the light neutrino masses given by the formula $m_\nu \simeq \frac{(h_\ell v)^2}{h^M V_{BL}}$; if we approximate h_ℓ to be the same as the Yukawa coupling of the charged lepton of the corresponding generation, then $m_{\nu_e} \simeq m_e^2/2h^M v_{BL}$ and for $h^M \simeq 1$ and $v_{BL} \simeq$ few Tevs, we get neutrino masses below the laboratory bounds. Shortly after the proposal of the idea of the Majoron, it was also pointed out[69] that, $\nu_{\mu,\tau}$ can decay to $\nu_e + \chi$ making an eV-Kev-Mev type spectrum for neutrinos consistent with cosmological constraints. More refined analysis of this question carried out in the context of both $SU(2)_L \times U(1)_Y$[70] as well as $SU(2)_L \times SU(2)_R \times U(1)_{B-L}$ models[54] have confirmed this belief. Thus, an immediate advantage of the concept of Majoron is that an eV-keV-Mev type neutrino spectrum is theoretcally and cosmologically consistent. Turning this question around, any evidence for an eV-Kev-Mev type neutrino spectrum will also, at the same time, be an evidence for the existence of the Majoron.

It is worth emphasizing at this point that Majoron couples[9] to ν_L at the tree level in the diagonal basis with a coupling strength $\approx (m_{\nu_L}/v_{BL})$ which is expected to be of order $\sim 10^{-11} - 10^{-12}$. This leads to $\nu\nu \to \chi\chi$ annihilation cross section of order of 10^{-66} cm^2 for $E_\nu \simeq$ few Mev, which is utterly negligible. The coupling of the Majoron to charged fermions u, d, e arises at the one loop level via the exchange of W or Z bosons and has the magnitude $\approx \frac{G_F}{\pi} m_f m_\nu$, where $f = u, d, e$. This is of order $\approx 10^{-16}$ or so. As a result, Majoron emission contributes a negligible amount to the energy loss of red giants as well as other stars. We see from this that the singlet Majoron is highly invisible. As we see below, there exist modification of this idea introduced and analyzed by Gelmini and Roncadelli[71] and Georgi, Glashow and Nussinov,[72] following the work of ref. 9, which leads to a number of testable laboratory prediction of the Majoron idea.

The triplet Majoron model[71,72] extends the standard model by introducing a Higgs triplet $\vec{\Delta}_L$ with $Y = 2$ and lepton number $L = 2$.

$$\vec{\tau} \cdot \vec{\Delta}_L = \begin{pmatrix} \Delta_L^+ & \sqrt{2}\,\Delta_L^{++} \\ \sqrt{2}\,\Delta_L^o & \Delta_L^+ \end{pmatrix} . \tag{8.3}$$

The introduction of this multiplet allows for the existence of the coupling $h\,\psi_L^T c^{-1} \tau_2 \vec{\tau} \cdot \vec{\Delta}_L \psi_L$. This Yukawa coupling includes a direct coupling of type $\nu_L^T\, c^{-1}\, \nu_L \Delta_L^o$ in contrast with the singlet Majoron. The global $B - L$ symmetry of the model is spontaneously broken by a vacuum expectation value of $\langle \Delta_L^o \rangle = v_T$, which then leads to a neutrino Majorana mass m_ν $= h_m\, v_L$. Furthermore, since the Majoron field in this case is given by

$$\chi = \frac{\sqrt{2}\, v_T Im\varphi^o - v\, Im\, \Delta^o}{\sqrt{v^2 + 2v_T^2}} \tag{8.4}$$

the left-handed neutrinos are rather strongly coupled to the Majoron, leading to testable predictions in laboratory process such as $\mu \to e\chi\chi$, $\pi \to e\nu\chi$, $(N, Z) \to (N-2, Z+2) + 2e^-$ $+ \chi$, etc.

Let us now discuss the constraints on v_L and coupling h_m from observations. The neutral current interaction in general $SU(2)_L \times U(1)$ theories, includes a parameter ρ, whose tree level value in the standard model is given by unity in agreement with experiment to within one percent. In the presence of the Higgs triplet that acquire non-zero v.e.v., v_T, we get,

$$\rho \simeq 1 - 4\frac{v_T^2}{v^2} . \tag{8.5}$$

Neutral current observations therefore imply that $v_T \leq 25$ GeV. However, further constraints on v_T arise from the triplet Majoron (TM) coupling to u, d, e, which is given by

$$f_{uu\chi} \approx \frac{g\, m_u}{m_W} \times \left(\frac{v_T}{v}\right) < 10^{-12} \tag{8.6}$$

and leads to the bound $v_T \leq 10$ kev.[73] Laboratory bounds on the electron neutrino mass, then, implies, $h_{ee} \lesssim 10^{-3}$. Bounds of the same order of magnitude, also emerge from the study of $(\beta\beta)_{o\nu}$ decay[34] (see fig. 4) and studies of $\pi \to e\nu\chi$,[74] $\mu \to e\chi\chi$.[75] A precise measurement of the Z-width at LEP will be a crucial test of the triplet Majoron model since its contribution is equivalent to that of two extra neutrino species. An immediate theoretical question then arises as to what is responsible for the small symmetry breaking scale. One can speculate that this may be related to the existence of very high scale in the theory. It is, however, not easy to construct such a model.

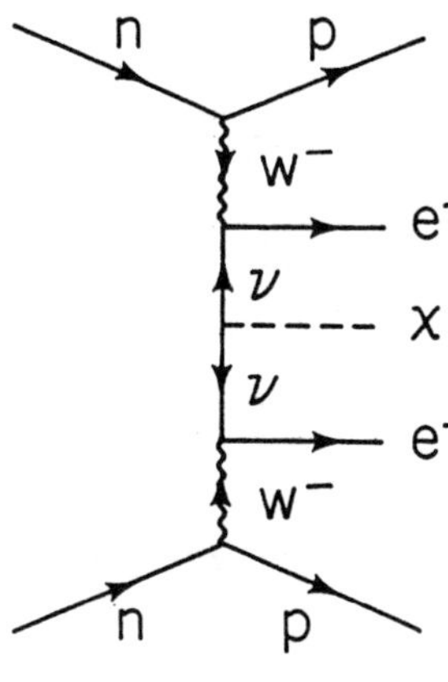

Fig. 4

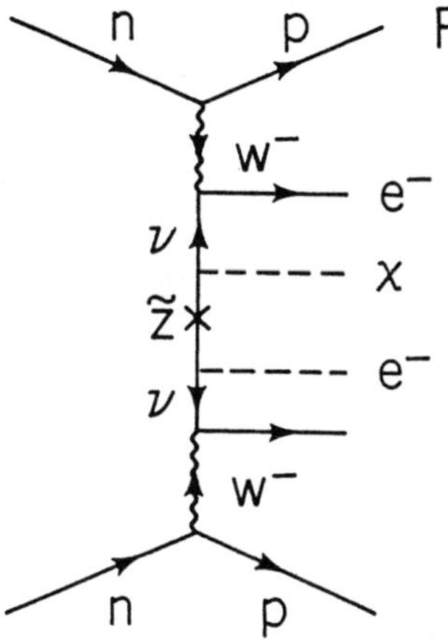

Fig. 5

Fig. 4. The Feynman diagram for single Majoron emission in neutrinoless double beta decay.

Fig. 5. The Feynman diagram that leads to neutrinoless double beta decay with two Majoron emission.

Finally, it is worth pointing out, in the context of supersymmetric models, one can spontaneously break lepton number symmetry by giving a non-zero v.e.v.[76] to the superpartner of the neutrino, which due to supersymmetry carries lepton number 1. This gives rise to the doublet Majoron. The doublet Majoron contribution to the Z-width is half the contribution of a single neutrino species. The constraints on the sneutrino v.e.v. coming from astrophysics are very similar to those of the triplet v.e.v. and imply that $\langle\tilde{\nu}\rangle \leq$ keV. An important test of the doublet Majoron model will be the new process of double Majoron emission shown in fig. 5. It appears with a strength which can be roughly estimated to be about $G_F^2 \cdot g^2/\tilde{M}_Z$ and can lead to a large contribution to neutrinoless double beta decay for $M_{\tilde{Z}} \leq 100$ GeV. It is, therefore, important to calculate the energy spectrum of the electron in this process.

9. Conclusion:

To summarize, we have given an extensive review of the basic ideas that go into understanding the smallness of neutrino masses in general gauge and superstring inspired models. We see that various kinds of mass spectra can emerge in different models. Testing between these various possibilities may be possible if the MSW solution to the solar neutrino puzzle is backed up by future experiments and other considerations. For instance, this would rule out a low mass m_{W_R} in an $SO(10)$ type model as well as the mechanisms (i) and (ii) in superstring inspired models.

An important question that we did not address in this article is the theoretical prediction of neutrino mixings. The existence of arbitrary parameters in the theories discussed precludes any reliable conclusion about this at the moment. We eagerly await new experimental developments regarding this as well as the whole question of neutrino masses.

References

[1] J.C. Pati and A. Salam, Phys. Rev. D**10**, 275 (1974); R.N. Mohapatra and J.C. Pati, Phys. Rev. D**11**, 566 (1975); G. Senjanović and R.N. Mohapatra, Phys. Rev. D**12**, 1502 (1975). For a review, see R.N. Mohapatra in "Quarks, Leptons and Beyond," ed. by H. Fritzsch et al, Plenum (N.Y.), p. 219.

[2] R.N. Mohapatra and G. Senjanović, Phys. Rev. Lett. **44**, 912 (1980); Phys. Rev. D**23**, 165 (1981).

[3] For a review, see "Superstring Theory" Vols. I and II by M. Green, J. Schwarz and E. Witten, Cambridge University Press (1986).

[4] For a review of standard model and grandunification see "Unification and Supersymmetry: The Frontiers of Quark-Lepton Physics," by R.N. Mohapatra, Springer-Verlag (N.Y.) (1986).

[5] R. Cowsik and J. Mclelland, Phys. Rev. Lett. **29**, 699 (1972).

[6] S.S. Gershtein and Ya B. Zeldovich, JETP Lett. **4**, 120 (1966).

[7] B.W. Lee and S. Weinberg, Phys. Rev. Lett. **39**, 165 (1977); P. Hut, Phys. Lett. **69B**, 85 (1977); K. Sato and M. Kobayashi, Prog. Theor. Phys. **58**, 1775 (1977); M.I. Vysotskii, A.D. Dolgov and Ya. B. Zeldovich, JETP Lett. **26**, 188 (1977).

[8] D. Dicus, E. Kolb and V. Teplitz, Phys. Rev. Lett. **39**, 169 (1977).

[8a] S. Sarkar and A.M. Cooper, Phys. Lett. **148B**, 347 (1984).

[9] Y. Chikashige, R.N. Mohapatra and R.D. Peccei, Phys. Lett. **98B**, 265 (1981).

[10] Y. Chikashige, R.N. Mohapatra and R.D. Peccei, Phys. Rev. Lett. **45**, 1926 (1980); G. Gelmini and J.W.F. Valle, **142B**, 181 (1984); A. Kumar and R.N. Mohapatra, Phys. Lett. **150B**, 191 (1985).

[11] M. Roncadelli and G. Senjanović, Phys. Lett. **107B**, 59 (1983); P. Pal, Nucl. Phys. B**227**, 237 (1983).

[12] R.N. Mohapatra and Palash Pal, Phys. Lett. **179B**, 105 (1986).

[13] J.K. Rowley, B.T. Cleveland and R. Davis, Jr. in "Solar Neutrino and Neutrino Astronomy," ed. M.L. Cherry (A.I.P. Conf. Proc. No. 126, New York, 1985).

[14] J.N. Bahcall, W.F. Huebner, S.H. Lubow, P.D. Parker and R.G. Ulrich, Rev. Mod. Phys. **54**, 767 (1982); J.N. Bahcall, B.T. Cleveland, R. Davis, Jr. and J.K. Rowley, Appl. J. **292**, 279 (1985), R. Bahcall and R.K. Ulrich, preprint (in preparation).

[15] B. Pontecorvo, Zh. Eksp. Teor. Fiz. **53**, 1717 (1967) Soviet. Phys. JETP **26**, 984 (1968); V. Gribov and B. Pontecorvo, Phys. lett. **28B**, 493 (1969).

[16] R.N. Mohapatra and J.W.F. Valle, Phys. Lett. 177B, 47 (1986).

[17] S.P. Mikheyev and A.Yu. Smirnov, Nuovo Cim. **9C**, 17 (1986); Yad. Fiz. **42**, 1441 (1985) [Soc. J. Nucl. Phys. **42**, 913 (1985)]; L. Wolfenstein, Phys. Rev. **D17**, 2369 (1978).

[18] For a recent review, see M. Fukugita, Kyoto Preprint, RIFP-718 (1987).

[19] J. Bahcall, N. Cabibbo and A. Yahil, Phys. Rev. Lett. **28**, 316 (1972).

[20] J. Bahcall, S. Petcov, S. Toshev and J.W.F. Valle, Phys. Lett. **181B**, 369 (1986).

[21] A. Cisneros, Applied Space Science **10**, 87 (1971).

[22] L.B. Okun, M.B. Voloshin and M.I. Vysotskii, Zh. Eksp. Teor. Fiz. **91**, 754 (1986) [Sov. Phys. JETP **64**, 446 (1986)].

[23] P. Sutherland, J.N. Ng, E. Flowers, M. Ruderman and C. Inman, Phys. Rev. **D13**, 2700 (1976).

[24] J.P. Morgan, Phys. Let. **102B**, 247 (1981).

[25] K. Hirata, et al., Phys. Rev. Lett. **58**, 1490 (1987).

[26] R.M. Bionta, et al., Phys. Rev. Lett. **58**, 1494 (1987).

[27] We give a very incomplete list of references on this subject: J.N. Bahcall and S.L. Glashow, Nature, **326**, 476 (1987); W.D. Arnett and J.L. Rosner, Phys. Rev. Lett. **58**, 1906 (1987); E.W. Kolb, A.J. Stebbing and M.S. Turner, Phys. Rev. **D35**, 3598 (1987).

[28] R. Barbieri and R.N. Mohapatra, Univ. of MD Preprint (1988); J. Cooperstein and J. Lattimer, Brookhaven Preprint (1988); S. Nussinov, G. Alexander, I. Goldman and A. Aharanov, Tel Aviv Preprint (1987); A. Dar, Tel Aviv Preprint (1987).

[29] A. Dar and S. Dado, Phys. Rev. Lett. **59**, 2368 (1987).

[30] For a recent review of all radiative neutrino decays, see, M. Roos, Proceedings of "international Conference on Neutrino Physics," ed. by H.V. Klapdor, Springer-Verlag (1988).

[31] Particle Data Table, Phys. Lett. **B** , April (1987).

[32] S. Petcov, Phys. Lett. B**1101**, 245 (1982); J.W.F. Valle, Phys. Rev. D**27**, 1672 (1983); M. Doi, T. Kotani, M. Kenmoku, and E. Takasugi, Phys. Rev. D**30**, 626 (1984). For a review, see B. Kayser, Comments in Nuc. and Part. Physics **14**, 69 (1985).

[33] M. Gell-Mann, P. Ramond and R. Slansky, in "Supergravity," ed. by D.Z. Freedman and P. Vanniuenhuizen, (North Holland 1980); T. Yanagida, KEK Lectures, (1979).

[34] For a recent review, see D. Caldwell, Proceedings of "XXIIIrd International Conference on High Energy Physics," ed. S. Loken, World Scientific Publishing Co. (1986), p. 951; H. Ejiri, Proceedings of INS Symposium on "Neutrino Mass and Related Topics," ed. by T. Oshima (1988). A recent report of the discovery of the Majoron in neutrinoless double β-decay by F. Avignone et al., Proceedings of the APS-DPF meeting, Salt Lake City, Utah (1987) has been disputed by other expts on $(\beta\beta)_{o\nu}$-decay, see for instance E. Fiorini, Proceedings of "VIIIth Workshop on Grandunification," ed. by K.C. Wali (World Scientific), 1988.

[35] R.N. Mohapatra and G. Senjanović, Phys. Rev. D**23**, 165 (1981).

[36] L. Durkin and P. Langacker, Phys. Lett. B**166**, 436 (1986); R. Robinett and J. Rosner, Phys. Rev. D**25**, 3036 (1982); V. Barger, E. Ma and K. Whisnant, Phys. Rev. D**26**, 2378 (1982).

[37] M.A.B. Bég, R.V. Budny, R.N. Mohapatra and A. Sirlin, Phys. Rev. Lett. **38**, 1252 (1977). For further discussion, see J. Maalampi, K. Mursula and M. Roos, Nuc. Phys. B**207**, 233 (1982).

[38] J. Carr **et al**, Phys. Rev. Lett. **51**, 627 (1983); D.P. Stoker, **et al.**, Phys. Rev. Lett. **54**, 1887 (1985).

[39] P. Herczeg, Los Alamos preprint LA-UR-85-2761 (1985).

[40] R.N. Mohapatra, New Frontiers in High Energy Physics, ed. A. Perlmutter and L. Scott (Plenum, New York, 1978), p. 337.

[41] J.F. Donoghue and B. Holstein, Phys. Lett. B**113**, 382 (1982).

[42] G. Beall, M. Bander and A. Soni, Phys. Rev. Lett. **48**, 848 (1982); R.N. Mohapatra, G. Senjanović and M.D. Tran, Phys. Rev. **D28**, 546 (1983); G. Ecker, W. Grimus and H. Neufeld, Phys. Lett. **127B**, 365 (1983); F.J. Gilman and M.H. Reno, Phys. Rev. **D29**, 937 (1984). Inclusion of QCD corrections improve this bound; see G. Ecker and W. Grimus, Z. Phys. C**30**, 293 (1986).

[43] I.I. Bigi and J.M. Frere, Phys. Lett. **110B**, 255 (1982).

[44] H. Abramowicz, **et al.**, Z. Phys. C**12**, 225 (1982).

[45] L. Wolfenstein, Phys. Rev. D**29**, 2130 (1984).

[46] E. Masso, Phys. Rev. Lett. **52**, 1956 (1984).

[47] D.O. Caldwell, **et al.**, Phys. Rev. D**33**, 2737 (1986); H. Ejiri, **et al.**, Nucl. Phys. A**448**, 27 (1986); E. Firoini, **et al.**, Phys. Lett. B**146**, 450 (1984); F.T. Avignone III, **et al.**, Phys. Rev. Lett. **54**, 2309 (1985).

[48] R.N. Mohapatra, Phys. Rev. D**34**, 909 (1986). For limits on m_{ν_R} from other considerations, see M. Gronau, C. Leung and J. Rosner, Phys. Rev. D**29**, 2539 (1984).

[49] For reviews, see, W. Haxton and G. Stephenson, Jr., Progress in Nuclear and Particle Physics, **12**, 409 (1984); J. Vergados, Phys. Rep. **133**, 1 (1986); M. Doi, T. Kotani and E. Takasugi, Prog. Theor. Phys. Supp. **83**, 1 (1985).

[50] R.N. Mohapatra, Proceedings of "The International Symposium on Weak and Electromagnetic Interactions in Nuclei," ed. H.V. Klapdor, Springer-Verlag (1986); S. Bertolini and J. Liu, CMU-HEP 87-09 (1987).

[51] D. Chang and R.N. Mohapatra, Phys. Rev. D**32**, 1248 (1985).

[52] D. Chang, R.N. Mohapatra and M.K. Parida, Phys. Rev. Lett. **52**, 1072 (1984).

[53] U. Chattopadhyay and P.B. Pal, Phys. Rev. D**34**, 3444 (1986).

[54] A. Kumar and R.N. Mohapatra, Phys. Lett. **150B**, 191 (1985); R.N. Mohapatra and P. Pal, Univ. of Massachusetts Preprint (1987).

[55] J.E. Kim, Phys. Rev. D**14**, 3000 (1976); M.A.B. Bég, W. Marciano and M. Ruderman, Phys. Rev. D**17**, 1395 (1978); J. Liu, Phys. Rev. **D35**, 3447 (1987); M.J. Duncan et al., Phys. Lett. **191B**, 304 (1987); G. Branco and J. Liu, Carnegie-Mellon Preprint (1988).

[56] L.B. Okun, M.B. Voloshin and M. Vysotski, Zh. Eksp. Teor. Fiz. **91**, 754 (1986); [Societ Phys. JETP **64**, 446 (1986)].

[57] D. Chang and R.N. Mohapatra, Phys. Rev. Lett. **58**, 1600 (1987); R.N. Mohapatra, Phys. Lett. **B** (to appear) (1988).

[58] R.N. Mohapatra and G. Senjanović, Zeit. für Phys. C**17**, 53 (1983); F. Stecker and Q. Shafi, Phys. Rev. Lett. **50**, 928 (1983); P. Langacker, R.D. Peccei and T. Yanagida, Mod. Phys. Lett. **A1**, 541 (1986).

[59] For reviews of strong CP-problem, see J.E. Kim, Phys. Rep. **150**, 1 (1987); H.Y. Cheng, Indiana Preprint IUHET-125 (1986).

[60] R.N. Mohapatra, Proceedings of "Eighth Workshop on Grandunification," ed. by K.C. Wali, (World Scientific Publishing, Singapore, 1987).

[61] R.N. Mohapatra, Phys. Rev. Lett. **56**, 561 (1986).

[62] J.W.F. Valle, private communication (1987).

[63] E. Witten, Nucl. Phys. **B268**, 79 (1986).

[64] R.N. Mohapatra and J.W.F. Valle, Phys. Rev. **D34**, 1642 (1986).

[65] S. Nandi and U. Sarkar, Phys. Rev. Lett. **56**, 564 (1986).

[66] A. Masiero, D.V. Nanopoulos and A.I. Sanda, Phys. Rev. Lett. **57**, 663 (1986); G.C. Branco and C.Q Geng, Phys. Rev. Lett. **58**, 969 (1987); E. Ma, Phys. Rev. Lett. **58**, 1047 (1987).

146

[67] M. Doi, T. Kotani, T. Kurimoto, H. Nishiura and E. Takasugi, Phys. Rev. **D37**, (1988).

[68] For a recent review of the properties of the Majoron, see, R.N. Mohapatra, Proceedings of the XXIII International Conference on High Energy Physics, ed. by S. Loken (World Scientific) p.

[69] Y. Chikashige, R.N. Mohapatra and R.D. Peccei, ref. 10. For further discussion of $\nu_{\mu,\tau} \rightarrow \nu_e + \chi$ decay, see J. Schecter and J.W.F. Valle, Phys. Rev. **D25**, 774 (1982).

[70] S.L. Glashow, Phys. Lett. **187B**, 367 (1987).

[71] G. Gelmini and M. Roncadelli, Phys. Lett. **99B**, 411 (1981).

[72] H. Georgi, S.L. Glashow and S. Nussinov, Nuc. Phys. **B193**, 297 (1981).

[73] M. Fukugita, S. Watamura and M. Yoshimura, Phys. Rev. Lett. **48**, 1522; D. Dearborn, D. Schramm and G. Steigman, Phys. Rev. Lett. **56**, 26 (1985).

[74] V. Barger, W.Y. Keung and S. Pakvasa, Phys. Rev. **D25**, 907 (1982).

[75] T. Goldman, E. Kolb and G. Stevenson, Phys. Rev. **D26**, 2503 (1982).

[76] C.S. Aulakh and R.N. Mohapatra, Phys. Lett. **119B**, 136 (1982). For further analysis of this model, see R. Santamaria and J.W.F. Valle, Phys. Rev. Lett. **59**, (1987). For a non-supersymmetric model with doublet Majoron, see, S. Bertolini and R. Santamaria, Carnegie-Mellon Preprint (1988).

Eur. Phys. J. C 9, 389–408 (1999)
Digital Object Identifier (DOI) 10.1007/s100529900001

THE EUROPEAN
PHYSICAL JOURNAL C

Neutrino textures in light of Super-Kamiokande data and a realistic string model

J. Ellis[1], G.K. Leontaris[1,2], S. Lola[1], D.V. Nanopoulos[3,4,5]

[1] Theory Division, CERN, CH-1211 Geneva 23, Switzerland
[2] Theoretical Physics Division, Ioannina University, GR-45110 Ioannina, Greece
[3] Center for Theoretical Physics, Department of Physics, Texas A&M University, College Station, TX 77843 4242, USA
[4] Astroparticle Physics Group, Houston Advanced Research Center (HARC), The Mitchell Campus, Woodlands, TX 77381, USA
[5] Academy of Athens, Chair of Theoretical Physics, Division of Natural Sciences, 28 Panepistimiou Ave., Athens GR-10679, Greece

Received: 6 November 1998 / Published online: 18 June 1999

Abstract. Motivated by the Super-Kamiokande atmospheric neutrino data, we discuss possible textures for Majorana and Dirac neutrino masses within the see-saw framework. There are two main purposes of this paper: first, to gain intuition into this area from a purely phenomenological analysis, and second, to explore to what extent it may be realized in a specific model. We comment initially on the simplified two-generation case, emphasizing that large mixing is not incompatible with a large hierarchy of mass eigenvalues. We also emphasize that renormalization-group effects may amplify neutrino mixing, and we present semi-analytic expressions for estimating this amplification. Several examples are then given of three-family neutrino mass textures, which may also accommodate the persistent solar neutrino deficit, with different assumptions for the neutrino Dirac mass matrices. We comment on a few features of neutrino mass textures arising in models with a $U(1)$ flavour symmetry. Finally, we discuss the possible pattern of neutrino masses in a "realistic" flipped $SU(5)$ model derived from string theory, illustrating how a desirable pattern of mixing may emerge. Both small- or large-angle MSW solutions are possible, while a hierarchy of neutrino masses appears more natural than near-degeneracy. This model contains some unanticipated features that may be relevant in other models also: The neutrino Dirac matrices may not be related closely to the quark mass matrices, and the heavy Majorana states may include extra gauge-singlet fields.

1 Introduction

There have recently been reports from the Super-Kamiokande Collaboration [1] and others [2] indicating that the atmospheric neutrino deficit is due to neutrino oscillations. The data on electron events with visible energy greater than 200 MeV are very consistent with standard model expectations. On the other hand, the number of events with muons is about half of the expected number, and the deficit becomes more acute for larger values of L/E, indicating that neutrino oscillations dilute the abundance of atmospheric ν_μ. The possibility that $\nu_\mu \to \nu_e$ oscillations dominate is disfavoured by both Super-Kamiokande [1] and CHOOZ data [3]. A fit to $\nu_\mu \to \nu_\tau$ oscillations, with $\Delta m^2 = 5 - 50 \times 10^{-4}$ eV^2 and $\theta \sim \pi/4$ matches the data very well, but an admixture of $\nu_\mu \to \nu_e$ oscillations cannot be excluded.

One intriguing feature of this scenario is the large mixing angle that is required, and the question of how one could achieve this in theoretically motivated models arises. Large mixing angles in the neutrino sector do arise naturally in a sub-class of GUT models with flavour symmetries, as in [4], where they were used to explain what was then only an "atmospheric neutrino anomaly" [5]. Many models with a single $U(1)$ symmetry predict small mixings [6], principally because of the constrained form of the Dirac mass matrices. However, this is not a generic feature, and textures with large $\nu_\mu \to \nu_\tau$ mixing have also been presented in [7]. Moreover, string-derived models may well have a richer structure, with three or four $U(1)$ symmetries.

However, models where the large neutrino mixing arises from the Dirac mass matrix may have a problem with quark masses. In many GUTs, for example, $SO(10)$, the neutrinos and up-type quarks couple to the same Higgs and are in the same multiplets, so their couplings arise from identical GUT terms[1]. Thus, in these cases one would simultaneously generate large mixing in the u-quark sector. Then, in order to obtain small mixing in V_{CKM}, one needs to invoke some cancellation with mixing in the d-quark sector. One way to overcome these difficulties may

[1] Fermion mass hierarchies in this class of models have been discussed in [8].

PHYSICAL REVIEW D, VOLUME 61, 113012

Zee neutrino mass model in a SUSY framework

Kingman Cheung
Department of Physics, University of California, Davis, California 95616

Otto C. W. Kong
Institute of Physics, Academia Sinica, Nankang, Taipei, Taiwan 11529
(Received 6 December 1999; published 10 May 2000)

We study the Zee model of neutrino mass in the framework of R-parity-violating supersymmetry. Within the matter content of the minimal supersymmetric standard model, any one of the three right-handed sleptons could be a suitable candidate for the charged-singlet scalar of the Zee model, and one of the Higgs doublets provides the extra necessary vacuum expectation value. A combination of one bilinear and two trilinear R-parity-violating couplings then completes the model. In this framework, we also discuss other various contributions to neutrino masses and derive the conditions for the dominance of the contribution from the Zee model, and hence maintain the successfully Zee mass texture. However, this model within the minimal supersymmetric standard model is shown to be only marginally feasible. More general versions of supersymmetrization of the Zee model are also discussed. A particularly interesting example that has extra Higgs doublets while the slepton, especially the selectron, keeps the role of the Zee scalar is illustrated.

PACS number(s): 14.60.Pq, 12.60.Jv

I. INTRODUCTION

Within the standard model (SM), the *V-A* nature of the weak interaction dictates a zero mass for all three families of neutrinos. The discovery of neutrino mass(es) or oscillation(s) will certainly push for new physics. Evidence for neutrino oscillations has been collected in a number of solar neutrino and atmospheric neutrino experiments. The most impressive results were the recent ν_μ neutrino deficit and the asymmetric zenith-angle distribution observed by the Super-Kamiokande Collaboration [1].

The atmospheric neutrino deficit and zenith-angle distribution can be explained by the ν_μ-ν_τ or ν_μ-ν_s oscillations (ν_s is a sterile neutrino that has negligible coupling to W or Z boson and the former has a slightly better fit). The oscillation parameters with $\nu_\mu \rightarrow \nu_\tau$ at 90% C.L. are [1]

$$\Delta m^2_{\rm atm} \simeq (2-6)\times 10^{-3}\ \ {\rm eV}^2,\ \ \sin^2 2\theta_{\rm atm} \gtrsim 0.85.$$

On the other hand, the solar neutrino deficit admits more than one solution. With $\nu_e \rightarrow \nu_\tau$ the solutions at 95% C.L. are [2]

$$\text{vacuum oscillation:}\quad \Delta m^2_{\rm sol} \simeq (5-8)\times 10^{-11}\ \ {\rm eV}^2,$$

$$\sin^2 2\theta_{\rm sol} \simeq 0.6-1.0,$$

$$\text{small angle MSW:}\quad \Delta m^2_{\rm sol} \simeq (4-9)\times 10^{-6}\ \ {\rm eV}^2,$$

$$\sin^2 2\theta_{\rm sol} \simeq (3.5-13)\times 10^{-3},$$

$$\text{large angle MSW:}\quad \Delta m^2_{\rm sol} \simeq (8-30)\times 10^{-6}\ \ {\rm eV}^2,$$

$$\sin^2 2\theta_{\rm sol} \simeq 0.4-0.8.$$

The above two neutrino-mass differences can be accommodated by the three species of neutrinos that we know from the SM. There is also another indication for neutrino oscillation from the accelerator experiment at the Liquid Scintillation Neutrino Detector (LSND) [3], which requires an oscillation of ν_μ into another neutrino with

$$\Delta m^2_{\rm LSND} \simeq 0.2-2\ \ {\rm eV}^2,\quad \sin^2 2\theta_{\rm LSND} \simeq 0.003-0.03.$$

To as well accommodate this data it requires an additional species of neutrino beyond the usual neutrinos. Nevertheless, further evidence from the next round of neutrino experiments is required to confirm the neutrino mass and oscillation. As neutrino is favored to be massive it is desirable to understand the generation of neutrino masses from physics beyond the SM, especially, to see if the new physics can give a neutrino mass pattern that can explain the atmospheric and solar neutrino data, and perhaps the LSND as well.

An economical way to generate small neutrino masses with a phenomenologically favorable texture is given by the Zee model [4–6], which generates masses via one-loop diagrams. The model consists of a charged gauge singlet scalar h^-, the Zee scalar, which couples to lepton doublets ψ_{Lj} via the interaction

$$f^{ij}(\psi^\alpha_{Li} C \psi^\beta_{Lj})\epsilon_{\alpha\beta} h^-, \tag{1}$$

where α, β are the SU(2) indices, i,j are the generation indices, C is the charge-conjugation matrix, and f^{ij} are Yukawa couplings antisymmetric in i and j. Another ingredient of the Zee model is an extra Higgs doublet (in addition to the one that gives masses to charged leptons) that develops a vacuum expectation value (VEV) and thus provides mass mixing between the charged Higgs boson and the Zee scalar boson. The corresponding coupling, together with the f^{ij}'s, enforces

PHYSICAL REVIEW D, VOLUME 62, 015007

Fermion masses and neutrino oscillations in $SO(10)$ supersymmetric grand unified theory with $D_3 \times U(1)$ family symmetry

Radovan Dermíšek and Stuart Raby
Department of Physics, The Ohio State University, 174 W. 18th Ave., Columbus, Ohio 43210
(Received 10 November 1999; published 30 May 2000)

Discrete non-Abelian gauge symmetries appear to be the most advantageous candidates for a family symmetry. We present a predictive $SO(10)$ SUSY GUT model with $D_3 \times U(1)$ family symmetry (D_3 is the dihedral group of order 6). The hierarchy in fermion masses is generated by the family symmetry breaking $D_3 \times U(1) \rightarrow Z_N \rightarrow$ nothing. This model fits the low energy data in the charged fermion sector quite well and naturally provides large angle ν_μ-ν_τ mixing describing atmospheric neutrino oscillation data and small angle ν_e-ν_s mixing consistent with the small mixing angle MSW solution to the solar neutrino data. In addition, the non-Abelian family symmetry D_3 is sufficient to suppress large flavor violations.

PACS number(s): 14.60.Pq, 11.30.Hv, 12.15.Ff

I. INTRODUCTION

The origin of the fermion mass hierarchy is one of the most challenging problems in elementary particle physics. In the standard model fermion masses and mixing angles are free parameters. Even though these 13 parameters [9 charged fermion masses; 3 angles and 1 phase in the Cabibbo-Koboyashi-Maskawa (CKM) matrix] are well known experimentally, the standard model does not offer any explanation. Supersymmetric (SUSY) grand unified theories (GUTs), in addition to gauge coupling unification, also provide relations between quark and lepton masses within generations. However, the understanding of the hierarchy between generations is still missing. A possible solution to the fermion mass hierarchy problem is to introduce a new symmetry—family symmetry—acting horizontally between generations. The hierarchy is then generated by sequential spontaneous breaking of this symmetry. Furthermore, acting differently on different generations, family symmetries can provide a solution to the problem of large flavor changing neutral currents (FCNCs) in SUSY [1].

A variety of models [2–9] with family symmetries were proposed. Among these, models with $U(2)$ (or its subgroups) family symmetry [5–9] appear to be very promising candidates for the theory of flavor. The reason for this is twofold: the top quark is the only fermion with mass of order the weak scale, thus distinguishing the third generation from the others; and by placing the first and second generations into a two dimensional irreducible representation of the family group the degeneracy of squarks in these two generations can be achieved, which is necessary to suppress FCNCs. Thus non-Abelian family symmetries, especially $U(2)$ or its subgroups, are naturally suggested.

We would like to focus here on a particular model presented in [7]. It is an $SO(10)$ SUSY GUT with family symmetry $U(2) \times U(1)$.[1] This model is "predictive" by which we mean that it is "natural"—the Lagrangian contains all terms consistent with the symmetries and particle content of the theory; and the number of arbitrary parameters is less than the number of observables. This model fits the low energy data in the charged fermion sector quite well and naturally provides large angle ν_μ-ν_τ mixing describing atmospheric neutrino oscillation data and small angle ν_e-ν_s mixing consistent with the small mixing angle Mikheyev-Smirnov-Wolfenstein (MSW) solution to solar neutrino data.

There are however complications associated with a $U(2)$ family symmetry in supersymmetric theories. It is believed that global symmetries do not arise in string theory and also these are thought to be violated by quantum gravity effects [10]. On the other hand, with continuous gauge symmetries there are associated D-term contributions to scalar masses which can lead to unacceptably large FCNCs [11]. As a result, we should consider discrete family gauge symmetries. Discrete gauge symmetries are not violated by quantum gravity effects [12] and can arise in spontaneous breaking of continuous gauge symmetries or directly in compactifications of string theory.

In this paper we present an $SO(10)$ SUSY GUT with $D_3 \times U(1)$ family gauge symmetry which does not suffer from the problems mentioned in the previous paragraph. This model provides exactly the same operators generating Yukawa matrices as model [7]. Thus it fits the low energy data in the charged lepton sector equally well and provides the same neutrino solution. In addition, the field content of this model is simpler than [7] and can naturally provide an explanation for sequential family symmetry breaking by the vacuum expectation values (VEVs) of "flavon" fields.

The rest of the paper is organized as follows. In Sec. II we briefly review possible discrete family symmetries, provide a motivation for $D_3 \times U(1)$ as a family symmetry and discuss anomalies associated with gauging of this symmetry. In Sec. III we construct the $SO(10) \times D_3 \times U(1)$ invariant superspace potential which, after family symmetry breaking, generates the quark and lepton Yukawa matrices. Our conclusions are in Sec. IV. For convenience, in Appendix A we summarize properties of the group D_3 and its representations, and calculate invariants used in Sec. III. In Appendix

[1]The model [7] is a modification of the $SO(10) \times U(2)$ model suggested in [6]. The modification only affects the results in the neutrino sector.

PHYSICAL REVIEW D, VOLUME 60, 113001

Neutrino oscillations in a predictive SUSY GUT

T. Blažek*
Department of Physics and Astronomy, Northwestern University, Evanston, Illinois 60208

S. Raby and K. Tobe
Department of Physics, The Ohio State University, 174 W. 18th Ave., Columbus, Ohio 43210
(Received 2 April 1999; published 28 October 1999)

In this paper we present a predictive SO(10) supersymmetric grand unified theory with the family symmetry U(2)×U(1) which has several nice features. We are able to fit fermion masses and mixing angles, including recent neutrino data, with nine parameters in the charged fermion sector and four in the neutrino sector. The family symmetry plays a preeminent role. (i) The model is ''natural''—we include all terms allowed by the symmetry. It restricts the number of arbitrary parameters and enforces many zeros in the effective mass matrices. (ii) Family symmetry breaking from U(2)×U(1)→U(1)→ nothing generates the family hierarchy. It also constrains squark and slepton mass matrices, thus ameliorating flavor violation resulting from squark and slepton loop contributions. (iii) It naturally gives large angle $\nu_\mu-\nu_\tau$ mixing describing atmospheric neutrino oscillation data and small angle $\nu_e-\nu_s$ mixing, consistent with the small mixing angle Mikheyev-Smirnov-Wolfenstein (MSW) solution to solar neutrino data. (iv) Finally, in this paper we assume minimal family symmetry-breaking vacuum expectation values (VEV's). As a result we cannot obtain a three neutrino solution to both atmospheric and solar neutrino oscillations. In addition, the solution discussed here cannot fit liquid scintillation neutrino detector (LSND) data even though this solution requires a sterile neutrino ν_s. It is important to note, however, that with nonminimal family symmetry-breaking VEV's, a three neutrino solution is possible with the small mixing angle MSW solution to solar neutrino data and large angle $\nu_\mu-\nu_\tau$ mixing describing atmospheric neutrino oscillation data. In the four neutrino case, nonminimal family VEV's may also permit a solution for LSND. The results with nonminimal family breaking are still under investigation and will be reported in a future paper. [S0556-2821(99)00619-0]

PACS number(s): 14.60.Pq, 12.15.Ff

I. INTRODUCTION

Solar [1], atmospheric [2], and accelerator [3] neutrino data strongly suggest that neutrinos have small masses and nonvanishing mixing angles. This hypothesis is also constrained by reactor [4] based experiments. In the near future, many more experiments will test the hypothesis of neutrino masses. In addition, a neutrino mass necessarily implies new physics beyond the standard model. Thus there is great excitement, both experimental and theoretical, in this field.

Phenomenological neutrino mass models [5] are designed to reproduce the best fits to all or some of the neutrino data. These models are only constrained by how much of the neutrino data one wants to fit. Three neutrino models with three active neutrinos (ν_e, ν_μ and ν_τ) are consistent with solar [1] and atmospheric [2] neutrino oscillations, while four neutrino models, including a sterile (or electroweak singlet) neutrino (ν_s), are consistent with solar, atmospheric, and liquid scintillation neutrino detector (LSND) [3] neutrino experiments. There are also six neutrino models, with three active and three sterile neutrinos, motivated by complete family symmetry [6].

It is important to address the theoretical question; to what extent can this new data on neutrino masses and mixing angles constrain the physics beyond the standard model; in particular, theories of fermion masses. Since any number of sterile neutrinos may mix with the three active neutrinos, even in a grand unified theory, it may always be possible to fit neutrino data without ever constraining the charged fermion sector of the theory. This would be an unfortunate circumstance. It is the purpose of this paper, however, to show that in any ''predictive'' theory of charged fermion masses, the neutrino sector is severely constrained.

By a ''predictive'' model of fermion masses we mean the following.

The Lagrangian is ''natural'' containing all terms consistent with the symmetries and particle content of the theory.

In addition there are necessarily grand unified theory (GUT) gauge symmetries as well as family symmetries which restrict the form of the Yukawa matrices [7–9]; thereby greatly reducing the number of arbitrary parameters.

In supersymmetric (SUSY) theories, these same family symmetries can usefully constrain the form of soft SUSY breaking squark and slepton masses as well [7–9]; thus ameliorating the problem of large flavor violation in SUSY theory [10].

In this paper, we demonstrate that these same family symmetries greatly restrict the form of neutrino masses and mixing. Hence neutrino data can greatly constrain any predictive theory of fermion masses.

We show this in the context of a particular SO(10) SUSY GUT which fits charged fermion masses and mixing angles well. SUSY GUT's are very attractive. They successfully predict the unification of gauge couplings observed at the

*On leave of absence from Faculty of Mathematics and Physics, Comenius Univ., Bratislava, Slovakia.

RAPID COMMUNICATIONS

PHYSICAL REVIEW D VOLUME 46, NUMBER 7 1 OCTOBER 1992

Neutrino masses within the minimal supersymmetric standard model

Mirjam Cvetič and Paul Langacker
Department of Physics, University of Pennsylvania, Philadelphia, Pennsylvania 19104-6396
(Received 20 May 1992)

We investigate the possibility of accommodating neutrino masses compatible with the Mikheyev-Smirnov-Wolfenstein study of the solar neutrino deficit within the minimal supersymmetric standard model. The "gravity-induced" seesaw mechanism based on an interplay of nonrenormalizable and renormalizable terms in the superpotential allows neutrino masses $m_\nu \propto m_u^2/M_I$, with m_u the corresponding quark mass and $M_I \simeq 4\times 10^{11}$ GeV, while at the same time ensuring the grand desert with the gauge coupling unification at $M_U \simeq 2\times 10^{16}$ GeV. The proposed scenario may be realized in a class of string vacua, i.e., large radius ($R^2/\alpha' \sim 20$) (0,2) Calabi-Yau spaces. In this case $M_U^2 = M_C^2/O(2R^2/\alpha')$ and $M_I = O(e^{-R^2/\alpha'})M_C$. Here $M_C = g\times 5.2\times 10^{17}$ GeV is the scale of the tree-level (genus-zero) gauge coupling (g) unification.

PACS number(s): 12.10.Dm, 11.17.+y, 12.15.Ff, 14.60.Gh

Precise data from the experiments at the CERN e^+e^- collider LEP indicate that the gauge couplings of the standard model meet at $M_U \simeq (1-4)\times 10^{16}$ GeV in the minimal supersymmetric extension of the standard model [1]. Another set of intriguing data arises from the solar neutrino experiments. The deficit of solar neutrinos can most efficiently be explained through the Mikheyev-Smirnov-Wolfenstein (MSW) [2] mechanism of matter-enhanced neutrino oscillations. In particular, current data favor [3] the mass splitting of the electron and muon neutrinos to be $\Delta m^2 \equiv |m_{\nu_\mu}^2 - m_{\nu_e}^2| \simeq (2\text{–}5)\times 10^{-7}$ eV^2 if the mixing angle $\theta_{\nu_\mu\nu_e} = O(\theta_C)$, where θ_C is the Cabibbo angle. For arbitrary mixing angles the nonadiabatic MSW solution favors $\Delta m^2 \simeq (1\text{–}16)\times 10^{-7}$ eV^2.

In many grand unified theory (GUT) models the lepton mixing matrix V_l and the Cabibbo-Kobayashi-Maskawa matrix $V_{\rm CKM}$ are predicted to be approximately equal [4]. However, these same models predict $m_e/m_\mu \simeq m_d/m_s$ ($\simeq 1/20$), which fails by an order of magnitude. Small perturbations on the models which rescue this mass relation [5] may also modify the mixing-angle predictions. In theories with no explicit intrafamily unification $V_{\rm CKM}$ and V_l are not expected to be equal, but could well be of the same order of magnitude. We will assume $\sin^2 2\theta_{\nu_\mu\nu_e} \sim \sin^2 2\theta_C \sim 0.18$ for the central value in our discussion, but will consider the entire range 4×10^{-3}–1 allowed by the nonadiabatic MSW solution. Assuming $m_{\nu_\mu} \gg m_{\nu_e}$, the corresponding values of m_{ν_μ} are $(5\text{–}7)\times 10^{-4}$ eV for $\theta_{\nu_\mu\nu_e} \sim \theta_C$ and $(3\text{–}40)\times 10^{-4}$ eV for general $\theta_{\nu_\mu\nu_e}$.

In the GUT seesaw scenario [6] masses of light neutrinos are given by $m_{\nu_{e,\mu}} \simeq c m_{u,c}^2/M_I$, where $m_{u,c}$ are the corresponding quark masses and $c \simeq 0.05\text{–}0.09$ is a factor due to the renormalization down to the low-energy scale [3]. This implies that $M_I \simeq (4\pm 3)\times 10^{11}$ GeV, the central value corresponding to $\theta_{\nu_\mu\nu_e} \sim \theta_C$. If the same scale applies to the third family, then $m_{\nu_\tau} \simeq c' m_t^2/M_I$ could be in the cosmologically interesting 10-eV range, with ν_μ-ν_τ oscillations observable in the laboratory.

Each of the two sets of experimental data has an elegant theoretical explanation. Unfortunately, the two theoretical models are mutually exclusive at first glance. In the minimal supersymmetric standard model there is a "grand desert" up to $M_U \simeq (1\text{–}4)\times 10^{16}$ GeV. Within this theory, the implementation of a naive seesaw mechanism would indicate that $m_{\nu_{e,\mu}} \simeq c m_{u,c}^2/M_I$, with $M_I \sim (10^{-2}\text{–}1) M_U \sim (1\text{–}4)\times 10^{15\pm 1}$ GeV, which is too small to be compatible with the favored experimental data and the MSW scenario [3]. Actually, in GUT models, in order to obtain a nonzero M_I with $M_I \sim M_U$ one has to introduce large Higgs representations [e.g., the **126** of SO(10)].

The aim of this note is to implement the neutrino masses in the minimal supersymmetric standard model in such a way that there is still a grand desert with the gauge coupling unification at $M_U \sim (1\text{–}4)\times 10^{16}$ GeV, while the effective scale M_I governing the neutrino masses is in the range of $(4\pm 3)\times 10^{11}$ GeV. We are proposing a "gravity-induced" seesaw mechanism (an extension of a mechanism proposed by Nandi and Sarkar [7]), realized through an interplay between the nonrenormalizable and renormalizable terms in the superpotential, as the origin of the neutrino masses. The essence of the idea is based on a supersymmetric theory with an extended gauge symmetry, which contains an additional sterile neutrino (a standard model singlet), and a *restricted* representation of the Higgs fields. Such fields break the extended gauge symmetry at the scale M_U. However, they cannot give the sterile neutrino a large Majorana mass proportional to M_U through the renormalizable (cubic) terms in the superpotential. On the other hand, through the nonrenormalizable (e.g., quartic) terms of the superpotential, which are suppressed by a scale $M_{\rm NR} > M_U$, such Higgs fields can give a Majorana mass of order $M_U^2/M_{\rm NR} < M_U$.

The origin of the nonrenormalizable terms is best motivated in theories which include gravity, e.g., Kaluza-Klein theories and superstring theory. In this case the exchange of heavy modes with masses of order the Planck mass $M_{\rm Pl}$ in general induce nonrenormalizable terms with $M_{\rm NR}=O(M_{\rm Pl})$. Thus, we shall call the proposed scenario the gravity-induced seesaw. In the case of $M_U \sim 2\times10^{16}$ GeV and $M_{\rm NR}\sim M_{\rm Pl}/\sqrt{8\pi} \sim 2\times10^{18}$ GeV one has $M_I \sim 2\times10^{14}$ GeV, which is about two to three orders of magnitude too large to be compatible with the favored MSW data. However, if such nonrenormalizable terms are *suppressed* by an additional factor 10^{-2}–10^{-3}, one can obtain the desired $M_I \sim 10^{11}$–10^{12} GeV.

Such a scenario can be accommodated within a GUT theory with restricted Higgs field representations; e.g., the SO(10) gauge group without 126-plets of Higgs fields [8]. For a broad class of simple GUT models [4], $V_{\rm CKM}\simeq V_l$. On the other hand, extended gauge symmetries based on a product of simple groups and U(1) factors [e.g., the standard model with additional U(1)'s and/or left-right-symmetric gauge symmetry] with restricted representations of the Higgs sector can also accomodate the gravity-induced seesaw mechanism. However, in this case, the relation between $V_{\rm CKM}$ and V_l is less obvious.

One can demonstrate the gravity-induced seesaw in an explicit (minimal) model with all the essential features. We choose the enhanced gauge symmetry $SU(3)_C\times SU(2)_L\times U(1)_Y\times U(1)_{Y'}$, where Y is the ordinary weak hypercharge. The matter consists of the particle content of the minimal standard model as well as of the standard model singlets L_i, S_1, and $\bar S_1$. L_i supermultiplets with $i=(1,2,3)$ contain a sterile neutrino which accompanies each of the three families. S_1 and $\bar S_1$ contain the Higgs fields which break the enhanced gauge symmetry with vacuum expectation values (VEV's) of order M_U.

Consistent with the anomaly constraint we choose the following values for the Y' charges: quark $SU(2)_L$ doublets, u_L^c quarks, and e_L^c leptons have (-1), lepton doublets and d_L^c quarks have $(+3)$, L_i and S_1 have (-5), $\bar S_1$ has $(+5)$, while Higgs doublets $H_{(1,2)}$ have (-2) and $(+2)$, respectively [9]. In the neutrino sector, the only renormalizable terms allowed in the superpotential are of the type $W=L_i\nu_i H_2$. Terms of the type $L_i\nu_i S_1$ or $L_iL_iS_1$ are not allowed by the quantum numbers. These constraints yield the following contribution to the neutrino mass matrix:

$$\begin{bmatrix} 0 & m \\ m & 0 \end{bmatrix}, \tag{1}$$

where m is proportional to the VEV of the Higgs doublet H_2. Since H_2 gives mass to the quarks as well, m is of the order of the corresponding quark masses unless there is a large difference in the magnitude of the Yukawa couplings. On the other hand, the only allowed nonrenormalizable term in the superpotential with a leading contribution to the neutrino mass matrix is of the type $W_{\rm NR}=L_iL_i\bar S_1\bar S_1/M_{\rm NR}$ [10]. This modifies the neutrino mass matrix:

$$\begin{bmatrix} 0 & m \\ m & M_I \end{bmatrix}, \tag{2}$$

where $M_I=M_U^2/M_{\rm NR}$. The quantum numbers prevent the contribution of any nonrenormalizable term to the $\nu\nu$ and νL masses that would be of the order of $M_U^K/M_{\rm NR}^{K-1}$ for any $K>1$.

As seen in the above model, the gravity-induced seesaw can be accommodated by an interplay of the renormalizable and nonrenormalizable terms in the superpotential, which takes place because of a restricted representation of the Higgs sector. While such a scenario is appealing on its own terms, we would like to motivate its origin. Study of the effective Lagrangian of superstring vacua provides a natural framework for the restricted representation of the chiral supermultiplets. One can also shed light on the origin of the nonrenormalizable terms in the superpotential, which in string theory arise due to the exchange of massive modes.

Let us first illustrate the neutrino mass pattern in the example discussed by Nandi and Sarkar [7] with a gauge group $G\in E_6$ and all the chiral supermultiplets (i.e., those which contain quarks, leptons and Higgs particles) arising from 27-plets of E_6. The renormalizable superpotential is of the type $W_R=\mathbf{27}_i\mathbf{27}_j\mathbf{27}_K$, where quarks and leptons arise from $\mathbf{27}_{i,j}$ and Higgs vacuum expectation values (VEV's) from $\mathbf{27}_K$. If one assumes that all the exotic quarks acquire large masses due to the large VEV's of the standard model singlets S_1 and S_2, while the Higgs doublets in $\mathbf{27}_K$ account for the masses of the ordinary quarks and leptons, this constrains [7] the mass matrix of the five neutral fermions (ν,N,N^c,ν^c,L) to be of the form

$$\begin{bmatrix} 0 & 0 & M_{U_1} & m_1 & 0 \\ 0 & 0 & M_{U_2} & 0 & m_1 \\ M_{U_1} & M_{U_2} & 0 & m_2 & m_3 \\ m_1 & 0 & m_2 & 0 & 0 \\ 0 & m_1 & m_3 & 0 & 0 \end{bmatrix}. \tag{3}$$

Under the $SU(3)_L\times SU(3)_R\times SU(3)_C$ subgroup of E_6, the neutral fields (ν,N,N^c,ν^c,L) are a part of a $(3,\bar 3,1)$ multiplet with the following entries:

$$\begin{bmatrix} N & & \\ & N^c & \nu \\ & \nu^c & L \end{bmatrix}.$$

M_{U_1} and M_{U_2} are masses due to the VEV's of the standard model singlets and $m_{1,2,3}$ are the light masses, related to the quark masses, which are due to the standard model Higgs doublets. The mass matrix (3) does not have a large Majorana mass along the diagonal of the lower two components. There is a heavy sector with large masses M_U [N^c and $(\nu+N)/\sqrt2$], a light sector with masses of order m [$(\nu-N)/\sqrt2$ and $(\nu^c-L)/\sqrt2$], and an ultralight standard model singlet [$(\nu^c+L)/\sqrt2$] with mass of order [7] m^2/M_U; this pattern is clearly incompatible with experiment.

However, nonrenormalizable terms could provide (along with the renormalizable ones) a derived seesaw pattern. Within the E_6 gauge group there may be terms in the superpotential of the type

$W_{NR}=27_i27_j\overline{27}_K\overline{27}_L/M_{NR}$, which can account for the heavy Majorana masses [7]. Namely, the mass matrix (3) becomes

$$\begin{bmatrix} 0 & 0 & M_{U_1} & m_1 & 0 \\ 0 & 0 & M_{U_2} & 0 & m_1 \\ M_{U_1} & M_{U_2} & 0 & m_2 & m_3 \\ m_1 & 0 & m_2 & M_{I_1} & M_{I_2} \\ 0 & m_1 & m_3 & M_{I_3} & M_{I_4} \end{bmatrix} \quad (4)$$

with $M_I=M_U^2/M_{NR}$.

Corrections to the pattern (4) are scaled by [7] $M_I/M_U \sim M_U/M_{NR}$ and have been neglected [11]. Such terms are expected to appear due to the exchange of heavy modes. Again M_I is somewhat too large for, say, $M_U \sim 2\times 10^{16}$ GeV and $M_{NR}\sim 10^{18}$ GeV. As we point out later, in superstring theory one can shed light on the magnitude of the nonrenormalizable terms of the superpotential in a quantitative way [12,13], and thus account for an additional suppression factor.

The above example is based on the constraints of the E_6 gauge group and the **27** representations. In order for this scenario to be compatible with the grand-desert scenario of the minimal supersymmetric standard model, a number of other constraints have to be satisfied: (i) the gauge couplings have to meet at $M_U \simeq (1\text{–}4)\times 10^{16}$ GeV, (ii) below M_U the gauge group has to be the standard model group [14], and (iii) the particle content contributing to the running of the gauge couplings has to be that of the minimal supersymmetric standard model.

In superstring theories these constraints place conditions on the string vacuum. Perhaps the most difficult to satisfy (with no existing example available) is (iii). Generically, (2,2) string vacua, e.g., Calabi-Yau manifolds with gauge and spin connection identified, possess a large number of additional multiplets. In particular, for vacua without Wilson lines, the gauge group is E_6, with **27**'s, $\overline{\mathbf{27}}$'s, and **1**'s of E_6. Some of the particles in these multiplets acquire large masses if there are flat directions in the space of specific string vacua. Finding flat directions [15,16] allows one to give large VEV's to fields in a particular set of **27**'s and $\overline{\mathbf{27}}$'s, which in turn can give mass to some of the unwanted massless multiplets. At the same time, E_6 is broken down to SO(10) or SU(5). Explicit examples of such directions have been found in blown-up orbifolds [17,18] as well as for a class of Calabi-Yau manifolds [19] based on Gepner's [20] construction.

Flat directions of (2,2) vacua provide one with a new class of [(0,2)] string vacua. In such vacua a large number of unwanted modes become heavy; however, the gauge group is still a simple GUT group [SO(5) or SU(5)]. Since the matter supermultiplets are in the fundamental representation or singlets of the gauge group [21], this prevents one from breaking the simple GUT groups down to the standard model, thus contradicting the constraint (ii).

This problem can be remedied by the introduction of Wilson lines on the compactified space, allowing a breakdown of the simple gauge group [E_6, SO(10), or SU(5)] to a direct product of simple groups and U(1)'s. It is in general possible [22,19] to introduce Wilson lines which break the gauge group down to the standard model. At the same time, this procedure decouples a large number of unwanted modes. Since there is no grand unification in the four-dimensional theory one does not expect observable proton decay, and the relationship between V_{CKM} and V_l is lost.

Thus, a viable scenario which could satisfy constraints (ii) and (iii) is to construct (2,2) string vacua with flat directions as well as Wilson lines. However, there exists no explicit construction of such a supersymmetric string vacuum which would contain only the minimal standard model particle spectrum.

The next issue to be addressed is the value of M_C, which is the scale at which the gauge couplings g, as determined at the tree level of the string theory, are equal. M_C is determined [23] in the dimensional regularization ($\overline{\text{DR}}$) scheme by the value of the Planck mass M_{Pl} and of the gauge coupling g in the following way:

$$M_C=\frac{e^{(1-\gamma)/2}\sqrt{2}}{3^{3/4}\sqrt{\pi\alpha'}}=g\frac{e^{(1-\gamma)/2}}{3^{3/4}4\pi}M_{Pl}$$
$$=g0.043M_{Pl}=g\times 5.2\times 10^{17}\text{ GeV}\ , \quad (5)$$

where $\gamma=0.57722$ is the Euler constant, $g^2=32\pi/(\alpha' M_{Pl}^2)$, with g defined according to the GUT convention [24] and $M_{Pl}=1.2\times 10^{19}$ GeV.

For the expected value $g\sim 0.7$ this is one order of magnitude too large compared with $M_U\sim(1\text{–}4)\times 10^{16}$ GeV, which is the scale of the gauge coupling unification of the minimal supersymmetric standard model. However, threshold effects, i.e., genus-one corrections to the gauge couplings, can split the gauge couplings at M_C, thus, in principle allowing for an effective unification scale $M_U<M_C$. Explicit calculations of the threshold corrections for a class of orbifolds are given in Refs. [23,25]. Extensive study [26] of threshold corrections in orbifolds indicate that $M_U<M_C$ if the massless spectrum satisfies certain constraints compatible with the target space one-loop modular anomaly. In such examples one would obtain $M_U=O(e^{(-cR^2/\alpha')})M_C$ when the orbifold radius is large ($R^2/\alpha' \gg 1$). The positive coefficient c depends [26] on the modular weights of the massless states. As we shall see later, the heavy Majorana mass turns out to be $M_I=O(e^{(-c'R^2/\alpha')})M_C$. In order to ensure $M_U\sim 10^{16}$ GeV and $M_I\sim 10^{12}$ GeV, this in turn involves detailed constraints on coefficients c and c'.

In the following, we shall pursue a different approach, i.e., study of smooth Calabi-Yau spaces. For simply connected (2,2) Calabi-Yau manifolds the nature of threshold corrections is different. Such spaces possess the $E_6\times E_8$ gauge group, and the compactified space corresponds to the smooth Calabi-Yau manifolds with the radius of compactification being large, i.e., in the conformal field theory language this corresponds to the (2,2) string vacuum with large VEV's for *all* the moduli. In this case the massive modes of the string theory do not contribute to the threshold corrections [27], and thus the gauge couplings do not have corrections proportional to the powers of the moduli VEV's. Instead, such corrections are milder, only logarithmic in the VEV's of moduli. The threshold corrections for Calabi-Yau spaces (with only one modulus [27] as well as for arbitrary number of moduli [28]) are of the form

$$\Delta\left[\frac{16\pi^2}{g^2}\right]\sim -b_G^{N=1}\ln(T+T^*)\ , \tag{6}$$

where the real part of the T field corresponds to an overall value of large moduli, i.e., $T+T^*=O(2R^2/\alpha')\gg 1$, where R is the radius of the compactification and α' is the string tension. $b_G^{N=1}$ is related to the one-loop $N=1$ beta function β_G for $G=E_6$ or E_8 as $\beta_G=b_G g^3/16\pi$. Since threshold corrections (6) to the gauge coupling of each of the gauge groups are proportional to the corresponding $N=1$ β function, this implies that the slope of the running gauge couplings is not changed. However the effective gauge coupling unification scale is lowered:

$$M_U^2=\frac{M_C^2}{T+T^*}=O\left[\frac{M_C^2}{2R^2/\alpha'}\right]. \tag{7}$$

The above results apply only to simply connected Calabi-Yau spaces. The gauge group E_6 can be broken if Wilson lines are introduced. In this case the contribution of the massless states to the threshold corrections has not been studied, yet. We proceed with the assumption that the nature of the threshold corrections is still of the type (6). From (7) one then sees that for $R^2/\alpha'\sim 20$ the gauge unification scale is lowered to $M_U\sim 6\times 10^{16}$ GeV, which is slightly too large. However, Eq. (7) relates M_U^2 to R^2/α' only by orders of magnitude. Thus, an additional factor of 2 in the relation of an overall modulus ReT to R^2/α' enables one to obtain M_U in the preferred range 4×10^{16} GeV.

We turn now to neutrino masses. In particular, we would like to address the size of the nonrenormalizable terms. In string theory the magnitude of the coefficient $M_{\rm NR}$ is proportional to M_C. However, one can prove explicitly [13] that for all (0,2) string vacua the nonrenormalizable terms are suppressed by an additional factor $e^{-R^2/\alpha'}$, i.e., the origin of the nonrenormalizable terms is due only to world-sheet instanton effects. This is a general stringy result, proven explicitly on (blown-up) orbifolds [12] as well as in σ-model perturbations [13] of Calabi-Yau manifolds. Therefore,

$$\frac{1}{M_{\rm NR}}=\frac{O(e^{-R^2/\alpha'})}{M_C}\ . \tag{8}$$

By choosing vacuum expectation values along the flat direction to be M_C (the only natural scale in the four-dimensional string vacuum) nonrenormalizable terms of the type (4) yield the heavy Majorana mass:

$$M_I=\frac{M_C^2}{M_{\rm NR}}=O(e^{-R^2/\alpha'})M_C\ . \tag{9}$$

It follows from (7) that we need $R^2/\alpha'\sim 20$ in order to achieve $M_U\sim 10^{16}$ GeV. In this case $M_I\sim 10^{-8}M_C\sim 10^{10}$ GeV. Although these are only order of magnitude statements, it is instructive to set the coefficients in (7)–(9) equal to unity. In that case the range $M_I\sim(4\pm 3)\times 10^{11}$ GeV suggested by the solar neutrino deficit implies $R^2/\alpha'\sim 13$–15, yielding a slightly too large $M_U\sim 7\times 10^{16}$ GeV.

To summarize, the desired scale of the gauge coupling unification $M_U\sim 2\times 10^{16}$ GeV and the scale of Majorana neutrino masses $M_I\sim 4\times 10^{11}$ GeV, may be achieved [29] for a superstring vacuum, corresponding to a (0,2) Calabi-Yau space obtained by deforming a (2,2) smooth large radius Calabi-Yau space along the exactly flat directions. The radius of the compactification has to be in the range $R^2/\alpha'\simeq(20)$, allowing for $M_U^2=M_C^2/O(2R^2/\alpha')$ and $M_I=O(e^{-R^2/\alpha'})M_C$. We believe it is intriguing that one can obtain M_U and M_I in the desired range on the basis of fairly general stringy arguments.

We would like to thank R. Brustein, B. Greene, and V. Kaplunovsky for useful discussions and comments. This research was supported in part by the U.S. DOE Grant No. DE-AC02-76-ERO-3071 and by the Texas National Research Laboratory Commission (M.C.).

[1] P. Langacker and M. Luo, Phys. Rev. D **44**, 817 (1991); U. Amaldi, W. de Boer, and H. Furstenau, Phys. Lett. B **260**, 447 (1991); J. Ellis, S. Kelley, and D. V. Nanopoulos, *ibid.* **249**, 441 (1990).

[2] S. P. Mikheyev and A. Yu. Smirnov, Yad. Fiz. **42**, 1441 (1985) [Sov. J. Nucl. Phys. **42**, 913 (1985)]; Nuovo Cimento **9C**, 17 (1986); L. Wolfenstein, Phys. Rev. D **17**, 2369 (1968); **20**, 2634 (1979).

[3] For a recent discussion, see S. Bludman, D. Kennedy, and P. Langacker, Phys. Rev. D **45**, 1810 (1992); Nucl. Phys. (to be published).

[4] This occurs in simple GUT seesaw models for a wide range of mass matrices for the heavy Majorana neutrinos [3].

[5] For a review, see P. Langacker, Phys. Rep. **72**, 185 (1981).

[6] M. Gell-Mann, P. Ramond, and R. Slansky, in *Supergravity*, Proceedings of the Workshop, Stony Brook, New York, 1979, edited by P. van Nieuwenhuizen and D. Freedman (North-Holland, Amsterdam, 1979), p. 315; T. Yanagida, Prog. Theor. Phys. B **135**, 66 (1978).

[7] S. Nandi and U. Sarkar, Phys. Rev. Lett. **56**, 564 (1986).

[8] Radiative corrections due to two-loop effects, as proposed by E. Witten, Phys. Lett. **91B**, 81 (1980), do not work in supersymmetric theory: there are no linear terms of the type $M\phi_{10}\phi_{16}\phi_{16}$ [where **10**, **16**, **16**, are representations of the corresponding scalar multiplets of SO(10)] because they break supersymmetry. They can be generated by soft supersymmetry-breaking terms, but this would then yield a Majorana mass of order $M_I\sim\alpha M_W$.

[9] The choice of the particle spectrum and the $U(1)_{Y'}$ quantum numbers are motivated by SO(10) gauge symmetry. Namely, the family particle content (including L_i's) corre-

sponds to full 16-plets of SO(10), the two Higgs doublets are parts of a 10-plet, while the standard model singlets S_1 and $\bar{S}_1$ are parts of a 16-plet and a $\overline{16}$-plet, respectively.

[10] A scenario that ensures large VEV's $\langle S \rangle = \langle \bar{S} \rangle = M_U$ is based on an interplay of soft supersymmetry-breaking mass terms of order M_W, i.e., terms of the type $-M_W^2(|S_1|^2+|\bar{S}_1|^2)$ and nonrenormalizable terms of the type $W=(27_I\overline{27}_J)^K/M_{\rm NR}^{2K-3}$, with $K \geq 4$. This allows for $M_U = O(M_W M_{\rm NR}^{2K-3})^{1/2K-2} \geq O(M_U) \sim 10^{16}$ GeV. In order to prevent the terms in the $(S_1, \bar{S}_1)$ sector with $K<4$ one has to impose a discrete symmetry; e.g., Z_4 symmetry with L_i, S_1, and $\bar{S}_1$ transforming with the phases $2\pi/4$, $4\pi/4$, and $2\pi/4$, respectively, ensures that the first nonzero term in the $(S_1, \bar{S}_1)$ sector corresponds to $K=4$ while at the same time allowing for the term $L_i L_i \bar{S}_1 \bar{S}_1 / M_{\rm NR}$. Note, that this scenario for the breakdown of the gauge symmetry at M_U is again motivated from the properties of string vacua.

[11] The physical light doublet neutrino of mass $m \sim cm^2/M_I$ is $\nu \cos\theta + N \sin\theta$ where $\tan\theta = M_{U_1}/M_{U_2}$, and ν and N are both doublets, while the orthogonal combination has a mass $\sim M_U$. This does not cause any problems with universality because there is exactly the same mixing between the charged doublet partners.

[12] M. Cvetič, Phys. Rev. Lett. **59**, 1795 (1987).

[13] M. Cvetič, Phys. Rev. D **37**, 2366 (1987); M. Dine and C. Lee, Phys. Lett. B **203**, 371 (1988).

[14] There could be additional group factors which commute with the standard model.

[15] E. Witten, Nucl. Phys. **B268**, 79 (1986).

[16] This is one of the possible constructions of (0,2) superstring vacua.

[17] M. Cvetič, Phys. Rev. Lett. **59**, 2829 (1987).

[18] A. Font, L. Ibañez, H. P. Nilles, and F. Quevedo, Nucl. Phys. **B307**, 105 (1988).

[19] B. Greene, Phys. Rev. D **40**, 1645 (1989).

[20] D. Gepner, Phys. Lett. B **199**, 380 (1987).

[21] This is the case for all level-1 Kač-Moody algebra constructions of string vacua.

[22] A. Font, L. Ibañez, F. Quevedo, and A. Sierra, Nucl. Phys. **B331**, 421 (1990).

[23] V. Kaplunovsky, Nucl. Phys. **B307**, 145 (1988); and erratum (to be published).

[24] In the literature, there are quoted different values of M_C, differing by factors of 2 or $\sqrt{2}$. As explained in the erratum of Ref. [23], using the correct numerical form of the original formula [23], and the tree-level relation $g^2 = 32\pi/(\alpha' M_{\rm Pl}^2)$, with g defined according to the GUT convention, yields the quoted result, which is the same as the original numerical value in Ref. [23]. Note that the gauge coupling g as defined in the effective string Lagrangian, e.g., P. Ginsparg, Phys. Lett. B **197**, 139 (1987), is by a factor of $\sqrt{2}$ smaller than the gauge coupling g defined according to the GUT convention. Namely, in the effective string Lagrangian the trace over the generators of the vector representation of SO(2N) gauge group is chosen to be $\mathrm{Tr}(T^a T^b) = -2\delta^{ab}$, while in the GUT theories the convention is $\mathrm{Tr}(T^a T^b) = -\delta^{ab}$, thus rendering the gauge coupling for a factor of $\sqrt{2}$ bigger in the latter case.

[25] L. Dixon, V. Kaplunovsky, and J. Louis, Nucl. Phys. **B355**, 649 (1991).

[26] L. Ibañez, D. Lüst, and G. Ross, Phys. Lett. B **272**, 251 (1991); L. Ibañez and D. Lüst, CERN Report No. CERN-TH.6380/92.

[27] L. Dixon (unpublished).

[28] M. Cvetič, Report No. UPR-489-T, 1991 (unpublished).

[29] There is a less appealing possibility in superstring theory, which for completeness we wish to discuss. In this case one would stick to (2,2) Calabi-Yau spaces with Wilson lines and $R^2/\alpha' \sim 1$. The threshold corrections are small and thus the gauge coupling unification scale is $M_C \sim 4 \times 10^{17}$ GeV. Because of the different matter representations, the gauge couplings of the semisimple group [e.g., SU(3)^3] factors run *differently* from M_C to $M_U \sim 10^{16}$ GeV, where the semisimple gauge group is broken down to the standard one. Thus, the gauge couplings are not equal at M_U anymore. In particular, for SU(3)^3 with $9q$ ($6\bar{q}$) and $7l$ ($4\bar{l}$) the difference in gauge couplings is below current experimental uncertainties if $|M_C/M_U| < 10$. The known scenario for obtaining large M_U is through an interplay of soft supersymmetry-breaking mass terms (of order M_W) and nonrenormalizable terms of the type $W=(27_I\overline{27}_J)^K/M_{\rm NR}^{2K-3}$, which allow for a large VEV of the standard model singlets $\langle\phi\rangle = O(M_W M_{\rm NR}^{2K-3})^{1/2K-2} \geq O(M_U) \sim 10^{16}$ GeV with $1/M_{\rm NR} = O(e^{-R^2/\alpha'})/M_C$ and $K \geq 4$. In addition, it is very difficult [as explored by B. Green, K. Kirklin, P. Miron, and G. Ross, Nucl. Phys. **B278**, 667 (1986); **B292**, 606 (1987)] to decouple all the unwanted particles at the same scale M_U. Barring these problems, in this scenario one has $1/M_{\rm NR} = O(e^{-R^2/\alpha'})/M_C$ and $M_I = (e^{-R^2/\alpha'}) M_U^2/M_C$. For $R^2/\alpha' \sim 5$ and $M_U \sim 10^{16}$ GeV one obtains $M_I \sim 10^{12}$ GeV.

hep-ph/0004129
April 2000

NEW DIRECTIONS FOR NEW DIMENSIONS: FROM STRINGS TO NEUTRINOS TO AXIONS TO... [a]

Keith R. Dienes

Theory Division, CERN, CH-1211 Geneva 23, Switzerland
Department of Physics, University of Arizona, Tucson, AZ 85721 USA [b]

In this talk, I discuss recent developments concerning the possibility of large extra spacetime dimensions. After briefly reviewing how such dimensions can lower the fundamental GUT, Planck, and string scales, I then outline how these scenarios lead to a new higher-dimensional seesaw mechanism for generating neutrino oscillations — perhaps even without neutrino masses. I also discuss how extra dimensions lead to new mechanisms contributing to the "invisibility" of the QCD axion. This talk reports on work done in collaboration with Emilian Dudas and Tony Gherghetta.

1 Introduction: Lowering the fundamental scales of physics

The possibility of large extra spacetime dimensions has recently received considerable attention. This is clearly an exciting prospect. One of the earliest proponents of TeV-scale extra dimensions was Antoniadis [1], who attempted to use such extra dimensions to explain supersymmetry breaking. Later, Witten [2] pointed out that extra large dimensions could lower the string scale below its usual value near 10^{18} GeV, and subsequently Lykken [3] proposed that Witten's idea could be extended to lower the string scale all the way to the TeV-range. Finally, in March 1998, it was proposed that extra dimensions could also be used to lower the fundamental Planck scale [4] as well as the fundamental GUT scale [5]. Thus, combining these different proposals, it becomes possible to contemplate a self-consistent scenario in which all high fundamental energy scales (GUT, Planck, and string scales) are eliminated in favor of large extra spacetime dimensions!

It is important to distinguish two different types of extra spacetime dimensions. First, there are so-called "universal" extra dimensions. These extra

[a] Invited plenary talk given at *PASCOS '99: 7th International Symposium on Particles, Strings, and Cosmology* (held at Lake Tahoe, California, 10–16 December 1999). To appear in the Proceedings.

[b] Current address. E-mail: dienes@physics.arizona.edu

PHYSICAL REVIEW D VOLUME 52, NUMBER 11 1 DECEMBER 1995

Reconciling present neutrino puzzles: Sterile neutrinos as mirror neutrinos

Zurab G. Berezhiani
Istituto Nazionale di Fisica Nucleare, Sezione di Ferrara, 44100 Ferrara, Italy
and Institute of Physics, Georgian Academy of Sciences, 380077 Tbilisi, Georgia

Rabindra N. Mohapatra
Department of Physics, University of Maryland, College Park, Maryland 20742
(Received 1 June 1995)

We suggest that recent puzzles in neutrino physics, i.e., the solar and atmospheric neutrino deficits, possible neutrino oscillations reported by the LSND, and neutrinos as hot dark matter, can all be naturally explained by assuming the existence of a mirror world described by an "electroweak" gauge symmetry $[SU(2)\times U(1)]'$, with the breaking scale larger by about a factor of 30 than the scale of the standard $SU(2)\times U(1)$ model. Such models are motivated by the superstring $E_8\times E_8'$ theories. The model also suggests that the bulk of the dark matter in the universe may be warm dark matter consisting of KeV mirror particles rather than the conventional cold dark matter.

PACS number(s): 14.60.Pq, 12.60.Cn, 14.60.St, 96.60.Kx

The present situation in neutrino physics is rather intriguing. On the one hand, direct measurements show no evidence for any of the neutrinos to be massive. As is well known the standard model (SM) renders neutrinos massless, since in the absence of the right-handed (RH) states lepton number conservation arises as an accidental global symmetry of the theory. However, with lepton number being a global symmetry, its conservation need not be respected by nonperturbative gravitational effects. In that case, even if the minimal SM is a true theory up to the Planck scale $M_{\rm Pl}$, at $M_{\rm Pl}$ nonrenormalizable operators of the type $\frac{1}{M_{\rm Pl}}l_il_jHH$ can emerge with $l_{1,2,3}$ being lepton doublets and H the Higgs doublet. These can then induce small neutrino Majorana masses, at most of the order of $\mathring{m}=\langle H\rangle^2/M_{\rm Pl}=3\times10^{-6}$ eV [1,2]. On the other hand, there have been indirect "positive" signals for neutrino masses and mixing accumulating during the past years. In particular, if any of the following hints will prove to be true, this would point towards neutrino masses much larger than $\mathring{m}$. These hints include the following:

(a) *The solar neutrino problem* (SNP). It appears that the deficit of solar neutrino fluxes cannot be explained by astrophysical reasons thereby implying new neutrino properties, the most popular and natural solution being the Mikheyev, Smirnov, and Wolfenstein (MSW) oscillation [3] mechanism based on ν_e-ν_x conversion in the solar medium. The required parameter range corresponds to $\delta m^2_{ex}\sim10^{-5}$ eV^2 and, in the preferable "small mixing angle" scenario, $\sin^2 2\theta_{ex}\sim10^{-3}-10^{-2}$.

(b) *The atmospheric neutrino problem* (ANP). There is evidence for a significant depletion of the atmospheric ν_μ flux by almost a factor of 2. This points to $\nu_\mu-\nu_x$ oscillation, with $\delta m^2_{\mu x}\sim10^{-2}$ eV^2 and $\sin^2 2\theta_{\mu x}\sim1$.

(c) *Dark matter problem.* The COBE measurements of the cosmic microwave background anisotropy suggests that cosmological dark matter consists of two components, cold dark matter (CDM) being a dominant component ($\Omega_{\rm CDM}\simeq0.7$) and hot dark matter (HDM) being a smaller admixture ($\Omega_{\rm HDM}\simeq0.25$) [4]. The latter role can be naturally played by neutrinos with mass of few eV's. As for the CDM, e.g., in supersymmetric theories with conserved R parity it could be the lightest neutralino. However, it can be of interest to think of cold dark matter as also consisting of neutrinos, this time heavier (keV range), so called warm dark matter (WDM) [5].

(d) *LSND result.* Direct evidence of $\bar\nu_\mu$-$\bar\nu_e$ oscillation from the recent Los Alamos experiment [6], with $\delta m^2_{e\mu}\geq0.3$ eV^2 and $\sin^2 2\theta_{e\mu}=10^{-3}-10^{-2}$.

These hints if proved correct would imply nontrivial physics beyond the SM at energies below $M_{\rm Pl}$. In particular, they would imply that the neutrino masses are induced by the effective operators of the type $\frac{h_{ij}}{\Lambda}l_il_jHH$ with the cutoff scale $\Lambda\ll M_{\rm Pl}$.

Furthermore, if we take all these results seriously, they have very interesting implications for the texture of neutrino masses [7]. In particular, it appears that only one possible neutrino mass texture is compatible with all the above mentioned data. It requires an extra light, sterile neutrino ν_s beyond the three known neutrinos ν_e, ν_μ, and ν_τ. In this scenario, assuming that $m_{\nu_{e,s}}\ll m_{\nu_{\mu,\tau}}$, the SNP is explained as a consequence of ν_e-ν_s oscillation while the ANP is explained by ν_μ-ν_τ oscillation. The ν_μ and ν_τ with $m_{\nu_\mu}\simeq m_{\nu_\tau}\simeq2.4$ eV provide the cosmological HDM and can also explain the LSND result. The detailed mass matrix for the neutrinos in this case is essentially a block matrix with each block being 2×2 and with a small mixing between the ν_e and ν_μ to account for the LSND result.

Indeed, the small mixing angle MSW oscillation ν_e-ν_s is the only place where the extra sterile neutrino could show up. Obviously, ν_s cannot constitute HDM, neither can it be applied to ANP, due to primordial nucleosynthesis bound on the effective number of neutrino species $N_\nu<3.1$ [8].

Concerning the sterile neutrinos, an immediate question that arises is "where do they all come from, and where do they all belong?": in particular, why can they be so light when their masses are allowed by the gauge symmetry? In the SM the role of sterile neutrinos can be played by the RH components of the usual left-handed (LH) ones. However, in this case it is difficult to understand why they in combination with LH neutrinos do not form the Dirac particles as heavy as the charged fermions, or why they do not have large Majorana masses in the spirit of the seesaw scenario. Many people have constructed models where this extra particle is included in the model by hand, but they involve *ad hoc* assumptions.

Here we suggest that the sterile neutrinos are in fact the neutrinos of a mirror world which is the mirror duplicate of our visible world except that its "electroweak" scale v' is scaled by some factor ζ relative to the standard electroweak scale v. Such models are motivated by superstring type fundamental theories which at the compactification imply the existence of an $E_8 \times E_8'$, where E_8 after symmetry breaking leads to our world where the E_8' could lead to the mirror world. The two worlds interact only through the gravity. In such a world, the mirror neutrinos $\nu'_{e,\mu,\tau}$ should be light by the same reason as the ordinary ones $\nu_{e,\mu,\tau}$—their direct mass terms are forbidden by the gauge symmetry. We show that this framework can provide a plausible solution to all present neutrino puzzles, in particular the MSW resolution of the solar neutrino puzzle will turn out to be purely gravitational.

To get the detailed pattern for neutrinos, we will assume separate lepton number ($L_{e,\mu,\tau}$) conservation in both sectors, and assume that these lepton numbers are broken in two stages: (a) the dominant entries ($\sim$ eV) in the neutrino mass matrix have origin in some intermediate scale physics which respects a global ZKM-type lepton number ($\bar{L} = L_e + L_\mu - L_\tau$ in our model); this fixes the skeleton of neutrino mass matrix with specific texture needed for reconciling the LSND oscillation and the HDM, while the ν_e and ν_e' states are left massless; (b) the conservation of $\bar{L}$ is violated by the Planck scale effects, which can then explain SNP and ANP. In particular, the Planck scale terms induce the oscillation of ν_e into its mirror partner ν_e' and moreover, for ζ of about 30 properly reproduce the orders of magnitudes of the parameters required by the small mixing angle MSW scenario.

An interesting prediction of our model is that the $\nu'_{\mu,\tau}$ will play the role of warm dark matter of the universe, since the neutrino mass matrix in the mirror sector will have a pattern similar to that of the usual neutrinos except that its entries should be scaled up by a factor of ζ^2. As far as $\nu_{\mu,\tau}$ with $\sim$ eV mass play a HDM role, then $\nu'_{\mu,\tau}$ being $\zeta^2 \sim 1000$ times heavier fall in the keV range and thus can constitute WDM of the universe. We also address their implications for nucleosynthesis.

WITHOUT THE MIRROR WORLD

We start by assuming that in the basis of the flavor eigenstates $\nu_{e,\mu,\tau}$ the neutrino mass matrix has a texture obeying the $\bar{L} = L_e + L_\mu - L_\tau$ conservation:

$$\hat{m}_\nu = \begin{pmatrix} 0 & 0 & a \\ 0 & 0 & b \\ a & b & 0 \end{pmatrix} . \tag{1}$$

This matrix has one massless (ν_1) and two massive ($\nu_{2,3}$) degenerated eigenstates, with mass $m_{2,3} = m = (a^2 + b^2)^{1/2}$. The latter can play a role of HDM provided that m is of few eV's (e.g., $m \simeq 2.4$ eV, according to Ref. [4]). On the other hand, ν_e and ν_μ are mixed by the angle $\theta_{e\mu} (\tan\theta_{e\mu} = a/b)$, and $\delta m^2_{e\mu} = m^2 \simeq 6$ eV^2 which can explain the LSND oscillation if $\sin^2 2\theta_{e\mu} \simeq 2 \times 10^{-3}$ [6].

Small explicit violation of $\bar{L}$ would induce nonzero entries in the matrix (1) and thus trigger the ν_μ-ν_τ oscillation. For example, if the small nonzero entry $\varepsilon \ll m$ appears in its (3,3) element, then we have $\delta m^2_{\mu\tau} \approx 2\varepsilon m$ and almost maximal mixing, $\sin 2\theta_{\mu\tau} \approx 1$. Then the ANP solution requires that $\delta m^2_{\mu\tau} \simeq 10^{-2}$ eV^2, which for $m \simeq 2.4$ eV in turn implies $\varepsilon \simeq 2 \times 10^{-3}$ eV. Thus only the SNP remains unresolved.

CONNECTING THE TWO WORLDS

As mentioned already, we assume that besides the world of the usual particles of the standard $G = SU(3) \times SU(2) \times U(1)$ model, there exists an analogous mirror world of particles belonging to the mirror gauge group $G' = [SU(3) \times SU(2) \times U(1)]'$. In particular, the ordinary and mirror lepton states (all in the LH basis) and Higgs doublets transform as

$$l_i = (\nu_i, e_i) \sim (\tfrac{1}{2}, -1; 0, 0) \ , \quad E_i^c \sim (0, 2; 0, 0) \ ,$$

$$H \sim (\tfrac{1}{2}, 1; 0, 0)$$

and

$$l_i' = (\nu_i', e_i') \sim (0, 0; \tfrac{1}{2}, -1) \ , E_i'^c \sim (0, 0; 0, 2) \ ,$$

$$H' \sim (0, 0; \tfrac{1}{2}, 1) \ ,$$

where the weak isospins I, I' and hypercharges Y, Y' are shown explicitly, and index $i = 1, 2, 3$ denotes the electron, muon, and taon families, respectively. Quarks can be included in a straightforward way: the ordinary quarks transform as triplets of SU(3) while the mirror ones are triplets of the mirror gauge group SU(3)$'$. One can impose also the invariance under discrete transformation $P(G \leftrightarrow G')$ simultaneously interchanging all corresponding particles of the ordinary and mirror worlds [9]. We assume that P is spontaneously broken, e.g., by the vacuum expectation value (VEV) of some P-odd singlet scalar η coupled to H and H'. As a result, H and H' can have different nonzero VEV's: $v' \gg v = 174$ GeV. Hence the mirror world is completely analogous to ours, but with all particle masses just scaled by the factor $\zeta = v'/v$. In particular, since the charged leptons get masses [illegible] the P-invariant Yukawa couplings $g_i l_i E_i^c \tilde{H} +$ [illegible], we obtain $m'_{e,\mu,\tau} = \zeta m_{e,\mu,\tau}$. The same applies to the mirror quark masses. The gauge boson masses are also scaled as $M_{W',Z'} = \zeta M_{W,Z}$ while photons and gluons remain massless in both of these worlds.

Let us suppose that the two worlds communicate only through gravity and possibly also via some superheavy gauge singlet matter, like the P-odd scalar η. We do not consider the possibility of the mixed representations. Concerning thermodynamics of the two worlds in the early Universe, we assume that at the inflationary reheating temperatures they are already decoupled from each other. If the inflaton couplings violate P invariance, then one can imagine the situation when the visible and mirror particles are "reheated" with different rates, so that after inflation the effective temperatures of the ordinary and mirror thermal bath are different. In this way, the present cosmological abundance of mirror particles (including lightest "charged" states e', u', d', mirror neutrinos $\nu'_{e,\mu,\tau}$, and photons γ') can be suppressed as compared to that of their visible partners. The contribution of ν' and γ' to the universe expansion rate at the nucleosynthesis epoch $t \sim 1$ should be very small: $\Delta N_\nu < 0.1$ [8], which for ζ varied from 10 to 100 can be translated into the upper limit on today's relative abundance $r = n_{\nu'}/n_\nu < 0.04 - 0.02$ [10].

With the minimal particle content given above, the neutrinos of both sectors stay massless unless one appeals to the gravity induced Planck scale effects which explicitly violate the global lepton number, and also can mix the neutrino states of the two worlds [2]. The relevant higher order operators are (here and in the following the C matrix is omitted)

$$\frac{\alpha_{ij}}{M_{\rm Pl}}(l_iH)(l_jH) + \frac{\alpha_{ij}}{M_{\rm Pl}}(l'_iH')(l'_jH') + \frac{\beta_{ij}}{M_{\rm Pl}}(l_iH)(l'_jH') + \text{H.c.}\ , \quad (2)$$

with the constants $\alpha, \beta \sim 1$. More in general, an order of magnitude less values of α, β should not come as a surprise, while it is also possible (e.g., in the string context) that the actual cutoff scale of the operators (2) is an order of magnitude less than $M_{\rm Pl}$, so that one can let these constants range, say, from 0.1 to 10. Obviously, these operators generate the neutrino mass terms in both worlds, as well as their mixing term:

$$m_{\nu_e} = \alpha\frac{v^2}{M_{\rm Pl}} = \alpha\mathring{m}\ , \quad m_{\nu'_e} = \alpha\frac{v'^2}{M_{\rm Pl}} = \zeta^2\alpha\mathring{m}\ ,$$

$$m_{\nu_e\nu'_e} = \beta\frac{vv'}{M_{\rm Pl}} = \zeta\beta\mathring{m}\ . \quad (3)$$

Hence the $\nu_e - \nu'_e$ oscillation emerges with parameters in the range $\delta m^2 = \alpha^2(\zeta/30)^4 \times 8 \times 10^{-6}$ eV^2, $\sin^2 2\theta = \frac{\beta^2}{\alpha^2}(30/\zeta)^2 \times 4.5 \times 10^{-3}$. For $\alpha, \beta \sim 1$, the choice $\zeta \sim 30$ perfectly fits the parameter range of the "small mixing angle" MSW solution to the SNP [3]. More in general, by taking into account the solar model uncertainties, the relevant parameter range can vary within $\sin^2 2\theta = 6 \times 10^{-4}$–$2 \times 10^{-2}$ and $\delta m^2 = (4\text{–}10) \times 10^{-6}$ eV^2 [3]. By assuming $\alpha \sim \beta$, this range for $\sin^2 2\theta$ corresponds to $\zeta = 10$–100, while for the mass difference we have $\delta m^2 \sim \alpha^2(5 \times 10^{-3}/\sin^2 2\theta)^2 \times 6 \times 10^{-6}$ eV^2, which still agrees with the MSW range for a proper α within 0.1–10.

Certainly, the dominant entries in the matrix (1) needed for the explanation of the other neutrino puzzles cannot emerge from the Planck scale effects. For this purpose we invoke operators, respectively, bilinear in H and H' and cutoff by the scale $\Lambda \ll M_{\rm Pl}$. Such operators can be effectively induced by several mechanisms and we will discuss one such mechanism in brief in the next section. But on general grounds, by assuming that these operators respect P parity [11], one can write

$$\frac{h_{ij}}{\Lambda}(l_iH)(l_jH) + \frac{h_{ij}}{\Lambda}(l'_iH')(l'_jH') + \text{H.c.}\ , \quad (4)$$

where h_{ij} are some $O(1)$ "Yukawa" constants with the $\bar{L}$-conserving texture (1). In doing so, the $\nu_1 \approx \nu_e$ state and its mirror partner $\nu'_1 \approx \nu'_e$ are rendered massless so that the SNP solution can be ascribed to the Planck scale effects in a manner demonstrated above, provided that $\zeta \sim 30$ or so. As for the states $\nu_{2,3}$ and $\nu'_{2,3}$, their masses are scaled as $m' = \zeta^2 m$. Therefore, for m in the eV range as it is needed for explaining the origin of HDM, $\nu'_{2,3}$ emerge in the keV range and thus can constitute the WDM. This can be expressed as

$$2m = \Omega_{\rm HDM}92h^2 \text{ eV}\ , \quad 2rm' = \Omega_{\rm WDM}92h^2 \text{ eV}\ , \quad (5)$$

where h is the Hubble constant in units 100 km s^{-1} Mpc^{-1}. Cosmological density of the visible baryonic matter corresponds to $\Omega_B \simeq 0.05$. We also recall that in our model $\Omega_{B'} < \Omega_B$ due to lower inflationary reheating, unless baryon asymmetry in the mirror world is much larger than that of normal baryons. Thus by taking rather conservatively $\Omega_B + \Omega'_B \leq 0.1$, the remaining cosmological density can be shared between HDM and WDM components as, say, $\Omega_{\rm HDM} \simeq 0.2$ and $\Omega_{\rm HDM} \simeq 0.7$. Then for the needed abundance of the mirror neutrinos relative to the active ones we obtain $r \simeq 3.5/\zeta^2$, which for $\zeta = 10 - 100$ is obviously consistent with the nucleosynthesis bound $r < 0.04$–0.02 derived above.

Let us remark that the ANP remains unresolved unless one introduces the explicit mechanism of $\bar{L}$ breaking including the nonzero entries $\varepsilon \sim 10^{-3}$ eV in the mass matrix (1). The Planck scale operators (2) can provide at most $\sim \mathring{m}$ contribution which is far below the needed value. However, in the next section we demonstrate a model in which APN solution can be also related to the Planck scale physics.

ORIGIN OF THE DOMINANT NEUTRINO MASSES

Let us suppose that the theory obeys conservation of individual global lepton numbers $L_i = L_{e,\mu,\tau}$, which is spontaneously broken by the VEV's of the gauge singlet scalars with mixed lepton numbers: $\Phi_{e\mu}$, $\Phi_{e\tau}$, and $\Phi_{\mu\tau}$. In particular, we assume that only $\langle\Phi_{e\tau}, \Phi_{\mu\tau}\rangle \sim V$ are nonzero, while $\langle\Phi_{e\mu}\rangle = 0$. In this way, the ZKM-type lepton number $\bar{L} = L_e + L_\mu - L_\tau$ remains conserved. We also introduce the additional neutral fermions $N_i(-1), S_i(+1)$ and $N_i'(-1), S_i'(+1)$, where brackets show the individual lepton charges L_i. Then the $G \times G' \times P$ invariant couplings, also respecting the individual lepton numbers, are the following [12]:

$$\mathcal{L}_{\rm Yuk} = f_i(l_i N_i H + l_i' N_i' H') + M_i(N_i S_i + N_i' S_i') + \lambda\Phi_{ij}(S_i S_j + S_i' S_j') \ . \qquad (6)$$

For simplicity we assume that the Dirac mass terms are closely degenerated: $M_{1,2,3} \sim M \gg v'$, while the Yukawa couplings f_i have approximately the same pattern as that of the charged fermions, say, $f_3 \sim 1$, $f_2 \sim 3 \times 10^{-2}$, and $f_1 \sim 10^{-3}$.

Then the total 9×9 mass matrices of the neutral states in each sector have the form

$$\mathcal{M} = \begin{pmatrix} 0 & \hat{f}v & 0 \\ \hat{f}v & 0 & \hat{M} \\ 0 & \hat{M} & \hat{\mu} \end{pmatrix} , \quad \mathcal{M}' = \begin{pmatrix} 0 & \hat{f}v' & 0 \\ \hat{f}v' & 0 & \hat{M} \\ 0 & \hat{M} & \hat{\mu} \end{pmatrix} , \qquad (7)$$

where the 3×3 Majorana mass matrix μ has a texture conserving $\bar{L} = L_e + L_\mu - L_\tau$ as in Eq. (1) with $a \sim b \sim \lambda V$. Clearly, at energies below M the theory effectively reduces the operators (4). Let us now take into account also the Planck scale effects explicitly violating the lepton numbers. Among the plethora of possible Planck scale operators relevant are only the ones given by Eq. (2) $\frac{\gamma_{ij}}{M_{\rm Pl}}(S_i S_j + S_i' S_j')\phi^2$, where ϕ^2 stands for the gauge invariant bilinear of all scalars involved and obviously it is dominated by the Higgs particles Φ_{ij} having the largest VEV's in our model [13]. Thus the mass matrix of the S states becomes $\hat{\mu} + \hat{\gamma}\mathring{\varepsilon}$, where $\mathring{\varepsilon} = V^2/M_{\rm Pl}$. As far as the latter operator violates $\bar{L}$ too, the matrix γ in general has no zeros. Thus, after decoupling the heavy states the 6×6 mass matrix of the light neutrinos reads

$$\hat{m}_\nu = \begin{pmatrix} \hat{s}(\hat{\mu} + \hat{\gamma}\mathring{\varepsilon})\hat{s} + \mathring{m}\hat{\alpha} & \zeta\mathring{m}\hat{\beta} \\ \zeta\mathring{m}\hat{\beta}^T & \zeta^2\hat{s}(\hat{\mu} + \hat{\gamma}\mathring{\varepsilon})\hat{s} + \zeta^2\mathring{m}\hat{\alpha} \end{pmatrix} ; \quad \hat{s} = {\rm diag}(s_1, s_2, s_3) \ , \qquad (8)$$

where $s_i = f_i v/M_i$. Clearly, in the absence of the Planck scale induced corrections the mass matrix $\hat{m}_\nu$ of the active neutrinos has the $\bar{L}$ conserving texture (1) with $a = s_1 s_3 A$ and $b = s_2 s_3 B$, while the mass matrix of their mirror partners is just scaled by factor of ζ^2: $\hat{m}_\nu' = \zeta^2\hat{m}_\nu$. Planck scale induced corrections violate $\bar{L}$, giving rise to small nonzero entries in neutrino mass matrix. Thus our model solves all present neutrino puzzles.

HDM+WDM: In order to represent the cosmological HDM component, the mass m of the degenerated states $\nu_{2,3}$ should be in the eV range, say, $m = 2.4$ eV [4]. Then for $\zeta \sim 30$ masses of their mirror partners $m' = \zeta^2 m$ are in the keV range and thus constitute WDM provided that their present abundance has a proper value.

LSND: Oscillation ν_e-ν_μ occurs with $\delta m^2_{e\mu} = m^2 \simeq 6$ eV^2 and $\sin\theta_{e\mu} = f_1 A/f_2 B$. For the assumed pattern of the Yukawa constants f_i we obtain $\sin^2 2\theta_{e\mu} \sim 10^{-3}$, in the range needed for the LSND oscillation.

ANP: The Planckian terms induce nonzero entries in the matrix $\hat{\mu}$ of the order of $\mathring{\varepsilon} \doteq V^2/M_{\rm Pl}$. This in turn induces in $\hat{m}_\nu$ nonzero (3,3) entry $\varepsilon = s_3^2\gamma\mathring{\varepsilon}$ which removes the ν_2-ν_3 degeneracy and thus triggers ν_μ-ν_τ oscillation with $\sin 2\theta \approx 1$ and $\delta m^2_{\mu\tau} = 2\varepsilon m$. Since $m = s_2 s_3 \lambda V \simeq 2.4$ eV, then the atmospheric neutrino oscillation range requires $\varepsilon \simeq 2 \times 10^{-3}$ eV. Thus, confronting the values of m and ε, we obtain the following estimates for the lepton number breaking scale V and the Dirac mass M:

$$V = \frac{\lambda f_2 \varepsilon}{\gamma f_3 m} M_{\rm Pl} \sim \frac{\lambda}{\gamma} 3 \times 10^{14} \ {\rm GeV} \ ,$$
$$M = \frac{\gamma^{1/2} f_3 v}{\sqrt{\varepsilon M_{\rm Pl}}} V \sim \frac{\lambda}{\gamma^{1/2}} 10^{13} \ {\rm GeV} \ , \qquad (9)$$

so that both scales are larger for reasonable values of λ and γ.

SNP: Obviously, the operators make negligible contribution to the electron neutrino mass. Indeed, their contribution to the (1,1) entry in the matrix $\hat{m}_\nu$ is $(f_1/f_3)^2\varepsilon \ll \mathring{m}$. Thus, for the mass terms of ν_e and ν_e' the relevant contributions come dominantly from the operators (2), leading to MSW oscillation with the parameter range given above.

In conclusion, we suggest that if all the present neutrino data are confirmed, an attractive resolution can be provided by postulating the existence of a mirror world. A crucial prediction is that the dark matter of the universe be warm rather than cold. The validity of this picture also requires the concept of an asymmetric inflation between the standard and the mirror world.

Note added. When our work was finished, we came across a paper by R. Foot and R. Vokas, Phys. Rev. D **52**, 6595 (1995), in which the explanation of recent neutrino puzzles is attempted with the mirror model with exact parity. Our approach and results, however, are drastically different from those obtained in this paper.

ACKNOWLEDGMENTS

Z.G.B. thanks V. Berezinsky, A. Dolgov, G. Fiorentini, and U. Miscili for valuable discussions. The work of R.N.M. was supported by the National Science Foundation under Grant No. PHY9421385.

[1] R. Barbieri, J. Ellis, and M. K. Gaillard, Phys. Lett. **90B**, 249 (1980).

[2] E. Akhmedov, Z. Berezhiani, and G. Senjanović, Phys. Rev. Lett. **69**, 3013 (1992).

[3] S. P. Mikheyev and A. Yu. Smirnov, Yad. Fiz. **42**, 1441 (1985); L. Wolfenstein, Phys. Rev. D **17**, 2369 (1978).

[4] J. R. Primack, J. Holtzman, A. Klypin, and D. O. Caldwell, Phys. Rev. Lett. **74**, 2160 (1995), and references therein.

[5] S. Dodelson and L. Widrow, Phys. Rev. Lett. **72**, 17 (1994); R. A. Malaney, G. D. Strakman, and L. Widrow, Phys. Rev. D **52**, 5480 (1995).

[6] C. Athanassopoulos *et al.*, Phys. Rev. Lett. **75**, 2650 (1995).

[7] D. Caldwell and R. N. Mohapatra, Phys. Rev. D **48**, 3259 (1993).

[8] T. Walker *et al.*, Astrophys. J. **376**, 51 (1991); P. Kernan and L. Krauss, Phys. Rev. Lett. **72**, 3309 (1994); K. Olive and G. Steigman, Phys. Lett. B **354**, 357 (1995).

[9] The concept of a hidden mirror world of particles has been considered in several earlier papers, e.g., Y. Kobzarev, L. Okun, and I. Pomeranchuk, Yad. Fiz. **3**, 1154 (1966); B. Holdom, Phys. Lett. **166B**, 196 (1985); S. L. Glashow, *ibid.* **167B**, 35 (1986); R. Foot, H. Lew, and R. Volkas, Phys. Lett. B **272**, 67 (1991); S. Barr, D. Chang, and G. Senjanović, Phys. Rev. Lett. **67**, 2725 (1991), and references therein.

[10] These estimates depend on the abundance of ν' relative to γ' at $t \sim 1$ s. The neutrino decoupling temperature in the mirror bath is related to one of the ordinary neutrinos T_D=3–5 MeV as $T'_D \sim \zeta^{4/3} T_D$. E.g., for $\zeta \sim 10$, ν' decouples from γ' after the QCD$'$ phase transition (due to P invariance, the "mirror" confinement scale cannot be substantially larger than that of the visible world: $\Lambda'_{\rm QCD} < 1$ GeV). Then for their effective temperatures at $t \sim 1$ s we obtain $T_{\nu'}/T_{\gamma'} = (4/11)^{1/3}$. On the contrary, for $\zeta \sim 100$ the decoupling temperature $T'_D \sim 2$ GeV and thus the contribution of light mirror quarks u', d' has also to be taken into account, which implies $T_{\nu'}/T_{\gamma'} = (4/53)^{1/3}$.

[11] This seems natural if $\Lambda \gg \langle\eta\rangle$. Indeed, in order to induce $\sim$ eV entries in the mass matrix $\hat{m}_\nu$ (1), Λ should be about 10^{13} GeV.

[12] R. N. Mohapatra and J. W. F. Valle, Phys. Rev. D **24**, 1642 (1986); M. C. Gonzalez-Garcia and J. W. F. Valle, Phys. Lett. B **216**, 360 (1989). Notice, however, that not all terms allowed by the symmetry are introduced. Their absence can be ensured by imposing additional symmetries.

[13] Once the individual global lepton numbers $L_{e,\mu,\tau}$ are broken spontaneously, there should exist multiflavor analogues of the singlet majoron. The Planck scale effects applied to these majorons induce their masses $\sim (V/M)^{1/2} V$, which, however, do not lead to any cosmological problems; see, e.g., E. Akhmedov, Z. Berezhiani, R. N. Mohapatra, and G. Senjanović, Phys. Lett. B **299**, 90 (1993).

Paper presented at the First Int. Symp. on Lepton and Baryon Number Violation, Trento, 1998

Exotic mechanisms for neutrino masses

Zurab Berezhiani [1]

Università di Ferrara, Via Paradiso 12, I-44100 Ferrara, Italy,
Institute of Physics, Georgian Academy of Sciences, Tbilisi, Georgia

Abstract. The problem of neutrino masses and mixings is discussed in the context of the standard model and its extensions. The recent hints of neutrino oscillations are reviewed as are the solar and atmospheric neutrino deficits and the LSND anomaly. If all these will be confirmed by future statistics, this would require existence of extra sterile neutrino. The origin of the latter can be naturally explained if it is identified to the neutrino of the hypothetical mirror world which is a complete duplicate to the visible one, except that its weak scale v' could be somewhat different from the ordinary weak scale. The mirror neutrinos can mix the ordinary ones through the Planck scale induced higher order operators, which gives rise to the oscillation of ν_e into its mirror partner ν'_e with parameters naturally in the Just-so range if $v' \sim v$, or in the MSW range if $v' \sim 30v$. We also discuss the cosmological and astrophysical implications of of mirror neutrinos and mirror baryons.

1. Neutrino masses in the standard model and beyond

1.1. In the standard model

The Standard Model (SM), being an internally consistent renormalizable gauge theory, has been extremely successfull in overcoming all experimental precission tests. It accomodates the observed quarks and leptons in a consistent way. Three families sharing the same quantum numbers under

[1] E-mail: berezhiani@fe.infn.it

PHYSICAL REVIEW D VOLUME 22, NUMBER 9 1 NOVEMBER 1980

Neutrino masses in SU(2) ⊗ U(1) theories

J. Schechter and J. W. F. Valle
Physics Department, Syracuse University, Syracuse, New York 13210
(Received 30 June 1980)

We analyze SU(2)×U(1) theories, denoted by (n,m), in which there are n neutrinos belonging to isodoublets and m neutrino isosinglets. The charged-current weak interactions are described by a rectangular matrix K which we explicitly parametrize. The neutral-current neutrino interactions are described by a square matrix $P = K^+K$. This has the consequences that neutrinos may decay into three lighter ones and that neutrino oscillations involving neutral-current interactions should exist. The general formalism for the latter situation is given. Associated material on parametrization of unitary matrices and the quantum theory of Majorana particles is also briefly discussed.

I. INTRODUCTION

Recently there has been a great deal of interest in the possibility that neutrinos may in fact be massive particles. On the experimental side this is in part due to the work of Reines *et al.*[1] on neutrino oscillations. Actually the earlier experiments of Davis on solar-neutrino flux were also interpreted as evidence for neutrino oscillations. The whole subject is nicely reviewed by Bilenky and Pontecorvo.[2] On the theoretical side this interest is due to the fact that many of the symmetry groups which unify $SU(2)_L \times U(1)$ with strong interactions require massive neutrinos for self-consistency.[3]

In the present note we will discuss the question of how the weak interactions involving massive neutrinos should be parametrized. That this is a nontrivial question can be seen by referring to the Kobayashi-Maskawa (KM) parametrization[4] of weak interactions involving massive quarks. There (in addition to the quark masses) four mixing angles are needed. In a certain sense we may think of these mixing angles as representing the "kinematics" or "geometry" of the theory. Now it is immediately clear that the parametrization depends on the particular model adopted. Since almost all theories of present interest are considered to reduce to $SU(2)_L \times U(1)$ (Ref. 5) effectively at low energies it seems reasonable to work in an $SU(2)_L \times U(1)$ framework. To give a logical structure to our presentation we will demand that the theory be *natural*,[6] in a sense to be spelled out precisely. As we shall see, the lepton mixings are inevitably more complicated than the KM scheme.

We consider a natural theory to be one in which, once the particle content is specified, the Lagrangian is the most general local one consistent with proper Lorentz invariance and renormalizability. The latter requirement includes the cancellation of anomalies and hence rules out certain particle assignments. It should be stressed that no assumption about the P, C, and T symmetries is to be made at the beginning. Whether or not and to what degree these symmetries hold should emerge from the theory itself. This is a sense in which the parametrization of the theory is related to its "geometry." Note that by the initial assumption the CPT theorem[7] will hold, so CPT is automatically a good symmetry. We are aware that if $SU(2)_L \times U(1)$ is embedded in a larger unifying group G, the criterion of renormalizability for the $SU(2)_L \times U(1)$ subgroup by itself is too restrictive. Nevertheless, it seems to be the most reasonable first approach. In any event, the main part of our analysis is independent of this assumption.

Now let us discuss in a step-by-step way how the usual $SU(2)_L \times U(1)$ model of leptons can be modified to include massive neutrinos. The usual theory[5] contains a complex $SU(2)_L$ Higgs doublet with weak hypercharge $Y=1$, n two-component fermion $SU(2)_L$ doublets

$$\psi_a = \begin{pmatrix} N_a \\ E_a \end{pmatrix}_L \tag{1.1}$$

[N_a is neutral, E_a is negatively charged, and L means $(1+\gamma_5)/2$ projection] with $Y=-1$, and n two-component (right-handed) fermion $SU(2)_L$ singlets E_{aR} with $Y=-2$. n is the number of "generations" which we will allow to be arbitrary. Note that all fermion fields are two-component spinors which can be considered to be van der Waerden spinors.[8] Introducing four-component Dirac fields, while convenient for computation, is something of a "mystification" in a theory where no assumptions about P, C, and T symmetries are made *a priori*. If one wishes to get massive neutrinos without introducing any new fermion fields in the theory, it is necessary to add a complex Higgs triplet with $Y=2$. The triplet may be put into a 2×2 matrix,

$$h = \begin{pmatrix} h^{(+)} & h^{(++)} \\ h^{(0)} & -h^{(+)} \end{pmatrix}, \tag{1.2}$$

Rabindra N. Mohapatra

Unification and Supersymmetry

Springer-Verlag
New York Berlin
Heidelberg Tokyo
London Paris

The Frontiers of Quark–Lepton Physics

Contents

S. P. ROSEN
T Division,
Los Alamos National Laboratory,
Los Alamos, New Mexico 87545

On Seeing the First Double Beta Decay

Key Words: *beta decay, lepton conservation, Majorana neutrino, Majoron MSW-effect*

At the end of August, 1987 Elliott, Hahn and Moe,[1] working at the University of California, Irvine announced the observation of the decay of ^{82}Se into ^{82}Kr plus two electrons and two neutrinos with a half-life of 10E(20) years. This is the first time that the extremely rare process of double beta decay[2] has ever been seen in the laboratory, and the work of the Irvine group constitutes a landmark achievement of experimental science.

Previous observations of double beta decay had relied upon the so-called "geochemical" method in which one takes ancient ores of the parent element and looks for traces of the daughter; one must then eliminate all conceivable background processes as the source of the daughter nucleus, and make a reliable estimate of the age of the ore in order to calculate the half-life for the decay. The beauty of the Irvine experiment is that it actually records the tracks of the electrons as they are emitted by the ^{82}Se nucleus in a time projection chamber (TPC). Backgrounds are an extremely severe problem, just as they are in geochemical experiments, and it is only through the heroic efforts of understanding the properties of the background and cleaning up the chamber that Elliott, Hahn and Moe were able to overcome them. In fact they had to go

Comments Nucl. Part. Phys.
1988, Vol. 18, No. 1, pp. 31–54
Photocopying permitted by license only

World Scientific Lecture Notes in Physics Vol. 25

THE PHYSICS OF MASSIVE NEUTRINOS

BORIS KAYSER
National Science Foundation

with **FRANÇOISE GIBRAT-DEBU**, CEN-Saclay

and **FREDERIC PERRIER**, Stanford Linear Accelerator Center

PHYSICAL REVIEW D VOLUME 25, NUMBER 11 1 JUNE 1982

Neutrinoless double-β decay in SU(2)$\times$U(1) theories

J. Schechter and J. W. F. Valle
Department of Physics, Syracuse University, Syracuse, New York 13210
(Received 14 December 1981)

It is shown that gauge theories give contributions to neutrinoless double-β decay [$(\beta\beta)_{0\nu}$] which are not covered by the standard parametrizations. While probably small, their existence raises the question of whether the observation of $(\beta\beta)_{0\nu}$ implies the existence of a Majorana mass term for the neutrino. For a "natural" gauge theory we argue that this is indeed the case.

The associated questions of neutrinoless double-β decay [denoted $(\beta\beta)_{0\nu}$] and neutrino mass have again become of general interest. A comprehensive recent discussion is given by Doi *et al.*[1] and a concise summary by Rosen.[2]

The classical analysis of the $(\beta\beta)_{0\nu}$ process, which of course predated gauge theories by many years, assumed that it arose as a result of neutrino exchange between two effective four-fermion vertices. This is shown in modern language in Fig. 1(a), which illustrates the process at the quark level. Thus, the parametrization of the $(\beta\beta)_{0\nu}$ process was considered to be given fundamentally by the parametrization of the four-fermion single-β-decay interaction.

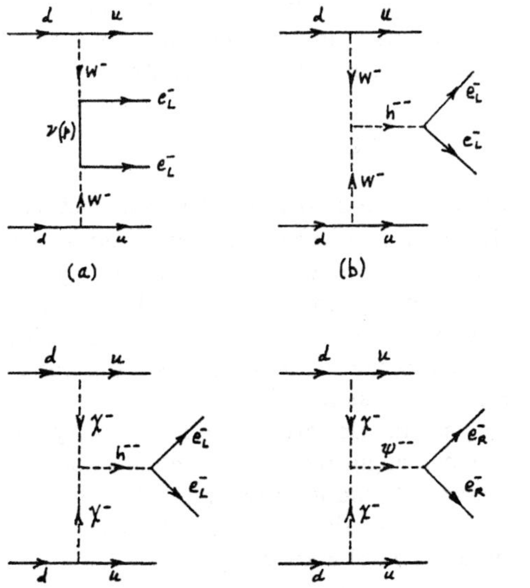

FIG. 1. Diagrams for neutrinoless double-β decay in an SU(2)$\times$U(1) gauge theory. The standard diagram is Fig. 1(a). It is the only one which contains a virtual neutrino (of four-momentum p). d and u are the down and up quarks.

We point out here that the situation is more complicated[3] if the weak interactions are described by a gauge theory. For definiteness we consider $(\beta\beta)_{0\nu}$ in the framework of the standard SU(2)$\times$U(1) gauge theory. Larger gauge groups usually contain SU(2)$\times$U(1) as a subgroup and composite models are usually contrived to also display this symmetry. Thus there is not much loss of generality in doing so.

A rather minimal way[4] to naturally include lepton-number violation in the theory is to add to the complex $Y=1, I=\frac{1}{2}$ Higgs doublet a complex $Y=2, I=1$ isotriplet H:

$$\phi=\begin{bmatrix}\phi^+\\ \phi^0\end{bmatrix},\quad H=\begin{bmatrix}\frac{H^+}{\sqrt{2}} & h^{++}\\ h^0 & \frac{-h^+}{\sqrt{2}}\end{bmatrix}. \tag{1}$$

The theory need not contain any more than one two-component Weyl neutrino field for each generation.[5] In a natural theory (no special adjustment of parameters) both ϕ^0 and h^0 will develop nonzero vacuum expectation values giving the relations[4]

$$\begin{aligned}&\langle\phi^0\rangle\equiv\lambda,\ \ \langle h^0\rangle\equiv y\ ,\\ &m^2(W)=\frac{g^2}{4}(\lambda^2+2y^2)\ ,\\ &m^2(Z)=\frac{g^2}{4\cos^2\theta_W}(\lambda^2+4y^2)\ ,\\ &\frac{G_F}{\sqrt{2}}=\frac{g^2}{8m^2(W)}\ ,\end{aligned} \tag{2}$$

in standard notation. The vacuum value y is expected to be small compared to λ, but the experimental constraints are not very tight.[6] For our purposes it is important to note that the singly charged field χ^- which is *not* absorbed by the gauge field is the linear combination

$$\chi^- \approx h^- + \omega\phi^-, \quad \omega = -\sqrt{2}\frac{y}{\lambda}\,, \tag{3}$$

where we have assumed ω to be small. Notice also that the "Yukawa" interaction of H allows a virtual h^{--} to annihilate into two left-handed electron fields (e_L^-) with strength m_ν/y (m_ν is a characteristic neutrino mass). Then there are a variety of diagrams for $(\beta\beta)_{0\nu}$ which involve no virtual neutrino line and hence cannot be parametrized by the standard β-decay interaction vertex. These diagrams will require a six-fermion amplitude for their description: The effective Lagrangian will be characterized by a parameter of dimension $(\text{mass})^{-5}$.

A basic new diagram is shown in Fig. 1(b). The trilinear $W^-W^-h^{++}$ vertex is due to a term in the fundamental Lagrangian:

$$\tfrac{1}{2}g^2\langle \bar{h}^0\rangle W_\mu^- W_\mu^- h^{++} + \text{H.c.} \tag{4}$$

derived from the kinetic term $-\frac{1}{2}\,\text{Tr}[(D_\mu H)^\dagger D_\mu H]$. The strength of the amplitude for $(\beta\beta)_{0\nu}$ from Fig. 1(b) is the product of the four trilinear coupling constants and the three propagators:

$$\frac{g^4 m_\nu}{m^2(h^{--})m^4(W)}\,. \tag{5}$$

Note that the triplet vacuum value has canceled out of Eq. (5). For comparison the "standard" diagram given in Fig. 1(a) is characterized by a strength[7]

$$\frac{g^4 m_\nu}{m^4(W)\langle p^2\rangle}\,. \tag{6}$$

Here $\langle p^2\rangle$ represents a suitable average of the squared four-momentum carried by the virtual neutrino, say about $(10\ \text{MeV})^2$. Of course the amplitude for the underlying quark process must be suitably[1,2] folded into the description of the real nucleus. The ratio

$$\frac{\text{amp for 1(b)}}{\text{amp for 1(a)}} = \frac{\langle p^2\rangle}{m^2(h^{--})} \tag{7}$$

is thus expected to be of order of 10^{-8}. Thus the contribution of Fig. 1(b) seems negligible.

Additional new diagrams involve a trilinear Higgs interaction [see Fig. 1(c)]. The relevant terms in the Higgs Lagrangian are

$$c_1\phi^\dagger(HH^\dagger - H^\dagger H)\phi + c_2\,\text{Tr}(HH)\,\text{Tr}(H^\dagger H^\dagger) + (d\phi^\dagger H\phi_c + \text{H.c.})\,, \tag{8}$$

where

$$\phi_c \equiv \begin{pmatrix} \bar{\phi}^0 \\ -\phi^- \end{pmatrix}.$$

Taking into account the mixing between the charged components of the Higgs doublet and the triplet, given by Eq. (3), results in the trilinear vertex

$$g_3\chi^-\chi^-h^{++} + \text{H.c.}\,, \qquad g_3 \approx \sqrt{2}\omega c_1\langle\phi^0\rangle + 2c_2\langle h^0\rangle - \omega^2 d\,. \tag{9}$$

The strength parameter associated with Fig. 1(c) is then

$$\frac{\omega^2 m_q{}^2 m_\nu g_3}{y\lambda^2 m^2(h^{--})m^4(\chi^-)} \tag{10}$$

($m_q \approx 10$ MeV is a light-quark mass). Equation (10) involves unknown parameters characterizing the effective Higgs sector of the theory. Let us first estimate the contribution to Eq. (10) coming from the c_1 and c_2 terms. These are dimensionless parameters and can be reasonably expected to be of order unity, at most. Further estimating $m^2(h^{--}) \approx m^2(\chi^-) \approx \lambda^2$ the ratio of this amplitude to the standard one in Eq. (6) is roughly

$$\frac{y^2}{\lambda^2}\frac{m_q{}^2\langle p^2\rangle}{\lambda^4} \approx 10^{-13}\omega^2\,, \tag{11}$$

which is quite negligible. The contribution to Eq. (10) from the d term (d has dimensions of mass) is potentially more interesting. If one imposes lepton-number conservation on the theory d will be zero and there will be no contribution. Then lepton number breaks spontaneously and one has a massless "Majoron."[8] In such a case ω will also be very small. However, since d does carry a dimension one might regard it as an indicator of a new mass scale[9] and let it remain. Then the ratio of the d contribution to Eq. (10) to the standard amplitude Eq. (6) is roughly

$$\frac{\omega^3 m_q{}^2\langle p^2\rangle}{\lambda^4}\frac{d}{\lambda} \approx 10^{-13}\omega^3\frac{d}{\lambda}\,. \tag{12}$$

This indicates that (even neglecting the suppression due to the ω^3 factor) d would have to be of the order of the grand unification mass scale for the diagram of Fig. 1(c) to play an important role.

The presence of the Higgs field χ^- in the theory

also results in new diagrams of the standard form 1(a) in which one or both of the W's is replaced by a χ^-. These diagrams, which also modify the $V-A$ structure of the single-β-decay interaction, are quite small.

Thus we reach the conclusion that for the SU(2)×U(1) theory defined by the Higgs content (1), the effect of the new diagrams is quite small if one considers only mass scales lower than that of grand unification.

One type of neutrinoless diagram which may conceivably be relatively strong without superheavy masses is shown in Fig. 1(d). Here a new $Y=-4$ isosinglet Higgs field ψ^{--} is introduced in addition to the doublet and the triplet. The virtual ψ^{--} decays into two e_R's rather than two e_L's as in the previous cases. In this case the $\psi^{++}e_R^-e_R^-$ Yukawa interaction is not proportional to the neutrino mass (as was required previously since the $h^{++}e_L^-e_L^-$ Yukawa term is related by an isospin transformation to the $h^0\nu\nu$ term which generates neutrino mass) and thus may be of order unity.[10] The term in the Higgs Lagrangian which generates the trilinear $\chi^-\chi^-\psi^{++}$ coupling in Fig. 1(d) is

$$\phi^\dagger H^\dagger \phi_c \psi^{++} + \text{H.c.} \quad (13)$$

The amplitude for Fig. 1(d) would then roughly be of order $\omega^2 m_q{}^2 y/\lambda^8$. The ratio of this to the usual amplitude, which is suppressed by a factor of m_ν, is about

$$\frac{\omega^2 y m_q{}^2\langle p^2\rangle}{m_\nu \lambda^4} . \quad (14)$$

This could be comparable to one if m_ν is exceptionally small.

Other models with extra Higgs fields can also boost the new contribution. For example, suppose that we add to (1) another complex doublet ϕ', as one might have in an axion scheme. Then there will be two physical singly charged fields and there is in general no need to have a suppression[10] of their Yukawa couplings to the quarks for small y. The ratio of the d contribution to the standard one [see Eq. (12)] is now roughly

$$10^{-13}\frac{\langle p^2\rangle^{1/2}}{y}\frac{d}{\lambda} \approx 10^{-6}\frac{d}{\lambda} ,$$

where we have taken $y \approx 1$ eV. Thus an intermediate scale $d \approx 10^8$ GeV could make the new diagrams important.

To sum up we can say that while neutrinoless diagrams might not be dominant, a careful analysis of $(\beta\beta)_{0\nu}$ decays should really take into account

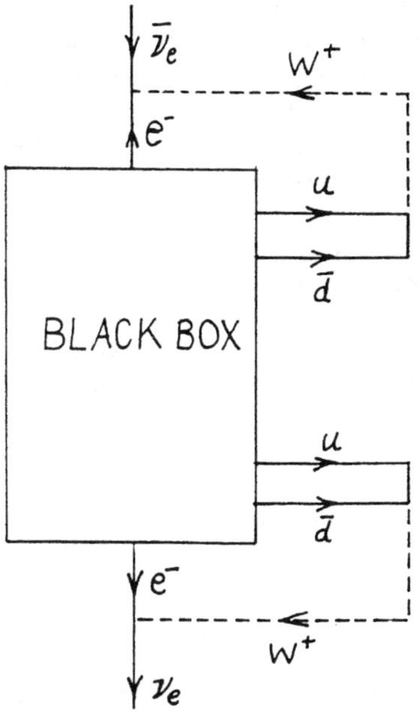

FIG. 2. Diagram showing how any neutrinoless double-β decay process induces a $\bar{\nu}_e$-to-ν_e transition, that is, an effective Majorana mass term.

their possible existence. This is because the general structure (as opposed to detailed predictions) of gauge theories seems to be the safest guide to the parametrization of weak-interaction amplitudes. It would be desirable to develop criteria[11] based on angular distributions of the decay products for distinguishing these diagrams from the usual ones.

We will conclude this paper with a brief discussion of the relation between the $(\beta\beta)_{0\nu}$ process and nonzero neutrino mass. After noticing the existence of neutrinoless diagrams one might be tempted to try to construct models without massive neutrinos and which would still give $(\beta\beta)_{0\nu}$. However, such a search would be in vain. For the model based on the Higgs content (1) this result is obvious since Eqs. (5) and (10) are proportional to m_ν. It is also true for the model with ψ^{--}: Although this model gives an amplitude with no m_ν factor there is an overall factor of $y=\langle h^0\rangle$. Now in a natural theory H will couple to the basic lepton doublet so that a nonzero value of y will generate a neutrino mass.

Still one might think that a yet more clever choice of the Higgs-representation content could do the job. Rather than attempt an enumeration of all possible Higgs structures we will give a general and yet very simple proof that the existence of $(\beta\beta)_{0\nu}$ implies that the electron neutrino has nonzero mass. Essentially all one needs is to assume that the weak interactions are described by a local gauge theory. In this framework crossing symmetry will hold so the existence of $(\beta\beta)_{0\nu}$ implies a nonzero amplitude for the virtual process $0\rightarrow uu\bar{d}\bar{d}\bar{e}\bar{e}$. This is shown in Fig. 2, where the "black box" may contain any mechanism whatsoever for generating $(\beta\beta)_{0\nu}$. Now any realistic gauge theory will include the ordinary W-gauge-field interaction with the left-handed electron and neutrino and with the u and d quarks. Using four of these vertices and connecting the lines together as in Fig. 2 shows that we develop an amplitude which gives a nonzero Majorana mass for the electron neutrino.

One might object that some other diagram might precisely cancel[12] Fig. 2, but this would clearly involve fine tuning of parameters and would be unnatural.

The converse question is whether a nonzero neutrino mass implies the existence of $(\beta\beta)_{0\nu}$. If the massive neutrino is of Majorana type, Fig. 1(a) shows that $(\beta\beta)_{0\nu}$ will occur. If the neutrino is of Dirac type the $(\beta\beta)_{0\nu}$ will not occur.[13] One may, however, exclude the possibility of neutrinos being of Dirac type if one postulates a "strong naturality" in which no global conservation laws are assumed *a priori*. In such a case massive neutrinos will be[14] of Majorana type.

This work was supported in part by the U. S. Department of Energy under Contract No. DE-AC02-76ER03533. The work of J. V. was supported by CNPq, Brazil.

---

[1]M. Doi, T. Kotani, H. Nishiura, K. Okuda, and E. Takasugi, Prog. Theor. Phys. 66, 1739 (1981); 66, 1765 (1981); see also H. Nishiura, Kyoto University Report No. RIFP-453, 1981 (unpublished).

[2]S. P. Rosen, in *Neutrino 81*, proceedings of the International Conference on Neutrino Physics and Astrophysics, Maui, Hawaii, edited by R. J. Cence, E. Ma, and A. Roberts (University of Hawaii High Energy Physics Group, Honolulu, 1981).

[3]After the work described in this paper was completed we received a paper [Phys. Rev. Lett. 47, 1713 (1981)] by R. N. Mohapatra and J. D. Vergados which also points out the need for the intrinsic six-fermion interactions. However, the discussions of the two papers are somewhat different.

[4]See, for example, J. Schechter and J. W. F. Valle, Phys. Rev. D 22, 2227 (1980); T. P. Cheng and L. F. Li, *ibid.* 22, 2860 (1980); G. B. Gelmini and M. Roncadelli, Phys. Lett. 99B, 411 (1981); H. M. Georgi, S. L. Glashow, and S. Nussinov, Harvard Report No. HUTP-81/A026 (unpublished); P. B. Pal and L. Wolfenstein, Phys. Rev. D 25, 766 (1982).

[5]For simplicity we will consider the theory to contain only one generation.

[6]See Cheng and Li, Ref. 4 above.

[7]The lepton-number-violating Majorana-neutrino propagator in Fig. 1(a) is responsible for the factor $m_\nu/\langle p^2\rangle$.

[8]See, for example, Y. Chikashige, R. N. Mohapatra, and R. D. Peccei, Phys. Lett. 98B, 265 (1981); Gelmini and Roncadelli and Georgi *et al.*, Ref. 4 above; J. Schechter and J. W. F. Valle, Phys. Rev. D 25, 774 (1982); V. Barger, W. Y. Keung, and S. Pakvasa, *ibid.* 25, 907 (1982).

[9]See, for example, R. N. Mohapatra and G. Senjanovic, Phys. Rev. D 21, 165 (1981).

[10]If the leptonic Yukawa coupling constant in Fig. 1(c) is of order unity, y will be approximately m_ν and one will have enormous suppression of this diagram due to the factor ω^2 in Eq. (10).

[11]See Refs. 1 and 2, for example.

[12]Also one would have to cancel all variations on the basic Fig. 2, which is extremely unlikely. For example, a photon may be exchanged between each u and $\bar{d}$ line.

[13]A Dirac neutrino can be conceived of as a pair of mass-degenerate Majorana neutrinos with opposite *CP* assignments. As pointed out by L. Wolfenstein [Carnegie-Mellon Report No. COO-3066-180 (unpublished)], their contributions to the $(\beta\beta)_{0\nu}$ rate cancel due to their relative *CP* phase. With several Majorana neutrinos with different *CP* eigenvalues the neutrinoless diagrams discussed here may not be entirely negligible.

[14]See, for example, the discussion in Sec. 1 of the first of Refs. 4 above.

RAPID COMMUNICATIONS

PHYSICAL REVIEW D VOLUME 27, NUMBER 7 1 APRIL 1983

Neutrinoless double-β decay with quasi-Dirac neutrinos

J. W. F. Valle*
Department of Physics, Syracuse University, Syracuse, New York 13210
(Received 2 July 1982)

A mechanism to generate the neutrinoless double-β decay process with quasi-Dirac neutrinos and no right-handed currents is described. Relatively heavy neutrinos can easily be made consistent with the constraints on the $(\beta\beta)_{0\nu}$ rate. The proposed scheme for the violation of lepton number can in principle be distinguished from a Majorana mass in $\Delta l=2$ oscillation phenomena.

Historically the first scheme proposed by Primakoff and Rosen[1] to generate the $(\beta\beta)_{0\nu}$ process was based on the existence of a right-handed charged leptonic current (RHC mechanism). More recently[2] a new scheme was found based on a nonvanishing Majorana neutrino mass (mass mechanism). The common denominator of both of these schemes is the assumption that the neutrino is described by a Majorana field operator.[3] In this Communication we stress that such an assumption is not necessary in general to generate the $(\beta\beta)_{0\nu}$ process.

Within a gauge framework both current and mass mechanisms naturally arise. For example, the standard $SU(2)\times U(1)$ theory with massive neutrinos naturally gives Majorana neutrinos so that the mass mechanism is operative. Left-right-symmetric extensions of the standard model can accomodate both mechanisms.[4] While Majorana neutrinos do naturally arise in many gauge models, their mixing pattern is exceedingly complex.[5] From this point of view it is desirable to seek models which use Dirac neutrinos, at least as an approximation, while incorporating the possibility of lepton-number violation.[6]

First we describe the mechanism to generate a $(\beta\beta)_{0\nu}$ transition mediated by a quasi-Dirac neutrino. We take the simplest $V-A$ effective charged-current weak Hamiltonian

$$H=\sqrt{2}G_F j_L\cdot J_L^\dagger+\text{H.c.}\ , \tag{1}$$

in which J_L is an hadronic current and j_L is the leptonic current,

$$j_{L\mu}=\bar{e}\gamma_\mu\frac{(1+\gamma_5)}{2}\left[\frac{\nu+\epsilon\nu^c}{(1+\epsilon^2)^{1/2}}\right]\ . \tag{2}$$

Here ν is a four-component massive Dirac field and $\nu^c=C\bar{\nu}^T$. The presence of the second term, proportional to a small parameter ϵ, breaks lepton-number conservation by two units. In terms of $SL(2,C)$ mass-eigenstate neutrinos ν_1 and ν_2,

$$\nu_L\cong\frac{\nu_2+i\nu_1}{\sqrt{2}}\ ,\quad (\nu^c)_L\cong\frac{\nu_2-i\nu_1}{\sqrt{2}}\ , \tag{3}$$

one rewrites the leptonic current as

$$j_{L\mu}=\sum_{\alpha=1}^{2}\bar{e}_L\gamma_\mu K_\alpha\nu_\alpha\ ,\qquad K_2\cong\frac{1+\epsilon}{\sqrt{2}},\quad K_1\cong i\frac{1-\epsilon}{\sqrt{2}}\ , \tag{4}$$

where ν_1 and ν_2 are characterized by a *common* mass so as to make up jointly a Dirac spinor ν. Equations (1) and (2) imply, in addition to the lepton-number-conserving β-decay process of Fig. 1(a), the lepton-number-violating process depicted in 1(b). It is then possible to Wick-contract the neutrino field operators and obtain Fig. 1(c) by joining the neutrino lines of 1(a) and 1(b). The $(\beta\beta)_{0\nu}$ amplitude will be proportional to

$$m_\nu\sum_{\alpha=1}^{2}K_\alpha{}^2=\frac{2m_\nu\epsilon}{1+\epsilon^2}\simeq 2m_\nu\epsilon\ , \tag{5}$$

so that a complete cancellation[7] of the ν_1 and ν_2 contributions is avoided even when $m_1=m_2=m_\nu$ (Dirac limit). We call ν a quasi-Dirac neutrino because, in a gauge framework, it will naturally develop a Majorana mass from radiative corrections.[8,9] One therefore sees from Eq. (5) that the present scheme requires not only an explicit $\Delta l=2$ interaction ($\epsilon\neq 0$) but

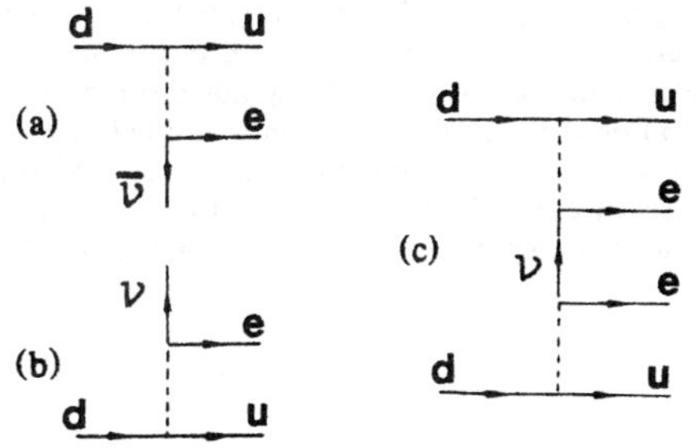

FIG. 1. (a) Ordinary β-decay process ($\Delta l=0$). (b) "Wrong" β decay ($\Delta l=2$). (c) Dirac-neutrino-mediated $(\beta\beta)_{0\nu}$ decay.

also a nonzero neutrino mass. Since the relevant combination is the product $m_\nu \epsilon$ the present mechanism can tolerate a wide range of neutrino masses even for a small $(\beta\beta)_{0\nu}$ rate. This sharply contrasts with the RHC mechanism. Taking over the recent experimental bound[3] one would expect, roughly,

$$\frac{m_\nu}{m_e}\epsilon \lesssim 10^{-5} \ . \tag{6}$$

For $\epsilon \simeq 10^{-2}$, a neutrino mass as large as $m_\nu \simeq 1$ keV would be allowed on the basis of this mechanism alone. The general predictions of this scheme regarding the angular correlation of electrons emitted in a $0^+ \rightarrow 0^+$ neutrinoless transition are indistinguishable from those of the mass mechanism but distinct from those of the RHC mechanism. Similarly $0^+ \rightarrow 2^+$ transitions would be forbidden both here and in the mass mechanism, in contrast with the RHC mechanism.

While the present mechanism avoids completely a RHC, notice that the $(\beta\beta)_{0\nu}$ decay mode will also be generated if the lepton number is broken minimally by means of an explicit RHC.[10]

The present scheme (lepton number broken minimally by a left-handed interaction) can in principle be distinguished from the usual mass mechanism (lepton number broken by Majorana masses) in $\Delta l = 2$ oscillations. The process is depicted in Figs. 2(a) and 2(b). If the present scheme holds, the neutrino produced in 2(a) from a charged lepton e_a propagates as a neutrino and instigates in 2(b) a charged-current reaction that produces an antilepton $\bar{e}_b$. The overall amplitude factor for the combined process is

$$\text{amp}(e_a \rightarrow \bar{e}_b, t) = \epsilon \frac{1}{E} \sum_c K_{ac} e^{-iE_c t} m_c K'_{bc} \ , \tag{7}$$

where K and K' are appropriate mixing matrices and m_c/E accounts for a helicity suppression, E is the neutrino energy, and $t \cong$ distance from the neutrino source at 2(a). Consider now the case in which lepton number is broken by Majorana neutrino masses. Then the neutrinos emitted in 2(a) would oscillate during their flight to 2(b) into antineutrinos so as to trigger, at the position of the second target, an inverse-β-decay reaction. Clearly this process will depend on the fraction of neutrinos emitted in 2(a) which have oscillated into antineutrinos by the time they reach 2(b). Then, via the usual $(\epsilon = 0)$ weak interactions they would produce antileptons $\bar{e}_b$. The

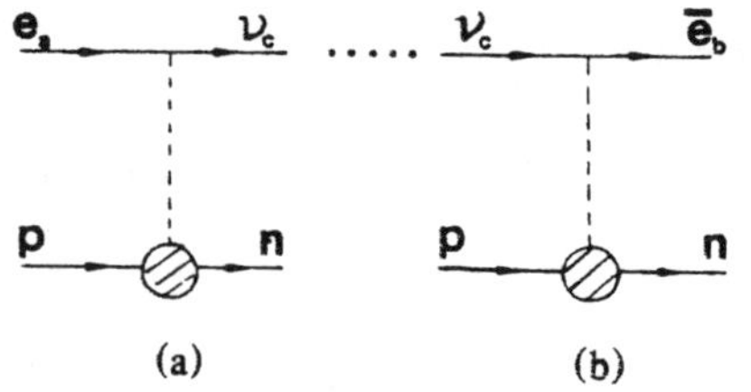

FIG. 2. (a) Ordinary charged-current reaction, in which a charged lepton e_a produces a mass-eigenstate Dirac neutrino ν_c. (b) After time evolution, ν_c triggers a $\Delta l = 2$ charged-current interaction, producting an antilepton $\bar{e}_b$.

overall amplitude factor is

$$\text{amp}(e_a \rightarrow \bar{e}_b, t) = \frac{1}{E} \sum_c K_{ac} e^{-iE_c t} m_c K_{bc} \ . \tag{8}$$

Comparing these two situations one sees that they are qualitatively distinct: In the first case the process is distance independent (lepton number broken in the weak-interaction vertex), while in the case of pure mass mechanism it is distance dependent.[11]

Finally we note that the interaction (1) and (2) can be derived from a gauge-theory framework. It suffices that the fields e_L, ν_L, and ν_L^c are assigned to the fundamental representation of the gauge group. It is not always true, however, that the two-component spinors ν_L and ν_L^c will amalgamate into a Dirac neutrino. A model that accomplishes this to lowest order in perturbation theory is discussed in Ref. 6. It is an extension of the standard model in which the new energy scale is associated with the *local* breakdown of lepton number.[12] The lepton-number-violating parameter ϵ is then essentially given by the ratio of the appropriate energy scales.

To conclude, we note that the ongoing analysis of the neutrinoless nuclear double-β decay should be sharpened by taking into account the possible existence of the mechanism described here.

I would like to thank Professor Henry Primakoff for stimulating discussions and comments, and Professor Arthur Halprin for the hospitality extended to me at the Lewes Center for Physics. My thanks also to Professor Carl Rozensweig and Robert Shrock and my colleagues at Syracuse University for their encouragment and discussions. I also want to thank Professor Peter Rosen for stressing to me the helicity factor in Eq. (7). This work was sponsored by the National Research Council (CNPq), Brazil.

*Present address: Theory Division, Rutherford Appleton Laboratory, Chilton, Didcot, Oxfordshire OX11 0QX, England.

[1]H. Primakoff and S. P. Rosen, Phys. Rev. 184, 1925 (1969).

[2]A. Halprin, P. Minkowski, H. Primakoff, and S. P. Rosen, Phys. Rev. D 13, 2569 (1976).

[3]H. Primakoff and S. P. Rosen, Annu. Rev. Nucl. Part. Sci. 31, 145 (1981); S. P. Rosen, in *Neutrino '81*, proceedings of the International Conference on Neutrino Physics and Astrophysics, Wailea, Hawaii, 1981, edited by R. J. Cence, E. Ma, and A. Roberts (Department of Physics, University of Hawaii, Honolulu, 1981).

[4]For a recent analysis carried out within this framework, see H. Nishiura, Kyoto Report No. RIFP-453, 1981 (unpublished). See also M. Doi *et al.*, Prog. Theor. Phys. 66, 1739 (1981); 66, 1765 (1981); W. Haxton, G. Stephenson, and D. Strottman, Phys. Rev. Lett. 47, 153 (1981); Phys. Rev. D 25, 2360 (1982). For a recent review see Ref. 3.

[5]J. Schechter and J. W. F. Valle, Phys. Rev. D 22, 2227 (1980); 23, 1666 (1981); M. Doi *et al.*, Phys. Lett. 102B, 323 (1981).

[6]J. W. F. Valle and M. Singer (in preparation).

[7]L. Wolfenstein, Phys. Lett. 107B, 77 (1981); J. Schechter and J. W. F. Valle, Phys. Rev. D 24, 1883 (1981); 25, 283 (1982).

[8]J. Schechter and J. W. F. Valle, Phys. Rev. 25, 2951 (1982).

[9]One can also talk of quasi-Dirac neutrinos in the context of a model with $\epsilon = 0$ and 2 nearly degenerate Majorana neutrinos: In this case the interaction (1) and (2) with $\epsilon = 0$ gives a $(\beta\beta)_{0\nu}$ amplitude proportional to their mass difference. This should be contrasted with the situation described in the text.

[10]See S. P. Rosen, in *Particles and Fields—1971*, proceedings of the Annual Meeting of Division of Particles and Fields of the APS, Rochester, New York, edited by A. Melissinos and P. Slattery (AIP, New York, 1971), p. 226.

[11]One can also contemplate a general situation in which "neutrino-antineutrino" oscillations also occur while the "neutrino" travels towards the second target. For that it would suffice to break lepton-number conservation by Majorana masses in addition to the present scheme.

[12]The possibility of spontaneous breakdown of *global* lepton number leads to the so-called Majoron schemes, analyzed in J. Schechter and J. W. F. Valle, Phys. Rev. D 25, 774 (1982) and references therein.

2.1.3 Double Beta Decay, Gauge Theories and Neutrino Mass:

The Double Beta Half Life and the Neutrino Mass

Progress of Theoretical Physics, Vol. 66, No. 5, November 1981

Neutrino Mass, the Right-Handed Interaction and the Double Beta Decay. I

—— *Formalism* ——

Masaru DOI, Tsuneyuki KOTANI,* Hiroyuki NISHIURA,* Kazuko OKUDA* and Eiichi TAKASUGI*

Osaka College of Pharmacy, Matsubara, Osaka 580
**Institute of Physics, College of General Education Osaka University, Toyonaka 560*

(Received March 23, 1981)

In order to shed light on the important question whether neutrinos are Dirac or Majorana particles, the double β decay is investigated within a general form of weak interaction Hamiltonian. The systematic study is made on the $0^+ \to J^+$ nuclear transitions for the two-neutrino and neutrinoless modes both in the two-nucleon- and N^*-mechanism. It is shown that for the neutrinoless mode, only the $0^+ \to 0^+$ transition in the two-nucleon mechanism is allowed if there is no right-handed interaction. When the right-handed interaction gives a sizable contribution, the role of the $0^+ \to 2^+$ transition becomes as important as the $0^+ \to 0^+$ transition. The comparison of our results with the previous ones is also presented.

§ 1. Introduction

The question whether neutrinos are massive or massless has become one of the recent important topics. This is motivated by the recent theoretical development of the grand unified theories where neutrinos are likely to be massive because leptons and quarks are treated on the equal basis. If neutrinos are massive, there arises an important question whether neutrinos are Dirac or Majorana particles. In models like $SO(10)$, neutrinos are assigned to Majorana particles to explain the small masses of the observed neutrinos.

In this paper, we investigate the double β decay which reveals directly the difference between Dirac and Majorana neutrinos. This seems to be the only experiment presently available for this purpose. There are two decay modes, i.e., the two-neutrino mode $(\beta\beta)_{2\nu}$,

$$N_A(A, Z) \to N_B(A, Z+2) + e^- + e^- + \bar{\nu}_e + \bar{\nu}_e\,, \tag{1·1}$$

and the neutrinoless mode $(\beta\beta)_{0\nu}$,

$$N_A(A, Z) \to N_B(A, Z+2) + e^- + e^-\,. \tag{1·2}$$

The $(\beta\beta)_{0\nu}$ mode is interesting because this takes place only when neutrinos are Majorana particles, while the $(\beta\beta)_{2\nu}$ mode occurs both for Dirac and Majorana neutrinos.

We analyze the $0^+ \to J^+$ nuclear transitions for these two modes in two mechanisms: One is the two-nucleon ($2n$)-mechanism where the successive transitions of two neutrons (n_1 and n_2) trigger the double β decay as shown in Fig. 1. The other is the N^*-mechanism[1)] where the double β decay occurs through the transitions*) of the nuclei involving $\Delta(1232)$ as shown in Fig. 2. The detailed discussion for these mechanisms is given in Appendix A.

In the following, we use the weak interaction Hamiltonian,

$$H_W(x) = \frac{G_F}{\sqrt{2}}[j_{L\mu}^-(x) J_L^{+\mu}(x) + \lambda j_{R\mu}^-(x) J_R^{+\mu}(x)] + \text{h.c.}, \qquad (1\cdot 3)$$

where $J_{L(R)}^{+\mu}(x)$ is the left (right)-handed hadronic current and the leptonic currents $j_{L\mu}^-(x)$ and $j_{R\mu}^-(x)$ are expressed as follows:

$$j_{L\mu}^-(x) = \bar{e}\gamma_\mu(1-\gamma_5)\nu_{eL}, \qquad j_{R\mu}^-(x) = \bar{e}\gamma_\mu(1+\gamma_5)\nu'_{eR}. \qquad (1\cdot 4)$$

The current neutrinos ν_{eL} and ν'_{eR} are assumed to be the superposition of the mass eigenstate neutrino N_j's with the corresponding mass m_j's,[2)]

$$\nu_{eL} = \sum_{j=1}^{2n} U_{ej} N_{jL}, \qquad \nu'_{eR} = \sum_{j=1}^{2n} V_{ej} N_{jR}, \qquad (1\cdot 5)$$

where n is the number of generations and U_{ej} (V_{ej}) is the left (right)-mixing matrix.**)

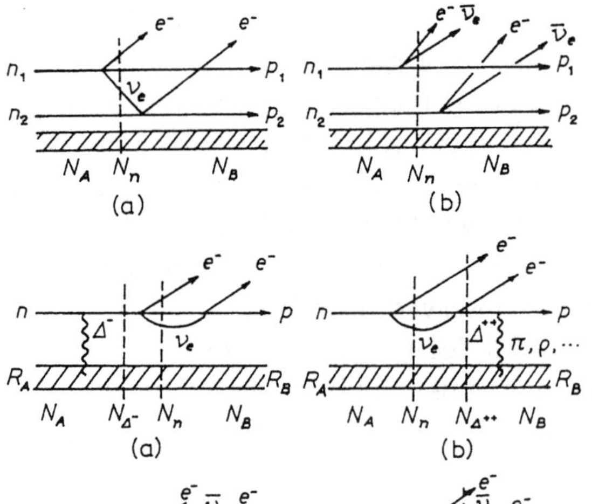

Fig. 1. The schematic diagrams for the $(\beta\beta)_{0\nu}$ mode (a) and for the $(\beta\beta)_{2\nu}$ mode (b) in *the 2n-mechanism*. The N_A, N_B and N_n are the parent, the daughter and the intermediate nucleus, respectively.

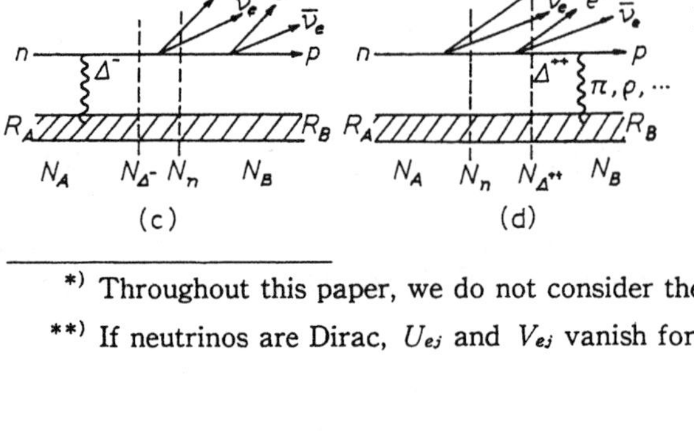

Fig. 2. The diagrams for the $(\beta\beta)_{0\nu}$ mode ((a) and (b)) and for the $(\beta\beta)_{2\nu}$ mode ((c) and (d)) in *the N^*-mechanism*. The N_{Δ^-} and $N_{\Delta^{++}}$ denote the intermediate nuclear states including Δ^- and Δ^{++}, respectively.

*) Throughout this paper, we do not consider the radial and orbital excitations of $\Delta(1232)$.

**) If neutrinos are Dirac, U_{ej} and V_{ej} vanish for $j > n$.

In the $2n$-mechanism, the hadronic currents may be written as

$$J_L^{+\mu}(x)=\bar{\psi}_N\tau^+\gamma^\mu(g_V-g_A\gamma_5)\psi_N\,,\qquad J_R^{+\mu}(x)=\bar{\psi}_N\tau^+\gamma^\mu(g_V'+g_A'\gamma_5)\psi_N\,,\qquad (1\cdot6)$$

where $\psi_N{}^T=(p,\ n)$ and τ^+ is the isospin raising matrix. Here $g_V=\cos\theta$ and $g_V=\cos\theta'$, where $\theta(\theta')$ is the left (right)-mixing angle between u and d quarks. We expect $\theta\simeq\theta_c$ where θ_c is the Cabbibo angle. Also,

$$g_A/g_V=g_A'/g_V'\simeq1.24\qquad (1\cdot7)$$

is expected. This deviation from unity is due to the strong interaction renormalization.

In the N^*-mechanism, the hadronic currents are considered to act on quarks in a hadron and may be obtained by replacing $\psi_N{}^T$ with $\psi_q{}^T=(u,\ d)$ and taking $g_A/g_V=g_A'/g_V'=1$. The effect from the strong interaction will be taken into account by evaluating the matrix elements in the $SU(6)$ quark model.

Many works have been made on the double β decay.[1),3)~9)] However, the structure of H_W used here is somewhat different from the ones previously used. Our results will be compared with those obtained by others only in some special limits like zero neutrino mass ($m_\nu=0$) or $\lambda=0$. Since there are some disagreements between our results and the previous theoretical estimates, we shall show the derivations in some detail. Also, in Appendix C, the decay formulae are presented for the conventionally used Hamiltonian to show how different they are in two H_W's.

In § 2, the $(\beta\beta)_{0\nu}$ mode is investigated for the $0^+\to J^+$ transitions both in the $2n$- and N^*-mechanism. The $(\beta\beta)_{2\nu}$ mode is analyzed in § 3. The concluding remarks are given in § 4.

§ 2. The $(\beta\beta)_{0\nu}$ mode

In the $2n$-mechanism, the double β decay occurs through the transition of two neutrons in N_A to two protons in N_B as shown in Fig. 1(a). On the other hand, there are two possible processes in the N^*-mechanism: (i) The strong interaction first creates the nuclear state N_{Δ^-} which includes Δ^-. Subsequently N_{Δ^-} makes the weak decay as shown in Fig. 2(a). (ii) The parent nucleus N_A makes the weak decay to the nuclear state $N_{\Delta^{++}}$ which is converted into the daughter nucleus N_B as shown in Fig. 2(b). Since the strong interaction operates as internal force, $N_{\Delta^-}(N_{\Delta^{++}})$ has the same J^P as $N_A(N_B)$ has. Therefore, we are able to single out the second order weak interaction part from Figs. 2(a) and (b), and make a unified treatment both for the $2n$- and N^*-mechanism.

Let us consider the second order weak interaction parts in Figs. 1(a), 2(a) and (b) which are collectively expressed by the $N_\alpha\to N_\beta+2e^-$ transition. The weak interaction patterns of our Hamiltonian are given in Fig. 3. If two lepton

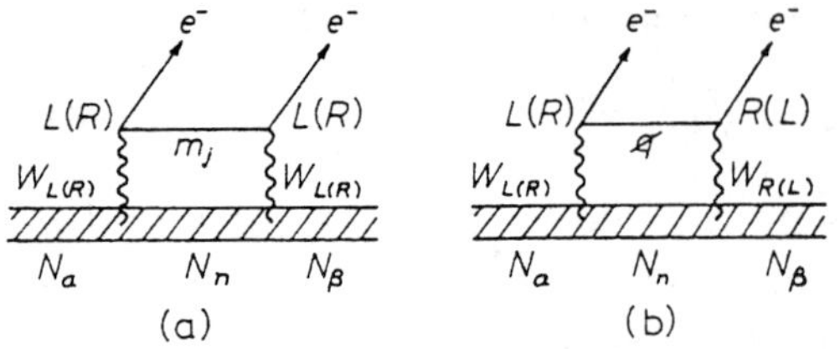

Fig. 3. The diagrams for the $(\beta\beta)_{0\nu}$ mode corresponding to the second order weak interactions. The wavy line represents the weak intermediate boson which controls the left- or right-handed weak interaction.

vertices are either combination of (L, L) or (R, R) as shown in Fig. 3(a), the contribution from this diagram is proportional to the mass m_j of the intermediate Majorana neutrino. When two vertices are (L, R) or (R, L) as in Fig. 3(b), the contribution is proportional to the neutrino four momentum q and the relative strength λ. This situation can be easily seen from the neutrino propagators given in Appendix B. Thus, the $(\beta\beta)_{0\nu}$ mode takes place only if neutrinos are Majorana and at least one of two parameters, m_j and λ, does not vanish.*)

From the above consideration, we can write the R-matrix element for $N_a \to N_\beta + e^-(p_1) + e^-(p_2)$,

$$R_W = \frac{1}{\sqrt{2}}\left(\frac{G_F}{\sqrt{2}}\right)^2 [(2\pi)^{-6}(p_1{}^0 p_2{}^0)^{-1} F(Z+2, p_1{}^0) F(Z+2, p_2{}^0)]^{1/2}$$
$$\times 2\sum_j \{ m_j [U_{ej}^2 t_{\mu\nu}^L K_{LL}^{\mu\nu} + \lambda^2 V_{ej}^2 t_{\mu\nu}^R K_{RR}^{\mu\nu}] + \lambda U_{ej} V_{ej} [u_{\mu\nu\rho}^L L_{LR}^{\mu\nu\rho} + u_{\mu\nu\rho}^R L_{RL}^{\mu\nu\rho}]\}, \tag{2·1}$$

where, by using $u^c = C\bar{u}^T$ with the charge conjugation matrix C,

$$t_{\mu\nu}^{L,R} = \bar{u}(p_1)\gamma_\mu(1 \mp \gamma_5)\gamma_\nu u^C(p_2), \tag{2·2}$$

$$u_{\mu\nu\rho}^{L,R} = \bar{u}(p_1)\gamma_\mu(1 \mp \gamma_5)\gamma_\rho\gamma_\nu u^C(p_2), \tag{2·3}$$

$$K_{ab}^{\mu\nu} = \int d\boldsymbol{x} d\boldsymbol{y} e^{-i(\boldsymbol{p}_1\boldsymbol{x} + \boldsymbol{p}_2\boldsymbol{y})} \int \frac{d\boldsymbol{q}}{2(2\pi)^3 q^0} e^{i\boldsymbol{q}(\boldsymbol{x}-\boldsymbol{y})}$$
$$\times \langle N_\beta | \sum_n \left\{ \frac{J_a^{+\mu}(\boldsymbol{x})|N_n\rangle\langle N_n|J_b^{+\nu}(\boldsymbol{y})}{q^0 + E_n - E_a + p_2{}^0} + \frac{J_b^{+\nu}(\boldsymbol{y})|N_n\rangle\langle N_n|J_a^{+\mu}(\boldsymbol{x})}{q^0 + E_n - E_a + p_1{}^0} \right\} |N_a\rangle, \tag{2·4}$$

$$L_{ab}^{\mu\nu\rho} = \int d\boldsymbol{x} d\boldsymbol{y} e^{-i(\boldsymbol{p}_1\boldsymbol{x} + \boldsymbol{p}_2\boldsymbol{y})} \int \frac{d\boldsymbol{q}}{2(2\pi)^3 q^0} q^\rho$$
$$\times \langle N_\beta | \sum_n \left\{ e^{i\boldsymbol{q}(\boldsymbol{x}-\boldsymbol{y})} \frac{J_a^{+\mu}(\boldsymbol{x})|N_n\rangle\langle N_n|J_b^{+\nu}(\boldsymbol{y})}{q^0 + E_n - E_a + p_2{}^0} \right.$$
$$\left. - e^{-i\boldsymbol{q}(\boldsymbol{x}-\boldsymbol{y})} \frac{J_b^{+\nu}(\boldsymbol{y})|N_n\rangle\langle N_n|J_a^{+\mu}(\boldsymbol{x})}{q^0 + E_n - E_a + p_1{}^0} \right\} |N_a\rangle. \tag{2·5}$$

---

*) In the $m_j = 0$ limit, R_W is proportional to $\lambda \sum U_{ej} V_{ej}$. If we take the gauge theories seriously, we should take $U_{e1} = V_{e,n+1} = 1$ and zero for others. Thus, the $(\beta\beta)_{0\nu}$ mode does not take place. However we keep the possibility $V_{e1} \neq 0$ on the phenomenological basis in order to compare our results with the previous works.

Here the first $1/\sqrt{2}$ in Eq. (2·1) is the statistical factor for the emitted two electrons, $a(b)$ takes L and R, and N_n is the intermediate nuclear state with the energy E_n. The Fermi factor for the emitted electrons is approximated by

$$F(Z, p^0)=(p^0/|\boldsymbol{p}|)2\pi\alpha Z[1-\exp(-2\pi\alpha Z)]^{-1}\,. \tag{2·6}$$

It should be noted that in the $2n$-mechanism the R-matrix is obtained by taking $N_\alpha=N_A$ and $N_\beta=N_B$ in the R_W-matrix. In the N^*-mechanism, the R_W-matrix which corresponds to the 2nd order perturbation of H_W is a part of the R-matrix, as given in Eq. (A·3) of Appendix A.

Now we adopt the following approximations: (i) The energy of the intermediate nucleus E_n is replaced by the average value $\langle E_n\rangle$. (ii) The non-relativistic impulse approximation is used for the hadronic currents $J_L^{+\mu}(\boldsymbol{x})$ and $J_R^{+\mu}(\boldsymbol{x})$. (iii) The first two terms of the multipole expansion for the lepton wave function are kept; $\exp[-i(\boldsymbol{p}_1\boldsymbol{x}+\boldsymbol{p}_2\boldsymbol{y})]\simeq 1-i(\boldsymbol{p}_1\boldsymbol{x}+\boldsymbol{p}_2\boldsymbol{y})$.

Under the approximation (i), the intermediate nuclear states can be summed by closure. By the approximation (ii), the hadronic currents may be expressed as follows:

$$J_L^{+\mu}(\boldsymbol{x})=\sum_n \tau_n{}^+(g_V g^{\mu 0}1_n+g_A g^{\mu j}\sigma_n{}^j)\delta(\boldsymbol{x}-\boldsymbol{r}_n), \tag{2·7}$$

where the subscript n implies that the operators act on the n-th nucleon in the $2n$-mechanism or the n-th quark in the N^*-mechanism. Note that for the parity conserving $0^+\to J^+$ transitions, the first term in the multipole expansion contributes to $K_{ab}^{\mu\nu}$ and $L_{ab}^{\mu\nu 0}$ terms, while the dipole term to $L_{ab}^{\mu\nu k}$.

Within these approximations, the q-integrations in $K_{ab}^{\mu\nu}$ and $L_{ab}^{\mu\nu\rho}$ can be formally performed and the results are

$$K_{LL}^{\mu\nu}=\frac{1}{8\pi}[\langle H_1(r,m_j)\rangle+\langle H_2(r,m_j)\rangle]$$
$$\times\langle N_\beta|\sum_{n,m}\tau_n{}^+\tau_m{}^+(g_V g^{\mu 0}+g_A\sigma_n{}^j g^{\mu j})(g_V g^{\nu 0}+g_A\sigma_m{}^k g^{\nu k})|N_\alpha\rangle\,, \tag{2·8}$$

$$L_{LR}^{\mu\nu 0}=\frac{1}{8\pi}[A_1\langle H_1(r,m_j)\rangle-A_2\langle H_2(r,m_j)\rangle]$$
$$\times\langle N_\beta|\sum_{n,m}\tau_n{}^+\tau_m{}^+(g_V g^{\mu 0}+g_A\sigma_n{}^j g^{\mu j})(g_V' g^{\nu 0}-g_A'\sigma_m{}^k g^{\nu k})|N_\alpha\rangle\,, \tag{2·9}$$

$$L_{LR}^{\mu\nu k}=\frac{-1}{8\pi}[\langle rH_1'(r,m_j)\rangle+\langle rH_2'(r,m_j)\rangle]$$
$$\times\langle N_\beta|\sum_{n,m}\tau_n{}^+\tau_m{}^+\frac{1}{2}[(p_1-p_2)^l\hat{r}_{nm}^l+(p_1+p_2)^l\hat{r}_{+nm}^l]\hat{r}_{nm}^k$$
$$\times(g_V g^{\mu 0}+g_A\sigma_n{}^p g^{\mu p})(g_V' g^{\nu 0}-g_A'\sigma_m{}^q g^{\nu q})|N_\alpha\rangle$$
$$-\frac{i}{8\pi}[\langle H_1'(r,m_j)\rangle+\langle H_2'(r,m_j)\rangle]$$
$$\times\langle N_\beta|\sum_{n,m}\tau_n{}^+\tau_m{}^+(g_V g^{\mu p}D_n{}^p+g_A g^{\mu 0}C_n)(g_V' g^{\nu q}D_m{}^q-g_A' g^{\nu 0}C_m)|N_\alpha\rangle\,. \tag{2·10a}$$

We note that only the first term of the multipole expansion is taken into account for $K_{LL}^{\mu\nu}$ and $L_{LR}^{\mu\nu 0}$, and the dipole term is used to obtain the first term of $L_{LR}^{\mu\nu k}$. In order to maintain the consistency of the approximation, the relativistic correction of the hadronic current should be included. The second term of $L_{LR}^{\mu\nu k}$ (2·10a) is due to this correction. Here C_n and $\boldsymbol{D}_n$ are defined by

$$C_n = \boldsymbol{\sigma}_n \cdot (\boldsymbol{q}_n - 2\boldsymbol{P}_n)/2M\ , \tag{2·10b}$$

$$\boldsymbol{D}_n = [(\boldsymbol{q}_n - 2\boldsymbol{P}_n) + i(\boldsymbol{\sigma}_n \times \boldsymbol{q}_n)]/2M\ , \tag{2·10c}$$

where $\boldsymbol{P}$ and $\boldsymbol{q}$ are the momentum and the momentum transfer of the nucleon, respectively.[11] In Eqs. (2·8)~(2·10), $H_i(r, m_j)$ is the potential-like term due to the exchange of neutrino and is defined as

$$H_i(|\boldsymbol{r}_n - \boldsymbol{r}_m|, m_j) = \int \frac{d\boldsymbol{q}}{2\pi^2} \frac{e^{i\boldsymbol{q}\cdot(\boldsymbol{r}_n - \boldsymbol{r}_m)}}{q^0(q^0 + A_i)}\ , \tag{2·11}$$

where $q^0 = \sqrt{|\boldsymbol{q}|^2 + m_j^2}$ and $A_i = \langle E_n \rangle - M_a + p_i^0$. Also, $H_i' = dH_i/dr$, $\boldsymbol{r}_{nm} = \boldsymbol{r}_n - \boldsymbol{r}_m$, $\hat{\boldsymbol{r}}_{nm} = \boldsymbol{r}_{nm}/|\boldsymbol{r}_{nm}|$, and $\hat{\boldsymbol{r}}_{+nm} = (\boldsymbol{r}_n + \boldsymbol{r}_m)/|\boldsymbol{r}_{nm}|$. The terms like $\langle H_i \rangle$ and $\langle rH_i' \rangle$ represent the average values of "potentials" with the weight of nuclear tensor operators.*) Note that the potential $H_i(r, m_j)$ behaves like $1/r$ for $m_j \lesssim O(\text{MeV})$ and $e^{-m_j r}/r$ for $m_j \gtrsim O(\text{GeV})$. The replacement E_n by $\langle E_n \rangle$ (the approximation (i)) is not crucial because the main contribution to the potentials comes from $|\boldsymbol{q}| \geq 20$ MeV which is much larger than A_i ($\simeq$a few MeV). The other terms, $K_{RR}^{\mu\nu}$, $L_{RL}^{\mu\nu 0}$ and $L_{RL}^{\mu\nu k}$, are obtained by taking the interchanges$(g_V \leftrightarrow g_V')and(g_A \leftrightarrow -g_A')$in the expressions of $K_{LL}^{\mu\nu}$, $L_{LR}^{\mu\nu 0}$ and $L_{LR}^{\mu\nu k}$, respectively.

The product of the leptonic and hadronic parts can be easily calculated and the results are as follows:

$$\begin{Bmatrix} t_{\mu\nu}^L & K_{LL}^{\mu\nu} \\ t_{\mu\nu}^R & K_{RR}^{\mu\nu} \end{Bmatrix} = \frac{1}{8\pi}[\langle H_1 \rangle + \langle H_2 \rangle] \bar{u}(p_1)(1 \pm \gamma_5) u^C(p_2) \begin{Bmatrix} g_A^2 \\ g_A'^2 \end{Bmatrix} \left[\left(\frac{g_V}{g_A} \right)^2 M_F - M_{GT} \right], \tag{2·12}$$

$$u_{\mu\nu 0}^L L_{LR}^{\mu\nu 0} + u_{\mu\nu 0}^R L_{RL}^{\mu\nu 0} = \frac{1}{8\pi}[A_1 \langle H_1 \rangle - A_2 \langle H_2 \rangle] \times \bar{u}(p_1) \gamma^0 u^C(p_2) g_A g_A' [(g_V/g_A)^2 M_F - M_{GT}], \tag{2·13}$$

$$\begin{aligned} u_{\mu\nu k}^L L_{LR}^{\mu\nu k} + u_{\mu\nu k}^R L_{RL}^{\mu\nu k} = &\frac{-1}{8\pi}[\langle rH_1' \rangle + \langle rH_2' \rangle] \\ &\times g_A g_A' \bar{u}(p_1)[\gamma^k (p_1 - p_2)^l P^{lk} + \gamma_5 \gamma^0 (p_1 + p_2)^l Q^l] u^C(p_2) \\ &+ \frac{i}{2\pi}[\langle H_1' \rangle + \langle H_2' \rangle] g_A g_A' \bar{u}(p_1) \gamma_5 \gamma^k u^C(p_2) R^k\ , \end{aligned} \tag{2·14}$$

---

*) The average is defined as $\langle f \rangle = \langle N_B | \Sigma_{n,m} f O_{nm} | N_A \rangle / \langle N_B | \Sigma_{n,m} O_{nm} | N_A \rangle$, where O_{nm} is the nuclear tensor operator.

where

$$M_F = \langle N_\beta | \sum_{n,m} \tau_n^+ \tau_m^+ | N_\alpha \rangle \,, \tag{2·15}$$

$$M_{GT} = \langle N_\beta | \sum_{n,m} \tau_n^+ \tau_m^+ \boldsymbol{\sigma}_n \cdot \boldsymbol{\sigma}_m | N_\alpha \rangle \,, \tag{2·16}$$

$$Q^l = \langle N_\beta | \sum_{n,m} \tau_n^+ \tau_m^+ \hat{r}^l_{+nm} \left[2\left(\frac{g_V}{g_A}\right)(\hat{r}_{nm} \cdot \boldsymbol{\sigma}_n) - i \hat{r}_{nm} \cdot (\boldsymbol{\sigma}_n \times \boldsymbol{\sigma}_m) \right] | N_\alpha \rangle \,, \tag{2·17}$$

$$P^{lk} = \langle N_\beta | \sum_{n,m} \tau_n^+ \tau_m^+ \hat{r}^l_{nm} \{ \hat{r}^k_{nm} [(g_V/g_A)^2 + \boldsymbol{\sigma}_n \cdot \boldsymbol{\sigma}_m]$$
$$+ 2i(g_V/g_A)(\hat{r}_{nm} \times \boldsymbol{\sigma}_n)^k - 2\sigma_m^k(\hat{r}_{nm} \cdot \boldsymbol{\sigma}_n) \} | N_\alpha \rangle \,, \tag{2·18a}$$

$$R^k = \langle N_\beta | \sum_{n,m} \tau_n^+ \tau_m^+ \left\{ i(\hat{r}_{nm} \times \boldsymbol{\sigma}_m)^k C_n + i\left(\frac{g_V}{g_A}\right)^2 (\hat{r}_{nm} \times \boldsymbol{D}_n)^k \right.$$
$$\left. + \left(\frac{g_V}{g_A}\right) [\hat{r}^k_{nm}(C_n - \boldsymbol{\sigma}_n \cdot \boldsymbol{D}_m) + \sigma_n^k (\hat{r}_{nm} \cdot \boldsymbol{D}_m) + (\hat{r}_{nm} \cdot \boldsymbol{\sigma}_n) D_m^k \right\} | N_\alpha \rangle \,. \tag{2·18b}$$

It is clear from the nuclear matrix elements given above that the $0^+ \to J^+$ ($J \geqq 3$) transitions are forbidden within our approximations.

In the following, a further simplification is made by replacing A_i in the potentials with their average value, i.e.,

$$\mu_0 = (A_1 + A_2)/2m_e = [\langle E_n \rangle - (M_A + M_B)/2]/m_e \,. \tag{2·19}$$

Table I. Allowed transitions and relative order of magnitudes. Each parenthesis under the interaction indicates the possible transition, and m_ν and λ represent the typical neutrino mass and the relative strength of the right-handed interaction, respectively.

$(\beta\beta)_{0\nu}$ mode					
interaction patterns	2*n*-mechanism		N^*-mechanism		Feynman diagrams
	$m_\nu \lesssim O$(ev)	$m_\nu \gtrsim O$(GeV)	$m_\nu \lesssim O$(eV)	$m_\nu \gtrsim O$(GeV)	
$m_\nu K^{\mu\nu}_{LL}$ $(0^+ \to 0^+)$	$m_\nu \left\langle \frac{1}{r} \right\rangle$	$m_\nu \left\langle \frac{e^{-m_\nu r}}{r} \right\rangle$	no contribution		Fig. 3(a)
$\lambda^2 m_\nu K^{\mu\nu}_{RR}$ $(0^+ \to 0^+)$	$\lambda^2 m_\nu \left\langle \frac{1}{r} \right\rangle$	$\lambda^2 m_\nu \left\langle \frac{e^{-m_\nu r}}{r} \right\rangle$	no contribution		
$\lambda L^{\mu\nu 0}_{ab}$ $(0^+ \to 0^+)$	$\lambda(p_1{}^0 - p_2{}^0) \left\langle \frac{1}{r} \right\rangle$	$\lambda(p_1{}^0 - p_2{}^0) \left\langle \frac{e^{-m_\nu r}}{r} \right\rangle$	no contribution		Fig. 3(b)
$\lambda L^{\mu\nu k}_{ab}$ $(0^+ \to 0^+, 1^+, 2^+)$	$\lambda \lvert \boldsymbol{p}_1 \pm \boldsymbol{p}_2 \rvert \left\langle \frac{1}{r} \right\rangle$	$\lambda \lvert \boldsymbol{p}_1 \pm \boldsymbol{p}_2 \rvert \left\langle \frac{e^{-m_\nu r}}{r} \right\rangle$	$\lambda \lvert \boldsymbol{p}_1 \pm \boldsymbol{p}_2 \rvert \left\langle \frac{1}{r} \right\rangle_\Delta$	$\lambda \lvert \boldsymbol{p}_1 \pm \boldsymbol{p}_2 \rvert \left\langle \frac{e^{-m_\nu r}}{r} \right\rangle_\Delta$	

As a consequence, we can write H_i's in the single form,

$$H_1(r, m_j) = H_2(r, m_j) \equiv H(r, m_j, \mu_0). \tag{2·20}$$

(2-a) *The 2n-mechanism*

The R-matrix is obtained from R_W in Eq. (2·1) by replacing N_α and N_β with N_A and N_B, respectively. It should be noted that the M_F and M_{GT} terms are of rank 0 with respect to the angular momentum, and the Q^l term of rank 1. On the other hand, the P^{lk} term consists of irreducible tensor operators of ranks 0, 1 and 2. Consequently, the terms $t^L_{\mu\nu}K^{\mu\nu}_{LL}$, $t^R_{\mu\nu}K^{\mu\nu}_{RR}$ and $u^L_{\mu\nu 0}L^{\mu\nu 0}_{LR} + u^R_{\mu\nu 0}L^{\mu\nu 0}_{RL}$ contribute only to the $0^+ \to 0^+$ transition. While, $u^L_{\mu\nu k}L^{\mu\nu k}_{LR} + u^R_{\mu\nu k}L^{\mu\nu k}_{RL}$ contributes to the $0^+ \to 0^+$, 1^+, and 2^+ transitions. These features and the relative order of magnitudes are listed in Table I for various types of terms.

(i) The $0^+ \to 0^+$ transition

The nuclear matrix elements M_F, M_{GT} and $P^{ii}/3$ contribute to this transition. We obtain

$$\begin{aligned} d\Gamma^{2n}_{0\nu}(0^+ \to 0^+) = {} & (a_{0\nu}/m_e^7)\, g_A^4 \\ & \times \{(p_1^0 p_2^0 - \mathbf{p}_1 \cdot \mathbf{p}_2)[|X_1 + X_4|^2 + |X_2 + X_4|^2] \\ & + \frac{1}{2m_e^2}(p_1^0 - p_2^0)^2 (p_1^0 p_2^0 + \mathbf{p}_1 \cdot \mathbf{p}_2 - m_e^2)|X_3 - X_4|^2 \\ & - 2m_e^2 \,\mathrm{Re}(X_1 + X_4)(X_2 + X_4)^* \\ & - (p_1^0 - p_2^0)^2 \,\mathrm{Re}(X_1 + X_2 + 2X_4)(X_3 - X_4)^*\} \\ & \times p_1^0 p_2^0 \delta(p_1^0 + p_2^0 + M_B - M_A)\, d\cos\theta\, dp_1^0\, dp_2^0 , \end{aligned} \tag{2·21}$$

where θ is the angle between two emitted electrons,

$$a_{0\nu} = \frac{G_F^2 m_e^9}{2(2\pi)^5} \left\{ \frac{2\pi\alpha(Z+2)}{1 - \exp[-2\pi\alpha(Z+2)]} \right\}^2 , \tag{2·22}$$

$$X_1 = \sum_j (m_j/m_e)\, U^2_{ej} \langle H \rangle [(g_V/g_A)^2 M_F - M_{GT}], \tag{2·23}$$

$$X_2 = \lambda^2 \sum_j (m_j/m_e)\, V^2_{ej} \langle H \rangle (g_V'/g_V)^2 [(g_V/g_A)^2 M_F - M_{GT}], \tag{2·24}$$

$$X_3 = \lambda \sum_j U_{ej} V_{ej} \langle H \rangle (g_V'/g_V) [(g_V/g_A)^2 M_F - M_{GT}], \tag{2·25}$$

$$X_4 = \lambda \sum_j U_{ej} V_{ej} \langle rH' \rangle (g_V'/3g_V) \left[(g_V/g_A)^2 M_F + \frac{1}{3} M_{GT} - 2M_T \right]. \tag{2·26}$$

Here M_F and M_{GT} are defined in Eqs. (2·15) and (2·16), and

$$M_T = \langle N_B | \sum_{n,m} \tau_n^+ \tau_m^+ \left[(\hat{r}_{nm} \cdot \boldsymbol{\sigma}_n)(\hat{r}_{nm} \cdot \boldsymbol{\sigma}_m) - \frac{1}{3} \boldsymbol{\sigma}_n \cdot \boldsymbol{\sigma}_m \right] | N_A \rangle . \tag{2·27}$$

The decay rate for the $0^+\to 0^+$ transition in the $2n$-mechanism is

$$\Gamma_{0\nu}^{2n}(0^+\to 0^+)=(a_{0\nu}/m_e^2)g_A^4\{G_{01}(T)[|X_1+X_4|^2+|X_2+X_4|^2]$$
$$+G_{02}(T)|X_3-X_4|^2-G_{03}(T)\mathrm{Re}(X_1+X_4)(X_2+X_4)^*$$
$$-G_{04}(T)\mathrm{Re}(X_1+X_2+2X_4)(X_3-X_4)^*\}, \qquad (2\cdot28)$$

where

$$G_{01}(T)=\frac{1}{15}T(T^4+10T^3+40T^2+60T+30), \qquad (2\cdot29)$$

$$G_{02}(T)=\frac{1}{210}T^4(T^3+14T^2+77T+70), \qquad (2\cdot30)$$

$$G_{03}(T)=\frac{2}{3}T(T^2+6T+6), \qquad (2\cdot31)$$

$$G_{04}(T)=\frac{1}{15}T^3(T^2+10T+10). \qquad (2\cdot32)$$

Here T is the maximum kinetic energy release,

$$T=(M_A-M_B-2m_e)/m_e\,. \qquad (2\cdot33)$$

Let us compare our results with those obtained previously. In the limit of $\lambda=0$ and $U_{ej}=\delta_{j1}$, we found that the overall normalization by Greuling and Whitten[5)] is twice as large as ours.[*)] In the other limit of $m_j=0$, our results can be compared with those by Primakoff and Rosen[4)] who used the quite general form for H_W.[**)] They assumed $M_T=0$ and their result is twice as large as ours.[*)]

(ii) The $0^+\to 2^+$ transition

Only the rank 2 part of the P^{lk} term contributes to this transition. The final result is

$$d\Gamma_{0\nu}^{2n}(0^+\to 2^+)=(a_{0\nu}/m_e^9)(g_A^2g_A'^2/30)|\lambda\sum_j U_{ej}V_{ej}\langle rH'\rangle|^2(N_2^{pq},N_2^{pq})$$
$$\times\{3(\boldsymbol{p}_1\cdot\boldsymbol{p}_2)^2-\boldsymbol{p}_1\cdot\boldsymbol{p}_2[10(p_1^0p_2^0+m_e^2)+|\boldsymbol{p}_1|^2+|\boldsymbol{p}_2|^2]$$
$$+5(p_1^0p_2^0+m_e^2)(|\boldsymbol{p}_1|^2+|\boldsymbol{p}_2|^2)-|\boldsymbol{p}_1|^2|\boldsymbol{p}_2|^2\}$$
$$\times p_1^0p_2^0\delta(p_1^0+p_2^0+M_B-M_A)d\cos\theta dp_1^0dp_2^0\,, \qquad (2\cdot34)$$

where $a_{0\nu}$ is defined in Eq. (2·22), $(N_2^{pq},N_2^{pq})\equiv\sum_{J_z}\sum_{p,q=1}^3 N_2^{pq*}N_2^{pq}$ with

*) It seems that they did not take account of the statistical factor. See, e.g., Eq. (20) of Ref. 5) and Eq. (33) of Ref. 4).

**) The correspondence between our notation and theirs in Ref. 4) is as follows: $C_a=D_a=G_F(g_a+\lambda g_a')/2$, $C_a\delta_a=D_a\delta_a=-G_F(g_a-\lambda g_a')/2$, for $a=V$ and A, and $\Gamma_V=\gamma_\mu$, $\Gamma_A=\gamma_\mu\gamma_5$, $U_{ej}=V_{ej}=\delta_{j1}$. It is also reminded that the result by Molina and Pascual[9)] is four times as large as ours.

$$N_2^{pq}=\langle N_B(2^+)|\sum_{n,m}\tau_n^+\tau_m^+\,\hat{r}_{nm}^p\{\hat{r}_{nm}^q[(g_V/g_A)^2+\boldsymbol{\sigma}_n\cdot\boldsymbol{\sigma}_m]$$
$$+2i(g_V/g_A)(\hat{r}_{nm}\times\boldsymbol{\sigma}_n)^q-2\sigma_m^q(\hat{r}_{nm}\cdot\boldsymbol{\sigma}_n)\}|N_A(0^+)\rangle\,. \qquad (2\cdot35)$$

The decay rate is

$$\Gamma_{0\nu}^{2n}(0^+\to 2^+)=(a_{0\nu}/m_e^2)(g_Ag_A')^2|\lambda\sum_j U_{ej}V_{ej}\langle rH'\rangle|^2\,(N_2^{pq},N_2^{pq})\,G_{21}(T), \qquad (2\cdot36)$$

where

$$G_{21}(T)=\frac{2^3}{7!}T^2(4T^5+56T^4+343T^3+1050T^2+1540T+840). \qquad (2\cdot37)$$

(iii) The $0^+\to1^+$ transition

The terms Q^l, $\varepsilon_{lkj}P^{kj}$ and R^l contribute to this transition. After small calculations, the decay formula is

$$d\Gamma_{0\nu}^{2n}(0^+\to1^+)=(a_{0\nu}/m_e^9)(g_A^2g_A'^2/6)|\lambda\sum_j U_{ej}V_{ej}|^2$$
$$\times[\{(\boldsymbol{p}_1+\boldsymbol{p}_2)^2(p_1^0p_2^0+\boldsymbol{p}_1\cdot\boldsymbol{p}_2+m_e^2)(N_1^q,N_1^q)$$
$$+2[(\boldsymbol{p}_1-\boldsymbol{p}_2)^2(p_1^0p_2^0+m_e^2)+(\boldsymbol{p}_1\cdot\boldsymbol{p}_2-|\boldsymbol{p}_1|^2)(\boldsymbol{p}_1\cdot\boldsymbol{p}_2-|\boldsymbol{p}_2|^2)](N_2^q,N_2^q)$$
$$+4(\boldsymbol{p}_1\times\boldsymbol{p}_2)^2\,\mathrm{Im}(N_1^q,N_2^q)\}\langle rH'\rangle^2$$
$$-8\{2[(p_1^0+p_2^0)\boldsymbol{p}_1\cdot\boldsymbol{p}_2-(p_1^0|\boldsymbol{p}_2|^2+p_2^0|\boldsymbol{p}_1|^2)\mathrm{Re}(N_2^q,R^q)$$
$$-[(p_1^0+p_2^0)\boldsymbol{p}_1\cdot\boldsymbol{p}_2+(p_1^0|\boldsymbol{p}_2|^2+p_2^0|\boldsymbol{p}_1|^2)\mathrm{Im}(N_1^q,R^q)\}\langle H'\rangle\langle rH'\rangle$$
$$+16\{3(p_1^0p_2^0-m_e^2)-\boldsymbol{p}_1\cdot\boldsymbol{p}_2\}(R^q,R^q)\langle H'\rangle^2]$$
$$\times p_1^0p_2^0\delta(p_1^0+p_2^0+M_B-M_A)\,d\cos\theta\, dp_1^0\,dp_2^0\,, \qquad (2\cdot38)$$

where $(N_i^q,N_j^q)\equiv\sum_{J_z}\sum_{q=1}^3 N_i^{q*}N_j^q$ with

$$N_1^q=\langle N_B(1^+)|\sum_{n,m}\tau_n^+\tau_m^+\,[2(g_V/g_A)(\hat{r}_{nm}\cdot\boldsymbol{\sigma}_n)-i\hat{r}_{nm}\cdot(\boldsymbol{\sigma}_n\times\boldsymbol{\sigma}_m)]\,\hat{r}_{+nm}^q|N_A(0^+)\rangle\,, \qquad (2\cdot39)$$

$$N_2^q=\langle N_B(1^+)|\sum_{n,m}\tau_n^+\tau_m^+$$
$$\times\{i(g_V/g_A)[\hat{r}_{nm}\times(\hat{r}_{nm}\times\boldsymbol{\sigma}_n)]^q-(\hat{r}_{nm}\cdot\boldsymbol{\sigma}_n)(\hat{r}_{nm}\times\boldsymbol{\sigma}_m)^q\}|N_A(0^+)\rangle\,. \qquad (2\cdot40)$$

The terms N_1^q and N_2^q come from Q^q and $\varepsilon_{qlk}P^{lk}$, respectively.

The decay rate is

$$\Gamma_{0\nu}^{2n}(0^+\to1^+)=(a_{0\nu}/m_e^2)(g_Ag_A')^2|\lambda\sum_j U_{ej}V_{ej}|^2$$
$$\times[\{G_{11}(T)[(N_1^q,N_1^q)+4(N_2^q,N_2^q)+4\,\mathrm{Im}(N_1^q,N_2^q)]$$

$$+G_{12}(T)[(N_1^q, N_1^q)+2(N_2^q, N_2^q)]\}\langle rH'\rangle^2$$
$$+G_{13}(T)(R^q, R^q)\langle H'\rangle^2/m_e^2$$
$$+G_{14}(T)[2\,\mathrm{Re}(N_2^q, R^q)+\mathrm{Im}(N_1^q, R^q)]\langle rH'\rangle\langle H'\rangle/m_e], \tag{2·41}$$

where

$$G_{11}(T)=\frac{2^3}{3\cdot 7!}T^3(3T^4+42T^3+210T^2+420T+280), \tag{2·42}$$

$$G_{12}(T)=\frac{2^3}{7!}T^2(4T^5+56T^4+343T^3+1050T^2+1540T+840), \tag{2·43a}$$

$$G_{13}(T)=\frac{2^6}{5!}T^2(T^3+10T^2+35T+30), \tag{2·43b}$$

$$G_{14}(T)=\frac{2^5}{3\cdot 5!}T^2(T^4+12T^3+55T^2+100T+60). \tag{2·43c}$$

(2-b) *The N^*-mechanism*

The R-matrix for the N^*-mechanism is obtained by substituting the R_W-matrices corresponding to the $N_{\Delta^-}\rightarrow N_B+2e^-$ and $N_A\rightarrow N_{\Delta^{++}}+2e^-$ transitions into Eq. (A·3) in Appendix A.

The nuclear matrix elements in the R_W-matrix are given by Eqs. (2·15) ~(2·18). Since there is no standard method to treat the relativistic correction in the quark model, this correction (the second term in $L_{LR}^{\mu\nu k}$) is discarded. It should be noted that the operators*) $\tau_{n(m)}^{+}$, $\sigma_{n(m)}^{j}$, $\hat{r}'_{nm}$ and $\hat{r}'_{+nm}$ act on quarks and change the intrinsic part of the hadron. The nucleon $N(\frac{1}{2}^+)$ and $\Delta(\frac{3}{2}^+)$ are assigned in the $SU(6)$ quark model to the states $l=0$, i.e., the zero orbital angular momentum states around the center of hadron. From these considerations, we conclude that M_F, M_{GT} and Q^l defined in Eqs. (2·15)~(2·17) turn out to be zero, and only the term P^{lk} in Eq. (2·18) contributes. (See Appendix A for the detailed discussion.) Consequently, the term $u_{\mu\nu k}^{L}L_{LR}^{\mu\nu k}+u_{\mu\nu k}^{R}L_{RL}^{\mu\nu k}$ contributes to the $(\beta\beta)_{0\nu}$ mode in the N^*-mechanism, while $t_{\mu\nu}^{L}K_{LL}^{\mu\nu}$, $t_{\mu\nu}^{R}K_{RR}^{\mu\nu}$ and $u_{\mu\nu 0}^{L}L_{LR}^{\mu\nu 0}+u_{\mu\nu 0}^{R}L_{RL}^{\mu\nu 0}$ vanish.

In summarizing the above discussion, the $(\beta\beta)_{0\nu}$ mode in the N^*-mechanism takes place only when $\lambda\neq 0$, whether neutrinos are massive or massless. These results are listed in Table I. Halprin et al.[7] have derived the bounds of the neutrino mass both in the $2n$- and N^*-mechanism. However, the bounds they obtained in the N^*-mechanism seem to be meaningless, because there is no contribution from the m_ν-term within the N^*-mechanism adopted in the present paper. We would like to emphasize here that the above discussions are independent of the "factorization hypothesis" which will be used later.

*) $\hat{r}'_{nm}$ and $\hat{r}'_{+nm}$ are defined in terms of the position operators of quarks measured from the center of the hadron.

The non-vanishing product of the leptonic and hadronic parts for the $N_{\Delta^-}\rightarrow N_B+2e^-$ transition is

$$u^L_{\mu\nu k}L^{\mu\nu k*}_{LR}+u^R_{\mu\nu k}L^{\mu\nu k*}_{RL}=\frac{1}{6\pi}\langle rH'\rangle_\Delta(g_Ag_A')$$

$$\times(p_1-p_2)^l\bar{u}(p_1)\gamma^k u^C(p_2)[i\epsilon_{lkj}(g_V/g_A)M^j+M^{lk}], \qquad (2\cdot44)$$

where

$$M^j=\langle N_B|\sum_{n,m}\tau_n{}^+\tau_m{}^+\sigma_n{}^j|N_{\Delta^-}\rangle\,, \qquad (2\cdot45)$$

$$M^{jk}=\langle N_B|\sum_{n,m}\tau_n{}^+\tau_m{}^+\sigma_n{}^j\sigma_m{}^k|N_{\Delta^-}\rangle\,. \qquad (2\cdot46)$$

The nuclear tensor operators in Eqs. (2·45) and (2·46) only change the spin and isospin of Δ^- and leave the remainder unchanged. Therefore, M^j and M^{jk} represent essentially the matrix elements between Δ^- and p. Also, the expectation value of the "potential" $\langle rH'\rangle$ should be taken between Δ^- and p. Note that $g_A=g_V$ and $g_A'=g_V'$ should be taken in the N^*-mechanism as explained in § 1.

Now we use the "factorization hypothesis"[2)] (see Eq. (A·10)) and evaluate the decay formula,[*)]

$$d\Gamma^{N^*}_{0\nu}=(a_{0\nu}/m_e{}^9)(2^6/3^3)(g_Vg_V')^2$$

$$\times|\lambda\sum_j U_{ej}V_{ej}\langle rH'\rangle_\Delta|^2P(\Delta)|\langle\Phi_f|\Phi_i\rangle|^2$$

$$\times\{2(\boldsymbol{p}_1\cdot\boldsymbol{p}_2)^2-\boldsymbol{p}_1\cdot\boldsymbol{p}_2[(p_1{}^0+p_2{}^0)^2+4(p_1{}^0p_2{}^0+m_e{}^2)]$$

$$+3(p_1{}^0p_2{}^0+m_e{}^2)(|\boldsymbol{p}_1|^2+|\boldsymbol{p}_2|^2)\}$$

$$\times p_1{}^0p_2{}^0\delta(p_1{}^0+p_2{}^0+M_B-M_A)d\cos\theta dp_1{}^0dp_2{}^0\,. \qquad (2\cdot47)$$

Here $P(\Delta)$ is the probability of producing Δ per nucleon inside the nucleus, and $\langle\Phi_f|\Phi_i\rangle$ represents the overlap between the initial and final nuclei. The $N_A\rightarrow N_{\Delta^{++}}+2e^-$ transition is also included in the above formula. We refer to Appendix A for the detailed discussion.

The decay rate is

$$\Gamma^{N^*}_{0\nu}=(a_{0\nu}/m_e{}^2)(g_Vg_V')^2$$

$$\times|\lambda\sum_j U_{ej}V_{ej}\langle rH'\rangle_\Delta|^2P(\Delta)|\langle\Phi_f|\Phi_i\rangle|^2G^{N^*}(T), \qquad (2\cdot48)$$

*) Here we have used the following results: $\sum\tilde{M}^{j*}\tilde{M}^k=(16/3)\delta_{jk}$, $\sum\tilde{M}^{j*}\tilde{M}^{kl}=0$, $\sum\tilde{M}^{jk*}\tilde{M}^{lm}=16(\delta_{jl}\delta_{km}+\delta_{jm}\delta_{kl}-\frac{2}{3}\delta_{jk}\delta_{lm})$, where $\tilde{M}^j$ and $\tilde{M}^{jk}$ are defined from M^j and M^{jk} by replacing N_{Δ^-} and N_B with Δ^- and p, respectively. The same relations hold for the matrix elements between n and Δ^{++}.

where

$$G^{N^*}(T)=\frac{2^{10}}{3^3\cdot 7!}T^2[39T^5+546T^4+3297T^3+9870T^2+14140T+7560]. \tag{2·49}$$

The above decay formula is applicable to all $0^+\to 0^+$, 1^+ and 2^+ transitions by the appropriate choice of $P(\Delta)|\langle\Phi_f|\Phi_i\rangle|^2$.

§ 3. The $(\beta\beta)_{2\nu}$ mode

In a similar way to the case of the $(\beta\beta)_{0\nu}$ mode, the $0^+\to J^+$ transitions are investigated. In our Hamiltonian in Eq. (1·3), the $(\beta\beta)_{2\nu}$ mode takes place through the process,

$$N_A(p_A)\to N_B(p_B)+e^-(p_1)+e^-(p_2)+\bar{N}_i(k_1)+\bar{N}_j(k_2). \tag{3·1}$$

The contribution from the right-handed interaction is suppressed by λ ($\lambda\ll 1$) so that this is neglected here.

The R_W-matrix due to the $V-A$ interaction for the $N_\alpha\to N_\beta+2e^-+\bar{N}_i+\bar{N}_j$ transition is expressed by

$$R_{Wij}=\frac{\epsilon_{ij}}{\sqrt{2}}\left(\frac{G_F}{\sqrt{2}}\right)^2 U_{ei}U_{ej}$$
$$\times[(2\pi)^{-12}(p_1^0p_2^0k_1^0k_2^0)^{-1}F(Z+2,p_1^0)F(Z+2,p_2^0)]^{1/2}$$
$$\times[E_{\mu\nu}J^{\mu\nu}-(p_1\leftrightarrow p_2)], \tag{3·2}$$

where

$$E_{\mu\nu}=\bar{u}(p_1)\gamma_\mu(1-\gamma_5)u^C(k_1)\bar{u}(p_2)\gamma_\nu(1-\gamma_5)u^C(k_2), \tag{3·3}$$

$$J^{\mu\nu}=\int dxdye^{-i[(p_1+k_1)x+(p_2+k_2)y]}$$
$$\times\langle N_\beta|\sum_n\left\{\frac{J_L^{+\mu}(x)|N_n\rangle\langle N_n|J_L^{+\nu}(y)}{E_n-E_\alpha+p_2^0+k_2^0}+\frac{J_L^{+\nu}(y)|N_n\rangle\langle N_n|J_L^{+\mu}(x)}{E_n-E_\alpha+p_1^0+k_1^0}\right\}|N_\alpha\rangle. \tag{3·4}$$

Here the term $\varepsilon_{ij}/\sqrt{2}$ is the statistical factor for the final two electrons and two neutrinos, i.e., $\varepsilon_{ij}=1/\sqrt{2}$ for $i=j$ and $=1$ for $i\neq j$. The full R-matrix for the $(\beta\beta)_{2\nu}$ mode can be readily obtained in the same manner as for the $(\beta\beta)_{0\nu}$ mode.

Now we use the approximations (i), (ii) and (iii) introduced in § 2. Under these assumptions, the nuclear part $J^{\mu\nu}$ can be simplified. Note that the $0^+\to J^+$ ($J\geqq 3$) transitions are forbidden.

(3-a) *The 2n-mechanism*

The R-matrix is obtained from the R_W-matrix by the replacements, $N_\alpha \to N_A$ and $N_\beta \to N_B$.

(i) The $0^+ \to 0^+$ transition

After straightforward calculations, we obtain

$$d\Gamma_{2\nu}^{2n}(0^+ \to 0^+) = (a_{2\nu}/m_e{}^9)\frac{1}{8}[p_1{}^0 p_2{}^0 C - \boldsymbol{p}_1 \cdot \boldsymbol{p}_2 D]$$
$$\times (k_1{}^0 k_2{}^0)^2 p_1{}^0 p_2{}^0 \delta(p_1{}^0 + p_2{}^0 + k_1{}^0 + k_2{}^0 + M_B - M_A)$$
$$\times d\cos\theta dp_1{}^0 dp_2{}^0 dk_1{}^0 dk_2{}^0 , \qquad (3\cdot5)$$

where

$$a_{2\nu} = (2\sum_{i\geq j}{}' \epsilon_{ij}^2 |U_{ei} U_{ej}|^2)\frac{1}{4}\frac{G_F{}^4 m_e{}^9}{2\pi^7}\left\{\frac{2\pi\alpha(Z+2)}{1-\exp[-2\pi\alpha(Z+2)]}\right\}^2 , \qquad (3\cdot6)$$

$$C = g_V{}^4(K^2 - KL + L^2)|M_F|^2$$
$$-2g_V{}^2 g_A{}^2 KL \,\mathrm{Re}(M_F M_{GT}^*) + \frac{1}{3} g_A{}^4(K^2 + KL + L^2)|M_{GT}|^2 , \qquad (3\cdot7)$$

$$D = g_V{}^4 KL|M_F|^4 - \frac{2}{3} g_V{}^2 g_A{}^2 (K^2 + KL + L^2)\mathrm{Re}(M_F M_{GT}^*)$$
$$+\frac{1}{9} g_A{}^4(2K^2 + 5KL + 2L^2)|M_{GT}|^2 . \qquad (3\cdot8)$$

Here M_F and M_{GT} are defined in Eqs. (2·15) and (2·16), and

$$K = [\langle E_n\rangle - M_A + p_1{}^0 + k_1{}^0]^{-1} + [\langle E_n\rangle - M_A + p_2{}^0 + k_2{}^0]^{-1} , \qquad (3\cdot9)$$
$$L = [\langle E_n\rangle - M_A + p_2{}^0 + k_1{}^0]^{-1} + [\langle E_n\rangle - M_A + p_1{}^0 + k_2{}^0]^{-1} . \qquad (3\cdot10)$$

The primed sum in Eq. (3·6) should extend over all energetically allowed neutrinos in the final state. Rigorously speaking, the neutrino masses m_j in $k_1{}^0$ and $k_2{}^0$ should be taken into account in this primed sum. If all neutrinos are allowed to contribute and the replacement of $k_i{}^0$ by $|\boldsymbol{k}_i|$ is permissible, then $(2\sum'_{i\leq j}\epsilon_{ij}^2|U_{ei}|^2|U_{ej}|^2) = 1$. The factor 1/4 in $a_{2\nu}$ is to represent the statistical factor for the case $U_{ej} = \delta_{j1}$.

To perform the phase space integration, we neglect the masses of neutrinos and assume the following replacement (within a few % errors): $p_i{}^0 + k_j{}^0 \to \langle p_i{}^0 + k_j{}^0\rangle = (M_A - M_B)/2$. Then $K \simeq L \simeq 2(\mu_0 m_e)^{-1}$.

Now the straightforward calculations lead to

$$d\Gamma_{2\nu}^{2n}(0^+\to 0^+)=(a_{2\nu}/m_e^{11})(g_A^4/60)|(g_V/g_A)^2M_F-M_{GT}|^2\mu_0^{-2}$$
$$\times p_1{}^0p_2{}^0(p_1{}^0p_2{}^0-\boldsymbol{p}_1\cdot\boldsymbol{p}_2)(M_A-M_B-p_1{}^0-p_2{}^0)^5$$
$$\times d\cos\theta dp_1{}^0dp_2{}^0\,. \tag{3·11}$$

The decay rate is

$$\Gamma_{2\nu}^{2n}(0^+\to 0^+)=a_{2\nu}g_A^4|(g_V/g_A)^2M_F-M_{GT}|^2\mu_0^{-2}F_0(T), \tag{3·12}$$

where

$$F_0(T)=\frac{2^4}{11!}T^7(T^4+22T^3+220T^2+990T+1980). \tag{3·13}$$

Primakoff and Rosen have derived Eq. (3·5) in the limit of $U_{ej}=\delta_{j1}$. Note that their result has a few misprints and also is four times as large as ours.*) Concerning this overall normalization, our result in Eq. (3·11) agrees in this limit with that by Konopinski.[6]

(ii) The $0^+\to 2^+$ transition

For this transition, we have

$$d\Gamma_{2\nu}^{2n}(0^+\to 2^+)=(a_{2\nu}/m_e^9)(g_A^4/8)$$
$$\times(M_2^{pq},\,M_2^{pq})(K-L)^2(p_1{}^0p_2{}^0+\frac{1}{3}\boldsymbol{p}_1\cdot\boldsymbol{p}_2)$$
$$\times(k_1{}^0k_2{}^0)^2p_1{}^0p_2{}^0\delta(p_1{}^0+p_2{}^0+k_1{}^0+k_2{}^0+M_B-M_A)$$
$$\times d\cos\theta dp_1{}^0dp_2{}^0dk_1{}^0dk_2{}^0\,, \tag{3·14}$$

where $(M_2^{pq},\,M_2^{pq})\equiv\sum_{J_z}\sum_{p,q=1}^3M_2^{pq*}M_2^{pq}$ with

$$M_2^{pq}=\langle N_B(2^+)|\sum_{n,m}\tau_n{}^+\tau_m{}^+\sigma_n{}^p\sigma_m{}^q|N_A(0^+)\rangle\,. \tag{3·15}$$

To simplify the term $(K-L)^2$, we use the approximation $p_i{}^0+k_j{}^0\simeq\langle p_i{}^0+k_j{}^0\rangle=(M_A-M_B)/2$ only in the denominator of $K-L$ and obtain $K-L=2(p_1{}^0-p_2{}^0)(k_1{}^0-k_2{}^0)(\mu_0m_e)^{-3}$. This approximation is valid within several % errors for $\mu_0\geq 4$. After the phase space integration, we get

$$d\Gamma_{2\nu}^{2n}(0^+\to 2^+)=(a_{2\nu}/m_e^{15})(g_A^4/420)(M_2^{pq},\,M_2^{pq})\mu_0^{-6}p_1{}^0p_2{}^0(p_1{}^0-p_2{}^0)^2$$
$$\times\left(p_1{}^0p_2{}^0+\frac{1}{3}\boldsymbol{p}_1\cdot\boldsymbol{p}_2\right)(M_A-M_B-p_1{}^0-p_2{}^0)^7d\cos\theta dp_1{}^0dp_2{}^0\,. \tag{3·16}$$

*) See the footnotes on page 1747 and Eq. (60) of Ref. 4). Greuling and Whitten[5] give results four times as large as ours.

The decay rate is given by

$$\Gamma_{2\nu}^{2n}(0^+\to 2^+)=a_{2\nu}g_A{}^4(M_2^{pq},\ M_2^{pq})\mu_0^{-6}F_2(T), \tag{3·17}$$

where

$$F_2(T)=\frac{2^6 3^2}{15!}T^{11}(T^4+30T^3+420T^2+1820T+2730). \tag{3·18}$$

(iii) The $0^+\to 1^+$ transition

Similarly, we obtain

$$d\Gamma_{2\nu}^{2n}(0^+\to 1^+)=(a_{2\nu}/m_e{}^9)(g_A{}^2g_V{}^2/4)$$
$$\times(M_1{}^p,\ M_1{}^p)(K-L)^2\left(p_1{}^0p_2{}^0+\frac{1}{3}\boldsymbol{p}_1\cdot\boldsymbol{p}_2\right)$$
$$\times(k_1{}^0k_2{}^0)^2p_1{}^0p_2{}^0\delta(p_1{}^0+p_2{}^0+k_1{}^0+k_2{}^0+M_B-M_A)$$
$$\times d\cos\theta dp_1{}^0dp_2{}^0dk_1{}^0dk_2{}^0\ , \tag{3·19}$$

where $(M_1{}^p,\ M_1{}^p)\equiv\sum_{J_z}\sum_{p=1}^{3}M_1^{p^*}M_1{}^p$ with

$$M_1{}^p=\langle N_B(1^+)|\sum_{n,m}\tau_n{}^+\tau_m{}^+\sigma_n{}^p|N_A(0^+)\rangle\ . \tag{3·20}$$

Note that the transition formula given above is exactly the same as the one for the $0^+\to 2^+$ transition given in Eq. (3·14), aside from the overall normalization. Therefore, the decay rate can be read off from the one for the $0^+\to 2^+$ transition.

Before closing this subsection, we would like to mention the work by Molina and Pascual[9] who estimated the $0^+\to J^+$ transitions. We found several errors in their formulae: (i) They showed that $\Gamma_{2\nu}^{2n}(0^+\to 1^+)=0$, while we get the non-vanishing rate as given in Eq. (3·19). (ii) Their decay rates of the $0^+\to 0^+$, 2^+ transitions for the $(\beta\beta)_{2\nu}$ mode are twice as large as ours.

(3-b) *The N^*-mechanism*

The R-matrix element for the N^*-mechanism is obtained from Eq. (A·3) in Appendix A by substituting the R_W-matrices corresponding to the $N_{\Delta^-}\to N_B$ and $N_A\to N_{\Delta^{++}}$ transitions shown in Figs. 2(c) and (d). The hadronic part of the amplitude for $N_{\Delta^-}\to N_B+2e^-+\bar{N}_i+\bar{N}_j$ is expressed as follows:

$$(J^{\mu\nu})_{N^*}=K\left\{g_Vg_A(g^{\mu 0}g^{\nu j}+g^{\mu j}g^{\nu 0})M^j+\frac{1}{2}g_A{}^2(g^{\mu j}g^{\nu k}+g^{\mu k}g^{\nu j})M^{jk}\right\}, \tag{3·21}$$

where K, M^j and M^{jk} are defined in Eqs. (A·9), (2·45) and (2·46). In order to express the contributions from the M^j and M^{jk} terms clearly, we retain g_V and g_A explicitly although $g_A=g_V$ should be taken in the N^*-mechanism.

By using the factorization hypothesis, the decay formula is obtained in the following form:

$$d\Gamma_{2\nu}^{N^*}=(a_{2\nu}/m_e{}^9)g_V{}^4 2^2\cdot 3P(\Delta)|\langle\Phi_f|\Phi_i\rangle|^2(K-L)^2\left(p_1{}^0p_2{}^0+\frac{1}{3}\boldsymbol{p}_1\cdot\boldsymbol{p}_2\right)$$
$$\times(k_1{}^0k_2{}^0)^2p_1{}^0p_2{}^0\delta(p_1{}^0+p_2{}^0+k_1{}^0+k_2{}^0+M_B-M_A)$$
$$\times d\cos\theta dp_1{}^0dp_2{}^0dk_1{}^0dk_2{}^0\,, \qquad (3\cdot22)$$

where both the $N_{A^-}\to N_B$ and $N_A\to N_{A^{++}}$ transitions are included according to the argument given in Appendix A. It is amusing to observe that the above formula is exactly the same as the one for the $0^+\to2^+$ transition in the $2n$-mechanism aside from the overall normalization.

The decay rate is

$$\Gamma_{2\nu}^{N^*}=a_{2\nu}g_V{}^4 2^5\cdot 3P(\Delta)|\langle\Phi_f|\Phi_i\rangle|^2\mu_0^{-6}F_2(T), \qquad (3\cdot23)$$

where $F_2(T)$ is defined in Eq. (3·18) and $a_{2\nu}$ is in Eq. (3·6).

We would like to note that this formula is completely different from the one obtained by Picciotto.[8] This is due to the fact that he used a crucial approximation*) for the R-matrix, instead of taking the spin sum explicitly. However, his approximation cannot be regarded as reasonable. Note also that he has neglected the M^{jk} term ($g_A{}^4$ term) in Eq. (3·22) which turns out to be dominant for the $(\beta\beta)_{2\nu}$ mode.

§ 4. Concluding remarks

In this paper, we have presented the formulae for parity conserving $0^+\to J^+$ transitions within the weak interaction Hamiltonian in Eq. (1·3) which is motivated by the grand unified theories. The emphasis has been made on the transitions to the excited states which become important in analyzing the data on the half-life, especially the data obtained by the geological method.

We have found several interesting "selection rules" as given in Table I: (i) If $\lambda=0$ (no right-handed interaction), the N^*-mechanism does not contribute to the $(\beta\beta)_{0\nu}$ mode, whether neutrinos are massive or massless. (ii) If $\lambda=0$, the $0^+\to1^+$ and 2^+ transitions of the $(\beta\beta)_{0\nu}$ mode are forbidden and the $0^+\to0^+$ transition is only allowed. We emphasize that these selection rules for the N^*-mechanism do not depend on the factorization hypothesis. The selection rule (i) seems to nullify the neutrino mass bounds derived by Halprin et al. in the N^*-mechanism.[7]

It may be worth while to mention that some care is necessary to use the previous theoretical estimates numerically, because there are various errors as discussed in § 3 for the $(\beta\beta)_{2\nu}$ mode and in § 2 and Appendix C for the $(\beta\beta)_{0\nu}$ mode.

*) Picciotto[8] used the approximation $\{E_{\mu\nu}J^{\mu\nu}-(p_1\leftrightarrow p_2)\}\sim E_{\mu\nu}g_Vg_A(g^{\mu0}g^{\nu k}+g^{\mu k}g^{\nu 0})M^k\mu_0^{-1}$. We have evaluated the spin sum exactly.

Acknowledgements

The authors would like to thank Professor M. Kenmoku and Professor H. Ohtsubo for helpful discussions. One of the authors (H.N.) would like to thank Iue-Kinenkai for financial support.

Appendix A

—— Brief Description of the 2n- and N-Mechanisms ——*

(a) *The general description*

In order to deal with the N^*-mechanism, the following effective Hamiltonian is considered,

$$H_{\rm Int} = H_W + H_S\ , \tag{A·1}$$

where H_S represents the effective interaction for the transition $N+N \leftrightarrow \Delta + N$ by the exchange of π, ρ, $\cdots$.

In the $2n$-mechanism, the double β decay takes place through the 2nd order perturbation in H_W and the 0th order in H_S as shown in Fig. 1 and the R-matrix is

$$R^{2n} = \langle N_B,\ l|R_W|N_A\rangle\ , \tag{A·2}$$

where l stands for either $2e^-$ or $2e^- + 2\bar{\nu}_e$ depending on the $(\beta\beta)_{0\nu}$ or $(\beta\beta)_{2\nu}$ mode, respectively. The R_W represents the R-matrix due to the 2nd order weak interaction.

In the N^*-mechanism, the double β decay occurs through the 2nd order in H_W and the 1st order in H_S as shown in Fig. 2.*) Then the R-matrix element for the N^*-mechanism may be expressed in the following forms:

$$R^{N^*} = \sum_{N_\Delta} \Big\{ \langle N_B,\ l|R_W|N_{\Delta^-}\rangle \frac{1}{M_A - E_{\Delta^-}} \langle N_{\Delta^-}|H_S|N_A\rangle + \langle N_B|H_S|N_{\Delta^{++}}\rangle \frac{1}{M_B - E_{\Delta^{++}}} \langle N_{\Delta^{++}},\ l|R_W|N_A\rangle \Big\}, \tag{A·3}$$

where M_A and M_B are the masses of the parent (N_A) and daughter (N_B) nuclei, respectively, and E_Δ is the energy of the intermediate nucleus which includes $\Delta(1232)$. Note that the nuclear state $N_{\Delta^-}(N_{\Delta^{++}})$ has the same J^P as $N_A(N_B)$ has.

Let us consider, for definiteness, the $N_{\Delta^-} \to N_B + 2e^-$ transition. The nuclear states N_{Δ^-} and N_B are expressed in the following forms:

$$|N_{\Delta^-}\rangle = |\Delta^-\rangle_S \otimes |\Delta^-\rangle_L \otimes |R_{\Delta^-}\rangle\ , \tag{A·4}$$

*) There may be the third possible combination of H_W and H_S such as the sequence H_W-H_S-H_W in contrast to the H_W-H_W-H_S in Fig. 2(a). Since this contribution is expected to be small, this is not considered in this paper.

$$|N_B\rangle = |p\rangle_s \otimes |p\rangle_L \otimes |R_B\rangle\,, \qquad (\text{A}\cdot 5)$$

where s and L in the hadronic states stand for the intrinsic (spin and isospin) part and the orbital angular momentum part with respect to the center of the nucleus. Here, $|R_{\Delta^-}\rangle$ and $|R_B\rangle$ represent the remainders of the nuclear states. It should be understood that there is some appropriate sum with respect to the angular momenta. The nuclear matrix elements in Eqs. (2·15)~(2·18) can be written in this notation as follows:

$$\langle N_B|O|N_{\Delta^-}\rangle = \langle R_B|\otimes_L\langle p|\otimes_s\langle p|\,O|\Delta^-\rangle_s\otimes|\Delta^-\rangle_L\otimes|R_{\Delta^-}\rangle\,, \qquad (\text{A}\cdot 6)$$

where O represents one of the nuclear tensor operators appearing in M_F, M_{GT}, Q^l and P^{lk}.

Let us discuss what kinds of nuclear tensor operators change quark states inside the hadron. Obviously, the operators $\tau^+_{n(m)}$ and $\sigma^j_{n(m)}$ act on quarks. As for $\hat{r}_{nm}$ and $\hat{r}_{+nm}$, some caution is necessary. Consider the following decomposition of the position operator for the n-th quark; $\boldsymbol{r}_n = \boldsymbol{r}_G + \boldsymbol{r}_n'$ where $\boldsymbol{r}_G$ is the position operator of Δ^- measured from the center of N_{Δ^-}. The relative coordinate $\boldsymbol{r}_n'$ changes the orbital angular momentum of quarks around the center of the hadron. Thus we conclude that the relevant operators for quarks in Δ^- are $\tau^+_{n(m)}$, $\sigma^j_{n(m)}$, $\boldsymbol{r}'_{nm}$ and $\boldsymbol{r}'_{+nm}$. With this caution, the nuclear matrix elements are calculated in the $SU(6)$ quark model where $\Delta(\frac{3}{2}^+)$ and the nucleon $N(\frac{1}{2}^+)$ are assigned to $l=0$. The nuclear tensor operators contributing to the transition $\Delta(\frac{3}{2}^+)\to N(\frac{1}{2}^+)$ should be of rank 0 with respect to $\boldsymbol{r}'_{nm}$ and $\boldsymbol{r}'_{+nm}$ and of rank 1 or 2 with respect to the spin part. We conclude from Eqs. (2·15)~(2·18) that M_F and M_{GT} do not contribute to this transition. The Q^l and P^{lk} take the following forms:

$$Q^l = \frac{1}{3}\,{}_s\langle p|\sum_{n,m}\tau_n^+\tau_m^+ \times[(\sigma_n^l-\sigma_m^l)-i(\boldsymbol{\sigma}_n\times\boldsymbol{\sigma}_m)^l](\hat{r}'_{nm}\cdot\hat{r}'_{+nm})|\Delta^-\rangle_s\cdot{}_L\langle p|\Delta^-\rangle_L\langle R_B|R_{\Delta^-}\rangle\,, \qquad (\text{A}\cdot 7)$$

$$P^{lk} = -\frac{2}{3}\,{}_s\langle p|\sum_{n,m}\tau_n^+\tau_m^+[i\epsilon_{lkj}\sigma_n^j+\sigma_n^k\sigma_m^l]|\Delta^-\rangle_s\cdot{}_L\langle p|\Delta^-\rangle_L\langle R_B|R_{\Delta^-}\rangle\,. \qquad (\text{A}\cdot 8)$$

In the $SU(6)$ quark model, we get $Q^l=0$ which may be understood from the following argument. Note that the spin operators in Q^l are antisymmetric under the interchange of quarks so that it is expected that the spin part of ${}_s\langle p|\sum_{nm}\tau_n^+\tau_m^+[(\sigma_n^l-\sigma_m^l)-i(\boldsymbol{\sigma}_n\times\boldsymbol{\sigma}_m)^l]$ is also antisymmetric. Since the spin wave function of $|\Delta^-\rangle_s$ is symmetric, $Q^l=0$ is concluded. Therefore, only P^{lk} contributes to the $(\beta\beta)_{0\nu}$ mode. The same argument also holds for the $N_A\to N_{\Delta^{++}}+2e^-$ transition.

A similar argument applies to the $(\beta\beta)_{2\nu}$ mode. Of course, it should be noted that K and L in Eq. (3·22) should be modified as follows:

$$K=[\langle E_n\rangle-E_{\Delta^-}+p_1{}^0+k_1{}^0]^{-1}+[\langle E_n\rangle-E_{\Delta^-}+p_2{}^0+k_2{}^0]^{-1}, \qquad (\text{A}\cdot 9)$$

and similarly for L.

(b) *The factorization hypothesis*

As we have seen in the previous subsection (a), the R_W-matrix may be written in the following form:

$$\langle N_B, l|R_W|N_{\Delta^-}\rangle = {}_s\langle p, l|R_W|\Delta^-\rangle_s\ {}_L\langle p|\Delta^-\rangle_L\langle R_B|R_{\Delta^-}\rangle\,. \qquad (\text{A}\cdot 10)$$

This is valid under the approximations (i), (ii) and (iii) introduced in § 2 and in the framework of the $SU(6)$ quark model. The "factorization hypothesis" means the approximation that the amplitude ${}_s\langle p, l|R_W|\Delta^-\rangle_s$ is modified by the following replacement,

$${}_s\langle p, l|R_W|\Delta^-\rangle_s \Rightarrow e^{i\varphi}[\sum_{S_{\Delta z}, S_{pz}}{}' |{}_s\langle p, l|R_W|\Delta^-\rangle_s|^2]^{1/2}\,, \qquad (\text{A}\cdot 11)$$

where the primed sum means the spin average with respect to Δ. Under this replacement, the R-matrix is rewritten as follows:

$$R^{N^*}=[\sum_{S_{\Delta z}, S_{pz}}{}' |{}_s\langle p, l|R_W|\Delta^-\rangle_s|^2]^{1/2}\sqrt{2}P(\Delta)^{1/2}\langle\Phi_f|\Phi_i\rangle\,, \qquad (\text{A}\cdot 12)$$

where

$$\sqrt{2}P(\Delta)^{1/2}\langle\Phi_f|\Phi_i\rangle = {}_L\langle p|\Delta^-\rangle_L\langle R_B|R_{\Delta^-}\rangle e^{i\varphi}\frac{\langle N_{\Delta^-}|H_S|N_A\rangle}{M_A-E_{\Delta^-}}$$
$$+\frac{\langle N_B|H_S|N_{\Delta^{++}}\rangle}{M_B-E_{\Delta^{++}}}e^{i\varphi'}\langle R_{\Delta^{++}}|R_A\rangle_L\langle\Delta^{++}|n\rangle_L\,. \qquad (\text{A}\cdot 13)$$

Here we have used the relation

$$\sum_{S_{\Delta z}, S_{pz}}{}' |{}_s\langle p, l|R_W|\Delta^-\rangle_s|^2 = \sum_{S_{\Delta z}, S_{nz}}{}' |{}_s\langle\Delta^{++}, l|R_W|n\rangle_s|^2\,. \qquad (\text{A}\cdot 14)$$

The factors $P(\Delta)$ and $\langle\Phi_f|\Phi_i\rangle$ are introduced to give some physical image of the N^*-mechanism. Let us assume the decomposition

$$\langle N_{\Delta^-}|H_S|N_A\rangle \simeq \langle\Delta^-|H_S|n\rangle\langle R_{\Delta^-}|R_A\rangle\,. \qquad (\text{A}\cdot 15)$$

Now the probability admixture $P(\Delta)$ may be defined as

$$P(\Delta)=\frac{1}{N_n}\sum_{n,\Delta^-}\left|\frac{\langle\Delta^-|H_S|n\rangle}{E_{\Delta^-}-M_A}\right|^2, \qquad (\text{A}\cdot 16)$$

where N_n is the number of neutrons which actively participate in the double β decay and the sum of n extends over all those neutrons. In other words, $P(\Delta)$

is the probability to make Δ^- per neutron. Now $\langle\Phi_f|\Phi_i\rangle$ may be written

$$\langle\Phi_f|\Phi_i\rangle\simeq{}_L\langle p|\Delta^-\rangle_L\langle R_B|R_{\Delta^-}\rangle\langle R_{\Delta^-}|R_A\rangle\simeq{}_L\langle p|\Delta^-\rangle_L\langle R_B|R_A\rangle\,. \tag{A·17}$$

Here we have used $\langle R_{\Delta^-}|R_A\rangle\simeq\delta_{R_\Delta\text{-}R_A}$. In this way, the $\langle\Phi_f|\Phi_i\rangle$ may be interpreted as the overlap between the initial and final nuclear wave functions.

Appendix B

—— *Majorana Neutrino Propagators* ——

The quantization of the Majorana field is rather complicated. It can be done straightforwardly by decomposing Majorana field into two-component field. The quantized form of the Majorana field is derived by substituting the quantized two-component field into it.[10] In this way, we obtain

$$N(x)=\frac{1}{(2\pi)^{3/2}}\int\frac{d\boldsymbol{p}}{\sqrt{p^0}}\sum_s\{a(p,s)u(p,s)e^{-ipx}+a^+(p,s)u^C(p,s)e^{ipx}\}, \tag{B·1}$$

where $u^C=C\bar{u}^T$. The creation and annihilation operators $a^+(p,s)$ and $a(p,s)$ satisfy the canonical commutation relation,

$$[a(p,s),a^+(p',s')]_+=\delta_{ss'}\delta(\boldsymbol{p}-\boldsymbol{p}'). \tag{B·2}$$

The Majorana propagator can be calculated straightforwardly from Eqs. (B·1) and (B·2),

$$\langle 0|\,T[N(x)\bar{N}(y)]|0\rangle=iS_F(x-y). \tag{B·3}$$

Note that we also obtain

$$\langle 0|\,T[N(x)N^T(y)]|0\rangle=iS_F(x-y)C^T\,. \tag{B·4}$$

This is only possible for the (self-conjugate) Majorana field and makes the $(\beta\beta)_{0\nu}$ mode possible.

As defined in Eq. (1·5), current neutrinos ν_{eL} and ν'_{eR} are the superpositions of massive Majorana neutrinos N_j. We obtain from Eqs. (1·5) and (B·4),

$$\begin{aligned}\langle 0|\,T[\nu_{eL}(x)\nu_{eL}^T(y)]|0\rangle &=i\sum_j U_{ej}^2\Big(\frac{1-\gamma_5}{2}\Big)S_F(x-y)C^T\Big(\frac{1-\gamma_5{}^T}{2}\Big)\\ &=i\sum_j m_jU_{ej}^2\int\frac{d^4q}{(2\pi)^4}\frac{e^{-iq(x-y)}}{q^2-m_j{}^2+i\epsilon}\Big(\frac{1-\gamma_5}{2}\Big)C^T\,,\end{aligned} \tag{B·5}$$

$$\begin{aligned}\langle 0|\,T[\nu_{eL}(x)\nu_{eR}'^T(y)]|0\rangle &=i\sum_j U_{ej}V_{ej}\Big(\frac{1-\gamma_5}{2}\Big)S_F(x-y)C^T\Big(\frac{1+\gamma_5{}^T}{2}\Big)\\ &=i\sum_j U_{ej}V_{ej}\int\frac{d^4q}{(2\pi)^4}\frac{\not{q}e^{-iq(x-y)}}{q^2-m_j{}^2+i\epsilon}\Big(\frac{1+\gamma_5}{2}\Big)C^T\,.\end{aligned} \tag{B·6}$$

The other propagators are similarly obtained.

Appendix C

—— The Double β Decay in the Weak Interaction Hamiltonian Previously Used ——

The weak interaction Hamiltonian used by many others[1),5),7)~9)] is

$$H_W=\left(\frac{G_F}{\sqrt{2}}\right)\bar{e}\gamma_\mu[(1-\gamma_5)+\eta(1+\gamma_5)]\nu_e\bar{p}\gamma^\mu(g_V-g_A\gamma_5)n+\text{h.c.} \tag{C·1}$$

It should be emphasized that the parameters η in Eq. (C·1) and λ in Eq. (1·3) have different physical meanings. The parameter η represents the admixture of $V+A$ interaction in the leptonic current, while λ is the relative strength of the right-handed to the left-handed weak interactions. For the $(\beta\beta)_{2\nu}$ mode, the decay formulae are the same as the ones given in § 4 because the contribution due to the η term can be neglected. Here we only present the results for the $(\beta\beta)_{0\nu}$ mode.

We obtain the following R_W-matrix corresponding to Eqs. (2·1), (2·12) ~(2·18),

$$R_W=\frac{1}{\sqrt{2}}\left(\frac{G_F}{\sqrt{2}}\right)^2[(2\pi)^{-6}(p_1{}^0p_2{}^0)^{-1}F(Z+2,p_1{}^0)F(Z+2,p_2{}^0)]^{-1/2}$$

$$\times 2\{m_\nu(t^L_{\mu\nu}+\eta^2t^R_{\mu\nu})K^{\mu\nu}_{LL}+\eta u^{L+R}_{\mu\nu\rho}L^{\mu\nu\rho}_{LL}\}, \tag{C·2}$$

where $u^{L+R}_{\mu\nu\rho}\equiv u^L_{\mu\nu\rho}+u^R_{\mu\nu\rho}$ and*)

$$(t^L_{\mu\nu}+\eta^2t^R_{\mu\nu})K^{\mu\nu}_{LL}=\frac{1}{4\pi}\langle H\rangle g_A{}^2$$

$$\times\bar{u}(p_1)[(1+\gamma_5)+\eta^2(1-\gamma_5)]u^C(p_2)[(g_V/g_A)^2M_F-M_{GT}], \tag{C·3}$$

$$u^{L+R}_{\mu\nu0}L^{\mu\nu0}_{LL}=\frac{1}{4\pi}\langle H\rangle g_A{}^2(p_1{}^0-p_2{}^0)$$

$$\times\bar{u}(p_1)\{\gamma^0[(g_V/g_A)^2M_F+M_{GT}]+2(g_V/g_A)\gamma^j\mathcal{R}^j\}u^C(p_2), \tag{C·4}$$

$$u^{L+R}_{\mu\nu k}L^{\mu\nu k}_{LL}=\frac{-1}{4\pi}\langle rH'\rangle g_A{}^2\bar{u}(p_1)\{(p_1-p_2)^l[\gamma^k\mathcal{P}^{lk}+\gamma^0\mathcal{P}^l]$$

$$+(p_1+p_2)^l[\gamma_5\gamma^0\mathcal{Q}^l+\gamma_5\gamma^k\mathcal{Q}^{lk}]\}u^C(p_2). \tag{C·5}$$

Here M_F and M_{GT} are defined in Eqs. (2·15) and (2·16), and

*) The relativistic correction terms (v/c) in the hadron current are not taken into account in this appendix.

$$\mathcal{R}^l=\langle\sigma_n{}^l\rangle,\quad \mathcal{P}^l=2(g_V/g_A)\langle\hat{r}^l_{nm}(\hat{r}_{nm}\cdot\boldsymbol{\sigma}_m)\rangle,$$
$$\mathcal{P}^{lk}=\langle\hat{r}^l_{nm}\{\hat{r}^k_{nm}[(g_V/g_A)^2-(\boldsymbol{\sigma}_n\cdot\boldsymbol{\sigma}_m)]+2\sigma_m{}^k(\hat{r}_{nm}\cdot\boldsymbol{\sigma}_n)\}\rangle,$$
$$\mathcal{Q}^l=i(g_V/g_A)^2\langle\hat{r}^l_{+nm}\hat{r}_{nm}\cdot(\boldsymbol{\sigma}_n\times\boldsymbol{\sigma}_m)\rangle,\quad \mathcal{Q}^{lk}=-2i(g_V/g_A)\langle\hat{r}^l_{+nm}(\hat{r}_{nm}\times\boldsymbol{\sigma}_m)^k\rangle, \tag{C·6}$$

where abbreviation $\langle O\rangle\equiv\langle N_\beta|\sum_{n,m}\tau_n{}^+\tau_m{}^+O|N_\alpha\rangle$ is used.

The $m_\nu K^{\mu\nu}_{LL}$ term contributes only to the $0^+\to0^+$ transition, while $\eta L^{\mu\nu 0}_{LL}$ contributes to $0^+\to0^+$, 1^+ and $\eta L^{\mu\nu k}_{LL}$ to $0^+\to0^+$, 1^+, 2^+, cf., Table I.

(a) *The 2n-mechanism*

For the $0^+\to0^+$ transition, the decay formula is expressed as follows:

$$\begin{aligned}d\Gamma^{2n}_{0\nu}(0^+\to0^+)=&(a_{0\nu}/m_e{}^7)g_A{}^4\\&\times\{(p_1{}^0p_2{}^0-\boldsymbol{p}_1\cdot\boldsymbol{p}_2)[|Y_1+Y_4-Y_5|^2+|Y_2+Y_4+Y_5|^2]\\&+\frac{1}{2m_e{}^2}(p_1{}^0-p_2{}^0)^2(\alpha_-+\boldsymbol{p}_1\cdot\boldsymbol{p}_2)|Y_3-Y_4|^2\\&+\frac{1}{2m_e{}^2}(p_1{}^0+p_2{}^0)^2(\alpha_++\boldsymbol{p}_1\cdot\boldsymbol{p}_2)|Y_5|^2\\&-2m_e{}^2\,\mathrm{Re}(Y_1+Y_4-Y_5)(Y_2+Y_4+Y_5)^*\\&-(p_1{}^0-p_2{}^0)^2\,\mathrm{Re}(Y_1+Y_2+2Y_4)(Y_3-Y_4)^*\\&+(p_1{}^0+p_2{}^0)^2\,\mathrm{Re}(Y_1-Y_2-2Y_5)Y_5{}^*\}\\&\times p_1{}^0p_2{}^0\,\delta(p_1{}^0+p_2{}^0+M_B-M_A)d\cos\theta dp_1{}^0dp_2{}^0\,,\end{aligned}\tag{C·7}$$

where

$$\alpha_\pm=p_1{}^0p_2{}^0\pm m_e{}^2 \tag{C·8}$$

and

$$\begin{aligned}&Y_1=(m_\nu/m_e)\langle H\rangle[(g_V/g_A)^2M_F-M_{GT}],\quad Y_2=\eta^2Y_1\,,\\&Y_3=\eta\langle H\rangle[(g_V/g_A)^2M_F+M_{GT}],\\&Y_4=(\eta/3)\langle rH'\rangle[(g_V/g_A)^2M_F-M_{GT}/3+2M_T],\\&Y_5=(\eta/3)(-2ig_V/g_A)\langle rH'\rangle\langle\hat{r}_{+nm}\cdot(\hat{r}_{nm}\times\boldsymbol{\sigma}_m)\rangle\,.\end{aligned}\tag{C·9}$$

Note that if $\eta=0$, the above formula agrees with Eq. (2·21) with $\lambda=0$. If $\eta\neq0$, there arise the following differences: (i) A new nuclear matrix element Y_5 appears. (ii) There are some sign differences in Y_3 and Y_4 in comparison with X_3 and X_4. (iii) If Y_5, M_F and M_T are discarded, both decay formulae are similar because only the difference comes from $Y_3 \simeq -X_3$ and $Y_4 \simeq -X_4$ whose contributions are minor. Thus, it is rather difficult to distinguish these two $H_{W'}$'s.

In the formula derived by Primakoff and Rosen,[1] the terms Y_5 as well as M_T are not included. Also, Greuling and Whitten[5] did not include Y_5 either. Note that their formulae are two times as large as ours.

For the $0^+ \to 2^+$ transition, the decay formula is

$$d\Gamma_{0\nu}^{2n}(0^+ \to 2^+) = (a_{0\nu}/m_e^9)(g_A^4/30)|\eta\langle rH'\rangle|^2$$
$$\times [x_+(\mathcal{N}_1^{pq}, \mathcal{N}_1^{pq}) + x_-(\mathcal{N}_2^{pq}, \mathcal{N}_2^{pq})]$$
$$\times p_1^0 p_2^0 \delta(p_1^0 + p_2^0 + M_B - M_A) d\cos\theta dp_1^0 dp_2^0 , \qquad (\mathrm{C}\cdot 10)$$

where

$$x_\pm = 3(\boldsymbol{p}_1 \cdot \boldsymbol{p}_2)^2 - \boldsymbol{p}_1 \cdot \boldsymbol{p}_1 (10\alpha_\pm + |\boldsymbol{p}_1|^2 + |\boldsymbol{p}_2|^2)$$
$$+ 5\alpha_\pm (|\boldsymbol{p}_1|^2 + |\boldsymbol{p}_2|^2) - |\boldsymbol{p}_1|^2 |\boldsymbol{p}_2|^2 ,$$
$$\mathcal{N}_1^{pq} = \langle \hat{r}_{nm}^p \{ \hat{r}_{nm}^q [(g_V/g_A)^2 - (\boldsymbol{\sigma}_n \cdot \boldsymbol{\sigma}_m)] + 2\sigma_m^q (\hat{r}_{nm} \cdot \boldsymbol{\sigma}_n) \} \rangle ,$$
$$\mathcal{N}_2^{pq} = -2i(g_V/g_A) \langle \hat{r}_{+nm}^p (\hat{r}_{nm} \times \boldsymbol{\sigma}_m)^q \rangle . \qquad (\mathrm{C}\cdot 11)$$

In a similar way to Eq. (2·34), the $0^+ \to 2^+$ transition occurs only if $\eta \neq 0$. In the place of the tensor operator $\hat{r}_{nm}^p (\hat{r}_{nm} \times \boldsymbol{\sigma}_m)^q$, a new operator $\hat{r}_{+nm}^p (\hat{r}_{nm} \times \boldsymbol{\sigma}_m)^q$ enters here. It is interesting to note that if $m_e = 0$, both decay formulae agree with each other except for the overall normalization.

For the $0^+ \to 1^+$ transition, we obtain

$$d\Gamma_{0\nu}^{2n}(0^+ \to 1^+) = (a_{0\nu}/m_e^9)(g_A^4/6)\eta^2 \sum_{a,b=1}^{5} D_{ab}(\mathcal{N}_a^p, \mathcal{N}_b^p)$$
$$\times p_1^0 p_2^0 \delta(p_1^0 + p_2^0 + M_B - M_A) d\cos\theta dp_1^0 dp_2^0 , \qquad (\mathrm{C}\cdot 12)$$

where

$$\mathcal{N}_1^j = (g_V/g_A)\langle H \rangle \langle \sigma_n^j \rangle , \quad \mathcal{N}_2^j = (g_V/g_A)\langle rH' \rangle \langle \hat{r}_{nm}^j (\hat{r}_{nm} \cdot \boldsymbol{\sigma}_m) \rangle ,$$
$$\mathcal{N}_3^j = \langle rH' \rangle \langle \hat{r}_{+nm}^j \hat{r}_{nm} \cdot (\boldsymbol{\sigma}_n \times \boldsymbol{\sigma}_m) \rangle , \quad \mathcal{N}_4^j = \langle rH' \rangle \langle (\hat{r}_{nm} \times \boldsymbol{\sigma}_m)^j (\hat{r}_{nm} \cdot \boldsymbol{\sigma}_n) \rangle ,$$
$$\mathcal{N}_5^j = (g_V/g_A)\langle rH' \rangle \langle [\hat{r}_{+nm} \times (\hat{r}_{nm} \times \boldsymbol{\sigma}_m)]^j \rangle . \qquad (\mathrm{C}\cdot 13)$$

The non-zero coefficients $D_{ab} = D_{ba}$ are as follows:

$$D_{11} = 4(p_1^0 - p_2^0)^2 (3\alpha_+ - \boldsymbol{p}_1 \cdot \boldsymbol{p}_2), \quad D_{12} = -4(p_1^0 - p_2^0)^2 (\alpha_+ + \boldsymbol{p}_1 \cdot \boldsymbol{p}_2),$$
$$D_{15} = 4(p_1^0 - p_2^0)^2 (\alpha_+ - \boldsymbol{p}_1 \cdot \boldsymbol{p}_2), \quad D_{22} = 4(\boldsymbol{p}_1 - \boldsymbol{p}_2)^2 (\alpha_- + \boldsymbol{p}_1 \cdot \boldsymbol{p}_2),$$
$$D_{25} = -2D_{34} = 8[|\boldsymbol{p}_1|^2 |\boldsymbol{p}_2|^2 - (\boldsymbol{p}_1 \cdot \boldsymbol{p}_2)^2],$$
$$D_{33} = (\boldsymbol{p}_1 + \boldsymbol{p}_2)^2 (\alpha_+ + \boldsymbol{p}_1 \cdot \boldsymbol{p}_2),$$

$$D_{44}=2[(\boldsymbol{p}_1\times\boldsymbol{p}_2)^2+(\boldsymbol{p}_1-\boldsymbol{p}_2)^2(\alpha_+-\boldsymbol{p}_1\cdot\boldsymbol{p}_2)],$$

$$D_{55}=-2[(\boldsymbol{p}_1\times\boldsymbol{p}_2)^2-(\boldsymbol{p}_1+\boldsymbol{p}_2)^2(\alpha_--\boldsymbol{p}_1\cdot\boldsymbol{p}_2)]. \tag{C·14}$$

In contrast to Eq. (2·38), the above formula is complicated. Instead of two first terms in $N_1{}^q$ and $N_2{}^q$ in Eqs. (2·39) and (2·40), the new terms $\mathcal{N}_1{}^j$, $\mathcal{N}_4{}^j$ and $\mathcal{N}_5{}^j$ now enter into the decay formula. Morina and Pascual[9] have also evaluated the $0^+\to2^+$ and 1^+ transitions. Their results are four times as large as ours in addition to some errors.

(b) *N^*-mechanism*

We have found that both $\eta L_{LL}^{\mu\nu0}$ and $\eta L_{LL}^{\mu\nu k}$ terms in Eqs. (C·4) and (C·5) contribute to the $(\beta\beta)_{0\nu}$ mode in contrast to the case in Table I, while the term $m_\nu K_{LL}^{\mu\nu}$ in Eq. (C·3) does not again. That is, whether neutrinos are massive or massless, the N^*-mechanism has no contribution to the $(\beta\beta)_{0\nu}$ mode if $\eta=0$.

By using the factorization hypothesis explained in Appendix A, we obtain

$$d\Gamma_{0\nu}^{N^*}=(a_{0\nu}/m_e{}^9)(2^5g_A{}^2/3^3)\eta^2P(\Delta)|\langle\Phi_f|\Phi_i\rangle|^2$$
$$\times[(\boldsymbol{p}_1\cdot\boldsymbol{p}_2)^2(3g_A{}^2-g_V{}^2)\langle rH'\rangle^2-\boldsymbol{p}_1\cdot\boldsymbol{p}_2\beta_1+\beta_2]$$
$$\times p_1{}^0p_2{}^0\delta(p_1{}^0+p_2{}^0+M_B-M_A)d\cos\theta dp_1{}^0dp_2{}^0\,, \tag{C·15}$$

where

$$\beta_1=\left\{g_A{}^2(10\alpha_++|\boldsymbol{p}_1|^2+|\boldsymbol{p}_2|^2)+g_V{}^2\left[\alpha_--\frac{1}{2}(|\boldsymbol{p}_1|^2+|\boldsymbol{p}_2|^2)\right]\right\}\langle rH'\rangle_\Delta{}^2$$
$$+(9g_V{}^2/2)(p_1{}^0-p_2{}^0)^2\langle H\rangle_\Delta{}^2+3g_V{}^2(p_1{}^0-p_2{}^0)^2\langle rH'\rangle_\Delta\langle H\rangle_\Delta\,, \tag{C·16}$$

$$\beta_2=\left\{g_A{}^2[5\alpha_+(|\boldsymbol{p}_1|^2+|\boldsymbol{p}_2|^2)-|\boldsymbol{p}_1|^2|\boldsymbol{p}_2|^2]-\frac{1}{2}g_V{}^2\alpha_-(|\boldsymbol{p}_1|^2+|\boldsymbol{p}_2|^2)\right\}\langle rH'\rangle_\Delta{}^2$$
$$+(27g_V{}^2/2)\alpha_+(p_1{}^0-p_2{}^0)^2\langle H\rangle_\Delta{}^2-3g_V{}^2\alpha_+(p_1{}^0-p_2{}^0)^2\langle rH'\rangle_\Delta\langle H\rangle_\Delta\,. \tag{C·17}$$

Here $g_A=g_V$ is understood. New terms proportional to $\langle H\rangle_\Delta$ which do not exist in Eq. (2·47) come from the $\eta L_{LL}^{\mu\nu0}$ term. The $g_A{}^4$ term in (C·15) gives exactly the same behavior as the one in (2·47) but the $g_A{}^2g_V{}^2$ term differs. However, since the $g_A{}^4$ term is expected to contribute dominantly, both decay formulae give similar results. Primakoff and Rosen[1] have worked out the decay formula by taking only the $g_A{}^2g_V{}^2$ term into account.

References

1) H. Primakoff and S. P. Rosen, Phys. Rev. **184** (1969), 1925.
2) B. Pontecorvo, Soviet Phys.-JETP **6** (1957), 429; **7** (1958), 172.
Z. Maki, M. Nakagawa and S. Sakata, Prog. Theor. Phys. **28** (1962), 870.

V. Gribov and B. Pontecorvo, Phys. Letters **28B** (1969), 493.
S. M. Bilenky, J. Hosek and S. T. Petcov, Phys. Letters **94B** (1980), 495.
T. Yanagida and M. Yoshimura, Prog. Theor. Phys. **64** (1980), 1870.
M. Doi, T. Kotani, H. Nishiura, K. Okuda and E. Takasugi, Phys. Letters **102B** (1981), 323.
3) M. G. Mayer, Phys. Rev. **48** (1935), 512.
W. H. Furry, Phys. Rev. **56** (1939), 1184.
D. Smith, C. Picciotto and D. Bryman, Phys. Letters **46B** (1973), 157.
D. Bryman and C. Picciotto, Rev. Mod. Phys. **50** (1978), 11.
4) H. Primakoff and S. P. Rosen, Reports Prog. Phys. **22** (1959), 121; Proc. Phys. Soc. **78** (1961), 464.
5) E. Greuling and R. C. Whitten, Ann. of Phys. **11** (1960), 510.
6) E. J. Konopinski, *Theory of Beta Radioactivity* (Oxford Univ. Press, London, England, 1966).
7) A. Halprin, P. Minkowski, H. Primakoff and S. P. Rosen, Phys. Rev. **D13** (1976), 2567.
8) C. Picciotto, Can. J. Phys. **56** (1978), 399.
9) A. Molina and P. Pascual, Nuovo Cim. **41** (1977), 756.
10) M. Kenmoku, private communication.
11) R. J. Blin-Stoyle, *Fundamental Interactions and the Nucleus* (North-Holland, Amsterdam, 1973), p. 29.

1765

Progress of Theoretical Physics, Vol. 66, No. 5, November 1981

Neutrino Mass, the Right-Handed Interaction and the Double Beta Decay. II

—— *General Properties and Data Analysis* ——

Masaru DOI, Tsuneyuki KOTANI,* Hiroyuki NISHIURA,* Kazuko OKUDA* and Eiichi TAKASUGI*

Osaka College of Pharmacy, Matsubara, Osaka 580
**Institute of Physics, College of General Education Osaka University, Toyonaka 560*

(Received March 30, 1981)

Based on the formulae for the double β decay obtained in the previous paper, the general properties of $0^+ \to J^+$ transitions are discussed and the analysis of the experimental data is presented. It is found that, for the two neutrino mode, the $0^+ \to 0^+$ transition in the two nucleon ($2n$)-mechanism dominates over the $0^+ \to 2^+$ transition as well as the contribution from the N^*-mechanism. For the neutrinoless mode, only the $0^+ \to 0^+$ transition in the $2n$-mechanism is allowed if there is no right-handed interaction. When the right-handed interaction gives a sizable contribution, the role of the $0^+ \to 2^+$ transition becomes as important as the $0^+ \to 0^+$ transition in this mode. It is concluded that the experimental data on the ratio of the ^{128}Te to ^{130}Te half-lives by Hennecke et al. suggest that neutrinos are Majorana particles, if we take the Vergados estimation of the nuclear matrix elements. Moreover, we find that the weighted average of neutrino masses is around 34 eV if there is no right-handed interaction.

§ 1. Introduction

The interest in the double β decay has been revived recently. This is because the double β decay offers one of the best tools to tell whether neutrinos are Dirac or Majorana particles. More precisely, the neutrinoless double β decay (the $(\beta\beta)_{0\nu}$ mode) involves the change of the total lepton number and occurs only if neutrinos are Majorana, while the two neutrino double β decay (the $(\beta\beta)_{2\nu}$ mode) occurs independently of their nature. Interestingly enough, in 1970, the Columbia group[1)] found one event which might be from the $(\beta\beta)_{0\nu}$ mode for ^{48}Ca. From the ratio of the ^{128}Te to ^{130}Te half-lives obtained geologically by Hennecke et al.,[2)] we suggested[3)] that the sizable contribution from the $(\beta\beta)_{0\nu}$ mode exists in the double β decay of ^{128}Te, i.e., neutrinos are Majorana.

In the previous paper[4)] (hereafter referred to as Part I), the theoretical analysis of the double β decay has been presented. The decay formulae have been derived for the parity conserving $0^+ \to J^+$ transitions of the $(\beta\beta)_{0\nu}$ and $(\beta\beta)_{2\nu}$ modes both in the two nucleon ($2n$)-and N^*-mechanism. In these calculations, we used the weak interaction Hamiltonian motivated by the grand unified theories. Our interaction includes one parameter λ which represents the relative

Progress of Theoretical Physics, Vol. 69, No. 2, February 1983

Double Beta Decay

Masaru DOI,* Tsuneyuki KOTANI, Hiroyuki NISHIURA**,*) and Eiichi TAKASUGI

Institute of Physics, College of General Education
Osaka University, Toyonaka 560
**Osaka College of Pharmacy, Matsubara, Osaka 580*
***Research Institute for Fundamental Physics*
Kyoto University, Kyoto 606

(Received August 16, 1982)

The double β decay is investigated by the use of the relativistic Coulomb wave function including the finite nuclear size effect. A general form of the $V \pm A$ interactions is considered. It is shown that the decay rate is enhanced considerably in comparison with the case of the plane wave multiplied by the non-relativistic Fermi factor.

It is confirmed that the selection rule derived on the basis of the plane wave is still valid in the relativistic Coulomb wave case. The neutrino mass and the $V+A$ current parameters are estimated from the analysis of experimental data.

§1. Introduction

The recent development of the grand unified theories has revived the interest in the double beta ($\beta\beta$) decay, which appears to be one of the most sensitive tests for the lepton-quark symmetry. Recently many works on this decay have been made in order to explore the lepton number non-conservation, the mass of neutrino and the existence of the right-handed current.[1)~9)]

In this paper, we present a detailed analysis of the $\beta\beta$ decay by treating the Coulomb distortion of the electron wave function relativistically with inclusion of the nuclear finite size effect.[10)] Also a general form of the weak interaction Hamiltonian is used. The necessity of the relativistic treatment is motivated from the observation by Haxton, Stephenson and Strottman[6)] and Nishiura[4)] that it gives the considerable enhancement of the decay rate in comparison with the use of the plane wave function multiplied by the non-relativistic form of Fermi factor. The latter function has been used in our previous papers[1)] (hereafter referred to as I).

As in I, two mechanisms are considered: One is the two-nucleon ($2n$)-mechanism[11),12)] where the successive transitions of two neutrons trigger the $\beta\beta$ decay. The other is the N^*-mechanism[13)] where the $\beta\beta$ decay occurs through the transition of $\Delta(1232)$ inside the nuclei.

---

*) Fellow of the Japan Society for the Promotion of Science.

Progress of Theoretical Physics, Vol. 70, No. 5, November 1983

The Energy Spectrum and the Angular Correlation in the $\beta\beta$ Decay

Masaru DOI, Tsuneyuki KOTANI,* Hiroyuki NISHIURA* and Eiichi TAKASUGI*

Osaka College of Pharmacy, Matsubara 580
**Institute of Physics, College of General Education*
Osaka University, Toyonaka 560

(Received June 11, 1983)

In the $\beta\beta$ decay, there are in general three parameters: the neutrino mass (m_ν) and two right-handed weak current parameters (λ and η). It is shown how to distinguish their contrib utions experimentally. Two useful measurable quantities are the single electron kinetic energy spectrum and the angular correlation between two emitted electrons. The numerical results are given for the ^{76}Ge case.

§ 1. Introduction

In the $\beta\beta$ decay, it is possible to draw direct information on the neutrino mass (m_ν) and two right-handed weak current parameters (λ and η), if neutrinos are the Majorana particles.[*),1)~3)] These three parameters are assumed to be zero in the present standard electroweak theory.

In this paper, we shall explore the essential features of terms proportional to these three parameters, and discuss how we are able to distinguish them experimentally.

The right-handed coupling parameter λ comes in through the existence of the right-handed weak gauge boson (W_R), and is determined as $\lambda=(m_{W_L}/m_{W_R})^2$ where m_{W_L} and m_{W_R} stand for the masses of the left- and right-handed weak gauge bosons, respectively.

If W_L and W_R are the mixing between two mass eigenstates with masses m_{W_1} and m_{W_2}, it is possible in general to have two parameters λ and η, which are different combinations of m_{W_1} and m_{W_2}, and a mixing angle ζ.[4)] In this mixing scheme, if λ becomes zero, η should be zero.[5)] As the smallness of the W_L-W_R mixing has been argued from the low energy data,[6)] it may be expected to be $|\lambda|\gg|\eta|$. However, it is possible to construct another scenario, i.e., the mixing between the left-handed particles and their mirror particles (the right-handed particles interacting with W_L).[7)] In this mixing scheme, we have another set of parameters, η and x. The parameter x has no measurable contribution to the $\beta\beta$ decay, as discussed in Eq. (2·10) of our previous paper,[1)] which will be referred to as I. We shall treat λ and η as two independent parameters, because of these two mixing schemes.

Let us explain briefly how the m_ν, λ and η terms show different characters. For this purpose, we shall consider the $0^+\to0^+$ transition in the neutrinoless $\beta\beta$ decay mode, the $(\beta\beta)_{0\nu}$ mode, and assume that it occurs in the second order perturbation.

---

*) The parameters λ and η are defined phenomenologically through the following weak interaction Hamiltonian,

$$H_W=(G\cos\theta_C/\sqrt{2})[j_{L\mu}J_L^{\mu\dagger}+\lambda j_{R\mu}J_R^{\mu\dagger}+\eta j_{R\mu}J_L^{\mu\dagger}+x j_{L\mu}J_R^{\mu\dagger}]+\text{h.c.} \qquad (1\cdot1)$$

Here j_μ and J_μ are the weak leptonic and hadronic currents, respectively.[1)]

Progress of Theoretical Physics Supplement No. 83, 1985 1

Double Beta Decay and Majorana Neutrino

Masaru DOI, Tsuneyuki KOTANI* and Eiichi TAKASUGI*

Osaka College of Pharmacy, Matsubara, Osaka 580
**Institute of Physics, College of General Education, Osaka University, Toyonaka 560*

(Received August 15, 1985)

This review consists of three parts: Various properties of the quantized neutrino fields are summarized in part I from the viewpoint that a Dirac neutrino consists of two Majorana neutrinos with a degenerate mass but with opposite *CP* signs. It is shown why the Dirac neutrino has a freedom of the phase transformation to guarantee the lepton number conservation, while the Majorana neutrino does not.

Part II is devoted to the theoretical investigation of the double beta decay and the analysis on the experimental data. Various corrections due to the vanishing terms in the closure approximation and the higher spherical waves of leptons are estimated. The constraints on the effective neutrino mass and the effective right-handed weak interaction parameters are derived from the present experimental data. In part III, the above bounds are compared with those from other processes like the ^{3}H decay, the neutrino oscillation, the π-μ-e decay, etc. The possible mixing schemes of the neutrino system are explored.

Contents

PHYSICAL REVIEW D VOLUME 37, NUMBER 9 1 MAY 1988

Neutrinoless double-beta decay with Majoron emission

M. Doi
Osaka University of Pharmaceutical Sciences, Matsubara 580, Japan

T. Kotani* and E. Takasugi
Institute of Physics, College of General Education, Osaka University, Toyonaka 560, Japan
(Received 18 September 1987)

The neutrinoless $\beta\beta$ decay with Majoron emission is examined in detail for the $0^+ \to 0^+$ and $0^+ \to 2^+$ transitions. Especially, the quantitative analysis is made on the correction term which was not dealt with properly before. We found that the correction term does not change the fundamental characters of the energy spectra of emitted electrons and the angular correlation obtained from the main term. The dominant contribution comes from the main term through two light neutrino propagations, except some special cases which have the large mixings between light and heavy neutrinos. Also it is shown that the $0^+ \to 2^+$ transition occurs only through the correction term, and it is by at least $\sim 10^{-7}$ smaller than the $0^+ \to 0^+$ transition.

I. INTRODUCTION

Recently, the Battelle–South Carolina (BS) group reported[1] the observation of the possible Majoron-emitting $\beta\beta$ decay of ^{76}Ge. The reported half-life was $T_{0\nu\beta}(0^+ \to 0^+) = (6\pm1)\times10^{20}$ yr. This aroused widespread interest, because the detection of this mode means the direct observation of the Nambu-Goldstone boson and the Majorana neutrinos. Subsequently, the Caltech-SIN-Neuchatel[2] and Santa Barbara–LBL (SBL) groups[3] gave bounds $T_{0\nu\beta}(0^+ \to 0^+) > 1.2\times10^{21}$ and 1.4×10^{21} yr, respectively, for the half-life of the ^{76}Ge decay, while the Irvine group[4] gave a bound $T_{0\nu\beta}(0^+ \to 0^+) > 4.4\times10^{20}$ yr for the ^{82}Se decay. All of them seem to exclude the data

TABLE I. The data of the half-lives of the neutrinoless double-beta decay with Majoron emission for various nuclei and the bounds on $\langle g_B \rangle$. The bounds on $\langle g_B \rangle$ are obtained by using the nuclear matrix elements, $M_{GT}^{(0\nu)}(1-\chi_F)$, in Refs. 8, 14, 24, 25, and 26. In order to derive the bounds indicated by r (Ref. 26), we adopted the values of nuclear matrix elements explicitly given in Ref. 27. Concerning the ratio of half-lives of ^{128}Te to ^{130}Te obtained by the Heidelberg group, its central value contradicts the theoretical estimates of the ratio for the $(\beta\beta)_{2\nu}$ mode by Refs. 14, 24, and 26. Therefore, the uncertainty of the two standard deviation is taken into account, i.e., $(R_T)^{-1} < 3.29\times10^{-4}$. Since the ratio estimated in Ref. 26 is still outside this extended limit, the bound on $\langle g_B \rangle$ is zero if both results are taken seriously. R_T is the ratio of half-lives of ^{128}Te to ^{130}Te.

	Experimental group	Half-life (yr)	$10^4 \mid \langle g_B \rangle \mid$
^{76}Ge	Osaka[a]	$>2\times10^{20}$	$<(5.0^m, 13^n, 15^p, 34^q, 100^r)$
	Moscow (ITEP)[b]	$>2\times10^{20}$	$<(5.0^m, 13^n, 15^p, 34^q, 100^r)$
	Batelle–South Carolina[c]	$(6\pm1)\times10^{20}$	$(2.9^m, 7.4^n, 8.6^p, 19^q, 60^r)$
	Caltech-SIN-Neuchatel[d]	$>12\times10^{20}$	$<(2.0^m, 5.3^n, 6.1^p, 14^q, 42^r)$
	Santa Barbara–LBL[e]	$>14\times10^{20}$	$<1.9^m, 4.8^n, 5.6^p, 13^q, 39^r)$
^{82}Se	Irvine[f]	$>4.4\times10^{20}$	$<(1.5^m, 3.7^n, 3.8^p, 9.8^q, 33^r)$
^{100}Mo	Osaka[g]	$>6\times10^{18}$	$<(7.2^m, 130^r)$
	Irvine[h]	$>7.5\times10^{18}$	$<(6.4^m, 120^r)$
^{128}Te/^{130}Te	Missouri[i]	$R_T = 2\times10^3$	$<(0.44^m, 2.4^n, 1.3^r)$
	Heidelberg[j]	$R_T^{-1} = (1.03\pm1.13)\times10^{-4}$	$<(0.28^m, 1.5^n)$
^{136}Xe	Milano[k]	$>1.6\times10^{19}$	$<(15^m, 110^r)$
	Moscow (INR)[l]	$>1.0\times10^{20}$	$<(5.9^m, 42^r)$
^{150}Nd	Moscow (INR)[s]	$>7.0\times10^{19}$	

[a]Reference 16.
[b]Reference 17.
[c]Reference 1.
[d]Reference 2.
[e]Reference 3.
[f]Reference 4.
[g]Reference 18.
[h]Reference 19.
[i]Reference 20.
[j]Reference 21.
[k]Reference 22.
[l]Reference 23.
[m]Reference 24.
[n]Reference 14.
[p]Reference 25.
[q]Reference 8.
[r]Reference 26.
[s]Reference 29.

Progress of Theoretical Physics, Vol. 89, No. 1, January 1993

Neutrinoless Modes of Double Beta Decay

Masaru DOI and Tsuneyuki KOTANI*

Osaka University of Pharmaceutical Sciences, Matsubara 580
**Osaka Gakuin University, Suita 564*

(Received August 17, 1992)

Three types of nuclear double beta decays without emission of neutrinos are compared; (1) capture of an atomic electron and emission of a positron, (2) emission of two positrons and (3) emission of two electrons. Captures from the K and L_I shells are taken into account. Half-lives for the $0^+\to 0^+$ transitions are calculated by using relativistic wave functions. They are shorter than those obtained by using solutions of the Schrödinger equation. For the case of one positron emission, the ^{124}Xe decay is expected to be the shortest half-life, if there is no anomalous cancellation of the nuclear matrix elements. It is also discussed that the transition rate for the capture of double atomic electrons is smaller than the one for the single atomic electron capture.

§ 1. Introduction

The possibility of the non-vanishing neutrino mass has been suggested from the ıark-lepton symmetry in a model such as GUTS. If the neutrino has a finite mass, ere arises another question whether it is the Dirac or Majorana type. All present ꞇperimental data indicate no deviation from the minimum standard model of the ectroweak interaction in which both the zero neutrino mass and the left-handed eak $(V-A)$ interaction are assumed. However, why does the nature restrict the ft-right symmetry to only the $(V-A)$ interaction? These have long been main ıbjects in particle physics, because they provide us with issues for the new physics ɜyond the standard model. In spite of development of the gauge theory in the last vo decades, these questions have not yet been settled.

The double beta decay without emission of neutrinos (the 0ν mode) is a sensitive robe to the above questions: The 0ν mode takes place only when the neutrino is the Iajorana type with the finite mass. In addition, if the 0ν mode is observed experi-ıentally, some information is obtained on both the right-handed weak $(V+A)$ ıteraction and the existence of right-handed heavy Majorana neutrinos. But no ecisive experimental indication of this mode has been reported yet.

So far, the most attention has been paid to the $(\beta^-\beta^-)_{0\nu}$ mode,

$$\text{the}\quad (\beta^-\beta^-)_{0\nu}\quad \text{mode;}\quad (A, Z-2)\to(A, Z)+2\beta^-, \qquad (1\cdot1)$$

ʼhere A and Z denote mass and atomic numbers of the daughter nuclei, respectively. ɔn the other hand, there is another decay mode,

$$\text{the}\quad (\beta^+\beta^+)_{0\nu}\quad \text{mode;}\quad (A, Z+2)\to(A, Z)+2\beta^+. \qquad (1\cdot2)$$

One of purposes of this paper is to make the difference between these decay rates lear by using the relativistic Coulomb wave functions for the case of a uniform harge distribution inside the nucleus. In the neutrino emitting mode (the 2ν mode), he decay rate for the $(\beta^+\beta^+)_{2\nu}$ mode is smaller by the order of 10^{-7} than for the

DOUBLE BETA DECAY*

W. C. HAXTON

Department of Physics, Princeton University, Princeton, NJ 08540 and
Theoretical Division, Los Alamos National Laboratory, Los Alamos, NM 87545, U.S.A.†

and

G. J. STEPHENSON, Jr.

Physics and Theoretical Divisions, Los Alamos National Laboratory, Los Alamos, NM 87545, U.S.A.

PERGAMON PRESS

Progress in
PARTICLE AND NUCLEAR PHYSICS
Volume 12
Edited by
Sir Denys Wilkinson, F.R.S.
Vice-Chancellor, University of Sussex

CONTENTS

* Work performed under the auspices of the U.S. Department of Energy and the National Science Foundation.
† Permanent address.

Are massive neutrinos Majorana particles?

Tsuneyuki Kotani

Institute of Physics, College of General Education
Osaka University
Toyonaka, Osaka 560, Japan

ABSTRACT

One Dirac field can be constructed from two Majorana fields theoretically. The origin of the different properties between the Dirac and Majorana neutrinos is explained. The theory of the neutrinoless double beta decay is briefly reviewed, and the data analysis and the ambiguities due to the nuclear matrix elements are discussed. Consistencies are considered for various experimental bounds from the neutrinoless double beta decay, the tritium beta decay and the neutrino oscillations. A consistent solution for these bounds is shown for the SO(10) model with three generations.

§ 1. Introduction

Among several assumptions in the minimum standard model of electroweak interaction, two followings will be discussed;

(1) Are all neutrinos massless, $m_\nu = 0$? If $m_\nu \neq 0$, are neutrinos Dirac or Majorana particles?

(2) Is the gauge group only $SU(2)_L \times U(1)$, i.e., the V-A interaction?

At present, there is no definite result for the finite mass of neutrino, except the experiment on the tritium β decay by the Moscow group. However, there are two indirect evidences indicating the finite neutrino mass; the Mikheyev-Smirnov-Wolfenstein enhancement of oscillation as a possible solution to the solar-neutrino problem and the dark matter problem in the astrophysics.

If neutrinos are massive, we can say that the Majorana neutrino is theoretically more fundamental than the Dirac neutrino, because one massive Dirac field consists of two Majorana fields (§ 2 and § 3). As

---

in Proc. 12th International Conference on "Neutrino Physics and Astrophysics", Sendai, Japan, June 3 - 8, 1986, eds *T. Kitagaki and H. Yuta, World Scientific, Singapore* (1986) 114 – 134

Double Beta Decay, Neutrino Mass and Nuclear Structure

K. Muto† *and H.V. Klapdor*

Max-Planck-Institut für Kernphysik, Saupferchweg 1,
D-6900 Heidelberg, Fed. Rep. of Germany

in "Neutrinos", Graduate texts in contemporary physics", ed. *H. V. Klapdor, Berlin, Germany: Springer* (1988) 183 – 238

1. Introduction

Nuclear double beta ($\beta\beta$) decay is one of the rarest processes in nature, with a half-life of the order of $\sim 10^{20}$ years or longer [Kir 67,68,69, Pri 81, Hax 84, Doi 85, Kir 86, Ver 86, Gro 86, Ell 87, Mut 88a]. It can become observable only when a nucleus cannot undergo ordinary single β decay because of energy conservation, or of very strong suppression of energetically allowed transitions by a large spin difference between the parent and daughter states (for example, ^{48}Ca). All $\beta\beta$ emitters and their daughter nuclei have even numbers of protons and neutrons, since the pairing force, which is a prominent feature of the nucleon-nucleon interaction acting between like nucleons, makes even-even nuclei more bound relative to the neighbouring odd-odd nuclei. As an example the mass spectrum for A = 76 isobars is shown in Fig. 1.1. For ^{76}Ge, β^- decay to ^{76}As and β^+/EC to ^{76}Ga are both forbidden, and consequently $\beta\beta$ decay to ^{76}Se is the only possible decay mode. In heavy-mass regions, even if a nucleus is stable against $\beta^\pm$ decay, the α-particle emission channel is usually open and $\beta\beta$ decay is hardly observable. (This will affect nuclei beyond ^{204}Hg in Table 5.3.)

A $\beta\beta$ decay is expected to take place mainly via the two decay modes of the two-nucleon mechanism (for other mechanisms, see sect. 3.4): the two-neutrino (2ν) mode in which two neutrinos are emitted together with two electrons, and the neutrinoless (0ν) mode in which no neutrino leaves the nucleus,

$$2\nu \text{ mode}: \quad (A,Z) \to (A,Z+2) + 2e^- + 2\bar{\nu} \tag{1.1a}$$

$$0\nu \text{ mode}: \quad (A,Z) \to (A,Z+2) + 2e^-. \tag{1.1b}$$

The 2ν mode (Fig. 1.2a) resembles two successive single β^- transitions with the exception that the intermediate states are virtual (see Fig. 3.1). It is described by second-order perturbation of the weak interaction within the standard electroweak model [Gla 61, Wei 67, Sal 68] irrespective of the nature of the neutrino, Dirac ($\nu \neq \nu^c$) or Majorana ($\nu = \nu^c$), massless or massive (ν^c denotes the charge-conjugate state (antiparticle)). On the other hand, the 0ν mode (Fig. 1.2b) is a lepton-number non-conserving process, and is forbidden in the Standard Model. It occurs when the neutrino emitted at one vertex can be absorbed by the nucleon at the other vertex. Therefore two conditions must be satisfied: (i) the neutrino is a Majorana particle, and (ii) the neutrino has a finite mass and/or there exists a right-handed component in the charged weak interaction. It should be noted that observation of 0ν $\beta\beta$ decay *always* implies that the electron neutrino is a massive Majorana particle [Sch 82, Nie 84, Tak 84]. Variants of 0ν $\beta\beta$ decay other than eq.(1.1b), among them decays implying the emission of a majoron, are discussed in sect. 3.4.

Historically, $\beta\beta$ decay was regarded as a sensitive test of the Dirac/Majorana character of the neutrino. The 2ν mode was first proposed by Goeppert-Mayer in 1935 [Goe 35], who estimated, on the basis of Fermi's β decay theory [Fer 34], half-lives of the order of 10^{20} years. In 1937, Majorana developed a new theory of the neutrino [Maj 37], a two-component theory in contrast to Dirac's

† On leave of absence from Department of Physics, Tokyo Institute of Technology, Japan.

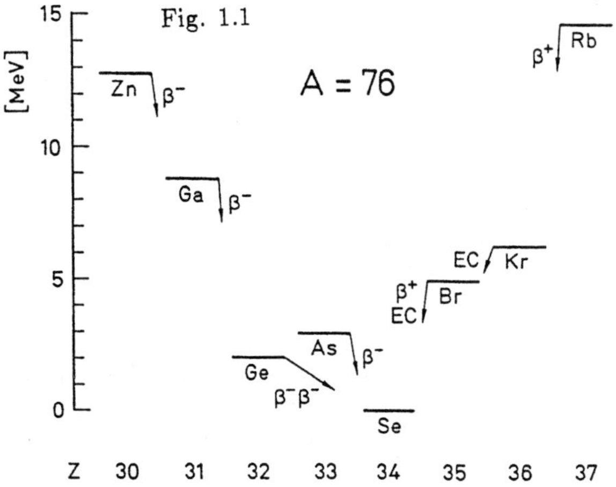

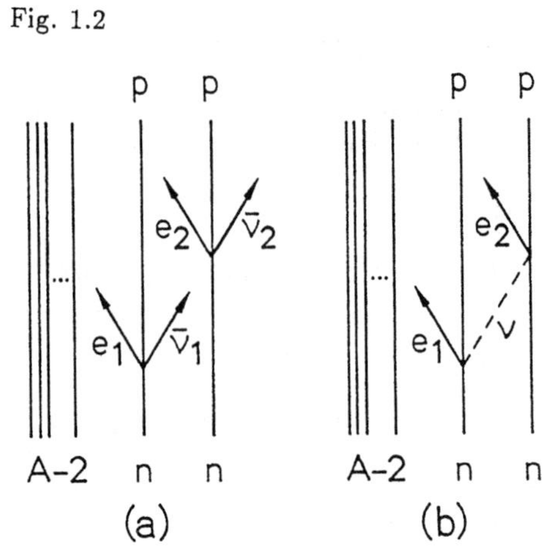

Fig. 1.1 Mass spectrum of isobars with mass number A = 76. Decay modes of the ground states are given. The $\beta^-\beta^-$ decay is the only possible mode for ^{76}Ge.

Fig. 1.2 The two modes of $\beta\beta$ decay possible in the two-nucleon mechanism; (a) the two-neutrino (2ν) mode, and (b) the neutrinoless (0ν) mode.

four-component theory. Furry suggested in 1939 the possibility of the 0ν mode [Fur 39] in which a Majorana neutrino is exchanged between two nucleons. The half-lives, calculated for various possible weak interactions, were shorter than the 2ν decay half-lives by large factors, 10^5 or more. Experimental efforts in the late 1940's and early 1950's (the first experiment devoted to search for the 0ν mode was performed just 40 years ago [Fir 49]) to look for the favoured 0ν decay resulted in negative evidence. The lower limits on half-lives exceeded by far the theoretical values for the 0ν mode. This result implied that the electron neutrino is a Dirac particle, and suggested the introduction of lepton number, which distinguishes the neutrino from the antineutrino as a conserved quantity. In 1955, Davis tried a neutrino capture experiment [Dav 55], using antineutrinos $\bar{\nu}$ from fission products in a reactor,

$$\text{(Reactor) } n \to p + e^- + \bar{\nu}$$
$$\nu + {}^{37}\text{Cl} \to {}^{37}\text{Ar} + e^- . \tag{1.2}$$

If the antineutrino $\bar{\nu}$ and the neutrino ν are, in fact, identical, the second reaction should be possible and ^{37}Ar should have been observed. However, this reaction did not show up, apparently confirming the Dirac character of the neutrino.

Soon after the Davis experiment, it was found that parity is maximally violated in the weak interaction, and its V-A form was established [Lee 56, Wu 57]. This implies that, as a massless (anti)neutrino has a definite helicity, the reaction of eq.(1.2) is forbidden even if the neutrino is a Majorana particle, i.e., if $\nu = \nu^c$. The neutrino emitted in the first reaction has positive helicity (right-handed) and the neutrino to be absorbed in the second reaction should have negative helicity (left-handed). This is also the case for the 0ν $\beta\beta$ decay. Thus, the interest in $\beta\beta$ decay waned after the discovery of parity violation and the establishment of lepton-number conservation. At the same time the question about the Dirac or Majorana character of the neutrino was reopened.

The development of modern gauge theories has revitalised the study of $\beta\beta$ decay. The standard electroweak theory unifies the electromagnetic and weak interactions in the framework of a $SU(2)_L \times U(1)$ gauge theory, and is quite successful in describing the known experimental data. This model contains no right-handed neutrinos, and the left-handed neutrino in the theory must be massless. The same is true for the minimal SU(5) model. However, massive neutrinos naturally appear in grand unified theories based on the SO(10) gauge group and in some superstring models (see [Lan 81,86,88, Moh 86,88]). They also imply the existence of both left-handed and right-handed Majorana neutrinos and of right-handed components of the weak interaction, but provide little information on precise values of the neutrino masses. The predicted values for light neutrinos lie anywhere between 10^{-11} eV and $\sim$ eV (see Langacker, Mohapatra in this volume).

The 0ν $\beta\beta$ decay is the most sensitive test of the existence of Majorana neutrinos, and sets, at present, the most stringent limits on the (electron) neutrino mass. Thus 0ν $\beta\beta$ decay experiments belong to the most important present-day underground physics experiments. Observation of 0ν $\beta\beta$ decay would be equivalent to uncovering new physics beyond the standard theory. The mass scales of grand unified theories probed by a finite neutrino mass are far beyond direct reach by present accelerators. On the other hand, reliable deductions of neutrino masses from measured 0ν $\beta\beta$ decay rates depend decisively on reliable calculations of nuclear matrix elements.

2. Majorana neutrinos and grand unified theories

Let us consider left-handed and right-handed fermion fields, ψ_L and ψ_R. They transform under the proper Lorentz transformation as (0,1/2) and (1/2,0), respectively, where (j_1,j_2) is the irreducible representation of the Lorentz group SO(1,3) [Wei 64]. For the sake of transparency, we define[1] the charge-conjugate fields as $(\psi_L)^c = C\bar{\psi}_L^T$ and $(\psi_R)^c = C\bar{\psi}_R^T$, with the charge-conjugation matrix C. Here, ψ^T denotes the transpose of ψ, and $\bar{\psi} = \psi^\dagger\gamma^0$. In the Weyl representation of γ matrices,

$$\gamma^0 = \begin{bmatrix} 0 & 1 \\ 1 & 0 \end{bmatrix}, \qquad \gamma^k = \begin{bmatrix} 0 & -\sigma_k \\ \sigma_k & 0 \end{bmatrix}, \qquad \gamma_5 = \begin{bmatrix} 1 & 0 \\ 0 & -1 \end{bmatrix},$$
$$C = i\gamma^2\gamma^0 = \begin{bmatrix} -i\sigma_2 & 0 \\ 0 & i\sigma_2 \end{bmatrix}. \tag{2.1}$$

The charge-conjugate field $(\psi_L)^c$ ($(\psi_R)^c$) behaves like ψ_R (ψ_L), in the sense that it is a right-handed (left-handed) field and transforms as (1/2,0) ((0,1/2)) under a Lorentz transformation. Therefore ψ_L can mix with $(\psi_R)^c$, and ψ_R with $(\psi_L)^c$. In terms of these four fields, the most general mass Lagrangian which is Hermitian and Lorentz invariant is

$$L_m = -\tfrac{1}{2}(\bar{\psi_L})^c m_L^M \psi_L - \tfrac{1}{2}\bar{\psi}_R m_R^M (\psi_R)^c - \bar{\psi}_R m^D \psi_L + h.c. \tag{2.2}$$

($h.c.$ denotes the hermitian conjugate of the three preceding terms) or in matrix form

$$L_m = -\tfrac{1}{2}\left[(\bar{\psi_L})^c \quad \bar{\psi}_R\right] M \begin{bmatrix} \psi_L \\ (\psi_R)^c \end{bmatrix} + h.c. \tag{2.3}$$

$$M = \begin{bmatrix} m_L^M & m^D \\ m^D & m_R^M \end{bmatrix}. \tag{2.4}$$

A Dirac mass m^D results from the coupling of the independent left-handed and right-handed fields, whereas Majorana masses m_L^M and m_R^M come from the coupling of fields with their charge-conjugate fields. The mass coupling scheme is illustrated in Fig. 2.1.

Charged fermions are forbidden to couple via the Majorana mass because of charge conservation, and have only Dirac masses. For example (Fig. 2.1b), the electron with a negative charge does not couple with the positively charged positron. Coupling is allowed only between the left-handed and right-handed components of the electron (positron) by the Dirac mass term,

$$L_m = -\bar{e}_R m^D e_L + h.c. = -m^D \bar{e}e. \tag{2.5}$$

The Dirac mass term is invariant with respect to the global gauge transformation $e_{L,R} \to e_{L,R}$ $\exp(i\phi)$. This invariance implies conservation of the lepton number: the electron can be distinguished from its charge-conjugate field, namely the electron is a Dirac particle.

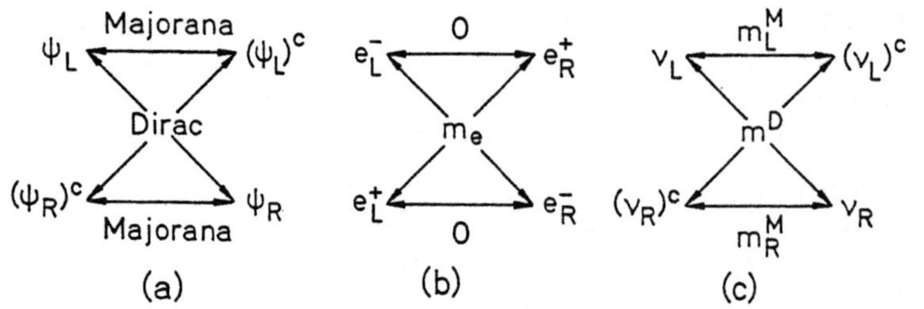

Fig. 2.1 Coupling schemes of fermion fields through Majorana and Dirac masses. (a) The general coupling scheme for the left- and right-handed fields and their charge conjugate fields. (b) The coupling scheme for electrons as an example of charged fermions. (c) The coupling scheme for neutrinos. Only neutrinos can have the general coupling via both Majorana and Dirac masses.

The neutrino is the only fermion that can have the general mass matrix consisting of Dirac as well as of Majorana mass terms (Fig. 2.1c), and can have the Majorana self-conjugate character, $\psi = \psi^c$. Mass eigenstates are obtained by diagonalizing the mass matrix, eq.(2.4). For simplicity, we assume here that CP is conserved. Then the mass matrix is real (symmetric), and is diagonalized by the orthogonal matrix O as

$$\begin{bmatrix} \nu_L \\ (\nu_R)^c \end{bmatrix} = O \begin{bmatrix} N'_{1L} \\ N'_{2L} \end{bmatrix} \tag{2.6a}$$

$$O^T M O = \begin{bmatrix} s_1 m_1 & 0 \\ 0 & s_2 m_2 \end{bmatrix} \tag{2.6b}$$

where $m_j > 0$ and s_j is a sign factor which can be $+1$ or -1. The sign factors can be absorbed in the fields and the mixing matrix [Sch 80, Doi 83b] by introducing an additional phase factor which is i for $s_j = -1$ (CP odd) and 1 for $s_j = +1$ (CP even). When $s_1 = -1$ and $s_2 = +1$,

$$\begin{bmatrix} N_{1L} \\ N_{2L} \end{bmatrix} = \begin{bmatrix} i & 0 \\ 0 & 1 \end{bmatrix} \begin{bmatrix} N'_{1L} \\ N'_{2L} \end{bmatrix} \tag{2.7}$$

$$U_\nu = O \begin{bmatrix} -i & 0 \\ 0 & 1 \end{bmatrix} = \begin{bmatrix} U \\ V^* \end{bmatrix} \tag{2.8}$$

$$\begin{bmatrix} \nu_L \\ (\nu_R)^c \end{bmatrix} = U_\nu \begin{bmatrix} N_{1L} \\ N_{2L} \end{bmatrix}. \tag{2.9}$$

The mass Lagrangian is then written as

$$L_m = -\frac{1}{2} \sum_{j=1}^{2} m_j \bar{N}_j N_j \tag{2.10}$$

with

$$N_j = N_{jL} + (N_{jL})^c. \tag{2.11}$$

Obviously, the mass eigenstate N_j has the mass m_j and satisfies the Majorana self-conjugate condition, $N_j = (N_j)^c$. This mass term is not invariant under any global gauge transformation of the neutrino fields, and can change the lepton number by two units.

One can have the following special cases in eq.(2.4):

a) The "see-saw" mechanism [Yan 79, Gel 79], see also [Moh 80, Doi 85] and Langacker in this volume.

Assuming $m_L^M = 0$, $m_R^M \gg m^D$, with m^D of the order of the charged lepton or quark masses, i.e $m_L^M \ll m^D \ll m_R^M$, diagonalization of the mass matrix leads to the following mass eigenvalues

$$m_1 \sim (m^D)^2/m_R^M \qquad \text{(light neutrino)} \qquad (2.12a)$$

$$m_2 \sim m_R^M \qquad \text{(heavy neutrino).} \qquad (2.12b)$$

In general, the right-handed Majorana mass m_R^M is assumed to be produced when the left-right symmetry is broken, while the Dirac mass m^D is induced when the $SU(2)_L \times U(1)$ symmetry is broken. The "see-saw" mechanism proves to be useful in reducing the large Dirac mass of the neutrino automatically present in the SO(10) model (see below) by an interplay between the large Dirac mass terms and an even larger Majorana mass term. It explains "automatically" the small mass of the observed left-handed neutrino in comparison with the charged fermion masses. The condition $m_R^M \geq 10^3$ GeV would lead to $m_1 \leq 1$ eV. The neutrino in this case is a Majorana particle.

b) Pseudo-Dirac neutrino [Wol 81, Pet 82, Doi 83b, Leu 83, Val 83a,83b, Wyl 83].
For a pseudo-Dirac neutrino, $m_L^M = m_R^M \ll m^D$. Diagonalization of the mass matrix yields two Majorana neutrinos with slightly different masses, $m_{1,2} = m^D \pm m^M$ and opposite CP signs. In this case the interaction neutrino is predominantly a Dirac particle with a very slight admixture of Majorana character. Even if the masses are not small, the pseudo-Dirac neutrinos could suppress $0\nu\ \beta\beta$ decay rates by a possible cancellation between the contributions from the almost degenerate masses. Therefore, this suppression mechanism could become attractive, if the effective neutrino mass deduced from $0\nu\ \beta\beta$ decay differs considerably from that measured in tritium decay. In these two processes, we see effective neutrino masses corresponding to different superpositions of the mass eigenstates (see v. Feilitzsch in this volume).

c) Dirac neutrino
In the limit of $m_L^M = m_R^M = 0$, a Dirac neutrino appears. It can be constructed with a pair of Majorana neutrinos which have the degenerate masses, $m_1 = m_2 = m^D$,

$$\psi_L = (N_{1L} + iN_{2L})/\sqrt{2} \qquad (2.13a)$$

$$\psi_R = ((N_{1L})^c + i(N_{2L})^c)/\sqrt{2} \qquad (2.13b)$$

where

$$N_k = \begin{bmatrix} -i\sigma_2\eta_k^* \\ \eta_k \end{bmatrix} \quad \text{or} \quad \begin{bmatrix} \xi_k \\ i\sigma_2\xi_k^* \end{bmatrix}. \qquad (2.14)$$

The two-component fields η_k and ξ_k are left-handed and right-handed fields, respectively.

The one flavour mass Lagrangian, eq.(2.4), can easily be generalized to a system consisting of n left-handed neutrinos ν_{lL} and n right-handed neutrinos ν_{lR} (l = e, μ, ...), corresponding to n flavours. In this case ν_{lL} and ν_{lR} are n-component column vectors. Correspondingly, the mass matrix is 2n $\times$ 2n (and symmetric),

$$M = \begin{bmatrix} m_L^M & (m^D)^T \\ m^D & m_R^M \end{bmatrix} \qquad (2.15)$$

(each submatrix is n $\times$ n). With the following transformation and mass eigenstates,

$$\begin{bmatrix} \nu_L \\ (\nu_R)^c \end{bmatrix} = U_\nu N_L \qquad (2.16)$$

$$N = N_L + (N_L)^c \qquad (2.17)$$

the mass term is expressed as

$$L_m = -\tfrac{1}{2}(\bar{N_L})^c D_\nu N_L + h.c. = -\tfrac{1}{2}\bar{N} D_\nu N \qquad (2.18)$$

where D_ν is a diagonal matrix with real, positive mass eigenvalues.

The existing species of neutrinos and their fundamental properties depend on the choice of the gauge model (see Langacker, Mohapatra in this volume, and also [Gro 88]). The standard electroweak theory of Glashow, Weinberg and Salam [Gla 61, Wei 67, Sal 68] is based on $SU(2)_L \times U(1)$ gauge invariance. The multiplets of leptons belonging to the first generation are shown in Fig. 2.2a. The left-handed components of the leptons form a doublet of the group $SU(2)_L$, while the right-handed component of the charged lepton forms a singlet. The left-handed and right-handed quarks are assumed to form multiplets separately from the leptons, and as a result the lepton and baryon numbers are conserved separately. There is no right-handed neutrino in the standard theory. This means that the left-handed neutrino must be massless. One can see from Fig. 2.1 that the Dirac mass coupling does not exist in the absence of right-handed neutrinos, and the Majorana mass should vanish because it violates lepton number conservation.

In spite of the great success of the standard theory, it seems incomplete and inadequate in many respects. The theory has in its minimal version 21 free parameters which have to be fixed to specify the theory. The global symmetry implying separate lepton- and baryon-number conservation cannot be considered as fundamental in the context of gauge theories.

The grand unified theories (GUT's) represent an attempt to unify the electroweak and strong interactions of leptons and quarks in the framework of gauge theories. The simplest GUT is the minimal SU(5) theory of Georgi and Glashow [Geo 74]. The fermions of each generation are assigned to the $(\bar{5}+10)$ representation of the gauge group, Fig. 2.2b. In this theory, superheavy gauge bosons, X and Y, are predicted, in addition to the vector bosons W and Z of the standard theory. In the SU(5) model, baryon number B and lepton number L are not conserved separately, but the combination $B-L$ is. However, as in the standard theory, the right-handed neutrino is absent, and the neutrino is massless. In the minimal SU(5) model, a right-handed neutrino again does not exist, leading to a vanishing Dirac mass; on the other hand, SU(5) invariant Majorana couplings are not possible with the Higgs content of the minimal model. An interesting consequence of the B non-conservation is the possible decay of the proton. Recent theoretical estimates based on the SU(5) models [Lan 81,86] lead to a half-life of $\tau_p \simeq 10^{28} - 10^{32}$ years, the decay mode $p \to e^+ + \pi^0$ being predicted as the strongest decay branch with a branching ratio of 40-60 %. The experimental value $\tau(p \to e^+ + \pi^0) > 4 \times 10^{32}$ years [Mey 86] rules out the minimal SU(5) model.

Massive neutrinos appear quite naturally in the SO(10) GUT's [Geo 75, Fri 75]. This is the minimal left-right symmetric grand unified model that gauges the $B-L$ symmetry (see [Moh 86]). All fermions of one generation form a 16-dimensional spinor representation of SO(10), which consists of the $\bar{5}$- and 10-dimensional SU(5) representations and one additional neutral fermion, which has to be interpreted as the right-handed neutrino ν_R. Fig. 2.2c shows the multiplet of the first generation. The Dirac mass term shows up here, since both left-handed and right-handed neutrinos exist. In addition, since $B-L$ is not necessarily conserved, the Majorana mass terms do not vanish. Thus, in the SO(10) theories, neutrinos are expected to have the general mass Lagrangian of eq. (2.3). Assuming the right-handed neutrino ν_R represented by the SU(5) singlet to

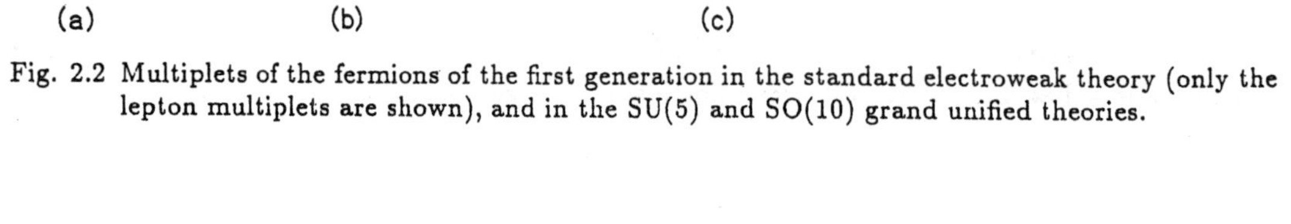

Fig. 2.2 Multiplets of the fermions of the first generation in the standard electroweak theory (only the lepton multiplets are shown), and in the SU(5) and SO(10) grand unified theories.

be very heavy — which is consistent with its non-observation — the see-saw mechanism leads in the way described above to the phenomenologically desired small mass of the observed left-handed neutrino. This mechanism predicts, neglecting for simplicity the possible mixing between different generations, that the mass of the left-handed (light) neutrino is proportional to the square of the fermion (quark or lepton) mass from the same generation, $m_1 \simeq (m^D)^2/m_R^M$ (see eq.(2.12a)),

$$m_{\nu_e} : m_{\nu_\mu} : m_{\nu_\tau} = m_u^2 : m_c^2 : m_t^2 \quad \text{or} \quad = m_e^2 : m_\mu^2 : m_\tau^2. \tag{2.19}$$

In some models, typical values of the electron neutrino mass are around 10^{-1} eV and 1 eV (Langacker, Mohapatra in this volume, and [Moh 81a, Cha 85]); in general, however, depending on the type of the see-saw mechanism applied (GUT see-saw, intermediate see-saw, $SU(2)_L\times SU(2)_R\times U(1)$ see-saw, ...), i.e. on the mass of the right-handed heavy neutrino, the predicted masses range between 10^{-11} eV and about 1 eV.

It should be noted that spontaneous breaking of the global $B-L$ symmetry would — in addition to allowing for neutrinoless $\beta\beta$ decay — require the existence of a new particle, the majoron [Gel 81, Chi 81, Geo 81], which should be emitted in one of the $\beta\beta$ decay modes (see sect. 3.4).

Finally, left-right symmetric models involve, as well as left-handed gauge bosons, right-handed gauge bosons which mediate right-handed (V+A) weak interactions.

(footnote 1)
This is not the usual charge conjugation, because under the so defined operation the handedness of the field changes. The usual definition of charge conjugation for the handed fields would be: $(\psi_{L/R})^c = C\,(\bar{\psi}_{R/L})^T$.

3. Decay rates of 2ν and 0ν $\beta\beta$ decay

3.1. Effective charged weak interaction

We adopt the following effective weak interaction consisting of the vector (V) and axial-vector (A) currents,

$$H_W = \frac{G}{\sqrt{2}}[j_L \cdot J_L^\dagger + \kappa\, j_L \cdot J_R^\dagger + \eta\, j_R \cdot J_L^\dagger + \lambda\, j_R \cdot J_R^\dagger] + h.c. \tag{3.1}$$

Here, G is the Fermi coupling constant, and κ, η and λ are mixing parameters of the right-handed currents. The leptonic currents are defined as

$$j_L^\mu = \bar{e}\gamma^\mu(1-\gamma_5)\nu_{eL}, \qquad j_R^\mu = \bar{e}\gamma^\mu(1+\gamma_5)\nu'_{eR} \tag{3.2}$$

$$\nu_{eL} = \textstyle\sum_j U_{ej}N_{jL}, \qquad \nu'_{eR} = \sum_j V_{ej}N_{jR} \tag{3.3}$$

where N_j is a Majorana neutrino with the mass eigenvalue m_j. This expression includes the case of Dirac neutrinos, since a Dirac neutrino can be composed of a pair of Majorana neutrinos with degenerate masses (see sect. 2). The hadronic currents are expressed in terms of quarks in a form similar to eq.(3.2), and can be reduced to currents for nucleons. The hadronic currents are modified by higher-order perturbations from the strong and electroweak interactions; as a consequence the A part is renormalized (the effective coupling constant is denoted by g_A), and in addition weak magnetism (g_W) and pseudo-scalar (g_P) terms are induced. In the non-relativistic impulse approximation, the hadronic currents in the form of operators acting on the nucleons can be written as

$$J_{L/R}^{0\dagger}(\mathbf{x}) = \sum_n t_{-n}(g_V \mp g_A C_n)\delta(\mathbf{x}-\mathbf{r}_n) \tag{3.4a}$$

$$J_{L/R}^{k\dagger}(\mathbf{x}) = \sum_n t_{-n}(\pm g_A \sigma_n^k - g_V D_n^k)\delta(\mathbf{x}-\mathbf{r}_n). \tag{3.4b}$$

The isospin operator t_- converts a neutron into a proton, σ and $\mathbf{r}$ are spin and position operators, respectively, and the sum n runs over all nucleons in a nucleus. The nucleon recoil terms, which are first-order terms of the non-relativistic expansion, have the following forms,

$$C_n = [(\mathbf{P}_n + \mathbf{P}'_n)\cdot\sigma_n - \frac{g_P}{g_V}(E_n - E'_n)(\mathbf{P}_n - \mathbf{P}'_n)\cdot\sigma_n]/(2M) \tag{3.5a}$$

$$\mathbf{D}_n = [(\mathbf{P}_n + \mathbf{P}'_n) - (1 - 2M\frac{g_W}{g_V})i\sigma_n \times (\mathbf{P}_n - \mathbf{P}'_n)]/(2M) \tag{3.5b}$$

where $(E_n, \mathbf{P}_n)$ and $(E'_n, \mathbf{P}'_n)$ are the initial and final nucleon four-momenta, and M the nucleon mass. Values of the coupling constants are $G = 1.16637 \times 10^{-5}$ GeV^{-2}, $g_V = 0.9737$, $g_A/g_V = -1.254$, $\mu_\beta = 1 - 2Mg_W/g_V = 4.706$ [PDG 86] and $g_P/g_A = 2M/m_\pi^2$ with m_π being the pion mass.

3.2. Decay rate formulae for the 2ν mode

The 2ν mode $\beta\beta$ decay can be described in second order perturbation of the weak interaction as successive Gamow-Teller transitions via virtual intermediate 1^+ states (see Fig. 3.1). Since the energy released by the decay is of the order of 1 MeV (the largest Q-value T_0 for $\beta\beta$ decay is that of ^{48}Ca with $T_0 = 4.27$ MeV) and is shared by the four leptons in the continuum, the electrons and neutrinos can be assumed to be emitted in S-wave states. Consequently higher-order terms can be neglected: the higher partial waves of the leptons, and finite wave-length effects. Furthermore, in what follows, we neglect the nucleon recoil terms and contributions from the right-handed currents.

Let us first consider $0_i^+ \to 0_f^+$ transitions, which have been extensively studied both experimentally and theoretically. In second order perturbation theory, the inverse half-life is given by [Gro 83a, Kla 84, Doi 85] (for a derivation see [Gro 88])

$$[T_{1/2}^{2\nu}(0^+ \to 0^+)]^{-1} = \frac{a^{2\nu}}{\ln 2}\int_{m_e}^{T_0+m_e} F_0(Z,e_1)k_1e_1\,de_1 \int_{m_e}^{T_0+2m_e-e_1} F_0(Z,e_2)k_2e_2\,de_2 \times \int_0^{T_0+2m_e-e_1-e_2} \nu_1^2\nu_2^2\,d\nu_1 \sum_{a,a'} A_{aa'} \tag{3.6}$$

$$(T_0 = e_1 + e_2 - 2m_e + \nu_1 + \nu_2)$$

$$a^{2\nu} = \frac{G^4 g_A^4}{32\pi^7 m_e} \tag{3.7}$$

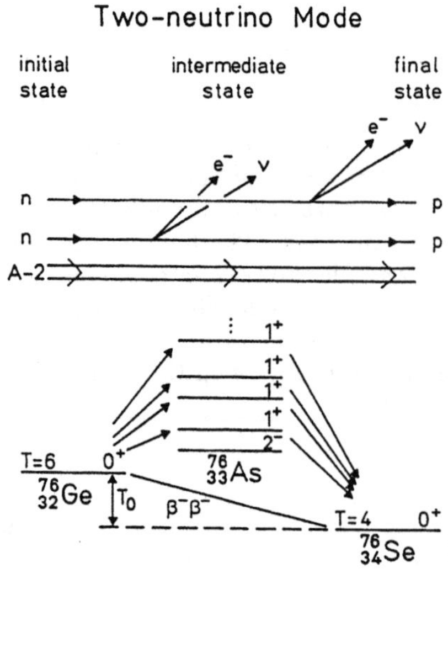

Fig. 3.1 Diagram and illustration of the 2ν $\beta\beta$ decay of ^{76}Ge.

where $k = \sqrt{e^2 - m_e^2}$ is the electron momentum, and $F_0(Z, e)$ is the Fermi function which takes into account the distortion of the electron wave functions due to the charge Z of the daughter nucleus. (The energies of the leptons are denoted here simply by e_i, ν_i.) The nuclear Gamow-Teller transition matrix elements and the energy denominators in the perturbation expression appear in $A_{aa'}$,

$$A_{aa'} = \langle 0_f^+ \| t_- \sigma \| 1_a^+ \rangle \langle 1_a^+ \| t_- \sigma \| 0_i^+ \rangle \langle 0_f^+ \| t_- \sigma \| 1_{a'}^+ \rangle \langle 1_{a'}^+ \| t_- \sigma \| 0_i^+ \rangle \times \frac{1}{3}(K_a K_{a'} + L_a L_{a'} + \frac{1}{2} K_a L_{a'} + \frac{1}{2} L_a K_{a'}) \tag{3.8}$$

where

$$K_a = \frac{1}{E_a + e_1 + \nu_1 - E_i} + \frac{1}{E_a + e_2 + \nu_2 - E_i} \tag{3.9a}$$

$$L_a = \frac{1}{E_a + e_1 + \nu_2 - E_i} + \frac{1}{E_a + e_2 + \nu_1 - E_i}. \tag{3.9b}$$

The quantity $A_{aa'}$ depends on the energies of both the intermediate nuclear states and the leptons. (For the reduced matrix elements we adopt the Edmonds convention [Edm 57].)

When defining the nuclear matrix element $M_{GT}^{2\nu}$ as

$$M_{GT}^{2\nu} = \sum_a \frac{\langle 0_f^+ \| t_- \sigma \| 1_a^+ \rangle \langle 1_a^+ \| t_- \sigma \| 0_i^+ \rangle}{E_a + T_0/2 + m_e - E_i} \tag{3.10}$$

replacing the lepton energy $e + \nu$ in eq.(3.9) by $T_0/2 + m_e$, the nuclear part of eq.(3.6) is separated from kinematical factors, and considerations of nuclear structure effects can focus on the nuclear matrix element $M_{GT}^{2\nu}$. The half-life is then

$$T_{1/2}^{2\nu} = [G^{2\nu} |M_{GT}^{2\nu}|^2]^{-1} \tag{3.11}$$

where $G^{2\nu}$ is the phase space integral. The replacement of the lepton energy by $T_0/2 + m_e$ is in general a good approximation, since the integrand of eq.(3.6) becomes large when an electron and neutrino pair carries a half of the maximum available kinetic energy T_0. However, the approximation of eq.(3.11) fails when an almost complete cancellation between terms in eq.(3.10) occurs. In this case, $M_{GT}^{2\nu}$ is merely a measure of the nuclear transition strength, and the integration over the lepton energies must be carried out explicitly for each intermediate nuclear state.

The closure approximation, which was sometimes employed in earlier calculations, introduces an additional approximation. That is to replace the intermediate state energies E_a by some average value $\langle E_a \rangle$, and then complete the sum by closure, $\sum_a |1_a^+\rangle\langle 1_a^+| = 1$. The nuclear matrix element, eq.(3.10), is then reduced to

$$M_{GT}^{2\nu} = -\frac{[M_{GT}^{2\nu}]_c}{\langle E_a \rangle + T_0/2 + m_e - E_i} \tag{3.12}$$

with the closure matrix element

$$[M_{GT}^{2\nu}]_c = \sum_{m,n} \langle 0_f^+ \| t_{-m} t_{-n} \sigma_m \cdot \sigma_n \| 0_i^+ \rangle. \tag{3.13}$$

In this approximation only wave functions of the initial and final 0^+ states are needed, and a tedious calculation of the many intermediate states is avoided. The validity of the closure approximation depends simply on the choice of $\langle E_a \rangle$. It is clear that it fails badly, if the matrix elements on the right side of eq.(3.10) have different signs and there is a significant cancellation in the sum. This is usually more or less the case (see e.g. Figs. 5.1 and 5.4).

The vector part (V) of the weak interaction gives no (or an extremely small) contribution to the 2ν mode. The vector part of the hadronic current, eq.(3.4a), contains only the isospin operator t_- of the Fermi transition, and the sum over the nucleons results in the isospin lowering operator for the nucleus, $\sum_n t_{-n} = T_-$. The operator T_- cannot change the isospin of nuclear states, and isospin is normally a good quantum number. The initial and final states of $\beta\beta$ decay have, however, isospins different by two units (see e.g. Fig. 3.1). A finite matrix element for the Fermi transition results mainly from isospin violating terms which are brought into the nuclear structure model by simplification (i.e., approximation) of the calculation.

In even-even nuclei, a 2^+ state appears at low excitation energy, usually as the first excited state, because of strong proton-neutron quadrupole correlations. When the 2^+ state in the daughter nucleus lies energetically lower than the parent state, $0_i^+ \to 2^+$ $\beta\beta$ decay may occur. The decay rate can be obtained from that of the $0_i^+ \to 0_f^+$ transition, eq.(3.6), by small changes; the maximum available kinetic energy T_0 should be interpreted as that of the $0_i^+ \to 2^+$ decay, and $A_{aa'}$ is replaced by

$$A_{aa'} = \frac{1}{3}\langle 2^+\|t_-\sigma\|1_a^+\rangle\langle 1_a^+\|t_-\sigma\|0_i^+\rangle\langle 2^+\|t_-\sigma\|1_{a'}^+\rangle\langle 1_{a'}^+\|t_-\sigma\|0_i^+\rangle \times (K_a - L_a)(K_{a'} - L_{a'}). \tag{3.14}$$

The energy denominators K_a and L_a are defined in eq.(3.9). In contrast to the $0_i^+ \to 0_f^+$ decay, differences of K_a and L_a appear. The leading terms cancel each other,

$$K_a - L_a \simeq \frac{2(\nu_2 - \nu_1)(e_2 - e_1)}{(E_a + T_0/2 + m_e - E_i)^3}. \tag{3.15}$$

The cubic factor in the denominator and the asymmetry in the electron and neutrino energies indicate strong suppression of the $0_i^+ \to 2^+$ decay, and moreover the smaller T_0 implies additional kinematical hindrance of the decay. As before the nuclear matrix element

$$M_{GT}^{2\nu}(0_i^+ \to 2^+) = \sum_a \frac{\langle 2^+\|t_-\sigma\|1_a^+\rangle\langle 1_a^+\|t_-\sigma\|0_i^+\rangle}{(E_a + T_0/2 + m_e - E_i)^3}. \tag{3.16}$$

is introduced and the half-life factorized as

$$T_{1/2}^{2\nu}(0_i^+ \to 2^+) = [G_2^{2\nu}|M_{GT}^{2\nu}(0_i^+ \to 2^+)|^2]^{-1} \tag{3.17}$$

where $G_2^{2\nu}$ is the phase space integral. (The lepton energy factor $2(\nu_2 - \nu_1)(e_2 - e_1)$ in eq.(3.15) should be included in the integral.)

In the closure approximation, the nuclear matrix element is reduced to

$$M_{GT}^{2\nu}(0_i^+ \to 2^+) \simeq \frac{[M_{GT}^{2\nu}(0_i^+ \to 2^+)]_c}{(\langle E_a\rangle + T_0/2 + m_e - E_i)^3} \tag{3.18}$$

with the closure matrix element

$$[M_{GT}^{2\nu}(0_i^+ \to 2^+)]_c = \sum_{m,n}\langle 2^+\|t_{-m}t_{-n}[\sigma_m \times \sigma_n]^{(2)}\|0_i^+\rangle \tag{3.19}$$

where the square bracket denotes the coupling of the two spin operators (spherical tensors of rank 1) to a spherical tensor operator of rank 2.

It is known that the Gamow-Teller strength is quenched by the excitation of subnuclear degrees of freedom in β decay (see. e.g., [Gro 83b,86]). To take into account virtual excitations of the Δ resonance, one can include the ΔN^{-1} space in a simple quark model treatment of the nucleons [Boh 81, Gro 83b,86]. Examples of $\beta\beta$ transition amplitudes involving ΔN^{-1} configurations instead of nucleonic particle-hole excitations are shown in Fig. 3.2. The effect on the $\beta\beta$ transition rates is of the order of 30 % and is discussed in detail in [Gro 86], see also Table 5.4.

3.3. Decay rate formulae for the 0ν mode

Fig. 3.3 shows a schematic diagram of the 0ν mode in the two-nucleon mechanism. (Here again there is also the possibility of ΔN^{-1} excitations, see Fig. 3.2.) An antineutrino is created at

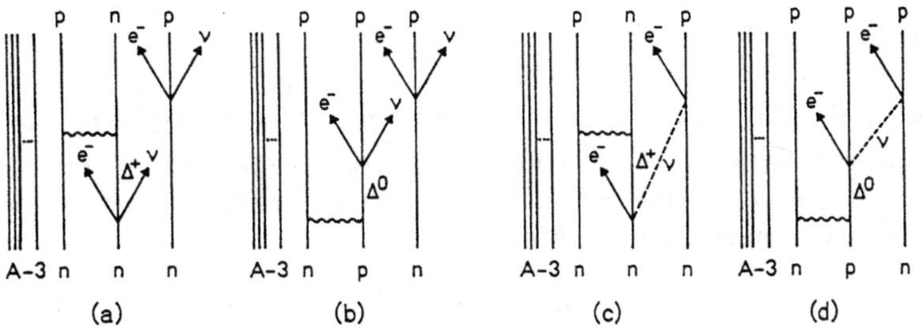

Fig. 3.2 Diagrams involving ΔN^{-1} excitations; (a) in the intermediate state, (b) in the initial state for the 2ν mode, (c) and (d) the corresponding diagrams for the 0ν mode. The wavy line represents the strong interaction.

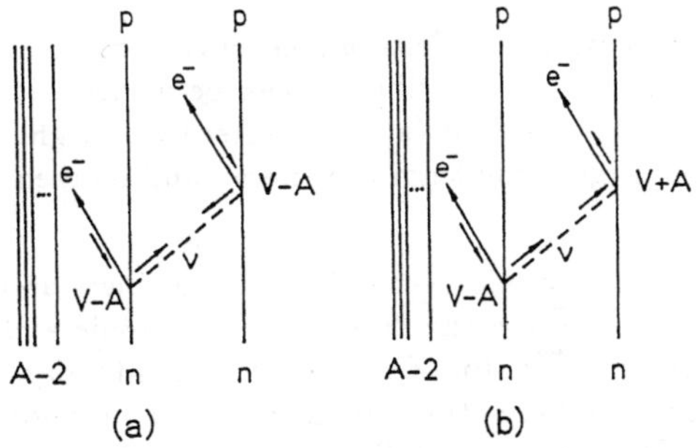

Fig. 3.3 Diagrams of the neutrinoless $\beta\beta$ decay in the two-nucleon mechanism. At each vertex are indicated the handedness of the weak interaction, V-A (left-handed) or V+A (right-handed), and the main helicity components of the leptons by short arrows.

one vertex, $n \to p + e^- + \bar{\nu}$, and a neutrino is to be absorbed at the other vertex, $\nu + n \to p + e^-$. If the (anti)neutrino has zero mass, helicity is a good quantum number. The antineutrino has positive helicity (right-handed), while the neutrino has negative helicity (left-handed), and consequently the 0ν $\beta\beta$ decay is forbidden if only a left-handed (V-A) weak interaction is acting. For massive neutrinos (Fig. 3.3a), the opposite helicity component can be mixed, and the small right-handed (left-handed) component of the neutrino (antineutrino) matches the main right-handed (left-handed) component of the antineutrino (neutrino), resulting in the occurrence of 0ν $\beta\beta$ decay. The helicity mismatch between the main components for the left-handed interaction is also avoided by a possible right-handed (V+A) interaction mediated by right-handed weak gauge bosons (Fig. 3.3b).

Since the formalism for deriving the 0ν decay rate is rather complex we refer for completeness to [Doi 81,83a,85, Gro 88] and concentrate here on some essential points.

The leptonic part of the 0ν transition operator is

$$
\begin{aligned}
T_{lep} &\sim \bar{e}(x)\gamma_\rho \frac{1 \pm \gamma_5}{2} N_j(x) \bar{e}(y) \gamma_\sigma \frac{1 \pm \gamma_5}{2} N_k(y) \\
&= -\bar{e}(x)\gamma_\rho \frac{1 \pm \gamma_5}{2} N_j(x) N_k(y)^T \frac{1 \pm \gamma_5}{2} \gamma_\sigma^T \bar{e}(y)^T \\
&= -i\delta_{jk} \int \frac{d^4q}{(2\pi)^4} \frac{e^{-iq(x-y)}}{q^2 - m_j^2 + i\epsilon} \bar{e}(x)\gamma_\rho \frac{1 \pm \gamma_5}{2} (q^\mu \gamma_\mu + m_j) \frac{1 \pm \gamma_5}{2} \gamma_\sigma e^c(y) \qquad (3.20)
\end{aligned}
$$

where a contraction between the neutrino operators N_j and N_k^T has been made, which is allowed for a Majorana neutrino, and in the last step the neutrino propagator $S_F(x-y)$ substituted,

$$S_F(x-y) = \int \frac{d^4q}{(2\pi)^4} \frac{e^{-iq(x-y)}}{q^2 - m_j^2 + i\epsilon} (q^\mu \gamma_\mu + m_j) \tag{3.21}$$

Here, $q^2 = \omega^2 - \mathbf{q}^2$ with $\mathbf{q}$ being the three-momentum of the neutrino and $\omega = \sqrt{\mathbf{q}^2 + m_j^2}$ the neutrino energy. By performing the integration over the neutrino energy, the factor $(q^2 - m_j^2)^{-1}$ in eq.(3.20) is replaced by its residue π/ω. Therefore the leptonic part of the transition operator includes the integration over the virtual neutrino momentum,

$$T_{lep} \sim \frac{1}{(2\pi)^3} \int \frac{d\mathbf{q}}{\omega} e^{-\mathbf{q}\cdot(\mathbf{x}-\mathbf{y})} \bar{e}(\mathbf{x}) \gamma_\rho \frac{1 \pm \gamma_5}{2} (\omega\gamma^0 - \mathbf{q}\cdot\gamma + m_\nu) \frac{1 \pm \gamma_5}{2} \gamma_\sigma e^c(\mathbf{y}). \tag{3.22}$$

Using the relations,

$$(1-\gamma_5)(\omega\gamma^0 - \mathbf{q}\cdot\gamma + m_\nu)(1-\gamma_5) = 2m_\nu(1-\gamma_5) \tag{3.23a}$$

$$(1-\gamma_5)(\omega\gamma^0 - \mathbf{q}\cdot\gamma + m_\nu)(1+\gamma_5) = 2(\omega\gamma^0 - \mathbf{q}\cdot\gamma)(1+\gamma_5) \tag{3.23b}$$

we can see that only the neutrino mass term contributes when the left-handed current acts at both vertices (or the right-handed current at both vertices but its contribution can be neglected), eq.(3.23a). On the other hand, the interference between left-handed (at one vertex) and right-handed (at the other vertex) currents allows a transition originating from the energy and momentum of the virtual neutrino, eq.(3.23b).

The integration over the neutrino momentum in the leptonic sector leads to a "neutrino potential" (eq.(3.39)) which acts on the nuclear wave function. (In the 2ν mode the leptonic and hadronic parts are almost separate, except for the energy denominator of the perturbation theory.) The potential, which represents the exchange of the neutrino between the two nucleons, is a function of the momentum transferred by the neutrino, or, in coordinate space, of distance $|\mathbf{r}_1 - \mathbf{r}_2|$ between the two nucleons. Such a potential induces transitions leading to states of various multipolarity (J^π) in the intermediate odd-odd nucleus. This can be seen by a multipole decomposition of the potential into products of two operators which, respectively, act on the two nucleons. For instance, in the case of the Gamow-Teller matrix element eq.(3.37a),

$$H(|\mathbf{r}_1 - \mathbf{r}_2|)\sigma_1 \cdot \sigma_2 = \sum_{J,L} (-1)^{L+1-J}(2L+1) \int_0^\infty dq q^2 v(q)$$
$$\times (j_L(qr_1)[\sigma_1 \times Y_L(\hat{\mathbf{r}}_1)]^{(J)}) \cdot (j_L(qr_2)[\sigma_2 \times Y_L(\hat{\mathbf{r}}_2)]^{(J)}) \tag{3.24}$$

where $j_L(qr)$ is the spherical Bessel function of rank L, $Y_L(\hat{\mathbf{r}})$ the spherical harmonic of rank L, and $v(q)$ the Fourier transform of the potential $H(|\mathbf{r}_1 - \mathbf{r}_2|)$. When we consider a matrix element of eq.(3.24) between the initial 0_i^+ and final 0_f^+ states of $\beta\beta$ decay, and insert intermediate states $\sum_{J,M} |JM\rangle\langle JM| = 1$ between the operator with index 1 and that with index 2, the operator $j_L(qr)[\sigma \times Y_L(\hat{\mathbf{r}})]^{(J)}$ leads to intermediate states with the angular momentum J and parity $\pi = (-1)^L$.

In the presence of a neutrino potential, wave functions at short internucleon distances must be treated carefully, in particular for the zero-range potential of the nucleon recoil matrix element (see, eqs. (3.45) and (3.46)). In the calculation of the nuclear matrix elements, nuclear wave functions consisting of (antisymmetrized) products of single-particle wave functions have usually been used, in which the short-range correlations are not taken into account. These correlations arise from the strong, short-range, repulsive components of the nucleon-nucleon interaction, which prevent two nucleons from coming close to each other. If the short-range correlations are taken into account, a matrix element of the zero-range potential vanishes completely. However, if we take into account the fact that the nucleon has a finite extention in space, the matrix element is not affected drastically by the short-range correlations. One can introduce the short-range correlations by multiplying the uncorrelated two-nucleon wave function with a function $f(|\mathbf{r}_1 - \mathbf{r}_2|)$

$$|\phi(\mathbf{r}_1)\phi(\mathbf{r}_2)\rangle \longrightarrow f(|\mathbf{r}_1-\mathbf{r}_2|)|\phi(\mathbf{r}_1)\phi(\mathbf{r}_2)\rangle, \tag{3.25}$$

for which the functional form

$$f(r) = 1 - e^{-ar^2}(1-br^2) \tag{3.26}$$

with $a = 1.1$ fm^{-2} and $b = 0.68$ fm^{-2} [Mil 76] has usually been used. The finite nucleon-size effects are taken into account by replacing the vector and axial-vector coupling constant by dipole form factors in momentum space [Ver 81,83],

$$g_V \longrightarrow g_V(\frac{\Lambda^2}{\Lambda^2+q^2})^2, \qquad g_A \longrightarrow g_A(\frac{\Lambda^2}{\Lambda^2+q^2})^2 \tag{3.27}$$

with typically $\Lambda = 850$ MeV. This amounts to smearing out the delta function in eq.(3.4) with the replacement

$$\delta(\mathbf{x}-\mathbf{r}_n) \longrightarrow \frac{\Lambda^3}{8\pi}\exp(-\Lambda|\mathbf{x}-\mathbf{r}_n|). \tag{3.28}$$

In 0ν $\beta\beta$ decay one can apply the closure approximation which in this case is good, in contrast to 2ν $\beta\beta$ decay. The reason is that the average energy of intermediate states appears in the neutrino potential, eq.(3.39), in the combination $\omega+\bar{E}$, where $\bar{E} = \langle E_a\rangle + T_0/2 + m_e - E_i$; this is typically ~ 10 MeV, while the energy of the virtual neutrino is typically $\omega \sim 1/\bar{r} \sim 100$ MeV (with $\bar{r}$ being the mean distance between two nucleons), i.e. much larger than $\bar{E}$.

3.3.1. $0^+ \to 0^+$ transitions

The inverse half-life of the 0ν mode $0_i^+ \to 0_f^+$ $\beta\beta$ decay is given by [Doi 85]

$$\begin{aligned}[T_{1/2}^{0\nu}(0_i^+ \to 0_f^+)]^{-1} &= C_{mm}(\frac{\langle m_\nu\rangle}{m_e})^2 + C_{\eta\eta}\langle\eta\rangle^2 + C_{\lambda\lambda}\langle\lambda\rangle^2 \\ &+ C_{m\eta}\langle\eta\rangle\frac{\langle m_\nu\rangle}{m_e} + C_{m\lambda}\langle\lambda\rangle\frac{\langle m_\nu\rangle}{m_e} + C_{\eta\lambda}\langle\eta\rangle\langle\lambda\rangle.\end{aligned} \tag{3.29}$$

Effective values of the neutrino mass and right-handed parameters are defined by

$$\langle m_\nu\rangle = \textstyle\sum_j m_j U_{ej}^2 \tag{3.30a}$$
$$\langle\eta\rangle = \eta\textstyle\sum_j U_{ej}V_{ej} \tag{3.30b}$$
$$\langle\lambda\rangle = \lambda\textstyle\sum_j U_{ej}V_{ej}. \tag{3.30c}$$

The sum extends over light neutrino mass eigenstates, $m_j < 10$ MeV. Terms including κ (see eq.(3.1)) are neglected, since κ always appears in the 0ν $\beta\beta$ decay amplitude in the combination $1\pm\kappa$ and we expect $\kappa \ll 1$. The coefficients C_{xy} include the nuclear matrix elements and phase space integrals (index 0ν on the matrix elements is suppressed),

$$C_{mm} = (M_{GT}-M_F)^2G_1 \tag{3.31a}$$
$$C_{\eta\eta} = M_{2+}^2G_2 + \frac{1}{9}M_{1-}^2G_4 - \frac{2}{9}M_{1-}M_{2+}G_3 + M_P^2G_8 - M_PM_RG_7 + M_R^2G_9 \tag{3.31b}$$
$$C_{\lambda\lambda} = M_{2-}^2G_2 + \frac{1}{9}M_{1+}^2G_4 - \frac{2}{9}M_{1+}M_{2-}G_3 \tag{3.31c}$$
$$C_{m\eta} = (M_{GT}-M_F)(M_{2+}G_3 - M_{1-}G_4 - M_PG_5 + M_RG_6) \tag{3.31d}$$
$$C_{m\lambda} = -(M_{GT}-M_F)(M_{2-}G_3 - M_{1+}G_4) \tag{3.31e}$$
$$C_{\eta\lambda} = -2M_{2+}M_{2-}G_2 + \frac{2}{9}(M_{1+}M_{2+} + M_{1-}M_{2-})G_3 - \frac{2}{9}M_{1-}M_{1+}G_4 \tag{3.31f}$$

with the combinations

$$M_{1\pm} = M'_{GT} \pm 3M'_F - 6M_T, \tag{3.32a}$$
$$M_{2\pm} = M_{GT\omega} \pm M_{F\omega} - \frac{1}{9}M_{1\pm}. \tag{3.32b}$$

The phase space integrals are

$$G_i = \frac{a^{0\nu}}{(m_eR)^2 \ln 2} \int_{m_e}^{T_0+m_e} F_0(Z,e_1)k_1e_1F_0(Z,e_2)k_2e_2g_i \, de_1 \tag{3.33}$$

$$(e_1 + e_2 - 2m_e = T_0)$$

$$a^{0\nu} = \frac{G^4 g_A^4 m_e^4}{32\pi^5} \tag{3.34}$$

with

$$g_1 = 1 \tag{3.35a}$$

$$g_2 = \frac{1}{2}(\frac{e_1 - e_2}{m_e})^2 \frac{e_1e_2 - m_e^2}{e_1e_2} \tag{3.35b}$$

$$g_3 = \frac{(e_1 - e_2)^2}{e_1e_2} \tag{3.35c}$$

$$g_4 = \frac{2}{9}\frac{e_1e_2 - m_e^2}{e_1e_2} \tag{3.35d}$$

$$g_5 = \frac{4}{3}[x_P \frac{m_e(T_0 + 2m_e)}{2e_1e_2m_eR} - \frac{e_1e_2 + m_e^2}{e_1e_2}] \tag{3.35e}$$

$$g_6 = \frac{4}{m_eR}\frac{T_0m_e + 2m_e^2}{e_1e_2} \tag{3.35f}$$

$$g_7 = \frac{16}{3}\frac{1}{m_eR}[x_P\frac{e_1e_2 + m_e^2}{2e_1e_2m_eR} - \frac{m_e(T_0 + 2m_e)}{e_1e_2}] \tag{3.35g}$$

$$g_8 = \frac{2}{9}\frac{1}{(m_eR)^2}([x_P^2 + (2m_eR)^2]\frac{e_1e_2 + m_e^2}{e_1e_2} - 4m_eRx_P\frac{m_e(T_0 + 2m_e)}{e_1e_2}) \tag{3.35h}$$

$$g_9 = (\frac{4}{m_eR})^2\frac{e_1e_2 + m_e^2}{2e_1e_2}. \tag{3.35i}$$

The quantity which characterizes the P-wave emission of the electron is denoted by x_P,

$$x_P = 3\alpha Z + R(T_0 + 2m_e) \tag{3.36}$$

where α is the fine structure constant and R the nuclear radius.

The nine types of nuclear matrix elements of the $0_i^+ \to 0_f^+$ 0ν $\beta\beta$ decay are summarized in Table 3.1. The first two elements originate from the neutrino mass term with left-handed interaction at both vertices, two electrons being emitted in S-wave states. The double Fermi transition (M_F) is possible in 0ν decay, in contrast to 2ν decay, because natural parity states with $\pi = (-1)^J$ are allowed for the nuclear intermediate states in the presence of the neutrino potential. The next two matrix elements in Table 3.1 are from the ω-term of the left-right interference, (V-A) interaction at one vertex and (V+A) at the other vertex. The scalar property of the neutrino energy results in a similar structure for the transition operator to that of the previous elements, with a different neutrino potential. On the other hand, the neutrino three-momentum is a vector and has negative parity. This odd parity can be compensated, for the $0^+ \to 0^+$ transition with no parity change, either by P-wave emission of one of the two electrons (M'_{GT}, M'_F, M_T and M_P), or by the nuclear part, i.e., the nucleon recoil term M_R.

Explicit forms of the closure matrix elements are

$$M_{GT}^{0\nu} = \langle H(r)\sigma_m \cdot \sigma_n \rangle \tag{3.37a}$$

$$M_F^{0\nu} = \langle H(r)\rangle(\frac{g_V}{g_A})^2 \tag{3.37b}$$

$$M_{GT\omega}^{0\nu} = \langle H_\omega(r)\sigma_m \cdot \sigma_n \rangle \tag{3.37c}$$

Table 3.1. Nature of the transition operators for the neutrino, electron and nuclear parts of $0_i^+ \to 0_f^+$ 0ν $\beta\beta$ decay. In the second column, L•L denotes the left-handed interaction at both vertices, and L•R left-handed interaction at one vertex and right-handed interaction at the other. The parity of each operator is presented in parentheses.

Matrix Element	Vertex	Neutrino (π)		Electron (π)		Nucleus	(π)
M_{GT}	L•L	m_ν	(+)	SS	(+)	$\sigma_1 \cdot \sigma_2$	(+)
M_F						$1_1 \cdot 1_2$	(+)
$M_{GT\omega}$	L•R	ω	(+)	SS	(+)	$\sigma_1 \cdot \sigma_2$	(+)
$M_{F\omega}$						$1_1 \cdot 1_2$	(+)
M'_{GT}		q	(-)	SP	(-)	$\sigma_1 \cdot \sigma_2$	(+)
M'_F						$1_1 \cdot 1_2$	(+)
M_T						S_{12}	(+)
M_P						$i(\sigma_1 - \sigma_2) \cdot (\hat{r} \times \hat{r}_+)$	(+)
M_R				SS	(+)	$\hat{r} \cdot (\sigma_1 \times D_2 + D_1 \times \sigma_2)$	(-)

$S_{12} = (\sigma_1 \cdot \hat{r})(\sigma_2 \cdot \hat{r}) - \frac{1}{3}(\sigma_1 \cdot \sigma_2)$

$$M_{F\omega}^{0\nu} = \langle H_\omega(r) \rangle (\frac{g_V}{g_A})^2 \tag{3.37d}$$

$$M_{GT}^{'0\nu} = \langle H'(r) \sigma_m \cdot \sigma_n \rangle \tag{3.37e}$$

$$M_F^{'0\nu} = \langle H'(r) \rangle (\frac{g_V}{g_A})^2 \tag{3.37f}$$

$$M_T^{0\nu} = \langle H'(r)[(\sigma_m \cdot \hat{\mathbf{r}})(\sigma_n \cdot \hat{\mathbf{r}}) - \frac{1}{3}\sigma_m \cdot \sigma_n] \rangle \tag{3.37g}$$

$$M_P^{0\nu} = \langle H'(r) \frac{r_+}{2r} i(\sigma_m - \sigma_n) \cdot (\hat{\mathbf{r}} \times \hat{\mathbf{r}}_+) \rangle (\frac{g_V}{g_A}) \tag{3.37h}$$

$$M_R^{0\nu} = \langle H'(r) \frac{R}{2r} \hat{\mathbf{r}} \cdot (\sigma_m \times \mathbf{D}_n + \mathbf{D}_m \times \sigma_n) \rangle (\frac{g_V}{g_A}) \tag{3.37i}$$

where

$$\mathbf{r} = \mathbf{r}_m - \mathbf{r}_n, \qquad r = |\mathbf{r}|, \qquad \hat{\mathbf{r}} = \mathbf{r}/r,$$
$$\mathbf{r}_+ = \mathbf{r}_m + \mathbf{r}_n, \qquad r_+ = |\mathbf{r}_+|, \qquad \hat{\mathbf{r}}_+ = \mathbf{r}_+/r_+.$$

The simplified form

$$\langle O_{mn} \rangle = \sum_{m,n} \langle 0_f^+ || t_{-m} t_{-n} O_{mn} || 0_i^+ \rangle \tag{3.38}$$

is used in the expressions of eq.(3.37). The scalar recoil term C_n of eq.(3.5a) does not contribute under the assumption that finite wave-length effects can be neglected.

The neutrino potentials, which are functions of r, are defined by

$$H(r) = \frac{R}{2\pi^2}\int \frac{d\mathbf{q}}{\omega}\frac{1}{\omega+\bar{E}}e^{i\mathbf{q}\cdot\mathbf{r}} \tag{3.39a}$$

$$H_\omega(r) = H(r) - \bar{E}\frac{R}{2\pi^2}\int \frac{d\mathbf{q}}{\omega}\frac{1}{(\omega+\bar{E})^2}e^{i\mathbf{q}\cdot\mathbf{r}} \tag{3.39b}$$

$$H'(r) = -r\frac{d}{dr}H(r). \tag{3.39c}$$

The factor ω^{-1} comes from the neutrino propagator (see eq.(3.22)), and $(\omega+\bar{E})^{-1}$ is the energy denominator of the perturbation theory. The radial form of the potential $H(r)$ is shown in Fig. 3.4 for some values of the neutrino mass. It behaves as $\sim 1/r$ for light neutrinos with $m_\nu < 10$ MeV, or as a Yukawa potential $e^{-m_\nu r}/r$ for heavy neutrinos. The dependence of the potential on the average energy of intermediate states is shown in Fig. 3.5. (The nuclear radius R has been introduced in the definition of the neutrino potentials, eq.(3.39), in order to make the potentials dimensionless. This is compensated by a factor R^{-2} in the phase space integrals, eq.(3.33).)

When there are no right-handed components of the weak interaction, the 0ν $\beta\beta$ decay originates only from the finite neutrino mass. The 0ν decay rate is then proportional to the square of the effective neutrino mass,

$$[T^{0\nu}_{1/2}(0^+_i \to 0^+_f)]^{-1} = (M^{0\nu}_{GT} - M^{0\nu}_F)^2 \left(\frac{\langle m_\nu\rangle}{m_e}\right)^2 G_1. \tag{3.40}$$

The phase space factor G_1 is defined by eqs. (3.33), (3.34) and (3.35a). The two nuclear matrix elements relevant to the neutrino mass, eqs. (3.37a) and (3.37b), are more explicitly expressed as

$$M^{0\nu}_{GT} = \sum_{m,n}\langle 0^+_f \| t_{-m}t_{-n}H(|\mathbf{r}_m - \mathbf{r}_n|)\sigma_m\cdot\sigma_n\|0^+_i\rangle \tag{3.41a}$$

$$M^{0\nu}_F = \sum_{m,n}\langle 0^+_f \| t_{-m}t_{-n}H(|\mathbf{r}_m - \mathbf{r}_n|)\|0^+_i\rangle \left(\frac{g_V}{g_A}\right)^2 \tag{3.41b}$$

with the neutrino potential defined by eq.(3.39a). Knowing the values of the phase space factor and the nuclear matrix elements, one can deduce from an observed 0ν decay half-life the effective value of the neutrino mass

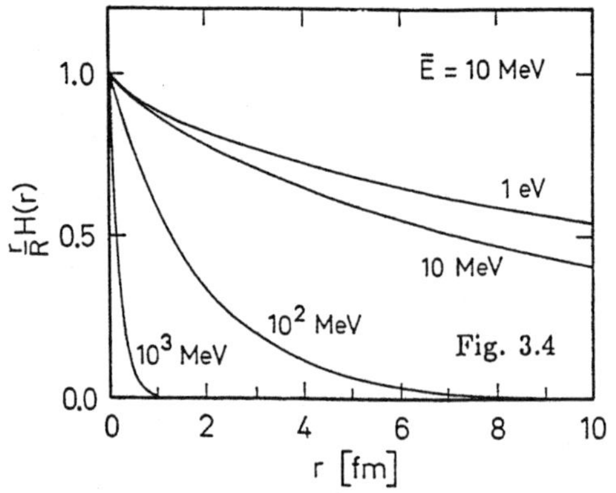

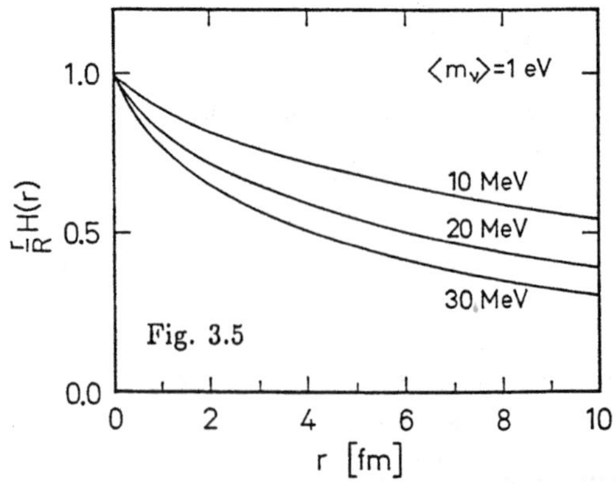

Fig. 3.4 Radial shape of the neutrino potential $H(r)$ for neutrino masses 1 eV, 10 MeV, 10^2 MeV and 10^3 MeV. An average energy of the intermediate states $\bar{E} = 10$ MeV is assumed. The neutrino potential is almost completely identical for neutrino masses $0 < m_\nu < 1$ MeV.

Fig. 3.5 Radial shape of the neutrino potential $H(r)$ for average energies of the intermediate states $\bar{E}$ = 10, 20 and 30 MeV. The neutrino mass is assumed to be 1 eV.

$$\langle m_\nu \rangle = \sum_j m_j U_{ej}^2 \tag{3.42}$$

where the sum runs over light neutrinos with mass eigenvalues $m_j < 10$ MeV. This factorization of the effective neutrino mass is possible for light neutrinos. Although the radial form of the neutrino potential $H(r)$ depends on the mass of the neutrino which is exchanged between two nucleons, the dependence is negligibly small for light neutrinos with masses $0 < m_j < \bar{E}$, where the average energy of intermediate nuclear states $\bar{E}$ is typically ~ 10 MeV, see Fig. 3.4.

If a heavy neutrino contributes to the 0ν decay, one must take into account the dependence of the neutrino potential on the neutrino mass. Then the nuclear matrix elements have an m_j dependence, and the expression for the half-life should be modified to

$$[T_{1/2}^{0\nu}(0_i^+ \to 0_f^+)]^{-1} = [\sum_j m_j U_{ej}^2 (M_{GT}^{0\nu}(m_j) - M_F^{0\nu}(m_j))]^2 \frac{G_1}{m_e^2}. \tag{3.43}$$

For two Majorana neutrinos with masses m_1 and m_2, one being light and the other heavy, $m_1 \ll \bar{E} < m_2$, the effective value of the neutrino mass which one deduces from an experimental half-life is expressed as [Hal 83, Ver 83]

$$\langle m_\nu \rangle = |m_1 \cos^2\theta - m_2 \sin^2\theta \frac{M_{GT}^{0\nu}(m_2) - M_F^{0\nu}(m_2)}{M_{GT}^{0\nu}(0) - M_F^{0\nu}(0)}| \tag{3.44}$$

where θ is the angle of mixing between the two neutrinos. The ratio of the nuclear matrix elements in the second term of eq.(3.44) depends sensitively on the short-range correlations (see the discussion above, eq.(3.25)), since the contributions to $M_{GT}^{0\nu}(m_2)$ and $M_F^{0\nu}(m_2)$ come from the short-distance region $|\mathbf{r}_1 - \mathbf{r}_2| < 1/m_2$. This ratio also depends on the mass number, and consequently the values of $\langle m_\nu \rangle$ deduced should be different for various $\beta\beta$ emitters (see, e.g., the application in [Gro 86]).

Among the nuclear matrix elements which originate from the right-handed currents, i.e., the V+A interaction at one vertex and the V-A interaction at the other vertex, the matrix element M_R would give the largest contribution to the 0ν decay rate. This matrix element comes from the three-momentum part of the neutrino propagator (see eq.(3.22) and Table 3.1). The negative parity of the momentum is compensated by the negative parity operator on the nuclear side, the nucleon recoil term, eq.(3.5b), which is the small relativistic amplitude of the nucleon wave function. (The negative parity can also be compensated on the electron side [Hax 84], but the corresponding matrix element is smaller than M_R by two orders of magnitude [Tom 86].) The transition operator corresponds to a forbidden single β decay. Such a forbidden transition, where the neutrino is a real particle, is highly hindered compared to an allowed Gamow-Teller β decay due to the small factor $q/Mc \leq 1/100$, where $Mc = 940$ MeV/c. However, the neutrino is virtual in the 0ν $\beta\beta$ decay, and the neutrino momentum is only limited by the requirement that the nucleons are confined in the nucleus, i.e., $q \leq 2k_F \sim 560$ MeV/c, with k_F being the Fermi momentum of a nucleon in the nucleus. The ratio q/Mc is then of the order of unity and is not small any more. The recoil matrix element is enhanced also by the large coupling constant $\mu_\beta = 4.7$, and together with the large phase factor G_9 it dominates the coefficient $C_{\eta\eta}$. Therefore, the matrix element $M_R^{0\nu}$ is the most important one for a determination of the mixing parameter η of the right-handed leptonic current.

The nucleon recoil term $\mathbf{D}_n$ of eq.(3.5b) involves $\mathbf{P}_n' - \mathbf{P}_n$, the difference between the nucleon momenta before and after the weak interaction vertex. The factor can be replaced by the momentum $\mathbf{q}$ which the virtual neutrino carries, by neglecting the small momentum of the emitted electron, $\mathbf{q} = \mathbf{P}_n - \mathbf{P}_n'$. The replacement gives rise to an additional $\mathbf{q}$ which is to be included in the momentum integral of the neutrino potential. With the resulting potential, the recoil matrix element $M_R^{0\nu}$ of eq.(3.37i) is rewritten as [Tom 86],

$$M_R^{0\nu} = \langle H_{RC}(r) \sigma_m \cdot \sigma_n \rangle + (\text{non}-\text{central} \quad \text{terms}) \tag{3.45}$$

$$H_{RC}(r) = \frac{2R^2}{3M}\mu_\beta[4\pi\delta(\mathbf{r}) - \frac{2}{\pi}\frac{\bar{E}}{r^2} + \bar{E}^2 H(r)]. \tag{3.46}$$

The zero-range potential singular at the origin yields a leading contribution to the recoil matrix element $M_R^{0\nu}$, and the matrix element has to be evaluated by taking account of both short-range correlations and nucleon finite size effects, see eqs. (3.25) and (3.27).

3.3.2. $0^+ \to 2^+$ transitions

As for the 2ν decay, a $0^+ \to 2^+$ transition could occur for the 0ν decay. In this case, the neutrino mass term does not contribute, and the main contribution of the right-handed (V+A) current arises when one electron is emitted in S-wave and the other in P-wave (j=3/2) because of angular momentum conservation.

The inverse half-life is

$$[T_{1/2}^{0\nu}(0_i^+ \to 2^+)]^{-1} = C_{2\eta\eta}\langle\eta\rangle^2 + C_{2\lambda\lambda}\langle\lambda\rangle^2 + C_{2\eta\lambda}\langle\eta\rangle\langle\lambda\rangle \tag{3.47}$$

The coefficients C_{2xy} contain the nuclear matrix elements and phase space integrals,

$$C_{2\eta\eta} = M_\eta^2 G_+ + M_\eta'^2 G_- \tag{3.48a}$$

$$C_{2\lambda\lambda} = M_\lambda^2 G_+ \tag{3.48b}$$

$$C_{2\eta\lambda} = -2M_\eta M_\lambda G_+. \tag{3.48c}$$

The nuclear matrix elements are given by

$$M_\eta = -\frac{2}{3}M_1 - \frac{\sqrt{6}}{3}M_2 - \frac{\sqrt{6}}{9}M_3 + \frac{\sqrt{14}}{3}M_4 \tag{3.49a}$$

$$M_\eta' = \sqrt{2}M_6 \tag{3.49b}$$

$$M_\lambda = -\frac{2}{3}M_1 + \frac{\sqrt{6}}{3}M_2 + \frac{\sqrt{6}}{9}M_3 + \frac{\sqrt{14}}{3}M_4 + M_5 \tag{3.49c}$$

$$M_1 = \langle H'(r)[\sigma_m \times \sigma_n]^{(2)}\rangle \tag{3.50a}$$

$$M_2 = \langle H'(r)C_2(\hat{\mathbf{r}})\rangle \tag{3.50b}$$

$$M_3 = \langle H'(r)C_2(\hat{\mathbf{r}})(\sigma_m \cdot \sigma_n)\rangle \tag{3.50c}$$

$$M_4 = \langle H'(r)[C_2(\hat{\mathbf{r}})[\sigma_m \times \sigma_n]^{(2)}]^{(2)}\rangle \tag{3.50d}$$

$$M_5 = \langle H'(r)[C_2(\hat{\mathbf{r}}) \times (\sigma_m + \sigma_n)]^{(2)}\rangle \tag{3.50e}$$

$$M_6 = \langle H'(r)\frac{r_+}{r}[\hat{\mathbf{r}}_+ \times [\hat{\mathbf{r}}\times(\sigma_m - \sigma_n)]^{(1)}]^{(2)}\rangle \tag{3.50f}$$

where $C_2(\hat{\mathbf{r}}) = \sqrt{4\pi/5}Y_2(\hat{\mathbf{r}})$ is the modified spherical harmonics of rank 2. The phase space integrals $G_\pm$ are obtained by replacing g_i in eq.(3.33) by $g_\pm$,

$$g_\pm = \frac{e_1 e_2 \pm m_e^2}{6e_1 e_2}[(\frac{p_1}{m_e})^2\frac{F_1(Z,e_1)}{F_0(Z,e_1)} + (\frac{p_2}{m_e})^2\frac{F_1(Z,e_2)}{F_0(Z,e_2)}]. \tag{3.51}$$

The first (second) term of eq.(3.51) corresponds to P-wave emission of the first (second) electron the other electron being emitted in S-wave.

3.4. Non-nucleonic mechanisms

We have considered in the previous sections 2ν and 0ν $\beta\beta$ transitions only in terms of two nucleon processes in the sense that two different nucleons decay — including the possibility of virtual Δ excitations. In this subsection we briefly discuss other possible mechanisms.

3.4.1. The N^* mechanism

The N^* mechanism is characterized by the fact that the same particle undergoes two successive decay steps, in contrast to the two-nucleon process in which two different nucleons decay (see Figs. 3.1, 3.2 and 3.3). Since the charge has to be changed by two units, this process involves necessarily the Δ resonance, but it should not be confused with Δ excitations in the two-nucleon process. The N^* mechanism was first introduced by Primakoff and Rosen for the 0ν mode [Pri 69] and later for the 2ν mode by Smith, Picciotto and Bryman [Smi 73]. Diagrams are shown in Fig. 3.6. The $\beta\beta$ decay takes place through the Δ isobar (intrinsic spin-parity $J^\pi = 3/2^+$, isospin 3/2) which is admixed by the strong interaction into the initial and final nuclear states. The N^* mechanism could be expected to be important for the 0ν mode, since the neutrino is emitted and absorbed by the same hadron and therefore the range of neutrino propagation is much shorter than for the two-nucleon mechanism. In the N^* mechanism, however, $0^+ \to 0^+$ transitions are forbidden [Doi 83a, Hax 81]. The hadronic current acts only on the intrinsic parts of the Δ and the nucleon in the static approximation, and the $\Delta J = 0$ operator of the $0^+ \to 0^+$ transition cannot connect the nucleon with $J^\pi = 1/2^+$ and the Δ with $J^\pi = 3/2^+$. (This transition occurs if higher partial waves ($l \geq 2$) of the lepton are taken into account, but is highly suppressed.) The selection rules of the two-nucleon and N^* mechanisms are summarized in Table 3.2.

Recently, Tomoda calculated the $0^+ \to 2^+$ 0ν $\beta\beta$ decay of ^{76}Ge, and showed the importance of the N^* mechanism particularly for the nuclear matrix element relevant to the determination of the right-handed current mixing parameter λ (the term of $j_R \cdot J_R$, see eq.(3.1)) [Tom 88].

Haxton and Stephenson pointed out another diagram of the N^* mechanism, Fig. 3.7, which allows the $0^+ \to 0^+$ transition through both the m_ν and V+A parts [Hax 84]. However, this process is of second order in the coupling between Δ and nucleon, and the decay rate is suppressed by the

Table 3.2. Selection rules for $\beta\beta$ decay in the two-nucleon mechanism and the N* mechanism (in the static quark model). Circles, crosses and triangles denote dominant, forbidden and suppressed transitions, respectively.

	2ν mode		0ν mode			
			m_ν		ω, q	
	2n	N*	2n	N*	2n	N*
$0^+ \to 0^+$ transition	o	×	o	×	o	×
$0^+ \to 2^+$ transition	Δ	Δ	×	×	o	o

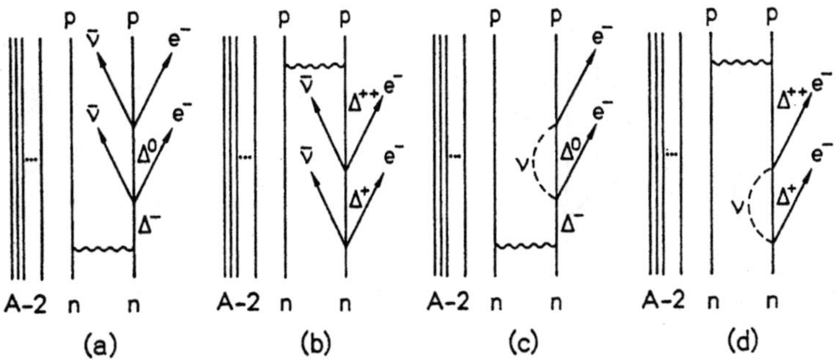

Fig. 3.6 Diagrams of the $\beta\beta$ decay amplitudes in the N^* mechanism for the 2ν mode, (a) and (b), and for the 0ν mode, (c) and (d). The wavy line represents the strong interaction.

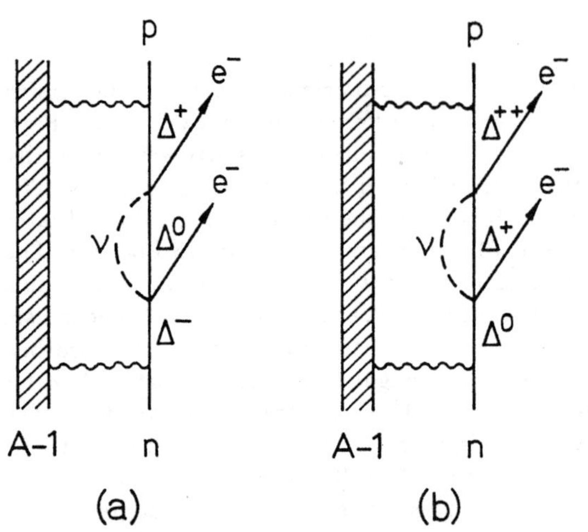

Fig. 3.7 The 0ν mode $\beta\beta$ decay in the higher-order N^* mechanism. The two wavy lines, which represent the strong interaction, couple to either one nucleon or two nucleons of the A-1 nucleons.

factor $P(\Delta)^2 \sim 10^{-4}$, where $P(\Delta)$ is the probability of finding a Δ in a nucleus. The $0^+ \to 0^+$ transition is also possible through various higher nucleon resonances with $J^\pi = 1/2$ [Hax 81].

3.4.2. The pion exchange mechanism

Since the nuclear interaction can be described in terms of pions and heavier mesons, these mesons could explicitly participate in the $\beta\beta$ decay. Diagrams of the pion exchange mechanism are shown in Fig. 3.8. Vergados considered the diagram 3.8c (0ν mode) and showed that, if the

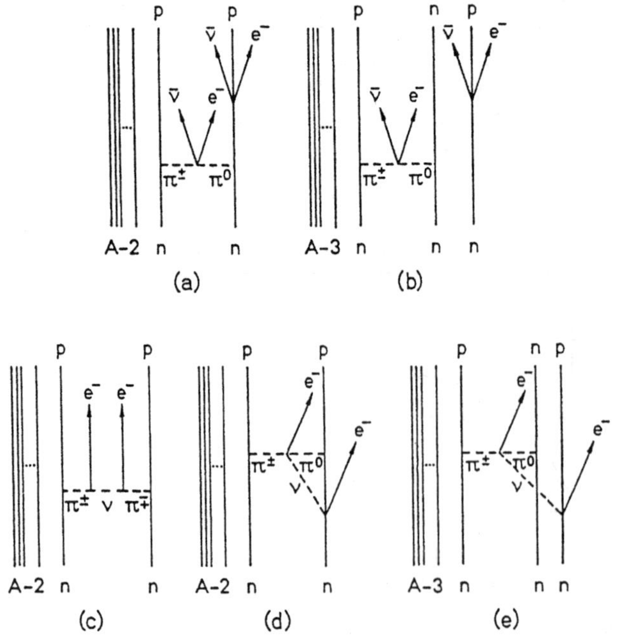

Fig. 3.8 Diagrams of the $\beta\beta$ decay in the pion exchange mechanism; (a) and (b) for the 2ν mode; (c) (d) and (e) for the 0ν mode.

mixing between light and heavy neutrinos is not negligible, the double charge exchange of pions in flight between two nucleons makes an important contribution to the 0ν $\beta\beta$ decay mediated by a heavy Majorana neutrino [Ver 82]. There are no estimates for the other meson exchange processes.

3.4.3. 0ν decay through Higgs bosons

Mohapatra and Vergados proposed a 0ν mode via Higgs bosons without virtual neutrinos [Moh 81b,86], as shown in Fig. 3.9. This process does not involve a Majorana neutrino, but proceeds by decay of a doubly charged Higgs boson to electrons. Contributions of this mechanism are, however, negligible because of the weak coupling of the physical Higgs particle to quarks [Sch 82] and furthermore because of the suppressed nuclear matrix element relevant for this process [Hax 82b].

3.4.4. 0ν mode with majoron and familon emission

Georgi, Glashow and Nussinov pointed out the 0ν mode associated with majoron emission [Geo 81, Gel 82]. The majoron is a light pseudo-scalar Goldstone boson arising from the spontaneous breaking of the global $B-L$ symmetry [Chi 80,81, Gel 81]. The diagram is shown in Fig. 3.10. For the effective interaction Hamiltonian

$$H_W = \frac{G}{\sqrt{2}} j_L \cdot J_L^\dagger + \frac{i}{2} \sum_{j,k} g_{jk} \bar{N}_j \gamma_5 N_k \phi_M \tag{3.52}$$

the inverse half-life for a $0_i^+ \to 0_f^+$ transition is given by [Doi 85]

$$[T_{1/2}^{0\nu,B}(0_i^+ \to 0_f^+)]^{-1} = |\langle g_B \rangle|^2 F^{0\nu,B} |M_{GT}^{0\nu} - M_F^{0\nu}|^2. \tag{3.53}$$

Here, $\langle g_B \rangle$ is the effective value of the coupling between the majoron and neutrinos,

$$\langle g_B \rangle = \sum_{j,k} g_{jk} U_{ej} U_{ek} \tag{3.54}$$

and the two nuclear matrix elements in eq.(3.53) are the same ones that appear in the 0ν decay of the two-nucleon mechanism, eqs. (3.37a) and (3.37b). The phase space integral is given by

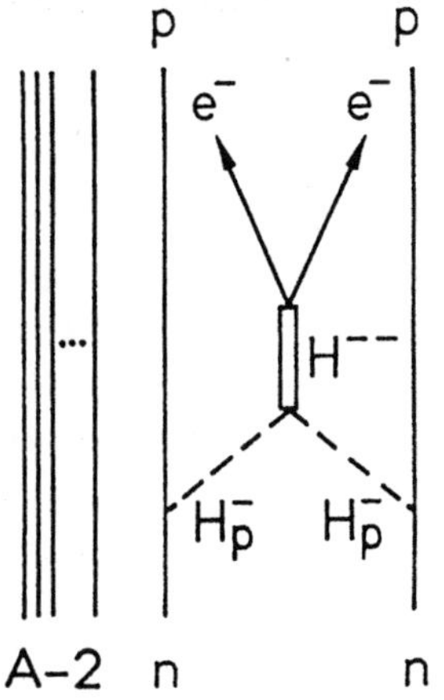

Fig. 3.9

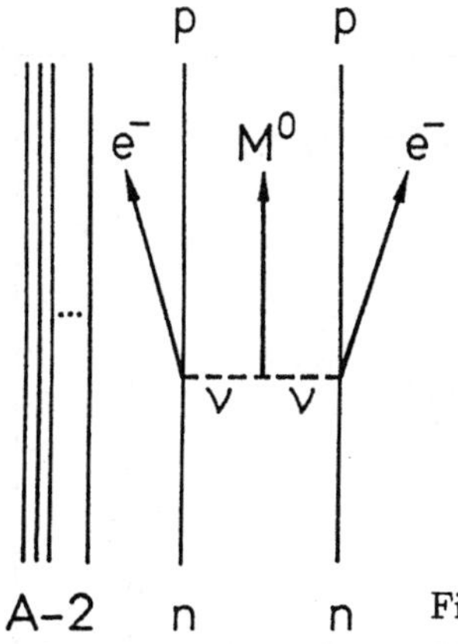

Fig. 3.10

Fig. 3.9 $\beta\beta$ decay through Higgs bosons. H_p^- represents the physical Higgs boson. The neutrino does not participate in this mechanism.

Fig. 3.10 0ν $\beta\beta$ decay with majoron emission. The virtual Majorana neutrino in flight emits the majoron.

$$F^{0\nu,B} = \frac{G^4 g_A^4}{128\pi^7 R^2} \int_{m_e}^{T_0+m_e} F(Z,e_1)k_1 e_1\, de_1 \int_{m_e}^{T_0+2m_e-e_1} F(Z,e_2)k_2 e_2 k_M\, de_2 \qquad (3.55)$$

where $k_M = T_0 - (e_1 + e_2 - 2m_e)$ is the momentum of the emitted majoron. A detailed examination of this process for $0^+ \rightarrow 0^+$ as well as for $0^+ \rightarrow 2^+$ transitions is given by [Doi 87]. The latter process is found to be suppressed by of the order of 10^{-7} compared to $0^+ \rightarrow 0^+$ transitions.

This process may become an important background in the electron spectrum in $\beta\beta$ decay counter experiments. The electron sum-energy spectra are illustrated in Fig. 3.11. Compared to the 2ν decay which leads to five-body final states (two electrons, two neutrinos and the residual nucleus), the sum-energy spectrum of the four-body majoron emission is shifted to higher energies. The sum-energy of the 0ν electrons should be identical to the $\beta\beta$ decay Q value, $e_1 + e_2 - 2m_e = T_0$.

A similar situation and similar results are expected for the familon, the Nambu-Goldstone boson associated with spontaneous breaking of the horizontal (family) symmetry [Rei 82, Wil 82, Gel 83].

4. Nuclear structure models used in calculations of $\beta\beta$ decay

We have seen in the preceding sections that for the calculation of 2ν and 0ν $\beta\beta$ decay rates and particularly for the deduction of the neutrino mass or of information on right-handed currents from measured 0ν decay rates, reliable calculations of the contributing nuclear matrix elements are the essential prerequisite.

Several approaches with different nuclear structure models have been used in the calculation of $\beta\beta$ decay nuclear matrix elements. In spite of many efforts, until recently there was a long-standing problem namely that calculations systematically overestimated most $\beta\beta$ decay rates of the 2ν mode. It was argued that the nuclear theory could have a poor predicticable power also for the 0ν mode, and that therefore the upper limits on the neutrino mass and the mixing parameters of right-handed currents, eqs. (3.29) and (3.47), deduced from experimental half-lives of the 0ν mode might have considerable uncertainties. Non-nuclear explanations of the problem were investigated [Gro 84], but ultimately a nuclear structure solution was found. The suppression mechanism of the 2ν decay rates has been clarified in recent studies [Kla 84, Gro 85,86, Vog 86, Civ 87, Mut 88a,88c] by recognizing the importance of ground state correlations for the calculated decay rates and by using a nuclear model based on the quasiparticle random phase approximation (QRPA). In the

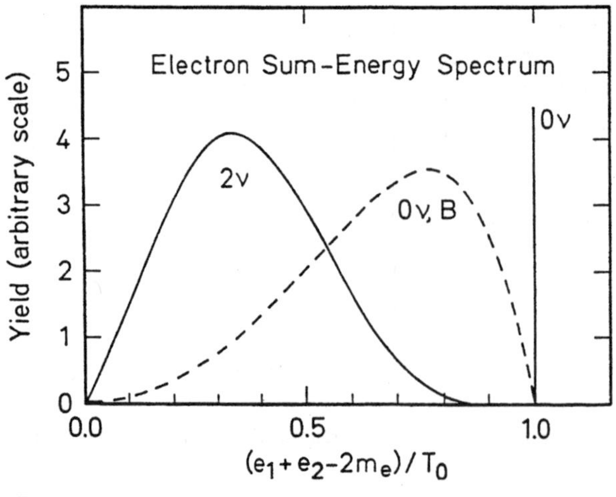

Fig. 3.11 Sum-energy spectra of the two electrons emitted in the $\beta\beta$ decay for the 2ν and 0ν modes and for the 0ν,B mode (0ν mode associated with majoron emission).

next section we discuss the general features of the different types of correlations which play a decisive role in $\beta\beta$ decay. In sections 4.2 - 4.4 we present general characteristic features of nuclear models used in the calculation of $\beta\beta$ decay. In sect. 5 we then give the results of calculations of $\beta\beta$ decay rates based on these models.

4.1. Ground state correlations

Nuclear states are not well described by a single Slater determinant in a microscopic model, even for the 0^+ ground state of an even-even nucleus. Fig. 4.1 shows the configuration for the ground states of parent and daughter nuclei for the $\beta\beta$ decay of ^{76}Ge in the simplest shell model. The 2ν $\beta\beta$ decay is forbidden for these ground state wave functions, since Gamow-Teller transitions are allowed only between orbits with the same orbital angular momentum belonging to the same oscillator major shell. The situation is the same for most nuclei which undergo $\beta\beta$ decay, except for some light nuclei such as ^{48}Ca. In the latter case wave functions of the simplest shell model configuration yield on the other hand a much too large nuclear matrix element of $\beta\beta$ decay. In order to obtain a better description of $\beta\beta$ transitions, configurations including particle-hole excitations have to be taken into account in the nuclear wave functions. Such admixtures in the ground states are in general called ground state correlations. There exist different types of correlations.

The shell model automatically takes into account any type of correlation, if the model space is large enough and a realistic residual interaction is used. But such a shell model calculation is practically impossible in most cases, and the model space must be truncated drastically, see sect. 4.2. It is, therefore, reasonable to include explicitly correlations which result from prominent features of the nucleon-nucleon interaction and are important for the transition under consideration.

The pairing force is the most prominent part of the residual nuclear interaction acting between like nucleons (between two protons or between two neutrons). This force makes two nucleons in the same orbit form a pair coupled to $J^\pi = 0^+$. These correlations are called pairing correlations. Because of the attractive character of the pairing force (the matrix element is negative), all even-even nuclei have the ground state spin-parity $J^\pi = 0^+$ and are energetically lower than the neighbouring nuclei with odd numbers of protons and neutrons.

The pairing correlations are taken into account in the BCS formalism, see sect. 4.3. Fig. 4.2 compares nucleon distributions among orbits for the cases without and with pairing correlations. When there are no correlations (Fig. 4.2a), low-lying orbits are completely filled and high-lying

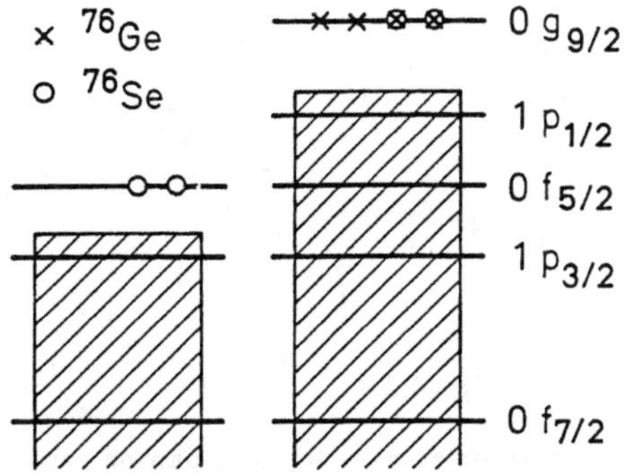

Fig. 4.1 Configurations in the simplest shell model for the ground states of ^{76}Ge and ^{76}Se (the initial and final states of ^{76}Ge $\rightarrow^{76}$Se $\beta\beta$ decay, respectively). The hatched area indicates fully occupied orbits.

orbits are empty (one orbit nearest to the Fermi energy can be partly filled). On the other hand, the pairing interaction smears out the distribution of nucleons (Fig. 4.2b). The occupation probability changes from 1 to 0 in orbits near the Fermi energy. All nucleons are paired to $J^\pi = 0^+$. This state is the vacuum for quasiparticles, which are defined by the Bogoliubov transformation, eq.(4.4). The pairing correlations give rise to a finite decay rate for 2ν $\beta\beta$ decay which is forbidden for the simplest shell model configurations (Fig. 4.1). They are therefore the origin of the most important collective effect to enhance $\beta\beta$ decay rates. The calculated half-lives, however, are much shorter than the experimental values by up to several orders of magnitude, when including *only* the pairing correlations (see Table 5.4 and Fig. 5.4).

The RPA (random phase approximation) formalism takes into account the correlations in the ground state which are directly connected with the nuclear excitation considered. There are several versions of RPA, depending on the mode of excitation under consideration and/or on the nucleus [Rin 80]. Here we mention the proton-neutron quasiparticle RPA (QRPA), see sect. 4.4. This RPA model describes charge-changing transitions which are associated with a change of the atomic number by one unit, such as Gamow-Teller transitions, and is expressed in terms of quasiparticle degrees of freedom. The ground state is the vacuum for the QRPA phonon, eq.(4.11). The ground state wave function is illustrated in Fig. 4.3. The main component is the BCS ground state with no quasiparticles (Fig. 4.2b), and the leading admixtures are four-quasiparticle states. When the RPA equation (eq.(4.12)) is solved for a J^π mode, the four-quasiparticle state consists of two pairs of proton- and neutron-quasiparticles which both have spin-parity J^π as indicated in Fig. 4.3,

$$|RPA\rangle \sim |-\rangle + \sum_{pnp'n'} \alpha_{pn,p'n'} |pn(J^\pi), p'n'(J^\pi); 0^+\rangle + \cdots \tag{4.1}$$

where $|-\rangle$ denotes the BCS ground state. Since, by the charge-changing transition, a creation of a pn pair from the main component and a destruction of one of the pairs in the four-quasiparticle components can lead to the same pn pair state in the odd-odd nucleus, the transition strength could be strongly affected by the inclusion of the ground state correlations through interference between the two transition amplitudes, see Fig. 4.6. For the calculation of 2ν $\beta\beta$ decay, which proceeds by successive Gamow-Teller transitions through 1^+ intermediate states, the RPA equation is solved for $J^\pi = 1^+$. In this case the wave function admixtures given in eq.(4.1), the spin-isospin correlations, play an important role for the suppression of the decay rate. The particle-particle interaction (see

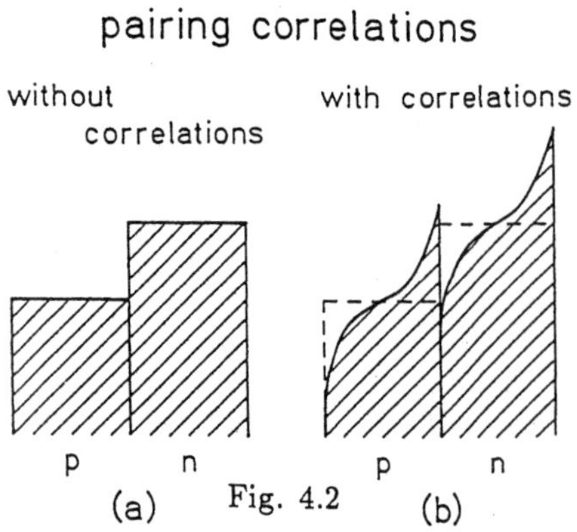

Fig. 4.2

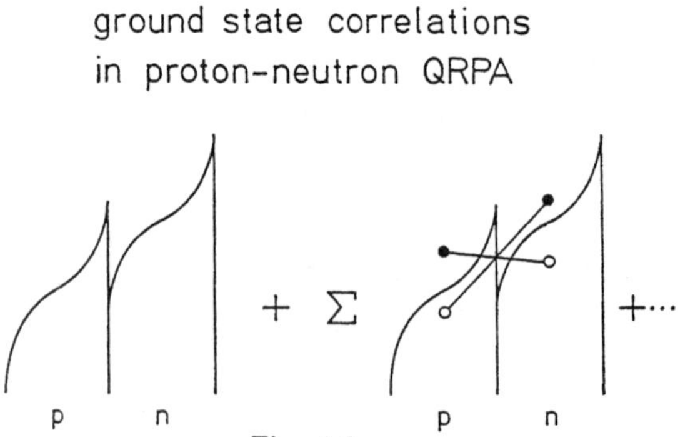

Fig. 4.3

Fig. 4.2 Distributions of nucleons among single-particle orbits in a nucleus; (a) without pairing correlations (the simplest shell model), (b) with pairing correlations.

Fig. 4.3 Ground-state wave function in the proton-neutron quasiparticle RPA. The line connecting circles, which denote quasiparticles, indicates angular momentum coupling of a proton-neutron pair. Both pairs have the same spin-parity J^π.

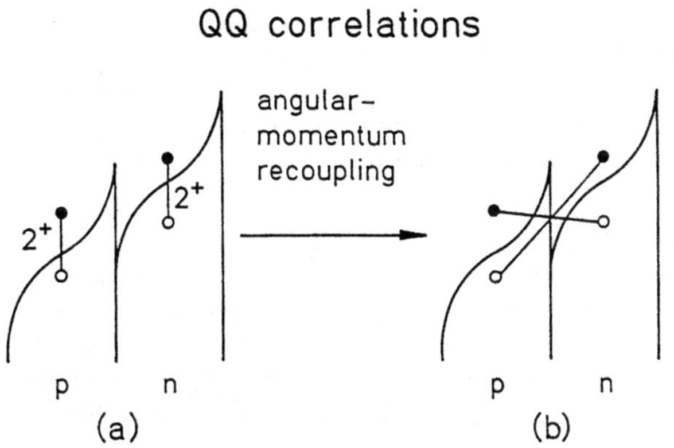

Fig. 4.4 Quadrupole-quadrupole (QQ) correlations in the ground-state wave function. (a) The dominant four-quasiparticle components of the QQ correlations consist of two protons coupled to 2^+ and two neutrons coupled to 2^+. (b) The quadrupole coupling can be expressed, by angular momentum recoupling, as a linear combination of two proton-neutron pairs which both have the same J^π.

eq.(4.14) and the following discussion) enhances the spin-isospin correlations, namely enhances the amplitudes $\alpha_{pn,p'n'}$ in eq.(4.1).

The quadrupole-quadrupole (QQ) force, which is the strongest component of the proton-neutron interaction, yields the QQ correlations in nuclear wave functions. Because of the QQ interaction, even-even nuclei have a low-lying 2^+ state, usually as the first excited state. The RPA ground state for the 2^+ mode excitation in the same nucleus (without a change of the atomic number, in contrast to charge-changing transitions) has the structure

$$|RPA\rangle \sim |-\rangle + \sum_{pp'nn'} \alpha'_{pp',nn'} |pp'(2^+), nn'(2^+); 0^+\rangle + \cdots \tag{4.2}$$

(cf. eq.(4.1) for the charge-changing mode). Klapdor and Grotz showed that the QQ correlations in the ground states suppress $\beta\beta$ decay rates of the 2ν mode [Kla 84, Gro 86]. This can be understood by the angular momentum recoupling, see also Fig. 4.4,

$$\begin{aligned} &|pp'(2^+), nn'(2^+); 0^+\rangle \\ &= \sum_J (-1)^{p+n'-2-J} \sqrt{5(2J+1)} W(pp'nn'; 2J) |pn(J^\pi), p'n'(J^\pi); 0^+\rangle. \end{aligned} \tag{4.3}$$

The QQ correlations thus include components which affect Gamow-Teller transitions ($J^\pi = 1^+$).

4.2. Shell model

All microscopic models are based on the shell model. Nucleons in a nucleus are assumed to move in a single-particle (s.p.) potential, which is generated by the nuclear and Coulomb forces acting between them, and interact with each other via the residual interaction. A nuclear wave function of the many-particle system is expressed as a linear combination of antisymmetrized, orthonormal basis states which are products of the s.p. wave functions. Amplitudes of the nuclear wave function are computed by diagonalizing the effective Hamiltonian consisting of the s.p. term and the residual interaction. A set of basis states is determined by specifying an inert core and valence s.p. orbits. The s.p. orbits in the core are fully occupied by nucleons, and its nucleons do not participate in nuclear transitions. The remaining nucleons are distributed among the valence orbits in accordance with the Pauli exclusion principle.

Application of conventional shell model techniques is limited to light $\beta\beta$ emitters. As already mentioned, for heavy nuclei the model space and configurations have to be seriously truncated and the collective effects important in $\beta\beta$ decay can only partly be included. It is difficult or even impossible to treat explicitly the intermediate 1^+ spectrum, which could be compared with results of (p,n) experiments, within this approach. This has only been done for ^{48}Ca which, besides ^{46}Ca, is the lightest $\beta\beta$ emitter.

For the states of parent (^{48}Ca), daughter (^{48}Ti) and intermediate (^{48}Sc) nuclei, ^{40}Ca is expected to be a good inert-core nucleus, and the additional eight nucleons can be restricted to an $f_{7/2}$ - $p_{3/2}$ - $p_{1/2}$ - $f_{5/2}$ space [Hax 84]. The smallest model space $(f_{7/2})^8$ was assumed in earlier calculations. Zamick and Auerbach used the pure configuration for qualitative discussion on the hindrance of 2ν decay by the K selection rule in the Nilsson-pairing scheme [Zam 82]. It was shown later, however, that a few nucleons have to be excited from $f_{7/2}$ to the higher-lying orbits [Sko 83, Tsu 84, Wu 85, Mut 86].

In its application to the $\beta\beta$ decay of heavier nuclei, the shell model approach has the difficulty that the Hamiltonian matrix becomes much too large to be handled. For instance, if four protons and fourteen neutrons are distributed among the valence orbits $p_{3/2}$, $p_{1/2}$, $f_{5/2}$ and $g_{9/2}$ (this situation corresponds to ^{76}Ge), the number of linearly independent basis states with $J^\pi = 0^+$ is 210,772; and for a larger model space consisting of full $3\hbar\omega$ and $4\hbar\omega$ oscillator major shells, the number exceeds 10^{21}. Thus, one is forced to reduce the number of basis states very drastically but preserving as much as possible the components important for the phenomena under consideration.

Haxton, Stephenson and Strottman [Hax 82a,84] chose the weak-coupling model in the calculation of nuclear matrix elements for the $\beta\beta$ decay of ^{76}Ge, ^{82}Se and ^{128,130}Te. They first solved eigenvalue equations for proton and neutron systems separately, and took from each system about fifty states with the smallest energies. Then the ground-state wave functions of both parent and daughter nuclei of the $\beta\beta$ decay were calculated by diagonalizing the proton-neutron residual interaction in the proton-neutron coupled states. The number of basis states with $J^\pi = 0^+$ is reduced to the order of 100. These authors had the problem of treating properly and explicitly the intermediate 1^+ spectrum. The closure approximation with its inherent problems in the case of 2ν decay (see sect. 3.2) was used in these calculations. In addition, they did not include the spin-orbit partners, for example $f_{7/2}$ for $f_{5/2}$, which play an important role for Gamow-Teller transitions.

4.3. Quasiparticle models

The quasiparticle picture is attractive, since it defines a new vacuum for quasiparticles and nuclear wave functions are described in terms of a small number of quasiparticle degrees of freedom (for the formalism see, e.g., [Sol 76, Rin 80]). Quasiparticles are introduced by the Bogoliubov transformation

$$a_k^\dagger = u_k c_k^\dagger - v_k c_{\bar{k}}$$
$$a_{\bar{k}}^\dagger = u_k c_{\bar{k}}^\dagger + v_k c_k \qquad (4.4)$$

where $a_k^\dagger$ and $c_k^\dagger$ are quasiparticle and particle (nucleon) creation operators, respectively, and $\bar{k}$ is a conjugate (usually time-reversed) state of k. The quasiparticle operators obey the fermion anti-commutation relations as the particle operators do. The occupation amplitudes u_k and v_k satisfy $u_k^2 + v_k^2 = 1$, v_k^2 being the occupation probability of a nucleon in the orbits k and $\bar{k}$. These coefficients of the Bogoliubov transformation can be calculated in the BCS approximation by the constrained variational principle

$$\delta\langle BCS|H - \lambda\hat{N}|BCS\rangle = 0 \qquad (4.5a)$$
$$\langle BCS|\hat{N}|BCS\rangle = N. \qquad (4.5b)$$

The Fermi energy λ is introduced as a Lagrange multiplier, N is the nucleon number and $\hat{N}$ the number operator. The variational principle yields a set of equations

$$\tilde{\epsilon}_k = \bar{\epsilon}_k + \sum_{k'>0} \frac{1}{2} v_{k'}^2 (\langle kk'|V|kk'\rangle + \langle \bar{k}k'|V|\bar{k}k'\rangle) - \lambda \tag{4.6}$$

$$\Delta_k = -\sum_{k'>0} u_{k'} v_{k'} \langle k\bar{k}|V|k'\bar{k}'\rangle \tag{4.7}$$

$$v_k^2 = \frac{1}{2}(1 - \frac{\tilde{\epsilon}_k}{\sqrt{\tilde{\epsilon}_k^2 + \Delta_k^2}}) \tag{4.8}$$

where $\tilde{\epsilon}_k$ and $\epsilon_k = \sqrt{\tilde{\epsilon}_k^2 + \Delta_k^2}$ are single-particle and quasiparticle energies, respectively. These equations are solved self-consistently with the particle-number condition, eq.(4.5b),

$$2\sum_{k>0} v_k^2 = N. \tag{4.9}$$

The BCS ground state defined by $|BCS\rangle = \prod_k a_k a_{\bar{k}}|\rangle$ is obviously the vacuum for the quasiparticles, $a_k|BCS\rangle = 0$ (here, $|\rangle$ is the vacuum for nucleons). In terms of nucleons, it is a superposition of states with different nucleon numbers,

$$|BCS\rangle \sim |\rangle + \sum_{k\geq 0} \frac{v_k}{u_k} c_k^\dagger c_{\bar{k}}^\dagger |\rangle + \frac{1}{2} \sum_{kk'\geq 0} \frac{v_k v_{k'}}{u_k u_{k'}} c_k^\dagger c_{\bar{k}}^\dagger c_{k'}^\dagger c_{\bar{k}'}^\dagger |\rangle + \cdots \tag{4.10}$$

and each state consists of pairs of nucleons coupled to $J^\pi = 0^+$ (see also Fig. 4.2). The quasiparticle model thus takes into account the pairing correlations which result from the prominent, strong pairing forces acting between like nucleons. The ground state of an even-even nucleus is the quasiparticle vacuum, and excited states are described to the lowest order by two-quasiparticle states.

The program VAMPIR (or MONSTER) of the Tübingen group [Sch 84a,84b] is a generalized version of the quasiparticle model. The authors developed these computer codes on the basis of the Hartree-Fock-Bogoliubov formalism; i.e., the s.p. states are calculated self-consistently and quasiparticles are defined by a Bogoliubov transformation more general than eq.(4.4). The nucleon number is restored by the particle-number projection, and the rotational invariance which is violated by the use of a deformed basis is restored by angular-momentum projection. The wave functions generated by this method have been used in the calculation of $\beta\beta$ decay rates [Tom 85,86,88].

Klapdor and Grotz [Kla 84, Gro 85a,85b,86] calculated half-lives of both 2ν and 0ν (the mass term) $\beta\beta$ decay for all possible $\beta\beta$ emitters with A $\geq$ 70 (see Table 5.3). They adopted two types of models. One is a QRPA calculation with a schematic Gamow-Teller force, which is the part of the proton-neutron interaction most important for Gamow-Teller transitions (the particle-particle component is neglected, see eq.(4.14)). The other is an extended nucleon-number-projected BCS calculation with a residual Gamow-Teller force, in a model space consisting of zero- and four-quasiparticle states, as well as an additional quadrupole-quadrupole (QQ) force, which is another important part of the proton-neutron interaction. These calculations included further virtual excitations of the Δ resonance, the Gamow-Teller force being extended to a generalized spin-isospin force in a simple quark model. Therefore, wave functions of the 0^+ ground states are linear combinations of eq.(4.2) and the second term of eq.(4.1) with $J^\pi = 1^+$ including Δ-quasiparticle components. These authors showed that the QQ ground state correlations reduce the 2ν decay rates by up to an order of magnitude compared with those of the QRPA calculation, and lead to a remarkable improvement in the agreement of the calculated 2ν $\beta\beta$ decay rates with experimental results (see Fig. 5.2). These calculations led to two important results. First they demonstrated the large sensitivity of the calculated $\beta\beta$ decay rates to various types of ground state correlations. Second, more generally, these calculations showed the way to the solution of the above mentioned long-standing problem in the calculation of 2ν $\beta\beta$ rates: The QQ force, which had a strongly suppressing effect on the 2ν $\beta\beta$ matrix elements, enhances the component in the ground state wave functions with $J^\pi = 1^+$ (second term in eq.(4.1), see eq.(4.3)). The QQ force, on the other hand, is contained as a constituent in the particle-particle interaction of the $J^\pi = 1^+$ mode, which enhances further the amplitude $\alpha_{pn,p'n'}$ of the spin-isospin correlations (see sects. 4.4 and 5.1.2).

It was, therefore, natural to investigate the particle-hole and particle-particle forces (eq.(4.14)) separately. This is possible and has been done in the most recent calculations [Vog 86, Civ 87, Tom 87, Mut 88a,88c] in which a particle-particle component was included in the QRPA calculations in addition to the particle-hole component (which was the only one considered in previous RPA calculations). This led to a final clarification of the suppression mechanism of 2ν $\beta\beta$ decay rates. The formalism of the QRPA including the particle-particle interaction, therefore, is described in some detail in the next section.

4.4. Quasiparticle random phase approximation (QRPA) including the particle-particle interaction

The proton-neutron QRPA was developed for the description of charge-changing excitations, such as Gamow-Teller transitions, from the 0^+ ground state of an even-even nucleus (A, Z) to states in the odd-odd nucleus $(A, Z \pm 1)$ [Hal 67, Mut 88b]. In the derivation of QRPA formulae one makes two main assumptions as in the usual RPA: (i) one assumes boson commutation relations for creation and annihilation operators of a proton-neutron pair, (ii) one neglects $a^\dagger a$ terms in the matrix elements of the QRPA ground state, $\langle RPA|a_k^\dagger a_{k'}|RPA\rangle = 0$.

The QRPA phonons with angular momentum J and its z-component M are defined by

$$A_\omega^\dagger(JM) = \sum_{pn}(X_\omega^{pn,J}[a_p^\dagger a_n^\dagger]_M^{(J)} - Y_\omega^{pn,J}(-1)^{1+J+M}([a_p^\dagger a_n^\dagger]_{-M}^{(J)})^\dagger) \tag{4.11}$$

where indices p and n distinguish between proton and neutron states, and the square bracket denotes angular-momentum coupling. The energy eigenvalues ω and forward- and backward-going amplitudes X and Y (see Fig. 4.6) are obtained by solving the QRPA equation

$$\begin{bmatrix} A & B \\ -B & -A \end{bmatrix} \begin{bmatrix} X \\ Y \end{bmatrix} = \omega \begin{bmatrix} X \\ Y \end{bmatrix} \tag{4.12}$$

with the normalization

$$\sum_{pn}((X_\omega^{pn,J})^2 - (Y_\omega^{pn,J})^2) = 1. \tag{4.13}$$

Elements of the submatrices are explicitly

$$\begin{aligned} A_{pn,p'n'} = {} & \delta(pn,p'n')(\epsilon_p + \epsilon_n) \\ & +(u_p v_n u_{p'} v_{n'} + v_p u_n v_{p'} u_{n'})\langle pn^{-1}|V|p'n'^{-1}\rangle_J \\ & +(u_p u_n u_{p'} u_{n'} + v_p v_n v_{p'} v_{n'})\langle pn|V|p'n'\rangle_J \end{aligned} \tag{4.14a}$$

$$\begin{aligned} B_{pn,p'n'} = {} & -(u_p v_n v_{p'} u_{n'} + v_p u_n u_{p'} v_{n'})\langle pn^{-1}|V|p'n'^{-1}\rangle_J \\ & +(u_p u_n v_{p'} v_{n'} + v_p v_n u_{p'} u_{n'})\langle pn|V|p'n'\rangle_J. \end{aligned} \tag{4.14b}$$

The quasiparticle energies ϵ_j and the occupation amplitudes u_j and v_j are obtained in the BCS calculation. The matrix elements of particle-particle $\langle pn|V|p'n'\rangle$ and particle-hole $\langle pn^{-1}|V|p'n'^{-1}\rangle$ interactions are related to each other by the Pandya transformation,

$$\begin{aligned} & \langle pn^{-1}|V|p'n'^{-1}\rangle_J \\ & = -(-1)^{p+n+p'+n'}\textstyle\sum_{J'}(2J'+1)W(pnp'n';JJ')\langle pn'|V|p'n\rangle_{J'}. \end{aligned} \tag{4.15}$$

The particle-particle and particle-hole interactions are reduced in the limit of the nucleon picture, $u_j v_j \to 0$, to interactions acting between two-particle (or two-hole) states and between particle-hole states, respectively. This can be seen from eq.(4.14) (see also Fig. 4.5). In the nucleon picture of a doubly closed shell nucleus, the orbits occupied by nucleons have $v_j = 1$ ($u_j = 0$) and the orbits with no nucleons have $u_j = 1$ ($v_j = 0$). The former can be called hole orbits and the latter particle orbits, because when a particle-hole pair is excited, the particle occupies a particle orbit and the hole is created in a hole orbit. The particle-particle interaction is always associated with the product $u_p u_n$ (or $v_p v_n$), which means that both proton and neutron occupy particle (or hole) orbits. On the other hand, the particle-hole interaction is associated with $u_p v_n$ or $v_p u_n$, namely one of the proton and neutron is in a particle orbit and the other in a hole orbit.

The QRPA ground state is the vacuum with respect to the phonons, $A_\omega(JM)|RPA\rangle = 0$, and is expressed as a superposition of states with quasiparticle number 0, 4, 8, ...

$$|RPA\rangle = n_0 \exp[\textstyle\sum_{pn,p'n'} Z_{pn,p'n'} [a_p^\dagger a_n^\dagger]^{(J)} \cdot [a_{p'}^\dagger a_{n'}^\dagger]^{(J)}] \, |BCS\rangle \tag{4.16}$$

where the normalization factor n_0 and the matrix of the ground state correlations Z are given by

$$n_0 = |\det X|^{-1/2} \tag{4.17}$$

$$Z = YX^{-1}. \tag{4.18}$$

The main component of the ground state has no quasiparticles (BCS ground state), and the next leading terms are four-quasiparticle states, which, in the nucleon picture, are two particle - two hole states. The probability of the (RPA-type) ground state correlations is given by $1 - \langle RPA|BCS\rangle^2 = 1 - n_0^2$.

Reduced matrix elements of charge-changing transition operators $t_\pm f_M^{(J)}$, from the 0^+ ground state (the RPA vacuum) of an even-even nucleus to states with J^π (one-phonon state) in the odd-odd nucleus, are given by

$$\langle \omega J \| t_- f^{(J)} \| 0^+ \rangle = \textstyle\sum_{pn} \langle p \| f^{(J)} \| n \rangle (X_\omega^{pn,J} u_p v_n - Y_\omega^{pn,J} v_p u_n) \tag{4.19a}$$

$$\langle \omega J \| t_+ f^{(J)} \| 0^+ \rangle = \textstyle\sum_{pn} \langle p \| f^{(J)} \| n \rangle (X_\omega^{pn,J} v_p u_n - Y_\omega^{pn,J} u_p v_n). \tag{4.19b}$$

For example, in case of the $0_i^+ \to 0_f^+$ 2ν $\beta\beta$ decay (see Fig. 4.6), the former corresponds to the Gamow-Teller transition from the parent 0_i^+ state to an intermediate 1^+ state, and the latter to the inverse of the Gamow-Teller transition from the intermediate state to the daughter 0_f^+ state. The X term of eq.(4.19a) ((4.19b)) is associated with $u_p v_n$ ($v_p u_n$), implying excitation of a proton particle - neutron hole (neutron particle - proton hole) pair. On the other hand, the amplitude Y is defined in eq.(4.11) with two quasiparticle annihilation operators. Therefore, the Y terms of eq.(4.19) originate from destruction of one of the particle-hole pairs in the RPA ground state, and depend on the degree of ground state correlations.

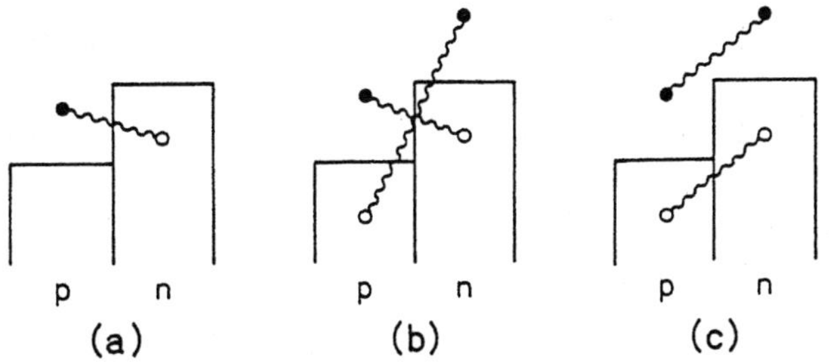

Fig. 4.5 The particle-hole (ph) interaction acting between proton particle - neutron hole (or proton hole - neutron particle) states, and particle-particle (pp) interaction between two-particle states or between two-hole states. (a) The ph interaction for a particle-hole state. (b) The ph interaction for a two particle - two hole state, and (c) the pp interaction for the same state.

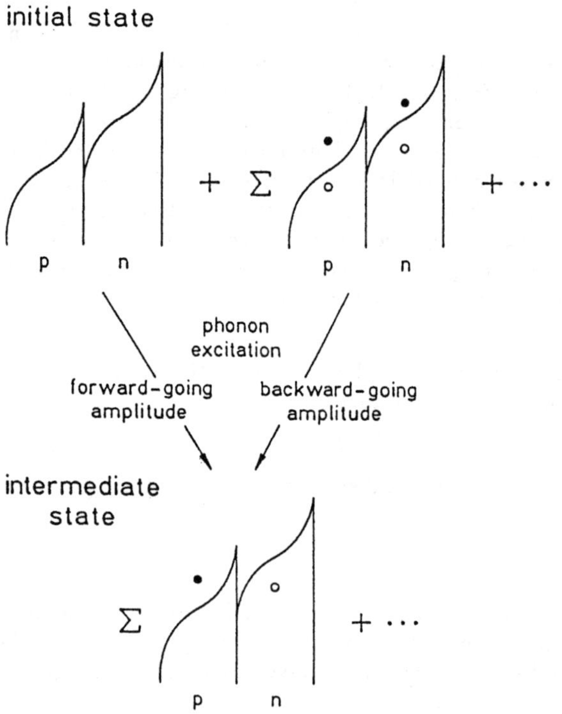

Fig. 4.6 A phonon excitation from the ground state (QRPA vacuum) to a one-phonon state in the proton-neutron QRPA. The dominant contribution of the forward-going amplitude term (from the zero-quasiparticle state to a proton-neutron pair state) interferes with the backward-going amplitude term (from four-quasiparticle states to the pair state). In the case of $\beta\beta$ decay, this phonon excitation corresponds to the transition from the initial (final) 0^+ state to an intermediate state.

In the QRPA formalism, the nuclear matrix element of a $0_i^+ \to 0_f^+$ 2ν $\beta\beta$ decay is expressed as (see eq.(3.10))

$$M_{GT}^{2\nu} = \sum_{a,b} \frac{\langle 0_f^+ \| t_- \sigma \| 1_b^+ \rangle \langle 1_b^+ | 1_a^+ \rangle \langle 1_a^+ \| t_- \sigma \| 0_i^+ \rangle}{E_a + T_0/2 + m_e - E_i}. \tag{4.20}$$

The amplitudes of Gamow-Teller transitions are obtained from eq.(4.19) by substituting $f^{(J)} = \sigma$ ($J^\pi = 1^+$),

$$\langle 1_a^+ \| t_- \sigma \| 0_i^+ \rangle = \sum_{p,n} \langle p \| \sigma \| n \rangle (X_a^{pn} u_p v_n - Y_a^{pn} v_p u_n) \tag{4.21a}$$

$$\langle 0_f^+ \| t_- \sigma \| 1_b^+ \rangle = \sum_{p,n} \langle p \| \sigma \| n \rangle (\bar{X}_b^{pn} \bar{v}_p \bar{u}_n - \bar{Y}_b^{pn} \bar{u}_p \bar{v}_n). \tag{4.21b}$$

The overlap of intermediate 1^+ states is given by

$$\langle 1_b^+ | 1_a^+ \rangle = \sum_{pn} (\bar{X}_b^{pn} X_a^{pn} - \bar{Y}_b^{pn} Y_a^{pn}). \tag{4.22}$$

The quantities without (with) bar are defined with respect to the initial 0_i^+ (final 0_f^+) state.

The 0ν $\beta\beta$ matrix elements of a $0_i^+ \to 0_f^+$ decay (see eqs. (3.37), (3.38) and (3.41)) are calculated with the following formulae,

$$\langle O_{mn}\rangle = \sum_{pnp'n'J\pi J'} \langle 0_f^+ \| [c_{p'}^\dagger \times \tilde{c}_{n'}]^{(J)\dagger} \| J_b^\pi \rangle \langle J_b^\pi | J_a^\pi \rangle \langle J_a^\pi \| [c_p^\dagger \times \tilde{c}_n]^{(J)} \| 0_i^+ \rangle$$
$$\times (-1)^{p+n+J+J'} (2J'+1) W(pnp'n'; JJ') \langle pp' | t_{-1} t_{-2} O_{12} | nn' \rangle_{J'} \qquad (4.23)$$

$$\langle J_a^\pi \| [c_p^\dagger \times \tilde{c}_n]^{(J)} \| 0_i^+ \rangle = \sqrt{2J+1} (X_a^{pn} u_p v_n - Y_a^{pn} v_p u_n) \qquad (4.24a)$$

$$\langle 0_f^+ \| [c_{p'}^\dagger \times \tilde{c}_{n'}]^{(J)\dagger} \| J_b^\pi \rangle = \sqrt{2J+1} (\bar{X}_b^{pn} \bar{v}_p \bar{u}_n - \bar{Y}_b^{pn} \bar{u}_p \bar{v}_n) \qquad (4.24b)$$

where $\tilde{c}_{jm} = (-1)^{j+m} c_{j-m}$ and the overlap is defined by eq.(4.22).

We make now an estimation of the order of the effects of the particle-particle interaction on the ground state correlations and charge-changing transitions. This is important because as mentioned the success of the QRPA calculations of 2ν $\beta\beta$ decay was achieved basically by including the particle-particle interaction. (It is noted that the particle-particle interaction is not an extra one being added to the QRPA model. It appears naturally by the Bogoliubov transformation, in addition to the particle-hole counterpart. But it was usually neglected in earlier calculations of $\beta\beta$ decay.) First of all, for nuclei with a large neutron excess, we expect that u_p and v_n are normally significantly larger than v_p and u_n,

$$u_p \sim v_n \sim l, \qquad v_p \sim u_n \sim s, \qquad s^2 \ll l^2. \qquad (4.25)$$

Then, the elements of QRPA equation (4.14) are expressed as

$$A_{pn,p'n'} \sim \delta(pn,p'n')(\epsilon_p + \epsilon_n)$$
$$+(l^4 + s^4) \langle pn^{-1} | V | p'n'^{-1} \rangle + 2l^2 s^2 \langle pn | V | p'n' \rangle \qquad (4.26a)$$

$$B_{pn,p'n'} \sim -2l^2 s^2 \langle pn^{-1} | V | p'n'^{-1} \rangle + 2l^2 s^2 \langle pn | V | p'n' \rangle. \qquad (4.26b)$$

The submatrix A is a matrix between main forward-going amplitudes X (or between backward-going amplitudes Y), and is therefore responsible for the mixing of various particle-hole states in the odd-odd nucleus reached by the charge-changing transition. For example, for the $J^\pi = 1^+$ mode the mixing leads to a concentration of β^- Gamow-Teller strength to the giant resonance (see Figs. 5.4 and 5.7). The elements of A are dominated by the particle-hole interaction and quasiparticle energies. The contribution of the particle-particle interaction is much smaller, by the factor s^2/l^2, implying that this interaction has little effect on the mixing of particle-hole states in the intermediate nucleus. On the other hand, the submatrix B is important for the ground state correlations since it connects X and Y. An increase of the backward-going amplitudes Y means enhancement of the ground state correlations (see eqs. (3.18) and (3.16)). The elements of B consist of both interaction terms of the same order, $\sim l^2 s^2$. This indicates that the particle-particle interaction, being added coherently to the particle-hole interaction, can significantly enhance the ground state correlations. The degree of enhancement depends on the nature of particle-hole and particle-particle interactions in the J^π channel.

The estimation of the effect on the transition amplitudes, eq.(4.19), clarifies a substantial difference between transitions induced by $t_- f^{(J)}$ and $t_+ f^{(J)}$. Let us take $\beta^\pm$ Gamow-Teller transitions to 1^+ states as an example. (Gamow-Teller strength distributions calculated in the QRPA model are compared in Fig. 5.7 for different strengths of the particle-particle interaction.) With the large factor l and the small factor s, the transition amplitudes are simplified such that

$$\beta^- : \langle \omega, 1^+ \| t_- \sigma \| 0^+ \rangle \sim X l^2 - Y s^2 \qquad (4.27a)$$

$$\beta^+ : \langle \omega, 1^+ \| t_+ \sigma \| 0^+ \rangle \sim X s^2 - Y l^2. \qquad (4.27b)$$

For β^- transitions ($n \to p$), the amplitude is dominated by the X term, which results from excitation of a proton particle - neutron hole pair. The Y term gives a very small contribution suppressed by the factor s^2 and furthermore by the ratio of Y over X. Normally the forward-going amplitudes X are much larger than the backward-going amplitudes Y, eq.(4.13). Therefore β^- transitions are insensitive to the ground state correlations, i.e., insensitive to the particle-particle interaction. The factors multiplied with X and Y are exchanged in β^+ transitions ($p \to n$).

When the particle-particle interaction is switched off, the X term dominates the transition, even if one considers the enhancement factor l^2 of the Y term since the ground state correlations are negligibly small. But, the particle-particle interaction enhances the ground state corrlations appreciably, and makes the Y term more important. The two terms are of the same order for $J^\pi = 1^+$, and β^+ transitions are strongly suppressed. Generally, in the 1^+ channel, the particle-hole interaction acting between particle-hole pairs is repulsive, while the particle-particle interaction is attractive. This characteristic feature results in a destructive interference between the X and Y terms. Thus, the particle-particle interaction plays a fundamental role in the suppression of β^+ transition strength. This conclusion can be understood also by the following simple argument. In a nucleus with a large neutron excess, most s.p. β^+ transitions ($p \to n$) are blocked by the Pauli exclusion principle in the absence of ground state correlations (the neutron orbits which can be reached by the Gamow-Teller transitions are fully occupied). When ground state correlations exist, vacancies in the neutron hole orbits can be found, as well as occupations in proton particle orbits, with the consequent occurrence of β^+ amplitudes which by destructive interference then reduce the total β^+ strength.

5. Results of calculations of $\beta\beta$ decay rates

5.1. Two-neutrino mode

5.1.1. Calculations till 1986

The main points of the development till 1986 were

(1) the observation [Gro 83,86, Kla 84, Tsu 84] that the closure approximation which was used in most earlier work such as that of [Hax 82a,84] cannot give reliable predictions of 2ν $\beta\beta$ decay rates. Figs. 5.1 and 5.4 are examples showing the importance of the signs of the transition matrix elements (see discussion in sect. 3). The closure approximation is, however, expected to be reliable in the calculation of 0ν $\beta\beta$ decay.

(2) the observation that the ground state correlations play the decisive role in the prediction of $\beta\beta$ decay rates of the 2ν mode, while the neutrino mass term of the 0ν mode is much less

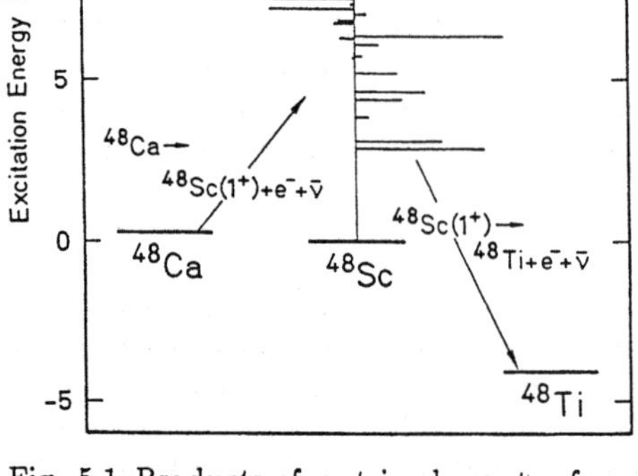

Fig. 5.1 Products of matrix elements of successive Gamow-Teller transitions through 1^+ intermediate states for the 2ν $\beta\beta$ decay, $^{48}Ca \to ^{48}Ti$, calculated in the shell model (from [Tsu 84]).

affected by the inclusion of the correlations [Kla 84, Gro 85,86]. The reason why details of the nuclear structure affect $M^{2\nu}_{GT}$ and $M^{0\nu}_{GT}$ ($M^{0\nu}_{F}$) so differently is discussed in sect. 5.2.2.

The unreliability of the closure approximation can be seen in the 2ν $\beta\beta$ decay of ^{48}Ca. The shell model approach is expected to work well for this doubly magic nucleus. Table 5.1 shows results of shell model calculations by [Hax 82a, Zam 82, Sko 83, Tsu 84]. The four calculations yielded similar values of the closure matrix element (eq.(3.13)). The difference in the half-life between [Tsu 84] and [Hax 82a] is traced to the use of the closure approximation by the latter authors. These authors estimated the average energy of intermediate states from the strength distribution of ^{48}Ca $\rightarrow$ ^{48}Sc Gamow-Teller excitations. On the other hand, for the matrix elements shown in Fig. 5.1, Tsuboi, Muto and Horie [Tsu 84] performed the phase space integral, eq.(3.6), explicitly for each intermediate state without assuming the closure approximation. The signs of the matrix elements for high-lying intermediate states are generally opposite to those for low-lying states. The sum in $M^{2\nu}_{GT}$, eq.(3.10), contains the energy denominator, while the closure matrix element, eq.(3.13) is a sum with equal weight (=1). The two sums involve different interference of those matrix elements. Klapdor and Grotz [Kla 84] used the same procedure as Tsuboi et al. for heavier nuclei and obtained for the average energies, also about half the values of Haxton et al. [Hax 84].

Fig. 5.2 and Table 5.2 show the most intensively investigated experimental and theoretical (till end of 1986) half-lives of 2ν $\beta\beta$ decay. The theoretical values are taken from systematic calculations, with the shell model [Hax 82a], the number-projected BCS [Kla 84, Gro 86], and the QRPA with a particle-hole Gamow-Teller force [Gro 85,86, Vog 85]. It is seen that for the heavier nuclei the discrepancy between experiment and calculation is rather large, up to two orders of magnitude for the calculations of [Hax 82a] and [Vog 85]. The calculations of [Kla 84, Gro 86] lead here to a remarkable improvement, but obviously cannot yet reproduce the full degree of observed suppression of the 2ν $\beta\beta$ decay rates.

A comment has to be made here concerning the experimental half-lives given in Table 5.2. The 2ν character is clear for the half-lives obtained by counter experiments for ^{76}Ge, ^{82}Se, ^{100}Mo, ^{136}Xe and ^{150}Nd. The recent counter measurement by Elliott, Hahn and Moe [Ell 87] yielded a half-life for ^{82}Se consistent with geochemical measurements [Kir 86, Man 86]. The values for ^{128,130}Te are, on the other hand, only from geochemical determination [Kir 86, Man 86], which cannot differentiate between different decay modes. However, according to the phase space argument discussed below, the $\beta\beta$ decay of ^{130}Te is clearly dominated by the 2ν mode.

Table 5.1. Half-lives, closure matrix elements and average energies of intermediate states of the 2ν $\beta\beta$ decay of ^{48}Ca, calculated in the shell model. The experimental lower limit of the half-life is given in the last line.

References	$T^{2\nu}_{1/2}$ (10^{19} y)	$(M^{2\nu}_{GT})_c$	$\bar{E}$ (MeV)
Calculation			
Hax 82	2.70	0.444	7.72
Zam 82		0.360	
Sko 83	4.1	0.250	4.6
Tsu 84	0.61	0.464	3.80
Experiment			
Bar 70	>3.6		

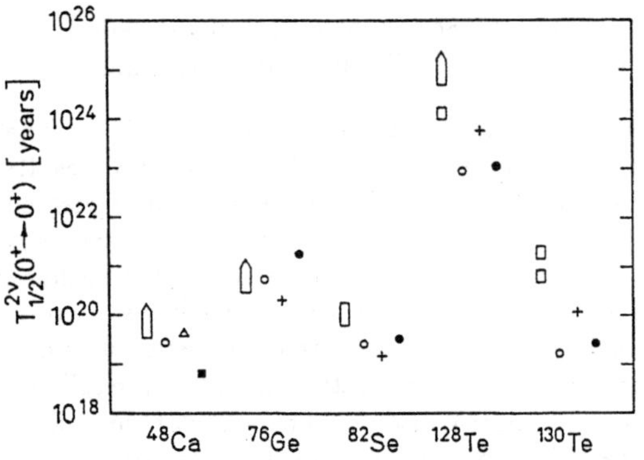

Fig. 5.2 Comparison of experimental and calculated half-lives of 2ν $\beta\beta$ decay. Experimental values are denoted by an open square, or an open arrow when only the lower limit is known. The theoretical values are from [Hax 84] (open circle), [Sko 83] (triangle), [Tsu 84] (filled square), [Gro 86] PBCS (cross) and [Vog 85] (filled circle).

Table 5.2. Comparison of $0^+_i \to 0^+_f$ 2ν $\beta\beta$ half-lives calculated with different nuclear structure models (see text). Recent experimental values are also presented.

References	^{76}Ge	^{82}Se	^{128}Te	^{130}Te	^{100}Mo	^{136}Xe	^{150}Nd
	10^{21}y	10^{20}y	10^{24}y	10^{21}y	10^{18}y	10^{19}y	10^{19}y
Calculation							
Hax 82	0.415	0.262	0.088	0.017	--	--	--
Vog 85	1.8	0.33	0.11	0.023	2.1	2.3	0.22
Gro 86							
QRPA (without pp force)	0.11	0.048	0.12	0.019	1.8	6.0	0.05
PBCS (GT)	0.13	0.051	0.062	0.015			
(GT + QQ)	0.22	0.15	0.57	0.12		3.3	
Experiment							
	>0.3[a]	1.30±0.05[b]	>5[b]	1.5-2.75[b]	>6.2[e]	>2.1[f]	>2.4[g]
		1.0 ± 0.4[c]	1.4±0.4[c]	0.7±0.2[c]			
		$1.1^{+0.8}_{-0.3}$[d]					

[a]Avi 86, [b]Kir 86, [c]Man 86, [d]Ell 87, [e]Smo 88, [f]Bar 87, [g]Kli 86.

The phase space argument for the branching ratio of ^{130}Te decay between 2ν and 0ν modes is as follows: The main assumption is that the nuclear matrix elements are assumed to be almost identical for ^{128}Te and ^{130}Te, which seems reasonable, since these nuclei have the same number of protons and the numbers of neutrons differ only by two units. Under this assumption, the half-life

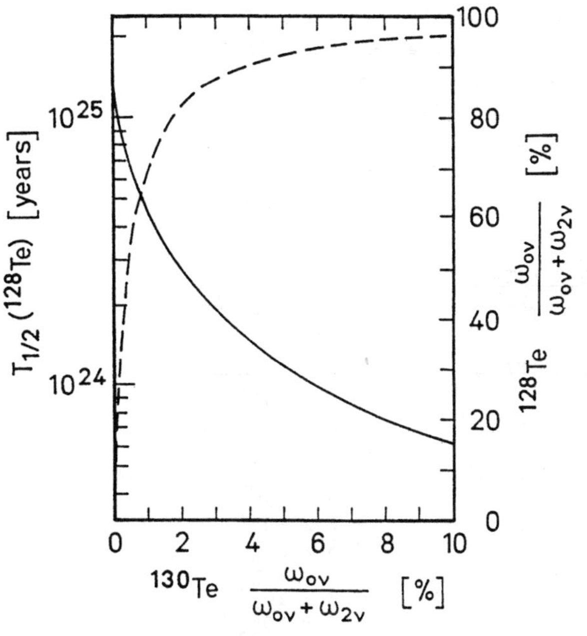

Fig. 5.3 The half-life (solid line) and the branching ratio (dashed line) of the $^{128}Te \rightarrow {}^{128}Xe$ $\beta\beta$ decay as function of the branching ratio of the ^{130}Te decay under the assumption that the corresponding 0ν and 2ν matrix elements are equal to each other for these isotopes. The experimental half-life $T_{1/2}(^{130}Te) = 2.55 \times 10^{21}$ y from the geochemical measurement [Kir 86] is used.

and branching ratio of ^{128}Te and ^{130}Te are related by the phase space integrals. Fig. 5.3 shows the relation between the branching ratio of ^{130}Te and the half-life and branching ratio of ^{128}Te, for $T_{1/2}(^{130}Te) = 2.55 \times 10^{21}$ years [Kir 86]. The observed half-lives of ^{128}Te, $T_{1/2}(^{128}Te) > 5 \times 10^{24}$ years [Kir 86] and $T_{1/2}(^{128}Te) = (1.4 \pm 0.4) \times 10^{24}$ years [Man 86], set the limit on the branching ratio of ^{130}Te. The $\beta\beta$ decay of ^{130}Te is therefore dominated by the 2ν mode, the branching ratio of the 0ν mode being of the order of 1% or smaller. However, the branching ratio of ^{128}Te is not restricted by this phase space argument. (Generally, the 0ν mode is more favoured for nuclei with a smaller $Q_{\beta\beta}$ value (T_0) because of the different dependence on T_0 for the 0ν and 2ν modes, respectively.)

Klapdor and Grotz [Kla 84, Gro 85a,85b,86] calculated half-lives of 2ν and 0ν $\beta\beta$ decay for all possible $\beta\beta$-decaying nuclei with $A \geq 70$, using two types of models. Details of the models are explained in sect. 4.3. The calculated half-lives are listed in Table 5.3. Although being systematically (in particular the QRPA results) somewhat too short, they may still provide a useful guide line for experimental planning. These authors also investigated the sensitivity of the calculated $\beta\beta$ decay rates to various types of ground state correlations and increasing complexity of the wave functions. Table 5.4 (from [Gro 86]) shows the sensitivity of the matrix elements to various types of correlations in the case of 2ν $\beta\beta$ decay (see also Table 5.2 for the half-lives). Table 5.6 demonstrates the insensitivity of the 0ν $\beta\beta$ decay rates to the ground state correlations. Using the projected BCS approach, these authors first took into account the quadrupole-quadrupole correlations and ΔN^{-1} excitations in the initial and final ground states of $0_i^+ \rightarrow 0_f^+$ transitions in addition to a residual Gamow-Teller interaction. As an example, transition amplitudes for the $\beta\beta$ decay of ^{130}Te are shown in Fig. 5.4. This investigation suggested a clue for solving the problem of the significant discrepancies between theoretical and experimental half-lives of the 2ν mode. The largest values of the closure matrix elements are obtained for the BCS wave functions. Significant reductions of the

Table 5.3. Calculated half-lives for 2ν and 0ν double beta decay for double beta emitters with A ≥ 70 (from [Gro 86]). The QRPA calculations were performed neglecting the pp force. The 0ν half-lives are neutrino mass dependent and the given figures correspond to $\langle m_\nu \rangle = 1$ eV. The $T_{1/2}^{2\nu *}$ and $T_{1/2}^{0\nu *}$ are calculated using the PBCS model including spin-isospin and quadrupole-quadrupole forces (see text). The $T_{1/2}^{0\nu}$ are relatively rough estimates, for their derivation see eqs. (44), (45) in [Gro 86]. T_o denotes the $\beta\beta$ Q value.

	T_0	QRPA $T_{1/2}^{2\nu}$	BCS (GT+QQ) $T_{1/2}^{2\nu *}$	$T_{1/2}^{0\nu} \times \langle m_\nu \rangle^2$
	(MeV)	(y)	(y)	(y × eV^2)
^{70}Zn	1.00	1.9×10^{22}		7.6×10^{23}
^{76}Ge	2.04	1.1×10^{20}	2.2×10^{20}	2.6×10^{23}*
^{80}Se	0.136	9.0×10^{28}		6.7×10^{25}
^{82}Se	3.01	4.5×10^{18}	1.5×10^{19}	9.5×10^{22}*
^{86}Kr	1.25	3.5×10^{22}		5.0×10^{24}
^{94}Zr	1.15	4.1×10^{21}		6.2×10^{23}
^{96}Zr	3.35	5.2×10^{17}		1.6×10^{22}
^{98}Mo	0.11	1.6×10^{29}		7.3×10^{25}
^{100}Mo	3.03	1.8×10^{18}		3.3×10^{22}
^{104}Ru	1.30	1.8×10^{21}		5.0×10^{23}
^{110}Pd	2.01	5.0×10^{19}		1.3×10^{23}
^{114}Cd	0.54	2.7×10^{24}		1.7×10^{25}
^{116}Cd	2.81	8.3×10^{18}		1.7×10^{23}
^{122}Sn	0.36	1.4×10^{26}		3.6×10^{25}
^{124}Sn	2.28	9.3×10^{19}		3.1×10^{23}
^{128}Te	0.87	1.2×10^{23}	5.7×10^{23}	9.8×10^{23}*
^{130}Te	2.53	1.9×10^{19}	1.2×10^{20}	4.6×10^{22}*
^{134}Xe	0.84	5.1×10^{22}	2.5×10^{23}	8.7×10^{23}*
^{136}Xe	2.48	6.0×10^{19}	3.3×10^{19}	3.0×10^{23}*
^{142}Ce	1.41	2.8×10^{21}	4.1×10^{20}	4.7×10^{23}*
^{146}Nd	0.06	2.9×10^{30}		5.6×10^{25}
^{148}Nd	1.93	2.5×10^{19}		1.1×10^{23}
^{150}Nd	3.37	4.8×10^{17}		2.4×10^{22}
^{154}Sm	1.25	9.5×10^{20}		2.4×10^{23}
^{160}Gd	1.73	4.4×10^{19}		6.4×10^{22}
^{170}Er	0.66	6.6×10^{22}		9.1×10^{23}
^{176}Yb	1.08	1.1×10^{21}		2.5×10^{23}
^{186}W	0.49	3.2×10^{23}		1.2×10^{24}
^{192}Os	0.41	1.7×10^{24}		1.6×10^{24}
^{198}Pt	1.04	1.2×10^{22}		1.6×10^{24}
^{204}Hg	0.41	4.6×10^{25}		2.6×10^{25}
^{232}Th	0.85	1.6×10^{20}		3.8×10^{22}
^{238}U	1.15	2.2×10^{19}		2.4×10^{22}

values are found when the spin-isospin force (without the particle-particle component) is switched on and later the QQ force. The spin-isospin force, which is diagonalized in a zero- plus four-quasiparticle space, induces admixtures of four-quasiparticle components in the ground state, i.e., ground state correlations of the spin-isospin mode $|pn(1^+), p'n'(1^+); 0^+\rangle$ (eq.(4.1)), which suppress the decay rates. The calculated half-lives are quite consistent with those of the QRPA calculation with the same force (see Table 5.2), in which the four-quasiparticle admixtures are taken into account in the RPA formalism. The inclusion of the QQ force further reduces the decay rates by large factors of up to an order of magnitude. The QQ force gives rise to the QQ correlations of eq.(4.2), and as a consequence, as one can see by the angular-momentum recoupling of eq.(4.3), enhances the spin-isospin ground state correlations. This implies a large sensitivity of the calculated 2ν $\beta\beta$

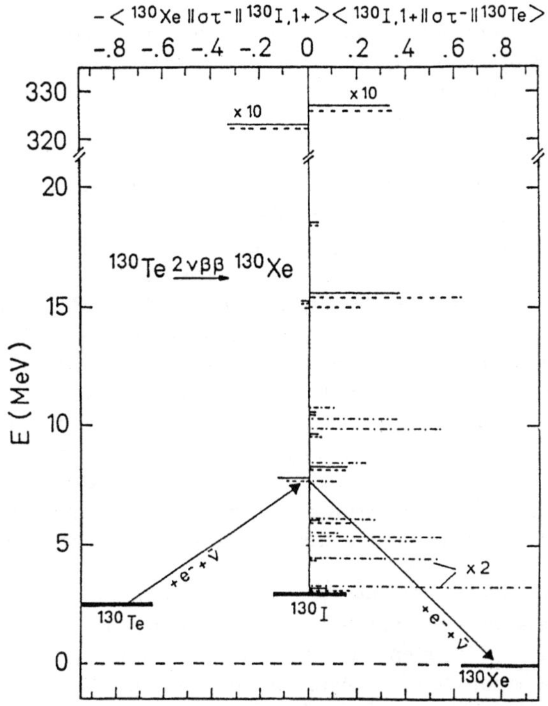

Fig. 5.4 Amplitudes for the 2ν $\beta\beta$ transition ^{130}Te $\rightarrow$ ^{130}Xe calculated with pairing force (dashed-dotted lines) and pairing plus Gamow-Teller forces including Δ-hole mixing (dashed lines). The solid lines show the final result including also quadrupole-quadrupole forces (from [Gro 86]).

Table 5.4. Closure matrix elements $(M_{GT}^{2\nu})_c$ calculated in the PBCS treatment in various approaches of increasing complexity from left to right (from [Gro 86]).

	Only pairing	Pairing + $H_{\sigma\tau}$ (no Δh mixing)	Pairing + $H_{\sigma\tau}^{\Delta}$ (with Δh mixing)	Pairing + $H_{\sigma\tau}^{\Delta}$ + H_{QQ}, without higher order phonon contributions	Pairing + $H_{\sigma\tau}^{\Delta}$ + H_{QQ} including higher order phonon contributions
^{128}Te	6.34	2.00	1.77	0.61	0.52
^{130}Te	5.86	1.75	1.55	0.56	0.48
^{82}Se	5.52	2.83	2.33	1.84	1.38
^{76}Ge	7.05	3.76	3.26	2.64	1.93

The first column gives the large matrix elements in the simple pairing model. These are reduced step by step including the spin-isospin force first in the nucleonic and then also in the Δ nucleon hole space and further by including the 2^+ phonon correlations. The matrix elements in the last column include also higher order g.s. correlations.

half-lives to the ground state correlations of the spin-isospin mode. The ground state correlations of the spin-isospin mode are naturally enhanced by including the particle-particle component of a residual interaction. This is discussed in the next section.

5.1.2. QRPA calculations including the particle-particle interaction

The suppression mechanism of 2ν $\beta\beta$ matrix elements was finally clarified by recent calculations by Vogel and Zirnbauer [Vog 86], Civitarese, Faessler and Tomoda [Civ 87], and Muto and Klapdor [Mut 88a]. All of these calculations are based on the QRPA approach, but with different effective interactions; zero-range force [Vog 86], a realistic force derived from the Bonn one-boson-exchange potential [Civ 87], and another realistic force from the Paris potential [Mut 88a]. The conclusion of these calculations was that the particle-particle interaction enhances the ground state correlations beyond those considered earlier [Kla 84, Gro 86] and as a result, the 2ν matrix elements are further strongly suppressed. For a reliable prediction of the half-life, the strength of the particle-particle interaction is crucial. Muto and Klapdor [Mut 88a], and later Muto, Bender and Klapdor [Mut 88c] taking a similar procedure, fixed the strength to the Gamow-Teller strengths observed in β^+ decay of a number of nuclei. These authors showed that the half-lives calculated with the fixed interaction strength are consistent with the experimental values, with a remarkable improvement over the earlier calculations. Therefore we will follow the calculation of [Mut 88a,88c].

Half-lives of $0_i^+ \to 0_f^+$ transitions were calculated for the $\beta\beta$ emitters ^{76}Ge, ^{82}Se, ^{100}Mo, ^{128,130}Te, ^{136}Xe, ^{150}Nd. The model space is assumed to include two oscillator major shells, for example full $3\hbar\omega$, $4\hbar\omega$ shells for $A = 76$ and $A = 82$ systems and full $4\hbar\omega$, $5\hbar\omega$ and $i_{13/2}$, $i_{11/2}$ shells for $A = 150$. Single-particle energies are computed by solving the eigenvalue equation for a nucleon in the Coulomb-corrected Woods-Saxon potential [Boh 69]. The depth of the Woods-Saxon central potential is modified by adding a term depending on the orbital angular momentum, $-0.05l(l+1)$ MeV, in order to better reproduce observed single (quasi)particle spectra. The occupation amplitudes, u_j and v_j, which define the quasiparticle basis are calculated in the BCS approximation. Then the QRPA equation for $J^\pi = 1^+$ is solved for both 0_i^+ and 0_f^+ ground states. The G-matrix of the Paris potential [Vin 79,Lac 80] fitted to a sum of Yukawa terms [Ana 83] is adopted as nucleon-nucleon residual interaction. The same interaction is used consistently for the BCS and RPA calculations. The Paris potential and the Bonn potential [Hol 81] are constructed from a phase-shift analysis of nucleon-nucleon scattering.

The realistic interaction could be renormalized in a nucleus, and additionally in the restricted model space. The renormalization could differ in the BCS and RPA approximations. In order to adjust the interaction strengths, four renormalization factors which multiply the corresponding interaction matrix elements are introduced. These factors, however, are not expected to deviate much from unity. In the BCS calculation, the renormalization factor of the pairing interaction, g^p_{pair} or g^n_{pair}, which multiplies the matrix elements of eq.(4.7), is fixed for proton and neutron systems separately so as to reproduce experimental even-odd mass differences [Wap 85]. The values of g^p_{pair} and g^n_{pair} thus fixed are close to unity, as expected, with deviations of at most 30%. In the RPA two factors are introduced, g_{ph} multiplying the particle-hole interaction, which is mainly responsible for mixing of particle-hole pairs in the odd-odd intermediate nucleus, and g_{pp} for the particle-particle interaction, which is important for the ground state correlations of the parent and daughter nuclei. By multiplying these renormalization factors, eq.(4.14) becomes

$$\begin{aligned} A_{pn,p'n'} &= \delta(pn,p'n')(\epsilon_p+\epsilon_n) \\ &\quad + g_{ph}(u_pv_nu_{p'}v_{n'} + v_pu_nv_{p'}u_{n'})\langle pn^{-1}|V|p'n'^{-1}\rangle_J \\ &\quad + g_{pp}(u_pu_nu_{p'}u_{n'} + v_pv_nv_{p'}v_{n'})\langle pn|V|p'n'\rangle_J \end{aligned} \tag{5.1a}$$

$$\begin{aligned} B_{pn,p'n'} &= -g_{ph}(u_pv_nv_{p'}u_{n'} + v_pu_nu_{p'}v_{n'})\langle pn^{-1}|V|p'n'^{-1}\rangle_J \\ &\quad + g_{pp}(u_pu_nv_{p'}v_{n'} + v_pv_nu_{p'}u_{n'})\langle pn|V|p'n'\rangle_J . \end{aligned} \tag{5.1b}$$

The particle-hole interaction strength is first adjusted so that the RPA calculation reproduces the excitation energy of the Gamow-Teller giant resonance observed in charge-exchange (p, n) reactions on the $\beta\beta$ emitters [Mad 87]. The fitted mass-number-dependent strength $g_{ph} = 1 + 0.002 \cdot A$ is consistently used in the calculation of $\beta\beta$ decay and also in β^+ decay for the adjustment of g_{pp}.

The nuclear matrix elements of $\beta\beta$ decay are calculated with the fixed interaction strengths but as a function of g_{pp}, which is not so far adjusted. Fig. 5.5 shows the behaviour of the matrix elements. As g_{pp} is increased, the matrix element decreases and crosses through zero at around $g_{pp} = 1$. Since g_{pp} is not expected to be far from unity, it is a nice feature of the calculation with the realistic interaction that the matrix element is strongly suppressed at $g_{pp} \sim 1$, which implies a long half-life consistent with experiment. However, it is seen that the matrix elements decrease so rapidly near the crossing point that the value of g_{pp} is critical in the prediction of half-lives with satisfactory reliability. The rapid decrease of the matrix elements comes from the enhancement of the ground state correlations with increasing particle-particle interaction strength. Fig. 5.6 shows the percentage of ground state correlations in the ground state wave functions as function of g_{pp}. When the particle-particle interaction is switched off ($g_{pp} = 0$), the percentage is smaller than 1%. It increases exponentially as a function of g_{pp} and reaches about 10% at $g_{pp} \sim 1$.

The particle-particle interaction strength g_{pp} is fixed from β^+ decay observed in nuclei with $A = 60 - 170$. β^+ decay, corresponding to the inverse of the second transition in $\beta\beta$ decay, $1^+ \to 0_f^+$, is sensitive to g_{pp}, while, as discussed in sect. 4.4, β^- transitions are much less sensitive to the ground state correlations, since most of the β^- transitions come from the excess neutrons or from spin-flip transitions which are allowed even if there are no ground state correlations. As an example of this difference in the sensitivity to the particle-particle interaction in β^- and β^+ decay, the transitions relevant for the $^{82}Se \to {}^{82}Kr$ $\beta\beta$ decay are shown in Fig. 5.7. The transitions

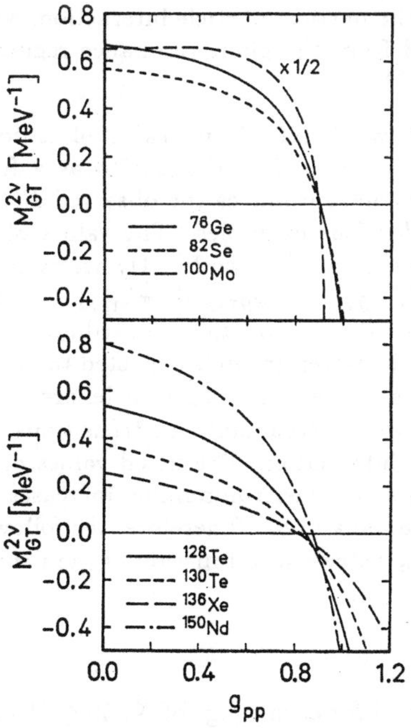

Fig. 5.5

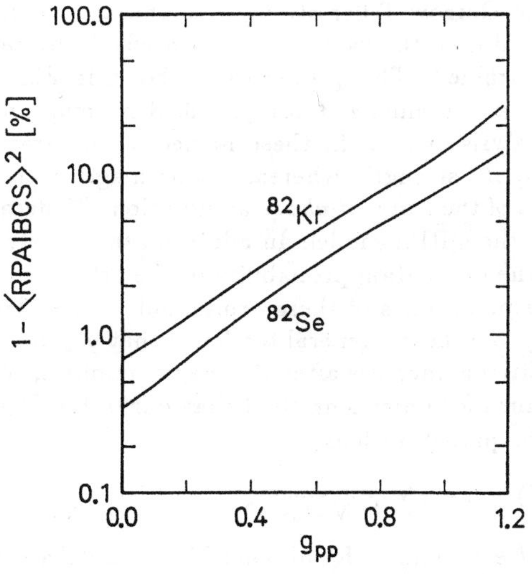

Fig. 5.6

Fig. 5.5 Nuclear matrix elements of the $0^+ \to 0^+$ 2ν $\beta\beta$ decay calculated as a function of the particle-particle interaction strength of the proton-neutron quasiparticle RPA.

Fig. 5.6 Percentages of the spin-isospin correlations in the QRPA ground-state wave functions as function of g_{pp} for ^{82}Se and ^{82}Kr.

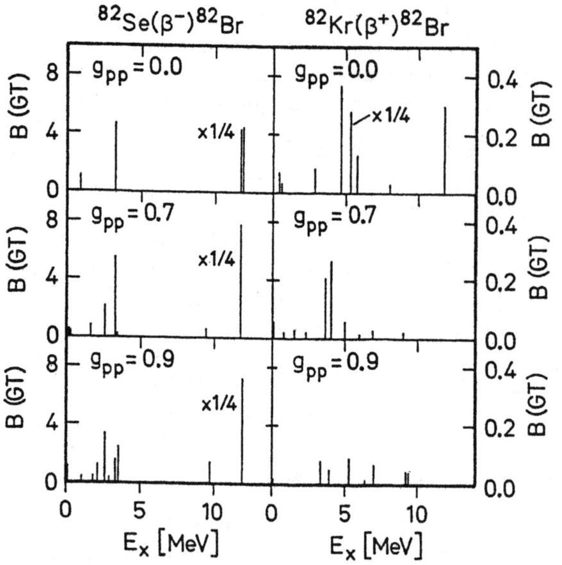

Fig. 5.7 Comparison of Gamow-Teller strength distributions for β^- and β^+ excitations relevant for $\beta\beta$ decay of ^{82}Se, for three values of the particle-particle interaction strength.

shown there are the ones occuring in the $\beta\beta$ nuclear matrix element. The Gamow-Teller strength distributions are plotted for three values of g_{pp}; $g_{pp} = 0$ (the particle-particle interaction is switched off), $g_{pp} = 0.7$, and $g_{pp} = 0.9$, where the 2ν $\beta\beta$ nuclear matrix element vanishes almost completely. It is obvious that the β^+ strength is drastically quenched by the particle-particle interaction, while in the β^- sector only a small fraction of strength is shifted from the giant resonance region to low-lying states.

For some thirty β^+-unstable even-even nuclei with A = 60-170 (from Table of Isotopes [Led 78]), the Gamow-Teller strength in the Q_β window is calculated in the QRPA as a function of g_{pp}, and then the value of g_{pp} with which the calculation reproduces the observed $B(GT)$ values is determined. This procedure is shown in Fig. 5.8 for four examples. The values of g_{pp} thus obtained are summarized in Fig. 5.9. Semi-magic nuclei such as ^{94}Ru and ^{148}Dy are excluded from this analysis of g_{pp}. In these nuclei, a s.p. transition $\pi g_{9/2} \rightarrow \nu g_{7/2}$ or $\pi h_{11/2} \rightarrow \nu h_{9/2}$ dominates the β^+ strength, whereas in non-magic nuclei various s.p. transitions contribute. From the viewpoint of the *random phase* assumption, β^+ decay of the latter group is expected to be well described by the QRPA model. In addition, the β^+ strengths of the semi-magic nuclei are rather sensitive to the occupation probability of the orbits relevant to the dominant s.p. transitions, i.e., sensitive to s.p. energies of these orbits and to the effective interaction. The fitted values of g_{pp} shown in Fig. 5.9 show a general tendency that g_{pp} decreases as the neutron number increases and show an additional increase after the magic numbers, N = 28, 50 and 82. Therefore, the following form is assumed for describing the behaviour of the fitted g_{pp} values, as a function of the neutron number of the parent nucleus,

$$g_{pp}(N) = a - bN + \frac{c}{N-d} \tag{5.2}$$

where a and b are common for all nuclei but c and d are different for each range of N, $30 \leq N < 50$, $50 < N < 82$, $82 < N \leq 98$. Values of the constants are determined using a least-squares fitting procedure. The fitted curves are plotted in Fig. 5.9; the standard deviation is $\delta g_{pp} = 0.039$.

For the calculated half-lives of 2ν $\beta\beta$ decay, we show two numbers; the lower limit of the half-life in the 1σ range of g_{pp}, and the average in the same range. The average half-life is defined

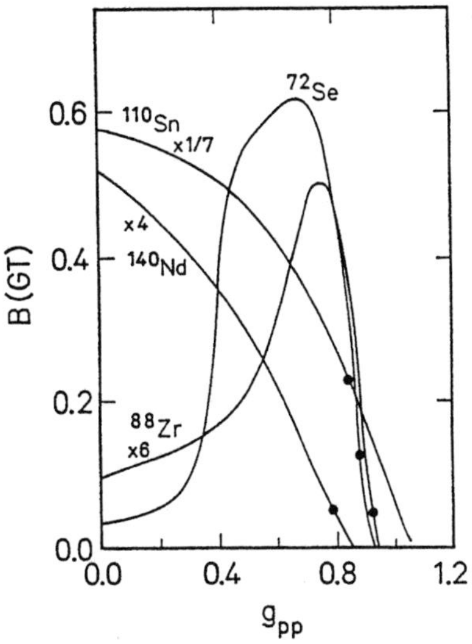

Fig. 5.8

Fig. 5.9

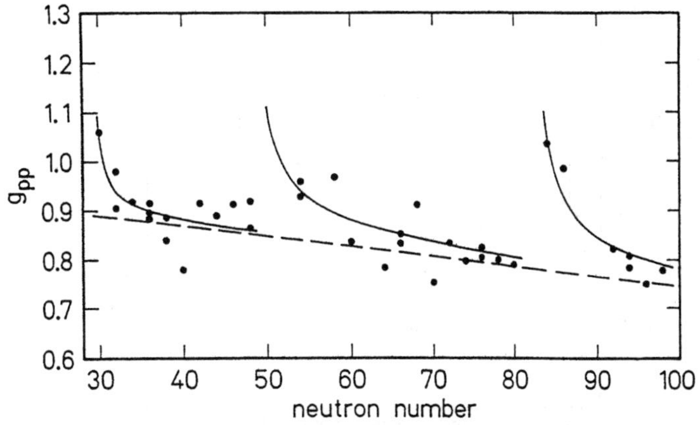

Fig. 5.8 The fitting procedure of the particle-particle interaction strength g_{pp} to the experimental Gamow-Teller strength of single β^+ decay. The experimental value is presented by a dot on the theoretical curve calculated as a function of g_{pp}.

Fig. 5.9 The values of the particle-particle interaction strength, with which a QRPA calculation reproduces experimental Gamow-Teller β^+-decay strength. The solid line denotes the function of eq.(5.2) fitted to the g_{pp} values presented by dots, the dashed line the linear term of the function.

by $T_{1/2}^{2\nu} = ln2/\bar{\omega}_{2\nu}$, where $\bar{\omega}_{2\nu}$ is obtained by simply averaging the decay rate over the 1σ range,

$$\bar{\omega}_{2\nu} = \frac{1}{2\delta g_{pp}} \int_{g_{pp}(N)-\delta g_{pp}}^{g_{pp}(N)+\delta g_{pp}} \omega_{2\nu}(g_{pp}) dg_{pp} \tag{5.3}$$

It should be noted that the nuclear matrix element of 2ν decay goes through zero at a g_{pp} lying within the range of one standard deviation from the optimum value corresponding to the neutron number of the daughter nucleus, implying a possible infinite half-life. The half-lives thus calculated are listed in Table 5.5, with the experimental values in the last row, and are also given in Fig. 5.10. The theoretical half-lives are consistent with the existing experimental data, and the large discrepancies of the earlier calculations are removed.

To conclude, a solution of the long-standing problem of understanding 2ν $\beta\beta$ decay half-lives is found in the nuclear structure calculation. The success is based on the fine tuning of the strength of the particle-particle interaction. This interaction is automatically taken into account, for instance, in a shell model calculation. However, although being simplified compared to such a sophisticated model, the QRPA approach gives insight into and clarifies the role of the particle-particle interaction for the suppression of the 2ν $\beta\beta$ decay rates precisely because of this simplification.

It is noted, however, that we have taken into account only the leading term in the calculation of 2ν $\beta\beta$ decay rates. When the leading term is strongly suppressed, higher order terms and other mechanisms could become more important.

5.1.2.1. Limitations of the QRPA approach

The calculations described in section 5.1.2. revealed limitations of the QRPA approach [Mut 88c]. While the results obtained for the nuclei ^{76}Ge, ^{82}Se, ^{128,130}Te and ^{136}Xe are considered to be reliable, such limitations can be clearly seen in the behaviour of the nuclear matrix

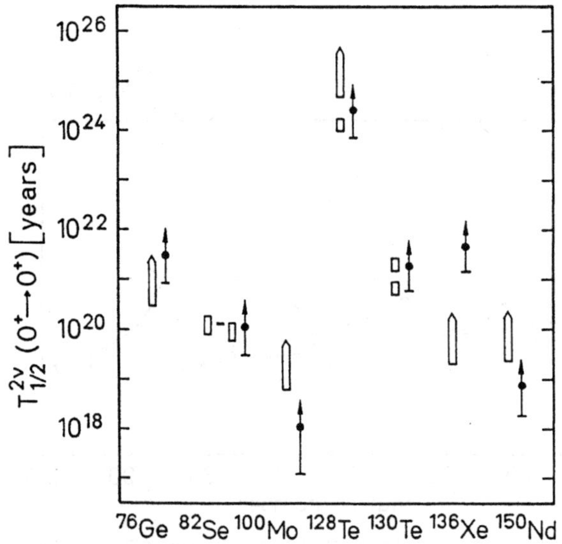

Fig. 5.10 Comparison of experimental and calculated half-lives of $0^+ \to 0^+$ 2ν $\beta\beta$ decay. The experimental values are given by open squares or open arrows for lower limits. The theoretical half-lives, calculated in the proton-neutron QRPA model [Mut 88c], are presented by arrows. The arrow starts from the lower limit and the circle on it denotes the average (see text).

Table 5.5. Comparison of calculated [Mut 88c] and experimental half-lives of $0^+ \to 0^+$ 2ν $\beta\beta$ decay. The half-lives evaluated at $g_{pp} = 0$ (no particle-particle interaction), and the lower limit and average in the 1σ range of g_{pp} (see text) are presented.

	^{76}Ge	^{82}Se	^{100}Mo	^{128}Te	^{130}Te	^{136}Xe	^{150}Nd
	10^{21}y	10^{20}y	10^{18}y	10^{24}y	10^{21}y	10^{21}y	10^{19}y
Calculation							
$g_{pp} = 0$	0.070	0.029	0.22	0.020	0.0069	0.015	0.0061
lower limit	0.822	0.306	0.125	0.723	0.594	1.43	0.189
average	2.99	1.09	1.13	2.63	1.84	4.64	0.737
Experiment	>0.3[a]	1.30±0.05[b]	>6.2[e]	>5[b]	1.5-2.75[b]	>0.021[f]	>2.4[g]
		1.0 ±0.4[c]		1.4±0.4[c]	0.7±0.2[c]		
		$1.1^{+0.8}_{-0.3}$[d]					

[a]Avi 86, [b]Kir 86, [c]Man 86, [d]Ell 87, [e]Smo 88, [f]Bar 87, [g]Kli 86

element for the decay of ^{100}Mo, for which the calculated half-lives are somewhat uncertain. The matrix element, as g_{pp} is increased, first increases and then very rapidly decreases, passing through zero (see Fig. 5.5), and after the crossing the RPA equation tends to a collapse. The collapse occurs, when the pp interaction enhances the ground-state correlations too much and pushes down the energy of the lowest eigenstate such that the depressed eigenvalue finally becomes imaginary.

The ground-state correlations of, say, more than 50% obviously contradict the main assumption of the RPA formalism. Therefore, even if the RPA equation is stable, the probable values of g_{pp} could enter an unreasonable region of the RPA solution. The lower limit of the half-life for ^{100}Mo is evaluated at a point just before the collapse, and consequently both lower-limit and 'average' would likely be underestimated.

The calculated half-lives of ^{100}Mo (and ^{150}Nd) also suffer from another origin of uncertainties. This nucleus is located in a region where the $\beta\beta$ decay is dominated by few strong single-particle transitions in $g_{9/2} - g_{7/2}$ ($h_{11/2} - h_{9/2}$) orbits. It means that the half-lives are affected rather sensitively by a change of occupation amplitudes of the relevant orbits, i.e., by a change of single-particle energies and/or that of the pairing interactions. These uncertainties would probably also result in an overestimation of the decay rate for the transitions strongly suppressed by the destructive interference between terms with opposite signs.

Finally, we point out that, in general, the 2ν $\beta\beta$ decay rate should not vanish, although the 2ν decay nuclear matrix element of the QRPA calculation vanishes completely at a g_{pp} value in its 1σ range. The matrix element, with $F_\mu = t_-\sigma_\mu$,

$$M_{GT}^{2\nu} = -\langle 0_f^+|\sum_\mu(-1)^\mu F_\mu \frac{1}{H-(M_i+M_f)/2}F_{-\mu}|0_i^+\rangle$$

can be expressed in two alternative forms

$$\begin{aligned} M_{GT}^{2\nu} &= \langle 0_f^+|\sum_\mu(-1)^\mu \left\{F_\mu[F_{-\mu}, \frac{1}{H-(M_i+M_f)/2}] + \frac{1}{(M_i-M_f)/2}F_\mu F_{-\mu}\right\}|0_i^+\rangle \\ &= \langle 0_f^+|\sum_\mu(-1)^\mu \left\{-[F_{-\mu}, \frac{1}{H-(M_i+M_f)/2}]F_\mu - \frac{1}{(M_i-M_f)/2}F_\mu F_{-\mu}\right\}|0_i^+\rangle \end{aligned}$$

where relations $H|0_i^+\rangle = M_i|0_i^+\rangle$ and $H|0_f^+\rangle = M_f|0_f^+\rangle$ are used for the second terms, which have the same magnitude and opposite signs. Therefore, the nuclear matrix element is given by

$$M_{GT}^{2\nu} = \frac{1}{2}\langle 0_f^+|\sum_\mu(-1)^\mu[F_\mu,[F_{-\mu}, \frac{1}{H-(M_i+M_f)/2}]]|0_i^+\rangle \tag{5.4}$$

This expression holds model independently. The Hamiltonian contains spin-isospin dependent components and they yield non-vanishing contributions through the double commutator to the nuclear matrix element. In the QRPA calculation, the crossing of $M_{GT}^{2\nu}$ through zero happens because of the artificial change of g_{pp} as a free parameter, resulting in a violation of the Pandya transformation, eq. (4.15), which connects pp and ph interaction matrix elements.

5.2. Neutrinoless mode

5.2.1. Calculations of nuclear matrix elements

The 0ν $\beta\beta$ decay nuclear matrix elements have been calculated using different nuclear models. Estimations of 0ν decay half-lives for all possible nuclei with $A \geq 70$ [Gro 86] are given in Table 5.3. Table 5.6 compares half-lives calculated with different nuclear models and assuming a neutrino mass $\langle m_\nu\rangle = 1$ eV (contributions of the right-handed currents are neglected). These half-lives are also shown in Fig. 5.11, in which $\langle m_\nu\rangle = 2.1$ eV is used. When this value is chosen the experimental lower limit of the half-life of ^{76}Ge, $T_{1/2}^{0\nu}(^{76}\text{Ge}) > 4.7 \times 10^{23}$ years [Cal 86], is reproduced with the nuclear matrix elements of the QRPA calculation with both particle-hole and particle-particle interactions [Tom 87]. Comparing the half-lives shown in Table 5.6 and Fig. 5.11, the four shell model calculations for ^{48}Ca yield half-lives quite consistent with each other. The shell model calculation for heavier nuclei by [Hax 84] gives half-lives similar to those of the QRPA calculation [Tom 87] for ^{76}Ge and ^{82}Se. The discrepancies for ^{128,130}Te are about a factor of three. The half-lives calculated by Klapdor and Grotz [Kla 84, Gro 86] are systematically shorter than those of the QRPA and shell model calculations, probably because only Gamow-Teller and quadrupole-quadrupole correlations are taken into account.

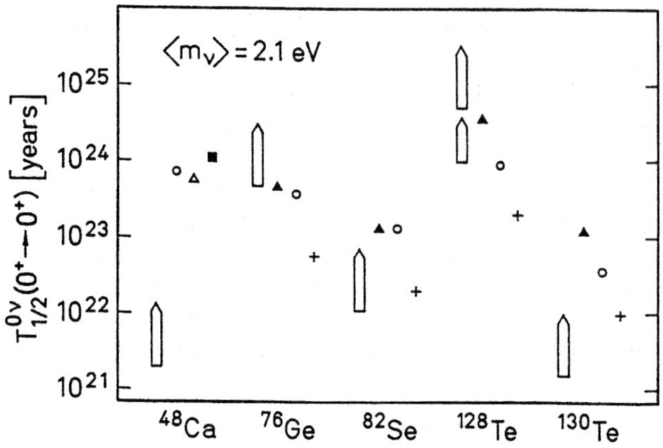

Fig. 5.11 Comparison of experimental and theoretical half-lives of $0^+ \to 0^+$ 0ν $\beta\beta$ decay. The experimental lower limits are given by open arrows. For the theoretical values $\langle m_\nu \rangle = 2.1$ eV is assumed, and contributions of right-handed currents are neglected. Nuclear matrix elements are taken from [Hax 84] (open circle), [Sko 83] (open triangle), [Mut 86] (filled square), [Tom 87] (filled triangle), [Gro 86] PBCS (cross).

Table 5.6. Calculated half-lives of 0ν $\beta\beta$ decay for an effective neutrino mass $\langle m_\nu \rangle$ = 1 eV. Contributions of right-handed currents are neglected. Present experimental half-lives are given in the last line.

References	^{48}Ca	^{76}Ge	^{82}Se	^{128}Te	^{130}Te	^{134}Xe	^{136}Xe	^{142}Ce
	10^{24}y	10^{24}y	10^{23}y	10^{24}y	10^{23}y	10^{23}y	10^{23}y	10^{23}y
Sko 83	2.6							
Wu 85	3.2							
Mut 86	5.4							
Hax 84	3.2	1.8	5.8	4.0	1.6			
Gro 86								
pairing		0.17	0.68	0.62	0.27			
pairing + $H_{\sigma\tau}+H_{QQ}$		0.26	0.95	0.98	0.46	8.7	3.0	4.7
Tom 86		0.91						
Tom 87		2.2	6.0	9.6	5.2			
Experiment	>0.002[a]	>0.47[b]	>0.11[c]	>5[d]	>0.015[d]		>0.012[e]	

[a]Bah 70, [b]Cal 86, [c]Ell 86, [d]Kir 86, [e]Bar 87,

Tomoda and Faessler [Tom 87] performed a QRPA calculation of the 0ν decay, using a realistic interaction, the G-matrix of the Bonn one-boson exchange potential [Hol 81], and consequently taking into account all possible modes of ground state correlations. (Some details of the results are discussed in sect. 5.2.2.) However, they did not fix the strength of the particle-particle interaction. Comparing the behaviour of the 2ν $\beta\beta$ decay matrix elements, which are calculated with the same interaction [Civ 87], with those of [Mut 88a,88c], in which the interaction strength is fixed to β^+

decay strength of many nuclei with $60 \leq A \leq 170$ (see sect. 5.1.2), we estimate the most probable value of g_{pp} for the Bonn potential to be very close to unity. The coefficients C_{xy} of the 0ν decay rate (eqs. (3.29) and (3.31)) evaluated at g_{pp} =1 are listed in Table 5.7.

Vogel and his collaborators have calculated $\beta\beta$ decay rates of the 2ν and 0ν (neutrino mass term) modes in the QRPA with schematic zero-range forces [Vog 86, Eng 88]. The results are qualitatively consistent with those of the QRPA calculations with realistic interactions by [Civ 87, Tom 87, Mut 88a,88c], but show a considerable suppression of 0ν decay rates, compared to [Tom 87], of up to two orders of magnitude. Such long half-lives have two main origins. First, these authors obtained a stronger particle-particle interaction, by fixing its strength to β^+ decay strengths for a few semi-magic nuclei. We have seen in sect. 5.1.2 that larger g_{pp} values are required for semi-magic nuclei than for open-shell nuclei (most $\beta\beta$ emitters belong to the latter class). The 0ν decay matrix element $M_{GT}^{0\nu}$ of Fig. 5.12 calculated with the Bonn potential, for which g_{pp} =1.0 seems to be optimum, would vanish at a slightly larger value $g_{pp} \sim 1.15$. Secondly, the use of zero-range forces has various consequences. It leads to a set of occupation amplitudes u_j and v_j (see sect. 4.3) which is significantly different from that of a realistic interaction, and as a result the calculated half-lives are long already at $g_{pp} = 0$. The adjustment of the zero-range force strengths, which was done for the $J^\pi = 1^+$ mode, does not necessarily mean correct force strengths for the other multipoles, which are also important for the calculation of the 0ν $\beta\beta$ decay matrix element. With such adjustment the zero-range force could only reproduce the short-range character of a residual interaction, but would underestimate long-range parts. Furthermore, these authors adopted a quenched axial-vector coupling constant, $g_A = 1.0$, which additionally suppresses decay rates by $(1.254)^4 = 2.5$.

The theoretical $\beta\beta$ decay half-lives of the 2ν [Mut 88a,88c] and 0ν [Tom 87] modes are consistent with the phase space argument for the branching ratios of ^{128}Te and ^{130}Te (see sect. 5.1.1).

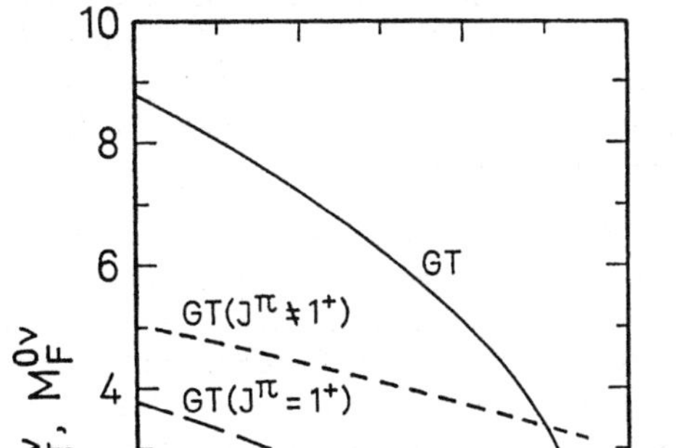

Fig. 5.12 The nuclear matrix elements (neutrino mass term) of the 0ν $\beta\beta$ decay of ^{76}Ge calculated in the proton-neutron QRPA as function of the particle-particle interaction strength (solid lines). Contributions through 1^+ (0^+) intermediate states for $M_{GT}^{0\nu}$ ($M_F^{0\nu}$) and those through the other states are shown separately by long-dashed and short-dashed lines, respectively (from [Tom 87]).

228

Table 5.7. The coefficients [y^{-1}] of $0^+ \to 0^+$ 0ν $\beta\beta$ decay rates calculated in the QRPA model including the particle-particle interaction with $g_{pp} = 1$ (from [Tom 87]).

	^{76}Ge	^{82}Se	^{128}Te	^{130}Te
C_{mm}	1.21×10^{-13}	4.36×10^{-13}	2.73×10^{-14}	5.05×10^{-13}
$C_{\eta\eta}$	3.69×10^{-9}	1.33×10^{-8}	1.18×10^{-9}	1.92×10^{-8}
$C_{\lambda\lambda}$	1.74×10^{-13}	1.31×10^{-12}	7.27×10^{-15}	1.16×10^{-12}
$C_{m\eta}$	2.08×10^{-11}	5.96×10^{-11}	7.58×10^{-12}	8.18×10^{-11}
$C_{m\lambda}$	-6.04×10^{-14}	-2.68×10^{-13}	-6.48×10^{-15}	-2.88×10^{-13}
$C_{\eta\lambda}$	-1.14×10^{-13}	-9.46×10^{-13}	-3.32×10^{-15}	-7.70×10^{-13}

Also the partial half-life $T^{0\nu}_{1/2}(^{130}\text{Te}) \sim 1 \times 10^{23}$ years evaluated at $\langle m_\nu \rangle = 2.1$ eV indicates the branching ratio of the 0ν mode to be at most a few percent of the total half-life from the geochemical measurement [Kir 86]. The ratio could be smaller since the effective value of the neutrino mass is taken from the experimental 0ν decay half-life of ^{76}Ge [Cal 86], which is only a lower limit.

5.2.2. Effects of ground state correlations

In sect. 5.1.2 it is shown that the QRPA calculation with carefully adjusted strengths of the residual interaction yields 2ν $\beta\beta$ half-lives quite consistent with the existing experimental data, with a remarkable improvement over the earlier calculations. The 2ν decay rates are strongly suppressed by the ground state correlations which are enhanced by the particle-particle interaction. This finding raises the question whether the matrix elements of the 0ν mode are also drastically affected by the particle-particle interaction and more generally by ground state correlations.

The recent QRPA calculation by Tomoda and Faessler [Tom 87] of the 0ν decay using a realistic interaction with both particle-hole and particle-particle components led to a negative answer to the question. Figs. 5.12 and 5.13 show the dependence of the matrix elements on the strength of the particle-particle interaction, for the 0ν $\beta\beta$ decay of ^{76}Ge. $M^{0\nu}_{GT}$ and $M^{0\nu}_{F}$ are relevant for the extraction of the neutrino mass $\langle m_\nu \rangle$, and $M^{0\nu}_{R}$ is most important for the determination of the mixing parameter $\langle \eta \rangle$ of the right-handed leptonic current. (These figures are reproduced from Fig. 1 of [Tom 87] by taking into account differences in the definitions of these matrix elements.)

The matrix elements $M^{0\nu}_{GT}$ and $M^{0\nu}_{F}$ (see Fig. 5.12) are not affected as much as $M^{2\nu}_{GT}$ by the inclusion of the particle-particle interaction. (A similar tendency was obtained for the quadrupole-quadrupole interaction, see sect. 5.2.1 and Table 5.6.) The reason why details of the nuclear structure affect $M^{0\nu}_{GT}$ and $M^{2\nu}_{GT}$ so differently is found in the different radial dependence of the involved transition operators:

$$M^{0\nu}_{GT} = \sum_{m,n} \langle 0^+_f \| t_{-m} t_{-n} \sigma_m \cdot \sigma_n H(|\mathbf{r}_m - \mathbf{r}_n|) \| 0^+_i \rangle \tag{5.5a}$$

$$M^{2\nu}_{GT} = \sum_{m,n} \langle 0^+_f \| t_{-m} t_{-n} \sigma_m \cdot \sigma_n \| 0^+_i \rangle \tag{5.5b}$$

Due to the neutrino potential $H(|\mathbf{r}_m - \mathbf{r}_n|)$ occuring in the 0ν matrix element with its radial dependence, $\sim |\mathbf{r}_m - \mathbf{r}_n|^{-1}$, the decay of two neutrons being close together is favoured. Alternatively, the presence of the neutrino potential in the 0ν decay makes transitions possible through intermediate

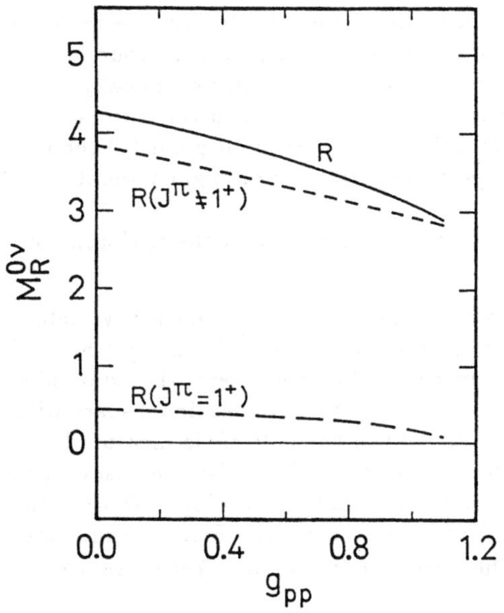

Fig. 5.13 The nucleon recoil matrix element for the ^{76}Ge 0ν decay calculated in the proton-neutron QRPA as function of the particle-particle interaction strength (solid line). This element is dominated by transitions through intermediate states with $J^\pi \neq 1^+$ (short-dashed line) while contributions through 1^+ states are small (long-dashed line) (from [Tom 87]).

states with various multipoles ($J^\pi = 0^+$ is forbidden for $M_{GT}^{0\nu}$, and $\pi = (-1)^{J+1}$ are forbidden for $M_F^{0\nu}$), while in the case of 2ν decay only $J^\pi = 1^+$ is allowed for the intermediate states. It is therefore interesting to decompose the 0ν matrix element $M_{GT}^{0\nu}$ ($M_F^{0\nu}$) into contributions through 1^+ (0^+) intermediate states and those through the other states. These contributions are separately plotted in Fig. 5.12 by dashed and dotted lines, respectively. The dashed line behaves quite similarly to the 2ν matrix element, passing through zero at $g_{pp} \sim 1$. On the other hand, the dotted line is rather constant against a change of g_{pp}. Compared with the values at $g_{pp} = 0$, the matrix elements at $g_{pp} = 1$ are reduced by a factor of 2 - 3, i.e., the 0ν decay rate (when right-handed current contributions are neglected) is suppressed by a factor of about 6, which is much less than in case of the 2ν decay rates.

The insensitivity of $M_{GT}^{0\nu}$ to the strength of the particle-particle interaction, compared to the drastic suppression effects on the 2ν decay rates, indicates negative evidence for the sometimes used procedure of scaling of 0ν matrix elements. The scaling method, which was applied in order to set a conservative limit on the neutrino mass [Hax 84, Doi 85], had been used before the suppression mechanism of the 2ν mode was found, and at a time when the calculations systematically overestimated the 2ν $\beta\beta$ decay rates.

The behaviour of the matrix element $M_R^{0\nu}$ as a function of g_{pp} is shown in Fig. 5.13. This matrix element is much less affected by the increase of the particle-particle interaction strength. In this case, the decomposition clarifies that the contribution through 1^+ intermediate states is much smaller than that through the other multipoles. The main contribution to $M_R^{0\nu}$ comes from the zero-range part of the neutrino potential $H_{RC}(r)$, eq.(3.46). The zero-range potential favours transitions through intermediate states with high multipoles, compared to a finite range potential. —— Generally, as the range of a neutrino potential becomes large, components of high momentum-transfer are suppressed, and low-multipole transitions are favoured. As the range tends to infinite, i.e. no radial dependence, which corresponds to the 2ν transition and is not realized in the 0ν decay, only the lowest possible multipole is allowed. —— The observation that higher-multipole transitions dominate the 0ν decay matrix elements and are almost not affected by the increase of g_{pp}, in contrast to the 2ν decay matrix element, indicates that corresponding ground state correlations are not affected by the particle-particle interactions. It thus reflects basically some well-known feature of the fundamental nature of the nucleon-nucleon (proton-neutron) interaction, namely the particle-particle interaction in the $J^\pi = 1^+$ (the 0^+ and highest multipolarity) channel is strongly attractive and that in other multipole channels is less attractive or slightly repulsive.

6. Present limits on the neutrino mass, right-handed current parameters and the majoron coupling constant from $\beta\beta$ decay

The 0ν $\beta\beta$ decay rate of the $0^+ \to 0^+$ transition is a quadratic function of the effective values of three quantities — the neutrino mass $\langle m_\nu \rangle$ and the mixing parameters of the right-handed currents $\langle \eta \rangle$ and $\langle \lambda \rangle$ (see eq.(3.29)). If the decay rate is measured by experiment, the admissible solutions for these quantities form the surface of a closed volume in a three dimensional space with one of the above quantities for one axis. Since only a lower limit of the half-life is measured, the solutions lie inside the volume. Fig. 6.1 shows cross sections of the volume for the cases when one quantity is put to zero, for the nuclear matrix elements of ^{76}Ge decay [Tom 87] with the experimental half-life $T_{1/2}^{0\nu}(^{76}\mathrm{Ge}) > 4.7 \times 10^{23}$ years [Cal 86]. As an example, Fig. 6.1a indicates the allowed region for $\langle m_\nu \rangle$ and $\langle \eta \rangle$ when $\langle \lambda \rangle = 0$. The upper limits of the effective values are summarized in Table 6.1.

Table 6.2 compares the upper limits of the neutrino mass for the nuclear matrix elements calculated with various nuclear structure models. Here, contributions of the right-handed currents are neglected (as a result the values of [Tom 87] in this table are slightly smaller than those in Table 6.1). For a given nucleus, a smaller neutrino mass indicates a larger nuclear matrix element. Among the present experimental results, the $0^+ \to 0^+$ 0ν decay of ^{76}Ge and ^{128}Te sets the smallest upper limits on the neutrino mass $\langle m_\nu \rangle \sim 2$ eV. It has to be kept in mind that the deduced neutrino masses are effective masses, i.e. linear combinations of mass eigenvalues of Majorana neutrinos (see eq.(3.30a)). This means that the true (electron) neutrino mass might in principle still be substantially larger than the limit for the effective mass deduced from $\beta\beta$ decay (see sect. 2). Possible interference effects between light and heavy neutrinos have been discussed in [Gro 86] (see also Langacker in this volume). In most Majorana neutrino models, however, one expects $\langle m_{\nu e} \rangle \approx m_{\nu e}$.

Haxton et al. calculated contributions of the right-handed currents [Hax 84]. Their nuclear matrix elements, for the experimental half-lives shown in Table 6.2, give effective values of $\langle \lambda \rangle$ (the

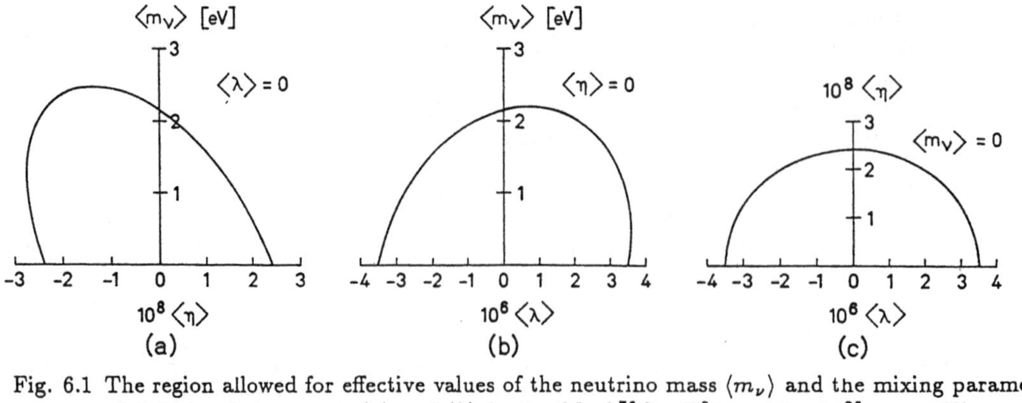

Fig. 6.1 The region allowed for effective values of the neutrino mass $\langle m_\nu \rangle$ and the mixing parameters of right-handed currents $\langle \eta \rangle$ and $\langle \lambda \rangle$ for 0ν $\beta\beta$ of ^{76}Ge, $T_{1/2}^{0\nu} > 4.7 \times 10^{23}$ years. The nuclear matrix elements are taken from [Tom 87]. The cross cuts are shown for (a) $\langle \lambda \rangle = 0$, (b) $\langle \eta \rangle = 0$, and (c) $\langle m_\nu \rangle = 0$.

Table 6.1. The upper limits of the effective values of the neutrino mass and right-handed current coupling constants deduced from the experimental lower limits of $0^+ \to 0^+$ 0ν $\beta\beta$ decay half-lives given in the last line (from [Tom 87]).

	^{76}Ge	^{82}Se	^{128}Te	^{130}Te
$\langle m_\nu \rangle$ [eV]	<2.5	<8.2	<1.9	<21
$\langle \eta \rangle$	$<2.8 \times 10^{-8}$	$<9.0 \times 10^{-8}$	$<1.8 \times 10^{-8}$	$<2.1 \times 10^{-7}$
$\langle \lambda \rangle$	$<3.6 \times 10^{-6}$	$<8.5 \times 10^{-6}$	$<5.5 \times 10^{-6}$	$<2.4 \times 10^{-5}$
$T^{0\nu}_{1/2}$[y]	$>4.7 \times 10^{23}$ [a]	$>1.1 \times 10^{22}$ [b]	$>5 \times 10^{24}$ [c]	$>1.5 \times 10^{21}$ [c]

[a]Cal 86, [b]Ell 86, [c]Kir 86

Table 6.2. Upper limits of the effective neutrino mass [eV] deduced from experimental $0^+ \to 0^+$ 0ν $\beta\beta$ decay half-life limits (last line) for the nuclear matrix elements of various nuclear structure calculations. Contributions of right-handed currents are neglected.

Refs.	^{48}Ca	^{76}Ge	^{82}Se	^{128}Te	^{130}Te	^{100}Mo	^{136}Xe	^{150}Nd
Sko 83	<36							
Wu 85	<40							
Mut 86	<52							
Hax 84	<40	<1.8	<7.3	<0.90	<10			
Gro 86		<0.7	<2.9	<0.43	<5.4	<8.66*	<13.3	<3.22*
Tom 86		<1.4						
Tom 87		<2.0	<7.4	<1.4	<19			
$T^{0\nu}_{1/2}$[y]	$>2\times10^{21}$ [a]	$>5.\times10^{23}$ [b]	$>1.1\times10^{22}$ [c]	$>5\times10^{24}$ [d]	$>1.5\times10^{21}$ [d]	$>4.4\times10^{20}$ [e]	$>1.7\times10^{21}$ [f]	$>2.3\times10^{21}$ [g]

[a]Bah 70, [b]Cal 87, [c]Ell 86, [d]Kir 86, [e]Smo 88, [f]Ale 88, [g]Kli 86.
*rough estimate, see Table 5.3.

parameter for the coupling of right-handed leptonic current and right-handed hadronic current $j_R \cdot J_R^\dagger$, see eq.(3.1)) of the same order as the corresponding values deduced by [Tom 87]. However, the effective values of $\langle \eta \rangle$ (the parameter for $j_R \cdot J_L^\dagger$) of [Tom 87], which are of the order of 10^{-8}, are smaller than those of [Hax 84] by two orders of magnitude, because the latter authors did not take into account the nucleon recoil matrix element $M_R^{0\nu}$ which is most important for the determination of $\langle \eta \rangle$.

As for 0ν $\beta\beta$ decay, no evidence exists at present for neutrinoless $\beta\beta$ decay associated with majoron emission (see Figs. 3.10 and 3.11). In Table 6.3 the observed lower limits of the half-lives for this process are given together with the deduced upper limits on the coupling constant $\langle g_B \rangle$ defined in eq.(3.54). The phase space factors G_B used are from [Doi 87], the 0ν decay matrix elements $(M_{GT}^{0\nu} - M_F^{0\nu})^2$ are taken from [Mut 86, Tom 87, Gro 86]. The deduced upper limits of the coupling constant are of the order $10^{-5} - 10^{-4}$.

Table 6.3. Effective values of the coupling constant $<g_B>$ of the majoron to Majorana neutrinos, deduced from experimental limits on $0\nu,B$ $\beta\beta$ decay half-lives.
The nuclear matrix elements and the phase space factors G_B are also presented.

	^{48}Ca	^{76}Ge	^{82}Se	^{128}Te	^{130}Te	^{100}Mo	^{150}Nd
$T_{1/2}^{0\nu,B}$[y]	$>8.6\times10^{20}$ [a]	$>1.4\times10^{21}$ [b]	$>4.4\times10^{20}$ [c]	$>5\times10^{24}$ [d]	$>1.5\times10^{21}$ [d]	$>3.3\times10^{20}$ [e]	$>1.3\times10^{20}$ [f]
$(M_{GT}^{0\nu}-M_F^{0\nu})^2$	0.746 [g]	19.1 [h]	15.4 [h]	14.9 [h]	11.2 [h]	173 [i]	51.9 [i]
$G_B[y^{-1}]$ [j]	3.97×10^{-15}	1.21×10^{-16}	9.98×10^{-16}	9.92×10^{-18}	1.31×10^{-15}	1.74×10^{-15}	1.03×10^{-14}
$\vert<g_B>\vert$	$<6.3\times10^{-4}$	$<5.6\times10^{-4}$	$<3.8\times10^{-4}$	$<3.7\times10^{-5}$	$<2.1\times10^{-4}$	$<1.0\times10^{-4}$	$<1.2\times10^{-4}$

[a]Bar 87, [b]Cal 87, [c]Ell 87, [d]Kir 86, [e]Als 88, [f]Kli 86, [g]Mut 86, [h]Tom 87, [i]Gro 86, [j]Doi 87

Bounds on neutrino masses have been estimated from various kinds of experiments — single β decay of tritium, electron capture of ^{163}Ho, neutrino oscillations, the solar neutrinos, heavy neutrino searches — and from cosmology. Among them, the tritium decay experiments have set the smallest upper limits on the electron neutrino mass, $m_{\nu e} < 18$ eV [Fri 86], $m_{\nu e} < 29$ eV [Bow 86]. The ITEP group reported a finite neutrino mass from the tritium decay experiment, 17 eV $< m_{\nu e} <$ 40 eV [Bor 85,87], but critical comments were made on the analysis of this experiment [Ber 85]. The time-spectrum analyses of neutrinos from the recently observed supernova explosion [Hir 87, Bio 87] resulted in bounds on the electron neutrino mass around 20 eV (see also [Abb 88]). Thus at present, the 0ν $\beta\beta$ decay experiments set the most stringent limit on the electron neutrino mass. As for the right-handed currents, the effective values of the mixing parameters deduced from the 0ν $\beta\beta$ decay rates are more stringent by several orders of magnitude, compared to constraints obtained from other sources such as single nuclear β decay, or decays of muons and kaons (for a detailed discussion see [Doi 85]).

7. Conclusion

The discovery of a non-zero neutrino mass would be an extemely profound example of 'new physics' beyond the simplest theories of grand unification. Neutrinoless $\beta\beta$ decay is at present the most sensitive test of the existence and mass of Majorana neutrinos. The most sensitive detector experiments currently yield an upper limit for the (electron) neutrino mass of about 1 eV. However, reliable deductions of neutrino masses from measured 0ν $\beta\beta$ decay rates depend decisively on reliable calculations of nuclear matrix elements. In this paper the present status of such calculations is reviewed. It is shown that the long-standing discrepancy between experimental and calculated 2ν $\beta\beta$ decay rates can be resolved by including particle-particle forces in a QRPA treatment. But, this progress has to be paid for by severe limitations in the accuracy with which 2ν $\beta\beta$ decay rates now can be predicted. Only lower limits can still be reliably calculated. Also, for some nuclei there are limitations of the QRPA approach. We have, furthermore, given some general commutator relation argument that the 2ν decay matrix elements should not vanish and that the vanishing of the 2ν matrix element in the QRPA at a certain choice of g_{pp} is an artifact of that model. Calculated rates for 0ν $\beta\beta$ decay are found — in contrast to the 2ν case — to be comparably insensitive to details of nuclear structure (various types of ground state correlations). This is of great importance: The search for 0ν $\beta\beta$ decay remains the most powerful tool for investigating the mass of the (electron) neutrino.

In view of the important role neutrinos play in the framework of theories of grand unification and for astrophysics and cosmology, this should encourage further experimental efforts along this line. Next generation $\beta\beta$ decay experiments, particularly those using enriched ^{76}Ge [Kla 86,87, Cal 86a] and ^{136}Xe [Ale 88, Boe 88], could explore the neutrino mass down to $\sim 10^{-1}$ eV, corresponding to a see-saw model with a right-handed Majorana mass term of about 1 TeV which may be realized in some models based on $SU(2)_L \times SU(2)_R \times U(1)$.

References

Abb 88 L.F. Abbott, A. de Rújula and T.P. Walker Nucl. Phys. **B299** (1988) 734.

Ale 88 A. Alessandrello, A. Giuliani, E. Bellotti, D. Camin, O. Cremonesi, E. Fiorini, C. Liguori, S. Ragazzi, L. Rossi, P.P. Sverzellati and L. Zanotti, Nucl. Phys. **A478** (1988) 453c.

Als 88 M. Alston-Garnjost, B. Dougherty, R. Kenney, J. Krivicich, R. Tripp, H. Nicholson, S. Sutton, B. Dieterle, J. Kang and C. Leavitt, Phys. Rev. Lett. **60** (1988) 1928.

Ana 83 N. Anantaraman, H. Toki and G.F. Bertsch, Nucl. Phys. **A398** (1983) 269.

Avi 86 F.T. Avignone, III, R.L. Brodzinski, J.C. Evans, Jr., W.K. Hensley, H.S. Miley and J.H. Reeves, Phys. Rev. **C34** (1986) 666.

Bar 70 R.K. Bardin, P.J. Gollon, J.D. Ullman and C.S. Wu, Nucl. Phys. **A158** (1970) 337.

Bar 87 A.S. Barabash, V.M. Lobashev, V.V. Kuz'minov, V.M. Novikov, B.M. Ovchinnikov and A.A. Pomansky, Pisma w JETP **45** (1987) 171; A.S. Barabash, preprint ITEP, Moscow, 1987.

Ber 85 K.E. Bergkvist, Phys. Lett. **154B** (1985) 224; Phys. Lett. **159B** (1985) 408.

Bio 87 R.M. Bionta, G. Blewitt, C.B. Bratton, D. Casper, A. Ciocio, R. Claus, B. Cortez, M. Crouch, S.T. Dye, S. Errede, G.W. Foster, W. Gajewski, K.S. Ganezer, M. Goldhaber, T.J. Haines, T.W. Jones, D. Kielczewska, W.R. Kropp, J.G. Learned, J.M. Losecco, J. Matthews, R. Miller, M.S. Mudan, H.S. Park, L.R. Price, F. Reines, J. Schultz, S. Seidel, E. Schumard, D. Sinclair, H.W. Sobel, J.L. Stone, L.R. Sulak, R. Svoboda, G. Thornton, J.C. van der Velde and C. Wuest, Phys. Rev. Lett. **58** (1987) 1494.

Boe 88 F. Boehm, P.H. Fisher, H.E. Henrikson, M.Z. Iqbal, B.M.G. O'Callahgan, J.H. Thomas, H.T. Wong, E. Bovet, L.W. Mitchell, M. Treichel, J.-C. Vuilleumier, J.-L. Vuilleumier and K. Gabathuler, Proposal E88-01.1, 1988.

Boh 69 A. Bohr and B.R. Mottelson, Nuclear Structure (Benjamin, New York, 1969) Vol. I.

Boh 81 A. Bohr and B.R. Mottelson, Phys. Lett. **100B** (1981) 10.

Bor 85 S.D. Boris, A.I. Golutvin, L.P. Laptin, V.A. Lyubimov, V.V. Nagovitsyn, E.G. Novikov, V.Z. Nozik, V.A. Soloshchenko, I.N. Tikhomirov, E.F. Tret'yakov and N.F. Myasoedov, Pisma w JETP **42** (1985) 107; S. Boris, A. Golutvin, L. Laptin, V. Lubimov, V. Nagovizin, E. Novikov, V. Nozik, V. Soloshenko and E. Tretjakov, Phys. Lett. **159B** (1985) 217.

Bor 87 S.D. Boris, A.I. Golutvin, L.P. Laptin, V.A. Lyubimov, N.F. Myasoedov, V.V. Nagovitsyn, V.Z. Nozik, E.G. Novikov, V.A. Soloshchenko, I.N. Tikhomirov and E.F. Tret'yakov, Pisma w JETP **45** (1987) 267.

Bow 86 T.J. Bowler, J.F. Wilkinson, J.C. Browne, M.P. Maley, R.G.H. Robertson, P.A. Knapp and J.A. Helffrich, Proc. Int. Symp. on Weak and Electromagnetic Interactions in Nuclei (WEIN '86), Heidelberg, July 1-5, 1986, ed.

H.V. Klapdor (Springer, Heidelberg, 1986) p. 782.
Cal 86 D.O. Caldwell, R.M. Eisberg, D.M. Grumm, D.L. Hale, M.S. Witherell, F.S. Goulding, D.A. Landis, N.W. Madden, D.F. Malone, R.H. Pehl and A.R. Smith, Phys. Rev. **D33** (1986) 2737.
Cal 86a D.C. Caldwell, Proc. 12th Int. Conf. on Neutrino Physics and Astrophysics, Sendai, 3-8 June, 1986, eds. T. Kitagaki and H. Yuta, (World Scientific, Singapore, 1986) p. 77.
Cal 87 D.O. Caldwell, R.M. Eisberg, D.M. Grumm, M.S. Witherell, Phys. Rev. Lett. **59** (1987) 419.
Cha 85 D. Chang and R.N. Mohapatra, Phys. Rev. **D31** (1985) 1718.
Chi 80 Y. Chikashige, R.N. Mohapatra and R.D. Peccei, Phys. Rev. Lett. **45** (1980) 1926.
Chi 81 Y. Chikashige, R.N. Mohapatra and R.D. Peccei, Phy. Lett. **98B** (1981) 265.
Civ 87 O. Civitarese, A. Faessler and T. Tomoda, Phys. Lett. **194B** (1987) 11.
Dav 55 R. Davis, Phys. Rev. **97** (1955) 76.
Doi 81 M. Doi, T. Kotani, H. Nishiura, K. Okuda and E. Takasugi, Phys. Lett. **103B** (1981) 219; Prog. Theor. Phys. **66** (1981) 1739, 1765.
Doi 83a M. Doi, T. Kotani, H. Nishiura and E. Takasugi, Prog. Theor. Phys. **69** (1983) 602; **70** (1983) 1353.
Doi 83b M. Doi, M. Kenmoku, T. Kotani, H. Nishiura and E. Takasugi, Prog. Theor. Phys. **70** (1983) 1331.
Doi 85 M. Doi, T. Kotani and E. Takasugi, Prog. Theor. Phys. Suppl. **83** (1985) 1.
Doi 87 M. Doi, T. Kotani and E. Takasugi, preprint OS-GE-87-07, Osaka, 1987.
Edm 57 A.R. Edmonds, Angular Momentum in Quantum Mechanics (Princeton University Press, 1957).
Ell 87 S.R. Elliott, A.A. Hahn and M.K. Moe, Phys. Rev. Lett. **59** (1987) 2020.
Eng 88 J. Engel, P. Vogel and M.R. Zirnbauer, Phys. Rev. **C37** (1988) 731.
Fer 34 E. Fermi, Z. Phys. **88** (1934) 161.
Fir 49 E.L. Fireman, Phys. Rev. **75** (1949) 323.
Fri 75 H. Fritzsch and P. Minkowski, Ann. Phys. (N.Y.) **93** (1975) 193.
Fri 86 M. Fritschi, E. Holzschuh, W. Kündig, J.W. Petersen, R.E. Pixley and H. Stüssi, Phys. Lett. **173B** (1986) 485.
Fur 39 W.H. Furry, Phys. Rev. **56** (1939) 1184.
Gel 79 M. Gell-Mann, P. Ramond and R. Slansky, in Supergravity, ed. van Nieuvenhuizen and Freeman (North-Holland, Amsterdam, 1979).
Gel 81 G.B. Gelmini and M. Roncadelli, Phys. Lett. **99B** (1981) 411.
Gel 82 G.B. Gelmini, S. Nussinov and M. Roncadelli, Nucl. Phys. **B209** (1982) 157.
Gel 83 G.B. Gelmini, S. Nussinov and T. Yanagida, Nucl. Phys. **B219** (1983) 31.
Geo 74 H. Georgi and S.L. Glashow, Phys. Rev. Lett. **32** (1974) 438.
Geo 75 H. Georgi, in Particles and Fields, ed. C.E. Carlson (AIP, New York, 1975), p. 575.
Geo 81 H. Georgi, S.L. Glashow and S. Nussinov, Nucl. Phys. **B193** (1981) 297.
Gla 61 S.L. Glashow, Nucl. Phys. **22** (1961) 579.
Goe 35 M. Goeppert-Mayer, Phys. Rev. **48** (1935) 512.
Gro 83a K. Grotz, H.V. Klapdor and J. Metzinger, J. Phys. **G9** (1983) L 169.
Gro 83b K. Grotz, H.V. Klapdor and J. Metzinger, Phys. Lett. **132B** (1983) 22.
Gro 84 K. Grotz and H.V. Klapdor, Phys. Rev. **D30** (1984) 2095.
Gro 85a K. Grotz and H.V. Klapdor, Phys. Lett. **153B** (1985) 1.
Gro 85b K. Grotz and H.V. Klapdor, Phys. Lett. **157B** (1985) 242.
Gro 86 K. Grotz and H.V. Klapdor, Nucl. Phys. **A460** (1986) 395.
Gro 88 K. Grotz and H.V. Klapdor, Die Schwache Wechselwirkung in Kern-, Teilchen- und Astrophysik, Teubner Studienbuch, 1988.

Hal 67 J.A. Halbleib and R.A. Sorensen, Nucl. Phys. **A98** (1967) 542.

Hal 83 A. Halprin, S.T. Petcov and S.P. Rosen, Phys. Lett. **125B** (1983) 335.

Hax 81 W.C. Haxton, G.J. Stephenson, Jr. and D. Strottman, Phys. Rev. Lett. **47** (1981) 153.

Hax 82a W.C. Haxton, G.J. Stephenson, Jr. and D. Strottman, Phys. Rev. **D25** (1982) 2360.

Hax 82b W.C. Haxton, S.P. Rosen and G.J. Stephenson, Jr., Phys. Rev. **D26** (1982) 1805.

Hax 84 W.C. Haxton and G.J. Stephenson, Jr., Prog. Part. Nucl. Phys. **12** (1984) 409.

Hir 87 K. Hirata, T. Kajita, M. Koshiba, M. Nakahata, Y. Oyama, N. Sato, A. Suzuki, M. Takita, Y. Totsuka, T. Kifune, T. Suda, K. Takahashi, T. Tanimori, K. Miyano, M. Yamada, E.W. Beier, L.R. Feldscher, S.B. Kim, A.K. Mann, F.M. Newcomer, R. van Berg, W. Zhang and B.G. Cortez, Phys. Rev. Lett. **58** (1987) 1490.

Hol 81 K. Holinde, Phys. Rep. **68** (1981) 121.

Kir 67 T. Kirsten, W. Gentner and O.A. Schaeffer, Z. Phys. **202** (1967) 273.

Kir 68 T. Kirsten, O.A. Schaeffer, E. Norton and R.W. Stoenner, Phys. Rev. Lett. **20** (1968) 1300.

Kir 69 T. Kirsten and H.W. Müller, Earth Planet. Sci. Lett. **6** (1969) 271.

Kir 86 T. Kirsten, E. Heusser, D. Kaether, J. Oehm, E. Pernicka and H. Richter, Proc. Int. Symp. on Nuclear Beta Decays and Neutrino, eds. T. Kotani, H. Ejiri and E. Takasugi (World Scientific, Singapore, 1986) p. 81.

Kla 84 H.V. Klapdor and K. Grotz, Phys. Lett. **142B** (1984) 323.

Kla 86 H.V. Klapdor, Proc. Int. Symp. on Nuclear Beta Decays and Neutrino, eds. T. Kotani, H. Ejiri and E. Takasugi, (World Scientific, Singapore, 1986) p. 251.

Kla 87 H.V. Klapdor, Proposal, Internal Report, MPI H - 1987 - V17.

Kli 86 A.A. Klimenko, S.B. Osetrov, A.A. Pomansky, A.A. Smolnikov and S.I. Vasilyev, Proc. Int. Symp. on Weak and Electromagnetic Interactions in Nuclei (WEIN '86), Heidelberg, July 1-5, 1986, ed. H.V. Klapdor (Springer, Heidelberg, 1986) p. 701.

Lac 80 M. Lacombe, B. Loiseau, J.M. Richard, R. Vinh Mau, J. Cote, P. Pires and R. de Tourreil, Phys. Rev. **C21** (1980) 861.

Lan 81 P. Langacker, Phys. Rep. **72** (1981) 185.

Lan 86 P. Langacker, Proc. Int. Symp. on Weak and Electromagnetic Interactions in Nuclei (WEIN '86), Heidelberg, July 1-5, 1986, ed. H.V. Klapdor (Springer, Heidelberg, 1986) p. 879.

Lan 88 P. Langacker, Proc. Workshop on 'Neutrino Physics', Heidelberg, 20-22 Oct. 1987, eds. H.V. Klapdor and B. Povh (Springer, Heidelberg, 1988).

Led 78 Table of Isotopes, 7th ed., eds. C.M. Lederer and V.S. Shirley (Wiley, New York, 1978).

Lee 56 T.D. Lee and C.N. Yang, Phys. Rev. **104** (1956) 254.

Leu 83 C.N. Leung and S.T. Petcov, Phys. Lett. **125B** (1983) 461.

Mad 87 R. Madey, B.S. Flanders, B.D. Anderson, A.R. Baldwin, J.W. Watson, S.M. Austin, C.C. Foster, H.V. Klapdor and K. Grotz, preprint, 1987.

Maj 37 E. Majorana, Nuovo Cimento **14** (1937) 171.

Man 86 O.K. Manuel, Proc. Int. Symp. on Nuclear Beta Decays and Neutrino, eds. T. Kotani, H. Ejiri and E. Takasugi, (World Scientific, Singapore, 1986) p. 71.

Mey 86 H. Meyer, Proc. Int. Symp. on Weak and Electromagnetic Interactions in Nuclei (WEIN '86), Heidelberg, July 1-5, 1986, ed. H.V. Klapdor (Springer, Heidelberg, 1986) p. 846.

Mil 76 J.E. Miller and J.E. Spencer, Ann. Phys. (N.Y.) **100** (1976) 562.

Moh 80 R.N. Mohapatra and G. Senjanovic, Phys. Rev. Lett. **44** (1980) 912.
Moh 81a R.N. Mohapatra and G. Senjanovic, Phys. Rev. **D23** (1981) 165.
Moh 81b R.N. Mohapatra and J.D. Vergados, Phys. Rev. Lett. **47** (1981) 1713.
Moh 86 R.N. Mohapatra, Unification and Supersymmetry (Springer, New York, 1986).
Moh 88 R.N. Mohapatra, Proc. Workshop on 'Neutrino Physics', Heidelberg, 20-22 Oct. 1987, eds. H.V. Klapdor and B. Povh (Springer, Heidelberg, 1988).
Mut 86 K. Muto, Proc. Int. Symp. on Nuclear Beta Decays and Neutrino, eds. T. Kotani, H. Ejiri and E. Takasugi, (World Scientific, Singapore, 1986) p. 177; Proc. Int. Symp. on Weak and Electromagnetic Interactions in Nuclei (WEIN '86), Heidelberg, July 1-5, 1986, ed. H.V. Klapdor (Springer, Heidelberg, 1986) p. 668.
Mut 88a K. Muto and H.V. Klapdor, Phys. Lett. **201B** (1988) 420.
Mut 88b K. Muto, E. Bender and H.V. Klapdor, submitted to Phys. Rev. C.
Mut 88c K. Muto, E. Bender and H.V. Klapdor, to be published.
Nie 84 J.F. Nieves, Phys. Lett. **147B** (1984) 375.
PDG86 Particle Data Group, Phys. Lett. **170B** (1986) 1.
Pet 82 S.T. Petcov, Phys. Lett. **110B** (1982) 245.
Pri 69 H. Primakoff and S.P. Rosen, Phys. Rev. **184** (1969) 1925.
Pri 81 H. Primakoff and S.P. Rosen, Ann. Rev. Nucl. Part. Sci. **31** (1981) 145.
Rei 82 D.B. Reix, Phys. Lett. **115B** (1982) 217.
Rin 80 P. Ring and P. Schuck, The Nuclear Many-Body Problem (Springer, New York, 1980).
Sal 68 A. Salam, Proc. of the Eighth Nobel Symp., ed. N. Svartholm (Almqvist and Wiksell, Stockholm, 1968) p. 367.
Sch 80 J. Schechter and J.W.F. Valle, Phys. Rev. **D22** (1980) 2227.
Sch 82 J. Schechter and J.W.F. Valle, Phys. Rev. **D25** (1982) 2951.
Sch 84a K.W. Schmid, F. Grümmer, A. Faessler, Phys. Rev. **C29** (1984) 308.
Sch 84b K.W. Schmid, F. Grümmer, A. Faessler, Nucl. Phys. **A431** (1984) 205.
Sko 83 L.D. Skouras and J.D. Vergados, Phys. Rev. **C28** (1983) 2122.
Smi 73 D. Smith, C.E. Picciotto and D. Bryman, Phys. Lett. **46B** (1973) 157.
Smo 88 A.A. Smolnikov, private communication, May 1988.
Sol 76 V.G. Soloviev, Theory of Complex Nuclei (Pergamon, Oxford, 1976).
Tak 84 E. Takasugi, Phys. Lett. **149B** (1984) 372.
Tom 85 T. Tomoda, A. Faessler, K.W. Schmid and F. Grümmer, Phys. Lett. **157B** (1985) 4.
Tom 86 T. Tomoda, A. Faessler, K.W. Schmid and F. Grümmer, Nucl. Phys. **A452** (1986) 591.
Tom 87 T. Tomoda and A. Faessler, Phys. Lett. **199B** (1987) 475.
Tom 88 T. Tomoda, preprint, 1988, submitted to Nucl. Phys. **A**.
Tsu 84 T. Tsuboi, K. Muto and H. Horie, Phys. Lett. **143B** (1984) 293.
Val 83a J.W.F. Valle, Phys. Rev. **D27** (1983) 1672.
Val 83b J.W.F. Valle and M. Singer, Phys. Rev. **D28** (1983) 540.
Ver 81 J.D. Vergados, Phys. Rev. **C24** (1981) 640.
Ver 82 J.D. Vergados, Phys. Rev. **D25** (1982) 914.
Ver 83 J.D. Vergados, Nucl. Phys. **B218** (1983) 109.
Ver 86 J.D. Vergados, Phys. Rep. **133** (1986) 1.
Vin 79 R. Vinh Mau, Mesons in Nuclei, Vol. I., eds. M. Rho and D.H. Wilkinson (North-Holland, Amsterdam, 1979) p. 151.
Vog 85 P. Vogel and P. Fisher, Phys. Rev. **C32** (1985) 1362.
Vog 86 P. Vogel and M.R. Zirnbauer, Phys. Rev. Lett. **57** (1986) 3148.
Wap 85 A.H. Wapstra and G. Audi, Nucl. Phys. **A432** (1985) 1.
Wei 64 S. Weinberg, Phys. Rev. **133B** (1964) 1318; Phys. Rev. **134B** (1964) 882.

Wei 67 S. Weinberg, Phys. Rev. Lett. **19** (1967) 1264.

Wil 82 F. Wilczek, Phys. Rev. Lett. **49** (1982) 1549.

Wol 81 L. Wolfenstein, Nucl. Phys. **B185** (1981) 147.

Wu 57 C.S. Wu, E. Ambler, R.H. Hayward, D.D. Hopper and R.P. Hudson, Phys. Rev. **105** (1957) 1413.

Wu 85 H.F. Wu, H.Q. Song, T.T.S. Kuo, W.K. Chang and D. Strottman, Phys. Lett. **162B** (1985) 227.

Wyl 83 D. Wyler and L. Wolfenstein, Nucl. Phys. **B218** (1983) 205.

Yan 79 T. Yanagida, Proc. Workshop on Unified Theory and Baryon Number in the Universe, ed. Sawada and K. Sugimoto (KEK 1979).

Zam 82 L. Zamick and N. Auerbach, Phys. Rev. **C26** (1982) 2185.

Rep. Prog. Phys. **54** (1991) 53–126. Printed in the UK

Double beta decay

T Tomoda

Paul Scherrer Institute, CH-5232 Villigen PSI, Switzerland

Abstract

Recent developments in the theoretical investigation of nuclear double-beta decay are reviewed. In particular, the neutrinoless mode is discussed in detail, since it is sensitive to lepton number violation as predicted by gauge theories beyond the standard model and it is expected to give important information on the nature of the neutrinos and the weak interaction. Various approximations made in the theoretical treatment of neutrinoless and two-neutrino double beta decay are examined, and the present limits on the effective Majorana mass of the electron neutrino as well as the coupling constants of the right-handed leptonic current are presented.

This review was received in July 1990.

Contents

0034-4885/91/010053+74$14.00

Die schwache Wechselwirkung in Kern-, Teilchen- und Astrophysik

Eine Einführung

Von Dr. rer. nat. Klaus Grotz
und Prof. Dr. rer. nat. Hans Volker Klapdor
Max-Planck-Institut für Kernphysik, Heidelberg

THE WEAK INTERACTION IN NUCLEAR, PARTICLE AND ASTROPHYSICS

Translated by S S WILSON

Слабое взаимодействие в физике ядра, частиц и астрофизике

Перевод с немецкого
Ю. А. Данилова

под редакцией
д-ра физ.-мат. наук Ю. В. Гапонова

核、粒子和天体物理中的弱相互作用

[德] K.Grotz H.V.Klapdor 著
朱允伦 袁照明 译
庆承瑞 校

Adam Hilger

B. G. Teubner Stuttgart 1989

ADAM HILGER
BRISTOL, PHILADELPHIA and NEW YORK

山东科学技术出版社

Москва «Мир» 1992

DOUBLE BETA DECAY

By H. PRIMAKOFF and S. P. ROSEN

Department of Physics, Washington University, St. Louis, Missouri, U.S.A.

CONTENTS

Abstract. The phenomenon of double beta decay is discussed rather fully both from the experimental and from the theoretical point of view, the relation between double beta decay and the possible kinds of inverse beta decay being also briefly treated. A summary of the available experimental data is given, all methods of detection of the various double beta decay processes being considered. The theory of double beta decay, both in the no-neutrino and in the two-neutrino cases, is worked out, *ab initio*, on the basis of a nucleon–lepton interaction without conservation of parity. It is particularly emphasized that, with a ' two-component neutrino ' type nucleon–lepton interaction without conservation of parity, absence of no-neutrino $\beta\beta$ decay does not by itself uniquely imply that neutrino and anti-neutrino are distinguishable and that the total lepton charge is conserved.

Comparison of the experimental limits on the double beta decay half-lives with the corresponding theoretical values indicates that two neutrinos are emitted together with two electrons in each double beta decay process—this conclusion is, however, not yet certain and must be confirmed by further experimental work. A treatment is given of the implications of the actual occurrence of two-neutrino rather than no-neutrino double beta decay for the problems of neutrino–anti-neutrino identity and conservation of total lepton charge. It is concluded, on the basis of (i) the available double beta decay, inverse beta decay *and* muon decay experimental data and (ii) the provisional assumption of the universal applicability of ' two-component neutrino ' type coupling, that a verdict may be tentatively reached in favour of a ' Dirac ' neutrino, operationally distinguishable from a ' Dirac ' anti-neutrino, and with conservation of total lepton charge valid in all neutrino interactions.

§ 1. Introduction

In the process of double beta decay, the (very long-lived) nucleus (A, Z) disintegrates spontaneously to the isobaric nucleus $(A, Z \pm 2)$ with the simultaneous emission of two electrons—negatrons or positrons—accompanied by either (*a*) *two* anti-neutrinos or neutrinos, or (*b*) *no* anti-neutrinos or neutrinos. Alternative to the emission of two positrons, emission of one positron and absorption of one orbital negatron, or absorption of two orbital negatrons, is always possible energetically. The existence of double beta decay is expected on general theoretical

Rep. Progr. Phys. **22** (1959) 121 – 166

ANNALS OF PHYSICS: **11**, 510–533 (1960)

Lepton Conservation and Double Beta-Decay*†

EUGENE GREULING AND R. C. WHITTEN‡

Department of Physics, Duke University, Durham, North Carolina

Double beta-decay is investigated without assuming lepton conservation. If the lepton part of the universal V–A current operator is linear in the massless Majorana neutrino field, leptons are conserved. The intermediate state neutrino is completely polarized and may not be reabsorbed to produce a neutrinoless double beta-decay final state. Two slight modifications of the interaction, (1) finite neutrino mass, and/or (2) a deviation from exact equality of the effective V and A lepton currents, result in the creation of an incompletely polarized intermediate state neutrino which may then be reabsorbed. Neutrinoless double beta-decay, in violation of lepton conservation, is then possible. These neutrinoless modes of double beta-decay are computed for several nuclei and are compared with the usual lepton conserving two-neutrino mode. Measureable competition between these decay modes may exist. The abandonment of lepton conservation does not alter significantly any of the predictions of the universal V–A Fermi interaction for all first order beta-processes.

1. INTRODUCTION AND SUMMARY

(A) LEPTON CONSERVATION AND EXPERIMENTAL EVIDENCE

According to the universal V–A Fermi interaction (*1*) double beta-decay without emission of neutrinos is absolutely forbidden. Leptons are conserved by the universal Fermi interaction[1] (UFI) independent of the nature of the neutrino field if the intrinsic neutrino mass is zero. Case (*2*) has pointed out that if the lepton current operators,[2]

$$J_\mu(\bar{\nu},e) = \bar{\phi}\gamma_\mu P_+\psi \qquad \text{and} \qquad J_\mu(\bar{e},\nu) = \bar{\psi}\gamma_\mu P_+\phi \tag{1}$$

* Supported in part by the National Science Foundation. A condensed paper, prepared by E.G. while a summer (1959) research participant at the Oak Ridge National Laboratory, was privately circulated.

† Based in part on a thesis submitted by R.C.W. in partial fulfillment of the requirements for the Ph.D. degree to Duke University, (1959).

‡ Now at Stanford Research Institute, Menlo Park, California.

[1] Hereafter the Feynman and Gell-Mann Universal V–A Fermi Interaction (*1*) will be abbreviated as the UFI.

[2] In our representation the $\gamma_\mu (\mu = 1,2,3,4,5)$ are anticommuting Hermitian matrices. $P_\pm = (\frac{1}{2})(1 \pm \gamma_5)$ are Feynman and Gell-Mann's a or $\bar{a}$ (*1*) and Ψ, Φ, ψ, and ϕ are the proton, neutron, electron, and neutrino fields, respectively.

Ann. Rev. Nucl. Part. Sci. 1981. 31: 145–192

BARYON NUMBER AND LEPTON NUMBER CONSERVATION LAWS

✖ 5630

Henry Primakoff
Department of Physics, University of Pennsylvania, Philadelphia, Pennsylvania 19104

S. Peter Rosen
Department of Physics, Purdue University, West Lafayette, Indiana 47907

CONTENTS

1 GENERALITIES

Recent years have witnessed a remarkable change of attitude among physicists toward baryon-number and lepton-number conservation. With the advent of the grand unified theories of strong, electromagnetic, and weak interactions (Langacker 1980, Ellis 1980), the old belief that baryon number and lepton number were exactly conserved has given way to a new belief that, because of the intrinsic structure of most grand unified theories, baryon-number and lepton-number conservation is only approximate. In part motivated by this new belief, active searches are now underway for baryon-number and lepton-number nonconserving processes such

0163-8998/81/1201-0145$01.00

6 May 1999

PHYSICS LETTERS B

ELSEVIER

Physics Letters B 453 (1999) 194–198

Towards a superformula for neutrinoless double beta decay

H. Päs [1], M. Hirsch [2], H.V. Klapdor-Kleingrothaus [3], S.G. Kovalenko [4]

Max-Planck-Institut für Kernphysik, P.O. Box 10 39 80, D-69029 Heidelberg, Germany

Received 18 November 1998; received in revised form 1 February 1999
Editor: W. Haxton

Abstract

A general Lorentz-invariant parameterization for the long-range part of the $0\nu\beta\beta$ decay rate is derived. Combined with the short range part this general parameterization in terms of effective $B-L$ violating couplings will allow it to extract the $0\nu\beta\beta$ limits on arbitrary lepton number violating theories. Several new nuclear matrix elements appear in the general formalism compared to the standard neutrino mass mechanism. Some of these new matrix elements have never been considered before and are calculated within pn-QRPA. Using these, limits on lepton number violating parameters are derived from experimental data on ^{76}Ge. © 1999 Elsevier Science B.V. All rights reserved.

Keywords: Double beta decay; Neutrino; QRPA

1. Introduction

Double beta decay has been proven to be one of the most powerful tools to constrain $B-L$ violating physics beyond the standard model [1]. In recent years, besides the most restrictive limit on the effective neutrino Majorana mass [1,2], stringent constraints on several theories beyond the Standard Model such as R-parity violating [3–5] as well as conserving [6] SUSY, leptoquarks [8], left-right symmetric models [9] and compositeness [10,11] have been derived (for a review see [1]).

While the neutrino mass limit is based on the well-known mechanism exchanging a massive Majorana neutrino between two standard model $V-A$ vertices, the effective vertices appearing in the new contributions involve non-standard currents such as scalar, pseudoscalar and tensor currents.

Thus we felt motivated to consider the neutrinoless double beta decay rate in a general framework, parameterizing the new physics contributions in terms of all effective low-energy currents allowed by Lorentz-invariance. Such an ansatz allows one to separate the nuclear physics part of double beta decay from the underlying particle physics model, and derive limits on arbitrary lepton number violating theories. The first step of this work, treating the long-range part, is presented here. Although the general decay rate is independent of the underlying nuclear physics model, to extract quantitative limits values for nuclear matrix elements are needed. First

[1] E-mail: Heinrich.Paes@mpi-hd.mpg.de
[2] Present Address: Inst. de Fisica Corpuscular – C.S.I.C. – Dept. de Fisica Teorica, Univ. de Valencia, 46100 Burjassot, Valencia, Spain; E-mail: mahirsch@flamenco.ific.uv.es
[3] E-mail: klapdor@mickey.mpi-hd.mpg.de
[4] On leave from Joint Institute for Nuclear Research, Dubna, Russia; E-mail: kovalen@nusun.jinr.dubna.su

0370-2693/99/$ - see front matter
PII: S0370-2693(99)00330-5

limits are derived using matrix elements calculated in proton-neutron (pn) QRPA, partially already available in the literature and partially calculated here for the first time.

2. General formalism

We consider the long-range part of neutrinoless double beta decay with two vertices, which are pointlike at the Fermi scale, and exchange of a light neutrino in between. The general Lagrangian can be written in terms of effective couplings ϵ_β^α, which correspond to the pointlike vertices at the Fermi scale so that Fierz rearrangement is applicable:

$$\mathscr{L} = \frac{G_F}{\sqrt{2}} \left\{ j_{V-A}^{\mu} J_{V-A,\mu}^{\dagger} + {\sum_{\alpha,\beta}}' \epsilon_\alpha^\beta j_\beta J_\alpha^\dagger \right\} \qquad (1)$$

with the combinations of hadronic and leptonic Lorentz currents $J_\alpha^\dagger = \bar{u} \mathscr{O}_\alpha d$ respectively $j_\beta = \bar{e} \mathscr{O}_\beta \nu$ of defined helicity. The operators $\mathscr{O}_{\alpha,\beta}$ are defined as

$$\mathscr{O}_{V-A} = \gamma^\mu (1 - \gamma_5), \quad \mathscr{O}_{V+A} = \gamma^\mu (1 + \gamma_5),$$
$$\mathscr{O}_{S-P} = (1 - \gamma_5), \quad \mathscr{O}_{S+P} = (1 + \gamma_5),$$
$$\mathscr{O}_{T_L} = \frac{i}{2} [\gamma_\mu, \gamma_\nu] (1 - \gamma_5),$$
$$\mathscr{O}_{T_R} = \frac{i}{2} [\gamma_\mu, \gamma_\nu] (1 + \gamma_5). \qquad (2)$$

The prime indicates the sum runs over all contractions allowed by Lorentz-invariance, except for $\alpha = \beta = V - A$. Note that all currents have been scaled relative to the strength of the ordinary $(V - A)$ interaction.

The effective Lagrangian given in Eq. (1) represents the most general low-energy 4-fermion charged-current interaction allowed by Lorentz invariance. The interpretation of the effective couplings ϵ_β^α, however, depend on the specific particle physics model. Nevertheless one realizes the following general feature. Using only the SM fermion fields [5] and working in the Majorana basis for the neutrinos $(\nu := \nu_L + \nu_L^C)$ it is easily seen that all currents involving operators proportional to $(1 + \gamma_5)$ violate lepton number by two units, i.e. the corresponding ϵ_β^α must also be lepton-number violating. (Such LNV ϵ_β^α are easily found, an example is given by R-parity violating supersymmetry treated in Refs. [3–5].)

The double beta decay amplitude is proportional to the time-ordered product of two effective Lagrangians (see Fig. 1):

$$T(\mathscr{L}_{(1)} \mathscr{L}_{(2)}) = \frac{G_F^2}{2} T \{ j_{V-A} J_{V-A}^\dagger j_{V-A} J_{V-A}^\dagger + \epsilon_\alpha^\beta j_\beta J_\alpha^\dagger j_{V-A} J_{V-A}^\dagger + \epsilon_\alpha^\beta \epsilon_\gamma^\delta j_\beta J_\alpha^\dagger j_\delta J_\gamma^\dagger \}. \qquad (3)$$

While the first term (contribution (a) in Fig. 1) corresponds to the standard model (SM) like neutrino exchange, and the 3rd term (contribution (c) in Fig. 1), which is quadratic in ϵ can be neglected, only the 2nd term (contribution (b)) is phenomenologically interesting. For this term one has to consider two general cases:

1) The leptonic SM $V - A$ current meets a left-handed non SM current j_β with $\beta = S - P, T_L$. For this contribution the neutrino propagator is

$$\sim P_L \frac{q^\mu \gamma_\mu + m_\nu}{q_\mu q^\mu - m_\nu^2} P_L = \frac{m_\nu}{q_\mu q^\mu - m_\nu^2} \qquad (4)$$

with the usual left- and right-handed projectors $P_{L/R} = \frac{1 \mp \gamma_5}{2}$. This expression is proportional to the unknown neutrino Majorana mass $m_\nu \leq \sim 0.5$ eV, for

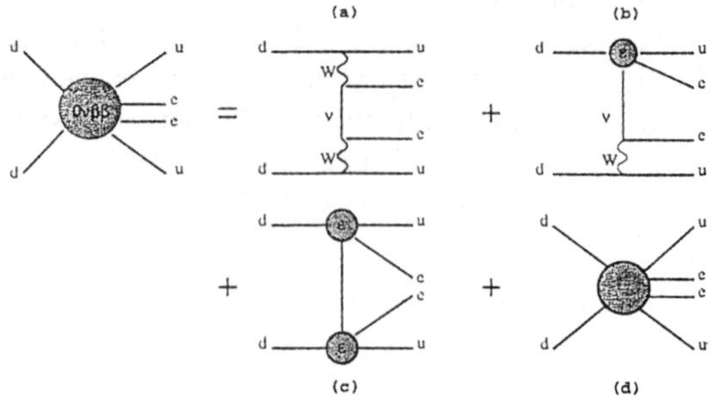

Fig. 1. Feynman graphs of the general double beta rate: The contributions (a)–(c) correspond to the long range part, the contribution d) is the short range part (to be discussed elsewhere).

[5] The use of only SM neutrinos does not really imply a loss of generality here, see the discussion on $V + A$ currents below.

which no lower bound exists. Therefore no limits on the corresponding parameters ϵ_α^β can be derived.

2) The leptonic SM $V-A$ current meets a right-handed non SM current j_β with $\beta = S+P, V+A, T_R$. For this contribution the neutrino propagator is

$$\sim P_L \frac{q^\mu \gamma_\mu + m_\nu}{q_\mu q^\mu - m_\nu^2} P_R = \frac{q^\mu \gamma_\mu}{q_\mu q^\mu - m_\nu^2} \tag{5}$$

which is proportional to the neutrino momentum $q_\mu \simeq p_F \simeq 100$ MeV with the nuclear Fermi momentum p_F, and thus will produce stringent limits on corresponding ϵ_α^β.

Taking these considerations into account, we are left with three interesting contributions discussed in the following section. With the present half-life limit of the Heidelberg-Moscow experiment $T_{1/2}^{0\nu\beta\beta} > 1.2 \cdot 10^{25} y$ [1] and considering only one ϵ_α^β at a time (evaluation "on axis")

$$\left[T_{1/2}^{0\nu\beta\beta}\right]^{-1} = \left(\epsilon_\alpha^\beta\right)^2 G_{0k} |ME|^2 \tag{6}$$

where G_{0k} denotes the phase space factors given in [12] and $|ME|$ the nuclear matrix elements discussed in the following. Note that evaluating "on axis", compared to the arbitrary evaluation, neglects interference terms of the different contributions. Although this is expected to be a small effect (see below), this remains to be discussed in the next step.

3. Calculational details and limits

3.1. SM meets $j_{V+A} J_{V+A}^\dagger$ and $j_{V+A} J_{V-A}^\dagger$

This contribution has been considered already in the context of left-right symmetric models [12,15,9]. For sake of completeness we repeat the updated results of [15] here in our notation: $\epsilon_{V+A}^{V+A} < 7.9 \cdot 10^{-7}$, $\epsilon_{V-A}^{V+A} < 4.9 \cdot 10^{-9}$ for the full calculation with arbitrary evaluation. Using s-wave approximation and "on axis" evaluation as was done in this work the limits are reduced to $\epsilon_{V+A}^{V+A} < 7.0 \cdot 10^{-7}$, $\epsilon_{V-A}^{V+A} < 4.4 \cdot 10^{-9}$. This example confirms the expectation that these assumptions will only slightly (less than 10%) affect the result.

It is worthwhile discussing the following subtlety in interpretation when comparing our ansatz and the work of [12]. Doi et al. [12] calculate the decay amplitude, writing down the Lagrangian of an explicitly left-right symmetric model. Therefore, their calculations do not only treat SM neutrinos, but contain also right-handed neutrinos. Nevertheless, the derivation of the decay amplitude for $0\nu\beta\beta$ decay is the same in both calculations, only some care is required when going from one notation to the other. For example, in the lepton-number violating $\langle \lambda \rangle$, defined in [12] as

$$\langle \lambda \rangle = \lambda {\sum}'_j U_{ej} V_{ej} \tag{7}$$

where $\lambda \approx (m_{W_L}/m_{W_R})^2$, LNV is due to the product of mixing matrices $\Sigma'_j U_{ej} V_{ej}$, which vanishes identically if LNV goes to zero. Thus, our (LNV) ϵ_{V+A}^{V+A} corresponds to $\langle \lambda \rangle$ of Doi et al. [12], (and *not* to λ). This example shows, that the use of only SM neutrinos in the derivation does not imply a loss of generality in our results, if the source of LNV of the particle physics model under consideration is carefully identified.

3.2. SM meets $j_{S+P} J_{S+P}^\dagger$ and $j_{S+P} J_{S-P}^\dagger$

Using s-wave approximation for the outgoing electrons and some assumptions according to [13,4] one gets

$$ME_{S+P}^{S+P} = -ME_{S-P}^{S+P} = -\frac{F_P^{(3)}(0)}{R m_e G_A}\left(\mathcal{M}_{T'} + \tfrac{1}{3}\mathcal{M}_{GT'}\right). \tag{8}$$

The phase space factor is defined

$$G_{01} = \frac{(G_F G_A)^4 m_e^4}{32\pi^5 (m_e R)^2 \ln 2} \times \int F_0(Z,\epsilon_1) p_1 \epsilon_1 F_0(Z,\epsilon_2) p_2 \epsilon_2 \times \delta(\epsilon_1 + \epsilon_2 + M_f - M_i) d\epsilon_1 d\epsilon_2 \tag{9}$$

and the matrix elements (summation over nucleons a,b is suppressed) are

$$\mathcal{M}_{GT'} = \langle 0_f^+ | h_R (\boldsymbol{\sigma}_a \boldsymbol{\sigma}_b) \tau_a^+ \tau_b^+ | 0_i^+ \rangle \tag{10}$$

$$\mathcal{M}_{T'} = \langle 0_f^+ | h_{T'} \{ (\boldsymbol{\sigma}_a \hat{\boldsymbol{r}}_{ab})(\boldsymbol{\sigma}_b \hat{\boldsymbol{r}}_{ab}) - \tfrac{1}{3}(\boldsymbol{\sigma}_a \boldsymbol{\sigma}_b) \} \times \tau_a^+ \tau_b^+ | 0_i^+ \rangle . \tag{11}$$

Table 1
Relevant Nuclear Matrix Elements calculated in pn-QRPA

$\mathcal{M}_{GT'}$	2.95
$\mathcal{M}_{F'}$	−0.663
$\mathcal{M}_{GT''}$	8.78
$\mathcal{M}_{T'}$	0.224
$\mathcal{M}_{T''}$	1.33

h_R and $h_{T'}$ are neutrino potentials defined as

$$h_R = \frac{2}{\pi}\frac{R^2}{m_P}\int_0^\infty dq q^4 \frac{j_0(qr_{ab})f^2(q^2)}{\omega(\omega+\bar{E})}, \tag{12}$$

$$h_{T'} = \frac{2}{\pi}\frac{R^2}{m_P}\int_0^\infty dq q^2 \frac{f^2(q^2)}{\omega(\omega+\bar{E})} \times\left\{q^2 j_0(qr_{ab}) - 3\frac{q}{r_{ab}}j_1(qr_{ab})\right\}. \tag{13}$$

Here R denotes the nuclear radius, m_P the proton mass, ϵ_i and p_i are electron energies and momenta, $F_0(Z,\epsilon_i)$ and is the Fermi function. Further $\omega = \sqrt{q^2+m_\nu^2}$, $q=|\boldsymbol{q}|$, $\hat{r}=\boldsymbol{r}/r$ and $j_k(qr)$ are spherical Bessel functions. $(\omega+\bar{E})^{-1}$ is the energy denominator of the perturbation theory. The form factors $F_i^a(0) = F_i^a(q^2)/f(q^2)$ with $f(q^2) = (1 + q^2/m_A^2)^{-2}$ ($m_A^2 = 0.85$ GeV) have been calculated in the MIT bag model in [14], $G_A \simeq 1.26$ and $G_V \simeq 1$.

Inserting the numerical value of the matrix elements $\mathcal{M}_{GT'}$ and $\mathcal{M}_{T'}$ (see [4] and Table 1), one derives $\epsilon_{S+P}^{S+P}, \epsilon_{S-P}^{S+P} < 1.1\cdot 10^{-8}$.

3.3. SM meets $j_{T_R}J_{T_R}^\dagger$ and $j_{T_R}J_{T_L}^\dagger$

In the tensor part the decay rate depends on the phase space G_{01} and new matrix elements not considered in the literature.

For the hadronic T_R contribution one gets under the assumptions used above

$$ME_{T_R}^{T_R} = -\alpha_1\tfrac{2}{3}\mathcal{M}_{GT'} + \alpha_1\mathcal{M}_{T'} \tag{14}$$

with

$$\alpha_1 = \frac{4T_1^{(3)}(0)G_V(1-2m_P(G_W/G_V))}{G_A^2 R m_e}. \tag{15}$$

Again $T_1^{(3)}(0)$ has been taken from [14] and the strength of the induced weak magnetism

$$(G_W/G_V) = \frac{\mu_P-\mu_n}{2m_P} \simeq \frac{-3.7}{2m_P}$$

is obtained by the CVC hypothesis. The involved nuclear matrix elements have been calculated in the QRPA-approach of [15,16]. Inserting the values obtained for the special case of ^{76}Ge (see Table 1) yields $\epsilon_{T_R}^{T_R} < 1.7\cdot 10^{-9}$.

For the hadronic T_L contribution in leading order of $(1/m_P)$ one finds

$$ME_{T_L}^{T_R} = \alpha_2\mathcal{M}_{F'} - \alpha_3\left(\mathcal{M}_{T''} + \tfrac{1}{3}\mathcal{M}_{GT''}\right) \tag{16}$$

with

$$\mathcal{M}_{F'} = \langle 0_f^+|h_R\tau_a^+\tau_b^+|0_i^+\rangle \tag{17}$$

$$\mathcal{M}_{GT''} = \langle 0_f^+|h_{R\omega}(\boldsymbol{\sigma}_a\boldsymbol{\sigma}_b)\tau_a^+\tau_b^+|0_i^+\rangle \tag{18}$$

$$\mathcal{M}_{T''} = \langle 0_f^+|\omega h_{T''}\{(\boldsymbol{\sigma}_a\hat{\boldsymbol{r}}_{ab})(\boldsymbol{\sigma}_b\hat{\boldsymbol{r}}_{ab}) - \tfrac{1}{3}(\boldsymbol{\sigma}_a\boldsymbol{\sigma}_b)\} \times\tau_a^+\tau_b^+|0_i^+\rangle. \tag{19}$$

The neutrino potentials are

$$h_R\omega = \frac{2}{\pi}\frac{R^3}{m_P}\int_0^\infty dq q^4\frac{j_0(qr_{ab})f^2(q^2)}{(\omega+\bar{E})}, \tag{20}$$

$$h_{T''} = \frac{2}{\pi}\frac{R^3}{m_P}\int_0^\infty dq q^2\frac{f^2(q^2)}{(\omega+\bar{E})} \times\left\{q^2 j_0(qr_{ab}) - 3\frac{q}{r_{ab}}j_1(qr_{ab})\right\}. \tag{21}$$

and

$$\alpha_2 = \frac{4\left(2\hat{T}_2^{(3)}(0) - T_1^{(3)}(0)\right)G_V}{G_A^2 R m_e}, \tag{22}$$

$$\alpha_3 = \frac{4T_1^{(3)}(0)(G_P/G_A)}{G_A R^2 m_e}. \tag{23}$$

Here $\hat{T}_2^{(3)}(0)$ has been taken from [14], $G_P/G_A = 2m_P/m_\pi^2$ and ω is the neutrino energy. Again the

Table 2
Limits on effective $B-L$ violating couplings evaluated "on axis"

ϵ_{V-A}^{V+A}	$4.4\cdot 10^{-9}$
ϵ_{V+A}^{V+A}	$7.0\cdot 10^{-7}$
ϵ_{S-P}^{S+P}	$1.1\cdot 10^{-8}$
ϵ_{S+P}^{S+P}	$1.1\cdot 10^{-8}$
ϵ_{TL}^{TR}	$6.4\cdot 10^{-10}$
ϵ_{TR}^{TR}	$1.7\cdot 10^{-9}$

matrix elements have been calculated in the model of [15,16]. A limit of $\epsilon_{T_L}^{T_R} < 7.3 \cdot 10^{-10}$ has been obtained. Factors R have been arbitrarily absorbed into the definition $\mathcal{M}_{GT''}$ and $\mathcal{M}_{T''}$ to get dimensionless quantities. One should notice that the corresponding factor included in α_3 compensates this choice.

4. Conclusion

We have presented a general parameterization for the long range part of the neutrinoless double beta decay rate in terms of effective couplings. The resulting bounds are summarized in Table 2. Combined with the short range part and contributions of derivative couplings, this parameterization will give the double beta decay constraints for arbitrary lepton number violating theories beyond the SM. The next step should include these contributions and discuss interference terms.

Acknowledgements

M.H. would like to acknowledge support by the European Union's TMR program under grant ERBFMBICT983000.

References

[1] H.V. Klapdor-Kleingrothaus, in [7]
[2] Heidelberg-Moscow Collaboration, L. Baudis, Phys. Lett. B 407 (1997) 219.
[3] M. Hirsch, H.V. Klapdor-Kleingrothaus, S.G. Kovalenko, Phys. Rev. Lett. 75 (1995) 17; Phys. Rev. D 53 (1996) 1329.
[4] M. Hirsch, H.V. Klapdor-Kleingrothaus, S.G. Kovalenko, Phys. Lett. B 372 (1996) 181; B 381 (1996) 488 (E).
[5] H. Päs, M. Hirsch, H.V. Klapdor-Kleingrothaus, in preparation.
[6] M. Hirsch, H.V. Klapdor-Kleingrothaus, S.G. Kovalenko, Phys. Lett. B 398 (1997) 311; Phys. Rev. D 57 (1998) 1947; contribution in [7]
[7] H.V. Klapdor-Kleingrothaus, H. Päs (Eds.), in: Proc. Int. Conf. Beyond the Desert - Accelerator- and Non-Accelerator Approaches, Castle Ringberg, Germany, 1997.
[8] M. Hirsch, H.V. Klapdor-Kleingrothaus, S.G. Kovalenko, Phys. Lett. B 378 (1996) 17; Phys. Rev. D 54 (1996) R4207.
[9] M. Hirsch, H.V. Klapdor-Kleingrothaus, O. Panella, Phys. Lett. B 374 (1996) 7.
[10] O. Panella, in [7]
[11] E. Takasugi, in [7]
[12] M. Doi, T. Kotani, E. Takasugi, Progr. Theor. Phys. (Suppl.) 83 (1985) 1.
[13] T. Tomoda, Rep. Progr. Phys. 54 (1991) 53.
[14] S. Adler, Phys. Rev. D 11 (1975) 3309.
[15] K. Muto, E. Bender, H.V. Klapdor, Z. Phys. A 334 (1989) 177, 187.
[16] A. Staudt, K. Muto, H.V. Klapdor-Kleingrothaus, Europhys. Lett. 13 (1990) 31; M. Hirsch, K. Muto, T. Oda, H.V. Klapdor-Kleingrothaus, Z. Phys. A 347 (1994) 151.

A superformula for neutrinoless double beta decay II: The short range part

H. Päs [1], M. Hirsch [2], H.V. Klapdor–Kleingrothaus [1], S. G. Kovalenko [3]

[1] *Max–Planck–Institut für Kernphysik, P.O.Box 10 39 80, D–69029 Heidelberg, Germany*

[2] *Inst. de Fisica Corpuscular - C.S.I.C. - Dept. de Fisica Teorica, Univ. de Valencia, 46100 Burjassot, Valencia, Spain*

[3] *Departamento de Fisica, Universidad Tecnica Federico Santa Maria, Casilla 110-V, Valparaiso, Chile*

Abstract

A general Lorentz–invariant parameterization for the short-range part of the $0\nu\beta\beta$ decay rate is derived. Combined with the long range part already published this general parameterization in terms of effective $B-L$ violating couplings will allow it to extract the $0\nu\beta\beta$ limits on arbitrary lepton number violating theories.

Keywords: Double beta decay; Neutrino; QRPA

1 Introduction

The search for neutrinoless double beta decay has been proven to be one of the most powerful tools to constrain $B-L$ violating physics beyond the standard model [1]. In recent years, besides the most restrictive limit on the effective neutrino Majorana mass [1,3], stringent constraints on several theories beyond the Standard Model such as R–parity violating [6,7] as well as conserving [8] SUSY, leptoquarks [10], left–right symmetric models [11] and compositeness [12,13] have been derived (for a review see [1]). While the neutrino mass limit is based on the well-known mechanism exchanging a massive Majorana neutrino between two standard model $V-A$ vertices, the effective vertices appearing in the new contributions involve non–standard currents such as scalar, pseudoscalar and tensor currents. Moreover, besides contributions with a light neutrino being exchanged between two separated vertices, the so-called long-range part, additional contributions from the short range part are expected, e.g. in SUSY without R-parity [5,4].

to be Publ. in 2000

PHYSICAL REVIEW D VOLUME 25, NUMBER 9 1 MAY 1982

Lepton-number conservation and the double-β decay of ^{128}Te and ^{130}Te

W. C. Haxton
Theoretical Division, Los Alamos National Laboratory, Los Alamos, New Mexico 87545 and Department of Physics, Purdue University, West Lafayette, Indiana 47907

G. J. Stephenson, Jr. and D. Strottman
Theoretical Division, Los Alamos National Laboratory, Los Alamos, New Mexico 87545
(Received 24 November 1981)

We consider the classic argument of Pontecorvo that the ratio of $\beta\beta$-decay half-lives for ^{130}Te and ^{128}Te may indicate violation of lepton number. We believe that the theoretical support for the assumption underlying this argument, equality of the nuclear matrix elements mediating the 2ν $\beta\beta$ decay of these isotopes, is weak due to a significant disparity between the calculated and geochemical absolute rates. If one ignores this inconsistency and attributes the breaking of the γ_5 invariance of the weak leptonic current to a Majorana mass, faithful treatment of the radial dependence of the 0ν-$\beta\beta$-decay operators yields $\langle m^{\mathrm{Maj}}\rangle_\nu \cong 10$ eV.

Current theoretical attempts to unify the electroweak and strong interactions may succeed only if certain symmetries of the Glashow-Weinberg-Salam model, such as exactly conserved baryon and lepton numbers and massless neutrinos, are abandoned. Sensitive experimental tests of such symmetries are thus being pursued with great urgency. One powerful probe of lepton-number conservation, of the mass and charge-conjugation properties of the neutrino, and of possible right-handed admixtures in the weak leptonic current is nuclear double-β decay, $(A,Z-2)\rightarrow(A,Z)$.[1] This process can be observed in a number of even-even nuclei where, due to the pairing interaction, the competing decay $(A,Z-2)\rightarrow(A,Z-1)$ is energetically inaccessible. The key issue is whether $\beta\beta$ decay proceeds, as in the standard model, entirely by those second-order weak processes in which two electrons and two neutrinos are produced [see Fig. 1(a)], or whether additional mechanisms involving violation of lepton number also contribute. In particular, if the Majorana mass of the electron neutrino is nonzero, the neutrinoless $\beta\beta$ decay shown in Fig. 1(b) will occur.[2] It has also been suggested that a first-order process of a superweak interaction may produce a neutrinoless final state[3,4] and Georgi, Glashow, and Nussinov have recently considered a $\beta\beta$-decay mechanism in which the outgoing electrons are accompanied by a light scalar boson.[5]

Currently, the issue of lepton-number conservation in $\beta\beta$ decay must be argued in the context of limited experimental information. Laboratory bounds on the neutrinoless decay of Fig. 1(b) have been used to restrict the Majorana mass of the electron neutrino to $\langle m^{\mathrm{Maj}}\rangle_\nu \lesssim 15$ eV, assuming that there is no explicit right-handed coupling of the Majorana neutrino.[6] Somewhat less stringent limits can be obtained by attributing total geochemical $\beta\beta$-decay rates to neutrinoless decay.[7]

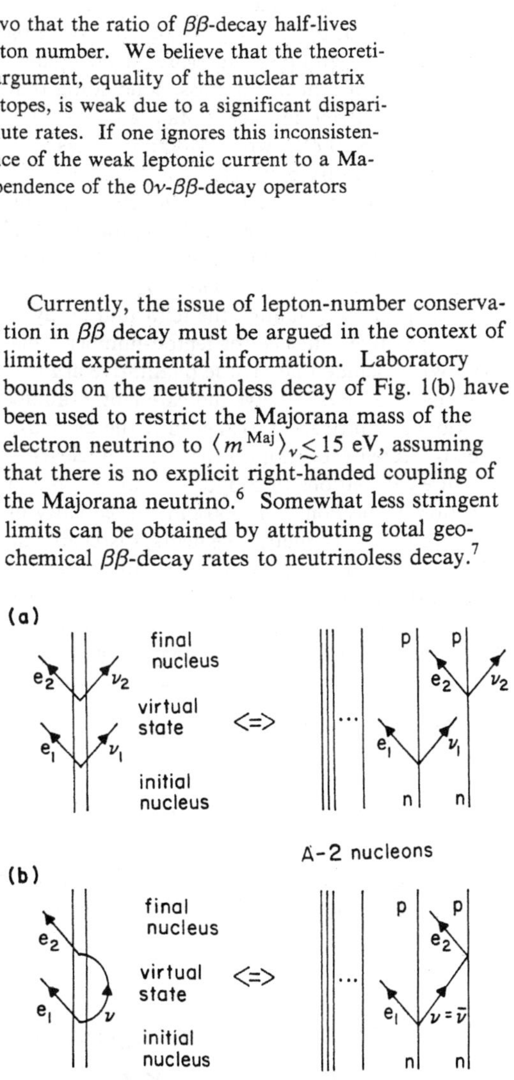

FIG. 1. Two-nucleon mechanisms for two-neutrino (a) and no-neutrino (b) $\beta\beta$ decay.

However, in both these approaches the limits on lepton-number violation can only be determined to within the accuracy permitted by nuclear-matrix-element calculations.

It is clearly of great importance to have a test of lepton-number violation which is less sensitive to structure uncertainties. One possibility, originally suggested by Pontecorvo,[3] is the ratio of the total $\beta\beta$-decay rates for ^{128}Te and ^{130}Te. The Pontecorvo argument has been used by Vergados[7] and by Bryman and Picciotto[8] to derive a nonzero value for a lepton-number-violating coupling in a phenomenological $\beta\beta$-decay Lagrangian. More recently the Osaka group has interpreted this coupling in terms of a neutrino mass, concluding that the electron neutrino may be a Majorana particle with $\langle m^{\rm Maj}\rangle_\nu \cong 30$ eV,[9] while Minkowski has found the much different value $\langle m^{\rm Maj}\rangle_\nu \cong 1$ eV.[10] It is thus unclear whether the mass determined in this manner is consistent[5] with the lepton-number-violating mechanism of Fig. 1(b), given the laboratory 0ν-$\beta\beta$-decay mass limit $\langle m^{\rm Maj}\rangle_\nu \lesssim 15$ eV.

In this paper we examine the tellurium decays and the putative evidence for lepton-number violation. We conclude that (1) the qualitative argument that the Te $\beta\beta$ decays indicate lepton-number violation is significantly weakened if the 2ν phase space is treated properly; (2) the use of nuclear-structure calculations to support the underlying assumption of the Pontecorvo argument, equality of the ^{128}Te and ^{130}Te matrix elements, disregards an existing discrepancy between theory and experiment in the absolute rates; and (3) as demonstrated by the significant discrepancy in the results of Refs. 9 and 10, a faithful treatment of the radial behavior of 0ν matrix elements is necessary for the determination of $\langle m^{\rm Maj}\rangle_\nu$ from the Te ratio. We present the mass estimate resulting from such a treatment.

The $\beta\beta$-decay amplitude for $J^\pi=0^+\rightarrow 0^+$ nuclear transitions is given, in the standard model, by the two-nucleon mechanism of Fig. 1(a). [A single-hadron mechanism where the $\Delta(1232)$ decays to a neutron and four leptons has been discussed.[11] However, the nuclear operator for this process is a spin vector and cannot contribute to $\Delta J=0$ transitions.] If we evaluate each nucleon β decay in the allowed approximation, the nuclear $\beta\beta$-decay amplitudes in time-dependent perturbation theory are

$$\sum_M \frac{\langle F|\sum_{i=1} O(i)|M\rangle\langle M|\sum_{i=1} O(i)|I\rangle}{E_M^N+E^L-E_I^N}, \qquad (1)$$

where the sum extends over a complete set of states M of the intermediate nucleus $(A,Z-1)$. Nuclear energies are denoted by E^N, and E^L is the sum of the lepton energies emitted in the first β decay. The $O(i)$ represent either the Fermi or Gamow-Teller operators, $\tau_+(i)$ or $\sigma(i)\tau_+(i)$.

The principal approximation we make is to complete the sum in Eq. (1) by closure after replacing E_M^N by an average value, either $\langle E_{\rm F}^N\rangle$ or $\langle E_{\rm GT}^N\rangle$. Our method of calculating these values is discussed in Ref. 6 and, more extensively, in Ref. 12. The resulting 2ν-$\beta\beta$-decay rate then can be written

$$\omega_{2\nu}=f_{\rm GT}|M_{\rm GT}|^2+f_{\rm F}|M_{\rm F}|^2 + f_{\rm F-GT}{\rm Re}(M_{\rm F}M_{\rm GT}^*),$$

where

$$M_{\rm F}=\left\langle F\left|\left|\tfrac{1}{2}\sum_{ij}\tau_+(i)\tau_+(j)\right|\right|I\right\rangle,$$

$$M_{\rm GT}=\left\langle F\left|\left|\tfrac{1}{2}\sum_{ij}\vec\sigma(i)\cdot\vec\sigma(j)\tau_+(i)\tau_+(j)\right|\right|I\right\rangle,$$

$$f_i=\frac{G^4}{8\pi^7}\int_{m_e}^{T_0+m_e}\mathcal{F}(Z,\epsilon_1)|\vec{k}_1|\epsilon_1 d\epsilon_1\int_{m_e}^{T_0+2m_e-\epsilon_1}\mathcal{F}(Z,\epsilon_2)|\vec{k}_2|\epsilon_2 d\epsilon_2 \times\int_0^{T_0-\epsilon_1-\epsilon_2+2m_e}\nu_1^2\nu_2^2 d\nu_1 A_i,\quad i=\text{GT, F or F}-\text{GT}$$

with

$$A_{\rm GT}=F_A^4(K_{\rm GT}^2+L_{\rm GT}^2+K_{\rm GT}L_{\rm GT})/3,\quad A_{\rm F}=F_1^4(K_{\rm F}^2+L_{\rm F}^2-K_{\rm F}L_{\rm F}),$$

$$A_{\rm F-GT}=-F_1^2F_A^2(K_{\rm GT}L_{\rm F}+K_{\rm F}L_{\rm GT}),$$

$$K=\frac{1}{\langle E^N\rangle+\epsilon_1+\nu_1-E_I}+\frac{1}{\langle E^N\rangle+\epsilon_2+\nu_2-E_I}, \qquad (2)$$

$$L=\frac{1}{\langle E^N\rangle+\epsilon_1+\nu_2-E_I}+\frac{1}{\langle E^N\rangle+\epsilon_2+\nu_1-E_I},$$

with F_A and F_1 the axial-vector and vector couplings, and where K_{GT} is obtained by assigning the appropriate subscript to $\langle E^N \rangle$, etc. In the above $(\epsilon, \vec{k})$ denotes an electron four-momentum, $T_0 = E_I - E_F - 2m_e$ is the total kinetic energy of the outgoing leptons, ν_1 is the first neutrino's energy, and $\nu_2 = E_I - E_F - \epsilon_1 - \epsilon_2 - \nu_1$. The factors $\mathscr{F}$ accounting for Coulomb distortion of the outgoing electrons in the field of the daughter nucleus of charge Z are taken from Ref. 13.

In early treatments of $\beta\beta$ decay by Primakoff and Rosen[1] and by Konopinski,[1] an approximation to this formula was often used. The Fermi matrix element vanishes in the limit of exact isospin; if Coulomb effects are included, we find typically $|M_F| < 0.02 << |M_{GT}|$. Thus M_F can be ignored for all but extremely suppressed $\beta\beta$-decay transitions. In addition, usually $\langle E^N \rangle - E_I >> \epsilon + \nu$, so we can approximate $\epsilon + \nu \cong T_0/2 + m_e$ without changing the energy denominators too much. Finally, a nonrelativistic point Coulomb correction is used in describing the distortion of the outgoing electrons,

$$\mathscr{F}(Z,\epsilon) = \frac{\epsilon}{|\vec{k}|} \mathscr{F}^{PR}(Z),$$
$$\mathscr{F}^{PR}(Z) = \frac{2\pi\alpha Z}{1-\exp(-2\pi\alpha Z)} . \quad (3)$$

Although this approximation fails badly for heavy nuclei, its use in the older papers of Ref. 1 was not inappropriate: estimates of $\beta\beta$-decay rates were then typically assigned uncertainties of 2 orders of magnitude. However, in the context of modern efforts to perform accurate calculations of $\beta\beta$-decay matrix elements, this approximation may lead to serious underestimation of rates. We shall return to this point in our discussion of the Te ratio.

These approximations permit an exact evaluation of the phase-space integrals of Eq. (2), yielding

$$\omega_{2\nu}^{PR} = f_{GT}^{PR} |M_{GT}|^2$$

with

$$f_{GT}^{PR} = \frac{16 G^4 F_A{}^4}{\pi^7} [\mathscr{F}^{PR}(Z)]^2 \frac{1}{(\langle E_N \rangle + \frac{1}{2} T_0 + m_e - E_I)^2} \frac{m_e{}^{11}}{8!} \tilde{T}_0{}^7 \left[1 + \frac{\tilde{T}_0}{2} + \frac{\tilde{T}_0{}^2}{9} + \frac{\tilde{T}_0{}^3}{90} + \frac{\tilde{T}_0{}^4}{1980} \right] \quad (4)$$

and with $\tilde{T}_0 = T_0/m_e$. This result agrees with that of Konopinski.[1]

We now turn to the evaluation of the lepton-number-violating two-nucleon process of Fig. 1(b). We use a phenomenological Lagrangian in which the weak leptonic current

$$\bar{\psi}_e(x)\gamma_\mu \left[(1-\gamma_5) + \eta(1+\gamma_5)\psi_{\nu_e}(x) \right] \quad (5)$$

couples to the usual hadronic current. The γ_5 invariance is broken by an explicit right-handed coupling η of the electron to the Majorana neutrino field ψ_ν and by the Majorana mass $\langle m^{Maj} \rangle_\nu$.[14] We again restrict our consideration to $\Delta J = 0$ normal-parity transitions, we make the allowed approximation, and we complete the sum over intermediate nuclear states by closure. (We use $\langle E^N \rangle \cong \langle E_{GT}^N \rangle$ as the 0ν rate is rather insensitive to the value of $\langle E^N \rangle$.) In addition, we specialize to neutrino mass components $m_i^{Maj} << \langle p_\nu \rangle$, with $\langle p_\nu \rangle$ the average momentum of the exchanged neutrino. (Physically this restricts us to $m_i^{Maj} << 1/R_0$, where R_0 is the nuclear radius, and thus to m_i^{Maj}, not more than a few tens of MeV.) We also make one further approximation. The matrix elements mediating 0ν decay have a radial dependence which varies as $[g(r_{ij}\xi_1) + g(r_{ij}\xi_2)]/2r_{ij}$,

$$g(x) = \frac{2}{\pi} \left\{ \sin x \, \mathrm{Ci}(x) + \cos x \left[\frac{\pi}{2} - \mathrm{Si}(x) \right] \right\} \quad (6)$$

with $\xi_{1,2} = \langle E_N \rangle + \epsilon_{1,2} - E_I$ and r_{ij} the separation between nucleons i and j. For a heavy nucleus, $g(x)$ may vary from 1.0 to 0.4 as r_{ij} goes from 0 to $R_0 \cong 1.2A^{1/3}$. Thus it has been customary to approximate

$$\frac{g(r_{ij}\xi_1) + g(r_{ij}\xi_2)}{2r_{ij}} \cong \frac{1}{r_{ij}} \cong \frac{1}{R_0} ,$$

which of course is only a very rough guess of the effect of this radial dependence on the nuclear matrix elements. We instead make the approximation

$$g(r_{ij}\xi_{1,2}) \cong g(r_{ij}\xi_0) ,$$

with

$$\xi_0 = \langle E_N \rangle + \frac{T_0}{2} + m_e - E_I \cong \xi_{1,2} ,$$

and explicitly evaluate matrix elements with such radial dependences. As $\langle E_N \rangle - E_I >> \epsilon$, we are making a small change in an argument of a slowly varying function, and thus our approximation should be quite reliable.

The resulting rate for the 0ν two-nucleon process of Fig. 1(b) then becomes, for the $\langle m^{\rm Maj}\rangle_\nu \neq 0$ and $\eta=0$ terms,

$$\omega_{0\nu,\eta=0}=\left[\frac{\langle m^{\rm Maj}\rangle_\nu}{m_e}\right]^2 f_{xx}\,|X|^2$$

with

$$f_{xx}=\frac{G^4 m_e^2}{8\pi^5}\int_{m_e}^{T_0+m_e}\mathcal{F}(Z,\epsilon_1)\mathcal{F}(Z,\epsilon_2)\times\epsilon_1\epsilon_2\,|\vec{k}_1|\,|\vec{k}_2|\,d\epsilon_1 ,$$

where

$$X=F_A^2 M_2'' - F_1^2 M_1'' ,$$

$$M_1''=\left\langle F\left|\tfrac{1}{2}\sum_{ij}\frac{g(r_{ij}\xi_0)}{r_{ij}}\tau_+(i)\tau_+(j)\right|I\right\rangle , \qquad (7)$$

$$M_2''=\left\langle F\left|\tfrac{1}{2}\sum_{ij}\frac{g(r_{ij}\xi^0)}{r_{ij}}\vec{\sigma}(i)\cdot\vec{\sigma}(j)\times\tau_+(i)\tau_+(j)\right|I\right\rangle .$$

This agrees, in the proper limit, with the result of Doi *et al.* (The result for $\eta \neq 0$ is given in Ref. 12. We concentrate on the $\langle m^{\rm Maj}\rangle_\nu$ contribution because of the keen interest in Majorana mass limits, but caution the reader that the $\eta\langle m^{\rm Maj}\rangle_\nu$ interference terms occur in the general expression.)

We now consider the additional approximations which are often made and which, in contrast to those discussed above, may introduce appreciable uncertainties in estimates of lepton-number violations. If we take $\mathcal{F}(Z,\epsilon)=\mathcal{F}^{\rm PR}(Z)$, the phase-space integrals can be done exactly, yielding

$$\omega_{0\nu,\eta=0}^{\rm GW}=f_{xx}^{\rm GW}\left[\frac{\langle m^{\rm Maj}\rangle_\nu}{m_e}\right]^2|X|^2 \qquad (8)$$

with

$$f_{xx}^{\rm GW}=\frac{G^4}{16\pi^5}\left[\mathcal{F}^{\rm PR}(Z)\right]^2 m_e^7\times\left[\frac{\tilde{T}_0^5}{15}+\frac{2\tilde{T}_0^4}{3}+\frac{8\tilde{T}_0^3}{3}+4\tilde{T}_0^2+2\tilde{T}_0\right] .$$

(We use the superscript GW as the first to derive the 0ν rate in this manner appear to be Grueling and Whitten.[15] Their $\eta=0$ result, however, is a factor of 2 too large becasue of their omission of the statistical factor for identical particles in the final state.) In addition, the radial dependence of the matrix elements is approximated as

$$\frac{g(r_{ij}\xi_0)}{r_{ij}}\cong\frac{1}{r_{ij}}\cong\frac{1}{R_0} . \qquad (9)$$

As M_1'' then becomes proportional to $M_{\rm F}$, the isospin-forbidden matrix element, it is ignored, leading to the simple result

$$\omega_{0\nu}^{\rm Approx}=f_{xx}^{\rm GW}\frac{F_A^4}{R_0^2}\left[\frac{\langle m^{\rm Maj}\rangle_\nu}{m_e}\right]^2|M_{\rm GT}|^2 . \qquad (10)$$

We now consider the Te decay ratio. This ratio is of interest because of the differing nuclear energy releases in ^{128}Te and ^{130}Te, $\tilde{T}_0=1.701$ and 4.957, respectively. As the leading-order contribution to $\omega_{0\nu,\eta=0}^{\rm GW}$ varies as $\tilde{T}_0^5$, while that for $\omega_{2\nu}^{\rm PR}$ varies as $\tilde{T}_0^{11}$, the ratio of these decay rates may depend sensitively on the mechanism for $\beta\beta$ decay. In particular, the ^{128}Te decay rate could be affected appreciably by 0ν decay proceeding via small Majorana mass terms $\langle m^{\rm Maj}\rangle_\nu/m_e$.

The half-life for the decay of ^{130}Te has been determined from the concentrations of ^{130}Xe in telluride ore samples by Inghram and Reynolds,[16] by Takaoka and Ogata,[17] by Kirsten *et al.*,[18] by Alexander *et al.*,[19] and by Srinivasan *et al.*[20] The more recent geochemical measurements[18–20] yield values in the range $(2.03-3.09)\times10^{21}$ yr. The half-life ratio $\tau_{1/2}^{128}/\tau_{1/2}^{130}=(1.59\pm0.05)\times10^3$ has been determined by Hennecke *et al.*[21] from the ratios of ^{130}Xe/^{132}Xe and ^{128}Xe/^{132}Xe in old ore; an earlier, smaller value for this ratio is given in Ref. 17, indicating that background ^{128}Xe may have contributed in that measurement.

If we assume that the nuclear matrix elements mediating 2ν $\beta\beta$ decay are equal for ^{128}Te and ^{130}Te, then Eq. (2) yields $(\tau^{128}/\tau^{130})_{2\nu}=5.07\times10^3$. The corresponding ratio employing the approximate result of Eq. (4) is $(\tau^{128}/\tau^{130})_{2\nu}^{\rm approx}=5.69\times10^3$. Likewise, the ratios for 0ν $\beta\beta$ decay are, with the same assumption, $(\tau^{128}/\tau^{130})_{0\nu,\eta=0}=25.0$ and 29.6, using Eq. (7) and the approximate result of Eq. (10), respectively. As these values bracket the result of Hennecke *et al.*, the discrepancy between the 2ν and experimental ratios may indicate the presence of 0ν decay.

The reliability of this conclusion, that a lepton-number-violating mechanism is contributing to these decays, depends on the correctness of the Hennecke *et al.* measurement and on the assumption of equal matrix elements for ^{128}Te and ^{130}Te. One expects the geochemical result to be accurate as uncertainties in the ore age and some effects of Xe diffusion will cancel in the ratio. However, one point we find worrisome in the results of Ref. 21: the Hennecke *et al.* half-lives for ^{130}Te, 0.89×10^{21} and 1.05×10^{21} yr, are considerably outside the range $(2.03-3.09)\times 10^{21}$ yr of other recent measurements. Absolute half-lives are determined from the ratio of ^{130}Xe to ^{132}Xe, which is not radiogenic, and thus some part of this discrepancy could be attributed to an error in dating the Hennecke *et al.* ore samples, which would then not affect the ^{128}Xe/^{130}Xe ratio. Yet it should be noted that the apparent difference in the ^{130}Xe/^{132}Xe ratio from other recent measurements is similar to the anomaly ^{128}Xe/^{130}Xe on which the argument for lepton-number violation rests.

If we accept the result of Hennecke *et al.*, the conclusion that lepton number is violated follows only after assuming the matrix elements for ^{128}Te and ^{130}Te are roughly equal. The support for this assumption, as cited in the work of Bryman and Picciotto,[8] Doi *et al.*,[9] and Minkowksi,[10] is the nuclear-matrix-element calculation of Vergados,[7] in which

$$\left|\frac{M_{\rm GT}(128)}{M_{\rm GT}(130)}\right| = \frac{0.284}{0.248} = 1.15 \cong 1 \ .$$

(A value of 1.78 would yield a ratio of 2ν-$\beta\beta$-decay rates in agreement with experiment.) It has been noted that the Vergados matrix element for ^{130}Te, whose decay proceeds predominantly by the 2ν mechanism for any reasonable $\langle m^{\rm Maj}\rangle_\nu$, yields $\tau_{1/2}^{2\nu,\rm PR} = 2.73\times 10^{21}$ yr, in good agreement with the general geochemical range. This has been construed as circumstantial evidence for the validity of the Vergados nuclear-structure approximations.

However, one must consider carefully the implications of the approximate treatment of phase space in Eq. (4). We illustrate this point in Table I, where $f_{\rm GT}$ and $f_{\rm GT}^{\rm PR}$ are compared for a variety of $\beta\beta$-decay nuclei and where $|M_{\rm GT}|$ is extracted directly from experiment under the assumption of purely GT 2ν $\beta\beta$ decay [$|M|_F \underset{\rm set}{=} 0$ in Eq. (2)]. Noting that $f_{\rm GT}^{\rm PR}$ underestimates the 2ν phase space by a factor of 5 (see Appendix), we find that the upper bound on $|M_{\rm GT}(130)|$ of 0.104−0.129 can be obtained from the recent geochemical estimates of $\tau_{1/2}^{130}$. Thus the Vergados matrix element underestimates $\tau_{1/2}^{130}$ by a factor of at least 3.7−5.7. Effectively, if one uses the Vergados calculation to argue in favor of lepton-number violation in the decay of ^{128}Te, one chooses to exacerbate an existing discrepancy in the theoretical rates in order to reproduce a ratio of rates. (That is, the Vergados estimate of $\tau_{1/2}^{128}$, roughly correct in the limit of lepton-number conservation, also becomes too small once the 0ν mechanism is introduced to improve the ratio of rates.) On more general grounds, one also sees that a correct treatment of phase space shows, directly from experiment, that $|M_{\rm GT}(130)|$ could be nearly a factor of 2 smaller than the celebrated suppressed matrix element of ^{48}Ca.[22] Certainly with such suppression, it is not obvious that the matrix element ratio will be unity.

One possible escape from this predicament is to find a consistent theoretical description of both the absolute rates and the ratio of rates. Clearly there are fundamental difficulties with the Vergados shell-model approximation. The neutron holes are restricted to the $h_{11/2}$ subshell in *both* initial and

TABLE I. The approximate and full phase-space factors $f_{\rm GT}^{\rm PR}$ and $f_{\rm GT}$ for 2ν $\beta\beta$ decay are compared. Average excitation energies $\langle E_{\rm GT}^N\rangle$, calculated in a statistical model (Refs. 6 and 12), and our shell-model results for the double Gamow-Teller matrix elements $|M_{\rm GT}|_{\rm th}$ are also shown. In the last column maximum experimental values for $|M_{\rm GT}|$ are derived by dividing total $\beta\beta$-decay rates by $f_{\rm GT}$.

	$f_{\rm GT}^{\rm PR}$ (sec^{-1})	$f_{\rm GT}$ (sec^{-1})	$\langle E_{\rm GT}^N\rangle$ (MeV)	$\lvert M_{\rm GT}\rvert_{\rm th}$	$\lvert M_{\rm GT}\rvert_{\rm exp}$
^{130}Te	1.31×10^{-28}	6.48×10^{-28}	11.50	1.483	0.104−0.129 (Refs. 18, 29, 20)
^{128}Te	2.34×10^{-32}	1.29×10^{-31}	11.59	1.474	0.185−0.230 (Ref. 21)
^{82}Se	5.82×10^{-28}	1.08×10^{-27}	8.07	0.938	1.430 (Ref. 27)
					0.272 (Ref. 28)
^{76}Ge	2.23×10^{-29}	3.67×10^{-29}	7.88	1.278	
^{48}Ca	1.45×10^{-26}	1.65×10^{-26}	5.07	0.222	<0.193 (Ref. 22)

final states, so that the *only two-body density matrix elements* which contribute in his first-order calculation are of the form $(h_{11/2})^{-2}J;n \rightarrow (h_{11/2})^2 J;p$. Yet the $(h_{11/2})^2$ configuration is expected to be a minor proton configuration, the $g_{7/2}$ and $d_{5/2}$ subshells being primarily occupied. *A priori* one might expect contributions of the form $(d_{3/2})^{-2}J;n \rightarrow (d_{5/2})^2 J;p$ connecting a minor neutron configuration to a major proton component to be of similar importance. More generally, to properly estimate $M_{\rm GT}$, one should insist that all Pauli-allowed configurations which can be reached by operating on the initial state with the double Gamow-Teller operator be included in the final wave function.

We have performed such a calculation, allowing both proton and neutron holes to range throughout the model space $(2d_{5/2}, 1g_{7/2}, 3s_{1/2}, 2d_{3/2}, 1h_{11/2})$. (We considered extending the model space beyond the magic numbers 50 and 82 in order that remaining spin partners $1h_{9/2}$ and $1g_{9/2}$ could be treated. Inclusion of these subshells, however, introduces wave-function spuriosity which could have serious effects as our effective interaction is not translationally invariant.) Details of these calculations, in which we use a potential derived by Baldridge and Vary from the Kuo bare G matrix,[23] are given in Ref. 12. Although we employ weak coupling, we do not take the extreme approach of Vergados, where the ground state is the simple product of the lowest neutron and proton state. Instead we take the lowest 50 proton and 50 neutron states from the full, identical-particle shell model calculations, form a weak-coupling basis of definite spin and parity by combining these states in all allowable combinations, and diagonalize the proton-neutron interaction in this basis. The two-body density matrix is then calculated and the various $\beta\beta$-decay matrix elements evaluated.

The results exacerbate the discrepancy between theory and experiment in the absolute rates. The new matrix elements $|M_{\rm GT}|$ (see Table I) are 5–6 times *larger* than those calculated by Vergados due to a coherent addition of many components of the density matrix. The ratio $|M_{\rm GT}(128)/M_{\rm GT}(130)|$ is unity to better than 1%. Again, then, this provides theoretical support for matrix element equality only if one is willing to ignore an alarming problem with the theoretical rates. (We believe at least two points should be studied before this improved theoretical treatment is accepted at face value. We are concerned, in view of recent experimental evidence for considerable collectivity in Gamow-Teller strengths,[24] that the approximate treatment of the pn interaction implicit in the weak-coupling approximation may be inadequate. We are also concerned about many-body modifications of the β-decay vertices. The quenching of F_A via Δ-hole excitations has been discussed frequently in the context of β-decay systematics of heavy nuclei.[25])

If both the result of Hennecke *et al.* and the assumption of equal matrix elements are accepted, the Te ratio indicates lepton-number violation. If the γ_5 invariance of the weak current is broken by a Majorana mass, rather than by a right-handed coupling of the Majorana neutrino, then a calculation of the 0ν mechanism of Fig. 1(b) will place an upper limit on $\langle m^{\rm Maj}\rangle_\nu$. (In the limit where only a single Majorana mass eigenstate contributes to ψ_{ν_e}, $\langle m^{\rm Maj}\rangle_\nu = C^2 m^{\rm Maj}$, with C the mixing amplitude and $m^{\rm Maj}$ the mass eigenvalue. See the note in Ref. 14 for the constraints $\beta\beta$ decay can place on more general mass matrices.) If, in addition, we assume that no other mechanism for lepton violation contributes (no superweak decay, no 0ν $\beta\beta$ decay with light scalar particles, etc.), that upper bound is the neutrino mass. We now consider the calculation of that mass.

In the conventional approach, we find from Eqs. (4) and (10),

$$\frac{\tau_{1/2}^{128}}{\tau_{1/2}^{130}} = \frac{\omega_{2\nu}^{\rm PR}(130)+\omega_{0\nu}^{\rm approx}(130)}{\omega_{2\nu}^{\rm PR}(128)+\omega_{0\nu}^{\rm approx}(128)} = \frac{f_{\rm GT}^{\rm PR}(130)+f_{xx}^{\rm GW}(130)F_A{}^4/R_0{}^2(130)(\langle m^{\rm Maj}\rangle_\nu/m_e)^2}{f_{\rm GT}^{\rm PR}(128)+f_{xx}^{\rm GW}(128)F_A{}^4/R_0{}^2(128)(\langle m^{\rm Maj}\rangle_\nu/m_e)^2}\,, \tag{11}$$

where we have assumed $|M_{\rm GT}(128)| = |M_{\rm GT}(130)|$. If we take $F_A=-1.25$, $R_0=1.2A^{1/3}f$, and $\tau_{1/2}^{128}/\tau_{1/2}^{130}=(1.59\pm0.05)\times10^3$, we find

$$\langle m^{\rm Maj}\rangle_\nu^{\rm approx}=(21.6\pm0.4)\ {\rm eV} \tag{12}$$

with the error reflecting only the uncertainty in the geochemical ratio. Note that in the work of Doi *et al.*, the GT matrix element ratio and $\langle E_N \rangle$ are taken from Vergados and $g(r_{ij})/r_{ij} \cong 0.6/R_0$, leading to $\langle m^{\rm Maj} \rangle_\nu^{\rm Approx} = 30$ eV, a value with which we agree given these approximations.

Now we consider the effect of using the more correct expressions of Eqs. (2) and (7). We find for equal GT matrix elements,

$$\frac{t_{1/2}^{128}}{t_{1/2}^{130}} = \frac{f_{\rm GT}(130) + f_{xx}(130) F_A{}^4/\tilde{R}_0{}^2(130)[1-\kappa(130)]^2(\langle m^{\rm Maj}\rangle_\nu/m_e)^2}{f_{\rm GT}(128) + f_{xx}(128) F_A{}^4/\tilde{R}_0{}^2(128)[1-\kappa(128)]^2(\langle m^{\rm Maj}\rangle_\nu/m_e)^2}$$

with

$$\tilde{R}_0 = |M_{\rm GT}| / |M_2''| \ , \qquad \kappa = F_1{}^2 M_1'' / F_A{}^2 M_2'' \ . \tag{13}$$

This differs from Eq. (11) to the extent that κ and $\tilde{R}_0$ differ from 0 and R_0, respectively, and due to the more careful treatment of phase space, though the substantial phase-space effects discussed earlier largely cancel in taking the ratio. In Table II we show the numerical results for κ and $\tilde{R}_0$ for the Te isotopes and for several $\beta\beta$-decay nuclei we have discussed elsewhere.[6,12] As κ is not negligible, we see that the factorization of the radial dependence from M_1'', yielding $\langle 1/r \rangle$ times the isospin-forbidden matrix element M_F, is unjustified.

However, the difference in various earlier estimates of $\tilde{R}_0$ account primarily for the broad range of $\langle m^{\rm Maj} \rangle_\nu$ values extracted from the Te ratio. Doi *et al.*[9] adopted the Primakoff-Rosen approximation that $\langle 1/r_{ij} \rangle \approx 1/R_0$ and then estimated that proper account of $g(r_{ij})$ at such large distance weakened this matrix element by a factor of 0.6, yielding $\tilde{R}_0^{\rm Doi} = 1.67\, R_0$. By explicit calculation we find that, for independent-particle wave functions and with $g(r_{ij}) = 1$, $\tilde{R}_0 = 0.37\, R_0$, which is not surprising as $1/r_{ij}$ preferentially weights smaller separations. If we then include the effects of $g(r_{ij})$ on the matrix element, $\tilde{R}_0$ becomes $0.47\, R_0$. [Of course, this does not imply that contributions at larger separations are enhanced. On the contrary, $g(r_{ij})$ reduces those contributions, but we absorb some average reduction factor for the numerator $g(r_{ij})$ into an effective $\tilde{R}_0$.] The effect of $g(r_{ij})$ is smaller than that estimated by Doi *et al.* because $g(r_{ij})$ is closer to unity for $r_{ij} \sim \tilde{R}_0 < R_0$. Finally, we correct for the absence of short-ranged correlations in our shell-model wave functions by modifying the two-body densities

$$\psi_{1,2}(\vec{r}_i,\vec{r}_j) \to \left[1-\beta(r_{ij})\right] \psi_{1,2}(\vec{r}_i,\vec{r}_j) \ , \qquad \beta(r) = e^{-ar^2}(1-br^2) \ , \tag{14}$$

with $a = 1.1$ fm^{-2} and $b = 0.68$ fm^{-2}.[26] This yields $\tilde{R}_0 = 0.59\, R_0$, which then is used in Eq. (13). We note that Minkowski[10] has recently advocated using a much smaller $\tilde{R}_0$, reducing the Doi *et al.* mass estimate by a factor of 13 to 20. Minkowski maintains that $\tilde{R}_0$ must be so drastically scaled to account for the $1/r_{ij}$ weighting of small separations and the loss of pair-spin correlations for $r_{ij} \geq 1$ fm. The explicit calculations we have performed do not support such drastic renormalizations of the Primakoff-Rosen estimate. Clearly long-range components of the nuclear force also in-

TABLE II. The matrix-element ratios $\kappa = F_1{}^2 M_1''/F_A{}^2 M_2''$ and $\tilde{R}_0 = |M_{\rm GT}/M_2''|$ resulting from shell-model calculations. We take $F_1 = 1$ and $F_A = -1.25$, and we give $\tilde{R}_0$ in units of $R_0 = 1.2A^{1/3}$. The third and fourth columns give values for $\tilde{R}_0$ with $g = 1$ and with the full g of Eq. (6) employed in M_2''. The fifth column shows the effects of modifying the shell-model two-particle densities by the correlation function of Eq. (14). It is this value that is employed in the present calculations.

	κ	$\tilde{R}_0 \left[\langle \frac{1}{r} \rangle\right]$	$\tilde{R}_0 \left[\langle \frac{g}{r} \rangle\right]$	$\tilde{R}_0 \left[\langle \frac{g}{r} \rangle_{\rm corr}\right]$
^{130}Te	−0.23	0.37	0.47	0.59
^{128}Te	−0.23	0.38	0.47	0.59
^{82}Se	−0.18	0.36	0.43	0.54
^{76}Ge	−0.20	0.41	0.49	0.61
^{48}Ca	−0.15	0.26	0.29	0.44

duce strong spin-singlet correlations in $T=1$ pairs.

Our numerical values for κ and $\tilde{R}_0$ then yield

$$\langle m^{\text{Maj}} \rangle_\nu = 10.1 \pm 0.2 \text{ eV} \tag{15}$$

with the error again reflecting only the uncertainty in the geochemical ratio. This can be compared with the mass limit obtained from laboratory bounds on the neutrinoless decay of ^{76}Ge and ^{82}Se, $\langle m^{\text{Maj}} \rangle_\nu \lesssim 15$ eV.[6] [If the prescription outlined here is followed precisely in calculating the 0ν decay rates of these nuclei, the values of Ref. 6 are changed slightly to $\langle m^{\text{Maj}} \rangle_\nu \lesssim 17$ eV and $\lesssim 13$ eV, respectively, for ^{76}Ge and ^{82}Se.] Previously, noting that the Doi *et al.*[9] estimate of $\langle m^{\text{Maj}} \rangle_\nu$ was in apparent conflict with these limits, Georgi, Glashow, and Nussinov[5] had pointed out that masses derived from the Te ratio cannot be compared directly to the results of laboratory searches for the 0ν mechanism of Fig. 1(b). In particular, in the context of a model suggested by Gelmini and Roncadelli,[5] in which neutrino masses are generated by virtue of the spontaneous breakdown of global $B-L$ symmetry with the appearance of a massless Goldstone particle, they discuss a second 0ν-$\beta\beta$-decay mode in which the emitted electrons are accompanied by a very light boson. This process would contribute to the inclusive test of lepton-number violation provided by the Te ratio, but would not be seen in laboratory searches for the two-electron mode. Thus, if the Doi result and the two-electron mass limits are accepted at face value, this might suggest the existence of such additional 0ν modes. Our result of $\langle m^{\text{Maj}} \rangle_\nu \simeq 10$ eV no longer requires any such adjustment to accommodate the two-electron mode mass limit. We also recall that the derivation of the 15 eV mass limit, which requires a calculation of the matrix element X, was made plausible by the agreement of the calculated 2ν rate with the laboratory measurement of Moe and Lowenthal.[27] However, that laboratory rate disagrees with the geochemical rate.[28] As the relative constancy of κ and $\tilde{R}_0$ in Table II indicates a rough scaling of 2ν and 0ν matrix elements, a more conservative mass limit could be obtained by renormalizing all matrix elements so that the geochemical result is reproduced. This would yield $\langle m^{\text{Maj}} \rangle_\nu \lesssim 52$ eV. We believe a second, more definitive laboratory measurement of the ^{82}Se 2ν decay rate is extremely important in clarifying whether systematic difficulties exist in theoretical half-life estimates.

In summary, we have argued that the evidence for lepton-number violation in the Te $\beta\beta$-decay ratio may not be as strong as commonly believed. As a proper treatment of phase space demonstrates that the 2ν matrix elements for ^{128}Te and ^{130}Te must be suppressed to account for the absolute geochemical rates, the argument that the matrix element ratio is near unity is not obvious *a priori*. While the present careful theoretical calculations of these matrix elements do support this argument, an alarming discrepancy in predicting the absolute geochemical rates remains troublesome. Stipulating that the matrix element ratio is unity, we have faithfully treated the radial dependence of the 0ν operators in order to determine the scaling of 0ν and 2ν matrix elements. This yields $\langle m^{\text{Maj}} \rangle_\nu \cong 10$ eV if we assume no explicit right-handed coupling of Majorana neutrinos to electrons; this mass is not inconsistent with the limits derived from laboratory searches for the two-electron mode. We emphasize again that the reliability of this result is difficult to assess until the origin of the discrepancy between theoretical and geochemical estimates of absolute rates is uncovered.

We thank T. Goldman, P. Herczeg, T. Kotani, and S. P. Rosen for helpful discussions. We also thank T. Kotani for his careful reading of the manuscript. This work was partially supported by the U. S. Department of Energy and the National Science Foundation (Grant No. PHY-8021272).

APPENDIX

We briefly compare the relativistic and nonrelativistic treatments of the Coulomb functions $\mathscr{F}(Z,\epsilon)$ which correct the electron plane waves for distortion due to the nuclear charge.

The nonrelativistic point Coulomb correction is derived by taking the square of the ratio of the Schrödinger scattering solution for a point charge Z to a plane wave, evaluated at the origin. For an electron,

$$\mathscr{F}^{\text{NR}}(Z,\epsilon) = \frac{2\pi\eta}{1-e^{-2\pi\eta}} = e^{\pi\eta} \, | \Gamma(1+i\eta) |^2 \,, \tag{A1}$$

with $\eta = \alpha Z/\beta$ and $\beta = k/\epsilon$, $k = |\vec{k}|$ the electron momentum. In much of the $\beta\beta$-decay literature this formula has been further simplified by setting $\beta = 1$ in the exponent, yielding Eq. (3), as this permits analytic integration over the $\beta\beta$-decay phase space.

In the present work we take $\mathcal{F}(Z,\epsilon)$ from Ref. 13, where the Coulomb correction is derived by numerical solution of the Dirac equation for an extended nuclear charge. We can write this correction factor as

$$\mathcal{F}(Z,\epsilon)=\mathcal{F}^R(Z,\epsilon)L_0 \tag{A2}$$

with $\mathcal{F}^R(Z,\epsilon)$ the square of the ratio of the Dirac scattering solution for a point charge Z to a plane wave, evaluated at the nuclear surface (the relativistic density for a point nucleus is infinite at the origin). The numerical factor L_0, generally small, takes into account the finite charge distribution and screening corrections, while

$$\mathcal{F}^R(Z,\epsilon)=2(1+\gamma)(2kR_0)^{2(\gamma-1)} \times e^{\pi\eta}\left|\frac{\Gamma(\gamma+i\eta)}{\Gamma(2\gamma+1)}\right|^2, \tag{A3}$$

with $\gamma=[1-(\alpha Z)^2]^{1/2}$ and R_0 the nuclear radius. Thus

$$\frac{\mathcal{F}^{\rm NR}(Z,\epsilon)}{\mathcal{F}(Z,\epsilon)}\cong\frac{\mathcal{F}^{\rm NR}(Z,\epsilon)}{\mathcal{F}^R(Z,\epsilon)} =\frac{\Gamma(3+2\xi)^2}{2(2+\xi)}\frac{1}{(2kR_0)^{2\xi}} \times\left|\frac{\Gamma(1+i\eta)}{\Gamma(1+\xi+i\eta)}\right|^2, \tag{A4}$$

with each of the three terms in this expression approaching unity in the limit of small Z, $\xi=\gamma-1\to 0$.

To illustrate the behavior of this ratio we expand the first and third terms in Eq. (A4) to $O(\xi)$ [note $\xi(^{130}\text{Xe})=-0.081$]. We find

$$\frac{\Gamma(3+2\xi)^2}{2(2+\xi)}=1+3.191\xi+O(\xi^2)\,,$$

$$\frac{\Gamma(1+i\eta)}{\Gamma(1+i\eta+\xi)}=1-\xi\psi^0(1+i\eta)+O(\xi^2)\,.$$

The digamma function ψ^0 is tabulated in Ref. 29. Using $R_0=1.2A^{1/3}f$, we obtain for ^{130}Xe with $k/m_e=0.2$

$$\frac{\mathcal{F}^{\rm NR}(Z,\epsilon)}{\mathcal{F}^R(Z,\epsilon)}\cong(0.742)(0.440)(1.131) =0.369$$

and with $k/m_e=3.0$

$$\frac{\mathcal{F}^{\rm NR}(Z,\epsilon)}{\mathcal{F}^R(Z,\epsilon)}\cong(0.742)(0.682)(0.939) =0.475\,.$$

Such factors generate the factor-of-5 enhancement in the ^{130}Te $\beta\beta$-decay phase space which we discussed in the main text.

---

[1]For general discussions, see H. Primakoff and S. P. Rosen, Rep. Prog. Phys. 22, 121 (1950); E. J. Konopinski, *Theory of Beta Radioactivity* (Oxford University Press, London, 1966); talks presented by C. S. Wu, S. P. Rosen, and G. J. Stephenson, Jr., in *Proceedings of the Neutrino Mass Miniconference, Telemark, Wisconsin, 1980*, edited by V. Barger and D. Cline (University of Wisconsin, Madison, 1981); S. P. Rosen, talk presented at Neutrino 81, proceedings of the International Conference on Neutrino Physics and Astrophysics, Maui, Hawaii, edited by R. J. Cence, E. Ma, and A. Roberts (University of Hawaii High Energy Physics Group, Honolulu, 1981); Nuovo Cimento 2, 1 (1971); Ref. 8.

[2]A. Halprin. P. Minkowski, H. Primakoff, and S. P. Rosen, Phys. Rev. D 13, 2567 (1976).

[3]B. Pontecorvo, Phys. Lett. 26B, 630 (1968).

[4]L. Wolfenstein, Phys. Rev. Lett. 13, 562 (1964).

[5]H. M. Georgi, S. L. Glashow, and S. Nussinov, Harvard University Report No. HUTP-81/A026, 1981 (unpublished); G. B. Gelmini and M. Roncadelli, Phys. Lett. 99B, 411 (1981).

[6]W. C. Haxton, G. J. Stephenson Jr., and D. Strottman, Phys. Rev. Lett. 46, 698 (1981); see also Ref. 9.

[7]J. D. Vergados, Phys. Rev. C 12, 865 (1976). Vergados attempted to subtract the 2ν rate from the geochemical rate prior to extracting a lepton-number-violation parameter. Our observation later in this paper on the size of the Vergados rates when the 2ν phase space is properly treated indicates that this procedure will lead to inconsistencies.

[8]D. Bryman and C. Picciotto, Rev. Mod. Phys. 50, 11 (1978).

[9]M. Doi, T. Kotani, H. Nishiura, K. Okuda, and E. Takasugi, Phys. Lett. 102B, 219 (1981).

[10]P. Minkowski, Univ. of Bern Report No. BUTP-15, 1981 (unpublished).

[11]C. Picciotto, Can. J. Phys. 56, 399 (1978).

[12]W. C. Haxton, G. J. Stephenson, Jr., and D. Strottman (in preparation).

[13]H. Behrens and J. Jännecke, *Numerical Tables for Beta-Decay and Electron Capture* (Springer, Berlin,

1969).

[4]We stated in Ref. 6 that $\langle m^{\rm Maj}\rangle_\nu = \sum_i C_i^2 m_i^{\rm Maj}$, with C_i^2 the probability of the electron neutrino being in the eigenstate with Majorana mass $m_i^{\rm Maj}$, $\sum_i C_i^2 \leq 1$. Lincoln Wolfenstein has recently pointed out that this generalization is not correct in *CP*-invariant theories in which two (or more) massive Majorana neutrinos having opposite *CP* values couple to the electron [Carnegie-Mellon University Report No. COO-30550180 (unpublished)]. In this case $\langle m^{\rm Maj}\rangle_\nu = \sum_i C_i^2 N_i m_i^{\rm Maj}$, with $N_i = \pm 1$ the appropriate *CP* eigenvalue. Thus, for two neutrinos, it may be the weighted mass difference that is constrained by $\beta\beta$-decay. In particular, the case $C_1^2 = C_2^2$ and $N_1 = -N_2$ corresponds to the limit of the two Majorana neutrinos merging to form a Dirac neutrino, with the $\beta\beta$-decay rate then vanishing. M. Doi *et al.* [Phys. Lett. 102B, 323 (1981)] have considered *CP*-noninvariant theories in which the $\beta\beta$-decay mass then becomes $\sum_i C_i C_i m_i^{\rm Maj}$, with the C_i complex, though their interpretation of this result should be tempered by Wolfenstein's observations.

[15]E. Grueling and R. Whitten, Ann. Phys. (N.Y.) 11, 510 (1960).

[16]M. G. Inghram and J. H. Reynolds, Phys. Rev. 76, 1265 (1949); 78, 822 (1950).

[17]N. Takaoka and K. Ogata, Z. Naturforsch. 21a, 84 (1966).

[18]T. Kirsten, W. Gentner, and O. A. Schaeffer, Z. Phys. 202, 273 (1967); T. Kirsten, W. Gentner, and O. Müller, Z. Naturforsch. 22a, 1703 (1967); T. Kirsten, A. O. Schaeffer, E. Norton, and R. W. Stoenner, Phys. Rev. Lett. 20, 1300 (1968).

[19]E. C. Alexander, Jr., B. Srinivasan, and O. K. Manuel, Earth Planet. Sci. Lett. 5, 478 (1969).

[20]B. Srinivasan, E. C. Alexander, Jr., and O. K. Manuel, Econ. Geol. 67, 592 (1972); J. Inorg. Nucl. Chem. 34, 2381 (1972).

[21]E. W. Hennecke, O. K. Manuel, and D. D. Sabu, Phys. Rev. C 11, 1378 (1975).

[22]R. K. Bardin, P. J. Gollon, J. D. Ullman, and C. S. Wu, Nucl. Phys. A158, 337 (1970); V. A. Khodel, Phys. Lett. B32, 503 (1970).

[23]T. T. S. Kuo and G. E. Brown, Nucl. Phys. A114, 241 (1968); W. T. Baldridge and J. P. Vary, Phys. Rev. C 14, 2246 (1976).

[24]D. E. Bainum *et al.*, Phys. Rev. Lett. 44, 1751 (1980); C. D. Goodman *et al.*, *ibid.* 44, 1755 (1980); B. D. Anderson *et al.*, *ibid.* 45, 699 (1980); D. J. Horen *et al.*, Phys. Lett. 95B, 27 (1980).

[25]See, e.g., *Mesons in Nuclei,* edited by M. Rho and D. H. Wilkinson (North-Holland, Amsterdam, 1979); A. Bohr and B. R. Mottelson, Phys. Lett. 100B, 10 (1981).

[26]G. A. Miller and J. E. Spencer, Ann. Phys. (N.Y.) 100, 562 (1976).

[27]M. K. Moe and D. Lowenthal, Phys. Rev. C 22, 2186 (1980).

[28]B. Srinivasan, E. C. Alexander, Jr., R. D. Beaty, D. E. Sinclair, and O. K. Manuel, Econ. Geol. 68, 252 (1973).

[29]M. Abramowitz and I. A. Stegun, *Handbook of Mathematical Functions,* (Dover, New York, 1972).

PHYSICAL REVIEW VOLUME 184, NUMBER 5 25 AUGUST 1969

Nuclear Double-Beta Decay and a New Limit on Lepton Nonconservation*

H. PRIMAKOFF
Department of Physics, University of Pennsylvania, Philadelphia, Pennsylvania 19104

AND

S. P. ROSEN
Department of Physics, Purdue University, Lafayette, Indiana 47907
(Received 21 April 1969)

Nuclear double-β decay is discussed on the assumptions (1) that the process, treated as second-order weak, occurs predominantly without neutrino emission and so violates lepton convervation, and (2) that the matrix element for the process is mainly due to the no-neutrino double-β decay of an isospin-$\frac{3}{2}$ nucleon resonance present, albeit with a small ($\approx 1\%$) probability, in the parent or daughter nucleus. The theoretical lifetimes obtained on this basis are compared with the corresponding experimental lifetimes; the various lifetime ratios are correctly predicted, and a limit $\approx 10^{-4}$ is imposed on the relevant "lepton-non-conservation" parameter. A discussion is also given of the convergence problems associated with the matrix elements for second-order weak processes in general. In addition it is shown that in spite of the presence of virtual-neutrino closed loops, divergences never arise in the matrix elements for no-neutrino double-β decay if the basic nuclear constituents (i.e., quarks or nucleons) have isospin not exceeding $\frac{1}{2}$ and mean separation greater than zero.

1. INTRODUCTION

NOW that nuclear double-β decay has definitely been observed,[1] we can begin to reexamine some of the theoretical questions associated with the phenomenon. Nuclear double-β decay has always been anticipated as a second-order effect of the usual weak ($\Delta Q=\pm 1$, $\Delta S=0$) interaction which gives rise, in first order, to nuclear single-β decay; there is, however, no *a priori* reason to prevent it from being, in part and even predominantly, a first-order effect of some hitherto unrecognized "superweak" ($\Delta=\pm 2$, $\Delta S=0$) interaction.[2] According to the various known conservation laws, it is always possible for two neutrinos to be emitted in the nuclear double-β decay; however, it has yet to be determined whether there occur nuclear double-β events in which no neutrinos are emitted and where, as a consequence, lepton conservation is not valid. These problems are intimately related to the problem of the appropriate description of the free neutrino and of its coupling to other particles, and it is from this point of view that we wish to analyze the present experimental situation. We shall also discuss questions concerning the convergence of second-order weak matrix elements arising in nuclear double-β decay and in certain related processes.

The experiments in which nuclear double-β decay was detected involved mass-spectrometric analyses of tellurium ores of known age.[1] Xenon occluded in the ores was found to contain a relative abundance of Xe^{130} far in excess of that contained in atmospheric xenon. A careful analysis showed that the excess of Xe^{130} could have resulted only from the double-β decay of Te^{130}, and the half-life of Te^{130} was deduced to be $10^{21.34\pm0.12}$ years.[1] In view of the ensuing discussion, it should also be noted that no excess of Xe^{128} was found to accompany the excess of Xe^{130}, and that as a consequence, the half-life for the double-β decay of Te^{128} could be estimated as greater than $10^{23.3}$ years.[1]

Because it is possible to detect only the daughter nucleus by the mass-spectrometric method, no direct conclusion can be drawn from the tellurium ore experiments about the presence or absence of neutrinos in the double-β decay final state. The measured half-life of Te^{130} is, however, in good agreement with a theoretical estimate of $10^{22.5\pm2.0}$ years obtained on the assumption that the nuclear double-β decay is a second-order weak process in which two neutrinos are always emitted so that lepton conservation holds.[3] Thus the measured half-life of Te^{130} seems to provide evidence against the occurrence of any lepton-nonconserving no-neutrino double-β decay, viewed either as a "superweak" or as a second-order weak process.

Pontecorvo has pointed out that this somewhat tentative conclusion may not be valid.[4] In one of the tellurium ore experiments,[5] an excess of Xe^{128} was found to accompany the excess of Xe^{130}, and if this excess of Xe^{128} is attributed to the double-β decay $Te^{128}\rightarrow Xe^{128}$, the corresponding Te^{128} half-life is $10^{22.5\pm0.5}$ years. Now if one makes the reasonable assumption that the nuclear matrix elements for $Te^{128}\rightarrow Xe^{128}$ and $Te^{130}\rightarrow$

* Supported in part by the National Science Foundation and the U. S. Atomic Energy Commission.

[1] T. Kirsten, O. A. Schaeffer, E. Norton, and R. W. Stoenner, Phys. Rev. Letters **20**, 1300 (1968); T. Kirsten, W. Gentner, and O. A. Schaeffer, Z. Physik **202**, 273 (1967); O. A. Schaeffer (private communication).

[2] R. G. Winter, Phys. Rev. **83**, 1070 (1951).

[3] S. P. Rosen and H. Primakoff, in *Alpha-, Beta-, and Gamma-Ray Spectroscopy*, edited by K. Siegbahn (North-Holland Publishing Co., Amsterdam, 1965), Vol. II, p. 1499; Rept. Progr. Phys. **22**, 121 (1959); Proc. Phys. Soc. (London) **78**, 464 (1961); E. J. Konopinski, in *Theory of Beta Radioactivity* (Oxford University Press, London, 1966), p. 245; E. Greuling and R. C. Whitten, Ann. Phys. (N. Y.) **11**, 510 (1960); T. D. Lee and C. S. Wu, Ann. Rev. Nucl. Sci. **15**, 381 (1965).

[4] B. Pontecorvo, Phys. Letters **26B**, 630 (1968).

[5] N. Takaoka and K. Ogata, Z. Naturforsch. **21a**, 84 (1966).

Xe^{130} are approximately equal, then these matrix elements cancel from the expression for the ratio of the corresponding double-β decay rates, and this ratio reduces to the ratio of the available phase spaces. For two-neutrino double-β decay, the phase space available to the (four) emitted leptons is roughly proportional to the eighth through 11th power of the energy release,[3] and, since the energy release for Te^{130} is three times that for Te^{128}, one would expect a lifetime ratio

$$T_{1/2}(\mathrm{Te}^{128})/T_{1/2}(\mathrm{Te}^{130})\cong 3^{8.4}=10^{4.0},$$

in contrast to the measured lifetime ratio

$$10^{22.5\pm0.5}/10^{21.34\pm0.12}=10^{1.2\pm0.6}.$$

One obvious explanation for this large discrepancy is that most of the excess Xe^{128} found in the experiments of Ref. 5 does not originate from double-β decay; in fact, the argument in favor of the double-β decay origin of this excess Xe^{128} is by no means as strong as the corresponding argument for the excess of Xe^{130},[1,5] particularly since no excess Xe^{128} is found in the experiments of Ref. 1. An alternative and more intriguing explanation noted by Pontecorvo[4] is based on the assumption that both Te^{128} and Te^{130} predominantly undergo no-neutrino double-β decay; in this case the phase space available to the (two) emitted leptons is roughly proportional to the fourth through fifth power of the energy release,[3] and one would estimate that

$$T_{1/2}(\mathrm{Te}^{128})/T_{1/2}(\mathrm{Te}^{130})\cong 3^{4.6}=10^{2.2},$$

in much better agreement with the measured lifetime ratio.

On the basis of this suggestion, Pontecorvo proposed that the decays of Te^{130} and Te^{128} are, predominantly, the first-order effect of a new superweak ($\Delta Q=\pm2$, $\Delta S=0$) interaction which mediates no-neutrino double-β decay, rather than the second-order effect of the usual weak ($\Delta Q=\pm1$, $\Delta S=0$) interaction. In fact, Pontecorvo drew a possible parallel between this superweak ($\Delta Q=\pm2$, $\Delta S=0$) interaction which violates lepton conservation and the Wolfenstein superweak ($\Delta Q=0$, $\Delta S=\pm2$) interaction which violates CP conservation.[6] Rather than considering further the implications of the Pontecorvo superweak interaction, we wish instead, in the present paper, to reexamine an alternative and much older point of view,[3] namely, that while the nuclear double-β decay is predominantly no-neutrino and therefore violates lepton conservation, it is nevertheless a second-order weak process. According to this view, the usual weak ($\Delta Q=\pm1$, $\Delta S=0$) interaction which gives rise in first order to nuclear single-β decay must also violate lepton conservation, though we expect that the violation is in some sense small (see below). The measured half-life of Te^{130} then enables us to set an upper limit on this violation.

[6] L. Wolfenstein, Phys. Rev. Letters **13**, 562 (1964).

In order to introduce a violation of lepton conservation in the usual weak ($\Delta Q=\pm1$, $\Delta S=0$) interaction Hamiltonian, we recall that when the neutrino mass $m_\nu=0$ and the lepton weak current L_λ is of pure "two-component" variety, i.e., when[7]

$$L_\lambda=\psi_e^\dagger\gamma_4\gamma_\lambda(1+\gamma_5)\psi_\nu, \quad (1)$$

the helicity projection operator $(1+\gamma_5)$ ensures that a neutrino emitted together with an electron by one hadron cannot be reabsorbed with simultaneous emission of a second electron by another hadron. Nuclear no-neutrino double-β decay is then forbidden and lepton conservation is valid to all orders in the weak ($\Delta Q=\pm1$, $\Delta S=0$) interaction, irrespective of whether ψ_ν describes a "Majorana" ($\psi_{\bar\nu}\equiv\bar\psi_\nu^\dagger=\psi_\nu$) or a "Dirac" ($\psi_{\bar\nu}\equiv\bar\psi_\nu^\dagger\neq\psi_\nu$) neutrino. If we now modify the lepton current of Eq. (1) by including a term with the helicity projection operator $(1-\gamma_5)$ and suppose that ψ_ν describes a "Majorana" neutrino, viz.,[3]

$$L_\lambda'=\psi_e^\dagger\gamma_4\gamma_\lambda[(1+\gamma_5)+\eta(1-\gamma_5)]\psi_\nu,\quad \bar\psi_\nu^\dagger=\psi_\nu \quad (2)$$

then a neutrino emitted together with an electron by one hadron can be reabsorbed with simultaneous emission of a second electron by another hadron with a probability amplitude proportional to the "lepton-nonconservation" parameter η.[8] It follows that the corresponding probability for nuclear no-neutrino double-β decay is proportional to η^2. The lepton current L_λ' parametrizes lepton nonconservation in a manner complementary to that of parity nonconservation, with $\eta=0$ corresponding to lepton conservation and maximal parity nonconservation, and $\eta=1$ corresponding to parity conservation and maximal lepton nonconservation. Thus for example, L_λ' predicts a longitudinal spin polarization for an electron emitted in a single-β decay, equal to $(v_{el}/c)(1-\eta^2)/(1+\eta^2)$, and this varies from v_{el}/c to 0 as η varies from 0 to 1.

The emission and reabsorption of the virtual neutrino in the lepton-nonconserving double-β decay of a nucleus corresponds to a sum of second-order weak diagrams, each with a virtual-neutrino closed loop (see Fig. 1), and so we might expect that the associated second-order weak matrix element will be divergent. In the standard

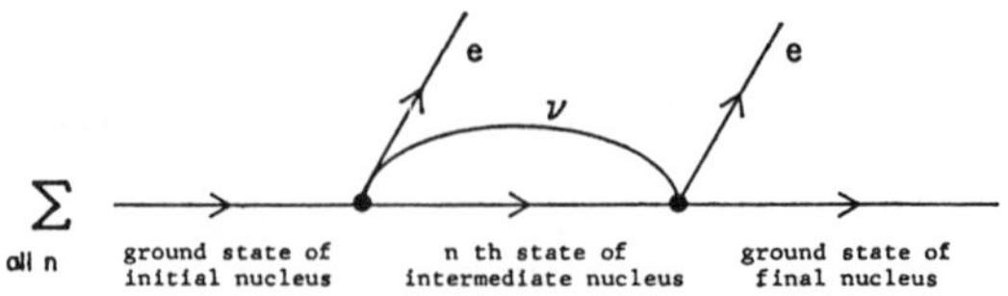

FIG. 1. Second-order weak diagrams for lepton-nonconserving double-β decay.

[7] We use the "Majorana" form of the γ_λ so that $\gamma_\lambda^*=\tilde\gamma_\lambda=\gamma_\lambda\times(1-2\delta_{\lambda4})$.

[8] We suppose in this paper that η is real: $\eta=\eta^*$. If η should differ from η^* the replacement of L_λ by L_λ' in the $\mathcal{H}_{\text{weak}}$ of Eq. (5) below will yield a weak interaction which not only violates lepton conservation but also violates CP conservation.

calculation,[3] however, the virtual neutrino is emitted together with an electron by one neutron in the nucleus, and is reabsorbed with simultaneous emission of a second electron by another neutron in the same nucleus. The reciprocal of the mean separation between the two neutrons then serves as an effective cutoff for the energy of the virtual neutrino, and the matrix element in question is actually finite. Underlying this result is the fact that the only charge states of the nucleon are zero and one; thus it is not possible for a single nucleon to emit and reabsorb the virtual neutrino with simultaneous emission of two electrons. In other words, there is a direct connection between (i) the convergence of the matrix element for lepton-nonconserving nuclear double-β decay, and (ii) the isospin and mean separation of the basic nuclear constituents, i.e., the nucleons in the standard picture: As long as this isospin does not exceed $\frac{1}{2}$ and the mean separation is greater than zero, the matrix element is finite.

This situation is to be contrasted with other second-order weak processes which can however all be considered in the limit of lepton conservation. Thus, in two-neutrino double-β decay, there are no virtual-neutrino closed loops, so that the associated matrix element is finite, and would in fact remain so even if the mean separation between the basic nuclear constiuents were taken as zero.[3] Further, in processes where an electron-positron pair is emitted, e.g., $O^{16*} \to O^{16} + e^- + e^+$, the virtual neutrino can be emitted and reabsorbed by the same basic constituent of the nuclues, i.e., the same nucleon in the standard picture. There is then, no mean separation available to provide a natural cutoff for the energy of the virtual neutrino in the second-order weak matrix element, and this matrix element diverges quadratically.

An immediate limit on the "lepton-nonconservation" parameter η of Eq. (2) can be obtained by comparing the measured half-life of Te^{130} with a theoretical estimate obtained on the basis of the standard two-nucleon mechanism; this yields the limit $\eta \approx 10^{-3}$ [see Eq. (27) below]. The theoretical lifetimes are inversely proportional to η^2, directly proportional to the square of the mean separation of a typical pair of nucleons in the parent or daughter nucleus $(\cong A^{1/3}/m_\pi)^2$, and directly proportional to the square of the reciprocal overlap between the wave functions of these nuclei; for the double-β decay $Te^{130} \to Xe^{130}$, each of these last two factors is quite large. It follows that if the expression for the theoretical lifetime can be reasonably modified in such a way as to reduce these last two factors, then a sharper limit on η can be set.

One way of achieving this modification is to take advantage of a paper by Kerman and Kisslinger, wherein it is argued rather convincingly that the deuteron contains about a 1% probability admixture of the nucleon resonance with $J^{(P)} = \frac{5}{2}^+$, $I = \frac{1}{2}$, and $m = 1688$ MeV, namely $N^*(1688)$.[9] Generalizing this to other nuclei and to other nucleon resonances, we contemplate the possibility that the various parent and daughter nuclei involved in double-β decay contain about 1% probability admixture of the nucleon resonance with $J^{(P)} = \frac{3}{2}^+$, $I = \frac{3}{2}$, and $m = 1236$ MeV [$N^*(1236)$]. Since the $N^*(1236)$ has isospin $\frac{3}{2}$ it can emit and reabsorb the virtual neutrino with simultaneous emission of two electrons, viz.,

$$N^{*-}(1236) \to p + e^- + e^-, \quad n \to N^{*++}(1236) + e^- + e^-. \quad (3)$$

Thus, if the processes of Eq. (3) contribute to nuclear double-β decay, the change of two units of charge within the nucleus can involve one particle instead of two, and the overlap between the wave functions of the parent and daughter nuclei is larger than in the standard two-nucleon mechanism. This situation will always occur if the nucleus contains particles with isospin greater than $\frac{1}{2}$. To avoid the divergence problem that will arise if the nucleon resonances and the nucleons are treated as "elementary," i.e., as point particles, we estimate the lifetimes of the processes of Eq. (3) on the assumption that each of these nucleon resonances and nucleons is a composite of three isospin-$\frac{1}{2}$ quarks. Then instead of the mean separation between nucleons in a nucleus, the lifetime estimate will contain the mean separation between quarks in a three-quark composite, and the latter mean separation is much smaller than the former. Taking all factors into account, we expect the corresponding limit on η to be an order of magnitude smaller than the limit obtained on the basis of the two-nucleon mechanism.

Certain aspects of the general convergence problem of second-order weak matrix elements are studied in the next section, while the limit on η obtained on the basis of the $N^*(1236)$ mechanism is described in Sec. 3. Section 4 summarizes our results.

2. CONVERGENCE OF SECOND-ORDER WEAK MATRIX ELEMENTS

As we have already indicated, the matrix elements for a second-order weak process between two hadron states in which a virtual neutrino is emitted and then reabsorbed may be divergent. To study this problem, we consider those processes in which a pair of charged leptons emerges, and we divide them into two clasess. In the first class, the lepton pair carries off zero charge ($\Delta Q = 0$; e.g., $O^{16*} \to O^{16} + e^- + e^+$, or $\pi^0 \to e^- + e^+$, or $K_L{}^0 \to \mu^- + \mu^+$), and in the second, the lepton pair carries off two units of charge ($\Delta Q = \pm 2$; e.g., no-neutrino double-β decay). As we have already noted, the matrix elements for the $\Delta Q = 0$ processes diverge, while the matrix elements for the $\Delta Q = \pm 2$ processes converge in any composite hadron model where the basic hadronic

[9] A. K. Kerman and L. S. Kisslinger, Phys. Rev. **180**, **1483** (1969); see also H. Arenhövel and M. Danos, Phys. Letters **28B**, 299 (1968); L. Kisslinger, Phys. Rev. **29B**, 211 (1969).

constituents have isospin not exceeding $\frac{1}{2}$ and mean separation greater than zero.

We begin by considering the second-order weak processes for which $\Delta Q=0$ and $\Delta S=0$, namely,

$$\alpha \begin{matrix} \nearrow \text{any } \gamma_n^{(+)}+\nu+e^- \searrow \\ \searrow \text{any } \gamma_n^{(-)}+\nu+e^+ \nearrow \end{matrix} \beta+e^-+e^+, \qquad (4)$$

where α, $\gamma_n^{(+)}$, $\gamma_n^{(-)}$, and β are hadrons; e.g., $\alpha=O^{16*}$, $\gamma_n^{(+)}=F^{16}$, $F^{16*}\cdots$, $\gamma_n^{(-)}=N^{16}$, $N^{16*}\cdots$, $\beta=O^{16}$ or $\alpha=\pi^0$, $\gamma_n^{(+)}=\pi^+$, $\rho^+\cdots$, $\gamma_n^{(-)}=\pi^-$, $\rho^-\cdots$, and β=vacuum. From a practical point of view, it is of course true that when strangeness is conserved the decay $\alpha\to\beta+e^-+e^+$ can be engendered by electromagnetic interactions alone, and that contributions to it arising from Eq. (4) (associated with, e.g., parity violation effects) will be extremely hard to detect. The process of Eq. (4) has no bearing on the matter of lepton conservation, and so we shall neglect the (in any case, small) parameter η of Eq. (2).

The current-current weak-interaction Hamiltonian density for strangeness-conserving semileptonic processes can be written

$$\mathcal{H}_{\text{weak}}(x)=(G/\sqrt{2})[L_\lambda(x)J_\lambda^{(+)}(x)+J_\lambda^{(-)}(x)L_\lambda^\dagger(x)], \qquad (5)$$

$$J_\lambda^{(-)}(x)=[J_\lambda^{(+)}(x)]^\dagger, \quad J_\lambda^{(\pm)}(x)=V_\lambda^{(\pm)}(x)+A_\lambda^{(\pm)}(x)$$

where G is the weak-coupling constant, $L_\lambda(x)$ is the lepton weak current given in Eq. (1), and $V_\lambda^{(\pm)}(x)$ and $A_\lambda^{(\pm)}(x)$ are polar and axial-vector hadron weak currents specified by CVC (conserved vector current), PCAC (partially conserved axial-vector current), and appropriate equal-time commutation relations. $\mathcal{H}_{\text{weak}}(x)$ is normally used to treat first-order weak processes but we shall use it here to treat second-order weak processes as well. The matrix element for the process of Eq. (4) is then given by

$$\mathfrak{M}(\alpha\to\beta+e^-+e^+)=\left(\frac{G}{\sqrt{2}}\right)^2\int\int d\mathbf{x}d\mathbf{y}\sum_{\mathbf{p}_\nu,S_\nu}\sum_{n;\mathbf{p}_n,S_n}\Bigg[\frac{\langle e^+|L_\mu^\dagger(\mathbf{x})|\nu\rangle\langle\beta|J_\mu^{(-)}(\mathbf{x})|\gamma_n^{(+)}\rangle\langle e^-\nu|L_\lambda(\mathbf{y})|0\rangle\langle\gamma_n^{(+)}|J_\lambda^{(+)}(\mathbf{y})|\alpha\rangle}{E_\nu+E_{e^-}+E_n^{(+)}-E_\alpha}$$
$$+\frac{\langle e^-|L_\lambda(\mathbf{y})|\nu\rangle\langle\beta|J_\lambda^{(+)}(\mathbf{y})|\gamma_n^{(-)}\rangle\langle e^+\nu|L_\mu^\dagger(\mathbf{x})|0\rangle\langle\gamma_n^{(-)}|J_\mu^{(-)}(\mathbf{x})|\alpha\rangle}{E_\nu+E_{e^+}+E_n^{(-)}-E_\alpha}\Bigg], \qquad (6)$$

where

$$S_\nu\equiv\mathbf{S}_\nu\cdot\hat{p}_\nu, \quad S_n\equiv\mathbf{S}_n\cdot\hat{p}_n,$$
$$E_\alpha=m_\alpha, \quad E_n^{(\pm)}=[(\mathbf{p}_n)^2+(m_n^{(\pm)})^2]^{1/2},$$
$$E_{e^\pm}=[(\mathbf{p}_\pm)^2+m_e^2]^{1/2}, \quad E_\nu=|\mathbf{p}_\nu|,$$

and where the intermediate hadron states $|\gamma_n^{(\pm)}\rangle$ form a complete set. Thus, summing over this complete set,

$$\mathfrak{M}(\alpha\to\beta+e^-+e^+)$$
$$=\left(\frac{G}{\sqrt{2}}\right)^2\int\int d\mathbf{x}d\mathbf{y}\sum_{\mathbf{p}_\nu,S_\nu}\Bigg[\frac{\langle e^+|L_\mu^\dagger(\mathbf{x})|\nu\rangle\langle e^-\nu|L_\lambda(\mathbf{y})|0\rangle}{E_\nu+E_{e^-}+\langle E_n^{(+)}\rangle-E_\alpha}$$
$$\times\langle\beta|J_\mu^{(-)}(\mathbf{x})J_\lambda^{(+)}(\mathbf{y})|\alpha\rangle$$
$$+\frac{\langle e^-|L_\lambda(\mathbf{y})|\nu\rangle\langle e^+\nu|L_\mu^\dagger(\mathbf{x})|0\rangle}{E_\nu+E_{e^+}+\langle E_n^{(-)}\rangle-E_\alpha}$$
$$\times\langle\beta|J_\lambda^{(+)}(\mathbf{y})J_\mu^{(-)}(\mathbf{x})|\alpha\rangle\Bigg], \qquad (7)$$

with the $\langle E_n^{(\pm)}\rangle$ being appropriate average values. Using Eq. (1) for L_λ, remembering that the field operators for e^+ and e^- anticommute, and summing over S_ν, we obtain from Eq. (7)

$$\mathfrak{M}(\alpha\to\beta+e^-+e^+)=\left(\frac{G}{\sqrt{2}}\right)^2\int\int d\mathbf{x}d\mathbf{y}$$
$$\times\int\frac{d\mathbf{p}_\nu}{(2\pi)^3}\left(l_{\lambda\mu}+i\mathbf{l}_{\lambda\mu}'\cdot\frac{\mathbf{p}_\nu}{E_\nu}\right)e^{-i(\mathbf{p}_+\cdot\mathbf{x}+\mathbf{p}_-\cdot\mathbf{y})}$$
$$\times\Bigg[\frac{e^{i\mathbf{p}_\nu\cdot(\mathbf{x}-\mathbf{y})}}{E_\nu+E_{e^-}+\langle E_n^{(+)}\rangle-E_\alpha}\langle\beta|J_\mu^{(-)}(\mathbf{x})J_\lambda^{(+)}(\mathbf{y})|\alpha\rangle$$
$$-\frac{e^{i\mathbf{p}_\nu\cdot(\mathbf{y}-\mathbf{x})}}{E_\nu+E_{e^+}+\langle E_n^{(-)}\rangle-E_\alpha}\langle\beta|J_\lambda^{(+)}(\mathbf{y})J_\mu^{(-)}(\mathbf{x})|\alpha\rangle\Bigg], \qquad (8)$$

where

$$l_{\lambda\mu}\equiv(u_e^\dagger(\mathbf{p}_-)\gamma_4\gamma_\lambda\gamma_4\gamma_\mu(1+\gamma_5)u_e^*(\mathbf{p}_+)),$$
$$\mathbf{l}_{\lambda\mu}'\equiv(u_e^\dagger(\mathbf{p}_-)\gamma_4\gamma_\lambda\gamma_4\boldsymbol{\gamma}\gamma_4\gamma_\mu(1+\gamma_5)u_e^*(\mathbf{p}_+)).$$

We now recall that the $\sum_{n;\mathbf{p}_n,S_n}\cdots$ in Eq. (6) is dominated by contributions from those $|\gamma_n^{(\pm)}\rangle$ which have a small four-momentum difference and a small mass difference *vis-à-vis* $|\alpha\rangle$; such $|\gamma_n^{(\pm)}\rangle$ are characterized by $|\mathbf{p}_n|\approx[(m_n^{(\pm)})^2-m_\alpha^2]/2m_\alpha$ and $m_n^{(\pm)}\approx m_\alpha$, i.e., by $E_n^{(\pm)}\approx[(m_n^{(\pm)})^2+m_\alpha^2]/2m_\alpha\approx m_\alpha$. Thus

$$E_\nu+E_{e\mp}+\langle E_n^{(\pm)}\rangle-E_\alpha$$
$$\approx E_\nu+(m_\beta-m_\alpha)+m_\alpha-m_\alpha\cong E_\nu, \qquad (9)$$

so that, inserting Eq. (9) into Eq. (8), and carrying

out the integration over $\mathbf{p}_\nu$,

$$\mathfrak{M}(\alpha\to\beta+e^-+e^+)$$
$$=-\left(\frac{G}{\sqrt{2}}\right)^2\left\{\left(\frac{1}{2\pi^2}\right)l_{\lambda\mu}\langle\beta|\int\int d\mathbf{x}d\mathbf{y}[J_\lambda^{(+)}(\mathbf{y}),J_\mu^{(-)}(\mathbf{x})]_-\right.$$
$$\times\frac{e^{-i(\mathbf{p}_-\cdot\mathbf{y}+\mathbf{p}_+\cdot\mathbf{x})}}{|\mathbf{y}-\mathbf{x}|^2}|\alpha\rangle-\left(\frac{1}{4\pi}\right)\mathbf{l}_{\lambda\mu}'\cdot\langle\beta|\int\int d\mathbf{x}d\mathbf{y}$$
$$\left.\times[J_\lambda^{(+)}(\mathbf{y}),J_\mu^{(-)}(\mathbf{x})]_+\frac{e^{-i(\mathbf{p}_-\cdot\mathbf{y}+\mathbf{p}_+\cdot\mathbf{x})}}{|\mathbf{y}-\mathbf{x}|^3}(\mathbf{y}-\mathbf{x})|\alpha\rangle\right\},\quad(10)$$

where $[\ \ ]_-$ and $[\ \ ]_+$ denote a commutator and an anticommutator, respectively. Eq. (10) shows that $\mathfrak{M}(\alpha\to\beta+e^-+e^+)$ receives a quadratically divergent contribution from the term containing $[J_\lambda^{(+)}(\mathbf{y}), J_\mu^{(-)}(\mathbf{x})]_-$ if, as predicted on the basis of any of the proposed equal-time current-current commutation relations, $[J_\lambda^{(+)}(\mathbf{y}),J_\mu^{(-)}(\mathbf{x})]_-=f_{\lambda\mu}^{(-)}(\mathbf{x})\delta^{(3)}(\mathbf{y}-\mathbf{x})+\cdots$. On the other hand, a term of the form $f_{\lambda\mu}^{(+)}(\mathbf{x})\delta^{(3)}(\mathbf{y}-\mathbf{x})$ in the expression for $[J_\lambda^{(+)}(\mathbf{y}),J_\mu^{-}(\mathbf{x})]_+$ contributes nothing to $\mathfrak{M}(\alpha\to\beta+e^-+e^+)$ because of the factor $(\mathbf{y}-\mathbf{x})$ in the second integral of Eq. (10). To investigate these matters in more detail and to specify explicitly the operators $f_{\lambda\mu}^{(\mp)}(\mathbf{y})$, we consider the nonrelativistic impulse approximation for $J_\lambda^{(\pm)}(\mathbf{y})$. In this approximation, $J_\lambda^{(\pm)}(\mathbf{y})$ is written as a sum of terms each of which acts only on one of the basic constituents (i.e., quarks or nucleons) of the hadrons α and β, viz.,

$$J_\lambda^{(\pm)}(\mathbf{y})=\sum_n\tau_n^{(\pm)}(\Gamma_\lambda)_n\delta^{(3)}(\mathbf{y}-\mathbf{r}_n),$$
$$[\tau_n^{(+)},\tau_m^{(-)}]_-=\delta_{nm}\tau_n^{(3)},$$
$$\tau_n^{(+)}\tau_n^{(+)}=\tau_n^{(-)}\tau_n^{(-)}=0\quad(11)$$
$$(\Gamma_\lambda)_n=g_V\delta_{\lambda4}+g_Ai(\sigma_\lambda)_n(1-\delta_{\lambda4}),$$
$$i\sigma=\gamma_4\gamma\gamma_5,\quad g_V=1.0,\quad g_A\cong1.2$$

where the $\tau_n^{(\pm)}$ appear because these basic constituents are assumed to have isospin $\frac{1}{2}$. Equation (11) yields

$$[J_\lambda^{(+)}(\mathbf{y}),J_\mu^{(-)}(\mathbf{x})]_-=\{\tfrac{1}{2}\sum_n(\tau_n^{(3)}[(\Gamma_\lambda)_n,(\Gamma_\mu)_n]_++[(\Gamma_\lambda)_n,(\Gamma_\mu)_n]_-)\delta^{(3)}(\mathbf{x}-\mathbf{r}_n)\}\delta^{(3)}(\mathbf{y}-\mathbf{x}),$$

$$[J_\lambda^{(+)}(\mathbf{y}),J_\mu^{(-)}(\mathbf{x})]_+=\{\tfrac{1}{2}\sum_n(\tau_n^{(3)}[(\Gamma_\lambda)_n,(\Gamma_\mu)_n]_-+[(\Gamma_\lambda)_n,(\Gamma_\mu)_n]_+)\delta^{(3)}(\mathbf{x}-\mathbf{r}_n)\}\delta^{(3)}(\mathbf{y}-\mathbf{x})$$
$$+2\sum_{n,m}\tau_n^{(+)}\tau_m^{(-)}(\Gamma_\lambda)_n(\Gamma_\mu)_m(1-\delta_{nm})\delta^{(3)}(\mathbf{y}-\mathbf{r}_n)\delta^{(3)}(\mathbf{x}-\mathbf{r}_m),\quad(12)$$

$$[(\Gamma_\lambda)_n,(\Gamma_\lambda)_n]_\pm=[g_V^2\delta_{\lambda4}-g_A^2(1-\delta_{\lambda4})](1\pm1),\quad[(\Gamma_\lambda)_n,(\Gamma_4)_n]_\pm=g_Vg_Ai(\sigma_\lambda)_n(1\pm1),\quad\lambda\neq4$$
$$[(\Gamma_\lambda)_n,(\Gamma_\mu)_n]_\pm=-g_A^2\epsilon_{\lambda\mu\nu}i(\sigma_\nu)_n(1\mp1),\quad\lambda\neq4,\ \mu\neq4,\ \lambda\neq\mu$$
$$(\Gamma_\lambda)_n(\Gamma_\mu)_m=g_V^2\delta_{\lambda4}\delta_{\mu4}-g_A^2(\sigma_\lambda)_n(\sigma_\mu)_m(1-\delta_{\lambda4})(1-\delta_{\mu4})$$
$$+ig_Vg_A[(\sigma_\mu)_m\delta_{\lambda4}(1-\delta_{\mu4})+(\sigma_\lambda)_n\delta_{\mu4}(1-\delta_{\lambda4})],\quad n\neq m$$

so that

$$f_{\lambda\mu}^{(\mp)}(\mathbf{x})=\tfrac{1}{2}\sum_n\{\tau_n^{(3)}[(\Gamma_\lambda)_n,(\Gamma_\mu)_n]_\pm+[(\Gamma_\lambda)_n,(\Gamma_\mu)_n]_\mp\}\delta^{(3)}(\mathbf{x}-\mathbf{r}_n);\quad(13)$$

$f_{\lambda\mu}^{(\mp)}(\mathbf{x})$ is different from zero since either $[(\Gamma_\lambda)_n, (\Gamma_\mu)_n]_+$ or $[(\Gamma_\lambda)_n,(\Gamma_\mu)_n]_-$ is different from zero. The combination of Eqs. (12), (13), and (10) gives

$$\mathfrak{M}(\alpha\to\beta+e^-+e^+)=-\left(\frac{G}{\sqrt{2}}\right)^2\left\{\left(\frac{1}{2\pi^2}\right)l_{\lambda\mu}\langle\beta|\int d\mathbf{x}f_{\lambda\mu}^{(-)}(\mathbf{x})e^{-i(\mathbf{p}_-+\mathbf{p}_+)\cdot\mathbf{x}}|\alpha\rangle\int d\mathbf{y}\frac{\delta^{(3)}(\mathbf{y}-\mathbf{x})}{|\mathbf{y}-\mathbf{x}|^2}\right.$$
$$-\left(\frac{1}{4\pi}\right)\mathbf{l}_{\lambda\mu}'\cdot\langle\beta|\int d\mathbf{x}f_{\lambda\mu}^{(+)}(\mathbf{x})e^{-i(\mathbf{p}_-+\mathbf{p}_+)\cdot\mathbf{x}}|\alpha\rangle\int d\mathbf{y}\frac{\delta^{(3)}(\mathbf{y}-\mathbf{x})}{|\mathbf{y}-\mathbf{x}|^3}(\mathbf{y}-\mathbf{x})$$
$$\left.-\left(\frac{1}{4\pi}\right)\mathbf{l}_{\lambda\mu}'\cdot\langle\beta|2\sum_{n,m}\tau_n^{(+)}\tau_m^{(-)}(\Gamma_\lambda)_n(\Gamma_\mu)_m(1-\delta_{nm})\frac{e^{-i(\mathbf{p}_-\cdot\mathbf{r}_n+\mathbf{p}_+\cdot\mathbf{r}_m)}}{|\mathbf{r}_n-\mathbf{r}_m|^3}(\mathbf{r}_n-\mathbf{r}_m)|\alpha\rangle\right\},\quad(14)$$

and since

$$\int dy \frac{\delta^{(3)}(\mathbf{y}-\mathbf{x})}{|\mathbf{y}-\mathbf{x}|^2} = \lim_{a\to 0} \int \frac{d\mathbf{r}}{r^2}\left(\frac{e^{-r^2/a^2}}{\pi^{3/2}a^3}\right) = \lim_{a\to 0}\left(\frac{2}{a^2}\right) = \infty\,, \quad (15)$$

$$\int dy \frac{\delta^{(3)}(\mathbf{y}-\mathbf{x})}{|\mathbf{y}-\mathbf{x}|^3}(\mathbf{y}-\mathbf{x}) = \lim_{a\to 0} \int \frac{d\mathbf{r}}{r^3}\left(\frac{e^{-r^2/a^2}}{\pi^{3/2}a^3}\right)\mathbf{r} = 0\,,$$

the quadratic divergence in $\mathfrak{M}(\alpha\to\beta+e^-+e^+)$ indeed arises from the equal-time commutation relation $[J_\lambda^{(+)}(\mathbf{y}),J_\mu^{(-)}(\mathbf{x})]_- = f_{\lambda\mu}^{(-)}(\mathbf{x})\delta^{(3)}(\mathbf{y}-\mathbf{x})$ with nonzero $f_{\lambda\mu}^{(-)}(\mathbf{x})$. Physically, the fact that $f_{\lambda\mu}^{(-)}(\mathbf{x})\neq 0$ corresponds to the fact that the same basic constituent of the hadron can emit and reabsorb the virtual neutrino when an electron-positron pair is emitted so that no natural cutoff for the energy of this virtual neutrino is available. We remark that had we approached the calculation of $\mathfrak{M}(\alpha\to\beta+e^-+e^+)$ using covariant rather than noncovariant perturbation theory, and had we retained only its leading part by means of the Bjorken procedure, we would have found just the quadratically divergent term in Eq. (10) containing $[J_\lambda^{(+)}(\mathbf{y}), J_\mu^{(-)}(\mathbf{x})]_-$, and thus drawn conclusions identical to those reached above. Finally, we note the last term on the right-hand side of Eq. (14) which gives a finite contribution to $\mathfrak{M}(\alpha\to\beta+e^-+e^+)$ and which arises from $[J_\lambda^{(+)}(\mathbf{y}),J_\mu^{(-)}(\mathbf{x})]_+$ [see Eqs. (12) and (10)]; it may well turn out that this term will constitute the main part of a physically sensible, second-order weak calculation of $\mathfrak{M}(\alpha\to\beta+e^-+e^+)$.

We proceed to consider the second-order weak processes with $\Delta Q=\pm 2$ and $\Delta S=0$, namely, the no-neutrino double-β decay

$$N_i \begin{matrix}\nearrow \text{any } N_n+e_1^-+\nu \searrow \\ \searrow \text{any } N_n+e_2^-+\nu \nearrow\end{matrix} N_f+e_1^-+e_2^-\,, \quad (16)$$

where N_i, N_n, and N_f are hadrons, e.g., $N_i=\mathrm{Te}^{130}$, $N_n=\mathrm{I}^{130}, \mathrm{I}^{130*}, \cdots, N_f=\mathrm{Xe}^{130}$. By means of an analysis analogous to that given in Eqs. (5)–(9), but with the $\mathcal{H}_{\text{weak}}$ of Eq. (5) containing the L_λ' of Eq. (2) instead of the L_λ of Eq. (1), we obtain an expression for $\mathfrak{M}(N_i\to N_f+e_1^-+e_2^-)$ similar to that in Eqs. (10) and (8), but with the essential difference that $J_\mu^{(-)}(\mathbf{x})$ is replaced by $J_\mu^{(+)}(\mathbf{x})$ and $(1+\gamma_5)$ in $l_{\lambda\mu}$ and $\mathbf{l}'_{\lambda\mu}$ is replaced by 2η. Since $[J_\lambda^{(+)}(\mathbf{y}),J_\mu^{(+)}(\mathbf{x})]_-=0$, it follows that $\mathfrak{M}(N_i\to N_f+e_1^-+e_2^-)$ is convergent and dependent only on $[J_\lambda^{(+)}(\mathbf{y}), J_\mu^{(+)}(\mathbf{x})]_+$. In addition we now have, using Eq. (11),

$$\begin{aligned}[J_\lambda^{(+)}(\mathbf{y}),J_\mu^{(+)}(\mathbf{x})]_+ &= 2J_\lambda^{(+)}(\mathbf{y})J_\mu^{(+)}(\mathbf{x}) = 2\sum_{n,m}\tau_n^{(+)}\tau_m^{(+)}(\Gamma_\lambda)_n(\Gamma_\mu)_m \\ &\quad\times\delta^{(3)}(\mathbf{y}-\mathbf{r}_n)\delta^{(3)}(\mathbf{x}-\mathbf{r}_m) \quad (17)\\ &= 2\sum_{n,m}\tau_n^{(+)}\tau_m^{(+)}(\Gamma_\lambda)_n(\Gamma_\mu)_m(1-\delta_{nm}) \\ &\quad\times\delta^{(3)}(\mathbf{y}-\mathbf{r}_n)\delta^{(3)}(\mathbf{x}-\mathbf{r}_m)\,,\end{aligned}$$

where the last equality is a consequence of $\tau_n^{(+)}\tau_n^{(+)}=0$. Thus, $\mathfrak{M}(N_i\to N_f+e_1^-+e_2^-)$ is proportional to

$$\begin{aligned}&\langle N_f|2\sum_{n,m}\tau_n^{(+)}\tau_m^{(+)}(\Gamma_\lambda)_n(\Gamma_\mu)_m \\ &\quad\times(1-\delta_{nm})\frac{e^{-i(\mathbf{p}_1\cdot\mathbf{r}_n+\mathbf{p}_2\cdot\mathbf{r}_m)}}{|\mathbf{r}_n-\mathbf{r}_m|^3}(\mathbf{r}_n-\mathbf{r}_m)|N_i\rangle \\ &\cong -\tfrac{1}{3}i(\mathbf{p}_1-\mathbf{p}_2)\langle N_f|2\sum_{n,m}\tau_n^{(+)}\tau_m^{(+)}(\Gamma_\lambda)_n(\Gamma_\mu)_m \\ &\quad\times(1-\delta_{nm})\left(\frac{1}{|\mathbf{r}_n-\mathbf{r}_m|}\right)|N_i\rangle\,, \quad (18)\end{aligned}$$

where the approximate equality follows since in all double-β decays of interest $|N_i\rangle$ and $|N_f\rangle$ are characterized by zero spin and the same parity. We note for the sake of completeness that a term proportional to

$$\begin{aligned}&(E_1-E_2)\langle N_f|2\sum_{n,m}\tau_n^{(+)}\tau_m^{(+)}(\Gamma_\lambda)_n(\Gamma_\mu)_m \\ &\quad\times(1-\delta_{nm})\left(\frac{1}{|\mathbf{r}_n-\mathbf{r}_m|}\right)|N_i\rangle \quad (19)\end{aligned}$$

also contributes to $\mathfrak{M}(N_i\to N_f+e_1^-+e_2^-)$; this term arises if one does not totally neglect E_1 and E_2 compared to E_ν in the energy denominators of the second-order perturbation expression for $\mathfrak{M}(N_i\to N_f+e_1^-+e_2^-)$.[3]

3. ESTIMATES OF THE DEGREE OF LEPTON NONCONSERVATION

For an initial estimate of the degree of lepton nonconservation in weak interactions, we use the standard two-nucleon mechanism for no-neutrino double-β decay.[3] Two neutrons within the nucleus undergo a second-order weak transition into two protons and simultaneously emit an electron pair. Since the decay rate is proportional to η^2 [see Eq. (2)] the measured half-life of Te^{130} enables us to place a limit on the size of η.

In order to sharpen the limit, we consider an alternative mechanism in which an $N^*(1236)$ within the nucleus undergoes no-neutrino double-β decay directly [Eq. (3)]. Even though we assume that the probability of finding the $N^*(1236)$ inside the nucleus is only about 1%, we gain more than enough from other factors to render the associated nuclear matrix element appreciably greater than in the two-nucleon mechanism. As a result, we improve the limit on η by an order of magnitude.

A. Two-Nucleon Mechanism

The half-life of the no-neutrino double-β decay of a nucleus on the basis of the two-nucleon mechanism can be expressed as

$$\frac{\ln 2}{T_{1/2}} = 2\pi\int\int\frac{d\epsilon_1}{(2\pi)^3}\frac{d\epsilon_2}{(2\pi)^3}\delta(\epsilon_0-\epsilon_1-\epsilon_2) \times(\epsilon_1+1)^2(\epsilon_2+1)^2P(\epsilon_1,\epsilon_2)\,, \quad (20)$$

where ϵ_1 and ϵ_2 are the kinetic energies of the electrons and ϵ_0 is the energy release, all in units of m_e. The most general form of $P(\epsilon_1,\epsilon_2)$ can be calculated along the lines described in Sec. 2 using the nuclear matrix elements specified in Eqs. (18) and (19) with $(\Gamma_\lambda)_n(\Gamma_\mu)_m$ given in Eq. (12). Making the further assumption that pairs of like nucleons in the nucleus are predominantly in singlet spin states, so that

$$\langle N_f|\sum_{n,m=1}^{A}\frac{\tau_n^{(+)}\tau_m^{(+)}}{r_{nm}}(1-\delta_{nm})|N_i\rangle$$

$$\cong-\tfrac{1}{3}\langle N_f|\sum_{n,m=1}^{A}\frac{\tau_n^{(+)}\tau_m^{(+)}}{r_{nm}}\boldsymbol{\sigma}_n\cdot\boldsymbol{\sigma}_m(1-\delta_{nm})|N_i\rangle, \quad (21)$$

we find, specializing some general results obtained in an earlier paper by the present authors,[3] that

$$P(\epsilon_1,\epsilon_2)\cong\eta^2G^4(2\pi Z/137)^2(1-e^{-2\pi Z/137})^{-2}$$
$$\times\{(16/E_1E_2)(E_1-E_2)^2(E_1E_2-1)\}$$
$$\times[(g_A{}^4/9R^2)|\langle N_f|(\mathbf{Y}^{(+)})^2|N_i\rangle|^2],$$
$$E_i\equiv\epsilon_1+1,\quad i=1,2$$

$$\left\langle\frac{1}{r_{nm}}\right\rangle\equiv\langle N_f|\sum_{n,m=1}^{A}\frac{\tau_n^{(+)}\tau_m^{(+)}}{r_{nm}}\boldsymbol{\sigma}_n\cdot\boldsymbol{\sigma}_m(1-\delta_{nm})|N_i\rangle\Big/$$
$$\langle N_f|\sum_{n,m=1}^{A}\tau_n^{(+)}\tau_m^{(+)}\boldsymbol{\sigma}_n\cdot\boldsymbol{\sigma}_m(1-\delta_{mn})|N_i\rangle, \quad (22)$$
$$\langle r_{nm}\rangle\cong R,$$
$$\sum_{n,m=1}^{A}\tau_n^{(+)}\tau_m^{(+)}\boldsymbol{\sigma}_n\cdot\boldsymbol{\sigma}_m(1-\delta_{nm})=(\sum_{n=1}^{A}\tau_n^{(+)}\boldsymbol{\sigma}_n)^2\equiv(\mathbf{Y}^{(+)})^2,$$

where $\langle r_{nm}\rangle$ is the mean separation between any two nucleons in the initial or the final nucleus, $R=(1.2\times10^{-13}$ cm) $A^{1/3}$ is the nuclear radius, and $\tau_n^{(+)}\tau_n^{(+)}=0$ (since the nucleons have isospin $\frac{1}{2}$). Thus, combining Eqs. (20) and (22), we obtain the half-life of the no-neutrino double-β decay of a nucleus calculated on the basis of the two-nucleon mechanism[3]

$$T_{1/2}\cong[10^{19.9}/\eta^2 f(\epsilon_0)](137/2\pi Z)^2(1-e^{-2\pi Z/137})^2$$
$$\times(A/130)^{2/3}|\langle N_f|(\mathbf{Y}^{(+)})^2|N_i\rangle|^{-2}\text{ years}, \quad (23)$$

where

$$f(\epsilon_0)\equiv\epsilon_0{}^2(\epsilon_0{}^5+13\epsilon_0{}^4+77\epsilon_0{}^3+70\epsilon_0{}^2). \quad (24)$$

We also record the corresponding electron-electron angular correlation function[3]

$$C(\hat{p}_1\cdot\hat{p}_2)\cong1+[p_1p_2/(E_1E_2-1)]\hat{p}_1\cdot\hat{p}_2\cong1+\hat{p}_1\cdot\hat{p}_2. \quad (25)$$

The nucleus Te^{130} has $\epsilon_0=5.0$ and so its half-life is predicted from Eqs. (23) and (24) to be $\cong10^{13.4}\eta^{-2}|\langle N_f|(\mathbf{Y}^{(+)})^2|N_i\rangle|^{-2}$ years. If all of the measured half-life $T_{1/2}(Te^{130})=10^{21.34}$ years is attributed to no-neutrino double-β decay, then

$$\eta\cong10^{-4.0}/|\langle N_f|(\mathbf{Y}^{(+)})^2|N_i\rangle|, \quad (26)$$

whence, if the nuclear matrix element $\langle N_f|(\mathbf{Y}^{(+)})^2|N_i\rangle$ is estimated to have a magnitude ≈0.1,

$$\eta\approx10^{-3}. \quad (27)$$

This limit on η is considerably lower than the limit obtained in other ways; for example, the nonobservation of the process $\bar{\nu}+Cl^{37}\rightarrow Ar^{37}+e^-$ yields $\eta\cong0.2$, while the measured longitudinal spin polarization of electrons emitted in single-β decay yields $\eta\cong0.1$.

With the value of $\eta|\langle N_f|(\mathbf{Y}^{(+)})^2|N_i\rangle|$ given in Eq. (26), we can now estimate the half-lives of Te^{128} and Ca^{48}, for which $\epsilon_0=1.7$ and 8.6, respectively. From Eqs. (23) and (24) we obtain

$$T_{1/2}(Te^{128})\cong T_{1/2}(Te^{130})f(5.0)/f(1.7)$$
$$=(10^{21.34}\text{ years})10^{2.40}=10^{23.7}\text{ years} \quad (28a)$$

and

$$T_{1/2}(Ca^{48})\cong T_{1/2}(Te^{130})\left(\frac{54}{22}\right)^2\left(\frac{1-e^{-2\pi\times22/137}}{1-e^{-2\pi\times54/137}}\right)^2$$
$$\times\left(\frac{48}{130}\right)^{2/3}\frac{f(5.0)}{f(8.6)} \quad (28b)$$
$$=(10^{21.34}\text{ years})10^{-0.75}=10^{20.6}\text{ years}.$$

This estimate of the Te^{128} half-life is much closer to the experimental value of $10^{22.5\pm0.5}$ years than is the estimated Te^{128} half-life for two-neutrino double-β decay: $\cong10^{25.3}$ years $[\cong T_{1/2}(Te^{130})\times10^{4.0}=(10^{21.34}$ years) $\times10^{4.0}$, see Sec. 1]. The estimate of the Ca^{48} half-life is not inconsistent with the experimental lower limit of $10^{21.3}$ years[10] since it may well be that the nuclear matrix element $\langle N_f|(\mathbf{Y}^{(+)})^2|N_i\rangle$ is actually some two to three times smaller in $Ca^{48}\rightarrow Ti^{48}$ than in $Te^{130}\rightarrow Xe^{130}$.

B. N*(1236) Mechanism

As we have already mentioned, it has been suggested that the deuteron and presumably most other nuclei contain about a 1% probability admixture of the $N^*(1688)$.[9] If this is indeed the case, then it is likely that the various nuclei involved in double-β decay also contain a similar probability admixture of $N^*(1236)$. This particle, having isospin $\frac{3}{2}$, can undergo no-neutrino double-β decay directly, as set down in Eq. (3), and its decay amplitude will therefore contribute to the amplitude for the nuclear no-neutrino double-β decay. In fact, the contribution of the $N^*(1236)$ may well dominate the nuclear amplitude, as we shall now show.

If the $N^*(1236)$ were treated as "elementary," i.e., as a point particle, then, as noted above, the emission and reabsorption of a virtual neutrino in the processes of Eq. (3) would give rise to a divergence in the associated second-order weak-matrix element. To overcome this difficulty, we adopt a quark model for the $N^*(1236)$ and the nucleon. Because the isospin of the (nonstrange)

[10] The limit quoted is from R. K. Bardin, P. J. Gollon, J. D. Ullman, and C. S. Wu, Phys. Letters **26B**, 112 (1967), where references to earlier experimental work are also given.

quarks is $\frac{1}{2}$, no single quark can emit and reabsorb the virtual neutrino and simultaneously emit two electrons, and so Eq. (3) represents a two-quark process. The mean separation between the two quarks involved then serves to cut off the divergence in the same way as in the two-nucleon case.

On the average, the mean separation between any two quarks within the $N^*(1236)$ or within the nucleon is about equal to the charge radius of the proton, i.e., about 0.1 the mean separation between any two nucleons in medium-to-heavy nuclei. This property compensates for the small probability of finding an $N^*(1236)$ inside the nucleus. In addition, because only one particle of the nucleus is now involved in the double-β decay, the overlap between initial and final nuclear wave functions is larger than in the two-nucleon mechanism so that, in general, the nuclear matrix element for no-neutrino double-β decay is estimated to be larger on the $N^*(1236)$ mechanism than on the two-nucleon mechanism.

In the $N^*(1236)$ mechanism, using the quark model, the matrix element for the first process of Eq. (3), $\mathfrak{M}(S_{N^*},S_p)$, can be calculated by the same method as is used in the two-nucleon mechanism.[3] The general form of this matrix element is given by

$$\mathfrak{M}(S_{N^*},S_p)=(\eta G^2/6\pi)(u_e^\dagger(\mathbf{p}_2)i\gamma_4\boldsymbol{\gamma}u_e^*(\mathbf{p}_1))\cdot(\mathbf{p}_1-\mathbf{p}_2)\times\mathbf{M}(S_{N^*},S_p),$$

$$\mathbf{M}(S_{N^*},S_p)\equiv g_Vg_A\langle p;S_p|\sum_{n,m=1}^{3}\frac{\tau_n^{(+)}\tau_m^{(+)}}{r_{nm}}\times\tfrac{1}{2}(\boldsymbol{\sigma}_n+\boldsymbol{\sigma}_m)(1-\delta_{nm})|N^{*-};S_{N^*}\rangle$$
$$=g_Vg_A\langle 1/r_{nm}\rangle\times\langle p;S_p|I^{(+)}\mathbf{Y}^{(+)}|N^{*-};S_{N^*}\rangle,\quad(29)$$

$$\sum_{n,m=1}^{3}\tau_n^{(+)}\tau_m^{(+)}\boldsymbol{\sigma}_m(1-\delta_{nm})=\left(\sum_{n=1}^{3}\tau_n^{(+)}\right)\left(\sum_{n=1}^{3}\tau_n^{(+)}\boldsymbol{\sigma}_n\right)\equiv I^{(+)}\mathbf{Y}^{(+)},$$

where $\langle r_{nm}\rangle$ is the mean separation between any two quarks within the $N^*(1236)$ or the nucleon, $\tau_n^{(+)}\tau_n^{(+)}=0$ (since the quarks have isospin $\frac{1}{2}$), S_{N^*} and S_P are the z components of the spin of N^{*-} and p, and $\mathbf{M}$ (S_{N^*},S_p) does not contain terms proportional to g_V^2 or to g_A^2 because the $N^*(1236)\leftrightarrow$ nucleon transition involves a spin change of one unit. Then, evaluating $\mathbf{M}(S_{N^*},S_p)$ by means of $SU(6)$-type spin-isospin wave functions with perfect radial overlap for the quarks in the $N^*(1236)$ and in the nucleon,[11] we get

$$[M(S_{N^*},S_p)]_{x\pm iy}=\mu C_{S_{N^*}\pm1S_p}{}^{\frac{3}{2}1\frac{1}{2}},$$
$$[M(S_{N^*},S_p)]_z=\mu C_{S_{N^*}0S_p}{}^{\frac{3}{2}1\frac{1}{2}},\quad(30)$$
$$\mu\equiv g_Vg_A\langle 1/r_{nm}\rangle$$

where the $C_{m_1m_2m_3}{}^{j_1j_2j_3}$ are Clebsch-Gordan coefficients. Using Eqs. (29) and (30), we obtain the half-life of the no-neutrino double-β decay of a nucleus calculated on the basis of the $N^*(1236)$ mechanism

$$T_{1/2}\cong\frac{10^{17.5}}{\eta^2g(\epsilon_0)}\left(\frac{137}{2\pi Z}\right)^2(1-e^{-2\pi Z/137})^2\left(\frac{\langle r_{nm}\rangle}{0.7\times10^{-13}\text{cm}}\right)^2\times(P(N^*))^{-1}|\langle\Phi_f|\Phi_i\rangle|^{-2}\text{ years},\quad(31)$$

where

$$g(\epsilon_0)\equiv\epsilon_0^2(\epsilon_0^5+14\epsilon_0^4+81\epsilon_0^3+221\epsilon_0^2+228\epsilon_0+140),\quad(32)$$

and where $P(N^*)$ is the probability of finding an $N^*(1236)$ in the nucleus, and $|\langle\Phi_f|\Phi_i\rangle|^2$ is an overlap factor between the initial and final nuclear wave functions. This overlap factor is normalized to unity for identical momentum distributions (1) of the N^* and the nucleon into (out of) which the N^* transforms, and (2) of the other $A-1$ nucleons. We also obtain the corresponding electron-electron angular-correlation function

$$(C\hat{p}_1\cdot\hat{p}_2)=1-\left(\frac{p_1p_2}{E_1E_2}\right)\left(\frac{(E_1+E_2)^2E_1E_2}{(E_1E_2+1)(p_1^2+p_2^2)+p_1^2p_2^2}\right)\hat{p}_1\cdot\hat{p}_2+\left(\frac{p_1p_2}{E_1E_2}\right)^2\frac{E_1^2E_2^2}{(E_1E_2+1)(p_1^2+p_2^2)+p_1^2p_2^2}(\hat{p}_1\cdot\hat{p}_2)^2$$
$$\cong(1-\hat{p}_1\cdot\hat{p}_2)(1-\tfrac{1}{3}\hat{p}_1\cdot\hat{p}_2).\quad(33)$$

Equations (31) and (32) with $\langle r_{nm}\rangle\cong0.7\times10^{-13}$ cm yield a half-life $\cong10^{10.8}\eta^{-2}[P(N^*)]^{-1}|\langle\Phi_f|\Phi_i\rangle|^{-2}$ years for $\text{Te}^{130}(\epsilon_0=5)$; comparison of this with the measured half-life of $10^{21.34}$ years yields

$$\eta\cong10^{-5.3}/[P(N^*)]^{1/2}|\langle\Phi_f|\Phi_i\rangle|,\quad(34)$$

whence, taking $P(N^*)\approx0.01$ and estimating (rather conservatively) $|\langle\Phi_f|\Phi_i\rangle|\approx(0.1)^{1/2}$,

$$\eta\approx10^{-4}.\quad(35)$$

This limit on η is lower than the limit on η in Eq. (27) by an order of magnitude. The half-lives of Te^{128} $(\epsilon_0=1.7)$ and Ca^{48} $(\epsilon_0=8.6)$ predicted on the basis of Eqs. (31) and (32) are

$$T_{1/2}(\text{Te}^{128})\cong T_{1/2}(\text{Te}^{130})g(5.0)/g(1.7)=(10^{21.34}\text{ years})10^{2.13}=10^{23.5}\text{ years},\quad(36\text{a})$$

$$T_{1/2}(\text{Ca}^{48})\cong T_{1/2}(\text{Te}^{130})\left(\frac{54}{22}\right)^2\left(\frac{1-e^{-2\pi\times22/137}}{1-e^{-2\pi\times54/137}}\right)^2\frac{g(5.0)}{g(8.6)}=(10^{21.34}\text{ years})10^{-0.82}=10^{20.5}\text{ years},\quad(36\text{b})$$

in essential agreement with Eqs. (28a) and (28b) and with the experimental data. Finally, it is instructive to record the half-life for the double-β decay of a free N^* as calculated from Eqs. (31), (32), and (35)—we find $T_{1/2}(N^{*-}\rightarrow p+e^-+e^-)\approx10^{17.5}/(10^{-4})^2(600)^7$ years

---

[11] R. H. Dalitz, in *High Energy Physics*, edited by C. DeWitt and M. Jacob (Gordon and Breach Science Publishers, Inc., New York, 1965), p. 253.

$=4\times10^{13}$ sec; this is greater by a factor of 10^{23} than the half-life for the single-β decay of a free N^* and by a factor of 10^{37} than the half-life for the pionic decay of a free N^*.

4. SUMMARY AND DISCUSSION

We have shown that if the transition $Te^{130}\rightarrow Xe^{130}$ is assumed to be largely due to lepton-nonconserving no-neutrino double-β decay, and if the dominant mechanism in this decay involves an $N^*(1236)\leftrightarrow$ nucleon transition within the nucleus, then, granting the validity of the above estimate of $P(N^*)|\langle\Phi_f|\Phi_i\rangle|^2$, the limit on the "lepton-nonconservation" parameter is $\eta\approx10^{-4}$. This limit is an order of magnitude smaller than the one obtained in an analogous way from the standard two-nucleon mechanism and is very much smaller than the limits on η which are obtained from the analysis of first-order weak processes. We also note that the electron-electron angular-correlation functions of Eqs. (33) and (25) are quite different so that a measurement of the electron-electron angular correlation could conceivably decide whether the $N^*(1236)$ mechanism is indeed dominant over the two-nucleon mechanism.

Although the measure half-life[1] for $Te^{130}\rightarrow Xe^{130}$ is in good agreement with theoretical estimates for the lepton-conserving two-neutrino double-β decay,[3] we have chosen, pursuing a suggestion of Pontecorvo,[4] to interpret this process as a lepton-nonconserving no-neutrino double-β decay because of an apparent anomaly in the case of Te^{128}. If the excess of Xe^{128} found in a recent mass-spectrometric experiment[5] can be ascribed to the double-β decay $Te^{128}\rightarrow Xe^{128}$, it corresponds to a half-life for Te^{128} whose ratio to the measured half-life of Te^{130} is consistent with the no-neutrino process, but is three orders of magnitude smaller than that expected for the two-neutrino process. At present, the arguments in favor of ascribing the excess Xe^{128} to $Te^{128}\rightarrow Xe^{128}$ are not conclusive, and they may not survive further experimentation. In the event that they do survive, however, we will be forced to accept the existence of lepton-nonconserving no-neutrino double-β decay, since it is extremely unlikely that the nuclear matrix elements for the lepton-conserving two-neutrino double-β decays of Te^{128} and Te^{130} differ in magnitude by a factor as large as $10^{3/2}$.[12] We emphasize, however, that by far the best way to settle the question of the presence or absence of neutrinos in double-β decay is to observe the electron-electron energy-sum spectrum.[3] This spectrum will exhibit two peaks if the half-lives of no-neutrino and two-neutrino double-beta decay are not too different: a broad relatively low peak in the vicinity of $\frac{1}{2}\epsilon_0$ from the two-neutrino process and a narrow relatively high peak at ϵ_0 from the no-neutrino process. A search for these peaks and a determination of the corresponding branching ratio is obviously of crucial interest.

In the above discussion we parametrized a possible lepton nonconservation by inclusion in the lepton weak current of an "opposite-helicity" term: $\eta\psi_e^\dagger\gamma_4\gamma_\lambda(1-\gamma_5)\psi_\nu$; however, we kept our "Majorana" neutrino massless: even though a nonvanishing η in general implies a nonvanishing m_ν proportional to η. Alternatively, we could have parametrized lepton nonservation by supposing that $\eta=0$ and $m_\nu\neq0$; lifetime calculations of no-neutrino double-β decay would then lead essentially to the formulas derived above but with η replaced by $(m_\nu\langle E_\nu^{\text{virtual}}\rangle^{-1})\approx(m_\nu\langle r_{nm}\rangle)$. The experimental limit $m_\nu<2\times10^{-4}m_e$[13] and the estimate $\langle r_{mn}\rangle\cong0.7\times10^{-13}\text{ cm}=\frac{1}{2}m_\pi^{-1}$ then yield $(m_\nu\langle r_{nm}\rangle)<4\times10^{-7}$ which is some 250 times smaller than the limit required to account for the measured lifetime of Te^{130} [see the limit on η in Eq. (35)]. Thus the effect of any possible nonzero m_ν can be safely neglected in the treatment of lepton-nonconserving no-neutrino double-β decay.

Still another possible parametrization of lepton nonconservation involves inclusion in the lepton weak current of an "opposite-helicity" scalar or tensor term $\eta_S\psi_e^\dagger\gamma_4(1-\gamma_5)\psi_\nu$ or $\eta_T\psi_e^\dagger\gamma_4\sigma_{\lambda\mu}(1-\gamma_5)\psi_\nu$, instead of the above-used "opposite-helicity" vector term $\eta\psi_e^\dagger\gamma_4\gamma_\lambda\times(1-\gamma_5)\psi_\nu$. The limits that one can then set on η_S or η_T are similar numerically to those set on η [Eq. (35) or (27)] and are very much smaller than the limits presently available on any $1+\gamma_5$ and/or $1-\gamma_5$ scalar or tensor weak currents as determined by an analysis of single-β decay.

As a final remark, we would like to emphasize that, in spite of the presence of virtual-neutrino closed loops, divergences never arise in second-order weak matrix elements with $\Delta Q=\pm2$ if the basic hadronic constituents (i.e., quarks or nucleons) have isospin not exceeding $\frac{1}{2}$ and a mean separation greater than zero; this circumstance permits a quantitative estimation of lifetimes in no-neutrino double-β decay in terms of a single "lepton-nonconservation" parameter. Unfortunately, divergences arising from virtual-neutrino closed loops are present in the second-order weak matrix elements with $\Delta Q=0$, and this necessitates the introduction of auxiliary "cutoff" or "cancellation" procedures in order to obtain quantitative estimates of the second-order weak contributions to the rates of processes such as $O^{16*}\rightarrow O^{16}+e^-+e^+$, $\pi^0\rightarrow e^-+e^+$, $K_L{}^0\rightarrow\mu^-+\mu^+$, etc.

ACKNOWLEDGMENT

We wish to thank Dr. O. A. Schaeffer for a very helpful communication on the lower limit of the half-life of Te^{128} as determined from the experiments of Ref. 1.

[12] These nuclear matrix elements are of the form

$$M(Te^{128}\rightarrow Xe^{128})\cong\sum_n\langle Xe^{128}|\mathbf{Y}^{(+)}|I_n{}^{128}\rangle\cdot\langle I_n{}^{128}|\mathbf{Y}^{(+)}|Te^{128}\rangle/[E(I_n{}^{128})-E(Te^{128})]$$

and

$$M(Te^{130}\rightarrow Xe^{130})\cong\sum_n\langle Xe^{130}|\mathbf{Y}^{(+)}|I_n{}^{130}\rangle\cdot\langle I_n{}^{130}|\mathbf{Y}^{(+)}|Te^{130}\rangle/[E(I_n{}^{130})-E(Te^{130})],$$

and, on the basis of any reasonable nuclear model, are expected to have about the same magnitude.

[13] K. E. Bergkvist, in *Proceedings of the CERN Topical Weak Interaction Conference* (CERN, Geneva, 1969), p. 91.

VOLUME 47, NUMBER 3 PHYSICAL REVIEW LETTERS 20 JULY 1981

Double Beta Decay and the Majorana Mass of the Electron Neutrino

W. C. Haxton
Department of Physics, Purdue University, West Lafayette, Indiana 47907,[(a)] *and Theory Division, Los Alamos Scientific Laboratory, Los Alamos, New Mexico 87545*

and

G. J. Stephenson, Jr., and D. Strottman
Theory Division, Los Alamos Scientific Laboratory, Los Alamos, New Mexico 87545

(Received 6 April 1981)

We have calculated the two-neutrino and no-neutrino, lepton-number-nonconserving double beta decay rates for ^{76}Ge and ^{82}Se. Our result for the two-neutrino decay of ^{82}Se is in agreement with a recent cloud-chamber measurement and thus in disagreement with the geochemical determination. We find that a constraint on the Majorana mass of the electron neutrino of $\langle m\rangle_\nu \gtrsim 15$ eV follows from the comparison of our calculations and the experimental limits on the neutrinoless decays of ^{76}Ge and ^{82}Se.

PACS numbers: 23.40.Bw, 14.20.Cg

The prejudice for conservation of lepton number is based in part on the early recognition that a Majorana electron neutrino would lead to neutrinoless double beta decay at rates much faster than found experimentally.[1] Later, with the discovery of parity nonconservation in the weak interaction, it was realized that exact γ_5 invariance of the weak leptonic current would forbid neutrinoless $\beta\beta$ decay, irrespective of the behavior of the neutrino wave function under charge conjugation. Recently, however, evidence is mounting that this γ_5 invariance may be only approximate, broken either by a small neutrino mass or by some explicit mixing of left- and right-handed components in the weak current. If this proves true, then again the presence or absence of neutrinoless $\beta\beta$ decay can place fundamental constraints on possible descriptions of the neutrino.

A recent cloud-chamber search for the two-neutrino $\beta\beta$ decay of ^{82}Se has also renewed interest in this conventional, lepton-number-conserving process. Moe and Lowenthal[2] obtained, from limited statistics, a two-neutrino decay rate which exceeds a preexisting total $\beta\beta$ decay rate, determined geochemically,[3] by a factor of 28. It is imperative that we understand the origin of this discrepancy as other geochemical results, particularly the $\beta\beta$ decay lifetimes of ^{130}Te and ^{128}Te, have been used in limiting possible violations of lepton-number conservation.[1] In this Letter we present what we believe to be the first calculations of sufficient precision to allow a meaningful comparison between theory and these two conflicting measurements. We then calculate the no-neutrino $\beta\beta$ decay rates for ^{82}Se and ^{76}Ge, taking the neutrino mass from a recent measurement of the triton β decay spectrum. The laboratory limits on the neutrinoless decays of these nuclei then indicate that the electron neutrino is not a Majorana mass eigenstate.

The lepton-number-conserving $\beta\beta$ decay of a nucleus occurs as a second-order weak process in which a final state of two electrons and two antineutrinos is produced

$$(A,Z)\to(A,Z+2)+2e^-+2\bar{\nu}_e .$$

This differs from two ordinary β decays in that the intermediate state $(A,Z+1)+e^-+\bar{\nu}_e$ is virtual. Thus the single β decay to the intermediate nucleus must be forbidden energetically (or at least strongly hindered), a situation not uncommon due to the pairing force which lowers the energy of the initial and final even-even nuclei relative to the odd-odd intermediate nucleus. Consequently, most interesting candidates for $\beta\beta$ decay involve initial and final states with $J^\pi=0^+$. We specialize our treatment to this case.

The $\beta\beta$ decay amplitude is given by the two-nucleon mechanism of Fig. 1(a). (A single-hadron mechanism where the Δ decays to a neutron and four leptons has been discussed[4]; however, the nuclear operator for this process is a spin vector and thus cannot contribute to $J=0\to0$ transitions.[5,6]) We evaluate each nucleon β decay in the allowed approximation. The nuclear $\beta\beta$ decay amplitudes in time-dependent perturbation theory then become

$$\sum_M \frac{\langle F|\sum_{i=1}^{A} O(i)|M\rangle\langle M|\sum_{i=1}^{A} O(i)|I\rangle}{E_M^N+E^L-E_I^N}, \tag{1}$$

where the sum extends over a complete set of

VOLUME 47, NUMBER 3 PHYSICAL REVIEW LETTERS 20 JULY 1981

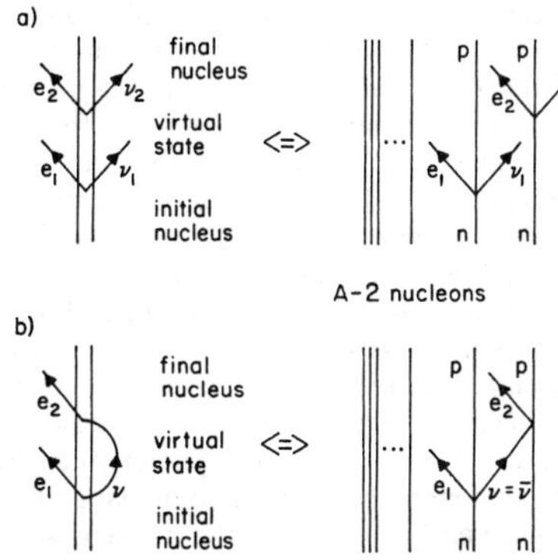

FIG. 1. Two-nucleon mechanisms for (a) two-neutrino and (b) no-neutrino $\beta\beta$ decay.

intermediate nuclear states M. Nuclear energies are denoted E^N, and E^L is the sum of the lepton energies emitted in the first β decay. The $O(i)$ represent either the Fermi or Gamow-Teller operators, $\tau_+(i)$ or $\vec{\sigma}(i)\tau_+(i)$.

The sum over intermediate nuclear states is performed by closure after replacing $E_M{}^N$ by an average value, either $\langle E_F{}^N\rangle$ or $\langle E_{GT}{}^N\rangle$. These are taken from the $(E_M{}^N+E^L)^{-1}$ moments of the Fermi and $T_<$ Gamow-Teller strengths for β^- decay of the initial nucleus, evaluated in a statistical model which incorporates the effects of pairing on the Fermi surface.[5] We find $\langle E_{GT}{}^N\rangle = 8.07$ (7.88) MeV and $\langle E_F{}^N\rangle = 9.17$ (8.79) MeV for ^{82}Se (^{76}Ge). (These are nuclear, not atomic, energies.) The resulting differential $\beta\beta$ decay rate $d\omega/d\epsilon_1 d\epsilon_2 d\nu_1$ then can be expressed in terms of the double Fermi and Gamow-Teller matrix elements[5]

$$M_F = \langle F|\tfrac{1}{2}\sum_{\substack{ij\\ i\neq j}} \tau_+(i)\tau_+(j)|I\rangle$$
$$M_{GT} = \langle F|\tfrac{1}{2}\sum_{\substack{ij\\ i\neq j}} \vec{\sigma}(i)\cdot\vec{\sigma}(j)\tau_+(i)\tau_+(j)|I\rangle \qquad (2)$$

which must be evaluated with realistic wave functions.

We describe the ground states of the initial and final nuclei of interest, ^{82}Se, ^{82}Kr, ^{76}Ge, and ^{76}Se, in weak coupling. First proton (neutron) wave functions are determined separately in full shell model calculations for four, six, and eight valence particles (two, four, six, and eight valence holes) in a model space consisting of the

TABLE I. Calculated and experimental $\beta\beta$ decay half-lives τ, in units of 10^{19} yr. The neutrino mass parameter $\eta = \langle m\rangle_\nu/m_e$. Experimental references are given in the text.

	$\tau_{calc}{}^{2\nu}$	$\tau_{expt}{}^{2\nu}$	$\tau_{calc}{}^{0\nu}$	$\tau_{expt}{}^{0\nu}$
^{82}Se	1.69[a]	1.0 ± 0.4	$1.59\times10^{-7}/\eta^2$	≥ 310
	2.35	27.6[b]		
^{76}Ge	37.3		$4.42\times10^{-7}/\eta^2$	≥ 500

[a] Explicit summation over intermediate states below 5 MeV in ^{82}Br makes this the more reliable theoretical value.
[b] This is the total geochemical half-life.

$2p_{1/2}$-$2p_{3/2}$-$1f_{5/2}$-$1g_{9/2}$ shells. This space is nonspurious.[7] The fifty proton states with the lowest energies are combined with a like number of neutron states to form a weak coupling basis for $J^\pi = 0^+$. These basis states are then allowed to mix through the proton-neutron interaction, and the resulting matrix is diagonalized to yield nuclear wave functions and eigenvalues.

In such calculations both the shell model effective interaction and the single particle energies must be specified. Potential matrix elements were taken from Kuo,[8] modified as discussed below, with the Coulomb interaction between protons added. The Kuo matrix elements are designed for the mass region near ^{56}Ni; it proved necessary to weaken this interaction somewhat to obtain a systematic fit to level spacings of nuclei near the middle of the fpg shell, reflecting the larger mean nucleon separations in these heavier nuclei. This reduction factor (= 0.8) and the proton single particle energies were determined by fitting thirteen levels in the closed-shell nuclei ^{84}Se, ^{85}Br, ^{86}Kr, ^{87}Rb, and ^{88}Sr. A similar fit to fourteen levels in the closed-shell nuclei ^{62}Ni, ^{63}Ni, and ^{65}Ni yielded the neutron energies. The mean square deviation from the experimental levels included in these fits is approximately 270 keV. Further details are given in Ref. 5.

The two-body density matrices determined from these weak coupling wave functions immediately yield $|M_{GT}| = 0.94$ (1.28) for ^{82}Se (^{76}Ge). The Fermi operator can be rewritten in terms of T_+T_+, where T_+ is the isospin raising operator, and thus cannot change the total isospin. Its matrix element differs from zero only to the extent that charge-dependent terms in the shell model potential mix the analog of the initial nuclear state (with $T = T_i = T_f + 2$, $M_T = -T_f$) into the ground state of the final nucleus. As the weak coupling approximation itself yields states without exact isospin, the Coulomb mixing of these states was calculated in perturbation theory. We find $|M_F| < 0.02 \ll |M_{GT}|$.

The results are shown in Table I. The calculated two-neutrino half-life for ^{82}Se, $\tau_{1/2}{}^{2\nu} = 2.35 \times 10^{19}$ yr, is in reasonable agreement with the result of Moe and Lowenthal,[2] $(1.0 \pm 0.4) \times 10^{19}$ yr, and more than an order of magnitude shorter than the total half-life determined from geochemical measurements of Srinivasan, Alexander, and Manuel,[3] 2.76×10^{20} yr. Calculations have also been performed[5] for the $\beta\beta$ decay transitions ^{128}Te$\rightarrow$^{128}Xe and ^{130}Te$\rightarrow$^{130}Xe, and rates much faster than the geochemical values are again found. Although we feel that the difficulty of the Te and Xe structure calculations render those results somewhat less reliable than the ^{82}Se results reported here, the suggestion of a systematic discrepancy between theory and the geochemical measurements is clear.

In view of this, we emphasize that the calculated ^{82}Se $\beta\beta$ decay amplitude is largely determined by a few dominant two-body density matrix elements which add coherently. Thus minor wave function adjustments are unlikely to alter our present result. Also, although no Gamow-Teller transition involving the intermediate nucleus ^{82}Br is known, a recent measurement of the $\frac{3}{2}^- \rightarrow \frac{1}{2}^-$ (0.19 MeV) transition from ^{81}Br$_{\rm g.s.}$ to ^{81}Kr does provide an opportunity for checking our weak coupling techniques. Use of proton bases identical to those employed here yields[9] $\log(ft) = 4.99$, in good agreement with the experimental value[10] $\log(ft) = 4.88 \pm 0.11$.

However, the most troublesome theoretical point may be in the derivation, and not the evaluation, of the matrix elements. The terms in Eq. (1) are not positive definite, so in principle the non-energy-weighted sum evaluated by closure may differ significantly from the exact result, particularly if terms with small E_M are important. To check the sum rule we generated the weak coupling 1^+ states lying within 5 MeV of the ground state of ^{82}Br and performed the summation of Eq. (1) explicitly for those states. Treating the higher excitations as before, this yields $\tau_{1/2}{}^{^{82}\mathrm{Se}} = 1.69 \times 10^{19}$ yr. Thus this more reliable estimate of $\tau_{1/2}{}^{^{82}\mathrm{Se}}$ is in good agreement with our previous value.

With these checks, in addition to the agreement with Moe and Lowenthal, encouraging some confidence in the present calculations, we turn to the neutrinoless $\beta\beta$ decay of ^{82}Se and ^{76}Ge. Primakoff and Rosen[1] have stressed that the nuclear matrix elements mediating the neutrinoless decay are expected to track those for two-neutrino decay; thus we expect, and find by explicit calculation, strong matrix elements for these cases as well.

The neutrinoless $\beta\beta$ decay amplitude is taken from the two-nucleon process shown in Fig. 1(b). (The comment made earlier on the absence of Δ-to-nucleon $\beta\beta$ decay transitions applies equally here.[5,6]) Details of the general evaluation of the decay rate (i.e., allowing both massive neutrinos and left- and right-handed admixtures in the current) are given in Ref. 5. In the present work we assume a Majorana electron neutrino and break the γ_5 invariance of the weak current by the neutrino mass. As in the two-neutrino case, the transition amplitude is taken from time-dependent perturbation theory; the integral over the intermediate neutrino momentum can be evaluated exactly for small m_ν, and the sum over nuclear states is performed by closure. The resulting differential decay rate $d\omega/d\epsilon_1$ is given in terms of nuclear matrix elements of the form[5]

$$M = \langle F | \tfrac{1}{2} \sum_{\substack{ij \\ i \neq j}} f(r_{ij}) [O(i) \times O(j)]_{00} | I \rangle$$

with $r_{ij} = |\vec{r}_i - \vec{r}_j|$, with $f(r_{ij})$ not too different from $1/r_{ij}$, and with $O(i) = \tau_+(i)$ or $\vec{\sigma}(i)\tau_+(i)$. [The third rotational invariant, $O = \vec{\sigma}(i) \cdot \hat{r}_{ij} \tau_+(i)$, arises only for admixed left- and right-handed currents.]

Our weak coupling wave functions then yield

$$\tau_{1/2}{}^{0\nu} = 1.59 \times 10^{12}\ \mathrm{yr}/\eta^2,\quad ^{82}\mathrm{Se},$$

$$\tau_{1/2}{}^{0\nu} = 4.42 \times 10^{12}\ \mathrm{yr}/\eta^2,\quad ^{76}\mathrm{Ge},$$

where $\eta = \langle m \rangle_\nu / m_e$ for this lepton-number- (and $B-L$)-nonconserving process. (Here $\langle m \rangle_\nu = \sum_i C_i{}^2 m_i$, with $C_i{}^2$ the probability of finding the electron neutrino in the eigenstate with Majorana mass m_i.) The laboratory limits for ^{82}Se and ^{76}Ge, $\tau_{1/2}{}^{0\nu} \geq 3.1 \times 10^{21}$ yr (Ref. 11) and $\tau_{1/2}{}^{0\nu} \geq 5 \times 10^{21}$ yr (Ref. 12), then yield $\langle m \rangle_\nu \leq 12$ eV and $\langle m \rangle_\nu \leq 15$ eV, respectively, at the 68% C.L. Clearly these constraints also apply separately to each $C_i{}^2 m_i$. Assuming a definite mass $m_1 = \langle m_\nu \rangle$, a recent measurement of the triton endpoint β spectrum gives $14 \leq m_1 \leq 46$ eV at the 99% C.L.[13] Thus if these experiments and the present calculations are correct, the electron neutrino is not a Majorana mass eigenstate.[14] The open question of the more general implications of our

$\langle m \rangle_\nu$ limit should encourage less restrictive analyses of the triton β spectrum.

It is apparent from this discussion that, given $\langle m \rangle_\nu \geq 14$ eV, existing limits on 0ν $\beta\beta$ decay are at the threshold of testing lepton-number conservation unambiguously. We thus believe it imperative to proceed with the new generation of experiments currently under discussion, including very sensitive searches for the neutrinoless decays of ^{76}Ge,[15] ^{82}Se,[2] and ^{136}Xe.[16] We also strongly urge a definitive laboratory measurement of the 2ν $\beta\beta$ decay of ^{82}Se,[2,17] as this rate provides a crucial test of the nuclear theory and of the assumptions implicit in geochemical determinations.

We thank A. Arima, F. T. Avignone, P. Herczeg, G. S. Hurst, L. S. Kisslinger, M. K. Moe, and S. P. Rosen for helpful discussions. This work was partially supported by the U. S. Department of Energy and by the National Science Foundation under Grant No. PHY-8021272.

[(a)]Present address.

[1]For general reviews, see H. Primakoff and S. P. Rosen, Rep. Prog. Phys. 22, 121 (1959); E. J. Konopinski, *Theory of Beta Radioactivity* (Oxford Univ. Press, London, 1966); D. Bryman and C. Picciotto, Rev. Mod. Phys. 50, 11 (1978). Also, see E. Grueling and R. C. Whitten, Ann. Phys. 11, 510 (1960); A. Halprin *et al.*, Phys. Rev. D 13, 2567 (1976); J. D. Vergados, Phys. Rev. C 13, 865 (1976), and University of Ioannina Report No. 149, 1980 (to be published).

[2]M. K. Moe and D. D. Lowenthal, Phys. Rev. C 22, 2186 (1980).

[3]B. Srinivasan, B. C. Alexander, and O. K. Manuel, Econ. Geol. 67, 592 (1972).

[4]C. Picciotto, Can. J. Phys. 56, 399 (1978).

[5]W. C. Haxton, G. J. Stephenson, and D. Strottman, to be published.

[6]A more general selection rule was first pointed out to us by P. Herczeg. Also, S. P. Rosen, in Proceedings of Orbis Scientiae 1981, University of Miami (to be published); M. Doi *et al.*, Osaka University Report No. OS-GE 80-27 (to be published); L. S. Kisslinger, private communication.

[7]Inclusion of either of the missing spin partners will introduce spuriosity. We discuss in Ref. 5 the possible effects of this truncation on nuclear matrix elements.

[8]T. T. S. Kuo, private communication; T. T. S. Kuo and G. E. Brown, Nucl. Phys. A114, 241 (1968).

[9]W. C. Haxton, to be published.

[10]C. L. Bennett *et al.*, Phys. Rev. C 22, 2245 (1980).

[11]B. T. Cleveland *et al.*, Phys. Rev. Lett. 35, 737 (1975).

[12]E. Fiorini *et al.*, Nuovo Cimento A13, 747 (1973).

[13]V. A. Lubimov *et al.*, Phys. Lett. 94B, 266 (1980).

[14]This argument is incomplete if the γ_5 invariance is also broken by explicit left- and right-handed admixtures. See Ref. 5.

[15]F. T. Avignone, III, R. L. Brodzinski, and N. A. Wogman, to be published.

[16]H. H. Chen and P. J. Doe, University of California at Irvine UCI-Neutrino Report No. 40, unpublished.

[17]G. S. Hurst, private communication.

Rejection of Evidence for Nonzero Neutrino Rest Mass from Double Beta Decay

T. Kirsten, H. Richter, and E. Jessberger
Max-Planck-Institut für Kernphysik, D-6900 Heidelberg, Germany

(Received 1 October 1982)

Previous evidence from double beta decay (DBD) of ^{128}Te and ^{130}Te required neutrinoless DBD involving a neutrino rest mass of $\hat{m}_\nu \sim 34$ eV (~ 10 eV in the most recent theoretical treatment) or else a strong violation of lepton-number conservation ($\eta \neq 0$) due to a $(V+A)$ admixture. The DBD rate ratio of ^{128}Te and ^{130}Te has been redetermined. The present result is consistent with $\hat{m}_\nu = \eta = 0$. The experimental limits still allow Majorana decay at levels of $\hat{m}_\nu \leq 5.6$ eV or $\eta \leq 2.4\times 10^{-5}$ (95% confidence). The DBD half-life for ^{128}Te is $T_{1/2}(^{128}Te) > 8\times 10^{24}$ yr (2σ).

PACS numbers: 23.40.Bw, 14.60.Gh, 27.60.+j

Double beta decay (DBD) may occur with neutrino emission (2ν, Dirac decay) or without neutrinos (0ν, Majorana decay). Neutrinoless DBD is made possible either by a right-handed leptonic current admixture[1] (violation of lepton-number conservation, amplitude η) or by an implicit helicity breaking due to a nonzero mass of the electron neutrino ($\hat{m}_\nu \neq 0$).[2-4] The presence or absence of measurable decay rates $\lambda_{0\nu}$ puts stringent limits on η and $\hat{m}_\nu$. Even very small values of $\eta, \hat{m}_\nu$ lead to effectively enhanced decay rates since Majorana decay is promoted by a phase-space factor $\sim 10^6$ relative to Dirac decay. DBD is thus considered to be "the most sensitive test for lepton-quark symmetry."[5]

DBD has been experimentally observed so far by the geochemical method.[6-11] This method yields effective decay rates $\lambda_\Sigma = \lambda_{0\nu} + \lambda_{2\nu}$ and one cannot distinguish between Majorana and Dirac decay. Limits on η have nevertheless been inferred from $\lambda_{0\nu} < \lambda_\Sigma$. These limits suffered, however, from the uncertainties of the nuclear matrix elements and hence the theoretical decay-rate predictions. Pontecorvo[12] pointed out that the situation is much improved if decay-rate ratios of pairs of similar nuclei such as ^{128}Te-^{130}Te (or ^{80}Se-^{82}Se) are considered because the ratio of their respective relevant nuclear matrix elements should be near unity. In addition, the ratio $\rho_{0\nu} = {}^{128}\lambda_{0\nu}/{}^{130}\lambda_{0\nu}$ becomes much larger than

VOLUME 50, NUMBER 7 PHYSICAL REVIEW LETTERS 14 FEBRUARY 1983

$\rho_{2\nu} = {}^{128}\lambda_{2\nu}/{}^{130}\lambda_{2\nu}$. This is because the decay energy of ^{128}Te (869 keV) is much lower than that of ^{130}Te (2533 keV). Even very small Majorana-decay contributions cause a drastic increase of the measurable ratio $\rho = \rho_\Sigma = {}^{128}\lambda_\Sigma/{}^{130}\lambda_\Sigma$ above the base level $\rho_{2\nu}$. Consequently, $\rho_{\rm exp} = \rho_\Sigma > \rho_{2\nu}$ if Majorana decay occurs at all.

Equally important are the experimental advantages in determining a decay-rate ratio rather than absolute rates. The geochemical method of DBD detection is based on the measurement of DBD product nuclei which have accumulated during long geological times in natural minerals rich in the parent nuclei. Tellurium minerals contain both ^{128}Te and ^{130}Te, and DBD produces ^{128}Xe* and ^{130}Xe*. (Throughout this paper an asterisk denotes excess over the atmospheric abundance of the same isotope in the sample.) The ratio $\rho = {}^{128}\lambda_\Sigma/{}^{130}\lambda_\Sigma$ is directly deduced from isotope ratio measurements. Errors in the effective accumulation time (absolute age error, or incomplete gas retention over geological time) or in the absolute Xe calibration will cancel.

Previous determinations of ρ gave the following results:

$\rho \leq 295\times10^{-4}$ (1σ) (Ref. 9);

$\rho \leq 194\times10^{-4}$ (1σ) (Ref. 10);

$\rho \leq 111\times10^{-4}$ (1σ) (Ref. 10);

$\rho \leq 5\times10^{-4}$ (1σ) (Ref. 13);

$\rho = (6.29\pm0.20)\times10^{-4}$ (Ref. 11);

$\rho = (6.37\pm0.41)\times10^{-4}$ (Ref. 11).

TABLE I. Xenon extracted from 4.494 g native tellurium, Goodhope Mine. (Melt fraction; 2 h at 480°C.)

	Sample	Atmosphere
^{124}Xe	0.3556 ± 0.0280[a]	0.3537
^{126}Xe	0.3208 ± 0.0208	0.3300
^{128}Xe	7.152 ± 0.036	7.136
^{129}Xe	146.9 ± 0.6	98.32
^{130}Xe	424.86 ± 1.69	15.136
^{131}Xe	113.36 ± 0.48	78.9
^{132}Xe	100	100
^{134}Xe	39.11 ± 0.48	38.79
^{136}Xe	33.39 ± 0.46	32.94
	After fission correction	
^{128}Xe$_{fc}$	7.175 ± 0.043	7.136
^{130}Xe$_{fc}$	426.25 ± 1.7	15.136

[a]All isotopes are measured relative to ^{132}Xe; errors stated are 1σ. ^{132}Xe component of sample is 460.5 $\times 10^{-14}$ cm^3/g.

The positive result of Hennecke, Manuel, and Sabu[11] is far above the theoretically expected ratio for Dirac decay ($\rho_{2\nu} \sim 2\times10^{-4}$).[5] It seemed to indicate the occurrence of Majorana decay and was interpreted by the authors in terms of $\eta \neq 0$. [The theoretical expectation values for $\rho_{0\nu}$ alone are between 62×10^{-4} ($\hat{m}_\nu = 0$) and 420×10^{-4} ($\eta = 0$).[5]]

With the advent of a quantitative theoretical formalism concerning the implicit helicity breaking by $m_\nu \neq 0$,[2-5,14] values of $\hat{m}_\nu \sim 34$ eV,[2] or, most recently, of $\hat{m}_\nu \sim 10$ eV,[4,5] have been inferred from the ρ of Ref. 11 (for $\eta = 0$).

In view of another observation (tritium end-point energy determination) calling for a nonvanishing neutrino rest mass of that same order[15] these data became extremely important and made an experimental confirmation desirable. We have redetermined ρ (and $^{130}\lambda_\Sigma$) in a mass spectrometric investigation of xenon in a sample from the same Precambrian tellurium ore which we used fourteen years ago, when DBD occurrence in nature was first unambiguously demonstrated for ^{130}Te.[6] The results are summarized in Tables I and II. Some experimental details are given

TABLE II. Partitioning of Xe components [units of 10^{-14} (cm^3 STP)/g] and deduced DBD quantities.

	Melt extraction	Total[b]
^{129}Xe$_{\rm Exc}$	223.6±2.7	251.6±3.1
^{131}Xe$_{\rm Exc}$	158.7±2.2	178.9±2.5
^{129}Xe$_{\rm Exc}$/^{131}Xe$_{\rm Exc}$	1.409±0.026	1.406±0.026
^{132}Xe$_{\rm measured}$	460.5	614.7
^{132}Xe$_{\rm Fission}$	1.47 ±1.47	
^{132}Xe$_{\rm F.corr.}$	459 ±18[a]	
^{130}Xe*	1887 ±74[a]	2113 ±83[a])
^{128}Xe*	0.18 (±) 0.20[a]	0.20 (±) 0.22[a]
$^{128}\lambda_\Sigma$ (10^{-26} y^{-1})[c]	$2.75^{+3.0}_{-2.75}$	
$^{130}\lambda_\Sigma$ (10^{-22} y^{-1})[c]	2.67 ± 0.29	
$R_{\rm exp} \equiv {}^{128}$Xe*/^{130}Xe*	(0.95 (±) 1.04)·10^{-4}	
$\rho = {}^{128}\lambda_\Sigma/{}^{130}\lambda_\Sigma$ [d]	(1.03 (±) 1.13)·10^{-4}	
$^{130}T_{1/2} = (2.60\pm0.28)\cdot10^{21}$ y;		
$^{128}T_{1/2}/{}^{130}T_{1/2} = \rho^{-1} =$ 9710 (mean)		
> 4425 (+ 1σ)		
> 3040 (+ 2σ)		

[a]Inclusive absolute calibration error.
[b]Inclusive sum of five preheating steps.
[c]Age = $(1.31\pm0.14)\times10^9$ yr.
[d]Obeying ^{128}Te/^{130}Te = 0.922.

VOLUME 50, NUMBER 7 PHYSICAL REVIEW LETTERS 14 FEBRUARY 1983

later.

Our result for the absolute half-life of ^{130}Te is $^{130}T_{1/2} = (2.60 \pm 0.28) \times 10^{21}$ yr. This agrees well with earlier measurements from different laboratories.[6,10] For $\rho = {}^{128}\lambda_\Sigma / {}^{130}\lambda_\Sigma$ we find $[1.03 \pm 1.13] \times 10^{-4}$ and cannot confirm the results of Ref. 11. In terms of $^{128}T_{1/2}$ our result $^{128}T_{1/2} \geqslant 8 \times 10^{24}$ yr (2σ) conflicts with $^{128}T_{1/2} = (1.54 \pm 0.17) \times 10^{24}$ yr.[11]

Our results are fully compatible with Dirac decay alone. Any possible contribution from Majorana decay is limited to values $\hat{m}_\nu \leqslant 5.6$ eV and $\eta \leqslant 2.4 \times 10^{-5}$ (95% confidence).

The allowed region for combinations of $\hat{m}_\nu \neq 0$, $\eta \neq 0$ is shown in Fig. 1 both for the result of Ref. 11 and for ours. The two results are incompatible. Also shown in Fig. 1 are the respective limits obtained from the search for neutrinoless DBD in a coincidence experiment with ^{82}Se.[4,5,16] The deduction of limits on m_ν and η from the data benefits from recent advances in our theoretical understanding of the relation between 0ν and 2ν matrix elements[4] and in the treatments of the DBD phase space.[4,5] The present analysis has been taken from a report by Doi *et al.*[5] that incorporates these improvements. The major remaining theoretical uncertainty is the value of the ratio $\xi = |M_{GT}^{0\nu}/M_{GT}^{2\nu}|$. The ξ value used $[\xi(128) \approx \xi(130) = 1.69]^4$ is conservative (see, e.g., Ref. 17) unless there are unrecognized large cancellations in the nuclear theory, a possibility which could explain a disagreement of theoretical and experimental matrix elements for ^{82}Se and ^{130}Te.[18] This caveat remains and cannot be resolved at present.

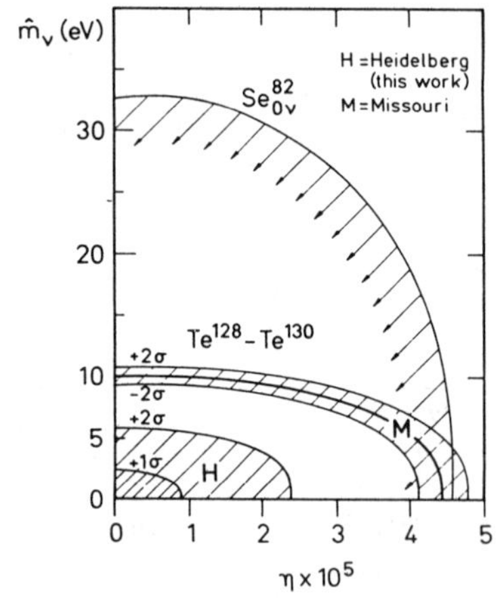

FIG. 1. Allowed regions in the $(\hat{m}_\nu, \eta)$ plane deduced from the measured ratio ρ of this work after theoretical treatment following Ref. 5. The neutrino rest mass is $\hat{m}_{\nu e} \leqslant 5.6$ eV (2σ). Note that the results of this work are incompatible with those of Ref. 11 ("Missouri") treated the same way. The ^{82}Se curve is calculated from the limit for $^{82}\lambda_{0\nu}$ set by Cleveland *et al.* (Ref. 16) in their direct counting experiment (1σ).

For details of the $\rho(\hat{m}_\nu, \eta)$ dependence we refer to Refs. 4, 5, and 19. In this Letter we give for better perception a numerically simplified approximation:

$$\rho \approx 1.97 + \hat{m}_\nu^2 + (\tfrac{1}{2}\eta)^2,$$

with $\hat{m}_\nu$ in units of $10^{-5} m_e$ (= 5.11 eV), η in units of 10^{-5}, and ρ in units of 10^{-4}. The approximation is valid for $\hat{m}_\nu \times \eta \leqslant 10$ in the same units. Note that the inferred neutrino rest mass (for $\eta = 0$) equals the square root of the difference of the experimentally determined ratio ρ and the respective ratio expected for Dirac decay alone ($\hat{m}_\nu = \eta = 0$). This illustrates the intrinsic power of the Te-ratio system. Deduced values are independent of absolute calibrations, ore ages, or gas retention properties.

In summary, from the results of our study we exclude a Majorana mass of the electron neutrino in excess of 5.6 eV at the 95% confidence level.

Experimental details and the complete results will be reported in a more comprehensive paper.[19] In this Letter we confine ourselves to the essentials.

The available sample was a compact piece of freshly appearing native tellurium from the Goodhope mine (Colorado). It was preserved from our earlier study.[6] The chemical analysis confirmed its purity [Te, (100 ± 0.6)%]. The K-Ar age of the ore had been determined earlier to be $(1.31 \pm 0.14) \times 10^9$ yr.[6] It should be mentioned here that both the K-Ar and the Te-Xe systems are based on rare-gas retention. In our context the time of gas retention is more relevant than the geological mineralization age if they should differ at all.

We have used our standard procedures for high-sensitivity mass spectrometric rare-gas analysis. The sample was preheated in a high-vacuum extraction line in five steps; 4 h each at 100, 200, 250, 300, and 330 °C. The Xe was analyzed in each step. Afterwards the sample was melted for 2 h at 480 °C ("melt extraction"). 89.3% of the total radiogenic ^{130}Xe* was released in the melt run (Table II) which gave ^{130}Xe*/^{130}Xe$_{atm}$ = 27. Table I contains the results. The errors stated are 1σ statistical errors from n subse-

VOLUME 50, NUMBER 7 PHYSICAL REVIEW LETTERS 14 FEBRUARY 1983

quent ratio determinations with ^{132}Xe serving as reference isotope ($n = 80$ for ^{128}Xe, ^{130}Xe, and ^{132}Xe; $n = 40$ for the other isotopes). Before and after the sample numerous air xenon standards of widely differing quantities [$(70-14\,000)\times 10^{-14}$ cm^3 STP ^{132}Xe] were analyzed. This served to check the reproducibility of isotope ratio determinations, to determine the mass fractionation $(0.32\pm 0.02)\%/$u, and to perform absolute sensitivity calibration.

Table II contains the results of the decomposition of the measured Xe into its components. As known from earlier studies of Xe in tellurium ores[7,9,11] these are as follows. (i) Xe of atmospheric composition: ^{132}Xe, ^{134}Xe, and ^{136}Xe are entirely atmospheric except for a small amount of U-fission Xe. (ii) Resonance-neutron-capture Xe: $^{129}Xe_{exc}$, $^{131}Xe_{exc}$. The observed ratio $^{129}Xe_n/{}^{131}Xe_n \sim 1.4$ is typical.[7,20] (iii) Radiogenic xenon (DBD): $^{128}Xe^*$, $^{130}Xe^*$. (iv) Fission xenon from ^{238}U: minor amounts of ^{136}Xe, ^{134}Xe, ^{132}Xe, ^{131}Xe; no ^{128}Xe, ^{130}Xe (shielded isotopes). The relevance of U-fission Xe is limited to the slight reduction of $^{132}Xe_{atm}$. This is the reference isotope needed to infer the atmospheric quantities of $^{128}Xe_{atm}$ and $^{130}Xe_{atm}$ above which $^{128}Xe^*$ and $^{130}Xe^*$ appear as excess. The fission contributions are deduced from the known isotopic compositions of Xe_{fiss} and of Xe_{atm} for ^{132}Xe, ^{134}Xe, and ^{136}Xe. The quantities are in accordance with a very low U-concentration of the ore (≤ 40 parts per 10^9).

The excess $^{130}Xe^*$ is undisputably due to DBD (see Refs. 6–8 for exclusion of other sources of $^{130}Xe_{exc}$).

The possible excess of $^{128}Xe^*$ is limited to $(0.20^{+0.22}_{-0.20})\times 10^{-14}$ (cm^3 STP)/g (1σ). Before assigning this excess to DBD it would be necessary in principle to exclude all other possible sources of excess ^{128}Xe. Examples are neutron capture on traces of ^{127}I in the ore or experimental artifacts such as the "memory" effect. These potential sources of larger apparent ^{128}Te decay constants must be definitely excluded before a positive result on $^{128}Xe^*$ is assigned to DBD. However, all these potential sources of error can only increase $^{128}Xe^*$. In our case the measured quantity leads only to an upper limit of the ^{128}Te decay constant, and hence these effects could only reduce this upper limit even further. Therefore we assume $^{128}Xe^* = {}^{128}Xe_{DBD}$. This is the most conservative assumption with respect to the inferred half-lives and theoretical consequences (Table II).

We acknowledge stimulating discussions with W. Hampel, W. Haxton, T. Kotani, O. Manuel, P. Minkowski, and S. Rosen.

[1]H. Primakoff and S. Rosen, Rep. Prog. Phys. 22, 121 (1959); for a recent review see H. Primakoff and S. Rosen, Ann. Rev. Nucl. Sci. 31, 145 (1981).

[2]M. Doi, T. Kotani, H. Nishimura, K. Okuda, and E. Takasugi, Phys. Lett. 103B, 219 (1981), and 113B, 513 (1982).

[3]A. Halprin, P. Minkowski, H. Primakoff, and S. Rosen, Phys. Rev. D 13, 2567 (1976).

[4]W. Haxton, G. Stephenson, and D. Strottman, Phys. Rev. D 25, 2360 (1982), and Phys. Rev. Lett. 47, 153 (1981).

[5]M. Doi, T. Kotani, H. Nishimura, and E. Takasugi, Osaka University Report No. Os-GE 82-43, 1982 (to be published).

[6]T. Kirsten, O. Schaeffer, E. Norton, and R. Stoenner, Phys. Rev. Lett. 20, 1300 (1968).

[7]T. Kirsten, W. Gentner, and O. Müller, Z. Naturforsch. 22a, 1783 (1967); T. Kirsten, W. Gentner, and O. A. Schaeffer, Z. Phys. 202, 273 (1967); T. Kirsten and H. W. Müller, Earth Planet. Sci. Lett. 6, 271 (1969); T. Kirsten, H. Müller, and E. Norton, in *Recent Developments in Mass Spectroscopy*, edited by K. Ogata (Univ. of Tokyo Press, Tokyo, 1970), p. 501.

[8]T. Kirsten, Fortsch. Phys. 18, 449 (1970).

[9]N. Takaoka and K. Ogata, Z. Naturforsch. 21a, 84 (1966).

[10]E. Alexander, B. Srinivasan, and O. Manuel, Earth Planet. Sci. Lett. 5, 478 (1969); B. Srinivasan, E. Alexander, and O. Manuel, J. Inorg. Nucl. Chem. 34, 2381 (1972).

[11]E. Hennecke, O. Manuel, and D. Sabu, Phys. Rev. C 11, 1378 (1975); E. Hennecke, Phys. Rev. C 17, 1168 (1978).

[12]B. Pontecorvo, Phys. Lett. 26B, 630 (1968).

[13]T. Kirsten, E. Jessberger, and J. Kiko, Max-Planck-Institut für Kernphysik, Heidelberg, Annual Report, 1972 (unpublished), p. 199; made known to the authors of Ref. 11 and cited by them therein.

[14]S. Rosen, in *Proceedings of the International Conference on Neutrino Physics and Astrophysics: Neutrino 81, Maui, Hawaii, 1981*, edited by R. J. Cence, E. Ma, and A. Roberts (Univ. of Hawaii Press, Honolulu, 1982), Vol. 2, p. 76.

[15]V. Lubimov, E. Novikov, V. Nozik, E. Tretyakov, and V. Kozik, Phys. Lett. 94B, 266 (1980).

[16]B. Cleveland, W. Leo, C. Wu, L. Kasday, A. Rushton, P. Gollon, and J. Ullman, Phys. Rev. Lett. 35, 757 (1975).

[17]P. Minkowski, Nucl. Phys. B 201, 269 (1982).

[18]T. Kirsten, in "Science Underground—1982," edited by M. M. Nieto, AIP Conference Proceedings No. 96 (American Institute of Physics, New York, to be published).

[19]T. Kirsten, H. Richter, E. Jessberger, Max-Planck-Institut Report No. MPI H 82 V19, 1982 (to be published).

[20]J. Browne and B. Berman, Phys. Rev. C 8, 2405 (1973).

2.2 The Nuclear Physics Side - Nuclear Matrix Elements

2.2.1 A Breakthrough to the Understanding of the Matrix Elements

J. Phys. G: Nucl. Phys. **9** (1983) L169–L173. Printed in Great Britain

LETTER TO THE EDITOR

Double beta decay of ^{128}Te/^{130}Te and the question of a neutrino Majorana mass

K Grotz, H V Klapdor and J Metzinger
Max-Planck-Institut für Kernphysik, Heidelberg, West Germany

Received 3 May 1983

Abstract. The 2ν double beta ($\beta\beta$) decay of ^{130}Te and ^{128}Te is calculated in the PBCS pairing model. A correct treatment of the phase-space integral shows that the Pontecorvo argument leads to an incorrect half-life ratio. This has severe impact on the discussion of neutrinoless β decay and a possible neutrino mass.

In the discussion of the question of neutrino mass, double beta ($\beta\beta$) decay (Primakoff and Rosen 1950, Konopinski 1966), especially of the isotopes ^{128}Te and ^{130}Te, has been attributed an important role. If the electron neutrino possesses a Majorana mass and, as a consequence, lepton number is not conserved, neutrinoless $\beta\beta$ decay should be possible (Halprin *et al* 1976). This additional decay mode would shorten the half-life of the decaying nucleus. Because of the uncertainties in calculations of absolute half-lives, pairs of nuclei differing by only two nucleons were thought to provide a much better test. The argument, first made by Pontecorvo (1968), was, that in such cases, differences in the nuclear matrix elements can be neglected and if one compared the two decay rates, the ratio should only depend on the nuclear mass differences. Most experimental information is available for the pair ^{128}Te and ^{130}Te (Inghram and Reynolds 1949, 1950, Takaoka and Ogata 1966, Kirsten *et al* 1967, 1983, Alexander *et al* 1969, Srinivasan *et al* 1972a, b, Hennecke *et al* 1975). The measurement of Hennecke *et al* (1975) leads to the half-life ratio $\tau_{128}/\tau_{130} = (1.6 \pm 0.05) \times 10^3$, whereas the simple Pontecorvo argument combined with the assumption that transitions other than ground-state ones are negligible gives $\tau_{128}/\tau_{130} \approx 6 \times 10^3$ for pure 2ν $\beta\beta$ decay. The same result is obtained in a simple shell-model calculation by Vergados (1976) and also in an investigation with an extended model space by Haxton *et al* (1982). From the factor three between the Hennecke measurement and the calculations mentioned above, several authors calculated a non-vanishing neutrino mass of $m_{\nu_e} \approx 34$ eV (Doi *et al* 1981), $m_{\nu_e} \approx 10$ eV (Haxton *et al* 1982) and $m_{\nu_e} > 0.8$ eV (Minkowski 1982). However, the recent measurement of Kirsten *et al* (1983) yields $\tau_{128}/\tau_{130} > 4.2 \times 10^3$ (1σ) or $> 3.0 \times 10^3$ (2σ), contradictory to Hennecke *et al.* Clearly, with this result the above argumentation leads only to an upper limit $m_{\nu_e} \leqslant 5.6$ eV (2σ) (Kirsten *et al* 1983). However, we are not going to discuss the correctness of the experimental methods, but restrict ourselves to theoretical aspects.

The most sophisticated calculation up to now by Haxton *et al* (1982) led to a ^{130}Te decay rate a factor of 60–160 too large, in comparison with the different experimental values. It will be shown that the presently existing theoretical approaches not only lead to wrong absolute 2ν $\beta\beta$ decay rates, but also the calculated 2ν rate ratio is much more uncertain than commonly assumed.

The 2ν $\beta\beta$ decay of a nucleus (Z, N) is described in perturbation theory by the transition amplitude

$$\sum_M \frac{\langle \mathrm{F}|\, \Sigma\, O(i)|M\rangle\langle M|\, \Sigma\, O(i)|\mathrm{I}\rangle}{E_M - E_\mathrm{I} + E_\mathrm{L}} \tag{1}$$

with the summation extending over a complete set of intermediate states M in the nucleus $(Z+1, N-1)$. E_I and E_M are the nuclear energies of the initial and intermediate states and E_L is the total energy of the leptons emitted in the first stage. $O(i)$ denotes the Fermi or Gamow–Teller operators $\tau_-(i)$ and $\sigma(i)\tau_-(i)$. Contributions from the Fermi operator are assumed to be negligibly small. The usual procedure to calculate $\beta\beta$ rates, also applied by Haxton *et al*, was to avoid the summation in (1) by replacing the energy denominator by some mean value and to complete the sum by closure. The evaluation of the operator $\Sigma_{ij}\, \frac{1}{2}\sigma(i)\tau_-(i)\sigma(j)\tau_-(j)$ between the 0^+ states then leads to matrix elements typically of order one.

In a previous investigation, Haxton *et al* (1981) found a 30% shorter half-life for the 2ν $\beta\beta$ decay of ^{82}Se when they performed an explicit summation over the 1^+ states within 5 MeV excitation energy in ^{82}Br. Although this result cannot be regarded as a firm basis for the closure approximation, they used the latter exclusively when calculating half-life *ratios*, also for the Te isotopes. We feel, however, that the justification for this approximation still has to be shown (see also Klapdor 1983) and studied it in this paper for the Te isotopes by performing the sum over intermediate 1^+ states in (1) for excitation energies up to about 20 MeV.

Obeying the commutation rules for the final lepton states (Konopinski 1966), one can get the 2ν rate of $0^+ \to 0^+$ transitions:

$$\omega_{2\nu} = \frac{F_\mathrm{A}^4 G^4}{32\pi^7} \int_{m_e}^{t_0+m_e} F(Z, \varepsilon_1)|K_1|\varepsilon_1\, \mathrm{d}\varepsilon_1 \int_{m_e}^{t_0+2m_e-\varepsilon_1} F(Z, \varepsilon_2)|K_2|\varepsilon_2\, \mathrm{d}\varepsilon_2$$
$$\times \int_0^{t_0-\varepsilon_1-\varepsilon_2+2m_e} \nu_1^2 \nu_2^2\, \mathrm{d}r_1 \sum_{MM'} A_{MM'} \tag{2}$$

with

$$A_{MM'} = \langle \mathrm{F}\|\sigma\tau_-\|M\rangle\langle M\|\sigma\tau_-\|\mathrm{I}\rangle\langle \mathrm{F}\|\sigma\tau_-\|M'\rangle\langle M'\|\sigma\tau_-\|\mathrm{I}\rangle$$
$$\times \tfrac{1}{3}(K_M K_{M'} + L_M L_{M'} + \tfrac{1}{2}K_M L_{M'} + \tfrac{1}{2}L_M K_{M'})$$

and

$$K_M = \frac{1}{E_M - E_\mathrm{I} + \varepsilon_1 + \nu_1} + \frac{1}{E_M - E_\mathrm{I} + \varepsilon_2 + \nu_2}$$
$$L_M = \frac{1}{E_M - E_\mathrm{I} + \varepsilon_1 + \nu_2} + \frac{1}{E_M - E_\mathrm{I} + \varepsilon_2 + \nu_1}$$
$$F_\mathrm{A} = 1.25.$$

Equation (2) agrees with the expression given by Haxton *et al* when approximating the energies E_M by some mean value.

We describe the nuclear states in the BCS pairing model as in Grotz *et al* (1983a, b) with additional particle number projection (PBCS) and evaluate equation (2) by numerical integration. The Fermi function was taken from Feister (1950). The gap parameters Δ_n have been calculated separately for initial and final 0^+ states using $G_n = 0.167$ MeV. The influence of Δ_p on the result was studied using various values for G_p. This is important because of the nearby $Z = 50$ shell closure and the resulting uncertainty for Δ_p. The matrix elements $\langle 0_F^+ \| \sigma\tau_- \| 1^+ \rangle$ occurring in (2) are replaced by the time-reversed process $\langle 1^+ \| \sigma\tau_+ \| 0_F^+ \rangle$ and the $\beta\beta$ decay can be thought of as a product of β^- and β^+ decay amplitudes starting from the 0^+ PBCS ground states. The intermediate 1^+ states are strongly influenced by the Gamow–Teller force and have been diagonalised in a configuration space including $1f_{7/2}$ to $2f_{5/2}$ neutron and $1f_{5/2}$ to $1i_{11/2}$ proton states. The calculations have been performed for two sets of single-particle levels (SPL). The first set SPL I was achieved by reproducing the energy levels in the ^{132}Sn region, and orbits not deducible by this method were taken from an oscillator potential with $\hbar\omega = 8.14$ MeV. A second set SPL II was fixed by the restriction to reproduce the measured (Madey 1983) ^{128}Te(p, n)^{128}I and ^{130}Te(p, n)^{130}I spectra. For details see Grotz *et al* (1983a, b). The β^- strength distribution in ^{128}I for SPL II is shown in figure 1. Resulting $\beta\beta$ half-lives for different proton gap parameters Δ_p are given in table 1. In a recent investigation of $\beta\beta$ decay of ^{76}Ge in the Nilsson pairing model, Zamick and Auerbach (1982) claimed that contributions from different shell transitions add coherently to give a large transition rate because matrix elements of $\sigma_1\sigma_2$ all have the same sign. However, there are at least two effects which lead to destructive interference. If one takes into account the mixing of the intermediate states, the summands $\langle 0^+ \| \sigma\tau_- \| M \rangle \langle M \| \sigma\tau_- \| 0^+ \rangle$ in equation (1) no longer are all of the same sign. Secondly, the sign of some $\sigma_1\sigma_2$ contributions changes after inclusion of four-quasiparticle admixtures of the form $|(\mathrm{np}(1^+); \mathrm{n'p'}(1^+))0^+\rangle$ to the ground state which have been shown by Randrup (1973) to strongly reduce β^+ amplitudes. Indeed, after including these effects, we find that while transitions via high-energetic states around the GTGR in ^{130}I (and also in ^{128}I) add coherently, some lower 1^+ states contribute

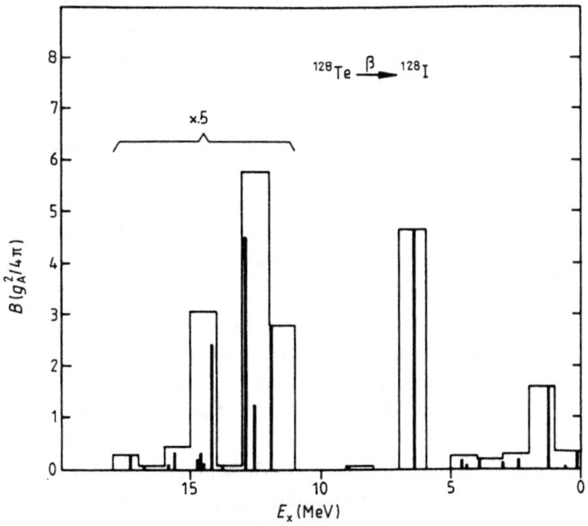

Figure 1. The calculated β^- strength distribution in ^{128}I (with SPL II and $\Delta_p = 1.22$ MeV).

Table 1. Calculated half-lives and half-life ratios for 2ν $\beta\beta$ decay with different gap parameters Δ_p in Te/Xe. Values in parentheses correspond to SPL II. The experimental numbers correspond to the sum of 2ν and 0ν decay rates.

This work Δ_p (MeV) in Te/Xe	Half-life of ^{130}Te/10^{19} yr	Half-life of ^{128}Te/10^{23} yr	$(\tau_{128}/\tau_{130}) \times 10^{-3}$
0.2/0.2	4.4 (1.4)	1.8 (0.59)	4.0 (4.2)
0.4/0.4	2.6 (0.54)	0.68 (0.23)	2.6 (4.3)
0.6/0.9	0.74 (0.35)	0.25 (0.15)	3.4 (4.2)
0.8/1.1	0.63 (0.34)	0.25 (0.14)	4.0 (4.1)
1.1/1.5	0.54 (0.37)	0.19 (0.16)	3.5 (4.3)
Haxton *et al* (1982)	1.8	0.89	5.0
Experiment:			
Hennecke *et al* (1975)	90–110	15.4 ± 1.7	1.6 ± 0.05
Kirsten *et al* (1983)	260 ± 28	> 80	> 4.6 (1σ) > 3.0 (2σ)

with the opposite sign. The number and positions of these states depend very much on details in the wavefunctions. However, still the calculated transition rates are much higher than the experimental ones (see table 1), at least by one order of magnitude even for an unrealistically small proton pairing gap Δ_p. The strong dependence of the calculated half-lives on Δ_p can easily be understood by the fact that partial occupation of high proton orbits opens more β^+ channels. Our decay rates assuming a realistic pairing strength ($\Delta_p = 0.8$ MeV for Te in table 1) are even higher than in the calculation of Haxton *et al* (1982). This might be explained by the larger configuration space used in our treatment.

So, if one assumes that the half-lives determined in the geochemical measurements are correct (which seems somewhat questionable when noting that equation (2) already yields the observed rates with the *experimental* matrix elements of the 1^+ ground state of ^{128}I, neglecting all other transitions completely), one is forced to look for alternative mechanisms which can strongly reduce the calculated 2ν $\beta\beta$ rates. One effect not included in the calculation is the influence of Δ–h admixtures in the 1^+ states. The estimate of Zamick and Auerbach (1982), which leads only to a small effect, is probably too simple. The Δ–h admixtures are expected to reduce the single β^- transition by about 30%. However, the influence on β^+ amplitudes, and therefore also on $\beta\beta$ amplitudes, has also to be considered. A strong effect on β^+ GT transitions was recently pointed out by Brown *et al* (1983) in the context of (n, p) experiments.

The second result of table 1 is the significant variation of the half-life ratio for the different parameters used. The range of values obtained in the previous and in this investigation shows that the theoretical uncertainty for this quantity is much bigger than commonly assumed. The results also indicate the unreliability of the closure approximation.

Although the ratio seems to be relatively stable when we use SPL II, with which the measured β^- GT strength function is well reproduced (except for the fact that the GT giant resonance carries only part of the calculated strength), we are not going to draw conclusions from this result. The half-lives are somewhat smaller than in the calculation with SPL I, which yields a somewhat different β^- GT strength function within the first few MeV of excitation energy.

In conclusion, according to our results and the results of previous investigations, we do not believe that the problem of double beta decay at present can be solved consistently. Provided that the underlying basic $\beta\beta$-decay theory used in *all* theoretical approaches up to now is correct, the fact that using only the experimental β-decay matrix elements of the 1^+ GS of ^{128}I is already sufficient to reproduce the observed 2ν $\beta\beta$-decay rate casts doubt on the experimental determinations of the $\beta\beta$-decay rates. On the other hand, it is necessary to include additional effects in the calculation which can lead to a reduction of 2ν $\beta\beta$ transition rates. One important extension would be the inclusion of subnuclear degrees of freedom.

References

Alexander E C Jr, Srinivasan B and Manuel O K 1969 *Earth Planet Sci. Lett.* **5** 478
Brown V R *et al* 1983 *Phys. Rev. Lett.* **50** 658
Doi M *et al* 1981 *Phys. Lett.* **103B** 219
Feister I 1950 *Phys. Rev.* **78** 375
Grotz K *et al* 1983a *Phys. Lett.* in press
Grotz K, Klapdor H V and Metzinger J 1983b to be published
Halprin A, Minkowski P, Primakoff H and Rosen S P 1976 *Phys. Rev.* D **13** 2567
Haxton W C, Stephenson G J Jr and Strottmann D 1981 *Phys. Rev. Lett.* **47** 153
—— 1982 *Phys. Rev.* D **25** 2360
—— 1983 *Comm. Nucl. Part. Phys.* **11** 41
Hennecke E W, Manuel O K and Sabu D D 1975 *Phys. Rev.* C **11** 1378
Inghram M G and Reynolds J H 1949 *Phys. Rev.* **76** 1265
—— 1950 *Phys. Rev.* **78** 822
Kirsten T, Gentner W and Schaeffer O 1967 *Z. Phys.* **202** 273
Kirsten T, Richter H and Jessberger E 1983 *Phys. Rev. Lett.* **50** 474
Klapdor H V 1983 *Prog. Part. Nucl. Phys.* in press
Konopinski E J 1966 *Theory of Beta Radioactivity* (London: Oxford University Press)
Madey R 1983 Private communication
Minkowski P 1982 *Nucl. Phys.* B **201** 269
Pontecorvo B 1968 *Phys. Lett.* **26B** 630
Primakoff H and Rosen S P 1950 *Rep. Prog. Phys.* **22** 121
Randrup J 1973 *Nucl. Phys.* A **207** 209
Srinivasan B, Alexander E C Jr and Manuel O K 1972 *Econ. Geol.* **67** 592
—— 1972 *J. Inorg. Nucl. Chem.* **34** 2381
Takaoka N and Ogata K 1966 *Z. Naturf.* **21a** 84
Vergados J 1976 *Phys. Rev.* C **13** 865
Zamick L and Auerbach N 1982 *Phys. Rev.* C **26** 2185

Volume 132B, number 1,2,3 PHYSICS LETTERS 24 November 1983

THE QUENCHING OF LOW-ENERGETIC NUCLEAR GAMOW–TELLER STRENGTH BY Δ-EXCITATIONS

K. GROTZ, H.V. KLAPDOR and J. METZINGER
Max-Planck-Institut für Kernphysik, Heidelberg, Germany

Received 19 July 1983

It is found in shell model calculations including ground state correlations, that Δ–nucleon hole (Δ–h) mixing affects the low-lying β^- Gamow–Teller transitions much less than the Gamow–Teller giant resonance (GTGR). This can be understood in a simple perturbation treatment as a destructive interference of first and second order contributions. It is therefore *not* possible to describe the Δ-quenching mechanism by a constant renormalization of the axial vector current.

Since systematic studies of the Gamow–Teller giant resonance (GTGR) in (p,n) charge exchange experiments have been performed, it is well-known that only about 33–51% [1–3] of the sum rule strength is experimentally found, the main part concentrated in the GTGR. One possible mechanism which reduces the Gamow–Teller (GT) strength was found in the Δ_{33} resonance [4]. Admixtures of Δ–nucleon hole (Δh) configurations lead to a reduction of the GT-strength in the region of nucleonic excitations of the order of 30% [5,6] compared to the sum rule.

It will be shown in this paper, that the application of this reduction factor to *low*-energetic GT-states overestimates the influence of the Δ–h excitations by roughly a factor of 2. This means that this quenching mechanism can *not* be described by a constant renormalization of the axial vector current. It will be further shown, that the small quenching of the low-lying strength is mainly arising from a destructive interference of the coupling of the low-lying states to the collective Δ–h excitations and to the Δ–h admixture in the GTGR.

We follow Bohr and Mottelson [4] and replace the operators $\sigma(i)\,\tau(i)$ acting on the nucleon i by

$$F(i) = \frac{3}{5} \sum_{q\in i} \sigma(q)\,\tau(q)\,, \qquad (1)$$

in which the spin and isospin operators act now on the quarks q in the nucleon i. Two collective $|\Delta h\rangle$ states

$$|\Delta h, T=1, T_z=+1\rangle$$
$$= [2/(Z+3N)]^{1/2}(\sqrt{3N}\,|\Delta^- n^{-1}\rangle - \sqrt{Z}\,|\Delta^0 p^{-1}\rangle)\,,$$
$$|\Delta h, T=1, T_z=-1\rangle$$
$$= [2/(3Z+N)]^{1/2}(\sqrt{3Z}\,|\Delta^{++} p^{-1}\rangle - \sqrt{N}\,|\Delta^+ n^{-1}\rangle)\,, \qquad (2)$$

evolve from the action of the new Gamow–Teller operators, defined by

$$Y^\pm = \sum_i F^\pm(i) = \frac{3}{5} \sum_i \sum_{q\in i} \sigma(q)\,\tau^\pm(q)\,, \qquad (3)$$

on the parent ground state. The excitation energy of these states is assumed to be 300 MeV plus a contribution from the collective interaction. The splitting of different orbits is neglected in this treatment.

The GT-strength function is calculated in a BCS shell model basis, diagonalizing a residual GT-interaction in TDA and including ground state correlations (gsc) in the parent nucleus explicitly to lowest order as described in ref. [7]. In addition, particle number projection is performed and the long range GT force [7] is extended to include (Δh) excitations (see fig. 1):

$$V_{GT} = 2\chi[F^+(i)\,F^-(j) + F^-(i)\,F^+(j)]\,. \qquad (4)$$

Volume 132B, number 1,2,3 PHYSICS LETTERS 24 November 1983

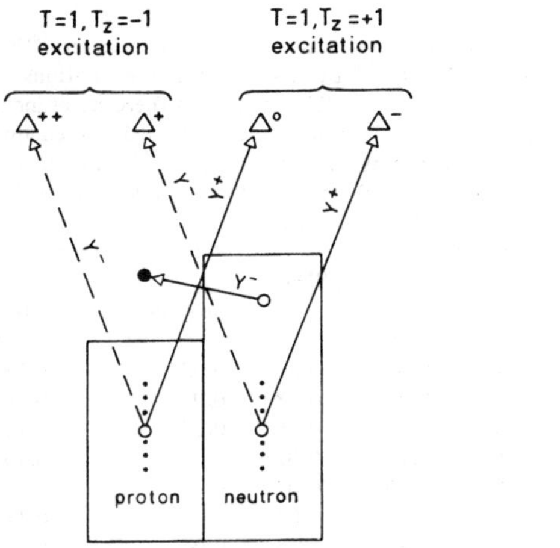

Fig. 1. Schematic picture of excitations involving the Δ-resonances: The dashed arrows correspond to a collective Δ–nucleon hole excitation by the action of the operator Y^-, carrying the quantum numbers $T = 1$, $T_z = -1$ and $L = 0$, $S = 1$. This excitation mixes with the nucleonic β^- excitations. The full arrows show an excitation mixing with the parent ground state (gsc). This involves a Δh excitation with $T = 1$, $T_z = +1$, mediated by Y^+ and a nucleonic ph β^- excitation.

In contrast to the purely nucleonic part of V_{GT}, where matrix elements are exactly evaluated, we retain only the direct term in the interaction of ph and Δh states (see fig. 2a):

$$\langle \Delta h, T = 1, T_z = -1 | V_{GT}^{dir} | j_n j_p \rangle$$

$$= \chi'(N + 3Z)^{1/2} u_p v_p \langle j_n \| \sigma \| j_p \rangle ,$$

$$\langle \Delta h, T = 1, T_z = +1 | V_{GT}^{dir} | j_n j_p \rangle$$

$$= \chi'(Z + 3N)^{1/2} v_p u_p \langle j_n \| \sigma \| j_p \rangle , \qquad (5)$$

$$\chi' = \tfrac{4}{5}(\tfrac{2}{3})^{1/2} \chi .$$

Only this one is of collective character and the exchange term is therefore much smaller.

The Δ-resonance is also present in the ground state due to gsc (see fig. 2b). The coupling is

$$\langle \Delta h, T_z = +1; j_p j_h, T_z = 1 | V_{GT}^{dir} | 0 \rangle$$

$$= \sqrt{3}\chi'(Z + 3N)^{1/2} u_p v_n \langle j_n \| \sigma \| j_p \rangle . \qquad (6)$$

An analogous expression holds for the $\langle \Delta h, T_z = -1 |$

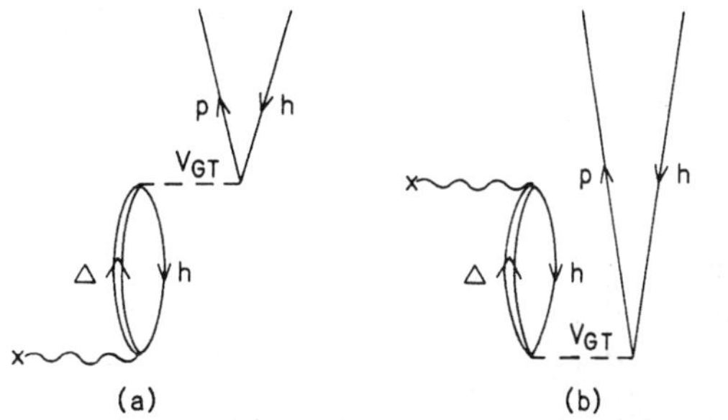

Fig. 2. (a) Final state interaction with Δ–h excitations. (b) Ground state correlation graph.

admixture, but only the $T_z = +1$ component is important for β^- transitions, because only this one can be annihilated by Y^-.

Numerical results of the above effects will be discussed for ^{208}Pb and ^{90}Zr. Diagonalization of V_{GT} was performed in a space including the collective Δh states and $\pi 1g_{9/2} - \pi 1j_{13/2}$, $\nu 1g_{9/2} - \nu 3d_{5/2}$ nucleonic single particle orbits in ^{208}Pb. Single particle energies as in ref. [7] were used. For ^{90}Zr the calculation was done in a harmonic oscillator basis with $\hbar\omega = 9.2$ MeV including $1f_{7/2} - 1h_{9/2}$ orbits. The coupling constant χ was taken to be 0.11 MeV and 0.17 MeV for ^{208}Pb and ^{90}Zr, respectively. These values resulted from a fit to the position of the GTGR in the mass region $A = 76 - 208$. Calculated β^- strength functions are shown in fig. 3a for ^{208}Pb and fig. 3b for ^{90}Zr. The Δ–h admixtures reduce the transition strength of *all* GT-states. However, the quenching of the GTGR is *markedly stronger* than of the other states. Tables 1 and 2 show the quenching factors, which we define as $q_\Delta = 1 - B_{GT}^{\Delta}/B_{GT}^{0}$, where B_{GT}^{Δ} and B_{GT}^{0} are the reduced transition probabilities, the index Δ or 0 referring to the calculations with and without Δ included, respectively. B_{GT}^{0} involves already nucleonic gsc, which lead also to a reduction of a few percent, however, can not quench the β strength below the sum rule value $3(N-Z)$. While the 34% quenching of the GTGR in ^{208}Pb is significantly stronger than in ^{90}Zr with only 23%, the quenching of strength below the GTGR is only ~14% in both cases and indicates no mass-dependence. We find also that the quenching of low-lying states is *not* very sensitive to the actual wave function varying between 10% and 20% with only few exceptions.

Volume 132B, number 1,2,3 PHYSICS LETTERS 24 November 1983

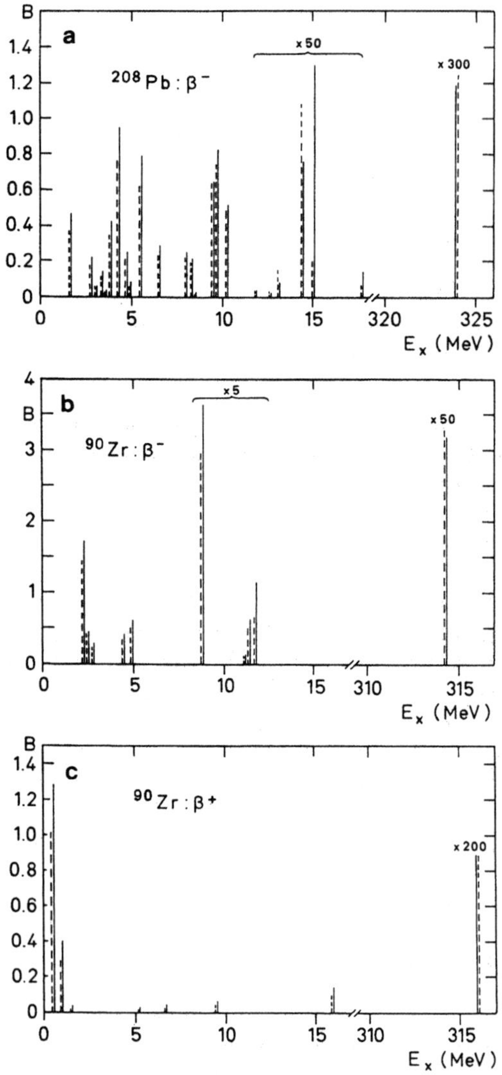

Fig. 3. (a) Calculated reduced β^- Gamow–Teller transition probability (in units of $g_A^2/4\pi$) for decay of ^{208}Pb without coupling to Δ–h excitations (solid lines) and including the interaction graphs fig. 2(a) and (b) (dashed lines). (b) Same as (a) for β^- decay of ^{90}Zr. (c) Strength calculated for β^+ decay of ^{90}Zr. Notation as in (a).

In tables 1 and 2 (forth column) quenching factors q_Δ ignoring gsc completely, as well in B_{GT}^{Δ} as in B_{GT}^{0}, are also given. From these figures it follows that there is *no* quenching of the low-lying GT strength in ^{208}Pb by final state interaction with Δ–h excitations (fig. 2a),

Table 1
Calculated energy and strength of GT states in ^{208}Bi (β^- decay of ^{208}Pb to ^{208}Bi) neglecting Δ–h excitations are given in columns 1 and 2. The strength of these states including the effect of Δ–h admixtures and the resulting quenching q_Δ is shown in columns 3 and 5. The quenching factors in the forth column result from a calculation without including gsc graphs [c.f. fig. 2(b)].

E	B_{GT}^{0}	B_{GT}^{Δ}	q_Δ no gs corr.	q_Δ with gs corr.
1.71	0.46	0.37	0.02	0.19
2.82	0.22	0.18	0.01	0.18
3.09	0.06	0.05	0.02	0.17
3.36	0.14	0.02	0	0.19
3.50	0.03	0.02	0	0.19
3.67	0.03	0.02	−0.02	0.19
3.83	0.42	0.35	−0.01	0.18
4.28	0.95	0.77	0.01	0.20
4.74	0.25	0.21	0.02	0.18
4.88	0.08	0.05	0.03	0.30
5.52	0.79	0.62	0.01	0.21
6.54	0.29	0.23	−0.01	0.18
7.18	0.01	0.01	0.06	−0.18
8.02	0.25	0.22	−0.06	0.10
8.29	0.21	0.19	−0.06	0.10
8.41	0.03	0.03	−0.05	0.13
9.49	0.64	0.64	−0.15	0.01
9.66	0.82	0.74	−0.11	0.10
10.26	0.51	0.49	−0.17	0.05
11.9	1.10	1.35	0.17	0.34
12.74	0.86	1.47	0.17	0.34
13.166	4.07	7.27	0.17	0.34
14.49	38.04	54.16	0.17	0.34
15.08	65.87	10.0	0.17	0.34
17.74	7.09	2.83	0.17	0.34

Table 2
Same as table 1 for β^- decay of ^{90}Zr to ^{90}Nb.

E	B_{GT}^{0}	B_{GT}^{Δ}	q_Δ no gs corr.	q_Δ with gs corr.
2.2	1.73	1.44	0.08	0.17
2.45	0.45	0.42	−0.07	0.03
2.86	0.26	0.21	0.10	0.20
4.41	0.40	0.34	0.03	0.16
4.86	0.61	0.52	0.06	0.16
8.81	18.25	14.57	0.12	0.23
11.15	0.40	0.42	0.12	0.23
11.31	3.09	2.44	0.12	0.23
11.71	5.63	3.30	0.12	0.23

Volume 132B, number 1,2,3 PHYSICS LETTERS 24 November 1983

but the total reduction is due to Δ–h admixtures to the ground state. For the GTGR we find equal contributions from *both* effects. In ^{90}Zr the final state interaction gives a contribution of 5% to the quenching of strength below the GTGR, but reduces the GTGR itself by 12%.

To get a better insight into the situation, it is instructive to look at a simple 3 state model, including a GT-state at low excitation energy, carrying only a small fraction of the total transition strength, the GTGR and the collective Δ–h state. Noting that the reduced transition strength to a configuration $|j_n j_p;1^+\rangle$ is given by

$$\langle j_n j_p;\, 1^+ \| Y^- \| 0\rangle = u_p v_n \langle j_n \| \sigma \| j_p\rangle\,, \qquad (7)$$

eq. (5) leads to

$$\langle \Delta h;\, T=1,\, T_z=-1\, |\, V_{\mathrm{GT}}^{\mathrm{dir}} |\, \mathrm{GT}\rangle = \chi'(N+3Z)^{1/2} \langle \mathrm{GT} \| Y^- \| 0\rangle\,, \qquad (8)$$

where $|\mathrm{GT}\rangle$ may be any mixed p–h GT state.

Treating the influence of the Δ–h state perturbatively leads to the mixing amplitude α with the GTGR

$$\alpha = \chi'(N+3Z)^{1/2}\langle \mathrm{GTGR} \| Y^- \| 0\rangle / [E(\mathrm{GTGR}) - E_\Delta] \sim -\chi'(N+3Z)^{1/2}[3(N-Z)]^{1/2}/300\ \mathrm{MeV}\,. \qquad (9)$$

For ^{208}Pb this estimate gives $\alpha \sim -0.056$, the reduction of the strength is

$$\{\alpha \cdot \tfrac{4}{5}\,\tfrac{3}{2}(N+3Z)^{1/2}/[3(N-Z)]^{1/2} + 1\}^2 \sim 0.83\,, \qquad (10)$$

in good agreement with the above calculation. For the low-lying state $\langle \mathrm{GT}\ 1|$ first order perturbation theory leads to the same reduction factor. However, the presence of the Δ–h admixture in the $\widetilde{\mathrm{GTGR}}$ leads to an interaction between $\langle \mathrm{GT}\ 1|$ and $\langle \widetilde{\mathrm{GTGR}}|$. The tilde denotes the GTGR including the Δ–h admixture. From this it follows that second order correlations have to be taken into account. The corrected $\langle \widetilde{\mathrm{GT}\ 1}|$ reads then:

$$\langle \widetilde{\mathrm{GT}\ 1}| = -\frac{\chi'(N+3Z)^{1/2}\langle \mathrm{GT}\ 1 \| Y^- \| 0\rangle}{300\ \mathrm{MeV}}\langle \Delta h| + \frac{\chi'^2(N+3Z)\langle \mathrm{GT}\ 1 \| Y^- \| 0\rangle \langle \mathrm{GTGR} \| Y^- \| 0\rangle}{300\ \mathrm{MeV}[E(\mathrm{GTGR}) - E(\mathrm{GT}\ 1)]}\langle \mathrm{GTGR}| + \beta\langle \mathrm{GT}\ 1|\,,$$
$$\beta \sim 1\,, \qquad (11)$$

leading to the correction factor for the transition amplitude for $\langle \widetilde{\mathrm{GT}\ 1}|$

$$\frac{N+3Z}{300\ \mathrm{MeV}}\left(-\chi'\tfrac{4}{5}\sqrt{3} + \chi'^2 \frac{\langle \mathrm{GTGR} \| Y^- \| 0\rangle^2}{E(\mathrm{GTGR}) - E(\mathrm{GT}\ 1)}\right) + 1\,. \qquad (12)$$

For ^{208}Pb and a typical low energetic GT state at 5 MeV excitation energy first and second order contributions are +0.085 and −0.085, respectively, cancelling each other. This explains the weak influence of the final state Δ-mixing on the low-lying states calculated above.

An analogous estimate leads to a mixing amplitude of 0.034 for the $\langle \Delta h,\, T=1,\, T_z=0|$ component in the M1 giant resonance in ^{208}Pb in agreement with the investigation of Knüpfer et al. [8]. The quenching q is related to the factor γ in ref. [8] by $\gamma^2 = 1-q$. For low-lying M1 transitions first and second order contributions are also found to be roughly of the same magnitude, resulting also in partial cancellation of quenching effects. However, one has to be more careful than for β^-GT transitions, because nucleonic gsc are larger here, and also pionic current processes have to be included [9].

We also show the β^+ transitions for ^{90}Zr in fig. 3c and table 3. Here the situation is different. Since there exists no GTGR for β^+ transitions in sufficiently neutron-rich nuclei, there appears also no similar cancellation of final state Δ–h interactions as described for β^- transitions. So we find an almost constant reduction of strength for all daughter states of 18% from 2.0 to 1.65. Nucleonic gsc are included in these num-

Table 3
Same as in table 1 for β^+ decay of ^{90}Zr to ^{90}Y.

E	B^0_{GT}	B^Δ_{GT}	q_Δ no gs corr.	q_Δ with gs corr.
0.51	1.28	1.01	0.13	0.21
1.04	0.41	0.29	0.18	0.28
1.50	0.02	0.01	0.22	0.33
5.20	0.02	0.01	0.19	0.43
6.62	0.04	0.03	0.33	0.36
9.49	0.06	0.04	0.17	0.31
10.04	0.03	0.02	0.13	0.37
10.87	0.03	0.02	0.21	0.44
12.98	0.02	0.01	0.18	0.39
15.81	0.14	0.09	0.18	0.35

Volume 132B, number 1,2,3 PHYSICS LETTERS 24 November 1983

bers. As is well-known they are important for β^+ transitions and reduced the strength already from 3.6 to 2.6. We find a further 8% reduction as a result of Δ–gsc. The total quenching including Δ–h admixtures as well as nucleonic and Δ-gsc is found to be 58%. A quarter of this total effect can be ascribed to the Δ_{33} resonance.

It is known that inclusion of the Δ excitations does not solve the missing strength problem completely. But the fraction of GT-strength seen in (p,n) experiments will very likely rise by treating the background below the GTGR properly [3]. Bertsch and Hamamoto [10] showed that there exists also effective purely nucleonic quenching by the coupling to 2p–2h configurations which they probably somewhat overestimated by their perturbative treatment. However, because of the density distribution of 2p–2h states this effect should hardly affect the low-lying GT states.

The conclusion from our results is that it is not possible to describe the quenching of GT strength by Δh mixing by a constant renormalization of the axial vector current and also not by a variable reduction of the single particle transition matrix elements. By such a procedure it was tried [11] to describe the quenching observed experimentally in light nuclei. On the other hand, the effect of gsc alone, in β^- mainly arising from Δ–h involving excitations, leads to a nearly constant reduction of the single particle matrix elements. So *this part* of the quenching can be described by a constant renormalization of the axial vector current, resulting in a constant reduction for all GT states including the GTGR. Altogether the influence of the Δ_{33} resonance leads to a reduction of the transition strength by only about 15% for low energetic GT-states, which are the decisive ones for the nuclear beta decay properties [12,13].

The authors acknowledge with pleasure fruitful discussions with R. Madey and E. Richter.

References

[1] G. Bertsch, Comments Nucl. Part. Phys. 10 (1981) 91.
[2] C. Gaarde, Trieste Nuclear Physics Workshop (Oct. 5–30, 1981), and references therein.
[3] F. Osterfeld, Phys. Rev. C26 (1982) 762.
[4] A. Bohr and B.R. Mottelson, Phys. Lett. 100B (1981) 10.
[5] H. Sagawa and Nguyen van Giai, Phys. Lett. 113B (1982) 119.
[6] H.R. Fiebig and J. Wambach, Nucl. Phys. A386 (1982) 381.
[7] K. Grotz et al., Phys. Lett. B126 (1983) 417; J. Phys. G9, No. 8 (1983).
[8] W. Knöpfer, M. Dillig and A. Richter, Phys. Lett. 122B (1983) 7.
[9] M. Kohno and D.W.L. Sprung, Phys. Rev. C26 (1982) 297.
[10] G.F. Bertsch and I. Hamamoto, Phys. Rev. C26 (1982) 1323.
[11] B.A. Brown, W. Chung and B.H. Wildenthal, Phys. Rev. Lett. 40 (1978) 1631.
[12] K.V. Klapdor, J. Metzinger and T. Oda, report MPI H-1981-V47 and At. Data Nucl. Data, to be published.
[13] H.V. Klapdor, Proc. Intern. School-Seminar on Heavy ion physics (Alushta, USSR, April 1983); Prog. Part. Nucl. Phys. 10 (1983) 131.

Volume 142B, number 5,6 PHYSICS LETTERS 2 August 1984

CALCULATION OF DOUBLE BETA DECAY OF ^{76}Ge, ^{82}Se, ^{128,130}Te

H.V. KLAPDOR and K. GROTZ

Max-Planck-Institut für Kernphysik, Heidelberg, Germany

Received 29 September 1983
Revised manuscript received 19 April 1984

Calculations of double beta decay rates of ^{76}Ge, ^{82}Se, ^{128,130}Te, including Δ–h excitations and collective effects arising from spin–isospin and quadrupole–quadrupole forces are presented. The serious discrepancy between previous shell model calculations and experimental rates for ^{128,130}Te is markedly reduced. The persisting discrepancy of a factor 10 for these nuclei and for ^{82}Se could raise the question of a possible non-nuclear structure origin of the disagreement.

In the context of grand unified theories (GUT's) besides baryon number (B) violation (proton decay) also lepton number (L) violation should occur. One of the most sensitive tests of lepton number violation is double beta ($\beta\beta$) decay [1]. The existence of a neutrinoless (0ν) decay mode would be equivalent to a $B - L$ violation and could either be interpreted in terms of a finite neutrino Majorana mass or in terms of a right-handed admixture η in the weak lepton current [2]. Although the 0ν mode should demonstrate its existence by a signal with a very distinctive signature, it is very difficult to establish this mode by a direct counting measurement, because of the extremely small decay rates. Experiments up to now yielded only upper limits for such a decay mode [3–5].

Because of the higher sensitivity achievable, the main interest has been focused on geochemical experiments [6–9] in which the total $\beta\beta$-decay rates resulting from the contributions of the 2ν and the probably existing 0ν mode are determined. By neglecting contributions from the 2ν mode already upper bounds for m_ν can be obtained. It is believed [9–12] that more stringent conclusions can be drawn by comparing the ratio of the measured [8,9] half lives of ^{128}Te and ^{130}Te with the theoretical expectation for the ratio of the 2ν mode. In fact, *only* this procedure has in principle the power to *establish* the existence of a 0ν mode and thus a finite ν mass or right-handed admixture. However, all such analyses up to now, as pointed out already in ref. [13] are questionable due to the fact that present shell model calculations [10] predict a 2ν $\beta\beta$-rate for ^{130}Te a factor $\sim$160 higher than one deduces from the geochemical result [‡1] for the *total* decay rate, while the calculated rate ratio for both isotopes has to be known to an accuracy of the order of 50%. But not only for getting the 0ν $\beta\beta$-rate out of the geochemical results the problems in the shell model calculations existing so far are a severe shortcoming, but also in relating the 0ν $\beta\beta$-rate to the lepton number violating parameters $\langle m_\nu \rangle_{\rm Maj}$ and η. Here enters [12] the *absolute* value of the nuclear matrix element $M^{0\nu}$, which is usually calculated according to $M^{0\nu} = (\xi/R_0)M^{2\nu}_{\rm exp}$ with $\xi = 0.6$–20 [10–12] and which thus cannot be expected to be better known from the calculations than the matrix element for the 2ν $\beta\beta$-decay.

We present in this paper calculations of $2\nu\beta\beta$-rates including for the first time quantitatively Δ–h excitations and 2^+ phonon admixtures in the initial and final ground states. Extending a recent study [15] we investigate the $2\nu\beta\beta$-decay of ^{128,130}Te, ^{82}Se and ^{76}Ge avoiding the closure approximation, which has been used in most other calculations up to now. In our more exact treatment not simply the matrix element of $\sigma\sigma\tau^-\tau^-$ between initial and final 0^+ states is

‡1 As geochemical values for ^{130}Te and ^{82}Se we take the weighted mean values from all available data, given in ref. [14].

0.370-2693/84/$ 03.00 © Elsevier Science Publishers B.V.
(North-Holland Physics Publishing Division)

Volume 142B, number 5,6 PHYSICS LETTERS 2 August 1984

calculated, but matrix elements $\langle M\|\sigma\tau^-\|I\rangle$ and $\langle F\|\sigma\tau^-\|M\rangle$ are calculated for the whole spectrum of intermediate 1^+ states M. Steps in this direction were made in refs. [10] and [16]. In ref. [10] an explicit summation restricted to 1^+ states within the first 5 MeV excitation energy in ^{82}Br has been performed and in ref. [16] the 1^+ spectrum in ^{128,130}I was calculated in an RPA approach, however, treating the lepton phase space approximately. In this work the spectrum of intermediate 1^+ states is taken into account up to ~300 MeV.

Starting from time dependent perturbation theory the $2\nu\beta\beta$ $0^+ \to 0^+$ decay rate neglecting contributions from Fermi decay is given by:

$$\omega_{2\nu} = \frac{F_A^4 G^4}{32\pi^7} \int_{m_e}^{t_0+m_e} F(Z,\epsilon_1)|K_1|\epsilon_1 \, d\epsilon_1 \times \int_{m_e}^{t_0+2m_e-\epsilon_1} F(Z,\epsilon_2)|K_2|\epsilon_2 \, d\epsilon_2 \times \int_0^{t_0-\epsilon_1-\epsilon_2+2m_e} \nu_1^2\nu_2^2 \, d\nu_1 \sum_{MM'} A_{MM'} \,, \qquad (1)$$

with

$$A_{MM'} = \langle F\|\sigma\tau^-\|M\rangle\langle M\|\sigma\tau^-\|I\rangle \times \langle F\|\sigma\tau^-\|M'\rangle\langle M'\|\sigma\tau^-\|I\rangle \times \tfrac{1}{3}(K_M K_{M'} + L_M L_{M'} + \tfrac{1}{2}K_M L_{M'} + \tfrac{1}{2}L_M K_{M'}),$$

with the same notation as in refs. [10] and [15].

The absolute values of $\langle M\|\sigma\tau^-\|I\rangle$ can be compared to (p, n) charge exchange experiments, from which the β^- Gamow–Teller (GT) strength function $B_{GT}^-(E)$ can be extracted. The matrix elements $\langle F\|\sigma\tau^-\|M\rangle$ are related to the β^+ strength B_{GT}^+ of the final 0^+ ground state by $B_{GT}^+ = |\langle F\|\sigma\tau^-\|M\rangle|^2$, which could in principle be extracted from an (n, p) experiment with the daughter nucleus as target.

β^- as well as β^+ matrix elements are calculated using PBCS wave functions with single-particle orbits taken from a Nilsson potential with $\delta = 0$. In this approximation it is not necessary to truncate the shell model basis seriously. The shell model space was extended beyond the magic numbers to include $\beta\beta$ matrix elements down to the 1% level. Particle number projection has been done and is necessary because of the large overlap of initial and final 0^+ ground states. To account for the high collectivity of GT transitions a residual GT interaction $V_{GT} = \chi \cdot \sigma\sigma\tau^+\tau^-$ (spin–isospin force) is diagonalized in a 2 quasi particle (qp) basis for the 1^+ states and a 4 qp basis for the 0^+ states. First order ground state correlations (gsc) of the GT interaction are included in this treatment.

As can be seen from table 1 (see also ref. [15]) the discrepancy between calculated and observed decay rates for ^{128}Te and ^{130}Te cannot be removed in this treatment. The same is true for ^{82}Se. (The direct counting experiment [17] quotes a much smaller half life of ^{82}Se (table 1) than the geochemical ex-

Table 1
$2\nu\,\beta\beta$ half lives calculated in this work with pairing and GT-forces (first column) and including (second column) Δ–h excitations. The third column shows the results obtained with pairing, GT-force, Δ–h mixing and first order 2^+ phonon corrections. In the fourth columns are given the results from a calculation inserting the known ground state matrix elements as described in the text. The experimental values in the last column correspond to the sum of 2ν and 0ν decay in the case of the geochemical results [8,9,14], to 2ν decay for ref. [17] and ^{76}Ge.

Nucleus	Calculated half lives for $2\nu\,\beta\beta$-decay (yr)				Experimental $\beta\beta$-half lives (yr)
	pairing + V_{GT}	pairing + V_{GT} + Δ–h mixing	pairing + V_{GT} + Δ–h mixing + V_{QQ}	from 1^+ g.s. decay	
^{128}Te	6.2×10^{22}	8.7×10^{22}	5.7×10^{23}	1.6×10^{25}	$>8 \times 10^{24} (2\sigma)$ a) $(1.5 \pm 0.2) \times 10^{24}$ b)
^{130}Te	1.5×10^{19}	2.0×10^{19}	1.2×10^{20}	2.6×10^{21}	$(2.55 \pm 0.20) \times 10^{21}$ c)
^{82}Se	5.1×10^{18}	7.2×10^{18}	1.5×10^{19}	1.8×10^{20}	$(1.45 \pm 0.15) \times 10^{20}$ c) $(1.0 \pm 0.4) \times 10^{19}$ d)
^{76}Ge	1.3×10^{20}	1.9×10^{20}	2.2×10^{20}	–	$>2.8 \times 10^{19}$ c)

a) Ref. [9]. b) Ref. [8]. c) Ref. [14]. d) Ref. [17].

periments, but suffers, however, from severe background problems.)

The GT strength in the "nuclear" region of excitation is quenched to some extent compared to the sum rule value by Δ–h excitations, which shift part of the strength up to an energy region around 300 MeV. Of course, not only the β^--part, but also the β^+-part of the matrix elements in eq. (1) is affected, with excitation energy dependent quenching factors [18]. We repeated therefore the above calculation in the quark model including Δ–h excitations in the form of two collective states $\langle \Delta - h, T = 1, T_Z = +1 |$ and $\langle \Delta - h, T = 1, T_Z = -1 |$, neglecting the energy splitting of the different Δ-orbits (see also ref. [18]). Gsc as well as final-state interaction graphs have been included (see fig. 1). The calculated 2ν $\beta\beta$-amplitudes entering in eq. (1) are shown in fig. 2. The transitions directly via the collective Δ–h states cancel each other, so that there is no direct contribution to the $\beta\beta$-decay from the 300 MeV region. The total decrease of the 2ν $\beta\beta$-rates is only 25–32% (see table 1) and therefore surprisingly small compared to the ~30% decrease already obtained for the total β^- strength of the parent nuclei (and smaller than the estimate by ref. [19]). The reasons for this are the following: It was shown [18] that

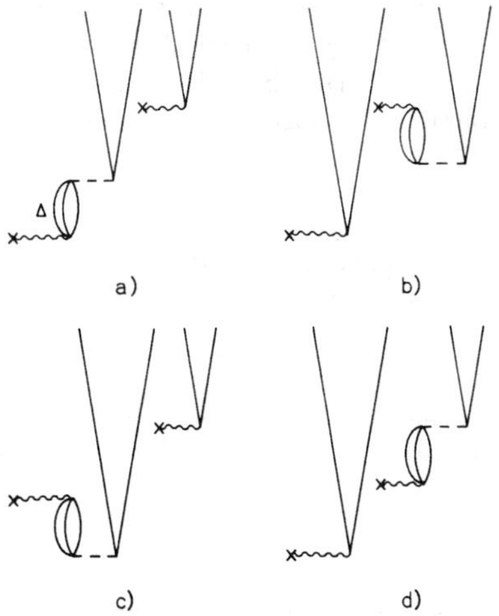

Fig. 1. Interaction graphs including a Δ-excitation, which give a sizeable contribution to the $\beta\beta$-amplitudes. Note that in the final state the 4 quasi particles have to couple to the 0^+ ground state, which is eqivalent to the quasi particle vacuum. Hence graphs (c) and (d) correspond to ground state correlations in the initial and final nucleus, respectively.

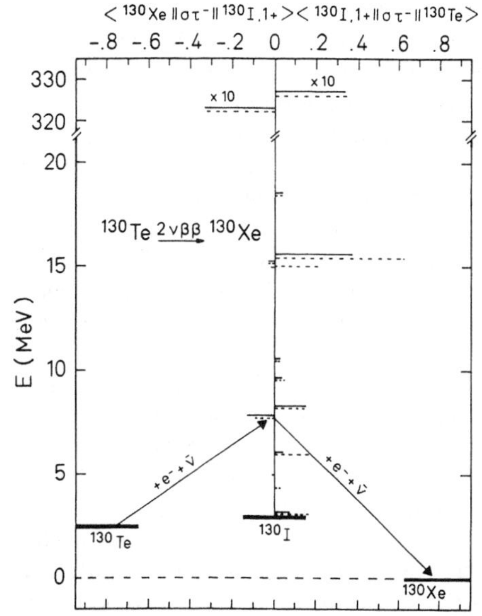

Fig. 2. 2ν $\beta\beta$-amplitudes for the transition $^{130}Te \xrightarrow{\beta\beta} {}^{130}Xe$ calculated with pairing plus Gamow–Teller forces and including Δ–H mixing (dashed line). The solid li es show the final result including also quadrupole–quadrupole forces.

the quenching of β^- strength below the GT giant resonance (GTGR) is much smaller than for the GTGR itself due to a destructive interference of first and second order effects. Secondly, it is found that the quenching of β^+ strength, responsible for the second decay step, is generally weaker. This is because the β^+ strength is already strongly quenched by nucleonic gsc and switching on the Δ–h excitations reduces the effect at the same time.

As was shown already in ref. [15] the $\beta\beta$-rates depend quite strongly on the pairing interaction, decreasing with decreasing pairing correlations. In the present calculation we used G_p = 15 MeV/A and G_N = 16 MeV/A yielding Δ_p = 0.4 MeV and Δ_n = 0.9 (0.8) MeV for ^{128}Te (^{130}Te). These are lower limits compatible with stripping [20] and pick-up [21] reaction data. Using standard values $G_p = G_N$ = 20 MeV/A the calculated 2ν $\beta\beta$ rates for ^{128,130}Te would be increased by a factor 1.8. To get agreement with the geochemical rates we would have to assume Δ_p, $\Delta_n \lesssim 0.08$ MeV, which means an extremely small deviation from a pure shell model configuration in clear contradiction to the experiment.

Volume 142B, number 5,6 PHYSICS LETTERS 2 August 1984

Therefore other effects have to be responsible for the strong suppression of the $2\nu\beta\beta$-rates. Besides the pairing and Gamow–Teller forces another important part of the residual interaction not taken into account in the above presented calculations can be parametrized by a quadrupole–quadrupole force. This attractive force leads to low-lying collective 2^+ states (phonons) and strong ground state correlations. The first order corrections of the $\beta\beta$-decay are given by the decay of the two-phonon 0^+ states mixed with the parent ground state into p–h configurations in the intermediate nucleus and also by the decay of the p–h intermediate configurations into two-photon 0^+ states mixed with the final ground state. These corrections are proportional to the s-coefficients resulting from a quadrupole–quadrupole RPA of parent and daughter nucleus, respectively. Besides these there are also higher order contributions involving phonon admixtures in both the initial and final nucleus which are proportional to a product of at least two s-coefficients. Moreover, for strong ground state correlations the renormalization of the zeroth order matrix elements cannot be neglected.

We calculated the first order corrections and the renormalization of the zeroth order resulting from a quadrupole–quadrupole RPA for the whole spectrum of $\beta\beta$-amplitudes. A modified version of the RPA equations was used to take into account the Pauli principle [22] and the coupling strength was adjusted to reproduce the position of the one-phonon states.

For the Te isotopes we find large destructive corrections from the d and $h_{11/2}$ shells, leading to a reduction of the decay rate by about a factor of 5. For ^{82}Se and ^{76}Ge the first order corrections are less important (see table 1). The second order corrections which are not small at least for ^{82}Se and ^{76}Ge may give a further reduction, but up to now have only been calculated in the closure approximation (see table 2). Their contribution to the closure matrix elements $\langle 0_F | \Sigma_{ij} \sigma(i)\tau^-(i)\sigma(j)\tau^-(j) | 0_I \rangle$ is -0.461 for ^{82}Se (our definition of the closure matrix element differs by a factor of two from the M_{GT} of ref. [10]). Inserting the closure energies obtained from an E^{-1} weighting of the β^- Gamow–Teller strength (as was done also in ref. [10]) the experimental matrix elements would be only a factor of two smaller than the calculated closure matrix elements. However, the explicit summation over intermediate 1^+ states, including all corrections (except the higher order phonon contribution) indicates, that closure energies obtained in this way are too high. This is the consequence of the fact, that low-lying 1^+ states within the first few MeV excitation energy contribute more to the $2\nu\beta\beta$ decay than the GTGR.

Since such low-lying states may vary appreciably between neighbouring odd–odd nuclei we would not suggest to apply the ratio argument for theoretical interpretation of the geochemical results in terms of m_ν and η. More reliable bounds for $B-L$ violation may be obtained by neglecting the contribution from

Table 2
Calculated and experimental closure matrix elements. The calculated matrix elements including first order phonon corrections correspond to the half lives of column three in table 1. The "experimental" matrix elements have been deduced using the closure energies given in the last column.

Nucleus	Closure matrix elements				
	This work		ref. [10]	deduced from experiments	
	with first order phonon corrections	including higher order phonon corrections		$\langle M^{2\nu}\rangle_{exp}$	$\langle E\rangle$ (MeV)
^{128}Te	0.61	0.52	2.95	<0.33 a)	11.5
^{130}Te	0.56	0.48	2.97	0.24 b)	10.7
^{82}Se	1.84	1.38	1.88	0.69 b)	6.7
^{76}Ge	2.64	1.93	2.56	<4.7 b)	8.12

a) Ref. [9]. b) Ref. [14].

Volume 142B, number 5,6 PHYSICS LETTERS 2 August 1984

2ν decay in ^{128}Te completely. The correspoinding limit for m_ν is 9 eV using the Kirsten data [9]. The main uncertainty then still entering in such an analysis of geochemical data as well as of direct counting results is the calculated value for the 0ν-matrix element $M^{0\nu}$ connecting the 0ν-rate with the parameters m_ν and η.

We conclude, that although the refined treatment presented here reduces the discrepancy between the geochemical experiments and previous calculations [10,15] considerably for the Te isotopes, the calculated rates are still a factor of ~10 too high in all three nuclei ^{92}Se, ^{128}Te and ^{130}Te, where there exist geochemical measurements. It is extremely unlikely that this can be explained by experimental errors, because of various consistency checks of the used methods [14]. The higher order phonon contributions neglected in table 1 we expect to affect predominantly the $\beta\beta$-amplitudes of the higher energetic 1^+ states, because they involve 2p–2h configurations in the intermediate nucleus. The overall contributions, however, are small for ^{128,130}Te, as can be seen from table 2. Inclusion of the electromagnetic interaction of the two outgoing electrons, which is neglected in all papers on this subject up to now, could lead to a further reduction. According to a very simple energetic argument the spin–spin interaction should lead sizeable corrections for transitions via the GTGR. The spin–spin interaction energy for two electrons separated by a distance $r = \hbar c/\Delta E$ corresponding to a virtual excitation of 10 MeV is of the same order as the Coulomb interaction with the nucleus and predominantly of opposite sign as can be seen from angular correlations.

From these arguments one could expect a reduction of the calculated $2\nu\,\beta\beta$ decay rates, mainly because of smaller contributions from the GTGR. However, it can hardly be expected that these corrections are large enough to confirm by the calculations what is indicated by the experimentally known matrix elements for the $1^+ \xrightarrow{\beta^\pm} 0^+$ ground state decays of ^{128}I and ^{80}Br, namely that already the lowest intermediate 1^+ state dominates the total $\beta\beta$-decay (table 1).

It is an interesting question, whether the still unexplained discrepancy is founded completely in nuclear structure or if there are elementary effects of non-nuclear structure origin responsible. Do we describe weak interactions properly? $\beta\beta$-decay is besides $K^0 \leftrightarrow \overline{K^0}$ oscillations the only experimentally accessible weak second order process. The calculation of the matrix element describing the K_1^0, K_2^0 mass difference is up to now a very delicate problem and involves cancellation of various gauge group dependent higher order terms [23]. In the case of $\beta\beta$-decay, however, a generalization of the phenomenological first order theory to second order seems to be straightforward. On the other hand, general limits of the applicability of the simple time dependent perturbation treatment such as nonexponential decay and the Zenon effect have been discussed recently [24,25].

The authors acknowledge with pleasure fruitful discussions with Professor B. Stech, Professor T. Kirsten, Professor W. Bühring, Professor H.A. Weidenmüller and Professor U. Schmidt-Rohr.

References

[1] H. Primakoff and S.P. Rosen, Rep. Prog. Phys. 22 (1959) 1;
B. Pontecorvo, Phys. Lett. 26B (1968) 630.
[2] A. Halprin et al., Phys. Rev. D 13 (1976) 2567.
[3] B.T. Cleveland et al., Phys. Rev. Lett. 35 (1975) 757.
[4] R.K. Bardin et al., Nucl. Phys. A158 (1970) 337.
[5] F.T. Avignone et al., Phys. Rev. Lett. 50 (1983) 721.
[6] T. Kirsten and H.W. Müller, Earth Planet Sci. Lett. 6 (1969) 271.
[7] B. Srinivasan et al., Econ. Geol. 68 (1973) 252.
[8] E.W. Hennecke, O.K. Manuel and D.D. Sabu, Phys. Rev. C11 (1975) 1378.
[9] T. Kirsten, H. Richter and E. Jessberger, Phys. Rev. Lett. 50 (1983) 474; Z. Phys. C16 (1983) 189.
[10] W.C. Haxton, G.J. Stephenson Jr. and D. Strottmann, Phys. Rev. Lett. 47 (1981) 153;
W.C. Haxton, Phys. Rev. D 25 (1982) 2360; Comm. Nucl. Part. Phys. 11 (1983) 41.
[11] P. Minkowski, Nucl. Phys. B201 (1982) 269.
[12] M. Doi et al., Prog. Theor. Phys. 69 (1983) 602.
[13] H.V. Klapdor, Prog. Part. Nucl. Phys. 10 (1983) 131.
[14] T. Kirsten, in: Proc. Workshop on Science underground (Los Alamos, 1982).
[15] K. Grotz, H.V. Klapdor and J. Metzinger, J. Phys. G9 (1983) L 169.
[16] A.H. Huffmann, Phys. Rev. C2 (1970) 741.
[17] M. Moe and D. Löwenthal, Phys. Rev. C22 (1980) 2186.
[18] K. Grotz, H.V. Klapdor and J. Metzinger, Phys. Lett. 132B (1983) 22.
[19] L. Zamick and N. Auerbach, Phys. Rev. C26 (1982) 2185.
[20] W.H. Moore, G.K. Schlegel and S. O'Dell, Nucl. Phys. A104 (1967) 327.

[21] R.L. Auble, J.P. Ball and Y.B. Fulmer, Nucl. Phys. A116 (1968) 14.
[22] K. Ideda, T. Udagawa and H. Yamaura, Prog. Theor. Phys. 33 (1965) 22.
[23] R.N. Mohapatra, J.C. Pati and P. Vinciarelli, Phys. Rev. D8 (1973) 2652.
[24] H.V. Klapdor and K. Grotz, Report MPI H-1983-V34; K. Grotz and H.V. Klapdor, Report MPI-H-1983-V35; and submitted to Phys. Rev. C.
[25] H.V. Klapdor, in: Proc. Intern. Symp. on Electromagnetic properties of atomic nuclei (Tokyo, November 1983); and Proc. Intern. Symp. on Nuclear spectroscopy and nuclear interactions (Osaka, March 1984).

Volume 153B, number 1,2 PHYSICS LETTERS 21 March 1985

THE NEUTRINO MASS FROM DOUBLE BETA DECAY

K. GROTZ and H.V. KLAPDOR

Max-Planck Institut für Kernphysik, D-6900 Heidelberg, West Germany

Received 5 November 1984; revised manuscript received 15 January 1985

New and more stringent limits for the mass of a Majorana neutrino are deduced from the existing experimental data on double beta decay by new calculations of the matrix elements for neutrinoless double beta decay.

Recently calculations of the two-neutrino (2ν) double beta ($\beta\beta$) decay were presented [1] reducing the existing discrepancy between the geochemically measured [2] decay rates of $^{128,130}Te$ and the previous calculations of Haxton et al. [3] by one order of magnitude. Such calculations of the so to say parameter-free 2ν decay are of importance not only for the analysis of geochemical $\beta\beta$ decay results but are at the same time a critical test for nuclear structure calculations with respect to the neutrinoless (0ν) mode. This decay branch, which has been experimentally not yet observed, is related to a finite neutrino mass m_ν and to right-handed current admixtures. Experimental data on $\beta\beta$ decay from direct counting as well as geochemical experiments exist at present for a considerable number of nuclei: ^{46}Ca, ^{48}Ca, ^{70}Zn, ^{76}Ge, ^{82}Se, ^{96}Zr, ^{100}Mo, ^{110}Pd, ^{124}Sn, ^{128}Te, ^{130}Te, ^{148}Nd, ^{150}Nd, ^{186}W, ^{192}Os and ^{198}Pt. For an overview and references see ref. [3]. New limits for the neutrinoless decay of ^{76}Ge have also been reported by the Guelph [4] and the Osaka [5] groups. New results exist further for ^{150}Nd [6].

In this paper we concentrate on the calculation of the nuclear matrix elements for neutrino mass induced 0ν $\beta\beta$ decay of ^{76}Ge, ^{82}Se and $^{128,130}Te$, the nuclei for which the most constraining experimental data exist. We simplify the discussion by neglecting the influence of possible right-handed current admixtures and consider only $0^+ \to 0^+$ transitions in the case $\langle m_\nu \rangle_{\mathrm{Majorana}} = |U_{\mathrm{e}i}^2 m_i| > 0$ (for a discussion of the mixing matrix U and the general formalism including also right-handed currents see for example Doi et al. [7]). To derive limits on right-handed parameters the study of $0^+ \to 1^+, 2^+$ transitions in principle is more powerful, because these are not induced by a mass term. However, because of the complexity nuclear structure calculations for this latter process are not yet available. The upper limits for the neutrino mass deduced in this paper may only very slightly be increased by taking into account right-handed current admixtures.

Neglecting right-handed currents the 0ν decay rate $\omega_{0\nu}$ for $0^+ \to 0^+$ transitions is related to an effective Majorana neutrino mass $\langle m_\nu \rangle = |U_{\mathrm{e}i}^2 m_i|$ by

$$\omega_{0\nu} = \langle m_\nu \rangle^2 / m_\mathrm{e}^2 f^{0\nu} |1 - x_\mathrm{F}|^2 |M^{0\nu} R_0|^2 . \tag{1}$$

$M^{0\nu}$ is the Gamow–Teller (GT)-like matrix element

$$M^{0\nu} = \langle 0_\mathrm{F}^+ | \sum_{ij} H_{ij} \boldsymbol{\sigma}_i \boldsymbol{\sigma}_j \tau_i^- \tau_j^- | 0_\mathrm{I}^+ \rangle ,$$

and x_F the fraction of Fermi-like transitions

$$x_\mathrm{F} = (g_\mathrm{V}/g_\mathrm{A})^2 \langle 0_\mathrm{F}^+ | \sum_{ij} H_{ij} \tau_i^- \tau_j^- | 0_\mathrm{I}^+ \rangle / M^{0\nu} .$$

In the derivation of eq. (1) the closure approximation was applied. This is expected to be appropriate for the 0ν mode *in contrast* to the 2ν decay (for the latter problem see ref. [1]) because of the energy dominance of the virtual neutrino in the intermediate state appearing in $M^{0\nu}$ through the neutrino propagator H_{ij}. The phase space factor $f^{0\nu}$ in eq. (1) is related to the quantity G_{01} defined in ref. [7] simply by $f^{0\nu} = \ln 2\, G_{01}$.

In previous analyses [2,3,7] of experimental data in applying eq. (1) the calculated $M_{\mathrm{calc}}^{0\nu}$ was scaled

0370-2693/85/$ 03.30 © Elsevier Science Publishers B.V.
(North-Holland Physics Publishing Division)

Volume 153B, number 1,2 PHYSICS LETTERS 21 March 1985

down to take care of the existing discrepancy between calculation and experiment concerning the 2ν rates in the following way:

$$M^{0\nu} = M^{0\nu}_{\mathrm{calc}} |M^{2\nu}_{\mathrm{exp}}/M^{2\nu}_{\mathrm{calc}}| \,. \qquad (2)$$

$M^{2\nu}$ is the corresponding closure matrix element for the 2ν mode. It was believed that the uncertainties in the nuclear structure calculations should cancel out in the dimensionless quantity $\xi = M^{0\nu}R_0/M^{2\nu}$ which in a zeroth order estimate is expected to be of order unity. The calculation of Haxton et al. [3] yielding ξ = 1.6–2.3 for the discussed nuclei indicated no strong deviations from this simple estimate.

We shall show, however, that this is no longer the case, when collective effects arising from spin–isospin forces and phonon correlations are included in the nuclear structure calculations.

In the present investigation it is found that such correlations affect the 0ν matrix elements only weakly. $M^{0\nu}$, ξ and x_{F} were calculated in the formalism of particle-number projected BCS with an additional residual interaction consisting of a spin–isospin and a quadrupole–quadrupole part. The total hamiltonian consists then of four parts

$$H = H_{\mathrm{s.p.}} + H_{\mathrm{pair}} + H_{\sigma\tau} + H_{QQ} \,,$$

$$H_{\mathrm{pair}} = \sum_{\eta} G_{\eta} \sum_{kk'} a^{+}_{\eta k'} a^{+}_{\eta \bar{k}'} a_{\eta k} a_{\eta \bar{k}} \,,$$

$$\eta = \text{proton, neutron} \,,$$

$$H_{\sigma\tau} = C_0 g_0' \frac{m_\pi^3}{4\pi} \sum_{ij} \frac{\exp(-m_\pi r_{ij})}{m_\pi r_{ij}} \boldsymbol{\sigma}_i \boldsymbol{\sigma}_j \tau_i^{+} \tau_j^{-} \,,$$

$$H_{QQ} = -\tfrac{1}{2}\kappa \sum_{\eta,\eta'} Q_\eta Q_{\eta'} \,.$$

The 0^+ wave functions contain therefore pairing, spin–isospin and quadrupole–quadrupole correlations and are essentially the same as in ref. [1] (except that Δ–N^{-1} excitations are neglected). $H_{\mathrm{s.p.}}$ was taken to be a Nilsson potential with zero deformation. The force strength parameters G, g_0' and κ where adjusted to reproduce stripping and pick up data [8], the position of the GTGR (known from (p,n) experiments) and the position of the low-lying collective 2^+ states. This lead to pairing strengths somewhat smaller than the standard value of 20 MeV/A and to values for g_0' close to the corresponding parameter used in the theory of finite Fermi systems [9]. For example for ^{130}Te we used G_{N} = 0.12 MeV, G_{p} = 0.12 MeV, g_0' = 0.98 and $\kappa\langle r^2\rangle^2$ = 1.2 MeV. With effective charges of 1.5 (0.5) for protons (neutrons) the experimental B(E2) values of the parent as well as the daughter nuclei are reproduced.

Defining

$$\epsilon_{1/2} = \langle E_{\mathrm{M}}\rangle + e_{1/2} - E_{\mathrm{I}} \,, \qquad \epsilon_0 = \langle E_{\mathrm{M}}\rangle + T_0/2 - E_{\mathrm{I}} \,,$$

with the nuclear closure energy $\langle E_{\mathrm{M}}\rangle$, the electron energies $e_{1/2}$ and the energy E_{I} of the initial nuclear state, the neutrino propagator for $m_\nu \ll |p_\nu|$ has the form

$$H_{ij} = [g(r_{ij}\epsilon_1) + g(r_{ij}\epsilon_2)]/2r_{ij} \approx g(r_{ij}\epsilon_0)/r_{ij} = \tilde{H}_{ij}(r_{ij}) \,.$$

Because the function $g(x)$ defined in ref. [10] varies only moderately for the range of arguments of importance, we apply a numerical approximation which is more easy to handle:

$$\tilde{H}_{ij}(r_{ij}) \approx \exp(-1.5 r_{ij}\epsilon_0)/r_{ij} \,,$$

with ϵ_0 = 12 MeV for ^{128,130}Te and ϵ_0 = 9.5 MeV for ^{76}Ge and ^{82}Se. The resulting values for $R_0 M^{0\nu}$, x_{F} and ξ are given in table 1. $R_0 M^{0\nu}$ is found to be surprisingly large for all nuclei. This is in contrast to the matrix elements for 2ν decay. For the latter strong cancellations arising from GT and QQ ground state correlations [1] reduce the closure matrix elements $M^{2\nu}$ for the Te isotopes by more than a factor of 10 compared to the values obtained using pure pairing wave functions (BCS-vacuum) (see table 1). But these correlations affect $M^{0\nu}$ only weakly. GT correlations reduce for example the no-neutrino matrix element of ^{128}Te by 18% and 2^+ phonons are practically negligible. The result is that $M^{0\nu}$ is not dramatically suppressed below the pairing value. The fact, that the calculated $M^{0\nu}$ values are even larger than those of the calculation of ref. [3], can at least partly be explained by the enlarged model space used here. (Corrections for short-range correlations would only moderately reduce our matrix elements.) The different behaviour of the two kinds of matrix elements expresses itself in ξ-values much larger than 1. This means that the scaling procedure commonly used is not an appropriate method and underestimates $M^{0\nu}$.

The reason, why details of the nuclear structure affect $M^{0\nu}$ and $M^{2\nu}$ so differently is founded in the dif-

Volume 153B, number 1,2 PHYSICS LETTERS 21 March 1985

Table 1
Calculated matrix elements for 0ν and 2ν $\beta\beta$ decay. Columns 2–4 give the calculated values for x_F and $R_0|M^{0\nu}|$ from this work. Besides the final result for $R_0|M^{0\nu}|$ the values obtained using pure BCS-wave functions are given in column two. 2ν matrix elements $|M^{2\nu}|$ are given in columns five and six (taken from the calculations of ref. [1]). ξ is the ratio of columns four and six.

Nucleus	x_F	$R_0\|M^{0\nu}\|$		$\|M^{2\nu}\|$		ξ
		pairing	final	pairing	final	
^{76}Ge	−0.24	12.45	10.40	7.05	1.93	5.4
^{82}Se	−0.23	9.68	8.20	5.52	1.38	5.9
^{128}Te	−0.24	12.30	10.03	6.34	0.52	19.3
^{130}Te	−0.24	11.86	9.44	5.86	0.48	19.7

ferent radial dependence of the involved operators. Due to the $1/r_{ij}$ factor in $M^{0\nu}$ the decay of two nucleons being close together is favored and long-range correlations as phonon correlations are unimportant. Minkowski [11] made also a rather qualitative argument in this direction. He argues, that to the 2ν matrix elements only the outermost shells contribute and that pair spin correlations should get lost for these. However, we find the situation not to be that simple. Not only the 0ν but also the 2ν matrix elements are collective in the sense, that a large number of shells contribute substantially. To what extent the long-range forces destroy the pairing correlations depends very much on the considered nucleus (see table 1).

Recently Bogdan et al. [12] calculated the 0ν decay of ^{76}Ge in the pairing model including deformation and rotation and avoiding the closure approximation. But they took into account only 1^+ intermediate states, whereas $M^{0\nu}$ involves also transitions with L-transfer. According to our closure calculation, which includes all possible intermediate J-couplings, they neglected a significant contribution to $M^{0\nu}$. Taking this into account, their result for $M^{0\nu}$ might be rather close to ours. However, they neglected completely neutron–

Table 2
Experimental limits for neutrinoless $\beta\beta$-decay and deduced limits for $\langle m_\nu \rangle_{\text{Majorana}}$. The limits deduced in this work are compared with those obtained from the calculations of ref. [3]. The limits given in the columns with heading "unscaled" follow directly from the calculated $M^{0\nu}$, whereas in the scaled values the observed discrepancy between experimental and calculated 2ν rates for ^{82}Se and ^{130}Te is taken into account as discussed in the text.

Nucleus	Experimental $T_{1/2}^{0\nu}$ (yr)	Ref.	Deduced m_ν (eV)			
			this work		ref. [3]	
			unscaled	scaled	unscaled	scaled
^{48}Ca	$>2 \times 10^{21}$ (80% CL)	[14]			<40	
^{76}Ge	$>7.2 \times 10^{22}$ (1σ)	[13]	< 1.9	< 5.8	< 4.8	< 11
	$>3.4 \times 10^{22}$ (2σ)		< 2.7	< 8.5	< 7.0	< 17
^{82}Se	$>1.30 \times 10^{20}$ (1σ) (G) [a]	[15]	<27	<83	<67	<159
	$>1.15 \times 10^{20}$ (2σ)		<28	<89	<71	<168
	$>3.1 \times 10^{21}$ (1σ)	[16]	< 5.5	<17	<14	< 32
^{130}Te	$>2.35 \times 10^{21}$ (1σ) (G)	[15]	< 4.3	<20	< 8.2	<103
	$>2.15 \times 10^{21}$ (2σ)		< 4.5	<21	< 8.6	<108
^{128}Te	$>1.2 \times 10^{25}$ (1σ) (G)	[2]	< 0.28	< 1.3	< 0.57	< 7.2
	$>8 \times 10^{24}$ (2σ)		< 0.35	< 1.6	< 0.70	< 8.8

[a] G denotes geochemical results.

Volume 153B, number 1,2 PHYSICS LETTERS 21 March 1985

proton correlations in the 0^+ ground states. GT neutron–proton correlations would greatly reduce their extremely large 2ν matrix element. In the case of $M^{0\nu}$ in principle again all possible J-coupled neutron–proton correlations should contribute. In the calculation presented here we took into account the leading contribution, namely 1^+ coupled neutron–proton ground state correlations (GT-correlations). These reduce $M^{0\nu}$ only moderately by ~20% and therefore the neglect of other J-couplings should not be a serious point.

In deriving the limits on m_ν given in table 2 we used two different sets of $M^{0\nu}$, firstly the calculated absolute values from table 1 and secondly the same values multiplied by a scaling factor $(\omega_{2\nu,\mathrm{exp}}/\omega_{2\nu,\mathrm{calc}})^{1/2}$ taken from the results for ^{82}Se and ^{130}Te of ref. [1]. This scaling in fact differs from eq. (2) to the extent the explicitly calculated 2ν rates differ from the closure approach. According to the arguments given above, the difference between the two $\langle m_\nu \rangle$ limits obtained with and without scaling reflects somehow the order of uncertainty of such an analysis. If the not yet removed discrepancy for the 2ν rates could be removed by an improved treatment of long-range correlations, however, the unscaled values and therefore the *lower* limits on $\langle m_\nu \rangle$ should be the more realistic ones.

We have analyzed the geochemical results for ^{128,130}Te and ^{82}Se with the new $M^{0\nu}$ values under the most conservative assumption:

$$\omega_{0\nu} \leqslant \omega_{\mathrm{geochemical}} .$$

The limit from the ^{128}Te data does, therefore, not rely on the often used "ratio argument" for the 2ν rates, which would give in principle more stringent but in fact less reliable limits (see the discussion in ref. [1]). The new obtained limit of 1.6 eV using the data of ref. [2] and the (more conservative) scaled value for the $M^{0\nu}$ is already much smaller than the previous limit of 8.8 eV (5.6 eV using the 2ν ratio argument) [2,3]. With the absolute unscaled value for $M^{0\nu}$ we obtain the more stringent limit of 0.35 eV. Also for the other investigated nuclei the limits on $\langle m_\nu \rangle$ are improved. The lowest limit from direct counting experiments is obtained for ^{76}Ge with the data from the Milano group [13]. This is $\langle m_\nu \rangle < 8.5$ eV (< 2.7 eV unscaled). On the other hand, by assuming the geochemically observed ^{130}Te decay to be totally caused by neutrinoless decay, the limit on $\langle m_\nu \rangle$ is of the same order of magnitude as presently obtained from the direct counting experiments.

Concluding, the correlations, that were found [1] to reduce the 2ν rates substantially, have only minor effects on the 0ν decay. Therefore the sensitivity for $B-L$ violating parameters in $\beta\beta$-decay experiments remains large, resulting in very stringent limits for $\langle m_\nu \rangle$. The presented new 0ν decay matrix elements and the new analysis of existing $\beta\beta$-decay data lead to improved limits for a Majorana neutrino mass.

It is a pleasure for the authors to acknowledge fruitful discussions with T. Kirsten.

References

[1] H.V. Klapdor and K. Grotz, Phys. Lett. 142B (1984) 323.
[2] T. Kirsten, H. Richter and E. Jessberger, Phys. Rev. Lett. 50 (1983) 474.
[3] W.C. Haxton, G.J. Stephenson Jr. and D. Strottman, Phys. Rev. Lett. 47 (1981) 153;
W.C. Haxton, Phys. Rev. D25 (1982) 2360;
W.C. Haxton and G.J. Stephenson Jr., preprint submitted to Prog. Part. Nucl. Phys.
[4] J.J. Simpson et al., Phys. Rev. Lett. 53 (1984) 141.
[5] H. Ejiri et al., Intern. Symp. on Nuclear spectroscopy and nuclear interactions (Osaka, 1984).
[6] A.A. Klimenko, A.A. Pomansky and A.A. Smolnikov, Proc. XIth Intern. Conf. on Neutrino physics and astrophysics (Neutrino '84) (Nordkirchen, June 1984).
[7] M. Doi et al., Prog. Theor. Phys. 69 (1983) 602.
[8] W.H. Moore, G.K. Schlegel and S. O'Dell, Nucl. Phys. A104 (1967) 327;
A.L. Auble, J.P. Ball and Y.B. Falmer, Nucl. Phys. A116 (1968) 14.
[9] J. Speth, V. Klempt, J. Wambach and G.E. Brown, Nucl. Phys. A343 (1980) 382.
[10] E. Greuling and R.C. Whitten, Ann. Phys. 11 (1960) 510.
[11] P. Minkowski, Nucl. Phys. B201 (1982) 269.
[12] D. Bogdan, A. Faessler, A. Petrovici and S. Holan, preprint (1984).
[13] E. Bellotti et al., Proc. XIth Intern. Conf. on Neutrino physics and astrophysics (Neutrino '84) (Nordkirchen, June 1984).
[14] R.K. Bardin, P.J. Gollon, J.D. Ullmann and C.S. Wu, Nucl. Phys. A158 (1970) 337.
[15] T. Kirsten, Proc. Workshop on Science underground (Los Alamos, 1982).
[16] B.T. Cleveland et al., Phys. Rev. Lett. 35 (1975) 757.

Volume 157B, number 4 PHYSICS LETTERS 18 July 1985

PREDICTIONS OF 2ν AND 0ν DOUBLE BETA DECAY RATES FOR NUCLEI WITH $A \geqslant 70$

K. GROTZ and H.V. KLAPDOR

Max-Planck-Institut für Kernphysik, D-6900 Heidelberg, Germany

Received 13 March 1985; revised manuscript received 26 April 1985

Microscopic calculations of double beta ($\beta\beta$) decay matrix elements are presented for the first time for all potential $\beta\beta$ decaying nuclei with $A \geqslant 70$. This replaces previous phase-space estimates for most of these nuclei. It is found that nuclear structure effects considerably influence the $\beta\beta$ decay rates. The best candidates for 2ν as well as 0ν $\beta\beta$ decay experiments are selected.

In recent years, double beta ($\beta\beta$) experiments have gained much attraction [1–8]. The main reason for this is to get improved limits for the neutrino mass [9] by searching for the neutrinoless (0ν) decay mode. The other point is that up to now no double beta decay event – also not of the conventional two neutrino (2ν) type – has been identified by a direct observation. Double beta decay has only been proven by the indirect geochemical method, which is limited to very few isotopes.

There exist 35 potential double beta emitters throughout the nuclidic chart. However, nuclear structure calculations needed for the theoretical prediction of decay rates in terms of lepton number violation, taking into account explicitly the spectrum of intermediate 1^+ states in the case of 2ν decay, up to now only existed for a very limited number of isotopes [10–12]. Predictions for the other $\beta\beta$ emitters were simply based on phase-space estimates [13]. As discussed in detail in ref. [11] for calculating the 2ν rates the intermediate 1^+ spectrum has to be treated properly, while the widely used closure approach [13–16] is not sufficient. It is no contradiction to this that closure is well justified for the 0ν decay, in which case the energy of the intermediate states is dominated by the virtual neutrino [9].

Following this line, in this paper we predict for the first time $\beta\beta$ half-lives for a nearly complete list of $\beta\beta$ emitters, based on microscopic nuclear structure calculations.

Most of these nuclei are either very heavy ($A \gtrsim 140$) or strongly deformed or both. For these reasons and because $\beta\beta$ decay shows collective features [9–11, 13,15], large model spaces have to be used. Such large basis calculations can be performed in the framework of the quasiparticle random phase approximation (QRPA), introduced for Gamow–Teller forces by ref. [17]. In this paper a method similar to the one used by Huffmann [10] for $^{128,130}Te$ but improved in some points is developed. In the QRPA treatment the lepton contribution to the virtual excitation energy in the intermediate state (see eq. (1) of ref. [11]) is replaced by $T_0/2 + m_e$ with T_0 being the $\beta\beta$ Q-value. In this way the virtual excitation energies of the nuclear states can be treated properly avoiding closure. The decay rate then is

$$\omega_{2\nu} = f^{2\nu} \Big| \sum_{M} \langle 0_F^+ | \sum_i \boldsymbol{\sigma}(i)\tau^-(i) | 1_M^+ \rangle \times \langle 1_M^+ | \sum_j \boldsymbol{\sigma}(j)\tau^-(j) | 0_I^+ \rangle (E_M - E_I + T_0/2 + m_e)^{-1} \Big|^2 \equiv f^{2\nu}\epsilon \,, \quad (1)$$

where $f^{2\nu}$ is the lepton phase space factor; the nuclear structure part is condensed in the quantity ϵ defined by eq. (1). The intermediate 1^+ and 0^+ states are described in the QRPA formalism with a residual p–n interaction

0370-2693/85/$ 03.30 © Elsevier Science Publishers B.V. (North-Holland Physics Publishing Division)

Volume 157B, number 4 PHYSICS LETTERS 18 July 1985

$$H_{\sigma\tau} = \chi \sum_{ij} \boldsymbol{\sigma}(i)\boldsymbol{\sigma}(j)\tau^-(i)\tau^+(j)\ . \tag{2}$$

The QRPA calculations were performed in a deformed Nilsson–BCS basis. The values for Δ_p, Δ_n and the U and V coefficients were separately optimized for parent and daughter nuclei (however, using the same deformation parameters). Because the QRPA calculation consistently treats the 0^+ ground state and 1^+ states of the neighbouring nuclei it was necessary to perform separately QRPA calculations based first on the parent and then on the daughter nucleus ground state. The summation in eq. (1) then involves products of the transition matrix elements of *both* QRPA calculations. One problem arises: the 1^+ states of the intermediate nucleus resulting from the two different QRPA calculations are mathematically not identical (of course, physically they should be). In order not to violate orthogonality relations the 1^+ eigenstates $|\bar{a}\rangle$ resulting from the daughter nucleus QRPA have to be expanded in the basis defined by the parent nucleus QRPA eigenstates labelled by $|a\rangle$. ϵ is then given by

$$\epsilon = \Big| \sum_{a\bar{a}} \langle 0_F^+ | \sum_i \boldsymbol{\sigma}(i)\tau^-(i) |\bar{a}\rangle\langle\bar{a}|a\rangle$$
$$\times \langle a| \sum_j \boldsymbol{\sigma}(j)\tau^-(j)|0_I^+\rangle (E_a - E_I + T_0/2 + m_e) \Big|^2 . \tag{3}$$

In eq. (3) the physical 1^+ states are assumed to be given by the set $|a\rangle$. One could, however, equally well choose the set $|\bar{a}\rangle$. This seems to be rather arbitrary, but in fact is not a serious point. Although the two sets of eigenstates do not fulfill the relation

$$\langle\bar{a}_k|a_l\rangle = \delta_{kl}\ ,$$

exactly, they fulfill it approximately in the sense

$$\langle\bar{a}|a\rangle \simeq 0\ , \quad \text{for } |E_{\bar{a}} - E_a| > \Delta E\ ,$$

and

$$\sum_{|E_{\bar{a}} - E_a| < \Delta E} |\langle\bar{a}|a\rangle|^2 \simeq 1\ ,$$

with ΔE being of the order of 1 MeV. This means that the strength of the states which are eigenstates in one calculation is fragmented over an energy interval of ~1 MeV in the other calculation. By this, in eq. (3) there is introduced an uncertainty in the energy denominator which is, however, small compared to the mean value $\langle E\rangle$.

The final point is to express the transition matrix elements in terms of r and s-QRPA expansion coefficients. The expansion coefficients we use are related to those of ref. [17] by

$$r_{np} = (-)^{J_n + J_p}\, a(p,n)\ , \quad s_{np} = (-)^{J_n + J_p + 1}\, b(p,n)\ .$$

Retaining terms up to linear order in s, we obtain:

$$\epsilon = \Big| \Big(\sum_{\substack{a\bar{a}\\ np}} O(n,p)\overline{r^a_{np}}\, r^a_{np} \sum_\mu (-)^\mu \sum_{\substack{n'p'\\ n''p''}} \langle n'|\sigma_\mu|p'\rangle$$
$$\times \langle n''|\sigma_{-\mu}|p''\rangle (\overline{r^a_{n''p''}}\, r^a_{n'p'} \bar{U}_{n''}\bar{V}_{p''} V_{n'} U_{p'}$$
$$- s^a_{n''p''}\overline{r^a_{n'p'}}\, V_{n''}\bar{U}_{p''}\bar{V}_{n'} U_{p'}$$
$$- \overline{r^a_{n''p''}}\, s^a_{n'p'} \bar{U}_{n''}\bar{V}_{p''} U_{n'} V_{p'}) \Big)$$
$$\times (E_a - E_I + T_0/2 + m_e)^{-1} \Big|^2 . \tag{4}$$

In the above expression all quantities with a bar correspond to the daughter nucleus. $O(n, p)$ is an overlap function actually depending on the orbits involved in the transition

$$O(n,p) = \prod_{n' \neq n} (\bar{U}_{n'} U_{n'} + \bar{V}_{n'} V_{n'})$$
$$\times \prod_{p' \neq p} (\bar{U}_{p'} U_{p'} + \bar{V}_{p'} V_{p'})\ . \tag{5}$$

In the actual calculation for simplicity this orbital dependence, which is rather weak, was skipped by including also the orbits n and p in the product eq. (5). If the barred quantities in eq. (4) are replaced by unbarred, the nuclear matrix part of eq. (4) simplifies (up to linear order in s coefficients) to the expression given for $M_{\beta\beta}$ in ref. [10]. Numerically the matrix elements obtained with this simplified expression are found to be smaller for almost all nuclei (up to a factor 2) than deduced from eq. (4). The reason is that the refined treatment does take into account the change in proton and neutron numbers in $\beta\beta$ decay. This increases the leading contributions in eq. (4) involving the factor $\bar{U}_n \bar{V}_p V_n U_p$. This factor is in general larger than the factor $U_n V_p V_n U_p$ in the approximate treatment, because mostly one has $\bar{U}_n^2 + V_n^2 > 1$ and $\bar{V}_p^2 + U_p^2 > 1$.

The parameters used in the calculations were chosen as follows: The single particle orbits were calculated

Volume 157B, number 4 PHYSICS LETTERS 18 July 1985

from a Nilsson potential [19] with deformation parameter δ fixed individually for each nucleus (see table 1 and ref. [18]). The number of orbits included in the calculation was taken as large as necessary to take into account all transitions down to the percent level and is typically 50 for medium heavy ($A \simeq 100$) and up to 70 for the very heavy ($A \gtrsim 180$) nuclei with approximately the same number of levels on both sides of the Fermi levels. It is observed that the influence of orbits far away from the Fermi level decreases rather slowly. If the model space is increased from 40 to 60 single particle orbits in the heavy nucleus ^{160}Gd this results still in an increase of ϵ of ~20%.

The BCS parameters were calculated from the gap equation with the pairing strength $G_n = G_p = 19$ MeV/A for $A < 150$ and $G_n = 20$ MeV/A, $G_p = 22$ MeV/A for $A > 150$. For the interaction constant χ

Table 1
Results of the present calculations of $\beta\beta$ decay. Given are the calculated closure matrix elements $M^{2\nu}$, the quantity ϵ and the calculated half-lives for the 2ν and 0ν decay modes. The 0ν half-lives are neutrino mass dependent and the given figures correspond to $\langle m_\nu \rangle = 1$ eV. $T_{1/2}^{2\nu *}$ are half-lives obtained with the method described in ref. [11]. The 0ν half-lives calculated as in ref. [9] are also marked with an asterisk.

Nucleus	T_0 (MeV)	QRPA				$T_{1/2}^{2\nu *}$ (yr)	$T_{1/2}^{0\nu} \times \langle m_\nu \rangle^2$ (yr × eV^2)
		δ	$M^{2\nu}$	ϵ (MeV^{-2})	$T_{1/2}^{2\nu}$ (yr)		
^{70}Zn	1.00	0.	6.79	0.720	1.9×10^{22}		7.6×10^{23}
^{76}Ge	2.04	0.2	5.54	0.278	1.1×10^{20}	2.2×10^{20}	2.6×10^{23}*
^{80}Se	0.136	0.2	5.33	0.273	9.0×10^{28}		6.7×10^{25}
^{82}Se	3.01	0.2	4.47	0.200	4.5×10^{18}	1.5×10^{19}	9.5×10^{22}*
^{86}Kr	1.25	0	1.63	0.039	3.5×10^{22}		5.0×10^{24}
^{94}Zr	1.15	−0.1	4.63	0.450	4.1×10^{21}		6.2×10^{23}
^{96}Zr	3.35	−0.12	4.83	0.490	5.2×10^{17}		1.6×10^{22}
^{98}Mo	0.111	−0.19	4.33	0.308	1.6×10^{29}		7.3×10^{25}
^{100}Mo	3.03	−0.24	3.81	0.258	1.8×10^{18}		3.3×10^{22}
^{104}Ru	1.30	−0.26	3.81	0.258	1.8×10^{21}		5.0×10^{23}
^{110}Pd	2.01	−0.23	3.56	0.219	5.0×10^{19}		1.3×10^{23}
^{114}Cd	0.54	0.14	1.80	0.111	2.7×10^{24}		1.7×10^{25}
^{116}Cd	2.81	0	1.63	0.065	8.3×10^{18}		1.7×10^{23}
^{122}Sn	0.36	0	1.75	0.036	1.4×10^{26}		3.6×10^{25}
^{124}Sn	2.28	0	1.65	0.030	9.3×10^{19}		3.1×10^{23}
^{128}Te	0.87	0.15	2.36	0.044	1.2×10^{23}	5.7×10^{23}	9.8×10^{23}*
^{130}Te	2.53	0.10	2.55	0.050	1.9×10^{19}	1.2×10^{20}	4.6×10^{22}*
^{134}Xe	0.84	0	3.18	0.110	5.1×10^{22}	2.5×10^{23}	8.7×10^{23}*
^{136}Xe	2.48	0	1.22	0.019	6.0×10^{19}	3.3×10^{19}	3.0×10^{23}*
^{142}Ce	1.41	0	1.21	0.024	2.8×10^{21}	4.1×10^{20}	4.7×10^{23}*
^{146}Nd	0.061	0	3.38	0.193	2.9×10^{30}		5.6×10^{25}
^{148}Nd	1.93	0.18	2.63	0.192	2.5×10^{19}		1.1×10^{23}
^{150}Nd	3.37	0.24	2.17	0.082	4.8×10^{17}		2.4×10^{22}
^{154}Sm	1.25	0.28	3.10	0.127	9.5×10^{20}		2.4×10^{23}
^{160}Gd	1.73	0.29	3.46	0.170	4.4×10^{19}		6.4×10^{22}
^{170}Er	0.66	0.27	2.80	0.139	6.6×10^{22}		9.1×10^{23}
^{176}Yb	1.08	0.26	2.69	0.161	1.1×10^{21}		2.5×10^{23}
^{186}W	0.49	0.20	2.46	0.131	3.2×10^{23}		1.2×10^{24}
^{192}Os	0.41	−0.15	2.32	0.076	1.7×10^{24}		1.6×10^{24}
^{198}Pt	1.04	−0.10	0.78	0.008	1.2×10^{22}		1.6×10^{24}
^{204}Hg	0.41	0	0.47	0.002	4.6×10^{25}		2.6×10^{25}
^{232}Th	0.85	0.23	3.57	0.311	1.6×10^{20}		3.8×10^{22}
^{238}U	1.15	0.24	2.87	0.245	2.2×10^{19}		2.4×10^{22}

Volume 157B, number 4 PHYSICS LETTERS 18 July 1985

the following A dependence was assumed (moderately modified A^{-1} dependence)

$$\chi = 20[1 + \beta(A - 130)]/A \text{ MeV}, \quad (6)$$

with $\beta = 0.001$ for $A < 130$ and $= 0.0035$ for $A > 130$. This parametrization was obtained from energy systematics of the Gamow–Teller giant resonance (GTGR) known from (p, n) experiments [20].

The results of the calculations are given in table 1. Besides the calculated 2ν half-life, also the so-called closure matrix elements

$$M^{2\nu} = \langle 0_F^+ | \sum \boldsymbol{\sigma}(i)\tau^-(i)\boldsymbol{\sigma}(j)\tau^-(j)|0_I^+\rangle \quad (7)$$

($M^{2\nu}$ is related to M_{GT} of ref. [13] by $M^{2\nu} = 2M_{GT}$) and the quantity ϵ defined in eq. (1) is given. The latter two quantities are used to parametrize the results in a convenient way. ϵ reflects the nuclear structure influence for the 2ν decay and below we relate $M^{2\nu}$ to the 0ν decay, which is insensitive to the intermediate 1^+ spectrum.

It is of interest to compare the results of the RPA approach presented here with the more refined treatment described in ref. [11] (called method 1 throughout this paper) which includes also the coupling of the 0^+ ground states to 2^+ phonons. The 2ν $\beta\beta$ half-lives obtained in ref. [11] are included in table 1. In addition the nuclei ^{134,136}Xe and ^{142}Ce have been calculated using method 1. This shows that inclusion of phonon correlations leads to considerably larger half-lives in the $A \simeq 130$ region than obtained in the QRPA calculation (for a discussion see ref. [11]). One exception is ^{136}Xe. Although in this nucleus, too, we find strong cancellations in the closure matrix element, a large low-energetic component dominates the transition rate and leads to an even smaller half-life than the QRPA. We would expect the QRPA results to be more reliable for the well-deformed and especially for the very heavy nuclei where $\beta\beta$ decay is a more collective effect involving a large number of single particle orbits.

However, overestimation of the 2ν rates by nuclear structure calculations seems to be a general trend (compare tables 1, 2), and was a systematic and even larger problem in earlier calculations (see, e.g., ref. [13] and the discussion in ref. [11]). The problem has been considerably reduced for ^{128,130}Te, however, not completely solved in the work of ref. [11].

Table 2
Present experimental data on $\beta\beta$ decay. The experiments are divided into two classes, epxeriments using counter techniques to observe directly decay events (marked D) and geochemical determinations (G). The given half-lives hold for the decay mode indicated.

Nucleus	$T_{1/2}$ a) (yr)	Selected mode	Method b)	Ref.
^{46}Ca	$>4 \times 10^{11}$		D	[23]
^{48}Ca	$>2 \times 10^{21}$ (80%CL)	0ν	D	[24]
	$>3.6\times 10^{19}$ (80%CL)	2ν	D	[24]
^{70}Zn	$>5 \times 10^{14}$		D	[23]
^{76}Ge	$>3.4\times 10^{22}$ (2σ)	0ν	D	[1] c)
^{82}Se	$>3.1\times 10^{21}$	0ν	D	[7]
	$>(1.45\pm 0.15)\times 10^{20}$		G	[8]
^{96}Zr	$>(1.4-6.5)\times 10^{19}$	0ν	D	[5]
^{100}Mo	$>2.1\times 10^{21}$	0ν	D	[5]
^{114}Cd	$>5 \times 10^{15}$		D	[23]
^{116}Cd	$>6 \times 10^{16}$	0ν	D	[25]
^{124}Sn	$>1.5\times 10^{17}$	0ν	D	[26]
^{128}Te	$>8 \times 10^{24}$ (2σ)		G	[3]
^{130}Te	$>1.2\times 10^{21}$	0ν	D	[5]
	$(2.6\pm 0.2)\times 10^{21}$		G	[8]
^{150}Nd	$>1.3\times 10^{19}$ (95%CL)	2ν	D	[6]
	$>1.2\times 10^{21}$ (95%CL)	0ν	D	[6]
^{186}W	$>1 \times 10^{15}$		D	[23]
^{192}Os	$>1 \times 10^{13}$		D	[23]
^{198}Pt	$>5 \times 10^{14}$		D	[23]

a) Unless otherwise stated, limits are based on one standard deviation.
b) D = direct counter experiment, G = geochemical determination.
c) Similar limits are also obtained by Caldwell et al. and by the Caltech, Osaka, Guelph-APTEC and Battelle groups [2].

Besides the QRPA half-lives for ^{82}Se, ^{128,130}Te the value obtained for ^{150}Nd is also definitely too short. ^{48}Ca is a further candidate (this light nucleus is not suitable for our calculational approach) where there is some indication that the 2ν rate is smaller than nuclear structure predictions [12]. On the other hand, up to now there does not exist any contrary case, the half-life of a $\beta\beta$ emitter being *shorter* than calculated (some speculative discussion connected with this point was given in ref. [21], see, however, also ref. [22]). In this respect, using the results of our calculation as a guideline for experimental planning, one should be conservative.

Concerning the neutrinoless decay mode induced by a possible Majorana mass term, in table 1 we give

Volume 157B, number 4 PHYSICS LETTERS 18 July 1985

estimates based on the following assumption. The 0ν decay rate $\omega_{0\nu}$ is calculated according to the relation (see ref. [9])

$$\omega_{0\nu} = f^{0\nu} |1 - x_{\rm F}|^2 \xi^2 |M^{2\nu}|^2 \langle m_\nu \rangle^2 / m_{\rm e}^2 \,. \qquad (8)$$

Besides the phase space factor $f^{0\nu}$ eq. (8) contains the nuclear structure parameters $x_{\rm F}$ and ξ, not obtained in the RPA approach, for which we use the values −0.24 [9] and 2.9, respectively. The value for ξ is obtained by dividing the matrix elements $|R_0 M^{0\nu}|$ for the four nuclei calculated in ref. [9] by the QRPA closure results $M^{2\nu}$ (yielding 1.8 to 4.3) and averaging. (We do not use the larger values for ξ calculated directly in ref. [9] because the QRPA calculation as discussed above does not include all effects leading to the smaller $M^{2\nu}$ values calculated in ref. [9].) The 0ν half-lives calculated in this way are given in table 1, last column. The given figures correspond to an assumed neutrino mass of 1 eV.

Concluding, we have presented for the first time nuclear structure calculations for a practically complete list (except the Ca isotopes) of $\beta\beta$ emitters. These are based on the collective treatment of Gamow–Teller transitions with the unlike QRPA formalism using large model spaces. Static deformation is included. This approach is therefore especially suitable for heavy complex nuclei.

The quantity ϵ introduced is a measure whether nuclear structure favours 2ν $\beta\beta$ decay or not. It is seen that ϵ is very small for nuclei around the closed shells $Z = 50, N = 82, Z = 82$.

Looking for nuclei with simultaneously favourable nuclear structure and phase space (large T_0), the best candidates for the 2ν mode are ^{82}Se, ^{96}Zr, ^{100}Mo and ^{150}Nd (the latter because of the large T_0). ^{136}Xe, which has received some attention for future experiments, unfortunately does not seem to be a very good candidate.

Concerning the more interesting 0ν decay, phase space considerations are less important because of the more moderate T_0 dependence. The nuclear structure influence is expressed by $\Sigma |M^{2\nu}|^2$ rather than by ϵ, the intermediate spectrum of GT states being less important for this decay process. The best candidates are here ^{82}Se, ^{96}Zr, ^{100}Mo, ^{110}Pd, ^{130}Te, ^{148}Nd, ^{150}Nd, ^{160}Gd, ^{232}Th and ^{238}U.

[1] E. Bellotti et al., Proc. XIth Intern. Conf. on Neutrino physics and astrophysics (Neutrino 84) (Nordkirchen, June 1984).
[2] D.O. Caldwell et al., Phys. Rev. Lett. 54 (1985) 281; A. Forster et al., Phys. Lett. 138B (1984) 301; E. Ejiri et al., Intern. Symp. on Nuclear spectroscopy and nuclear interactions (Osaka, March 1984); J.J. Simpson et al., Phys. Rev. Lett. 53 (1984) 141; F.T. Avignone et al., Phys. Rev. Lett. 50 (1983) 721.
[3] T. Kirsten, H. Richter and E. Jessberger, Phys. Rev. Lett. 50 (1983) 474.
[4] M. Moe and D. Löwenthal, Phys. Rev. C 22 (1980) 2186.
[5] Yu.G. Zdesenko, V.N. Kutz, I.A. Mytsyk and A.S. Nikolaiko, Izv. Akad. Nauk. (SSSR), Ser. Fiz. 45 (1981) 1856; Yu.G. Zdesenko, JETP Lett. 32 (1980) 62.
[6] A.A. Klimenko, A.A. Pomansky and A.A. Smolnikov, INR Academy of Science preprint.
[7] B.T. Cleveland et al., Phys. Rev. Lett. 35 (1975) 757.
[8] T. Kirsten, Proc. Workshop on Science underground (Los Alamos, 1982).
[9] K. Grotz and H.V. Klapdor, Phys. Lett. 153B (1985) 1.
[10] A.H. Huffmann, Phys. Rev. C 2 (1970) 741.
[11] H.V. Klapdor and K. Grotz, Phys. Lett. 142B (1984) 323.
[12] T. Tsuboi, K. Muto and H. Horie, Phys. Lett. 143B (1984) 293.
[13] W.C. Haxton, G.J. Stephenson Jr. and D. Strottman, Phys. Rev. Lett. 50 (1983) 474; W.C. Haxton and G.J. Stephenson Jr., Prog. Part. Nucl. Phys. 12 (1984) 409; W.C. Haxton, C.A. Gowan and M. Goldhaber, Phys. Rev. C 28 (1983) 467.
[14] V.A. Khodel, Phys. Lett. 32B (1970) 583.
[15] L. Zamick and N. Auerbach, Phys. Rev. C 26 (1982) 2185.
[16] T. Tomoda, A. Faessler, K.W. Schmid and F. Grümmer, preprint.
[17] J.A. Halbleib and R.A. Sorensen, Nucl. Phys. A 98 (1967) 542.
[18] V.G. Soloviev, Theory of complex nuclei (Pergamon, London, 1976) p. 244ff; J.G. Rohozinski, J. Dobaczewski, B. Nerlo-Pomorska, U. Pomorski and J. Srebrny, Nucl. Phys. A 292 (1977) 66; K. Kumar and M. Baranger, Nucl. Phys. A 110 (1968) 529.
[19] I. Ragnarsson, Proc. 6th Intern. Conf. on Atomic masses and fundamental constants (1979).
[20] C. Gaarde et al., Proc. 4th Intern. Conf. on Nuclei far from stability (Helsingor, 1981) p. 281.
[21] H.V. Klapdor and K. Grotz, Report MPI H-1983-V34.
[22] K. Grotz and H.V. Klapdor, Phys. Rev. C 30 (1984) 2098.
[23] J.H. Fremlin and M.C. Walters, Proc. Phys. Soc. A 65 (1952) 911.
[24] R.K. Bardin et al., Nucl. Phys. A 158 (1970) 337.
[25] J.F. Detoeuf and R. Moch, J. Phys. Radium 16 (1955) 897.
[26] J.A. McCarthy, Phys. Rev. 90 (1953) 853.

Nuclear Physics **A460** (1986) 395–436
North-Holland, Amsterdam

CALCULATION OF DOUBLE BETA DECAY RATES AND THE NEUTRINO MASS

K. GROTZ and H.V. KLAPDOR

Max-Planck-Institut für Kernphysik, Heidelberg, Germany

Received 29 January 1986

Abstract: Predictions for 2ν and 0ν double beta decay rates are given for all nuclei with $A \geqslant 70$, for which double beta decay is energetically allowed. These predictions are based on detailed nuclear structure studies of the beta strength distribution and replace earlier estimates basing mostly on phase space considerations. New and more stringent limits on the Majorana neutrino mass are deduced from existing double beta decay experiments. Since the collective effects arising from spin–isospin as well as quadrupole–quadrupole forces are found to lead to a strong reduction of the nuclear matrix elements for two-neutrino double beta decay, but to have only minor influence on the matrix elements $M^{0\nu}$ for the neutrinoless decay mode, the smaller limits for m_ν result mainly from the fact that the widely used scaling procedure underestimates the 0ν matrix elements. It is further discussed to what extent interference between different neutrinos affects the obtained mass limits.

1. Introduction: neutrino masses and double beta decay

1.1. NEUTRINO MASSES

The exploration of the nature of the neutrino at the lower end of the mass hierarchy of elementary particles (fig. 1) is a great challenge. Nuclear physics can contribute to the solution of the question of neutrino masses and mixing by the investigation of neutrinoless double beta decay (0ν $\beta\beta$ decay). The latter is a sensitive probe for a Majorana mass of the neutrino [1]), which might be a unique property of the neutrino among the elementary fermions.

Fermion fields are in general described by four linear independent basic spinor fields, two of which are related to the others by charge conjugation. A possible basis for the description of a fermion field is, therefore, given by ψ_L, ψ_R, $(\psi_L)^C$, and $(\psi_R)^C$ ($\psi^C = C(\bar{\psi})^T$)*. As a consequence of fermion masses being described as coupling between right- and left-handed field components, arises the distinction between Dirac and Majorana masses (fig. 2): Dirac masses result from the coupling of two independent fields (independent in the sense that they are not related by charge conjugation) $m^D\bar{\psi}_L\psi_R$. In contrast Majorana masses correspond to the

* For the sake of transparency we define the charge conjugation for the handed field components according to $(\psi_{L/R})^C = C(\bar{\psi}_{L/R})^T$ with $C = i\gamma_2\gamma_0$ (ψ^T denotes the transposed of ψ). This is not the usual charge conjugation, because under the so defined operation the handedness of the field changes. The usual definition of charge conjugation for the handed fields would be: $(\psi_{L/R})^C = C(\bar{\psi}_{R/L})^T$.

0375-9474/86/$03.50 © Elsevier Science Publishers B.V.
(North-Holland Physics Publishing Division)

2.2.2 QRPA Calculations

Including the pp–Force

VOLUME 57, NUMBER 25 PHYSICAL REVIEW LETTERS 22 DECEMBER 1986

Suppression of the Two-Neutrino Double-Beta Decay by Nuclear-Structure Effects

P. Vogel
Physics Department, California Institute of Technology, Pasadena, California 91125

M. R. Zirnbauer
W. K. Kellogg Radiation Laboratory, California Institute of Technology, Pasadena, California 91125
(Received 14 July 1986)

The nuclear matrix element for 2ν double-beta decay is calculated within the quasiparticle random-phase approximation. It is shown that the decay matrix element passes through zero as a function of the strength g^{pp} of the particle-particle component of the spin-isospin polarization force, neglected previously. The analysis of electron capture/β^+ decay rates for semimagic neutron-deficient nuclei suggests values for g^{pp} in the very vicinity of this zero, which gives rise to long lifetimes. The qualitative features of nuclear $\beta\beta$ decay are illustrated with the example of an exactly soluble model.

PACS numbers: 23.40.Bw, 21.10.Pc, 23.40.Hc

There are two modes of $\beta\beta$ decay, one involving the emission of two neutrinos (2ν mode) and another associated with no neutrino emission (0ν mode) which requires the existence of massive Majorana neutrinos. The 2ν mode has two electrons and two antineutrinos in the final state and occurs in the standard theory in second order. In order to interpret the results of experiments on $\beta\beta$ decay, we need to understand the nuclear-structure part of the decay process. The 2ν mode, being independent of the neutrino properties, offers a sensitive test of our ability to deal with this problem.

Calculation of the nuclear matrix elements for the 2ν $\beta\beta$ decay is the topic of this Letter. The general theory has been reviewed recently by Haxton and Stephenson[1] In all previous investigations the calculated decay rates were faster than the experimental ones. The discrepancy was particularly disturbing in the case of ^{130}Te, where experiment and theory differ by a factor of 100. We offer here a possible solution to this problem.

The nuclear matrix element for the $2\nu\,\beta\beta$ decay connecting the 0^+ ground states of two even-even nuclei is, in the standard approximation, given by

$$M_{\rm GT}=\sum_m \frac{\langle f|\sigma\tau^+|m\rangle\cdot\langle m|\sigma\tau^+|i\rangle}{E_m-(M_i+M_f)/2}. \qquad (1)$$

Like Vogel and Fisher,[2] we use the quasiparticle random-phase approximation (QRPA) to calculate the matrix elements connecting both the initial and the final 0^+ state with each of the intermediate 1^+ states. The energies of the intermediate states, E_m, are obtained at the same time.

The QRPA equations, and the QRPA matrices A and B, are of the form

$$\begin{bmatrix} A & B \\ -B & -A \end{bmatrix}\begin{bmatrix} x \\ y \end{bmatrix}=\omega\begin{bmatrix} x \\ y \end{bmatrix}, \qquad (2)$$

$$A_{pn,p'n'}=(\varepsilon_p+\varepsilon_n)\delta_{pn,p'n'}+g^{ph}\bar V_{pn,p'n'}(u_pv_nu_{p'}v_{n'}+v_pu_nv_{p'}u_{n'})+g^{pp}V_{pn,p'n'}(u_pu_nu_{p'}u_{n'}+v_pv_nv_{p'}v_{n'}), \qquad (3a)$$

$$B_{pn,p'n'}=g^{ph}\bar V_{pn,p'n'}(v_pu_nu_{p'}v_{n'}+u_pv_nv_{p'}u_{n'})-g^{pp}V_{pn,p'n'}(u_pu_nv_{p'}v_{n'}+v_pv_nu_{p'}u_{n'}). \qquad (3b)$$

Here, ε_i are the quasiparticle energies and u,v the usual occupation amplitudes. There are two types of two-body interaction matrix elements in (3a) and (3b); the particle-hole matrix element $\bar V$ and the particle-particle matrix element V. We parametrize both of them by the respective matrix elements of a δ-function interaction $[V=\sigma_1\cdot\sigma_2\tau_1\cdot\tau_2\delta(\mathbf r_1-\mathbf r_2)]$. In a first-principles calculation these two matrix elements would be related by angular-momentum recoupling. However, our use of an *effective* shell-model space requires the use of an *effective* two-body interaction, and we are forced to take the two interaction constants g^{ph} and g^{pp} as independent. The particle-hole coupling constant g^{ph} is determined from the requirement that the energy of the giant Gamow-Teller (GT) resonance be correctly reproduced; we here obtain $g^{ph}=-333$ MeV fm^3, with an uncertainty of $\sim 10\%$.

The particle-particle interaction has been neglected in all previous treatments of $\beta\beta$ decay which use the QRPA and in most papers dealing with the β^- and β^+ decays. This interaction indeed has little effect on the giant GT resonance. However, as we shall see, it plays an essential role when one tries to calculate strongly suppressed quantities such as the matrix element $M_{\rm GT}$ in (1) or the β^+ decay rates in nuclei with a neutron excess.

In order to determine the value of the particle-particle coupling constant g^{pp}, we use the following procedure.

VOLUME 57, NUMBER 25 PHYSICAL REVIEW LETTERS 22 DECEMBER 1986

First, we note that the β^+ strength, which is suppressed by Pauli blocking, depends sensitively on g^{pp}. There is no experimental information on this strength from (n,p) reactions. However, in several semimagic neutron-deficient nuclei the electron capture/β^+ decay is expected to contain essentially the whole β^+ strength. We thus use the experimental ft values for the decays of ^{94}Ru, ^{96}Pd, ^{148}Dy, ^{150}Er, and ^{152}Yb (from Kleinheinz *et al.*,[3] Kurcewicz *et al.*,[4] and Nolte *et al.*[5]) to fix g^{pp}. In these nuclei the extreme shell model overestimates the β^+ strength by a factor of ~ 7. If we assume, following Towner,[6] that the interactions not accounted for (Δ isobar, etc.) can be included approximately by use of an axial-vector-current coupling constant $g_A = 1.0$ instead of 1.25, we obtain a good description of the experimental ft values with g^{pp} in a relatively narrow interval 128–144 MeV fm^3. The actual value of g^{pp} depends to some extent on the number of spherical subshells used in the calculation. (Addition of a pair of subshells changes the effective g^{pp} typically by $\sim 10\%$.) In our calculation we used all subshells within 10 MeV of the Fermi energy plus their spin-orbit partners. The single-particle energies were taken from a Wood-Saxon Coulomb-corrected potential, and the same number of subshells was used for neutrons and protons. QRPA calculations of $M_{\rm GT}$ are carried out for both the initial and final nuclei and the resulting matrix elements are averaged. The BCS equations were solved with a δ-function interaction of strength $g^{\rm pair} \sim -300$ MeV fm^3. It is interesting to note that the effective coupling constants $-3g^{pp}$ and $g^{\rm pair}$ are very close to each other.

We now turn to the evaluation of the 2ν $\beta\beta$ decay matrix element $M_{\rm GT}$. The matrix elements in Eq. (1) are obtained from

$$\langle 1_m^+ | \sigma\tau^+ | 0_i^+ \rangle = \sum_{pn} \langle p \| \sigma \| n \rangle (u_p v_n x_{pn}^m + v_p u_n y_{pn}^m), \quad (4a)$$

$$\langle 0_f^+ | \sigma\tau^+ | 1_m^+ \rangle = \sum_{pn} \langle p \| \sigma \| n \rangle (v_p u_n x_{pn}^m + u_p v_n y_{pn}^m). \quad (4b)$$

For nuclei with a neutron excess the complex conjugate of the β^+ matrix element, Eq. (4b), is small and very sensitive to the amount of pairing and ground-state correlations, characterized by the factors u, v, and the amplitudes y_{pn}^m, respectively. The terms proportional to x and y on the right-hand side of (4b) typically have opposite signs, a feature that can be understood from the following argument. We write the second of the QRPA equations (2) in the form $-Bx = (A+\omega)y$. Using the premise that the matrix $(A+\omega)$ is positive definite, we see that the relative sign of x and y is controlled by the sign of B. The latter matrix is largely positive, as a result of the repulsive nature of the particle-hole interaction, the attractive nature of the particle-particle interaction, and the way in which these interactions enter into the expression (3b). We note further that the two terms $v_p u_n x_{pn}^m$ and $u_p v_n y_{pn}^m$ are of the same order because the intrinsic smallness of y is compensated for by the difference in occupation amplitudes. As the value of g^{pp} increases, the second term in fact overpowers the first one, and each matrix element on the left-hand side of (4b) *passes through zero.* (For very large values of g^{pp} the QRPA becomes unstable and gives rise to complex frequencies.) As the β^- matrix elements (4a) are only weakly dependent on g^{pp}, $M_{\rm GT}$ passes through zero, too. Figure 1 illustrates this point. (The decay rate is proportional to the square of $M_{\rm GT}$; see Refs. 1 and 2 for the coefficient of proportionality.) We observe that the 2ν $\beta\beta$ decay matrix element $M_{\rm GT}$, and the decay rate, vanish for $g^{pp} \sim 110$ MeV fm^3 and vary very fast with g^{pp} in the vicinity of this value. This does not mean that the terms in Eq. (1) vanish all at once. Rather, the individual terms in the sum over intermediate states cancel each other. The closure approximation in which the E_m in Eq. (1) are replaced by a constant is of dubious value at this point, and the whole sum can be as small as the contribution of a single term, a feature noted[7] in the case of ^{128}Te. Quantitative prediction of the $\beta\beta$ decay lifetime now becomes very difficult. On the other hand, one can easily understand very long lifetimes in some nuclei, such as ^{130}Te.

The extreme sensitivity to g^{pp} may explain why the suppression of $M_{\rm GT}$ observed here went unnoticed in the shell-model work of Haxton and Stephenson as described in Ref. 1. Their G matrix definitely contains the particle-particle interaction that we find to be of crucial importance, but, because of the considerable uncertainty in present effective interactions, it is not clear whether the relevant interaction matrix elements are correctly

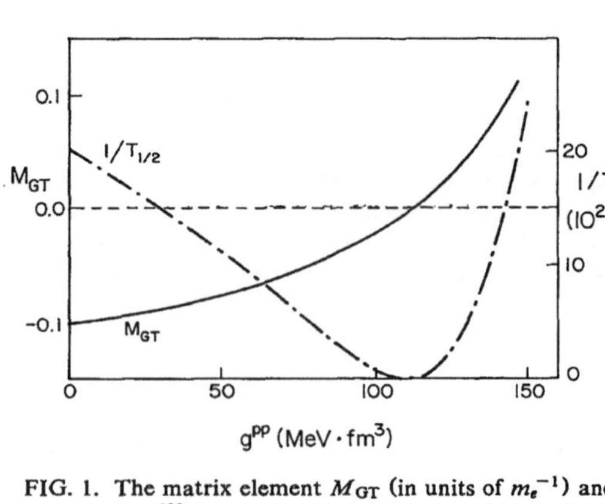

FIG. 1. The matrix element $M_{\rm GT}$ (in units of m_e^{-1}) and the decay rate for ^{130}Te as a function of the particle-particle coupling constant g^{pp}.

represented.

Is the described behavior of M_{GT} a genuine property of nuclei or just an artifact of the QRPA? To be able to give a convincing answer to this question, we have constructed a schematic model which can be solved exactly and yet contains most of the qualitative features of a more realistic description. We consider a set of single-particle states with the associated creation and time-reversed annihilation operators, $a^\dagger$ and $\tilde{a}$, which we label by their orbital angular momentum, spin, isospin, and the respective z components. Using LS coupling we then define sixteen particle-hole operators and twelve pair operators with definite spin and isospin; in all of them the orbital angular momenta are coupled to zero. These operators close under commutation; they form, in fact, the Lie algebra of SO(8).

From these operators we construct the following Hamiltonian:

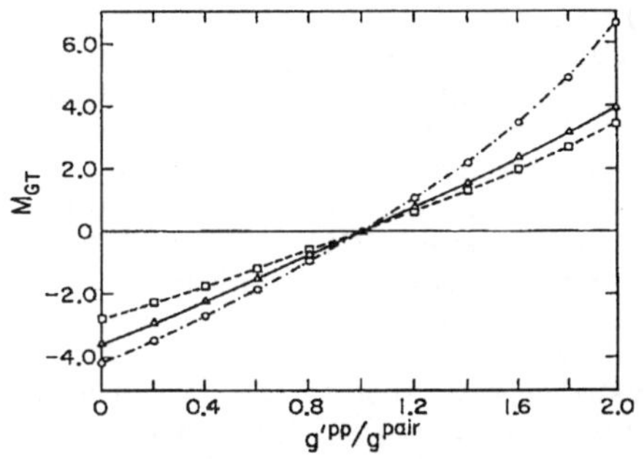

FIG. 2. Dependence of M_{GT} (in arbitrary units) on g'^{pp}/g^{pair} for the exactly soluble model described in the text. The full line connecting triangles is the exact solution, and the dashed line (dot-dashed line) connecting squares (circles) is the QRPA result for the initial (final) nucleus.

$$H = g^{pair}(S^+_{nn}S^-_{nn} + S^+_{pn}S^-_{pn} + S^+_{pp}S^-_{pp}) + g'^{pp}P^+\cdot P^- + \tfrac{1}{2}g'^{ph}\sum_{\mu\nu}(-1)^{\mu+\nu}F^\nu_\mu F^{-\nu}_{-\mu}, \tag{5}$$

where S^+ is the pair creation operator with spin 0 and isospin 1 (S^- is the adjoint operator), P is the pair operator with spin 1 and isospin 0, and F^ν_μ is the particle-hole operator with spin 1, projection μ, and isospin 1, projection ν (Gamow-Teller operator). Note that this Hamiltonian is properly isoscalar and rotationally invariant. Moreover, it contains all of the features of the nuclear residual interaction directly relevant to $\beta\beta$ decay: pairing, the particle-hole interaction in the $J^\pi = 1^+$ channel, and the corresponding particle-particle interaction. What is missing from the Hamiltonian (5) is the nondegeneracy of the single-particle subshells in actual nuclei and especially the splitting of spin-orbit partners by the spin-orbit force.

The virtue of the Hamiltonian (5) is that its eigenvalues and eigenfunctions can be worked out *exactly*. In Fig. 2 we show the resulting $\beta\beta$ decay matrix element as a function of the ratio g'^{pp}/g^{pair}. The parameters we use are $g^{pair} = -0.30$, $g'^{ph} = +1.5$, and we take for the initial nucleus four proton particles and six neutron holes in a major shell with $l = 0,2,4$. Figure 2 is a clear and indisputable demonstration of the effect of increasing the value of g'^{pp}: It forces M_{GT} to pass through zero!

In the present model, the vanishing of M_{GT} has a clear-cut explanation. For $g^{pair} = g'^{pp}$ and arbitrary g'^{ph}, H acquires a dynamical SU(4) symmetry (Wigner's supermultiplet symmetry), which leads to new conserved quantum numbers in addition to those of spin and isospin. Because of the pair superstructure of our model, it is useful to reinterpret SU(4) as SO(6), by use of the known isomorphism between these two algebras. It can then be shown that the ground state of H for $g^{pair} = g'^{pp}$ is the highest-weight state of a totally symmetric irreducible representation of SO(6). Because the Gamow-Teller operator is one of the raising operators of

TABLE I. Lifetimes of the 2ν $\beta\beta$ decay. Lifetimes are given in years, coupling constants in megaelectronvolts per cubic femtometers.

Interaction parameters	$T^{exper\,a}_{1/2}$	$T^{calc}_{1/2}$ Schematic interaction $\chi = 23/A$	δ interaction $g^{pp} = 0$	δ interaction $g^{pp} = 144$	δ interaction $g^{pp} = 136$	δ interaction $g^{pp} = 128$
^{76}Ge	$> 8\times10^{19}$	1.8×10^{21}	1.6×10^{21}	5.0×10^{20}	1.8×10^{21}	6.3×10^{21}
^{82}Se	$> 1\times10^{20}$	3.3×10^{19}	7.4×10^{19}	5.5×10^{19}	2.5×10^{20}	6.3×10^{20}
^{96}Zr	...	...	2.8×10^{18}	5×10^{17} b	3.2×10^{17}	8.5×10^{18}
^{100}Mo	...	2.1×10^{18}	4.1×10^{18}	8×10^{17} b	2.8×10^{18}	3.7×10^{19}
^{116}Cd	...	...	7.5×10^{19}	9.3×10^{19}	3.4×10^{20}	1.9×10^{21}
^{128}Te	$> 7\times10^{24}$	1.1×10^{23}	1.5×10^{23}	1.0×10^{23}	2.7×10^{23}	8.2×10^{23}
^{130}Te	2.6×10^{21}	2.3×10^{19}	5.0×10^{19}	4.4×10^{19}	1.0×10^{20}	2.5×10^{20}
^{136}Xe	$> 2\times10^{19}$	2.3×10^{19}	1.5×10^{20}	2.0×10^{20}	5.1×10^{20}	1.5×10^{21}

[a]Experimental data from the following sources: Ge, Ref. 8; Se, Ref. 9, see also Ref. 10; Te, Ref. 10; Xe, Ref. 11.

[b]Unstable solution for the final nucleus; only the initial nucleus is used.

VOLUME 57, NUMBER 25 PHYSICAL REVIEW LETTERS 22 DECEMBER 1986

SO(6) $\simeq$ SU(4), it follows that the β^+ strength of the final nucleus vanishes identically, and that M_{GT} must vanish, too. It is gratifying to see that the QRPA reproduces the location of this zero exactly. We have investigated several different parameter sets and find that the qualitative behavior is always similar to that seen in Fig. 2. In heavy nuclei the SU(4) symmetry is badly broken by the spin-orbit splitting. Nevertheless, we believe that the exact solution of the schematic model is instructive for two reasons. It shows that the QRPA yields a reasonable approximation to M_{GT}, and it demonstrates the existence of a zero in M_{GT} in more general situations, as any perturbation to the Hamiltonian (5) can only change details of the curve in Fig. 2 but not its qualitative behavior.

In Table I we show the calculated $2\nu\,\beta\beta$ decay lifetimes for a number of spherical $\beta\beta$ decay candidates. For comparison we show in column 3 the lifetimes obtained with a schematic interaction,[2] which are very similar to those in column 4. The last three columns give lifetimes for three closely spaced values of g^{pp} in the predetermined interval. The extreme sensitivity is obvious. All lifetimes (except for column 3 where the effect of the Δ isobar was taken into account explicitly) were calculated with $g_A = 1.0$, as mentioned earlier. The experimental values are within, or close to, the boundaries of the calculated interval of lifetimes. It should be clear that the observed suppression makes M_{GT} susceptible to corrections arising from other components of the wave function such as quadrupole correlations, not considered here.

In conclusion, we have shown that the particle-particle interaction causes a large suppression of M_{GT}. The existence of a zero in this matrix element is a consequence of general properties of the nuclear residual interaction. Is the same sensitivity to g^{pp} found here present for the 0ν mode as well? Detailed analysis will be necessary to answer this question.

It is a pleasure to thank S. E. Koonin, H. A. Weidenmüller, and A. Winther for valuable discussions. This work was supported by the U. S. Department of Energy under Contract No. DE-AT03-81ER40002 and by the National Science Foundation, under Grants No. PHY82-07332 and PHY85-05682.

[1]W. C. Haxton and G. J. Stephenson Prog. Part. Nucl. Phys. **12**, 409 (1984).

[2]P. Vogel and P. Fisher, Phys. Rev. C **32**, 1362 (1985).

[3]P. Kleinheinz *et al.*, Phys. Rev. Lett. **55**, 2664 (1985).

[4]W. Kurcewicz *et al.*, Z. Phys. A **308**, 21 (1982).

[5]E. Nolte *et al.*, Z. Phys. A **306**, 223 (1983), and **309**, 33 (1982).

[6]I. S. Towner, Nucl. Phys. A **444**, 402 (1985).

[7]H. V. Klapdor and K. Grotz, Phys. Lett. **142B**, 323 (1984).

[8]D. O. Caldwell *et al.*, Phys. Rev. D **33**, 2737 (1986).

[9]S. R. Elliott, A. A. Hahn, and M. K. Moe, Phys. Rev. Lett. **56**, 2582 (1986).

[10]T. Kirsten, in *Science Underground—1983,* edited by M. N. Nieto, W. C. Haxton, C. M. Hoffman, E. W. Kolb, V. D. Sandberg, and J. W. Toevs, AIP Conference Proceedings No. 96 (American Institute of Physics, New York, 1983).

[11]A. S. Barabash *et al.*, contribution to the International Symposium on Nuclear Beta Decays and Neutrino, Osaka, June 1986 (unpublished).

PHYSICAL REVIEW C VOLUME 37, NUMBER 2 FEBRUARY 1988

Nuclear structure effects in double-beta decay

J. Engel and P. Vogel
Physics Department, California Institute of Technology, Pasadena, California 91125

M. R. Zirnbauer
W. K. Kellogg Radiation Laboratory, California Institute of Technology, Pasadena, California 91125
(Received 8 June 1987)

Using the quasiparticle random phase approximation, we calculate the nuclear matrix elements governing two-neutrino and neutrinoless double-beta decay. We show that a consistent treatment, including both particle-hole and particle-particle interactions, helps to resolve the longstanding discrepancy between experimental and calculated two-neutrino decay rates. The particle-particle force, which allows us to bring calculated EC/β^+ decay rates in semimagic nuclei into closer agreement with experiment, is in large part responsible for suppressing calculated two-neutrino decay rates that are otherwise too fast. We test the validity of our procedure by comparing quasiparticle random phase approximation results with exact solutions for a solvable model, in which the suppression of two-neutrino decay by the particle-particle interaction is confirmed. We then extend our approach to the neutrinoless decay associated with a finite Majorana neutrino mass and, conceivably, with majoron emission, and demonstrate that the nuclear matrix elements governing these processes are also suppressed. We present predicted half-lives for both two-neutrino and neutrinoless double-beta decay in several candidate nuclei, and discuss the difficulties associated with the calculation of such highly suppressed quantities.

I. INTRODUCTION

Double-beta decay is the process by which a nucleus with neutron and proton number (N,Z) undergoes a transition to the nucleus $(N-2,Z+2)$, when the transition to the intermediate nucleus $(N-1,Z+1)$ is energetically forbidden or highly retarded. There are two modes of double-beta decay, one involving the emission of two antineutrinos and two electrons (2ν mode), and another in which only electrons are emitted (0ν mode). The 2ν mode of double-beta decay occurs in second order in the standard theory of weak interactions and its rate is independent of a possible small neutrino mass. On the other hand, the 0ν mode violates the law of lepton number conservation and requires the existence of massive Majorana neutrinos. Observation of 0ν double-beta decay would therefore signal physics beyond the standard minimal electroweak model. For this reason, considerable experimental effort is being devoted to the study of this mode of nuclear decay. In order to plan and correctly interpret the results of experiments on double-beta decay, one has to know the various nuclear matrix elements associated with the decay process. The calculation of these matrix elements is the subject of our paper.

The theory and the experimental status of the two modes of double-beta decay have been reviewed by Doi *et al.*[1] and by Haxton and Stephenson,[2] and we refer to these papers for details. Even though a careful analysis in the framework of general gauge theories[3] shows that the 0ν mode indeed always requires the existence of a massive Majorana neutrino, one should distinguish between two mechanisms for 0ν decay: one in which the decay rate involves the effective neutrino Majorana mass $\langle m_\nu \rangle$ (mass mechanism), and a second one in which the rate contains the strength of a weak interaction involving right-handed leptonic currents (RHC mechanism). In the present work, we consider the nuclear structure aspects of the 2ν mode and of the mass mechanism for the 0ν mode. We hope to treat the RHC mechanism for the 0ν mode elsewhere.

A third, somewhat exotic mechanism for 0ν double-beta decay involves the emission of a hypothetical scalar particle, the majoron, along with the two electrons. This decay is believed to be governed by the same nuclear matrix elements that describe the mass mechanism for the 0ν mode.[1] Hence our calculations are applicable to majoron emission as well. On the other hand, we do not consider the decay mediated by very heavy neutrinos.[5]

Half-lives for the 2ν double-beta decay have recently been calculated by Vogel and Zirnbauer[6] (this paper will hereafter be referred to as I), who used the quasiparticle random phase approximation (QRPA) to deal with the nuclear structure aspects of the process. It was shown in I that the inclusion of the particle-particle component of the residual nucleon-nucleon interaction results in a considerable suppression of the decay rate. This suppression, in turn, provides a possible resolution of the longstanding discrepancy between calculated and measured 2ν double-beta decay rates.[1,2] In the present work, we elaborate on this result and extend the treatment to the nuclear matrix elements of the 0ν mode, showing that they are also strongly affected by the residual nucleon-nucleon force.

Z. Phys. A – Atomic Nuclei 334, 177–186 (1989)

Zeitschrift für Physik A
Atomic Nuclei

Effects of Ground-State Correlations on $2\nu\beta\beta$ Decay Rates and Limitations of the QRPA Approach

K. Muto*, E. Bender, and H.V. Klapdor
Max-Planck-Institut für Kernphysik, Heidelberg, Federal Republic of Germany

Received April 10, 1989

Half-lives of the $2\nu\beta\beta$ decay are calculated in the proton-neutron QRPA for ^{76}Ge, ^{82}Se, ^{100}Mo, ^{128,130}Te, ^{136}Xe and ^{150}Nd. The strength of the particle-particle interaction, which plays a decisive role for a reliable evaluation of the half-lives, is determined from a QRPA calculation of single β^+ decays. The 2ν decay rates calculated with the interaction strength fitted in this way are strongly suppressed and found to be consistent with the existing experimental data. Effects of the ground-state correlations on the suppression are investigated. On the other hand, the present calculation indicates limitations of the QRPA approach.

PACS: 23.40-s; 23.40.Hc; 21.60.Jz

1. Introduction

Nuclear double beta ($\beta\beta$) decay is expected to take place mainly via two decay modes of the two-nucleon mechanism (see [1, 2]), which are illustrated in Fig. 1; the two-neutrino (2ν) mode in which two neutrinos are emitted together with two electrons, and the neutrinoless (0ν) mode in which no neutrino leaves the nucleus. The 0ν mode is of particular interest. Observation of the 0ν decay always implies [3] that the electron neutrino is a massive Majorana particle, and also gives information on the existence of right-handed components of the charged weak interaction. This process is forbidden in the standard $SU(2)_L \times U(1)$ gauge model of the electroweak interaction [4], in which neutrinos are massless and there exist no right-handed currents. (The distinction between Dirac and Majorana character disappears for a massless neutrino, if there are no right-handed currents.) Massive neutrinos naturally appear in most grand unified theories [5, 6]. The $0\nu\beta\beta$ decay sets, at present, the most stringent limits on the mass of neutrino and also the coupling constants of the right-handed currents (see [2]).

The neutrino masses and the right-handed current coupling constants, which are deduced from an experimental 0ν decay half-life, depend decisively on the relevant nuclear matrix elements which have to be evaluated theoretically. A sensitive test of the nuclear structure calculation is provided by the $2\nu\beta\beta$ decay, since this process occurs within the standard model and the half-life can be calculated free from unknown parameters on the particle physics side. However, till 1986, large discrepancies existed between experimental half-lives and those calculated with various nuclear structure models; in particular for ^{128}Te and ^{130}Te the discrepancies were up to two orders of magnitude. The experimental half-lives (or the lower limits) imply that the 2ν decay nuclear matrix elements are suppressed, and the calculations could not reproduce the strong suppression. It was not clear at that time

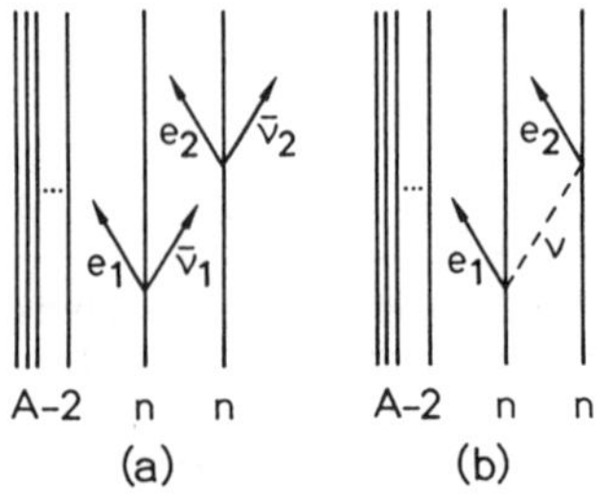

Fig. 1a and b. Two modes of $\beta\beta$ decay possible in the two-nucleon mechanism; (**a**) 2ν mode and (**b**) 0ν mode

* Permanent address: Department of Physics, Tokyo Institute of Technology, Ohokayama, Meguro, Tokyo 152, Japan

Z. Phys. A – Atomic Nuclei 334, 187–194 (1989)

Zeitschrift für Physik A
Atomic Nuclei

Nuclear Structure Effects on the Neutrinoless Double Beta Decay

K. Muto [1], **E. Bender** [2], **and H.V. Klapdor** [2]

[1] Department of Physics, Tokyo Institute of Technology, Tokyo, Japan
[2] Max-Planck-Institut für Kernphysik, Heidelberg, Federal Republic of Germany

Received April 20, 1989

The nuclear matrix elements of the $0\nu\,\beta\beta$ decay of ^{76}Ge, ^{82}Se, ^{100}Mo, ^{128,130}Te, ^{136}Xe and ^{150}Nd are calculated in the proton-neutron quasiparticle RPA with the *G*-matrix of the Paris potential. It is shown that the matrix elements are not sensitive to details of nuclear structure, in contrast to the $2\nu\,\beta\beta$ decay. We investigate effects of ground-state correlations and those of short-range correlations on the suppression of the nuclear matrix elements. We also discuss effective values of the neutrino mass which are deduced from experimental $0\nu\,\beta\beta$ decay half-lives.

PACS: 23.40.-s; 23.40.Hc; 21.60.Jz

1. Introduction

The neutrinoless double beta ($0\nu\,\beta\beta$) decay is most sensitive to the existence of massive Majorana neutrinos, practically the only possibility to distinguish between Majorana and Dirac neutrinos, and sets the most stringent limits on the neutrino mass and the coupling constants of right-handed currents [1, 2]. This lepton-number non-conserving process is forbidden in the standard model of the electroweak interaction [3], in which neutrinos are massless and there are no right-handed currents. In most grand unified theories neutrinos are naturally predicted to be massive Majorana particles [4]. Observation of the $0\nu\,\beta\beta$ decay would therefore uncover physics beyond the standard model. On the other hand, the mass of neutrinos has important consequences to astrophysics and cosmology [5].

In order to obtain quantitative answers for the mass of neutrinos and the properties of the weak interaction from the $0\nu\,\beta\beta$ decay, we have to understand the mechanism of nuclear transitions and then evaluate the nuclear matrix elements responsible for the decay with high reliability.

The suppression mechanism of the $2\nu\,\beta\beta$ decay was found in nuclear structure [2, 6–8]. The calculations in the proton-neutron quasiparticle random phase approximation (pnQRPA [9]) have clarified that the particle-particle (pp) interaction, which is the counterpart of the particle-hole interaction, enhances the spin-isospin correlations in the ground-state wave functions and the correlations suppress the 2ν decay rate. We determined [2, 8] the pp interaction strength from β^+ decays of a number of nuclei and showed that the 2ν decay rates calculated with the interaction strength fixed in this way are consistent with the existing experimental data.

The finding of the suppression mechanism of the 2ν decay raises the question whether the nuclear matrix elements of the $0\nu\,\beta\beta$ decay are also affected by the ground-state correlations, i.e., by the pp interaction. Tomoda and Faessler [10] and Engel, Vogel and Zirnbauer [11] obtained opposite conclusions by similar QRPA calculations; the former concluded that the 0ν elements are much less sensitive to the pp interaction than the 2ν element, whereas the latter showed that the 0ν elements are rather strongly suppressed by the ground-state correlations and also by the short-range correlations. In this paper we show that the nuclear matrix elements of the $0\nu\,\beta\beta$ decay are much less sensitive to details of nuclear structure than 2ν decay rates and can be predicted with much higher accuracy in the QRPA approach.

2. Calculations – the Model

We assume the effective weak interaction hamiltonian density

$$H_W=\frac{G}{\sqrt{2}}(j_{L\mu}J_L^{\mu\dagger}+\kappa j_{L\mu}J_R^{\mu\dagger}+\eta j_{R\mu}J_L^{\mu\dagger}+\lambda j_{R\mu}J_R^{\mu\dagger}) + \text{h.c.} \tag{1}$$

The left- and right-handed leptonic currents are

$$j_L^\mu = \bar{e}\,\gamma^\mu(1-\gamma_5)\,\nu_{eL}, \qquad j_R^\mu = \bar{e}\,\gamma^\mu(1+\gamma_5)\,\nu'_{eR} \tag{2}$$

and the current electron neutrinos are superpositions of mass-eigenstate Majorana neutrinos N_j with the mass m_j,

$$\nu_{eL} = \sum_j U_{ej} N_{jL}, \qquad \nu'_{eR} = \sum_j V_{ej} N_{jR}. \tag{3}$$

Hadronic currents, which are expressed in terms of quarks similarly to (2), are converted into the currents of nucleons. The nuclear currents are, in the non-relativistic impulse approximation,

$$(J_{L/R}^{0\dagger}(\mathbf{x}), \mathbf{J}_{L/R}^\dagger(\mathbf{x})) = \sum_{n=1}^{A} t_{-n}(g_V \mp g_A C_n, \mp g_A \boldsymbol{\sigma}_n + g_V \mathbf{D}_n)\,\delta(\mathbf{x}-\mathbf{r}_n) \tag{4}$$

with the nucleon recoil terms

$$C_n = [(\mathbf{p}_n + \mathbf{p}'_n)\cdot\boldsymbol{\sigma}_n - (g_P/g_A)(E_n - E'_n)(\mathbf{p}_n - \mathbf{p}'_n)\cdot\boldsymbol{\sigma}_n]/(2M), \tag{5a}$$

$$\mathbf{D}_n = [(\mathbf{p}_n + \mathbf{p}'_n) - i\,\mu_\beta\,\boldsymbol{\sigma}_n \times (\mathbf{p}_n - \mathbf{p}'_n)]/(2M) \tag{5b}$$

where $(E_n, \mathbf{p}_n)$ and $(E'_n, \mathbf{p}'_n)$ are the initial and final four-momenta of the n-th nucleon, respectively, and M the nucleon mass. We take the values of the coupling constants, $G = 1.16637 \times 10^{-5}\,\mathrm{MeV}^{-2}$, $g_V = 0.9737$, $g_A/g_V = 1.254$ and $\mu_\beta = 4.706$ [12]. The scalar recoil term C_n does not contribute under the assumption that finite de Broglie wave length corrections of electrons are neglected. Terms proportional to κ will be neglected, since it always appears in the combination $1 \pm \kappa$ and $|\kappa| \ll 1$ is expected.

The inverse half-life of the $0\nu\,\beta\beta$ decay, $0_i^+ \to 0_f^+$, can be written as

$$[T_{1/2}^{0\nu}]^{-1} = C_{mm}\left(\frac{\langle m_\nu\rangle}{m_e}\right)^2 + C_{\eta\eta}\langle\eta\rangle^2 + C_{\lambda\lambda}\langle\lambda\rangle^2 + C_{m\eta}\frac{\langle m_\nu\rangle}{m_e}\langle\eta\rangle + C_{m\lambda}\frac{\langle m_\nu\rangle}{m_e}\langle\lambda\rangle + C_{\eta\lambda}\langle\eta\rangle\langle\lambda\rangle \tag{6}$$

where effective values of the neutrino mass and coupling strengths of the right-handed currents are

$$\langle m_\nu\rangle = \sum_j m_j U_{ej}^2, \qquad \langle\eta\rangle = \eta \sum_j U_{ej} V_{ej}, \qquad \langle\lambda\rangle = \lambda \sum_j U_{ej} V_{ej}. \tag{7}$$

Coefficients C_{xy} consist of products of electron phase-space integrals G_k ($k = 0$–9) and the nuclear matrix elements M_α ($\alpha = GT, F, GT\omega, F\omega, GTq, Fq, T, P$ and R),

$$\begin{aligned}
C_{mm} &= (M_{GT} - M_F)^2 G_1, \\
C_{\eta\eta} &= M_{2+}^2 G_2 + \tfrac{1}{9} M_{1-}^2 G_4 - \tfrac{2}{9} M_{1-} M_{2+} G_3 + M_P^2 G_8 - M_P M_R G_7 + M_R^2 G_9, \\
C_{\lambda\lambda} &= M_{2-}^2 G_2 + \tfrac{1}{9} M_{1+}^2 G_4 - \tfrac{2}{9} M_{1+} M_{2-} G_3, \\
C_{m\eta} &= (M_{GT} - M_F)(M_{2+} G_3 - M_{1-} G_4 - M_P G_5 + M_R G_6), \\
C_{m\lambda} &= -(M_{GT} - M_F)(M_{2-} G_3 - M_{1+} G_4), \\
C_{\eta\lambda} &= -2 M_{2+} M_{2-} G_2 + \tfrac{2}{9}(M_{1+} M_{2+} + M_{1-} M_{2-}) G_3 - \tfrac{2}{9} M_{1-} M_{1+} G_4.
\end{aligned} \tag{8}$$

The nine nuclear matrix elements are

$$M_\alpha = \sum_{m,n} \langle 0_f^+ \| t_{-m} t_{-n} O_{mn}^\alpha \| 0_i^+ \rangle \tag{9}$$

with the following two-body transition operators

$$\begin{aligned}
O_{12}^{GT} &= \boldsymbol{\sigma}_1 \cdot \boldsymbol{\sigma}_2 H_m(r), \qquad O_{12}^{GT\omega} = \boldsymbol{\sigma}_1 \cdot \boldsymbol{\sigma}_2 H_\omega(r), \\
O_{12}^{GTq} &= \boldsymbol{\sigma}_1 \cdot \boldsymbol{\sigma}_2 \frac{r}{R} H_q(r), \qquad O_{12}^{F} = H_m(r)\left(\frac{g_V}{g_A}\right)^2, \\
O_{12}^{F\omega} &= H_\omega(r)\left(\frac{g_V}{g_A}\right)^2, \qquad O_{12}^{Fq} = \frac{r}{R} H_q(r)\left(\frac{g_V}{g_A}\right)^2, \\
O_{12}^{T} &= [(\boldsymbol{\sigma}_1 \cdot \hat{\mathbf{r}})(\boldsymbol{\sigma}_2 \cdot \hat{\mathbf{r}}) - \tfrac{1}{3}\boldsymbol{\sigma}_1 \cdot \boldsymbol{\sigma}_2] \frac{r}{R} H_q(r), \\
O_{12}^{P} &= i(\boldsymbol{\sigma}_1 - \boldsymbol{\sigma}_2) \cdot \left(\hat{\mathbf{r}} \times \frac{\mathbf{r}_+}{R}\right) H_q(r)\left(\frac{g_V}{g_A}\right), \\
O_{12}^{R} &= \boldsymbol{\sigma}_1 \cdot \boldsymbol{\sigma}_2 H_R(r) \frac{\mu_\beta}{3}\left(\frac{g_V}{g_A}\right),
\end{aligned} \tag{10}$$

where $\mathbf{r} = \mathbf{r}_1 - \mathbf{r}_2$, $r = |\mathbf{r}|$, $\hat{\mathbf{r}} = \mathbf{r}/r$, $\mathbf{r}_+ = (\mathbf{r}_1 + \mathbf{r}_2)/2$, and the nuclear radius R is introduced to make the matrix elements dimensionless (Neutrino potentials are also defined to be dimensionless, (12)). We retain the dominant central part of the recoil matrix element M_R [13]. The combinations $M_{1\pm}$ and $M_{2\pm}$ are

$$\begin{aligned}
M_{1\pm} &= M_{GTq} \pm 3 M_{Fq} - 6 M_T, \\
M_{2\pm} &= M_{GT\omega} \pm M_{F\omega} - \tfrac{1}{9} M_{1\mp}.
\end{aligned} \tag{11}$$

The four types of "neutrino potentials" are defined by the integral of the neutrino momentum,

$$\begin{aligned}
H_m(r) &= \frac{2R}{\pi}\frac{1}{r}\int_0^\infty \frac{q\sin(qr)}{\omega(\omega+\bar{E})}\,dq, \\
H_\omega(r) &= \frac{2R}{\pi}\frac{1}{r}\int_0^\infty \frac{q\sin(qr)}{(\omega+\bar{E})^2}\,dq, \\
H_q(r) &= -R\frac{d}{dr} H_m(r), \\
H_R(r) &= -\frac{R}{M}\frac{d^2}{dr^2} H_m(r)
\end{aligned} \tag{12}$$

where $\omega=\sqrt{q^2+m_\nu^2}$ and $\bar{E}$ is the average energy of intermediate nuclear states. In the $0\nu\,\beta\beta$ decay, one can apply the closure approximation, although it fails badly in case of the 2ν decay, since $\bar{E}\sim 10$ MeV is much smaller than the neutrino energy ω which is typically ~ 100 MeV for a light neutrino.

Because of the two-body character of the transition operators, O^α_{12}, the nuclear matrix elements can be reduced to a sum of products of two-body transition densities and transition matrix elements for two-particle states,

$$M_\alpha=\sum_{p'pn'nJ_2J\pi} Z(p'p, n'n; J_2 J^\pi) \cdot \langle p'p; J_2 | t_{-1} t_{-2} O^\alpha_{12} | n'n; J_2\rangle. \quad (13)$$

The $\beta\beta$ decay occurs by transitions of two nucleons whose angular momenta are coupled to J_2, $n'n\to p'p$; two successive single-particle transitions, $n\to p$ from the 0_i^+ parent state to an intermediate state with spin-parity J^π and $n'\to p'$ from the intermediate state to the 0_f^+ daughter state. The short-range correlations between two nucleons are taken into account by multiplying the two-particle wave functions by the correlation function [14]

$$1-f(r)=1-e^{-ar^2}(1-b\,r^2) \quad (14)$$

with $a=1.1\ \mathrm{fm}^{-2}$ and $b=0.68\ \mathrm{fm}^{-2}$. The finite nucleon size effects are introduced by the monopole form factors in momentum space [15]

$$g_{V/A}\to g_{V/A}\left(\frac{\Lambda^2}{\Lambda^2+q^2}\right)^2 \quad (15)$$

with $\Lambda=850$ MeV, which are inserted in (12).

In the pnQRPA model, the transition densities are expressed as

$$\begin{aligned}&Z(p'p, n'n; J_2 J^\pi)\\ &=(-1)^{p'+n+J_2+J}(2J_2+1)(2J+1)\,W(p'pn'n; J_2 J)\\ &\quad\cdot\sum_{a'a}(\bar{X}^{p'n',J\pi}_{a'}\,\bar{v}_{p'}\,\bar{u}_{n'}-\bar{Y}^{p'n',J\pi}_{a'}\,\bar{u}_{p'}\,\bar{v}_{n'})\\ &\quad\cdot\langle J^\pi_{a'}|J^\pi_a\rangle\,(X^{pn,J\pi}_a\,u_p\,v_n-Y^{pn,J\pi}_a\,v_p\,u_n)\end{aligned} \quad (16)$$

with the overlap

$$\langle J^\pi_{a'}|J^\pi_a\rangle=\sum_{pn}(\bar{X}^{pn,J\pi}_{a'}X^{pn,J\pi}_a-\bar{Y}^{pn,J\pi}_{a'}Y^{pn,J\pi}_a). \quad (17)$$

Here, u and v are occupation amplitudes which are calculated in the BCS approximation, X and Y are eigenfunctions of the QRPA equation, and a (a') distinguishes the eigenstates. The QRPA equation is solved for each J^π (excitation from the 0^+ ground state to intermediate states with spin-parity J^π) for both parent and daughter states. The quantities without (with) a bar are defined with respect to the 0_i^+ (0_f^+) state.

3. Results and Discussion

The nuclear matrix elements are calculated in pnQRPA with a realistic interaction, the G-matrix of the Paris potential [16], for ^{76}Ge, ^{82}Se, ^{100}Mo, ^{128,130}Te, ^{136}Xe and ^{150}Nd. The $0\nu\,\beta\beta$ decays of these nuclei have extensively been studied experimentally. We take the value of g_{pp}, the renormalization factor of the pp interaction, which is determined from β^+/EC decays of nuclei with the mass number $60\leqq A\leqq 170$ [2, 8]. With the interaction strength, the $2\nu\,\beta\beta$ decay half-lives consistent with the experimental data are obtained by the QRPA calculation. The coefficients of (8) are then evaluated with the calculated nuclear matrix elements and the phase-space integrals taken from [1]. The obtained values are listed in Table 1.

The nuclear matrix elements of the present calculation are in remarkably good agreement with those obtained by Tomoda and Faessler [10], except the tensor matrix element, M_T. These authors did not calculate this matrix element, since a negligibly small value of M_T was obtained [13] in the Hartree-Fock-Bogoliubov calculation with the angular-momentum and nucleon number projections. Our values of M_T, however, are not so small. The omission of this matrix element does not significantly affect upper bounds on the coupling strength $\langle\lambda\rangle$. The coupling strength of the right-handed weak current, $\langle\eta\rangle$, is determined mostly by the recoil matrix element M_R, as was pointed out in [13].

In the $0\nu\,\beta\beta$ decay, the neutrino is virtual and its momentum integration gives rise to neutrino potentials, (12), which act on nuclear wave functions. Due to the neutrino potential the 0ν decay proceeds through intermediate nuclear states with various multipolarities J^π. By contrast, only 1^+ states are allowed in the case of 2ν decay in which real neutrinos are emitted, and the 2ν decay rates are strongly suppressed by the spin-isospin ($J^\pi=1^+$) ground-state correlations. This distinctive feature of the 0ν decay results in different effects of nuclear structure, compared to the 2ν decay. Taking the $0\nu\,\beta\beta$ decay of ^{76}Ge $\to{}^{76}$Se as a typical example, we will see the nuclear structure effects on the nuclear matrix elements $M_{GT}-M_F$ and M_R, which are responsible for the determination of the neutrino mass $\langle m_\nu\rangle$ and the coupling strength $\langle\eta\rangle$, respectively.

Figure 1 shows the behaviour of the ground-state correlations in the daughter state (upper), to which

Table 1. The nuclear matrix elements of the $0\nu\beta\beta$ decay and the coefficients (in y^{-1}) of the decay rate formula. The nuclear matrix elements are calculated in pnQRPA with the most probable value of the pp interaction strength [8]. The electron phase-space integrals are taken from [1]

	^{76}Ge	^{82}Se	^{100}Mo	^{128}Te	^{130}Te	^{136}Xe	^{150}Nd
M_{GT}	3.014	2.847	0.763	3.103	2.493	1.120	4.254
M_F	−1.173	−1.071	−1.356	−1.184	−0.977	−0.461	−1.821
$M_{GT\omega}$	2.912	2.744	1.330	3.011	2.442	1.172	4.206
$M_{F\omega}$	−1.025	−0.939	−1.218	−1.047	−0.867	−0.411	−1.630
M_{GTq}	1.945	1.886	−1.145	1.999	1.526	0.480	2.485
M_{Fq}	−1.058	−0.966	−1.161	−1.054	−0.860	−0.389	−1.592
M_T	−0.612	−0.789	−0.823	−0.583	−0.574	−0.482	−1.148
M_P	−0.530	−0.500	1.182	−0.483	−0.387	−0.360	0.998
M_R	3.594	3.343	4.528	4.371	3.736	2.193	7.005
C_{mm}	1.12×10^{-13}	4.33×10^{-13}	2.05×10^{-13}	3.36×10^{-14}	5.34×10^{-13}	1.18×10^{-13}	7.74×10^{-12}
$C_{m\eta}$	2.19×10^{-11}	6.37×10^{-11}	6.48×10^{-11}	9.46×10^{-12}	9.10×10^{-11}	2.61×10^{-11}	1.02×10^{-9}
$C_{m\lambda}$	-4.11×10^{-14}	-1.60×10^{-13}	-1.61×10^{-13}	-4.86×10^{-15}	-2.17×10^{-13}	-2.80×10^{-14}	-3.57×10^{-12}
$C_{\eta\eta}$	4.44×10^{-9}	1.54×10^{-8}	3.50×10^{-8}	1.50×10^{-9}	2.25×10^{-8}	8.27×10^{-9}	2.95×10^{-7}
$C_{\eta\lambda}$	-4.99×10^{-14}	-3.84×10^{-13}	7.03×10^{-13}	-1.87×10^{-15}	-4.13×10^{-13}	-6.22×10^{-14}	-8.39×10^{-12}
$C_{\lambda\lambda}$	1.36×10^{-13}	1.01×10^{-12}	1.05×10^{-12}	7.39×10^{-15}	1.05×10^{-12}	2.04×10^{-13}	2.68×10^{-11}

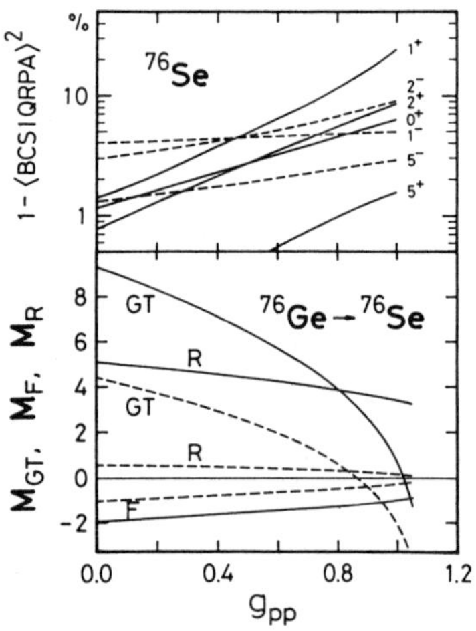

Fig. 1. Upper: Percentages of the ground-state correlations of ^{76}Se calculated in pnQRPA for some J^π values. Lower: The nuclear matrix elements M_{GT}, M_F and M_R of the $0\nu\beta\beta$ decay $^{76}Ge\rightarrow{}^{76}Se$, calculated as function of g_{pp}. Contributions to M_{GT} and M_R (M_F) through intermediate states with $J^\pi=1^+$ (0^+) are shown by a dashed line

the nuclear matrix elements are more sensitive than to those of the parent state, and the nuclear matrix elements (lower) as function of g_{pp}. For this decay, $g_{pp}=0.875$ is the most probable value [8]. In the probable range of g_{pp}, the percentage of the 1^+ correlations increases rapidly and is more than twice that of the other multipolarities. This is reflected in the behaviour of the nuclear matrix elements, in particular M_{GT}. The decrease of M_{GT} with g_{pp} comes mostly from that of the 1^+ contribution, and the other contributions remain almost constant. The 1^+ contribution, which is about a half of M_{GT} at $g_{pp}=0$, decreases and almost completely vanishes at $g_{pp}=0.875$, as the 2ν decay matrix element does. Consequently, M_{GT} is suppressed by the pp interaction by a factor of 3. This suppression is much smaller than that of the $2\nu\,\beta\beta$ decay, $M_{GT}^{2\nu}$, which is suppressed by up to two orders of magnitude [8]. In case of M_F (only natural-parity states are allowed for the intermediate states), the decrease with g_{pp} is mainly due to the 0^+ contribution. The recoil matrix element, M_R, is much less sensitive to a change of g_{pp}, although its transition operator is similar to that of M_{GT}, because of the short-range character of the neutrino potential,

$$H_R(r) \sim \frac{\delta(r)}{r^2} - \frac{2\bar{E}}{\pi}\frac{1}{r^2} + \bar{E}^2 H_m(r). \qquad (18)$$

The zero-range potential favours transitions via high-multipole intermediate states.

A decomposition of the nuclear matrix elements is shown in Fig. 2. In the mass term, M_{GT} and $-M_F$ have the same sign in each J^π mode, and all components of different J^π are added constructively, except for $J^\pi=1^+$, in which the calculation involves the largest uncertainties. The recoil matrix element comprises all multipole contributions with the same sign. There exist sizable contributions up to the highest multipolarities, the $J^\pi=9^+$ component comes only from two successive $\nu g_{9/2}\rightarrow\pi g_{9/2}$ transitions.

Figure 3 shows an alternative decomposition of $M_{GT}-M_F$ and M_R into components of two-particle transitions in which the angular momenta of the two nucleons are coupled to J_2^π (see (13)). For example, the component with $J_2^\pi=0^+$ arises from transitions of 0^+-coupled two neutrons to 0^+-coupled two protons. All components of $J_2^\pi \neq 0^+$ have negative sign and the sum of them cancels the positive contribution of $J_2^\pi = 0^+$, particularly in $M_{GT} - M_F$. This means that a too large nuclear matrix element is obtained by nuclear wave functions with only pairing correlations, and it is reduced by ground-state correlations of various modes, for example, the quadrupole correlations, $J_2^\pi = 2^+$. It is also noted that the $\beta\beta$ decay occurs exclusively via transitions with $\pi = (-1)^{J_2}$, implying transitions between spin-singlet states.

Thus, the nuclear matrix elements of the $0\nu\,\beta\beta$ decay are much less sensitive to details of nuclear structure than the $2\nu\,\beta\beta$ decay matrix element, due to the fact that the neutrino is virtual. Consequently they can be calculated in the QRPA approach with high accuracy. Figure 4 illustrates the insensitivity for the $\beta\beta$ decay of ^{128}Te. In the vicinity of the most probable value of g_{pp}, the calculated half-life of the 0ν mode increases gradually with g_{pp}, whereas the 2ν half-life changes by orders of magnitude.

Figure 5 compares, for the seven nuclei under consideration, the experimental lower limit of the 0ν half-life with the theoretical value calculated with the assumption $m_\nu = 1\,$eV, neglecting contributions of the right-handed currents. It should be kept in mind, however, that the nuclear structure calculation is somewhat uncertain in the $J^\pi = 1^+$ channel, i.e., the spin-isospin ground-state correlations, because of the rapid decrease of the 1^+ contribution (see Fig. 1). Uncertainties of $M_{GT} - M_F$ corresponding to the 1σ range of g_{pp} [8] are shown in Fig. 6. The recoil matrix element, M_R, can be calculated more accurately, since it involves a much smaller 1^+ contribution (see Fig. 1). The choice of g_{pp} which yields 2ν decay rates consistent with experiment is important for the reliable evaluation of the 0ν decay nuclear matrix elements.

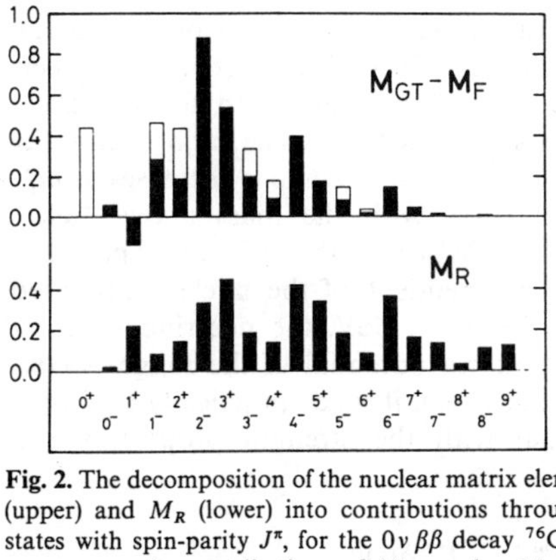

Fig. 2. The decomposition of the nuclear matrix elements $M_{GT} - M_F$ (upper) and M_R (lower) into contributions through intermediate states with spin-parity J^π, for the $0\nu\,\beta\beta$ decay $^{76}\mathrm{Ge} \to {}^{76}\mathrm{Se}$. In the upper part, the contributions of M_{GT} and $-M_F$ are denoted by the filled and open histograms, respectively

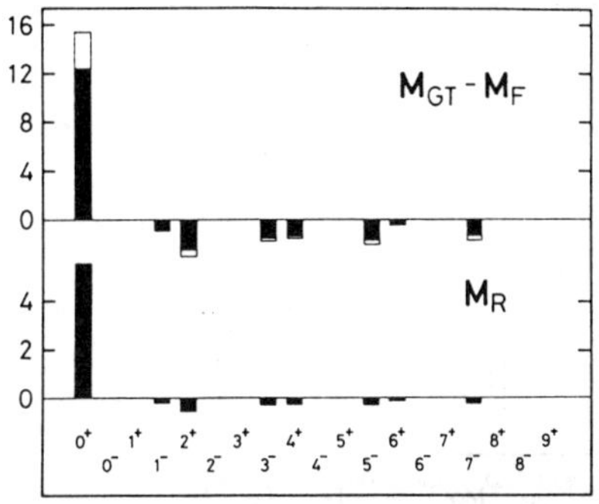

Fig. 3. The decompositon of $M_{GT} - M_F$ (upper) and M_R (lower) into contributions arising from transitions of two nucleons coupled to the spin-parity J^π, for the $0\nu\,\beta\beta$ decay $^{76}\mathrm{Ge} \to {}^{76}\mathrm{Se}$. The contributions of M_{GT} and $-M_F$ are denoted by the filled and open histograms, respectively

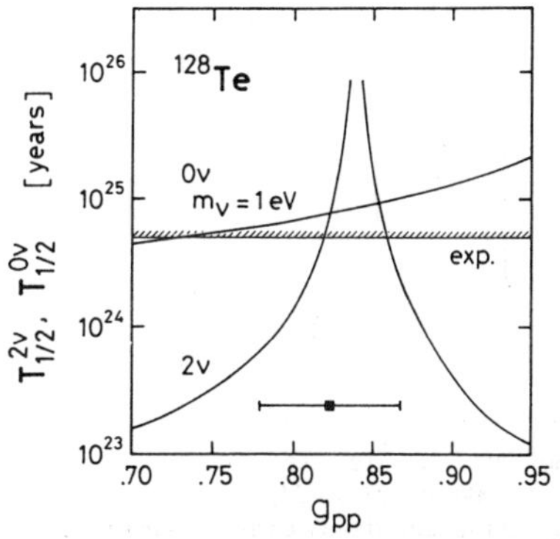

Fig. 4. The 2ν and $0\nu\,\beta\beta$ decay half-lives of ^{128}Te calculated as function of g_{pp}. In the 0ν decay, $m_\nu = 1\,$eV is assumed and contributions of the right-handed currents are neglected. The square and bar denote the most probable value of g_{pp} and its uncertainty (1σ) which were obtained in the least-squares fit of the g_{pp} values determined by single β^+ decays of a number of nuclei [8]. The experimental lower limit of geochemical measurements [20] is indicated by a horizontal line with hatch

Upper bounds on the neutrino mass and right-handed current coupling strengths are deduced from experimental half-lives of $0\nu\,\beta\beta$ decay [17–22] by using the coefficients C_{xy} of Table 1. The maximum and the value on axis are listed in Table 2, together with the experimental half-lives. The decays of ^{76}Ge and ^{128}Te set the sharpest limits; $|\langle m_\nu \rangle| \lesssim 2\,$eV, $|\langle \eta \rangle| \lesssim 2 \times 10^{-8}$, and $|\langle \lambda \rangle| \lesssim 4 \times 10^{-6}$. These are more stringent than those deduced from other experimental sources (see [1, 2]).

The upper bounds which are deduced by using the coefficients obtained by the similar pnQRPA cal-

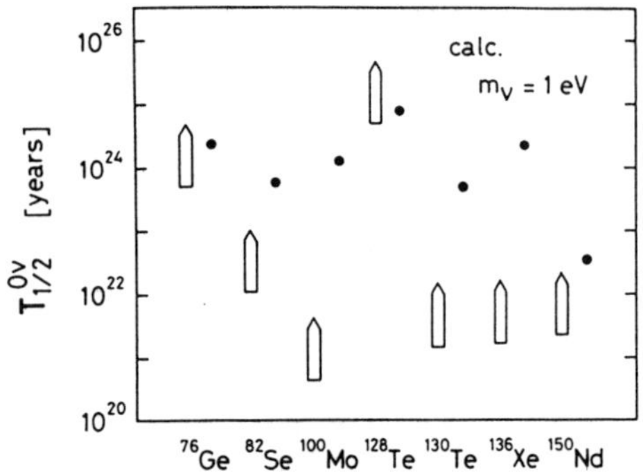

Fig. 5. Comparison of the experimental [17–22] and theoretical half-lives of the $0\nu\,\beta\beta$ decay. The experimental lower limits are denoted by an open arrow. The theoretical values are obtained by assuming $m_\nu = 1$ eV and neglecting right-handed current contributions

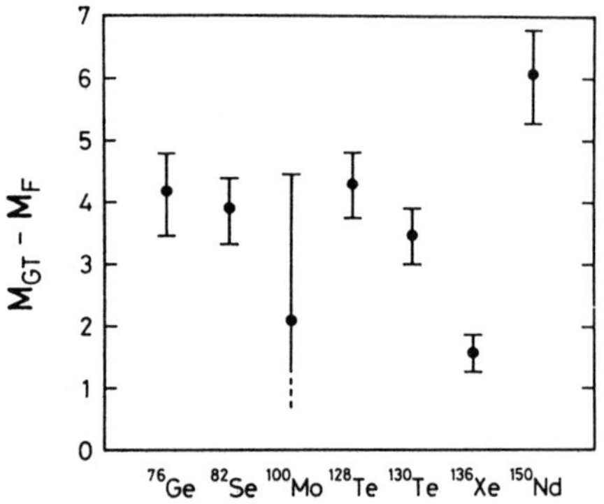

Fig. 6. The nuclear matrix element $M_{GT}-M_F$ calculated at the most probable value of g_{pp} and its uncertainty corresponding to the 1σ range of g_{pp} denoted by the "error" bar (see caption of Fig. 4)

culations [10, 11] are also given in Table 2. The agreement of our result with that by [10] is naturally good, in particular for the neutrino mass. Tomoda and Faessler took another realistic interaction, the G-matrix of the Bonn potential [23], and chose $g_{pp}=1$.

On the other hand, the bounds on the neutrino mass of [11] (They calculated only the nuclear matrix elements relevant to the neutrino mass, and consequently the deduced masses are values on axis.) are systematically larger than the values we obtained, i.e., their nuclear matrix elements are more suppressed. This strong suppression of the 0ν matrix element could be attributed to coherent effects of some origins. First, these authors adopted zero-range interactions and estimated a probable g_{pp} window from β^+ decays of semimagic nuclei. Their g_{pp} window corresponds to a range higher than ours. At a larger value of g_{pp}, the 1^+ contribution is negative and large in the absolute value, and the negative contribution cancels positive contributions of the other multipolarities to more extent, leading to a stronger suppression of M_{GT}. Since the nuclear matrix element of the 2ν decay behaves as the 1^+ contribution of the 0ν decay as a function of g_{pp}, the 2ν decay rates of [11] are systematically overestimated. Secondly, M_{GT} is more reduced by the short-range correlations (SRC) at a higher g_{pp}. We have calculated M_{GT} with and without SRC, and the result is shown in Fig. 7. The nuclear matrix element is reduced by an almost same amount independently of g_{pp}. As a result, the reduction at $g_{pp}=0$ is only 13%, and the corresponding values at $g_{pp}=0.875$ and $g_{pp}=1.0$ are 27% and 60%, respectively. The SRC suppress transitions of two nucleons coupled to $J_2^\pi=0^+$, in which angular momenta of the two nucleons are anti-parallel and the nucleons have a high possibility to come close to each other. Thirdly, the use of zero-range interactions would enhance the SRC effects. A zero-range interaction favours two nucleons to be at short distances, and transitions arising from these components are most strongly affected by SRC.

Finally, we will discuss on the effective value of neutrino mass, $\langle m_\nu \rangle$. It is a weighted average of masses of Majorana neutrinos. The definition of (7) is not correct, if a (left-handed) heavy neutrino contributes to the $0\nu\,\beta\beta$ decay. In this case, the effective value that is deduced from the 0ν decay is [1]

$$\langle m_\nu \rangle = \sum_j m_j U_{ej}^2 R(m_j) \tag{19}$$

where

$$R(m_j) = \frac{M_{GT}(m_j) - M_F(m_j)}{M_{GT}(0) - M_F(0)}. \tag{20}$$

Here, the neutrino-mass dependence of the nuclear matrix elements is indicated explicitly. The dependence of the ratio $R(m_\nu)$ calculated for the decay $^{76}Ge \rightarrow {}^{76}Se$ is shown in Fig. 8. The ratio is unity for light neutrinos, $m_\nu \leqq 1$ MeV, while it decreases as m_ν^{-2} for heavy neutrinos with the mass above 1 GeV, $R(m_\nu) = 3.2 \times (10^8\ \mathrm{eV}/m_\nu)^2$ with m_ν in eV. The factor 3.2 is almost independent of the nuclei considered. In the region of $m_\nu > 1$ GeV, the neutrino potential resembles a Yukawa potential of the range $\sim 1/m_\nu$, and, as m_ν is increased, it tends practically to a zero-range potential with the strength proportional to m_ν^{-2},

$$H_m(r) \rightarrow R\,\frac{e^{-m_\nu r}}{r} \rightarrow \frac{R}{m_\nu^2}\,\frac{\delta(r)}{r^2}. \tag{21}$$

Table 2. Upper bounds on the effective values of the neutrino mass $\langle m_\nu \rangle$ (in eV) and the coupling strengths of the right-handed currents $\langle \eta \rangle$ and $\langle \lambda \rangle$. The experimental lower limits of the $0\nu\,\beta\beta$ decay half-life (in y) are given in the last line

	^{76}Ge	^{82}Se	^{100}Mo	^{128}Te	^{130}Te	^{136}Xe	^{150}Nd
This work							
Maximum							
$\langle m_\nu \rangle$	2.6	8.1	59	1.7	20	40	4.1
$\langle \eta \rangle\ 10^{-8}$	2.5	8.3	28	1.6	19	29	4.1
$\langle \lambda \rangle\ 10^{-6}$	4.0	9.6	47	5.3	25	54	4.1
On axis							
$\langle m_\nu \rangle$	2.2	7.4	54	1.2	18	36	3.8
$\langle \eta \rangle\ 10^{-8}$	2.2	7.7	25	1.2	17	27	3.8
$\langle \lambda \rangle\ 10^{-6}$	4.0	9.5	47	5.2	25	54	4.0
Ref. [10]							
Maximum							
$\langle m_\nu \rangle$	2.5	8.2		1.9	21		
$\langle \eta \rangle\ 10^{-8}$	2.8	9.0		1.8	21		
$\langle \lambda \rangle\ 10^{-6}$	3.6	8.5		5.5	24		
On axis							
$\langle m_\nu \rangle$	2.1	7.4		1.4	19		
$\langle \eta \rangle\ 10^{-8}$	2.4	8.3		1.3	19		
$\langle \lambda \rangle\ 10^{-6}$	3.5	8.3		5.2	24		
Ref. [11]							
$\langle m_\nu \rangle^{a}$	7.6	32	66	2.2	29	61	
^{b}	17	64		3.0	39	80	
$(T^{0\nu}_{1/2})_{exp}$	4.7×10^{23}	1.1×10^{22}	4.4×10^{20}	5×10^{24}	1.5×10^{21}	1.7×10^{21}	2.3×10^{21}
[Ref.]	[17]	[18]	[19]	[20]	[20]	[21]	[22]

The two values of [11] correspond to the strength of the zero-range pp interaction, $^{a}\ \alpha'_1 = -390$ MeV fm^3 and $^{b}\ \alpha'_1 = -432$ MeV fm^3

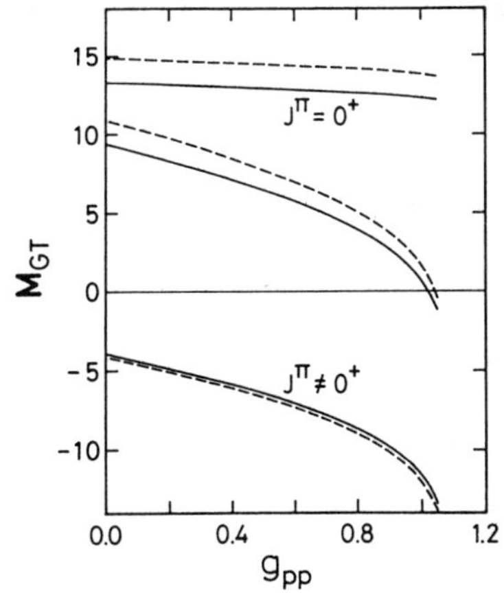

Fig. 7. The nuclear matrix element M_{GT} calculated with (solid line) and without (dashed line) the correlation function, (14), which takes into account the short-range correlations between the two nucleons which make a transition. The contribution of two-particle transitions with $J^\pi = 0^+$ and the sum of the other transitions ($J^\pi \neq 0^+$) are separately shown

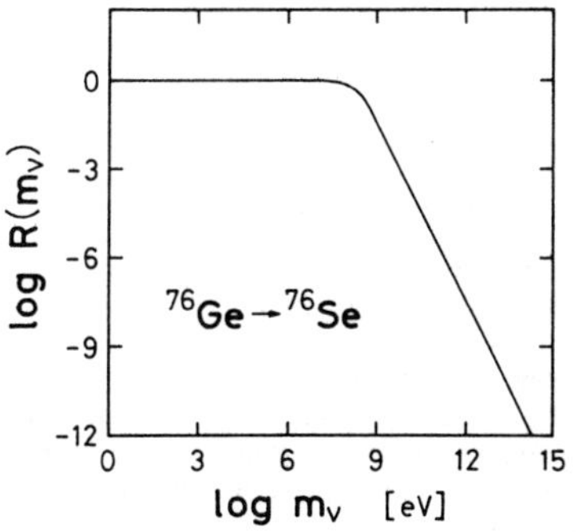

Fig. 8. The ratio $R(m_\nu) = (M_{GT}(m_\nu) - M_F(m_\nu))/(M_{GT}(0) - M_F(0))$ calculated as a function of the mass of the Majorana neutrino m_ν for the $0\nu\,\beta\beta$ decay $^{76}Ge \to {}^{76}Se$. When a heavy Majorana neutrino contributes to the decay, in addition to a light neutrino, the effective value of the neutrino mass which is deduced from an experimental half-life is defined by including the ratio, $\langle m_\nu \rangle = \sum_j m_j U_{ej}^2 R(m_j)$, where m_j is the mass of the Majorana neutrino N_j, and U_{ej} is the mixing amplitude in the 'electron' neutrino, $\nu_{eL} = \sum_j U_{ej} N_{jL}$

Consequences of mixing of a heavy Majorana neutrino and implications on the neutrino mass deduced from the $0\nu\ \beta\beta$ decay will be discussed elsewhere.

4. Conclusion

We have calculated the nuclear matrix elements of the $0\nu\ \beta\beta$ decay in pnQRPA with a realistic interaction, and carried out investigations of nuclear structure effects. It is shown that, compared to the 2ν decay, the 0ν nuclear matrix elements are not so much suppressed by the ground-state correlations, i.e., much less sensitive to the pp interaction, and therefore they can be reliably calculated in the QRPA approach. To evaluate the nuclear matrix elements with high accuracy, it is important to take the pp interaction strength which yields the 2ν decay half-life consistent with the experimental value. We have also discussed briefly the case of mixing of a heavy Majorana neutrino in the 'electron' neutrino.

The present calculation supports the result of Tomoda and Faessler [10]. The detailed investigation of nuclear structure effects indicates some plausible reasons why Engel et al. [11] have suggested the opposite conclusion that the 0ν matrix elements are rather strongly suppressed by the ground-state correlations and also by the short-range correlations.

Numerical calculations were performed by using the FACOM M-780 computer system at the Institute for Nuclear Study, University of Tokyo.

References

1. Doi, M., Kotani, T., Takasugi, E.: Prog. Theor. Phys. Suppl. **83**, 1 (1985)
2. Muto, K., Klapdor, H.V.: In: Neutrinos. Klapdor, H.V. (ed.) Berlin, Heidelberg, NewYork: Springer 1988
3. Glashow, S.L.: Nucl. Phys. **22**, 579 (1961);
Weinberg, S.: Phys. Rev. Lett. **19**, 1264 (1967);
Salam, A.: Proceedings of the Eighth Nobel Symposium. Svartholm, N. (ed.), p. 367. Stockholm: Almquist and Wiksell 1968
4. Langacker, P.: In: Neutrinos. Klapdor, H.V. (ed.) Berlin, Heidelberg, New York: Springer 1988;
Mohapatra, R.N.: *ibid*
5. Gelmini, G.B.: In: Neutrinos. Klapdor, H.V. (ed.) Berlin, Heidelberg, New York: Springer 1988
6. Vogel, P., Zirnbauer, M.R.: Phys. Rev. Lett. **57**, 3148 (1986)
7. Civitarese, O., Faessler, A., Tomoda, T.: Phys. Lett. **194**B, 11 (1987)
8. Muto, K., Klapdor, H.V.: Phys. Lett. **201** B, 420 (1988);
Muto, K., Bender, E., Klapdor, H.V.: Z. Phys. A – Atomic Nuclei **334**, 177 (1989)
9. Halbleib, J.A., Sorensen, R.A.: Nucl. Phys. A**98**, 542 (1967)
10. Tomoda, T., Faessler, A.: Phys. Lett.**199**B, 475 (1987)
11. Engel, J., Vogel, P., Zirnbauer, M.R.: Phys. Rev. C**37**, 731 (1988)
12. Particle Data Group, Phys. Lett. **170**B, 1 (1986)
13. Tomoda, T., Faessler, A., Schmid, K.W., Grümmer, F.: Nucl. Phys. A**452**, 591 (1986)
14. Miller, G.A., Spencer, J.E.: Ann. Phys. **100**, 562 (1976)
15. Vergados, J.D.: Phys. Rev. C**24**, 640 (1981); Nucl. Phys. B**218**, 109 (1983)
16. Vinh Mau, R.: Mesons in nuclei. Rho, M., Wilkinson, D.H. (eds.), Vol. I, p. 151. Amsterdam: North-Holland 1979
Anantaraman, N., Toki, H., Bertsch, G.F.: Nucl. Phys. A**398**, 269 (1983)
17. Caldwell, D.O., Eisberg, R.M., Grumm, D.M., Witherell, M.S.: Phys. Rev. Lett. **59**, 419 (1987)
18. Elliott, S.R., Hahn, A.A., Moe, M.K.: Proceedings of the International Symposium on Weak and Electromagnetic Interactions in Nuclei (WEIN '86). Klapdor, H.V. (ed.), p. 692. Berlin, Heidelberg, New York: Springer 1986
19. Smolnikov, A.A.: Private communication 1988
20. Kirsten, T., Heusser, E., Kaether, D., Oehm, J., Pernicka, E., Richter, H.: Proceedings of the International Symposium on Nuclear Beta Decays and Neutrino. Kotani, T., Ejiri, H., Takasugi, E. (eds.), p. 81. Singapore: World Scientific 1986
21. Alesandrello, A., Giuliani, A., Bellotti, E., Camin, D., Cremonesi, O., Fiorini, E., Liguori, C., Ragazzi, S., Rossi, L., Sverzellati, P.P., Zanotti, L.: Nucl. Phys. A**478**, 453c (1988)
22. Klimenko, A.A., Osetrov, S.B., Pomansky, A.A., Smolnikov, A.A., Vasilyev, S.I.: Proceedings of the International Symposium on Weak and Electromagnetic Interactions in Nuclei (WEIN '86). Klapdor, H.V. (ed.), p. 701. Berlin, Heidelberg, New York: Springer 1986
23. Holinde, K.: Phys. Rep. **68**, 121 (1981)

K. Muto
Department of Physics
Tokyo Institute of Technology
Tokyo 152
Japan

E. Bender
H.V. Klapdor
Max-Planck-Institut für Kernphysik
Postfach 103980
D-6900 Heidelberg 1
Federal Republic of Germany

EUROPHYSICS LETTERS 1 September 1990

Europhys. Lett., **13** (1), pp. 31-36 (1990)

Calculation of 2ν and 0ν Double-Beta Decay Rates.

A. STAUDT, K. MUTO (*) and H. V. KLAPDOR-KLEINGROTHAUS

Max-Planck-Institut für Kernphysik - D-6900 Heidelberg, F.R.G.

(received 11 June 1990; accepted 5 July 1990)

PACS. 23.40 – Beta decay; electron and muon capture.
PACS. 23.40H – Nuclear matrix elements and nuclear structure inferred from beta decay.
PACS. 21.60J – Hartree-Fock and random-phase approximations.

Abstract. – 2ν and 0ν double-beta decay rates for all potential ββ emitters with $A \geqslant 70$ are predicted. The nuclear transition matrix elements are calculated within the QRPA with a realistic effective nucleon-nucleon interaction. The results for neutrinoless decays are rather insensitive to details of the nuclear structure, except for the cases of ^{70}Zn, ^{100}Mo and ^{148}Nd.

Neutrinos are still one of the most puzzling particles in elementary-particle physics. Nuclear physics can contribute to the solution of the question of the mass and the nature of the neutrino by the investigation of neutrinoless double-beta decay [1, 2].

Nuclear double-beta decay is expected to proceed mainly via two decay modes. The two-neutrino mode (2νββ) occurs in second-order perturbation of the standard theory of weak interaction irrespective of the nature of the neutrino. The zero-neutrino mode (0νββ) violates the law of lepton-number conservation and is forbidden in the standard model and in the minimal SU(5) model. One usually distinguishes between two mechanisms of the 0ν mode. One which involves the effective neutrino mass $\langle m_\nu \rangle$ and a second involving right-handed charged weak currents. However, in the framework of general gauge theories neutrinoless double-beta decay always requires the existence of a massive Majorana neutrino [3].

The rate of the 0ν mode is related to the neutrino mass and the right-handed mixing parameters by nuclear transition matrix elements that depend on the complicated structure of the nuclei involved. Therefore it is of upmost importance to calculate these matrix elements as accurately as possible. Since 0νββ decay provides at present the most sensitive possibility of distinguishing between Majorana and Dirac neutrinos and sets the most stringent limits on the mass and the right-handed mixing amplitudes, much effort has been devoted to this problem. In the present paper we will present the results of a QRPA calculation of both 2ν and 0ν double-beta decay rates. For details of the calculation we refer to a future publication.

Recent QRPA calculations have led to a considerable progress concerning the understanding of the suppression mechanism of 2ν matrix elements [4-7]. It had been shown earlier already [8] that ground-state correlations play a decisive role. The inclusion of the particle-particle interaction considerably enhances these ground-state correlations. As a

(*) Permanent address: Department of Physics, Tokyo Institute of Technology, Tokyo, Japan.

result the $2\nu\beta\beta$ matrix elements are strongly suppressed, consistent with experimental data [7]. The nuclear matrix element of 2ν decay is [2]

$$M_{\rm GT}^{2\nu} = \sum_{a,b} \frac{\langle 0_f^+ \| t_- \sigma \| 1_b^+ \rangle \langle 1_b^+ | 1_a^+ \rangle \langle 1_a^+ \| t_- \sigma \| 0_i^+ \rangle}{E_a + Q_{\beta\beta}/2 + m_e - E_i} . \tag{1}$$

The various quantities in the above equation are readily calculated from the pnQRPA energies and amplitudes as described in detail in ref. [2]. We have applied a realistic effective nucleon-nucleon potential, *i.e.* the G matrix of the Paris potential simulated by a sum of Yukawa terms [9]. Pairing is treated in the BCS approach, single-particle energies are obtained from a Coulomb-corrected Woods-Saxon potential. The details are the same as described earlier [6, 7]. The value of the strength of the particle-particle interaction $g_{\rm pp}$ is most critical for a reliable evaluation of the decay rates since $M_{\rm GT}^{2\nu}$ largely depends on $g_{\rm pp}$. For the purpose of the present calculation, systematics of the pp-force are investigated by an analysis of single β^+/EC decays [7] which are sensitive to the same type of interaction. The fitted $\bar{g}_{\rm pp}$-values slightly decrease as a function of the neutron number N, apart from sudden increases at magic neutron numbers. The standard deviation is given by $\delta g_{\rm pp} = 0.044$. The strength of the particle-hole interaction is fixed to excitation energies of β^- Gamow-Teller giant resonances [7]. The half-life for $2\nu\beta\beta$ decay can be given in a factorized form as

$$[T_{1/2}^{2\nu}]^- = F^{2\nu} |M_{\rm GT}^{2\nu}|^2 , \tag{2}$$

where $F^{2\nu}$ is a lepton phase-space integral. However, in the calculation of the decay rates the integrations over lepton energies are carried out explicitly for each intermediate nuclear state.

The matrix elements $M_{\rm GT}^{2\nu}$ decrease as $g_{\rm pp}$ is increased and pass through zero in the vicinity of the fitted value $\bar{g}_{\rm pp}(N)$. Thus a small change of the strength of the pp-force could cause a drastic change of the calculated half-life. This explains the difficulties in the prediction of precise $2\nu\beta\beta$ decay half-lives. Therefore we have calculated an average half-life $\bar{T}_{1/2} = \ln 2/\bar{\omega}_{2\nu}$, where $\bar{\omega}_{2\nu}$ is a simple average over the 1σ range of $g_{\rm pp}$,

$$\bar{\omega}_{2\nu} = \frac{1}{2\,\delta g_{\rm pp}} \int_{g_{\rm pp}(N) - \delta g_{\rm pp}}^{g_{\rm pp}(N) + \delta g_{\rm pp}} \omega_{2\nu}(g_{\rm pp})\, {\rm d}g_{\rm pp} . \tag{3}$$

In table I these average half-lives as well as the half-lives calculated at $g_{\rm pp} = 0$ and $\bar{g}_{\rm pp}$ are listed for all $\beta\beta$ emitters with $A \geq 70$. In addition we give the lower limit of the half-life corresponding to the largest nuclear matrix element within the probable range of $g_{\rm pp}$.

An increase of $g_{\rm pp}$ finally causes the RPA equation to collapse, *i.e.* the eigenvalues become complex. For the isotopes ^{104}Ru, ^{110}Pd, ^{146,148}Nd this collapse occurs already at $g_{\rm pp} < \bar{g}_{\rm pp}$ (denoted by «C» in table I). Therefore we are not able to determine the decay rate at these points. In some cases (denoted by (*)) the half-lives are evaluated close to the collapse. Consequently, these values are probably somewhat underestimated. In general, the parameters of the Woods-Saxon potential are taken from ref. [10]. In some cases the $\beta\beta$ transitions are dominated by only a few single-particle transitions (*e.g.* $g_{9/2}$-$g_{7/2}$ in ^{100}Mo). Thus the half-lives are rather sensitively affected by a change of the occupation amplitudes, for example by a change of the single-particle energies which causes some uncertainties in the predictions. For the isotopes ^{94,96}Zr, ^{98}Mo and ^{114}Cd the calculated behaviour of the matrix elements $M_{\rm GT}^{2\nu}$ seems not to be physical, probably because of the above-mentioned large sensitivity to the BCS amplitudes. However, when we apply the Woods-Saxon

TABLE I. – *2ν and 0ν half-lives are given for all ββ emitters with* $A \geq 70$. $T^{2\nu}_{1/2}(0)$ *is the half-life evaluated at* $g_{pp}=0$, $T^{2\nu}_{1/2}(\bar{g})$ *is that calculated at* $\bar{g}_{pp}$, $T^{2\nu}_{1/2}(\mathrm{l.l.})$ *is the lower limit within the* 1σ *range of* g_{pp} *and* $\bar{T}^{2\nu}_{1/2}$ *is the average half-life defined in eq.* (3). (*) *indicates that the* 2ν *decay rates are evaluated close to the collapse of the* RPA, $^{(B)}$ *refers to the parametrization of the Woods-Saxon potential according to* [11] *and «C» indicates the* RPA *collapse. Finally* $^{(D)}$ *denotes isotopes whose* 0ν *matrix elements are evaluated before the occurrence of the collapse at* $g_{pp} < \bar{g}_{pp}$.

Nucleus	$T^{2\nu}_{1/2}(0)$ (y)	$T^{2\nu}_{1/2}(\mathrm{l.l.})$ (y)	$T^{2\nu}_{1/2}(\bar{g})$ (y)	$\bar{T}^{2\nu}_{1/2}$ (y)	$T^{0\nu}_{1/2}\langle m_\nu\rangle^2$ (y · eV²)
$^{70}Zn^{(*)}$	$1.51 \cdot 10^{22}$	$4.47 \cdot 10^{21}$	$3.60 \cdot 10^{24}$	$3.99 \cdot 10^{22}$	$9.83 \cdot 10^{25}$
^{76}Ge	$7.01 \cdot 10^{19}$	$8.22 \cdot 10^{20}$	$2.66 \cdot 10^{21}$	$2.99 \cdot 10^{21}$	$2.33 \cdot 10^{24}$
^{80}Se	$7.52 \cdot 10^{28}$	$5.05 \cdot 10^{29}$	$1.92 \cdot 10^{30}$	$1.18 \cdot 10^{30}$	$1.14 \cdot 10^{27}$
^{82}Se	$2.90 \cdot 10^{18}$	$3.06 \cdot 10^{19}$	$5.85 \cdot 10^{21}$	$1.09 \cdot 10^{20}$	$6.03 \cdot 10^{23}$
^{86}Kr	$1.58 \cdot 10^{22}$	$1.57 \cdot 10^{23}$	$2.19 \cdot 10^{24}$	$6.93 \cdot 10^{23}$	$2.78 \cdot 10^{25}$
$^{94}Zr^{(B)}$	$1.70 \cdot 10^{22}$	$4.29 \cdot 10^{22}$	$7.38 \cdot 10^{22}$	$6.93 \cdot 10^{22}$	$3.97 \cdot 10^{25}$
$^{96}Zr^{(B)}$	$3.00 \cdot 10^{18}$	$1.94 \cdot 10^{18}$	$1.94 \cdot 10^{19}$	$1.08 \cdot 10^{19}$	$5.30 \cdot 10^{23}$
$^{98}Mo^{(B)}$	$3.36 \cdot 10^{29}$	$1.25 \cdot 10^{30}$	$3.57 \cdot 10^{30}$	$2.96 \cdot 10^{30}$	$1.05 \cdot 10^{27}$
$^{100}Mo^{(*)}$	$2.25 \cdot 10^{17}$	$1.25 \cdot 10^{17}$	$1.10 \cdot 10^{20}$	$1.13 \cdot 10^{18}$	$1.27 \cdot 10^{24}$
$^{104}Ru^{(*)}$	$3.24 \cdot 10^{20}$	$6.14 \cdot 10^{21}$	*C*	$6.29 \cdot 10^{21}$	$4.24 \cdot 10^{24(D)}$
$^{110}Pd^{(*)}$	$1.13 \cdot 10^{19}$	$8.52 \cdot 10^{18}$	*C*	$1.16 \cdot 10^{19}$	$1.96 \cdot 10^{24(D)}$
$^{114}Cd^{(B)}$	$7.32 \cdot 10^{23}$	$1.40 \cdot 10^{24}$	$1.75 \cdot 10^{24}$	$1.78 \cdot 10^{24}$	$5.07 \cdot 10^{25}$
^{116}Cd	$2.85 \cdot 10^{18}$	$1.94 \cdot 10^{19}$	$1.20 \cdot 10^{20}$	$6.31 \cdot 10^{19}$	$4.87 \cdot 10^{23}$
^{122}Sn	$4.85 \cdot 10^{25}$	$1.33 \cdot 10^{27}$	$1.58 \cdot 10^{28}$	$5.44 \cdot 10^{27}$	$1.27 \cdot 10^{26}$
^{124}Sn	$3.64 \cdot 10^{19}$	$1.28 \cdot 10^{21}$	$1.58 \cdot 10^{22}$	$5.25 \cdot 10^{21}$	$1.36 \cdot 10^{24}$
^{128}Te	$1.98 \cdot 10^{22}$	$7.23 \cdot 10^{23}$	$3.35 \cdot 10^{27}$	$2.63 \cdot 10^{24}$	$7.77 \cdot 10^{24}$
^{130}Te	$6.87 \cdot 10^{18}$	$5.94 \cdot 10^{20}$	$1.72 \cdot 10^{24}$	$1.84 \cdot 10^{21}$	$4.89 \cdot 10^{23}$
^{134}Xe	$5.38 \cdot 10^{22}$	$1.46 \cdot 10^{24}$	$1.93 \cdot 10^{25}$	$6.09 \cdot 10^{24}$	$1.69 \cdot 10^{25}$
^{136}Xe	$1.50 \cdot 10^{19}$	$1.43 \cdot 10^{21}$	$2.11 \cdot 10^{22}$	$4.64 \cdot 10^{21}$	$2.21 \cdot 10^{24}$
^{142}Ce	$2.15 \cdot 10^{20}$	$4.03 \cdot 10^{20}$	$4.21 \cdot 10^{21}$	$1.58 \cdot 10^{21}$	$2.77 \cdot 10^{24}$
^{146}Nd	$1.08 \cdot 10^{30}$	*C*	*C*	*C*	$4.36 \cdot 10^{26(D)}$
$^{148}Nd^{(*)}$	$1.40 \cdot 10^{19}$	$1.14 \cdot 10^{19}$	*C*	$(5.17 \cdot 10^{19})$	$1.36 \cdot 10^{24(D)}$
^{150}Nd	$6.10 \cdot 10^{16}$	$1.89 \cdot 10^{18}$	$3.86 \cdot 10^{20}$	$7.37 \cdot 10^{18}$	$3.37 \cdot 10^{22}$
^{154}Sm	$2.72 \cdot 10^{20}$	$5.91 \cdot 10^{20}$	$2.47 \cdot 10^{21}$	$1.80 \cdot 10^{21}$	$1.39 \cdot 10^{24}$
^{160}Gd	$4.93 \cdot 10^{18}$	$1.05 \cdot 10^{20}$	$9.86 \cdot 10^{20}$	$4.21 \cdot 10^{20}$	$8.56 \cdot 10^{23}$
^{170}Er	$1.65 \cdot 10^{23}$	$6.40 \cdot 10^{24}$	$1.57 \cdot 10^{26}$	$2.68 \cdot 10^{25}$	$1.37 \cdot 10^{25}$
^{176}Yb	$1.32 \cdot 10^{20}$	$2.60 \cdot 10^{21}$	$3.70 \cdot 10^{24}$	$9.83 \cdot 10^{21}$	$1.36 \cdot 10^{24}$
^{186}W	$7.05 \cdot 10^{23}$	$4.78 \cdot 10^{24}$	$1.15 \cdot 10^{25}$	$1.07 \cdot 10^{25}$	$6.35 \cdot 10^{24}$
^{192}Os	$1.26 \cdot 10^{24}$	$1.04 \cdot 10^{25}$	$2.02 \cdot 10^{25}$	$1.94 \cdot 10^{25}$	$4.08 \cdot 10^{24}$
^{198}Pt	$4.76 \cdot 10^{21}$	$1.16 \cdot 10^{23}$	$4.78 \cdot 10^{23}$	$3.53 \cdot 10^{23}$	$4.70 \cdot 10^{23}$
^{204}Hg	$3.08 \cdot 10^{25}$	$6.37 \cdot 10^{26}$	$2.51 \cdot 10^{27}$	$1.87 \cdot 10^{27}$	$8.22 \cdot 10^{24}$
^{232}Th	$4.24 \cdot 10^{21}$	$7.98 \cdot 10^{22}$	$1.79 \cdot 10^{23}$	$1.60 \cdot 10^{23}$	$3.07 \cdot 10^{23}$
^{238}U	$9.70 \cdot 10^{20}$	$5.20 \cdot 10^{22}$	$3.78 \cdot 10^{23}$	$1.53 \cdot 10^{23}$	$2.60 \cdot 10^{23}$
^{244}Pu	$9.33 \cdot 10^{21}$	$3.76 \cdot 10^{22}$	$6.87 \cdot 10^{22}$	$6.54 \cdot 10^{22}$	$5.72 \cdot 10^{23}$

parameters of ref. [11], then the results are reasonable, the behaviour being similar to that of the other $\beta\beta$ emitters.

The calculated upper limits of the 2ν half-lives are infinite since the matrix elements in most cases cross zero within the probable range of g_{pp}. However, one should note that the $2\nu\beta\beta$ decay rates should not vanish. In the QRPA the crossing of $M^{2\nu}_{GT}$ through zero is due to the artificial change of g_{pp} as a free parameter independently of g_{ph}, violating the Pandya transformation (see ref. [7]). Compared to the only to date existing complete calculation for all double-beta emitters by Grotz and Klapdor [8] our averaged 2ν half-lives are generally considerably larger (about one order of magnitude) because of the inclusion of the particle-particle interaction.

The 0ν transition matrix elements are known to be less sensitive to details of the nuclear structure. This was first observed by Klapdor and Grotz [8]. They are much less affected by the particle-particle interaction compared to the 2ν mode [12-14]. For the 0ν mode reliable results are obtained with the commonly used closure approximation which is also applied in the present work (see also [15]). Staudt *et al.* [16] recently investigated the influence of different renormalized effective interactions on the nuclear transition matrix elements. The 0ν matrix elements were shown to vary only moderately with respect to the choice of the effective interaction which leads to the conclusion that the $0\nu\beta\beta$ decay matrix elements in general can reliably be evaluated within the framework of the QRPA within a factor of 2 or 3.

The inverse half-life for this decay mode for the $0_i^+ \to 0_f^+$ transition is expressed as [2]

$$[T^{0\nu}_{1/2}]^{-1} = C_{mm}\left(\frac{\langle m_\nu\rangle}{m_e}\right)^2 + C_{\eta\eta}\langle\eta\rangle^2 + C_{\lambda\lambda}\langle\lambda\rangle^2 + + C_{m\eta}\frac{\langle m_\nu\rangle}{m_e}\langle\eta\rangle + C_{m\lambda}\frac{\langle m_\nu\rangle}{m_e}\langle\lambda\rangle + C_{\eta\lambda}\langle\eta\rangle\langle\lambda\rangle\,. \tag{4}$$

For a definition of the effective values of the neutrino mass $\langle m_\nu\rangle$, the coupling strengths of the right-handed currents $\langle\eta\rangle$, $\langle\lambda\rangle$ and the coefficients C_{xy} we refer to ref. [13]. The coefficients C_{xy} consist of products of electron phase-space integrals $G^{0\nu}_k$ and the nuclear matrix elements $M^{0\nu}_\alpha$. In the closure approximation the latter are given by

$$M^{0\nu}_\alpha = \sum_{m,n}\langle 0_f^+\|t_{-m}t_{-n}\,\mathcal{O}^\alpha_{mn}\|0_i^+\rangle\,, \tag{5}$$

where $\mathcal{O}^\alpha_{mn}$ are the relevant two-body transition operators. In this letter we shall concentrate on the mass mechanism of neutrinoless double-beta decay. The matrix elements corresponding to right-handed currents will be discussed elsewhere. Thus neglecting right-handed contributions we end up with

$$T^{0\nu}_{1/2}\langle m_\nu\rangle^2 = \frac{m_e^2}{(M^{0\nu}_{GT} - M^{0\nu}_F)^2 G^{0\nu}_1}\,. \tag{6}$$

The relevant transition operators are defined by

$$\mathcal{O}^{GT}_{12} = \boldsymbol{\sigma}_1\cdot\boldsymbol{\sigma}_2 H_m(r)\,, \tag{7a}$$

$$\mathcal{O}^{F}_{12} = H_m(r)\left(\frac{g_V}{g_A}\right)^2, \tag{7b}$$

where $H_m(r)$ is the «neutrino potential» [13] which represents the exchange of a virtual neutrino between two nucleons.

Because of the two-body character of the operators $\mathcal{O}^{\alpha}_{12}$ the nuclear matrix elements can be reduced to a sum of products of two-particle transition densities Z and two-particle matrix elements of the above operators. From the solutions of the pnQRPA equation the densities Z are readily evaluated [2, 13]. Because of the neutrino potential the 0ν transitions proceed through intermediate states of all possible multipolarities—in contrast to the 2ν transitions which exclusively proceed through 1^+ states. This feature together with the nature of the nucleon-nucleon interaction explains the relative insensitivity to the strength of the particle-particle interaction g_{pp} [13]. The electron phase-space integral $G_1^{0\nu}$ in eq. (6) is taken from ref. [1] as far as available. In the remaining cases it is determined from the Fermi function calculated according to the prescription of [17]. We assume $g_A/g_V = -1.254$ [18].

The results for the 0ν mode are listed in the last column of table I. For the isotopes where the collapse of the RPA for the 1^+ component occurs at values of g_{pp} smaller than $\bar{g}_{pp}$ (labelled by $^{(D)}$), the 0ν matrix elements are evaluated before the collapse. This seems permissable because the variation of the 0ν matrix elements as a function of g_{pp} is rather moderate. In general, the variation of the quantity $T^{0\nu}_{1/2}\langle m_\nu \rangle^2$ within the 1σ range of g_{pp} allows a reliable deduction of the neutrino mass within a factor of about 2 or 3 (see fig. 1 and also [16]). More details will be presented in a future publication.

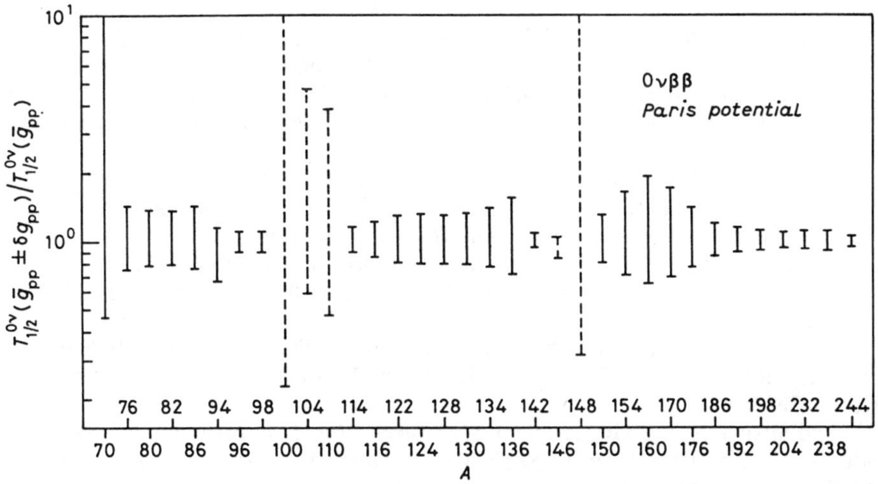

Fig. 1. – Illustration of the variation of the 0ν decay rates within the probable range of the strength of the particle-particle interaction g_{pp}. The dashed «error bars» indicate isotopes for which the matrix elements are evaluated close to the collapse of the RPA or at $g_{pp} < \bar{g}_{pp}$.

For a consistent evaluation of the neutrino mass $\langle m_\nu \rangle$ and the right-handed mixing amplitudes $\langle \lambda \rangle$ and $\langle \eta \rangle$ the knowledge of all coefficients C_{xy} in eq. (4) is required. In table II these are given as example for the isotope ^{116}Cd. The resulting upper bounds on $\langle m_\nu \rangle$, $\langle \lambda \rangle$ and $\langle \eta \rangle$ extracted from the recently measured lower limit of the half-life $T^{0\nu}_{1/2} \geq 1.3 \cdot 10^{21}$ y [19] are $\langle m_\nu \rangle \leq 21.8$ eV, $\langle \eta \rangle \leq 2.8 \cdot 10^{-7}$ and $\langle \lambda \rangle \leq 2.5 \cdot 10^{-5}$. The most stringent limits of all direct (detector) experiments are obtained from ^{76}Ge for which the values are given in ref. [13].

TABLE II. – *Values of the coefficients* C_{xy}, *defined in eq.* (4), *for the isotope* ^{116}Cd.

C_{mm} (y^{-1})	$C_{m\eta}$ (y^{-1})	$C_{m\lambda}$ (y^{-1})	$C_{\eta\eta}$ (y^{-1})	$C_{\lambda\lambda}$ (y^{-1})	$C_{\eta\lambda}$ (y^{-1})
$5.36 \cdot 10^{-13}$	$6.79 \cdot 10^{-11}$	$-3.00 \cdot 10^{-13}$	$1.18 \cdot 10^{-8}$	$1.31 \cdot 10^{-12}$	$-8.71 \cdot 10^{-13}$

The last point we want to make concerns the Reid SC potential [20]. Comparing the behaviour of the 2ν matrix elements calculated with the Paris and the Reid potential, respectively, one finds that the value of g_{pp} at which $M_{GT}^{2\nu}$ crosses zero is larger for the Reid potential by about 0.042, nearly independent of the mass number. If we add this constant shift to the fitted $\bar{g}_{pp}(\bar{g}_{pp}(\text{Reid}) = \bar{g}_{pp}(\text{Paris}) + 0.042)$, then the Reid potential yields results for both two-neutrino and zero-neutrino decay rates which are very close to those calculated from the Paris potential.

Concluding, we present in this paper a complete list of both two-neutrino and zero-neutrino decay rates for all potential ββ-emitters with $A \geq 70$.

REFERENCES

[1] DOI M., KOTANI T. and TAKASUGI E., *Prog. Theor. Phys. Suppl.*, **83** (1985) 1.
[2] MUTO K. and KLAPDOR H. V., in *Neutrinos, Graduate Texts in Contemporary Physics*, edited by H. V. KLAPDOR (Springer, Heidelberg, New York, N.Y.) 1988, p. 183.
[3] SCHECHTER J. and VALLE J. W. F., *Phys. Rev. D*, **25** (1982) 2951; NIEVES J. F., *Phys. Lett. B*, **147** (1984) 375; TAKASUGI E., *Phys. Lett. B*, **149** (1984) 372; KAYSER B., in *Nuclear Beta Decays and Neutrino*, edited by T. KOTANI, H. EJIRI and E. TAKASUGI (World-Scientific, Singapore) 1986, p. 473.
[4] VOGEL P. and ZIRNBAUER M. R., *Phys. Rev. Lett.*, **57** (1986) 3148.
[5] CIVITARESE O., FAESSLER A. and TOMODA T., *Phys. Lett. B*, **194** (1987) 11.
[6] MUTO K. and KLAPDOR H. V., *Phys. Lett. B*, **201** (1988) 420.
[7] MUTO K., BENDER E. and KLAPDOR H. V., *Z. Phys. A*, **334** (1989) 177.
[8] KLAPDOR H. V. and GROTZ K., *Phys. Lett. B*, **142** (1984) 323; GROTZ K. and KLAPDOR H. V., *Nucl. Phys. A*, **460** (1986) 395.
[9] ANANTARAMAN N., TOKI H. and BERTSCH G. F., *Nucl. Phys. A*, **398** (1983) 269; LACOMBE M., LOISEAU B., RICHARD J. M., VINH MAU R., CÔTÉ J., PIRÈS P. and DE TOURREIL R., *Phys. Rev. C*, **21** (1980) 861.
[10] BOHR A. and MOTTELSON B. R., *Nuclear Structures*, Vol. **1** (Benjamin, New York, N.Y.) 1969.
[11] BERTSCH G. F., *The Practitioner's Shell Model* (North-Holland, Amsterdam) 1972.
[12] TOMODA T. and FAESSLER A., *Phys. Lett. B*, **199** (1987) 475.
[13] MUTO K., BENDER E. and KLAPDOR H. V., *Z. Phys. A*, **334** (1989) 187.
[14] ENGEL J., VOGEL P., JI X. and PITTEL S., *Phys. Lett. B*, **225** (1989) 5.
[15] SUHONEN J., KHADKIKAR S. B. and FAESSLER A., *Phys. Lett. B*, **237** (1990) 8.
[16] STAUDT A., KUO T. T. S. and KLAPDOR-KLEINGROTHAUS H. V., *Phys. Lett. B*, **242** (1990) 17.
[17] GOVE N. B. and MARTIN M. J., *Nucl. Data Tables*, **10** (1971) 205.
[18] PARTICLE DATA GROUP, *Phys. Lett. B*, **170** (1986) 1.
[19] DANEVICH F. A., ZDESENKO YU. G., NIKOLAIKO A. S. and TRETYAK V. I., *JETP Lett.*, **49** (1989) 476.
[20] REID R. V., *Ann. Phys. (N.Y.)*, **50** (1968) 411.

Volume 242, number 1 PHYSICS LETTERS B 31 May 1990

ββ DECAY OF ^{76}Ge WITH RENORMALIZED EFFECTIVE INTERACTION DERIVED FROM PARIS, BONN AND REID POTENTIALS

A. STAUDT, T.T.S. KUO [1,2] and H.V. KLAPDOR-KLEINGROTHAUS
Max-Planck-Institut für Kernphysik, D-6900 Heidelberg, FRG

Received 5 March 1990

Two-neutrino and neutrinoless double beta decays of ^{76} Ge are calculated using renormalized effective interactions derived from the Paris, Bonn-A Reid potentials. Core polarization and folded diagrams are considered in our effective interaction. The dependence of the two-neutrino nuclear matrix elements on the strength of the particle–particle interaction in the QRPA for the various potentials are compared. For the neutrinoless decay our calculated values for $T^{0\nu}_{1/2}\langle m_\nu\rangle^2$ are typically 3×10^{23}yr eV^2. The present extensive study shows that the neutrino mass should be reliably extractable from ββ-decay experiments within a factor of three.

Nuclear double beta decay is expected to proceed via two decay modes, the two-neutrino mode (2ν ββ) and the neutrinoless mode (0ν ββ). The latter is of particular interest, both experimentally and theoretically, as it is this decay process which is sensitive to the existence of massive Majorana neutrinos and sets the most stringent limits on the neutrino mass and the coupling constants of the right-handed components of the charged weak interaction [1,2]. This lepton-number non-conserving process is forbidden in the standard $SU(2)_L\times U(1)$ gauge model of the electroweak interaction, in which the neutrinos are massless and there exist no right-handed currents. Massive neutrinos naturally appear in grand unified theories (GUTs) [3]. The nature (Dirac or Majorana particle) and the mass of the neutrino could yield important boundary conditions for these models.

Reliable deductions of the ν-mass from measured 0ν ββ decay rates are, however, sensitively dependent on predictions of nuclear matrix elements. Therefore it is of upmost importance to calculate these matrix elements as accurately as possible. Since the nuclear wave functions and the underlying theory for treating the 0ν ββ and the 2ν ββ decay are the same, one commonly first calculates the matrix elements for the latter in order to test the nuclear structure theory. This is because there exist experimental results for the 2ν ββ processes. Recently considerable progress has been made concerning the understanding of the suppression mechanism of the 2ν transition matrix elements [4–7].

^{76}Ge is one of the most promising ββ emitters for the study of 0ν ββ decay. Among the direct (non-geochemical) experiments it gives at present the smallest limit of the neutrino mass, $\langle m_\nu\rangle<2.6$ eV [8] corresponding to the half-life limit of $T^{0\nu}_{1/2}>4.7\times10^{23}$yr [9]. Our interest in this nucleus is further motivated by the ongoing Heidelberg–Moscow ββ collaboration [10]. This project employs enriched ^{76}Ge as both source and detectors, instead of the commonly used natural Ge, and thus could substantially improve the deduced life-time limits, exploring the half-life of the 0ν ββ decay up to the 10^{25} yr range.

Calculation of nuclear ββ-decay matrix elements is nevertheless a rather sensitive matter and one is naturally concerned about the choice of the effective interaction ($V_{\rm eff}$) to be used in such calculations. In other words one would want to know what is the variation of these matrix elements due to the use of various effective interactions. In the present work we shall consider the three realistic nucleon–nucleon

1 Work supported in part by US DOE Grant DE-FG02-88ER40388.

2 Permanent address: Physics Department, State University of New York at Stony Brook, Stony Brook, NY 11794-3800, USA.

Volume 242, number 1 PHYSICS LETTERS B 31 May 1990

(NN) interactions – Paris [11], Bonn-A [12,13] and Reid [14]. These three interactions all reproduce the two-nucleon data at low energy ($\lesssim$300 MeV) remarkably well, the quality of their fits to the data being practically indistinguishable. But there is one important intrinsic difference among them.

The strength of the NN tensor force is among the most unsettled issues in NN interactions. The recent Bonn-A potential, which is derived from relativistic meson theory and it is not parametrized in terms of local functions as is usually adopted by other NN potentials, has a considerably weaker tensor force than the Paris and Reid potentials. The predicted deuteron D-state probability P_D, a commonly used measure for the tensor force strength, is 6.5%, 5.8% and 4.4% for the Reid, Paris and Bonn-A potentials respectively. Tensor force plays a highly important role in nuclear structure calculations, as it is well known. Thus it should be of interest as well as useful to investigate the connection between the tensor force strength and the predicted nuclear ββ-decay matrix elements.

To proceed we need first to derive from these NN potentials the respective V_{eff} matrix elements for ββ-decay calculations. This is, however, a long and rather difficult task. Here we would want to put up a "best possible" effort in carrying out this derivation. The Pauli exclusion operator in the G matrix will be treated in an essentially exact way in the present work. In earlier works this has been treated only approximately. In addition we shall also consider the renormalization contributions to V_{eff} from core polarization and folded diagrams.

Nuclear structure calculations are almost always carried out in a restricted model space commonly denoted by P. The free NN potentials such as those mentioned above are, however, designed to be used in the full space $P+Q$, Q being the complement of P. Hence a first step is to convert the full-space secular equation $H\Psi=E\Psi$ to a model-space secular equation $H_{eff}P\Psi=EP\Psi$. Here the interaction part of H is V, the free NN interaction, while that of H_{eff} is V_{eff}, the model space effective interaction. Starting from a given V and P, there exist a number of formally exact theories for carrying out the calculation of V_{eff}. One such theory is the folded-diagram method formulated by Kuo, Lee and Ratcliff [15,16]. This method is relatively convenient and systematic for actual calculations and is adopted in the present work. A first step here is the calculation of the reaction matrix G defined as

$$G(\omega)=V+VQ_{2p}\frac{1}{\omega-Q_{2p}T(2p)Q_{2p}}Q_{2p}G(\omega)\,, \qquad (1)$$

where $T(2p)$ denotes the two-particle kinetic energy.

The model space we employed for our ββ-decay calculations of ^{76}Ge consists of nine proton orbits and nine neutron orbits, each ranging from $0f_{7/2}$ to $2s_{1/2}$, outside a closed ^{40}Ca core. We must avoid double counting, in the sense that the correlations included in the intermediate states of the G matrix must not be repeated by those generated by solving the nuclear secular equation within the above model space. The exclusion operator Q_{2p} in the above equation is mainly for this purpose and we must treat this exclusion operator, which projects onto all the two-particle states outside the model space, accurately. This is, however, a rather difficult requirement to enact, a major difficulty being that the G-matrix equation is usually solved in the two-nucleon relative and center-of-mass frame but on this representation the operator Q_{2p} is not diagonal and is very complicated. This problem is further complicated because our present calculation involves a rather large model space.

There exists, however, an essentially exact method in correctly treating the above exclusion operator [17,18]. This method involves the inversion of rather large P-space matrices and in the past it has been only feasible to apply it to light nuclei such as those in the sd shell [18]. We have now extended this method to the ^{76}Ge region. With this method the solution of the above model space G-matrix equation is given exactly as

$$G_T(\omega)=G_{TF}(\omega)+\Delta G(\omega)\,, \qquad (2)$$

where G_{TF} is the free G matrix defined as

$$G_{TF}(\omega)=V+V\frac{1}{\omega-T(2p)}G_{TF}(\omega)\,. \qquad (3)$$

$\Delta G(\omega)$ is a correction term defined entirely within the P-space, given as

$$\Delta G(\omega) = -G_{\rm TF}(\omega)\,\frac{1}{e}\,P_{\rm 2p}$$
$$\times \frac{1}{P_{\rm 2p}[(1/e)+(1/e)G_{\rm TF}(\omega)(1/e)]P_{\rm 2p}}$$
$$\times P_{\rm 2p}\,\frac{1}{e}\,G_{\rm TF}(\omega)\,, \qquad (4)$$

where $e \equiv \omega - T(2\mathrm{p})$. $P_{\rm 2p}$ is the complement of $Q_{\rm 2p}$ in the two-particle subspace. Thus we see that the G matrix is now expressed as the sum of two terms; the first term is the free G matrix while the second is calculated using only some simple model-space matrix operations involving the free G matrix. The free G matrix no longer contains the troublesome Q-space projection operator and may be calculated via momentum-space matrix-inversion methods [19]. Indeed the above formalism provides a convenient and essentially exact method for calculating the model-space G matrix.

The model-space used in the present work is specified with (n_1, n_2, n_3) given by (6, 15, 28) [18]. (In this notation the nine orbits mentioned above are denoted as orbits 7–15). This is a rather sizeable model space and requires a considerable amount of computing effort in carrying out the respective G-matrix calculation. (For comparison, the calculation performed by Krenciglowa et al. [18] for the sd shell nuclei involved a much smaller P-space of size $(n_1, n_2, n_3) = (3, 10, 21)$.) We have performed G-matrix calculations as outlined above for the Paris, Bonn-A and Reid potentials for several values for the energy variables ω.

Based on the above folded-diagram theory, our effective interaction is then expressed as a folded diagram series, grouped according to the number of folds. Namely

$$V_{\rm eff} = F_0 + F_1 + F_2 + F_3 + \ldots \qquad (5)$$

where F_n denotes a $(n+1)$ $\hat{Q}$-box term connected with n sets of folded lines. For example, the three-time folded term has the form

$$F_3 = -\hat{Q}\int\hat{Q}\int\hat{Q}\int\hat{Q}\,, \qquad (6)$$

where $\int$ stands for a generalized folding operation.

In the present calculation we have included in the $\hat{Q}$-box all the irreducible diagrams first and second order in G, namely the same set of seven diagrams included in the sd shell calculation of Shurpin et al. [20]. We shall present the details of our calculation in a future publication; for the present let us just mention the various second order diagrams are calculated with respect to a ^{40}Ca core in consistence with our chosen model space. In fact we have used three schemes to evaluate the effective interactions:

(1) G: Here the effective interaction $V_{\rm eff}$ is taken as the bare G matrix only.

(2) $G+G^2(G^{(2)})$: Here we do not include any folded diagrams in $V_{\rm eff}$ and take it as composed of just the three two-body G-matrix diagrams G, $G_{\rm 3p1h}$ – the core polarization diagram – and $G_{\rm 2h}$. This is in analogy to the choice made sometime ago by Kuo and Brown [21].

(3) Lesu: This is the "full" calculation where the folded diagram series is summed up to all orders using the Lee–Suzuki iteration method [22]. In the $\hat{Q}$-box we have included the seven diagrams as mentioned above.

The nuclear matrix elements for the ββ decays are then calculated within the framework of the proton–neutron quasiparticle random phase approximation (pnQRPA) [23]. The G matrix is dependent on the energy variable ω. When using the effective interaction prescribed by (1) and (2) there is the ambiguity about what value of ω one should use for G. (A value of -10 MeV is used in the present work.) This ambiguity is removed in the Lesu case as here the derived $V_{\rm eff}$ is energy independent [15,16,22].

The nuclear matrix element of 2ν decay is [2]

$$M^{2\nu}_{\rm GT}(7) = \sum_{a,b} \frac{\langle 0_{\rm f}^+ \| t_-\sigma \| 1_b^+ \rangle \langle 1_b^+ | 1_a^+ \rangle \langle 1_a^+ \| t_-\sigma \| 0_{\rm i}^+ \rangle}{E_a + Q_{\beta\beta}/2 + m_{\rm e} - E_i}\,, \qquad (7)$$

where we have replaced the lepton energy in the intermediate state by $Q_{\beta\beta}/2 + m_{\rm e}$. The various quantities in the above equation are readily calculated from the pnQRPA energies and amplitudes as described in detail in ref. [2]. We note that the interaction terms in the pnQRPA secular matrix are written as $g_{\rm pp} V^{\rm pp}_{\rm pn,p'n'}$ for the particle–particle interaction and $g_{\rm ph} V^{\rm ph}_{\rm pn,p'n'}$ for the particle–hole interaction. The strength parameters $g_{\rm pp}$ and $g_{\rm ph}$ are introduced in order to study the sensitivity of the calculated ββ matrix elements to these two types of interactions.

Volume 242, number 1 PHYSICS LETTERS B 31 May 1990

Turning to the $0\nu\beta\beta$ mode, we write its inverse half-life as

$$[T^{0\nu}_{1/2}]^{-1}=C_{mm}\left(\frac{\langle m_\nu\rangle}{m_e}\right)^2+C_{\eta\eta}\langle\eta\rangle^2+C_{\lambda\lambda}\langle\lambda\rangle^2$$

$$+C_{m\eta}\frac{\langle m_\nu\rangle}{m_e}\langle\eta\rangle+C_{m\lambda}\frac{\langle m_\nu\rangle}{m_e}\langle\lambda\rangle$$

$$+C_{\eta\lambda}\langle\eta\rangle\langle\lambda\rangle\,. \tag{8}$$

For a definition of the effective values of the neutrino mass $\langle m_\nu\rangle$, the coupling strengths of the right-handed currents $\langle\eta\rangle$, $\langle\lambda\rangle$ and the coefficients C_{xy} we refer to ref. [8]. The coefficients C_{xy} consist of products of electron phase-space integrals G_k and the nuclear matrix elements M_α

$$M_\alpha=\sum_{m,n}\langle 0_f^+\|t_{-m}t_{-n}\mathcal{O}^\alpha_{mn}\|0_i^+\rangle\,, \tag{9}$$

where $\mathcal{O}^\alpha_{mn}$ are the relevant two-body transition operators. Neglecting contributions of right-handed currents, only the matrix elements $M^{0\nu}_{GT}$ and $M^{0\nu}_F$

$$\mathcal{O}^{GT}_{12}=\sigma_1\cdot\sigma_2 H_m(r)\,,\quad \mathcal{O}^F_{12}=H_m(r)\left(\frac{g_V}{g_A}\right)^2 \tag{10a,b}$$

contribute, where $H_m(r)$ is the "neutrino potential" [8] which represents the exchange of a virtual neutrino between two nucleons. The inter-nucleon short-range correlations and the nucleon finite size effects are taken into account in the standard way [8,24].

Because of the two-body character of the operators $\mathcal{O}^\alpha_{12}$ the nuclear matrix elements can be reduced to a sum of products of two-particle transition densities Z and two-particle matrix elements of the above operators. From the solutions of the pnQRPA equation the densities Z are readily evaluated [8].

The single-particle energies used in our calculation are obtained from a Coulomb corrected Woods–Saxon potential [25], where the depth of the central potential is modified by adding the l-dependent term $-0.05l(l+1)$MeV. In the subsequent BCS calculation the self-energy term μ is set to zero because this shift in the single-particle energies is presumably already taken into account by the use of the appropriate N- and Z-dependent Woods–Saxon potential. The strengths of the pairing interaction of the BCS calculation are adjusted to experimental even–odd mass differences [26] for proton and neutron systems separately, by multiplying the respective pp and nn pairing forces in the gap equation with the renormalization factors g_p^{pair} and g_n^{pair}. These and other details concerning the BCS and pnQRPA calculations are the same as described earlier [2,7,8].

In table 1 we present some calculated proton and neutron pairing gaps with $g_{p,n}^{pair}=1$ for some shells near the Fermi surface for different effective interactions. It is clearly seen that the Bonn force gives generally about 15% more pairing than the Paris and Reid po-

Table 1
Proton and neutron pairing gaps calculated from various effective interactions and the strength parameters needed to reproduce the experimental odd–even mass differences.

	Δ_p				Δ_n				g_p^{pair}	g_n^{pair}
	$f_{5/2}$	$p_{3/2}$	$p_{1/2}$	$f_{9/2}$	$f_{5/2}$	$p_{3/2}$	$p_{1/2}$	$g_{9/2}$		
Bonn-bare	1.483	1.643	1.724	1.366	1.753	1.428	1.454	1.177	0.956	1.157
Bonn-$G^{(2)}$	3.074	2.766	2.816	2.223	2.889	2.629	2.517	2.030	0.675	0.873
Bonn–Lesu	1.820	1.754	1.777	1.343	1.422	1.578	1.348	1.036	0.909	1.211
Paris-bare	1.247	1.448	1.530	1.189	1.539	1.189	1.222	0.957	1.050	1.283
Paris-$G^{(2)}$	2.502	2.298	2.361	1.823	2.358	2.106	2.005	1.597	0.761	0.991
Paris–Lesu	1.466	1.432	1.473	1.079	1.075	1.254	1.033	0.771	1.032	1.380
Paris–Yukawa	1.219	1.414	1.516	1.195	1.622	1.145	1.190	0.891	1.062	1.319
Reid-bare	1.252	1.443	1.525	1.193	1.547	1.197	1.227	0.954	1.048	1.285
Reid-$G^{(2)}$	2.427	2.242	2.315	1.774	2.303	2.038	1.939	1.544	0.773	1.009
Reid–Lesu	1.425	1.398	1.463	1.092	1.136	1.201	1.201	0.828	1.047	1.350
Reid–Yukawa	1.209	1.404	1.503	1.181	1.605	1.131	1.177	0.872	1.066	1.331

Volume 242, number 1 PHYSICS LETTERS B 31 May 1990

tentials. For example the neutron $g_{9/2}$ pairing gaps given by the bare G matrix derived from these three interactions are respectively 1.177, 0.957 and 0.954 MeV. This observation is also seen from the tabulated $g_{\rm p,n}^{\rm pair}$ values, those for the Bonn cases being clearly smaller.

For comparison, the results obtained with the G matrix of the Paris potential simulated by sum of Yukawa potentials [27] are listed in addition. This interaction was applied in the work of refs. [7,8]. The same procedure may also be used to derive an effective NN interaction from the Reid potential. We note that the pairing gaps given by the simulated G matrices are quite similar to those by our bare G matrices, where the exclusion operator has been treated in a more accurate way.

In general, the nn-pairing force seems to be too small, as indicated by the tabulated $g_{\rm n}^{\rm pair}$ values, in all cases except for the second order G matrices ($G^{(2)}$) of Paris and Bonn potentials. Comparing with the bare-G results, $G^{(2)}$ largely enhances the pairing and this is mainly due to the contribution from the core-polarization diagrams $G_{\rm 3p1h}$. We feel that $G^{(2)}$ may have over-enhanced the pairing force. A main effect of the folded diagrams included in the Lesu cases appears to be the reducing of this enhancement, rendering the resulting pairing gaps close to the bare-G values. One may also note that the results of the Reid and Paris NN-potentials in table 1 are remarkably similar to each other.

Our results for the 2ν matrix elements are presented in figs. 1 and 2. Fig. 1 illustrates the dependence of $M_{\rm GT}^{2\nu}$ on $g_{\rm pp}$ for different effective interactions derived from the Paris potential. We find a similar behaviour as in earlier QRPA studies, namely the matrix elements stay more or less constant for small $g_{\rm pp}$ values but decrease towards zero rapidly near $g_{\rm pp}=1$. A further increase of $g_{\rm pp}$ will cause the QRPA to collapse – the occurrence of complex QRPA eigenvalues. It is of interest that for the Lesu $V_{\rm eff}$ the collapse of the QRPA equation occurs at significantly larger values of $g_{\rm pp}$ compared to the other forces. If we set $g_{\rm p,n}^{\rm pair}=1$ then $M_{\rm GT}^{2\nu}$ is substantially more reduced for the bare G matrix and the Lesu $V_{\rm eff}$ whereas it is enhanced for $G^{(2)}$. For a more detailed discussion we refer to a future publication. Compared to the results of Muto et al. [7] we obtain somewhat less suppressed matrix elements and the slope is much

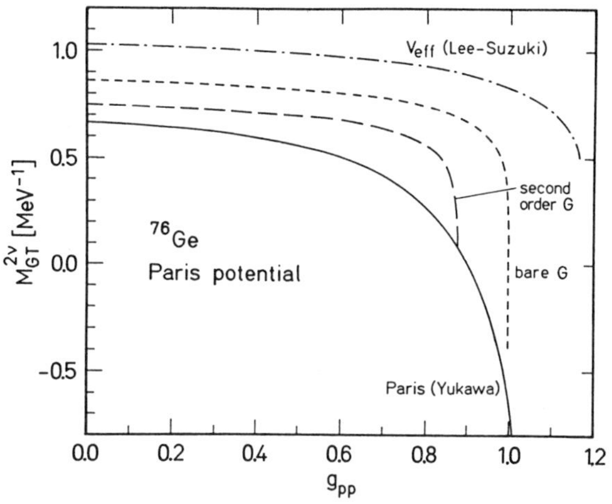

Fig. 1. 2ν matrix elements calculated with different choices for the effective interaction.

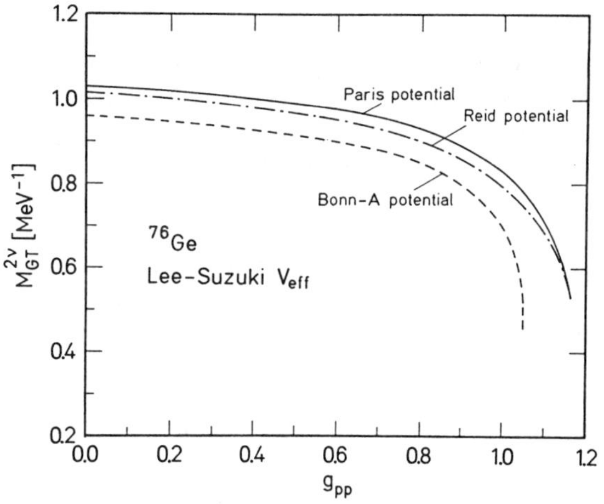

Fig. 2. 2ν matrix elements calculated from renormalized effective interactions derived from Paris, Bonn-A and Reid potentials.

steeper, especially for the bare G matrix.

Fig. 2 compares the 2ν matrix elements using the Lesu $V_{\rm eff}$ derived from different NN potentials. The results for the Paris and Reid potentials are quite similar, while the Bonn-A potential yields a considerably smaller matrix element.

The 0ν transition matrix elements are known to be less sensitive to details of nuclear structure [8,24,28]. Table 2 contains the calculated values for $M_{\rm F}$ and $M_{\rm GT}$ for different values of the strength of the particle–

Volume 242, number 1 PHYSICS LETTERS B 31 May 1990

Table 2
0ν transition matrix elements and half-lives for different effective interactions and particle–particle interaction strengths (blanks indicate the collapse of the QRPA equation).

	$g_{pp}=0.875$			$g_{pp}=0.90$			$g_{pp}=1.00$		
	M_{GT}	M_F	$T^{0\nu}_{1/2}\langle m_\nu\rangle^2$[yr eV^2]	M_{GT}	M_F	$T^{0\nu}_{1/2}\langle m_\nu\rangle^2$[yr eV^2]	M_{GT}	M_F	$T^{0\nu}_{1/2}\langle m_\nu\rangle^2$[yr eV^2]
Bonn-bare	7.77	−1.36	4.89×10^{23}	7.57	−1.33	5.15×10^{23}			
Bonn-$G^{(2)}$									
Bonn–Lesu	10.28	−1.62	2.88×10^{23}	10.19	−1.58	2.94×10^{23}	9.70	−1.41	3.31×10^{23}
Paris-bare	8.25	−1.49	4.31×10^{23}	8.11	−1.46	4.46×10^{23}	4.60	−1.33	1.16×10^{24}
Paris-$G^{(2)}$	6.79	−1.13	6.51×10^{23}						
Paris–Lesu	10.91	−1.83	2.51×10^{23}	10.82	−1.80	2.56×10^{23}	10.44	−1.66	2.79×10^{23}
Paris–Yukawa	3.00	−1.17	2.34×10^{24}	2.65	−1.14	2.84×10^{24}	0.67	−1.01	1.46×10^{25}
Reid-bare	8.26	−1.41	4.36×10^{23}	8.13	−1.39	4.51×10^{23}	7.32	−1.26	5.53×10^{23}
Reid-$G^{(2)}$	7.61	−1.13	5.34×10^{23}	4.38	−1.08	1.37×10^{24}			
Reid–Lesu	10.33	−1.72	2.81×10^{23}	10.24	−1.69	2.87×10^{23}	9.85	−1.56	3.13×10^{23}
Reid–Yukawa	3.36	−1.12	2.03×10^{24}	3.06	−1.09	2.37×10^{24}	1.53	−0.97	6.58×10^{24}

particle interaction. g_{pp} is expected to be close to unity provided that an appropriate force is chosen. Therefore we compare the present calculation at $g_{pp}=1$ with that of ref. [8] at $g_{pp}=0.875$ which was determined from a QRPA calculation of single β^+ decays. We also present the results for an effective interaction simulated by a sum of Yukawa potentials derived from the Reid potential for which we expect a g_{pp} value slightly larger than 0.875 ($g_{pp}\approx0.9$ [29]). Table 2 indicates that the Lesu V_{eff} yields values of $T^{0\nu}_{1/2}\langle m_\nu\rangle^2$ which are smaller by about one order of magnitude compared to the QRPA results of refs. [8,24] which would result in about three times smaller limits of $\langle m_\nu\rangle$ at a given half-life. It may be noted that the corresponding value given by Grotz and Klapdor [4] sometime ago based on a PBCS approach and diagonalisation of a spin–isospin and quadrupole–quadrupole force is 2.6×10^{23} yr eV^2. In contrast to refs. [8,29], we obtain in all cases relatively small values of the tensor matrix element M_T, which was also found by Tomoda et al. [30] in a Hartree–Fock–Bogoliubov calculation with angular-momentum and nucleon number projections.

In conclusion we would like to mention the following: The Bonn-A potential yields significantly larger pairing gaps than the Paris and Reid potentials. Although the pairing gaps given by the bare G matrix and those calculated from the Lesu V_{eff} are comparable, the latter interaction apparently has an important property in hindering the QRPA collapse. Our calculated 2ν matrix elements seemed to be less suppressed compared with earlier calculations. For the 0ν matrix elements our results show rather little variation with respect to the choice of the effective interaction. The main result of the present extensive study is that the neutrinò mass can be reliably deduced from 0ν ββ-decay experiments within a factor of 2–3.

The G-matrix programs were developed together with Dan Strottman and Brian Stout. Their help is gratefully acknowledged. The authors would like to thank also Dr. K. Muto for useful discussions.

References

[1] M. Doi, T. Kotani and E. Takasugi, Prog. Theor. Phys. Suppl. 83 (1985) 1.
[2] K. Muto and H.V. Klapdor, in: Neutrinos. Graduate Texts in Contemporary Physics, ed. H.V. Klapdor (Springer, Berlin, 1988) p. 183.
[3] P. Langacker, in: Neutrinos. Graduate Texts in Contemporary Physics, ed. H.V. Klapdor (Springer, Berlin, 1988) p. 71.
[4] K. Grotz and H.V. Klapdor, Nucl. Phys. A 460 (1986) 395.
[5] P. Vogel and M.R. Zirnbauer, Phys. Rev. Lett. 57 (1986) 3148.
[6] O. Civitarese, A. Faessler and T. Tomoda, Phys. Lett. B 194 (1987) 11.

Volume 242, number 1 PHYSICS LETTERS B 31 May 1990

[7] K. Muto and H.V. Klapdor, Phys. Lett. B 201 (1988) 420; K. Muto, E. Bender and H.V. Klapdor, Z. Phys. A 334 (1989) 177.
[8] K. Muto, E. Bender and H.V. Klapdor, Z. Phys. A 334 (1989) 187.
[9] D.O. Caldwell, R.M. Eiseberg, D.M. Grumm and M.S. Witherell, Phys. Rev. Lett. 59 (1987) 419.
[10] H.V. Klapdor, A. Piepke, G. Heusser, E. Buchner, A. Müller, U. Schmidt-Rohr, H. Strecker, S.T. Belyaev, A. Balish, A. Gurov, A. Demehin, I. Kondratenko and V.I. Lebedev, Proc. Intern. Symp. on Weak and electromagnetic interactions in nuclei, (Montreal, May 1989).
[11] M. Lacombe, B. Loiseau, J.M. Richard, R. Vinh Mau, J. Côté, P. Pirès and R. de Tourreil, Phys. Rev. C 21 (1980) 861.
[12] R. Machleidt, K. Holinde and C. Elster, Phys. Rep. 149 (1987) 1.
[13] R. Machleidt, Adv. Nucl. Phys. 19 (1989) 189.
[14] R.V. Reid, Ann. Phys. (NY) 50 (1968) 411.
[15] T.T.S. Kuo, S.Y. Lee and K.F. Ratcliff, Nucl. Phys. A 176 (1971) 65.
[16] T.T.S. Kuo, Lecture Notes in Physics, Vol. 144 (Springer, Berlin, 1981) p. 248.
[17] S.F. Tsai and T.T.S. Kuo, Phys. Lett. B 39 (1972) 427.
[18] E.M. Krenciglowa, C.L. Kung, T.T.S. Kuo and E. Osnes, Ann. Phys. (NY) 101 (1976) 154.
[19] G.E. Brown, A.D. Jackson and T.T.S. Kuo, Nucl. Phys. A 133 (1969) 481.
[20] J. Shurpin, T.T.S. Kuo and D. Strottman, Nucl. Phys. A 408 (1983) 310.
[21] T.T.S. Kuo and G.E. Brown, Nucl. Phys. A 85 (1966) 40.
[22] S.Y. Lee and K. Suzuki, Phys. Lett. B 91 (1980) 173; Prog. Theo. Phys. 64 (1980) 2091.
[23] J.A. Halbleib and R.A. Sorensen, Nucl. Phys. A 98 (1967) 542;
K. Muto, E. Bender and H.V. Klapdor, Z. Phys. A 333 (1989) 125.
[24] T. Tomoda and A. Faessler, Phys. Lett. B 199 (1987) 475.
[25] A. Bohr and B.R. Mottelson, Nuclear structure, Vol. 1, (Benjamin, New York, 1969).
[26] A.H. Wapstra and G. Audi, Nucl. Phys. A 432 (1985) 1.
[27] N. Anantaraman, H. Toki and G.F. Bertsch, Nucl. Phys. A 398 (1983) 269.
[28] J. Engel, P. Vogel, Xiangdong JI and S. Pittel, Phys. Lett. B 225 (1989) 5.
[29] A. Staudt, K. Muto and H.V. Klapdor, to be published.
[30] T. Tomoda, A. Faessler, K.W. Schmid and F. Grümmer, Nucl. Phys. A 452 (1986) 591.

PHYSICAL REVIEW C VOLUME 46, NUMBER 3 SEPTEMBER 1992

$\beta\beta$ decay of ^{128}Te, ^{130}Te, and ^{76}Ge with renormalized effective interactions derived from Paris and Bonn potentials

A. Staudt, T. T. S. Kuo,* and H. V. Klapdor-Kleingrothaus
Max-Planck-Institut für Kernphysik, D-6900 Heidelberg, Germany
(Received 5 December 1991)

We perform $2\nu\beta\beta$ and $0\nu\beta\beta$ calculations for ^{76}Ge, ^{128}Te, and ^{130}Te using effective interactions derived from the Paris and Bonn-A potentials. Extended model spaces are employed in setting up the quasiparticle random phase approximation (QRPA) equations, on which our present calculations are based. For ^{76}Ge the model space consists of nine orbits, the two major shells from $0f_{7/2}$ to $2s_{1/2}$, and for the tellurium case we include eleven orbits spanning three major shells from $1p_{3/2}$ to $1f_{7/2}$. The bare-*G*-matrix elements are first calculated, with the Pauli exclusion operator carefully treated with a matrix inversion method, so that double counting between the calculated effective interaction and the above model spaces is strictly avoided. We then calculate the renormalized effective interaction, including corrections from core polarizations and folded diagrams. The effect of core polarization is found to be highly significant, especially for ^{76}Ge. There appears to be a compensating effect from the folded diagrams; the net results with core polarizations and folded diagrams both included become rather close to the bare-*G* results. Unlike earlier QRPA calculations, our calculated $M_{\rm GT}$ matrix elements for $2\nu\beta\beta$ do not seem to exhibit strong dependence on $g_{\rm pp}$, the particle-particle interaction strength parameter, in the vicinity of $g_{\rm pp}=1.0$. For $0\nu\beta\beta$ decays, our calculated values for $T^{0\nu}_{1/2}\langle m_\nu\rangle^2$ are typically 5×10^{23} yr eV^2 for ^{76}Ge, 2×10^{24} yr eV^2 for ^{128}Te, and 9×10^{22} yr eV^2 for ^{130}Te. Although the Bonn-A potential gives generally more pairing force, the final results for $\beta\beta$ decays given by the Paris and Bonn-A potentials are rather close to each other.

PACS number(s): 23.40.Hc, 21.30.+y, 27.60.+j, 27.50.+e

I. INTRODUCTION

A new generation of nuclear double-beta- ($\beta\beta$-) decay experiments, using direct-counter methods, are rapidly progressing [1–3], and it may be of interest to carry out further theoretical studies of nuclear $\beta\beta$ decays. Traditionally nuclear $\beta\beta$ experiments were mostly performed using geochemical methods. This is an inclusive measurement, including both the neutrinoless ($0\nu\beta\beta$) modes and the 2ν ones. An advantage of direct-counting experiments is their ability in differentiating the above two modes. Thus the moment of directly observing the fundamentally important $0\nu\beta\beta$ decays of nuclei in the laboratory seems to have finally "arrived." In fact a number of this type of experiments are in progress, such as the Heidelberg-Moscow experiment using enriched ^{76}Ge [2,3]. As of now, the $0\nu\beta\beta$ mode of ^{76}Ge seems to have not been observed, remaining to be elusive. However, a number of experiments have reported direct observations of the $2\nu\beta\beta$ modes. For instance, Moe and collaborators [4] have observed the $2\nu\beta\beta$ decay of ^{82}Se. Ejiri and collaborators [5] have observed the $2\nu\beta\beta$ decay of ^{100}Mo. And two groups, a Russian and an American group, have observed the $2\nu\beta\beta$ decay of ^{76}Ge [6,7].

The above suggests that it may be worthwhile to carry out further calculations for the $2\nu\beta\beta$ decays, as they can be compared with the latest experimental results. The main purpose of the present work is to perform $2\nu\beta\beta$ and $0\nu\beta\beta$ calculations for nuclei ^{128}Te, ^{130}Te and ^{76}Ge, using renormalized effective interactions derived from realistic nucleon-nucleon (*NN*) potentials—the Paris [8] and Bonn-A [9,10] potentials. A brief account of our ^{76}Ge work, where the Reid *NN* potential was also used, has been reported [11].

The study of tellurium is of particular interest since there are two neighboring isotopes, ^{128}Te and ^{130}Te, which are expected to disintegrate by $\beta\beta$ decay. One may argue that these two isotopes, which only differ by two neutrons, are likely to exhibit similar nuclear structure effects in the calculation of the relevant transition amplitudes. The ratio of the $\beta\beta$ half-lives of both isotopes have recently been determined in geochemical measurements [12,13]. This ratio provides a test of theoretical matrix elements and together with the latter can also help to decide whether the decays are dominated by the 2ν or the 0ν mode [14]. However, there exists a disagreement between the measurements by Kirsten *et al.* [15] and those of [12,13,16]. The reason for this discrepancy seems not yet to be fully understood.

Nuclei where $\beta\beta$ decays take place are usually rather far from the closed shells. And calculations for these nuclei using full-fledged shell-model approaches become generally prohibitive. One has to adopt some approximation scheme. A commonly adopted one is the *pn* QRPA (proton neutron quasiparticle random phase approximation) (see, e.g., [14,17,18,28], and references quoted therein). Although it is the $0\nu\beta\beta$ mode which provides

*Permanent address: Physics Department, State University of New York at Stony Brook, Stony Brook, NY 11794-3800.

Teilchenphysik ohne Beschleuniger

Non-accelerator Particle Physics

Неускорительная физика элементарных частиц

Von Prof. Dr. rer. nat. Hans Volker Klapdor-Kleingrothaus
Max-Planck-Institut für Kernphysik, Heidelberg
und Dr. rer. nat. Andreas Staudt
Zentrale Forschung Bayer AG, Leverkusen

Mit zahlreichen Abbildungen und Tabellen

Translated by S S Wilson

Перевод с немецкого В. А. БЕДНЯКОВА

Inhalt

B. G. Teubner Stuttgart 1995

Institute of Physics Publishing
Bristol and Philadelphia

МОСКВА
НАУКА • ФИЗМАТЛИТ
1997

VOLUME 75, NUMBER 3 PHYSICAL REVIEW LETTERS 17 JULY 1995

Renormalized Proton-Neutron Quasiparticle Random-Phase Approximation and Its Application to Double Beta Decay

J. Toivanen and J. Suhonen

Department of Physics, University of Jyväskylä, P.O. Box 35, SF-40351, Jyväskylä, Finland

(Received 21 February 1995)

A self-consistent method of treating excitations of the proton-neutron quasiparticle random-phase approximation is presented. The non-self-consistent methods violate the Pauli exclusion principle and lead to an eventual collapse of the ground state. This behavior renders a reliable calculation of the nuclear matrix elements, relevant for the prediction of double-beta-decay half-lives, difficult. The present formalism promotes the Pauli exclusion principle and avoids the collapse of the double-beta-decay matrix elements. We have applied this formalism to the double beta decay of ^{100}Mo.

PACS numbers: 21.60.Jz, 23.40.Hc, 27.60.+j

At the moment, the neutrinoless double beta decay ($0\nu\beta\beta$) is the best probe for physics beyond the standard model of electroweak interactions. Its existence connects with many fundamental aspects of neutrino physics [1]. To date, only lower limits on half-lives of different nuclei have been obtained experimentally. These limits are used to deduce upper limits on the Majorana-neutrino mass, the right-handed-current admixture parameters, the Majoron–Majorana-neutrino coupling constant, etc. One particular source of uncertainty in the above analyses is the evaluation of the nuclear matrix elements involved. For the decay to the ground state of the daughter nucleus, the mass mechanism is the dominant one [1], leading to the need of evaluating the associated double Gamow-Teller matrix element (DGT) as reliably as possible.

It is important to note that in connection with the $0\nu\beta\beta$ decay, the detection of double beta decay with the emission of two neutrinos ($2\nu\beta\beta$), which is an allowed process of second order in the standard model, enables experimental determination of the magnitude of the DGT. This is because both modes of double beta decay require the same basic many-body nuclear-structure wave functions leading to the need of developing theoretical schemes of calculation in connection with both the $2\nu\beta\beta$ decays and the $0\nu\beta\beta$ decays. The improved theoretical methods yield better predictions for double-beta half-lives and stimulate new and better-focused double-beta experiments.

After its successful application to $2\nu\beta\beta$ decay [2], the proton-neutron quasiparticle random-phase approximation (pnQRPA) has been the most powerful tool in studies of double beta decay of medium-heavy and heavy open-shell nuclei. In these studies both schematic [2] and realistic [3] two-body matrix elements have been used. A shortcoming of these calculations is that they violate the Pauli exclusion principle and lead to an eventual collapse of the pnQRPA ground state. This collapse is caused by generation of too many ground-state correlations with increasing strength of the proton-neutron interaction. This means that in the region of realistic pn-interaction strengths, i.e., near $g_{pp} = 1.0$, where g_{pp} is the strength parameter of the particle-particle part of the pn interaction [2], the DGT may be very unstable, changing its magnitude very fast as a function of g_{pp}. Therefore, a reliable calculation of the nuclear matrix elements, relevant for the prediction of double-beta-decay half-lives, becomes difficult. In particular, this instability has been speculated [4,5] to hinder a reliable calculation of the DGT in connection with the $2\nu\beta\beta$ transition ^{100}Mo(g.s.) $\rightarrow$ ^{100}Ru(g.s.).

Several attempts [4,6–8] have been made to shift the collapse of the pnQRPA ground state to higher values of g_{pp}, possibly beyond the physically acceptable region. In Refs. [4,7] the method of particle-number-projected pnQRPA equations was used and in Ref. [6] the BCS self-energy corrections were included in the calculations. The authors of Ref. [8] tried to achieve a consistent determination of g_{pp} by invoking partial restoration of isospin and Wigner SU(4) symmetries. All these methods, however, disregard the main source of ground-state instability, namely, the increasing violation of the Pauli principle with building up of an excessive amount of ground-state correlations.

In this Letter we present an easy-to-apply method with which one can avoid the instability of the DGT and produce more reliable predictions for the double-beta half-lives. This method we call the renormalized pnQRPA, RQRPA for short. It is not restricted to any particular choice of the single-particle basis or form of the two-body interaction. To our knowledge, the RQRPA is the first attempt to retain the Pauli exclusion principle in the correlated ground state of the pnQRPA with increasing strength of the proton-neutron interaction. Below we summarize its basic ingredients and demonstrate its application by comparing its results with the results of the ordinary pnQRPA in the case of the previously mentioned ^{100}Mo decay. On one hand, this comparison is stimulated by the suggested [4,5] instability of the DGT, and on the other hand, by the present experimental efforts [9] being invested in the study of the ^{100}Mo double beta decay.

The first step in the RQRPA calculation, as in the ordinary pnQRPA, is the BCS calculation of the even-even

VOLUME 75, NUMBER 3 PHYSICAL REVIEW LETTERS 17 JULY 1995

ground state leading to the quasiparticle representation of the nuclear Hamiltonian. After that, one can derive the RQRPA equations in a straightforward way by using the equations-of-motion (EOM) method [10]. In the EOM one starts from the creation and annihilation operators of excited states, which, in the case of odd-odd nuclei, consist of proton-neutron two-quasiparticle components

$$Q^\dagger(m;JM) = \sum_{pn} [X^m_{pn} A^\dagger(pn;JM) - Y^m_{pn} \tilde{A}(pn;JM)], \tag{1}$$

where the notation is from [11]. Using the machinery of the EOM theory [10], one arrives at the matrix equation

$$\begin{pmatrix} A & B \\ B & A \end{pmatrix} \begin{pmatrix} X^m \\ Y^m \end{pmatrix} = \hbar\omega_m \begin{pmatrix} U & 0 \\ 0 & -U \end{pmatrix} \begin{pmatrix} X^m \\ Y^m \end{pmatrix}, \tag{2}$$

where the matrices A and B are defined as

$$A(pnp'n'J) = \langle \mathrm{RPA}| [A(pn;JM), \hat{H}, A^\dagger(p'n';JM)] \times |\mathrm{RPA}\rangle, \tag{3a}$$

$$B(pnp'n'J) = \langle \mathrm{RPA}| [A(pn;JM), \hat{H}, A(p'n';JM)] \times |\mathrm{RPA}\rangle, \tag{3b}$$

Here the double commutators are defined as $[A,B,C] = \frac{1}{2}[A,[B,C]] + \frac{1}{2}[[A,B],C]$. The matrices A and B in (2) carry contributions from one-body and two-body ground-state densities [10]. These contributions go beyond the ordinary pnQRPA level and take into account the Pauli exclusion principle [10]. They are nonzero because in the EOM one takes the intrinsic fermion structure of the pair operators $A(pn;JM)$ and $A^\dagger(pn;JM)$ fully into account when deriving Eq. (2.)

Keeping only the diagonal parts of the one-body densities, the overlap matrix U on the right-hand side of (2) becomes

$$U(pnp'n') = \langle \mathrm{RPA}| [A(pn;JM), A^\dagger(p'n';JM)] |\mathrm{RPA}\rangle = \delta_{pp'}\delta_{nn'}(1 - \hat{j}_p^{-1}\langle \mathrm{RPA}| [a_p^\dagger \tilde{a}_p]_{00}|\mathrm{RPA}\rangle - \hat{j}_p^{-1}\langle \mathrm{RPA}| [a_n^\dagger \tilde{a}_n]_{00}|\mathrm{RPA}\rangle) \equiv \delta_{pp'}\delta_{nn'}D_{pn}, \tag{4}$$

where $\hat{j} = \sqrt{2j+1}$. This form of overlap matrix is exact if the single-particle valence space contains at most two major oscillator shells (which is the case in the present calculation). This approximation leads to a diagonal overlap matrix U, and Eq. (2) can be transformed to standard RPA form [1]. After this transformation, and omitting the two-body densities (supposed to act incoherently with random phases, see Ref. [10]), one obtains the RQRPA equations with new matrices $\overline{A}$ and $\overline{B}$

$$\begin{pmatrix} \overline{A} & \overline{B} \\ \overline{B} & \overline{A} \end{pmatrix} \begin{pmatrix} \overline{X}^m \\ \overline{Y}^m \end{pmatrix} = \hbar\omega_m \begin{pmatrix} \overline{X}^m \\ -\overline{Y}^m \end{pmatrix}, \tag{5}$$

where now the matrices $\overline{A}$ and $\overline{B}$ read

$$\overline{A}(pnp'n'J) = (E_p + E_n)\delta_{pp'}\delta_{nn'} - D_{pn}^{1/2}[2G(pnp'n'J)(u_p u_n u_{p'} u_{n'} + v_p v_n v_{p'} v_{n'}) + 2F(pnp'n'J)(u_p v_n u_{p'} v_{n'} + v_p u_n v_{p'} u_{n'})]D_{p'n'}^{1/2},$$

$$\overline{B}(pnp'n'J) = D_{pn}^{1/2}[2G(pnp'n'J)(u_p u_n v_{p'} v_{n'} + v_p v_n u_{p'} u_{n'}) - 2F(pnp'n'J)(u_p v_n v_{p'} u_{n'} + v_p u_n u_{p'} v_{n'})]D_{p'n'}^{1/2}. \tag{6}$$

The amplitudes $\overline{X}^m_{pn} = D_{pn}^{1/2} X^m_{pn}$ and $\overline{Y}^m_{pn} = D_{pn}^{1/2} Y^m_{pn}$ fulfill the usual RPA orthogonality relations, as one can easily check using Eqs. (1) and (4).

So far nothing has been said about the calculation of the overlap matrix elements D_{pn}. We adopt the method of Ref. [12] where the one-body densities are expressed in terms of fermion-pair creation and annihilation operators $A^\dagger$ and A. One can expand the product of the fermion-pair operators [12] by using completeness of the intermediate states and then apply Eq. (4) to yield

$$D_{pn} = 1 - \hat{j}_p^{-2} \sum_{n'} D_{pn'} \left(\sum_{Jm} \hat{J}^2 |\overline{Y}^{Jm}_{pn'}|^2 \right) - \hat{j}_n^{-2} \sum_{p'} D_{p'n} \left(\sum_{Jm} \hat{J}^2 |\overline{Y}^{Jm}_{p'n}|^2 \right). \tag{7}$$

The above equation cannot be converted into an ordinary matrix-inversion problem and must therefore be solved by iteration. Equations (5) and (7) together form a doubly iterative problem. When the coefficients D_{pn} on the right-hand side of Eq. (7) are put equal to unity, we obtain the usual pnQRPA expression of ground-state correlations, which is valid as long as ground-state correlations are small. In this case no double iteration is needed. To access the difference between these two approximations of D_{pn}, we have used them both in our calculations.

To be able to calculate beta-decay and double-beta-decay transition amplitudes, one has to know the renormalized charge-changing transition densities of the RQRPA, namely,

$$(\mathrm{RPA}|| [c_p^\dagger \tilde{c}_n]_L || J_m) = \hat{L}\delta_{LJ}[v_p u_n \overline{X}^{Jm}_{pn} + u_p v_n \overline{Y}^{Jm}_{pn}]D_{pn}^{1/2}, \tag{8a}$$

$$(\mathrm{RPA}|| [c_n^\dagger \tilde{c}_p]_L || J_m) = \hat{L}\delta_{LJ}[u_p v_n \overline{X}^{Jm}_{pn} + v_p u_n \overline{Y}^{Jm}_{pn}]D_{pn}^{1/2}\eta_{J_{pn}}, \tag{8b}$$

VOLUME 75, NUMBER 3 PHYSICAL REVIEW LETTERS 17 JULY 1995

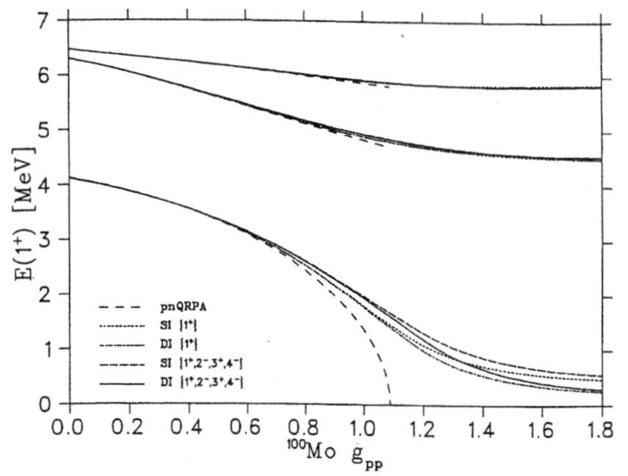

FIG. 1. Energies of the three lowest 1^+ states of ^{100}Tc calculated using the *pn*QRPA and four different approximations of the RQRPA as a function of the particle-particle interaction strength g_{pp}. Short-dashed line: single-iterative (SI) with intermediate 1^+ states only. Dot-dashed line: double-iterative (DI) with intermediate 1^+ states only. Long-dashed line: SI using intermediate 1^+, 2^-, 3^+, and 4^- states. Solid line: DI using intermediate 1^+, 2^-, 3^+, and 4^- states.

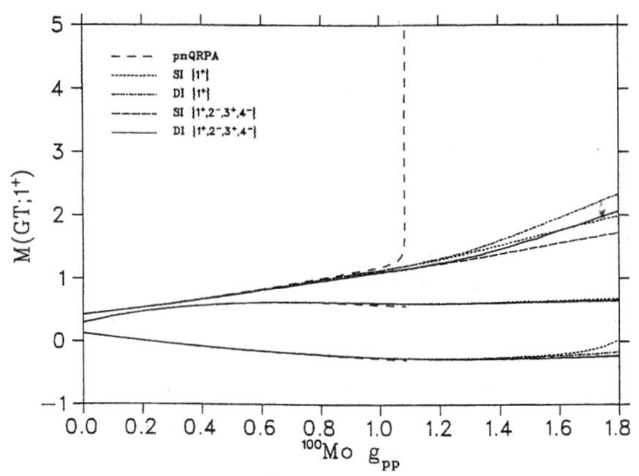

FIG. 2. β^- transition amplitudes $M(\mathrm{GT})$ for the decay of ^{100}Mo(g.s.) to the three lowest 1^+ states of ^{100}Tc as functions of the particle-particle interaction strength g_{pp}. Highest group of lines: $M(\mathrm{GT})$ for the state 1_1^+. Middle group of lines: $M(\mathrm{GT})$ for the state 1_2^+. Lowest group of lines: $M(\mathrm{GT})$ for the state 1_3^+. For the meaning of the various line types, see the caption of Fig. 1.

where $\eta_{J_{pn}} \equiv (-1)^{j_p+j_n+J+1}$. These transition matrix elements can be inserted into the standard expressions of the Gamow-Teller beta-decay [11] and double-beta-decay [1–3] amplitudes.

We demonstrate the use of the RQRPA by calculating the $2\nu\beta\beta$ transition amplitude ^{100}Mo(g.s.) $\rightarrow^{100}$ Ru(g.s.). Both the *pn*QRPA and RQRPA calculations follow the scheme of Ref. [5]. Our single-particle valence space consists of major shells $3\hbar\omega$, $4\hbar\omega$, and the $0h_{11/2}$ intruder orbital from the $5\hbar\omega$ major shell. The adopted single-particle energies and the two-body G-matrix interaction are the ones of Ref. [5].

Our results are summarized in Figs. 1–3. Figures 1 and 2 show the three lowest energies of the intermediate 1^+ states and the β^- transition amplitudes to these states from ^{100}Mo(g.s.) as functions of the proton-neutron interaction strength g_{pp}. Corresponding results for ^{100}Ru are qualitatively similar. Finally, in Fig. 3 we summarize the calculated results for the $2\nu\beta\beta$ decay ^{100}Mo(g.s.) $\rightarrow^{100}$ Ru(g.s.) as a function of g_{pp}.

In all figures we call those results where Eq. (7) is simplified noniterative as "SI" and those results where we take Eq. (7) fully iterative as "DI". The full RQRPA calculation needs fairly big computational effort because in each step of the singly or doubly iterative procedure of solving Eqs. (5) and (7), for a given g_{pp}, we must, in principle, find eigenstates for all intermediate multipolarities J^π. To study the effects of different numbers of intermediate multipolarities, we have performed the calculations using two sets of J^π: either only 1^+ states (denoted $\{1^+\}$ in the figures) or states 1^+, 2^-, 3^+, and 4^- ($\{1^+, 2^-, 3^+, 4^-\}$ in the figures) which, in this case, are the only important multipoles affecting the solution of Eq. (7).

As can be seen from the figures, for part of the physically acceptable values of particle-particle interaction strength, e.g., $0.8 \le g_{pp} \le 1.0$, the results of different approximation schemes differ only little from each other. For larger values of g_{pp} the lowest 1^+ state vector, which in ordinary *pn*QRPA collapses at some critical particle-particle interaction strength, in this case at $g_{pp} = 1.03$, behaves smoothly in the RQRPA. Thus the *pn*QRPA collapses in the physically acceptable region of g_{pp} and may lead

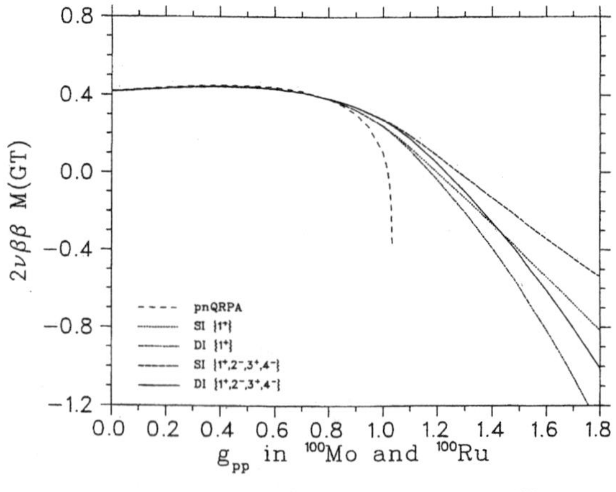

FIG. 3. The $2\nu\beta\beta$ transition amplitude ^{100}Mo(g.s.) $\rightarrow^{100}$ Ru(g.s.) as a function of the particle-particle interaction strength g_{pp}. For the meaning of the various line types, see the caption of Fig. 1.

to ambiguous determination of the beta-decay and double-beta-decay Gamow-Teller matrix elements. This is further demonstrated in Fig. 2 where one observes that near $g_{pp} = 1.0$ the decay amplitude to the lowest intermediate 1^+ state becomes unphysically large for the pnQRPA, whereas in the RQRPA the corresponding amplitude remains stable. Figure 3, in turn, indicates that the $2\nu\beta\beta$ amplitude remains rather stable at $g_{pp} = 1.0$ for the pnQRPA and does not differ drastically from the result of the RQRPA, establishing the validity of the results of Ref. [5]. In some other cases [3], especially for some double β^+ and electron capture decays, the situation may not be that fortunate and severe difficulties can arise in the pnQRPA.

For higher 1^+ states both energies and beta-decay amplitudes are well described by the pnQRPA, even beyond the physical region of g_{pp}. As a general observation, the RQRPA amplitudes calculated with Eq. (7) tend to diminish faster than the corresponding amplitudes calculated using the noniterative approximation of D_{pn}. This is in keeping with the observations of Ref. [12]. However, for all shown values of g_{pp} the different RQRPA approaches yield amplitudes of the same order of magnitude.

In conclusion, we have developed a straightforward method, the renormalized proton-neutron QRPA, for a reliable calculation of beta and double-beta transition matrix elements in a region of two-body interactions where the ordinary pnQRPA becomes unstable. This method is not restricted to any particular choice of the single-particle basis or form of the two-body interaction. It simply promotes the Pauli exclusion principle violated by the excessive ground-state correlations of the pnQRPA in the physically interesting region of the proton-neutron interaction. This aspect has been neglected in all previous attempts to overcome the eventual collapse of the pnQRPA. We demonstrate the RQRPA method by comparing its results with the pnQRPA results in the case of the $2\nu\beta\beta$ transition ^{100}Mo(g.s.) $\rightarrow^{100}$ Ru(g.s.).

This work was supported by the Academy of Finland, Contract No. 15152.

[1] T. Tomoda, Rep. Prog. Phys. **54**, 1 (1991); M. Moe and P. Vogel, Annu. Rev. Nucl. Part. Sci. **44**, 247 (1994).

[2] P. Vogel and M.R. Zirnbauer, Phys. Rev. Lett. **57**, 3148 (1986); J. Engel, P. Vogel, and M.R. Zirnbauer, Phys. Rev. C **37**, 731 (1988).

[3] O. Civitarese, A. Faessler, and T. Tomoda, Phys. Lett. B **194**, 11 (1987); K. Muto, E. Bender, and H.V. Klapdor, Z. Phys. A **334**, 177 (1989); **334**, 187 (1989); A. Staudt, T.T.S. Kuo, and H.V. Klapdor-Kleingrothaus, Phys. Lett. B **242**, 17 (1990); J. Suhonen, Phys. Rev. C **48**, 574 (1994).

[4] O. Civitarese, A. Faessler, J. Suhonen, and X.R. Wu, J. Phys. G **17**, 943 (1991).

[5] A. Griffiths and P. Vogel, Phys. Rev. C **46**, 181 (1992); J. Suhonen and O. Civitarese, Phys. Rev. C **49**, 3055 (1994).

[6] D.B. Stout and T.T.S. Kuo, Phys. Rev. Lett. **69**, 1900 (1992); S.S. Hsiao, Y. Tzeng, and T.T.S. Kuo, Phys. Rev. C **49**, 2233 (1994).

[7] F. Krmpotić, A. Mariano, T.T.S. Kuo, and K. Nakayama, Phys. Lett. B **319**, 393 (1993).

[8] F. Krmpotić and S.S. Sharma, Nucl. Phys. **A572**, 329 (1994).

[9] A.S. Barabash *et al.*, Phys. Lett. B **249**, 186 (1990); H. Ejiri *et al.*, Phys. Lett. B **258**, 17 (1991); M. Alston-Garnjost *et al.*, Phys. Rev. Lett. **71**, 831 (1993); N. Kudomi *et al.*, Phys. Rev. C **46**, R2132 (1992); NEMO Collaboration (to be published).

[10] D.J. Rowe, Rev. Mod. Phys. **40**, 153 (1968).

[11] J. Suhonen, T. Taigel, and A. Faessler, Nucl. Phys. **A486**, 91 (1988).

[12] D. Karadjov, V.V. Voronov, and F. Catara, Phys. Lett. B **306**, 197 (1993); F. Catara, N. Dinh Dang, and M. Sambataro, Nucl. Phys. **A579**, 1 (1994).

PHYSICAL REVIEW C VOLUME 47, NUMBER 2 FEBRUARY 1993

Gamow-Teller matrix elements for two-neutrino double β decay within a second quasi-random-phase approximation

S. Stoica
Institute of Atomic Physics, Bucharest, P.O. Box MG-6, Romania

W. A. Kaminski
Institute of Physics, M. Curie-Sklodowska University, PL-20-031 Lublin, Poland

(Received 8 May 1992)

A second quasi-random-phase approximation (QRPA) procedure is employed to perform calculations of the $M_{\rm GT}^{2\nu}$ matrix elements for several $\beta\beta$ emitters. It is found that higher-order QRPA (HQRPA) corrections display a weak dependence on particle-particle (pp) strength $g_{\rm pp}$ and, in the physical region (around $g_{\rm pp}=1.0$), they become important by comparison with QRPA predictions. It turns out that a further investigation of the HQRPA corrections could be illuminating to obtain more stable and reliable values for $M_{\rm GT}^{2\nu}$ matrix elements in this region, where they are very sensitive to the pp strength.

PACS number(s): 23.40.Hc, 21.60.Jz, 23.90.+w

In the last decade the double-β-decay phenomenon has been the subject of many experimental as well as theoretical investigations [1–7]. The interest in this field is mainly related to the neutrinoless double beta decay ($0\nu\beta\beta$) mode since its discovery would imply that (i) lepton number is not conserved; (ii) ν_e is a Majorana particle with a nonvanishing mass and/or there is a small right-handed component in the weak interaction. Reliable predictions for upper limits of neutrino mass and right handedness of the weak interaction are possible only if we have reliable calculations for the nuclear matrix elements (ME) which enter the half-life formulas. Since there are no experimental data to confirm the existence of the $0\nu\beta\beta$ decay mode, a good test for nuclear ME would be to use them for computing the two-neutrino double beta decay ($2\nu\beta\beta$) rates and then to compare the predictions to the existing data [8–11]. One of the many-body methods which were successfully used to compute the nuclear ME in double β decay is the proton-neutron (QRPA) (*pn* QRPA) [12–16]. In the frame of this approach the agreement with the experiment, for the predicted $\beta\beta$ half-lives, was achieved only when pp interactions are taken into account [14–17]. However, the drawback of these calculations is the fact that the Gamow-Teller (GT) ME ($M_{\rm GT}^{2\nu}$) are functions very sensitive on $g_{\rm pp}$, especially in the physical region. An attempt to improve the calculations has been to use a projected-QRPA procedure in order to restore the particle number and spin symmetries [17]. Another one has been to go beyond QRPA and to include HQRPA corrections in computing the GT ME [18].

The present work reports the results of the computation of $M_{\rm GT}^{2\nu}$ ME for several $\beta\beta$ emitters, performed in the frame of a second QRPA-type approach. This means that we used improved QRPA wave functions (by including HQRPA corrections) but we kept the HFB vacuum unchanged. The procedure is described in more detail in references [18]. Here we have extended the calculations to several $\beta\beta$ emitters in order to reach a final conclusion about the influence of HQRPA corrections upon the GT ME. We have shown that these corrections are functions slightly sensitive to the $g_{\rm pp}$ parameter and, in the physical region, they become important by comparison with QRPA predictions. To describe the properties of the parent, daughter, and intermediate nuclei involved in the $\beta\beta$ decay process we used a many-body Hamiltonian with a one body term describing the independent motion of the nucleons in a Woods-Saxon potential including Coulomb corrections and a nucleon-nucleon residual interaction taken as G matrices $\langle (ab)J|G|(cd)J\rangle$, $(J=0,1,2)$ calculated with Bonn OBEP. In our numerical calculations the single-particle space is restricted to two full shells. This truncation requires a renormalization of the two-body matrix elements which can be achieved by multiplying the G matrices

$$\langle (ab)J=0|G_\tau|(cd)J=0\rangle\ ,$$

$\tau=p,n$ by strength constants as follows: (i) for pairing interaction $(J=0)$ by $g_{\rm pp}^{\tau}$; (ii) for dipole interaction $(J=1)$ by $g_{\rm ph}^{(1)}$ if b, d are hole states and by $g_{\rm pp}^{(1)}$ when b,d are particle states; (iii) for quadrupole interaction $(J=2)$ by $g_{\rm ph}^{(2)}$ and $g_{\rm pp}^{(2)}$ both for the proton and neutron system. Accordingly to the previous experience for such type of calculations these constants are not far from unity and are fixed in the following way: $g_{\rm pp}^{(\tau)}$ are chosen so that the experimental odd-even mass differences are reproduced. $g_{\rm ph}^{(1)}$ is fitted so that the position of the GT resonance for the intermediate odd-odd nucleus be reproduced and $g_{\rm ph}^{(2)}$ is determined by fitting the experimental energy of the first 2^+ state for the parent and daughter nuclei. $g_{\rm pp}^{(2)}$ is set equal to unity while $g_{\rm pp}^{(1)}$ is varied from 0.0 to 1.15. In the frame of the QRPA approach phonon states of a nucleus are described by applying the corresponding operators:

$$\Gamma_{s\mu}^{+}(l)=\sum_k [x_l^s(k)A_{s\mu}^{\dagger}(k)-Y_l^s(K)A_{s-\mu}(k)(-1)^{s+\mu}] \quad (1)$$

to the QRPA vacuum state. The index s takes on the

values 1 or 2 for particle-unlike or particle-like QRPA approaches. X and Y are the upwards and backwards QRPA amplitudes while $A^{\dagger}_{s\nu}(k)$ are the pair operators and are defined as two quasiparticle tensors of rank s. The summation index k stands for the (p,n) pair states or for (pp) and (nn) pair states for particle-unlike or particle-like QRPA approaches, respectively. As is well known, QRPA can be viewed as the first term in a boson expansion of certain quasiparticle operator combinations. To introduce HQRPA corrections we have developed the quasiparticle pair operators $A^{\dagger}_{s\mu}(k)$ in terms of QRPA boson operators $\Gamma^{\dagger}_{s\mu}(k)$ and retained the next order beyond QRPA in the series expansion. In this way the properties of the initial, intermediate and final nuclei involved in the $\beta\beta$ decay are described by the following multiphonon states.

(1) Initial (i):

$$|0\rangle_i\,,\quad {}_i\Gamma^{\dagger}_{2\mu}(k)|0\rangle_i\,; \tag{2}$$

(2) intermediate (int):

$${}_i\Gamma^{\dagger}_{1\mu}(k)|0\rangle_i\,,\quad ({}_i\Gamma^{\dagger}_{1\mu_1}(k_1)\,{}_i\Gamma^{\dagger}_{2\mu_2}(k_2))_{1\mu}|0\rangle_i\,; \tag{3a}$$

$${}_f\Gamma^{\dagger}_{1\mu}(k)|0\rangle_f\,,\quad ({}_f\Gamma^{\dagger}_{1\mu_1}(k_1)\,{}_f\Gamma^{\dagger}_{2\mu_2}(k_2))_{1\mu}|0\rangle f\,; \tag{3b}$$

(3) final (f):

$$|0\rangle_f\,,\quad {}_f\Gamma^{\dagger}_{2\mu}(k)|0\rangle_f\,. \tag{4}$$

The two sets of multiphonon states which describe the intermediate nucleus come from two different QRPA procedures applied for the initial and final nuclei, respectively. The $2\nu\beta\beta$ decay rate for the transition $0^+_i\to0^+_f$ has the expression:

$$M_{\rm GT}=\sum_{k,m}\frac{{}_f\langle 0\|\beta^{\dagger}\|km\rangle_f\langle km|km'\rangle_i\langle km'\|\beta^{\dagger}\|0\rangle_i}{E_{\rm int}-E_f+mc^2+Q_{\beta\beta}/2}=\sum_{k=1,2}M^{(k)}_{\rm GT}\,, \tag{5}$$

where $k=1$ and 2 means one- and two-boson states, respectively. In order to calculate the ME of the transition operators $\beta^{\dagger}$ from (5), we have also to expand their QRPA expressions in terms of Γ boson operators. Thus, for the $\beta^{\dagger}$ operator one obtains:

$$\beta^{\dagger}_{\mu}=\sum_k[B^{(10)}_k\Gamma^{\dagger}_{1\mu}(k)+B^{(01)}_k\Gamma_{1-\mu}(k)(-1)^{1-\mu}]+\sum_{k_1,k_2}[B^{(20)}_{k_1k_2}(\Gamma^{\dagger}_1(k_1)\Gamma^{\dagger}_2(k_2))_{1\mu}+B^{(02)}_{k_1k_2}(\Gamma_1(k_1)\Gamma_2(k_2))_{1\mu}]$$
$$+\sum_{k_1,k_2}[B^{(12)}_{k_1k_2}(\Gamma^{\dagger}_1(k_1)\Gamma_2(k_2))_{1\mu}+B^{(21)}_{k_1k_2}(\Gamma^{\dagger}_2(k_2)\Gamma_1(k_1))_{1\mu}]\,. \tag{6}$$

The coefficients B were obtained by requiring that the ME of the $\beta^{\dagger}_{\mu}$ operators, in the boson basis, be identical to those corresponding to the rhs of Eq. (6) [20]. If, in the expansion (6), one keeps only the linear terms, one find the QRPA boson image of the fermion $\beta^{\dagger}_{\mu}$ operators. We have neglected in (6) the terms of the type $(\Gamma^{\dagger}_1\Gamma_2)_{1\mu}$ since they have a vanishing effect on the transition $0^+_i\to0^+_f$. We have also neglected all three boson state contributions for the following reasons: (i) if one takes into account only two-boson-state contributions to the HQRPA corrections, the sum rules which single beta strengths must fulfill very accurate in the frame of QRPA $\sum_l[\langle\beta^+\rangle^2_l-\langle\beta^-\rangle^2_l]=3(N-Z)$, are violated only about 1%. (ii) One can easily show [21] that the expression (6) of the β operators still commutes with the expansions of particle number and isospin operators taken in the same approximation as (6). If we add three boson state contributions, the communtation relations mentioned above no longer hold and we could have an additional source of spurious contributions during the computation. Accordingly with the previous results from boson expansion theory these three boson state contributions should be at most of the same order of magnitude as two boson state ones and hence do not modify qualitatively our conclusion about the influence that HQRPA corrections could have on GT ME. Our numerical results refer to six $\beta\beta$ transitions: ^{76}Ge$\to^{76}$Se, ^{110}Cd$\to^{110}$Pd, ^{116}Cd$\to^{116}$Sn, ^{128}Te$\to^{128}$Xe, ^{130}Te$\to^{130}$Xe, ^{136}Ba$\to^{136}$Xe. The $M^{2\nu}_{\rm GT}(g_{\rm pp})$ functions are plotted in Figs. 1–3: the upper curves (which drop steeply with respect to $g_{\rm pp}$) represent the pnQRPA result of the computation of these ME while the lower curves represent HQRPA corrections as functions of $g_{\rm pp}$. One can see that these corrections are functions more stable on this parameter. In spite of the small

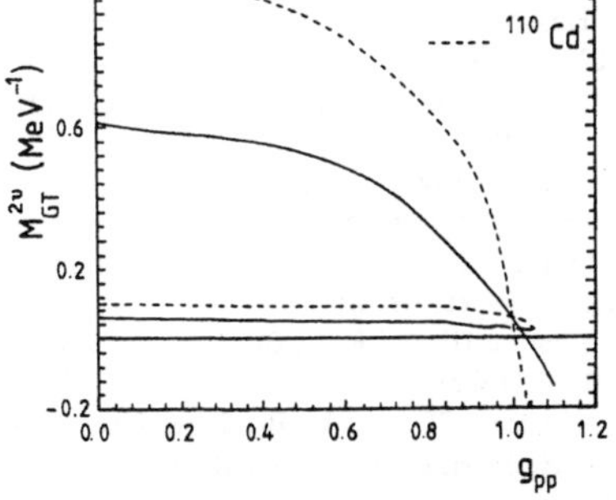

FIG. 1. The GT ME $M_{\rm GT}$ are plotted as functions of $g^{(1)}_{\rm pp}$ for ^{76}Ge (solid lines) and ^{110}Cd (dashed lines) cases. The upper curves (which drop steeply with respect to $g_{\rm pp}$) represent the pnQRPA result while the lower curves represent the HQRPA corrections. One can see that these corrections are functions more stable on this parameter and become important around $g_{\rm pp}=1$. The inclusion of them shifts the zero point of the $M^{2\nu}_{\rm GT}$ functions, computed in the frame of QRPA, to the region where the QRPA procedure collapses.

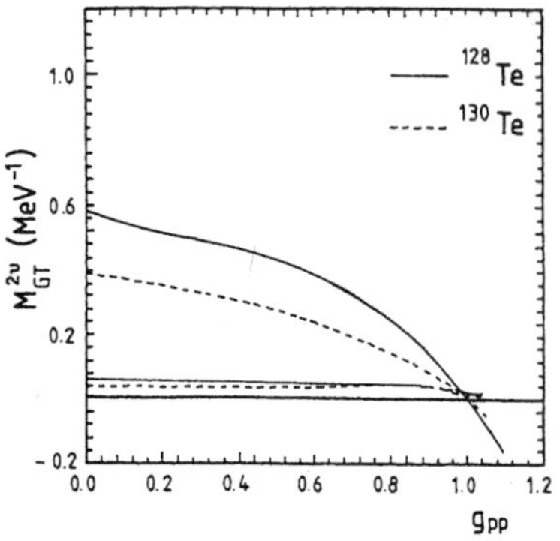

FIG. 2. The same as Fig. 1 but for ^{128}Te and ^{130}Te cases.

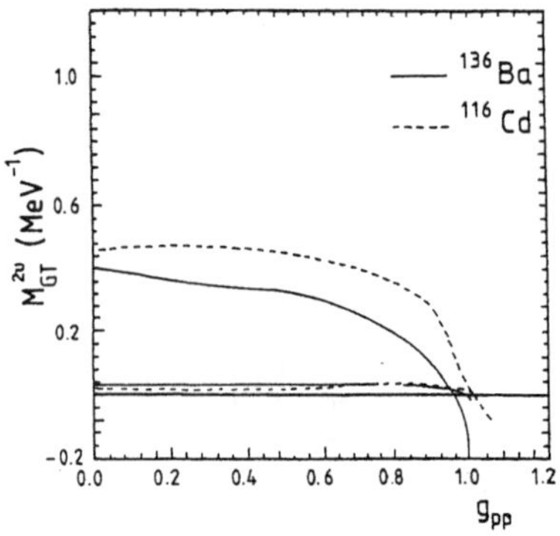

FIG. 3. The same as Fig. 1 but for ^{136}Ba and ^{116}Cd cases.

values of the HQRPA corrections, they become important by comparison with QRPA values just around $g_{pp}=1.0$, where the desired suppression for M_{GT} is acquired, for all the transitions which are studied. By including them one gains stability in the computation of the GT ME since the zero point of the function $M_{GT}^{2\nu}(g_{pp})$ is shifted to the region where QRPA procedure collapses. This happens because in our formalism are included not only proton-neutron interactions (as in the standard *pn*QRPA) but also proton-proton and neutron-neutron interactions (through HQRPA corrections). These particle-like interactions are also present in extended QRPA-type theories [19]. Concluding, a further investigation of the HQRPA corrections could be illuminating for reliable predictions of the GT ME. For example as an extension of our formalism one can take also into account three boson state contributions by using a projection procedure in order to avoid the spurious states contributions which violate the single beta sum rules by about 20%. Some work is in progress on this line. On the other hand, additional meaningful correlations could be included if an extended QRPA approach [19] would be employed where, beside the higher-order correlations to the QRPA wave functions, an improved HFB vacuum is also used.

One of us (S.S.) has deeply appreciated the generous amount of computer time provided by the Institute of Physics of the University M. Curie-Sklodowska. This work was supported in part by the Scientific Research Committee (Poland) under Contract No. 2 034291.01.

[1] H. Primakov and S. P. Rosen, Rep. Prog. Phys. **22**, 121 (1959).
[2] W. C. Haxton and G. J. Stephensen, Prog. Part. Nucl. Phys. **12**, 409 (1984).
[3] M. Doi, T. Kotani, and E. Takasugi, Prog. Theor. Phys. Suppl. **83**, 1 (1985).
[4] J. D. Vergados, Phys. Rep. **133**, 1 (1986).
[5] A. Faessler, Prog. Part. Nucl. Phys. **21**, 183 (1988).
[6] K. Muto and H. V. Klapdor, *Neutrinos*, edited by H. V. Klapdor (Springer, Berlin, 1988).
[7] F. T. Avignone III and R. L. Brodzinski, Prog. Part. Nucl. Phys. **21**, 99 (1988).
[8] F. T. Avignone III, R. L. Brodzinski, J. C. Evans, K. Hensley, H. S. Miley, and J. H. Reeves, Phys. Rev. C **34**, 666 (1986).
[9] D. O. Caldwell *et al.*, Phys. Rev. D **33**, 2737 (1986).
[10] T. Kirsten, in *Proceedings of the International Symposium on Nuclear Beta Decay and Neutrinos*, Osaka, Japan, 1986, edited by T. Kotani, H. Fujiri, and E. Takasugi (World Scientific, Singapore, 1986), p. 81.
[11] S. R. Elliott, A. A. Hahn, and M. K. Moe, Phys. Rev. Lett. **59**, 2020 (1987).
[12] J. A. Halbleib and R. A. Sorensen, Nucl. Phys. **A98**, 542 (1967).
[13] D. Cha, Phys. Rev. C **27**, 2269 (1987).
[14] P. Vogel and M. R. Zirnbauer, Phys. Rev. Lett. **57**, 3148 (1986).
[15] T. Tomoda and A. Faessler, Phys. Lett. B **199**, 475 (1987).
[16] K. Muto and H. V. Klapdor, Phys. Lett. B **201**, 420 (1988); K. Muto, E. Bender, and H. V. Klapdor, Z. Phys. A **334**, 177 (1989).
[17] O. Civitarese, A. Faessler, J. Suhonen, and X. R. Wu, Nucl. Phys. **A524**, 404 (1991); J. Phys. G **17**, 943 (1991); J. Suhonen, T. Taigel, and A. Faessler, Nucl. Phys. **A486**, 91 (1988).
[18] A. A. Raduta, A. Faessler, S. Stoica, and W. A. Kaminski, Phys. Lett. B **254**, 7 (1991); A. A. Raduta, A. Faessler, and S. Stoica, Nucl. Phys. **A534**, 149 (1991).
[19] K. Takayamagi, K. Shimizu, and A. Arima, Nucl. Phys. **A477**, 205 (1988); L. Zhao and A. Sustich, Michigan State University report, 1991 (unpublished).
[20] T. Marumori, M. Yamamura, and A. Tokunaga, Prog. Theor. Phys. **31**, 1009 (1964).
[21] A. Huffman, Phys. Rev. C **2**, 742 (1970).

Second-QRPA Calculations for Two Neutrino Double-beta Decay

S. Stoica

Department of Theoretical Physics, Institute of Atomic Physics 76900-Bucharest, P.O. Box MG-6, Romania

Abstract. A second-QRPA method suitable for charge-changing processes is developed. It goes beyond pnQRPA using an improved wave function but keeping unchanged the HFB vacuum. Applications of this method to the calculation of the $2\nu\beta\beta$ decay half-lives for transitions to the final ground as well as excited states, are further presented. It is shown that the higher-order corrections positively influence the pnQRPA results making the Gamow-Teller matrix elements less sensitive to the variation of the particle-particle strength parameter and, consequently, the calculations of them more reliable.

in Proc. "Double Beta Decay and Related Topics", Trento, Italy, April-May, 1995, eds. *H. V. Klapdor-Kleingrothaus and S. Stoica, Singapore: World Scientific* (1996) 382 – 398

1 Introduction

The proton-neutron Quasi Random Phase Approximation (pnQRPA) [1,2] has proved to be a very popular method to compute the reduce matrix elements of the $\beta^{\pm}$ transition operators [3-9].

The quenching of the strength in the allowed β^{+}/EC decay as well as the quenching of the Gamow-Teller (GT) matrix elements (ME) entering the double-beta decay half-lives have been succesfully explained in the framework of this method, in cases of the medium-heavy nuclei, when the particle-particle (PP) interaction, besides the particle-hole (PH) one, was taken into account in building the pnQRPA matrices [4-9].

However, the calculations are very sensitive to the PP interaction strength parameter (g_{pp}) and, morever, in the region around $g_{pp} \simeq 1.0$, there is an instability of the pnQRPA numerical procedure. Some attemps have been done to overcome this drawback:

- i) the projection tehnique has been used to restore particle number conservation both at the level of BCS quasiparticle (qp) and RPA phonon excitations by developing a projected pnQRPA method [11].

Double-beta decay matrix elements for ^{76}Ge

S. Stoica[a], H.V. Klapdor-Kleingrothaus[b]
a) *National Institute of Physics and Nuclear Engineering, P.O. Box MG-6, 76900-Bucharest, Romania*
b) *Max-Planck-Institut für Kernphysik, W-6900 Heidelberg, Germany*

June 6, 2000

Abstract

Double-beta decay matrix elements (ME) for ^{76}Ge are calculated with different quasi random phase approximation (QRPA)-based methods. First, the ME for the two-neutrino mode are computed using two choices for the single particle (s.p.) basis: i) $3-4\hbar\omega$ full shells and ii) $2-4\hbar\omega$ full shells. When calculated with the renormalized QRPA (RQRPA) and full-RQRPA their values are rather dependent on the size of the single particle basis used, while calculated with proton-neutron QRPA (pnQRPA) and second-QRPA approaches such a dependence was found to be small. The Ikeda sum rule was well fulfilled within pnQRPA for both choices of the s.p. basis and with a good approximation within second-QRPA, while the RQRPA and full-RQRPA methods give deviations up to 21%. Further, the ME for the neutrinoless mode are calculated with the pn-QRPA, RQRPA and full-RQRPA methods. They all give close results for the calculation with the smaller basis (i), while for the larger basis (ii), the results differ significantly either from one method to another or within the same method. Finally, using the most recent experimental limit for the $0\nu\beta\beta$ decay half-life of ^{76}Ge a critical discussion on the upper limits for the neutrino mass parameter obtained with different theoretical approaches is given.

Pacs: 21.60.Jz (Hartree-Fock and random-phase approximations).
Pacs: 23.40.-s (....).
Pacs: 23.40.Hc (Relation with nuclear matrix elements and nuclear structure)

---

to be Publ. in 2000

Physics of Atomic Nuclei, Vol. 61, No. 7, 1998, pp. 1211–1217. From Yadernaya Fizika, Vol. 61, No. 7, 1998, pp. 1311–1317.

NUCLEON AND NUCLEAR STRUCTURE
IN SEARCHING FOR NEW PHYSICS

QRPA and RQRPA Calculations of Neutrinoless Double-Beta Decay beyond the Point of Collapse*

G. Pantis[1] **, F. Šimkovic[2], and A. Faessler[3]

Abstract—Calculations of neutrinoless double beta decay ($0\nu\beta\beta$ decay) by the methods relying on standard quasiparticle random-phase approximation (QRPA) show a sensitivity of nuclear matrix elements to the particle–particle component of the residual interaction of the nuclear Hamiltonian. Extensions of the standard QRPA, such as renormalization of particle–particle forces that takes into account the Pauli exclusion principle (renormalized QRPA, abbreviated as RQRPA by convention), and inclusion of proton–neutron pairing (full renormalized QRPA, commonly known as full RQRPA) offer a stable solution in the physically acceptable region of particle–particle strengths. In particular, calculations of the $0\nu\beta\beta$ decay of ^{76}Ge, ^{100}Mo, ^{116}Cd, ^{128}Te, and ^{136}Xe show that, if a large model space is used, the sensitivity of nuclear matrix elements to the choice of particle–particle force is eliminated, in which case more accurate values on the effective neutrino mass can be deduced.

1. INTRODUCTION

It is well known (see, for example, [1] and references therein) that the transition

$$(A, Z) \longrightarrow (A, Z+2) + 2e^- \quad (1)$$

in which a nucleus (A, Z) emits of two electrons, going over to a nucleus $(A, Z+2)$ (that is, neutrinoless double-beta decay denoted by $0\nu\beta\beta$), violates the lepton number L_e by two units and is forbidden in the Standard Model. It has not hitherto been observed experimentally, and its possible detection would undoubtedly give a signal of new physics beyond the Standard Model. So far, experiments have set lower limits constraining severely the effective neutrino mass, the coupling constants of supersymmetric particles, and the parameters of right-handed currents. These limits are very important because it can be stated, among other things, that the neutrino will be a Majorana particle (that is, a particle coincident with its antiparticle) if it is found to be heavy.

The other neutrino mode, the two-neutrino double-beta decay ($2\nu\beta\beta$ decay) is allowed in the Standard Model. It produces two antineutrinos and two electrons in the final state; therefore, it does not violate the law of lepton-number conservation. In addition, it is a much simpler process from the particle-physics point of view and can be used as a basis for obtaining deeper insight into problems associated with nuclear structure—in particular, two-neutrino double-beta decay can furnish valuable information in cases where nuclear matrix elements that are relevant to the transition being considered cannot be estimated to a sufficiently high degree of precision. This is so for almost all-double beta transitions because nuclei undergoing $\beta\beta$ decay are generally away from closed shells.

Several methods have been proposed for performing nuclear-structure calculations associated with $\beta\beta$-decay transitions. One of the most advanced methods is based on the proton–neutron quasiparticle random-phase approximation [2–7] (*pn*-QRPA), referred thereon to as the standard QRPA. This method, relying on the quasiboson approximation, has already produced interesting results for 0ν and $2\nu\beta\beta$ decays. However, some of these calculations—in particular, those for $2\nu\beta\beta$ decay—revealed a high sensitivity of the nuclear matrix elements to the particle–particle strength of the nuclear Hamiltonian. In fact, it turned out that the matrix elements collapse for physical values of the particle–particle strength. Some other shortcomings, such as particle number nonconservation, the violation of the Pauli exclusion principle, and disregard of proton–neutron pairing, also became obvious. This revealed that the standard QRPA is not sufficiently accurate to make definitive predictions. To remedy this situation, several extensions and modifications have been proposed in recent years. Toivanen and Suhonen [8] introduced a proton–neutron renormalized QRPA, which takes into account the Pauli exclusion principle in an approximate way and which will be referred to below as RQRPA. Schwieger *et al.* [9] extended it by including proton–neutron pairing. The method extended in this way is commonly known as the full renormalized QRPA (full RQRPA), because it allows for proton–neutron pairing in addition to the renormalization taking the Pauli exclusion principle into account.

---

* This article was submitted by the authors in English.

[1] Department of Physics, University of Ioannina, PO Box 1186, GR-45110 Ioannina, Greece.
[2] Department of Nuclear Physics, Comenius University, Mlynská dolina, SK-84215 Bratislava, Slovakia.
[3] Institut für Theoretische Physik der Universität Tübingen, Auf der Morgenstelle 14, D-72076 Tübingen, Germany.
** e-mail: gpantis@cc.uoi.gr

J. Phys. G: Nucl. Part. Phys. **24** (1998) 2139–2178. Printed in the UK PII: S0954-3899(98)55035-9

TOPICAL REVIEW

Double beta decay

Amand Faessler†§ and Fedor Šimkovic‡||

† Institute für Theoretische Physik der Universität Tübingen Auf der Morgenstelle 14, D-72076 Tübingen, Germany

‡ Department of Nuclear Physics, Comenius University, Mlynská dol., pav. F1, SK-842 15 Bratislava, Slovakia

Received 21 April 1998

Abstract. We review the recent developments in the field of nuclear double beta decay, which is presently an important topic in both nuclear and particle physics. The mechanism of lepton number violation within the neutrinoless double beta decay ($0\nu\beta\beta$-decay) is discussed in the context of the problem of neutrino mixing and the R-parity violating supersymmetric extensions of the standard model. The problem of reliable determination of the nuclear matrix elements governing both two-neutrino and neutrinoless modes of the double beta decay is addressed. The validity of different approximation schemes in the considered nuclear structure studies is analysed and the role of the Pauli exclusion principle for a correct treatment of nuclear matrix elements is emphasized. The constraints on different lepton-number violating parameters such as effective electron neutrino mass, effective right-handed weak interaction parameters, effective Majoron coupling constant and R-parity violating SUSY parameters are derived from the best presently available experimental limits on the half-life of $0\nu\beta\beta$-decay.

1. Introduction

Double beta ($\beta\beta$) decay has attracted the attention of both experimentalists and theoreticians for a long period and remains of major importance both for particle and nuclear physics. It is a second-order process of weak interaction and there are many nuclear systems [1], which offer an opportunity to study it. $\beta\beta$-decay can be observed because the pairing force renders the even–even nuclei with an even number of protons and neutrons more stable than the odd–odd nuclei with broken pairs. Thus, the single beta decay transition from the even–even parent nucleus (A, Z) to the neighbouring odd–odd nucleus $(A, Z+1)$ is forbidden energetically and the $\beta\beta$-decay to the daughter nucleus $(A, Z+2)$ is the only possible decay channel.

There are different possible modes of the $\beta\beta$-decay, which differ from each other by the light particles accompanying the emission of two electrons. We distinguish the $\beta\beta$-decay modes with and without lepton-number violation.

The two-neutrino double beta decay ($2\nu\beta\beta$-decay), which involves the emission of two electrons and two antineutrinos,

$$(A, Z) \rightarrow (A, Z+2) + 2e^- + 2\overline{\nu}_e \qquad (1)$$

is a process fully consistent with the standard model (SM) of electroweak interaction formulated by Glashow [2], Weinberg [3] and Salam [4]. This decay mode with obvious

§ E-mail address: amand.faessler@uni-tuebingen.de

|| E-mail address: simkovic@fmph.uniba.sk

ЯДЕРНАЯ ФИЗИКА, 1999, том 62, № 2, с.

ЯДРА

A FIELD THEORY APPROACH TO TWO-NEUTRINO DOUBLE BETA DECAY

© 1999 г. F. Šimkovic[1)], G. Pantis[2)]

Received May 20, 1997

A detailed analysis of the two-neutrino-double-beta-decay amplitude has been performed by using a field theory approach. It is shown that the summation over the intermediate nuclear states in the present two-neutrino-double-beta-decay studies corresponds to a summation over a class of meson-exchange diagrams. We offer some arguments which suggest that the two-nucleon mechanism considered at present does not provide the whole contribution to the two-neutrino-double-beta-decay amplitude and thus we propose additional new mechanisms for this process.

1. INTRODUCTION

Nuclear double beta decay is a subject of the current intensive activity, both experimentally and theoretically. The neutrinoless mode of this process ($0\nu 2\beta$-decay) is related to physics beyond the standard model and requires the existence of the lepton number violating massive Majorana neutrinos [1–5]. Up to now, experimentally only lower half-life limits for $0\nu 2\beta$-decay have been reported in the literature. Growing interest is also paid to the two-neutrino mode ($2\nu 2\beta$-decay). It is the rarest process observed so far in nature and this mode being independent of the neutrino properties offers a sensitive test of nuclear structure calculations.

The usual strategy has been first to try to reproduce the observed $2\nu 2\beta$-decay half-life times in order to gain confidence in the calculated $0\nu 2\beta$-decay nuclear matrix elements. In most calculations of $2\nu 2\beta$-decay only the Gamow–Teller nuclear matrix element is considered,

$$M_{\rm GT} = \sum_n \frac{\langle 0_f^+|A_k|1_n^+\rangle\langle 1_n^+|A_k|0_i^+\rangle}{E_n - E_i + \Delta}, \qquad (1)$$

where $|0_i^+\rangle$, $|0_f^+\rangle$ and $|1_n^+\rangle$ are, respectively, the wave functions of the initial, final and intermediate nuclei with corresponding energies E_i, E_f and E_n. Δ denotes the average energy $\Delta = \overline{(E_i - E_f)}/2$. A_k is the Gamow–Teller transition operator $A_k = \sum_i \tau_i^+(\sigma_i)_k$, $k = 1, 2, 3$.

The proton–neutron quasiparticle random phase approximation (*pn*-QRPA) is the nuclear structure method most widely used to calculate $M_{\rm GT}$. However, the results are extremely sensitive to the details of the nuclear hamiltonian, in particular, to the factor g_{pp}, which is introduced to renormalize the particle–particle interaction strength [6–9]. The range of the magnitude of g_{pp} consistent with the calculation of β^+ decay is not broad, nevertheless $M_{\rm GT}$ as a function of g_{pp} crosses zero within this range. Thus this sensitivity of *pn*-QRPA to g_{pp} renters it difficult to make definite rate predictions. Several modifications of *pn*-QRPA have been proposed towards overcoming this difficulty, e. g., higher-order RPA corrections [10], nuclear deformation [11] and particle number projection [12, 13], however with a limited success since none of these modifications inhibits the matrix element $M_{\rm GT}$ to pass through zero near the natural value of $g_{pp} = 1$.

Recent calculations of Toivanen and Suhonen [14] within the renormalized *pn*-QRPA (*pn*-RQRPA) and Schwieger, Šimkovic and Faessler [15] within the *pn*-RQRPA with proton–neutron pairing (full-RQRPA) suggest that the violation of the Pauli principle could be the origin of this peculiar behavior of the $M_{\rm GT}$ as a function of g_{pp}. This argument is supported further by a large Hilbert space full-RQRPA calculation of the neutrinoless double beta decay [16]. It has been also found there that the RQRPA approximation offers advantages over the QRPA. However, the RQRPA has still a few shortcomings which we still do not know how they affect the evaluation of $M_{\rm GT}$. To mention only a few of these we refer to the fact that the Pauli exclusion principle is considered only in an approximate way, the Ikeda sum rule is violated [17], and the two sets of intermediate nuclear states generated from the initial and final nuclei are different. The common origin of these shortcomings is the particle number non-conservation. Therefore it is important to develop alternative approaches for the calculation of the nuclear many-body Green function, which might produce more reliable results than that of the renormalized QRPA.

In the present work we will adopt this point of view and we will try to analyze the $2\nu 2\beta$-decay amplitude in the field-theory approach. Our aim will be to show that the calculation of the many-body Green function in $M_{\rm GT}$ of equation (1) is equivalent to the calculation of

1) Department of Nuclear Physics, Comenius University, Bratislava, Slovakia; e-mail: simkovic@fmph.uniba.sk.
2) Theoretical Physics Section, University of Ioannina, Greece; e-mail: gpantis@cc.uoi.gr

2.2.3 Shell Model Calculations

VOLUME 77, NUMBER 10 PHYSICAL REVIEW LETTERS 2 SEPTEMBER 1996

Shell Model Studies of the Double Beta Decays of ^{76}Ge, ^{82}Se, and ^{136}Xe

E. Caurier,[1] F. Nowacki,[1] A. Poves,[2] and J. Retamosa[2]
[1]Centre de Recherches Nucleaires/Université Louis Pasteur, BP20 F-67037, Strasbourg Cedex-2, France
[2]Departamento de Física Teórica C-XI, Universidad Autónoma de Madrid, E-28049 Madrid, Spain
(Received 10 January 1996)

The double beta decays of ^{76}Ge, ^{82}Se, and ^{136}Xe are studied in very large shell model spaces. The dimensions of the shell model spaces used in these calculations reach $O(10^8)$. The two neutrino matrix elements obtained are in good agreement with the available experimental data. Assuming that the mass of the neutrino is smaller than 1 eV, we get the following upper bounds to the half-lives for the neutrinoless decays: $T_{1/2}^{(0\nu)}(\mathrm{Ge}) > 1.85 \times 10^{25}$ yr, $T_{1/2}^{(0\nu)}(\mathrm{Se}) > 2.36 \times 10^{24}$ yr, and $T_{1/2}^{(0\nu)}(\mathrm{Xe}) > 1.21 \times 10^{25}$ yr. [S0031-9007(96)00984-2]

PACS numbers: 23.40.Bw, 21.60.Cs, 27.50.+e, 27.60.+j

The importance of the nuclear double beta decay ($\beta\beta$) is well established. The two neutrino mode ($\beta\beta_{(2\nu)}$), very sensitive to the nuclear correlations, provides a severe test of the nuclear wave functions. The neutrinoless mode ($\beta\beta_{(0\nu)}$) is one of the best probes for the physics beyond the standard model. It is particularly suitable to explore intrinsic properties of the neutrino such as its mass and the existence of right-handed weak currents [1–4]. Reasonable values of these lepton-number-violating parameters can be extracted from the experiment provided that accurate nuclear matrix elements are used.

Before this work, large shell model (SM) calculations of the $\beta\beta$ decay had been only possible in ^{48}Ca [1,5]. In order to describe heavier nuclei different approximations had to be devised. Haxton, Stephenson, and Strottman implemented a weak coupling approximation in order to calculate the decays of ^{76}Ge, ^{82}Se, and ^{128,130}Te [1]. They also used the closure approximation to circumvent the calculation of the 1^+ states in the intermediate nuclei. The uncertainty in their results is difficult to estimate because the choice of the energy denominator in the two neutrino matrix element is not unique. Using a statistical method to determine the energy denominators and setting $g_a = 1$, they gave reasonable values for the lifetimes in all cases, except for the tellurium isotopes.

The quasiparticle random-phase approximation (QRPA) has also been used to study the decay of medium and heavy nuclei [6]. A shortcoming of this approach [7,8] is that the matrix elements of the Gamow-Teller and double Gamow-Teller operators are very sensitive to the particle-particle interaction. The strength of this interaction (g_{pp}) is treated as a parameter and fitted to the available $\beta_\pm$ data. The introduction of g_{pp} as a phenomenological factor makes it possible to reproduce the observed half-lives. However, the predictive power of the QRPA approach is limited because of the large variation of the relevant matrix element $M_{\mathrm{GT}}^{(2\nu)}$ in the physical window for g_{pp}. The zero neutrino mode is less sensitive to the correlations; still, different predictions for $M_{\mathrm{GT}}^{(0\nu)}$ may differ by a factor of 3 [9–13].

Very recently the shell model Monte Carlo method (SMMC) [14] has been applied to the description of the two neutrino channel. For the decay $^{48}\mathrm{Ca} \rightarrow {}^{48}\mathrm{Ti}$ the SMMC matrix element 0.15 ± 0.07 MeV^{-1} can be compared to the result of the exact diagonalization using the same interaction, 0.08 MeV^{-1}. For the decay of ^{76}Ge a matrix element 0.13 ± 0.05 MeV^{-1} is obtained.

In this Letter we study the $\beta\beta$ decays of ^{76}Ge, ^{82}Se, and ^{136}Xe in the shell model framework. For ^{76}Ge and ^{82}Se the Schrödinger equation is solved in the valence space spanned by the orbits $p_{3/2}$, $f_{5/2}$, $p_{1/2}$, and $g_{9/2}$. For ^{136}Xe the valence space contains the $d_{5/2}$, $s_{1/2}$, $g_{7/2}$, $d_{3/2}$, and $h_{11/2}$ shells. For each valence space we define a truncation scheme labeled by t: the number of particles that are allowed to jump from the pf subshells to the $g_{9/2}$ orbital ($A = 76, 82$) or the number of particles that can be excited from the lower $g_{7/2}$ and $d_{5/2}$ subshells to the remaining three orbitals ($A = 136$). Later on we shall analyze the convergence of the relevant matrix elements as a function of the truncation. We can perform full calculations for the $A = 82$ (maximum dimension 70 757 366) and $A = 136$ nuclei. In $A = 76$ we are limited to $t = 4$ (Ge and Se) and $t = 5$ (As). These calculations have been made possible by the SM code ANTOINE [15].

The effective interactions used in the calculations are realistic G matrices whose monopole part has been phenomenologically adjusted [16]. For $A = 76, 82$ we use a G matrix calculated by Kuo [17]. To fix the interaction the monopole parameters are fitted to 60 energy levels of the Ni isotopes and $N = 50$ isotones [18]. The main monopole changes the amount to weaken the interaction among the pf orbits and the $g_{9/2}$ shell. For ^{136}Xe, the interaction is the G matrix obtained from the Bonn potential [19] with monopole modifications in order to reproduce the excitation energies of 150 states of the $N = 82$ isotones [18]. In both cases the χ^2 value is about 250 keV. Table I shows our predictions for the 2^+ energies around $A = 82$. The electromagnetic properties are also well reproduced by our calculation.

 0031-9007/96/77(10)/1954(4)$10.00

TABLE I. Selected $E_x(2^+)$.

		^{80}Ge	^{82}Ge	^{82}Se	^{82}Kr	^{84}Kr	^{84}Se
$E_x(2^+)$	Th	0.620	1.578	0.612	0.685	0.773	1.814
	Exp	0.659	1.348	0.655	0.776	0.882	1.455

The weak processes in the nucleus are described by the effective Hamiltonian proposed by Doi *et al.* [2] that consists of V and A currents and that is compatible with $SU(2)_L \otimes SU(2)_R \otimes U(1)$ grand unification models:

$$H_w = \frac{G}{\sqrt{2}}[j_L^\mu(J_{L\mu} + \chi J_{R\mu})^+ + j_R^\mu(\eta J_{L\mu} + \lambda J_{R\mu})^+] + \text{H.c.} \quad (1)$$

In addition to the right-left and right-right coupling constants, the neutrino mass (implicit in the leptonic currents) also breaks the maximal parity violation of the standard theory. This Hamiltonian leads us to a scenario without maximal parity violation and without lepton number conservation. The neutrinoless double beta decay is mediated by (virtual) massive Majorana neutrinos. Moreover, the description of the two neutrino mode is essentially the same as in the standard model. This is so because the relevant contributions come from the standard left-left coupling. In the hadronic currents, all the terms up to order v/c are included [20].

Results for the $(\beta\beta)_{2\nu}$ *decays.*—The partial 2ν half-life can be approximated as

$$[T_{1/2}^{(2\nu)}(0^+ \to 0^+)]^{-1} = G|M_{\text{GT}}^{(2\nu)}|^2, \quad (2)$$

where G is an integral kinematical factor and $M_{\text{GT}}^{(2\nu)}$ is the energy weighted double Gamow-Teller matrix element. We adopt the effective value $g_A = 1$ that takes into account the reduction of the Gamow-Teller strength at low energies due to correlations outside the valence space. The operator in the definition of the matrix element is just $\boldsymbol{\sigma} \cdot \boldsymbol{\tau}$; the constants g are included in the integral kinematic factor. The precise definition of this matrix element as well as the algorithm that we use to calculate it can be found in Ref. [5]. By means of this algorithm we can get a reliable approximation to the Gamow-Teller strength in the intermediate nucleus.

Partial $(\beta\beta)_{2\nu}$ lifetimes have been measured for several emitters. We summarize our results in Table II. Two sets of matrix elements are listed. The first one is obtained using the theoretical energies of the 1^+ states. In the second, the spectrum of these states is globally shifted in order to place the first 1^+ at its experimental energy. As it can be seen in Table II the matrix elements increase by 20% when the experimental energies are used. The agreement with the experimental data is, in both cases, reasonably good. In the $A = 76, 82$ region the matrix element increases very slowly as the valence space is enlarged. Consequently, the half-life decreases and there is a factor of 2 between the $t = 2$ and the final predictions. This behavior is different from that found in the decay of ^{136}Xe, where the matrix element is almost constant.

Our closure matrix elements $[(M_{\text{GT}}^{2\nu})_c]$ are very different from those of Ref. [1]; 0.68 compared to 2.56 (^{76}Ge) and 0.74 compared to 1.876 (^{82}Se). These discrepancies also lead to quite different effective 1^+ centroids ($\overline{E}$). Table II lists also the total Gamow-Teller strengths $S_{+/-}$. For the ^{76}Ge decay the SM result, 0.14, is very close to the SMMC extrapolation, 0.13 ± 0.05. Nevertheless, other relevant quantities ($(M_{\text{GT}}^{2\nu})_c$, S_+, $\overline{E}$) are quite different. Notice, however, that, although the valence space is the same, the effective interaction is different.

There are two sources of uncertainty in our calculation: the first one is the absence in the valence spaces of some of the spin orbit partners, leading to total Gamow-Teller strengths that are smaller than the $3(N - Z)$ values ($S_-^{(\text{Ge})} = 17.14$, $S_-^{(\text{Se})} = 21.66$, and $S_-^{(\text{Xe})} = 52.30$

TABLE II. Two neutrino matrix elements and half-lives. $M_{\text{GT}}^{(2\nu)}$ in MeV^{-1} and $T_{1/2}$ in years. The experimental values are taken from Refs. [24–26].

		$M_{\text{GT}}^{(2\nu)}$			$T_{1/2}^{2\nu}$					
Decay	t	ΔE(exp)	ΔE(th)	$(M_{\text{GT}}^{(2\nu)})_{\text{exp}}$	ΔE(exp)	ΔE(th)	$(T_{1/2}^{2\nu})_{\text{exp}}$	$(M_{\text{GT}}^{(2\nu)})_c$	S_-	S_+
	0	0.000	0.000					0.000	16.73	0.000
^{76}Ge → ^{76}Se	2	0.112	0.088		5.678×10^{21}	9.197×10^{21}		0.465	17.13	0.146
	4	0.180	0.140		2.198×10^{21}	3.634×10^{21}		0.676	17.14	0.258
	Full			0.22			1.80×10^{21}			
	0	0.000	0.000					0.000	21.66	0.000
^{82}Se → ^{82}Kr	2	0.128	0.102		1.312×10^{20}	2.065×10^{20}		0.483	21.61	0.121
	4	0.198	0.155		5.482×10^{19}	8.946×10^{19}		0.745	21.56	0.209
	Full	0.208	0.164	0.14	4.968×10^{19}	7.991×10^{19}	1.08×10^{20}	0.799	21.55	0.226
	0	0.026	0.028		2.487×10^{21}	2.455×10^{21}		0.106	52.75	0.004
^{136}Xe → ^{136}Ba	2	0.036	0.039		1.485×10^{21}	1.265×10^{21}		0.178	52.37	0.007
	4	0.032	0.035		1.879×10^{21}	1.571×10^{21}		0.146	52.30	0.008
	Full	0.031	0.034	<0.06	2.003×10^{21}	1.665×10^{21}	5.50×10^{20}	0.143	52.30	0.008

VOLUME 77, NUMBER 10 PHYSICAL REVIEW LETTERS 2 SEPTEMBER 1996

compared to 36, 42, and 84, respectively). However, the influence of the excluded orbits in $M_{\rm GT}^{(2\nu)}$ is expected to be small because (i) the missing one-particle–one-hole (1p-1h) intermediate states are outside the energy window relevant for the matrix element (~6 MeV) and (ii) the 2p-2h correlations in the ground states are hindered by the large effective energy denominators (≃20 MeV) involved. We have calculated the $M_{\rm GT}^{(2\nu)}$ for the decay of ^{60}Ni in a model space that includes the $1f7/2$ and $1g7/2$ orbits and the effect amounts to less than 30%. The second one is related to possible defects in the effective interaction. A good test of the combined effect of truncation and interaction is provided by the comparison of the predicted $\beta_\pm$ strength functions with the experimental ones, measured in (p,n) and (n,p) reactions. Unfortunately, only the former is available in some cases. Our predictions for the Gamow-Teller strength functions compare fairly well with the (p,n) results of Ref. [21], in the energy range relevant for the 2ν decay. As an example, the calculated B(GT)'s to the first 1^+ states in ^{76}As and ^{82}Br are 0.15 and 0.28, compared to the measured 0.15 and 0.34. These two states contribute more than 50% to the 2ν matrix element. Another estimation of the quality of the interaction is given by its predictions for the binding energies. Recent SMMC calculations show that they are in reasonable agreement with the experimental results [22]. All put together, it would be safe to affect our $M_{\rm GT}^{(2\nu)}$ values of a "theoretical" error of 50%.

Results for the $(\beta\beta)_{0\nu}$ decays.—In the closure approximation the half-life of the $0^+ \to 0^+$ decay can be written as

$$[T_{1/2}^{(0\nu)}(0^+ \to 0^+)]^{-1} = |M_{\rm GT}^{(0\nu)}|^2\left\{C_{mm}\left(\frac{\langle m_\nu\rangle}{m_e}\right)^2 + C_{\lambda\lambda}\langle\lambda\rangle^2 + C_{\eta\eta}\langle\eta\rangle^2 + C_{\lambda m}\langle\lambda\rangle\frac{\langle m_\nu\rangle}{m_e}\cos\psi_1 + +C_{\eta m}\langle\eta\rangle\frac{\langle m_\nu\rangle}{m_e}\cos\psi_2 + C_{\lambda\eta}\langle\lambda\rangle\langle\eta\rangle\cos(\psi_1-\psi_2)\right\}, \tag{3}$$

where $\langle m_\nu\rangle$, $\langle\lambda\rangle$, and $\langle\eta\rangle$ are the effective lepton violating parameters, $\psi_{1(2)}$ are the *CP* phases, and the C_{xy} coefficients are linear combinations of the nine matrix elements and nine integral factors. A clear and comprehensive definition of them all can be found in Ref. [2].

The nine matrix elements shown in Table III are calculated in the light neutrino approximation (see Doi [2]) and using $g_A/g_V = 1.25$. Since this mode has not been observed yet, we assume a half-life of 10^{25} yr, which is very close to the expected experimental limits [23], to obtain the upper bounds to the three lepton violating parameters $\langle m_\nu\rangle$, $\langle\lambda\rangle$ and $\langle\eta\rangle$. The results are compiled in Table IV. For $\langle m_\nu\rangle < 1$ eV we find the following lower bounds: $T_{1/2}^{(0\nu)}$(Ge) $> 1.85\times10^{25}$ yr, $T_{1/2}^{(0\nu)}$(Se) $> 2.36\times10^{24}$ yr, and $T_{1/2}^{(0\nu)}$(Xe) $> 1.21\times10^{25}$ yr. Although the result for ^{76}Ge has been obtained in a $t=4$ calculation, we expect the matrix element and therefore the half-life to be close to convergence, as it is the case in ^{82}Se. Among the three nuclei studied, ^{82}Se is the best candidate for the detection of the zero neutrino mode as it has the smallest half-life and the largest ratio $T_{1/2}^{(2\nu)}/T_{1/2}^{(0\nu)}$.

Table V compares the $M_{\rm GT}^{(0\nu)}$ values predicted by several authors. Notice that there is at least a factor of 2 among the different predictions. Our matrix elements are similar to those of Refs. [11,13].

TABLE III. Nuclear matrix elements (ME) for the $(\beta\beta)_{(0\nu)}$ mode. $\chi_{\rm GT}^{\omega} = 2 - \chi_{\rm GT}^{\prime}$, $\chi_F^{\omega} = 2\chi_F - \chi_F^{\prime}$.

ME		$t=0$	$t=2$	$t=4$	Full
$M_{\rm GT}^{(0\nu)}$	^{76}Ge	0.721	1.294	1.568	
	^{82}Se	0.505	1.340	1.846	1.970
	^{136}Xe	0.484	0.630	0.649	0.651
χ_F	^{76}Ge	−0.068	−0.098	−0.106	
	^{82}Se	−0.101	−0.107	−0.107	−0.108
	^{136}Xe	−0.172	−0.156	−0.157	−0.158
$\chi_{\rm GT}^{\prime}$	^{76}Ge	1.074	1.107	1.115	
	^{82}Se	1.069	1.114	1.119	1.120
	^{136}Xe	1.103	1.106	1.099	1.097
$\chi_{F'}$	^{76}Ge	−0.060	−0.102	−0.109	
	^{82}Se	−0.104	−0.112	−0.112	−0.112
	^{136}Xe	−0.184	−0.166	−0.167	−0.167
χ_T	^{76}Ge	0.186	0.043	0.017	
	^{82}Se	0.156	0.049	0.031	0.028
	^{136}Xe	0.039	−0.006	−0.031	−0.031
χ_P	^{76}Ge	−1.435	−0.832	−0.544	
	^{82}Se	1.710	0.852	0.574	0.494
	^{136}Xe	0.898	0.411	0.280	0.256
χ_R	^{76}Ge	0.761	0.707	0.684	
	^{82}Se	0.873	0.706	0.683	0.680
	^{136}Xe	0.780	0.872	0.942	0.955

TABLE IV. Shell model predictions for the upper bounds to the three lepton violating parameters (LPV): the effective mass $\langle m_\nu\rangle$ and the coupling constants $\langle\lambda\rangle$ and $\langle\eta\rangle$ for $T_{1/2} > 10^{25}$ yr.

		$t=0$	$t=2$	$t=4$	Full
^{76}Ge	$\langle m_\nu\rangle$ (eV)	2.98	1.60	1.32	
	$\langle\eta\rangle\ 10^8$	2.78	1.83	1.65	
	$\langle\lambda\rangle\ 10^6$	4.31	2.65	2.24	
^{82}Se	$\langle m_\nu\rangle$ (eV)	1.93	0.72	0.52	0.49
	$\langle\eta\rangle\ 10^8$	3.03	1.22	0.86	0.79
	$\langle\lambda\rangle\ 10^6$	1.93	0.82	0.61	0.57
^{136}Xe	$\langle m_\nu\rangle$ (eV)	1.47	1.14	1.10	1.10
	$\langle\eta\rangle\ 10^8$	3.10	1.77	1.53	1.49
	$\langle\lambda\rangle\ 10^6$	1.88	1.53	1.50	1.49

TABLE V. SM versus QRPA $M_{GT}^{(0\nu)}$ matrix elements.

	This work	Haxton [1]	Tomoda [8]	Muto [10]	Engel [11]	Staut [12]	Pantis [13]
^{76}Ge	1.57	4.18	3.36	3.01	1.75	10.91	1.85
^{82}Se	1.97	3.45	3.06	2.85	1.29		1.15
^{136}Xe	0.65		1.12		1.25		1.35

In summary, we have presented the first large scale SM calculations of the $\beta\beta$ decays of ^{76}Ge, ^{82}Se, and ^{136}Xe. The reasonable agreement between theory and experiment for the spectroscopy of the regions around these $\beta\beta$ emitters, as well as for the $\beta\beta_{(2\nu)}$ half-lives, strongly support our predictions for the zero neutrino mode.

We thank M. Hjorth-Jensen and T.T. Kuo for making their G matrix available to us. We thank D. Dean, P. Vogel, and A. Zuker for their help. This work has been partly supported by the IN2P3 (France)-CICYT(Spain) agreements and by a grant of DGICYT (Spain) PB93-263.

[1] W.C. Haxton and G.J. Stephenson, Jr., Prog. Part. Nucl. Phys. **12**, 409 (1984).
[2] M. Doi, T. Kotani, and E. Takasugi, Prog. Theor. Phys. Suppl. **83**, 1 (1985).
[3] F. Boehm and P. Vogel, *Physics of Massive Neutrinos* (Cambridge University Press, Cambridge, 1992).
[4] M. Moe and P. Vogel, Annu. Rev. Nucl. Part. Sci. **44**, 247 (1994).
[5] E. Caurier, A.P. Zuker, A. Poves, and G. Martínez-Pinedo, Phys. Rev. C **50**, 225 (1994); J. Retamosa, E. Caurier, and F. Nowacki, Phys. Rev. C **51**, 371 (1995); J. Engel, W.C. Haxton, and P. Vogel, Phys. Rev. C **46**, R2153 (1992).
[6] P. Vogel and M.R. Zimbauer, Phys. Rev. Lett. **57**, 3148 (1986).
[7] O. Civitarese, A. Faessler, and T. Tomoda, Phys. Lett. B **194**, 11 (1987).
[8] K. Muto and H.V. Klapdor, Phys. Lett. B **201**, 420 (1988); K. Muto, E. Bender, and H.V. Klapdor, Z. Phys. A **334**, 177 (1989).
[9] T. Tomoda and A. Faessler, Phys. Lett. B **199**, 475 (1987).
[10] K. Muto, E. Bender, and H.V. Klapdor, Z. Phys. A **334**, 187 (1989).
[11] J. Engel, P. Vogel, JI Xiandong, and S. Pittel, Phys. Lett. B **225**, 5 (1989).
[12] A. Staudt, T.T.S. Kuo, and H.V. Klapdor, Phys. Lett. B **242**, 17 (1990).
[13] G. Pantis, F. Simkovic, J.D. Vergados, and Amand Faessler, Phys. Rev. C **53**, 695 (1996).
[14] P.B. Radha *et al.*, Phys. Rev. Lett. **76**, 2642 (1996).
[15] E. Caurier, SM code ANTOINE (1989).
[16] A. Abzouzi, E. Caurier, and A.P. Zuker, Phys. Rev. Lett. **66**, 1134 (1991).
[17] T.T. Kuo (private communication).
[18] F. Nowacki, Ph.D. thesis, ULP Strasbourg, 1995 (unpublished).
[19] M. Hjorth-Jensen, T.T.S. Kuo, and E. Osnes, Phys. Rep. **261**, 125 (1995).
[20] T. Tomoda, A. Faessler, K.W. Schmid, and F. Grummer, Nucl. Phys. **A452**, 591 (1986).
[21] R. Madey *et al.*, Phys. Rev. C **40**, 540 (1989).
[22] D. Dean (private communication).
[23] NEMO 3 Proposal, LAL 94-29 (1994).
[24] H.V. Klapdor-Kleingrothaus, in *Proceedings of the International Workshop on Double Beta Decay and Related Topics, Trento, 1995,* edited by H.V. Klapdor-Kleingrothaus and S. Stoica (World Scientific, Singapore, 1995).
[25] S.R. Elliot *et al.*, Phys. Rev. C **46**, 1535 (1992).
[26] G. Gervasic *et al.*, in *Proceedings of the International Workshop on Double Beta Decay, European Centre for Theoretical Studies, Trento, Italy 1995,* edited by H.V. Klapdor-Kleingrothaus and S. Estoica (World Scientific, Singapore, 1995).

PHYSICAL REVIEW C VOLUME 42, NUMBER 3 SEPTEMBER 1990

Shell-model calculation for two-neutrino double beta decay of ^{48}Ca

Liang Zhao and B. Alex Brown
National Superconducting Cyclotron Laboratory and Department of Physics and Astronomy, Michigan State University, East Lansing, Michigan 48824

W. A. Richter
Physics Department, University of Stellenbosch, Stellenbosch 7600, South Africa
(Received 12 February 1990)

The $2\nu\beta\beta$ decay matrix element of ^{48}Ca has been studied in a large basis shell-model space. The theoretical and experimental β^- and β^+ spectra and their relation to $2\nu\beta\beta$ are discussed. A new empirical effective interaction is found to give the best agreement to β^- and β^+ spectra with the effective Gamow-Teller operator $\tilde{\sigma}t=0.77\sigma t$. The predicted $T_{1/2}=1.9\times10^{19}$ yr differs by a factor of 2 from the present experimental limit of $T_{1/2}>3.6\times10^{19}$ yr. We also compare the matrix elements obtained at various levels of the truncation in the shell-model space.

Double beta ($\beta\beta$) decay is a rare transition between two nuclei of the same mass number having a change of two units of nuclear charge. In cases of interest, ordinary single beta decay is forbidden because of the energy conservation or angular momentum mismatch. There are two modes of double beta decay. One involving the emission of two antineutrinos and two electrons (2ν mode), occurs in second order of the standard weak interaction theory and is independent of a possible small neutrino mass. The other involving no neutrinos and two electrons (0ν mode) violates the lepton number conservation and requires the neutrino to have a nonzero mass.[1-3,7] Analysis of the experimental result to determine the character of the neutrino in $\beta\beta$ decay strongly depends on the precise calculation of the nuclear matrix elements. In particular, agreement between experiment and theory for the standard 2ν mode is one of the prerequisites for a reliable interpretation of the more exotic 0ν mode. In this paper, we study the $2\nu\beta\beta$ decay of ^{48}Ca which has the largest double beta decay Q value of any nucleus.

There are several difficulties with previous shell-model calculations for the $2\nu\beta\beta$ decay of ^{48}Ca. In cases where intermediate states in ^{48}Sc were considered explicitly the fp shell-model space was highly truncated,[4-7] in other cases where the truncation was less severe the intermediate states were not calculated and the closure approximation was used instead.[1,7,8] Also the effective interactions used were not always well tested with regard to the nuclear spectra. In a more recent calculation (Ref. 9) a new method was used to implicitly take into account the spectrum of the intermediate 1^+ states exactly. However, we will emphasize below the importance of testing the interactions with respect to the explicit intermediate spectrum.

The quasiparticle random phase approximation (QRPA) is another widely used method to calculate β and $\beta\beta$ decay matrix elements.[10-12] But there are many shortcomings with the QRPA. The equations violate the particle number conservation, and some approximations must be made to match the excited states of odd-odd nuclei based on the different initial and final even-even nuclei ground states. There are also uncertainties in the interactions used, and the QRPA equations become unstable with the interaction strengths just above realistic values.

In this paper, we calculate the nuclear matrix elements for $2\nu\beta\beta$ decay of ^{48}Ca and the related β^- and β^+ decay in the $0f_{7/2},1p_{3/2},0f_{5/2},1p_{1/2}(fp)$ shell-model space with a much larger basis than previously used and with a new and more reliable effective interaction. The truncation in the fp shell is defined by the set of partitions $f_{7/2}^{8-n}(p_{3/2}f_{5/2}p_{1/2})^n$. In this work, the partitions assumed for ^{48}Ca$(0^+,T=4)$, ^{48}Sc$(1^+,T=3)$, and ^{48}Ti$(0^+,T=2)$ are $(n\le4)$, $(n\le5)$, and $(n\le4)$, respectively. The $n\le n_{max}$ means that $n=0,\ldots,n_{max}$ are allowed. The corresponding J-scheme dimensions are 133, 5599, and 3613, respectively. This is an order of magnitude larger basis than has been used in previous calculations. Our calculations were carried out with the shell-model code OXBASH (Ref. 13) on a VAX computer. The most complete fp shell calculation should be based on the full-basis space $(n\le8)$, but at present this is impossible because of the large dimensions involved. For example, the J-scheme dimension for the ^{48}Ti ground state is 10 872 in the full-basis space. It is at the edge of our current computer capability. In later discussions, we will argue that our truncation is a good approximation to the full-basis space.

The model space for the intermediate nucleus (^{48}Sc) should include all states reachable by a one-body operator from the initial and final nuclei. Thus for our initial (^{48}Ca) and final (^{48}Ti) states which have $n\le4$, we include $n\le5$ configurations in the intermediate system. Then the B(GT) (where GT denotes Gamow-Teller) from the ^{48}Ti or ^{48}Ca ground states satisfy the sum rule,

$$\sum B(\mathrm{GT}^-)-\sum B(\mathrm{GT}^+)=3(N-Z)\ , \qquad (1)$$

where

$$B(\mathrm{GT})=(\langle f|\sigma t|i\rangle)^2=(\langle f\|\sigma t\|i\rangle)^2/(2J_i+1)\ .$$

VOLUME 76, NUMBER 15 PHYSICAL REVIEW LETTERS 8 APRIL 1996

Shell Model Monte Carlo Method for Two-Neutrino Double Beta Decay

P. B. Radha,[1] D. J. Dean,[1,2] S. E. Koonin,[1] T. T. S. Kuo,[3] K. Langanke,[1] A. Poves,[1,4] J. Retamosa,[4] and P. Vogel[5]
[1]*W. K. Kellogg Radiation Laboratory, California Institute of Technology, Pasadena, California 91125*
[2]*Physics Division, Oak Ridge National Laboratory, P.O. Box 2008, Oak Ridge, Tennessee 37381*
[3]*Physics Department, SUNY at Stony Brook, Stony Brook, New York 11794*
[4]*Departmento de Física Teórica C-XI, Universidad Atuónoma de Madrid, E-28049 Madrid, Spain*
[5]*Division of Physics, Mathematics and Astronomy, California Institute of Technology, Pasadena, California 91125*
(Received 24 October 1995)

Shell model Monte Carlo techniques are used to calculate two-neutrino double beta decay matrix elements. We test the approach against direct diagonalization for ^{48}Ca in the complete *pf* shell using the KB3 interaction. The method is then applied to the decay of ^{76}Ge in the $(0f_{5/2}, 1p, 0g_{9/2})$ model space using a newly calculated realistic interaction. Our result for the matrix element is 0.12 ± 0.05 MeV^{-1}, in reasonable agreement with the experimental value.

PACS numbers: 21.60.Cs, 21.60.Ka, 27.40.+z, 23.40.–s

The double beta ($\beta\beta$) decay of a nucleus is a rare second order weak process [1,2]. The as-yet unobserved neutrinoless mode is of fundamental interest, as it would signal a neutrino mass, lepton number nonconservation, or admixtures of right handed weak currents. In contrast, the existence of the 2ν mode has been firmly established (see the review in Ref. [2]). The ability to accurately describe this latter process is an important element in the interpretation of limits on neutrinoless decays. Unfortunately, it seems that 2ν matrix elements are highly suppressed and so depend sensitively on small, poorly determined parts of the nuclear wave functions.

Most recent calculations of $2\nu\beta\beta$ matrix elements for nuclei heavier than ^{48}Ca rely on the quasiparticle random phase approximation (QRPA) [2]. While this approach is computationally simple and includes many features of the two-body interaction known to be relevant for $\beta\beta$ decay, the calculated matrix elements are uncertain because of their great sensitivity to the $J = 1^+, T = 0$ particle-particle interaction [3]. The interacting shell model offers a more microscopic approach to the problem. Complete $0\hbar\omega$ shell model calculations [4] not only recover more quenching of Gamow-Teller (GT) transitions than QRPA calculations, but also are in agreement with observations (after the universal renormalization of g_A to 1.0). However, computational limitations have restricted shell model calculations of the $2\nu\beta\beta$ decay matrix element to ^{48}Ca, the lightest of all $\beta\beta$ candidates.

In this Letter we show how shell model Monte Carlo (SMMC) methods can be used to calculate $2\nu\beta\beta$ decay matrix elements. We first calculate the decay of ^{48}Ca in the complete *pf* shell and validate our method against direct diagonalization. We then present results for the decay of ^{76}Ge, one of the few nuclei where the $2\nu\beta\beta$ decay has been precisely measured and where the best limits on the 0ν decay have been obtained [5]. Our calculation in the complete $(0f_{5/2}, 1p, 0g_{9/2})$ orbitals, which is impractical using traditional shell model methods, is the first of $2\nu\beta\beta$ decay in such a large model space.

The $2\nu\beta\beta$ matrix elements between the 0^+ ground states of the initial and final even-even nuclei is given by [6]

$$M^{2\nu} = \sum_m \frac{\langle 0_f^+ |\mathbf{G}| 1_m^+\rangle \cdot \langle 1_m^+ |\mathbf{G}| 0_i^+\rangle}{E_m - (E_i^0 + E_f^0)/2}. \tag{1}$$

Here, $|0_i^+\rangle$ ($|0_f^+\rangle$) is the ground state of the initial (final) nucleus with energy E_i^0 (E_f^0); $|1_m^+\rangle$ are the 1^+ states of the intermediate odd-odd nucleus with energies E_m; and $\mathbf{G}$ is the GT operator $\sum_l \boldsymbol{\sigma}_l \tau_l^-$, where $\boldsymbol{\sigma}_l$ is the Pauli spin operator for nucleon l and τ_l^- is the isospin lowering operator that changes a neutron into a proton.

Previous shell model calculations for nuclei heavier than ^{48}Ca ([1], and references therein) have invoked the so-called closure approximation, where the matrix element is written as $M^{2\nu} = M_c/\tilde{E}$, with $M_c = \langle 0_f^+|\mathbf{G}\cdot\mathbf{G}|0_i^+\rangle$ and $\overline{E}$ an average energy denominator. As there is no prescription for choosing the average energy denominator (and even the closure matrix element is usually calculated in a severely truncated basis), the uncertainty in this approximation is difficult to estimate.

To calculate the exact $2\nu\beta\beta$ matrix element, Eq. (1), we consider the function

$$\phi(\tau,\tau') = \frac{\mathrm{Tr}[e^{-(\beta-\tau-\tau')H}\mathbf{G}^\dagger \cdot \mathbf{G}^\dagger e^{-\tau H}\mathbf{G}e^{-\tau' H} \cdot \mathbf{G}]}{\mathrm{Tr}[e^{-\beta H}]}, \tag{2}$$

where H is the many-body Hamiltonian and the trace is over all states of the initial nucleus. The quantities $\beta - \tau - \tau'$ and τ play the role of the inverse temperature in the parent and daughter nucleus, respectively. A spectral expansion of ϕ shows that large values of these parameters guarantee cooling to the parent and daughter ground states. In these limits, we note that $\phi(\tau,\tau' = 0)$ approaches $e^{-\tau Q}|M_c|^2$, where $Q = E_i^0 - E_f^0$ is the energy release, so that a calculation of $\phi(\tau,0)$ leads

 0031-9007/96/76(15)/2642(4)$10.00

VOLUME 76, NUMBER 15 PHYSICAL REVIEW LETTERS 8 APRIL 1996

directly to the closure matrix element. If we then define

$$\eta(T,\tau) \equiv \int_0^T d\tau' \, \phi(\tau,\tau') e^{-\tau' Q/2} \tag{3}$$

and

$$M^{2\nu}(T,\tau) \equiv \frac{\eta(T,\tau) M_c^*}{\phi(\tau,0)}, \tag{4}$$

it is easy to see that in the limit of large τ, $\beta - \tau - \tau'$, and $T, M^{2\nu}(T,\tau)$ becomes independent of these parameters and is equal to the matrix element in Eq. (1).

We use SMMC methods [7] to calculate $\phi(\tau,\tau')$, and hence $M^{2\nu}$. These techniques scale more gently than direct diagonalization with the number of valence nucleons and single particle orbits and so allow calculations larger then possible otherwise. They are based on the discretization of the many-body propagator $e^{-\beta H}$ into a finite number of "time" slices N_t, each of duration $\Delta\beta = \beta/N_t$. At each time slice the many-body propagator is linearized via the Hubbard-Stratonovich transformations [8]; observables are then calculated as expectation values in the canonical ensemble of nuclear states.

To circumvent the "sign problem" encountered in the SMMC calculations with realistic interactions, we use the extrapolation procedure outlined in [9]. One defines a set of Hamiltonians $H(g,\chi) = [1 - (1-g)/\chi]H_G + gH_B$ such that $H(g=1,\chi) = H$ is the physical Hamiltonian and $H_{G,B}$ are the "good" and "bad" parts of the Hamiltonian, respectively. For $g \le 0, H(g,\chi)$ is free of the sign problem; the matrix elements are therefore calculated for several values $g \le 0$ and extrapolated to $g = 1$. The value of χ is chosen to make the linear g extrapolation as smooth as possible.

To validate our method, we calculated the matrix elements for ^{48}Ca in the complete *pf* shell with the KB3 interaction [10] for six equally spaced g values between -1.0 and 0.0 using $\chi = 4$ and extrapolated to the physical result at $g = 1$. Each calculation involved 2500–3500 Monte Carlo samples and was performed at $\beta = 2$ MeV^{-1} with $N_t = 48$. The direct diagonalization calculations for ^{48}Ca with which we compare our results were performed using an implementation of the Lanczos algorithm [11]. We calculated both the closure and exact matrix elements for the same Hamiltonians $H(g,\chi)$ as used in the SMMC.

We found the slope of $\ln[\phi(\tau,0)]$ to be in good agreement with that expected from the difference of the energies for ^{48}Ti and ^{48}Ca (Fig. 1) and extracted $|M_c|$ from the intercept. The SMMC closure matrix elements for $g \le 0$ are in very good agreement with the direct diagonalization results (Fig. 1) indicating that our temperatures are sufficiently low to correctly calculate the closure matrix element from the ground state of ^{48}Ca to ^{48}Ti. However, the direct diagonalization calculations show a small curvature near $g = 1.0$ that the extrapolation cannot reproduce. Our linear extrapolation of the closure matrix element, which

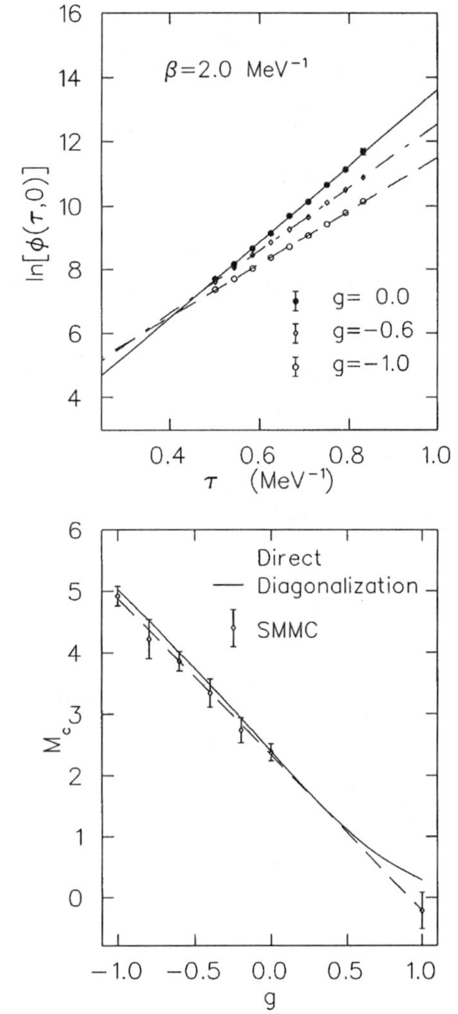

FIG. 1. Upper: $\ln[\phi(\tau,0)]$ for ^{48}Ca calculated at $\beta = 2.0$ MeV^{-1} with $N_t = 48$. The lines are best fits. Lower: SMMC and direct diagonalization closure matrix elements for ^{48}Ca. The SMMC points are linearly extrapolated to $g = 1.0$.

takes place over almost a factor of 20, therefore underestimates the physical ($g = 1.0$) calculation. We obtain -0.21 ± 0.29 for the closure matrix element to be compared with the direct diagonalization result of 0.29. As the natural scale for M_c is given by the sum rule [2] as ≈ 21, we may conclude that the SMMC successfully reproduces the shell model suppression of a factor of 70.

The calculation of the function $\phi(\tau,\tau')$ was performed for $\tau = 0.5$ MeV^{-1} and for thirteen τ' values spaced equally between 0.0 and 0.5 MeV^{-1}. This combination of parameters was checked to give converged results for the matrix element. We then calculated $\eta(T,\tau)$ for thirteen values of $T \leq 0.5$ MeV^{-1}; the upper limit of T is sufficiently large for the integral in Eq. (3) to converge. From these, we obtained $M^{2\nu}(T,\tau)$ [Eq. 6] as shown in Fig. 2 for some representative values of g [12]. In Fig. 2, we show the good agreement between the SMMC matrix elements and direct diagonalization for $g \leq 0$. Even though the value of $\chi = 4.0$ was chosen to make the linear extrapolation as smooth as possible, the direct diagonalization results still have a small curvature. For the exact 2ν matrix element we obtain an extrapolated

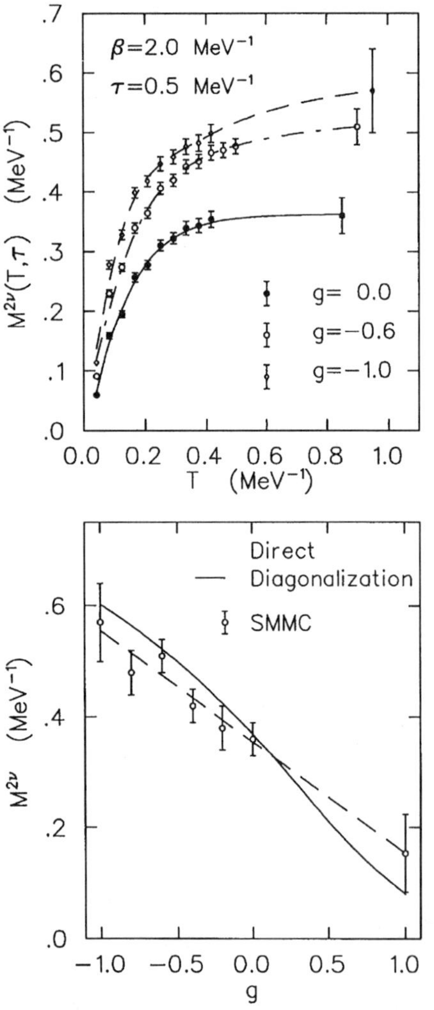

FIG. 2. Upper: $M^{2\nu}(T,\tau)$ for ^{48}Ca calculated at $\beta =$ 2.0 MeV^{-1} with $N_t = 48$. The points at large T show the asymptotic value ($T \rightarrow \infty$) of the matrix elements obtained according to [12]; lines are drawn to guide the eye. Lower: SMMC exact matrix elements and the direct diagonalization results for ^{48}Ca. The SMMC matrix elements are linearly extrapolated to $g = 1.0$.

value of 0.15 ± 0.07 MeV^{-1} whereas the calculation of Caurier, Poves, and Zuker [13] (including the erratum in [10]) gives 0.08 MeV^{-1}. There is thus agreement within the uncertainty.

We now apply the SMMC method to a heavier nucleus, where direct diagonalization is not possible. In particular, we calculate the 2ν matrix element for ^{76}Ge using an effective interaction based on the Paris potential in the $(0f_{5/2}, 1p, 0g_{9/2})$ orbitals, with the single particle energies taken from the levels of ^{57}Ni relative to the ^{56}Ni core [14]. This interaction has been constructed using a G-matrix folded-diagram method, in close analogy with the calculations carried out by Shurpin, Kuo, and Strottman [15] and by Dean *et al.* [4]. The model space comprises some 10^8 configurations, so that our SMMC calculation is significantly larger than previous shell model treatments of ^{76}Ge [1]. While it avoids spurious excitations of the center of mass, it does not include all spin-orbit pairs of orbitals and thus does not obey the Ikeda sum rule for GT strengths. However, this model space (with the choice of an appropriate effective interaction) should adequately describe those low-lying states expected to be the most important for $2\nu\beta\beta$ decay [16].

We performed the ^{76}Ge calculation at $\beta = 2.5$ MeV^{-1} with $N_t = 60$. The effective interaction used reproduces the experimental mass splitting of ^{76}Ge and ^{76}Se well: 21.35 ± 0.30 MeV compared to the experimental splitting of 20.72 MeV (the Coulomb energy was calculated following Ref. [17]). The mass splitting of ^{76}Ge and other $A =$ 76 nuclei (^{76}Zn, ^{76}Kr, ^{76}Sr) compares favorably with the Coulomb corrected experimental values or value from systematics in the case of ^{76}Sr. Our value for the β^- strength of ^{76}Ge is $B(\mathrm{GT}_-) = \langle \mathbf{G}^\dagger \cdot \mathbf{G} \rangle = 19.09 \pm 0.39$, and we find an energy centroid of 6.3 ± 0.2 MeV, while the experimental values are 19.9 and 9.1 MeV, respectively [18]; the apparent near consistency of this total strength with experiment is misleading as we have not employed the renormalization of $g_A = 1.0$ and we have missing strength in our model space corresponding mainly to the transitions between the $g_{9/2}$ and $g_{7/2}$ orbitals. This missing high-energy strength should not affect the low-lying states of ^{76}As that are important for $2\nu\beta\beta$ decay. Our value for the β^+ strength for ^{76}Se is $B(\mathrm{GT}_+) = 0.60 \pm 0.13$. This strength is identically zero in the independent particle model, and it is generated only by the smearing of the Fermi surface due to the interaction.

We performed two independent sets of calculations for both the closure and the exact matrix element using the $\chi = 4$ and $\chi = \infty$ families of $H(g,\chi)$. The best-fit extrapolations to $g = 1.0$ were linear for both the closure and the exact matrix elements in both cases. Our results for the closure matrix elements are -0.36 ± 0.34 and 0.08 ± 0.17 for $\chi = 4$ and $\chi = \infty$, respectively. These are to be compared with the truncated shell model calculation of Haxton and Stephenson [1] (using a different effective interaction) that resulted in a value of 2.56.

VOLUME 76, NUMBER 15 PHYSICAL REVIEW LETTERS 8 APRIL 1996

We find consistent $M^{2\nu}$ values for the $\chi = 4$ and $\chi = \infty$ cases (Fig. 3). Our results are 0.12 ± 0.07 and 0.12 ± 0.06 MeV^{-1}, respectively (a combined value of 0.12 ± 0.05 MeV^{-1}), while the experimental value of this matrix element (using $g_A = 1.26$) is 0.14 ± 0.01 MeV^{-1} [5]. However, shell model calculations of ordinary β decay consistently suggest that g_A is renormalized to 1.0 in the nuclear medium [4], in which case the experimental matrix element is 0.22 ± 0.01 MeV^{-1}.

There has been no previous shell model calculation of $M^{2\nu}$. Haxton and Stephenson [1] obtained an estimate in the closure approximation by taking the average energy denominator to be the position of the β^- GT resonance in ^{76}Ge (9.4 MeV). We find significantly smaller values of $\overline{E} \equiv M_c/M^{2\nu}$ (-3.0 ± 3.3 and 0.57 ± 1.26 MeV for $\chi = 4$ and ∞, respectively), in agreement with other $2\nu\beta\beta$ decay candidates such as ^{48}Ca, ^{100}Mo, and ^{128}Te where $\overline{E}$ is known to be significantly smaller than the position of the β^- GT resonance [16].

In this Letter, we have demonstrated an SMMC method to calculate $2\nu\beta\beta$ decay matrix elements and test it for the ^{48}Ca decay against direct diagonalization. We have also calculated the matrix element for ^{76}Ge in a model space significantly larger than previous calculations, and obtain a result that is in reasonable agreement with experiment. Although our extrapolations of the matrix elements may introduce systematic errors in our physical value, the fact that two independent calculations give consistent values enhances the confidence in our result. The dependence of $M^{2\nu}$ on the effective interaction and single-particle energies remains to be investigated. A more detailed description of these calculations will be given elsewhere, and work is in progress to calculate the matrix elements for several other, heavier nuclei.

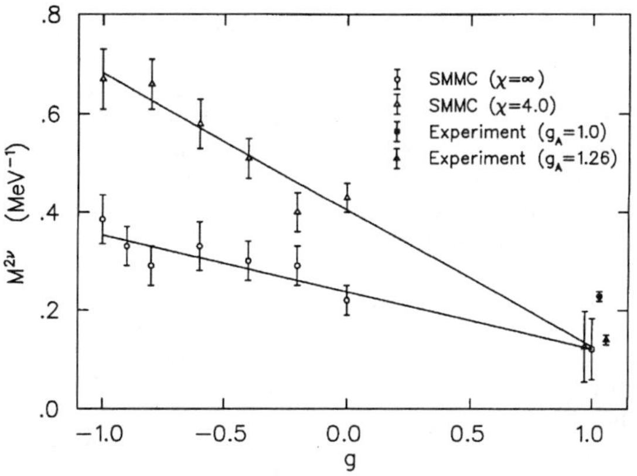

FIG. 3. SMMC exact matrix elements for ^{76}Ge calculated using the Hamiltonians $H(g, \chi)$ with $\chi = 4$ and $\chi = \infty$. The lines are linear fits to the points in both cases. The extrapolated values and the experimental result of Ref. [5] are shown staggered around $g = 1.0$ for clarity.

This work was carried out under grants from the NSF and the DOE. Computational cycles were provided by the Concurrent Supercomputing Consortium on the Intel Touchstone Delta and the Intel Paragon and on the IBM SP2 at the Maui High Performance Computing Center.

[1] W. C. Haxton and G. J. Stephenson, Jr., Prog. Part. Nucl. Phys. **12**, 409 (1984).

[2] M. Moe and P. Vogel, Ann. Rev. Nucl. Part. Sci. **44**, 247 (1994).

[3] P. Vogel and M. R. Zirnbauer, Phys. Rev. Lett. **57**, 3148 (1986).

[4] B. A. Brown and B. H. Wildenthal, At. Data Nucl. Data Tables **33**, 347 (1985); K. Langanke, D. J. Dean, P. B. Radha, Y. Alhassid, and S. E. Koonin, Phys. Rev. C **52**, 718 (1995); D. J. Dean, S. E. Koonin, T. T. S. Kuo, K. Langanke, and P. B. Radha, Phys. Lett. B **367**, 17 (1996).

[5] A. Balysh *et al.*, Phys. Lett. B **322**, 176 (1994).

[6] F. Boehm and P. Vogel, *Physics of Massive Neutrinos* (Cambridge University Press, Cambridge, England, 1992), 2nd ed.

[7] G. H. Lang, C. W. Johnson, S. E. Koonin, and W. E. Ormand, Phys. Rev. C **48**, 1518 (1993).

[8] J. Hubbard, Phys. Rev. Lett. **3**, 77 (1959); R. L. Stratonovich, Dokl Akad. Nauk. SSSR [Sov. Phys. Dokl.] **115**, 1097 (1957).

[9] D. J. Dean, S. E. Koonin, K. Langanke, P. B. Radha, and Y. Alhassid, Phys. Rev. Lett. **74**, 2909 (1995).

[10] E. Caurier, A. P. Zuker, A. Poves, and G. Martinez-Pinedo, Phys. Rev. C **50**, 225 (1994).

[11] E. Caurier, code ANTOINE, Strasbourg (1989).

[12] To determine the asymptotic ($T \to \infty$) value of $M^{2\nu}(\tau, T)$, we construct an ensemble of Gaussian-distributed $\phi(\tau, \tau')$ with mean and standard deviations as calculated in SMMC. Each $M^{2\nu}(\tau, T)$ is then fit by the form $a - be^{-cT}$ between $T = 0.17$ and 0.5 MeV^{-1}, and the average and standard deviation of a gives the required matrix element and its error (shown at large T in the upper panel of Fig. 2).

[13] E. Caurier, A. Poves, and A. P. Zuker, Phys. Lett. B **252**, 13 (1990).

[14] A. Bohr and B. Mottelson, *Nuclear Structure* (Benjamin, New York, 1969), Vol. 1.

[15] J. Shurpin, T. T. S. Kuo, and D. Strottman, Nucl. Phys. **A408**, 310 (1983).

[16] M. Ericson, T. Ericson, and P. Vogel, Phys. Lett. B **328**, 259 (1994).

[17] W. D. Myers and W. J. Swiatecki, Nucl. Phys. **81**, 1 (1966).

[18] R. Madey *et al.*, Phys. Rev. C **40**, 540 (1989).

Z. Phys. A - Hadrons and Nuclei 339, 435-444 (1991)

Zeitschrift für Physik A Hadrons and Nuclei

A comparative study of double beta decay by shell model and quasiparticle RPA

K. Muto[1], E. Bender[1,*], and H.V. Klapdor-Kleingrothaus[2]

[1] Department of Physics, Tokyo Institute of Technology, Oh-okayama, Meguro, Tokyo, 152 Japan
[2] Max-Planck-Institut für Kernphysik, W-6900 Heidelberg, Federal Republic of Germany

Received June 7, 1990; revised version December 27, 1990

Nuclear matrix elements of the two-neutrino and neutrinoless $\beta\beta$ decays of $^{48}Ca(0^+_{g.s.}) \to {}^{48}Ti(0^+_{g.s.})$ are calculated by shell model and QRPA. The two-neutrino matrix element $M^{2\nu}_{GT}$ is rather reliably evaluated in the QRPA approach by a careful fit of the particle-particle interaction strength in the 1^+ channel, which governs the spin-isospin ground-state correlations. The shell-model value of $M^{2\nu}_{GT}$ depends not only on the 1^+ interaction but largely on the pairing and quadrupole interactions. Concerning the neutrinoless-mode nuclear matrix elements, the shell model gives generally smaller values than the QRPA. A detailed analysis indicates that the discrepancies originate mainly from the truncation of shell-model configurations (fp-space). The QRPA calculation in a larger model space well takes into account transitions from/to single-particle orbits far from the Fermi surface, and those transitions give rise to sizable contributions because of large momentum transfers due to the exchange of a virtual neutrino.

PACS: 21.60-n; 23.40-s; 23.40.Hc

1. Introduction

Nuclear double beta ($\beta\beta$) decay [1–3] is expected to occur via two decay modes, the two-neutrino (2ν) mode and the neutrinoless (0ν) mode. The latter is of particular interest, since this lepton-number non-conserving process is forbidden in the standard electroweak theory, and observation of the 0ν $\beta\beta$ decay always implies that the neutrino is a massive Majorana particle. The 0ν decay is practically the only possibility to distinguish between Dirac and Majorana neutrinos, and sets the most stringent limits on the neutrino mass and the coupling constants of the right-handed charged weak currents. The deduction of the limits from experimental 0ν decay rates requires a reliable theoretical estimation of the relevant nuclear matrix elements.

The $\beta\beta$ decay has been studied from the nuclear structure point of view mostly by using the shell model and the proton-neutron quasiparticle random phase approximation (QRPA). Much progress has been achieved by recent QRPA studies [4–7] in understanding the suppression mechanism of the 2ν decay rates. An important finding is that β^+ Gamow-Teller transitions in neutron-rich nuclei are very sensitive to ground-state correlations of the spin-isospin type and the correlations are enhanced by the strongly attractive proton-neutron interaction in the $J^\pi = 1^+$ channel. Muto et al. [6, 7] determined the interaction strength by fitting the QRPA predictions of β^+-decay strengths to experimental values for some forty nuclei with $A = 60 - 170$, and obtained 2ν decay rates consistent with existing experimental data. A solution was thus found in the QRPA approach to the long-standing problem of large discrepancies up to two orders of magnitude between experimental half-lives and theoretical values which were obtained by shell-model calculations [1].

The QRPA model has also been applied to the 0ν $\beta\beta$ decay [8–10], and it was found that the 0ν decay nuclear matrix elements are not so sensitive to details of nuclear structure as the 2ν matrix elements. It is significant that the QRPA results calculated with different effective interactions are consistent with each other. The agreement is excellent between two calculations with the G-matrix of realistic nucleon-nucleon interactions; Muto, Bender and Klapdor [10] with the Paris potential [11], and Tomoda and Faessler [8] with the Bonn potential [12]. A QRPA calculation with zero-range forces by Engel, Vogel and Zirnbauer [9] yielded nuclear matrix elements and effects of short-range correlations which are considerably different from those of the former calculations. However, Muto et al. [10] pointed out that the deviations arise from their choice of a stronger interaction strength and partly from the use of schematic forces.

* Visiting scholar of Faculty of Science, Tokyo Institute of Technology
Present address: Institut für Theoretische Physik, Universität Tübingen, W-7400 Tübingen, Federal Republic of Germany

The progress in understanding the transition mechanisms has been achieved in the QRPA approach by the detailed analysis of the $\beta\beta$-decay nuclear transitions in connection with ground-state correlations and nuclear interactions, especially in the spin-isospin channel, and also that of the multipole decomposition of the 0ν nuclear matrix elements into contributions through intermediate states with different J^π values. However, such an extensive analysis has not been attempted in shell-model studies of the $\beta\beta$ decay. Therefore it is not clear to what extent shell-model prediction of the nuclear matrix elements involve uncertainties, and furthermore possible origins of uncertainties have not been investigated.

The two practical approaches so far employed, shell model and QRPA, can be regarded as different types of approximations to an ideal, complete shell-model calculation. Such a complete calculation is impossible in a model space large enough for the $\beta\beta$-decay nuclear matrix elements. For example, the number of linearly independent basis states in the full $3\hbar\omega$ and $4\hbar\omega$ oscillator major shells is more than 10^{21} for the ground state of ^{76}Ge. One is forced to reduce the number of basis states very drastically in a certain way by preserving as much as possible the components important for the $\beta\beta$ decay. Haxton et al. [1, 13] assumed a small model space (a $p_{3/2}-p_{1/2}-f_{5/2}-g_{9/2}$ space for the decay of ^{76}Ge, but this model space does not include the spin-orbit partners $f_{7/2}$ and $g_{7/2}$ which play an important role in Gamow-Teller transitions), and further truncated by taking a weak-coupling scheme. The number of basis states is thus reduced to the order of hundred. On the other hand, the QRPA model simplifies a nuclear structure calculation in a different sense. It introduces quasiparticles and defines a vacuum for the quasiparticles, which has many particle - many hole components in the shell-model picture, and then the RPA equation is solved for proton-neutron quasiparticle pairs. The number of quasiparticle pairs with $J^\pi=1^+$ in the full $3\hbar\omega$ and $4\hbar\omega$ space is 23. Nuclear excitations are described in terms of such a small number of quasiparticle degrees of freedom on the vacuum. This simplicity enables to perform a QRPA calculation in a much larger model space and makes it possible to take into account ground-state correlations which are most important for the nuclear transitions. It is interesting and of importance to compare results of the very different approximations to an exact solution of the many-body problem. The two approaches are expected to be complementary to each other for a better understanding of nuclear transition mechanisms and providing more reliable nuclear matrix elements of the $\beta\beta$ decay.

In this paper, we calculate nuclear matrix elements of the $\beta\beta$ decay, $^{48}\mathrm{Ca}\,(0^+_{\mathrm{g.s.}})\rightarrow{}^{48}\mathrm{Ti}\,(0^+_{\mathrm{g.s.}})$, by shell model and QRPA in the standard procedures which have been used in the previous calculations. The main purpose is, through a detailed analysis of the calculated nuclear matrix elements, to clarify fundamental features of the $\beta\beta$ nuclear transitions, and also uncertainties and shortcomings involved in the standard procedures of the nuclear structure calculations. For this purpose the decay of ^{48}Ca is the best example, since ^{48}Ca is one of the lightest $\beta\beta$ emitters and the shell model is expected to describe structure of the initial, final and intermediate (^{48}Sc) states presumably well. The procedures of the model calculations are explained in Sect. 2. The calculated results are shown and discussed for the 2ν decay in Sect. 3 and for the 0ν decay in Sect. 4. Concluding remarks are made in Sect. 5.

2. Models

2.1. QRPA

Detailed procedures of the QRPA calculation of $\beta\beta$-decay nuclear matrix elements are given in [3, 6, 7, 9, 10]. Here, we focus on the basic ideas and some important points of the QRPA formalism in comparison with a shell-model calculation.

The proton-neutron QRPA was developed [14] (see also [15]) in order to describe change-changing excitations by RPA in the quasiparticle picture. The transformation from a nucleon basis to a quasiparticle basis is defined by a special Bogoliubov transformation. The transformation coefficients u_j and v_j, which are occupation amplitudes of single-particle orbits satisfying $u_j^2+v_j^2=1$, are determined in the BCS approximation, together with single quasiparticle energies ε_j. In the basis of proton-neutron quasiparticle pairs coupled to a spin-parity J^π, the RPA equation for the J^π-mode is solved,

$$\begin{bmatrix} A & B \\ -B & -A \end{bmatrix} \begin{bmatrix} X \\ Y \end{bmatrix} = \omega \begin{bmatrix} X \\ Y \end{bmatrix}, \tag{1}$$

and energy eigenvalues ω and wave functions X and Y are obtained. Elements of the submatrices are

$$\begin{aligned} A^{J\pi}_{pn,p'n'} &= \delta(j_p j_n, j'_p j'_n)(\varepsilon_{j_p}+\varepsilon_{j_n}) \\ &\quad + g_{\mathrm{pp}}(u_{j_p}u_{j_n}u_{j'_p}u_{j'_n}+v_{j_p}v_{j_n}v_{j'_p}v_{j'_n}) \\ &\quad \times\langle j_p j_n | V | j'_p j'_n\rangle_J \\ &\quad + g_{\mathrm{ph}}(u_{j_p}v_{j_n}u_{j'_p}v_{j'_n}+v_{j_p}u_{j_n}v_{j'_p}u_{j'_n}) \\ &\quad \times\langle j_p j_n^{-1} | V | j'_p j'^{-1}_n\rangle_J, \end{aligned} \tag{2a}$$

$$\begin{aligned} B^{J\pi}_{pn,p'n'} &= g_{\mathrm{pp}}(u_{j_p}u_{j_n}v_{j'_p}v_{j'_n}+v_{j_p}v_{j_n}u_{j'_p}u_{j'_n}) \\ &\quad \times\langle j_p j_n | V | j'_p j'_n\rangle_J \\ &\quad - g_{\mathrm{ph}}(u_{j_p}v_{j_n}v_{j'_p}u_{j'_n}+v_{j_p}u_{j_n}u_{j'_p}v_{j'_n}) \\ &\quad \times\langle j_p j_n^{-1} | V | j'_p j'^{-1}_n\rangle_J. \end{aligned} \tag{2b}$$

Subscripts p and n distinguish between proton and neutron single-(quasi)particle states. We have introduced renormalization factors g_{ph} and g_{pp} multiplying particle-hole (ph) and particle-particle (pp) matrix elements, respectively. The ph elements are defined by the Pandya transformation,

$$\begin{aligned} &\langle j_p j_n^{-1} | V | j'_p j'^{-1}_n\rangle_J \\ &\quad = -(-1)^{j_p+j_n+j'_p+j'_n}\sum_{J'}(2J'+1) \\ &\qquad W(j_p j_n j'_n j'_p; JJ')\langle j_p j'_n | V | j'_p j_n\rangle_{J'}. \end{aligned} \tag{3}$$

It is noted that the J^π-mode QRPA equation has proton-neutron interaction matrix elements for only J^π-coupled pairs, and the QRPA ground state contains correlations in which each proton-neutron pair has the same spin-parity J^π. By contrast, a shell-model wave function of the ground state has various modes of correlations. A typical example is the quadrupole correlation which is induced by a quadrupole-quadrupole component of the proton-neutron interaction. The leading terms of the correlation can be expressed in the quasiparticle picture as $|j_p j'_p(2^+), j_n j'_n(2^+); 0^+\rangle$. By the angular-momentum recoupling, the like-nucleon coupling is transformed to the proton-neutron coupling,

$$|j_p j'_p(2^+), j_n j'_n(2^+); 0^+\rangle = \sum_J (-1)^{j_p+j'_n-J} \sqrt{5(2J+1)} \times W(j_p j'_p j_n j'_n; 2J) | j_p j_n(J^\pi), j'_p j'_n(J^\pi); 0^+\rangle . \quad (4)$$

The RPA calculation for $J^\pi = 1^+$ takes into account the $J^\pi = 1^+$ term of (4). Also, other modes of shell-model correlations involve 1^+-coupled components in the proton-neutron coupling scheme. Thus the idea of the QRPA formalism can be understood, in comparison with shell model, that correlations of various modes present in a shell-model wave function are mapped onto a restricted space comprising proton-neutron quasiparticle pairs coupled to the definite spin-parity to be considered. This mapping is done on the operator basis.

The RPA equation of $J^\pi = 1^+$ is solved for the calculation of the 2ν decay nuclear matrix element, since the 2ν decay is described by successive Gamow-Teller transitions through virtual intermediate nuclear states with $J^\pi = 1^+$. On the other hand, in case of the 0ν decay, many spin-parity values are allowed for the intermediate states due to the exchange of a virtual neutrino and the resulting "neutrino potential" in the nuclear transition operators. Consequently, the RPA calculation is repeated for all J^π which are possible in the model space, and the 0ν decay matrix elements are obtained by summing up contributions through the intermediate states.

Numerical calculations are made in a harmonic oscillator basis with the oscillator constant from [16]. The model space consists of nine single-particle orbits, $1s0d - 1p0f - 0g$. The single-particle energies are taken from one-nucleon transfer reactions on ^{48}Ca and ^{48}Ti [17], when available, and the others are calculated with a Coulomb-corrected Woods-Saxon potential [18, 7]. The G-matrix of the Paris potential [11] is used consistently for both BCS and RPA calculations. The strength of the pairing interaction is determined in the BCS calculation, for proton and neutron systems separately, so as to reproduce experimental gap energies which are evaluated from masses given in the compilation by Wapstra et al. [19]. The renormalization factor of the ph interaction is fixed to $g_{\text{ph}} = 1.096$ [7]. The most probable value of g_{pp} for the decay of ^{48}Ca is estimated to be $g_{\text{pp}} = 0.920$ by extending the analysis of neutron-number dependence of g_{pp} [7] to $N < 30$.

2.2. Shell model

The procedure of a shell-model calculation is well established [20, 1]. Hamiltonian matrix elements for n-particle states with definite isospin and angular momentum TJ are

$$\begin{aligned}\langle n\alpha_1 TJ | H | n\alpha_2 TJ\rangle &= \delta(\alpha_1, \alpha_2) \sum_j \langle n\alpha_1 TJ | \varepsilon_j \hat{N}_j | n\alpha_1 TJ\rangle \\ &+ \sum_{\bar{\alpha}\bar{T}\bar{J}T'J'(j_1 j'_1)(j_2 j'_2)} \binom{n}{2} \\ &\times \langle n\alpha_1 TJ \{ | n-2\bar{\alpha}\bar{T}\bar{J}, j_1 j'_1 T'J'\rangle \\ &\times \langle j_1 j'_1 | V | j_2 j'_2\rangle_{T'J'} \\ &\times \langle n-2\bar{\alpha}\bar{T}\bar{J}, j_2 j'_2 T'J' | \} n\alpha_2 TJ\rangle , \end{aligned} \quad (5)$$

where α represents additional quantum numbers necessary to specify the basis state, ε_j is the single-particle energy and $\hat{N}_j$ the number operator of the j-orbit. Matrix elements of a two-body interaction are expressed in terms of those for two-particle states and coefficients of fractional parentage (see [20, 21]).

It is clear from (5) that two-particle matrix elements $\langle j_1 j'_1 | V | j_2 j'_2\rangle_{T'J'}$ with various $T'J'$ values contribute to the hamiltonian matrix elements in calculating eigenstates with TJ. Therefore, the calculated eigenstate contains accordingly various modes of correlations.

The renormalization of the interaction strength by multiplying g_{pp} (g_{ph}) in the QRPA calculation would correspond, in the shell-model calculation, to that of the pp (ph) matrix elements of the proton-neutron interaction,

$$\langle j_p j_n | V | j'_p j'_n\rangle_J = \frac{\sqrt{(1+\delta(j_p, j_n))(1+\delta(j'_p, j'_n))}}{2} \times [\langle j_p j_n | V | j'_p j'_n\rangle_{T=0,J} + \langle j_p j_n | V | j'_p j'_n\rangle_{T=1,J}] . \quad (6)$$

However, the change of the proton-neutron matrix elements is not unique. Also, a renormalization of pp matrix elements automatically affects ph elements which are connected by the Pandya transformation. Therefore, we introduce the renormalization factors g_{ph} and g_{pp} only in the $J^\pi = 1^+$ channel by multiplying both $T=0$ and $T=1$ elements with the same factor. When ph elements are renormalized, we first multiply the elements with g_{ph} and then use the inverse transformation of (3) to obtain the corresponding pp elements, which are put into the shell-model calculation. A simultaneous change of many ph (pp) elements violates the Pandya transformation very badly in the shell-model calculation. On the contrary, in the QRPA calculation, the change in a J^π channel can be carried out without affecting interactions in the other channels, since only elements with one J^π value contribute to the RPA equation (2) and the ph and pp interactions appear separately in the equation.

We assume the following isospin-scheme configurations on an inert ^{40}Ca core,

$$\sum_{m=0,1,2} f_{7/2}^{8-m}(p_{3/2}, p_{1/2}, f_{5/2})^m .$$

The isospin scheme guarantees that isospin is always a good quantum number irrespective of the truncation of model space. The eight valence nucleons occupy mostly the lowest-lying orbital $f_{7/2}$ in the 0^+ ground states. Then, charge-changing transitions, such as Gamow-Teller ones, convert an $f_{7/2}$ neutron (proton) into a proton (neutron) in one of the four orbits, leading to $m=0$ and $m=1$ configurations. The two-particle excited configurations ($m=2$) constitute main components of ground-state correlations and give rise to first-order corrections through interference with the dominant transition amplitudes. Higher configurations ($m>2$) would have little effects on the $\beta\beta$ decay. In this model space, the numbers of basis states are 11, 291 and 201 for $T; J^\pi = 4; 0^+$, $2; 0^+$ and $3; 1^+$, respectively. We take the same effective interaction, the G-matrix of the Paris potential, as in the QRPA calculation. It is normally assumed that $g_{\rm ph}=g_{\rm pp}=1$, i.e., no renormalization, in shell-model calculations. Single-particle energies ε_j for the ^{40}Ca core are adjusted such that experimental single-hole ($A=47$) and single-particle ($A=49$) energies with respect to ^{48}Ca [17] are reproduced with the two-body interaction.

3. Two-neutrino mode

The nuclear part of the $2\nu\,\beta\beta$ decay is described by successive Gamow-Teller transitions, $^{48}{\rm Ca}(0_i^+)\to{}^{48}{\rm Sc}(1^+)\to{}^{48}{\rm Ti}(0_f^+)$. In a good approximation, the nuclear part is separated from the phase-space factor of the emitted leptons, and the nuclear matrix element is defined by

$$M_{\rm GT}^{2\nu}=\sum_a \frac{\langle 0_f^+\|t_-\sigma\|1_a^+\rangle\langle 1_a^+\|t_-\sigma\|0_i^+\rangle}{E_a-(M_i+M_f)/2}, \qquad (7)$$

where E_a are energies of 1^+ intermediate states, and M_i and M_f are masses of the initial and final nuclei, respectively. In the QRPA model, since phonons calculated with respect to the initial (0_i^+) and final (0_f^+) states are not identical to each other, we define the nuclear matrix element as [5–7]

$$M_{\rm GT}^{2\nu}=\sum_{a,b} \frac{\langle 0_f^+\|t_-\sigma\|1_b^+\rangle\langle 1_b^+|1_a^+\rangle\langle 1_a^+\|t_-\sigma\|0_i^+\rangle}{E_a-(M_i+M_f)/2}, \qquad (8)$$

with overlaps

$$\langle 1_b^+|1_a^+\rangle=\sum_{pn}(X_b^{pn}X_a^{pn}-Y_b^{pn}Y_a^{pn}). \qquad (9)$$

Figure 1a shows the nuclear matrix elements calculated as function of $g_{\rm pp}$, the strength of the pp interaction in the 1^+ channel. The most important issue is the crossing through zero of the nuclear matrix element. $M_{\rm GT}^{2\nu}$ of QRPA behaves as those of the heavier $\beta\beta$ emitters calculated with QRPA [7], and vanishes near the most probable value, $g_{\rm pp}=0.92$. A vanishing of $M_{\rm GT}^{2\nu}$ is obtained also by the shell model near $g_{\rm pp}=1.0$. The decrease of $M_{\rm GT}^{2\nu}$ with $g_{\rm pp}$ is correlated with that of the sum of the β^+ strength,

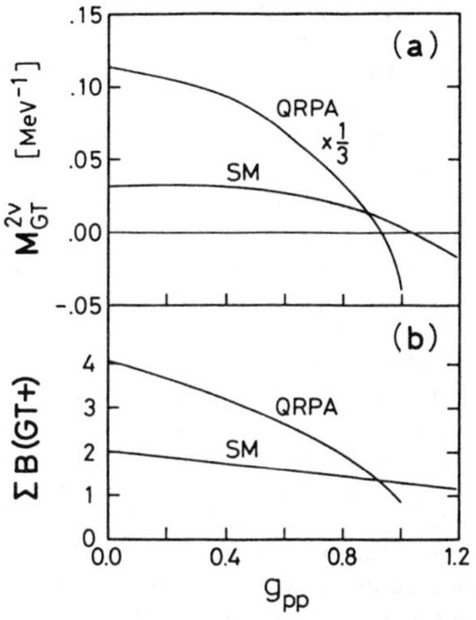

Fig. 1. a The nuclear matrix element of the $2\nu\,\beta\beta$ decay and **b** the strength sum of Gamow-Teller excitation $^{48}{\rm Ti}(0_{\rm g.s.}^+)\to{}^{48}{\rm Sc}(1^+)$, calculated by the shell model (SM) and QRPA as function of $g_{\rm pp}$, the strength of the particle-particle interaction in $J^\pi=1^+$ channel

$$\sum B({\rm GT}+)=\sum_k |\langle 1_k^+\|t_+\sigma\|0_f^+\rangle|^2, \qquad (10)$$

as shown in Fig. 1b. The sum $\sum B({\rm GT}+)=1.34$ of the QRPA calculation at $g_{\rm pp}=0.92$ agrees well with the shell-model value $\sum B({\rm GT}+)=1.30$ at $g_{\rm pp}=1.0$. The consistent results of the shell-model and QRPA calculations confirm the conclusions obtained by the QRPA approach [4–7] on the suppression mechanism of the $2\nu\,\beta\beta$ decay rate that the proton-neutron pp interaction enhances the spin-isospin ground-state correlations and the correlations suppress the 2ν decay rate, mainly by the suppression of β^+ transitions from the daughter 0_f^+ state to intermediate 1^+ states. The suppression mechanism, which was found in the QRPA approach, is now understood to be fundamental for the $2\nu\,\beta\beta$ decay.

The two curves in Fig. 1a have to be compared with caution. $M_{\rm GT}^{2\nu}$ of QRPA is large at $g_{\rm pp}=0$, where the ground-state wave function only slightly deviates from the quasiparticle vacuum and has negligibly small spin-isospin correlations. It decreases rapidly with $g_{\rm pp}$ by the enhancement of the correlations and crosses through zero. (The RPA equation tends to a collapse after the crossing.) Thus, the QRPA prediction of $M_{\rm GT}^{2\nu}$ is very sensitive to $g_{\rm pp}$, but, on the other hand, the nuclear matrix element can rather reliably be predicted by a careful fit of the interaction strength, as was done in our previous paper [7]. On the other hand, the shell-model wave function always contains a considerable amount of spin-isospin correlations arising from various modes of correlations (see Subsect. 2.2), even if the proton-neutron 1^+ interaction is switched off ($g_{pp}=0$). Consequently, $M_{GT}^{2\nu}$ of shell model starts with an already suppressed value at $g_{pp}=0$, and the decrease with g_{pp} is moderate. However, this does not mean that the shell-model prediction is more reliable than that of QRPA. While the shell-model value of $M_{GT}^{2\nu}$ is less sensitive to the strength of $J^\pi=1^+$ interaction, it is expected to depend also on some other J^π interactions and on some important combinations of them. This is discussed in the following.

Among various components of a nucleon-nucleon interaction, the most prominent feature is a strongly attractive pairing component which acts between like nucleons. In general, the pairing component is relatively weak in a bare *G*-matrix, and is most enhanced by renormalizing core-polarization diagrams. Recently, Staudt et al. [22] performed a QRPA calculation of the $\beta\beta$ decay of ^{76}Ge with a variety of *G*-matrix interactions, and in the course of the calculation it was found that the renormalization seems to result in an overestimation of the pairing strength. This implies possible uncertainties involved in this important component of the nucleon-nucleon interaction. In the QRPA calculation, the pairing strength is fixed to experimental BCS gap energies. Besides the pairing component, the proton-neutron quadrupole interaction is expected to play a significant role for the suppression of the 2ν decay rate. This was pointed out by Grotz and Klapdor [23] in a number-projected BCS calculation with separable Gamow-Teller and quadrupole-quadrupole forces. The latter force induces ground-state correlations of the type given in (4), for example, an admixture of $|\pi f_{7/2}^2(2^+), \nu f_{7/2}^{-2}(2^+); 0^+\rangle$ in the ground state of ^{48}Ti, and it is clear from (4) that the spin-isospin correlations are affected by the quadrupole interaction.

Motivated by these arguments, we have repeated the shell-model calculation by multiplying the strength of the pairing and quadrupole components with a factor of 0.8 or 1.2. For a realistic nucleon-nucleon interaction, the quadrupole component can be defined as the $k=2$ term in the multipole expansion of the proton-neutron interaction,

$$\langle j_p j_n | V | j_p' j_n' \rangle_J = \sum_k (-1)^{j_p+j_n'-J}(2k+1)\, W(j_p j_n j_p' j_n'; Jk) \times f^{(k)}(j_p j_n; j_p' j_n'), \tag{11a}$$

where the multipole strengths are, by the inverse transformation,

$$f^{(k)}(j_p j_n; j_p' j_n') = \sum_J (-1)^{j_p+j_n'-J}(2J+1)\, W(j_p j_n j_p' j_n'; Jk) \times \langle j_p j_n | V | j_p' j_n' \rangle_J . \tag{11b}$$

The calculated nuclear matrix elements are shown in Fig. 2. The pairing and quadrupole interactions have effects on $M_{GT}^{2\nu}$ in the opposite directions; $M_{GT}^{2\nu}$ is more

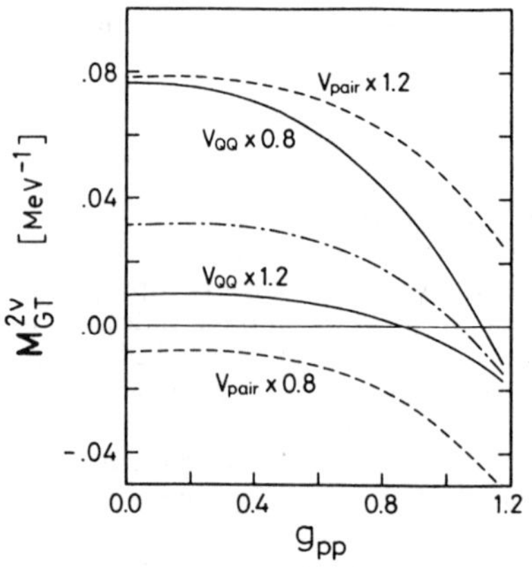

Fig. 2. The 2ν nuclear matrix element calculated by the shell model with different strengths of the pairing (V_{pair}) and quadrupole (V_{QQ}) interactions. The dash-dotted line denotes the matrix element calculated with no change of the interaction strength

(less) suppressed by a stronger (weaker) quadrupole interaction, and consequently it crosses through zero at a smaller (larger) value of g_{pp}. This agrees qualitatively with the result by a shell-model calculation with a schematic quadrupole-quadrupole force [24]. A change of the pairing interaction affects the nuclear matrix element more largely. If the pairing strength is enlarged by only 20%, $M_{GT}^{2\nu}$ becomes more than two times larger at $g_{pp}=0$ and is much less suppressed at $g_{pp}\approx 1$. It may therefore be attributed to an inappropriate strength of the pairing component (and to that of the quadrupole component) of the renormalized *G*-matrix interaction that the earlier shell-model calculations [25–27] of the $2\nu\ \beta\beta$ decay of ^{48}Ca could not reproduce the experimentally inferred suppression of the nuclear matrix element [28].

The half-life of the 2ν decay is obtained by performing the phase-space integrals for each intermediate state with the calculated energy in the perturbation energy denominator. Because of the rapid decrease of $M_{GT}^{2\nu}$ of QRPA in the vicinity of the crossing through zero, we present, following the procedure of [7], the lower limit and average of the $2\nu\ \beta\beta$ decay half-life of ^{48}Ca

$$[T_{1/2}^{2\nu}]^{QRPA}_{\text{lower-limit}} = 4.85\times 10^{19}\,\text{y},$$

$$[T_{1/2}^{2\nu}]^{QRPA}_{\text{average}} = 1.49\times 10^{20}\,\text{y},$$

which are consistent with the experimental limit $(T_{1/2}^{2\nu})^{exp} > 3.6\times 10^{19}$ y [28]. The reliability of the prediction may be evidenced by the fact that the standard QRPA calculation [7] gives 2ν decay "average" half-lives for ^{82}Se (1.09×10^{20} y) and for ^{130}Te (1.84×10^{21} y) in good agreement with experiment. These are, at present, the only two nuclides for which experiments have set an upper limit of the half-life; experimental half-lives for ^{82}Se by counter ($1.1^{+0.8}_{-0.3}\times 10^{20}$ y) [29] and geochemical ((1.30 ± 0.05) $\times 10^{20}$ y) [30] measurements are consistent with each other, and the geochemical method gave a finite half-life of ^{130}Te (($1.5-2.75)\times 10^{21}$ y) [30].

440

The present shell-model calculation for ^{48}Ca yields (at $g_{pp}=1.0$) a much longer half-life of

$$[T_{1/2}^{2\nu}]^{SM}=6.41\times 10^{21}\,y.$$

However, half-lives calculated with the slightly different strengths of the pairing and quadrupole components,

$$T_{1/2}^{2\nu}(V_{\rm pair}\times 0.8)=8.75\times 10^{19}\,y,$$

$$T_{1/2}^{2\nu}(V_{\rm pair}\times 1.2)=4.57\times 10^{19}\,y,$$

$$T_{1/2}^{2\nu}(V_{QQ}\times 0.8)=2.77\times 10^{20}\,y,$$

$$T_{1/2}^{2\nu}(V_{QQ}\times 1.2)=2.81\times 10^{21}\,y,$$

imply that a shorter half-life would be obtained by fixing the important interaction strengths to, for example, experimental β^+ strengths.

4. Neutrinoless mode

The $0\nu\,\beta\beta$ decay is expected to occur by the exchange of a virtual neutrino between two nucleons in a nucleus. This is possible if the neutrino is a massive Majorana particle ($\nu=\bar\nu$) [1, 31], and then possible right-handed components of the charged weak current would also contribute to the decay. The inverse half-life of the $0\nu\,\beta\beta$ decay is given by

$$[T_{1/2}^{0\nu}]^{-1}=C_{mm}\left(\frac{\langle m_\nu\rangle}{m_e}\right)^2+C_{\eta\eta}\langle\eta\rangle^2+C_{\lambda\lambda}\langle\lambda\rangle^2 + C_{m\eta}\frac{\langle m_\nu\rangle}{m_e}\langle\eta\rangle+C_{m\lambda}\frac{\langle m_\nu\rangle}{m_e}\langle\lambda\rangle + C_{\eta\lambda}\langle\eta\rangle\langle\lambda\rangle, \quad (12)$$

where $\langle m_\nu\rangle$ is the effective neutrino mass, $\langle\eta\rangle$ and $\langle\lambda\rangle$ the effective coupling constants of the right-handed currents, and C_{xy} are expressed by nuclear matrix elements and phase-space integrals. In evaluating the nuclear matrix elements, the closure approximation is used, and the inter-nucleon short-range correlations [32] and nucleon finite size effects [33] are taken into account in the standard way. For details see [2, 3, 10].

Among the nine nuclear matrix elements, a combination of $M_{\rm GT}^{0\nu}-M_{\rm F}^{0\nu}$ is relevant for the determination of $\langle m_\nu\rangle$, and the recoil matrix element $M_R^{0\nu}$ dominates the term multiplying $\langle\eta\rangle$, the coupling strength of the right-handed leptonic current relative to that of the left-handed current. These matrix elements are explicitly

$$M_{\rm GT}^{0\nu}=\sum_{m,n}\langle 0_f^+\,\|(t_-\sigma)_m \times(t_-\sigma)_n H_m(|\mathbf{r}_m-\mathbf{r}_n|)\|0_i^+\rangle, \quad (13a)$$

$$M_{\rm F}^{0\nu}=\sum_{m,n}\langle 0_f^+\,\|(t_-)_m(t_-)_n \times H_m(|\mathbf{r}_m-\mathbf{r}_n|)\|0_i^+\rangle\left(\frac{g_V}{g_A}\right)^2, \quad (13b)$$

$$M_R^{0\nu}=\sum_{m,n}\langle 0_f^+\,\|(t_-\sigma)_m \times(t_-\sigma)_n H_R(|\mathbf{r}_m-\mathbf{r}_n|)\|0_i^+\rangle\frac{\mu_\beta}{3}\frac{g_V}{g_A}, \quad (13c)$$

where $H_m(r)$ and $H_R(r)$ are "neutrino potentials" which represent the exchange of a neutrino. Because of the neutrino potential, the $0\nu\,\beta\beta$ decay proceeds via intermediate states with various J^π-values.

A decomposition of the nuclear matrix elements into contributions of different J^π states enables a detailed comparison of the calculated matrix elements. The decomposition has been used in the QRPA studies [8–10]. It is also possible in the shell model. Because of the two-body character of the $\beta\beta$ transition operators, the nuclear matrix elements are reduced to a sum of products of two-body transition densities and matrix elements for (non-antisymmetrized) two-particle states,

$$M_\alpha^{0\nu}=\sum_{m,n}\langle 0_f^+\,\|O_{m,n}^\alpha\|0_i^+\rangle = \sum_{j_pj'_pj_nj'_nJ'} Z(j_pj'_p,j_nj'_n;J')\langle j_pj'_pJ'\|O_{12}^\alpha\|j_nj'_nJ'\rangle. \quad (14)$$

The transition operator can be expressed in a form of momentum integral

$$O_{12}=N\int_0^\infty dq\,q^2v(q)\sum_{J\pi}[F^{(J\pi)}(q)]_1 \cdot[F^{(J\pi)}(q)]_2, \quad (15)$$

where N represents constant factors, $v(q)$ the Fourier transform of the neutrino potential and $[F^{(J\pi)}(q)]_{1\,(2)}$ is a spherical tensor operator of rank J and parity π acting on the particle 1 (2). The two-particle matrix elements are then

$$\langle j_pj'_pJ'\|O_{12}\|j_nj'_nJ'\rangle = N\int_0^\infty dq\,q^2v(q)\sum_{J\pi}(-1)^{j_p+j'_n-J'}\sqrt{2J'+1} \times W(j_pj_nj'_pj'_n;JJ') \times\langle j_p\|F^{(J\pi)}(q)\|j_n\rangle\langle j'_p\|F^{(J\pi)}(q)\|j'_n\rangle. \quad (16)$$

Here, J^π is the spin-parity of intermediate states in the $0^+\to 0^+$ $\beta\beta$ decay. Using the orthogonality of Racah coefficients, we can decompose the two-particle matrix elements,

$$\langle j_pj'_pJ'\|O_{12}\|j_nj'_nJ'\rangle = (-1)^{J'}\sqrt{2J'+1}\sum_J(2J+1) \times W(j_pj_nj'_pj'_n;JJ') \times\sum_{J''}(-1)^{J''}\sqrt{2J''+1}\,W(j_pj_nj'_pj'_n;JJ'') \times\langle j_pj'_pJ''\|O_{12}\|j_nj'_nJ''\rangle. \quad (17)$$

Figures 3 and 4 compare the decompositions of $M^{0\nu}_{\rm GT}-M^{0\nu}_{\rm F}$ and $M^{0\nu}_{R}$, respectively. A striking difference between the shell-model and QRPA results arises from the model space. The shell model in the fp-space gives contributions through only positive-parity ($J^{\pi}=0^{+}-7^{+}$) intermediate states. On the other hand, the QRPA in the larger space yields negative-parity contributions with comparable magnitudes. The same tendency is seen in both $M^{0\nu}_{\rm GT}$ and $M^{0\nu}_{R}$. The negative-parity contributions arise from $1\,\hbar\omega$ excitations, i.e., transitions between $1s0d(0g)$ and $1p0f$ shells. The BCS calculation well reproduces experimentally observed small vacancies in the positive-parity proton shells below $Z=20$ [17]. The QRPA result therefore implies that, even if vacancies (u_j^2) of hole orbits or occupations (v_j^2) of particle orbits seem to be so small that those orbits would have little effects on low-lying transitions, they give rise to substantial contributions to the $0\nu\,\beta\beta$ decay. The transitions from or to those single-particle orbits far from the Fermi energy are enhanced by large momenta of the virtual neutrino, which are typically $q\sim 100$ MeV/c. By contrast, the Gamow-Teller operator which mediates the 2ν decay carries no momentum. Thus, the calculation of the 0ν decay nuclear matrix elements requires a larger model space than the one which is presumably large enough for describing nuclear states at low excitation energies.

Of particular importance is that in $M^{0\nu}_{\rm GT}$ the relative magnitudes of the positive-parity contributions differ between the shell-model and QRPA calculations, especially of the 1^{+} contribution. The strong correlation of the 1^{+} contribution of the 0ν decay with the nuclear matrix element of the 2ν decay has been suggested in the QRPA calculations for heavier $\beta\beta$ nuclei [8–10]. Namely, the 1^{+} contribution decreases with $g_{\rm pp}$ and is strongly suppressed at probable values of $g_{\rm pp}$, as $M^{2\nu}_{\rm GT}$ is. Figure 5 shows the $g_{\rm pp}$-dependence of $M^{0\nu}_{\rm GT}$ and its 1^{+} contribution. The behaviour mentioned above is seen in the QRPA result. In contrast, the shell-model calculation indicates that the 1^{+} contribution and consequently $M^{0\nu}_{\rm GT}$ are more or less constant in the broard range of $g_{\rm pp}$. This is probably also a result of the truncation of the shell-model configurations. We have performed a QRPA calculation in the fp-space, and the result shows no suppression of the 1^{+} contribution (and no suppression of $M^{0\nu}_{\rm GT}$). This is traced to the rigid Fermi surface of the proton system in ^{48}Ca. It is noted, however, that there exist two main differences between $M^{2\nu}_{\rm GT}$ and the 1^{+}contribution to

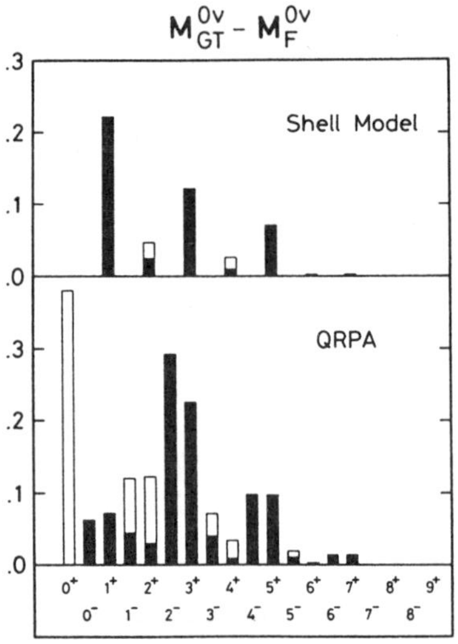

Fig. 3. Decomposition of the combination of the 0ν nuclear matrix elements $M^{0\nu}_{\rm GT}-M^{0\nu}_{\rm F}$ into contributions through intermediate states with different spin-parities, calculated by the shell model ($g_{\rm pp}=1.0$) and QRPA ($g_{\rm pp}=0.92$). Contributions of $M^{0\nu}_{\rm GT}$ and $-M^{0\nu}_{\rm F}$ are separately shown by filled and open histograms, respectively

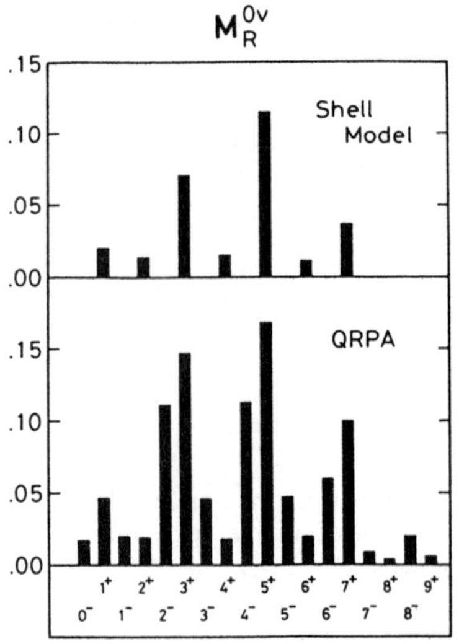

Fig. 4. Decomposition of the 0ν recoil matrix element $M^{0\nu}_{R}$ into contributions through intermediate states with different spin-parities, calculated by the shell model ($g_{\rm pp}=1.0$) and QRPA ($g_{\rm pp}=0.92$)

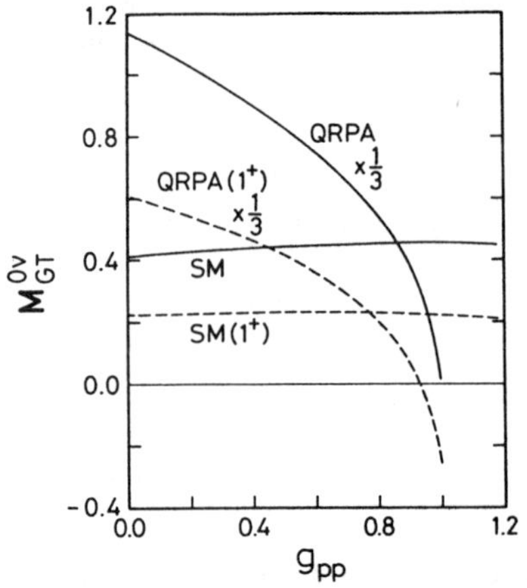

Fig. 5. Comparison of the 0ν nuclear matrix elements $M^{0\nu}_{\rm GT}$ (solid lines) calculated by the shell model (SM) and QRPA as function of $g_{\rm pp}$. Contributions through 1^{+} intermediate states are also shown by dashed lines

$M^{0\nu}_{GT}$, although both are successive Gamow-Teller type transitions through virtual 1^+ intermediate states. Compared to the Gamow-Teller operator $t_- \sigma$ of the 2ν decay, the transition operator for the 0ν decay consists of $t_- j_0(qr)\sigma$ and $t_- j_2(qr)[\sigma \times Y_2(\hat{r})]^{(1)}$, both of them have finite momentum transfers and the latter can induce a transition with a change of orbital angular momentum. The other difference lies in the perturbation energy denominator. The closure approximation can be applied to the 0ν decay and transition amplitudes through 1^+ intermediate states are summed up with equal weight. On the other hand, in $M^{2\nu}_{GT}$, the amplitudes are added with different weigths depending on the energy of the intermediate states (see (7)).

A considerable difference is found in the 0^+ component of $M^{0\nu}_{F}$ between the two model calculations (see Fig. 3); the QRPA value is large, while the corresponding component of the shell model is almost vanishing. The 0^+ component comes from the $L=0$ term of the multipole expansion of the neutrino potential,

$$H_m(|\mathbf{r}_1-\mathbf{r}_2|)=\sum_L (2L+1)\int dq q^2 v(q)[j_L(qr_1)\, Y_L(\hat{r}_1)] \cdot [j_L(qr_2)\, Y_L(\hat{r}_2)], \tag{18}$$

where $v(q)$ is the Fourier transform of $H_m(r)$, and the $L=0$ operator acting on a nucleon wave function resembles the isospin-lowering operator. The large QRPA value of the 0^+ component is probably a result of isospin breaking involved in the QRPA calculation, while the transition is almost completely forbidden in the isospin-scheme shell-model calculation. The isospin symmetry is violated by approximations; for example, through single-particle energies which are not consistent with the two-body interaction, and also by the model space truncation in a proton-neutron scheme shell-model calculation. Recently, Hirsch and Krmpotić discussed [34] restoration of isospin symmetry in the context of a role of $SU(4)$ symmetry for the suppression of the $\beta\beta$ decay. However, the isospin breaking would not be a serious problem, because $M^{0\nu}_{F}$ appears in the decay rate formula always in the combination of $M^{0\nu}_{GT}-M^{0\nu}_{F}$ and $M^{0\nu}_{GT}$ dominates over $M^{0\nu}_{F}$, and the breaking is expected to be smaller in nuclei with a larger neutron excess, such as most $\beta\beta$ emitters.

Table 1 lists nuclear matrix elements for the 0ν decay of ^{48}Ca and a typical nuclide ^{76}Ge, calculated with different nuclear structure models and effective interactions. The nuclear structure calculations given in the table are, in addition to the present QRPA and shell-model (SM) calculations; for ^{48}Ca: the shell model [35] in the same model space with a modified Kuo-Brown interaction [36, 37]; and for ^{76}Ge: our previous QRPA calculation [10], the shell model calculation in the proton-neutron weak coupling scheme [1] with the Kuo-Brown interaction [36], and a quasiparticle model based on Hartree-Fock-Bogoliubov with nucleon-number and angular-momentum projections (VAMPIR) [38, 39]. The eight matrix elements, except $M^{0\nu}_{GT}$, are shown by the ratio to $M^{0\nu}_{GT}$,

$$\chi_\alpha = M^{0\nu}_\alpha / M^{0\nu}_{GT}. \tag{19}$$

In each calculation, the Gamow-Teller type elements, $M^{0\nu}_{GT}$, $M^{0\nu}_{GT\omega}$ and $M^{0\nu}_{GTq}$, have similar values to each other, and this is also the case for the Fermi type elements, $M^{0\nu}_{F}$, $M^{0\nu}_{F\omega}$ and $M^{0\nu}_{Fq}$. However, the relative magnitudes of the Fermi type elements by QRPA are generally larger, for both ^{48}Ca and ^{76}Ge, than those by the shell-model and VAMPIR calculations. The same feature is seen in the tensor matrix element M_T and the recoil matrix element M_R. The discrepancy is partly caused by the isospin symmetry breaking of the QRPA calculation, and would partly be attributed to the model space truncation and to the different choices of the effective interaction. The recent QRPA study of the ^{76}Ge $\beta\beta$ decay with bare and renormalized G-matrices derived from Bonn, Paris and Reid potentials showed that the Fermi type elements are

Table 1. Nuclear matrix elements of the $0\nu\,\beta\beta$ decay of ^{48}Ca and ^{76}Ge calculated with different nuclear models and effective interactions, and limits on the neutrino mass and the coupling strengths of the right-handed currents deduced from experimental half-life limits (in the last row). The eight matrix elements are given by the ratio $\chi_\alpha = M^{0\nu}_\alpha / M^{0\nu}_{GT}$. See text for the nuclear structure calculations

	^{48}Ca			^{76}Ge		
	QRPA this work	SM this work	SM [35]	QRPA [10]	SM [1]	VAMPIR [38]
$M^{0\nu}_{GT}$	1.00	0.45	0.77	3.01	4.17	5.72
χ_F	− 0.62	− 0.09	− 0.14	−0.39	−0.20	−0.22
$\chi_{GT\omega}$	0.94	0.87	0.91	0.97	1.00	0.86
$\chi_{F\omega}$	− 0.52	− 0.09	− 0.14	−0.34	−0.20	−0.18
χ_{GTq}	0.67	0.89	1.09	0.65	1.14	1.14
χ_{Fq}	− 0.62	− 0.05	− 0.14	−0.35	−0.23	−0.26
χ_T	− 0.39	− 0.14	− 0.21	−0.20	−0.01	−0.03
χ_P	− 0.15	0.04	0.10	−0.18	0.27	−0.22
χ_R	1.02	0.63	0.81	1.19	−	0.42
$\langle m_\nu \rangle$ [eV]	< 30	< 96	< 54	< 2.6	< 2.2	< 1.6
$\langle \eta \rangle$	$< 3.5\cdot10^{-7}$	$< 1.3\cdot10^{-6}$	$< 6.0\cdot10^{-7}$	$< 2.5\cdot10^{-8}$	$< 7.5\cdot10^{-6}$	$< 3.8\cdot10^{-8}$
$\langle \lambda \rangle$	$< 2.6\cdot10^{-5}$	$< 1.0\cdot10^{-4}$	$< 5.5\cdot10^{-5}$	$< 4.0\cdot10^{-6}$	$< 4.9\cdot10^{-6}$	$< 2.7\cdot10^{-6}$
$[T^{0\nu}_{1/2}]^{exp}$ [y]		$> 2\cdot10^{21}$ [28]			$> 4.7\cdot10^{23}$ [40]	

more or less constant irrespective of the variety of interactions, whereas the Gamow-Teller elements are dependent on the effective interaction [22].

In the shell-model calculation, the change of the pairing and quadrupole interaction strengths induces a simultaneous change of all nuclear matrix elements in such a way that the ratios χ_α are approximately preserved. The limits of the neutrino mass and the right-handed current coupling strengths, for the change of V_{pair} and V_{QQ} by 20%, are in the ranges of $\langle m_\nu \rangle < (70-153)$ eV, $\langle \eta \rangle < (1.1-1.6)\times 10^{-6}$, and $\langle \lambda \rangle < (0.7-1.6)\times 10^{-4}$. The less sharp limits, compared to those of the QRPA calculation, will be reduced by taking a larger model space.

The present QRPA calculation gives a half-life of the $0\nu\,\beta\beta$ decay of ^{48}Ca,

$$[T_{1/2}^{0\nu}]^{QRPA} \approx 2\times 10^{24}\,\mathrm{y}$$

by assuming $\langle m_\nu \rangle = 1$ eV and neglecting possible contributions from the right-handed currents. The predicted half-life is longer than the present experimental limit [28] by three orders of magnitude.

5. Conclusion

We have calculated nuclear matrix elements of the $\beta\beta$ decay of ^{48}Ca by shell model and QRPA, which have been employed in previous calculations. The two models are different types of approximations to an exact solution of the many-body problem. These complementary calculations lead, by a detailed comparison and analysis of the calculated results, to a deeper insight into the understanding of the transition mechanism of $\beta\beta$ decay, and show a guide line along which the so far used model should be improved in future calculations of the $\beta\beta$ decay nuclear matrix elements.

The results for the 2ν matrix element $M_{GT}^{2\nu}$ obtained by both models are consistent with each other. This confirms the suppression mechanism of the 2ν decay suggested by the previous QRPA studies for heavier $\beta\beta$ nuclei; namely, the spin-isospin correlations in the ground-state wave function give rise to transitions which interfere destructively with those coming from 0^+-paired components of the wave function, especially the destructive interference results in a very strong suppression of β^+ transitions in neutron-rich nuclei, such as the inverse of ^{48}Sc$(1^+)\to{}^{48}$Ti$(0^+_{g.s.})$. The nuclear matrix element $M_{GT}^{2\nu}$ crosses through zero in both models in the reasonable range of the strength of the particle-particle interaction in the $J^\pi = 1^+$ channel, which is most responsible for the enhancement of the spin-isospin correlations. The suppression mechanism of the 2ν-decay nuclear matrix element is understood to be fundamental.

$M_{GT}^{2\nu}$ of QRPA is very sensitive to the 1^+ interaction. This is because the ground-state correlations are mapped onto the proton-neutron quasiparticle pairs with $J^\pi = 1^+$, and the correlations in this space are governed by the 1^+ interaction. The pairing correlations, which are also important, are taken into account in a separate calculation (the first step of the QRPA calculation), where the strength of the pairing interaction is fixed to experimental even-odd mass differences. Therefore, a careful choice of the 1^+ interaction strength enables a rather reliable evaluation of the 2ν decay nuclear matrix element.

On the other hand, all components of a nucleon-nucleon interaction are put together in the shell-model eigenvalue problem, and consequently the shell-model value of $M_{GT}^{2\nu}$ depends on various components of the effective interaction. The present study indicates that the pairing interaction (and the quadrupole interaction) plays a decisive role as well as the 1^+ interaction for the evaluation of $M_{GT}^{2\nu}$. The strength of the pairing interaction possibly involves significant uncertainties; for example, it is most affected by the renormalization orginating from the model-space truncation, and is dependent on the choice of diagrams which are included in the renormalization procedure. A fit of the pairing strength in a certain way will be required in the shell-model approach.

The result for the 0ν decay are not consistent in some aspects between the two model calculations. The discrepancies are traced mainly to the truncation of the shell-model configurations. The *fp*-space of the shell model does not seem to be large enough for the 0ν decay, although the shell-model calculation in the same model space gives consistent results with QRPA for the 2ν decay. The $0\nu\,\beta\beta$ decay occurs by the exchange of a virtual neutrino between two nucleons which make a $\beta\beta$ transition. The exchange of a neutrino brings about two main differences, compared to the 2ν decay; one is a resulting neutrino potential in the nuclear transition operators, and the other lies in finite momenta of typically ~ 100 MeV/c carried by the neutrino. The neutrino potential allows various spin-parities for the nuclear intermediate states, and the QRPA calculation shows that transitions from and/or to single-particle orbits far from the nucleon Fermi surface are enhanced by the large momentum transfers. This suggests that a shell-model calculation of a manageable size should include simple configurations in a large number of active orbits rather than complicated configurations in a small space.

A systematic discrepancy between the shell-model and QRPA calculations is found in the magnitudes of the Fermi type matrix elements, such as $M_F^{0\nu}$, relative to those of the Gamow-Teller type elements. The QRPA calculation violates the isospin symmetry to some extent, resulting in an overestimation of the Fermi type matrix elements, but only a part of the discrepancy would be accounted for by isospin breaking.

In spite of the shortcomings of the nuclear structure models, the limits on the important quantities, the neutrino mass and right-handed current strengths, deduced from the experimental half-life agree within a factor of about three. Uncertainties in the predicted nuclear matrix elements, and consequently those of the deduced limits, will be reduced in future calculations which are improved along the line suggested in the present work.

One of the authors (E.B.) is grateful for financial support to Professor E. Arai of the Research Laboratory for Nuclear Reactors,

444

Tokyo Institute of Technology, and to the Centennial Commemoration Fund of Tokyo Institute of Technology. Numerical calculations were performed with FACOM M780/M380 computer systems of the Institute for Nuclear Study, University of Tokyo.

References

1. Haxton, W.C., Stephenson, G.J., Jr. :Prog. Part. Nucl. Phys. **12**, 409 (1984)
2. Doi, M., Kotani, T., Takasugi, E.: Prog. Theor. Phys. [Suppl.] **83**, 1 (1985)
3. Muto, K., Klapdor, H.V.: In :Klapdor, H.V. (ed.), Neutrinos. Berlin Heidelberg New York: Springer 1988
4. Vogel, P., Zirnbauer, M.R.: Phys. Rev. Lett. **57**, 3148 (1986)
5. Civitarese, O., Faessler, A., Tomoda, T.: Phys. Lett. **194**B, 11 (1987)
6. Muto, K., Klapdor, H.V.: Phys. Lett. **201**B, 420 (1988)
7. Muto, K., Bender, E., Klapdor, H.V.: Z. Phys. A – Atomic Nuclei **334**, 177 (1989)
8. Tomoda, T., Faessler, A.: Phys. Lett. **199**B, 475 (1987)
9. Engel, J., Vogel, P., Zirnbauer, M.R.: Phys. Rev. C**37**, 731 (1988)
10. Muto, K., Bender, E., Klapdor, H.V.: Z. Phys. A – Atomic Nuclei **334**, 187 (1989)
11. Vinh Mau, R.: Mesons in Nuclei, Vol. 1, p. 151. Rho, M., Wilkinson, D.H. (eds.). Amsterdam: North-Holland 1979; Lacombe, M., Loiseau, B., Richard, J.M., Vinh Mau, R., Côté, J., Pirès, P., Tourreil, R. de: Phys. Rev. C**21**, 861 (1980); Anantaraman, N., Toki, H., Bertsch, G.F.: Nucl. Phys. A**398**, 269 (1983)
12. Holinde, K.: Phys. Rep. **68**, 121 (1981)
13. Haxton, W.C., Stephenson, G.J., Jr., Strottman, D.: Phys. Rev. D**26**, 1805 (1982)
14. Halbleib, J.A., Sorensen, R.A.: Nucl. Phys. A**98**, 542 (1967)
15. Muto, K., Bender, E., Klapdor, H.V.: Z. Phys. A – Atomic Nuclei **333**, 125 (1989)
16. Bertsch. G.F.: The practitioners shell model. Amsterdam: North-Holland 1972
17. Burrows, T.W.: Nucl. Data Sheets **48**, 1, 569 (1986)
18. Bohr, A., Mottelson, B.R.: Nuclear Structure, Vol. I. New York: Benjamin 1969
19. Wapstra, A.H., Audi, G., Hoekstra, R.: At. Data Nucl. Data Tables **39**, 281 (1988)
20. Shalit, A. de, Talmi, I.: Nuclear Shell Theory. New York: Academic Press 1963
21. Horie, H.: J. Phys. Soc. Jpn. **19**, 1783 (1964); Horie, H., Oda, T.: Prog. Theor. Phys. [Suppl.], 315 (1968)
22. Staudt, A., Kuo, T.T.S., Klapdor, H.V.: Phys. Lett. **242**B, 17 (1990)
23. Grotz, K., Klapdor, H.V.: Nucl. Phys. A**460**, 395 (1986); Klapdor, H,V., Grotz, K.: Phys. Lett. **142**B, 323 (1984)
24. Engel, J., Vogel, P., Civitarese, O., Zirnbauer, M.R.: Phys. Lett. **208**B, 187 (1988)
25. Haxton, W.C., Stephenson, G.J., Jr., Strottman, D.: Phys. Rev. D**25**, 2360 (1982)
26. Tsuboi, T., Muto, K., Horie, H.: Phys. Lett. **143**B, 293 (1984)
27. Wu, H.F., Song, H.Q., Kuo, T.T.S., Chang, W.K., Strottman, D.: Phys. Lett. **162**B 227 (1985)
28. Bardin, R.K., Gollon, P.J., Ullman, J.D.-, Wu, C.S.: Nucl. Phys. A**158**, 337 (1970)
29. Elliott, S.R., Hahn, A.A., Moe, M.K.: Phys. Rev. Lett. **59**, 2020 (1987)
30. Kirsten, T., Heusser, E., Kaether, D., Oehm, J., Pernicka, E., Richter, H.: In: Kotani, T., Ejiri, H., Takasugi, E. (eds.). Proc. Int. Symp. on Nuclear Beta Decays and Neutrino, p. 81. Singapore: World Scientific 1986
31. Schechter, J., Valle, J.W.F.: Phys. Rev. D**25**, 2951 (1982); Nieves, J.F.: Phys. Lett. **147**B, 375 (1984); Takasugi, E.: Phys. Lett. **149**B, 372 (1984)
32. Miller, G.A., Spencer, J.E.: Ann. Phys. **100**, 562 (1976)
33. Vergados, J.D.: Phys. Rev. C**24**, 640 (1981); Nucl. Phys. B**218**, 109 (1983)
34. Hirsch, J., Krmpotić, F.: Phys. Rev. C**41**, 792 (1990)
35. Muto, K.: In: Kotani, T., Ejiri, H., Takasugi, E. (eds.). Proc. Int. Symp. on Nuclear Beta Decays and Neutrino, p. 177. Singapore: World Scientific 1986
36. Kuo, T.T.S., Brown, G.E.: Nucl. Phys. A**114**, 241 (1968)
37. McGrory, J.B., Wildenthal, B.H.: Phys. Lett. **103**B, 173 (1981)
38. Tomoda, T., Faessler, A., Schmid, K.W., Grümmer, F.: Nucl. Phys. A**452**, 591 (1986)
39. Schmid, K.W., Grümmer, F., Faessler, A.: Nucl. Phys. A**431**, 205 (1984)
40. Caldwell, D.O., Eisberg, R.M., Grumm, D.M., Witherell, M.S.: Phys. Rev. Lett. **59**, 419 (1987)

2.2.4 The Operator Expansion Method

Commun. in Theor. Phys. (Beijing, China) Vol.10, No.1, 1988 45-51

AN ALTERNATIVE APPROACH IN NUCLEAR DOUBLE BETA DECAY THEORY

C.R. CHING(庆承瑞) and T.H. HO(何祚庥)

Centre of Theoretical Physics, CCAST (World Laboratory)
and the Institute of Theoretical physics, Academia Sinica
Beijing, China

Received December 22,1987

Abstract

A new formula as a series of commutators of two axial vector currents and the nuclear Hamiltonian H_S is derived for estimating the 2β-decay nuclear matrix element without using explicitly the closure approximation. With a simple assumption of the nuclear wave function and the Hamiltonian H_S, it is shown that the leading term in 2ν-2β decay matrix element vanishes and thus the smallness of 2ν-2β decay probability seems to be understood.

I. Intorduction

It has been noted that in order to keep pace with the longer and longer nuclear 2ν-2β decay lifetime obtained from the experiments, the theorists have to try to reduce the nuclear matrix elements of 2β-decay by introducing different pairing forces and, indeed, there are some indications that these efforts have met with success, bringing the theoretical estimates in agreement with the experiment ones[1,2]. However, since the experimental results obtained from the direct laboratory measurements are the lower limits of the lifetime[3] it is therefore, very probable that the agreement might be temporary. Nevertheless, one thing is certain: the nuclear matrix element of 2β-decay of these most studied nuclei, namely ^{48}Ca, ^{76}Ge, ^{82}Se, and in particular, the nuclear matrix elements of 2ν-2β decay are very small. It is very desirable to have an additional argument which can illustrate the smallness of the nuclear matrix elements without going to the detail of the nuclear structure, and provide at the same time a new method of estimating the nucler matrix elements without using the closure approximation[4].

In this paper we first derived a new formula as series of commutators for calculating the nuclear matrix element, of both 2ν-2β and 0ν-2β transitions. Then we compared our formulas with those used in the literatures[4] And finally with a simple assumption of the nuclear wave functions we showed that the leading term in the 2ν-2β decay matrix element vanishes and the approach hopefully provides an alternative way to calculate the 2β-decay probability and it will be interesting to compare quantitatively the estimates with thoes obtained by the conventional method.

II. Derivation

The basic assumptions we adopted for estimating the 2β-decay probability are as the same as that used in the conventional approach. They are:

i). Second order V-A weak interaction;

ii). Two-nucleon mechanism;

iii). The nuclear energy states are the eigenstates of the isospin, and;

iv). The nonrelativistic approximation.

With these assumptions the standard perturbation calculation leads to the following expression for 2ν-2β transition matrix element:

$$M_{fi}^{2\nu 2\beta}=i\left(\frac{G_F}{\sqrt{2}}\right)^2\sum_n\left\{\frac{\langle f|A_k|n\rangle\langle n|A_i|i\rangle j_k(e_1\nu_1)j_i(e_2\nu_2)}{E_n+\nu_{20}+p_{20}-E_i}\right.$$

$$+\frac{\langle f|A_i|n\rangle\langle n|A_k|i\rangle j_i(e_2\nu_2)j_k(e_1\nu_1)}{E_n+\nu_{10}+p_{10}-E_i}-\frac{\langle f|A_k|n\rangle\langle n|A_i|i\rangle j_k(e_2\nu_1)j_i(e_1\nu_2)}{E_n+\nu_{20}+p_{10}-E_i}$$

$$\left.-\frac{\langle f|A_i|n\rangle\langle n|A_k|i\rangle j_i(e_1\nu_2)j_k(e_2\nu_1)}{E_n+\nu_{10}+p_{20}-E_i}\right\}, \quad (1)$$

where A_i is the space component of the nuclear axial vector current, and

$$A_i=\sum_{\ell=1}^{A}\sigma_i^{(\ell)}\tau_\ell^+ \qquad i=1,2,3, \quad (2)$$

with σ_i the usual pauli matrix, and τ^+ the charge raising operator. $j_i(e_j,\nu_k)$ is the corresponding lepton current composed of the wave functions of the j-th electron and k-th antineutrino with j,k=1,2. E_n and E_i are, respectively, the total energies of the intermediate and the initial nuclear states, ν_{j0} and p_{k0} are the total energies of the j-th neutrino and k-th electron respectively.

Now using the energy conservation condition

$$E_i-\nu_{j0}-p_{k0} = E_f+\nu_{k0}+p_{j0}.$$

We transform the denominators of the second and fourth term in Eq. (1) to the corresponding expression involving only E_n and E_f . And introducing an energy parameter Δ as

$$E_i - E_f \equiv 2\Delta \qquad (3)$$

and taking an average $\nu_{j0}+p_{k0}=\Delta$ approximately, we expand formally Eq. (1) in Taylor series with respect to Δ:

$$M_{fi}^{2\nu 2\beta}=i\left(\frac{G_F}{\sqrt{2}}\right)^2 \left\{ \frac{1}{\Delta}\sum_n <f|A_k|n>\left(1-\frac{E_n-E_i}{\Delta}+\frac{(E_n-E_i)^2}{\Delta^2}-\ldots\right)<n|A_i|i>\right.$$
$$\left. -\frac{1}{\Delta}\sum_n <f|A_i|n>\left(1-\frac{E_f-E_n}{\Delta}+\frac{(E_f-E_n)^2}{\Delta^2}-\ldots\right)<n|A_k|i>\right\} [j_k(e_1\nu_1)j_i(e_2\nu_2)$$
$$- (e_1 \leftrightarrows e_2)]. \qquad (4)$$

Now we define a total nonrelativistic Hamiltonian of the A nucleons' system H_s so that E_i, E_f and E_n, apart from A nucleon's rest mass, are the eigenvalues of the corresponding eigenstates

$$H_s|n>= \varepsilon_n|n>,$$
$$H_s|i>= \varepsilon_i|i>,$$
$$H_s|f>= \varepsilon_f|f>,$$

where ε is the binding energy of the nuclear system. Then it is easy to show that

$$M_{fi}^{2\nu 2\beta}= i\left(\frac{G_F}{\sqrt{2}}\right)^2\frac{1}{\Delta} \{ <f|[A_k,A_i]|i>+ \frac{1}{\Delta}<f|[A_k,[H_s,A_i]]|i>$$
$$+ \frac{1}{\Delta^2}<f|[A_k,[H_s,[H_s,A_i]]]|i>+\ldots\ldots\} [j_k(e_1\nu_1)j_i(e_2\nu_2)-(e_1\leftrightarrows e_2)]. \qquad (5)$$

The same method can also be applied to 0ν-2β decay with the only difference that in the $o\nu$-2β case the expansion is performed with respect to the virtual neutrino energy E_ν, and the transition matrix element now reads

$$M_{fi}^{0\nu 2\beta}\propto\left(\frac{G_F}{\sqrt{2}}\right)^2\int\frac{d^3\vec{\nu}}{(2\pi)^3}\frac{m_\nu}{E_\nu^2}\left\{<f|\{J_\mu,J_\lambda\}|i> + \frac{<f|\{J_\mu,[H_s,[H_s,J_\lambda]]\} \quad |i>}{E_\nu^2}\right.$$
$$\left.+ O\left(\frac{1}{E_\nu^4}\right) +\ldots\ldots\right\} [\bar{u}(\vec{p}_1)\gamma_\mu (1-\gamma_5) \gamma_\lambda\upsilon(\vec{p}_2)], \qquad (6)$$

where J_μ is the nuclear weak current. If only the axial vector current is retained, then instead of J_μ and J_μ we shall have the space components A_i and A_j.

It is interesting to note that the structures of the transition matrix element of 2ν-2β and 0ν-2β decay are quite different. In the former case the series contains the commutators of two axial currents, while in the latter case only anticommutators of two axial currents appear.

III. A Discussion of the Validity of the Expansion

The expressions of Eq.(4) and Eq.(5) we have just obtained rely entirely on the validity of Taylor expansion. Generally it is believed that in 0ν-2β decay the virtual neutrino energy is high and therefore a Taylor expansion with respect to E_ν can be justified. In fact, it easily sees that the first term in Eq.(6) is just the usual expression for 0ν-2β decay matrix element obtained by using the closure approximation. The additional term may give rise to an important correction especially as far as light neutrino is concerned. For 2ν-2β decay the situation is more complicated. Since a typical magnitude of the energy parameter Δ is only 1-2MeV, and the ground state of the intermediate nucleus lies higher than both the initial and final nuclear systems. Therefore it seems that the Taylor expansion as in Eq.(5) is illegal. However, as in every term of this expansion (see Eq.(4)) one must sum over the complete set of the intermediate states, one does not know a priori how strongly the cancellation occurs between different states. In other words, we cannot prove either the convergence or the divergence of Eq.(5). Therefore the way we shall follow is to try to estimate the nuclear matrix element given by the series (5) by using a reasonable Hamiltonian and wave functions.

First of all, the nuclear Hamiltonian H_s we are using has the following expression:

$$H_s = \sum_i^A - \frac{-\nabla_i^2}{2m_N} + V_c + V_s \ , \qquad (7)$$

where m_N is the nuclear mass, V_c is the Coulomb energy:

$$V_c = \frac{e^2}{8} \sum_{i\neq j}^A (1+\tau_3^{(i)})(1+\tau_3^{(i)}) \frac{1}{\gamma_{ij}} \ , \qquad (8)$$

where $\vec{\tau}$ is the Pauli matrices in the isospin space. The nuclear part of the effective potential energy V_s which we shall use is of the following form[5]

$$V_s = V_0 + V_\sigma + V_\tau + V_{\sigma\tau} , \tag{9}$$

where V_0 only depends upon the nucleon space coordinates, and

$$\left.\begin{aligned} V_\sigma &= \frac{g'_\sigma}{2} \sum_{i \neq j} (\vec{\sigma}_i \cdot \vec{\sigma}_j)\, V_\sigma(\gamma_{ij}) , \\ V_\tau &= \frac{g'_\tau}{2} \sum_{i \neq j} (\vec{\tau}_i \cdot \vec{\tau}_j)\, V_\tau(\gamma_{ij}) , \\ V_{\sigma\tau} &= \frac{g'_{\sigma\tau}}{2} \sum_{i \neq j} (\vec{\tau}_i \cdot \vec{\tau}_j)(\vec{\sigma}_i \cdot \vec{\sigma}_j)\, V_{\sigma\tau}(\gamma_{ij}) . \end{aligned}\right\} \tag{10}$$

The radial parts of all these potentials V_σ , V_τ and $V_{\sigma\tau}$ are, of course, of short range, and for simplicity we assume all of them are δ -force:

$$V(\gamma_{ij}) = \delta(\gamma_i - \gamma_j), \tag{11}$$

and g'_σ, g'_τ and $g'_{\sigma\tau}$ are the corresponding effective coupling strength.

We still need some general information of the wave functions. Let us first discuss an idealized case, where the two neutrons that transform to two protons during 2β-decay are in the same j-state outside a closed shell. Then the symmetry consideration shows that for both j-j or L-S coupling L=0, S=0, and L=1, S=1 are possible in the ground state of even-even nucleus. But since the pairing forces are of short range, and in a limiting case of δ -force Eq.(11), the odd L-component vanishes[6] so that it is a good approximation to consider the two-nucleon pair always in the spin-singlet state.

Now we are able to estimate the 2ν-2β nuclear matrix element by evaluating the commutators in Eq.(5). First of all we note that the kinetic energy as well as V_0 in V_s (see Eq.(9)) does not contribute to the commutators in Eq.(5). And we also neglect the Coulomb interactions V_c in accordance with one of the basic assumptions we made that the isospin is a good quantum number. Therefore we need only to consider $V_{\sigma\tau}$, V_σ and V_τ in Eq.(10).

The first commutator in Eq.(5)

$$\langle f|\, [A_i, A_k]\, |i\rangle = 0 \tag{12}$$

due to the fact that the two particles cannot be emitted from the same nucleon. And a direct calculation shows

$$\langle S=0\, |[A_i,[V_{\sigma\tau},A_k]]|\, S=0\rangle = 0 . \tag{13}$$

Furthermore, if use is made of Eq.(13) and Jacobian identity, one also has

$$\langle S=0\,|[A_i,[V_{\sigma\tau},[V_{\sigma\tau},A_k]]]|\,S=0\rangle=0 \quad . \tag{14}$$

If we go further to higher and higher order commutators, then upon neglect of terms involving more than two-particle correlations, one can show

$$\langle S=0\,|[A_i,\ [V_{\sigma\tau}^{(1)}\ldots\ldots[V_{\sigma\tau}^{n},A\ ]\ldots\ldots]|\,S=0\rangle=0 \ , \tag{15}$$

if all (n-1) commutators equal to zero.

But unfortunately it is not the case for V_σ and V_τ separately. However, it is easy to realize that $V_{\sigma\tau}$-commutator contains both the V_σ-commutator and V_τ-commutator with equal weight. Therefore the same conclusions Eq.(13)-Eq.(15) hold for $V_\tau+V_\sigma$ if

$$g'_\sigma \approx g'_\tau \ . \tag{16}$$

From the analysis given in reference[7] the strength of $V_{\sigma\tau}$, i.e. $g'_{\sigma\tau}$, is insensitive to energy, while for the coupling strength g'_τ a strong energy dependence has been found[8]. The general trend is: apart from very small incoming particle energy in the (p,n) reaction, $g'_{\sigma\tau}$ always prevails over g'_τ in a wide energy range. As for the nucleon-nucleon interaction inside a neucleus we don't know exactly what the energy range one should take, but if judging from the relative velocity of nucleons in nucleus, the energy cannot be very small, therefore it is reasonable to think $V_{\sigma\tau}$ in Eq.(9) as the dominant part. And in comparing with $V_{\sigma\tau}$ and V_τ, V_σ is less known, and the conclusions seem disputable[9,10]. But for not very high energy V_σ is also small[5] and some authors prefer[9,10]

$$g'_\sigma \gtrsim 0.3\ g'_{\sigma\tau} \quad .$$

If we insist that interior the nuclear matter the following relation holds:

$$g'_\sigma \approx g'_\tau \quad .$$

Then the contributions arisen from V_σ and V_τ cancel with each other, and consequently we have

$$\langle S=0|\ [A_i,[H_s^{(1)}\ldots\ldots[H_s^{n},A_k]\ldots\ldots]|\ S=0\rangle = 0 \tag{17}$$

and the nuclear matrix element under the approximations we adopted equals to zero even for the most favorable transition where the initial and final nuclei have the same configuration. In practice it would mean that the commutator series Eq.(5) might converge rather rapidly. And if so, Eq.(5) provides an interesting way for estimating 2ν-2β decay matrix element without using the closure approximation. It seems also that we have a new argument to understand why the 2ν-2β decay probability is so small. The same method can also be applied in 0ν-2β decay, giving an additional correction (see Eq. (6)).

References

[1] O. Civitarese, et al., Phys. Lett. B194(1987)11.

[2] P. Vogel and M.R. Zirnbauer, Phys. Rev. Lett. 57(1986)3148.

[3] D.O. Caldwell, et al., Phys. Rev. D33(1986)2739; S.R. Elliot, et al., Phys. Rev. Lett. 56(1986)2582; H. Ejiri, et al., J. Phys. G: Nucl. Phys. 13(1987) 839; A.S. Barabash, et al., "Nuclear Beta Decay and Neutrions," Eds. T. Kotani, H.Ejiri and E. Takasugi, World Scientific (1986), Singapore.

[4] See, for instance, W.C. Haxton and G.J. Stephenson, Jr, Progress in Part. and Nucl. Phys. 12(1984)409.

[5] S.O. Backman and G.E. Brown, Phys. Rep. 124(1985)1.

[6] L.C. Bildenharn and J.D. Louck, "Angular Momentum in Quantum Physics: Theory and Application" (Encyclopadia of Mathematics and its Applications V.8), Addison-Wesley Publishing Company (1981).

[7] G.E. Brown, J. Speth and J. Wamback, Phys. Rev. Lett. 46(1981)1057.

[8] C.D. Goodman et al., Phys. Rev. Lett. 44(1980)1755; B.D. Anderson et al., Phys. Rev. Lett. 45(1980)699; D. Bainum et al., Phys. Rev. Lett. 44(1980)1751.

[9] S. Krewald and J. Speth, Phys. Rev. Lett. 45(1980)417.

[10] J. Speth et al., Nucl Phys. A343(1980)382.

[11] S.O. Backman, O. Sjoberg and A.D.Jackson, Nucl. Phys. A321(1979)10.

Co..mmun. Theor. Phys. 11(1989)495–497

A Discussion of the Question in Connexion with the Expansion of a Divergent Series Encountered in Calculating Nuclear 2ν–2β Decay Matrix Element

C.R. CHING and T.H. HO
Centre for Theoretical Physics, CCAST (World Laboratory) and
Institute of Theoretical Physics, Academia Sinica
P. O. Box 2735, Beijing 100080, China

(Received February 14, 1989)

Abstract

It is shown that the expansion and summation of the divergent series occurring in 2ν–2β decay calculation[1,2] can be well-defined without mathematical flaw.

Recently we suggested a new method for calculating the nuclear 2ν–2β decay without using the closure approximation[1,2]. This approach, however, has met with some criticisms. The main objection is: we expanded the energy denominators occurring in any second-order perturbation calculation into power series which are far beyond the convergence radius.

In this note, we would like to discuss this question more concretely. In 2ν–2β transition matrix element there appears a pair of energy denominators

$$\frac{1}{E_n - E_i + \nu_{20} + p_{20}} \quad \text{and} \quad \frac{1}{E_n - E_f - \nu_{20} - p_{20}} .$$

Now let Δ_{22} stand for $\nu_{20} + p_{20}$, the total energy of an electron and a neutrino. Then we expand these two denominators into power series by using binomial theorem or performing the division:

$$\begin{aligned} \frac{1}{E_n - E_i + \Delta_{22}} &= \frac{1}{\Delta_{22}}\left(1 - \frac{E_n - E_i}{\Delta_{22}} + \frac{(E_n - E_i)^2}{\Delta_{22}^2} - \cdots\right) , \\ \frac{1}{E_n - E_f - \Delta_{22}} &= \frac{-1}{\Delta_{22}}\left(1 - \frac{E_f - E_n}{\Delta_{22}} + \frac{(E_f - E_n)^2}{\Delta_{22}^2} - \cdots\right) . \end{aligned} \tag{1}$$

It is well known that Δ_{22} is approximately 1∼ 2 MeV in most nuclear 2ν–2β decay, and the average $(E_n - E_i)$, or $(E_n - E_f)$, as a generally accepted fact, is about 7∼10 MeV[3]. The natural question is: do we have the right to expand these denominators and then resume them in the same spirit?

In fact, such a kind of questions had been studied extensively long before by many mathematicians and was summarised in a book "*Divergent Series*" written by G.H. Hardy[4]. What we did is just to follow the recipe given in this book.

A sum of a series (including divergent series) S is defined as satisfying the following axioms[5]:

$$\text{If}\quad \sum_n a_n = S\,, \qquad \text{then}\quad \sum_n k a_n = kS\,. \tag{I}$$

$$\text{If}\quad \sum_n a_n = S\,, \qquad \text{and}\quad \sum_n b_n = t\,, \qquad \text{then}\quad \sum_n (a_n + b_n) = S + t\,. \tag{II}$$

$$\text{If}\quad a_0 + a_1 + a_2 + \cdots = S, \text{ then }\quad a_1 + a_2 + \cdots = S - a_0\,, \text{ and } \textit{vice versa.} \tag{III}$$

Then using (I)—(III), one easily sums the right-hand side series of Eq. (1) as

$$\begin{aligned} S &= \left[1 - \frac{E_n - E_i}{\Delta_{22}}\left(1 - \frac{E_n - E_i}{\Delta_{22}} + \frac{(E_n - E_i)^2}{\Delta_{22}^2} - \cdots\right)\right] \\ &= \left(1 - \frac{E_n - E_i}{\Delta_{22}} S\right)\,, \end{aligned} \tag{2}$$

and hence

$$S = \frac{1}{1 + (E_n - E_i)/\Delta_{22}}\,.$$

and conversely it is also true. This derivation, of course, is formalistic. However, using the method of Euler, we are led to the same result, and the meaning of the sum of S can be defined more rigorously. This method states as follows.

If $\sum a_n x^n$ is convergent for small x, and we can define a function $f(x)$ of the complex variable x, which is one-valued and regular in an open and connected region containing the origin and the point $x = 1$, and $f(1) = S$, then we call S the sum of $\sum a_n$[6]. The value of S may naturally depend on the chosen region.

Thus let us examine $M_{fi}^{2\nu-2\beta}(x)$ instead of $M_{fi}^{2\nu-2\beta}$, where

$$\begin{aligned} M_{fi}^{2\nu-2\beta}(x) = i\left(\frac{G_F}{\sqrt{2}}\right)^2 \sum_n \Big\{ &\frac{\langle f \mid A_k \mid n\rangle\langle n \mid A_i \mid i\rangle}{\Delta_{22} + (E_n - E_i)x} j_k(e_1\nu_1) j_i(e_2\nu_2) \\ &- \frac{\langle f \mid A_i \mid n\rangle\langle n \mid A_k \mid i\rangle}{\Delta_{22} + (E_f - E_n)x} j_i(e_2\nu_2) j_k(e_1\nu_1) - (e_1 \rightleftharpoons e_2)\Big\} \end{aligned} \tag{3}$$

and

$$M_{fi}^{2\nu-2\beta} = \lim_{x=1} M_{fi}^{2\nu-2\beta}(x)\,, \tag{4}$$

here all the notations are the same as in Ref. [2].

Then for very small x, $M_{fi}^{2\nu-2\beta}(x)$ can be expanded in terms of the binomial theorem. So that we can get all the equations through Eq. (4) to Eq. (17) of Ref. [2] with x standing before every operator H_s . The resummation is also performed with x standing before every potential functions ν_σ, ν_τ and $\nu_{\sigma\tau}$. And after performing the limiting procedure $x \to 1$ we are led to Eq. (17) of Ref. [2].

The essential point here lies in that how to take the order of limit. What we did can be simplified as follows. Suppose, for example, that

$$M(n,\ x) = \sum_{i=0}^{n} a_i x^i\,.$$

let

$$BM = \lim_{x=1} \sum_{i=0}^{n} a_i x^i = \sum_{i=0}^{n} a_i$$

and

$$PM = \lim_{n=\infty} \lim_{x=1} \sum_{0}^{n} a_i x^i$$

if and only if $\sum_{0}^{\infty} a_i$ is convergent. However, we can also have

$$AM = \lim_{n=\infty} \sum_{0}^{n} a_i x^i = \sum_{0}^{\infty} a_i x^i = f(x)$$

whenever the last series is convergent for $x \ll 1$. And now let

$$QM = \lim_{x=1} AM \equiv M .$$

In many cases PM does not exist while QM does. Then we may take QM as the definition[7]. In this case $f(x)$ defines an analytic function regular at the origin and M is obtained as its analytic continuation. And there is no ambiguity in our calculation since the generating function we have started from is well-defined and regular in x-complex plane. And the resummation is performed in exactly the same way but reversed after some approximations are taken.

Acknowledgments

we are grateful to Prof. H.W. PENG for beneficial discussions.

References

[1] C.R. CHING and T.H. HO, Commun. Theor. Phys. 10(1988)45.
[2] C.R. CHING and T.H. HO, preprint AS-ITP-88-004.
[3] See, for instance, W.C. Haxton and G.J. Stephenson, Jr, Progress in Particle and Nuclear Physics 12(1984)409.
[4] G.H. Hardy, *Divergent Series*, Oxford at the Clarendon Press, 1949.
[5] ibid, p.6.
[6] ibid, p.7.
[7] ibid, p.89.

Краткие сообщения ОИЯИ №6[39]-89 *JINR Rapid Communications No.6[39]-89*
УДК 539.165

A NEW METHOD OF THE CALCULATION OF THE TWO-NEUTRINO DOUBLE BETA DECAY AMPLITUDES

F.Šimkovic *

A new method is proposed of the calculation of the two-neutrino double-beta decay amplitude, which does not use the spectrum of intermediate nuclear states. The method is based on evaluation of a series of commutators of the nuclear Hamiltonian and weak nuclear hadron current. As a result, two-neutrino double-beta decay amplitude with a nuclear matrix element of a simple form of the two-nucleon operator has been obtained.

The investigation has been performed at the Laboratory of Theoretical Physics, JINR.

Новый метод вычисления амплитуды двухнейтринного двойного бета-распада

Ф.Шимковиц

Предложен новый метод вычисления скорости двухнейтринного двойного бета-распада, основанный на коммутационных соотношениях ядерного гамильтониана и слабых ядерных адронных токов. Метод не требует построения спектра "промежуточного" ядра. Получено замкнутое выражение для амплитуды данного процесса.

Работа выполнена в Лаборатории теоретической физики ОИЯИ.

Introduction

Recent first experimental observation of the two-neutrino mode of the double beta decay of ^{82}Se /1/ has revived interest in this process both among theoreticians and èxperimentalists.

The two-neutrino emitting mode of the double beta decay ($2\nu 2\beta$):

$$(A,Z) \to (A,Z+2) + 2e^- + 2\tilde{\nu}_e \,, \qquad (1)$$

* *Permanent address: Comenius University, Bratislava, Czechoslovakia.*

ИЗВЕСТИЯ АКАДЕМИИ НАУК СССР
Т. 54, № 9 СЕРИЯ ФИЗИЧЕСКАЯ 1990

УДК 539.165.2

М. ГМИТРО, Ф. ШИМКОВИЦ

НОВОЕ В ТЕОРИИ ДВОЙНОГО β-РАСПАДА *

Рассмотрены расчеты периода полураспада двухнейтринного двойного β-распада методом случайных фаз и прямым вычислением коммутаторов слабых адронных токов.

Введение

Теорией слабого взаимодействия предсказывается существование нескольких типов двойного β-распада. Экспериментально подтверждено лишь существование двухнейтринного двойного β-распада ($2\nu2\beta$):

$$A(Z) \to A(Z+2) + 2e^- + 2\tilde{\nu}_e. \quad (1)$$

Эллиотт, Хан и Мо [1] определили в изящном эксперименте для реакции

$$^{82}Se \to {}^{82}Kr + 2e^- + 2\tilde{\nu}_e \quad (2)$$

период полураспада

$$T_{1/2}(^{82}Se;\ 2\nu2\beta) = \left(1{,}1\ {}^{+0{,}8}_{-0{,}3}\right)\cdot 10^{20}\ \text{лет}. \quad (3)$$

Это значение придает вес ранним геохимическим данным.

Для реакции безнейтринного двойного β-распада ($0\nu2\beta$)

$$A(Z) \to A(Z+2) + 2e^- \quad (4)$$

экспериментально определена лишь нижняя граница [2]:

$$T_{1/2}(^{76}Ge;\ 0\nu2\beta) \geqslant (5-10)\cdot 10^{23}\ \text{лет}. \quad (5)$$

Среди более экзотических типов 2β-распада наиболее часто обсуждается двойной β-распад с испусканием нейтральной частицы «майорона»:

$$A(Z) \to A(Z+2) + 2e^- + M. \quad (6)$$

Сообщение об экспериментальном обнаружении этого распада не нашло подтверждения в последующих работах.

Двухнейтринный двойной β-распад [3—7] представляет собою очень редкий процесс. Однако существование такого распада не нарушает представлений теории Вайнберга — Салама о физике слабого взаимодействия. Процесс в современном понимании интересен сам по себе не только ввиду своей исключительной редкости, но и потому, что дает возможность проверить адекватность методов теории ядра, применяемых в аналогичных ситуациях, в частности для извлечения физической информации из данных по ($0\nu2\beta$)-распаду.

Двойной β-распад с испусканием майорона описывается в некоторых специальных моделях несохранения лептонных чисел [3—6]. Экспериментальное обнаружение такого процесса привело бы к достаточно глубоким изменениям наших представлений о природе слабого взаимодействия. Наиболее интенсивно в настоящее время исследуется, однако, ($0\nu2\beta$)-распад [3—7]. Его наблюдение в природе равносильно следующим утверждениям: 1) лептонные числа не сохраняются, 2) хотя бы один вид нейтрино является майорановской частицей ($\bar{\nu}=\nu$) с ненулевой массой.

* Обзорный доклад.

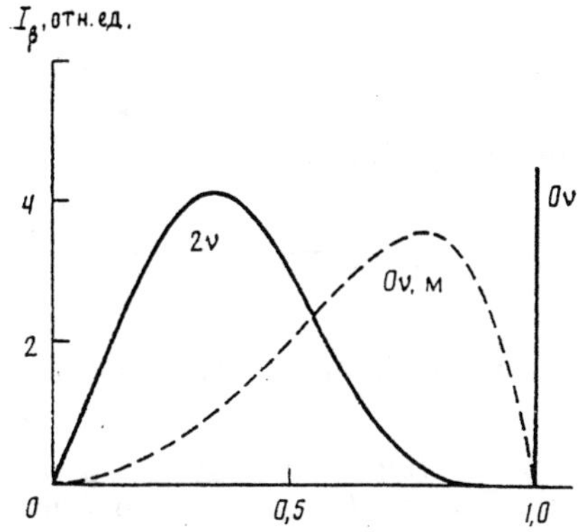

Рис. 1

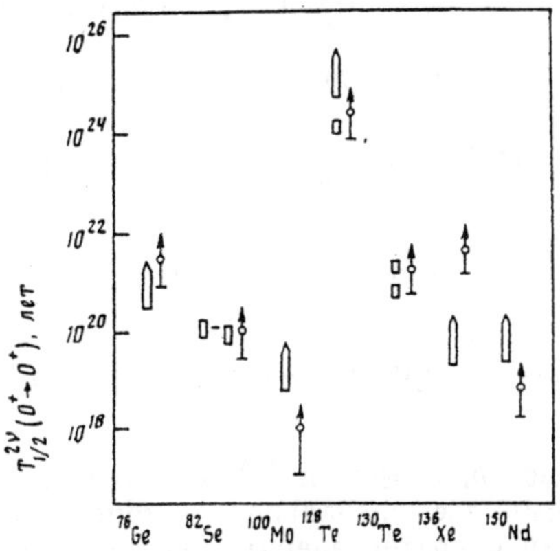

Рис. 3

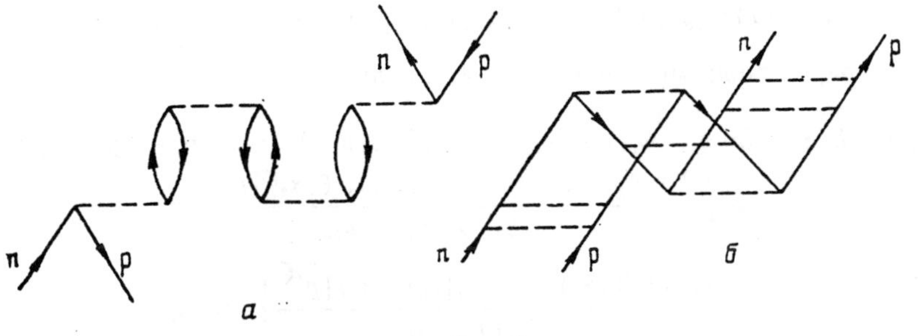

Рис. 2

Рис. 1. Суммарная энергия электронов, испускаемых при двойном β-распаде для 2ν- и 0ν-распадов и распада, связанного с испусканием майорона. По оси абсцисс $(e_1+ +e_2-2m_e)/T_0$

Рис. 2. Частично(нейтрон)-дырочное(протон) (*а*) и частично-частичное взаимодействие (*б*). Знак минус, связанный со вторым членом, указывает на притягивающий характер частично-частичного взаимодействия, приводящий к компенсации вклада отталкивающего частично-дырочного члена

Рис. 3. Сравнение экспериментальных и теоретических периодов полураспада для (2ν2β)-распада. Экспериментальным результатам соответствуют пустые стрелки и прямоугольники. Теоретические значения (стрелки с точкой) получены в квазичастичном приближении случайных фаз [7]

Здесь мы не будем касаться вопроса о существовании так называемых правых токов.

Таким образом, наблюдение (0ν2β)-распада представляло бы первый выход за рамки стандартной модели. Для иллюстрации сказанного отметим, что определение нижней границы $T_{1/2}$ (5) позволяет установить пределы массы (майорановского) нейтрино. Расчеты дают для верхней границы массы значение порядка нескольких эВ.

Экспериментальный сигнал, указывающий на наблюдение одной из трех упомянутых разновидностей 2β-распада, достаточно однозначен (см. рис. 1): суммарная энергия двух испущенных электронов образует непрерывный спектр, в котором положение максимума характеризует (2ν2β)-, (0ν2β*M*)-процесс или для (0ν2β)-распада выделяет отдельную линию, соответствующую максимальной кинематически разрешенной энергии. Огромные экспериментальные трудности, сопровождающие работы по 2β-распаду, связаны с тем, что приходится выделять исключительно редкий сигнал на фоне разнообразных достаточно быстро протекающих реакций.

Цель настоящей работы — обсудить современные теоретические подходы для оценки скорости (2ν2β)-распада.

1. Метод случайных фаз в расчетах $(2\nu 2\beta)$-распада

Амплитуда двухнейтринного $\beta\beta$-распада имеет вид

$$\langle f|S^{(2)}|i\rangle=\frac{i}{2}\left(\frac{G_F}{\sqrt{2}}\right)^2\frac{1}{(2\pi)^6}(16p_{10}p_{20}k_{10}k_{20})^{-1/2}\bar{u}(p_1)\gamma_\alpha(1+\gamma_5)u^c(k_1)\times$$

$$\times\bar{u}(p_2)\gamma_\beta(1+\gamma_5)u^c(k_2)J_{\alpha\beta}-(p_1\leftrightarrow p_2)-(k_1\leftrightarrow k_2)+(p_1\leftrightarrow p_2)(k_1\leftrightarrow k_2), \quad (7)$$

где

$$J_{\alpha\beta}=i\int\exp[-i(p_1+k_1)x_1]\exp[-i(p_2+k_2)x_2]\langle p_f|T(J_\alpha(x_1)J_\beta(x_2))|p_i\rangle dx_1\,dx_2. \quad (8)$$

Здесь p_1 и p_2(k_1 и k_2) 4-импульсы электронов (антинейтрино), p_i и p_f — 4-импульсы начального и конечного ядра, $T(J_\alpha(x_1)J_\beta(x_2))$ — хронологический оператор ядерных адронных слабых токов J_α и J_β. Используя определение

$$T[J_\alpha(x)J_\beta(y)]=\Theta(x_0-y_0)J_\alpha(x)J_\beta(y)+\Theta(y_0-x_0)J_\beta(y)J_\alpha(x) \quad (9)$$

и интегрируя по временной координате, получаем

$$J_{\alpha\beta}=2\pi\delta(E_f-E_i+p_{10}+k_{10}+p_{20}+k_{20})\int\exp[-i(\mathbf{p}_1+\mathbf{k}_1)\mathbf{x}_1]\exp[-i(\mathbf{p}_2+\mathbf{k}_2)\mathbf{x}_2]\times$$
$$\times\sum_n\left\{\frac{\langle p_f|J_\alpha(0,\mathbf{x}_1)|p_n\rangle\langle p_n|J_\beta(0,\mathbf{x}_2)|p_i\rangle}{E_n-E_i+p_{20}+k_{20}}+\right.$$
$$\left.+\frac{\langle p_f|J_\beta(0,\mathbf{x}_2)|p_n\rangle\langle p_n|J_\alpha(0,\mathbf{x}_i)|p_i\rangle}{E_n-E_i+p_{10}+k_{10}}\right\}d\mathbf{x}_1\,d\mathbf{x}_2. \quad (10)$$

Многочисленные попытки использования полноты $\left(\sum_n|p_n\rangle\langle p_n|=1\right)$ системы состояний промежуточного ядра (после выбора некоторой средней энергии $\bar{E}$ в энергетических знаменателях в формуле (10)) привели к выводу о невозможности ответственного выбора $\bar{E}$, так как результат очень чувствителен к этой величине.

Большинство теоретических работ по (2β)-распаду, опубликованных в последние годы, см., например, [8—11], посвящено вычислению спектра $\{E_n, |p_n\rangle\}$ в рамках одной из ядерных моделей. Для $(2\nu 2\beta)$-распада, как правило, ограничиваются вычислением спектра состояний с $J^\pi=1^+$. Сумма по состояниям промежуточного ядра в уравнении (10) после этого проводится численно. Типичные результаты, полученные в рамках оболочечной модели (модели случайных фаз), приводят к оценкам периода полураспада в 10—100 раз ниже экспериментальных нижних пределов [12, 13].

Анализ такой неблагоприятной ситуации [8—11] показал, что приведенная схема вычисления ядерного матричного элемента 2β-распада предельно чувствительна к выбору остаточного нуклон-нуклонного взаимодействия. Оказалось, что наряду с привычным взаимодействием протона и нейтронной дырки (рис. 2, *а*) $\langle p_1 n_1^{-1}|V|p_2 n_2^{-1}\rangle_J$ необходимо учитывать также частично-частичное взаимодействие (рис. 2, *б*) $\langle p_1 n_1|V|p_2 n_2\rangle_J$, которым в других задачах структуры ядра можно пренебречь ввиду его малости. Обратимся к более подробному описанию возникающей ситуации.

В расчетах [10, 11] использовались реалистические взаимодействия в форме боннского и парижского потенциалов. Такие потенциалы характеризуют взаимодействие свободных нуклонов и для применения в задачах ядерной структуры нуждаются в перенормировке, учитывающей эффекты ядерной среды, а также перенормировке, связанной с выбором конечного (а фактически очень ограниченного) модельного пространства, в котором проводится расчет ядерных волновых функций. Конкретно для этого проводят замену

$$\langle p_1 n_1^{-1} | V | p_2 n_2^{-1} \rangle_J \to g_{ph} \langle p_1 n_1^{-1} | V | p_2 n_2^{-1} \rangle_J, \qquad (11)$$

$$\langle p_1 n_1 | V | p_2 n_2 \rangle_J \to g_{pp} \langle p_1 n_1 | V | p_2 n_2 \rangle_J, \qquad (12)$$

а величины g_{ph} и g_{pp} интерпретируют как свободные параметры. Можно ожидать, что g_{ph} и g_{pp} будут лишь ненамного отличаться от единицы.

Величину g_{ph} фиксируют [8—11] таким образом, чтобы рассчитанные в рамках приближения случайных фаз возбуждения гамов-теллеровского гигантского резонанса воспроизводили экспериментальные данные для реакции перезарядки (pn) на материнском ядре 2β-распада.

Ядерный матричный элемент (2ν2β)-распада, если его рассчитывать, используя только что зафиксированное значение константы g_{ph}, является быстро убывающей функцией g_{pp}. Значение матричного элемента меняет знак в окрестности $g_{pp}=1$. Такое поведение дает возможность понять большие значения периода полураспада, наблюдаемые на эксперименте для (2ν2β)-распада. Потребовав, например, правильное описание гамов-теллеровской силовой функции β^+-распада $p \to n$ из дочернего в промежуточное ядро, удается зафиксировать некоторый интервал допустимых значений g_{pp} (в окрестности $g_{pp}=1$). Интервал этот заключает, однако, во всех рассмотренных случаях также значение g_{pp}, зануляющее матричный элемент (2ν2β)-распада.

Видно, что хотя мы и прояснили природу проблемы теоретических оценок вероятности (2ν2β)-распада, убеждаемся одновременно, что теория на этом пути теряет предсказательную силу и позволяет вычислить лишь нижний предел $T_{1/2}$(2ν2β)-распада. Примеры теоретических и экспериментальных результатов, взятые из работы [7], приведены на рис. 3.

2. Новый метод вычисления амплитуды (2ν2β)-распада

Вычислительная схема может быть переформулирована в таком виде, в котором построение спектра состояний промежуточного ядра не требуется [14, 15]. Для этого запишем уравнение (9) в эквивалентном виде

$$T(J_\alpha(x) J_\beta(y)) = J_\alpha(x) J_\beta(y) + \Theta(y_0 - x_0)\,[J_\beta(y), J_\alpha(x)] \qquad (13)$$

и получим вместо (10) выражение

$$\begin{aligned} J_{\alpha\beta} = i \sum_n 2\pi\delta(E_f - E_n + p_{10} + k_{10})\, 2\pi\delta(E_n - E_i + p_{20} + k_{20}) \int \exp[-i(\mathbf{p}_1 + \mathbf{k}_1)\mathbf{x}_1] \times \\ \times \exp[-i(\mathbf{p}_2 + \mathbf{k}_2)\mathbf{x}_2] \langle p_f | J_\alpha(0, \mathbf{x}_1) | p_n \rangle \langle p_n | J_\beta(0, \mathbf{x}_2) | p_i \rangle d\mathbf{x}_1\, d\mathbf{x}_2 + \qquad (15) \\ + i2\pi\delta(E_f - E_i + p_{10} + k_{10} + p_{20} + k_{20}) \int \exp[-i(\mathbf{p}_1 + \mathbf{k}_1)\mathbf{x}_1] \exp[-i(\mathbf{p}_2 + \mathbf{k}_2)\mathbf{x}_2] \times \\ \times \int_0^\infty \exp[i(p_{20} + k_{20})t] \langle p_f | [J_\beta(t, \mathbf{x}_2),\ J_\alpha(0, \mathbf{x}_1)] | p_i \rangle\, dt\, d\mathbf{x}_1\, d\mathbf{x}_2. \qquad (14) \end{aligned}$$

Можно видеть, что первый член в выражении (14) соответствует двум последовательным β-распадам. Двойной β-распад может наблюдаться лишь в том случае, когда простой β-распад сильно подавлен или запрещен по энергии. Формально это отражено δ-функцией с положительным аргументом $E_n - E_i + p_{20} + k_{20} > 0$. Таким образом, в уравнении (14) ненулевой вклад возникает лишь от второго члена, содержащего коммутатор ядерных токов J_α и J_β. Легко увидеть, что

$$\begin{aligned} \mathrm{Im}\, J_{\alpha\beta} = \frac{1}{2} \int \exp[-i(p_1 + k_1)x_1] \exp[-i(p_2 + k_2)x_2] \langle p_f | [J_\beta(x_2),\ J_\alpha(x_1)] | p_i \rangle \times \\ (15) \\ \times dx_1\, dx_2 = 2\pi\delta(E_f - E_i + p_{10} + k_{10} + p_{20} + k_{20}) \int \exp[-i(\mathbf{p}_1 + \mathbf{k}_1)\mathbf{x}_1] \times \\ \times \exp[-i(\mathbf{p}_2 + \mathbf{k}_2)\mathbf{x}_2] \sum_n \{ \langle p_f | J_\beta(0, \mathbf{x}_2) | p_n \rangle \langle p_n | J_\alpha(0, \mathbf{x}_1) | p_i \rangle \times \end{aligned}$$

$$\times \pi\delta(E_n-E_i+p_{10}+k_{10})-\langle p_f|J_\alpha(0,\mathbf{x}_1)|p_n\rangle\langle p_n|J_\beta(0,\mathbf{x}_2)|p_i\rangle\times$$
$$\times\pi\delta(E_n-E_i+p_{20}+k_{20})\}d\mathbf{x}_1\,d\mathbf{x}_2 \qquad (16)$$

тоже равна нулю из-за запрета по энергии β-перехода на промежуточные ядерные состояния ($E_n>E_i$). Это означает, что ядерный матричный элемент $J_{\alpha\beta}$ веществен, что важно для следующих рассуждений.

Рассмотрим наиболее интересный случай 2β-переходов $0^+\to0^+$ и предположим, что испускаемые электроны и антинейтрино находятся в $s_{1/2}$-состоянии.

Ядерный адронный ток в нерелятивистском импульсном приближении имеет вид

$$J_\alpha(0,\mathbf{x})=\sum_n \tau_n{}^+(\delta_{\alpha4}+ig_A\delta_{\alpha k}(\sigma_n)_k)\,\delta(\mathbf{x}-\mathbf{x}_n), \qquad (17)$$

где суммирование проводится по всем нуклонам ядра. Уравнение (14) перепишем в виде

$$J_{\alpha\beta}=2\pi\delta(E_f-E_i+p_{10}+k_{10}+p_{20}+k_{20})\left(\delta_{\alpha4}\delta_{\beta4}M_{VV}-\frac{1}{3}g_A{}^2\delta_{\alpha k}\delta_{\beta k}M_{AA}\right), \qquad (18)$$

где

$$M_{VV}=i\int_0^\infty \exp\left[i\left(p_{20}+k_{20}+\frac{E_f-E_i}{2}\right)t\right]\langle 0_f{}^+|[V(t/2),V(-t/2)]|0_i{}^+\rangle dt, \qquad (19)$$

$$M_{AA}=i\int_0^\infty \exp\left[i\left(p_{20}+k_{20}+\frac{E_f-E_i}{2}\right)t\right]\langle 0_f{}^+|[A_l(t/2),A_l(-t/2)]|0_i{}^+\rangle dt, \qquad (20)$$

где

$$V(t/2)=e^{iHt/2}V(0)\,e^{-iHt/2}, \qquad (21)$$

$$A_l(t/2)=e^{iHt/2}A_l(0)\,e^{-iHt/2}, \qquad (22)$$

$$V(0)=\sum_n \tau_n{}^+, \qquad (23)$$

$$A_l(0)=\sum_n \tau_n{}^+(\sigma_n)_l. \qquad (24)$$

Здесь H — ядерный гамильтониан. Ядерные матричные элементы M_{VV} и M_{AA} могут быть вычислены в замкнутом виде для гамильтониана H общего вида, содержащего тензорные и спин-орбитальные силы. Чтобы избежать написание громоздких формул, продемонстрируем способ их вычисления для более простого ядерного гамильтониана

$$H=H_0+V_C+V_S, \qquad (25)$$

где H_0 — гамильтониан свободных нуклонов, V_C — оператор потенциальной энергии кулоновского взаимодействия нуклонов,

$$V_C=\frac{1}{2}\sum_{i\neq j} g_C(r_{ij})\,O_\tau, \qquad (26)$$

$$O_\tau={}^1/_4(1+\tau_i{}^0)(1+\tau_j{}^0), \qquad (27)$$

V_S — оператор центрального взаимодействия,

$$V_S=\frac{1}{2}\sum_{i\neq j}(g_W(r_{ij})+g_B(r_{ij})P_\sigma-g_H(r_{ij})P_\tau-g_M(r_{ij})P_\tau P_\sigma). \qquad (28)$$

Здесь P_σ и P_τ — проекционные операторы

$$P_\sigma={}^1/_2(1+\sigma_i\cdot\sigma_j), \qquad (29)$$

$$P_\tau = {}^1/_2(1+\boldsymbol{\tau}_i\cdot\boldsymbol{\tau}_j), \tag{30}$$

$g_a(r_{ij})$, $a=W$, B, H, M представляют скалярные функции относительной координаты r_{ij} взаимодействующих нуклонов; $g_C(r_{ij})=e^2/r_{ij}$.

Рассмотрим матричный элемент

$$M_{AA}(t)=\langle 0_f^+|[A_l(t/2),\ A_l(-t/2)]|0_i^+\rangle. \tag{31}$$

Предположим, что вкладами 3-, 4- и многочастичных операторов можно пренебречь. Получим

$$M_{AA}(t)=\langle 0_f^+|\frac{1}{2}\sum_{n\neq m}[e^{iG\,t/2}(\mathbf{A}_{nm})_l e^{-iG\,t/2},\ e^{-iG\,t/2}(\mathbf{A}_{nm})_l e^{iG\,t/2}]|0_i^+\rangle, \tag{32}$$

где

$$G=g_C(r_{nm})O_\tau+g_B(r_{nm})P_\sigma-g_H(r_{nm})P_\tau-g_M(r_{nm})P_\sigma P_\tau, \tag{33}$$

$$\mathbf{A}_{nm}=\tau_n^+\boldsymbol{\sigma}_n+\tau_m^+\boldsymbol{\sigma}_m. \tag{34}$$

При этом было учтено, что

$$\left[A_l(0),H_0+\frac{1}{2}\sum_{i\neq j}g_W(r_{ij})\right]=0. \tag{35}$$

Двухчастичный оператор в матричном элементе (32) можно переписать в более простом виде, для чего вычислим выражение

$$\mathbf{B}_{nm}=\exp(ig_C O_\tau\,t/2)\,\mathbf{A}_{nm}\exp(-ig_C O_\tau\,t/2)= \tag{36}$$

$$=\sum_{k=0}^{\infty}\frac{(ig_C\,t/2)^k}{k!}[\overbrace{O_\tau\ldots[O_\tau}^{k\ \text{раз}},\mathbf{A}_{nm}]\ldots], \tag{37}$$

используя равенство $O_\tau^2=O_\tau$, что позволяет просуммировать ряд коммутаторов (37).

Получим

$$\mathbf{B}_{nm}=\mathbf{A}_{nm}-O_\tau\mathbf{A}_{nm}-\mathbf{A}_{nm}O_\tau+2O_\tau\mathbf{A}_{nm}O_\tau+$$
$$+\cos(g_Ct/2)\,(O_\tau\mathbf{A}_{nm}+\mathbf{A}_{nm}O_\tau-2O_\tau\mathbf{A}_{nm}O_\tau)+i\sin(g_Ct/2)\,[O_\tau,\ \mathbf{A}_{nm}]. \tag{38}$$

Аналогично имеем

$$\exp(ig_BP_\sigma\,t/2)\,\mathbf{B}_{nm}\exp(-ig_BP_\sigma\,t/2)=\sum_{k=0}^{\infty}\frac{(ig_B\,t/2)^k}{k!}[\overbrace{P_\sigma\ldots P_\sigma}^{k\ \text{раз}},\mathbf{B}_{nm}]\ldots]. \tag{39}$$

Используя равенство $P_\sigma^2=1$, можем просуммировать ряд коммутаторов (39). При этом получим

$$\exp(ig_BP_\sigma t/2)\,\mathbf{B}_{nm}\exp(-ig_BP_\sigma t/2)={}^1/_2\{\mathbf{B}_{nm}+P_\sigma\mathbf{B}_{nm}P_\sigma+$$
$$+\cos(g_Bt)\,(\mathbf{B}_{nm}-P_\sigma\mathbf{B}_{nm}P_\sigma)+i\sin(g_Bt)\,[P_\sigma,\ \mathbf{B}_{nm}]\}. \tag{40}$$

Повторное применение (40) с заменой g_BP_σ на g_HP_τ и $g_MP_\sigma P_\tau$ приводит к результату

$$M_{AA}(t)=2i\langle 0_f^+|\sum_{n,m}\tau_n^+\tau_m^+\{+3\sin(2g_Ht+2g_Bt-g_Ct/2)\,\pi_s^\sigma+$$
$$+\lfloor 2\sin(g_Ct/2)+\sin(2g_Ht-2g_Bt-g_Ct/2)\rfloor\pi_t^\sigma|0_i^+\rangle, \tag{41}$$

где

$$\pi_s^\sigma={}^1/_2(1-P_\sigma),\quad \pi_t^\sigma={}^1/_2(1+P_\sigma). \tag{42}$$

Аналогичным путем можно получить

$$M_{VV}(t)=\langle 0_f^+|[V(t/2),\ V(-t/2)]|0_i^+\rangle=2i\langle 0_f^+|\sum_{n,m}\tau_n^+\tau_m^+\sin(g_C\,t/2)\,|0_i^+\rangle. \tag{43}$$

Подставляя (41) и (43) в уравнения (19) и (20), проведем интегрирование по временной координате. Для этого используем стандартную операцию выключения взаимодействия при $t\to\infty$

$$\int_0^\infty e^{iat}\sin(bt)\,dt \to \lim_{\varepsilon\to 0}\int_0^\infty e^{i(a+i\varepsilon)t}\sin(bt)\,dt = \lim_{\varepsilon\to 0}\frac{b}{b^2-a^2-i\varepsilon} =$$

$$= \mathscr{P}\frac{b}{b^2-a^2} + i\pi\delta(b^2-a^2), \tag{45}$$

где $\mathscr{P}$ символ главного значения. Учитывая снова $\operatorname{Im} J_{\alpha\beta}=0$, получаем после интегрирования

$$M_{AA} = \langle 0_f^+|\sum_{n,m}\tau_n^+\tau_m^+\left\{\frac{-6(2g_H+2g_B-g_C/2)}{(2g_H+2g_B+g_C/2)^2-\Delta_{22}^2}\pi_s^\sigma + \right.$$

$$\left. +\left(\frac{-2g_C}{(g_C/2)^2-\Delta_{22}^2} - \frac{2(2g_H-2g_B-g_C/2)}{(2g_H-2g_B-g_C/2)^2-\Delta_{22}^2}\right)\pi_t^\sigma|0_i^+\rangle\right\}, \tag{46}$$

$$M_{VV} = \langle 0_f^+|\sum_{n,m}\tau_n^+\tau_m^+\frac{-g_C(r_{nm})}{(g_C(r_{nm})/2)^2-\Delta_{22}^2}|0_i^+\rangle, \tag{47}$$

где

$$\Delta_{22} = -\Delta_{11} = \left(p_{20}+k_{20}+\frac{E_f-E_i}{2}\right). \tag{48}$$

Интегрирование по переменной r_{nm} в ядерных матричных элементах проводится в смысле главного значения.

Ядерные матричные элементы M_{AA} и M_{VV} соответствуют в методе случайных фаз матричным элементам M_{GT} и M_F

$$M_{GT} = 2\sum_n \langle 0_f^+|\sum_i \tau_i^+(\sigma_i)_l|1_n^+\rangle\langle 1_n^+|\sum_j \tau_j^+|(\sigma_j)_l|0_i^+\rangle\times$$

$$\times\frac{E_n-(E_i+E_f)/2}{\left(E_n-\dfrac{E_i-E_f}{2}\right)^2-\Delta_{22}^2}, \tag{49}$$

$$M_F = 2\sum_n \langle 0_f^+|\sum_i \tau_i^+|1_n^+\rangle\langle 1_n^+|\sum_j \tau_j^+|0_i^+\rangle\frac{E_n-(E_i+E_f)/2}{\left(E_n-\dfrac{E_i+E_f}{2}\right)^2-\Delta_{22}^2}.$$

В работе [14] были получены другим методом формулы, аналогичные (46) и (47), в них, однако, не учтено кулоновское взаимодействие нуклонов в гамильтониане H, роль которого может оказаться существенной. Тем не менее, числовые результаты, полученные в работе [14], показывают, что для матричных элементов ($2\nu2\beta$)-распада метод приводит к 2—6-кратному уменьшению по сравнению с вычислениями по методу случайных фаз.

Значительный процесс экспериментальных исследований $\beta\beta$-распада и изложенные выше идеи нового теоретического подхода позволяют надеяться на существенное продвижение вперед в нашем понимании редчайшего процесса уже в ближайшем будущем.

СПИСОК ЛИТЕРАТУРЫ

1. *Elliot S. R., Hahn A. A., Moe M. K.* // Phys. Rev. Lett. 1987. V. 59. P. 2020.
2. *Caldwell D. O. et al.* // Phys. Rev. 1986. V. D33. P. 2737.
3. *Haxton W. C., Stephenson G. J.* Jr. // Progr. Part. Nucl. Phys. 1984. V. 12. P. 409.
4. *Doi M., Kotani T., Takasugi E.* // Prog. Theor. Phys. 1985. Supplement № 83. P. 1.
5. *Vergados J. D.* // Phys. Rep. 1986. V. 133. P. 1.
6. *Щепкин М. Г.* // УФН. 1984. Т. 143. С. 513.
7. *Muto K., Klapdor H. V.* // Neutrinos/Ed Klapdor H. V. Berlin; Heidelberg; New York: Springer Verlag. 1988. P. 183.
8. *Vogel P., Zirnbauer M. R.* // Phys. Rev. Lett. 1986. V. 57. P. 3148.
9. *Engel J., Vogel P., Zirnbauer M. R.* // Phys. Rev. 1988. V. C37. P. 731.
10. *Muto K., Klapdor H. V.* // Phys. Lett. 1988. V. B201. P. 420.
11. *Civitarese O., Faessler A., Tomoda T.* // Phys. Lett. 1987. B. 194. P. 11.
12. *Vogel P., Fisher P.* // Phys. Rev. 1985. V. C32. P. 1362.
13. *Klapdor H. V., Grotz K.* // Phys. Lett. 1984. V. B142. P. 323.
14. *Ching C. R., Ho T. H., Wu X. R.* // Phys. Rev. 1989. V. C40. P. 304.
15. *Simkovic F.* // JINR Rapid Communications. 1989. V. 39. P. 21.

Physics Letters B 272 (1991) 169–172
North-Holland

PHYSICS LETTERS B

Two-neutrino double beta decay with operator expansion method

X.R. Wu, A. Staudt, H.V. Klapdor-Kleingrothaus
Max-Planck-Institut für Kernphysik, W-6900 Heidelberg, FRG

Cheng-rui Ching and Tso-hsiu Ho
Centre for Theoretical Physics, CCAST (World Laboratory), Beijing, China
and Institute of Theoretical Physics, Academia Sinica, P.O. Box 2735, Beijing 100 039, China

Received 9 July 1991; revised manuscript received 20 September 1991

The 2νββ-decay matrix elements of ^{76}Ge, ^{82}Se, ^{100}Mo and ^{130}Te are calculated by use of a new approach – the operator expansion method. This new method does not require any information about the intermediate states and avoids limitations unavoidable in approaches calculating the full spectrum of intermediate states explicitly. It thus overcomes, e.g., the extreme dependence of QRPA on the strength of the particle–particle interaction, and seems to be a major step beyond this latter approach.

The theory of ββ decay has attracted enormous interest because not only the ββ decay process is related to physics beyond the standard model but also the calculation of ββ decay matrix elements provides a test of our understanding of nuclear theory [1–8]. There exist two widely used approaches for the calculation of ββ transition matrix elements: the shell model method [1] and the quasi-particle random phase approximation (QRPA) [2–5]. Except for the lightest ββ nucleus ^{48}Ca, the calculations with shell model methods are forced to use the so-called closure approximation which seems reasonable for 0νββ but doubtable for 2νββ decay where the theoretical decay rates are 1–2 orders of magnitude larger than the experimental data for several nuclei. The QRPA calculation can provide a suppressed 2νββ decay rate consistent with the experiment. However, the results given by QRPA are very sensitively dependent on the strength of the particle–particle interaction g_{pp} when $g_{pp} \sim 1$. In this letter, we present an approach which goes beyond the usual QRPA approximation and does not explicitly use the calculated intermediate energy spectrum. In this way the sensitivity to the strength of the particle–particle force can significantly be reduced.

In order to avoid the closure approximation and to treat the sum over infinite intermediate states without the explicit knowledge of the intermediate states, an alternative method – the operator expansion method (OEM) – for the calculation of ββ matrix elements has been suggested by Ching and Ho [6] and applied to the calculation of ββ decay of ^{48}Ca [7]. The results for ββ decay of ^{48}Ca seem quite encouraging [7]. As a further application of OEM, this paper is devoted to the study of 2νββ decays of ^{76}Ge, ^{82}Se, ^{100}Mo and ^{130}Te. Here we would like to note that a similar idea has been pursued by ref. [8].

First we briefly describe OEM of 2νββ decay [6,7]. The elementary 2νββ decay matrix element is

$$M_{\mathrm{GT}} = \tfrac{1}{2} \sum_{\mathrm{N}} \left(\frac{\langle \mathrm{F}|A^{\alpha}|\mathrm{N}\rangle \langle \mathrm{N}|A^{\beta}|\mathrm{I}\rangle}{\Delta + (E_{\mathrm{N}} - E_{\mathrm{I}})} - \frac{\langle \mathrm{F}|A^{\beta}|\mathrm{N}\rangle \langle \mathrm{N}|A^{\alpha}|\mathrm{I}\rangle}{\Delta + (E_{\mathrm{F}} - E_{\mathrm{N}})} \right). \tag{1}$$

For $0^+ \to 0^+$ transitions, the matrix element has the well-known form

$$M^{2\nu}_{\mathrm{GT}} = \sum_{\mathrm{N}} \frac{\langle 0^+_{\mathrm{F}} \,|A^{\alpha}|\, 1^+_{\mathrm{N}} \rangle \langle 1^+_{\mathrm{N}} \,|A^{\alpha}|\, 0^+_{\mathrm{I}} \rangle}{\Delta + (E_{\mathrm{N}} - E_{\mathrm{I}})}, \tag{2}$$

where $|\mathrm{I}\rangle$, $|\mathrm{N}\rangle$ and $|\mathrm{F}\rangle$ are wave functions of the initial, intermediate and final nuclear states with corresponding energies E_{I}, E_{N}, and E_{F}, respectively. Δ denotes the average energy of an electron and a neu-

Volume 272, number 3,4 PHYSICS LETTERS B 5 December 1991

trino, $\Delta=\frac{1}{2}(E_{\rm I}-E_{\rm F})$. A^α is the Gamow–Teller transition operator, $A^\alpha=\sum_i \tau_i^- \sigma_i^\alpha$ with nucleonic index i.

The nuclear state and the corresponding energy can be considered as the eigenstate and eigenvalue of a nuclear hamiltonian $H_{\rm s}$,

$$H_{\rm s}|{\rm I}\rangle=E_{\rm I}|{\rm I}\rangle,\quad H_{\rm s}|{\rm F}\rangle=E_{\rm F}|{\rm F}\rangle\,,$$
$$H_{\rm s}|{\rm N}\rangle=E_{\rm N}|{\rm N}\rangle\,. \tag{3}$$

We can expand the energy denominator in eq. (1) into an infinite series, so that all nuclear energies appear in the numerator instead of the energy denominator:

$$\frac{1}{\Delta+(E_{\rm N}-E_{\rm I})}=\frac{1}{\Delta}\left(1-\frac{E_{\rm N}-E_{\rm I}}{\Delta}+\frac{(E_{\rm N}-E_{\rm I})^2}{\Delta^2}-...\right). \tag{4}$$

Then we can replace the energies $E_{\rm I}$, $E_{\rm N}$ and $E_{\rm F}$ by the hamiltonian operator $H_{\rm s}$ according to eq. (3), namely

$$(E_{\rm N}-E_{\rm I})\langle{\rm N}|A^\beta|{\rm I}\rangle=\langle{\rm N}|[H_{\rm s},A^\beta]|{\rm I}\rangle\,,$$
$$(E_{\rm N}-E_{\rm I})^2\langle{\rm N}|A^\beta|{\rm I}\rangle=\langle{\rm N}|[H_{\rm s},[H_{\rm s},A^\beta]]|{\rm I}\rangle\,,$$
$$\vdots \tag{5}$$

Eq. (1) now reads

$$M_{\rm GT}=\frac{1}{2\Delta}\langle{\rm F}|\Big([A^\alpha,A^\beta]-\frac{1}{\Delta}[A^\alpha,[H_{\rm s},A^\beta]]+\frac{1}{\Delta^2}[A^\alpha,[H_{\rm s},[H_{\rm s},A^\beta]]]-...\Big)|{\rm I}\rangle\,. \tag{6}$$

As pointed out in refs. [6,7], eq. (6) should not be regarded as a perturbative series and it should be treated without truncation. In order to calculate the commutators in the series eq. (6), one needs to specify the nuclear hamiltonian $H_{\rm s}$. As generally accepted, $H_{\rm s}$ consists of the hamiltonian of A free nucleons H_0, the Coulomb interaction $V_{\rm c}$ and the strong nuclear force $V_{\rm s}$ arising from the nucleon–nucleon interaction. H_0 and $V_{\rm c}$ play less important roles here [6–8]. As a first approximation, the strong nuclear force $V_{\rm s}$ can be expressed by [6–8]

$$V_{\rm s}=\frac{1}{2}\sum_{i\neq j}[v_0(r_{ij})+v_\tau(r_{ij})\boldsymbol{\tau}_i\cdot\boldsymbol{\tau}_j+v_\sigma(r_{ij})\boldsymbol{\sigma}_i\cdot\boldsymbol{\sigma}_j+v_{\sigma\tau}(r_{ij})\boldsymbol{\sigma}_i\cdot\boldsymbol{\sigma}_j\,\boldsymbol{\tau}_i\cdot\boldsymbol{\tau}_j]\,, \tag{7}$$

which is the most general central static force. Here $v_0(r_{ij})$, $v_\sigma(r_{ij})$, $v_\tau(r_{ij})$ and $v_{\sigma\tau}(r_{ij})$ denote the radial parts of the corresponding interactions; r_{ij} denote the relative coordinates of the nucleons.

After very tedious deviation, where many-particle scattering terms are ignored, one can obtain the general formula of commutators [6,7] and sum up the series (6). Finally the $0^+\to0^+$ transition matrix element (2) can be expressed by

$$M_{\rm GT}\equiv\langle 0_{\rm F}^+|\sum_{i\neq j}\mathcal{M}_{ij}\tau_i^-\tau_j^-|0_{\rm I}^+\rangle\,, \tag{8}$$

$$\mathcal{M}_{ij}=\frac{12[v_\sigma(r)-v_\tau(r)]\Omega_0(ij)}{\Delta^2-16[v_\sigma(r)-v_\tau(r)]^2}+\frac{4[2v_{\sigma\tau}(r)-v_\sigma(r)-v_\tau(r)]\Omega_1(ij)}{\Delta^2-16[2v_{\sigma\tau}(r)-v_\sigma(r)-v_\tau(r)]^2}, \tag{9}$$

where the spin singlet and spin triplet projection operators are

$$\Omega_1(ij)=\tfrac{1}{4}(3+\boldsymbol{\sigma}_i\cdot\boldsymbol{\sigma}_j),$$
$$\Omega_0(ij)=\tfrac{1}{4}(1-\boldsymbol{\sigma}_i\cdot\boldsymbol{\sigma}_j)\,. \tag{10}$$

In eq. (9) we do not have to care about the problem that the bare NN interaction becomes very large at small distance, because $\mathcal{M}_{ij}$ itself stays finite. Thus after summing over infinite intermediate states, the matrix element (2) has been converted into a matrix which is not explicitly dependent on intermediate states. The operator $\mathcal{M}_{ij}$ can be determined by using the effective nucleon–nucleon force. Thus OEM requires only the knowledge of the wave functions of the initial and final states. In principle, one can use the shell model to calculate the wave functions of the initial and final states. However, it is very difficult to derive wave functions of heavy nuclei such as ^{76}Ge, ^{82}Se, ^{100}Mo and ^{130}Te by using standard shell model techniques. Alternatively, we employ the method used in QRPA calculations of ββ decay matrix elements [3,5]. However, we note that here QRPA is only used to obtain the knowledge of the wave functions of the initial and final states, while in usual QRPA also the wave functions and energy spectrum of the intermediate states have to be calculated.

The QRPA expression for the ββ matrix element is

Volume 272, number 3,4 PHYSICS LETTERS B 5 December 1991

$$M_{\mathrm{GT}} = \sum_{\mathrm{pnp'n'}Jab} O^J_{\mathrm{pnp'n'}}$$
$$\times [\bar{X}^{\mathrm{p'n'}J}_b \bar{v}_{\mathrm{p'}} \bar{u}_{\mathrm{n'}} - \bar{Y}^{\mathrm{p'n'}J}_b \bar{u}_{\mathrm{p'}} \bar{v}_{\mathrm{n'}}] \xi_{ab}$$
$$\times [X^{\mathrm{pn}J}_a u_{\mathrm{p}} v_{\mathrm{n}} - Y^{\mathrm{pn}J}_a v_{\mathrm{p}} u_{\mathrm{n}}] , \qquad (11)$$

where ξ_{ab} is the overlap integral of intermediate states calculated with respect to the parent (0_{I}^+) and daughter (0_{F}^+) states, respectively. u and v are BCS occupation amplitudes, X and Y denote the eigenfunctions of the QRPA equation. The indices a and b distinguish between the two different sets of calculated eigenstates.

For the 2νββ mode one has to substitute

$$O^J_{\mathrm{pnp'n'}} = \delta_{J,1} \frac{\langle \mathrm{p}|\sigma\tau^-|\mathrm{n}\rangle \langle \mathrm{p'}|\sigma\tau^-|\mathrm{n'}\rangle}{\Delta + (E_a - E_{\mathrm{I}})} , \qquad (12)$$

whereas for the 0νββ one has for the GT matrix element (in closure approximation)

$$O^J_{\mathrm{pnp'n'}} = (2J+1)$$
$$\times \sum_{J'} (2J'+1)(-1)^{j_{\mathrm{p'}}+j_{\mathrm{n}}+J+J'} W(\mathrm{p'pn'n}; J'J)$$
$$\times \langle \mathrm{p'p}; J'| H(r)\, \boldsymbol{\sigma}_1 \cdot \boldsymbol{\sigma}_2\, \tau_1^- \tau_2^- |\mathrm{n'n}; J'\rangle . \qquad (13)$$

The OEM leads to an expression for 2νββ decay (when expressing the wave functions of initial and final states in QRPA) of the form of eq. (11) with $O^J_{\mathrm{pnp'n'}}$ of eq. (13), but with the operator $H(r)\boldsymbol{\sigma}_1 \cdot \boldsymbol{\sigma}_2$ replaced by $\mathscr{M}_{12}$ which is defined by eq. (9). The operator $\mathscr{M}_{12}$ which depends on the radial parts of nucleon–nucleon interaction $v_\sigma(r)$, $v_\tau(r)$ and $v_{\sigma\tau}(r)$ can be determined from the Paris potential. The monopole interaction for the BCS calculation and G-matrix in the QRPA equation are all taken from the Paris potential. The Woods–Saxon potential with Bohr–Mottelson's parameters is used as the self-consistent potential (if one uses Woods–Saxon potential with Bertsch's parameters, the matrix elements are a little (~10%) smaller). The model spaces of active shells are chosen as in refs. [3,5].

We have calculated the 2νββ decay matrix elements of ^{76}Ge, ^{82}Se, ^{100}Mo, and ^{130}Te by using particle–hole and particle–particle interaction strengths, both $g_{\mathrm{ph}} = g_{\mathrm{pp}} = 1$ and the empirical values obtained by fitting single beta decay [3,5]. As shown in fig. 1 for the 2νββ decay of ^{100}Mo, the matrix element turns out to be insensitive to the particle–particle interac-

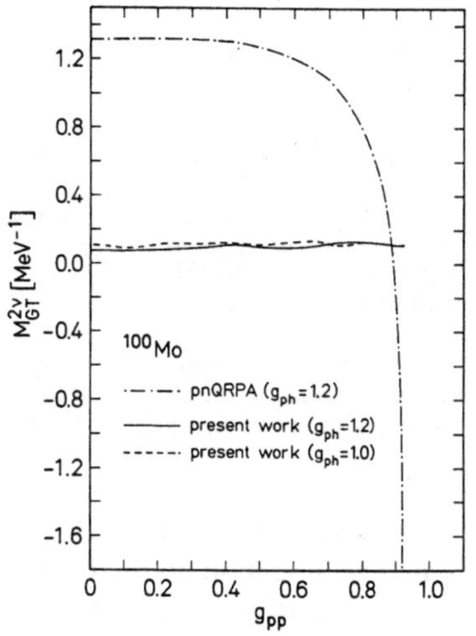

Fig. 1.

tion strength g_{pp}, independent of the choice of the strength of the particle–hole force. Therefore we can use either the physical value $g_{\mathrm{pp}} = g_{\mathrm{ph}} = 1$ (except for ^{100}Mo, where the RPA equation collapses already at $g_{\mathrm{pp}} = 0.9185$) or the values of g_{pp} and g_{ph} determined by the fit to the single β$^+$ data. The theoretical matrix elements M_{GT} half-lives (for $g_A = 1.254$), as well as the experimental half-lives are listed in table 1. In general, good agreement between calculation and experiment is obtained, only in the case of ^{130}Te the predicted half-life is shorter than the lower limit given by the geochemical method by a factor of ~7.

The feature that M_{GT} is insensitive to g_{pp} can be understood by the fact that the matrix operator $\mathscr{M}_{ij}$ in eq. (9) is not explicitly dependent on the intermediate energy spectrum. The intermediate states simply occur as a complete set of mathematical functions, whereas in usual QRPA calculation [2], the Gamow–Teller matrix element explicitly depends on both energies and the wave functions of the intermediate states. Thus, the present approach is a step beyond the QRPA approach. It seems likely that the very strong dependence of nuclear matrix elements on the strength of the particle–particle channel of the proton–neutron interaction is a feature of RPA-type models and can be avoided in more general approaches.

An extension of OEM to include the tensor force term and the nonlocal spin–orbit force term in the

Volume 272, number 3,4 PHYSICS LETTERS B 5 December 1991

Table 1
The calculated $2\nu\beta\beta$ Gamow–Teller matrix elements and half-lives. Case A is with g_{ph} and g_{pp} from refs. [3,5]; case B is with $g_{ph}=g_{pp}=1$ ($g_{pp}=0.9$ for ^{100}Mo).

Nuclei	M_{GT} (MeV^{-1})		Half-life$_{theor}$		Half-life$_{exp}$	
	case A	case B	case A	case B		
^{76}Ge	0.3355	0.3351	2.76×10^{20}	2.77×10^{20}	$(9.2^{+0.7}_{-0.4})\times 10^{20}$ a)	
^{82}Se	0.1096	0.1019	0.759×10^{20}	0.878×10^{20}	$(1.1^{+0.8}_{-0.3})\times 10^{20}$ b)	
^{100}Mo	0.1063	0.1065	3.37×10^{19}	3.37×10^{19}	$(1.15^{+0.30}_{-0.20})\times 10^{19}$ c)	1.16×10^{19} d)
^{130}Te	0.09425	0.1006	1.065×10^{20}	0.937×10^{20}	$(1.5\text{–}2.8)\times 10^{21}$ e),f)	$(0.75\pm 0.03)\times 10^{21}$ g),f)

a) Ref. [10]. b) Ref. [11]. c) Ref. [12]. d) Ref. [13]. e) Ref. [14]. f) Geochemical data (total $\beta\beta$ decay half-life). g) Ref. [15].

nuclear hamiltonian (7) is under way. As suggested by ref. [8] it is also possible to retain the Coulomb interaction in the nuclear hamiltonian, which accounts for the difference in the ground state energies of initial and final nuclei.

References

[1] W.C. Haxton and G.J. Stephenson Jr., Prog. Part. Nucl. Phys. 12 (1984) 409;
M. Doi, T. Kotani and E. Takasugi, Prog. Theor. Phys. Suppl. 83 (1985) 1;
H.V. Klapdor and K. Grotz, Phys. Lett. B 142 (1984) 323;
J.D. Vergados, Phys. Rep. 133 (1986) 1;
T. Tomoda, Rep. Prog. Phys. 54 (1991) 53.

[2] K. Grotz and H.V. Klapdor, Phys. Lett. B 153 (1985) 1; B 157 (1985) 242; Nucl. Phys. A 460 (1986) 395;
P. Vogel and M.R. Zirnbauer, Phys. Rev. Lett. 57 (1986) 3148;
O. Civitarese, A. Faessler and T. Tomoda, Phys. Lett. B 194 (1987) 11;
K. Muto and H.V. Klapdor, Phys. Lett. B 201 (1988) 420.

[3] K. Muto, E. Bender and H.V. Klapdor, Z. Phys. A 334 (1989) 177, 187.

[4] A. Staudt, T.T.S. Kuo and H.V. Klapdor-Kleingrothaus, Phys. Lett. B 242 (1990) 17.

[5] A. Staudt, K. Muto and H.V. Klapdor-Kleingrothaus, Europhys. Lett. 13 (1990) 31.

[6] C.R. Ching and T.H. Ho, Commun. Theor. Phys. 11 (1989) 433.

[7] C.-R. Ching, T.-H. Ho and X.-R. Wu, Phys. Rev. C 40 (1989) 304; Commun. Theor. Phys. 12 (1989) 167; in: Proc. BIMP Symp. on Heavy flavor physics, eds. K. Chao et al. (World Scientific, Singapore, 1989) p. 359.

[8] F. Simkovic, JINR Rapid Commun. 39 (1989) 21;
M. Gmitro and F. Simkovic, Izv. Akad. Nauk SSSR 54 (1990) 1780.

[9] A. Bohr and B.R. Mottelson, Nuclear structure, Vol. 1 (Benjamin, New York, 1969);
G.F. Bertsch, The practitioner's shell model (North-Holland, Amsterdam, 1972).

[10] F.T. Avignone III et al., Phys. Lett. B 256 (1991) 559.

[11] S.R. Elliot, A.A. Hahn and M.K. Moe, Phys. Rev. Lett. 59 (1987) 2020.

[12] H. Ejiri et al., Phys. Lett. B 258 (1991) 17.

[13] M.K. Moe, Proc. 14th EPS Conf. (Bratislava, October 1990), J. Phys. G, to be published.

[14] T. Kirsten, Proc. Intern. Symp. on Nuclear beta decays and neutrino (Osaka, June 1986), eds. T. Kotani, H. Ejiri and E. Takasugi (World Scientific, Singapore, 1986) p. 81.

[15] W.J. Lin et al., Nucl. Phys. A 481 (1988) 477.

Physics Letters B 276 (1992) 274–278
North-Holland

PHYSICS LETTERS B

Tensor force and operator expansion method for nuclear double beta decay

X.R. Wu, A. Staudt, T.T.S. Kuo [1,2] and H.V. Klapdor-Kleingrothaus
Max-Planck-Institut für Kernphysik, W-6900 Heidelberg, FRG

Received 9 September 1991; revised manuscript received 25 November 1991

Using a general effective interaction containing $\sigma\cdot\sigma$, $\tau\cdot\tau$, $\sigma\cdot\sigma\,\tau\cdot\tau$ and tensor terms, we have derived an approximate formula for calculating the M_{GT} matrix element for nuclear $2\nu\beta\beta$ decays. Our method is an extension of the operator expansion method of Ching and Ho which is for an effective interaction without tensor force. With the present formula the calculation of M_{GT} does not require any information about the nuclear intermediate states, and our result is obtained *without* invoking closure approximation.

Nuclear double beta decays, both the neutrinoless ($0\nu\beta\beta$) mode and the two-neutrino ($2\nu\beta\beta$) mode, have received much attention in recent years. The $0\nu\beta\beta$ mode is of course very important as its existence would be directly, or conclusively, related to the fundamental question of whether neutrino is a Dirac or Majorana particle and whether it is massive or massless. A growing interest is, however, also being paid to the 2ν mode of double-beta decay. There are primarily two reasons for this trend, one experimental and the other theoretical: Previously double-beta decays were, by and large, observed using geochemical methods. The past few years have, however, witnessed a new and very successful experimental development, the direct-counter method for measuring double-beta decays. As a result, there now exists an abundance of high-accuracy experimental results about the $2\nu\beta\beta$ mode [1–3].

The $2\nu\beta\beta$ nuclear matrix element can be written as

$$M_{GT}(Z)=\tfrac{1}{2}\sum_N\left(\frac{\langle F|A^\alpha|N\rangle\langle N|A^\beta|I\rangle}{\Delta+(E_N-E_I)Z}-\frac{\langle F|A^\beta|N\rangle\langle N|A^\alpha|I\rangle}{\Delta+(E_F-E_N)Z}\right), \qquad (1)$$

where $|I\rangle$, $|N\rangle$ and $|F\rangle$ are respectively the wave functions of the initial, intermediate and final nuclei with the corresponding energies E_I, E_N, and E_F. Δ denotes the average energy $\Delta=\frac{1}{2}(E_I-E_F)$. A^α is the Gamow–Teller transition operator $A^\alpha=\sum_i\tau_i^-\sigma_i^\alpha$. Z is a parameter to be set equal to unity. For $0^+\to0^+$ transitions, the above matrix element takes the familiar form (with $Z=1$)

$$M_{GT}^{2\nu}=\sum_N\frac{\langle 0_F^+|A^\alpha|1_N^+\rangle\langle 1_N^+|A^\alpha|0_I^+\rangle}{\Delta+(E_N-E_I)}. \qquad (2)$$

The above equations appear to be simple, merely a second-order perturbation theory, and it would seem to be a simple task to evaluate M_{GT}. It has, however, turned out to be a highly difficult and even somewhat "tricky" undertaking. In fact the evaluation of M_{GT} has in the last few years attracted the interest of many theorists, leading to a number of formalisms and calculations [4–15], some based on a shell model and some based on QRPA (quasi particle random phase approximation).

In this letter we would like to propose a new approximation method for evaluating M_{GT}. It is an extension of

[1] Work supported in part by US DOE Grant DE-FG02-88ER40388.
[2] Permanent address: Physics Department, State University of New York at Stony Brook, Stony Brook, NY 11794-3800, USA.

Volume 276, number 3 PHYSICS LETTERS B 13 February 1992

the operator-expansion method (OEM) proposed by Ching and Ho [16,17]. We shall refer to their method as OEM1, while our present method as OEM2. Within the commutator-series formalism of OEM1, it would be very difficult to include the nucleon–nucleon (NN) tensor force and it is mainly for this reason that the NN tensor force was not considered in OEM1. Simkovic and Gmitro [18] have given a different derivation for OEM1 and in addition extended it to the inclusion of the Coulomb force. They also did not include the tensor force. In this letter, we would like to generalize OEM1 so that the NN tensor force can also be included in the calculation of M_{GT} with an OEM-type expression. We have found that this can in fact be attained in a fairly straightforward way.

A major difficulty in evaluating the M_{GT} matrix element has been the treatment of the nuclear intermediate states in eqs. (1) or (2). Nuclei where double-beta decays may take place are generally all far from the closed shells. Hence shell-model calculations for them are prohibitively difficult unless some truncation schemes are imposed, so as to make the calculation manageable. It is nevertheless difficult to assess their accuracy. In QRPA we treat the nuclear system as a system of quasi-particles and include only a restricted type of quasi-particle excitations. In this way the treatment of the intermediate states becomes largely simplified. It is an appealing method, but it also has its drawbacks. For instance QRPA calculations often depend sensitively on the particle–particle interaction strength g_{pp} in the vicinity of $g_{pp}=1$ [11–14].

Thus it would seem to be of use to have a method where explicit information about the intermediate states are not needed for the calculation of M_{GT}. OEM1 is such a method. Let us briefly describe below some main features of it, so as to prepare us to generalize it to the case with tensor force.

In OEM1 the NN effective interaction V_S is taken to have the form

$$V_S=\frac{1}{2}\sum_{i\neq j}[v_0(r_{ij})+v_\tau(r_{ij})\tau_i\cdot\tau_j+v_\sigma(r_{ij})\sigma_i\cdot\sigma_j+v_{\sigma\tau}(r_{ij})\sigma_i\cdot\sigma_j\,\tau_i\cdot\tau_j]\,, \tag{3}$$

where r_{ij} is the internucleon distance, and $v_0(r_{ij})$, $v_\sigma(r_{ij})$, $v_\tau(r_{ij})$, $v_{\sigma\tau}(r_{ij})$ are functions of r_{ij} only. Empirical NN effective interactions are frequently of this form. With the above effective interaction, OEM1 gives the following result for the $0^+\to0^+$ $2\nu\beta\beta$ matrix element:

$$M_{GT}\equiv\langle 0_F^+|\sum_{i\neq j}\mathcal{M}_{ij}\tau_i^-\tau_j^-|0_I^+\rangle\,, \tag{4}$$

$$\mathcal{M}_{ij}=\frac{6[g_H(r)+g_B(r)]\Omega_0(ij)}{\Delta^2-4[g_H(r)+g_B(r)]^2}+\frac{2[g_H(r)-g_B(r)]\Omega_1(ij)}{\Delta^2-4[g_H(r)-g_B(r)]^2}\,. \tag{5}$$

The various radial functions are

$$g_W(r)=v_0(r)-v_\sigma(r)-v_\tau(r)+v_{\sigma\tau}(r),\quad g_M(r)=-4v_{\sigma\tau}(r)\,,$$
$$g_B(r)=2v_\sigma(r)-2v_{\sigma\tau}(r),\quad g_H(r)=-2v_\tau(r)+2v_{\sigma\tau}(r)\,, \tag{6}$$

and the spin singlet and triplet projection operators are

$$\Omega_1(ij)=\tfrac{1}{4}(3+\sigma_i\cdot\sigma_j),\quad \Omega_0(ij)=\tfrac{1}{4}(1-\sigma_i\cdot\sigma_j)\,. \tag{7}$$

We have used the abbreviated notation r for r_{ij}. We note that the above result was obtained with certain approximations [16,17], namely in the treatment of the commutators involved one has neglected certain multiple scattering terms and approximated the unperturbed hamiltonian by its averaged value. The accuracy of these approximations remains to be investigated, although they are expected [16,17] to be reasonable. We shall come back to discuss these approximations later. The above OEM1 result is rather interesting and appealing, and it would seem to be worthwhile to further study this approach.

To evaluate the $2\nu\beta\beta$ matrix element with the above method we only need to know the effective NN interaction and the wave functions for the initial state 0_I^+ and final state 0_F^+. The intermediate-state wave functions and energies of eq. (2) are never needed here. The above OEM1 method has been applied to the $\beta\beta$ decays of

^{48}Ca [17,18], and to ^{76}Ge, ^{82}Se, ^{100}Mo and ^{130}Te [19], both with rather encouraging results. The OEM1 method has also been generalized and extensively applied to double-charge-exchange nuclear reactions. (See ref. [20] and references quoted therein.)

The effective interaction of eq. (3) above does not have a tensor force component, which is well known to be important for describing a number of nuclear properties. It would seem to be of interest to consider a more general effective interaction which contains a tensor force term. To proceed, we start from the OEM1 expression for eq. (1), namely

$$M_{GT}(Z)=\frac{1}{2\Delta}\langle F|\left([A^\alpha,A^\beta]-\frac{Z}{\Delta}[A^\alpha,[H_S,A^\beta]]+\frac{Z^2}{\Delta^2}[A^\alpha,[H_S,[H_S,A^\beta]]]-\ldots\right)|I\rangle$$
$$\equiv\tfrac{1}{2}\sum_{n=0}^{\infty}\frac{(-Z)^n}{\Delta^{n+1}}[A^\alpha,\overbrace{[H_S,\ldots[H_S}^{n\text{ times}},A^\beta]\ldots]]\,. \tag{8}$$

By way of the relations

$$\int_0^\infty dt\,\exp(i\Delta t)=\lim_{\epsilon\to 0}\int_0^\infty dt\,\exp[(i\Delta-\epsilon)t]=\frac{i}{\Delta+i\epsilon}=\frac{i}{\Delta}, \tag{9}$$

$$\int_0^\infty dt\,\exp(i\Delta t)\,t^n=\lim_{\epsilon\to 0}\frac{i^{n+1}\,n!}{(\Delta+i\epsilon)^{n+1}}, \tag{10}$$

and

$$\exp(iOt)\,A\exp(-iOt)=A+it[O,A]+\frac{(it)^2}{2!}[O,[O,A]]+\ldots=\sum_{n=0}^{\infty}\frac{(it)^n}{n!}[O,\overbrace{[O,\ldots[O}^{n\text{ times}},A]\ldots]]\,; \tag{11}$$

eq. (8) can be rewritten as

$$M_{GT}(Z)=-\tfrac{1}{2}i\langle F|\int_0^\infty dt\,\exp(i\Delta t)\,[A^\alpha,\exp(iH_SZt)\,A^\beta\exp(-iH_sZt)]|I\rangle\,. \tag{12}$$

In the above H_S denotes the nuclear hamiltonian H_0+V_S, the former being the single-particle part. We use the symbols E_I, E_F, I, F to denote the eigenvalues and eigenfunctions of H_S. Using a field theory method, Simkovic and Gmitro [18] have also derived eq. (12). We like to point out that the inclusion of tensor force directly using eq. (8) is apparently most complicated, if not impossible. But things are much facilitated with eq. (12) which expresses M_{GT} in a time integral form.

We shall consider the following effective interaction:

$$V_S=\tfrac{1}{2}\sum_{i\neq j}[g'_W(r_{ij})+g'_B(r_{ij})P_\sigma(ij)-g_H(r_{ij})P_\tau(ij)-g_M(r_{ij})P_\sigma(ij)P_\tau(ij)+g_T(r_{ij})S_T(ij)] \tag{13}$$

$$=\tfrac{1}{2}\sum_{i\neq j}[g_W(r_{ij})+g_B(r_{ij})P_\sigma(ij)-g_H(r_{ij})P_\tau(ij)-g_M(r_{ij})P_\sigma(ij)P_\tau(ij)+g_T(r_{ij})S(ij)]\,, \tag{14}$$

with

$$P_\sigma(ij)=\tfrac{1}{2}(1+\sigma_i\cdot\sigma_j),\quad P_\tau(ij)=\tfrac{1}{2}(1+\tau_i\cdot\tau_j),\quad S_T(ij)=\sigma_i\cdot\hat{r}\sigma_j\cdot\hat{r}-\tfrac{1}{3}\sigma_i\cdot\sigma_j,\quad S(ij)=\sigma_i\cdot\hat{r}\,\sigma_j\cdot\hat{r}\,. \tag{15}$$

It is a useful step in rewriting eq. (13) as eq. (14), absorbing the $\sigma_i\cdot\sigma_j$ term of the tensor force S_T into g_W and g_B, with $g_W=g'_W+\frac{1}{3}g_T$ and $g_B=g'_B-\frac{2}{3}g_T$. The radial functions g may be determined empirically by fitting certain experimental data or deduced from some calculated G-matrix elements. An example for the latter is the well-known M3Y effective interaction [21].

For an operator O with $O^2=1$, one has

$$\exp(\mathrm{i}gO)\,B\exp(-\mathrm{i}gO)=\tfrac{1}{2}\{B+OBO+\cos(2g)(B-OBO)+\mathrm{i}\sin(2g)[O,B]\}\,. \tag{16}$$

We want to use the above relation to simplify the interaction parts of eq. (12). The two-body operator S satisfies the condition $O^2=1$. (Note that the operator S_T does not have this property.) Then we have

$$\exp(\mathrm{i}g_T S)\,\sigma_i^\alpha\exp(-\mathrm{i}g_T S)=[1-\cos(2g_T t)]\hat{r}^\alpha\,\sigma_i\cdot\hat{r}+\cos(2g_T t)\sigma_i^\alpha-\sin(2g_T t)(\sigma_i\wedge\hat{r})^\alpha\,\sigma_j\cdot\hat{r}\,. \tag{17}$$

All terms in V_S of eq. (14) commute with each other. Hence we can write

$$\exp(\mathrm{i}V_S t)\,A^\beta\exp(-\mathrm{i}V_S t)=\exp[\mathrm{i}(g_B P_\sigma-g_H P_\tau-g_M P_\sigma P_\tau)]\exp(\mathrm{i}g_T S)\,A^\beta\exp(-\mathrm{i}g_T S)$$
$$\times\exp[-\mathrm{i}(g_B P_\sigma-g_H P_\tau-g_M P_\sigma P_\tau)]\,. \tag{18}$$

Using $P_\sigma^2=1$ and $P_\tau^2=1$ one can derive, similar to eq. (16), the following relation:

$$\exp[\mathrm{i}(g_B P_\sigma-g_H P_\tau-g_M P_\sigma P_\tau)]=\tfrac{1}{2}[\cos(g_B t+g_H t)(1-P_\sigma P_\tau)+\mathrm{i}\sin(g_B t+g_H t)(P_\sigma-P_\tau)]\exp(\mathrm{i}g_M t)$$
$$+\tfrac{1}{2}[\cos(g_B t-g_H t)(1+P_\sigma P_\tau)+\mathrm{i}\sin(g_B t-g_H t)(P_\sigma+P_\tau)]\exp(-\mathrm{i}g_M t)\,. \tag{19}$$

With the above relations, M_{GT} of eq. (12) may now be evaluated. But before proceeding, there is an important step, namely H_S in eq. (12) is approximated by V_S. Recall that H_S is actually equal to H_0+V_S. This approximation was also used in refs. [16–18] and discussed there, in the context that the average value of the kinetic energy is generally significantly smaller than the interaction potential energy and hence the above approximation is expected to be reasonable. We take a slightly different viewpoint.

Suppose we use a small, in terms of energy spread, model space for the calculation, then within this model space H_0 is nearly degenerate and we may replace it as a first approximation by its average $H_0^{\deg}$, which is a constant. Clearly $H_0^{\deg}$ does not contribute to eq. (12). The difference $H_0^{\deg}-H_0$ is small compared to V_S, for the case of a small model space. Thus in this case we are perhaps allowed to replace H_S in eq. (12) by V_S. We emphasize that the above step is a main approximation in OEM. There is another type of approximation imposed, concerning the multiple scattering terms [16,17]. Briefly speaking, relations (17) and (19) treat the two-body scattering terms exactly for an interaction of the form of eq. (14). For multiple scattering terms the above relations are only approximately true, but the error caused is probably small. The reason is that a major part of the multiple scattering terms, which are not handled correctly in eqs. (17) and (19), cancel among themselves when one evaluates the commutators of eq. (8), as in refs. [16,17], or the commutator of eq. (12).

Adopting the above approximations, the commutator in eq. (12) can be evaluated and the time integral performed. A fair amount of algebraic manipulation is needed, but compared with OEM1 the present derivation is still perhaps simpler. Finally we obtain the OEM2 result for the $0^+\to0^+ M_{GT}$ transition matrix element as

$$M_{GT}\equiv\langle 0_F^+|\sum_{i\neq j}\mathcal{M}_{ij}\tau_i^-\tau_j^-|0_I^+\rangle\,, \tag{20}$$

with

$$\mathcal{M}_{ij}=\frac{1}{4}\left(\frac{2g_B(r)+2g_H(r)}{\Delta^2-[2g_B(r)+2g_H(r)]^2}(1-\sigma_i\cdot\sigma_j)+\frac{2g_B(r)-2g_H(r)}{\Delta^2-[2g_B(r)-2g_H(r)]^2}(2\sigma_i\cdot\hat{r}\,\sigma_j\cdot\hat{r}-\sigma_i\cdot\sigma_j-1)\right.$$
$$+\frac{2g_T(r)}{\Delta^2-[2g_T(r)]^2}(6\sigma_i\cdot\hat{r}\,\sigma_j\cdot\hat{r}-2\sigma_i\cdot\sigma_j)+\frac{2g_B(r)+2g_H(r)+2g_T(r)}{\Delta^2-[2g_B(r)+2g_H(r)+2g_T(r)]^2}(2-2\sigma_i\cdot\sigma_j)$$
$$\left.+\frac{2g_B(r)-2g_H(r)+2g_T(r)}{\Delta^2-[2g_B(r)-2g_H(r)+2g_T(r)]^2}(-2\sigma_i\cdot\hat{r}\,\sigma_j\cdot\hat{r}-2)\right)\,. \tag{21}$$

When setting the tensor force equal to zero, the above reduces to the OEM1 result of eqs. (4) and (5) as it should. For certain radii the denominators in the above may vanish. This is, however, not a problem. Recall

Volume 276, number 3 PHYSICS LETTERS B 13 February 1992

that we have introduced an adiabatic converging factor starting from eq. (9). Thus our M_{GT} of eq. (20) has actually two parts, a principal-value integral and an imaginary delta-function part. By definition, see eqs. (1) and (2), M_{GT} in real. Hence in evaluating eqs. (20) and (21) we retain only the principal-value integrals.

It may be noted that the central force g_W does not enter the calculation of M_{GT}, as seen in eqs. (20) and (21), except through its role in determining the initial and final wave functions 0_F^+ and 0_I^+. This is a general result as can be seen from eq. (12).

In summary, we have derived an approximate formula for calculating the M_{GT} matrix element for nuclear $2\nu\beta\beta$ decays, using a general effective interaction with spin–spin, isospin–isospin and tensor components. Our formula does not need any information about the intermediate states, and this is achieved *without* using closure approximation. Our formula seems to be rather convenient for calculation; aside from the effective NN interaction it only needs to have the initial and final ground-state wave functions, which may be easier to calculate, comparing with those of the excited states. For example we may calculate them using either variation methods or Lanczos shell-model methods. These ground-state wave functions may also be calculated with QRPA. Such a calculation was recently reported [19] and the resulting M_{GT} values calculated according to eqs. (4) and (5) were rather insensitive to the particle–particle interaction strength g_{pp} and in general, good agreement between calculation and experiments was obtained [19].

References

[1] S.R. Elliott, A.A. Hahn and M.K. Moe, Phys. Rev. Lett. 59 (1987) 2020.

[2] H. Ejiri, K. Fushimi, T. Kamada, H. Kinoshita, H. Kobiki, H. Ohsumi, K. Okada, H. Sano, T. Shibata, T. Shima, N. Tanabe, J. Tanaka, T. Taniguchi, T. Watanabe and N. Yamamoto, Phys. Lett. B 258 (1991) 17.

[3] A.A. Vasenko, I.V. Kirpichnikov, V.A. Kuznetsov, A.S. Starostin, A.G. Djanyan, V.S. Pogosov, S.P. Shachysisyan and A.G. Tamanyan, Mod. Phys. Lett. A 5 (1990) 1299;
H.S. Miley, F.T. Avignone III, R.L. Brodzinski, J.I. Collar and J.H. Reeves, Phys. Rev. Lett. 65 (1990) 3092;
F.T. Avignone III, R.L. Brodzinski, C.K. Guerard, I.V. Kirpichnikov, H.S. Miley, V.S. Pogosov, J.H. Reeves, A.S. Starostin and A.G. Tamangan, Phys. Lett. B 256 (1991) 559.

[4] W.C. Haxton, G.J. Stephenson Jr. and D. Strottman, Phys. Rev. D 25 (1982) 2360.

[5] L. Zamick and N. Auerbach, Phys. Rev. C 26 (1982) 2185.

[6] L.D. Skouras and J.D. Vergados, Phys. Rev. C 28 (1983) 2122.

[7] T. Tsuboi, K. Muto and H. Horie, Phys. Lett. B 143 (1984) 293.

[8] J. Engel, P. Vogel, O. Civitarese and M.R. Zirnbauer, Phys. Lett. B 208 (1988) 187.

[9] J. Sinatkas, L.D. Skouras and J.D. Vergados, Phys. Rev. C 37 (1988) 1229.

[10] Liang Zhao, B.A. Brown and W.A. Richter, Phys. Rev. C 42 (1990) 1120.

[11] K. Muto and H.V. Klapdor, in: Neutrinos, graduate texts in contemporary physics, ed. H.V. Klapdor (Springer, Berlin, 1988).

[12] P. Vogel and M.R. Zirnbauer, Phys. Rev. Lett. 57 (1986) 3148;
J. Engel, P. Vogel and M.R. Zirnbauer, Phys. Rev. C 37 (1988) 731.

[13] O. Civitarese, A. Faessler and T. Tomoda, Phys. Lett. B 194 (1987) 11.

[14] K. Grotz and H.V. Klapdor, Nucl. Phys. A 460 (1986) 395.

[15] A. Staudt, T.T.S. Kuo and H.V. Klapdor-Kleingrothaus, Phys. Lett. B 242 (1990) 17.

[16] C.R. Ching and T.H. Ho, Commun. Theor. Phys. 11 (1989) 433.

[17] C.-R. Ching, T.-H. Ho and X.-R. Wu, Phys. Rev. C 40 (1989) 304; Commun. Theor. Phys. 12 (1989) 167; in: Proc. BIMP Symp. on Heavy flavor physics, eds. Chao Kuangta et al. (World Scientific, Singapore, 1989) p. 359.

[18] F. Simkovic, JINR Rapid Commun. 39 (1989) 21;
M. Gmitro and F. Simkovic, Izv. AN SSSR 54 (1990) 1780.

[19] X.R. Wu, A. Staudt, H.V. Klapdor-Kleingrothaus, C.R. Ching and T.H. Ho, Phys. Lett. B 272 (1991) 169.

[20] C.-R. Ching, T.-H. Ho and B. Zou, Nucl. Phys. A 510 (1990) 630; A 513 (1990) 697.

[21] N. Anantaraman, H. Toki and G. Bertsch, Nucl. Phys. A 398 (1983) 269.

Commun. Theor. Phys. 20(1993)453–460
© International Academic Publishers Vol. 20, No. 4

New Theoretical Results of $2\nu\beta\beta$ Decay with the Operator Expansion Method[1]

X.R. WU, M. Hirsch, A. Staudt and H.V. Klapdor-Kleingrothaus
Max-Planck-Institut für Kernphysik, D-6900 Heidelberg, Germany
Cheng-Rui CHING and Tso-Hsiu HO
Centre for Theoretical Physics, CCAST (World Laboratory) and
Institute of Theoretical Physics, Academia Sinica, Beijing 100080, China
(Received April 9, 1992)

Abstract

The half-lives for $2\nu\beta\beta$ decay for all the potential $\beta\beta$-emitters with A$\geq$ 70 are calculated by the operator expansion method. Compared with the directly measured half-lives of ^{76}Ge, ^{82}Se, ^{100}Mo and ^{238}U the theoretical values are in excellent agreement with experimental ones.

Recently, more than half a century after Goeppert-Mayer first discussed $\beta\beta$ decay under Wigner's suggestion in 1935[1], direct counter experiments reported the observation of $2\nu\beta\beta$ decay for the nuclei ^{76}Ge[2,3], ^{82}Se[4], ^{100}Mo[5,6] and ^{238}U[7]. These measurements now supplement the earlier geochemical experiments for the $\beta\beta$ candidates ^{82}Se, ^{128}Te and ^{130}Te[8–11].

For the more interesting $0\nu\beta\beta$ decay mode, however, no evidence has been found so far, and only lower limits for the half-lives have been quoted in the literature[2–7]. At present new experiments like that of the Heidelberg-Moscow collaboration have been started[12,13], in a new attempt to look for $0\nu\beta\beta$ decay with increased sensitivity.

The amount of information which can be extracted from such experiments depends in a decisive way on the reliability of the theoretical estimates for the $0\nu\beta\beta$ decay matrix elements. The calculation of $2\nu\beta\beta$ decay is a valuable test for the accuracy which can be expected for the $0\nu\beta\beta$ transition matrix elements, since both calculations involve essentially the same nuclear physics. Here, we would like to report the calculation of $2\nu\beta\beta$ decay half-lives for all potential $\beta\beta$-emitters with $A \geq 70$, within a new model, the operator expansion method[14–15].

The calculation of the $\beta\beta$ nuclear transition matrix elements between the initial even-even parent nucleus (A, Z) and the final state in the daughter nucleus $(A, Z+2)$ includes a sum of — in principle — infinite number of intermediate states in the adjacent odd-odd nucleus $(A, Z+1)$. The determination of these intermediate states is a difficult task and their treatment in shell model calculations[16,17] as well as calculations within the usual quasiparticle random-phase approximation (QRPA)[18–20] shows some weakness. As is well known, in shell model calculations the energies of the intermediate states are replaced by an average value (closure approximation) which has been shown to be doubtable for $2\nu\beta\beta$-decay. In QRPA the matrix elements for $2\nu\beta\beta$ decay are very sensitive to the particle-particle interaction parameter g_{pp} when g_{pp} is near its physical value of $g_{pp} = 1$. Therefore, recently an alternative method, the operator expansion method (OEM), which can treat the sum over the infinite intermediate states in a more elegant way, has been proposed by Ching and Ho and applied to the calculation of ^{48}Ca, ^{76}Ge, ^{82}Se, ^{100}Mo and ^{130}Te with rather promising results[14–15]. Encouraged by these successes and as a further application of OEM, we have calculated the $2\nu\beta\beta$ decay half-lives for all the possible $\beta\beta$-emitters with A$\geq$ 70.

[1]The project supported partly by National Natural Science Foundation of China.

If one assumes that the sum of the energies for each pair of emitted electron and neutrino can be replaced by the average value of $\Delta = (E_I - E_F)/2 = \frac{1}{2}Q_{\beta\beta} + m_e$, (for a discussion see for example Ref. [21]) then, as is well known, the half-life for $0^+ \to 0^+$ $2\nu\beta\beta$ decay can be expressed in a factorized form as

$$\left[T^{2\nu}_{1/2}\right]^{-1} = F^{2\nu}\left|M_{GT}\right|^2 , \tag{1}$$

where $F^{2\nu}$ is a lepton phase-space integral. The matrix element is given by

$$M^{2\nu}_{GT} = \sum_N \frac{\langle 0^+_F|A^\alpha|1^+_N\rangle\langle 1^+_N|A^\alpha|0^+_I\rangle}{\Delta + (E_N - E_I)} . \tag{2}$$

where $M^{2\nu}_{GT}$ is a special case for $0^+ \to 0^+$ transition. The basic idea of OEM now is to transfer the energy denominator to the numerator by the following mathematical procedure. We take instead of $M^{2\nu}_{GT}$ the following matrix element

$$M_{GT}(Z) = \frac{1}{2}\sum_N \left\{ \frac{\langle F|A^\alpha|N\rangle\langle N|A^\beta|I\rangle}{\Delta + (E_N - E_I)Z} - \frac{\langle F|A^\beta|N\rangle\langle N|A^\alpha|I\rangle}{\Delta + (E_F - E_N)Z} \right\} , \tag{3}$$

where $|I\rangle$, $|N\rangle$ and $|F\rangle$ denote the wave functions of the initial, intermediate and final nuclear states, respectively and E_I, E_N, and E_F are the corresponding energies and Z is a complex variable. Mathematically $M_{GT}(Z)$ is a single-valued, regular function of Z in an open, connected region containing the origin and $Z = 1$, except some possible poles along the real axis. For small Z, $M_{GT}(Z)$ can be expanded using the binomial theorem:

$$\frac{1}{\Delta + (E_N - E_I)Z} = \frac{1}{\Delta}\left\{1 - \frac{Z(E_N - E_I)}{\Delta} + \frac{Z^2(E_N - E_I)^2}{\Delta^2} - \dots\right\} . \tag{4}$$

Then, after introducing the nuclear Hamiltonian H_S the following $M_{GT}(Z)$ is obtained

$$\begin{aligned} M_{GT} = \frac{1}{2\Delta}\Big\langle F\Big|\Big\{&[A^\alpha, A^\beta] - \frac{Z}{\Delta}[A^\alpha, [H_S, A^\beta]] \\ &+ \frac{Z^2}{\Delta^2}\Big[A^\alpha, \big[H_S, [H_S, A^\beta]\big]\Big] - \cdots\Big\}\Big|I\Big\rangle , \end{aligned} \tag{5}$$

where the summation over the intermediate states has been carried out. It is important to note that equation (5) is an exactly equivalent formulation of the matrix element of Eq. (3). On the other hand, mathematically equation (5) is a divergent series for $2\nu\beta\beta$ with $Z = 1$ and thus one has to sum up all the terms up to infinity for the calculation of the matrix element (see for example G.M. Hardy, in: Divergent Series[22]).

For the summation of the infinite series, however, we have to assume a specific form of the nuclear Hamiltonian, and take

$$H_S = \langle H_0\rangle + V_S , \tag{6}$$

where

$$V_S = \frac{1}{2}\sum_{i\neq j}\left\{v_0(r_{ij}) + v_\tau(r_{ij})\tau_i\cdot\tau_j + v_\sigma(r_{ij})\sigma_i\cdot\sigma_j + v_{\sigma\tau}(r_{ij})\sigma_i\cdot\sigma_j\tau_i\cdot\tau_j\right\} , \tag{7}$$

which is the most general central static force. $v_0(r_{ij})$, $v_\sigma(r_{ij})$, $v_\tau(r_{ij})$ and $v_{\sigma\tau}(r_{ij})$ denote the radial parts of the corresponding interactions, respectively, and $r_{ij} = |\mathbf{r}_i - \mathbf{r}_j|$ stands for the relative coordinates of the nucleons.

The assumption that the unperturbed nuclear Hamiltonian can be replaced by its average value $\langle H_0\rangle$, which is constant and thus does not contribute to the commutator series (5), is the main approximation of the present OEM. The accuracy of this approximation has still to be checked. However, one can argue[23] that fot the case of a small model space H_0 is nearly degenerate, i.e., $\langle H_0\rangle - H_0$ is small compared with V_S and thus the approximation should be reasonable.

Adopting this approximation, and after a fair amount of algebraic operations, one can sum up the infinite series (5) and simplify the $0^+ \to 0^+$ transition matrix element (2) into

$$M_{GT} \equiv \left\langle 0_F^+ \middle| \sum_{i\neq j} \mathcal{M}_{ij} + \sum_{i\neq j\neq k} \mathcal{M}_{ijk} + \cdots \middle| 0_I^+ \right\rangle, \tag{8}$$

where there are totally $(A-1)$ operators from 2-body operator $\mathcal{M}_{ij}$ to A-body operator.

Finally for the two-body operator OEM gives the matrix element M_{GT} as the analytical continuation $M_{GT} = M_{GT}(Z)|_{Z=1}$,

$$\begin{aligned}\mathcal{M}_{ij}(Z) = & \frac{12\big(v_\sigma(r) - v_\tau(r)\big)Z}{\Delta^2 - 16Z^2\big(v_\sigma(r) - v_\tau(r)\big)^2}\,\Omega_0(ij) \\ & + \frac{4Z\big(2v_{\sigma\tau}(r) - v_\sigma(r) - v_\tau(r)\big)}{\Delta^2 - 16Z^2\big(2v_{\sigma\tau}(r) - v_\sigma(r) - v_\tau(r)\big)^2}\,\Omega_1(ij)\,,\end{aligned} \tag{9}$$

where we have introduced the spin singlet and spin triplet operators

$$\Omega_1(ij) = \frac{3+\sigma_i\cdot\sigma_j}{4}, \qquad \Omega_0(ij) = \frac{1-\sigma_i\cdot\sigma_j}{4}. \tag{10}$$

The mathematical procedure just outlined is the so-called Euler's method of summation. We note that Simkovic and Gmitro[24] starting from the time-ordered product of two operators rederived the OEM result for the two-body operator (9). They obtained exactly the same expression for $\mathcal{M}_{ij}$.

In principle, all the $A-1$ terms can be given a compact form, but due to the technical difficulty to calculate the matrix elements of the many-body operators, only the two-body terms are taken into account. This is our second approximation. We expect this approximation to be reasonable, since for $\beta\beta$ decay the transition operator is of two-body character. However, we plan to investigate the influence of the three-body terms in a future publication.

We note that if $v_\sigma = v_\tau = v_{\sigma\tau}$ in the effective interaction (7), the Hamiltonian (7) would be of [0] irreducible tensor form of Wigner's SU(4) multiplet, and the two-body operator $\mathcal{M}_{ij}$ in equation (9) is exactly zero. The nonzero contributions come from differences of the interaction strengths v_σ, v_τ and $v_{\sigma\tau}$. We note that in Ref. [25], an $H' = V(\tau_i\cdot\tau_j - \sigma_i\cdot\sigma_j)$ SU(4) symmetry-breaking Hamiltonian was taken into account and the M_{GT} given there by making use of the SU(4) algebra perhaps corresponds to the contribution of the $\Omega_0(ij)$ term (the first term) in Eq. (9). But the Hamiltonian used in the present work is more realistic than that used in the SU(4) algebra techniques and thus more effects of SU(4) symmetry-breaking are picked up here.

To calculate the $2\nu\beta\beta$ matrix element (8) we only need to know the effective interactions for evaluating $\mathcal{M}_{ij}$ and the ground state wave functions of the initial and final states. This is an delighting advantage of the present approach.

The operator $\mathcal{M}_{ij}$ can be determined by the effective interaction, for which we take the Paris-potential[26]. In principle, one should take the eigenfunctions of H_S with the correct eigenvalues E_I and E_F, respectively. However, since most of the potential $\beta\beta$-emitters are located far away from closed shells, it is impossible at present to use shell model wave functions for the initial and final states. Thus, for the determination of $|0_I^+\rangle$ and $|0_F^+\rangle$ QRPA wave functions are used in the present work. For a description of the QRPA we refer to the original literature for brevity[18-20], for the combination of OEM and QRPA wavefunctions to Ref. [15].

We describe the initial $|0_I^+\rangle$ and final state $|0_F^+\rangle$ by the QRPA vacuum based on BCS states. Then we insert two sets of complete and orthogonal mathematical functions $|a\rangle$ and $|b\rangle$ into Eq. (9). We construct these two sets of mathematical functions by the phonon operators acting on the QRPA vacuum, these phonons consisting of the quasiparticle proton-neutron-pair. Then we can directly use the QRPA techniques to calculate the transition matrix element M_{GT}.

The main advantage of the present approach lies in the fact that the calculated matrix element M_{GT} is only weakly sensitive to the choice of the particle-particle interaction parameter g_{pp}. This was demonstrated in Fig. 1 for the example of Ref. [15] for ^{100}Mo. The similar behaviour of the matrix element holds also for ^{238}U. The OEM leads to matrix elements which are always smaller than the QRPA calculation for $g_{pp} = 0$, but do not exhibit a strong dependence on g_{pp}.

It is a natural feature of the present approach that M_{GT} is insensitive to g_{pp}. Recall that the two-body operator $\mathcal{M}_{ij}$ in Eq. (9) is not explicitly dependent on the intermediate energy spectrum and that the wave functions of the intermediate states are only used as a complete and orthogonal set of mathematical functions. The numerical results of M_{GT} should only care for whether these two sets of functions are complete and orthogonal, and thus $M_{GT}^{2\nu}$ should be constant no matter how g_{pp} changes. On the other hand, our present calculation uses the QRPA vacuum for the description of the initial $|0_I^+\rangle$ and final state $|0_F^+\rangle$ and these depend on g_{pp}. However, the QRPA ground state is

$$|\,\mathrm{QRPA}\,\rangle = |\,\mathrm{BCS}\,\rangle + YX^{-1}|\,\mathrm{BCS}\,\rangle + \cdots . \tag{11}$$

Clearly, the $|\,\mathrm{BCS}\,\rangle$ state does not depend on g_{pp} and since in QRPA the coefficient necessarily fulfills the condition $YX^{-1} \ll 1$, in the first order, the QRPA vacuum and thus our calculated M_{GT} should only weakly depend on g_{pp}. On the other hand, in usual QRPA calculations the wave functions of the intermediate states should be the real wave functions of the odd-odd nucleus. The energy eigenvalues of these intermediate states depend on g_{pp}, and consequently in usual QRPA the results are more sensitive to g_{pp}.

We think that the strong dependence of the $2\nu\beta\beta$ decay half-lives on g_{pp} is a major disadvantage of QRPA calculations, as we would like to discuss for the special case of ^{238}U. As can be seen from Fig. 1 the $2\nu\beta\beta$ matrix element crosses zero for a certain value of g_{pp}, translating into an infinite half-life. For the parameter choice of Ref. [20] one obtains $T_{1/2}^{2\nu} = 1.5 \cdot 10^{23}$ y, which is a factor of 100 larger than the experimental result of Ref. [7] but could easily be adapted to the experimental value by a small change of g_{pp}. On the other hand, the OEM result of $T_{1/2}^{2\nu} = 0.9 \cdot 10^{21}$ y is relatively stable against variations on g_{pp}, and also in good agreement with the experimental data.

Since the results of the present approach are not sensitive to a specific choice of the parameters g_{ph} and g_{pp} we can use the physical values $g_{pp} = g_{ph} = 1$ directly. However, in some cases (as indicated in Table 1) the QRPA equation collapses for values below $g_{pp} = 1$. For these isotopes the g_{pp} values before the collapse points are used. The calculated matrix elements M_{GT} and corresponding half-lives for all the potential $\beta\beta$-emitters are given in Table 1. The half-lives given are for $g_A/g_V = -1.254$ and the phase-space factors of Ref. [21]. Note, however, that for the heaviest isotopes the phase space factor of Ref. [21] differs from the one used in the QRPA calculation of Ref. [20] by a factor of ~ 3. On the other hand, for medium heavy isotopes, such as ^{76}Ge, the phase factors of Refs. [21] and [20] are essentially equal.

Table 1. The calculated Gamow-Teller matrix elements and the corresponding half-lives of all potential $\beta\beta$-emitters with $A \geq 70$. The experimental half-lives for some nuclei are included. For all isotopes, except those marked by †, $g_{pp} = 1$ is taken. For isotopes marked by † the QRPA equation collapses for $g_{pp} = 1$, and thus g_{pp} before the collapse point is used. [a] Ref. [2], [b] Ref. [3], [c] Ref. [4], [d] Ref. [9], [e] Ref. [5], [f] Ref. [6], [g] Ref. [10], [h] Ref. [11], [i] Ref. [27], [j] Ref. [28], [k] Ref. [29] and [l] Ref. [7].

Nuclei	M_{GT}	Half life (yrs)	Experimental half life (years)
^{70}Zn	0.0918†	1.44×10^{24}	
^{76}Ge	0.3351	2.61×10^{20}	$(1.1^{+0.6}_{-0.3}) \times 10^{21}$ [a], $(9.2^{+0.7}_{-0.4}) \times 10^{20}$ [b]
^{80}Se	0.3424	2.68×10^{29}	
^{82}Se	0.1019	0.848×10^{20}	$(1.1^{+0.8}_{-0.3}) \times 10^{20}$ [c], $(1.30 \pm 0.05) \times 10^{20}$ [d]
^{86}Kr	0.0580	3.42×10^{23}	
^{94}Zr	0.0117†	1.68×10^{24}	
^{96}Zr	0.0314†	2.02×10^{20}	
^{98}Mo	0.0800	6.16×10^{30}	
^{100}Mo	0.1065†	3.58×10^{19}	$(1.15^{+0.3}_{-0.2}) \times 10^{19}$ [e], $(1.16^{+0.34}_{-0.22}) \times 10^{19}$ [f]
^{104}Ru	0.1162†	3.09×10^{22}	
^{110}Pd	0.0879†	1.24×10^{21}	
^{114}Cd	0.1642†	9.84×10^{24}	
^{116}Cd	0.0171†	1.57×10^{22}	
^{122}Sn	0.1712	1.25×10^{26}	
^{124}Sn	0.0391	1.49×10^{21}	
^{128}Te	0.1462	2.11×10^{23}	$> 5 \times 10^{24}$ [g], $(1.8 \pm 0.7) \times 10^{24}$ [h]
^{130}Te	0.1006	0.787×10^{20}	$(1.5 - 2.8) \times 10^{21}$ [d], $(7.5 \pm 0.3) \times 10^{20}$ [h]
^{134}Xe	0.1286	2.69×10^{23}	
^{136}Xe	0.0280	1.01×10^{21}	$> 8.4 \times 10^{19}$ [i], $> 1.6 \times 10^{20}$ [j]
^{142}Ce	0.0400	3.30×10^{22}	
^{146}Nd	0.3285†	7.31×10^{30}	
^{148}Nd	0.0548†	1.19×10^{21}	
^{150}Nd	0.0441	1.66×10^{19}	$\geq 1.8 \times 10^{19}$ [k]
^{154}Sm	0.0793	1.49×10^{22}	
^{160}Gd	0.0454	2.81×10^{21}	
^{170}Er	0.1685	2.46×10^{23}	
^{176}Yb	0.1593†	4.92×10^{21}	
^{186}W	0.1506†	1.30×10^{24}	
^{192}Os	0.1777	2.40×10^{24}	
^{198}Pt	0.0741	1.14×10^{22}	
^{204}Hg	0.0510	1.81×10^{25}	
^{232}Th	0.1263	4.03×10^{21}	
^{238}U	0.0785	0.914×10^{21}	$(2.0 \pm 0.6) \times 10^{21}$ [l]

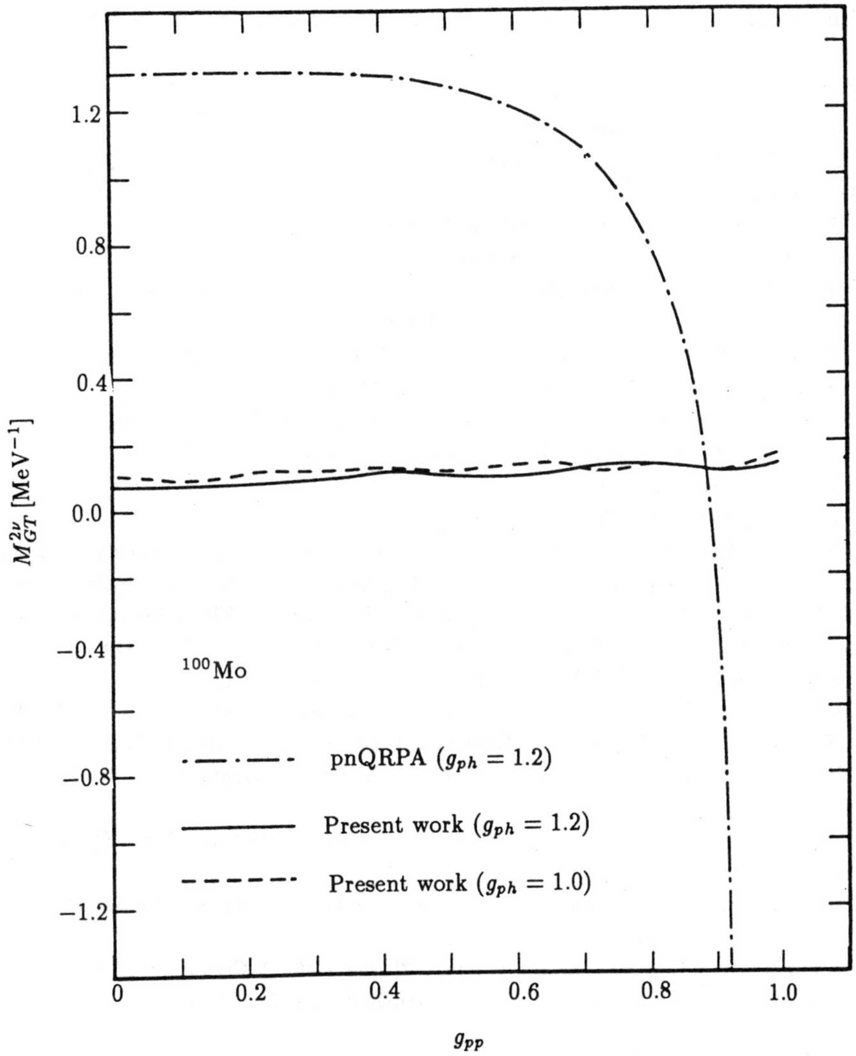

Fig. 1. Calculated Gamow-Teller matrix element M_{GT} for ^{238}U as a function of the particle-particle interaction parameter g_{pp} for the present work as well as for a usual QRPA calculation. It is shown that within the OEM, M_{GT} is relatively insensitive to the choice of g_{pp}.

For a comparison we listed also the experimental data available to us. For all nuclides which have been measured by direct counter experiments (^{76}Ge, ^{82}Se, ^{100}Mo and ^{238}U), the theoretical half-lives are in quite well agreement with the experimental ones. However for the Te isotopes our calculation gives a larger decay rate than the geochemical experiments.

To summarize, we have calculated $2\nu\beta\beta$ decay half-lives for all the potential $\beta\beta$ emitters with $A \geq 70$. Our results are not sensitive to a specific choice for the particle-particle interaction parameter g_{pp}. The present approach thus overcomes problems of earlier QRPA calculations.

Acknowledgments

We thank Dr. K. Muto for many helpful discussions and for his QRPA code[2].

References

[1] M. Goeppert-Mayer, Phys. Rev. **48**(1935)512.

[2] H.S. Miley, F.T. Avignone, III, R.L. Brodzinski, J.I. Collar and J.H. Reeves, Phys. Rev. Lett. **65**(1990)3092.

[3] F.T. Avignone III *et al.*, Phys. Lett. **B256**(1991)559.

[4] S.R. Elliott, A.A. Hahn and M.K. Moe, Phys. Rev. Lett. **59**(1987)2020.

[5] H. Ejiri *et al.*, Phys. Lett. **B258**(1991)17.

[6] K. Okada *et al.*, Nucl. Phys. B (Proc. Suppl.) 19(1991).

[7] A.L. Turkevich, T.E. Economou and G.A. Cowan, Phys. Rev. Lett. **67**(1991)3211.

[8] T. Kirsten, W. Gentner and O.A. Schaeffer, Z. Phys. **202**(1967)273.

[9] T. Kirsten, Proc. Int. Symp on Nuclear Beta Decays and Neutrino, June 1986, Osaka, eds. T. Kotani, H. Ejiri and E. Takasugi, World Scientific, Singapore (1986) p. 81.

[10] O.K. Manuel, as Ref. [9] but p. 103.

[11] W.J. Lin *et al.*, Nucl. Phys. **A481**(1988)477.

[12] H.V. Klapdor-Kleingrothaus, in: Proc. Second Int. Workshop on Theoretical and Phenomenological Aspects of Underground Physics (TAUP 91), Sep. 9–13 1991, Tolédo Spain, in press.

[13] M. Beck, J. Bockholt, J. Echternach, G. Heusser, H.V. Klapdor-Kleingrothaus, B. Maier, F. Petry, A. Piepke, U. Schmidt-Rohr, H. Strecker, R. Zuber, A. Balysh, S.T. Belyaev, A. Demehin, A. Gurov, I. Kondratenko, V.I. Lebedev, submitted to Phys. Lett. B.

[14] C.R. CHING and T.H. HO, Commun. Theor. Phys. **11**(1989)433; C.R. CHING, T.H. HO and X.R. WU, Phys. Rev. **C40**(1989)304; Commun. Theor. Phys. **12**(1989)167; in "*Proceedings of the BIMP Symposium on Heavy Flavor Physics*, eds. Chao Kuangta *et al.*, World Scientific, Singapore (1989), p. 359.

[15] X.R. WU, S. Staudt, H.V. Klapdor-Kleingrothaus, C.R. CHING and T.H. HO, Phys. Lett. **B272**(1991)169.

[16] W.C. Haxton and G.J. Stephenson, Jr., Progress in Particle and Nuclear Physics **12**(1984)409.

[17] J.D. Vergados, Phys. Rep. **133**(1986)1.

[18] P. Vogel and M.R. Zirnbauer, Phys. Rev. Lett. **57**(1986)3148; O. Civitarese, A. Faessler and T. Tomoda, Phys. Lett. **B194**(1987)11; K. Muto and H.V. Klapdor, Phys. Lett. **B201**(1988)420;

[2] The present calculation was completed while one of the authors, X.R. WU was visiting M.P.I. in the beginning of 1992. Recently we have learned that in a paper appearing in Phys. Rev. C46 (1992)R2153 the authors J. Engel, W. C. Haxton and P. Vogel criticized strongly our approach in double beta decay calculation. First, they criticized the "power series expansion beyond the radius of convergence"; second is the neglect of kinetic energy operator T in the commutators; and third, the neglect of many-body scattering terms. Since the improvement of the OEM calculation is still under way, and a detailed analysis on the questions raised by these authors will be given elsewhere. Here we only brief our main arguments by pointing out 1), the first criticism is groundless, and this question is also explained in the present paper; 2), the neglect of operator T in the commutators certainly brings error in the matrix element, but it is tolerable for a realistic nuclear force according to the standard of nuclear physics, moreover, the claim that this approximation always leads to smaller matrix element (comparing to the exact one) is wrong. It is potential-dependent; 3), the neglect of many-body scattering terms is a real problem in OEM. Howevr, part of this effect has been absorbed in the wave functions. And the validity can be evaluated by including the three-body terms, which can be shown to be summable as in the case of two-body case. The results presented in this paper are interesting not only because they are not so easily obtained using method different from OEM, but also the OEM+QRPA wave functions is no less better than other results which are inevitably based upon a certain approximation

T. Tomoda, Rep. Prog. Phys. **54**(1991)53.

[19] H.V. Klapdor and K. Grotz, Phys. Lett. **B142**(1984)323; K. Grotz and H.V. Klapdor, Phys. Lett. **B153**(1985)1, Phys. Lett. **B157**(1985)242; Nucl. Phys. **A460**(1986)395.

[20] A. Staudt, K. Muto and H.V. Klapdor-Kleingrothaus, Europhys. Lett. **13**(1990)31; A. Staudt, T.T.S. Kuo and H.V. Klapdor-Kleingrothaus, Phys. Lett. **B242**(1990)17; K. Muto, E. Bender and H.V. Klapdor, Z. Phys. **A334**(1989)177, 187.

[21] M. Doi, T. Kotani and E. Takasugi, Prog. Theor. Phys. Suppl. **83**(1985)1.

[22] G.M. Hardy, Divergent Series, Clarendon, Oxford, 1949.

[23] X.R. WU, S. Staudt, T.T.S. Kuo and H.V. Klapdor-Kleingrothaus, Phys. Lett. (in press).

[24] F. Simkovic, JINR Rapid Commun. **39**(1989)21 and M. Gmitro and F. Simkovic, Izv. AN SSSR **54**(1990)1780.

[25] J. Bernabeu *et al.*, Z. Phys. **C46**(1990)792; K.T. Hecht, S.C. Pang, J. Math. Phys. **8**(1969)1571.

[26] M. Lacombe *et al.*, Phys. Rev. **C21**(1980)861.

[27] A.S. Barabash, V.V. Kuzminov *et al.*, Sov. J. Nucl. Phys. **51**(1990)1.

[28] E. Bellotti *et al.*, Phys. Lett. **266B**(1989)209.

[29] A.A. Klimenko *et al.*, Nucl. Instrum. and Meth. **B16**(1986)446.

Z. Phys. A 345, 163–169 (1993)

ZEITSCHRIFT
FÜR PHYSIK A

Matrix elements for $0\nu\beta\beta$ decay calculated with the operator expansion method and QRPA wave functions

M. Hirsch [1], X.R. Wu [1], H.V. Klapdor-Kleingrothaus [1], Ching Cheng-rui [2,*], Ho Tso-hsiu [2,*]

[1] Max-Planck-Institut für Kernphysik, Postfach 103980, W-6900 Heidelberg, Germany
[2] Centre for Theoretical Physics, CCAST (World Laboratory) and Institute of Theoretical Physics, Academia Sinica, Beijing, PR China

Received: 18 August 1992 / Revised version: 21 October 1992

Abstract. Matrix elements for $0\nu\beta\beta$ decay of ^{76}Ge, ^{82}Se, ^{116}Cd, ^{128,130}Te, ^{136}Xe and ^{150}Nd are calculated combining proton-neutron quasiparticle RPA wave functions with the operator expansion method (OEM). The differences between OEM and earlier QRPA calculations, using the closure approximation, are investigated in detail. Special emphasis is put on the discussion on the differences between the $2\nu\beta\beta$ and the $0\nu\beta\beta$ mode, since currently 0ν matrix elements can only indirectly be tested. By a comparison with experimental data upper limits on the effective neutrino mass are derived.

PACS: 23.40.

1. Introduction

Neutrinoless double beta decay ($0\nu\beta\beta$) has attracted considerable attention during recent years [1–3], since the observation of $0\nu\beta\beta$ decay could answer some of the most longstanding questions in particle physics: is the neutrino a Dirac or a Majorana particle? And has the neutrino a finite rest mass? In the standard model of electroweak interaction [4] neutrinos are assumed to be massless, but massive Majorana neutrinos naturally arise in most grand unified theories (see for example [5]), with predicted masses in the range of 10^{-11} to 10 eV. Thus much effort has been devoted from both, experimental [6–13] and theoretical side [14–20] to explore this exotic phenomenon (see also the review in [21]).

Up to now, experimentally only lower half life limits for $0\nu\beta\beta$ decay have been reported in the literature. The main motivation for the detailed theoretical studies, however is that reliable estimates for the nuclear transition matrix elements are necessary to convert the experimental half life limits into upper limits for the effective neutrino mass.

* Project supported by the Natural Science Foundation of China

If $0\nu\beta\beta$ decay ever would be observed, for the analysis of the data, in addition to the matrix elements related to the mass mechanism, theoretical information on transitions involving right-handed weak currents would be required. In the present work we consider only the matrix elements relevant for the neutrino mass (inclusion of right-handed weak currents would lead to slightly less stringent bounds by $\sim$10–20%).

Calculations of $\beta\beta$ decay matrix elements involve, in principle, the determination of a full set of intermediate states in the adjacent odd-odd nucleus. Since most of the $\beta\beta$ candidates are located far away from closed shells, it is surely this part of the calculation which poses most difficulties for the theoretical studies of $\beta\beta$ decay. Many attempts thus have introduced the so-called closure approximation to circumvent this complicated problem [2, 15–20]. The use of closure has been shown to be doubtable for $2\nu\beta\beta$ decay, but can be argued to be reasonable for the 0ν mode (see [19]). On the other hand, recently a novel method has been proposed for the calculation of $\beta\beta$ decay matrix elements which avoids both, the closure approximation as well as a direct calculation of the intermediate spectrum, the operator expansion method (OEM) [22–24]. OEM has also recently been combined with a state of the art calculation for the nuclear ground state wave functions for $2\nu\beta\beta$ decay [25]. The results of this work [25], using the proton-neutron quasiparticle RPA model of [16, 20] for the calculation of the nuclear wave functions, were rather surprising: matrix elements of this model seemed to be only weakly dependent on a special model parameter, the strength of the particle-particle interaction g_{pp}, which played a crucial role in all earlier studies – using QRPA – of $2\nu\beta\beta$ decay [3, 17, 18, 20]. We thus felt encouraged to combine the OEM also for $0\nu\beta\beta$ decay matrix elements with QRPA wave functions. In this work $0\nu\beta\beta$ decay matrix elements for ^{76}Ge, ^{82}Se, ^{116}Cd, ^{128,30}Te, ^{136}Xe and ^{150}Nd are calculated by this combined OEM + QRPA model, as well as in the common closure approximation. The differences between both calculations are examined in detail and found to be small, indicating that closure works

ELSEVIER

Physics Reports 242 (1994) 403–422

PHYSICS REPORTS

Operator expansion method and nuclear ββ decay

M. Hirsch[a], X.R. Wu[a, 1], H.V. Klapdor-Kleingrothaus[a], Ching Cheng-rui[b, 2], Ho Tso-hsiu[b, 2]

[a] *Max-Planck-Institut für Kernphysik, D-6900 Heidelberg, Germany*
[b] *Centre for Theoretical Physics, CCAST (World Laboratory) and Institute of Theoretical Physics, Academia Sinica, Beijing, China*

Abstract

Recent developments in the theoretical description of double beta decay matrix elements are outlined. We present the formalism of the operator expansion method (OEM) and discuss its physical basis. The OEM has been combined with ground state wave functions obtained in a usual QRPA calculation and applied to the problem of evaluating nuclear matrix elements for double beta decay in the 2νββ as well as in the 0νββ decay mode. For 2νββ decay we have calculated the half-lives for all possible $\beta^-\beta^-$ decay candidates with $A \geq 70$. We confront our results with the existing experimental data and discuss the quality of the model. 0νββ decay matrix elements are then used to extract upper limits for the effective neutrino mass from published experimental lower half-life limits. The presently sharpest limits come from experiments using isotopically enriched ^{76}Ge. A short discussion is devoted to a recent measurement using ^{238}U in a radiochemical experiment. Our conclusion is that the experimental result should be interpreted as being due to 2νββ decay, and lepton number violation is not necessary to explain the experimental finding.

1. A short introduction to nuclear double beta decay

Today, 60 years after Pauli proposed the existence of the neutrino [1] in order to explain the energy spectra observed in nuclear β decays, the nature and most of the properties of this elusive particle remain a miracle. For example, there exists no convincing argument based on theoretical grounds that neutrinos should have a vanishing mass, although in the standard model of the electroweak interaction [2] they are assumed to be massless. Moreover, in many extensions of the standard model [3] neutrino masses arise quite naturally. The observation of a finite neutrino mass could thus prove or disfavour the different models and would be a clear sign of physics beyond the

[1] Talk presented by X.R. Wu.
[2] Project supported by the Natural Science Foundation of China.

SSDI 0370-1573(94)00033-Y

Z. Phys. A 352, 33–45 (1995)

ZEITSCHRIFT
FÜR PHYSIK A

Extended operator expansion method for neutrinoless double beta decay

M. Hirsch,[1] O. Kadowaki,[1] H.V. Klapdor-Kleingrothaus,[2] K. Muto,[1] T. Oda[1]

[1]Department of Physics, Tokyo Institute of Technology, Oh-okayama, Meguro, Tokyo 152, Japan
[2]Max-Planck-Institut für Kernphysik, D-69029 Heidelberg, Germany

Received: 2 September 1994/Revised version: 22 December 1994

Abstract. Reliable calculations of nuclear matrix elements are a prerequisite for the determination of the effective neutrino mass and other particle physics parameters from neutrinoless double beta decay. Here, the operator expansion method is improved by including Coulomb, tensor and central interactions simultaneously. Furthermore, the formalism of the OEM is extended to those matrix elements necessary to extract the right-handed parameters $\langle \lambda \rangle$ and $\langle \eta \rangle$ from $0\nu\beta\beta$ decay. OEM includes the dependence of the nuclear matrix elements on the intermediate states implicitly and can therefore be understood as a step beyond the closure approximation. Numerical studies are carried out for the isotope ^{76}Ge combining the OEM expressions with ground-state wave functions calculated within a proton-neutron quasiparticle Random Phase Approximation (pn-QRPA) model. The influence and relative importance of central, tensor and Coulomb interactions is investigated. Within the OEM, contributions from the Coulomb force are found to be negligible in $0\nu\beta\beta$ decay, while the tensor force leads to a moderate change of the results, of the order of (10–30) %, giving a better agreement between sets of calculations which employ different NN-interactions. Generally, results of the OEM + QRPA calculation are similar to previous calculations of $0\nu\beta\beta$ decay matrix elements, indicating that $0\nu\beta\beta$ decay is not sensitive to model approximations and might therefore be more accurately calculated than the strongly suppressed $2\nu\beta\beta$ decay matrix elements.

PACS: 21.60; 23.40

1. Introduction

One of the fascinating consequences of the observation of neutrinoless double beta ($0\nu\beta\beta$) decay would be that the neutrino has to be a massive Majorana particle. Based on this motivation growing interest has focused on the study of $\beta\beta$ decay – both theoretical [1–3] and experimental [4–11] – during the past years. Massive Majorana neutrinos naturally appear in many grand unified theories (for reviews, see for example [12,13]), but predicted mass scales are essentially arbitrary.

No positive evidence for any $0\nu\beta\beta$ decay has been reported to date, but experiments have improved half-life limits [4–11] steadily and data taking continues. To convert experimental half-life limits on $0\nu\beta\beta$ decay into upper limits of the neutrino mass (and other particle physics parameters), however, nuclear matrix elements need to be known. Calculations of these were carried out by different groups [3, 14–21] using different nuclear structure models and/or approximations. On the other hand, $0\nu\beta\beta$ decay matrix elements can at present only indirectly be checked; tests of the models – or any approximations used – are therefore important.

Purpose of the present work is to improve an interesting idea for the calculation of $\beta\beta$ decay matrix elements, which has been recently proposed, the Operator Expansion Method (OEM) [22, 23]. OEM has the advantage to avoid the explicit sum over intermediate states and instead describes $\beta\beta$ decay matrix elements by an (infinite) series of multiple commutators of the nuclear Hamiltonian and transition operators. OEM accounts for intermediate states implicitly and is therefore an improvement of calculations based on the closure calculation. In principle, OEM provides an equivalent expression to the more common definitions of nuclear matrix elements. Published results of the OEM for $0\nu\beta\beta$ decay are, however, based on two main assumptions [22–24]: (i) from the nuclear Hamiltonian only the central components have been taken into account, and (ii) many-body scattering terms are neglected. Numerical results [24] combining the OEM $0\nu\beta\beta$ decay formula [23] with ground-state wave functions calculated by pn-QRPA [14,15], obtained 0ν matrix elements similar to those of previous studies [15, 17, 18, 20]. On the other hand, in a recent paper Muto included additionally tensor and Coulomb forces into the two-neutrino (2ν) $\beta\beta$ decay OEM formula, concluding the latter to be of essential importance [25]. To investigate the influence of these interactions also in the case of $0\nu\beta\beta$ decay, we derived OEM expressions including central, tensor and Coulomb terms

34

simultaneously. However, we still take the two-body approximation.

As is well-known, $0\nu\beta\beta$ decay can occur not only by a finite neutrino mass but also if the weak interaction has a right-handed component. Matrix elements for the latter are different from those of the mass mechanism. Numerical calculations of $0\nu\beta\beta$ decay within OEM [24] have up to now been restricted to only two matrix elements. These are the Gamow-Teller and Fermi matrix elements, which are necessary to determine the neutrino mass. The second purpose of the present work is to extend the OEM formalism also to the matrix elements for the right-handed current (RHC) mode. In addition, numerical calculations for all nine matrix elements of $0\nu\beta\beta$ decay are carried out.

Generally, values for all OEM + QRPA $0\nu\beta\beta$ decay matrix elements are relatively close to previous calculations. This is quite different from the situation for $2\nu\beta\beta$ reported in [25], which showed the OEM not to give matrix elements as suppressed as the earlier QRPA calculations [14]. It seems that $0\nu\beta\beta$ decay matrix elements are not so seriously affected by model approximations as are the $2\nu\beta\beta$ decay matrix elements. Therefore, it can be concluded that sharp limits on $\langle m_\nu \rangle$, $\langle \lambda \rangle$ and $\langle \eta \rangle$ can be derived from the nonobservation of $0\nu\beta\beta$ decay.

This paper is organized as follows. Section 2 deals with theory. After a brief summary of definitions (Sect. 2.1), the basic idea of the Operator Expansion Method will be described (Sect. 2.2), while OEM potentials for the different matrix elements are derived in the following. Section 3 then is devoted to the numerical results. The OEM is combined with ground-state wave functions of a pn-QRPA model [14,15]. The relative importance of central, tensor and Coulomb interactions for the nuclear matrix elements is discussed. While the inclusion of the tensor components lead to a moderate change of the calculated matrix elements, Coulomb force is found to have a negligible effect for the OEM $0\nu\beta\beta$ decay calculation, in contrast to the situation reported for $2\nu\beta\beta$ decay [25].

2. Theoretical considerations

2.1. Fundamental definitions for $0\nu\beta\beta$ decay

Before deriving the OEM expressions, it might be useful to give a brief summary of basic definitions for $0\nu\beta\beta$ decay matrix elements. For more details about the derivation of the decay rate formula we refer to the work of Doi et al. [2].

$0\nu\beta\beta$ decay can proceed either via the mass mechanism or if the weak interaction has a right-handed component. As the starting point of the calculation, therefore, the following effective Hamiltonian is used,

$$H_W = \frac{G}{\sqrt{2}} (j_{L_\mu} J_L^{\mu\dagger} + \kappa j_{L_\mu} J_R^{\mu\dagger} + \eta j_{R_\mu} J_L^{\mu\dagger} + \lambda j_{R_\mu} J_R^{\mu\dagger}) + \text{h.c.}, \tag{II.1}$$

where right-handed parameters κ, η and λ are defined such that the ordinary left-handed weak interaction has a relative strength of unity. $j_{L_\mu} = \bar{e}\gamma^\mu(1-\gamma_5)\,\nu_{e_L}$ and $j_{R_\mu} = \bar{e}\gamma^\mu(1+\gamma_5)\,\nu_{e_R}$ are left- and right-handed leptonic currents, respectively. Within the non-relativistic impulse approximation hadronic currents are given by

$$J_L^{\mu\dagger} = \sum_n t_n^- \left[(g_V - g_A C_n) g^{\mu 0} + (g_A \sigma_n^k - g_V D_n^k) g^{\mu k}\right] \delta(x - r_n), \tag{II.2a}$$

$$J_R^{\mu\dagger} = \sum_n t_n^- \left[(g_V + g_A C_n) g^{\mu 0} + (-g_A \sigma_n^k - g_V D_n^k) g^{\mu k}\right] \delta(x - r_n), \tag{II.2b}$$

where the sum runs over all nucleons in the nucleus and the recoil terms are

$$C_n = \left[(P_n + P_n')\cdot\sigma_n - \left(\frac{g_P}{g_A}\right)(E_n - E_n')(P_n - P_n')\cdot\sigma_n\right] \Big/ (2M), \tag{11.3a}$$

$$D_n = \left[(P_n + P_n') - \left(1 - 2M\frac{g_W}{g_V}\right) i\sigma_n (P_n - P_n')\right] \Big/ (2M), \tag{II.3b}$$

Here, (E_n, P_n) and (E_n', P_n') are initial and final four momenta of the n-th nucleon. M in (II.3) is the nucleon mass, $g_P/g_A = 2M/m_\pi^2$ and $g_W/g_V = -3.7/(2M)$.

With the above definitions the following expression for the $0\nu\beta\beta$ decay rate is obtained,

$$\begin{aligned}[T_{1/2}^{0\nu\beta\beta}(0_i^+ \to 0_f^+)]^{-1} &= C_{mm}\left(\frac{\langle m_\nu\rangle}{m_e}\right)^2 + C_{\eta\eta}\langle\eta\rangle^2 + C_{\lambda\lambda}\langle\lambda\rangle^2 + C_{m\eta}\frac{\langle m_\nu\rangle}{m_e}\langle\eta\rangle \\ &+ C_{m\lambda}\frac{\langle m_\nu\rangle}{m_e}\langle\lambda\rangle + C_{\eta\lambda}\langle\eta\rangle\langle\lambda\rangle. \end{aligned} \tag{II.4}$$

Terms proportional to κ do not appear in (II.4), since they always can be combined as $(1 \pm \kappa)$ and $\kappa \ll 1$ is expected. Effective values of the neutrino mass and right-handed parameters are defined by

$$\langle m_\nu \rangle = \sum_j{}' U_{ej}^2 m_j, \tag{II.5a}$$

$$\langle \eta \rangle = \eta \sum_j{}' U_{ej} V_{ej}, \tag{II.5b}$$

$$\langle \lambda \rangle = \lambda \sum_j{}' U_{ej} V_{ej}, \tag{II.5c}$$

where U_{ej} and V_{ej} are neutrino mixing coefficients for left-handed and right-handed Majorana neutrinos. In order to obtain the factorized expression for the inverse half-life it is assumed that only light neutrino mass eigenstates ($m_j < \mathcal{O}$(MeV)) contribute to the decay rate.

Coefficients C_{xy} are products of leptonic phase space integrals G_k ($k = 1$–9) and nine different nuclear matrix elements M_β ($\beta = GT, F, GT\omega, F\omega, GTq, Fq, T, P$ and R),

$$C_{mm} = (M_{GT} - M_F)^2 G_1, \tag{II.6a}$$

$$C_{m\lambda} = (M_{GT} - M_F)[-M_{2-}G_3 + M_{1+}G_4], \tag{II.6b}$$

$$C_{m\eta} = (M_{GT} - M_F)[M_{2+}G_3 - M_{1-}G_4 - M_P G_5 + M_R G_6], \tag{II.6c}$$

$$C_{\lambda\lambda} = M_{2-}^2 G_2 - \tfrac{1}{9}[2M_{1+}M_{2-}G_3 - M_{1+}^2 G_4], \tag{II.6d}$$

$$C_{\eta\eta} = M_{2+}^2 G_2 - \tfrac{1}{9}[2M_{1-}M_{2+}G_3 - M_{1-}^2 G_4] - M_P M_R G_7 + M_P^2 G_8 + M_R^2 G_9, \tag{II.6e}$$

$$C_{\eta\lambda} = -2\{M_{2+}M_{2-}G_2 - \tfrac{1}{9}[(M_{1+}M_{2+} + M_{1-}M_{2-})G_3 - M_{1+}M_{1-}G_4]\}. \tag{II.6f}$$

Shorthands $M_{1\pm}$ and $M_{2\pm}$ are defined by

$$M_{1\pm} = M_{GTq} - 6M_T \pm 3M_{Fq}, \tag{II.7a}$$

$$M_{2\pm} = M_{GT\omega} \pm M_{F\omega} - \tfrac{1}{9} M_{1\mp}. \tag{II.7b}$$

Definitions and numerical values of the phase space integrals G_k can be found in [2,26].

As will be shown below, matrix elements for OEM $0\nu\beta\beta$ decay can be written in a form very similar to the matrix elements of the common closure approximation. Therefore we recall that the nine matrix elements of (II.6) which do not vanish in the closure approximation can be expressed as

$$M_\alpha = \sum_{i \neq j} \langle F | t_i^- t_j^- \mathcal{O}_{ij}^\alpha | I \rangle, \tag{II.8}$$

where the two-body operators are

$$\mathcal{O}_{ij}^{GT} = \sigma_i \cdot \sigma_j H_m(r_{ij}), \tag{II.9a}$$

$$\mathcal{O}_{ij}^{F} = H_m(r_{ij}) \left(\frac{g_V}{g_A}\right)^2, \tag{II.9b}$$

$$\mathcal{O}_{ij}^{GT\omega} = \sigma_i \cdot \sigma_j H_\omega(r_{ij}), \tag{II.9c}$$

$$\mathcal{O}_{ij}^{F\omega} = H_\omega(r_{ij}) \left(\frac{g_V}{g_A}\right)^2 \tag{II.9d}$$

$$\mathcal{O}_{ij}^{GTq} = \sigma_i \cdot \sigma_j H_q(r_{ij}), \tag{II.9e}$$

$$\mathcal{O}_{ij}^{Fq} = H_q(r_{ij}) \left(\frac{g_V}{g_A}\right)^2, \tag{II.9f}$$

$$\mathcal{O}_{ij}^{T} = [(\sigma_i \cdot \hat{r}_{ij})(\sigma_j \cdot \hat{r}_{ij}) - \tfrac{1}{3}\sigma_i \cdot \sigma_j] H_q(r_{ij}), \tag{II.9g}$$

$$\mathcal{O}_{ij}^{P} = i(\sigma_i - \sigma_j) \cdot (\hat{r}_{ij} \times R_{ij}) H_q(r_{ij}) \left(\frac{g_V}{g_A}\right), \tag{II.9h}$$

$$\mathcal{O}_{ij}^{R} = \sigma_i \cdot \sigma_j H_R(r_{ij}) \frac{\mu_\beta}{3} \left(\frac{g_V}{g_A}\right). \tag{II.9i}$$

$\mu_\beta = 4.7$ and $r_{ij} = r_i - r_j$, $r_{ij} = |r_{ij}|$,

$\hat{r}_{ij} = r_{ij}/r_{ij}$, $R_{ij} = (r_i + r_j)/2$.

Two-body operators $\mathcal{O}_{ij}^\alpha$ consist of products of nuclear operators and "neutrino potentials". The latter are momentum integrals which appear due to the virtual nature of the neutrino in $0\nu\beta\beta$ decay. They are given by [2,3]

$$H_m(r_{ij}) = \frac{2R}{\pi}\frac{1}{r_{ij}} \int_0^\infty \frac{q \sin(q r_{ij})}{\omega(\omega + \langle E \rangle)} dq, \tag{II.10a}$$

$$H_\omega(r_{ij}) = \frac{2R}{\pi}\frac{1}{r_{ij}} \int_0^\infty \frac{q \sin(q r_{ij})}{(\omega + \langle E \rangle)^2} dq, \tag{II.10b}$$

$$H_q(r_{ij}) = -r_{ij} \frac{d}{dr_{ij}} H_m(r_{ij}), \tag{II.10c}$$

$$H_R(r_{ij}) = -\frac{R}{M} \frac{d^2}{d^2 r_{ij}} H_m(r_{ij}), \tag{II.10d}$$

where R is the nuclear radius. Within closure approximation, there appears $\omega + \langle E \rangle$ in the denominators of the neutrino potentials (from the second order perturbation matrix elements), with $\langle E \rangle$ being the average intermediate state energy, usually assumed to be of the order of $\langle E \rangle = 10$ MeV. The modifications necessary for evaluation of the momentum integration within OEM will be given in appendix C.

2.2. *Basic machinery of the OEM*

$0\nu\beta\beta$ decay is a second-order process of the weak interaction. Different nuclear operators appear in the nine matrix elements of the decay rate formula, but generally the hadronic part of the matrix elements can be written in the following form,

$$M_x = \frac{1}{2}\sum_N \frac{\langle F|\mathcal{A}^\alpha|N\rangle\langle N|\mathcal{A}^\beta|I\rangle}{\omega + \varepsilon_1 + E_N - E_I} \pm (e_1 \leftrightarrow e_2)$$
$$= \frac{1}{2}\sum_N \frac{\langle F|\mathcal{A}^\alpha|N\rangle\langle N|\mathcal{A}^\beta|I\rangle}{\omega - \varepsilon_2 + E_N - E_F} \pm (e_1 \leftrightarrow e_2), \tag{II.11}$$

where the operators $\mathcal{A}^{\alpha/\beta} = \sum_i t_i^- \mathcal{A}_i^{\alpha/\beta}$ stand for any part of the hadronic currents in (II.2). (For Gamow-Teller transitions, for example, $\mathcal{A}_i^\alpha = \sigma_i^\alpha$.) $|I\rangle$, $|N\rangle$ and $|F\rangle$ are wave functions of the initial, intermediate and final nuclear states, with energies E_I, E_N and E_F, respectively. The latter are eigenstates of the nuclear Hamiltonian,

$$H|I\rangle = E_I|I\rangle, \quad H|N\rangle = E_N|N\rangle, \quad H|F\rangle = E_F|F\rangle. \tag{II.12}$$

ω and ε_i in (II.11) are energies of the neutrino and the i-th electron.

Before proceeding a few comments on (II.11) might be in order. Firstly, it has to be kept in mind that in $0\nu\beta\beta$ decay one has to integrate over the neutrino four-momentum, which has been suppressed here for convenience. Energy conservation is written as $E_I - E_F = \varepsilon_1 + \varepsilon_2$, which leads to the equal sign between the two lines of (II.11). However, as will be explained in appendix B, both terms will be kept in the following for a simple inclusion of the Coulomb force into the OEM formalism.

36

Moreover, one has to be careful about the exchange terms ($e_1 \leftrightarrow e_2$). For the matrix elements of the mass mechanism these are added constructively. Assuming $\varepsilon_1 \simeq \varepsilon_2$, which is a good approximation since $\varepsilon_i \ll \omega$, exchange terms will not be important for M_{GT} and M_F. (Simply evaluate one term and multiply it by a factor of 2.) However, from the neutrino propagator one also has matrix elements proportional to the neutrino four-momentum $q_\nu = (\omega, q)$. As is explained in [2], negative signs in (II.11) have to be taken for the ω-terms. These terms will therefore be suppressed by $\varepsilon_{12} = \varepsilon_1 - \varepsilon_2$ and would vanish, if $\varepsilon_1 = \varepsilon_2$ is assumed. Exchange terms, on the other hand, can easily be retained in the OEM formalism. In the derivation they are kept, regardless of the sign.

Next, energy denominators are expanded into an infinite series[1],

$$\frac{1}{\omega + \varepsilon_1 + E_N - E_I} = \frac{1}{\omega + \varepsilon_1} \sum_{n=0}^{\infty} \left(-\frac{1}{\omega + \varepsilon_1} \right)^n (E_N - E_I)^n, \tag{II.13a}$$

$$\frac{1}{\omega - \varepsilon_2 + E_N - E_F} = \frac{1}{\omega - \varepsilon_2} \sum_{n=0}^{\infty} \left(-\frac{1}{\omega - \varepsilon_2} \right)^n (E_N - E_F)^n. \tag{II.13b}$$

Applying the eigenvalue equation (II.12) repeatedly, it is easy to show that the matrix elements can be written as an infinite series of multiple commutators [22, 23],

$$M_x = \frac{1}{\omega + \varepsilon_1} \langle F | \sum_{n=0}^{\infty} \left(-\frac{1}{\omega + \varepsilon_1} \right)^n (\mathscr{A}^\alpha [H^{(n)}, \mathscr{A}^\beta]) | I \rangle, \tag{II.14a}$$

$$M_x = \frac{1}{\omega - \varepsilon_2} \langle F | \sum_{n=0}^{\infty} \left(-\frac{1}{\omega - \varepsilon_2} \right)^n ([\mathscr{A}^\alpha, H^{(n)}] \mathscr{A}^\beta) | I \rangle, \tag{II.14b}$$

where

$$[H^{(n)}, \mathscr{A}^\alpha] = \overbrace{[H, [\cdots, [H, \mathscr{A}^\alpha] \cdots]}^{n\text{-times}}, \tag{II.15}$$

and $\sum_N |N\rangle\langle N| = 1$ has been used. An equivalent expression, which may be more convenient for evaluation of the commutators under the two-body approximation (see below), is obtained by applying the following relations [27],

$$e^{iHt} \mathscr{A}^\alpha e^{-iHt} = \sum_{n=0}^{\infty} \frac{(it)^n}{n!} [H^{(n)}, \mathscr{A}^\alpha], \tag{II.16}$$

$$e^{-iHt} \mathscr{A}^\alpha e^{iHt} = \sum_{n=0}^{\infty} \frac{(it)^n}{n!} [\mathscr{A}^\alpha, H^{(n)}], \tag{II.17}$$

and

$$\int_0^\infty t^n e^{i\Delta t} dt = \lim_{\varepsilon \to 0} \frac{i^{n+1} n!}{(\Delta + i\varepsilon)^{n+1}}. \tag{II.18}$$

The matrix elements are then written as

$$M_x = \frac{i}{4} \langle F | \int_0^\infty dt \{ \mathscr{A}^\alpha e^{iHt} \mathscr{A}^\beta e^{-iHt} (e^{i\Delta_1 t} \pm e^{i\Delta_2 t}) + e^{-iHt} \mathscr{A}^\alpha e^{iHt} \mathscr{A}^\beta (e^{i\Delta_1' t} \pm e^{i\Delta_2' t}) \} | I \rangle, \tag{II.19}$$

where

$$\Delta_1 = \omega + \varepsilon_1, \qquad \Delta_2 = \omega + \varepsilon_2, \tag{II.20a}$$

$$\Delta_1' = \omega - \varepsilon_2, \qquad \Delta_2' = \omega - \varepsilon_1. \tag{II.20b}$$

Within the OEM formalism, the influence of the intermediate states on the nuclear matrix elements is implicitly taken into account in the commutators. Expressions of (II.14) and (II.19) are equivalent to the full calculation incorporating all intermediate states. The advantage is to avoid the tedious calculation of the full set of intermediate states. However, certain approximations to the exact expressions are necessary for numerical computation.

2.3. Derivation of OEM potentials for M_{GT} and M_F

The evaluation of the commutators is quite similar for all nine matrix elements of (II.6). In this section potentials for M_{GT} and M_F are derived, while Sect. 2.4 briefly discusses the differences of the formalism between these and the other seven matrix elements.

The derivation of the OEM potentials is based on two main assumptions. The first approximation is to assume that the Hamiltonian can be represented by

$$H \simeq V = V_C + V_{CN} + V_{TN}. \tag{II.21}$$

Coulomb (V_C), central (V_{CN}) and tensor (V_{TN}) parts can be written as (Tensor potentials are partly shifted to the central part for an easier handling of the commutators, see appendix A.)

$$V_C = \frac{1}{2} \sum_{i \neq j} v_C(r_{ij}) \Xi_{ij}, \tag{II.22a}$$

$$V_{CN} = \frac{1}{2} \sum_{i \neq j} \{ g_W(r_{ij}) + g_B(r_{ij}) P^\sigma_{ij} - g_H(r_{ij}) P^\tau_{ij} - g_M(r_{ij}) P^\tau_{ij} P^\sigma_{ij} \}, \tag{II.22b}$$

$$V_{TN} = \frac{1}{2} \sum_{i \neq j} \{ g_{TW}(r_{ij}) \Phi_{ij} + g_{TH}(r_{ij}) \Phi_{ij} P^\tau_{ij} \}, \tag{II.22c}$$

where the four different operators are defined by

$$\Xi_{ij} = \tfrac{1}{4}(1 - \tau^z)_i (1 - \tau^z)_j, \tag{II.23a}$$

$$P^\tau_{ij} = \tfrac{1}{2}(1 + \tau_i \cdot \tau_j), \tag{II.23b}$$

$$P^\sigma_{ij} = \tfrac{1}{2}(1 + \sigma_i \cdot \sigma_j), \tag{II.23c}$$

$$\Phi_{ij} = (\sigma_i \cdot \hat{r}_{ij})(\sigma_j \cdot \hat{r}_{ij}). \tag{II.23d}$$

[1] Since on average $\omega > E_N - E_I$, the expansion for $0\nu\beta\beta$ decay is in principle convergent. However, within the two-body approximation it is easy to keep terms of all orders, as will be done here

Radial functions $g_X(r_{ij})$ can be taken from some realistic NN-interaction and $v_C(r_{ij}) = e^2/r_{ij}$. The Hamiltonian of (II.21) is more general [25] than that considered previously [22–24], which accounted only for the central part of the NN interaction. However, we still neglect the one-body part of H.

The second simplification is the so-called two-body approximation. The multiple commutators of OEM lead to a sum of many-body operators. Provided the Hamiltonian consists of one-body and two-body interactions, the highest operator for a system of A nucleons is A-body. It would however be difficult to calculate all many-body operators and corresponding many-body transition densities. This difficulty might be avoided by taking only the two-body terms, the lowest order terms for a process where the atomic number changes by two units. Within the two-body approximation, we can write,

$$\frac{1}{2}\sum_{i\neq j} v_\alpha(r_{ij})\frac{1}{2}\sum_{k\neq l} v_\beta(r_{kl}) \rightarrow \frac{1}{2}\sum_{i\neq j} v_\alpha(r_{ij})v_\beta(r_{ij}). \tag{II.24}$$

In principle, this approximation is of the same order as RPA (the quasi-boson approximation) [25]. Nevertheless, one could argue it to be the main approximation in the current OEM, since it is applied many times for the multiple commutators [25]. Note that in the following sums over nucleons will be suppressed for convenience.

Using (II.19), (II.21) and (II.24) the derivation becomes straightforward. First of all, all operators of (II.23) commute with each other. As shown in appendix B, the inclusion of the Coulomb force can be performed rather simply. We will therefore concentrate on the treatment of the nuclear part of the Hamiltonian $V_S = V_{CN} + V_{TN}$. The projection operators as well as the tensor operator Φ_{ij} have the following useful property,

$$(P^\tau_{ij})^2 = 1, \quad (P^\sigma_{ij})^2 = 1, \quad (\Phi_{ij})^2 = 1. \tag{II.25}$$

Equation (II.25) allows to reduce products of two or more operators into mere sums of operators. Furthermore, considering that for any operator $\mathcal{O}$ with $\mathcal{O}^2 = 1$

$$e^{ig_x\mathcal{O}t} = \cos(g_x t) + i\sin(g_x t)\mathcal{O}, \tag{II.26}$$

it is easy to rewrite

$$\mathcal{A}^\alpha_i t^-_i \, e^{i(V_{CN}+V_{TN})t} \mathcal{A}^\beta_j t^-_j \, e^{-i(V_{CN}+V_{TN})t}$$
$$= \sum_{k=1}^{4}\sum_{l=1}^{4} a(k,l)\mathcal{A}^\alpha_i\mathcal{O}(k)\mathcal{A}^\beta_j\mathcal{O}(l)t^-_i t^-_j, \tag{II.27}$$

where

$$\mathcal{O}(1) = 1, \tag{II.28a}$$

$$\mathcal{O}(2) = P^\sigma_{ij}, \tag{II.28b}$$

$$\mathcal{O}(3) = \Phi_{ij}, \tag{II.28c}$$

$$\mathcal{O}(4) = P^\sigma_{ij}\Phi_{ij} = \Phi_{ij} - P^\sigma_{ij} + 1. \tag{II.28d}$$

Altogether, only (2 ×) 16 noncommutable products have to be evaluated for each matrix element.[2] The coefficients $a(k, l)$ consist merely of products of sine and cosine of the radial parts of the interaction, the calculation of which is simple algebra.[3]

Note that the isospin projection operator P^τ_{ij} does no longer occur in (II.27). This is because hadronic currents always involve isospin lowering operators and

$$t^-_i P^\tau_{ij} t^-_j = t^-_i t^-_i P^\tau_{ij} = P^\tau_{ij} t^-_j t^-_j = 0, \tag{II.29a}$$

$$t^-_j P^\tau_{ij} t^-_j = t^-_j t^-_i P^\tau_{ij} = P^\tau_{ij} t^-_i t^-_j = t^-_i t^-_j. \tag{II.29b}$$

Evaluating the 16 terms of (II.27), with the help of the rules given in appendix A, and rearranging with respect to spin operators, one finds for the Gamow-Teller matrix element

$$\sigma_i t^-_i \, e^{i(V_{CN}+V_{TN})t} \sigma_j t^-_j \, e^{-i(V_{CN}+V_{TN})t}$$
$$= \tfrac{1}{4}\{\alpha^{GT}(1) + \alpha^{GT}(2)P^\sigma_{ij} + \alpha^{GT}(3)\Phi_{ij}\} t^-_i t^-_j, \tag{II.30}$$

where

$$\begin{aligned}\alpha^{GT}(1) = 1 &- e^{2i(g_B-g_M)t} + e^{2i(g_H+g_M-g_{TH})t} - e^{2i(g_B+g_H+g_{TH})t} \\ &- e^{2i(g_H-g_B-g_{TW})t} + e^{2i(g_H+g_M+g_{TW})t} \\ &+ e^{-2i(g_{TH}+g_{TW})t} - e^{2i(g_M-g_B-g_{TH}-g_{TW})t} \\ &- e^{2i(g_B+g_H+g_{TW})t} - 2e^{2i(g_B-g_M+g_{TH}+g_{TW})t},\end{aligned} \tag{II.31a}$$

$$\begin{aligned}\alpha^{GT}(2) = {} & e^{2i(g_B-g_M)t} + e^{2i(g_M-g_B)t} \\ &- e^{2i(g_H-g_B+g_{TH})t} + e^{2i(g_B+g_H+g_{TH})t} \\ &+ 2e^{2i(g_B+g_H+g_{TW})t} + 2e^{2i(g_H+g_M+g_{TW})t} \\ &+ 2e^{2i(g_{TH}+g_{TW})t} + 2e^{2i(g_B-g_M+g_{TH}+g_{TW})t},\end{aligned} \tag{II.31b}$$

$$\begin{aligned}\alpha^{GT}(3) = 1 &+ e^{2i(g_M-g_B)t} + e^{2i(g_H+g_M-g_{TH})t} + e^{2i(g_H-g_B+g_{TH})t} \\ &- e^{2i(g_H-g_B-g_{TW})t} + e^{2i(g_H+g_M-g_{TW})t} \\ &+ e^{-2i(g_{TH}+g_{TW})t} - e^{2i(g_M-g_B-g_{TH}-g_{TW})t} \\ &- 2e^{2i(g_H+g_M+g_{TW})t} - 2e^{2i(g_{TH}+g_{TW})t},\end{aligned} \tag{II.31c}$$

and for the Fermi matrix element

$$t^-_i \, e^{i(V_{CN}+V_{TN})t} t^-_j \, e^{-i(V_{CN}+V_{TN})t}$$
$$= \tfrac{1}{4}\{\alpha^F(1) + \alpha^F(2)P^\sigma_{ij} + \alpha^F(3)\Phi_{ij}\} t^-_i t^-_j, \tag{II.32}$$

where

$$\alpha^F(1) = 2 + e^{2i(g_H+g_M-g_{TH})t} + e^{2i(g_H-g_M+g_{TH})t}, \tag{II.33a}$$

$$\alpha^F(2) = e^{2i(g_H+g_M+g_{TH})t} - e^{2i(g_H-g_M+g_{TH})t}, \tag{II.33b}$$

$$\alpha^F(3) = e^{2i(g_H+g_M-g_{TH})t} - e^{2i(g_H+g_M+g_{TH})t}. \tag{II.33c}$$

The terms of the sum over nucleons where i equals j, when evaluated, lead to the same form as (II.30) and (II.32). The corresponding $\alpha(k)$'s have the same combinations of nuclear potentials in the exponents and differ only in some signs between terms from the $\alpha(l)$'s given above. They will therefore not be repeated here.

Finally, the time integration is carried out along the lines of (II.18) and matrix elements are rewritten as

$$M_\alpha = \langle F| \sum_{i\neq j} t^-_i t^-_j \mathcal{M}^\alpha_{ij} |I\rangle, \tag{II.34}$$

[2] In the sums over nucleons the lable i does or does not equal j. This is the origin of terms proportinal to 1 and $\exp(iq\cdot r_{ij})$, see below

[3] While individual $a(k, l)$ consist of many terms, most of them cancel each other, when the sums over l and k are performed

recalling the sum over nucleons. The two-body operators $\mathcal{M}^{\alpha}_{ij}$ are given in the following form,

$$\mathcal{M}^{\alpha}_{ij} = \mathcal{I}_\beta(v^{\text{singlet}}_\alpha(r_{ij}))\,\Omega^0_{ij} + \mathcal{I}_\beta(v^{\text{triplet}}_\alpha(r_{ij}))\,\Omega^1_{ij} + \mathcal{I}_\beta(v^{\text{tensor}}_\alpha(r_{ij}))\,S_{ij}. \quad \text{(II.35)}$$

$\mathcal{I}_\beta(v)$ stands for the integration over the virtual neutrino momentum, from which the neutrino potentials are obtained. They are given explicitly in Appendix C. For M_{GT} and M_F the index $\beta = m$ or $\beta = m'$ (for m = mass), see (C.1) and (C.2). In (II.35) the spin singlet operator Ω^0_{ij}, the spin triplet operator Ω^1_{ij} and the tensor operator $S_{ij} = \Phi_{ij} - (1/3)\,\sigma_i \cdot \sigma_j$ have been introduced. Definitions and relations between the different operators in (II.22) and (II.35) (and corresponding radial functions) can be found in appendix A.[4] Note that, in order to account for Coulomb force, radial functions $v(r_{ij})$ have to be actually calculated with the help of

$$v(r_{ij}) = \tfrac{1}{2}\,\{v_L(r_{ij}, \Delta_L) + v_R(r_{ij}, \Delta_R)\}. \quad \text{(II.36)}$$

The difference between v_L and v_R, as shown in Appendix B, accounts for the Coulomb force, and

$$\Delta_L = \omega + \tfrac{1}{2}\,Q_{\beta\beta} + m_e, \quad \text{(II.37a)}$$

$$\Delta_R = \omega - (\tfrac{1}{2}\,Q_{\beta\beta} + m_e) - v_C(r_{ij}). \quad \text{(II.37b)}$$

In the following the indices L/R will not be given explicitly, keeping in mind that for all expression containing Δ actually two functions have to be evaluated. Note that in writing down (II.37) we assumed $\varepsilon_1 \simeq \varepsilon_2 \simeq \frac{1}{2}Q_{\beta\beta} + m_e$ which is a good approximation numerically, since $\varepsilon_i \ll \omega$.

By defining the following abbreviations,

$$\mathcal{F}[\pm V_k] = \frac{1}{\Delta \pm V_k}, \quad \text{(II.38a)}$$

$$\mathcal{F}(0) = \frac{1}{\Delta}, \quad \text{(II.38b)}$$

the radial functions $v(r_{ij})$ for the Gamow-Teller and Fermi matrix elements are then given by

$$v^{\text{singlet}}_{GT}(r_{ij}) = \{-(\mathcal{F}[V_1] + \tfrac{1}{2}\mathcal{F}[V_2]) \times (1 + e^{iq\cdot r_{ij}}) + (\mathcal{F}[V_3] + \tfrac{1}{2}\mathcal{F}[V_4]) \times (1 - e^{iq\cdot r_{ij}})\}, \quad \text{(II.39a)}$$

$$v^{\text{triplet}}_{GT}(r_{ij}) = \tfrac{1}{3}\,\{(\mathcal{F}(0) - \mathcal{F}[V_5] - \tfrac{1}{2}\mathcal{F}[V_6] + \mathcal{F}[V_{10}] + \mathcal{F}[-V_{10}]) \times (1 + e^{iq\cdot r_{ij}}) + (\mathcal{F}[-V_3] + \tfrac{1}{2}\mathcal{F}[-V_4] - \mathcal{F}[V_7] - \mathcal{F}[V_8] - \mathcal{F}[V_9]) \times (1 - e^{iq\cdot r_{ij}})\}, \quad \text{(II.39b)}$$

$$v^{\text{tensort}}_{GT}(r_{ij}) = \tfrac{1}{4}\,\{(\mathcal{F}(0) - 2\mathcal{F}[V_{10}] + \mathcal{F}[-V_{10}] - \mathcal{F}[V_5] + \mathcal{F}[V_6]) \times (1 + e^{iq\cdot r_{ij}}) + (\mathcal{F}[-V_3] - \mathcal{F}[-V_4] - \mathcal{F}[V_7] - \mathcal{F}[V_8] + 2\mathcal{F}[V_9]) \times (1 - e^{iq\cdot r_{ij}})\}, \quad \text{(II.39c)}$$

[4] The derivation of OEM potentials using (A.1a) instead of (II.22) and the method outlined in [25] leads to the same final expression

Table 1. The twelve different combinations of radial parts of the nucleon-nucleon potential which appear in the OEM formalism for $V = V_{CN} + V_{TN} + V_C$

V_k	
V_1	$v_{TE} - v_{SE} + \frac{2}{3}v_{TNE}$
V_2	$v_{TE} - v_{SE} - \frac{4}{3}v_{TNE}$
V_3	$v_{TO} - v_{SE} + \frac{2}{3}v_{TNO}$
V_4	$v_{TO} - v_{SE} - \frac{4}{3}v_{TNO}$
V_5	$v_{SO} - v_{TO} - \frac{2}{3}v_{TNO}$
V_6	$v_{SO} - v_{TO} + \frac{4}{3}v_{TNO}$
V_7	$v_{TE} - v_{TO} + \frac{2}{3}v_{TNE} - \frac{2}{3}v_{TNO}$
V_8	$v_{TE} - v_{TO} - \frac{4}{3}v_{TNE} - \frac{2}{3}v_{TNO}$
V_9	$v_{TE} - v_{TO} + \frac{2}{3}v_{TNE} + \frac{4}{3}v_{TNO}$
V_{10}	$2v_{TNO}$
V_{11}	$v_{SO} - v_{SE}$
V_{12}	$v_{TE} - v_{TO} - \frac{4}{3}v_{TNE} + \frac{4}{3}v_{TNO}$

and

$$v^{\text{singlet}}_F(r_{ij}) = \tfrac{1}{2}\{\mathcal{F}(0) \times (1 + e^{iq\cdot r_{ij}}) - \mathcal{F}[V_{11}] \times (1 - e^{iq\cdot r_{ij}})\}, \quad \text{(II.40a)}$$

$$v^{\text{triplet}}_F(r_{ij}) = \tfrac{1}{3}\,\{\tfrac{3}{2}\,\mathcal{F}(0) \times (1 + e^{iq\cdot r_{ij}}) - (\mathcal{F}[V_7] + \tfrac{1}{2}\mathcal{F}[V_{12}]) \times (1 - e^{iq\cdot r_{ij}})\}, \quad \text{(II.40b)}$$

$$v^{\text{tensor}}_F(r_{ij}) = \tfrac{1}{4}(-\mathcal{F}[V_7] + \mathcal{F}[V_{12}]) \times (1 - e^{iq\cdot r_{ij}}). \quad \text{(II.40c)}$$

Altogether 12 different combinations of radial functions of the nuclear NN potential appear in the OEM potentials. These are summarized in Table 1.

Note that putting the tensor and Coulomb terms equal to zero, (II.39) and (II.40) reduce to the OEM equations derived previously [23,24], as they naturally should.

2.4. OEM matrix elements for right-handed currents

The remaining seven matrix elements can be calculated within OEM with only moderate modifications of the procedure outlined above. Therefore, their derivation will be discussed briefly.

As mentioned below (II.11), negative signs for the exchange terms have to be taken for the matrix elements proportional to the neutrino energy ω. However, the negative sign leads to

$$i\int_0^\infty dt\, e^{iV_k t}\,\{e^{i\Delta_1 t} - e^{i\Delta_2 t}\} = i\int_0^\infty dt\, e^{iV_k t}\,\{e^{i(\omega+\frac{1}{2}(\varepsilon_1+\varepsilon_2+\varepsilon_{12}))t} - e^{i(\omega+\frac{1}{2}(\varepsilon_1+\varepsilon_2-\varepsilon_{12}))t}\} = \frac{\varepsilon_{12}}{(\omega + \frac{1}{2}(\varepsilon_1+\varepsilon_2) + V_k)^2 - (\varepsilon_{12}/2)^2}, \quad \text{(II.41)}$$

where $\Delta_1 = \omega + \varepsilon_1$, $\Delta_2 = \omega + \varepsilon_2$ and $\varepsilon_{12} = \varepsilon_1 - \varepsilon_2$. Nuclear operators of $M_{GT\omega}$ and $M_{F\omega}$, on the other hand, are exactly the same as for M_{GT} and M_F, respectively. By introducing[5]

$$\mathscr{G}[\pm V_k] = \frac{1}{(\Delta \pm V_k)^2 - (\varepsilon_{12}/2)^2}, \tag{II.42a}$$

$$\mathscr{G}(0) = \frac{1}{\Delta^2 - (\varepsilon_{12}/2)^2}, \tag{II.42b}$$

$M_{GT\omega}$ and $M_{F\omega}$ can then be calculated in exactly the same manner as M_{GT} and M_F, by simply replacing the neutrino potentials $\mathscr{I}_{m/m'}$ by $\mathscr{I}_{\omega/\omega'}$, corresponding to a replacement of all $\mathscr{F}$'s in the radial functions $v(r_{ij})$ by $\mathscr{G}$'s.

Also for the next two matrix elements M_{GTq} and M_{Fq} the nuclear operators are the same as for M_{GT} and M_F, respectively. Due to the parity-odd character of q, however, for M_{GTq} and M_{Fq} one of the emitted electrons will be in a $P_{1/2}$ state, which introduces a factor of r_{ij} into the formalism. Since r_{ii} vanishes identically, M_{GTq} and M_{Fq} (and for the same reason M_T and M_P) do not have terms proportional to 1. Neutrino potentials are calculated by (C.5) and the various radial functions for the two-body operators are given by

$$v_{GTq}^{\text{singlet}}(r_{ij}) = -\{\mathscr{F}[V_1] + \tfrac{1}{2}\mathscr{F}[V_2] + \mathscr{F}[V_3] + \tfrac{1}{2}\mathscr{F}[V_4]\}\,\mathrm{e}^{iq\cdot r_{ij}}, \tag{II.43a}$$

$$\begin{aligned} v_{GTq}^{\text{triplet}}(r_{ij}) = \tfrac{1}{3}\{&\mathscr{F}(0) - \mathscr{F}[-V_3] - \tfrac{1}{2}\mathscr{F}[V_4] - \mathscr{F}[V_5] \\ &- \tfrac{1}{2}\mathscr{F}[V_6] + \mathscr{F}[V_7] + \mathscr{F}[V_8] + \mathscr{F}[V_9] \\ &+ \mathscr{F}[V_{10}] + \mathscr{F}[-V_{10}]\}\,\mathrm{e}^{iq\cdot r_{ij}}, \end{aligned} \tag{II.43b}$$

$$\begin{aligned} v_{GTq}^{\text{tensor}}(r_{ij}) = \tfrac{1}{4}\{&\mathscr{F}(0) - \mathscr{F}[-V_3] + \mathscr{F}[-V_4] - \mathscr{F}[V_5] \\ &+ \mathscr{F}[V_6] - \mathscr{F}[V_7] - \mathscr{F}[V_8] + 2\mathscr{F}[V_9] \\ &- 2\mathscr{F}[V_{10}] + \mathscr{F}[-V_{10}]\}\,\mathrm{e}^{iq\cdot r_{ij}}, \end{aligned} \tag{II.43c}$$

and

$$v_{Fq}^{\text{singlet}}(r_{ij}) = \tfrac{1}{2}\{\mathscr{F}(0) + \mathscr{F}[V_{11}]\}\,\mathrm{e}^{iq\cdot r_{ij}}, \tag{II.44a}$$

$$v_{Fq}^{\text{triplet}}(r_{ij}) = \tfrac{1}{3}\{\tfrac{3}{2}\mathscr{F}(0) + \mathscr{F}[V_7] + \tfrac{1}{2}\mathscr{F}[V_{12}]\}\,\mathrm{e}^{iq\cdot r_{ij}}, \tag{II.44b}$$

$$v_{Fq}^{\text{tensor}}(r_{ij}) = \tfrac{1}{4}\{\mathscr{F}[V_7] - \mathscr{F}[V_{12}]\}\,\mathrm{e}^{iq\cdot r_{ij}}. \tag{II.44c}$$

For the tensor matrix element the nuclear currents differ from those discussed previously. Consequently, also different combinations of nuclear potentials appear in the two-body operator $\mathscr{M}_{ij}^T$. Evaluating the commutators with the help of (II.27), $\mathscr{M}_{ij}^T$ however can also be written in the form of (II.35) with radial functions given by

$$v_T^{\text{singlet}}(r_{ij}) = \tfrac{1}{3}\{\mathscr{F}[V_1] - \mathscr{F}[V_2] + \mathscr{F}[V_3] - \mathscr{F}[V_4]\}\,\mathrm{e}^{iq\cdot r_{ij}}, \tag{II.45a}$$

$$\begin{aligned} v_T^{\text{triplet}}(r_{ij}) = \tfrac{1}{9}\{&2\mathscr{F}(0) + \mathscr{F}[-V_3] \\ &- \mathscr{F}[-V_4] + \mathscr{F}[V_5] - \mathscr{F}[V_6] \\ &+ 2\mathscr{F}[V_7] - \mathscr{F}[V_8] - \mathscr{F}[V_9] - \mathscr{F}[V_{10}] \\ &- \mathscr{F}[-V_{10}]\}\,\mathrm{e}^{iq\cdot r_{ij}}, \end{aligned} \tag{II.45b}$$

$$\begin{aligned} v_T^{\text{tensor}}(r_{ij}) = \tfrac{1}{12}\{&2\mathscr{F}(0) + \mathscr{F}[-V_3] + 2\mathscr{F}[-V_4] \\ &+ \mathscr{F}[V_5] + 2\mathscr{F}[V_6] + 2\mathscr{F}[V_7] \\ &- \mathscr{F}[V_8] + 2\mathscr{F}[V_9] + 2\mathscr{F}[V_{10}] \\ &- \mathscr{F}[-V_{10}]\}\,\mathrm{e}^{iq\cdot r_{ij}}. \end{aligned} \tag{II.45c}$$

An exact calculation for the recoil matrix element would be very complicated, since the nuclear recoil terms contain the nuclear momenta. These should be understood actually as operators, which do not exactly commute with H. However, as shown by Tomoda et al. [28],

$$\begin{aligned} &\tfrac{1}{2}\hat{r}_{ij}\cdot(\sigma_i \times D_j + D_i \times \sigma_j) \\ &= \frac{1}{M}\{-\tfrac{1}{4}\sigma_i\cdot(\hat{r}_{ij}\times(P_j + P_j')) + \tfrac{1}{4}\sigma_j\cdot(\hat{r}_{ij}\times(P_i + P_i')) \\ &\quad + i\mu_\beta\tfrac{1}{2}[\sigma_i\times\sigma_j]^{(2)}\cdot[\hat{r}_{ij}\times Q_{ij}]^{(2)} \\ &\quad + i\mu_\beta\tfrac{1}{4}(\sigma_i\times\sigma_j)\cdot(\hat{r}_{ij}\times P_{ij}) \\ &\quad - i\left(\frac{\mu_\beta}{3}\right)\sigma_i\cdot\sigma_j(\hat{r}_{ij}\cdot Q_{ij})\} \end{aligned} \tag{II.46}$$

where

$$P_{ij} = \tfrac{1}{2}[(P_i - P_i') + (P_j - P_j')], \tag{II.47a}$$

$$Q_{ij} = \tfrac{1}{2}[(P_i - P_i') - (P_j - P_j')]. \tag{II.47b}$$

A numerical calculation by the same authors [29] then concluded the last term to be the by far dominating one. Later evaluations of the recoil matrix element therefore have been restricted to only this term. Since it is now a good approximation to assume $Q_{ij} \simeq -q$, due to $m_e \ll m_p \simeq m_n$, the two-body operator $\mathscr{M}_{ij}^R$ contains the same combination of radial functions of V_{NN}'s as M_{GTq}. Note, however, that in the calculation of the matrix element the neutrino potential $\mathscr{I}_R$ (C.6) has to be taken.

Lastly, the P-wave matrix element is the only one which cannot be expressed in terms of Ω_{ij}^0, Ω_{ij}^1 and S_{ij}. This is due to the fact that, while all other matrix elements incorporate either products of two σ's or none, the P-wave matrix element is of the "mixed" Gamow-Teller-Fermi type. After evaluation of the commutators it is, however, possible to classify all potentials with only two different transition operators, as:

$$\begin{aligned} \mathscr{M}_{ij}^P = (&i v_P^{(\sigma_i-\sigma_j)}(r_{ij})(\sigma_i - \sigma_j) \\ &+ v_P^{(\sigma_i\times\sigma_j)}(r_{ij})(\sigma_i\times\sigma_j))\cdot(\hat{r}_{ij}\times\hat{R}_{ij}) \end{aligned} \tag{II.48}$$

with radial functions $v_P(r_{ij})$ given by

$$\begin{aligned} v_P^{(\sigma_i-\sigma_j)}(r_{ij}) = \tfrac{1}{8}\{&2\mathscr{F}(0) + \mathscr{F}[V_1] + \mathscr{F}[V_3] + \mathscr{F}[-V_3] \\ &+ \mathscr{F}[V_5] + \mathscr{F}[V_7] + \mathscr{F}[V_{11}]\}\,\mathrm{e}^{iq\cdot r_{ij}}, \end{aligned} \tag{II.49a}$$

$$\begin{aligned} v_P^{(\sigma_i-\sigma_j)}(r_{ij}) = \tfrac{1}{8}\{&\mathscr{F}[V_1] + \mathscr{F}[V_3] - \mathscr{F}[-V_3] \\ &- \mathscr{F}[V_5] - \mathscr{F}[V_7] + \mathscr{F}[V_{11}]\}\,\mathrm{e}^{iq\cdot r_{ij}}. \end{aligned} \tag{II.49b}$$

[5] The factor ε_{12} is absorbed into the definition of the phase space integrals

40

3. Numerical results

The OEM provides expressions for the "intermediate part" of the nuclear matrix elements. For the numerical calculation also ground-state wave functions are needed, which are taken from the pn-QRPA model of [14,15], here. For the effective interaction the G-matrix interaction [30] of the Paris potential [31] (Paris-G) is taken. Calculations with and without Coulomb and tensor interactions in the OEM formalism are carried out, in order to discuss their relative importance for the different matrix elements.

In the numerical calculation the short-range correlations between two nucleons are taken into account by multiplying the two-particle relative wave functions by the correlation function [34]

$$1 + f(r_{ij}) = 1 - e^{-ar_{ij}^2}(1 - br_{ij}^2), \qquad \text{(III.1)}$$

with $a = 1.1\ \text{fm}^{-2}$ and $b = 0.68\ \text{fm}^{-2}$. The finite nucleon size effects are introduced by the nuclear form factor [35]

$$g_{V/A} \rightarrow g_{V/A}\left(\frac{\Lambda^2}{\Lambda^2 + q^2}\right)^2, \qquad \text{(III.2)}$$

where we take $\Lambda = 850$ MeV.

Since it is known [15–21] that the strength of the particle-particle interaction g_{pp} plays a key role in the determination of $\beta\beta$ decay matrix elements, the influence of g_{pp} on the OEM calculation has been investigated and results are displayed in Fig. 1. The figure shows $M_{GT} - M_F$, by which the neutrino mass is determined, for the isotope ^{76}Ge for different cases. The full line is an OEM calculation including $V_{CN} + V_{TN} + V_C$, whereas the dashed line is the result when accounting only for the central parts of the interaction. For comparison, the dash-dotted line shows a usual closure calculation, with the average intermediate state energy fixed at $\langle E_N \rangle = 10$ MeV. The G-matrix of the Paris potential is used consistently for both transition densities and radial functions. All calculations are relatively close to each other and show a similar dependence on g_{pp}. Increasing values of g_{pp} lead to decreasing matrix elements, in agreement with previous QRPA calculations [15–21] and OEM + QRPA for $0\nu\beta\beta$ decay [24]. The inclusion of the tensor force leads only to a moderate change of the matrix elements, probably in the right direction when compared to a calculation where intermediate states are calculated within QRPA [32].

It is interesting to note that the inclusion of the Coulomb force has a negligible effect (see also Table 2) on $0\nu\beta\beta$ decay matrix elements. This is very different to the situation reported for $2\nu\beta\beta$ decay within OEM [25], where Coulomb force was found to be essential. We think, however, this difference can be explained as follows. At relatively small distances the OEM potentials are dominated by the nuclear part of the interaction. Beyond approximately $r_{ij} \simeq 3$ fm, on the other hand, the situation reverses: nuclear potentials have exponentially damped off and the main contribution comes from the Coulomb force, which has the weaker radial dependence $\sim 1/r_{ij}$. In $2\nu\beta\beta$ decay OEM potentials change the sign at $r_{ij} \simeq 2$ fm (for ^{76}Ge), and there is a substantial cancellation of those parts coming from small distances. As a result contributions to the matrix elements from relatively large distances are important, say up to the nuclear radius R (~ 5 fm for ^{76}Ge). In $0\nu\beta\beta$ decay, on the other hand, both decay vertices are connected by the virtual neutrino. Due to the momentum integration the main contribution to the matrix elements comes from a region around $r_{ij} \simeq (1-2)$ fm, where V_C plays only a minor role.

To investigate the influence of the choice of the interaction on the numerical results, an OEM calculation using the Paris potential [31] has been carried out. The results are compared to the OEM using Paris-G and a closure calculation in Fig. 2, again for different cases.

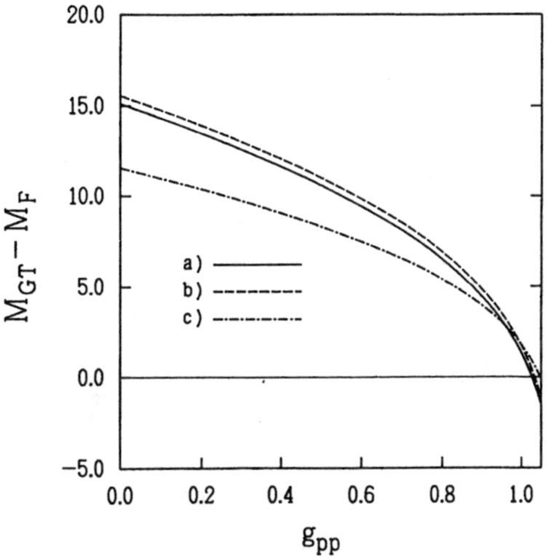

Fig. 1. Comparison of matrix elements $M_{GT} - M_F$ for different calculations. **a** is an OEM + QRPA calculation using the Paris-G potential, accounting for central, tensor and Coulomb interaction, while **b** is OEM + QRPA for the central parts of the interaction only. For comparison **c** results of a closure calculation are shown

Table 2. The different nuclear matrix elements of $0\nu\beta\beta$ decay for the isotope ^{76}Ge for different calculations. a) Result of a closure calculation, assuming $\langle E_N \rangle = 10$ MeV, b) OEM calculation including $V = V_{CN} + V_{TN} + V_C$, c) OEM: $V = V_{CN} + V_{TN}$, d) OEM: $V = V_{CN} + V_C$, e) OEM: $V = V_{CN}$. All calculations have been carried out employing the G-matrix effective interaction of the Paris potential [30]. Note that the values given for the closure calculation do not coincide numerically with those given in [15]. This is due to two reasons. First, we have updated the physical constants by [36]. Second, in [37] a small change of the strength of the spin-orbit force in the Woods-Saxon potential was introduced, which gave a better description of single-particle levels in the $A = 100$ region

M_x	a)	b)	c)	d)	e)
M_{GT}	3.01	3.35	3.36	3.77	3.76
M_F	− 1.26	− 1.56	− 1.55	− 1.57	− 1.55
$M_{GT\omega}$	2.94	3.35	3.37	4.27	4.26
$M_{F\omega}$	− 1.10	− 1.61	− 1.58	− 1.62	− 1.59
M_{GTq}	1.91	1.62	1.62	1.60	1.60
M_{Fq}	− 1.16	− 1.16	− 1.16	− 1.16	− 1.16
M_T	− 0.66	− 0.84	− 0.84	− 0.76	− 0.75
M_P	− 1.03	− 1.14	− 1.13	− 1.10	− 1.10
M_R	3.59	3.65	3.64	3.56	3.55

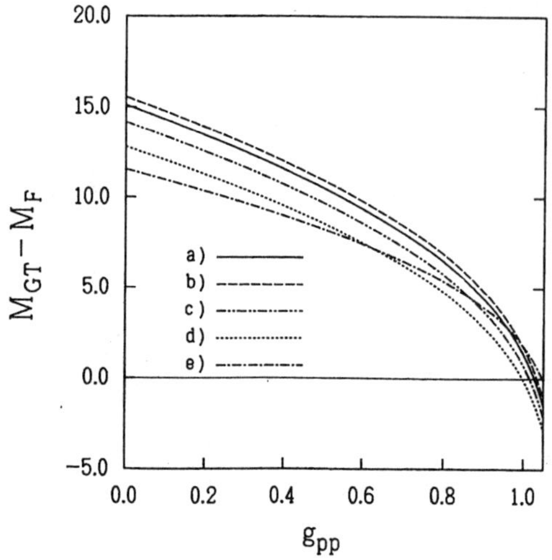

Fig. 2. $M_{GT} - M_F$ employing either the Paris potential or Paris-G in the OEM calculation. Different cases are compared: **a** Paris-G, $V = V_{CN} + V_{TN} + V_C$; **b** Paris-G, $V = V_{CN}$; **c** Paris, $V = V_{CN} + V_{TN} + V_C$; **d** Paris, $V = V_{CN}$; **e** closure. Inclusion of tensor force decreases the differences between the calculations using Paris-G and Paris potentials

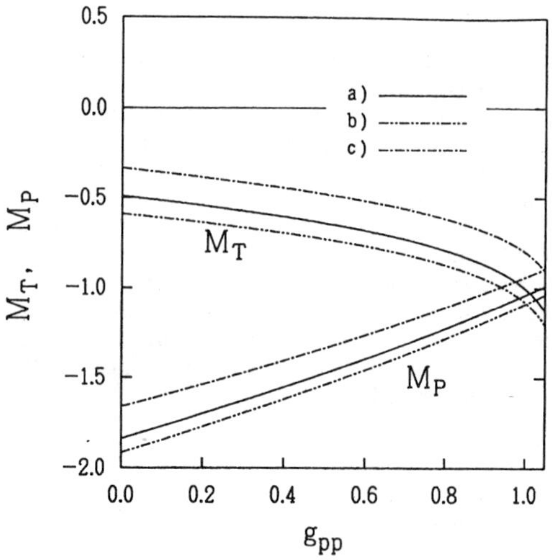

Fig. 4. OEM + QRPA and closure results for M_T and M_P. For OEM $V = V_{CN} + V_{TN} + V_C$ is taken. Shown are: **a** Paris-G; **b** Paris; *c* closure. For discussion see text

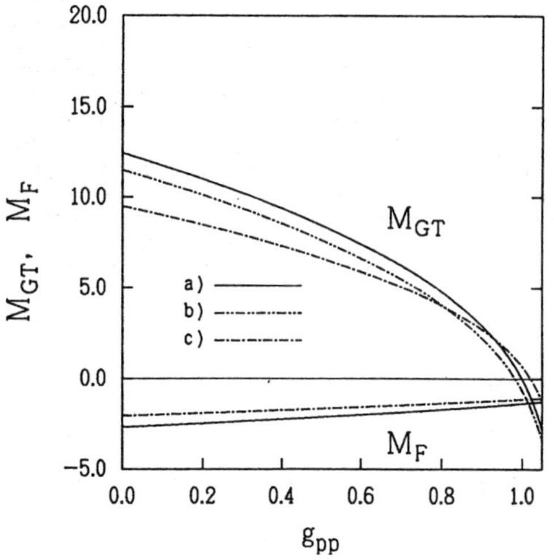

Fig. 3. Results for M_{GT} and M_F for: **a** Paris-G, $V = V_{CN} + V_{TN} + V_C$; **b** Paris, $V = V_{CN} + V_{TN} + V_C$; **c** closure. In case of M_F the OEM results using Paris or Paris-G are too close to each other to be distinguished in the plot

Generally, the use of the Paris potential leads to smaller values of $M_{GT} - M_F$ than does the Paris-G. The difference between both calculations is mainly due to differences in M_{GT}, as can be seen from Fig. 3. It is interesting to note, however, that the inclusion of the tensor force leads to an increase of the matrix element for Paris, whereas a decrease is found in the Paris-G. Inclusion of tensor force therefore leads to smaller differences between calculations employing different NN-potentials, especially in the region of the most probable value of g_{pp}.

Calculations for $M_{GT\omega}$, $M_{F\omega}$, M_{GTq} and M_{Fq} show that these matrix elements depend in a very similar manner on the particle-particle force as M_{GT} and M_F, respectively, as could be expected. From Table 2 it is found that also for these matrix elements values comparable to the closure values.

The matrix elements M_T, M_P and M_R depend only rather weakly on the particle-particle force in all calculations. Figure 4 shows M_T and M_P for different cases. As for other matrix elements, the OEM leads to a small correction, differences between the various calculations are of the order of (*some*) %.

The recoil matrix element is of special interest, since its contribution can easily dominate the $\langle \eta \rangle$ coefficients of the decay rate formula. As is shown in Fig. 5, however, also for the recoil matrix element, OEM leads to results similar to those obtained previously. The inclusion of the tensor force again leads to a decrease of the differences between the Paris and Paris-G calculations.

To summarize this discussion, it might be stated that the inclusion of the tensor force into the OEM formalism leads to a moderate change of the nuclear matrix elements, but it is nevertheless important for the agreement between calculations employing different NN-interactions. The similarity between closure and OEM might be explained by the fact that $0\nu\beta\beta$ decay matrix elements are dominated by the virtual exchange of a neutrino with a typical energy of $R^{-1} > 50$ MeV [24]. Intermediate nuclear states, which have typically $\langle E_N \rangle \sim 10$ MeV, will only lead to small corrections to the matrix elements, similar to the numerical results using OEM.

Coefficients of the decay rate formula have been derived from the matrix elements of Table 2, and the results are summarized in Table 3. Compared are the coefficients

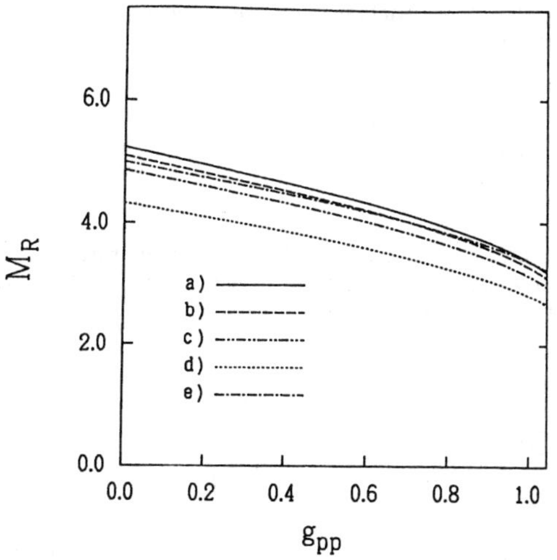

Fig. 5. OEM + QRPA and closure results for M_R. The following different calculations are shown: **a** Paris-G, $V = V_{CN} + V_{TN} + V_C$ **b** Paris-G, $V = V_{CN}$; **c** Paris, $V = V_{CN} + V_{TN} + V_C$; **d** Paris, $V = V_{CN}$; **e** closure. As found for M_{GT}, inclusion of the tensor force decreases the differences between the calculations using Paris-G and Paris potentials.

Table 3. Coefficients of the decay rate formula in units of inverse years, derived from the matrix elements of Table 2. Compared are a) OEM + QRPA calculation for $V = V_{CN} + V_{TN} + V_C$ and b) closure calculation

C_{xy}	a)	b)
C_{mm}	1.61×10^{-13}	1.22×10^{-13}
$C_{m\lambda}$	-6.23×10^{-14}	-4.55×10^{-14}
$C_{m\eta}$	2.78×10^{-11}	2.37×10^{-11}
$C_{\lambda\lambda}$	2.23×10^{-13}	1.51×10^{-13}
$C_{\eta\eta}$	4.98×10^{-9}	4.78×10^{-9}
$C_{\eta\lambda}$	-3.23×10^{-14}	-4.25×10^{-14}

of the OEM + QRPA calculation using Paris-G, including central, tensor and Coulomb terms, and results obtained in closure. Due to the similarity of the matrix elements, naturally also the decay rate coefficients are found to be very close to each other. Taking the lower limit of the $0\nu\beta\beta$ decay half-life for ^{76}Ge [33], the following upper limits for $\langle m_\nu \rangle$, $\langle \lambda \rangle$ and $\langle \eta \rangle$ can be derived:

OEM + QRPA:

$$|\langle m_\nu \rangle| < 1.22 \text{ eV}, \qquad \text{(III.3a)}$$

$$|\langle \lambda \rangle| < 1.76 \cdot 10^{-6}, \qquad \text{(III.3b)}$$

$$|\langle \eta \rangle| < 1.34 \cdot 10^{-8}, \qquad \text{(III.3c)}$$

closure + QRPA:

$$|\langle m_\nu \rangle| < 1.40 \text{ eV}, \qquad \text{(III.4a)}$$

$$|\langle \lambda \rangle| < 2.15 \cdot 10^{-6}, \qquad \text{(III.4b)}$$

$$|\langle \eta \rangle| < 1.37 \cdot 10^{-8}, \qquad \text{(III.4c)}$$

for arbitrary variation of the other parameters.

4. Summary

The Operator Expansion Method avoids the explicit calculation of the full set of intermediate states and instead takes their influence implicitly into account by rewriting the nuclear matrix elements into a commutator series. OEM can therefore be viewed as a step beyond the closure approximation. We have extended the formalism of the OEM for neutrinoless double beta decay by two means. First, for the Hamiltonian we included Coulomb and tensor forces in addition to the central components of the nucleon-nucleon interaction. Furthermore, OEM expressions for the matrix elements of the right-handed current mode of $0\nu\beta\beta$ decay have been derived. A simple method to deal with the commutators of the OEM has been outlined and the differences between the formalism of the various matrix elements have been discussed. However, we still take the two-body approximation, which we think is the main approximation of the current work.

A numerical study has been carried out, combining the OEM with ground-state wave functions of a pn-QRPA model. For $0\nu\beta\beta$ decay, the OEM including tensor force leads to results close to the closure approximation (for an assumed intermediate state energy of $\langle E_N \rangle = 10$ MeV). The results of OEM are also similar to those obtained, when the intermediate states are calculated within QRPA [32, 38]. This confirms the conclusion of earlier calculations of OEM [24], using the central parts of the nucleon-nucleon interaction only, that OEM seems to work rather well in case of $0\nu\beta\beta$ decay. This is different to the application of OEM to the calculation of $2\nu\beta\beta$ decay matrix elements according to [25], which showed deficits of the OEM, when compared to a "full" calculation of intermediate states within QRPA. In $0\nu\beta\beta$ decay Coulomb force was found to give a negligible contribution, while the tensor force changes matrix elements by $\sim$(10–30) %. The inclusion of the tensor force considerably decreases the differences between OEM calculations employing different NN-interactions and is therefore an important improvement of the OEM.

OEM aims at an improvement over calculations using the closure approximation, by implicitly taking into account the intermediate state energies. On the other hand, the two-body approximation has been assumed. Both approximations will yield certain deviations from the "exact" calculation employing the same transition densities. Concerning the numerical results, the improved OEM seems to give corrections to the closure result in the right direction, but a more definite conclusion must await a detailed comparison with an exact calculation. The two-body approximation is therefore still an open problem of the OEM.

The coefficients of the decay rate formula derived from the OEM + QRPA calculation are similar to those obtained previously [15], and, combined with experimental data [33], lead to sharp limits on the effective neutrino mass and right-handed parameters. It seems after all that neutrinoless double beta decay remains a sensitive tool to probe physics beyond the standard model.

This study is supported in part by a Grant-in-Aid for Scientific Research (05243204) from the Ministry of Education, Science and

Culture (Monbusho). M.H. would like to thank the Japanese Ministry of Education, Science and Culture for financial support. Numerical studies were performed by using the FACOM M780 computer system at the Institute for Nuclear Study, University of Tokyo.

Appendix A: Relations between various operators and radial functions

Nucleon-nucleon potentials can be defined in different conventions. The Paris potential [31], for example, is formulated in terms of (A.1a), while (A.1b) allows a simpler derivation of the OEM expressions according to (II.19). This appendix gives relations between the two different formulations used in the text and a brief summary of definitions.

The two-body part of the nuclear interaction, accounting for central and tensor components, can be written as

$$V_S = \frac{1}{2}\sum_{i\neq j}\{v_{\rm SE}(r_{ij})\Pi^1_{ij}\Omega^0_{ij} + v_{\rm TO}(r_{ij})\Pi^1_{ij}\Omega^1_{ij} + v_{\rm SO}(r_{ij})\Pi^0_{ij}\Omega^0_{ij} + v_{\rm TE}(r_{ij})\Pi^0_{ij}\Omega^1_{ij} + v_{\rm TNO}(r_{ij})\Pi^1_{ij}S_{ij} + v_{\rm TNE}(r_{ij})\Pi^0_{ij}S_{ij}\} \tag{A.1a}$$

$$= \frac{1}{2}\sum_{i\neq j}\{g_W(r_{ij}) + g_B(r_{ij})P^\sigma_{ij} - g_H(r_{ij})P^\tau_{ij} - g_M(r_{ij})P^\sigma_{ij}P^\tau_{ij} + g_{TW}(r_{ij})\Phi_{ij} + g_{TH}(r_{ij})\Phi_{ij}P^\tau_{ij}\}, \tag{A.1b}$$

where the different operators are defined by

$$\Pi^1_{ij} = \tfrac{1}{4}(3 + \tau_i\cdot\tau_j), \tag{A.2a}$$

$$\Pi^0_{ij} = \tfrac{1}{4}(1 - \tau_i\cdot\tau_j), \tag{A.2b}$$

$$\Omega^1_{ij} = \tfrac{1}{4}(3 + \sigma_i\cdot\sigma_j), \tag{A.2c}$$

$$\Omega^0_{ij} = \tfrac{1}{4}(1 - \sigma_i\cdot\sigma_j), \tag{A.2d}$$

$$S_{ij} = (\sigma_i\cdot\hat{r}_{ij})(\sigma_j\cdot\hat{r}_{ij}) - \tfrac{1}{3}\sigma_i\cdot\sigma_j, \tag{A.2e}$$

$$P^\tau_{ij} = \tfrac{1}{2}(1 + \tau_i\cdot\tau_j), \tag{A.2f}$$

$$P^\sigma_{ij} = \tfrac{1}{2}(1 + \sigma_i\cdot\sigma_j), \tag{A.2g}$$

$$\Phi_{ij} = (\sigma_i\cdot\hat{r}_{ij})(\sigma_j\cdot\hat{r}_{ij}). \tag{A.2h}$$

Radial functions of (A.1b) are related to those of (A.1a) by

$$g_W(r_{ij}) = \tfrac{1}{4}(v_{\rm TE}(r_{ij}) + v_{\rm SE}(r_{ij}) + v_{\rm TO}(r_{ij}) + v_{\rm SO}(r_{ij})) + \tfrac{1}{6}(v_{\rm TNO}(r_{ij}) + v_{\rm TNE}(r_{ij})), \tag{A.3a}$$

$$g_B(r_{ij}) = \tfrac{1}{4}(v_{\rm TE}(r_{ij}) - v_{\rm SE}(r_{ij}) + v_{\rm TO}(r_{ij}) - v_{\rm SO}(r_{ij})) - \tfrac{1}{3}(v_{\rm TNO}(r_{ij}) + v_{\rm TNE}(r_{ij})), \tag{A.3b}$$

$$g_H(r_{ij}) = \tfrac{1}{4}(v_{\rm TE}(r_{ij}) - v_{\rm SE}(r_{ij}) - v_{\rm TO}(r_{ij}) + v_{\rm SO}(r_{ij})) + \tfrac{1}{6}(v_{\rm TNE}(r_{ij}) - v_{\rm TNO}(r_{ij})), \tag{A.3c}$$

$$g_M(r_{ij}) = \tfrac{1}{4}(v_{\rm TE}(r_{ij}) + v_{\rm SE}(r_{ij}) - v_{\rm TO}(r_{ij}) - v_{\rm SO}(r_{ij})) + \tfrac{1}{3}(v_{\rm TNO}(r_{ij}) - v_{\rm TNE}(r_{ij})), \tag{A.3d}$$

$$g_{TH}(r_{ij}) = \tfrac{1}{2}(v_{\rm TNO}(r_{ij}) - v_{\rm TNE}(r_{ij})), \tag{A.3e}$$

$$g_{TW}(r_{ij}) = \tfrac{1}{2}(v_{\rm TNO}(r_{ij}) + v_{\rm TNE}(r_{ij})). \tag{A.3f}$$

As is easy to show, nuclear spin operators are related to each other by

$$P^\sigma_{ij} = \Omega^1_{ij} - \Omega^0_{ij}, \tag{A.4a}$$

$$1 = \Omega^1_{ij} + \Omega^0_{ij}, \tag{A.4b}$$

$$\Phi_{ij} = S_{ij} + \tfrac{1}{3}\Omega^1_{ij} - \Omega^0_{ij}. \tag{A.4c}$$

Similar relations can be given for isospin operators. To evaluate the non-commutable products of (II.27) the following rules for the spin operators σ and $(\sigma_i\cdot\hat{r}_{ij})$ are needed,

$$\sigma_i\cdot\sigma_i = 3, \tag{A.5a}$$

$$\sigma_i\cdot\sigma_j = 2P^\sigma_{ij} - 1, \tag{A.5b}$$

$$\sigma_i P^\sigma_{ij} = P^\sigma_{ij}\sigma_j, \tag{A.5c}$$

$$(\sigma_i\cdot\hat{r}_{ij})(\sigma_i\cdot\hat{r}_{ij}) = 1, \tag{A.5d}$$

$$(\sigma_i\cdot\hat{r}_{ij})P^\sigma_{ij} = P^\sigma_{ij}(\sigma_j\cdot\hat{r}_{ij}), \tag{A.5e}$$

$$\sigma_i\Phi_{ij}\sigma_i = -\Phi_{ij}, \tag{A.5f}$$

$$\sigma_i\Phi_{ij}\sigma_j = 2P^\sigma_{ij} - \Phi_{ij}. \tag{A.5g}$$

Equation (A.5) gives the rules necessary for all matrix elements except for the P-wave matrix element. For the P-wave matrix element one has to calculate terms such as

$$\sigma_i P^\sigma_{ij} = \tfrac{1}{2}(\sigma_i + \sigma_j - i\sigma_i\times\sigma_j), \tag{A.6a}$$

$$\sigma_i\Phi_{ij} = \hat{r}_{ij}(\sigma_j\cdot\hat{r}_{ij}) - i(\sigma_i\times\hat{r}_{ij})(\sigma_j\cdot\hat{r}_{ij}), \tag{A.6b}$$

$$\sigma_i\Phi_{ij}P^\sigma_{ij} = \tfrac{1}{2}(\sigma_i - \sigma_j) + \frac{i}{2}(\sigma_i\times\sigma_j) + \hat{r}_{ij}(\sigma_j\cdot\hat{r}_{ij}) - i(\sigma_i\times\hat{r}_{ij})(\sigma_j\cdot\hat{r}_{ij}), \tag{A.6c}$$

$$\Phi_{ij}\sigma_i\Phi_{ij} = \sigma_i + 2i(\sigma_i\times\hat{r}_{ij})(\sigma_i\cdot\hat{r}_{ij}). \tag{A.6d}$$

Appendix B. Inclusion of Coulomb force

For the Coulomb force

$$V_C = \frac{1}{2}\sum_{i\neq j} v_C(r_{ij})\Xi_{ij}, \tag{B.1}$$

$$\Xi_{ij} = \frac{1}{4}(1 - \tau^z)_i(1 - \tau^z)_j, \tag{B.2}$$

the following relation holds, due to $\Xi^2_{ij} = \Xi_{ij}$,

$$e^{\pm i v_C \Xi_{ij} t} = 1 - \Xi_{ij} + e^{\pm i v_C t}\,\Xi_{ij}. \tag{B.3}$$

In double beta decay the nuclear currents always incorporate (two) isospin (lowering) operators. Trivially,

$$t^-_i t^-_j \Xi_{ij} = t^-_i t^-_i \Xi_{ij} = t^-_i \Xi_{ij} t^-_j = t^-_j \Xi_{ij} t^-_i = 0, \tag{B.4a}$$

$$\Xi_{ij} t^-_i t^-_j = t^-_i t^-_j, \tag{B.4b}$$

44

since the Coulomb force acts only between two proton states. Furthermore, Ξ_{ij} commutes with all other operators of H (II.21). Therefore,

$$\begin{aligned}&\mathscr{A}_i^\alpha t_i^- e^{iVt} \mathscr{A}_j^\alpha t_j^- e^{-iVt}\\ &= \mathscr{A}_i^\alpha t_i^- e^{iV_{st}}(1-\Xi_{ij}+e^{iv_{ct}}\Xi_{ij})\mathscr{A}_j^\beta t_j^-(1-\Xi_{ij}\\ &\quad + e^{-iv_{ct}}\Xi_{ij})e^{-iV_{st}} = \mathscr{A}_i^\alpha t_i^- e^{iV_{st}}\mathscr{A}_j^\beta t_j^- e^{-iV_{st}}.\end{aligned} \tag{B.5}$$

On the other hand,

$$\begin{aligned}&e^{-iVt}\mathscr{A}_i^\alpha t_i^- e^{iVt}\mathscr{A}_j^\beta t_j^-\\ &= e^{-iV_{st}}(1-\Xi_{ij}+e^{-iv_{ct}}\Xi_{ij})\mathscr{A}_i^\alpha t_i^-(1-\Xi_{ij}\\ &\quad + e^{iv_{ct}}\Xi_{ij})e^{iV_{st}}\mathscr{A}_j^\beta t_j^-\\ &= e^{-iV_{st}}\mathscr{A}_i^\alpha t_i^- e^{iV_{st}}\mathscr{A}_j^\beta t_j^- e^{-iv_{ct}}.\end{aligned} \tag{B.6}$$

It is straightforward, although a bit lengthy, to show for the nuclear part

$$\mathscr{A}_i^\alpha t_i^- e^{iV_{st}}\mathscr{A}_j^\beta t_j^- e^{-iV_{st}} = e^{-iV_{st}}\mathscr{A}_i^\alpha t_i^- e^{iV_{st}}\mathscr{A}_j^\beta t_j^-. \tag{B.7}$$

Considering that within OEM one has to calculate

$$\tfrac{1}{2}\{\mathscr{A}_i^\alpha t_i^- e^{iVt}\mathscr{A}_j^\beta t_j^- e^{-iVt} + e^{-iVt}\mathscr{A}_i^\alpha t_i^- e^{iVt}\mathscr{A}_j^\beta t_j^-\}, \tag{B.8}$$

Coulomb force can be accounted for simply by replacing all radial functions $v(r_{ij})$ by

$$v(r_{ij}) = \tfrac{1}{2}\{v_L(r_{ij},\Delta_L) + v_R(r_{ij},\Delta_R)\}, \tag{B.9}$$

where the only difference between v_L and v_R is due to

$$\Delta_L = \omega + \tfrac{1}{2}Q_{\beta\beta} + m_e, \tag{B.10a}$$

$$\Delta_R = \omega - (\tfrac{1}{2}Q_{\beta\beta} + m_e) - v_C(r_{ij}), \tag{B.10b}$$

and the various combinations of nuclear potentials for the radial functions $v(r_{ij})$ are given in the text. Note that in (B.10) we assumed $\varepsilon_1 \simeq \varepsilon_2 \simeq \frac{1}{2}Q_{\beta\beta} + m_e$ in order to separate the phase space integration from the evaluation of nuclear matrix elements. It should however be mentioned that matrix elements do not sensitively depend on this approximation, due to $\varepsilon_i \ll \omega$.

Appendix C. Momentum integrals

In the calculation of the OEM potentials, there appear six types of momentum integrals. Two integrals are required for the nuclear matrix elements of the mass mechanism. One of them involves a factor of $\exp(iq\cdot r_{ij})$, while the other does not. Similarly, there are two integrals for the nuclear matrix elements associated with the neutrino energy ω. Nuclear matrix elements arising from the neutrino momentum require only the integral involving $\exp(iq\cdot r_{ij})$ (M_{GTq}, M_{Fq}, M_T, M_P and M_R). Integrals are distinguished by suffices m (mass), ω (energy), R (recoil) and q (momentum), respectively. The six integrals are defined by

$$\begin{aligned}\mathscr{I}_m(\mathscr{F},\Lambda;r_{ij}) &= \frac{R}{2\pi^2}\int\frac{1}{\omega}\mathscr{F}\left(\frac{\Lambda^2}{q^2+\Lambda^2}\right)^4 e^{iq\cdot r_{ij}}dq\\ &= \frac{2R}{\pi}\int_0^\infty\frac{q^2}{\omega}\mathscr{F}j_0(qr_{ij})\left(\frac{\Lambda^2}{q^2+\Lambda^2}\right)^4 dq,\end{aligned} \tag{C.1}$$

$$\begin{aligned}\mathscr{I}_{m'}(\mathscr{F},\Lambda) &= \frac{R}{2\pi^2}\int\frac{1}{\omega}\mathscr{F}\left(\frac{\Lambda^2}{q^2+\Lambda^2}\right)^4 dq\\ &= \frac{2R}{\pi}\int_0^\infty\frac{q^2}{\omega}\mathscr{F}\left(\frac{\Lambda^2}{q^2+\Lambda^2}\right)^4 dq,\end{aligned} \tag{C.2}$$

$$\begin{aligned}\mathscr{I}_\omega(\mathscr{G},\Lambda;r_{ij}) &= \frac{R}{2\pi^2}\int\mathscr{G}\left(\frac{\Lambda^2}{q^2+\Lambda^2}\right)^4 e^{iq\cdot r_{ij}}dq\\ &= \frac{2R}{\pi}\int_0^\infty q^2\mathscr{G}j_0(qr_{ij})\left(\frac{\Lambda^2}{q^2+\Lambda^2}\right)^4 dq,\end{aligned} \tag{C.3}$$

$$\begin{aligned}\mathscr{I}_{\omega'}(\mathscr{G},\Lambda) &= \frac{R}{2\pi^2}\int\mathscr{G}\left(\frac{\Lambda^2}{q^2+\Lambda^2}\right)^4 dq\\ &= \frac{2R}{\pi}\int_0^\infty q^2\mathscr{G}\left(\frac{\Lambda^2}{q^2+\Lambda^2}\right)^4 dq,\end{aligned} \tag{C.4}$$

$$\begin{aligned}\mathscr{I}_q(\mathscr{F},\Lambda;r_{ij}) &= \frac{R}{2\pi^2}\int\frac{-ir_{ij}\cdot q}{\omega}\mathscr{F}\left(\frac{\Lambda^2}{q^2+\Lambda^2}\right)^4 e^{iq\cdot r_{ij}}dq\\ &= \frac{2Rr_{ij}}{\pi}\int_0^\infty\frac{q^3}{\omega}\mathscr{F}j_1(qr_{ij})\left(\frac{\Lambda^2}{q^2+\Lambda^2}\right)^4 dq,\end{aligned} \tag{C.5}$$

$$\begin{aligned}\mathscr{I}_R(\mathscr{F},\Lambda;r_{ij}) &= \frac{R}{2\pi^2 M}\int\frac{q\cdot q}{\omega}\mathscr{F}\left(\frac{\Lambda^2}{q^2+\Lambda^2}\right)^4 e^{iq\cdot r_{ij}}dq\\ &= \frac{2R^2}{\pi M}\int_0^\infty\frac{q^4}{\omega}\mathscr{F}j_0(qr_{ij})\left(\frac{\Lambda^2}{q^2+\Lambda^2}\right)^4 dq.\end{aligned} \tag{C.6}$$

Here, $\mathscr{F}$ and $\mathscr{G}$ stand for any of the combination of nuclear potentials in the radial functions $v(r_{ij})$. Λ is the nuclear form factor taken to be $\Lambda = 850$ MeV, which is introduced to account for the finite extension of nucleons. In the definitions the neutrino energy has been denoted by $\omega = \sqrt{m^2+q^2}$, where m is the neutrino mass.

Note that the integrals without $\exp(iq\cdot r_{ij})$ can be solved analytically, if $\omega = q$ is assumed, whereas for the other ones a numerical integration has to be performed.

References

1. W.C. Haxton and G.J. Stephenson, Progr. Part. Nucl. Phys. **12**, 409 (1984)
2. M. Doi, T. Kotani and E. Takasugi, Progr. Theor. Phys. Suppl. **83**, 1 (1985)
3. K. Muto and H.V. Klapdor, in: Neutrinos, ed. H.V. Klapdor, Springer, Berlin, Heidelberg, New York, 1988, p. 183
4. A. Balysh, M. Beck, S.T. Belyaev, J. Bockholt, A. Demehin, J. Echternach, A. Gurov, G. Heusser, M. Hirsch, H.V. Klapdor-Kleingrothaus, I. Kondratenko, V.I. Lebedev, B. Maier, A. Müller, F. Petry, A. Piepke, U. Schmidt-Rohr, H. Strecker, K. Zuber, Phys. Lett **B283**, 32 (1992)
5. F.T. Avignone, R.L. Brodzinski, J.T. Collar, C.K. Guerard, H.S. Miley and J.H. Reeves, Phys. Lett. **B 256**, 559 (1991)
6. A. Balysh, M. Beck, S.T. Belyaev, F. Bensch, J. Bockholt, A. Demehin, A. Gurov, G. Heusser, M. Hirsch, H.V. Klapdor-Kleingrothaus, I. Kondratenko, V.I. Lebedev, B. Maier, A. Müller, F. Petry, A. Piepke, H. Strecker, M. Völlinger and K. Zuber, Proc. 26th Int. Conf. on High Energy Physics, Dallas, USA, August 1992, AIP Conf. Proc. **272** (1993) p. 1141; Phys. Rev. Lett. **70**, 2853 (1993)
7. S.R. Elliott, A.A. Hahn and M.K. Moe, Phys. Rev. Lett. **59**, 2020 (1987)

8. H. Ejiri, K. Fushimi, M. Kawasaki, H. Kinoshita, H. Ohsumi, K. Okada, H. Sano, T. Shima, E. Takasugi, J. Tanaka, T. Watanabe, J. Phys. **G 17** Proc. Suppl. (1991) p.155
9. T. Kirsten, G. Heusser, D. Kaether, J. Oehm, E. Pernicka, H. Richter, in: Proc. Int. Symp. on Nuclear Beta Decays and Neutrino, eds T. Kotani, H. Ejiri, E. Takasugi (World Scientific, Singapore, 1986) p.81
10. T. Bernatowicz, J. Brannon, R. Brazzle, R. Cowsik, C. Hohenberg and F. Podosek, Phys. Rev. Lett. **69**, 2341 (1992)
11. A.L. Turkevich, T.E. Economou and G.A Cowan, Phys. Rev. Lett. **67**, 3211 (1991)
12. P. Langacker, in: Neutrinos, ed. H.V. Klapdor, Springer, Berlin, Heidelberg, New York, 1988, p. 71
13. R.N. Mohapatra and P. Pahl, Massive Neutrinos in Physics and Astrophysics (World Scientific, Singapore, 1990)
14. K. Muto, E. Bender and H.V. Klapdor, Z. Phys. **A334**, 177 (1989)
15. K. Muto, E. Bender and H.V. Klapdor, Z. Phys. **A334**, 187 (1989)
16. P. Vogel and M.R. Zirnbauer, Phys. Rev. Lett. **57**, 3148 (1986)
17. O. Civitarese, A. Faessler and T. Tomoda, Phys. Lett. **B 194**, 11 (1987)
18. T. Tomoda, A. Faessler, Phys. Lett. **B199**, 475 (1987)
19. J. Engel, P. Vogel, M.R. Zirnbauer, Phys. Rev. **C37**, 731 (1988)
20. J. Suhonen, T. Taigel and A. Faessler, Nucl Phys. **A 486**, 91 (1988)
21. A. Staudt, K. Muto, H.V. Klapdor-Kleingrothaus, Europhys. Lett. **13**, 31 (1990)
22. C.R. Ching, T.H. Ho and X.R. Wu, Commun. Theor. Phys. **12**, 167 (1989); C.R. Ching and T.H. Ho, Commun. Theor. Phys. **11**, 433 (1989)
23. C.R. Ching, T.H. Ho and X.R. Wu, Phys. Rev. **C40**, 304 (1989)
24. M. Hirsch, X.R. Wu, H.V. Klapdor-Kleingrothaus, C.R. Ching and T.H. Ho, Z. Phys. **A 345**, 163 (1993)
25. K. Muto, Phys. Rev. **C 48**, 402 (1993)
26. M. Doi and T. Kotani, Progr. Theor. Phys. **89**, 139 (1993)
27. X.R. Wu, A. Staudt, T.T.S. Kuo and H.V. Klapdor-Kleingrothaus, Phys. Lett. **B276**, 274 (1992)
28. T. Tomoda, A. Faessler, K.W. Schmid and F. Grümmer, Phys. Lett. **B 157**, 4 (1985)
29. T. Tomoda, A. Faessler, K.W. Schmid and F. Grümmer, Nucl. Phys. **A 452**, 591 (1986)
30. N. Anantaraman, H. Toki and G.F. Bertsch, Nucl. Phys. **A 398**, 269 (1983)
31. M. Lacombe, B. Loiseau, J.M. Richard, R. Vinh Mau, J. Côté, P. Pires and R. de Tourreil, Phys. Rev. **C21**, 861 (1980)
32. K. Muto, in: Proceedings of the International Symposium on Spin-Isospin Responses and Weak Processes in Hadrons and Nuclei, March, 8-10, 1994, Osaka, Japan, Eds. H. Ejiri, Y. Mizuno and T. Suzuki, Nucl. Phys. **A577**, 415c (1994)
33. A. Balysh, M. Beck, S.T. Belyaev, F. Bensch, J. Bockholt, A. Demehin, A. Gurov, G. Heusser, H.V. Klapdor-Kleingrothaus, I. Kondratenko, D. Kotel'nikov, V.I. Lebedev, B. Maier, A. Müller, F. Petry, A. Piepke, A. Pronsky, H. Strecker, M. Völlinger and K. Zuber, Phys. Lett. **B 322**, 176 (1994)
34. G.A. Miller and J.E. Spencer, Ann. Phys. **100**, 562 (1976)
35. J.D. Vergados, Phys. Rev. **C 24**, 640 (1981)
36. Particle Data Group, Phys. Rev. **D 45**, S1(1) (1992)
37. M. Hirsch, K. Muto, T. Oda and H.V. Klapdor-Kleingrothaus, Z. Phys. **A 347**, 151 (1994)
38. J. Suhonen, S.B. Khadkikar and A. Faessler, Nucl. Phys. **A535**, 509 (1991); Nucl. Phys. **A529**, 727 (1991); Phys. Lett. **B237**, 8 (1990)

Physics of Atomic Nuclei, Vol. 61, No. 7, 1998, pp. 1218–1228. From Yadernaya Fizika, Vol. 61, No. 7, 1998, pp. 1318–1328.

NUCLEON AND NUCLEAR STRUCTURE IN SEARCHING FOR NEW PHYSICS

Two-Neutrino Double Beta Decay: Critical Analysis*

F. Šimkovic[1), **], G. Pantis[2)], and A. Faessler[3)]

Abstract—A critical analysis of various approximate schemes for calculating the matrix elements for two-neutrino double-beta decay ($2\beta2\nu$ decay) is performed. The time integral representation of the $2\beta2\nu$-decay matrix element is used for this purpose. It is shown that, within the single-particle approximation of the nuclear Hamiltonian, the $2\beta2\nu$-decay matrix element is equal to zero because of the mutual cancellation of the direct and cross terms. The quasiboson approximation (QBA) and renormalized QBA (RQBA) schemes imply that the $2\beta2\nu$-decay transition operator is constant if the initial and final quasiparticle-random-phase-approximation (QRPA) and renormalized-QRPA (RQRPA) Hamiltonians are required to be equivalent. This means that $2\beta2\nu$-decay is a higher order process in the boson expansion of the nuclear Hamiltonian and that its higher order boson approximations are important. The equivalence of the initial and final QRPA and RQRPA Hamiltonians is discussed within the QBA and the RQBA, respectively. It is found that the mismatch of the two Hamiltonians becomes worse with increasing strength of particle–particle interaction, especially in the case of QRPA Hamiltonians. It is assumed to be one of the factors responsible for extreme sensitivity of the $2\beta2\nu$-decay matrix element to the residual interaction appearing in explicit calculations involving an intermediate nucleus. Further, the operator expansion method (OEM) is reconsidered, and new $2\beta2\nu$-decay transition operators are rederived in a consistent way. The validity of the OEM approximation is discussed in respect to the other approximation schemes. The OEM, combined with QRPA or RQRPA ground-state wave functions, reflects sensitively instabilities incorporated in the ground states being considered. Therefore, the predictive power of the OEM must be studied with aid of the other ground-state wave functions—for example, shell-model ones.

1. INTRODUCTION

Two-neutrino double-beta decay ($2\beta2\nu$ decay) is a second-order weak-interaction process, which is allowed by the Standard model [1]. In $2\beta2\nu$ decay, a nucleus (A, Z) undergoes a transition to the nucleus $(A, Z+2)$ through the emission of two electrons and two antineutrinos. This rare process has already been well established experimentally for a couple of isotopes. The inverse half-live with respect to $2\beta2\nu$ decay is free of unknown parameters on the particle-physics side and is expressed as the product of a phase-space factor and the relevant $2\beta2\nu$-decay nuclear matrix element. Since the phase-space factor can be calculated to a desired precision, experimental investigation of $2\beta2\nu$ decay directly yields the value of the $2\beta2\nu$-decay nuclear matrix element. Thereby, $2\beta2\nu$ decay offers a sensitive test of nuclear structure calculations. The calculation of the $2\nu\beta\beta$-decay nuclear transition continues to be challenging because the predicted nuclear matrix elements are small and because the mechanism suppressing these matrix elements has yet to be completely understood.

The proton–neutron quasiparticle random phase approximation (*pn*-QRPA) is the most frequently used nuclear-structure method for evaluating $2\beta2\nu$-decay rates because of the remarkable success achieved in revealing the suppression mechanism for $2\beta2\nu$-decay matrix elements [2–4]. However, the extreme sensitivity of the $2\beta2\nu$-decay matrix elements to the *pn* 1^+ particle–particle matrix element and the collapse of the QRPA solution in the physically acceptable region of the particle–particle strength in the nuclear Hamiltonian render it difficult to make definitive rate predictions.

Some attempts have been made to overcome the above drawbacks, for example, by including higher order RPA corrections [5], particle-number projection [6, 7], and proton–neutron pairing [8] in the theory. However, none of these modifications of the QRPA prevents the collapse or inhibits the vanishing of the nuclear matrix element near the physical value of the particle–particle force. Toivanen and Suhonen proposed a proton–neutron renormalized QRPA (*pn*-RQRPA) [9], which goes beyond the QRPA, taking into account the Pauli exclusion principle in an approximate way. It was shown that the above phenomena could be associated with the limitation of the QRPA approach, the quasiboson approximation (QBA), which violates the Pauli exclusion principle. The renormalized QBA, which underlies *pn*-RQRPA, inhibits the collapse of the *pn*-RQRPA solution for a physical value of the strength of particle–particle interaction. In addition, the $2\beta2\nu$-decay nuclear matrix elements calculated via the *pn*-RQRPA were found to be significantly less sensitive to the particle–particle force within its physical values in relation to those obtained by the *pn*-QRPA method [9–

* This article was submitted by the authors in English.

1) Department of Nuclear Physics, Comenius University, Mlynská dolina, SK-84215 Bratislava, Slovakia.
2) Department of Physics, University of Ioannina, PO Box 1186, GR-45110 Ioannina, Greece.
3) Institute für Theoretische Physik der Universität Tübingen, Auf der Morgenstelle 14, D-72076 Tübingen, Germany.
** e-mail: simkovic@fmph.uniba.sk

1063-7788/98/6107-1218$15.00 © 1998 МАИК Наука/Interperiodica Publishing

2.2.5 $\beta^+\beta^+$, EC/EC, β^+/EC decay

Physics Letters B 268 (1991) 312–316
North-Holland

PHYSICS LETTERS B

Nuclear matrix elements for double positron emission

A. Staudt [a], K. Muto [b] and H.V. Klapdor-Kleingrothaus [a]

[a] *Max-Planck-Institut für Kernphysik, W-6900 Heidelberg, FRG*
[b] *Tokyo Institute of Technology, Tokyo, Japan*

Received 21 June 1991

Nuclear matrix elements for double positron emissions are calculated for both two-neutrino and zero-neutrino decay modes within the pnQRPA model with a realistic effective nucleon–nucleon interaction. This is the first complete microscopic calculation for all potential double positron emitters. The predicted lifetimes are several orders of magnitude larger than those for $\beta^-\beta^-$ decays and difficult to reach in present days experiments.

Investigations of neutrinoless double beta decay provide very sensitive tests of lepton number conservation and of a possible Majorana mass of the neutrino. In this way they also yield the possibility to solve one of the long-standing questions about the nature of the neutrino, namely to decide whether the neutrino is a Dirac or a Majorana particle.

One usually distinguishes between two mechanisms of the 0ν mode. One which involves the effective neutrino mass and a second involving right-handed charged weak currents. However, in the framework of general gauge theories neutrinoless double beta decay always requires a nonvanishing Majorana mass of the neutrino [1]. Thus, the observation of the 0ν decay would constitute convincing proof that there exists a massive Majorana neutrino which couples to the electron. For this reason, considerable experimental effort is being devoted to the search for this rare process [2–4]. However, the extraction of parameters related to the properties of the neutrino and the structure of weak interaction requires the knowledge of nuclear matrix elements.

Two-neutrino double beta decay is allowed in the standard theory of the electroweak interaction irrespective of the nature of the neutrino and has been seen in the laboratory [5–7]. Since both decay modes are related to each other, the ability to reproduce the observed 2ν decay rates in nuclear structure calculations can give some indication of the accuracy with which the neutrino mass and possible right-handed mixing parameters can be extracted from limits on 0ν decay lifetimes. Therefore, a reliable calculation of nuclear matrix elements governing both modes of ββ decay is of considerable importance.

Most experimental and theoretical work up to now has been devoted to the $\beta^-\beta^-$ decay. The corresponding nuclear matrix elements have been calculated by several authors [8–15] and were shown to be reliable within a factor of about three for the 0ν decay mode [14,16].

From an experimental point of view double positron emission and electron–positron conversion have the advantage of a clear signature due to the emitted positrons. This fact may help to compensate for a kinematical retardation of the decays. In the following we will focus on double positron emission,

$$(Z,A)\rightarrow(Z-2,A)+2e^++2\nu_e \quad (2\nu\beta\beta)\,, \tag{1a}$$

$$(Z,A)\rightarrow(Z-2,A)+2e^+ \quad (0\nu\beta\beta)\,, \tag{1b}$$

which is energetically allowed for the following seven nuclides: ^{78}Kr, ^{96}Ru, ^{106}Cd, ^{124}Xe, ^{130}Ba, ^{136}Ce and ^{148}Gd. One should note that electron–positron conversion and double electron capture processes are always accompanying the double positron emission and may also occur when $\beta^+\beta^+$ decay is energetically forbidden.

At present there only exists a phase space estimation of $2\nu\,\beta^+\beta^+$ half-lives for nuclei with $A=78$–136 [8]. The authors assume a typical "closure" Ga-

Volume 268, number 3,4 PHYSICS LETTERS B 17 October 1991

mow–Teller matrix element to be valid for all nuclides. Lifetimes for two-neutrino and neutrinoless $\beta^+\beta^+$ transitions of the $A=96$ system have been predicted in ref. [17]. In the latter work the nuclear matrix elements are calculated in a shell model approach.

In the present letter we provide for the first time a complete set of microscopic predictions for half-lives of all potential double positron emitters. The ββ-decay matrix elements are evaluated in the QRPA model which has successfully been applied to the calculation of ββ matrix elements as well as single β^-- and β^+-decay properties [9–14,16, 18–21]. In particular, after the strong sensitivity of 2νββ decay rates to ground-state correlations had first been shown in ref. [9], the suppression mechanism of 2ν and β^+ matrix elements could be traced back to spin–isospin ground-state correlations in refs. [10–13,18,19,21]. In this context the inclusion of the particle–particle (pp) force, which considerably enhances correlations in the ground state, was shown to be of decisive importance.

In QRPA one describes pairing interaction of like nucleons in a BCS calculation. The proton–neutron residual interaction is accounted for in the RPA. The realistic effective interaction to be used in the BCS and RPA calculations is derived from the Paris potential [22,23]. Single-particle energies are calculated from a Coulomb-corrected Woods–Saxon potential. The choice of the model parameters is the same as in our previous work on $\beta^-\beta^-$ decay [13,14]. Renormalization effects in finite nuclei are taken into account in the same way as described in refs. [13,16]. The strengths of the pairing interaction of the BCS calculation are adjusted to experimental even–odd mass differences for proton and neutron systems separately. Most important, the strength of the pp force is determined by a fit to single β^+-decay data as described in refs. [13,14].

As in the case of $\beta^-\beta^-$ emitters, the reduction of $\beta^+\beta^+$ amplitudes due to RPA correlations is particularly pronounced in the case of 2νββ decay where the corresponding amplitude vanishes at a certain value of the strength g_{pp} of the particle–particle component of the proton–neutron interaction. This cancellation feature remains essentially unchanged when particle-number symmetries both for protons and neutrons are restored before the solution of the RPA equation [24] (see also ref. [9]). Therefore, we neglect the particle-number projection in the present letter.

The nuclear matrix element of 2ν decay is defined by

$$M_{\mathrm{GT}}^{2\nu} = \sum_{a,b} \frac{\langle 0_f^+ \| t_+\sigma \| 1_b^+ \rangle \langle 1_b^+ | 1_a^+ \rangle \langle 1_a^+ \| t_+\sigma \| 0_i^+ \rangle}{E_a + \frac{1}{2}Q_{\beta\beta} + m_e - E_i} . \tag{2}$$

This expression involves transition amplitudes to the intermediate 1^+ states which are determined in two separate calculations, one based on the initial (A, Z) and the other based on the final $(A, Z-2)$ nuclei. The overlap between the two different sets of 1^+ states are accounted for as described in refs. [9,13]. The various quantities in the above equation are readily calculated from the pnQRPA energies and amplitudes.

The half-life for the 2νββ decay can be written in a factorized form [18,25],

$$T_{1/2}^{2\nu} = (F^{2\nu} |M_{\mathrm{GT}}^{2\nu}|^2)^{-1} , \tag{3}$$

where $F^{2\nu}$ is a leptonic phase space integral, which is independent of the nuclear structure.

The 2ν half-lives for the abovementioned potential double positron emitters with $A \geqslant 78$ are given in table 1. Since the calculated matrix elements cross zero in the vicinity of the most probable values of the strength of the pp force ($\bar{g}_{pp}$), it is difficult to predict a precise 2ν half-life. Therefore, we have also calculated a mean half-life by simply averaging the decay probability over the 1σ range of g_{pp} given by the fit to

Table 1
Calculated 2ν half-lives for all potential double positron emitters. $T_{1/2}^{2\nu}(\bar{g}_{pp})$ denotes the half-life evaluated at the fitted strength of the particle–particle interaction, $T_{1/2}^{2\nu}$(l.l.) is the lower limit within the 1σ range of g_{pp}. $\bar{T}_{1/2}^{2\nu}$ is obtained by averaging the decay rate over the 1σ range of the strength parameter g_{pp}.

Nuclide	$T_{1/2}^{2\nu}$(l.l.) (yr)	$T_{1/2}^{2\nu}(\bar{g}_{pp})$ (yr)	$\bar{T}_{1/2}^{2\nu}$ (yr)
^{78}Kr	4.07×10^{25}	4.52×10^{29}	1.93×10^{26}
^{96}Ru	5.10×10^{26}	5.49×10^{26}	5.31×10^{26}
^{106}Cd	3.48×10^{25}	6.93×10^{25}	4.94×10^{25}
^{124}Xe	4.30×10^{25}	8.44×10^{25}	8.17×10^{25}
^{130}Ba	5.00×10^{28}	1.68×10^{29}	1.37×10^{29}
^{136}Ce	1.03×10^{31}	3.84×10^{32}	4.51×10^{31}
^{148}Gd	2.15×10^{26}	collapse	5.81×10^{26}

Volume 268, number 3,4 PHYSICS LETTERS B 17 October 1991

single β^+-decay data. These average half-lives are denoted by $\bar{T}^{2\nu}_{1/2}$. In addition, we give a lower limit for the half-life which corresponds to the largest nuclear matrix element within the 1σ range of g_{pp} [$T^{2\nu}_{1/2}$(l.l.)]. In the case of ^{148}Gd the collapse of the RPA equation, that is the occurrence of complex eigenvalues, is already found at a strength smaller than that expected from the fit to single β^+ data. The nucleus $^{148}_{64}Gd_{84}$ lies just above the $N=82$ magic number where the QRPA model has the largest uncertainties. The collapse of the RPA equation indicates that the 1^+ component is ambiguous and thus also limits somewhat the reliability of the result on the neutrinoless decay of this particular nuclide.

The shell model prediction of ref. [17] for ^{96}Ru is $T^{2\nu}_{1/2}=1.6\times10^{28}$ yr which is more than one order of magnitude larger than our value. This difference is probably due to the choice of the model space. In the approach of ref. [17] the model space consists of six protons in the $1p_{1/2}$ and $0g_{9/2}$ shells and two neutrons in the $1d_{5/2}$ and $2s_{1/2}$ shells. Neutrons in the $0g_{7/2}$ orbit are only included in perturbation theory. Therefore one might expect to get too small Gamow–Teller matrix elements. In the present calculation we have chosen the active valence space to consist of the oscillator shells $3\hbar\omega$ and $4\hbar\omega$ and the $0h_{9/2}$ and $0h_{11/2}$ orbits for both neutrons and protons. The BCS calculation yields an occupation amplitude $v=0.26$ for the neutron in the $0g_{7/2}$ shell and only 0.14 in the $2s_{1/2}$ shell. This clearly underlines the importance of the $g_{7/2}$ states which are only treated perturbatively in ref. [17].

The 0ν transition matrix elements are known to be less sensitive to details of the nuclear structure. In the following we will restrict ourselves to the discussion of the mass mechanism of the 0ν mode. If we neglect the contribution from right-handed admixtures to charged weak currents, the nuclear structure information is contained in the matrix elements

$$M_{GT}=\sum_{m,n}\langle 0_f^+\|t_{+m}t_{+n}\sigma_m\cdot\sigma_n H(r)\|0_i^+\rangle\,, \qquad (4a)$$

$$M_F=\left(\frac{g_V}{g_A}\right)^2\sum_{m,n}\langle 0_f^+\|t_{+m}t_{+n}H(r)\|0_i^+\rangle\,, \qquad (4b)$$

which are obtained by the closure approximation. They have similar structure to those in (2) for the 2ν mode. However, they involve the neutrino potential $H(r)$ that represents the effect of the neutrino propagation between the two nucleons and thus introduces a dependence of the transition operators on the relative distance r.

The inverse half-life for neutrinoless double beta decay is of the form

$$[T^{0\nu}_{1/2}(0^+\to0^+)]^{-1}=G_1^{0\nu}\frac{\langle m_\nu\rangle^2}{m_e^2}\,|M_{GT}-M_F|^2\,, \qquad (5)$$

where $G_1^{0\nu}$ is the two-positron phase space integral, which can be found in ref. [26]. The 0ν matrix elements are calculated in the pnQRPA model following the work of refs. [13,14]. The inter-nucleon short range correlations are treated in the standard way [12,13], effects of the finite nucleon size are accounted for by replacing the vector and axial-vector coupling constants by dipole form factors in momentum space [17]. In table 2 the results on neutrinoless double positron emission are summarized. Only the matrix elements for the decay induced by a nonvanishing Majorana mass of the exchanged neutrino are given here.

The decay rates for both decay modes are considerably suppressed in comparison with $\beta^-\beta^-$ decay. This is mainly due to the nuclear Coulomb repulsion on positrons and the unfavourable phase space. With the calculated nuclear matrix elements upper bounds on the effective neutrino mass $\langle m_\nu\rangle$ can be deduced from measured lower limits on $0\nu\,\beta\beta$ half-lives.

While much effort has been devoted to searches for double β^- decay, relatively little experimental work has been done on double β^+ decay. These experiments obviously suffer from the small energy avail-

Table 2
Results on $0\nu\,\beta^+\beta^+$ decay. Upper bounds on the effective value of the neutrino mass are given in the last column.

Transition	$Q_{\beta^+\beta^+}$ (MeV)	$T^{0\nu}_{1/2}\langle m_\nu\rangle^2$ (yr eV^2)	$\langle m_\nu\rangle$ (eV)
^{78}Kr→ ^{78}Se	0.833	9.32×10^{27}	–
^{96}Ru→^{96}Mo	0.677	4.86×10^{28}	$<1.3\times10^6$
^{106}Cd→^{106}Pd	0.734	3.20×10^{28}	$<3.5\times10^5$
^{124}Xe→^{124}Te	0.822	6.58×10^{28}	$<1.3\times10^5$
^{130}Ba→^{130}Xe	0.538	2.03×10^{29}	–
^{136}Ce→^{136}Ba	0.365	5.17×10^{30}	–
^{148}Gd [a)]→^{148}Sm	1.024	1.63×10^{28}	–

[a)] α-emitter with an α half-life of about 75 yr.

Volume 268, number 3,4 PHYSICS LETTERS B 17 October 1991

able in the decay. However, one great advantage is the possibility to utilize the fact that if a $\beta^+\beta^+$ transition occurred within a thick sample of material, the positrons would stop and annihilate within the sample. Subsequently, four coincident 511 keV annihilation γ-rays would be emitted, giving a typical signature for a real event.

The most recent measurements yield the following limits on $\beta^+\beta^+$ half-lives: ^{96}Ru: $T_{\beta^+\beta^+}(2\nu+0\nu) > 3.1\times10^{16}$ yr [27], ^{106}Cd: $T_{\beta^+\beta^+}(2\nu+0\nu) > 2.6\times10^{17}$ yr [28], ^{124}Xe: $T_{\beta^+\beta^+}(2\nu) > 2\times10^{14}$ yr and $T_{\beta^+\beta^+}(0\nu) > 4.2\times10^{17}$ yr [29]. These values are still far below the theoretical predictions presented in this letter. The predicted 2ν half-life for ^{124}Xe is more than 10 orders of magnitude larger than the present experimental limit.

The extracted upper bounds on the effective neutrino mass are also given in table 2. They are by far less stringent than those obtained from studies of double β^- decay, where the present experiments yield an upper limit of the effective Majorana mass of the neutrino in the range of 1–2 eV [2–4].

In conclusion, it seems unlikely that double positron emission is observable with the sensitivity of present days experiments. The advantage of the larger decay energy available in processes involving electron capture instead of positron emission is most pronounced for the 2ν mode but less important for the 0ν decay mode. Since the neutrinoless mode of double electron capture also appears to be hopelessly slow as pointed out in ref. [17], the only observable neutrinoless process may be electron–positron conversion. This process is kinematically favoured in comparison with double positron emission and still has the advantage of a clear signature due to the emitted positron.

References

[1] J. Schechter and J.W.F. Valle, Phys. Rev. D 25 (1982) 2951.
[2] D.O. Caldwell, Intern. J. Mod. Phys. A 4 (1989) 1851.
[3] H.V. Klapdor, A. Piepke, G. Heusser, E. Buchner, A. Müller, U. Schmidt-Rohr, H. Strecker, S.T. Belyaev, A. Balish, A. Gurov, A. Demehin, I. Kondratenko and V.I. Lebedev, in: Proc. Intern. Symp. on Weak and electromagnetic interactions in nuclei (Montreal, May 1989), ed. P. Depommier (Editions Frontières, Gif-sur-Yvette, 1989) p. 701.
[4] H.V. Klapdor-Kleingrothaus, Proc. 14th Europhys. Conf. on Nuclear physics (Bratislava, October 1990), J. Phys. G, to be published.
[5] S.R. Elliott, A.A. Hahn and M.K. Moe, Phys. Rev. Lett. 59 (1987) 2020.
[6] A.A. Vasenko, I.V. Kirpichnikov, V.A. Kuznetsov, A.S. Starostin, A.G. Djanyan, V.S. Pogosov, S.P. Shachysisyan and A.G. Tamanyan, Mod. Phys. Lett. A 5 (1990) 1299;
H.S. Miley, F.T. Avignone III, R.L. Brodzinski, J.I. Collar and J.H. Reeves, Phys. Rev. Lett. 65 (1990) 3092;
F.T. Avignone III, R.L. Brodzinski, C.K. Guerard, I.V. Kirpichnikov, H.S. Miley, V.S. Pogosov, J.H. Reeves, A.S. Starostin and A.G. Tamangan, Phys. Lett. B 256 (1991) 559.
[7] H. Ejiri, K. Fushimi, T. Kamada, H. Kinoshita, H. Kobiki, H. Ohsumi, K. Okada, H. Sano, T. Shibata, T. Shima, N. Tanabe, J. Tanaka, T. Taniguchi, T. Watanabe and N. Yamamoto, Phys. Lett. B 258 (1991) 17.
[8] W.C. Haxton and G.J. Stephenson Jr., Prog. Part. Nucl. Phys. 12 (1984) 409.
[9] H.V. Klapdor and K. Grotz, Phys. Lett. B 142 (1984) 323;
K. Grotz and H.V. Klapdor, Phys. Lett. B 153 (1985) 1; B 157 (1985) 242.
[10] K. Grotz and H.V. Klapdor, Nucl. Phys. A 460 (1986) 395.
[11] P. Vogel and M.R. Zirnbauer, Phys. Rev. Lett. 57 (1986) 3148;
J. Engel, P. Vogel and M.R. Zirnbauer, Phys. Rev. C 37 (1988) 731.
[12] O. Civitarese, A. Faessler and T. Tomoda, Phys. Lett. B 194 (1987) 11;
T. Tomoda and A. Faessler, Phys. Lett. B 199 (1987) 475.
[13] K. Muto and H.V. Klapdor, Phys. Lett. B 201 (1988) 420;
K. Muto, E. Bender and H.V. Klapdor, Z. Phys. A 334 (1989) 177, 187.
[14] A. Staudt, K. Muto and H.V. Klapdor-Kleingrothaus, Europhys. Lett. 13 (1990) 31.
[15] J. Bernabeu, B. Desplanques, J. Navarro and S. Noguera, Z. Phys. C 46 (1990) 323.
[16] A. Staudt, T.T.S. Kuo and H.V. Klapdor-Kleingrothaus, Phys. Lett. B 242 (1990) 17.
[17] J.D. Vergados, Nucl. Phys. B 218 (1983) 109.
[18] K. Muto and H.V. Klapdor, in: Neutrinos, ed. H.V. Klapdor (Springer, Berlin, 1988) p. 183.
[19] J. Suhonen, T. Taigel and A. Faessler, Nucl. Phys. A 486 (1988) 91.
[20] A. Staudt, E. Bender, K. Muto and H.V. Klapdor-Kleingrothaus, At. Data Nucl. Data Tables 44 (1990) 79.
[21] A. Staudt, M. Hirsch K. Muto and H.V. Klapdor-Kleingrothaus, Phys. Rev. Lett. 65 (1990) 1543.
[22] M. Lacombe, B. Loiseau, J.M. Richard, R. Vinh Mau, J. Côté, P. Pirès and R. de Tourreil, Phys. Rev. C 21 (1980) 861.
[23] N. Anantaraman, H. Toki and G.F. Bertsch, Nucl. Phys. A 398 (1983) 269.
[24] O. Civitarese, A. Faessler, J. Suhonen and X.R. Wu, Phys. Lett. B 251 (1990) 333.

Volume 268, number 3,4 PHYSICS LETTERS B 17 October 1991

[25] M. Doi, T. Kotani and E. Takasugi, Prog. Theor. Phys. Suppl. 83 (1985) 1.

[26] F. Boehm and P. Vogel, Physics of massive neutrinos (Cambridge U.P., Cambridge, 1987).

[27] E.B. Norman, Phys. Rev. C 31 (1985) 1937.

[28] E.B. Norman and M.A. DeFaccio, Phys. Lett. B 148 (1984) 31.

[29] A.S. Barabash, V.V. Kuzminov, V.M. Lobashev, V.M. Novikov, B.M. Ovchinnikov and A.A. Pomansky, Phys. Lett. B 223 (1989) 273.

Z. Phys. A 347, 151–160 (1994)

ZEITSCHRIFT
FÜR PHYSIK A

Nuclear structure calculation of $\beta^+\beta^+$, β^+/EC and EC/EC decay matrix elements

M. Hirsch [1], **K. Muto** [1], **T. Oda** [1], **H.V. Klapdor-Kleingrothaus** [2]

[1] Department of Physics, Tokyo Institute of Technology, Meguro, Tokyo, 152 Japan
[2] Max-Planck-Institut für Kernphysik, Postfach 103980, D-69029 Heidelberg, Germany

Received: 6 September 1993

Abstract. Nuclear matrix elements for double positron emisson ($\beta^+\beta^+$), positron emission/electron capture (β^+/EC) and double electron capture (EC/EC) in the $2\nu\beta\beta$ decay mode and for $\beta^+\beta^+$ and β^+/EC decay in the $0\nu\beta\beta$ mode are calculated for the experimentally most promising isotopes ^{58}Ni, ^{78}Kr, ^{96}Ru, ^{106}Cd, ^{124}Xe, ^{130}Ba and ^{136}Ce within pn-QRPA. We point out that the matrix element for the $2\nu\beta^+$/EC decay differs from the $2\nu\beta^+\beta^+$ matrix element, an effect not considered previously. For the neutrino accompanied decays our calculation predicts for the β^+/EC and the EC/EC mode half lives which are shorter typically by 4–7 orders of magnitude than those for the double positron emission. However, even for the best candidates typical values for $2\nu\beta^+$/EC (2νEC/EC) are still in the range of $\sim 10^{22}$ ((*some*) 10^{21}) years. For $0\nu\beta\beta$ decay we have calculated all matrix elements relevant for both, the mass mechanism and the right-handed currents for the first time complete. A detailed discussion of the differences between the $0\nu\beta^+\beta^+$, the $0\nu\beta^+$/EC and $0\nu\beta^-\beta^-$ decay is given.

PACS: 21.60; 23.40.

1. Introduction

Double beta decay is known as a sensitive tool to explore neutrino properties ever since Furry [1] suggested the neutrinoless double beta decay mode ($0\nu\beta\beta$) as a possibility to decide whether the neutrino is a Dirac or a Majorana particle. Though only experiments can finally prove which description of the neutrino is correct, double beta decay has also been subject of much theoretical work since quantitative information about, for example, the effective neutrino mass requires a reliable calculation of nuclear matrix elements.

Lower limits on half lives for $0\nu\beta\beta$ decay have been improved considerably [2–7], but up to now no conclusive evidence for $0\nu\beta\beta$ decay has been reported. On the other hand, experimental progress has finally led to the observation of $2\nu\beta\beta$ decay for the isotopes ^{76}Ge [2, 3], ^{82}Se [4], ^{100}Mo [5], ^{128,130}Te [6, 7] and ^{238}U [8] – all of which are $\beta^-\beta^-$ decays. However, there exists no positive evidence for any $\beta^+\beta^+$ decay.

The main reason why $\beta^-\beta^-$ decay has attracted much more interest is simple: Even the largest $Q-$value for double positron emission, $Q_{\beta^+\beta^+}(^{124}\mathrm{Xe})\simeq 1.02$ MeV, is low compared to those of the best $\beta^-\beta^-$ decay candidates. This results – due to the strong dependence of the phase space on the available energy – in half life estimations far out of reach of present day experiments [9].

In any case, if double positron emission is possible energetically, electron capture always acts as a competitive process and in double beta plus decay we therefore have to deal with three different possible decay modes,

$$2\nu\beta^+\beta^+:\quad (Z,A)\rightarrow(Z-2,A)+2e^+ +2\nu_e, \qquad (1.1a)$$

$$2\nu\beta^+/\mathrm{EC}:\quad e^- +(Z,A)\rightarrow(Z-2,A)+e^+ +2\nu_e, \qquad (1.1b)$$

$$2\nu\mathrm{EC/EC}:\quad 2e^- +(Z,A)\rightarrow(Z-2,A)+2\nu_e. \qquad (1.1c)$$

In the following we will refer to the β^+/EC decay more briefly as the mixed mode. The most important difference between processes (1.1a)–(1.1c) is perhaps their different Q-values,

$$Q_{\beta^+\beta^+}=M(A,Z)-M(A,Z-2)-4m_e c^2, \qquad (1.2a)$$

$$Q_{\beta^+/\mathrm{EC}}=M(A,Z)-M(A,Z-2)-2m_e c^2, \qquad (1.2b)$$

$$Q_{\mathrm{EC/EC}}=M(A,Z)-M(A,Z-2). \qquad (1.2c)$$

Therefore, even for ^{124}Xe the Q-value for EC/EC decay is larger than the one for double positron emission by a factor of ~ 3. Shorter half lives for processes (1.1b) and (1.1c) are thus expected.

$2\nu\beta\beta$ decay half lives for double positron emission have been calculated earlier by phase space estimations [9]. Later on, Staudt et al. [10] calculated $2\nu\beta^+\beta^+$ half lives within the *pn*-QRPA model of [11, 12], essentially confirming the pessimistic estimations of Haxton and Stephenson [9]. Kim and Kubodera [13] estimated half lives also for the electron capture modes within a relatively simple model, assuming a constant nuclear matrix element. Recently more advanced phase space estimations for the mixed mode and the EC/EC mode have been published by Doi and Kotani [14]. Though in the work of [14] the important nuclear matrix elements have not been calculated, their results indicated that the half lives for $2\nu\beta^+$/EC and 2νEC/EC decay might be much shorter than those for double positron emission and probably be within reach of advanced $\beta\beta$ decay experiments. However, in order to draw firm conclusions we have to calculate the nuclear structure matrix elements. This is the first purpose of this paper.

In this context we would like to mention that recently we received a preprint by Suhonen [15], dealing with the calculation of matrix elements for 2νEC/EC decay. However, our present work differs from this one in two respects. The first one is that, while [15] deals only with the neutrino emitting mode of EC/EC decay, we consider also the mixed mode as well as the neutrinoless modes. More important is that our results for the 2νEC/EC decay differ considerably from those given by Suhonen [15], so that we arrive at a somewhat different conclusion.

As mentioned above, one of the most interesting decay modes of double beta decay is the $0\nu\beta\beta$ decay. From $0\nu\beta\beta$ decay not only information on the effective neutrino mass can be obtained, but it could also be possible if the weak interaction has a right-handed component (RHC) [9, 16, 17]. Matrix elements and phase spaces for the RHCs differ from those for the mass mechanism. To obtain complete information it is necessary to calculate various nuclear matrix elements.

Also $0\nu\beta^+\beta^+$ decay matrix elements for the mass mechanism have been calculated in [10], but no calculation for the matrix elements for the RHCs exists up to now in the literature. Moreover, for the more interesting $0\nu\beta^+$/EC decay mode to the best of our knowledge there exists no calculation. The complete calculation of the decay rate coefficients for these both modes is the second purpose of the present paper.

Though experimental limits on $0\nu\beta^+\beta^+$ and $0\nu\beta^+$/EC decay are much weaker than those of the best $\beta^-\beta^-$ decay candidates (and hence are derived limits on the effective neutrino mass) one important result of the present work is an interesting enhancement effect in the $\langle\lambda\rangle$-terms of the RHCs in $0\nu\beta^+$/EC decay. A detailed discussion will be given in Sect. 3.3. However, the situation can be summarized by stating that the $0\nu\beta^+$/EC decay is – similarly to the $0\nu\beta^-\beta^-(0^+\to 2^+)$ transitions – much more sensitive to the RHC mode than to the mass mechanism, but having much shorter half lives than those expected for $0^+\to 2^+$ transitions. On the other hand, as discussed in Sect. 3.3, the same enhancement effect will make it markedly difficult to determine exact limits on the neutrino mass and right-handed parameters from the nonobservation of $0\nu\beta^+$/EC decay alone; additional information from $0\nu\beta^+\beta^+$ or $0\nu\beta^-\beta^-$ decay will be required.

This paper is organized as follows: in Sect. 2 we consider the 2ν decay mode. Matrix elements are calculated within the *pn*-QRPA formalism [11, 12]. Predictions of $2\nu\beta\beta$ decay half lives are given for all three possible neutrino emitting modes for the isotopes ^{58}Ni, ^{78}Kr, ^{96}Ru, ^{106}Cd, ^{124}Xe, ^{130}Ba and ^{136}Ce. By the special example of ^{106}Cd, however, we would also like to discuss the limitations of present $2\nu\beta\beta$ decay calculations. Section 3 is then concerned with the calculation of the matrix elements for the neutrinoless modes, followed by a discussion of the implications of our results on the determination of the neutrino mass and the right-handed parameters. We then close with a short summary and outlook.

2. $2\nu\beta\beta$ decay

2.1. General considerations

For $2\nu\beta\beta$ decay the inverse half life can be expressed as [16, 17],

$$[T_{1/2}^{2\nu}]^{-1}=\frac{a_{2\nu}}{\ln(2)}\int d\Omega_{2\nu}^{\alpha}\sum_{a,a'}A_{a,a'}^{\alpha}, \tag{2.1}$$

where $a_{2\nu}=(Gg_A)^4/(32\pi^7\hbar)$ and the relevant nuclear structure information has been confined into the coefficients $A_{a,a'}$, see below. $d\Omega_{2\nu}^{\alpha}$ stands for the leptonic phase space and α distinguishes between the different decay modes $\alpha=\beta^+\beta^+$, β^+/EC or EC/EC. Phase spaces can, in principle, be calculated unambigously and also very accurately [14, 16]. We follow essentially the description of Doi and coworkers and refer for brevity to their original papers [14, 16].

Here instead we would like to discuss one aspect of the nuclear structure calculation in some detail; this is the matrix elements of the $2\nu\beta^+$/EC and the $2\nu\beta^+\beta^+$ decay are different contrary to previous belief. The coefficients $A_{a,a'}$ in (2.1) can be expressed as

$$\begin{aligned}A_{a,a'}^{\alpha}=&\langle 0_f^+\|t_+\sigma\|1_a^+\rangle\langle 1_a^+\|t_+\sigma\|0_i^+\rangle\\&\times\langle 0_f^+\|t_+\sigma\|1_{a'}^+\rangle\langle 1_{a'}^+\|t_+\sigma\|0_i^+\rangle\\&\times\tfrac{1}{3}(K_aK_{a'}+L_aL_{a'}+\tfrac{1}{2}K_aL_{a'}+\tfrac{1}{2}L_aK_{a'}),\end{aligned} \tag{2.2}$$

where $|0_i^+\rangle$ and $\langle 0_f^+|$ are the wave functions of the initial and final nuclear states and one has to sum up over all intermediate states $|1_a^+\rangle$. The factors K_a and L_a are the typical energy denominators of the second-order perturbation and for double positron emission given by

$$K_a=\frac{1}{E_a-E_i+e_1+\nu_1}+\frac{1}{E_a-E_i+e_2+\nu_2} \tag{2.3a}$$

$$L_a=\frac{1}{E_a-E_i+e_1+\nu_2}+\frac{1}{E_a-E_i+e_2+\nu_1}. \tag{2.3b}$$

For a captured electron, however, we have to replace one (continuum) positron energy by the energy of the captured electron, $e_x\to -e_C$. This leads to a smaller energy denominator for the mixed mode than for the other two modes and consequently to an enhancement of the matrix element for β^+/EC decay, as can be best seen with the help of the following consideration.

Assume that the lepton energies in the denominators can be replaced by $e+v\simeq\frac{1}{2}Q+m_e$, irrespectively of the indices. Then one can separate the phase space integration and the calculation of the nuclear matrix elements, leading to the well-known expression

$$[T_{1/2}^{2\nu}]^{-1}=F^{2\nu}(M_{GT})^2, \tag{2.4}$$

where

$$M_{GT}=\sum_a \frac{\langle 0_f^+ \| t_+\sigma \| 1_a^+\rangle\langle 1_a^+ \| t_+\sigma \| 0_i^+\rangle}{E_a-E_i+\frac{1}{2}Q+m_e} \tag{2.5}$$

and $F^{2\nu}$ as the leptonic phase space integral. Note that by this assumption the four possible combinations of denominators have become equal. For $(E_a-E_i)\gg e_x+v_x$ we expect that the half life calculated by (2.4) approaches that of the exact (2.1). On the other hand, especially if there is a large cancellation among the different terms in (2.5) deviations between the results of (2.1) and (2.5) have to be expected. Moreover, for the mixed mode this approximation will not be well satisfied since the energy denominators are asymmetric with respect to the *number of leptons in continuum states.* In order to give an equivalent expression also for the mixed mode it might be possible to assume that the available energy is shared equally between the three emitted leptons. (Then the denominator containing the captured electron will be smaller ($\frac{1}{3}Q$) than the one with the positron in the continuum state ($\frac{2}{3}Q$).) Numerically, however, we found that though this approximation better reproduces the result of the exact calculation (2.1) there still remain some differences. In the calculation of the half lives we therefore carried out the integration over leptonic energies in (2.1) for each calculated intermediate state exactly.

2.2. *Matrix elements calculated in* QRPA

Nuclear matrix elements and the resulting half lives have been calculated within *pn*-QRPA (proton-neutron quasiparticle Random phase approximation). *pn*-QRPA has extensively been applied to the calculation of $2\nu\beta^-\beta^-$ decay half lives throughout the past few years [11, 12, 17–25]. We will therefore not repeat the details of the model here but briefly summarize some main results.

The RPA of charge-changing transitions has been developed by Halbleib and Sorensen [26]. Early calculations, however, mainly based on the shell model (for a review of theoretical $\beta\beta$ decay calculations predating 1984 see [9]), overestimated the 2ν decay rates by quite large factors. Vogel and Zirnbauer [18] then could show that the inclusion of the particle-particle force, first considered by Cha [27], leads to a strong reduction of the $2\nu\beta\beta$ decay matrix element in better agreement with experimental data. In this work [18] a zero-range interaction was taken, however, Civitarese et al. [9], using the Bonn potential [28], proofed that the suppression mechanism persisted also for more realistic nucleon-nucleon interactions. Muto et al. [11] then, by a careful fit of the strength of the particle-particle interaction to single β^+ decay data, were able to calculate $2\nu\beta^-\beta^-$ decay half lives for the experimentally most interesting isotopes consistent with existing measurements.

The strong dependence of calculated $2\nu\beta\beta$ decay matrix elements on the particle-particle interaction initiated several subsequent studies employing various refinements. For example, in [23] the effects of a particle-number projection on the QRPA calculation has been studied, whereas Staudt et al. [25] calculated $2\nu\beta^-\beta^-$ decay matrix elements including effects of core polarization and the so-called folded diagrams in the effective interactions. This latter work [25] showed that the use of different effective interactions does not drastically affect calculated 2ν matrix elements. However, the principle features of the earlier 2ν calculations remained basically unchanged in these studies [23, 25].

In our present work we follow essentially the description of Muto et al. [11, 17]. Single particle energies are obtained from a Coulomb-corrected Woods-Saxon potential, with the parameters taken from Bohr and Mottelson [29]. Typically we account for two major oscillator shells in the numerical calculation. A realistic interaction, the Paris potential [30, 31], is used consistently in both the BCS and the RPA calculations. We adjust the strengths of the pairing interactions g^n_{pair} and g^p_{pair} such that the experimental pairing gaps [32] are reproduced. For the strength of the particle-hole interaction g_{ph} we take the empirical formula $g_{\text{ph}}=1+0.002\,A$ [11], which leads to a better reproduction of the excitation energy of the Gamow-Teller giant resonance. The fit of g_{ph}, however, influences the final 2ν matrix element only very weakly [11, 17]. For the choice of the more important particle-particle strength parameter g_{pp} see the discussion below.

Since the earlier $\beta^-\beta^-$ decay calculations [11, 12, 17–25] have shown that the particle-particle interaction plays a crucial role for the calculation of $2\nu\beta\beta$ decay half lives, the influence of g_{pp} on the $\beta^+\beta^+$ decay matrix elements has been investigated in detail. A typical result is displayed in Fig. 1 for the example of ^{130}Ba. For the plot g_{pp} has been varied between 0 and 1. Note however, that such a free variation has no physical meaning; it should be purely understood as a study of the model parameter dependence of the final result. For the sepa-

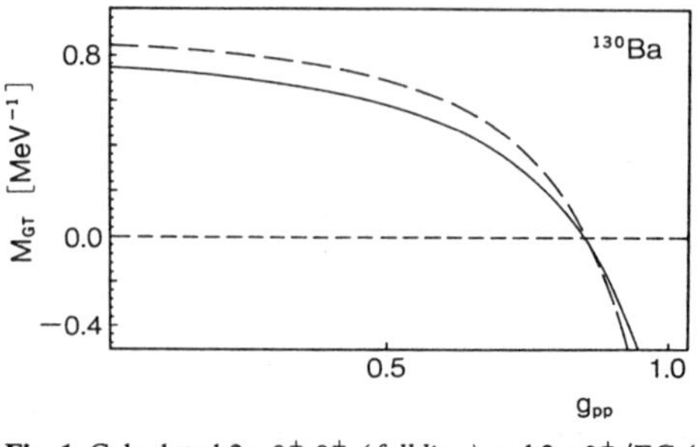

Fig. 1. Calculated $2\nu\beta^+\beta^+$ *(full line)* and $2\nu\beta^+$/EC *(dashed line)* decay matrix elements for the representative example of ^{130}Ba as a function of the particle-particle strength parameter g_{pp}. The results obtained are similar to those of previous $2\nu\beta^-\beta^-$ decay calculations

ration of the nuclear matrix elements and the leptonic phase space integration we assumed for the $\beta^+\beta^+$ matrix elements the common $(\frac{1}{2}Q_{\beta\beta}+m_e)$-approximation for the average leptonic energies (full line), whereas for the mixed mode the available kinetic energy was distributed equally among the 3 emitted leptons (dashed line).

In agreement with the expectation, matrix elements are decreasing functions of g_{pp}. At a certain value of g_{pp} the matrix elements vanishes completely. In the most probable range of g_{pp} calculated matrix elements are strongly suppressed, in agreement with the results of earlier $2\nu\beta^-\beta^-$ decay calculations.

Moreover, Fig. 1 shows the relative enhancement of the mixed mode matrix element compared to the $\beta^+\beta^+$ decay matrix element. From the discussion in Sect. 2.1 it is clear that both matrix elements would become equal in the limit $(E_a-E_i)\to\infty$. On the other hand, the lower the average excitation energy in the intermediate nucleus is the more enhanced will be the mixed mode matrix element. The enhancement is therefore different for different isotopes and dependent on g_{pp}, typical values are in the range of (10–30)%.

While for the isotopes ^{58}Ni, ^{78}Kr, ^{124}Xe, ^{130}Ba and ^{136}Ce we consider the *pn*-QRPA calculation to be reliable, the nuclei ^{96}Ru and ^{106}Cd revealed also some limitations of the present approach. The case of ^{106}Cd shall be discussed in some detail now (quite similar arguments can be applied to ^{96}Ru).

Recall that within QRPA a β-transition is described as the creation of a QRPA phonon out off the ground state. The double beta decay matrix element can be decomposed into β^- and β^+ transitions from the ground states of the mother and daughter isotopes to the excited states in the intermediate nucleus. In QRPA the relevant Gamow-Teller transition matrix elements are then given by

β^--transitions:

$$\langle 1_a^+|t_-\sigma|0^+\rangle = \sum_{p,n}\langle p\|\sigma\|n\rangle(X_a^{pn}u_pv_n - Y_a^{pn}v_pu_n), \qquad (2.6a)$$

β^+-transitions:

$$\langle 1_a^+|t_+\sigma|0^+\rangle = \sum_{p,n}\langle p\|\sigma\|n\rangle(X_a^{pn}v_pu_n - Y_a^{pn}u_pv_n), \qquad (2.6b)$$

where $\langle p\|\sigma\|n\rangle$ are the reduced matrix elements of σ, X_a^{pn} and Y_a^{pn} are the forward and backward going amplitudes of the RPA phonons, whereas u and v are the occupation amplitudes determined by the BCS calculation.

The RPA assumptions then ensures us that if there are many individual terms contributing with a comparable magnitude to the sums in (2.6) the calculation should be reliable and moreover relatively stable against small variations of model parameters. The importance of the ground state correlations (i.e. the Y-terms) for the $\beta\beta$ decay matrix element can also be understood easily. Though in RPA we assume $X\gg Y$, there will be some contributions from terms where $u_pv_n\gg v_pu_n$, so that products Xv_pu_n and Yu_pv_n can be of comparable magnitude. The enhancement of the ground state correlations by the particle-particle force can therefore lead to a strong suppression of the $2\nu\beta\beta$ decay matrix element.

On the other hand, the application of RPA might sensitively depend on the choice of model parameters if the sum in (2.6) is dominated by only a few (or even only one) terms. An example for such a case is the isotope ^{106}Cd. For ^{106}Cd the $2\nu\beta\beta$ matrix element is dominated by a few strong single-particle transitions in $g_{9/2}-g_{7/2}$ orbitals. We have therefore calculated 2ν matrix elements for ^{106}Cd for two sets of input parameters for the Woods-Saxon potential. The first calculation has been carried out with the standard choice of parameters described above, while for the second one only the strength of the spin-orbit force of the Woods-Saxon potential has been changed by ~9% from its standard value [29]. While such a change generally only leads to a small shift in the calculated single-particle levels, the main effect of this variation for ^{106}Cd is to decrease the energy difference between the $\nu d_{5/2}$ and the $\nu g_{7/2}$ by about ~0.3 MeV. The results of these calculations are shown in Fig. 2. Taking the standard Woods-Saxon parameters the $2\nu\beta\beta$ decay matrix element of ^{106}Cd stays nearly constant over a wide range of g_{pp}, but then starts to *increase* just before the QRPA calculation collapses. On the other hand, by the small change of the strength of the spin-orbit force one gets a sharp decrease of the 2ν matrix element, leading to a strong suppression of M_{GT} around $g_{pp}\simeq 0.85$. The explanation for this unusually strong model parameter dependence is found along the lines discussed above. Unfortunately, no experimental information on the energies of single-particle levels in this mass region are available, which at least in the case of ^{106}Cd would be very helpful. However, as can be seen from Fig. 2, by lowering the strength of the spin-orbit force the calculation yields a suppressed $2\nu\beta\beta$ decay matrix element in better agreement with the expectation. The numerical instability of the calculation makes the prediction of a $2\nu\beta\beta$ decay half life for ^{106}Cd very difficult; the half life of ^{106}Cd (and ^{96}Ru) given in Table 1 should therefore be understood only as a lower limit.

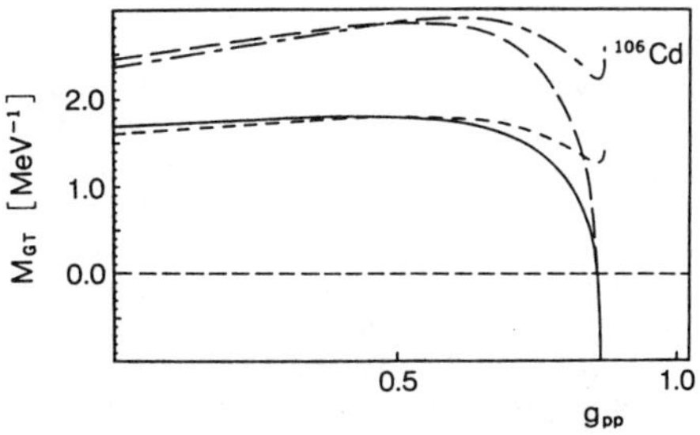

Fig. 2. Calculated $2\nu\beta^+\beta^+$ and $2\nu\beta^+$/EC decay matrix elements for ^{106}Cd as a function of g_{pp} for two different sets of input parameters. For a discussion see text

Table 1a, b. Predicted $2\nu\beta\beta$ decay half lives for the three possible decay modes. Half lives are given in years. Note that the half lives for ^{96}Ru and ^{106}Cd should be taken as lower limits, as is discussed in the text. For completness also existing experimental limits at 68% c.l. are given. With the exception of ^{124}Xe the experimental numbers refer to the sum of all decay modes of double beta decay

a $2\nu\beta^+\beta^+$ decay half lives

Isotope	$T_{1/2}^{\text{cal}}$	$T_{1/2}^{\text{exp}} >$	Ref. (exp)
^{78}Kr	2.3×10^{26}		
^{96}Ru	5.8×10^{26}	3.1×10^{16}	[33]
^{106}Cd	4.2×10^{26}	2.6×10^{17}	[34]
^{124}Xe	1.4×10^{27}	2.0×10^{14}	[35]
^{130}Ba	1.7×10^{29}		
^{136}Ce	5.2×10^{31}		

b $2\nu\beta^+$/EC and 2νEC/EC decay half lives

Isotope	$T_{1/2}^{\text{cal}}(\beta^+/\text{EC})$	$T_{1/2}^{\text{cal}}(\text{EC/EC})$	$T_{1/2}^{\text{exp}}(\beta^+/\text{EC}) >$	Ref. (exp)
^{58}Ni	5.5×10^{25}	3.9×10^{24}	6.2×10^{19}	[34]
^{78}Kr	5.3×10^{22}	3.7×10^{22}		
^{96}Ru	1.2×10^{22}	2.1×10^{21}	6.7×10^{16}	[33]
^{106}Cd	4.1×10^{21}	8.7×10^{20}	5.7×10^{17}	[34]
^{124}Xe	3.0×10^{22}	2.9×10^{21}	4.8×10^{16}	[35]
^{130}Ba	1.0×10^{23}	4.2×10^{21}		
^{136}Ce	9.2×10^{23}	1.7×10^{22}		

The dependence of the calculated matrix elements on g_{pp} makes predictions of $2\nu\beta\beta$ decay half lives somewhat uncertain. The similarity between the $\beta^+\beta^+$ decay matrix elements of our present work and those of [11], on the other hand, gives us some confidence that the following procedure might be applicable.[1] We take the most probable values of g_{pp} from the fit to single β^+ decay data [17] and calculate average half lives in the spirit of [11] by summing over calculated decay rates within the (1σ)-interval of the most probable value of g_{pp}.

Calculated half lives for the 3 different neutrino-accompanied decay modes are given in Table 1. Though, as discussed above, we have to expect a considerable uncertainty for the $2\nu\beta\beta$ decay calculation there are some interesting facts which can be learned from the quoted values. In the case of double positron emission expected half lives are very large, confirming the earlier calculations [9, 10]. Note, that our values do not coincide with those given in [10], which can be traced back mainly to the different phase space treatment.

More interesting, however, is that the other two decay modes have half lives which are shorter by many orders of magnitude, but even for the best candidates we expect the mixed mode half life to be still in the range of 10^{22} years. These numbers should be compared to existing experimental limits [33–35], for completeness also given in Table 1. On the other hand, advanced $\beta\beta$ decay experiments might improve these limits by many orders of magnitude. Just for example, Ohsumi and Ejiri [36] estimated the sensitivity of the ELEGANTS V detector [37] with which a limit of $\sim3.5\times10^{20}$ years for the mixed mode of ^{106}Cd might be reached. Nevertheless, it seems that much experimental work remains to be done before the $2\nu\beta^+$/EC decay will be detected.

As given in Table 1, half lives for double electron capture are even shorter than those of the mixed mode; with the difference increasing with increasing proton number Z. The latter effect can be explained simply by the higher probability to find an electron inside the nucleus for heavier isotopes. Experimentally, however, the detection of 2νEC/EC decay has also some disadvantages. While there are two 511 keV γ-rays as a signal in $2\nu\beta^+$/EC decay, there are only two X-rays from the deexcitation of the orbital electrons in 2νEC/EC decay as observable. On the other hand, [37] quotes a sensitivity of $\sim7\times10^{20}$ years for the double electron capture decay of ^{106}Cd, which is not too far away from the expectation. There might therefore be some hope that the detection of 2νEC/EC decay becomes possible in the not too far future.

Finally, we would like to compare our results with those of Suhonen [15]. In this work [15] 2νEC/EC decay half lives of the following 4 isotopes have been calculated. ^{58}Ni($T_{1/2}\simeq3.9\times10^{23}$y), ^{96}Ru($T_{1/2}\simeq2.8\times10^{21}$y), ^{106}Cd ($T_{1/2}\simeq2.6\times10^{20}$ y) and ^{136}Ce ($T_{1/2}\simeq6.4\times10^{19}$ y). Suhonen [15] also made use of the *pn*-QRPA approach (in the particle-number projected version [23]) but took the Bonn potential [28] for the effective interaction. Compared to our calculation, however, these half lives appear to be surprisingly short. For the isotopes ^{96}Ru and ^{106}Cd this might be explained with the unusual behaviour of the $2\nu\beta\beta$ decay matrix elements which has been discussed for the case of ^{106}Cd above. A similar feature can be found in the calculation of Suhonen (see Fig. 1 of [15]), the matrix elements for these isotopes are – using the standard Woods-Saxon parameters – nearly unsupressed by the inclusion of the particle-particle force, in contrast to the results of previous $2\nu\beta^-\beta^-$ decay calculations [11, 17–25].

To summarize the results on $2\nu\beta\beta$ decay we state that in agreement with the expectation half lives for $2\nu\beta^+$/EC and 2νEC/EC decay are shorter than those for double positron emission. Half lives for the experimentally most promising isotopes are given in Table 1, which we hope may be useful as a guideline for future $\beta^+\beta^+$ decay experiments.

3. Neutrinoless decay modes

3.1. Decay rate formulas

In the case of $0\nu\beta^+\beta^+$ and $0\nu\beta^+$/EC decay the theoretical description is quite similar to that of the more common $0\nu\beta^-\beta^-$ decay [38]. As in the neutrino-accompanied decays, in principle also (neutrinoless) double electron capture could occur. However, as pointed out by Vergados [39] and discussed in some detail by Doi and Kotani [38], the 0νEC/EC decay mode appears to be hopelessly slow; the relative order of magnitude of its decay rate compared to the neutrinoless mixed mode is

[1] The application of this procedure to the calculation of the half life of ^{76}Ge led to the prediction of $T_{1/2}\simeq3\times10^{21}$ y [11] in good agreement with the most recent measurement $T_{1/2}\simeq1.4\times10^{21}$ y [3]

expected to be at least $< 10^{-4}$ [38]. Therefore, we will consider only the $0\nu\beta^+\beta^+$ and the $0\nu\beta^+/EC$ decay here.

We adopt the following effective weak hamiltonian,

$$H_w = \frac{G}{\sqrt{2}} (j_{L_\mu} J_L^{\mu\dagger} + \kappa j_{L_\mu} J_R^{\mu\dagger} + \eta j_{R_\mu} J_L^{\mu\dagger} + \lambda j_{R_\mu} J_R^{\mu\dagger}) + \text{h.c.}, \tag{3.1}$$

where small (capital) letters denote leptonic (hadronic) currents and the subscripts L and R indicate left-handed and right-handed couplings, respectively. Note that in this convention the right-handed parameters κ, η and λ are chosen such that the ordinary left-left handed interaction has a relative strength of unity.

The derivation of the decay rate(s) follows closely that outlined for the $\beta^-\beta^-$ decay [12, 16, 17, 38]. Neglecting terms proportional to κ, which should be a good approximation since they always appear as $1 \pm \kappa$ and $\kappa \ll 1$ is expected, and considering only light neutrino eigenstates, the inverse half life for $0\nu\beta\beta$ decay can be expressed in a factorized form as:

$$[T_{1/2}^{0\nu\beta\beta,\alpha}(0_i^+ \to 0_f^+)]^{-1} = C_{mm}^{\alpha} \left(\frac{\langle m_\nu \rangle}{m_e}\right)^2 + C_{\eta\eta}^{\alpha} \langle \eta \rangle^2 + C_{\lambda\lambda}^{\alpha} \langle \lambda \rangle^2 + C_{m\eta}^{\alpha} \frac{\langle m_\nu \rangle}{m_e} \langle \eta \rangle + C_{m\lambda}^{\alpha} \frac{\langle m_\nu \rangle}{m_e} \langle \lambda \rangle + C_{\eta\lambda}^{\alpha} \langle \eta \rangle \langle \lambda \rangle , \tag{3.2}$$

where α is a discriminator ($\alpha = \beta^-\beta^-, \beta^+\beta^+$ and β^+/EC), indicating that the coefficients C_{xy}^{α} are different for the different decay modes.

In writing down (3.2) we made use of the following definitions for the *effective* neutrino mass and right-handed parameters

$$\langle m_\nu \rangle = \sum_j{}' U_{ej}^2 m_j , \tag{3.3a}$$

$$\langle \eta \rangle = \eta \sum_j{}' U_{ej} V_{ej} , \tag{3.3b}$$

$$\langle \lambda \rangle = \lambda \sum_j{}' U_{ej} V_{ej} , \tag{3.3c}$$

where the prime indicates that the sums are restricted to light neutrino mass eigenstates ($m_j < 10$ MeV).

The coefficients C_{xy} contain products of nuclear matrix elements and leptonic phase spaces. Within the non-relativistic impulse approximation there appear 9 different matrix elements M_β ($\beta = GT, F, GT\omega, F\omega, GTq, Fq, T, P$ and R) and also 9 phase space integrals,

$$C_{mm}^{\alpha} = (M_{GT} - M_F)^2 G_1^{\alpha} , \tag{3.4a}$$

$$C_{m\lambda}^{\alpha} = (M_{GT} - M_F)[-S_e M_{2-} G_3^{\alpha} + M_{1+} G_4^{\alpha}] , \tag{3.4b}$$

$$C_{m\eta}^{\alpha} = (M_{GT} - M_F)[S_e M_{2+} G_3^{\alpha} - M_{1-} G_4^{\alpha} - S_\beta (M_P G_5^{\alpha} - S_e M_R G_6^{\alpha})] , \tag{3.4c}$$

$$C_{\lambda\lambda}^{\alpha} = M_{2-}^2 G_2^{\alpha} - \tfrac{1}{9} f_e [2 S_e M_{1+} M_{2-} G_3^{\alpha} - M_{1+}^2 G_4^{\alpha}] , \tag{3.4d}$$

$$C_{\eta\eta}^{\alpha} = M_{2+}^2 G_2^{\alpha} - \tfrac{1}{9} f_e [2 S_e M_{1-} M_{2+} G_3^{\alpha} - M_{1-}^2 G_4^{\alpha}] - S_e M_P M_R G_7^{\alpha} + M_P^2 G_8^{\alpha} + M_R^2 G_9^{\alpha} , \tag{3.4e}$$

$$C_{\eta\lambda}^{\alpha} = -2\{M_{2+} M_{2-} G_2^{\alpha} - \tfrac{1}{9} f_e [S_e (M_{1+} M_{2+} + M_{1-} M_{2-}) G_3^{\alpha} - M_{1+} M_{1-} G_4^{\alpha}]\} . \tag{3.4f}$$

For shorthand in (3.4) we made use of the following relations,

$$M_{1\pm} = M_{GTq} - 6 M_T \pm 3 M_{Fq} , \tag{3.5a}$$

$$M_{2\pm} = M_{GT\omega} \pm M_{F\omega} - \tfrac{1}{9} M_{1\mp} . \tag{3.5b}$$

The 9 nuclear matrix elements in (3.4) are (within the closure approximation) the same for the different decay modes but appear in different combinations. They involve 5 different kinds of transition operators and 4 types of "neutrino potentials". A complete definition of them is given in [12]. The factor S_β takes $S_\beta = +1$ for the $\beta^-\beta^-$ and $S_\beta = -1$ for the $\beta^+\beta^+$ and the β^+/EC mode, because of the different character of the Coulomb potential for emitted electrons and positrons. The factor S_e is

$$S_e = \begin{cases} +1, & \text{for } \beta^-\beta^- \text{ and } \beta^+\beta^+ \\ -1, & \text{for } \beta^+/\text{EC}, \end{cases} \tag{3.6}$$

which appears due to the different number of fermions in the final states [38]. Finally,

$$f_e = \begin{cases} +1, & \text{for } \beta^-\beta^- \text{ and } \beta^+\beta^+ \\ 1 - \dfrac{3\alpha}{2 m_e R}, & \text{for } \beta^+/\text{EC}, \end{cases} \tag{3.7}$$

which comes from the fact that a captured electron and an emitted positron feel different nuclear charges [38]. α in (3.7) is the fine structure constant and R is the nuclear radius.

3.2. *Numerical calculations*

We have calculated the various nuclear matrix elements appearing in the decay rate formula (3.2) within *pn*-QRPA [12]. For the evaluation we performed the closure approximation [9]. Though closure is known to fail badly in the case of $2\nu\beta\beta$ decay, the comparatively large momentum transfer by the virtual neutrino in $0\nu\beta\beta$ decays ensures that in the neutrinoless modes closure should be a good approximation. We can further strengthen this statement with the results of two recent numerical studies [40, 41], which showed that the use of closure introduces only minor corrections to the matrix elements for neutrinoless decays.

For the determination of the coefficients C_{xy} matrix elements have been calculated with the most probable values of the particle-particle interaction strength g_{pp} [11]. Also the other model parameters have been chosen in the same way as for the $2\nu\beta\beta$ decay calculation discussed in Sect. 2. This procedure is mainly motivated by the attempt to calculate $0\nu\beta\beta$ decay matrix elements in an as similar model as possible to that for the $2\nu\beta\beta$ decay calculation.

It should be noted, however, that there are some differences between the 2ν and the 0ν calculations due to the virtual nature of the neutrino in the latter. One important difference is that while in 2ν decay, due to the selection rules for Gamow-Teller transitions, the intermediate states are $|1^+\rangle$-states exclusively, in 0ν decay also higher multipolarities contribute to the total matrix element. It is known [12] that the particle-particle interaction is attractive mainly in the 1^+ channel, while the higher multipolarities are only weakly affected. The suppression of the $0\nu\beta\beta$ decay matrix elements by the inclusion of the particle-particle interaction is therefore less strong than that of the $2\nu\beta\beta$ decay matrix elements. $0\nu\beta\beta$ decay matrix elements are consequently less affected by uncertainties in the best values of g_{pp} and might more reliably be calculable than $2\nu\beta\beta$ decay half lives.

$0\nu\beta^+\beta^+$ decay matrix elements show a very similar model parameter dependence than that reported in earlier $0\nu\beta^-\beta^-$ decay calculations [12]. We will therefore not repeat the discussion here. However, we would like to mention that a reasonable change of g_{pp} changes the calculated matrix elements usually by less than $\sim 50\%$.

Consistent with the expectation and the discussion presented above even for the special case of ^{106}Cd, $0\nu\beta\beta$ decay matrix elements are only weakly affected by a variation of g_{pp} and other model parameters. A change of the strength of the spin-orbit force, as discussed in Sect. 2.2, would change the results of the $0\nu\beta\beta$ decay calculation by at most a few percent. Also a variation of g_{pp} within 2σ of the most probable value changes the different matrix elements by only (10–40)% from the mean.

The various matrix elements resulting from the *pn*-QRPA calculation are given in Table 2. With the help of the phase space integrals of [33] we are then able to determine the coefficients of the decay rate formula. They are summarized in Table 3. A detailed discussion of these results will be given in the next section.

3.3. Discussion of the results for the neutrinoless modes

The coefficients C_{xy} summarized in Table 3 are the main result of the present work for the neutrinoless modes; their implications for the effective neutrino mass and right-handed parameters shall be discussed briefly.

In agreement with simple phase space considerations the coefficients for the neutrinoless mixed mode are larger than those for the $0\nu\beta^+\beta^+$ decays. The relative enhancement of the different coefficients for a given isotope is, however, not a common factor as could naively be

Table 2. Calculated $0\nu\beta\beta$ decay matrix elements for the 7 most interesting isotopes

Isotope	^{58}Ni	^{78}Kr	^{96}Ru	^{106}Cd	^{124}Xe	^{130}Ba	^{136}Ce
M_{GT}	1.36	3.28	2.62	3.34	3.92	4.02	2.44
M_F	−0.30	−1.42	−0.98	−1.22	−1.35	−1.50	−1.02
$M_{GT\omega}$	1.20	3.19	2.47	3.14	3.72	3.86	2.44
$M_{F\omega}$	−0.27	−1.23	−0.88	−1.09	−1.19	−1.32	−0.91
M_{GTq}	1.19	2.13	1.81	2.35	2.82	2.70	1.40
M_{Fq}	−0.26	−1.31	−0.84	−1.05	−1.21	−1.34	−0.89
M_T	−0.026	−0.67	−0.31	−0.38	−0.67	−0.87	−0.66
M_P	0.99	−0.89	0.82	1.43	−0.61	−0.63	−0.51
M_R	0.92	3.83	3.24	4.10	4.73	5.44	3.90

Table 3a. Coeffients in the decay rate formula for the double positron emission mode

Isotope	^{78}Kr	^{96}Ru	^{106}Cd	^{124}Xe	^{130}Ba	^{136}Ce
C_{mm}	1.6×10^{-16}	3.0×10^{-17}	5.4×10^{-17}	8.7×10^{-17}	1.6×10^{-17}	1.1×10^{-18}
$C_{m\eta}$	-4.3×10^{-14}	-9.2×10^{-15}	-1.6×10^{-14}	-2.1×10^{-14}	-4.8×10^{-15}	-4.0×10^{-16}
$C_{m\lambda}$	-9.3×10^{-18}	-1.3×10^{-18}	-2.6×10^{-18}	-5.8×10^{-19}	7.3×10^{-19}	7.2×10^{-20}
$C_{\eta\eta}$	6.0×10^{-12}	1.4×10^{-12}	2.4×10^{-12}	2.8×10^{-12}	6.5×10^{-13}	5.9×10^{-14}
$C_{\lambda\lambda}$	1.9×10^{-17}	2.0×10^{-18}	4.4×10^{-18}	8.5×10^{-18}	5.5×10^{-19}	1.8×10^{-20}
$C_{\eta\lambda}$	-3.3×10^{-18}	-6.5×10^{-19}	-1.6×10^{-18}	-4.9×10^{-18}	-6.0×10^{-19}	-4.2×10^{-20}

Table 3b. Coefficients in the decay rate formula for the electron capture positron emission mode

Isotope	^{58}Ni	^{78}Kr	^{96}Ru	^{106}Cd	^{124}Xe	^{130}Ba	^{136}Ce
C_{mm}	9.1×10^{-18}	4.0×10^{-16}	3.5×10^{-16}	7.7×10^{-16}	1.6×10^{-15}	1.5×10^{-15}	5.6×10^{-16}
$C_{m\eta}$	1.2×10^{-15}	7.4×10^{-14}	7.1×10^{-14}	1.5×10^{-13}	2.9×10^{-13}	2.8×10^{-13}	1.1×10^{-13}
$C_{m\lambda}$	3.9×10^{-17}	2.4×10^{-15}	2.0×10^{-15}	4.5×10^{-15}	9.2×10^{-15}	8.0×10^{-15}	2.9×10^{-15}
$C_{\eta\eta}$	1.2×10^{-13}	8.9×10^{-12}	9.4×10^{-12}	2.0×10^{-11}	3.3×10^{-11}	3.5×10^{-11}	1.6×10^{-11}
$C_{\lambda\lambda}$	6.2×10^{-17}	5.9×10^{-15}	4.8×10^{-15}	1.1×10^{-14}	2.2×10^{-14}	1.7×10^{-14}	5.8×10^{-15}
$C_{\eta\lambda}$	-6.2×10^{-17}	-2.5×10^{-15}	-2.8×10^{-15}	-6.6×10^{-15}	-1.4×10^{-14}	-9.8×10^{-15}	-2.9×10^{-15}

expected. Just to give an illustrate example, for ^{124}Xe one finds for the ratio $C_{mm}^{\beta^+/\mathrm{EC}}/C_{mm}^{\beta^+\beta^+} \simeq 18$, whereas $C_{\lambda\lambda}^{\beta^+/(\mathrm{EC}}/C_{\lambda\lambda}^{\beta^+\beta^+} \simeq 2600$! Generally, all coefficients involving $\langle\lambda\rangle$ have much larger enhancement factors than those for $\langle\eta\rangle$ or $\langle m_\nu\rangle$.

The large enhancement factors of the coefficients $C_{\lambda\lambda}$, $C_{m\lambda}$ and $C_{\eta\lambda}$ in the mixed mode can be explained as follows. The (possible) contributions of a right-handed interaction are proportional to the neutrino four-momentum $q=(\omega, \mathbf{q})$. It is convenient to divide it into two parts, the ω-terms and the $\mathbf{q}$-terms (with their associated matrix elements given in Table 2). Usually one would expect the ω-terms to be the dominant ones, since $\mathbf{q}$ acts as a parity-odd operator between two parity-even nuclear states (remember that in $\beta\beta$ decay mother and daughter isotopes always have $|0^+\rangle$ ground states), which requires an additional operator vanishing in the usual approximation for allowed transitions[2]. However, it is not so for double lepton emission since for the ω-terms there appears a large cancellation among different combinations of energy denominators, as is explained in detail in [16]. The associated kinematical factors G_2 and G_3 therefore carry factors of $(\varepsilon_{12}/m_e)^2$ and (ε_{12}/m_e) respectively, where $\varepsilon_{12}=\varepsilon_1-\varepsilon_2$ is the energy difference between the two emitted leptons. These factors lead to a strong damping of G_2 and G_3 in double lepton emission, since ε_{12} is a very small quantity *on average*. On the other hand, in the mixed mode $\varepsilon_{12}=\varepsilon+\varepsilon_C$, where ε_C is the energy of the captured electron, is nearly equal to $Q_{\beta^+/\mathrm{EC}}$ which cannot be called "small". Therefore, while G_1 and G_4-G_9 are larger in the mixed mode by moderate factors, mainly due to the larger available energy, both G_2 and G_3 are enhanced strongly. In combination with the nuclear matrix elements this enhancement acts then constructively for $C_{\lambda\lambda}$, $C_{m\lambda}$ and $C_{\eta\lambda}$ [3].

Combining the coefficients C_{xy} with available experimental limits on $0\nu\beta^+/\mathrm{EC}$ and $0\nu\beta^+\beta^+$ decay half lives [33-35] the upper bounds on $\langle m_\nu\rangle$, $\langle\lambda\rangle$ and $\langle\eta\rangle$ summarized in Table 4 are obtained. Presently existing experimental limits on $0\nu\beta^+\beta^+$ and $0\nu\beta^+/\mathrm{EC}$ decays are much weaker than those for the best $\beta^-\beta^-$ decay candidates [2-7]. Consequently also the limits on $\langle m_\nu\rangle$, $\langle\lambda\rangle$ and $\langle\eta\rangle$ are far less stringent. Just for comparison we mention the currently best limits deduced from the non-observation of ^{76}Ge $0\nu\beta\beta$ decay [3]: $\langle m_\nu\rangle < 1.4(1.2)$ eV, $\langle\lambda\rangle < 2.2(2.2)\times 10^{-6}$ and $\langle\eta\rangle < 1.4(1.2)\times 10^{-8}$ for maximum ("on axis", which means that the other two parameters are assumed to be zero), with coefficients taken from [12]. Note, however, that the use of the coefficients calculated by the Tübingen group [42] would only lead to minor changes for the limits from ^{76}Ge.

While from the discussion above it is obvious that the $0\nu\beta^+/\mathrm{EC}$ decay is relatively more sensitive to the $\langle\lambda\rangle$-mechanism, Table 4 on the other hand shows that from the *non-observation* of $0\nu\beta^+/\mathrm{EC}$ decay alone more stringent *limits* on the $\beta\beta$ decay parameters are difficult to get; while values "on axis" from $0\nu\beta^+/\mathrm{EC}$ decays are smaller by large factors compared to those of $0\nu\beta^+\beta^+$ decays it is not so for an arbitrary variation of the other two parameters ("maximum"). The reason is that the ellipsoids of the mixed mode are highly deformed, since not only the coefficients $C_{\lambda\lambda}$ but also the $C_{m\lambda}$'s and $C_{\eta\lambda}$'s are enhanced – by comparable factors. (Only if $C_{\lambda\lambda}$ is scaled by ζ^2 and $C_{m\lambda}$ and $C_{\eta\lambda}$ by ζ also, the maximum allowed value of $\langle\lambda\rangle$ would be more restrictive by (exactly) a factor of ζ.) The situation is therefore comparable to that in $0\nu\beta\beta\,(0^+\to 2^+)$ decays, which are more sensitive to $\langle\lambda\rangle$ and $\langle\eta\rangle$ than to the mass mechanism, such that the existence of right-handed weak currents would be the most likely explanation if $0\nu\beta\beta\,(0^+\to 2^+)$ decay ever would be observed[4].

Finally we would like to add some speculations about the usefulness of $0\nu\beta^+/\mathrm{EC}$ decay data. While it is difficult to get valuable information on the $\beta\beta$ decay parameters from the experimental data at present, the situation would change completely if neutrinoless double beta decay ever would be observed. (For the following discussion we have to assume some definite numbers, however, the qualitative features of the arguments presented would not be changed if $0\nu\beta\beta$ decay would be observed with another value of the half life. Simply scale all quoted numbers by a common factor.) Assume, for example, $\langle m_\nu\rangle = 1.0$ eV, then we would expect that the

Table 4a, b. Limits on $\langle m_\nu\rangle$, $\langle\lambda\rangle$ and $\langle\eta\rangle$ deduced from the non-observation of $0\nu\beta^+\beta^+$ and $0\nu\beta^+/\mathrm{EC}$ decay. Experimental data taken from [33-35]. Upper values given are those obtained letting the other two parameters vary arbitrarily, whereas the lower ones are "on axis" assuming the other two parameters to be equal to zero

a $\beta^+\beta^+$ decays

Isotope	$\langle m_\nu\rangle <$ [eV]	$\langle\lambda\rangle <$	$\langle\eta\rangle <$
^{96}Ru	7.8×10^5	4.1×10^0	7.1×10^{-3}
	5.4×10^5	4.0×10^0	4.9×10^{-3}
^{106}Cd	2.0×10^5	9.5×10^{-1}	1.8×10^{-3}
	1.4×10^5	9.4×10^{-1}	1.3×10^{-3}
^{124}Xe	1.2×10^5	5.3×10^{-1}	1.2×10^{-3}
	8.5×10^4	5.3×10^{-1}	9.2×10^{-4}

b β^+/EC decays

Isotope	$\langle m_\nu\rangle <$ [eV]	$\langle\lambda\rangle <$	$\langle\eta\rangle <$
^{58}Ni	1.2×10^6	7.4×10^{-1}	1.2×10^{-2}
	2.2×10^4	1.6×10^{-2}	3.7×10^{-4}
^{96}Ru	5.6×10^6	2.4×10^0	4.2×10^{-2}
	1.1×10^5	5.6×10^{-2}	1.3×10^{-3}
^{106}Cd	3.7×10^6	1.5×10^0	2.8×10^{-2}
	2.5×10^4	1.3×10^{-2}	3.0×10^{-4}
^{124}Xe	6.6×10^5	2.7×10^{-1}	5.7×10^{-3}
	1.2×10^4	6.2×10^{-3}	1.6×10^{-4}

[2] This is the origin of the *P*-wave and recoil terms which appear in the coefficients $C_{\eta\eta}$ and $C_{m\eta}$, compare (3.4)

[3] The enhancement of G_2 and G_3 is not important for $C_{\eta\eta}$ and $C_{m\eta}$ since these are dominated by the contribution from the nuclear recoil terms, which are multiplied by the large G_9 and G_6 respectively

[4] Even the observation of $0\nu\beta\beta\,(0^+\to 2^+)$ decay would not be a *proof* of the existence of right-handed weak currents, as has sometimes been stated [43], since the contribution from the mass mechanism vanishes only in first order. Contributions from the mass mechanism become possible, though suppressed, for example, allowing the outgoing two leptons to be in $S_{1/2}-D_{3/2}$ or $P_{1/2}-P_{3/2}$ states [16]

$0\nu\beta\beta$ decay half life of ^{76}Ge would be equal to 2.3×10^{24} years. On the other hand, this value could also be obtained for a nearly arbitrarily small value of $\langle m_\nu\rangle$, but $\langle\lambda\rangle=1.8\times10^{-6}$ or $\langle\eta\rangle=1.0\times10^{-8}$. For a complete analysis further information would be required.

There are, in principle, different possibilities to decide whether the decay is dominated by the mass mechanism or by right-handed current interaction. One is to measure the angular distribution of the outgoing leptons, in addition to the half life [16] (the angular distribution for the $\langle\lambda\rangle$-terms differ from the ones for the mass mechanism). However, such an experiment would require quite large a statistics and moreover can not be done within an experiment using semiconductors, such as ^{76}Ge. Also the observation of $0\nu\beta\beta\,(0^+\to2^+)$ decays might be helpful, as mentioned above. However, in that case very long half lives need to be measured. For example, taking the coefficients of [43] for ^{76}Ge $0\nu\beta\beta\,(0^+\to2^+)$ decay the following half lives can be estimated: $\langle\lambda\rangle=1.8\times10^{-6}$ ($\langle\eta\rangle=1.0\times10^{-8}$) leads to $T_{1/2}(0^+\to2^+)\simeq2\times10^{29}$ years ($T_{1/2}(0^+\to2^+)\simeq1.2\times10^{32}$ years).

In principle, a third possibility would be to measure the $0\nu\beta\beta$ decay half lives of different isotopes, since the values of, for example, $\langle m_\nu\rangle$ and $\langle\lambda\rangle$ which lead to the same half life for ^{76}Ge give slightly different half lives for ^{136}Xe (or any other isotope). However, in such an analysis one would have to account for both, the experimental error and the theoretical uncertainties in the calculation of the matrix elements. Accounting for these uncertainties unfortunately limits the usefulness of this idea as long as one considers only $0\nu\beta^-\beta^-$ decays, as is shown in Fig. 3a. For the plot we assumed a $0\nu\beta\beta$ decay half life of $(1.5\pm0.5)\times10^{24}$ y for both ^{76}Ge and ^{136}Xe (larger or smaller values for the half lives again lead only to a scaling factor). The region between the full lines (dashed lines) are allowed by ^{76}Ge (^{136}Xe), and the shaded area is allowed by *both* experiments.

On the other hand, if we insert the values for $\langle m_\nu\rangle$, $\langle\lambda\rangle$ and $\langle\eta\rangle$ quoted above we get estimations of mixed mode and $\beta^+\beta^+$ decay half lives ("on axis") presented in Table 5. For ^{124}Xe, for example, we expect the mixed mode half life to be shorter by more than an order of magnitude, if the ^{76}Ge decay is dominated by $\langle\lambda\rangle$ than if it is dominated by the mass mechanism. While a full 3-dimensional analysis need to be done in principle, even the non-observation of the mixed mode decay could be helpful (still assuming that at least one $0\nu\beta\beta$ decay has been observed), see below. The expected "on axis" half lives are quite large compared to existing limits, however shorter than those expected for the ^{76}Ge$(0^+\to2^+)$ decay by 4 (7) orders of magnitude for the $\langle\lambda\rangle$-terms (the $\langle\eta\rangle$-terms).

A graphical representation of the large differences between the neutrino mass induced neutrinoless mixed mode decay and that caused by the $\langle\lambda\rangle$-contributions is given in Fig. 3b, c. Assuming again $T_{1/2}^{0\nu\beta\beta}=(1.5\pm0.5)\times10^{24}$ y for ^{76}Ge and $T_{1/2}^{0\nu\beta\beta}=(1.5\pm0.5)\times10^{25}$ (10^{26}) y for ^{124}Xe in Fig. 3b (Fig. 3c), one finds that while the shorter value used in Fig. 3b would still be marginally consistent with an arbitrarily small effective neutrino mass large regions of parameter space could be excluded if the

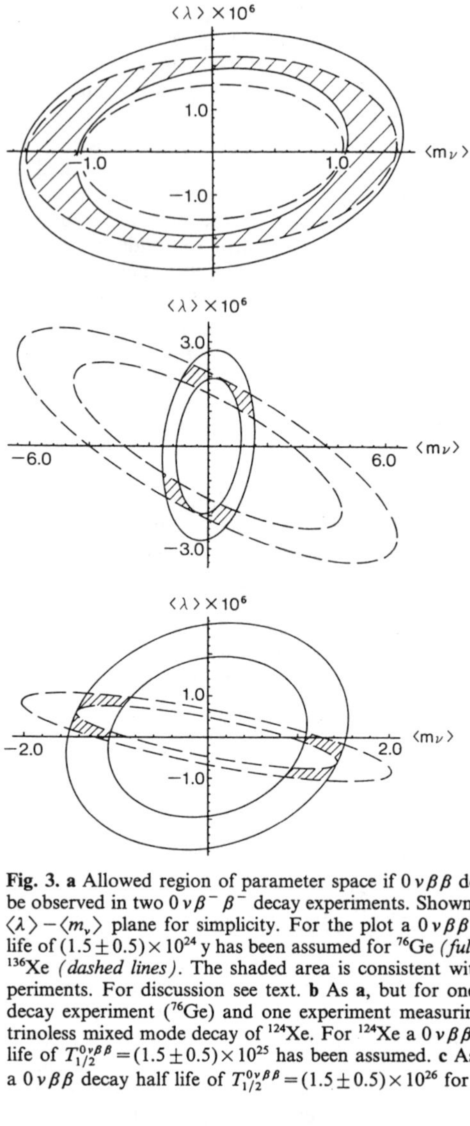

Fig. 3. **a** Allowed region of parameter space if $0\nu\beta\beta$ decay would be observed in two $0\nu\beta^-\beta^-$ decay experiments. Shown is only the $\langle\lambda\rangle-\langle m_\nu\rangle$ plane for simplicity. For the plot a $0\nu\beta\beta$ decay half life of $(1.5\pm0.5)\times10^{24}$ y has been assumed for ^{76}Ge *(full lines)* and ^{136}Xe *(dashed lines)*. The shaded area is consistent with both experiments. For discussion see text. **b** As **a**, but for one $0\nu\beta^-\beta^-$ decay experiment (^{76}Ge) and one experiment measuring the neutrinoless mixed mode decay of ^{124}Xe. For ^{124}Xe a $0\nu\beta\beta$ decay half life of $T_{1/2}^{0\nu\beta\beta}=(1.5\pm0.5)\times10^{25}$ has been assumed. **c** As **b**, but for a $0\nu\beta\beta$ decay half life of $T_{1/2}^{0\nu\beta\beta}=(1.5\pm0.5)\times10^{26}$ for ^{124}Xe

Table 5. Expected "on axis" half lives in years for $0\nu\beta^+\beta^+$ and $0\nu\beta^+$/EC decays for $\langle m_\nu\rangle=1.0$ eV, $\langle\lambda\rangle=1.8\times10^{-6}$ and $\langle\eta\rangle=1.0\times10^{-8}$. The table shows the different sensitivity of these modes to the mass mechanism and the right-handed current parameter $\langle\lambda\rangle$. For discussion see text

Isotope	Mode	$\langle m_\nu\rangle$	$\langle\lambda\rangle$	$\langle\eta\rangle$
^{78}Kr	$0\nu\beta^+\beta^+$	1.6×10^{27}	1.6×10^{28}	1.7×10^{27}
	$0\nu\beta^+$/EC	6.5×10^{26}	5.2×10^{25}	1.1×10^{27}
^{96}Ru	$0\nu\beta^+\beta^+$	8.7×10^{27}	1.5×10^{29}	7.1×10^{27}
	$0\nu\beta^+$/EC	7.5×10^{26}	6.4×10^{25}	1.1×10^{27}
^{106}Cd	$0\nu\beta^+\beta^+$	4.8×10^{27}	7.0×10^{28}	4.2×10^{27}
	$0\nu\beta^+$/EC	3.4×10^{26}	2.8×10^{25}	5.0×10^{26}
^{124}Xe	$0\beta^+\beta^+$	3.0×10^{27}	3.6×10^{28}	3.6×10^{27}
	$0\nu\beta^+$/EC	1.6×10^{26}	1.4×10^{25}	3.0×10^{26}
^{130}Ba	$0\nu\beta^+\beta^+$	1.6×10^{28}	5.6×10^{29}	1.5×10^{28}
	$0\nu\beta^+$/EC	1.7×10^{26}	1.8×10^{25}	2.8×10^{26}
^{136}Ce	$0\nu\beta^+\beta^+$	2.4×10^{29}	1.7×10^{31}	1.7×10^{29}
	$0\nu\beta^+$/EC	4.7×10^{26}	5.3×10^{25}	6.3×10^{26}

160

larger half life would be correct. Moreover, if $0\nu\beta\beta$ decay would ever be observed in a $0\nu\beta^-\beta^-$ decay experiment even the non-observation of the neutrinoless mixed mode at some given level would allow similar conclusions (limits allow consistent solutions only inside the ellipsoids, compare Fig. 3, therefore also larger limits can be used to exclude regions in parameter space).

To summarize the discussion on the neutrinoless $\beta^+\beta^+$ and β^+/EC decays we state that while presently existing half life limits lead to only rather loose bounds on $\langle m_\nu \rangle$, $\langle \lambda \rangle$ and $\langle \eta \rangle$, an interesting enhancement effect for the $\langle \lambda \rangle$-terms has been found for the neutrinoless mixed mode. If $0\nu\beta\beta$ decay ever would be observed, an experiment on the neutrinoless mixed mode decay therefore might offer an additional possibility to decide whether the decay is dominated by contributions from a finite neutrino mass or right-handed weak currents.

4. Summary

We have calculated $2\nu\beta\beta$ decay half lives for the experimentally most promising isotopes for all three possible decay modes ($\beta^+\beta^+$, β^+/EC and EC/EC). While the calculated $2\nu\beta^+\beta^+$ decay half lives are very large, confirming earlier studies [9, 10], the predicted half-lives for the mixed mode and the double electron capture are shorter by many orders of magnitude.

For $0\nu\beta\beta$ decay we have calculated all matrix elements for both the mass mechanism and the right-handed weak current terms for the first time complete. For $\beta^+\beta^+$ and β^+/EC decay coefficients of the decay rate for the isotopes ^{58}Ni, ^{78}Kr, ^{96}Ru, ^{106}Cd, ^{124}Xe, ^{130}Ba and ^{136}Ce are tabulated.

While limits on the effective neutrino mass and right-handed parameters deduced from experimental limits are at present not very restrictive when compared to those of the best $\beta^-\beta^-$ decay candidates, an interesting enhancement effect for the $\langle \lambda \rangle$-terms in the neutrinoless mixed mode has been pointed out. We think that if neutrinoless double beta decay ever would be observed the consideration of neutrinoless mixed mode decays offers a possibility to decide whether the observed decay is dominated by the mass mechanism or by right-handed weak currents.

Let us finally mention that while the pessimistic estimates for the 2ν double positron emission half lives might have discouraged experimentalists, we hope that the considerably shorter half lives for the other two decay modes leads to renewed interest in double beta plus decay experiments.

This work is supported in part by a Grant-in-Aid for Scientific Research (05243204) from the Ministry of Education, Science and Culture. One of us (M.H.) would like to thank the Japanese Ministry of Education, Science and Culture (Monbusho) for financial support. He also acknowledges valuable discussions with S.S.

References

1. Furry, W.H.: Phys. Rev. **56**, 1184 (1939)
2. Avignone, F.T., Brodzinski, R.L., Collar, J.T., Guerard, C.K., Miley, H.S., Reeves, J.H.: Phys. Lett. B**256**, 559 (1991)
3. Balysh, A., Beck, M., Belyaev, S.T., Bensch, F., Bockholt, J., Demehin, A., Gurov, A., Heusser, G., Hirsch, M., Klapdor-Kleingrothaus, H.V., Kondratenko, I., Lebedev, V.I., Maier, B., Müller, A., Petry, F., Piepke, A., Strecker, H., Völlinger, M., Zuber, K.: Proc. 26th Int. Conf. on High Energy Physics, Dallas 1992. AIP Conf. Proc. **272**, 1141 (1993); Phys. Rev. Lett. **70**, 2853 (1993)
4. Elliott, S.R., Hahn, A.A., Moe, M.K.: Phys. Rev. Lett. **59**, 2020 (1987)
5. Ejiri, H., Fushimi, K., Kawasaki, M., Kinoshita, H., Ohsumi, H., Okado, K., Sano, H., Shima, T., Takasugi, E., Tanaka, J., Watanabe, T.: J. Phys. G [Proc. Suppl.] **17**, 155 (1991)
6. Kirsten, T., Heusser, G., Kaether, D., Oehm, J., Pernicka, E., Richter, H.: In: Proc. Int. Symp. on Nuclear Beta Decays and Neutrino. Kotani, T., Ejiri, H., Takasugi, E. (eds.), p. 81. Singapore: World Scientific 1986
7. Bernatowicz, T., Brannon, J., Brazzle, R., Cowsik, R., Hohenberg, C., Podosek, F.: Phys. Rev. Lett. **69**, 2341 (1992)
8. Turkevich, A.L., Economou, T.E., Cowan, G.A.: Phys. Rev. Lett. **67**, 3211 (1991)
9. Haxton, W.C., Stephenson, G.J.: Progr. Part. Nucl. Phys. **12**, 409 (1984)
10. Staudt, A., Muto, K., Klapdor-Kleingrothaus, H.V.: Phys. Lett. B**268**, 312 (1991)
11. Muto, K., Bender, E., Klapdor, H.V.: Z. Phys. A**334**, 177 (1989)
12. Muto, K., Bender, E., Klapdor, H.V.: Z. Phys. A**334**, 187 (1989)
13. Kim, C.W., Kubodera, K.: Phys. Rev. D**27**, 2765 (1983)
14. Doi, M., Kotani, T.: Progr. Theor. Phys. **87**, 1207 (1992)
15. Suhonen, J.: University of Jyväskylä Preprint No. 30/1992
16. Doi, M., Kotani, T., Takasugi, E.: Progr. Theor. Phys. [Suppl.] **83**, 1 (1985)
17. Muto, K., Klapdor, H.V.: In: Neutrinos. Klapdor, H.V. (ed.), p. 183. Berlin, Heidelberg, New York: Springer 1988
18. Vogel, P., Zirnbauer, M.R.: Phys. Rev. Lett. **57**, 3148 (1986)
19. Civitarese, O., Faessler, A., Tomoda, T.: Phys. Lett. B**194**, 11 (1987)
20. Tomoda, T., Faessler, A.: Phys. Lett. B**199**, 475 (1987)
21. Engel, J., Vogel, P., Zirnbauer, M.R.: Phys. Rev. C**37**, 731 (1988)
22. Suhonen, J., Taigel, T., Faessler, A.: Nucl. Phys. A**486**, 91 (1988)
23. Civitarese, O., Faessler, A., Suhonen, J., Wu, X.R.: Phys. Lett. B**251**, 333 (1990)
24. Staudt, A., Muto, K., Klapdor-Kleingrothaus, H.V.: Europhys. Lett. **13**, 31 (1990)
25. Staudt, A., Kuo, T.T.S., Klapdor-Kleingrothaus, H.V.: Phys. Rev. C**46**, 871 (1992)
26. Halbleib, J.A., Sorensen, R.A.: Nucl. Phys. A**98**, 542 (1967)
27. Cha, D.: Phys. Rev. C**27**, 2269 (1983)
28. Holinde, K.: Phys. Rep. **68**, 121 (1981)
29. Bohr, A., Mottelson, B.R.: Nuclear structure. Vol. 1. New York: Benjamin 1969
30. Lacombe, M., Loiseau, B., Richard, J.M., Vinh Mau, R., Côté, J., Pires, P., Tourreil, R. de: Phys. Rev. C**21**, 861 (1980)
31. Anantaraman, N., Toki, H., Bertsch, G.F.: Nucl. Phys. A**398**, 269 (1983)
32. Wapstra, A.H., Audi, G., Hoekstra, R.: At. Data Nucl. Data Tables **39**, 281 (1988)
33. Norman, E.B., DeFaccia, M.A.: Phys. Lett. B**148**, 31 (1984)
34. Norman, E.B.: Phys. Rev. C**31**, 1937 (1985)
35. Barabash, A.S., Kuzminov, V.V., Lobashev, V.M., Novikov, V.M., Ovchinikov, B.M., Pomansky, A.A.: Phys. Lett. B**223**, 273 (1989)
36. Ohsumi, H., Ejiri, H.: Private communication 1993
37. Ejiri, H. et al.: Nucl. Instrum. Methods A**302**, 304 (1991)
38. Doi, M., Kotani, T.: Progr. Theor. Phys. **89**, 139 (1993)
39. Vergados, J.D.: Nucl. Phys. B**218**, 109 (1983)
40. Suhonen, J., Khadkikar, S.B., Faessler, A.: Phys. Lett. B**237**, 8 (1990)
41. Hirsch, M., Wu, X.R., Klapdor-Kleingrothaus, H.V., Ching, C.R., Ho, T.H.: Z. Phys. A**345**, 163 (1993)
42. Suhonen, J., Khadkikar, S.B., Faessler, A.: Nucl. Phys. A**535**, 509 (1991)
43. Tomoda, T.: Nucl. Phys. A**484**, 635 (1988)

2.2.6 Matrix Elements for Exchange of Heavy Particles

International Journal of Modern Physics A, Vol. 11, No. 9 (1996) 1591–1605

$e^-e^- \to W^-W^-$ MEDIATED BY HEAVY NEUTRINO FLAVORS

PETER MINKOWSKI
Institute for Theoretical Physics, University of Bern
CH - 3012 Bern, Switzerland

The lepton flavor violating reaction $e^-e^- \to W^-W^-$, mediated by a set of heavy neutrino flavors extending minimally the standard model, is dominated at high energy in Born approximation by a single channel amplitude, scattering two left-handed incoming electrons into two outgoing longitudinally polarized standard W bosons. This amplitude depends universally on the Fermi constant and specifically on mass and mixing parameters of heavy neutrino flavors. It satisfies unitarity constraints and, if observed, is capable of revealing the CP-violating character of these parameters as interference patterns between heavy flavor contributions, provided their total number exceeds two. Cross section estimates are presented.

1. The Born Amplitude Dominant Far Above Threshold

We work in the minimal framework of the standard model extended by an a priori arbitrary set of (heavy) neutrino flavors with general $SU2_L$ quantum numbers and Yukawa couplings compatible with renormalizability. We further restrict this framework, assuming that only three neutrino flavors exhibit a small mass, which is neglected, while the remaining flavors are heavy — much heavier than the top quark — to be specific. We call this setting 'minimal mass generating case' (MMC) following Ref. 1.

Given the MMC setting we consider the reaction $e^-e^- \to W^-W^-$ at center of mass energies much in excess of the 2 W threshold. Then the dominant Born amplitude, denoted by T, connects the left-handed helicity states of incoming electrons to their longitudinally polarized W counterparts. We adopt standard s, t, u notation for the squares of center of mass energy and momentum transfers appropriate for 2–2 body scattering.

T corresponds to the t–u symmetric sum of heavy neutrino exchanges in the respective channels, given by the Feynman diagrams in Fig. 1 Heavy neutrino flavors are labelled $\mathcal{N}_\Lambda; \Lambda = 1 \cdots N'$, while M_Λ denotes their respective physical masses. We denote the total number of heavy flavors by N'.

We adopt as mass parameter the quantity

$$v_{ch} = \left(2\sqrt{2}G_F\right)^{-1/2} = 174.1 \text{ GeV} \qquad (1)$$

International Journal of Modern Physics A, Vol. 13, No. 14 (1998) 2363–2381

HEAVY MAJORANA NEUTRINOS IN e^-e^- COLLISIONS *

CHRISTOPH GREUB AND PETER MINKOWSKI
Institute for Theoretical Physics,
University of Berne, Berne, Switzerland
E-mail: greub@itp.unibe.ch, mink@itp.unibe.ch

We discuss the process $e^-e^- \to W^-W^-$ mediated by heavy Majorana neutrino exchange in the t- and u channel. In our model the cross section for this reaction is a function of the masses (m_N) of the heavy Majorana neutrinos and mixing parameters (U_{eN}) originating from mixing between the ordinary left-handed standard model neutrinos and additional singlet right-handed neutrino fields. Taking into account the standard model background and constraints from low energy measurements, we present discovery limits in the (m_N, U_{eN}^2) plane. We also discuss how to measure in principle the CP violating phases, i.e., the relative phases between the mixing parameters.

1. Introduction

The question why the masses of the observed neutrinos are much lighter than those of the charged leptons is a central unsolved problem in particle physics. An attractive scenario to understand this is the following: Adding right-handed singlet neutrino fields to the standard model, the resulting mass spectrum can be such that there are 3 essentially zero mass Majorana neutrinos and additional heavy Majorana neutrinos. While in the generic case the masses of these heavy neutrinos are too large to detect observable consequences at present and future colliders, it is not excluded however, that their masses are in the range of a few TeV and that their coupling strenghs to the charged leptons are rather large. For the latter case, the phenomenology for e^+e^- collisions where heavy Majorana neutrinos can be directly produced via t- channel W and s- channel Z^0 exchange, has been worked out in detail in ref. 1. The main conclusion is that the cross sections and designed luminosities are large enough to find heavy Majorana neutrinos essentially up to the kinematical limit, i.e., up to $m_N \approx \sqrt{s}$, provided that the relevant mixing angles are near the present bounds inferred from low energy data. A similar conclusion also holds for single Majorana neutrino production in $e\gamma$ collisions. Another attractive reaction to be discussed in detail in this paper is $e^-e^- \to W^-W^-$, which is mediated by t- and u- channel Majorana neutrino exchange (see Fig. 1). As the neutrinos do not have to be produced in this reaction, one is potentally sensitive to neutrino masses which exceed $\sqrt{s}$. Therefore, the aim of this paper is to work out the discovery region for such neutrinos in the parameter space (masses and mixing angles), taking into account standard model background reactions.

*Work supported in part by Schweizerischer Nationalfonds.

On nuclear matrix element uncertainties in short range $0\nu\beta\beta$ decay

H.V. Klapdor–Kleingrothaus and H. Päs

Max–Planck–Institut für Kernphysik
P.O. Box 103980, D–69029 Heidelberg, Germany

Abstract

The evaluation of short range contributions to neutrinoless double beta decay has been challenged due to critics of the ansatz of the nuclear matrix element calculations. We comment on the critics and uncertainties of these calculations and the effect on the derived limits.

Neutrinoless double beta decay

$$^A_ZX \rightarrow ^A_{Z+2} X + 2e^- \tag{1}$$

has been proven to be one of the most sensitive tools to search for particle physics beyond the standard model. Besides the most stringent limit on the effective neutrino Majorana mass and neutrino-mediated contributions from R-parity violating SUSY and leptoquarks also contributions due to heavy particle exchange (superheavy neutrinos and SUSY partners) have been discussed, and extremely stringent constraints on the effective superheavy neutrino mass $\langle m_H \rangle$ and R-parity violating coupling λ'_{111} have been derived (for an overview and recent limits see [1, 2]):

$$\langle m_H \rangle = \left(\sum_j \frac{U_{ej}^2}{m_j} \right)^{-1} > 9 \cdot 10^7 GeV \tag{2}$$

$$\lambda'_{111} \leq 4 \times 10^{-4} \Big(\frac{m_{\tilde{q}}}{100GeV}\Big)^2 \Big(\frac{m_{\tilde{g}}}{100GeV}\Big)^{1/2}. \tag{3}$$

For comparison, a future linear collider with a center of mass energy of 1 TeV would be sensitive to 250 TeV $< \langle m_H \rangle <$ 5000 TeV, only [3] (for a serious discussion of possibilities to observe inverse neutrinoless double beta decay at future colliders due to finetuned cancellations of mass eigenstates in the double beta decay observable see [4]). The conclusions from the $0\nu\beta\beta$ decay half life limit have been challenged by critics [5] concerning the matrix element calculations at short distances. The standard ansatz for nuclear matrix element calculations treats double beta decay in terms of nucleons of finite size with a hard core. The finite nucleon size effect is taken into account by nucleon form factors in momentum space [6]

$$F(q^2) = F(0)\left(1 - \frac{q^2}{m_A^2}\right)^{-2} \tag{4}$$

with $m_A = 0.85$ GeV. The form factors $F(0)$ used have been calculated treating the quarks in the MIT bag model [7].

The nucleon-nucleon repulsion at short distances is considered in two ways. First, the repulsion effect is included in the nucleon potential. In addition, to be conservative, the nucleon hard core is simulated with introducing a cutoff by multiplying the two particle wave functions by the correlation function [8]

$$1 - f(r) = 1 - e^{-ar^2}(1 - br^2). \tag{5}$$

The parameters a and b can be related to each other so that effectively, there is one free parameter, the correlation length

$$l_c = -\int_0^\infty ds([1 + f(r)]^2 - 1. \tag{6}$$

The standard value of $l_c \simeq 0.7$ fm fits experimental data from nucleon-nucleon scattering. In this approach the total suppression of short range matrix elements compared to long range matrix elements with the same transition operator equals 1/20-1/30.

The dependence of short range nuclear matrix element calculations in the pn-QRPA model on the quantities m_A and l_C has been discussed extensively in [9]. It has been shown that in this approach the main contribution to the matrix element comes from nuclear distances larger than 1 fm. The matrix elements are stable to variations of m_A and l_C, changes up to 50 % of the standard values yield only comparable variations of the nuclear matrix elements. Although no guarantee - in the sense the nucleon can not be derived from QCD and no direct experimental test apart from comparison with data from nucleon-nucleon scattering is possible - exists that this approach is applicable for the case of heavy particle exchange, it was successful in predicting the matrix element of the standard model mode of double beta decay (two neutrino emitting decay) with an accuracy of $\sqrt{2}$ (compare refs. [10] and [11]).

The criticism of ref. [5] is based on the argument that for intermediate particle masses as heavy as discussed here the correct picture would be the quark rather than the nucleon picture. One should keep in mind, however, that the heavy exchanged particles are virtual and that the momenta transferred are much smaller. The total suppression of short range transitions compared to long range transitions due to the quark-quark repulsion has been estimated in ref. [5] to yield a suppression by a factor of 1/40 or less. This estimation is based on a spin singlet requirement to achieve an overall antisymmetric wave function ($\simeq 2/3$), the color Coulomb repulsion of the involved d-quarks ($\simeq 1/3$ estimated by a WKB evaluation of the color Coulomb barrier) and a similar factor from the interaction of the remaining two quarks in the nucleus, which is justified by the picture that each of the two decaying d-quarks is "pulled on by a u- and a d-quark from its own nucleon", the latter being estimated to be $\simeq (1/3)^2$ or less. Whether attracting interactions between quarks belonging to the other nucleon change this picture is not discussed in ref. [5]. Also effects of the nuclear environment may change this picture and are totally ignored in this estimation. While this estimation is not based on an approach which is generally accepted (the preprint ref. [5] from 1996 is not published yet), the total suppression factor 1/40 argued, confirms the order of magnitude of the suppression of short range matrix elements compared to long range matrix elements in the pn-QRPA approach 1/20-1/30. However, ref. [5] incorrectly applied this suppression factor to the limits derived with the pn-QRPA short range matrix elements and this way considered the suppression factor two times. Moreover, old experimental limits have been used in the comparison of double beta decay and the inverse process.

In fact the to our knowledge only serious attempt of a calculation based on a relativistic quark model, see ref. [12], confirms matrix element calculations in the standard approach with an accuracy of a factor of three. We therefore assume it to be rather premature, to classify (as in [3]) all matrix elements calculated for heavy particle exchange as "old" in the sense of no more valid.

If one in spite of these facts assumes (incorrectly) the estimated suppression of short range matrix elements from ref. [3], the limit on the superheavy Majorana neutrino becomes 2000 TeV, still being competitive to a 1 TeV linear collider. For supersymmetric contributions in addition one has to take into account that the bound on the coupling scales with the square root of the nuclear matrix element, so that the estimated suppression would lead to a limit on λ'_{111} being worse only by a factor of order 5.

Summarizing, we commented on the critics of short range matrix element calculations for neutrinoless double beta decay. Since a real alternative based on a treatment in the quark picture is missing and in view of the lack of any reasonable estimation leading to considerably worse limits (i.e. more than a factor of three) we find it useful to present as limits furthermore the results of the calculations in the nucleon picture. Moreover, even if one assumes the - clearly incorrect - estimation of ref. [5], limits on SUSY are only worse by a factor of five and limits on superheavy neutrinos are still compatible to what could be obtained at future linear colliders. It should be stressed further, that these critics do not concern the neutrinoless double beta decay contributions with light particle exchange yielding limits on light neutrino masses [13], R-parity violating SUSY [14], leptoquarks [15] as well as violations of the equivalence principle and Lorentz invariance [16].

Acknowledgement

We thank M. Hirsch for useful discussions.

References

[1] H.V. Klapdor-Kleingrothaus, H. Päs, hep-ph/0002109, in Proc. Cosmo'99, Trieste, Italy

[2] H.V. Klapdor-Kleingrothaus, Springer Tracts in Modern Physics, Vol. 163 (2000), p. 69-104

[3] C. Greub, P. Minkowski, Int. J. Mod. Phys. A 13 (1998) 2363-2381

[4] G. Belanger, Proc. *Lepton and Baryon Number Violation, Trento/Italy 1998*, Eds. H.V. Klapdor–Kleingrothaus, I.V. Krivosheina, IOP Bristol & Philadelphia 1999; G. Belanger, F. Boudjema, D. London, H. Nadeau, Phys. Rev. D 53 (1996) 6292

[5] C.A. Heusch, P. Minkowski, hep-ph/9611353

[6] J.D. Vergados, Phys. Rev. C 24 (1981) 640; Nucl. Phys. B 218 (1983) 109

[7] S. Adler *et al.*, Phys. Rev. **D 11**, (1975) 3309

[8] G.A. Miller, J.E. Spencer, Ann. Phys. (N.Y.) 100 (1976) 562

[9] M. Hirsch, H.V. Klapdor–Kleingrothaus, S.G. Kovalenko, Phys. Rev. D 53 (1996) 1329-1348

[10] K. Muto, E. Bender, H.V. Klapdor, Z. Phys. A 334 (1989) 177

[11] Heidelberg–Moscow Collab., Phys. Lett. B 322 (1994) 176

[12] J. Suhonen, S.B. Khadkikar, A. Faessler, Phys. Lett. B 237 (1990) 8; J. Suhonen, S.B. Khadkikar, A. Faessler, Nucl. Phys. A 529 (1991) 727; J. Suhonen, S.B. Khadkikar, A. Faessler, Nucl. Phys. A 535 (1991) 509.

[13] Heidelberg–Moscow Collab., Phys. Rev. Lett. 83 (1999) 42

[14] H. Päs, M. Hirsch, H.V. Klapdor–Kleingrothaus, Phys. Lett. B 459 (1999) 450

[15] M. Hirsch, H.V. Klapdor–Kleingrothaus, S.G. Kovalenko, Phys. Rev. D 54 (1996) R4207

[16] H.V. Klapdor–Kleingrothaus, H. Päs, U.Sarkar, Eur. Phys. J. A 5 (1999) 3

2.3 Effective Neutrino Masses from Double Beta Decay, Neutrino Mass Models and Cosmological Parameters – Present Status and Prospects

PHYSICAL REVIEW D VOLUME 48, NUMBER 7 1 OCTOBER 1993

Neutrino mass explanations of solar and atmospheric neutrino deficits and hot dark matter

David O. Caldwell
Department of Physics, University of California, Santa Barbara, California 93106-9530

Rabindra N. Mohapatra
Department of Physics, University of Maryland, College Park, Maryland 20742
(Received 17 March 1993)

If the solar and atmospheric neutrino deficits and the apparent need for a hot dark matter component all result from neutrino mass, one of three neutrino mass patterns is required. Of these, two appear quite unlikely, leading to a unique solution which points to further experimental tests. We briefly outline possible theoretical models which could generate these three neutrino mass matrices without adjustment of parameters.

PACS number(s): 14.60.Gh, 12.15.Ff, 95.35.+d, 96.60.Kx

Current indications for neutrino mass are [1] (1) the deficit of solar electron neutrinos, (2) the depletion of atmospheric muon neutrinos relative to electron neutrinos, and (3) the apparent need for some hot dark matter. Conventional wisdom is that (1) results from $\nu_e \to \nu_\mu$ oscillations, so on the basis of a seesaw model ν_τ could have the ~ 7 eV mass desired for (3). If (2) is evidence for neutrino mass, however, the small $\nu_\mu - \nu_\tau$ mass difference needed makes this scenario wrong, and there are three different possibilities for the pattern of neutrino masses and mixings.

These possible scenarios of neutrino mass, if the above three observations are indeed manifestations of neutrino mass, are (a) ν_e, ν_μ, and ν_τ all are $\sim 2-3$ eV, (b) the ν_e, ν_μ, and ν_τ are all very light, and a sterile neutrino ν_s supplies the hot dark matter, or (c) the ν_e and ν_s are light and the ν_μ and ν_τ are $\sim 3-4$ eV. The consequences of each of these alternatives will be examined below after a short review of the experimental input.

1. Solar neutrinos. Evidence for a solar neutrino deficit comes from four experiments [2]. If the relative values of the Kamiokande and Homestake results are correct, the problem is not astrophysical but must relate to neutrino properties [3], most likely neutrino mass. This would result in ν_e's produced in the Sun oscillating on the way to Earth to another neutrino species, which could be ν_μ, ν_τ, or ν_s (a sterile neutrino). The required mass difference and mixing are [3,4] (a) $\Delta m^2_{ei} = (0.3-1.2)\times 10^{-5}$ eV^2, $\sin^2 2\theta = (0.4-1.5)\times 10^{-2}$ [nonadiabatic Mikheyev-Smirnov-Wolfenstein (MSW) mechanism], (b) $\Delta m^2_{ei} = (0.3-3)\times 10^{-5}$ eV^2, $\sin^2 2\theta = 0.6-0.9$ (large-mixing MSW mechanism), or (c) $\Delta m^2_{ei} = (0.5-1.1)\times 10^{-10}$ eV^2, $\sin^2 2\theta = 0.8-1.0$ (vacuum oscillations).

2. Atmospheric neutrinos. Depletions of ν_μ relative to ν_e produced in the upper atmosphere mainly by pion and subsequent muon decay have been observed by three experiments. The ratio of observed μ events to e events normalized to the flux ratio expected from calculations are $R(\mu/e) = 0.60\pm 0.07\pm 0.05$ (Kamiokande) [5], $0.54\pm 0.05\pm 0.12$ (IMB) [6], $0.69\pm 0.19\pm 0.09$ (Soudan II, preliminary) [7]. Combining these results with the observations of upward-going muons by the Kamiokande [5], IMB [6], and Baksan [8] groups, and the negative Fréjus [9] results, Frati *et al.* [10] have concluded that ν_μ could be oscillating into either ν_e, ν_τ, or ν_s with $\Delta m^2 \approx 0.5$ to 0.005 eV^2 and $\sin^2 2\theta \approx 0.5$.

3. Dark matter neutrinos. Data on the extent of structure in the Universe are now available on a wide range of distance scales. Evidence from the Cosmic Background Explorer (COBE) results on the anisotropy of the cosmic microwave background radiation, galaxy-galaxy angular correlations, large-scale velocity fields, and correlations of galactic clusters can all be fit [11] by a model [12] of the Universe containing 70% cold dark matter and 30% hot dark matter contributed by ~ 7 eV in neutrino mass. Such a model provides a consistent explanation of not only the shape of the density fluctuation spectrum, but also the observed estimates of the absolute density on small and large scales. While the fits have been made with a single neutrino, dividing the mass among more than one neutrino would work even better [13].

4. Other constraints on neutrinos. Two constraints are important here. The first is the nonobservation of neutrinoless double β decay ($\beta\beta_{0\nu}$), which provides an important limit on an effective Majorana mass, $\langle m_\nu \rangle \lesssim 1-2$ eV [14]. Second, since the possibility of four light neutrinos is considered, their effect on nucleosynthesis is of concern, as the 95% C.L. limit on the number of light neutrinos is 3.3 from the primordial ^{4}He abundance [15]. To keep ν_s out of equilibrium at the time of nucleosynthesis requires a limit on its mixing angle with one of the three active species which depends on Δm^2_{si}. In the atmospheric neutrino case, even the most favorable $\Delta m^2_{\mu s}$ (≈ 0.005 eV^2) would have to have $\sin^2 2\theta_{\mu s}$, almost an order of magnitude smaller than the required 0.5 value in order for the $\nu_\mu \to \nu_s$ oscillation not to contribute 0.3 of an effective neutrino at that time, ruling out this possibility [16].

(a) *Case of three similar neutrino masses.* If the solar ν_e deficit is solved by $\nu_e \to \nu_\mu$, the atmospheric ν_μ deficit by $\nu_\mu \to \nu_\tau$, and only these three light neutrinos exist, then having hot dark matter requires the three to be nearly degenerate, with a mass ~ 2.5 eV each. The necessary Δm^2 values make it impossible for fewer neutrinos to supply the needed dark matter mass. In this case a mass matrix of the following type would fit all the constraints:

$$M = \begin{bmatrix} m & \delta s_2^2 s_1 & -\delta s_1 s_2 \\ \delta s_2^2 s_1 & m+\delta s_2^2 & -\delta s_2 \\ -\delta s_1 s_2 & -\delta s_2 & m+\delta \end{bmatrix}, \tag{1}$$

where the columns refer, in order, to ν_e, ν_μ, and ν_τ and the rows to the mass eigenstates m_1, m_2, and m_3. For the nonadiabatic MSW solution of the solar neutrino problem, typical parameters of the matrix are $m \approx 2.5$ eV, $\delta \approx 0.08$ eV, $s_1 \approx 0.05$, and $s_2 \approx 0.35$. This highly fine-tuned mass matrix is difficult to obtain naturally in gauge theories.

In this case, ν_e-ν_τ mixing was made negligible by choice, but the matrix can be generalized to allow for nonzero ν_e-ν_τ mixing angle. Unitarity constraints, however, imply that the maximum value of this mixing angle would be 0.05. No experimental limit comes close to even this maximum value of $\sin^2 2\theta_{e\tau} = 10^{-2}$. The smallness of this angle and also of $\theta_{e\mu}$, or more exactly the smallness of the matrix elements M_{e2} and M_{e3} above, make it impossible to arrange a cancellation of the effective neutrino mass in $\beta\beta_{0\nu}$, $\langle m_\nu \rangle \approx \sum_j \xi_j M_{ej}^2 m_j$, by choosing suitable CP eigenvalues, $\xi_j = \pm 1$. Instead, $\langle m_\nu \rangle \approx 2.5$ eV. Thus the likelihood that this case represents reality is marginal at best, since Majorana masses are involved. To avoid experimental limits [14], one must rely on improbable nuclear matrix element suppression in $\beta\beta_{0\nu}$, the neutrino mass needed for hot dark matter to be much less than 7 eV, or the theoretically unlikely introduction of Dirac neutrino masses. Viewed more positively, if this were the correct scenario, $\beta\beta_{0\nu}$ experiments could soon demonstrate its existence.

(b) *Case of three very light active neutrinos and one heavier sterile neutrino.* For this logical possibility [17], the matrix in Eq. (1) could apply here as well, with m now very small or even zero. The sterile neutrino, which could be completely decoupled from the active ones, must have a reduced contribution to the energy density at the era of nucleosynthesis to respect the constraint [15] that $\delta N_\nu < 0.3$. The number density of ν_s now must satisfy the inequality $n_{\nu_s} < n_{\nu_e} (\delta N_\nu^{\max})^{3/4}$. If ν_s is to contribute 30% of dark matter, this implies $m_{\nu_s} = 27.6 (\delta N_\nu)^{-3/4} h^2$. For $\delta N_\nu = 0.3$, $h = \frac{1}{2}$ (the Hubble constant in units of 100 km s^{-1} Mpc^{-1}), $m_{\nu_s} \approx 17$ eV. It is more likely that ν_s decouples before the quark-hadron transition temperature ($T \sim 200$–400 MeV), making its contribution $\delta N_\nu = 0.1$ and requiring $m_{\nu_s} \approx 39$ eV. Such large values of ν_s are unlikely to be acceptable as hot dark matter [18]. That this is a very improbable scenario is fortunate, since direct detection of such ν_s dark matter appears impossible, and its coupling to active neutrinos is forced by the nucleosynthesis constraint to be so small ($\sin^2 2\theta_{sl} \lesssim 10^{-4}$) as to make neutrino disappearance detection extremely difficult.

(c) *Case of very light ν_e, ν_s and heavier ν_μ, ν_τ.* A much more attractive scheme is to have $\nu_e \to \nu_s$ solve the solar neutrino problem [19], the parameters being almost the same as for $\nu_e \to \nu_\mu$, and have ν_μ and ν_τ both account for the atmospheric ν_μ deficit and be the hot dark matter. For the Δm_{es}^2 and $\sin^2 2\theta_{es}$ involved, the nucleosynthesis constraint is obeyed for the nonadiabatic MSW or vacuum oscillation cases, but most likely violated by the large-mixing MSW case. The simplest mass matrix attaining these goals is

$$M = \begin{bmatrix} \mu_1 & \mu_3 & 0 & 0 \\ \mu_3 & \mu_2 & 0 & 0 \\ 0 & 0 & m & \delta/2 \\ 0 & 0 & \delta/2 & m+\delta \end{bmatrix}, \tag{2}$$

where the columns refer in order to ν_s, ν_e, ν_μ, and ν_τ, $\mu_{1,2}$ are of order 10^{-2} eV, μ_3 is of order 5×10^{-4} eV, and $m = 3.5$ eV and $\delta = 0.07$ to 0.0007 eV for the numbers given above. A similar mass matrix was advocated [20] to account for the solar neutrino and atmospheric ν_μ deficits and for the 17-keV neutrino, but with $\delta \gg m$ in Ref. [20].

For the sake of simplicity we have "disconnected" the ν_e-ν_s sector from the ν_μ-ν_τ sector. One could, for example, add a nonzero $\nu_e \nu_\mu$ element leading to ν_e-ν_μ oscillations, important to determining the reality of this scenario. While $\nu_e \to \nu_s$ can be tested vs $\nu_e \to \nu_\mu$ by observing neutral current interactions at SNO or BOREXINO, and $\nu_\mu \to \nu_\tau$ can be checked in proposed long-baseline oscillation experiments, the hope for seeing a mass difference between the ν_e-ν_s and ν_μ-ν_τ sectors lies mainly in $\nu_\mu \to \nu_e$ oscillation experiments. The LSND experiment at Los Alamos by 1994 could have sufficient data to reduce the present limit in this $\Delta m_{e\mu}^2$ region of $\sin^2 2\theta_{e\mu} = 2 \times 10^{-3}$ by an order of magnitude. It would be very difficult to attain such sensitivity in a $\nu_e \to \nu_\tau$ experiment.

Another model-dependent issue is the nature of the ν_e-ν_s and ν_μ-ν_τ mixings. While splitting of a light Dirac particle into Majorana ν_e and ν_s can be accomplished to give other than maximal mixing (although that would be appropriate in the vacuum oscillation case), that is very difficult to do in the heavy nearly degenerate ν_μ-ν_τ case. However, we present later an acceptable model based on an $S_3 \times Z_4$ symmetry having this property.

Theoretical model for case (a). The scenario in which all three active neutrinos have the same mass of about 2.5 eV cannot be realized in the conventional seesaw mechanisms generated by the mass matrix

$$\begin{bmatrix} 0 & m_D \\ m_D & M \end{bmatrix}, \tag{3}$$

where m_D is a Dirac mass and M is a large Majorana mass, since neutrino masses scale quadratically with the charged fermion masses (e.g., $m_u^2 : m_c^2 : m_t^2$). On the other hand, it has been noted [21] that both in left-right-symmetric models, as well as SO(10) models, the ν_L-N_R

mass matrix that emerges naturally has the same form as Eq. (3) with 0 replaced by fv_L and M by fv_R, where $v_L=(4m_{W_L}^2\gamma)/(g^2v_R)$. Here, v_R represents the scale at which the local $B-L$ symmetry [subgroup of SO(10)] breaks, and v_L is the vacuum expectation value of the $SU(2)_L$ triplet field [contained in the {126}-dimensional Higgs multiplet in the case of SO(10) theory] that couples to the lepton doublets. For $v_R\approx 10^{12}-10^{13}$ GeV and $f\approx(0.1-1)$, $\gamma\sim 1$, we get contributions to all neutrino masses of order 4 eV. The off-diagonal contributions would be dominated by the usual seesaw terms and are of order 10^{-9} eV, 10^{-3} eV, and a few eV, respectively. If we demanded that, by some symmetry, f be a unit matrix in the generation space, then indeed we would get $m_{\nu_e}\approx m_{\nu_\mu}\approx m_{\nu_\tau}\approx 2.5$ eV without unnatural fine-tuning of parameters. We would also get $\Delta m_{e\mu}^2\approx 5\times 10^{-3}$ eV^2 and $\Delta m_{\mu\tau}^2\approx$few eV^2, values which are too large, but at least close to those desired.

Possible model for case (b). With the solar and atmospheric neutrino puzzles accounted for by the ν_e, ν_μ, and ν_τ, the dark matter would be provided by one or more sterile neutrinos. Three sterile neutrinos, each with mass ≈ 13 eV, is very suggestive of an E_6 grand unified theory, where the {27}-dimensional representation (to which the fermions are usually assigned) contains one sterile neutrino, leading to three for three generations. Under the $SO(10)\times U(1)_X$ subgroup of E_6 we have, $\{27\}=\{16\}_{+1}+\{10\}_{-2}+\{1\}_{+4}$, where the subscript denotes the $U(1)_X$ quantum number. The seesaw mechanism in this case is implemented by the Higgs boson in the {351'} representation of E_6, which could be used to generate the light masses for (ν_e,ν_μ,ν_τ), as usual. (The Dirac masses arise from a {27}-dimensional Higgs boson.) However, the sterile SO(10)-singlet fermion remains massless at the tree level in this minimal scenario. It picks up mass at the one-loop level from a combination of $m_0[\phi\{27\}]^3$ and $b\Psi\{27\}\Psi\{27\}\phi\{27\}$ couplings, where Ψ are fermions and ϕ are Higgs bosons. A typical magnitude of this contribution is $m_{\nu_s}\approx(b^2m_0\langle S\rangle^2)/(16\pi^2M_H^2)$, where S is the SO(10) singlet in $\phi\{27\}$. For $b\approx 10^{-2}$, $m_0\approx$TeV, we need $\langle S\rangle/M_H\approx 10^{-7}$ to get $m_{\nu_s}\approx 13$ eV. If M_H is of the order of the grand unified theory (GUT) scale ($\approx 10^{16}$ GeV), this implies $\langle S\rangle\approx\frac{1}{2}\times 10^{10}$ GeV, which is a plausible intermediate scale.

Theoretical model for case (c). The mass matrix in Eq. (2) for this case can be generated by a combination of seesaw and loop mechanisms without adjustment of parameters. The model is presented as an argument to show that, while the mass matrix may appear highly fine tuned, its existence may have some theoretical basis in terms of hidden symmetries of nature. Consider an extension of the standard model with three sterile neutrinos ν_s, N_μ, and N_τ. Demand that this theory be invariant under an $S_3\times Z_4$ permutation symmetry. The S_3 group has one {2}-dimensional and two {1}-dimensional representations, denoted here by 1^- and 1^+, respectively. The fields then have the representations shown in Table I.

The most general gauge-, S_3-, and Z_4-invariant Yukawa coupling is

$$\begin{aligned}\mathcal{L}_Y=&M(N_\mu^TC^{-1}N_\mu+N_\tau^TC^{-1}N_\tau)+h_0(\bar\Psi_\mu\phi_uN_\mu+\bar\Psi_\tau\phi_uN_\tau)+h_e\bar\Psi_e\phi_{d_3}e_R+h_e'\bar\Psi_e(\phi_{d_1}\mu_R+\phi_{d_2}\tau_R)\\&+h_\mu[(\bar\Psi_\mu\phi_{d_2}+\bar\Psi_\tau\phi_{d_1})\mu_R+(\bar\Psi_\mu\phi_{d_1}-\bar\Psi_\tau\phi_{d_2})\tau_R]+f_1\Psi_\mu^TC^{-1}\tau_2\Psi_\tau\eta_2^++f_2\nu_s^TC^{-1}e_R\eta_2^+\\&+f_4k_2^{++}e_R^TC^{-1}e_R+f_3k_1^{++}(\mu_R^TC^{-1}\mu_R+\tau_R^TC^{-1}\tau_R)+\text{H.c.}\end{aligned}\tag{4}$$

Similarly, the potential, in addition to the usual terms, consists of

$$\begin{aligned}V'=&\mu_1k_1^{++}\eta_1^-\eta_1^-+\lambda_2k_2^{++}\eta_1^-\eta_2^-\sigma+\mu_3k_1^{++}\eta_2^-\eta_2^-\\&+\mu_4\phi_{d_3}^T\tau_2\phi_{d_4}\eta_1^-+\mu_s\eta_1\eta_2^*\sigma'\ ,\end{aligned}\tag{5}$$

where σ is a real S_3-odd singlet with a vacuum expectation value.

The first two terms in Eq. (4) lead to degenerate ν_μ,ν_τ Majorana states with masses $m_\nu\approx h_0^2\kappa_u^2/M$, where $\kappa_u=\langle\phi_u^0\rangle$. For $\kappa_u\approx 100$ GeV, $h_0\approx\frac{1}{2}\times 10^{-4}$, and $M\approx 7$ TeV, we get $m_\nu\approx 3.5$ eV, as required. The purpose of explicit numbers is to show that the choice of parameters is not too different in order of magnitude from those of the standard model. At the tree level, $m_{\nu_e}=m_{\nu_s}=0$. At the two-loop level, these masses will arise. Note that even with S_3 symmetry, m_e, m_μ, and m_τ can be very different from each other.

The two-loop graphs are similar to those in Ref. [22] and are shown in Figs. 1(a)–1(c). The $\nu_\mu\nu_\tau$ and the $\nu_\mu\nu_\mu$ elements are roughly [in terms of the parameters in Eqs. (4) and (5)]

$$m_{\nu_\mu\nu_\tau}\approx\frac{2f_1^2}{(16\pi^2)^2}\frac{m_\tau m_{\mu\tau}\mu_3}{M_H^2}\ ,\tag{6}$$

$$m_{\nu_\mu\nu_\mu}\approx\frac{f_1^2}{(16\pi^2)^2}\frac{\mu_3}{M_H^2}(m_\tau^2+m_{\mu\tau}^2)\ .\tag{7}$$

M_H is a typical Higgs boson mass in the loop, chosen ≈ 100 GeV; $m_{\mu_2}\approx h_\mu\langle\phi_{d_1}^0\rangle$. For $f_1\approx 10^{-2}$, $\mu_3\approx M_H$, $m_{\nu_\mu\nu_\tau}\approx 3\times 10^{-2}$ eV $\approx m_{\nu_\mu\nu_\mu}$, which is roughly of the right order of magnitude. Equations (6) and (7) together predict the ν_μ-ν_τ mixing angle to be approximately $\tan 2\theta_{\mu\tau}\approx 2m_{\mu\tau}/m_\tau$, avoiding maximal mixing.

In the $\nu_e\nu_s$ sector we can select parameters with reasonable values and obtain $m_{\nu_e\nu_s}\approx m_{\nu_s\nu_s}\approx 10^{-2}-10^{-3}$ eV, if we choose $f_2\approx 10^{-3}$, $f_4\approx 10^{-1}$ and $\langle\sigma\rangle\approx\langle\phi_{d_4}\rangle\approx M_H$. This explains the solar neutrino puzzle via the nonadiabatic MSW mechanism. There are

TABLE I. Gauge and discrete symmetry transformation properties of the fermions and bosons in model (c).

Fields		$SU(2)\times U(1)_Y$	S_3	Z_4
Fermions:	$\begin{bmatrix}\Psi_\mu\\ \Psi_\tau\end{bmatrix}$	(2,−1)	2	+1
	$\begin{bmatrix}N_\mu\\ N_\tau\end{bmatrix}$	(1,0)	2	−1
	$\begin{bmatrix}\mu_R\\ \tau_R\end{bmatrix}$	(1,−2)	2	+1
	Ψ_e	(2,−1)	1^+	+1
	e_R	(1,−2)	1^+	$-i$
	ν_s	(1,0)	1^-	$+i$
Higgs bosons:	ϕ_u	(2,−1)	1^+	−1
	$\begin{bmatrix}\phi_{d_1}\\ \phi_{d_2}\end{bmatrix}$	(2,+1)	2	+1
	ϕ_{d_3}	(2,+1)	1^+	i
	ϕ_{d_4}	(2,+1)	1^+	$-i$
	η_1^+	(1,+2)	1^+	+1
	η_2^+	(1,2)	1^-	+1
	k_1^{++}	(1,4)	1^+	+1
	k_2^{++}	(1,4)	1^+	−1
	σ	(1,0)	1^-	−1
	σ'	(1,0)	1^-	+1

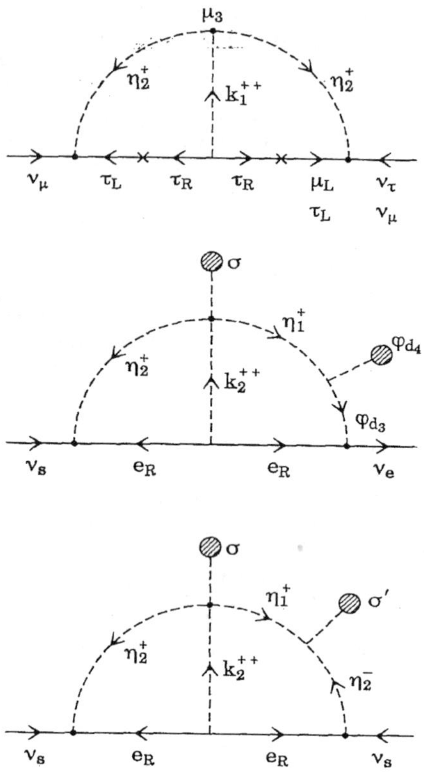

FIG. 1. Two-loop graphs contributing to $m_{\nu_\mu\nu_\mu}$, $m_{\nu_\mu\nu_\tau}$, $m_{\nu_e\nu_s}$, $m_{\nu_s\nu_s}$ in model (c).

also small contributions connecting the ν_e-ν_s and ν_μ-ν_τ sectors proportional to h'_e. The mixing angle cannot be predicted because it depends on the unknown parameter h'_e.

The strength of the ν_s-e interaction in this model is given by $f_2^2/4M_{\eta_2}^2$. For $M_{\eta_2}\simeq 100$ GeV and the above quoted value of $f_2(\approx 10^{-3})$, the strength of ν_s-e scattering is $\simeq 2.5\times 10^{-6}G_F$. This implies that in the early Universe the sterile neutrinos go out of equilibrium around $T\simeq 10$ GeV, resulting in their contribution to energy density at the epoch of nucleosynthesis being an effective $\delta N_\nu < 0.05$.

In summary, if present indications of neutrino mass from solar and atmospheric neutrino experiments, as well as indirect evidence for a component of hot dark matter, are correct, then one of three scenarios for the masses of neutrinos must be realized. The first, in which ν_e, ν_μ, and ν_τ are all ~2–3 eV, is either ruled out now, or soon could be, by neutrinoless double β decay experiments. The second, having very light ν_e, ν_μ, and ν_τ, with dark matter being supplied by one or more sterile neutrinos ν_s probably does not provide fast enough dark matter. The third, and by far most likely, in which light $\nu_e\to\nu_s$ solves the solar neutrino deficit and the heavier ν_μ-ν_τ sector accounts for hot dark matter and the deficiency of atmospheric ν_μ's, is testable by SNO and BOREXINO (for $\nu_e\to\nu_s$), by long-base-line oscillation experiments (for $\nu_\mu\to\nu_\tau$), but particularly by detecting $\nu_\mu\to\nu_e$ or $\nu_e\to\nu_\tau$ oscillations to see the mass difference between the two sectors. Theoretical possibilities for each of the three scenarios are given to demonstrate that such arrangements of neutrino mass are not impossible to achieve in gauge theories and to give indications for future theoretical work.

This work was supported by the National Science Foundation (R.N.M.) and the Department of Energy (D.O.C.).

[1] For neutrino masses, if in addition to these three indications the 17-keV neutrino exists, see D. O. Caldwell, Phys. Lett. B **289**, 389 (1992); E. Kh. Akhmedov *et al.*, Phys. Rev. D **47**, 3245 (1993).

[2] R. Davis, Jr. *et al.*, in *Proceedings of the 21st International Cosmic Ray Conference*, Adelaide, Australia, 1989, edited by R. J. Protheroe (University of Adelaide Press, Adelaide, 1990), Vol. 12, p. 143; K. S. Hirata *et al.*, Phys. Rev.

Lett. **65**, 1297 (1990); **65**, 1301 (1990); **66**, 9 (1990); A. I. Abazov *et al.*, *ibid.* **67**, 3332 (1991); P. Anselmann *et al.*, Phys. Lett. B **285**, 376 (1992).

[3] S. Bludman, D. Kennedy, and P. Langacker, Nucl. Phys. **B374**, 373 (1992).

[4] L. Krauss, E. Gates, and M. White, Phys. Lett. B **299**, 94 (1993); P. I. Krastev and S. T. Petcov, *ibid.* **299**, 99 (1993); S. Bludman *et al.*, Phys. Rev. D **47**, 2220 (1993).

[5] K. S. Hirata *et al.*, Phys. Lett. B **280**, 146 (1992).

[6] R. Becker-Szendy *et al.*, Phys. Rev. D **46**, 3720 (1992).

[7] P. J. Litchfield, in International Europhysics Conference on High Energy Physics, Marseille, France, 1993 (unpublished).

[8] M. M. Boliev *et al.*, in *Proceedings of the 3rd International Workshop on Neutrino Telescopes,* Venice, Italy, 1991, edited by M. Baldo-Ceolin (Istitute Nazionale di Fisica Nucleare, Padova, 1991), p. 235.

[9] Ch. Berger *et al.*, Phys. Lett. B **245**, 305 (1990); **227**, 489 (1989).

[10] W. Frati *et al.*, Phys. Rev. D **48**, 1140 (1993).

[11] E. L. Wright *et al.*, Astrophys. J. **396**, L13 (1992); M. Davis, F. J. Summers, and D. Schlegel, Nature **359**, 393 (1992); A. N. Taylor and M. Rowan-Robinson, *ibid.* **359**, 396 (1992); R. K. Schaefer and Q. Shafi, Report No. BA-92-28, 1992 (unpublished); J. A. Holtzman and J. R. Primack, Astrophys. J. **405**, 428 (1993).

[12] Q. Shafi and F. W. Stecker, Phys. Rev. Lett. **53**, 1292 (1984).

[13] M. Davis and J. Primack (private communication).

[14] D. O. Caldwell *et al.*, in *Weak Interactions and Neutrinos,* Proceedings, Ginosar, Israel, 1989, edited by P. Singer and B. Gad Eilam [Nucl. Phys. B (Proc. Suppl.) **13**, 547 (1990)]; A. Balysh *et al.*, Phys. Lett. B **283**, 32 (1992).

[15] T. P. Walker *et al.*, Astrophys. J. **51**, 376 (1991).

[16] K. Enqvist, K. Kainulainen, and M. Thomson, Phys. Lett. B **280**, 245 (1992).

[17] Suggested to one of us (D.O.C.) by J. T. Peltoniemi.

[18] J. R. Primack and R. K. Schaefer (private communication).

[19] D. O. Caldwell and P. Langacker, Phys. Rev. D **44**, 823 (1991), and references therein.

[20] K. S. Babu, R. N. Mohapatra, and I. Rothstein, Phys. Rev. D **45**, 5 (1992).

[21] R. N. Mohapatra and G. Senjanović, Phys. Rev. D **23**, 165 (1981).

[22] K. S. Babu and R. N. Mohapatra, Phys. Rev. D **46**, 374 (1992).

NEUTRINO MASSES AND OSCILLATIONS

Alexei Yu. Smirnov

International Centre for Theoretical Physics
Strada Costiera 11, 34100 Trieste, Italy

Institute for Nuclear Research, Russian Academy of Sciences,
117312 Moscow, Russia

New effects related to refraction of neutrinos in different media are reviewed and implication of the effects to neutrino mass and mixing are discussed. Patterns of neutrino masses and mixing implied by existing hints/bounds are described. Recent results on neutrino mass generation are presented. They include neutrino masses in SO(10) GUT's and models with anomalous $U(1)$, generation of neutrino mass via neutrino-neutralino mixing, models of sterile neutrino.

1 Introduction

1.1 Hints

A number of results testifies for non-zero neutrino masses and mixing:

- Solar neutrino spectroscopy.
- Results on atmospheric neutrinos.
- Large scale structure of the Universe. (Its formation may imply some amount of the hot dark matter (HDM)).
- LSND results.
- Hydrogen ionization in the Universe.
- Peculiar velocities of pulsars.
- Excess of events in tritium spectrum.

First four items are reviewed by Y. Suzuki [1], the fifth item was discussed in [2], and the last two will be presented in sect. 2 .

1.2 Upper bounds

Majority of results gives just upper bounds on neutrino masses and mixing. The most strong bounds relevant for the discussion come from:
- reactor oscillation experiments BUGEY [3], Krasnoyarsk [4] ($\bar{\nu}_e$ - ν_x oscillations);
- meson factory oscillation experiments (KARMEN [5], LSND [6]);
- accelerator experiments E531 ($\nu_\mu \to \nu_\tau$) [7] and E776 (ν_μ - ν_τ) [8];
- direct kinematic searches of neutrino mass: tritium experiment in Troitzk on $\bar{\nu}_e$ [9], (see also Maiz experiment [10]), PSI experiment on the mass of ν_μ [11], LEP ALEPH result on ν_τ [12] (see also OPAL result [13]; new possibility to measure the mass of ν_τ has been suggested in [14]);
- searches for the neutrinoless double beta decay in Heidelberg-Moscow experiment [15], (see also IGEX [16]);
- supernova 1987A data [17,18,19]; dynamics of supernovas [17];
- nucleosyntesis in supernovas [20];
- primordial Nucleosyntesis [21,22,23];
- cosmology [24];
- structure formation in the Universe [25,26,27,28].

These results give important restrictions on possible pattern of neutrino masses and mixing.

1.3 Lower bounds on neutrino mass?

Neutrinos are the only fermions for which the Standard model predicts masses. It prdicts that neutrino masses are zero. This follows from the content of the model, namely, from the fact that in the model there is

- no right handed neutrino components,
- no Higgs triplets which can give the Majorana mass for the left handed neutrinos.

The absence of the ν_R gives an explanation of strong upper bounds on the neutrino masses. However the absence of ν_R looks rather anesthetic.

The Standard Model is not the end of the story and we know that at least there is the gravity. The gravity can questioned both above items:

THE HEIDELBERG-MOSCOW DOUBLE BETA DECAY EXPERIMENT WITH ENRICHED ^{76}GE: STATUS AND PERSPECTIVES

H.V. Klapdor-Kleingrothaus*

Max-Planck-Institut für Kernphysik, Heidelberg, Germany

Abstract:

Double beta decay probes beyond standard model physics. Status and perspectives of the HEIDELBERG-MOSCOW double beta decay experiment which is undertaken in the GRAN SASSO laboratory, are presented. Of the 16.9 kg of enriched (86%) ^{76}Ge in hands of the cooperation, ~ 12 kg have been converted into detectors (or crystals). At present 6 kg of detectors are in operation for data taking. The significance of the experiment at present is 2649 kg · d. Present limits for $0\nu\beta\beta$ decay to the g.s. and first excited state of ^{76}Se, respectively, are:
$T_{1/2}^{0\nu}(0^+ \to 0^+) > 1.9 \times 10^{24}$ y and $T_{1/2}^{0\nu}(0^+ \to 2^+) > 8.0 \times 10^{23}\,y$ with 90% c.l. These values are the most stringent directly measured $0\nu\beta\beta$ half life limits known up to now. They correspond to an effective neutrino mass limit of $< m_\nu >< 1.1$ eV. On the other hand we seem to observe a systematic excess of counts beyond the expected background precisely at the $0\nu\beta\beta$ $(0^+ \to 0^+)$ transition energy. For the neutrinoless decay with majoron emission a half life of 1.7 x 10^{22} y can be excluded leading to an upper limit for the neutrino-majoron coupling of $< g_{\nu\chi} > < 1.8 \cdot 10^{-4}$ (90% c.l.). The background around 2 MeV is b = 0.2 counts/kg y keV for the total array. The experiment opens new perspectives for the investigation of exotic processes like electron decay, solar axions and dark matter. It was, e.g., possible to improve the existing cross section limits for WIMP masses above ~ 150 GeV and to exclude Dirac neutrinos in the range 26 GeV to 4.7 TeV as the dominant component of the dark halo. The new $\beta\beta$-technology has found also application in high resolution balloon and satellite gamma-ray astronomy.
Concerning the matrix elements for $\beta\beta$ decay a major step beyond the frequently used QRPA model has been made by applying the Operator Expansion Method (OEM).

1. INTRODUCTION

The neutrino mass m_ν is one of the key quantities for the structure of grand unified

* for the HEIDELBERG-MOSCOW collaboration:

M. Beck, J. Bockholt, G. Heusser, M. Hirsch, H.V. Klapdor-Kleingrothaus#, B. Maier, F. Petry , A. Piepke, H. Strecker, M. Völlinger, K. Zuber (Max-Planck-Institut für Kernphysik, Heidelberg)

A. Balysh, S.T. Belyaev#, A. Demehin, A. Gurov, I. Kondratenko, V. I. Lebedev (Kurchatov Institute, Moscow)

A. Müller (INFN, Gran Sasso)

#) speakers of the collaboration

in Proc. "Particles and Nuclei", Perugia, Italy, June-July, 1993, *World Scientific, Singapore* (1994) 283 – 289

PHYSICAL REVIEW D VOLUME 50, NUMBER 5 1 SEPTEMBER 1994

Accommodating solar and atmospheric neutrino deficits, hot dark matter, and a double β decay signal

David O. Caldwell
Department of Physics, University of California, Santa Barbara, California 93106

Rabindra N. Mohapatra
Department of Physics, University of Maryland, College Park, Maryland 20742
(Received 8 February 1994)

Neutrino mass explanations of the solar and atmospheric neutrino deficits and a hot dark matter component require one of three patterns of those masses, as already pointed out by us. Recently there have been indications of a nonvanishing amplitude for neutrinoless double β decay. If this additional hint of neutrino mass is true, it would make even less likely the one unfavored pattern (a sterile neutrino giving warm, rather than hot dark matter), which would alter another by making the ν_e a contributor to the hot dark matter, and would make the third (ν_e, ν_μ, and ν_τ approximately degenerate) much more likely than previously. For this third case we construct a gauge model consistent with other weak interaction data. This model utilizes a more general version of the seesaw mechanism, which is very likely to be the source of neutrino mass, if this degenerate pattern is correct. A new supernova constraint is utilized, and implications and tests of the different mass matrices are noted.

PACS number(s): 14.60.Pq, 23.40.Bw, 95.35.+d, 96.60.Kx

I. INTRODUCTION

There are several different observations involving neutrinos which receive a plausible and satisfactory explanation if the neutrinos are massive, which they are not in the standard model. First is the well-known solar neutrino deficit [1], observed by four different experiments [2]. Second is the deficit of muon neutrinos relative to electron neutrinos produced in the atmosphere, as measured by three experiments [3]. Third is the likely need for a neutrino component of the dark matter of the Universe to understand the structure and density on all distance scales [4]. We showed [5] that, consistent with particle physics and cosmological constraints, there are only three conceivable patterns of neutrino mass which could explain these three phenomena. Of the three, one gave warm dark matter, rather than the favored hot component, and one other also appeared dubious for Majorana masses because of the limitation on electron neutrino mass from neutrinoless double β decay, $\beta\beta_{0\nu}$.

The $\beta\beta_{0\nu}$ situation has now changed, however, and leads us to emphasize this formerly less favored pattern of neutrino masses. What was previously [6] a limit on effective neutrino mass has, after another year and a half of data taking, become about a 2-SD indication for that mass [7]. With current matrix element calculations, this effective Majorana mass from the enriched ^{76}Ge experiment is $\langle m_\nu \rangle \sim$ 1–2 eV. Adding interest to this possibility is the observation in a ^{130}Te experiment [8] of a similar 2-SD effect. In this case the even more uncertain matrix element calculations would favor $\langle m_\nu \rangle \sim 4$ eV. Clearly much more experimental work is needed before this hint of neutrino mass is even at the level of believability of the three mentioned above, but the uncertainties in the nuclear matrix elements make it possible that the ^{76}Ge and ^{130}Te experiments are observing the same real effect. In case this is true, we wish to point out how all of these results could be accommodated theoretically.

First, however, we review the constraints on neutrino mass from experiments, cosmology, and astrophysics.

II. INPUT INFORMATION

A. Solar neutrino deficit

For massive neutrinos which can oscillate from one species to another, the solar electron neutrino observations [2] can be understood if the neutrino mass differences and mixing angles fall into one of the following ranges [9], where the Mikheyev-Smirnov-Wolfenstein (MSW) mechanism is included [10]:

$$\begin{aligned}
&\text{(a) small-angle MSW,}\quad \Delta m^2_{ei} \sim 6\times 10^{-6}\ \text{eV}^2\ ,\\
&\qquad \sin^2 2\theta_{ie} \sim 7\times 10^{-3}\ ,\\
&\text{(b) large-angle MSW,}\quad \Delta m^2_{ei} \sim 9\times 10^{-6}\ \text{eV}^2\ ,\\
&\qquad \sin^2 2\theta_{ei} \sim 0.6\ ,\\
&\text{(c) vacuum oscillation,}\quad \Delta m^2_{ei} \sim 10^{-10}\ \text{eV}^2\ ,\\
&\qquad \sin^2 2\theta_{ei} \sim 0.9\ .
\end{aligned} \tag{1}$$

Perspectives of Double Beta and Dark Matter Search as Windows to New Physics

H.V. Klapdor-Kleingrothaus

Max-Planck-Institut für Kernphysik, P.O. Box 10 39 80, D-69029 Heidelberg, Germany

Abstract. Nuclear double-beta decay provides an extraordinarily broad potential to search for beyond-standard-model physics, which is already probing the TeV scale, on which new physics should manifest itself. The following possibilities are reviewed here. First, the results of present-generation experiments are presented. The most sensitive one of them – the Heidelberg–Moscow experiment in the Gran Sasso – probes the electron mass in the sub-eV region and has recently reached a limit of ~0.1 eV. This limit has striking influence on presently discussed neutrino-mass scenarios. Based to a large extent on the theoretical work of the Heidelberg Double Beta Group in the last two years, results are obtained also for SUSY models (R-parity breaking, sneutrino mass), leptoquarks (leptoquark–Higgs coupling), compositeness, right-handed W boson mass, test of special relativity and the equivalence principle in the neutrino sector and others. These results are comfortably competitive to corresponding results from high-energy accelerators like TEVATRON, HERA, etc. One of the enriched ^{76}Ge detectors also yields the most-stringent limits for cold dark matter (WIMPs) to date by using raw data. Second, future perspectives of $\beta\beta$ research are discussed. A new Heidelberg experimental proposal (GENIUS) will allow us to increase the sensitivity for Majorana neutrino masses from the present level of at best 0.1 eV down to 0.01 or even 0.001 eV. Its physical potential would be a breakthrough into the multi-TeV range for many beyond-standard models. Its sensitivity for neutrino-oscillation parameters would be larger than all present terrestrial neutrino-oscillation experiments and those planned for the future. It could probe directly the large-angle, and for almost-degenerate neutrino-mass scenarios even the small-angle solution of the solar-neutrino problem. It would further, already in a first step using only 100 kg of natural Ge detectors, cover almost the full MSSM parameter space for prediction of neutralinos as cold dark matter, making the experiment competitive to LHC in the search for supersymmetry. Finally GENIUS could be used as the first real-time detector of solar *pp* neutrinos.

1 Motivation for the Search for Double-Beta Decay and a Future Perspective: GENIUS

Double-beta decay yields – besides proton decay – the most promising possibilities to probe beyond-standard-model physics above accelerator energy scales.

The potential of double-beta decay includes information on the neutrino and sneutrino mass, SUSY models, compositeness, leptoquarks, right-handed

Springer Tracts in Modern Physics, Vol. 163
Symmetries in Intermediate and High Energy Physics
Eds.: Faessler, Kosmas, Leontaris

VOLUME 83, NUMBER 1 PHYSICAL REVIEW LETTERS 5 JULY 1999

Limits on the Majorana Neutrino Mass in the 0.1 eV Range

L. Baudis, A. Dietz, G. Heusser, H. V. Klapdor-Kleingrothaus,* I. V. Krivosheina,† St. Kolb, B. Majorovits, V. F. Melnikov,† H. Päs, F. Schwamm, and H. Strecker

Max-Planck-Institut für Kernphysik, P.O. Box 10 39 80, D-69029 Heidelberg, Germany

V. Alexeev, A. Balysh, A. Bakalyarov, S. T. Belyaev,* V. I. Lebedev, and S. Zhukov

Russian Science Centre, Kurchatov Institute, 123 182 Moscow, Russia

(Received 26 January 1999)

The Heidelberg-Moscow experiment gives the most stringent limit on the Majorana neutrino mass. After 24 kg yr of data with pulse shape measurements, we set a lower limit on the half-life of the $0\nu\beta\beta$ decay in ^{76}Ge of $T_{1/2}^{0\nu} \geq 5.7 \times 10^{25}$ yr at 90% C.L. (after PDG98 [C. Caso *et al.*, Eur. Phys. J. **C3**, 1 (1998)]), the sensitivity of the experiment being $T_{1/2}^{0\nu} \geq 1.6 \times 10^{25}$ yr at 90% C.L. We thus exclude an effective Majorana neutrino mass greater than 0.2 eV (0.39 eV sensitivity), using the matrix elements of A. Staudt, K. Muto, and H. V. Klapdor-Kleingrothaus, Europhys. Lett. **13**, 31 (1990). This limit sets strong constraints on degenerate neutrino mass models.

PACS numbers: 14.60.Pq, 23.40.Bw

Neutrinoless double beta ($0\nu\beta\beta$) decay is an extremely sensitive tool to probe theories beyond the standard model (see [1]). While the standard model exactly conserves $B - L$, $0\nu\beta\beta$ decay violates lepton number, and $B - L$, by two units. The simplest mechanism which can induce $0\nu\beta\beta$ decay is the exchange of a Majorana neutrino between the decaying neutrons. Alternatively, any theory that contains lepton number violating interactions can lead to the process. Independently of the underlying mechanism, an observation of the $0\nu\beta\beta$ decay would be an evidence for a nonzero Majorana neutrino mass [2]. There are several indications for nonzero neutrino masses, the most stringent ones come from solar and atmospheric neutrino experiments. In particular, the confirmation by Super-Kamiokande of the atmospheric neutrino deficit [3] provides strong evidence for neutrino oscillations, although also other solutions are possible [4]. If a neutrino as a hot dark matter (HDM) component is taken into account, then fitting the atmospheric, solar, and HDM scales with three neutrinos is possible only in the degenerate mass scenario, where all neutrinos have nearly the same mass, of the order of $\mathcal{O}$(eV) [5]. This would lead to an amplitude for $0\nu\beta\beta$ decay mediated by the neutrino mass which is accessible by the present sensitivity of the Heidelberg-Moscow experiment.

The Heidelberg-Moscow experiment operates five *p*-type HPGe detectors in the Gran Sasso Underground Laboratory. The Ge crystals were grown out of 19.2 kg of 86% enriched ^{76}Ge material. The total active mass of the detectors is 10.96 kg, corresponding to 125.5 mol of ^{76}Ge, presently the largest source strength of all double beta experiments. Four detectors are placed in a common 30 cm thick lead shielding in a radon-free nitrogen atmosphere, surrounded by 10 cm of boron-loaded polyethylene and with two layers of 1 cm thick scintillators on top. The remaining detector is situated in a separate box with 27 cm electrolytical copper and 20 cm lead shielding, flushed with gaseous nitrogen and with 10 cm of boron-loaded polyethylene below the box. A detailed description of the experiment and its background is given in [6]. For a further reduction of the already very low background of the experiment, a pulse shape analysis (PSA) method was developed [7]. The analysis distinguishes between multiple scattered interaction in the Ge crystal, so called multiple site events (MSE), and pointlike interactions, i.e., single site events (SSE). Since double beta decay events belong to the SSE category, the method allows one to effectively reduce the background of multiple Compton scattered photons. The probability of correct detection for a SSE is 75%, and 74% for a MSE [7].

Figure 1 shows the total spectrum of the five enriched detectors of the Heidelberg-Moscow experiment, with a statistical significance of 41.55 kg yr. Because of the good energy resolution of the detectors, the γ activities can be easily identified via their specific lines. The neutron background can be estimated from measurements with and without the neutron shielding, while the effect of muons can be deduced from the scintillator measurements in coincidence with Ge detectors. In the region of interest for the $0\nu\beta\beta$ decay in ^{76}Ge (Q value $= 2038.56 \pm 0.32$ keV [8]) the dominant background originates from the Compton continuum of the ^{208}Tl line (2615 keV), the summed ^{60}Co line (2506 keV), and some ^{214}Bi lines (2119, 2204, 2248, and 2293 keV) (all together about 60% of the total background), from neutron-induced (about 30%) and muon-induced (about 10%) events.

For the evaluation of the $0\nu\beta\beta$ decay we consider both data sets, with and without pulse shape analysis. We see in none of them an indication for a peak at the Q value of the $0\nu\beta\beta$ decay. The total spectrum of the five detectors with a statistical significance of 41.55 kg yr contains all the data with the exception of the first 200 d of measurement of each detector. The interpolated energy resolution at the energy of the hypothetical $0\nu\beta\beta$ peak is 3.85 ± 0.16 keV. To estimate the expected background in the $0\nu\beta\beta$ region, we take the energy interval from 2000 to 2080 keV. The expected background in the 3σ peak interval, centered at 2038.56 keV, is 78 ± 3 events, whereas the number of measured events in the same peak region is 68. This number is below the expectation and might be due to peaks in the background interval, which are however too weak to be identified. To extract a half-life limit for the $0\nu\beta\beta$ decay we make the assumption, as recommended by [9], that the number of measured events equals the background expectation. With the achieved energy resolution, the number of excluded events in the 3σ peak region is 15.9 (9.52) with 90% C.L. (68% C.L.), resulting in a half-life limit of (for the $0^+ \rightarrow 0^+$ transition)

$$T_{1/2}^{0\nu} \geq 1.3 \times 10^{25} \text{ yr}, \quad 90\% \text{ C.L.}$$
$$T_{1/2}^{0\nu} \geq 2.1 \times 10^{25} \text{ yr}, \quad 68\% \text{ C.L.}$$

We consider now the data with pulse shape measurements, with a statistical significance of 24.16 kg yr and an energy resolution at 2038.56 keV of 4.2 ± 0.17 keV. The expected number of events from the background left and right of the peak is 13 ± 1 events, the measured number of events in the 3σ peak region is 7. Considering again the number of expected events instead of the number of measured ones [9], we can exclude 7.17 (4.17) events with 90% C.L. (68% C.L.). The limit on the half-life is

$$T_{1/2}^{0\nu} \geq 1.6 \times 10^{25} \text{ yr}, \quad 90\% \text{ C.L.}$$
$$T_{1/2}^{0\nu} \geq 2.8 \times 10^{25} \text{ yr}, \quad 68\% \text{ C.L.}$$

Obviously the pulse shape data are now not only competitive with the complete data set, but they deliver more stringent lower limits on the half-life of the $0\nu\beta\beta$ decay. The pulse shape analysis reduces the background in the interesting energy region by a factor of 3 [background index without PSA: (0.18 ± 0.02) events/(kg yr keV), with PSA: (0.06 ± 0.02) events/(kg yr keV)]. This reduction factor is due to the large fraction of multiple Compton scattered events in the $0\nu\beta\beta$ decay region.

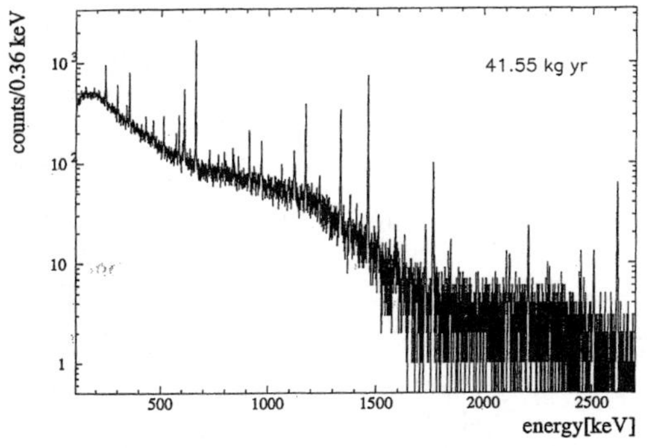

FIG. 1. Sum spectrum of all detectors of the Heidelberg-Moscow experiment with 41.55 kg yr for the energy region between 100 and 2700 keV.

Figure 2 shows the combined spectrum of the five detectors after 41.55 kg yr and the SSE spectrum, corrected for the detection efficiency, after 24.16 kg yr. The solid lines represent the exclusion limits for the two spectra at the 90% C.L.

In addition we report the limits obtained by evaluating the SSE data with the new method proposed by the Particle Data Group 98 [10]. The number of excluded events for an observation of seven events and a background expectation of 13 is 2.07 (0.47) with 90% C.L. (68% C.L.) (see Tables III and V in [11]). The resulting upper limits for the half-life of the $0\nu\beta\beta$ decay are the following:

$$T_{1/2}^{0\nu} \geq 5.7 \times 10^{25} \text{ yr}, \quad 90\% \text{ C.L.}$$
$$T_{1/2}^{0\nu} \geq 2.5 \times 10^{26} \text{ yr}, \quad 68\% \text{ C.L.}$$

The sensitivity of the experiment, as defined in [11], is again obtained by setting the measured number of events equal to the expected background. With 7.51 (4.71) excluded events at 90% C.L. (68% C.L.), the half-life limit is

$$T_{1/2}^{0\nu} \geq 1.6 \times 10^{25} \text{ yr}, \quad 90\% \text{ C.L.}$$
$$T_{1/2}^{0\nu} \geq 2.5 \times 10^{25} \text{ yr}, \quad 68\% \text{ C.L.}$$

Figure 3 shows the SSE spectrum and the excluded peaks (upper limit and sensitivity) with 90% C.L.

Using the matrix elements of [12] and neglecting right-handed currents, we can convert the lower half-life limit into an upper limit on the effective Majorana neutrino mass. The obtained limits are shown in Table I. The Heidelberg-Moscow experiment is thus, at present, giving the most stringent upper limit on the Majorana neutrino

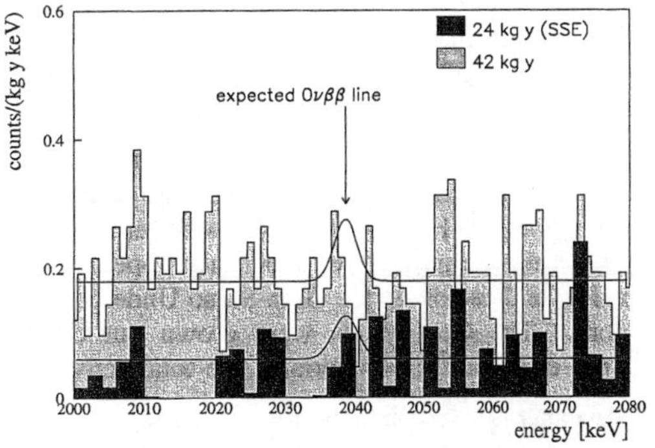

FIG. 2. Sum spectrum of all five detectors with 41.55 kg yr and SSE spectrum with 24.16 kg yr in the region of interest for the $0\nu\beta\beta$ decay. The curves correspond to the excluded signals with $T_{1/2}^{0\nu} \geq 1.3 \times 10^{25}$ yr (90% C.L.) and $T_{1/2}^{0\nu} \geq 1.6 \times 10^{25}$ yr (90% C.L.)

VOLUME 83, NUMBER 1 PHYSICAL REVIEW LETTERS 5 JULY 1999

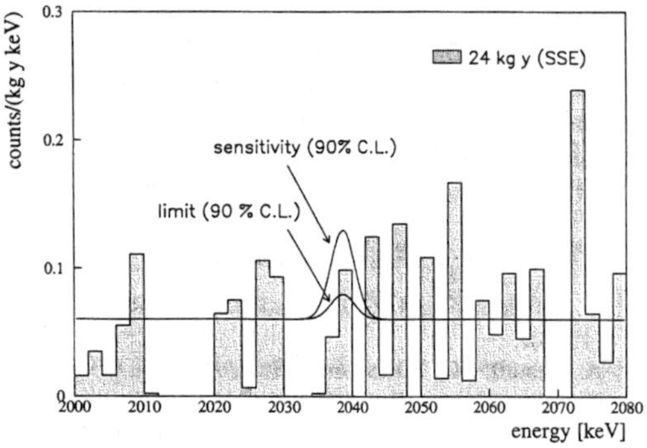

FIG. 3. SSE spectrum after 24.16 kg yr and the corresponding excluded signal and sensitivity of the experiment, at 90% C.L. after [10].

mass, of 0.2 eV at 90% C.L. (0.1 eV at 68% C.L.) (after PDG98 [10]). The sensitivity of the experiment is 0.38 eV at 90% C.L. [10]. This is equal to the limit obtained with the method proposed by PDG96 [9] (see Table I).

Table II gives a list of the limits from the presently most sensitive experiments on double beta decay and a comparison between different matrix element calculations. It should be mentioned that the calculations of [22] do not use a realistic nucleon-nucleon force, those of [21] use too small a configuration space (leaving out the important spin-orbit partners), and the recent calculation of [24] does not fulfill the Ikeda sum rule. This favors the matrix elements of [12,23] (for a discussion, see also [1]). Some further confidence in the matrix elements of [12] might be derived from the fact that the $2\nu\beta\beta$ decay half-life of ^{76}Ge [27] has been predicted correctly [28] within a factor of 2 (the $2\nu\beta\beta$ decay matrix element within $\sqrt{2}$).

The limit on the effective Majorana neutrino mass given by the Heidelberg-Moscow experiment is about an order of magnitude lower than for other double beta experiments. This is the result of the high source strength, good energy resolution, high material purities combined with the efficiency of the pulse shape analysis, and, last, but not least, the excellent long-term stability of the experiment.

TABLE I. Limits on the effective Majorana neutrino mass from the $0\nu\beta\beta$ decay of ^{76}Ge for the matrix elements from [12].

	$T_{1/2}^{0\nu}$ (yr)	$\langle m \rangle$ (eV)	C.L. (%)
Full data set	$\geq 1.3 \times 10^{25}$	≤ 0.43	90
	$\geq 2.1 \times 10^{25}$	≤ 0.33	68
SSE data after [9]	$\geq 1.6 \times 10^{25}$	≤ 0.38	90
	$\geq 2.8 \times 10^{25}$	≤ 0.29	68
SSE data after [10]	$\geq 5.7 \times 10^{25}$	≤ 0.20	90
	$\geq 2.5 \times 10^{26}$	≤ 0.10	68
Sensitivity	$\geq 1.6 \times 10^{25}$	≤ 0.38	90
	$\geq 2.5 \times 10^{25}$	≤ 0.30	68

After about another five years of measurement, assuming no other improvement in the present background index, the Heidelberg-Moscow experiment will be able to explore the half-life of the $0\nu\beta\beta$ decay up to some 10^{26} yr in the best case. For a significant improvement of the experimental sensitivity, a much lower background counting rate and a higher source strength are required.

A major step forward in this field would bring the new project proposed by our group, GENIUS (Germanium in liquid Nitrogen underground setup) [1,29], which would operate enriched Ge crystals directly in liquid nitrogen. The goal is to reduce the background by about 3 orders of magnitude by removing essentially all materials from the vicinity of the measurement crystals. It was shown that Ge detectors work reliably in liquid nitrogen [29,30], also detailed Monte Carlo simulations of the various expected background components [30] confirm the possibility of obtaining a counting rate of 0.3 events/(ton yr keV) in the $0\nu\beta\beta$ region. GENIUS is conceived to test the neutrino mass down to the level of 0.01 eV and lower.

In conclusion, already in the present stage, the Heidelberg-Moscow experiment is setting the most stringent limit on the Majorana neutrino mass and is testing the predictions of degenerate neutrino mass models. For example, in models which try to accommodate the solar

TABLE II. Limits on the half-lives of the $0\nu\beta\beta$ decay and on the effective Majorana neutrino mass for different matrix element calculations for the best existing experiments with half-life limits $>10^{21}$ yr.

Isotope	$T_{1/2}^{0\nu}$ (yr)	C.L. (%)	$\langle m \rangle$ (eV)
^{48}Ca	$\geq 9.5 \times 10^{21}$	76 [13]	$\leq$13 [12]
^{76}Ge limit	$\geq 5.7 \times 10^{25}$	90	$\leq$0.20 [12], 0.56 [21], 0.52 [22], 0.19 [23], 0.4 eV [24], 0.17 eV [25], 0.22 eV [26]
^{76}Ge sensitivity	$\geq 1.6 \times 10^{25}$	90	$\leq$0.38 [12], 1.07 [21], 0.93 [22], 0.36 [23], 0.77 [24], 0.32 eV [25], 0.41 eV [26]
^{82}Se	$\geq 9.5 \times 10^{21}$	90 [14]	$\leq$7.9 [12], 15.7 [21], 24 [22], 8 [23], 7.8 eV [25], 8.6 eV [26]
	$\geq 2.7 \times 10^{22}$	68 [15]	$\leq$4.7 [12], 9.4 [21], 14.4 [22], 4.75 [23], 4.6 eV [25], 5.1 eV [26]
^{100}Mo	$\geq 5.2 \times 10^{22}$	68 [16]	$\leq$4.9 [12], 4.34 [22], 2.2 [23], 2.2 [24]
^{116}Cd	$\geq 3.2 \times 10^{22}$	90 [17]	$\leq$3.9 [12], 4.26 eV [26]
^{130}Te	$\geq 5.6 \times 10^{22}$	90 [18]	$\leq$2.9 [12], 3.43 [22], 3.1 [23], 2.88 [24], 1.69 [25], 3.59 [26]
^{136}Xe	$\geq 4.4 \times 10^{23}$	90 [19]	$\leq$2.2 [12], 5.2 [21], 2.7 [22], 1.8 [23], 3.12 [26]
^{150}Nd	$\geq 1.22 \times 10^{21}$	90 [20]	$\leq$5.2 [12], 6.7 [26]

and atmospheric neutrino problems while considering the neutrino as a hot dark matter candidate with a mass of a few eV, the small angle Mikheyev, Smirnov, Wolfenstein (MSW) solution is practically ruled out [31]. For the large angle MSW solution, an effective Majorana neutrino mass smaller than 0.2 eV would need an unnatural fine-tuning to account for a relevant neutrino mass in a mixed hot and cold dark matter cosmology [32].

This, as well as other implications for physics beyond the standard model, such as left-right symmetric models, supersymmetry, leptoquarks, and compositeness will be discussed in detail elsewhere.

The Heidelberg-Moscow experiment was supported by the Bundesministerium für Forschung und Technologie der Bundesrepublik Deutschland, the State Committee of Atomic Energy of Russia, and the Istituto Nazionale di Fisica Nucleare of Italy. L. B. was supported by the Graduiertenkolleg of the University of Heidelberg.

*Author to whom correspondence should be addressed.
†On leave from the Radiophysical Research Institute (NIRFI), Nishnij Novgorod, Russia.

[1] H. V. Klapdor-Kleingrothaus, Int. J. Mod. Phys. A **13**, 3953–3992 (1998); H. V. Klapdor-Kleingrothaus, in *Proceedings of Beyond the Desert 97, First International Conference on Particle Physics Beyond the Standard Model, Castle Ringberg, Germany, 1997*, edited by H. V. Klapdor-Kleingrothaus and H. Päs (IOP, Bristol, 1998), pp. 485–531; in *Proceedings of Neutrino98, Takayama, Japan, 1998* (World Scientific, Singapore, 1999).
[2] J. Schechter and J. W. F. Valle, Phys. Rev. D **25**, 2951 (1982).
[3] The Super-Kamiokande Collaboration, Y. Fukuda *et al.*, Phys. Rev. Lett. **81**, 1562-1567 (1998).
[4] M. C. Gonzalez-Garcia *et al.*, hep-ph/9809531.
[5] D. O. Caldwell and R. N. Mohapatra, Phys. Rev. D **48**, 3259 (1993); R. N. Mohapatra, in *Proceedings of the 17th International Conference on Neutrino Physics and Astrophysics, Neutrino96, Helsinki, 1996* (World Scientific, Singapore, 1997); A. Yu. Smirnov, in *Proceedings of the 28th International Conference on High Energy Physics, Warsaw, 1996* (World Scientific, Singapore, 1996).
[6] Heidelberg-Moscow Collaboration, M. Günther *et al.*, Phys. Rev. D **55**, 54 (1997).
[7] Heidelberg-Moscow Collaboration, L. Baudis *et al.*, Phys. Lett. B **407**, 219 (1997).
[8] J. G. Hykawy *et al.*, Phys. Rev. Lett **67**, 1708 (1991).
[9] Particle Data Group (PDG 96), R. M. Barnett *et al.*, Phys. Rev. D **54**, 1 (1996).
[10] Particle Data Group (PDG 98), C. Caso *et al.*, Eur. Phys. J. **C3**, 1 (1998).
[11] G. J. Feldman and R. D. Cousins, Phys. Rev. D **57**, 3873 (1998).
[12] A. Staudt, K. Muto, and H. V. Klapdor-Kleingrothaus, Europhys. Lett. **13**, 31 (1990).
[13] Y. Ke *et al.*, Phys. Lett. B **265**, 53 (1991).
[14] R. Arnold *et al.*, Nucl. Phys. **A636**, 209–233 (1998).
[15] S. R. Elliott *et al.*, Phys. Rev. C **46**, 1535 (1992).
[16] N. Kudomi *et al.*, Nucl. Phys. **A629**, 527c–530c (1998).
[17] F. A. Danevich *et al.*, Nucl. Phys. B (Proc. Suppl.) **70**, 246–248 (1999).
[18] A. Alessadrello *et al.*, Phys. Lett. B **433**, 156–162 (1998).
[19] R. Luescher *et al.*, Nucl. Phys. B (Proc. Suppl.) **66**, 195–198 (1998).
[20] A. De Silva *et al.*, Phys. Rev. C **56**, 2451–2467 (1997).
[21] E. Caurier *et al.*, Phys. Rev. Lett. **77**, 1954–1957 (1996).
[22] J. Engel *et al.*, Phys. Rev. C **37**, 731–746 (1988).
[23] T. Tomoda, Rep. Prog. Phys. **54**, 53 (1991).
[24] F. Simkovic *et al.*, Phys. Lett. B **393**, 267–273 (1997).
[25] W. C. Haxton *et al.*, Nucl. Phys. B (Proc. Suppl.) **31**, 82 (1993).
[26] X. R. Wu *et al.*, Phys. Lett. B **272**, 169 (1991); **276**, 274 (1992).
[27] A. Balysh *et al.*, Phys. Lett. B **322**, 176 (1994).
[28] K. Muto, E. Bender, and H. V. Klapdor, Z. Phys. A **334**, 177 (1989).
[29] H. V. Klapdor-Kleingrothaus, J. Hellmig, and M. Hirsch, J. Phys. G **24**, 483 (1998).
[30] L. Baudis, G. Heusser, B. Majorovits, Y. Ramachers, H. Strecker, and H. V. Klapdor-Kleingrothaus, Nucl. Instrum. Methods Phys. Res., Sect. A **426**, 425–435 (1999).
[31] H. Minakata and O. Yasuda, Phys. Rev. D **56**, 1692 (1997); R. Adhikari and G. Rajasekaran, hep-ph/98123361.
[32] H. Minakata and O. Yasuda, Nucl. Phys. **B523**, 597–610 (1998); R. N. Mohapatra, hep-ph/9808284.

Double Beta Decay of ^{76}Ge: New Results from the Heidelberg-Moscow Experiment[1]

H.V. Klapdor-Kleingrothaus[2], L. Baudis, A. Dietz, I.V. Krivosheina, G. Heusser, St. Kolb, B.Majorovits, H. Päs, H. Strecker and H. Tu

Max-Planck-Institut für Kernphysik, Heidelberg, Gernmany

V. Alexeev, A. Balysh, A. Bakalyarov, S.T. Belyaev, V.I. Lebedev and S. Zhukov

Russian Science Center Kurchatov Institute, 123 182 Moscow, Russia

Abstract

New results for the double beta decay in ^{76}Ge are presented. They are extracted from the Heidelberg-Moscow experiment, which operates five enriched ^{76}Ge detectors in an extreme low-level environment in the Gran Sasso underground laboratory. The two-neutrino accompanied double beta decay is evaluated for the first time for all five detectors with a statistical significance of 47.7 kg y, resulting in a half life of $T^{2\nu}_{1/2} = [1.55 \pm 0.01(stat)^{+0.19}_{-0.15}(syst)] \times 10^{21}$ years. No evidence for a Majoron emitting decay mode or for the neutrinoless mode is observed. The half-life of the $0\nu\chi\beta\beta$ -decay is $T^{0\nu}_{1/2} > 6.4 \times 10^{22}$ yr with 90% C.L. The inferred upper limit on the effective Majoron neutrino coupling of 8×10^{-5} is the most stringent limit obtained so far in a direct measurement. The lower limit on the half-life of the $0\nu\beta\beta$ - decay after 35.5 kg y of measurement with pulse shape analysis is $T^{0\nu}_{1/2} > 1.9 \times 10^{25}$ $(1.3 \times 10^{25}$ years with 90% C.L. (68% C.L.). Using the matrix elements of A. Staudt, K. Muto, H.V. Klapdor-Kleingrothaus, Europhys. Lett. **13**, 31 (1990), an effective Majorana neutrino mass greater that 0.35 eV (0.27eV) is excluded.

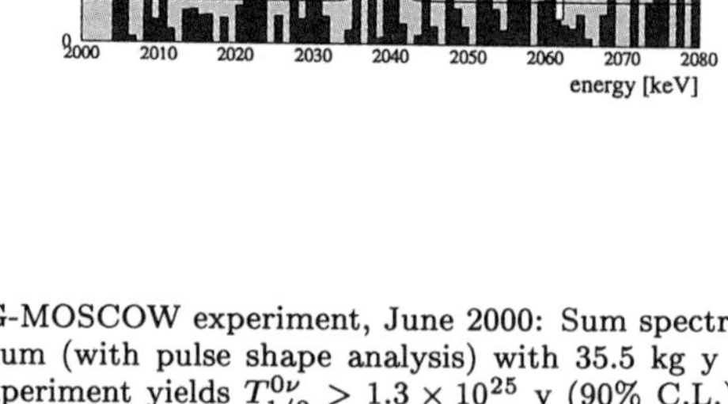

Fig. 4. HEIDELBERG-MOSCOW experiment, June 2000: Sum spectrum of all five detectors with 53.9 kg y and SSE spectrum (with pulse shape analysis) with 35.5 kg y in the region of interest for the $0\nu\beta\beta$ - decay. The experiment yields $T^{0\nu}_{1/2} > 1.3 \times 10^{25}$ y (90% C.L.), $T^{0\nu}_{1/2} > 2.2 \times 10^{25}$ y (68% C.L.), for the full spectrum and $T^{0\nu}_{1/2} > 1.9 \times 10^{25}$ y (90% C.L.), $T^{0\nu}_{1/2} > 3.1 \times 10^{25}$ y (68% C.L.) for the SSE spectrum. The curves correspond to the excluded signals at 90% C.L. The corresponding Majorana neutrino mass limits are (for the SSE spectrum) 0.35 eV (90% C.L.) and 0.27 eV (68% C.L.)

[1]Preprint submitted to Elsevier Preprint

[2]spokesman of the collaboration

RAPID COMMUNICATIONS

PHYSICAL REVIEW D, VOLUME 61, 031301(R)

Constraints on mixing angles of Majorana neutrinos

Rathin Adhikari* and G. Rajasekaran†

The Institute of Mathematical Sciences, C.I.T Campus, Taramani, Chennai-600 113, India

(Received 14 December 1998; revised manuscript received 28 May 1999; published 21 December 1999)

By combining the inputs from the neutrinoless double beta decay and the fits of cosmological models of dark matter with solar and atmospheric neutrino data, we obtain constraints on two of the mixing angles of Majorana neutrinos, which become stronger when coupled with the reactor neutrino data. These constraints are strong enough to rule out Majorana neutrinos if the small angle solution of solar neutrino puzzle is borne out.

PACS number(s): 14.60.Pq, 14.60.St, 23.40.Bw, 26.65.+t

It is well-known [1,2] that unless neutrinos are very massive and nonrelativistic, or interact through both left- and right-handed currents, experimental data on neutrino-induced reactions cannot distinguish between Dirac and Majorana neutrinos. Neutrinoless double beta decay remains as the only feasible tool to probe this question. Although experiments [3–5] have so far provided only upper limits on the rate of this decay, recent limits [5] combined with other inputs on neutrino physics might already lead to important information on whether neutrinos are Dirac or Majorana particles. These other inputs are indications from the analysis of the cosmic microwave background for the presence of a hot dark matter component, which is presumably neutrinos of mass ≈ 1 eV [6,8,9], and the indications [10–12] from the analysis of solar and atmospheric neutrinos that neutrinos do oscillate and that the mass differences among the three neutrinos are much smaller than this scale of 1 eV.

We show in this paper that if neutrinos are Majorana particles, a combined study of all of the above pieces of data leads to rather stringent restrictions on two of the mixing angles that occur for three flavors of neutrinos. If these results are then confronted with the values of these mixing angles allowed by solar, atmospheric, and reactor neutrino data, the allowed regions are further narrowed and in fact, in a few cases, one is already close to contradiction, thus leading to the conclusion that neutrinos are not Majorana particles.

In the three flavor mixing scheme the neutrino flavor eigenstates $\nu_\alpha = \nu_{e,\mu,\tau}$ are related to the mass eigenstates $\nu_i = \nu_{1,2,3}$ by

$$\nu_\alpha = \sum_i U_{\alpha i} \nu_i , \tag{1}$$

where $U_{\alpha i}$ are the elements of the unitary mixing matrix U. We note that for the Majorana neutrino [2,13] there are three CP-violating phases in contrast with the case of the Dirac neutrino which has only one phase. We use the parametrization [13]

$$U = \begin{pmatrix} c_\omega c_\phi & s_\omega c_\phi e^{-i\delta_1} & s_\phi e^{-i\delta_2} \\ -s_\omega c_\psi e^{i\delta_1} - c_\omega s_\psi s_\phi e^{i(\delta_2+\delta_3)} & c_\omega c_\psi - s_\omega s_\psi s_\phi e^{i(\delta_3+\delta_2-\delta_1)} & s_\psi c_\phi e^{i\delta_3} \\ s_\omega s_\psi e^{i(\delta_1-\delta_3)} - c_\omega c_\psi s_\phi e^{i\delta_2} & -c_\omega s_\psi e^{-i\delta_3} - s_\omega c_\psi s_\phi e^{i(\delta_2-\delta_1)} & c_\psi c_\phi \end{pmatrix}, \tag{2}$$

where δ_1, δ_2, and δ_3 are the three CP-violating phases and c and s stand for sine and cosine of the associated mixing angle ω, ϕ, or ψ with the angle placed as subscript.

The rate for the neutrinoless double beta decay depends on the following combination of the neutrino parameters [2]:

$$m_{0\nu\beta\beta} = \left| \sum_{i=1,2,3} \eta_i U_{ei}^2 m_{\nu_i} \right| , \tag{3}$$

where m_{ν_i} are the Majorana neutrino masses, U_{ei} the elements of the first row of the mixing matrix given in Eq. (2), $\eta_i = (1/i)\eta_i^{CP} = \pm 1$, and η_i^{CP} is the CP parity of the Majorana neutrino ν_i. Although neutrinoless double beta decay has not yet been seen experimentally, the experimental upper limits on this rate have recently improved to a significant extent. In particular one may refer to the results of the Tellurium [3] and Germanium experiments [4,5]. The strongest upper limit so far comes from the Germanium experiment [5,14] and it is

$$m_{0\nu\beta\beta} < 0.56 \text{ eV (99\% C.L.)}$$
$$< 0.46 \text{ eV (90\% C.L.)}. \tag{4}$$

These numbers have been obtained using the nuclear matrix elements calculated in [15]. We shall take into account the

*Email address: rathin@imsc.ernet.in
†Email address: graj@imsc.ernet.in

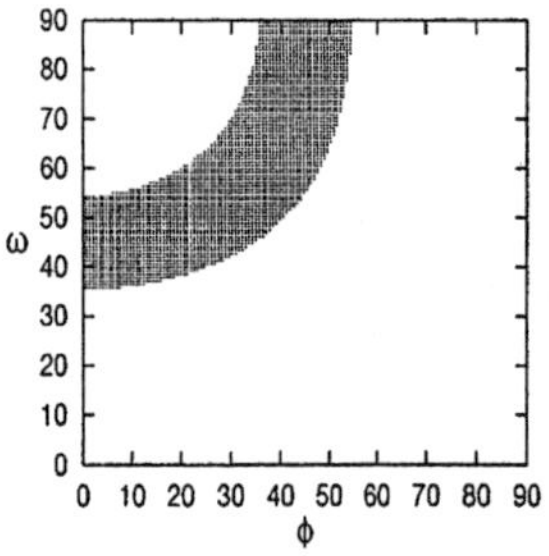

FIG. 1. The allowed region in (ω,ϕ) is shown shaded for $m_\nu = 1.7$ eV, $\eta_1 = -\eta_2 = \eta_3 = \pm 1$, and $\delta_1 = \delta_2 = 0$.

uncertainties in this calculation (see below). Using the conservative number 0.56 eV we have

$$\left| \sum_{i=1,2,3} \eta_i U_{ei}^2 m_{\nu_i} \right| < 0.56 \text{ eV}. \tag{5}$$

Next we consider the fits to the recent data on the anisotropies of the cosmic microwave background radiation [16] and the large scale structure of the universe [17]. The best fit [9,8] requires a mixture of 10% ordinary baryonic matter, 70% cold, and 20% hot dark matter with $\Omega_m = 1$. If the hot component is identified with neutrinos, the model implies [6]

$$\sum_{i=1,2,3} m_{\nu_i} \approx 5 \text{ eV}. \tag{6}$$

The right-hand side of Eq. (6) is not expected to be less than 3 eV for $\Omega_m = 1$ [7] as otherwise there is too much small scale power [18,19]. On the other hand, solar and atmospheric neutrino data suggest that the two mass-squared differences among the three neutrinos are very small [10–12]; $m_2^2 - m_1^2 \approx 10^{-5}$ eV^2 or smaller and $m_3^2 - m_1^2 \approx 10^{-3} - 10^{-2}$ eV^2. Hence we take all three neutrinos as almost degenerate in mass and using Eq. (6) we obtain

$$m_{\nu_i} \approx m_\nu \approx 1.7 \text{ eV}. \tag{7}$$

We shall allow m_ν to vary over a range around 1.7 eV. This will take care of the uncertainties of cosmological models as well as those of the calculations of the nuclear matrix elements in double beta decay, since only the ratio $0.56/m_\nu$ enters into our analysis. Any possible improvement in the neutrinoless double beta decay limit can also be incorporated by scaling m_ν appropriately.

Combining all the inputs, we have the basic inequality

$$|(\eta_1 \cos^2\omega + \eta_2 \sin^2\omega e^{-i2\delta_1})\cos^2\phi + \eta_3 \sin^2\phi e^{-i2\delta_2}| < \frac{0.56}{m_\nu}, \tag{8}$$

where m_ν is expressed in eV. One can rewrite this inequality in terms of two effective phases by combining η_i and δ_i. However, to make our discussion on *CP* conserving and *CP*-violating cases more transparent we have kept both η_i and δ_i above. We proceed to extract the bounds on the mixing angles ω and ϕ implied by this inequality for various choices of η_i, δ_i, and typical values of m_ν favored by the cosmological models. It is to be noted that in contrast to the usual oscillation phenomena studied in neutrino physics, *CP* violation plays an important role in neutrinoless double beta decay.

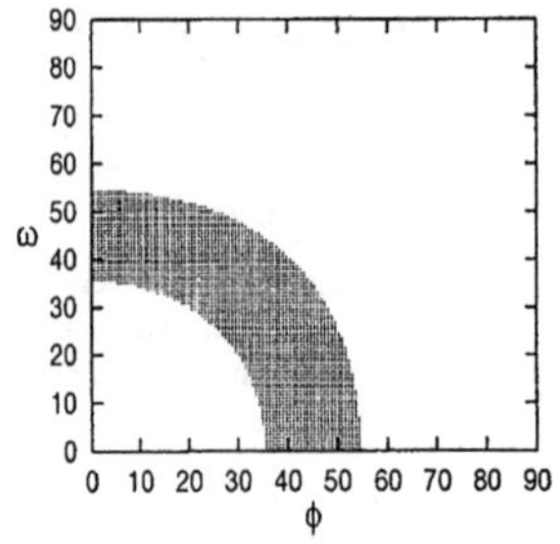

FIG. 2. Same as Fig. 1 but $-\eta_1 = \eta_2 = \eta_3 = \pm 1$.

We shall first consider $\delta_1 = \delta_2 = 0$ in Eq. (8). Out of eight possible combinations for different values of η_i in Eq. (8), four combinations are equivalent to the other four, as only the overall magnitude in the left-hand side of this inequality is constrained. So we shall analyze Eq. (8) on the basis of four cases: case I, $\eta_1 = \eta_2 = \eta_3 = \pm 1$; case II, $\eta_1 = -\eta_2 = \eta_3 = \pm 1$; case III, $-\eta_1 = \eta_2 = \eta_3 = \pm 1$; and case IV, $\eta_1 = \eta_2 = -\eta_3 = \pm 1$. Case I is the natural choice for η_i, if there exists a symmetry linking the three generations. But then the left-hand side of Eq. (8) is unity and so unless $m_\nu \leq 0.56$ eV, the inequality cannot be satisfied [20]. Since such low values of m_ν are not expected in the cosmological models, we conclude that the case of equal intrinsic *CP* parities for the three Majorana neutrinos is not favored. The allowed regions in ω and ϕ for cases II, III, and IV are plotted in Figs. 1, 2, and 3, respectively, for $m_\nu = 1.7$ eV. For both cases II and III, small ϕ $(\phi \approx 0)$ requires ω to be in the region of 45°. Whereas small ω in case II is not allowed at all, small ω in case III requires large ϕ. Case IV leads to a constraint condition on ϕ that is independent of ω and requires ϕ to be in the region of 45°. Figures 4 and 5

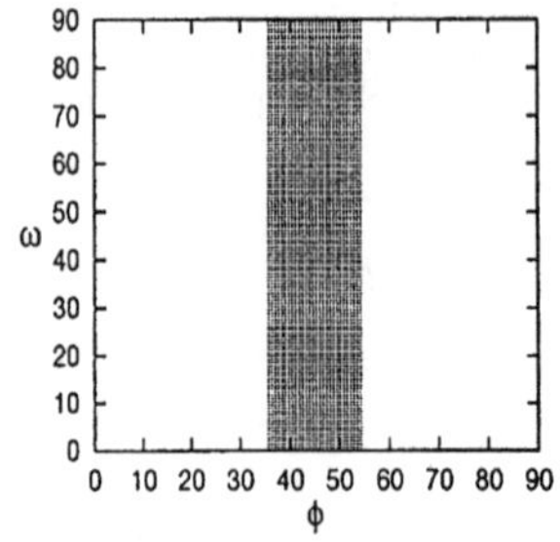

FIG. 3. Same as Fig. 1 but $\eta_1 = \eta_2 = -\eta_3 = \pm 1$.

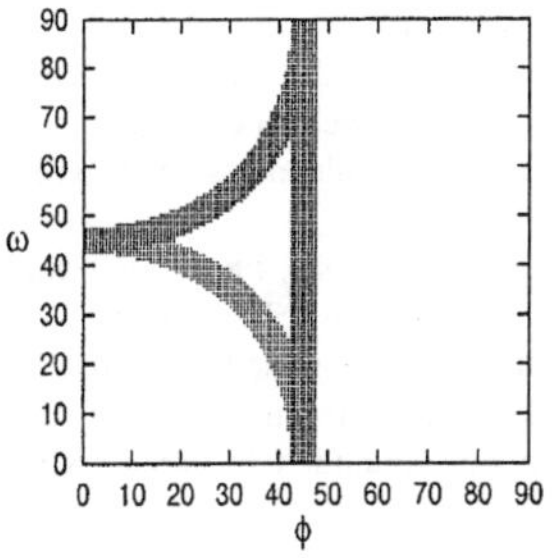

FIG. 4. Allowed region for all possible combinations of η_i, $m_\nu=6$ eV and $\delta_1=\delta_2=0$.

show the total allowed regions for all possible combinations of η_i (i.e., all the cases I, II, III, and IV) for $m_\nu=6$ eV and 0.64 eV, respectively. One may note that for $m_{0\nu\beta\beta}/m_\nu \to 0$, the allowed values of ω and ϕ are constrained to lie on the three curves $\tan^2\phi=\pm\cos 2\omega,1$. Figure 4 (where $m_{0\nu\beta\beta}/m_\nu=0.093$) gives a small width to these curves.

We next consider the *CP*-violating case. Now the choice $\eta_1=\eta_2=\eta_3=\pm 1$ is *not* ruled out. Thus *CP* violation is capable of changing the conclusions dramatically. Figures 6–8 show the allowed regions in ω and ϕ for a few choices of the parameters δ_1 and δ_2.

We have not considered "maximal" *CP* violation $\delta_1 = \pi/2$ and/or $\delta_2=\pi/2$ since these cases are subsumed by the choices of relative negative η_i, as far as inequality (8) is concerned. Therefore, it is important to keep in mind that Figs. 1–5 are relevant for such maximal *CP*-violating cases also. In particular, Fig. 1 and Fig. 2 can be considered to be the maximally *CP*-violating case of $\delta_1=\pi/2$ and $\delta_2=0$.

Let us now compare the above bounds on the mixing angles with the results of the analysis of data from reactor, solar, and atmospheric neutrino experiments. Before we make this comparison, we must comment on the role of *CP* violation in these experiments. We note that even for Majorana neutrinos, the oscillation phenomena are controlled by a single *CP*-violating phase δ [21].

The CHOOZ reactor experiment [22] interpreted within a three flavor framework [23] leads to the constraint

$$\phi \leq 12.5°. \tag{9}$$

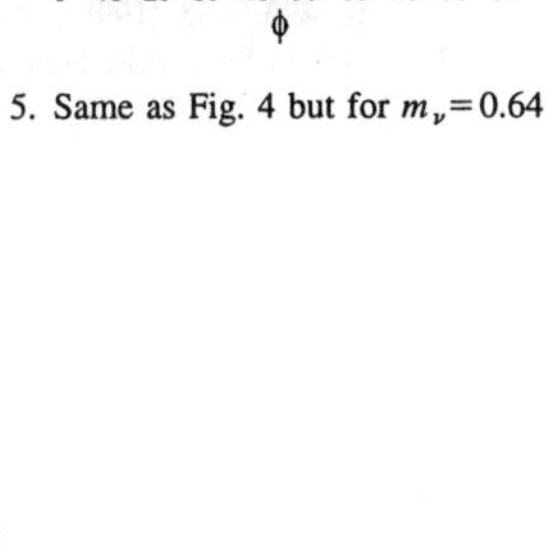

FIG. 5. Same as Fig. 4 but for $m_\nu=0.64$ eV.

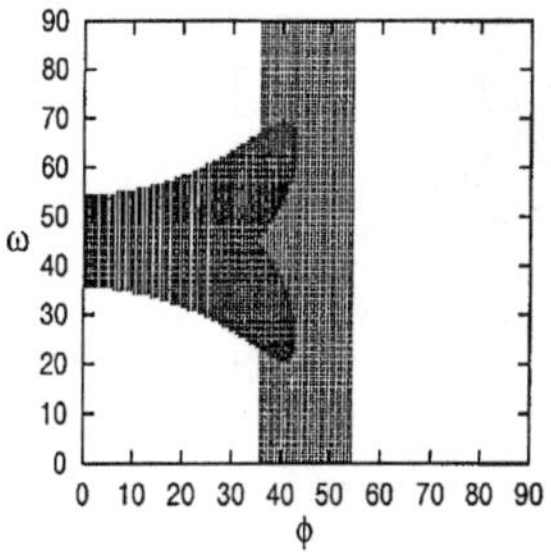

FIG. 6. Allowed region for all possible η_i, $m_\nu=1.7$ eV and $\delta_1=0$, $\pi/8\leq\delta_2\leq 3\pi/8$.

Although *CP* violation was neglected in reference [23], it is easy to see that Eq. (9) is valid even if *CP* is violated. Comparing Eq. (9) with Fig. 3, we see that case IV is ruled out, while comparisons with Figs. 1 and 2 rule out all values of ω except those in the region of 45°. These are for $m_\nu = 1.7$ eV. The results are weakened for $m_\nu=0.64$ eV (Fig. 5) and are further strengthened for $m_\nu=6$ eV (Fig. 4). If *CP* is violated, with $\delta_1=0$ and $\delta_2\neq 0$ (Fig. 6), the above conclusion is still valid. But for $m_\nu=1.7$ eV, $\pi/8 \gtrsim \delta_1 \lesssim 3\pi/8$, and $\delta_2=0$ (Fig. 7), as well as $\pi/8 \gtrsim \delta_1$, $\delta_2 \lesssim 3\pi/8$ (Fig. 8) there is no allowed region at all after using Eq. (9).

It is important to note that as a consequence of Eq. (9), the effect of *CP* violation in all the neutrino-oscillation phenomena is much reduced since the *CP*-violating phase factors $e^{\pm i\delta}$ always occur in combination with $\sin\phi$, as $\sin\phi e^{\pm i\delta}$. Hence it is legitimate to compare with the results of analyses of solar and atmospheric neutrinos even though *CP* violation is usually ignored in those analyses.

Analysis of solar neutrino data on the basis of the Mikheyev-Smirnov-Wolfenstein (MSW) effect [24] leads to the allowed ω values of either 2°–3° or 20°–40° for $\phi \leq 12.5°$. So comparing with the above result, we conclude that the result of the present analysis contradicts the small ω MSW solution ($\omega \approx 2° - 3°$), but there exists some overlap with the large ω MSW solution ($\omega \approx 20° - 40°$). Also, there is no contradiction with the vaccuum oscillation as a solution of the solar neutrino problem since this also requires large ω [27]. Finally there is no contradiction with the results of the

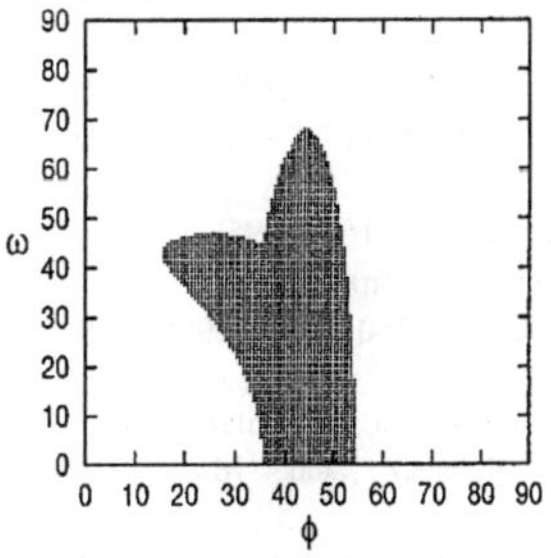

FIG. 7. Allowed region for all possible η_i, $m_\nu=1.7$ eV and $\pi/8\leq\delta_1\leq 3\pi/8$, $\delta_2=0$.

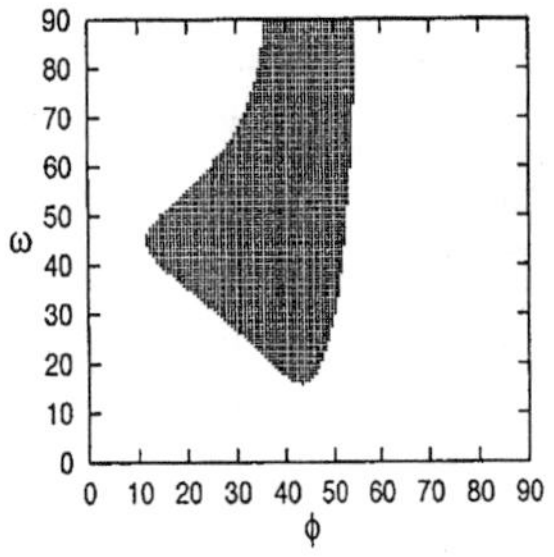

FIG. 8. Allowed region for all possible η_i, $m_\nu = 1.7$ eV and $\pi/8 \leqslant \delta_1$, $\delta_2 \leqslant 3\pi/8$.

atmospheric neutrino analysis [24] since this does not involve ω.

So far we have used the information on m_ν from cosmology to get results on the mixing angles which were then compared with the results of reactor, solar, and atmospheric neutrinos. In view of the uncertainties of cosmological models, one can ask what kind of information on the quasidegenerate mass for Majorana neutrinos can be obtained from our analysis, if we drop the cosmological input completely. It is clear from Fig. 5 that for the small ω MSW solution of the solar neutrino problem, we get an upper bound on m_ν of about 0.7 eV and we have checked that this upper bound becomes 4 eV for the large ω MSW solution.

The quantitative results of our analysis are contained in Figs. 1–8 in the form of restrictions in ω and ϕ. We may also state two qualitative conclusions that emerge from our analysis.

(1) If neutrinos are Majorana fermions and *CP* is conserved, all three neutrinos cannot have the same *CP* parities. This conclusion (which perhaps is well-known and is included here only for the sake of completeness) may be important for model building.

(2) If neutrinos are Majorana fermions, the mixing angle ω cannot be small. Hence, if the small ω solution turns out to be the only correct solution of the solar neutrino problem, then neutrinos cannot be Majorana fermions. This will have serious consequences for models intended to explain small neutrino masses.

The above results and conclusions are based on the present indications that the experiments on the neutrinoless double beta decay require $m_{0\nu\beta\beta}$ to be less than a fraction of an eV while models with neutrinos as the hot component of dark matter require m_ν to be higher than about 1 eV. The restrictions become more severe and the conclusions become stronger if the upper limit on $m_{0\nu\beta\beta}$ decreases [25,26] and/or m_ν increases.

Note added. After the first submission of our manuscript, we came across the papers of Georgi and Glashow [28] and of Branco, Rebelo, and Silva-Marcos [29], whose contents have partial overlap with our work. Our analysis is more general than both of these works in which ϕ is put as zero and in addition we have considered the *CP*-violating cases in detail.

We thank Raj Gandhi, Mohan Narayan, and S. Uma Sankar for discussions and Sandip Pakvasa for a useful communication. We thank Rahul Sinha for an important discussion that clarified the role of *CP* violation in neutrino oscillations.

[1] B. Kayser and R. Shrock, Phys. Lett. **112B**, 137 (1982).
[2] R.N. Mohapatra and P.B. Pal, *Massive Neutrinos in Physics and Astrophysics* (World Scientific, Singapore, 1991); B. Kayser *et al.*, *The Physics of Massive Neutrinos* (World Scientific, Singapore, 1989).
[3] T. Bernatowicz *et al.*, Phys. Rev. Lett. **69**, 2341 (1992); Phys. Rev. C **47**, 806 (1993).
[4] Heidelberg-Moscow Experiment, A. Balysh *et al.*, Phys. Lett. B **356**, 450 (1995).
[5] M. Günther *et al.*, Phys. Rev. D **55**, 54 (1997); Heidelberg-Moscow Experiment, L. Baudis *et al.*, Phys. Lett. B **407**, 219 (1997).
[6] J.R. Primack, Phys. Rev. Lett. **74**, 2160 (1995).
[7] J.R. Primack (private communication).
[8] E. Gawiser and J. Silk, Science **280**, 1405 (1998); J.R. Primack, *ibid.* **280**, 1398 (1998).
[9] J.R. Primack, presented in the Jerusalem Winter School, 1996, astro-ph/9707285; J.R. Primack and M.A.K. Gross, astro-ph/9810204.
[10] B.T. Cleveland *et al.*, Nucl. Phys. B (Proc. Suppl.) **38**, 47 (1995); Y. Fukuda *et al.*, Phys. Rev. Lett. **77**, 1683 (1996).
[11] Super-Kamiohande Collaboration, Y. Fukuda *et al.*, Phys. Lett. B **433**, 9 (1998); **436**, 33 (1998); Phys. Rev. Lett. **81**, 1158 (1998); **81**, 4279 (1998); **81**, 1562 (1998).
[12] N. Hata *et al.*, Phys. Rev. D **56**, 6107 (1997); P. Langacker, Nucl. Phys. B (Proc. Suppl.) **77**, 241 (1999).
[13] K. Zuber, Phys. Rep. **305**, 295 (1998) we have corrected some printing errors in the mixing matrix given in this report.
[14] H. Päs (private communication).
[15] K. Muto, A. Staudt, and H.V. Klapdor-Kleingrothaus, Europhys. Lett. **13**, 31 (1990).
[16] K.M. Gorski *et al.*, Astrophys. J. Lett. **430**, L89 (1994).
[17] G. Efstathiou, J.R. Bond, and S.D.M. White, Mon. Not. R. Astron. Soc. **258**, P1 (1992).
[18] Although the observations of high redshift supernovae have been interpreted [19] in terms of a nonvanishing cosmological constant Λ, cosmological models with $\Omega_m + \Omega_\Lambda = 1$ do not provide good fit to the structure data (CMBR anisotropies and the large-scale structure). Leaving this puzzle to be resolved by the cosmologists, we shall stick to Eq. (6) and deduce the consequences for neutrino physics.
[19] S. Perlmutter *et al.*, Nature (London) **391**, 51 (1998); P.M. Garnavich *et al.*, Astrophys. J. Lett. **493**, L53 (1998); A.G. Riess *et al.*, Astrophys. J. **116**, 1009 (1998); B.P. Schmidt *et al.*, astro-ph/9805200.

[20] Although specific value for m_ν is mentioned here as well as below, it must be scaled suitably if the calculated value of the nuclear matrix element in neutrinoless double beta decay changes or if the experimental limit on this decay improves, since only the ratio of the upper limit of $m_{0\nu\beta\beta}$ to m_ν enters into our analysis.

[21] M. Doi *et al.*, Phys. Lett. **102B**, 323 (1981); S.M. Bilenky *et al.*, *ibid.* **94B**, 495 (1980).

[22] CHOOZ Collaboration, M. Appolonio *et al.*, Phys. Lett. B **420**, 397 (1998).

[23] Mohan Narayan, G. Rajasekaran, and S. Uma Sankar, Phys. Rev. D **58**, 031301 (1998).

[24] M. Narayan *et al.*, Phys. Rev. D **53**, 2809 (1996); G.L. Fogli *et al.*, *ibid.* **54**, 2048 (1996); M. Narayan *et al.*, *ibid.* **56**, 437 (1997); G.L. Fogli *et al.*, Phys. Lett. B **434**, 333 (1998).

[25] L. Baudis *et al.*, Phys. Rev. Lett. **83**, 41 (1999).

[26] In fact this has already happened; according to Reference [25], $m_{0\nu\beta\beta} < 0.2$ eV at 90% C.L.

[27] S.L. Glashow *et al.*, Phys. Lett. B **190**, 199 (1987); B. Faid *et al.*, Phys. Rev. D **55**, 1353 (1997); P. Osland *et al.*, Phys. Lett. B **438**, 129 (1998); S.L. Glashow *et al.*, *ibid.* **445**, 412 (1999).

[28] H. Georgi and S.L. Glashow, Phys. Rev. D (to be published), hep-ph/9808293.

[29] G.C. Branco *et al.*, Phys. Rev. Lett. **82**, 683 (1999).

PHYSICAL REVIEW D VOLUME 56, NUMBER 3 1 AUGUST 1997

Constraining almost degenerate three-flavor neutrinos

Hisakazu Minakata
Department of Physics, Tokyo Metropolitan University, Minami-Osawa, Hachioji, Tokyo 192-03, Japan and Institute for Nuclear Theory, University of Washington, Seattle, Washington 98195-1550

Osamu Yasuda
Department of Physics, Tokyo Metropolitan University, Minami-Osawa, Hachioji, Tokyo 192-03, Japan

(Received 5 December 1996)

We discuss constraints on a scenario of almost degenerate three-flavor neutrinos imposed by the solar and the atmospheric neutrino anomalies, hot dark matter, and in particular by the neutrinoless double β decay experiments. It is found that in the Majorana version of the model the region with relatively large θ_{13} is favored and the model is not compatible with the popular small-θ_{12} Mikheyev-Smirnov-Wolfenstein solution of the solar neutrino problem. A constraint on the CP-violating phases including the one characteristic to Majorana neutrinos is also obtained. The stability of our conclusion against the uncertainty of the nuclear matrix elements of double β decay is briefly addressed. [S0556-2821(97)02015-8]

PACS number(s): 14.60.Pq, 23.40.−s, 26.65.+t, 95.35.+d

There exist several experimental hints which indicate that most probably neutrinos have tiny masses and flavor mixings. The first is the solar neutrino deficit observed in four different experiments: the chlorine, the Kamiokande II-III, GALLEX, and SAGE [1–4]. It became highly unlikely that the data of various experiments can be reconciled with any sensible modifications of the standard solar model. The second is the atmospheric neutrino anomaly, the large deviation in the observed ratio ν_μ/ν_e from the expectation of the Monte Carlo simulations [5,6]. While the anomaly was not observed within the statistics of the NUSEX and the Frejus experiments [7,8], the evidence of the Kamiokande and IMB detectors are so impressive that they force us to seriously consider the anomaly. The presence of the anomaly is also supported by the newest tracking detector, Soudan 2 [9].

The possible third hint for neutrino masses comes from the cosmological model with cold and hot dark matter (CHDM). Neutrinos are the only known candidates for the hot component. They could be responsible for the large-scale structure formation in a way consistent with the Cosmic Background Explorer (COBE) observation of anisotropy of cosmic microwave background [10–12]. While less direct compared with the first and the second hints, it provides a good motivation for examining the possibility of neutrino masses of a few eV range.

It has been pointed out by various authors that if at least one of the neutrino states has mass of the dark matter scale and if there is a hierarchy in two Δm^2, the difference in squared masses, the accelerator, and the reactor experiments put powerful constraints on mixing angles [13–15]. It is very remarkable that the mixing pattern of neutrinos is determined to be essentially unique [13] if one imposes the additional constraints that come from the requirement of solving either the solar neutrino problem or the atmospheric neutrino anomaly, together with that from neutrinoless double β decays [16,17].

The problem with the above framework with only three-flavor neutrinos (i.e., without sterile neutrinos) is that one cannot account for the solar neutrino deficit, the atmospheric neutrino anomaly, and the hot dark matter simultaneously. The only known possibility that can accommodate these two phenomena as well as supplying neutrino masses appropriate for hot dark matter within the standard three-flavor framework is the case of almost degenerate neutrinos (ADN's). An incomplete list of earlier references on ADN's is in [20].

In this paper we discuss the constraints that can be imposed on such an almost degenerate neutrino scenario from the solar and atmospheric neutrino observations as well as the terrestrial neutrino experiments. We will point out that, in the case of Majorana neutrinos, the neutrinoless double β decay experiment is of key importance. In particular, the solar neutrino and the double β decay experiments constrain the mixing angle θ_{13} not to be small.

Let us start by defining more precisely what we mean by the almost degenerate neutrinos. Due to the requirement of solving the solar and the atmospheric neutrino problems the two Δm^2 should have values $\lesssim 10^{-5}$ and $\sim 10^{-2}$ eV ^{2}, respectively. This implies that three neutrino states are degenerate up to the accuracy of 0.1 eV. Then, the requirement from the hot dark matter hypothesis implies that they must have masses of the order of a few to several eV [10–12]. Then, the degeneracy in the masses is better than 0.01 eV, hence the name of almost degenerate neutrinos (ADN's).

For definiteness, we assign the smaller Δm^2 to $\Delta m^2_{12} \equiv m^2_2 - m^2_1$ and the larger to Δm^2_{13}. It should be noticed that this can be done without loss of generality. Despite the almost degeneracy in neutrino masses there is a hierarchy in $\Delta m^2; \Delta m^2_{13} \simeq \Delta m^2_{23} \gg \Delta m^2_{12}$. It allows us to simplify greatly formulas for the oscillation probabilities. With neutrino mixing matrix $U_{\alpha i}$ they read

$$P(\nu_\beta \to \nu_\alpha) = 4|U_{\alpha 3}|^2 |U_{\beta 3}|^2 \sin^2\left(\frac{\Delta m^2_{13} L}{4E}\right), \quad (1)$$

Constraining Degenerate Neutrino Mass Models and Implications

Osamu Yasuda[1]

Department of Physics, Tokyo Metropolitan University, Minami-Osawa, Hachioji, Tokyo 192-0397, Japan

Abstract. A scenario of almost degenerate three flavor neutrinos which have masses suggested by the mixed dark matter model is reexamined with the recent data of the solar, atmospheric neutrinos and neutrinoless double β decay experiments. It is found that the Majorana version of the model with the MSW type solution for solar neutrinos is excluded at least at 2.5σ confidence level. On the other hand, this scenario with the vacuum oscillation solar solution is consistent with all the constraints and the best fit is close to the set of parameters for the bi-maximal mixing model.

1. Introduction

There have been several experimental data which suggest that neutrinos have tiny masses and flavor mixings. One of them is the solar neutrino deficit observed in four different experiments [1, 2, 3, 4, 5]. The other evidence comes from the atmospheric neutrino anomaly [6, 7, 8, 9, 10]. One another possible evidence may be the data of the LSND experiment [11]. While the KARMEN2 experiment [12] has been unable to exclude the LSND results, we have to wait for other experiments which may or may not confirm their results in the future. Here we assume that the solar and atmospheric neutrino anomalies are accounted for by neutrino oscillations among three neutrinos.

On the other hand, the so called mixed dark matter scenario has been proposed in which hot dark matter component as well as cold dark matter

[1] Email: yasuda@phys.metro-u.ac.jp

in Proc. "Beyond the Desert'99" eds.
H. V. Klapdor–Kleingrothaus and I. V. Krivosheina,
Tegernsee, Germany, June 1999, *IOP* (2000)

may be responsible for the large-scale structure formation [13, 14, 15]. It has been pointed out recently [16] that the mixed dark matter scenario does not necessarily reproduce a realistic power spectrum when the new supernova data are taken into account. Since arguments based on cosmological observations are always accompanied by systematic errors, it is not so clear how much unlikely the mixed dark matter scenario has become because of the new observational data. The purpose of this talk is to discuss the almost degenerate neutrino scenario of a few eV mass using only constraints of particle physics, independent of cosmological arguments.

The only known way to explain simultaneously the solar neutrino deficit, the atmospheric neutrino anomaly and the hot dark matter is the case of almost degenerate neutrinos (ADN) [17]. This scenario with the MSW type solar neutrino solutions [18] has been analyzed in [19] along with constraints of the solar and atmospheric neutrino observations as well as the terrestrial neutrino experiments. Here we will reexamine this scenario with all possible solar neutrino solutions including the vacuum oscillation solution, using slightly modified graphs and the most up-to-date data of atmospheric neutrinos, neutrinoless double β decay, and the reactor experiments [2]. In the case of Majorana neutrinos, the neutrinoless double β decay experiment is of key importance. The recent data by Heidelberg-Moscow group is so stringent that the almost degenerate neutrino scenario of Majorana type with the MSW type solar solution turns out to be inconsistent with other constraints.

2. The almost degenerate neutrino scenario

To solve the solar and the atmospheric neutrino problems the two mass squared differences have to have values $\Delta m^2_{\rm smaller} \sim 10^{-5}{\rm eV}^2$ or $10^{-10}{\rm eV}^2$ and $\Delta m^2_{\rm larger} \sim 10^{-2.5}{\rm eV}^2$, respectively. On the other hand, the requirement from the hot dark matter hypothesis implies that they must have masses of the order of a few to several eV [13, 14, 15]. Without loss of generality we assume $\Delta m^2_{21} \equiv m^2_2 - m^2_1$ to be the smallest and Δm^2_{31} to be the largest. Notice that there is a hierarchy in Δm^2, i.e., $\Delta m^2_{31} \simeq \Delta m^2_{32} \gg \Delta m^2_{21}$ despite the almost degeneracy in neutrino masses.

We use the standard parametrization [23] of the MNS mixing matrix [24]:

$$U = \begin{pmatrix} c_{12}c_{13} & s_{12}c_{13} & s_{13}e^{-i\delta} \\ -s_{12}c_{23} - c_{12}s_{23}s_{13}e^{i\delta} & c_{12}c_{23} - s_{12}s_{23}s_{13}e^{i\delta} & s_{23}c_{13} \\ s_{12}s_{23} - c_{12}c_{23}s_{13}e^{i\delta} & -c_{12}s_{23} - s_{12}c_{23}s_{13}e^{i\delta} & c_{23}c_{13} \end{pmatrix}. \quad (1)$$

[2] This talk is based on the work [20]. Similar analyses have been given, e.g., in [21, 22].

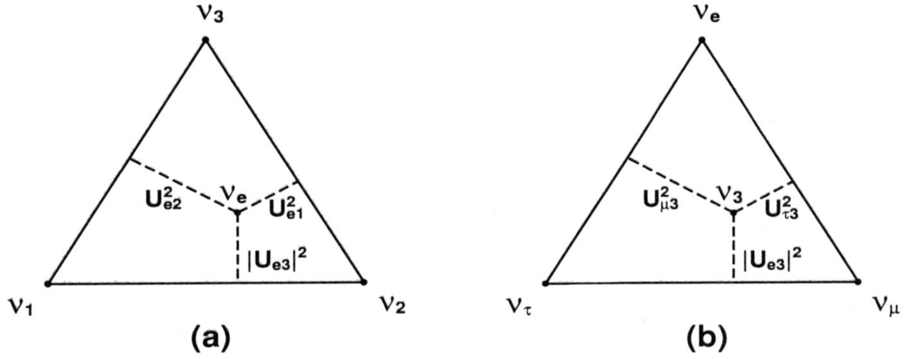

Figure 1. Triangles of unit height for (a) ν_e and (b) ν_3.

We will adopt the triangle representation which has been introduced by Fogli, Lisi, and Scioscia [25]. Fig. 1 (a), which will be useful in discussions of the solar neutrinos and the reactor data, shows how the ν_e state can be written as a linear combination of the three mass eigenstates with the coefficients $|U_{ej}|^2$ $(j = 1, 2, 3)$. Fig. 1 (b), which will play a role in the analysis of atmospheric neutrino data, represents how the most massive state ν_3 mixes with three flavor eigenstates with the coefficients $|U_{\alpha 3}|^2$ $(\alpha = e, \mu, \tau)$. In either case, the vertical position of the state in each triangle graph is s_{13}^2, which indicates deviation from the two flavor mixings.

3. Constraints from neutrinoless double β decay experiments

Let us first discuss the constraint from neutrinoless double β decays, which applies only to the Majorana neutrinos. Although there has been no experimental evidence so far, it is natural from the theoretical viewpoint that neutrinos are of Majorana type, as the see-saw mechanism offers a nice explanation for small neutrino masses [26]. Observation of no neutrinoless double β decay implies the upper bound on $\langle m_{\nu e}\rangle$, which can be written in our notation of the mixing matrix as

$$\langle m_{\nu e}\rangle = \left| c_{12}^2 c_{13}^2 m_1 e^{-i(\beta+\gamma)} + s_{12}^2 c_{13}^2 m_2 e^{i(\beta-\gamma)} + s_{13}^2 m_3 e^{2i(\gamma-\delta)} \right|, \quad (2)$$

where β and γ are the extra CP-violating phases characteristic to Majorana neutrinos [27, 28].

3.1. CP invariant case

We will first consider the CP-invariant case $(e^{2i\beta}, e^{i(\beta+3\gamma-2\delta)} = \pm 1)$. In this case the phase factors in (2) can be reduced to the CP parities η_j of mass eigenstates j with masses m_j [29]. Under the condition of almost degeneracy with which we are working we can approximate the expressions of $\langle m_{\nu e}\rangle$ by ignoring the mass differences. Let us take the convention that $\eta_1 = +1$ and denote them collectively as $(\eta_1, \eta_2, \eta_3) \equiv (1, e^{2i\beta}, e^{i(\beta+3\gamma-2\delta)}) = (++-)$ etc. Then we have

$$r \equiv \frac{\langle m_{\nu e}\rangle}{m} = \begin{cases} 1 & \text{for} \quad (+++) \\ |1-2s_{13}^2| & \text{for} \quad (++-) \\ |1-2s_{12}^2 c_{13}^2| & \text{for} \quad (+-+) \\ |1-2c_{12}^2 c_{13}^2| & \text{for} \quad (+--) \end{cases} . \tag{3}$$

We will refer to the ratio $\langle m_{\nu e}\rangle/m$ as r hereafter.

We assume that 20-30% of the universe is shared by the hot dark matter. We note that the neutrino contribution to the Ω parameter is $\Omega_\nu = (\sum m_i/91.5\text{eV})h^{-2}$ [30]. The mixed dark matter model with three kinds of neutrinos has been analyzed by Pogosyan and Starobinsky [14] and they concluded that the allowed region is given by $0.55 \le h \le 0.7$, $0.25 \le \Omega_\nu \le 0.3$. If we take these values, we obtain 2.3 eV$\le m_j \simeq m_{\text{HDM}} \le 4.5$eV as masses of almost degenerate neutrinos. We will use m_{HDM}=2.3 eV and 4.5 eV for neutrino masses as reference values in the following analysis. We impose the experimental bound on $\langle m_{\nu e}\rangle$ obtained by negative results of neutrinoless double β decays. The most stringent one to date is $\langle m_{\nu e}\rangle < 0.2$ eV derived in the ^{76}Ge experiment done by Heidelberg-Moscow group [31]. The bound is based on the calculation of the nuclear matrix elements in [32]. It implies the bounds on the r parameter $r \le 0.04$ and 0.09 for neutrino masses m_{HDM}=2.3 eV and m_{HDM}=4.5 eV, respectively.

From (3) we immediately see that the first pattern of the CP parity, (+++), is excluded. Other patterns are not immediately excluded but their parameters are subject to the constraint. From (3) the allowed regions for the cases of $(++-)$, $(+-+)$ and $(+--)$, which are given by $|1-2|U_{e3}|^2| \le r_{\text{max}}$, $|1-2|U_{e2}|^2| \le r_{\text{max}}$, $|1-2|U_{e1}|^2| \le r_{\text{max}}$ where r_{max} stands for the possible maximum value that r can have, are those bounded by the two straight lines which are parallel to $|U_{ej}|^2 = (1 \pm r_{\text{max}}^2)/2$, where j = 3, 2, 1 for the cases of $(++-)$, $(+-+)$ and $(+--)$, respectively (See the lighter shadowed regions in Fig. 2).

3.2. CP violating case

In general CP-noninvariant cases, we have to keep the two CP-violating phases β and γ in (2). Namely, we have

$$\begin{aligned} r_{\max} > \frac{\langle m_{\nu e}\rangle}{m_{\rm HDM}} &= \left| c_{13}^2 (e^{-i\beta} c_{12}^2 + e^{i\beta} s_{12}^2) + e^{i(3\gamma - 2\delta)} s_{13}^2 \right| \\ &\geq \left| c_{13}^2 (1 - \sin^2\beta \sin^2 2\theta_{12})^{1/2} - s_{13}^2 \right|, \end{aligned} \tag{4}$$

where the equality in the second line holds when

$$\arg(e^{-i\beta} c_{12}^2 + e^{i\beta} s_{12}^2) = 3\gamma - 2\delta + (2n+1)\pi, \tag{5}$$

where n is an integer. Note that the constraint from neutrinoless double β decays becomes even more stringent if the CP violating phases β, γ, δ do not satisfy the relation (5). Now (4) can be rewritten as

$$\begin{aligned} & (1 - |U_{e3}|^2)^2 - (|U_{e3}|^2 + r_{\max}^2)^2 \\ \leq\ & 4|U_{e1}|^2 |U_{e2}|^2 \sin^2\beta \\ \leq\ & (1 - |U_{e3}|^2)^2 - (|U_{e3}|^2 - r_{\max}^2)^2. \end{aligned} \tag{6}$$

By noting $4|U_{e1}|^2|U_{e2}|^2 \sin^2\beta \geq 0$ in the second inequality in (6), we get

$$|U_{e3}|^2 \leq \frac{1 + r_{\max}^2}{2}. \tag{7}$$

On the other hand, using the facts $4\sin^2\beta \leq 4$ and $1 - |U_{e3}|^2 = |U_{e1}|^2 + |U_{e2}|^2$ in the first inequality in (6), we have

$$(|U_{e1}|^2 - \frac{1 + r_{\max}^2}{2})(|U_{e2}|^2 - \frac{1 + r_{\max}^2}{2}) \geq 0. \tag{8}$$

Since $|U_{e1}|^2 + |U_{e2}|^2 = 1 - |U_{e3}|^2 \leq 1$, we obtain

$$|U_{e1}|^2 \leq \frac{1 + r_{\max}^2}{2}, \qquad |U_{e2}|^2 \leq \frac{1 + r_{\max}^2}{2}. \tag{9}$$

(7) and (9) show that the allowed region is the inside of the hexagon bounded by the straight lines $|U_{ej}|^2 \leq (1 + r_{\max}^2)/2$ (j=1,2,3) and $|U_{ej}|^2 \geq 0$ (j=1,2,3), which is the region inside of the thick solid lines in Fig. 2.

4. Constraints from the solar neutrino data

For the MSW type solutions the three flavor analysis has been given by Fogli, Lisi, and Montanino [33] with the assumption of mass hierarchy. The darker shadowed regions in Fig. 2 are the allowed regions for various

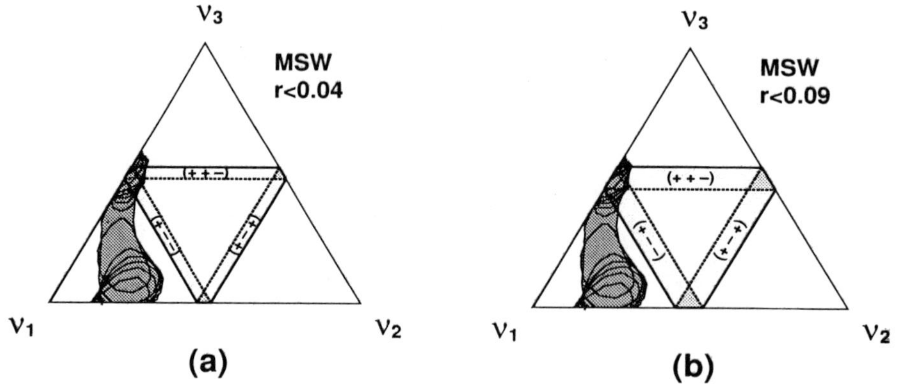

Figure 2. The darker shadowed areas are allowed region by the MSW type solar solutions. The area bounded by the straight lines are the allowed region by neutrinoless double β decay for (a) $\langle m_{\nu e}\rangle$ <0.2eV, $m_{\rm HDM}$ =4.5eV and (b) $\langle m_{\nu e}\rangle$ <0.2eV, $m_{\rm HDM}$ =2.3eV.

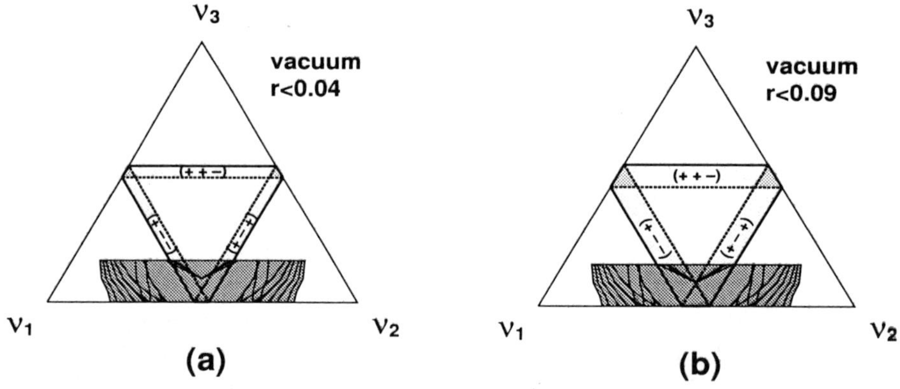

Figure 3. The allowed region for the vacuum oscillation solar solutions and the allowed region by neutrinoless double β decay for (a) $\langle m_{\nu e}\rangle$ <0.2eV, $m_{\rm HDM}$ =4.5eV and (b) $\langle m_{\nu e}\rangle$ <0.2eV, $m_{\rm HDM}$ =2.3eV.

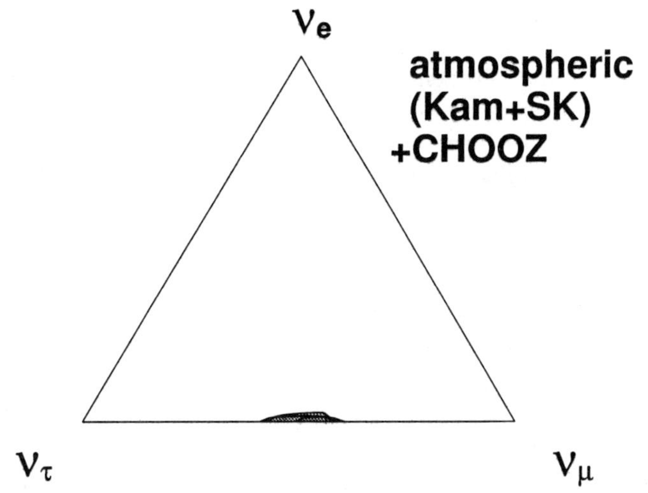

Figure 4. The allowed regions at 90%CL for various Δm_{32}^2 by the constraints of atmospheric neutrino data of the Kamiokande contained events, the Superkamiokande contained and upward going μ events, and the CHOOZ reactor data. All the shadowed regions are located near the $\nu_\mu - \nu_\tau$ line.

values of the mass squared difference ($2.2 \times 10^{-6}\text{eV}^2 \leq \Delta m_{21}^2 \leq 1.5 \times 10^{-4}\text{eV}^2$), projected on the triangle, where the small and the large mixing angle solutions at $\theta_{13} = 0$ are connected in the (θ_{12}, θ_{13}) plane.

For the vacuum oscillation solutions, the analysis has been done by Osland and Vigdel [34]. The allowed regions for the vacuum oscillation solar solutions are the darker shadowed areas in Fig. 3, where each allowed region for different mass squared difference ($0.4 \times 10^{-10}\text{eV}^2 \leq \Delta m_{21}^2 \leq 3.2 \times 10^{-10}\text{eV}^2$) is projected onto the (θ_{12}, θ_{13}) plane. For some Δm_{21}^2 the area near $\theta_{12} = \pi/4$ is not contained but for other Δm_{21}^2 it is allowed [34].

To give the allowed region in the three flavor framework one needs a three dimensional plot in the (θ_{12}, θ_{13}, Δm_{21}^2) space, and such intuitive figures are given for both MSW type and vacuum oscillation solutions in [35].

5. Constraints from atmospheric neutrino anomaly

The atmospheric neutrino data of Superkamiokande have been analyzed by [36, 37] in the three flavor framework. The two flavor analysis of the most up-to-date data by the Superkamiokande group shows that the allowed region of the mass squared difference is $1 \times 10^{-3}\text{eV}^2 < \Delta m^2 < 7 \times 10^{-3}\text{eV}^2$

at 90% CL [9]. The analyses in [36, 37] are strictly speaking different from the original one in [9], since the full data which are binned with respect to the energy as well as the zenith angle are not used in [36, 37]. The analysis in [36] has been updated with the recent data, where the upward going μ data [10] have also been incorporated. It was found that the region of the mass squared difference which is as small as $5 \times 10^{-4}\text{eV}^2$ is allowed at 90%CL.

On the other hand, the CHOOZ group has updated their result on $P(\bar{\nu}_e \to \bar{\nu}_e)$ in the reactor disappearance experiment [38], and the mass squared difference is limited to $\Delta m^2 < 7 \times 10^{-4}\text{eV}^2$. In our ADN scenario with mass hierarchy, the disappearance probability is given by

$$P(\bar{\nu}_e \to \bar{\nu}_e) = 1 - \sin^2 2\theta_{13} \sin^2 \left(\frac{\Delta m_{32}^2 L}{4E} \right), \tag{10}$$

so if $\Delta m_{32}^2 > 7 \times 10^{-4}\text{eV}^2$ then $\sin^2 2\theta_{13}$ has to be small. However, for $\Delta m_{32}^2 < 7 \times 10^{-4}\text{eV}^2$ there is no constraint by the CHOOZ data and relatively large θ_{13} is still allowed, as can been seen in the allowed region given in [36, 37] [3].

To obtain more stringent bound on $\sin^2 2\theta_{13}$, we include the three flavor analysis [39] of the contained events of the Kamiokande atmospheric neutrino data [6], for which the value of the mass squared difference in the allowed region tends to be higher than that of Superkamiokande. In fact $\Delta m_{32}^2 < 2 \times 10^{-3}\text{eV}^2$ is excluded at 90%CL by including the Kamiokande data.

Combining the atmospheric neutrino data of Superkamiokande, Kamiokande and the CHOOZ reactor data, we obtain the allowed region (the shadowed area in Fig. 4), which is quite narrow as far as θ_{13} is concerned. Fig. 4 shows that $\sin^2 2\theta_{13} < 0.1$ has to be satisfied, which is basically the consequence from the CHOOZ data.

6. Combined analysis of the solar, atmospheric and neutrinoless double β decay experiments

6.1. MSW type solutions for solar neutrinos

Let us now combine the allowed region of the solar neutrino data and that of the neutrinoless double β decay data. We have plotted the allowed regions of the MSW type solar neutrino solutions and the neutrinoless double β

[3] In [22] it is argued that θ_{13} has to be small because of absence of the deviation of the e-like data from the Monte Carlo predictions. However the numerical results of [36, 37] show that the Superkamiokande data alone are not sufficient to constrain θ_{13} to be small.

decay constraint for $r < 0.04$ ($\langle m_{\nu e}\rangle$ <0.2eV, $m_{\rm HDM}$ =4.5eV, Fig. 2 (a)) and $r < 0.09$ ($\langle m_{\nu e}\rangle$ <0.2eV, $m_{\rm HDM}$ =2.3eV, Fig. 2 (b)), respectively. For $r < 0.04$, the solution exists only if $s_{13}^2 > 0.40$, and for $r < 0.09$ only if $s_{13}^2 > 0.37$. These values are obviously inconsistent with the atmospheric neutrino and the CHOOZ data, which tell us that $s_{13}^2 \leq 0.025$ (See Fig. 4). In fact we find that the region of $s_{13}^2 > 0.37$ is excluded at 4.3 σ (99.998%) CL by the atmospheric neutrino and the CHOOZ data. So we conclude that the ADN scenario of Majorana type with the MSW type solar solution is ruled out.

6.2. Vacuum oscillation solutions for solar neutrinos

The allowed regions of the vacuum oscillation solar solutions and the neutrinoless double β decay constraint are depicted for $r < 0.04$ ($\langle m_{\nu e}\rangle$ <0.2eV, $m_{\rm HDM}$ =4.5eV, Fig. 3 (a)) and $r < 0.09$ ($\langle m_{\nu e}\rangle$ <0.2eV, $m_{\rm HDM}$ =2.3eV, Fig. 3 (b)), respectively. In this case the set of parameters (θ_{12}, θ_{13})=($\pi/4$, 0) is included by both of the allowed regions, and this parameter set is also consistent with the atmospheric neutrino data plus the CHOOZ data, provided that $\theta_{23} \simeq \pi/4$. In fact the set of parameters (θ_{12}, θ_{13}, θ_{23}) =($\pi/4$, 0, $\pi/4$) corresponds to the case of the bi-maximal mixing model [40], which has the following MNS matrix:

$$U = \begin{pmatrix} \frac{1}{\sqrt{2}} & \frac{1}{\sqrt{2}} & 0 \\ -\frac{1}{2} & \frac{1}{2} & \frac{1}{\sqrt{2}} \\ \frac{1}{2} & -\frac{1}{2} & \frac{1}{\sqrt{2}} \end{pmatrix}. \tag{11}$$

7. Theoretical uncertainties

Let us address the issue of theoretical uncertainties of the bound on $\langle m_{\nu e}\rangle$ on which our discussion has heavily relied. It is difficult to estimate the theoretical uncertainties in the calculation of the nuclear matrix elements [41]. In [31] 5 different calculations of the nuclear matrix elements are quoted. Depending on the nuclear matrix elements, the limit on the Majorana neutrino mass becomes $\langle m_{\nu e}\rangle < 0.20$ eV [32], 0.56 eV [42], 0.52 eV [43], 0.19 eV [44], 0.4 eV [45].

We take more conservative bounds $\langle m_{\nu e}\rangle < 0.4$ eV and $\langle m_{\nu e}\rangle < 0.56$ eV as reference values for $\langle m_{\nu e}\rangle$ and analyze the constraints again. In Figs. 5 (a) and 5 (b) we present the allowed regions of the MSW type solar neutrino solution and the neutrinoless double β decay for $r < 0.17$ and $r < 0.24$. In these figures we only plot the case of $m_{\rm HDM} = 2.3$ eV which leads to $r < 0.17$ and $r < 0.24$. In the case of $r < 0.17$, $s_{13}^2 \simeq 0.09$ or $s_{13}^2 > 0.27$ is obtained and this is still inconsistent with the atmospheric neutrino plus the CHOOZ data. In fact the region $s_{13}^2 \simeq 0.09$ is excluded

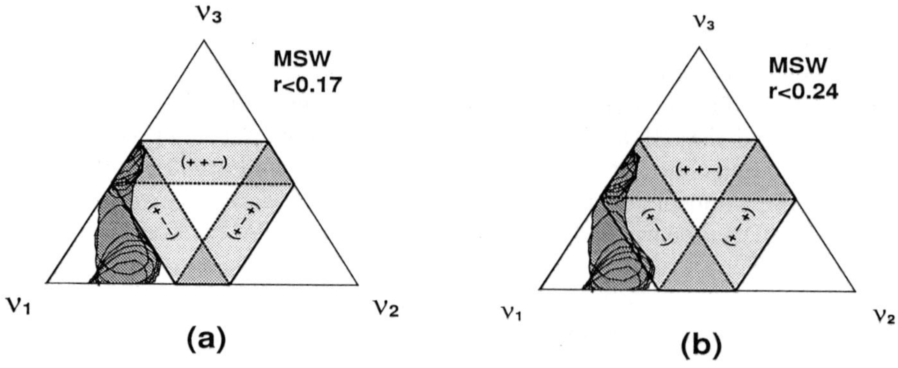

Figure 5. The allowed region for the MSW type solar solution and the allowed region by neutrinoless double β decay (a) for $\langle m_{\nu e}\rangle$ <0.4eV, $m_{\rm HDM}$ =2.3eV and (b) for $\langle m_{\nu e}\rangle$ <0.56eV, $m_{\rm HDM}$ =2.3eV.

at 2.9σ (99.6%) CL by the atmospheric neutrino and the CHOOZ data, and if we combine the solar neutrino data it is excluded at 3.2σ (99.9%) CL. The confidence level of the region of $s_{13}^2 > 0.27$ is 3.6σ (99.97%) by the atmospheric neutrino plus the CHOOZ data, and 3.8σ (99.99%) if the solar neutrino data are also included. In the case of $r < 0.24$, the constraint becomes weaker. We have $s_{13}^2 > 0.04$ which corresponds to 2.5σ (98.7%) CL by the atmospheric neutrino plus the CHOOZ data, and 2.9σ (99.6%) CL by combining the solar neutrino data. We conclude that the ADN scenario with the MSW type solar neutrino solution is excluded at least 2.5σ (98.7%) CL even with the most conservative estimate of the nuclear matrix elements.

8. Conclusions

We have discussed the almost degenerate three flavor neutrino scenario as a simultaneous solution to the solar, the atmospheric and the dark matter problems. We have shown that the ADN scenario of Majorana type with the MSW type solar solution is excluded at least at 2.5σ (98.7%) CL by the constraints from neutrinoless double β decays as well as these observational data of solar and atmospheric neutrinos. The ADN scenario with the vacuum oscillation solar solutions, on the other hand, is perfectly consistent with all these constraints and the best fit parameter set is close to the bi-maximal mixing case.

9. Acknowledgement

The author would like to thank Prof. H. V. Klapdor-Kleingrothaus and other organizers for invitation and hospitality during the conference, and H. Minakata for collaboration and discussions. He would also like to thank S. D. H. Hsu and other members of Institute of Theoretical Science, University of Oregon for hospitality during part of this work. This work has been supported in part by Grant-in-Aid for Scientific Research of the Ministry of Education, Science and Culture under #10640280 and #09045036.

References

[1] Cleveland B T *et al* 1995 *Nucl. Phys.* B (Proc. Suppl.) **38** 47
Lande K 1996 *Proc. 17th International Conference on Neutrino Physics and Astrophysics* (Singapore: World Scientific) p 25

[2] Hirata K S *et al* 1991 *Phys. Rev.* **D44** 2241; 1989 *Phys. Rev. Lett.* **63** 16; 1990 *Phys. Rev. Lett.* **65** 1297; 1990 *Phys. Rev. Lett.* **65** 1301; 1991 *Phys. Rev. Lett.* **66** 9;
Fukuda Y *et al* 1996 *Phys. Rev. Lett.* **77** 1683;
Suzuki Y 1995 *Nucl. Phys.* B (Proc. Suppl.) **38** 54

[3] Fukuda Y *et al* 1998 *Phys. Rev. Lett.* **81** 1158, 4279 (E); 1999 *Phys. Rev. Lett.* **82** 1810; 1999 *Phys. Rev. Lett.* **82** 2430;
Suzuki Y 1999 *Nucl. Phys.* B (Proc. Suppl.) **77** 35

[4] Anselmann P *et al* 1992 *Phys. Lett.* **B285** 376; 1993 *Phys. Lett.* **B314** 445; 1994 *Phys. Lett.* **327** 377; 1995 *Phys. Lett.* **B342** 440; 1995 *Phys. Lett.* **B357** 237; 1996 *Phys. Lett.* **B388** 384; 1999 *Nucl. Phys.* B (Proc. Suppl.) **70** 284; 1999 *Phys. Lett.* **B447** 127; 1999 *Nucl. Phys.* B (Proc. Suppl.) **77** 26

[5] Abazov A I *et al* 1991 *Phys. Rev. Lett.* **67** 3332;
Abdurashitov J N *et al* 1995 *Nucl. Phys.* B (Proc. Suppl.) **38** 60; 1999 *Nucl. Phys.* B (Proc. Suppl.) **77** 20

[6] Hirata K S *et al* 1988 *Phys. Lett.* **B205** 416; 1992 *Phys. Lett.* **B280** 146;
Fukuda Y *et al* 1994 *Phys. Lett.* **B335** 237

[7] Casper D *et al* 1991 *Phys. Rev. Lett.* **66** 2561;
Becker-Szendy R *et al* 1992 *Phys. Rev.* **D46** 3720

[8] Peterson E 1999 *Nucl. Phys.* B (Proc. Suppl.) **77** 111;
Allison W W M 1999 *Phys. Lett.* **B449** 137

[9] Fukuda Y *et al* 1998 *Phys. Lett.* **B433** 9; 1998 *Phys. Lett.* **B436** 33; 1998 *Phys. Rev. Lett.* **81** 1562;
Kajita T 1999 *Nucl. Phys.* B (Proc. Suppl.) **77** 123; 1999 these proceeding;
Scholberg K 1999 hep-ex/9905016

[10] Fukuda Y *et al* 1999 *Phys. Rev. Lett.* **82** 2644

[11] Athanassopoulos *et al* 1995 *Phys. Rev. Lett.* **75** 2650; 1996 *Phys. Rev. Lett.* **77** 3082; 1996 *Phys. Rev.* **C54** 2685; 1998 *Phys. Rev. Lett.* **81** 1774; 1998 *Phys. Rev.* **C58** 2489;
Smith D 1999 these proceeding

[12] Eitel K 1999 *Proc. Lake Louise Winter Institute*, http://www-ik1.fzk.de/www/karmen/ps/llwi.ps-gz;
Jannakos T 1999 *Proc. LES RENCONTRES DE MORIOND*, http://www-ik1.fzk.de/www/karmen/ps/moriond99_2.ps;
Steidl M 1999 *Proc. LES RENCONTRES DE PHYSIQUE DE LA VALLEE AOSTE*, http://www-ik1.fzk.de/www/karmen/ps/thuile99.ps;
Drexlin G J 1999 these proceeding

[13] Holtzman J A 1989 Astrophys. Suppl. J **71** 1;
Holtzman J A and Primack J R 1993 Astrophys. J. **405** 428;
Primack J R, Holtzman J, Klypin A, and Caldwell D O 1995 *Phys. Rev. Lett.* **74** 2160;
Caldwell D O 1999 *Nucl. Phys.* B (Proc. Suppl.) **77** 420

[14] Pogosyan D and Starobinsky A 1995 astro-ph/9502019

[15] Babu K S, Schaefer R K, and Shafi Q 1996 *Phys. Rev.* **D53** 606

[16] Primack J R and Gross M A K 1998 astro-ph/9810204

[17] Caldwell D and Mohapatra R N 1993 *Phys. Rev.* **D48** 3259; 1994 *Phys. Rev.* **D50** 3477;
Petcov S T and Smirnov A Yu 1994 *Phys. Lett.* **B322** 109;
Joshipura A S 1994 *Z. Phys.* **C64** 31; 1995 *Phys. Rev.* **D51** 1321;
Bamert P and Burgess C P 1994 *Phys. Lett.* **B329** 289;
Lee G D and Mohapatra R N 1994 *Phys. Lett.* **B329** 463;
Ioannissyan A and Valle J F W 1994 *Phys. Lett.* **B332** 93;
Mohapatra R N and Nussinov S 1995 *Phys. Lett.* **B346** 75

[18] Mikheyev S P and Smirnov A 1986 *Nuovo Cimento* **9C** 17;
Wolfenstein L 1978 *Phys. Rev.* **D17** 2369

[19] Minakata H and Yasuda O 1997 *Phys. Rev.* **D56** 1692

[20] Minakata H and Yasuda O 1999 to appear

[21] Vissani F 1997 hep-ph/970843;
Barger V and Whisnant K 1999 *Phys. Lett.* **B456** 194.

[22] Georgi H and Glashow S L 1998 hep-ph/9808293

[23] Particle Data Group 1998 Euro. Phys. J. **C3** 1

[24] Maki Z , Nakagawa M and Sakata S 1962 Prog. Theor. Phys. **28** 870

[25] Fogli G L, Lisi E and Scioscia G 1995 *Phys. Rev.* **D52** 5334

[26] Yanagida T 1979 *Proc. Workshop on Unified Theory and Baryon Number in the Universe* (Tsukuba: KEK) p 95;
Gell-Mann M, Slansky and Ramond P 1979 *Supergravity* (Amsterdam: North Holland) p 7

[27] Schechter J and J F W Valle 1980 *Phys. Rev.* **D22** 2227;
Bilenky S M, Hosek J, and Petcov S T 1980 *Phys. Lett.* **B94** 495;
Doi M *et al* 1981 *Phys. Lett.* **B102** 323

[28] Fukugita M and Yanagida T 1994 *Physics and Astrophysics of Neutrinos* (Tokyo: Springer-Verlag)

[29] Wolfenstein L 1981 *Phys. Lett.* **B107** 77

[30] Kolb E W and Turner M S 1990 *The Early Universe* (California: Addison-Wesley Publishing Co.)

[31] Baudis L *et al* 1999 *Phys. Rev. Lett.* **83** 41

[32] Staudt A, Muto K, Klapdor-Kleingrothaus H V 1990 Europhys. Lett. **13** 31

[33] Fogli G L, Lisi E, and Montanino D 1996 *Phys. Rev.* **D54** 2048

[34] Osland P and Vigdel G 1998 *Phys. Lett.* **B438** 129

[35] Lisi E 1999 *New Era in Neutrino Physics* (Tokyo: Universal Academic Press) p 153;
Fogli G L 1998 *Proc. NEUTRINO OSCILLATION WORKSHOP* http://www.nikhef.nl/pub/conferences/now98/presentations.html/#fogli

[36] Yasuda O 1999 *New Era in Neutrino Physics* (Tokyo: Universal Academic Press) p 165; 1998 *Phys. Rev.* **D58** 091301

[37] Fogli G L, Lisi E, Montanino D and Scioscia G 1999 *Phys. Rev.* **D59** 033001; 1999 hep-ph/9904465;
Fogli G L 1999 these proceeding

[38] Apollonio M *et al* 1998 *Phys. Lett.* **B338** 383; 1999 hep-ex/9907037;
Declais Y 1999 these proceeding

[39] Yasuda O 1997 hep-ph/9706546

[40] Barger V, Pakvasa S, Weiler T J and Whisnant K 1998 *Phys. Lett.* **B437** 107;
Nomura Y and Yanagida T 1999 *Phys. Rev.* **D59** 017303;
Davidson S and King S F 1998 *Phys. Lett.* **B445** 191;
Mohapatra R N and Nussinov S 1998 *Phys. Lett.* **B441** 299;
Giunti C 1999 *Phys. Rev.* **D59** 077301;
Kang S K and Kim C S 1999 *Phys. Rev.* **D59** 091302;
Jarlskog C, Matsuda M, Skadhauge S and Tanimoto M 1999 *Phys. Lett.* **B449** 240;
Jezabek M and Sumino Y 1999 *Phys. Lett.* **B457** 139

[41] Vergados J 1999 these proceeding

[42] Caurier E *et al* 1996 *Phys. Rev. Lett.* **77** 1954

[43] Engel J, Vogel P and Zirnbauer M R 1988 *Phys. Rev.* **C37** 731

[44] Tomoda T 1991 Rep. Prog. Phys. **54** 53

[45] Simkovic F *et al* 1997 *Phys. Lett.* **B393** 267

TMUP-HEL-0007

Answering the Sphinx's Questions on Neutrinos

Hisakazu Minakata*

Department of Physics, Tokyo Metropolitan University

1-1 Minami-Osawa, Hachioji, Tokyo 192-0397, Japan, and

Research Center for Cosmic Neutrinos, Institute for Cosmic Ray Research,

University of Tokyo, Kashiwa, Chiba 277-8582, Japan†

Abstract

In answering the difficult questions on neutrinos asked by Sphinx I argue that search for proton decay is the most important experiment in coming 5-10 years. I also emphasize the crucial importance of the neutrinoless double beta decay with sensitivity of $\langle m_{\nu e} \rangle \sim 0.01$ eV level as the unique feasible way of directly detecting neutrinos of atmospheric mass scale in laboratories. I point out that, if observed at this level, it means not only that neutrinos are Majorana particle but also that they must obey an inverted mass hierarchy.

---

*Talk presented at Workshop on Neutrino Oscillations and Their Origin, Fujiyoshida, Japan, February 11-13, 2000, to appear in Proceedings published by Universal Academy Press, Tokyo.

†Address correction requested

PHYSICAL REVIEW D, VOLUME 61, 097301

Neutrinos on Earth and in the heavens

Howard Georgi* and S. L. Glashow†
Laboratory of Physics, Harvard University, Cambridge, Massachusetts 02138
(Received 28 May 1999; published 4 April 2000)

Recent data suggest a simple and intriguing form of the neutrino mass matrix. We show how the data may constrain solar neutrino oscillations to be nearly maximal [and rule out the small-angle Mikheyev-Smirnov-Wolfenstein (MSW) explanation of solar neutrino observations] if relic neutrinos comprise at least three percent of the critical mass density of the universe.

PACS number(s): 14.60.Pq

Cosmologists differ on whether or not neutrinos play an essential role in the evolution of the large-scale structure of the universe [1]. In this paper, we assume that they do, and that the sum of their masses is several electron volts so that relic neutrinos comprise several percent of the critical mass density. Under this hypothesis, we demonstrate how a wide range of observations and deductions relating to neutrinos can be explained in terms of a specific effective neutrino mass matrix $\mathcal{M}$ involving a suggestive pattern of neutrino masses and mixings. Although many of our arguments may be found elsewhere in part or in other contexts [2], a cogent synthesis may be useful.

The literature is rife with both experimental and theoretical claims regarding neutrino properties, many of them in conflict with one another. Below is a somewhat arbitrary selection of neutrino "facts" we shall accept and describe. They are consistent with one another and are suggested by current data, but they are not decisively established. We do not intend to argue that these particular facts should be accepted as true [3], but rather that, if true, they constrain the neutrino mass matrix to have a form that we find both fascinating and a bit bizarre [4].

Fact 1. There exist precisely three chiral neutrino states with Majorana masses, m_1, m_2, and m_3 (taken to be real and non-negative). In particular we do not consider the existence of additional neutrinos, sterile or otherwise.

Fact 2. Atmospheric neutrinos rarely oscillate into electron neutrinos. This is a plausible, but not inescapable [5], interpretation of recent data from the Super-Kamiokande Collaboration [6] and from CHOOZ [7].

Fact 3. Atmospheric muon neutrinos suffer maximal, or nearly maximal, two-flavor oscillations into tau neutrinos,

$$\sin^2 2\theta > 0.82, \tag{1}$$

and the required neutrino mass-squared difference Δ_a satisfies [6]. The relevant mixing angle satisfies [8]

$$10^{-3}\ \mathrm{eV}^2 < \Delta a < 7\times 10^{-3}\ \mathrm{eV}^2. \tag{2}$$

Fact 4. Oscillations are needed to resolve the discrepancy between the observed and computed solar neutrino fluxes [9,10]. A relevant neutrino squared-mass different in the range

$$6\times 10^{-11}\ \mathrm{eV}^2 < \Delta_s < 2\times 10^{-5}\ \mathrm{eV}^2 \tag{3}$$

provide Mikheyev-Smirnov-Wolfenstein (MSW) explanations for larger values of Δ_s, and just-so explanations for smaller values. It has been suggested [11] that the solar neutrino deficit may result from maximal time-averaged vacuum oscillations. If so, the bound $\Delta_s < 10^{-3}\ \mathrm{eV}^2$ is obtained from reactor experiments [7]. It is premature and unnecessary for us to choose among these proposed solutions to the solar neutrino puzzle.

Fact 5. Here we assume that neutrino masses are large enough to play a significant cosmological role and take

$$\frac{m_1+m_2+m_3}{3} \equiv M \sim 2\ \mathrm{eV}. \tag{4}$$

This is the least well established fact in our list, but it is crucial to our discussion.

Fact 6. Careful studies of many nuclear species have failed to detect neutrinoless double beta decay. These experiments provide bounds on $\mathcal{M}_{ee}$, the ee component of the Majorana neutrino mass matrix in the charged lepton flavor basis—a weighted average of neutrino masses. In this paper, we adopt the strongest published bound, $\mathcal{M}_{ee} < B = 0.46$ eV [12].

These "facts" dramatically constrain the form of the neutrino mass matrix. In addition, they compel solar neutrino oscillations to be maximal, thereby ruling out the small-angle MSW solution of the solar neutrino puzzle. We begin by considering the well-known implications of fact 1. The most general mass matrix involving three chiral neutrinos is a 3×3 complex symmetric matrix $\mathcal{M}$. It may be written

$$\mathcal{M} = e^{i\eta} U_0^* D_0 U_0^\dagger, \tag{5}$$

where U_0 is an element of SU(3) and D is a diagonal matrix with real non-negative entries m_i. The mass matrix would be real were CP conserved, but it is not. Consequently $\mathcal{M}$ involves nine convention-independent parameters. Judicious choice of the phases of the flavor eigenstates allows us to rewrite Eq. (5) as

$$\mathcal{M} = U^* D U^\dagger, \tag{6}$$

where U is a unitary "Kobayashi-Maskawa" matrix (involving three angles θ_i and a complex phase δ) expressing flavor eigenstates in terms of mass eigenstates. In standard notation,

*Email address: georgi@physics.harvard.edu
†Email address: glashow@physics.harvard.edu

$$\begin{pmatrix} \nu_e \\ \nu_\mu \\ \nu_\tau \end{pmatrix} = \begin{pmatrix} c_2c_3 & c_2s_3 & s_2e^{-i\delta} \\ -c_1s_3-s_1s_2c_3e^{i\delta} & +c_1c_3-s_1s_2s_3e^{i\delta} & s_1c_2 \\ +s_1s_3-c_1s_2c_3e^{i\delta} & -s_1c_3-c_1s_2s_3e^{i\delta} & c_1c_2 \end{pmatrix} \begin{pmatrix} \nu_1 \\ \nu_2 \\ \nu_3 \end{pmatrix}, \tag{7}$$

with s_i and c_i standing for sines and cosines of θ_i. The remaining five parameters appear in the diagonal matrix D, which may be written

$$D = \begin{pmatrix} m_1e^{i\phi} & 0 & 0 \\ 0 & m_2e^{i\phi'} & 0 \\ 0 & 0 & m_3 \end{pmatrix}. \tag{8}$$

Each of the phase factors $(e^{i\delta}, e^{i\phi}, e^{i\phi'})$, if not real, is *CP* violating.

The amplitude for atmospheric muon neutrinos with energy E_a to oscillate into ν_e over a distance R_a is

$$\sum_j U_{\mu j} U_{ej}^* e^{im_j^2 R_a/2E_a}. \tag{9}$$

According to fact 2, this amplitude must be small over the range of R_a and E_a relevant to atmospheric neutrinos, around $2E_a/R_a \approx 10^{-3}$ eV^2. It follows that $|m_j^2 - m_k^2| R_a/2E_a$ must be small for some pair of neutrino mass eigenstates j and k. To prove this, we assume the contrary. It follows that the amplitude (9) is small for a range R_a and E_a if and only if $U_{\mu j} U_{ej}^*$ is small for each j. But fact 3 requires that ν_μ is not close to a mass eigenstate. Thus $U_{\mu j} U_{ej}^*$ can be small for each j only if ν_e is close to a mass eigenstate. This would lead us to the so-called small-angle MSW solution that requires $\Delta_s < 2\times 10^{-5}$ eV^2: small compared to $2E_a/R_a$ and contrary to the hypothesis—QED.

Thus the neutrino mass eigenstates associated with atmospheric oscillations must have a squared-mass difference $\Delta_a > 5\times 10^{-4}$ eV^2, while those associated with solar oscillations must have a much smaller squared-mass difference, $\Delta_s \ll \Delta_a$. Without loss of generality, we take $\Delta_a \equiv |m_3^2 - m_2^2|$ and $\Delta_s \equiv |m_2^2 - m_1^2|$. This may all sound rather obvious, and indeed it is the standard wisdom. But notice that the argument of the previous paragraph rules out the possibility of the time-averaged solar neutrino solution with Δ_s near the CHOOZ bound.

We assume that all differences of squares of neutrino masses are less than 7×10^{-3} eV^2. Yet according to fact 5 [and Eq. (4) in particular] the sum of the neutrino masses must be several eV.[1] Thus the three neutrinos must have equal mass to a precision of at least 10^{-3} if they are to play a role in large-scale structure formation. From the point of view of particle theory, this is truly a bizarre result and not at all what one would expect from the simplest seesaw mechanisms. It is nonetheless an immediate consequence of the facts we have accepted. We now proceed to a more detailed discussion of the mass matrix (6), from which additional constraints can be found.

We may express the *in vacua* energy-dependent survival probabilities for solar and atmospheric neutrinos in terms of the parameters so defined. Because the path length R_s of a solar neutrino is roughly an astronomical unit, Eq. (2) yields $\Delta_a R_s/E \gg 1$. Using this relation, we obtain

$$P(\nu_e \to \nu_e)|_{\text{solar}} \simeq 1 - \frac{\sin^2 2\theta_2}{2} - \cos^4\theta_2 \sin^2 2\theta_3 \sin^2(\Delta_s R_s/4E). \tag{10}$$

Because the path length of an atmospheric neutrino R_a can be no greater than Earth's diameter, Eq. (3) yields $\Delta_s R_a/E \ll 1$. Using this relation we obtain

$$P(\nu_\mu \to \nu_\mu)|_{\text{atmospheric}} = 1 - 4\sin^2\theta_1 \cos^2\theta_2 (1 - \sin^2\theta_1 \cos^2\theta_2) \sin^2(\Delta_a R_a/4E). \tag{11}$$

These oscillations produce electron or tau neutrinos in the ratio

$$\left.\frac{P(\nu_\mu \to \nu_e)}{P(\nu_\mu \to \nu_r)}\right|_{\text{atmospheric}} \simeq \frac{\tan^2\theta_2}{\cos^2\theta_1}. \tag{12}$$

Note that none of Eqs. (10), (11), and (12) involve the *CP*-violating parameter δ.

We turn to the consequences of our other tentatively accepted facts. Fact 2 and Eq. (12) yield

$$\theta_2 \simeq 0, \tag{13}$$

expressing the absence of oscillations of atmospheric oscillations into electron neutrinos. This result greatly simplifies Eqs. (10) and (11), which become

$$P|_{\text{solar}} \simeq 1 - \sin^2 2\theta_3 \sin^2(\Delta_s R_s/4E),$$
$$P|_{\text{atmospheric}} \simeq 1 - \sin^2 2\theta_1 \sin^2(\Delta_a R_a/4E). \tag{14}$$

Thus θ_1 is the parameter controlling atmospheric neutrino oscillations, and we conclude from fact 3 and Eq. (1) that

$$\sin 2\theta_1 \simeq 1, \tag{15}$$

expressing the observation that these oscillations are nearly maximal.

[1]The bound $M < 4.4$ eV follows from a recent measurement of the tritium beta spectrum [13].

Finally, we must address fact 6, the observed suppression of neutrinoless double beta decay. The amplitude for this process is proportional to the quantity $\mathcal{M}_{ee}$, on which the bound is

$$\mathcal{M}_{ee} \equiv |m_1 c_2^2 c_3^2 e^{i\phi} + m_2 c_2^2 s_3^2 e^{i\phi'} + m_3 s_2^2 e^{i2\delta}| < B. \quad (16)$$

We have seen that fact 5 requires neutrino masses to be equal to a precision sufficient to neglect their differences in Eq. (16). Furthermore, Eq. (13) lets us put Eq. (16) into the simple form

$$|\cos^2\theta_3 e^{i\phi} + \sin^2\theta_3 e^{i\phi'}| < B/M. \quad (17)$$

If we set $M=2$ eV and $B=0.46$ eV (the current experimental upper bound), we find that Eq. (16) can be satisfied if $\phi + \phi' \simeq \pi$ and $|\cos 2\theta_3| < 0.23$, or

$$\sin^2 2\theta_3 > 0.95. \quad (18)$$

If $M \geqslant 2$ eV, our six facts are mutually consistent if and only if solar neutrino oscillations are nearly maximal. Somewhat stronger bounds on neutrinoless double beta decay could strengthen Eq. (18) enough to leave just-so oscillations [14] as the only viable explanation of the solar neutrino data [10]. Conversely, if the small angle MSW description of solar neutrino oscillations is correct, the sum of the neutrino masses is bounded above by $3B \simeq 1.4$ eV. In this case, future double beta-decay experiments may exclude the cosmological relevance of relic neutrinos.

The neutrino matrix we are led to has approximately the following form:

$$\mathcal{M} = M \begin{pmatrix} 0 & \frac{1}{\sqrt{2}} & \frac{1}{\sqrt{2}} \\ \frac{1}{\sqrt{2}} & \frac{1}{2} & -\frac{1}{2} \\ \frac{1}{\sqrt{2}} & -\frac{1}{2} & \frac{1}{2} \end{pmatrix}. \quad (19)$$

This has several intriguing properties: $\mathcal{M}\mathcal{M}^\dagger$ is approximately a multiple of the unit matrix and $\mathcal{M}_{ee} \simeq 0$. To the extent that these relations are satisfied, the following processes are forbidden at all orders in perturbation theory: neutrinoless double beta decay; muon decay into $e+\gamma$ or into $e+e+\bar{e}$; muonium-antimuonium transitions; muon-electron conversion via capture; the induction of an electron electric dipole moment. This point is academic because the detection of any of the above processes (except neutrinosless double beta decay) would require a radical revision of the standard model.

Neutrino astronomy is a new science. Future observations of neutrinos from nearby supernovae, or among ultrahigh energy cosmic rays, are likely sources of new information about particles and the universe. These neutrinos, having traversed great distances, will experience time-averaged oscillations so that their composition at detection will not coincide with their composition at birth. Let D_l be the number of detected neutrinos with identities ν_l, and B_l their numbers at birth. With the above mixing parameters we find

$$\begin{pmatrix} D_e \\ D_\mu \\ D_\tau \end{pmatrix} = \frac{1}{8} \begin{pmatrix} 4 & 2 & 2 \\ 2 & 3 & 3 \\ 2 & 3 & 3 \end{pmatrix} \begin{pmatrix} B_e \\ B_\mu \\ B_\tau \end{pmatrix}. \quad (20)$$

Half of the ν_e burst from a supernovae reach Earth as ν_e, while cosmic ν_μ's are seen as 25% ν_e's and 37.5% ν_τ's.

Let us summarize our results. There are nine parameters in the neutrino mass matrix, all but one of which are severely constrained by the facts we have accepted. The three neutrino masses are nearly (but not quite) the same. The angles relating flavor and mass eigenstates take the following simple values: $\theta_1 \simeq \theta_3 \simeq \pi/4$ and $\theta_2 \simeq 0$. These simple relations may indicate a deeper underlying truth.

In this connection, note that just one of the three *a priori* *CP*-violating parameters in $\mathcal{M}$ is unconstrained by our analysis: We must have $\phi - \phi' \simeq \pi$ to suppress neutrinoless double beta decay and the parameter δ is *hors de combat* because it always occurs multiplied by $\sin\theta_2$, which nearly vanishes.

One of us (S.L.G.) thanks Maurice Goldhaber, Lawrence Krauss, and Lisa Randall for stimulating discussions. We are also grateful to John Bahcall, Ed Kearns, and Stephen Parke for comments. This work was supported in part by the National Science Foundation under Grant No. NSF-PHY/98-02709.

[1] For a recent review, see J. R. Primack, *SCIPP-96-59-REV*, in the Proceedings of Midrasha Mathematicae in Jerusalem: Winter School in Dynamical Systems, Jerusalem, Israel, 1997, astro-ph/9707285.

[2] A good recent review is R. N. Mohapatra, Nucl. Phys. B (Proc. Suppl.) **77**, 376 (1999). Other recent papers on the subject include: J.W.F. Valle, hep-ph/9509306; H. Minakata and O. Yasuda, Nucl. Phys. **B523**, 597 (1997); F. Vissani, hep-ph/9708483; H. Minakata and O. Yasuda, Phys. Rev. D **56**, 1692 (1997).

[3] For a recent discussion of the full range of possibilities, see A. Y. Smirnov, "Lepton mixing: Small, large, maximal?," hep-ph/9907296.

[4] The neutrino mass matrix we are led to coincides with the "bimaximal mixing matrix" described by V. Barger, S. Pakvasa, T. J. Weiler, and K. Whisnant, Phys. Lett. B **437**, 107 (1998).

[5] See R. Barbieri, J. High Energy Phys. **12**, 017 (1998).

[6] Super-Kamiokande Collaboration, Y. Fukuda *et al.*, Phys. Rev. Lett. **81**, 1562 (1998).

[7] M. Appollonio *et al.*, Phys. Lett. B **420**, 397 (1998).

[8] Kate Scholberg for the Super-Kamiokande Collaboration, K. Scholberg, hep-ex/9905016. We have taken this limit from Fig. 10.

[9] For example, Y. Suzuki, in Proceedings of the XVIII Interna-

tional Conference on Neutrino Physics and Astrophysics, Takayama, Japan, 1998.
[10] For example, J. N. Bahcall, Phys. Rev. D **58**, 096016 (1998).
[11] S. Nussinov, Phys. Lett. **63B**, 201 (1976); see [10] for other references and for a critique of this possibility.
[12] L. Baudis *et al.*, Phys. Lett. B **407**, 219 (1997).
[13] A. I. Belesev *et al.*, Phys. Lett. B **350**, 263 (1995).
[14] V. Barger, K. Whisnant, and R. J. N. Phillips, Phys. Rev. D **24**, 538 (1981); S. L. Glashow and L. M. Krauss, Phys. Lett. B **190**, 199 (1987).

VOLUME 83, NUMBER 13 PHYSICAL REVIEW LETTERS 27 SEPTEMBER 1999

Simple Connection Between Atmospheric and Solar Neutrino Vacuum Oscillations

Ernest Ma
Department of Physics, University of California, Riverside, California 92521
(Received 18 February 1999)

Extending the minimal standard model of particle interactions (without right-handed singlet neutrinos) to include a heavy scalar triplet ξ to obtain nonzero Majorana neutrino masses, I derive the following simple realistic connection between atmospheric and solar neutrino vacuum oscillations: $(\Delta m^2)_{\rm sol}(\Delta m^2)_{\rm atm}/m_\nu^4(\sin^2 2\theta)_{\rm atm} = 2I^2$, where m_ν is the assumed common approximate mass of each neutrino (which may be suitable for hot dark matter) and $I = (3/16\pi^2)(G_F/\sqrt{2})m_\tau^2 \ln(m_\xi^2/m_W^2)$ comes from the radiative splitting of the degeneracy due to the charged leptons.

PACS numbers: 14.60.Pq, 26.65.+t, 96.40.Tv

There is now a vast literature on models of neutrino oscillations [1]. Most try to understand why atmospheric neutrino oscillations [2] of $\nu_\mu(\bar\nu_\mu)$ to $\nu_\tau(\bar\nu_\tau)$ require near-maximal mixing [3]. Many also suggest that solar neutrino oscillations [4] of ν_e to a linear combination of ν_μ and ν_τ should have near-maximal mixing as well [5]. Both are possible in the context of three nearly mass-degenerate neutrinos [6,7] which could then be considered as candidates for hot dark matter [8].

Recently it has been pointed out [9] that if all three neutrinos obtain equal Majorana masses of order 1 eV from the canonical seesaw mechanism [10], then their splitting due to the different charged-lepton masses from the two-loop exchange of two W bosons [11] is of the right magnitude for solar neutrino vacuum oscillations. However, the inclusion of atmospheric neutrino oscillations has to be rather *ad hoc* in this case. In fact, it is rare indeed that any bona fide model of neutrino masses even gets a relationship between the mass difference of one oscillation and that of another. [One exception is the recently proposed model [12] of radiative masses for ν_e, ν_μ, ν_τ, plus a singlet (sterile) neutrino ν_s, which explains atmospheric and solar neutrino oscillations as well as the $\bar\nu_\mu(\nu_\mu)$ to $\bar\nu_e(\nu_e)$ data of the LSND (Liquid Scintillator Neutrino Detector) experiment [13]. It has the successful relationship $(\Delta m^2)_{\rm atm} \simeq 2[(\Delta m^2)_{\rm sol}(\Delta m^2)_{\rm LSND}]^{1/2}$, where $(\Delta m^2)_{\rm sol}$ refers to the matter-enhanced solution [14] of the solar neutrino deficit.]

In this Letter I present the most economical model to date of neutrino masses which has the following simple realistic connection between atmospheric and solar neutrino vacuum oscillations:

$$\frac{(\Delta m^2)_{\rm sol}(\Delta m^2)_{\rm atm}}{m_\nu^4(\sin^2 2\theta)_{\rm atm}} = 2I^2 = 4.9 \times 10^{-13}\left[\ln\frac{m_\xi^2}{m_W^2}\right]^2, \tag{1}$$

where m_ν is the assumed common approximate mass of each neutrino, m_ξ is the mass of a heavy scalar triplet, and

$$I = \frac{3G_F m_\tau^2}{16\pi^2\sqrt{2}} \ln\frac{m_\xi^2}{m_W^2} \tag{2}$$

comes from the one-loop radiative splitting of the degeneracy due to the charged leptons, as explained below. Numerically, let $m_\nu = 0.6$ eV, $(\sin^2 2\theta)_{\rm atm} = 1$, and $m_\xi = 1$ TeV, then Eq. (1) is satisfied with the best fit values of $(\Delta m^2)_{\rm sol} = 4.0 \times 10^{-10}$ eV^2 and $(\Delta m^2)_{\rm atm} = 4.0 \times 10^{-3}$ eV^2.

To start with, the minimal standard model (without right-handed singlet neutrinos) is extended to include a heavy scalar triplet $\xi = (\xi^{++}, \xi^+, \xi^0)$, where $m_\xi^2 \gg m_W^2$ is assumed. This provides the three neutrinos ν_e, ν_μ, ν_τ with small Majorana masses [15]. As emphasized recently [16], such an alternative is as simple and natural as the canonical seesaw mechanism [10] which was used in Ref. [9]. Now let there be a discrete S_3 symmetry (which has irreducible representations $\underline{2}$, $\underline{1}$, and $\underline{1}'$) such that ξ is a $\underline{1}$ and the standard Higgs doublet $\Phi = (\phi^+, \phi^0)$ is also a $\underline{1}$, whereas two of the lepton doublets form a $\underline{2}$ and the third is a $\underline{1}$ or $\underline{1}'$. The relevant terms in the interaction Lagrangian are then given by

$$\mathcal{L}_{\rm int} = \xi^0[f_0(\nu_1\nu_2 + \nu_2\nu_1) + f_3\nu_3\nu_3] + \mu\bar\xi^0\phi^0\phi^0 + \text{H.c.} \tag{3}$$

The field ξ^0 acquires a naturally small vacuum expectation value [15] $u \simeq -\mu\langle\phi^0\rangle^2/m_\xi^2$ and the 3×3 Majorana neutrino mass matrix is of the form

$$\mathcal{M}_\nu = \begin{pmatrix} 0 & m_0 & 0 \\ m_0 & 0 & 0 \\ 0 & 0 & m_3 \end{pmatrix}, \tag{4}$$

where $m_0 = 2f_0 u$ and $m_3 = 2f_3 u$. Actually, the difference between m_0 and m_3 will be assumed small compared to either m_0 or m_3 in the following; i.e., each neutrino is accorded an approximate common mass m_ν.

The neutrinos are now identified with their charged-lepton partners as follows:

$$\nu_1 = \nu_e, \qquad \nu_2 = c\nu_\mu - s\nu_\tau, \qquad \nu_3 = c\nu_\tau + s\nu_\mu, \tag{5}$$

where $s \equiv \sin\theta$ and $c \equiv \cos\theta$. This construction is made to accommodate the atmospheric data [2] as $\nu_\mu - \nu_\tau$

 0031-9007/99/83(13)/2514(4)$15.00

oscillations with $\sin^2 2\theta = 4s^2c^2$ and $\Delta m^2 = m_0^2 - m_3^2$. At this point, the eigenvalues of $\mathcal{M}_\nu$ of Eq. (4) are $-m_0$, m_0, and m_3. However, since the charged-lepton masses break the assumed S_3 symmetry, the twofold degeneracy of the $\nu_1 - \nu_2$ sector is broken radiatively in one loop. There are two effects. One is a finite correction to the mass matrix, as shown in Fig. 1. The other is a renormalization of the coupling matrix [17] from the shift in mass scale from m_ξ to m_W. As expected, the dominant contribution comes from the τ Yukawa coupling. The two contributions are naturally of the same texture and are easily calculated to be $4I/3$ and $-I/3$, respectively, where I is already given by Eq. (2). The mass matrix $\mathcal{M}_\nu$ is now corrected to read

$$\mathcal{M}_\nu = \begin{pmatrix} 0 & m_0(1 + s^2I) & -scm_0I \\ m_0(1 + s^2I) & 0 & -scm_3I \\ -scm_0I & -scm_3I & m_3(1 + 2c^2I) \end{pmatrix}. \tag{6}$$

The twofold degeneracy of the $\nu_1 - \nu_2$ sector is then lifted, with the following mass eigenvalues:

$$\begin{aligned} &-m_0(1 + s^2I) - \frac{s^2c^2(m_0 - m_3)^2I^2}{2(m_0 + m_3)}, \\ &m_0(1 + s^2I) + \frac{s^2c^2(m_0 + m_3)^2I^2}{2(m_0 - m_3)}, \end{aligned} \tag{7}$$

where $I^2 \ll (m_0 - m_3)^2/(m_0 + m_3)^2$ has been used, being justified numerically. Hence their mass-squared difference is

$$\begin{aligned} \Delta m^2 &\simeq s^2c^2m_0I^2\left[\frac{(m_0 + m_3)^2}{m_0 - m_3} - \frac{(m_0 - m_3)^2}{m_0 + m_3}\right] \\ &\simeq \frac{8s^2c^2I^2m_\nu^4}{m_0^2 - m_3^2}, \end{aligned} \tag{8}$$

where $m_\nu \simeq m_0 \simeq m_3$ has been used. Identifying this with solar neutrino vacuum oscillations then yields Eq. (1).

In the above, the choice $\nu_1 = \nu_e$ leads to $(\sin^2 2\theta)_{\rm sol} = 1$. The eigenstates of $\mathcal{M}_\nu$ from Eq. (4) or Eq. (6) are the same to first order,

$$\frac{1}{\sqrt{2}}(\nu_e - c\nu_\mu + s\nu_\tau), \qquad \frac{1}{\sqrt{2}}(\nu_e + c\nu_\mu - s\nu_\tau), \qquad s\nu_\mu + c\nu_\tau. \tag{9}$$

For $s = c = 1/\sqrt{2}$, the so-called bimaximal mixing solution [5] of neutrino oscillations is obtained. With the assumed form of Eq. (4), it is also worth noting that renormalization effects due to the τ and μ Yukawa couplings do not affect the degeneracy of the $\nu_1 - \nu_2$ sector to first order. This is why $(\Delta m^2)_{\rm sol}$ can be small enough here to be suitable for vacuum oscillations. The zero $\nu_e - \nu_e$ entry in the neutrino mass matrix is crucial for the validity of Eq. (7) and has been chosen to avoid neutrinoless double beta decay [18]. This is an important constraint as long as m_ν is greater than about 1 eV, which used to be a desirable feature as a component of dark matter [8]. However, with the recent observation of a nonzero cosmological constant [19], whereas m_ν is probably still needed for large-scale structure formation in the universe, its magnitude can be much smaller. In general, ν_1 may be a linear combination of ν_e, ν_μ, and ν_τ, but it has to be predominantly ν_e. Otherwise, m_τ (and m_μ) radiative contributions would appear in the diagonal entries of Eq. (6) and modify Eqs. (7) and (8). For illustration, the values $m_\nu = 0.6$ eV and $m_\xi = 1$ TeV have been used. It may be argued that m_ξ is naturally of order 10^{13} GeV or greater [15], in which case m_ν should be somewhat smaller. More precisely,

$$m_\nu \sim 1.3\ \text{eV}\left[\ln\frac{m_\xi^2}{m_W^2}\right]^{-1/2}. \tag{10}$$

For $m_\xi = 10^{13}$ GeV, the required m_ν is then about 0.18 eV.

The charged-lepton mass matrix which accompanies $\mathcal{M}_\nu$ of Eq. (4) is not uniquely defined, because only the left-handed fields are correlated with it. Nevertheless, S_3 is clearly violated. So far, I have not identified the origin of this violation. It may simply be explicit, or it may be spontaneous, in the sense that it occurs only when the electroweak gauge symmetry is broken. An example of the latter is the following model. Under S_3, let

$$\left[\begin{pmatrix}\nu_1\\ l_1\end{pmatrix}_L, \begin{pmatrix}\nu_2\\ l_2\end{pmatrix}_L\right] \sim 2, \quad \begin{pmatrix}\nu_3\\ l_3\end{pmatrix}_L \sim 1, \qquad [l_{1L}^c, l_{2L}^c] \sim 2, \quad l_{3L}^c \sim 1, \tag{11}$$

$$\left[\begin{pmatrix}\phi_1^0\\ \phi_1^-\end{pmatrix}, \begin{pmatrix}\phi_2^0\\ \phi_2^-\end{pmatrix}\right] \sim 2, \quad \begin{pmatrix}\phi_3^0\\ \phi_3^-\end{pmatrix} \sim 1, \qquad (\xi^{++}, \xi^+, \xi^0) \sim 1. \tag{12}$$

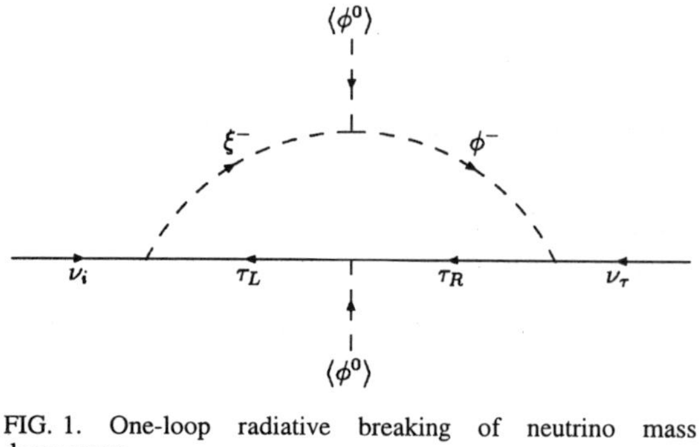

FIG. 1. One-loop radiative breaking of neutrino mass degeneracy.

VOLUME 83, NUMBER 13 PHYSICAL REVIEW LETTERS 27 SEPTEMBER 1999

With $\langle \xi^0 \rangle \neq 0$, $\mathcal{M}_\nu$ of Eq. (4) is obtained, whereas $\mathcal{M}_l$ is now given by

$$\mathcal{M}_l = \begin{bmatrix} h_1\langle\phi_1^0\rangle & h_2\langle\phi_3^0\rangle & h_3\langle\phi_2^0\rangle \\ h_2\langle\phi_3^0\rangle & h_1\langle\phi_2^0\rangle & h_3\langle\phi_1^0\rangle \\ h_4\langle\phi_2^0\rangle & h_4\langle\phi_1^0\rangle & h_5\langle\phi_3^0\rangle \end{bmatrix}, \quad (13)$$

where $h_{1,2,3,4,5}$ are the couplings of all possible Yukawa terms invariant under S_3. Before electroweak symmetry breaking, charged-lepton masses as well as neutrino masses are zero, as in the standard model. After electroweak symmetry breaking, let $\langle\phi_2^0\rangle \ll \langle\phi_1^0\rangle$, then S_3 is also broken in $\mathcal{M}_l$ at tree level but not in $\mathcal{M}_\nu$. Radiative corrections then break S_3 in $\mathcal{M}_\nu$ as shown in this paper. In the limit $h_2 \to 0$ and $\langle\phi_2^0\rangle \to 0$, e_L is indeed separated from the $\mu_L - \tau_L$ sector and Eq. (5) holds as desired. In terms of fine-tuning, this model is no worse than the standard model which also requires arbitrary Yukawa couplings to fix the charged-lepton masses.

In conclusion, I have presented in this Letter a new and economical extension of the minimal standard model, where a heavy scalar triplet ξ is added to provide the three known neutrinos with nonzero Majorana masses. This replaces the usual method of adding three heavy right-handed neutrino singlets. A discrete S_3 symmetry is then assumed so that two neutrinos are degenerate in mass, with their splitting controlled by one-loop radiative corrections. This results in a simple realistic connection between atmospheric and solar neutrino vacuum oscillations as given by Eq. (1). It is consistent with the present data and will get tested further as more data become available in the near future from planned experiments in neutrino oscillations, neutrinoless double beta decay, neutrino mass, searches for dark matter and for new particles in high-energy accelerators.

This work was supported in part by the U.S. Department of Energy under Grant No. DE-FG03-94ER40837.

[1] For a brief review, see, for example, M. Tanimoto, in Proceedings of the 17th International Workshop on Weak Interactions and Neutrinos (WIN99), Cape Town, South Africa, 1999 [hep-ph/9903505].

[2] Y. Fukuda *et al.,* Phys. Lett. B **433**, 9 (1998); **436**, 33 (1998); Phys. Rev. Lett. **81**, 1562 (1998); **82**, 2644 (1999).

[3] C.H. Albright, K.S. Babu, and S.M. Barr, Phys. Rev. Lett. **81**, 1167 (1998); B.C. Allanach, Phys. Lett. B **450**, 182 (1999); S.F. King, Phys. Lett. B **439**, 350 (1998); J.K. Elwood, N. Irges, and P. Ramond, Phys. Rev. Lett. **81**, 5064 (1998); R. Barbieri, L.J. Hall, D. Smith, A. Strumia, and N. Weiner, J. High Energy Phys. **9812**, 017 (1998); G. Altarelli and F. Feruglio, Phys. Lett. B **439**, 112 (1998); J. High Energy Phys. **9811**, 021 (1998); Phys. Lett. B **451**, 388 (1999); E. Ma, Phys. Lett. B **442**, 238 (1998); N. Haba, Phys. Rev. D **59**, 035011 (1999); K. Oda, E. Takasugi, M. Tanaka, and M. Yoshimura, Phys. Rev. D **59**, 055001 (1999); J. Ellis, G.K. Leontaris, S. Lola, and D.V. Nanopoulos, Eur. Phys. J. C **9**, 389 (1999); R.N. Mohapatra and S. Nussinov, Phys. Lett. B **441**, 299 (1998); R. Barbieri, L.J. Hall, and A. Strumia, Phys. Lett. B **445**, 407 (1999); Y. Grossman, Y. Nir, and Y. Shadmi, J. High Energy Phys. **9810**, 007 (1998); M. Fukugita, M. Tanimoto, and T. Yanagida, Phys. Rev. D **59**, 113016 (1999); E. Ma, D.P. Roy, and U. Sarkar, Phys. Lett. B **444**, 391 (1998); E. Malkawi, hep-ph/9810542; E. Ma and D.P. Roy, Phys. Rev. D **59**, 097702 (1999); L.J. Hall and N. Weiner, Phys. Rev. D **60**, 033005 (1999); Z. Berezhiani and A. Rossi, J. High Energy Phys. **9903**, 002 (1999); K.S. Babu, J.C. Pati, and F. Wilczek, hep-ph/9812538; S. Lola and G.G. Ross, hep-ph/9902283.

[4] R. Davis, Prog. Part. Nucl. Phys. **32**, 13 (1994); P. Anselmann *et al.,* Phys. Lett. B **357**, 237 (1995); **361**, 235 (1996); J.N. Abdurashitov *et al.,* Phys. Lett. B **328**, 234 (1994); Y. Fukuda *et al.,* Phys. Rev. Lett. **77**, 1683 (1996); **81**, 1158 (1998); **82**, 1810 (1999); **82**, 2430 (1999).

[5] V. Barger, S. Pakvasa, T.J. Weiler, and K. Whisnant, Phys. Lett. B **437**, 107 (1998); A.J. Baltz, A.S. Goldhaber, and M. Goldhaber, Phys. Rev. Lett. **81**, 5730 (1998); Y. Nomura and T. Yanagida, Phys. Rev. D **59**, 017303 (1999); M. Tanimoto, Phys. Rev. D **59**, 017304 (1999); M. Jezabek and Y. Sumino, Phys. Lett. B **440**, 327 (1998); H. Fritzsch and Z. Xing, Phys. Lett. B **440**, 313 (1998); S.L. Glashow, P.J. Kernan, and L.M. Krauss, Phys. Lett. B **445**, 412 (1999); S. Davidson and S.F. King, Phys. Lett. B **445**, 191 (1998); A.S. Joshipura and S.D. Rindani, hep-ph/9811252; C. Jarlskog, M. Matsuda, S. Skadhauge, and M. Tanimoto, Phys. Lett. B **449**, 240 (1999).

[6] D. Caldwell and R.N. Mohapatra, Phys. Rev. D **48**, 3259 (1993); A.S. Joshipura, Z. Phys. C **64**, 31 (1994); Phys. Rev. D **51**, 1321 (1995); P. Bamert and C.P. Burgess, Phys. Lett. B **329**, 289 (1994); D.-G. Lee and R.N. Mohapatra, Phys. Lett. B **329**, 463 (1994); A. Ioannisian and J.W.F. Valle, Phys. Lett. B **332**, 93 (1994); A. Ghosal, Phys. Lett. B **398**, 315 (1997); A.K. Ray and S. Sarkar, Phys. Rev. D **58**, 055010 (1998); C.D. Carone and M. Sher, Phys. Lett. B **420**, 83 (1998); U. Sarkar, Phys. Rev. D **59**, 037302 (1999); G.C. Branco, M.N. Rebelo, and J.I. Silva-Marcos, Phys. Rev. Lett. **82**, 683 (1999).

[7] F. Vissani, hep-ph/9708483; H. Georgi and S.L. Glashow, hep-ph/9808293; R.N. Mohapatra and S. Nussinov, Phys. Rev. D **60**, 013002 (1999); Y.L. Wu, hep-ph/9810491; hep-ph/9901245; hep-ph/9901320; C. Wetterich, Phys. Lett. B **451**, 397 (1999); R. Barbieri, L.J. Hall, G.L. Kane, and G.G. Ross, hep-ph/9901228.

[8] E. Gawiser and J. Silk, Science **280**, 1405 (1998); J.R. Primack and M.A.K. Gross, astro-ph/9810204; K.S. Babu, R.K. Schaefer, and Q. Shafi, Phys. Rev. D **53**, 606 (1996).

[9] E. Ma, Phys. Lett. B **456**, 48 (1999).

[10] M. Gell-Mann, P. Ramond, and R. Slansky, in *Supergravity,* edited by P. van Nieuwenhuizen and D.Z. Freedman (North-Holland, Amsterdam, 1979), p. 315; T. Yanagida, in *Proceedings of the Workshop on the Unified Theory and the Baryon Number in the Universe,* edited by O. Sawada and A. Sugamoto, KEK Report No. 79-18 (KEK, Tsukuba, Japan, 1979), p. 95; R.N. Mohapatra and G. Senjanovic, Phys. Rev. Lett. **44**, 912 (1980).

VOLUME 83, NUMBER 13 PHYSICAL REVIEW LETTERS 27 SEPTEMBER 1999

[11] K. S. Babu and E. Ma, Phys. Rev. Lett. **61**, 674 (1988); Phys. Lett. B **228**, 508 (1989). See also S. T. Petcov and S. T. Toshev, Phys. Lett. **143B**, 175 (1984).
[12] N. Gaur, A. Ghosal, E. Ma, and P. Roy, Phys. Rev. D **58**, 071301 (1998).
[13] C. Athanassopoulos *et al.,* Phys. Rev. Lett. **75**, 2650 (1995); **77**, 3082 (1996); **81**, 1774 (1998).
[14] L. Wolfenstein, Phys. Rev. D **17**, 2369 (1978); S.P. Mikheyev and A. Yu. Smirnov, Sov. J. Nucl. Phys. **42**, 913 (1986).
[15] E. Ma and U. Sarkar, Phys. Rev. Lett. **80**, 5716 (1998).
[16] E. Ma, Phys. Rev. Lett. **81**, 1171 (1998).
[17] J. Ellis and S. Lola, hep-ph/9904279; J. A. Casas, J. R. Espinosa, A. Ibarra, and I. Navarro, hep-ph/9904395.
[18] For a review, see, for example, H. V. Klapdor-Kleingrothaus, hep-ex/9901021.
[19] A. G. Reiss *et al.,* astro-ph/9805201; S. Perlmutter *et al.,* astro-ph/9812133.

PHYSICAL REVIEW D, VOLUME 61, 036002

Neutrinoless double-β decay with three or four neutrino mixing

Carlo Giunti
INFN, Sezione di Torino, and Dipartimento di Fisica Teorica, Università di Torino, Via P. Giuria 1, I-10125 Torino, Italy
(Received 8 June 1999; published 3 January 2000)

Considering the scheme with mixing of three neutrinos and a mass hierarchy that can accommodate the results of solar and atmospheric neutrino experiments, it is shown that the results of solar neutrino experiments imply a lower bound for the effective Majorana mass in neutrinoless double-β decay, under the natural assumptions that massive neutrinos are Majorana particles and there are no unlikely fine-tuned cancellations among the contributions of the different neutrino masses. Considering the four-neutrino schemes that can accommodate also the results of the LSND experiment, it is shown that one of them is favored by the results of neutrinoless double-β decay experiments and the measurement of the abundances of primordial elements produced in big-bang nucleosynthesis. It is shown that in this scheme, under the assumptions that massive neutrinos are Majorana particles and there are no cancellations among the contributions of the different neutrino masses, the results of the LSND experiment imply a lower bound for the effective Majorana mass in neutrinoless double-β decay.

PACS number(s): 23.40.−s, 14.60.Pq, 14.60.St

I. INTRODUCTION

Neutrino oscillation [1–3] is one of the most intriguing phenomena of present day high-energy physics and one of the most promising ways to explore physics beyond the standard model. At present there are three experimental indications in favor of neutrino oscillations that have been obtained in solar neutrino experiments (Homestake [4], Kamiokande [5], GALLEX [6], SAGE [7], and Super-Kamiokande [8]), in atmospheric neutrino experiments (Kamiokande [9], IMB [10], Super-Kamiokande [11], Soudan-2 [12], and MACRO [13]), and in the Liquid Scintillation Neutrino Detector (LSND) experiment [14,15]. On the other hand, neutrinoless double-β decay ($\beta\beta_{0\nu}$) experiments [16–19] and the experiments on the direct measurement of neutrino masses [20] have not obtained any positive result so far (see [21]). The connection between the properties of neutrinos that determine neutrino oscillations (mass squared differences and neutrino mixing) and neutrinoless double-β decay has been discussed in many papers [22–33]. In this paper we discuss some implications of the latest results of neutrino oscillation experiments for neutrinoless double-β decay and we show that under reasonable assumptions there is a lower bound for the effective Majorana neutrino mass measured in $\beta\beta_{0\nu}$ decay experiments.

If massive neutrinos are Majorana particles, the matrix element of $\beta\beta_{0\nu}$ decay is proportional to the effective Majorana mass

$$|\langle m\rangle|=\left|\sum_k U_{ek}^2 m_k\right|, \tag{1.1}$$

where U is the mixing matrix that connects the flavor neutrino fields $\nu_{\alpha L}$ ($\alpha=e,\mu,\tau$) to the fields ν_{kL} of neutrinos with masses m_k through the relation

$$\nu_{\alpha L}=\sum_k U_{\alpha k}\nu_{kL}. \tag{1.2}$$

The present experimental upper limit for $|\langle m\rangle|$,

$$|\langle m\rangle|_{\text{expt}}\leq 0.2-0.4 \text{ eV} \ \ (90\% \text{ C.L.}), \tag{1.3}$$

has been obtained from the measurement of the half-life of ^{76}Ge in the Heidelberg-Moscow experiment [$T_{1/2}^{0\nu}(^{76}\text{Ge}) \geq 5.7\times 10^{25}$ yr at 90% C.L.] [18]. The range of the upper bound (1.3) is due to the uncertainty of the theoretical calculation of the nuclear matrix element and has been obtained from the results of different calculations using the quasiparticle random phase approximation (QRPA) [34,35]. In particular, the recent QRPA calculation in [35] yields the rather stringent upper bound $|\langle m\rangle|_{\text{expt}}\leq 0.27$ eV. On the other hand, the shell model calculation in [36] yields the loser bound $|\langle m\rangle|_{\text{expt}}\leq 0.56$ eV. However, the calculation of the nuclear matrix element for the neutrinoless double-β decay of ^{76}Ge presented in [36] has been truncated before reaching convergence and the full calculation is expected to yield a more stringent upper bound for $|\langle m\rangle|_{\text{expt}}$. Therefore, in the following we will consider the range in Eq. (1.3) as a reliable estimate of the uncertainty of the experimental upper bound for the effective Majorana mass $|\langle m\rangle|$.

The next generation of $\beta\beta_{0\nu}$ decay experiments is expected to be sensitive to values of $|\langle m\rangle|$ in the range $10^{-2}-10^{-1}$ eV [17]. Values of $|\langle m\rangle|$ as small as about 10^{-3} eV may be reachable not far in the future [19].

After the measurement in the Super-Kamiokande experiment of an up-down asymmetry of μ-like events induced by atmospheric muon neutrinos, the experimental evidence in favor of oscillations of atmospheric neutrinos is widely considered to be beyond reasonable doubts (see, for example, [2,37,3]). There are also convincing arguments in favor of a neutrino oscillation explanation of the solar neutrino problem (see, for example, [38,2,39]). Therefore, in this paper we will consider first, in Sec. II, the implications for $\beta\beta_{0\nu}$ decay in the scheme with mixing of three neutrinos and a mass hierarchy that can accommodate the results of atmospheric and solar neutrino experiments. In Sec. III we consider the schemes with four massive neutrinos that can accommodate all neutrino oscillation data, including the LSND results in favor of $\nu_\mu \to \nu_e$ and $\bar{\nu}_\mu \to \bar{\nu}_e$ oscillations that wait for independent confirmations by other experiments [40,41].

II. THREE NEUTRINOS WITH A MASS HIERARCHY

The results of solar and atmospheric neutrino experiments indicate the existence of a hierarchy of mass-squared differences (see [38,42–45,11]):

$$\Delta m^2_{\rm sun} \lesssim 10^{-4}\ {\rm eV}^2 \ll 10^{-3}\ {\rm eV}^2 \lesssim \Delta m^2_{\rm atm} \lesssim 10^{-2}\ {\rm eV}^2, \tag{2.1}$$

where $\Delta m^2_{\rm sun}$ and $\Delta m^2_{\rm atm}$ are the mass-squared differences relevant for solar and atmospheric neutrino oscillations, respectively. A natural scheme that can accommodate this hierarchy is the one with three neutrinos and a mass hierarchy,

$$\underbrace{\overbrace{m_1 \ll m_2}^{\rm sun} \ll m_3}_{\rm atm}\ . \tag{2.2}$$

In the framework of the hierarchical spectrum (2.2) the mass-squared differences relevant for the oscillations of solar and atmospheric neutrinos are

$$\Delta m^2_{\rm sun} = \Delta m^2_{21} \equiv m_2^2 - m_1^2 \simeq m_2^2,$$
$$\Delta m^2_{\rm atm} = \Delta m^2_{31} \equiv m_3^2 - m_1^2 \simeq m_3^2. \tag{2.3}$$

The mass hierarchy (2.2) is predicted by the seesaw mechanism [46], which predicts also that the tree light massive neutrinos are Majorana particles. In this case neutrinoless double-β decay is possible.

It has been shown in [25–27] that the results of neutrino oscillation experiments imply a rather stringent upper bound (about 6×10^{-3} eV) for the effective Majorana mass in neutrinoless double-β decay in the scheme with mixing of three neutrinos and a mass hierarchy. In principle the effective Majorana mass (1.1) can be vanishingly small because of cancellations among the contributions of the different mass eigenstates. However, since the neutrino masses and the elements of the neutrino mixing matrix are independent quantities, if there is a hierarchy of neutrino masses, such a cancellation would be the result of an unlikely fine-tuning, unless some unknown symmetry is at work. Here we consider the possibility that no such symmetry exists and *no unlikely fine-tuning operates to suppress the effective Majorana mass* (1.1). In this case we have

$$|\langle m\rangle| \simeq \max_k |U_{ek}|^2 m_k. \tag{2.4}$$

Let us define the absolute value of the contribution of the neutrino mass m_k to $|\langle m\rangle|$ as

$$|\langle m\rangle|_k \equiv |U_{ek}|^2 m_k,. \tag{2.5}$$

In the following we will estimate the value of $|\langle m\rangle|$ using the largest $|\langle m\rangle|_k$ obtained from the results of neutrino oscillation experiments.

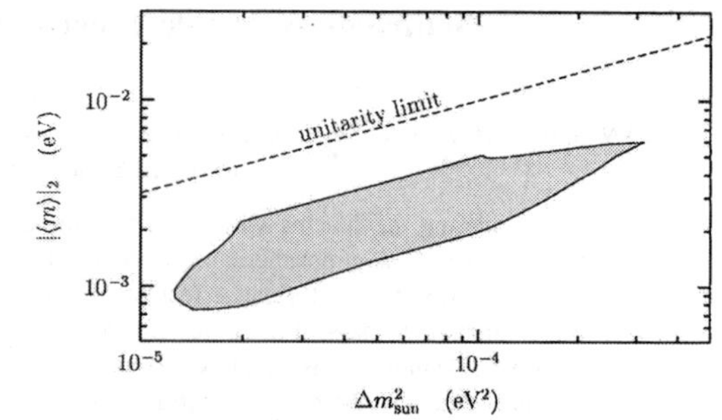

FIG. 1. The shadowed area shows the allowed range for $|\langle m\rangle|_2$ as a function of $\Delta m^2_{\rm sun}$ in the scheme with mixing of three neutrinos with a mass hierarchy discussed in Sec. II, in the case of the LMA-MSW solution of the solar neutrino problem. The shadowed area has been obtained using Eq. (2.6) and the allowed range for $\sin^2 2\vartheta_{\rm sun}$ given by the LMA-MSW region (99% C.L.) in the $\sin^2 2\vartheta_{\rm sun}$–$\Delta m^2_{\rm sun}$ plane presented in Fig. 2 of Ref. [42]. The dashed line represents the unitarity limit $|\langle m\rangle|_2 \leq \sqrt{\Delta m^2_{\rm sun}}$.

The results of the CHOOZ experiment [47] and the Super-Kamiokande atmospheric neutrino data [11] imply that $|U_{e3}|^2$ is small ($|U_{e3}|^2 \lesssim 5\times 10^{-2}$) and there is an upper limit of about 6×10^{-3} eV for the contribution $|\langle m\rangle|_3$ to the effective Majorana mass in $\beta\beta_{0\nu}$ decay [25–27]. Since there is no lower bound for $|U_{e3}|^2$ from experimental data, $|\langle m\rangle|_3$ could be very small.

Hence, the largest contribution to $|\langle m\rangle|$ could come from $|\langle m\rangle|_2 \equiv |U_{e2}|^2 m_2$. Since in the framework of the scheme with mixing of three neutrinos and a mass hierarchy $\Delta m^2_{\rm sun} \simeq m_2^2$ and $|U_{e2}|^2 \simeq \frac{1}{2}(1-\sqrt{1-\sin^2 2\vartheta_{\rm sun}})$ [48], where $\vartheta_{\rm sun}$ is the two-neutrino mixing angle used in the analysis of solar neutrino data, we have

$$|\langle m\rangle|_2 \simeq \tfrac{1}{2}\,(1-\sqrt{1-\sin^2 2\vartheta_{\rm sun}})\,\sqrt{\Delta m^2_{\rm sun}}. \tag{2.6}$$

Solar neutrino data imply bounds for $\sin^2 2\vartheta_{\rm sun}$ and $\Delta m^2_{\rm sun}$. In particular the large mixing angle (LMA) Mikheyev-Smirnov-Wolfenstein (MSW) [49] solution (LMA-MSW) of the solar neutrino problem, which seems to be favored by recent data [50], implies that [42]

$$1.2\times 10^{-5}\ {\rm eV}^2 \leq \Delta m^2_{\rm sun} \leq 3.1\times 10^{-4}\ {\rm eV}^2,$$
$$0.58 \leq \sin^2 2\vartheta_{\rm sun} \leq 1.00 \tag{2.7}$$

at 99% C.L., taking into account the total rates measured in solar neutrino experiments and the day-night variations observed in the Super-Kamiokande experiment [8]. Hence, for the contribution of m_2 to the effective Majorana mass we obtain

$$6\times 10^{-4}\ {\rm eV} \lesssim |\langle m\rangle|_2 \lesssim 9\times 10^{-3}\ {\rm eV}. \tag{2.8}$$

This estimate does not take into account the correlation between $\Delta m^2_{\rm sun}$ and $\sin^2 2\vartheta_{\rm sun}$. The precise allowed range for $|\langle m\rangle|_2$ as a function of $\Delta m^2_{\rm sun}$ obtained with Eq. (2.6) from the LMA-MSW region (99% C.L.) in Fig. 2 of Ref. [42] is shown in Fig. 1. The dashed line in Fig. 1 represents the unitarity limit $|\langle m\rangle|_2 \leq \sqrt{\Delta m^2_{\text{sun}}}$. From Fig. 1 one can see that the LMA-MSW solution of the solar neutrino problem implies that[1]

$$7.4\times 10^{-4}\ \text{eV} \lesssim |\langle m\rangle|_2 \lesssim 6.0\times 10^{-3}\ \text{eV}. \quad (2.9)$$

Assuming the absence of fine-tuned cancellations among the contributions of the three neutrino masses to the effective Majorana mass, if $|U_{e3}|^2$ is very small and $|\langle m\rangle|_3 \ll |\langle m\rangle|_2$, from Eqs. (2.4) and (2.9) we obtain

$$7\times 10^{-4}\ \text{eV} \lesssim |\langle m\rangle| \lesssim 6\times 10^{-3}\ \text{eV}. \quad (2.10)$$

Hence, we see that, *assuming the absence of an unlikely fine-tuned suppression of* $|\langle m\rangle|$, *the results of solar neutrino experiments give an indication of the value of the effective Majorana mass in* $\beta\beta_{0\nu}$ *decay, with a lower bound of about* 7×10^{-4} *eV in the case of the LMA-MSW solution of the solar neutrino problem*. This bound is rather small, but values of $|\langle m\rangle|$ of the order of 10^{-3} eV, indicated by the range (2.10), may be measurable in a not too far future [19].

Also the small mixing angle MSW (SMA-MSW) and the vacuum oscillation (VO) solutions of the solar neutrino problem imply allowed ranges for $|\langle m\rangle|_2$, but their values are much smaller than in the case of the LMA-MSW solution. Using the 99% C.L. allowed regions obtained in [38] from the analysis of the total rates measured in solar neutrino experiments we have

$$5\times 10^{-7}\ \text{eV} \lesssim |\langle m\rangle|_2 \lesssim 10^{-5}\ \text{eV (SMA-MSW)}, \quad (2.11)$$

$$10^{-6}\ \text{eV} \lesssim |\langle m\rangle|_2 \lesssim 2\times 10^{-5}\ \text{eV (VO)}. \quad (2.12)$$

III. FOUR NEUTRINOS

If, in addition to the solar and atmospheric neutrino data, also the results of the accelerator LSND experiment are taken into account, at least three independent mass-squared differences are needed. This can be seen by considering the general expression for the probability of $\nu_\alpha \to \nu_\beta$ transitions in vacuum [1–3], which can be written as

$$P_{\nu_\alpha\to\nu_\beta} = \left|\sum_k U^*_{\alpha k} U_{\beta k} \exp\left(-i\frac{\Delta m^2_{kj} L}{2E}\right)\right|^2, \quad (3.1)$$

where $\Delta m^2_{kj} \equiv m^2_k - m^2_j$, j is any of the mass-eigenstate indices, L is the distance between the neutrino source and detector, and E is the neutrino energy. The range of L/E probed by each type of experiment is different: $L/E \gtrsim 10^{10}\ \text{eV}^{-2}$ for solar neutrino experiments, $L/E \sim 10^2$–$10^3\ \text{eV}^{-2}$ for atmospheric neutrino experiments, and $L/E \sim 1\ \text{eV}^{-2}$ for the LSND experiment. From Eq. (3.1) it is clear that neutrino oscillations occur in an experiment only if there is at least one mass-squared difference Δm^2_{kj} such that

$$\frac{\Delta m^2_{kj} L}{2E} \gtrsim 0.1 \quad (3.2)$$

(the precise lower bound depends on the sensitivity of the experiment) in a significant part of the energy and source-detector distance intervals of that experiment [if the condition (3.2) is not satisfied, $P_{\nu_\alpha\to\nu_\beta} \simeq |\Sigma_k U^*_{\alpha k} U_{\beta k}|^2 = \delta_{\alpha\beta}$]. Since the range of L/E probed by the LSND experiment is the smaller one, a mass-squared difference is needed for LSND oscillations:

$$\Delta m^2_{\text{LSND}} \gtrsim 10^{-1}\ \text{eV}^2. \quad (3.3)$$

Specifically, the maximum likelihood analysis of the LSND data in terms of two-neutrino oscillations gives [15]

$$0.20\ \text{eV}^2 \leq \Delta m^2_{\text{LSND}} \leq 2.0\ \text{eV}^2. \quad (3.4)$$

Furthermore, from Eq. (3.1) it is clear that a dependence of the oscillation probability from the neutrino energy E and the source-detector distance L is observable only if there is at least one mass-squared difference Δm^2_{kj} such that

$$\frac{\Delta m^2_{kj} L}{2E} \sim 1. \quad (3.5)$$

Since a variation of the transition probability as a function of neutrino energy has been observed both in solar and atmospheric neutrino experiments and the range of L/E probed by each type of experiment is different, two more mass-squared differences with different scales are needed:

$$\Delta m^2_{\text{sun}} \sim 10^{-10}\ \text{eV}^2\ \ \text{(VO)}, \quad (3.6)$$

$$\Delta m^2_{\text{atm}} \sim 10^{-3}\text{–}10^{-2}\ \text{eV}^2. \quad (3.7)$$

The condition (3.6) for the solar mass-squared difference Δm^2_{sun} has been obtained under the assumption of vacuum oscillations. If the disappearance of solar ν_e's is due to the MSW effect [49], the condition

$$\Delta m^2_{\text{sun}} \lesssim 10^{-4}\ \text{eV}^2\ \text{(MSW)} \quad (3.8)$$

must be fulfilled in order to have a resonance in the interior of the Sun. Hence, in the MSW case Δm^2_{sun} must be at least one order of magnitude smaller than Δm^2_{atm}.

The existence of three different scales of neutrino mass-squared differences[2] implies that at least four light massive neutrinos must exist in nature. Here we consider the schemes with four light and mixed neutrinos [52–58], which constitute the minimal possibility that allows to explain all the data of neutrino oscillation experiments. In this case, in the flavor basis the three active neutrinos ν_e, ν_μ, ν_τ are accompanied by a sterile neutrino ν_s that does not take part in standard

---

[1]The upper limit $|\langle m\rangle|_2 \lesssim 3\times 10^{-3}$ eV presented in [27] has been obtained from the 90% C.L. LMA-MSW region in Fig. 8a of Ref. [43], using Eq. (2.6). The 99% C.L. LMA-MSW region in Fig. 8a of Ref. [43] gives $|\langle m\rangle|_2 \lesssim 6\times 10^{-3}$ eV, in agreement with the upper bound in Eq. (2.9).

weak interactions.

The existence of four light massive neutrinos is a low-energy manifestation of physics beyond the standard model (see, for example, [59]). In most theories beyond the standard model neutrinos are naturally Majorana particles and neutrinoless double-β decay is allowed. Therefore, we see that the experimental evidence in favor of neutrino oscillations indicates that neutrinos may be Majorana particles and neutrinoless double-β decay is a concrete possibility.

It has been shown [55] that there are only two schemes with four-neutrino mixing that can accommodate the results of all neutrino oscillation experiments:

$$\text{(A)} \quad \underbrace{\overbrace{m_1 < m_2}^{\text{atm}} < \overbrace{m_3 < m_4}^{\text{sun}}}_{\text{LSND}}, \qquad \text{(B)} \quad \underbrace{\overbrace{m_1 < m_2}^{\text{sun}} < \overbrace{m_3 < m_4}^{\text{atm}}}_{\text{LSND}}. \tag{3.9}$$

These two spectra are characterized by the presence of two pairs of close masses separated by a gap of about 1 eV which provides the mass-squared difference $\Delta m^2_{\text{LSND}} = \Delta m^2_{41}$ responsible of the oscillations observed in the LSND experiment. In the scheme A, $\Delta m^2_{\text{atm}} = \Delta m^2_{21}$ and $\Delta m^2_{\text{sun}} = \Delta m^2_{43}$, whereas in scheme B, $\Delta m^2_{\text{atm}} = \Delta m^2_{43}$ and $\Delta m^2_{\text{sun}} = \Delta m^2_{21}$.

The results of the short-base-line $\bar{\nu}_e$ disappearance experiment Bugey [51], in which no indication in favor of neutrino oscillations was found, imply that the mixing of ν_e with the two "heavy" neutrinos ν_3 and ν_4 is large in scheme A and small in scheme B [55,2]:

$$1 - (|U_{e3}|^2 + |U_{e4}|^2) \lesssim 3 \times 10^{-2} \ \text{(A)}, \tag{3.10}$$

$$|U_{e3}|^2 + |U_{e4}|^2 \lesssim 3 \times 10^{-2} \ \text{(B)}, \tag{3.11}$$

for Δm^2_{LSND} in the LSND-allowed range (3.4). Therefore, if scheme A is realized in nature, the effective Majorana mass in $\beta\beta_{0\nu}$ decay can be as large as $m_3 \simeq m_4 \simeq \sqrt{\Delta m^2_{\text{LSND}}} \simeq 0.45\text{–}1.4$ eV [24–27]. On the other hand, in scheme B neutrinoless double-β decay is strongly suppressed [25–27]. In the following two subsections we discuss some connections between the results of neutrino oscillation experiments and neutrinoless double-β decay in the schemes A and B.

[2]It is possible to ask if three different scales of neutrino mass-squared differences are needed even if the results of the Homestake solar neutrino experiment [4] is neglected, allowing an energy-independent suppression of the solar ν_e flux. The answer is that still the data cannot be fitted with only two neutrino mass-squared differences because an energy-independent suppression of the solar ν_e flux requires large $\nu_e \to \nu_\mu$ or $\nu_e \to \nu_\tau$ transitions generated by Δm^2_{atm} or Δm^2_{LSND}. These transitions are forbidden by the results of the Bugey [51] and CHOOZ [47] $\bar{\nu}_e$ disappearance experiments and by the nonobservation of an up-down asymmetry of e-like events in the Super-Kamiokande atmospheric neutrino experiment [11]. I would like to thank S.T. Petcov for useful discussions about this point.

A. Scheme A

In the four-neutrino scheme A, from Eq. (3.10) and the unitarity of the mixing matrix we have $|U_{e1}|^2 + |U_{e2}|^2 \lesssim 3 \times 10^{-2}$. Therefore, the contribution of the two light masses m_1 and m_2 to the effective Majorana mass in $\beta\beta_{0\nu}$ decay can be neglected and we have

$$|\langle m \rangle| \simeq |U_{e3}^2 m_3 + U_{e4}^2 m_4|, \tag{3.12}$$

which implies the limits

$$\| U_{e3}|^2 m_3 - |U_{e4}|^2 m_4| \lesssim |\langle m \rangle| \lesssim |U_{e3}|^2 m_3 + |U_{e4}|^2 m_4. \tag{3.13}$$

Neglecting the small difference between m_3 and m_4 and taking into account that $m_3 \simeq m_4 \simeq \sqrt{\Delta m^2_{\text{LSND}}}$, we have

$$\| U_{e3}|^2 - |U_{e4}|^2 | \sqrt{\Delta m^2_{\text{LSND}}} \lesssim |\langle m \rangle| \lesssim (|U_{e3}|^2 + |U_{e4}|^2) \sqrt{\Delta m^2_{\text{LSND}}}. \tag{3.14}$$

Since the quantity $|U_{e3}|^2 + |U_{e4}|^2$ is large in scheme A [see Eq. (3.10)], we obtain

$$\| U_{e3}|^2 - |U_{e4}|^2 | \sqrt{\Delta m^2_{\text{LSND}}} \lesssim |\langle m \rangle| \lesssim \sqrt{\Delta m^2_{\text{LSND}}}. \tag{3.15}$$

Furthermore, from the inequality (3.10) one can see that the contribution of the mixing of ν_e with ν_1 and ν_2 to the survival probability of solar electron neutrinos is negligible and $|U_{e3}|^2$ and $|U_{e4}|^2$ are related to the mixing angle ϑ_{sun} obtained from the two-generation fit of solar neutrino experiments by

$$|U_{e3}|^2 \simeq \cos^2 \vartheta_{\text{sun}}, \quad |U_{e4}|^2 \simeq \sin^2 \vartheta_{\text{sun}}. \tag{3.16}$$

Hence, the range (3.15) can be written as

$$\sqrt{(1 - \sin^2 2\vartheta_{\text{sun}}) \Delta m^2_{\text{LSND}}} \lesssim |\langle m \rangle| \lesssim \sqrt{\Delta m^2_{\text{LSND}}}, \tag{3.17}$$

Let us emphasize that this allowed range for $|\langle m\rangle|$ in scheme A depends only on the assumption that massive neutrinos are Majorana particles.

In the case of the SMA-MSW solution of the solar neutrino problem (for both $\nu_e\to\nu_\tau$ or $\nu_e\to\nu_s$ transitions) $\sin^2 2\vartheta_{\rm sun}$ is very small ($\sin^2 2\vartheta_{\rm sun}\lesssim 10^{-2}$) and we have

$$|\langle m\rangle|\simeq\sqrt{\Delta m^2_{\rm LSND}}\simeq 0.45-1.4 \text{ eV} \quad \text{(SMA-MSW)}. \tag{3.18}$$

Hence, the experimental upper bound (1.3) indicates that *the SMA-MSW solution of the solar neutrino problem is disfavored in scheme A.*

Furthermore, the upper bound $N_\nu^{\rm BBN}<4$ for the effective number of neutrinos in big-bang nucleosynthesis (BBN) (see, for example, [60]) implies that [54,57,58]

$$|U_{s1}|^2+|U_{s2}|^2\lesssim 10^{-2}, \tag{3.19}$$

in scheme A. The analysis of recent astrophysical data yields the upper bound[3] $N_\nu^{\rm BBN}\leq 3.2$ at 95% C.L. [61], although the issue is still rather controversial (see [62,63]). The inequalities (3.10) and (3.19), together with the unitarity of the mixing matrix, imply that the oscillations of solar neutrinos occur mainly in the $\nu_e\to\nu_s$ channel [57,58]. In this case, the analysis of solar neutrino data in terms of two-generation $\nu_e\to\nu_s$ oscillations is valid in the four-neutrino scheme A if the usual two-generation mixing parameters Δm^2 and ϑ are identified, respectively, with $\Delta m^2_{\rm sun}=\Delta m^2_{43}$ and $\vartheta_{\rm sun}$ defined in Eq. (3.16) [from Eqs. (3.10), (3.16), (3.19) and the unitarity of the mixing matrix we obtain $|U_{s3}|^2\simeq\sin^2\vartheta_{\rm sun}$ and $|U_{s4}|^2\simeq\cos^2\vartheta_{\rm sun}$]. The results of the analyses of solar neutrino data in terms of two-generation $\nu_e\to\nu_s$ oscillations show that only the SMA-MSW solution is allowed [38,42,43]. Therefore, comparing Eqs. (1.3) and (3.18) we conclude that *scheme A is disfavored by the experimental upper bound $|\langle m\rangle|$ and the BBN bound $N_\nu^{\rm BBN}<4$.*

Summarizing, the data from oscillation experiments, from neutrinoless double-β decay experiments and from the measurement of the abundances of primordial elements indicate that, *among all the possible four-neutrino schemes there is one preferred, scheme B.* Let us recall, however, that the validity of the $\beta\beta_{0\nu}$ bound (1.3) and the validity of the BBN bound $N_\nu^{\rm BBN}<4$ are controversial. Hopefully, future experimental and theoretical research will clarify this issue.

B. Scheme B

In scheme B, the BBN upper bound $N_\nu^{\rm BBN}<4$ implies that [57,58]

$$|U_{s3}|^2+|U_{s4}|^2\lesssim 10^{-4}. \tag{3.20}$$

From this inequality, Eq. (3.11), and the unitarity of the mixing matrix it follows that the oscillations of solar neutrinos occur mainly in the $\nu_e\to\nu_s$ channel. Therefore, the analysis of solar neutrino data in terms of two-generation $\nu_e\to\nu_s$ oscillations is valid in the four-neutrino scheme B if the usual two-generation mixing parameters Δm^2 and ϑ are identified, respectively, with $\Delta m^2_{\rm sun}=\Delta m^2_{21}$ and $\vartheta_{\rm sun}$ defined by

$$|U_{e1}|^2\simeq\cos^2\vartheta_{\rm sun}, \quad |U_{e2}|^2\simeq\sin^2\vartheta_{\rm sun} \tag{3.21}$$

[from Eqs. (3.11), (3.20), (3.21) and the unitarity of the mixing matrix we obtain $|U_{s1}|^2\simeq\sin^2\vartheta_{\rm sun}$ and $|U_{s2}|^2\simeq\cos^2\vartheta_{\rm sun}$]. Since the results of the analyses of solar neutrino data in terms of two-generation $\nu_e\to\nu_s$ oscillations [38,42,43] show that only the SMA-MSW solution is allowed, with $10^{-3}\lesssim\sin^2 2\vartheta_{\rm sun}\lesssim 10^{-2}$, we have $2.5\times10^{-4}\lesssim|U_{e2}|^2\lesssim 2.5\times10^{-3}$. Therefore, in scheme B we have

$$|U_{e1}|^2\simeq 1, \quad |U_{e2}|^2,|U_{e3}|^2,|U_{e4}|^2\ll 1. \tag{3.22}$$

In the scheme B there are two possibilities: a quasidegenerate mass spectrum

$$\underbrace{\overbrace{m_1<m_2}^{\rm sun}\lesssim\overbrace{m_3<m_4}^{\rm atm}}_{\rm LSND} \qquad \text{(BD)} \tag{3.23}$$

or a mass hierarchy

$$\underbrace{\overbrace{m_1<m_2}^{\rm sun}\ll\overbrace{m_3<m_4}^{\rm atm}}_{\rm LSND} \qquad \text{(BH)} \tag{3.24}$$

If the quasidegenerate mass spectrum BD is realized in nature, it is clear that from Eqs. (1.1) and (3.22) we have

$$|\langle m\rangle|\simeq m_1. \tag{3.25}$$

In this case, the experimental upper bound (1.3) implies that

$$m_1\lesssim 0.2\text{–}0.4 \text{ eV}. \tag{3.26}$$

The observation of neutrinoless double-β decay by the next generation of experiments, which are sensitive to values of $|\langle m\rangle|$ in the range 10^{-2}–10^{-1} eV [17], together with the confirmation of the four-neutrino scheme B by neutrino oscillation experiments, will provide an evidence in favor of the quasidegenerate scheme BD.

If the hierarchical mass spectrum BH is realized in nature, the absence of unlikely fine-tuned cancellations between the contributions of m_1,m_2 and m_3,m_4 to the effective Majorana mass (1.1) implies that

$$|\langle m\rangle|\simeq\max[|\langle m\rangle|_{12},|\langle m\rangle|_{34}], \tag{3.27}$$

with

$$|\langle m\rangle|_{12}\equiv|U_{e1}^2 m_1+U_{e2}^2 m_2|, \tag{3.28}$$

$$|\langle m\rangle|_{34}\equiv|U_{e3}^2 m_3+U_{e4}^2 m_4|. \tag{3.29}$$

From Eq. (3.22), if $m_1\simeq m_2$, we have $|\langle m\rangle|_{12}\simeq m_2$ and the contribution of $|\langle m\rangle|_{12}$ to $|\langle m\rangle|$ could be sizable. On the

[3]The bound $N_\nu^{\rm BBN}\leq 3.2$ [61] implies that $|U_{s1}|^2+|U_{s2}|^2\lesssim 5\times10^{-4}$ [57,58] in scheme A.

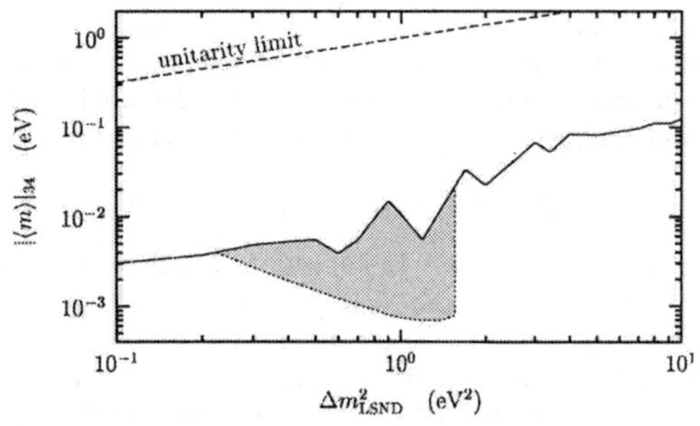

FIG. 2. The shadowed region shows the allowed range for $|\langle m\rangle|_{34}$ as a function of $\Delta m^2_{\rm LSND}$ in the four-neutrino scheme BH [see Eq. (3.24)]. The solid line represents the upper bound in Eq. (3.36) and the dotted line represents the lower bound in Eq. (3.39). The dashed line represents the unitarity limit $|\langle m\rangle|_{34} \leq \sqrt{\Delta m^2_{\rm LSND}}$.

other hand, if $m_1 \ll m_2$, we have $|\langle m\rangle|_{12} \ll m_2$ and the contribution of $|\langle m\rangle|_{12}$ to $|\langle m\rangle|$ is strongly suppressed. In this case, if there are no unlikely fine-tuned cancellations between the contributions of m_1 and m_2 to $|\langle m\rangle|_{12}$, we have $|\langle m\rangle|_{12} \simeq |\langle m\rangle|_2$ with $|\langle m\rangle|_2$ in the range (2.11). In any case, at present it is not known if $m_1 \simeq m_2$ or $m_1 \ll m_2$ and we cannot infer the contribution of $|\langle m\rangle|_{12}$ to $|\langle m\rangle|$.

Let us consider now the contribution $|\langle m\rangle|_{34}$ to the effective Majorana mass. In principle it is possible that $|\langle m\rangle|_{34} = 0$ if $U_{e3}^2 m_3 + U_{e4}^2 m_4 = 0$, i.e.,

$$|U_{e3}|^2 m_3 = |U_{e4}|^2 m_4 \tag{3.30}$$

and

$$|\arg(U_{e3}) - \arg(U_{e4})| = \pi/2 \tag{3.31}$$

[the condition (3.31) is satisfied if CP is conserved and ν_3 and ν_4 have opposite CP parities [64,24,2,27]]. However, even if $m_3 \simeq m_4$, since $|U_{e3}|^2 + |U_{e4}|^2 \ll 1$ [see Eq. (3.11)], there is no reason to have $|U_{e3}|^2 \simeq |U_{e4}|^2$. On the other hand, the explanation of the atmospheric neutrino data with $\nu_\mu \to \nu_\tau$ oscillations [9–13], which is favored by the latest Super-Kamiokande data [65], requires a large mixing in the ν_μ, ν_τ-ν_3, ν_4 sector which could be related to the fact that $|U_{\mu 3}|^2 + |U_{\mu 4}|^2$ and $|U_{\tau 3}|^2 + |U_{\tau 4}|^2$ are large (close to 1) and $m_3 \simeq m_4$.

Therefore, in the following we will assume that $|U_{e3}|^2$ and $|U_{e4}|^2$ have different orders of magnitude. In this case, the contribution of m_3 and m_4 to the effective Majorana mass is given by

$$|\langle m\rangle|_{34} \simeq d_e \sqrt{\Delta m^2_{\rm LSND}}, \tag{3.32}$$

where we have taken into account that $m_3 \simeq m_4 \simeq \sqrt{\Delta m^2_{41}} = \sqrt{\Delta m^2_{\rm LSND}}$ and we have defined

$$d_\alpha \equiv \sum_{k=3,4} |U_{\alpha k}|^2 \quad (\alpha = e, \mu, \tau, s). \tag{3.33}$$

It has been shown in [55] that d_e is small in scheme B:

$$d_e \leq a_e^{\rm Bugey}, \tag{3.34}$$

with

$$a_e^{\rm Bugey} = \frac{1}{2}\left(1 - \sqrt{1 - \sin^2 2\vartheta_{\rm Bugey}}\right). \tag{3.35}$$

Here $\sin^2 2\vartheta_{\rm Bugey}$ is the upper value of the two-neutrino mixing parameter $\sin^2 2\vartheta$ obtained from the Bugey exclusion curve [51] as a function of $\Delta m^2 = \Delta m^2_{\rm LSND}$, where Δm^2 is the two-neutrino mass-squared difference used in the analysis of the Bugey data [the upper bound (3.11) has been obtained from the inequality (3.34) restricting $\Delta m^2_{\rm LSND}$ in the LSND-allowed range (3.4)].

From Eqs. (3.32) and (3.34), for $|\langle m\rangle|_{34}$ we obtain the upper bound [25–27]

$$|\langle m\rangle|_{34} \lesssim a_e^{\rm Bugey} \sqrt{\Delta m^2_{\rm LSND}}. \tag{3.36}$$

The numerical value of this upper bound as a function of $\Delta m^2_{\rm LSND}$ is depicted by the solid line in Fig. 2. The dashed line in Fig. 2 represents the unitarity limit $|\langle m\rangle|_{34} \leq \sqrt{\Delta m^2_{\rm LSND}}$.

The amplitude $A_{\mu e} = 4|\Sigma_{k=3,4} U_{ek} U_{\mu k}^*|^2$ of short-base-line $\bar{\nu}_\mu \to \bar{\nu}_e$ oscillations in scheme B is bounded by [55]

$$A_{\mu e} \leq 4 d_e d_\mu. \tag{3.37}$$

Since d_μ is large in scheme B [55], we have

$$d_e \geq \frac{A_{\mu e}^{\rm min}}{4}, \tag{3.38}$$

where $A_{\mu e}^{\rm min}$ is the minimum value of $A_{\mu e}$ measured in the LSND experiment. The physical reason of this lower bound for d_e is that ν_e must have some mixing with ν_3 and/or ν_4 in order to generate the oscillations observed in the LSND experiment.

Taking into account Eq. (3.32), the inequality (3.38) leads to the lower bound

$$|\langle m\rangle|_{34} \gtrsim \frac{A_{\mu e}^{\rm min}}{4} \sqrt{\Delta m^2_{\rm LSND}}. \tag{3.39}$$

The numerical value of this lower bound as a function of $\Delta m^2_{\rm LSND}$ is shown in Fig. 2 by the dotted curve that, together with the solid line obtained from the upper bound (3.36), defines an allowed region in the $\Delta m^2_{\rm LSND}$–$|\langle m\rangle|_{34}$ plane (shadowed area). From Fig. 2 one can see that

$$6.9 \times 10^{-4}\ {\rm eV} \lesssim |\langle m\rangle|_{34} \lesssim 2.1 \times 10^{-2}\ {\rm eV}. \tag{3.40}$$

Summarizing, in the framework of the scheme BH in Eq. (3.24) we have made three assumptions: (i) massive neutrinos are Majorana particles, (ii) there is no unlikely fine-tuned cancellations between the contributions of m_1, m_2 and m_3, m_4 to the effective Majorana mass $|\langle m\rangle|$, and (iii) the two small elements U_{e3} and U_{e4} of the neutrino mixing matrix have different orders of magnitude. Under these reasonable assumptions, we have obtained the following allowed range for the effective Majorana mass in $\beta\beta_{0\nu}$ decay:

$$7\times10^{-4}\ \text{eV}\lesssim|\langle m\rangle|\lesssim2\times10^{-2}\ \text{eV}, \tag{3.41}$$

from the contribution of m_3 and m_4 alone (the effective Majorana mass could be even larger than 2×10^{-2} eV if $m_1 \simeq m_2 \gtrsim 2\times10^{-2}$ eV). Such values of the effective Majorana mass could be measured by future $\beta\beta_{0\nu}$ decay experiments [17,19].

IV. CONCLUSIONS

We have derived lower limits for the effective Majorana mass in neutrinoless double-β decay in the scheme with mixing of three neutrinos and a mass hierarchy [Eq. (2.2)] under the natural assumptions that massive neutrinos are Majorana particles and there are no large cancellations among the contributions of the different neutrino masses. If there is a hierarchy of neutrino masses, large cancellations are unlikely (unless an unknown symmetry is at work), because they require a fine-tuning among the values of the neutrino masses and the elements of the neutrino mixing matrix, which are independent quantities.

Under the only assumption that massive neutrinos are Majorana particles, we have shown that, among all the possible four-neutrino schemes that can accommodate the results of solar and atmospheric experiments and the results of the LSND experiment, the scheme B [Eq. (3.9)] is favored by the experimental results on neutrinoless double-β decay and the measurements of the cosmic abundances of elements produced in big-bang nucleosynthesis. In the scheme B there are two possibilities: the quasidegenerate mass spectrum BD [Eq. (3.23)] and the hierarchical mass spectrum BH [Eq. (3.24)].

If the quasi-degenerate four-neutrino scheme BD is realized in nature, neutrinoless double-β decay should be observed by the next generation of experiments, which will be sensitive to $|\langle m\rangle|\sim10^{-2}-10^{-1}$ eV.

In the framework of the hierarchical four-neutrino scheme BH, we have shown that there is a lower bound for the effective Majorana mass in $\beta\beta_{0\nu}$ decay, under the assumptions that massive neutrinos are Majorana particles, there are no large cancellations among the contributions of m_1,m_2 and m_3,m_4, and the two small elements U_{e3} and U_{e4} of the neutrino mixing matrix have different orders of magnitude.

We hope that the indications presented here in favor of a lower bound, albeit small, for the effective Majorana mass in neutrinoless double-β decay will encourage the development of future $\beta\beta_{0\nu}$ experiments.

[1] S.M. Bilenky and B. Pontecorvo, Phys. Rep. **41**, 225 (1978); S.M. Bilenky and S.T. Petcov, Rev. Mod. Phys. **59**, 671 (1987); C.W. Kim and A. Pevsner, *Neutrinos in Physics and Astrophysics*, Contemporary Concepts in Physics (Harwood Academic, Chur, Switzerland, 1993), Vol. 8.

[2] S.M. Bilenky, C. Giunti, and W. Grimus, Prog. Part. Nucl. Phys. **43**, 1 (1999).

[3] P. Fisher, B. Kayser, and K.S. McFarland, hep-ph/9906244.

[4] B.T. Cleveland *et al.*, Astrophys. J. **496**, 505 (1998).

[5] Y. Fukuda *et al.*, Phys. Rev. Lett. **77**, 1683 (1996).

[6] W. Hampel *et al.*, Phys. Lett. B **388**, 384 (1996); **447**, 127 (1999).

[7] J.N. Abdurashitov *et al.*, Phys. Rev. Lett. **77**, 4708 (1996); Phys. Rev. C **59**, 2246 (1999).

[8] Y. Fukuda, Phys. Rev. Lett. **81**, 1158 (1998); **81**, 4279(E) (1998); **82**, 2430 (1999); Super-Kamiokande Collaboration, M.B. Smy, hep-ex/9903034.

[9] Y. Fukuda *et al.*, Phys. Lett. B **335**, 237 (1994).

[10] R. Becker-Szendy *et al.*, Nucl. Phys. B (Proc. Suppl.) **38**, 331 (1995).

[11] Y. Fukuda *et al.*, Phys. Rev. Lett. **81**, 1562 (1998); **82**, 2644 (1999); Super-Kamiokande Collaboration, A. Habig, hep-ex/9903047; Super-Kamiokande Collaboration, K. Scholberg, hep-ex/9905016.

[12] W.W.M. Allison *et al.*, Phys. Lett. B **449**, 137 (1999).

[13] M. Ambrosio *et al.*, Phys. Lett. B **434**, 451 (1998); MACRO Collaboration, F. Ronga, hep-ex/9905025; MACRO Collaboration, A. Surdo, hep-ex/9905028.

[14] C. Athanassopoulos *et al.*, Phys. Rev. Lett. **75**, 2650 (1995); **77**, 3082 (1996); **81**, 1774 (1998).

[15] LSND Collaboration, G. Mills, talk presented at the XXXIVth Rencontres de Moriond "Electroweak Interactions and Unified Theories," Les Arcs, 1999 (scanned transparencies available at http://moriond.in2p3.fr/EW/transparencies).

[16] M. Moe and P. Vogel, Annu. Rev. Nucl. Part. Sci. **44**, 247 (1994); A. Morales, Nucl. Phys. B (Proc. Suppl.) **77**, 335 (1999).

[17] C.E. Aalseth *et al.*, Nucl. Phys. B (Proc. Suppl.) **70**, 236 (1999); NEMO Collaboration, X. Sarazin, *ibid.* **70**, 239 (1999); V.D. Ashitkov *et al.*, *ibid.* **70**, 233 (1999); E. Fiorini, Phys. Rep. **307**, 309 (1998).

[18] L. Baudis *et al.*, Phys. Rev. Lett. **83**, 41 (1999).

[19] J. Hellmig and H.V. Klapdor-Kleingrothaus, Z. Phys. A **359**, 351 (1997); H.V. Klapdor-Kleingrothaus and M. Hirsch, *ibid.* **359**, 361 (1997); H.V. Klapdor-Kleingrothaus, J. Hellmig, and M. Hirsch, J. Phys. G **24**, 483 (1998).

[20] A.I. Belesev, Phys. Lett. B **350**, 263 (1995); H. Barth *et al.*, Prog. Part. Nucl. Phys. **40**, 353 (1998).

[21] Particle Data Group, C. Caso *et al.*, Eur. Phys. J. C **3**, 1 (1998).

[22] S.T. Petcov and A.Yu. Smirnov, Phys. Lett. B **322**, 109 (1994).

[23] S.M. Bilenky, A. Bottino, C. Giunti, and C.W. Kim, Phys. Lett. B **356**, 273 (1995); Phys. Rev. D **54**, 1881 (1996).

[24] S.M. Bilenky, C. Giunti, C.W. Kim, and S.T. Petcov, Phys. Rev. D **54**, 4432 (1996).

[25] S.M. Bilenky, C. Giunti, C.W. Kim, and M. Monteno, Phys. Rev. D **57**, 6981 (1998).

[26] S.M. Bilenky, C. Giunti, and W. Grimus, Nucl. Phys. B (Proc. Suppl.) **77**, 151 (1999); S.M. Bilenky and C. Giunti, talk presented at WIN99, Cape Town, South Africa, 1999, hep-ph/9904328.

[27] S.M. Bilenky, C. Giunti, W. Grimus, B. Kayser, and S.T. Petcov, hep-ph/9907234.
[28] F. Vissani, hep-ph/9708483; hep-ph/9904349; J. High Energy Phys. **906**, 022 (1999).
[29] H. Minakata and O. Yasuda, Phys. Rev. D **56**, 1692 (1997); Nucl. Phys. **B523**, 597 (1998).
[30] T. Fukuyama, K. Matsuda, and H. Nishiura, Phys. Rev. D **57**, 5844 (1998); Mod. Phys. Lett. A **13**, 2279 (1998).
[31] R. Adhikari and G. Rajasekaran, hep-ph/9812361.
[32] G.C. Branco, M.N. Rebelo, and J.I. Silva-Marcos, Phys. Rev. Lett. **82**, 683 (1999).
[33] V. Barger and K. Whisnant, Phys. Lett. B **456**, 194 (1999).
[34] A. Staudt, K. Muto, and H.V. Klapdor-Kleingrothaus, Europhys. Lett. **13**, 31 (1990); T. Tomoda, Rep. Prog. Phys **54**, 53 (1991); W. Haxton *et al.*, Nucl. Phys. B (Proc. Suppl.) **31**, 82 (1993); X.R. Wu *et al.*, Phys. Lett. B **272**, 169 (1991); **276**, 274 (1992); F. Simkovic *et al.*, *ibid.* **393**, 267 (1997).
[35] F. Simkovic *et al.*, Phys. Rev. C **60**, 055502 (1999).
[36] E. Caurier *et al.*, Phys. Rev. Lett. **77**, 1954 (1996).
[37] P. Lipari, hep-ph/9904443.
[38] J.N. Bahcall, P.I. Krastev, and A.Yu. Smirnov, Phys. Rev. D **58**, 096016 (1998).
[39] V. Berezinsky, hep-ph/9904259.
[40] KARMEN Collaboration, K. Eitel, Proceedings of Lake Louise Winter Institute, 1999, Lake Louise, 1999; KARMEN Collaboration, T. Jannakos, Proceedings of Les Rencontres De Moriond 1999, Les Arc, 1800, 1999; KARMEN WWW page: http://www-ik1.fzk.de/www/karmen/karmen_e.html.
[41] Booster Neutrino Experiment (BooNE), http://nu1.lampf.lanl.gov/BooNE; I-216 $\nu_\mu \to \nu_e$ proposal at CERN, http://chorus01.cern.ch/~pzucchel/loi/; Oak Ridge Large Neutrino Detector (ORLaND), http://www.phys.subr.edu/orland/; Neutrinos at the European Spallation Source (NESS), http://www.isis.rl.ac.uk/ess/neut%5Fess.htm.
[42] Y. Fukuda *et al.*, Phys. Rev. Lett. **82**, 1810 (1999).
[43] M.C. Gonzalez-Garcia *et al.*, hep-ph/9906469.
[44] M.C. Gonzalez-Garcia *et al.*, Phys. Rev. D **58**, 033004 (1998); Nucl. Phys. **B543**, 3 (1999).
[45] G.L. Fogli, E. Lisi, A. Marrone, and G. Scioscia, Phys. Rev. D **59**, 033001 (1999).
[46] M. Gell-Mann, P. Ramond, and R. Slansky, in *Supergravity*, edited by F. van Nieuwenhuizen and D. Freedman (North-Holland, Amsterdam, 1979), p. 315; T. Yanagida, in Proceedings of the Workshop on Unified Theory and the Baryon Number of the Universe, KEK, Japan, 1979; R.N. Mohapatra and G. Senjanović, Phys. Rev. Lett. **44**, 912 (1980).
[47] M. Apollonio *et al.*, Phys. Lett. B **420**, 397 (1998).
[48] S.M. Bilenky and C. Giunti, Phys. Lett. B **444**, 379 (1998).
[49] S.P. Mikheyev and A.Yu. Smirnov, Yad. Fiz. **42**, 1441 (1985) [Sov. J. Nucl. Phys. **42**, 913 (1985)]; Nuovo Cimento C **9**, 17 (1986); L. Wolfenstein, Phys. Rev. D **17**, 2369 (1978); **20**, 2634 (1979).
[50] J.N. Bahcall, P.I. Krastev, and A.Yu. Smirnov, Phys. Rev. D **60**, 093001 (1999).
[51] B. Achkar *et al.*, Nucl. Phys. **B434**, 503 (1995).
[52] J.T. Peltoniemi, D. Tommasini, and J.W.F. Valle, Phys. Lett. B **298**, 383 (1993); E.J. Chun *et al.*, *ibid.* **357**, 608 (1995); S.C. Gibbons *et al.*, *ibid.* **430**, 296 (1998); B. Brahmachari and R.N. Mohapatra, *ibid.* **437**, 100 (1998); S. Mohanty, D.P. Roy, and U. Sarkar, *ibid.* **445**, 185 (1998); J.T. Peltoniemi and J.W.F. Valle, Nucl. Phys. **B406**, 409 (1993); Q.Y. Liu and A.Yu. Smirnov, *ibid.* **B524**, 505 (1998); D.O. Caldwell and R.N. Mohapatra, Phys. Rev. D **48**, 3259 (1993); E. Ma and P. Roy, *ibid.* **52**, R4780 (1995); A.Yu. Smirnov and M. Tanimoto, *ibid.* **55**, 1665 (1997); N. Gaur *et al.*, *ibid.* **58**, 071301 (1998); E.J. Chun, C.W. Kim, and U.W. Lee, *ibid.* **58**, 093003 (1998); K. Benakli and A.Yu. Smirnov, Phys. Rev. Lett. **79**, 4314 (1997); Y. Chikira, N. Haba, and Y. Mimura, hep-ph/9808254; C. Liu and J. Song, Phys. Rev. D **60**, 036002 (1999); W. Grimus, R. Pfeiffer, and T. Schwetz, hep-ph/9905320.
[53] J.J. Gomez-Cadenas and M.C. Gonzalez-Garcia, Z. Phys. C **71**, 443 (1996); S. Goswami, Phys. Rev. D **55**, 2931 (1997); S.M. Bilenky, C. Giunti, C.W. Kim, and S.T. Petcov, *ibid.* **54**, 4432 (1996); V. Barger, Y.B. Dai, K. Whisnant, and B.L. Young, *ibid.* **59**, 113010 (1999); V. Barger, T.J. Weiler, and K. Whisnant, Phys. Lett. B **427**, 97 (1998).
[54] N. Okada and O. Yasuda, Int. J. Mod. Phys. A **12**, 3669 (1997).
[55] S.M. Bilenky, C. Giunti, and W. Grimus, Eur. Phys. J. C **1**, 247 (1998); in *Proceedings of "Neutrino '96,"* Helsinki, 1996, edited by K. Enqvist *et al.*, World Scientific, 1997, p. 174; S.M. Bilenky, C. Giunti, W. Grimus, and T. Schwetz, Phys. Rev. D **60**, 073007 (1999).
[56] S.M. Bilenky, C. Giunti, and W. Grimus, Phys. Rev. D **57**, 1920 (1998); **58**, 033001 (1998).
[57] S.M. Bilenky, C. Giunti, W. Grimus, and T. Schwetz, Astropart. Phys. **11**, 413 (1999).
[58] S.M. Bilenky, C. Giunti, W. Grimus, and T. Schwetz, in *Proceedings of the Ringberg Euroconference*, Tergernsee, Germany, 1998, edited by B.A. Kniehl, G.G. Raffelt, and N. Schmitz (World Scientific, Singapore, 1999), p. 117; in *New Era in Neutrino Physics*, edited by H. Minakata and O. Yasuda (Universal Academy Press, Tokyo, Japan, 1999).
[59] A. Yu. Smirnov, hep-ph/9901208; R.N. Mohapatra, hep-ph/9903261.
[60] D.N. Schramm and M.S. Turner, Rev. Mod. Phys. **70**, 303 (1998).
[61] S. Burles, K.M. Nollett, J.N. Truran, and M.S. Turner, Phys. Rev. Lett. **82**, 4176 (1999).
[62] K.A. Olive, astro-ph/9903309.
[63] E. Lisi, S. Sarkar, and F.L. Villante, Phys. Rev. D **59**, 123520 (1999); S. Sarkar, astro-ph/9903183.
[64] S.M. Bilenky, N.P. Nedelcheva, and S.T. Petcov, Nucl. Phys. **B247**, 61 (1984); B. Kayser, Phys. Rev. D **30**, 1023 (1984).
[65] J. Learned, talk presented at the 23rd Johns Hopkins Workshop on Current Problems in Particle Theory "Neutrinos in the Next Millennium," Baltimore, 1999.

Neutrino Mass Spectrum and Neutrinoless Double Beta Decay

H.V. Klapdor–Kleingrothaus, H. Päs
Max–Planck–Institut für Kernphysik,
P.O. Box 103980, D-69029 Heidelberg, Germany

A. Yu. Smirnov
The Abdus Salam International Center of Theoretical Physics,
Strada Costiera 11, Trieste, Italy
Institute for Nuclear Research, RAS, Moscow, Russia

Abstract

The relations between the effective Majorana mass of the electron neutrino, m_{ee}, responsible for neutrinoless double beta decay, and the neutrino oscillation parameters are considered. We show that for any specific oscillation pattern m_{ee} can take any value (from zero to the existing upper bound) for normal mass hierarchy and it can have a minimum for inverse hierarchy. This means that oscillation experiments cannot fix in general m_{ee}. Mass ranges for m_{ee} can be predicted in terms of oscillation parameters with additional assumptions about the level of degeneracy and the type of hierarchy of the neutrino mass spectrum. These predictions for m_{ee} are systematically studied in the specific schemes of neutrino mass and flavor which explain the solar and atmospheric neutrino data. The contributions from individual mass eigenstates in terms of oscillation parameters have been quantified. We study the dependence of m_{ee} on the non-oscillation parameters: the overall scale of the neutrino mass and the relative mass phases. We analyze how forthcoming oscillation experiments will improve the predictions for m_{ee}. On the basis of these studies we evaluate the discovery potential of future $0\nu\beta\beta$ decay searches. The role $0\nu\beta\beta$ decay searches will play in the reconstruction of the neutrino mass spectrum is clarified. The key scales of m_{ee}, which will lead to the discrimination among various schemes are: $m_{ee} \sim 0.1$ eV and $m_{ee} \sim 0.005$ eV.

1 Introduction

The goal of the search for neutrinoless double beta decay ($0\nu\beta\beta$ decay) is to establish the violation of (total) lepton number L and to measure the Majorana mass of the electron neutrino, thus identifying the nature of the neutrino [1, 2]. Both issues are related: Even if the main mechanism of $0\nu\beta\beta$ decay may be induced by e.g. lepton number violating right-handed currents, R-parity violation in SUSY models, leptoquark-Higgs couplings (for an overview see e.g. [3]), the observation of $0\nu\beta\beta$ decay implies always a non-vanishing effective neutrino Majorana mass ($0\nu\beta\beta$-mass) at loop level [4].

If $0\nu\beta\beta$ decay is induced dominantly by the exchange of a light Majorana neutrino ($m < 30$ MeV), the decay rate is proportional to the Majorana mass of the electron neutrino m_{ee} squared:

$$\Gamma \propto m_{ee}^2. \tag{1}$$

Thus, in absence of lepton mixing the observation of $0\nu\beta\beta$ decay would provide an information about the absolute scale of the Majorana neutrino mass.

The situation is changed in presence of neutrino mixing when the electron neutrino is not a mass eigenstate but turns out to be a combination of several mass eigenstates, ν_i, with mass eigenvalues m_i:

$$\nu_e = \sum_i U_{ei}\nu_i \;, \qquad i = 1, 2, 3, \ldots \;. \tag{2}$$

Here U_{ej} are the elements of the mixing matrix relating the flavor states to the mass eigenstates. In this general case the mass parameter ($0\nu\beta\beta$-mass) which enters the $0\nu\beta\beta$ decay rate is not the physical mass of the neutrino but the combination m_{ee} of physical masses:

$$|m_{ee}| = \left| \sum_j |U_{ej}|^2 e^{i\phi_j} m_j \right| . \tag{3}$$

Apart from the absolute values of masses m_j and mixing matrix elements, the effective Majorana mass depends also on new parameters: phases ϕ_j which originate from a possible complexity of the mass eigenvalues and from the mixing matrix elements. Thus searches for double beta decay are sensitive not only to masses but also to mixing matrix elements and phases ϕ_j.

Notice that in the presence of mixing m_{ee} is still the ee-element of the neutrino mass matrix in the flavor basis [1]. In this sense it gives the scale of elements of the neutrino mass matrix. However, in general, m_{ee} does not determine the scale of the physical masses. If $0\nu\beta\beta$ decay will be discovered and if it will be proven to proceed via the Majorana neutrino mass mechanism, then the m_{ee} extracted from the decay rate will give a lower bound on some physical masses. As it is easy to see from Eq. (3)), at least one physical mass, m_j, should be

$$m_j \geq \frac{m_{ee}}{3} \tag{4}$$

for the three-neutrino case.

Can m_{ee} be predicted? According to (3) the mass m_{ee} depends on absolute values of masses, mixings and phases ϕ_j. Certain information about masses and mixing can be obtained from (i) oscillation searches, (ii) direct kinematical measurements and (iii) cosmology. Let us comment on these issues in order.

[1]In general the experimental value of m_{ee} depends on the process being considered. It coincides with the theoretical m_{ee} of eq. (3) if all masses $m_i \ll Q$, where Q is the energy release of a given process. This fact may become important for comparing heavy neutrino contributions in $0\nu\beta\beta$ decay and inverse neutrinoless double beta decay at colliders, see *e.g.* [5].

1). The oscillation pattern is determined by mass squared differences, moduli of elements of the mixing matrix, and (for three neutrino mixing) only one complex phase which leads to CP violating effects in neutrino oscillations:

$$\Delta m_{ij}^2 \equiv |m_i|^2 - |m_j|^2, \qquad |U_{ej}|^2, \qquad \delta_{CP}. \tag{5}$$

(We indicated here only mixing elements which enter m_{ee}.) In what follows we will call (5) the *oscillation parameters.*

Neutrino oscillations and neutrinoless double beta decay, however, depend on different combinations of neutrino masses and mixings. In terms of the oscillation parameters the mass (3) can be rewritten as

$$|m_{ee}| = \left| \sum_j |U_{ej}|^2 e^{i\phi_j} \sqrt{\Delta m_{j1}^2 + m_1^2} \right|, \tag{6}$$

where we assumed for definiteness m_1 to be the smallest mass. We also put $\phi_1 = 0$ and consider the other ϕ_j as *the relative phases.*

According to (6) the oscillation parameters do not allow one to determine uniquely m_{ee}. Apart from these parameters, the mass m_{ee} depends also on the absolute value of the first mass (absolute scale) and on the relative phases:

$$m_1, \qquad \phi_j, \qquad j = 2, 3, ... \tag{7}$$

These parameters can not be determined from oscillation experiments and we will call them *non-oscillation parameters.*

The mass squared difference gives the absolute value of the mass only in the case of strong mass hierarchy: $m_j \gg m_1$, when $|m_j| \approx \sqrt{\Delta m_{j1}^2}$. However, even in this case the lightest mass (which can give a significant or even dominant contribution to m_{ee}) is not determined.

The relative phases ϕ_j which appear in m_{ee} (eq. (3)) differ from δ_{CP} and can not be determined from oscillation experiments, since the oscillation pattern is determined by moduli $|m_i|^2$. On the other hand, the phase relevant for neutrino oscillations does not enter m_{ee} or can be absorbed in phases of masses.

2). Apart from neutrino oscillations, informations on neutrino masses and especially on the absolute scale of masses can be obtained from direct kinematical searches and cosmology.

There is still some chance that future kinematical studies of the tritium beta decay will measure the electron neutrino mass, and thus will allow us to fix the absolute scale of masses. Projects are under consideration which will have a sensitivity of about 1 eV and less (ref. [6]).

3). The expansion of the universe and its large scale structure are sensitive to neutrinos with masses larger than about 0.5 eV. The status of neutrinos as the hot dark matter (HDM) component of the universe is rather uncertain now: it seems that the present cosmological observations do not require a significant Ω_ν contribution and therefore a large $\mathcal{O}(1\text{eV})$ neutrino mass. However in some cases massive neutrinos may help to get a better fit of the data on density perturbations.

In order to predict m_{ee} one should not only determine the oscillation parameters but make additional assumptions which will fix the non-oscillation parameters. If the oscillation parameters are known, then, depending on these assumptions, one can predict m_{ee} completely or get certain bounds on m_{ee}.

What are these assumptions?

It was pointed out in [8] that predictions on m_{ee} significantly depend on two points:

- The level of degeneracy of the neutrino mass spectrum, which is related to the absolute scale of neutrino masses.

- The solution of the solar neutrino problem; this solution determines to a large extent the distribution of the electron neutrino flavor in the mass eigenstates, that is, $|U_{ej}|^2$.

The assumptions about the level of degeneracy allow one to fix the absolute scale of the neutrino mass. In fact, at present even the oscillation parameters are essentially unknown, so that further assumptions are needed. Evidences of neutrino oscillations (atmospheric, solar neutrino problems, LSND result) allow us in principle to determine the oscillation parameters up to a certain ambiguity related, in particular, to the existence of several possible solutions of the solar neutrino problem.

A number of studies of the $0\nu\beta\beta$-mass have been performed, using various assumptions about the hierarchy/degeneracy of the spectrum which remove the ambiguity in interpretations of existing oscillation data. In fact, these assumptions allow one to construct the neutrino mass and mixing spectrum, and some studies have been performed for specific neutrino spectra. Most of the spectra considered so far explain the atmospheric neutrino problem and the solar neutrino problem assuming one of the suggested solutions. Some results have also been obtained for schemes with 4 neutrinos which also explain the LSND result. Let us summarize the main directions of these studies.

(1) Three-neutrino schemes with normal mass hierarchy which explain the solar and atmospheric neutrino data have been studied in [8, 9, 10, 11, 12, 13]. Various solutions of the $\nu_\odot$-problem were assumed. These schemes give the most stringent constraints on $0\nu\beta\beta$-mass in terms of oscillation parameters.

(2) The $0\nu\beta\beta$-mass in three-neutrino schemes with inverse mass hierarchy has been considered in [14, 12, 13]. These schemes favor m_{ee} to be close to the present experimental bound.

(3) Three-neutrino schemes with *partial* degeneracy of the spectrum and various solutions of the $\nu_\odot$-problem were discussed in [8, 13]. In these schemes m_{ee} can also be close to the present experimental bound.

(4) Large attention was devoted to the three-neutrino schemes with complete degeneracy [8, 15, 16, 17, 18, 19, 20, 21, 22, 23, 13] since they can explain solar and atmospheric neutrino data and also give a significant amount of the HDM in the universe. In these schemes the predictions of m_{ee} depend mainly on the absolute mass scale and on the mixing angle relevant for the solar neutrinos.

Some intermediate situations between hierarchical and degenerate spectra have been discussed in [19, 13].

5. The $0\nu\beta\beta$-mass in scenarios with 4 neutrinos which can accommodate also the LSND result have been analyzed in Ref. [14, 10, 11].

Some general bounds on the $0\nu\beta\beta$-mass under various assumptions have been discussed in [24, 25, 26, 19, 27, 28].

In a number of papers an inverse problem has been solved: using relations between the $0\nu\beta\beta$-mass and oscillation parameters which appear in certain schemes restrictions on oscillation parameters have been found from existing bounds on m_{ee}. In particular the 3ν-schemes with mass degeneracy [15] and mass hierarchy [12] have been discussed.

An important ingredient for the prediction of m_{ee} are the phases (see eq. 7). Unfortunately, there is no theory or compelling assumptions which allow to determine these phases.

In this paper we will analyze the discovery potential of future $0\nu\beta\beta$ decay searches in view of existing and forthcoming oscillation experiments. We will clarify the role $0\nu\beta\beta$ decay searches will play in the identification of the neutrino mass spectrum.

For this we first (sect. 2) consider the general relations between the effective Majorana mass of the electron neutrino and the oscillation parameters. We will study the dependence of m_{ee} on the non-oscillations parameters. The crucial assumptions which lead to predictions for m_{ee} are identified.

In sects. 3 - 7 we present a systematic and updated study of predictions for m_{ee} for possible neutrino mass spectra. In contrast with most previous studies using oscillation data we quantify the contributions from *individual* mass eigenstates and we keep explicitly the dependence on unknown relative mass phases. The dependence of predictions on non-oscillation parameters – the absolute mass value and the phases ϕ_i is studied in detail. We consider 3ν–schemes with mass hierarchy (section 3), partial degeneracy (section 4), inverse hierarchy (section 5), total degeneracy (section 6) and schemes with sterile neutrinos (section 7). We analyse how forthcoming and planned oscillation experiments will sharpen the predictions for m_{ee}. In sect. 8, comparing predictions of m_{ee} from different schemes we clarify the role future searches for $0\nu\beta\beta$ decay can play in the identification of the neutrino mass spectrum.

2 Neutrino oscillations and neutrinoless double beta decay

As has been pointed out in the introduction, the prediction of m_{ee} depends on oscillation ($|U_{ei}|$, Δm^2) and non-oscillation (m_1 and ϕ_j) parameters. In this section we will consider general relations between m_{ee} and the oscillation parameters. We analyse the dependence of these relations on non-oscillation parameters. We quantify ambiguities which exist in predictions of m_{ee}. Our results will be presented in a way which will be convenient for implementations of future oscillation results.

2.1 Effective Majorana mass and oscillation parameters

The oscillation pattern is determined by the effective Hamiltonian (in the flavor basis):

$$H = \frac{1}{2E} M M^{\dagger} + V , \tag{8}$$

where E is the neutrino energy, M is the mass matrix and V is the (diagonal) matrix of effective potentials which describe the interaction of neutrinos in a medium.

The oscillation pattern is not changed if we add to H a term proportional to the unity matrix:

$$M M^{\dagger} \rightarrow M M^{\dagger} \pm m_0^2 I . \tag{9}$$

Indeed, the additional term does not change the mixing, it leads just to a shift of the mass eigenstates squared by the same value without affecting Δm_{ij}^2:

$$m_i^2 \rightarrow m_i^2 \pm m_0^2 . \tag{10}$$

(we consider $m_i^2 - m_0^2 \geq 0$ for all i to keep the Hermiticity of the Hamiltonian). The additional term changes, however, the $0\nu\beta\beta$-mass . Thus, for a given oscillation pattern there is a freedom in m_{ee}, associated with m_0^2.

Let us study how arbitrary m_{ee} can be for a given oscillation pattern. According to (9,10) for the three-neutrino case we get

$$m_{ee} = |m_{ee}^{(1)}| + e^{i\phi_2} |m_{ee}^{(2)}| + e^{i\phi_3} |m_{ee}^{(3)}| , \tag{11}$$

where $m_{ee}^{(i)} \equiv |m_{ee}^{(i)}| \exp{(i\phi_i)}$ $(i = 1, 2, 3)$ are the contributions to m_{ee} from individual mass eigenstates which can be written in terms of oscillation parameters as:

$$|m_{ee}^{(1)}| = |U_{e1}|^2 m_1 , \tag{12}$$

$$|m_{ee}^{(2)}| = |U_{e2}|^2 \sqrt{\Delta m_{21}^2 + m_1^2} , \tag{13}$$

$$|m_{ee}^{(3)}| = |U_{e3}|^2 \sqrt{\Delta m_{31}^2 + m_1^2} . \tag{14}$$

and ϕ_i are the relative phases of the contributions from masses m_i and m_j (the mass m_0^2 has been absorbed in definition of m_1^2). The contributions $m_{ee}^{(i)}$ can be shown as vectors in the complex plane (fig. 1).

Without loss of generality we assume $m_3 > m_2 > m_1 \geq 0$, so that m_1 is the lightest state. Then *normal* mass hierarchy corresponds to the case when the electron flavor prevails in the lightest state: $|U_{e1}|^2 > |U_{e2}|^2, |U_{e3}|^2$. We will refer to *inverse* hierarchy as to the case when $|U_{e1}|^2 < |U_{e2}|^2$ or/and $|U_{e3}|^2$, *i.e.* when the admixture of the electron neutrino flavor in the lightest state is not the largest one.

Let us consider the dependence of m_{ee} on *non-oscillation parameters* $m_{ee} = m_{ee}(m_1, \phi_j)$. It is obvious that due to the freedom in the choice of m_1 there is no upper bound for m_{ee}. However, in some special cases lower bounds on m_{ee} exist.

Let us start with the two neutrino case which would correspond to zero (or negligibly small) ν_e admixture in one of mass eigenstates, *e.g.* $|U_{e3}|$. We consider first the case of

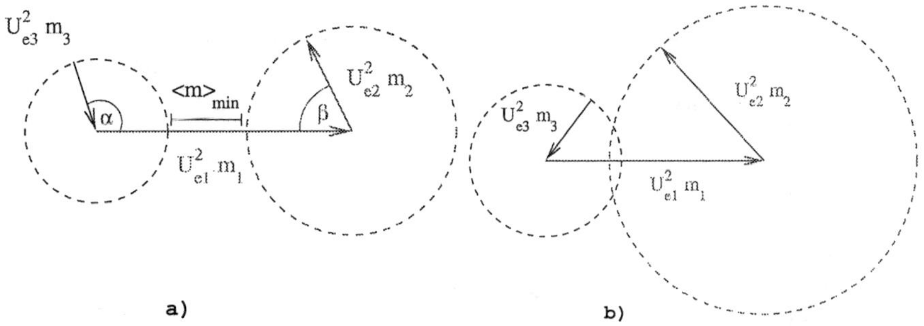

Figure 1: The effective Majorana mass m_{ee} in the complex plane. Vectors show contributions to m_{ee} from individual eigenstates. The total m_{ee} appears as the sum of the three vectors. Allowed values of m_{ee} correspond to modulies of vectors which connect two points on the circles. Here $\alpha = \phi_3 - \pi$, $\beta = \pi - \phi_2$. a). $|m_{ee}^{(1)}| > |m_{ee}^{(2)}| + |m_{ee}^{(3)}|$: the vectors $\vec{m}_{ee}^{(i)}$ can not form a triangle and no complete cancellation occurs. b) $|m_{ee}^{(1)}| \leq |m_{ee}^{(2)}| + |m_{ee}^{(3)}|$: in this case complete cancellation occurs in the intersection points of the circles, so that $m_{ee} = 0$.

normal hierarchy $U_{e1}^2 > U_{e2}^2$. For $m_1 = 0$ the effective mass m_{ee} is uniquely fixed in terms of oscillation parameters:

$$m_{ee}^0 = |U_{e2}|^2\sqrt{\Delta m_{21}^2} \equiv \sin^2\theta\sqrt{\Delta m_{21}^2}\ , \tag{15}$$

where $\sin\theta \equiv U_{e2}$. For non-zero m_1, the maximal and minimal values of m_{ee} correspond to $\phi_2 = 0$ and $\phi_2 = \pi$.

The upper bound ($\phi_2 = 0$) on m_{ee} increases with m_1 monotonously from m_{ee}^0 at $m_1 = 0$ and approaches the asymptotic dependence $m_{ee} = m_1$ for large m_1 (see fig. 2 a). The lower limit ($\phi_2 = \pi$) decreases monotonously with increase of m_1 starting by m_{ee}^0. It reaches zero at

$$m_1 = \frac{\sin^2\theta}{\sqrt{|\cos 2\theta|}}\sqrt{\Delta m_{21}^2}, \tag{16}$$

and approaches the asymptotic dependence $m_{ee} = |\cos 2\theta|\ m_1$ at large m_1 (see fig. 2 a). Thus, for arbitrary values of oscillation parameters, no bound on $|m_{ee}|$ exists.

(2) In the case of inverse hierarchy, $|U_{e2}| > |U_{e1}|$, the function $m_{ee}(m_1)$ has a minimum which differs from zero,

$$m_{ee}^{min} = \sqrt{|\cos 2\theta|\Delta m_{21}^2} \tag{17}$$

at

$$m_1 = \frac{\cos^2\theta}{\sqrt{\cos 2\theta}}\sqrt{\Delta m_{21}^2}. \tag{18}$$

At large m_1 it has the asymptotics $m_{ee} = |\cos 2\theta| m_1$. (fig. 2 b). As we will see, the existence of a minimal value of $|m_{ee}|$ can play an important role in the discrimination of various scenarios.

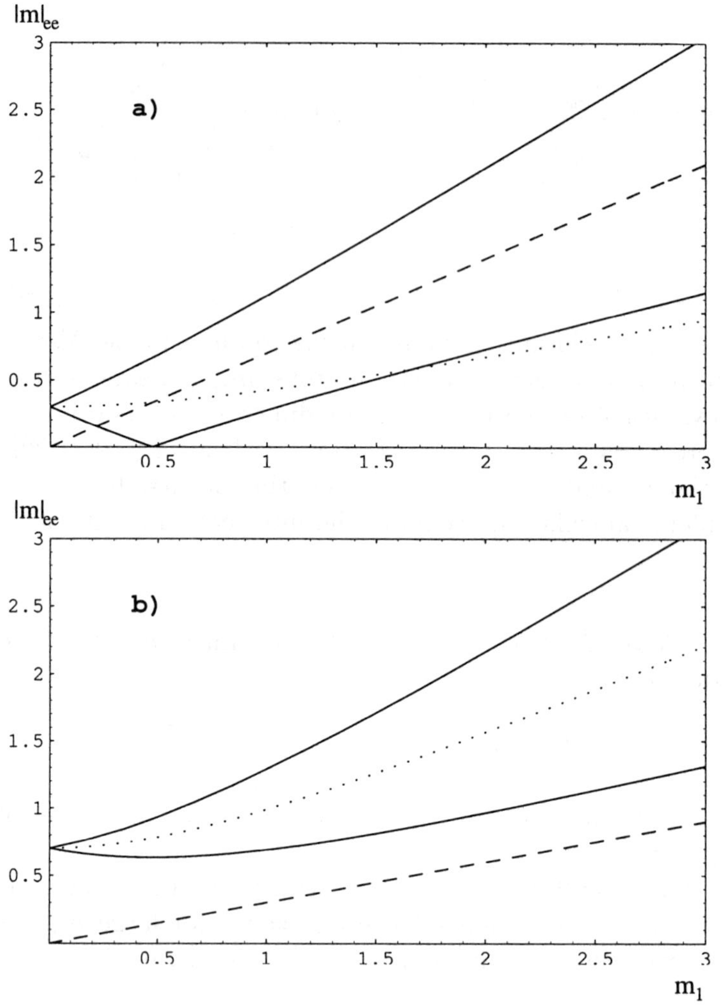

Figure 2: Qualitative dependence of the effective Majorana mass m_{ee} on m_1 in the two neutrino mixing case. a) corresponds to the case of normal mass hierarchy, b) to the case of inverse hierarchy. Shown are the contributions $m_{ee}^{(1)}$ (dashed) and $m_{ee}^{(2)}$ (dotted). The solid lines give m_{ee}^{max} and m_{ee}^{min}, which correspond to $\phi_{21} = 0$ and $\phi_{21} = \pi$, respectively. While in the case of normal hierarchy a complete cancellation is possible, so that $m_{ee} = 0$, for inverse hierarchy a minimal value $|m_{ee}^{min}| > 0$ exists.

Let us consider now the three-neutrino case. The mass m_{ee} is given by the sum of three vectors $\vec{m}_{ee}^{(1)}$, $\vec{m}_{ee}^{(2)}$ and $\vec{m}_{ee}^{(3)}$ in the complex plane (see fig. 1), so that a complete cancellation corresponds to a closed triangle. The sufficient condition for having a minimal value of $|m_{ee}|$ which differs from zero for arbitrary non-oscillation parameters is

$$|m_{ee}^{(i)}| > \sum_{j \neq i} |m_{ee}^{(j)}|. \tag{19}$$

That is, one of contributions $m_{ee}^{(i)}$ should be larger than the sum of the moduli of the two others.

Let us prove that this condition can not be satisfied for the *normal* hierarchy case. Indeed, in eq. (19) i can not be 1. For $m_1 = 0$ we have $m_{ee}^{(1)} = 0$, at the same time the right-handed side of eq. (19) is larger than zero as long as $U_{e1}^2 \neq 1$, (*i.e.* any non-zero mixing of ν_e exists). The condition (19) can not be satisfied for $i \neq 1$ either. In this case for a large enough m_1, so that $m_1 \simeq m_2 \simeq m_3$, we get $m_{ee}^{(i)} < m_{ee}^{(1)}$ since $U_{e1} > U_{ei}$. This proof holds also for schemes with more than three neutrinos. Thus, one can conclude that neither an upper nor a lower bound on $|m_{ee}|$ exists for any oscillation pattern and normal mass hierarchy. Any value $|m_{ee}| \geq 0$ can be obtained by varying the non-oscillation parameters m_1 and ϕ_{ij}.

For inverse mass hierarchy we find that condition eq. (19) can be fulfilled for $i = 3$. Since now both $m_3 > m_2, m_1$ and $U_{e3} > U_{e2}, U_{e1}$ one can get

$$m_3|U_{e3}|^2 > m_2|U_{e2}|^2 + m_1|U_{e1}|^2 \qquad (20)$$

for any set of values of non-oscillation parameters provided that the mixing of the heaviest state fulfills

$$|U_{e3}|^2 > 0.5. \qquad (21)$$

Indeed, for large enough m_1, such that $m_1 \simeq m_2 \simeq m_3$, the condition (20) reduces to $|U_{e3}|^2 > |U_{e2}|^2 + |U_{e1}|^2 \equiv 1 - |U_{e3}|^2$, and the latter is satisfied for (21). For smaller values of m_1 the relative difference of masses $m_3 > m_2, m_1$ increases and the inequality of contributions in eq. (20) becomes even stronger. Thus, the inequality (21) is the sufficient condition for all values of m_1. This statement is true also for any number of neutrinos. It is also independent of the relative size of U_{e2} and U_{e1}.

Summarizing we conclude that

- No upper bound on $|m_{ee}|$ can be derived from oscillation experiments.
- A lower bound exists only for scenarios with inverse mass hierarchy when the heaviest state (ν_3) mixes strongly with the electron neutrino: $|U_{e3}|^2 > 0.5$.
  For normal mass hierarchy certain values of the non-oscillation parameters m_1, ϕ_j exist for which $m_{ee} = 0$.

The cases of normal and inverse mass hierarchy (they differ by signs of Δm^2 once the flavor of the states is fixed) can not be distinguished in vacuum oscillations. However, it is possible to identify the type of the hierarchy in studies of neutrino oscillations in matter, since matter effects depend on the relative signs of the potential V and Δm_{ij}^2. This will be possible in future atmospheric neutrino experiments, long base-line experiments and also studies of properties of the neutrino bursts from supernova [37].

2.2 Effective Majorana mass and the degeneracy of the spectrum

As follows from fig. 2, predictions for m_{ee} can be further restricted under assumptions about the absolute scale of neutrino masses m_1. With increase of m_1 the level of degeneracy of the neutrino spectrum increases and we can distinguish three extreme cases:

- $m_1^2 \ll \Delta m_{21}^2 \ll \Delta m_{31}^2$, in this case the spectrum has a strong *mass hierarchy.*
- $\Delta m_{21}^2 \ll m_1^2 \ll \Delta m_{31}^2$, this is the case of *partial* degeneracy;
- Inequality $\Delta m_{21}^2 \ll \Delta m_{31}^2 \ll m_1^2$ corresponds to *strong* degeneracy.

There are also two transition regions when $m_1^2 \sim \Delta m_{21}^2$ and $m_1^2 \sim \Delta m_{31}^2$. In what follows we will consider all these cases in order.

2.3 Effective Majorana mass and present oscillation data

Present oscillation data do not determine precisely all oscillation parameters. The only conclusion that can be drawn with high confidence level is that the muon neutrino has large (maximal) mixing with some non-electron neutrino state. The channel $\nu_\mu \leftrightarrow \nu_\tau$ is the preferable one, and it is the only possibility, if no sterile neutrino exists. Thus, in 3ν schemes the atmospheric neutrino data are described by $\nu_\mu \leftrightarrow \nu_\tau$ oscillations as dominant mode with

$$\Delta m_{atm}^2 = (2 \div 6) \cdot 10^{-3} \mathrm{eV}^2 , \quad \sin^2 2\theta_{atm} = 0.84 - 1, \tag{22}$$

and the best fit point

$$\Delta m_{atm}^2 = 3.5 \cdot 10^{-3} \mathrm{eV}^2 , \quad \sin^2 2\theta_{atm} = 1.0, \tag{23}$$

[29], see also [30]. A small contribution of the $\nu_\mu \leftrightarrow \nu_e$ mode is possible and probably required in view of an excess in the e -like events in the Super-K experiment.

As it was realized some time ago [8], predictions for m_{ee} depend crucially on the solution of the solar neutrino problem. The solution of the solar neutrino problem determines the distribution of the ν_e-flavor in the mass eigenstates, and this affects considerably expectations for the $0\nu\beta\beta$-mass . Up to now the unique solution is not yet identified and there are several possibilities [31], see also [32]:

1. Small mixing angle MSW solution with

$$\Delta m_\odot^2 = (0.4 \div 1) \cdot 10^{-5} \mathrm{eV}^2 , \quad \sin^2 2\theta_\odot = (0.2 \div 1.2) \cdot 10^{-2} \tag{24}$$

2. Large mixing angle MSW solution with

$$\Delta m_\odot^2 = (0.1 \div 1.5) \cdot 10^{-4} \mathrm{eV}^2 , \quad \sin^2 2\theta_\odot = (0.53 \div 1) \tag{25}$$

3. Low mass MSW (LOW) solution with

$$\Delta m_\odot^2 = (0.3 \div 2.5) \cdot 10^{-7} \mathrm{eV}^2 , \quad \sin^2 2\theta_\odot = (0.8 \div 1) \tag{26}$$

4. Several regions of vacuum oscillation (VO) solutions exist with

$$\Delta m^2_{\odot} < 10^{-9}\text{eV}^2 \ , \quad \sin^2 2\theta_{\odot} > 0.7 \ . \tag{27}$$

There is a good chance that before the new generation of double beta decay experiments starts operation studies of the solar neutrino fluxes by existing and forthcoming experiments will allow us to identify the solution of the solar neutrino problem. The key measurements include the day-night effect, the zenith angle dependence of the signal during the night, seasonal variations, energy spectrum distortions and the neutral current event rate.

The LSND result [33] which implies

$$\Delta m^2_{LSND} = (0.2 \div 2)\ \text{eV}^2, \quad \sin^2 2\theta_{LSND} = (0.2 \div 4) \cdot 10^{-2} \tag{28}$$

is considered as the most ambiguous hint for neutrino oscillations. The KARMEN [34] experiment does not confirm the LSND result but it also does not fully exclude this result (see [33]). The oscillation interpretation of the LSND result will be checked by the MINIBOONE [35] experiment. A simultaneous explanation of the LSND result and of the solutions of the solar and atmospheric neutrino problems in terms of neutrio mass and mixing requires the introduction of a forth neutrino. We will discuss the 4ν schemes in section 7.

Summarizing, there is a triple uncertainty affecting predictions of m_{ee}:

1. An uncertainty in oscillation parameters. The oscillation pattern does not determine uniquely the $0\nu\beta\beta$-mass . Moreover, not all relevant oscillation parameters are known, so that additional assumptions are needed.

2. An uncertainty in the absolute scale m_1. Some information on m_1 can be obtained from cosmology and may be from direct kinematical measurements.

3. An uncertainty in the relative phases. Clearly, the dependence on the phases is small in the case if one of the eigenstates gives a dominating contribution to m_{ee}.

In what follows we will consider predictions for the $0\nu\beta\beta$-mass in schemes of neutrino masses and mixings which explain the solar and the atmospheric neutrino data. The schemes differ by the solution of the solar neutrino problem, the type of the hierarchy and the level of degeneracy. Relative phases are considered as free parameters.

3 Schemes with normal mass hierarchy

In the case of strong mass hierarchy,

$$m_1^2 \ll \Delta m^2_{21} \ll \Delta m^2_{31} \ , \tag{29}$$

the absolute mass values of the two heavy neutrinos are completely determined by the mass squared differences:

$$m_3^2 = \Delta m^2_{31} = \Delta m^2_{atm}; \quad m_2^2 = \Delta m^2_{21} = \Delta m^2_{\odot}. \tag{30}$$

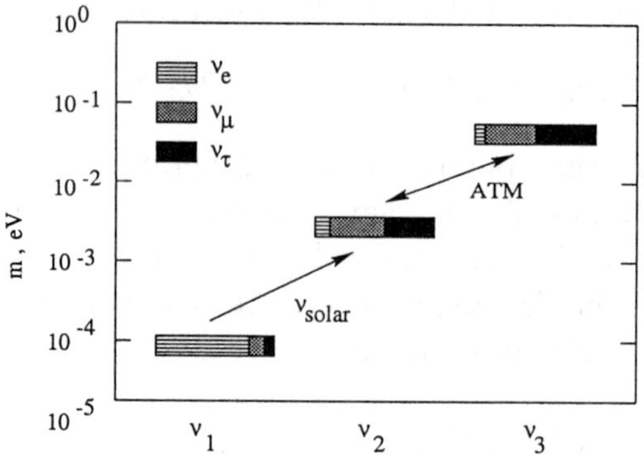

Figure 3: Pattern of neutrino masses and mixing in the scenario with mass hierarchy and SMA solution of the solar neutrino problem. The boxes correspond to mass eigenstates, the sizes of different regions in the boxes show admixtures of different flavors. Weakly hatched regions correspond to the electron flavor, strongly hatched regions depict the muon flavor, black regions present the tau flavor.

The only freedom left is the choice of the value of m_1. In this case the $0\nu\beta\beta$-mass is to a large extent determined by the oscillation parameters.

Since the heaviest neutrino has a mass $m_3 \leq 0.1$ eV, the neutrino contribution to the Hot Dark Matter component of the universe is small: $\Omega_\nu < 0.01$. This neutrino contribution cannot be seen with present and future experimental sensitivity in the CMB radiation, unless a large lepton asymmetry exists [36]. Oberservational evidence of a significant amount of the HDM component $\Omega_\nu \gg 0.01$ would testify against this scenario.

3.1 Single maximal (large) mixing

In this scheme ν_μ and ν_τ are mixed strongly in ν_2 and ν_3 (see fig. 3). The electron flavor is weakly mixed: it is mainly in ν_1 with small admixtures in the heavy states. The solar neutrino data are explained by $\nu_e \to \nu_\mu, \nu_\tau$ resonance conversion inside the Sun. (Notice that ν_e converts to ν_μ and ν_τ in comparable portions.) A small admixture of ν_e in ν_3 can lead to resonantly enhanced oscillations of ν_e to ν_τ in the matter of the Earth.

Let us consider the contributions to m_{ee} from individual mass eigenstates. The contribution from the third state, $m_{ee}^{(3)}$, can be written in terms of oscillation parameters as

$$m_{ee}^{(3)} \simeq \frac{1}{4}\sqrt{\Delta m_{atm}^2} \sin^2 2\theta_{ee}, \tag{31}$$

where the mixing $\sin^2 2\theta_{ee} \approx 4|U_{e3}|^2$ determines the oscillations of ν_e driven by the atmospheric Δm_{atm}^2. The parameter $\sin^2 2\theta_{ee}$ immediately gives the depth of oscillation of the ν_e -survival probability and it is severely constrained by the CHOOZ experiment. In fig. 4 the iso-mass lines of equal $m_{ee}^{(3)}$ in the oscillation parameter space are shown, together with various bounds from existing and future reactor and accelerator oscillation experiments. The shaded region shows the favored range of Δm_{13}^2 from the atmospheric neutrino data. As follows from fig. 4 in the Super-K favored region the CHOOZ bound leads to

$$m_{ee}^{(3)} < 2 \cdot 10^{-3}\text{eV}. \tag{32}$$

For the best fit value of the atmospheric neutrinos the bound is slightly stronger: $m_{ee}^{(3)} < 1.5 \cdot 10^{-3}$ eV.

The mixing $\sin^2 2\theta_{ee}$ and therefore m_{ee} can be further restricted by searches of $\nu_\mu \leftrightarrow \nu_e$ oscillations in the long baseline (LBL) experiments (K2K, MINOS, CERN-Gran-Sasso). The effective mixing parameter measured in these experiments equals $\sin^2 2\theta_{e\mu} = 4|U_{e3}|^2|U_{\mu 3}|^2$ so that

$$\sin^2 2\theta_{ee} = \frac{\sin^2 2\theta_{e\mu}}{|U_{\mu 3}|^2}, \tag{33}$$

where the matrix element $|U_{\mu 3}|^2$ is determined by the dominant mode of the atmospheric neutrino oscillations. Using Eq. (33), the value $|U_{\mu 3}|^2 = 1/2$ and the expected sensitivity to $\sin^2 2\theta_{e\mu}(\Delta m^2)$ of K2K and MINOS experiments, we have constructed corresponding bounds in fig. 4. According to fig. 4, these experiments will be able to improve the bound on $m_{ee}^{(3)}$ by a factor of 2 - 5 depending on Δm^2 and reach 2×10^{-4} eV for a value of Δm^2_{atm} at the present upper bound. For smaller values of $|U_{\mu 3}|^2$ the bound on m_{ee} will be weaker. Taking the smallest value $|U_{\mu 3}|^2 = 0.3$ allowed by the atmospheric neutrino data, we get that the bound on m_{ee} will be 1.7 times weaker. In any case, future LBL experiments will be able to probe the whole region of sensitivity of even the second stage of the GENIUS experiment.

A much stronger bound on $m_{ee}^{(3)}$ can be obtained from studies of neutrino bursts from Supernovae [37]. A mixing parameter as small as $\sin^2 2\theta_{ee} = 10^{-4}$ can give an observable effect in the energy spectra of supernova neutrinos. This corresponds to $m_{ee}^{(3)} \sim 2 \cdot 10^{-6}$ eV.

The contribution from the second mass eigenstate is completely determined by the parameters being responsible for the solution of the solar neutrino problem:

$$m_{ee}^{(2)} \simeq \frac{1}{4}\sqrt{\Delta m^2_{\odot}} \sin^2 2\theta_{\odot}. \tag{34}$$

Taking the 99 % C.L. region of solution (24) we obtain

$$m_{ee}^{(2)} = (5 \cdot 10^{-7} \div 10^{-5}) \text{ eV}, \tag{35}$$

and in the best fit point

$$m_{ee}^{(2)} = 4 \cdot 10^{-6}\text{eV}. \tag{36}$$

The contribution from the lightest state is

$$m_{ee}^{(1)} = m_1 \cos^2 \theta_{\odot} \simeq m_1 \ll m_2 < 2 \cdot 10^{-3}\text{eV}, \tag{37}$$

which can be even larger than $m_{ee}^{(2)}$: if the hierarchy between the masses of the first and the second state is not too strong, $m_1/m_2 > 10^{-2}$ (for comparison $m_e/m_\mu = 5 \cdot 10^{-3}$), we

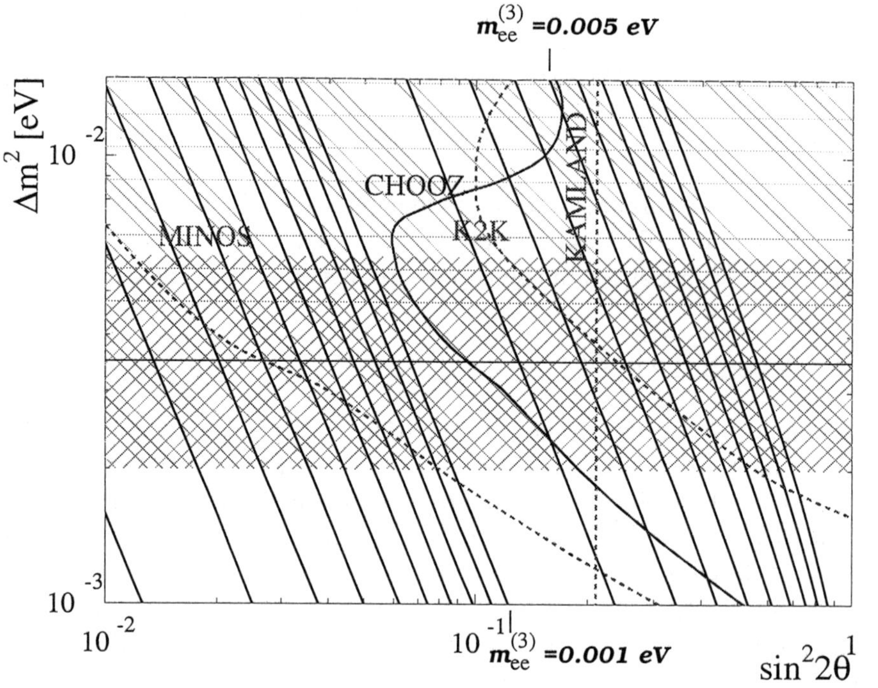

Figure 4: Iso-mass ($|m_{ee}^{(3)}|$) lines (solid lines) in the scheme with hierarchical mass spectrum. From the upper right downward $|m_{ee}^{(3)}|$ decreases from to 0.0001 eV. Also shown are the regions favored by the Super–Kamiokande atmospheric neutrino data with current bestfit (solid horizxontal line) and Kamiokande (lower and upper shaded areas, respectively, according to [29]) and the borders of the regions excluded by CHOOZ (solid line) [38] as well as the expected final sensitivity of KAMLAND and K2K (dashed) [39] as well as of MINOS [40].

get $m_{ee}^{(1)} > 2 \cdot 10^{-5}$ eV, with a typical interval $m_{ee}^{(1)} \sim (0.2 - 2) \cdot 10^{-4}$ eV.

Summing up the contributions (see fig. 5) one finds a maximal value for the $0\nu\beta\beta$-mass

$$m_{ee}^{max} = (2 - 3) \cdot 10^{-3}\text{eV} \tag{38}$$

which is dominated by the third mass eigenstate. No lower bound on m_{ee} can be obtained from the present data. Indeed, U_{e3} and therefore $m_{ee}^{(3)}$ can be zero. The same statement is true for $m_{ee}^{(1)}$, since no lower bound for m_1 exists. The only contribution bounded from below is $m_{ee}^{(2)} > 10^{-6}$eV. However, cancellations with the two other states can yield a zero value for the total m_{ee} (see fig. 5).

The following conclusions on future double beta experiments and neutrino oscillations can be drawn

1). If future experiments will detect neutrinoless beta decay with a rate corresponding to $m_{ee} > 2{\cdot}10^{-3}$ eV, the scenario under consideration will be excluded, unless contributions

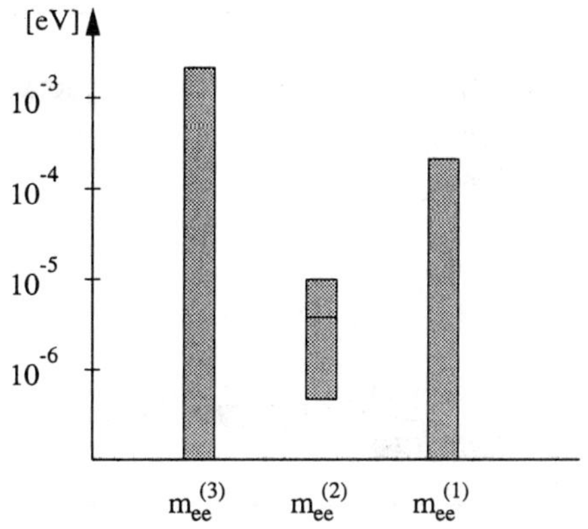

Figure 5: Contributions to m_{ee} from the individual mass eigenstates for the single maximal mixing scheme with mass hierarchy. The bars correspond to allowed regions.

to $0\nu\beta\beta$ decay from alternative mechanisms exist.

2). As we have pointed out, future long-baseline oscillation experiments on $\nu_\mu \to \nu_e$ oscillations (MINOS) may further improve the bound on U_{e3}^2 and therefore on m_{ee}^{max} by a factor of $\sim 2-5$. A much stronger bound may be obtained from supernovae studies [37]. As follows from fig. 4 and from the fact the SMA solution is realized, KAMLAND should give a zero-result in this scheme.

3). An important conclusion can be drawn if future LBL and atmospheric neutrino experiments will observe ν_e-oscillations near the present upper bound. In particular, an up-down asymmetry of the e-like events at Super-Kamiokande is one of the manifestations of these oscillations [41]. In this case the ν_3 contribution to m_{ee} dominates, no significant cancellation is expected and the dependence on the relative phases is weak. One predicts then the result

$$m_{ee} \approx m_{ee}^{(3)} \sim U_{e3}^2 \sqrt{\Delta m_{atm}^2}. \tag{39}$$

The observation of $0\nu\beta\beta$ decay with $m_{ee}^{exp} \sim m_{ee}^{(3)}$ would provide a strong evidence of the scheme, provided that the SMA solution will be established. On the other hand it will be difficult to exclude the scheme if $0\nu\beta\beta$ decay will not be observed at the level which corresponds to m_{ee} (39). In this case the scheme will be disfavored. However one should take into account also possible cancellations of $m_{ee}^{(3)}$ and $m_{ee}^{(1)}$, if the mass hierarchy is weak.

3.2 Bi-large mixing

The previous scheme can be modified in such a way that the solar neutrino data are explained by the large angle MSW conversion. Now the ν_e flavor is strongly mixed in ν_1 and ν_2 (see fig. 6).

The contribution from the third state is the same as in the previous scheme (see eq. (31)) with the upper bound $m_{ee}^{(3)} < 2 \cdot 10^{-3}$ eV (32).

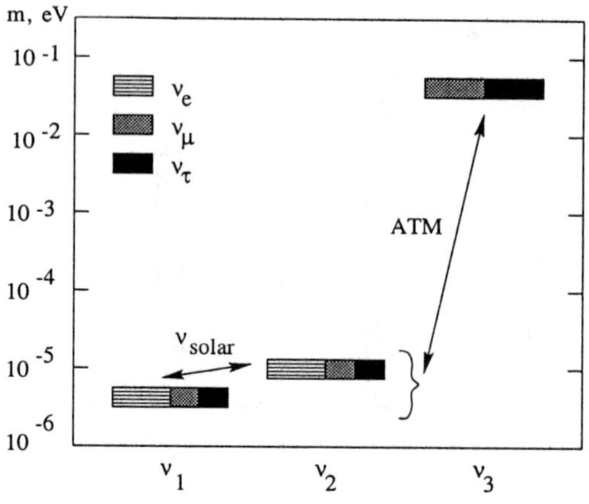

Figure 6: Neutrino masses and mixing pattern of the bi-maximal mixing scheme with mass hierarchy.

The contribution from the second level,

$$m_{ee}^{(2)} = \frac{1}{2}\left(1 - \sqrt{1 - \sin^2 2\theta_{\odot}}\right)\sqrt{\Delta m_{\odot}^2} \ , \tag{40}$$

can be significant: both the mixing parameter and the mass are now larger. According to fig. 7, in the region of the LMA solution of the solar neutrino problem, the contribution can vary in the interval

$$m_{ee}^{(2)} = (0.5 - 4) \cdot 10^{-3}\mathrm{eV} \ . \tag{41}$$

In the best fit point we get $m_{ee}^{(2)} \simeq 1.4 \cdot 10^{-3}$eV. Notice that a lower bound on $m_{ee}^{(2)}$ exists in this scheme, provided that $sin^2 2\theta < 1$. Notice that a day-night asymmetry of about 6 % indicated by the Super-Kamiokande experiment would correspond to $m_{ee}^{(3)} = (1-3) \cdot 10^{-3}$ eV.

The contribution $m_{ee}^{(1)}$:

$$m_{ee}^{(1)} \simeq \cos^2\theta_{\odot} m_1 \ , \tag{42}$$

where $\cos^2\theta \simeq 0.5 - 0.84$, is smaller than in the previous scheme of sect. 3.1, since now ν_ϵ is not purely ν_1 and m_1 can be as large as $1 \cdot 10^{-3}$ eV for $m_1/m_2 < 0.1$. Due to the mass hierarchy $m_{ee}^{(1)}$ is much smaller than $m_{ee}^{(2)}$ (see fig. 8).

Summing up the contributions, we get a maximal value of $m_{ee}^{max} \simeq 7 \cdot 10^{-3}$ eV. The typical expected value for m_{ee} is in the range of several 10^{-3} eV. However, no lower bound on m_{ee} can be obtained on the basis of the present data, although values of m_{ee} being smaller than 10^{-3} eV require some cancellation of the contributions $m_{ee}^{(2)}$ and $m_{ee}^{(3)}$.

Let us consider possible implications of future results from oscillations and $0\nu\beta\beta$ decay searches:

- The observation of $m_{ee} > (few) \cdot 10^{-2}$ eV will exclude the scheme.

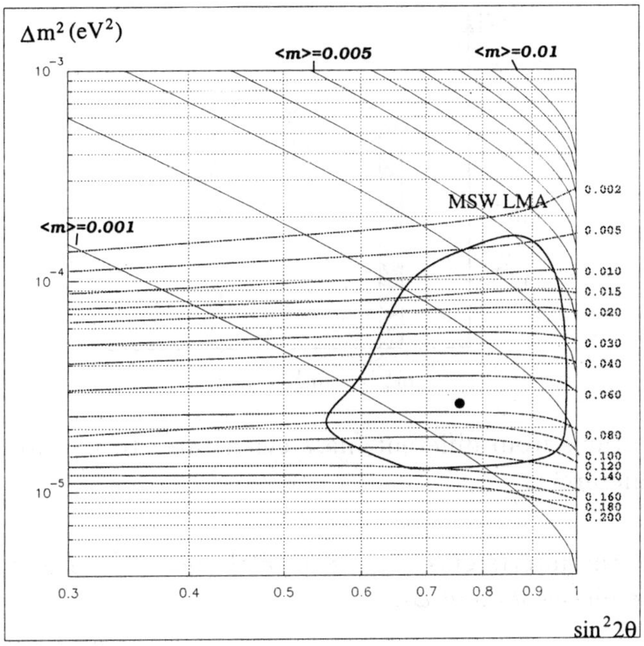

Figure 7: The iso-mass $m_{ee}^{(2)}$ lines, determining the contribution of the second state in the $\Delta m_{12}^2 - \sin^2 2\theta_{12}$ plane for the hierarchical scheme with the LMA MSW solution. From the upper right downward: $|m_{ee}^{(2)}|$ decreases from 0.01 to 0.001 eV. Also shown is the MSW LMA 99 % C.L. allowed region from the combined analysis of the Homestake, Gallex, Sage and Super-Kamiokande rates and the Super-Kamiokande and the day-night asymmetry at Super-Kamiokande. The point indictes the best fit value parameters. The horizontal lines correspond to contours of constant day-night assymmetry [42]. KAMLAND should observe a disappearance signal in this model.

- The non-observation of $0\nu\beta\beta$ decay will not exclude the scheme due to possible cancellations.

The situation can, however, change in the future, if oscillation experiments restrict strongly one of the contributions $m_{ee}^{(2)}$ or $m_{ee}^{(3)}$. Let us discuss possible developments in this direction:

- Within several years solar neutrino experiments will check the LMA-solution. In particular, further measurements of the day-night asymmetry and zenith angle distribution at Super-K and SNO could give a decisive identification of the solution of the solar neutrino problem (see fig. 7). Notice that precise measurements of the day/night asymmetry can sharpen the predictions of $m_{ee}^{(2)}$. Moreover, the LBL reactor experiment KAMLAND should observe an oscillation effect thus providing an independent check of the LMA MSW solution.

- If MINOS or atmospheric neutrino studies will fix $m_{ee}^{(3)}$ near the present upper bound,

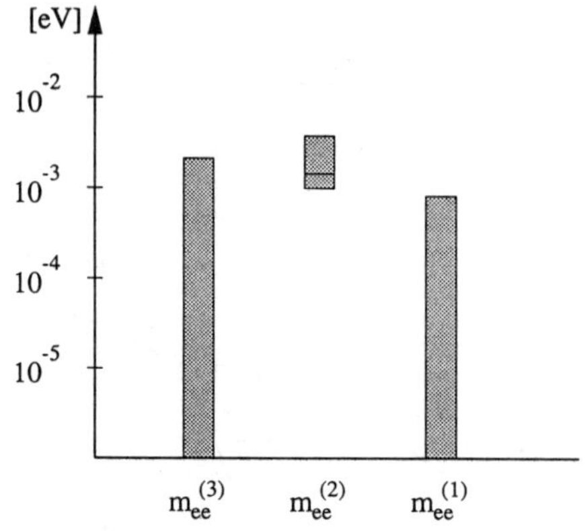

Figure 8: Contributions from different mass eigenstates to m_{ee} for the bi-large mixing scheme with mass hierarchy.

one can study the interference effects of $m_{ee}^{(2)}$ and $m_{ee}^{(3)}$ in $0\nu\beta\beta$ decay determined by the relative phases ϕ_2 and ϕ_3.

3.3 Scheme with Vacuum oscillation solution

The solar electron neutrinos ν_e oscillate in vacuum into comparable mixtures of ν_μ and ν_τ. The fit to the data indicates several disconnected regions in the $\Delta m^2 - \sin^2 2\theta$ - plot. We consider the large Δm^2 region, $\Delta m^2 = (4-9)\cdot 10^{-10}\mathrm{eV}^2$, and $\sin^2 2\theta > 0.8$, where oscillations allow one to explain an excess of the e-like events in the recoil electron spectrum indicated by Super-Kamiokande. (Obviously a small Δm^2 will give even smaller contributions to the effective Majorana mass). In this case

$$m_{ee}^{(2)} = \sqrt{\Delta m^2}\sin^2\theta_\odot < 2\cdot 10^{-5}\mathrm{eV}. \tag{43}$$

Due to the mass hierarchy and large mixing, the lightest mass eigenstate gives an even smaller contribution: $m_{ee}^{(1)} \ll m_{ee}^{(2)}$. The contribution from the third state is the same as in Eq. (31) and in fig. 4. For the sum we get

$$m_{ee} \simeq m_{ee}^{(3)} < 2\cdot 10^{-3}\mathrm{eV}\ , \tag{44}$$

and clearly, $m_{ee}^{(3)}$ can be the dominant contribution (fig. 9).

The following conclusions may be drawn:

1). The observation of $m_{ee} > 10^{-2}$ eV will exclude the scheme. On the other hand there is no minimal value for m_{ee} according to the present data, so that negative results of searches for $0\nu\beta\beta$ decay will have no serious implications for this scheme.

2) A positive signal for atmospheric ν_e oscillations or in the MINOS experiment will allow us to predict uniquely the value of m_{ee}. Then searches for m_{ee} will give a crucial check of the scheme. The absolute scale of the neutrino mass will be fixed.

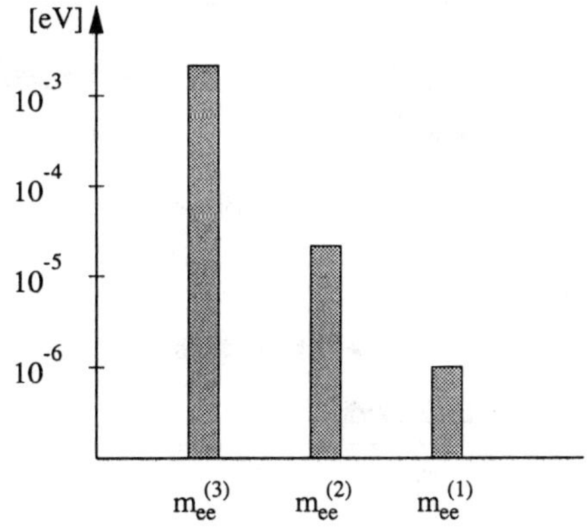

Figure 9: Contributions from different mass eigenstates to m_{ee} for the bi-maximal mixing scheme with mass hierarchy.

Similar results can be obtained for the LOW MSW solution. Here the mass squared difference $\Delta m_{21}^2 = 3 \cdot 10^{-7} \mathrm{eV}^2$ implies

$$m_{ee}^{(2)} < 3 \cdot 10^{-4} \mathrm{eV}. \tag{45}$$

Again $m_{ee}^{(1)} \ll m_{ee}^{(2)}$, and the main contribution may arise from the third state.

Thus, models with normal mass hierarchy lead to rather small values of m_{ee}, certainly below 10^{-2}eV. Moreover, the largest value can be obtained in the scheme with the LMA MSW solution of the solar neutrino problem. The lower bound is of the order $\sim 10^{-3}$eV, unless cancellation (which looks rather unnatural) occurs. Clearly, only the second stage of the GENIUS experiment can obtain positive results.

3.4 Triple maximal mixing scheme

In the scheme of [43] all elements of the mixing matrix are assumed to be equal: $|U_{ij}| = 1/\sqrt{3}$ (see fig. 10). The $0\nu\beta\beta$-mass is dominated by the contribution from the third state:

$$m_{ee}^{(3)} = \frac{1}{3}\sqrt{\Delta m_{atm}^2} \ . \tag{46}$$

The best fit of the atmospheric neutrino data in this scheme implies that $\Delta m_{atm}^2 \simeq 8 \cdot 10^{-4}$ eV^2 and thus

$$m_{ee} \approx 10^{-2} \mathrm{eV} \ . \tag{47}$$

The scheme has rather definite predictions for solar and atmospheric neutrinos. It does not give a good fit of the data and will be tested by forthcoming experiments.

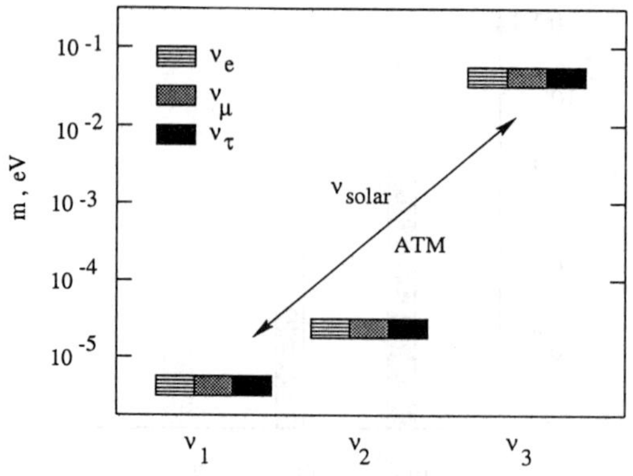

Figure 10: Neutrino masses and mixing in the scheme with threefold maximal mixing.

4 Schemes with partial degeneracy

In the case of partial mass degeneracy,

$$\Delta m_{21}^2 \ll m_1^2 \ll \Delta m_{31}^2 \ ; \tag{48}$$

the masses of the two light neutrinos are approximately equal to m_1 and the heaviest mass is determined by the atmospheric mass squared difference:

$$m_1 \approx m_2; \quad m_3 \approx \sqrt{\Delta m_{31}^2} = \sqrt{\Delta m_{atm}^2} \ . \tag{49}$$

The interval of masses implied by the condition of partial degeneracy (48) is rather narrow especially for the LMA and SMA solutions of the solar neutrino problem, when Δm_{31}^2 and Δm_{21}^2 differ by two orders of magnitude only. A mass value of $m_1 > 3 \cdot 10^{-2}$ eV will shift m_3 to larger values, and therefore influence the contribution from the third eigenstate. We will consider this "transition" case separately in sect. 6.

The contribution from the third state is the same as in hierarchical schemes (see fig. 4). For the two light states, the contribution can be written as

$$m_{ee}^{(1)} + m_{ee}^{(2)} \simeq m_1(\cos^2\theta_\odot + e^{i\phi_2}\sin^2\theta_\odot) \ , \tag{50}$$

and depending on the relative phase ϕ_2 it varies in the interval

$$m_{ee}^{(1)} + m_{ee}^{(2)} = m_1(\cos 2\theta_\odot - 1). \tag{51}$$

This contribution can be further restricted, if the solution of the solar neutrino problem will be identified. In the case of the SMA MSW solution $m_{ee}^{(1)}$ dominates; the dependence on the phase practically disappears and one gets

$$m_{ee}^{(1)} + m_{ee}^{(2)} \simeq m_1. \tag{52}$$

The condition of partial degeneracy implies that the mass m_1 should be in the interval:

$$0.5 \cdot 10^{-2}\text{eV} < m_1 < 3 \cdot 10^{-2}\text{eV},$$

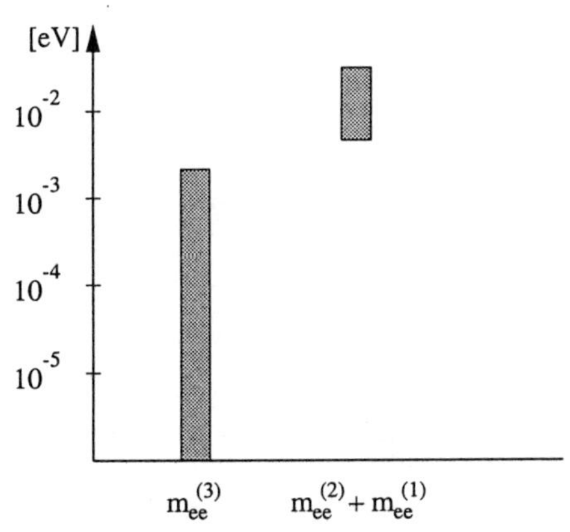

Figure 11: Contributions from different mass eigenstates to m_{ee} for partially degenerate scenarios with MSW SMA solution.

and therefore, m_1 can reach $3 \cdot 10^{-2}$ eV at most (fig. 11).

Summing up all contributions we expect m_{ee} between 10^{-3} and $3 \cdot 10^{-2}$ eV. Notice that a lower bound on m_{ee} exists here. Near the upper bound the mass m_{ee} is dominated by the contribution from the lightest states and therefore the $0\nu\beta\beta$ decay rate will give a direct measurement of m_1: $m_1 \approx m_{ee}$.

Observations of m_{ee} larger than $m_{ee}^{(3)} = U_{e3}^2 m_3$ ($m_{ee}^{(3)}$ can be determined from oscillation experiments) by oscillation experiments) would favor the scheme, although will not allow one to identify it unambigously.

Future observations of $0\nu\beta\beta$ decay with $m_{ee} > 3 \cdot 10^{-2}$ eV will exclude the scheme testifying for spectra with complete degeneracy or inverse hierarchy (see sect. 5 or 7).

For the LMA solution the typical Δm_{21}^2 is bigger than in the SMA case and the condition of partial degeneracy implies an even narrower interval $m_1 = (1-3) \cdot 10^{-2}$eV. Moreover, for m_1 at the lower limit of this interval, the difference of light mass eigenvalues can give a substantial correction to formula (50). In the lowest approximation of $\frac{\Delta m^2}{m_1^2}$ we get:

$$m_{ee}^{(1)} + m_{ee}^{(2)} \simeq m_1(\cos^2\theta_\odot + e^{i\phi_2}\sin^2\theta_\odot) + e^{i\phi_2}\frac{\Delta m_\odot^2}{2m_1}\sin^2\theta_\odot. \qquad (53)$$

The correction (last term in this equation) can be as big as 10^{-3} eV and may turn out to be important when a cancellation of $m_{ee}^{(1)}$ and $m_{ee}^{(2)}$ occurs.

Summing up the contributions we find, that the maximal value of m_{ee} can be about $3 \cdot 10^{-2}$ eV as in the case of the SMA solution with similar implications for future $0\nu\beta\beta$ decay searches. In contrast with the SMA case, now due to possible strong cancellations of the contributions no lower bound on m_{ee} can be obtained from the present data (fig.12).

Future oscillation results will allow to sharpen the predictions of m_{ee}. In particular, the solar neutrino experiments will allow to measure a deviation of mixing from the maximal value. The bound $1 - \sin^2 2\theta_\odot > 0.1$ would imply that $m_{ee}^{(1)} + m_{ee}^{(2)} > 3 \cdot 10^{-3}$ eV. In this case no complete cancellation in m_{ee} is possible and a minimum value $m_{ee} \geq 10^{-3}$

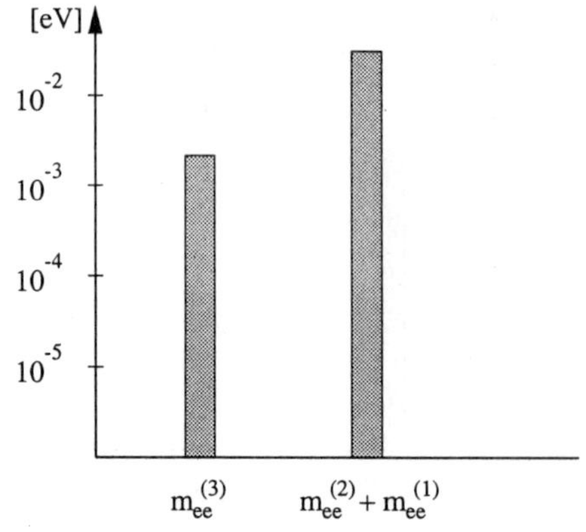

Figure 12: Contributions to m_{ee} from different mass eigenstates in schemes with partial degeneracy and LMA, LOW or VO solutions of the solar neutrino problem. The degenerate states give the main contribution. A complete cancellation of contributions is possible.

eV appears. The searches for ν_e-oscillations driven by Δm^2_{atm} will further restrict (or measure) $m^{(3)}_{ee}$.

Future studies of the $0\nu\beta\beta$ decay can have the following implications: (i) A measurement of $m_{ee} > 2 \cdot 10^{-2}$ eV will exclude the scheme. (ii) The non-observation of m_{ee} at the level of 10^{-3} eV (second stage of GENIUS) can exclude the scheme if future oscillation experiments will lead to a determination of the sum $m^{(1)}_{ee} + m^{(2)}_{ee}$ and a lower bound on m_{ee} will be derived. (iii) If m_{ee} will be observed at the level $(0.3 - 2)10^{-2}$ eV (and alternative schemes which yield a prediction in this interval will be rejected by other observations), then m_{ee} measurements will imply a certain bound in the $m_1 - \phi_2$ plane.

In the case of the LOW solution $\Delta m^2_{\odot}$ is much smaller than for the LMA solution and m_1 can be in the interval $m_1 = (10^{-3} - 3 \cdot 10^{-2})$ eV. Correspondingly, the contribution from the two lightest states can be in the wider range $(10^{-4} - 3 \cdot 10^{-2})$ eV. The maximal value for m_{ee} can reach $3 \cdot 10^{-2}$ eV. However it will be impossible to establish a lower bound on m_{ee} even if the solar mixing angle $\theta_{\odot}$ will be measured.

Notice that the LOW solution can be identified by a specific enhancement of the regeneration effects (in particular the day/night asymmetry) in the lower energy part of the solar neutrino spectrum. An especially strong effect is expected on the ^7Be-line.

For vacuum oscillations the situation is similar to the LOW case.

5 Schemes with complete mass degeneracy

In schemes with a degenerate neutrino mass spectrum the common mass m_1 is much larger than the mass splittings:

$$\Delta m^2_{21} \ll \Delta m^2_{31} \ll m^2_1. \tag{54}$$

This can be realized as long as $m_1 > 0.1$ eV.

Already the present bound $m_{ee} < 0.2 - 0.4$ eV [44] implies not too strong degeneracy unless a substantial cancellation of the contributions in m_{ee} occurs. Indeed, if $\Delta m^2_{atm} = 3 \cdot 10^{-3}$ eV^2, and $m_1 \sim 0.2$ eV, we get

$$\frac{\Delta m_{23}}{m_1} \approx \frac{\Delta m^2_{atm}}{2m_2^2} = 4 \cdot 10^{-2}. \qquad (55)$$

For $m_1 = 0.1$ eV, the ratio equals 0.15.

In the case of the SMA solution the ν_e flavor is mainly concentrated in ν_1 and

$$m_{ee}^{(1)} \simeq m_1 \cos^2\theta_\odot \gg m_{ee}^{(2)} \gg m_{ee}^{(3)}. \qquad (56)$$

Numerically, we get $m_{ee} \sim m_{ee}^{(1)} \sim m_1 > 0.1$ eV; *i.e.* close to the present bound. Basically one measures m_1 by measuring the $0\nu\beta\beta$-mass (see fig. 13).

Important conclusions follow from a comparison of the $0\nu\beta\beta$ decay results with the cosmological bounds on the neutrino mass [20], as well as from bounds which follow from observations of the large scale structure of the Universe. In this scenario one expects

$$\sum m_i = 3m_{ee} \; , \qquad (57)$$

and therefore

$$\Omega_\nu = \frac{3m_{ee}}{91.5\text{eV}} h^{-2}. \qquad (58)$$

Thus the effective Majorana mass, Ω_ν and the Hubble constant are related. This relation may have the following implications:

1). A significant deviation from equation (58) will exclude the scheme. The present bound on m_{ee} implies $3m_1 \leq (0.6 - 1)$ eV. If, *e.g.*, data on the large scale structure of the universe will require $\sum m_i \simeq 1$ eV, this scheme will be excluded [15, 17].

2). The discovery of $0\nu\beta\beta$ decay at the level of the present bound, 0.1 - 0.2 eV, will give $\sum m_i \simeq 0.3 - 0.6$ eV. This range can be probed by MAP and Planck. If these experiments will put a bound on the sum of neutrino masses below 0.3 eV the scheme will be excluded.

3). This scheme will also be excluded, if cosmological observations will require $\sum m_i > 0.3$ eV, but $0\nu\beta\beta$ decay searches will give a bound below $m_{ee} < 0.1$ eV, the scheme will be excluded.

Apart from a confirmation of the SMA solution future oscillation experiments will not influence predictions of m_{ee} in this scheme.

Observations of m_{ee} at the level $0.1 - 0.4$ eV will be in favor of the scheme.

For the LMA solution a significant cancellation of the contributions from the first and the second state may occur, resembling the situation in the partially degenerate case. We have

$$m_{ee}^{(1)} + m_{ee}^{(2)} \simeq m_1(\cos^2\theta_\odot + e^{i\phi_2}\sin^2\theta_\odot) \; , m_{ee}^{(3)} = m_1|U_{e3}|^2 \; . \qquad (59)$$

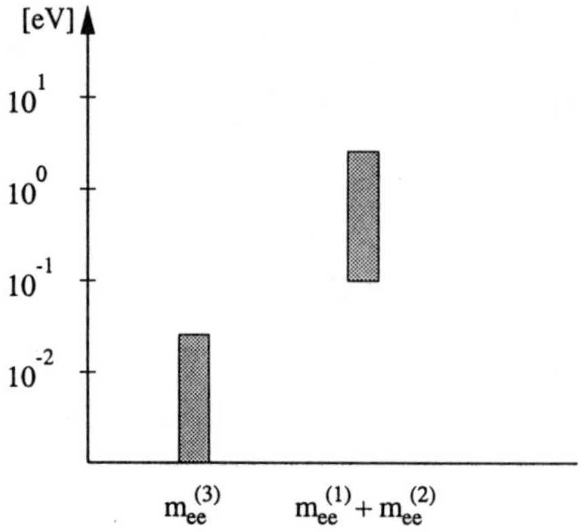

Figure 13: Contributions from different mass eigenstates to m_{ee} for the scenario with complete mass degeneracy and the SMA solution of the solar neutrino problem.

Since $U_{e3}^2 \leq 0.03$, the contribution from the first two states dominates (fig. 15), unless strong mixing, which has an extremely small deviation from maximal mixing, is introduced: $|1 - \sin^2 2\theta_\odot| < 10^{-3}$, which leads to strong cancellation. Thus

$$m_{ee} \approx m_{ee}^{(1)} + m_{ee}^{(2)} = (\cos 2\theta_\odot - 1) \cdot m_1 = (0.2 - 1)m_1 \ , \tag{60}$$

and since $m_1 > 0.1$ eV, we expect for $\sin^2 2\theta = 0.96$:

$$m_{ee} \geq 2 \cdot 10^{-2}\text{eV}. \tag{61}$$

Notice that some recent studies show that even exact maximal mixing is allowed by the present data, so that the cancellation can be complete.

Precise measurements of $\theta_\odot$ in future oscillation experiments will play a crucial role for predictions of the mass m_{ee}.

If the scheme will be identified, measurements of m_{ee} will provide a bound in the $m_1 - \phi$-plane.

The same results hold also for the LOW solution.

For the vacuum oscillations of the $\nu_\odot$-problem, the situation is similar to the one with LMA MSW. In the strict bi-maximal scheme $U_{e3}^2 = 0$ and $U_{e1}^2 = U_{e2}^2$, so that $m_{ee} \equiv 0$ in the limit of equal masses. Small deviations from zero can be related to the mass difference of m_1 and m_2:

$$m_{ee} \simeq \frac{1}{2}(m_1 - m_2) = \frac{1}{4}\frac{\Delta m_\odot^2}{m_1} \simeq 10^{-10}\text{eV}\frac{1\text{eV}}{m_1}. \tag{62}$$

Thus, no unique prediction for m_{ee} exists. Although a large value $m_{ee} > 0.1$ eV would favor degenerate scenarios, the non-observation of $0\nu\beta\beta$ decay at the level of 0.1 eV will not rule out the scheme.

The identification of the scenario will require (i) a strong upper bound on U_{e3}, (ii) the confirmation of the vacuum oscillation solution and (iii) a large $m_{ee} > 0.1$ eV. This would

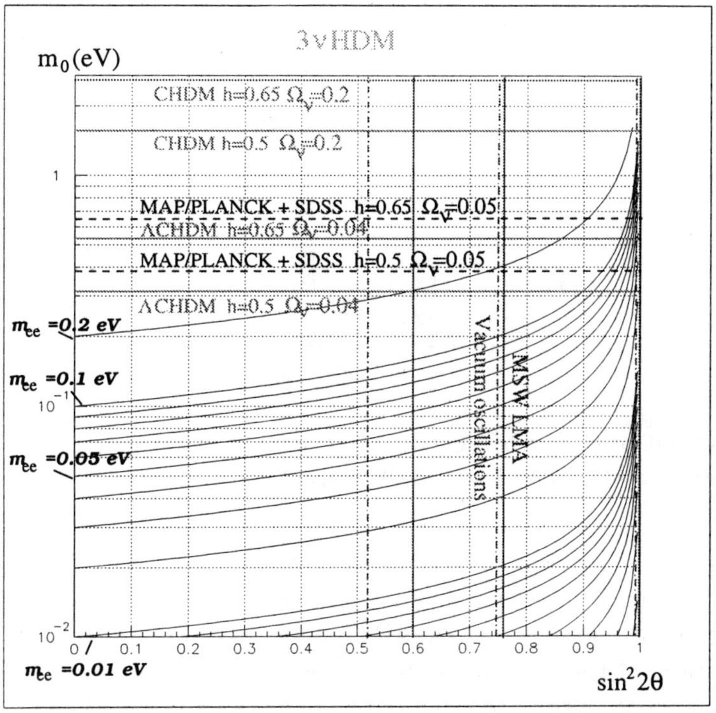

Figure 14: Plotted are iso-mass lines in the $m_1 - \sin^2 2\theta$ plane for the case of cancellation between the contributions m_1 and m_2 in degenerate scenarios. Mass splitting is neglected, since $m_2 - m_1 \ll m_1$ and $m_3 - m_1 < 0.1 m_1$ Shown are the bestfit values of m_0 for cold+hot dark matter (CHDM) (according to [45]) and cosmological-constant+cold+hot dark matter (ΛCHDM) (according to [46]) scenarios for different values of the Hubble constant. Also shown is the sensitivity to m_0 of MAP/Planck combined with SDSS according to [47]. Also shown are the regions of the LMA and VO solutions with best fit line according to [48].

testify for the case of addition of the ν_1 and ν_2 contributions. An upper bound on m_1 can be obtained from cosmology.

If however $0\nu\beta\beta$ decay will not be discovered it will be practically impossible to exclude the scenario (and distinguish it from the hierarchical cases), unless cosmology will be able to measure neutrino masses down to $m_1 \simeq 0.1$ eV. The key element is the precise determination of $\theta_\odot$ and its deviation from the maximal mixing.

Let us consider how deviations from the exact bi-maximal case affect the predictions for the rate in double beta decay experiments. The lower bound on the effective mass turns out to be

$$m_{ee} \simeq m_1 \sqrt{1 - \sin^2 2\theta} \; . \tag{63}$$

We show this result in fig. 14 as lines of minimal values of m_{ee} in the $m_1 - \sin^2 2\theta$ plane. These lines give the lower bound on values of m_1 and the upper bound on values of $\sin^2 2\theta$

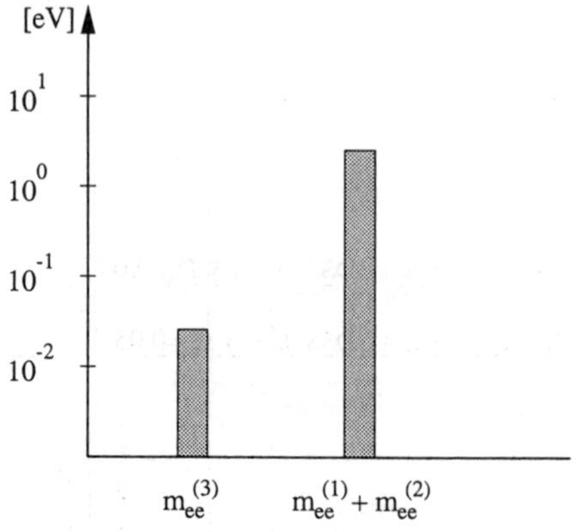

Figure 15: Contributions from different mass eigenstates to m_{ee} for the scenario with complete degeneracy and LMA, LOW or VO solution.

for which a given value m_{ee} can be reproduced. We have shown also the favored regions of the solar MSW large mixing angle solution as well as the "Just-so" vacuum oscillation solution. For the bestfit value of the ΛCHDM model $m_1 = 0.3$, a Hubble constant $h = 0.5$ and the best fit value of the mixing angle $\sin^2 2\theta = 0.76$ we get from the fig 14 $m_{ee} \simeq 0.2$ eV close to the present experimental bound. Larger mixing allows for smaller values of m_{ee}. In fig. 14 also shown is the sensitivity of CMB studies with MAP and Planck combined with the Sloan Digital Sky Survey. For not too large mixing already the present $0\nu\beta\beta$ decay bound ontained from the Heidelberg-Moscow experiment [44] is close to the sensitivity of these cosmological observations.

6 Transition regions

There are two intermediate regions of m_1:

1) The region with $m_1 \sim \sqrt{\Delta m^2_{\odot}}$, where the transition between the hierarchical case and the case with partial degeneracy occurs. Here $m_1 \sim m_2 \ll m_3$.

2). The region with $m_1 \sim \sqrt{\Delta m^2_{atm}}$ which corresponds to the transition between partial degeneracy to complete degeneracy: $m_1 \approx m_2 \sim m_3$. Here the two lightest states are strongly degenerate and their contributions are described by eq. (51) in sect. 4. with the only difference that m_1 now can be larger. For $m_1^2 \sim \Delta m^2_{atm}$, m_1 and therefore m_{ee} can reach 0.1 eV.

The contribution from the third state is also modified:

$$m_{ee}^{(3)} = |U_{e3}|^2 \sqrt{\Delta m^2_{atm}} \left[1 + \frac{m_1^2}{\Delta m^2_{atm}}\right]^{1/2}. \tag{64}$$

Thus now for the same values of the oscillation parameters the contribution $m_{ee}^{(3)}$ can be $[1 + m_1^2/\Delta m^2_{atm}]^{1/2} \sim (1-2)$ times larger.

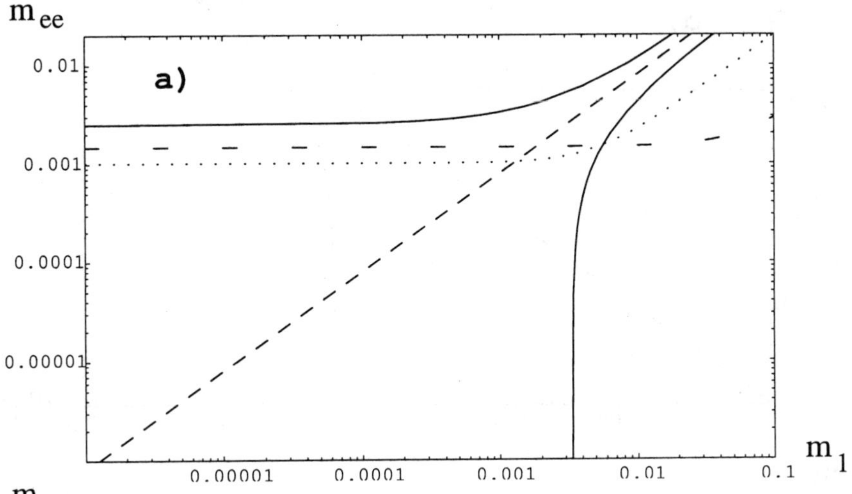
m ee
a)
0.01
0.001
0.0001
0.00001
0.00001
0.0001
0.001
0.01
0.1
m 1

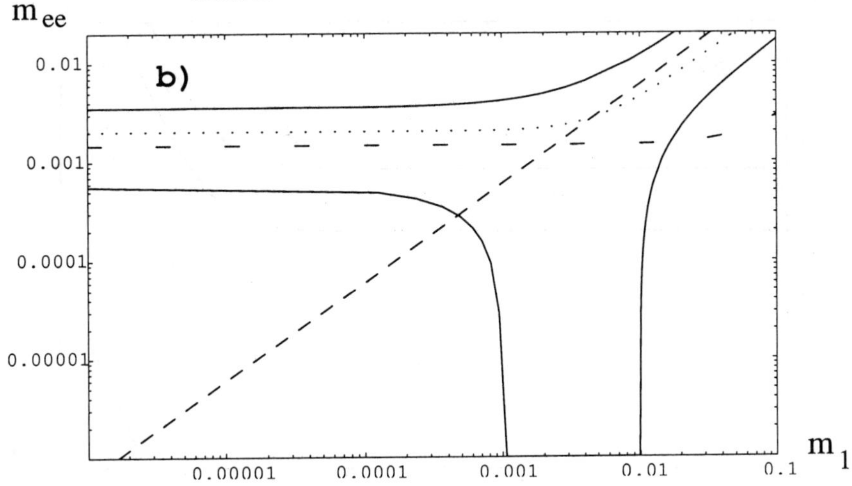
m ee
b)
0.01
0.001
0.0001
0.00001
0.00001
0.0001
0.001
0.01
0.1
m 1

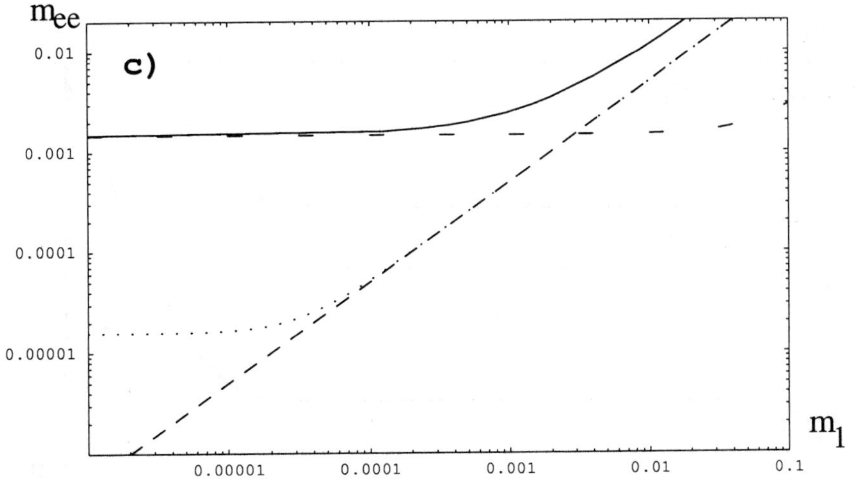
m ee
c)
0.01
0.001
0.0001
0.00001
0.00001
0.0001
0.001
0.01
0.1
m 1

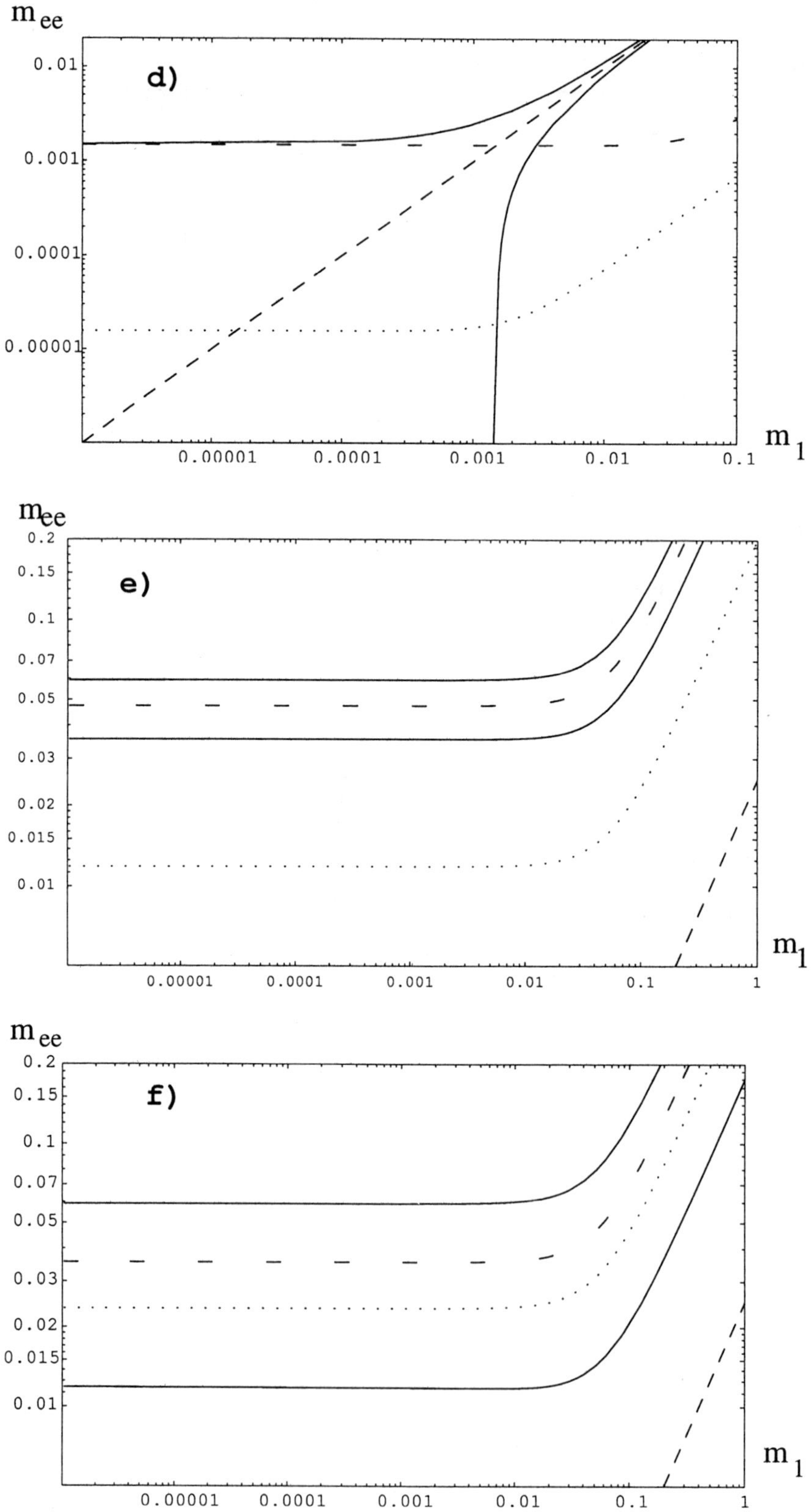
m ee
0.01
0.001
0.0001
0.00001
d)
0.00001
0.0001
0.001
0.01
0.1
m 1
m ee
0.2
0.15
0.1
0.07
0.05
0.03
0.02
0.015
0.01
e)
0.00001
0.0001
0.001
0.01
0.1
1
m 1
m ee
0.2
0.15
0.1
0.07
0.05
0.03
0.02
0.015
0.01
f)
0.00001
0.0001
0.001
0.01
0.1
1
m 1

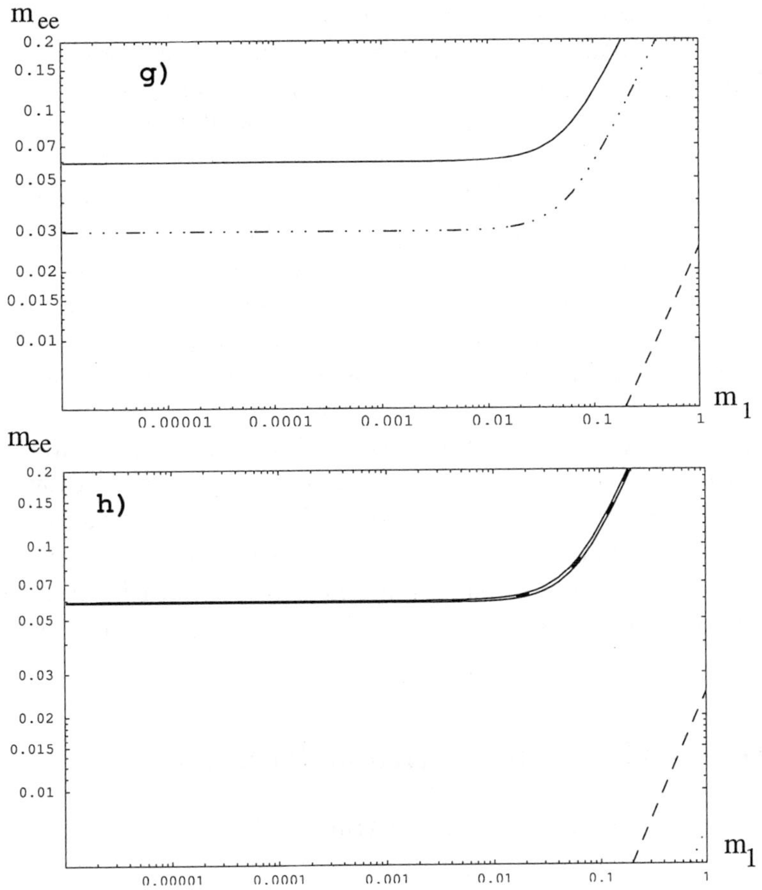

Figure 16: m_{ee} (eV) as a function of m_1 (eV) for three-neutrino mixing. Shown are the contributions $m_{ee}^{(1)}$ (dashed), $m_{ee}^{(2)}$ (dotted) and $m_{ee}^{(3)}$ (interrupted dashes). The solid lines correspond to m_{ee}^{max} and m_{ee}^{min} and show the allowed region for m_{ee}. Panels a)-d) correspond to the case for normal hierarchy, panels e)-h) – inverse hierarchy. Shown are the cases a) and e) LMA MSW with $U_{e2}^2 = 0.2$, b) and f) LMA MSW with $U_{e2}^2 = 0.4$, c) and g) vacuum oscillation with $U_{e2}^2 = 0.5$ and d) and h) SMA MSW with $U_{e2}^2 = 7 \cdot 10^{-3}$. The mixing of the third state is varied from zero to its upper bound, $U_{e3}^2 = 2.5 \cdot 10^{-2}$.

In fig.16 we show the dependence of the individual contributions to m_{ee} on m_1. For $m_{ee}^{(3)}$ only the upper bound is used; the two other lines represent possible values of $m_{ee}^{(1)}$ and $m_{ee}^{(2)}$ for certain neutrino mixing parameters. We show also the maximal and the minimal possible values of m_{ee}.

The position of the first transition region is determined by the specific solution of the solar neutrino problem: According to fig. 16, $m_1 = (2 - 15) \cdot 10^{-3}$ eV for the LMA solution, $m_1 = (1 - 9) \cdot 10^{-3}$ eV for the SMA MSW solution and $m_1 = (1 - 10) \cdot 10^{-5}$ eV for the VO solution.

The position of the second transition region, $m_1 = (3 - 20) \cdot 10^{-2}$ eV, is similar in all cases.

The upper bounds on m_{ee} as functions of m_1 have a similar dependence for all the cases. The lower bounds are different and depend on specific values of oscillation parameters.

Thus, for the LMA solutions a lower bound exists in the range of mass hierarchy ($m_1 < 10^{-3}$ eV) if the solar mixing angle is sufficiently large (see fig. 16 a)). In this case the contribution from ν_2 dominates and no cancellation is possible even for maximal possible $m_{ee}^{(3)}$. In contrast, for a lower $\sin^2 2\theta_\odot$ the cancellation can be complete so that no lower bound appears (see fig. 16 a)).

In the first transition region all states contribute with comparable portions to m_{ee}, thus cancellation is possible and no lower bound exists.

In the second transition region as well as in the completely degenerate case the first and the second state give the dominating contributions to m_{ee} and the increase of m_3 does not influence significantly the total m_{ee}. The mass m_{ee} is determined by m_1 and $\theta_\odot$. Moreover, a larger $\sin^2 2\theta_\odot$ implies a larger possible range of m_{ee} for a given m_1 (fig. 16 a,b)).

Let us consider the SMA MSW solution (fig. 16 d)). In the mass hierarchy region the third state gives the main contribution and no lower bound exists. A lower bound on m_{ee} appears at $m_1 > 1.5 \cdot 10^{-3}$ eV and at $m_1 > 10^{-2}$ eV, where the mass m_{ee} is given by m_1.

In the case of the VO solution (fig. 16 c) the upper bound on m_{ee} is given by $m_{ee}^{(3)}$ up to $m_1 \sim 2 \cdot 10^{-4}$ eV. In the range of partial degeneracy the contribution from the first and the second states become important. No lower bound on $m_{ee}^{(3)}$ can be established from the present data in the whole range of m_1.

7 Scheme with inverse mass hierarchy

Let us consider the partially degenerate spectrum with

$$m_3^2 \approx m_2^2 = \Delta m_{atm}^2, \quad m_1^2 \ll m_2^2, \quad \Delta m_{23}^2 = \Delta m_\odot^2 \ , \tag{65}$$

so that the mass of the second and third neutrino are determined from the atmospheric neutrino data. The ν_e flavor is concentrated in the heavy states (inverse mass hierarchy). A small admixture of ν_e in the lightest state can exist (fig. 17).

The contribution to m_{ee} from the first state equals

$$m_{ee}^{(1)} = m_1 U_{e1}^2. \tag{66}$$

The inequality $m_1^2 \ll \Delta m_{atm}^2$ implies $m_1 < 2 \cdot 10^{-2}$ eV for $m_1^2/m_2^2 < 0.1$. Using then the CHOOZ result which restricts (in schemes with inverse hierarchy) U_{e1}^2: $U_{e1}^2 < 2.5 \cdot 10^{-2}$, we get

$$m_{ee}^{(1)} < 5 \cdot 10^{-4} \text{ eV} \ . \tag{67}$$

The sum of the contributions from the two heavy degenerate states can be written as

$$m_{ee}^{(2)} + m_{ee}^{(3)} \simeq \sqrt{\Delta m_{atm}^2}(\sin^2\theta_\odot + e^{i\phi_{23}}\cos^2\theta_\odot) \ , \tag{68}$$

where $\phi_{23} \equiv \phi_2 - \phi_3$. For the SMA solution we get from eq. (68)

$$m_{ee} \approx m_{ee}^{(2)} + m_{ee}^{(3)} \approx \sqrt{\Delta m_{atm}^2} = (4-8) \cdot 10^{-2}\text{eV} \tag{69}$$

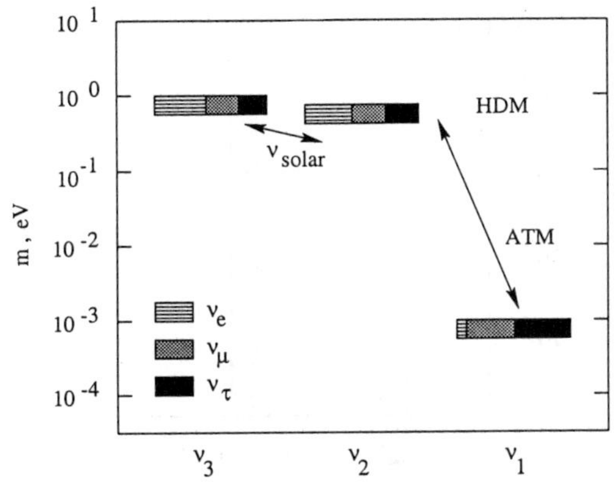

Figure 17: The neutrino mass and mixing pattern in the bi-large mixing scheme with inverse mass hierarchy.

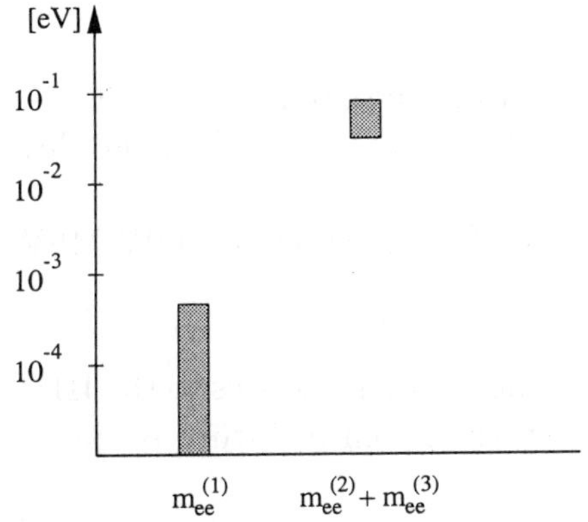

Figure 18: Contributions to m_{ee} from different mass eigenstates in the scheme with inverse mass hierarchy and SMA solution. The degenerate states give the main contribution, implying a unique prediction for m_{ee}.

and in the bestfit point of the atmospheric neutrino data: $m_{ee} \approx 6 \cdot 10^{-2}$eV. This means, that the predicted value of m_{ee}^2 coincides with Δm_{atm}^2 (fig. 18). This coincidence provides a unique possibility to identify the scheme (see also, e.g. [12]).

The relation $m_{ee}^2 = \Delta m_{atm}^2$ applies also for the case of the LMA solution as long as $\phi_2 = 0$ [2].

For the LMA solution the sum of the contributions from the two heavy states lies in the interval

$$m_{ee}^{(2)} + m_{ee}^{(3)} = (\cos 2\theta_{\odot} - 1) \cdot \sqrt{\Delta m_{atm}^2}. \tag{70}$$

For $\sin^2 2\theta_{\odot} < 0.98$ we get $m_{ee} > 4 \cdot 10^{-3}$ eV which is still much larger than $m_{ee}^{(1)}$. The compensation is complete if the mixing is maximal. The value $m_{ee}^{(2)} + m_{ee}^{(3)} < 2 \cdot 10^{-3}$ eV

[2]Notice that if the mass degeneracy originates from some flavor blind interactions one may indeed expect that masses of ν_2 and ν_3 have the same phase.

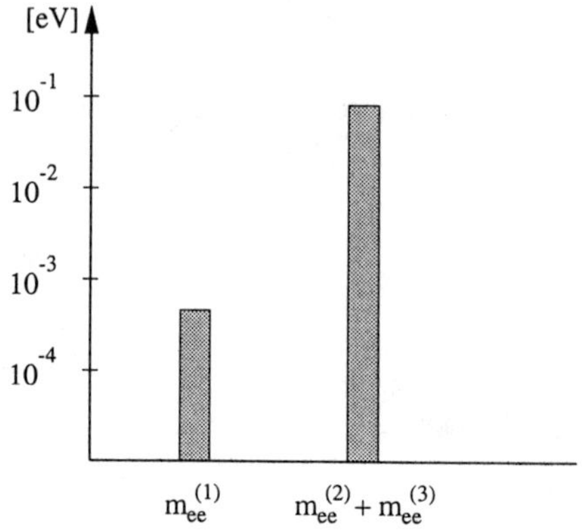

Figure 19: Contributions to m_{ee} from different mass eigenstates in the schemes with inverse hierarchy and LMA, LOW or VO solutions. Now cancellation between the degenerate states is possible leading to a wide range of values allowed for m_{ee}.

requires a very small deviation from maximal mixing: $1 - \sin^2 2\theta_\odot < 2 \cdot 10^{-3}$. Thus, the lower bound on m_{ee} can be further strengthened, if the deviation from maximal mixing will be established.

A similar consideration holds for the cases of LOW MSW or vacuum oscillation solutions (69) (see fig. 19).

The contribution of the two heavier eigenstates to the HDM, $\Omega_\nu = (2m_1)/(91.5\,\text{eV}h^2) \sim 0.01$, is rather small and below the reach of future projects on measurements of cosmological parameters.

If the ν_e admixture in the lightest state is non-zero, so that the ν_e-oscillations driven by Δm^2 exist, the scheme can be identified by studying matter effects in atmospheric and supernova neutrinos as well as in the long-baseline experiments.

Indeed, in the case of inverse mass hierarchy the $\nu_e - \nu_3'$ level crossing (in matter) occurs in the *antineutrino* channel, so that in supernovae the antineutrinos $\bar{\nu}_e$ will be strongly converted into a combination of $\bar{\nu}_\mu$, $\bar{\nu}_\tau$ and vice versa. This leads to a hard $\bar{\nu}_e$'s spectrum at the Earth detector which coincides with the original $\bar{\nu}_\mu$ spectrum [37].

In atmospheric neutrinos the identification of the type of mass hierarchy will be possible if the sensitivity will be enough to detect oscillation effects in e-like events (electron neutrinos and antineutrinos). It will be also important to measure the sign of the electric charge of the lepton, since the matter effects are different in the neutrino and antineutrino channels and this difference depends on the type of mass hierarchy.

These matter effects can be studied in LBL experiments [39, 40] with neutrinos from neutrino factories where beams of neutrinos and antineutrinos are well controlled.

Let us consider the dependence of the predictions for m_{ee} on m_1. In the schemes with inverse hierarchy there is only one transition region: $m_1 \sim \sqrt{\Delta m^2_{atm}}$, that is $m_1 \sim m_2 \approx m_3$ or $m_1 = (1-8) \cdot 10^{-2}$ eV. The sum of the contributions from the second and the third states dominates in the whole range of m_1. It is determined by the "solar" mixing angle

$\theta_\odot$ and $m_2 \approx m_3$. The latter changes from $\sqrt{\Delta m^2_{atm}}$ in the hierarchical region to m_1 in the region of complete degeneracy (see fig. 16 e-h). The mass m_{ee} is completely predicted in terms of m_2 for the SMA solution (fig. 16 h).

No lower bound on m_{ee} appears when the (solar) mixing parameter is maximal or close to maximal.

8 Four neutrino scenarios

The introduction of new ("sterile") neutrinos mixed with the usual SU(2) doublet neutrinos opens new possibilities for the construction of the neutrino mass spectrum and for the explanation of the data. It also modifies predictions of m_{ee}. Here we will consider several scenarios which are motivated both by phenomenology and theory. All scenarios we will discuss contain one or two (degenerate) states in the range relevant for structure formation in the universe and/or for the LSND oscillations.

8.1 Scenario with small flavor mixing and mass hierarchy

The scheme (fig. 20) is characterized by a mass hierarchy:

$$m_4 = m_{HDM}; \quad m_3 \approx \sqrt{\Delta m^2_{ATM}}; \; m_2 \approx \sqrt{\Delta m^2_\odot}; \quad m_1 \ll m_2 \; . \tag{71}$$

The states ν_μ and ν_s are strongly mixed in the second and fourth mass eigenstates, so that $\nu_\mu \leftrightarrow \nu_s$ oscillations solve the atmospheric neutrino problem. All other mixings are small. In particular, the solar neutrino problem is solved by small mixing MSW conversion $\nu_e \to \nu_\mu, \nu_s$.

The main motivation for this scheme is to avoid the introduction of large mixing between flavor states and to keep in this way as much as possible correspondence with the quark sector.

A clear signature of the scheme is the $\nu_\mu \leftrightarrow \nu_s$ oscillation solution of the atmospheric neutrino problem. The solution can be tested by (i) studies of the neutral current interactions in atmospheric neutrinos, in particular, $\nu N \to \nu N \pi^0$ (with $N = n, p$), which gives the main contribution to the sample of the so called π^0 events (the rate should be lower in the $\nu_\mu \to \nu_s$ case); (ii) studies of the zenith angle distribution of the upward going muons (stopping and through-going); (iii) detection of the τ leptons produced by converted ν_τ.

Recent Super-Kamiokande data do not show a deficit of π^0 events, and moreover the $\nu_\mu \to \nu_\tau$ oscillations give a better fit (of about $2-3\sigma$) of the zenith angle distribution thus favoring the $\nu_\mu \to \nu_\tau$ interpretation. However, more data are needed to draw a definite conclusion (see [30]).

The novel element of this scheme (compared with the 3ν - schemes discussed in the previous sections) is the existence of a heavy state in the HDM range. Its contribution to m_{ee} equals:

$$m_{ee}^{(4)} = U_{e4}^2 m_4 \; . \tag{72}$$

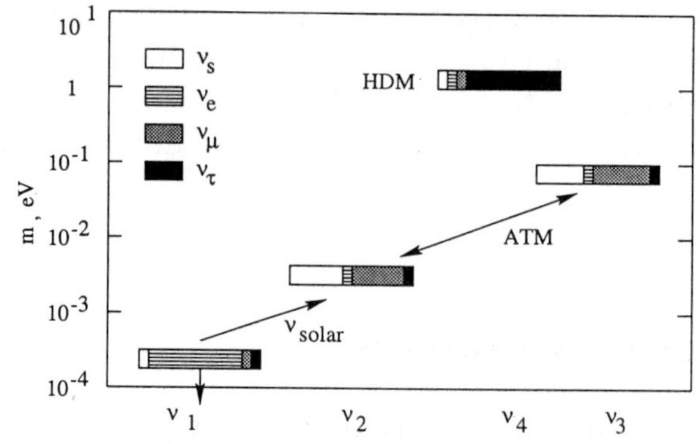

Figure 20: Neutrino masses and mixing in the 4 ν scenario with small flavor mixing and mass hierarchy. Here the white parts of the boxes correspond to admixtures of the sterile state.

The relevant parameters, U_{e4} and m_4, can be determined from studies of the short range $\nu_e \leftrightarrow \nu_e$ oscillations (disappearance) driven by the largest mass splitting $m_4 \approx \sqrt{\Delta m^2} \approx m_{HDM}$. For this channel the effective mixing angle equals

$$\sin^2 2\theta_{ee} = 4|U_{e4}|^2(1 - |U_{e4}|^2) \approx 4|U_{e4}|^2, \tag{73}$$

so that

$$m_{ee}^{(4)} \approx \frac{1}{4}\sqrt{\Delta m^2}\sin^2 2\theta_{ee}. \tag{74}$$

The corresponding iso-mass lines in the $\Delta m^2 - \sin^2 2\theta$ plot together with various oscillation bounds are shown in fig. 21.

In the cosmologically interesting range, $\sqrt{\Delta m^2} \approx m_{HDM} \simeq (0.5 - 5)$ eV, the mixing is constrained by the BUGEY experiment: $\sin^2 2\theta_{ee} = (2-4)\cdot 10^{-2}$. Therefore we get the upper bound

$$m_{ee}^{(4)} \simeq (2-5)\cdot 10^{-2}\text{eV} . \tag{75}$$

There is no strict relation between m_{ee} and the parameters of the $\nu_e \leftrightarrow \nu_\mu$ oscillations since both relevant mixing elements U_{e3} and $U_{\mu 3}$ are small. Indeed, now the effective depth of oscillations is determined by $\sin^2 2\theta_{e\mu} = 4|U_{e3}|^2|U_{\mu 3}|^2$, so that

$$m_{ee}^{(4)} = \frac{\sin^2 2\theta_{e\mu} m_{HDM}}{4|U_{\mu 4}|^2} . \tag{76}$$

It is impossible to infer useful information from this unless the $U_{\mu 4}$ will be determined from other experiments. Taking the bound $|U_{\mu 4}|^2 < 0.25$ from the 3ν - analysis of the atmospheric neutrino data, we get from eq. (76) the lower bound

$$m_{ee}^{(4)} > \sin^2 2\theta_{e\mu} m_{HDM} . \tag{77}$$

To get an estimation we assume $\sin^2 2\theta_{e\mu} \sim 10^{-3}$ which corresponds to upper bounds on the elements U_{e4} and $U_{\mu 4}$ from the BUGEY experiment and searches for $\nu_\mu \leftrightarrow \nu_\tau$

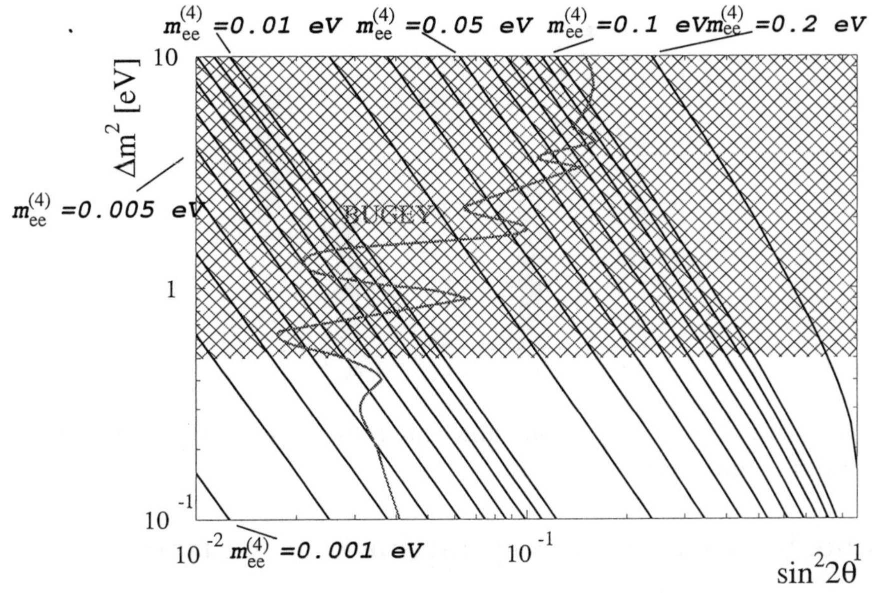

Figure 21: Iso-mass $|m_{ee}^{(4)}|$ lines in the four-neutrino scenario with small flavor mixing and mass hierarchy. The shadowed area shows the region for neutrino masses of cosmological interest as HDM. Also shown are the regions excluded by the reactor experiment BUGEY (from [49]).

oscillations. (Notice that LSND result can not be completely explained in this scheme.) This leads to

$$m_{ee}^{(4)} \gtrsim 10^{-3}\text{eV} . \qquad (78)$$

The contributions from the three light states are similar to the contributions in the 3ν single maximal mixing scheme with mass hierarchy (sect. 3.1). In particular, the largest contribution may come from the third mass eigenstate: $m_{ee}^{(3)} = \sqrt{\Delta m_{atm}^2} U_{e3}^2 < 2 \cdot 10^{-3}$ eV (see eq. (32)). The contributions from the two lightest states can be estimated as $m_{ee}^{(2)} = (5 \cdot 10^{-7} - 10^{-5})$ eV and $m_{ee}^{(1)} < 2 \cdot 10^{-3}$ eV.

Thus, the $0\nu\beta\beta$-mass (see fig. 22) can be dominated by the contribution of the heaviest state which can reach $m_{ee} \approx m_{ee}^{(4)} \sim 5 \cdot 10^{-2}$ eV.

The contribution depends strongly on the mixing angle $\sin^2 2\theta_{e\tau}$. Short baseline experiments (such as the rejected short baseline neutrino oscillation proposal TOSCA) could in principle test the region of large masses $m_{HDM} \simeq 10$ eV down to $\sin^2 2\theta_{e\tau} = 10^{-3}$, which correspond to an improvement of the upper bound on m_{ee} by 1 - 2 orders of magnitude. Due to possible cancellations between the contributions no lower bound on m_{ee} can be obtained from the present data.

Notice that the MINIBOONE experiment will probe the mixing angle $\sin^2 2\theta_{e\mu}$ down to $4 \cdot 10^{-4}$ eV and thus will check the LSND result. A confirmation of the LSND result will exclude this scheme.

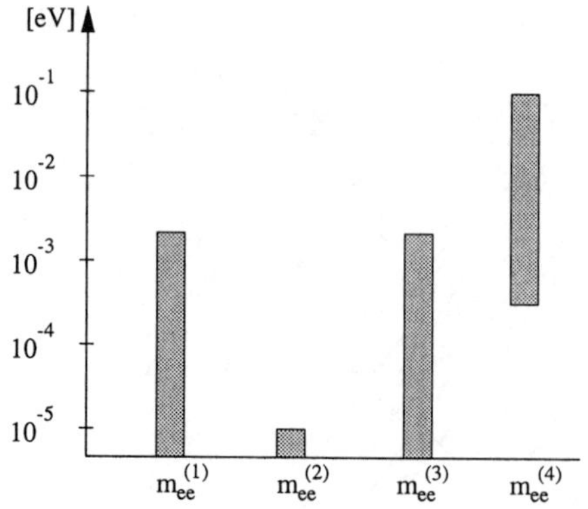

Figure 22: Contributions to $m_{\epsilon\epsilon}$ from different mass eigenstates in the 4ν scheme with small flavor mixing and mass hierarchy.

8.2 Scenario with two heavy degenerate neutrinos

The main motivation for this scenario (see fig. 23) is to explain the LSND result along with oscillation solutions of the solar and atmospheric neutrino problems [50, 51]. The masses are determined as

$$m_3 \approx m_4 \approx \sqrt{\Delta m^2_{LSND}}, \qquad m_2 \approx \sqrt{\Delta m^2_{\odot}}, \qquad m_1 \ll m_2 \; . \tag{79}$$

The neutrinos ν_μ and ν_τ are strongly mixed in the two heavy mass eigenstates ν_2 and ν_3, so that $\nu_\mu \leftrightarrow \nu_\tau$ oscillations solve the atmospheric neutrino problem. The two other neutrinos, ν_ϵ and ν_s, are weakly mixed in the two lightest mass states and the resonance $\nu_\epsilon \to \nu_s$ conversion solves the solar neutrino problem.

The two heavy neutrinos with masses $m_3 \approx m_4$ can be relevant for cosmology, their contribution to a hot dark matter component equals:

$$m_{HDM} = 2m_3 = 2\sqrt{\Delta m^2_{LSND}} \; . \tag{80}$$

In this scheme the new element is the existence of two heavy degenerate states. Let us consider in details their contribution to $m_{\epsilon\epsilon}$ (the effect of the two lightest states is small). Using relations (79) we can write this contribution as

$$m^{(3)}_{\epsilon\epsilon} + m^{(4)}_{\epsilon\epsilon} \approx (|U_{\epsilon 3}|^2 + |U_{\epsilon 4}|^2 e^{i\phi_{34}})\sqrt{\Delta m^2_{LSND}} \tag{81}$$

and $\phi_{34} \equiv \phi_4 - \phi_3$ is the relative phase of the ν_3 and ν_4 masses. Let us express the masses in eq. (81) in terms of oscillation parameters. In short base-line experiments the only oscillation phases which enter are the ones between heavy states and light states. One can neglect the oscillation phase between the two light states which is determined by $\Delta m^2_{\odot}$ and the phase between the two heavy states which is determined by Δm^2_{atm}. In this case the oscillations are reduced to two neutrino oscillations with a phase determined by Δm^2_{LSND} and the width for the $\nu_\epsilon \leftrightarrow \nu_\mu$ channel:

$$\sin^2 2\theta_{\epsilon\mu} = 4|U^*_{\epsilon 3}U_{\mu 3} + U^*_{\epsilon 4}U_{\mu 4}|^2 \; . \tag{82}$$

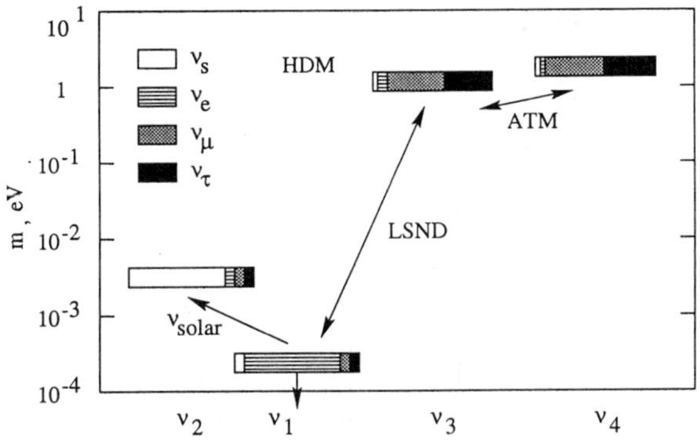

Figure 23: The pattern of the neutrino mass and mixing in the scheme with two degenerate neutrinos and one sterile component.

Let us consider two extreme situations: suppose an admixture of the ν_e flavor in one of the heavy states is much larger than in the other one, *e.g.* $|U_{e3}| \gg |U_{e4}|$, then $\sin^2 2\theta_{e\mu} = |U_{e3}|^2|U_{\mu3}|^2$, and therefore $|U_{e3}|^2 = \sin^2 2\theta_{e\mu}/|U_{\mu3}|^2$. In this case we get from eq. (81) $m_{ee}^{(3)} + m_{ee}^{(4)} \approx |U_{e3}|^2 m_3$, and consequently,

$$m_{ee}^{(3)} + m_{ee}^{(4)} \approx \frac{\sin^2 2\theta_{e\mu}}{|U_{\mu3}|^2}\sqrt{\Delta m_{LSND}^2}\ . \tag{83}$$

Since $|U_{\mu3}|^2 \sim 0.5$ is determined by atmospheric neutrino oscillations, taking $\sin^2 2\theta_{e\mu} = 2\cdot 10^{-3}$ and $\sqrt{\Delta m_{LSND}^2} = 1$ eV we find $m_{ee}^{(3)} + m_{ee}^{(4)} \approx 10^{-3}$ eV.

Let us now take $U_{e3}^* U_{\mu3} \approx U_{e4}^* U_{\mu4}$, then $\sin^2 2\theta_{e\mu} = 16|U_{e3}^* U_{\mu3}|$ and $m_{ee}^{(3)} + m_{ee}^{(4)} \approx \sin^2 2\theta_{e\mu}\sqrt{\Delta m_{LSND}^2}/2$, provided that the two contributions are in phase. This result is two times smaller than the result in the previous case.

For the $\nu_e \leftrightarrow \nu_e$ channel we find the depth of oscillations

$$\sin^2 2\theta_{ee} = 4U_+(1 - U_+) \approx 4U_+, \tag{84}$$

where $U_+ \equiv |U_{e3}|^2 + |U_{e4}|^2$. At the same time $m_{ee} \leq U_+\sqrt{\Delta m_{LSND}^2}/2$, so that

$$m_{ee}^{(3)} + m_{ee}^{(4)} \approx \frac{1}{4}\sin^2 2\theta_{ee}\sqrt{\Delta m_{LSND}^2}. \tag{85}$$

Using the BUGEY bound on $\sin^2 2\theta_{ee}$ we get $m_{ee}^{(3)} + m_{ee}^{(4)} \lesssim 10^{-2}$ eV.

Since cancellations may show up, no lower bound can be obtained.

The same combination of neutrino mixing matrix elements (84) determines the ν_e mode of oscillations in atmospheric neutrinos. It will lead to an overall suppression of the number of the e-like events.

The contributions from the light states are similar to those in the 3ν case (see sect. 3.1): $m_{ee}^{(2)} = (5\cdot 10^{-7} - 10^{-5})$ eV and $m_{ee}^{(1)} \ll 2\cdot 10^{-3}$ eV.

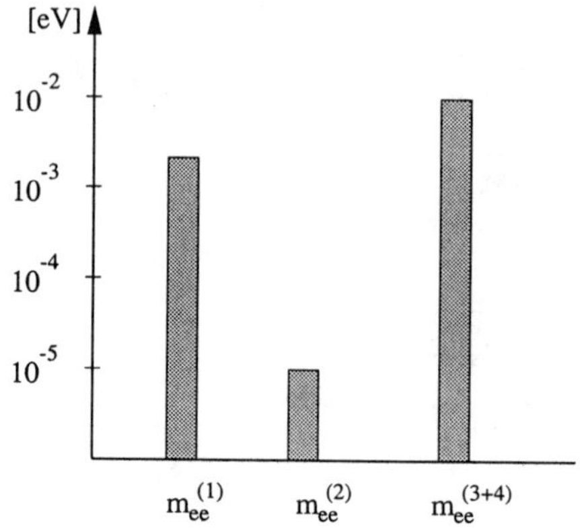

Figure 24: Contributions to m_{ee} from different mass eigenstates for the 4ν scheme with two degenerate pairs of neutrinos and normal mass hierarchy.

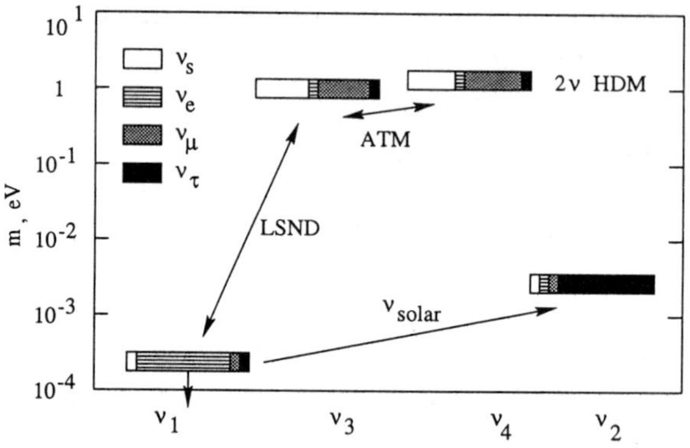

Figure 25: The neutrino masses and mixing in the "Grand Unification" scenario.

Summing up all the contributions we get that the $0\nu\beta\beta$-mass can be at most $(few) \times 10^{-2}$ eV being dominated by the contribution of the heavy states at the upper bound (fig. 24). A coincidence of a $0\nu\beta\beta$ decay signal in this range with a confirmation of the LSND oscillations by MINIBOONE can be considered as a hint for this scheme. At the same time, since cancellation between different contributions can show up, no lower bound on m_{ee} exists. Thus, a non-observation of $0\nu\beta\beta$ decay of the order of magnitude $(few) \cdot 10^{-2}$ eV does not rule out the scheme.

A similar situation appears in the "Grand Unification" scenario [52, 53] which is characterized by strong mixing of ν_μ and ν_s in the two heavy states and mixing of ν_e and ν_τ in the two light states (fig. 25). Here the atmospheric neutrino problem is solved by $\nu_\mu \leftrightarrow \nu_s$ oscillations whereas the solar neutrino data are explained by $\nu_e - \nu_\tau$ conversion.

8.3 Scenario with inverse mass hierarchy

The mass hierarchy in the two schemes with two pairs of states with small splitting can be inverse. In the first case, ν_e and ν_s flavors are concentrated in the two heavy states ν_3 and ν_4, whereas ν_μ and ν_τ are in the two light states. The dominating contribution comes from the third state which almost coincides with ν_e:

$$m_{ee} \approx m_{ee}^{(3)} \approx \sqrt{\Delta m^2_{LSND}} \approx 0.4 - 1\mathrm{eV}. \tag{86}$$

Thus in the context of this scheme the double beta decay searches check immediately the LSND result, and in fact, already existing data disfavor the scheme.

Another possibility of the inverse hierarchy is that the ν_e and ν_τ flavors are concentrated in the heavy states, whereas ν_μ and ν_s are in the pair of light mass states whose splitting leads to the atmospheric neutrino oscillations. The situation is similar to that for the 3ν scheme with inverse hierarchy (see sect. 7) with the only difference that Δm^2_{atm} should be substituted by Δm^2_{LSND}:

$$m_{ee}^{(3)} + m_{ee}^{(4)} \simeq \sqrt{\Delta m^2_{LSND}}(\sin^2\theta_\odot + e^{i\phi_{23}}\cos^2\theta_\odot)\ . \tag{87}$$

The third and the fourth mass eigenstates give the dominating contributions. Thus the expected interval for the total effective mass is

$$m_{ee} \approx \sqrt{\Delta m^2_{LSND}}(\cos 2\theta_\odot - 1). \tag{88}$$

This interval can be probed already by existing experiments, although for large mixing angle solutions of the solar neutrino problem (LMA, LOW, VO) strong cancellation can occur.

9 Discussion and Conclusions

In fig. 26 we summarize the predictions for m_{ee} in various schemes considered in this paper. We also show the present upper bound of $0\nu\beta\beta$ decay experiments [44] and regions of sensitivity which can be reached in future double beta decay experiments. Future double beta decay projects such as GENIUS [54, 13, 55], CUORE [56], MOON [57] will lead to a significant improvement of the sensitivity. The most ambitious and at the same time most realistic project, GENIUS, will test m_{ee} down to $2 \cdot 10^{-2}$ eV in the one ton version with one year of measurement time and down to $2 \cdot 10^{-3}$ eV in the 10 ton version with 10 years of measurement time.

According to figure 26 there are two key scales of m_{ee}, which will allow one to discriminate among various schemes: $m_{ee} \sim 0.1$ eV and $m_{ee} = 0.005$ eV.

1). If future searches will show that $m_{ee} > 0.1$ eV, then the schemes which will survive are those with neutrino mass degeneracy or 4ν schemes with inverse mass hierarchy. All other schemes will be excluded.

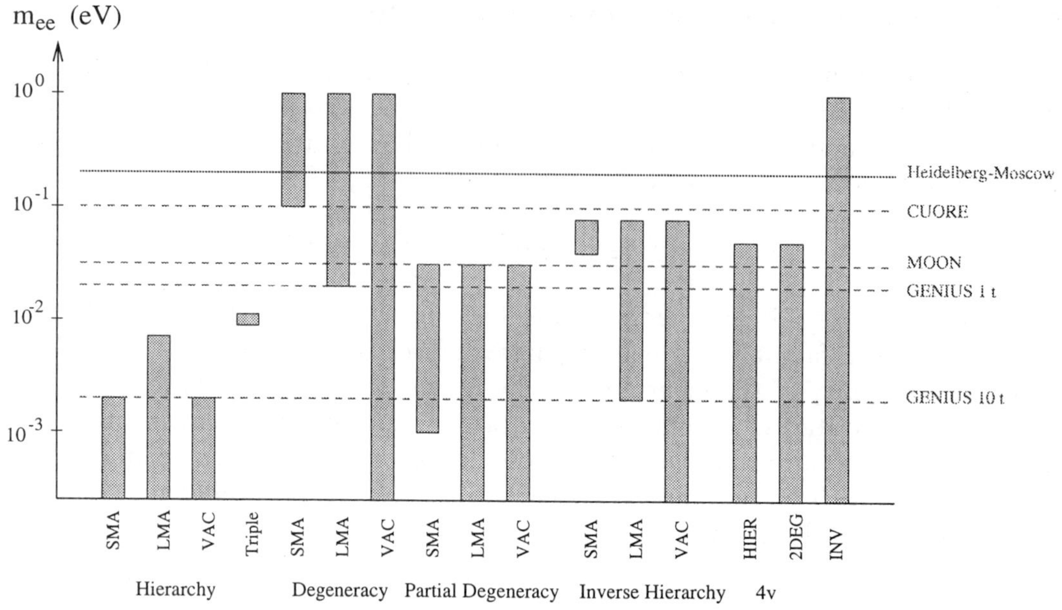

Figure 26: Summary of expected values for m_{ee} in the different schemes discussed in this paper. The expectations are compared with the recent neutrino mass limits obtained from the Heidelberg–Moscoscow [44] experiment as well as the expected sensitivities for the CUORE [56], MOON [57] proposals and the 1 ton and 10 ton proposal of GENIUS [13].

2). For masses in the interval $m_{ee} = 0.005 - 0.1$ eV, possible schemes include: 3ν schemes with partial degeneracy, triple maximal scheme, 3ν schemes with inverse mass hierarchy and 4ν scheme with one heavy (O (1 eV)) neutrino.

3). If $m_{ee} < 0.005$ eV, the schemes which survive are 3ν schemes with mass hierarchy, schemes with partial degeneracy, and the 4ν schemes with normal hierarchy. The schemes with degenerate spectrum and inverse mass hierarchy will be excluded, unless large mixing allows for strong cancellations. For $m_{ee} < 0.001$ eV this applies also for schemes with a partial degenerate spectrum.

Future oscillation experiments will significantly reduce the uncertainty in predictions for m_{ee} and therefore modify implications of $0\nu\beta\beta$ decay searches. Before a new generation of $0\nu\beta\beta$ decay experiments will start to operate we can expect that

- The solution of the solar neutrino problem will be identified. Moreover, $\Delta m^2_{\odot}$ and $\sin^2 2\theta_{\odot}$ will be determined with better accuracy. In particular, in the case of the solutions with large mixing (LMA, LOW, VO) the deviation of $1 - \sin^2 2\theta_{\odot}$ from zero can be established.

- The dominant channel of the atmospheric neutrino oscillation ($\nu_\mu - \nu_\tau$ or $\nu_\mu - \nu_s$) will be identified. The mass Δm^2_{atm} will be measured with better precision.

- A stronger bound on the element U_{e3} will be obtained or it will be measured if oscillations of electron neutrinos to tau neutrinos driven by Δm^2_{atm} will be discovered in the atmospheric neutrino or LBL experiments.
- The LSND result will be checked by MINIBOONE.

In the following we summarize possible consequences of these oscillation results.

1). Let us first comment on how the identification of the solution of the solar neutrino problem will modify implications of the $0\nu\beta\beta$ decay searches.

- If the SMA solution of the solar neutrino problem turns out to be realized in nature, a value of $m_{ee} > 0.2$ eV will imply a completely degenerate neutrino mass spectrum or schemes with inverse mass hierarchy. The measured value of m_{ee} will coincide with m_1 and will fix the absolute mass scale in the neutrino sector. A confirmation of this conclusion can be obtained from the CMB experiments MAP and Planck, if the degenerate neutrino mass is larger than $\simeq 0.4$ eV. However this value is already disfavored by the recent limits obtained from the Heidelberg-Moscow experiment.

  For lower values, $m_{ee} = 2 \cdot 10^{-3} - 10^{-2}$ eV, a scheme with partially degenerate spectrum will be favored. Again, we have $m_{ee} = m_1$ and the mass scale can be fixed.

  For even lower mass values: $m_{ee} < 2 \cdot 10^{-3}$ eV, or, after MINOS improved the bound on $m^{(3)}_{ee}$, $m_{ee} \lesssim 4 \cdot 10^{-4}$ eV, with the contribution $m^{(3)}_{ee}$ a new parameters enters, which for larger m_{ee} could be neglected. Thus it will be impossible to quantify the contribution of each single state to m_{ee}, unless $m^{(3)}_{ee}$ will be fixed in atmospheric or LBL oscillations.

- If the LMA solution of the solar neutrino deficit turns out to be realized in nature, a value $m_{ee} > 2 \cdot 10^{-2}$ eV will testify for a scenario with degenerate mass spectrum. A confirmation of this result will be obtained from the CMB experiments MAP and Planck, if the degenerate neutrino mass is larger than $\simeq 0.4$ eV. Using the mixing angle determined in solar neutrino experiments the range for the absolute mass scale can be determined from m_{ee} according to fig. 14. A value of $m_{ee} < 2 \cdot 10^{-2}$ eV will favor schemes with partial degeneracy or hierarchical spectrum. As soon as $m_{ee} < 2 \cdot 10^{-3}$ eV $m^{(3)}_{ee}$ becomes important and enters as a new parameter and it will be difficult to reconstruct the type of hierarchy.

- If the LOW or VO solution is the solution of the solar neutrino problem, the situation is similar to the MSW LMA case. The only difference is, that an observed $0\nu\beta\beta$-mass $m_{ee} > 2 \cdot 10^{-3}$ eV will imply a partially or completely degenerate scheme. Below this value the type of hierarchy can not be identified until bounds on $m^{(3)}_{ee}$ will be improved.

For schemes with inverse mass hierarchy the situation can be more definite:

- If the MSW SMA solution turns out to be true, a value of $m_{ee} = \sqrt{\Delta m^2_{atm}} = (5-8)\cdot 10^{-2}$ eV is expected. This value coincides with $m_1 \simeq m_2$ and therefore will give the absolute mass scale. For larger masses: $m_{ee} > 8\cdot 10^{-2}$ eV the transition to a completely degenerate spectrum occurs.

- If the MSW LMA, MSW LOW or vacuum oscillation solution is realized, a value of $m_{ee} = (0.02-8)\cdot 10^{-2}$ eV will testify for inverse mass hierarchy. The interval of expected values of m_{ee} can be narrower once the deviation of $1-\sin^2 2\theta$ from zero will be measured in solar neutrino experiments.

  For larger values of masses: $m_{ee} > 8\cdot 10^{-2}$ eV the scheme approaches the degenerate case.

2). The discovery of a sterile neutrino will have significant impact on the implications of the double beta decay searches.

The existence of a sterile neutrino can be established by a confirmation of the LSND result in MINIBOONE, or by a proof of the $\nu_e \to \nu_s$ oscillations solution of the solar neutrino problem by SNO, or by studies of the atmospheric neutrinos.

For 4 ν scenarios the interpretation of the $0\nu\beta\beta$ decay results is rather ambiguous. A value of $m_{ee} > (few)\times 10^{-2}$ eV will favor the intermediate mass scale scenario, while a value of $m_{ee} < 10^{-3}$ eV will favor a scenario with two degenerate pairs of neutrinos and normal mass hierarchy. A value of $m_{ee} > 10^{-1}$ eV will clearly favor an inverse mass hierarchy scheme. In all cases it will be difficult to disentangle the single contributions and to identify a specific spectrum. Important input in this case may come from the CMB experiments MAP and Planck by fixing the mass of the heaviest state.

3). U_{e3}: further searches for ν_e oscillations in atmospheric neutrinos, LBL and reactor experiments will allow one to measure or further restrict this mixing element. This, in turn, will be important for sharpening the predictions for m_{ee} especially in the schemes with strong mass hierarchy.

4). Matter effects and hierarchy: Studies of matter effects on neutrino oscillations will allow to establish the type of mass hierarchy, which in turn is of great importance for predictions of m_{ee}.

We can conclude from this summary, that in 3ν scenarios any measurement of $m_{ee} > 2\cdot 10^{-3}$ eV in $0\nu\beta\beta$ decay (corresponding to the final sensitivity of the 10 ton version of GENIUS) will provide informations about the character of hierarchy of the neutrino mass spectrum and in some cases also to fix the absolute mass scale of neutrinos. For values of $m_{ee} < 2\cdot 10^{-3}$ eV no reconstruction of the spectrum is possible until the contribution $m^{(3)}_{ee}$ will be fixed or bounded more stringent in atmospheric or LBL neutrino oscillations. For four-neutrino scenarios it will be not that easy to fix the mass scale of the neutrino sector. Crucial informations can be obtained from tests of the LSND signal and cosmology.

As has been mentioned before, a non-zero $0\nu\beta\beta$ decay rate always implies a non-vanishing neutrino Majorana mass [4]. Let us comment finally on possible ambiguities in the interpretation of a positive signal in neutrinoless double beta decay in terms of m_{ee}, in view of the existence of different alternative mechanisms, which could induce neutrinoless double beta decay, such as R-parity violating SUSY, right-handed currents, or leptoquarks. While no absolute unique method to identify the mechanism being responsible for neutrinoless double beta decay exists, the following remarks can be done:

1). Many of the possible alternative contributions require new particles, *e.g.* SUSY partners, leptoquarks, right-handed W bosons or neutrinos having masses in or below the TeV range, which to date not have been observed. Thus one expects to observe effects of new particles at future high energy colliders as the LHC or the NLC, giving independent informations on possible contributions to $0\nu\beta\beta$ decay (keeping in mind an uncertainty in nuclear matrix elements of about a factor of $\mathcal{O}(2)$). Notice that the same new interactions mentioned here may induce effects in neutrino oscillations and imply ambiguities in the interpretation of the data also there (see e.g. [58, 59]).

2). Using different source isotopes in different experiments and figuring out the values of $0\nu\beta\beta$ decay nuclear matrix elements for different contributions may help to identify the dominant one. Also a future experiment being sensitive to angular correlations of outgoing electrons could be useful in the discrimination of different contributions. Observing a positive signal in $0\nu\beta\beta$ decay should encourage new experimental efforts to confirm the results.

3). Last but not least and as discussed in this paper, a non-zero $0\nu\beta\beta$ decay signal can be related to some experimental results (both positive and negative) in neutrino oscillations and cosmology. A coincident and non-contradictory identification of a single neutrino mass scheme from the complementary results in such different experiments thus should be respected as a strong hint for this scheme.

In conclusion, after Super-Kamiokande has established large mixing in atmospheric neutrinos, the simplest neutrino spectrum with strong mass hierarchy and small flavor mixing (which typically predicts an undetectable m_{ee}) is excluded. Now the neutrino mass spectrum can exhibit any surprise: it can have a normal or inverse mass hierarchy, be partially or completely degenerate. More than three mass eigenstates can be involved in the mixing. In view of this more complicated situation a detection of a positive signal in future double beta decay searches seems to be rather plausible. We have shown that for a given oscillation pattern any value of m_{ee} is possible, which is still not excluded by the experimental bounds on the neutrinoless double beta decay half life limit. (A lower bound on m_{ee} appears in the case of inverse mass hierarchy.) This means that even after all oscillations parameters will be measured no unique prediction of m_{ee} can be derived. On the other hand this means that double beta decay searches provides informations being independent on informations obtained from oscillation experiments. Combining the results of double beta decay and oscillation searches offers a unique possibility to shed some light on the absolute scale of the neutrino mass, the type of hierarchy and the level of degeneracy of the spectrum. If we want eventually to reconstruct the neutrino mass and flavor spectrum, further searches for neutrinoless double beta decay with increased sensitivity seem to be unavoidable.

Note added: When this paper has been prepared for submission the papers of ref. [60] appeared which discuss similar topics.

References

[1] M. Doi, T. Kotani, E. Takasugi, Progr. Theor. Phys. Suppl. 83 (1985) 1;

[2] T. Tomoda, Rep. Progr. Phys. 54 (1991) 53

[3] H.V. Klapdor–Kleingrothaus, H. Päs, hep-ph/0002109, talk given at COSMO99, Trieste/Italy; H.V. Klapdor–Kleingrothaus, hep-ex/9901021, Proc. Int. Conf. on Lepton- and Baryon Number Non-Conservation, Trento, Italy, April 20-25 1998, IOP Bristol (1999) ed. H.V. Klapdor-Kleingrothaus and I. Krivosheina, p.251

[4] J. Schechter, J.W.F. Valle, Phys. Rev. D 25 (1982) 2951

[5] D. London, hep-ph/9907419, talk given at Beyond the Desert 99, Castle Ringberg, Tegernsee, Germany, June 1999, IOP Bristol 2000, ed. H.V. Klapdor-Kleingrothaus and I. Krivosheina

[6] V. Lobashev et al. Phys. Lett B. 460 (1999) 227; Ch. Weinheimer et al., Phys. Lett B. 460 (1999) 219

[7] A. A. Klimenko, A. A. Pomansky, A. A. Smolnikov, Proceedings of the Moriond-1985 Workshop on Perspectives in Electroweak Interactions. v. 2 (1985) 303.

[8] S.T. Petcov, A.Y. Smirnov, Phys. Lett. B. 322 (1994) 109

[9] S. Bilenky, A. Bottino, C. Giunti, C. Kim, Phys. Rev. D54 (1996) 1881-1890

[10] S.M. Bilenki, C. Giunti, C.W. Kim, M. Monteno, Phys. Rev. D57 (1998) 6981-6988; hep-ph/9904328

[11] C. Giunti, Phys. Rev. D 61 (2000) 036002 (hep-ph/9906275); S. Bilenky, C. Giunti, W. Grimus, hep-ph/9809368 Neutrino '98, Nucl. Phys. B 77 (Proc. Suppl.) (1999) ed. Y. Suzuki and Y. Totsuka; S. Bilenky, C. Giunti, hep-ph/9904328 Proc. 'WIN99', Cape Town, south Africa, 24-30 January (1999)

[12] S.M. Bilenky, C. Giunti, W. Grimus, B. Kayser, S.T. Petcov, hep-ph/9907234, Phys.Lett. B465 (1999) 193-202

[13] J. Hellmig, H.V. Klapdor–Kleingrothaus, Z. Phys. A 359 (1997) 351; H.V. Klapdor–Kleingrothaus, M. Hirsch, Z. Phys. A 359 (1997) 361; H.V. Klapdor–Kleingrothaus, J. Hellmig, M. Hirsch, J. Phys. G 24 (1998) 483;

[14] S. Bilenky, C. Giunti, C. Kim, S. Petcov, Phys. Rev. D54 (1996) 4432-4444

[15] R. Adhikari, G. Rajasekaran, hep-ph/9812361, Phys. Rev. D 61 (2000) 031301

[16] H. Minakata, O. Yasuda, Phys. Rev. D 56 (1997) 1692-1697

[17] H. Minakata, O. Yasuda, Nucl. Phys. B 523 (1998) 597-610

[18] F. Vissani, hep-ph/9708483

[19] F. Vissani, hep-ph/9904349, Proc. of the "6th Tropical Seminar on Neutrino and Astroparticle Physics", May 17-21 1999, San Miniato, Italy

[20] V. Barger, K. Whisnant, Phys. Lett. B 456 (1999) 194-200

[21] J. Ellis, S. Lola, hep-ph/9904279 Phys. Lett. B 458 (1999) 310-321

[22] G.C. Branco, M.N. Rebelo, J.I. Silva-Marcos, Phys. Rev. Lett. 82 (1999) 683-686

[23] H. Georgi, S. L. Glashow, hep-ph/9808293

[24] T. Fukuyama, K. Matsuda, H. Nishiura hep-ph/9708397

[25] T. Fukuyama, K. Matsuda, H. Nishiura, hep-ph/9708397; Mod. Phys. Lett. A13 (1998) 2279

[26] T. Fukuyama, K. Matsuda, H. Nishiura; Phys. Rev. D57 (1998) 5844

[27] F. Vissani, JHEP 9906 (1999) 022, hep-ph/9906525

[28] M. Czakon, J. Gluza, M. Zralek, hep-ph/9906381 Phys. Lett. B 465 (1999) 211-218

[29] T. Kajita, talk given at Beyond the Desert 99, Castle Ringberg, Tegernsee, Germany, June 1999, IOP Bristol 2000, ed. H.V. Klapdor-Kleingrothaus and I. Krivosheina

[30] N. Fornengo, M.C. Gonzalez-Garcia, J.W.F. Valle, hep-ph/0002147,

[31] J. N. Bahcall, P. I. Krastev, A. Yu. Smirnov, to be published

[32] M.C. Gonzalez-Garcia, P.C. de Holanda, C. Pena-Garay, J.W.F. Valle, hep-ph/9906469

[33] C. Athanassopoulos et al., Phys. Rev. Lett. 81 (1998) 1774-1777 K. Eitel, hep-ex/9909036;

[34] T.E. Jannakos, hep-ex/9908043

[35] A.O. Bazarko, hep-ex/9906003

[36] J. Lesgourgues, S. Pastor, S. Prunet, hep-ph/9912363

[37] A.S. Dighe, A.Yu. Smirnov, hep-ph/9907423

[38] M. Apollonio et al. (CHOOZ collab.), hep-ex/9907037, Phys. Lett. B 466 (1999) 415-430

[39] K. Zuber, hep-ex/9810022

[40] MINOS detectors technical design report, version 1.0, October 1998, *http : //www.hep.anl.gov/ndk/hypertext/minos$_t$dr.html*; we thank Kai Zuber for bringing this reference to our attention

[41] E.Kh. Akhmedov, A. Dighe, P. Lipari, A.Yu. Smirnov, Nucl. Phys. B 542 (1999) 3

[42] J. N. Bahcall, P. I. Krastev, A. Yu. Smirnov, Phys. Rev. D 60 (1999) 093001

[43] P.F. Harrison, D.H. Perkins, W.G. Scott, Phys. Lett. B 196 (1997) 186

[44] Heidelberg–Moscow Collab., Phys. Rev. Lett 83 (1999) 41

[45] E. Gawiser, J. Silk, Science 280 (1998) 1405-1411

[46] J.R. Primack, M.A.K. Gross, astro-ph/9810204, Proc. of the Xth Recontres de Blois "The birth of galaxies", June 28 - July 4 1998

[47] D.J. Eisenstein, W. Hu, M. Tegmark astro-ph/9807130, Submitted to ApJ.

[48] J.N. Bahcall, P.I. Krastev, A.Yu. Smirnov, Phys. Rev. D 58 (1998) 096016

[49] L. di Lella, hep-ex/9912010

[50] J.T. Peltoniemi, J.W.F. Valle, Nucl. Phys. B 406 (1993) 409

[51] D.O. Caldwell, R.N. Mohapatra, Phys. Rev. D 48(1993) 3259

[52] J.T. Peltoniemi, D. Tommasini, J.W.F. Valle, Phys.Lett.B298:383-390,1993

[53] A.S. Joshipura, A.Yu. Smirnov, Phys.Lett.B439:103-111,1998

[54] H.V. Klapdor–Kleingrothaus, in *Beyond the Desert '97 – Accelerator and Non-Accelerator Approaches* (Eds. H.V. Klapdor–Kleingrothaus, H. Päs), Proc. Int. Workshop on Particle Physics beyond the Standard Model, Castle Ringberg, June 8-14, 1997, IOP Publ., Bristol, Philadelphia, p. 485 and Int. J. Mod. Phys. A 13 (1998) 3953

[55] H. V. Klapdor-Kleingrothaus, L. Baudis, G. Heusser, B. Majorovits, H. Päs, hep-ph/9910205

[56] E. Fiorini et al., Phys. Rep 307 (1998) 309

[57] H. Eijiri et al., nucl-ex/9911008

[58] S. Bergmann, Nucl.Phys. B515 (1998) 363-383

[59] N. Fornengo, M. C. Gonzalez-Garcia, J. W. F. Valle, hep-ph/9906539

[60] K.Matsuda, N.Takeda, T.Fukuyama, H.Nishiura, hep-ph/0003055; W. Rodejohann, hep-ph/0003149 M. Czakon, J. Gluza, M. Zralek, hep-ph/0003161.

Preprint typeset in JHEP style. - HYPER VERSION

Signal of neutrinoless double beta decay, neutrino spectrum and oscillation scenarios

Francesco Vissani

Deutsches Elektronen-Synchrotron, DESY
Notkestraße 85, D-22603 Hamburg, Germany, and
International Centre for Theoretical Physics, ICTP
Strada Costiera 11, 34100 Trieste, Italy
E-mail: vissani@ictp.trieste.it

ABSTRACT: The lower and upper bounds on the neutrinoless double beta ($0\nu2\beta$) decay rate are obtained, as functions of the parameters of neutrino oscillations and of the lightest neutrino mass. The constraints on these parameters from the search for the $0\nu2\beta$ transition, as well as from the interpretation of solar and atmospheric neutrino data in terms of oscillations, can be conveniently represented in one unitarity triangle. This representation helps to clarify the cases when the $0\nu2\beta$ rate is small; the crucial dependence on the scenarios assumed for solar neutrino oscillations and on the neutrino spectrum is emphasized. We consider hierarchical and non-hierarchical neutrino spectra, and discuss their interest in view of future searches of the $0\nu2\beta$ decay.

KEYWORDS: Neutrino Physics, Solar and Atmospheric Neutrinos.

Contents

1. Informations on neutrino parameters

1.1 Massive neutrinos and $0\nu2\beta$ decay

Atmospheric neutrino data can be interpreted in terms of a dominant $\nu_\mu - \nu_\tau$ oscillation channel, although a sub-dominant channel $\nu_\mu - \nu_e$ is not excluded [1]. The latter may be due to a ν_e component of the heaviest (lightest) neutrino state ν_3 (ν_1) for spectra with "normal" ("inverted") hierarchy–our definition of "hierarchy" is discussed in section 3. Several possibilities are open for the interpretation of the solar neutrino data, depending on the frequencies of oscillation and mixings.

Hence, the indications for massive neutrinos are strong. However, there is quite a limited knowledge on the *neutrino mass spectrum itself*, and particularly on the lightest neutrino mass. The search for $0\nu2\beta$ decay can shed light on this important issue. The bound of 0.2 eV obtained [2] on the parameter

$$\mathcal{M}_{ee} = |\sum_i U^2_{ei}\, m_i| \tag{1.1}$$

is sensibly smaller than the mass scales probed by present studies of β-decay, or those inferred in cosmology [3]. In eq. (1.1), the non-negative quantities m_i, $i = 1, 2, 3...N$ are the neutrino masses ($m_{i+1} \geq m_i$); the complex quantities $U_{\ell i}$, $\ell = \mathrm{e}, \mu, \tau...$, are the elements of the mixing matrix, which relates the flavor eigenstates to the mass eigenstates: $\nu_\ell(x) = \sum_i U_{\ell i}\, \nu_i(x)$. Hence, $\mathcal{M}_{\mathrm{ee}}$ can be thought of as (the absolute value of) the ee–entry of the neutrino mass matrix. Let us recall that, beside the $(N-1)(N-2)/2$ phases relevant to neutrino oscillations, there are still $N-1$ physical phases in the lepton sector that have no analogy in the quark sector, and arise from the Majorana structure of the neutrino mass matrix. Notice that both the amplitudes *and* the phases of the elements of the mixing matrix $U_{\mathrm{e}i}$ are relevant in determining the size of $\mathcal{M}_{\mathrm{ee}}$.

1.2 Extremal values of $\mathcal{M}_{\mathrm{ee}}$ for $0\nu 2\beta$ decay

We obtain in this section the extremal values of $\mathcal{M}_{\mathrm{ee}}$ under arbitrary variations of the phases, keeping fixed the neutrino masses m_i and the "mixing elements"[1] $|U_{\mathrm{e}i}^2|$. The maximum value of $\mathcal{M}_{\mathrm{ee}}$ is simply:

$$\mathcal{M}_{\mathrm{ee}}^{max} = \sum_i |U_{\mathrm{e}i}^2|\; m_i. \tag{1.2}$$

The minimum value can be written as:

$$\mathcal{M}_{\mathrm{ee}}^{min} = \max\{\ 2\ |U_{\mathrm{e}i}^2|\ m_i - \mathcal{M}_{\mathrm{ee}}^{max},\ \ 0\ \}. \tag{1.3}$$

To demonstrate this formula, let us consider the absolute value of the sum of three complex numbers: $r = |z_1 + z_2 + z_3|$. We want to minimize r by keeping fixed $|z_i|$, namely, by varying the phases. Let us define the quantities $r_{1,2,3}$ and $q_{1,2,3}$ as: $r_1 = |z_1| - |z_2| - |z_3|$, $q_1 = |z_1| - |z_2 + z_3|$, and similar eqs., but permuting the indices for $r_{2,3}$ and $q_{2,3}$. Notice that *at most* one of the r_i's is positive. Assuming that $r_1 > 0$, it is simple to show that $r^{min} = r_1$; in fact, using twice the Schwartz inequality, we get $r \geq |q_1| = q_1 \geq r_1$. Similar considerations if $r_2 > 0$, or $r_3 > 0$. The last case has $r_i \leq 0$ for $i = 1, 2, 3$. If one of the r_i's is zero, then $r^{min} = 0$, hence we need to consider the case when $r_i < 0$ for all i's. In this case, the quantity q_1 goes from negative, when the phases of z_2 and z_3 are equal, to positive, when these phases are opposite. By continuity, a phase choice exists such that $q_1 = 0$. Since by proper choice of the phase of z_1 we can get $r = |q_1|$ we conclude that, again, $r^{min} = 0$. In conclusion, the general case is covered by the formula: $r^{min} = \max\{r_i,\ 0\}$. This is equivalent to eq. (1.3), after noticing that $r_i = 2|z_i| - \sum_{i=1}^{3} |z_i|$. The generalization of these results to N neutrinos is quite simple: Just limit the sum in eq. (1.2) to $N = 3$. However, we will be concerned only with the case of three neutrinos in the rest of the work.

[1] In the following, we will always refer with the term "mixing elements" to the absolute value of the elements of the mixing matrix.

The previous two equations give the extremal values of $\mathcal{M}_{ee}$, once the neutrino spectrum *and* the mixing elements are known. Such extremal values are important, being independent of the complex phases. The information we get from the experimental upper bound is $\mathcal{M}_{ee}^{bound} \geq \mathcal{M}_{ee}^{min}$; the informations we could get from a positive signal, instead, is $\mathcal{M}_{ee}^{signal} \in [\mathcal{M}_{ee}^{min}, \mathcal{M}_{ee}^{max}]$. In the following it will be shown how to use and represent $\mathcal{M}_{ee}^{min}$ and $\mathcal{M}_{ee}^{max}$, and what we can learn on them assuming specific neutrino spectra, and scenarios of neutrino oscillations.

2. Representation of $\mathcal{M}_{ee}^{min}$ and $\mathcal{M}_{ee}^{max}$

We introduce and discuss in this section a graphical representation of the values of $\mathcal{M}_{ee}^{min}$ and $\mathcal{M}_{ee}^{max}$. For this purpose we will make reference to fig. 1, where the representation of $\mathcal{M}_{ee}^{min}$ is displayed, for an illustrative choice of the neutrino spectrum: $m_3 = 2\,m_2$ and $m_2 = 2\,m_1$. In order to fix the ideas, we point out from the beginning the two essential features of fig. 1: (1) the value of $\mathcal{M}_{ee}$ at the vertices, namely the masses of the neutrinos m_i; (2) the position of the inner triangle (also determined by the masses of the neutrinos).

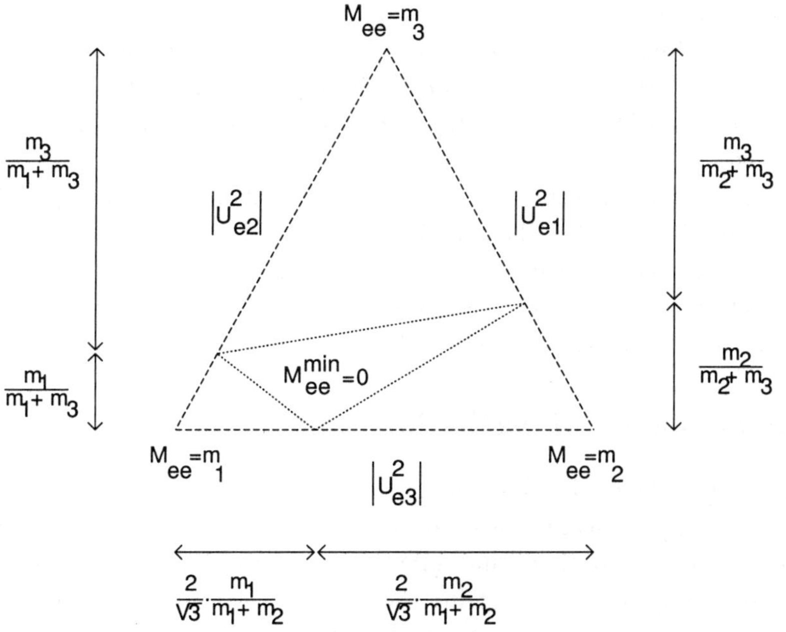

Figure 1: Representation of the minimum value of $\mathcal{M}_{ee}$, eq. (1.3), in the unitarity triangle. For a given internal point, the distance from the side labelled by $|U_{ei}^2|$ represents the size of the corresponding mixing element. The inner triangle encloses the region where $\mathcal{M}_{ee}^{min} = 0$. The position of the vertices of the inner triangle relative to the vertices of the unitarity triangle (corresponding to eq. (2.1)) is indicated by the arrows with labels.

Let us begin by recalling some basic facts. The three mixing elements $|U_{ei}^2|$ are constrained by the unitarity condition $\sum_i |U_{ei}^2| = 1$. This condition can be represented by using the inner region of one equilateral triangle with unit height, where the distance from the ith side represents the value of $|U_{ei}^2|$, see fig. 1 (this triangle was first used in [4], to analyze solar neutrino oscillations). To exemplify the use of the triangle, let us consider two special cases: (a) When ν_e is an equal admixture of the three mass eigenstates, we have $|U_{ei}^2| = 1/3$. This point is represented by the barycentre of the equilateral triangle of fig. 1. (b) When ν_e coincides with the mass eigenstate ν_1, we have $|U_{e1}^2| = 1$, and the other two mixing elements are zero. This point is represented by the 1st-vertex (by definition, the 1st vertex is opposite to the 1st side, denoted with the label $|U_{e1}^2|$ in fig. 1, *etc.*).

From eq. (1.3), $\mathcal{M}_{ee}^{min}$ is *zero* in the inner triangular region represented in fig. 1. The vertices of this inner triangle are given by:

$$|U_{e1}^2|/|U_{e2}^2| = m_2/m_1 \text{ when } |U_{e3}^2| = 0, \tag{2.1}$$

and by the two additional equations obtained by the replacement $3 \leftrightarrow 1$, and $3 \leftrightarrow 2$. The condition $|U_{e3}^2| = 0$ in eq. (2.1) tells us that we are on the 3rd (lower) side of the unitarity triangle of fig. 1.

At the ith vertex of the unitarity triangle $\mathcal{M}_{ee}^{min} = \mathcal{M}_{ee} = m_i$, as is clear from eq. (1.3), and as illustrated in fig. 1. The value of $\mathcal{M}_{ee}^{min}$ decreases linearly when moving from one vertex toward the inner triangle. In fact, $\mathcal{M}_{ee}^{min}$ is non-zero only close to the vertices of the unitarity triangle (assuming $m_1 > 0$). This concludes the illustration of fig. 1.

The unitarity triangle can also be used to represent the maximum possible value $\mathcal{M}_{ee}$. Quite simply, $\mathcal{M}_{ee}^{max}$ is the function of the mixing elements $|U_{ei}^2|$ that interpolates linearly among the values $\mathcal{M}_{ee} = m_i$ taken at the vertices of the unitarity triangle, as clear from eq. (1.2). However, since $\mathcal{M}_{ee}^{max}$ is just the sum of positive contributions (eq. (1.2)), the analysis of $\mathcal{M}_{ee}^{max}$ is nearly trivial.

3. Phenomenology of oscillations and $0\nu2\beta$

We discuss now the $0\nu2\beta$ signal assuming some specific spectra, and scenarios of oscillation, using the graphical representation introduced above. We take advantage of the indications from atmospheric and solar neutrinos, that can be accounted in terms of two different frequencies of neutrino oscillations, related to the mass differences squared Δm^2_{atm} and $\Delta m^2_{\odot}$ ($\Delta m^2_{atm} \gg \Delta m^2_{\odot}$). We consider the following three cases:

- Case $[\mathcal{N}]$: "normal" hierarchy, $m_1 \ll (\Delta m^2_{atm})^{1/2}$;
- Case $[\mathcal{I}]$: "inverted" hierarchy, $m_1 \ll (\Delta m^2_{atm})^{1/2}$;
- Case $[\mathcal{D}]$: "normal" and "inverted" hierarchies, $m_1 \gg (\Delta m^2_{atm})^{1/2}$;

from these cases, it will be easy to understand also the "intermediate" situations when $m_1 \sim (\Delta m^2_{atm})^{1/2}$. With the term "hierarchy" (either "normal" or "inverted") we refer to *the mass differences squared* (see eqs. (3.1) and (3.4) below)[2]. We assume that the electronic admixture in atmospheric neutrinos is sub-dominant [1], and use for the mass splittings Δm^2_{atm} and $\Delta m^2_{\odot}$ the values suggested by the phenomenology. For solar neutrino solutions we use the terminology of [5], that we will recall in the following. A similar study has been performed in reference [6], with the goal to extract informations on the mixing angles, knowing $\mathcal{M}_{ee}$ and the neutrino spectrum. For other recent works oriented toward the phenomenology, see [7].

3.1 Case [$\mathcal{N}$]: "normal" hierarchy, $m_1 \ll (\Delta m^2_{atm})^{1/2}$

What is the expected value of $\mathcal{M}_{ee}$ for a neutrino spectrum with "normal" hierarchy:

$$m_3^2 - m_2^2 = \Delta m^2_{atm} \gg m_2^2 - m_1^2 = \Delta m^2_{\odot}, \tag{3.1}$$

assuming, to begin with, that m_1 is negligible? For the values of $\Delta m^2_{\odot}$ suggested by the MSW [8] small mixing angle solution of the solar neutrino problem (SMA) or vacuum oscillation (VO), the only important contribution to $0\nu2\beta$ decay rate comes from the heaviest eigenstate: $\mathcal{M}_{ee} \approx |U^2_{e3}| m_3$. It is *possible* to have a comparable contribution from the second eigenstate assuming MSW solutions of the solar neutrino problem with large mixing angle (LMA) $\delta\mathcal{M}_{ee}|_{\odot} = |U^2_{e2}|\ m_2 \approx 4 \times 10^{-3}$ eV (using $\Delta m^2_{\odot} \approx 10^{-4}$ eV^2 and $|U^2_{e2}| \approx 0.4$). This is of the same size of the contribution from the heaviest eigenstate, $\delta\mathcal{M}_{ee}|_{atm} = |U^2_{e3}|\ m_3$, if $|U^2_{e3}| \approx 0.1$ and $\Delta m^2_{atm} \approx 2 \times 10^{-3}$ eV^2. We conclude that, if future experiments searching for the $0\nu2\beta$ transition will prove that

$$\mathcal{M}_{ee} > 10^{-2} \text{ eV}, \tag{3.2}$$

the hypothesis of a spectrum with "normal" hierarchy and very small m_1 will be disfavoured [9][3].

The function $\mathcal{M}^{min}_{ee}$ is represented in fig. 2 for two different values of $\Delta m^2_{\odot}$: 10^{-4} eV^2 in the 1st plot, and 10^{-5} eV^2 in the 2nd (we assumed $\Delta m^2_{atm} = 2 \times 10^{-3}$ eV^2). Notice that assuming $m_1 = 0$ the inner triangle of fig. 1 degenerates into a line (for much smaller values of $\Delta m^2_{\odot}$, say for VO, the line practically coincides with the side $U_{e3} = 0$). Recalling that the inner triangle corresponds to the region where $\mathcal{M}^{min}_{ee} = 0$, we appreciate from fig. 2 the crucial dependence on the parameter $|U^2_{e3}|$ of the $0\nu2\beta$ transition rate.

---

[2]In order to simplify the connection with the phenomenology, we use a definition of "hierarchy" that is relevant to neutrino oscillations, which involves *just* the mass differences squared. Notice that sometimes in the literature, "hierarchy" is used in reference to the neutrino spectrum itself.

[3]Alternatively, one should postulate a different origin of the $0\nu2\beta$ decay.

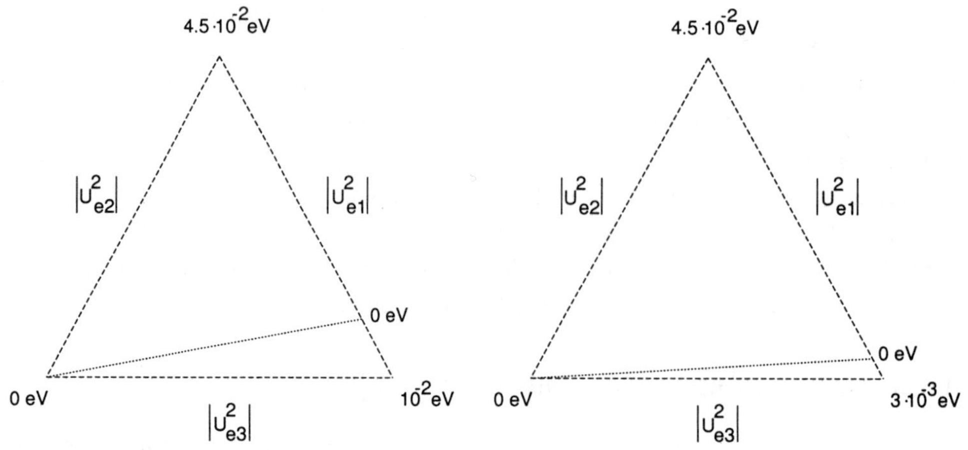

Figure 2: Same as fig. 1, for spectra with $m_1 = 0$ and "normal" hierarchy. For "inverted" hierarchy, the value at the 2nd vertex increases from the value 10^{-2} in the first plot (3×10^{-3} in the second plot) up to $\approx 4.5 \times 10^{-2}$ eV, and the internal line approaches closely the bisector $|U_{e2}^2| = |U_{e3}^2|$.

Let us increase now the size of m_1, keeping $m_1 \ll (\Delta m^2_{atm})^{1/2} \approx m_3$. The (degenerate) inner triangle in fig. 2 becomes an *obtuse* isosceles triangle when $m_1 \approx m_2 \gtrsim (\Delta m^2_{\odot})^{1/2}$; the base being parallel to the 3rd side, where $U_{e3} = 0$. A complete suppression of the $0\nu2\beta$ transition can take place for those solutions of the solar neutrino problem that fall in this inner triangle, and for this reason, the most important conclusion is unchanged: The size of $|U_{e3}^2|$ is very important in determining whether the case $\mathcal{M}_{ee}^{min} = 0$ is possible or not. More precisely, this mixing element has to be compared with the height of the triangle, $\sim m_1/m_3$ (see fig. 1). Incidentally, we notice the simple formula

$$\mathcal{M}_{ee} \approx |m_1 + |U_{e3}^2|\, m_3\, e^{i\varphi}| \quad \text{where } \varphi = \arg[\, U_{e3}^2/U_{e1}^2\,] \tag{3.3}$$

valid for the SMA case, which illustrates that $\mathcal{M}_{ee} \approx 0$ is possible when $m_1/m_3 \approx |U_{e3}^2|$ ($m_3 \approx (\Delta m^2_{atm})^{1/2}$ in present hypotheses) and the phases of U_{e3}^2 and U_{e1}^2 are opposite.

3.2 Case [$\mathcal{I}$]: "inverted" hierarchy, $m_1 \ll (\Delta m^2_{atm})^{1/2}$

Let us assume a spectrum with "inverted" hierarchy, namely

$$m_3^2 - m_2^2 = \Delta m^2_{\odot} \ll m_2^2 - m_1^2 = \Delta m^2_{atm}, \tag{3.4}$$

and suppose, to begin with, that m_1 is negligible. In this case, since the sub-dominant mixing element is $|U_{e1}^2|$, we can obtain large maximum values [12]:

$$\mathcal{M}_{ee}^{max} \approx (\Delta m^2_{atm})^{1/2} = (3 \text{ to } 9) \times 10^{-2} \text{ eV}. \tag{3.5}$$

This could be close to the present bound [2], if also the nuclear matrix elements take the highest values allowed by present uncertainties, $\sim 2-3$ [11].

In these hypotheses, $\mathcal{M}_{ee}^{min}$ can be (close to) zero only if $|U_{e2}^2|$ is very close to $|U_{e3}^2|$; the contribution from $|U_{e1}^2|$ being irrelevant. In a graphical representation like in fig. 2, this corresponds to the fact that the inner triangle almost coincides with the bisector $|U_{e2}^2| = |U_{e3}^2|$ (the "small" mixing element $|U_{e1}^2|$ is represented by the distance from the 1st–right–side).

Let us increase the size of m_1, keeping $m_1 \ll (\Delta m_{atm}^2)^{1/2} \approx m_3$. The inner triangle is, in this assumption, *acute* isosceles, the base being parallel to the side $U_{e1} = 0$, and with length $\sim m_1/m_3 \times 2/\sqrt{3}$. Hence, only those solutions of the solar neutrino problem which have almost maximal mixing angles (VO, averaged oscillations and perhaps LMA) fall in the region where the $0\nu 2\beta$ transition rate may be strongly suppressed. In the case of SMA, since $|U_{e3}^2|$ is small by assumption (and $|U_{e1}^2|$ is not large) we have simply:

$$\mathcal{M}_{ee} \approx m_2 \approx (m_1^2 + \Delta m_{atm}^2)^{1/2}. \tag{3.6}$$

Hence, $\mathcal{M}_{ee} \approx 0$ is impossible if the SMA solution is correct. Quite generally, in the case of "inverted" hierarchy, it is less likely that $\mathcal{M}_{ee}^{min}$ is zero.

3.3 Case [$\mathcal{D}$]: "nearly degenerate" spectrum, $m_1 \gg (\Delta m_{atm}^2)^{1/2}$

Largest values of $\mathcal{M}_{ee}$ (up to the experimental bound) can be taken for a "nearly degenerate" neutrino spectrum [10, 13]. The maximum value is simply $\mathcal{M}_{ee}^{max} = m_1 + \mathcal{O}(\Delta m^2/m_1)$, m_1 playing the role of mass spectrum offset.

The corresponding minimum value, $\mathcal{M}_{ee}^{min}/m_1 = \max\{2|U_{ei}^2| - 1, 0\}$ is represented in fig. 3 assuming "normal" hierarchy of the mass differences (eq. (3.1)); $\mathcal{O}(\Delta m^2/m_1^2)$ terms have been neglected. From this figure it is visible that, to interpret properly the results of $0\nu 2\beta$ decay studies (and possibly, to exclude the inner region in the 1st plot, the one where $\mathcal{M}_{ee} \ll m_1$ is *possible*) we need precise information on the mixing elements. This requires distinguishing among oscillation scenarios. The plots also illustrate the importance to quantify the size of $|U_{e3}^2|$ [10, 13], [1]. Similar considerations apply when the mass differences have "inverted" hierarchy, eq. (3.4) with $|U_{e1}^2|$ playing the role of $|U_{e3}^2|$. Notice in particular that with approximate mass degeneracy the role of the sub-dominant mixing is almost the same for "normal" and "'inverted" hierarchy; this should be contrasted with the conclusions for the cases [$\mathcal{N}$] and [$\mathcal{I}$], when $m_1 \ll (\Delta m_{atm}^2)^{1/2}$.

In the particular case of SMA solution, eq. (3.3) is still valid, with $m_1 \approx m_3$ (and $U_{e3} \to U_{e1}$ for "inverted" hierarchy); hence, up to sub-dominant mixing terms $\mathcal{M}_{ee} \approx \mathcal{M}_{ee}^{max} \approx m_1$, and a complete cancellation is impossible.

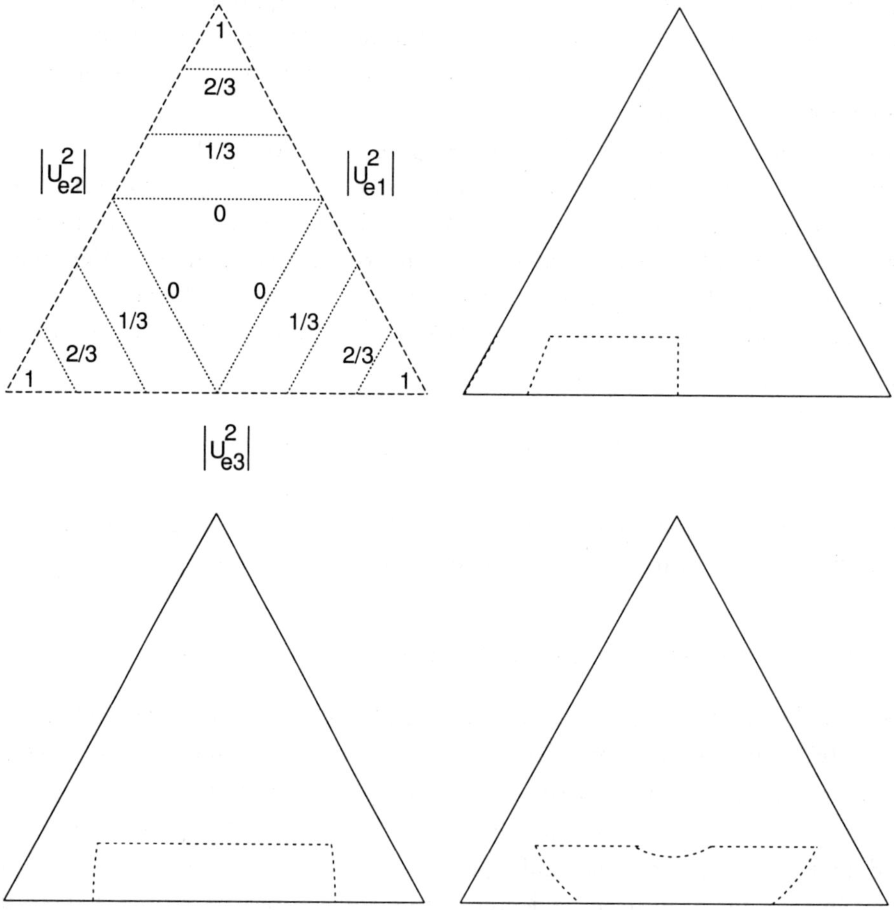

Figure 3: Case of "nearly degenerate" neutrino spectrum ($m_1 \approx m_2 \approx m_3$). From up to down, left to right: 1st plot, the minimum value of $\mathcal{M}_{ee}/m_1$ represented in the unitarity triangle. Reference numerical values of 1, 2/3, 1/3 and 0 are indicated. 2nd plot, indicative allowed regions for MSW enhanced transitions; the SMA solution is almost superimposed to the left–second–side, where $U_{e2} = 0$; 3rd, allowed region for vacuum oscillations; 4th, allowed region for averaged oscillations. We assume $|U_{e3}^2| < 0.15$ [1] and "normal" hierarchy. "Inverted" hierarchy corresponds approximatively to a 120° rotation of the last three plots.

3.4 A complementary representation

In order to recapitulate and confirm the results obtained in this section, we present a complementary graphical representation. Supposing that the mixing elements are known with good precision, we can plot the range of values of $\mathcal{M}_{ee}$ as a function of the only residual parameter: The mass of the lightest neutrino[4]. This is done in fig. 4, where we assume the mass splittings $\Delta m^2_{atm} = 2\times 10^{-3}$ eV^2 and $\Delta m^2_{\odot} = 10^{-4}$ eV^2 for "normal" and "inverted" hierarchy. The mixing $|U^2_{e3}|$ (resp. $|U^2_{e1}|$) with the heaviest (resp. lightest) state is 0, 2, 4 and 6 $\times 10^{-2}$ in the 4 types of curves, going from inner to outer ones. We fixed $|U^2_{e2}| = 0.4$ (resp. $|U^2_{e3}| = 0.4$), which corresponds roughly to an LMA solution. The figure confirms the conclusions obtained in section 3.1 for the case $[\mathcal{N}]$, about the importance of $\Delta m^2_{\odot}$, and of the small mixing element $|U^2_{e3}|$. For the case $[\mathcal{I}]$, instead, $|U^2_{e1}|$ and $\Delta m^2_{\odot}$ are less important in agreement with the discussion in section 3.2.

This representation emphasizes that also a *null* experimental result may be a very important information on the massive neutrino parameters: In fact, $\mathcal{M}^{min}_{ee} \lesssim 10^{-2}$ eV could rule out the assumption of "inverted" hierarchy, see the second plot of fig. 4; or, a bound on $\mathcal{M}^{min}_{ee}$ at the 10^{-3} level could amount to a measurement of the lightest neutrino mass, see the first plot of the same figure. Unfortunately, the value of m_1 determined in this way depends strongly on the parameters of oscillation, since:

$$\mathcal{M}^{min}_{ee} = \Big|\ |U^2_{e2}|\ (\Delta m^2_{\odot})^{1/2} - |U^2_{e3}|\ (\Delta m^2_{atm})^{1/2}\ \Big| \quad \text{for } m_1 = 0; \tag{3.7}$$

so that, even in the LMA case we are considering, it will be a real challenge to prove that $m_1 \neq 0$.

4. Concluding remarks

4.1 On the case $\mathcal{M}_{ee} \approx 0$

We regarded $\mathcal{M}_{ee}$ as a function of several parameters: the mixing elements, the squared mass splittings, the mass of the lightest neutrino and the complex phases. Following this approach, one may be led to wonder whether the cases when the rate is small as a consequence of cancellations among the various parameters are (in some sense) "natural".

We show here how the smallness can arise in a "natural" manner. Let us postulate that the neutrino mass matrix has a hierarchical structure, analogous to the structure of the Yukawa couplings of the charged fermions.

[4]In practice, this representation will be useful when the parameters of oscillation will be known reliably.

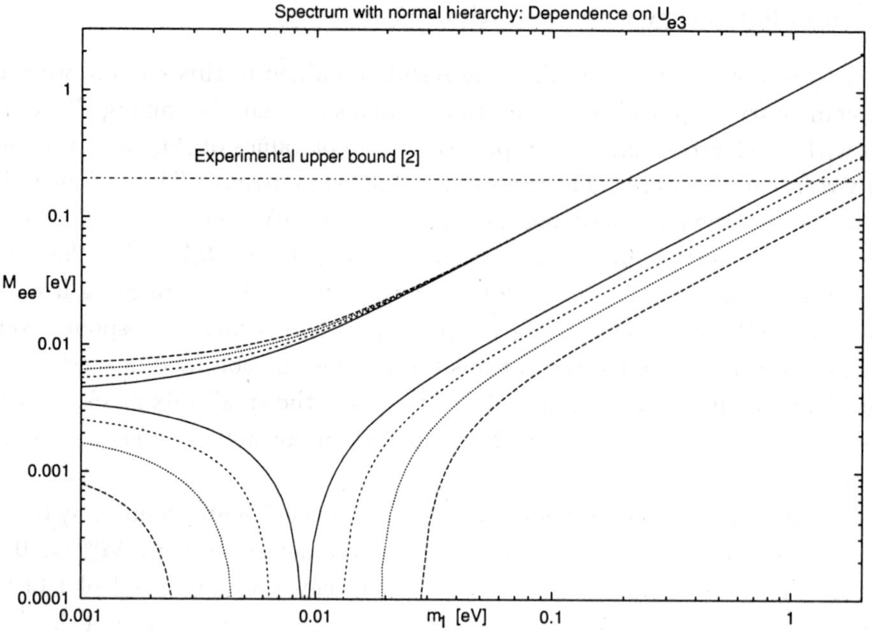

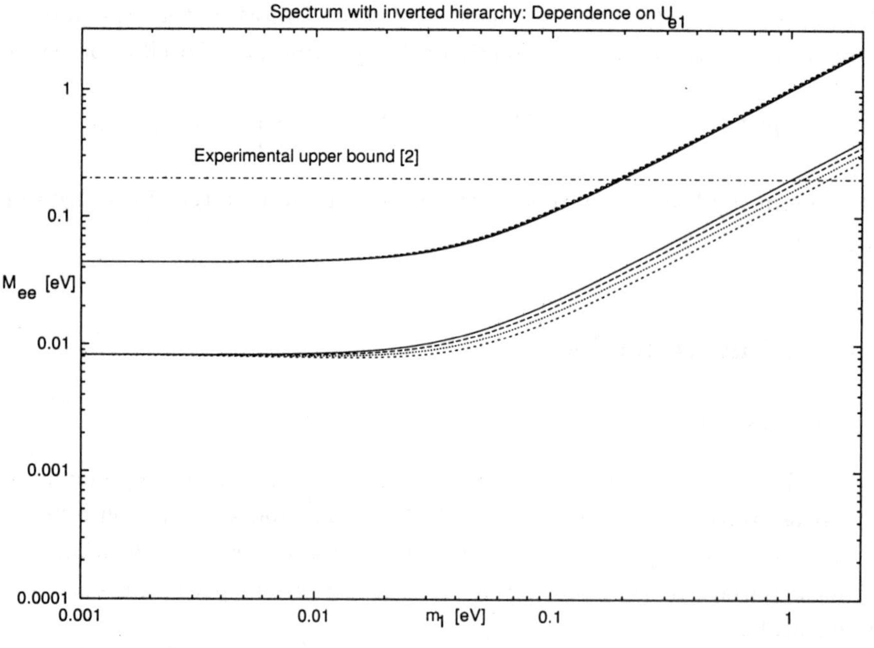

Figure 4: Range of values of $\mathcal{M}_{ee}$. The four upper curves in each plot represent $\mathcal{M}_{ee}^{max}$, the four lower ones $\mathcal{M}_{ee}^{min}$ — eqs. (1.2), (1.3). The four types of curve differ for the values of the sub-dominant mixing element: For "normal" hierarchy, $|U_{e3}^2| = 0, 2, 4$ and 6×10^{-2}, going from inner curves (continuous line) to outer ones (long-dashed); same values, but for $|U_{e1}^2|$, in the case of "inverted" hierarchy.

In this case, we can expect that the "ee-entry" of the neutrino mass matrix ($= \mathcal{M}_{ee}$) is the smallest one, and also $\mathcal{M}_{ee} \ll (\Delta m^2_{atm})^{1/2}$. This is what happens in the two models of references [14], where:

$$\mathcal{M}_{ee} \approx (\Delta m^2_{atm})^{1/2} \times (\sin\theta_C)^{2n}; \tag{4.1}$$

θ_C is the Cabibbo angle, and $n = 2, 3$ in the two models respectively. The value of $\mathcal{M}_{ee}$ in these models is rather small (see also [15]). Although the contribution from third family is modest, LMA solutions with relatively large mass splittings are possible in this type of models [16], which *a priori* may imply much larger values of $\mathcal{M}_{ee}$, as remarked for the case of section 3.1. Thus, these models provide examples of cases when $\mathcal{M}_{ee}$ is small as a consequence of cancellations among the various contributions.

In another sense, the statement $\mathcal{M}_{ee} \approx 0$ is surely "natural" in a standard model framework, since at one loop level the radiative corrections are tiny: $\sim y_e^2/(4\pi)^2 \sim 5 \times 10^{-14}$, where y_e is the electron Yukawa coupling.

4.2 What is the maximum value of $\mathcal{M}_{ee}$?

Let us briefly summarize the results of section 3, about an aspect of importance for experimental search: The maximum value of $\mathcal{M}_{ee}$ that we can *a priori* expect.

For given mixing elements, $\mathcal{M}_{ee}^{max}$ *increases* passing from the cases discussed in sections 3.1 (case $[\mathcal{N}]$) to section 3.2 (case $[\mathcal{I}]$), and finally to section 3.3 (case $[\mathcal{D}]$). Indeed, $\mathcal{M}_{ee}^{max}$ reaches at most the 10^{-2} eV level in case $[\mathcal{N}]$, depending on the subdominant mixing element $|U_{e3}^2|$ and on the scenario of oscillation (eq. (3.2)); it can be of the order of 3 to 9×10^{-2} eV in the case $[\mathcal{I}]$, depending on the size of Δm^2_{atm} (eq. (3.5)); finally, $\mathcal{M}_{ee}^{max}$ can be as large as the experimental upper limit of 0.2 eV in the case $[\mathcal{D}]$. In this sense, the *a priori* hope of a positive experimental result increases when going from $[\mathcal{N}]$ to $[\mathcal{I}]$, and from $[\mathcal{I}]$ to $[\mathcal{D}]$[5].

4.3 Studies of neutrino oscillations and search for $0\nu2\beta$ decay

We have shown that the parameters of oscillations are strictly related to the possible value of the $0\nu2\beta$ decay rate. However, the dependence on the type of spectrum is also essential. We summarize here some results of special interest (making reference for details to the previous section):

• For the small angle MSW solution, $\mathcal{M}_{ee}$ is quite large for "inverted" hierarchy in the case $m_1 \ll (\Delta m^2_{atm})^{1/2}$, see eq. (3.6); for "normal" hierarchy, we have instead eq. (3.3), which is smaller than the previous case by a factor of $|U_{e3}^2|$ when m_1 is small, and possibly even smaller (eq. (3.3)).

[5]On the contrary, one might argue that the case $[\mathcal{N}]$ is more likely than $[\mathcal{I}]$, and this latter more likely than $[\mathcal{D}]$, again on the basis of an analogy between the neutrino spectrum and the spectra of the charged fermions.

• For the large mixing angle MSW solution, contributions from "solar" frequency, order $(\Delta m^2_{\odot})^{1/2}$ are *not* negligible, and they may lead to cancellations (or enhancements) depending on the size of $|U^2_{e3}|$ in the case of "normal" hierarchy (sections 3.1 and 3.4).
• For VO solution, and "normal" hierarchy, the dependence of $\mathcal{M}^{min}_{ee}$ on $|U^2_{e3}|$ in quite appreciable (section 3.1).
• For "inverted" hierarchy, cancellations are not easy to obtain if m_1 is small in comparison with $(\Delta m^2_{atm})^{1/2}$, except for solutions of the solar neutrino problem with almost maximal mixing angles (section 3.2).
• Largest values of $\mathcal{M}_{ee}$ are taken in the case of "nearly degenerate" spectrum, $m_1 \gg (\Delta m^2_{atm})^{1/2}$ (section 3.3). In this extreme case, cancellations are possible especially for quite large mixing angle solutions, with relevant dependence on the size of the sub-dominant mixing, for both "normal" and "inverted" hierarchies.

4.4 Conclusions and perspectives

In this work, we discussed the interplay between the studies of neutrino oscillations and the search for $0\nu2\beta$ decay. We introduced new graphical representations, aimed at clarifying the relations between the neutrino spectra, the scenarios of oscillations and the rate of the neutrinoless double beta decay. For the perspectives, it has to be noticed that the present information on massive neutrinos is compatible with quite different oscillations scenarios and neutrino spectra. Future experiments aiming at a signal of the $0\nu2\beta$ process above the 10^{-2} eV level [17] will have an important role in deciding among the alternative possibilities.

Acknowledgments

I thank R. Barbieri, C. Giunti, M. Maris and A. Yu. Smirnov for useful discussions, and the Referee of the work for having suggested important improvements. Earlier accounts were presented in [18].

References

[1] G.L. Fogli, E. Lisi, A. Marrone and G. Scioscia, *Phys. Rev.* **D 59** (1999) 033001 [hep-ph/9808205];
G.L. Fogli, talk at the *VIII* Int. Workshop on "Neutrino Telescopes", Venice, Feb. 99.

[2] For most recent results (Heidelberg-Moskow experiment):
L. Baudis et al., *Limits on the Majorana neutrino mass in the 0.1 eV range*, hep-ex/9902014.

[3] See for instance C. Caso *et al*, *Eur. Phys. J.* **C 3** (1998) 1; PDG internet site at: http://pdg.lbl.gov/.

[4] G. L. Fogli, E. Lisi and D. Montanino, *Phys. Rev.* **D 54** (1996) 2048 [hep-ph/9605273].

[5] J. N. Bahcall, P. I. Krastev and A. Yu. Smirnov, *Phys. Rev.* **D 58** (1998) 096016 [hep-ph/9807216].

[6] T. Fukuyama, K. Matsuda and H. Nishiura, *Phys. Rev.* **D 57** (1998) 5844 [hep-ph/9807216]; *Mod. Phys. Lett.* **A 13** (1998) 2279 [hep-ph/9804262].

[7] V. Barger and K. Whisnant, *Majorana Neutrino Masses from Neutrinoless Double Beta Decay and Cosmology*, hep-ph/9904281;
C. Giunti, *Neutrinoless double-beta decay with three or four neutrino mixing,* hep-ph/9906275.

[8] L. Wolfenstein, *Phys. Rev.* **D 17** (1978) 2369;
S. P. Mikheyev and A. Yu. Smirnov, *Yad. Fiz.* **42** (1985) 1441 [*Sov. J. Nucl. Phys.* **42** (1985) 913]; *Nuovo Cim.* **C9** (1988) 17.

[9] S. M. Bilenkii, A. Bottino, C. Giunti and C. W. Kim, *Phys. Rev.* **D 54** (1996) 1881 [hep-ph/9602216]; see also second paper in [10].

[10] D. O. Caldwell and R. Mohapatra, *Phys. Rev.* **D 48** (1993) 3259;
S. T. Petcov and A. Yu. Smirnov, *Phys. Lett.* **B 322** (1994) 109 [hep-ph/9311204];
A. S. Joshipura, *Z. Physik* **C 64** (1994) 31.

[11] A. Faessler and F. Šimkovic, *J. Phys.* **G 24** (1998) 2139, reviewed the nuclear physics aspects of the $0\nu2\beta$ decay. Calculations of the matrix elements are compared in table V therein.

[12] S. M. Bilenkii, C. Giunti, C. W. Kim and S. T. Petcov, *Phys. Rev.* **D 54** (1996) 4432 [hep-ph/9604364].

[13] H. Minakata and O. Yasuda, *Phys. Rev.* **D 56** (1997) 1692 [hep-ph/9609276];
F. Vissani, *A study of the scenario with nearly degenerate Majorana neutrinos,* hep-ph/9708483 (compare also with H. Georgi and S. L. Glashow, *Neutrinos on Earth and in the Heavens*, hep-ph/9808293).

[14] J. Sato and T. Yanagida, *Phys. Lett.* **B 430** (1998) 127 [hep-ph/9710516];
N. Irges, S. Lavignac and P. Ramond, *Phys. Rev.* **D 58** (1998) 035003 [hep-ph/9802334].

[15] W. Buchmüller and T. Yanagida, *Phys. Lett.* **B 445** (1999) 399 [hep-ph/9810308].

[16] F. Vissani, *J. High Energy Phys.* **11** (1998) 025 [hep-ph/9810435].

[17] J. Hellmig and H. V. Klapdor-Kleingrothaus, *Z. Physik* **A 359** (1997) 351 [nucl-ex/9801004];
H. V. Klapdor-Kleingrothaus, *Prog. Part. Nucl. Phys.* **40** (1998) 265.

[18] A. Dighe, S. Pastor and A. Yu. Smirnov, *The physics of relic neutrinos*, hep-ph/9812244, proceedings of the "Relic Neutrino" workshop, Trieste, Sep. 98; F. Vissani, IC/99/36, contributed work for "Sixth Topical Seminar on Neutrino and Astro-Particle Physics", San Miniato, May 99.

HIP-2000-11/TH

Neutrinoless double beta decay in four-neutrino models

Anna Kalliomäki [1] and Jukka Maalampi [2]

Department of Physics, Theoretical Physics Division
University of Helsinki, Finland

ABSTRACT

The most stringent constraint on the so-called effective electron neutrino mass from the present neutrinoless double beta decay experiments is $|M_{ee}| < 0.2$ eV, while the planned next generation experiment GENIUS is anticipated to reach a considerably more stringent limit $|M_{ee}| < 0.001$ eV. We investigate the constraints these bounds set on the neutrino masses and mixings of neutrinos in four-neutrino models where there exists a sterile neutrino along with the three ordinary neutrinos. We find that the GENIUS experiment would be sensitive to the electron neutrino masses down to the limit $m_{\nu_e} \lesssim 0.024$ eV in such a scenario.

[1]E-mail address: amkallio@pcu.helsinki.fi
[2]E-mail address: maalampi@pcu.helsinki.fi

The Quest for the Neutrino Mass Spectrum

H.V. Klapdor–Kleingrothaus, H. Päs

1. Introduction

Recently the particle physics community was shocked with breathtaking news from the neutrino sector: Neutrino oscillations have been confirmed finally in the Super-Kamiokande [1] experiment. Now for the first time, ongoing and future experiments in neutrino oscillations (Super-Kamiokande, Borexino, SNO, MINOS, KAMLAND, MINIBOONE,...) and double beta decay (Heidelberg–Moscow, GENIUS,...) together can aim to solve the neutrino mass puzzle.

It was in 1930, when Wolfgang Pauli (fig. 1) wrote his famous letter adressed as *Liebe radioaktive Damen und Herren (Dear radioactive Ladies and Gentlemen)*, where he informed the participants of a nuclear physics workshop in Tübingen about his absence (he preferred to participate in a dance party) and postulated the neutrino to solve the problem of energy nonconservation in the nuclear beta decay. In 1956 the neutrino was observed for the first time by Clyde Cowan and Fred Reines in Los Alamos, who originally planned to explode a nuclear bomb for their experiment [2]. Finally, two years ago, the Super-Kamiokande experiment, a 50,000 ton water tank viewed by more than 11,000 photo multipliers 1,000 meter underground below a holy mountain in Japan, announced a significant signal for neutrino oscillations and established a non-vanishing mass of the neutrino as the first experimental signal of physics beyond the standard model. However, in spite of these successes, entering a new millenium the neutrino is still the most mysterious of the known particles. Alternatingly compared with spaceships travelling through the universe, ghosts penetrating solid rocks and vampires missing a mirror image [3], it still inspires the phantasy of hundreds of adventurous particle, nuclear and astro physicists being motivated by the hope, the neutrino could act as a key to the old human dream of a final theory, describing all particles and forces in a unified framework, and to a deeper understanding of the fate of the universe.

The attributes, making the neutrino this kind of outlaw among the known particles, are the following:

PHYSICAL REVIEW D, VOLUME 59, 022001

New limits on dark-matter weakly interacting particles from the Heidelberg-Moscow experiment

L. Baudis, J. Hellmig, G. Heusser, H. V. Klapdor-Kleingrothaus,* S. Kolb, B. Majorovits, H. Päs, Y. Ramachers, and H. Strecker
Max-Planck-Institut für Kernphysik, P.O. Box 10 39 80, D-69029 Heidelberg, Germany

V. Alexeev, A. Bakalyarov, A. Balysh, S. T. Belyaev,* V. I. Lebedev, and S. Zhukov
Russian Science Center Kurchatov Institute, 123182 Moscow, Russia
(Received 5 June 1998; published 24 December 1998)

New results after 0.69 kg yr of measurement with an enriched ^{76}Ge detector of the Heidelberg-Moscow experiment with an active mass of 2.758 kg are presented. An energy threshold of 9 keV and a background level of 0.042 cts/(kg d keV) in the energy region between 15 keV and 40 keV was reached. The derived limits on the weakly interacting massive particles–nucleon cross section are the most stringent limits on spin-independent interactions obtained to date by using essentially raw data without background subtraction. [S0556-2821(98)05424-1]

PACS number(s): 95.35.+d, 14.80.Ly

INTRODUCTION

The nature of dark matter in the Universe remains a challenging question. Even if new measurements will confirm that we live in a low Ω_{matter} universe ($\Omega_{\text{matter}}\sim 0.3-0.4$) [1,2] a considerable amount of nonbaryonic dark matter is needed. WIMPs (weakly interacting massive particles) are among the most discussed candidates [3], being well motivated from early universe physics [4] and supersymmetry [5].

WIMP detection experiments can decide whether WIMPs dominate the halo of our Galaxy. For this reason, considerable effort is made towards direct WIMP search experiments which look for energy depositions from elastic WIMP-nucleus scattering [6]. Germanium experiments designed for the search for neutrinoless double beta decay were among the first to set such kinds of limits [7,8]. The Heidelberg-Moscow experiment gave the most stringent upper limits on spin-independent WIMP interactions [9] until recently. The present best limits on the WIMP-nucleon cross section come from the DAMA NaI Experiment [10].

The Heidelberg-Moscow experiment operates five enriched ^{76}Ge detectors with an active mass of 10.96 kg in the Gran Sasso Underground Laboratory. It is optimized for the search for the neutrinoless double beta decay of ^{76}Ge in the energy region of 2038 keV. For a detailed description of the experiment and latest results see [12].

In this paper we report on results from one of the enriched Ge detectors, which took data in a period of 0.249 years in a special configuration developed for low energy measurements. A lower energy threshold and a smaller background counting rate has been achieved with the same detector as used in 1994 [9], mainly due to the lower cosmogenic activities in the Ge crystal and in the surrounding copper after four years without activation.

*Spokesmen of the collaboration.

EXPERIMENTAL SETUP

The utilized detector is a coaxial, intrinsic p-type HPGe detector with an active mass of 2.758 kg. The enrichment in ^{76}Ge is 86%. The sensitivity to spin-dependent interactions becomes negligible, since ^{73}Ge, the only stable Ge isotope with nonzero spin, is deenriched to 0.12% (7.8% for natural Ge). The detector has been in the Gran Sasso Underground Laboratory since September 1991; a detailed description of its background can be found in [12].

The data acquisition system allows an event-by-event sampling and pulse shape measurements. The energy output of the preamplifier is divided and amplified with two different shaping time constants, 2 μs and 4 μs. The fast 2 μs signal serves as a stop signal for the 250 MHz flash analogue to digital converter (ADC) which records the pulse shape of each interaction. The best energy resolution and thus lowest energy threshold is obtained with the 4 μs shaped signal. A third branch of the energy output is shaped with 3 μs and amplified to record events up to 8 MeV in order to identify radioactive impurities contributing to the background. The spectra are measured with 13bit ADCs, which also release the trigger for an event by a peak detect signal. Further triggers are vetoed until the complete event information, including the pulse shape, has been recorded. To record the pulse shape (for details see [13]) the timing output of the preamplifier is divided into four branches, each signal being integrated and differentiated in filtering amplifiers (TFAs) with different time constants. The TFAs are used since the charge current is integrated within the preamplifier. The signals are amplified to record low as well as high-energetic pulses.

DATA ANALYSIS

An energy threshold of 9 keV has been reached. This rather high value is due to the large detector size and a 50 cm distance between FET and detector. Both effects lead to a higher system capacitance and thus to an enhancement of the baseline noise.

We calibrate the detector with a standard ^{152}Eu-^{228}Th

2.4 Other Beyond Standard Model Physics: From SUSY and Leptoquarks to Compositeness and Quantum Foam

2.4.1. General

44

Neutrinoless Double Beta Decay and Physics Beyond The Standard Model

Rabindra N. Mohapatra*

Department of Physics, University of Maryland, College Park, MD 20742, U. S. A.

Abstract

Neutrinoless double beta decay is a sensitive probe of new physics beyond the standard model. In this review, we begin by describing the various mechanisms for this process and the kind of new physics scenarios where these mechanisms can arise. The present experimental lower bound on the lifetime for $\beta\beta_{0\nu}$ is then used to set limits on the parameters of these new physics scenarios. We then consider the positive indications for neutrino masses present in various experiments such as those involving the solar and atmospheric neutrinos as well as the LSND result. Coupled with other astrophysical and cosmological constraints on neutrino masses and mixings, they restrict the allowed profiles for the Majorana neutrino mass matrices considerably. We then show how ongoing searches for $\beta\beta_{0\nu}$ decay can confirm or rule out the various scenarios for neutrino masses. In the last section, we present the outlook for observable $\beta\beta_{0\nu}$ amplitude in some specific grand unified theories.

in Proc. Int. Workshop "Double Beta Decay and Related Topics", Trento, Italy, April-May, 1995, eds. *H.V. Klapdor–Kleingrothaus and S. Stoica: Singapore: World Scientific* (1996) 44 – 68

*Work supported by the National Science Foundation Grant No.PHY-9119745 and in part by the Distinguished Faculty Research Fellowship Award by the University of Maryland

Introduction

In the standard electroweak model of Glashow, Weinberg and Salam, the absence of the right-handed neutrinos and the existence of an exact accidental global $B-L$ symmetry guarantees that the neutrinos are massless to all orders in perturbation theory. Any experimental evidence for a non-zero neutrino mass therefore constitutes evidence for new physics beyond the standard model and will be major step towards a deeper understanding of nature[1]. There are at this moment many experiments under way searching directly or indirectly (e.g. via neutrino oscillations) for neutrino masses, one of the most important ones being the search for neutrinoless double beta decay if the neutrino happens to be its own antiparticle (Majorana neutrino) as is implied by many extensions of the standard model. However, Majorana mass of the neutrino is not the only way to get an observable amplitude for neutrinoless double beta decay ($\beta\beta_{0\nu}$), as will be made clear in this article. Since $\beta\beta_{0\nu}$ decay changes lepton number (L_e) by two units any time there is violation of electron lepton number L_e in a theory, one can in principle expect this process to turn on. This therefore reflects the tremendous versatility of $\beta\beta_{0\nu}$ as a probe of all kinds of new physics beyond the standard model. Indeed we will see that already very stringent constraints on new physics scenarios such as the left-right symmetric models with the see-saw mechanism[2] and supersymmetric models with R-parity violation[3], scales of possible compositeness of leptons etc are implied by the existing experimental limits[4] on this process.

This talk is organized as follows: In part I, I discuss the basic mechanisms for neutrinoless double beta decay ; in part II, I go on to discuss the kind of new physics scenarios that can be probed by $\beta\beta_{0\nu}$ decay and the kind of constraints on the parameters of the new physics scenarios implied by the already existing experimental data; In part III, I address the question of the theoretical and phenomenological outlook for $\beta\beta_{0\nu}$ decay being observable given our present information about neutrinos; in part IV, the question of observability of $\beta\beta_{0\nu}$ in some popular grand unified scenarios for neutrino masses is addressed.

Part I

Mechanisms for $\beta\beta_{0\nu}$ decay

Before starting the discussion of the various mechanisms for $\beta\beta_{0\nu}$, let us write down the basic Four-Fermi V-A interaction resposible for known weak phenomena involving only the first generation:

$$H_{wk} = \frac{G_F}{\sqrt{2}}[\overline{u}\gamma^{\mu}(1-\gamma_5)d\overline{e}\gamma_{\mu}(1-\gamma_5)\nu_e + \ h.c.] \tag{1}$$

Note that in the second order in G_F, the above Hamiltonian leads to the two neutrino double beta decay process which has now been observed in many nuclei [4, 5]. Eq.1

immediately implies that if the neutrino is its own antiparticle, then the two neutrinos from the two weak Hamiltonians in 1 can annihilate into vacuum leading to neutrinoless double beta decay. Since in the standard model, neutrino is not its own antiparticle, $\beta\beta_{0\nu}$ decay probes physics beyond the standard model. It could of course be that there are effective four-Fermi interactions which involve heavier fermions in scenarios beyond the standard models. If such particles are their own antiparticles, again similar arguments as above could also lead to $\beta\beta_{0\nu}$ decay. Examples of such particles abound in literature: right-handed neutrino, photino, gluino to mention a few popular ones. One could therefore give an arbitrary classification of the mechanisms for $\beta\beta_{0\nu}$ decay into two kinds: (A) one class that involves the exchange of light neutrinos; and (B) the second class that involves heavy fermions or bosons.

I.A: Light neutrino exchange:

As already mentioned, if the neutrino is considered as a Majorana particle, the process $\beta\beta_{0\nu}$ can arise. The four-Fermi interaction involving the neutrino however need not be purely $V-A$ type as in Eq.1 once we entertain physics beyond the standard model. One can therefore contemplate several kinds of mechanisms involving the light neutrino exchange. Before presenting them, it is important to remark that these kind of light neutrino exchange diagrams always lead to a long range neutrino potential inside the nucleons and therefore, crudely speaking the two nucleons "far" from each other can lead to double beta decay. This has important implications for the evaluation of the nuclear matrix element[6]

I.A.1: Helicity flip neutrino mass mechanism:

If both the four-Fermi interactions involve $V-A$ currents, the $\beta\beta_{0\nu}$ will be non-vanishing only if there is a flip of neutrino helicity; this can happen only if the neutrino has a mass m_ν[7]. The diagram of Fig.1 can then lead to $\beta\beta_{0\nu}$ decay. Neglecting nuclear physics complications, one can write down the amplitude $A_{\beta\beta}$ for neutrinoless double beta decay for this case to be:

$$A^{(m)}_{\beta\beta} \simeq \frac{G_F^2}{2}\langle\frac{m_\nu}{k^2}\rangle_{Nucl.} \qquad (2)$$

and

The nuclear average in general consists of a three-momentum integral which roughly converts this amplitude to: $A^{(m)}_{\beta\beta} \simeq G_F^2 m_\nu p_F$. The width for the decay can then be written as (again ignoring detailed nuclear factors):

$$\Gamma_{\beta\beta} \simeq \frac{Q^5|A|^2}{60\pi^3} \qquad (3)$$

Here, Q is the available energy for the two electrons. The factors of π can also be easily seen from Fermi golden rule combined with the appropriate phase space factors (e.g. a factor 2π from the golden rule; $(2\pi)^{-6}$ from the phase space for two electrons and

$(4\pi)^2$ from the integration of the two electron solid angles). To get a feeling for the kind of restrictions they imply on m_ν and the η parameter, let us use the present bound on the $\Gamma_{\beta\beta}$ indicated by the present Heidelberg-Moscow ^{76}Ge experiment at Gran Sasso i.e. $\Gamma_{\beta\beta} \leq 3.477 \times 10^{-57}$ GeV; using for a rough estimate $Q \simeq 2$ MeV and $p_F \simeq 50$ MeV, we see very easily that one gets an upper limit of .7 eV for the neutrino mass. Of course these limits are very crude; but they do indicate the severity of the constraints on the parameters beyond the standard model from the neutrinoless double beta decay process. A more careful treatment of the particle physics part of the calculation implies that the m_{ν_e} in Eq.2 should be replaced by $\Sigma_i U_{ei}^2 \zeta_i m_{\nu_i}$ where the U_{ei} denotes the mixing angle of the electron neutrino with other neutrinos and ζ_i denotes the CP-phase of the i-th neutrino. So in principle, if different neutrinos had different CP-phase, then the effective mass that appears in $\beta\beta_{0\nu}$ amplitude could be small while keeping the individual masses bigger.

I.A.2 Helicity nonflip vector-vector mechanism:

If in addition to the usual $V - A$ type four-Fermi interaction, there exist neutrino interactions involving admixture of $V + A$ type leptonic currents[8], then a helicity conserving mechanism for $\beta\beta_{0\nu}$ (and hence without the need for a neutrino mass) emerges (Fig.2). This for instance can happen, when one replaces $\bar{e}\gamma_\mu(1 - \gamma_5)\nu_e$ in Eq.1 by $\bar{e}\gamma_\mu[(1 - \gamma_5) + \eta(1 + \gamma_5)]\nu_e$. The amplitude for $\beta\beta_{0\nu}$ (ignoring nuclear physics factors) can then be written as:

$$A^{\eta}_{\beta\beta} \simeq \frac{G_F^2}{2}\langle\frac{\eta}{\gamma \cdot k}\rangle_{Nucl.} \tag{4}$$

Such contributions depend on the value of η and are nonzero as long as the neutrino is a Majorana particle regardless of how big its mass is. It is sometimes called the left-right mixing contribution and leads to $0^+ \to 2^+$ type of transition. Order of magnitude arguments of the type just given for the mass mechanism leads to an upper bound for η of about 10^{-8} or so. More careful nuclear physics arguments also lead to similar bounds[9]

I.A.3 Helicity nonflip vector-scalar mechanism:

A completely new class of contributions to $\beta\beta_{0\nu}$ involving the exchange of light neutrinos has been pointed out recently[10]. The new contributions arise from the combination of two effective four-Fermi interactions of the following type:

$$\begin{aligned} H_{eff} &= \frac{G_F}{\sqrt{2}}(\bar{e}\gamma_\mu(1-\gamma_5)\nu_e\bar{u}\gamma^\mu(1-\gamma_5)d + \epsilon_1^{ee}\bar{d}(1-\gamma_5)u\nu_e^T C^{-1}(1-\gamma_5)e \; + \\ &\quad \epsilon_2^{ee}\bar{d}(1-\gamma_5)\nu_e u^T C^{-1}(1-\gamma_5)e\;) \; + \; h.c. \end{aligned} \tag{5}$$

In the above, the first term is the usual (V-A) interaction, the other two are effective lepton number violating terms. The $V - A$ term in the above equation in collaboration with either of the last two terms can lead to $\beta\beta_{0\nu}$ decay via a Feynman diagram which

48

is similar to Fig.2 . In order to evaluate the matrix elements between nuclear states, we need to do Fierz reordering of the ϵ_2^{ee} term, which casts it in the form:

$$\frac{G_F}{2\sqrt{2}}\epsilon_2^{ee}\left(\bar{d}(1-\gamma_5)u\overline{e^c}(1-\gamma_5)\nu_e + \frac{1}{2}\bar{d}\sigma^{\mu\alpha}(1-\gamma_5)u\overline{e^c}(1-\gamma_5)\sigma_{\mu\alpha}\nu_e\right) .$$

The resulting effective double beta amplitude can be written in momentum space as:

$$A_{\beta\beta}^{vs} \simeq G_F^2(\epsilon_i^{ee})\langle\frac{1}{\gamma\cdot k}\rangle \tag{6}$$

As a crude estimate, we assume an average value of k as before to be equal to the Fermi momentum p_F of the nucleons in the nucleus ($\approx$ 50 MeV). The present upper limits on m_ν of about 1 eV then translates to an upper limit on the new interaction parameter ϵ as follows:

$$\epsilon_{1,2}^{ee} \leq 1\times10^{-8}. \tag{7}$$

I.B: Heavy particle exchange:

The second class of mechanisms consists of exchange of heavy particles (such as majorana fermions) which often arise in physics scenarios beyond the standard model. In the low energy limit, the effective Hamiltonian that leads to $\beta\beta_{0\nu}$ decay in these cases requires point interaction between nucleons; as a result, in general the nuclear matrix elements in thses cases are expected to be smaller; nevertheless, a lot of extremely useful information have been extracted about new physics where these mechanisms operate. Symbolically, such contributions can arise from effective Hamiltonians of the following type(we have suppressed all gamma matrices as well as color indices):

$$H^{(1)} = G_{eff}\,\bar{u}\Gamma d\bar{e}\Gamma F + h.c. \tag{8}$$

or

$$H^{(2)} = \lambda_\Delta\left(\frac{1}{M^3}\,\bar{u}\Gamma d\bar{u}\Gamma d + e^-e^-\right)\Delta^{++} + h.c. \tag{9}$$

Here F represents a neutral majorana fermion such as the right-handed neutrino (N)[11] or gluino $\tilde{G}$ or photino $\tilde{\gamma}$ and Δ^{++} represents a doubly charged scalar or vector particle. In the above equations, the coupling G_{eff} has dimension of M^{-2} and λ_Δ is dimensionless. The possibility of the doubly charged scalar contribution to $\beta\beta_{0\nu}$ was first noted in [12] and have been discussed subsequently in [13]. The contributions to neutrinoless double beta decay due to the above interactions arise from diagrams in Fig. 3 and 4 and lead to $\beta\beta_{0\nu}$ amplitudes as follows:

$$A(^F)_{\beta\beta} \simeq G_{eff}^2\frac{1}{M_F}(p^{eff})^3 \tag{10}$$

and

$$A^{\Delta}_{\beta\beta} \simeq \left(\frac{\lambda^2_{\Delta}}{M^3 M^2_{\Delta}}\right)(p^{eff})^3 \tag{11}$$

Here again we have crudely replaced all nuclear effects by the effctive momentum parameter p^{eff}. If we choose $p^{eff} \simeq 50$ MeV, then the present lower limit on the lifetime for ^{76}Ge decay leads to a crude upper limit on the effective couplings as follows:

$$G_{eff} \leq 10^{-7}\left(\frac{M_F}{100\ GeV}\right)^{\frac{1}{2}} \tag{12}$$

and

$$\lambda_{\Delta} \leq 10^{-3}\left(\frac{M}{100\ GeV}\right)^{\frac{5}{2}} \tag{13}$$

In the second equation above, we have set $M = M_{\Delta}$. Note that these limits are rather stringent and therefore have the potential to provide useful constraints on the new physics scenarios that lead to such pictures.

Part II

Implications for physics beyond the standard model:

Let us now discuss what kind of new physics scenarios beyond the standard model where the above mechanisms can be realized. Let us first consider the the neutrino mass mechanism. Any theory which gives the electron neutrino a significant ($\simeq$ eV) Majorana mass or any other species (e.g. ν_μ or ν_τ) a large enough mass and mixing angle with the ν_e so that $U^2_{ei} m_{\nu_i}$ is of order of an electron volt will make itself open to testability by the $\beta\beta_{0\nu}$ decay experiment. There are many theories with such expectations for neutrinos. Below I described two examples: (i) the singlet majoron model and (ii) the left-right symmetric model. This is intimately connected with ways to understand the small neutrino mass in gauge theories.

II.A: The singlet majoron model:

This model[14] is the simplest extension of the standard model that provides a naturally small mass for the neutrinos by employing the the see-saw mechanism[15]. It extends the standard model by the addition of three right-handed neutrinos and the addition of a single complex Higgs field Δ which is an $SU(2)_L \times U(1)_Y$ singlet but with a lepton number +2. There is now a Dirac mass for the neutrinos and a Majorana mass for the right handed neutrinos proportional to the vacuum expectation value (vev) $\langle\Delta\rangle \equiv v_R$. This leads to a mass matrix for the neutrinos with the usual see-saw form:

$$M = \begin{pmatrix} 0 & m_D \\ m_D^T & f v_R \end{pmatrix} \tag{14}$$

50

This leads to both the light and heavy (right-handed) neutrinos being Majorana particles with the mutual mass relation being given by the see-saw formula:

$$m_{\nu_i} \simeq m_{iD}(M_{iR}^{-1})m_{iD}^T \tag{15}$$

where we have ignored all mixings and $M_{iR} \simeq f_{ii}v_R$ denote the masses of the heavy right-handed neutrinos . It is clear that the electron neutrino mass can be in the electron-volt range if the values of m_{1D} are chosen to be of similar order of magnitude to the electron mass. In fact, for $m_{1D} = m_e$, and $m_{1R} = 250$ GeV, one gets $m_{\nu_e} = 1$ eV which is the range of masses being probed by the ongoing and proposed $\beta\beta_{0\nu}$ experiments. This model would predict a hierarchical pattern for neutrino masses with an eV-KeV-MeV masses for the three neutrinos. Cosmological consistency for such a spectrum has been studied in several papers[16] and we do not go into details. We simply mention two recent arguments which have brought attention to such scenarios (especially the tau neutrino mass being in the MeV range)

The first point has to do with a recent analysis of the constraints on the number of neutrino flavors imposed by the big bang nucleosynthesis[17] (BBN). According to this analysis, the present data on ^3He, D and ^4He abundances in the universe combined with theoretical models for chemical evolution of ^3He and D, implies that the total number of neutrino species N_ν at the epoch of BBN must be $\simeq 2$. Since LEP data has confirmed the existence of ν_τ along with ν_μ and ν_e, the ν_τ somehow must not contribute to BBN. The only way it can happen is if it has decayed by the BBN time. In the singlet majoron model, such short decay lifetimes have been shown possible within the other existing constraints on the model[16].

A second argument arises from considerations of structure formation. Apparently all known data for structure in the universe can be accomodated by assuming the existence of of cold dark matter in conjunction with an MeV ν_τ decaying with a lifetime of 10-100 sec[18], a lifetime value in the same range as required by the BBN argument.

II.B: Left-right symmetric models:

Let us now consider the minimal left-right symmetric model with a see-saw mechanism for neutrino masses as described in [2]. Below, we provide a brief description of the structure of the model. The three generations of quark and lepton fields are denoted by $Q_a^T \equiv (u_a, d_a)$ and $\Psi_a^T \equiv (\nu_a,\ e_a)$ respectively, where $a\ =\ 1,\ 2,\ 3$ is the generation index. Under the gauge group $SU(2)_L \times SU(2)_R \times U(1)_{B-L}$, they are assumed to transform as $\Psi_{a\ L} \equiv (1/2,\ 0,\ -1)$ and $\Psi_{a\ R} \equiv (0,\ 1/2,\ -1)$ and similarly for the quarks denoted by $Q^T \equiv (u,\ d)$. In this model, there is a right-handed counterpart to the $W_L^{\pm}$ to be denoted by $W_R^{\pm}$. Their gauge interactions then lead to the following expanded structure for the charged weak currents in the model for one generation prior to symmetry breaking (for our discussion , the quark mixings and the higher generations are not very important; so we will ignore them in what follows.)

$$L_{wk} = \frac{g}{2\sqrt{2}}[W^{-}_{\mu L}\left(\bar{d}\gamma^{\mu}(1-\gamma_5)u + \bar{e}\gamma^{\mu}(1-\gamma_5)\nu_e\right) + L \to R, -\gamma_5 \to +\gamma_5, \nu_e \to N_e\ ] \tag{16}$$

The Higgs sector of the model consists of the bi-doublet field $\phi \equiv (1/2,\ 1/2,\ 0)$ and triplet Higgs fields: $\Delta_L(1,\ 0,\ +2) \oplus \Delta_R(0,\ 1,\ +2)$.

The Yukawa couplings which are invariant under gauge and parity symmetry can be written as:

$$\begin{aligned}\mathcal{L}_Y &= \overline{\Psi}_L h^{\ell}\phi\Psi_R + \overline{\Psi}_L\tilde{h}^{\ell}\tilde{\phi}\Psi_R + \overline{Q}_L\phi h^q Q_R + \overline{Q}_L\tilde{h}^q\tilde{\phi}Q_R + \\ &\quad \Psi_L^T f\tau_2\vec{\tau}\cdot\vec{\Delta}_L C^{-1}\Psi_L + L \to R + h.c.\end{aligned} \tag{17}$$

where h, $\tilde{h}$ are hermitian matrices while f is a symmetric matrix in the generation space. Ψ and Q here denote the leptonic and quark doublets respectively.

The gauge symmetry is spontaneously broken by the vacuum expectation values: $< \Delta^0_R >= V_R$; $< \Delta^0_L >= 0$; and $< \phi >= \begin{pmatrix} \kappa & 0 \\ 0 & \kappa' \end{pmatrix}$. As usual, $< \phi >$ gives masses to the charged fermions and Dirac masses to the neutrinos whereas $< \Delta^0_R >$ leads to the see-saw mechanism for the neutrinos in the standard way[2]. For one generation the see-saw matrix is in the form given in Eq.14 and leads as before to a light and a heavy state as discussed in the previous section. For our discussion here it is important to know the structure of the light and the heavy neutrino eigenstates:

$$\begin{aligned}\nu &\equiv \nu_e + \xi N_e \\ N &\equiv N_e - \xi\nu_e\end{aligned} \tag{18}$$

where $\xi \simeq \sqrt{m_{\nu_e}/m_N}$ and is therefore a small number. Substituting these eigenstates into the charged current Lagrangian in Eq.16, we see that the right-handed W_R interaction involves also the light neutrino with a small strength proportional to ξ. To second order in the gauge coupling g, the effective weak interaction Hamiltonian involving both the light and the heavy neutrino becomes:

$$\begin{aligned}H_{wk} &= \frac{G_F}{\sqrt{2}}\left(\bar{u}\gamma^{\mu}(1-\gamma_5)d[\bar{e}\gamma_{\mu}[(1-\gamma_5) + \xi(\frac{m^2_{W_L}}{m^2_{W_R}})(1+\gamma_5)]\nu + \xi\bar{e}(1-\gamma_5)N\right) \\ &\quad + \frac{G_F}{\sqrt{2}}\left(\frac{m^2_{W_L}}{m^2_{W_R}}\right)(\bar{u}\gamma^{\mu}(1+\gamma_5)d\bar{e}\gamma_{\mu}(1+\gamma_5)N) + h.c.\end{aligned} \tag{19}$$

From Eq. (19), we see that there are several contributions to the $\beta\beta_{0\nu}$. Aside from the usual neutrino mass diagram (Fig.1), there is a contribution due to the wrong helicity admixture with $\eta \simeq \xi\left(\frac{m^2_{W_L}}{m^2_{W_R}}\right)$ and there are contributions arising from the exchange of heavy right-handed neutrinos (Fig.3). This last contribution is given by :

$$A^{(R)}_{\beta\beta} \simeq \frac{G_F^2}{2}\left(\frac{m^4_{W_L}}{m^4_{W_R}} + \xi^2\right)\frac{1}{m_N} \tag{20}$$

52

The present limits on neutrinoless double beta decay lifetime then imposes a correlated constraint on the parameters m_{W_R} and m_N[20]. This is shown in Fig.5 and a correlated constraint on ξ and m_{W_R} shown in Fig.6 due to Hirsch[21]. It is clear from fugure that if we combine the theoretical constraints of vacuum stability then, the present ^{76}Ge data provides a lower limit on the masses of the right handed neutrino (N_e) and the W_R of 1 TeV, which is a rather stringent constraint. The limits on ξ on the other hand are not more stringent than what would be expected from the structure of the theory. We have of course assumed that the leptonic mixing angles are small so that there is no cancellation between the parameters.

Finally, the Higgs sector of the theory generates two types of contributions to $\beta\beta_{0\nu}$ decay. One arises from the coupling of the doubly charged Higgs boson to electrons (see Fig.4). The amplitude for the decay is same as in Eq.11 except we have $\lambda_\Delta = f_{11}$ and

$$\frac{\lambda_\Delta}{M^3} = 2^{7/4} G_F^{3/2} \left(\frac{m_{W_L}}{M_{W_R}}\right)^3 \tag{21}$$

Using this expression, we find that the present ^{76}Ge data implies that (assuming $m_{W_R} \geq 1$ TeV)

$$M_{\Delta^{++}} \geq \sqrt{f_{11}}\ \ 80GeV \tag{22}$$

A second type Higgs induced contribution arises from the mixing among the charged Higgs fields in ϕ and Δ_L which arise from the couplings in the Higgs potential, such as $\mathrm{Tr}(\Delta_L \phi \Delta_R^\dagger \phi^\dagger)$ after the full gauge symmetry is broken down to $U(1)_{em}$. Let us denote this mixing term by an angle θ. This will contribute to the four-Fermi interaction of the form given by the ϵ_1^{ee} term through the diagram shown in Fig.7 with

$$\epsilon_1^{ee} \simeq \frac{h_u f_{11} sin2\theta}{4\sqrt{2} G_F M_{H+}^2}\ , \tag{23}$$

where we have assumed that H^+ is the lighter of the two Higgs fields. We get $h_u f_{11} \sin 2\theta \leq 6 \times 10^{-9} (M_{H^+}/100\ GeV)^2$, which is quite a stringent constraint on the parameters of the theory. To appreciate this somewhat more, we point out that one expects $h_u \approx m_u/m_W \approx 5 \times 10^{-5}$ in which case, we get an upper limit for the coupling of the Higgs triplets to leptons $f_{11}\sin 2\theta \leq 10^{-4}$ (for $m_{H^+} = 100\ GeV$). Taking a reasonable choice of $\theta \sim M_{W_L}/M_{W_R} \sim 10^{-1}$ would correspond to a limit $f_{11} \leq 10^{-3}$. Limits on this parameters from analysis[19] of Bhabha scattering is only of order .2 or so for the same value of the Higgs mass.

II.C: MSSM with R-parity violation:

The next class of theories we will consider is the supersymmetric stamdard model. As is well-known, the minimal supersymmetric standard model can have explicit[3] violation of the R-symmetry (defined by $(-1)^{3B+L+2S}$), leading to lepton number violating interactions in the low energy Lagrangian. The three possible types of couplings in the superpotential are :

$$W' \;=\; \lambda_{ijk}L_iL_jE^c_k + \lambda'_{ijk}L_iQ_jD^c_k + \lambda''_{ijk}U^c_iD^c_jD^c_k \;. \tag{24}$$

Here L, Q stand for the lepton and quark doublet superfields, E^c for the lepton singlet superfield and U^c, D^c for the quark singlet superfields. i, j, k are the generation indices and we have $\lambda_{ijk} = -\lambda_{jik}$, $\lambda''_{ijk} = -\lambda''_{ikj}$. The $SU(2)$ and color indices in Eq. (24) are contracted as follows: $L_iQ_jD^c_k = (\nu_i d^\alpha_j - e_i u^\alpha_j)D^c_{k\alpha}$, etc. The simultaneous presence of all three terms in Eq. (24) will imply rapid proton decay, which can be avoided by setting the $\lambda'' = 0$. In this case, baryon number remains an unbroken symmetry while lepton number is violated.

There are two types of to $\beta\beta_{0\nu}$ decay in this model. One class dominantly mediated by heavy gluino exchange[20] falls into the class of type II contributions discussed in the previous section. The dominant diagram of this class is ahown in Fig.8. Detailed evaluation of the nuclear matrix element for this class of models has recently been carried out by Hirsch et. al.[22] and they have found that a very stringent bound on the following R-violating parameter can be given:

$$\lambda'_{111} \leq 3.9 \times 10^{-4}\left(\frac{m_{\tilde q}}{100GeV}\right)^2\left(\frac{m_{\tilde g}}{100GeV}\right)^{1/2} \tag{25}$$

The second class of contributions fall into the light neutrino exchange vector-scalar type[10] and the dominant diagram of this type is shown in Fig.9.(where the exchanged scalar particles are the $\tilde b - \tilde b^c$ pair). This leads to a contribution to ϵ^{ee}_2 given by

$$\epsilon^{ee}_2 \;\simeq\; \left(\frac{(\lambda'_{113}\lambda'_{131})}{2\sqrt{2}G_F M^2_{\tilde b}}\right)\left(\frac{m_b}{M^2_{\tilde b^c}}\right)(\mu \tan\beta + A_b m_0) \;. \tag{26}$$

Here A_b, m_0 are supersymmetry breaking parameters, while μ is the supersymmetric mass of the Higgs bosons. $\tan\beta$ is the ratio of the two Higgs vacuum expectation values and lies in the range $1 \leq \tan\beta \leq m_t/m_b \approx 60$. For the choice of all squark masses as well as μ and the SUSY breaking mass parameters being of order of 100 GeV, $A_b = 1$, $\tan\beta = 1$,the following bound on R-violating couplings is obtained:

$$\lambda'_{113}\lambda'_{131} \leq 3 \times 10^{-8} \tag{27}$$

This bound is a more stringent limit on this parameter than the existing ones [23]. The present limits on these parameters are $\lambda'_{113} \leq 0.03, \lambda'_{131} \leq 0.26$, which shows that the bound derived here from $\beta\beta_{0\nu}$ is about five orders of magnitude more stringent on the product $\lambda'_{113}\lambda'_{131}$. If the exchanged scalar particles in Fig.9 are the $\tilde s - \tilde s^c$ pair, one obtains a limit

$$\lambda'_{121}\lambda_{112} \leq 1 \times 10^{-6} \tag{28}$$

which also is more stringent by about four orders of magnitude than the existing limits ($\lambda'_{121} \leq 0.26, \lambda'_{112} \leq 0.03$).

54

II.D: Models with heavy sterile neutrinos

The models we have discussed so far are very strongly motivated by independent physics considerations (other than understanding small neutrino masses). There is however a class of models which one can construct simply to use the see-saw mechanism (or variations of it) to understand the small neytrino mass. A simple example such models can be constructed by taking the singlet Majoron model and eliminating the lepton number carrying scalar boson Δ and instead adding an explicit heavy majorana mass for the three right handed neutrinos. The see-saw mechanism still operates so that small neutrino masses come out naturally. Let us ask if these models have any interesting implication for $\beta\beta_{0\nu}$ decay other than the usual neutrino mass contribution. The only possible new contribution would arise from the exchange of the heavy majorana right-handed neutrino- but as will be clear soon, this contribution is suppressed due to the see-saw mechanism. The point is clear if we look at Eq. and realize that the mixing parameter ξ between the light ν_e and the heavy N_e is given by $\sqrt{m_\nu/m_N}$ and generic see-saw formula for neutrinos require m_N in the range of tens of GeV. This makes $\xi \leq 10^{-5}$ or so. Therefore, the double beta amplitude contributed by the N exchange is at most of order $G_F^2 \times 10^{-10}/m_N$ which is much too small to be observable.

Since the heavy sterile sector is largely unknown, a possibility to consider is to have two heavy sterile leptons which participate in a 3×3 see-saw with the light neutrino to make m_ν small. The analog of the mixing parameter ξ is then not constrained to be small[24] by the see-saw considerations and also a larger range of masses for the heavy sterile particles are then admissible. Such models are however subject to a variety of cosmological and astrophysical constraints. These constraints have been analyzed in detail in [24] and it is found that there is a large range of the parameter space for the sterile particles which can be probed by the ongoing neutrinoless double beta experiments (see Fig.10 from [24]).

II.E: Limits on the scale of lepton compositeness

If the quarks and leptons are composite particles, it is natural to expect excited leptons which will interact with the electron via some effective interaction involving the W_L boson. If the excited neutrino is a majorana particle, then there will be contributions to $\beta\beta_{0\nu}$ decay mediated by the excited neutrinos (ν^*). The effctive interaction responsible for this is obtained from the primordial interaction:

$$H_{eff}^{\nu^*} = g\frac{\lambda_W^{(\nu^*)}}{2m_{\nu^*}}\bar{e}\sigma^{\mu\nu}(\eta_L^*(1-\gamma_5)+\eta_R^*(1+\gamma_5))\nu^* W_{\mu\nu} + \ h.c. \tag{29}$$

Here L and R denote the left and right chirality states. This contribution falls into our type II heavy particle exchange category and has been studied in detail in two recent papers[25] and have led to the conclusion that it leads to a lower bound

$$m_{\nu^*} \geq 5.9 \times 10^4 TeV \tag{30}$$

for $\lambda_W^{(\nu^*)} \geq 1$. This is a rather stringent bound on the compositeness scale.

Part III

Outlook for $\beta\beta_{0\nu}$ decay given present data on neutrinos:

The present situation in the neutrino physics is rather intriguing. On the one hand, the direct measurements show no evidence for any of the neutrinos to be massive, providing only the upper bounds $m_{\nu_e} < 4.5\,\mathrm{eV}$ ($< 0.7\,\mathrm{eV}$ $[\beta\beta_{0\nu}]$) $m_{\nu_\mu} < 160\,\mathrm{keV}$ and $m_{\nu_\tau} < 20\,\mathrm{MeV}$. The neutrinos could therefore be massless as far as these experiments are concerned. On the other hand there are several other experiments which provide strong indications in favor of neutrino masses and mixings. Let us describe them now.

III.A: Solar Neutrino Deficit

For massive neutrinos which can oscillate from one species to another, the solar electron neutrino observations[26] can be understood if the neutrino mass differences and mixing angles fall into one of the following ranges [28], where the Mikheyev-Smirnov-Wolfenstein (MSW) mechanism is included[27]: a) Small-angle MSW, $\Delta m^2_{ei} \simeq 6\times10^{-6}eV^2, sin^2 2\theta_{ei} \simeq 7\times10^{-3}$; b) Large-angle MSW, $\Delta m^2_{ei} \simeq 9\times10^{-6}eV^2, sin^2 2\theta_{ei} \simeq 0.6$; c) Vacuum oscillation, $\Delta m^2_{ei} \simeq 10^{-10}eV^2, sin^2 2\theta_{ei} \simeq 0.9$.

III.B: Atmospheric Neutrino Deficit

The second set of experiments indicating non-zero neutrino masses and mixings has to do with atmospheric ν_μ's and ν_e's arising from the decays of π's and K's and the subsequent decays of secondary muons produced in the final states of the π and K decays. In the underground experiments the ν_μ and $\bar{\nu}_\mu$ produce muons and the ν_e and $\bar{\nu}_e$ lead to $e^\pm$. Observations of $\mu^\pm$ and $e^\pm$ indicate a far lower value for ν_μ and $\bar{\nu}_\mu$ than suggested by naïve counting arguments which imply that $N(\nu_\mu+\bar{\nu}_\mu) = 2N(\nu_e+\bar{\nu}_e)$. More precisely, the ratio of μ events to e-events can be normalized to the ratio of calculated fluxes to reduce flux uncertainties, giving $R(\mu/e) = 0.60\pm0.07\pm0.05(Kamiokande); \ = 0.54\pm 0.05\pm0.12(IMB) and \ = 0.69\pm0.19\pm0.09(SoudanII)$

Combining these results with observations of upward going muons by Kamiokande[29], IMB[30], and Baksan[34] and the negative Fréjus[32] and NUSEX[33] results leads to the conclusion[29] that neutrino oscillations can give an explanation of these results, provided $\Delta m^2_{\mu i} \approx 0.005 \ to \ 0.5eV^2, \ sin^2 2\theta_{\mu i} \approx 0.5$.

III.C: Hot Dark Matter

There is increasing evidence that more than 90% of the mass in the universe must be detectable so far only by its gravitational effects. This dark matter is likely to be a mix of $\sim$ 20 to 30% of particles which were relativistic at the time of freeze-out from equilibrium in the early universe (hot dark matter) and $\sim$ 70% of particles which were non-relativistic (cold dark matter). Such a mixture[35] gives the best fit of any available model to the

56

structure and density of the universe on all distance scales, such as the anisotropy of the microwave background, galaxy-galaxy angular correlations, velocity fields on large and small scales, correlations of galaxy clusters, etc. A very plausible candidate for hot dark matter is one or more species of neutrinos with total mass of $m_{\nu_H} = 93h^2 F_H \Omega = 5$ eV, if $h = 0.5$ (the Hubble constant in units of 100 km·s^{-1}·Mpc^{-1}), $F_H = 0.2$ (the fraction of dark matter which is hot), and $\Omega = 1$ (the ratio of density of the universe to closure density). We shall use the frequently quoted 5 eV below, but different determinations give $h = 0.45 \pm 0.09$[36] or $h = 0.80 \pm 0.11$[37] (a value giving difficulties with $\Omega = 1$), making $m_{\nu_H} = 2$ or 21 eV.

It is usually assumed that the ν_τ would supply the hot dark matter. This is justified on the basis of an appropriately chosen see-saw model[15] and a $\nu_e \to \nu_\mu$ MSW explanation of the solar ν deficit. However, if the atmospheric ν_μ deficit is due to $\nu_\mu \to \nu_\tau$, the ν_τ alone cannot be the hot dark matter, since the ν_μ and ν_τ need to have close to the same mass. It is interesting that instead of a single $\simeq 5$ eV neutrino, sharing that $\simeq 5$ eV between two or among three neutrino species provides a better fit to the universe structure and particularly a better understanding of the variation of matter density with distance scale[38].

III.D: Indications of $\overline{\nu}_\mu \to \overline{\nu}_e$ oscillation from LSND

There appear to be some indications in favor of a possible oscillation of $\overline{\nu}_\mu \to \overline{\nu}_e$ from the recent LSND experiment[39]. While these results are not completely conclusive, taken at face value a $\Delta m^2 \approx 1 - 6\ eV^2$ and $sin^2\theta \approx 10^{-2}$ appears to be the preferred range of parameters needed to explain observations.

III.E: Nucleosynthesis Limits on Neutrino Species

While the Z^0 width limits the number of weakly interacting neutrino species to three, the nucleosynthesis limit[40] on the number of light neutrinos (denoted by N_ν) is more useful here, since it is independent of the neutrino interactions with the Z-boson. Until a few months ago, the limit on N_ν was 3.3. However, a recent analysis by Hata et al.[41] concludes that after one includes the evolutionary effects on the ^3He and Deuterium abundances, the present primordial ^4He abundance rules out $N_\nu = 3$ at 99.7% confidence level and favors a value close to $N_\nu = 2$. Thus one would have trouble understanding present helium abundance using three light neutrinos in the framework of the standard model. One possibility is to have a unstable tau neutrino with mass in the MeV range with a life time of the order of a few seconds decaying to ν_e+ majoron.

Within this set of constraints for example, the atmospheric ν_μ problem cannot be explained by $\nu_\mu \to \nu_s$ because it requires a large mixing angle and in that case, for the $\Delta m^2_{\mu s}$ involved, ν_s would have contributed as one extra neutrino species. On the other hand, the solar ν_e problem can be explained by $\nu_e \to \nu_s$ for either the small-angle MSW or the vacuum oscillation solutions, but not for the less favored large-angle MSW solution

III.F:Supernova r-Process Constraint

Another set of constraints on neutrino mixings have been derived from the assumption that heavy elements in the universe are produced in the neutron rich environment around the supernovae by rapid neutron capture known as r-process. It has been shown that unless ν_e–ν_μ and ν_e–ν_τ mixing angles are severely restricted ($\sin^2 2\theta \leq 4 \times 10^{-4}$) for $\Delta m^2 \geq 4$ eV^2 (with a rapid decrease in $\sin^2 2\theta$ for larger Δm^2)[42], the energetic ν_μ and ν_τ ($\langle E \rangle \approx 25$ MeV) can convert to ν_e's which have much higher energy than the thermal ν_e's ($\langle E \rangle \approx 11$ MeV). The higher energy ν_e's, having a larger cross section, will reduce the neutron density via $\nu_e + n \rightarrow e^- + p$, diminishing heavy element formation.

III.G: Possible patterns of Neutrino masses consistent with the above constraints

A. *Patterns Required by Solar and Atmospheric Neutrino Deficits and Hot Dark Matter*

With the above input information, if we stay within the minimal three neutrino picture, then the solar neutrino puzzle can be resolved by $\nu_e \rightarrow \nu_\mu$ oscillations and the atmospheric neutrino deficit by $\nu_\mu \rightarrow \nu_\tau$ oscillations. Note that these observables are controlled only by the mass square difference; on the other hand, the required hot dark matter implies that at least one or more of the neutrinos must have mass in the few eV range. It was pointed out[43] in 1993 that, in the minimal picture, this leads to the following scenario, labelled (A):

All three neutrinos are nearly degenerate, with $m_{\nu_e} \approx m_{\nu_\mu} \approx m_{\nu_\tau} \approx 2$ eV, since $\nu_e \rightarrow \nu_\mu$ and $\nu_\mu \rightarrow \nu_\tau$ both require small mass differences, but the required dark matter mass can be shared. The mass matrix for this case in the ν_e, ν_μ, ν_τ basis is given by:

$$M = \begin{pmatrix} m + \delta_1 s_1^2 & -\delta_1 c_1 c_2 s_1 & -\delta_1 c_1 s_1 s_2 \\ -\delta_1 c_1 c_2 s_1 & m + \delta_1 c_1^2 c_2^2 + \delta_2 s_2^2 & \delta_1 c_1^2 c_2 s_2 - \delta_2 c_2 s_2 \\ -\delta_1 c_1 s_1 s_2 & \delta_1 c_1^2 s_2 c_2 - \delta_2 c_2 s_2 & m + \delta_1 c_1^2 s_2^2 + \delta_2 c_2^2 \end{pmatrix} \tag{31}$$

where $c_i = \cos\theta_i$ and $s_i = \sin\theta_i$, $m = 2$ eV; $\delta_1 \simeq 1.5 \times 10^{-6}$ eV; $\delta_2 \simeq .2$ to $.002$ eV; $s_1 \simeq 0.05$; and $s_2 \simeq 0.4$ for the small-angle MSW solution.

In this case, the LSND results cannot be accomodated. However, there will be an observable amplitude for neutrinoless double beta decay mediated by the neutrino mass mechanism. In fact, if the limit on $\langle m_\nu \rangle$ goes below 1 eV (without any nuclear matrix element uncertainty), then this model will fail to provide a viable hot dark matter candidate.

B. *Mass matrix Accomodating the solar, atmospheric and LSND data and HDM:*

In this case, an additional light sterile neutrino (to be called ν_s) is essential[43]. The ν_e and ν_s[44] are assumed to be quite light to take care of the solar neutrino problem while the ν_μ and ν_τ share the dark matter role, being ~ 2.4 eV each, and explain the

58

atmospheric ν_μ deficit. Recently, several interesting gauge models realizing this texture have been constructed[45].

In this case as in case A, one will have to use the small-angle MSW solution, since the ν_s has to be very weakly mixed to satisfy the nucleosynthesis bound. Nucleosynthesis also eliminates the large-angle MSW solution and in the vacuum oscillation case forces the ν_s to be mixed strongly only with the ν_e. The form of the Majorana mass matrix for this case is given by[43]: (in the basis $(\nu_s, \nu_e, \nu_\mu, \nu_\tau)$,

$$M = \begin{pmatrix} \mu_1 & \mu_3 & 0 & 0 \\ \mu_3 & \mu_2 & \epsilon_{21} & 0 \\ 0 & \epsilon_{21} & m & \delta/2 \\ 0 & 0 & \delta/2 & m+\delta \end{pmatrix} \tag{32}$$

As is clear, model can also accomodate the LSND $\nu_e \to \nu_\mu$ oscillation. Needless to say that the neutrinoless double beta decay will be unobservable in this case.

C. *Inverted Mass Hierarchy for Solar Neutrino Puzzle, HDM and LSND*

Since the atmospheric neutrino anomaly is perhaps on a somewhat weaker footing due to some experiments (e.g. Frejus and NUSEX) not showing this anomaly, it may be interesting to see what kind of mass pattern is allowed by eliminating it as a constraint. This has been studied in several papers recently[46] and it has been noted that in this case, there is again no need for a sterile neutrino and one can write the following 3×3 mass matrix for the three Majorana neutrinos $(\nu_e, \nu_\mu, \nu_\tau)$.

$$M = \begin{pmatrix} -m\beta-\delta & -\mu_1 & m+\delta \\ -\mu_1 & \mu & -\mu_1 \\ m+\delta & -\mu_1 & m\beta-\delta \end{pmatrix} \tag{33}$$

We assume that $\mu \ll m \simeq 2.4$ eV; μ_1/m denotes the $\nu_e - \nu_\mu$ mixing angle responsible for the LSND results and $\delta \simeq 10^{-5}$ eV to fit the solar neutrino data. The key new implication here is that the solar neutrino puzzle must be resolved by the large angle MSW solution, a result which can be tested at SuperKamiokande by looking for the day-night variation in the neutrino flux. Note the presence of an exact $L_e - L_\tau$ symmetry of the mass matrix in the limit of vanishing β, δ, μ_1. In this case also the neutrinoless double beta decay is unobservable. It must however be said that if the dominant neutrino masses in this case came by putting the entry m above along the diagonal rather along the antidiagonal positions, then there would be an observable neutrinoless double beta decay amplitude in the ongoing experiments.

Part IV

Theoretical scenarios

Let us note some important qualitative points about the mass matrices described here. Generically, they indicate two kinds of scales: one corresponding to the mass differences which are typically of order $\approx 10^{-3}$ eV or so and another corresponding to a mass of order of a few eV. In the canonical see-saw models one has $m_{\nu_i} \simeq \frac{m^2_{u_i}}{f v_R}$ which leads to a hierarchical neutrino mass pattern i.e. $m_{\nu_e} \ll m_{\nu_\mu} \ll m_{\nu_\tau}$ with $m_{\nu_\tau} \approx 2 - 4$ eV. While this simple picture can very easily accomodate a solution to the solar neutrino puzzle and an HDM, it has room neither for the LSND result nor for the atmospheric neutrino data. One has to go beyond this picture to understand all existing observations. It turns out that the canonical see-saw picture is not realized[2] in many popular unified models and the correct see-saw matrix (to be called type II see-saw matrix here) that arises for instance in the SO(10) models helps in generating the matrix texture (A).

IV.A Type II See-Saw and Degenerate Neutrinos in SO(10) GUT

In the early days of the discussion of the see-saw formula for neutrino masses, it was pointed out[2] that implementing it in the simplest left-right or SO(10) models resulted in a ν_L-ν_R mass matrix of the modified see-saw form:

$$\begin{pmatrix} f v_L & m_{\nu_D} \\ m^T_{\nu_D} & f v_R \end{pmatrix} \tag{34}$$

where f and m_{ν_D} are 3×3 matrices and $v_L \approx \lambda M^2_{W_L}/v_R$. The light neutrino mass matrix that follows from diagonalizing the above mass matrix is (type II see-saw formula)

$$m_\nu \approx f \lambda M^2_{W_L}/v_R - m_{\nu_D} f^{-1} m^T_{\nu_D}/v_R + \ldots \tag{35}$$

While both terms vanish as $v_R \to \infty$, the first term always dominates over the second one for neutrino masses. This negates the usual quadratic formula (i.e., the second term) for neutrino masses.

Within the type II see-saw formula, it is clear that if a symmetry dictates that $f_{ab} = f\delta_{ab}$, then the neutrino masses are degenerate to leading order. For $v_R \approx 10^{13.5}$ GeV, $f\lambda \approx 1$, we get $m_{\nu_e} = m_{\nu_\mu} = m_{\nu_\tau} \approx 1.5$eV, $m^2_{\nu_\mu} - m^2_{\nu_e} \approx 3m^2_c/(10 f v_R) \approx 10^{-4}/f$ eV^2, and $m^2_{\nu_\tau} - m^2_{\nu_\mu} \approx 3m^2_t/(10 f v_R) \approx (2/f)(m_t/150 \text{ GeV})^2$ eV^2. These mass differences are of the right order of magnitude to explain the solar neutrino (via the MSW mechanism) and the atmospheric neutrino puzzles, while the sum of all the neutrino masses roughly give the needed amount of hot dark matter.

It is also interesting to note that the B-L breaking scale of $v_R \sim 10^{13}$ GeV emerges naturally from constraints of $\sin^2 2\theta_W$ and α_s in non-supersymmetric SO(10) grandunified theories , enhancing the reason for an SO(10) scenario. To guarantee the neutrino degeneracy (i.e., $f_{ab} = f\delta_{ab}$), an extra family symmetry is imposed on the model. This family

60

symmetry will be broken softly by terms in the Lagrangian of dimension two, so that departures from the degeneracy in the neutrino sector are naturally small. An explicit example with an S_4 horizontal symmetry was worked out in [47], where it was possible to predict the complete neutrino mixing matrix:

$$V^l = \begin{pmatrix} -.9982 & .05733 & .01476 \\ .05884 & .9334 & .3541 \\ -.00652 & -.3544 & .9351 \end{pmatrix} \tag{36}$$

This mixing matrix can be tested by the proposed long baseline experiments such as the Fermilab , Brookhaven and KEK experiments. If future experiments bear out a degenerate light neutrino spectrum, this detailed SO(10) model may or may not be the appropriate description of the physics. However, its essential ingredient, the type II see-saw formula, will almost surely be required to fit those data.

IV.B: Inverted Mass Hierarchy from an $L_e - L_\tau$ symmetric Left-right Model

We saw in the last section that to get three degenerate neutrinos requires a very elaborate horizontal symmetry structure. On the other hand, it turns out that if only two Majorana neutrinos are to be degenerate with opposite CP properties, it suffices to have a simpler $U(1)$ symmetry involving the two leptons. In the present case, the relevant symmetry is $L_e - L_\tau$ as has been noted in the second paper of [46]. Then one can use the type II see-saw formula in the context of a left-right symmetric model to generate the above mass matrix.

In conclusion, neutrinoless double beta decay provides a very versatile way to probe scenarios of physics beyond the standard model. In this review, we have focussed only on the 0ν mode; there can also be single and multi majoron modes which test for the possibility of lepton number being a spontaneously broken global symmetry. The theory and phenomenology of this type of modes have been discussed in this volume by C. Burgess[48]. The 0ν mode acquires special interest in view of certain SO(10) models predicting such spectra without contradicting the solar and atmospheric neutrino data.

Acknowledgement

I am grateful to K. S. Babu, D. Caldwell, C. Burgess, H. Klapdor-Kleingrothaus, M. Hirsch, S. Kovalenko, E. Takasugi for many discussions on the subject of this review.

References

[1] R.N. Mohapatra and P.B. Pal, "Massive Neutrinos in Physics and Astrophysics", World Scientific, Singapore, 1991.

[2] R. N. Mohapatra and G. Senjanović, Phys. Rev. Lett. **44**, 912 (1980); Phys. Rev. **D23**, 165 (1981).

[3] C. S. Aulakh and R. N. Mohapatra, Phys. Lett. **119B**, 136 (1983); F. Zwirner, Phys. Lett. **132B**, 103 (1983); L. Hall and M. Suzuki, Nucl. Phys. **B231**, 419 (1984); G. G. Ross and J. W. F. Valle, Phys. Lett. **B151**, 375 (1985).

[4] H. Klapdor-Kleingrothaus, Prog. in Part. and Nucl. Phys., **32**, 261 (1994); A. Balysh et. al., Phys. Lett. (to appear).

[5] M. Moe and P. Vogel, Ann. Rev. Nucl. Sc. **44**, 247 (1994).

[6] see the talks by P. Vogel, K. Muto, S. Stoica and others in this volume.

[7] M. Doi, T. Kotani, E. Takasugi, Prog. Theor. Phys. Suppl. **83**, 1 (1985).

[8] H. Primakoff and S. P. Rosen, Rep. Prog. Phys. **22**, 121 (1959)

[9] M. Doi et. al.[7]; W. C. Haxton and G. Stephenson, Prog. in Part. and Nucl. Phys. **12**, 409 (1984); H. Grotz and H. Klapdor, The Weak Interactions in Nuclear, Particle and Astrophysics, Adam Hilger, Bristol, (1990); D. Caldwell, Nucl. Phys. Proc. Suppl. **B 13**, 547 (1990).

[10] K. S. Babu and R. N. Mohapatra, hep-ph/9506354.

[11] A. Halprin, P. Minkowski, S. P. Rosen and H. Primakoff, Phys. Rev. **D13**, 2567 (1976).

[12] R.N. Mohapatra and J. Vergados, Phys. Rev. Lett. **47**, 1713 (1981).

[13] J. Schecter and J.W.F. Valle, Phys. Rev. **D25**, 2951 (1982); W.C. Haxton, S.P. Rosen and G.J. Stephenson, *ibid.*, **D26**,1805 (1982); L. Wolfenstein, *ibid.*, **D26**, 2507 (1982).

[14] Y. Chikashige, R. N. Mohapatra and R. D. Peccei, Phys. Lett. **98B**, 265 (1981).

[15] M. Gell-Mann, P. Ramond and R. Slansky, *in* "Supergravity", Ed. D.Freedman et al. (North-Holland, Amsterdam, 1979); T. Yanagida, Prog. Th. Phys. **B135** (1978) 66; R.N. Mohapatra and G. Senjanović, Phys. Rev. Lett. **44** (1980) 912.

[16] R. N. Mohapatra and S. Nussinov, Phys. Rev. **D51**, 3843 (1995).

[17] N. Hata et al., Ohio State Univ. preprint, hep-ph/9505319

[18] S. Dodelson, G. Gyuk and M. Turner, Phys. Rev. Lett. **72**, 3578 (1995).

[19] M. Schwarz, Phys. Rev. **D40**, 1521 (1989).

[20] R.N. Mohapatra, Phys. Rev. **D34**, 909 (1986).

[21] M. Hirsch, H. Klapdor-Kleingrothaus and S. Kovalenko, Heidelberg preprint (1995).

[22] M. Hirsch, H. V. Klapdor-Kleingrothaus and S. G. Kovalenko, Heidelberg Preprint (1995).

[23] V. Barger, G. Giudice and T. Han, Phys. Rev. **D40**, 2987 (1989).

[24] P. Bamert, C. Burgess and R. N. Mohapatra, Nucl. Phys. **B438**, 3 (1995).

[25] O. Panella and Y. N. Srivastava, College de France Preprint, LPC 94-39; E. Takasugi, hep-ph/9506379.

[26] GALLEX Collaboration, Phys. Lett. **B327**, 377 (1994); SAGE Collaboration, Phys. Lett. **B328**, 234 (1994); Kamiokande Collaboration, Nucl. Phys. **B38** (Proc. Suppl.), 55 (1995); Homestake Collaboration, *ibid.* 47.

[27] S.P. Mikheyev and A.Yu. Smirnov, Yad. Fiz. **42**, 1441 (1985); Nuovo Cim. **9C**, 17 (1986); L. Wolfenstein, Phys. Rev. **D17**, 2369 (1978).

[28] For the recent status see P.I. Krastev and A.Yu. Smirnov, Phys. Lett. **B338**, 882 (1994); V.S. Berezinsky, G. Fiorentini and M. Lissia, Phys. Lett. **B341**, 38 (1994); N. Hata and P. Langacker, Phys. Rev. **D50**, 632 (1994); V. Castellani, S. Degl'Innocenti and G. Fiorentini, Astron. Astrophys. **271** (1993) 601; Phys. Lett. **B303**, 68 (1993); J.N. Bahcall and H.A. Bethe, Phys. Rev. Lett. **65**,2233 (1993); S.A. Bludman, N. Hata and P. Langacker, Phys. Rev. **D 49**, 3622 (1994); V.S. Berezinsky, Comments Nucl. Part. Phys. **21**, 249 (1994); A.Yu. Smirnov, preprint DOE/ER/40561-136-INT94-13-01 (1994).

[29] Kamiokande Collaboration, Y. Fukuda et al., Phys. Lett. **B335**, 237 (1994).

[30] R. Becker-Szendy et al., Phys. Rev. **D 46**, 3720 (1992).

[31] P. J. Lichtfield et al., in International Europhysics Conference on High Energy Physics, Marseille, France (1993).

[32] K. Daum et al., Zeit. für Phys. **C 66**, 417 (1995).

[33] M. Aglieta et al., Europhys. Lett. **8**, 611 (1989).

[34] M. M. Boliev et al., in Proc. of Third International Workshop on Neutrino telescopes, Venice ed by M. Baldo-ceolin (1991), p.235.

[35] R. Shaefer and Q. Shafi, Nature **359**, 199 (1992); M. Davis. F.J. Summers and D. Schagel, *ibid.* 396; A.N. Taylor and M. Rowan-Robinson, *ibid.* 396; E.L. Wright et al., Ap. J. **396**, L13 (1992); J.A. Holtzman and J.R. Primack, Ap. J. **405**, 428 (1993); A. Klypin et al., Ap. J. **416**,1 (1993).

[36] A. Sandage et. al., Astrophys. J. **401**, L7 (1992).

[37] M.J. Pierce et al., Nature **371** (1994) 385;

[38] J.R. Primack, J. Holtzman, A. Klypin and D.O. Caldwell, Phys. Rev. Lett. **74**, 2160 (1995).

[39] C. Athanassopoulos et al., preprint LA-UR-95-1238 (nucl-ex/9504002).

[40] T. Walker et al., Ap. J. **376** (1991) 51; P. Kernan and L. Krauss, Phys. Rev. Lett. **72** (1994) 3309; K. Olive and G. Steigman, preprint OSU-TA-2/95.

[41] Recently N. Hata et al., preprint OSU-TA-6/95 (May 1995), reconsidered the nucleosynthesis bound and found the upper limit $N_\nu < 2.5$ at For a recent review, see K. Olive and S. T. Scully, UMN-TH-1341/95.

[42] Y-Z Qian and G. Fuller, Phys. Rev. D (in press).

[43] D. Caldwell and R. N. Mohapatra, Phys. Rev. **D48**, 3259 (1993).

[44] See J. Peltoniemi, Sissa Preprint (1995).

[45] D. Caldwell and R. N. Mohapatra, Ref.[43]; J. Peltoniemi and J. W. F. Valle, Nucl. Phys. **B406**, 409 (1993); E. Ma and P. Roy, UCRHEP-T145; E. J. Chun, A. Joshipura and A. Smirnov, hep-ph/9505275; Z. Berezhiani and R. N. Mohapatra, hep-ph/9505385.

[46] G. Raffelt and J. Silk, Berkeley preprint, 1995; D. Caldwell and R. N. Mohapatra, Phys. Lett. B (to appear).

[47] D. G. Lee and R. N. Mohapatra, Phys. Lett. **B329**, 463 (1994).

[48] C. Burgess, this volume.

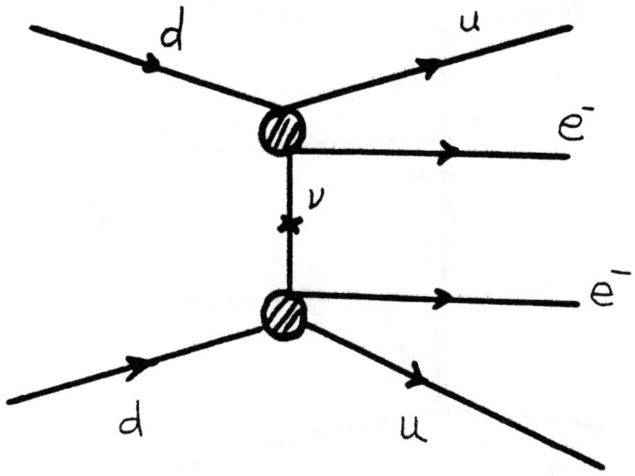

Figure 1. Feynman diagram involving neutrino majorana mass that contributes to $\beta\beta_{0\nu}$ decay.

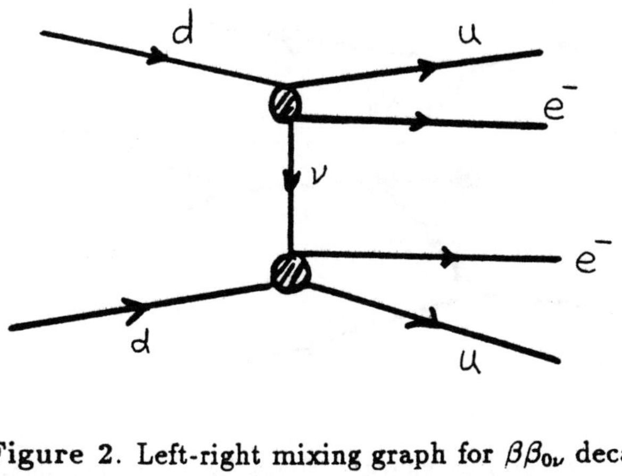

Figure 2. Left-right mixing graph for $\beta\beta_{0\nu}$ decay.

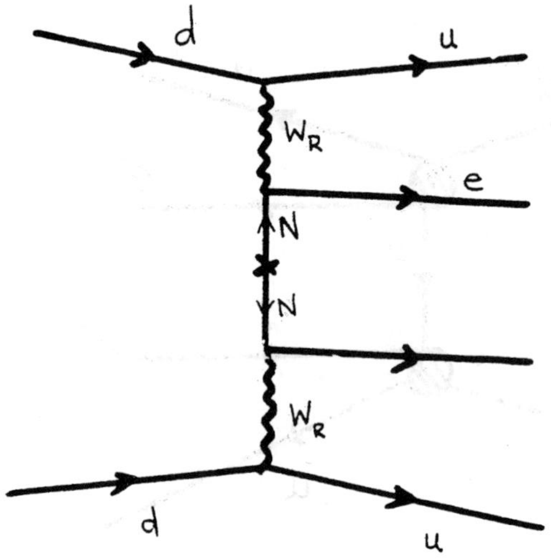

Figure 3. Heavy right handed neutrino contribution to $\beta\beta_{0\nu}$ in the left-right symmetric model.

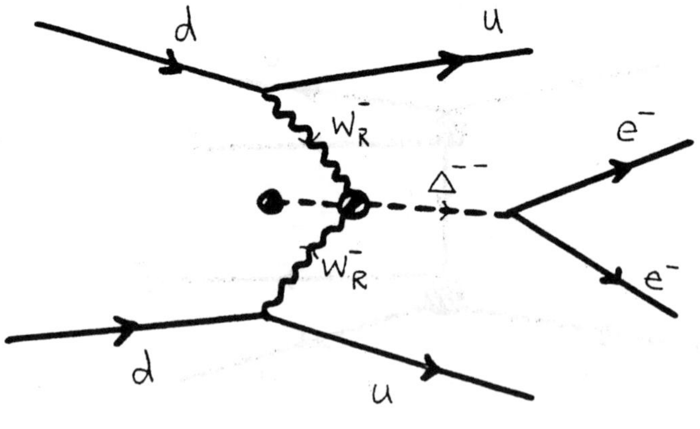

Figure 4. Contribution of the doubly charged Higgs boson in the left-right symmetric model.

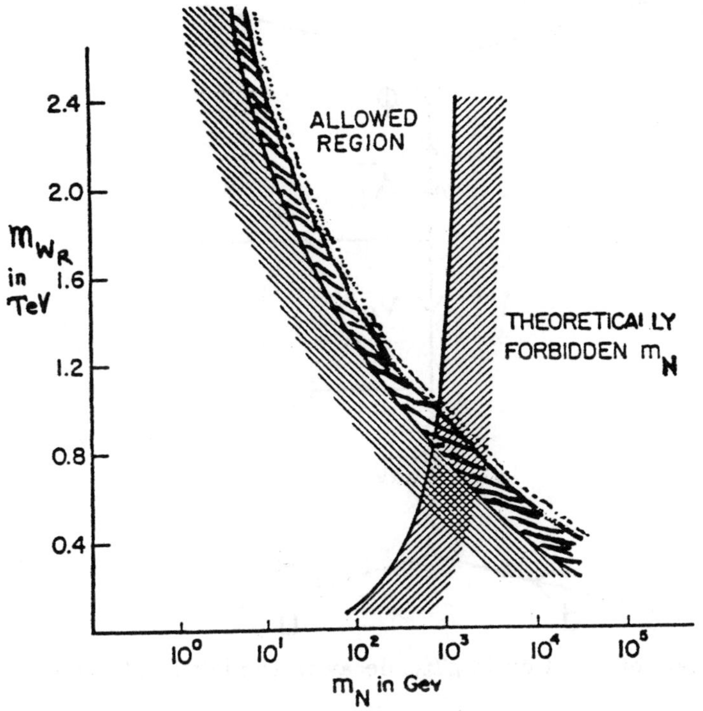

Figure 5. Bounds on the masses of m_{W_R} and m_N from $\beta\beta_{0\nu}$ lifetime and theoretical arguments of vacuum stability [20].

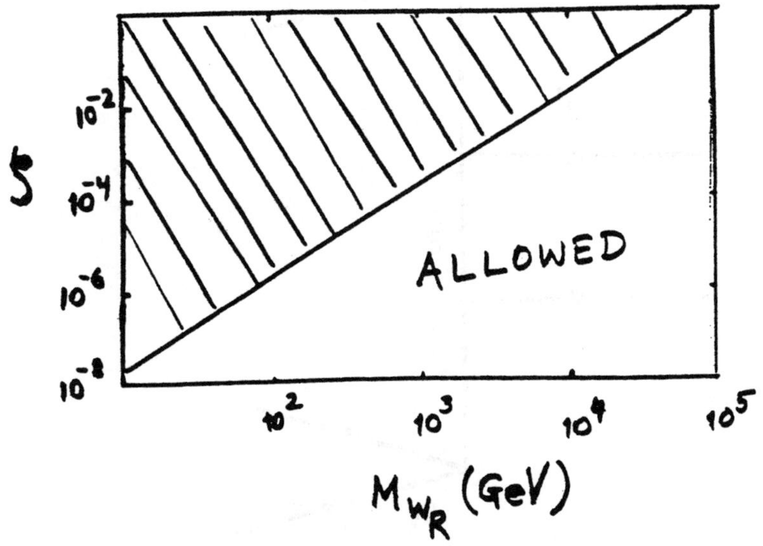

Figure 6. Bounds on the light and heavy neutrino mixing parameter in the left-right model from ^{76}Ge data.

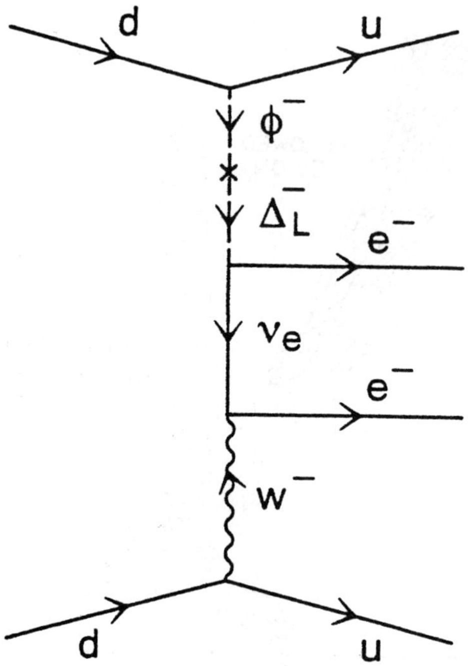

Figure 7. Vector-scalar contribution to $\beta\beta_{0\nu}$ decay in the left-right symmetric model.

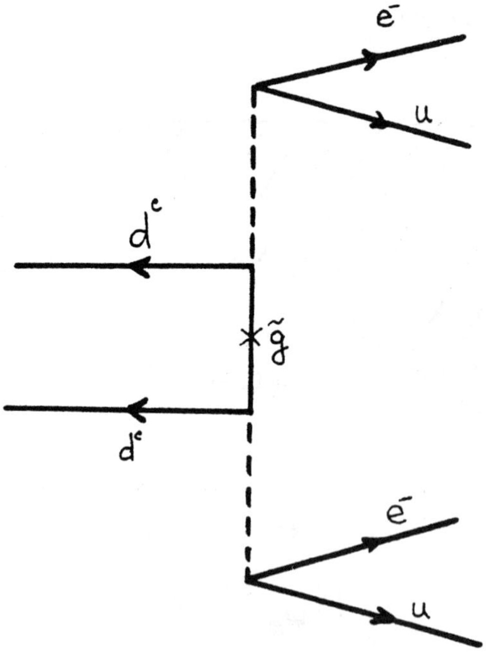

Figure 8. Gluino mediated contribution in MSSM with R-parity violation.

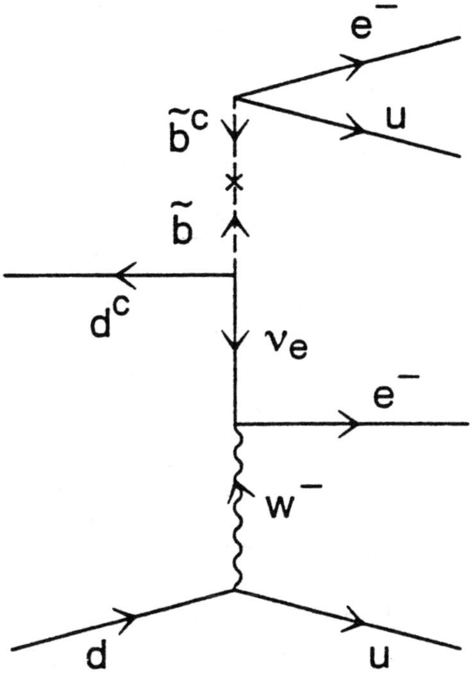

Figure 9. Vector-scalar contribution in MSSM with R-parity violation.

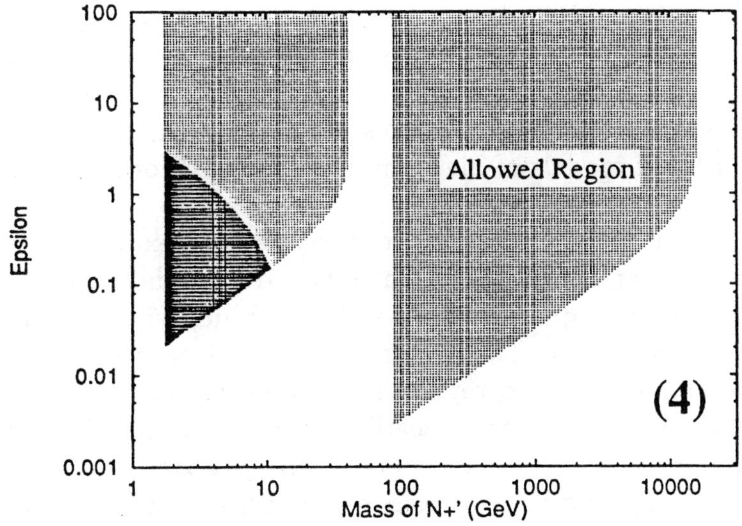

Figure 10. The shaded area in the figure represents the range of mixing and mass parameters of a specific heavy sterile neutrino which are allowed by low energy and cosmological bounds and where the $\beta\beta_{0\nu}$ amplitude is in the observable range.

Double Beta Decay – Physics Beyond the Standard Model

H.V. Klapdor–Kleingrothaus
Max–Planck–Institut für Kernphysik
P.O.Box 10 39 80, D–69029 Heidelberg, Germany

ABSTRACT

Double beta decay yields – besides proton decay – the most promising possibilities to probe beyond standard model physics at beyond accelerator energies. The possibilities include the neutrino mass, SUSY models, compositeness, leptoquarks, right–handed W bosons and others. We discuss the status and the future perspectives of $\beta\beta$ research, including applications some double beta technology can find in the search for dark matter. It is found that $0\nu\beta\beta$ decay probes already now the TeV scale on which new physics should manifest itself according to present theoretical expectations.

in Proc. "Neutrino Physics and Astrophysics", NEUTRINO'96, Helsinki, Finland, June 1996, eds. *K. Enqvist, K. Huitu and J. Maalampi, World Scientific, Singapore,* (1997) 317 – 341

1. Introduction

Many central questions of particle physics are beyond the capabilities of modern accelerators. They can, however, to some extent be investigated via non–accelerator experiments (see, e.g.[Kla95]). LHC, for example, the main enterprise of High Energy Physics in the next decade, will cover physics up to scale of a few TeV and may search for the Higgs particle, some SUSY particles and others.[Zer93,Nuc86] In general, however, accelerator physicists at present are forced to search in the extreme 'low–energy' range of the parameter spaces of models of 'new physics' like SUSY or leptoquark signatures and others (see e.g.[Buc91]).

This explains the increasing trend to non–accelerator experiments in numerous underground laboratories and elsewhere. Double beta decay, and proton decay, to mention the most prominent examples, are among those, which yield the most promising possibilities to probe beyond the standard model (SM) physics at beyond accelerator energies. Propagator physics has to replace direct observations. That this method is very effective, is obvious from important earlier research work and has been stressed, e.g. by.[Rub96] Examples are the proporties of W and Z bosons derived from neutral weak currents and β–decay, and the top mass deduced from LEP electroweak radiative corrections. Also for accelerators, the search for new bosons or compositeness, for example, can be to some degree extended beyond kinematical production limits through the study of *indirect* effects from virtual particle exchange (see, e.g.[H195]).

Doppelbetazerfall – Physik jenseits des Standardmodells

Hans Volker Klapdor-Kleingrothaus

Der Doppelbetazerfall bietet gegenwärtig eine der vielversprechendsten Möglichkeiten, Physik jenseits des Standardmodells der Elementarteilchenphysik und jenseits der direkten Möglichkeiten gegenwärtiger und zukünftiger Hochenergiebeschleuniger zu sondieren.

Im Jahre 1935, fünf Jahre nach Postulierung des Neutrinos durch Wolfgang Pauli und nur ein Jahr nach Enrico Fermis Theorie der schwachen Wechselwirkung, untersuchte Maria Goeppert-Mayer einen neuen ungewöhnlichen Prozeß, den Doppelbetazerfall. Dieser Effekt, in dem der zerfallende Atomkern zwei Elektronen und zwei Neutrinos emittiert (Abbildung 1), wurde erst 50 Jahre später nachgewiesen, zunächst über geochemische Experimente, und 1987 auch erstmalig in einem direkten Zählerexperiment in Irvine, Kalifornien. Der Doppelbetazerfall gibt heute interessante Aufschlüsse zur Struktur der Atomkerne.

Doppelbetazerfall und Neutrinomasse

Der Doppelbetazerfall ist sehr selten: Die Lebensdauer eines Doppelbeta-Emitters übertrifft das Alter des Universums um das Billionenfache. Heute suchen wir vornehmlich nach einem anderen, noch selteneren Prozeß, dem sogenannten neutrinolosen Doppelbetazerfall (Abbildung 1). Hierbei wandeln sich zwei Neutronen im Kern unter Austausch eines Neutrinos und unter Abstrahlung zweier Elektronen um. Zudem wird möglicherweise ein masseloses Teilchen, ein sogenanntes Goldstone-Boson, frei, das Majoron genannt wird.

Voraussetzung für diesen Prozeß ist jedoch eine Verletzung der Leptonenzahlerhaltung oder der Größe Baryonenzahl minus Leptonenzahl (B-L). Außerdem muß das Neutrino ein sogenanntes Majorana-Teilchen sein. Dies bedeutet, daß Neutrino und Antineutrino identisch sind, und es muß eine nichtverschwindende Ruhemasse haben.

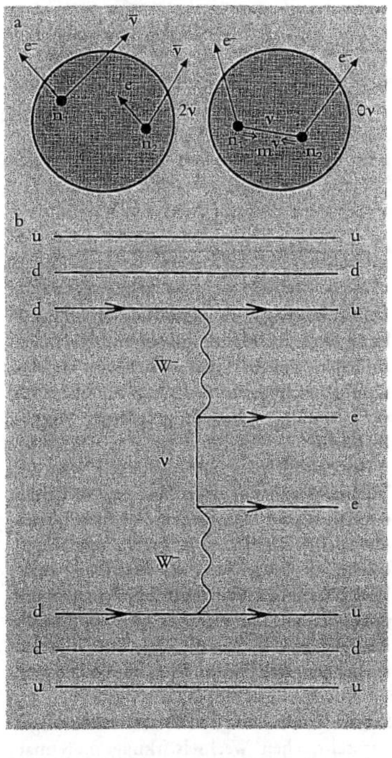

Abb. 1. (a) Schematische Darstellung des Doppelbetazerfalls von Atomkernen. Links der Zerfall unter Emission zweier Antineutrinos, rechts der neutrinolose Zerfall. Hier wird ein massives Neutrino zwischen den zerfallenden Neutronen ausgetauscht. (b) Feynman-Graph für neutrinolosen Doppelbetazerfall durch Austausch eines linkshändigen leichten oder schweren Neutrinos (ν). Up- und Down-Quarks sind mit u und d bezeichnet.

Tatsächlich sagen die meisten Theorien der Großen Vereinigung der Elementarteilchenphysik im Gegensatz zum Standardmodell vorher, daß das Neutrino ein Majorana-Teilchen ist und eine Masse besitzt. Die erwartete Rate für den Zerfallsprozeß ist proportional dem Quadrat der Neutrinomasse.

Die Frage nach der Natur der Neutrinos ist in vielfältiger Weise mit ungeklärten Problemen der Teilchenphysik verknüpft. Insbesondere mit möglichen Erweiterungen des heutigen Standardmodells. Im folgenden werden wir einige Aspekte schildern.

Neutrinos als Sonden für „neue Physik"

Die Neutrinomasse ist eine der Schlüsselgrößen für die Großen Vereinigungstheorien der Elementarteilchenphysik und gleichzeitig von außerordentlicher Bedeutung für die Kosmologie. Doppelbetaexperimente sind derzeit in eine Phase besonderer Aktualität eingetreten. Es werden nämlich gegenwärtig mögliche Hinweise auf „neue Physik", das heißt Physik jenseits des Standardmodells der Elementarteilchenphysik, diskutiert, die am einfachsten zu verstehen sind, wenn Neutrinos eine Ruhemasse besitzen (vgl. *Physik in unserer Zeit* **26**, 129 (3/1995)).

Hierzu zählt das solare Neutrinoproblem, wonach nur etwa die Hälfte der erwarteten Neutrinos von der Sonne gemessen werden (vgl. *Physik in unserer Zeit* **23**, 246 (6/1992)). Ein anderes Problem ist die Anomalie des zu geringen atmosphärischen Myonneutrino Flusses. In der Atmosphäre entstehen durch energiereiche kosmische Teilchen Myon- und Elektronneutrinos mit einem theoretischen

 0031-9252/98/0305-0123 $ 17.50 + .50/0

Häufigkeitsverhältnis $\nu_\mu/\nu_e = 2$. Gemessen werden jedoch etwa gleich viele Myon- und Elektronneutrinos. Die Ergebnisse der solaren und atmosphärischen Neutrinos ließen sich durch Flavour-Oszillationen von Neutrinos erklären, die allerdings nur möglich sind, wenn diese eine Ruhemasse besitzen. Überdies sagen einige Modelle von der Entwicklung des Universums die Existenz unsichtbarer Materie voraus. Neutrinos bilden hierfür mögliche Kandidaten.

Es wurde kürzlich gezeigt, daß alle drei Probleme in einem erweiterten Modell der Großen Vereinheitlichungstheorien, GUT, verständlich werden [1]. Dieses einfachste Szenario für die drei bekannten leichten Neutrinos, welches das solare und atmosphärische Neutrinodefizit und außerdem die Vorstellungen von der Dunklen Materie erklärt, fordert für die drei bekannten Neutrinoflavours Massen zwischen 1 und 2 eV – im Gegensatz zu GUT-Modellen, die eine ausgeprägte Massenhierarchie der Neutrinos voraussetzen. Auch die bislang noch umstrittenen Ergebnisse des Beschleunigerexperiments in Los Alamos, das auf Neutrino-Oszillationen hindeutet, lassen sich mit weitgehend entarteten Massen der Neutrinoflavours erklären [2].

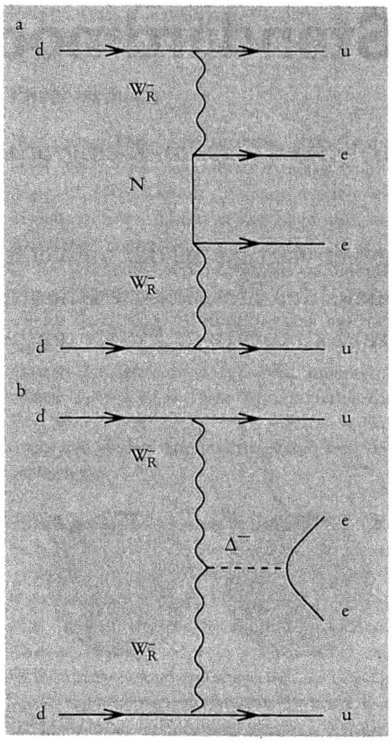

Abb. 2. **Beitrag zum neutrinolosen Doppelbetazerfall in rechts-links-symmetrischen Modellen durch Austausch eines rechtshändigen schweren Neutrinos N oder eines rechtshändigen doppelt geladenen Higgs-Bosons Δ^-**

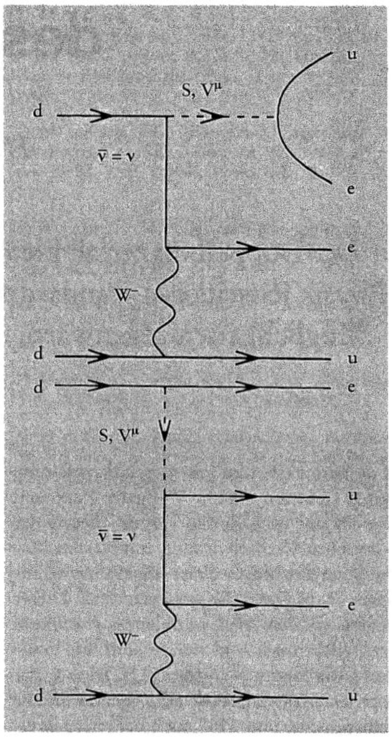

Abb. 3. Beispiele für Feynman-Graphen neutrinoloser Doppelbetazerfälle unter Austausch von Leptoquarks. S und V^μ stehen für skalare und vektorielle Leptoquarks [6].

Die Aktualität des laufenden Doppelbetaexperiments besteht nun darin, daß es die Existenz einer Majorana-Masse um 1 eV in naher Zukunft bestätigen oder verwerfen kann. Bislang konnte der neutrinolose Doppelbetazerfall nicht nachgewiesen werden. Wie wir weiter unten darlegen werden, ergibt sich derzeit aus der gemessenen unteren Grenze für die Halbwertszeit des neutrinolosen Doppelbetazerfalls von $1{,}1 \cdot 10^{25}$ Jahren eine obere Grenze für die Elektronneutrinomasse von 0,5 eV. Die Unsicherheit dieser Zahl, die durch das Kernmatrixelement hereinkommt, beträgt etwa einen Faktor 2. In den nächsten Jahren ist das Ziel die Sondierung der Neutrinomasse bis hinab zu 0,1 eV (fast zwei Größenordnungen unterhalb der gegenwärtigen Grenzen des besten Tritium-Experiments).

Majorana- oder Dirac-Neutrinos?

Der Doppelbetazerfall liefert ferner die einzigartige Möglichkeit, die Frage zu klären, ob Neutrinos Dirac- oder Majorana-Teilchen sind. Unter einem Antiteilchen verstand man historisch ein Teilchen mit gleicher Masse, aber entgegengesetzter elektrischer Ladung. Man fragt sich natürlich sofort, ob ein solcher „Vorzeichenwechsel" bei einem neutralen Teilchen überhaupt einen Sinn ergibt. Erhält man bei dieser Transformation etwas Neues oder nicht? Dies ist, etwas vereinfacht ausgedrückt, die Frage nach der Majorana- oder Dirac-Natur des Neutrinos. Ein Teilchen, das unter Ladungskonjugation in sich selbst übergeht, würde man als Majorana-Teilchen bezeichnen. Das Photon ist dafür ein Beispiel.

Seit der Entdeckung der Paritätsverletzung in der schwachen Wechselwirkung muß man, um vom Teilchen zum Antiteilchen zu gelangen, aber auch noch eine Raumspiegelung durchführen. Diese führt linkshändige Teilchen in rechtshändige über und trägt damit der Tatsache Rechnung, daß nur linkshändige Teilchen und rechtshändige Antiteilchen an der schwachen Wechselwirkung teilnehmen. Damit ist die Frage, ob das Neutrino nun ein Majorana- oder ein Dirac-Teilchen ist, etwas komplizierter geworden. Unterscheiden sich das linkshändige Neutrino und das rechtshändige Antineutrino nur durch ihre Händigkeit (Majorana-Fall) oder gibt es noch ein weiteres Unterscheidungsmerkmal (Dirac-Fall)?

Nur falls Neutrinos Majorana-Teilchen sind, erwarten wir die Existenz des neutrinolosen Doppelbetazerfalls, kurz $0\nu\beta\beta$-Zerfall. Die gegenwärtige Halbwertzeitgrenze läßt sich dann in eine Obergrenze für die Majorana-Masse der Neutrinos umrechnen. Die berühmten solaren Neutrinoexperimente, wie Gallex (vgl. *Physik in unserer Zeit* **22**, 246 (6/1992)) machen keine direkten Aussagen für diese Größe, da sie lediglich Differenzen von Massen verschiedener Neutrinotypen liefern.

Trotz jahrzehntelanger Suche konnte bislang kein einziger $0\nu\beta\beta$-Zerfall nachgewiesen werden. Aber auch dieser „Nicht-Nachweis" ermöglicht wichtige Schlüsse im Bereich der

Teilchenphysik. Grund ist, daß der Doppelbetazerfall nämlich auch durch den Austausch *anderer* Teilchen als das Neutrino ausgelöst werden kann, etwa durch Austausch supersymmetrischer Teilchen wie Gluinos oder Sneutrinos (Abbildungen 2, 3, 4) (siehe „Supersymmetrie und R-Parität" S. 129).

Links-rechts-symmetrische Modelle und Doppelbetazerfall

Experimentell scheinen an der schwachen Wechselwirkung nur linkshändige Teilchen teilzunehmen. Die Natur scheint links und rechts unterscheiden zu können. Die Ursache hierfür ist bisher nicht ausreichend verstanden. Das Standardmodell gibt für diese Tatsache keine Begründung, sie ist einfach „von Hand" eingebaut.

Links-rechts-symmetrische Modelle der schwachen Wechselwirkung sind ein möglicher Erklärungsversuch. Bei der Ausarbeitung dieser Modelle macht man sich eine einfache Grundtatsache zunutze. Die Austauschteilchen der schwachen Wechselwirkung haben eine sehr große Masse, die schwache Wechselwirkung damit eine sehr kurze Reichweite. Stellt man sich nun vor, es existiere neben den W- und Z-Bosonen des Standardmodells eine weitere Ausgabe dieser Teilchen, die nur an die rechtshändigen Fermionen koppeln, so ließe sich die beobachtete Linkshändigkeit der schwachen Wechselwirkung einfach erklären, wenn man diesen neuen Bosonen eine größere Masse als den gewöhnlichen W und Z zuordnen würde.

Auf den ersten Blick erscheint es vielleicht so, als hätte man die Frage „Warum ist die schwache Wechselwirkung linkshändig?" lediglich durch die neue Frage „Warum sind die Bosonen, die an die rechtshändigen Fermionen koppeln, schwerer?" ersetzt. Die Suche nach experimentellen Hinweisen für links-rechts-symmetrische Modelle ist aber sinnvoll, wenn man bedenkt, daß diese neben der genannten „Erklärung" der Linkshändigkeit der schwachen Wechselwirkung auch noch andere Fragestellungen angehen könnten. Das prominenteste Beispiel dafür ist der sogenannte Seesaw-Mechanismus, der die Kleinheit von Neutrinomassen zwanglos erklären könnte. Dies jedoch nur, wenn Neutrinos Majorana-Teilchen sind.

Jedoch machen links-rechts-symmetrische Modelle keine Aussagen, bei welcher Masse die neuen „rechtshändigen" W-Bosonen zu finden sind. Experimentell sind bis heute nur Untergrenzen für ihre Massen bekannt. Die Untersuchung des $0\nu\beta\beta$-Zerfalls (Abbildung 2) und speziell die Heidelberger Untersuchungen liefern nach heutigem Stand bereits die schärfste Massengrenze dieser hypothetischen W-Bosonen.

Leptoquarks und Doppelbetazerfall

Leptoquarks sind hypothetische Teilchen, skalarer oder vektorieller Natur, die an jeweils ein Lepton und ein Quark koppeln sollten. Solche Teilchen könnten nicht nur die Gleichheit der Ladung von Elektron und Proton zwanglos erklären, sie werden auch von einer Reihe von Großen Vereinheitlichungstheorien gefordert.

Koppeln Leptoquarks außer an ein Quark und ein Lepton auch an zwei Quarks gleichzeitig, können sie im allgemeinen den Zerfall des Protons induzieren. Solche Teilchen müssen dann außerordentlich schwer, um 10^{15} GeV oder mehr, sein. Koppeln sie nicht an zwei Quarks gleichzeitig, so sind ihre Massengrenzen sehr viel schwächer. Um für Beschleunigerexperimente zugänglich zu sein, muß man von Leptoquarks noch eine weitere Eigenschaft fordern. Sie sollten nur an links- oder an rechtshändige Fermionen koppeln, nicht aber an beide (chirale Leptoquarks).

Leptoquarks können andererseits auch an das im Standardmodell enthaltene Higgs-Boson koppeln [3]. Eine solche Kopplung führt dann automatisch dazu, daß die Chiralitätsforderung verletzt wird. Experimente wie der Doppelbetazerfall können dann dazu benutzt werden, Eigenschaften dieser Leptoquarks einzuschränken (Abbildung 3). Die bisherige Nichtbeobachtung des $0\nu\beta\beta$-Zerfalls erweist sich hier als das beste Instrument.

Supersymmetrie und Doppelbetazerfall

Supersymmetrie ist eine gedachte Operation, die Fermionen mit Bosonen verknüpft. Supersymmetrie, oder kurz SUSY, die zunächst ein rein theoretisches Konstrukt war, hat in den letzten Jahren zunehmend an Bedeutung gewonnen. Die sogenannte minimale supersymmetrische Erweiterung des Standardmodells (MSSM) gilt heutzutage als führender Kandidat für eine Physik jenseits des Standardmodells. Die Liste der Gründe dafür ist recht lang, sie reicht von rein theoretischen Überlegungen, wie der Einbeziehung der Gravitation in die Großen Vereinigungstheorien, über indirekte experimentelle Hinweise, wie die Vereinigung der drei Standardmodell-Kopplungskonstanten, bis hin zu Präzisionsfits der neuesten experimentellen Beschleunigerdaten, die unter Annahme des MSSM bestens erklärt werden können.

Andererseits gibt es bis heute keinen experimentellen Nachweis eines supersymmetrischen Teilchens. Demnach muß SUSY gebrochen sein, wenn sie existiert. Man erwartet Massen von supersymmetrischen Teilchen typischerweise in der Größenordnung von 100 GeV bis maximal 1 TeV.

Doppelbetazerfall läßt sich auf zwei sehr verschiedenen Wegen mit SUSY verknüpfen (Abbildung 4). Da ist zum einen die Frage nach der Erhaltung der sogenannten R-Parität (vgl. „Supersymmetrie und R-Parität", S. 129). Anders als das Standardmodell läßt SUSY in der allgemeinen Form Wechselwirkungen zu, die zu Leptonenzahlverletzungen führen. Dies rührt daher, daß die supersymmetrischen Partner der Quarks und Leptonen in gewöhnliche Teilchen zerfallen könnten. Ordnet man den SUSY-Teilchen eine hypothetische Quantenzahl $R_P = -1$ und allen gewöhnlichen Teilchen $R_P = +1$ zu, so hat man ein Mittel an der Hand, solche Zerfälle über R-Paritätserhaltung zu verbieten.

Da dies ohne eine zugrunde liegende Symmetrie geschieht, können letztendlich nur Experimente darüber entscheiden, ob diese Quantenzahl tatsächlich erhalten ist. Doppelbetazerfall liefert gegenwärtig von allen Experimenten die schärfsten Grenzen für R-Paritätsverletzung von Teilchen der ersten Generation [4].

Die Verknüpfung von Doppelbetazerfall mit SUSY ist aus einem weiteren Grunde von größtem Interesse, wenn man von der Existenz eines Majorana-Neutrinos ausgeht. Zunächst scheint die Frage nach Majorana- oder Dirac-Neutrinos völlig unabhängig von der Existenz oder Nichtexistenz von SUSY zu sein. Zieht man aber die Existenz eines skalaren Partnerteilchens des Neutrinos, des Sneutrinos, in Betracht, dann ist die Frage erlaubt, ob dieses Sneutrino nun – ähnlich wie das Neutrino – Majorana-Eigenschaften haben kann. Und das ist tatsächlich der Fall. Mehr noch, ist das Neutrino ein Majorana-Teilchen, so muß auch das Sneutrino

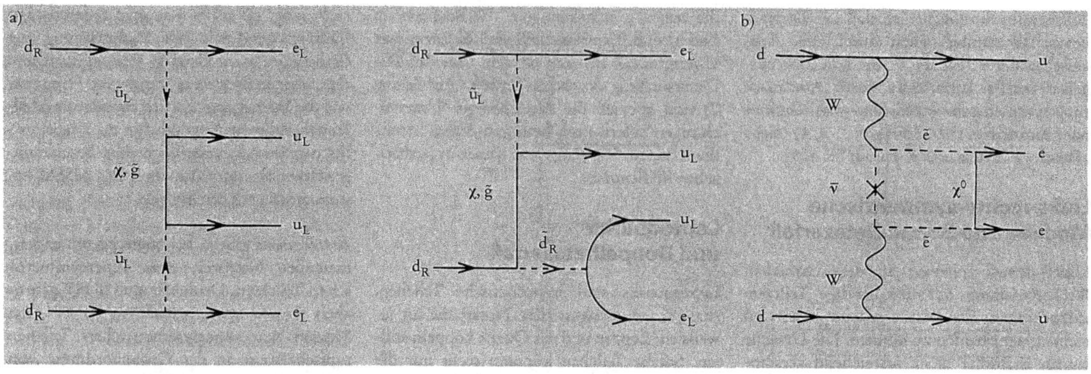

Abb. 4. Beispiele für neutrinolose Doppelbetazerfälle innerhalb R-paritätsverletzender SUSY-Modelle. Die Signatur ist identisch zu dem in Abbildung 1b gezeigten Prozeß: (a) keine Neutrinos nehmen teil. Anstelle des Neutrinos werden Gluinos ($\tilde{g}$) oder Neutralinos (χ) ausgetauscht. Die Zerfallsamplitude ist proportional zu $\lambda'^2_{111} m^4_{\tilde{q}}, m_{\tilde{g}\chi}$ M, wobei $m_{\tilde{q},\tilde{g},\chi}$ die Massen von Squarks, Gluinos, Neutralinos bezeichnen. λ'_{111} ist eine die Stärke einer R-paritätsbrechenden Wechselwirkung im Superpotential des MSSM beschreibende Yukawa-Kopplung und M ein Kernmatrixelement. (b) Eines der dominierenden Box-Diagramme zum Doppelbetazerfall in R-paritätserhaltenden SUSY-Modellen. Der Prozeß erlaubt Aussagen zur Masse des ausgetauschten skalaren Neutrinos.

eine Majorana-Masse besitzen und umgekehrt.

Diese direkte Verknüpfung erlaubt es nicht nur, aus dem Doppelbetazerfall Schlüsse auf die Majorana-Masse des Sneutrinos zu ziehen (Abbildung 4b). Sie verbindet darüber hinaus auch den Doppelbetazerfall mit Beschleunigerexperimenten, die nach der Existenz von SUSY-Teilchen oder an zukünftigen Beschleunigern nach Sneutrino-Antisneutrino-Oszillationen suchen [5]. Sollte man jemals SUSY-Teilchen finden, so könnte man durch Untersuchung der Eigenschaften des Sneutrinos die Frage nach der Majorana-Natur des Neutrinos indirekt klären.

Compositeness und Doppelbetazerfall

Obwohl bislang kein Experiment auf eine Substruktur von Quarks und Leptonen hinweist, gibt es Spekulationen, daß bei Energien oberhalb von etwa 1 TeV eine solche Substruktur sichtbar werden könnte. Beschleuniger ergeben als untere Grenze für diese sogenannte Compositeness bereits einen Wert von 1,6 TeV. Die Massen angeregter Leptonen und Quarks sollten nicht kleiner sein als dieser Wert, untere Grenzen aus Beschleunigerexperimenten liegen bei 100 bis 500 GeV. Aus sogenannten Präonen zusammengesetzte angeregte (Majorana-) Neutrinos könnten auch neutrinolosen Doppelbetazerfall auslösen. Die gegenwärtigen experimentellen Grenzen für letzteren liefern andererseits eine untere Grenze für die Compositeness-Skala [6].

Untersuchung von Doppelbetazerfällen

Die Suche nach Doppelbetazerfällen ist über den langen Zeitraum von 40 Jahren, in dem die getesteten Halbwertszeiten von 10^{16} auf 10^{25} Jahre erhöht wurden, immer wieder aktuell gewesen und hat – wie ausgeführt – besonders in den letzten Jahren an Aktualität gewonnen. Nicht der früher beobachtete Zerfall unter gleichzeitiger Emission von zwei Elektronen und zwei Neutrinos, sondern der Zerfall unter alleiniger Emission von zwei Elektronen ist die Motivation für die aktuellen Experimente.

Beim $0\nu\beta\beta$-Zerfall ist die Summenenergie der beiden emittierten Elektronen durch die Energiedifferenz von Ausgangs- und Endkern bestimmt und sollte daher als monochromatische Linie am oberen Ende des Doppelbetazerfall-Spektrums erscheinen (Abbildung 5). Eine solche Linie wurde bisher in keinem Experiment beobachtet.

Stattdessen mißt man einen kontinuierlichen Untergrund von radioaktiven Prozessen. Dessen Reduzierung innerhalb und außerhalb des Detektors ist also von entscheidender Bedeutung für ein erfolgreiches Experiment, zumal eine extrem geringe Zerfallsrate von nur wenigen Ereignissen pro Jahr erwartet wird. Ebenso wichtig ist eine möglichst große Quellstärke des $\beta\beta$-Emitters.

Im Jahre 1988 wurde eine Kollaboration des Max-Planck-Instituts für Kernphysik in Heidelberg mit dem Kurtschatow-Institut in Moskau ins Leben gerufen. Hoch angereichertes ^{76}Ge aus der damaligen UdSSR sollte auf seinen Doppelbetazerfall hin untersucht werden. In diesem Experiment werden zu 86 % angereicherte hochreine Detektoren aus ^{76}Ge eingesetzt, die gleichzeitig als Quelle der Strahlung und als Detektor dienen. Die im Zeitraum von 1990 bis 1995 gebauten fünf Detektoren (Abbildung 6) haben insgesamt eine Masse von 11,5 kg. Dieses Experiment besitzt die bei weitem größte Quellstärke aller gegenwärtigen $\beta\beta$-Experimente [7].

Die Empfindlichkeit des Experiments entspricht der eines Experiments mit über 1,2 Tonnen an natürlichem Germanium. Die italienische Forschungsorganisation INFN (Istituto Nazionale di Fisica Nucleare) ließ für das Experiment im Gran-Sasso-Labor bei Rom ein Gebäude errichten, dessen Wände aus aktivitätsarmem Beton hergestellt sind, um den durch Neutronen verursachten radioaktiven Untergrund zu minimieren (Abbildung 7). Das Experiment liegt in unmittelbarer Nachbarschaft des solaren Neutrino-Ex-

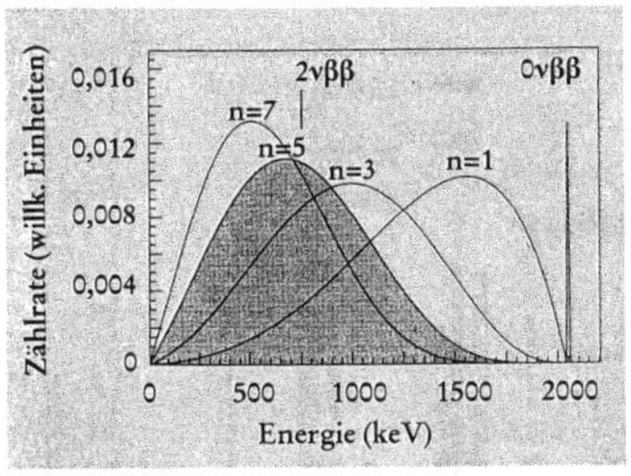

Abb. 5. Erwartete Spektren beim Doppelbetazerfall. Nur für den neutrinolosen Zerfall erwartet man eine scharfe Linie, die im Falle des Emitters ^{76}Ge bei einer Energie von 2038,5 keV liegt.

Abb. 7. Das Experimentiergebäude des Heidelberg-Moskau-Experiments und des Dunkle-Materie-Experiments (HDMS) im Gran-Sasso-Untergrundlaboratorium.

Abb. 6. Einer der angereicherten ^{76}Ge-Kristalle. Er war 1990 der größte je gezogene Germaniumkristall überhaupt.

8

Abb. 8. Drei der insgesamt fünf angereicherten Germaniumdetektoren beim Einbau in die Blei-Abschirmung.

periments GALLEX und des vorwiegend für die Suche nach magnetischen Monopolen gebauten MACRO-Experiments.

Jahrelange Anstrengungen haben es ermöglicht, das Niveau der Untergrundereignisse auf einen extrem geringen Wert von 0,07 Ereignisse pro Jahr und Kilogramm Detektor und pro Energiebereich von 1 keV im relevanten Energiebereich abzusenken. Zunächst bieten die 1400 m Felsabschirmung, entsprechend 3500 m Wasseräquivalent, einen vollständigen Schutz vor der hadronischen Komponente der kosmischen Strahlung und reduzieren den Myonenfluß um sechs Größenordnungen. Extreme Selektion der Materialien für den Bau der Detektoren, Halterungen und Kryostaten lieferten eine weiteren entscheidenden Beitrag für einen weitgehend untergrundfreien Meßbetrieb.

Die Detektoren sind in dem Meßbunker nach dem Zwiebelschalenprinzip von verschiedenen Schichten passiver Abschirmung umgeben (Abbildung 8). Die innerste Schicht ist das Kupfer des Kristallhalters und Kryostatsystems. Darauf folgen zwei Schichten aus hochreinem Blei. Insgesamt wiegt der experimentelle Aufbau, der vier Detektoren aus angereichertem Material enthält, sieben Tonnen. Dieser Bleiwürfel befindet sich in einer luftdichten Stahlkiste, die mit Stickstoff gespült wird. Durch die Verdrängung des Radons aus der Luft verringert sich die Zählrate in den Detektoren um das Zehnfache. Die Stahlkiste wiederum ist von einer 10 Zentimeter starken Borpolyethylenabschirmung ummantelt. Durch die Protonen im Polyethylen werden die Neutronen aus der Umgebung moderiert und von den Borkernen mit einem großen Einfangquerschnitt absorbiert. Oberhalb des Aufbaus befinden sich schließlich zwei Lagen Szintillatorplatten, die durchgehende Myonen in Koinzidenz nachweisen.

Der fünfte Detektor befindet sich in einem separaten Aufbau, der ausschließlich eine innere Abschirmung aus hochreinem Kupfer enthält. Die Zählrate in allen Detektoren zusammen beträgt über das gesamte Energiespektrum 4 Ereignisse in 10 Minuten. Der natürlichen Umgebungsradioaktivität, deren Ursachen hauptsächlich ^{40}K und die Zerfallsreihen von Uran und Thorium sind, würde eine um den Faktor 10^6 höhere Zählrate entsprechen. Durch den Unterschied von sechs Größenordnungen in der Aktivität wird ersichtlich, wie katastrophal sich die kleinsten Verunreinigungen auf das Meßergebnis auswirken würden.

Eine weitere Reduzierung des Untergrundes wird durch ein neu entwickeltes Verfahren zur Pulsformanalyse erreicht. Dieses erlaubt, Doppelbeta-Ereignisse von mehrfachgestreuten Gamma-Ereignissen zu trennen. Das Verfahren ist umgekehrt auch geeignet, die Nachweisempfindlichkeit von Ge-Detektoren für den Nachweis schwacher γ-Aktivitäten zu erhöhen und somit möglicherweise ökonomisch interessant. Es wurde daher zur Patentierung angemeldet.

Ergebnisse des Heidelberg-Moskau-Experiments

Das Heidelberg-Moskau-Experiment hat die in es gesetzten Erwartungen in vieler Hinsicht weit übertroffen [8].

Standardmodell. Für den im Standardmodell erlaubten, neutrinobegleiteten Zerfall ergibt sich eine Halbwertszeit von (1,77 ± 0,13) 10^{21} Jahre. Die Bestimmung dieser Halbwertszeit stützt sich auf circa 20 000 Ereignisse, im Vergleich zu den 40 Ereignissen, aus denen im Jahre 1987 Mike Moe die erste nicht-geochemisch bestimmte Halbwertszeit ableitete. Doppelbetaspektroskopie ist damit erstmalig in den Bereich „normaler" Kern-

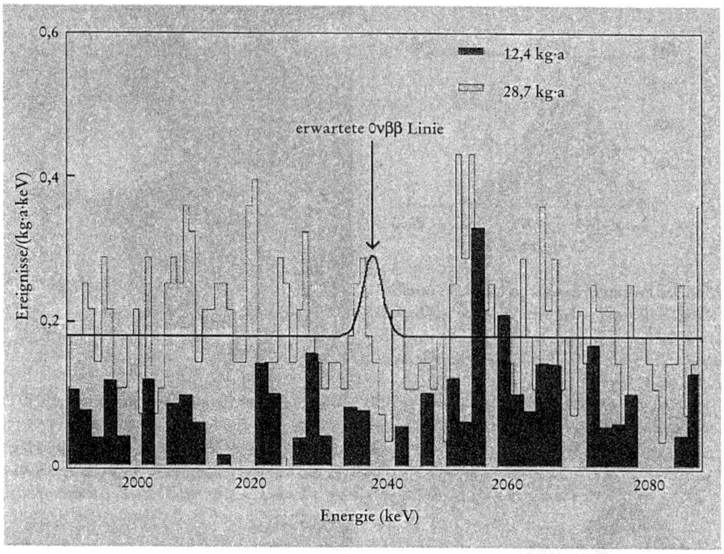

Abb. 9 Gemessenes Spektrum nach 28,7 Kilogrammjahren im Bereich der erwarteten Linie aus dem neutrinolosen Doppelbetazerfall. Die gestrichelte Kurve entspricht dem mit 90 % Vertrauensbereich (2σ) ausgeschlossenen Signal. Ihm entspricht eine Halbwertszeit von mehr als $1{,}1 \cdot 10^{25}$ Jahren. Das dunkle Histogramm entspricht den in 12,4 Kilogrammjahren erhaltenen Spektrum bei weiterer Untergrundreduktion mittels Pulsformanalyse.

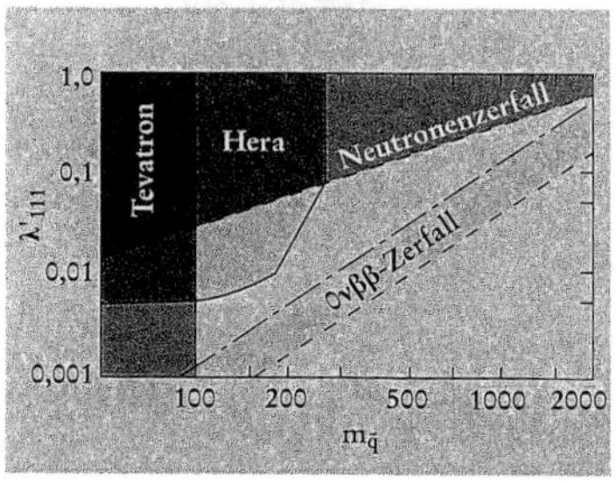

Abb. 10. Grenzen aus dem Doppelbetazerfall für die Verletzung der R-Parität in Abhängigkeit von der Masse des Squark, m_q. Hierbei wurden für das Gluino jeweils eine Masse von 1 TeV und 0,1 TeV angenommen. Ebenfalls angegeben sind Grenzen aufgrund von Beschleunigerexperimenten am Tevatron und HERA sowie aus dem Neutronenzerfall. Ausgeschlossen sind die Bereiche m_q < 100 GeV sowie oberhalb der Kurven.

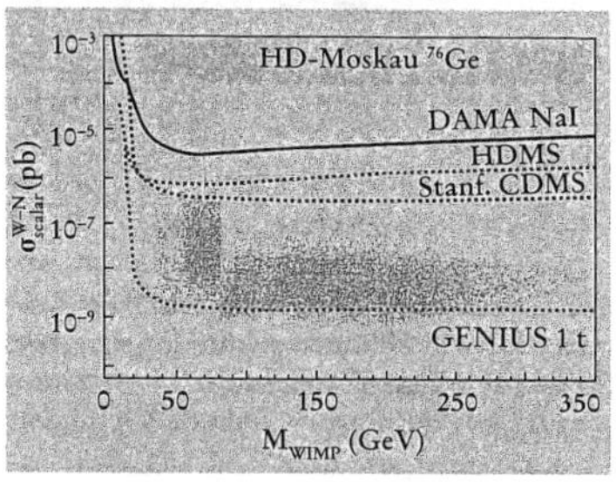

Abb. 11. Gegenwärtig beste Grenzen für den Wirkungsquerschnitt von Neutralinos als Kandidaten für WIMPS (durchgezogene Linie). Die Bereiche oberhalb der Linien sind ausgeschlossen. Mit den im Aufbau befindlichen Experimenten CDMS (Berkeley) und HDMS (Heidelberg) will man in den Bereich der Erwartungen für Neutralinos vorstoßen. Das geplante GENIUS-Experiment wird in der Lage sein, den vollen Parameterraum der SUSY-Vorhersagen zu überdecken. Die grauen Punkte entsprechen verschiedenen SUSY-GUT-Modellrechnungen.

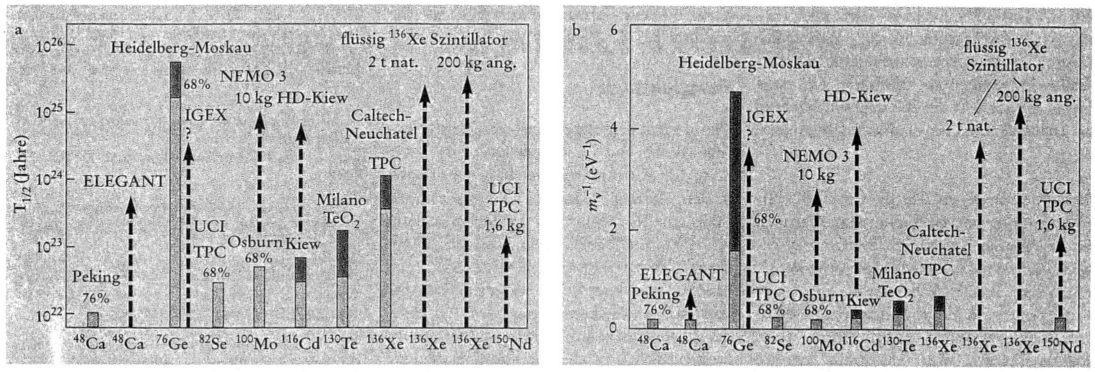

Abb. 12. Status und Perspektiven von Doppelbeta-Experimenten für die erreichbaren Halbwertszeiten und daraus folgenden Neutrinomassen. Die hellgrauen Balken entsprechen dem gegenwärtigen Stand, dunkelgraue Balken sicheren Extrapolationen für die kommenden Jahre; gestrichelte Linien zeigen unsichere Extrapolationen oder derzeit diskutierte Experimente.

spektroskopie eingetreten. Dies erlaubt nun, nach Abweichungen vom reinen 2ν-Spektrum zu suchen, die von der gleichzeitigen Emission zusätzlicher Teilchen, wie Majoronen, herrühren könnten.

Jenseits des Standardmodells. Das Heidelberg-Moskau-Experiment liefert gegenwärtig die weltweit beste Grenze für die Halbwertszeit des ^{76}Ge für neutrinolosen Doppelbetazerfall von $1{,}1 \cdot 10^{25}$ Jahren (Abbildung 9). Hieraus ergibt sich für die Masse des (Elektron-) Neutrinos eine obere Grenze von 0,48 eV – sofern es sich um eine Majorana-Teilchen handelt. Für die Masse eines linkshändigen superschweren Neutrinos liefert das Experiment eine untere Grenze von $7 \cdot 10^7$ GeV.

Für die weiteren oben erwähnten Modelle jenseits des Standardmodells ergibt sich zusammenfassend folgendes.

Die untere Massengrenze für ein in rechts-links-symmetrischen Modellen auftretendes rechtshändiges W-Boson liegt bei 1,1 TeV, diejenige für Leptoquarks beträgt 10 – 14 TeV für eine Leptoquark-Higgs-Kopplung von der Größenordnung 1. Um Konsistenz des Doppelbeta-Experiments mit den jüngsten HERA-Ergebnissen zu erhalten, muß andererseits für eine Leptoquark-Masse von 200 GeV die Leptoquark-Higgs-Kopplung von der Größenordnung von einigen 10^{-6} sein. Unter der Annahme einer Substruktur von Leptonen und Quarks ergibt sich für die Compositeness-Skala eine untere Grenze von 0,1 TeV.

Ein besonders interessantes Ergebnis ergibt sich auch für SUSY-Modelle [4]. Abbildung 10 zeigt die aus dem Heidelberger Experiment abgeleiteten Grenzen (Bereiche oberhalb der Linien sind ausgeschlossen) für R-Paritätsverletzung (beschrieben durch den Parameter λ'_{111}) von Teilchen der ersten Generation, als Funktion der Squarkmasse $m_{\tilde{q}}$, zusammen mit Grenzen, die von den Hochenergiebeschleunigern Tevatron und HERA ermittelt wurden oder noch werden sollen. Deutlich ist die energetische Begrenzung der Beschleuniger zu erkennen. Das Doppelbetaexperiment ergibt für Squarkmassen oberhalb

Supersymmetrie und R-Parität

Die Theorie der Supersymmetrie sagt eine Symmetrie zwischen Fermionen und Bosonen voraus. Sie kann nur dann realisiert sein, wenn jedes Teilchen mit Spin j einen supersymmeterischen Partner mit Spin j-1/2 besitzt. Zu jedem fundamentalen Fermion (Lepton oder Quark) gibt es also einen skalaren, bosonischen Partner mit Spin 0. Er bekommt den Vorsatz „S", also Slepton oder Squark. Ebenso gibt es zu jedem Boson (Photon, Gluon, Z, W, Higgs) ein entsprechendes Fermion, das die Endung -ino erhält. Dadurch würde sich die Anzahl der Elementarteilchen also verdoppeln. Jedes GUT-Modell kann zu SUSY-Versionen erweitert werden. Bis heute ließ sich kein einziges supersymmetrisches Teilchen nachweisen. Lokal supersymmetrische Modelle enthalten die Gravitation und heißen Supergravitations-Modelle.

Die R-Parität ist eine multiplikative Quantenzahl und verknüpft mit der Baryonenzahl B, der Leptonenzahl L und dem Spin S gemäß $R_p = -1^{3B+L+2S}$. Dadurch ist $R_p = 1$ für normale Teilchen und $R_p = -1$ für SUSY-Partner. Erhaltung der R-Parität bedeutet dann beispielsweise, daß das leichteste SUSY-Teilchen, das Neutralino, stabil sein muß.

Teilchen	SUSY-Partner	Symbol	Spin
Quark	Squark	$\tilde{q}$	0
Lepton	Slepton	$\tilde{e}$	0
Gluon	Gluino	$\tilde{g}$	1/2
W-Boson	Wino	$\tilde{W}$	1/2
Photon	Photino	$\tilde{\gamma}$	1/2
Higgs	Higgsino	$\tilde{h}$	1/2
Graviton	Gravitino		3/2

von 100 GeV die erheblich schärferen Grenzen. Auch das Ergebnis für λ'_{111} ist von Bedeutung für die Interpretation der kürzlichen HERA-Ergebnisse im Hinblick auf neue Physik. Es schließt die Bildung von Squarks der ersten Generation in diesen Ereignissen aus.

Aus der gemessenen Grenze für die $0\nu\beta\beta$-Halbwertszeit wurden ferner obere Grenzen für die Majorana-Masse von Sneutrinos abgeleitet [9]. Demnach ist das Elektron-Sneutrinos leichter als 22 MeV. Es wurde gezeigt, daß somit zwar zukünftige Beschleuniger unter Umständen wertvolle Informationen über Majorana-Massen von Sneutrinos zweiter und dritter Generation erwarten lassen, sie aber für Sneutrinos erster Generation nicht mit Doppelbetazerfall werden konkurrieren können.

Dunkle Materie

Beobachtungen der Rotation von Spiralgalaxien haben zu der Überzeugung geführt, daß diese Sternsysteme von einem ausgedehnten Halo aus unsichtbarer, sogenannter Dunkler Materie umgeben sind (*Physik in unserer Zeit* **28**, 16 (1/1997)). Unklar ist, woraus dieser Halo besteht. Eine Möglichkeit sind noch unbekannte, schwach wechselwirkende Teilchen, sogenannte WIMPs (Weakly Interacting Massive Particles). Wenn wir annehmen, daß WIMPs den dunklen Halo unserer Galaxie bilden, sollten sie in Laborexperimenten nachweisbar sein. Bislang geben Experimente nur Grenzen für Masse und elastischen Streuquerschnitt von WIMPs an.

Die besten Laborgrenzen erzielt unser Heidelberg-Moskau-Experiment (Abbildung 11) [10]. Dirac-Neutrinos konnten hiermit als dominante Komponente des dunkles Halos im Massenbereich von 26 GeV bis 4,7 TeV ausgeschlossen werden, ebenso wie schwere skalare Neutrinos in Szenarien des minimalen supersymmetrischen Standardmodells [11]. Die berechneten Erwartungswerte aus SUSY-GUT-Szenarien für Neutralinos, den leichtesten supersymmetrischen Teilchen, als dunkle Materie [12] sind ebenfalls in Abbildung 11 gezeigt.

Es ist ersichtlich, daß erst Experimente größerer Empfindlichkeit in diesen Bereich vordringen können. Verschiedene Experimente sind geplant oder befinden sich im Bau. Auch unsere Gruppe beginnt soeben ein solches Experiment (HDMS) mit speziell konstruierten Germaniumkristallen als Detektoren [13].

Perspektiven

Den derzeitigen Stand und die Perspektiven einer Reihe von Experimenten zeigt Abbildung 12. Mit unserem Detektor werden wir in den nächsten Jahren in der Lage sein, die Majorana-Neutrinomasse bis hinab zu etwa 0,1 eV zu sondieren. Im Hinblick auf die moderne Neutrinophysik nimmt das Experiment damit eine Schlüsselposition ein. Bezüglich verschiedener Fragen zu Theorien jenseits des Standardmodells wie Supersymmetrie oder Leptoquarks konkurrieren die Ergebnisse des Doppelbetazerfalls bereits jetzt mit denen der leistungsfähigsten Beschleuniger.

Die Zukunft wird uns noch weitere Möglichkeiten eröffnen. Insbesondere soll eine weitere geplante Ausbaustufe, genannt GENIUS (Germanium Nitrogen Underground Setup), eine Tonne an angereicherten „nackten" Germaniumdetektoren in einem Tank mit flüssigem Stickstoff einsetzen. Mit ihm könnte es möglich sein, $\beta\beta$-Halbwertszeiten von 10^{28} Jahren zu bestimmen, wodurch sich der Bereich für die Neutrinomasse bis hinab zu 10^{-2} eV sondieren ließe. In einer zweiten Ausbaustufe sollte das Experiment überdies einen wesentlichen Beitrag zur direkten Prüfung der Oszillationshypothese als Erklärung für das solare Neutrinoproblem liefern [16].

GENIUS wird auch einen weiteren (möglicherweise sogar den vollen) Vorhersagebereich von SUSY-Modellen für Dunkle Materie sondieren können (Abbildung 11), der in Kolliderexperimenten zum Teil nur schwer abzudecken ist. Dies wäre bereits mit nur 100 kg an natürlichem Germanium zu erreichen [16]. Auch wenn man mit dem LHC supersymmetrische Teilchen zuerst finden würde, wäre es immer noch faszinierend, die Existenz von Neutralinos als Dunkler-Materie-Teilchen direkt nachzuweisen [14,15].

Literatur

[1] R.N. Mohapatra, Progr. Part. Nucl. Phys. 32 (1994) 187, und Proc. Neutrino '96, Helsinki, Juni 1996; S. Petcov and Yu. Smirnov, Phys. Lett. B **322**, 109 (1994); A. Ioannissyan, J.W.F. Valle, Phys. Lett. B **332**, 93 (1994); D.G. Lee, R.N. Mohapatra, Phys. Lett. B **329**, 963 (1994).

[2] G. Raffelt, J. Silk, Phys. Lett. B **366**, 429 (1996); J.R. Primack et al. Phys. Rev. Lett. **74**, 2160 (1995).

[3] M. Hirsch, H.V. Klapdor-Kleingrothaus, S. Kovalenko, Phys. Lett. B **378**, 17 (1996); Phys. Rev. D **54**, R4207 (1996); St. Kolb, M. Hirsch, H.V. Klapdor-Kleingrothaus, Phys. Lett. B **391**, 131 (1997).

[4] M. Hirsch, H.V. Klapdor-Kleingrothaus, S. Kovalenko, Phys. Rev. Lett. **75**, 17 (1995); Phys. Rev. D **53**, 1329 (1996); Phys. Lett. B **372**, 181 (1996).

[5] M. Hirsch, H.V. Klapdor-Kleingrothaus, S. Kovalenko, Phys. Lett. B **398**, 311 (1997); Phys. Lett. B **403**, 291 (1997).

[6] O. Panella, in: Double Beta Decay and Related Topics, H.V. Klapdor-Kleingrothaus, S. Stoica (Hrsg.), World Scientific, Singapore, 1996.

[7] Heidelberg-Moskau-Kollaboration, Phys. Rev. D **55**, 54 (1997); Phys. Lett. **407**, 219 (1997).

[8] H.V. Klapdor-Kleingrothaus, Proc., Neutrino '96, Helsinki, Juni 1996; Proc., Beyond the Desert, Schloß Ringberg, Juni 1997.

[9] M. Hirsch, H.V. Klapdor-Kleingrothaus, S. Kovalenko, Phys. Rev. D **57**, 2020 (1998).

[10] Heidelberg-Moskau-Kollaboration, Phys. Lett. B **336**, 141 (1994).

[11] T. Falk, K. Olive, M. Srednicki, Phys. Lett. B **339**, 248 (1994).

[12] V. Bednyakov, H.V. Klapdor-Kleingrothaus, S. Kovalenko, Phys. Rev. D **55**, 503 (1997); Z. Phys. A **357**, 339 (1997).

[13] L. Baudis et al. Nucl. Instr. Meth. A **385**, 265 (1997).

[14] H.V. Klapdor-Kleingrothaus, A. Staudt, Teilchenphysik ohne Beschleuniger, Teubner-Verlag, Stuttgart, 1995; Non-Accelerator Particle Physics, IOP, Bristol, 1995.

[15] H.V. Klapdor-Kleingrothaus, K. Zuber, Teilchenastrophysik, Teubner-Verlag, Stuttgart, 1997; Particle Astrophysics, IOP, Bristol, 1997.

[16] H.V. Klapdor-Kleingrothaus, J. Helmig, M. Hirsch, J. Phys G **24**, 483 (1998).

Hans Volker Klapdor-Kleingrothaus, geboren 1942, promovierte an der Universität Hamburg im Bereich der Kernphysik und arbeitet seit 1969 am Max-Planck-Institut für Kernphysik. Er ist Initiator und Sprecher der Heidelberg-Moskau Doppelbeta-Kollaboration. Er ist Verfasser mehrerer Bücher.

Anschrift

Prof. Dr. Hans Volker Klapdor-Kleingrothaus, Max-Planck-Institut für Kernphysik, Postfach 10 39 80, 69029 Heidelberg.

NEUTRINOLESS DOUBLE BETA DECAY AND NEW PHYSICS IN THE NEUTRINO SECTOR

H.V. KLAPDOR-KLEINGROTHAUS, H. PÄS
Max-Planck-Institut für Kernphysik, P.O. Box 103980, D-69029 Heidelberg
E-mail: klapdor@gustav.mpi-hd.mpg.de
E-mail: Heinrich.Paes@mpi-hd.mpg.de

Neutrinoless double beta decay belongs to the most sensitive tools for the search of new physics beyond the standard model. The recent half life limit of the Heidelberg–Moscow experiment implies restrictive bounds on the absolute mass scale in the neutrino sector. Possible improvements by the GENIUS project provide a unique possibility to reconstruct the neutrino mass spectrum. Further constraints on new interactions in the neutrino sector are given in a model-independent way. Consequences for neutrino anomalies and theories beyond the standard model such as left-right symmetric models, R-parity violating SUSY and leptoquarks are discussed. The potential of double beta decay experiments in the search for WIMP dark matter is reviewed.

1 Introduction

Double beta decay [1,2] corresponds to two single beta decays occuring in one nucleus and converts a nucleus (Z,A) into a nucleus (Z+2,A). While even the standard model (SM) allowed process emitting two antineutrinos

$$^A_ZX \rightarrow ^A_{Z+2} X + 2e^- + 2\overline{\nu}_e \tag{1}$$

is one of the rarest processes in nature with half lives in the region of 10^{21-24} years, more interesting is the search for the neutrinoless mode ($0\nu\beta\beta$),

$$^A_ZX \rightarrow ^A_{Z+2} X + 2e^- \tag{2}$$

which violates lepton number by two units and thus implies physics beyond the SM [3].

The most sensitive experiment so far, the Heidelberg–Moscow experiment [1,4] is searching for the $0\nu\beta\beta$ decay of ^{76}Ge. The results after 31 kg y measuring time using digital pulse shape analysis correspond to a conservative half life limit of [4]

$$\begin{aligned} T_{1/2}^{0\nu\beta\beta} &> 1.8 \cdot 10^{25} y \quad (90\% C.L.), \\ T_{1/2}^{0\nu\beta\beta} &> 3.0 \cdot 10^{25} y \quad (68\% C.L.). \end{aligned} \tag{3}$$

To render possible a further breakthrough in search for neutrino masses and physics beyond the SM, GENIUS, an experiment operating a large amount

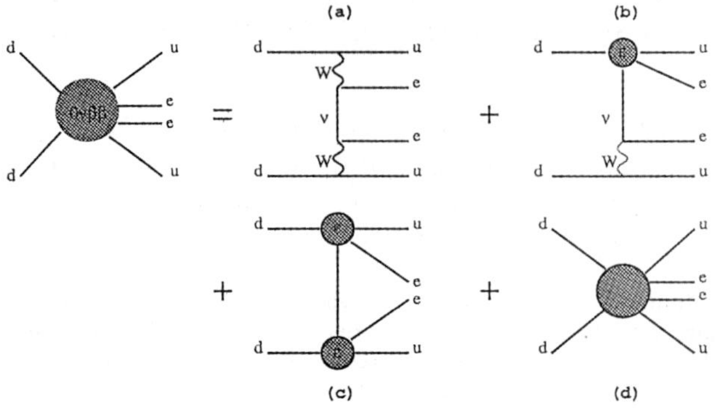

Figure 1. *Feynman graphs of the general double beta rate: The contribution a) corresponds to the neutrino mass mechanism with SM interactions and is discussed in the context of neutrino oscillations in section 2. The contributions b) - d) include new neutrino interactions and are discussed in section 3).*

of naked Ge–detectors in a liquid nitrogen shielding, has been proposed [5,6]. Operating 288 enriched ^{76}Ge detectors with a total mass of 1 ton inside a nitrogen tank of $\sim$ 12 m height and diameter, one could access half lifes of $T_{1/2}^{0\nu\beta\beta} = 6 \cdot 10^{27} y$ after one year of measurement. A ten ton version would reach a final sensitivity of $T_{1/2}^{0\nu\beta\beta} = 6 \cdot 10^{29} y$ within 10 years of measurement time.

2 Neutrino masses and oscillations

The search for $0\nu\beta\beta$ decay exchanging a massive left–handed Majorana neutrino between two SM vertices (contribution a) in fig. 1) at present provides the most sensitive approach to determine an absolute neutrino mass and also a unique possibility to distinguish between the Dirac or Majorana nature of the neutrino. With the recent half life limit of the Heidelberg–Moscow experiment [4] the following conservative limit on the effective neutrino mass

$$\langle m_\nu \rangle = \sum_i U_{ei}^2 m_i \leq 0.36 eV \quad (90\% C.L.)$$
$$\langle m_\nu \rangle = \sum_i U_{ei}^2 m_i \leq 0.28 eV \quad (68\% C.L.) \qquad (4)$$

can be deduced. Here the sum extends over light mass eigenstates m_i only. The GENIUS project could access effective neutrino masses down to 10^{-2} eV or even 10^{-3} eV in the 1 ton or 10 ton version, respectively. Since in specific models of neutrino masses the quantity $\langle m \rangle$ can be related to the oscillation parameters Δm^2 and $\sin^2 2\theta$, these bounds imply restrictive bounds on the neutrino mass spectrum, which in most cases are more stringent than the bounds from precision measurements of the CMB by MAP and Planck. The extreme cases discussed here are degenerate and hierarchical models (for a detailed discussion see [7]).

Degenerate models: Such models have been proposed to get a large mass scale for neutrinos acting as hot dark matter and at the same time accomodate two of the three neutrino anomalies (solar, atmospheric & LSND neutrinos). In this framework neutrino masses up to (few) eV are predicted. Thus large mixing (vacuum or MSW LMA oscillations) and strong cancellations are required to be consistent with eq. 4. E.g., maximal cancellation, the MSW LMA bestfit and a ΛCHDM model with a Hubble constant of $h \simeq 0.5$ implies a value of $\langle m \rangle = 0.15$ eV just below the present limit.

Hierarchical models: Such models give less optimistic predictions for $0\nu\beta\beta$ decay. However, assuming the MSW LMA solution, still sizable contributions up to $\langle m \rangle =$ (few) 10^{-2} in the reach of GENIUS are possible. Since $0\nu\beta\beta$ decay is most sensitive in the large Δm^2 region of the LMA solution, it may provide complementary informations to the search for day-night effects in solar neutrinos.

For a *super-heavy left-handed neutrino* a bound of $\langle m_H \rangle = \left(\sum_j \frac{U_{ej}^2}{m_j} \right)^{-1} > 9 \cdot 10^7 GeV$ can be deduced, with heavy mass eigenstates m_j. This constraint makes a heavy neutrino unobservable at linear colliders except in the most contrived scenarios [8].

3 New interactions

Besides the exchange of massive Majorana neutrinos between two SM vertices, a variety of theories beyond the SM predicts new lepton number violating interactions contributing to neutrinoless double beta decay, leading to the idea to construct the general double beta decay rate allowed by Lorentz-invariance [9,11]. This approach allows to constrain lepton number violating parameters in arbitrary models.

For the long range part of the decay rate with two separable vertices and light neutrino exchange in between (contributions b) and c) in fig. 1), one has to consider the Lorentz-invariant contractions of six projections with defined helicity both for the leptonic (j_α) and hadronic (J_α) current. The general Lagrangian can be written in terms of effective couplings ϵ_β^α, which correspond to the pointlike vertices at the Fermi scale so that Fierz rearrangement is applicable:

$$\mathcal{L} = \frac{G_F}{\sqrt{2}}\{j^{\mu}_{V-A}J^{\dagger}_{V-A,\mu} + \sum_{\alpha,\beta}{}' \epsilon^{\beta}_{\alpha} j_{\beta} J^{\dagger}_{\alpha}\} \tag{5}$$

with the combinations of hadronic and leptonic Lorentz currents of defined helicity $\alpha, \beta = V-A, V+A, S-P, S+P, T_L, T_R$. The prime indicates that the sum runs over all contractions allowed by Lorentz–invariance, except for $\alpha = \beta = V - A$. Here $\epsilon^{\beta}_{\alpha}$ denotes the strength of the non–SM couplings. For the helicity suppressed terms proportional to the (from below) unconstrained neutrino mass no limit can be derived and terms proportional $(\epsilon^{\beta}_{\alpha})^2$ can be neglected. The limits on the remaining non–SM couplings derived in s-wave approximation and evaluated "on axis" are [9] (here and in the following $90\% C.L.$): $\epsilon^{V+A}_{V+A} < 6\cdot 10^{-7}$, $\epsilon^{V+A}_{V-A} < 4\cdot 10^{-9}$, $\epsilon^{S+P}_{S+P} < 9\cdot 10^{-9}$, $\epsilon^{S+P}_{S-P} < 9\cdot 10^{-9}$, $\epsilon^{T_R}_{T_R} < 1\cdot 10^{-9}$, $\epsilon^{T_R}_{T_L} < 6\cdot 10^{-10}$. These bounds e.g. exclude the possibility to fake the LSND anomaly (and this way accomodate for all neutrino anomalies with only three neutrinos) via the lepton number violating reaction $\nu_e u_L \to d_R e^+$ in a model-independent way [10].

For the short range part the hadronic currents have to be contracted with leptonic currents $j_\alpha = \bar{e}\mathcal{O}_\alpha e^C$, where $\mathcal{O}_\alpha$ denotes the operators of defined helicty discussed above. In this case the general Lagrangian is

$$\mathcal{L} = \frac{G_F^2}{2} m_P^{-1}\{\epsilon_1 JJj + \epsilon_2 J^{\mu\nu}J_{\mu\nu}j + \epsilon_3 J^{\mu}J_{\mu}j + \epsilon_4 J^{\mu}J_{\mu\nu}j^{\nu} + \epsilon_5 J^{\mu}Jj_{\mu}$$
$$+\epsilon_6 J^{\mu}J^{\nu}j_{\mu\nu} + \epsilon_7 JJ^{\mu\nu}j_{\mu\nu} + \epsilon_8 J_{\mu\kappa}J^{\nu\kappa}j^{\mu}_{\nu}\}, \tag{6}$$

where indices α have been suppressed. Since no fundamental tensors exist in renormalizable theories and since the leptonic tensor current vanishes in the s-wave approximation, the contributions proportional to ϵ_4, ϵ_6, ϵ_7, ϵ_8 can be neglected. The remaining terms are constrained as follows [11]: $\epsilon_1 < 3\cdot 10^{-7}$, $\epsilon_2 < 2\cdot 10^{-9}$, $\epsilon_3 < 4\cdot 10^{-8}/1\cdot 10^{-8}$ $(V\mp A\ V\mp A/V\mp A\ V\pm A)$, $\epsilon_5 < 2\cdot 10^{-7}$.

4 Left-right symmetry, R-parity violation, Leptoquarks

In this section we apply the general discussion above to specific theories of physics beyond the SM.

Left-Right-Symmetric Models: In left–right symmetric models the left–handedness of weak interactions is explained as due to the effect of different symmetry breaking scales in the left– and in the right–handed sector. $0\nu\beta\beta$ decay proceeds through exchange of the heavy right–handed partner of the ordinary neutrino between right-handed W vertices, leading to a limit of

$$m_{W_R} \geq 1.4\Big(\frac{m_N}{1TeV}\Big)^{-(1/4)} TeV. \tag{7}$$

Including a theoretical limit obtained from considerations of vacuum stability [12] one can deduce an absolute lower limit on the right–handed W mass of [13]

$$m_{W_R} \geq 1.4TeV. \tag{8}$$

Supersymmetry: While in the minimal supersymmetric extension (MSSM) R–parity is assumed to be conserved, there are no theoretical reasons for R_p conservation and several GUT and Superstring models require R–parity violation in the low energy regime. In this case $0\nu\beta\beta$ decay can occur through Feynman graphs involving the exchange of superpartners as well as $\not R_P$-couplings λ' [14,16,17,18,19]. The half–life limit of the Heidelberg–Moscow experiment leads to bounds in a multidimensional parameter space [14,17]

$$\lambda'_{111} \leq 4 \times 10^{-4}\Big(\frac{m_{\tilde{q}}}{100GeV}\Big)^2\Big(\frac{m_{\tilde{g}}}{100GeV}\Big)^{1/2} \tag{9}$$

(for $m_{\tilde{d}_R} = m_{\tilde{u}_L}$), which are the sharpest limits on $\not R_P$-SUSY.

$0\nu\beta\beta$ decay is not only sensitive to λ'_{111}. Taking into account the fact that the SUSY partners of the left- and right–handed quark states can mix with each other, new diagrams appear in which the neutrino-mediated double beta decay is accompanied by SUSY exchange in the vertices [15,18]. A calculation of previously neglected tensor contributions to the decay rate allows to derive improved limits on different combinations of λ' [19]. Assuming the supersymmetric mass parameters of order 100 GeV, the half life limit of the Heidelberg–Moscow Experiment implies: $\lambda'_{113}\lambda'_{131} \leq 3 \cdot 10^{-8}$, $\lambda'_{112}\lambda'_{121} \leq 1 \cdot 10^{-6}$

In addition, stringent bounds on coupling products can be derived directly from the effective mass bound eq. 4, since R-parity violating interactions will produce neutrino Majorana masses on loop level. It implies [21] $\lambda'_{133}\lambda'_{133} < 5 \cdot 10^{-8}$, $\lambda'_{132}\lambda'_{123} < 1 \cdot 10^{-6}$, $\lambda'_{122}\lambda'_{122} < 3 \cdot 10^{-5}$, $\lambda_{133}\lambda_{133} < 9 \cdot 10^{-7}$, $\lambda_{132}\lambda_{123} < 2 \cdot 10^{-5}$, $\lambda_{122}\lambda_{122} < 2 \cdot 10^{-4}$.

In the case of R–parity conserving SUSY, based on a theorem proven in [20], the $0\nu\beta\beta$ mass limits can be converted in sneutrino Majorana mass term limits being more restrictive than what could be obtained in inverse neutrinoless double beta decay and single sneutrino production at future linear colliders (NLC) [20].

Leptoquarks: Leptoquarks are scalar or vector particles coupling both to leptons and quarks, which appear naturally in GUT, extended Technicolor or Compositeness models. The mixing of different multiplets by introducing a leptoquark–Higgs coupling would lead to a contribution to $0\nu\beta\beta$ decay

[22]. Combined with the half–life limit of the Heidelberg–Moscow experiment bounds on effective couplings can be derived [23]. Assuming only one lepton number violating $\Delta L = 2$ LQ–Higgs coupling unequal to zero and the leptoquark masses not too different, one can derive from this limit either a bound on the LQ–Higgs coupling

$$Y_{LQ-Higgs} = (few) \cdot 10^{-6} \tag{10}$$

or a limit excluding leptoquarks with masses in the range of $\mathcal{O}(200GeV)$. Assuming $Y_{LQ} \sim \mathcal{O}(1)$ leptoquark masses should be larger than (few) 10 TeV.

5 Violations of the equivalence principle and Lorentz invariance

Special relativity and the equivalence principle can be considered as the most basic foundations of the theory of gravity. However, string theories may allow for or even predict the violation of these laws. Such effects in the neutrino sector have been extensively studied in the framework of neutrino oscillations [24]. A typical feature of the violation of Lorentz invariance (VLI) is that different species of matter may have characteristic maximal attainable velocities. The quantity δv provides an observable for VLI. The corresponding quantity describing violations of the equivalence principle (VEP) is the difference of characteristic couplings δg to the gravitational potential ϕ. While previous studies of neutrino oscillations are restricted to the region of large mixing of velocity/gravitational and flavor eigenstates, $0\nu\beta\beta$ decay provides a bound in the previously unconstrained region of zero mixing [25]: $\delta v < 4 \cdot 10^{-16}$, $\phi\delta g < 4 \cdot 10^{-16}$.

6 WIMP Dark Matter Search with Double Beta Experiments

Weakly interacting masssive particles (WIMPs) such as the lightest supersymmetric particle (LSP) are major candidates for the cold component of nonbaryonic dark matter in the universe. Due to its low background properties double beta technology can also find applications in the search for direct detection of WIMPs. The Heidelberg–Moscow Experiment, without being specially designed for this purpose, gave the most stringent limits on WIMPs for several years [26]. New results with 0.69 kg y of measurement reached a background level of 0.042 cts/(kg d keV) in the region between 15 keV and 40 keV. The derived limit excludes WIMPS with masses greater than 13 GeV and cross sections as low as $1.12 \cdot 10^{-5}$ pb. These are the most stringent limits on spin-independent interactions using only raw data [27]. The GENIUS experiment would allow to test almost the entire MSSM parameter space already in a first step using only 100 kg of enriched or even natural Ge [1,5].

7 Summary

Neutrinoless double beta decay and dark matter search belong to the most sensitive approaches with great perspectives to test particle physics beyond the SM.

The possibilities to use $0\nu\beta\beta$ decay (and the most sensitive Heidelberg–Moscow experiment) for constraining neutrino masses, new interactions beyond the standard model, and violations of Lorentz invariance and the equivalence principle have been reviewed. Experimental limits on $0\nu\beta\beta$ decay are not only complementary to accelerator experiments, neutrino oscillations and cosmological precision measurements, but at least in some cases competitive or superior to the best existing or planned approaches. Direct WIMP detection experiments can compete with recent and future accelerator experiments in the search for SUSY and experiments using double beta technology belong to the most promising approaches in this field of research. A further large breakthrough, both for double beta decay and dark matter search, will be possible realizing the GENIUS proposal, which would improve the obtained limits by up to 1-2 orders of magnitude.

References

1. H.V. Klapdor–Kleingrothaus, Int. J. Mod. Phys. A 13 (1998) 3953; Proc. Lepton and Baryon Number Violation, IOP Bristol & Philadelphia 1999, Eds. H.V. Klapdor–Kleingrothaus, I. Krivosheina, 251-301
2. W.C. Haxton, G.J. Stephenson, Progr. Part. Nucl. Phys. 12 (1984) 409; M. Moe, P. Vogel, Annual Review of Nucl. Part. Science 44 (1994) 247; M. Doi, T. Kotani, E. Takasugi, Progr. Theor. Phys. Suppl. 83 (1985) 1
3. H.V. Klapdor–Kleingrothaus, H. Päs (Eds.): *Beyond the Desert – Accelerator and Non-Accelerator Approaches*, Proc. Int. Workshop on Particle Physics beyond the Standard Model, Castle Ringberg, June 8-14, 1997, IOP Publ., Bristol, Philadelphia
4. HEIDELBERG–MOSCOW collab., Phys. Rev. Lett. 83 (1999) 41; priv. comm.
5. H.V. Klapdor–Kleingrothaus, in [3]; J. Hellmig, H.V. Klapdor–Kleingrothaus, Z. Phys. A 359 (1997) 351; H.V. Klapdor–Kleingrothaus, M. Hirsch, Z. Phys. A 359 (1997) 361; H.V. Klapdor–Kleingrothaus,

J. Hellmig, M. Hirsch, J. Phys. G 24 (1998) 483; H.V. Klapdor-Kleingrothaus, Y. Ramachers, Eur. Phys. J. A 3 (1998) 85
6. H.V. Klapdor–Kleingrothaus, L. Baudis, G. Heusser, B. Majorovits, H. Päs, hep-ph/9910205
7. H.V. Klapdor–Kleingrothaus, H. Päs, A.Y. Smirnov, to be published
8. G. Belanger, Proc. Lepton and Baryon Number Violation, see [1]
9. H. Päs, M. Hirsch, S.G. Kovalenko, H.V. Klapdor–Kleingrothaus, Phys. Lett. B 453 (1999) 194
10. S. Bergmann, H.V. Klapdor–Kleingrothaus, H. Päs, to be published
11. H. Päs, M. Hirsch, H.V. Klapdor–Kleingrothaus, S.G. Kovalenko, to be published
12. R.N. Mohapatra, Phys. Rev D 34 (1986) 3457
13. M. Hirsch, H.V. Klapdor–Kleingrothaus, O. Panella, Phys. Lett. B 374 (1996) 7
14. M. Hirsch, H.V. Klapdor–Kleingrothaus, S.G. Kovalenko, Phys. Rev. Lett. 75 (1995) 17
15. K. S. Babu, R. N. Mohapatra, Phys.Rev.Lett. 75 (1995) 2276
16. M. Hirsch, H.V. Klapdor–Kleingrothaus, S. Kovalenko, Phys. Lett. B 352 (1995) 1
17. M. Hirsch, H.V. Klapdor–Kleingrothaus, S. Kovalenko, Phys. Rev. D 53 (1996) 1329
18. M. Hirsch, H.V. Klapdor–Kleingrothaus, S.G. Kovalenko, Phys. Lett. B 372 (1996) 181
19. H. Päs, M. Hirsch, H.V. Klapdor–Kleingrothaus, Phys. Lett. B 459 (1999) 450
20. M. Hirsch, H.V. Klapdor–Kleingrothaus, S.G. Kovalenko, Phys. Lett. B 398 (1997) 311 and 403 (1997) 291; M. Hirsch, H.V. Klapdor-Kleingrothaus, S.G. Kovalenko, Phys.Rev. D 57 (1998) 1947;
21. G. Bhattacharyya, H.V. Klapdor-Kleingrothaus, H. Päs, Phys.Lett. B463 (1999) 77-82
22. M. Hirsch, H.V. Klapdor–Kleingrothaus, S.G. Kovalenko, Phys. Lett. B 378 (1996) 17
23. M. Hirsch, H.V. Klapdor–Kleingrothaus, S.G. Kovalenko Phys. Rev. D 54 (1996) R4207
24. C. Leung, hep-ph/0002073 and references therein
25. H.V. Klapdor–Kleingrothaus, H. Päs, U. Sarkar, Eur. Phys. J A5 (1999) 3
26. HEIDELBERG–MOSCOW collab., Phys. Lett. B 336 (1994) 141
27. HEIDELBERG–MOSCOW collab., Phys. Rev. D 59 (1998) 022001

World Scientific Lecture Notes in Physics – Vol. 41

World Scientific
Singapore • New Jersey • London • Hong Kong

MASSIVE NEUTRINOS IN PHYSICS AND ASTROPHYSICS

Rabindra N Mohapatra
University of Maryland
College Park

Palash B Pal
University of Oregon
Eugene

Teilchenastrophysik

Particle Astrophysics

Von Prof. Dr. rer. nat. Hans Volker Klapdor-Kleingrothaus
Max-Planck-Institut für Kernphysik, Heidelberg
und Dr. rer. nat. Kai Zuber
Universität Dortmund

Mit zahlreichen Abbildungen und Tabellen

Translated by S M Foster and B Foster

Перевод с немецкого В. А. БЕДНЯКОВА

Inhalt

МОСКВА
НАУКА • ФИЗМАТЛИТ
2000

B. G. Teubner Stuttgart 1997

Institute of Physics Publishing
Bristol and Philadelphia

2.4.2. Doubly Charged Higgs and Pion Double Charge Exchange, and Double Beta Decay

VOLUME 47, NUMBER 24 PHYSICAL REVIEW LETTERS 14 DECEMBER 1981

New Contribution to Neutrinoless Double Beta Decay in Gauge Models

R. N. Mohapatra

Department of Physics, City College of the City University of New York, New York, New York 10031

and

J. D. Vergados

Department of Physics, University of Ioannina, Ioannina, Greece

(Received 28 September 1981)

It is pointed out that in a general class of gauge models, there exist new contributions to neutrinoless double β transitions, that do not involve a Majorana neutrino but the decay of a doubly charged Higgs boson to electrons. Explicit calculations for the case $^{48}Ca \to {}^{48}Ti$ indicate that for reasonable choice of parameters, this new contribution may dominate over that involving light or heavy Majorana neutrinos.

PACS numbers: 23.40.Bw, 11.10.Np, 12.30.-s

All observed weak-interaction processes to date appear to respect a $U(1)_L$ global symmetry associated with lepton number.[1] Lacking any dynamical reason for the existence of $U(1)_L$ it is important to ask whether lepton number is indeed a good or an approximate quantum number? In the context of modern gauge theories, if neutrinos are taken massive, satisfactory understanding of the smallness of their mass[2] requires that the leptonic $U(1)_L$ symmetry (local[2] or global[3]) be spontaneously broken. Phenomenological tests of this idea, of course, depend on the scale of the $U(1)_L$-symmetry breakdown. If this scale is low[4] ($\approx$TeV), an outstanding signature for this breaking is the observation of neutrinoless double beta decay[5] $[(\beta\beta)_{0\nu}]$

$$(N, Z) \to (N-2, Z+2) + e^- + e^-$$

with a lifetime $\tau_{(\beta\beta)_{0\nu}} \approx 10^{22}-10^{24}$ yr, not much higher than the existing experimental lower limits[5] on $\tau_{(\beta\beta)_{0\nu}} \gtrsim 10^{21}$ yr.

The traditional theoretical analysis of the $(\beta\beta)_{0\nu}$ process had to suppose the existence of a Majorana neutrino ν_m, which could be ν_e itself or could couple to the electron, and one had to make one (or both) of the following assumptions: (i) There exists a small right-handed component in the predominantly left-handed current. (ii) The Majorana neutrino ν_m has a mass not much less than a few eV.

These mechanisms have been analyzed in the past[5-9] and limits on the neutrino mass have been extracted.[7-9] In this note, we point out that a new contribution to $(\beta\beta)_{0\nu}$ involving no neutrinos, but the decay of doubly charged Higgs boson to electrons, can exist in gauge models where neutrinos are Majorana particles. For reasonable values of parameters, we estimate this contribution for the transition $^{48}Ca \to {}^{48}Ti$ and find that it may dominate over the neutrino-exchange graphs.

We first point out the origin of the new Higgs exchange contribution to $(\beta\beta)_{0\nu}$, in a gauge model based on the group $SU(2)\otimes U(1)$ with only left-handed Majorana neutrinos. It is only relevant for our purpose to display the leptonic and Higgs content of our model,[10]

$$\psi_L \equiv \begin{pmatrix} \nu_L \\ e_L^- \end{pmatrix} \ (\tfrac{1}{2}, -1), \qquad e_R^- \ (0, -2); \tag{1}$$

$$\phi \equiv \begin{pmatrix} \varphi^+ \\ \varphi^0 \end{pmatrix} \ (\tfrac{1}{2}, 1), \qquad \vec{\Delta} \equiv \begin{pmatrix} \Delta^{++} \\ \Delta^+ \\ \Delta^0 \end{pmatrix} \ (1, 2). \tag{2}$$

Gauge invariance allows the following Yukawa-type couplings among the above fields:

$$\mathcal{L}_Y = h_1 \bar{\psi}_L \phi e_R^- + i h_2 \psi_L^T \tau_2 \vec{\tau} \psi_L \vec{\Delta} + \text{H.c.} \tag{3}$$

As in the standard Glashow-Weinberg-Salam model, we assume that the ϕ also couples to the quarks. Spontaneous breakdown of the $SU(2)_L \otimes U(1)$ gauge symmetry is caused by a $\langle \phi \rangle = \binom{0}{v}$.

The existence of a trilinear Higgs coupling of the form $\mathcal{L}_3 \equiv \lambda M \phi^T \tau_2 \vec{\tau} \phi \cdot \vec{\Delta}$ implies that $\langle \Delta^0 \rangle = \kappa \neq 0$ leading to a Majorana mass term for the left-handed neutrino, with $m_\nu = h_2 \kappa$. If we assume that $h_2 \approx 1$, this implies $\kappa \lesssim 3.5 \times 10^{-8}$ GeV. Since $v \simeq 10^3$ GeV and we choose $\lambda \approx 1$ and $M \approx v$, we need a fine tuning of the mass term of $\vec{\Delta}$ to one part in 10^{-11}. This is obviously unsatisfactory but no worse than in other models with vastly disparate mass scales such as grand unified ones.

VOLUME 47, NUMBER 24 PHYSICAL REVIEW LETTERS 14 DECEMBER 1981

The presence of the $\mathcal{L}_3$ term breaks lepton number by two units and contributes to $(\beta\beta)_{0\nu}$ decay without exchange of Majorana neutrinos as shown in Fig. 1. It leads to an effective double β-decay Hamiltonian of the form

$$\mathcal{H}_{\Delta L=2} \simeq G_2 \bar{d}_L u_R \bar{d}_L u_R e_L{}^T c^{-1} e_L + \text{H.c.} \tag{4}$$

with

$$G_2 \simeq \frac{h_q{}^2 h_2 \lambda M}{m_\varphi{}^4 - m_\Delta{}^2}. \tag{5}$$

The strength of this interaction is estimated to be $\simeq G_F{}^2 \times 10^{-8}$ GeV^{-1}, if we assume $h_q{}^2 \simeq 10^{-9}$ (since it is related to the quark masses) and $m_\varphi \approx m_\Delta \simeq 100$ GeV. This value is of the same order as the present limits on the interaction.[11]

We consider next the $SU(2)_L \otimes SU(2)_R \otimes U(1)_{B-L}$ model.[4] We remind the reader about the leptonic multiplets of the model: left-handed doublet

$$\psi_L \equiv (\nu_L, e_L{}^-): \ (\tfrac{1}{2}, 0, -1);$$

right-handed doublet

$$\psi_R \equiv (N_R, e_R{}^-): \ (0, \tfrac{1}{2}, -1);$$

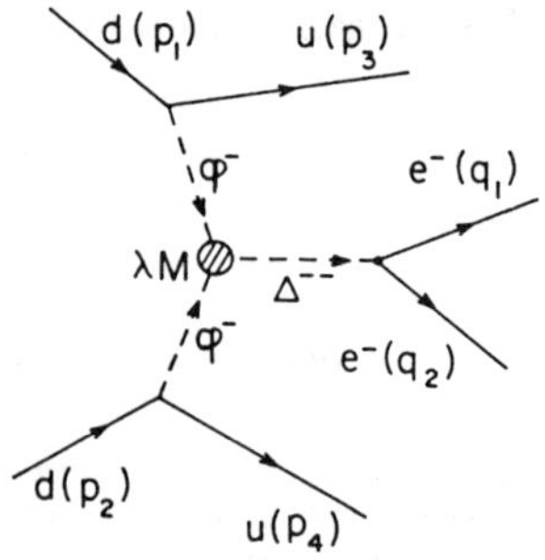

FIG. 1. New Higgs exchange contribution to neutrinoless double β decay.

Higgs multiplets

$$\phi: \ (\tfrac{1}{2}, \tfrac{1}{2}, 0); \quad \Delta_L: \ (1, 0, +2); \quad \Delta_R: \ (0, 1, +2).$$

It is then straightforward to observe that the existence of the scalar coupling $\text{Tr}(\vec{\tau}\cdot \Delta_L \phi \vec{\Delta}_R{}^\dagger \cdot \vec{\tau}\phi^\dagger)$ upon substituting $\langle \Delta_R{}^0 \rangle = v_R \neq 0$ leads to the Feynman diagram in Fig. 1, with M replaced by v_R. Since v_R is bigger than M_w, the value G_2 is expected to be bigger in this model.

The amplitude associated with the Feynman diagram of Fig. 1 can be written down. Assuming that all Higgs particles involved are quite heavy, we obtain

$$\mathfrak{M} = G_2 \bar{u}(p_3)(1-\gamma_5) d(p_1) \bar{u}(p_4)(1-\gamma_5) d(p_2)(1-P_{12}) \bar{e}(q_2)(1+\gamma_5) e^c(q_1), \tag{6}$$

where P_{12} is the permutation operator (interchanging spins and momenta of the two electrons). At the nucleon level the above amplitude can be written as

$$\mathfrak{M} = G_2 \bar{u}(p_3)[F_S{}^{(3)} - F_P{}^{(3)}\gamma_5]\tau_+ u(p_1) \bar{u}(p_4)[F_S{}^{(3)} - F_P{}^{(3)}\gamma_5]\tau_+ u(p_2)(1-P_{12}) \bar{e}(q_2)(1+\gamma_5) e^c(q_1), \tag{7}$$

where $F_S{}^{(3)}$ and $F_P{}^{(3)}$ are the isovector scalar and pseudoscalar form factors. We use the parametrization for them suggested by Adler *et al.*[12]

$$F_P{}^{(3)}(k^2) = \frac{F_P{}^{(3)}(0)}{(1-k^2/M_A{}^2)(1-k^2/M_\pi{}^2)}, \tag{8}$$

$$F_S{}^{(3)}(k^2) = \frac{F_S{}^{(3)}(0)}{(1-k^2/M_A{}^2)^2}, \tag{9}$$

where $M_A = 0.85$ GeV$/c^2$.

For the reasons discussed in Ref. 9, the momentum dependence of the form factors cannot be ignored; namely, if we neglect such dependence, the transition operator in coordinate space would behave like a δ function in the separation of two interaction nucleons. Because of the Pauli principle, it has the following consequences: First, the transition rate is greatly suppressed; and second, the results obtained depend crucially on not so well understood physics like short-range correlations. It has been argued in Ref. 9 that introduction of the form factor cures both these problems.

It is now tedious but straightforward to make a nonrelativistic reduction of Eq. (8) and obtain the following result in coordinate space[9]:

$$\mathfrak{M} = \tfrac{1}{2} G_2 \frac{\hbar c}{m_p c^2 R_0} m_e m_p{}^2 f_A{}^2 [(1-P_{12})\bar{e}(q_2)(1+\gamma_5)e^c(q_1)] \langle f | \Omega_\Delta | i \rangle, \tag{10}$$

$$\Omega_\Delta = \frac{m_p}{m_e} \sum_{i \neq j} \tau_+(i)\tau_+(j) \frac{R_0}{r_{ij}} \{\alpha_S F_1(r_{ij}) + \alpha_P [\vec{\sigma}_i \cdot \vec{\sigma}_j F_2(r_{ij}) + (\tfrac{24}{5}\pi)^{1/2} Y^2(\hat{r})(\vec{\sigma}_i \times \vec{\sigma}_j)^2 F_3(r_{ij})]\}, \tag{11}$$

with $R_0 = r_0 A^{1/3}$ ($r_0 = 1.1$ fm), the nuclear radius; $\vec{r}_{ij} = \vec{r}_i - \vec{r}_j$, $\hat{r} = \vec{r}_{ij}/|\vec{r}_{ij}|$,

$$\alpha_S = (1/3f_A^2)[F_S^{(3)}(0)]^2, \quad \alpha_P = \tfrac{1}{12}(M_\pi/m_p)^4[F_P^{(3)}(0)]^2, \tag{12a}$$

$$F_1(r_{ij}) = \tfrac{1}{48}(M_A/m_p)^2 x_A(x_A^2 + 3x_A + 3)e^{-x_A}, \tag{12b}$$

$$F_2(r_{ij}) = 2[(M_\pi/M_A)^2 e^{-x_\pi} - e^{-x_A}] + \tfrac{1}{2}[(x_\pi - 2)e^{-x_\pi} + (x_A - 2)e^{-x_A}], \tag{12c}$$

$$F_3(r_{ij}) = 2[(M_\pi/M_A)^2 Z(x_A)] + \tfrac{1}{2}[(x_\pi + 1)e^{-x_\pi} + (x_A + 1)e^{-x_A}], \tag{12d}$$

$$x_A = (M_A c^2/\hbar c)|\vec{r}_{ij}|; \quad x_\pi = (M_\pi c^2/\hbar c)|\vec{r}_{ij}|; \quad Z(x) = (x^2 + 3x + 3)x^{-2}e^{-x}. \tag{12e}$$

We notice that the structure of the operator Ω_Δ is not very different from the one involved in the heavy-Majorana-neutrino–mediated process[9] except that it contains a tensor component.

If one now defines a lepton-number–nonservation parameter $\eta_\Delta = (G_2 m_p/G_F^2)$, the transition probability associated with Fig. 1 takes a familiar form[5,7,9]:

$$W_{0\nu}(i \to f e^- e^-) = K_2 |\chi(\alpha(z+2))|^2 A^{-2/3} f_2(\epsilon_0) |\eta_\Delta \langle f|\Omega_\Delta|i\rangle|^2, \tag{13}$$

where

$$K_2 = c\,\frac{m_e c^2}{\hbar c}\,\frac{(G_F m_p^2)^4}{(2\pi)^3}\left(\frac{\hbar c}{m_p c^2 r_0}\right)^2 \left(\frac{m_e}{m_p}\right)^6 f_A^4$$
$$= 2.6 \times 10^{-15}\ \mathrm{yr}^{-1}, \tag{14}$$

$$f_2(\epsilon_0) = \tfrac{2}{15}\epsilon_0[\epsilon_0^4 + 10\epsilon_0^3 + 40\epsilon_0^2 + 60\epsilon_0 + 30] \tag{15a}$$

(ϵ_0 is the available energy of double β decay in units of $m_e c^2$),

$$\chi(\alpha z) = \frac{2\pi\alpha Z}{1 - \exp(-2\pi\alpha Z)} \tag{15b}$$

represents the distortion of the electron wave function. In order to estimate the rate given by Eq. (13), we taken the values of α_s and $\alpha_p = 0.40$; $\eta_\Delta \approx 10^{-8}$. We then compute[9] the nuclear matrix element for the simplest and experimentally most interesting process: ^{48}Ca → ^{48}Ti, in the shell model where we find $\langle f|\Omega_\Delta|i\rangle = 530$. This value is approximately reduced by 20% when a simple step-function short-range correlation is used. From Eqs. (14), (16), and (17), we find

$$(T_{1/2})_{0\nu} \simeq 3 \times 10^2\ \mathrm{yr} \tag{16}$$

which almost coincides with the present experimental lower limit. Of course, it must be pointed out that although we believe $M_\phi \approx M_\Delta \approx 100$ GeV is quite a reasonable choice for Higgs masses, $T_{1/2}$ depends on these masses rather sensitively (i.e., $T_{1/2} \sim M_\phi^8 M_\Delta^4$), and a slight increase in the values of these masses could increase the half-lives significantly.

We remark that in the presence of the proposed mechanism, the full double β-decay experiments impose a constraint of the form

$$|\eta_\nu \langle f|\Omega_\nu|i\rangle + \eta_N \langle f|\Omega_N|i\rangle + \eta_\Delta \langle f|\Omega_\Delta|i\rangle|^2 \leq b, \tag{17}$$

where η_ν and η_N correspond to the lepton-number–nonconservation parameters associated with the conventional light and heavy Majorana neutrinos, which in the notation of Ref. 9 can be written as $\eta_\nu = m_\nu/m_e$ and $\eta_N = \beta^2 m_p/m_N$ where β is the effective coupling of N to electrons. It has been shown in Ref. 9 that $\langle f|\Omega_\nu|i\rangle \approx 1$ and $\langle f|\Omega_N|i\rangle \approx 72$. From the data on ^{48}Ca → ^{48}Ti decay, we get $b = 0.4 \times 10^{-10}$, which from Eq. (16) leads to useful constraints on gauge models. For instance, if $\eta_N = \eta_\Delta = 0$, this implies $\eta_\nu \leq 0.6 \times 10^{-5}$ or $m_\nu \leq 3$ eV, etc.

In conclusion, we have pointed out a new contribution to double β decay, which does not involve Majorana neutrinos and makes a measurable contribution to it, provided that η_Δ is not smaller than 10^{-8}. Our results are based on the ^{48}Ca → ^{48}Ti decay but we see no reason why it should not hold in general.

Both the authors would like to acknowledge the hospitality of the CERN theory group where this work was done. This work was supported in part by National Science Foundation Grant No. PHY-78-2488 and by CUNY-PSC-BHE Faculty Research Award.

[1]R. E. Marshak, Riazuddin, and C. P. Ryan, *Theory of Weak Interactions in Particle Physics* (Wiley, New York, 1969). For a recent review and references, see H. Primakoff and S. P. Rosen, to be published.

[2]M. Gell-Mann, P. Ramond, and R. Slansky, in *Supergravity*, edited by D. Z. Freedman *et al.* (North-Holland, Amsterdam, 1979); T. Yanagida, unpublished; R. Barbieri, G. Morchio, D. Nanopoulos, and F. Strocci, Phys. Lett. 90B, 81 (1980); R. N. Mohapatra and G. Senjanović, Phys. Rev. Lett. 44, 912 (1980), and Phys. Rev. D 23, 165 (1981).

[3]Y. Chikashige, R. N. Mohapatra, and R. D. Peccei, Phys. Lett. 98B, 265 (1981), and Phys. Rev. Lett. 45,

VOLUME 47, NUMBER 24 PHYSICAL REVIEW LETTERS 14 DECEMBER 1981

1926 (1980). For further considerations along this approach, see G. Gelmini and M. Roncadelli, Phys. Lett. 99B, 411 (1981); H. Georgi, S. L. Glashow, and S. Nussinov, to be published.

[4]R. N. Mohapatra and G. Senjanović, Ref. 2; T. Rizzo and G. Senjanović, Phys. Rev. Lett. 46, 1315 (1981); R. Barbieri, R. N. Mohapatra, and A. Masiero, Phys. Lett. 105B, 369 (1981).

[5]For early theoretical treatment of double β decay, see H. Primakoff and S. P. Rosen, Phys. Rev. 184, 1925 (1969), and references therein.

[6]For useful reviews, see, e.g., D. Bryman and C. Picciotto, Rev. Mod. Phys. 50, 11 (1978); Yu. G. Zdesenko, Fiz. Elem. Chastits At. Yadra 11, 1369 (1980) [Sov. J. Part. Nucl. 11, 542 (1980)].

[7]M. Doi, T. Kotani, H. Nishiura, K. Okuda, and E. Takashugi, Phys. Lett. 103B, 819 (1981), and Osaka Univ. Report No. OS-GE 81-28, 1981 (unpublished).

[8]W. C. Haxton, G. J. Stephenson, and D. Strottman, Phys. Rev. Lett. 47, 153 (1981).

[9]J. D. Vergados, to be published, and Phys. Rev. C 13, 865 (1976).

[10]Riazuddin, R. E. Marshak, and R. N. Mohapatra, Phys. Rev. D 24, 1310 (1981); T. P. Cheng and L.-F. Li, Phys. Rev. D 22, 2680 (1980).

[11]B. J. Cleveland *et al.*, Phys. Rev. Lett. 35, 757 (1975); B. Srinivasan, E. Alexander, and O. Manuel, Econ. Geol. 67, 592 (1972); M. Moe and D. Lowenthal, University of California, Irvine, Report No. 10P19-143, 1979 (unpublished); T. Kirsten *et al.*, Phys. Rev. Lett. 20, 1300 (1968).

[12]S. L. Adler *et al.*, Phys. Rev. D 11, 3309 (1975).

PHYSICAL REVIEW D VOLUME 25, NUMBER 3 1 FEBRUARY 1982

Pion-double-charge-exchange contribution to neutrinoless double-β decay

J. D. Vergados
Physics Department, University of Ioannina, Ioannina, Greece

(Received 13 October 1981)

It is shown that the double charge exchange of pions in flight between two nucleons makes a substantial contribution to neutrinoless nuclear double-β decay if it is mediated by heavy Majorana neutrinos.

Neutrinoless double-β decay has recently become[1–6] a very interesting process due to recent developments in gauge theories[7–12] which predict that lepton charge (number) must be broken at some level. In this Communication we explore a new mechanism for neutrinoless double-β decay which involves the double charge exchange of pions in flight between two nucleons.

In a previous paper,[13] which will be referred to as I, we have examined the neutrinoless double-β decay

$$(A,Z)\rightarrow(A,Z+2)+e^-+e^- , \tag{1}$$

concentrating solely on a mechanism which involved only nucleons. We showed that even if the mediating Majorana neutrino is very massive the two-body transition operator behaves as e^{-mr}/r with $m\approx m_p$ provided that the nucleon form factor is taken into account. Thus, this operator does not suffer from severe short-range pathologies and the two-nucleon mechanism will substantially contribute to neutrinoless double-β decay even in the presence of the nuclear hard core.[13] It is tempting, however, to consider the contribution of other particles which can undergo a change of charge by two units and have a reasonable chance to be found in the nuclear medium. Such candidates are the nucleon isobars[14] and in particular

$$n\rightarrow\Delta^{++}+e^-+e^- . \tag{2}$$

Such a process, however, does not contribute to the $0^+\rightarrow0^+$ nuclear decay due to angular momentum selection rules.[15] Other processes, such as

$$\Delta^0\rightarrow\Delta^{++}+e^-+e^-,\quad \Delta^-\rightarrow\Delta^++e^-+e^- , \tag{3}$$

may contribute to second order but their contribution will be small since they are proportional to $P_\Delta{}^2\approx10^{-4}$ (P_Δ is the probability of finding such an isobar inside the nucleus). More recently[16] Halprin has investigated the contribution of processes like $n\rightarrow\Delta(1910)+e^-+e^-$ to reaction (1). Such a process is indeed allowed from the point of view of angular momentum. Kotani, however, has argued[16] that even this mechanism will not substantially contribute to $0^+\rightarrow0^+$ nuclear transitions since the $\Delta(1910)$ must predominantly have $l=2$ (around its center of mass). Furthermore, the probability of finding such a particle inside the nucleus is not known. Kotani argues again that, due to the heavy mass of $\Delta(1910)$, this probability must be very small.[16] Thus the situation at this point is not completely clear.

In the present paper we will investigate another mechanism which does not suffer from the above ambiguities, namely, the double charge exchange of pions in flight between the two nucleons:

$$\pi^-\rightarrow\pi^++e^-+e^- . \tag{4}$$

The relevant Feynman diagrams which give rise to (4) are shown in Fig. 1. The amplitude associated with Fig. 1(a) takes the form

$$\mathfrak{M}=[\bar N(p_3)g_r\gamma_5\tau_+N][\bar N(p_4)g_r\tau_+\gamma_5N(p_2)]\frac{1}{(p_1-p_3)^2-m_\pi{}^2}\frac{1}{(p_4-p_2)^2-m_\pi{}^2}\mathfrak{M}_1 . \tag{5}$$

All the essential new physics is contained in the amplitude $\mathfrak{M}_1$ associated with Fig. 1(b). In evaluating this diagram we will follow a procedure analogous to that described previously[13] in I. The charged current in the leptonic vertex is taken to be of the form

$$\begin{pmatrix}\nu_e+\beta N_0\\ e^-\end{pmatrix}_L , \tag{6}$$

where ν_e and N_0 are the light and heavy components of the intermediate Majorana neutrino ν_m given by

$$\nu_e=\sum_j X_{ej}^{(1)}\nu_j,\quad N_0=\sum_j X_{ej}^{(2)}N_j , \tag{7}$$

where ν_j (N_j) are the (Majorana) mass eigenstates in the light (heavy) sector and β is a convenient parametrization of the mixing between the two groups.

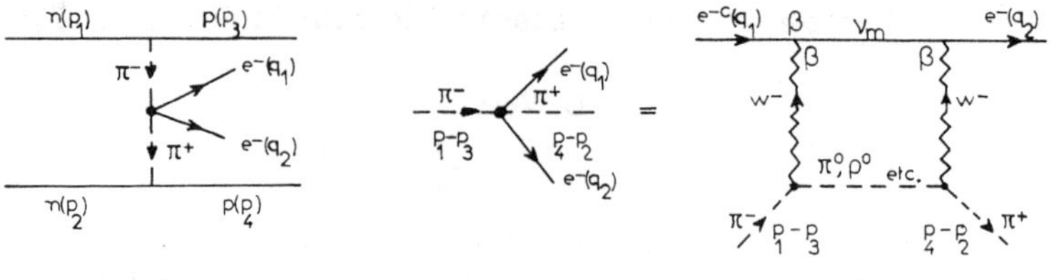

(a) (b)

FIG. 1. (a) Lepton-number-violating diagrams associated with the double charge exchange of a pion in flight between the two nucleons. (b) The elementary process $\pi^- \to e^- + e^- + \pi^+$ is exhibited.

The amplitude associated with the heavy sector takes the form

$$\mathfrak{M}_1 = \frac{G_F}{m_p}\eta_\sigma I_{\lambda\mu}(1-P_{12}) \times [\bar{e}^-(q_2)\gamma^\lambda\gamma^\mu(1+\gamma_5)e^{-c}(q_2)] \ , \quad (8a)$$

where G_F is the Fermi coupling constant,

$$\eta_\sigma = \beta^2\frac{m_p}{m_\sigma}, \quad \frac{1}{m_\sigma} = \sum_j (X_{ej}^{(2)})^2\frac{1}{M_j} \ , \quad (8b)$$

P_{12} is the permutation operator exchanging the spins and momenta of the two electrons, and M_j are the heavy-mass eigenvalues.

The quantity $I_{\lambda\mu}$ will be calculated in two ways.

Case (i). The pions will be treated as "elementary." Only the π^0 will be included in the intermediate mesonic states. The vertex functions are taken[17] to be $\sqrt{2}(p_1-p_2+q)_\lambda$ and $\sqrt{2}(p_4-p_2+q)_\mu$, each multiplied by a dipole shape form factor[13]

$$F(k^2) = \frac{1}{(1-k^2/m_A{}^2)^2}, \quad m_A = 0.85 \text{ GeV}/c^2 \ , \quad (9)$$

where k is the momentum transfer at the vertex. Ignoring the external momenta in front of m_A and employing standard loop-integral techniques[18] we obtain

$$I_{\lambda\mu} \approx 2\int \frac{d^4q}{(2\pi)^4}\frac{(p_1-p_3+q)_\lambda}{q^2-m_\pi{}^2}\frac{(p_4-p_2+2)_\mu}{(1-q^2/m_A{}^2)^4} \cong g_{\lambda\mu}\frac{M_A{}^4}{192\pi^2} \ , \quad (10)$$

$$\mathfrak{M}_1 = \tfrac{1}{2}(G_F m_p)^2\eta_\sigma 4m_\pi\xi_\pi(1-P_{12}) \times [\bar{e}^-(q_2)(1+\gamma_5)e^{-c}(q_1)] \ , \quad (11)$$

with

$$\xi_\pi = \frac{1}{96\pi^2}\frac{m_A}{m_\pi}\left(\frac{m_A}{m_p}\right)^3 \ . \quad (12)$$

It is now pretty straightforward to make a nonrelativistic reduction of the amplitude of Eq. (5). Furthermore, since the momenta of the electrons are negligible in front of m_π we can easily transform this amplitude to coordinate space to obtain

$$\mathfrak{M} = \tfrac{1}{2}(G_F m_p)^2\eta_\sigma\frac{m_e}{m_p}f_A{}^2\frac{\hbar c}{m_p c^2 R_0}\langle f|\Omega_\pi|i\rangle \times (1-P_{12})[\bar{e}^-(q_2)(1+\gamma_5)e^{-c}(q_1)] \ , \quad (13)$$

where $R_0 = r_0 A^{1/3}$ = nuclear radius, $f_A = 1.24$, and

$$\Omega_\pi = \alpha_\pi\frac{m_p}{m_e}\sum_{i\neq j}\tau_+(i)\tau_+(j)\frac{R_0}{r_{ij}} \times [F_1(x_\pi)\vec{\sigma}_i\cdot\vec{\sigma}_j + F_2(x_\pi)(3\vec{\sigma}_i\cdot\hat{r}\,\vec{\sigma}_j\cdot\hat{r} - \vec{\sigma}_i\cdot\vec{\sigma}_j)] \ , \quad (14a)$$

$$\hat{r} = \frac{\vec{r}_{ij}}{r_{ij}}, \quad \vec{r}_{ij} = \vec{r}_i - \vec{r}_j, \quad x_\pi = r_{ij}\frac{m_\pi c^2}{\hbar c} \ , \quad (14b)$$

$$F_1(x) = (x-2)e^{-x}, \quad F_2(x) = (x+1)e^{-x} \ , \quad (14c)$$

$$\alpha_\pi = \frac{1}{72\pi}\left(\frac{m_A}{m_p}\right)^2\left(\frac{m_A}{m_\pi}\right)^2\frac{(f_{\pi NN})^2}{f_A{}^2} = 0.0072 \ , \quad (14d)$$

$(f_{\pi NN})^2 = 0.08$ . .

Case (ii). The pions are considered as composite particles described in terms of a bound quark-antiquark system. The bound wave function can be treated in the fashion of Adler *et al.*[19] or in the quark-bag-model approach.[20] For our purposes it is adequate to consider the quarks as nonrelativistic with a quark density corresponding to the phenomenological form factor of Eq. (9). The operator at the quark level can be obtained as in I. Employing closure for the intermediate meson states we finally obtain

$$I_{\lambda\mu} \approx \frac{m_A{}^4}{96\pi^2}\langle\pi^+|\omega_{\lambda\mu}|\pi^-\rangle \ , \quad (15a)$$

where

$$\omega_{\lambda\mu} = \sum_{i\neq j}\tau_+(i)\tau_+(j)L_\lambda(i)L_\mu(j) \ ,$$

$$L_\lambda(i) = \begin{cases} 1, & \lambda = 0 \\ -\vec{\sigma}(i), & \lambda \neq 0 \ . \end{cases} \quad (15b)$$

In the above expression unlike Eq. (14a), the indices i and j label the quarks and $|\pi^+\rangle$, $|\pi^-\rangle$ describe the spin-isospin part of the quark-antiquark wave function. Combining Eqs. (15a), (15b), and (8) we once again obtain (3) with

$$\alpha_\pi = 0.06 \quad . \tag{16}$$

The larger value is attributed to the fact that, by invoking closure, we included here intermediate states other than π^0. Equation (13) is the same with that obtained in I except for the nuclear matrix element which in I was found to be[13] $\langle f|\Omega_\sigma|i\rangle$ with

$$\Omega_\sigma = \sum_{i\neq j} \tau_+(i)\tau_+(j)\left(\frac{f_V^2}{f_A^2} - \vec{\sigma}_i\cdot\vec{\sigma}_j\right)\frac{R_0}{r_{ij}}F(x_A) \quad ,$$

$$x_A = \frac{m_A c^2 r_{ij}}{\hbar c}, \quad F(x) = \tfrac{1}{48}x(x^2+3x+3)e^{-x} \quad .$$

Thus one can apply the analysis of double-β decay given in I with the only modification that

$$\langle f|\Omega_\sigma|i\rangle \rightarrow \langle f|\Omega_\sigma + \Omega_\pi|i\rangle \quad .$$

In order to estimate the pionic contribution to double-β decay we will examine the $^{48}Ca \rightarrow {}^{48}Ti$ transition.[5,13,19] We will employ a nuclear model which has been described in some detail elsewhere.[6,13] For the reader's convenience we summarize its basic features here. The $N = 0, 1, 2$ harmonic-oscillator shells are assumed completely filled (inert core) and the active nucleons are distributed in the $1f_{7/2}$ shell. Thus the initial ^{48}Ca state is a closed $1f_{7/2}$ neutron shell, while the final state contains two protons and six neutrons in the $1f_{7/2}$ shell. The precise final nuclear wave function depends on the effective two-nucleon interaction[13] used, but the final results reported here are not too sensitive to such variations. Our numerical results were obtained using

$$^{48}\mathrm{Ti}(\mathrm{g.s.}) = 0.8451|1\rangle - 0.5233|2\rangle + 0.1094|3\rangle - 0.0009|4\rangle \quad ,$$

with

$$|i\rangle = |f^6_{7/2}(n)J_i; f^2_{7/2}(p)J_i;0^+\rangle, \quad J_i = 0,2,4,6 \quad .$$

With the above nuclear wave function we obtained

$$\langle f|\Omega_\pi|i\rangle = 14 \quad [\,(\mathrm{i})\,]$$

and

$$\langle f|\Omega_\pi|i\rangle = 94 \quad [\,(\mathrm{ii})\,] \quad .$$

This must be compared with the matrix element of the two-nucleon mode

$$\langle f|\Omega_\sigma|i\rangle = 72 \quad .$$

With the above pionic contribution included, the limits obtained in I change by approximately a factor of 2. Thus using the experimental lower limit[21] on the lifetime of $^{48}Ca \rightarrow {}^{48}Ti$ we obtain

$$\beta^2\frac{m_p}{m_\sigma} \lesssim 5\times 10^{-8} \quad ,$$

$$m_\sigma \gtrsim (2\times 10^4)-(2\times 10^5)\ \mathrm{GeV}/c^2 \quad .$$

Analogous calculation for the light-neutrino sector shows that the pionic matrix element is less than 1% of that involving only nucleons. So in this case it can safely be neglected.

In conclusion, we have shown that if η_σ is not negligible, the heavy component of the Majorana neutrino will make an important contribution to neutrinoless double-β decay. A substantial part of this contribution may come from the pionic mode discussed in this work.

---

[1]See, e.g., H. Primakoff and S. P. Rosen, Phys. Rev. 184, 1925 (1969), and references therein.

[2]For a review, particularly of the experimental situation, and a complete list of references, see D. Bryman and C. Picciotto, Rev. Mod. Phys. 50, 11 (1978); Yu. G. Zdesenko, Fiz. Elem. Chastits At. Vatra 11, 1369 (1980) [Sov. J. Part. Nucl. 11, 542 (1980)]; see also, H. Primakoff and S. P. Rosen, Purdue University report 1981 (unpublished).

[3]W. C. Haxton, G. J. Stephenson, and S. Strottman, Phys. Rev. Lett. 47, 153 (1981).

[4]M. Doi, T. Kotani, H. Nishiura, K. Otuda, and E. Takusugi, Phys. Lett. 30B, 219 (1981).

[5]E. Fiorini *et al.*, Nuovo Cimento A 13, 747 (1973).

[6]J. D. Vergados, Phys. Rev. C 13, 865 (1976).

[7]In the standard Weinberg-Salam model [S. Weinberg, Phys. Rev. Lett. 19, 1264 (1967); A. Salam, in *Elementary Particle Theory, Relativistic Groups and Analyticity (Nobel Symposium No. 8)*, edited by N. Svartholm (Almqvist and Wiksell, Stockholm, 1968), p. 367; S. L. Glashow, Nucl. Phys. 22, 579 (1961)], lepton number is conserved.

[8]Grand unification schemes, with the one exception of the "minimal" SU(5), contain $\Delta L = 2$ terms. For SU(5) see, e.g., H. Georgi and S. L. Glashow, Phys. Rev. Lett. 32, 438 (1974); for SO(10) see, e.g., H. Fritzsch and P. Minkowski, Ann. Phys. 93, 193 (1975); H. Georgi and D. V. Nanopoulos, Nucl. Phys. B155, 52 (1979).

[9]M. Gell-Mann, P. Ramond, and R. Slansky, in *Supergravity*, edited by P. van Nieuwenhuizen and D. Z. Freedman (North-Holland, Amsterdam, 1979), p. 315.

[10]R. Barbieri, D. V. Nanopoulos, G. Morchio, and F. Strocchi, Phys. Lett. 90B, 91 (1980).

[11]E. Witten, Phys. Rev. Lett. 91B, 81 (1980). For a clear review on the Majorana mass see, e.g., E. Witten, Harvard University Report No. HUTP-80/A051 (unpublished).

[12]R. N. Mohapatra and G. Senjanovic, Phys. Rev. D 23, 165 (1981).
[13]J. D. Vergados, Phys. Lett. (to be published).
[14]A. Halprin, P. Minkowski, H. Primakoff, and S. P. Rosen, Phys. Rev. D 13, 2567 (1976); J. D. Vergados and M. Ericson, Nucl. Phys. (to be published).
[15]M. Doi, T. Kotani, H. Nishiura, K. Okuda, and E. Takasugi, Osaka University Report No. OS-GE 81-28, 1981 (unpublished).
[16]A. Halprin, Phys. Rev. D 24, 2988 (1981); T. Kotani (private communication).
[17]E. D. Commins, *Introduction to Weak Interactions* (McGraw-Hill, New York, 1973), p. 92.
[18]G. 't Hooft and M. Veltman, Report No. CERN 73-9, 1973 (unpublished).
[19]S. L. Adler, E. W. Colglazier, Jr., J. B. Healy, T. Karliner, J. Lieberman, Y. J. Ng, and H. S. Tsao, Phys. Rev. D 11, 3309 (1975).
[20]A. Chodos, R. L. Jaffe, K. Johnson, and C. B. Thorn, Phys. Rev. D 10, 2599 (1974); A. Chodos *et al.*, *ibid.* 9, 3471 (1974); W. A. Bardeen *et al.*, 11, *ibid.* 1094 (1975).
[21]B. J. Cleveland, W. R. Leo, C. S. Wu, P. J. Callon, and J. D. Ullman, Phys. Rev. Lett. 35, 757 (1975); R. K. Burdin, P. J. Callon, J. D. Ullman, and C. S. Wu, Nucl. Phys. A158, 1337 (1970); E. Mateosian and M. Goldhaber, Phys. Rev. 146, 810 (1966).

2.4.3 Superstring — Inspired Models, *R*-Parity Breaking Supersymmetry and Double Beta Decay

Z. Phys. C – Particles and Fields 41, 623–629 (1989)

Zeitschrift für Physik C Particles and Fields

Lepton phenomenology and neutrinoless double β-decay in superstring inspired theories

G.K. Leontaris[1,*] and J.D. Vergados[2]

[1] CERN 7, CH-1211 Geneva 23, Switzerland
[2] Department of Physics, University of Ioannina, GR-45110 Ioannina, Greece

Received 5 April 1988

Abstract. The properties of neutrinos, s-leptons, s-quarks as well as the coloured isosinglet $\tilde{D}$ and $\tilde{D}^c$ are studied in superstring inspired models. Furthermore the neutrinoless double β-decay process is investigated, including light and heavy majorana neutrinos, as well as the contribution of the exotic Higgs scalars of the theory. From the current experimental limits on $o\nu$–$\beta\beta$-decay, useful constraints are imposed on the masses and Yukawa couplings of the theory.

1 Introduction

The neutrinoless double β-decay [1–4]

$$(A, Z) \rightarrow (A, Z \pm 2) + e^{\mp} + e^{\mp} \tag{1}$$

which can occur in the presence of lepton number violating interactions is known to impose rather severe constraints on the parameters of gauge models (for a recent review see [4]). Such are e.g. the Majorana neutrino mass, the mass of the right-handed neutrino, the couplings of Higgs scalars to leptons and even the masses of the supersymmetric partners of quarks and leptons, i.e., s-quarks and s-leptons [5]. In the present paper, we will examine the neutrinoless double β-decay in the context of superstring-inspired models [6–14]. This process is ideal for such purposes since it only involves particles of the first generation.

Superstring theories [7–14] appear to be the only candidates for a unified theory of all known interactions. The first attempt to obtain a realistic theory involved the compactification of the field theory limit of the $E_8 \times E_8$ ten-dimensional heterotic string compactified on a Calabi–Yau manifold [7]. A second attempt has been made through orbifold compactifications [9]. The advantage of the latter attempt is that it directly provides us with string theories with four uncompactified dimensions. Furthermore [13] there are now other ways, using the fermionic representations, of directly constructing string models in four dimensions. However, although four-dimensional string theories seem to be promising, as far as we know, no model which is completely realistic [14] has yet been found. In fact, four-dimensional string theories do not preclude a priori the existence of phenomenological models, consistent with $SU(3) \times SU(2) \times U(1)$ theory, like those arising from the ten-dimensional heterotic string compactified on Calabi–Yau manifolds.

The field theory limit of the latter contains fundamental superfields in the adjoint 248-dimensional representation of E_8 which under $SU(3) \times E_6$ decomposes as follows:

$$248 = (1, 78) + (3, 27) + (\bar{3}, 27) + (8.1). \tag{2}$$

The chiral fields belong to the 27-dimensional representation of E_6 which under $SO(10)$ and $SU(5)$ decomposes as follows:

$$\begin{array}{l} 27 = (u, d, d^c, e^+) + \nu^c + (d^c, \nu, e) + (D, \tilde{H}) + (D^c, \tilde{\tilde{H}}) + \eta \\ \quad\;\; (16, 10) \qquad (16, 1)\ (16, \bar{5}) \quad (10, \bar{5}) \quad (10, 5) \quad (1, 1). \end{array} \tag{3}$$

In realistic unified string theories, it is known that the compactification scale has to be of the order of the string tension [10]. This, however, has important constraints on superstring model building. Here we will concentrate our attention on a model which is allowed under the aforementioned restrictions. For illustrative purposes we study the model which arises when E_6 breaks down to the six-rank group:

* Permanent address: Physics Dept., University of Ioannina, Ioannina GR-45110 Greece

PHYSICAL REVIEW D VOLUME 34, NUMBER 11 1 DECEMBER 1986

New contributions to neutrinoless double-beta decay in supersymmetric theories

Rabindra N. Mohapatra
Department of Physics and Astronomy, University of Maryland, College Park, Maryland 20742
(Received 9 June 1986)

In supersymmetric theories with R-parity violation, there are new contributions to neutrinoless double-beta [$(\beta\beta)_{0\nu}$] decay that do not involve the exchange of Majorana neutrinos. Experimental information on $(\beta\beta)_{0\nu}$ decay can therefore be used to constrain parameters of supersymmetric theories. We also discuss neutrino mass in these theories. We then discuss how R-parity-violating interactions can be induced in the low-energy sector of a theory that respects R-parity conservation prior to symmetry breaking.

I. INTRODUCTION

Neutrinoless double-beta decay[1] [$(\beta\beta)_{0\nu}$ decay] is believed to be a sensitive probe of physics beyond the standard model. In the standard $SU(2)_L\times U(1)$ model since $B-L$ conservation is automatic, $(\beta\beta)_{0\nu}$ vanishes to all orders in the weak interactions. However, once the possibility of $B-L$ violation (either spontaneous or explicit) is accepted, the above process, which violates the $B-L$ quantum number by two units, can occur in a variety of ways. The simplest and the most well-known way[2] is via the exchange of Majorana neutrinos or the presence of $(B-L)$-violating right-handed currents, and in fact, present limits on the lifetime for $(\beta\beta)_{0\nu}$ decay are generally interpreted to give an upper limit on the Majorana mass of the neutrino. It has since been pointed out that, in left-right-symmetric theories with Majorana neutrinos,[3] there exist new contributions to $(\beta\beta)_{0\nu}$ decay that involve the exchange of heavy right-handed neutrinos[3,4] as well as the exchange of doubly charged Higgs bosons.[5] Experimental information on $(\beta\beta)_{0\nu}$ decay can, therefore, be used to constrain parameters of the left-right-symmetric models. In this paper, we point out that general supersymmetric extensions of the standard model give rise to lepton-number-violating interactions,[6] which, in turn, can lead to completely new contributions to $(\beta\beta)_{0\nu}$ decay that involve exchange of gluinos and photinos (and not Majorana neutrinos). Existing low-energy data constrain the parameters that control the magnitude of the double-beta decay rate but are such that the new contribution can be large enough to be visible in on-going experiments. Turning the question around, experimental data on neutrinoless double-beta decay can be used to constrain the parameters of the supersymmetric model. We also study the neutrino masses in these models and then discuss the simplest extensions of the standard model where $B-L$ conservation is automatically restored. In such models, R parity can be broken by the vacuum leading to an effective theory at low energies that has induced R-parity-violating interactions. This kind of embedding may help to fix the strength of the R-parity-violating interactions.

II. DESCRIPTION OF THE MODEL

The minimal supersymmetric extension of the standard model[7] consists of the following superfields, with their $SU(2)_L\times U(1)\times SU(3)_c$ transformation indicated within parentheses next to the fields: quarks Q $(2,\frac{1}{3},3)$; U^c $(1,-\frac{4}{3},3^*)$; D^c $(1+\frac{2}{3},3^*)$; leptons L $(2,-1,1)$; E^c $(1,+2,1)$; Higgs bosons H_u $(2,1,0)$; H_d $(2,-1,0)$. We have suppressed the generation index for fermions and assume that supersymmetry (SUSY) breaking is dictated by $N=1$ supergravity.[7] The nongauge interactions of this model are specified by the following general form of the superpotential where we have imposed global baryon-number symmetry to avoid rapid proton decay:

$$W=h_uQH_uU^c+h_dQH_dD^c+h_eLH_dE^c+\mu_1H_uH_d +\mu_2H_uL+f_{pqr}Q_pL_qD_r^c+\tilde{h}_{[p,q]r}L_pL_qE_r^c\ , \tag{1}$$

where p,q,r stand for generation indices and $[p,q]$ implies antisymmetry in the two indices; h_u, h_d, and h_e are matrices in generation space. Supersymmetry-breaking terms in the Lagrangian dictated by a super-Higgs effect in an $N=1$ supergravity model have the form

$$\mathscr{L}_{SB}=m_{3/2}\int d^2\theta\,\theta^2W+\int d^2\theta\,\theta^2\mu_aW_a^\alpha W_{\alpha,a}+\text{H.c.}\ , \tag{2}$$

where a goes over all the gaugino fields of the model such as the W gaugino ($\tilde{W}$) Z gaugino ($\tilde{Z}$), photino ($\tilde{\gamma}$), and gluino ($\tilde{g}$); W^α represents the supersymmetric representation of the gauge field tensor $F_{\mu\nu}$; and θ is the two-component fermionic coordinate.

It is clear from Eq. (1) that R parity[8] defined by $(-1)^{3B+L+2S}$ (where B, L, and S stand for baryon number, lepton number, and spin, respectively) and lepton-number symmetry are violated by the μ_2, f, and $\tilde{h}$ terms in the superpotential. They will, therefore, lead to processes that violate lepton-number conservation. Many of these processes have already been studied previously.[9] Here, we wish to focus on the neutrino mass and neutrinoless double-beta decay. Before proceeding to this applica-

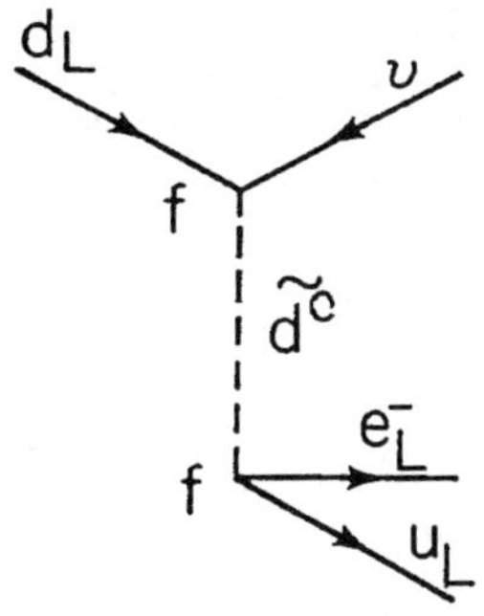

FIG. 1. A new contribution to neutron beta decay in a supersymmetric extension of the standard model.

tion, we would like to briefly comment on the allowed ranges of the couplings f and $\tilde{h}$. The most stringent constraints on the couplings come from $\mu \to 3e$ and $K_L \to \mu\bar{e}$ decay;[1] but both these constraints can be satisfied by constraining $\tilde{h}_{[\mu,e]e^c}$ and $\tilde{h}_{[\mu,e]\mu^c}$ to be tiny; but since these couplings do not play any role in neutrinoless double-beta decay, we will not concern ourselves with the details and simply set $\tilde{h}=0$. Also for simplicity, let us set $\mu_2=0$ and assume that f is diagonal in the quark indices. The part of the f coupling relevant in our discussion is given by

$$W_{\Delta L\neq 0}=f_{ude}Q_1L_1D_1^c+\text{H.c.}+\cdots\ , \tag{3}$$

where the subscript 1 stands for fermions of the first generation.

Let us now turn to the magnitude of the coupling. To study this, we note that Eq. (3) contributes to β decay of neutrino via the Feynman diagram in Fig. 1. The strength of this contribution is given by $f_{ude}^2/M_{\tilde{d}^c}^2$ and has pure $V-A$ form, as can be easily seen after Fierz reshuffling. As a result, it adds to the usual Fermi coupling. This contribution must therefore be less than 1% of the Fermi coupling from observations on β decay and $\pi \to e\bar{\nu}$ decay leading to the constraint

$$f_{ude}^2/M_{\tilde{d}^c}^2 \leq 2\sqrt{2}G_F\times 10^{-2}\ . \tag{4}$$

III. MAGNITUDE FOR $(\beta\beta)_{0\nu}$ DECAY

Coupled with the Majorana masses for the Z gaugino, photino, and gluino, the R-parity-violating interaction in Eq. (3) will contribute to $(\beta\beta)_{0\nu}$ decay via the Feynman diagrams in Figs. 2(a)—2(c). The contribution of the gluino exchange diagram in Figs. 2(a) and 2(c) can be estimated to be

$$M^{(\tilde{g})}\simeq 2f_{ude}^2\left[\frac{1}{M_{\tilde{u}}^4}+\frac{1}{M_{\tilde{d}^c}^4}\right]\times\frac{M_{\tilde{g}}4\pi\alpha_s}{p^2+M_{\tilde{g}}^2}\bar{u}_L^{\alpha}d_{R\beta}\bar{u}_L^{\beta}d_{R,\alpha}e_L^{\dagger}Ce_L^{*} \tag{5}$$

(α,β denote the color index), whereas those involving the exchange of $\tilde{\gamma}$, $\tilde{Z}$ from Figs. 2(a) and 2(b) are

$$M^{(\tilde{Z})}\simeq\frac{f_{ude}^2g^2}{\cos^2\theta_W}\left[\frac{\frac{1}{2}-\frac{2}{3}\sin^2\theta_W}{M_{\tilde{u}}^4}+\frac{-\frac{1}{2}+\sin^2\theta_W}{M_{\tilde{e}}^4}+\frac{\frac{1}{3}\sin^2\theta_W}{M_{\tilde{d}^c}^4}\right]\bar{u}_L^{\alpha}d_{R\alpha}\bar{u}_L^{\beta}d_{R\beta}e_L^{\dagger}Ce_L^{*}\frac{M_{\tilde{Z}}}{p^2+M_{\tilde{Z}}^2}\ , \tag{6}$$

$$M^{(\tilde{\gamma})}\simeq f_{ude}^2 4\pi\alpha\left[\frac{2}{3M_{\tilde{u}}^4}-\frac{1}{M_{\tilde{e}}^4}+\frac{1}{3M_{\tilde{d}^c}^4}\right]\bar{u}_L^{\alpha}d_{R\alpha}\bar{u}_L^{\beta}d_{R\beta}e_L^{\dagger}Ce_L^{*}\frac{M_{\tilde{\gamma}}}{p^2+M_{\tilde{\gamma}}^2}\ . \tag{7}$$

In order to estimate these contributions we first consider the gluino-exchange diagram which has the strongest couplings, among all these graphs. Although it involves color-octet operators if the gluino is heavy, the net interaction is pointlike; as a result, we can do a Fierz reshuffling and make both the operators into color-singlet ones with an extra factor $\frac{4}{3}$. For further estimation, let us assume $M_{\tilde{u}}=M_{\tilde{d}^c}$; since the present collider result seems to indicate[10] that $M_{\tilde{u}}=M_{\tilde{d}^c}\equiv M_{\tilde{q}}\geq 60$ GeV, with $M_{\tilde{g}}>100$ GeV; using this, we can predict the strength for neutrinoless double-beta transition decay to be

$$A_{(\beta\beta)_{0\nu}}^{(\tilde{g})}\simeq\frac{16}{3}\frac{f_{ude}^2}{M_{\tilde{u}}^4}4\pi\alpha_s M_{\tilde{g}}A_{\text{nuc}}(M_{\tilde{g}})\ , \tag{8}$$

where $A_{\text{nuc}}(M_{\tilde{g}})$ represents the nuclear matrix element of the hadronic part of M.

However, we wish to emphasize that, crucial to our conclusion is the assumption that we can do Fierz reshuffling leading to color-singlet operators which is, of course, valid in the strict point-interaction limit. If this is not allowed, then, the matrix element will be extremely suppressed since we will need colored intermediate states. These objections do not apply to $\tilde{\gamma}$ or $\tilde{Z}$ exchange, which we consider now. Again, for simplicity, we assume $M_{\tilde{u}}=M_{\tilde{e}}=M_{\tilde{d}^c}\gtrsim 60$ GeV for the $\tilde{Z}$ exchange and we get for the strength of the $(\beta\beta)_{0\nu}$ amplitude

$$A_{(\beta\beta)_{0\nu}}^{(\tilde{Z})}\simeq\frac{f_{ude}^2 7\times 10^{-2}}{M_{\tilde{q}}^4}M_{\tilde{Z}}A_{\text{nuc}}(M_{\tilde{Z}})\ . \tag{9}$$

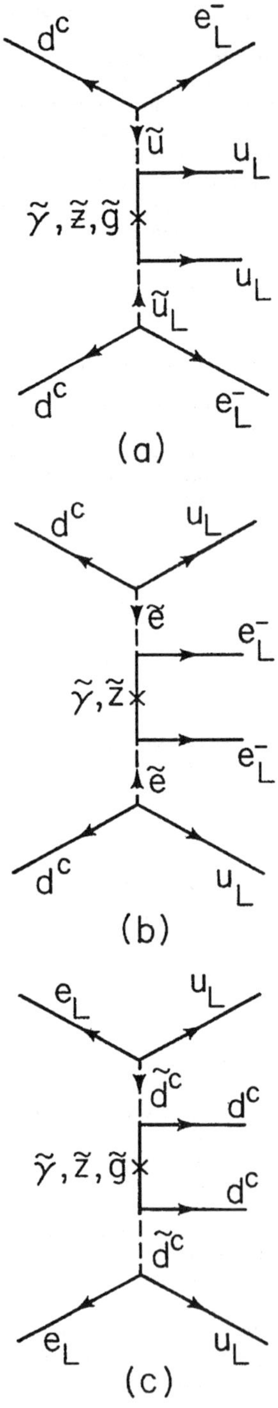

FIG. 2. New contributions to neutrinoless double-beta decay in supersymmetric extension of the standard model mediated by (a) photino ($\tilde{\gamma}$), (b) Z-gaugino ($\tilde{Z}$), and (c) gluino ($\tilde{g}$) exchange.

As far as the photino contribution goes if we make the same assumption as above, the amplitude vanishes; however, experimentally, the lower bound on $M_{\tilde{e}}$ is about 50 GeV, whereas as mentioned before collider results[10] put a lower bound on $M_{\tilde{q}} \geq 60$ to 75 GeV. As a result, we expect the cancellation to be imperfect and if we define $\Delta m^2 = M_{\tilde{q}}^2 - M_{\tilde{e}}^2$, we find

$$A^{(\tilde{\gamma})}_{(\beta\beta)_{0\nu}} \simeq \frac{f_{ude}^2}{M_{\tilde{q}}^4}\left[\frac{\Delta m^2}{M_{\tilde{e}}^2}\right] 9\times 10^{-2} M_{\tilde{\gamma}} A_{\rm nuc}(M_{\tilde{\gamma}})\left[1+\frac{m_{\tilde{q}}^2}{m_{\tilde{e}}^2}\right]. \tag{10}$$

The photino and Z-gaugino exchange diagrams could be comparable, because while the present lower limit on $M_{\tilde{\gamma}}$ is about[10] 5 GeV and that on $M_{\tilde{Z}}$ could be of order[11] 33 GeV, we assume the factor $(1+m_{\tilde{q}}^2/m_{\tilde{e}}^2)\Delta m^2/M_{\tilde{e}}^2 = \frac{1}{5}$ to $\frac{1}{10}$. From now on we will assume the effective $\tilde{Z}$ strength to be comparable.

Let us now turn to discussion of the nuclear matrix element in order to put bounds on scalar-quark and scalar-lepton masses from experimental limits on double-beta decay. For this purpose, we will consider the gluino diagram separately from the photino and Z-gaugino diagram. In the case of the former, we assume that we can make a Fierz transform.

The nuclear matrix elements here are Fermi type and, as has been noted,[12] they are expected to be about 1% of the corresponding Gamow-Teller (GT) matrix elements, due to the fact that there are no nuclear states in the same isomultiplet as the ones that undergo double-beta transition. Using a finite-size approximation for the nucleus[13] and following the discussion of Haxton and Stephenson[1] (their Fig. 14b) and using the suppression of the Fermi-type matrix element, $M_F = 10^{-2} M_{\rm GT}$, we estimate that the constraint of $(\beta\beta)_{0\nu}$ decay of ^{76}Ge implies (for a typical gluino, photino, or Z-gaugino mass of 100 GeV)

$$\frac{f_{ude}}{M_{\tilde{u}}^4} a_x \lesssim G_F^2 \times 10^{-2.5}, \tag{11}$$

TABLE I. We summarize the lower bounds on the scalar-quark masses for various values of the gaugino mass coming from a lower bound on ^{76}Ge 0ν half-life of 2.5×10^{23} yr. The gluino case and the $\tilde{Z}+\tilde{\gamma}$ case are given separately.

Gaugino mass	Lower bound on $m_{\tilde{u}}$ for the gluino case	Lower bound on $m_{\tilde{u}}$ for the $\tilde{Z}+\tilde{\gamma}$ case
100 GeV	3 TeV	440 GeV
10^3 GeV	950 GeV	150 GeV
10^4 GeV	300 GeV	40 GeV
10^5 GeV	105 GeV	

where

$$a_{\tilde{g}} \approx \frac{16}{3}$$

or

$$a_{\tilde{Z}+\tilde{\gamma}} = 1.6\times 10^{-1} , \qquad (12)$$

where we have assumed

$$M_{\tilde{Z}} A_{\text{nuc}}(M_{\tilde{Z}}) \approx (\Delta m^2/M_{\tilde{Z}}^2) M_{\tilde{\gamma}} A_{\text{nuc}}(M_{\tilde{\gamma}}) .$$

Clearly, using the maximum value in the constraint in Eq. (4), we find, for the gluino case

$$M_{\tilde{q}} \geq 2 \text{ TeV} \qquad (13)$$

and for the photino, Z-gaugino case

$$M_{\tilde{q}} \geq 380 \text{ GeV} .$$

In Table I, we give the constraints for various values of the gaugino masses using Fig. 14 of Haxton *et al.*[1] as rescaled by Caldwell *et al.* using the ^{76}Ge 0ν half-life[14] of $\geq 2.5\times 10^{23}$ yr.

IV. NEUTRINO MASS

The dominant contribution to neutrino masses in this class of theories will come from Feynman diagrams depicted in Fig. 3. They involve two electroweak symmetry-breaking doublet vacuum expectation values (VEV's) and also a supersymmetry-breaking parameter $m_{3/2}$. If we ignore the contribution of higher generations to the electron neutrino mass, then we estimate that the induced Majorana mass for the neutrino from Fig. 3 is given by

$$m_\nu \simeq \frac{f_{ude}{}^2}{16\pi^2}\frac{m_d{}^2}{m_{3/2}} . \qquad (14)$$

For a scalar-quark mass of 500 GeV, $f_{ude}{}^2 \leq 7.5\times 10^{-2}$; choosing the SUSY-breaking scale $m_{3/2} \simeq 100-1000$ GeV, we find $m_\nu \leq 0.5-0.05$ eV. This is interesting since it implies that contribution of the neutrino mass to $(\beta\beta)_{0\nu}$ decay is about 10—100 times smaller than the contributions of direct R-parity-violating interactions. As a result, detailed characteristics of $(\beta\beta)_{0\nu}$ decay must be studied before one can conclude a value for the Majorana mass for the neutrino from the $(\beta\beta)_{0\nu}$ lifetime, once it is observed.

V. R-PARITY CONSERVATION IN EXTENDED TO ELECTROWEAK MODELS

We now wish to point out that the existence of an R-parity- (and L-) violating interaction depends on the kind of electroweak gauge group one works with. As an illustrative example, consider the SUSY electroweak group to be $SU(2)_L\times U(1)_{I_{3R}}\times U(1)_{B-L}$. In this case, there exist no interactions[15] in the superpotential that violate lepton number. To see this, we assign the quarks and leptons to the gauge group as follows:[15]

for quarks, $Q(2,0,\frac{1}{3})$, $U^c(1,-\frac{1}{2},-\frac{1}{3})$, $D^c(1,+\frac{1}{2},-\frac{1}{3})$;

for leptons, $L(2,0,-1)$, $E^c(1,+\frac{1}{2},+1)$, $\nu^c(1,-\frac{1}{2},+1)$;

for Higgs bosons, $H_u(2,\frac{1}{2},0)$, $H_d(2,-\frac{1}{2},0)$.

The most general gauge-invariant superpotential in this model can be written as

$$W = h_u Q H_u U^c + h_d Q H_d D^c + h_e L H_d E^c + h_\nu L H_u \nu^c + \mu H_u H_d . \qquad (15)$$

We see that all terms in Eq. (15) conserve the lepton number (and R parity). Of course, the h_ν term in Eq. (15) may induce a large Dirac mass for the neutrino, which in turn may require R-parity violation[15] or lepton-number violation via new Higgs bosons to understand the small neutrino mass. R-parity- (and lepton-number) violation is not intrinsic to the theory (for instance, we could set $h_\nu = 0$ and, we will not need any lepton-number violation at all). Thus, it may be more natural to work with this kind of extended gauge models to understand the origin of lepton-number violation should it be observed. In order to obtain R-parity-violating interactions from this picture, we can give a nonzero VEV to the scalar partner of the right-handed neutrino,[15,16] i.e., $\langle\tilde{\nu}^c\rangle \neq 0$. Then, via the Feynman diagram shown in Fig. 4, interactions of type QLD^c and LLE^c are introduced. The strengths of these interactions is estimated to be the following:

$$\text{for } QLD^c,\ h_d h_\nu \langle\tilde{\nu}^c\rangle/\mu ;$$
$$\text{for } LLE^c,\ h_e h_\nu \langle\tilde{\nu}^c\rangle/\mu . \qquad (16)$$

Since h_ν contributes to Dirac masses of neutrinos, we expect $h_\nu \simeq (m_D/M_W)$, which for the first-generation fermions is $\approx 10^{-4}$ or so. Similarly, we expect $h_d \approx 10^{-4}$; as

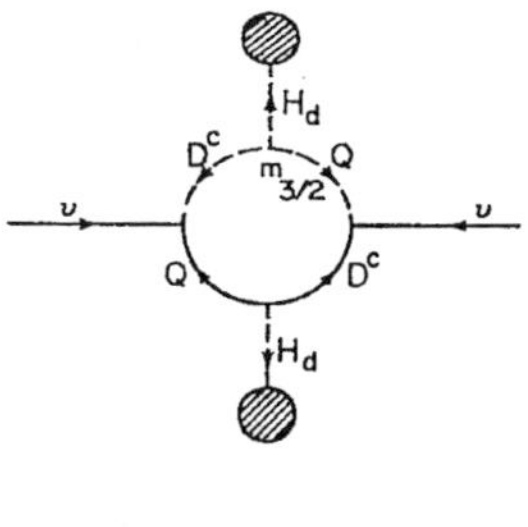

FIG. 3. One-loop diagram for neutrino mass in the presence of R-parity-violating interactions.

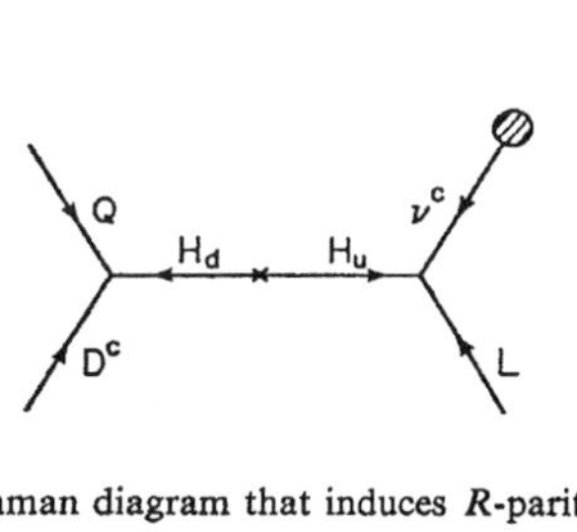

FIG. 4. Feynman diagram that induces R-parity violation at low energies.

a result, even if $\langle\tilde{\nu}^c\rangle \approx \mu$, the strength of effective R-parity-violating interactions is very weak. It is, however, easy to extend this theory further by duplicating the Higgs superfields to (H_u^i, H_d^i), $i=1,2$ with the same quantum numbers as H_u and H_d such that the strength of induced R-parity-violating coupling becomes much bigger. To see this, we write the new superpotential as

$$W=\sum_{i=1,2}(h_u^i QH_u^i U^c + h_d^i QH_d^i D^c + h_e^i LH_d^i E^c + h_\nu^i LH_u^i \nu^c) . \quad (17)$$

It is now phenomenologically acceptable for the Higgs fields $H_u{}^2$ and $H_d{}^2$ to have no VEV; the Yukawa couplings $h_u{}^2$, $h_d{}^2$, $h_e{}^2$, and $h_\nu{}^2$ can then be of order 1 leading to large induced R-parity-violating interactions at low energies.

In conclusion, we point out that in the most general supersymmetric extension of the standard model, there exist new contributions to neutrinoless double-beta decay which may be used to provide constraints on the parameters of the model. We also comment on the possible extensions of the model where such contributions are absent prior to spontaneous symmetry breaking and can arise with arbitrary strength in the low-energy sector after symmetry breaking.

ACKNOWLEDGMENTS

I would like to thank S. P. Rosen and D. Caldwell for valuable discussions. This work was supported by a grant from the National Science Foundation.

---

[1]For recent reviews, see M. Doi, T. Kotani, and E. Takasugi, Prog. Theor. Phys. (Suppl.) 83, 1 (1985); W. Haxton and G. Stevenson, Prog. Part. Nucl. Phys. 12, 409 (1984); J. D. Vergados, Phys. Rep. 133, 1 (1986).

[2]H. Primakoff and S. P. Rosen, Proc. Phys. Soc. 78, 464 (1961).

[3]R. N. Mohapatra and G. Senjanovic, Phys. Rev. D 23, 165 (1981).

[4]Riazuddin, R. E. Marshak, and R. N. Mohapatra, Phys. Rev. D 24, 1310 (1981).

[5]R. N. Mohapatra and J. D. Vergados, Phys. Rev. Lett. 47, 1713 (1981); C. Piccioto and M. S. Zahir, Phys. Rev. D 26, 2320 (1982).

[6]C. S. Aulakh and R. N. Mohapatra, Phys. Lett. 119B, 136 (1982); 121B, 147 (1983); L. J. Hall and M. Suzuki, Nucl. Phys. B231, 419 (1984); G. G. Ross and J. W. F. Valle, Phys. Lett. 151B, 375 (1985); S. Dawson, Nucl. Phys. B261, 297 (1985).

[7]For reviews, see H. P. Nilles, Phys. Rep. 110, 1 (1984); H. Haber and G. Kane, *ibid.* 117, 75 (1985); R. N. Mohapatra, *Unification and Supersymmetry* (Springer, New York, 1986), Chap. 13.

[8]P. Fayet and G. Farrar Phys. Lett. 76, 575 (1978).

[9]Hall and Suzuki (Ref. 6); Dawson (Ref. 6).

[10]For a recent review, see R. M. Barnett, Report No. LBL-20492, 1986 (unpublished); I. Hinchliffe, Report No. LBL-20747, 1986 (unpublished).

[11]JADE Collaboration, W. Bartel *et al.*, Phys. Lett. 155B, 288 (1985); Mark J Collaboration, B. Adeva *et al.*, Phys. Rev. Lett. 53, 1806 (1984).

[12]W. Haxton, S. P. Rosen, and G. J. Stephenson, Jr., Phys. Rev. D 26, 1805 (1982).

[13]J. D. Vergados, Phys. Rev. C 24, 640 (1981).

[14]D. O. Caldwell *et al.*, Phys. Rev. D 33, 2737 (1986); T. Ejiri *et al.*, Nucl. Phys. A448, 27 (1986); E. Fiorini *et al.*, Phys. Lett. 121B, 72 (1983); F. Avignone *et al.*, Phys. Rev. Lett. 50, 721 (1983).

[15]R. N. Mohapatra, Phys. Rev. Lett. 56, 561 (1986).

[16]M. Hayashi and A. Murayama, Phys. Lett. 153B, 251 (1985).

VOLUME 75, NUMBER 1 PHYSICAL REVIEW LETTERS 3 JULY 1995

New Constraints on R-Parity-Broken Supersymmetry from Neutrinoless Double Beta Decay

M. Hirsch* and H. V. Klapdor-Kleingrothaus[†]
Max-Planck-Institut für Kernphysik, P.O. Box 10 39 80, D-69029, Heidelberg, Germany

S. G. Kovalenko[‡]
Joint Institute for Nuclear Research, Dubna, Russia

(Received 20 March 1995)

New constraints on the parameters of the minimal supersymmetric standard model with explicit R-parity violation ($\not{R}_p$MSSM) are obtained from the current experimental lower bound on the half-life of ^{76}Ge $0\nu\beta\beta$ decay. These constraints are shown to be more stringent than those from other low-energy processes and are competitive with or even more stringent than constraints expected from accelerator searches.

PACS numbers: 12.60.Jv, 11.30.Er, 23.40.Bw

Neutrinoless double beta decay ($0\nu\beta\beta$) has long been recognized as a sensitive tool to put theories beyond the standard model (SM) to the test (for reviews see Refs. [1–3]). A variety of mechanisms which may cause $0\nu\beta\beta$ decay has been studied in the past. The simplest and the most well-known possibilities are via the exchange of a massive Majorana neutrino between the decaying neutrons or due to $(B-L)$-violating right-handed currents. Another mechanism was found within supersymmetric (SUSY) models with $R_p = (-1)^{3B+L+2S}$ violation ($\not{R}_p$) [4]. In Refs. [5] and [6] this SUSY mechanism has been investigated within the minimal supersymmetric standard model (MSSM) with explicit R_p violation in the superpotential ($\not{R}_p$MSSM). In this Letter we extract constraints on the $\not{R}_p$MSSM parameter space from the current experimental lower bound on the ^{76}Ge $0\nu\beta\beta$ decay half-life.

Recently, impressive progress has been achieved in experiments with this isotope, in both the $2\nu\beta\beta$ and the $0\nu\beta\beta$ decay mode [7,8]. We will use, in our analysis, the experimental half-life limit of ^{76}Ge, recently measured by the Heidelberg-Moscow Collaboration [8],

$$T_{1/2}^{0\nu\beta\beta}(^{76}\mathrm{Ge}, 0^+ \to 0^+) > 5.6 \times 10^{24}\ \mathrm{yr}\ (90\%\ \mathrm{C.L.}). \tag{1}$$

The $\not{R}_p$MSSM has the MSSM field content and is completely specified by the standard SU(3) × SU(2) × U(1) gauge couplings, as well as by the low-energy superpotential and "soft" SUSY breaking terms [9]. The superpotential can be written as $W = W_{R_p} W_{\not{R}_p}$. The R_p-conserving term, W_{R_p}, coincides with the MSSM superpotential [9]. The most general gauge-invariant form of the R_p-violating term is [10]

$$W_{\not{R}_p} = \lambda_{ijk} L_i L_j \overline{E}_k + \lambda'_{ijk} L_i Q_j \overline{D}_k + \lambda''_{ijk} \overline{U}_i \overline{D}_j \overline{D}_k . \tag{2}$$

We use notations L and Q for lepton and quark doublet superfields and use $\overline{E}$, $\overline{U}$, and $\overline{D}$ for lepton and *up* and *down* quark singlet superfields. Indices i, j, k denote generations. The coupling constants λ (λ'') are antisymmetric in the first (last) two indices. The first two terms lead to lepton number violation, while the last one violates baryon number conservation. Proton stability forbids simultaneous presence of lepton- and baryon-number-violating terms in the superpotential [11] (unless the couplings are very small). Therefore, either λ, λ' *or* λ'' Yukawa couplings can be nonzero. Neutrinoless double beta decay, which is the subject of the present paper, requires lepton-number-violating interactions. Therefore we bind ourselves to the $\not{R}_p$MSSM with lepton number violation ($\lambda \neq 0$, $\lambda' \neq 0$) and baryon number conservation ($\lambda'' = 0$). Apparently, $0\nu\beta\beta$ decay can probe only the first-generation lepton-number-violating Yukawa coupling λ'_{111}, because only the first-generation fermions u, d, e are involved in this process.

Let us write down explicitly the $\not{R}_p$ interaction terms of the $\not{R}_p$MSSM Lagrangian relevant for $0\nu\beta\beta$ decay. We use the four-component Dirac bispinor notation for fermion fields. The lepton-number-violating part of the Lagrangian obtained directly from the $W_{\not{R}_p}$ superpotential part [Eq. (2)] has the form

$$\mathcal{L}_{\not{R}_p} = -\lambda'_{111}\Bigg[(\overline{u}_L\ \overline{d}_R)\begin{pmatrix} e_R^c \\ -\nu_R^c \end{pmatrix}\overline{d}_R + (\overline{e}_L\ \overline{\nu}_L)d_R\begin{pmatrix} \tilde{u}_L^* \\ -\tilde{d}_L^* \end{pmatrix} + (\overline{u}_L\ \overline{d}_L)d_R\begin{pmatrix} \tilde{e}_L^* \\ -\tilde{\nu}_L^* \end{pmatrix} + \mathrm{H.c.}\Bigg]. \tag{3}$$

To construct diagrams contributing to $0\nu\beta\beta$ decay one also needs the MSSM R_p-conserving gluino $\tilde{g}$ and neutralino χ_i interactions with quarks, electrons, and their superpartners. The corresponding terms of the MSSM Lagrangian are well known and can be taken from Ref. [9].

Having specified all necessary interaction terms, one can construct diagrams describing the $\not{R}_p$MSSM contribution to the $0\nu\beta\beta$ decay. The complete set of these diagrams presented in Fig. 1 has been found in Ref. [6]. The supersymmetric mechanism of $0\nu\beta\beta$ decay was first proposed in Ref. [4] and later studied in more detail in Ref. [5]. However, only a subset

ELSEVIER

8 June 1995

Physics Letters B 352 (1995) 1–7

PHYSICS LETTERS B

New contributions to supersymmetric mechanism of neutrinoless double beta decay

M. Hirsch, H.V. Klapdor-Kleingrothaus, S.G. Kovalenko [1]

Max-Planck-Institut für Kernphysik, P.O. 10 39 80, D-69029, Heidelberg, Germany

Received 15 February 1995; revised manuscript received 13 April 1995
Editor: C. Mahaux

Abstract

The neutrinoless double beta ($0\nu\beta\beta$) decay is analyzed within the Minimal Supersymmetric Standard Model with explicit R-parity violation ($\not{R}_p$MSSM). We have found new supersymmetric contributions to this process and give the complete set of relevant Feynman diagrams. Operators describing $0^+ \to 0^+$ nuclear transitions induced by the supersymmetric interactions of the $\not{R}_p$MSSM are derived. These operators can be used for calculating the $0\nu\beta\beta$ decay rate applying any specific nuclear model wave functions.

The observation of neutrinoless double beta ($0\nu\beta\beta$) decay would be a clear signal for physics beyond the standard model, since it violates lepton number by two units (for reviews see [1,2]). No definite observation of $0\nu\beta\beta$ decay has been reported to date, but recent experimental progress has pushed the existing half-life limits of $0\nu\beta\beta$ decay beyond 10^{24} years and further progress can be expected in the near future [3].

Half-life limits on $0\nu\beta\beta$ decay are usually interpreted as limits on the effective Majorana neutrino mass (see Fig. 1). However, it is known for some time that there exist also other mechanisms which might induce $0\nu\beta\beta$ decays, like for example in left-right symmetric extensions of the standard model gauge group [2]. In this paper we study contributions to $0\nu\beta\beta$ decay within supersymmetric (SUSY) theories with explicit R-parity breaking. R-parity (R_p) is a discrete, multiplicative symmetry defined as $R_p = (-1)^{3B+L+2S}$, where S, B and L are the spin, the baryon and the lepton quantum number. Conservation of R-parity in turn implies baryon number (B) and lepton number (L) conservation. This symmetry has been imposed on the minimal supersymmetric standard model (MSSM) (for a review see [4]). However, neither gauge invariance nor supersymmetry re-

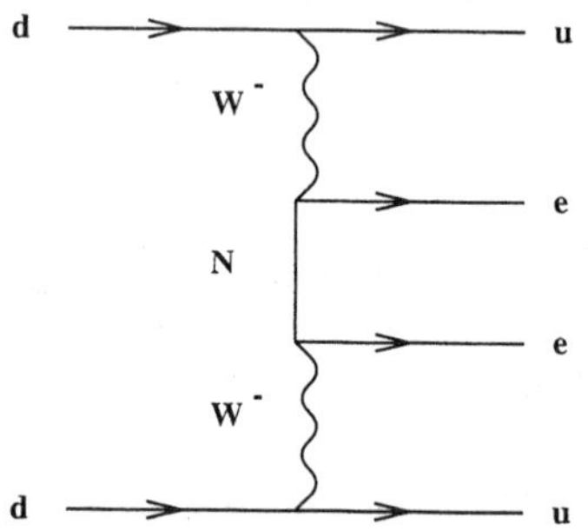

Fig. 1. Feynman graphs for the conventional mechanism of $0\nu\beta\beta$ decay by massive Majorana neutrino exchange.

[1] Joint Institute for Nuclear Research, Dubna, Russia.

Elsevier Science B.V.
SSDI 0370-2693(95)00460-2

Aus der Wissenschaft

Test von Supersymmetrie: Doppelbetazerfall und Teilchenbeschleuniger

Supersymmetrie („SUSY") gilt heute als der wahrscheinlichste Kandidat für die Theorie jenseits des Standardmodells der Elementarteilchenphysik. SUSY bezeichnet eine Verallgemeinerung des Symmetriebegriffs, die Bosonen mit Fermionen verknüpft (s. z. B. *J. Wess*, Phys. Bl. 43 (1987) 2). Neue Abschätzungen aus dem Doppelbetazerfall liefern nun die schärfsten bisher bekannten Grenzen für die Stärke der R-Paritätsverletzung.

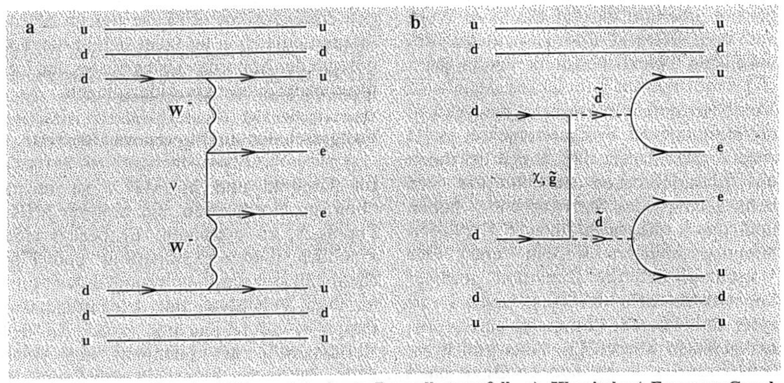

Abb. 1: Feynman-Graphen für neutrinolosen Doppelbetazerfall. a) ‚Klassischer' Feynman-Graph unter Neutrinoaustausch, b) SUSY-Feynman-Graph in R-Paritätsverletzenden Theorien. In beiden Fällen wandeln sich zwei d-Quarks in zwei u-Quarks um, unter Aussendung zweier Elektronen. Im ‚klassischen' Fall zerfallen die beiden W-Bosonen unter Neutrinoaustausch. Aus der Untersuchung dieses Prozesses lassen sich die derzeit besten Grenzen für die Majorana-Masse des Neutrinos ableiten. Im supersymmetrischen Fall dagegen treten anstelle der W-Bosonen und des Neutrinos supersymmetrische Teilchen auf, wie skalare Quarks, Gluinos (SUSY-Partner des Gluons) oder auch Neutralinos. $0\nu\beta\beta$-Zerfall liefert also auch Aussagen über supersymmetrische Theorien.

Die wohl bekannteste Vorhersage supersymmetrischer Modelle ist die Existenz sogenannter supersymmetrischer Partnerteilchen. Die Massen dieser SUSY-Teilchen sollten 1 TeV nicht überschreiten. Nach solchen Teilchen wurde in vielen Beschleunigerexperimenten – leider bisher erfolglos – gesucht. Gegenwärtige Untergrenzen für Massen von SUSY-Teilchen liegen in der Gegend von 20–100 GeV.

SUSY-Teilchen unterscheiden sich von gewöhnlichen Teilchen nicht nur in ihren Massen, sondern auch in ihrer R-Parität. (R-Parität ist eine Quantenzahl, die $R_P = +1$ für normale Teilchen und $R_P = -1$ für SUSY-Teilchen annimmt.) Ist die R-Parität erhalten, so können SUSY-Teilchen nur paarweise erzeugt werden, und das leichteste supersymmetrische Teilchen ist absolut stabil. R-Paritätserhaltung führt zu besonders einfachen SUSY-Modellen, weshalb früher hauptsächlich solche Theorien untersucht wurden. Seit einiger Zeit hat man sich aber verstärkt sogenannten R_P-verletzenden SUSY-Modellen zugewandt, da es für R_P-Erhaltung keine überzeugenden Argumente gibt. Bisher existieren Grenzen für die Stärke einer R_P-verletzenden Wechselwirkung λ aus dem Neutronzerfall und von Beschleunigerexperimenten, z. B. am TEVATRON. Geplant ist auch, Daten von HERA auf diese Größe hin zu untersuchen.

Was hat das Ganze, wie im Titel angedeutet, mit Doppelbetazerfall zu tun? – Bisher beruhte die Motivation zum Studium des $\beta\beta$-Zerfalls hauptsächlich auf einem speziellen Zerfallsmodus, dem neutrinolosen $\beta\beta$-Zerfall, bzw. dem entsprechenden Feynmangraphen, s. Abb. 1a, in dem ein Majorana-Neutrino zwischen zwei Nukleonen ausgetauscht wird. Einen analogen Zerfall kann man aber auch durch Austausch von supersymmetrischen Teilchen bewirken (Abb. 1b). Im Diagramm 1.a gehen zwei Neutronen (Quarkgehalt: (*udd*)) eines Atomkerns in zwei Protonen (*uud*) und zwei Elektronen über. Offensichtlich ist dieser Zerfall im Standardmodell verboten, da er die Leptonenzahlerhaltung um zwei Einheiten verletzt. Es ist nun leicht zu zeigen, daß die resultierende Halbwertszeit für diesen Prozeß invers proportional zum Quadrat der effektiven Neutrinomasse ist, wenn beide W-Bosonen linkshändig sind. (In links-rechtssymmetrischen Erweiterungen des Standardmodells können auch Terme auftreten, die Rückschlüsse auf Eigenschaften bisher nicht beobachteter, aber von solchen Modellen geforderter, *rechtshändiger* W-Bosonen erlauben.)

Eine ganze Reihe von Experimenten hat deshalb mit verschiedenen Methoden nach $0\nu\beta\beta$-Zerfall gesucht, aber bisher nur Grenzen angeben können. Aus der Nichtbeobachtung des $0\nu\beta\beta$-Zerfalls von ^{76}Ge kann man gegenwärtig eine Obergrenze von $m_\nu < 0{,}65$ eV ableiten. Dies ist der schärfste bekannte Test auf den Majorana-Charakter des Neutrinos.

Abb. 1b zeigt ein Beispiel eines Abb. 1a entsprechenden SUSY-Feynmangraphen für $0\nu\beta\beta$-Zerfall. Es gibt insgesamt sechs solcher SUSY-Graphen, die alle recht ähnlich aussehen [1]. Frühere Untersuchungen [2] gingen von nur drei, einfacheren Graphen aus. Auch hier gehen zwei Neutronen unter Aussendung von Elektronen in zwei Protonen über. Wie man sieht, ist die experimentelle Signatur, die man für diesen Beitrag erwartet, völlig analog zum gewöhnlichen $0\nu\beta\beta$-Zerfall unter Neutrinoaustausch. Jedoch werden in diesem Beispiel an Stelle des Neutrinos und der W-Bosonen nun supersymmetrische Teilchen, wie etwa ein Gluino $\tilde{g}$ (der supersymmetrische Partner des Gluons) und zwei skalare d-Quarks ($\tilde{d}$ ausge-

0031–9279/95/0505–0418 $ 5.00+.25/0 – © VCH, D-69451 Weinheim, 1995 Phys. Bl. 51 (1995) Nr. 5

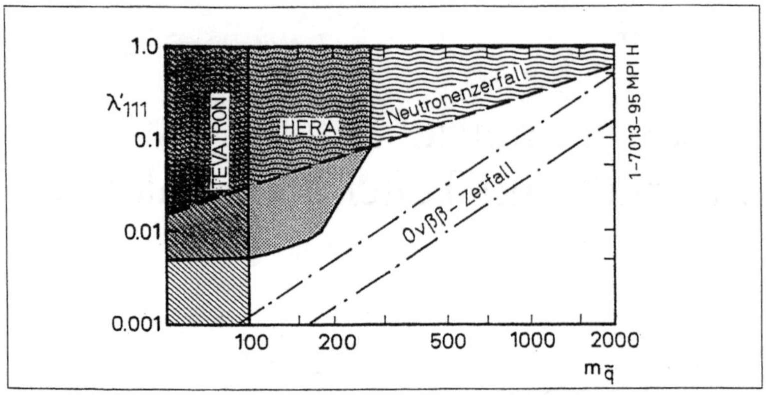

Abb. 2: Grenzen für die Stärke der R-paritätsverletzenden Wechselwirkung λ gegen die Masse skalarer Quarks (aus [1]). Bereiche links der Kurven sind von den Experimenten ausgeschlossen. Die senkrechte Linie ist die Untergrenze für Massen skalarer Quarks aus Daten am Tevatron, die dicke Linie kennzeichnet den Bereich, der von Experimenten bei HERA untersucht werden könnte [4]. Für die Berechnung dieser Grenzen aus Beschleunigerexperimenten muß man annehmen, daß das skalare Quark in ein leichteres, supersymmetrisches Teilchen (Photino: $\tilde{\gamma}$) zerfallen kann, d. h. daß die Masse des Photinos kleiner als die des skalaren Quarks ist. Die gestrichelte Linie ist das bisher beste Limit aus Niederenergieexperimenten [5], während die strichpunktierten Linien die Grenze aus $0\nu\beta\beta$-Zerfall – dem Heidelberg-Moskau-Experiment – angeben (links für eine Gluinomasse von 1 TeV, rechts für eine Gluinomasse von 100 GeV. Für letztere erhielte man schärfere Grenzen). Die gemessene Halbwertszeitgrenze für $0\nu\beta\beta$-Zerfall liefert also für Massen skalarer Quarks größer 100 GeV die schärfsten Grenzen (selbst wenn das Gluino 1 TeV schwer sein sollte, was man gemeinhin als größte ‚natürliche' Masse in SUSY-Modellen ansieht). Die Obergrenze für die Verletzung der R-Parität, die sich hieraus ergibt, wirft nun die Frage auf, ob R-Parität nicht doch absolut erhalten ist.

tauscht, d. h. auch für supersymmetrische Theorien liefert $0\nu\beta\beta$-Zerfall interessante Aussagen.

Die detaillierte Berechnung dieser supersymmetrischen Beiträge zum $0\nu\beta\beta$-Zerfall [1] ergab nun, zusammen mit der Halbwertszeitgrenze des Heidelberg-Moskau-$\beta\beta$-Experiments [3], ein erstaunliches Ergebnis: $0\nu\beta\beta$-Zerfall liefert Grenzen für die R-paritätsverletzende Wechselwirkung λ, die für Massen supersymmetrischer Teilchen größer als 100 GeV sogar schärfer als die der größten, existierenden Beschleuniger sind. Dies ist in Abb. 2 veranschaulicht. Demnach kann die Stärke der R-paritätsverletzenden Wechselwirkung nicht größer als

$$\lambda \leq 3{,}9 \times 10^{-4} \left(\frac{m\tilde{q}}{100\ \mathrm{GeV}} \right)^2 \times \left(\frac{m\tilde{g}}{100\ \mathrm{GeV}} \right)^{1/2}$$

sein. Hierbei steht $m_{\tilde{q}}$ bzw. $m_{\tilde{g}}$ für die Massen der skalaren Quarks bzw. des Gluinos, der supersymmetrischen Partner der Quarks und des Gluons. Abb. 2 zeigt Grenzen aus $0\nu\beta\beta$-Zerfall für zwei Werte der Gluinomasse $m_{\tilde{g}}$ = 100 GeV bzw. 1 TeV. Die der Gluinomasse von 1 TeV entsprechende Kurve ist eine konservative Grenze, da nach allen bekannten SUSY-Modellen das Gluino nicht schwerer als 1 TeV sein sollte. Selbst diese Grenze ist fast eine Größenordnung schärfer als die besten bisherigen und gegenwärtig machbaren Beschleunigergrenzen.

Der erhaltene kleine Wert von λ wirft nun die Frage auf: Ist R-Parität doch erhalten oder liegen die Massen von SUSY-Teilchen höher als erwartet? Vielleicht werden erst Experimente am LHC in der Lage sein, diese Frage zu beantworten.

M. Hirsch, H. V. Klapdor-Kleingrothaus,
MPI für Kernphysik Heidelberg
S. Kovalenko, JINR, Dubna

Literatur

[1] *M. Hirsch, H. V. Klapdor-Kleingrothaus, S. G. Kovalenko*, eingereicht bei Phys. Rev. D.

[2] *R. Mohapatra*, Phys. Rev. **D34** (1986) 3457, *J. D. Vergados*, Phys. Lett. **D184** (1987) 55.

[3] *A. Balysh* et al., Proceed. XXVII. Internat. Conf. on High Energy Physics, Glasgow, Ang. 1994; *H. V. Klapdor-Kleingrothaus*, Progr. Part. Nucl. Phys. **32** (1994) 261.

[4] *J. Butterworth* u. *H. Dreiner*, Nucl. Phys. **B397 (1993).**

[5] *V. Barger* et al., Phys. Rev. **D40** (1989) 2987.

PHYSICAL REVIEW D VOLUME 53, NUMBER 3 1 FEBRUARY 1996

Supersymmetry and neutrinoless double beta decay

M. Hirsch* and H. V. Klapdor-Kleingrothaus†
Max-Plank-Institut für Kernphysik, P.O. 10 39 80, D-69029 Heidelberg, Germany

S. G. Kovalenko‡
Joint Institute for Nuclear Research, Dubna, Russia

(Received 6 March 1995)

Neutrinoless double beta decay ($0\nu\beta\beta$) induced by superparticle exchange is investigated. Such a supersymmetric (SUSY) mechanism of $0\nu\beta\beta$ decay arises within SUSY theories with R-parity nonconservation ($\not R_p$). We consider the minimal supersymmetric standard model (MSSM) with explicit $\not R_p$ terms in the superpotential ($\not R_p$ MSSM). The decay rate for the SUSY mechanism of $0\nu\beta\beta$ decay is calculated. Numerical values for nuclear matrix elements for the experimentally most interesting isotopes are calculated within the proton-neutron quasiparticle random phase approximation. Constraints on the $\not R_p$ MSSM parameter space are extracted from current experimental half-life limits. The most stringent limits are derived from data on ^{76}Ge. It is shown that these constraints are more stringent than those from other low-energy processes and are competitive with or even more stringent than constraints expected from accelerator searches.

PACS number(s): 12.60.Jv, 11.30.Er, 23.40.Bw

I. INTRODUCTION

In the standard model (SM), since B-L conservation is exact, neutrinoless double β decay ($0\nu\beta\beta$), which violates lepton number by two units, is forbidden. On the other hand B-L and L violation is expected in theories beyond the SM. That is why $0\nu\beta\beta$ decay has long been recognized as a sensitive tool to put theories beyond the SM to the test (for reviews, see [1–3]). A variety of mechanisms which may cause $0\nu\beta\beta$ have been studied in the past. The simplest and the most well-known possibility is via the exchange of a Majorana neutrino between the decaying neutrons or due to (B-L)-violating right-handed currents. $0\nu\beta\beta$ decay has not yet been seen, but limits on various model parameters can be deduced (see [2,3] and references therein) from its nonobservation.

Recently impressive progress has been achieved in the experimental investigation of double beta decay, both in the $2\nu\beta\beta$ and the $0\nu\beta\beta$ decay mode [4–6]. In the near future essential advances in this direction are expected. Experimental lower bounds on $0\nu\beta\beta$ decay half-lives are often represented in terms of an upper limit on the Majorana neutrino mass $\langle m_\nu \rangle$. At least one experiment currently in operation will reach a final sensitivity of about $\langle m_\nu \rangle = (0.1\text{–}0.2)$ eV and has already pushed the existing limit below 1 eV [6].

In view of the rising experimental sensitivity it is of great interest to pursue a more comprehensive theoretical study of the possible mechanisms of $0\nu\beta\beta$ decay.

In this work we investigate contributions to $0\nu\beta\beta$ decay within supersymmetric (SUSY) theories with explicit R-parity breaking. R-parity (R_p) is a discrete, multiplicative symmetry defined as $R_p = (-1)^{3B+L+2S}$ [7], where S, B, and L are the spin, the baryon, and the lepton quantum number. The SM fields, including additional Higgs boson field appearing in the extended gauge models, have $R_p = +1$ while their superpartners have $R_p = -1$. This symmetry has been imposed on the minimal supersymmetric standard model (MSSM) (for a review see [8]) to ensure baryon number (B) and lepton number (L) conservation. However, neither gauge invariance nor supersymmetry require R_p conservation. The question of whether or not R_p is a good symmetry of the supersymmetric theory is a dynamical problem which might be related to more fundamental physics at the Planck scale. In general R_p can be either broken explicitly [9–11] or spontaneously [12,13] by the expectation value of the scalar superpartner of the R_p-odd isosinglet lepton field [9].

Supersymmetric models with R_p nonconservation ($\not R_p$) have been extensively discussed in the literature not only because of their great theoretical interest, but also because they have interesting phenomenological and cosmological implications.

Existing constraints on $\not R_p$ SUSY theories are either direct from collider experiments [14] or indirect from low-energy processes [15], matter stability [16–18], and cosmology [19–22]. We will discuss the former two constraints later in more detail in comparison with the bounds from $0\nu\beta\beta$ decay.

Recently $\not R_p$ SUSY models have been analyzed in connection with current and forthcoming collider experiments. The $\not R_p$ SUSY gives rise to spectacular signatures

*Electronic address: MAHIRSCH@ENULL.MPI-HD.MPG.DE
†Electronic address: KLAPDOR@ENULL.MPI-HD.MPG.DE
‡Electronic address: KOVALEN@NUSUN.JINR.DUBNA.SU

[23,24] of events in collider detectors which would yield a very clean signal for supersymmetry. Consequently, expected sensitivities of experiments at the DESY *ep* collider HERA [25,26], the Fermilab Tevatron [27], the CERN e^+e^+ collider LEP 200 [28], and the CERN Large Hadron Collider (LHC) [29] have been analyzed recently. In the present paper we pay special attention to comparing the capability of $0\nu\beta\beta$ decay experiments with the collider experiments in establishing better constraints on $\not{R}_p$ SUSY models.

The rest of the paper is organized as follows. In the next section, we specify the minimal supersymmetric standard model with R_p nonconservation. Section III discusses the basic diagrams inducing $0\nu\beta\beta$ decay in the $\not{R}_p$ MSSM. The effective low-energy Lagrangian as well as the different lepton number violating parameters are defined. Section IV outlines the procedure for obtaining nucleon matrix elements from the quark currents in the nonrelativistic impulse approximation. Section V then deals with the numerical calculation of the relevant nuclear structure matrix elements. We briefly summarize the main features of the nuclear structure model before presenting numerical results for those isotopes which are currently the most promising experimentally. Special attention is paid to a discussion of the theoretical uncertainties of the nuclear structure calculation. In Sec. VI, on the basis of the current experimental limit on the half-life of ^{76}Ge [6], we analyze constraints on the supersymmetric parameter space imposed by the nonobservation of $0\nu\beta\beta$ decay. We have found that these limits are more stringent than those from other low-energy processes and also more stringent than those expected from experiments with the ZEUS detector at HERA [25]. We then close with a short summary and outlook.

II. MINIMAL SUPERSYMMETRIC STANDARD MODEL WITH R-PARITY NONCONSERVATION

In the following we will use the MSSM extended by inclusion of the explicit R-parity nonconserving terms ($\not{R}_p$) into the superpotential. This model has the MSSM field content and is completely specified by the standard SU(3)×SU(2)×U(1) gauge couplings as well as by the low-energy superpotential and "soft" SUSY breaking terms [8]. The most general gauge invariant form of the superpotential is

$$W = W_{R_p} + W_{\not{R}_p} . \tag{1}$$

The R_p conserving part has the standard MSSM form

$$W_{R_p} = h_L H_1 L \bar{E} + h_D H_1 Q \bar{D} - h_U H_2 Q \bar{U} - \mu H_1 H_2 . \tag{2}$$

We use notations L, Q for lepton and quark doublet superfields and $\bar{E}$, $\bar{U}$, $\bar{D}$ for lepton and *up*, *down* quark singlet superfields; H_1 and H_2 are the Higgs doublet superfields with a weak hypercharge $Y = -1, +1$, respectively. Summation over generations is implied. For simplicity generation indices of fields and Yukawa coupling constants h_L, h_U, h_D are suppressed. The mass-mixing parameter μ is a free parameter describing mixing between the Higgs bosons H_1-H_2 as well as between Higgsinos $\tilde{H}_1$-$\tilde{H}_2$.

The R_p violating part of the superpotential (1) can be written as [9,10]

$$W_{\not{R}_p} = \lambda_{ijk} L_i L_j \bar{E}_k + \lambda'_{ijk} L_i Q_j \bar{D}_k + \lambda''_{ijk} \bar{U}_i \bar{D}_j \bar{D}_k , \tag{3}$$

where indices i, j, k denote generations, and the fields have been defined so that the bilinear lepton-number-violating operators $L_i H_2$ [10] have been rotated away. The coupling constants $\lambda(\lambda'')$ are antisymmetric in the first (last) two indices. The first two terms lead to lepton number violation, while the last one violates baryon number conservation.

Proton stability forbids the simultaneous presence of lepton and baryon number violating terms in the superpotential [16] (unless the couplings are very small). Therefore, only λ, λ', or λ'' type interactions can be present. There may exist an underlying discrete symmetry in the theory which allows either the first or the second set of couplings [17,30,31]. An example of such a symmetry, which forbids baryon number violating couplings but allows lepton violating ones, is given by the transformation rules [24]

$$(Q, \bar{U}, \bar{D}) \to -(Q, \bar{U}, \bar{D}), (L, \bar{E}, H_{1,2}) \to +(L, \bar{E}, H_{1,2}) . \tag{4}$$

This discrete symmetry can be justified on a more fundamental level of Planck scale physics. It has been shown to be compatible with the ordinary SU(5) [10] and "flipped" SU(5)×U(1) [32] grand unified theory (GUT) scenarios as well as with phenomenologically viable superstring theories [33].

Neutrinoless double beta decay, which is the main subject of the present paper, requires lepton number violating interactions. Therefore we bind ourselves to the $\not{R}_p$ MSSM with lepton number violation ($\lambda \neq 0$, $\lambda' \neq 0$) and baryon number conservation ($\lambda'' = 0$). The Lagrangian of this model possesses the discrete symmetry Eq. (4). Apparently, $0\nu\beta\beta$ can probe only the first generation lepton number violating coupling λ'_{111} because only the first generation fermions u, d, e are involved in this process.

In addition to proton decay constraints on $\not{R}_p$ couplings there are also constraints which follow from cosmological arguments, requiring that the baryon asymmetry generated at the GUT scale is not washed out by B-L violating interactions present in Eq. (3). These cosmological constraints have been thought to affect all $\not{R}_p$ couplings $\lambda, \lambda', \lambda'' \ll 10^{-7}$, making these models phenomenologically uninteresting. These arguments, however, were proved strongly model dependent [21]; bounds can be evaded in perfectly reasonable scenarios of matter genesis [22].

The effect of "soft" supersymmetry breaking can be parametrized at the Fermi scale as a part of the scalar potential:

$$V_{\rm soft} = \sum_{i={\rm scalars}} m_i^2|\phi_i|^2 + h_L A_L H_1 \tilde{L}\tilde{\bar{E}} + h_D A_D H_1 \tilde{Q}\tilde{\bar{D}}$$
$$-h_U A_U H_2 \tilde{Q}\tilde{\bar{U}} - \mu B H_1 H_2 + \text{H.c.} \quad (5)$$

and a "soft" gaugino mass term

$$\mathcal{L}_{\rm GM} = -\frac{1}{2}[M_1\tilde{B}\tilde{B} + M_2\tilde{W}^k\tilde{W}^k + M_3\tilde{g}^a\tilde{g}^a] - \text{H.c.} \quad (6)$$

As usual, $M_{3,2,1}$ are the masses of the SU(3)×SU(2)×U(1) gauginos $\tilde{g}$, $\tilde{W}$, $\tilde{B}$, and m_i are the masses of scalar fields. A_L, A_D, A_U, and B are trilinear and bilinear "soft" supersymmetry breaking parameters. These quantities are free SUSY model parameters which, due to the renormalization effect, depend on the energy scale Λ.

Considering a GUT scenario within the MSSM one can claim the following unification conditions at the GUT scale $\Lambda \sim M_X$:

$$A_U(M_X) = A_D(M_X) = A_L(M_X) = A_0\,, \quad (7)$$

$$m_L(M_X) = m_E(M_X) = m_Q(M_X) = m_U(M_X)$$
$$= m_D(M_X) = m_0\,, \quad (8)$$
$$M_1(M_X) = M_2(M_X) = M_3(M_X) = m_{1/2}\,, \quad (9)$$

$$g_1(M_X) = g_2(M_X) = g_3(M_X) = g_{\rm GUT}\,, \quad (10)$$

where g_3, g_2, g_1 are the SU(3)×SU(2)×U(1) gauge coupling constants equal to $g_{\rm GUT}$ at the unification scale M_X.

At the Fermi scale $\Lambda \sim M_W$ these parameters can be evaluated on the basis of the MSSM renormalization group equations (RGE's) [34,35]. We assume that the $R\!\!\!/_p$ Yukawa coupling constants λ, λ', λ'' are small enough to be neglected in these equations. Equation (9) implies, at $\Lambda \sim M_W$,

$$M_1 = \tfrac{5}{3}\tan^2\theta_W M_2, \;\; M_2 \simeq 0.3 m_{\tilde{g}}\,. \quad (11)$$

Here $m_{\tilde{g}} = M_3$ is the gluino mass.

Now the model is completely specified and we can deduce the interaction terms of the $R\!\!\!/_p$ MSSM Lagrangian relevant for neutrinoless double beta decay.

Write down these interaction terms explicitly. Note that in the following we use for fermion fields the 4 component Dirac bispinor notation.

First, let us write down an ordinary charged current interaction

$$\mathcal{L}_{CC} = -\frac{g_2}{\sqrt{2}}[W_\mu^+(\bar{u}_L\gamma^\mu d_L + \bar{\nu}_L\gamma^\mu e_L)$$
$$+W_\mu^-(\bar{d}_L\gamma^\mu u_L + \bar{e}_L\gamma^\mu \nu_L)]\,. \quad (12)$$

The lepton number violating part of the Lagrangian can be obtained directly from the superpotential (3). It has the form

$$\mathcal{L}_{R\!\!\!/_p} = \lambda'_{111}\Bigg[(\bar{u}_L\bar{d}_R)\begin{pmatrix} e_R^c \\ -\nu_R^c \end{pmatrix}\tilde{d}_R + (\bar{e}_L\bar{\nu}_L)d_R\begin{pmatrix} \tilde{u}_L^* \\ -\tilde{d}_L^* \end{pmatrix}$$
$$+(\bar{u}_L\bar{d}_L)d_R\begin{pmatrix} \tilde{e}_L^* \\ -\tilde{\nu}_L^* \end{pmatrix} + \text{H.c.}\Bigg]\,. \quad (13)$$

The Lagrangian terms corresponding to gluino $\mathcal{L}_{\tilde{g}}$ and neutralino $\mathcal{L}_\chi$ interactions with fermions $\psi = \{u, d, e\}$, $q = \{u, d\}$ and their superpartners $\tilde{\psi} = \{\tilde{u}, \tilde{d}, \tilde{e}\}$, $\tilde{q} = \{\tilde{u}, \tilde{d}\}$ are [8]

$$\mathcal{L}_{\tilde{g}} = -\sqrt{2}g_3\frac{\lambda^{(a)}_{\alpha\beta}}{2}(\bar{q}_L^\alpha\tilde{g}^a\tilde{q}_L^\beta - \bar{q}_R^\alpha\tilde{g}^a\tilde{q}_R^\beta) + \text{H.c.}\,, \quad (14)$$

$$\mathcal{L}_\chi = \sqrt{2}g_2\sum_{i=1}^{4}[\epsilon_{Li}(\psi)\bar{\psi}_L\chi_i\tilde{\psi}_L + \epsilon_{Ri}(\psi)\bar{\psi}_R\chi_i\tilde{\psi}_R] + \text{H.c.} \quad (15)$$

Here $\lambda^{(a)}$ are 3×3 Gell-Mann matrices ($a = 1, \ldots, 8$). Superscripts α, β in Eq. (14) are color indices. Neutralino coupling constants are defined as [8]

$$\epsilon_{Li}(\psi) = -T_3(\psi)\mathcal{N}_{i2} + \tan\theta_W[T_3(\psi) - Q(\psi)]\mathcal{N}_{i1}\,, \quad (16)$$

$$\epsilon_{Ri}(\psi) = Q(\psi)\tan\theta_W\mathcal{N}_{i1}\,. \quad (17)$$

Here $Q(\psi)$ and $T_3(\psi)$ are the electric charge and weak isospin of the field ψ. The coefficients $\mathcal{N}_{ij}$ are elements of the orthogonal neutralino mixing matrix which diagonalizes the neutralino mass matrix. In the $R\!\!\!/_p$ MSSM the neutralino mass matrix is identical to the MSSM one [8] and in the basis of fields $(\tilde{B}, \tilde{W}^3, \tilde{H}_1^0, \tilde{H}_2^0)$ has the form

$$M_\chi = \begin{pmatrix} M_1 & 0 & -M_Z s_W c_\beta & M_Z s_W s_\beta \\ 0 & M_2 & M_Z c_W c_\beta & -M_Z c_W s_\beta \\ -M_Z s_W c_\beta & M_Z c_W c_\beta & 0 & -\mu \\ M_Z s_W s_\beta & -M_Z c_W s_\beta & -\mu & 0 \end{pmatrix}, \quad (18)$$

where $c_W = \cos\theta_W$, $s_W = \sin\theta_W$, $s_\beta = \sin\beta$, $c_\beta = \cos\beta$. The angle β is defined as $\tan\beta = \langle H_2^0\rangle/\langle H_1^0\rangle$. Here $\langle H_2^0\rangle$ and $\langle H_1^0\rangle$ are vacuum expectation values of the neutral components H_2^0 and H_1^0 of the Higgs doublet fields with weak hypercharges $Y(H_2^0) = +1$ and $Y(H_1^0) = -1$, respectively. The mass parameters M_1, M_2 are related to the gluino mass $m_{\tilde{g}}$ according to Eq. (11).

By diagonalizing the mass matrix (18) one can obtain

four Majorana neutralinos χ_i with masses m_{χ_i} and the field content

$$\chi_i = \mathcal{N}_{i1}\tilde{B} + \mathcal{N}_{i2}\tilde{W}^3 + \mathcal{N}_{i3}\tilde{H}_1^0 + \mathcal{N}_{i4}\tilde{H}_2^0 . \quad (19)$$

Recall again that we use notation $\tilde{W}^3$, $\tilde{B}$ for neutral $SU(2)_L \times U(1)$ gauginos and $\tilde{H}_2^0$, $\tilde{H}_1^0$ for Higgsinos which are the superpartners of the two neutral Higgs boson fields H_1^0 and H_2^0.

We apply a diagonalization by means of a real orthogonal matrix $\mathcal{N}$. Therefore the coefficients $\mathcal{N}_{ij}$ are real and masses m_{χ_i} are either positive or negative. The sign of the mass coincides with the CP parity of the corresponding neutralino mass eigenstate χ_i. If necessary, a negative mass can always be made positive by a redefinition [36] of the neutralino field χ_i. It leads to a redefinition of the relevant mixing coefficients $\mathcal{N}_{ij} \to i\mathcal{N}_{ij}$.

The lightest neutralino is commonly assumed to be the lightest supersymmetric particle (LSP). That is true in almost all phenomenologically viable SUSY models with R_p conservation. For $\not{R}_p$ SUSY models this is a very nontrivial assumption, since the cosmological constraints [37] requiring the LSP to be color and electrically neutral no longer apply [20]. *A priori*, the LSP could be any superparticle in $\not{R}_p$ SUSY models. However, the RGE analysis in minimal supergravity models suggests that the LSP is a neutralino if $\not{R}_p$ couplings are reasonably small [34]. Therefore, we also assume that the LSP is the lightest neutralino. If R_p is conserved the LSP is a stable particle. Otherwise it decays into ordinary matter.

Squarks $\tilde{u}_{L,R}$, $\tilde{d}_{L,R}$ and selectron $\tilde{e}_{L,R}$ in Eqs. (14), (15) are with a good precision mass eigenstates. Possible $\tilde{f}_L$-$\tilde{f}_R$ mixing for the first generation of squarks and sleptons are negligible due to the smallness of the relevant Yukawa couplings. In this case the MSSM mass formulas can be written as [8,34,35]

$$m^2_{\tilde{e}_L} = m_0^2 + 0.07m^2_{\tilde{g}} + \tfrac{1}{2}\cos2\beta M_Z^2(2\sin^2\theta_W - 1)\,, \quad (20)$$

$$m^2_{\tilde{e}_R} = m_0^2 + 0.02m^2_{\tilde{g}} - \cos2\beta M_Z^2\sin^2\theta_W\,, \quad (21)$$

$$m^2_{\tilde{u}_L} = m_0^2 + 0.83m^2_{\tilde{g}} + \tfrac{1}{2}\cos2\beta M_Z^2(1 - \tfrac{4}{3}\sin^2\theta_W)\,, \quad (22)$$

$$m^2_{\tilde{d}_L} = m_0^2 + 0.83m^2_{\tilde{g}} - \tfrac{1}{2}\cos2\beta M_Z^2(1 - \tfrac{2}{3}\sin^2\theta_W)\,, \quad (23)$$

$$m^2_{\tilde{u}_R} = m_0^2 + 0.77m^2_{\tilde{g}} + \cos2\beta M_Z^2(\tfrac{2}{3})\sin^2\theta_W\,, \quad (24)$$

$$m^2_{\tilde{d}_R} = m_0^2 + 0.76m^2_{\tilde{g}} - \cos2\beta M_Z^2(\tfrac{1}{3})\sin^2\theta_W\,. \quad (25)$$

Here, $m_{\tilde{g}}$ is the gluino mass and m_0 is the common sfermion mass at the unification scale [see (8)]. From Eqs. (20)–(25) one can estimate in the region $m^2_{\tilde{u}_L}$, $m^2_{\tilde{d}_R} \lesssim 300$ GeV that $m^2_{\tilde{u}_L} \approx m^2_{\tilde{d}_R}$. This approximate relation will be helpful for understanding the results of our numerical analysis in Sec. VI.

Having specified the Lagrangian interaction terms (13)–(15) and the mass eigenstates $\tilde{g}$, χ_i, $\tilde{q}$, $\tilde{e}$ involved in the interactions we can construct diagrams describing the $\not{R}_p$ MSSM contribution to the neutrinoless double beta decay.

In principle, in the $\not{R}_p$ MSSM a small neutrino mass of Majorana type may arise radiatively [10,38]. On the other hand in some GUT scenarios based on large gauge groups like SO(10) there might also be heavy Majorana fermions having non-negligible $SU(2)_L$ components. It is often identified with the heavy Majorana neutrino N. In the presence of either the light Majorana neutrinos or the heavy ones (or both) $0\nu\beta\beta$ decay can be induced by the conventional neutrino mass mechanism. In the present paper we concentrate on the $\not{R}_p$ SUSY mechanism of $0\nu\beta\beta$ decay. The effect of additional contributions from the neutrino mass mechanism is investigated in the case of heavy Majorana neutrino exchange. (Inclusion of the light Majorana neutrino contribution will not change the results of our analysis.)

III. LOW-ENERGY $\Delta L_e = 2$ EFFECTIVE LAGRANGIAN

The basic diagram corresponding to the neutrinoless double beta decay at the nucleon level is presented in Fig. 1(a). Two neutrons from the initial nucleus A_i, after interaction, transform into two protons of the final nucleus A_f, emitting two electrons. Apparently this process breaks the electron lepton number by two units, $\Delta L_e = 2$. At the quark level it can be induced by the subprocess with two initial d quarks and two final u quarks accompanied by two electrons as shown in Fig. 1(b). In the case of $0\nu\beta\beta$ decay when momenta of external particles are much smaller than intermediate particle masses one can treat interactions in Fig. 1(b) as pointlike.

Typical momenta of particles inside the nucleus are of order of the nucleon Fermi momentum $p_F \sim 100$ MeV

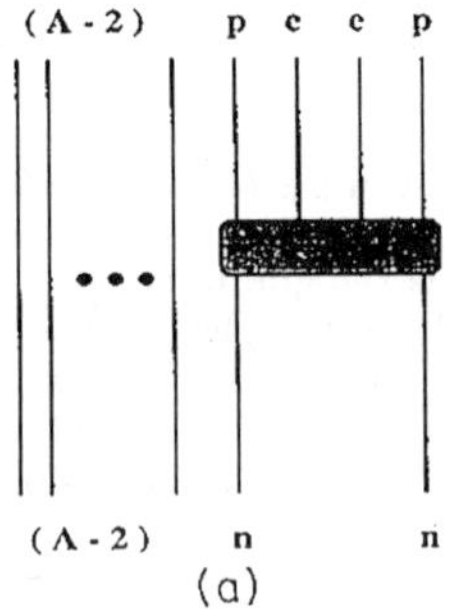

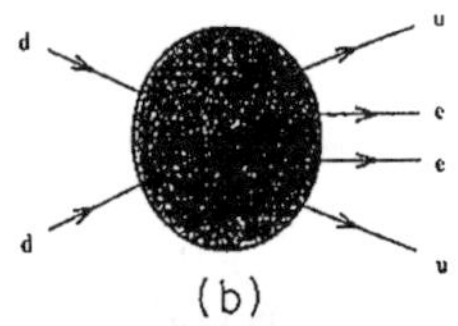

FIG. 1. Basic diagram for neutrinoless double beta decay. (a) at the nucleon level, (b) at the quark level.

while intermediate particles $\tilde{g}$, $\tilde{q}$, χ, N, $W^{\pm}$ have masses $m_i \sim 100$ GeV$\gg p_F$. A suitable formalism in this case is the effective Lagrangian approach. It gives a natural framework for relating quark and nuclear structure levels in calculating the matrix element of the $0\nu\beta\beta$ decay. Formally, it allows one to interpret the vertex in Fig. 1(b) in terms of nucleon variables instead of quark ones and then to use the nonrelativistic impulse approximation for derivation of nuclear transition operators.

A. General formalism

Define the low-energy effective Lagrangian $\mathcal{L}_{\text{eff}}^{\Delta L_e=2}(x)$ as

$$\begin{aligned}\mathcal{R}_{0\nu\beta\beta} &= \langle (A,Z+2),2e^-|S-1|(A,Z)\rangle \\ &= \langle (A,Z+2),2e^-|T\exp\left(i\int d^4x\mathcal{L}_{\text{int}}(x)\right)|(A,Z)\rangle_{m_i\gg p_F} \\ &\approx i\int d^4x\langle (A,Z+2),2e^-|\mathcal{L}_{\text{eff}}^{\Delta L_e=2}(x)|(A,Z)\rangle\,. \end{aligned} \tag{26}$$

Thus, the matrix element of $0\nu\beta\beta$ decay $\mathcal{R}_{0\nu\beta\beta}$ can be approximated as a first-order process in terms of the effective Lagrangian $\mathcal{L}_{\text{eff}}^{\Delta L_e=2}(x)$ while it is of higher order in terms of the fundamental interaction Lagrangian of the $R\!\!\!/_p$ MSSM. The term relevant to the vertex in Fig. 1(b) is given by

$$\mathcal{L}_{\text{int}} = \mathcal{L}_{CC} + \mathcal{L}_{R\!\!\!/_p} + \mathcal{L}_{\tilde{g}} + \mathcal{L}_{\chi}\,, \tag{27}$$

where $\mathcal{L}_{CC}$, $\mathcal{L}_{R\!\!\!/_p}$, $\mathcal{L}_{\tilde{g}}$, and $\mathcal{L}_{\chi}$ are specified in Eqs. (12)–(15). In Eq. (27) we have omitted terms giving negligible contributions to Eq. (26), such as interactions with Higgs fields.

The effective Lagrangian $\mathcal{L}_{\text{eff}}^{\Delta L_e=2}(x)$ describing the interaction in Fig. 1(b) at low energies can be written in the general form as

$$\mathcal{L}_{\text{eff}}^{\Delta L_e=2} = \sum_{\alpha,\beta,\gamma,\delta} \mathcal{C}^{\sigma,\omega}_{\alpha,\beta,\gamma,\delta}\bar{u}^{\alpha}d^{\beta}\cdot\bar{u}^{\gamma}d^{\delta}\cdot\bar{e}_{\sigma}e^c_{\omega} \tag{28}$$

$$= \sum_k \eta_k\cdot J^{\mu}_{(k)}J_{(k)\mu}\cdot\bar{e}\Gamma_k e^c + \text{color octet contribution}\,, \tag{29}$$

where the charge conjugate electron field is $e^c = C(\bar{e})^T$ with $C = i\gamma^2\gamma^0$ being charge conjugation matrix.

In the first part of this equation $\alpha,\beta,\gamma,\delta$ are collective indexes of quark fields corresponding to Lorentz and internal symmetries while σ and ω are Lorentz indexes of electron fields. The coefficients $\mathcal{C}^{\alpha,\cdots}$ can be obtained as a perturbation series from the definition of the effective Lagrangian Eq. (26) in terms of the fundamental one $\mathcal{L}_{\text{int}}$. The procedure implies expansion of the T exponent in Eq. (26) using Wick's theorem with

$$\begin{aligned}\langle 0|T(\psi_i(x)\bar{\psi}_j(y)|0\rangle &= -\delta^4(x-y)\frac{i}{m_i}\delta_{ij}\,, \\ \langle 0|T(\phi_i(x)\phi^*_j(y)|0\rangle &= -\delta^4(x-y)\frac{i}{m_i^2}\delta_{ij}\,,\end{aligned} \tag{30}$$

as propagators for heavy fermions $\psi_i = \chi, \tilde{g}$ and scalars $\phi_i = \tilde{q}, \tilde{e}$.

In the second part of Eq. (28) we introduced so-called lepton number violating parameters η_k and color-singlet quark currents $J^{\mu}_{(k)} = \bar{u}^{\alpha}\mathcal{O}^{\mu}_{(k)}d_{\alpha}$ (α is a color index). Here $\mathcal{O}^{\mu}_{(k)}$ and in Eq. (28) Γ_k are certain combinations of Dirac gamma matrices. As discussed in Sec. IV, nucleon matrix elements of these currents $\langle P(p)|J^{\mu}_{(k)}|N(p')\rangle$ can be parametrized in terms of nucleon form factors. In order to transform the first part of Eq. (28) to the second one, expressed in terms of color-singlet quark currents, one should use a Fierz rearrangement procedure followed by extraction of the color-singlet component of the relevant quark field operators. The latter is based on the following formulas

$$\lambda^{(k)}_{\alpha\beta}\cdot\lambda^{(k)}_{\gamma\delta} = \frac{16}{9}\delta_{\alpha\delta}\delta_{\gamma\beta} - \frac{1}{3}\lambda^{(k)}_{\alpha\delta}\cdot\lambda^{(k)}_{\gamma\beta}\,, \tag{31}$$

$$\delta^{(k)}_{\alpha\beta}\cdot\delta^{(k)}_{\gamma\delta} = \frac{1}{3}\delta_{\alpha\delta}\delta_{\gamma\beta} + \frac{1}{2}\lambda^{(k)}_{\alpha\delta}\cdot\lambda^{(k)}_{\gamma\beta}\,, \tag{32}$$

following from the fact that the system $\{I,\lambda^{(k)}\}$ is a basis in the space of 3×3 matrices. In the right-hand side of these equations the product of two δ's leads to the product of the two color singlet currents we are looking for, while the product of two λ's corresponds to the product of two color octet currents. The latter do not contribute to the nucleon matrix elements because the nucleon is a colorless state. Henceforth, we neglect the corresponding octet contributions to $\mathcal{L}_{\text{eff}}^{\Delta L_e=2}$ indicated in Eq. (29).

B. Leading order $R\!\!\!/_p$ MSSM effective Lagrangian

Now let us apply the above described formalism and derive the leading order contribution to the low-energy effective Lagrangian $\mathcal{L}_{\text{eff}}^{\Delta L_e=2}$ in the $R\!\!\!/_p$ MSSM. Feynman graphs corresponding to leading order contributions are presented in Fig. 2.

The conventional mass mechanism with Majorana neutrino exchange is presented in Fig. 2(a). In the following we consider the $R\!\!\!/_p$ MSSM contribution to $0\nu\beta\beta$ decay. Starting from the fundamental interactions (13)–(15) we have found [39] the complete set of diagrams, presented in Figs. 2(b) and 2(c), which contribute to this subprocess.

The supersymmetric mechanism of $0\nu\beta\beta$ decay was first proposed by Mohapatra [40] and later studied in more detail by Vergados [41]. In these papers [40,41] only three diagrams similar to those in Fig. 2(b) were considered. Instead of neutralinos χ_i, which are actual mass eigenstates in the MSSM, the consideration of Refs. [40,41] used Z-ino ($\tilde{Z}$) and photino ($\tilde{\gamma}$) fields in intermediate states. $\tilde{Z}$ and $\tilde{\gamma}$ can be mass eigenstates only at special values of parameters of the neutralino mass matrix. In general, these fields are not mass eigenstates. Furthermore, it was recently realized that a cosmologically viable lightest neutralino is very likely B-ino dominant [42]. Using in the intermediate states such fields that are not mass eigenstates leads to neglecting diagrams with mixed intermediate states when, for instance, $\tilde{Z}$ turns to $\tilde{\gamma}$ due to mixing proportional to the relevant entry of the neutralino mass matrix. The effect of mixing is taken into account completely in the set of diagrams displayed in Figs. 2(b) and 2(c) with all neutralino mass eigenstates χ_i involved.

Using the above-described formalism it is straightforward to find the operators in the effective Lagrangian which correspond to the diagrams in Figs. 2(b) and 2(c). The result is [39]

$$\begin{aligned}
\mathcal{L}_{\text{eff}}^{\Delta L_e=2}(x) = {} & 8\pi\alpha_2\lambda'^2_{111}\sum_{i=1}^{4}\frac{1}{m_{\chi_i}}\left[\frac{\epsilon^2_{Li}(e)}{m^4_{\tilde{e}_L}}(\bar{u}^\alpha_L d_{R\alpha})(\bar{u}^\beta_L d_{R\beta})(\bar{e}_L e^c_R)\right. \\
& +\frac{\epsilon^2_{Li}(u)}{m^4_{\tilde{u}_L}}(\bar{u}^\alpha_L u^c_{R\beta})(\bar{e}_L d_{R\alpha})(\bar{e}_L d^\beta_R)+\frac{\epsilon^2_{Ri}(d)}{m^4_{\tilde{d}_R}}(\bar{u}^\alpha_L e^c_R)(\bar{u}^\beta_L e^c_R)(\overline{d^c}_{L\alpha} d_{R\beta}) \\
& \left.+\left(\frac{\epsilon_{Li}(u)\epsilon_{Ri}(d)}{m^2_{\tilde{u}_L}m^2_{\tilde{d}_R}}+\frac{\epsilon_{Li}(u)\epsilon_{Li}(e)}{m^2_{\tilde{u}_L}m^2_{\tilde{e}_L}}+\frac{\epsilon_{Li}(e)\epsilon_{Ri}(d)}{m^2_{\tilde{e}_L}m^2_{\tilde{d}_R}}\right)(\bar{u}^\alpha_L d^\beta_R)(\bar{u}_{L\beta} e^c_R)(\bar{e}_L d_{R\alpha})\right] \\
& +\lambda'^2_{111}\frac{8\pi\alpha_s}{m_{\tilde{g}}}\frac{\lambda^{(a)}_{\alpha\beta}}{2}\frac{\lambda^{(a)}_{\gamma\delta}}{2}\left[\frac{1}{m^4_{\tilde{u}_L}}(\bar{u}^\alpha_L u^{c\gamma}_R)(\bar{e}_L d^\beta_R)(\bar{e}_L d^\delta_R)\right. \\
& \left.+\frac{1}{m^4_{\tilde{d}_R}}(\bar{u}^\alpha_L e^c_R)(\bar{u}^\gamma_L e^c_R)(\overline{d^c}^\beta_L d^\delta_R)-\frac{1}{m^2_{\tilde{d}_R}m^2_{\tilde{u}_L}}(\bar{u}^\alpha_L d^\delta_R)(\bar{u}^\gamma_L e^c_R)(\bar{e}_L d^\beta_R)\right].
\end{aligned} \tag{33}$$

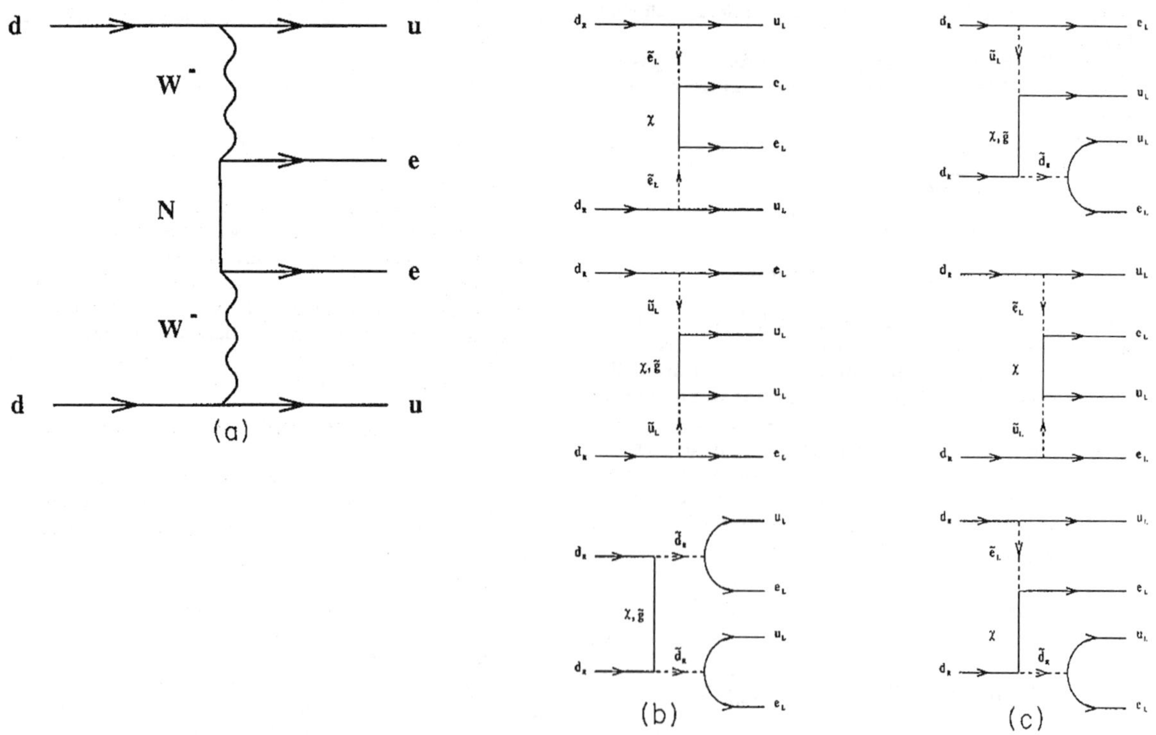

FIG. 2. Feynman graphs for (a) the conventional mechanism of $0\nu\beta\beta$ decay by massive Majorana neutrino exchange, (b) "diagonal," and (c) new "nondiagonal" [39] supersymmetric contributions to $0\nu\beta\beta$ decay.

This expression takes the form of the first part of the general Eq. (28) after applying the following relations $\psi^c = C(\bar{\psi})^T$, $\overline{\psi^c} = \psi^T C$, $\bar{\psi} = (\psi^c)^T C$.

Gauge coupling constants $\alpha_2 = g_2^2/(4\pi)$ and $\alpha_s = g_3^2/(4\pi)$ are running coupling constants which should be estimated at the proper energy scale Λ. One can see from the diagrams in Fig. 2 that the typical scale at which the $\tilde{g}$-$\tilde{q}$-q and χ-$\tilde{q}$-q interactions occur is of the order of the gluino or neutralino mass. In the mass region which will be analyzed numerically in the subsequent sections we may take approximately these couplings at the Z-boson pole and use their values from Ref. [43]:

$$\alpha_s(M_Z) = 0.127, \quad \alpha_2(M_Z) = 0.0337. \tag{34}$$

The Lagrangian (33) has terms in a form which do not allow a direct application of the nonrelativistic impulse approximation for further calculation of the $0\nu\beta\beta$ reaction matrix element. As explained at the beginning of this section, one should rearrange the right-hand side of Eq. (33) to the form of a product of two color-singlet quark currents and the leptonic current. It can be accomplished by a Fierz rearrangement procedure and subsequent extraction of color singlets from the product of two color-triplet and color-antitriplet quark fields using formulas (31) and (32). The final result is

$$\mathcal{L}_{\text{eff}}^{\Delta L_e=2}(x) = \frac{G_F^2}{2} \cdot m_P^{-1} \left[\left((\eta_{\tilde{g}} + \eta_\chi)(J_{PS}J_{PS} - \frac{1}{4} J_T^{\mu\nu} J_{T\mu\nu} \right) + (\eta_{\chi\tilde{e}} + \eta'_{\tilde{g}} - \eta_{\chi\tilde{f}}) J_{PS}J_{PS} + \eta_N J_{VA}^{\mu} J_{VA\mu} \right] [\bar{e}(1+\gamma_5)e^c]. \tag{35}$$

The last term corresponds to the heavy Majorana neutrino contribution described by Fig. 2(a) which we also consider at the end of Sec. II.

The lepton number violating parameters are defined as

$$\eta_{\tilde{g}} = \frac{2\pi\alpha_s}{9} \frac{\lambda'^2_{111}}{G_F^2 m_{\tilde{d}_R}^4} \frac{m_P}{m_{\tilde{g}}} \left[1 + \left(\frac{m_{\tilde{d}_R}}{m_{\tilde{u}_L}} \right)^4 \right], \tag{36}$$

$$\eta_\chi = \frac{\pi\alpha_2}{6} \frac{\lambda'^2_{111}}{G_F^2 m_{\tilde{d}_R}^4} \sum_{i=1}^{4} \frac{m_P}{m_{\chi i}} \left[\epsilon_{Ri}^2(d) + \epsilon_{Li}^2(u) \left(\frac{m_{\tilde{d}_R}}{m_{\tilde{u}_L}} \right)^4 \right], \tag{37}$$

$$\eta_{\chi\tilde{e}} = 2\pi\alpha_2 \frac{\lambda'^2_{111}}{G_F^2 m_{\tilde{d}_R}^4} \left(\frac{m_{\tilde{d}_R}}{m_{\tilde{e}_L}} \right)^4 \sum_{i=1}^{4} \epsilon_{Li}^2(e) \frac{m_P}{m_{\chi i}}, \tag{38}$$

$$\eta'_{\tilde{g}} = \frac{4\pi\alpha_s}{9} \frac{\lambda'^2_{111}}{G_F^2 m_{\tilde{d}_R}^4} \frac{m_P}{m_{\tilde{g}}} \left(\frac{m_{\tilde{d}_R}}{m_{\tilde{u}_L}} \right)^2, \tag{39}$$

$$\eta_{\chi\tilde{f}} = \frac{\pi\alpha_2}{3} \frac{\lambda'^2_{111}}{G_F^2 m_{\tilde{d}_R}^4} \left(\frac{m_{\tilde{d}_R}}{m_{\tilde{e}_L}} \right)^2 \sum_{i=1}^{4} \frac{m_P}{m_{\chi i}} \left[\epsilon_{Ri}(d)\epsilon_{Li}(e) + \epsilon_{Li}(u)\epsilon_{Ri}(d) \left(\frac{m_{\tilde{e}_L}}{m_{\tilde{u}_L}} \right)^2 + \epsilon_{Li}(u)\epsilon_{Li}(e) \left(\frac{m_{\tilde{d}_R}}{m_{\tilde{u}_L}} \right)^2 \right], \tag{40}$$

$$\eta_N = \frac{m_P}{\langle m_N \rangle}, \tag{41}$$

where $\langle m_N \rangle$ is the *effective* heavy Majorana neutrino mass (for the definition see [3]).

Color-singlet hadronic currents have the form

$$J_{PS} = \bar{u}^\alpha (1+\gamma_5) d_\alpha, \tag{42}$$

$$J_T^{\mu\nu} = \bar{u}^\alpha \sigma^{\mu\nu} (1+\gamma_5) d_\alpha, \tag{43}$$

$$J_{AV}^{\mu} = \bar{u}^\alpha \gamma^\mu (1-\gamma_5) d_\alpha, \tag{44}$$

where $\sigma^{\mu\nu} = \frac{i}{2}[\gamma^\mu, \gamma^\nu]$.

Now we can consider nucleon matrix elements of these quark currents. These matrix elements provide us with certain information about quark states inside the nucleon. Further we will use in Sec. V the proper nuclear wave functions to describe nucleon states in the nucleus. Applying this standard two-step procedure we will obtain the reaction matrix element for the $0\nu\beta\beta$ decay.

IV. FROM QUARK TO NUCLEAR LEVEL

Adopting the above mentioned two-step procedure let us write down the $0\nu\beta\beta$ decay matrix element $\mathcal{R}_{0\nu\beta\beta}$ corresponding to the effective Lagrangian Eq. (35). Using the general formula Eq. (A3) one can write

$$\mathcal{R}_{0\nu\beta\beta} = \frac{G_F^2}{\sqrt{2}} m_P^{-1} C_{0\nu}^{-1} \{\bar{e}(1+\gamma_5)e^c\} [(\eta_{\tilde{g}} + \eta_\chi)\langle F|\Omega_{\tilde{q}}|I\rangle + (\eta_{\chi\tilde{e}} + \eta'_{\tilde{g}} - \eta_{\chi\tilde{f}})\langle F|\Omega_{\tilde{f}}|I\rangle + \eta_N \langle F|\Omega_N|I\rangle]. \tag{45}$$

The heavy Majorana neutrino exchange contribution has been included in the last term of this equation. It corresponds to the last term of Eq. (35).

We have introduced transition operators Ω_i [see Eq. (A4) in the Appendix] which are useful for separating the particle physics part of the calculation from the nuclear physics one. The transition operators contain information about the underlying interactions at the quark level (35) and quark states inside the nucleon. They are independent of the initial $|I\rangle$ and the final $\langle F|$ nuclear states. To calculate the nuclear matrix elements in Eq. (45) one should use nuclear model wave functions.

Nuclear matrix elements of the transition operators $\Omega_{\tilde{q}}$, $\Omega_{\tilde{f}}$, and Ω_N describe transitions induced by quark color singlet currents (42)–(44) in the first, second, and third terms of Eq. (35). Diagrams in Fig. 2 with the intermediate states $\{W - N - W\}$, $\{\tilde{u}(\tilde{d}) - \chi, \tilde{g} - \tilde{u}(\tilde{d})\}$, and $\{\tilde{u}(\tilde{d}) - \chi, \tilde{g} - \tilde{d}(\tilde{u}); \tilde{e} - \chi - \tilde{e}, \tilde{q}\}$ contribute to operators Ω_N, $\Omega_{\tilde{q}}$, and $\Omega_{\tilde{f}}$, respectively.

The relevant formula for calculating the transition operators in the nonrelativistic impulse approximation (NRIA) [see Eq. (A6)] require the nucleon matrix elements of these currents.

Now we turn to the derivation of nucleon matrix elements of the color singlet quark currents (42)–(44) using the results of Ref. [44]. The relevant matrix elements are

$$\langle P(p)|\bar{u}d|N(p')\rangle = F_S^{(3)}(q^2)\cdot \bar{\mathcal{N}}(p)\tau_+\mathcal{N}(p'),\tag{46}$$

$$\langle P(p)|\bar{u}\gamma_5 d|N(p')\rangle = F_P^{(3)}(q^2)\cdot \bar{\mathcal{N}}(p)\gamma_5\tau_+\mathcal{N}(p'),$$

$$\langle P(p)|\bar{u}\sigma^{\mu\nu}(1+\gamma_5)d|N(p')\rangle = \bar{\mathcal{N}}(p)\left(J^{\mu\nu}+\frac{i}{2}\epsilon^{\mu\nu\rho\sigma}J_{\rho\sigma}\right)\tau_+\mathcal{N}(p'),$$

$$\langle P(p)|\bar{u}\gamma^{\mu}(1-\gamma_5)d|N(p')\rangle = \bar{\mathcal{N}}(p)\gamma^{\mu}[F_V(q^2)-F_A(q^2)\gamma_5+F_W(q^2)i\sigma^{\mu\nu}q_\nu+F_P/(q^2)q_\mu\gamma_5]\tau_+\mathcal{N}(p'),$$

where $\mathcal{N}=\binom{P}{N}$ is a nucleon isodoublet, $q=p-p'$. The tensor structure is defined as

$$J^{\mu\nu} = T_1^{(3)}(q^2)\sigma^{\mu\nu}+\frac{iT_2^{(3)}}{m_P}(\gamma^\mu q^\nu-\gamma^\nu q^\mu) +\frac{T_3^{(3)}}{m_P^2}(\sigma^{\mu\rho}q_\rho q^\nu-\sigma^{\nu\rho}q_\rho q^\mu).\tag{47}$$

For all form factors [$F_W(q^2)$ and $F_P(q^2)$, will be neglected in the following] we take following Ref. [45] a dipole form

$$\frac{F_{V,A}(q^2)}{f_{V,A}}=\frac{F_{S,P}^{(3)}(q^2)}{F_{S,P}^{(3)}(0)}=\frac{T_i^{(3)}(q^2)}{T_i^{(3)}(0)}=\left(1-\frac{q^2}{m_A^2}\right)^{-2}\tag{48}$$

with $m_A=0.85$ GeV and $f_V\approx 1$, $f_A\approx 1.261$. Form factor normalization values $F_{S,P}^{(3)}(0), T_i^{(3)}(0)$ were calculated in Ref. [44] and are given in Table I.

Using formulas (46) and (47) we may derive the nonrelativistic limit $m_P \gg |\vec{p}|$ for nucleon matrix elements of the three combinations of the quark currents in (35). Keeping all terms up to order q^2 in the nonrelativistic expansion we find from Eq. (A6) the relevant transition operators $\Omega_{\tilde{q}}, \Omega_{\tilde{f}}, \Omega_N$ for two outgoing electrons in the S-wave state:

$$\Omega_{\tilde{q}}=\frac{m_P}{m_e}\{\alpha_V^{(0)}\Omega_{F,N}+\alpha_A^{(0)}\Omega_{\mathrm{GT},N}+\alpha_V^{(1)}\Omega_{F'} +\alpha_A^{(1)}\Omega_{\mathrm{GT}'}+\alpha_T\Omega_{T'}\},\tag{49}$$

$$\Omega_{\tilde{f}}=\Omega_{\tilde{q}}(T_i=0),\tag{50}$$

$$\Omega_N=\frac{m_P}{m_e}\left\{\left(\frac{f_V}{f_A}\right)^2\Omega_{F,N}-\Omega_{\mathrm{GT},N}\right\},\tag{51}$$

where partial transition operators are

$$\Omega_{\mathrm{GT},N}=\sum_{i\neq j}\tau_+^{(i)}\tau_+^{(j)}\boldsymbol{\sigma}_i\cdot\boldsymbol{\sigma}_j\left(\frac{R_0}{r_{ij}}\right)F_N(x_A),\tag{52}$$

$$\Omega_{F,N}=\sum_{i\neq j}\tau_+^{(i)}\tau_+^{(j)}\left(\frac{R_0}{r_{ij}}\right)F_N(x_A),\tag{53}$$

$$\Omega_{\mathrm{GT}'}=\sum_{i\neq j}\tau_+^{(i)}\tau_+^{(j)}\boldsymbol{\sigma}_i\cdot\boldsymbol{\sigma}_j\left(\frac{R_0}{r_{ij}}\right)F_4(x_A),\tag{54}$$

$$\Omega_{F'}=\sum_{i\neq j}\tau_+^{(i)}\tau_+^{(j)}\left(\frac{R_0}{r_{ij}}\right)F_4(x_A),\tag{55}$$

$$\Omega_{T'}=\sum_{i\neq j}\tau_+^{(i)}\tau_+^{(j)}\{3(\boldsymbol{\sigma}_i\cdot\hat{\boldsymbol{r}}_{ij})(\boldsymbol{\sigma}_j\cdot\hat{\boldsymbol{r}}_{ij})-\boldsymbol{\sigma}_i\cdot\boldsymbol{\sigma}_j\} \times\left(\frac{R_0}{r_{ij}}\right)F_5(x_A).\tag{56}$$

Here, R_0 is the nuclear radius, introduced to make the matrix elements dimensionless (compensating factors have been absorbed into the phase space integrals [3]). The following notation is used:

$$\boldsymbol{r}_{ij}=(\vec{r}_i-\vec{r}_j),\quad r_{ij}=|\boldsymbol{r}_{ij}|,\quad \hat{\boldsymbol{r}}_{ij}=\boldsymbol{r}_{ij}/r_{ij},\quad x_A=m_A r_{ij}.$$

Here $\vec{r}_i$ is the coordinate of the ith nucleon. The above matrix elements have been written in the closure approximation, which is well satisfied due to the large masses of the intermediate particles.

The nucleon structure coefficients in (49) are given by

$$\alpha_V^{(0)}=\left(\frac{F_S^{(3)}}{f_A}\right)^2,\quad \alpha_A^{(0)}=-\left(\frac{T_1^{(3)}}{f_A}\right)^2,\tag{57}$$

$$\alpha_V^{(1)}=-\left(\frac{m_A}{m_P}\right)^2\alpha_A^{(0)}\left[\frac{1}{4}+\left(\frac{T_2^{(3)}}{T_1^{(3)}}\right)^2-\frac{T_2^{(3)}}{T_1^{(3)}}\right],\tag{58}$$

$$\alpha_A^{(1)}=-\left(\frac{m_A}{m_P}\right)^2\left[\alpha_A^{(0)}\left(\frac{1}{6}-\frac{2}{3}\frac{T_2^{(3)}}{T_1^{(3)}}+\frac{4}{3}\frac{T_3^{(3)}}{T_1^{(3)}}\right) +\frac{1}{12}\left(\frac{F_P^{(3)}}{f_A}\right)^2\right],\tag{59}$$

$$\alpha_T=-\left(\frac{m_A}{m_P}\right)^2\left[\alpha_A^{(0)}\left(\frac{1}{12}-\frac{1}{3}\frac{T_2^{(3)}}{T_1^{(3)}}+\frac{2}{3}\frac{T_3^{(3)}}{T_1^{(3)}}\right) -\frac{1}{12}\left(\frac{F_P^{(3)}}{f_A}\right)^2\right].\tag{60}$$

TABLE I. Nucleon form factor normalizations (at $q^2=0$) as calculated in [44].

Set	$F_S^{(3)}$	$F_P^{(3)}$	$T_1^{(3)}$	$T_2^{(3)}$	$T_3^{(3)}$
(A) Bag model	0.48	4.41	1.38	−3.30	−0.62
(B) Nonrelativistic quark model	0.62	4.65	1.45	−1.48	−0.66

Here $F^{(3)}_{S,P} \equiv F^{(3)}_{S,P}(0)$, $T^{(3)}_i \equiv T^{(3)}_i(0)$.

Three different structure functions F_i appear in Eqs. (52)–(56) (we use notations of Ref. [41]). They are given by the integrals over the momentum q transferred between two nucleons [see Eq. (A6)]:

$$F_N(x_A) = 4\pi m_A^6 r_{ij} \int \frac{d^3q}{(2\pi)^3} \frac{1}{(m_A^2+q^2)^4} e^{iqr_{ij}} , \qquad (61)$$

$$F_4(x_A) = 4\pi m_A^4 r_{ij} \int \frac{d^3q}{(2\pi)^3} \frac{q^2}{(m_A^2+q^2)^4} e^{iqr_{ij}} , \qquad (62)$$

$$F_5(x_A) = 2\pi m_A^4 r_{ij} \int \frac{d^3q}{(2\pi)^3} \frac{q^2 - \frac{1}{3}(q\cdot\hat{r}_{ij})^2}{(m_A^2+q^2)^4} e^{iqr_{ij}} . \qquad (63)$$

These functions are analogous to the "neutrino potentials" for the case of the light Majorana neutrino exchange. Nuclear energy denominators typical for such momentum integrals can be neglected safely from Eqs. (61)–(63) in the case of heavy intermediate particles, such as SUSY particles and the heavy Majorana neutrino N, with masses much larger than the characteristic nuclear energy and are therefore suppressed in (61)–(63). The analytic solutions of the integrals in Eqs. (61)–(63) are

$$F_N(x) = \frac{x}{48}(3+3x+x^2)e^{-x} ,$$
$$F_4(x) = \frac{x}{48}(3+3x-x^2)e^{-x} , \qquad (64)$$
$$F_5(x) = \frac{x^3}{48}e^{-x} .$$

At this stage we point out that our formulas for the coefficients (57)–(60) of the transition operators (49) and (50) disagree with the corresponding formulas derived in Ref. [41]. Particularly, in our case $\alpha_A^{(0)} \le 0$ while in Ref. [41] this coefficient is positive. This sign difference has an important consequence if one considers simultaneously the supersymmetric and the heavy Majorana neutrino contributions. It will be seen in the next section that in both the neutrino exchange and in the $\not{R}_p$ SUSY mechanisms the dominant contributions correspond to the $\Omega_{\mathrm{GT},N}$ transition operator, initiating Gamow-Teller $0^+ \to 0^+$ nuclear transitions. Comparing Eq. (49) with Eq. (51) one can see that our formulas correspond to a constructive interference between the dominant $\Omega_{\mathrm{GT},N}$ terms of these two mechanisms while formulas from Ref. [41] correspond to a destructive one. In the latter case both contributions can cancel each other and by a proper choice of λ'_{111} in (36)–(40) the matrix element of $0\nu\beta\beta$ decay can be set to zero at any values of particle masses involved in the formulas. As a result the $0\nu\beta\beta$ decay half-life limit would constrain neither supersymmetric particle masses nor that for Majorana neutrinos. Our formulas (49), (50), and (57)–(60) always lead to certain constraints on these masses. We return to a detailed discussion of this point in Sec. VI.

V. NUCLEAR MATRIX ELEMENTS IN THE *pn* QRPA APPROACH

While the formalism of the SUSY $0\nu\beta\beta$ decay outlined in the previous sections is independent of the nuclear structure model used to generate the wave functions, numerical values of matrix elements are model dependent. We therefore separate this part of our work from the rest of the formalism in order to split up nuclear and particle physics uncertainties in a most distinctive way.

A. Basic summary of the nuclear model and its parameters

Nuclear matrix elements have been calculated within *pn* QRPA (proton-neutron quasiparticle random phase approximation). *pn* QRPA, the RPA for charge-changing transitions has been developed by Halbleib and Sorenson [46], and during recent years widely been applied to double beta decay calculations [47–53]. In the present work we follow essentially the description of [50,51,53].

Further we just briefly summarize the numerically most important features.

We account for typically two major oscillator shells, for example, $3\hbar\omega$ and $4\hbar\omega$ in the case of ^{76}Ge. Single-particle energies are calculated from a Coulomb-corrected Woods-Saxon potential with Bohr-Mottelson parameters [54], except that the strength of the spin-orbit force has been reduced according to [53] by $\sim 9\%$ for a better agreement with experimental data. For the pairing [in the Bardeen-Cooper-Schriefer (BCS) approximation] and RPA calculation we used consistently the nucleon-nucleon potential of Lacombe *et al.* [55]. The strengths of the pairing interactions in the BCS calculation have been adjusted to reproduce the experimentally observed pairing gaps [56]. In the RPA calculation, for the strength of the particle-hole interaction a simple mass number dependence was used, $g_{ph} = 1 + 0.002A$ [50]. For the strength of the particle-particle interaction numerical values of matrix elements are given for values of g_{pp}, which have been fitted to experimentally measured β^+/EC decay strengths [50].

As mentioned in the previous sections, in the SUSY $0\nu\beta\beta$ decay the virtually exchanged particles are assumed to be heavy. Thus, the corresponding operators will be short ranged. For this reason the short-range part of the nuclear wave functions has to be treated carefully. In addition to the finite nucleon size effects, which are taken into account by the nucleon form factors in momentum space [see Eq. (48)], the nucleon-nucleon repulsion at short distances must be accounted for. These short-range correlations are treated by multiplying the two particle wave functions by the correlation function [57]

$$1 - f(r) = 1 - e^{-ar^2}(1 - br^2) . \qquad (65)$$

The two parameters a and b can be related to each other, so that effectively there is only one free parameter, the so-called correlation length defined by

$$l_c = -\int_0^\infty ds([1+f(r)]^2 - 1) \simeq \frac{0.748}{\sqrt{a}}, \quad (66)$$

with the standard value of $l_c = 0.7$ fm.

B. Numerical results and discussion

In the SUSY nuclear transition operators (49) and (50) there appear five basic structures (52)–(56). These, combined with the nuclear wave functions, give rise to the following five nuclear structure matrix elements[1]

$$\mathcal{M}_{\mathrm{GT},N} = \langle F|\Omega_{\mathrm{GT},N}|I\rangle = \left\langle F \left| \sum_{i\neq j} \tau_+^{(i)}\tau_+^{(j)} \boldsymbol{\sigma}_i\cdot\boldsymbol{\sigma}_j \left(\frac{R_0}{r_{ij}}\right) F_N(x_A) \right| I \right\rangle, \quad (67)$$

$$\mathcal{M}_{F,N} = \langle F|\Omega_{F,N}|I\rangle = \left\langle F \left| \sum_{i\neq j} \tau_+^{(i)}\tau_+^{(j)} \left(\frac{R_0}{r_{ij}}\right) F_N(x_A) \right| I \right\rangle, \quad (68)$$

$$\mathcal{M}_{\mathrm{GT}'} = \langle F|\Omega_{\mathrm{GT}'}|I\rangle = \left\langle F \left| \sum_{i\neq j} \tau_+^{(i)}\tau_+^{(j)} \boldsymbol{\sigma}_i\cdot\boldsymbol{\sigma}_j \left(\frac{R_0}{r_{ij}}\right) F_4(x_A) \right| I \right\rangle, \quad (69)$$

$$\mathcal{M}_{F'} = \langle F|\Omega_{F'}|I\rangle = \left\langle F \left| \sum_{i\neq j} \tau_+^{(i)}\tau_+^{(j)} \left(\frac{R_0}{r_{ij}}\right) F_4(x_A) \right| I \right\rangle, \quad (70)$$

$$\mathcal{M}_{T'} = \langle F|\Omega_{T'}|I\rangle = \left\langle F \left| \sum_{i\neq j} \tau_+^{(i)}\tau_+^{(j)} \{3(\boldsymbol{\sigma}_i\cdot\hat{\boldsymbol{r}}_{ij})(\boldsymbol{\sigma}_j\cdot\hat{\boldsymbol{r}}_{ij}) - \boldsymbol{\sigma}_i\cdot\boldsymbol{\sigma}_j\} \left(\frac{R_0}{r_{ij}}\right) F_5(x_A) \right| I \right\rangle. \quad (71)$$

We have calculated these five basic matrix elements for the experimentally most interesting isotopes. The results are given in Table II. As expected $\mathcal{M}_{\mathrm{GT},N}$ and $\mathcal{M}_{F,N}$ have the largest numerical values, whereas the other three matrix elements give only minor corrections. This can be simply understood, if one notes that matrix elements $\mathcal{M}_{\mathrm{GT}'}, \mathcal{M}_{F'}, \mathcal{M}_{T'}$ have additional q^2 factors in the integrals (62) and (63) over the momentum q, transferred between two decaying neutrons. The effective nuclear cutoff for momentum transfer is $q \lesssim p_F$, where $p_F \sim 100$ MeV is the momentum of the nucleon Fermi motion inside the nucleus. This then results in the relative suppression of the nuclear matrix elements of the last three operators in Eqs. (69)–(71) by factors of $(p_F/m_P)^2 \simeq$(few)%.

The data of Table II, however, show also another interesting fact: Corresponding matrix elements for different isotopes are relatively similar to each other, with a spread of about a factor of 2 from the mean only. As is known, much larger differences are usually found in calculations of $2\nu\beta\beta$ decay matrix elements as one goes from one isotope to another. The similarity of the SUSY $0\nu\beta\beta$ decay matrix element is, on the other hand, not really a surprise. Recall that due to the large masses of the intermediate particles the transition operators are short ranged. Therefore only the part of the wave functions at short distances contribute appreciably to the matrix elements, and nuclear structure effects, which are so important in $2\nu\beta\beta$ decay, play a less dominant role in SUSY $0\nu\beta\beta$ decay. (See also the discussion in the next section.)

According to Eqs. (49) and (50) the five basic matrix elements (67)–(71) can then be combined to define the following nuclear matrix elements describing contributions to $0\nu\beta\beta$ decay which correspond to the $\not{R}_p$ SUSY mechanism

$$\mathcal{M}_{\tilde q} = \langle F|\Omega_{\tilde q}|I\rangle \quad (72)$$
$$= \frac{m_P}{m_e}\Big\{\alpha_V^{(0)}\mathcal{M}_{F,N} + \alpha_A^{(0)}\mathcal{M}_{\mathrm{GT},N} + \alpha_V^{(1)}\mathcal{M}_{F'} + \alpha_A^{(1)}\mathcal{M}_{\mathrm{GT}'} + \alpha_T\mathcal{M}_{T'}\Big\},$$

$$\mathcal{M}_{\tilde f} = \langle F|\Omega_{\tilde f}|I\rangle = \mathcal{M}_{\tilde q}|_{\forall T_i=0} \quad (73)$$

and the conventional mass mechanism due to heavy Majorana neutrino exchange

$$\mathcal{M}_N = \langle F|\Omega_N|I\rangle = \frac{m_P}{m_e}\left\{\left(\frac{f_V}{f_A}\right)^2 \mathcal{M}_{F,N} - \mathcal{M}_{\mathrm{GT},N}\right\}, \quad (74)$$

Table III shows $\mathcal{M}_{\tilde q}$ and $\mathcal{M}_{\tilde f}$ for the isotopes considered for the two sets of input parameters for the nucleon structure constants $\alpha^{(i)}$ as calculated from[2] Table I.

[1]Note that, compared to the definitions of Fermi-type matrix elements for the light neutrino case [51,53], we took out the factor $(f_V/f_A)^2$ here; all coupling-constant combinations are absorbed into the coefficients $\alpha^{(i)}$.

[2]Although $\mathcal{M}_{\tilde q}$ and $\mathcal{M}_{\tilde f}$ are negative, positive values are given in Table III. This was done to adjust the signs of matrix elements to the notation used in our previous publications on $0\nu\beta\beta$ decay in left-right-symmetric models [51,53].

TABLE II. Nuclear matrix elements for SUSY $0\nu\beta\beta$ decay for the experimentally most interesting isotopes calculated within *pn* QRPA. $\mathcal{M}_\alpha \times 10^x$ implies that the matrix element should be divided by 10^x to get the correct numerical value. These factors are introduced here, just to show the differences in magnitude of the different matrix elements.

^AY	$\mathcal{M}_{\mathrm{GT},N} \times 10^1$	$\mathcal{M}_{F,N} \times 10^2$	$\mathcal{M}_{\mathrm{GT}'} \times 10^3$	$\mathcal{M}_{F'} \times 10^3$	$\mathcal{M}_{T'} \times 10^3$
^{76}Ge	1.13	−4.07	−7.70	3.06	−3.09
^{82}Se	1.02	−3.60	−7.13	2.76	−2.76
^{100}Mo	1.29	−4.89	−7.88	3.52	−4.93
^{116}Cd	0.75	−2.71	−5.49	2.19	−2.65
^{128}Te	1.19	−4.19	−7.95	3.11	−3.85
^{130}Te	1.05	−3.69	−6.97	2.73	−3.53
^{136}Xe	0.58	−2.03	−3.81	1.49	−1.81
^{150}Nd	1.65	−5.91	−10.4	4.32	−7.51

Again, it is noteworthy that both sets of parameters lead to rather similar results. This is due to a partial cancellation of the differences in the input values for the $\alpha^{(i)}$, at least in the case of $\mathcal{M}_{\tilde{q}}$.

Table III moreover shows that the $\mathcal{M}_{\tilde{f}}$'s are much smaller than the $\mathcal{M}_{\tilde{q}}$'s. This difference can be simply understood because only scalar and pseudoscalar currents $J_{S,P}$ contribute to $\mathcal{M}_{\tilde{f}}$, but no tensor current J_T. Since in this case [see (73)] $\alpha_A^{(0)} \sim (T_1^{(3)})^2 = 0$, there is no contribution from the large $\mathcal{M}_{\mathrm{GT},N}$ to $\mathcal{M}_{\tilde{f}}$ and consequently $\mathcal{M}_{\tilde{f}} \ll \mathcal{M}_{\tilde{q}}$. This result has important implications in the numerical analysis of Sec. VI.

C. Uncertainties of the nuclear structure matrix elements

We have investigated the dependence of the numerical values of the matrix elements on our choice of nuclear model parameters. Besides the numerical uncertainties discussed here, there will also be deviations from the "true" matrix element due to model approximations. These, however, can not be quantified exactly. Some confidence in the model might be derived from the fact that the $2\nu\beta\beta$ decay half-life of ^{76}Ge [58] has been *predicted* correctly [50] within a factor of 2 (the $2\nu\beta\beta$ decay matrix element within $\sqrt{2}$).

TABLE III. Nuclear matrix elements for SUSY $0\nu\beta\beta$ decay. Shown are $\mathcal{M}_{\tilde{q}}$ and $\mathcal{M}_{\tilde{f}}$ for the two sets of input values of coefficients for the $\alpha^{(i)}$ of Table I.

^AY	(A) $\mathcal{M}_{\tilde{q}}$	(B) $\mathcal{M}_{\tilde{q}}$	(A) $\mathcal{M}_{\tilde{f}}$	(B) $\mathcal{M}_{\tilde{f}}$
^{76}Ge	283	304	13.2	20.7
^{82}Se	253	272	11.3	17.9
^{100}Mo	328	356	23.6	33.5
^{116}Cd	190	205	11.0	16.2
^{128}Te	298	323	16.7	24.8
^{130}Te	262	284	15.4	22.5
^{136}Xe	143	155	7.91	11.8
^{150}Nd	416	456	34.4	47.0

In the following the three most important parameters are discussed in some detail. Other parameters, like for example g_{ph} or the use of another nucleon-nucleon potential (Reid soft-core potential [59]), instead of the Paris force, have been found to have negligible effects.

In calculations of $2\nu\beta\beta$ decay [47,48,50] it has been shown that the $2\nu\beta\beta$ decay matrix elements calculated within *pn* QRPA strongly depend on the strength of the particle-particle interaction g_{pp}. We have therefore calculated the SUSY $0\nu\beta\beta$ decay matrix elements as a function of g_{pp} for all isotopes considered in the present work. A typical example is shown in Fig. 3. Compared to the

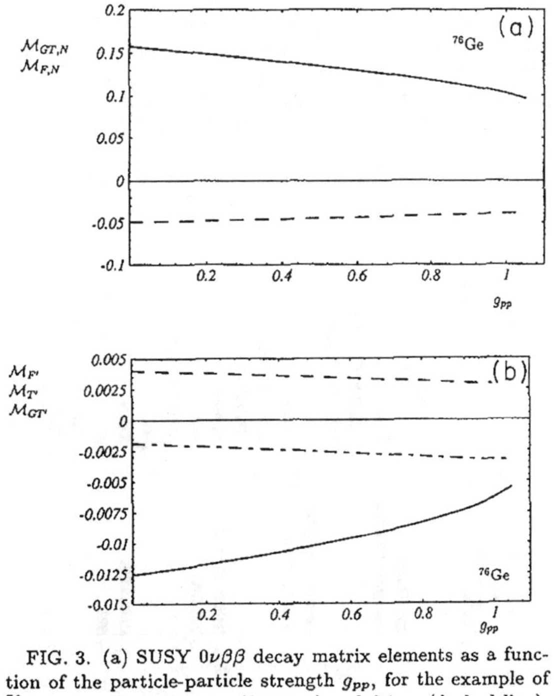

FIG. 3. (a) SUSY $0\nu\beta\beta$ decay matrix elements as a function of the particle-particle strength g_{pp}, for the example of ^{76}Ge. Shown are $M_{\mathrm{GT},N}$ (full line) and $M_{F,N}$ (dashed line). (b) As for (a), but for $\mathcal{M}_{\mathrm{GT}'}$ (full line), $\mathcal{M}_{F'}$ (dashed line), and $\mathcal{M}_{T'}$ (dash-dotted line).

results of $2\nu\beta\beta$ decay calculations only a very weak dependence of the matrix elements on g_{pp} is found. A variation of g_{pp} within ± 2 standard deviations from its fitted value [50] changes the SUSY matrix elements only by about 10%.

The explanation of this weak sensitivity of matrix elements on g_{pp} is similar to the one discussed in $0\nu\beta\beta$ decay calculations for light neutrino exchange [51]. Because of the selection rules for the Gamow-Teller operator, in $2\nu\beta\beta$ decay the intermediate nuclear states are 1^+ states, exclusively. On the other hand, higher multipoles also contribute to the $0\nu\beta\beta$ matrix elements in Eqs. (67)–(71) similarly to the above-mentioned case of the light neutrino exchange mechanism. A typical example of a multipole decomposition for $\mathcal{M}_{\mathrm{GT},N}$ and $\mathcal{M}_{F,N}$ is shown in Fig. 4. Quite large contributions up to high multipoles are found. Since the particle-particle force mainly affects the 1^+ states [51], SUSY $0\nu\beta\beta$ decay matrix elements are not very sensitive to the actual choice of g_{pp}.

Figure 5 shows the nuclear matrix elements for the example of ^{130}Te as a function of the momentum cutoff factor m_A. All matrix elements vanish in the limit $m_A \to \infty$ corresponding to pointlike nucleons. This is in agreement with the expectation since the calculation takes into account short range correlations, i.e., interactions at very small distances are cut off, and, for pointlike nucleons, matrix elements then approach zero. As seen from Fig. 5 matrix elements $\mathcal{M}_{\mathrm{GT},N}$, $\mathcal{M}_{F,N}$, and $\mathcal{M}_{T'}$ are rather stable against variations of m_A, within the region 650 MeV$\leq m_A \leq 1.5$ GeV, or so.

A larger sensitivity, however, is found for $\mathcal{M}_{\mathrm{GT}'}$ and $\mathcal{M}_{F'}$. Moreover, the sign change at around $m_A \simeq 650$ MeV is most likely an artifact and indicates that improvements in the treatment of nucleon finite size effects are desirable. Since $\mathcal{M}_{\tilde{q}}$ is dominated by $\mathcal{M}_{\mathrm{GT},N}$, it will not be very sensitive to variations of m_A as well. $\mathcal{M}_{\tilde{f}}$, on the other hand, has to be expected to depend more strongly on m_A.

The reason for the different dependence of the various nuclear matrix elements on m_A can only be understood considering in some detail the radial integrals, which need to be done in calculating SUSY $0\nu\beta\beta$ decay matrix elements. As usual in $\beta\beta$ decay calculations [2,50,53,47,48] the radial part of nuclear matrix elements is obtained using harmonic oscillator wave functions, $\Psi_{n,l}(r)$ multiplied by the short range correlation function, i.e., $\Psi_{n,l}(r) \to [1 + f(r)]\Psi_{n,l}(r)$. In addition, in performing the radial integration at each step, radial wave functions have to be "weighted" by the integrals over the momenta of virtual particles, $F_N(x_A)$, $F_4(x_A)$, or $F_5(x_A)$ for the corresponding matrix elements. Examples of such radial functions are shown in Fig. 6 for $n = l = 0$. In this figure (a) shows the calculation including $F_N(x_A)$ while (b) is the calculation containing $F_4(x_A)$. [The case for $F_5(x_A)$ is similar to $F_N(x_A)$ and will therefore not be discussed separately.] Figure 6 is obtained using the standard value of l_C and three values of m_A (see figure caption). Radial functions including $F_N(x_A)$ are reduced by increasing m_A, but always stay positive. This is to be contrasted with the situation encountered using $F_4(x_A)$, which is not positively defined, see (64). As can be seen from Fig. 6(b) the integrals over r contain a delicate cancellation between positive and negative contributions. It is exactly this cancellation, which results in the sensitivity of $\mathcal{M}_{\mathrm{GT}'}$ and $\mathcal{M}_{F'}$ on the parameter m_A, that makes the exact prediction of these two matrix elements so difficult.

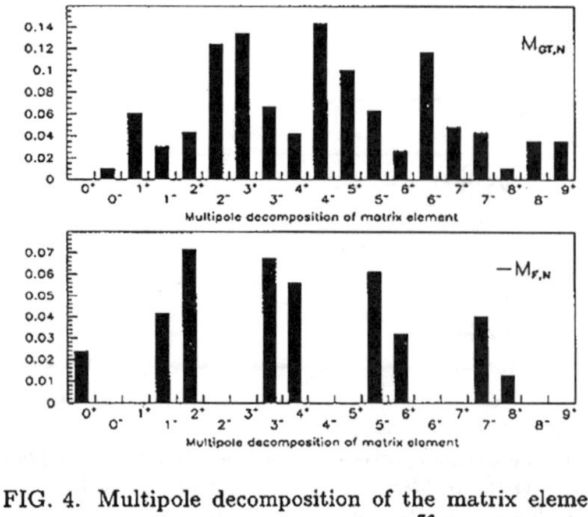

FIG. 4. Multipole decomposition of the matrix elements $\mathcal{M}_{\mathrm{GT},N}$ and $\mathcal{M}_{F,N}$ for the example of ^{76}Ge. Plotted are $10 \times \mathcal{M}_{\mathrm{GT},N}$ and $-10 \times \mathcal{M}_{F,N}$ for convenience.

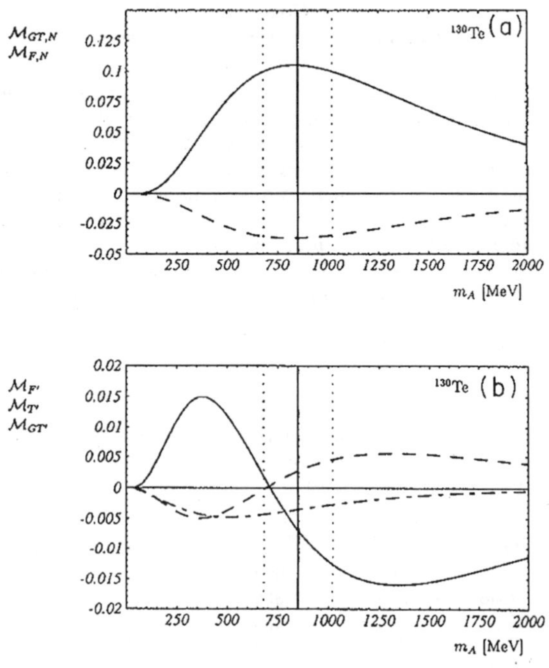

FIG. 5. (a) Matrix elements $\mathcal{M}_{\mathrm{GT},N}$ (full line) and $\mathcal{M}_{F,N}$ (dashed line) for ^{130}Te as a function of the momentum cutoff factor m_A (MeV). (b) As for (a), but for $\mathcal{M}_{\mathrm{GT}'}$ (full line), $\mathcal{M}_{F'}$ (dashed line), and $\mathcal{M}_{T'}$ (dash-dotted line).

Figure 7 then plots matrix elements as a function of the correlation length l_c for the example of ^{150}Nd. Increasing the correlation length decreases matrix elements, which have to vanish in the limit $l_C \to \infty$. As can be read off, a 20% change in l_c from our adopted standard value of $l_C = 0.7$ fm typically changes calculated $\mathcal{M}_{\mathrm{GT},N}$ and $\mathcal{M}_{F,N}$ by about (20–30)%. Similar to the case when the cutoff factor is varied, $\mathcal{M}_{\mathrm{GT}'}$ and $\mathcal{M}_{F'}$ are found to be more sensitive to changes in l_c. Even though we do not think a correlation length of smaller than, say, $l_C \leq 0.5$ fm is reasonable, again Fig. 7 shows that $\mathcal{M}_{\mathrm{GT}'}$ and $\mathcal{M}_{F'}$ are much more difficult to predict than the other matrix elements. [The cancellation of negative and positive parts in the radial integrals for matrix elements including $F_4(x_A)$ is also found when varying l_C.]

Since changes in all three parameters can lead to both larger and smaller matrix elements, the influence of a simultaneous change of these cannot be estimated by simply adding up the individual errors. To estimate the total uncertainty, we have therefore calculated $\mathcal{M}_{\tilde{q}}$ and $\mathcal{M}_{\tilde{f}}$ at 125 grid points in this three-dimensional parameter space. (Parameter variations are within ±20% of their standard values for l_c and m_A and within 2 standard deviations for g_{pp}.) From this set we then calculated the "mean" matrix elements, the corresponding standard deviations and extreme values.

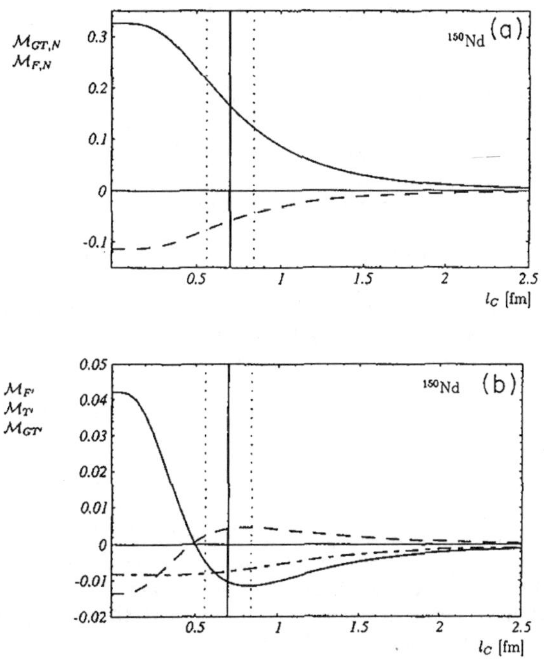

FIG. 7. (a) Matrix elements $\mathcal{M}_{\mathrm{GT},N}$ (full line) and $\mathcal{M}_{F,N}$ (dashed line) for ^{150}Nd as a function of the correlation length l_C (fm). (b) As for (a), but for $\mathcal{M}_{\mathrm{GT}'}$ (full line), $\mathcal{M}_{F'}$ (dashed line), and $\mathcal{M}_{T'}$ (dash-dotted line).

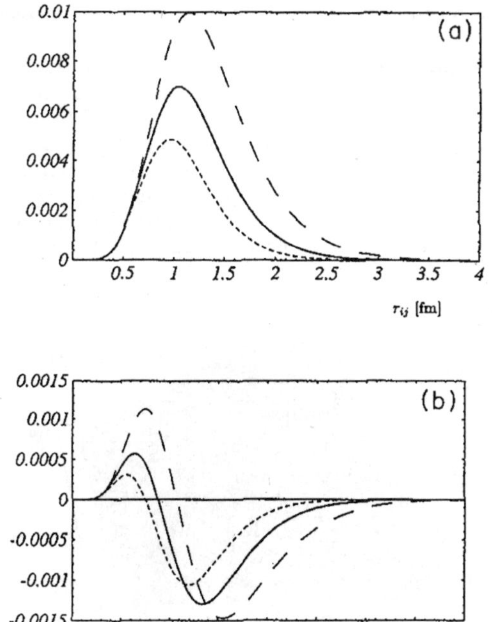

FIG. 6. Harmonic oscillator radial wave functions for $n = l = 0$ and three different values of the cutoff factor m_A, including the short range correlation function. The dashed line corresponds to $m_A = 0.7$ GeV, while the full line is for $m_A = 0.85$ GeV and the short-dashed line is obtained for $m_A = 1.0$ GeV. Normalization is arbitrary. (a) Calculated with $F_N(x_A)$ and (b) calculated with $F_4(x_A)$. For discussion see text.

In the case of $\mathcal{M}_{\tilde{q}}$ these mean matrix elements are always very similar to those calculated for the standard values of input parameters (typical differences up to a few %). The standard deviations for $\mathcal{M}_{\tilde{q}}$ and its extreme values deviate typically less than 20% and 50%, respectively, from the mean.

The situation is different in the case of $\mathcal{M}_{\tilde{f}}$. As discussed above, $\mathcal{M}_{\mathrm{GT}'}$ shows a larger sensitivity to model parameter variations than do other matrix elements. Moreover, in the case of $\mathcal{M}_{\tilde{f}}$ there occurs a destructive interference between contributions from $\mathcal{M}_{F,N}$, $\mathcal{M}_{T'}$, and $\mathcal{M}_{\mathrm{GT}'}$. In certain regions of the model parameter space $\mathcal{M}_{\mathrm{GT}'}$ tends to cancel the contributions from the other two matrix elements. The conclusion therefore unfortunately must be that $\mathcal{M}_{\tilde{f}}$ is considered to carry a large uncertainty: At extreme combinations of parameters $\mathcal{M}_{\tilde{f}}$ is smaller than (larger than) in the standard calculation by up to a factor of 5 (2).

To summarize this discussion, it can be stated that while $\mathcal{M}_{\tilde{q}}$ can be reliably calculated, the numerical value of $\mathcal{M}_{\tilde{f}}$ must be considered rather uncertain. Fortunately we will see in the next section that only the $\mathcal{M}_{\tilde{q}}$ matrix element is relevant for the extraction of the R_p MSSM parameters from experimental data on $0\nu\beta\beta$ decay.

VI. CONSTRAINTS ON THE $R\!\!\!/_p$ MSSM PARAMETER SPACE

In the following we will analyze constraints on the $R\!\!\!/_p$ MSSM parameter space using the experimental half-life limit of ^{76}Ge, recently measured by the Heidelberg-Moscow Collaboration [6] (for constraints derived from experiments on other isotopes, see below):

$$T_{1/2}^{0\nu\beta\beta}(^{76}\mathrm{Ge}, 0^+ \to 0^+) > 5.6 \times 10^{24}\,\mathrm{yr},\quad 90\%\ \mathrm{C.L.} \tag{75}$$

The theoretical expression for the inverse $0\nu\beta\beta$ half-life [see (A9)] corresponding to the reaction matrix element $\mathcal{R}_{0\nu\beta\beta}$ in Eq. (45) can be written as

$$[T_{1/2}^{0\nu\beta\beta}(0^+ \to 0^+)]^{-1} = G_{01}|(\eta_{\tilde{g}} + \eta_\chi)\mathcal{M}_{\tilde{q}} + (\eta_{\chi\tilde{e}} + \eta'_{\tilde{g}} - \eta_{\chi\tilde{f}})\mathcal{M}_{\tilde{f}} + \eta_N \mathcal{M}_N|^2 . \tag{76}$$

In the numerical analysis we will employ for definiteness the matrix elements for set A of the coefficients $\alpha^{(i)}$ (see Table I). Matrix elements of set B would only yield slightly more stringent bounds, but not change any of the arguments presented below.

Neglecting for the moment the contribution from Majorana neutrino exchange [the last term in Eq. (76)] we will first analyze the supersymmetric contribution alone.

Equations (75) and (76) define an excluded area within a five-dimensional $R\!\!\!/_p$ MSSM parameter space

$$\{\tan\beta, \mu, m_{\tilde{g}}, m_0, \lambda'_{111}\} . \tag{77}$$

These parameters, as explained in Sec. II, completely define the supersymmetric particle spectrum and their interactions initiating $0\nu\beta\beta$. However, there are too many free parameters to be defined from only one constraint [Eq. (75)]. Fortunately, the analysis gets considerably simplified realizing that the contributions from the five different lepton number violating parameters $\eta_{\tilde{g}}$, η_χ, $\eta_{\chi\tilde{e}}$, $\eta_{\chi\tilde{f}}$, $\eta'_{\tilde{g}}$ enter Eq. (76) with very different magnitudes. It can be seen from Eqs. (36)–(40) that

$$\eta_{\tilde{g}}, \eta'_{\tilde{g}} \gg \eta_\chi, \eta_{\chi\tilde{e}}, \eta_{\chi\tilde{f}} \ \text{ if } m_{\tilde{q}} \sim m_{\tilde{e}}, m_{\chi_i} \gtrsim 0.02 m_{\tilde{g}} \tag{78}$$

with the values (34) of the gauge coupling constants α_2 and α_s and for any field composition of the neutralino states χ_i [see Eq. (19)]. The last mass inequality in (78) is well satisfied even for a gluino mass $m_{\tilde{g}}$ as large as 500 GeV if $m_{\chi_i} \gtrsim 20$ GeV. The latter is guaranteed by the present experimental lower bound on the lightest neutralino $\chi \equiv \chi_1$ from the LEP experiments [60] $m_\chi \gtrsim 20$ GeV. Later on we consider the gluino dominance more carefully and will see that it holds for $m_{\tilde{g}}$ up to 1 TeV. Another interesting fact can be derived from the nuclear matrix elements $\mathcal{M}_{\tilde{q},\tilde{f}}$ of Table III. It can be seen that the value of $\mathcal{M}_{\tilde{q}}$ is much larger than the one of $\mathcal{M}_{\tilde{f}}$. Therefore, taking into account gluino dominance Eq. (78), it follows that the combination $\eta_{\tilde{g}}\mathcal{M}_{\tilde{q}}$ absolutely dominates in the half-life formula Eq. (76). The advantage of this fact is twofold. First, only three $R\!\!\!/_p$ MSSM parameters $m_{\tilde{q}}$, $m_{\tilde{g}}$, and λ'_{111} are involved in the numerical analysis. Second, as stated at the end of Sec. V, the nuclear model calculations for the matrix element $\mathcal{M}_{\tilde{q}}$ are much more reliable than those for $\mathcal{M}_{\tilde{f}}$. Thus, we are lucky to have a theoretically well controlled nuclear structure dependence in the dominant term of Eq. (76). Having this in mind we proceed with the numerical analysis.

Two limiting cases are interesting to discuss: (a) $m_{\tilde{d}_R} = m_{\tilde{u}_L} \equiv m_{\tilde{q}}$ and (b) $m_{\tilde{u}_L} \gg m_{\tilde{d}_R} \equiv m_{\tilde{q}}$ (the case $m_{\tilde{d}_R} \gg m_{\tilde{u}_L}$ is equivalent). The following bounds are obtained:

case (a)

$$\lambda'_{111} \leq 3.9 \times 10^{-4} \left(\frac{m_{\tilde{q}}}{100\ \mathrm{GeV}}\right)^2 \left(\frac{m_{\tilde{g}}}{100\ \mathrm{GeV}}\right)^{1/2} , \tag{79}$$

case (b)

$$\lambda'_{111} \leq 5.6 \times 10^{-4} \left(\frac{m_{\tilde{q}}}{100\ \mathrm{GeV}}\right)^2 \left(\frac{m_{\tilde{g}}}{100\ \mathrm{GeV}}\right)^{1/2} . \tag{80}$$

These two extreme cases differ just by a factor of $\sqrt{2}$, and for all other ratios of $(m_{\tilde{d}_R}/m_{\tilde{u}_L})$ limits between case (a) and case (b) are found. Motivated by the MSSM mass formulas, Eqs. (20)–(25), in the following discussion we will always assume $m_{\tilde{d}_R} = m_{\tilde{u}_L}$ which corresponds to case (a).

A graphical representation of this bound [Eq. (79)] is shown in Fig. 8. The area on the backside of the surface is forbidden by the $0\nu\beta\beta$ decay constraint (75). Shown is the mass range between 10 GeV and 1 TeV and a coupling range between 10^{-3} and 2×10^{-1}. It is interesting to note that employing the upper bounds $m_{\tilde{q},\tilde{g}} \lesssim 1$ TeV, motivated by the SUSY naturalness argument, one can obtain from (79) an upper bound for the $R\!\!\!/_p$ Yukawa

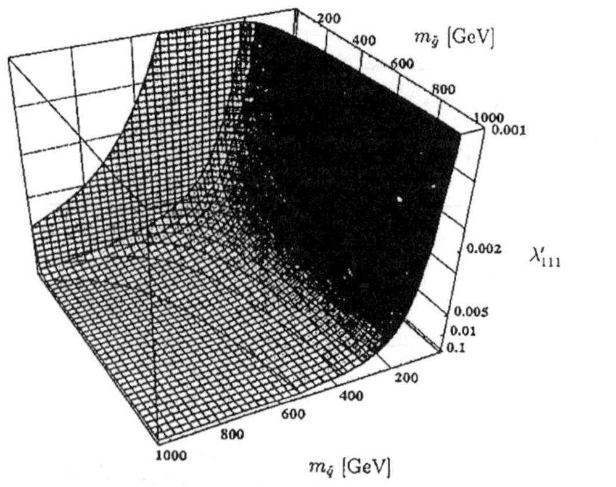

FIG. 8. Constraints from $0\nu\beta\beta$ decay on the squark $m_{\tilde{q}}$ and gluino $m_{\tilde{g}}$ masses and the R-parity-violating Yukawa coupling constant λ'_{111}. Values on the backside of the surface are forbidden by nonobservation of $0\nu\beta\beta$ decay. $m_{\tilde{d}_R} = m_{\tilde{u}_L} = m_{\tilde{q}}$ is assumed and the matrix element $\mathcal{M}_{\tilde{q}}$ for set A of the coefficient $\alpha^{(i)}$ has been used.

coupling constant

$$\lambda'_{111} \lesssim 0.124\,.$$

As mentioned before we have analyzed how the presence of a nonzero neutralino contribution to Eq. (76), proportional to η_χ,[3] affects the bounds derived from Eq. (79). From Eq. (76), one can expect that bounds derived from $\eta_{\tilde{g}}$ would be sharpened when η_χ is switched on. The large difference in magnitude, however, leads to the result that only for very light neutralinos will this effect be of importance. This is displayed in Fig. 9, where the excluded area in the ($m_{\tilde{q}}$, $m_{\tilde{g}}$, m_χ) space is shown for the example of $\lambda'_{111} = 10^{-2}$. For a neutralino heavier than a few GeV, the squark mass and gluino mass bounds are independent of the actual value of m_χ. As can be shown, this is true for *any* neutralino composition. Since such a light neutralino mass eigenstate is already excluded by the experiments at LEP [60], which tell us that $m_\chi \geq 20$ GeV, the conclusion is that the limits of Fig. 8 are practically independent of the neutralino mass.

Now another important question arises. How much will the obtained limits on superparticle masses and the $R\!\!\!/_p$ Yukawa coupling be affected by uncertainties of the nuclear structure calculation? From the definition of the lepton number violating parameters η Eqs. (36)–(40) and the $0\nu\beta\beta$ decay constraint Eqs. (75) and (76) the following conclusion can be made. Any shift of the nuclear matrix element $\mathcal{M}_{\tilde{q}}$ by the value $\Delta\mathcal{M}_{\tilde{q}}$, associated with the theoretical uncertainties, would change the extracted limits on λ'_{111}, gaugino $m_{\tilde{g},\chi}$, and squark $m_{\tilde{q},\tilde{e}}$ masses as

$$\frac{\Delta m_{\tilde{g},\chi}}{m_{\tilde{g},\chi}} \sim \frac{\Delta\mathcal{M}_{\tilde{q}}}{\mathcal{M}_{\tilde{q}}}\,, \tag{81}$$

$$\frac{\Delta\lambda'_{111}}{\lambda'_{111}} \sim 1 - \left(1 + \frac{\Delta\mathcal{M}_{\tilde{q}}}{\mathcal{M}_{\tilde{q}}}\right)^{-1/2}, \tag{82}$$

$$\frac{\Delta m_{\tilde{q}}}{m_{\tilde{q}}} \sim \left(1 + \frac{\Delta\mathcal{M}_{\tilde{q}}}{\mathcal{M}_{\tilde{q}}}\right)^{1/4} - 1\,. \tag{83}$$

From these equations one can see that the limits on squark masses $m_{\tilde{q}}$ and the $R\!\!\!/_p$ Yukawa coupling constant λ'_{111}, deduced from $0\nu\beta\beta$ decay depend only very weakly on the nuclear physics uncertainties. For instance, a change of $\mathcal{M}_{\tilde{q}}$ by even a factor of 2 changes the squark mass limit by less than 20%.

Up to now, we have not considered the influence of a possible contribution from Majorana neutrinos to $0\nu\beta\beta$ decay. In this paper we bound ourselves to the heavy Majorana neutrino (N) contribution, as discussed at the end of Sec. II. In Sec. IV it was shown that contributions to $0\nu\beta\beta$ decay from the SUSY mechanism and the ordinary neutrino mass mechanism add up coherently.

[3]Although in general all four neutralino mass eigenstates contribute to the $0\nu\beta\beta$ decay rate, it is sufficient to consider only the lightest neutralino with mass m_χ.

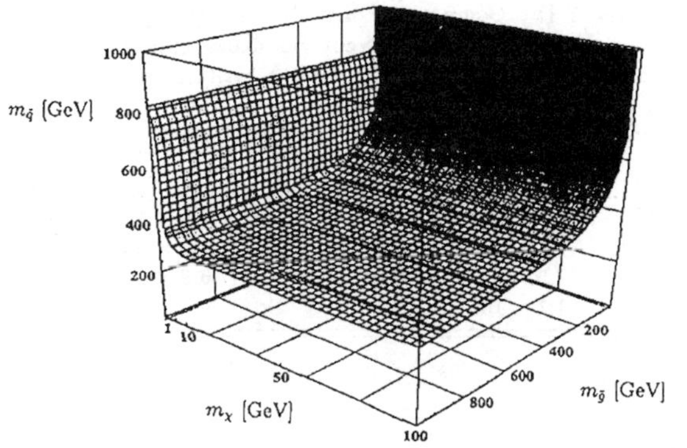

FIG. 9. Dependence of the squark and gluino mass limits derived from $0\nu\beta\beta$ decay on the actual value of the neutralino mass m_χ for $\lambda'_{111} = 10^{-2}$.

This is opposite of the results of [41], where a destructive interference was derived. The combined constraint in the ($m_{\tilde{q}}$, $m_{\tilde{g}}$, $\langle m_N\rangle$) space, where $\langle m_N\rangle$ is the *effective* heavy Majorana neutrino mass [see Eq. (41)], is shown in Fig. 10, for the case of $\lambda'_{111} = 10^{-2}$. It can be seen that limits on $\langle m_N\rangle$ also remain valid if one allows contributions from supersymmetric theories to $0\nu\beta\beta$ decay (or, vice versa $0\nu\beta\beta$ decay limits on the $R\!\!\!/_p$ MSSM are not sensitive to the actual value of the neutrino mass). This is an important result of the present work, since for a destructive interference between the neutrino mass and the SUSY mechanisms of $0\nu\beta\beta$ decay, it would always be possible to find regions in parameter space where *no* constraints from $0\nu\beta\beta$ decay could be derived. As an interesting byproduct of this analysis, we note that the current limit (75) on the $0\nu\beta\beta$ decay half-life of ^{76}Ge implies

$$\langle m_N\rangle \gtrsim 5.1 \times 10^7 \text{ GeV}\,. \tag{84}$$

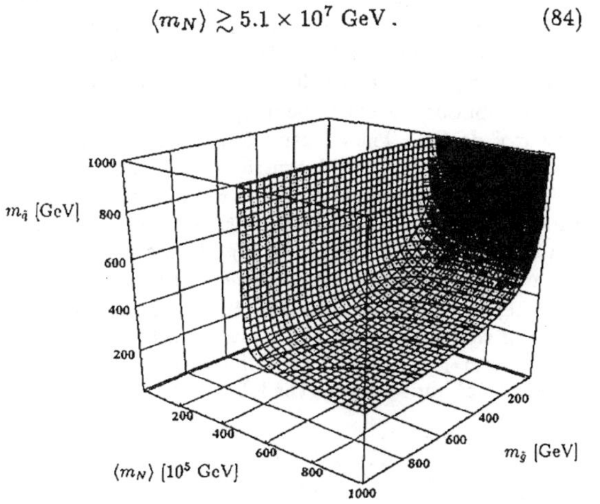

FIG. 10. Combined limits on supersymmetric particle masses $m_{\tilde{q},\tilde{g}}$ and the effective heavy Majorana neutrino mass $\langle m_N\rangle$ for $\lambda'_{111} = 10^{-2}$.

TABLE IV. Comparison of limits on supersymmetric parameters derived from different $\beta\beta$ decay experiments (see quoted references for half-life limits), written in the form of Eq. (79): $\lambda'_{111} \leq \epsilon(m_{\tilde{q}}/100\ \mathrm{GeV})^2(m_{\tilde{g}}/100\ \mathrm{GeV})^{1/2}$. Currently ^{76}Ge provides the most stringent limits.

^{A}Y	ϵ	Ref.	^{A}Y	ϵ	Ref.
^{76}Ge	3.9×10^{-4}	[6][a]	^{128}Te	4.9×10^{-4}	[61][b]
^{82}Se	1.1×10^{-3}	[62][b]	^{130}Te	1.1×10^{-3}	[63][a]
^{100}Mo	7.5×10^{-4}	[64][b]	^{136}Xe	6.8×10^{-4}	[65][a]
^{116}Cd	1.1×10^{-3}	[66][a]	^{150}Nd	9.7×10^{-4}	[67][a]

[a]90% C.L.
[b]68% C.L.

As mentioned above, currently ^{76}Ge provides the most stringent limits on $0\nu\beta\beta$ decay. For completeness, we have calculated the limits on supersymmetric parameters in the form of Eq. (79) for all isotopes considered in the present work. A summary of these limits, together with the references for the experimental half-life limits, are given in Table IV.

VII. COMPARISON OF CONSTRAINTS FROM $0\nu\beta\beta$ DECAY WITH OTHER EXPERIMENTS

It is interesting to compare limits on the $\not R_p$ MSSM parameters from $0\nu\beta\beta$ decay with those derived from other experiments. These constraints can be derived from low-energy processes involving virtual superparticles [15] or from direct accelerator searches for SUSY signals [14,25,26]. Since $0\nu\beta\beta$ decay is sensitive only to the first generation lepton number violating $\not R_p$ coupling, we will restrict the discussion to limits on λ'_{111}. Limits on other couplings might be found in the literature [14,15].

In Ref. [15] various low-energy processes have been analyzed. It was concluded that the most restrictive limit on λ'_{111} might be derived from charged current universality. The limitation follows from the fact that the existence of $\not R_p$ Yukawa coupling $\lambda'_{ijk}L_iQ_j\bar{D}_k$ gives an extra contribution to quark semileptonic decays (e.g., in nuclear β decay). The effective four-fermion interaction induced by the $\not R_p$ MSSM contribution in Fig. 11 has a $(V-A)\otimes(V-A)$ form identical to the one derived in the standard model. Therefore its contribution is equivalent to a shift in the Fermi constant G_F. If one assumes that only one $\not R_p$ operator has a sizable coupling constant, for instance, $\lambda'_{111}L_1Q_1D_1$ [15,9,26], then it leads to a violation of charged current (CC) universality [15] because the shift of G_F is different for different generations. The experimental limits impose the following bound at the 2σ level:

$$\lambda'_{111} \leq 0.03\left(\frac{m_{\tilde{d}_R}}{100\ \mathrm{GeV}}\right). \tag{85}$$

If one allows more than one $\not R_p$ operator to contribute the violation of CC universality is reduced and the above bound is weakened. Comparing (79) and (85) one can see that for masses in the range of 100 GeV, the bound (85) is less restrictive by nearly 2 orders of magnitude than the bound (79) derived from $0\nu\beta\beta$ decay.

Accelerator searches on supersymmetric particles are usually based on the assumption of R-parity conservation. In this case the LSP, which is assumed to be the neutralino χ, is stable and only weakly interacting. Therefore it escapes from the detector yielding the prominent missing transverse energy ($\not E_T$) signature of SUSY events. The Collider Detector at Fermilab (CDF) bounds on superparticle masses [68] rely essentially on $\not E_T$ signals and are not valid in the $\not R_p$ case. However, in Ref. [14] it was shown that limits on superparticle masses in the $\not R_p$ case can be derived using the CDF dilepton search data. The corresponding mass limit is

$$m_{\tilde{g},\tilde{q}} \geq 100\ \mathrm{GeV}, \tag{86}$$

independently of λ'_{111} if $\lambda'_{111} \geq 10^{-5}$ is assumed. For smaller values of λ'_{111}, the LSP has a negligible decay probability inside the detector and the $\not R_p$ signal has an $\not E_T$ pattern as in the R_p conserving case. Then the limits obtained in Ref. [68] can be applied for such small $\not R_p$ couplings. A recent calculation [27] shows that even larger masses than in Eq. (86) might be probed at the Tevatron in the near future.

There are in the literature other proposals for searching $\not R_p$ SUSY signals in future accelerator experiments. The isolated like-sign dilepton signature is proposed as a characteristic feature of the $\not R_p$ events at the LHC [29] and at LEP 200 [28].

Recently, searching for $\not R_p$ SUSY events in deep inelastic ep-scattering experiments with the ZEUS detector at HERA was proposed [25,26]. Two sets of signals could be identified with these events. They are the $\not R_p$ resonant squark production followed by the $\not R_p$ cascade decay of the neutralino [25] and the MSSM R_p conserving production of selectron and squark with their subsequent $\not R_p$ decay to ordinary matter [26]. It was advocated that searching for these signals provide HERA with a promising discovery potential for $\not R_p$ SUSY.

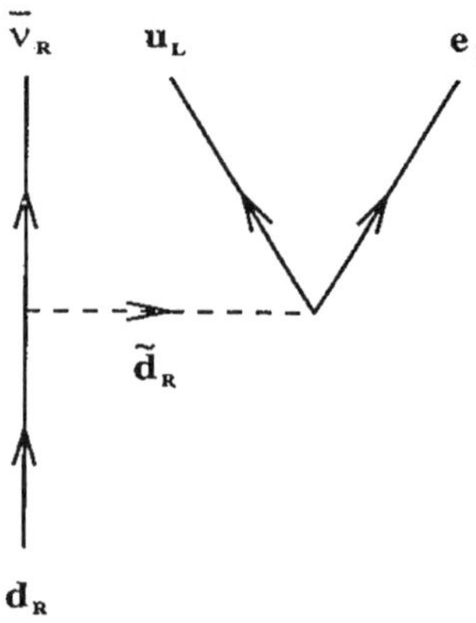

FIG. 11. Feynman graph for neutron decay in R_p-violating supersymmetric theories, see text.

A comparison of the bounds, which according to [25,26] might be reached with one year of HERA data, and the other limits discussed above with the limits we have obtained from the $0\nu\beta\beta$ decay constraint (75) is shown in Fig. 12 in the λ'_{111}-$m_{\tilde{q}}$ plane. The region enclosed by the thick line corresponds to the expected reach in the case of the $\not{R}_p$ resonant quark production [25], while the dotted vertical line is for $\not{R}_p$ decay [26] assuming $m_{\tilde{e},\tilde{\nu}} = 45$ GeV. (Limits in the mass reach of HERA are expressed as $m_{\tilde{e},\tilde{\nu}} + m_{\tilde{q}} \leq 205$ GeV in [26].) In the case of $0\nu\beta\beta$ decay, limits for two different values of the gluino mass are shown. It can be seen that even for a gluino mass as large as 1 TeV, which is marginal from the point of view of SUSY naturalness, the present double beta decay half-life limit yields the most restrictive bounds. (This result, however, does not touch upon $\not{R}_p$ couplings other than λ'_{111}.)

To summarize this discussion, $0\nu\beta\beta$ decay allows one to stringently restrict R_p-violating supersymmetric theories. We have shown that these limits are more stringent than those from other low-energy processes as well as those which can be derived from the HERA experiments by searching for resonant squark production [25] (see Fig. 12). Generally accelerator experiments allow one to probe even smaller values of the Yukawa couplings than indirect constraints, but are on the other hand restricted in the mass reached compared to $0\nu\beta\beta$ decay (and other low-energy constraints).

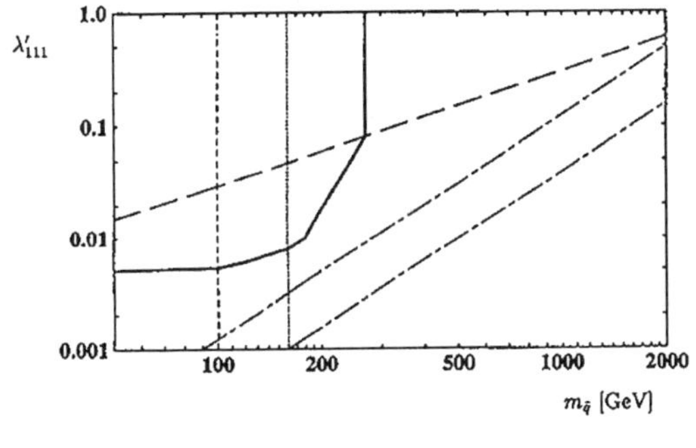

FIG. 12. Comparison of limits on R-parity violating supersymmetric theories from different experiments in the ($m_{\tilde{q}}$-λ'_{111}) plane. The dashed line is the limit from charged-current universality according to [15]. The dashed vertical line is the lower limit on squark masses in R_p-violating supersymmetric theories from the TEVATRON, according to [14]. The thick full line is the region which might be explored by HERA with about one year of data [25] in the study of resonant squark production, while the dotted vertical line is the mass reach of HERA probing $\not{R}_p$ violating decays [26]. (Note that the latter assumes $m_{\tilde{e},\tilde{\nu}} = 45$ GeV.) The two dash-dotted lines to the right are the limits obtained from nonobservation of the $0\nu\beta\beta$ decay of ^{76}Ge for gluino masses of (from left to right) $m_{\tilde{g}} = 1$ TeV, 100 GeV, respectively. The parameter regions to the upper left of the lines are forbidden by the different experiments. Note the double logarithmic scale of this plot.

VIII. CONCLUSIONS

We have investigated contributions to $0\nu\beta\beta$ decay from supersymmetric theories with explicit R-parity violation. The complete set of diagrams describing quark-lepton interactions at short distances has been considered. On this basis we have obtained the relevant low-energy effective Lagrangian in terms of nucleon and lepton currents. Then we have derived the transition operators describing $0^+ \to 0^+$ nuclear transitions induced by the $\not{R}_p$ MSSM interactions.

It has been found that contributions from the $\not{R}_p$ SUSY mechanism of $0\nu\beta\beta$ decay add coherently to the well-known neutrino mass mechanism. As a result, limits on the Majorana neutrino mass derived from $0\nu\beta\beta$ decay remain valid also if the supersymmetric contributions are taken into account.

We have calculated nuclear matrix elements of these transition operators within a realistic nuclear structure model for experimentally interesting isotopes. Special attention has been paid to the theoretical uncertainties of the nuclear matrix elements and to the question of how these uncertainties affect limits on the $\not{R}_p$ MSSM model parameters we extracted from $0\nu\beta\beta$. We were able to conclude that these limits depend only very weakly on nuclear physics uncertainties.

Using existing experimental lower bounds on the $0\nu\beta\beta$ decay half-life $T_{1/2}$, we then analyzed constraints on the $\not{R}_p$ MSSM parameters from $0\nu\beta\beta$ decay. The most restrictive limits are found from the current ^{76}Ge $0\nu\beta\beta$ decay experiment by the Heidelberg-Moscow Collaboration. We conclude that $0\nu\beta\beta$ decay imposes very restrictive bounds on the lepton-number-violating sector of $\not{R}_p$ SUSY models. We have presented these bounds as a three-dimensional exclusion plot in the space of the $\not{R}_p$ Yukawa coupling constant λ'_{111}, squark $m_{\tilde{q}}$, and gluino $m_{\tilde{g}}$ masses (Fig. 8) as well as a two-dimensional one in the λ'_{111}-$m_{\tilde{q}}$ plane (Fig. 12) which also shows bounds from other low and high energy processes.

We infer that the $0\nu\beta\beta$ bounds are able to compete with or are even more stringent than those derived from current and near future accelerator experiments. Particularly, they exclude the domain which is accessible for the experiments with the ZEUS detector at HERA in the $\not{R}_p$ resonant squark production mechanism [25].

ACKNOWLEDGMENTS

We thank V. A. Bednyakov, V. B. Brudanin, M. Lindner, and J. W. F. Valle for helpful discussions. We also thank H. Dreiner for bringing Ref. [26] to our attention. The research described in this publication was made possible in part by Grant No. GNTP 215 NUKLON from the Russian Ministry of Science and Technology (S.G.K.). M.H. would like to thank the Deutsche Forschungsgemeinschaft for financial support by Grants kl 253/8-1 and 446 JAP-113/101/0.

562 (1976).
[58] A. Balysh *et al.*, Phys. Lett. B **322**, 176 (1994).
[59] R. V. Reid, Ann. Phys. (N.Y.) **50**, 411 (1968).
[60] L. Roszkowski, Phys. Lett. B **252**, 471 (1990); K. Hidaka, Phys. Rev. D **44**, 927 (1991); Particle Data Group, K. Hikasa *et al.*, *ibid.* D **45**, S1 (1992).
[61] T. Bernatowicz *et al.*, Phys. Rev. Lett. **69**, 2341 (1992).
[62] S. R. Elliott *et al.*, Phys. Rev. C **46**, 1535 (1992).
[63] A. Alessandrello *et al.*, Phys. Lett. B **335**, 519 (1994).
[64] M. Alston-Garnjost *et al.*, Phys. Rev. Lett. **71**, 831 (1993).
[65] J.-L. Vuilleumier *et al.*, Phys. Rev. D **48**, 1009 (1993).
[66] F. A. Danevich *et al.*, Phys. Lett. B **344**, 72 (1995).
[67] M. K. Moe *et al.*, Prog. Part. Nucl. Phys. **32**, 247 (1994).
[68] D0 Collaboration, J. T. White, presented at the Ninth Topical Workshop on Proton-Antiproton Collider Physics, Tsukuba, Japan, 1993 (unpublished); CDF Collaboration, Y. Kato, *ibid.*

Z. Phys. C 74, 595–603 (1997)

ZEITSCHRIFT FÜR PHYSIK C

Theoretical Physics

Leptoquark/Squark interpretation of HERA events: Virtual effects in e^+e^- annihilation to hadrons

J. Kalinowski[1,2], R. Rückl[3,*], H. Spiesberger[4,*], P.M. Zerwas[1]

[1] Deutsches Elektronen-Synchrotron DESY, D-22607 Hamburg
[2] Institute of Theoretical Physics, Warsaw University, PL-00681 Warsaw
[3] Institut für Theoretische Physik, Universität Würzburg, D-97074 Würzburg
[4] Fakultät für Physik, Universität Bielefeld, D-33501 Bielefeld

Received: 12 March 1997

Abstract. In reference to the recently observed high Q^2, large x events in deep-inelastic positron–proton scattering at HERA, various leptoquark and supersymmetric scenarios are discussed. We study the impact of virtual leptoquark or R-parity breaking squark exchange as well as generic contact interaction on the production of quark–antiquark pairs in e^+e^- annihilation, in particular at LEP2.

1 Introduction

The recent observation of events in deep-inelastic positron–proton scattering with very high Q^2 and large x at HERA [1] has refuelled speculations on physics beyond the Standard Model, in particular on low-mass leptoquark-type particles. Such particles had been suggested a long time ago in a variety of physical scenarios: Pati–Salam SU(4) unification of quarks and leptons [2], grand unified theories such as SU(5) or E_6 [3], and composite models [4]. Moreover, in supersymmetric theories squarks couple to lepton–quark pairs if the R-symmetry is broken in the trilinear couplings of the superfields [5, 6]. Vector leptoquarks in grand unified theories with both lepton-quark and diquark couplings must be very heavy to suppress proton decay; certain scalar leptoquarks in GUT multiplets could nevertheless be relatively light [7] (disregarding the notorious hierarchy problem for the time being). Squarks in supersymmetric theories should naturally be expected in the mass range of a few hundred GeV.

A general classification of these novel states[1] has been presented in [8]. In this analysis the couplings of leptoquarks to lepton–quark pairs are assumed to be baryon- and lepton-number conserving in order to avoid rapid proton decay, family diagonal to exclude FCNC processes beyond the CKM mixing, and chiral to preserve the helicity suppression in leptonic pion decay. Moreover, the couplings are taken dimensionless and all interactions are assumed to respect the $SU(3)_C \times SU(2)_L \times U(1)_Y$ symmetry of the Standard Model. The allowed states can be classified according to spin, weak isospin and fermion number. We adopt the notation of [9] to conform with the notation generally employed in experimental papers: vector leptoquarks are denoted by V_I, scalar leptoquarks by S_I; isomultiplets with different hypercharges are distinguished by a tilde.

For convenience the nine possible states of scalar and vector leptoquarks are listed in Table 1. Leptoquarks in the upper (lower) part of Table 1 carry fermion number $F = 2$ ($F = 0$). The couplings are denoted generically by g_R or g_L with R, L refering to the chirality of the lepton. Each state can couple with different strength; for simplicity, the additional indices are suppressed. In principle the two scalar states S_0 and $S_{1/2}$ and the two vector states V_0 and $V_{1/2}$ could have both chiral g_R and g_L couplings at the same time; however, since the product of the two couplings is constrained very strongly by rare decays [10, 11, 12, 13], we assume only one of the two couplings to be non-zero. The special type of leptoquark that does not induce proton decay (as a result of its quantum numbers) and is compatible with the renormalization of the electroweak mixing angle from the symmetry value 3/8 at the GUT scale down to $\sin^2 \Theta_W \cong 0.23$ at the electroweak scale [7], is marked by an asterisk. Only a small subset of all these states is realized in supersymmetric theories with R-parity breaking. Moreover, supersymmetry requires these states to have universal left-handed couplings to leptons.

If leptoquarks or squarks in R-parity breaking supersymmetric theories exist, a large variety of phenomena are expected to be observed experimentally. In electron/positron–proton collisions, these particles are produced as single resonances, with the rate determined by the strength of the LQ-l-q Yukawa couplings [8, 14]. Pair production is an important production mechanism for leptoquarks in proton–(anti)proton [15], electron–positron [16] and photon–photon collisions [17]. In these reactions, the size of the cross section is determined (*modulo* anomalous couplings and form-factor effects) by the color, the electric and the electroweak charges of a given leptoquark (if the Yukawa couplings

* Supported by Bundesministerium für Bildung, Wissenschaft, Forschung und Technologie, Bonn, Germany, Contracts 05 7BI92P (9) and 05 7WZ91P (0)

[1] We shall generically denote leptoquarks and squarks in R-parity breaking scenarios by LQ

Paper presented at Beyond the Desert 1997: Accelerator and Non-Accelerator Approaches 159

High Q^2 DIS at HERA and Squark Production

H. Dreiner†, M. Krämer† and P. Morawitz‡

† Rutherford Laboratory, Oxfordshire, UK

‡ Imperial College, London, UK

Abstract. We discuss the high-Q^2 HERA anomaly in terms of resonance squark production via supersymmetric R-parity violating operators. Since the announcement by the HERA experiments several indirect bounds as well as the direct bounds from the Tevatron have been improved. We combine the CDF and D0 bounds and determine their gluino mass dependence. We use all the new bounds to update the possible solutions. We find the two-squark solution is no longer viable.

1. Introduction

Both experiments at HERA have reported an excess in their measured neutral current deep inelastic scattering (NC DIS) cross sections at high Q^2 [1]. Including the more recent 1997 data, the excess still persists [2], though slightly diminished. For $Q^2 > 20,000\,GeV^2$ the observed and the Standard Model combined H1 and ZEUS cross sections are $\sigma^{obs}(Q^2 > 20,000\,GeV^2) = 0.30\,^{+0.092}_{-0.076}\,pb$, $\sigma^{SM}(Q^2 > 20,000\,GeV^2) = 0.161\,pb$. The excess cross section is given by

$$\sigma^{ex}_{NC}(Q^2 > 20,000\,GeV^2) = (0.139\,^{+0.092}_{-0.076})\,pb. \tag{1}$$

In the following we shall assume this excess is due to the resonant production of scalar quarks via supersymmetric R-parity violating Yukawa couplings [3]. For these proceedings we shall consider the resonant masses

$$M_{\tilde{q}} = (200 \pm 10)\,GeV. \tag{2}$$

PHYSICAL REVIEW D, VOLUME 60, 115007

Grand unified theory constrained supersymmetry and neutrinoless double β decay

Andrzej Wodecki and Wiesław A. Kamiński
Department of Theoretical Physics, Maria Curie–Sklodowska University, PL-20 031 Lublin, Poland

Fedor Šimkovic
Department of Nuclear Physics, Comenius University, Mlynská dolina F1, SK-842 15 Bratislava, Slovakia
(Received 4 March 1999; published 5 November 1999)

We analyze the contributions to the neutrinoless double β decay ($0\nu\beta\beta$ decay) coming from the grand unified theory (GUT) constrained minimal supersymmetric standard model (MSSM) with trilinear R-parity breaking. We discuss the importance of two-nucleon and pion-exchange realizations of the quark-level $0\nu\beta\beta$-decay transitions. In this context, the questions of reliability of the calculated relevant nuclear matrix elements within the renormalized quasiparticle random phase approximation for several medium and heavy open-shell nuclei are addressed. The importance of gluino and neutralino contributions to $0\nu\beta\beta$ decay is also analyzed. We review the present experiments and deduce limits on the trilinear R-parity breaking parameter λ'_{111} from the nonobservability of $0\nu\beta\beta$ decay for different GUT constrained supersymmetry scenarios. In addition, a detailed study of limits on the MSSM parameter space coming from the $B\rightarrow X_s\gamma$ processes by using the recent CLEO and OPAL results is performed. Some studies in respect to the future $0\nu\beta\beta$-decay project GENIUS are also presented. [S0556-2821(99)02819-2]

PACS number(s): 12.60.Jv, 11.30.Er, 23.40.Bw

I. INTRODUCTION

The neutrinoless double beta decay ($0\nu\beta\beta$-decay) is forbidden in the standard model (SM) since it violates lepton number by two units ($\Delta L=2$). Therefore this decay is a sensitive probe for different aspects of physics beyond the SM. (For recent reviews see, e.g., Refs. [1,2].) Many generalization of the SM admit the violation of laws of the SM to a small extent. In this context, the nonobservability of $0\nu\beta\beta$ decay is used to constrain different extensions of the SM such as those with the left–right symmetry [3,4], leptoquarks [5], R-parity violating supersymmetric ($\not{R}_p$SUSY) models [6–12], and composite neutrinos [17,18].

The most widely discussed case in the literature has been the upper bound on the light effective electron neutrino Majorana mass $\langle m_\nu\rangle$ deduced from the experimental lower limit on the half-life of $0\nu\beta\beta$ decay [1,2]. Currently, the most restrictive limit on $\langle m_\nu\rangle$ is found from $0\nu\beta\beta$-decay in ^{76}Ge by the Heidelberg-Moscow Collaboration [19]: $\langle m_\nu\rangle\leq 0.4-1.3$ eV [1]. The uncertainty of this parameter is due to the ambiguity of $0\nu\beta\beta$-decay nuclear matrix elements. It is expected that the future double beta decay experiment GENIUS [20] based on 1 tonn of enriched ^{76}Ge would reach the sensitivity for $\langle m_\nu\rangle$ (0.01–0.001 eV).

In addition to the simplest and the best known mechanism of lepton number violation based on the mixing of massive Majorana neutrinos advocated by different variants of the grand unified theories (GUT) the R-parity violation proposed in the context of minimal supersymmetry extensions of the SM (MSSM) is becoming the most popular scenario for lepton number violation (see, e.g., Refs. [14,15]). The R parity is a multiplicative quantum number defined as $R=(-1)^{3B+L+2S}$ with B, L, and S being the baryon, lepton, and spin quantum numbers of the particle, respectively. Thus, all the SM particles have $R=+1$, while their superpartners have $R=-1$. We note that the R-parity conservation, which guarantee the baryon and lepton number conservation, is not required by gauge invariance or supersymmetry and might be broken explicitly or spontaneously at the Planck scale [16].

The R parity can be broken by involving the bilinear and trilinear terms in superpotential of the MSSM. The bilinear terms generate nonzero vacuum expectation value for the sneutrino field, which leads to the lepton number violating interactions based on the neutralino-neutrino and chargino-electron mixing [12]. The trilinear terms represent the interactions, which violate directly the lepton number and lepton flavor [8–12]. Both ways influence the low-energy phenomenology and therefore using some exotic processes such as the $0\nu\beta\beta$ decay one can impose limits on the parameters connected with the new physics.

Supersymmetric models with R-parity nonconservation ($\not{R}_p$MSSM) have been extensively discussed in the literature (see, e.g., Refs. [21,22]) and were also used for the study of R-parity violating trilinear term contribution to the $0\nu\beta\beta$ decay by Mohapatra [6] and Vergados [7]. The first calculations were concentrated only on the conventional two-nucleon mode of $0\nu\beta\beta$ decay, which assumes direct interactions between quarks of two-decaying neutrons [6–9]. A detailed study of this mechanism was carried out in Ref. [8]. By using viable phenomenological assumptions about some of the fundamental parameters of the $\not{R}_p$MSSM (e.g., the ansatz of universal sparticle masses and that the lightest neutralino is B-ino-like) it was found that the limit on the R-parity violating first generation Yukawa coupling λ'_{111} derived from the observed absence of the $0\nu\beta\beta$ decay is more stringent than the corresponding limit expected from the forthcoming accelerator experiment at the DESY ep collider HERA. The authors end up with the conclusion that the gluino exchange R-parity violating mechanisms of the $0\nu\beta\beta$ decay dominates over the neutralino ones [8].

Another scenario for reduction of a number of supersymmetric parameters associated with the limit on λ'_{111} has been outlined in Ref. [9]. The authors implemented relations

RAPID COMMUNICATIONS

PHYSICAL REVIEW C VOLUME 59, NUMBER 3 MARCH 1999

Limits on R-parity nonconservation from neutrinoless double β decay in the minimal supersymmetric standard model with gauge mediated breaking

A. Wodecki and Wiesław A. Kamiński*

Theoretical Physics Department, Maria Curie-Skłodowska University, Radziszewskiego 10, 20-031 Lublin, Poland

(Received 6 October 1998)

The R-parity nonconservation phenomenology within the minimal supersymmetric standard model with gauge mediated supersymmetry breaking has been studied. New constraints on the lepton number violating constant λ'_{111} were imposed by nonobservability of the neutrinoless double β decay. We have found that λ'_{111} depends strongly on the effective supersymmetry breaking scale Λ only and deduced limits which are much stronger than those previously found in literature. [S0556-2813(99)50103-9]

PACS number(s): 24.80.+y, 12.60.Jv, 11.30.Er, 23.40.Bw

Gauge-mediated theories of supersymmetry (SUSY) breaking (GMSB) have attracted a great deal of attention recently because of their high predictability, natural solution of the flavor problem, new phenomenology, and much less free parameters compared to the minimal supersymmetric standard model (MSSM), where SUSY breaking is mediated by the gravitational interaction [1–8]. In the GMSB models SUSY breaking is transmitted to the superpartners of quarks, leptons, and gauge bosons via the usual SU(3)×SU(2)×U(1) gauge interactions and occurs at the scale of the order $M_{\mathrm{SUSY}} \sim 10^5$ GeV. Gauginos and sfermions acquire their masses through interactions with the messenger sector at the one-loop and two-loop levels, respectively, resulting in different phenomenology of the low-energy world from the MSSM one. In these models flavor-diagonal sfermions mass matrices are induced in a rather low-energy scale, therefore they supply us with a very natural mechanism of suppressing flavor changing neutral currents (FCNC). Moreover, since the soft masses arise as gauge charges squared, the sizable hierarchy proportional to the gauge quantum numbers appears among the superpartner masses. As in MSSM the electroweak symmetry breaking (EWSB) is also driven by negative radiative corrections to the up-type Higgs mass resulting from the large top-quark Yukawa coupling and stop masses. Thus, with EWSB constraints the minimal GMSB models are highly constrained and strongly predictive. Recently renewed interest in GMSB [6,7] is then understood.

Motivated by the above features of the GMSB models we study the R-parity breaking phenomenology of MSSM and propose to use nonstandard processes like the neutrinoless double β decay ($0\nu\beta\beta$) for deducing the limits on some R-parity breaking constants. In the previous studies such estimates were performed in the framework of MSSM with supergravity mediated SUSY breaking by means of additional assumptions relating sfermion and gauginos masses [9,10] or using the GUT constraints [11]. Within the GMSB models one can find quantitatively new constraints, which may serve as a hint for further accelerator searches of R-parity breaking.

The R parity imposed on MSSM can be explicitly violated by the bilinear [13] and trilinear [12] terms in the superpotential. The trilinear terms lead to the lepton number and flavor violation, while the bilinear terms generate the nonzero vacuum expectation values for the sneutrino fields $\langle\tilde{\nu}_l\rangle$, causing neutrino-neutralino mixing and electron-chargino mixing. Because of the lepton number violation, supersymmetric models with the R parity broken may serve as a basis for the description of some exotic nuclear processes. Thus, from experimental bounds for such processes, e.g., from nonobservability of $0\nu\beta\beta$, one can impose limits on the R-parity breaking parameters. It has been found that this procedure is a powerful tool for such constraints [9–11].

To obtain the effective Lagrangian of the $0\nu\beta\beta$ decay for the discussed mechanism we have to write the complete superpotential W including R-parity conserving W_0 and breaking parts $W_{\not R}$:

$$W_0 = h^U_{ij}\hat{Q}_i\hat{H}_u\hat{u}^c_j + h^D_{ij}\hat{Q}_i\hat{H}_d\hat{d}^c_j + h^E_{ij}\hat{L}_i\hat{H}_d\hat{e}^c_j + \mu\hat{H}_d\hat{H}_u, \tag{1}$$

$$W_{\not R} = \lambda_{ijk}\hat{L}_i\hat{L}_j\hat{u}^c_k + \lambda'_{ijk}\hat{L}_i\hat{Q}_j\hat{d}^c_k + \lambda''_{ijk}\hat{u}^c_i\hat{d}^c_j\hat{d}^c_k + \mu_j\hat{L}_j\hat{H}_u. \tag{2}$$

In Eqs. (1) and (2) $\hat{Q},\hat{L}$ denote the quark and lepton SU(2) doublet superfields and $\hat{u}^c,\hat{d}^c,\hat{e}^c$ are corresponding SU(2) singlets. The Higgs superfields $\hat{H}_u,\hat{H}_d$ contain scalar components giving mass to the up-type and down-type quarks and leptons. In the R-parity breaking part we set $\lambda_{ijk}=\lambda''_{ijk}=0$ to avoid the unsuppressed proton decay. Since supersymmetry in the low-energy world is broken, the Lagrangian of the theory is supplemented with the "soft" supersymmetry breaking, both R-parity breaking and conserving, terms:

$$\begin{aligned}-\mathcal{L}_{\mathrm{soft}} = {} & (A^U_{(ij)}h^U_{ij}\tilde{Q}_iH_u\tilde{u}^c_j + A^D_{(ij)}h^D_{ij}\tilde{Q}_iH_d\tilde{d}^c_j + A^E_{(ij)}h^E_{ij}\tilde{L}_iH_d\tilde{e}^c_j \\ & + \mathrm{H.c.}) + B\mu(H_dH_u + \mathrm{H.c.}) + m^2_{H_d}|H_d|^2 + m^2_{H_u}|H_u|^2 \\ & + m^2_{\tilde{L}}|\tilde{L}|^2 + m^2_{\tilde{e}^c}|\tilde{e}^c|^2 + m^2_{\tilde{Q}}|\tilde{Q}|^2 + m^2_{\tilde{u}^c}|\tilde{u}^c|^2 + m^2_{\tilde{d}^c}|\tilde{d}^c|^2 \\ & + \left(\frac{1}{2}M_1\bar{\psi}_B\psi_B + \frac{1}{2}M_2\bar{\psi}^a_W\psi^a_W + \frac{1}{2}m_{\tilde{g}}\bar{\psi}^a_g\psi^a_g + \mathrm{H.c.}\right),\end{aligned} \tag{3}$$

*Fax: (48-81) 537 61 90. Electronic address: kaminski@neuron.umcs.lublin.pl

and

$$-\mathcal{L}_{\text{soft}_{R}}=\tilde{\lambda}_{ijk}\tilde{L}_i\tilde{L}_j\tilde{u}_k^c+\tilde{\lambda}'_{ijk}\tilde{L}_i\tilde{Q}_j\tilde{d}_k^c+\tilde{\lambda}''_{ijk}\tilde{u}_i^c\tilde{d}_j^c\tilde{d}_k^c+\tilde{\mu}_{2j}^2\tilde{L}_j\hat{H}_u+\tilde{\mu}_{1j}^2\tilde{L}_j\hat{H}_d, \quad (4)$$

where the tilde denotes the scalar partners of quark and lepton fields, while ψ_i's are the spin-$\frac{1}{2}$ partners of the gauge bosons.

In this Rapid Communication we concentrate on the role of the trilinear mechanism only, leaving treatment of both, i.e., bilinear and trilinear terms, simultaneously for the subsequent paper. It is done not only because of much more elaborated calculations but also due to some different physics connected with both R-parity breaking mechanisms (new spectrum of neutralinos and additional free parameters). Of course, dropping the bilinear term influences the obtained limits on the supersymmetry breaking parameters in forcing their upper values.

Using the R-breaking term W_{R} one can find the lepton number violating Lagrangian:

$$\mathcal{L}_{\lambda'_{111}}=-\lambda'_{111}\left[(\bar{u}_L,\bar{d}_L)\begin{pmatrix} e_R^c \\ -\nu_R^c\end{pmatrix}\tilde{d}_R^*+(\overline{e_L},\overline{\nu_L})d_R\begin{pmatrix}\tilde{u}_L^* \\ -\tilde{d}_L^*\end{pmatrix}+(\overline{u_L},\overline{d_L})d_R\begin{pmatrix}\tilde{e}_L^* \\ -\tilde{\nu}_L^*\end{pmatrix}+\text{H.c.}\right]. \quad (5)$$

Together with Lagrangians describing interactions among gluinos, neutralinos, fermions, and sfermions [8] we end up, after an integration out of heavy degrees of freedom, with the final effective Lagrangian:

$$\mathcal{L}_{\text{eff}}^{\Delta L_e=2}=\frac{G_F^2}{2m_p}\bar{e}(1+\gamma_5)e^c\left[\eta_{PS}J_{PS}J_{PS}-\frac{1}{4}\eta_T J_T^{\mu\nu}J_{T\mu\nu}\right], \quad (6)$$

where the color singlet hadronic currents[1] are $J_{PS}=\bar{u}^\alpha\gamma_5 d_\alpha+\bar{u}^\alpha d_\alpha$, $J_T^{\mu\nu}=\bar{u}^\alpha\sigma^{\mu\nu}(1+\gamma_5)d_\alpha$, with α as a color index and $\sigma^{\mu\nu}=(i/2)[\gamma^\mu,\gamma^\nu]$. The effective lepton-number violating parameters η_{PS} and η_T in Eq. (6) accumulate fundamental parameters of MSSM:

$$\eta_{PS}=\eta_{\chi\tilde{e}}+\eta_{\chi\tilde{f}}+\eta_\chi+\eta_{\tilde{g}}+7\eta'_{\tilde{g}}, \quad (7)$$

$$\eta_T=\eta_\chi-\eta_{\chi\tilde{f}}+\eta_{\tilde{g}}-\eta'_{\tilde{g}}, \quad (8)$$

where

$$\eta_{\tilde{g}}=\frac{\pi\alpha_s}{6}\frac{\lambda'^2_{111}}{G_F^2 m_{\tilde{d}_R}^4}\frac{m_P}{m_{\tilde{g}}}\left[1+\left(\frac{m_{\tilde{d}_R}}{m_{\tilde{u}_L}}\right)^4\right],$$

$$\eta_\chi=\frac{\pi\alpha_2}{2}\frac{\lambda'^2_{111}}{G_F^2 m_{\tilde{d}_R}^4}\sum_{i=1}^4\frac{m_P}{m_{\chi_i}}\left[\epsilon_{Ri}^2(d)+\epsilon_{Li}^2(u)\left(\frac{m_{\tilde{d}_R}}{m_{\tilde{u}_L}}\right)^4\right],$$

$$\eta_{\chi\tilde{e}}=2\pi\alpha_2\frac{\lambda'^2_{111}}{G_F^2 m_{\tilde{d}_R}^4}\left(\frac{m_{\tilde{d}_R}}{m_{\tilde{e}_L}}\right)^4\sum_{i=1}^4\epsilon_{Li}^2(e)\frac{m_P}{m_{\chi_i}},$$

$$\eta'_{\tilde{g}}=\frac{\pi\alpha_s}{12}\frac{\lambda'^2_{111}}{G_F^2 m_{\tilde{d}_R}^4}\frac{m_P}{m_{\tilde{g}}}\left(\frac{m_{\tilde{d}_R}}{m_{\tilde{u}_L}}\right)^2,$$

$$\eta_{\chi\tilde{f}}=\frac{\pi\alpha_2}{2}\frac{\lambda'^2_{111}}{G_F^2 m_{\tilde{d}_R}^4}\left(\frac{m_{\tilde{d}_R}}{m_{\tilde{e}_L}}\right)^2\sum_{i=1}^4\frac{m_P}{m_{\chi_i}}\left[\epsilon_{Ri}(d)\epsilon_{Li}(e)+\epsilon_{Li}(u)\epsilon_{Ri}(d)\left(\frac{m_{\tilde{e}_L}}{m_{\tilde{u}_L}}\right)^2+\epsilon_{Li}(u)\epsilon_{Li}(e)\left(\frac{m_{\tilde{d}_R}}{m_{\tilde{u}_L}}\right)^2\right]. \quad (9)$$

So, the effective Lagrangian depends on many supersymmetric parameters. In order to find their low-energy values in the GMSB model, we proceed as follows.

The minimal model of GMSB considered in this paper consists of messenger fields which transform as single flavors of $5+\bar{5}$ of SU(5), i.e., they are $SU(2)_L$ doublets l and $\bar{l}$ and SU(3) triplets q and $\bar{q}$. The SU(3)×SU(2)×U(1) gauge interactions of those fields communicate supersymmetry breaking from a hidden sector to the fields of the visible world via coupling to a gauge singlet S in the superpotential

$$W=\lambda_2 Sl\bar{l}+\lambda_3 Sq\bar{q}. \quad (10)$$

The lowest components and F components of the singlet superfield S acquire vacuum expectation values and set the overall scale for the messenger sector and SUSY breaking, respectively. For $F\neq 0$ the messenger sector loses its supersymmetry:

$$m_b=M\sqrt{1\pm\frac{\Lambda}{M}}, \quad (11)$$

$$m_f=M, \quad (12)$$

where $M=\lambda S, \Lambda=F/S$.

The messenger fields transmit the SUSY breaking to the visible sector through loops containing insertions of S and results with the gaugino and scalar masses

$$m_{\lambda_i}(M)=N_m g\left(\frac{\Lambda}{M}\right)\frac{\alpha_i(M)}{4\pi}\Lambda, \quad (13)$$

$$m^2(M)=2N_m f\left(\frac{\Lambda}{M}\right)\sum_{i=1}^3 k_i\left(\frac{\alpha_i(M)}{4\pi}\right)^2\Lambda^2. \quad (14)$$

In Eq. (14) we sum up over $SU(3)\times SU(2)_L\times U(1)_Y$ with $k_1=\frac{3}{5}(Y/2)^2, Y=2(Q-T_3), k_2=\frac{3}{4}$ for $SU(2)_L$ doublets and zero for singlets, and $k_3=\frac{4}{3}$ for $SU(3)_C$ triplets and zero for singlets. N_m is the number of messenger generations equal to 1 in the minimal model. The messenger threshold functions $g(x)$ and $f(x)$ are

[1]At this point we want to stress a need for a proper treatment of the color currents in Lagrangian (6) [11]. In the previous papers [9] this Lagrangian is erroneous.

$$g(x)=\frac{1+x}{x^2}\ln(1+x)+(x\to -x), \tag{15}$$

$$f(x)=g(x)-\frac{2(1+x)}{x^2}\left[Li_2\left(\frac{x}{1+x}\right)-\frac{1}{4}Li_2\left(\frac{x}{1+x}\right)\right]+(x\to -x). \tag{16}$$

It is worthwhile to note, that both functions $g(x)$ and $f(x)$ are closed to unity for $\Lambda \ll M$. They reach values 1.4 and 0.7 as $x\to 1.0$, respectively.

The formulas for gaugino and sfermion masses at the messenger scale M set up the boundary conditions for renormalization group equations (RGE's) used to calculate the low-energy spectrum. In our procedure we first evolve all the gauge and Yukawa couplings up to the scale M using the two-loop standard model RGE's below the mass threshold for SUSY particles (initially set up to $M_{\mathrm{SUSY}}=1$ TeV) and MSSM RGE's above that scale. We want to stress in this point that we do not use the full set of RGE's appropriate for the R-parity broken MSSM [14]. We estimate that an influence of the R-parity breaking constants on other quantities running from the messenger to the electroweak scale is marginal due to the smallness of λ's and the short "distance" between the M scale and m_Z. So, we limit our attention to RGE's for MSSM with R parity conserved [15].

When the couplings reach the scale M we find the values of gaugino and scalar masses and perform the RGE evolution of all the quantities back to m_Z. It is well known, that the running of $m^2_{H_u}$ is dominated by negative contribution from the top Yukawa coupling, which drives this parameter to a negative value at some scale and causes dynamic breaking of EWSB. This mechanism additionally allows us to express some GUT-scale free parameters in terms of low-energy ones.

The tree-level Higgs potential has the form

$$V_0=m_1^2|H_d^0|^2+m_2^2|H_u^0|^2+m_3^2(H_d^0H_u^0+\mathrm{H.c.})+\frac{g_1^2+g_2^2}{8}(|H_d^0|^2-|H_u^0|^2)^2, \tag{17}$$

where $m_{1,2}^2\equiv m^2_{H_{d,u}}+\mu^2$, $m_3^2\equiv B\mu$, and phases of fields are chosen so that $m_3^2<0$. Minimization of the full one-loop Higgs effective potential leads to the set of two equations:

$$|\mu|^2+\frac{m_Z^2}{2}=\frac{(m^2_{H_d}+\Sigma_d)-(m^2_{H_u}+\Sigma_u)\tan^2\beta}{\tan^2\beta-1}, \tag{18}$$

$$\sin 2\beta=\frac{-2B\mu}{(m^2_{H_u}+\Sigma_u)+(m^2_{H_d}+\Sigma_d)+2|\mu|^2}, \tag{19}$$

where Σ_u,Σ_d are given, e.g., in Ref. [16]. In order to minimize the stop contribution to the finite corrections we equal the minimization scale $Q_{\min}$ to the geometric mean of stop masses.

EWSB causes mixing among many particles. In particular, mixing in the gaugino sector results in four physical neutralinos $\chi_i (i=1,2,3,4)$,

$$\chi_i=\sum_{j=1}^{4} N_{ij}\psi_j. \tag{20}$$

In this equation the matrix N diagonalizing the matrix M_χ is real and orthogonal. Thus, for real N_{ij} neutralino masses are either positive or negative. If necessary, a negative mass can always be made positive by a redefinition of the relevant mixing coefficients $N_{ij}\to iN_{ij}$.

Similar mixing appears in the slepton and squark sector. After calculating the scalar particles masses and obtaining tree-level values for μ and $B\mu$, one is able to set up thresholds in RGE's for gauge and Yukawa couplings. Repeating the procedure by running everything up to the M scale and setting the M-scale conditions for soft parameters again, we proceed back to the m_Z scale. At this point we calculate new mass eigenstates, run everything up to $Q_{\min}$, minimize the one-loop corrected Higgs potential, and perform RGE's run of all the quantities to the m_Z scale. We iterate the procedure to obtain stable values of μ and $B\mu$.

Following the above scheme one can calculate the low-energy values of supersymmetric parameters appearing in the effective Lagrangian (6). Some of them can be restricted using experimental data for nonobservability of exotic nuclear processes. For example limits on λ'_{111} can be deduced from nonobservability of $0\nu\beta\beta$. In this case the half-life for the neutrinoless double β decay regarding all possibilities of hadronization of quarks can be written in the form

$$[T^{0\nu}_{1/2}(0^+\to 0^+)]^{-1}=G_{01}\left|\eta_T\mathcal{M}^{2N}_{\tilde{q}}+(\eta_{PS}-\eta_T)\mathcal{M}^{2N}_{\tilde{f}}+\frac{3}{8}\left(\eta_T+\frac{5}{8}\eta_{PS}\right)\mathcal{M}^{\pi N}\right|^2, \tag{21}$$

where $\mathcal{M}^{2N}_{\tilde{q}}$, $\mathcal{M}^{2N}_{\tilde{f}}$, and $\mathcal{M}^{\pi N}$ are defined in Ref. [10] and G_{01} is the standard phase space factor. The nuclear matrix elements used in these calculations were obtained for the bag model.

Using appropriate values for the nuclear matrix elements $\mathcal{M}^{2N}_{\tilde{q}}$, $\mathcal{M}^{2N}_{\tilde{f}}$, $\mathcal{M}^{\pi N}$ and the one-pion and two-pion contributions that dominate the 2-N mechanism [10] we could obtain the neutrinoless double β decay half-life as a function a few supersymmetric parameters entering formula (6) only. Our procedure limits the number of these free parameters to

Λ, M, $\tan\beta$, sgn(μ), and N_m. As the loop diagrams with messenger fields do not affect the A terms considerably, we can equal the common soft SUSY breaking parameter A_0 to zero at the M scale.

Setting the upper limit on the half-life of the neutrinoless double β decay in ^{76}Ge to 1.1×10^{25} y [17] we are able to study constraints imposed by experiment on the R-parity breaking parameter λ'_{111} as a function of the above-mentioned free parameters. We have found extremely weak dependence of λ'_{111} on M, $\tan\beta$ and sgn(μ). For example, a change of this parameter is less than 1% as one increases the value of $\tan\beta$ in the range 3–40. Contrary, a much more pronounced influence of the Λ parameter was detected. For the quantities commonly used in literature, λ'_{111}, $\lambda'_{111}/(m_{\tilde{q}}/100\text{ GeV})$ and $\lambda'_{111}/(m_{\tilde{q}}/100\text{ GeV})^2\ (m_{\tilde{g}}/100\text{ GeV})^{1/2}$, such a dependence is shown in Figs. 1(a)–(c). In the case of $\lambda'_{111}/(m_{\tilde{q}}/100\text{ GeV})^2(m_{\tilde{g}}/100\text{ GeV})^{1/2}$ one can observe the flat behavior for $\Lambda>70$ TeV that allows us to estimate the lepton number violating constant:

$$\frac{\lambda'_{111}}{(m_{\tilde{q}}/100\ \text{GeV})^2(m_{\tilde{g}}/100\ \text{GeV})^{1/2}}<3.2\times10^{-5}. \quad (22)$$

Comparison of this result with the constraints on the R-parity broken MSSM coming from other estimates [9,11] shows that limit (22) is approximately 1 order of magnitude stronger. The charged current universality violation combined with the gravity mediated MSSM imposes the bound $\lambda'_{111}\leq0.03(m_{\tilde{d}_R}/100\text{ GeV})$ [18] almost 2 order of magnitude weaker than the one obtained in the GMSB scenario. The difference comes from the different spectrums of superparticles in GMSB and gravity mediated MSSM. Because selectron masses compared to other sparticles masses are smaller within GMSB their contributions in Eq. (9) are increased. As a consequence, the limits on λ'_{111} are stronger. We also note that due to an increase of sparticles masses with the number of messenger generations N_m the strongest constraints come from the minimal model ($N_m=1$).

In conclusion, we have applied the experimental limits on the neutrinoless double β decay of ^{76}Ge to the analysis of the R-parity nonconservation phenomenology within the GMSB scenario. The presented approach serves as a consistent and free of *ad hoc* assumptions about mutual dependence of model parameters one. We have shown that the neutrinoless double β decay imposes strong limits on the lepton number violating constant λ'_{111}. We have also found that λ'_{111} depends significantly on Λ only. The obtained stringent limit on the λ'_{111} parameter may be helpful for planning future searches for R-parity breaking signatures in different experiments.

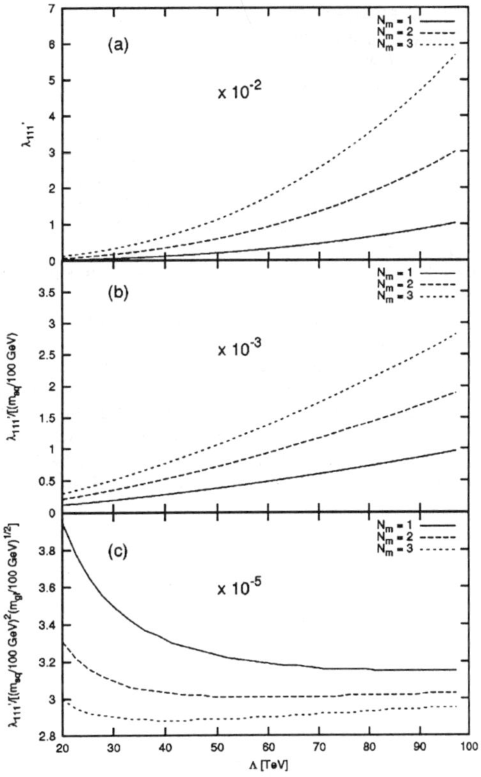

FIG. 1. Dependence of different representations of the λ'_{111} parameter on Λ. Three curves for different numbers of the messenger multiplets ($N_m=1,2,3$) are shown. Other free parameters are fixed as follows: $M=100$ TeV, $\tan\beta=20$, sgn(μ)$=+1$. For details see the text.

We are grateful to Dr. F. Simkovic for providing us with the nuclear matrix elements. This work was supported in part by the State Committee for Scientific Researches (Poland) Grant No. 2P03B00516.

[1] M. Dine, W. Fischler, and M. Srednicki, Nucl. Phys. **B189**, 575 (1981); S. Dimopoulos and S. Raby, *ibid.* **B192**, 353 (1981); M. Dine and W. Fischler, Phys. Lett. **110B**, 227 (1982); M. Dine and M. Srednicki, Nucl. Phys. **B202**, 328 (1982); L. Alvarez-Gaumé, M. Claudson, and M. Wise, *ibid.* **B207**, 96 (1982); C. Nappi and B. Ovrut, Phys. Lett. **113B**, 175 (1982).
[2] M. Dine and W. Fischler, Nucl. Phys. **B204**, 346 (1982); S. Dimopoulos and S. Raby, *ibid.* **B219**, 479 (1983).
[3] M. Dine and A. Nelson, Phys. Rev. D **48**, 1277 (1993); M. Dine, A. Nelson, and Y. Shirman, *ibid.* **51**, 1362 (1995); M. Dine, A. Nelson, Y. Nir, and Y. Shirman, *ibid.* **53**, 2658 (1996).
[4] S. Martin, Phys. Rev. D **55**, 3177 (1997).
[5] S. Dimopoulos, G. Giudice, and A. Pomarol, Phys. Lett. B **389**, 37 (1996); S. Dimopoulos, G. Dvali, and R. Ratazzi, *ibid.* **413**, 336 (1997).
[6] Z. Cahcko, B. Dutta, R. N. Mohapathra, and S. Nandi, Phys.

Rev. D **56**, 5466 (1997); A. Ghosal, A. Kundu, and B. Mukhopadhyaya, *ibid.* **56**, 504 (1997); Z. Tavartkiladze, Phys. Lett. B **427**, 65 (1998).

[7] S. Dimopoulos, S. Thomas, and J. D. Wells, Nucl. Phys. **B488**, 39 (1997).

[8] H. E. Haber and G. L. Kane, Phys. Rep. **117**, 75 (1985).

[9] M. Hirsch, H. V. Klapdor-Kleingrothaus, and S. G. Kovalenko, Phys. Rev. Lett. **75**, 17 (1995); M. Hirsch, H. V. Klapdor-Kleingrothaus, and S. G. Kovalenko, Phys. Rev. D **53**, 1329 (1996); A. Faessler, S. Kovalenko, F. Šimkovic, and J. Schwieger, Phys. Rev. Lett. **78**, 183 (1997).

[10] A. Faessler, S. Kovalenko, and F. Šimkovic, Phys. Rev. D **58**, 055004 (1998).

[11] A. Wodecki, W. A. Kamiński, and S. Pagerka, Phys. Lett. B **413**, 342 (1997).

[12] L. J. Hall and M. Suzuki, Nucl. Phys. **B231**, 419 (1984); G. G. Ross and J. W. F. Valle, Phys. Lett. **151B**, 375 (1985); R. Barbieri, D. E. Brahm, L. J. Hall, and S. D. Hsu, Phys. Lett. B **238**, 86 (1990); J. C. Ramao and J. W. F. Valle, Nucl. Phys. **B381**, 87 (1992); H. Dreiner and G. G. Ross, *ibid.* **B410**, 188 (1993); D. Comelli *et al.*, Phys. Lett. B **234**, 397 (1994); G. Bhattacharyya, D. Choudhury, and K. Sridhar, *ibid.* **355**, 193 (1995); G. Bhattacharyya and A. Raychaudhuri, *ibid.* **374**, 93 (1996); A. Y. Smirnov and F. Vissani, *ibid.* **380**, 317 (1996).

[13] M. A. Diaz, J. C. Romaõ, and J. V. Valle, Nucl. Phys. **B524**, 23 (1998); A. Akeroyd, M. A. Diaz, J. Ferrandis, M. A. Garcia-Jareno, and J. W. F. Valle, *ibid.* **B529**, 3 (1998); A. S. Joshipura and M. Nowakowski, Phys. Rev. D **51**, 2421 (1995); **51**, 5271 (1995); M. Nowakowski and A. Pilaftsis, Nucl. Phys. **B461**, 19 (1996).

[14] B. de Carlos and P. L. White, Phys. Rev. D **54**, 3427 (1996).

[15] V. Barger, M. S. Berger, and P. Ohmann, Phys. Rev. D **47**, 1093 (1993); G. L. Kane, C. Kolda, L. Roszkowski, and J. D. Wells, *ibid.* **49**, 6173 (1994); W. de Boer, Prog. Part. Nucl. Phys. **33**, 201 (1994).

[16] R. Arnowitt and P. Nath, Phys. Rev. D **46**, 3981 (1992); V. Barger, M. Berger, and P. Ohmann, *ibid.* **49**, 4908 (1994).

[17] Heidelberg-Moscow Collaboration, M. Gunther *et al.*, Phys. Rev. D **55**, 54 (1997); L. Baudis *et al.*, Phys. Lett. B **407**, 219 (1997).

[18] V. Barger, G. F. Guidice, and T. Han, Phys. Rev. D **40**, 2987 (1989).

VOLUME 75, NUMBER 12 PHYSICAL REVIEW LETTERS 18 SEPTEMBER 1995

New Vector-Scalar Contributions to Neutrinoless Double Beta Decay and Constraints on *R*-Parity Violation

K. S. Babu*
Bartol Research Institute, University of Delaware, Newark, Delaware 19716

R. N. Mohapatra
Department of Physics and Astronomy, University of Maryland, College Park, Maryland 20742
(Received 19 June 1995)

We show that in the minimal supersymmetric standard model (MSSM), with R-parity breaking as well as in the left-right symmetric model, there are new observable contributions to neutrinoless double beta decay ($\beta\beta_{0\nu}$) arising from hitherto overlooked diagrams involving the exchange of one W boson and one scalar boson. In particular, in the case of MSSM, the present experimental bounds on $\beta\beta_{0\nu}$ lifetime improves the limits on certain R-parity violating couplings by about 2 orders of magnitude. It is shown that similar diagrams also lead to enhanced rates for $\mu^- \to e^+$ conversion in nuclei, which are in the range accessible to ongoing experiments.

PACS numbers: 12.60.Jv, 11.30.Er, 12.60.Cn, 23.40.Bw

With the standard model of electroweak interactions brilliantly confirmed by a host of experiments, the search is on for the next level of physics at TeV or higher scales. There exist many interesting scenarios which address the various naturalness problems of the standard model framework. A generic feature of many of these scenarios is that conservation laws present in the standard model no longer remain valid. A typical conservation law that breaks down is the one corresponding to lepton-number symmetry. To make any theoretical headway, one needs to know the strength of the lepton-number-nonconserving interactions as well as any other kind of interactions that may accompany them.

It has been known for a long time [1] that neutrinoless double beta decay ($\beta\beta_{0\nu}$) is a very sensitive probe of lepton-number-violating terms in the Lagrangian such as the Majorana mass of the light neutrinos [2], right-handed weak couplings involving heavy Majorana nuetrinos [3,4], as well as Higgs [5] and other interactions such as those involving R-parity breaking in the supersymmetric model [6]. The reason why this observation is interesting is that the steadily improving experimental limits [7] on $\beta\beta_{0\nu}$ lifetime can then be translated into more stringent limits [8] on the parameters of these new physics scenarios. This is extremely valuable information to have in our search for physics beyond the standard model.

It is the goal of this Letter to point out another class of hitherto unnoticed contributions to $\beta\beta_{0\nu}$ decay in the following two classes of theories: (i) minimal supersymmetric models (MSSM) with R-parity violation and (ii) left-right symmetric models with a low mass W_R. These new contributions are of vector-scalar type in that they involve the exchange of a W_L together with a charged scalar boson with a virtual light neutrino. These contributions do not involve a helicity flip of the internal light neutrino, and, therefore, their amplitudes are enhanced relative to the ordinary contribution proportional to the light Majorana neutrino mass by roughly a factor of $p_F/m_\nu \sim (100\text{ MeV})/(1\text{ eV}) \sim 10^8$ (p_F is the Fermi momentum of the nucleons in the nuclei). Turning the argument around, the strength of such lepton-number-violating scalar boson interactions can be constrained to be $G_{\text{eff}} \leq 10^{-8} G_F$ from the $(\beta\beta)_{0\nu}$ process (modulo nuclear matrix element uncertainties), which indeed is a severe constraint. The effect turns out to be most dramatic on certain R-violating couplings in the MSSM in that it directly involves the R-violating couplings and superpartner masses. In the left-right models, although such contributions involve unknown Higgs boson mixings, they do lead to interesting bounds for a certain range of values for these parameters. We also point out that similar enhanced vector-scalar contributions exist in other rare processes such as $\mu^- \to e^+$ conversion in nuclei which turn out to be in the experimentally accessible range.

The new contributions arise from the combination of two effective four Fermi interactions of the following type which, as we will show later, can arise in several interesting classes of gauge models:

$$H_{\text{eff}} = \frac{G_F}{\sqrt{2}}[\bar{e}\gamma_\mu(1-\gamma_5)\nu_e\bar{u}\gamma^\mu(1-\gamma_5)d + \epsilon_1^{ee}\bar{d}(1-\gamma_5)u\nu_e^T C^{-1}(1-\gamma_5)e + \epsilon_2^{ee}\bar{d}(1-\gamma_5)\nu_e u^T C^{-1}(1-\gamma_5)e] + \text{H.c.} \quad (1)$$

In the above, the first term is the usual $(V-A)$ interaction; the other two are effective lepton-number-violating terms. In order to evaluate the matrix elements between nuclear states, we need to do Fierz reordering of the ϵ_2^{ee} term, which

 0031-9007/95/75(12)/2276(4)$06.00

casts it in the form

$$\frac{G_F}{2\sqrt{2}}\,\epsilon_2^{ee}(\overline{d}(1-\gamma_5)u\overline{e^c}(1-\gamma_5)\nu_e + \tfrac{1}{4}\overline{d}\sigma(1-\gamma_5)u\overline{e^c}(1-\gamma_5)\sigma_{\mu\alpha}\nu_e).$$

The parameters ϵ_i^{ee} ($i = 1,2$) characterize the new interactions that arise in a gauge model so that any limit on them translates into limits on the parameters on the theory leading to this interaction.

It is easy to see that the two interaction terms in Eq. (1) give contributions to $\beta\beta_{0\nu}$ decay which do not depend on the neutrino mass and involve a vector current at one hadronic vertex and a scalar current in the other [hence the name vector scalar; of course as just mentioned, the ϵ_2^{ee}-type scalar interaction of Eq. (1) after Fierz reordering generates also a tensor coupling]. The resulting effective $\Delta L = 2$ Hamiltonian is then given in momentum space by

$$H_{\Delta L=2} = G_F^2\Big(\Big(-\epsilon_1^{ee} + \frac{\epsilon_2^{ee}}{2}\Big)\overline{u}(1+\gamma_5)d\overline{u}\gamma_\mu(1-\gamma_5)d\overline{e}\gamma^\mu(1-\gamma_5)\frac{1}{\gamma\cdot q}C\overline{e}^T + \frac{\epsilon_2^{ee}}{8}\overline{u}\sigma^{\lambda\alpha}(1+\gamma_5)d\overline{u}\gamma_\mu(1-\gamma_5)d\overline{e}\gamma^\mu(1-\gamma_5)\frac{1}{\gamma\cdot q}\sigma_{\lambda\alpha}C\overline{e}^T\Big) + \text{H.c.} \tag{2}$$

In Eq. (2), q refers to the momentum of the internal light neutrino propagator. This effective Hamiltonian of course has to be evaluated between nuclear states. This gives rise to an effective neutrino potential as in the case of the neutrino mass contribution to $\beta\beta_{0\nu}$ decay with the difference that $\langle m_\nu/q^2\rangle_{\text{Nucl}}$ gets replaced by $\langle\gamma\cdot q/q^2\rangle_{\text{Nucl}}$ (and a different operator for the tensor coupling). To the best of our knowledge, such nuclear matrix elements for $\beta\beta_{0\nu}$ decay have not been evaluated in the literature. We therefore resort to a crude estimate and assume an average value of q to be equal to the Fermi momentum p_F of the nucleons in the nucleus ($\approx$100 MeV). The present upper limits on m_ν of about 1 eV then translates to an upper limit on the new interaction parameter ϵ as follows:

$$\epsilon_{1,2}^{ee} \leq 1\times 10^{-8}. \tag{3}$$

We realize that due to the crudeness of our estimate of the nuclear matrix element, the above upper limit is likely to be uncertain by perhaps a factor of 5 or so. Nevertheless, it is important to note that the limit on $\epsilon_{1,2}$, barring unforeseen nuclear suppression, is rather stringent and will imply important restrictions on the parameters of the gauge models leading to Eq. (1). Let us therefore proceed to the kind of gauge models where the last two terms in Eq. (1) can arise at low energies.

MSSM with R-parity violation.—As is well known, the MSSM can have explicit [9] violation of the R symmetry [defined by $(-1)^{3B+L+2S}$], leading to lepton-number-violating interactions in the low energy Lagrangian. The three possible types of couplings in the superpotential are [10]

$$W' = \lambda_{ijk}L_iL_jE_k^c + \lambda'_{ijk}L_iQ_jD_k^c + \lambda''_{ijk}U_i^cD_j^cD_k^c. \tag{4}$$

Here L, Q stand for the lepton and quark doublet superfields, E^c for the lepton singlet superfield, and U^c, D^c for the quark singlet superfields. i,j,k are the generation indices and we have $\lambda_{ijk} = -\lambda_{jik}$, $\lambda''_{ijk} = -\lambda''_{ikj}$. The SU(2) and color indices in Eq. (4) are contracted as follows: $L_iQ_jD_k^c = (\nu_i d_j^\alpha - e_i u_j^\alpha)D_{k\alpha}^c$, etc. The simultaneous presence of all three terms in Eq. (4) will imply rapid proton decay, which can be avoided by setting the $\lambda'' = 0$. In this case, the baryon number remains an unbroken symmetry while the lepton number is violated [11].

There are two types of vector-scalar contributions to $\beta\beta_{0\nu}$ and related $\Delta L = 2$ processes. These are shown in Figs. 1 and 2. The dominant contribution to $\beta\beta_{0\nu}$ in this model arises from Fig. 1, where the exchanged scalar particles are the $\tilde{b} - \tilde{b}^c$ pair. This leads to a contribution to ϵ_2^{ee} given by

$$\epsilon_2^{ee} \simeq \left(\frac{\lambda'_{113}\lambda'_{131}}{2\sqrt{2}\,G_F M_{\tilde{b}}^2}\right)\left(\frac{m_b}{M_{\tilde{b}^c}^2}\right)(\mu\tan\beta + A_b m_0). \tag{5}$$

Here A, m_0 are supersymmetry breaking parameters, while μ is the supersymmetric mass of the Higgs bosons. $\tan\beta$ is the ratio of the two Higgs vacuum expectation values and lies in the range $1 \leq \tan\beta \leq m_t/m_b \approx 60$. For the choice of all squark masses as well as μ and the supersymmetry breaking mass parameters being of order 100 GeV, $A = 1$, $\tan\beta = 1$, we find from Eq. (5) that $\lambda'_{113}\lambda'_{131} \leq 3\times 10^{-8}$, which is a more stringent limit on this parameter than the existing ones [12]. The present limits on these parameters are $\lambda'_{113} \leq 0.03$, $\lambda'_{131} \leq 0.26$, which shows that the bound derived here from $\beta\beta_{0\nu}$ is

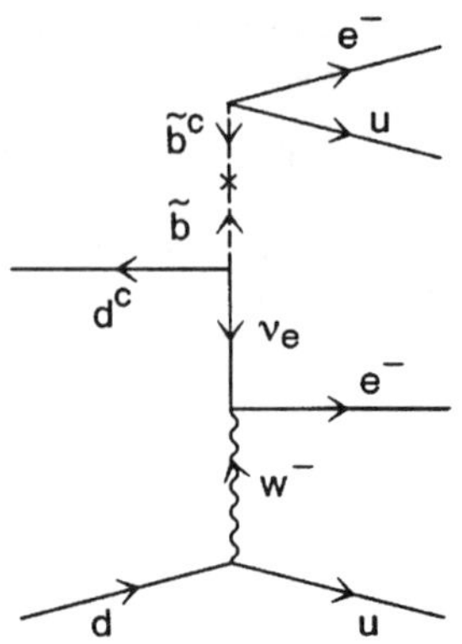

FIG. 1. The dominant diagram contributing to ϵ_2^{ee} type four Fermi term [Eq. (1)] in the supersymmetric model.

VOLUME 75, NUMBER 12 PHYSICAL REVIEW LETTERS 18 SEPTEMBER 1995

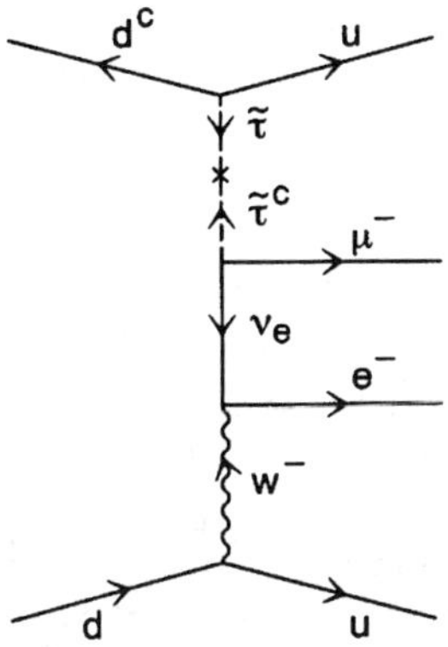

FIG. 2. The diagram contributing to $\epsilon_1^{e\mu}$ type four Fermi terms relevant for $\mu^- \to e^+$ conversion in the supersymmetric model.

about 5 orders of magnitude more stringent on the product $\lambda'_{113}\lambda'_{131}$. If the exchanged scalar particles in Fig. 1 are the $\tilde{s} - \tilde{s}^c$ pair, b gets replaced by s in Eq. (5) and one obtains a limit $\lambda'_{121}\lambda'_{112} \leq 1 \times 10^6$, which also is more stringent by about 4 orders of magnitude than existing limits ($\lambda'_{121} \leq 0.26, \lambda'_{112} \leq 0.03$). We note that the gluino exchange diagram [6,8] discussed in the context of $\beta\beta_{0\nu}$ constrains only the parameter λ'_{111}, while the vector-scalar exchange graphs constrain several other couplings.

The diagram in Fig. 2, due to the antisymmetry of λ_{ijk}, does not contribute to $\beta\beta_{0\nu}$ decay, but will be important for $\mu^- \to e^+$ conversion (see discussion below).

Let us note that there exist indirect limits on the λ and λ' couplings arising from the induced neutrino masses. The magnitudes of these masses are given generically by $m_{ij}^{\nu} \sim \lambda'_{ikl}\lambda'_{jkl}m_k m_l/16\pi^2 M_{\tilde{q}}$, where $m_{k,l}$ stand for the masses of the kth and lth down quarks (with a similar expression for the λ couplings). The coupling $\lambda'_{i33}\lambda'_{j33} \leq 6 \times 10^{-5}$ is the most severely constrained (being proportional to the b-quark mass squared) where the induced neutrino mass has been assumed to be $\leq$100 eV, which is cosmologically safe. The corresponding limit on $\lambda'_{131}\lambda'_{113} \leq 0.24$ (using $m_\nu \leq 1$ eV) is a trivial constraint to be compared with the $\beta\beta_{0\nu}$ decay limit derived here.

Constraints on the left-right symmetric model.—Let us consider the minimal left-right symmetric model with a seesaw mechanism for neutrino masses [13]. The gauge group of the model is $SU(3)_C \times SU(2)_L \times SU(2)_R \times U(1)_{B-L}$. The Higgs sector of the model consists of the bidoublet field $\phi \equiv (1/2, 1/2, 0)$ and triplet Higgs fields: $\Delta_L(1,0,+2) \oplus \Delta_R(0,1,+2)$. The Yukawa couplings which are invariant under gauge and parity symmetry can be written as

$$\mathcal{L}_Y = \overline{\Psi}_L h^\ell \phi \Psi_R + \overline{\Psi}_L \tilde{h}^\ell \tilde{\phi} \Psi_R + \overline{Q}_L \phi h^q Q_R + \overline{Q}_L \tilde{h}^q \tilde{\phi} Q_R + \Psi_L^T \tau_2 \vec{\tau} \cdot f \vec{\Delta}_L C^{-1} \Psi_L + L \to R + \text{H.c.}, \quad (6)$$

where h, $\tilde{h}$ are Hermitian matrices while f is a symmetric matrix in the generation space. Ψ and Q here denote the leptonic and quark doublets, respectively.

The gauge symmetry is spontaneously broken by the vacuum expectation values (VEVs): $\langle\Delta_R^0\rangle = V_R$, $\langle\Delta_L^0\rangle \simeq 0$, and $\langle\phi\rangle = \text{diag}(\kappa, \kappa')$. As usual, $\langle\phi\rangle$ gives masses to the charged fermions and Dirac masses to the neutrinos whereas $\langle\Delta_R^0\rangle$ leads to the seesaw mechanism for the neutrinos in the standard way [13].

The physics we are interested in comes from the left-handed triplet sector of the theory through its mixing with the bidoublet field which arises from the couplings in the Higgs potential, such as $\text{Tr}(\Delta_L \phi \Delta_R^\dagger \phi^\dagger)$ after the full gauge symmetry is broken down to $U(1)_{\text{em}}$. Specifically, there is a mixing between the singly charged components of ϕ and Δ_L [14] (we denote this mixing term by an angle θ). This will contribute to the four Fermi interaction of the form given by the ϵ_1^{ee} term through the diagram shown in Fig. 3 with

$$\epsilon_1^{ee} \simeq \frac{h_u f_{11} \sin 2\theta}{4\sqrt{2}\, G_F M_{H^+}^2}, \quad (7)$$

where we have assumed that H^+ is the lighter of the two Higgs fields. We get $h_u f_{11} \sin 2\theta \leq 6 \times 10^{-9}[M_{H^+}/(100\text{ GeV})]^2$, which is quite a stringent constraint on the parameters of the theory. To appreciate this somewhat more, we point out that one expects $h_u \approx m_u/m_W \approx 5 \times 10^{-5}$ in which case we get an upper limit for the coupling of the Higgs triplets to leptons $f_{11} \sin 2\theta \leq 10^{-4}$ (for $m_{H^+} = 100$. GeV). Taking a reasonable choice of $\theta \sim M_{W_L}/M_{W_R} \sim 10^{-1}$ would correspond to a limit $f_{11} \leq 10^{-3}$. Limits on this parameter from analysis [15] of Bhabha scattering is only of order 0.2 or so for the same value of the Higgs mass.

$\mu^- \to e^+$ conversion.—Another class of rare processes where the new vector-scalar contribution makes an important impact is the process of $\mu^- \to e^+$ conversion in nuclei which arises with an observable strength in the R-violating MSSM. This inolves the couplings λ_{ijk} which

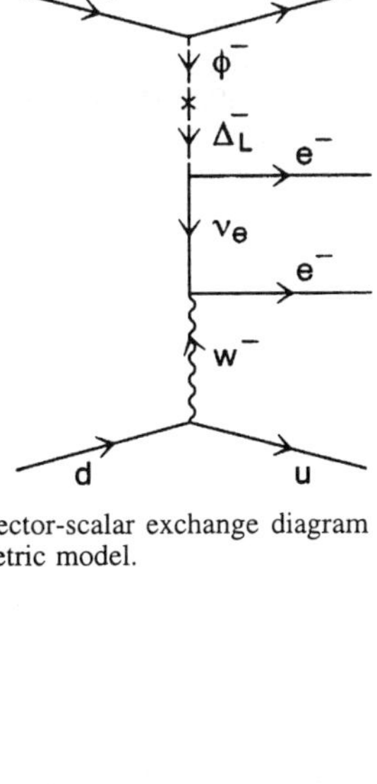

FIG. 3. The vector-scalar exchange diagram for $\beta\beta_{0\nu}$ in the left-right symmetric model.

were not constrained by the considerations of neutrinoless double beta decay due to anitsymmetry of the Yukawa couplings. Another contribution involves a different product of λ' couplings than what appeared in the $\beta\beta_{0\nu}$ process. We parametrize the effective four Fermi interaction for this process in analogy to Eq. (1) with ϵ_1^{ee} replaced by two terms $\epsilon_1^{e\mu}$ and $\epsilon_1^{\mu e}$ and similarly for ϵ_2. Here $\epsilon_1^{e\mu}$ is the coefficient of the term $G_F/\sqrt{2}[\bar{d}(1-\gamma_5)u\nu_e^T C^{-1}(1-\gamma_5)\mu]$, etc. The effective strengths of these couplings arising from Fig. 2 (as well as from Fig. 1 with e replaced by a μ) are found to be

$$\epsilon_1^{e\mu} = \frac{\lambda_{123}\lambda'_{311}}{2\sqrt{2}\,G_F m_{\tilde{\tau}}^2 m_{\tilde{\tau}^c}^2}\, m_\tau(\mu \tan\beta + A_\tau m_0),$$

$$\epsilon_2^{e\mu} = \frac{\lambda'_{213}\lambda'_{131}}{2\sqrt{2}\,G_F m_{\tilde{b}}^2 m_{\tilde{b}^c}^2}\, m_b(\mu \tan\beta + A_b m_0), \quad (8)$$

with similar expressions for $\epsilon_{1,2}^{\mu e}$. The existing limits on the λ' in Eq. (8) are $\lambda'_{213} \leq 0.09$, $\lambda'_{131} \leq 0.26$, so that for the quark masses of order 100 GeV, $\epsilon_2^{e\mu}$ can be as large as 10^{-1} or so. The branching ratio for $\mu^- \rightarrow e^+$ conversion relative to μ capture can be obtained by a naive scaling by the nucleon mass and we find $B(\mu^- \rightarrow e^+) \simeq 10^{-12}$ for this choice of $\epsilon_2^{e\mu}$. It is interesting that this is in the accessible range of the current experiments. The sensitivity of these experiments is expected to improve by another order of magnitude in the near future [16].

We note that the corresponding predictions for the left-right model is down by several orders of magnitude.

In summary, we have discussed a new class of contributions to the double lepton number violating processes which may arise in several extensions of the standard model. In particular, we find that the existing experimental limits on neutrinoless double beta decay lead to very stringent constraints on the R-violating couplings in MSSM as well as the lepton-number-violating Higgs couplings of the left-right symmetric model. Furthermore, we find the exciting possibility that for the presently allowed range of parameters in the MSSM, $\mu^- \rightarrow e^+$ conversion in nuclei is in the observable range. While we have focused on only two classes of models, our results are more general and should apply to other schemes, such as L-violating leptoquark models.

The work of K. S. B. was supported by Department of Energy Grant No. DE-FG02-91ER406267. The work of R. N. M. was supported by National Science Foundation Grant No. PHY-9119745.

*Present address: School of Natural Sciences, Institute for Advanced Study, Olden Lane, Princeton, New Jersey 08540.

[1] W. C. Haxton and G. Stephenson, Prog. Part. Nucl. Phys. **12**, 409 (1984); H. Grotz and H. Klapdor, *The Weak Interactions in Nuclear, Particle and Astrophysics* (Adam Hilger, Bristol, 1990); R. N. Mohapatra and P. B. Pal *Massive Neutrinos in Physics and Astrophysics* (World Scientific, Singapore, 1991); M. Moe and P. Vogel, Ann. Rev. Nucl. Part. Sci. **44**, 247 (1994).
[2] M. Doi, T. Kotani, H. Nishiura, and E. Takasugi, Prog. Theor. Phys. Suppl. **83**, 1 (1985).
[3] A. Halprin, P. Minkowski, H. Primakoff, and S. P. Rosen, Phys. Rev. D **13**, 2567 (1976).
[4] R. N. Mohapatra, Phys. Rev. D **34**, 909 (1986).
[5] R. N. Mohapatra and J. Vergados, Phys. Rev. Lett. **47**, 1713 (1981); J. Schecter and J. W. F. Valle, Phys. Rev. D **25**, 2951 (1982); W. C. Haxton, S. P. Rosen, and G. J. Stephenson, *ibid.* **26**, 1805 (1982); L. Wolfenstein, *ibid.* **26**, 2507 (1982).
[6] R. N. Mohapatra, Phys. Rev. D **34**, 3457 (1986).
[7] H. V. Klapdor-Kleingrothaus, Prog. Part. Nucl. Phys. **32**, 261 (1994); A. Balysh *et al.* (to be published).
[8] J. D. Vergados, Phys. Lett. B **184**, 55 (1987); M. Hirsch, H. V. Klapdor-Kleingrothaus, and S. G. Kovalenko (to be published).
[9] L. Hall and M. Suzuki, Nucl. Phys. **B231**, 419 (1984).
[10] In addition, we can write a term $\mu_i L_i H_u$ in W'. Although this term can be rotated away by a field redefinition, the VEVs of the scalar neutrinos cannot be set to zero in general then. In supergravity models, however, such VEVs get induced only be the effect of large Yukawa couplings, in which case the snuetrino VEVs will remain zero.
[11] There is also the possibility of setting $\lambda = \lambda' = 0$, with $\lambda'' \neq 0$, in which case B is violated, but L is not. We shall not pursue this alternative here.
[12] V. Barger, G. Giudice, and T. Han, Phys. Rev. D **40**, 2987 (1989).
[13] R. N. Mohapatra and G. Senjanović, Phys. Rev. Lett. **44**, 912 (1980); Phys. Rev. D **23**, 165 (1981).
[14] Note that in the limit $\kappa' = 0$, this coupling does not induce a $\langle\Delta_L\rangle$. See K. S. Babu and R. N. Mohapatra, Phys. Rev. Lett. **64**, 9 (1990).
[15] M. Schwarz, Phys. Rev. D **40**, 1521 (1989).
[16] W. Bertl (private communication).

ELSEVIER

11 April 1996

Physics Letters B 372 (1996) 181–186

PHYSICS LETTERS B

On the SUSY accompanied neutrino exchange mechanism of neutrinoless double beta decay

M. Hirsch, H.V. Klapdor-Kleingrothaus, S.G. Kovalenko [1]

Max-Planck-Institut für Kernphysik, P.O. 10 39 80, D-69029, Heidelberg, Germany

Received 6 December 1995
Editor: C. Mahaux

Abstract

The neutrinoless double beta decay ($0\nu\beta\beta$) induced by light Majorana neutrino exchange between decaying nucleons, accompanied by the squark exchange inside one nucleon, recently discussed by Babu and Mohapatra, is carefully analyzed both from the particle and nuclear physics sides. New nuclear matrix elements relevant to this mechanism are calculated. We extend the analysis to include mixing of light neutrinos with heavy and "sterile" neutrinos. It introduces another supersymmetric (SUSY) contribution to $0\nu\beta\beta$. We discuss constraints on the $\not{R}_p$MSSM parameters imposed by the current experimental limit on $0\nu\beta\beta$ decay half-life of ^{76}Ge.

PACS: 11.30; 12.30; 13.15; 23.40
Keywords: Supersymmetry; R-parity violation; Double beta decay; Neutrino; Weak interaction

Neutrinoless double beta ($0\nu\beta\beta$) decay is a sensitive probe of physics beyond the standard model, since it violates lepton number. Recent experimental progress has pushed the existing half-life limits of $0\nu\beta\beta$ decay beyond $T_{1/2}(0\nu\beta\beta) \geq 7.4 \times 10^{24}$ years and further progress can be expected in the near future [1]. This experimental result casts stringent constraints on new physics. Particularly, for the conventional mechanism of $0\nu\beta\beta$-decay with massive Majorana neutrino exchange between decaying nucleons (see Fig. 1) it implies an upper bound on the neutrino mass below 1 eV [1]. (There exist, however, other mechanisms which might induce $0\nu\beta\beta$ decays as well [2,3].)

In this paper we study contributions to $0\nu\beta\beta$ decay within the R-parity violating Minimal Supersymmetric Standard Model ($\not{R}_p$ MSSM). The $\not{R}_p$ MSSM has been extensively discussed in the literature since it has very interesting phenomenological [4] and cosmological [5] implications. It also gives a very natural framework for rare lepton number violating processes and particularly $0\nu\beta\beta$ decay [6–10].

The supersymmetric mechanism of $0\nu\beta\beta$ decay was first proposed by Mohapatra [6] and later studied in more details in Refs. [7,8]. In Ref. [9] it was shown that the gluino exchange contribution to $0\nu\beta\beta$-decay leads to a very stringent limit on the first generation $\not{R}_p$ Yukawa coupling $\lambda'_{111} \leq 3.9 \cdot 10^{-4}$. Recently, Babu and Mohapatra [10] found another contribution comparable in size with the gluino exchange. It allows one to set stringent limits on combinations of the intergeneration $\not{R}_p$-Yukawa couplings such as $\lambda'_{11i}\lambda'_{1i1}$, where i denotes generations.

However, in Ref. [10] limits were deduced using

[1] Joint Institute for Nuclear Research, Dubna, Russia.

PII S0370-2693(96)00050-0

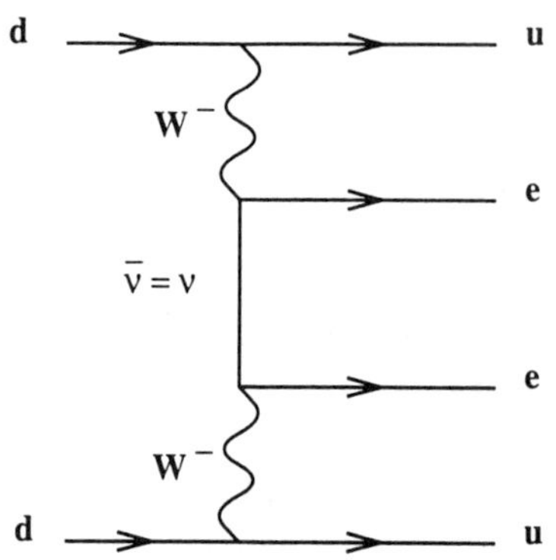

Fig. 1. Feynman graphs for the conventional mechanism of $0\nu\beta\beta$ decay by exchange of a massive Majorana neutrino.

a simplified estimation of the nuclear structure matrix elements. The approach used in [10] is based on a simple replacement of the virtual particle momenta and energies by the (estimated) Fermi momentum p_F and energy E_F of a nucleon inside the nucleus, therefore neglecting essentially all nuclear structure effects. Since the stringent constraints found in [10] may have important consequences for the $R\!\!\!/_p$ MSSM phenomenology, it is desirable to substantiate these results by detailed calculations.

In this letter we present the results of detailed particle and nuclear physics calculations relevant to the contribution of the diagrams in Fig. 2 to $0\nu\beta\beta$ decay. We extend the previous work to include mixing of the standard light neutrinos with exotic heavy neutrinos, neutralinos or some other neutral heavy particles or light "sterile" neutrino singlets [11]. Inclusion of mixing leads to the new slepton exchange diagram in Fig. 2b.

The $R\!\!\!/_p$MSSM is an extension of the MSSM which results from the inclusion of explicit R-parity violating ($R\!\!\!/_p$) terms $W_{R\!\!\!/_p}$ into the superpotential $W = W_{MSSM} + W_{R\!\!\!/_p}$, where

$$W_{R\!\!\!/_p} = \lambda_{ijk}L_iL_j\bar{E}_k + \lambda'_{ijk}L_iQ_j\bar{D}_k + \lambda''_{ijk}\bar{U}_i\bar{D}_j\bar{D}_k\,. \quad (1)$$

Indices i, j, k stand for generations. L, Q denote lepton and quark doublet superfields and $\bar{E}$, $\bar{U}$, $\bar{D}$ lepton and *up*, *down* quark singlet superfields. The first two terms in Eq. (1) lead to lepton number violation, while the last one violates baryon number. For $0\nu\beta\beta$ decay only the λ and λ' type couplings are of relevance.

Mixing between scalar superpartners $\tilde{f}_{L,R}$ of the left- and right-handed fermions $f_{L,R}$ will play a crucial role in our subsequent consideration. It occurs due to non-diagonality of the mass matrix which can be written as

$$\mathcal{M}^2_{\tilde{f}} = \begin{pmatrix} m^2_{\tilde{f}_L} + m^2_f - 0.42D_Z & -m_f(A_f + \mu\tan\beta) \\ -m_f(A_f + \mu\tan\beta) & m^2_{\tilde{f}_R} + m^2_f - 0.08D_Z \end{pmatrix} \quad (2)$$

Here, $f = d, s, b, e, \mu, \tau$ and $\tilde{f}$ are their superpartners. $D_Z = M_Z^2\cos 2\beta$ with $\tan\beta = \langle H_2^0\rangle/\langle H_1^0\rangle$ being the ratio of vacuum expectation values of the two Higgs doublets, $m_{\tilde{f}_{L,R}}$ are soft sfermion masses, A_f are soft SUSY breaking parameters describing the strength of trilinear scalar interactions, and μ is the supersymmetric Higgs(ino) mass parameter. Once sfermion mixing is included, the current eigenstates $\tilde{f}_L$, $\tilde{f}_R$ become superpositions of the mass eigenstates $\tilde{f}_i$ with the masses $m_{\tilde{f}_i}$ and the corresponding mixing angle θ^f is defined as

$$m^2_{\tilde{q}_{1,2}} = \tfrac{1}{2}[m^2_{LL} + m^2_{RR} \mp \sqrt{(m^2_{LL} - m^2_{RR})^2 + 4m^4_{LR}}\ ],$$

$$\sin 2\theta^f = \frac{2m^2_{(f)LR}}{m^2_{\tilde{f}_1} - m^2_{\tilde{f}_2}}, \quad (3)$$

where $m^2_{LR}, m^2_{LL}, m^2_{RR}$ denote the $(1,2), (1,1), (2,2)$ entries of the mass matrix (2).

Now it is straightforward to find the effective 4-fermion $\nu - u - d - e$ vertex induced by the sfermion exchange in the diagrams presented in Fig. 2. The corresponding effective Lagrangian, after a Fiertz rearrangement, takes the form

$$\begin{aligned}\mathcal{L}^{\rm eff}_{\rm SUSY}(x) = \frac{G_F}{\sqrt{2}}[&\tfrac{1}{4}(\eta^{nj}_{(q)LR} - 4\eta^{nj}_{(l)LR})U^*_{ni} \\ &\times (\bar{\nu}_i(1+\gamma_5)e^c_j)(\bar{u}(1+\gamma_5)d) \\ &- 2\eta^{nj}_{(l)LL}U_{ni}(\bar{\nu}_i(1-\gamma_5)e^c_j)(\bar{u}(1+\gamma_5)d) \\ &+ \tfrac{1}{2}\eta^{nj}_{(q)RR}U_{ni}(\bar{\nu}_i\,\gamma^\mu(1+\gamma_5)e^c_j)(\bar{u}\,\gamma_\mu(1-\gamma_5)d) \\ &+ \tfrac{1}{8}\eta^{nj}_{(q)LR}U^*_{ni}(\bar{\nu}_i\,\sigma^{\mu\nu}(1+\gamma_5)e^c_j) \\ &\times (\bar{u}\,\sigma_{\mu\nu}(1+\gamma_5)d)]\,. \end{aligned} \quad (4)$$

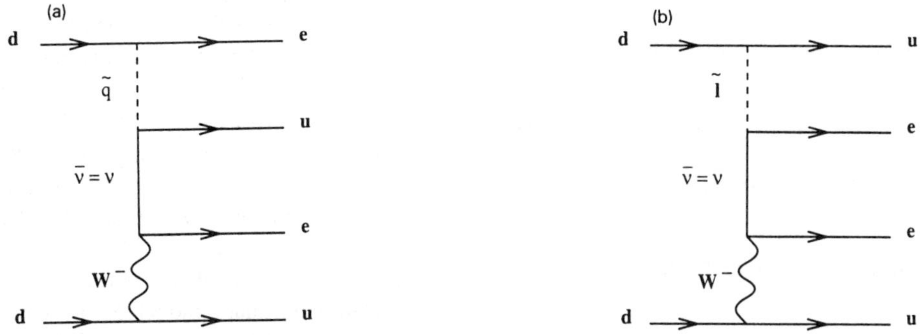

Fig. 2. Feynman graphs for the supersymmetry accompanied Majorana neutrino exchange mechanism of the $0\nu\beta\beta$ decay corresponding to (a) squarks and (b) sleptons contributions.

The $\not{R}_p$MSSM parameters η and neutrino mixing matrix U_{ij} are defined as follows

$$\eta^{nj}_{(q)LR} = \sum_k \frac{\lambda'_{j1k}\lambda'_{nk1}}{2\sqrt{2}G_F} \sin 2\theta^d_{(k)} \left(\frac{1}{m^2_{\tilde{d}_1(k)}} - \frac{1}{m^2_{\tilde{d}_2(k)}} \right), \tag{5}$$

$$\eta^{nj}_{(q)RR} = \sum_k \frac{\lambda'_{j1k}\lambda'_{n1k}}{2\sqrt{2}G_F} \left(\frac{\sin\theta^d_{(k)}}{m^2_{\tilde{d}_1(k)}} + \frac{\cos\theta^d_{(k)}}{m^2_{\tilde{d}_2(k)}} \right), \tag{6}$$

$$\eta^{nj}_{(l)LR} = \sum_k \frac{\lambda'_{k11}\lambda_{njk}}{2\sqrt{2}G_F} \sin 2\theta^e_{(k)} \left(\frac{1}{m^2_{\tilde{e}_1(k)}} - \frac{1}{m^2_{\tilde{e}_2(k)}} \right), \tag{7}$$

$$\eta^{nj}_{(l)LL} = \sum_k \frac{\lambda'_{k11}\lambda_{nkj}}{2\sqrt{2}G_F} \left(\frac{\cos\theta^e_{(k)}}{m^2_{\tilde{e}_1(k)}} + \frac{\sin\theta^e_{(k)}}{m^2_{\tilde{e}_2(k)}} \right), \tag{8}$$

$$\nu^0_i = \sum_j U_{ij}\nu_j . \tag{9}$$

Here $\eta_{(f)LR}$ denotes the contribution vanishing in the absence of $\tilde{f}_L - \tilde{f}_R$ - mixing while $\eta_{(f)LL}$ and $\eta_{(f)RR}$ in this limit correspond to the $\tilde{f}_L$ and $\tilde{f}_R$ exchange contribution in Fig. 2. We use the notations $d_{(k)} = d, s, b$ and $e_{(k)} = e, \mu, \tau$. Due to the antisymmetry of the Yukawa coupling λ_{njk} in nj it follows that $\eta^{nn}_{(l)LR} = 0$. This is an essential difference between the slepton $\tilde{l}_L - \tilde{l}_R$ and the squark $\tilde{q}_L - \tilde{q}_R$ contributions. The latter is not imposed to vanish at any combination of indexes.

The matrix element of the SUSY accompanied neutrino exchange mechanism can then be calculated according to the the diagrams in Fig. 2 with the point-like 4-fermion vertex described by the effective Lagrangian Eq. (4) with the sfermion exchange parts in the top. The bottom parts of these diagrams is the standard model charged current (SMCC) interaction. Applying the standard procedure (for details see [9]), one can get the matrix element $\mathcal{R}_{0\nu\beta\beta}(0^+ \to 0^+)$ of the $0\nu\beta\beta$ decay for $0^+ \to 0^+$ transitions. For two outgoing electrons in S-wave states it takes the form

$$\mathcal{R}_{0\nu\beta\beta}(0^+ \to 0^+) = C_{0\nu} f_A^2 \left[a \cdot \bar{e}(1+\gamma_5)e^c + b \cdot \bar{e}\gamma_0\gamma_5 e^c \right], \tag{10}$$

$$a = \left(4\eta^{n1}_{(l)LR} - \eta^{n1}_{(q)LR}\right) U^*_{ni} U_{ei} \mathcal{M}^{(i)}_1 (m_e R)^{-1} + \left(\eta^{n1}_{(q)RR} - \delta_{ne}\right) U_{ni} U_{ei} \frac{m_{\nu_i}}{m_e} \mathcal{M}^{(i)}_2 , \tag{11}$$

$$b = 4\, \eta^{n1}_{(l)LL} U_{ni} U_{ei} \frac{m_{\nu_i}}{m_e} \mathcal{M}^{(i)}_3 . \tag{12}$$

The term proportional to δ_{ne} corresponds to the ordinary neutrino exchange contribution with two standard model charged current vertices in Fig. 1. The normalization factor $C_{0\nu}$ is defined as $C_{0\nu} = (G_F^2 2m_e)/(8\sqrt{2}\pi R)$. We would like to stress that the terms proportional to b in Eq. (10) can, in principle, be discriminated from the terms proportional to a (particularly from the ordinary mass mechanism of $0\nu\beta\beta$-decay) by measuring the angular correlation between the two outgoing electrons.

The following nuclear matrix elements are involved in the calculation of $R_{0\nu\beta\beta}$ in Eq. (10)

$$\mathcal{M}^{(i)}_1 = \alpha_1 \left[\mathcal{M}^{(i)}_{T'} + \tfrac{1}{3}\mathcal{M}^{(i)}_{GT'} \right], \tag{13}$$

$$\mathcal{M}_2^{(i)} = \alpha_2 \mathcal{M}_F^{(i)} - \mathcal{M}_{GT}^{(i)}, \quad \mathcal{M}_3^{(i)} = \alpha_3 \mathcal{M}_F^{(i)}. \quad (14)$$

They are defined by (summation over nucleons is suppressed)

$$\mathcal{M}_F^{(i)} =< 0_f^+ || h_+(r_{ab}, m_{\nu_i}) \tau_a^+ \tau_b^+ || 0_i^+ >, \quad (15)$$

$$\mathcal{M}_{GT}^{(i)} =< 0_f^+ || h_+(r_{ab}, m_{\nu_i}) (\boldsymbol{\sigma}_a \boldsymbol{\sigma}_b) \tau_a^+ \tau_b^+ || 0_i^+ >, \quad (16)$$

$$\mathcal{M}_{GT'}^{(i)} =< 0_f^+ || h_R(r_{ab}, m_{\nu_i}) (\boldsymbol{\sigma}_a \boldsymbol{\sigma}_b) \tau_a^+ \tau_b^+ || 0_i^+ >, \quad (17)$$

$$\mathcal{M}_{T'}^{(i)} =< 0_f^+ || h_{T'}(r_{ab}, m_{\nu_i}) \times \left[(\boldsymbol{\sigma}_a \hat{\boldsymbol{r}}_{ab}) (\boldsymbol{\sigma}_b \hat{\boldsymbol{r}}_{ab}) - \tfrac{1}{3} (\boldsymbol{\sigma}_a \boldsymbol{\sigma}_b) \right] \tau_a^+ \tau_b^+ || 0_i^+ > . \quad (18)$$

Neutrino potentials can be written in the integral form

$$h_+(r_{ab}, m_{\nu_i}) = \frac{2}{\pi} R \int_0^\infty dq \cdot q^2 \frac{j_0(q r_{ab}) f^2(q^2)}{\omega(\omega + \bar{A})}, \quad (19)$$

$$h_R(r_{ab}, m_{\nu_i}) = \frac{2}{\pi} \frac{R^2}{m_P} \int_0^\infty dq \cdot q^4 \frac{j_0(q r_{ab}) f^2(q^2)}{\omega(\omega + \bar{A})}, \quad (20)$$

$$h_{T'}(r_{ab}, m_{\nu_i}) = \frac{2}{\pi} \frac{R^2}{m_P} \int_0^\infty dq \cdot q^4 \frac{j_0(q r_{ab}) - 3 j_1(q r_{ab})}{\omega(\omega + \bar{A})} f^2(q^2). \quad (21)$$

Here, $\omega = \sqrt{q^2 + m_{\nu_i}^2}$; $j_k(qr)$ are spherical Bessel functions and R_0 is the nuclear radius, introduced to make the matrix elements dimensionless. The following notations are used: $\boldsymbol{r}_{ab} = (\boldsymbol{r}_a - \boldsymbol{r}_b)$, $r_{ab} = |\boldsymbol{r}_{ab}|$, $\hat{\boldsymbol{r}}_{ab} = \boldsymbol{r}_{ab}/r_{ab}$. The above formulae have been written in the closure approximation which is well motivated for the $0\nu\beta\beta$ decay [2]. $\bar{A}$ in Eq. (19) is the average intermediate state energy.

Note that the nuclear matrix element $\mathcal{M}_{T'}^{(i)}$ has never been considered in the literature before. We have calculated the nuclear matrix elements relevant for our subsequent numerical analysis using the pn-QRPA model of [13]. We take into account both short-range correlations and finite nucleon size effects. The latter is described by introducing nucleon form factors in momentum space. In the present case these are the SMCC form factors $F_{(V,A)}^{CC}(q^2)$ and the isovector scalar and pseudoscalar current form factors $F_{P,S}^3(q^2)$. For all form factors we take, as usual, a dipole parameterization: $F_i^a(q^2)/F_i^a(0) = f(q^2) = \left(1 + q^2/m_A^2\right)^{-2}$ with $m_A = 0.85$ GeV. The q-dependent factor $f(q^2)$ is included in the definition of the neutrino potentials in Eqs. (19). The form factor normalizations are $F_{V(A)}^{CC}(0) = f_{V(A)} \approx 1(1.261)$. $F_{S,P}^3(0)$ can be calculated within the conventional non-relativistic quark model or the bag model. We take their numerical values from Ref. [14] $F_S^3(0) \approx 0.48$ $F_P^3(0) \approx 4.41$. These values correspond to the bag model calculations.

The nucleon structure coefficients in (13) are defined as $\alpha_1 = (F_P^{(3)}/(2f_A)$, $\alpha_2 = (f_V/f_A)^2$, $\alpha_3 = (f_V/f_A)(F_S^{(3)}/f_A)$.

Now having the $0\nu\beta\beta$ matrix element completely specified the inverse half life $T_{1/2}(0\nu\beta\beta)$ can be written as

$$T_{1/2}^{-1}(0\nu\beta\beta) = |a|^2 G_{01} + |b|^2 h_9 G_{09} + \mathrm{Re}(a^* b) h_6 G_{06}, \quad (22)$$

where $h_6 = m_e R/8$, $h_9 = (m_e R)^2/16$, and phase space factors G_{0i} are given in Ref. [2].

The formulae presented above describe the contribution of the diagrams in Fig. 2 for a general neutrino content (see Eq. (9)). We now turn to a particular case and assume that all neutrino mass eigenstates fall into two groups: light neutrinos $m_{\nu_i} < 10$ MeV and heavy neutrinos $m_{\nu_i} > 10$ GeV. We denote them as ν_i and N_i, respectively. There might be also a "sterile" neutrino ν_s with respect to the SM gauge group. Assume further, that the light neutrinos have non-negligible mixing with some heavy neutrinos N_i or probably with the "sterile" neutrino ν_s. Than the neutrino composition (9) and the unitarity relation for the mixing matrix U_{ij} can be written as

$$\nu_i^0 = \sum_j^{\prime} U_{ij} \nu_j + \sum_j^{\prime\prime} U_{ij} \nu_j + U_{is} \nu_s, \quad (23)$$

$$\sum_j U^*_{ij} U_{ei} = \sum_j{}' U^*_{ij} U_{ei} + \Delta_i ,$$

$$\Delta_i = \sum_j{}'' U^*_{ij} U_{ei} + U_{is}\nu_s . \tag{24}$$

Such a structure of the neutrino sector breaks contributions of each nuclear matrix elements in Eq. (10) into two pieces with a simple dependence on the neutrino mass. As an example of this effect consider the first term in Eq. (11)

$$\sum_i U^*_{ni} U_{ei} \mathcal{M}^{(i)}_1$$
$$= \sum_i{}' U^*_{ni} U_{ei} \mathcal{M}^{\nu}_1 + \sum_i{}'' \frac{U^*_{ni} U_{ei}}{M^2_{N_i}} \mathcal{M}^N_1 . \tag{25}$$

The "sterile" neutrino does not contribute being a SM singlet. The matrix elements are $\mathcal{M}^{\nu}_1 = \mathcal{M}^{(i)}_1 (m_\nu = 0)$ and $\mathcal{M}^N_1 = \lim_{M^2_N \to \infty} M_N \mathcal{M}^{(i)}_1 (M_N)$. They do not depend on neutrino masses. Proceeding in a similar way with the other matrix elements in Eq. (10) and substituting the result in the half-life formula Eq. (22) we obtain a polynomial in m_ν and M^{-1}_N. Assume light neutrinos ν to be very light while heavy neutrinos N to be very heavy so that one can neglect terms depending on these masses. Than, keeping only leading terms we retain

$$T^{-1}_{1/2}(0\nu\beta\beta) = G_{01} \mathcal{M}^{\nu}_1 (m_e R)^{-1}$$
$$\times \left(4\bar\eta_{(l)} - \bar\eta_{(q)} + \eta_{(q)}\right)^2 , \tag{26}$$

where Eq. (25) and the property $\eta^{nn}_{(l)LR} = 0$ have been used. We denoted $\eta_{(q)} = \eta^{11}_{(q)LR}$ and introduced the effective parameters as

$$\bar\eta_{(l,q)} = \sum_n \Delta_n \eta^{n1}_{(l,q)LR} . \tag{27}$$

For the $\bar\eta_{(l)}$ summation starts from $n = 2$. The nuclear matrix element $\mathcal{M}^{\nu}_1$ in Eq. (26) can be directly obtained from Eq. (13) for $\mathcal{M}^{(i)}_1$ as explained after Eq. (25). This matrix element has never been calculated in the literature before. Its value calculated in the pn-QRPA for the particular case of ^{76}Ge analyzed below is $\mathcal{M}^{(\nu)}_1(^{76}\text{Ge}) = 2.1$.[2]

[2] Recall that this numerical value corresponds to our dimensionless convention.

Now we are ready to discuss constraints on the parameters in Eq. (26) imposed by the current experimental lower half-life limit. We use the result from the Heidelberg-Moscow ^{76}Ge experiment $T^{0\nu\beta\beta}_{1/2}(^{76}Ge, 0^+ \to 0^+) > 7.4 \times 10^{24}$ years 90% c.l.

Disregarding a situation with unnatural fine-tuning between the different terms in Eq. (26) one can extract individual limits, numerically $\eta_{(q)} \le 2.9 \times 10^{-8}$, $\bar\eta_{(l)} \le 7.2 \times 10^{-9}$. These constraints lead, in principle, to a multidimensional exclusion curve. A simplified picture can be obtained under some reasonable assumptions. Assume all the MSSM mass parameters in Eqs. (5) and (2), (3) to be approximately equal to the "effective" SUSY breaking scale Λ_{SUSY}. Then we get a simplified set of constraints

$$\lambda'_{11i}\lambda'_{1i1} \le \epsilon'_i \left(\frac{\Lambda_{SUSY}}{100\ \text{GeV}}\right)^3 ,$$

$$\Delta_n \lambda'_{311}\lambda_{n13} \le \epsilon \left(\frac{\Lambda_{SUSY}}{100\ \text{GeV}}\right)^3 . \tag{28}$$

In the last equation we kept only the term corresponding to $\tilde\tau$ exchange. For the known values of SM quark and lepton masses in the following we use $m_{d_1} = m_d = 7.5$ MeV, $m_{d_2} = m_s = 150$ MeV, $m_{d_3} = m_b = 4.5$ GeV and $m_\tau = 1.8$ GeV. These masses are present in Eqs. (2), (3). Then, $\epsilon'_{1,2,3} = \{6.4\times10^{-5}, 3.2\times10^{-6}, 1.1\times10^{-7}\}$ and $\epsilon = 6.5 \times 10^{-8}$.

To obtain information on the $R\!\!\!/_p$ Yukawa couplings themselves one may use some reasonable values for Λ_{SUSY}. If one takes, following Ref. [10], the value $\Lambda_{SUSY} = 100$ GeV lying in the region of the current experimental lower bound for the SUSY particles, one gets $\lambda'_{113}\lambda'_{131} \le 1.1 \times 10^{-7}$, $\lambda'_{112}\lambda'_{121} \le 3.2 \times 10^{-6}$, $\lambda'^2_{111} \le 6.4 \times 10^{-5}$ and $\Delta_n\lambda'\lambda \le 6.5 \times 10^{-8}$. The former two limits correspond to those in Ref. [10], but are somewhat weaker than quoted by Babu and Mohapatra. This difference can be partly traced back to our numerical value of the matrix element, which turns out to be smaller than anticipated. The limit on λ'^2_{111} can be compared with the corresponding limit obtained from the gluino exchange diagram in Ref. [9]. The latter being $\lambda'^2_{111} \le 1.52 \cdot 10^{-7}$ is much more stringent then the one derived here. More conservative estimations can be derived from Eq. (28) implying $\Lambda_{SUSY} \sim 1$ TeV motivated by the SUSY naturalness arguments. Then, $\lambda'_{113}\lambda'_{131} \le 1.1 \times 10^{-4}$, $\lambda'_{112}\lambda'_{121} \le$

3.2×10^{-3}, $\lambda'^2_{111} \le 6.4 \times 10^{-2}$ and $\Delta_n \lambda' \lambda \le 6.5 \times 10^{-4}$.

Neglecting mixing between the light SM non-singlet neutrinos and the non-standard sector discussed above, one arrives at the case considered by Babu and Mohapatra [10]. In this case $\Delta_n = 0$ and in turn $\bar{\eta}(l,q) = 0$. Then the squark exchange diagram in Fig. 2a is the only contribution to Eq. (26). Introducing mixing one can obtain new information about λ type interactions. In this case, however, this information is accessible only in a form of some effective value of the R_p Yukawa couplings $\Delta\lambda'\lambda$. The upper limit for the couplings itself implies an extra uncertainty due to the unknown mixing factor Δ.

The new contributions analysed in this letter together with the previously discussed in the literature [6–9] complete the tree-level mechanisms of $0\nu\beta\beta$ decay within the R-parity violating Minimal Supersymmetric Standard Model (R_p MSSM). A detailed presentation of the analysis will be given elsewhere.

In conclusion we would like to stress that the current experimental limit on $0\nu\beta\beta$ decay half life allows one to establish rather stringent limits on R_p violating SUSY interactions. This might be an additional valuable motivation for present and forthcoming $0\nu\beta\beta$ decay experiments.

Acknowledgments

We thank V.A. Bednyakov and R.N. Mohapatra for helpful discussions. The research described in this publication was made possible in part (S.G.K.) by Grant GNTP 215NUCLON from the Russian ministry of science. M.H. would like to thank the Deutsche Forschungsgemeinschaft for financial support by grants kl 253/8-1 and 446 JAP-113/101/0.

References

[1] HEIDELBERG-MOSCOW Collaboration: A. Balysh et al., Phys. Lett. B 356 (1995) 450;
H.V. Klapdor-Kleingrothaus, Progr. Part. Nucl. Phys. 32 (1994) p. 261, and Proc. Workshop on Double Beta Decay and related topics, Trento, Italy (1995), World Scientific, Singapore, in press.

[2] M. Doi, T. Kotani and E. Takasugi, Prog. Theor. Phys. Suppl. 83 (1985) 1.

[3] J.D. Vergados, Phys. Report 133 (1986) 1;
J.W.F. Valle, Prog. Part. Nucl. Phys. 26 (1991) 91;
R.N. Mohapatra, Talk at the Workshop on Double Beta Decay and Related Topics, Trento, Italy (1995) (to appear in the Workshop Proceedings).

[4] V. Barger, G.F. Guidice and T. Han, Phys. Rev. D 40 (1989) 2987;
R. Barbieri and A. Masiero, Nucl. Phys. B 267 (1986) 679;
D.P. Roy, Phys. Lett. B 283 (1992) 270;
F. Zwirner, Phys. Lett. B 132 (1983) 103;
H. Dreiner and P. Morawitz, Nucl. Phys. B 428 (1994) 31.

[5] A. Bouquet and P. Salati, Nucl. Phys. B 284 (1987);
A. Nelson and S.M. Barr, Phys. Lett. B 258 (1991) 45;
B.A. Campbell, S. Davidson, J. Ellis and K. Olive, Phys. Lett. B 256 (1991) 457; CERN-preprint, CERN-TH-6208-91;
W. Fieschler, G. Guidice, R.G. Leigh and S. Paban, Phys. Lett. B 258 (1991) 45.

[6] R.N. Mohapatra, Phys. Rev. D 34 (1986) 3457.

[7] J.D. Vergados, Phys. Lett. B 184 (1987) 55.

[8] M. Hirsch, H.V. Klapdor-Kleingrothaus and S.G. Kovalenko, Phys. Lett. B 352 (1995) 1.

[9] M. Hirsch, H.V. Klapdor-Kleingrothaus and S.G. Kovalenko, Phys. Rev. Lett. 75 (1995) 17; Preprint MPI-H-V 6-1995, Heidelberg (1995), to appear in Phys. Rev. D (1995).

[10] K.S. Babu and R.N. Mohapatra, Phys. Rev. Lett. 75 (1995) 2276.

[11] D. Caldwell and R.N. Mohapatra, Phys. Rev. D 48 (1993) 3259;
J. Peltoniemi and J.W.F. Valle, Nucl. Phys. B 406 (1993) 409.

[12] L. Hall and M. Suzuki, Nucl. Phys. B 231 (1984) 419;
D. Braham and L. Hall, Phys. Rev. D 40 (1989) 2449;
M. Bento, L.J. Hall and G.G. Ross, Nucl. Phys. B 292 (1987) 400;
G. Lazarides, P.K. Mohapatra, C. Panagiotakopoulos and Q. Shafi, Nucl. Phys. B 323 (1989) 614.

[13] K. Muto, E. Bender and H.V. Klapdor, Z. Phys. A 334 (1989) 177, 187;
M. Hirsch, K. Muto, T. Oda and H.V. Klapdor-Kleingrothaus, Z. Phys. A 347 (1994) 151.

[14] S.L. Adler et al., Phys. Rev. D 11 (1975) 3309.

25 July 1996

PHYSICS LETTERS B

ELSEVIER

Physics Letters B 381 (1996) 488

Erratum

On the SUSY accompanied neutrino exchange mechanism of neutrinoless double beta decay [Phys. Lett. B 372 (1996) 181] ★

M. Hirsch, H.V. Klapdor-Kleingrothaus, S.G. Kovalenko

Fig. 2a should be replaced by the version given below. Our derivation is based on the correct Feynman graph, thus the formulas remain unaffected by this replacement.

Misprints were found in the formulas (6), (8) and (21). Eqs. (6) and (8) should be replaced by, respectively,

$$\eta^{nj}_{(q)RR} = \sum_k \frac{\lambda'_{j1k}\lambda'_{n1k}}{2\sqrt{2}G_F}\left(\frac{\sin^2\theta^d_{(k)}}{m^2_{\tilde{d}_1(k)}} + \frac{\cos^2\theta^d_{(k)}}{m^2_{\tilde{d}_2(k)}}\right),$$

$$\eta^{nj}_{(l)LL} = \sum_k \frac{\lambda'_{k11}\lambda_{nkj}}{2\sqrt{2}G_F}\left(\frac{\cos^2\theta^e_{(k)}}{m^2_{\tilde{e}_1(k)}} + \frac{\sin^2\theta^e_{(k)}}{m^2_{\tilde{e}_2(k)}}\right).$$

For the neutrino potential for the tensor matrix element ($\mathcal{M}^{(i)}_{T'}$), Eq. (21), the correct version reads

$$h_{T'}(r_{ab}, m_{\nu_i}) = \frac{2}{\pi}\frac{R^2}{m_P}\int_0^\infty dq \cdot q^2 \frac{f^2(q^2)}{\omega(\omega+\bar{A})}\{q^2 j_0(qr_{ab}) - 3\frac{q}{r_{ab}} j_1(qr_{ab})\}.$$

The numerical calculations were based on the correct formulas. Thus, our numerical results and the conclusion remain unchanged.

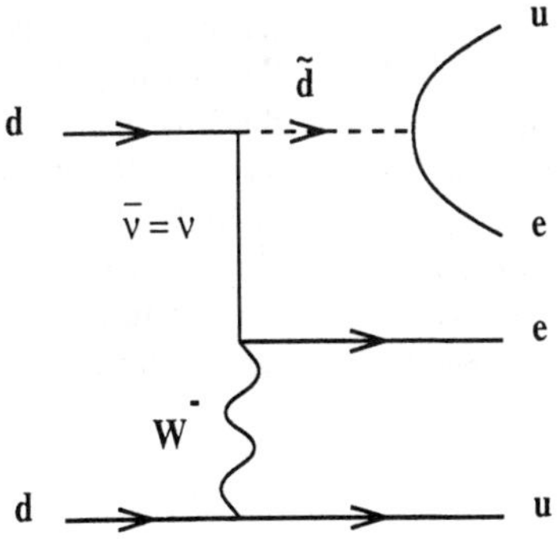

Fig. 2a.

★ PII of original article: S0370-2693(96)00050-0

PII S0370-2693(96)00689-2

ELSEVIER

29 July 1999

PHYSICS LETTERS B

Physics Letters B 459 (1999) 450–454

Improved bounds on SUSY accompanied neutrinoless double beta decay

H. Päs [1], M. Hirsch [2], H.V. Klapdor-Kleingrothaus [3]

Max-Planck-Institut für Kernphysik, P.O.Box 10 39 80, D-69029 Heidelberg, Germany

Received 30 October 1998; received in revised form 5 March 1999
Editor: W. Haxton

Abstract

Neutrinoless double beta decay ($0\nu\beta\beta$) induced by light Majorana neutrino exchange between two decaying nucleons with squark/slepton exchange inside one and W exchange inside the other nucleon (so-called vector-scalar exchange) gives stringent limits on R-parity violating interactions. We have extended previous work by including the tensor contribution to the transition rate. We discuss the improved limits on trilinear $R\!\!\!/_p$-MSSM couplings imposed by the current experimental limit on the $0\nu\beta\beta$ decay half-life of ^{76}Ge. © 1999 Published by Elsevier Science B.V. All rights reserved.

PACS: 13.15; 23.40; 21.60J; 14.80
Keywords: Supersymmetry; R-parity violation; Double beta decay; Neutrino; QRPA

In supersymmetric models, the new SUSY partners differ from the SM field content in a discrete multiplicative quantum number R-parity defined as

$$R_p = (-1)^{3B+L+2S}. \tag{1}$$

Here B denotes the baryon number, L the lepton number and S the spin of a particle leading to $R_p = +1$ for SM particles and $R_p = -1$ for superpartners. Thus in R_p conserving models superpartners can only be produced in pairs and the LSP is stable, leading to a natural WIMP dark matter candidate. While in the minimal supersymmetric extension (MSSM) of the standard model (SM) R-parity is assumed to be conserved, there are no theoretical reasons for R_p conservation and several GUT [1] and Superstring [2] models require R-parity violation in the low energy regime. Also the reports concerning an anomaly at HERA in the inelastic e^+p scattering at high Q^2 and x [3] have renewed the interest in $R\!\!\!/_p$-SUSY (see for example [4,5]). Generally, one can add the following trilinear R-parity violating terms to the superpotential [1]

$$W_{R_p} = \lambda_{ijk} L_i L_j \overline{E}_k + \lambda'_{ijk} L_i Q_j \overline{D}_k + \lambda''_{ijk} \overline{U}_i \overline{D}_j \overline{D}_k, \tag{2}$$

where i,j,k denote generation indices, L,Q denote lepton and quark doublet superfields and $\overline{E},\overline{U},\overline{D}$ lepton, up- and down quark singlet superfields. Terms

[1] E-mail: heinrich.paes@mpi-hd.mpg.de
[2] Present address: Inst. de Fisica Corpuscular, C.S.I.C., Dept. de Fisica Teorica, Univ. de Valencia, 46100 Burjassot, Valencia, Spain; E-mail: mahirsch@flamenco.ific.uv.es
[3] E-mail: klapdor@mickey.mpi-hd.mpg.de

PII: S0370-2693(99)00711-X

proportional to λ, λ' violate lepton number, those proportional λ'' violate baryon number. While simultaneous presence of both kinds of terms would lead to too fast proton decay and thus is forbidden, assuming the λ'' terms to be zero no constraints on λ and λ' terms can be derived from proton decay.

The search for neutrinoless double beta decay, converting a nucleus (Z,A) into a nucleus $(Z+2,A)$ under emission of two electrons, has been proven to belong to the most powerful tools to search for lepton number violating physics beyond the SM (for a review see [6,7]). Contributions occuring through Feynman graphs involving the exchange of superpartners as well as $\not{R}_P$-couplings λ' have been discussed in Refs. [8–12] and yield the most stringent bound on λ'_{111}. Taking into account the fact that the SUSY partners of the left and right-handed quark states can mix with each other, also diagrams appear in which the neutrino-mediated double beta decay is accompanied by SUSY exchange in the vertices (see Fig. 1). These contributions allow to constrain also combinations of couplings of higher generations $\lambda'_{11j}\lambda'_{1j1}$. They have been discussed in Ref. [13] and more extensively in Ref. [11], where however only scalar-pseudoscalar currents have been taken into account, whereas the tensor contribution to the decay rate has been neglected. On the other hand, in a recent work [14] the dominance of the tensor contribution in a general framework for neutrino mediated double beta decay has been proven. In the present letter we reanalyze SUSY-accompanied double beta decay and discuss the relative importance of the different nuclear matrix elements involved.

The mixing between scalar superpartners $\tilde{f}_{L,R}$ of the left and right-handed fermions $f_{L,R}$ occurs due to

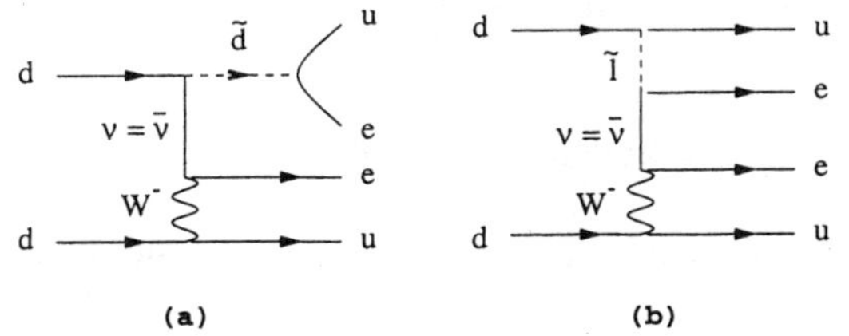

Fig. 1. Feyman graphs for the neutrino exchange mechanism of neutrinoless double beta decay accompanied by (a) squark and (b) slepton exchange.

non-diagonality of the mass matrix which can be written as

$$\mathcal{M}_{\tilde{f}}^2 = \begin{pmatrix} m_{\tilde{f}_L}^2 + m_f^2 - 0.42\,D_Z & -m_f(A_f + \mu\tan\beta) \\ -m_f(A_f+\mu\tan\beta) & m_{\tilde{f}_R}^2 + m_f^2 - 0.08\,D_Z \end{pmatrix}. \quad (3)$$

Here, $f = d,s,b,e,\mu,\tau$ and $\tilde{f}$ are their superpartners. $D_Z = M_Z^2\cos 2\beta$ with $\tan\beta = \langle H_2^0\rangle/\langle H_1^0\rangle$ being the ratio of vacuum expectation values of the two Higgs doublets, $m_{\tilde{f}_{L,R}}$ are soft sfermion masses, A_f are soft SUSY breaking parameters describing the strength of trilinear scalar interactions, and μ is the supersymmetric Higgs(ino) mass parameter. Once sfermion mixing is included, the current eigenstates $\tilde{f}_L, \tilde{f}_R$ become superpositions of the mass eigenstates $\tilde{f}_i$ with the masses $m_{\tilde{f}_i}$ and the corresponding mixing angle θ^f is defined as

$$\sin 2\theta^f = \frac{2m_{(f)LR}^2}{m_{\tilde{f}_1}^2 - m_{\tilde{f}_2}^2},$$

$$m_{\tilde{f}_{1,2}}^2 = \tfrac{1}{2}\left(m_{LL}^2 + m_{RR}^2 \mp \sqrt{\left(m_{LL}^2 - m_{RR}^2\right)^2 + 4m_{LR}^4} \right) \quad (4)$$

where $m_{LR}^2, m_{LL}^2, m_{RR}^2$ denote the (1,2),(1,1),(2,2) entries of the mass matrix (3).

Now it is straightforward to find the effective 4-fermion $\nu - u - d - e$ vertex induced by the sfermion exchange in the diagrams presented in Fig. 1. The corresponding effective Lagrangian, after a Fiertz rearrangement, takes the form

$$\begin{aligned} \mathcal{L}_{\text{SUSY}}^{\text{eff}}(x) = \frac{G_F}{\sqrt{2}} \Big[& \tfrac{1}{4}\left(\eta_{(q)LR}^{nj} - 4\eta_{(l)LR}^{nj}\right) \\ & \cdot U_{ni}^* \cdot \left(\bar{\nu}_i(1+\gamma_5)e_j^c\right)\left(\bar{u}(1+\gamma_5)d\right) \\ & - 2\eta_{(l)LL}^{nj} \cdot U_{ni} \cdot \left(\bar{\nu}_i(1-\gamma_5)e_j^c\right) \\ & \times \left(\bar{u}(1+\gamma_5)d\right) \\ & + \tfrac{1}{2}\eta_{(q)RR}^{nj} \cdot U_{ni}\left(\bar{\nu}_i\gamma^\mu(1+\gamma_5)e_j^c\right) \\ & \times \left(\bar{u}\gamma_\mu(1-\gamma_5)d\right) \\ & + \tfrac{1}{8}\eta_{(q)LR}^{nj} \cdot U_{ni}^* \cdot \left(\bar{\nu}_i\sigma^{\mu\nu}(1+\gamma_5)e_j^c\right) \\ & \times \left(\bar{u}\sigma_{\mu\nu}(1+\gamma_5)d\right)\Big]. \end{aligned} \quad (5)$$

The $\not{R}_p$MSSM parameters η and neutrino mixing matrix U_{ij} are defined as follows

$$\eta^{nj}_{(q)LR} = \sum_k \frac{\lambda'_{j1k}\lambda'_{nk1}}{2\sqrt{2}\,G_F}\sin 2\theta^d_{(k)}\left(\frac{1}{m^2_{\tilde{d}_1(k)}} - \frac{1}{m^2_{\tilde{d}_2(k)}}\right), \quad (6)$$

$$\eta^{nj}_{(q)RR} = \sum_k \frac{\lambda'_{j1k}\lambda'_{n1k}}{2\sqrt{2}\,G_F}\left(\frac{\sin^2\theta^d_{(k)}}{m^2_{\tilde{d}_1(k)}} + \frac{\cos^2\theta^d_{(k)}}{m^2_{\tilde{d}_2(k)}}\right), \quad (7)$$

$$\eta^{nj}_{(l)LR} = \sum_k \frac{\lambda'_{k11}\lambda_{njk}}{2\sqrt{2}\,G_F}\sin 2\theta^e_{(k)}\left(\frac{1}{m^2_{\tilde{e}_1(k)}} - \frac{1}{m^2_{\tilde{e}_2(k)}}\right), \quad (8)$$

$$\eta^{nj}_{(l)LL} = \sum_k \frac{\lambda'_{k11}\lambda_{nkj}}{2\sqrt{2}\,G_F}\left(\frac{\cos^2\theta^e_{(k)}}{m^2_{\tilde{e}_1(k)}} + \frac{\sin^2\theta^e_{(k)}}{m^2_{\tilde{e}_2(k)}}\right), \quad (9)$$

$$\nu^0_i = \sum_j U_{ij}\nu_j. \quad (10)$$

Here $\eta_{(f)LR}$ denotes the contribution vanishing in the absence of $\tilde{f}_L - \tilde{f}_R$ - mixing while $\eta_{(f)LL}$ and $\eta_{(f)RR}$ in this limit correspond to the $\tilde{f}_L$ and $\tilde{f}_R$ exchange contribution in Fig. 1. We use the notations $d_{(k)} = d,s,b$ and $e_{(k)} = e,\mu,\tau$. Due to the antisymmetry of the Yukawa coupling λ_{njk} in nj it follows that $\eta^{nn}_{(l)LR} = 0$. This is an essential difference between the slepton $\tilde{l}_L - \tilde{l}_R$ and the squark $\tilde{q}_L - \tilde{q}_R$ contributions. The latter is not imposed to vanish at any combination of indexes. Since the $\eta^{nj}_{(l)LL}$ and $\eta^{nj}_{(q)RR}$ contributions to the diagram Fig. 1 are helicity suppressed $\propto \langle m_\nu \rangle \simeq \mathcal{O}(0.5)$eV, we can restrict ourselves to the consideration of $\eta^{nj}_{(q)LR}$ and $\eta^{nj}_{(l)LR}$, which contributions are momentum enhanced $\propto q \simeq p_F \simeq 100$ MeV, where p_F denotes the Fermi momentum. Here $\langle m_\nu \rangle$ and q denote the effective neutrino mass and momentum entering the neutrino propagator. The first line in Eq. (5) corresponds to the scalar-pseudoscalar ($S+P$) contribution, the last line in (5) to the tensor contribution.

In the light neutrino case the $0\nu\beta\beta$ decay rate is given by

$$T^{-1}_{1/2}(0\nu\beta\beta) = G_{01}\{4\overline{\eta}_{(l)}\mathcal{M}_{S+P} + (-\overline{\eta}_{(q)} + \eta_{(q)})(\mathcal{M}_{S+P} + \mathcal{M}_T)\}^2. \quad (11)$$

Here G_{01} denotes the phase space factor defined in Ref. [15], $\eta_{(q)} = \eta^{11}_{(q)LR}$ and effective parameters are introduced as

$$\overline{\eta}_{(l,q)} = \sum_n \Delta_n \eta^{n1}_{(l,q)LR}. \quad (12)$$

For the $\overline{\eta}_{(l)}$ summation starts from $n=2$ and Δ_n denotes the combination of mixing matrices corresponding to heavy or sterile neutrinos

$$\Delta_i = \sum_j^{\prime\prime} U^*_{ij}U_{ei} + U^*_{is}U_{ei}, \quad (13)$$

where the sum extends over heavy mass eigenstates $m_{\nu_i} > 10$ GeV (see [11]). In s-wave approximation for the outgoing electrons and under some assumptions according to [16,11] (the s-wave approximation is expected to affect the result less than 10% [14]) the matrix element is

$$\mathcal{M}_{S+P} = \frac{F^{(3)}_P(0)}{4Rm_eG_A}\left(\mathcal{M}_{T'} + \tfrac{1}{3}\mathcal{M}_{GT'}\right), \quad (14)$$

$$\mathcal{M}_T = \alpha_1\left(\tfrac{2}{3}\mathcal{M}_{GT'} - \mathcal{M}_{T'}\right), \quad (15)$$

with (summation over nucleons a,b is suppressed)

$$\mathcal{M}_{GT'} = \langle 0^+_f | h_R(\boldsymbol{\sigma}_a\boldsymbol{\sigma}_b)\tau^+_a\tau^+_b|0^+_i\rangle \quad (16)$$

$$\mathcal{M}_{T'} = \langle 0^+_f | h_{T'}\{(\boldsymbol{\sigma}_a\hat{\boldsymbol{r}}_{ab})(\boldsymbol{\sigma}_b\hat{\boldsymbol{r}}_{ab}) - \tfrac{1}{3}(\boldsymbol{\sigma}_a\boldsymbol{\sigma}_b)\}\tau^+_a\tau^+_b|0^+_i\rangle, \quad (17)$$

$$\alpha_1 = \frac{T^{(3)}_1(0)G_V(1 - 2m_P(G_W/G_V))}{2G^2_A Rm_e}. \quad (18)$$

h_R and $h_{T'}$ are neutrino potentials defined as

$$h_R = \frac{2}{\pi}\frac{R^2}{m_P}\int_0^\infty dq q^4 \frac{j_0(qr_{ab})f^2(q^2)}{\omega(\omega+\overline{E})}, \quad (19)$$

$$h_{T'} = \frac{2}{\pi}\frac{R^2}{m_P}\int_0^\infty dq q^2 \frac{f^2(q^2)}{\omega(\omega+\overline{E})} \times\left\{q^2 j_0(qr_{ab}) - 3\frac{q}{r_{ab}}j_1(qr_{ab})\right\}. \quad (20)$$

Here R denotes the nuclear radius, m_P the proton mass.

Further $\omega = \sqrt{q^2 + m^2_\nu}$, $q = |\boldsymbol{q}|$, $\hat{\boldsymbol{r}} = \boldsymbol{r}/r$ and $j_k(qr)$ are spherical Bessel functions. $(\omega + \overline{E})^{-1}$ is the energy denominator of the perturbation theory.

The form factors $F_i^a(0) = F_i^a(q^2)/f(q^2)$ and $T_1^{(3)}(0) = T_1^{(3)}(q^2)/f(q^2)$ with $f(q^2) = (1 + q^2/m_A^2)^{-2}$ ($m_A^2 = 0.85$ GeV) have been calculated in the MIT bag model in Ref. [17], $G_A \simeq 1.26$, $G_V \simeq 1$ and the strength of the induced weak magnetism $(G_W/G_V) = \frac{\mu_p - \mu_n}{2m_p} \simeq \frac{-3.7}{2m_p}$ is obtained by the CVC hypothesis.

For comparison and to correct some details in Ref. [11] we first concentrate on the $S+P$ part i.e. $\mathcal{M}_T = 0$ in Eq. (11). Inserting the numerical value of the matrix elements $\mathcal{M}_{GT'} = 2.95$ and $\mathcal{M}_{T'} = 0.224$ [11] for the special case of ^{76}Ge and the half life limit obtained from the Heidelberg–Moscow experiment, $T_{1/2}^{0\nu\beta\beta} > 1.2 \cdot 10^{25} y$ [18,7], one derives $\eta_q \leq 4.5 \cdot 10^{-8}$, $\overline{\eta}_{(l)} \leq 1.1 \cdot 10^{-8}$, corresponding to limits of

$$\lambda'_{112}\lambda'_{121} \leq 5.0 \cdot 10^{-6} \left(\frac{\Lambda_{\text{SUSY}}}{100\,\text{GeV}} \right)^3,$$

$$\lambda'_{113}\lambda'_{131} \leq 1.6 \cdot 10^{-7} \left(\frac{\Lambda_{\text{SUSY}}}{100\,\text{GeV}} \right)^3,$$

$$\Delta_n \lambda'_{311}\lambda_{n13} \leq 9.9 \cdot 10^{-8} \left(\frac{\Lambda_{\text{SUSY}}}{100\,\text{GeV}} \right)^3. \tag{21}$$

Here in the last equation only the term corresponding to $\tilde{\tau}$ exchange has been kept. These limits differ from the ones given in Ref. [11] due to an erroneous factor of 2 in the definition of $\mathcal{M}_{S+P}$ in that reference as well as the improvement in the experimental bound.

Keeping the tensor part in Eq. (11) the half life limit of the Heidelberg–Moscow Experiment implies:

$$\lambda'_{112}\lambda'_{121} \leq 1.1 \cdot 10^{-6} \left(\frac{\Lambda_{\text{SUSY}}}{100\,\text{GeV}} \right)^3,$$

$$\lambda'_{113}\lambda'_{131} \leq 3.8 \cdot 10^{-8} \left(\frac{\Lambda_{\text{SUSY}}}{100\,\text{GeV}} \right)^3 \tag{22}$$

for supersymmetric mass parameters of order 100 GeV. For $\Lambda_{\text{SUSY}} \sim 1$ TeV, motivated by SUSY naturalness arguments, one obtains $\lambda'_{112}\lambda'_{121} \leq 1.1 \cdot 10^{-3}$, $\lambda'_{113}\lambda'_{131} \leq 3.8 \cdot 10^{-5}$. These are by more than a factor of four more stringent than the limits obtained from the scalar pseudoscalar part considered in Ref. [11]. The uncertainty of the nuclear matrix elements involved can be estimated to be less than a factor of two. This is motivated by the fact that the $2\nu\beta\beta$ halflife of ^{76}Ge has been predicted correctly within a factor of 2 (the $2\nu\beta\beta$ matrix element within a factor of $\sqrt{2}$) [19]. Since the uncertainty in both parts ($S+P$ and tensor) is expected to be about the same, the improvement can still be considered as substantial.

The obtained bound should be compared with the limits obtained from tree level $K^0 - \overline{K}^0$ and $B^0 - \overline{B}^0$ mixing, which yield $\lambda'_{i12}\lambda'_{i21} < 1 \cdot 10^{-9}$ and $\lambda'_{i13}\lambda'_{i31} < 8 \cdot 10^{-8}$ respectively, for $m_{\tilde{e}} = 100$ GeV [20]. While the second generation is bounded by about three orders of magnitude more stringent from the K system, for the third generation the double beta bound is most stringent. Moreover, since the masses of different exchanged particles (selectrons and squarks) enter, the limits are complementary in some sense.

In conclusion, we have performed a reanalysis of the SUSY-accompanied neutrino exchange mode of neutrinoless double beta decay. Contrary to the previous ansatz we included the tensor contribution to the decay rate, which has been shown to be the dominant contribution. This improves the limits on $\lambda'_{11j}\lambda'_{1j1}$ derived without the tensor contribution by a factor of four and thus provides the most stringent bound on $\lambda'_{113}\lambda'_{131}$.

Acknowledgements

M.H. would like to acknowledge support by the European Union's TMR program under grant ERBFMBICT983000.

References

[1] L. Hall, M. Suzuki, Nucl. Phys. B 231 (1984) 419; D. Brahm, L. Hall, Phys. Rev. D 40 (1989) 2449; K. Tamvakis Phys. Lett. B 382 (1996) 251; G.F. Guidice, R. Rattazzi, hep-ph/9604339; R. Barbieri, A. Strumia, Z. Berezhiani, hep-ph/9704275; K. Tamvakis, Phys. Lett B 383 (1996) 307; R. Hempfling, Nucl. Phys. B 478 (1996) 3; A.Y. Smirnov, F. Vissani, Nucl. Phys. B 460 (1996) 37.

[2] M.C. Bento, L. Hall, G.G. Ross, Nucl. Phys. B 292 (1987) 400; N. Ganoulis, G. Lazarides, Q. Shafi, Nucl. Phys. B 323 (1989) 374; A. Faraggi, Phys. Lett. B 398 (1997) 95.

[3] C. Adloff et al., H1 Collaboration, Z. Phys. C 74 (1997) 191; J. Breitweg et al., ZEUS Collaboration Z. Phys. C 74 (1997) 207.

[4] J. Kalinowski, R. Rueckl, H. Spiesberger, P.M. Zerwas, Z. Phys. C 74 (1997) 595.
[5] H. Dreiner, in: G.L. Kane (Ed.), Perspectives on Supersymmetry, World Scientific, Singapore; hep-ph/9707435.
[6] H.V. Klapdor-Kleingrothaus, in: H.V. Klapdor-Kleingrothaus, H. Päs (Eds.), Proc. Int. Conf. Beyond the Desert - Accelerator- and Non-Accelerator Approaches, Castle Ringberg, Germany, 1997, IOP, Bristol, 1998, p. 485.
[7] H.V. Klapdor-Kleingrothaus, H. Päs, in: Proc. of the 6th Int. Symposium on Particles, Strings and Cosmology (PASCOS98), Boston, MA, USA, March 22–27, 1998, to be publ. by World Scientific.
[8] R.N. Mohapatra, Phys. Rev. D 34 (1986) 3457; J.D. Vergados, Phys. Lett. B 184 (1987) 55.
[9] M. Hirsch, H.V. Klapdor-Kleingrothaus, S.G. Kovalenko, Phys. Rev. Lett. 75 (1995) 17, M. Hirsch, H.V. Klapdor-Kleingrothaus, S. Kovalenko, Phys. Rev. D 53 (1996) 1329.
[10] M. Hirsch, H.V. Klapdor-Kleingrothaus, S. Kovalenko, Phys. Lett. B 352 (1995) 1.
[11] M. Hirsch, H.V. Klapdor-Kleingrothaus, S.G. Kovalenko, Phys. Lett. B 372 (1996) 181; B 381 (1996) 488 (E).
[12] A. Faessler, S.G. Kovalenko, F. Simkovic, J. Schwieger, Phys. Rev. Lett. 78 (1997) 183.
[13] K.S. Babu, R.N. Mohapatra, Phys. Rev. Lett. 75 (1995) 2276.
[14] H. Päs, M. Hirsch, S.G. Kovalenko, H.V. Klapdor-Kleingrothaus, submitted to Phys. Lett. B
[15] M. Doi, T. Kotani, E. Takasugi, Progr. Theor. Phys. Suppl. 83 (1985) 1.
[16] T. Tomoda, Rep. Progr. Phys. 54 (1991) 53.
[17] S. Adler et al., Phys. Rev. D 11 (1975) 3309.
[18] Heidelberg-Moscow Collaboration, L. Baudis et al., Phys. Lett. B 407 (1997) 219.
[19] A. Balysh et al., Heidelberg-Moscow Collaboration, Phys. Lett. B 322 (1994) 176; K. Muto, E. Bender, H.V. Klapdor, Z. Phys. A 334 (1989) 177.
[20] G. Bhattacharyya, in Ref. [6]; we thank the author for bringing this bound to our attention.

2.4.4 *R*-Parity Conserving Supersymmetry and Double Beta Decay

17 April 1997

ELSEVIER

PHYSICS LETTERS B

Physics Letters B 398 (1997) 311–314

B-L-violating masses in softly broken supersymmetry

M. Hirsch [a], H.V. Klapdor-Kleingrothaus [a], S.G. Kovalenko [b]

[a] *Max-Planck-Institut für Kernphysik, P.O. Box 10 39 80, D-69029 Heidelberg, Germany*

[b] *Joint Institute for Nuclear Research, Dubna, Russia*

Received 16 January 1997
Editor: P.V. Landshoff

Abstract

We prove a general low-energy theorem establishing a generic relation between the neutrino Majorana mass and the superpartner sneutrino B-L-violating "Majorana"-like mass term. The theorem states that if one of these two quantities is non-zero the other one is also non-zero and, vice versa, if one of them vanishes the other vanishes, too. The theorem is a consequence of the underlying supersymmetry (SUSY) and valid for any realistic gauge model with weak scale softly broken SUSY. © 1997 Elsevier Science B.V.

PACS: 11.30; 12.10; 13.15
Keywords: Supersymmetry; Discrete symmetry; Majorana neutrino mass

Neutrinos are believed to be massive particles. Despite the lack of unambiguous experimental confirmation of this belief there are insisting indications for non-zero neutrino masses from cosmology, the solar and atmospheric neutrino puzzles (for recent review see [1]) as well as from recent LSND results on possible $\bar{\nu}_e - \bar{\nu}_\mu$ neutrino oscillations [2].

Among the known explanations for the extreme smallness of the neutrino mass compared to masses of the other fermions the most natural one is based on the see-saw mechanism [3]. It leads to a B-L-violating Majorana mass term for the neutrino. Various 1-loop contributions to the neutrino self-energy, widely discussed in the literature [4–9], also induce a small Majorana mass for neutrinos. Furthermore, the Grand Unification paradigm definitely prefers a Majorana mass for neutrinos. Due to these arguments, it has become a common trend to think of neutrinos as Majorana particles.

In supersymmetric (SUSY) models the neutrino ν has its scalar superpartner the sneutrino $\tilde{\nu}$. Given that they are components of the same superfield one may suspect a certain interplay between the neutrino and sneutrino properties at low energies as a relic of the underlying supersymmetry.

In the present note we prove a low-energy theorem establishing an intimate relation between the neutrino Majorana mass term and the B-L-violating as well as B-L-conserving sneutrino mass terms. Our consideration refers to the general structure of the low-energy effective Lagrangian assuming weak scale softly broken supersymmetry and stability of the ground state after electro-weak symmetry breaking. The proof of the low-energy theorem, consisting of three statements, is based on symmetry arguments and is of a general significance.

The effective Lagrangian of a generic model of weak scale supersymmetry contains after electro-weak

PII S0370-2693(97)00234-7

26 June 1997

PHYSICS LETTERS B

ELSEVIER

Physics Letters B 403 (1997) 291–296

Sneutrino oscillations and neutrinoless double beta decay

M. Hirsch [a], H.V. Klapdor-Kleingrothaus [a], S.G. Kovalenko [b]

[a] *Max-Planck-Institut für Kernphysik, P.O. 103 980, D-69029 Heidelberg, Germany*

[b] *Joint Institute for Nuclear Research, Dubna, Russia*

Received 4 March 1997; revised manuscript received 9 April 1997
Editor: P.V. Landshoff

Abstract

We prove that the neutrino Majorana mass, the B–L violating mass of the sneutrino ($\tilde{\nu}$) and the neutrinoless double beta decay amplitude are intimately related to each other such that if one of them is non-zero the other two are also non-zero and, vice versa, if one of them vanishes the other two vanish as well. The theorem is valid for any weak-scale softly broken supersymmetric model and independent of the mechanisms of neutrinoless double beta decay and (s)neutrino mass generation. The B–L-violating sneutrino mass leads to mass splitting in the sneutrino–antisneutrino ($\tilde{\nu} - \tilde{\nu}^c$) system and to the effect of lepton number violating $\tilde{\nu} - \tilde{\nu}^c$ oscillations. © 1997 Elsevier Science B.V.

Neutrinos are special among the known fermions in the sense that – being electrically neutral – they could be either Dirac or Majorana particles. Experimentally at present only upper limits on neutrino masses have been firmly established, but there are also an accumulating number of hints for non-zero neutrino masses from, for example, the solar and atmospheric neutrinos (for a recent review see Ref. [1]) as well as from recent LSND results [2]. While the discovery of any non-zero neutrino mass would present a major breakthrough, unfortunately none of the above experiments could tell whether the neutrino is a Dirac or a Majorana particle.

From a theoretical point of view Majorana neutrinos are clearly preferred. In particular, grand unified theories (GUTs) most naturally lead to Majorana neutrinos. A Majorana mass for the neutrinos could quite elegantly explain the observed smallness of neutrino masses via the famous see-saw mechanism. Also various 1-loop contributions to the neutrino self-energy, allowed in extensions of the SM [4], induce a small Majorana mass for neutrinos. Nevertheless, only experiments can finally settle the question about the nature of the neutrino. The importance of neutrinoless double beta ($0\nu\beta\beta$) decay derives from the fact that it is sensitive to the Majorana nature of neutrinos.

That there is a generic relation between the amplitude of neutrinoless double beta ($0\nu\beta\beta$) decay and the (B–L)-violating Majorana mass of the neutrino has been recognized about 15 years ago [3]. A general theorem relating these two observables has been proven in Ref. [3]. It states that if any of the two quantities – the Majorana neutrino mass or the neutrinoless double decay amplitude – vanishes the other one vanishes necessarily too and, vice versa, if one of them is non-zero the other one must also differ from zero. Recall, that $0\nu\beta\beta$-decay is strictly forbidden if the neutrino is a Dirac particle having only a (B–L)-conserving Dirac mass. This theorem is valid for any gauge model with spontaneously broken symmetry at the weak-scale, independent of the mechanism of $0\nu\beta\beta$-decay. The simple neutrino exchange

PII S0370-2693(97)00481-4

PHYSICAL REVIEW D VOLUME 57, NUMBER 3 1 FEBRUARY 1998

R-parity-conserving supersymmetry, neutrino mass, and neutrinoless double beta decay

M. Hirsch* and H. V. Klapdor-Kleingrothaus†
Max-Plank-Institut für Kernphysik, P.O. Box 10 39 80, D-69029, Heidelberg, Germany

S. G. Kovalenko‡
Joint Institute for Nuclear Research, Dubna, Russia
(Received 1 July 1997; published 22 December 1997)

We consider contributions of R-parity-conserving softly broken supersymmetry (SUSY) to neutrinoless double beta ($0\nu\beta\beta$) decay via the $(B-L)$-violating sneutrino mass term. The latter is a generic ingredient of any weak-scale SUSY model with a Majorana neutrino mass. The new R-parity-conserving SUSY contributions to $0\nu\beta\beta$ are realized at the level of box diagrams. We derive the effective Lagrangian describing the SUSY-box mechanism of $0\nu\beta\beta$ decay and the corresponding nuclear matrix elements. The one-loop sneutrino contribution to the Majorana neutrino mass is also derived. Given the data on the $0\nu\beta\beta$ decay half-life of ^{76}Ge and the neutrino mass we obtain constraints on the $(B-L)$-violating sneutrino mass. These constraints leave room for accelerator searches for certain manifestations of the second and third generation $(B-L)$-violating sneutrino mass term, but are most probably too tight for first generation $(B-L)$-violating sneutrino masses to be searched for directly. [S0556-2821(98)03401-8]

PACS number(s): 14.80.Ly, 12.60.Jv, 14.60.Pq

I. INTRODUCTION

Neutrinoless double beta ($0\nu\beta\beta$) decay [1,2] is a unique example of a nuclear process which allows one to probe for lepton number violation. Given the fact that the standard model (SM) conserves L at the classical level, $0\nu\beta\beta$ decay can proceed only via non-SM interactions. Therefore, taking into account the stringent experimental limits [3], $0\nu\beta\beta$ decay is extremely sensitive to physics beyond the standard model.

The simplest source of lepton number violation directly leading to $0\nu\beta\beta$ decay is a finite Majorana neutrino rest mass. It violates lepton number by two units $\Delta L=2$ precisely what is necessary for $0\nu\beta\beta$ decay. The corresponding contribution is described by the tree-level diagram of the fourth order in the weak coupling constant with one Majorana neutrino propagator as shown in Fig. 1(a).

Supersymmetric (SUSY) extensions of the SM bring in new sources of lepton number violation and, as a result, new mechanisms of $0\nu\beta\beta$ decay. Even with a minimal matter and Higgs field content there are renormalizable L- and B-violating terms in the superpotential and in the sector of soft-supersymmetry-breaking interactions which are not forbidden by gauge symmetry. These terms violate not only L and B but also R parity defined as $R_p=(-1)^{3B+L+2S}$ with S being a particle spin. Models containing such interaction terms are usually referred to as SUSY models with explicit R_P breaking [4]. (R_P can also be broken spontaneously via a nonzero vacuum expectation value of the sneutrino field $\langle\tilde{\nu}\rangle\neq 0$ [5].)

Formerly it was widely believed that supersymmetry can contribute to $0\nu\beta\beta$ decay only if R_P is broken [6–10]. The corresponding mechanisms have been comprehensively studied in the last few years [7–11]. In particular, it was shown

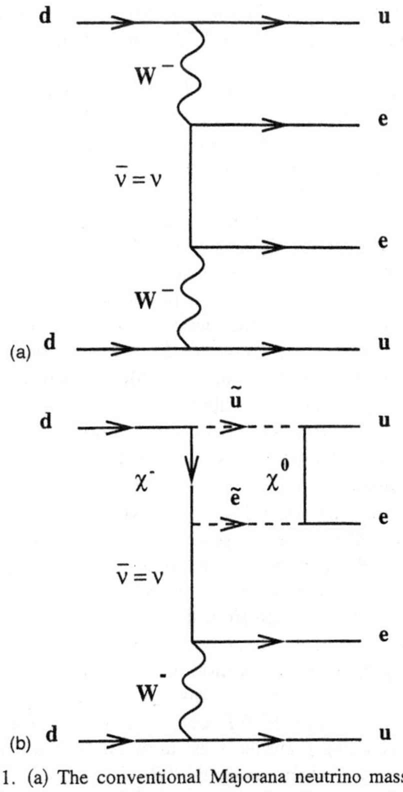

FIG. 1. (a) The conventional Majorana neutrino mass mechanism of $0\nu\beta\beta$ decay; (b) an example of an R_P-conserving SUSY contribution to the $0\nu\beta\beta$ decay.

*Email address: M.Hirsch@mpi-hd.mpg.de
†Email address: KLAPDOR@ENULL.MPI-HD.MPG.DE
‡Email address: KOVALEN@NUSUN.JINR.DUBNA.SU

that the current experimental limits from nonobservation of $0\nu\beta\beta$ decay sets upper bounds on certain $R\!\!\!/_p$ Yukawa coupling constants which are more stringent [7,8] than previously known from various accelerator and nonaccelerator experiments. Moreover, they turned out to be more stringent than those from some forthcoming accelerator experiments. This conclusion has put $0\nu\beta\beta$ decay forward as an interesting probe of supersymmetry.

Although there are no compelling theoretical arguments for R-parity conservation, there exist a number of well-known phenomenological drawbacks for supersymmetric models in which R_P is violated. Maybe the most serious one is the instability of the lightest SUSY particle. As a result, the supersymmetric solution for the dark matter problem is lost unless $R\!\!\!/_p$ Yukawa coupling constants λ become unnaturally small, typically $\lambda \leqslant 10^{-16}$.

In view of this and other problems for $R\!\!\!/_p$ SUSY there arises a natural question whether R_P violation is an inevitable condition for SUSY to contribute to $0\nu\beta\beta$ decay. The present paper addresses this question. We will demonstrate that there is a nontrivial R-parity-conserving SUSY contribution to $0\nu\beta\beta$ decay. In Fig. 1(b) we present an example of a diagram associated with the lowest-order contribution to $0\nu\beta\beta$ decay within the R-parity-conserving minimal supersymmetric standard model (MSSM) with a Majorana neutrino mass m_M^ν. This particular example gives an explicit answer to the above question: R-parity violation is not a necessary condition for a SUSY contribution to $0\nu\beta\beta$ decay.

It is also obvious that this SUSY contribution is strongly suppressed compared to the non-SUSY diagram in Fig. 1(a). This is because it is of higher order in perturbation theory, contains heavy sparticles in intermediate states, and receives a typical suppression due to the loop integration. Moreover, this diagram is proportional to the very small factor m_M^ν/p_F where $p_F \approx 80$ MeV is the nucleon Fermi momentum. The latter is also true for the simplest non-SUSY diagram in Fig. 1(a). The reason is common for both diagrams. In fact, both the SM and the MSSM interactions conserve lepton number L. Therefore, the only source for $\Delta L = 2$ violation, necessary for $0\nu\beta\beta$ decay to proceed, is the Majorana neutrino mass term. If $m_M^\nu = 0$, lepton number would be a conserved quantity and the $0\nu\beta\beta$-decay amplitude should vanish, $R_{0\nu\beta\beta} = 0$. Such a behavior corresponds to $R_{0\nu\beta\beta} \sim m_M^\nu/p_F$ in the limit of small m_M^ν.

Thus, the diagram in Fig. 1(b), given its very small contribution to $0\nu\beta\beta$ decay, provides just a principal demonstration of the fact that $0\nu\beta\beta$ decay can be triggered by R-parity-conserving supersymmetry. No practical consequences can be obtained from this new diagram in the sense of establishing new constraints either on SUSY parameters or on m_M^ν from nonobservation of $0\nu\beta\beta$ decay.

However, we will show that there are other R-parity-conserving SUSY contributions via the lepton-number-violating sneutrino mass term. As was shown in [12] the Majorana neutrino mass, the $(B-L)$-violating sneutrino mass and the $0\nu\beta\beta$-decay amplitude are generically connected to each other. Namely, nonvanishing of one of these three quantities implies nonzero values of the remaining two. Thus, the $0\nu\beta\beta$ amplitude should always contain a contribution corresponding to the $(B-L)$-violating sneutrino mass term.

We will study this contribution and extract constraints on the $(B-L)$-violating sneutrino mass term from the current experimental data on $0\nu\beta\beta$ decay of ^{76}Ge.

This paper is organized as follows. In Sec. II we give a short account of the structure of neutrino-sneutrino mass terms and the theorem which establishes the above-mentioned relations between the neutrino and sneutrino masses and $0\nu\beta\beta$ decay. Section III is devoted to some general properties of possible SUSY contributions to $0\nu\beta\beta$ decay. In this section we specify the box diagrams describing the R_P-conserving SUSY contribution. Our approach to the derivation of the corresponding $0\nu\beta\beta$-transition operators and nuclear matrix elements is outlined in Sec. IV. Section V deals with $0\nu\beta\beta$-decay constraints on the $(B-L)$-violating sneutrino mass. In Sec. VI we calculate the sneutrino contribution to the Majorana neutrino mass and derive then limits from experimental data on neutrino masses. We then close with a short summary.

II. STRUCTURE OF THE NEUTRINO-SNEUTRINO MASS TERMS

As shown in [12], the self-consistent form of the neutrino and sneutrino mass terms is

$$\mathcal{L}_{\text{mass}}^{\nu\tilde{\nu}} = -\tfrac{1}{2}(m_M^\nu \overline{\nu_L^c}\nu_L + \text{H.c.}) - \tfrac{1}{2}(\tilde{m}_M^2 \tilde{\nu}_L \tilde{\nu}_L + \text{H.c.}) - \tilde{m}_D^2 \tilde{\nu}_L^* \tilde{\nu}_L, \quad (1)$$

where $\nu = \nu^c$ is a Majorana field. The first two terms violate the global $(B-L)$ symmetry while the last one respects it. The first term is a Majorana mass term of the neutrino. We call the second term a "Majorana"-like mass, while the third one is referred to as a "Dirac"-like sneutrino mass term. This reflects an analogy with Majorana and Dirac mass terms for neutrinos. In the presence of the right-handed neutrino field ν_R the Dirac neutrino mass term $m_D^\nu(\bar{\nu}_L\nu_R + \bar{\nu}_R\nu_L)$ could also be included in Eq. 1 but it is not required by the self-consistency arguments. Note that $\tilde{m}_M^2$ is not a positively defined parameter.

Equation (1) is a generic consequence of weak-scale softly broken supersymmetry and does not depend on the specific mechanism of mass generation in the low-energy theory. For the sake of simplicity and without any loss of generality we ignore possible neutrino mixing.

The low-energy theorem proven in Ref. [12] relates the following three $(B-L)$-violating quantities: the neutrino Majorana mass m_M^ν, the "Majorana"-like sneutrino mass $\tilde{m}_M$, and the amplitude of $0\nu\beta\beta$ decay, $R_{0\nu\beta\beta}$. Here we shortly describe the proof of this theorem.

It is relatively easy to see that if at least one of the quantities is nonzero, the two others are generated in higher orders of perturbation theory as demonstrated in Fig. 2, where only dominant diagrams are shown. Internal lines in these diagrams are neutralinos χ_i, gluinos $\tilde{g}$, charginos $\chi^\pm$, selectron $\tilde{e}$, u squark $\tilde{u}$, and sneutrino $\tilde{\nu}$. The latter is to be identified with the $(B-L)$-violating "Majorana" propagator

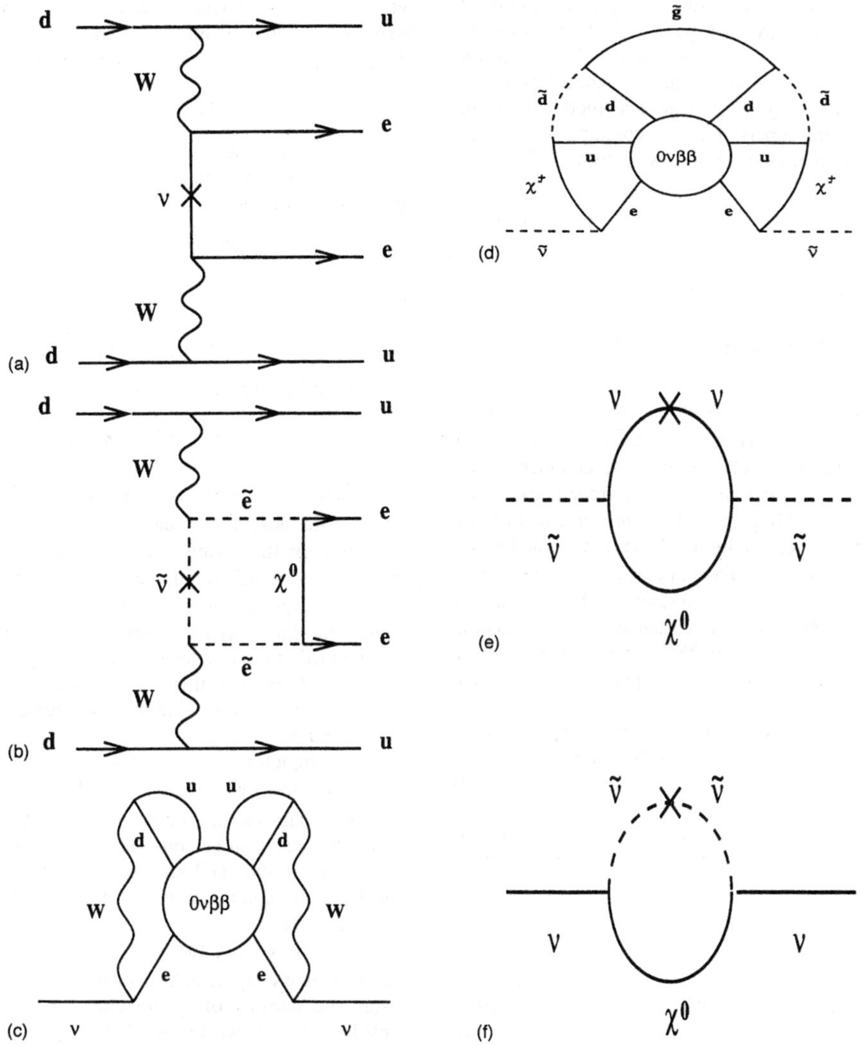

FIG. 2. Lowest-order perturbation theory diagrams representing the relation between the neutrino Majorana mass m_M^ν, the "Majorana"-like $(B-L)$-violating sneutrino mass $\widetilde{m}_M$, and the amplitude of neutrinoless double beta decay $R_{0\nu\beta\beta}$. (a) The neutrino and (b) an example of sneutrino contribution to the $0\nu\beta\beta$-decay amplitude $R_{0\nu\beta\beta}$, the $0\nu\beta\beta$-vertex contribution to (c) the neutrino Majorana mass and (d) to the "Majorana"-like sneutrino mass, (e) neutrino contribution to the sneutrino "Majorana"-like mass, and (f) sneutrino contribution to the neutrino Majorana mass. Crossed (s)neutrino lines correspond to the $(B-L)$-violating propagators.

proportional to $\widetilde{m}_M^2$. The sneutrino "Majorana" propagator was explicitly derived in Ref. [12] and will be given below.

The various diagrams lead to relations among the three $(B-L)$-violating observables, which we write down schematically

$$z_i = \sum_{i \neq j} a_{ij} \cdot z_j + \mathcal{A}_i . \tag{2}$$

Here, z_i can stand for $z_i = m_M^\nu, \widetilde{m}_M^2, R_{0\nu\beta\beta}$. The coefficients a_{ij} correspond to contributions of the diagrams in Figs. 2(a)–2(f) so that $i,j = a,b,c,d,e,f$. Terms $\mathcal{A}_i$ represent any other possible contributions. The explicit form of a_{ij} and $\mathcal{A}_i$ is not essential in the following. Important is only the presence of a correlation between m_M^ν, $\widetilde{m}_M^2$, and $R_{0\nu\beta\beta}$, expressed by Eq. 2.

Now we are going to prove that if $z_{i_1}=0$, then $z_{i_2}=z_{i_3}=0$ (the same will be true for any permutation). On the basis of Fig. 2 and Eqs. (2) one can expect such properties of the set of observables z_i. Indeed $z_{i_1}=0$ in the left-hand side of Eq. (2) strongly disfavors $z_{j_2}\neq 0$ and $z_{j_3}\neq 0$, because it requires either all the three terms in the right-hand sides to vanish or their net cancelation. The latter is "unnatural." Even if such a cancellation were done by hand, using (unnatural) fine-tuning of certain parameters, in some specific order of perturbation theory, it would be spoiled again in higher orders of perturbation theory. The cancellation of all terms in the right-hand side of Eqs. (2) in all orders of perturbation theory could only be guaranteed by a special unbroken symmetry. Let us envisage this possibility in details.

The effective Lagrangian of a generic model of weak-scale softly broken supersymmetry contains after electroweak symmetry breaking the following terms [13]:

$$\mathcal{L}=-\sqrt{2}g\epsilon_i\cdot\overline{\nu}_L\chi_i\tilde{\nu}_L-g\epsilon_i^-\cdot\overline{e}_L\chi_i^-\tilde{\nu}_L-g\epsilon_i^+\cdot\overline{\nu}_L\chi_i^+\tilde{e}_L$$
$$+\frac{g}{\sqrt{2}}(\overline{\nu}_L\gamma^\mu e_L+\overline{u}_L\gamma^\mu d_L)W_\mu^++g\cdot\overline{\chi}_i\gamma^\mu$$
$$\times(O_{ij}^LP_L+O_{ij}^RP_R)\chi_j^+W_\mu^-+\cdots+\text{H.c.}\tag{3}$$

The ellipsis denotes other terms which are not essential for our further considerations. Here, $\tilde{\nu}_L$ and $\tilde{e}_L$ represent scalar superpartners of the left-handed neutrino ν_L and electron e_L fields. The chargino $\chi_i^\pm$ and neutralino χ_i are superpositions of the gaugino and the Higgsino fields. The contents of these superpositions depend on the model. Note that the neutralino is a Majorana field $\chi_i^c=\chi_i$. The explicit form of the coefficients ϵ_i, $\epsilon_i^\pm$, and $O_{ij}^{L,R}$, is also unessential. For the case of the MSSM one can find them, for instance, in [13]. Equation (3) is a general consequence of the underlying weak-scale softly broken supersymmetry and the spontaneously broken electroweak gauge symmetry.

The Lagrangian (3) does not possess any continuous symmetry having nontrivial $(B-L)$ transformation properties. Recall, that $U(1)_{B-L}$ is assumed to be broken since we admit $(B-L)$-violating mass terms in Eq. (1). However, there might be an appropriate unbroken discrete symmetry. Let us specify this discrete symmetry group by the following field transformations:

$$\nu\to\eta_\nu\nu,\quad \tilde{\nu}\to\eta_{\tilde{\nu}}\tilde{\nu},\quad e_L\to\eta_e e_L,\quad \tilde{e}_L\to\eta_{\tilde{e}}\tilde{e}_L,$$
$$W^+\to\eta_W W^+,\quad \chi_i\to\eta_{\chi_i}\chi_i,\quad \chi^+\to\eta_{\chi^+}\chi^+,\quad q_L\to\eta_q q_L.\tag{4}$$

Here η_i are phase factors. Since the Lagrangian 3 is assumed to be invariant under these transformations, one obtains the following relations:

$$\eta_\nu^*\eta_{\tilde{\nu}}\eta_{\chi_i}=1,\quad \eta_e\eta_{\chi^+}\eta_{\tilde{\nu}}^*=1,\ldots,$$
$$\eta_e\eta_W\eta_\nu^*=1,\quad \eta_W^*\eta_{\chi^+}\eta_{\chi_i}^*=1,\quad \eta_d\eta_W\eta_u^*=1,\ldots.\tag{5}$$

The ellipsis denotes other relations which are not essential here. The complete set of these equations defines the admissible discrete symmetry group of the Lagrangian in Eq. (3).

Let us find the transformation property of the operator structure responsible for $0\nu\beta\beta$ decay under this group. At the quark level $0\nu\beta\beta$ decay implies the transition $dd\to uuee$, described by the effective operator

$$\mathcal{O}_{0\nu\beta\beta}=\alpha_i\cdot\overline{u}\Gamma_i^{(1)}d\cdot\overline{u}\Gamma_i^{(1)}d\cdot\overline{e}\Gamma_i^{(2)}e^c,\tag{6}$$

where α_i are numerical constants and $\Gamma_i^{(k)}$ are certain combinations of Dirac γ matrices. The $0\nu\beta\beta$-decay amplitude $R_{0\nu\beta\beta}$ is related to the matrix element of this operator,

$$R_{0\nu\beta\beta}\sim\langle 2e^-(A,Z+2)|\mathcal{O}_{0\nu\beta\beta}|(A,Z)\rangle,\tag{7}$$

where (A,Z) is a nucleus with the atomic weight A and the total charge Z. The operator in Eq. (6) transforms under the group (4) as follows:

$$\mathcal{O}_{0\nu\beta\beta}\to\eta_{0\nu\beta\beta}^2\mathcal{O}_{0\nu\beta\beta},\tag{8}$$

with

$$\eta_{0\nu\beta\beta}=\eta_d\eta_u^*\eta_e^*.\tag{9}$$

Solving Eqs. (5) and (9), one finds

$$\eta_\nu^2=\eta_{\tilde{\nu}}^2=(\eta_{0\nu\beta\beta}^*)^2.\tag{10}$$

This relation proves the statements 1 and 2. To see this we note that the observable quantity $z_i=(m_M^\nu,\ \tilde{m}_M^2,\ R_{0\nu\beta\beta})$ is forbidden by this symmetry if the corresponding discrete group factor is nontrivial, i.e., $\eta_i^2\neq 1$. On the contrary, if $\eta_i^2=1$, this quantity is not protected by the symmetry and appears in higher orders of perturbation theory, even if it is not included at the tree level. Relation (10) claims that if one of the z_i is forbidden, then the two others are also forbidden and, vice versa, if one of them is not forbidden, they are all not forbidden.

This completes the proof of the theorem relating the neutrino Majorana mass m_M^ν, the "Majorana"-like sneutrino mass $\tilde{m}_M$, and the amplitude of $0\nu\beta\beta$ decay, $R_{0\nu\beta\beta}$. The proven theorem can be considered as a supersymmetric generalization of the well-known theorem [14] relating only neutrino Majorana mass and the neutrinoless double beta decay amplitude.

Let us show that the "Dirac"-like $(B-L)$-conserving sneutrino mass $\tilde{m}_D$ should also be present in the theory to ensure the stability of the vacuum state. Towards this end consider the last two terms of Eq. (1) which we denote as $\mathcal{L}_{\text{mass}}^{\tilde{\nu}}$ and use the real field representation for the complex scalar sneutrino field,

$$\tilde{\nu}=(\tilde{\nu}_1+i\tilde{\nu}_2)/\sqrt{2},\tag{11}$$

where $\tilde{\nu}_{1,2}$ are real fields. Then

$$\mathcal{L}_{\text{mass}}^{\tilde{\nu}}=-\tfrac{1}{2}(\tilde{m}_M^2\tilde{\nu}_L\tilde{\nu}_L+\text{H.c.})-\tilde{m}_D^2\tilde{\nu}_L^*\tilde{\nu}_L$$
$$=-\tfrac{1}{2}\tilde{m}_1^2\tilde{\nu}_1^2-\tfrac{1}{2}\tilde{m}_2^2\tilde{\nu}_2^2,\tag{12}$$

where

$$\tilde{m}_{1,2}^2=\tilde{m}_D^2\pm|\tilde{m}_M^2|.\tag{13}$$

Assume the vacuum state is stable. Then $\tilde{m}_{1,2}^2\geqslant 0$, i.e., $\tilde{m}_D^2\geqslant|\tilde{m}_M^2|$; otherwise, the vacuum is unstable and subsequent spontaneous symmetry breaking occurs via nonzero vacuum expectation values of the sneutrino fields $\langle\tilde{\nu}_i\rangle\neq 0$. The broken symmetry in this case is the R parity. It is a discrete symmetry defined as $R_p=(-1)^{3B+L+2S}$, where S, B, and L are the spin, the baryon, and the lepton quantum number.

Therefore, as indicated at the beginning of this section the self-consistent structure of the mass terms of the neutrino-sneutrino sector are given by Eq. (1). The mass parameter $\tilde{m}_M$ gives a measure of sneutrino-antisneutrino mixing ($\tilde{\nu}-\tilde{\nu}^*$). This $(B-L)$-violating effect is an evident manifestation of the second term in Eq. (1). On the other hand, a finite $\tilde{m}_M$ gives rise to splitting the complex scalar field $\tilde{\nu}=(\tilde{\nu}_1+i\tilde{\nu}_2)/\sqrt{2}$ into two real mass eigenfields $\tilde{\nu}_{1,2}$ with the masses $\tilde{m}_{1,2}^2=\tilde{m}_D^2\pm|\tilde{m}_M^2|$. According to the above definition, $\tilde{\nu}_1$ is the CP-even state, while $\tilde{\nu}_2$ is the CP-odd one.

Let us write down the explicit form of the above-mentioned $(B-L)$-violating "Majorana" propagator $\Delta_{\tilde{\nu}}^M$ for the sneutrino [12] which is necessary for our subsequent considerations. It can be derived by the use of the real field representation as in Eq. (12). For comparison we also give the $(B-L)$-conserving "Dirac" $\Delta_{\tilde{\nu}}^D$ sneutrino propagator,

$$\Delta_{\tilde{\nu}}^D(x-y)=i\langle 0|T[\tilde{\nu}_L(x)\tilde{\nu}_L^\dagger(y)]|0\rangle = \tfrac{1}{2}[\Delta_{\tilde{m}_1}(x-y)+\Delta_{\tilde{m}_2}(x-y)], \quad (14)$$

$$\Delta_{\tilde{\nu}}^M(x-y)=i\langle 0|T[\tilde{\nu}_L(x)\tilde{\nu}_L(y)]|0\rangle = \tfrac{1}{2}[\Delta_{\tilde{m}_1}(x-y)-\Delta_{\tilde{m}_2}(x-y)], \quad (15)$$

where

$$\Delta_{\tilde{m}_i}(x)=\int\frac{d^4k}{(2\pi)^4}\frac{e^{-ikx}}{\tilde{m}_i^2-k^2-i\epsilon} \quad (16)$$

is the ordinary propagator for a scalar particle with mass $\tilde{m}_i$. Using the definition of $\tilde{m}_{1,2}$ as in Eq. (12) one finds

$$\Delta_{\tilde{\nu}}^D(x)=\int\frac{d^4k}{(2\pi)^4}\frac{\tilde{m}_D^2-k^2}{(\tilde{m}_1^2-k^2-i\epsilon)(\tilde{m}_2^2-k^2-i\epsilon)}e^{-ikx}, \quad (17)$$

$$\Delta_{\tilde{\nu}}^M(x)=-\tilde{m}_M^2\int\frac{d^4k}{(2\pi)^4}\frac{e^{-ikx}}{(\tilde{m}_1^2-k^2-i\epsilon)(\tilde{m}_2^2-k^2-i\epsilon)}. \quad (18)$$

It is seen that in absence of the $(B-L)$-violating sneutrino "Majorana"-like mass term $\tilde{m}_M^2=0$ the $(B-L)$-violating propagator $\Delta_{\tilde{\nu}}^M$ vanishes while the $(B-L)$-conserving one $\Delta_{\tilde{\nu}}^D$ becomes the ordinary propagator of a scalar particle with mass $\tilde{m}_1=\tilde{m}_2=\tilde{m}_D$.

Majorana neutrino fields can propagate in a virtual state conserving the $(B-L)$ quantum number as well as violating it. In the Majorana representation of the neutrino field,

$$\nu=P_L\nu_L+P_R(\nu_L)^c, \quad (19)$$

where $\nu^c=\nu$, the corresponding $(B-L)$-conserving (S^D) and $(B-L)$-violating (S^M) propagators can be written as

$$S^D(x-y)=i\langle 0|T[\nu_L(x)\bar{\nu}_L(y)]|0\rangle = iP_L\langle 0|T[\nu(x)\bar{\nu}(y)]|0\rangle P_R = \int\frac{d^4k}{(2\pi)^4}\frac{\gamma_\mu k^\mu}{m_\nu^2-k^2-i\epsilon}e^{-ik(x-y)}, \quad (20)$$

$$S^M(x-y)=i\langle 0|T[\nu_L(x)\bar{\nu_L^c}(y)]|0\rangle = P_L\langle 0|T[\nu(x)\bar{\nu}(y)]|0\rangle P_L = m_\nu\int\frac{d^4k}{(2\pi)^4}\frac{1}{m_\nu^2-k^2-i\epsilon}e^{-ik(x-y)}, \quad (21)$$

where $m_\nu\equiv m_M^\nu$ for simplicity. Therefore, the effect of $(B-L)$ violation originating from the neutrino propagator is proportional to the Majorana neutrino mass while a $(B-L)$-conserving contribution to $0\nu\beta\beta$ decay via neutrino propagation in the Dirac mode does essentially not depend on the neutrino mass and leads to a contribution proportional to the mean neutrino momentum in a nucleus. The latter is typically of the order of the Fermi momentum $\sim p_F\approx 80$ MeV. As a result such a contribution, if it exists, is greatly enhanced compared to the Majorana mass contribution.

III. MSSM CONTRIBUTION TO $0\nu\beta\beta$-DECAY: GENERAL PROPERTIES AND EFFECTIVE LAGRANGIAN

In this section we are considering the general properties of the R_P-conserving MSSM contribution to $0\nu\beta\beta$ decay and derive the corresponding effective Lagrangian in terms of color-singlet quark charged currents.

Within our supersymmetric framework there are two sources of lepton number violation, the Majorana neutrino mass m_M^ν and the (square of the) "Majorana"-like sneutrino mass $\tilde{m}_M^2$. Both of them violate—by construction—lepton number by two units. $0\nu\beta\beta$ decay also violates L by two units. The $0\nu\beta\beta$ amplitude $R_{0\nu\beta\beta}$ will therefore be proportional,

$$R_{0\nu\beta\beta}\sim m_M^\nu\mathcal{O}^{(1)}+\tilde{m}_M^2\mathcal{O}^{(2)}, \quad (22)$$

with $\mathcal{O}^{(i)}$ representing some matrix elements. Contributions proportional to m_M^ν can be classified by $\mathcal{O}^{(1)}=\mathcal{O}^{(1)}_{(\mathrm{SM})}+\mathcal{O}^{(1)}_{(\mathrm{SUSY})}$, where the first part stands symbolically for the usual mass mechanism diagram [see Fig. 1(a)], while the second part summarizes all kinds of diagrams involving virtual SUSY particles. An example of this type of contribution was given in the introduction [see Fig. 1(b)]. Clearly, all diagrams of $0\nu\beta\beta$ decay involving SUSY particles must have at least six basic vertices. They are thus of higher order compared to Fig. 1(a) and can be safely neglected: $\mathcal{O}^{(1)}=\mathcal{O}^{(1)}_{(\mathrm{SM})}+\mathcal{O}^{(1)}_{(\mathrm{SUSY})}\cong\mathcal{O}^{(1)}_{(\mathrm{SM})}$.

Let us consider $R_{0\nu\beta\beta}$ in more detail. At the quark level neutrinoless double beta decay is induced by the transition of two d quarks into two u quarks and two electrons. This process is schematically represented by the diagram in Fig. 3(a), encoding all possible contributions to $0\nu\beta\beta$ decay. In

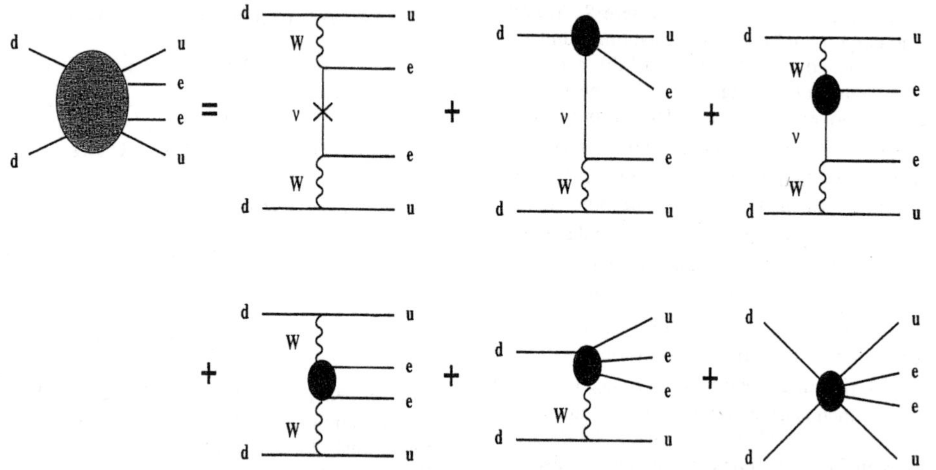

FIG. 3. The decomposition of the $0\nu\beta\beta$ effective vertex. (a) To the left: the $0\nu\beta\beta$ effective vertex. To the right the decomposition: (b)–(d) first line, neutrino-accompanied contributions to $0\nu\beta\beta$ decay; (e)–(g) second line, purely supersymmetric contributions without neutrinos.

the R_P-conserving supersymmetric model it is useful to decompose the basic diagram Fig. 3(a) as shown in Figs. 3(b)–3(g).

Given that the low-energy theory contains a light neutrino there are always contributions involving a long-distance interaction component associated with the light neutrino exchange. Let us use this fact for a decomposition of the quark-lepton $0\nu\beta\beta$ effective vertex in Fig. 3(a) into different parts explicitly showing the presence of the neutrino line, Figs. 3(b)–3(d). It is also instructive to isolate the SUSY contributions in the form of effective vertices induced by heavy SUSY particle exchange. This decomposition is indicated by the small black spots corresponding to the short-distance, approximately pointlike, effective SUSY-induced interactions, Figs. 3(c)–3(g). The first diagram, Fig. 3(b), corresponds to the conventional standard model Majorana neutrino exchange contribution mentioned above. The crossed neutrino line indicates the lepton-number-violating Majorana neutrino propagator S^M. Diagrams Figs. 3(c) and 3(d) correspond to the neutrino accompanied SUSY contributions. Here, neutrino lines are uncrossed and correspond to lepton-number-conserving neutrino propagators S^D. (As noted above we neglect SUSY diagrams with lepton-number-violating propagators proportional to the small neutrino mass m_M^ν.) The last three diagrams represent the purely supersymmetric contribution and are of short-ranged nature.

This decomposition allows one to apply, if necessary, a Fierz rearrangement to the approximately pointlike black-spot SUSY vertices in Fig. 3 and represent them in the form of a product of color-singlet quark charged currents and a leptonic part. Such a representation is crucial for the derivation of the $0\nu\beta\beta$-transition operators in the nonrelativistic impulse approximation and for the subsequent nuclear structure calculations discussed in Sec. IV.

Assuming the Fierz rearrangement applied to the pointlike SUSY vertices one can write down the general form of the effective Lagrangian reproducing the decomposition in Fig. 3 within fourth order of perturbation theory. It can be written in the form

$$\mathcal{L}_{0\nu\beta\beta}=\mathcal{L}_{W\bar{f}f}+\frac{\lambda_i^{(1)}}{m_{\rm SUSY}^{n_1}}j_i\cdot\bar{e}\Gamma_{i(1)}\nu_L^c+\frac{\lambda_i^{(2)}}{m_{\rm SUSY}^{n_2}}W_\mu^-\cdot\bar{e}\Gamma_{i(2)}^\mu\nu_L^c$$
$$+\frac{\lambda_i^{(3)}}{m_{\rm SUSY}^{n_3}}W_\mu^-W_\nu^-\cdot\bar{e}\Gamma_{i(3)}^{\mu\nu}e^c+\frac{\lambda_i^{(4)}}{m_{\rm SUSY}^{n_4}}j_i^\mu W_\mu^-\cdot\bar{e}\Gamma_{i(4)}e^c$$
$$+\frac{\lambda_{ij}^{(5)}}{m_{\rm SUSY}^{n_5}}j_ij_j\cdot\bar{e}\Gamma_{ij(5)}e^c, \tag{23}$$

where the SM term $\mathcal{L}_{W\bar{f}f}$ is also introduced [see Eq. (A2) in Appendix A]. Color-singlet local diquark operators are defined as

$$j_i=\bar{u}^\alpha\mathcal{O}_i d_\alpha, \tag{24}$$

with α being a color index. The objects $\Gamma_{i(k)}$ and $\mathcal{O}_i$ are constructed of Dirac gamma matrices as well as derivatives.

Effective couplings $\lambda^{(k)}$ are dimensionless constants. Different terms are scaled out by the characteristic SUSY-breaking mass scale $m_{\rm SUSY}$ with an appropriate degree n_i to accommodate the correct physical dimension of the corresponding term. As seen from the leptonic part of the effective Lagrangian (23), the first term conserves lepton number $(\Delta L=0)$ while the remaining five terms violate it by two units $(\Delta L=2)$.

In the effective Lagrangian (23) we neglected possible L-conserving terms with the lepton operator structure $\sim\bar{e}\Gamma_i\nu_L$. Their contributions to the $0\nu\beta\beta$ amplitude are strongly suppressed compared to the contributions of the similar L-violating terms $\sim\bar{e}\Gamma_i\nu_L^c$. To see this fact let us have a closer look at the corresponding leading order diagrams in Figs. 3(c) and 3(d). The bottom parts of these diagrams are the SM charged current (SMCC) interactions of the form $(\bar{u}\gamma_\mu P_L d)(\bar{e}\gamma^\mu P_L\nu_n)U_{en}$, while the top parts correspond to the effective SUSY vertices. If they are given by the second and third terms of the effective Lagrangian, Eq. (23), the resulting leptonic tensor $\mathcal{L}_{\rm SUSY}$ can then be written schematically as

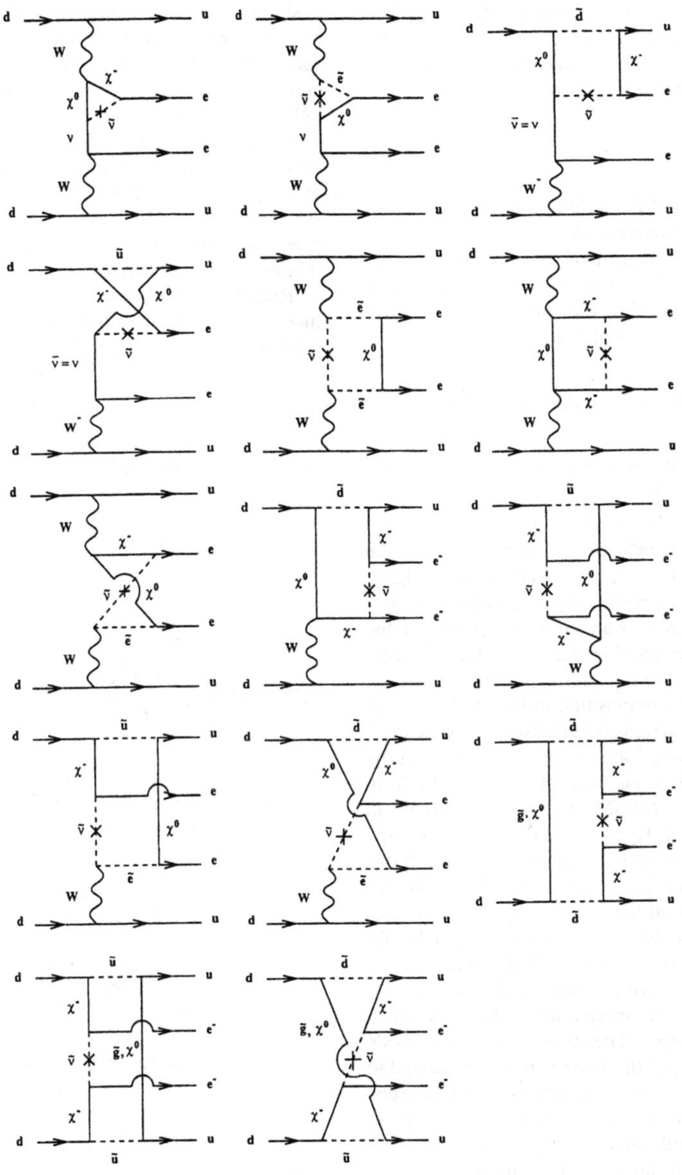

FIG. 4. R_P-conserving MSSM contributions to the $0\nu\beta\beta$-decay amplitude. Leading order diagrams; see also the text.

$$\mathcal{L}_{\Delta L=2} \sim \bar{e}\gamma_\mu P_L \langle 0|T(\nu_k \bar{\nu}_n)|0\rangle P_R \Gamma e^c \cdot U_{ek} U_{en}^* \sim \bar{e}\gamma_\mu \gamma_\nu P_R \Gamma e^c \cdot q^\nu / q^2, \tag{25}$$

where q is the neutrino momentum. On the right-hand side neutrino masses $m_M^{\nu_n}$ are neglected since $m_M^{\nu_n} \ll \langle q \rangle$, where $\langle q \rangle \sim p_F \approx 100$ MeV is the average momentum of a neutrino propagating in a nucleus (p_F is the nucleon Fermi momentum). The mixing matrix elements disappear on the right-hand side due to the unitarity relation $U_{ek} U_{en}^* \delta_{kn} = 1$. This should be compared with the contribution of the possible L-conserving terms $\sim \bar{e}\Gamma_i \nu_L$ which we neglected in the effective Lagrangian in Eq. (23). The leptonic tensor in this case takes the form

$$\mathcal{L}_{\Delta L=0} \sim \bar{e}\Gamma_i P_L \langle 0|T(\nu_k \bar{\nu}_n)|0\rangle P_L \gamma_\rho e^c \cdot U_{ek} U_{en} \sim \bar{e}\Gamma_i \gamma_\rho P_R e^c \cdot \langle m_\nu \rangle / q^2, \tag{26}$$

where $\langle m_\nu \rangle = m_M^{\nu_n} U_{en}^2$. The same structure appears in the standard neutrino mass mechanism with the SMCC at both ends of the virtual neutrino line.

Comparing Eqs. (25) and (26) one can see that the SUSY contribution corresponding to the $\Delta L=2$ operators receives from the leptonic sector a huge enhancement compared to the contribution of the $\Delta L=0$ operators. In fact

$$\mathcal{L}_{\Delta L=2}/\mathcal{L}_{\Delta L=0} \sim p_F/\langle m_\nu\rangle \sim 10^8 \times (1\ \mathrm{eV}/\langle m_\nu\rangle). \quad (27)$$

For this reason we neglected the SUSY-induced $\Delta L=0$ operators in the effective Lagrangian (23).

Now let us turn from the general consideration to the concrete case of the SUSY contribution to $0\nu\beta\beta$ decay within the MSSM. The following Lagrangian terms are relevant to $0\nu\beta\beta$-quark transitions in Fig. 3(a):

$$\mathcal{L}_{\rm int} = \mathcal{L}_{W\tilde{f}f} + \mathcal{L}_{W\tilde{f}\tilde{f}} + \mathcal{L}_{\chi^+ f\tilde{f}} + \mathcal{L}_{\chi f\tilde{f}} + \mathcal{L}_{\tilde{g}\tilde{q}q} + \mathcal{L}_{W\chi^+\chi}. \quad (28)$$

The Lagrangian terms in the right-hand side (RHS) are explicitly given in Appendix A.

Starting from this Lagrangian one can find 14 dominant diagrams proportional to $\widetilde{m}_M^2$ which contribute to the $0\nu\beta\beta$-quark transition in Fig. 3(a). They are listed in Fig. 4. (Note that in addition to the graphs shown, there exist several graphs corresponding simply to an exchange of two of the external momenta and are not shown for brevity.) As seen, all diagrams in Fig. 4 fall into five classes represented by the last five diagrams of the decomposition in Fig. 3. The supersymmetric part of these diagrams, as discussed above, can be parametrized by the effective Lagrangian $\mathcal{L}_{0\nu\beta\beta}$ given in the general form of Eq. (23). One can reconstruct a specific form of this Lagrangian in the MSSM comparing diagrams in Figs. 3 and 4 and separating the basic SUSY vertices denoted in Fig. 3 by the black spots representing five different terms of the effective Lagrangian $\mathcal{L}_{0\nu\beta\beta}$ in Eq. (23). The next step of the derivation is based on the standard approximate procedure relying on the fact that all intermediate particles involved in evaluation of the effective SUSY vertices are heavy SUSY particles with typical masses of order $m_{\rm SUSY}$. As a result these vertices can approximately be represented in the form of local operators. The local form of the SUSY operators allows one to apply the Fierz rearrangement and to collect the quark fields in the color-singlet quark charged currents. Straightforward realization of this strategy leads to an effective Lagrangian with the following leading order operators violating the lepton number L by 2 units:

$$\mathcal{L}_{\Delta L=2} = -(\eta^{(1)}_{WW} + \eta^{(2)}_{WW} + \eta^{(3)}_{WW})\frac{W^-_\mu W^{-\mu}}{m_{\rm SUSY}} \cdot \bar{e}(1+\gamma_5)e^c + (\eta_{\tilde{g}\tilde{u}} + \eta_{\tilde{g}\tilde{d}})\frac{j^\mu_{\rm AV} j_{\mu{\rm AV}}}{m^5_{\rm SUSY}} \cdot \bar{e}(1+\gamma_5)e^c. \quad (29)$$

The color-singlet quark charged currents are defined as usual:

$$j^\mu_{\rm AV} = \cos\theta_c \bar{u}\gamma^\mu(1-\gamma_5)d. \quad (30)$$

Note that since we take only the leading order contributions into account in Eq. (29), not all possible terms of the decomposition Eq. (23) are retained in Eq. (29). Naively one might have expected that the diagrams in Fig. 4 [and the corresponding terms in Eq. (23)] are ordered with respect to decreasing importance, since a larger number of heavy sparticles in the loops results in larger (loop) suppression factors. However, the explicit calculation shows that this is not the case. Terms corresponding in structure to the second, third, and the fifth terms in Eq. (23) are suppressed by the helicity structure of the basic MSSM interactions and/or typical suppression factors of order $m_e/m_{\rm SUSY}$ from left-right sfermion mixing or Higgsino-like interactions.

Retaining only leading contributions to the operator structures in Eq. (29), the dimensionless lepton-number-violating parameters take the form

$$\eta_{\tilde{g}\tilde{d}} = \frac{g_s^2 g^4}{72}\left(\frac{\widetilde{m}_M}{m_{\rm SUSY}}\right)^2 \sum_{i,j} U_{i1}V_{i1}U_{j1}V_{j1} \times \left(\frac{m_{\chi^\pm_j}}{m_{\rm SUSY}}\right)\left(\frac{m_{\chi^\pm_i}}{m_{\rm SUSY}}\right)\left(\frac{m_{\tilde{g}}}{m_{\rm SUSY}}\right)\mathcal{G}(m_{\tilde{g}}, m_{\chi^\pm_j}, m_{\chi^\pm_i}), \quad (31)$$

$$\eta_{\tilde{g}\tilde{u}} = \frac{g_s^2 g^4}{72}\left(\frac{\widetilde{m}_M}{m_{\rm SUSY}}\right)^2 \sum_{i,j} V^2_{i1}V^2_{j1}\left(\frac{m_{\tilde{g}}}{m_{\rm SUSY}}\right)\mathcal{F}(m_{\tilde{g}}, m_{\chi^\pm_j}, m_{\chi^\pm_i}), \quad (32)$$

$$\eta^{(1)}_{WW} = \frac{g^4}{4}\left(\frac{\widetilde{m}_M}{m_{\rm SUSY}}\right)^2 \sum_{i,j,k} V_{k1}V_{j1} \times \Bigg[\mathcal{O}^L_{ik}\mathcal{O}^L_{ij}\left(\frac{m_{\chi_i}}{m_{\rm SUSY}}\right)\mathcal{J}(m_{\chi_i}, m_{\chi^\pm_j}, m_{\chi^\pm_k}) + \mathcal{O}^R_{ik}\mathcal{O}^L_{ij}\left(\frac{m_{\chi^\pm_k}}{m_{\rm SUSY}}\right)\mathcal{J}(m_{\chi_i}, m_{\chi^\pm_j}, m_{\chi^\pm_k}) + \mathcal{O}^R_{ik}\mathcal{O}^R_{ij}\left(\frac{m_{\chi^\pm_j}}{m_{\rm SUSY}}\right)\left(\frac{m_{\chi^\pm_k}}{m_{\rm SUSY}}\right)\left(\frac{m_{\chi_i}}{m_{\rm SUSY}}\right) \times \mathcal{I}(m_{\chi_i}, m_{\chi^\pm_j}, m_{\chi^\pm_k})\Bigg], \quad (33)$$

$$\eta^{(2)}_{WW} = \frac{g^4}{4}\left(\frac{\widetilde{m}_M}{m_{\rm SUSY}}\right)^2 \sum_{i,j} \mathcal{J}(m_{\tilde{e}}, m_{\chi_i}, m_{\chi^\pm_j})\epsilon_{L_i}(e)V_{j1} \times \left[\mathcal{O}^R_{ij}\left(\frac{m_{\chi^\pm_j}}{m_{\rm SUSY}}\right) + \mathcal{O}^L_{ij}\left(\frac{m_{\chi_i}}{m_{\rm SUSY}}\right)\right], \quad (34)$$

$$\eta^{(3)}_{WW} = \frac{g^4}{4}\left(\frac{\widetilde{m}_M}{m_{\rm SUSY}}\right)^2 \sum_i \mathcal{J}(m_{\chi_i}, m_{\tilde{e}}, m_{\tilde{e}}) \times \epsilon^2_{L_i}(e)\left(\frac{m_{\chi_i}}{m_{\rm SUSY}}\right). \quad (35)$$

The dimensionless loop factors $\mathcal{F}(m_i)$, $\mathcal{G}(m_i)$, $\mathcal{J}(m_i)$, and $\mathcal{I}(m_i)$ are given in Appendix B. They depend on the sparticle masses m_i in the corresponding loop. Recall that g and g_s are the $SU(2)_L$ and $SU(3)_c$ coupling constants. Further definitions of couplings and mixing parameters can be found in Appendix A.

The next step of the calculation deals with reformulating the problem in terms of nucleon degrees of freedom instead of quark ones. This is relevant for the nuclear structure part of the calculations.

IV. FROM THE QUARK TO NUCLEAR LEVEL

So far the discussion has focused on particle physics aspects, deriving the low-energy effective Lagrangian in Eq. (29) formulated in terms of quark fields. However, our goal is the calculation of the amplitude $R_{0\nu\beta\beta}$ of $0\nu\beta\beta$ decay which is a nuclear process proceeding not at the level of quark degrees of freedom but at the level of nucleon ones. Formally one can write down

$$\mathcal{R}_{0\nu\beta\beta}=\langle (A,Z+2),2e^-|S-1|(A,Z)\rangle=\langle (A,Z+2),2e^-|T\,\exp\left[i\int d^4x\mathcal{L}_{0\nu\beta\beta}(x)\right]|(A,Z)\rangle, \tag{36}$$

where the effective Lagrangian $\mathcal{L}_{0\nu\beta\beta}=\mathcal{L}_{W\bar{f}f}+\mathcal{L}_{\Delta L=2}$ is given by Eqs. (A2) and (29). The nuclear structure is involved via the initial (A,Z) and the final $(A,Z+2)$ nuclear states having the same atomic weight A, but different electric charges Z and $Z+2$. The standard framework for the calculation of this nuclear matrix element is the nonrelativistic impulse approximation (NRIA). It implies the substitution of the quark current j^μ_{AV} in the effective Lagrangian $\mathcal{L}_{0\nu\beta\beta}$ in Eq. (36) by the nonrelativistic nucleon current

$$j^\mu_{AV}\overset{\text{NRIAJ}}{\longrightarrow} J^\mu.$$

The latter is an incoherent sum over individual nucleon currents of a nucleus and is given by the formula [15]

$$J^\mu(\mathbf{x})=\sum_i \tau_+^{(i)}[(f_V-f_AC_i)g^{\mu 0}-(f_A\sigma_i^k+f_VD_i^k)g^{\mu k}]\times\frac{m_A^3}{8\pi}e^{-m_A|\mathbf{x}-\mathbf{r}_i|}, \tag{37}$$

Here $f_V\approx 1$, $f_A\approx 1.261$, $g^{\mu\nu}$ is the metric tensor, $\tau_+^{(i)}$ is the isospin-raising operator, $\mathbf{r}_i$ is the position of the ith nucleon, and the superscript k stands for the spatial component. C_i and $\mathbf{D}_i$ are the well-known scalar and the vector nuclear recoil terms given by [2]

$$C_i=\frac{1}{2m_P}\left[(\mathbf{p}_i+\mathbf{p}_i')\cdot\boldsymbol{\sigma}_i-\frac{f_P}{f_A}(E_i-E_i')\mathbf{q}_i\cdot\boldsymbol{\sigma}_i\right], \tag{38}$$

$$\mathbf{D}_i=\frac{1}{2m_P}\left[(\mathbf{p}_i+\mathbf{p}_i')+i\left(1-2m_P\frac{f_W}{f_V}\right)\mathbf{q}_i\times\boldsymbol{\sigma}_i\right], \tag{39}$$

Here $(\mathbf{p}_i,E_i)$ and $(\mathbf{p}_i',E_i')$ are initial and final three-momentum and energy of the ith nucleon and the three-momentum transfer is $\mathbf{q}_i=\mathbf{p}_i-\mathbf{p}_i'$. The nucleon couplings obey the relations

$$f_W/f_V=-(\mu_p-\mu_n)/(2m_P)\approx 3.7/(2m_P),$$
$$f_P/f_A=2m_P/m_\pi^2, \tag{40}$$

where m_π is the pion mass and $\mu_{p(n)}$ is the proton (neutron) magnetic moment.

Since we are interested in the dominant contributions only, recoil terms will be neglected in the rest of this paper and have been given above for completeness.

The exponential factor in Eq. (37) is introduced instead of the local δ function in order to take into account the finite nucleon size. It is the Fourier transform of the nucleon form factor $F(\mathbf{q}^2)$ in the conventional parametrization a dipole form

$$F(\mathbf{q}^2)=\left(1+\frac{\mathbf{q}^2}{m_A^2}\right)^{-2}, \tag{41}$$

with $m_A=0.85$ GeV. The finite nucleon size effects are known [16] to be important for the short-distance contributions to the $0\nu\beta\beta$ amplitude such as those corresponding to the dominant terms in the effective Lagrangian (29).

Now, starting from Eq. (36), it is straightforward to calculate the $0\nu\beta\beta$ amplitude $R_{0\nu\beta\beta}$ within the nonrelativistic impulse approximation. The final result for the $0^+\to 0^+$ transition amplitude can be written as follows:

$$R_{0\nu\beta\beta}(0^+\to 0^+)=\frac{\eta^{\text{SUSY}}}{m_{\text{SUSY}}^5}\sqrt{2}\cdot C_{0\nu}^{-1}[\bar{e}(1+\gamma_5)e^c]\times\langle F|\Omega^{\text{SUSY}}|I\rangle. \tag{42}$$

The normalization factor is

$$C_{0\nu}=\frac{4\pi}{m_Pm_e}\frac{R_0}{f_A^2}. \tag{43}$$

Here, R_0 is the nuclear radius, and m_P and m_e are the proton and the electron masses.

The effective lepton-number-violating parameter is defined as

$$\eta^{\mathrm{SUSY}}=(\eta_{\tilde{g}\tilde{d}}+\eta_{\tilde{g}\tilde{u}})+g^2\left(\frac{m_{\mathrm{SUSY}}}{M_W}\right)^4 \times(\eta_{WW}^{(1)}+\eta_{WW}^{(2)}+\eta_{WW}^{(3)}). \quad (44)$$

In Eq. (42) we have introduced the transition operator Ω^{SUSY} in order to separate the nuclear physics part of the calculation from the particle physics one. Having the transition operator one can calculate the corresponding nuclear matrix element for any $0\nu\beta\beta$-decaying candidate isotope within any specific model of nuclear structure. From now on we define

$$\mathcal{M}^{\mathrm{SUSY}}=\langle F|\Omega^{\mathrm{SUSY}}|I\rangle. \quad (45)$$

This nuclear matrix element is found to be equal to those for heavy neutrino exchange [17]

$$\mathcal{M}^{\mathrm{SUSY}}=\{\mathcal{M}_{F,N}-\mathcal{M}_{\mathrm{GT},N}\}, \quad (46)$$

where

$$\mathcal{M}_{GT,N}=\left(\frac{m_A^2}{m_e m_p}\right)\langle F|\Omega_{\mathrm{GT},N}|I\rangle, \quad (47)$$

where

$$\langle F|\Omega_{\mathrm{GT},N}|I\rangle=\langle F|\sum_{i\neq j}\tau_+^{(i)}\tau_+^{(j)}\boldsymbol{\sigma}_i\cdot\boldsymbol{\sigma}_j\left(\frac{R_0}{r_{ij}}\right)F_N(x_A)|I\rangle \quad (48)$$

and

$$\mathcal{M}_{F,N}=\left(\frac{m_A^2}{m_e m_p}\right)\left(\frac{f_V}{f_A}\right)^2\langle F|\Omega_{F,N}|I\rangle, \quad (49)$$

where

$$\langle F|\Omega_{F,N}|I\rangle=\langle F|\sum_{i\neq j}\tau_+^{(i)}\tau_+^{(j)}\left(\frac{R_0}{r_{ij}}\right)F_N(x_A)|I\rangle. \quad (50)$$

Here, $F_N(x_A)$ is the short-ranged potential

$$F_N(x_A)=4\pi m_A^6 r_{ij}\int\frac{d^3\mathbf{q}}{(2\pi)^3}\frac{1}{(m_A^2+\mathbf{q}^2)^4}e^{i\mathbf{q}\cdot\mathbf{r}_{ij}}, \quad (51)$$

with $x_A=m_A r_{ij}$. This potential takes into account the finite nucleon size [see Eq. (41)]; its analytic solution is given by the formula

$$F_N(x)=\frac{x}{48}(3+3x+x^2)e^{-x}. \quad (52)$$

The above definitions are general in the sense that one can apply nuclear wave functions of any nuclear structure model for their calculation. For the following analysis, on the other hand, numerical values for the matrix elements are needed.

In our numerical analysis we will use the following value for the $0\nu\beta\beta$ decay of ^{76}Ge [17]:

$$\mathcal{M}^{\mathrm{SUSY}}=289. \quad (53)$$

This value is based on a proton-neutron-QRPA model [18], which has been discussed already several times in the literature [18] and has been applied previously to calculations of the R-parity-violating contributions to $0\nu\beta\beta$ decay [8,10], as well as to $0\nu\beta\beta$ decay in left-right symmetric models [17]. We will therefore not repeat the details of the calculation here, and refer for brevity to [18]. Uncertainties associated with the pn-QRPA have been discussed in [8] for $\mathcal{M}_{F,N}$ and $\mathcal{M}_{\mathrm{GT},N}$.

V. $0\nu\beta\beta$ CONSTRAINTS ON $(B-L)$-VIOLATING SNEUTRINO MASS

The $0\nu\beta\beta$-decay amplitude given in Eq. (42) leads to the following half-life formula:

$$[T_{1/2}^{0\nu\beta\beta}(0^+\to0^+)]^{-1}=G_{01}\frac{4m_P^2}{G_F^4}\left|\frac{\eta^{\mathrm{SUSY}}}{m_{\mathrm{SUSY}}^5}\mathcal{M}^{\mathrm{SUSY}}\right|^2. \quad (54)$$

Here G_{01} is the standard phase space factor tabulated for various nuclei in [2].

Equation (54) takes into account only the contributions from sneutrino exchange. There might be other contributions which we assume not to cancel the SUSY contribution. If there is no unnatural fine-tuning between different contributions, we may retain only the SUSY one in deriving upper bounds for the lepton-number-violating parameters.

The most stringent experimental lower limit on $0\nu\beta\beta$ decay has been obtained for ^{76}Ge [3]:

$$T_{1/2}^{0\nu\beta\beta\mathrm{expt}}(0^+\to0^+)\geq1.0\times10^{25}\ \mathrm{yr}\ \ 90\%\ \mathrm{C.L.} \quad (55)$$

Combining this bound with Eq. (54) and the numerical value of the nuclear matrix element $\mathcal{M}^{\mathrm{SUSY}}$ given in Eq. (53) we get the following constraint on the effective MSSM parameter:

$$\eta^{\mathrm{SUSY}}\leq1.0\times10^{-8}\left(\frac{m_{\mathrm{SUSY}}}{100\ \mathrm{GeV}}\right)^5. \quad (56)$$

Since we are interested in deriving constraints on the $(B-L)$-violating sneutrino mass $\tilde{m}_M$ from η^{SUSY}, we will adopt the following simplifying assumptions. Assume all SUSY particle masses to be equal to the effective SUSY-breaking scale m_{SUSY} introduced in Eq. (23) and consider two limiting cases for the lightest neutralino χ composition. In the first case it is assumed to be a pure B-ino as suggested by the SUSY solution of the dark matter problem [19], in the second a pure Higgsino. These two cases can be understood as extreme cases, and actual values for $\tilde{m}_M$ for other choices of the neutralino composition should therefore lie in between the two extreme values given below. In the Higgsino case essentially only the last three graphs in Fig. 4 with gluino lines contribute to $0\nu\beta\beta$ decay. As seen from Eq. (31) the corresponding lepton-number-violating parameter $\eta_{\tilde{q}}$ does not depend on the neutralino composition and survives in this limiting case. With the currently accepted [20] values of the gauge coupling constants and W-boson mass we derive

$$\tilde{m}_M\leq2\left(\frac{m_{\mathrm{SUSY}}}{100\ \mathrm{GeV}}\right)^{3/2}\ \mathrm{GeV},\quad \chi\sim\tilde{B}, \quad (57)$$

$$\tilde{m}_M \leq 11 \left(\frac{m_{\rm SUSY}}{100\ {\rm GeV}} \right)^{7/2} {\rm GeV}, \quad \chi \sim \tilde{H}. \tag{58}$$

VI. NEUTRINO MASS CONSTRAINTS

As already mentioned, the sneutrino contributes to the Majorana neutrino mass m_M^ν at the one-loop level via the $(B-L)$-violating propagator, Eq. (18), proportional to the sneutrino $(B-L)$-violating mass parameter $\tilde{m}_M^2$. The corresponding diagram given in Fig. 2(f) gives rise to an induced Majorana neutrino mass δm^ν. Thus, in the presence of a nonzero tree-level contribution $m_M^{\nu({\rm tree})}$ the total neutrino mass is

$$m_M^\nu = m_M^{\nu({\rm tree})} + \delta m^\nu. \tag{59}$$

As seen from Eq. (18) the $(B-L)$-violating sneutrino propagator in momentum space is a rapidly decreasing function of momentum q. In the ultraviolet limit it behaves as $\Delta_{\tilde{\nu}}^M(q) \sim 1/q^4$ unlike an ordinary scalar propagator decreasing only as $\sim 1/q^2$. As a result, the one-loop diagram in Fig. 2(f) leads to a finite loop integral. Hence, the sneutrino-induced Majorana neutrino mass δm^ν is a calculable object. It is given by the formula

$$\delta m^{\nu(i)} = g^2 \sum_{k=1}^{4} \epsilon_{L_k}^{\nu^2} m_{\chi_k} \tilde{m}_{M_{(i)}}^2 I(\tilde{m}_M^2, \tilde{m}_D^2, m_{\chi_k}), \tag{60}$$

where the subscript i stands for generation. The neutralino-neutrino-sneutrino coupling is $\epsilon_{L_k}^\nu = \tan\theta_W \mathcal{N}_{k1} - \mathcal{N}_{k2}$. The loop integral is defined as

$$I(\tilde{m}_M^2, \tilde{m}_D^2, m_{\chi_k})$$
$$= -i \int \frac{d^4 q}{(2\pi)^4} \frac{1}{(\tilde{m}_1^2 - q^2)} \frac{1}{(\tilde{m}_2^2 - q^2)} \frac{1}{(m_{\chi_k}^2 - q^2)}. \tag{61}$$

For an approximate numerical estimation we take all superpartner masses equal to the common mass scale of supersymmetry breaking $m_{\chi_k} \approx \tilde{m}_1 \approx \tilde{m}_2 \approx m_{\rm SUSY}$. Then in this approximation one gets for the lightest neutralino χ contribution the following constraint on the sneutrino mass splitting parameter:

$$\tilde{m}_{M(i)} \leq \frac{1.7 \times 10^{-2}}{|\tan\theta_W \mathcal{N}_{11} - \mathcal{N}_{12}|} \left(\frac{m_{\rm SUSY}}{100\ {\rm GeV}} \right)^{1/2} \left(\frac{m_{\nu_{(i)}}^{\rm expt}}{1\ {\rm eV}} \right)^{1/2} {\rm GeV}. \tag{62}$$

Here $m_{\nu_{(i)}} \leq m_{\nu_{(i)}}^{\rm expt}$ are the best laboratory limits on the neutrino masses which can be summarized as [20] $m_{\nu_{(e)}}^{\rm expt} = 15$ eV, $m_{\nu_{(\mu)}}^{\rm expt} = 170$ keV, and $m_{\nu_{(\tau)}}^{\rm expt} = 23$ MeV. Equation (62) assumes that there is no significant cancellation in Eq. (59) between the tree and one-loop level contributions.

If the neutralino is B-ino dominant, then we derive the following limits on the sneutrino mass splitting parameter:

$$\tilde{m}_{M(e)} \leq 120\ {\rm MeV}, \quad \tilde{m}_{M(\mu)} \leq 13\ {\rm GeV},$$
$$\tilde{m}_{M(\tau)} \leq 149\ {\rm GeV}. \tag{63}$$

Thus, for the second and third generation sneutrinos large splittings are not excluded by experimental data.

An interesting question to ask is whether there are certain loopholes in the constraints on the sneutrino mass splitting derived from the experimental upper limits on neutrino masses. It could happen, for instance, that the lightest neutralino is Higgsino dominated, in which case one would expect that the bound (62) might have to be relaxed considerably. However, one should remember that all neutralino states contribute to Eq. (60), so that even if there is no constraint from the lightest neutralino, the other mass eigenstates will provide a finite contribution to the neutrino mass.

In order to investigate this question a little bit more quantitatively we did a numerical scan of the SUSY parameter space, calculating upper bounds on $\tilde{m}_M$, taking into account all four neutralino states. In this case instead of Eq. (62) we have

$$\tilde{m}_{M(i)} \leq \frac{1.7 \times 10^{-2} {\rm GeV}}{\left[\sum_{i=1}^{4} (\tan\theta_W N_{i1} - N_{i2})^2 y_i C_3(x, y_i) \right]^{1/2}} \left(\frac{m_{\rm SUSY}}{100\ {\rm GeV}} \right) \times \left(\frac{m_{\nu_{(i)}}^{\rm expt}}{1\ {\rm eV}} \right)^{1/2}. \tag{64}$$

Here $x = \tilde{m}_D / m_{\rm SUSY}$ and $y_i = m_{\chi_i} / m_{\rm SUSY}$. $C_3(x, y_i)$ takes into account the fact that the masses of the particles in the loop are no longer taken to be equal. In the limit where $\tilde{m}_M \ll \tilde{m}_D$ the one-loop integral $C_3(x, y_i)$ is given by

$$C_3(x, y) = 2 \frac{x^2 + y^2 [\ln(y^2) - \ln(x^2) - 1]}{(y^2 - x^2)^2}. \tag{65}$$

$C_3(x, y)$ is normalized such that it approaches 1 in the limit where x and y approach 1. We let the parameters of the neutralino mass matrix vary from 0 to 1000 GeV, for μ and M_2, and $\tan\beta$ from 1 to 50, for both positive and negative μ. The unification condition $M_1 = (5/3) \tan^2\theta_W M_2$ has been assumed in this calculation. We required the lightest mass eigenstate to be heavier than 20 GeV, motivated by the LEP measurements. About 10^8 solutions were calculated. From these we calculated the "average constraint" and an "absolute" upper bound. These are

$$\tilde{m}_{M(i)} \leq 60(125) \left(\frac{m_{\nu_{(i)}}^{\rm expt}}{1\ {\rm eV}} \right)^{1/2} {\rm MeV}, \tag{66}$$

on average ("absolute"), if $\tilde{m}_D \approx m_{\rm SUSY} = 100$ GeV is assumed. These numbers are about a factor of 2 (4) less stringent than taking only the lightest neutralino (being B-ino) fixed at 100 GeV. This simply reflects the fact that within the above-mentioned parameter ranges many solutions exist where even the lightest neutralino mass state can be considerably heavier than 100 GeV. On the other hand, it seems that within the typical range of SUSY parameters, the constraint on $\widetilde{m}_M$ is essentially "stable" and has to be taken seriously. (This conclusion remains unchanged even if we drop the unification assumption on M_1, although the bounds might have to be slightly relaxed in some cases.)

Note that, in principle, more stringent limits on $\widetilde{m}_M$ than in Eq. (63) could be derived from the upper bounds on the neutrino mass given by nonobservation of $0\nu\beta\beta$ decay. However, in this case the situation is more complex, since $0\nu\beta\beta$ decay measures an effective neutrino mass $\langle m_M^\nu \rangle = \Sigma' U_{ej}^2 m_j$, where U_{ej} are mixing coefficients connecting the weak and the mass eigenstate basis for neutrinos. Thus limits on $\widetilde{m}_M$ derived from the neutrino mass limit of $0\nu\beta\beta$ decay will also depend on assumptions on neutrino mixing. If one assumes for simplicity $U_{ej} \approx \delta_{e1}$ one could derive $\widetilde{m}_{M(e)} \leq 22$ MeV from the data on ^{76}Ge [3].

VII. CONCLUSION

In summary, we have proven a low-energy theorem for weak-scale softly broken supersymmetry relating the $(B-L)$-violating mass terms of the neutrino and the sneutrino as well as the amplitude of neutrinoless double beta decay. This theorem can be regarded as a supersymmetric generalization of the well-known theorem [14] relating only the neutrino Majorana mass and the neutrinoless double beta decay amplitude.

According to Eq. (12) the parameter $\widetilde{m}_M$ describes a splitting in the sneutrino mass spectrum. This splitting leads to mixing in the sneutrino-antisneutrino ($\widetilde{\nu}-\widetilde{\nu}^c$) system and to the effect of lepton-number-violating $\widetilde{\nu}--\widetilde{\nu}^c$ oscillations [21,22].

The mass-splitting parameter $\widetilde{m}_M$ is constrained by the experimental data on neutrinoless double beta decay $0\nu\beta\beta$ and the neutrino mass discussed in the present paper. The neutrino mass constraint on $\widetilde{m}_M$ is found to be more stringent then the direct $0\nu\beta\beta$-decay constraint. This is opposite to the conclusion reached for R-parity-violating SUSY, for which the direct double beta decay constraints have been found to be more stringent than those derivable from the neutrino mass [7,8]. However, in contrast to the neutrino mass limits, the corresponding constraint on $\widetilde{m}_M$ from neutrinoless double beta decay is completely independent of assumptions about neutralino masses and mixings.

The constraints derived here, nevertheless, leave quite some room for accelerator searches for sneutrino-mediated $(B-L)$-violating effects for the second and third generations. The sneutrino-mass-splitting parameter $\widetilde{m}_M$ might be searched for at future colliders such as the NLC or a first muon collider [21,22]. Probably, dedicated searches for Majorana sneutrinos have a chance of detecting positive signal, within the above-discussed low-energy limits. For $\widetilde{\nu}_e$, on the other hand, these limits seem to be too stringent and accelerator experiments should not be expected to give positive signals.

ACKNOWLEDGMENTS

We thank V.A. Bednyakov, for helpful discussions. The research described in this publication was made possible in part (S.G.K.) by Grant No. GNTP 315 NUCLON from the Russian ministry of science. M.H. would like to thank the Deutsche Forschungsgemeinschaft for financial support by Grants Nos. kl 253/8-2 and 446 JAP-113/101/0.

APPENDIX A: SUPERSYMMETRIC LAGRANGIAN TERMS CONTRIBUTING TO THE $0\nu\beta\beta$-DECAY AMPLITUDE

In the presence of the $(B-L)$-violating (s)neutrino masses, given in Eq. (1), $0\nu\beta\beta$ decay is triggered by the following terms of the MSSM Lagrangian:

$$\mathcal{L}_{\rm MSSM} = \mathcal{L}_{W\bar{f}f} + \mathcal{L}_{W\bar{\tilde{f}}\tilde{f}} + \mathcal{L}_{\chi^+ f\tilde{f}} + \mathcal{L}_{\chi f\tilde{f}} + \mathcal{L}_{\tilde{g}\bar{\tilde{q}}q} + \mathcal{L}_{W\chi^+\chi}. \tag{A1}$$

The individual terms can be found in the standard sources like Refs. [13,23]. Let us list them explicitly.

(a) *Gauge boson-fermion-fermion term.* This is nothing but the usual standard model charged current Lagrangian:

$$\mathcal{L}_{W\bar{f}f} = -\frac{g}{\sqrt{2}}[W_\mu^+(\bar{u}_L\gamma^\mu d_L + \bar{\nu}_L\gamma^\mu e_L) + \text{H.c.}] \tag{A2}$$

$$\equiv \mathcal{L}_{W\bar{q}q} + \mathcal{L}_{W\bar{l}l}. \tag{A3}$$

(b) *Gauge boson-sfermion-sfermion term.* Only the charged current part of this type of the MSSM interactions is of interest in $0\nu\beta\beta$ decay:

$$\mathcal{L}_{W\bar{\tilde{f}}\tilde{f}} \equiv \mathcal{L}_{W\bar{\tilde{q}}\tilde{q}} + \mathcal{L}_{W\bar{\tilde{l}}\tilde{l}} \tag{A4}$$

$$= -\frac{ig}{\sqrt{2}} W_\mu^+(\tilde{u}_L^* \overleftrightarrow{\partial}^\mu \tilde{d}_L) - \frac{ig}{\sqrt{2}} W_\mu^+(\tilde{\nu}_L^* \overleftrightarrow{\partial}^\mu \tilde{e}_L) + \text{H.c.} \tag{A5}$$

Note that $\overleftrightarrow{\partial}^\mu$ is defined by $\overleftrightarrow{\partial}^\mu = \overrightarrow{\partial}^\mu - \overleftarrow{\partial}^\mu$.

(c) *Chargino-fermion-sfermion term:*

$$\begin{aligned}\mathcal{L}_{\chi^\pm f\tilde{f}} = {} & C_{LL}^u \cdot \bar{u}_L\chi_i^+ \tilde{d}_L + C_{LL}^d \cdot \bar{d}_L\chi_i^{c+} \tilde{u}_L + C_{LR}^u \cdot \bar{u}_L\chi_i^+ \tilde{d}_R \\ & + C_{RL}^u \cdot \bar{u}_R\chi_i^+ \tilde{d}_L + C_{RL}^d \cdot \bar{d}_R\chi_i^- \tilde{u}_L + C_{LR}^d \cdot \bar{d}_L\chi_i^- \tilde{u}_R \\ & + C_{LL}^\nu \cdot \bar{\nu}_L\chi_i^+ \tilde{e}_L + C_{LL}^e \cdot \bar{e}_L\chi_i^- \tilde{\nu}_L + \text{H.c.},\end{aligned} \tag{A6}$$

where the following shorthand notation has been defined:

$$C_{LL}^{u,\nu} = -gU_{i1}, \quad C_{LL}^{d,e} = -gV_{i1},$$

$$C_{LR}^u = \frac{gm_d}{\sqrt{2}m_W\cos\beta}U_{i2}, \quad C_{RL}^u = \frac{gm_u}{\sqrt{2}m_W\sin\beta}V_{i2}^*,$$

$$C_{RL}^d = \frac{gm_d}{\sqrt{2}m_W\cos\beta}U_{i2}^*, \quad C_{LR}^d = \frac{gm_u}{\sqrt{2}m_W\sin\beta}V_{i2}.$$

Coefficients C_{LL} and C_{RR} are fermion-sfermion couplings to the gaugino component of the chargino while C_{LR} and C_{RL} describe couplings to the Higgsino component. The latter are proportional to the fermion mass and, therefore, can be neglected in the fermion-sfermion sector as is done in the present paper.

The chargino-mixing matrices U_{ij} and V_{ij} are defined from the diagonalization of the chargino mass matrix $M_{\chi^\pm}$:

$$U^* \cdot M_{\chi^\pm} V^\dagger = \mathrm{Diag}\{m_{\chi^\pm}\}. \qquad \text{(A7)}$$

For details see Ref. [23].

(d) *Neutralino-fermion-sfermion term.* The neutralino interaction can be written as

$$\mathcal{L}_{\chi f \tilde{f}} = -\sqrt{2} g [\epsilon^{\psi}_{L(i)} \bar{\psi}_L \chi_i \tilde{\psi}_L + \epsilon^{\psi}_{R(i)} \bar{\psi}_R \chi_i \tilde{\psi}_R + \epsilon^{\psi}_{LR(i)} \bar{\psi}_L \chi_i \tilde{\psi}_R + \epsilon^{\psi}_{RL(i)} \bar{\psi}_R \chi_i \tilde{\psi}_L] - \mathrm{H.c.}, \qquad \text{(A8)}$$

where $\psi_L = u_L, d_L, e_L, \nu_L$ and their scalar superpartners $\tilde{\psi}_L = \tilde{u}_L, \tilde{d}_L, \tilde{e}_L, \tilde{\nu}_L$. The corresponding chiral coefficients are

$$\epsilon^{\psi}_{L(i)} = -T_3(\psi)\mathcal{N}_{i2} + \tan\theta_W [T_3(\psi) - Q(\psi)]\mathcal{N}_{i1},$$

$$\epsilon^{\psi}_{R(i)} = Q(\psi)\tan\theta_W \mathcal{N}_{i1}, \qquad \text{(A9)}$$

$$\epsilon^{u}_{LR(i)} = \frac{m_u}{m_W \sin\beta}\mathcal{N}^*_{j4}, \qquad \text{(A10)}$$

$$\epsilon^{u}_{RL(i)} = \frac{m_u}{m_W \sin\beta}\mathcal{N}_{j4}, \qquad \text{(A11)}$$

$$\epsilon^{d}_{LR(i)} = \frac{m_d}{m_W \cos\beta}\mathcal{N}^*_{j3}, \qquad \text{(A12)}$$

$$\epsilon^{d}_{RL(i)} = \frac{m_d}{m_W \cos\beta}\mathcal{N}_{j3}. \qquad \text{(A13)}$$

Coefficients N_{ij} are elements of the orthogonal neutralino-mixing matrix diagonalizing the neutralino mass matrix M_χ. In the $\not{R}_p$ MSSM the neutralino mass matrix is identical to the MSSM one [13] and in the basis of fields $(\tilde{B}, \tilde{W}^3, \tilde{H}^0_1, \tilde{H}^0_2)$ has the form

$$M_\chi = \begin{pmatrix} M_1 & 0 & -M_Z s_W c_\beta & M_Z s_W s_\beta \\ 0 & M_2 & M_Z c_W c_\beta & -M_Z c_W s_\beta \\ -M_Z s_W c_\beta & M_Z c_W c_\beta & 0 & -\mu \\ M_Z s_W s_\beta & -M_Z c_W s_\beta & -\mu & 0 \end{pmatrix}, \qquad \text{(A14)}$$

where $c_W = \cos\theta_W$, $s_W = \sin\theta_W$, $t_W = \tan\theta_W$, $s_\beta = \sin\beta$, and $c_\beta = \cos\beta$. The angle β is defined as $\tan\beta = \langle H^0_2\rangle / \langle H^0_1\rangle$. Here $\langle H^0_2\rangle$ and $\langle H^0_1\rangle$ are vacuum expectation values of the neutral components H^0_2 and H^0_1 of the Higgs doublet fields with weak hypercharges $Y(H^0_2) = +1$ and $Y(H^0_1) = -1$, respectively. The mass parameters M_1 and M_2 are the soft $SU(2)_L$ and $U(1)_Y$ gaugino masses. In grand unification scenarios they are related to each other as follows:

$$M_1 = \tfrac{5}{3}\tan^2\theta_W \cdot M_2. \qquad \text{(A15)}$$

By diagonalizing the mass matrix (A14) one can obtain four neutralinos χ_i with masses m_{χ_i} and the field content

$$\chi_i = \mathcal{N}_{i1}\tilde{B} + \mathcal{N}_{i2}\tilde{W}^3 + \mathcal{N}_{i3}\tilde{H}^0_1 + \mathcal{N}_{i4}\tilde{H}^0_2. \qquad \text{(A16)}$$

Recall again that we use notation $\tilde{W}^3$ and $\tilde{B}$ for neutral $SU(2)_L \times U(1)$ gauginos and $\tilde{H}^0_1$ and $\tilde{H}^0_2$ for Higgsinos which are the superpartners of the two neutral Higgs boson fields H^0_1 and H^0_2.

We apply a diagonalization by means of a real orthogonal matrix $\mathcal{N}$. Therefore the coefficients $\mathcal{N}_{ij}$ are real and masses m_{χ_i} are either positive or negative. The sign of the mass coincides with the CP parity of the corresponding neutralino mass eigenstate χ_i. If necessary, a negative mass can be always made positive by a redefinition [23] of the neutralino field χ_i. It leads to a redefinition of the relevant mixing coefficients $\mathcal{N}_{ij} \to i \cdot \mathcal{N}_{ij}$.

(e) *Gluino-squark-quark term.* The gluino interaction is given by

$$\mathcal{L}_{\tilde{g}} = -\sqrt{2} g_3 \frac{\lambda^{(a)}_{\alpha\beta}}{2} (\bar{q}^{\alpha}_L \tilde{g}^{(a)} \tilde{q}^{\beta}_L - \bar{q}^{\alpha}_R \tilde{g}^{(a)} \tilde{q}^{\beta}_R) + \mathrm{H.c.}, \qquad \text{(A17)}$$

Here $\lambda^{(a)}$ are 3×3 Gell-Mann matrices $(a = 1, \ldots, 8)$. Superscripts α, β in Eq. (A17) are color indices.

(f) *Gauge boson-chargino-neutralino term.* In the notation of Ref. [13],

$$\mathcal{L}_{W\chi^+\chi} = g W^-_\mu \bar{\chi}_i \gamma^\mu (\mathcal{O}^L_{ij} P_L + \mathcal{O}^R_{ij} P_R)\chi^+_j + \mathrm{H.c.}, \qquad \text{(A18)}$$

where

$$\mathcal{O}^L_{ij} = -\frac{1}{\sqrt{2}}\mathcal{N}_{i4} V^*_{j2} + \mathcal{N}_{i2} V^*_{j1}, \qquad \text{(A19)}$$

$$\mathcal{O}^R_{ij} = +\frac{1}{\sqrt{2}}\mathcal{N}^*_{i3} U_{j2} + \mathcal{N}^*_{i2} U_{j1}. \qquad \text{(A20)}$$

Further details and conventions on the definitions used can be found in the paper by Haber and Kane [13].

APPENDIX B: BOX INTEGRALS

In this appendix some relevant formulas for the calculation of the loop integrals are summarized. There are four kinds of integrals encountered in the Eqs. (31)–(35):

$$\mathcal{G}(m_1,m_2,m_3)=-i\int\frac{d^4k}{(2\pi)^4}\frac{m_{\mathrm{SUSY}}^{10}}{(\widetilde{m}_1^2-k^2)(\widetilde{m}_2^2-k^2)(m_{\tilde{d}}^2-k^2)^2(m_1^2-k^2)(m_2^2-k^2)(m_3^2-k^2)},\tag{B1}$$

$$\mathcal{F}(m_1,m_2,m_3)=-i\int\frac{d^4k}{(2\pi)^4}\frac{m_{\mathrm{SUSY}}^{8}k^2}{(\widetilde{m}_1^2-k^2)(\widetilde{m}_2^2-k^2)(m_{\tilde{u}}^2-k^2)^2(m_1^2-k^2)(m_2^2-k^2)(m_3^2-k^2)},\tag{B2}$$

$$\mathcal{I}(m_1,m_2,m_3)=-i\int\frac{d^4k}{(2\pi)^4}\frac{m_{\mathrm{SUSY}}^{6}}{(\widetilde{m}_1^2-k^2)(\widetilde{m}_2^2-k^2)(m_1^2-k^2)(m_2^2-k^2)(m_3^2-k^2)},\tag{B3}$$

$$\mathcal{J}(m_1,m_2,m_3)=-i\int\frac{d^4k}{(2\pi)^4}\frac{m_{\mathrm{SUSY}}^{4}k^2}{(\widetilde{m}_1^2-k^2)(\widetilde{m}_2^2-k^2)(m_1^2-k^2)(m_2^2-k^2)(m_3^2-k^2)}.\tag{B4}$$

All four integrals are finite and can be calculated analytically using standard methods. Simple solutions can be found for the case when the masses of all particles in the loops are assumed to be equal to m_{SUSY}. In this limit one finds

$$\mathcal{G}(m_{\mathrm{SUSY}})=(480\pi^2)^{-1},\quad \mathcal{F}(m_{\mathrm{SUSY}})=(960\pi^2)^{-1},$$

$$\mathcal{I}(m_{\mathrm{SUSY}})=\mathcal{J}(m_{\mathrm{SUSY}})=(192\pi^2)^{-1}.\tag{B5}$$

[1] W. C. Haxton and G. J. Stephenson, Prog. Part. Nucl. Phys. **12**, 409 (1984); J. D. Vergados, Phys. Rep. **133**, 1 (1986); K. Grotz and H. V. Klapdor-Kleingrothaus, *The Weak Interactions in Nuclear, Particle and Astrophysics* (Hilger, Bristol, New York, 1990); R. N. Mohapatra and P. B. Pal, *Massive Neutrinos in Physics and Astrophysics* (World Scientific, Singapore, 1991).

[2] M. Doi, T. Kotani, and E. Takasugi, Prog. Theor. Phys. Suppl. **83**, 1 (1985).

[3] Heidelberg-Moscow Collaboration, M. Günther *et al.*, Phys. Rev. D **55**, 54 (1997); J. Hellmig *et al.*, in *Proceedings of the International Workshop on Dark Matter in Astro and Particle Physics*, Heidelberg, 1997 (World Scientific, Singapore, in press).

[4] S. Dimopoulos and L. J. Hall, Phys. Lett. B **207**, 210 (1987); L. Hall and M. Suzuki, Nucl. Phys. **B231**, 419 (1984); E. Ma and D. Ng, Phys. Rev. D **41**, 1005 (1990).

[5] C. Aulakh and R. Mohapatra, Phys. Lett. **119B**, 136 (1983); G. G. Ross and J. W. F. Valle, *ibid.* **151B**, 375 (1985); J. Ellis *et al.*, *ibid.* **150B**, 142 (1985); A. Santamaria and J. W. F. Valle, Phys. Lett. B **195**, 423 (1987); Phys. Rev. Lett. **60**, 397 (1988); Phys. Rev. D **39**, 1780 (1989); M. C. Gonzalez-Garsia and J. W. F. Valle, Nucl. Phys. **B355**, 330 (1991); J. W. F. Valle, Phys. Lett. B **196**, 157 (1987); A. Masiero and J. W. F. Valle, *ibid.* **251**, 273 (1990).

[6] R. Mohapatra, Phys. Rev. D **34**, 3457 (1986); J. D. Vergados, Phys. Lett. B **184**, 55 (1987).

[7] M. Hirsch, H. V. Klapdor-Kleingrothaus, and S. G. Kovalenko, Phys. Rev. Lett. **75**, 17 (1995).

[8] M. Hirsch, H. V. Klapdor-Kleingrothaus, and S. G. Kovalenko, Phys. Lett. B **352**, 1 (1995); Phys. Rev. D **53**, 1329 (1996).

[9] K. S. Babu and R. N. Mohapatra, Phys. Rev. Lett. **75**, 2276 (1995).

[10] M. Hirsch, H. V. Klapdor-Kleingrothaus, and S. G. Kovalenko, Phys. Lett. B **372**, 181 (1996); **381**, 488(E) (1996).

[11] A. Faessler, S. Kovalenko, F. Simkovic, and J. Schwieger, Phys. Rev. Lett. **78**, 183 (1997).

[12] M. Hirsch, H. V. Klapdor-Kleingrothaus, and S. G. Kovalenko, Phys. Lett. B **398**, 311 (1997); Phys. Lett. B **403**, 291 (1997).

[13] H. E. Haber and G. L. Kane, Phys. Rep. **117**, 75 (1985); H. P. Nilles, *ibid.* **110**, 1 (1984).

[14] J. Schechter and J. W. F. Valle, Phys. Rev. D **25**, 2951 (1982); J. F. Nieves, Phys. Lett. **147B**, 375 (1984); E. Takasugi, Phys. Lett. **149B**, 372 (1984); B. Kayser, in *Proceedings of the XXIII International Conference on High Energy Physics*, edited by S. Loken (World Scientific, Singapore, 1987), p. 945; S. Petcov, in *Proceedings of '86 Massive Neutrinos in Astrophysics and in Particle Physics*, edited by O. Fackler and J. Trân Than Vân (Editions Frontieres, Gif-sur-Yvette, France, 1986), p. 187; S. P. Rosen, UTAPHY-HEP-4, hep-ph/9210202.

[15] K. Muto and H. V. Klapdor, *Neutrinos* (Springer, Heidelberg, 1988), p. 183.

[16] J. D. Vergados, Phys. Rev. C **24**, 640 (1981); Nucl. Phys. **B218**, 109 (1983).

[17] M. Hirsch, H. V. Klapdor-Kleingrothaus, and O. Panella, Phys. Lett. B **374**, (1996); M. Hirsch and H. V. Klapdor-Kleingrothaus, in *Proceedings of the International Workshop on Double Beta Decay and Related Topics*, Trento, Italy, 1995, edited by H. V. Klapdor-Kleingrothaus and S. Stoica (World Scientific, Singapore, 1996), p. 175.

[18] K. Muto, E. Bender, and H. V. Klapdor, Z. Phys. A **334**, 177 (1989); **334**, 187 (1989); M. Hirsch, K. Muto, T. Oda, and H. V. Klapdor-Kleingrothaus, *ibid.* **347**, 151 (1994).

[19] For a recent review see, for instance, G. Jungman, M. Kamion-

kowski, and K. Griest, Phys. Rep. **267**, 195 (1996).

[20] Particle Data Group, R. M. Barnett *et al.*, Phys. Rev. D **54**, 1 (1996), pp. 1–720.

[21] M. Hirsch, H. V. Klapdor-Kleingrothaus, St. Kolb, and S. G. Kovalenko, this issue, Phys. Rev. D **57**, 2020 (1998).

[22] Y. Grossman and H. E. Haber, Phys. Rev. Lett. **78**, 3438 (1997).

[23] J. F. Gunion, H. E. Haber, and G. L. Kane, Nucl. Phys. **B272**, 1 (1986).

PHYSICAL REVIEW D VOLUME 57, NUMBER 3 1 FEBRUARY 1998

Phenomenological implications of "Majorana" sneutrinos at future accelerators

M. Hirsch, H. V. Klapdor-Kleingrothaus, and St. Kolb
Max-Planck-Institut für Kernphysik, P.O. 10 39 80, D-69029, Heidelberg, Germany

S. G. Kovalenko
Joint Institute for Nuclear Research, Dubna, Russia
(Received 2 May 1997; published 12 January 1998)

It has recently been shown that in the framework of weak-scale softly broken supersymmetry the existence of a Majorana neutrino inevitably implies a finite (B−L)-violating "Majorana" mass $\tilde{m}_M$ for scalar neutrinos, with the effect of splitting the complex sneutrino field into two real fields separated in mass by $\tilde{m}_1^2-\tilde{m}_2^2=2|\tilde{m}_M^2|$. The existence of a finite Majorana sneutrino mass leads to new processes potentially observable at future colliders. Taking into account existing low-energy constraints, to find positive signals for electron sneutrinos will be very difficult, although not impossible. For muon and tau sneutrinos a dedicated search for Majorana masses has much better prospects. [S0556-2821(98)00105-2]

PACS number(s): 14.80.Ly, 12.60.Jv, 14.60.Pq

Whether or not neutrinos are Majorana or Dirac particles is an open question [1]. Theoreticians are clearly biased to believe in Majorana neutrinos, but an experimental verification would certainly be a major breakthrough in neutrino physics with profound implications not only for particle phenomenology and model building.

Currently only experiments searching for neutrinoless double beta ($0\nu\beta\beta$) decay are sensitive to the Majorana character of the neutrino [2]. Unfortunately $0\nu\beta\beta$ decay has not been observed yet, although quite stringent limits corresponding to $\langle m_M^\nu\rangle \leq \mathcal{O}(0.5\text{ eV})$ have been published [3].

Low-energy supersymmetry is an attractive candidate for physics beyond the standard model. Especially the minimal supersymmetric extension of the standard model (MSSM) [4] has been studied in great detail in recent years. However, in the MSSM neutrinos are kept exactly massless—as in the SM —by the minimal Higgs content and the assumed absence of right-handed neutrinos. Theoretically, on the other hand, it is easy to extend the MSSM to allow for Majorana neutrino masses. Then, an interesting question to ask is: Given that the neutrino and the scalar neutrino are components of the same superfield will there be any relation between the properties of the neutrino and sneutrino fields even after supersymmetry breaking?

A recent paper [5] has shown that this is indeed the case. Specifically, it has been shown [5] that if the neutrino has a finite Majorana mass then the sneutrino necessarily has a (B−L)-violating "Majorana" mass too, independent of the mechanism of mass generation. The inverse of this statement is also true. This observation [5] constitutes the basis for the present work, since a finite Majorana sneutrino mass should lead to potentially observable effects at future accelerators.

Here we will limit ourselves to a few examples how finite sneutrino Majorana masses might be searched for. None of the effects we have considered so far will be readily measurable but for anticipated parameters of future colliders, such as the next linear collider [6] or a first muon collider [7], with sufficient accumulated luminosity discovering Majorana sneutrinos could be possible, at least for the second and third generation sneutrinos.

We will work within the MSSM [4], minimally extended to allow for finite Majorana neutrino and sneutrino masses. Especially we will assume that R-parity is conserved in the following.

As proven in [5] the self-consistent form of the neutrino and sneutrino mass terms is

$$\mathcal{L}_{mass}^{\nu\tilde{\nu}} = -\frac{1}{2}(m_M^\nu \overline{\nu^c}\nu + \text{H.c.}) - \frac{1}{2}(\tilde{m}_M^2 \tilde{\nu}_L \tilde{\nu}_L + \text{H.c.}) - \tilde{m}_D^2 \tilde{\nu}_L^* \tilde{\nu}_L. \quad (1)$$

Here, $\nu=\nu^c$ denotes a Majorana neutrino field. m_M^ν and $\tilde{m}_M$ are (B−L) violating Majorana masses of the neutrino and sneutrino, while $\tilde{m}_D$ denotes the (B−L) conserving sneutrino Dirac mass. $\tilde{m}_M$ gives a measure of sneutrino-antisneutrino mixing. For finite $\tilde{m}_M$ the complex scalar field $\tilde{\nu}=(\tilde{\nu}_1+i\tilde{\nu}_2)/\sqrt{2}$ splits into two real mass eigenstate fields $\tilde{\nu}_{1,2}$ with masses $\tilde{m}_{1,2}^2=\tilde{m}_D^2\pm|\tilde{m}_M^2|$. According to the above definition, $\tilde{\nu}_1$ is the CP-even state while $\tilde{\nu}_2$ is the CP-odd one. Note that from stability of the ground state $\tilde{m}_D^2 \geq |\tilde{m}_M^2|$, otherwise one of the fields $\tilde{\nu}_i$ develops a nonzero vacuum expectation $\langle\tilde{\nu}_i\rangle \neq 0$ which spontaneously breaks the discrete R_P-symmetry [5].

The sneutrino-antisneutrino mixing leads to a (B−L) violating propagator for the sneutrino of the form

$$i\Delta_{\tilde{\nu}}^M(x-y) = \langle 0|T(\tilde{\nu}(x)\tilde{\nu}(y))|0\rangle = -\tilde{m}_M^2 \int \frac{d^4k}{(2\pi)^4} \times \frac{e^{-ik(x-y)}}{(\tilde{m}_1^2-k^2+i\epsilon)(\tilde{m}_2^2-k^2+i\epsilon)}. \quad (2)$$

Obviously $\Delta_{\tilde{\nu}}^M$ vanishes smoothly in the limit where $\tilde{m}_M$ goes to zero, as it should.

Since at 1-loop level the sneutrinos contribute to the Majorana neutrino mass via the (B−L)-violating propagator $\Delta_{\tilde{\nu}}^M$

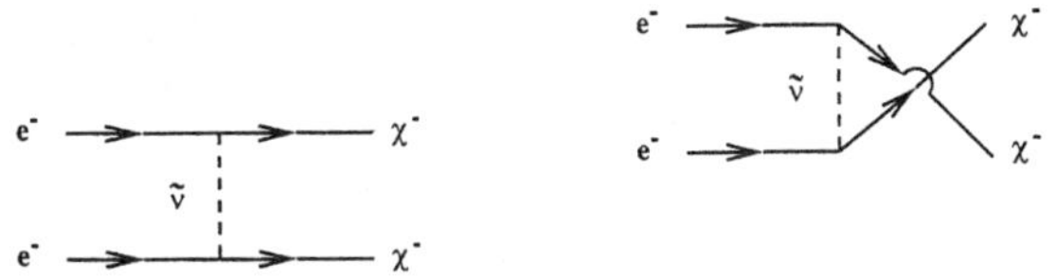

FIG. 1. Feynman graphs for the process $l^- l^- \to \chi^- \chi^-$. e can stand for either electron or muon.

[5] one can derive upper bounds on $\tilde{m}_{M(i)}$ from the experimental upper limits on neutrino masses m_{ν_i}, where i is a generation index. The final result is [8]

$$\tilde{m}_{M(i)} \leqslant \frac{1.7\times 10^{-2}}{|\tan\theta_W N_{11} - N_{12}|}\left(\frac{\Lambda_{SUSY}}{100\ \mathrm{GeV}}\right)^{1/2}\left(\frac{m^{exp}_{\nu(i)}}{1\ \mathrm{eV}}\right)^{1/2}\ \mathrm{GeV}. \tag{3}$$

Here, Λ_{SUSY} is the effective supersymmetry breaking scale which sets the typical scale of superpartner masses. It is assumed that $\Lambda_{SUSY} \simeq m_{\chi^0} \simeq \tilde{m}_D$. N_{kj} are elements of the orthogonal mixing matrix diagonalizing the neutralino mass matrix in the basis $\chi_k = N_{k1}\tilde{B} + N_{k2}\tilde{W}^{(3)} + N_{k3}\tilde{H}_1 + N_{k4}\tilde{H}_2$. For a B-ino dominant lightest neutralino, using the experimental limits of $m_{\nu_e} \leqslant 15$ eV, $m_{\nu_\mu} \leqslant 170$ keV, $m_{\nu_\tau} \leqslant 23$ MeV [9], and assuming $\Lambda_{SUSY} \sim 100$ GeV one obtains

$$\tilde{m}_{M(e)} \leqslant 120\ \mathrm{MeV},\quad \tilde{m}_{M(\mu)} \leqslant 13\ \mathrm{GeV},\quad \tilde{m}_{M(\tau)} \leqslant 149\ \mathrm{GeV}. \tag{4}$$

Thus, for the second and third generation sneutrinos large splittings are not excluded by the experimental data. The limits in Eq. (4) assume that there are no cancellations between the radiatively induced neutrino masses and (possible) tree-level Majorana neutrino masses. Strictly speaking they should thus be understood as order of magnitude estimates.

Now let us turn to a discussion of how $\tilde{m}_M$ might be searched for in collider experiments. We will limit the discussion to just three examples. First it is fairly obvious that for Majorana sneutrinos one has to expect that lepton number violating processes, such as $e^-e^- \to \chi^-\chi^-$ or $\mu^-\mu^- \to \chi^-\chi^-$, will occur. Feynman graphs for this process, which might be called the SUSY analog to inverse neutrinoless double beta decay, are shown in Fig. 1. With the definitions given above, the calculation of the cross section is rather straightforward. The final result can be written as,

$$\frac{d\sigma}{dt} = \frac{g^4 |V_{i1}|^4 \tilde{m}_M^4}{32\pi}\mathcal{F}(s,t), \tag{5}$$

where V_{i1} is an element of the (2×2) unitary matrix diagonalizing the chargino mass matrix. $\mathcal{F}(s,t)$ contains the dependence on the center-of-mass energy $\sqrt{s}$ and is defined by

$$\mathcal{F}(s,t) = \frac{1}{s^2}\left\{\frac{(m_\chi^2 - t)^2}{(t-\tilde{m}_1^2)^2(t-\tilde{m}_2^2)^2} + \frac{(m_\chi^2 - u)^2}{(u-\tilde{m}_1^2)^2(u-\tilde{m}_2^2)^2} - \frac{(m_\chi^2 - t)^2 + (m_\chi^2 - u)^2 - s(s-2m_\chi^2)}{(t-\tilde{m}_1^2)(t-\tilde{m}_2^2)(u-\tilde{m}_1^2)(u-\tilde{m}_2^2)}\right\}. \tag{6}$$

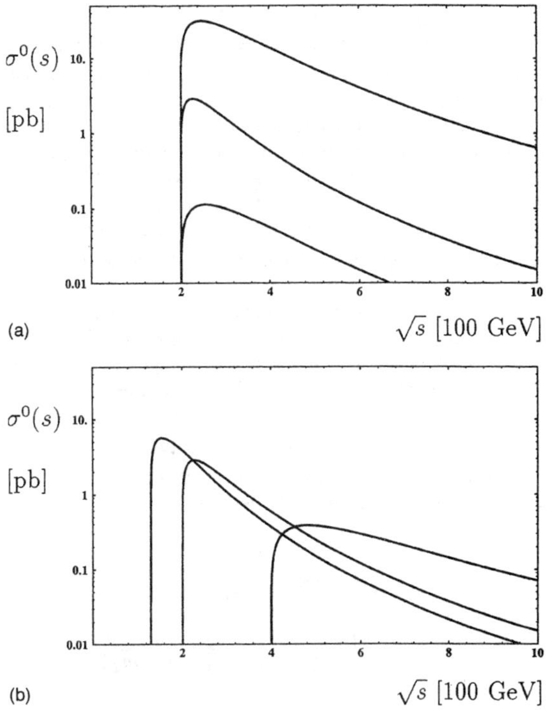

FIG. 2. Calculated cross section $\sigma^0(s)$ for different values of m_{χ^-} and $\tilde{m}_D$. (a) $m_{\chi^-} = 100$ GeV and (from top to bottom) $\tilde{m}_D = 45{,}100{,}200$ GeV; (b) $\tilde{m}_D = 100$ GeV and $m_{\chi^-} = 65{,}100{,}200$ GeV.

Here, m_{χ^-} is the mass of the chargino and $\tilde{m}_i$ are the masses of the sneutrino mass eigenstates as defined above. In leading order in the small parameter $\tilde{m}_M/\Lambda_{SUSY}$, the cross section is proportional to $\tilde{m}_M^4$. This simply reflects the effect of the (square of the) Majorana sneutrino propagator $\Delta^M_{\tilde{\nu}}$. Since $\mathcal{F}(s,t)$ depends only very weakly on $\tilde{m}_M$ one can write down the total cross section in the approximate form $\sigma^{tot}(s) = (\tilde{m}_M/100\ \mathrm{GeV})^4\sigma^0(s) + \mathcal{O}((\tilde{m}_M/100\ \mathrm{GeV})^8)$ and for $\tilde{m}_M \ll 100$ GeV only $\sigma^0(s)$ which is independent of $\tilde{m}_M$ has to be considered.

We have calculated the cross section $\sigma^0(s)$ and Fig. 2 shows examples for some typical choices of chargino and Dirac sneutrino masses. For masses of $m_{\chi^-} = \tilde{m}_D = 100$ GeV cross sections of the order of a few pb are found. The choices of $\tilde{m}_D = 45$ GeV and $m_{\chi^-} = 65$ GeV reflect approximately the current experimental limits on these masses [9,10]. [$\sigma^0(s)$ can be used for both the $\mu^-\mu^-$ and the e^-e^- mode, since $m_e, m_\mu \ll \Lambda_{SUSY}$.]

The design targets for the next linear (NLC) [6] or a first muon collider [7] give typically luminosities of $\mathcal{L} \sim 5 \times 10^{33}\,(1/\mathrm{cm}^2\mathrm{s})$, which could deliver an annual integrated luminosity of some 50 fb^{-1} yr^{-1}. For SUSY particle masses of $m_{\chi^-} = \tilde{m}_D = 100$ and $\tilde{m}_M = 13$ GeV (roughly the current limit for $\tilde{m}_{M(\mu)}$ from the muon neutrino mass) one would expect about 40 events per year. However, with the same parameters observing no event after one year one could give a limit on $\widetilde{m}_M$ of about $\widetilde{m}_M = 5$ GeV.

Because of the fourth power dependence of the cross section on $\widetilde{m}_M$ better sensitivity could be reached only with much larger luminosities. Thus it seems that for first generation sneutrinos a measurement of $e^-e^- \to \chi^-\chi^-$ cannot compete—except for some possible loopholes in the low-energy limit, as briefly noted above—with the limits from the electron neutrino mass.[1] The prospects for discovering $\mu^-\mu^- \to \chi^-\chi^-$ look much brighter given the essentially less stringent 2nd generation neutrino mass bound of $\widetilde{m}_{M(\mu)} \leq 13$ GeV.

A second possibility to probe for Majorana sneutrino masses are processes in which sneutrinos are singly produced. Such an experiment has the advantage that it could be done, in principle, at the NLC and would not require a muon collider. The basic idea is rather simple. Because a finite $\widetilde{m}_M$ splits the complex sneutrino field into two real fields, for $\widetilde{m}_M \neq 0$ cross sections for single sneutrino production show a characteristic double threshold behavior:

$$\sigma^{tot}(s) = \frac{1}{2}\sigma_1(s)\Theta\left[s - \left(\sum_f m_f + \widetilde{m}_1\right)^2\right] + \frac{1}{2}\sigma_2(s)\Theta\left[s - \left(\sum_f m_f + \widetilde{m}_2\right)^2\right]. \quad (7)$$

Here, $\Sigma_f m_f$ denotes symbolically the sum over all final state particle masses except the neutrino and $\widetilde{m}_{1,2}$ are again the two sneutrino mass eigenstates. It is therefore possible to search for Majorana sneutrino masses by measuring cross sections close to threshold.

Let us investigate this possibility a little bit more closely. Consider for example the associated production of sneutrino and charginos in electron-photon collisions $e^-\gamma \to \widetilde{\nu}_e \chi^-$. Formulas for this process in the limiting case $\widetilde{m}_M = 0$ can be found, for example, in Ref. [11]. We have calculated cross sections for $e^-\gamma \to \widetilde{\nu}_e\chi^-$ for assumed masses of $m_{\chi^-} = \widetilde{m}_D = 100$ and $\widetilde{m}_M = 0,5,10$ GeV. The results are shown in Fig. 3. The effect of a finite $\widetilde{m}_M$ is fairly obvious in the plot. To search for the "kink" in the cross section would, however, require that sufficient luminosity is accumulated. Nevertheless, the advantage of such a measurement is that it does not depend on any *explicit* power of $\widetilde{m}_M$. Moreover, an accurate measurement of the masses of the sneutrino and the chargino would, in principle, not be necessary to detect the effect, since event identification would be sufficient. With sufficient accumulated luminosity the sensitivity of these kinds of experiments would be finally limited by the uncertainty of the beam energy. Note, that an uncertainty in the center of mass energy of about ± 20 MeV corresponds to a final sensitivity on $\widetilde{m}_M$ of about ± 2 GeV. Although similar accuracy in the determination of the center of mass energy has been reached at the CERN e^+e^- collider LEP, for the NLC running in the $e\gamma$ option such a requirement looks unrealistic, because of the energy spread of the γ's [6].

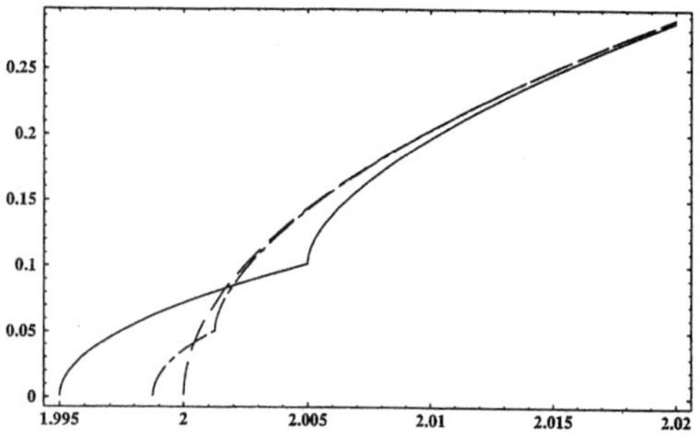

FIG. 3. Calculated cross sections for $e^-\gamma \to \widetilde{\nu}_e\chi^-$ assuming $\widetilde{m}_D = m_{\chi^-} = 100$ GeV and $\widetilde{m}_M = 0$ (dashed line), $\widetilde{m}_M = 5$ GeV (dot-dashed line) and $\widetilde{m}_M = 10$ GeV (full line). For finite $\widetilde{m}_M$ the cross section shows a characteristic double threshhold behavior.

Since for first generation sneutrinos the limit in Eq. (4) indicates that we have to expect even smaller mass splittings, again one should consider processes in which 2nd or 3rd generation sneutrinos are produced, such as $e^-e^+ \to \chi^-\widetilde{\nu}_j e_j$, where $j = 2,3$. However, with three particles in the final state it might be expected that the phase space suppression close to threshold is too severe to allow for an experiment with realistic parameters for future colliders. We are currently doing a quantitative investigation of these processes.

A third interesting way to search for Majorana sneutrinos, which we would like to discuss here for the sake of completeness, is to look for sneutrino-antisneutrino oscillations. This possibility has also been discussed in [8] and [13]. $(\widetilde{\nu} - \widetilde{\nu}^c)$ oscillations arise from the fact that $\widetilde{\nu}$ and $\widetilde{\nu}^c$ are the MSSM interaction eigenstates but not mass eigenstates. Therefore an initially produced pure $|\widetilde{\nu}\rangle$ state will evolve in time into a mixed $|\widetilde{\nu}\rangle - |\widetilde{\nu}^c\rangle$ state [8]. The theory of oscillations of unstable particles is well-known [12]. The time evolution is given by formula

$$|\widetilde{\nu}(t)\rangle = \frac{1}{\sqrt{2}}\left(e^{-i(m_1 + i\Gamma_1/2)t}|\widetilde{\nu}_1\rangle + ie^{-i(m_2 + i\Gamma_2/2)t}|\widetilde{\nu}_2\rangle\right) = A(t)_{\widetilde{\nu}-\widetilde{\nu}}|\widetilde{\nu}\rangle + A(t)_{\widetilde{\nu}-\widetilde{\nu}^c}|\widetilde{\nu}^c\rangle. \quad (8)$$

where Γ_i are the total decay widths of the $\widetilde{\nu}_i$ mass eigenstates $|\widetilde{\nu}_i\rangle$, $A(t)_{\widetilde{\nu}-\widetilde{\nu}(\widetilde{\nu}^c)}$ are transition amplitudes. The total time-integrated probability of finding the antisneutrino state $|\widetilde{\nu}^c\rangle$ in this superposition is given by $\mathcal{P}_{\widetilde{\nu}-\widetilde{\nu}^c} = x^2/2(1+x^2)$, where $x = (\widetilde{m}_1 - \widetilde{m}_2)/\Gamma$ with $\Gamma = (\Gamma_1 + \Gamma_2)/2$. For the case of interest $\widetilde{m}_M << \widetilde{m}_D$ one gets approximately $x \simeq \widetilde{m}_M^2/(\widetilde{m}_D\Gamma)$ [8].

Whether or not sneutrino-antisneutrino oscillations are observable is a strongly model-dependent question. For example, if the sneutrino is the lightest supersymmetric particle (LSP) it will be stable and oscillations are not detectable in any realistic experiment. The same conclusion is reached if

[1]The situation changes of course, if there is sizeable mixing between generations in the sneutrino sector.

$m_{\chi^0} < \tilde{m}_i < m_{\chi^-}$, since the sneutrino in this case decays only into $\chi^0 \nu$. As has been noted, however, in [13] if $m_{\chi^0} < m_{\chi^-} < \tilde{m}_i$ the lepton number of the sneutrino could be tagged by the lepton charge from the decay $\tilde{\nu} \to \chi^{\pm} e^{\mp}$. For a typical decay width of the sneutrino of about $\Gamma \approx 1$ GeV [4], $\tilde{m}_M \approx 2$ GeV and $\tilde{m}_D \approx 100$ GeV about $8 \times [BR(\tilde{\nu} \to \chi^{\pm} e^{\mp})]^2$ events out of 10^4 would appear to have the wrong sign. However, as has been pointed out in Ref. [13], if the decay width of the sneutrino is exceptionally small, searches for sneutrino-antisneutrino oscillations will be superior to the other possibilities discussed above.

In summary, in models of softly broken supersymmetry either both the sneutrino and the neutrino have a Majorana mass or none of them has [5]. A Majorana mass of the sneutrino might be searched for at future colliders such as the NLC [6] or a first muon collider [7]. We have discussed a few examples of Majorana sneutrino phenomenology to estimate the typical sensitivities accelerator experiments might reach. It seems that for second and third generation sneutrinos a dedicated search for Majorana sneutrinos has a chance of providing a positive signal, even after taking the existing low-energy limits into account. For $\tilde{\nu}_e$ the limits obtained from the electron neutrino mass seem to be too stringent—except for some possible loopholes—and accelerator experiments should not be expected to give positive signals.

M.H. would like to thank the Deutsche Forschungsgemeinschaft for financial support by grants kl 253/8-2 and 446 JAP-113/101/0. S.G.K. acknowledges support by Grant GNTP 315NUCLON from the Russian ministry of science.

[1] For a thorough review of the physics of massive neutrinos see, for example, R. N. Mohapatra and P. B. Pal, *Massive Neutrinos in Physics and Astrophysics* (World Scientific, Singapore, 1991).

[2] For recent reviews see, for example, *Proceedings of the International Workshop on Double Beta Decay and Related Topics*, Trento, Italy, 1995, edited by H. V. Klapdor-Kleingrothaus and S. Stoica (World Scientific, Singapore, 1996).

[3] Heidelberg-Moscow Collaboration, M. Günther *et al.*, Phys. Rev. D **55**, 54 (1997).

[4] H. E. Haber and G. L. Kane, Phys. Rep. **117**, 75 (1985); H. P. Nilles, *ibid.* **110**, 1 (1984).

[5] M. Hirsch, H. V. Klapdor-Kleingrothaus, and S. G. Kovalenko, Phys. Lett. B **398**, 311 (1997).

[6] *Sixth International Workshop on Linear Colliders (LC95)*, Proceedings of the International Workshop, Tsukuba, Japan, edited by J. Urakawa (KEK Report. No. 95-5, Tsukuba, 1995); *International Linear Collider Technical Review Committee Report 1995*, SLAC-R-95-471; *e^+e^- Collisions at 500 GeV: The Physics Potential*, Proceedings of the Workshop, Munich, Annecy, Hamburg, 1991, edited by P.M. Zerwas (DESY Report No. 93-123 C, Hamburg, 1993); H. Murayama and M. E. Peskin, SLAC-PUB-7149 and hep-ex/9606003.

[7] Proceedings of the First Workshop on the Physics Potential and Development of $\mu^+\mu^-$ Colliders, Napa, California (1992) [Nucl. Instrum. Methods Phys. Res. A **350**, 24 (1994)]; *Second Workshop on the Physics Potential and Development of $\mu^+\mu^-$ Colliders*, Proceedings, Sausalito, California, 1994, edited by D. Cline, AIP Conf. Proc. No. 352 (AIP, New York, 1994); Proceedings of the Symposium on Physics Potential and Development of *$\mu^+\mu^-$ Colliders*, San Francisco, California, edited by D. Cline [Nucl. Phys. B (Proc. Suppl.) **51A**, (1996)]; J. F. Gunion, Proceedings of the Rencontres de Physique de la Valle d'Aoste, 1996 and hep-ph/9605396.

[8] M. Hirsch, H. V. Klapdor-Kleingrothaus, and S. G. Kovalenko, Phys. Lett. B **403**, 291 (1997).

[9] Particle Data Group, R. M. Barnett *et al.*, Phys. Rev. D **54**, 1 (1996).

[10] OPAL Collaboration, K. Ackerstaff *et al.*, Phys. Lett. B **389**, 616 (1996); DELPHI Collaboration, P. Abreu *et al.*, *ibid.* **382**, 323 (1996).

[11] S. Hesselbach and H. Fraas, Phys. Rev. D **55**, 1343 (1997).

[12] P. J. Franzini, Phys. Rep. **173**, 1 (1989).

[13] Y. Grossman and H. E. Haber, Phys. Rev. Lett. **78**, 3438 (1997).

Light Lepton Number Violating Sneutrinos and the Baryon Number of the Universe

H.V. Klapdor-Kleingrothaus †, St. Kolb†,‡ and V.A. Kuzmin‡

† Max-Planck-Institut für Kernphysik, P.O. 10 39 80, D-69029 Heidelberg, Germany

‡ Institute for Nuclear Research of the Russian Academy of Sciences, 60th October Anniversary Prospect 7a, 117321 Moscow, Russia

Abstract

Recent results of neutrino oscillation experiments point to a non-vanishing neutrino mass. Neutrino mass models favour Majorana-type neutrinos. In such circumstances it is natural that the supersymmetric counterpart of the neutrino, the sneutrino, bears also lepton number violating properties. If the amount of lepton number violation is large enough the sneutrino may be the Cold Dark Matter in the universe. On the other hand, the fact that the universe exhibits an asymmetry in the baryon and antibaryon numbers poses constraints on the extent of lepton number violation in the light sneutrino sector if the electroweak phase transition is second or weak first order. From the requirement that the Baryon Asymmetry of the Universe should not be washed out by sneutrino induced lepton number violating interactions and sphalerons below the critical temperature of the electroweak phase transition we find that the mass-splitting of the light sneutrino mass states is compatible with the sneutrino Cold Dark Matter hypothesis only for heavy gauginos $M_1, M_2 \gtrsim 500GeV$ and opposite sign gaugino mass parameters.

I Introduction

There are hints from neutrino oscillation experiments that the neutrino is massive ([1] and refs. therein. For a recent overview see *e.g.* [2]). In most neutrino mass models the neutrino is of Majorana-type, *i.e.* it violates lepton number L. If this is indeed the case the next generation of experiments searching for neutrinoless double beta ($0\nu\beta\beta$) decay, which are the only experiments capable of deciding on the nature of the neutrino, possibly will be able to indeed observe a $0\nu\beta\beta$-decay signal (for a recent overview see *e.g.* [3]).

On the other hand, it has been shown in [4] that if the neutrino is a massive Majorana field the low energy effective theory of the supersymmetric extension of the Standard Model (for a phenomenological overview see *e.g.* [5]) will contain mass terms for the sneutrino which violate L too, regardless of the mechanism which is

2.4.5 Leptoquarks and

and Double Beta Decay

PHYSICAL REVIEW D VOLUME 54, NUMBER 7 1 OCTOBER 1996

New leptoquark mechanism of neutrinoless double β decay

M. Hirsch, H. V. Klapdor-Kleingrothaus, and S. G. Kovalenko*
Max-Planck-Institut für Kernphysik, P.O. 10 39 80, D-69029, Heidelberg, Germany
(Received 11 March 1996; revised manuscript received 31 May 1996)

A new mechanism for neutrinoless double β ($0\nu\beta\beta$) decay based on leptoquark exchange is discussed. Because of the specific helicity structure of the effective four-fermion interaction this contribution is strongly enhanced compared to the well-known mass mechanism of $0\nu\beta\beta$ decay. As a result the corresponding leptoquark parameters are severely constrained from nonobservation of $0\nu\beta\beta$ decay. These constraints are more stringent than those derived from other experiments. [S0556-2821(96)50119-0]

PACS number(s): 12.40.−j, 23.40.Bw

Neutrinoless double β decay ($0\nu\beta\beta$) is forbidden in the standard model (SM) of electroweak interactions since it violates lepton number conservation. Therefore, experimental observation of this exotic process would be an unambiguous signal of physics beyond the SM (see Refs. [1–3] for reviews).

Essential progress in the exploration of $0\nu\beta\beta$ decay both from theoretical and experimental sides has been achieved in the last few years (see, for instance [3], and references therein). The considerably improved experimental lower bounds on the half-lives of various isotopes enhance the potential of $0\nu\beta\beta$ experiments in testing different concepts of physics beyond the SM such as supersymmetry (SUSY) and leptoquarks (LQ's).

The SUSY mechanisms of $0\nu\beta\beta$ decay were comprehensively investigated in a series of papers [4–7]. It turned out that constraints on certain SUSY parameters from nonobservation of $0\nu\beta\beta$ decay [5] are stronger than those from current and near future accelerator and nonaccelerator experiments. Therefore, it is useful to investigate other possible contributions of physics beyond the SM to $0\nu\beta\beta$ decay to obtain $0\nu\beta\beta$ constraints on the corresponding parameters.

In this Rapid Communication we present a new mechanism of $0\nu\beta\beta$ decay associated with the leptoquark contribution to the effective low-energy charged current lepton-quark interactions. The diagrams describing this contribution are presented in Fig. 1.

The SM symmetries allow 5 scalar (S) and 5 vector (V^μ) LQ's with the following LQ[$SU(3)_c \otimes SU(2)_L \otimes U(1)_Y$] assignments: $S_0(3_c,1;-2/3)$, $\widetilde{S}_0(3_c,1;-8/3)$, $S_{1/2}(\overline{3}_c,2;-7/3)$, $\widetilde{S}_{1/2}(\overline{3}_c,2;-1/3)$, $S_1(3_c,3;-2/3)$, $V_0(\overline{3}_c,1;-4/3)$, $\widetilde{V}_0(\overline{3}_c,1;-10/3)$, $V_{1/2}(3_c,2;-5/3)$, $\widetilde{V}_{1/2}(3_c,2;1/3)$, $V_1(\overline{3}_c,3;-4/3)$, where $Y=2(Q_{em}-T_3)$.

The most general form of the renormalizable LQ-quark-lepton interactions consistent with $SU(3)_c \otimes SU(2)_L \otimes U(1)_Y$ gauge symmetry can be written as [8]

$$\mathcal{L}_{LQ-l-q}=\lambda^{(R)}_{S_0}\cdot\overline{u^c_R}e_R\cdot S^{R\dagger}_0+\lambda^{(R)}_{\widetilde{S}_0}\cdot\overline{d^c_R}e_R\cdot\widetilde{S}^\dagger_0+\lambda^{(R)}_{S_{1/2}}\cdot\overline{u_R}\ell_L\cdot S^{R\dagger}_{1/2}+\lambda^{(R)}_{\widetilde{S}_{1/2}}\cdot\overline{d_R}\ell_L\cdot\widetilde{S}^\dagger_{1/2}+\lambda^{(L)}_{S_0}\cdot\overline{q^c_L}i\tau_2\ell_L\cdot S^{L\dagger}_0+\lambda^{(L)}_{S_{1/2}}\cdot\overline{q_L}i\tau_2 e_R\cdot S^{L\dagger}_{1/2}$$
$$+\lambda^{(L)}_{S_1}\cdot\overline{q^c_L}i\tau_2\hat{S}^\dagger_1\ell_L+\lambda^{(R)}_{V_0}\cdot\overline{d_R}\gamma^\mu e_R\cdot V^{R\dagger}_{0\mu}+\lambda^{(R)}_{\widetilde{V}_0}\cdot\overline{u_R}\gamma^\mu e_R\cdot\widetilde{V}^\dagger_{0\mu}+\lambda^{(R)}_{V_{1/2}}\cdot\overline{d^c_R}\gamma^\mu\ell_L\cdot V^{R\dagger}_{1/2\mu}+\lambda^{(R)}_{\widetilde{V}_{1/2}}\cdot\overline{u^c_R}\gamma^\mu\ell_L\cdot\widetilde{V}^\dagger_{1/2\mu}$$
$$+\lambda^{(L)}_{V_0}\cdot\overline{q_L}\gamma^\mu\ell_L\cdot V^{L\dagger}_{0\mu}+\lambda^{(L)}_{V_{1/2}}\cdot\overline{q^c_L}\gamma^\mu e_R\cdot V^{L\dagger}_{1/2\mu}+\lambda^{(L)}_{V_1}\cdot\overline{q_L}\gamma^\mu\hat{V}^\dagger_{1\mu}\ell_L+\text{H.c.} \tag{1}$$

Here q and ℓ are the quark and the lepton doublets. Following [8,9] we distinguish $S(V)^{L,R}$ being LQ's coupled to the left-handed and right-handed quarks, respectively (see, however, the discussion on chiral couplings in [10]). For LQ triplets $\Phi_1=S_1, V^\mu_1$ the notation $\hat{\Phi}_1=\vec{\tau}\cdot\vec{\Phi}_1$ is used.

On the same footing the LQ fields couple to the SM Higgs doublet field H. A complete list of the renormalizable LQ-Higgs interactions is given in Ref. [10]. These new interactions are especially important for $0\nu\beta\beta$ decay, since after electroweak symmetry breaking they lead to mixing between different LQ multiplets. In turn this mixing generates the effective four-fermion interactions involving right-handed leptonic currents. In combination with the ordinary SM left-handed charged current interactions the latter produce the contribution to $0\nu\beta\beta$ decay shown in the diagrams of Fig. 1 with large enhancement factors. This type of contribution is absent in the case of decoupled LQ and Higgs sectors [10].

Under electroweak symmetry breaking the neutral component of the SM Higgs field acquires a nonzero vacuum expectation value $\langle H^0\rangle$, which creates via LQ-Higgs interaction terms nondiagonal mass matrices for LQ fields with the same electric charge but from different $SU(2)_L$ multiplets. To obtain observable predictions from the LQ-lepton-quark in-

*Permanent address: Joint Institute for Nuclear Research, Dubna, Russia.

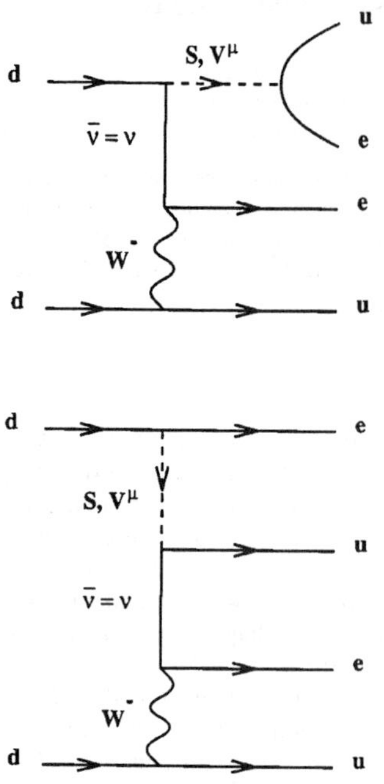

FIG. 1. Feynman graphs for the leptoquark-induced mechanism of $0\nu\beta\beta$ decay. S and V^{μ} stand symbolically for (a) $Q=-1/3$ (upper part) and (b) $Q=2/3$ (lower part) scalar and vector LQ's.

teraction Lagrangian in Eq. (1), the LQ fields ($I=S,V$) with nondiagonal mass matrices have to be rotated to the mass eigenstate basis I'. This can be done in the standard way: $I(Q)=\mathcal{N}^{(I)}(Q)\cdot I'(Q)$, where $\mathcal{N}^{(I)}(Q)$ are orthogonal matrices such that $\mathcal{N}^{(I)T}(Q_I)\cdot\mathcal{M}_I^2(Q)\cdot\mathcal{N}^{(I)}(Q)=\mathrm{Diag}\{M^2_{I_n}\}$, with the M_{I_n} being the mass of the relevant mass eigenstate field I'.

Now it is straightforward to derive the effective four-fermion ν-u-d-e interaction terms generated by the LQ exchange in the upper parts of the diagrams in Fig. 1. After Fierz rearrangement they take the form [10]

$$\mathcal{L}^{\mathrm{eff}}_{\mathrm{LQ}}=U^*_{en}(\bar{\nu}_nP_Re^c)\left[\frac{\epsilon_S}{M_S^2}(\bar{u}P_Rd)+\frac{\epsilon_V}{M_V^2}(\bar{u}P_Ld)\right] - U^*_{en}(\bar{\nu}_n\gamma^{\mu}P_Le^c)\left[\left(\frac{\alpha_S^{(R)}}{M_S^2}+\frac{\alpha_V^{(R)}}{M_V^2}\right)(\bar{u}\gamma_{\mu}P_Rd) - \sqrt{2}\left(\frac{\alpha_S^{(L)}}{M_S^2}+\frac{\alpha_V^{(L)}}{M_V^2}\right)(\bar{u}\gamma_{\mu}P_Ld)\right], \qquad (2)$$

where ν_n are the Majorana mass eigenstate neutrino fields related to the weak eigenstate ν_e^0 by the mixing matrix $\nu_e^0=U_{en}\nu_n$. The effective LQ couplings are defined as

$$\epsilon_I=2^{-\eta_I}\{\lambda^{(L)}_{I_1}\lambda^{(R)}_{\tilde{I}_{1/2}}[\theta^I_{43}(Q_I^{(1)})+\eta_I\sqrt{2}\,\theta^I_{41}(Q_I^{(2)})] - \lambda^{(L)}_{I_0}\lambda^{(R)}_{\tilde{I}_{1/2}}\theta^I_{13}(Q_I^{(1)})\}, \qquad (3)$$

$$\alpha_I^{(L)}=\frac{2}{3+\eta_I}\lambda^{(L)}_{I_{1/2}}\lambda^{(L)}_{I_1}\theta^I_{24}(Q_I^{(2)}),$$

$$\alpha_I^{(R)}=\frac{2}{3+\eta_I}\lambda^{(R)}_{I_0}\lambda^{(R)}_{\tilde{I}_{1/2}}\theta^I_{23}(Q_I^{(1)}). \qquad (4)$$

$\eta_{S,V}=1,-1$ for scalar and vector LQ's. $\theta^I_{kn}(Q)$ is a mixing parameter defined by

$$\theta^I_{kn}(Q)=\sum_l \mathcal{N}^{(I)}_{kl}(Q)\mathcal{N}^{(I)}_{nl}(Q)\left(\frac{M_I}{M_{I_l}(Q)}\right)^2, \qquad (5)$$

where $\mathcal{N}^{(I)}(Q)$ are mixing matrix elements for the scalar $I=S$ and vector $I=V$ LQ fields with electric charges $Q=-1/3,-2/3$. Common mass scales M_S of scalar and M_V of vector LQ's are introduced for convenience.

Following the well-known procedure [2] one can find the LQ contribution to the $0\nu\beta\beta$-decay matrix element for the diagrams in Fig. 1. The LQ exchange sectors of these diagrams are described by the pointlike four-fermion interactions specified by the effective Lagrangian in Eq. (2). Their bottom parts are the SM charged current (SMCC) interactions of the form $(\bar{u}\gamma_{\mu}P_Ld)(\bar{e}\gamma^{\mu}P_L\nu_n)\cdot U_{en}$. The leptonic part of the LQ contribution to the matrix element can be written schematically as

$$\bar{e}\gamma_{\mu}P_L\langle 0|T(\nu_k\bar{\nu}_n)|0\rangle P_RO^{(i)}e^c\cdot U_{ek}U^*_{en} \sim \bar{e}\gamma_{\mu}\gamma_{\nu}P_RO^{(i)}e^c\cdot q^{\nu}/q^2,$$

where q is the neutrino momentum $O^{(1,2)}=1,\gamma_{\rho}$ [see Eq. (2)]. In the right-hand side neutrino masses m_{ν_n} are neglected since $m_{\nu_n}\ll\langle q\rangle$, where $\langle q\rangle\sim p_F\approx 80$ MeV is the average momentum of a neutrino propagating in a nucleus (p_F is the nucleon Fermi momentum). The mixing matrix elements disappear on the right- hand side due to the unitarity relation $U_{ek}U^*_{en}\delta_{kn}=1$. This should be compared with the standard neutrino mass mechanism with the SMCC at both ends of the virtual neutrino line when $\bar{e}\gamma_{\mu}P_L\langle 0|T(\nu_k\bar{\nu}_n)|0\rangle P_L\gamma_{\rho}e^c\cdot U_{ek}U_{en}\sim\bar{e}\gamma_{\mu}\gamma_{\rho}P_Re^c\cdot\langle m_{\nu}\rangle/q^2$, where $\langle m_{\nu}\rangle=m_{\nu_n}U^2_{en}$. It is important that the LQ contribution has a huge enhancement factor $\sim\langle q\rangle/\langle m_{\nu}\rangle\sim 10^8\cdot(1\ \mathrm{eV}/\langle m_{\nu}\rangle)$ compared to the standard contribution. The final formula for the inverse half-life of $0\nu\beta\beta$ decay taking into account only the LQ contribution reads

$$T^{-1}_{1/2}(0\nu\beta\beta)=|M_{GT}|^2\frac{2}{G_F^2}[\widetilde{C}_1a^2+a(\widetilde{C}_2b_R-\sqrt{2}\widetilde{C}_3b_L) + C_4b_R^2+2C_5b_L^2-\sqrt{2}C_6b_Lb_R], \qquad (6)$$

with

$$a=\frac{\epsilon_S}{M_S^2}+\frac{\epsilon_V}{M_V^2},\qquad b_{L,R}=\left(\frac{\alpha_S^{(L,R)}}{M_S^2}+\frac{\alpha_V^{(L,R)}}{M_V^2}\right),$$

$$\widetilde{C}_n=C_1\left(\frac{\mathcal{M}_1^{(\nu)}/(m_eR)}{M_{GT}-\alpha_2M_F}\right)^{\kappa_n} \qquad (7)$$

where $\kappa_1=2$, $\kappa_{2,3}=1$; m_e and R are the electron mass and nuclear radius. Our final formulas Eqs. (6) and (7) do not explicitly depend on the neutrino mass m_ν. Nevertheless $m_\nu=0$ leads to $a=b_{L,R}=0$ and $\rightarrow T^{-1}_{1/2}(0\nu\beta\beta)=0$. This follows from the fact that the LQ interactions $\mathcal{L}^{\rm eff}_{\rm LQ}$ in Eq. (2) contribute to m_ν at the loop level and, therefore, vanish in the massless neutrino case. This property is a particular case of the general theorem relating the $0\nu\beta\beta$ matrix element $R_{0\nu\beta\beta}$ and neutrino mass m_ν [11]. This theorem is valid in any gauge theory. A quantitative estimation of the $R_{0\nu\beta\beta}-m_\nu$ relation in the specific case of left-right symmetric models was found in Ref. [12]. A similar quantitative application of the above-mentioned theorem to the LQ mechanism is beyond the scope of the present Rapid Communication.

We use notations of Ref. [2] for the nuclear structure coefficients C_i and the nuclear matrix elements $M_{GT,F}$. The new matrix element $\mathcal{M}_1^{(\nu)}$ was introduced and calculated in Ref. [7] within the pn quasiparticle random phase approximation (QRPA) framework [13]. Calculating C_i within the same approach for the particular case of ^{76}Ge we have a complete set of the nuclear structure coefficients in Eq. (6): $|M_{GT}|^2\widetilde{C}_1=1.70\times10^{-10}$, $|M_{GT}|^2\widetilde{C}_2=-1.72\times10^{-12}$, $|M_{GT}|^2\widetilde{C}_3=9.03\times10^{-10}$, $|M_{GT}|^2C_4=1.55\times10^{-13}$, $|M_{GT}|^2C_5=4.99\times10^{-9}$, and $|M_{GT}|^2C_6=-4.63\times10^{-14}$ (all in units of inverse years).

Now we are ready to derive constraints on the LQ parameters $a,b_{L,R}$ in Eq. (6). We use the result from the Heidelberg-Moscow ^{76}Ge experiment [14] $T^{0\nu\beta\beta}_{1/2}(^{76}{\rm Ge},0^+\rightarrow0^+)>7.4\times10^{24}$ yr 90% C.L.

Assuming that either scalar or vector leptoquarks contribute in Eq. (6) we derive the following constraints on the effective LQ parameters:

$$\epsilon_I\leqslant2.8(2.4)\times10^{-9}\left(\frac{M_I}{100\ {\rm GeV}}\right)^2, \tag{8}$$

$$\alpha_I^{(L)}\leqslant3.5(3.1)\times10^{-10}\left(\frac{M_I}{100\ {\rm GeV}}\right)^2, \tag{9}$$

$$\alpha_I^{(R)}\leqslant7.9(7.8)\times10^{-8}\left(\frac{M_I}{100\ {\rm GeV}}\right)^2. \tag{10}$$

Values in brackets are derived assuming only one parameter is nonzero at a time, while those without brackets are for arbitrary values of the other two parameters. Recall $I=S,V$.

It is interesting to compare these constraints with the corresponding constraints from other processes [9]. Consider the helicity-suppressed decay $\pi\rightarrow e\nu$, which is extremely sensitive to the first two scalar-pseudoscalar terms in Eq. (2), leading to a helicity-unsuppressed amplitude [9]. The following constraint from $\pi\rightarrow e\nu$-decay data was obtained in Ref. [10]: $\epsilon_I\leqslant5\times10^{-7}(M_I/100\ {\rm GeV})^2$. Apparently, the corresponding constraints from $0\nu\beta\beta$ decay in Eq. 8 are more stringent by about two orders of magnitude. This confirms that $0\nu\beta\beta$ decay is a powerful probe of physics beyond the standard model.

In summary, nonobservation of $0\nu\beta\beta$ decay can provide stringent bounds on parameters of extensions of the standard model. Moreover, the $0\nu\beta\beta$ decay bounds on some of these fundamental parameters can be much more stringent than those from other experiments. Previously such a conclusion was obtained for the case of the R-parity-violating supersymmetric contribution to $0\nu\beta\beta$ decay [5–7]. In this Rapid Communication we have shown that the leptoquark mechanism allows similar conclusions.

We thank V.A. Bednyakov and D.I. Kazakov for helpful discussions. M.H. would like to thank the Deutsche Forschungsgemeinschaft for financial support by Grants No. kl 253/8-1 and 446 JAP-113/101/0.

[1] W.C. Haxton and G.J. Stephenson, Prog. Part. Nucl. Phys. **12**, 409 (1994); K. Grotz and H.V. Klapdov-Kleingrothaus, *The Weak Interaction in Nuclear, Particle and Astrophysics* (Adam Hilger, Bristol, 1990); R.N. Mohapatra and P.B. Pal, *Massive Neutrinos in Physics and Astrophysics* (World Scientific, Singapore, 1991); M. Moe and P. Vogel, Annu. Rev. Nucl. Part. Sci. **44**, 247 (1994); J.D. Vergados, Phys. Rep. **133**, 1 (1986); J.W.F. Valle, Prog. Part. Nucl. Phys. **26**, 91 (1991).
[2] M. Doi, T. Kotani, and E. Takasugi, Prog. Theor. Phys. Suppl. **83**, 1 (1985).
[3] *Proccedings of the International Workshop on Double Beta Decay and Related Topics, Trento, Italy, 1995*, edited by H.V. Klapdor-Kleingrothaus and S. Stoica (World Scientific, Singapore, 1996).
[4] R.N. Mohapatra, Phys. Rev. D **34**, 3457 (1986); J.D. Vergados, Phys. Lett. B **184**, 55 (1987).
[5] M. Hirsch, H.V. Klapdor-Kleingrothaus, and S.G. Kovalenko, Phys. Lett. B **352**, 1 (1995); Phys. Rev. Lett. **75**, 17 (1995); Phys. Rev. D **53**, 1329 (1996).
[6] K.S. Babu and R.N. Mohapatra, Phys. Rev. Lett. **75**, 2276 (1995).
[7] M. Hirsch, H.V. Klapdor-Kleingrothaus, and S.G. Kovalenko, Phys. Lett. B **372**, 181 (1996).
[8] W. Buchmüller, R. Rückl, and D. Wyler, Phys. Lett. B **191**, 442 (1987).
[9] S. Davidson, D. Bailey, and A. Campbell, Z. Phys. C **61**, 613 (1994); M. Leurer, Phys. Rev. Lett. **71**, 1324 (1993); Phys. Rev. D **50**, 536 (1994).
[10] M. Hirsch, H.V. Klapdor-Kleingrothaus, and S.G. Kovalenko, Phys. Lett. B **378**, 17 (1996).
[11] J. Schechter and J.W.F. Valle, Phys. Rev. D **25**, 2951 (1982).
[12] J.F. Nieves, Phys. Lett. **147B**, 375 (1984); E. Takasugi, *ibid.* **149B**, 372 (1984); B. Kayser, in *Proceedings of the XXIII International Conference on High Energy Physics*, Berkeley, California, 1986, edited by S. Loken (World Scientific, Singapore, 1987), p. 945; S. Petcov, in *'86 Massive Neutrinos in Astrophysics and in Particle Physics*, Proceedings of the 21st Rencontre de Moriond, Tignes, France, edited by O. Fack-

ler and J. Trân Than Vân (Editions Frontieres, Gif-sur-Yvette, France, 1986), p. 187; S. P. Rosen, Reports No. UTAPHY-HEP-4 and hep-ph/9210202 (unpublished).
[13] K. Muto, E. Bender, and H.V. Klapdor, Z. Phys. A **334**, 177(1989); **334**, 187 (1989).
[14] HEIDELBERG-MOSCOW Collaboration, A. Balysh *et al.*, Phys. Lett. B **356**, 450 (1995); H.V. Klapdor-Kleingrothaus, in Ref. [3].

ELSEVIER

20 June 1996

Physics Letters B 378 (1996) 17–22

PHYSICS LETTERS B

New low-energy leptoquark interactions

M. Hirsch [a], H.V. Klapdor-Kleingrothaus [a], S.G. Kovalenko [a,b]

[a] *Max-Planck-Institut für Kernphysik, P.O. 10 39 80, D-69029 Heidelberg, Germany*

[b] *Joint Institute for Nuclear Research, Dubna, Russia*

Received 21 February 1996; revised manuscript received 1 April 1996
Editor: C. Mahaux

Abstract

We discuss an extension of the standard model (SM) with vector and scalar leptoquarks. The renormalizable leptoquark Lagrangian consistent with the SM gauge symmetry is presented including the leptoquark-Higgs interactions previously not considered in the literature. We discuss the importance of these new interactions for leptoquark phenomenology. After the electro-weak symmetry breaking they generate non-trivial leptoquark mass matrices. These lead to mixing between different $SU(2)_L$-multiplets of the leptoquarks and induce at low energies new effective 4-fermion lepton-quark vertices. The latter affect the standard leptoquark phenomenology. We discuss constraints on these interactions from the helicity-suppressed $\pi \rightarrow \nu + e$ decay.

PACS: 11.30; 12.30; 13.15; 14.80
Keywords: Leptoquarks; Higgs; Effective interaction; Non-chiral; Pion decay

The interest on leptoquarks (LQ) [1] has been renewed during the last few years since ongoing collider experiments have good prospects for searching these particles [2]. LQs are vector or scalar particles carrying both lepton and baryon numbers and, therefore, have a well distinguished experimental signature.

LQs can be quite naturally introduced in the low-energy theory as a relic of a more fundamental theory at some high-energy scale. In such a way LQs can emerge from grand unified theories (GUT) [3,4], including the superstring-inspired versions of GUT [4], models of extended technicolour [5] and composite models [6].

Possible LQ manifestations in various processes have been extensively investigated [2–7] (and references therein). Various constraints on LQ masses and couplings have been deduced from existing experimental data and prospects for the forthcoming experiments have been estimated.

Direct searches of LQs as s-channel resonances in deep inelastic ep-scattering at HERA experiments [10] placed lower limits on their mass $M_{LQ} \geq 140 - 235$ GeV [11] depending on the LQ type and couplings. With larger accumulated luminosity HERA will be able to cover almost the whole kinematical region in the LQ masses up to 296 GeV, for couplings to quarks and leptons above 10^{-2}. There are also bounds from other collider experiments. The LEP experiments exclude any LQ lighter then 45 GeV [12], the D0 collaboration rules out LQs lighter than 133 GeV if they couple to the first generation fermions [13], and the CDF collaboration sets a corresponding lower bound at 113 GeV [14].

Dramatic improvements of these constraints are expected in future collider experiments (see for instance [7] and references therein).

PII S0370-2693(96)00419-4

Table 1
The Standard model assignments of the scalar S and vector V_μ leptoquarks (LQ). ($Y = 2(Q_{em} - T_3)$)

LQ	$SU(3)_c$	$SU(2)_L$	Y	Q_{em}
S_0	**3**	**1**	$-2/3$	$-1/3$
$\tilde{S}_0$	**3**	**1**	$-8/3$	$-4/3$
$S_{1/2}$	**3***	**2**	$-7/3$	$(-2/3, -5/3)$
$\tilde{S}_{1/2}$	**3***	**2**	$-1/3$	$(1/3, -2/3)$
S_1	**3**	**3**	$-2/3$	$(2/3, -1/3, -4/3)$
V_0	**3***	**1**	$-4/3$	$-2/3$
$\tilde{V}_0$	**3***	**1**	$-10/3$	$-5/3$
$V_{1/2}$	**3**	**2**	$-5/3$	$(-1/3, -4/3)$
$\tilde{V}_{1/2}$	**3**	**2**	$1/3$	$(2/3, -1/3)$
V_1	**3***	**3**	$-4/3$	$(1/3, -2/3, -5/3)$

However, at present the most stringent limits on LQs come from low-energy experiments [8], [9]. Effective 4-fermion interactions, induced by virtual LQ exchange at energies much smaller than their masses, can contribute to atomic parity violation, flavour-changing neutral current (FCNC) processes, meson decays, meson-antimeson mixing and some rare processes. For instance, a typical bound on non-chirally coupled LQs imposed by the helicity-suppressed $\pi \to e\nu$ decay is $M_{LQ}^2/|g_L g_R| \geq (100\text{TeV})^2$, where $g_{L,R}$ are LQ couplings [8].

To consider LQ phenomenology in a model-independent fashion one usually follows some general principles in constructing the Lagrangian of the LQ interactions with the standard model (SM) fields. Generic principles are renormalizability **(p1)** and invariance **(p2)** under the SM gauge group $SU(3)_c \otimes SU(2)_L \otimes U(1)_Y$. In order to obey the stringent constraints from **(c1)** helicity-suppressed $\pi \to e\nu$ decay, from **(c2)** FCNC processes and **(c3)** proton stability, the following three assumptions are also commonly adopted: **(a1)** LQ couplings are "chiral", i.e. each type of LQs couples either to left-handed or to right-handed quarks only (call them left- and right-type LQ); **(a2)** LQ couplings are generation "diagonal", i.e. they couple only to a single generation of leptons and a single generation of quarks; **(a3)** LQ interactions conserve baryon (B) and lepton (L) numbers.

Of course, if these empirical assumptions (a1)–(a3) really determine the LQ interactions, they have to be explained in terms of an underlying theory predicting light LQs. We will show, however, that assumption (a1) does not solve problem (c1) since the LQ couplings with the SM Higgs doublet reintroduce the non-chiral interaction terms. Therefore, to obey (c1) one should not only claim chirality of the LQ-quark couplings (a1) but also absence of some LQ-Higgs couplings. It is unlikely that both requested properties can have the same origin in the underlying theory.

In the following we consider changes in LQ phenomenology caused by the LQ-Higgs interactions. We base our consideration on the general principles (p1), (p2) as well as on the assumptions (a1)–(a3).

The SM symmetries allow 5 scalar S and 5 vector V^μ LQs with the following $LQ(SU(3)_c \otimes SU(2)_L \otimes U(1)_Y)$ assignments: $S_0(3_c, 1; -2/3)$, $\tilde{S}_0(3_c, 1; -8/3)$, $S_{1/2}(\bar{3}_c, 2; -7/3)$, $\tilde{S}_{1/2}(\bar{3}_c, 2; -1/3)$, $S_1(3_c, 3; -2/3)$, $V_0(\bar{3}_c, 1; -4/3)$, $\tilde{V}_0(\bar{3}_c, 1; -10/3)$, $V_{1/2}(3_c, 2; -5/3)$, $\tilde{V}_{1/2}(3_c, 2; 1/3)$, $V_1(\bar{3}_c, 3; -4/3)$, where $Y = 2(Q_{em} - T_3)$.

In the literature only LQ-lepton-quark interaction terms have been considered. They have the following form [2]

$$
\begin{aligned}
\mathcal{L}_{LQ-l-q} = {} & \lambda_{S_0}^{(R)} \cdot \overline{u^c} P_R e \cdot S_0^{R\dagger} + \lambda_{\tilde{S}_0}^{(R)} \cdot \overline{d^c} P_R e \cdot \tilde{S}_0^\dagger + \lambda_{S_{1/2}}^{(R)} \cdot \overline{u} P_L l \cdot S_{1/2}^{R\dagger} \\
& + \lambda_{\tilde{S}_{1/2}}^{(R)} \cdot \overline{d} P_L l \cdot \tilde{S}_{1/2}^\dagger + \lambda_{S_0}^{(L)} \cdot \overline{q^c} P_L i\tau_2 l \cdot S_0^{L\dagger} + \lambda_{S_{1/2}}^{(L)} \cdot \overline{q} P_R i\tau_2 e \cdot S_{1/2}^{L\dagger} \\
& + \lambda_{S_1}^{(L)} \cdot \overline{q^c} P_L i\tau_2 \hat{S}_1^\dagger l + \lambda_{V_0}^{(R)} \cdot \overline{d} \gamma^\mu P_R e \cdot V_{0\mu}^{R\dagger} + \lambda_{\tilde{V}_0}^{(R)} \cdot \overline{u} \gamma^\mu P_R e \cdot \tilde{V}_{0\mu}^\dagger \\
& + \lambda_{V_{1/2}}^{(R)} \cdot \overline{d^c} \gamma^\mu P_L l \cdot V_{1/2\mu}^{R\dagger} + \lambda_{\tilde{V}_{1/2}}^{(R)} \cdot \overline{u^c} \gamma^\mu P_L l \cdot \tilde{V}_{1/2\mu}^\dagger + \lambda_{V_0}^{(L)} \cdot \overline{q} \gamma^\mu P_L l \cdot V_{0\mu}^{L\dagger}
\end{aligned}
\tag{1}
$$

$$+\lambda^{(L)}_{V_{1/2}}\cdot\overline{q^c}\gamma^\mu P_R e\cdot V^{L\dagger}_{1/2\mu}+\lambda^{(L)}_{V_1}\cdot\bar{q}\gamma^\mu P_L\hat{V}^\dagger_{1\mu}l+\text{h.c.}$$

Here $P_{L,R}=(1\mp\gamma_5)/2$; q and l are the quark and the lepton doublets; S^j_i and V^j_i are the scalar and vector LQs with the weak isospin $i=0,1/2,1$ coupled to left-handed ($j=L$) or right-handed ($j=R$) quarks respectively. (Our notations differ from those in Ref. [2]) For LQ triplets $\Phi_1=S_1,V^\mu_1$ we use the notation $\hat{\Phi}_1=\boldsymbol{\tau}\cdot\boldsymbol{\Phi}_1$.

There are no fundamental reasons forbidding LQ interactions with the standard model Higgs doublet H. The most general form of the LQ-Higgs interaction Lagrangian, consistent with (p1) and (p2) can be expressed as

$$\mathcal{L}_{LQ-H}=h^{(i)}_{S_0}Hi\tau_2\tilde{S}_{1/2}\cdot S^i_0+h^{(i)}_{V_0}Hi\tau_2\tilde{V}^\mu_{1/2}\cdot V^i_{0\mu}+h_{S_1}Hi\tau_2\hat{S}_1\cdot\tilde{S}_{1/2} \quad (2)$$

$$+h_{V_1}Hi\tau_2\hat{V}^\mu_1\cdot\tilde{V}_{1/2\mu}+Y^{(i)}_{S_{1/2}}\left(Hi\tau_2S^i_{1/2}\right)\cdot\left(\tilde{S}^\dagger_{1/2}H\right)+Y^{(i)}_{V_{1/2}}\left(Hi\tau_2V^{\mu(i)}_{1/2}\right)\cdot\left(\tilde{V}^\dagger_{1/2\mu}H\right)$$

$$+Y_{S_1}\left(Hi\tau_2\hat{S}^\dagger_1H\right)\cdot\tilde{S}_0+Y_{V_1}\left(Hi\tau_2\hat{V}^\dagger_{1\mu}H\right)\cdot\tilde{V}^\mu_0+\kappa^{(i)}_S\left(H^\dagger\hat{S}_1H\right)\cdot S^{i\dagger}_0$$

$$+\kappa^{(i)}_V\left(H^\dagger\hat{V}^\mu_1H\right)\cdot V^{i\dagger}_{0\mu}-\left(\eta_\Phi M^2_\Phi-g^{(i_1i_2)}_\Phi H^\dagger H\right)\Phi^{i_1\dagger}\Phi^{i_2}+\text{h.c.}$$

Here H is the SM $SU(2)_L$-doublet Higgs field. Φ^i is a cumulative notation for all leptoquark fields with $i=L,R$ (the same for $i_{1,2}$). We included diagonal mass terms $\eta_\Phi M^2_\Phi\Phi^\dagger\Phi$ of the scalar ($\eta_S=1$) and the vector ($\eta_V=-1$) LQ fields. These terms can be generated by spontaneous breaking of the underlying symmetry down to the electro-weak gauge group at some high-energy scale.

The subsequent electro-weak symmetry breaking at the Fermi scale produces additional non-diagonal LQ mass terms leading to non-trivial mixing between LQs from different $SU(2)_L$ multiplets as well as between left- and right-types of LQs, discussed above. The relevant LQ mass matrices can be read off from Eq. (2). There are 8 non-diagonal LQ mass matrices squared ($I=S,V$):

$$\mathcal{M}^2_I(Q^{(1)}_I)=\eta_I\cdot\begin{pmatrix}\eta_I\bar{M}^2_{I^L_0} & g^{LR}_{I_0}|v|^2 & h^{(L)}_{I_0}v & \kappa^{(L)}_I|v|^2\\ g^{LR}_{I_0}|v|^2 & \eta_I\bar{M}^2_{I^R_0} & h^{(R)}_{I_0}v & \kappa^{(R)}_I|v|^2\\ h^{(L)}_{I_0}v & h^{(R)}_{I_0}v & \eta_I\bar{M}^2_{\tilde{I}_{1/2}} & h_{I_1}v\\ \kappa^{(L)}_I|v|^2 & \kappa^{(R)}_I|v|^2 & h_{I_1}v & \eta_I\bar{M}^2_{I_1}\end{pmatrix}, \quad (3)$$

$$\mathcal{M}^2_I(Q^{(2)}_I)=\eta_I\cdot\begin{pmatrix}\eta_I\bar{M}^2_{\tilde{I}_{1/2}} & Y^L_{I_{1/2}}v^2 & Y^R_{I_{1/2}}v^2 & \sqrt{2}h_{I_1}v\\ Y^L_{I_{1/2}}v^2 & \eta_I\bar{M}^2_{I^L_{1/2}} & g^{(LR)}_{I_{1/2}}|v|^2 & 0\\ Y^R_{I_{1/2}}v^2 & g^{(LR)}_{I_{1/2}}|v|^2 & \eta_I\bar{M}^2_{I^R_{1/2}} & 0\\ \sqrt{2}h_{I_1}v & 0 & 0 & \eta_I\bar{M}^2_{I_1}\end{pmatrix}, \quad (4)$$

$$\mathcal{M}^2_I(Q^{(3)}_I)=\eta_I\cdot\begin{pmatrix}\eta_I\bar{M}^2_{\tilde{I}_0} & \sqrt{2}Y_{I_1}v^2\\ \sqrt{2}Y_{I_1}v^2 & \eta_I\bar{M}^2_{I_1}\end{pmatrix}, \quad (5)$$

$$\mathcal{M}^2_I(Q^{(4)}_I)=\eta_I\cdot\begin{pmatrix}\eta_I\bar{M}^2_{I^L_{1/2}} & -g^{(LR)}_{I_{1/2}}|v|^2\\ -g^{(LR)}_{I_{1/2}}|v|^2 & \eta_I\bar{M}^2_{I^R_{1/2}}\end{pmatrix}, \quad (6)$$

where $\bar{M}^2_I=M^2_I+\eta_Ig_I|v|^2$ is the "shifted" diagonal mass, $v^2=\langle H^0\rangle^2=(2\sqrt{2}G_F)^{-1}$ is the SM Higgs field vacuum expectation value, G_F is the Fermi constant and $\eta_{S,V}=1,-1$. The cumulative notation $\mathcal{M}^2_I(Q^{(k)}_I)$ encodes the mass matrices squared for the scalar ($I=S$) and vector ($I=V_\mu$) LQs with electric charges $Q^{(1)}_S=Q^{(2)}_V=-1/3$, $Q^{(2)}_S=Q^{(1)}_V=-2/3$, $Q^{(3)}_S=Q^{(4)}_V=-4/3$, $Q^{(3)}_V=Q^{(4)}_S=-5/3$ in the interaction eigenstate basises: ($k=1$) $I(Q^{(1)}_I)=(I^L_0,I^R_0,\tilde{I}^\dagger_{1/2},I_1)$; ($k=2$) $I(Q^{(2)}_I)=(\tilde{I}_{1/2},I^L_{1/2},I^R_{1/2},I^\dagger_1)$; ($k=3$) $I(Q^{(3)}_I)=(\tilde{I}_0,I_1)$; ($k=4$) $I(Q^{(4)}_I)=(I^L_{1/2},I^R_{1/2})$. Thus, there is a non-trivial mixing of LQs from different

$SU(2)_L$ multiplets as well as the $I^L - I^R$ mixing. The latter spoils chirality of the LQ-quark-lepton couplings (a1) and leads to reappearance of the problem with the constraint (c1). The mass matrices of all other LQ fields remain diagonal after electro-weak symmetry breaking.

To obtain observable predictions from the LQ-lepton-quark interaction Lagrangian in Eq. (1), fields with non-diagonal mass matrices have to be rotated to the mass eigenstate basis I'. This can be done in the standard way $I(Q) = \mathcal{N}^{(I)}(Q) \cdot I'(Q)$ where $\mathcal{N}^{(I)}$ are orthogonal matrices such that $\mathcal{N}^{(I)T}(Q) \cdot \mathcal{M}_I^2(Q) \cdot \mathcal{N}^{(I)}(Q) = Diag\{M_{I_n}^2\}$ with the M_{I_n} being the mass of the relevant mass eigenstate field I'. All phenomenological consequences of the LQ interactions in Eqs. (1)–(2) should be derived in terms of these fields I'.

In this letter we concentrate on the LQ induced 4-fermion lepton-quark effective interactions. For vanishing LQ-Higgs couplings (Eq. (2)) these interaction terms are listed in Ref. [8]. Mixing between different $SU(2)_L$ multiplets of LQs leads to new terms, vanishing in the limiting case of decoupled LQ and Higgs sectors. Below we list only those new terms which can be most stringently restricted from low-energy processes. After Fierz rearrangement they take the form

$$\mathcal{L}^{\rm eff}_{\rm mix} = (\bar{\nu} P_R e^c)\left[\frac{\epsilon_S}{M_S^2}(\bar{u}P_R d) + \frac{\epsilon_V}{M_V^2}(\bar{u}P_L d)\right] + (\overline{\nu^c} P_L e^c)\left[\frac{\omega_S}{M_S^2}(\bar{u}P_L d) + \frac{\omega_V}{M_V^2}(\bar{u}P_R d)\right]$$
$$- (\bar{\nu}\gamma^\mu P_L e^c)\left[\left(\frac{\alpha_S^{(R)}}{M_S^2} + \frac{\alpha_V^{(R)}}{M_V^2}\right)(\bar{u}\gamma_\mu P_R d) - \sqrt{2}\left(\frac{\alpha_S^{(L)}}{M_S^2} + \frac{\alpha_V^{(L)}}{M_V^2}\right)(\bar{u}\gamma_\mu P_L d)\right], \tag{7}$$

where

$$\epsilon_I = 2^{-\eta_I}\left[\lambda_{I_1}^{(L)}\lambda_{\tilde{I}_{1/2}}^{(R)}\left(\theta_{43}^I(Q_I^{(1)}) + \eta_I\sqrt{2}\theta_{41}^I(Q_I^{(2)})\right) - \lambda_{I_0}^{(L)}\lambda_{\tilde{I}_{1/2}}^{(R)}\theta_{13}^I(Q_I^{(1)})\right],$$
$$\omega_I = 2^{-\eta_I}\left[\lambda_{I_0}^{(L)}\lambda_{\tilde{I}_0}^{(R)}\theta_{12}^I(Q_I^{(1)}) + \lambda_{I_0}^{(R)}\lambda_{I_1}^{(L)}\theta_{42}^I(Q_I^{(1)}) + \lambda_{I_{1/2}}^{(L)}\lambda_{I_{1/2}}^{(R)}\theta_{32}^I(Q_I^{(2)})\right],$$
$$\alpha_I^{(L)} = \frac{2}{3+\eta_I}\lambda_{I_{1/2}}^{(L)}\lambda_{I_1}^{(L)}\theta_{24}^I(Q_I^{(2)}), \quad \alpha_I^{(R)} = \frac{2}{3+\eta_I}\lambda_{I_0}^{(R)}\lambda_{\tilde{I}_{1/2}}^{(R)}\theta_{23}^I(Q_I^{(1)}). \tag{8}$$

Here we introduced a mixing parameter

$$\theta_{kn}^I(Q) = \sum_l \mathcal{N}_{kl}^{(I)}(Q)\mathcal{N}_{nl}^{(I)}(Q)\left(\frac{M_I}{M_{I_l}(Q)}\right)^2, \tag{9}$$

where $Q = -1/3, -2/3$ and $I = S, V$. Common mass scales M_S of scalar and M_V of vector LQs were introduced for convenience.

The interaction terms Eq. (7) contribute to various low-energy processes. Using existing experimental data one can obtain constraints on the relevant coupling constants. Here we are not going to discuss this subject in detail but rather present only the most stringent bounds from the helicity-suppressed decay $\pi \to e\nu$. This process is especially sensitive to the first two scalar-pseudoscalar terms leading to a helicity-unsuppressed amplitude. Assuming no spurious cancellations between different contributions we derive on the basis of Ref. [8] the following severe constraints:

$$\epsilon_I, \omega_I \leq 5 \times 10^{-7}\left(\frac{M_I}{100\ \text{GeV}}\right)^2. \tag{10}$$

Other couplings in Eq. (7) are much weaker constrained by low-energy processes previously considered in connection with the LQ phenomenology [8,9]. We expect that new stringent constraints on these LQ couplings can be derived from neutrinoless double beta decay ($0\nu\beta\beta$). Consider the conventional mechanism of $0\nu\beta\beta$-decay based on the Majorana neutrino exchange between decaying nucleons. Let the neutrino propagator connect the ordinary SM charged current vertex with the LQ-generated one related to the ϵ or α couplings in Eq. (7). The corresponding amplitude is proportional to $P_L[q_\mu\gamma^\mu + m_\nu]P_{R(L)} \sim q(m_\nu)$, where m_ν and q are the neutrino mass and momentum. The case in the brackets () corresponds to both vertices being SM left-handed ones. It is seen that the LQ induced vertex in combination with the SM one produces an enormously large enhancement factor, $\langle q\rangle/m_\nu \sim p_F/m_\nu \sim 10^8$, compared to the pure SM case. (Here, $\langle q\rangle$ is the averaged momentum transfer between two nucleons in a nucleus approximately equal to the Fermi motion momentum $p_F \approx 100$MeV.) This enhancement should be compensated by the smallness of ϵ or α couplings. To determine actual constraints on these parameters requests a calculation of the nuclear matrix elements within some nuclear structure model. This is beyond the scope of this letter and will be published in a separate paper. A similar effect in $0\nu\beta\beta$-decay was recently found within the minimal supersymmetric standard model with R-parity violation [15].

In conclusion, we have derived the leptoquark interactions with the standard model Higgs field. We have shown that these interactions generate new 4-fermion lepton-quark couplings which contribute to various low-energy processes. Considering the helicity-suppressed $\pi \rightarrow e\nu$ decay, we have found that special combinations of the leptoquark couplings to the quarks, leptons and the Higgs fields (see Eq. (10)) are stringently constrained despite their chirality. This may indicate a dramatic impact of the LQ-Higgs interactions on the standard leptoquark phenomenology which should be reconsidered taking these new interactions into account. We stress that an underlying high-energy scale theory containing light leptoquarks must explain not only the chirality of leptoquark couplings to quarks and leptons (see (a1) at the beginning) but also the absence (or smallness) of at least those leptoquark-Higgs couplings which are suppressed by the constraints in Eq. (10).

Acknowledgments

We thank V.A. Bednyakov for helpful discussions. M.H. would like to thank the Deutsche Forschungsgemeinschaft for financial support by grants kl 253/8-1 and 446 JAP-113/101/0.

References

[1] J.C. Pati and A. Salam, Phys. Rev. D 10 (1974) 275.
[2] W. Buchmüller, R. Rückl and D. Wyler, Phys. Lett. B 191 (1987) 442.
[3] See for instance P. Langacker, Phys. Rep. 72 (1981) 185;
P.H. Frampton, Mod. Phys. Lett. A 7 (1992) 559.
[4] J. Hewett and T. Rizzo, Phys. Rep. 183 (1989) 193.
[5] S. Dimopoulos and J. Ellis, Nucl. Phys. B 182 (1981) 505.
[6] W. Buchmüller, Phys. Lett. B 145 (1984) 151;
B. Schrempp and F. Schrempp, Phys. Lett. B 153 (1985) 101.
[7] J. Wudka, Phys. Lett. B 167 (1986) 337;
J.L. Hewett and T.G. Rizzo, Phys. Rev. D 36 (1987) 3367;
J.E. Cieza Montalvo and O.J.P. Éboli, Phys. Rev. D 47 (1993) 837;
G. Bélanger, D. London and H. Nadeau, Phys. Rev. D 49 (1993) 3140;
J. Blümlein and R. Rückl, Phys. Lett. B 304 (1993) 337;
J. Blümlein and E. Boos, Nucl. Phys. B (Proc. Suppl.) B 37 (1994) 181;
J. Blümlein, E. Boos and A. Pukhov, Mod. Phys. Lett. A 9 (1994) 3007;
J. Ohnemus, S. Rudaz, T.F. Walsh and P. Zerwas, Phys. Lett. B 334 (1994) 203;
G. Bhattacharyya, J. Ellis and K. Sridhar, Phys. Lett. B 336 (1994) 100;
D. Choudhury, Phys. Lett. B 346 (1995) 291;
M.A. Doncheski and S. Godfrey, Phys. Rev. D 51 (1995) 1040.
[8] S. Davidson, D. Bailey and A. Campbell, Z. Phys. C 61 (1994) 613.
[9] M. Leurer, Phys. Rev. Lett. 71 (1993) 1324; Phys. Rev. D 49 (1994) 333; D 50 (1994) 536.
[10] H1 Collaboration, I. Abt et al., Nucl. Phys. B 396 (1993) 3;
ZEUS Collaboration, M. Derrick et al., Phys. Lett. B 306 (1993) 173.

[11] T. Ahmed et al., Z. Phys. C 64 (1994) 545.
[12] DELPHI Collaboration, P. Abreu et al., Phys. Lett. B 316 (1993) 620;
L3 Collaboration, B. Adeva et al., Phys. Lett. B 261 (1991) 169;
OPAL Collaboration, G. Alexander et al., Phys. Lett. B 263 (1991) 123.
[13] D0 Collaboration, S. Abachi et al., Phys. Rev. Lett. 72 (1994) 965.
[14] CDF Collaboration, F. Abe et al., Phys. Rev. D 48 (1993) 3939.
[15] K.S. Babu and R.N. Mohapatra, Phys. Rev. Lett., 75 (1995) 2276;
M. Hirsch, H.V. Klapdor-Kleingrothaus and S.G. Kovalenko, Phys. Lett. B (1996), hep-ph/9512237.

322 *Paper presented at Beyond the Desert 1997: Accelerator and Non-Accelerator Approaches*

Neutrinoless double beta decay and the HERA "anomaly"

M. Hirsch†[1], H.V. Klapdor-Kleingrothaus†[2] and S.G. Kovalenko‡[3]

† Max-Planck-Institut für Kernphysik, P.O. Box: 10 39 80, D-69029 Heidelberg, Germany

‡ Joint Institute for Nuclear Research, Dubna, Russia

Abstract. Recent results from the H1 and ZEUS experiments at HERA have renewed the interest in the phenomenology of both leptoquarks and R-parity violating supersymmetry. This paper discusses constraints on these extensions of the standard model which can be obtained from currently existing limits on neutrinoless double beta decay. In summary, neutrinoless double beta decay rules out first generation scalar quarks as an explanation of the HERA events. Moreover, absence of neutrinoless double beta decay strongly restricts the couplings of leptoquarks to the standard model Higgs bosons.

1. Introduction

As well known by now, both HERA experiments, H1 [1] and ZEUS [2] have reported an excess of events with respect to the standard model (SM) expectation in $e^+p \rightarrow e^+X$ at very large values of $Q^2 \geq 1.5 \times 10^4 GeV^2$. Although at present it is far from clear whether the observed effect is due to new physics or simply a statistical fluctuation these results have triggered a huge amount of theoretical activity [3].

Neutrinoless double beta ($0\nu\beta\beta$) decay is sensitive to any extension of the standard model containing lepton number violation. Therefore, any

[1] E-mail: M.Hirsch@mpi-hd.mpg.de
[2] E-mail: klapdor@enull.mpi-hd.mpg.de
[3] E-mail: kovalen@nusun.jinr.dubna.su

of the possible explanations of the HERA data which is built upon lepton number violating extensions of the standard model should also show up in $0\nu\beta\beta$ decay, and the current limits on this exotic process can be used to derive limits on the parameters of these models and then eventually be compared to the HERA data.

Essentially one can identify two candidates of models in which the occurence of $0\nu\beta\beta$ decay is expected. The first one is supersymmetry, or more concretely the R-parity violating extension of the MSSM. The second one are leptoquarks in models where the leptoquarks couple to the standard model Higgs boson. These two are discussed in some detail in the following sections.

Of course, there are other possible explanations for the HERA data [3], so absence of $0\nu\beta\beta$ decay alone does not imply that HERA can not see physics beyond the SM. However, it is remarkable that the current double beta decay experiments can in some parts of "model-space" compete in sensitivity even with large scale accelerators and lead to interesting constraints for model building.

2. $\not{R}_p$SUSY and $0\nu\beta\beta$ decay

Supersymmetry with R-parity violating interactions has been put forward as a possible explanation of the HERA events [3]. R-parity, defined as $R_P = (-1)^{3B+L+2S}$ [4], is assumed to be conserved exactly in the MSSM [5]. Although R-parity conserving models of SUSY have, among others, the advantage of providing a candidate for the cold dark matter, from a theoretical point of view alone $\not{R}_p$ SUSY models are as well motivated as R-parity conserving ones [6].

The R_p violating part of the superpotential can be written as [7], [8],

$$W_{\not{R}_p} = \lambda_{ijk} L_i L_j \bar{E}_k + \lambda'_{ijk} L_i Q_j \bar{D}_k + \lambda''_{ijk} \bar{U}_i \bar{D}_j \bar{D}_k, \tag{1}$$

where indices i, j, k denote generations. The coupling constants λ (λ'') are antisymmetric in the first (last) two indices. The first two terms lead to lepton number violation, while the last one violates baryon number conservation.

Proton stability forbids the simultaneous presence of lepton and baryon number violating terms in the superpotential [9] (unless the couplings are very small). Therefore, only λ, λ' *or* λ'' type interactions can be present. For a possible explanation of the HERA data as well as for $0\nu\beta\beta$ decay only the terms proportional to λ' are of interest.

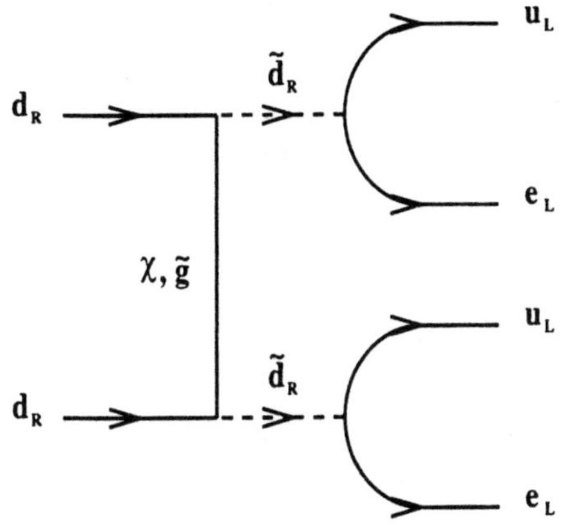

Fig. 1: *Example of $\not{R}_p SUSY$ contribution to $0\nu\beta\beta$ decay.*

Fig. 1 shows an example of a Feynman graph for $0\nu\beta\beta$ decay in $\not{R}_p$SUSY. Altogether there are six graphs of this kind contributing to $0\nu\beta\beta$ decay [10]. With the updated half life limit recently published by the Heidelberg-Moscow collaboration [12] absence of $0\nu\beta\beta$ decay at present implies [11]:

$$\lambda'_{111} \leq 3.3 \times 10^{-4} \Big(\frac{m_{\tilde{q}}}{100GeV}\Big)^2 \Big(\frac{m_{\tilde{g}}}{100GeV}\Big)^{(1/2)}, \tag{2}$$

for $m_{\tilde{d}_R} \approx m_{\tilde{u}_L}$.

Searching for $\not{R}_p$SUSY events in deep inelastic ep-scattering experiments with the ZEUS detector at HERA has been proposed [14, 15]. Two sets of signals could be identified with these events. They are the $\not{R}_p$resonant squark production followed by the $\not{R}_p$cascade decay of the neutralino [14] and the MSSM R_p conserving production of selectron and squark with their subsequent $\not{R}_p$decay to ordinary matter [15].

The recent H1 [1] and ZEUS [2] data could be explained by $\not{R}_p$SUSY if a scalar quark with a mass of approximately 210 GeV exists [3]. Typically one would need $\lambda'_{1j1} \sim 0.04$ for the observed number of events. For $j = 1$ this is clearly much larger than the bound from $0\nu\beta\beta$ decay. Therefore, first generation squarks can not possibly be a solution for the H1 and ZEUS observations. However, as has been emphasized in many papers [3], the cases $j = 2, 3$ are not restricted by double beta decay, and scalar top production in $\not{R}_p$ SUSY is not yet ruled out as an explanation of the experimental data at present.

Nevertheless, besides the purely supersymmetric contribution to $0\nu\beta\beta$ decay there are also supersymmetric diagrams with neutrino exchange [13]. $0\nu\beta\beta$ decay in this case is sensitive to products of two λ' of the form $\lambda'_{1j1} \cdot \lambda'_{11j}$. For superparticle masses of the order of 100 GeV one finds:

$\lambda'_{113}\lambda'_{131} \leq 1.1\times10^{-7}$, $\lambda'_{112}\lambda'_{121} \leq 3.2\times10^{-6}$, $\lambda'^{2}_{111} \leq 6.4\times10^{-5}$. Of course, if only one λ' is finite such products are zero by assumption. On the other hand, if the HERA data were due to scalar quarks of the third generation, thus implying $\lambda'_{131} \sim 0.04$, λ'_{113} must be smaller than $\lambda'_{113} \leq 2.2 \times 10^{-5}$ otherwise $0\nu\beta\beta$ decay should have been found.

3. Leptoquarks and the Higgs

The results of the HERA experiments have renewed the interests in the phenomenology of leptoquarks (LQs) [16, 17]. By definition leptoquarks are particles coupled to a quark and a lepton. Particles with such properties appear naturally, for example, in grand unified theories (GUTs) [18].

However, LQs can be of interest in accelerator physics only if their masses do not exceed a few 100 GeV. In order to be phenomenologically acceptable and still in this mass range, LQs have to satisfy the following 3 criteria [17, 19, 20]: *i)* LQs should not couple to quark pairs, *ii)* LQs should couple to one generation of fermions only and iii) LQ couplings have to be "chiral", i.e. each type of LQs couples either to left-handed or to right-handed quarks only. These requirements can be understood easily. If condition *i)* is violated, proton decay will occur, for LQ masses in the $0.1 - 1$ TeV range at an unacceptable rate. In fact, for LQs with diquark couplings of the order of gauge couplings the LQ masses better should be of the order of the GUT scale! Condition ii) has to be imposed since otherwhise flavour changing neutral currents (FCNCs) are induced. Finally condition *iii)* is needed to guarantee that helicity suppressed decays, such as $\pi \to e\bar{\nu}_e$, remain in agreement with the standard model expectation. It is important to realize that none of these assumptions is required by a deeply-hidden reason in some "leptoquark-theory", but instead imposed by hand to keep the low-energy limits on LQs worse enough that accelerators have a chance of finding them.

However, once we accept these assumptions one can construct a general LQ-lepton-quark Lagrangian [17]. Assuming further that the Lagrangian is renormalizable and invariant under the standard model gauge group for scalar LQs one gets,

$$\begin{aligned}\mathcal{L}_{S-l-q} &= \lambda^{(R)}_{S_0}\cdot\overline{u^c}P_R e\cdot S_0^{R\dagger} + \lambda^{(R)}_{\tilde{S}_0}\cdot\overline{d^c}P_R e\cdot\tilde{S}_0^{\dagger} + \\ &+ \lambda^{(R)}_{S_{1/2}}\cdot\overline{u}P_L l\cdot S_{1/2}^{R\dagger} + \lambda^{(R)}_{\tilde{S}_{1/2}}\cdot\overline{d}P_L l\cdot\tilde{S}_{1/2}^{\dagger} + \\ &+ \lambda^{(L)}_{S_0}\cdot\overline{q^c}P_L i\tau_2 l\cdot S_0^{L\dagger} + \lambda^{(L)}_{S_{1/2}}\cdot\overline{q}P_R i\tau_2 e\cdot S_{1/2}^{L\dagger} + \\ &+ \lambda^{(L)}_{S_1}\cdot\overline{q^c}P_L i\tau_2\hat{S}_1^{\dagger} l + h.c.\end{aligned} \qquad (3)$$

Table 1: The Standard model assignments of the five scalar S and vector V_μ leptoquarks (LQ). ($Y = 2(Q_{em} - T_3)$)

LQ	$SU(3)_c$	$SU(2)_L$	Y	Q_{em}
S_0	**3**	**1**	-2/3	-1/3
$\tilde{S}_0$	**3**	**1**	-8/3	-4/3
$S_{1/2}$	**3***	**2**	-7/3	(-2/3, -5/3)
$\tilde{S}_{1/2}$	**3***	**2**	-1/3	(1/3, -2/3)
S_1	**3**	**3**	-2/3	(2/3, -1/3,-4/3)
V_0	**3***	**1**	-4/3	-2/3
$\tilde{V}_0$	**3***	**1**	-10/3	-5/3
$V_{1/2}$	**3**	**2**	-5/3	(-1/3, -4/3)
$\tilde{V}_{1/2}$	**3**	**2**	1/3	(2/3, -1/3)
V_1	**3***	**3**	-4/3	(1/3, -2/3,-5/3)

and for the vector LQs

$$\begin{aligned}\mathcal{L}_{V-l-q} &= \lambda^{(R)}_{V_0} \cdot \bar{d}\gamma^\mu P_R e \cdot V^{R\dagger}_{0\mu} \lambda^{(R)}_{\tilde{V}_0} \cdot \overline{u}\gamma^\mu P_R e \cdot \tilde{V}^\dagger_{0\mu} + \\ &+ \lambda^{(R)}_{V_{1/2}} \cdot \overline{d^c}\gamma^\mu P_L l \cdot V^{R\dagger}_{1/2\mu} + \lambda^{(R)}_{\tilde{V}_{1/2}} \cdot \overline{u^c}\gamma^\mu P_L l \cdot \tilde{V}^\dagger_{1/2\mu} + \\ &+ \lambda^{(L)}_{V_0} \cdot \overline{q}\gamma^\mu P_L l \cdot V^{L\dagger}_{0\mu} + \lambda^{(L)}_{V_{1/2}} \cdot \overline{q^c}\gamma^\mu P_R e \cdot V^{L\dagger}_{1/2\mu} + \\ &+ \lambda^{(L)}_{V_1} \cdot \overline{q}\gamma^\mu P_L \hat{V}^\dagger_{1\mu} l + h.c. \end{aligned} \quad (4)$$

Here $P_{L,R} = (1 \mp \gamma_5)/2$; q and l are the quark and the lepton doublets; S_i^j and V_i^j are the scalar and vector LQs with the weak isospin i=0, 1/2, 1 coupled to left-handed ($j = L$) or right-handed ($j = R$) quarks respectively. The LQ quantum numbers are listed in Table 1. For LQ triplets $\Phi_1 = S_1, V_1^\mu$ we use the notation $\hat{\Phi}_1 = \vec{\tau} \cdot \vec{\Phi}_1$.

Limits one LQs obeying the restrictions of the above-given Lagrangian have been extensively reviewed in the literature [19, 20].

There is, however, one subtlety not discussed up to now. According to eqs. (3) and (4), LQs interact with quarks and leptons only. Can there be other interactions of LQs? Naturally everybody will agree that LQs should interact with the SM gauge bosons, since LQs carry SM charges. On the other hand, there is the SM Higgs boson. Are there any reasons to forbid the LQs to interact with the Higgs? The answer to this question is no [21]. In fact, one can construct a LQ-Higgs interaction Lagrangian which equally well satisfies the conditions i)-iii), renormalizability and invariance under the SM gauge group - prior to electro-weak symmetry breaking. In its most general form it is given by,

$$\mathcal{L}_{LQ-H} = h^{(i)}_{S_0} H i\tau_2 \tilde{S}_{1/2} \cdot S^i_0 + h^{(i)}_{V_0} H i\tau_2 \tilde{V}^\mu_{1/2} \cdot V^i_{0\mu} + \quad (5)$$

$$
\begin{aligned}
&+ \quad h_{S_1} H i\tau_2 \hat{S}_1 \cdot \tilde{S}_{1/2} + h_{V_1} H i\tau_2 \hat{V}_1^{\mu} \cdot \tilde{V}_{1/2\mu} + \\
&+ \quad Y^{(i)}_{S_{1/2}} \left(H i\tau_2 S^i_{1/2} \right) \cdot \left(\tilde{S}^{\dagger}_{1/2} H \right) + Y^{(i)}_{V_{1/2}} \left(H i\tau_2 V^{\mu(i)}_{1/2} \right) \cdot \left(\tilde{V}^{\dagger}_{1/2\mu} H \right) \\
&+ \quad Y_{S_1} \left(H i\tau_2 \hat{S}^{\dagger}_1 H \right) \cdot \tilde{S}_0 + Y_{V_1} \left(H i\tau_2 \hat{V}^{\dagger}_{1\mu} H \right) \cdot \tilde{V}^{\mu}_0 + \\
&+ \quad \kappa^{(i)}_S \left(H^{\dagger} \hat{S}_1 H \right) \cdot S^{i\dagger}_0 + \kappa^{(i)}_V \left(H^{\dagger} \hat{V}^{\mu}_1 H \right) \cdot V^{i\dagger}_{0\mu} + h.c. - \\
&- \quad \left(\eta_{\Phi} M^2_{\Phi} - g^{(i_1 i_2)}_{\Phi} H^{\dagger} H \right) \Phi^{i_1 \dagger} \Phi^{i_2}.
\end{aligned}
$$

Here $H = \begin{pmatrix} H^+ \\ H^0 \end{pmatrix}$ is the SM $SU(2)_L$-doublet Higgs field.

Φ^i is a cumulative notation for all leptoquark fields with $i = L, R$ (the same for $i_{1,2}$). Exactly as it is for the LQ-lepton-quark interaction, gauge invariance fixes the structure of this Lagrangian, but tells us nothing about the size of the couplings.

To see the effects of eq. (5) consider for the moment only one term $h_{S_1} H i\tau_2 \hat{S}_1 \cdot \tilde{S}_{1/2}$ and write it in components:

$$
\begin{aligned}
H i\tau_2 \hat{S}_1 \cdot \tilde{S}_{1/2} \quad &= \quad H^+ \Big(\sqrt{2} \tilde{S}_{1/2}(\tfrac{1}{3}) S_1(\tfrac{-4}{3}) - \tilde{S}_{1/2}(\tfrac{-2}{3}) S_1(\tfrac{-1}{3}) \Big) \\
&- \quad H^0 \Big(\tilde{S}_{1/2}(\tfrac{1}{3}) S_1(\tfrac{-1}{3}) - \sqrt{2} \tilde{S}_{1/2}(\tfrac{-2}{3}) S_1(\tfrac{2}{3}) \Big) \qquad (6)
\end{aligned}
$$

Now, after electro-weak symmetry breaking, the neutral component of the Higgs develops a vacuum expectation value and the second line of eq. (6) looks exactly like a mass term for the LQs. Taking into account this term only, one would find a mass matrix, which for the $Q = 2/3$ (scalar) states for example is given by

$$
\mathcal{M}^2_S(2/3) \quad = \quad \begin{pmatrix} \bar{M}^2_{\tilde{S}_{1/2}} & \sqrt{2} h_{S_1} v \\ \sqrt{2} h_{S_1} v^2 & \bar{M}^2_{S_1} \end{pmatrix}. \qquad (7)
$$

As seen from eq. (7), the interaction of the Higgs with the LQs leads to a mixing between LQs from different multiplets and of LQs with couplings to different chiralities of quarks. In other words, if LQs are allowed to couple to the SM Higgs, the LQ-lepton-quark Lagrangian automatically contains non-chiral terms!

In general there are 8 different mass matrices for LQs. Their exact definitions can be found in [21]. What is important here is that pion decay [21] and neutrinoless double beta decay [22] lead to restrictions on the allowed size of the LQ-Higgs couplings. [4] Although the details depend on

[4] $0\nu\beta\beta$ decay constraints numerically are usually (1-2) orders of magnitude more stringent that those from π-decays. However, pion decays and $0\nu\beta\beta$ decays are sensitive to partly different combinations of LQ-Higgs couplings.

which LQ-Higgs coupling is assumed to be non-zero etc., generally one can say that for masses of LQs in the range of $\mathcal{O}(200)$ GeV LQ-Higgs couplings have to be smaller than typically 10^{-5} to avoid the constraints from $0\nu\beta\beta$ decay, see fig. 2. Putting it the other way, LQs coupling with strengths of order one to the SM Higgs should be heavier than at least several TeV, otherwhise $0\nu\beta\beta$ decay should have been observed.

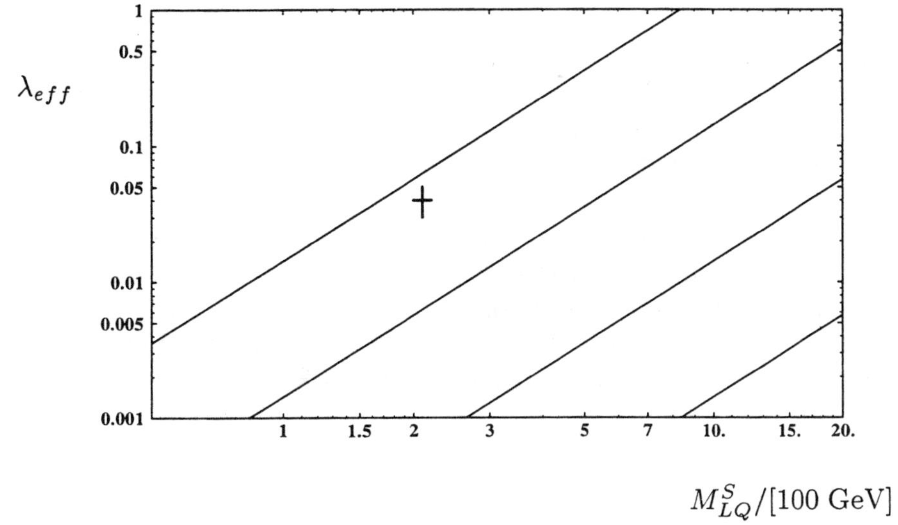

Fig. 2: *Constraints from $0\nu\beta\beta$ decay on the leptoquark parameter space. For the plot it has been assumed that only one LQ-Higgs coupling is different from zero (h_{S_1}). Furthermore, the masses of the two scalar leptoquarks which mix due to h_{S_1} are assumed to be about equal. λ_{eff} is defined as $\lambda_{eff} := \sqrt{\lambda^{(R)}_{\tilde{S}_{1/2}} \lambda^{(L)}_{S_1}}$, and h_{S_1} has been scaled relative to the vacuum expectation value of the SM Higgs field. Parameter ranges to the left of the lines are forbidden from absence of $0\nu\beta\beta$ decay. The 4 lines correspond to $(h_{S_1}/v) = 10^{-6}, 10^{-4}, 10^{-2}, 1$, respectively. The cross denotes the paramter combinations which could explain the HERA data by leptoquark production. Consistency between absence of $0\nu\beta\beta$ decay and the HERA data requires that the LQ-Higgs coupling is smaller than (few) $\times 10^{-6}$.*

There are, however, exceptions from this general feature. First, it could be that only one LQ state is light, while the others exist above the TeV scale. Second, if one allows only those LQ Higgs couplings which do not lead to non-chiral interactions, for example, $Y^{(R)}_{S_{1/2}} \left(H i\tau_2 S^R_{1/2}\right) \cdot \left(\tilde{S}^\dagger_{1/2} H\right)$, but forbids the presence of all other terms, neither $0\nu\beta\beta$ nor pion decay lead to any sensible constraint.

Summarizing this section, it can be stated that non-chiral interactions are a natural feature of LQs. Even if chirality is imposed by hand in the LQ-lepton-quark interaction it will reappear once LQs are allowed to interact with the Higgs. Such non-chiral LQ interactions are severly constrained by π and $0\nu\beta\beta$ decays and unless the LQ-Higgs coupling is very small, LQs can not exist on a mass scale of $\mathcal{O}(200)$ GeV.

4. Summary

Neutrinoless double beta decay is a sensitive probe for physics beyond the standard model. Unfortunately, $0\nu\beta\beta$ decay has up to now not been observed. Nevertheless, absence of $0\nu\beta\beta$ decay can be used to rule out some possible explanations for the recent HERA results [1, 2],which have been discussed in the literature.

In this paper, we have focussed on $R\!\!\!/_p$ SUSY and leptoquarks. From $0\nu\beta\beta$ decay one can learn that the HERA events can not be explained by scalar quarks of the first generation. Furthermore, if the HERA events are to be explained by leptoquarks, $0\nu\beta\beta$ decay tells us, that the leptoquark-Higgs coupling is (typically) smaller than (few) $\times 10^{-6}$.

Acknowledgments

This work was supported by the Deutsche Forschungsgemeinschaft by grants kl 253/8-2 and 446 JAP-113/101/0.

References

[1] Adloff C. et al., H1 Collaboration, 1997 Z. Phys. **C74** 191

[2] Breitweg J. et al., ZEUS Collaboration, 1997 Z. Phys. **C74** 207

[3] More comprehensive lists of theoretical papers on this subject can be found in, for example: Rückl R., these proceedings Dreiner H., these proceedings Rizzo T., these proceedings

[4] Fayet P., 1975 Nucl.Phys. **B90** 104; 1977 Phys.Lett. **B69** 489; Farrar G.R. and Fayet P., 1978 Phys.Let. **B76** 575.

[5] For reviews on the MSSM see, for example: Haber H.E. and Kane G.L., 1985 Phys.Rep. **117** 75; Nilles H.P., 1984 Phys.Report. **110** 1.

[6] For a recent introduction to $R\!\!\!/_p$ SUSY see, for example: Dreiner H., 1997 hep-ph/9707435

[7] Dimopoulos S. and Hall L.J., 1987 Phys.Lett. **B207** 210.

[8] Hall L. and Suzuki M., 1984 Nucl.Phys. **B231** 419.

[9] Zwirner F., 1983 Phys. Lett. **B132** 103.

330

[10] Hirsch M., Klapdor-Kleingrothaus H.V. and Kovalenko S.G., 1995 Phys. Lett. **B352** 1

[11] Hirsch M., Klapdor-Kleingrothaus H.V. and Kovalenko S.G., 1995 Phys. Rev. Lett. **75** 17 and 1996 Phys. Rev. **D53** 1329

[12] Heidelberg-Moscow collaboration, Günther M. et al., 1997 Phys.Rev. **D55** 54; and Hellmig J. et al., in: *Proc. Int. Workshop on Dark Matter in Astro and Particle Physics*, Heidelberg, Sept. 16-20, 1997, World Scientific, in press

[13] Hirsch M., Klapdor-Kleingrothaus H.V. and Kovalenko S.G., 1996 Phys. Lett. **B372** 181; Erratum, Phys. Lett. **B381** 488.

[14] Butterworth J. and Dreiner H., 1993 Nucl. Phys. **B397** 3;

[15] Dreiner H. and Morawitz P., 1994 Nucl. Phys. **B428** 31;

[16] Pati J.C. and Salam A., 1974 Phys.Rev. **D10** 275.

[17] Buchmüller W., Rückl R. and Wyler D., 1987 Phys.Lett. **B191** 442.

[18] see for instance Langacker P., 1981 Phys.Rep.**72** 185; Frampton P.H., 1992 Mod.Phys. Lett. **A7** 559.

[19] Davidson S., Bailey D. and Campbell A., 1994 Z.Phys. **C61** 613.

[20] Leurer M., 1993 Phys.Rev.Lett. **71** 1324; 1994 Phys.Rev. **D49** 333; 1994 Phys.Rev. **D50** 536;

[21] Hirsch M., Klapdor-Kleingrothaus H.V. and Kovalenko S.G., 1996 Phys. Lett. **B378** 17

[22] Hirsch M., Klapdor-Kleingrothaus H.V. and Kovalenko S.G., 1996 Phys. Rev. **D54** R4207

PHYSICAL REVIEW D VOLUME 56, NUMBER 9 1 NOVEMBER 1997

Much ado about leptoquarks: A comprehensive analysis

JoAnne L. Hewett and Thomas G. Rizzo
Stanford Linear Accelerator Center, Stanford, California 94309
(Received 9 April 1997)

We examine the phenomenological implications of a ~200 GeV leptoquark in light of the recent excess of events at DESY HERA. Given the relative predictions of events rates in e^+p versus e^-p, we demonstrate that classes of leptoquarks may be excluded, including those contained in E_6 grand unified theory (GUT) models. It is shown that future studies with polarized beams at HERA could reveal the chirality of the leptoquark fermionic coupling and that given sufficient luminosity in each $e^{\pm}_{L,R}$ channel the leptoquark quantum numbers could be determined. The implications of 200–220 GeV leptoquarks at the Fermilab Tevatron are examined. While present Tevatron data most likely exclude vector leptoquarks and leptogluons in this mass region, it does allow for scalar leptoquarks. We find that while leptoquarks have little influence on Drell-Yan production, further studies at the Main Injector may be possible in the single production channel provided the Yukawa couplings are sufficiently large. We investigate precision electroweak measurements as well as the process $e^+e^-\to q\bar{q}$ at CERN LEP II and find they provide no further restrictions on these leptoquark models. We then ascertain that cross section and polarization asymmetry measurements at the Next Linear Collider (NLC) provide the only direct mechanism to determine the leptoquark's electroweak quantum numbers. The single production of leptoquarks in γe collisions by both the back-scattered laser and Weisacker-Williams techniques at the NLC is also discussed. Finally, we demonstrate that we can obtain successful coupling constant unification in models with leptoquarks, both with or without supersymmetry. The supersymmetric case requires the GUT group to be larger than SU(5) such as flipped SU(5)$\times$U(1)$_X$. [S0556-2821(97)04019-8]

PACS number(s): 12.60.$-$i, 12.10.Kt, 14.80.$-$j

I. INTRODUCTION

The apparent symmetry between the quark and lepton generations is a mysterious occurrence within the standard model (SM) and has inspired many theories which go beyond the SM to relate them at a more fundamental level. As a result many of these models naturally contain leptoquarks, or particles that couple to a lepton-quark pair. Theories which fall in this category include, composite models with quark and lepton substructure [1], the strong coupling version of the SM [2], horizontal symmetry theories [3], extended technicolor [4], and grand unified theories (GUT's) based on the gauge groups SU(5) [5], SO(10) with Pati-Salam SU(4) color symmetry [6], SU(15) [7], and superstring-inspired E_6 models [8,9]. In all cases, the leptoquarks carry both baryon and lepton number and are color triplets under $SU(3)_C$. In models where baryon and lepton number are separately conserved, which includes most of the above cases, leptoquarks can be light (of order the electroweak scale) and still avoid conflicts with rapid proton decay. Their remaining properties, such as spin, weak isospin, electric charge, chirality of their fermionic couplings, and fermion number, depend on the structure of each specific model. If leptoquarks were to exist we would clearly need to determine these properties in order to ascertain their origin.

An excess of events at large values of Q^2 have recently been reported [10] by both the H1 and ZEUS Collaborations at the DESY ep collider HERA in their neutral current deep inelastic scattering (DIS) data. ZEUS has collected 20.1 pb^{-1} of integrated luminosity in e^+p collisions and observes five events with $Q^2>15\,000\ \mathrm{GeV}^2$ with $x>0.45$ and $y>0.25$, where x and y are the usual DIS scattering variables, while expecting two events from the SM in this region. H1 reports seven events in the kinematic region $m=\sqrt{xs}>180$ GeV and $y>0.4$, compared to a SM prediction of 1.83 ± 0.33 with 14.19 ± 0.32 pb^{-1} of integrated luminosity. Clearly, the statistical sample is too small at present to draw any conclusions and it is likely that this excess is merely the result of a statistical fluctuation. Another possibility is that this discrepancy is the result of deviations from current parton distribution parametrizations at large x. This case, however, has been examined by the H1 and ZEUS Collaborations [10] and is found to be unlikely. Also, such large modifications in the parton densities would most likely result in disagreement with the dijet data samples at the Tevatron [11,12]. It is also possible that this HERA data might signal the first hint of physics beyond the SM. Such an excess in event rate at large Q^2 is a classic signature for compositeness if the events show no specific kinematic structure. This scenario has recently been analyzed [13] in light of the HERA data, with the result that an $eeqq$ contact interaction with a right-left helicity structure (in order to avoid the constraints arising from atomic parity violation data discussed below) and a scale of ~3 TeV is consistent with the data. However, if instead, the events cluster in x while being isotropic in y, they would signal the production of a new particle. While the reconstructed values of the mass ($m=\sqrt{xs}$) of such a hypothetical particle show some spread between the two experiments, they are consistent within the evaluated errors, yielding a central value in the approximate range 200–220 GeV.

If this excess of events turns out to be the resonant production of a new particle, we will need to examine the possible classes of new lepton-hadron interactions which could give rise to such a signature. At present, there are three leading scenarios of this type: (i) models with leptogluons, (ii)

2.4.6 Superheavy Neutrinos and Double Beta Decay

RAPID COMMUNICATIONS

PHYSICAL REVIEW D VOLUME 34, NUMBER 3 1 AUGUST 1986

Limits on the mass of the right-handed Majorana neutrino

Rabindra N. Mohapatra
Department of Physics and Astronomy, University of Maryland, College Park, Maryland 20742
(Received 28 May 1986)

We derive limits on the mass of the right-handed Majorana neutrino (N_R) in terms of the mass of the right-handed W_R boson in left-right-symmetric theories of weak interactions from the recent results on lifetime for neutrinoless double-β decay of ^{76}Ge as well as from theoretical considerations of vacuum stability.

One of the new particles that is required for the consistency of the left-right-symmetric models[1] of weak interactions is the right-handed neutrino. Understanding of the small mass of the left-handed neutrino requires that the right-handed neutrino have large Majorana mass.[2,3] The value of the mass depends on two parameters: (a) the Yukawa coupling of the right-handed Higgs multiplets[3] Δ_R and (b) the mass of the right-handed W_R boson. Phenomenologically this value can be anywhere from a few Gev (Ref. 4) to as large as, or perhaps larger than, the mass of the right-handed W_R boson. An extensive study of the phenomenological constraints on the mass of the right-handed neutrino has been given by Gronau, Leung, and Rosner.[4] Our goal in this paper is more modest and should be taken as a supplement to the work of Ref. 4. We consider limits on the mass of the right-handed neutrino (N_R) from two considerations: (a) theoretical considerations of vacuum stability and (b) most recent data[5] on the neutrinoless double-β decay of ^{76}Ge.[6] In both cases, we correlate the mass of the right-handed neutrino with that of the W_R boson.

BOUND ON M_{N_R} FROM VACUUM STABILITY

It is well known from the works of Coleman and E. Weinberg[7] that one-loop corrections to the tree-level potential in a gauge theory can affect the picture of spontaneous symmetry breaking and lead to vacuum instability unless the parameters of the theory are restricted to a limited range. These considerations have been utilized to obtain lower bounds on the Higgs-boson mass[8] as well as to obtain upper bounds on the fermion masses[9] in the standard model. In this note, we use the same techniques to limit the mass of the right-handed neutrino.

To carry out our derivation, we remind the reader that the right-handed neutrino mass owes its origin to the Higgs mesons[3] $\Delta_L(3,1,2)+\Delta_R(1,3,2)$ with transformation properties under $SU(2)_L\times SU(2)_R\times U(1)_{B-L}$ group indicated within parentheses. Denoting the leptonic doublets by $\psi_L\equiv(\nu_L,e_L^-)$ and $\psi_R\equiv(N_R,e_R^-)$, the relevant Yukawa coupling can be denoted by

$$\mathcal{L}_Y=h[\psi_L^T C^{-1}\tau_2\Delta_L\psi_L+(L\leftrightarrow R)]+\text{H.c.}\ , \quad (1)$$

where

$$\Delta_L=\begin{pmatrix}\Delta^+ & \Delta^{++}\\ \Delta^0 & \Delta^+\end{pmatrix}_L ,$$

and similarly for $L\leftrightarrow R$. The $B-L$ as well as the $SU(2)_R$ symmetry are broken by $\langle\Delta_R^0\rangle=V_R\neq 0$. This gives mass to the W_R and Z_R boson, i.e., $M_{W_R}=gV_R$ as well as to the right-handed neutrino N_R: $M_{N_R}=hV_R$. The Higgs potential involving the Δ_L and Δ_R receives one-loop contributions from the gauge and Yukawa couplings as

$$V^{(1\text{ loop})}=\left\{\frac{3}{64\pi^2}[\tfrac{1}{2}g^4+(g^2+g'^2)^2]-\sum_{i=e,\mu,\tau}\frac{h_i^4}{64\pi^2}\right\}\times(\Delta_R^{0+}\Delta_R^0)^2\ln\Delta_R^{0+}\Delta_R^0+(R\to L)\ , \quad (2)$$

where we have suppressed the Δ^{++}, Δ_R^+ terms since they do not play any role in the symmetry breaking and g' represents the $U(1)_{B-L}$ coupling. We have also ignored the contribution of Higgs-boson self-couplings λ to $V^{1\text{ loop}}$. It is clear that vacuum stability requires that

$$\sum_{i=e,\mu,\tau}h_i^4\leq 3[\tfrac{1}{2}g^4+(g^2+g'^2)^2]\ . \quad (3)$$

Using the fact that $g=e\cos\theta_W$ and $g'=e\sqrt{\sec 2\theta_W}$, we

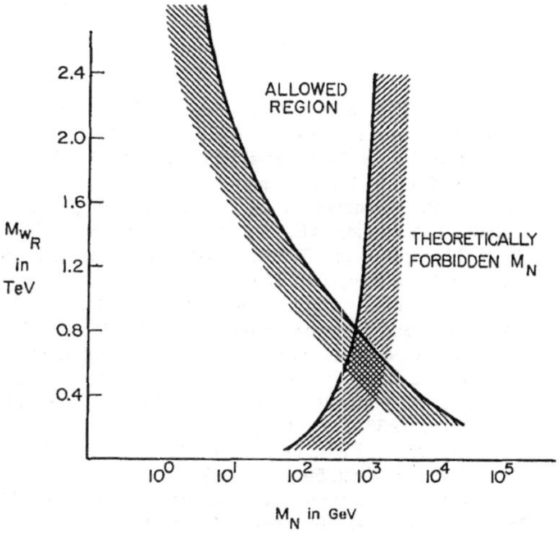

FIG. 1. Allowed values for the M_{W_R} and M_{N_R} from recent results on neutrinoless double-β decay and vacuum stability.

find that

$$\left[\sum_i M_{N_{R,i}}{}^4\right]^{1/4} \leq 0.95 M_{W_R} , \qquad (4)$$

where we used $\sin^2\theta_W = 0.22$. This is our main result. The main uncertainty in this inequality comes from lack of knowledge of the Higgs-boson self-coupling λ; if we assume it to be one, the bound becomes

$$\left[\sum_i M_{N_{R,i}}{}^4\right]^{1/4} \leq 1.18 M_{W_R} . \qquad (5)$$

Furthermore, if right-handed neutrinos of all three generations have the same mass, then we can bound that mass $M_N \lesssim 0.9 M_{W_R}$ using Eq. (5). In Fig. 1, we plot this upper limit on a semilogarithmic scale.

CONSTRAINTS FROM NEUTRINOLESS DOUBLE-β DECAY

It was pointed out in Refs. 3 and 10 that neutrinoless double-β decay[11] gets additional contributions from heavy right-handed neutrino intermediate states in left-right-symmetric models. There are basically two kinds of contributions:[12] one that involves two W_L exchanges and the other that involves the exchange of two W_R bosons. In the first case, the amplitude is proportional to

$$M^{LL} \approx G_F{}^2 \xi^2 M_N \left\langle \frac{e^{-M_N r}}{r} \right\rangle_{\text{nuc}}, \text{ where } \xi \simeq \left[\frac{M_e}{M_N}\right], \qquad (6)$$

whereas in the second case it is given by

$$M^{RR} \approx G_F{}^2 (M_{W_L}/M_{W_R})^4 M_N \left\langle \frac{e^{-M_N r}}{r} \right\rangle_{\text{nuc}} . \qquad (7)$$

For M_N in the GeV range, $\xi^2 \simeq 2.5\times10^{-8}/(M_N \text{ in GeV})^2$ and M^{LL} is then negligible compared to M^{RR} for $M_{W_R} \lesssim 20$ TeV. The estimate of the nuclear matrix element has been reviewed in Ref. 11. Using the latest experimental results we can limit the right-handed neutrino mass in terms of the W_R mass. This is done in Fig. 1. The region to the left of the left line (the shaded area) is forbidden by the latest neutrinoless double-β decay results.

There are also limits on the $0^+ \rightarrow 2^+$ transitions that arise in left-right-symmetric models. Their typical strength is of order

$$M_{LR} \approx G_F{}^2 \xi \left[\frac{M_{W_L}}{M_{W_R}}\right]^2 . \qquad (8)$$

Ejiri *et al.* have the best bound on this lifetime[6] of $0^+ \rightarrow 2^+$ transition in ^{76}Ge to be $> 4.6\times10^{22}$ years which implies $\xi(M_{W_L}/M_{W_R})^2 \lesssim 1.6\times10^{-5}$. For $M_N \gtrsim 2$ GeV and $M_{W_R} \gtrsim 0.8$ TeV, $\xi(M_{W_L}/M_{W_R})^2 \lesssim 0.2\times10^{-5}$. Therefore, no new constraints on M_{N_R} emerge from this decay mode.

One important implication of this analysis is that for lower M_{W_R} the right-handed neutrino must be heavier; for instance, $M_{W_R} \gtrsim 2$ TeV implies $M_{N_R} \gtrsim 17$ GeV or so, and $M_{W_R} \approx M_{N_R}$ at $M_{W_R} \simeq 0.8$ TeV.

In summary, we have obtained limits on the mass of the right-handed neutrino in terms of the mass of the right-handed W_R boson, which will be of phenomenological interest in the analysis of left-right-symmetric models. Our result is based on the theoretical input that uses vacuum stability arguments at the one-loop level and the latest experimental results on neutrinoless double-β decay. It may be noted that there exist other constraints on the M_{N_R} masses in the MeV range from cosmological considerations[13] as well as weak decay processes.[4] The new constraints of this paper are complementary to these.

This work was supported by a grant from the National Science Foundation.

[1] J. C. Pati and A. Salam, Phys. Rev. D **10**, 275 (1974); R. N. Mohapatra and J. C. Pati, *ibid.* **11**, 566, 2558 (1975); G. Senjanovic and R. N. Mohapatra, *ibid.* **12**, 1502 (1975).

[2] M. Gell-Mann, P. Ramond, and R. Slansky, *Supergravity*, edited by D. Freedman *et al.* (North-Holland, Amsterdam, 1980); T. Yanagida, KEK report, 1979 (unpublished).

[3] R. N. Mohapatra and G. Senjanovic, Phys. Rev. Lett. **44**, 912 (1980); Phys. Rev. D **23**, 165 (1981).

[4] J. Rosner, Nucl. Phys. **B248**, 503 (1984); M. Gronau, C. Leung, and J. Rosner, Phys. Rev. D **29**, 2539 (1984); F. J. Gilman, Comments Nucl. Part. Phys. (to be published); M. Gronau and R. Yahalom, Nucl. Phys. **B236**, 233 (1984).

[5] T. Ejiri *et al.*, Nucl. Phys. **A448**, 271 (1986); D. Caldwell *et al.*, Phys. Rev. D **33**, 2737 (1986).

[6] E. Fiorini *et al.*, Phys. Lett. **121B**, 72 (1983); F. Avignone *et al.*, Phys. Rev. Lett. **50**, 721 (1983).

[7] S. Coleman and E. Weinberg, Phys. Rev. D **7**, 1888 (1973).

[8] S. Weinberg, Phys. Rev. Lett. **36**, 294 (1976); A. Linde, Zh. Eksp. Teor. Fiz. **23**, 64 (1976) [JETP Lett. **23**, 73 (1976)].

[9] P. Q. Hung, Phys. Rev. Lett. **42**, 873 (1979); S. Wolfram and H. D. Politzer, California Institute of Technology report, 1979 (unpublished).

[10] Riazuddin, R. E. Marshak, and R. N. Mohapatra, Phys. Rev. D **24**, 1310 (1981).

[11] For recent reviews, see W. Haxton and G. Stephenson, Jr., Prog. Part. Nucl. Phys. **12**, 409 (1984); M. Doi, T. Kotani, and E. Takasugi, Prog. Theor. Phys. (Suppl.) **83**, 1 (1985); J. D. Vergado, Phys. Rep. **133**, 1 (1986).

[12] This point has also been reemphasized recently by B. Kayser, in *Proceedings of the Oregon Meeting*, the Annual Meeting of the Division of Particles and Fields of the American Physical Society, Eugene, 1985, edited by R. C. Hwa (World Scientific, Singapore, 1986).

[13] S. Sarkar, Rutherford Laboratory Report No. RAL-85-115 (unpublished).

ELSEVIER

2 May 1996

Physics Letters B 374 (1996) 7–12

PHYSICS LETTERS B

Double beta decay in left-right symmetric models

M. Hirsch [a,1], H.V. Klapdor-Kleingrothaus [a,2], O. Panella [b,3],

[a] *Max-Planck-Institut für Kernphysik, P.O. Box 10 39 80, D-69029 Heidelberg, Germany*

[b] *Istituto Nazionale di Fisica Nucleare, Sezione di Perugia, Via A. Pascoli, I-06123 Perugia, Italy and Laboratoire de Physique Corpusculaire, Collège de France, Place Marcelin Berhtelot, F-75231, Paris Cedex 05, France*

Received 30 January 1996
Editor: C. Mahaux

Abstract

Left-right symmetric models provide a natural framework for neutrinoless double beta ($0\nu\beta\beta$) decay. In the analysis of $0\nu\beta\beta$ decay in left-right symmetric models, however, it is usually assumed that all neutrinos are light. On the other hand, heavy *right-handed* neutrinos appear quite naturally in left-right symmetric models and should therefore not be neglected. Assuming the existence of at least one right-handed heavy neutrino, absence of $0\nu\beta\beta$ decay of ^{76}Ge currently provides the following limits on the mass and mixing angle of right-handed W-bosons: $m_{W_R} \geq 1.1$ TeV and $\tan(\zeta) \leq 4.7 \times 10^{-3}$ for a particular value of the effective right-handed neutrino mass, $\langle m_N^{(V)} \rangle = 1$ TeV, and in the limit of infinitely massive doubly charged Higgs (Δ^{--}). The effects of the inclusion of the Higgs triplet on $0\nu\beta\beta$ decay are also discussed.

PACS: 11.30; 12.30; 13.10; 13.15; 14.60; 14.80; 23.40
Keywords: Left-right symmetric models; Double beta decay; Neutrino mass; Right-handed W-bosons; Higgs triplet models

Left-right symmetric models (LR) [1] aim at explaining two of the most puzzling questions of the standard model (SM), both of which are intimately related to neutrinoless double beta decay ($0\nu\beta\beta$) [2,3]: i) The weak interaction violates parity, and ii) in the Standard Model neutrino masses are zero. Especially if current hints on finite neutrino masses are correct, LR models provide a very attractive explanation for their small values – when compared to those of the charged leptons – via the well-known see-saw mechanism [4].

Of course, $0\nu\beta\beta$ decay has been studied in connection with LR models by many theoretical groups before, see [5] for reviews. However, the analysis of $0\nu\beta\beta$ decay is usually either restricted to the case where all neutrinos are light [3,6,7] or simplified by considering only left-handed neutrinos [5]. Although both approximations look reasonable from a Standard Model point of view, the situation is very different in LR models in general. Actually, taking the see-saw mechanism as a motivation for LR models one has to expect the existence of some heavy, right-handed neutrino.

The importance of heavy right-handed neutrinos for $0\nu\beta\beta$ decay has been pointed out by Mohapatra [8], while Doi and Kotani [9] derived a quite general decay rate, keeping terms for both left- and right-handed neutrinos. Both of these papers, however, are not complete. While Mohapatra [8] considered only the contribution proportional to $(m_{W_L}/m_{W_R})^2$, Doi and Kotani

[1] E-mail: mahirsch@enull.mpi-hd.mpg.de.
[2] E-mail: klapdor@enull.mpi-hd.mpg.de.
[3] E-mail: panella@hppg04.pg.infn.it.

PII 0370-2693(96)00185-2

[9] did not calculate the relevant nuclear matrix elements. In view of the experimental progress on double beta decay [10,11] we therefore felt motivated to reconsider $0\nu\beta\beta$ decay in LR models and derive constraints on the various parameters of the decay rate in a more general way. For this purpose we have calculated matrix elements in the limit of heavy neutrino exchange in a pn-QRPA model [7,12].

Furthermore, we discuss modifications of the formalism once the contribution of the Higgs triplet is taken into account. Assuming the validity of the SM gauge group and simply adding an Higgs triplet to the particle content opens up new decay channels for $0\nu\beta\beta$ decay [13], which however were shown to be negligible by Schechter and Valle [14], Wolfenstein [15] and Haxton et al. [16]. Again, the situation is different in LR models. While an Higgs triplet is fairly exotic an extension of the SM, in LR models it could provide an attractive explanation of the Majorana nature of the neutrino [17]. Moreover, Rizzo [17] has argued that the contribution of the doubly-charged Higgs to the *inverse* $0\nu\beta\beta$ *decay* is a necessary ingredient to preserve the unitarity of the cross section. Thus, although of quite modest numerical importance for limits on W_R in usual $0\nu\beta\beta$ decay, as we will show at the end of this work, we felt the necessity to include the Higgs triplet in our analysis.

As a starting point for the calculation the following effective Hamiltonian (in the notation of [3]) is used:

$$\mathcal{H}_W^{CC} = \frac{G}{\sqrt{2}}\left\{ J^{\dagger}_{\mu L} j^{-}_{\mu L} + \kappa J^{\dagger}_{\mu R} j^{-}_{\mu L} + \eta J^{\dagger}_{\mu L} j^{-}_{\mu R} + \lambda J^{\dagger}_{\mu R} j^{-}_{\mu R} \right\}. \quad (1)$$

Here, $J^{\dagger}_{L/R}$ and $j^{-}_{L/R}$ are left- and right-handed hadronic and leptonic currents, respectively. κ, η and λ are the right-handed parameters, defined such that the SM charged weak current Hamiltonian is obtained in the limit when κ, η and λ approach zero [3].

Since $\lambda, \eta \ll 1$ one could think of deriving the decay rate considering only contributions of λ and η in lowest order. However, such a procedure leads to the neglection of important terms. In general, keeping also higher order terms, the decay rate can be written as a fourth-order polynomial in λ and η [4] as derived in [9]. However, to separate the particle from the nuclear physics part of the calculation, it is convenient to assume that there are no neutrinos with mass eigenstates in the range of $\mathcal{O}(10\text{–}1000)$ [MeV]. Using this well-motivated assumption, after some lengthy but straightforward calculation, we write the inverse half-life for $0\nu\beta\beta$ decay in a *factorized* form as [18],

$$\begin{aligned}
\left[T_{1/2}^{0\nu\beta\beta}(0^+ \to 0^+)\right]^{-1} &= \left(\frac{\langle m_\nu\rangle}{m_e}\right)^2 C^{LL}_{mm} \\
&+ \left(\frac{m_p}{\langle m_N^{(U)}\rangle}\right)^2 C^{NN}_{mm} + \langle\lambda\rangle^2 C^{LL}_{\lambda\lambda} + \langle\eta\rangle^2 C^{LL}_{\eta\eta} \\
&+ \langle\xi\rangle^2 C^{NN}_{mm} + \left(\frac{\langle m_\nu\rangle}{m_e}\right)\left(\frac{m_p}{\langle m_N^{(U)}\rangle}\right) C^{NL}_{mm} \\
&+ \left(\frac{\langle m_\nu\rangle}{m_e}\right)\langle\lambda\rangle C^{LL}_{m\lambda} + \left(\frac{\langle m_\nu\rangle}{m_e}\right)\langle\eta\rangle C^{LL}_{m\eta} \\
&+ \left(\frac{\langle m_\nu\rangle}{m_e}\right)\langle\xi\rangle C^{NL}_{m\xi} + \left(\frac{m_p}{\langle m_N^{(U)}\rangle}\right)\langle\lambda\rangle C^{NL}_{m\lambda} \\
&+ \left(\frac{m_p}{\langle m_N^{(U)}\rangle}\right)\langle\eta\rangle C^{NL}_{m\eta} + \left(\frac{m_p}{\langle m_N^{(U)}\rangle}\right)\langle\xi\rangle C^{NN}_{mm} \\
&+ \langle\lambda\rangle\langle\eta\rangle C^{LL}_{\lambda\eta} + \langle\lambda\rangle\langle\xi\rangle C^{NL}_{m\lambda} + \langle\eta\rangle\langle\xi\rangle C^{NL}_{m\eta}, \quad (2)
\end{aligned}$$

where $C^{\alpha\beta}_{xy}$ are products of nuclear matrix elements and phase space integrals. In the limit when all neutrinos are light, Eq. (2) reduces to the expression previously used [3,7]. Correspondingly, all coefficients with "LL" superscripts coincide with those of the light neutrino case, see [3,7]. Complete definitions for the coefficients for the heavy neutrino case are given in [18]. The particle physics parameters of the decay rate are defined as

$$\langle m_\nu\rangle = {\sum_j}' U^2_{ej} m_j, \quad (3)$$

$$\langle m_N^{(U)}\rangle^{-1} = {\sum_j}'' U^2_{ej} m_j^{-1}, \quad (4)$$

$$\langle\lambda\rangle = {\sum_j}' U_{ej} V_{ej} \lambda, \quad (5)$$

$$\langle\eta\rangle = {\sum_j}' U_{ej} V_{ej} \eta, \quad (6)$$

$$\langle\xi\rangle = \left[\lambda^2 + \eta^2 - 2\lambda\eta\left(\frac{M^N_{GT} + M^N_F}{M^N_{GT} - M^N_F}\right)\right] {\sum_j}'' V^2_{ej}\left(\frac{m_p}{m_j}\right). \quad (7)$$

[4] Terms proportional to κ can be safely neglected [3].

Table 1
Nuclear matrix elements for heavy neutrino exchange in $0\nu\beta\beta$ decay for the experimentally most interesting isotopes calculated within pn-QRPA

	^{A}Y							
	^{76}Ge	^{82}Se	^{100}Mo	^{116}Cd	^{128}Te	^{130}Te	^{136}Xe	^{150}Nd
M_{GT}^{N}	236	213	269	156	248	219	121	344
M_{F}^{N}	−53	−47	−64	−36	−55	−48	−27	−78

U_{ej} and V_{ej} are the elements of the neutrino mixing matrix for the left- and right-handed sectors, which satisfy the completeness ($\sum_j |U_{ej}|^2 = \sum_j |V_{ej}|^2 = 1$) and the orthogonality relation ($\sum_j U_{ej}V_{ej} = 0$) [3]. As usual the primed sum indicates [3] that the sums extend over light mass eigenstates only, whereas the double primed sums extend over the heavy mass eigenstates. $\langle\xi\rangle$ describes right-handed neutrino exchange and the first term in $\langle\xi\rangle$ corresponds to the one considered by Mohapatra [8]. Neglecting all other terms and assuming no mixing between the W-bosons, our decay rate reproduces the one considered by Mohapatra [8]. Note that in deriving Eq. (2) we have neglected light right-handed neutrinos, as well as terms proportional to $\sum_j'' U_{ej}V_{ej}\lambda m_j^{-2}$ and $\sum_j'' U_{ej}V_{ej}\eta m_j^{-2}$, since the latter are suppressed by additional powers of large neutrino masses.

We have calculated the matrix elements for heavy neutrino exchange within the pn-QRPA model of Muto et al. [7,12] Numerical results for the experimentally most interesting isotopes are given in Table 1. Corresponding matrix elements for light neutrino exchange can be found in Ref. [7]. Table 1 shows that, with the possible exception of the two heaviest isotopes, all matrix elements have rather similar numerical values, in agreement with the expectation.

With the calculated matrix elements at hand, using the half-life limit on ^{76}Ge $0\nu\beta\beta$ decay as recently measured by the Heidelberg-Moscow collaboration, $T_{1/2}^{0\nu\beta\beta}(^{76}\mathrm{Ge}) \geq 7.4 \times 10^{24}$ years (90% c.l.) [10] we are ready to derive quantitative constraints on the various LR model parameters. In principle, the experimental half-life limit and Eq. (2) define an excluded area in a 5-dimensional parameter space. However, since in LR models heavy neutrinos are expected to be right-handed, we will restrict the discussion to $\langle m_\nu\rangle$, $\langle\lambda\rangle$, $\langle\eta\rangle$ and $\langle\xi\rangle$.[5]

Constraints can be derived under the assumption that only one parameter contributes to the decay rate at a time ("on axis"), or for an arbitrary variation of all four parameters. Numerically we find: $\langle m_\nu\rangle = 0.66$ (0.56) [eV], $\langle\lambda\rangle = 1.1\ (1.0) \times 10^{-6}$, $\langle\eta\rangle = 6.4\ (5.5) \times 10^{-9}$ and $\langle\xi\rangle = 1.7\ (1.7) \times 10^{-8}$ for the "arbitrary" ("on axis") cases, respectively.

As is clear from Eqs. (5), (6), limits on $\langle\lambda\rangle$ and $\langle\eta\rangle$ cannot be converted into limits on the mass or mixing angle of right-handed W-bosons, without making assumptions about the size of the neutrino mixing matrix coefficients and their respective CP eigenvalues. Instead, for example, $\langle\lambda\rangle$ defines an excluded area in the plane $\{\sum_j' U_{ej}V_{ej}, m_{W_R}\}$, as is shown in Fig. 1. Although for large mixing very stringent limits would be obtained, for typical values of $\sum_j' U_{ej}V_{ej} \simeq \mathcal{O}(10^{-6})$ or so, only $m_{W_R} \leq m_{W_L}$ is excluded, certainly not a very stringent constraint.

Much more interesting in this sense is the limit on $\langle\xi\rangle$. From the completeness relation we know that there is at least one right-handed neutrino with $V_{ej}^2 \simeq \mathcal{O}(1)$, which in LR models should be quite heavy. Defining

$$\langle m_N^{(V)}\rangle^{-1} = \sum_j{}'' V_{ej}^2 m_j^{-1} \tag{8}$$

from the limit on $\langle\xi\rangle$ one can derive

$$m_{W_R} \geq 1.1\left(\frac{\langle m_N^{(V)}\rangle}{1\ \mathrm{TeV}}\right)^{(-1/4)} [\mathrm{TeV}], \tag{9}$$

$$\tan(\zeta) \leq 4.7 \times 10^{-3}\left(\frac{\langle m_N^{(V)}\rangle}{1\ \mathrm{TeV}}\right)^{(1/2)}. \tag{10}$$

To compare the limit on the mass of the W_R to the one derived by Mohapatra [8], we mention that

$$m_{W_R} \geq 1.23\left(\frac{\langle m_N^{(V)}\rangle}{1\ \mathrm{TeV}}\right)^{(-1/4)} [\mathrm{TeV}]$$

[5] Assuming only left-handed heavy neutrinos to contribute to $0\nu\beta\beta$ decay one could also derive the constraint $\langle m_N^{(U)}\rangle \geq 6.0 \times 10^7$ GeV. However, $\langle m_N^{(U)}\rangle$ incorporates $\left(\sum_j'' U_{ej}^2\right)^{(-1)}$ which has to be expected to be small. Moreover, since the *effective* masses include the unknown mixing coefficients, it has to be noted that $\langle m_N^{(U)}\rangle$ is not necessarily positive definite. Thus, by an extreme fine-tuning it is possible to cancel the contributions from light and heavy *left-handed* neutrinos. We disregard such an unlikely situation in the following.

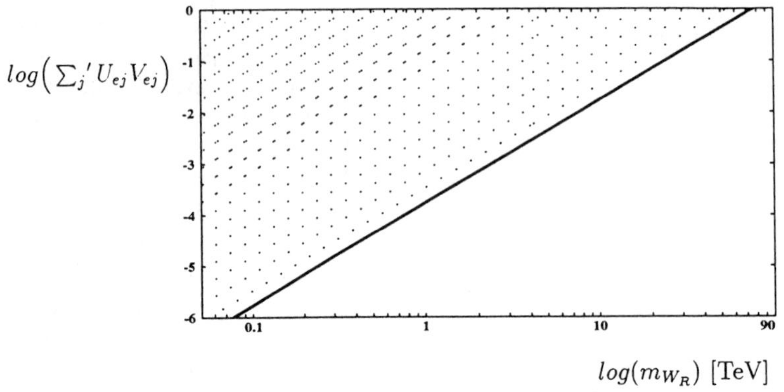

Fig. 1. Excluded area in the plane $\{\sum_j' U_{ej}V_{ej}, m_{W_R}\}$, for ^{76}Ge. Combinations to the upper left of the thick line are not allowed.

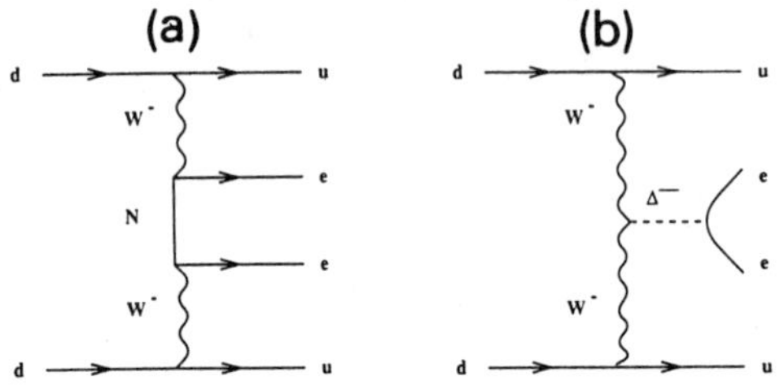

Fig. 2. a) Heavy neutrino exchange contribution to neutrinoless double beta decay in left right symmetric models, and b) Feynman graph for the virtual exchange of a doubly-charged Higgs boson, see text.

can be derived, if we take the limit $\tan(\zeta) \to 0$. Finally, following the argument [8] that vacuum stability requires $\langle m_N^{(V)} \rangle \leq g m_{W_R}$, where g is of order $\mathcal{O}(1)$, an absolute lower bound on m_{W_R} can be derived. The quantitative difference between our result and that of Ref. [8] is mainly due to the improved half-life limit used in our calculation. We stress that it is *not* due to errors or uncertainties in the matrix element calculation – uncertainties of limits on m_{W_R} scale only as the fourth square root of the uncertainties in the nuclear matrix elements.

Let us now turn to a brief discussion of the Higgs triplet contribution to $0\nu\beta\beta$ decay. The generation of Majorana masses in left-right symmetric models is achieved quite naturally if the Higgs sector of the theory contains two additional triplets, $\Delta_{L/R} = (\Delta^{--}, \Delta^{-}, \Delta^{0})_{L/R}$ [20]. This implies that $0\nu\beta\beta$ decay cannot only occur through the usual neutrino exchange diagram (Fig. 2a), but in addition also through the graph involving the exchange of a doubly-charged Higgs (Fig. 2b).[6] It is straightforward to show that the contribution of this graph to $0\nu\beta\beta$ decay is proportional to [20]

$$\frac{1}{m_{W_R}^4}\frac{m_N}{m_{\Delta_R^{--}}^2}.$$

(In general, also Δ_R and Δ_L can mix with each other as is the case for the W-bosons. The left-handed doubly-charged Higgs, however, has a negligible coupling strength proportional to the light neutrino mass. We neglect this possibility for simplicity.)

The inclusion of the graph in Fig. 3b therefore modifies the contribution of the λ^4-terms, which are then proportional to (neglecting mixing among neutrinos for simplicity)

$$\left(\frac{m_{W_L}}{m_{W_R}}\right)^4\left(\frac{1}{m_N} + \frac{m_N}{m_{\Delta_R^{--}}^2}\right). \tag{11}$$

Eq. (11) leads to a modified constraint on the mass of the right-handed W-bosons as shown in Fig. 3, where limits are shown as a function of the heavy, right-handed neutrino mass and various values of $m_{\Delta_R^{--}}$. Given that there is no *upper* bound on the mass of the right-handed Higgs triplet, however, no more stringent

[6] In addition, there is the possibility that the two W-bosons of Fig. 2a are replaced by the singly-charged component of the triplet. This contribution, however, is negligible due to the small coupling of the Higgs to quarks [14], and, in addition, due to the small relevant nuclear matrix elements [16].

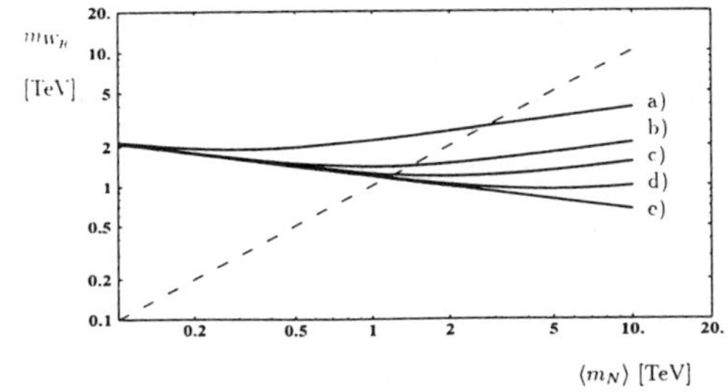

Fig. 3. Limits on the mass of the right-handed W-boson from neutrinoless double beta decay (full lines) and vacuum stability (dashed line). Combinations below the lines are forbidden. The five full lines correspond to the following masses of the doubly charged Higgs, $m_{\Delta^{--}}$: a) 0.3, b) 1.0, c) 2.0, d) 5.0 and e) ∞ [TeV].

constraints on m_{W_R} can be inferred from $0\nu\beta\beta$ decay, than the one quoted in Eq. (9).

To summarize, it is concluded that right-handed neutrino exchange in $0\nu\beta\beta$ decay leads to much more stringent limits on the mass and mixing angle of right-handed W-bosons, than the left-right mixing mechanism, usually expressed in terms of the *effective parameters* $\langle \lambda \rangle$ and $\langle \eta \rangle$. This is mainly due to the small mixing, which has to be expected between the left- and the right-handed neutrino sectors. Terms proportional to right-handed neutrinos cannot be neglected in the decay rate of $0\nu\beta\beta$ decay in left-right symmetric models. Although limits on the mass of the right-handed W-boson do depend only weakly on nuclear matrix elements, it therefore seems to be desirable also to reconsider the calculation of nuclear matrix elements for heavy particle exchange more carefully than has been done up to now.

We have also discussed the modifications of the formalism, due to the contribution of the right-handed Higgs triplet. Although such contribution turns out to be numerically small (unless Δ^{--} is very light) for $0\nu\beta\beta$ decay, as is shown in Fig. 3, it is necessary to include these terms if one wants to make a consistent comparison of the constraints on LR models as derived from $0\nu\beta\beta$ decay with those inferred from inverse neutrinoless double beta decay searched for at particle accelerators [21,22].

We would like to thank S.G. Kovalenko for useful discussions. M.H. is supported by the Deutsche Forschungsgemeinschaft (kl 253/8-1 and 446 JAP-113/101/0). O.P. acknowledges partial support from the E.U. program "Human Capital and Mobility" under contract No. CHRX-CT92-0026. He would also like to thank the Max-Planck-Institut für Kernphysik in Heidelberg for the very kind hospitality.

References

[1] J.C. Pati and A. Salam, Phys. Rev. D 10 (1974) 275;
R.N. Mohapatra and J.C. Pati, Phys. Rev. D 11 (1975) 566, 2558;
G. Senjanovic and R.N. Mohapatra, Phys. Rev. D 12 (1975) 1502.

[2] K. Grotz and H.V. Klapdor-Kleingrothaus, The Weak Interactions in Nuclear, Particle and Astrophysics, Adam Hilger, Bristol, New York (1990).

[3] M. Doi, T. Kotani and E. Takasugi, Progr. Theor. Phys. Suppl. 83 (1985) 1.

[4] M. Gell-Mann, P. Ramond and R. Slansky, in: Supergravity, eds. F. van Nieuwenhuizen and D. Freedman, North-Holland, Amsterdam (1979);
R.N. Mohapatra and G. Senjanovic, Phys. Rev. D 23 (1981) 165.

[5] W.C. Haxton and G.J. Stephenson, Progr. Part. Nucl. Phys. 12 (1984) 409;
J.D. Vergados, Phys. Report, 133 (1986) 1;
K. Muto and H.V. Klapdor, in: Neutrinos, ed. H.V. Klapdor, Springer, Heidelberg (1988) p. 183;
T. Tomoda, Rep. Prog. Phys. 54 (1991) 53.

[6] T. Tomoda, A. Faessler, K.W. Schmid and F. Grümmer, Nucl. Phys. A 452 (1986) 591;
O. Civitarese, A. Faessler and T. Tomoda, Phys. Lett. B 194 (1987) 11;
T. Tomoda, A. Faessler, Phys. Lett. B 199 (1987) 475;
J. Suhonen, T. Taigel and A. Faessler, Nucl. Phys. A 486 (1988) 91.

[7] K. Muto, E. Bender and H.V. Klapdor, Z. Phys. A 334 (1989) 177, 187.

[8] R.N. Mohapatra, Phys. Rev. D 34 (1986) 909.

[9] M. Doi and T. Kotani, Progr. Theor. Phys. 89 (1993) 139.

[10] HEIDELBERG-MOSCOW Collaboration: A. Balysh et al., Phys. Lett. B 356 (1995) 450;
H.V. Klapdor-Kleingrothaus, in: Proc. Int. Workshop on Double Beta Decay and Related Topics, Trento, Italy (April 1995), World Scientific, Singapore, in press.

[11] Recent reviews of the experimental situation may be found in: M. Moe, Int. J. Mod. Phys. E 2 (1993) 507;
M. Moe and P. Vogel, Ann. Rev. Nucl. Part. Sc.;
H.V. Klapdor-Kleingrothaus, Progr. Part. Nucl. Phys. 32 (1994) 261.

[12] M. Hirsch, K. Muto, T. Oda and H.V. Klapdor-Kleingrothaus, Z. Phys. A 347 (1994) 151.

[13] R.N. Mohapatra and J.D. Vergados, Phys. Rev. Lett. 47 (1981) 1713.

[14] J. Schechter and J.W.F. Valle, Phys. Rev. D 25 (1982) 2951.

[15] L. Wolfenstein, Phys. Rev. D 26 (1982) 2507.

[16] W. Haxton, S.P. Rosen and G.J. Stephenson, Phys. Rev. D 26 (1982) 1805.
[17] T.G. Rizzo, Phys. Lett. B 116 (1982) 23.
[18] M. Hirsch and H.V. Klapdor-Kleingrothaus, in: Proc. Int. Workshop on Double Beta Decay and Related Topics, Trento, Italy (April 1995), World Scientific, Singapore, in the press.
[19] J.D. Vergados, Phys. Rev. C 24 (1981) 640.
[20] T.G. Rizzo, Phys. Rev. D 25 (1982) 1355.
[21] G. Bélanger, F. Boudjema, D. London and H. Nadeau, hep-ph/9508317.
[22] T.G. Rizzo, Phys. Rev. D 50 (1994) 5602.

Paper presented at the First Int. Symp. on Lepton and Baryon Number Violation, Trento, 1998 605

Inverse neutrinoless double beta decay and $\Delta L = 2$ processes at linear colliders

Geneviève Bélanger

LAPTH,[1] Chemin de Bellevue, B.P. 110, F-74941 Annecy-le-Vieux, Cedex, France.

Abstract. Prospects for observation of the $\Delta L = 2$ lepton-number-violating process $e^-e^- \to W^-W^-$ at linear colliders are reviewed. In the standard model with additional Majorana neutrinos, constraints from neutrinoless double beta decay render the process unobservable at a linear collider of $\sqrt{s} < 2$ TeV except in the most contrived scenarios. In that respect, some models beyond the standard model fare better.

1. Introduction

The problem of the mass of the neutrino has been at the forefront of particle physics recently. What is even more intriguing than the value of the neutrino mass would be to know whether it has a Dirac or a Majorana ma
If the neutrino has a Majorana mass, then it will contribute to ΔL 2 lepton-number-violating processes such as neutrinoless double beta decay ($\beta\beta_{0\nu}$) through the subprocess $W^-W^- \to e^-e^-$. The neutrinoless double beta decay data constrains both the mass of a Majorana neutrino and its mixing with the electron neutrino. In the design of a future high energy linear e^+e^- collider, one option being considered seriously is to replace the positron beam by an electron beam in order to look at e^-e^- collisions. The inverse neutrinoless beta decay process $e^-e^- \to W^-W^-$ would then be accessible.

The process $e^-e^- \to W^-W^-$ has been looked at several times, by different authors, over the last decade or so [1]-[6]. In one recent analysis,

[1] URA 14-36 du CNRS, associée à l'Université de Savoie.

it was found that within the context of the standard model with additional Majorana neutrinos, once the constraint from $\beta\beta_{0\nu}$ is taken into account, the cross section for $e^-e^- \to W^-W^-$ is in general too small for it to be seen at a .5-1.5 TeV linear collider[7]. Here I will review the current situation taking into account the latest data[8] and considering the high luminosity version of a linear collider being discussed now[9]. Recently, some efforts were made to explore ways to evade the constraint from double beta decay [10]. While it was found that under specific conditions this could be done, in general, $e^-e^- \to W^-W^-$ is not the most suitable process to test the Majorana nature of the neutrino. Processes involving the single production of a Majorana neutrino have a much larger cross-section[10, 11]. I will as much as possible keep to a model independent approach not relying on any assumptions on neutrino masses and mixings.

When going beyond the standard model, the constraint from neutrinoless double beta decay still plays a role although it is possible to find models where the inverse neutrinoless double beta decay process can occur at an observable rate, such as composite neutrinos models. Furthermore, in extensions of the standard model, such as left-right symmetric models or supersymmetric models, there are other $\Delta L = 2$ processes involving the production of new particles that could be interesting to probe the Majorana nature of the neutrino. Of course these entail that new particles such as W_R or charginos are kinematically accessible[12, 13].

At a linear collider many options are under consideration. It is possible to backscatter laser light off one or both of the beams, creating an $e\gamma$ or $\gamma\gamma$ collider [14]. There are a number of $\Delta L = 2$ processes mediated by a Majorana electron neutrino that could in principle take place et these colliders, for example $e^-\gamma \to e^+W^+W^-$. Unfortunately these higher order processes are also subject to the neutrinoless double beta decay constraint. One would need energies in the multi TeV range to observe them, certainly a very remote possibility[7].

Clearly the limits from $\beta\beta_{0\nu}$ apply only to ν_e. Should the ν_μ have a Majorana mass, it will contribute to the processes $\mu^-\mu^- \to W^-W^-$ (and similarly for the ν_τ), with no constraints from low-energy processes. The best way to find evidence of the Majorana nature of the ν_μ would be with a $\mu^-\mu^-$ collider [15]. Lacking such a collider such lepton-number-violating processes cannot take place directly but they can occur as subprocesses in any of the proposed options of the linear collider. For example, the observation of $\gamma\gamma \to \mu^+\mu^+W^-W^-$ would be evidence for a Majorana ν_μ. However again a linear collider with a centre-of-mass energy of at least 4 TeV is necessary[7].

2. Inverse neutrinoless double beta decay, $e^-e^- \to W^-W^-$

2.1. Neutrino Mixing

Suppose that the ν_e mixes with an unspecified number of other neutrinos, after diagonalization of the mass matrix, ν_e can be expressed in terms of the mass eigenstates N_i:

$$\nu_e = \sum_i U_{ei} N_i \ , \tag{1}$$

where the mixing matrix U is unitary. Phenomenologically, we have observed two things. First, the ν_e does not mix much with other neutrinos [16]:

$$\sum_{i \neq 1} |U_{ei}|^2 < 6.6 \times 10^{-3} \ \ (90\% \ c.l.) \ . \tag{2}$$

This limit is essentially independent of the $SU(2)_L$ transformation properties of the neutrino(s) with which the ν_e mixes. Also, the limit is quite conservative – it allows for the possibility that the other charged fermions also mix with new, exotic charged particles [17]. If one assumes that the only new particles are neutrinos, then the above limit improves somewhat to 5.0×10^{-3}. Thus, the ν_e is mainly N_1, furthermore, from muon decay, we know that the N_1 is very light: $M_1 < 7$ eV [18].

2.2. Cross Section for $e^-e^- \to W^-W^-$

Assuming that the N_i are Majorana neutrinos, they will contribute to the process $e^-e^- \to W^-W^-$ through the diagrams of Fig. 1.

The cross section for unpolarized electrons is rather simple in the limit that $s \gg M_W^2$. This is a reasonable approximation for the linear collider,

$$\frac{d\sigma}{d\cos\theta} = \frac{g^4}{1024\pi M_W^4}\left(\sum_i M_i \left(U_{ei}\right)^2 \left[\frac{t}{(t - M_i^2)} + \frac{u}{(u - M_i^2)}\right]\right)^2 . \tag{3}$$

(The full expression given for example in [7] is used when presenting numerical results.)

In this energy regime the production of longitudinal W's dominates the process $e^-e^- \to W^-W^-$, in fact Eq. 3 is precisely what is obtained if one calculates the cross section for the process $e^-e^- \to W^-_{long}W^-_{long}$, where W^-_{long} is the longitudinal component of the W^-. The full helicity amplitudes are given for example in [7].

There are two limiting cases of Eq. 3 which will be useful in what follows. First, if $s \gg M_i^2$, the cross section becomes

$$\frac{d\sigma}{d\cos\theta} = \frac{g^4}{256\pi M_W^4}\left(\sum_i M_i \left(U_{ei}\right)^2\right)^2 . \tag{4}$$

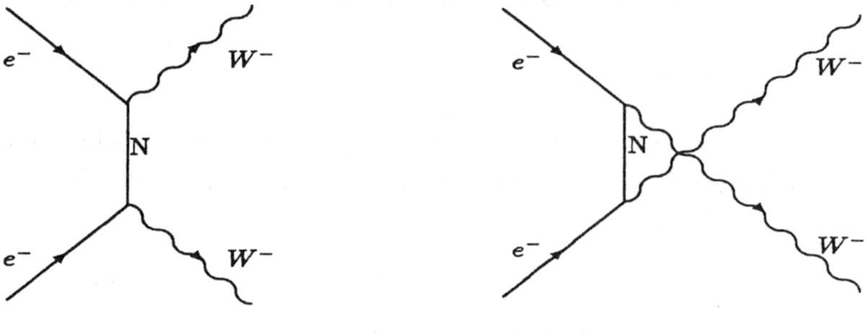

Figure 1. *Diagrams contributing to* $e^-e^- \to W^-W^-$.

Second, in the other limit, $M_i^2 \gg s$, the cross-section grows like s^2

$$\frac{d\sigma}{d\cos\theta} = \frac{g^4}{1024\pi M_W^4} s^2 \left(\sum_i \frac{(U_{ei})^2}{M_i} \right)^2 . \tag{5}$$

2.3. Unitarity Considerations

From Eq. 4, we see that, in the high-energy limit ($s \to \infty$), the cross section tends towards a constant:

$$\sigma_{s \to \infty} = \frac{g^4}{128\pi M_W^4} \left(\sum_i M_i (U_{ei})^2 \right)^2 . \tag{6}$$

In this particular case this indicates a violation of unitarity, since the amplitude (which is a pure s-wave) grows as $\sqrt{s}$.

There are basically two ways in which this unitarity violation can be cured. The first through the inclusion of a Higgs triplet. If the neutrinos with which the ν_e mixes are $SU(2)_L$ doublets, then they can acquire Majorana masses by giving the Higgs triplet a vacuum expectation value (v.e.v.). This Higgs triplet includes a doubly-charged Higgs, H^{--}. In this case unitarity is restored through the inclusion of a diagram in which the H^{--} is exchanged in the s-channel. However this solution is not favoured phenomenologically and this direction will not be pursued here.

In the absence of Higgs triplets, the only way to restore unitarity is to require that the neutrinos' masses and mixing angles satisfy

$$\sum_i (U_{ei})^2 M_i = 0 \ . \tag{7}$$

Although this relation may appear arbitrary at first sight, it is in fact automatically satisfied. It is straightforward to show that

$$\sum_i (U_{ei})^2 M_i = M_{ee}^* \ , \tag{8}$$

where M_{ee} is the Majorana mass of the ν_e. However, because there are no Higgs triplets, this mass is equal to zero, so that Eq. 7 holds. This is exactly what happens in the famous seesaw mechanism where one adds a right-handed neutrino N_R to the spectrum. This neutrino acquires a large Majorana mass M through the v.e.v. of a Higgs singlet, and the combination $\bar{N}_R \nu_{eL} + h.c.$ obtains a Dirac mass m once the ordinary Higgs doublet gets a v.e.v. The mass matrix looks like

$$\begin{pmatrix} 0 & m \\ m & M \end{pmatrix} . \tag{9}$$

The two mass eigenstates are N_1 and N_2, with masses $-m^2/M$ and M, respectively (the minus sign in front of M_1 can be removed by a γ_5 rotation). For m of the order of the electron mass and M about 1 TeV, one obtains a mass of about 1 eV for the lightest neutrino. (Thus, in such models, the large range of neutrino masses is explained in a natural way, unlike the Higgs triplet models.) The ν_e is a linear combination of these two physical neutrinos:

$$\nu_e = \cos\theta N_1 + \sin\theta N_2 \ , \qquad \sin\theta = \frac{m}{M} \ . \tag{10}$$

It is clear that, with these masses and mixing angles, the relation in Eq. 7 is automatically satisfied.

The downside of this particular solution is that the mixing of the ν_e with the N is tiny: for $m \sim m_e$ and $M \sim 1$ TeV, $\sin\theta \sim 10^{-6}$! This would make the cross section for $e^-e^- \to W^-W^-$ invisible, since $\sigma(e^-e^- \to W^-W^-) \propto \sin^4\theta$, and is typical of what happens in left-right symmetric models [13]. However, if the ν_e mixes with more neutrinos, it is, in principle, possible to satisfy Eq. 7 without having such small values of U_{ei}. (This is the assumed solution in Ref. [3].) This is perhaps a bit artificial, and probably requires some fine-tuning, but it is possible. If this is how unitarity restoration comes about, then Eq. 3 contains all the contributions to $e^-e^- \to W^-W^-$.

Although it is interesting to understand how unitarity is restored in different models, the above discussion demonstrates that the cross section for $e^-e^- \to W^-W^-$ is essentially unconstrained by such considerations –

the U_{ei} and M_i can take any values consistent with the phenomenological limits in Sec. 2.1. This is not the case when the experimental limits on neutrinoless double beta decay are taken into account.

2.4. Limits from $\beta\beta_{0\nu}$

As mentioned in the introduction, $e^-e^- \to W^-W^-$ is essentially the inverse of neutrinoless double beta decay. We might therefore expect that the limits on the latter process could constrain the former.

If some of the neutrinos have masses $M_i \ll 1$ GeV, then, for these neutrinos, the quantity which contributes to $\beta\beta_{0\nu}$ is

$$\langle m_\nu \rangle = \sum_i{}' \left(U_{ei}\right)^2 M_i \ , \tag{11}$$

where the sum is over the light neutrinos. (For simplicity, we have ignored factors corresponding to complications from the nuclear matrix elements – their inclusion does not change our conclusions. For more details we refer the reader to Ref. [19],[20].) At present, the best experimental limit on $\langle m_\nu \rangle$ is obtained by the Heidelberg-Moscow group[8],

$$\langle m_\nu \rangle \lesssim .45 \text{ eV} \ . \tag{12}$$

As for the neutrinos which are heavy, $M_i \gg 1$ GeV, they can still mediate $\beta\beta_{0\nu}$ decay. In this case the relevant quantity is

$$\langle m_\nu^{-1} \rangle_H = \sum_i{}'' \left(U_{ei}\right)^2 \frac{1}{M_i} \ , \tag{13}$$

where the sum is over the heavy neutrinos. Now the experimental limit on $\beta\beta_{0\nu}$ implies the following [21]:

$$\langle q^2 \rangle \sum_i{}'' \left(U_{ei}\right)^2 \frac{1}{M_i} \lesssim .45 \text{ eV} \ , \tag{14}$$

where q is an average nuclear momentum transfer. If one takes q to be roughly about 100 MeV, one obtains the right order-of-magnitude constraint. However, a more careful calculation, including all the nuclear effects, gives

$$\sum_i{}'' \left(U_{ei}\right)^2 \frac{1}{M_i} < 5 \times 10^{-5} \text{ TeV}^{-1} \ . \tag{15}$$

This limit has been arrived at by combining the nuclear matrix element calculation of Ref. [22] with the latest lower bound of 1.2×10^{24} yr on the half-life for ${}^{76}Ge \to {}^{76}Se + 2e^-$ [8]. While there are some variations in the matrix elements calculations, it is now generally accepted [20] that the uncertainty is at most a factor of 2 or 3. In fact the $\beta\beta_{0\nu}$ transition matrix elements are expected to be much less sensitive than the $\beta\beta_{2\nu}$ to the details of the nuclear structure. In particular they are much less affected by the particle-particle interactions.

Assuming no cancellations between contributions of different heavy Majorana neutrinos, the $\beta\beta_{0\nu}$ constraint implies a generic lower bound on the mass of the heavy neutrino:

$$M_i > 1.4 \times 10^4 \left(U_{ei}\right)^2 \text{ TeV} . \tag{16}$$

For $\left(U_{ei}\right)^2 \sim 5 \times 10^{-3}$, this gives $M_i > 100$ TeV. However, it is possible to evade this order-of-magnitude bound if one allows cancellations among the various terms, I will come back to that point below, for the moment I will assume that this constraint holds for all neutrinos.

As was reported earlier in this conference[23], the Heidelberg-Moscow experiment should improve by a factor of 5 their limit on the lifetime of neutrinoless double beta decay. There has also been a proposal, GENIUS, to probe half-life at the level of $5.8 \times 10^{27} yr$ or even $6.4 \times 10^{28} yr$ with a one ton detector[24]. The limit on neutrino mass improving only as the square root of the life-time, the proposed experiment could reach a sensitivity for the neutrino mass below the .01 eV level.

2.5. $e^-e^- \to W^-W^-$ at the linear collider

The constraint from $\beta\beta_{0\nu}$ gives us one of two conditions, depending on whether the new neutrinos are very light or very heavy relative to the energy scale of neutrinoless double beta decay. For the case of light neutrinos, the cross section for $e^-e^- \to W^-W^-$, which is independent of $\sqrt{s}$ is much too small to be observable at any future collider, $\sigma(e^-e^- \to W^-W^-) \leq 1.3 \times 10^{-17}\ fb$.

If no cancellations are allowed in Eq. 15, then $\beta\beta_{0\nu}$ constrains the neutrinos to be very massive (Eq. 16). For linear colliders with centre-of-mass energies in the TeV range, we have $M_i \gg \sqrt{s}$ and Eq. 5 can be used to calculate the cross section for $e^-e^- \to W^-W^-$. Using the limit in Eq. 15, we find that, at $\sqrt{s} = 500$ GeV,

$$\sigma(e^-e^- \to W^-W^-) < 4.9 \times 10^{-3}\ fb . \tag{17}$$

Even with a high luminosity collider, $\mathcal{L} = 500 fb^{-1}$, the process $e^-e^- \to W^-W^-$ would be unobservable.

In this section we assume that the cross section for $e^-e^- \to W^-W^-$ is dominated by the exchange of a single neutrino. (Of course, additional, heavier neutrinos must be present to satisfy the bound from unitarity.) Note that even if one assumes that more than one neutrino contributes to

$e^-e^- \to W^-W^-$, this will not change the cross section significantly, since the mixing angles of all the neutrinos must be correspondingly reduced in order to satisfy the constraint in Eq. 15.

In Ref.[7] it was shown that a linear collider of at least 2 TeV would be necessary to improve on the present neutrinoless double beta decay limits on massive Majorana neutrinos (see Fig. 2 in Ref. [7]). In particular it was found that with the luminosity, $80(\sqrt{s}/(1\ \text{TeV}))^2\ fb^{-1}$, for $\sqrt{s} = 500$ GeV and 1 TeV, the values of M_i and $(U_{ei})^2$ which produce an observable $e^-e^- \to W^-W^-$ cross section are already ruled out by neutrinoless double beta decay. For $\sqrt{s} = 2$ TeV, the discovery limit and the limit from $\beta\beta_{0\nu}$ are roughly equal. Although the 2 TeV linear collider opens a very small region of parameter space when polarized e^- beams are used and hence does slightly better than $\beta\beta_{0\nu}$.

Since that work was published, the linear collider design team has claimed that a luminosity of $\mathcal{L} = 500 fb^{-1}$ could be achieved with the 500 and 800 GeV linear collider[9]. This could improve the prospects for inverse neutrinoless double beta decay. Unfortunately the $\beta\beta_{0\nu}$ constraint is still too good to have any chance of seing the inverse process at those energies.

In Fig. 2, the discovery limit for $e^-e^- \to W^-W^-$ at the linear collider for several centre-of-mass energies as a function of M_i and $(U_{ei})^2$ are presented. We demand 10 events for discovery, and assume polarized e^- beams and a luminosity of $500 fb^{-1}$ for 500 and 800 GeV. For higher energies, $\sqrt{s} = 2, 4$ and 10 TeV, the luminosity $80(\sqrt{s}/(1\ \text{TeV}))^2\ fb^{-1}$ and unpolarized beams are assumed. Note that perfect polarization basically increases the cross-section by a factor of 4. The improvement that polarization provides at 2 TeV is also displayed in Fig. 2. Note that we have not included efficiencies for the detection of the W's, nor have we included any backgrounds. Our discovery limits are therefore as optimistic as they possibly could be. In this figure the phenomenological limit on $(U_{ei})^2$, as well as the constraints from $\beta\beta_{0\nu}$ are superimposed. By the time a linear collider is built the $\beta\beta_{0\nu}$ limits is expected to improve by a factor of 5 on the lifetime measurement, the limit achievable in this case is also shown as well as the limit from the proposed GENIUS experiments.

From Fig. 2 we conclude that in the context of the standard model with additional Majorana neutrinos, there are no chance of seing the inverse neutrinoless process at a collider of energy below 2TeV. Even a very high luminosity collider of 800 GeV with perfectly polarized beams is no match for the neutrinoless double beta decay experiment. Note that the bound on $U_{ei}^2 - M_i$ improves as the square root of the luminosity. A 2 TeV version of a linear collider with polarized beams can extend the discovery reach compare with present day low-energy experiments. However by the time such a collider is built, the low-energy constraint would have improved as displayed. In summary, prospects for 2 TeV collider to improve upon neutrinoless double beta decay are marginal at best.

Finally, for 4 TeV and 10 TeV linear colliders, there exists a sizeable region of M_i-$(U_{ei})^2$ parameter space, not ruled out by $\beta\beta_{0\nu}$, which produces an observable signal for $e^-e^- \to W^-W^-$. In Fig. 2, one can easily see that a 10 TeV collider is needed to do better than the next generation of neutrinoless double beta decay experiments for probing Majorana neutrinos.

In all this discussion the bound used for $\beta\beta_{0\nu}$ (Eq. 5) was obtained from the calculation in Ref. [22], it is clear that a factor of 2 or 3 uncertainty in the nuclear matrix element does not alter any of the conclusions. In fact, the matrix elements calculation quoted here is not the most restrictive and using the published results of various groups would only strengthen the conclusions[22].

2.6. Evading the $\beta\beta_{0\nu}$ constraint

The analysis just presented might be too restrictive and one should wonder if there is a way to evade the constraint from neutrinoless double beta decay and still obtain an observable cross-section for the inverse process. As the nuclear matrix element calculation is well under control[20], the only alternative to evade the $\beta\beta_{0\nu}$ constraint rests with particle physics. To obtain some cancellation between the contributions of different neutrinos in the bound in Eq.16 , the following conditions must be met: many Majorana neutrinos that mix with ν_e and some of the mixing matrix elements square are purely imaginary. In other words at least one neutrino must have a different intrinsic CP parity, η_{CP}. Some explicit examples of what is needed are given in Ref. [7].

First, consider the case where there are only two neutrinos. To have a cancellation the heavier neutrino must have a larger mixing with the electron neutrino than the lighter one, this is certainly possible but not very natural. Putting aside the naturality argument, it was shown by scanning over the parameter space of masses and mixing angles that when the neutrinoless double beta decay constraint is loosened, the $e^-e^- \to W^-W^-$ cross-section goes down as well and becomes unobservable at either 500 GeV or 1 TeV[10]. Perhaps what is even more interesting is that a neutrino of mass less than $\sqrt{s}$ which has a significant mixing could in principle be observed at a linear collider in the single production process $e^+e^- \to \nu_e N$. With the decay products of the heavier neutrino one could determine whether it is Majorana in nature[25]. After taking into account the neutrinoless double beta decay constraint Gluza and Zrałek showed recently that one could get an observable cross-section for $e^+e^- \to \nu N$ at 1 TeV[10, 11]. Large cross-sections ($\mathcal{O}(100 fb)$) are obtained when the two

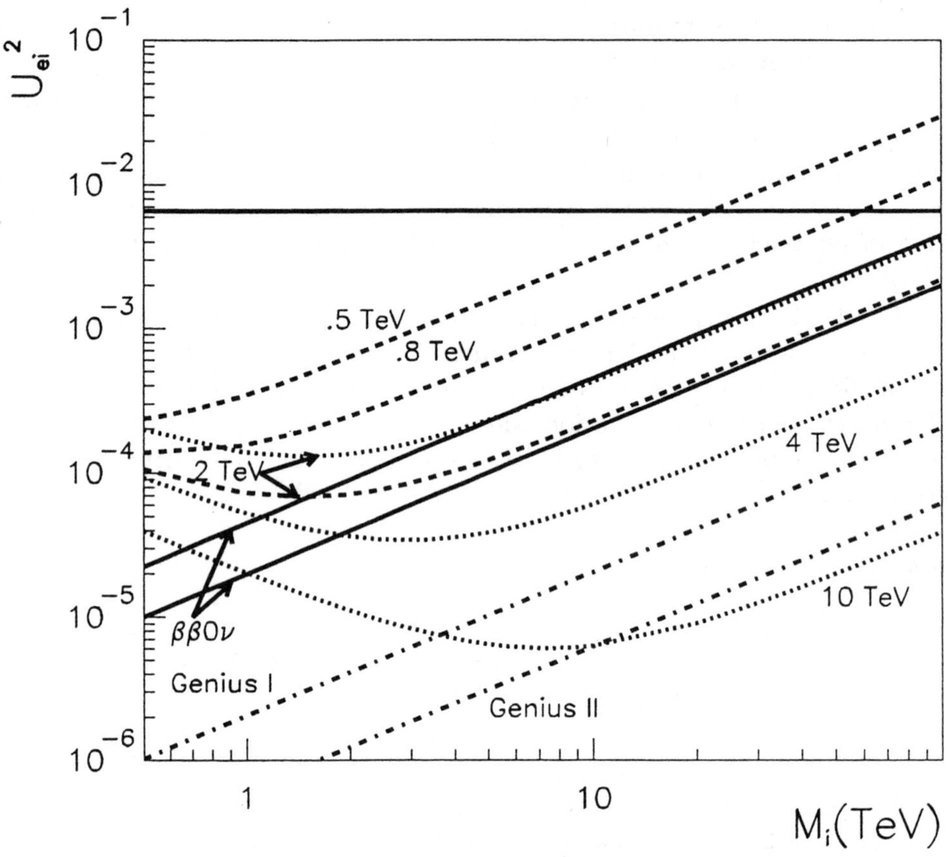

Figure 2. *Discovery limit for $e^-e^- \to W^-W^-$ at a linear collider as a function of M_i and $(U_{ei})^2$ for $\sqrt{s}$ = 500 GeV, 800 GeV, 2 TeV, 4 TeV and 10 TeV with luminosities of $500 fb^{-1}$ for energies below 1 TeV and $80(\sqrt{s}/(1\ TeV))^2\ fb^{-1}$ above. In all cases, the parameter space above the line corresponds to observable events. Dashed lines correspond to perfectly polarized beams while dotted ones to unpolarized beams. We also superimpose the present and future experimental limit from $\beta\beta_{0\nu}$ (diagonal solid line), from the Heidelberg Moscow experiment as well as the prospective limit from the future Genius experiment (dash-dot) The horizontal line represents the limit on neutrino mixing, $(U_{ei})^2$. Here, the parameter space above the line is ruled out.*

neutrinos are nearly degenerate, while neutrino masses ratio around 100 made for at most one fb.

As one adds more neutrinos, there is more freedom in the choice of parameters and one might be able to have some signal in inverse neutrinoless double beta decay. It is sufficient to look more closely at the three neutrinos case. As above one still needs one of the neutrinos with opposite intrinsic CP parity in order to evade the $\beta\beta_{0\nu}$ constraint. Assuming this being the neutrino with mass M_2, with $M_1 < M_2 < M_3$, Gluza and Zrałek [10] scanned the full parameter space for three neutrino masses. They found values for the masses and mixings where $\sigma(e^-e^- \to W^-W^-)$ exceeds one fb while the $\beta\beta_{0\nu}$ constraint is satisfied. In general, this implies that two neutrinos are nearly degenerate. However, when the Majorana neutrino mass is below the center of mass energy, the single production process is also allowed and the cross-section is quite large. This process is much favoured to discover a heavy Majorana neutrino, furthermore it takes place in the e^+e^- version of the linear collider. There exists a small region of parameter space where $e^-e^- \to W^-W^-$ is above the discovery limit at 1TeV and $\mathcal{L} = 80 fb^{-1}$(see Fig. 3 in Ref.[10]). Such is also the case at 500GeV if the high luminosity collider is available.

Clearly with more neutrinos there is more freedom and the same conclusion would be reached, it might be possible to observe the inverse neutrinoless double beta decay process at a linear collider with specific choices of masses, mixings and CP parity but in the major part of parameter space the single production process is the favoured one.

3. Other $\Delta L = 2$ Processes at the linear collider

For various reasons, it is conceivable that, even if a linear collider is built, the e^-e^- option may never be used while the $e\gamma$ or $\gamma\gamma$ option might be available. In the absence of an e^-e^- collider, the only real possibility for detecting a Majorana ν_e through $\Delta L = 2$ processes is the $2 \to 3$ process $e^-\gamma \to e^+W^-W^-$ which contains $e^-e^- \to W^-W^-$ as a subprocess as shown in Fig. 3. Being subject to the $\beta\beta_{0\nu}$ constraint one does not expect this process to do better than inverse neutrinoless double beta decay. Indeed it was shown that a few events could be observed in this process only at extremely high energies, beyond $\sqrt{s} = 4$ TeV[7]. A similar conclusion is reached when considering other processes such as $\gamma\gamma \to e^+e^+W^-W^-$.

Once again for the $e\gamma$ collider there exist a more direct way of seeing a Majorana neutrino: single production in $e\gamma \to WN$. The authors of Ref. [26] have shown recently, by scanning over all the possible mixing angles and masses that up to the kinematic limit one could get large cross-sections for single neutrino production. In particular in the case of

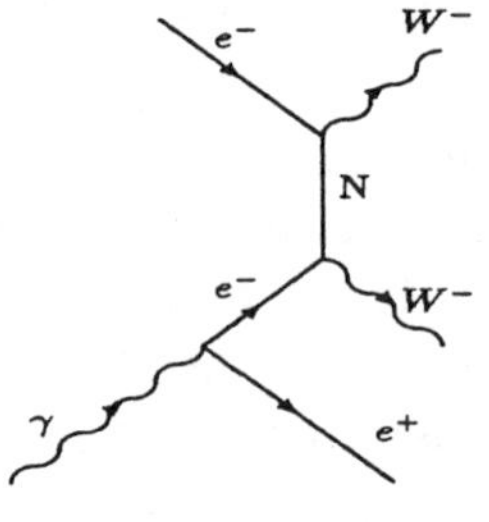

Figure 3. *The dominant diagram in $e^-\gamma \to e^+W^-W^-$.*

three neutrinos, cross sections above $100fb$ could be reached at a collider of 460GeV. Of course this presupposes that single neutrino production is kinematically accessible.

Although the constraint from neutrinoless double beta decay has proved to be very stringent in the context of the standard model, one should not forget that $\beta\beta_{0\nu}$ constrains only the ν_e –it says nothing about the ν_μ or the ν_τ. It therefore seems reasonable to ask about the possibilities for observing other $\Delta L = 2$ processes at the linear collider, specifically those involving a Majorana ν_μ or ν_τ. For this one could use the following options currently discussed in the context of the design of a linear collider: $e\gamma, \gamma\gamma$ or of a muon collider, $\mu^+\mu^-, \mu^-\mu^-$.

The most interesting option is obviously the $\mu^-\mu^-$ collider. A Majorana ν_μ will mediate processes such as $\mu^-\mu^- \to W^-W^-$. In order to observe $\Delta L = 2$ processes involving the ν_μ, this neutrino must mix with heavy Majorana neutrinos, just as was the case for the ν_e. Indeed even if the mass of the ν_μ lies at the upper allowed limit, $m_{\nu_\mu} < 0.27$ MeV but do not mix with heavy neutrinos, then the cross section for $\mu^-\mu^- \to W^-W^-$ is still unobservable. From Eq. 4, it is at most $O(10^{-3})\ fb$.

For a muon neutrino mixing with a heavy Majorana neutrino, the analysis is very similar to the one presented in the electron case, in particular the cross section formula and the numerical values for $\mu^-\mu^- \to W^-W^-$ are exactly the same. The curves in Fig. 2 then also represent the potential of a muon collider for probing the Majorana nature of the neutrino safe for the constraint from $\beta\beta_{0\nu}$ which is irrelevant here. The upper limit on the mixing angle of the ν_μ differs slightly from the ν_e case, $\sum_{i\neq 2}|U_{\mu i}|^2 < 6.0 \times 10^{-3}$[16]. As with the ν_e, this conservative limit is for the case where the other fermions also mix with new particles. If one assumes that only the neutrinos mix, then the limits improve to 1.8×10^{-3}.

In the absence of a muon collider, a muon Majorana neutrino can in principle also be looked for since the process $\mu\mu \to WW$ is also a subprocess in a number of 2 to 4 processes at linear colliders, from $\gamma\gamma \to \ell^+\ell^+W^-W^-$ to $e^-\gamma \to \nu_e\ell^-\ell^-W^+$ (here $l = \mu, \tau$ since these processes can also be used for a tau neutrino). Unfortunately again one needs to go to very high energies (4-10 TeV)to get a reasonable cross-section even if one does not have to deal with the $\beta\beta_{0\nu}$ constraint[7].

4. Beyond the standard model and heavy neutrinos

Beyond the context of the standard model the picture could be quite different as far as the prospect of observing the effect of a heavy Majorana neutrino mixing with the electron neutrino. In some models the constraint from $\beta\beta_{0\nu}$ is less stringent and in other models, additional processes with lepton number violation mediated by Majorana neutrinos are available. Composite models are example of the former and left-right symmetric models and supersymmetric models of the latter.

In composite models one could have an excited neutrino ν_e^* which can mix with a number of Majorana neutrinos. The coupling of the excited neutrino to leptons has the generic form

$$\mathcal{L}_{int} = g\frac{\lambda_W}{m_\nu^*}\bar{e}\sigma^{\mu\nu}(1 \pm \gamma_5)\nu_e^*\partial_\mu W_\nu^- + h.c. \tag{18}$$

where the mass dimension m_ν^* is taken to be of the order of the excited neutrino mass and the sign of the chirality operator depends on the chirality of the neutrino. The present limit on such an excited neutrino from accelerator searches is $m_{\nu^*} > 91GeV$[27]. With the $\sigma^{\mu\nu}$ type of coupling, the $\beta\beta_{0\nu}$ constraint is much less stringent in composite models [28]. In fact Kinuya and Takasagi[29] found that the constraint from neutrinoless double beta decay is basically irrelevant when considering the $e^-e^- \to W^-W^-$ cross-section, it can exceed the pb. On the other hand if one assumes that $\lambda_W \approx 1$ the cross -section is significantly reduced but remains in the observable range (i.e $>$ .1 fb) for masses $m^* \approx 20TeV$ even after taking into account the $\beta\beta_{0\nu}$ constraint. One characteristic feature of this class of models is that the W's produced are transverse rather than longitudinal.

In left-right symmetric models it is possible to choose parameters to evade both low-energy limits (including double beta decay) on Majorana neutrinos and lower bounds on right-handed W's in order to look for lepton-number violation in e^-e^-. However, the process of interest involves the production of a right-handed W either singly or in pair, see Ref. [13]. While certainly interesting, this scenario relies on the production of a new particle whose existence would first be probed in the e^+e^- version of the linear collider not to mention hadronic machines.

Other scenarios for lepton-number violation include those classes of supersymmetric models with R-parity violation, see Ref. [30]. These models have been shown to be very much constrained by double beta decay [30, 31]. In fact, in the context of the analysis presented in Section 2, the question is whether these new contributions could interfere destructively with those from a Majorana neutrino (that could also be present) with the result of evading those limits on the Majorana neutrino that we have used. Though such fine tuning seems unnatural, in this eventuality $e^-e^- \to W^-W^-$ could be used to single out the $\Delta L = 2$ lepton-number violation due to the Majorana neutrino. However, it is found from the analysis on double beta decay in Ref. [31] that these contributions add up coherently and that the limits on the Majorana neutrinos used above remain valid.

Another $\Delta L = 2$ process exists in supersymmetric models, $e^-e^- \to \chi^-\chi^-$ mediated by a Majorana sneutrino. Indeed it was shown in [32] that a Majorana sneutrino is tied to the existence of its standard superpartner, a Majorana neutrino. The cross-section for this process could be sizeable and should be easily observable [12].

5. Conclusions

I have reviewed the prospects for the observation of $e^-e^- \to W^-W^-$, mediated by a Majorana neutrino, at a high-energy e^-e^- collider . This process is essentially the inverse of neutrinoless double beta decay ($\beta\beta_{0\nu}$). Once the constraints from $\beta\beta_{0\nu}$ are taken into account, we have found that $e^-e^- \to W^-W^-$ is unobservable at an linear collider of $\sqrt{s} < 2$ TeV within the context of the standard model plus an extra Majorana neutrino . It is possible to evade the constraints, but this requires models which are extremely contrived and fine-tuned. In order to evade the constraint one needs at least three neutrinos mixing with the electron neutrino, the mixing of some heavier neutrinos larger than the one of the lighter ones, and at least one neutrino with an opposite intrinsic CP. Then for some values of the parameters $e^-e^- \to W^-W^-$ is observable. However in general the best way to probe the Majorana nature of the neutrino is through the single direct production process, if kinematically accessible.

Going beyond the standard model one can also find processes with lepton number violation mediated by Majorana neutrinos or other particles.

In most cases those involve the production of a new particle, such as a right-handed W or a chargino. Only in the case of composite Majorana neutrinos can one evade easily the constraint from $\beta\beta_{0\nu}$ and get a large cross-section in the inverse process.

Other $\Delta L = 2$ process mediated by ν_μ or ν_τ are not subject to the $\beta\beta_{0\nu}$ constraint. However even in this case subprocesses involving the μ or τ sector still require colliders of very high energies unless a $\mu^-\mu^-$ collider exists. In conclusion, the only way to find evidence of Majorana neutrinos in the standard model at a lepton collider of energies below 2 TeV without appealing to very specific models of Majorana neutrino masses would be at a $\mu^-\mu^-$ collider.

References

[1] T.G. Rizzo, *Phys. Lett.* **116B** (1982) 23.

[2] D. London, G. Bélanger and J.N. Ng, *Phys. Lett.* **188B** (1987) 155.

[3] D.A. Dicus, D.D. Karatas and P. Roy, *Phys. Rev.* **D44** (1991) 2033.

[4] J. Maalampi, A. Pietilä and J. Vuori, *Phys. Lett.* **297B** (1992) 327.

[5] T.G. Rizzo, *Phys. Rev.* **D50** (1994) 5602.

[6] C.A. Heusch and P. Minkowski, *Nucl. Phys.* **B416** (1994) 3.

[7] G. Bélanger , F. Boudjema, D. London,H. Nadeau, *Phys. Rev.* **D53** (1996) 6292.

[8] Heidelberg-Moscow Collaboration, *Phys. Lett.* **407B** (1997) 219.

[9] ECFA Study on Physics and Detectors for a linear electron-positron Collider.

[10] J. Gluza and M. Zrałek,*Phys. Lett.* **362B** (1995) 259, J. Gluza and M. Zrałek,*Phys. Lett.* **362B** (1995) 148.

[11] J. Gluza and M. Zrałek, *Phys. Rev.* **D55** (1997) 7030.

[12] M. Hirsch, H. V. Klapdor-Kleingrothaus, S. Kolb and S. G. Kovalenko, *Phys. Rev.* **D57** (1998) 2020.

[13] T. G. Rizzo, hep-ph/9510349, Oct. 95.

[14] I.F. Ginzburg, G.L. Kotkin, V.G. Serbo and V.I. Telnov, *Pis'ma ZhETF* **34** (1981) 514; *Sov. Yad. Fiz.* **38** (1983) 372; *Nucl. Instr. Methods* **205** (1983) 47; I.F. Ginzburg, G.L. Kotkin, S.L. Panfil, V.G. Serbo and V.I. Telnov, *Sov. Yad. Fiz.* **38** (1983) 1021; *Nucl. Instr. Methods* **219** (1984) 5.

[15] For a discussion of $\mu^-\mu^-$ colliders, see C.A. Heusch and F. Cuypers, hep-ph/9508230, August 1995.

[16] E. Nardi, E. Roulet and D. Tommasini, *Phys. Lett.* **344B** (1995) 225.

[17] For the formalism of fermion mixing, see P. Langacker and D. London, *Phys. Rev.* **D38** (1988) 886.

[18] *Review of Particle Properties,* Particle Data Group, *Phys. Rev.* **D50** (1994) 1173, Part I.

[19] For a review, see, for example, M. Moe and P. Vogel, *Ann. Rev. Nucl. Part. Sci.* **44**, (1994) 247.

[20] A. Staudt, K. Muto and H. V. Klapdor-Kleingrothaus, Europhys. Lett., **13** (1990) 31.

[21] For discussions of the contribution of heavy Majorana neutrinos to $\beta\beta_{0\nu}$, see, for example, J.D. Vergados, *Phys. Rep.* **C133** (1986) 1; R.N. Mohapatra and P.B Pal, *Massive Neutrinos in Physics and Astrophysics*, World Scientific Lecture Notes in Physics – Vol 41, Singapore, 1991, p. 204; G. Pantis, A. Faessler, W.A. Kaminski and J.D. Vergados, *J. Phys. G: Nucl. Part. Phys.* **18** (1992) 605.

[22] G. Pantis *et al*, Ref. [21].

[23] H. V. Klapdor-Kleingrothaus, these proceedings.

[24] H. V. Klapdor-Kleingrothaus, hep-ex/9802007.

[25] W. Buchmuller and C. Greub, *Nucl. Phys.* **B363** (1991) 345.

[26] J. Gluza, J.Maalampi, M. Raidal and M. Zrałek, *Phys. Lett.* **407B** (1997) 45.

[27] Aleph Collaboration, *Phys. Rep.* **C216** (1992) 253.

[28] O. Panella and Y. N. Srivastava, *Phys. Rev.* **D52** (1995) 5308.

[29] R. Kinyua and E. Takasugi, hep-ph/9804245.

[30] For a recent review on neutrinoless double beta decay which also discusses supersymmetric models with R-parity violation, see R. N. Mohapatra, University of Maryland Preprint, UMD-PP-95-147, hep-ph/9507234, July 95.

[31] M. Hirsch, H. V. Klapdor-Kleingrothaus and S. G. Kovalenko, *Phys. Lett.* **352B** (1995) 1, *ibid.*, *Phys. Lett.* **372B** (1996) 181, *ibid.*, *Phys. Rev.* **D53** (1996) 1329.

[32] M. Hirsch, H. V. Klapdor-Kleingrothaus and S. G. Kovalenko, *Phys. Lett.* **398B** (1997) 311

2.4.7 Compositeness

and Double Beta Decay

PHYSICAL REVIEW D VOLUME 56, NUMBER 9 1 NOVEMBER 1997

Neutrinoless double β decay with composite neutrinos

O. Panella,[1,2,*] C. Carimalo,[2] Y. N. Srivastava,[1,3] and A. Widom[3]
[1]*Istituto Nazionale di Fisica Nucleare, Sezione di Perugia, Via A. Pascoli, I-06123 Perugia, Italy*
[2]*Laboratoire de Physique Corpusculaire, Collège de France, 11 Place Marcelin Berthelot, F-75231, Paris Cedex 05, France*
[3]*Physics Department, Northeastern University, Boston, Massachusetts 02115*
(Received 18 December 1996)

We study in detail the contribution of heavy composite Majorana neutrinos to neutrinoless double beta decay ($0\nu\beta\beta$). Our analysis confirms the result of a previous estimate by two of the authors. Excited neutrinos couple to the electroweak gauge bosons through a magnetic-type effective Lagrangian. The relevant nuclear matrix element is related to matrix elements available in the literature and current bounds on the half-life of $0\nu\beta\beta$ are converted into bounds on the compositeness scale and/or the heavy neutrino mass. Our bounds are of the same order of magnitude as those available from accelerator experiments. [S0556-2821(97)03919-2]

PACS number(s): 12.60.Rc, 13.15.+g, 14.80.Mz, 23.40.Bw

I. INTRODUCTION

Neutrinoless double beta decay ($0\nu\beta\beta$) (see Fig. 1) is certainly one of the more interesting nonaccelerator processes that are presently being sought. The interest in this process stems from the fact that its observation would undoubtedly signal lepton number violation and at the same time would shed light onto the nature of the neutrino, one of the most elusive elementary particles. For these reasons it has received considerable attention from both the nuclear and the particle physics community [1].

In the standard model, $0\nu\beta\beta$ can only proceed if the neutrino is of Majorana type and has a *nonzero* mass. A number of mechanisms studied in models beyond the standard electroweak theory [2] have verified that neutrinoless double β decay is a very good probe of physics beyond the standard model. The experimental lower bound on the lifetime of the decay has been used to obtain constraints on the scale of new physics.

Recent work along these lines include (i) an investigation of supersymmetric contributions from the R-parity-violating minimal supersymmetric standard model [3] shows that constraints on parameters of the model from nonobservation of $0\nu\beta\beta$ are stronger than those available from accelerator experiments, (ii) a detailed analysis of the contribution from left-right symmetric models [4], and (iii) a study of the effective low-energy charged current lepton quark interactions due to the exchange of heavy leptoquarks [5]. The phenomenology of Majoron models has also been studied in detail [6].

Panella and Srivastava [7] showed that the compositeness scenario can give an additional contribution to $0\nu\beta\beta$ and derived bounds on the compositeness parameters from the nonobservation of $0\nu\beta\beta$. They explored phenomenologically the idea that the excited state of an ordinary neutrino might be a heavy Majorana neutral particle with a mass M_N ranging from a few hundred GeV up to 1 TeV. However, the nuclear aspect of the calculation was treated only approximately: Time ordering of the hadronic charged current was neglected and an upper bound for the nuclear matrix element was used in deriving constraints on the compositeness parameters. Use of the Heidelberg-Moscow $\beta\beta$ experiment lower bound [8] on the half-life of the decay $^{76}\mathrm{Ge}\to{}^{76}\mathrm{Se}+2e^-$ yielded the following constraint[1] on the scale (Λ_c), the heavy neutrino mass (M_N), and the dimensionless coupling constant f [7]:

$$|f|\leqslant 3.9\,\frac{\Lambda_c}{1\ \mathrm{TeV}}\left(\frac{M_N}{1\ \mathrm{TeV}}\right)^{1/2}. \tag{1}$$

Apart from the obvious desire to improve on the abovementioned approximations, the main motivation for the present work is twofold. First, after the completion of the work of Ref. [7], a related work by Takasugi [9] appeared, who considered the same problem ($0\nu\beta\beta$ via the exchange of a heavy composite Majorana neutral particle) but arrived at quite different conclusions [9]:[2]

$$|f|\leqslant 5.8\times 10^{-3}\,\frac{\Lambda_c}{1\ \mathrm{TeV}}\left(\frac{1\ \mathrm{TeV}}{M_N}\right)^{1/2}. \tag{2}$$

In view of this discrepancy it is, of course, mandatory to investigate further the problem in order to understand if reliable constraints on the compositeness scale from $0\nu\beta\beta$ are given by Eq. (1) or by Eq. (2).

Second, there is a recent claim by the Collider Detector at Fermilab (CDF) Collaboration of a possible signal of compositeness in high-energy proton-proton collisions [10]: The measured cross section for large transverse energy jets is significantly higher than predictions based on perturbative QCD calculations to $O(\alpha_s^3)$; a compositeness scale of $\Lambda_c=1.6$ TeV is suggested by the CDF study. Were this claim to withstand further data and analysis (such as angular distribution of dijets presently under way), the interest in new physics effects arising from a composite scenario will undoubtly increase enormously. If so, low-energy processes,

*Author to whom correspondence should be addressed. Electronic address: o.panella@pg.infn.it

[1]The numerical value used in Ref. [7] was $T_{1/2}^{0\nu\beta\beta}\geqslant 5.1\times 10^{24}$ yr.

[2]The numerical value used in Ref. [9] was $T_{1/2}^{0\nu\beta\beta}\geqslant 5.6\times 10^{24}$ yr.

PHYSICAL REVIEW D, VOLUME 62, 015013

Production of like sign dileptons in *p-p* collisions through composite Majorana neutrinos

O. Panella*
Istituto Nazionale di Fisica Nucleare, Sezione di Perugia, Via A. Pascoli, I-06123 Perugia, Italy

C. Carimalo
Laboratoire de Physique Nucléaire et de Hautes Energies, IN2P3-CNRS Universités Paris VI/VII, 4 place Jussieu, F-75252, Paris Cedex 05, France

Y. N. Srivastava
Dipartimento di Fisica dell' Università e INFN, Sezione di Perugia, Via A. Pascoli, I-06123 Perugia, Italy and Northeastern University, Physics Department, Boston, Massachusetts 02155

(Received 8 March 1999; revised manuscript received 4 February 2000; published 12 June 2000)

The production of like-sign dileptons (LSD) in the high-energy lepton-number-violating ($\Delta L=+2$) reaction $pp\rightarrow 2$ jets$+l^+l^+$ $(l=e,\mu,\tau)$, of interest for experiments to be performed at the forthcoming CERN Large Hadron Collider (LHC), is investigated in detail, taking up a composite model scenario in which the exchanged virtual *composite* neutrino is assumed to be a Majorana particle that couples to the light leptons via the $SU(2)\times U(1)$ gauge bosons through a magnetic type coupling ($\sigma_{\mu\nu}$). A helicity projection method is used to evaluate exactly the tree-level amplitudes of the contributing parton subprocesses ($2\rightarrow 4$), which allows one to take into account all exchange diagrams and occurring interferences. Numerical estimates of the corresponding signal cross section that implement kinematical cuts needed to suppress the standard model background are presented which show that in some regions of the parameter space the total number of LSD events is well above the background. Assuming nonobservation of the LSD signal it is found that LHC would exclude a composite Majorana neutrino up to ≈ 900 GeV (if one requires 10 events for discovery). The sensitivity of LHC experiments to the parameter space is then compared to that of the next generation of neutrinoless double beta decay ($\beta\beta_{0\nu}$) experiment, GENIUS, and it is shown that they will provide constraints of the same order of magnitude and will play a complementary role.

PACS number(s): 12.60.Rc, 13.15.+g, 13.85.Rm, 14.60.St

I. INTRODUCTION

Since the discovery of the Z^0 and $W^{\pm}$ gauge bosons [1] the standard model (SM) of electroweak interactions [2] based on the $SU(2)\times U(1)$ gauge group has scored an impressive record of experimental checks. However, some unexplained facts of the model, such as the mass hierarchy, the proliferation of elementary particles, and the total number of free parameters, have lead to the belief that it is only a low-energy manifestation of a yet unknown underlying fundamental theory, which would be free of the above theoretical difficulties. Therefore despite the enormous experimental success of the SM many alternative theories have been developed such as left-right symmetric models, composite models, supersymmetry, string theory, and grand unified models. The investigation of the effects predicted by the new theories that are absent in the standard theory is therefore very important since, were these effects to be experimentally observed, they would signal *new physics* unaccounted for by the SM. It is in this direction that a great portion of recent theoretical and experimental studies have been concentrated [3], and this is indeed the spirit of this work which deals with lepton-number-violating processes.

The conservation of the total lepton number (L) is one of the symmetries of the SM experimentally observed to hold true until now. In the SM with massless Dirac neutrinos processes with $\Delta L\neq 0$ are not possible. Violation of this symmetry is generally related to the existence of massive Majorana particles and many extensions of the SM contain L-violating interactions involving Majorana neutrinos. Left-right symmetric models for example contain right-handed Majorana neutrinos, with a mass that could be in the TeV range, and coupled to the light leptons via the right-handed gauge bosons (W_R,Z_R) [4]. Superstring generated E_6 models also have neutral Majorana leptons [5]. Finally Ref. [6] provides an example of a composite model with Majorana neutrals.

The effect which seems most promising with respect to showing violations of the lepton number is the neutrinoless double beta decay ($\beta\beta_{0\nu}$), a second order process where, in a nucleus, two protons (neutrons) undergo simultaneously a weak beta decay emitting two positrons (electrons) while the two neutrinos annihilate into the vacuum [3]:

$$A(Z+2)\rightarrow A(Z)+e^+e^+, \quad \Delta L=+2. \tag{1}$$

This process is only possible if the neutrino is a massive Majorana particle, and thus it is impossible within the SM. Experiments that search for such rare decay have long since been performed but always with negative results [7]. Currently the Heidelberg-Moscow $\beta\beta$ experiment at the Gran-

*Author to whom correspondence should be addressed. Electronic address: Orlando.Panella@PG.infn.it

Sasso laboratory in Italy provides the best experimental lower bound on the half-life of the process [8]

$$^{76}\text{Ge} \to {}^{76}\text{Se} + 2\,e^-,$$
$$T_{1/2}^{\beta\beta_{0\nu}} > 5.7 \times 10^{25}\ \text{yr}. \tag{2}$$

The proposed GENIUS double beta experiment (see Sec. VI), now under development, will either increase the lower bound on the half-life by two or three orders of magnitude or observe the decay. From the theoretical point of view, the strong bound on the half-life in Eq. (2) has been turned into a powerful tool to impose constraints on models of new physics which predict a nonzero amplitude for the $\beta\beta_{0\nu}$ decay [9]. Studies in this direction include: an investigation of new super-symmetric contributions from R-parity violating minimal supersymmetric standard model (MSSM) [10] which shows how constraints on parameters of the model from nonobservation of $\beta\beta_{0\nu}$ are stronger than those available from accelerator experiments; a detailed analysis of the contribution to $\beta\beta_{0\nu}$ from left-right symmetric models [11]; a study of the effective low-energy charged current lepton-quark interactions due to the exchange of heavy leptoquarks [12].

The present authors have, in a series of recent papers [13–15], investigated the contribution, to the neutrinoless double beta decay, of a heavy Majorana neutrino, arising from a composite model scenario in which the excited partner of the neutrino (the excited neutrino, $\nu^\star$) is assumed to be a Majorana particle. This study revealed that $\beta\beta_{0\nu}$ constraints are competitive, and in some regions of the parameter space, even more restrictive than those derived from high-energy direct search of excited particles [15,16]. This result led to consider the potential of the experiments to be performed at the forthcoming Large Hadron Collider (LHC) at CERN, with respect to the possibility of observing the production of like-sign dileptons (LSD) l^+l^+ or l^-l^-, $l = e,\mu,\tau$, in proton-proton collisions with an energy of 14 TeV in the center of mass frame

$$pp \to 2\text{jets} + \text{LSD}, \quad \Delta L = +2. \tag{3}$$

In hadronic collisions LSD can be produced in quark-quark (antiquark-antiquark) scattering, through the elementary subprocess $W^+W^+ \to l^+l^+$ (virtual W-boson fusion) as depicted in Fig. 1 where the dashed blob represents all contributing diagrams within a given model. As regards this mechanism of LSD production one can say that it is the high-energy analogue of the neutrinoless double beta decay which indeed proceeds through the same Feynman diagrams (see, for example, Ref. [15]). Figure 2(a) shows explicitly the Feynman diagram for the production of LSD through the exchange of a heavy Majorana neutrino (basic mechanism). In the case of quark-antiquark scattering in addition to the W fusion another mechanism must be considered that leads to LSD production: the direct production of a heavy Majorana neutrino via quark-antiquark annihilation $q\bar{q}' \to l^+N$ with the subsequent decay of the heavy neutrino $N \to l^+q\bar{q}'$ (annihilation mechanism). This is depicted in Fig. 2(b).

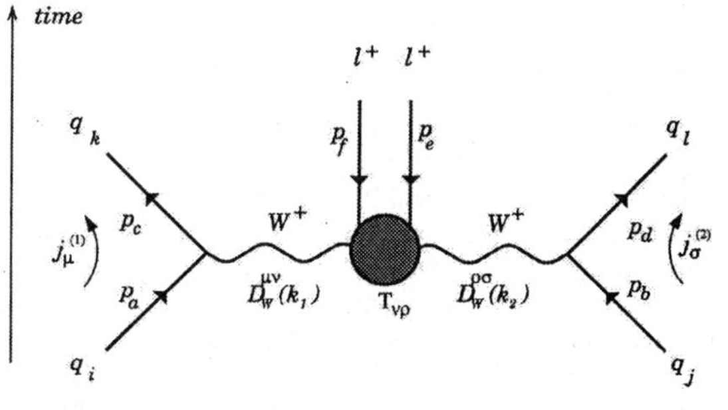

FIG. 1. Parton level mechanism for production of like-sign-dileptons (LSD) in high-energy hadronic collisions. The shaded blob contains all contributing diagrams for the virtual subprocess $W^+W^+ \to l^+l^+$.

Production of LSD has been considered in the past by several authors and within the context of different models. In the left-right symmetric model, Keung and Senjanović [17] already in 1983 realized that the associate production of LSD with two hadronic jets would signal the annihilation of quark-antiquark pairs into the right-handed gauge boson of

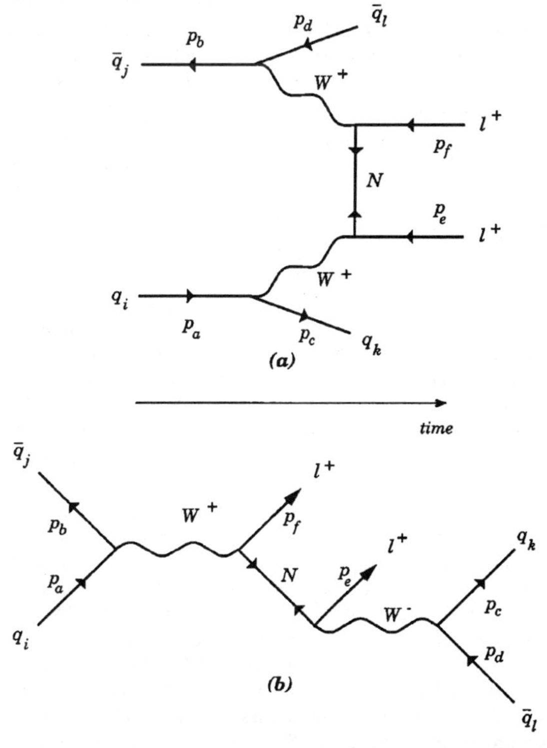

FIG. 2. Production of LSD through quark-antiquark scattering. There are here two interfering mechanisms to be considered (a) virtual W fusion, (b) l^+N_l production via quark-antiquark annihilation with subsequent hadronic decay of the heavy neutrino $N_l \to l^+q\bar{q}$.

the model (W_R). Estimates were given for pp collisions at $\sqrt{s}=800$ GeV. The study of this model was later taken up to higher energies [Superconducting Super Collider (SSC) and LHC] by Datta, Guchait, and Roy in Ref. [18] where the authors indicated how to effectively reduce the SM background. Dicus, Karatas, and Roy [19] have studied LSD production at high-energy hadron colliders through the exchange of heavy right-handed Majorana neutrinos, without commitment to a specific model (beyond the SM). They used a γ_μ-type coupling and found the LSD signal detectable at the SSC while at the LHC the SM background would probably preclude detection. Two of the present authors [20] provided a rough estimate of the signal cross sections for $pp \to 2\text{jets}+\text{LSD}$ at LHC within the context of composite models (exchange of a heavy composite Majorana neutrino with a $\sigma_{\mu\nu}$-type coupling) using an equivalent W-boson approximation [21] (similar to the Weiszäcker-Williams approximation for the photon field) and integrating over the complete phase-space of the subprocess $W^+W^+ \to l^+l^+$. The result was that the signal could be observable at the LHC.

One remark should be made at this point that applies to all works just cited that have investigated LSD production in pp collisions. None of them deals, *at the same time*, with the two mechanisms of LSD production, i.e., W^+W^+ fusion and $q\bar{q}'$ annihilation. Indeed when dealing with $q\bar{q}'$ scattering both mechanisms must be considered and the corresponding amplitudes should be added coherently. In order to do so one needs a way of efficiently computing the amplitudes. In this paper it is done precisely so, calculating analytically the helicity amplitudes of the occurring tree-level diagrams and accounting thus for the interference term between the W^+W^+ fusion and the $q\bar{q}'$ annihilation.

Thus the goal of this paper is twofold: (i) to address the sensitivity of LHC experiments with respect to the parameters of the composite model effective Lagrangian and compare this to that of the next generation of double beta decay experiments now under development (GENIUS); (ii) to present a calculation of LSD production in pp collisions (via the exchange of a heavy composite Majorana neutrino) which goes beyond the approximations of Ref. [20] and which, in the case of $q\bar{q}'$ annihilation, includes coherently the two competing mechanisms.

The rest of the paper is organized as follows. In Sec. II the reader is briefly reminded of the effective Lagrangian describing the coupling of the excited neutrino with the electron and a comparison between recent bounds on the parameters from the low-energy $\beta\beta_{0\nu}$ experiment by the Heidelberg-Moscow Collaboration and those from high energy experiments performed by the DELPHI Collaboration at the CERN Large Electron Positron (LEP) Collider is presented. In Sec. III the amplitudes of the L-violating parton subprocesses are presented. Section IV contains (i) a description of the kinematical cuts applied with a short discussion of the background; (ii) our numerical results for the signal cross sections. In Sec. V the sensitivity to the parameter space of LHC is compared to that of GENIUS. Finally, Sec. VI contains the conclusions.

II. COMPOSITENESS AND EXISTING $\beta\beta_{0\nu}$ CONSTRAINTS

It is well known that one possible scenario of physics beyond the SM is one in which quarks and leptons are not elementary particles but possess an internal structure; i.e., they are bound states of, yet unknown, new constituents, generally referred to as *preons*, bound together by a new dynamical interaction. Theories that follow this path are called *composite models* and although many have been proposed [22] none has emerged as a new dynamically consistent theory. However, there are some model-independent consequences of the idea of compositeness which can be addressed without commitment to any specific model. These are (i) contact interactions between ordinary fermions, (ii) the existence of excited partners for quarks and leptons with masses of the order of the compositeness scale Λ_c. Phenomenologically these ideas have been studied via effective interactions [14,23]. In particular in this work, the case of excited neutrinos (N), is taken up and only the relevant coupling with the light electron are reviewed. Effective couplings between the heavy and light leptons (or quarks) have been proposed, using weak isospin (I_W) and hypercharge (Y) conservation [24]. Assuming that such states are grouped in $SU(2)\times U(1)$ multiplets, since light fermions have $I_W = 0,1/2$ and electroweak gauge bosons have $I_W=0,1$, only multiplets with $I_W \leqslant 3/2$ can be excited in the lowest order perturbation theory. Also, since none of the gauge fields carry hypercharge, a given excited multiplet can couple only to a light multiplet with the same Y. Conservation of the electromagnetic current forces the transition coupling of heavy-to-light fermions to be of the magnetic moment type respect to any electroweak gauge bosons [24]. In fact, a vector (γ_μ) transition coupling between the ordinary electron e and its excited partner E mediated by the $\vec{W}^\mu$ and B^μ gauge fields, would result in an electromagnetic current of the type $j^\mu_{\text{e.m.}} \approx \bar{\psi}_E \gamma^\mu \psi_e$ which would not be conserved due to the different masses of excited and ordinary fermions (as it is expected that $m_E \gg m_e$). The $SU(2)\times U(1)$ symmetry forces then the tensor structure ($\sigma_{\mu\nu}$) to the transition coupling also in the charged and neutral weak currents.

Restrict here to the first family and consider spin-1/2 excited states grouped in multiplets with $I_W = 1/2$ and $Y=-1$ (the so called homodoublet model [23]),

$$L=\begin{pmatrix} N \\ E \end{pmatrix} \tag{4}$$

which can couple to the light left-handed multiplet

$$l_L=\begin{pmatrix} \nu_L \\ e_L \end{pmatrix} = \frac{1-\gamma_5}{2}\begin{pmatrix} \nu \\ e \end{pmatrix} \tag{5}$$

through the gauge fields $\vec{W}^\mu$ and B^μ. The relevant interaction is written [24] in terms of two *new* independent coupling constants f and f':

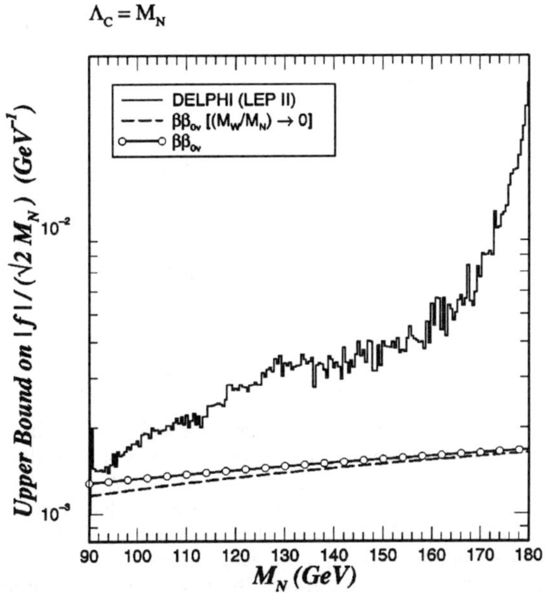

FIG. 3. Comparison between the $\beta\beta_{0\nu}$ and the LEP II upper bound on the quantity $|f|/(\sqrt{2}M_N)$ as a function of the heavy neutrino mass M_N, with the choice $\Lambda_c = M_N$. Regions above the curves are excluded. The dashed curve is the $\beta\beta_{0\nu}$ bound of Eq. (10), while the solid-circle curve includes numerical effects of terms of higher order in M_W/M_N as discussed in Ref. [16].

$$\mathcal{L}_{int} = \frac{gf}{\Lambda_c}\bar{L}\sigma_{\mu\nu}\frac{\vec{\tau}}{2}l_L \cdot \partial^\nu \vec{W}^\mu + \frac{g'f'}{\Lambda_c}\left(-\frac{1}{2}\bar{L}\sigma_{\mu\nu}l_L\right)\cdot\partial^\nu B^\mu + \text{H.c.}, \tag{6}$$

where $\vec{\tau}$ are the Pauli SU(2) matrices, g and g' are the usual SU(2) and U(1) gauge coupling constants, and the factor of $-1/2$ in the second term is the hypercharge of the U(1) current. This effective Lagrangian is widely used in the literature to predict production cross sections and decay rates of the excited particles [23,25,26]. In terms of the physical gauge fields the interaction Lagrangian describing the coupling of the heavy excited neutrino with the light electron is therefore

$$\mathcal{L}_{\text{eff}} = \left(\frac{gf}{\sqrt{2}\Lambda_c}\right)\left\{\left(\bar{N}\sigma^{\mu\nu}\frac{1-\gamma_5}{2}e\right)\partial_\nu W^+_\mu\right\} + \text{H.c.} \tag{7}$$

In the analysis carried out in Refs. [13,14] it was assumed that the excited neutrino is a Majorana particle with mass M_N, expected to be of the order of the compositeness scale Λ_c, which would then contribute to the neutrinoless double beta decay.

It should be emphasized how this scenario differs considerably from the more usual one of the left-right symmetric model where a SU(2) singlet Majorana neutrino couples to the charged leptons via a right-handed W gauge boson (W_R) and ($V+A$) coupling [27]. Here (and in Refs. [13,14]) instead the heavy composite Majorana neutrino belongs to a SU(2) doublet [see Eq. (4)] and interacts with the charged lepton via the standard model W gauge boson (left handed) with a tensor coupling $[\sigma_{\mu\nu}(1-\gamma_5)]$.

In Ref. [14] the half-life of the $\beta\beta_{0\nu}$ mediated by a heavy composite Majorana neutrino (coupling to the charged lepton via the SM W gauge boson and with tensor coupling) was calculated and found to be given by

$$T_{1/2}^{-1} = \left(\frac{f}{\Lambda_c}\right)^4 \frac{m_A^8}{M_N^2}|\mathcal{M}_{FI}|^2\frac{G_{01}}{m_e^2}, \tag{8}$$

where $m_A = 0.85$ GeV is a parameter entering the nuclear form factors, $\mathcal{M}_{FI} = -5.45\times10^{-2}$ is a nuclear matrix element, m_e is the electron mass, and $G_{01} = 6.4\times10^{-15}\ \text{yr}^{-1}$ is a phase space integral. Combining this result with the non-observation of the decay ($T_{1/2} > T_{1/2}^{\text{lower bound}}$) one obtains a constraint on the parameters of the model

$$\left|\frac{f}{\Lambda_c}\right| < M_N^{1/2}\left(\frac{m_e^2}{m_A^8}\right)^{1/4}\frac{[G_{01}T_{1/2}^{\text{lower bound}}]^{-1/4}}{|\mathcal{M}_{FI}|^{1/2}}. \tag{9}$$

Using the current experimental lower bound on the half-life of the ^{76}Ge decay provided by the Heidelberg-Moscow $\beta\beta$ experiment, the following constraint on the parameters f, Λ_c, M_N appearing in Eq. (7) is deduced:[1]

$$|f| \le 8.03\frac{\Lambda_c}{1\ \text{TeV}}\left(\frac{M_N}{1\ \text{TeV}}\right)^{1/2}. \tag{10}$$

This double beta bound on compositeness can be compared with bounds on the same parameters from high-energy experiments performed at the Large Electron Positron (LEP) collider, phase II. The DELPHI Collaboration has reported [28] on a search for excited leptons in e^+e^- collisions at $\sqrt{s} = 183$ GeV, where both the single and double production mode were studied. It should be emphasized that the analysis in Ref. [28] was carried out using the same effective Lagrangian that was considered in Refs. [13,15], cf. Eq. (7), so that it makes sense to compare the corresponding bounds. In Fig. 3 the bound of Eq. (10) is plotted against the exclusion curve of the DELPHI Collaboration [28], and one can see that for masses above ≈90 GeV the double beta bound is more constraining, i.e., it excludes a portion of parameter phase space still allowed by the DELPHI exclusion plot.[2]

---

[1]This is an updated constraint respect to that of Ref. [15] where a previous value of the half-life was used.

[2]It should be noted that the ALEPH Collaboration has also recently published results of a search for compositeness at LEP I. In Ref. [29] bounds on the compositeness scale, in particular regarding the same excited neutrino couplings discussed here, are reported. Choosing $f = f' = 1$ a neutrino mass-dependent lower bound on Λ_c is found which is about 16 TeV at $M_N = \mathcal{O}(10\ \text{GeV})$ while it drops down to 4 TeV at the maximum value of M_N explored of 80 GeV. This result is not directly comparable to Eq. (10) since this was derived within the hypothesis $M_N >> M_W$ [15]. Assuming $|f| = 1$, Eq. (10) gives $\Lambda_c \geq 0.12$ TeV at $M_N = 1$ TeV.

This result prompted the present authors to study the potential of the LHC with respect to the same type of lepton-number-violating processes, with an emphasis on comparing its sensitivity with that of the next generation of double beta decay experiments.

One comment is finally due on the the related issue of the detectability of the charged partner of the heavy composite Majorana neutrino, the so called excited electron E. The heavy electron E is currently being searched for in experiments performed at high-energy facilities which look for its direct production and/or indirect effects in electron-positron colliders (LEP at CERN) and/or electron-proton colliders (ep collider HERA at DESY). The most recent and stringent bounds on the mass of the heavy electron E is from studies of indirect (propagator) effects in $e^+e^-\to\gamma\gamma$ where the E is exchanged in the t channel. The ALEPH Collaboration reports $m_E>250$ GeV [30]. The single production in electron-proton collisions ($ep\to E+X$), gives also bounds of the same order of magnitude: $m_E>200$ GeV [31]. As regards the production of excited electrons E at hadron colliders it was realized some time ago [32] that these can be copiously produced via contact interactions (CTs), arising from a new *strong* preon dynamics, as opposed to the gauge interactions (G) being discussed here. One process that could be looked upon at the LHC to estimate E production through gauge interactions is $u\bar{d}\to W^*\to \nu E$ with the subsequent decay $E\to e\gamma$. This would give a signature $pp\to e\nu\gamma+X$ whose cross section is expected to be of the same order of magnitude of those described here. However, this goes beyond the scope of the present work, and should be the object of further investigation.

The following section deals with the calculation of the lepton number violating processes in pp collisions described by the diagrams of Figs. 1 and 2. They have been carried out with a choice of the parameters that satisfies the bounds from $\beta\beta_{0\nu}$ just discussed.

III. AMPLITUDES OF L-VIOLATING PARTON SUBPROCESSES

In the following the helicity amplitudes for parton subprocesses that contribute to production of LSD via the exchange (or production) of a heavy Majorana composite neutrino are presented. The effective interaction used is that of Eq. (7). Considering for the moment only the first family, three different types of processes should be distinguished:

$$\begin{aligned}&(\mathrm{i})\ uu\to dd+l^+l^+,\\&(\mathrm{ii})\ u\bar{d}\to d\bar{u}+l^+l^+,\\&(\mathrm{iii})\ \bar{d}\bar{d}\to\bar{u}\bar{u}+l^+l^+.\end{aligned}\tag{11}$$

The amplitudes are written using the following definitions of propagator factors:

$$\begin{aligned}1/A&=[(p_a-p_c)^2-M_W^2][(p_b-p_d)^2-M_W^2],\\1/B&=[(p_a-p_d)^2-M_W^2][(p_b-p_c)^2-M_W^2],\end{aligned}\tag{12}$$

$$\begin{aligned}1/\tilde{A}&=[(p_a+p_b)^2-M_W^2+iM_W\Gamma_W]\\&\quad\times[(p_c+p_d)^2-M_W^2+iM_W\Gamma_W],\\C&=(p_a-p_c-p_e)^2-M_N^2,\\D&=(p_a-p_c-p_f)^2-M_N^2,\\E&=(p_a-p_d-p_e)^2-M_N^2,\\F&=(p_a-p_d-p_f)^2-M_N^2,\\\tilde{C}&=(p_c+p_d+p_f)^2-M_N^2+iM_N\Gamma_N,\\\tilde{D}&=(p_c+p_d+p_e)^2-M_N^2+iM_N\Gamma_N.\end{aligned}\tag{13}$$

The width of the heavy composite neutrino Γ_N is of course a quantity which depends on the free parameters of the particular model that is being considered here, $|f|$, Λ_c, and M_N, and has been the object of discussion in the literature [32,26]. Typically the width of excited leptons (quarks) receives contributions from the gauge interactions of Eq. (7) and from contact terms arising from novel *strong* preon interactions [26].[3] In order to keep the numerical computations of cross-sections presented in the following reasonably simple, a constant value of $\Gamma_N=70$ GeV has been adopted, which is a somewhat average value in the mass range considered.

Define also the quantities

$$\begin{aligned}s(m,n)&=s(p_m,p_n)=\bar{u}_+(p_m)u_-(p_n),\\t(m,n)&=t(p_m,p_n)=\bar{u}_-(p_m)u_+(p_n),\end{aligned}\tag{14}$$

which are given by

$$\begin{aligned}s(m,n)&=-2\sqrt{E_mE_n}G_{mn},\\t(m,n)&=+2\sqrt{E_mE_n}F_{mn},\end{aligned}\tag{15}$$

with

$$\begin{aligned}G_{mn}&=\cos(\theta_m/2)\sin(\theta_n/2)\,e^{+i(\phi_m-\phi_n)/2}\\&\quad-\sin(\theta_m/2)\cos(\theta_n/2)\,e^{-i(\phi_m-\phi_n)/2},\\F_{mn}&=(G_{mn})^*.\end{aligned}\tag{16}$$

Let the tensor $T_{\mu\nu}$ describe the virtual sub-process $W^*W^*\to l^+l^+$ [Fig. 2(a)], while the tensor $\tilde{T}_{\mu\nu}$ describes the virtual subprocess $(W^*)^+\to l^+l^+(W^*)^-$ appearing in the diagram of Fig. 2(b). $J_{a,c}$ and $\bar{J}_{b,d}$, are the quark (antiquark) currents that couple in the t channel to the virtual gauge bosons of the standard model [Fig. 2(a)] while $\tilde{J}_{a,b}$ and $\tilde{J}^*_{c,d}$, are the in-

[3]It is to be noted, however, that these contact terms while contributing to the total width of the excited neutrino cannot contribute to the production of LSD via the diagrams discussed in this work.

coming and outgoing currents of the $q\bar{q}'$ pair that couples in the s channel to the W bosons [Fig. 2(b)].

$$J^{\mu}_{a,c}=\bar{u}(p_c)\,\gamma^{\mu}\frac{1-\gamma_5}{2}u(p_a),$$

$$\mathcal{J}^{\mu}_{b,d}=\bar{v}(p_b)\,\gamma^{\mu}\frac{1-\gamma_5}{2}v(p_d),$$

$$\bar{\mathcal{J}}^{\mu}_{a,b}=\bar{v}(p_b)\,\gamma^{\mu}\frac{1-\gamma_5}{2}u(p_a),$$

$$(\bar{\mathcal{J}}^{\mu}_{c,d})^{*}=\bar{u}(p_c)\,\gamma^{\mu}\frac{1-\gamma_5}{2}v(p_d). \tag{17}$$

The amplitudes are (unitary gauge)

(i) $U_iU_j\to D_kD_l+l^+l^+$

$$\mathcal{M}=\mathcal{K}\{V^{*}_{U_iD_k}V^{*}_{U_jD_l}A\,[J^{\mu}_{(a,c)}\,T_{\mu\nu}\,\mathcal{J}^{\nu}_{(b,d)}]$$
$$-V^{*}_{U_iD_l}V^{*}_{U_jD_k}B\,[\,(p_c\leftrightarrow p_d)\,]\}, \tag{18}$$

(ii) $U_i\bar{D}_j\to D_k\bar{U}_l+l^+l^+$

$$\mathcal{M}(WW-\text{fusion})=\mathcal{K}(V_{U_iD_k})^{*}(V_{U_lD_j})^{*}\,[A\,J^{\mu}_{(a,c)}\,T_{\mu\nu}\bar{\mathcal{J}}^{\nu}_{(b,d)}\,],$$

$$\mathcal{M}(q\bar{q}'-\text{annihilation})=\mathcal{K}(V_{U_iD_j})^{*}(V_{U_lD_k})^{*}\,[\tilde{A}\,\bar{\mathcal{J}}^{\mu}_{(a,b)}\,\tilde{T}_{\mu\nu}(\bar{\mathcal{J}}^{\nu}_{(c,d)})^{*}\,], \tag{19}$$

(iii) $\bar{D}_i\bar{D}_j\to\bar{U}_k\bar{U}_l+l^+l^+$

$$\mathcal{M}=\mathcal{K}\{V^{*}_{U_kD_i}V^{*}_{U_lD_j}A\,[\bar{\mathcal{J}}^{\mu}_{(a,c)}\,T_{\mu\nu}\,\bar{\mathcal{J}}^{\nu}_{(b,d)}\,]$$
$$-V^{*}_{U_lD_i}V^{*}_{U_kD_j}B[\,(p_c\leftrightarrow p_d)\,]\}, \tag{20}$$

where U_i denotes a positively charged quark (up-type) while D_i dentotes a negatively charged one (down-type). The quantities $V_{U_iD_j}$ are the elements of the CKM mixing matrix. Of course the annihilation diagram of Fig. 2(a) comes in only in quark-antiquark scattering. In processes (i) and (iii) the part of the amplitude depending on the factor B is due to the diagrams obtained exchanging the final state quarks. In the framework of the effective Lagrangian [see Eq. (7)] as discussed in Sec. II, it is found

$$T_{\mu\nu}=\bar{u}(p_e)\left[\frac{\sigma_{\mu\rho}\sigma_{\nu\sigma}}{C}+\frac{\sigma_{\nu\sigma}\sigma_{\mu\rho}}{D}\right]\frac{1-\gamma_5}{2}v(p_f)$$
$$\times(p_a-p_c)^{\rho}(p_b-p_d)^{\sigma},$$

$$\tilde{T}_{\mu\nu}=\bar{u}(p_e)\left[\frac{\sigma_{\mu\rho}\sigma_{\nu\sigma}}{\tilde{C}}+\frac{\sigma_{\nu\sigma}\sigma_{\mu\rho}}{\tilde{D}}\right]\frac{1-\gamma_5}{2}v(p_f)$$
$$\times(p_a+p_b)^{\rho}(p_c+p_d)^{\sigma},$$

$$\mathcal{K}=\frac{g^4}{4}\left(\frac{f}{\Lambda_c}\right)^2 M_N. \tag{21}$$

Because of the chiral nature of the couplings involved, the calculation is particularly simple if performed in the helicity basis [33]. In the massless approximation only one helicity amplitude is nonzero. The following result is found:

(i) $U_iU_j\to D_kD_l+l^+l^+$

$$\mathcal{M}=4\,\mathcal{K}\,s(a,b)\Bigg\{(V_{U_iD_k})^{*}(V_{U_jD_l})^{*}A\;t(a,c)t(d,b)$$
$$\times\left[\frac{s(e,a)s(b,f)}{C}-\frac{s(f,a)s(b,e)}{D}\right]$$
$$-(V_{U_iD_l})^{*}(V_{U_jD_k})^{*}B\;t(a,d)t(c,b)$$
$$\times\left[\frac{s(e,a)s(b,f)}{E}-\frac{s(f,a)s(b,e)}{F}\right]\Bigg\}, \tag{22}$$

(ii) $U_i\bar{D}_j\to D_k\bar{U}_l+l^+l^+$

$$(WW-\text{fusion}):\;\mathcal{M}=+4\,\mathcal{K}(V_{U_iD_k})^{*}(V_{U_lD_j})^{*}$$
$$\times A\;s(a,d)t(a,c)t(d,b)$$
$$\times\left\{\left[\frac{s(e,a)s(d,f)}{C}-\frac{s(f,a)s(d,e)}{D}\right]\right\}, \tag{23}$$

$(q\bar{q}'-\text{annihilation})$:

$$\mathcal{M}=-4\,\mathcal{K}(V_{U_iD_j})^{*}(V_{U_lD_k})^{*}\tilde{A}\;t(a,b)s(a,d)t(d,c)$$
$$\times\left\{\left[\frac{s(e,a)s(d,f)}{\tilde{C}}-\frac{s(f,a)s(d,e)}{\tilde{D}}\right]\right\}, \tag{24}$$

(iii) $\bar{D}_i\bar{D}_j\to\bar{U}_k\bar{U}_l+l^+l^+$

$$\mathcal{M}=4\,\mathcal{K}\,s(c,d)\Bigg\{(V_{U_kD_i})^{*}(V_{U_lD_j})^{*}A\;\;t(a,c)t(d,b)$$
$$\times\left[\frac{s(e,c)s(d,f)}{C}-\frac{s(f,c)s(d,e)}{D}\right]+(V_{U_lD_i})^{*}(V_{U_kD_j})^{*}$$
$$\times B\;\;t(a,d)t(c,b)\left[\frac{s(e,d)s(c,f)}{E}-\frac{s(f,d)s(c,e)}{F}\right]\Bigg\}. \tag{25}$$

The above simple analytic form of the amplitudes is also very easy to implement in a code for numerical applications, since the quantities $s(p_i,p_j)$ and $u(p_i,p_j)$ are just functions of the energies and angles of the particle's momenta; see Eqs. (15) and (16).

IV. DISCUSSION AND RESULTS

Before giving details of numerical calculations of the signal cross section and discussing the results one should be reminded that there are processes of the standard model that also lead to LSD production and are thus sources of background to the signal. This question was already considered in Refs. [18,19]. An immediate source of background comes from the subprocesses $uu \rightarrow ddW^+W^+$, $u\bar{d} \rightarrow d\bar{u}W^+W^+$, $\bar{d}\bar{d} \rightarrow \bar{u}\bar{u}W^+W^+$ and similar ones involving higher-generation quarks and antiquarks, each W subsequently decaying into $l\nu_l$. The corresponding overall reaction $pp \rightarrow 2\,\text{jets}\, l\nu_l l\nu_l$ can mimic the signal when the total missing P_T carried away by the neutrinos is small. As shown in Ref. [19], that background can be most efficiently reduced to a percent of fb in LHC conditions, which will be shown to be at the same level of the signal, in some regions of the parameter space, or even well below the latter in other regions. This background reduction is accomplished by limiting the missing P_T of neutrinos, that is, requiring a "P_T conservation" which is actually a characteristic of the signal.

As also observed in Refs. [18,19], a copious and more dangerous source of standard-model background seems to be due to $t\bar{t}$ production from gluon and quark initial states. In that process, one has the decay chains $t \rightarrow bW^+$, $W^+ \rightarrow l\nu_l$ on one side and $\bar{t} \rightarrow \bar{b}W^-$, $\bar{b} \rightarrow \bar{c}l\nu_l$, $W^- \rightarrow qq'$ on the other side. For LHC conditions, that reaction leads to a total production of about 4×10^6 LSD per year. Here again, a limitation of missing P_T together with the condition of large P_T leptons allow one to reduce substantially that background. The additional requirement of lepton isolation further reduces the background. But while the two requirements of missing P_T limitation and lepton isolation will certainly eliminate two other similar backgrounds coming from direct $c\bar{c}$ and $b\bar{b}$ production, that of $t\bar{t}$ production seems to remain, according to Refs. [18,19], at a level which might jeopardize measurement of the signal at LHC.

At this point, it is worth noticing that within the standard model one can observe in pp collisions not only events with like-sign dileptons of a given species ($e^\pm e^\pm$, $\mu^\pm\mu^\pm$, $\tau^\pm\tau^\pm$), but also events with "hybrid" like-sign dileptons (HLSD) such as $e^\pm\mu^\pm$, $e^\pm\tau^\pm$, $\mu^\pm\tau^\pm$, with practically the same production rate for all these events since the W's decay into any $l\nu_l$ final state at the same rate. Thus, one can get an idea on the amount of standard-model LSD background and eventually make appropriate subtraction by comparing, under given kinematical constraints, LSD production with HLSD production. At LHC, it would be most probably a comparison between $\mu^\pm\mu^\pm$, $\tau^\pm\tau^\pm$ production and $\mu^\pm\tau^\pm$ production. Said differently, once appropriate kinematical cuts performed, any significant difference between LSD production and HLSD production would signal lepton number violating processes such as those here considered. However, let us remark that a no-deviation result could not rule out new physics models allowing for lepton mixing.

In any case let us remark that an analysis of the background dedicated specifically to the LHC experimental conditions, and perhaps more complete than that presented in Refs. [18,19], is necessary (including in particular a detailed calculation of the amplitude of the processes involved), and will be the matter of a forthcoming work. Here the estimate of the background given in Ref. [19] is assumed:

$$\sigma_{\text{background}} = 3\times 10^{-2}\ \text{fb}. \tag{26}$$

In order to compare the signal cross-section with Eq. (26) kinematical cuts as discussed in Ref. [19] are used. The following selection criteria are needed in order to ensure lepton and jet identification:

$$\begin{aligned} |\eta_{\text{lep}}|<4 \quad & p_T(\text{lep})>5\ \text{GeV}, \\ |\eta_{\text{jet}}|<4 \quad & p_T(\text{jet})>20\ \text{GeV}. \end{aligned} \tag{27}$$

The signal cross sections are obtained by folding the square of the amplitudes with the four-particle phase-space and the parton distribution functions

$$\begin{aligned} d\sigma = \int dx_a dx_b \frac{1}{1+\delta_{ij}} [f_i(x_a,Q^2) f_j(x_b,Q^2) + x_a \leftrightarrow x_b] \\ \times \frac{1}{2\hat{s}} |\mathcal{M}|^2 (2\pi)^4 \delta^4\left(p_a + p_b - \sum_{m=1}^{4} p_m \right) \\ \times \frac{1}{2}\left(1 - \frac{\delta_{kl}}{2}\right) \prod_{n=1}^{4} \frac{d^3 p_n}{(2\pi)^3 2E_n}, \end{aligned} \tag{28}$$

where $\hat{s} = x_a x_b S$ is the squared center of mass energy of the parton collision and the factor $(1/2)(1-\delta_{kl}/2)$ accounts for the presence of the two identical fermions (l^+l^+) and the possibly identical quarks U_k, U_l $(\bar{U}_k, \bar{U}_l)$ in the final state. The distribution functions are those of set 1.1 of Duke-Owens (updated version of set 1) as described in Ref. [34] with $\Lambda_{QCD} = 177$ MeV/c. $\sqrt{S} = 14$ TeV has been used while the scale Q^2 is fixed at the value $Q^2 = \hat{s}$. With a proper choice of the transverse axis the phase-space reduces to a nine-dimensional integration that is performed with the well known VEGAS [35] routine which is based on a Monte Carlo algorithm. This allows easy implementation of kinematical cuts as described above. As regards the $u\bar{d}$ process the interference between the WW-fusion and annihilation mechanisms is naturally taken into account since the two (complex) amplitudes are summed before squaring.

In Figs. 4–6 the integrated cross section with the parameter $|f|=1$, i.e., $\sigma_1 = \sigma(|f|=1)$ is given. As $\mathcal{M} \propto f^2$, the total cross section for other values of $|f|$ can be easily recovered $(\sigma = |f|^4 \times \sigma_1)$. Keeping fixed $|f|=1$ there are other two parameters on which our signal rate is dependent: Λ_c and M_N. In order to sample different regions of the parameter space two cases have been considered. Case (a) $\Lambda_c = 1$ TeV, and case (b) $\Lambda_c = M_N$.

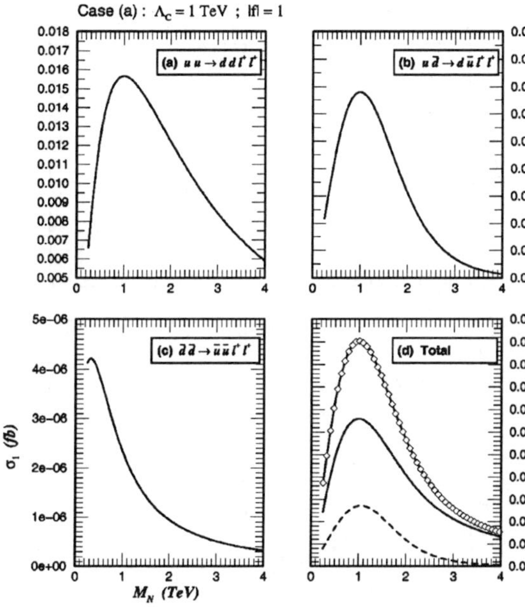

FIG. 4. Cross section normalized to $|f|=1$, i.e., $\sigma_1=\sigma/|f|^4$ with the choice $\Lambda_c=1$ TeV. (a)$uu\to dd+l^+l^+$, (b) $u\bar{d}\to d\bar{u}+l^+l^+$, (c) $\bar{d}\bar{d}\to\bar{u}\bar{u}+l^+l^+$, (d) the *solid* line is the sum of the contributions from Figs. 4(a)–4(c) including factorizable corrections according to Eqs. (A1),(A2); the *dashed* line is the process $u\bar{d}\to s\bar{c}+l^+l^+$ according to Eq. (A3). Finally the *solid-diamond* line in (d) is the total cross section σ_1 including the sum of the subleading contributions reported in Fig. 5.

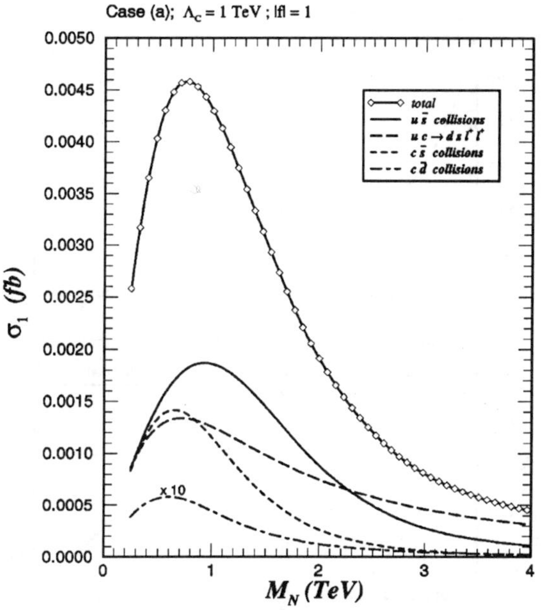

FIG. 5. Subleading processes: the *solid* line is the sum of $u\bar{s}$ collisions Eqs. (A4) and (A5); the long-dashed line is the process $uc\to ds+l^+l^+$; the dashed line is the sum of $c\bar{s}$ collisions, Eqs. (A6) and (A7); the dot-dashed line is the the sum of $c\bar{d}$ collisions, Eqs. (A8) and (A9), scaled by a factor of 10. Finally the *solid-diamond* line is the total contribution to σ_1 of the above processes.

Case (a) is shown in Fig. 4 where cross sections corresponding to the three subprocesses of the first quark family, cf. Eq. (11), are plotted versus the mass of the excited Majorana neutrino M_N [Figs. 4(a)–4(c)]. Since the subprocess $\bar{d}\bar{d}\to\bar{u}\bar{u}+l^+l^+$ is weighted by sea-quark distribution functions it is totally negligible relative to the other two. In Fig. 4(d) the *total cross section* is plotted versus M_N including contributions from other subprocesses with second generation quarks, (c,s) as described in the appendix. Some of these subprocesses are, however, weighted by off-diagonal elements of the CKM mixing matrix and therefore give only small corrections. The shape of the curves as a function of M_N is clearly understood since the only dependence on the new parameters is of the type[4]

$$\sigma\sim\left(\frac{|f|}{\Lambda_c}\right)^4 M_N^2\int\sum_K\frac{1}{(K^2-M_N^2)^2+\theta(K^2)(M_N\Gamma_N)^2}, \tag{29}$$

where K are different momenta flowing in the Majorana propagator. Thus in case (a), $\sigma\to 0$ as $M_N\to 0$ while $\sigma\sim M_N^{-2}$ as $M_N\to\infty$, and there is an intermediate region with a maximum. There is a mass interval from $M_N=250$ GeV up to $M_N\approx 3$ TeV where σ_1 is bigger than the lowest measurable cross section of 10^{-2} fb that corresponds to one event per year given the luminosity $\mathcal{L}_0=100\ \mathrm{fb}^{-1}$ (integrated over one year) planned at LHC. For example, the total signal cross section σ_1 is at most about 5×10^{-2} fb, which is at the same level (though bigger) of the background [Eq. 26], and would only give five events per year. It seems therefore that, *with this particular choice of parameters* [case (a) $\Lambda_c=1$ TeV, $|f|=1$], the lepton number violating signal due to the composite Majorana neutrino would hardly be measurable, unless a better set of kinematical cuts is found that enhances the absolute value of the signal rate while reducing still further the background. However one should keep in mind the dependence on the parameter $|f|$, which in Fig. 4 has been fixed to $|f|=1$. Since the signal cross section is proportional to $|f|^4$ even a slightly larger value of $|f|$ could increase sensibly the signal cross section.

Case (b) is shown in Fig. 6 with the same notation as in Fig. 4. Again the subprocess $\bar{d}\bar{d}\to\bar{u}\bar{u}+l^+l^+$ [Fig. 6(c)] is

[4]It should be remarked that this is only true within the approximation of a constant width Γ_N for the heavy neutrino. Taking into account the dependence of Γ_N with the new physics parameters $|f|$, M_N, and Λ_c (and those pertaining to contact terms) could modify, to some extent, the contibution of the quark-antiquark scattering. However, as pointed out in Ref. [26], Γ_N receives the largest contribution from contact terms which are independent of $|f|$.

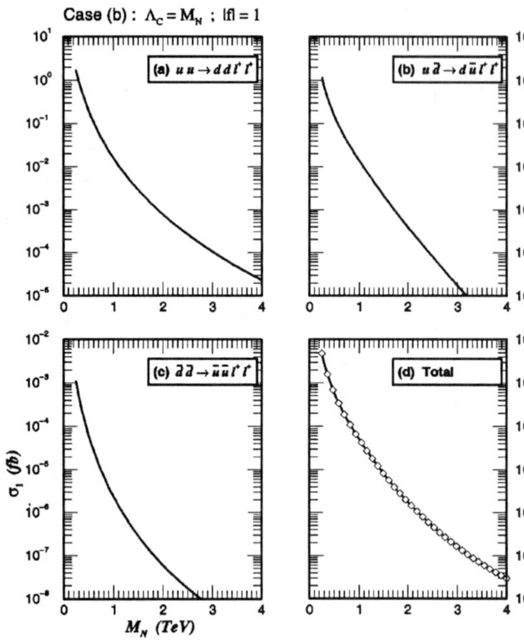

FIG. 6. Same as in Fig. 4 but with the choice $\Lambda_c = M_N$. As explained in the text the different shape of the cross section σ_1 as function of M_N with respect to Fig. 4 is because $\sigma_1 \propto \Lambda_c^{-4}$. Thus fixing $\Lambda_c = M_N$ gives of course a different function of M_N than choosing $\Lambda_c = 1$ TeV. The solid-diamond line in (d) again describes the total σ_1 as done in Fig. 4.

totally negligible relative to the other two. The different shape of the cross section σ_1 as a function of M_N is of course due to the choice $\Lambda_c = M_N$, which according to Eq. (29) gives roughly $\sigma_1 \sim M_N^{-6}$ as $M_N \to \infty$ while $\sigma_1 \sim M_N^{-2}$ as $M_N \to 0$. Thus σ_1 is strongly enhanced with respect to case (a) for values of $M_N < 1$ TeV, while for $M_N > 1$ TeV it will be severely decreased. The cross section σ_1 will be measurable in the mass interval $M_N = 250$ GeV (400 events/yr) up to $M_N \approx 1.4$ TeV (1 event/yr). This portion of the parameter space has therefore the potential of giving rise to a signal with a substantially higher number of events with respect to the background, at least up to $M_N = 850$ GeV (10 events/yr).

The sensitivity of numerical results on different parametrizations of the parton distribution functions (PDFs) has been studied in detail. Calculations concerning the dominant processes have been performed using three other, more recent, sets of PDFs, namely: CTEQ (2pM) [36], Martin-Roberts-Stirling (MRS) (G) [37] and Glück-Reya-Voyt 1994 harmonic oscillator (GRV94) HO [38]. In order to understand better possible dependencies on PDFs the scale of the average x probed by the processes under consideration has been estimated by evaluating:

$$\langle x_a x_b \rangle = \frac{\int dx_a dx_b (x_a x_b) \frac{d\sigma}{dx_a dx_b}}{\int dx_a dx_b \frac{d\sigma}{dx_a dx_b}} \tag{30}$$

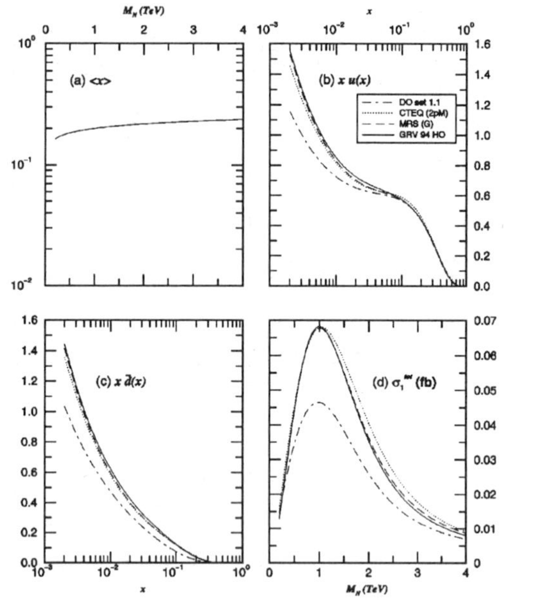

FIG. 7. (a) Average x as a function of M_N calculated for the process $uu \to dd + \text{LSD}$ with the parton distribution functions of DO set 1.1; (b) u quark distribution function at $Q = 1$ TeV for the different parametrizations chosen; (c) same as in (b) but for $\bar{d}$; (d) the total integrated cross section including the processes $uu \to dd + \text{LSD}$, $u\bar{d} \to d\bar{u} + \text{LSD}$, and $u\bar{d} \to s\bar{c} + \text{LSD}$. σ_{tot} is not including the subleading processes reported in Fig. 5. These were shown to give corrections of the order of 10%.

and then taking

$$\langle x \rangle \sim \sqrt{\langle x_a x_b \rangle}. \tag{31}$$

As shown in Fig. 7(a) the LSD signal probes x values in the region of $x \approx 0.2$ over the full range of heavy neutrino masses considered.

Figures 7(b) and 7(c) show a comparison of the parton distribution functions of DO, CTEQ, MRS, and GRV. The latter have a considerably larger $\bar{d}(x)$ relative to DO down to the region of $x \approx 0.2$ while the $u(x)$ distribution function differs, in this very same region, only by a few percent. One expects therefore that using CTEQ, MRS, and GRV will change only slightly the cross-section of the process $uu \to dd + \text{LSD}$ while that of $u\bar{d} \to d\bar{u} + \text{LSD}$ process will be enhanced by somewhat larger factors. This is indeed verified by the numerical calculation.

In Table I sample numerical results for the integrated cross section of $uu \to dd + \text{LSD}$ are shown at different values of the heavy neutrino mass M_N. One finds that the cross section of the uu process with CTEQ, MRS, and GRV, relative to the calculation performed with DO, varies up to $\approx +6\%$ for $M_N = 1$ TeV. The same comparison is shown in Table II for the process $u\bar{d} \to d\bar{u} + \text{LSD}$. In this case a much stronger sensitivity is found, as expected. At $M_N = 1$ TeV the

TABLE I. Sensitivity of numerical results with respect to the different parton distribution parametrizations. Comparing the cross section of the subprocess $uu \to dd + l^{\pm} l^{\pm}$. Case (a) $\Lambda_c = 1$ TeV. Cross sections are expressed in fb.

M_N (GeV)	DO set 1.1	CTEQ 2pM	MRS (G)	GRV 94 HO
500	0.12712×10^{-1}	0.13667×10^{-1}	0.13491×10^{-1}	0.12743×10^{-1}
1000	0.16176×10^{-1}	0.17333×10^{-1}	0.17254×10^{-1}	0.15994×10^{-1}
2000	0.12501×10^{-1}	0.13351×10^{-1}	0.13421×10^{-1}	0.12220×10^{-1}

average value obtained with CTEQ, MRS, and GRV is $\approx +65\%$ higher than that of D0.

In Fig. 7(d) the total integrated cross section (not including the subleading processes discussed in Fig. 5) is compared between the different set of PDFs. One finds that at $M_N = 1$ TeV (where the cross section reaches its maximum value) CTEQ, MRS, and GRV predict a value larger by about 55% with respect to the DO result. This difference of course is very important as far as one is interested in the total number of signal events but, as it will be shown, changes only by about 10% the bounds on the compositeness parameters discussed on Sec. V. One might say (since the CTEQ, MRS, and GRV predictions are all very close), that the DO prediction is rather conservative.

Finally it should be remarked that the discussion, so far, has been quite general with respect to the lepton flavor and applicable to all three of them but (LSD$= l^{\pm} l^{\pm}$, $l = e, \mu, \tau$) at the LHC, muons will be the leptons most easily detected while the other lepton flavors will be detectable but with lower efficiencies [39]. For this reason the numerical results presented here refer to only one lepton generation.

V. COMPARING THE LHC VS THE GENIUS POTENTIAL

This section contains a comparative discussion of the constraints on the parameters $|f|$, M_N, Λ_c that could be derived by the nonobservation of the L-violating signals discussed in the previous section at the high-energy LHC experiments as opposed to those deriving from the nonobservation of low-energy neutrinoless double beta decay experiments, present (Heidelberg-Moscow) and next-generation (GENIUS). The new $\beta\beta_{0\nu}$ GENIUS experiment, (GErmanium-detectors in liquid NItrogen as shielding in an Underground Setup) [40], has the potential to improve by orders of magnitude the lower bound on the $\beta\beta_{0\nu}$ decay half-life. Monte Carlo simulations have shown (for a 1 ton setup) that in one (four) year(s) of measurement the lower bound will be increased, respectively, to [40,41]

$$T^{0\nu}_{1/2} > 5.8 \times 10^{27} \text{ yr}, \quad [\text{one year}],$$

$$T^{0\nu}_{1/2} > 2.3 \times 10^{28} \text{ yr} \quad [\text{four years}].$$

Figures 8 and 9 show the upper bound on the parameter $|f|$ as function of the heavy neutrino mass M_N for the two cases (a) and (b) defined in the previous section. The curves concerning the $\beta\beta_{0\nu}$ bound are based on formulas that can be found in Ref. [16] which, relative to Eq. (10) above, include small correction terms of order $\mathcal{O}(M_W/M_N)$. The LHC curves are found using the numerical cross-sections presented in the previous section, requiring 10 events/yr as a criterion for discovery of the L-violating signal, and assuming an integrated luminosity of $\mathcal{L}_0 = 100$ fb^{-1} as before. Thus nonobservation of the signal at LHC means that $|f|^4 \sigma_1(M_N, \Lambda_c) \mathcal{L}_0 < 10$, which is translated into a constraint on $|f|$ that is the corresponding LHC upper bound to that in Eq. (10) from $\beta\beta_{0\nu}$:

$$|f| < \left(\frac{10}{\sigma_1 \mathcal{L}_0} \right)^{1/4}. \tag{32}$$

One remark is due here as regards the sensitivity to different sets of parton distribution functions. In Figs. 8 and 9 the LHC curve corresponding to the cross section σ_1^{tot} which includes the subleading terms discussed in Fig. 5 and found with the DO parametrization is the thick dot-dashed line. The other thin lines dot-dashed, dotted, long-dashed, and solid, are, respectively, the DO, CTEQ, MRS, and GRV bounds corresponding to the cross section σ_1^{tot} of Fig. 7(d) (all without subleading terms). Comparing the two dot-dashed lines one concludes that the inclusion of subleading terms gives corrections which are smaller than (or at most comparable to) the indeterminacy due the PDFs. The fact that the CTEQ, MRS, and GRV sets of PDFs give somewhat larger cross sections relative to DO, though important as regards the number of signal events and the signal to background ratio, is numerically less evident in Figs. 8 and 9 as here the LHC

TABLE II. Sensitivity of numerical results with respect to the different parton distribution parametrizations. Comparing the cross section of the subprocess $u\bar{d} \to d\bar{u} + l^{\pm} l^{\pm}$. Case (a) $\Lambda_c = 1$ TeV.

M_N (GeV)	DO set 1.1	CTEQ 2pM	MRS (G)	GRV 94 HO
500	0.11390×10^{-1}	0.15779×10^{-1}	0.12992×10^{-1}	0.13811×10^{-1}
1000	0.13617×10^{-1}	0.20878×10^{-1}	0.23133×10^{-1}	0.24700×10^{-1}
2000	0.65991×10^{-2}	0.12706×10^{-1}	0.11379×10^{-1}	0.10284×10^{-1}

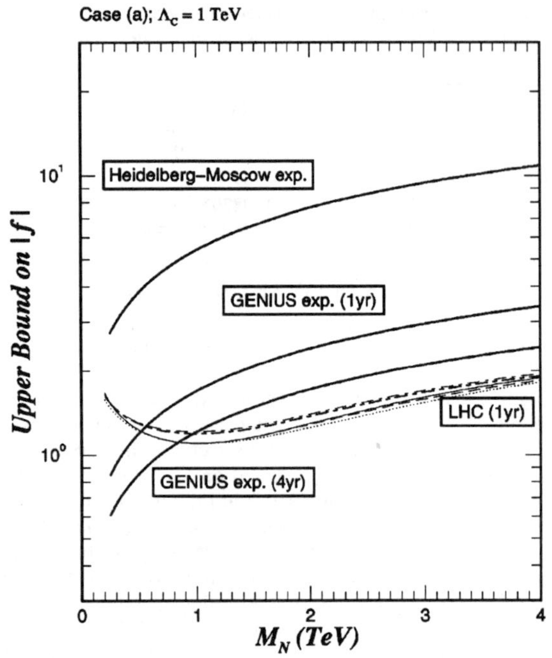

FIG. 8. Sensitivity of LHC vs current and next generation (GENIUS) double beta experiments to the compositeness parameters. Case (a) $\Lambda_c = 1$ TeV. Nonobservation of the signal excludes regions above the curves. If no signal will be observed both LHC and GENIUS will be able to get upper bounds on $|f|$ stronger by almost an order of magnitude with respect to the present Heidelberg Moscow bound. The thick solid lines are the $\beta\beta_{0\nu}$; the thick dot-dashed line is the LHC bound with DO set 1.1 parametrization and σ_1^{tot} of Fig. 4(d) which includes the subleading terms of Fig. 5; the thin lines are the bounds corresponding to Fig. 7(d), DO (dot-dashed), CTEQ (dotted), MRS (long dashed), and GRV (solid). There is a region where the LHC bound is weaker than the GENIUS-1 yr bound $M_N < 550$ GeV (DO), 500 GeV (CTEQ, MRS, and GRV) while for $M_N > 550 (500)$ GeV the LHC bound is stronger. The LHC bound is weaker than that of GENIUS-4 yr for $M_N < 1000$ GeV (DO), 820 GeV (CTEQ, MRS, and GRV) while for $M_N > 1000 (820)$ GeV the LHC bound is stronger.

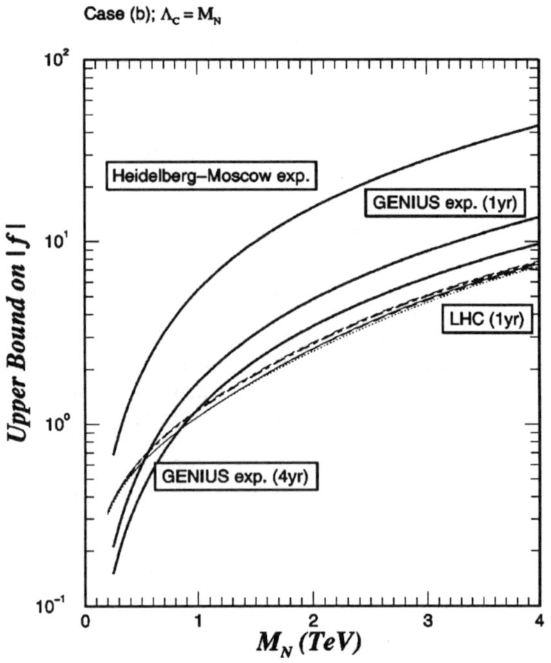

FIG. 9. Same as in Fig. 8 but with $\Lambda_c = M_N$. Also here regions above the curves are excluded. Here the shape of the LHC exclusion plot is similar to that of $\beta\beta_{0\nu}$. The values of M_N at which the LHC curves cross those of GENIUS are the same as in Fig. 8.

curves are obtained through Eq. (32) where σ_1 enters with a power of 1/4, and thus a change of a factor 1.5 in σ_1 gives a change in the bound on $|f|$ within 10%.

From Figs. 8 and 9 one can infer lower bounds on the composite neutrino mass (or equivalentely the compositeness scale) by assuming the dimensionless coupling $|f| \sim \mathcal{O}(1)$. For case (a), Fig. 8, one obtains the bounds shown in Table III while in Table IV the corresponding bounds for case (b), Fig. 9, are given. One comment is in order here. The LHC curve in Fig. 8 has a different behavior for $M_N < 1$ TeV as compared to those of the $\beta\beta_{0\nu}$. This is due to the fact that as $M_N \to 0$, $\sigma_1 \to 0$ and thus the LHC upper bound on $|f|$ becomes weaker and weaker. This does not happen in the $\beta\beta_{0\nu}$ whose squared amplitude behaves as $|\mathcal{M}_{\beta\beta_{0\nu}}|^2 \sim M_N^{-2}$ [15] and at lower masses gives a bigger effect and therefore a stronger constraint. It is for this reason that Table III, for case (a), does not show a lower bound on M_N for LHC. In case (b) if $|f| \sim \mathcal{O}(1)$ GENIUS (1 yr) can exclude Majorana composite neutrinos up to a mass of $M_N \sim 700$ GeV, while LHC (with DO) and GENIUS (4 yr) can go up to about 850 GeV (the LHC bound is 930 GeV if using CTEQ, MRS, or GRV). It is important to realize that the nonaccelerator, low-energy, GENIUS-4yr experiment has the potential to probe the compositeness scale into the TeV region.

At this point the reader should be made aware that investigations of the same type of effective Lagrangians for com-

TABLE III. Lower bound on M_N for case (a) [$\Lambda_c = 1$ TeV, $|f| = 1$]. The bounds are derived from the nonobservation of neutrinoless double beta decay ($\beta\beta_{0\nu}$) at the current (Heidelberg-Moscow) experiment and for the prospected GENIUS experiment after 1 and 4 years of running [40]. At LHC nonobservation of the LSD signal would not imply a lower bound on the composite neutrino mass because of the different shape of the exclusion plot. See Fig. 8.

Experiment	Exp. constraint	Lower bound on M_N (GeV)
Heidelberg-Moscow	$T_{1/2} > 5.7 \times 10^{25}$ yr	$M_N > \sim 10$
GENIUS 1 yr	$T_{1/2} > 6.0 \times 10^{27}$ yr	$M_N > \sim 350$
GENIUS 4 yr	$T_{1/2} > 2.3 \times 10^{28}$ yr	$M_N > \sim 700$
LHC	$N_{events} < 10$	

TABLE IV. Lower bound on M_N for case (b) $[\Lambda_c = M_N, |f| = 1]$. The bounds are derived from the nonobservation of neutrinoless double beta decay $(\beta\beta_{0\nu})$ at the current (Heidelberg-Moscow) experiment and for the prospected GENIUS experiment after 1 and 4 years of running [40] and from nonobservation of the LSD signal at LHC (less than 10 events in one year). See Fig. 9.

Experiment	Exp. constraint	Lower bound on M_N (GeV)
Heidelberg-Moscow	$T_{1/2} > 5.7 \times 10^{25}$ yr	$M_N > \sim 320$
GENIUS 1 yr	$T_{1/2} > 6.0 \times 10^{27}$ yr	$M_N > \sim 700$
GENIUS 4 yr	$T_{1/2} > 2.3 \times 10^{28}$ yr	$M_N > \sim 900$
LHC	$N_{\text{events}} < 10$	$M_N > \sim 850$ (DO) $> \sim 930$ (CTEQ,MRS,GRV)

positeness within the context of LHC experiments have already been reported in the literature. In particular while the production of excited quarks at LHC has been investigated both via magnetic type gauge (G) interactions and contact terms (CT) [32], the production of excited *leptons* has however been considered *only through* CT and a mass sensitivity of up to about 4–5 TeV is found [32]. This work is therefore the first report concerning excited leptons at LHC within the context of magnetic type gauge interactions, and, while the discovery limit derived for contact terms [14,32] cannot be directly compared with the constraints derived in Refs. [13–15] from the nonobservation of $\beta\beta_{0\nu}$ (that were based on gauge interactions G), the discovery limit for LHC reported here (M_N up to 850 GeV) can be directly compared with that of $\beta\beta_{0\nu}$ as done explicitly in Table IV and Figs. 8 and 9.

Finally it is worthwhile to note the complementary role that accelerator (LHC) and nonaccelerator experiments (GENIUS) can have. Figures 8 and 9 show explicitly that, in both cases (a) and (b), while for low masses the $\beta\beta_{0\nu}$ bound is more restrictive there is always a crossing point where the LHC constraint becomes stronger, though of the same order of magnitude.

VI. CONCLUSIONS

In this work the production of like-sign dileptons (LSD) via the exchange of a heavy composite Majorana neutrino in pp collisions has been studied in detail at LHC energies. The coupling of the Majorana neutrino is assumed to be a gauge interaction of the magnetic moment type $(\sigma_{\mu\nu})$. The helicity amplitudes have been presented and the resulting cross-sections within kinematical cuts, needed to suppress the SM background down to the fb level, are reported. Regions of the parameter space are pinned down where the signal is well above the estimated background $(\Lambda_c = M_N, |f| \sim 1, M_N < 850$ GeV). However, a study of the background specifically dedicated to the LHC experimental conditions would certainly be of help towards a better understanding of the lepton number violating processes discussed here. The comparison of the LHC potential with respect to observing L-violating processes with that of the new generation of the non-accelerator type $\beta\beta_{0\nu}$ experiment, GENIUS, shows how the two approaches, high- vs low-energy, do play a complementary role.

The approach developed here to discuss LSD production via composite Majorana neutrinos at LHC is being extended to other models of physics beyond the SM which provide L-violating interactions. The results of these analyses will be reported elsewhere.

One final remark is to be added concerning the interplay of low- vs high-energy facilities with respect to the study of lepton-number violation. The class of diagrams that give rise to $\Delta L = \pm 2$ processes discussed in this work could also trigger lepton-number-violating rare kaon decays such as $K^+ \to \pi^- e^+ e^+$. At the Frascati Φ factory, DAΦNE [42] (presently under commissioning), these decays could either be observed or, otherwise, the corresponding bounds on the branching ratios are susceptible to be strengthened. The current bound on the branching ratio for the $(\Delta L = -2)$ K^+ decay is $\text{Br}(K^+ \to \pi^- e^+ e^+) < 1.0 \times 10^{-8}$ [23], while the sensitivity of the KLOE experiment [43] to be performed at DAΦNE could reach the level of 10^{-9}; the KLOE experiment might thus provide insights on lepton-number-violating interactions beyond the standard model. Work along these lines is in progress.

ACKNOWLEDGMENTS

The authors would like to thank M. Espirito-Santo, of the DELPHI Collaboration, for providing the data histogram plotted in Fig. 3. O.P. would like to acknowledge useful discussions with the experimental colleagues of the Compact Muon Solenoid (CMS) Collaboration, L. Servoli, G. M. Bilei, and M. Biasimi. C.C. acknowledges support by a grant from the Istituto Nazionale di Fisica Nucleare of Italy (INFN), that allowed his stay in Perugia, where this work was completed. He is also indebted to Z. Ajaltouni (ALEPH Collaboration), F. Fleuret, and S. Jan (ATLAS Collaboration) for useful discussions. This project is partially supported by the EEC-TMR Program, Contract No. CT98-0169.

APPENDIX A: LIST OF PARTON SUBPROCESSES

A list of all subprocesses leading to the production of LSD within the first two families of quarks follows.

(i) Quark scattering, $U_i U_j \to D_k D_l + l^+ l^+$,

$(k=l)$	$(k \neq l)$
$uu \to dd[ss] + l^+ l^+$	$uu \to ds + l^+ l^+$
$cc \to ss\,[dd] + l^+ l^+$	$cc \to ds + l^+ l^+$
$uc \to ss\,[dd] + l^+ l^+$	$uc \to ds + l^+ l^+$.

(ii) Quark antiquark scattering $(U_i \bar{D}_j \to D_k \bar{U}_l + l^+ l^+)$:

$$u\bar{d} \to d\bar{u}[d\bar{c}, s\bar{u}, s\bar{c}] + l^+ l^+$$
$$u\bar{s} \to d\bar{c}[d\bar{u}, s\bar{u}, s\bar{c}] + l^+ l^+$$
$$c\bar{s} \to s\bar{c}\,[s\bar{u}, d\bar{c}, d\bar{u}] + l^+ l^+$$
$$c\bar{d} \to s\bar{u}\,[s\bar{c}, d\bar{c}, d\bar{u}] + l^+ l^+ .$$

(iii) Antiquark scattering ($\bar{D}_i\bar{D}_i \to \bar{U}_k\bar{U}_l + l^+l^+$):

$(k=l)$	$(k\neq l)$
$\bar{d}\bar{d}\to\bar{u}\bar{u}\ [\bar{c}\bar{c}]+l^+l^+$	$\bar{d}\bar{d}\to\bar{u}\bar{c}+l^+l^+$
$\bar{s}\bar{s}\to\bar{c}\bar{c}\ [\bar{u}\bar{u}]+l^+l^+$	$\bar{s}\bar{s}\to\bar{u}\bar{c}+l^+l^+$
$\bar{d}\bar{s}\to\bar{u}\bar{u}\ [\bar{c}\bar{c}]+l^+l^+$	$\bar{d}\bar{s}\to\bar{c}\bar{u}+l^+l^+$.

Numerical results reported in Figs. 4(d) and 6(d) contain contributions from some of the processes listed above.

The following equations (A1)–(A9) have been adopted to estimate the contribution of the subprocesses due to second family partons. Note that in this section, and in numerical computations, the complex phases of the elements of the CKM mixing matrix have been neglected, assuming $V_{ij}^* = V_{ij}$, as only the first two generations are being considered.

Processes initiated by two *sea* partons and not receiving contribution from the annihilation diagram have not been considered since Figs. 4(c) and 6(c) show that they are clearly negligible. As regards quark scattering only two cases have been considered; subprocesses initiated by uu and uc collisions, i.e., with at least one u-quark in the initial state.

uu initiated subprocesses: the processes $uu\to dd+l^+l^+$, $uu\to ds+l^+l^+$, and $uu\to ss+l^+l^+$ are factorized as follows

$$|\mathcal{M}_{uu\text{-initiated}}|^2=\left[1+2\left|\frac{V_{us}}{V_{ud}}\right|^2+\left|\frac{V_{us}}{V_{ud}}\right|^4\right]\times|\mathcal{M}_{uu\to dd+l^+l^+}|^2, \tag{A1}$$

the additional factor of 2, in the equation above, accounts for the fact that the process $uu\to ds+l^+l^+$ does not contain identical quarks in the final state as opposed to the processes $uu\to dd+l^+l^+$ and $uu\to ss+l^+l^+$ and thus for it Eq. (28) applies with $k\neq l$.

Quark-antiquark scattering subprocesses have been divided into

$u\bar{d}$ collisions: the processes $u\bar{d}\to[d\bar{u},s\bar{u},d\bar{c}]+l^+l^+$ are factorized as

$$|\mathcal{M}_{u\bar{d}\text{-initiated}}|^2=\left[1+\left|\frac{V_{us}}{V_{ud}}\right|^2+\left|\frac{V_{dc}}{V_{ud}}\right|^2\right]\times|\mathcal{M}_{u\bar{d}\to d\bar{u}+l^+l^+}|^2; \tag{A2}$$

the process $u\bar{d}\to s\bar{c}+l^+l^+$, does not factorize as above due to the fact that the WW fusion and the annihilation diagram come in with different factors of the elements of the CKM matrix

$$|\mathcal{M}_{u\bar{d}\to s\bar{c}+l^+l^+}|^2=\left|\frac{V_{us}V_{dc}}{V_{ud}^2}\mathcal{M}^{(WW\text{-fusion})}_{u\bar{d}\to d\bar{u}+l^+l^+}+\frac{V_{cs}}{V_{ud}}\mathcal{M}^{(u\bar{d}\text{-annihil})}_{u\bar{d}\to d\bar{u}+l^+l^+}\right|^2 \tag{A3}$$

[see Eq. (23]. It turns out to be numerically the most important subprocess, between those containing second family partons. It gives a contribution which is roughly equal to that of $uu\to dd+l^+l^+$ and $u\bar{d}\to d\bar{u}+l^+l^+$, accounting thus for about 30% of the total σ_1 reported in Fig. 4.

$u\bar{s}$ collisions: the processes $u\bar{s}\to[s\bar{u},d\bar{u},s\bar{c}]+l^+l^+$ can be factorized as

$$|\mathcal{M}_{u\bar{s}\text{-initiated}}|^2=\left(\frac{V_{us}}{V_{ud}}\right)^4\left[1+\left|\frac{V_{ud}}{V_{us}}\right|^2+\left|\frac{V_{cs}}{V_{us}}\right|^2\right]\times|\mathcal{M}_{u\bar{d}\to d\bar{u}+l^+l^+}|^2, \tag{A4}$$

and using the fact the within the set of parton densities used here (set 1.1 of Duke and Owens [34]), $\bar{u}(x)=\bar{d}(x)=\bar{s}(x)$, the cross section for $u\bar{s}$ initiated collisions can be simply obtained from $\sigma_1(u\bar{d}\to d\bar{u}+l^+l^+)$ by multiplying it with the above CKM factor which is $0.10\approx 10\%$; the process $u\bar{s}\to d\bar{c}+l^+l^+$ does not factorize as in the above equation and must be considered separately (it is shown in Fig. 5):

$$\mathcal{M}_{u\bar{s}\to d\bar{c}+l^+l^+}=\left|\frac{V_{cs}}{V_{ud}}\mathcal{M}^{(WW\text{-fusion})}_{u\bar{d}\to d\bar{u}+l^+l^+}+\frac{V_{us}V_{dc}}{(V_{ud})^2}\mathcal{M}^{(u\bar{d}\text{-annihil})}_{u\bar{d}\to d\bar{u}+l^+l^+}\right|^2. \tag{A5}$$

$c\bar{s}$ collisions: the processes $c\bar{s}\to[s\bar{c},s\bar{u},d\bar{c}]+l^+l^+$ can be factorized as

$$\begin{aligned}|\mathcal{M}_{c\bar{s}\text{-initiated}}|^2&=\left(\frac{V_{cs}}{V_{ud}}\right)^4\left[1+\left|\frac{V_{us}}{V_{cs}}\right|^2+\left|\frac{V_{cd}}{V_{cs}}\right|^2\right]\\&\quad\times|\mathcal{M}_{u\bar{d}\to d\bar{u}+l^+l^+}|^2,\\&=\left[1+\left|\frac{V_{us}}{V_{cd}}\right|^2+\left|\frac{V_{cs}}{V_{cd}}\right|^2\right]\times|\mathcal{M}_{c\bar{s}\to d\bar{c}+l^+l^+}|^2,\end{aligned} \tag{A6}$$

while the process $c\bar{s}\to d\bar{u}+l^+l^+$ has to be considered separately:

$$|\mathcal{M}_{c\bar{s}\to d\bar{u}+l^+l^+}|^2=\left|\frac{V_{cd}V_{us}}{V_{ud}^2}\mathcal{M}^{(WW\text{-fusion})}_{u\bar{d}\to d\bar{u}+l^+l^+}+\frac{V_{cs}}{V_{ud}}\mathcal{M}^{(u\bar{d}\text{-annihil})}_{u\bar{d}\to d\bar{u}+l^+l^+}\right|^2. \tag{A7}$$

$c\bar{d}$ collisions: the processes $c\bar{d}\to[s\bar{c},d\bar{u},d\bar{c}]+l^+l^+$ can be factorized as

$$\begin{aligned}|\mathcal{M}_{c\bar{d}\text{-initiated}}|^2&=\left(\frac{V_{cd}}{V_{ud}}\right)^2\left[1+\left|\frac{V_{cd}}{V_{ud}}\right|^2+\left|\frac{V_{cs}}{V_{ud}}\right|^2\right]\\&\quad\times|\mathcal{M}_{u\bar{d}\to d\bar{u}+l^+l^+}|^2,\\&=\left[1+\left|\frac{V_{cd}}{V_{cs}}\right|^2+\left|\frac{V_{ud}}{V_{cs}}\right|^2\right]\times|\mathcal{M}_{c\bar{s}\to d\bar{c}+l^+l^+}|^2;\end{aligned} \tag{A8}$$

and the process $c\bar{d}\to s\bar{u}+l^+l^+$ has to be considered separately:

$$|\mathcal{M}_{c\bar{d}\to s\bar{u}+l^+l^+}|^2=\left|\frac{V_{cs}}{V_{ud}}\mathcal{M}^{(WW\text{-fusion})}_{u\bar{d}\to d\bar{u}+l^+l^+}+\frac{V_{cd}V_{us}}{V_{ud}^2}\mathcal{M}^{(u\bar{d}\text{-annihil})}_{u\bar{d}\to d\bar{u}+l^+l^+}\right|^2. \quad \text{(A9)}$$

Finally the amplitude of the process $uc\to ds+l^+l^+$ although weighted by only one u-quark distribution function contains a graph multiplied by diagonal elements of the CKM matrix ($\propto V_{ud}^2V_{cs}^2$) see Eq. (22) and turns out to yield a contribution comparable to that of the $q\bar{q}'$ subprocesses described above (see Fig. 5). The contributions discussed in Eqs. (A4)–(A9) are reported in Fig. 5 together with the process $uc\to ds+l^+l^+$. The sum of these subprocesses accounts for about 10% of the total σ_1 reported in Fig. 4.

APPENDIX B: SQUARE OF AMPLITUDES

For the convenience of the reader interested in numerical applications the square of the amplitudes of the WW fusion mechanism is given here expressed in terms of the particles' momenta scalar products. In the numerical calculations it has been checked that one obtains an agreement of 1 part in 10^5 between this way of calculating the square of the amplitudes and the other consisting in writing down complex amplitudes and numerically taking the square of the absolute value.

Defining the quantities $K_i(i=1,2,3)$ by

$$\sum_{\text{pol}}|\mathcal{M}_i|^2=512\,\mathcal{F}^{(i)}_{CKM}\,\mathcal{K}^2\,K_i, \quad \text{(B1)}$$

they are explicitly (i) $U_iU_j\to D_kD_l+l^+l^+$:

$$\begin{aligned}K_i=p_a\cdot p_b\Bigg(&+A^2p_a\cdot p_c\,p_b\cdot p_d\left[+\frac{p_a\cdot p_e\,p_b\cdot p_f}{C^2}+\frac{p_a\cdot p_f\,p_b\cdot p_e}{D^2}-\frac{L(p_a,p_e,p_b,p_f)}{CD}\right]\\&+B^2p_a\cdot p_d\,p_b\cdot p_c\left[+\frac{p_a\cdot p_e\,p_b\cdot p_f}{E^2}+\frac{p_a\cdot p_f\,p_b\cdot p_e}{F^2}-\frac{L(p_a,p_e,p_b,p_f)}{EF}\right]\\&-AB\Bigg\{L(p_a,p_c,p_b,p_d)\left[\frac{p_a\cdot p_e\,p_b\cdot p_f}{CE}+\frac{p_a\cdot p_f\,p_b\cdot p_e}{DF}-\frac{1}{2}L(p_a,p_e,p_b,p_f)\left(\frac{1}{CF}+\frac{1}{DE}\right)\right]\\&-\frac{1}{2}\epsilon(p_a,p_b,p_c,p_d)\cdot\epsilon(p_a,p_b,p_e,p_f)\left(\frac{1}{CF}-\frac{1}{DE}\right)\Bigg\}\Bigg); \quad \text{(B2)}\end{aligned}$$

(ii) $U_i\bar{D}_j\to D_k\bar{U}_l+l^+l^+$:

$$K_{ii}=p_a\cdot p_d\,p_b\cdot p_d\,p_c\cdot p_a\,A^2\left\{+\frac{p_e\cdot p_a\,p_f\cdot p_d}{C^2}+\frac{p_f\cdot p_a\,p_e\cdot p_d}{D^2}-\frac{L(p_e,p_a,p_f,p_d)}{CD}\right\}; \quad \text{(B3)}$$

(iii) $\bar{D}_i\bar{D}_j\to\bar{U}_k\bar{U}_l+l^+l^+$:

$$\begin{aligned}K_{iii}=p_c\cdot p_d\Bigg\{&+A^2p_a\cdot p_c\,p_b\cdot p_d\left[+\frac{p_c\cdot p_e\,p_f\cdot p_d}{C^2}+\frac{p_c\cdot p_f\,p_e\cdot p_d}{D^2}-\frac{L(p_c,p_e,p_d,p_f)}{CD}\right]\\&+B^2p_a\cdot p_d\,p_b\cdot p_c\left[+\frac{p_c\cdot p_f\,p_e\cdot p_d}{E^2}+\frac{p_c\cdot p_e\,p_f\cdot p_d}{F^2}-\frac{L(p_c,p_e,p_d,p_f)}{EF}\right]\\&-AB\Bigg\{L(p_a,p_c,p_b,p_d)\left[\frac{p_c\cdot p_e\,p_f\cdot p_d}{CF}+\frac{p_c\cdot p_f\,p_e\cdot p_d}{ED}-\frac{1}{2}L(p_e,p_c,p_f,p_d)\left(\frac{1}{CE}+\frac{1}{DF}\right)\right]\\&-\frac{1}{2}\epsilon(p_a,p_b,p_c,p_d)\cdot\epsilon(p_e,p_f,p_c,p_d)\left(\frac{1}{CE}-\frac{1}{DF}\right)\Bigg\}\Bigg\} \quad \text{(B4)}\end{aligned}$$

with $L(p_a,p_b,p_c,p_d)=p_a\cdot p_b\,p_c\cdot p_d+p_a\cdot p_d\,p_b\cdot p_c-p_a\cdot p_c\,p_b\cdot p_d$.

[1] C. Rubbia, Rev. Mod. Phys. **57**, 699 (1985).

[2] S. Weinberg, Phys. Rev. Lett. **19**, 1264 (1967); A. Salam, in *Elementary Particle Theory: Relativistic Groups and Analyticity (Nobel Symposium No. 8)*, edited by Svartholm (Almqvist and Forlag, Stockholm, 1968).

[3] Proceedings of the First International Conference on *Particle Physics Beyond the Standard Model*, Castle Ringberg, Germany, 1997; *Beyond the Desert 1997*, edited by H. V. Klapdor-Kleingrothaus and H. Päs (Institute of Physics, Bristol, 1998).

[4] J. C. Pati and A. Salam, Phys. Rev. D **10**, 275 (1974).

[5] J. L. Hewett and T. G. Rizzo, Phys. Rep. **183**, 193 (1989).

[6] R. Barbieri, R. N. Mohapatra, and A. Masiero, Phys. Lett. **105B**, 369 (1981); **107B**, 455(E) (1981).

[7] For a review and further references see, for example, W. C. Haxton and G. J. Stephenson, Prog. Part. Nucl. Phys. **12**, 409 (1984).

[8] H. V. Klapdor-Kleingrothaus, in Ref. [3], p. 485. In Eq. (10) the most recent $\beta\beta_{0\nu}$ bound has been used as reported in L. Baudis *et al.*, Phys. Rev. Lett. **83**, 41 (1999); H. V. Klapdor-Kleingrothaus, hep-ph/9901021. In *Trento 1998, Lepton and Baryon Number Violation* 251-301. Proceedings of the International Symposium On Lepton And Baryon Number Violation, 1998, Trento, Italy, edited by H. V. Klapdor-Kleingrothaus and I. V. Krivosheina (IOP, Bristol, 1999), p. 760.

[9] Proceedings of the International Workshop *Double Beta Decay and Related Topics* held at the European Centre for Theoretical Studies (ECT), Trento, Italy, 1995, edited by H. V. Klapdor-Kleingrothaus and S. Stoica (World Scientific, Singapore, 1996).

[10] M. Hirsch, H. V. Klapdor-Kleingrothaus, and S. G. Kovalenko, Phys. Lett. B **352**, 1 (1995); Phys. Rev. Lett. **75**, 17 (1995); Phys. Rev. D **53**, 1329 (1996).

[11] M. Hirsch, H. V. Klapdor-Kleingrothaus, and O. Panella, Phys. Lett. B **374**, 7 (1996).

[12] M. Hirsch, H. V. Klapdor-Kleingrothaus, and S. G. Kovalenko, Phys. Rev. D **54**, 4207 (1996); Phys. Lett. B **378**, 17 (1996).

[13] O. Panella and Y. N. Srivastava, Phys. Rev. D **52**, 5308 (1995).

[14] O. Panella, in Proceedings of the Trento workshop, see Ref. [9].

[15] O. Panella, C. Carimalo, Y. N. Srivastava, and A. Widom, Phys. Rev. D **56**, 5766 (1997).

[16] O. Panella, C. Carimalo, Y. N. Srivastava, and A. Widom, in *Beyond the Desert 1997*, Ref. [3].

[17] Wai-Yee Keung and Goran Senjanović, Phys. Rev. Lett. **50**, 1427 (1983).

[18] Amitava Datta, Manoranjan Guchait, and D. P. Roy, Phys. Rev. D **47**, 961 (1993).

[19] D. A. Dicus, D. D. Karatas, and P. Roy, Phys. Rev. D **44**, 2033 (1991).

[20] O. Panella, Ph.D. thesis, Northeastern University, 1991; O. Panella, Y. N. Srivastava, and A. Widom, Northeastern University report, 1991 (unpublished).

[21] S. Dawson, Nucl. Phys. **B249**, 42 (1985).

[22] H. Terazawa, Y. Chikashige, and K. Akama, Phys. Rev. D **15**, 480 (1977); H. Harari, Phys. Lett. **86B**, 83 (1979); H. Fritzsch and G. Mandelbaum, *ibid.* **102B**, 319 (1981); O. Greenberg and J. Schuler, *ibid.* **99B**, 339 (1981); R. Barbieri, R. N. Mohapatra, and A. Masiero, *ibid.* **105B**, 369 (1981); for further references see, for example, H. Harari, Phys. Rep. **104**, 159 (1984); I. A. D'Souza and C. S. Kalman, *Preons, Models of Leptons, Quarks and Gauge Bosons as Composite Objects* (World Scientific, Singapore, 1992).

[23] Particle Data Group, C. Caso *et al.*, Euro. Phys. J. C **3**, 1 (1998; 1999 off-year partial update for the 2000 edition available on the PDG WWW pages (URL http://pdg.lbl.gov/).

[24] N. Cabibbo, L. Maiani, and Y. Srivastava, Phys. Lett. **139B**, 459 (1984).

[25] A. De Rujula, L. Maiani, and R. Petronzio, Phys. Lett. **140B**, 253 (1984).

[26] U. Baur, I. Hinchliffe, and D. Zeppenfeld, Int. J. Mod. Phys. A **2**, 1285 (1987).

[27] J. C. Pati and A. Salam, Phys. Rev. Lett. **31**, 661 (1973); R. N. Mohapatra and J. C. Pati, Phys. Rev. D **11**, 566 (1975); R. N. Mohapatra, *Unification and Supersymmetry* (Springer, New York, 1986).

[28] DELPHI Collaboration, Report No. CERN-EP/196, 1998.

[29] ALEPH Collaboration, R. Barate *et al.*, Eur. Phys. J. C **4**, 571 (1998).

[30] ALEPH Collaboration, R. Barate *et al.*, Phys. Lett. B **429**, 201 (1998).

[31] ZEUS Collaboration, Breitweg *et al.*, Z. Phys. C **76**, 631 (1997).

[32] A. Djouadi, J. Ng, and T. G. Rizzo (unpublished); in *Electroweak Symmetry Breaking and Beyond the Standard Model*, edited by T. Barklow, S. Dawson, H. E. Haber, and S. Siegrist (World Scientific, Singapore, in press), hep-ph/9504210; U. Baur, M. Spira, and P. M. Zerwas, Phys. Rev. D **42**, 815 (1990).

[33] R. Kleiss and W. J. Stirling, Nucl. Phys. **B262**, 235 (1985).

[34] J. F. Owens, Phys. Lett. B **266**, 126 (1991).

[35] G. P. Lepage, J. Comput. Phys. **27**, 192 (1978).

[36] CTEQ Collaboration, J. Botts *et al.*, Phys. Lett. B **304**, 159 (1993).

[37] A. D. Martin, R. G. Roberts, and W. J. Stirling, Phys. Lett. B **354**, 155 (1995).

[38] M. Glück, E. Reya, and A. Vogt, Z. Phys. C **67**, 433 (1995).

[39] CMS Collaboration, L. Servoli (private communication).

[40] J. Hellmig and H. V. Klapdor-Kleingrothaus, Z. Phys. A **359**, 351 (1997).

[41] M. Hirsch (private communication); see also H. V. Klapdor-Kleingrothaus and M. Hirsch, Z. Phys. A **359**, 361 (1997).

[42] The DAΦNE Project Team, "DAΦNE, Status and Plans," Proceedings of PAC95, 1995.

[43] The KLOE Collaboration, "Status of the KLOE experiment," LNF-97/033, 1997.

2.4.8 Sterile Neutrinos, Majorons and Double Beta Decay

Spontaneously Broken Lepton Number and Cosmological Constraints on the Neutrino Mass Spectrum

Y. Chikashige, R. N. Mohapatra,[a] and R. D. Peccei
Max-Planck-Institut für Physik und Astrophysik, D-8000 München 40, Federal Republic of Germany
(Received 7 October 1980)

The cosmological constraints on neutrino masses can be altered if lepton number is broken globally giving rise to a very weakly coupled Goldstone boson—the Majoron. Then heavy neutrinos can decay sufficiently rapidly by Majoron emission, thereby giving negligible contributions to the mass density of the universe. Specifically, if M is the mass scale associated with lepton number breakdown, for $M \lesssim 10^6$ GeV there are no constraints on neutrino masses while for M high enough ($M \gtrsim 10^9$–10^{10} GeV) the standard bounds remain.

PACS numbers: 14.60.Gh, 11.30.Qc, 14.80.Kx, 98.80.Bp

A topic of great current interest is the spectrum of neutrino masses and its implication for the nature of the weak interactions. The most stringent limits on this spectrum come from cosmological and astrophysical arguments. If neutrinos were stable, then considerations relating to the observed mass density in the present universe indicate that the sum of neutrino masses should either be less than[1,2] about 50 eV or, alternatively, that these masses should be heavier than a few gigaelectronvolts.[3] This forbidden gap can in principle be breached by neutrinos which can decay sufficiently rapidly. Dicus, Kolb, and Teplitz[4] obtained limits on the lifetime of the process $\nu_H \to \nu_L + \gamma$ which would allow for heavy-neutrino masses to be in the forbidden zone. However, the required lifetimes found in Ref. 4 are, in general, too short to be obtained in realistic weak-interaction models. Furthermore, if one studies the effects which the photons produced in these decays have on the element abundance of the present universe,[5] one finds that the lifetimes required for heavy-neutrino decay are even shorter than those obtained in Ref. 4, being typically of the order of hours. Hence, if these cosmological arguments are correct, neutrinos with conventional weak interactions cannot have masses in the 50 eV to the gigaelectronvolt range.

In this Letter we would like to discuss how the above limits can be obviated if lepton number is a spontaneously broken *global* symmetry. We indicate here briefly the physics scenario and proceed, further on, to the details. If lepton number is indeed spontaneously broken, there is necessarily a zero-mass Goldstone boson in the theory. It turns out that, in a large class of realistic models studied by us recently,[6] this Goldstone boson (the Majoron) couples dominantly, but very weakly, to neutrinos and essentially negligibly to matter. Hence the Majoron's existence is not ruled out by experiment. Heavy neutrinos thus can decay via Majoron emission to light neutrinos, with a rate which depends both on the mass of the neutrinos as well as on the scale which characterizes the spontaneous breakdown of lepton number. The typical lifetime of neutrinos with masses in the cosmological forbidden range, and for values of the lepton-number-nonconserving scale which are sensible, turn out to be short compared with the lifetime of the universe. The heavy nuetrinos, because they have mostly decayed, will only make a negligible contribution to the mass density of the universe. Therefore this removes the need to restrict their masses. However, since the lifetime of the heavy neutrinos scales with M—the lepton-number-nonconserving scale—eventually for large enough M one has stable (with respect to the lifetime of the universe) neutrinos and one recovers the old cosmological bounds.

We shall assume that the weak interactions of the neutrinos are governed by the standard SU(2) $\otimes$ U(1) theory.[7] In addition, however, we shall suppose that SU(2) $\otimes$ U(1)-singlet, right-handed, neutrino fields exist. Neutrinos of a given generation can then have both lepton-number-conserving Dirac masses m, as well as lepton-number-nonconserving Majorana masses M. We shall suppose that only the right-handed neutrinos have a Majorana mass term. Then if $M \gg m$ the physical neutrinos, which are Majorana fields, have masses M and $m_\nu = m(m/M)$. The observed neutrinos are the light neutrinos of mass m_ν, while the superheavy neutrinos are at scales beyond the range of present experimentation. If M arises from the vacuum expectation value of a singlet Higgs field Φ, which carries lepton number, then $\langle\Phi\rangle \neq 0$ implies the spontaneous breakdown of lepton number. Unless lepton number is gauged, there will appear in general a Goldstone boson associated with this spontaneous breakdown—the Majoron.

In a recent note[6] we discussed the properties of the Majoron—for a one-generation model—and we summarize the salient features of that analysis here. Let ν be the light neutrino of mass m_ν and η be the superheavy neutrino of mass M. Then the effective coupling of the Majoron field χ to neutrino and matter fields (quarks and leptons) is given by

$$\mathcal{L} = -\frac{ih}{\sqrt{2}}\,\chi\left\{\bar\eta\gamma_5\eta - \left(\frac{m_\nu}{M}\right)^{1/2}(\bar\eta\gamma_5\nu + \bar\nu\gamma_5\eta) + \frac{m_\nu}{M}\,\bar\nu\gamma_5\nu + \frac{G_F}{8\sqrt{2}\,\pi^2}\,m_\nu m_f g_f(\bar f\gamma_5 f)\right\}. \tag{1}$$

Here $g_f = +1$ for $f = e$ or u quarks, or $g_f = -1$ for d quarks, and h is a Higgs-fermion coupling constant. As can be seen from (1) the strength of the coupling of the Majoron to matter is extremely weak. Its coupling to neutrinos is also small, being suppressed by the factor m_ν/M. The Majoron only couples strongly to the superheavy neutrinos η. But these particles are unstable, decaying very rapidly into the light neutrinos via Majoron emission ($\tau_\eta \simeq 10^{-10}$ sec for $m_\nu \simeq 1$ eV, $h \simeq 10^{-2}$).

Because the Majoron's coupling to matter is so weak, terrestrial experiments cannot be used to rule out its existence. This point is discussed in some detail in Ref. 6, but it might be useful to

VOLUME 45, NUMBER 24 PHYSICAL REVIEW LETTERS 15 DECEMBER 1980

summarize here the principal results of that analysis. The pseudoscalar nature of the coupling of the Majoron to matter yields a spin-dependent $1/r^3$ potential between two fermions (quarks or leptons). The strength of this potential is characterized by a coupling constant $\lambda_f = (4\pi)^{-1}(hG_F m_\nu/16\pi^2)^2$ which, taking as typical parameters $h \simeq 10^{-2}$, $m_\nu \simeq 1$ eV, is of order 10^{-65} cm^2. This number is so small that the Majoron exchange force is only comparable to gravity at typical nuclear distances. Hence Eötvos-type experiments[8] are totally insensitive to Majorons. The analysis of Feinberg and Sucher,[9] of possible nonmagnetic spin-dependent forces, again is of little use for Majorons since the bounds found for λ_f by these authors are many orders of magnitude larger than 10^{-65} cm^2 (typically $\lambda_f \leq 10^{-32}$ cm^2). Similarly axion searches could not have uncovered the Majoron because its coupling to matter is so weak.[10]

The analysis of Ref. 6 is easily generalized to the case of many generations of neutrinos. In this case again an equation like (1) ensues, but now, in general, the Majoron has off-diagonal couplings with neutrinos of different generations. Again, the superheavy neutrinos decay rapidly by Majoron emission. But now it is also possible for the heavier of the light neutrinos ν_H to decay to the lightest state ν_L by Majoron emission: $\nu_H \to \nu_L + \chi$. The lifetime for such a process is easily estimated

$$\tau(\nu_H \to \nu_L + \chi) = \frac{32\pi}{h^2 \sin^2\theta}\left(\frac{M}{m_{\nu_H}}\right)^2 \frac{1}{m_{\nu_H}}, \qquad (2)$$

where $\sin\theta$ is an intrageneration mixing angle.

The lifetime τ can be sufficiently short on a cosmological scale to remove effectively the ν_H contribution to the mass density of the universe, provided that M is not too large. If we take as typical parameters $h \simeq 10^{-2}$, $\sin\theta \simeq 10^{-1}$, and $M \simeq 10^5$ GeV, we find that $\tau \simeq 2\times 10^7$ yr for $m_{\nu H} = 100$ eV. Thus we see that for "sensible" scales M, where one might expect new physics to arise, neutrino decay by Majoron emission can play an effective role in removing the heavy-neutrino contribution to the mass of the universe. Actually the above discussion can be sharpened. One can determine a range of values of M for which heavy neutrinos of mass $m_{\nu H}$ can exist without violating any cosmological bounds. We proceed to do this below.

Our discussion here will be analogous to the analysis of Dicus, Kolb, and Teplitz.[4] First we note that the dominant interaction of ν_H and ν_L is through the standard weak-interaction neutral-current process $\nu_H + \bar\nu_H \to \nu_L + \bar\nu_L$. Thus the calculation of when the heavy neutrinos decouple remains the same as that of Ref. 4 and we will adopt their values for the decoupling temperature T_D, decoupling time t_D, and the density $n_H(T_D)$ at that epoch. One can then calculate the contribution to the energy density of the Majorons coming from ν_H decay. [There is, of course, a remnant Majoron black-body energy density coming from Majorons which decoupled at temperatures of the order of M, but this component—like that of the photons—makes a totally negligible contribution to the present-day energy density.] The energy density of the Majorons which are decay by-products is given by[4]

$$\rho = 2n_H(T_D)\left(\frac{1.9\ \mathrm{°K}}{T_D}\right)^3 \int_{t_D}^{t_U} dt \left(\frac{m_{\nu H}}{2}\right)\left(\frac{t}{t_U}\right)^{1/2} \frac{1}{\tau} \exp\left(-\frac{t-t_D}{\tau}\right). \qquad (3)$$

Here t_U is the universe lifetime and t_D is the decoupling time for ν_H. The remaining factors in Eq. (3) are easily understood. The overall factor of 2 accounts for both heavy neutrino and antineutrino decays. The factor $(1.9\ \mathrm{°K}/T_D)^3$ takes into account of the volume expansion from the decoupling of the neutrinos to the present epoch, while $n_H(T_D)\exp[-(t-t_D)/\tau]$ is the ν_H density at the moment of decay. Finally the factor $\frac{1}{2}m_{\nu H}(t/t_U)^{1/2}$ accounts for the red shift of the Majoron's energy from its decay value $(\frac{1}{2}m_{\nu H})$, while τ^{-1} in Eq. (3) is just the probability that ν_H decay occurred.

Since $t_d \simeq 10^{-1}$ sec we have, in general, $\tau \gg t_d$. Furthermore, if $\tau \ll t_U$ we can approximate (3) by

$$\rho \simeq n_H(T_D)\left(\frac{1.9\ \mathrm{°K}}{T_D}\right)^3 \left(\frac{\sqrt{\pi}}{2}\right) m_{\nu H}\left(\frac{\tau}{t_U}\right)^{1/2}. \qquad (4)$$

So as to avoid conflicts with astrophysical bounds we must require that the energy density of the Majorons from ν_H decay be less than the critical density[5] $\rho_c \simeq 5\times 10^{-3}$ MeV/cm^3. This requirement for stable neutrinos [essentially $\tau = t_U$ in Eq. (4)] gives the cosmological forbidden zone for neutrino masses[1-3]: 50 eV $\leq m_\nu \leq$ 2 GeV. In our case, since we have a free parameter M—the scale of the lepton-number breakdown—the constraint $\rho \leq \rho_c$ will give for a given M allowed values for m_{ν_H}. Using again as typical parameters $h \simeq 10^{-2}$ and $\sin\theta \simeq 10^{-1}$, one obtains the graph of Fig. 1 for the allowed zone of neutrino masses. We note the important result than if $M \leq 10^6$ GeV there are *no* cosmological constraints on neutrino masses.

VOLUME 45, NUMBER 24 PHYSICAL REVIEW LETTERS 15 DECEMBER 1980

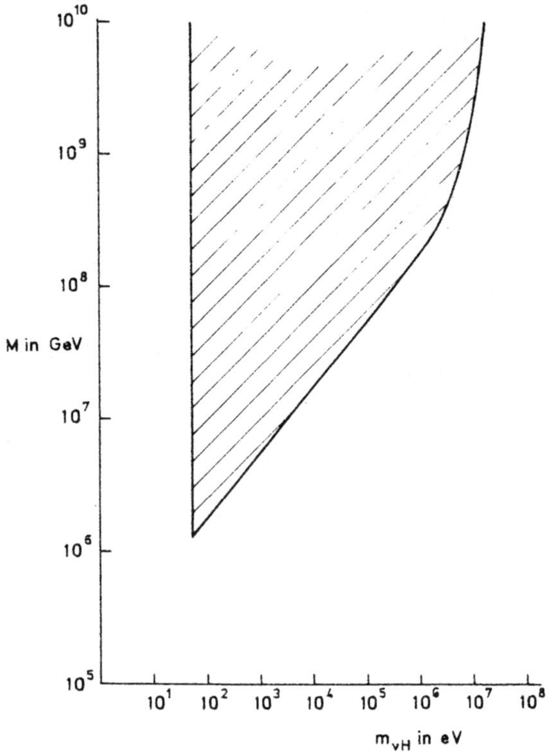

FIG. 1. The shaded area represents the forbidden domain of neutrino masses for a given range of the heavy Majorana lepton mass M.

On the other hand, if $M \gtrsim 10^9$–10^{10} GeV, the forbidden zone for neutrino masses essentially remains that of the standard analysis.[1-3] We remark that the straight-line portion of the boundary in Fig. 1 for $m_{\nu_H} \lesssim 1$ MeV arises because there $n_H(T_D)$ and T_D are fixed [see Ref. 4: $n_H(T_D) = 6 \times 10^{32}$ cm^{-3} and $T_D = 3.4 \times 10^{10}$ °K]. Then $\rho \simeq (M^2/m_{\nu_H})^{1/2}$. Above $m_{\nu_H} \simeq 1$ MeV the density factor $n_H(T_D)(1.9\ °\mathrm{K}/T_D)^3$ decreases rapidly and the bound $\rho \lesssim \rho_c$ begins to be ineffective.

We have shown that, if lepton number is a spontaneously broken global symmetry accompanied by a very weakly coupled Goldstone boson, it is possible to avoid cosmological constraints on the neutrino spectrum provided the scale $M \lesssim 10^6$ GeV. One may ask whether this is a reasonable scale for lepton-number breakdown. In terms of the vacuum expectation value of the Higgs field Φ which gives rise to the breakdown one has that $\langle\Phi\rangle \simeq M/h \lesssim 10^8$ GeV. Breakdown of lepton number at scales below these have been invoked previously in the context of left-right-symmetric models[11] and in connection with horizontal symmetries.[12] Thus in this sense the bound on $\langle\Phi\rangle$ seems an eminently reasonable one. However, we should point out that if the lepton-number breakdown occurs at a grand unified scale, one expects $\langle\Phi\rangle \simeq 10^{13}$–$10^{15}$ GeV and in that case even if Majorons exist, no lifting of the cosmological constraints is possible.

Two of us (Y.C. and R.N.M.) are recipients of Alexander von Humboldt Fellowships. This work was supported in part by the National Science Foundation, Professional Staff Congress-Board of Higher Education Research Award under Grant No. RF13406.

[(a)]Also at Physics Department, University of Munich, Munich, Federal Republic of Germany. On leave from Physics Department, City College of New York, New York, N. Y. 10031.

[1]R. Cowsik and J. McClelland, Phys. Rev. Lett. 29, 669 (1972).

[2]D. N. Schram and G. Steigman, Gen. Relat. Grav. (to be published).

[3]B. W. Lee and S. Weinberg, Phys. Rev. Lett. 39, 165 (1977).

[4]D. Dicus, E. Kolb, and V. Teplitz, Phys. Rev. Lett. 39, 168, 973(E) (1977).

[5]K. Sato and M. Kobayashi, Prog. Theor. Phys. 58, 1775 (1977); D. Dicus, E. Kolb, V. Teplitz, and R. Wagoner, Phys. Rev. D 17, 1529 (1978).

[6]Y. Chikashige, R. N. Mohapatra, and R. D. Peccei, Max Planck Institute Report No. MPI-PAE/PTh 36/80, 1980 (to be published).

[7]S. Weinberg, Phys. Rev. Lett. 19, 1264 (1967); A. Salam and J. C. Ward, Phys. Lett. 13, 168 (1964); S. L. Glashow, J. Iliopoulos, and L. Maiani, Phys. Rev. D 2, 1285 (1970).

[8]P. G. Roll *et al.*, Ann. Phys. (N.Y.) 26, 442 (1964); R. Spero *et al.*, Phys. Rev. Lett. 44, 1645 (1980).

[9]G. Feinberg and J. Sucher, Phys. Rev. D 20, 1717 (1979).

[10]R. D. Peccei, in *Proceedings of the Nineteenth International Conference on High Energy Physics, Tokyo, Japan, 1978*, eddited by S. Homma, M. Kawaguchi, and H. Miyazawa (Physical Society of Japan, Tokyo, 1979); see also D. Dicus *et al.*, Phys. Rev. D 22, 839 (1980).

[11]R. N. Mohapatra and G. Senjanović, Phys. Rev. Lett. 44, 912 (1980), and Fermilab Report No. 80/61 THY (to be published); R. N. Mohapatra and R. E. Marshak, Phys. Rev. Lett. 44, 1316 (1980).

[12]Y. Chikashige, G. Gelmini, R. D. Peccei, and M. Roncadelli, Phys. Lett. 94B, 499 (1980).

PHYSICAL REVIEW D VOLUME 49, NUMBER 11 1 JUNE 1994

New class of Majoron-emitting double-β decays

C. P. Burgess*
Physics Department, McGill University, 3600 University Street, Montréal, Québec, Canada H3A 2T8

J. M. Cline†
Theoretical Physics Institute, The University of Minnesota, Minneapolis, Minnesota 55455
(Received 22 July 1993)

Motivated by the excess events that have recently been found near the end points of the double-β decay spectra of several elements, we reexamine models in which double-β decay can proceed through the neutrinoless emission of massless Nambu-Goldstone bosons (Majorons). Noting that models proposed to date for this process must fine-tune either a scalar mass or a VEV to be less than 10 keV, we introduce a new kind of Majoron which avoids this difficulty by carrying lepton number $L=-2$. We analyze in detail the requirements that models of both the conventional and our new type must satisfy if they are to account for the observed excess events. We find (1) the electron sum-energy spectrum can be used to distinguish the two classes of models from one another, (2) the decay rate for the new models depends on different nuclear matrix elements than for ordinary Majorons, and (3) all models require a (pseudo) Dirac neutrino, having a mass of a several hundred MeV, which mixes with ν_e.

PACS number(s): 14.80.Mz, 12.60.−i, 14.60.Pq, 23.40.Bw

I. INTRODUCTION AND SUMMARY

Only six years ago the first direct observations of double-β decay were made in the laboratory [1]. The major interest of these experiments is in the search for deviations in the shape of the spectrum from that which is predicted for ordinary neutrino-emitting decays. While most of the attention has gone toward searches for neutrinoless decays that would indicate neutrino masses, a third decay mode has also been discussed, in which a massless Goldstone boson, the Majoron,[1] is emitted in lieu of neutrinos. The main purpose of this paper is to determine whether there is any theoretical hope for the last decay to occur at observable levels. Our conclusion is that there may be, but only if the Majoron has rather different properties than have previously been assumed.

Our more immediate motivation for studying this question is the recent observation of a mysterious excess of high-energy electrons in the electron spectrum for the double-β ($\beta\beta$) decay of several elements. This claim was first made in 1987 for the decay $^{76}\mathrm{Ge}\rightarrow{}^{76}\mathrm{Se}+2e^-$ by Avignone *et al.* [2], although the effect was discounted when they, as well as other groups, subsequently excluded a signal having the original strength [3]. The mysterious events reappeared, however, when the UC Irvine group found excess numbers of electrons near but below the end points for ^{100}Mo, ^{82}Se, and ^{150}Nd, with a statistical significance of 5σ [4]. Such events also persist in the ^{76}Ge data [5,6] at approximately one-tenth of the original rate.

Since these are difficult experiments, it is possible that the anomalous events will turn out to be due to systematic error or to a hitherto unsuspected nuclear physics effect.[2] But they may also be the fingerprint of the new fundamental interaction [8,9], of Majorons with neutrinos [10,11]. If so, these observations are of vital importance since they provide us with a glimpse of physics beyond the standard electroweak theory.

We assume for the sake of argument that any excess events which might be detected in these experiments are due to Majoron emission, denoted by $\beta\beta_M$. Our goal is to explore the implications of $\beta\beta_M$ taken together with the other known constraints on neutrino physics. In so doing, we have found that the candidate models capable of describing Majoron emission from nuclei fall into two broad classes.

In the first class of models for $\beta\beta_M$, which to our knowledge includes everything that has been proposed until recently [9,12], the Majoron is the Nambu-

*Electronic address: cliff@physics.mcgill.ca
†Electronic address: jcline@mnhep.hep.umn.edu

[1]The term "Majoron" was originally used for the Nambu-Goldstone boson associated with spontaneous breaking of lepton number, since the same lepton number breaking induced a Majorana mass for the neutrinos. We enlarge the meaning of the name in this paper by applying it even if the scalar is massive or if the model in question does not generate Majorana masses.

[2]After completing this work, we were informed of evidence that the anomalous events reported by the UC Irvine group may be due to resolution problems for the higher-energy electrons [7]. Even excluding events which can be explained in this way, however, there remains a smaller set of residual events whose magnitude is in agreement with observations of the Moscow-Heidelberg experiment.

0556-2821/94/49(11)/5925(20)/$06.00 49 5925

110

SCALAR-EMITTING MODES IN DOUBLE-BETA DECAY

C.P. BURGESS*

Physics Department, McGill University
3600 University St., Montréal, Québec, Canada, H3A 2T8.

ABSTRACT

The sum-energy spectrum of electrons emitted in double-beta decay is a well-known diagnostic for the nature of the physics which is responsible for the decay. Three types of spectra are usually considered when these experiments are analysed: one each for the standard two-neutrino ($\beta\beta_{2\nu}$) decay, neutrinoless ($\beta\beta_{0\nu}$) and a Majoron-emitting ($\beta\beta_{\varphi}$) decay. It has recently been shown that two other electron spectra can be possible for scalar-emitting modes, in addition to these traditional three. One of these is softer than the Standard-Model $\beta\beta_{2\nu}$ decay, while the other is intermediate between the $\beta\beta_{2\nu}$ and the usual $\beta\beta_{\varphi}$ spectra. The models which predict these new spectra are generically more natural than those which predict the traditional $\beta\beta_{\varphi}$ spectrum, in that they can accomodate the constraints following from the steadily improving limits on $\beta\beta_{0\nu}$ decay without requiring the fine-tuning that is endemic to the usual models. This article reviews the properties of the physics which can produce the new kinds of electron spectra.

1. Introduction

Double-beta ($\beta\beta$) decay is an extremely rare process in which two nuclear neutrons simultaneously convert into two protons and two electrons. Within the Standard Electroweak Model (SM) this decay occurs at second order in the charged-current weak interactions, and is accompanied by the emission of two antineutrinos, giving rise to a characteristic ($\beta\beta_{2\nu}$) electron spectrum. Despite the extremely long half-lives involved — typically 10^{20} yr or more — heroic efforts[1] over the past ten years have been rewarded by its experimental detection.

Because it is such a rare process, $\beta\beta$ decay experiments also furnish a unique window onto whatever new physics may replace the SM at energies very much higher than those that are directly accessible in today's accelerators. This is because the effects for these experiments of new interactions can in some circumstances compete with those of run-of-the-mill SM decays.

---

*Invited talk presented to the Workshop on Double Beta Decay and Related Topics, Trento Italy, April 1995.

VOLUME 59, NUMBER 15 PHYSICAL REVIEW LETTERS 12 OCTOBER 1987

Limit on Neutrinoless Double-Beta Decay with Majoron Emission in ^{82}Se

S. R. Elliott, A. A. Hahn, and M. K. Moe

Physics Department, University of California, Irvine, Irvine, California 92717

(Received 20 July 1987)

The University of California, Irvine, ^{82}Se double-beta-decay experiment has now accumulated 7960 h of live time. After imposition of kinematic cuts, thirteen candidate events remain in the sum-energy window between 2.0 and 3.0 MeV. If these are interpreted as neutrinoless double-beta decay with the emission of a Goldstone boson (known as the majoron), we obtain a half-life limit of 4.4×10^{20} yr (90% confidence level). This gives, dependent upon the various calculations of the nuclear matrix elements, a limit of the majoron coupling constant $g_{ee} \lesssim (2-30)\times10^{-4}$.

PACS numbers: 23.40.Bw, 14.60.Gh, 14.80.Gt

The observation of neutrinoless double-beta decay would signal the violation of lepton-number conservation in the weak interaction. In addition, it would provide evidence for the existence of neutrino mass and perhaps right-handed weak currents. Traditionally, experimental searches have concentrated on the decay mode into a daughter atom and two electrons (0ν), where because of the large mass of the daughter, the two electrons together share the total energy release of the decay. The experimental signature for the 0ν mode is quite dramatic (if it is ever seen): a spike in the sum-electron-energy spectrum at the end-point energy. Another recently proposed neutrinoless decay mode[1] is the decay into two electrons and a massless neutral scalar boson B, known as the majoron. The majoron would be the Goldstone boson resulting from the spontaneous breaking of the global symmetry of baryon-lepton number $(B-L)$.[1-3] The total energy release is shared now by the two electrons and the majoron. Therefore the sum energy of the two electrons (the majoron being undetected) is continuous. Figure 1 shows the theoretical spectra for the 0ν and $(0\nu,B)$ as well as the standard second-order two-neutrino decay modes.

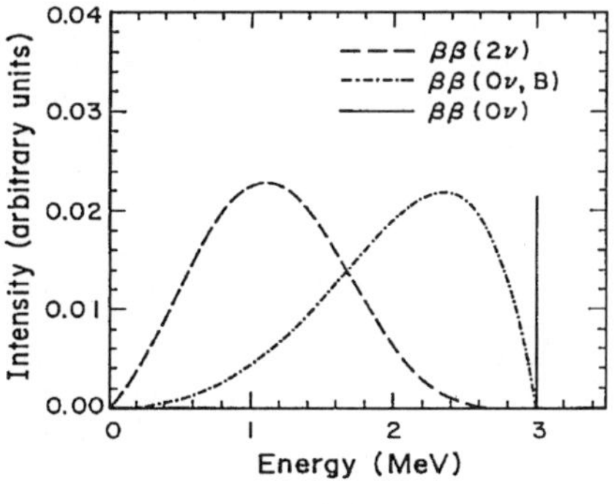

FIG. 1. Theoretical sum-energy spectra of the two electrons for three modes of double-beta decay in ^{82}Se. The normalizations are arbitrary.

The University of California, Irvine, double-beta-decay experiment is a time projection chamber (TPC) surrounding a thin aluminized Mylar foil which contains 14 g of 97% enriched ^{82}Se. The total source thickness is 7 mg/cm^2. Helmholtz coils provide a uniform magnetic field of 715 G perpendicular to the source plane. This apparatus, which has been described in more detail elsewhere,[4] is located in a basement laboratory of the Physical Sciences building at the University of California, Irvine. The principle advantage of the TPC over other double-beta-decay experiments lies in its ability to reconstruct the ionization tracks left by charged particles. Therefore a double-beta-decay candidate event is one in which two electron tracks are reconstructed emerging from the same point on the source plane. The energy of the individual electrons is determined by the radius and pitch of the helix.

The raw spectrum (7960 h of live time) of all double-beta-decay candidates is shown in Fig. 2. All events above 3.0 MeV, save one,[5] can be ascribed to Compton scattered electrons from cosmic-ray–induced gamma rays which also undergo Moeller scattering. Moeller scattering is characterized by both the opening angle between the two electrons, $\cos(\theta)=(T_1T_2/p_1p_2)$, and typically by a very low value of the ratio T_{lo}/T_{hi}, with T and

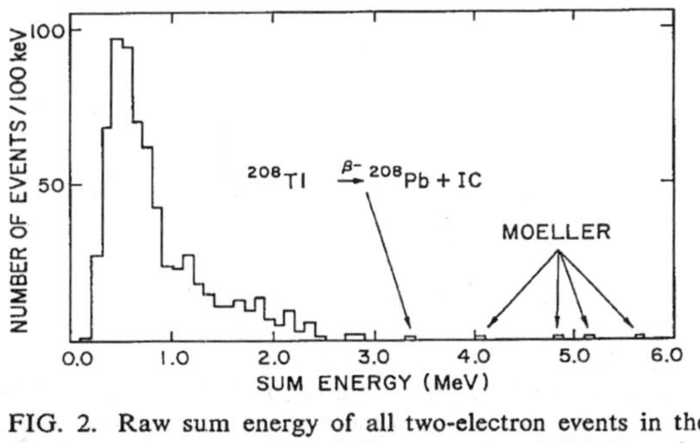

FIG. 2. Raw sum energy of all two-electron events in the TPC. The apparent sources of the events above 3 MeV are indicated. The large rate below 1 MeV is most likely due to the beta-decay, internal-conversion sequence in ^{214}Pb.

VOLUME 59, NUMBER 15 PHYSICAL REVIEW LETTERS 12 OCTOBER 1987

p being the electron kinetic energy and momentum, respectively, and lo/hi referring to the lower/higher-energy electron. However, the opening angle may be distorted by multiple scattering in the source. This is especially pronounced in Moeller scattering, because of the preponderance of low-energy (several hundred kiloelectronvolts) electrons. Indeed, of the events between 3 and 6 MeV, only one retains the Moeller angle.

In Fig. 3 the solid line shows the double-beta-decay candidates after our imposing both an energy threshold of 150 keV on each electron of an event and a sum-energy threshold of $T_1+T_2 \geq 800$ keV. The shaded area represents the candidate events with the sum energy $T_1+T_2 \geq 2.0$ MeV left after a cut is applied to remove Moeller events. The cut removes any event in which either the ratio of $T_{lo}/T_{hi} \leq 0.2$ and/or the measured opening angle is within 3° of the computed Moeller angle. This reduces the original 25 candidates between 2.0–3.0 MeV to thirteen candidates. The dotted-dashed curve is the theoretical spectrum of two-neutrino events assuming that the half-life determined from a geochemical experiment[6] is completely due to the two-neutrino mode of decay. The dashed curve is the known background from the beta-decay internal-conversion sequence in ^{214}Bi and ^{208}Tl decays. These two spectra are shown for illustrative purposes only and have not been subtracted from the data.

Conservatively, we assume for a limit that *all* thirteen events between 2.0–3.0 MeV which have survived the Moeller cut are due to the majoron decay mode of double-beta decay. Approximately 55% of all majoron decays in ^{82}Se will give sum-electron energies between 2.0–3.0 MeV. The efficiency of the TPC within this energy band is 16% when our energy thresholds are imposed on the theoretical majoron spectrum. This value has been based upon our measured/calculated efficiency

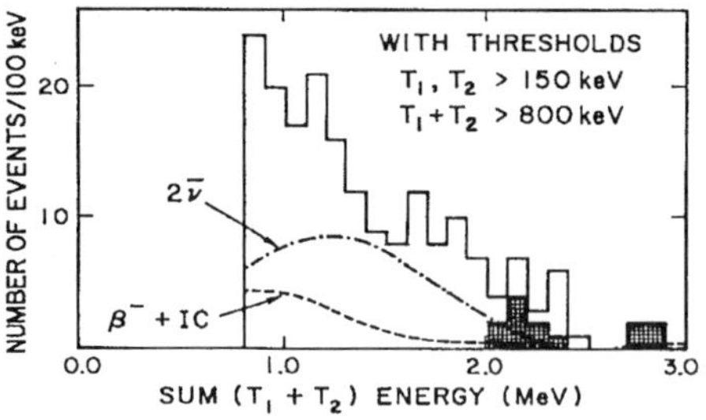

FIG. 3. Sum-energy spectrum between 0.8 and 3.0 MeV after a software threshold of 150 keV has been applied to each electron of an event. The dashed curve is the expected background due to sources of natural radioactivity within the TPC. The dotted-dashed curve is the rate of 2ν double-beta decay, normalized to a geochemical experiment (Ref. 6). The shaded area represents the events above 2 MeV remaining after a cut to remove Moeller scattering events has been applied.

in the two-neutrino branch.[4] The acceptance of the Moeller cut on the majoron spectrum between 2.0–3.0 MeV is 84%. The total efficiency is therefore $0.16\times0.84=0.13$. The half-life at the 90% confidence level (giving 19.0 events) implies a lower limit of $t_{1/2} \geq 4.4\times10^{20}$ yr.

In order to compare our results to those of a recent report by Avignone *et al.*[7] [University of South Carolina–Pacific Northwest Laboratory Collaboration (USC–PNL)] of positive evidence for the majoron mode of double-beta decay in ^{76}Ge, one must take into account the ratio of phase-space factors (8.2 in favor of ^{82}Se)[8] and the square of the different nuclear matrix elements. The ratio of $|M_{Se}^{0\nu}/M_{Ge}^{0\nu}|^2$ calculated by different authors for the zero-neutrino mode (with or without majoron) has values of 0.53,[9] 0.59,[10] 0.65,[11] and 0.78.[12] This ratio is remarkably insensitive to the calculation schemes, especially when one considers that the value of $|M^{0\nu}|^2$ varies by up to 2 orders of magnitude among these same authors. The net effect of both phase space and nuclear matrix elements is to make the half-life of ^{82}Se a factor of 4.3–6.4 times more sensitive than an equivalent half-life value or limit in ^{76}Ge. Therefore our limit (90% confidence level) would correspond to a ^{76}Ge limit of $t_{1/2} \geq (19-28)\times10^{20}$ yr, a factor of about 3–4 longer than the USC–PNL value of $6(1)\times10^{20}$ yr.[7]

Our result may also be expressed in terms of g_{ee}, the coupling of the majoron to the neutrino. Using the expression for g_{ee} from Doi, Kotani, and Takesugi,[8] we find that $g_{ee} \leq (2-30)\times10^{-4}$ depending on the value of $|M_{Se}^{0\nu}|$ chosen. Correspondingly using the USC–PNL half-life[7] we calculate $g_{ee}=(4-50)\times10^{-4}$. In both cases, the lower value of the range corresponds to the matrix element calculation from Ref. 10 and the higher one to Ref. 9.

We are in the process of analyzing all our candidate events for all three modes of double-beta decay. As one can see from Fig. 3, the tail of the 2ν energy distribution could account for up to three or four counts in the majoron region. Presumably (assuming the majoron does not exist), the remainder are due to gamma rays from both ^{208}Tl beta decay ($E_\gamma=2.6$ MeV) and cosmic rays. These gamma rays can produce two-electron events by the previously described Compton-Moeller process (and yet escape our Moeller cut) and/or by multiple Compton scattering.

We are now installing a new TPC in the same location at the University of California, Irvine. The new TPC has been constructed to eliminate the obvious sources of natural radioactivity. In addition we are planning to reduce the background even further by taking the experiment underground as soon as the new TPC has been debugged and commissioned. Even a modest 20-m depth will reduce the cosmic-ray flux a factor of 10. If the majoron is responsible for our remaining events between 2 and 3 MeV, the rate underground with the new detector will remain unchanged from the current surface level results. Furthermore, once the two-neutrino rate has been demonstrated, we will increase the source mass to 38 g of ^{82}Se. This will correspondingly increase our sensitivity to the zero-neutrino mode over the current experiment by a factor of 2.7.

We acknowledge the encouragement and advice of Professor F. Reines. A discussion with Dr. J. Engel has been very helpful. This project is supported by the U.S. Department of Energy under Contract No. DE-AT03-76ER71019.

Note added.— We have recently received a preprint[13] from the authors of Ref. 8, in which they correct an error in their phase-space calculations of double-beta decay with majoron emission. The effect of this change is to reduce all our calculated g_{ee} values by a factor of $\sqrt{2}$. This does not affect the comparison of our results in ^{82}Se to those in ^{76}Ge because this is a common factor in both cases.

[1]H. M. Georgi, S. L. Glashow, and S. Nussinov, Nucl. Phys. **B193**, 297 (1981).

[2]G. B. Gelmini and M. Roncadelli, Phys. Lett. **99B**, 411 (1981).

[3]Y. Chikashigi, R. N. Mohapatra, and R. D. Peccei, Phys. Lett. **98B**, 265 (1981).

[4]S. R. Elliott, A. A. Hahn, and M. K. Moe, Phys. Rev. Lett. **56**, 2582 (1986); M. K. Moe, A. A. Hahn, and H. E. Brown, in *The Time Projection Chamber—1983,* edited by J. A. MacDonald, AIP Conferences Proceedings No. 108 (American Institute of Physics, New York, 1984), p. 37.

[5]The lone remaining event at 3.3 MeV which is not attributable to Moeller scattering is most likely due to the beta-decay, internal-conversion sequence of ^{208}Th into ^{208}Pb (lead) on the source plane. It is identified from the energy of one of its two electrons which matches the energy of the ^{208}Pb internal-conversion line.

[6]T. Kirsten, in *Proceedings of the International Symposium on Nuclear Beta Decay and Neutrinos, Osaka, Japan, 1986,* edited by T. Kotani, H. Ejiri, and E. Takasugi (World Scientific, Singapore, 1986), p. 81.

[7]F. T. Avignone, R. L. Brodzinski, H. S. Miley, and J. H. Reeves, to be published.

[8]M. Doi, T. Kotani, and E. Takasugi, Prog. Theor. Phys., Suppl. **83**, 1 (1985).

[9]J. Engel, P. Vogel, and M. R. Zirnbauer, California Institute of Technology Report No. MAP-95, 1987 (to be published).

[10]K. Grotz and H. V. Klapdor, Phys. Lett. **153B**, 1 (1985).

[11]W. C. Haxton and G. J. Stephenson, Jr., Prog. Part. Nucl. Phys. **12**, 409 (1984).

[12]T. Tomoda and A. Faessler, to be published.

[13]M. Doi, T. Kotani, and E. Takasugi, Osaka University Report No. OS-GE-87-07, 1987 (to be published).

VOLUME 70, NUMBER 19 PHYSICAL REVIEW LETTERS 10 MAY 1993

Investigation of the Majoron-Accompanied Double-Beta Decay Mode of ^{76}Ge

M. Beck,[1] F. Bensch,[1] J. Bockholt,[1] G. Heusser,[1] M. Hirsch,[1] H. V. Klapdor-Kleingrothaus,[1,a] B. Maier,[1] F. Petry,[1] A. Piepke,[1] H. Strecker,[1] M. Völlinger,[1] K. Zuber,[1] A. Balysh,[2] S. T. Belyaev,[2,a] A. Demehin,[2] A. Gurov,[2] I. Kondratenko,[2] V. I. Lebedev,[2] and A. Müller[3]

[1]*Max-Planck-Institut für Kernphysik, P.O. Box 103980, 6900 Heidelberg, Germany*

[2]*Kurchatov Institute, Moscow, Russia*

[3]*LNGS, Istituto Nazionale di Fisica Nucleare Gran Sasso, Assergi, Italy*

(Received 24 August 1992; revised manuscript received 6 April 1993)

We have examined the double-beta decay mode accompanied by Majoron emission. After 223 days of measurement corresponding to about 615 kgd or 19.3 molyr of exposure we find a lower half-life limit for this decay channel of $T_{1/2}^{0\nu\chi} > 1.66(1.99)\times 10^{22}$ yr [90% (68%) C.L.] yielding a new limit for the Majoron-neutrino coupling of $\langle g_{\nu\chi}\rangle < 1.8\times 10^{-4}$ (90% C.L.).

PACS numbers: 23.40.Bw, 14.80.Gt

A test of the fundamental nature of neutrinos (Dirac or Majorana particle) can be done by double-beta decay research. Usually three possible decay modes are discussed:

$$^{Z}A \rightarrow {}^{Z+2}A + 2e^- + 2\bar{\nu}_e\,, \tag{1}$$

$$^{Z}A \rightarrow {}^{Z+2}A + 2e^-\,, \tag{2}$$

$$^{Z}A \rightarrow {}^{Z+2}A + 2e^- + \chi\,. \tag{3}$$

The first decay mode can be understood as a process of second-order Fermi theory, while the second one is not possible in the standard model. The observation of this decay would require massive Majorana neutrinos, if the forces are described by gauge theories. The Majorana term breaks lepton number conservation by two units, and therefore also baryon number minus lepton number, $B-L$. There are three possibilities to break $B-L$ in theory [1]: explicit $B-L$ breaking, meaning the Lagrangian contains $B-L$ violating expressions; spontaneous breaking of a local $B-L$ symmetry; spontaneous breaking of a global $B-L$ symmetry. Associated with the third possibility is the existence of a massless Nambu-Goldstone boson called the Majoron χ. There exist different possibilities to generate Majorana mass terms in extensions of the standard model and therefore also different Majoron models characterized by their weak isospin. By using an additional Higgs triplet Gelmini and Roncadelli [2,3] proposed a model leading to the so-called triplet Majoron. Because it should contribute [3] two neutrino flavors to the Z^0 width it seems to be ruled out by the recent measurements at LEP [4]. Also doublet Majoron models seem to be ruled out, which would contribute half a neutrino width. Nevertheless the existence of singlet Majorons [5] or a mixture of singlet and doublet Majorons is possible. Since Majorons couple to neutrinos it might occur in double-beta decay (see, e.g., [6]). Moreover the model of singlet Majorons has recently experienced growing interest due to the attempt to build possible neutrino mass schemes involving a 17 keV neutrino [7].

The corresponding diagram for Majoron emission in double-beta decay (hereafter the $0\nu\chi$ mode) is shown in Fig. 1. Since it is a three-body decay the energy spectrum is continuous. The sum energy spectrum of both electrons has a maximum around 1500 keV for ^{76}Ge. In deriving a half-life limit we have used the second enriched detector of the Heidelberg-Moscow $\beta\beta$ experiment [8] used to measure the electrons emitted in the $0\nu\chi$ decay. The mass of the detector is 2.88 kg where the fraction of the decaying isotope ^{76}Ge is 86% compared to only 7.8% in natural Ge. The measuring time was 223 d.

To obtain limits on the Majoron spectrum we use the following procedure. Out of the measured spectrum we make a cutoff of all observed spectral lines. The surrounding background is extrapolated into the line regions. This reduces the integral count rate by about 5%. In the region from 1.5–2.1 MeV only the weak 1764.5 keV line of ^{214}Bi is removed. In addition we subtract the Compton continua of the cosmogenically produced isotopes in the copper parts of the crystal holder (^{57}Co, ^{58}Co, ^{54}Mn, and ^{60}CO) and in the crystal (^{57}Co, ^{58}Co, ^{54}Mn, and ^{65}Zn), identified by their characteristic γ lines and the full absorption peaks shifted by the x-ray energy. In the energy interval from 2000 to 2880 keV the corresponding con-

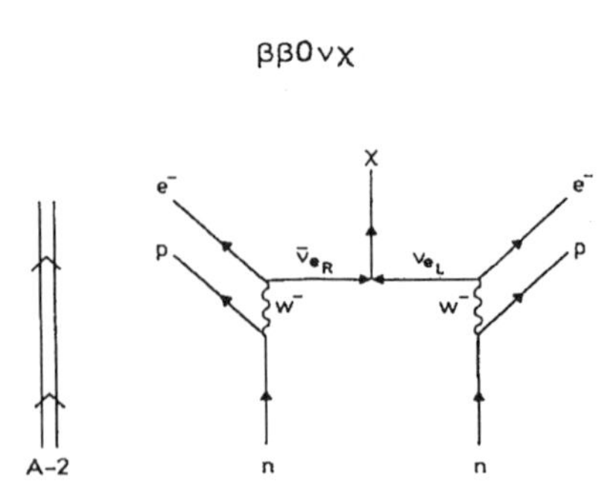

FIG. 1. Feynman diagram for double-beta decay with the emission of a Majoron.

VOLUME 70, NUMBER 19 PHYSICAL REVIEW LETTERS 10 MAY 1993

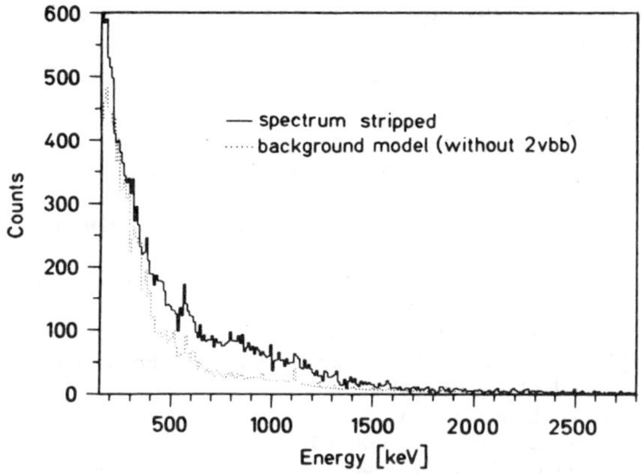

FIG. 2. Comparison of the measured spectrum without lines (solid curve) and the contribution of all the background components described in the text (dotted curve). It is clearly visible that it is not the case that the whole spectrum can be caused by these components.

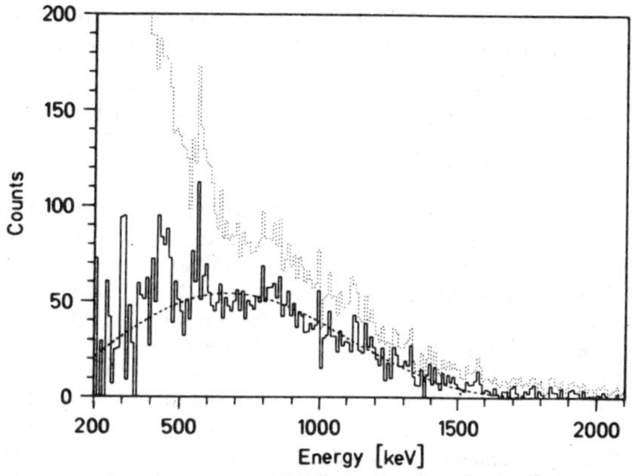

FIG. 3. Comparison of the measured (dotted curve) and background-subtracted (solid curve) spectrum. Especially the part below 800 keV experiences some modification. For comparison a calculated 2ν spectrum with a half-life of $T=1.43\times10^{21}$ yr is shown (dashed curve), which is the major background component.

tinuous contributions lead to a correction of the order of 14% and 1.5%, respectively. The continuous background of the natural decay chains ($4N,4N+2$) identified by the observed γ lines is also subtracted (10.2% correction). We further simulated the ^{210}Bi (a ^{210}Pb daughter) bremsstrahlung spectrum, dominating the energy range up to 1.1 MeV and contributing 33.8% to the measured background, but having a negligible effect on the Majoron spectrum. The ^{210}Pb content of the lead shielding was measured by low-level α spectroscopy to be 0.36 ± 0.03 Bq/kg. A quantitative treatment of the ^{210}Pb spectrum is important for the analysis of the $2\nu\beta\beta$ decay. All these background contributions are calculated with the help of the Monte Carlo code GEANT3 (version 3.14), which was checked experimentally with an x-y scan of the detector using collimated γ sources. Because not all of the background above 2 MeV originates from the discussed components, we determined the remaining averaged background between 2.1 and 2.5 MeV (0.23 count/keV) and subtracted a corresponding background over the whole energy range (in total a correction of 4%). This correction is a conservative limit, because at lower energies there will be additional contributions from Compton scattering. The sum of all these contributions in comparison with the measured spectrum is shown in Fig. 2. Two groups [9,10] have reported evidence for the $2\nu\beta\beta$ decay which was later confirmed [11]. The reported half-lives were all about 1×10^{21} yr. Figure 3 shows our measured spectrum with and without subtraction of the background model and also a theoretically calculated 2ν spectrum assuming $T_{1/2}^{2\nu}=[1.43\pm0.04(\text{stat})\pm0.13(\text{syst})]\times10^{21}$ yr. This is the value we derive for the $2\nu\beta\beta$ decay by a maximum likelihood fit. It will be discussed elsewhere [12].

The energy range from 1.1 to 2.05 MeV contains 74.5% of the theoretical $0\nu\chi$ spectrum. In this range there are still 208 remaining events corresponding to a difference from zero at a confidence level of 98%. Figure 4 shows the solid curve of Fig. 3 after further subtraction of the 2ν spectrum. However, the analysis of the logarithmic likelihood ratio shows that there is not good agreement with the expected form of a Majoron spectrum. Interpreting the result as an upper limit for this decay mode, we find $T_{1/2}^{0\nu\chi}>1.66(1.99)\times10^{22}$ yr 90% C.L. (68% C.L.). (This value would still increase when subtracting the continuous part of the background produced by ^{56}Co, ^{60}Co, and especially ^{68}Ge in the crystal.) This half-life limit can be converted into a limit for the neutrino-Majoron coupling constant

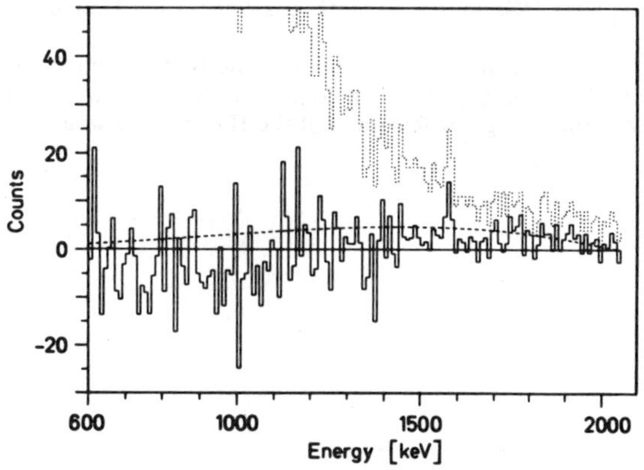

FIG. 4. Remaining spectrum (solid curve) after subtraction of the 2ν mode, and a calculated $0\nu\chi$ spectrum with a half-life of $T=1.66\times10^{22}$ yr (dashed curve). Also shown is the measured spectrum (dotted curve).

VOLUME 70, NUMBER 19 PHYSICAL REVIEW LETTERS 10 MAY 1993

TABLE I. Half-life limits for the $0\nu\chi$ decay and the corresponding limits for the neutrino-Majoron coupling constant for several isotopes.

Isotope	Experiment	$T_{1/2}$ (10^{21} yr)	$10^4\langle g_{\nu\chi}\rangle$	Ref.
^{76}Ge	MPIK-KIAE	16.6(90%)	1.8	
^{76}Ge	ITEP	10(68%)	2.2	[9]
^{76}Ge	UCSB-LBL	1.4(90%)	5.8	[16]
^{76}Ge	PNL-USC	6.0	2.8	[10]
^{76}Ge	Cal.PSI-Neu.	1.0(90%)	6.9	[17]
^{100}Mo	LBL-MHC-UNM	0.33(90%)	6.2	[18]
^{136}Xe	ITEP	0.19(68%)	12.5	[19]
^{136}Xe	Cal.-PSI-Neu.	7.2(90%)	2.0	[20]
^{82}Se	UCI	1.6(68%)	12.5	[21]
^{150}Nd	INR	0.07(68%)	1.9	[22]
^{48}Ca	ITEP	0.72	5.1	[23]
^{128}Te[a]	Washington Univ.-Tata	7700	0.3	[24]

[a]Geochemical experiment.

$$T_{1/2}^{-1}[\mathrm{yr}^{-1}] = |M_{GT} - M_F|^2 F^{0\nu\chi} \langle g_{\nu\chi}\rangle|^2 , \quad (4)$$

where

$$\langle g_{\nu\chi}\rangle = \sum_{ij} g_{\nu\chi} U_{ei} U_{ej} . \quad (5)$$

This results in a neutrino-Majoron coupling constant of $\langle g_{\nu\chi}\rangle < 1.75(1.60)\times 10^{-4}$ (90% and 68% C.L., respectively).

A comparison of experimental limits for Majoron emission half-lives in double-beta decay and coupling constants determined for several isotopes is given in Table I. Another value for the coupling constant of about 3×10^{-4} was discussed in [13]. All phase space factors $F^{0\nu\chi}$ [14] and all the matrix elements used in Table I are taken from [15]. Neglecting the contribution of heavy neutrino exchange the nuclear matrix element is identical to that of the neutrinoless double-beta decay (see also the discussion in [14]).

Concluding, we have examined the possibility of Majoron emission in the double-beta decay of ^{76}Ge. The Heidelberg-Moscow experiment setup to probe the neutrino mass down to about 0.1 eV yields, with only part of its final ^{76}Ge source strength, a stringent limit for $\beta\beta 0\nu\chi$ decay.

The experiment is supported by the Bundesministerium für Forschung und Technologie, the State Committee of Atomic Energy of Russia, and the INFN. We would like to thank especially Professor E. Bellotti and Professor N. Cabibbo for their continuous generous support.

(a)Spokesmen of the collaboration.

[1] R. N. Mohapatra and M. Pal, *Massive Neutrinos in Physics and Astrophysics* (World Scientific, Singapore, 1989).
[2] G. B. Gelmini and M. Roncadelli, Phys. Lett. **99B**, 411 (1981).
[3] H. M. Georgi, S. L. Glashow, and S. Nussinov, Nucl. Phys. **B193**, 297 (1981).
[4] J. Steinberger, Phys. Rep. **203**, 345 (1991).
[5] Y. Chikashige, R. N. Mohapatra, and R. D. Peccei, Phys. Rev. Lett. **45**, 1926 (1980).
[6] Z. G. Berezhiani, A. Y. Smirnov, and J. W. F. Valle, Phys. Lett. B (to be published).
[7] G. Gelmini, S. Nussinov, R. D. Peccei, Report No. UCLA-TEP 15-91, 1991 (unpublished).
[8] A. Balysh, M. Beck, S. T. Belyaev, F. Bensch, J. Bockholt, A. Demehin, J. Echternach, A. Gurov, G. Heusser, H. V. Klapdor-Kleingrothaus, I. Kondratenko, V. I. Lebedev, B. Maier, A. Müller, F. Petry, A. Piepke, U. Schmidt-Rohr, H. Strecker, and K. Zuber, Phys. Lett. B **283**, 32 (1992).
[9] A. A. Vasenko, *et al.*, Mod. Phys. Lett. A **5**, 1299 (1990).
[10] H. S. Miley *et al.*, Phys. Rev. Lett. **65**, 3092 (1990).
[11] F. T. Avignone *et al.*, Phys. Lett. B **256**, 559 (1991).
[12] Heidelberg-Moscow Collaboration, A. Balysh, *et al.* (to be published).
[13] M. Moe *et al.*, Report No. UCI-Neutrino 92-1 (to be published).
[14] M. Doi, T. Kotani, and E. Takasugi, Phys. Rev. D **37**, 2575 (1988).
[15] A. Staudt, K. Muto, and H. V. Klapdor-Kleingrothaus, Europhys. Lett. **13**, 31 (1990).
[16] D. O. Caldwell *et al.*, Phys. Rev. Lett. **59**, 419 (1987).
[17] P. Fisher *et al.*, Phys. Lett. B **218**, 257 (1989).
[18] M. Alston-Garnjost *et al.*, Phys. Rev. Lett. **60**, 1928 (1988).
[19] A. S. Barabash *et al.*, Yad. Fiz. **51**, 3 (1990) [Sov. J. Nucl. Phys. 51, 1 (1990)].
[20] J.-L. Vuilleumier (private communication).
[21] S. R. Elliot, A. A. Hahn, and M. K. Moe, Phys. Rev. Lett. **59**, 1649 (1987).
[22] A. A. Klimenko, A. A. Pomansky, and A. A. Smolnikov, in *Proceedings of the Eleventh International Conference on Neutrino Physics and Astrophysics* (World Scientific, Singapore, 1984).
[23] A. S. Barabash, Phys. Lett. B **216**, 257 (1989).
[24] T. Bernatowicz *et al.*, Phys. Rev. Lett. **69**, 2341 (1992).

4 April 1996

PHYSICS LETTERS B

ELSEVIER

Physics Letters B 372 (1996) 8–14

On the observability of Majoron emitting double beta decays

M. Hirsch, H.V. Klapdor-Kleingrothaus, S.G. Kovalenko [1], H. Päs

Max-Planck-Institut für Kernphysik, P.O. Box 10 39 80, D-69029 Heidelberg, Germany

Received 24 October 1995; revised manuscript received 8 January 1996
Editor: C. Mahaux

Abstract

Because of the fine-tuning problem in classical Majoron models in recent years several new models were invented. It is pointed out that double beta decays with new Majoron emission depend on new matrix elements, which have not been considered in the literature. A calculation of these matrix elements and phase space integrals is presented. We find that for new Majoron models extremely small decay rates are expected.

PACS: 13.15; 23.40; 21.60J; 14.80
Keywords: Majoron; Double beta decay; QRPA; Neutrino interactions

In many theories of physics beyond the standard model neutrinoless double beta decays can occur with the emission of new bosons, so-called Majorons [1–4]:

$$2n \rightarrow 2p + 2e^- + \phi, \tag{1}$$

$$2n \rightarrow 2p + 2e^- + 2\phi. \tag{2}$$

Since classical Majoron models [1,5] require severe fine-tuning in order to preserve existing bounds on neutrino masses and at the same time get an observable rate for Majoron emitting double beta decays in recent years several new Majoron models have been constructed [6–8], where the terminus Majoron means in a more general sense light or massless bosons with couplings to neutrinos. The main novel features of the "New Majorons" are that they can carry units of leptonic charge, that there can be Majorons which are no Goldstone bosons [6] and that decays with the emission of two Majorons [4,7] can occur. The latter can be scalar-mediated or fermion-mediated. In vector Majoron models the Majoron becomes the longitudinal component of a massive gauge boson [8] emitted in double beta processes. For simplicity we will call it Majoron, too.

In Table 1 the nine Majoron models we considered are summarized [7,8]. It is divided in the sections I for lepton number breaking and II for lepton number conserving models. The table shows also whether the corresponding double beta decay is accompanied by the emission of one or two Majorons.

The next three entries list the main features of the models: The third column lists whether the Majoron is a Goldstone boson or not (or a gauge boson in case of vector Majorons IIF). In column four the leptonic charge L is given. In column five the "spectral index" n of the sum energy of the emitted electrons is listed, which is defined from the phase space of the emitted particles, $G \sim (Q_{\beta\beta} - T)^n$, where $Q_{\beta\beta}$ is the energy

[1] On leave from Joint Institute for Nuclear Research, Dubna, Russia.

PII S0370-2693(96)00038-X

Table 1
Different Majoron models according to Bamert/Burgess/Mohapatra [9]. The case IIF corresponds to the model of Carone [10].

Case	Modus	Goldstone boson	L	n	Matrix element
IB	$\beta\beta\phi$	no	0	1	$M_F - M_{GT}$
IC	$\beta\beta\phi$	yes	0	1	$M_F - M_{GT}$
ID	$\beta\beta\phi\phi$	no	0	3	$M_{F\omega^2} - M_{GT\omega^2}$
IE	$\beta\beta\phi\phi$	yes	0	3	$M_{F\omega^2} - M_{GT\omega^2}$
IIB	$\beta\beta\phi$	no	−2	1	$M_F - M_{GT}$
IIC	$\beta\beta\phi$	yes	−2	3	M_{CR}
IID	$\beta\beta\phi\phi$	no	−1	3	$M_{F\omega^2} - M_{GT\omega^2}$
IIE	$\beta\beta\phi\phi$	yes	−1	7	$M_{F\omega^2} - M_{GT\omega^2}$
IIF	$\beta\beta\phi$	gauge boson	−2	3	M_{CR}

release of the decay and T the sum energy of the two electrons. The different shapes can be used to distinguish the different decay modes from each other and the double beta decay with emission of two neutrinos. In the last column we listed the nuclear matrix elements which will be defined in more detail later. Nuclear matrix elements are necessary to convert half-lives (or limits thereof) into values for the effective Majoron-neutrino coupling constant, using the approximate (see below) relations [4,9]:

$$[T_{1/2}]^{-1} = |\langle g_\alpha \rangle|^m \cdot |M_\alpha|^2 \cdot G_{BB_\alpha}, \tag{3}$$

with $m = 2$ for $\beta\beta\phi$-decays or $m = 4$ for $\beta\beta\phi\phi$-decays. The index α in Eq. (3) indicates that effective coupling constants g_α, nuclear matrix elements M_α and phase spaces G_{BB_α} differ for different models.

As shown in Table 1, several Majoron models with different theoretical motivation can lead to signals in double beta decays which are experimentally indistinguishable. The interpretation of experimental half-life limits in terms of the "effective Majoron-neutrino coupling constant" is therefore model dependent. Subsequently we give a brief summary of the theoretical background on which our conclusions on the different Majoron models are based.

Single Majoron emitting double beta decays ($0\nu\beta\beta\phi$) can be roughly divided into two classes, $n = 1$ (case IB, IC and IIB) and $n = 3$ (IIC and IIF) decays.

As has been noted in [7] as long as $0\nu\beta\beta$ decay has not been observed, the three $n = 1$ decays are indistinguishable from each other. We will call these Majorons "ordinary", since they contain the subgroup IC, which leads to the classical Majoron models [1,2,10]. For all ordinary Majorons the effective Majoron-neutrino interaction Lagrangian, leading to $0\nu\beta\beta\phi$ decay is [2,6]

$$L^{\mathrm{O.M.}}_{\phi\nu\nu} = -\tfrac{1}{2}\overline{\nu_i}(a_{ij}P_L + b_{ij}P_R)\nu_j\phi^* + \mathrm{h.c.} \tag{4}$$

Here, $P_{R/L} = 1/2(1 \pm \gamma_5)$. Using Eq. (4) the amplitude corresponding to the Feynman graph is, in the notation of [6]

$$\mathcal{A}^{\mathrm{O.M.}}(0\nu\beta\beta\phi) = 4\sqrt{2}\sum_{i,j} V_{ei}V_{ej} \times \int \frac{d^4q}{(2\pi)^4} \frac{m_i m_j a_{ij} + q^2 b_{ij}}{(q^2 - m_i^2 + i\epsilon)(q^2 - m_j^2 + i\epsilon)} \times (w_F - w_{GT}). \tag{5}$$

V_{ei}, V_{ej} are elements of the neutrino mixing matrix, m_i and m_j denote neutrino mass eigenvalues and $w_{F/GT}$ are nuclear matrix elements containing double Fermi and Gamow–Teller operators. To arrive at the factorized decay rate Eq. (3), the usual assumption $m_{i,j} \ll q \approx p_F \approx \mathcal{O}(100\ \mathrm{MeV})$, where p_F is the typical Fermi momentum of nucleons, is made. By this assumption the term proportional to a_{ij} can be dropped and the effective coupling constant is defined as

$$\langle g \rangle^{\mathrm{O.M.}} = \sum_{i,j} V_{ei}V_{ej}b_{ij}. \tag{6}$$

In this approximation matrix elements for ordinary Majoron decays coincide with the leading terms M_{GT} and M_F of the well-known mass mechanism of $0\nu\beta\beta$ decay.

Burgess and Cline advocated the so-called charged Majoron model IIC [6]. In this model the effective interaction Lagrangian is

$$L^{\mathrm{C.M.}}_{\phi\nu\nu} = -\frac{i}{2f}\overline{\nu}\gamma^\mu(A_L P_L + A_R P_R)\nu\partial_\mu\phi + \mathrm{h.c.} \tag{7}$$

Note that in the charged Majoron model the two additional powers of n in the phase space integrals originate from the derivative coupling of the Majoron in $L^{\mathrm{C.M.}}_{\phi\nu\nu}$. As shown in [6], for charged Majorons the contribution from the leading order matrix elements to the decay rate vanishes identically, so that one has to go to the next higher order in the non-relativistic impulse approximation of hadronic currents. The amplitude for $0\nu\beta\beta\phi$ decay is then given by

$$\mathcal{A}^{\mathrm{C.M.}}(0\nu\beta\beta\phi) = 8\sqrt{2}\sum_{i,j} V_{ei}V_{ej} \times \int \frac{d^4q}{(2\pi)^4} \frac{q^2 b_{ij}}{(q^2 - m_i^2 + i\epsilon)(q^2 - m_j^2 + i\epsilon)} \times (w_5 + w_6), \quad (8)$$

which leads to an effective coupling constant $\langle g\rangle^{\mathrm{C.M.}}$ as in the ordinary Majoron case, but with b_{ij} given by $b = \frac{i}{f}(A_L m^* + m^* A_R)$, with the neutrino mass matrix m, generator matrices $A_{L/R}$ and the decay constant f. The hadronic term w_6 is similar but not identical to the recoil matrix element of $0\nu\beta\beta$ decay induced by right-handed currents. This difference has turned out to be important. In the notation of [6]

$$w_5 = \frac{i\mu \boldsymbol{q}}{|q|^2(q_0^2 - \mu^2 + i\epsilon)} \langle F| e^{-iqr}[g_A^2(\boldsymbol{C}_n \boldsymbol{\sigma}_m - \boldsymbol{C}_m \boldsymbol{\sigma}_n) + g_V^2(\boldsymbol{D}_m - \boldsymbol{D}_n)]|I\rangle, \quad (9)$$

$$w_6 = \frac{\mu g_A g_V \boldsymbol{q}}{|q|^2(q_0^2 - \mu^2 + i\epsilon)} \langle F| e^{-iqr}[\boldsymbol{D}_n \times \boldsymbol{\sigma}_m + \boldsymbol{D}_m \times \boldsymbol{\sigma}_n]|I\rangle, \quad (10)$$

in which the summation over $\sum_{nm} \tau_n^+ \tau_m^+$ is suppressed. Here, C_n and $\boldsymbol{D}_n$ are nuclear recoil terms [9]

$$C_n = (\boldsymbol{P}_n + \boldsymbol{P}'_n)\cdot \boldsymbol{\sigma}_n/(2M_n) - (E_n - E'_n)(\boldsymbol{P}_n - \boldsymbol{P}'_n)\cdot \boldsymbol{\sigma}_n/(2m_\pi^2), \quad (11)$$

$$\boldsymbol{D}_n = [(\boldsymbol{P}_n + \boldsymbol{P}'_n) + i\mu_\beta(\boldsymbol{P}_n - \boldsymbol{P}'_n)\times \boldsymbol{\sigma}_n]/(2M_n); \quad (12)$$

$\boldsymbol{P}_n$ (E_n) and $\boldsymbol{P}'_n$ (E'_n) are momenta (energies) of initial and final state nucleons, m_π is the pion and M_n the nucleon mass and μ_β originates from the weak magnetism.

The terms of w_5 are neglected compared to w_6 due to the estimation $(\boldsymbol{P}_n+\boldsymbol{P}'_n) \leq (\boldsymbol{P}_n-\boldsymbol{P}'_n), (E_n-E'_n) \leq \mathcal{O}(Q_{\beta\beta})$ [9]. Following [11] we will also keep only the central part of the recoil term $\boldsymbol{D}$. Although both are approximations, which needs to be checked numerically, we do not expect it to affect any of our conclusions.

Finally for vector Majoron models (case IIF) [8]

$$L_{\phi\nu\nu}^{\mathrm{V.M.}} = -\frac{i}{2f}\overline{\nu}\gamma^\mu(c_{ij}P_L + d_{ij}P_R)\nu X^\mu + \text{h.c.}, \quad (13)$$

where X^μ is the emitted massive gauge boson. The effective coupling constant can be defined as in the ordinary Majoron model, with the replacement $b_{ij} = \frac{1}{2M}(c_{ij}m_j - m_i d_{ij})$, where M is the gauge boson mass. As discussed in [8], the vector Majoron amplitude approaches the charged Majoron one in the limit of vanishing gauge boson masses, which we assume in the phase space integration. They depend on the same nuclear matrix elements than the charged Majoron discussed above. We will therefore not repeat the definitions here.

Double Majoron emitting decays ($0\nu\beta\beta\phi\phi$), mediated by fermions, can have either spectral index $n = 7$ or $n = 3$, depending on whether the Majoron couples derivatively suppressed or not. [7]

In addition, in principle $0\nu\beta\beta\phi\phi$ decays could also be mediated by exotic scalars. The amplitude of scalar-mediated decays, however, is expected to be very much suppressed, since the scalars must have masses larger than about 50 GeV due to the LEP-measurements [7]. We will therefore concentrate on the fermion-mediated decays.

The Yukawa coupling of the Majoron to the neutrinos for the $n = 3$ decays (cases ID, IE, IID) is given as

$$L_{\phi\nu\nu}^{yuk} = -\overline{\nu_i}(A_{ia}P_L + B_{ia}P_R)N_a\phi + \text{h.c.} \quad (14)$$

where A_{ia} and B_{ia} represent arbitrary Yukawa-coupling matrices and N_a are sterile neutrinos. The corresponding amplitude for $0\nu\beta\beta\phi\phi$ decay is

$$\mathcal{A}^{\mathrm{D.M.}}(0\nu\beta\beta\phi\phi) = \sqrt{\frac{2}{3\pi^2}}\sum_{i,j,a} V_{ei}V_{ej}\int \frac{d^4q}{(2\pi)^4} \times \frac{\mathcal{N}_{ija}}{(q^2 - m_i^2 + i\epsilon)(q^2 - m_j^2 + i\epsilon)(q^2 - m_a^2 + i\epsilon)} \times (w_F - w_{GT}). \quad (15)$$

Although for $\mathcal{A}^{\mathrm{D.M.}}(0\nu\beta\beta\phi\phi)$ the same combination of nuclear operators appears ($w_F - w_{GT}$), note the additional $(q^2 - m^2)^{-1}$ compared to $\mathcal{A}^{\mathrm{O.M.}}(0\nu\beta\beta\phi)$. $\mathcal{N}_{ija}$ in (15) is given by

$$\mathcal{N}_{ija} = -q^2(A_{ia}B_{ja}m_{\nu_i} + A_{ja}B_{ia}m_{\nu_j} + B_{ia}B_{ja}m_{N_a}) + A_{ia}A_{ja}m_{\nu_i}m_{\nu_j}m_{N_a}. \quad (16)$$

In order to separate the particle physics parameters from the nuclear structure calculation, it is most convenient to neglect the last term in Eq. (16). This can be justified by considering that the mass eigenvalues

$m_{\nu_{i,j}} \ll p_F$ so that the last term in Eq. (16) for not too large m_{N_a} is suppressed compared to the first three by at least $m_{\nu_{i,j}}/p_F \simeq \mathcal{O}(10^{-5-6})$. Then, the q^2 is absorbed into the neutrino potentials and we redefine $\mathcal{N}_{ija}$ to obtain the effective coupling constant as

$$\langle g\rangle = (\frac{1}{m_e}\sum_{ij} V_{ei}V_{ej}[A_{ia}B_{ja}m_{\nu_i} + A_{ja}B_{ia}m_{\nu_j} + B_{ia}B_{ja}m_{N_a}])^{\frac{1}{2}}. \quad (17)$$

Note that we have arbitrarily absorbed a factor of m_e^{-1} into the definition of $\langle g\rangle$ here to get for the effective coupling a dimensionless quantity.

For the $n = 7$ $0\nu\beta\beta\phi\phi$ decays, the effective Lagrangian is (IIE/fermion mediated)

$$L_{\phi\nu\nu} = -i\overline{\nu_i}\gamma^\mu(X_{ia}P_L + Y_{ia}P_R)N_a\partial_\mu\phi + \text{h.c.} \quad (18)$$

Again, N_a denotes a sterile neutrino and the derivative coupling of ϕ accounts for the additional powers of n in the phase space integrals. The amplitude for $n = 7$ decays is the same as for the $n = 3$ case, discussed above, with the replacement: $\mathcal{N}_{ija} = X_{ia}Y_{ia}m_{N_a}$.

Note that X_{ia} and Y_{ia} have the dimension of an inverse mass. Therefore, also $\langle g\rangle$ has a dimension of an inverse mass. To define a dimensionless coupling constant in this case one would have to specify the symmetry breaking scale, which is however undetermined by the model.

For the Majoron models considered in this work there are five nuclear matrix elements to be calculated. Within the closure approximation they are defined as

$$M_F = (\frac{g_V^2}{g_A^2})\langle N_f\|h_{\text{mass}}(\mu, \boldsymbol{r})\tau_n^+\tau_m^+\|N_i\rangle, \quad (19)$$

$$M_{GT} = \langle N_f\|h_{\text{mass}}(\mu, \boldsymbol{r})\tau_n^+\tau_m^+\boldsymbol{\sigma}_n\boldsymbol{\sigma}_m\|N_i\rangle, \quad (20)$$

$$M_{CR} = (\frac{g_V}{g_A})\frac{f_W}{3}\langle N_f\|h_R(\mu, \boldsymbol{r})\tau_n^+\tau_m^+\boldsymbol{\sigma}_n\boldsymbol{\sigma}_m\|N_i\rangle, \quad (21)$$

$$M_{F\omega^2} = (\frac{g_V^2}{g_A^2})\langle N_f\|h_{\omega^2}(\mu, \boldsymbol{r})\tau_n^+\tau_m^+\|N_i\rangle, \quad (22)$$

$$M_{GT\omega^2} = \langle N_f\|h_{\omega^2}(\mu, \boldsymbol{r})\tau_n^+\tau_m^+\boldsymbol{\sigma}_n\boldsymbol{\sigma}_m\|N_i\rangle, \quad (23)$$

where h_α denote the neutrino potentials

$$h_{\text{mass}}(\mu, \boldsymbol{r}) = \frac{R}{2\pi^2}\int\frac{d^3\boldsymbol{q}}{\omega}\frac{e^{i\boldsymbol{qr}}}{\omega+\mu}, \quad (24)$$

$$h_R(\mu, \boldsymbol{r}) = \frac{1}{4\pi^2}(\frac{1}{M})\int\frac{d^3\boldsymbol{q}}{\omega}e^{i\boldsymbol{qr}}\frac{\mu+2\omega}{(\mu+\omega)^2}, \quad (25)$$

$$h_{\omega^2}(\mu, \boldsymbol{r}) = \frac{m_e^2 R}{16\pi^2}\int d^3\boldsymbol{q}q^2e^{i\boldsymbol{qr}}\frac{3\mu^2+9\mu\omega+8\omega^2}{\omega^5(\mu+\omega)^3}. \quad (26)$$

Here $\mu = \langle E_N - E_I\rangle$ denotes the average excitation energy of the intermediate nuclear states. $\omega = \sqrt{q^2+m^2}$ is the energy of the neutrino and since we assume all neutrinos to be light, the indices on neutrino masses have been dropped. Note that in order to define matrix elements dimensionless we follow the convention of [9]. That is $h_{\text{mass}}(\boldsymbol{r})$ and $h_{\omega^2}(\boldsymbol{r})$ are arbitrarily multiplied by the nuclear radius $R = r_0 A^{\frac{1}{3}}$ with $r_0 = 1.2$ fm, while $h_R(\boldsymbol{r})$ includes the nucleon mass. Compensating factors appear in the prefactors of the phase space integrals.

We have carried out a numerical calculation of these matrix elements within the pn-QRPA model of [12,13]. To estimate the uncertainties of the nuclear structure matrix elements the parameter dependence of the numerical results has been investigated. Since the matrix elements M_{GT} and M_F have been studied before [12], we will concentrate on M_{CR}, $M_{GT\omega^2}$ and $M_{F\omega^2}$. M_{GT} and M_F can be calculated with an accuracy of about a factor of 2 [12].

The matrix element M_{CR} shows a very similar behaviour as M_{GT}. This is in agreement with the expectation, since only the central part of the recoil terms is taken into account, so that apart from the different neutrino potential M_{CR} has the same structure as M_{GT}. Neither variations of the strength of the particle–particle force g_{pp} nor a change in the intermediate state energies significantly affects the numerical value of M_{CR}. We therefore conclude that M_{CR} should be accurate up to a factor of 2, as is expected for M_{GT}.

Unfortunately, in the case of the matrix elements $M_{GT\omega^2}$ and $M_{F\omega^2}$ the situation is very different. Both, variations of g_{pp} or μ, can change the numerical results drastically (Fig. 1). In fact, it is found that $M_{GT\omega^2}$ displays a very similar dependence on g_{pp} as has been reported in pn-QRPA studies of $2\nu\beta\beta$ decay matrix elements [12]. Especially important is that in the region of the most probable value of g_{pp} $M_{GT\omega^2}$ crosses zero.

Also for variations of the assumed average intermediate state energy a rather strong dependence of the

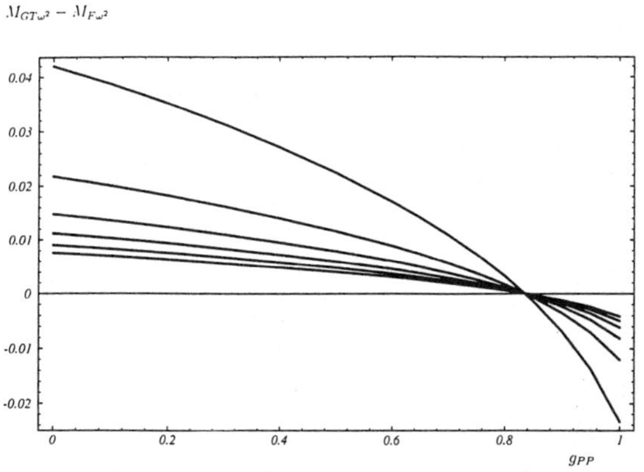

Fig. 1. $M_{\mathrm{GT}\omega^2} - M_{\mathrm{F}\omega^2}$ dependence of g_{PP} for different intermediate state energies $E_n = 4$ (top on the left), 8, 12, 16, 20, 24 (bottom on the left) MeV for ^{76}Ge.

results on the adopted value of μ has been found. As a consequence of this unpleasant strong dependence, for an accurate prediction of $M_{\mathrm{GT}\omega^2}$ and $M_{\mathrm{F}\omega^2}$ it seems necessary to go beyond the closure approximation.

The basic reason for the unusual sensitivity of $M_{\mathrm{GT}\omega^2}$ and $M_{\mathrm{F}\omega^2}$ on μ can be traced back to a certain difference in the neutrino potential of these matrix elements compared to $M_{\mathrm{GT/F}}$, $h_{\mathrm{mass}}(\mu, r) \sim \omega^{-2}$ while $h_{\omega^2}(\mu, r) \sim \omega^{-4}$. Contributions from very low momenta are therefore much preferred in $h_{\omega^2}(\mu, r)$ compared to $h_{\mathrm{mass}}(\mu, r)$. (Note that this leads also to a much smaller value for a typical ω than the naive expectation of $\omega \sim p_{\mathrm{F}} \sim \mathcal{O}(50\text{–}100)$ MeV!). With typical ω of only $\mathcal{O}(\text{few})$ MeV the strong dependence of $h_{\omega^2}(\mu, r)$ on μ becomes obvious.

Results of the calculation for various experimentally interesting isotopes are summarized in Table 2. Note that the matrix elements are valid for the limit of small intermediate particle masses, up to the order of 10 MeV. If any of the virtual particles in the Feynman graphs can have masses larger than 10 MeV, the matrix elements are no longer constant and the values in Table 2 should only be taken as upper limits for the analysis of data.

In comparison to the nuclear matrix elements phase space integrals can be calculated very accurately, so uncertainties of this calculation will not be discussed. We define the phase space integral as

Table 2
Dimensionless nuclear matrix elements of Majoron emitting modes calculated in this work.

Nucleus	$M_{\mathrm{F}} - M_{\mathrm{GT}}$	M_{CR}	$M_{\mathrm{F}\omega^2} - M_{\mathrm{GT}\omega^2}$
Ge 76	4.33	0.16	$\sim 10^{-3\pm 1}$
Se 82	4.03	0.14	$\sim 10^{-3\pm 1}$
Mo 100	4.86	0.16	$\sim 10^{-3\pm 1}$
Cd 116	3.29	0.10	$\sim 10^{-3\pm 1}$
Te 128	4.49	0.14	$\sim 10^{-3\pm 1}$
Te 130	3.90	0.12	$\sim 10^{-3\pm 1}$
Xe 136	1.82	0.05	$\sim 10^{-3\pm 1}$
Nd 150	5.29	0.15	$\sim 10^{-3\pm 1}$

$$G_{\mathrm{BB}_\alpha} = a_\alpha \cdot \int (Q_{\beta\beta} - \epsilon_1 - \epsilon_2)^n \prod_k p_k \epsilon_k f(\epsilon_k) d\epsilon_k, \tag{27}$$

where the prefactor a_α depends on the Majoron mode under consideration. A summary of the definitions is given in Table 3. $Q_{\beta\beta}$ is the maximum decay energy, ϵ_k and p_k are the energies and momenta of the outgoing electrons and $f(\epsilon_k)$ is the Fermi function calculated according to the description of [9]. Note the large difference in the phase space values of the old ($n = 1$) and new Majoron models.

Having calculated nuclear matrix elements and phase space integrals, it is straightforward to derive limits on the effective Majoron-neutrino coupling constants for the various Majoron models from experiment.

Although experimental half-life limits are comparable for all decay modes, as observed recently for ^{76}Ge decay [14,15], restrictive limits on the coupling constants of ordinary Majoron models contrast with limits on any of the new Majoron models, which will be weaker by 3–4 orders of magnitude.

The surprisingly weak limits which one obtains for the neutrino-Majoron coupling constant due to small matrix elements and phase spaces for all of the new Majoron models, require further explanation. (Note that the following discussion is independent of the isotope under consideration.) Consider, for example, ordinary and charged Majoron $0\nu\beta\beta\phi$ decays. Limits on the effective coupling constant for single Majoron emitting decays will scale as $\langle g \rangle \sim \mathcal{A}^{-1}(T_{1/2} \cdot G_{\mathrm{BB}})^{-1/2}$. Thus, the relative sensitivity of a double beta decay experiment on ordinary and charged Majoron decays can be expressed as

Table 3
Values of phase space integrals calculated in this work.

Nucleus	$\beta\beta\phi$		$\beta\beta\phi\phi$	
	$n=1$	$n=3$	$n=3$	$n=7$
	$a_\alpha = \frac{(G_F g_A)^4 \cdot 2 \cdot m_e^2}{256\pi^7 \ln(2)\hbar(m_e R)^2}$	$a_\alpha = \frac{(G_F g_A)^4 \cdot 2}{64\pi^7 \ln(2)\hbar}$	$a_\alpha = \frac{(G_F g_A)^4 \cdot 2}{12288\pi^9 \ln(2)\hbar(m_e R)^2}$	$a_\alpha = \frac{(G_F g_A)^4 \cdot 2}{215040\pi^9 m_e^4 \ln(2)\hbar(m_e R)^2}$
Ge 76	$1.25 \cdot 10^{-16}$	$2.07 \cdot 10^{-19}$	$6.32 \cdot 10^{-19}$	$1.21 \cdot 10^{-18}$
Se 82	$1.03 \cdot 10^{-15}$	$3.49 \cdot 10^{-18}$	$1.01 \cdot 10^{-17}$	$7.73 \cdot 10^{-17}$
Mo 100	$1.80 \cdot 10^{-15}$	$7.28 \cdot 10^{-18}$	$1.85 \cdot 10^{-17}$	$1.54 \cdot 10^{-16}$
Cd 116	$1.75 \cdot 10^{-15}$	$6.95 \cdot 10^{-18}$	$1.60 \cdot 10^{-17}$	$1.03 \cdot 10^{-16}$
Te 128	$1.02 \cdot 10^{-17}$	$5.96 \cdot 10^{-21}$	$1.28 \cdot 10^{-20}$	$1.20 \cdot 10^{-21}$
Te 130	$1.35 \cdot 10^{-15}$	$4.97 \cdot 10^{-18}$	$1.06 \cdot 10^{-17}$	$4.83 \cdot 10^{-17}$
Xe 136	$1.40 \cdot 10^{-15}$	$5.15 \cdot 10^{-18}$	$1.06 \cdot 10^{-17}$	$4.54 \cdot 10^{-17}$
Nd 150	$1.07 \cdot 10^{-14}$	$7.27 \cdot 10^{-17}$	$1.41 \cdot 10^{-16}$	$1.85 \cdot 10^{-15}$

$$\frac{\langle g\rangle^{\rm O.M.}}{\langle g\rangle^{\rm C.M.}} \sim \frac{\mathcal{A}^{\rm C.M.}}{\mathcal{A}^{\rm O.M.}} \Big(\frac{T_{1/2}^{\rm C.M.}}{T_{1/2}^{\rm O.M.}}\Big)^{\frac{1}{2}} \cdot (Q_{\beta\beta} - T).$$

Inserting the definitions of the corresponding amplitudes, it is clear that even if the half-life limit derived for the charged Majoron decay equals that of the ordinary Majoron mode, limits on the charged Majoron-neutrino coupling constant will be weaker by $M_n/(Q_{\beta\beta}-T) \simeq 1000$! (Note that this crude estimation is to first approximation independent of nuclear structure properties.)

A similar analysis can be easily done for double Majoron emitting decays. Again, very crudely, a reduced sensitivity of $(48\pi^2) \cdot p_F/(Q_{\beta\beta} - T) \simeq$ (few) $\times 10^4$ for $n = 3$ double Majoron decay, compared to ordinary Majoron decays, is expected. Here, the factor $(48\pi^2)$ is due to the phase space integration over the additional emitted particle, while the latter factor comes from the additional propagator.

One might think that since our definition of the effective coupling constant for the $n = 3$ $0\nu\beta\beta\phi\phi$ decays includes a factor m_{N_a}/m_e, where m_{N_a} is the sterile neutrino mass, one could get $\langle g\rangle$ easily as large as wanted, since the mass of the sterile neutrino is not bounded experimentally. However, matrix elements will fall off $M \sim m_{N_a}^{-2}$ as soon as m_{N_a} is larger than the typical momenta. While for the matrix elements $M_{\rm GT/F}$ for ordinary Majoron decays such a reduction occurs starting from masses of exchanged virtual particles in the region of 100–1000 MeV, for $M_{{\rm GT}\omega^2/{\rm F}\omega^2}$ the suppression will be important already for much smaller masses (see the E_n-dependence Fig. 1). Also, the contribution of the last term in Eq. (16) for larger neutrino masses can be at most as big as the terms we took into account. Since our conclusions are based on a very rough estimation of $M_{{\rm GT}\omega^2} - M_{{\rm F}\omega^2}$, they would not be affected by the omission of these terms.

Table 4
Comparison of half-lives calculated for different $\langle g\rangle$-values for the new Majoron models with experimental best fit values [16,18].

Model	$T_{1/2}(\langle g\rangle = 10^{-4})$	$T_{1/2}(\langle g\rangle = 1)$	$T_{1/2\ \rm exp}$
IB,IC,IIB	$4 \cdot 10^{22}$	$4 \cdot 10^{14}$	$5.38 \cdot 10^{22}$
ID,IE,IID	10^{38-42}	10^{22-26}	$1.67 \cdot 10^{22}$
IIC,IIF	$2 \cdot 10^{28}$	$2 \cdot 10^{20}$	$1.67 \cdot 10^{22}$
IIE	10^{38-42}	10^{22-26}	$3.37 \cdot 10^{22}$

Since the sensitivity of double beta decay experiments to the new Majoron models is so weak, it might be interesting to compare expected half-lives for the different models for different $\langle g\rangle$, $\langle g\rangle \approx 10^{-4}$ as a typical sensitivity in coupling constant for ordinary Majoron models and $\langle g\rangle = 1$ as an upper possible limit allowed by perturbation theory, with current experimental limits of $\mathcal{O}(10^{22})$ years (see Table 4). From this consideration it is very unlikely that any of the new Majoron models can produce an observable rate in planned or ongoing double beta decay experiments. Only the charged and the vector Majoron model [6,8] could produce an observable effect if $\sum_{ij} V_{ei}V_{ej}$ is not smaller than 0.1 and the real coupling constant of order $\mathcal{O}(1)$.

The authors would like to thank C.P. Burgess and E. Takasugi for several discussions on the theoretical aspects of Majoron models. The research described in this publication was made possible in part (M.H.) by the Deutsche Forschungsgemeinschaft (446 JAP-113/101/0 and Kl 253/8-1) and (S.G.K.) by Grant No. RFM300 from the International Science Foundation.

References

[1] Y. Chikashige, R.N. Mohapatra and R.D. Peccei, Phys. Lett. B 98 (1981) 265; Phys. Rev. Lett. 45 (1980) 265.
[2] G.B. Gelmini and M. Roncadelli, Phys. Lett. B 99 (1981) 411.
[3] H.M. Georgi, S.L. Glashow and S. Nussinov, Nucl. Phys. B 193 (1981) 297.
[4] R.N. Mohapatra and E. Takasugi, Phys. Lett. B 211 (1988) 192.
[5] Z.G. Berezhiani, A.Yu. Smirnov and J.W.F. Valle, Phys. Lett. B 291 (1992) 99.
[6] C.P. Burgess and J.M. Cline, Phys. Lett. B 298 (1993) 141; Phys. Rev. D 49 (1994) 5925.
[7] P. Bamert, C.P. Burgess and R.N. Mohapatra, Nucl. Phys. B 449 (1995) 25.
[8] C.D. Carone, Phys. Lett. B 308 (1993) 85.
[9] M. Doi, T. Kotani and E. Takasugi, Progr. Theor. Phys. Suppl. 83 (1985) 1.
[10] C.S. Aulakh and R.N. Mohapatra, Phys. Lett. B 119 (1982) 136.
[11] T. Tomoda, A. Faessler, K.W. Schmid and F. Grümmer, Nucl. Phys. A 452 (1986) 591.
[12] K. Muto, E. Bender and H.V. Klapdor, Z. Phys. A 334 (1989) 177, 187;
A. Staudt, K. Muto and H.V. Klapdor-Kleingrothaus, Europhys. Lett. 13 (1990) 31.
[13] M. Hirsch, K. Muto, T. Oda and H.V. Klapdor-Kleingrothaus, Z. Phys. A 347 (1994) 151.
[14] A. Balysh et al. (HEIDELBERG-MOSCOW collaboration), to be publ.
[15] M. Hirsch, H.V. Klapdor-Kleingrothaus, B. Maier and H. Päs, in Proc. Int. Workshop on Double Beta Decay and Related Topics, Trento, 24.4.-5.5.95, World Scientific Singapore, eds. H.V. Klapdor-Kleingrothaus and S. Stoica.
[16] M. Günther et al. (HEIDELBERG-MOSCOW collaboration), in Proc. Int. Workshop on Double Beta Decay and Related Topics, Trento, 24.4.-5.5.95, World Scientific Singapore, eds. H.V. Klapdor-Kleingrothaus and S. Stoica.

PHYSICAL REVIEW D VOLUME 54, NUMBER 5 1 SEPTEMBER 1996

BRIEF REPORTS

Brief Reports are accounts of completed research which do not warrant regular articles or the priority handling given to Rapid Communications; however, the same standards of scientific quality apply. (Addenda are included in Brief Reports.) A Brief Report may be no longer than four printed pages and must be accompanied by an abstract.

Bounds on new Majoron models from the Heidelberg-Moscow experiment

M. Günther, J. Hellmig, G. Heusser, M. Hirsch, H. V. Klapdor-Kleingrothaus,* B. Maier, H. Päs, F. Petry, Y. Ramachers, H. Strecker, and M. Völlinger
Max-Planck-Institut für Kernphysik, P.O. Box 10 39 80, D-69029 Heidelberg, Germany

A. Balysh, S. T. Belyaev,* A. Demehin, A. Gurov, I. Kondratenko, D. Kotel'nikov, and V. I. Lebedev
Russian Science Center Kurchatov Institute, 123 182 Moscow, Russia

A. Müller
Istituto Nazionale di Fisica Nucleare, I-67010 Assergi, Italy
(Received 13 November 1995)

In recent years several new Majoron models were invented to avoid the shortcomings of the ordinary models while leading to observable decay rates in double β experiments. We give the first experimental half-life bounds on double β decays with new Majoron emission and derive bounds on the effective neutrino-Majoron couplings from the data of the ^{76}Ge Heidelberg-Moscow experiment. While stringent half-life limits for all decay modes and the coupling constants of the ordinary models were obtained, small matrix elements and phase space integrals result in much weaker limits on the effective coupling constants of the new Majoron models. [S0556-2821(96)05217-4]

PACS number(s): 23.40.Bw, 14.80.Mz

In many theories of physics beyond the standard model neutrinoless double β decays can occur with the emission of new bosons, so-called Majorons [3–6]. While neutrinoless double β decay yields the most stringent limits on Majorana masses of neutrinos [7], the half-life bound for Majoron emitting modes yields limits on the effective Majoron-neutrino coupling:

$$2n \rightarrow 2p + 2e^- + \phi, \quad (1)$$

$$2n \rightarrow +2p + 2e^- + 2\phi. \quad (2)$$

In the Majoron model invented by Gelmini and Roncadelli in 1981 [4], the Majoron is the Nambu-Goldstone boson associated with the spontaneous breaking of the B-L symmetry and so generates Majorana masses of neutrinos. As pointed out by Georgi *et al.* [5], a sizable contribution to double β decay via Eq. (1) is expected for the Gelmini-Roncadelli Majoron. However, in this model the Majoron is an electroweak isospin triplet and therefore should contribute the equivalent of two neutrino species to the width of the Z^0, which was ruled out by the CERN e^+e^- collider LEP [8]. Also the doublet Majoron [9] was ruled out by this measurement.

On the other hand, ordinary Majoron models in which the Majoron is an electroweak isospin singlet [3,10] are still viable. The drawback of the singlet Majoron model is that in these models the Majoron couples to the neutrino at tree level with a coupling strength of roughly $g \simeq (m_{\nu_L}/v_{BL})$, where v_{BL} is the symmetry-breaking scale. In order to preserve existing bounds on neutrino masses and at the same time get an observable rate for Majoron emitting double β decays the singlet Majoron model requires severe fine tuning.

To avoid such an unnatural fine tuning in recent years several new Majoron models have been constructed; where the term Majoron means in a more common sense light or massless bosons with couplings to neutrinos. Since all these models were invented with the same intention of giving observable contributions to double β decays, we felt motivated to analyze the experimental data on ^{76}Ge to determine the experimentally allowed size of the effect.

The main features of the "new Majorons" (Figs. 1 and 2) are that they are not restricted to Goldstone bosons breaking a global lepton number symmetry. Majorons carrying leptonic charge appear in models where the Majoron is responsible for breaking down an extended symmetry group to the global lepton number symmetry [11]. In vector Majoron models one assumes this extended group to be gauged and the Majoron becomes the longitudinal component of a massive gauge boson [13] emitted in double β processes. For simplicity we will call it Majoron, too. Also Majorons which are no Goldstone bosons [11] are possible and decays with the emission of two Majorons can occur in models with Majoron fields carrying one unit of lepton number [12]. The latter is mediated by a sterile neutrino.

In Table I the nine Majoron models we considered are summarized [12,13]. It is divided in the Secs. I for lepton

*Spokesmen of the collaboration.

2.4.9 Lepton Number Violating Interactions, Nonexponential Decay and Time Dependence of the Weak Interaction

WIS-4/00/Apr-DPP
hep-ph/0004048

Lepton number violation interactions and their effects on neutrino oscillation experiments

Sven Bergmann [a], H.V. Klapdor–Kleingrothaus [b] and Heinrich Päs [b]

[a] *Department of Particle Physics
Weizmann Institute of Science, Rehovot 76100, Israel*

[b] *Max–Planck–Institut für Kernphysik
P.O. Box 103980, D-69029 Heidelberg, Germany*

Abstract

Mixing between bosons that transform differently under the standard model gauge group, but identically under its unbroken subgroup, can induce interactions that violate the total lepton number. We discuss four-fermion operators that mediate lepton number violating neutrino interactions both in a model-independent framework and within supersymmetry (SUSY) without R-parity. The effective couplings of such operators are constrained by: i) the upper bounds on the relevant elementary couplings between the bosons and the fermions, ii) by the limit on universality violation in pion decays, iii) by the data on neutrinoless double beta decay and, iv) by loop-induced neutrino masses. We find that the present bounds imply that lepton number violating neutrino interactions are not relevant for the solar and atmospheric neutrino problems. Within SUSY without R-parity also the LSND anomaly cannot be explained by such interactions, but one çannot rule out an effect model-independently. Possible consequences for future terrestrial neutrino oscillation experiments and for neutrinos from a supernova are discussed.

PHYSICAL REVIEW C VOLUME 30, NUMBER 6 DECEMBER 1984

Time scale of short time deviations from exponential decay

K. Grotz and H. V. Klapdor
Max-Planck Institut für Kernphysik, Heidelberg, Germany
(Received 24 April 1984)

It is discussed to what extent nonexponential decay or the so-called zenon effect may lead to a measurable slowing down of extremely rare decay processes as double beta decay or proton decay.

In recent papers,[1,2] we presented calculations of the two-neutrino double beta ($\beta\beta$) decay of ^{76}Ge, ^{82}Se, ^{128}Te, and ^{130}Te. Such calculations are needed for the analysis of geochemical $\beta\beta$-decay measurements[3] in terms of a B-L (baryon-lepton number) violation allowing for a neutrinoless decay mode, since it is not possible to distinguish in such experiments between the two decay modes.

The nuclear structure calculations of this decay, which include Δ-h excitations and collective effects arising from spin-isospin and quadrupole-quadrupole forces show a persisting discrepancy of a factor of ~ 10 with experiment. Because of the extraordinary large half-lives ($T_{1/2} > 10^{18}$ yr) as one among several possible explanations, the question was raised[4] whether here deviations from the simple perturbation treatment of a decay process in the form of nonexponential decay are observed. This question is also of interest and has been recently discussed with respect to proton decay.[5-8]

In this Brief Report we discuss critically the question of whether such quantum mechanically rigorously demanded deviations from the usual decay formulas may lead to observable effects and give estimates using the Heisenberg uncertainty relation.

It is easily seen that the exponential decay law following from a statistical ansatz is only an approximation in a quantum mechanical description. Consider an unstable state $|\Psi\rangle$ being prepared at an instant $t=0$. The probability of finding the system at some later time t still in the state $|\Psi\rangle$ is given by

$$P_0(t) = |\langle\Psi|e^{-iHt/\hbar}|\Psi\rangle|^2 \ . \tag{1}$$

Expanding this expression for small t up to order t^2 gives

$$\begin{aligned} P_0(t) &\cong \left[1+\frac{it}{\hbar}\langle\Psi|H|\Psi\rangle - \frac{t^2}{2\hbar^2}\langle\Psi|H^2|\Psi\rangle\right] \\ &\quad\times\left[1-\frac{it}{\hbar}\langle\Psi|H|\Psi\rangle - \frac{t^2}{2\hbar^2}\langle\Psi|H^2|\Psi\rangle\right] \\ &= 1-\frac{t^2}{\hbar^2}\langle\Psi|(H-\bar{E})^2|\Psi\rangle + 0(t^4) \ , \end{aligned} \tag{2}$$

with

$$\bar{E} = \langle\Psi|H|\Psi\rangle \ .$$

So for very small times, the decay rate is not constant as characteristic for an exponential decay law, but varies proportional to t. From Eq. (2), already the theorem[9]

$$\left.\frac{dP_0(t)}{dt}\right|_{t=0} = 0 \tag{3}$$

follows. From a detailed treatment, it can be seen that the mean square spread of the energy distribution $\Delta E^2 = \langle\Psi|(H-\bar{E})^2|\Psi\rangle$ appearing in Eq. (2) is the essential parameter which determines the onset of exponentiality. For the characteristic time t_0, before which strong nonexponential effects should occur, the following relation can be deduced:[7]

$$t_0 \cong \frac{\hbar^2}{\tau\Delta E^2} \ . \tag{4}$$

Here τ means the usual lifetime neglecting any deviations from exponential decay.

Equations (2) and (3) tell us that for sufficiently short times, the decay rate is whatever small. However, to make any quantitative estimate is very difficult. Peres[10] uses the threshold effect to get a quantitative estimate for the onset of the exponential decay:

$$t_0 = \frac{\hbar}{(\bar{E}-E_{\text{threshold}})} \ . \tag{5}$$

Applying this estimate to $\beta\beta$ decay yields $t_0 \cong 10^{-21}$ sec, which is much too small to give any measurable effect. On the other hand, this would require the rest mass distribution of the $\beta\beta$-decaying state $\langle E|\Psi\rangle$ to extend over a range of 10^{25} natural decay widths Γ in the case of ^{130}Te. The observed decay rates, which are roughly a factor of 10 smaller than calculated assuming exponential decay, would evolve, if $\Delta E = 10^4\ \Gamma$ for ^{130}Te, ^{82}Se and $10^6\ \Gamma$ for ^{128}Te, which means that the energy spectrum of $|\Psi\rangle$ had to be strongly suppressed[7] outside the region $|E-\bar{E}| < \Delta E^2/\Gamma$.

Now we show, that the Peres estimate already evolves from a time dependent perturbation treatment, as it is found in textbooks, if the transition to infinite times is not performed. As usual the total Hamiltonian H is divided in two parts, $H=H_0+V$, H_0 including all interactions which do not lead to a decay, and V having nonvanishing matrix elements between the decaying nucleus and the decay products. The eigenstates of H_0 with energy E_i will be denoted by $|\Psi_i\rangle$. Let us assume that the nucleus has not yet decayed at $t=0$. Then in the perturbation approximation it is described by an eigenstate $|\Psi_1\rangle$ of H_0 at $t=0$. If we look at some time t for decay products, their state vector is also approximated by an eigenstate $|\Psi_2\rangle$ of H_0. We define $|\Psi_2\rangle$ to also include the radiated particles. By doing this we avoid the introduction of a time dependent V. In the time interval $|0,t|$ we allow V to act on $|\Psi_1\rangle$. The time evolution for a first order process is then given by (there is no essential difference between first and second order effects for the following discussion)

$$\langle\Psi_2|e^{-iHt/\hbar}|\Psi_1\rangle = -\frac{i}{\hbar}e^{-iE_2t/\hbar}\int_0^t\langle\Psi_2|V|\Psi_1\rangle\exp[i(E_2-E_1)t'/\hbar]\,dt' = e^{-iE_2t/\hbar}\langle\Psi_2|V|\Psi_1\rangle\frac{\exp[i(E_2-E_1)t/\hbar]-1}{E_1-E_2} \ . \tag{6}$$

The decay probability $P_{1\to 2}(t)$ into the state $|\Psi_2\rangle$ is then

$$P_{1\to 2}(t) = |\langle \Psi_2 | e^{-iHt/\hbar} | \Psi_1 \rangle|^2 = \frac{4|\langle \Psi_2 | V | \Psi_1 \rangle|^2}{(E_1 - E_2)^2} \sin^2\left[(E_1 - E_2)\frac{t}{2\hbar}\right] . \quad (7)$$

This result seems to be rather unphysical for two reasons. First, $P_{1\to 2}(t)$ is a periodic function, which means that the probability of finding $|\Psi_1\rangle$ having decayed into $|\Psi_2\rangle$ after having reached a maximum goes back to zero at some later time. Secondly, there is a finite transition probability for $E_2 \neq E_1$, which seems to violate the conservation of energy. The reason for the latter is that in the initial $(t' < 0)$ and final $(t' > t)$ states the interaction V is assumed to be switched off. The switching on and off of V results in a contribution to the energy balance of the order $\langle \Psi_2 | V | \Psi_1 \rangle$. The appropriate exponential behavior of the decay process follows from an integration over final states: The *total* decay probability into any final state $P_{\text{int}}(t)$ is given by

$$P_{\text{int}}(t) = \int_{-\infty}^{+\infty} \rho(E_f) P_{1\to f}(t)\, dE_f \quad (8)$$

with the density of final states $\rho(E)$. Assuming $\rho(E_f) = \text{const}$ and $V(E) \equiv \langle \Psi_f | V | \Psi_1 \rangle = \text{const}$ leads to Fermi's golden rule

$$P_{\text{int}}(t) = |V(E_1)|^2 \rho(E_1) t 2\pi/\hbar \quad \text{for } t \ll \tau \quad (9a)$$

or in the usual differential form

$$\frac{dP_{\text{int}}(t)}{dt} = \frac{2\pi}{\hbar} |V(E_1)|^2 \rho(E_1) . \quad (9b)$$

Now let us look at the various approximations that have been made. First, $V(E)\rho(E)$ is by no means constant, but $\rho(E)$ varies with some power law for not too large E. The simplest assumption beyond $\rho(E) = \text{const}$ is a sharp cutoff at the threshold energy E_{th}, which is given by the rest mass of the decay products. In this approximation, $P_{\text{int}}(t)$ is calculated by inserting Eq. (7) into (8), and thus

$$P_{\text{int}}(t) = 4\rho(E_1)|V(E_1)|^2 \times \int_{E_{\text{th}}}^{+\infty} \left\{ \frac{\sin[(E_1 - E)t/2\hbar]}{E_1 - E} \right\}^2 dE . \quad (10)$$

It is seen that the missing part of the integral from $E = -\infty$ up to E_{th} involves the sin with frequencies higher than $(E_1 - E_{\text{th}})(4\pi\hbar)^{-1}$. Consequently deviations from Eq. (9a) occur for $t \lesssim \hbar/(E_1 - E_{\text{th}})$ as was found by Peres. The actual form of $V(E)\rho(E)$ may be quite different. However, from an evaluation of Eq. (8) for a definite time t, it can be seen that provided $V(E)\rho(E)$ is a "smooth" function (for example, some approximate power law), contributions from outside the region $|E_f - E_i| \lesssim \hbar/t$ can be neglected. This means that for times t orders of magnitude larger than t_0 in Eq. (5), $P_{\text{int}}(t)$ is determined by a small energy window around E_1, for which $V(E)\rho(E) = V(\bar{E})\rho(\bar{E})$ is a good approximation. So, in perturbation theory, the exponential decay law is a result of the smoothness of the function $V(E)\rho(E)$.

The crucial condition for the application of perturbation theory is that initial and final states can be described by eigenstates of the unperturbed Hamiltonian H_0 to a good approximation. Once this assumption is made, the dispersion ΔE of the parent state is fixed, and no room is left for a quantity as the preparation function introduced by Khalfin.[5] An *exact* treatment of the decay process, in contrast, includes the perturbing part V of H also in the initial and final state. But this involves the difficulty that then the initial state can no longer be described by an eigenstate of H and, in fact, is not a unique state, but its nature depends strongly on the process of formation. We want to discuss now the implications following from the uncertainty principle without going into details of the general very complicated quantum mechanical formalism of unstable systems.

As an illustrating example, we take a resonance scattering process. In a consistent description, without switching on and off additional fields, energy conservation must hold exactly. This means that the energy distribution of the resonance state is determined by the asymptotic $(t \to -\infty)$ distribution of the forming components and the dynamics of the reaction mechanism. Therefore in such a process, the energetic composition of the wave function of the resonance can be restricted to a small energy window by restricting the kinematics of the forming components. But an energy spread ΔE_{form} is connected to a time spread Δt via the Heisenberg relation. This means that the state describing the forming components with energy spread ΔE_{form} and henceforth, also the formation process, cannot be restricted to a time interval Δt less than $\hbar/\Delta E_{\text{form}}$. The resulting delay of exponentiality from a cutoff of the Breit-Wigner resonance shape at $|E - \bar{E}| = \Delta E_{\text{form}}$ extends according to Eq. (4) until $t_0 \cong \hbar/\Delta E_{\text{form}}$ (provided $\Delta E_{\text{form}} < \bar{E} - E_{\text{th}}$). This means that the nonexponential regime in the decay of the resonance cannot continue much longer than the formation process. If one would apply the uncertainty relation to the energy spread ΔE of the formed state itself instead of ΔE_{form}, an even sharper restriction evolves. The discussed situation can be realized in a resonance fluorescence experiment using Laser techniques. The energy spread ΔE_f of the Laser beam is connected with a minimum time $\Delta t = \hbar/\Delta E_f$, which is needed for switching on and off the Laser.

Generalizing the above considerations, by relating any suppression of ΔE in any formation process by the energy-time uncertainty to the minimum time extension Δt of the formation process, we conclude that the nonexponential behavior cannot persist longer than Δt.

This would mean that formation process and nonexponential decay cannot be separated in time. This result, of course, holds already in classical mechanics. As long as the formation process is going on, there is also no exponential behavior assuming purely statistical decay laws. If the above generalization holds, the implications for $\beta\beta$ decay as well as for proton decay clearly are, that deviations from exponentiality are negligible nowadays. To our knowledge, protons have been formed in the big bang within 10^{-6} sec and $\beta\beta$-decaying nuclei are the products of β-decay chains in the r process with typical decay constants of the order of minutes.

There has been given an argument by Fleming,[7] which seems to contradict this conclusion. According to Fleming a mass spectrum of the proton, as assumed by Peres, would lead to a "kinematical fragmentation" within, at maximum, 10^7 yr. (For $\beta\beta$ decay of ^{130}Te, the corresponding time would be 10^5 yr.) To avoid this fragmentation with its unacceptable consequences, Fleming concludes that t_0 *must* be of the order of 10^{26} yr. However, for this argument to work he has to assume free propagation of the proton wave packets. The interactions of the proton with other particles should destroy, however, the phase relations and hence mix the different mass components, and hence no fragmentation should occur.

Another effect, namely, the so-called zenon effect[11] has been discussed[12,6,8] as an amplification mechanism of the nonexponential behavior. The argument is that if during the nonexponential stage of the decay a measurement is performed probing the unstable system, the decay process is then restarted. If successive measurements are repeated in time intervals smaller than t_0, the system will never reach the exponential decay one would expect from usual perturbation theory. However, the zenon effect is, of course, automatically included in a complete quantum mechanical description of the whole system including also the measuring apparatus and all interactions. So the zenon effect, in principle, can only arise from the neglection of a part of the interactions. In this sense Valanju, Sudarshan, and Chiu[13] derived sizeable effects for pion production in hadron nucleus collisions. Recently Horwitz and Katznelson[6] suggested important corrections for the proton decay inside the nucleus from nucleon-nucleon interactions. However, their interpretation of nucleon-nucleon interactions as a measuring process in the sense of the zenon effect has been criticized.[14-16] We do not discuss here the question of how to define a measurement but give a simple argument using again the uncertainty principle. To have a significant influence on the decay process, the measuring process has to take place with a frequency of at least $1/t_0$. According to the Heisenberg principle, this frequency is related to an interaction energy $E_{\text{int}} \cong \hbar/t_0$ between the unstable system and the part which acts as measuring apparatus. If one accepts the Peres estimate, Eq. (5), this means the interaction energy has to be as large as the Q value of the decay. Of course, if there is such a strong interaction, it is nothing unexpected, that the decay is influenced. From this point one could consider the strong interactions of the nucleons inside a nucleus generating a zenon effect in β decay and also in $\beta\beta$ decay. However, such effects are, of course, automatically included in nuclear structure calculations. Interactions not completely included in such calculations are the electromagnetic interactions with the electron cloud. But these are according to the above criterion much too weak to lead to sizeable effects. In the case of proton decay already the nucleon-nucleon interaction energy is more than one order of magnitude smaller than the decay Q value. Consequently a slowing down of the decay by the zenon effect can hardly be expected either for $\beta\beta$ or proton decay.

[1]K. Grotz, H. V. Klapdor, and J. Metzinger, J. Phys. G **9**, L169 (1983).

[2]H. V. Klapdor and K. Grotz, Phys. Lett. **142B**, 323 (1984).

[3]T. Kirsten, H. Richter, and E. Jessberger, Phys. Rev. Lett. **50**, 474 (1983).

[4]H. V. Klapdor and K. Grotz, Report No. MPI H-1983-V34, 1983 (unpublished); K. Grotz and H. V. Klapdor, Report No. MPI H-1983-V35, 1983 (unpublished); H. V. Klapdor, Proceedings of the International Symposium on Nuclear Spectroscopy and Nuclear Interactions, Osaka, March 21–24 (1984).

[5]L. A. Khalfin, Phys. Lett. **112B**, 223 (1982).

[6]L. P. Horwitz and E. Katznelson, Phys. Rev. Lett. **50**, 1184 (1983).

[7]G. N. Fleming, Phys. Lett. **125B**, 287 (1983).

[8]L. Fonda, G. C. Chirardi, and T. Weber, Phys. Lett. **131B**, 309 (1983).

[9]L. A. Khalfin, Pis'ma Zh. Eksp. Teor. Fiz. **8**, 106 (1968) [JETP Lett. **8**, 65 (1968)].

[10]A. Peres, Ann. Phys. (N.Y.) **129**, 33 (1980).

[11]C. Chiu, E. C. G. Sudarshan, and B. Misra, Phys. Rev. D **16**, 520 (1977).

[12]A. Peres, Am. J. Phys. **48**, 931 (1980).

[13]P. Valanju, E. C. G. Sudarshan, and C. B. Chiu, Phys. Rev. D **21**, 1304 (1980).

[14]G. P. Lepage, T. M. Yan, and D. R. Yennie, Phys. Rev. Lett. **51**, 1599 (1983).

[15]K. Cahill, Phys. Rev. Lett. **51**, 1600 (1983).

[16]J. Wheater and R. Peierls, Phys. Rev. Lett. **51**, 1601 (1983).

Is weak interaction constant really constant?

A.S. Barabash

Institute of Theoretical and Experimental Physics, B. Cheremushkinskaya 25, 117259 Moscow, Russia, e-mail: barabash@vxitep.itep.ru

the date of receipt and acceptance should be inserted later

Abstract. A comparison is made of the probability of the process of two neutrino double beta decay for ^{82}Se and ^{96}Zr in direct (counter) and geochemical experiments. The experimental data for ^{130}Te are also analyzed. It is shown that the probability is systematically lower in geochemical experiments, which characterize the probability of $\beta\beta(2\nu)$ decay a few billions years ago. In addition geochemical measurements on young minerals give lower values of $T_{1/2}(^{130}\mathrm{Te})$ as compared to measurements on old minerals. It is proposed that this could be due to a change in the weak interaction constant with time. Possibilities of new, precise measurements be performed with the aid of counters and geochemical experiments are discussed. New geochemical experiment with ^{100}Mo is proposed.

PACS. 23.40.-s beta decay; double beta decay; electron and muon capture

1 Introduction

The question of the dependence of the fundamental constants on time was formulated by P. Dirac in 1937 - this is so-called Large Number Hypothesis [1]. This question was later discussed in Refs. [2-7]. Although Dirac's hypothesis was not confirmed in its initial form, interest in this problem gathered new strength in the 1980s, since a time dependence of the coupling constants appears in multidimensional Kaluza-Klein models [8,9] and in superstring theories [10] (see also Refs. [11] and [12]). These theories are formulated for a multidimensional space, which must then be compactified to the four observable dimensions of space-time. In these theories the fundamental coupling constants are associated with the radii of additional dimensions and these additional dimensions can manifest themselves through a time dependence of the coupling constants. The radii can shrink, increase, or even oscillate. It has not been ruled out that the compactification process is continuing at present. A time dependence of the fundamental constants also arises in models with a mass-

2.4.10 Test of Lorentz Invariance, Equivalence Principle and Quantum Foam

Eur. Phys. J. A **5**, 3–6 (1999)

THE EUROPEAN PHYSICAL JOURNAL A

Short note

Test of special relativity and equivalence principle from neutrinoless double beta decay

H.V. Klapdor–Kleingrothaus[1], H. Päs[1], U. Sarkar[2,3]

[1] Max–Planck–Institut für Kernphysik P.O. Box 103980, D–69029 Heidelberg, Germany
[2] DESY, Notkestrasse 85, D–22607 Hamburg, Germany
[3] Physical Research Laboratory, Ahmedabad, 380 009, India

Received: 19 January 1999
Communicated by B. Povh

Abstract. We generalize the formalism for testing Lorentz invariance and the weak equivalence principle in the neutrino sector. While neutrino oscillation bounds constrain the region of large mixing of the the weak and gravitational eigenstates, we obtain new constraints on violations of Lorentz invariance and the equivalence principle from neutrinoless double beta decay. These bounds apply even in the case of no mixing and thus probe a totally unconstrained region in the parameter space.

Special relativity and the equivalence principle can be considered as the most basic foundations of the theory of gravity. Many experiments already have tested these principles to a very high level of accuracy [1] for ordinary matter – generally for quarks and leptons of the first generation. These precision tests of local Lorentz invariance – violation of the equivalence principle should produce a similar effect [2] – probe for any dependence of the (non-gravitational) laws of physics on a laboratory's position, orientation or velocity relative to some preferred frame of reference, such as the frame in which the cosmic microwave background is isotropic.

A typical feature of the violation of local Lorentz invariance (VLI) is that different species of matter have a characteristical maximum attainable speed. This can be tested in various sectors of the standard model through vacuum Cerenkov radiation [3], photon decay [4], neutrino oscillations [5,8–11] and K–physics [6,7]. In this article we extend these arguments to derive new constraints from neutrinoless double beta decay.

The equivalence principle implies that spacetime is described by unique operational geometry and hence universality of the gravitational coupling for all species of matter. In the recent years there have been attempts to constrain a possible amount of violation of the equivalence principle (VEP) in the neutrino sector from neutrino oscillation experiments [8–11]. However, these bounds don't apply when the gravitational and the weak eigenstates have small mixing. In this article we present a generalized formalism of the neutrino sector to test the VEP and point out that neutrinoless double beta decay also constrains the VEP. VEP implies different neutrino species to suffer from different gravitational potentials while propagating through the nucleus and hence the effect of different eigenvalues doesn't cancel for the same effective momentum. Earlier results on neutrino oscillations come out as special case from our present formalism. The main result is that neutrinoless double beta decay can constrain the amount of VEP even when the mixing angle is zero, *i.e.*, when only the weak equivalence principle is violated, for which there does not exist any bound at present.

We shall first present our formalism for VLI and then for VEP. For sake of clarity we formulate the problem for a two generation scenario involving ν_e and ν_x with $x = \mu, \tau, s$. Neutrinos of different species may have different maximum attainable velocities if there is violation of local Lorentz invariance (VLI) and hence violation of special relativity [4]. We first assume that the weak eigenstates cannot be diagonalized simultaneously with the velocity eigenstates and the neutrinos are relativistic point particles. The effective Hamiltonian in the weak basis $[\nu_e \;\; \nu_x]$ is

$$H = U_m H_m U_m^{-1} + U_v H_v U_v^{-1}. \qquad (1)$$

In absence of VLI the neutrino mass matrix in the mass basis $[\nu_1 \;\; \nu_2]$ is given by

$$H_m = \frac{(M_m)^2}{2p} = \frac{1}{2p}\begin{pmatrix} m_1 & 0 \\ 0 & m_2 \end{pmatrix}^2 \qquad (2)$$

and the VLI part of the hamiltonian as

$$H_v = \begin{pmatrix} v_1 & 0 \\ 0 & v_2 \end{pmatrix} p, \tag{3}$$

to leading order in $\bar{m}^2/p^2$. Here p denotes the momentum and $\bar{m}$ the average mass, and for any quantity X we define $\delta X \equiv (X_1 - X_2)$, $\bar{X} = (X_1 + X_2)/2$.

In the absence of VLI, *i.e.*, when the special threory of relativity is valid, $v_i = 1$, and H_v simply becomes the momentum of the neutrinos. Here we are interested in a single neutrino beam (for neutrino oscillation experiments) or a single virtual neutrino propagating inside the nucleus with a particular momentum. For this reason we assume the momenta of both the neutrinos are p. Then v_i corresponds to the maximum attainable speed of the corresponding momentum eigenstates. Hence $v_1 - v_2 = \delta v$ is a measure of VLI in the neutrino sector. As typical or "standard" maximum attainable speed $\frac{v_1+v_2}{2} = 1$ is assumed. All previous bounds on this quantity δv in the neutrino sector were derived from neutrino oscillation experiments and for that reason these bounds are valid only for large gravitational mixing. As we shall point out, neutrinoless double beta decay can constrain δv even when the mixing angle vanishes.

We shall not consider any CP violation, and hence H_m and H_v are real symmetric matrices and U_m and U_v are orthogonal matrices $U^{-1} = U^T$. They can be parametrized as $U_i = \begin{pmatrix} \cos\theta_i & \sin\theta_i \\ -\sin\theta_i & \cos\theta_i \end{pmatrix}$, where θ_i represents weak mixing angle θ_m or velocity mixing angle θ_v. We can now write down the weak Hamiltonian H_w in the basis $[\nu_e \;\; \nu_x]$, in which the charged lepton mass matrix is diagonal and the charged current interaction is also diagonal, as

$$H = pI + \frac{1}{2p}\begin{pmatrix} M_+ & M_{12} \\ M_{12} & M_- \end{pmatrix}^2.$$

Here I is the identity matrix and

$$\begin{aligned} M_\pm &= \bar{m} \pm \frac{cos2\theta_m}{2}\delta m \\ &\pm \frac{p^2}{\bar{m}}\delta v \left(\frac{\cos 2\theta_v}{2} - \frac{\delta m}{4\bar{m}} \cos 2(\theta_m - \theta_v) \right) \\ M_{12} &= -\frac{\sin 2\theta_m}{2}\delta m \\ &- \frac{p^2}{\bar{m}}\delta v \left(\frac{\sin 2\theta_v}{2} + \frac{\delta m}{4\bar{m}} \sin 2(\theta_m - \theta_v) \right). \end{aligned} \tag{4}$$

In the case of exact Lorentz invariance, we usually write the mass matrix in the weak basis as $\begin{pmatrix} m_{ee} & m_{e\mu} \\ m_{e\mu} & m_{\mu\mu} \end{pmatrix}$. In the mass mechanism of neutrinoless double beta decay, the decay rate

$$[T_{1/2}^{0\nu\beta\beta}]^{-1} = \frac{\langle m \rangle^2}{m_e^2} G_{01} |ME|^2, \tag{5}$$

is proportional to the effective neutrino mass $\langle m \rangle = m_{ee} = M_+$. Here ME denotes the nuclear matrix element $ME = M_F - M_{GT}$, G_{01} corresponds to the phase space factor defined in [12] and m_e is the electron mass. The double beta observable can be written as

$$\begin{aligned} \langle m \rangle &= \sum_i U_{ei}^2 m_i = m_1 \cos^2\theta_w + m_2 \sin^2\theta_w \\ &= \bar{m} + \frac{1}{2}\delta m \cos 2\theta_w. \end{aligned} \tag{6}$$

If $m_{ee} = 0$, the two physical eigenstates with eigenvalues m_1 and m_2 will contribute to the neutrinoless double beta decay by an amount $U_{e1}^2 m_1$ and $U_{e2}^2 m_2$, respectively, which cancels each other. However, if these two physical states have different maximum attainable speed, corresponding to VLI, this cancellation will not be exact for the same cut–off effective momentum in the neutrino propagator. As a result, even when $m_{ee} = 0$, we can have neutrinoless double beta decay, which is proportional to the amount of VLI and the double beta observable is given by M_+ in (4). From (4) it can easily be seen that in the region of maximal mixing $\cos 2\theta_v = 0$, the double beta decay rate vanishes. Thus neutrinoless double beta decay doesn't constrain the amount of VLI for maximal mixing. However, when the mixing approaches zero, the most stringent bound from neutrinoless double beta decay is obtained. In this case $\delta v/2$ can be understood as derivation from the standard maximum attainable speed $\bar{v}$. As it is obvious, when there is no mixing the neutrino oscillation experiments cannot give any bound on the amount of VLI, since in absence of mixing only VLI cannot allow neutrino oscillations.

To give a bound on VLI in the small mixing region (including $\theta_v = \theta_m = 0$) we assume conservatively $\langle m \rangle \simeq 0$. We also assume $\delta m \leq \bar{m}$, and thus $\frac{\delta m}{4\bar{m}}$ may be neglected. Due to the p^2 enhancement the nuclear matrix elements of the mass mechanism have to be replaced by $\frac{m_p}{R} \cdot (M_F' - M_{GT}')$ with the nuclear radius R and the proton mass m_p, which have been calculated in [13]. Inserting the recent half life limit obtained from the Heidelberg–Moscow experiment [14], $T_{1/2}^{0\nu\beta\beta} > 1.2 \cdot 10^{25} y$, a bound on the amount of VLI as a function of the average neutrino mass $\bar{m}$ can be given. The most reliable assumption for $\bar{m}$ is obtained from the cosmological bound $\sum_i m_i < 40$ eV [15], i.e., $\bar{m} < 13$ eV for three generations, implying a bound of

$$\delta v < 4 \times 10^{-16} \quad \text{for} \quad \theta_v = \theta_m = 0.$$

However, combining the present experimental constraints from atmospheric and solar neutrino data as well as from tritium beta decay in a three neutrino framework and assuming a typical hierarchical mass pattern spectrum $m_3 \gg m_{1,2}$ or $m_3 \simeq m_2 \gg m_1$ implies $\bar{m} <\sim 0.08$ eV [16] and improves the bound to $\delta v < 2 \cdot 10^{-18}$ for $\theta_v = \theta_m = 0$.

In Fig. 1 the bound implied by double beta decay is presented for the entire range of $sin^2 2\theta_v$ and compared with bounds obtained from neutrino oscillation experiments in [10]. It should be stressed also that the GENIUS proposal of the Heidelberg group [17] could improve these bounds by about 1–2 orders of magnitude.

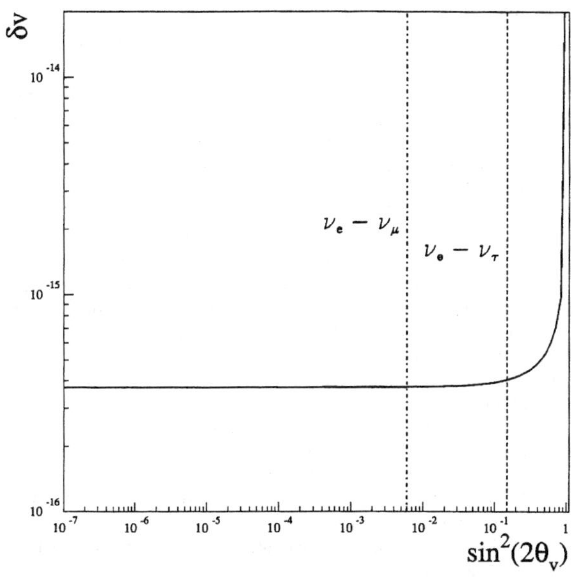

Fig. 1. Double beta decay bound (solid line) on violation of Lorentz invariance in the neutrino sector, excluding the region to the upper left. Shown is a double logarithmic plot in the δv–$\sin^2(2\theta)$ parameter space. The bound becomes most stringent for the small mixing region, which has not been constrained from any other experiments. For comparison the bounds obtained from neutrino oscillation experiments (from [10]) in the $\nu_e - \nu_\tau$ (dashed lines) and in the $\nu_e - \nu_\mu$ (dashed-dotted lines) channel, excluding the region to the right, are shown

For comparison, in the following the amount of VLI in neutrino oscillation experiments is calculated in this formalism. In the basis of the physical states ν_a and ν_b, the Hamiltonian becomes

$$H = \begin{pmatrix} p + \frac{m_a^2}{2p} & 0 \\ 0 & p + \frac{m_b^2}{2p} \end{pmatrix} = \begin{pmatrix} \bar{E} & 0 \\ 0 & \bar{E} \end{pmatrix} + \frac{1}{2}\begin{pmatrix} \Delta E & 0 \\ 0 & -\Delta E \end{pmatrix}, \qquad (7)$$

where $\bar{E} = (p + \frac{\bar{m}^2}{2p})$ and

$$\frac{p}{\bar{m}}\Delta E = m_a - m_b = \left[(\delta m)^2 + \left(\delta v \frac{p^2}{\bar{m}}\right)^2 + 2\delta m \delta v \frac{p^2}{\bar{m}} \cos(2(\theta_w - \theta_v))\right]^{1/2} \qquad (8)$$

The new mixing angle θ_{tot} is a function of θ_m and θ_v and the oscillation probability is now given by

$$P(\nu_e \to \nu_x) = \sin^2 2\theta_{tot} \sin^2 \frac{\pi L}{\lambda}, \qquad (9)$$

where $\lambda = \frac{\pi p}{\Delta m^2}$ and $\Delta m^2 = m_a^2 - m_b^2$. In the limit of vanishing neutrino masses, neutrino oscillations are implied only by VLI. In this case the oscillation probability becomes

$$P(\nu_e \to \nu_x) = \sin^2 2\theta_v \sin^2 pL\delta v,$$

which corresponds to the expression obtained earlier [5]. Here p denotes the total beam energy. From this expression it beomes clear that in the case of no mixing, $\theta_v = 0$, neutrino oscillation experiments don't constrain the size of VLI effects.

In the following we present the formalism for violation of the equivalence principle (VEP). While in the final expression the amount of VLI just will be replaced by VEP, the origin differs. In a linearized theory the gravitational part of the Lagrangian to first order in a weak gravitational field $g_{\mu\nu} = \eta_{\mu\nu} + h_{\mu\nu}$ ($h_{\mu\nu} = 2\frac{\phi}{c^2}diag(1,1,1,1)$) can be written as $\mathcal{L} = -\frac{1}{2}(1+g_i)h_{\mu\nu}T^{\mu\nu}$, where $T^{\mu\nu}$ is the stress-energy in the gravitational eigenbasis. In the presence of VEP the g_i may differ. Assuming only violation of the weak equivalence principle, the gravitational interaction is diagonal but the couplings differ. In this case there does not exist any bound on the amount of VEP. We point out that this region of the parameter region is most restrictively bounded by neutrinoless double beta decay.

The effective Hamiltonian in the weak basis again can be written as

$$H = pI + U_m H_m U_m^{-1} + U_G H_G U_G^{-1}, \qquad (10)$$

with H_m given in (2) and

$$H_G = \begin{pmatrix} G_1 & 0 \\ 0 & G_2 \end{pmatrix} = \begin{pmatrix} -2(1+g_1)\phi(p + \frac{\bar{m}^2}{2p}) & 0 \\ 0 & -2(1+g_2)\phi(p + \frac{\bar{m}^2}{2p}) \end{pmatrix} \qquad (11)$$

to first order in $\bar{m}^2/p^2$. In formalisms where only violation of the weak equivalence principle is assumed, one starts with U_G proportional to U_W, in which case there does not exist any bound from the neutrino oscillation experiments. The Hamiltonian in the weak basis is

$$\begin{aligned} M_\pm &= \bar{m} \pm \frac{\cos 2\theta_m}{2}\delta m \\ &\pm \frac{p}{\bar{m}}\delta G\left(\frac{\cos 2\theta_G}{2} - \frac{\delta m}{4\bar{m}}\cos(\theta_m - \theta_v)\right) \\ M_{12} &= -\frac{\sin 2\theta_m}{2}\delta m \\ &- \frac{p}{\bar{m}}\delta G\left(\frac{\sin 2\theta_G}{2} + \frac{\delta m}{4\bar{m}}\sin(\theta_m - \theta_v)\right). \end{aligned} \qquad (12)$$

Compared to VLI, the expressions for this case of VEP remain unchanged, except for replacing δv by $\frac{1}{p}\delta G$. Again, $\bar{g} = \frac{g_1+g_2}{2}$ can be considered as the standard gravitational coupling, for which the equivalence principle applies.

Thus the discussion of VLI can be directly translated to the VEP case and the bound from neutrinoless double beta decay for $\theta_v = \theta_m = 0$ is now given by

$$\begin{aligned} \phi\delta g &< 4 \times 10^{-16} \text{ (for } \bar{m} < 13\text{eV)} \\ \phi\delta g &< 2 \times 10^{-18} \text{ (for } \bar{m} < 0.08\text{eV)} \end{aligned} \qquad (13)$$

In this case, $\delta G = p\phi\delta g$, where ϕ is the background Newtonian gravitational potential on the surface of the earth. A natural choice for ϕ would be the earth's gravitational potential ($\sim 10^{-9}$), but another well motivated choice could be the potential due to all forms of distant matter. Unlike the case of VLI, the bound on the VEP will depend on what one chooses for the Newtonian potential ϕ. For this reason, here we only present the combined bound on $\phi\delta g$.

In summary, we presented a general formalism for the study of both VLI and VEP in the neutrino sector. We pointed out that neutrinoless double beta decay can constrain the amount of VLI or VEP. In particular, when the mixing of the gravitational eigenstates vanishes, the bounds from neutrinoless double beta decay become most stringent, while this region is not constrained by any other experiments.

One of us (US) would like to acknowledge the hospitality of the Theory Group, DESY, Hamburg, the Max-Planck-Institut für Kernphysik, Heidelberg and a fellowship from the Alexander von Humboldt Foundation.

References

1. V.W. Hughes, H.G. Robinson, and V. Beltran-Lopez, Phys. Rev. Lett. **4**, 342 (1960); R.W.P. Drever, Philos. Mag. **6**, 683 (1961); D. Newman, G.W. Ford, A. Rich and E. Sweetman, Phys. Rev. Lett. **40**, 1355 (1978); A. Brillet and J.L. Hall, Phys. Rev. Lett. **42**, 549 (1979); J.D. Prestage, J.J. Bollinger, W.M. Itano, and D.J. Wineland, Phys. Rev. Lett. **54**, 2387 (1985); S.K. Lamoureaux, J.P. Jacobs, B.R. Heckel, R.J. Raab, and E.N. Fortson, Phys. Rev. Lett. **57**, 3125 (1986)
2. C. M. Will, *Theory and Experiment in Gravitational Physics*, 2nd edition (Cambridge University Press, Cambridge, 1992)
3. M. Gasperini, Phys. Rev. Lett. **62**, 1945 (1989)
4. S. Coleman and S.L. Glashow, Phys. Lett. **B 405**, 249 (1997)
5. S.L. Glashow, A. Halprin, P.I. Krastev, C.N. Leung, and J. Panteleone, Phys. Rev. **D 56**, 2433 (1997)
6. T. Hambye, R.B. Mann and U. Sarkar, Phys. Lett. **B 421**, 105 (1998); Phys. Rev. **D 58**, 025003 (1998)
7. M.L. Good, Phys. Rev. **121**, 311 (1961); O. Nachtmann, Acta Physica Austriaca, Supp. VI *Particle Physics* ed. P. Urban, (1969) p. 485; S.H. Aronson, G.J. Bock, H-Y Cheng and E. Fishbach, **48**, 1306 (1982); Phys. Rev. **D28**, 495 (1983); I.R. Kenyon, Phys. Lett. **B237**, 274 (1990); R.J. Hughes, Phys. Rev. **D46**, R2283 (1992)
8. M. Gasperini, Phys. Rev. **D38**, 2635 (1988); *ibid.* **D39**, 3606 (1989)
9. A. Halprin and C.N. Leung, Phys. Rev. Lett. **67**, 1833 (1991); Nucl. Phys. **B28A** (Proc. Supp.), 139 (1992); J. N. Bahcall, P. I. Krastev, and C. N. Leung, Phys. Rev. **D 52**, 1770 (1995); R.B. Mann and U. Sarkar, Phys. Rev. Lett **76** (1996) 865;J.R. Mureika and R.B. Mann, Phys. Rev. **D 54** (1996) 2761-2778; R.B. Mann and J. Mureika, Phys. Lett. **B 368** (1996) 112–118
10. A. Halprin, C.N. Leung, J. Pantalone, Phys. Rev. **D 53** (1996) 5365
11. M. N. Butler *et al.*, Phys. Rev. **D47**, 2615 (1993); A. Halprin and C. N. Leung, Phys. Rev. Lett. **67**, 1833 (1991); J. Pantaleone, A. Halprin, and C. N. Leung, Phys. Rev. **D47**, R4199 (1993); K. Iida, H. Minakata and O. Yasuda, Mod. Phys. Lett. **A8** (1993) 1037
12. M. Doi, T. Kotani, E. Takasugi, Progr. Theor. Phys. Suppl. **83** (1985) 1
13. H. Päs, M. Hirsch, S.G. Kovalenko, H.V. Klapdor–Kleingrothaus, subm. to Phys. Rev. Lett. M. Hirsch, H.V. Klapdor–Kleingrothaus, S.G. Kovalenko, Phys. Lett. **B 372** (1996) 181, Erratum: Phys. Lett. **B 381** (1996) 488
14. H.V. Klapdor-Kleingrothaus, H. Päs, Proc. of the 6th Int. Symposium on Particles, Strings and Cosmology, (PASCOS98), Boston(MA), USA, March 22-27 1998, to be publ. by World Scientific; M. Günther et al, Phys. Rev. **D 55** (1997) 54; L. Baudis et al, Phys. Lett. **B 407** (1997) 219
15. G. Raffelt in Proc. TAUP97, Nucl. Phys. **B** (Proc. Suppl.) **70** (1998) 169
16. V. Barger, T.J. Weiler, K. Whisnant, hep-ph/9808367
17. H.V. Klapdor–Kleingrothaus, hep-ex/9802007, in: H.V. Klapdor–Kleingrothaus, H. Päs (Eds.), Proc. Int. Conf. "Beyond the Desert – Accelerator- and Non-Accelerator Approaches", Castle Ringberg, Germany, 1997; H.V. Klapdor–Kleingrothaus, M. Hirsch, Z. Phys. **A 359** (1997) 361; H.V. Klapdor–Kleingrothaus, J. Hellmig, M. Hirsch, J. Phys. **G 24** (1998) 483

ELSEVIER

6 April 2000

PHYSICS LETTERS B

Physics Letters B 478 (2000) 269–274

Effects of new gravitational interactions on neutrinoless double beta decay

H.V. Klapdor-Kleingrothaus [a], H. Päs [a], U. Sarkar [b]

[a] *Max-Planck-Institut für Kernphysik, P.O. Box 103980, D-69029 Heidelberg, Germany*
[b] *Physical Research Laboratory, Ahmedabad 380 009, India*

Received 30 December 1999; received in revised form 18 February 2000; accepted 25 February 2000
Editor: P.V. Landshoff

Abstract

It has recently been proposed that violations of Lorentz invariance or violations of the equivalence principle can be constrained from the non-observation of neutrinoless double beta decay. We generalize this analysis to all possible new gravitational interactions and discuss briefly the constraints for different cases. © 2000 Elsevier Science B.V. All rights reserved.

Although there is no evidence for the violation of gravitational laws, lots of work has been done to find out to what accuracy this is true. Many experiments have been performed to test the equivalence principle [1] for ordinary matter and to test local Lorentz invariance [2,3]. In recent times there has been some effort to test these laws for the gravitational couplings of neutrinos. Assuming that neutrinos of different generations have characteristic couplings to gravity with differing strength and that the gravitational eigenstates differ from the mass eigenstates, one can constrain the amount of violation of equivalence principle (VEP) in the neutrino sector from present neutrino oscillation results [4,5]. Similar bounds were also obtained for the amount of violation of local Lorentz invariance (VLI), assuming that neutrinos of different generations have characteristic maximum attainable velocities [6,7]. Recently we have pointed out that in both these cases it is possible to constrain some otherwise unconstrained region in the parameter space from neutrinoless double beta decay [8].

In this article we propose a general framework to study the effect of new gravitational interactions in the neutrinoless double beta decay. This formalism is similar to the one used in the study of K-system [9,10]. We classify all possible interactions as scalar, vector and tensor interactions. Since both the VEP and VLI are tensor interactions, it is expected that in both cases similar constraints should be obtained, as observed. On the other hand, a recent string motivated violation of the equivalence principle a la Damour and Polyakov [11] is a scalar interaction. Thus the constraint in this case is of different nature than in the cases of VEP or VLI considered previously. The possible fifth force [12] discussed in the literature is a vector interaction and thus also has a different phenomenology. Our analysis can be extended to study the effects of gravitational interactions in neutrino oscillation experiments.

PII: S0370-2693(00)00253-7

We write down the most general lagrangian for interactions of neutrinos with scalar, vector and tensor fields in the weak basis $[\nu_i]$ following the general framework developed for the K-system [9],

$$\mathscr{L} = G_{ij}\nu_i\nu_j + G_{ij}^{\mu}\nu_{i,\mu}\nu_j + G_{ij}^{\mu\nu}\nu_{i,\mu}\nu_{j,\nu}, \tag{1}$$

where i,j are generation indices and G_{ij}, G_{ij}^{μ} and $G_{ij}^{\mu\nu}$ are scalar, vector and tensor fields respectively. We shall not work beyond the external field approximation. These fields have some restrictions coming from the symmetry properties and by discarding the total divergence expressions from the lagrangian, which have been discussed in Ref. [9] in detail. We further assumed that the gravitational eigenstates could be different from the mass eigenstate as well as the weak eigenstate. For simplicity from now on we shall work in an two generation scenario, $i,j = e,x$ with $x = \mu,\tau,s$.

We now can write down the Feynmann diagrams and hence the self energy matrix in the same way as in Ref. [9], from which the contribution to the effective hamiltonian can be read off in the weak basis, given by

$$\mathscr{H}_{ij} = G_{ij} + iG_{ij}^{\mu}p_{\mu} + G_{ij}^{\mu\nu}p_{\mu}p_{\nu}. \tag{2}$$

This hamiltonian is related to the effective hamiltonian in the mass and gravitational bases through unitary rotations

$$H = U_m H_m U_m^{-1} + U_G H_G U_G^{-1}. \tag{3}$$

In absence of any new gravitational interactions the neutrino mass matrix in the mass basis $[\nu_1\,\nu_2]$ is given by

$$H_m = \frac{(M_m)^2}{2p} = \frac{1}{2p}\begin{pmatrix} m_1 & 0 \\ 0 & m_2 \end{pmatrix}^2 \tag{4}$$

and the gravitational interaction part of the hamiltonian is

$$H_G = pI + \frac{(M_G)^2}{2p} = pI + \frac{1}{2p}\begin{pmatrix} g_1^a & 0 \\ 0 & g_2^a \end{pmatrix}. \tag{5}$$

Here p denotes the momentum, I represents an unit matrix and $\overline{m}$ the average mass, and for any quantity X we define $\delta X \equiv (X_1 - X_2)$, $\overline{X} = (X_1 + X_2)/2$. $a = S,V,T$ represents scalar, vector and tensor interactions respectively.

The scalar, vector and tensor gravitational interactions can be written in the following forms so as to reproduce the correct dimensions of Eq. (2),

$$g_i^S = 2\alpha_i^S m_i^2, \quad g_i^V = 2\alpha_i^V m_i p, \quad g_i^T = 2\alpha_i^T p^2$$

In the absence of any gravitational interactions $\alpha_i^a = 0$, H_G simply becomes the momentum of the neutrinos. Here we are interested in a single virtual neutrino propagating inside the nucleus with a particular momentum. For this reason we assume the momenta of both the neutrinos are p in the absence of any new gravitational interactions. Hence $\alpha_1^a - \alpha_2^a = \delta\alpha^a$ is a measure of the new gravitational interactions in the neutrino sector. To compare our result with the neutrino oscillation experiments we further assume, $\alpha_1^a + \alpha_2^a = 0$, i.e., there is no mean deviation from the gravitational laws and there is only a relative violation given by the measure $\delta\alpha^a$. This approximation will reduce the the number of parameters so that we can compare the bounds from the neutrino oscillation experiments with the ones from neutrinoless double beta decay.

We tried to keep the above discussions as general as possible with the restrictions that we do not go beyond the perturbative regime. We assume that the corrections to gravity comes from interactions with some external scalar, vector or tensor fields only and there is no non-renormalizable higher dimensional operators which modifies gravity with inverse mass scales. Our general parametrization has one drawback that although we are working in the gravitational basis, the masses involved in the expressions for g_i are considered in the mass basis. This can be justified by assuming VEP to be a small effect. In the case of tensor interaction masses do not enter, only in the scalar and vector cases this problem appears. However, as we shall see, the final result for the scalar case comes out to be the same as the one derived from other approaches [19]. Moreover, in the case of some scalar interactions the gravitational basis is equal to the mass basis [1], and then this question will not arise.

We shall not consider any CP violation, and hence H_m and H_G are real symmetric matrices and

[1] this point will be discussed in a forthcoming article, where the dilaton-exchange gravity will be studied by the authors

U_m and U_G are orthogonal matrices $U^{-1} = U^T$. They can be parametrized as

$$U_i = \begin{pmatrix} \cos\theta_i & \sin\theta_i \\ -\sin\theta_i & \cos\theta_i \end{pmatrix},$$

where θ_i represents weak mixing angle θ_m or gravitational mixing angle θ_G. We can now write down the weak Hamiltonian H_w in the weak basis, in which the charged lepton mass matrix is diagonal and the charged current interaction is also diagonal, as

$$H = pI + \frac{1}{2p}\begin{pmatrix} M_+ & M_{12} \\ M_{12} & M_- \end{pmatrix}^2$$
$$= pI + \frac{1}{2p}\left(M_1 + \tfrac{1}{2}M_2^2 M_1^{-1}\right)^2. \quad (6)$$

where

$$M_1 = U_m M_m U_m^{-1}, \quad M_2 = U_G M_G U_G^{-1}$$

and we assumed $M_1^2 \gg M_2^2$, so that the gravitational effects are much smaller than the usual neutrino masses. Since no new gravitational effects have been observed so far, we use this formalism to constrain the parameters of the new gravitational interactions, for which this assumption is justified. We then obtain,

$$M_\pm = \overline{m} \pm \frac{\cos 2\theta_m}{2}\delta m \pm \left[\mp g^{\bar{a}}\overline{m} - \delta g^a \overline{m}\frac{\cos 2\theta_G}{2} \pm \frac{\delta m \delta g^a}{4}\cos 2(\theta_G - \theta_m) + \delta m g^{\bar{a}}\frac{\cos 2\theta_m}{2}\right] / \left[2(\delta m^2 - \overline{m}^2)\right],$$

$$M_{12} = -\frac{\sin 2\theta_m}{2}\delta m + \left[\delta g^a \overline{m}\frac{\sin 2\theta_G}{2} - \delta m g^{\bar{a}}\frac{\sin 2\theta_m}{2}\right] / 2(\delta m^2 - \overline{m}^2). \quad (7)$$

to a leading order in δg^a.

The decay rate for the neutrinoless double beta decay is given by

$$\left[T_{1/2}^{0\nu\beta\beta}\right]^{-1} = \frac{M_+^2}{m_e^2} G_{01} |ME|^2, \quad (8)$$

where ME denotes the nuclear matrix element, G_{01} corresponds to the phase space factor defined in [13] and m_e is the electron mass. The momentum dependence of M_+ must be absorbed into the nuclear matrix element, so that this quantity contains all the momentum dependence and the remaining part is estimated using zero momentum transfer approximation. Thus, if one ignores the nuclear matrix element, then obviously there cannot be any effect of the vector and tensor type gravitational interactions in neutrinoless double beta decay, which was mistaken in Ref. [14]. As it has been discussed earlier [8], the momentum dependence of the tensor type gravitational interactions enters the nuclear matrix element, which then is enhanced by a factor p^2 coming through M_+ in the above expression. We shall now present a more detail explanation of this analysis.

In Ref. [14] it is claimed that neither violations of Lorentz invariance nor violations of the equivalence principle may give sizable contributions to neutrinoless double beta decay. The argument discussed is the following: taking the neutrino propagator

$$\int d^4q \frac{e^{-iq(x-y)}\langle m\rangle c_a^2}{m^2 c_a^4 - q_0^2 c_a^2 + \boldsymbol{q}^2 c_a^2} \quad (9)$$

with the standard $0\nu\beta\beta$ observable $\langle m\rangle$, the neutrino four momentum q and the characteristic maximal velocity c_a. If one would neglect now q_0 and m in the denominator, c_a drops out and the decay rate is independent of c_a.

There are two mistakes in this argument:

(i) The propagator has been taken before performing the q_0-integration. After the integration one gets

$$\int d^3\boldsymbol{q} \frac{e^{iqc_a(x-y)}\langle m\rangle c_a^4}{2\sqrt{(m^2 c_a^4 + \boldsymbol{q}^2 c_a^2)}}. \quad (10)$$

The 3-integration now has the physical meaning of an integration over the neutrino momentum, which becomes the so-called "neutrino potential" in the nuclear structure calculation. Now the neutrino rest energy in the denominator can be neglected compared to the kinectic term and the whole expression becomes

$$\int d^3q \frac{e^{iqc_a(x-y)}\langle m\rangle c_a^3}{2q}, \tag{11}$$

which has an explicit dependence on c_a.

(ii) In [8] it has been shown starting from the Hamiltonian level that the propagator (or the $0\nu\beta\beta$ observable) is changed itself violating Lorentz number. Since

$$H = qc_a + \frac{m^2c_a^4}{2qc_a} = qI + \frac{m^{(*)2}c_a^4}{2qc_a} \tag{12}$$

with $c_a = I + \delta v$ and $m^{(*)2} = m^2 + 2q^2c_a\delta v$ an additional contribution is obtained $\propto q^2\delta v$. This mass-like term has a q^2 enhancement and is not proportional to the small neutrino mass. This consideration answers also the frequently asked question "What is the source of lepton number violation?" in this mechanism. Comparable to a usual mass term, which can be both of Majoran type as well as Dirac type the mass-like term $2q^2c_a\delta v$ can be of Majorana type and act as the source of lepton number violation in this context.

We shall now discuss the three different cases of scalar, vector and tensor interactions and their phenomenology. In the case of tensor interaction the constraint has already been discussed in Ref. [8]. The violation of local Lorentz invariance and the violation of the equivalence principle both fall under this category (their equivalence has been pointed out elsewhere [2]). In both these cases the effect of the new gravitational interactions have quadratic momentum dependence.

In case of the tensorial gravitational interaction we have, $g^{\overline{T}} = 2\alpha^{\overline{T}}p^2 = 0$ and $\delta g^T = 2\delta\alpha^T p^2$. In particular, for the violation of the equivalence principle we substitute $\delta\alpha^T = 4\delta g\phi$ (following the notation of Ref. [8]), where ϕ is the Newtonian gravitational potential on the surface of earth. On the other hand, for the violation of the local Lorentz invariance, we substitute instead, $\delta\alpha^T = 2\delta v$. In both these cases bounds were given in Ref. [8].

To give a bound on tensorial gravitational interactions in the small mixing region (including $\theta_v \sim \theta_m \sim 0$) conservatively $\langle m\rangle \simeq 0$ was assumed. It was also assumed that $\delta m \leq \overline{m}$, and thus $\delta m/4\overline{m}$ may be neglected. Due to the p^2 enhancement the nuclear matrix elements of the mass mechanism have to be replaced by $\frac{m_p}{R}\cdot(M'_F - M'_{GT})$ with the nuclear radius R and the proton mass m_p, which have been calculated in [15]. Inserting the recent half life limit obtained from the Heidelberg–Moscow experiment [16], a bound on the amount of tensorial gravitational interactions as a function of the average neutrino mass $\overline{m}$ was given [8]. It should be stressed also that the GENIUS proposal of the Heidelberg group [17] could improve these bounds by about 1–2 orders of magnitude.

For the vector type gravitational interactions there is a linear momentum dependence. In this case, $g^{\overline{V}} = 2\alpha^{\overline{V}}pm = 0$ and $\delta g^V = 2\delta\alpha^V pm$. The fifth force, as discussed by Fishbach et al. [12] in the context of K-physics is a vector type gravitational interaction. Since no studies of this type of forces exist for neutrino oscillation experiments, with which neutrinoless double beta decay results could be compared, we shall not study this case.

A similar generic structure was considered in a recent analysis of the atmospheric neutrino anomaly [18], where they used the power of momentum dependence as a parameter. From their analysis it becomes apparent that the atmospheric neutrino anomaly may not be explained by either tensorial or vectorial gravitational analysis alone [18].

Recently it has been argued by Damour and Polyakov [11] that string theory may lead to a new scalar type gravitational interaction through interaction of the dilaton field and subsequently its consequence to neutrino oscillation has been studied [19]. Damour and Polyakov have shown that the massless dilaton interaction modifies the gravitational potential energy and there is an additional contribution from an spin-0 exchange, which results in a scalar type gravitational interaction [11]. The resulting theory is of scalar-tensor type with the two particle static gravitational energy

$$V(r) = -G_N m_A m_B(1 + \alpha_A\alpha_B)/r, \tag{13}$$

where G_N is Newton's gravitational constant and α_j denotes the couplings of the dilaton field ϕ to the matter field ψ_j, leading to a gravitational energy of

$$L = m_j \alpha_j \overline{\psi_j} \psi_j \phi. \tag{14}$$

Thus the modified effective mass matrix of the neutrinos are now given by [19]

$$m^{(*)} = m - m\alpha\phi_c, \tag{15}$$

where the classical value of the dilaton field $\phi_c = \phi_N \alpha_{\text{ext}}$ is characterised by the α value of the bulk matter producing it and for a static matter distribution proportional to the Newtonian potential ϕ_N.

The effective mass squared difference

$$\Delta m^{(*)2} = -2m^2\phi_N \alpha_{\text{ext}} \delta\alpha \tag{16}$$

(for almost degenerate masses m) gives rise to neutrino oscillation. The corresponding effect for $0\nu\beta\beta$ decay is obtained by replacing $\delta g^S = 2\delta\alpha^S m^2$ (for almost degenerate mass $m_1 \sim m_2 \sim m$). Comparing the arguments in the oscillations probabilities we get

$$M_+ = m + m\alpha_{\text{ext}} \Phi_N \delta\alpha \frac{\cos(2\theta_G)}{2}. \tag{17}$$

In this case, it is difficult to obtain any bound from neutrino experiments since for α_{ext} only upper bounds exist. However, to get an idea of the constraints which can come from neutrino experiments in the future if α_{ext} is known, according to Ref. [19] we assume $\phi_N = 3 \cdot 10^{-5}$, $\alpha_{\text{ext}} = \sqrt{10^{-3}}$ and $m =$ 2.5 eV (as an upper bound obtained from tritium beta decay experiments [20]). In this case the quantity $\delta\alpha$ is not constrained from neutrinoless double beta decay.

In summary, we presented a general formalism for the study of effects of new gravitational interactions in neutrinoless double beta decay, which allows to constrain the amount of violation of the gravitational laws. Various scenarios discussed in the literature have been analyzed as special cases of the present formalism.

Acknowledgements

We thank G. Bhattacharyya for useful discussions.

References

[1] V.W. Hughes, H.G. Robinson, and V. Beltran-Lopez, Phys. Rev. Lett. 4, 342 (1960); R.W.P. Drever, Philos. Mag. 6 (1961) 683; D. Newman, G.W. Ford, A. Rich and E. Sweetman, Phys. Rev. Lett. 40 (1978) 1355; A. Brillet and J.L. Hall, Phys. Rev. Lett. 42 (1979) 549; J.D. Prestage, J.J. Bollinger, W.M. Itano, and D.J. Wineland, Phys. Rev. Lett. 54 (1985) 2387; S.K. Lamoureaux, J.P. Jacobs, B.R. Heckel, R.J. Raab, and E.N. Fortson, Phys. Rev. Lett. 57 (1986) 3125.

[2] C.M. Will, Theory and Experiment in Gravitational Physics, 2nd edition (Cambridge University Press, Cambridge, 1992).

[3] M.P. Haugan and C.M. Will, Phys. Today 40 (1987) 60; E. Fishbach, M.P. Haugan, D. Tadic and H.Y. Cheng, Phys. Rev. D 32 (1985) 154; G.L. Greene, M.S. Dewey, E.G. Kessler,Jr., and E. Fishbach, Phys. Rev. D 44 (1991) 2216.

[4] M. Gasperini, Phys. Rev. Lett. 62 (1989) 1945; Phys. Rev. D 38 (1988) 2635; ibid. D 39 (1989) 3606.

[5] A. Halprin and C.N. Leung, Phys. Rev. Lett. 67 (1991) 1833; Nucl. Phys. B28A (Proc. Supp.) (1992) 139; M.N. Butler et al., Phys. Rev. D 47 (1993) 2615; K. Iida, H. Minakata and O. Yasuda, Mod. Phys. Lett. A 8 (1993) 1037; J.N. Bahcall, P.I. Krastev, and C.N. Leung, Phys. Rev. D 52 (1995) 1770; R.B. Mann and U. Sarkar, Phys. Rev. Lett 76 (1996) 865; A. Halprin, C.N. Leung, J. Pantalone, Phys. Rev. D 53 (1996) 5365; J.R. Mureika and R.B. Mann, Phys. Rev. D 54 (1996) 2761.

[6] S. Coleman and S.L. Glashow, Phys. Lett. B 405 (1997) 249.

[7] S.L. Glashow, A. Halprin, P.I. Krastev, C.N. Leung, and J. Panteleone, Phys. Rev. D 56 (1997) 2433.

[8] H.V. Klapdor-Kleingrothaus, H. Päs and U. Sarkar, Eur. Phys. Jour. A 5 (1999) 3.

[9] O. Nachtmann, Acta Physica Austriaca, Supp. VI Particle Physics, ed. P. Urban (1969) p. 485.

[10] M.L. Good, Phys. Rev. 121 (1961) 311; T. Hambye, R.B. Mann and U. Sarkar, Phys. Lett. B 421 (1998) 105; Phys. Rev. D 58 (1998) 025003.

[11] T. Damour and A.M. Polyakov, Gen. Rel. Grav. 26 (1994) 1171; Nucl. Phys. B 423 (1994) 532.

[12] S.H. Aronson, G.J. Bock, H-Y Cheng and E. Fishbach, 48 (1982) 1306; Phys. Rev. D 28 (1983) 495.

[13] M. Doi, T. Kotani, E. Takasugi, Progr. Theor. Phys. Suppl. 83 (1985) 1.

[14] A. Halprin and R.R. Volkas, Phys. Lett. B 459 (1999) 183.

[15] H. Päs, M. Hirsch, S.G. Kovalenko, H.V. Klapdor-Kleingrothaus, Phys. Lett. B 453 (1999) 194; M. Hirsch, H.V. Klapdor-Kleingrothaus, S.G. Kovalenko, Phys. Lett. B 372 (1996) 181, Erratum: Phys. Lett. B 381 (1996) 488

[16] L. Baudis et al. (Heidelberg–Moscow Collab.), Phys. Rev. Lett. 83 (1999) 41.
[17] H.V. Klapdor-Kleingrothaus, hep-ex/9802007, in: H.V. Klapdor-Kleingrothaus, H. Päs (Eds.), Proc. Int. Conf. "Beyond the Desert – Accelerator- and Non-Accelerator Approaches", Castle Ringberg, Germany, 1997; H.V. Klapdor-Kleingrothaus, M. Hirsch, Z. Phys. A 359 (1997) 361; H.V. Klapdor-Kleingrothaus, J. Hellmig, M. Hirsch, J. Phys. G 24 (1998) 483; H. V. Klapdor-Kleingrothaus, L. Baudis, G. Heusser, B. Majorovits, H. Päs, hep-ph/9910205
[18] G.L. Fogli, E. Lisi, A. Marrone and G. Scioscia, hep-ph/9904248.
[19] A. Halprin and C.N. Leung, Phys. Lett. B 416 (1998) 361.
[20] V. Lobashev et al., Phys. Lett. B 460 (1999) 227; Ch. Weinheimer et al., Phys. Lett. B 460 (1999) 219.

February 2000

Comment on "Closing the neutrinoless double beta decay window into VEP and/or VLI"

H.V. Klapdor–Kleingrothaus[1], H. Päs[1] and U. Sarkar[2]

[1] *Max–Planck–Institut für Kernphysik, P.O. Box 103980, D–69029 Heidelberg, Germany*
[2] *Physical Research Laboratory, Ahmedabad 380 009, India*

Abstract

The constraints from the non-observation of neutrinoless double beta decay on the violations of Lorentz invariance (VLI) or violations of the equivalence principle (VEP) have recently been re-examined and it was claimed that the constraints are not valid [6]. In this reply we point out that this statement is not correct and prove that the arguments given are wrong.

In recent times there have been many attempts to find out to what accuracy gravitational laws are correct in the neutrino sector. To constrain the amount of violation of the equivalence principle (VEP) from neutrino oscillation experiments, one assumes that the neutrinos of different generations have different characteristic couplings to gravity [1, 2], while to constrain the amount of violation of local Lorentz invariance (VLI) one assumes that neutrinos of different generations have characteristic maximum attainable velocities [3, 4]. Some time back we pointed out that in both these cases it is possible to constrain some otherwise unconstrained region in the parameter space from neutrinoless double beta decay [5]. However, a recent paper [6] came to the contrary conclusions. In this comment we point out what went wrong in their arguments.

The decay rate for the neutrinoless double beta decay is given by,

$$[T_{1/2}^{0\nu\beta\beta}]^{-1} = \frac{M_+^2}{m_e^2} G_{01} |ME|^2, \tag{1}$$

where ME denotes the nuclear matrix element, G_{01} corresponds to the phase space factor defined in [7] and m_e is the electron mass. The momentum dependence of M_+ must be absorbed into the nuclear matrix element $|ME|$. It will be pointed out in the following, that the dominant contribution to neutrinoless double beta decay in the case of VLI or VEP results from the momentum dependence of the observable itself and has been missed in the ansatz in [6].

We shall now present a more detail explanation of this argument. Reference [6] starts with the neutrino propagator,

$$\int d^4q \frac{e^{-iq(x-y)}\langle m\rangle c_a^2}{m^2c_a^4 - q_0^2c_a^2 + \vec{q}^2c_a^2} \tag{2}$$

with the standard $0\nu\beta\beta$ observable $\langle m\rangle$, the neutrino four momentum q and the characteristic maximal velocity c_a. They neglect m in the denominator, so that c_a drops out and the decay rate becomes independent of c_a.

However, to derive the double beta decay rate correctly, one has to start from the Hamiltonian level. In the original paper [5] it has been shown that the propagator (or the $0\nu\beta\beta$ observable) itself is changed when the maximum attainable velocities of different neutrino species are different. Since

$$\begin{aligned} H &= \vec{q}c_a + \frac{m^2c_a^4}{2\vec{q}c_a} \\ &= \vec{q}I + \frac{m^{(*)2}c_a^4}{2\vec{q}c_a} \end{aligned} \tag{3}$$

with $c_a = I + \delta v$ and $m^{(*)2} = m^2 + 2\vec{q}^2c_a\delta v$ an additional contribution to the effective mass is obtained $\propto \vec{q}^2\delta v$. This mass-like term has a $\vec{q}^2$ enhancement and is not proportional to the small neutrino mass. This contribution was included through the nuclear matrix element $|ME|$. But in reference [6] the authors started with a propagator in the zero momentum transfer approximation and obviously did not get this additional important term. If the observable M_+ is assumed to be momentum-independent in the following (i.e. neglecting nuclear recoil), of course the momentum dependence has to be included in the nuclear matrix element.

In the original paper [5], a bound on VEP or VLI was presented in the small mixing region assuming conservatively $\langle m \rangle \simeq 0$ and $\delta m \leq \bar{m}$. Due to the q^2 enhancement the nuclear matrix elements of the mass mechanism were replaced by $\frac{m_p}{R} \cdot (M_F^{'} - M_{GT}^{'})$ with the nuclear radius R and the proton mass m_p, which have been calculated in [8]. Inserting the recent half life limit obtained from the Heidelberg–Moscow experiment [9], a bound on the amount of tensorial gravitational interactions as a function of the average neutrino mass $\bar{m}$ was presented [5]. **It is obvious that if one ignores the momentum enhancement of the nuclear matrix element (as it was done in ref [6]), the neutrinoless double beta decay cannot give any significant constraint.** In fact, including the momentum enhancement of the nuclear matrix elements not only these constraints are significant, they will further improve by 1–2 orders of magnitude with the GENIUS proposal of the Heidelberg group [10].

Acknowledgement We thank G. Bhattacharyya for useful discussions.

References

[1] M. Gasperini, Phys. Rev. Lett. **62**, 1945 (1989); Phys. Rev. **D38**, 2635 (1988); *ibid.* **D39**, 3606 (1989).

[2] A. Halprin and C.N. Leung, Phys. Rev. Lett. **67**, 1833 (1991); Nucl. Phys. **B28A** (Proc. Supp.), 139 (1992); M. N. Butler *et al.*, Phys. Rev. **D47**, 2615 (1993); K. Iida, H. Minakata and O. Yasuda, Mod. Phys. Lett. **A8** (1993) 1037; J. N. Bahcall, P. I. Krastev, and C. N. Leung, Phys. Rev. **D52**, 1770 (1995); R.B. Mann and U. Sarkar, Phys. Rev. Lett **76** (1996) 865; A. Halprin, C.N. Leung, J. Pantalone, Phys.Rev. **D 53** (1996) 5365; J.R. Mureika and R. B. Mann, Phys. Rev. **D 54** (1996) 2761.

[3] S. Coleman and S.L. Glashow, Phys. Lett. **B 405**, 249 (1997).

[4] S.L. Glashow, A. Halprin, P.I. Krastev, C.N. Leung, and J. Panteleone, Phys. Rev. **D 56**, 2433 (1997).

[5] H.V. Klapdor-Kleingrothaus, H. Päs and U. Sarkar, Eur. Phys. Jour. **A 5** (1999) 3.

[6] A. Halprin and R.R. Volkas, Phys. Lett. **B 459** (1999) 183.

[7] M. Doi, T. Kotani, E. Takasugi, Progr. Theor. Phys. Suppl. **83** (1985) 1.

[8] H. Päs, M. Hirsch, S.G. Kovalenko, H.V. Klapdor–Kleingrothaus, Phys. Lett. **B 453** (1999) 194; M. Hirsch, H.V. Klapdor–Kleingrothaus, S.G. Kovalenko, Phys. Lett. **B 372** (1996) 181, Erratum: Phys. Lett. **B 381** (1996) 488

[9] L. Baudis et al. (Heidelberg–Moscow Collab.), Phys. Rev. Lett. **83** (1999) 41

[10] H.V. Klapdor–Kleingrothaus, hep-ex/9802007, in: H.V. Klapdor–Kleingrothaus, H. Päs (Eds.), Proc. Int. Conf. "Beyond the Desert - Accelerator- and Non-Accelerator Approaches", Castle Ringberg, Germany, 1997, IOP Publishing, Bristol & Philadelphia 1998; H.V. Klapdor–Kleingrothaus, M. Hirsch, Z. Phys. **A 359** (1997) 361; H.V. Klapdor–Kleingrothaus, J. Hellmig, M. Hirsch, J. Phys. **G 24** (1998) 483; H. V. Klapdor-Kleingrothaus, L. Baudis, G. Heusser, B. Majorovits, H. Päs, hep-ph/9910205

PHYSICAL REVIEW D VOLUME 56, NUMBER 10 15 NOVEMBER 1997

Effect of violation of quantum mechanics on neutrino oscillation

Yong Liu, Liangzhong Hu, and Mo-Lin Ge
Theoretical Physics Division, Nankai Institute of Mathematics, Nankai University, Tianjin 300071, China
(Received 17 March 1997)

The effect of quantum mechanics violation due to quantum gravity on neutrino oscillation is investigated. It is found that the mechanism introduced by Ellis, Hagelin, Nanopoulos, and Srednicki through the modification of the Liouville equation can affect neutrino oscillation behavior and may be taken as a new solution of the solar neutrino problem. [S0556-2821(97)03222-0]

PACS number(s): 14.60.Pq, 03.65.Bz

I. INTRODUCTION

More than 20 years ago, Hawking found that quantum mechanics allows black holes to emit particles in a thermal spectrum [1,2]. Because the black hole creates particles in pairs, with one particle always falling into the hole and the other possibly escaping to infinity, part of the information about the state of the system is lost down the black hole and the final situation is represented by a density matrix rather than a pure quantum state [3]. Hawking proposed that if such a decay of a pure quantum state into a mixed state can occur with a macroscopic configuration such as a black hole, it also ought to occur on a microscopic elementary particle level because of quantum fluctuations of the metric which could be interpreted as virtual black holes which appear and disappear again [4]. Furthermore, Hawking introduced a new operator, called the superscattering operator, to describe the process. This operator can map the initial mixed states to final mixed states [3].

The evolution of pure states into mixed states has aroused considerable attention in physics. Page showed that any such dynamics can lead to conflict with *CPT* conservation [5]. Banks, Peskin, and Susskind found that, in such a theory which allows the evolution of pure states into mixed states, there is a serious conflict between energy-momentum conservation and locality [6]. Thereafter, Ellis, Hagelin, Nanopoulos, and Srednicki (EHNS) set up a modified Hamiltonian formalism for the time evolution of density matrices which includes violation of quantum mechanics such as the evolution of pure states into mixed states [7]. Following EHNS, Ellis, Mavromatos, and Nanopoulos reconsidered the analysis of EHNS for the K_0-$\overline{K_0}$ system. They suggested that this new source of *CPT* violation might fully account for the observed *CP* violation in the K_0-$\overline{K_0}$ system [8,9]. But Huet and Peskin using the classic results of the Carithers *et al.* [10] and CERN-Heidelberg experiments [11] and the results from CPLEAR [12] determined the two of the three new *CPT*-violation parameters α, β, and γ of EHNS. They argued that the *CP* violation observed in the K_0-$\overline{K_0}$ system is dominantly quantum mechanical in nature and of *CPT*-conserving origin [13]. Works along this direction are still going on [14,15].

On the other hand, the oscillation among neutrinos of different flavors such as ν_e-ν_μ is much like the strangeness oscillation phenomenon in the K_0-$\overline{K_0}$ system [16]. Because the neutrino oscillations in vacuum and/or in matter are in connection with the solar neutrino experiments and as a possible solution of the solar neutrino problem, it has caused a great interest in this subject for many years [17–19].

In this work, we use the EHNS mechanism to investigate the effect of the quantum mechanics violation proposed by Hawking on neutrino oscillations. To make this paper self-contained, we will introduce the modified Liouville equation of EHNS following Huet and Peskin [13] and the relative formulas about neutrino oscillation in Sec. II. Then we will list some numerical results in Sec. III. The conclusion and a discussion are given in Sec. IV.

II. FORMALISM

The neutrino weak eigenstates may not coincide with the eigenstates of its mass matrix. If such is the case, due to the different time evolution properties, oscillations will occur [16–19].

We consider the simplest case of two neutrinos. Let $|\nu_1\rangle$ and $|\nu_2\rangle$ be the mass eigenstates with masses m_1 and m_2. Suppose that neutrinos mix through a vacuum mixing angle θ; then, the weak eigenstates are

$$|\nu_e\rangle=\cos\theta|\nu_1\rangle+\sin\theta|\nu_2\rangle,$$

$$|\nu_\mu\rangle=-\sin\theta|\nu_2\rangle+\cos\theta|\nu_2\rangle. \tag{1}$$

The two states evolve differently; thus,

$$|\nu_e(t)\rangle=\cos\theta e^{-iE_1t}|\nu_1\rangle+\sin\theta e^{-iE_2t}|\nu_2\rangle,$$

$$|\nu_\mu(t)\rangle=-\sin\theta e^{-iE_1t}|\nu_2\rangle+\cos\theta e^{-iE_2t}|\nu_2\rangle. \tag{2}$$

As a result, a state originally $|\nu_e\rangle$ may oscillate into $|\nu_\mu\rangle$ with the probability

$$p[\nu_e\rightarrow\nu_\mu(t)]=\sin^2(2\theta)\sin^2[\tfrac{1}{2}(E_2-E_1)t] \tag{3}$$

and the probability for it to remain as itself is

$$p[\nu_e\rightarrow\nu_e(t)]=1-\sin^2(2\theta)\sin^2[\tfrac{1}{2}(E_2-E_1)t]. \tag{4}$$

Because of the smallness of neutrino masses, their energy and momentum are very close; hence, we can rewrite the probability as [20,21]

PHYSICAL REVIEW D VOLUME 56, NUMBER 4 15 AUGUST 1997

BRIEF REPORTS

Brief Reports are accounts of completed research which do not warrant regular articles or the priority handling given to Rapid Communications; however, the same standards of scientific quality apply. (Addenda are included in Brief Reports.) A Brief Report may be no longer than four printed pages and must be accompanied by an abstract.

Remarks on neutrino tests of special relativity

S. L. Glashow,[1] A. Halprin,[2] P. I. Krastev,[3] C. N. Leung,[2,4] and J. Pantaleone[5]
[1]*Lyman Laboratory of Physics, Harvard University, Cambridge, Massachusetts 02138*
[2]*Department of Physics and Astronomy, University of Delaware, Newark, Delaware 19716*
[3]*School of Natural Sciences, The Institute for Advanced Study, Princeton, New Jersey 08540*
[4]*Centro de Física das Interacções Fundamentais, Instituto Superior Técnico, Lisboa, Portugal*
[5]*Department of Physics and Astronomy, University of Alaska, Anchorage, Alaska 99508*
(Received 1 April 1997)

We point out that the assumption of Lorentz noninvariance examined recently by Coleman and Glashow leads to neutrino flavor oscillations which are phenomenologically equivalent to those obtained by assuming the neutrinos violate the principle of equivalence. We then comment on the limits on Lorentz noninvariance which can be derived from solar, atmospheric, and accelerator neutrino experiments. [S0556-2821(97)04916-3]

PACS number(s): 03.30.+p, 11.30.Cp, 14.60.Pq, 96.40.Tv

In a recent paper [1], Coleman and Glashow have proposed several interesting ways to test how well Lorentz invariance is obeyed in nature. If Lorentz invariance is violated, one possible consequence is that the propagation of a free particle will depend on its identity. In the case of massless neutrinos, this may lead to neutrino flavor oscillations because different neutrino species may have different maximum attainable velocities (which are no longer necessarily c). For this to happen, it is necessary that the neutrino flavor eigenstates be distinct from their velocity eigenstates, defined to be the energy eigenstates at infinite momentum, so that a flavor eigenstate is a linear superposition of the velocity eigenstates and vice versa. If one considers the case of two-neutrino mixing, say, ν_e and ν_μ, the ν_e survival probability is given by [1]

$$P(\nu_e \rightarrow \nu_e) = 1 - \sin^2(2\theta_v)\sin^2(\delta v EL/2), \tag{1}$$

where $\delta v = v_1 - v_2$ is the difference between the velocities of the velocity eigenstates ν_1 and ν_2, θ_v is the mixing angle,

$$\nu_e = \nu_1 \cos\theta_v - \nu_2 \sin\theta_v, \quad \nu_\mu = \nu_1 \sin\theta_v + \nu_2 \cos\theta_v, \tag{2}$$

E is the neutrino energy, and L is the distance traveled by the neutrino.

We would like to point out that the energy dependence described in Eq. (1) is exactly the same as what one will get if one assumes that neutrinos violate the principle of equivalence in a certain way [2,3]. This is interesting but not totally surprising because general coordinate invariance is violated in both cases. The phenomenology of the case of equivalence principle violation has been studied in some detail over the last several years [4–11]. The results of these studies can be straightforwardly translated to set limits on the possible violation of Lorentz invariance. This is what we will discuss in the remainder of this paper.

We shall use the notation of Ref. [10]. It is easy to see that [see, e.g., Eqs. (14)–(16) in Ref. [10]] the parameter $|\delta v|$, which measures the degree of violation of Lorentz invariance, should be compared with $2|\phi\Delta\gamma|$ in Ref. [10], where ϕ is the gravitational potential in which the neutrino propagates and $\Delta\gamma$ is a parameter which measures the degree of violation of the equivalence principle. The two are equivalent for the case of a constant gravitational potential. In addition, the mixing angle θ_v is equivalent to the mixing angle θ_G in Ref. [10]. Currently, there are positive indications of neutrino flavor mixing from solar and atmospheric neutrino experiments. These data have been used to obtain allowed regions for the mixing parameters $\sin^2(2\theta_G)$ and $|\phi\Delta\gamma|$ (see Refs. [4], [8], and [10]), which can be translated directly into allowed regions for $\sin^2(2\theta_v)$ and $|\delta v|$.

First of all, due to the specific energy dependence, the solar neutrino data cannot be explained by long-wavelength vacuum oscillations caused by the Lorentz invariance violation. This is because, if the mixing parameters are chosen such that enough ^{8}B neutrinos are suppressed, there will not be sufficient suppression for the lower energy solar neutrinos, in contradiction to the data. It is therefore necessary to invoke the Mikheyev-Smirnov-Wolfenstein mechanism [12] of matter-enhanced transitions in the Sun. If we assume δv to be constant inside the Sun, the situation will be equivalent to the case of constant ϕ analyzed in Ref. [10]. Using the solar model of Bahcall and Pinsonneault [13], one would find that a large portion of the mixing parameter space has already been excluded by the solar neutrino data, with two remaining allowed regions: a small mixing angle region for which

$$|\delta v| \sim 6\times 10^{-19}, \quad 0.002 < \sin^2(2\theta_v) < 0.003, \tag{3}$$

at 90% confidence level, and a large mixing angle region for which

$$4\times10^{-22}<|\delta v|<4\times10^{-21}, \quad 0.38<\sin^2(2\theta_v)<0.81, \tag{4}$$

also at 90% confidence level. Furthermore, the energy dependence implies that higher energy neutrinos have shorter oscillation length. As a consequence, the higher energy atmospheric neutrino data imply a violation of Lorentz invariance in a small but overlapping parameter region (see Fig. 4 in Ref. [10]). It is quite remarkable that the mixing of two neutrinos is sufficient to account for both the solar neutrino and the atmospheric neutrino data.

Aside from offering a possible resolution to the solar neutrino problem, velocity oscillations of neutrinos may provide the most sensitive tests of Lorentz invariance and the equivalence principle. Unlike conventional neutrino oscillations, velocity oscillations become more important at higher energies. Presently available accelerator data already provide useful constraints, as mentioned in Ref. [1] (see also Fig. 1 in Ref. [10]). In fact, part of the large angle region allowed by the solar neutrino data may be ruled out by the accelerator neutrino data. Planned long-base-line neutrino oscillation experiments will be able to push the limit on $|\delta v|$ lower by one to two orders of magnitude, thereby limiting possible departures from special relativity or the equivalence principle.

Whether neutrinos have observable masses is a central question of particle physics. It is often said that the observation of neutrino oscillations at accelerators (or their deduction from solar neutrino or cosmic ray experiments) would be conclusive evidence that at least one neutrino is massive. This is not true. Neutrino oscillations can also result from a tiny breakdown of Lorentz invariance and/or the principle of equivalence. More information than mere detection is needed to determine the underlying mechanism of neutrino oscillation. The differing energy dependence between the mass mechanism and the mechanism due to Lorentz noninvariance (or due to equivalence principle violation) suggests that an accurate spectral measurement is required. Super Kamiokande and SNO can accurately measure the solar neutrino spectrum, and, as the analysis in Ref. [8] shows, this measurement will test the viability of the small mixing region. The current atmospheric neutrino data favor the large mixing region and this possibility will be tested by new atmospheric neutrino data from Super Kamiokande which should be available in the very near future.

It is of course a distinct possibility that neutrinos have nondegenerate masses and, at the same time, Lorentz invariance and/or the principle of equivalence is violated. The phenomenology of neutrino oscillations in this case will be much more complicated. As Coleman and Glashow pointed out, and as also discussed in Sec. 2.3 of Ref. [10], for the case of two-neutrino mixing there is an additional phase parameter beside the doubling of mixing parameters. This will present a serious challenge to future neutrino experiments.

This work was supported in part by the U.S. Department of Energy under Grant No. DE-FG02-84ER40163 and by the National Science Foundation under Grant No. NSF-PHYS-92-18167. The work of P.K. was partially supported by NSF Grant No. PHY-9513835. C.N.L. would like to thank G. C. Branco for bringing Ref. [1] to his attention. He is also grateful for the hospitality extended to him by G. C. Branco, L. Ferreira, J. Pulido, M. Rebelo, E. Ribeiro, A. Rossi, V. Vieira, and the other members of the Centro de Física das Interacções Fundamentais at the Instituto Superior Técnico, where part of this work was carried out.

[1] S. Coleman and S. L. Glashow, ''Cosmic Ray and Neutrino Tests of Special Relativity,'' Harvard University Report No. HUTP-97/A008, hep-ph/9703240.

[2] M. Gasperini, Phys. Rev. D **38**, 2635 (1988); **39**, 3606 (1989).

[3] A. Halprin and C. N. Leung, Phys. Rev. Lett. **67**, 1833 (1991).

[4] J. Pantaleone, A. Halprin, and C. N. Leung, Phys. Rev. D **47**, R4199 (1993).

[5] K. Iida, H. Minakata, and O. Yasuda, Mod. Phys. Lett. A **8**, 1037 (1993).

[6] M. N. Butler, S. Nozawa, R. Malaney, and A. I. Boothroyd, Phys. Rev. D **47**, 2615 (1993).

[7] H. Minakata and H. Nunokawa, Phys. Rev. D **51**, 6625 (1995).

[8] J. N. Bahcall, P. I. Krastev, and C. N. Leung, Phys. Rev. D **52**, 1770 (1995).

[9] R. B. Mann and U. Sarkar, Phys. Rev. Lett. **76**, 865 (1996).

[10] A. Halprin, C. N. Leung, and J. Pantaleone, Phys. Rev. D **53**, 5365 (1996).

[11] H. Minakata and A. Yu Smirnov, Phys. Rev. D **54**, 3698 (1996).

[12] L. Wolfenstein, Phys. Rev. D **17**, 2369 (1978); **20**, 2634 (1979); S. P. Mikheyev and A. Yu Smirnov, Yad. Fiz. **42**, 1441 (1985) [Sov. J. Nucl. Phys. **42**, 913 (1985)]; Nuovo Cimento C **9**, 17 (1986).

[13] J. N. Bahcall and M. M. Pinsonneault, Rev. Mod. Phys. **64**, 885 (1992).

Effects of quantum space time foam in the neutrino sector

H.V. Klapdor-Kleingrothaus [1], H. Päs [1], U. Sarkar [2]

[1] Max-Planck-Institut für Kernphysik, P.O. Box 103980, D-69029 Heidelberg, Germany
[2] Physical Research Laboratory, Ahmedabad, 380 009, India

Abstract

We discuss violations of CPT and quantum mechanics due to interactions of neutrinos with space-time quantum foam. Neutrinoless double beta decay and oscillations of neutrinos from astrophysical sources (supernovae, active galactic nuclei) are analysed. It is found that the propagation distance is the crucial quantity entering any bounds on EHNS parameters. Thus, while the bounds from neutrinoless double beta decay are not significant, the data of the supernova 1987a imply a bound being several orders of magnitude more stringent than the ones known from the literature. Even more stringent limits may be obtained from the investigation of neutrino oscillations from active galactic nuclei sources, which have an impressive potential for the search of quantum foam interactions in the neutrino sector.

1 Introduction

While in the context of local quantum field theories CPT has to be conserved, CPT violating effects may show up in the framework of quantum gravity. As an example, Hawking radiation of black holes can be understood as a pair creation process near the event horizon, with one particle falling into the black hole and the other one escaping. Since with the particle falling into the black hole some phase information of the quantum state is lost, the thermic final state is a mixed state rather than a pure one. As Hawking has pointed out [1], such an evolution of a pure state into a mixed state violates the laws of conventional quantum mechanics (QMV). If the space time possesses a foamy structure at the Planck scale, including the creation and annihilation of black holes with Planck radius and Planck lifetime, such effects also may influence microscopical processes in the vacuum [2]. In the following Page [3] showed that such processes violate also CPT and the possibility of experimental tests in the $K_0 - \bar{K}_0$ sector was discussed by Eberhard [4]. Ellis, Hagellin, Nanopoulos and Srednicki independently developed an evolution equation formalism in the space of density matrices [5] containing three CPT violating (EHNS) parameters α, β, γ which have a dimension of mass and which might be expected to be of order $m_K^2/M_{Pl} \sim 10^{-20}$ GeV in the Kaon sector. Recently the topic has been reconsidered by Ellis, Mavromatos and Nanopoulos [6] and Huet and Peskin [7]. CPT violating processes in the neutrino sector have been discussed for the first time by Liu et al. [8] and in the following in [9], where neutrino oscillations due to CPT violation has been discussed as a solution to the solar neutrino problem. Recently another paper [11] explored the possibility of explaining the atmospheric neutrino anomaly with quantum foam effects and came to a negative conclusion. In this note we extend the discussion of quantum foam effects in the neutrino sector to the cases of neutrinoless double beta decay and oscillations of neutrinos from astrophysical sources, supernovae as well as active galactic nuclei. New, extremely stringent bounds are found improving constraints found in the literature by several orders of magnitude.

2 Density matrix formalism

For mixed states it is useful to work in the framework of the density matrix formalism, following the methodology as presented in ref. [8]. We start with the Schrödinger equation for the density matrix,

$$i\frac{d}{dt}\rho = [H, \rho]. \quad (1)$$

Here ρ is the density matrix of the system, which can be expanded in the Pauli matrix basis,

$$\rho = \rho^0 I + \rho^i \sigma^i, \quad (2)$$

where I is the unity matrix and σ^i are the Pauli matrices. In [9] a lepton number violating parametrization for the evolution equation of the components of the density matrix has been assumed:

$$\frac{d}{dt}\begin{pmatrix} \rho^0 \\ \rho^1 \\ \rho^2 \\ \rho^3 \end{pmatrix} = 2\begin{pmatrix} 0 & 0 & 0 & 0 \\ 0 & 0 & \Delta m^2/(4E) & 0 \\ 0 & -\Delta m^2/(4E) & -\alpha & -\beta \\ 0 & 0 & -\beta & -\gamma \end{pmatrix}\begin{pmatrix} \rho^0 \\ \rho^1 \\ \rho^2 \\ \rho^3 \end{pmatrix}. \tag{3}$$

Here $\beta \ll \alpha, \gamma$ [5, 7, 8]. In [8, 9] also an alternative, lepton-number conserving parametrization has been discussed. However this parametrization will not influence neither the double beta decay observable nor the oscillation probability in the asymptotics of large propagation distances compared to the standard case of neutrino masses [9]. Thus we will concentrate on the lepton-number violating case in the following. It should be mentioned however that a full analysis of the generalized dynamics requires six parameters [10].

Moreover, it should be stressed that this non-relativistic ansatz may not be suitable to describe ultra-relativisic particles such as neutrinos. However, while the covariant treatment of open quantum systems is still an unsolved problem, the density matrix ansatz has been successfully used in previous works to derive the "standard" mass mechanism neutrino oscillation probability also, see [11]. Thus, while future works should improve the present ansatz, this approach seems to be suitable to provide at least a possibility for the comparison of the sensitivity of different experiments and a rough estimation for the order of magnitude of the obtained bounds.

3 Neutrinoless double beta decay

Neutrinoless double beta decay is one of the most sensitive tools in neutrino physics. It corresponds to two single beta decays occuring simultaneous in one nucleus, with a virtual neutrino propagating between the vertices. Important impact of this process has been derived on the reconstruction of the neutrino mass spectrum, physics beyond the standard model as well as more exotic phenomena such as violations of the equivalence principle or Lorentz invariance (for an overview see [12, 13]). In the following we will study the potential of neutrinoless double beta decay for searches for CPT violations due to quantum foam interactions in the neutrino sector. The observable measured in neutrinoless double beta decay is the ee entry of the neutrino mass matrix in the flavor space,

$$m_{ee} = \bar{m} - \frac{\delta m}{2} cos(2\theta) \tag{4}$$

in a two neutrino scenario with $\bar{m} = (m_1 + m_2)/2$ and $\delta m = (m_2 - m_1)$ and $m_{1,2}$ being the mass eigenstates. This quantity will be modified in the presence of QMV. The recent experimental constraint is $m_{ee} < 0.3$ eV, obtained from the Heidelberg–Moscow experiment searching for double beta decays of ^{76}Ge [14]. The GENIUS project will be sensitive to $m_{ee} = 10^{-2} - 10^{-3}$ eV [15]. In the density matrix formalism the double beta decay observable can be expressed as follows:

$$\begin{aligned} Tr(\rho_{\nu_e}\mathcal{O}) &= Tr\begin{pmatrix} \rho_0 + \rho_3 & \rho_1 - i\rho_2 \\ \rho_1 + i\rho_2 & \rho_0 + \rho_3 \end{pmatrix} \cdot \begin{pmatrix} m_1 & 0 \\ 0 & m_2 \end{pmatrix} \\ &= (m_1 + m_2)\rho_0 + (m_2 - m_1)\rho_3. \end{aligned} \tag{5}$$

The propagation time of the neutrino

$$t = \frac{1}{4\pi\Delta E} \simeq 6 \cdot 10^{-24} s \tag{6}$$

can be estimated by taking its energy to be of the size of the nuclear Fermi momentum $p_F \simeq 100$ MeV for ^{76}Ge. Assuming $\beta \ll \alpha, \gamma$, eq. (5) yields

$$\frac{d}{dt}\rho_0 = 0 \tag{7}$$

$$\frac{d}{dt}\rho_3 = -2\gamma\rho_3 \tag{8}$$

and thus, using eq. (4)

$$\rho_0 = \frac{1}{2} \tag{9}$$

$$\rho_3 = e^{-2\gamma t}\frac{cos(2\theta)}{2}. \tag{10}$$

This implies

$$m_{ee}^{QMV} = \bar{m} + e^{-2\gamma t}\frac{\Delta m}{2}\cos 2\theta. \tag{11}$$

Due to the tiny propagation time (6) no significant variation of the double beta decay observable is obtained. However, from this analysis we realize that the distance plays a crucial role in constraining the QMV parameters, so we shall consider the bounds on the neutrino oscillation probability where neutrinos are propagating over large distances.

4 Oscillations of neutrinos from astrophysical sources

In the following we study the effect of quantum mechanics violation in neutrino oscillations from astrophysical sources. The most distant sources that have been discussed in the context of neutrino oscillations are supernovae (SN) and active galactic nuclei (AGN). While astrophysical sources have been discussed in the context of QMV effects on life time measurements [16, 17], they have not been considered for the case of QMV induced neutrino oscillations so far.

For the neutrino oscillation case we get the survival and disappearance oscillation probabilities [8, 9]

$$P(\nu_x \to \nu_x) = Tr[\rho_{\nu_x}(t)\rho_{\nu_x}] \tag{12}$$

$$P(\nu_x \to \nu_{x'}) = Tr[\rho_{\nu_x}(t)\rho_{\nu_{x'}}], \tag{13}$$

respectively. Here the density matrices can be parametrized as

$$\rho_{\nu_x} = \begin{pmatrix} \cos^2\theta & \cos\theta\sin\theta \\ \cos\theta\sin\theta & \sin^2\theta \end{pmatrix}, \tag{14}$$

$$\rho_{\nu_{x'}} = \begin{pmatrix} \sin^2\theta & -\cos\theta\sin\theta \\ -\cos\theta\sin\theta & \cos^2\theta \end{pmatrix}. \tag{15}$$

As initial condition we assume

$$\rho(t=0) = \rho(\nu_e) \tag{16}$$

and thus [8, 9]:

$$\rho_0 = \frac{1}{2} \tag{17}$$

$$\rho_1 = \frac{1}{2}\sin(2\theta) \tag{18}$$

$$\rho_2 = 0 \tag{19}$$

$$\rho_3 = \frac{1}{2}\cos(2\theta). \tag{20}$$

The interesting observable is the oscillation propability

$$P^{QMV}_{\nu_x\to\nu_{x'}} = Tr[\rho(t)\rho_x] = \frac{1}{2} - \frac{1}{2}e^{-\gamma L}\cos^2 2\theta - \frac{1}{2}e^{-\alpha L}\sin^2 2\theta\cos(\frac{\Delta m^2}{2E_\nu}L), \tag{21}$$

where $\beta \ll \alpha, \gamma$ has been assumed. For the n-flavour case the oscillation probability for large propagation distances is given by [9],

$$P^{QMV}_{\nu_x\to\nu_{x'}} = \frac{1}{n} - \frac{1}{n}e^{-\gamma L}, \tag{22}$$

where L is the propagation distance of the neutrinos.

This QMV oscillation probability can easily be distinguished from the asymptotics of the "standard" mass induced oscillation probability:

$$P^{mass}_{\nu_x \to \nu_{x'}} = \frac{\sin^2 2\theta}{2}. \tag{23}$$

The quantity P^{mass} is fixed experimentally to $P^{mass}_{\nu_\mu \to \nu_\tau} \simeq 0.5$ due to the maximal mixing in atmospheric neutrinos [21] and $P^{mass}_{\nu_e \to \nu_\tau} \lesssim 0.05$ due to the CHOOZ bound [22].

Supernovae 1987a: In supernovae strong neutrino oscillations will significantly distort the ν_e spectra at the earth, since the ν_e will aquire the spectra of the more energetic ν_μ and ν_τ. The distance is very large. As a result, the condition that QMV should satisfy the bound on the oscillation probability gives a very strong bound. In the case of supernova 1987a, $L \sim 50 Mpc \sim 7 \cdot 10^{39} GeV$, so that the observed constraint on the oscillation probability [18] $P^{exp}_{\nu_e \to \nu_{\mu,\tau}} < 0.2$ is satisfied for the three neutrino case when

$$\gamma < \frac{0.6}{L} \sim 10^{-40} GeV. \tag{24}$$

We assumed here that P^{exp} is the accuracy with which the deviations from the asymptotics $1/n = 1/3$ can be measured. Due to the unknown energy dependence of the EHNS parameters and the Lorentz non-invariant ansatz it is difficult to compare these bounds with the bound coming from K-physics. Following [8] we assume γ to be of the order E_ν^2/M_{Pl} and scale the obtained bound by the neutrino energy to the kaon mass squared,

$$\gamma_\nu \propto \frac{E_\nu^2}{m_K^2}; \tag{25}$$

implying $\gamma_K < 10^{-37} GeV$, which is an improvement of about 16 orders of magnitude. This disfavors strongly any solution of the solar or atmospheric neutrino problem by lepton number violating QMV effects. If one assumes that the same QMV parametrization is valid for the $K-$system, then any observational possibility in the $K-$system will also be excluded by the present constraints from the supernovae analysis. A relativistic treatment of the problem can modify this bound to some extent, but it is most unlikely that the modification is by several orders of magnitude.

Active Galactic Nuclei (AGN): AGN can be intense sources of high energy neutrinos ($E_\nu \sim \mathcal{O}$ (1 PeV)) [19]. According to representative models the flux of these neutrinos is flavor dependent and the ν_τ flux is reduced by at least two orders of magnitude compared to the ν_e, ν_μ fluxes. An unique appearance signal of high energy ν_τ neutrinos can be a double bang signal of the produced τ leptons: The first bang originates from the CC interaction of the τ neutrino and the second one from the hadronic decay of the τ lepton. Deep underwater or ice neutrino detectors have been estimated to be sensitive on neutrino oscillation probabilities of [19, 20]

$$P^{exp}_{\nu_{e,\mu} \to \nu_\tau} < 5 \times 10^{-3}. \tag{26}$$

Since the QMV effects become strong for large distances and higher energies, it is likely that when we have data from the active galactic nuclei on neutrino oscillations, these bounds will be modified by several orders of magnitude. Considering the distance to be $L \sim 100 Mpc$ and the average energy of the neutrinos to be around 1 PeV, a bound on the neutrino oscillation of $P_{\nu_e \to \nu_\mu} < 5 \times 10^{-3}$ will imply a corresponding bound on the QMV parameter

$$\gamma_\nu < 10^{-42} GeV. \tag{27}$$

Translation to the kaon mass scale yields

$$\gamma_K < 10^{-55} GeV, \tag{28}$$

which would imply the by far strongest bound on QMV parameters. This will provide a decisive test for any contribution of lepton number violating QMV effects in the neutrino sector.

5 Conclusions

We studied the effects of violation of quantum mechanics due to quantum space time foam interactions in neutrino experiments. While the non-observation of neutrinoless double beta decay does not give any significant constraint, the supernova 1987a implies a constraint being 16 orders of magnitude more stringent than the bounds known from the literature. This disfavors strongly any possibility of observable effects of lepton number violating QMV in any other experiments. The non-observation of QMV induced neutrino oscillations from active galactic nuclei will be able to improve this bound by many orders of magnitude. While the chosen non-relativistic ansatz might not be totally suitable for neutrinos, it should be at least useful to compare the sensitivity of different neutrino sources. Moreover the bounds obtained are that stringent, that, even in view of this ambiguity, they should be considered as the most restrictive ones.

Acknowledgements

We thank J. Ellis, E. Lisi, S. Pakvasa, A.Y. Smirnov, the referee and especially N.E. Mavromatos for comments and useful discussions.

References

[1] S.W. Hawking, Nature 248, (1974) 30

[2] S.W. Hawking, Commun. Math. Phys. 43 (1975) 199

[3] D.N. Page, Gen. Relativ. Gravit. 14 (1982) 1; Phys. Rev. Lett. 44 (1980) 301

[4] P.H. Eberhard, CERN report 72-1 (1972) unpubl.; W.C. Carithers, J.H. Christenson, P.H. Eberhard, D.R. Nygren, T. Modis, T.P. Pun, E.L. Schwartz, H. Sticker, Phys. Rev. D 14 (1976) 290

[5] J. Ellis, J.S. Hagellin, D.V. Nanopoulos, M. Srednicki, Nucl. Phys. B 241 (1984) 381

[6] J. Ellis, N.E. Mavromatos, D.V. Nanopoulos, Phys. Lett. B 293 (1992) 193

[7] P. Huet, M.E. Peskin, Nucl.Phys. B434 (1995) 3-38

[8] Y. Liu, L. Hu, M.-L. Ge, Phys. Rev. D 56 (1997) 6648

[9] Y. Liu, J.-L. Chen, M.-L. Ge, J. Phys. G 24 (1998) 2289; F. Ma, H. Hu, hep-ph/9805391; C.-H. Chang, W.-S. Dai, X.-Q. Li, Y. Liu, F.-C. Ma, Z.-J. Tao, Phys. Rev. D 60 (1999) 033006

[10] F. Benatti, R. Floreanini, JHEP 02 (2000) 032

[11] E. Lisi, A. Marrone, D. Montanino, hep-ph/0002053

[12] H.V. Klapdor-Kleingrothaus, H. Päs, hep-ph/0002109, in Proc. *Cosmo '99, Trieste, Italy*

[13] H.V. Klapdor-Kleingrothaus, H. Päs, A.Y. Smirnov, hep-ph/0003219

[14] HEIDELBERG-MOSCOW collab., Phys. Rev. Lett. 83 (1999) 41; priv. comm.

[15] H.V. Klapdor-Kleingrothaus, Proc. *Beyond the Desert '97*; J. Hellmig, H.V. Klapdor-Kleingrothaus, Z. Phys. A 359 (1997) 351; H.V. Klapdor-Kleingrothaus, M. Hirsch, Z. Phys. A 359 (1997) 361; H.V. Klapdor-Kleingrothaus, J. Hellmig, M. Hirsch, J. Phys. G 24 (1998) 483; H.V. Klapdor-Kleingrothaus, L. Baudis, G. Heusser, B. Majorovits, H. Päs, hep-ph/9910205

[16] J. Ellis, N.E. Mavromatos, D.V. Nanopoulos, G. Volkov, gr-qc/9911055

[17] O. Bertolami, C.S. Carvalho, gr-qc/9912117

[18] A.Y. Smirnov, D.N. Spergel, J.N. Bahcall, Phys. Rev. D 49 (1994) 1389

[19] J.G. Learned, S. Pakvasa, Astropart. Phys. 3 (1995) 267-274; S. Pakvasa, hep-ph/9503369; T.J. Weiler, W.A. Simmons, S. Pakvasa, J.G. Learned, hep-ph/9411432

[20] H. Minakata, A.Y. Smirnov, Phys.Rev. D54 (1996) 3698-3705

[21] Y. Fukuda et al. (SuperKamiokande Collab.), Phys. Lett. B 433 (1998) 9

[22] M. Apollonio et al. (CHOOZ collab.), hep-ex/9907037, Phys. Lett. B 466 (1999) 415-430

2.5 The Experimental Race: from the Late Eighties to the Future

2.5.1 General

Weak and Electromagnetic Interactions in Nuclei

Proceedings of the International Symposium
Heidelberg, July 1-5, 1986

Editor: H. V. Klapdor

Springer-Verlag Berlin Heidelberg New York
London Paris Tokyo

Neutrinos

Edited by H.V. Klapdor

With Contributions by
F. T. Avignone, III, R. L. Brodzinski, P. Depommier,
F. von Feilitzsch, G. Gelmini, W. Hillebrandt, H. V. Klapdor,
P. Langacker, S. P. Mikheyev, R. N. Mohapatra, K. Muto,
A. Y. Smirnov, K. Winter

Contents

Springer-Verlag Berlin Heidelberg New York
London Paris Tokyo

Neutrino Physics

Proceedings of an International Workshop Held in Heidelberg, October 20-22, 1987

Editors: H.V. Klapdor and B. Povh

Contents

VII VIII

Springer-Verlag Berlin Heidelberg New York
London Paris Tokyo

International Journal of Modern Physics A, Vol. 4, No. 8 (1989) 1851–1869

REVIEW OF DOUBLE BETA DECAY EXPERIMENTS

DAVID O. CALDWELL
Physics Department, University of California, Santa Barbara, CA 93106, USA

Received 12 August 1988

The type of double beta decay (two-neutrino) which must occur via a second order weak interaction has finally been observed in the laboratory, and the result agrees with previous geochemical determinations. Calculations can be made to agree with these experiments by adding a particle-particle force to decrease the rate, but the additional force may not have a large effect on predictions of no-neutrino double beta decay. The latter, which requires lepton-number violation and one other element of new physics beyond the standard model (such as neutrino mass) could occur in at least two forms. One of these, Majoron-induced decay, has definitely not been observed so far, contrary to an earlier experimental result. The other form of neutrinoless decay is also not yet observed, the lifetime limit being about 10^{24} years. This limit can be used to set constraints on effective Majorana neutrino mass (~ 2 eV), heavy Majorana neutrino mass, right-handed currents, and supersymmetric particles in theories with R parity violation.

1. Introduction

It is particularly fitting to review the status of double beta decay experiments at this time, because it was in 1948—just 40 years ago—that the first experimental result was reported. That experiment[1] yielded a positive result of a lifetime of $(4–9) \times 10^{15}$ years and was attributed to neutrinoless double beta decay, which is still sought, now with a lifetime limit 8 orders of magnitude longer.[2] However, another form of double-beta decay, that with the emission of neutrinos, has recently been observed in the laboratory,[3] confirming results obtained some time ago by geochemical means.[4,5] The significance of this new result will be discussed later, after a discussion of the reasons for searching for double beta decay and an introduction to the different types of decays.

2. Motivation for the Search for No-Neutrino Double Beta Decay

No-neutrino double beta decay is a low-energy process which tests mass scales beyond the reach of present accelerators. Thus, it is a promising way to search for physics beyond the standard model. For it to occur would require not only that lepton number not be conserved but also that there be one additional piece of new physics. The latter might be the existence of electron neutrino mass, of a heavy Majorana neutrino which mixes with the electron neutrino, of right-handed currents, of a Goldstone boson (such as the Majoron), or of supersymmetric particles violating R parity.

Limits set on the lifetime for no-neutrino double beta decay therefore give corresponding restrictions on all these areas of possible new physics. Since this second order weak process with potentially large phase space is so sensitive, these limits are

J. Phys. G: Nucl. Part. Phys. 17 (1991) S537–S543. Printed in the UK

SUMMARY TALK

H.V. Klapdor-Kleingrothaus

14th EUROPHYSICS CONFERENCE ON NUCLEAR PHYSICS
- Rare Nuclear Decays and Fundamental Processes -
22 - 26 Oct., 1990, Bratislava, Czecho-Slovakia

CONCLUDING REMARKS

H.V. Klapdor-Kleingrothaus
Max-Planck-Institut für Kernphysik, Heidelberg, Germany

Ladies and Gentlemen,

I am sure you all agree that the organizers of this first EPS Nuclear Physics Divisional conference held in Eastern Europe did a very fine job. Professor R. Bock from GSI asked me to transfer to them and in particular to the chairman of the conference, Prof. Pavel Povinec, on behalf of the Nuclear Physics board of the European Physical Society the thanks of the EPS.
The organizers succeeded in bringing together experts in the most exciting field of low-level radiation physics and kept their tradition in such conferences in the best possible way.
This conference was one in a line of earlier meetings on low-radioactivity physics held at Bratislava (or in the Tatra mountains) since 1975 in intervals of 5 years by the chairman of the present meeting. While the earlier conferences were more confined in topics to problems of low-level counting, material research, environmental protection, etc., there was at the present meeting - although these fields still played an important role in these days - a clear tendency of a phase transition to particle physics problems investigated by non-accelerator physics. Such a transitional state of nuclear physics could be observed already at some earlier conferences (e.g. WEIN '86 and '89 at Heidelberg and Montreal).

In fact a large part of the conference was devoted to topics related with the nature of the neutrino, as single and double beta decay looking for the neutrino mass, to solar neutrinos, to dark matter search with Ge or Si underground detectors, and proton decay - topics which are examples even of *beyond*-accelerator physics. The neutrino plays since the time of Pauli and Fermi a central role in our understanding of the weak interaction and plays nowadays a corresponding key role in the framework of Grand unification theories (GUTs, SUSYs,...). Neutrinos with a vanishing Majorana mass would correspond to the breaking of baryon-lepton number conservation and would rule out, for example, the minimal SU(5) model.

Double beta decay and proton decay in grand unified theories probe energy scales between 10^3 and 10^{15} GeV, far beyond the ranges which will become accessible even with the largest future accelerators (Fig. 1).

These energy scales - to stay with the example of double beta decay - correspond to the energies at which the left-right symmetry of the particular GUT model is broken, which can occur somewhere between the extremes $SO(10) \rightarrow SU(5) \rightarrow SU(3)_c \otimes SU(2)_L \otimes U(1)$ and $SO(10) \rightarrow SU(4)_{EC} \otimes$

Annu. Rev. Nucl. Part. Sci. 1994. 44:247–283

DOUBLE BETA DECAY

Michael Moe

Physics Department, University of California, Irvine, California 92717

Petr Vogel

Norman Bridge Laboratory of Physics, California Institute of Technology, Pasadena, California 91125

KEY WORDS: two-neutrino decay, neutrinoless decay, massive Majorana neutrinos, matrix elements, experiments

CONTENTS

1. INTRODUCTION

Double beta ($\beta\beta$) decay is a rare transition between two nuclei with the same mass number that changes the nuclear charge number by two units. It has been long recognized as a powerful tool for the study of lepton conservation in general and of neutrino properties in particular. Because the lifetimes of $\beta\beta$ decay are so long, the experimental study of $\beta\beta$ decay is particularly challenging and has spawned a whole field of experiments requiring very low background.

$\beta\beta$ decay has been and continues to be a popular topic. A superficial search of titles of the leading physics journals reveals that papers on

0163-8998/94/1201-0247$05.00

ATOMIC DATA AND NUCLEAR DATA TABLES **61**, 43–90 (1995)

TABLES OF DOUBLE BETA DECAY DATA

V. I. TRETYAK*,† and YU. G. ZDESENKO*

*Institute for Nuclear Research of the Ukrainian National Academy of Sciences
252028 Kiev, Ukraine
and
† Centre de Recherches Nucléaires, IN2P3-CNRS et Université Louis Pasteur
67037 Strasbourg, France

A compilation of data on double beta decay is presented. The tables contain the most stringent known experimental limits or positive results on half-lives for 2β transitions to ground and excited states of daughter nuclei for different channels ($2\beta^-$; $2\beta^+$, $\epsilon\beta^+$; 2ϵ) and modes (0ν; 2ν; 0νM) of decay. Theoretical estimates are given for comparison. Formulas for energy and angular distributions of electrons in various modes of 2β decay are presented as a supplement.

0092-640X/95 $12.00

Proceedings of the International Workshop held at European Centre for Theoretical Studies (ECT*)

World Scientific
Singapore • New Jersey • London • Hong Kong

Double-Beta Decay — and Related Topics —

Trento, Italy April 24–May 5, 1995

Editors

H V Klapdor-Kleingrothaus
Heidelberg

S Stoica
Bucharest

3

Double Beta Decay – Physics at Beyond Accelerator Energies

H.V. Klapdor–Kleingrothaus
Max–Planck–Institut für Kernphysik
P.O.Box 10 39 80, D–69029 Heidelberg, Germany

ABSTRACT

Double beta decay yields – besides proton decay – one of the most promising possibilities to probe beyond standard model physics at beyond accelerator energies. The possibilities include the neutrino mass, SUSY models, compositeness, leptoquarks, right–handed W bosons and others. We discuss the status and the future perspectives of $\beta\beta$ research, including applications some double beta technology can find in the search for dark matter and in high resolution balloon and satellite gamma–ray astronomy.

in Proc. "Double Beta Decay and Related Topics", International Workshop, Trento, Italy, April 24 – May 5, 1995, eds. *H.V. Klapdor–Kleingrothaus and S. Stoica*: *Singapore: World Scientific* (1996) 3 - 43

1. Introduction

Many central questions of particle physics are beyond the capabilities of modern accelerators. They can, however, to some extent be investigated via non–accelerator experiments (see, e.g.[1]). Table 1 and Fig.1 show some probable areas of research of future accelerators. LHC, for example, the main enterprise of High Energy Physics in the next decade, will cover physics up to scale of a few TeV and may search for the Higgs particle, some SUSY particles and others. In general, however, accelerator physicists at present are forced to search in the extreme 'low–energy' range of the parameter spaces of models of 'new physics' like SUSY or leptoquark signatures and others (see e.g.[3]). Many questions of modern physics have to be studied at higher energy scales: grand unification with or without supersymmetry, and with or without left–right symmetry, compositeness, leptoquarks, neutrino mass (e.g. see–saw–mechanism), Majorons,...

This explains the increasing trend to non–accelerator experiments in numerous underground laboratories and elsewhere. Double beta decay, and proton decay, to mention the most prominent examples, are among those, which yield the most promising possibilities to probe beyond the standard model (SM) physics at beyond accelerator energies and bridge the time gap to the occurence of future larger accelerators. Propagator physics has to replace direct observations. That this method is very effective, is obvious from important earlier research work and has been stressed, e.g. by.[4] Examples are the proporties of W and Z bosons derived from neutral weak currents and β–decay, and the top mass deduced from LEP electroweak radiative corrections. Also for accelerators, the search for new bosons or compositeness, for example, can be to some

2.5.2 Ionisation and Time Projection Chambers, and Combinations with Plastic Scintillators

VOLUME 59, NUMBER 18 PHYSICAL REVIEW LETTERS 2 NOVEMBER 1987

Direct Evidence for Two-Neutrino Double-Beta Decay in ^{82}Se

S. R. Elliott, A. A. Hahn, and M. K. Moe
Department of Physics, University of California, Irvine, Irvine, California 92717
(Received 31 August 1987)

The two-neutrino mode of double-beta decay in ^{82}Se has been observed in a time-projection chamber at a half-life of $(1.1^{+0.8}_{-0.3})\times 10^{20}$ yr (68% confidence level). This result from direct counting confirms the earlier geochemical measurements and helps provide a standard by which to test the double-beta-decay matrix elements of nuclear theory. It is the rarest natural decay process ever observed directly in the laboratory.

PACS numbers: 23.40.Bw

Double-beta decay was suggested by Wigner in the 1930's as a second-order weak transition between isobars differing by two units in atomic number. Assuming the emission of two electrons and two neutrinos (as required by the theory of Dirac), Geoppert-Mayer[1] in 1935 made the first theoretical estimate of the extremely low rates for this process [$\beta\beta(2\nu)$]. Four years later, Furry[2] invoked the theory of Majorana (in which the neutrino and antineutrino are identical) to propose an alternative mode of double-beta decay having no neutrinos in the final state [$\beta\beta(0\nu)$]. Experimentally the two modes can be distinguished by the spectra of their sum energies; that is, the sum of the kinetic energies of the two emitted electrons. For $\beta\beta(2\nu)$ a broadly distributed spectrum is expected since a portion of the energy is carried off by unseen neutrinos. Essentially all of the transition energy is available to the electrons in $\beta\beta(0\nu)$. The resulting sharp energy spike would be typically 100 times easier to detect. Furry pointed out that because of the greater phase space available to $\beta\beta(0\nu)$, this mode could proceed at a faster rate than $\beta\beta(2\nu)$ by many orders of magnitude. "According to the older theory," he remarked, "it seemed certain that double-beta disintegration could never be observed because of its extremely minute probability, but the Majorana theory indicates that this is by no means necessarily the case."

Ironically, it is $\beta\beta(2\nu)$ that has now been seen while Furry's potentially faster and experimentally much more distinctive $\beta\beta(0\nu)$ is yet to be detected.[3] Possibly neutrinos are indeed Dirac particles, but the Majorana alternative is not excluded by any known evidence. It is recognized today that parity nonconservation exerts a strong inhibiting influence, but need not exclude $\beta\beta(0\nu)$ entirely.

An observation of $\beta\beta(0\nu)$ would have profound consequences for particle physics. The two electrons appearing alone would constitute a violation of lepton-number conservation— a symmetry breakdown expected in some theories of grand unification. A neutrinoless decay would further indicate a nonzero mass for the neutrino. Under some circumstances, $\beta\beta(0\nu)$ would reveal a right-handed component in the weak leptonic current. The large phase-space enhancement for $\beta\beta(0\nu)$ and the unique character of the monoenergetic spectrum make this mode a very sensitive test for these neutrino properties. Intensive searches for $\beta\beta(0\nu)$ are underway.[4]

The neutrino mass and right-handed–current parameters are related to the $\beta\beta(0\nu)$ rate by matrix elements calculated from nuclear theory. Difficulties in making these calculations result in fairly large uncertainties. Unlike $\beta\beta(0\nu)$, the $\beta\beta(2\nu)$ mode is expected in the standard model. The matrix elements for this mode directly predict the $\beta\beta(2\nu)$ decay rate, independent of any assumptions about unknown properties of the neutrino. Thus a measurement of the $\beta\beta(2\nu)$ rate provides a test of the nuclear theory from which $\beta\beta(2\nu)$ and $\beta\beta(0\nu)$ matrix elements are calculated in similar ways. The theory of double-beta decay and the experimental history have been discussed extensively in the literature.[5-10]

Compelling indirect evidence for the existence of double-beta decay comes from geochemical measurements,[11-13] but until now the process had not been seen in the laboratory. The broad sum-energy spectrum for $\beta\beta(2\nu)$ makes this mode difficult to see without the background-rejecting capabilities of a tracking chamber. At the University of California, Irvine (UCI), a time-projection chamber (TPC) has been employed to search for $\beta\beta(2\nu)$ in ^{82}Se. The experiment is designed to look for the two electrons in the decay

$$^{82}\mathrm{Se} \rightarrow {}^{82}\mathrm{Kr} + 2e^- + 2\bar{\nu}.$$

The trajectories of the electrons are recorded by the TPC and are analyzed to measure their kinematic characteristics. The experimental setup and detector performance have been described elsewhere[14-16] and only the essentials will be summarized here. The source is 14 g of 97% enriched ^{82}Se deposited on a thin Mylar foil which forms the central electrode of the TPC. The source thickness is 7 mg/cm^2. The TPC is surrounded by a lead shield 10 to 15 cm thick. This lead house is enclosed within a 4π cosmic-ray veto, and the entire system is immersed in a 700-G magnetic field. The experiment is located in a basement laboratory in the physical sciences building at UCI.

The data reported here represent 7960 h of live time. Figure 1 shows the sum-energy spectrum with an 800-keV threshold. The energy of each individual electron was required to be larger than 150 keV. Also plotted in this figure is the theoretical sum-energy spectrum normalized to our best-fit value for the half-life. It is clear from this figure that the signal-to-background ratio is highest in the region 1.3 to 2.0 MeV.

There are several mechanisms which simulate $\beta\beta(2\nu)$ and thus must be considered as possible contributions to the data of Fig. 1. The most efficient of these are common beta decays which are followed by internal conversion (β+IC), and Moeller scattering. The origin of the electrons which undergo Moeller scattering can be either β rays of Compton electrons arising from γ rays impinging on the source plane.

The β+IC isotopes which cause the most trouble are members of the uranium and thorium series. Two of these isotopes, ^{214}Bi from the uranium series and ^{212}Bi from the thorium series, have daughters which are unstable against α decay. These β-α cascades are very easy to identify in the TPC. Thus, the contribution from most isotopes within these decay chains can be estimated from the observed decay rates of ^{214}Bi and ^{212}Bi. The contributions due to Moeller scattering and any other β+IC isotopes can be estimated by the examination of the spectrum of lone electrons (not members of pairs). A much more rigorous description of the background estimation can be found elsewhere.[15,16]

The result of the background studies shows that the β+IC of ^{208}Tl and Moeller scattering are the only significant background contributions to the data in the 1.3- to 2.0-MeV energy range, contributing 2.8 ± 0.7 and 9.3 ± 2.6 counts, respectively. Thus, we considered these contributions and a possible contribution of $\beta\beta(2\nu)$ as the explanations for the observed events between 1.3 and 2.0 MeV shown in Fig. 1. The histograms in Fig. 2 show the observed events which fall between these two sum-energy levels and which survive the 150-keV single-energy threshold. Furthermore, we required that the two electrons be emitted on opposite sides of the source plane, since the opening-angle bias of the detector was understood in this case. Figure 2 includes the sum- and single-energy spectra along with the opening-angle spectrum for the 46 observed events which survive these cuts. The opening-angle-distribution predictions include the effect of the detector bias.

The superimposed curves in Fig. 2 represent the shapes of the three previously mentioned contributions. The background spectra were all determined from measured data, and the $\beta\beta(2\nu)$ spectra are those determined from the calculations of Doi, Kotani, and Takasugi.[7]

The best fit was determined by a maximum-likelihood procedure for both the cases of a constrained and an unconstrained background. In the unconstrained fit, each of the three components was allowed to vary in integer steps between 0 and 140% of the observed integral number of counts (46). The likelihood value was calculated as follows: First, the admixture of the three components

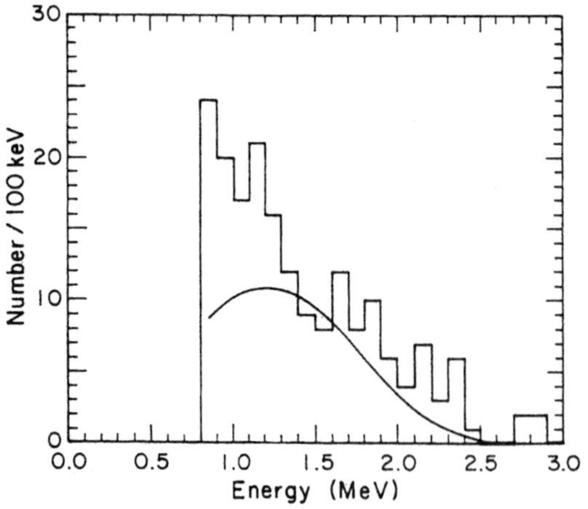

FIG. 1. The observed sum-energy spectrum of two-electron events. A threshold of 800 keV was imposed on the sum energy of the events, and a threshold of 150 keV was imposed on the single energy. The curve is the theoretical $\beta\beta(2\nu)$ sum-energy spectrum normalized to 1.1×10^{20} yr.

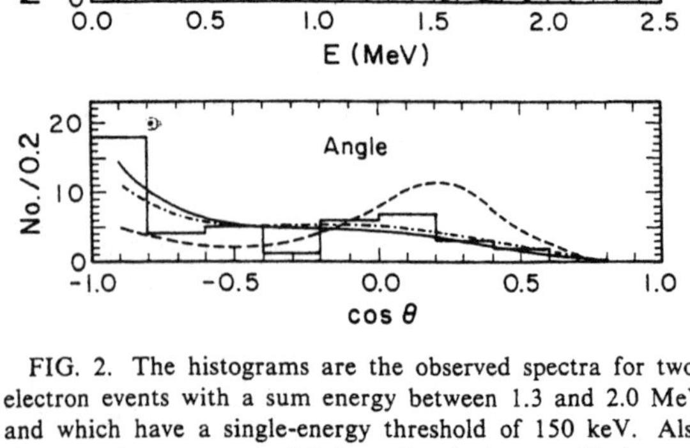

FIG. 2. The histograms are the observed spectra for two-electron events with a sum energy between 1.3 and 2.0 MeV and which have a single-energy threshold of 150 keV. Also these events were required to have the electrons emitted on opposite sides of the source plane. The curves show the shapes of the three assumed contributions to the data: $\beta\beta(2\nu)$ (solid curve), Moeller scattering (dashed curve), and ^{208}Tl (dot-dashed curve).

VOLUME 59, NUMBER 18 PHYSICAL REVIEW LETTERS 2 NOVEMBER 1987

was normalized to 46 counts. For each bin in each histogram, the number of counts expected was defined as the integral of the predicted spectrum over that bin. The probability of finding the number of counts actually observed in that bin was calculated with Poisson statistics. The product of these probabilities over all bins from all histograms was then multiplied by the probability of observing 46 counts given the predicted integral number before normalization. This then represented the unconstrained probability value for the particular mix of components. The constrained probability value was calculated by multiplication of the unconstrained value by the probability that the particular admixture of backgrounds had fluctuated from the 2.8 ± 0.7 ^{208}Tl and 9.3 ± 2.6 Moeller contributions mentioned earlier.

Figure 3 shows a mesh plot (and a contour plot) from the unconstrained fit for the case where the predicted integral number of counts equaled the observed number. The best fit (which lies in this particular slice) predicts that 78% or 35.9 of the 46 events are due to $\beta\beta(2\nu)$, 10.1 are due to Moeller scattering, and 0 are due to ^{208}Tl. Thus the freely floating ratios in the unconstrained maximum-likelihood fit settled at values closely duplicating the background contributions calculated independently.

In the unconstrained fit, the 68% confidence level allowed the two-neutrino mode to vary between 19 and 54 counts. The constrained fit tightened up the range to between 21 and 47 counts for the 68% confidence level, with the best fit at 35 counts. The efficiency for a $\beta\beta(2\nu)$ event to be detected in the TPC and to survive these cuts is $(6.2 \pm 0.5)\%$, and, hence, the best fit corresponds to a half-life of 1.1×10^{20} yr with a 68%-confidence-level range of $(0.8$ to $1.9) \times 10^{20}$ yr for the constrained case. As a check of the dependence of this result on the choice of energy range selected for analysis, this procedure was repeated for the much broader sum-energy limits of 0.8 and 2.5 MeV and gave a very similar conclusion.

Backgrounds in the regions above and below the 1.3–2.0-MeV range, although higher with respect to the signal, are also largely understood. A new radioactively cleaner TPC has been operating for a few months, and preliminary indications are that most of this background has disappeared, while the counting rate in the 1.3–2.0-MeV region is nearly unchanged.

We believe that the similarity of the measured spectra and angular distributions to the predicted shapes for $\beta\beta(2\nu)$, the agreement of the maximum-likelihood fit with background contributions calculated independently, the independence of the conclusion on the energy range of analysis, and the persistence of these counts in the

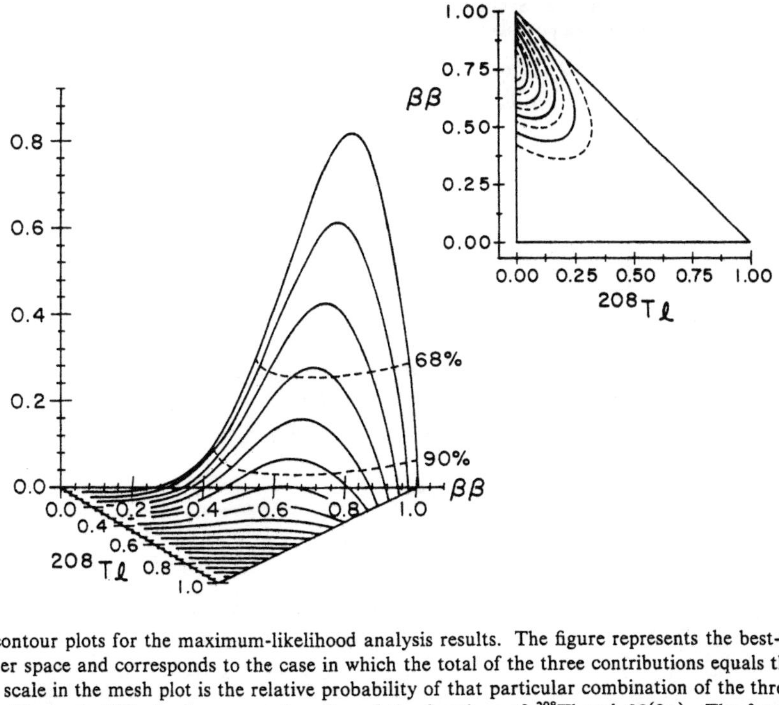

FIG. 3. Mesh and contour plots for the maximum-likelihood analysis results. The figure represents the best-fit slice through the unconstrained parameter space and corresponds to the case in which the total of the three contributions equals the number of events observed. The vertical scale in the mesh plot is the relative probability of that particular combination of the three contributions giving the observed data. The probability is shown as a function of the fractions of ^{208}Tl and $\beta\beta(3\nu)$. The fraction due to Moeller scattering can be deduced by the requirement that the sum of the three fractions be unity. The 68% and 90% curves indicate the regions which contain those percentages of the volume under the surface. The contour plot is of the same function.

VOLUME 59, NUMBER 18 PHYSICAL REVIEW LETTERS 2 NOVEMBER 1987

face of reduced backgrounds in the new chamber constitute strong evidence for the observation of $\beta\beta(2\nu)$. This result is consistent with the geochemical measurements of Kirsten,[11] $(1.30\pm0.05)\times10^{20}$ yr, and Manuel,[12] $(1.0\pm0.4)\times10^{20}$ yr, and the cosmochemical (meteorite) measurement of Marti and Murty,[13] $0.97^{+0.36}_{-0.45}\times10^{20}$ yr.

We hope that the other mode(s) of double-beta decay will ultimately be detected, and that this direct laboratory observation of $\beta\beta(2\nu)$ will contribute to the theoretical understanding of future results. The TPC experiment continues to operate in an attempt to reduce the uncertainty in the $\beta\beta(2\nu)$ half-life and to extend the search for neutrinoless decay in ^{82}Se.

We would like to thank Professor Frederick Reines for his advice and encouragement throughout this work. This project is supported by U.S. Department of Energy Contract No. DE-AT03-76ER71019.

[1]M. Goeppert-Mayer, Phys. Rev. **48**, 512 (1935).

[2]W. H. Furry, Phys. Rev. **56**, 1184 (1939).

[3]A second neutrinoless mode, involving a massless Goldstone boson or majoron, has been proposed but not observed. See, for example, Y. Chikashige, R. N. Mohapatra, and R. D. Pecci, Phys. Lett. **98B**, 265 (1981).

[4]A recent review of double-beta decay experiments is given by D. O. Caldwell, University of California, Santa Barbara, Report No. UCSB-HEP-87-10, 1987 (to be published).

[5]H. Primakoff and S. P. Rosen, Rep. Prog. Phys. **22**, 121 (1959), and Annu. Rev. Nucl. Part. Sci. **31**, 145 (1981).

[6]W. C. Haxton and G. J. Stephenson, Jr., Prog. Part. Nucl. Phys. **12**, 409 (1984).

[7]M. Doi, T. Kotani, and E. Takasugi, Prog. Theor. Phys. Suppl. **83**, 1 (1985), and Osaka University Report No. OS-GE 87-06, 1987 (to be published).

[8]J. D. Vergados, Phys. Rep. **133**, 1 (1986).

[9]P. Vogel and M. R. Zirnbauer, Phys. Rev. Lett. **57**, 3148 (1986); J. Engel, P. Vogel, and M. R. Zirnbauer, California Institute of Technology Report No. MAP-95, 1987 (to be published).

[10]O. Civitarese, Amand Faessler, and T. Tomoda, Phys. Lett. B **914**, 11 (1987).

[11]T. Kirsten, in *Proceedings of the International Symposium on Nuclear Beta Decay and Neutrinos, Osaka, Japan, 1986,* edited by T. Kotani, H. Ejiri, and E. Takasugi (World Scientific, Singapore, 1986), p. 81.

[12]O. K. Manuel, in *Proceedings of the International Symposium on Nuclear Beta Decay and Neutrinos, Osaka, Japan, 1986,* edited by T. Kotani, H. Ejiri, and E. Takasugi (World Scientific, Singapore, 1986), p. 71.

[13]K. Marti and S. V. S. Murty, Phys. Lett. B **163**, 71 (1985).

[14]M. K. Moe, A. A. Hahn, and H. E. Brown, in *The Time Projection Chamber,* edited by J. A. MacDonald, AIP Conference Proceedings No. 108 (American Institute of Physics, New York, 1984), p. 37.

[15]S. R. Elliott, A. A. Hahn, and M. K. Moe, Phys. Rev. Lett. **56**, 2582 (1986).

[16]S. R. Elliott, Ph.D dissertation, University of California, Irvine, 1987 (unpublished); S. R. Elliott, A. A. Hahn, and M. K. Moe, to be published.

342 in Proc. NEUTRINO'96, Helsinki, Finland, June 1996, eds. *K. Enqvist, K. Huitu and J. Maalampi, W.S., Singapore*, (1997) 342 – 346

DOUBLE BETA DECAYS AND DARK MATTERS STUDIED BY ELEGANT DETECTORS

H. EJIRI
Research Center for Nuclear Physics (RCNP), Ibaraki, Osaka, 567 JAPAN

Neutrinos (ν) and dark matters (DM) were studied with ELEGANT detectors by measuring nuclear double beta decays and nuclear elastic/inelastic scatterings of DM. Nuclear spin isospin responses associated with ν and DM were studied by charge-exchange spin-flip (^{3}He, t) reactions.

The present report aims at reporting briefly recent works on neutrino studies by measuring double beta decays ($\beta\beta$) and on spin coupled dark matters (DM) by measuring elastic and inelastic scatterings of DM from nuclei. The present work has been partly presented at the XIV RCNP symposium [1].

The neutrinoless double beta decay ($0\nu\beta\beta$), which violates the lepton number conservation law, is a very sensitive probe for studying such quantities as the Majorana neutrino mass $\langle m_\nu \rangle$, the right handed weak current $\langle$RHC$\rangle$, the R-parity violating coupling with SUSY (super symmetry) particles, and others. The neutrinoless $\beta\beta$ decay accompanied by Majoron ($0\nu\beta\beta$B) is sensitive to the Majoron-neutrino coupling $\langle g_B \rangle$. These are beyond the standard $\mathrm{SU(2)_L} \times$ U(1) theory.

The transition rate $T^{0\nu}$ for $0\nu\beta\beta$ with $\langle m_\nu \rangle$ and $\langle$RHC$\rangle$ is given as

$$\begin{aligned} T^{0\nu} &= G^{0\nu} \mid M^{0\nu} \mid^2 [\langle m_\nu \rangle^2 + C_{\lambda\lambda}\langle\lambda\rangle^2 + C_{\eta\eta}\langle\eta\rangle^2 + \\ & \quad C_{m\lambda}\langle m_\nu\rangle\langle\lambda\rangle + C_{m\eta}\langle m_\nu\rangle\langle\eta\rangle + C_{\lambda\eta}\langle\lambda\rangle\langle\eta\rangle], \end{aligned} \tag{1}$$

where $\langle\lambda\rangle$ and $\langle\eta\rangle$ are the $\langle$RHC$\rangle$ terms.

The $0\nu\beta\beta$, $0\nu\beta\beta B$ and $2\nu\beta\beta$ are studied by measuring sum energy spectra of $E_\beta + E_{\beta'}$. Individual terms of $\langle m_\nu\rangle$, $\langle\lambda\rangle$ and $\langle\eta\rangle$ of $0\nu\beta\beta$ are determined by studying energy and angular correlations of two β rays.

The transition rate $T^{2\nu}$ for the two neutrino double beta decay ($2\nu\beta\beta$), which is within the standard theory, is written as

$$T^{2\nu} = G^{2\nu} \mid M^{2\nu} \mid^2, \tag{2}$$

where $G^{2\nu}$ and $M^{2\nu}$ are the phase space factor and the matrix element, respectively. Thus observed halflives give $2\nu\beta\beta$ matrix elements $M^{2\nu}$. They are used to check nuclear structure calculations for $M^{2\nu}$ and $M^{0\nu}$.

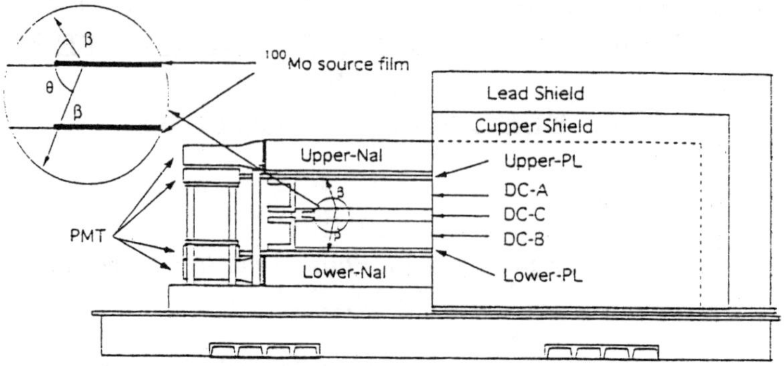

Figure 1: ELEGANT V, DC: drift chambers for β α. PL plastic scintillators [2].

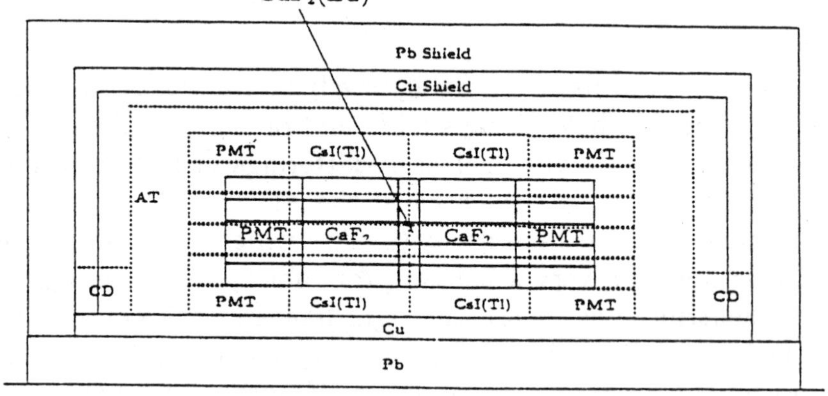

Figure 2: ELEGANT VI with CaF_2 for $\beta\beta$ and DM [6].

The high sensitive detector of ELEGANT [ELEctron GAmma-ray Neutrino Telescope] V (Fig. 1) was used for measuring $\beta\beta$ decays of ^{100}Mo and ^{116}Cd at the Kamioka underground lab [2]. The first ^{100}Mo run with 104 gr ^{100}Mo foils gave for the first time a finite $2\nu\beta\beta$ halflife of $1.15^{+0.3}_{-0.2} \cdot 10^{19}y$ for the ^{100}Mo $0^+ \rightarrow 0^+$ decay [2]. The ^{116}Cd run with 91 gr ^{116}Cd foils gave for the first time a finite $2\nu\beta\beta$ halflife of $2.6 \cdot ^{+0.9}_{-0.5} 10^{19}y$ for the ^{116}Cd $0^+ \rightarrow 0^+$ [3]. The $2\nu\beta\beta$ matrix elements in unit of $(m_e c^2)^{-1}$ are 0.09 for ^{100}Mo and 0.069 for ^{116}Cd.

The second ^{100}Mo run with 179 gr ^{100}Mo foils gave most stringent limits on the $0\nu\beta\beta$ and $0\nu\beta\beta$B processes [4]. The obtained halflife limits with 68 % confidence level(CL) for the ground state $\beta\beta$ transition are $5.2 \cdot 10^{22}y$ $3.9 \cdot 10^{22}y$ $5.1 \cdot 10^{22}y$ and $0.54 \cdot 10^{22}y$ for the $0\nu\beta\beta$ m_ν, λ, η modes and the $0\nu\beta\beta$B mode, respectively. These lead to the upper limits of $\langle m_\nu \rangle < 2.2$eV, $\langle \lambda \rangle < 3.7 \cdot 10^{-6}$, $\langle \eta \rangle < 2,5 \cdot 10^{-8}$ and $\langle g_B \rangle < 7.3 \cdot 10^{-5}$. Here the limits are the on-axis values with nuclear matrix elements in Ref. 5.

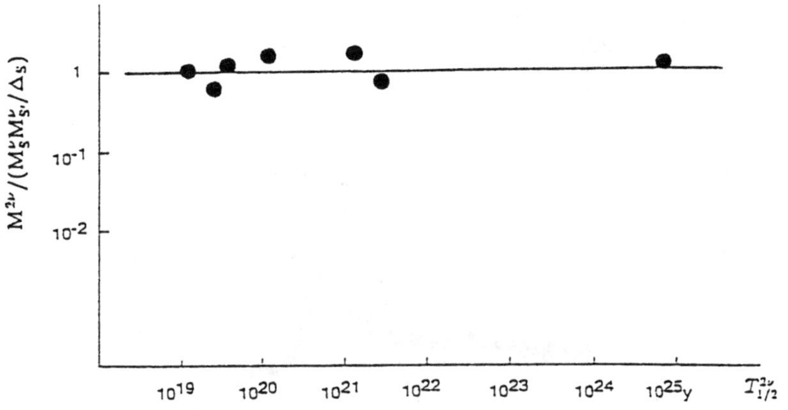

Figure 3: Ratio of the $2\nu\beta\beta$ matrix elements $M^{2\nu}$ and the product of the matrix elements of successive single β decays through the single particle-hole 1^+ state [7].

EL. VI consists of 25 modules of CaF_2 scintillators surrounded by 3π CsI detectors, as shown in Fig. 2 [6]. The CaF_2 module consists of a central CaF_2 crystal with $45 \times 45 \times$ 200mm^3 and a pair of side CaF_2 crystals with $45 \times 45 \times$ 75mm^3, each at both sides of the central crystal. The total ^{48}Ca content in the 25 central detectors is 31 gr. Because of the large Q value of 4.27 MeV, one may study well both $0\nu\beta\beta$ and $2\nu\beta\beta$ of ^{48}Ca.

EL. V and VI are installed at the new underground lab. of Oto Cosmo Observatory. Background levels there are small, $4 \cdot 10^{-3}/m^2/sec$ for cosmic rays, $4 \cdot 10^{-1}/m^2/sec$ for neutrons, and $10Bq/m^3$ for Rn.

The observed $2\nu\beta\beta$ matrix elements are approximately given by the product of the successive single-β matrix elements, M_S^ν and $M_{S'}^\nu$, through the single particle-hole 1^+ state at the low excitation region (Fig.3). They are

$$M^{2\nu} = \frac{M_S^\nu M_{S'}^\nu}{\Delta_S}. \tag{3}$$

Charged axial weak processes associated with nuclear isospin spin ($\tau\sigma$) responses are investigated by charge-exchange (τ) spin-flip (σ) nuclear reactions. Spin isospin responses for ^{100}Mo $\beta\beta$ decays were studied by the ^{100}Mo (^{3}He, t) ^{100}Tc. The 0.45 GeV ^{3}He beam obtained from the RCNP cyclotron was used. The $\tau\sigma$ strengths are concentrated, as shown in Fig. 4, into three states (resonances), the single particle-hole 1^+ state($\mid S\rangle$) at the low excitation region, the isobaric analogue state ($\mid$ IAS$\rangle$) and the GT giant resonance $\mid G\rangle$.

The $2\nu\beta\beta$ matrix element is analyzed, for simplicity, in terms of the coupling of two GT(1^+) states of $\mid S\rangle$ and $\mid G\rangle$ [7]. Then the $2\nu\beta\beta$ matrix element is written as $M^{2\nu} = M_S^\nu M_{S'}^\nu/\Delta_S + M_G^\nu M_{G'}^\nu/\Delta_G$, where the first and the second

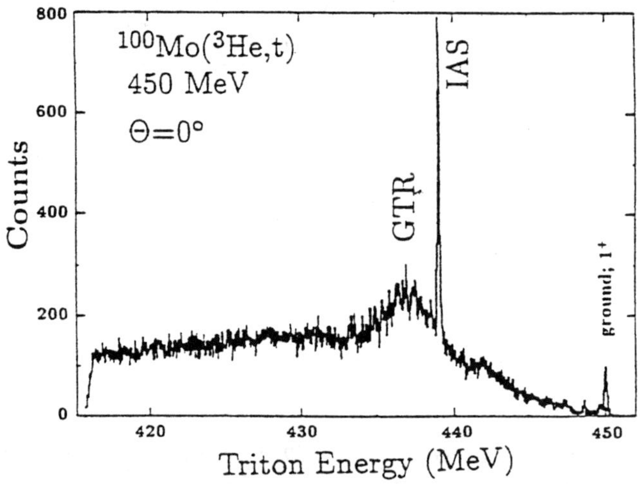

Figure 4: Energy spectrum for ^{100}Mo (^{3}He, t) ^{100}Tc. IAS and GTR are the isobaric analogue state and GT giant resonances, respectively [8].

terms correspond to the successive single β decays through $\mid S\rangle$ and $\mid G\rangle$ in the intermediate nucleus, respectively. The second term vanishes because of the cancellation between the contributions from the admixture of $\mid S\rangle$ into the $\mid G\rangle$ and the admixture of $\mid G'\rangle$ into $\mid 0_f\rangle$. Thus one gets eq.(3) as found empirically. M_S^ν derived from the (^{3}He, t) reaction, together with $M_{S'}^\nu$ from the log ft value, gives $M_S^\nu M_{S'}^\nu/\Delta_S = 0.1$, in agreement with $M^{2\nu} = 0.09$ obtained from the $2\nu\beta\beta$ halflife [8].

The high sensitive EL. V and VI are used also for studying WIMP's (weakly interacting massive particle) of the cold DM candidates through nuclear elastic (coherent) and inelastic scatterings [9]. The NaI detector of EL V is useful because of the large volume (0.76 tons) of the scintillator with 100% odd ^{23}Na($3/2^+$) and ^{127}I($5/2^+$) isotopes, and of the low noise level (~4 KeV). In particular, the inelastic scattering from the $5/2^+$ ground state to the $7/2^+$ excited state at E_x=58 KeV in ^{127}I is used to study spin coupled WIMP's because the spin response is known from the inverse M1 γ transition, and the sum energy signal of the nuclear recoil energy (E_R) and the γ-decay energy (E_X) is much larger than $E_R \approx 0$. The upper limits on the elastic and inelastic scatterings lead to upper limits on the vector and spin-coupled WIMP's, as shown in Fig. 5.

Search for the spin-coupled DM is now going on by means of CaF_2 (Eu) detectors of EL VI [6] at the new underground lab. of Oto Cosmo Observatory. Here the CaF_2 pure crystals used for studying ^{48}Ca $\beta\beta$ decays are replaced by the 25 modules of 45×45×45mm^3 CaF_2 (Eu) crystals in order to get large

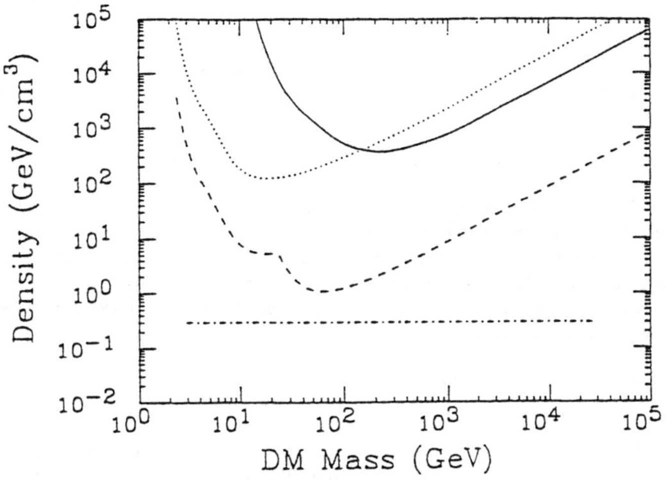

Figure 5: Upper limits on the DM densities. Dashed line: vector coupled Dirac ν by elastic scatterings. Dotted line: axial-vector coupled Majorana ν by elastic scatterings with SP matrix element. Solid line: axial-coupled Majorana ν by inelastic scatterings [9].

light outputs for DM. Elastic scattering of DM from the $1/2^+$ ^{19}F isotope with the large spin matrix element is studied.

References

1. H. Ejiri, *Proc. Int. Symp. Nuclear Reaction Dynamics of Nucleon Hadron Manybody System, 1995, Osaka*, (World Scientific 1996).
2. H. Ejiri et al., *Nucl. Instr. Meth.* **A302** (1991) 304, H. Ejiri, et al., *Phys. Lett.* **258B** (1991) 17; *J. Phys. G. Nucl. Phys.* **17** (1991) S155.
3. H. Ejiri et al., *J. Phys. Soc. Japan.* **64** (1995) 339, K. Kume et al., *Nucl. Phys.* **A577** (1994) 405c.
4. H. Ejiri, et al., *Nucl. Phys.* (1996) to be published;
   N. Kudomi et al., *Proc. WEIN'95*, ed. H. Ejiri et al., (World Scientific 1995) 204.
5. T. Tomoda, *Rep. Prog. Part. Phys.* **54** (1991) 53.
6. R. Hazama et al., *Proc. WEIN'95*, ed H. Ejiri, et al., (World Scientific 1995) 635.
7. H. Ejiri and H. Toki, *J. Phys. Soc. Japan-Letters*, 65 (1995) 7.
8. H. Akimune, H. Ejiri and M. Fujiwara,*Phys. Lett* **B** (1996).
9. K. Fushimi, et al., *Phys. Rev.* **C44** (1991) 502;
   H. Ejiri, et al., *Phys. Lett.* **B282** (1992) 281 **317** (1993) 14.

PHYSICAL REVIEW C, VOLUME 61, 035501

Results of a search for the two neutrino double β decay of ^{136}Xe with proportional counters

Ju. M. Gavriljuk, V. V. Kuzminov,* and N. Ya. Osetrova
Institute for Nuclear Research, Moscow 117312, Russia

S. S. Ratkevich†
Kharkov State University, Kharkov 310077, Ukraine
(Received 11 August 1999; published 14 February 2000)

Results of a search for ^{136}Xe double β decay with high-pressure multiwire wall-less proportional counters at the Baksan Neutrino Observatory are presented. The experimental method and the characteristics of the detectors are described. The detector background in the energy range 0.5–3.5 MeV has been reduced due to event position discrimination and pulse shape discrimination. Results of the analysis of background components for different event types and source positions are described. A new lower limit of $T_{1/2} \geqslant 0.81 \times 10^{21}$ yr (90% C.L.) is found for the ^{136}Xe $2\nu\beta\beta$ decay mode.

PACS number(s): 23.40.Bw, 27.60.+j, 29.40.Cs

I. INTRODUCTION

A search for the different modes of ^{136}Xe ($Q = 2.48$ MeV) double β decay is presented in Refs. [1–4]. Xenon was used as a working gas and double β decay source simultaneously in the different gas and liquid detectors. The types of detectors, working media, and results are summarized in Table I. A comparison of the results shows that the setup sensitivity to neutrinoless double β decay has been increased more than 200 times. This progress was achieved by lowering the background in the energy range 1.6–3.0 MeV by separating the two electron events [5]. The setup sensitivity to ^{136}Xe in the two neutrino double β decay mode has not changed substantially. Limits on the half-life of the ^{136}Xe $2\nu\beta\beta$ decay mode listed in Table I were established mainly by the analysis of spectra obtained after substraction of the natural Xe spectrum from the spectrum of xenon enriched with ^{136}Xe in the energy range of 0.8–2.0 MeV in Refs. [2,3]. In Ref. [5] the limit was obtained by the analysis of the background spectrum only in the energy range of 1.67–2.0 MeV.

Theoretical predictions for the half-life of the ^{136}Xe $2\nu\beta\beta$ decay mode are $T_{1/2} = 0.82 \times 10^{21}$ yr [7], $T_{1/2} = 1.0 \times 10^{21}$ yr [8], $T_{1/2} = 4.64 \times 10^{21}$ yr [9]. To reach the level of the theoretical half life, it is necessary to increase the experimental sensitivity by 2–10 times.

II. EXPERIMENTAL SETUP

The main features of the new experimental setup that enables one to obtain a sensitivity to ^{136}Xe $2\nu\beta\beta$ decay of better than 8×10^{20} yr are described in [10]. This increase in sensitivity is achieved by using a multidetector measurement scheme which allows testing samples of natural xenon and xenon enriched in ^{136}Xe simultaneously. This scheme makes it possible to reduce the systematic errors related to variations of experimental external conditions. The detector is a high-pressure multiwire wall-less proportional counter (MWPC) which consists of a central main counter (MC) and a surrounding protection ring counter (RC) in the same body. Detectors of similar construction are widely used for low background measurements with gaseous radioactive sources (see, for example, Refs. [11] and [12]). The diameters of the inner detector, RC anode grid, common MC and RC cathode grid are 122, 110, and 98 mm, respectively. The volumes of MC and RC are 4.44 and 2.57 l, respectively. The working pressure of xenon is 16.8 atm.

An increase of pulse amplitude due to gas amplification allows signals to be read from both ends of the MC anode (PC1 and PC2 signals). A parameter $\beta \equiv \mathrm{PC1} \times 100/(\mathrm{PC1}+\mathrm{PC2})$ determines the event coordinate along the detector. The selection of events inside the working anode length enables one to eliminate the background spectrum components related to microdischarges in the high voltage circuits and the insulator surfaces. The MC background component, related to the charged particles (electrons from radioactive impurities and external γ rays), is eliminated by an anticoincidence operation of the MC and RC (PAC signal). The MC event amplitude is formed as a sum of signals (PC1+PC2). A shaping amplifier with 26 μs integration and differentiation shaping times was used to obtain adequate energy resolution. The signals from the outputs of a nonshaping amplifier are joined together. The resulting signal, $P12$, is amplified in another shaping amplifier with $\tau_{\mathrm{int}} \simeq 0$ and $\tau_{\mathrm{dif}} = 0.75\ \mu$s.

The parameter $f \equiv P12 \times 100/(\mathrm{PC1}+\mathrm{PC2})$, depending on pulse rise time, is used to obtain its relative value. These data help to reduce the α-particle background.

Measurements were done in the underground laboratory of the Baksan Neutrino Observatory at 4900 m.w.e. depth. The experimental setup consists of three MWPC's installed in the low background shield (15 cm of lead, 8 cm of borated polyethylene and 11 cm of copper). The MWPC No. 1 was filled with a natural xenon without radioactive ^{85}Kr; MWPC Nos. 2 and 3 were filled with xenon isotopically enriched in ^{136}Xe to 93%. The gases in the counters No. 1 and No. 2 were periodically purified and replaced. MWPC No. 3 was used as a control detector to expose the possible sources of

*Electronic address: kuzminov@neutr.novoch.ru
†Electronic address: ratkevich@univer.kharkov.ua

VOLUME 67, NUMBER 10 PHYSICAL REVIEW LETTERS 2 SEPTEMBER 1991

New Limit on Neutrinoless Double β Decay in ^{136}Xe with a Time Projection Chamber

H. T. Wong,[(1),(a)] F. Boehm,[(1)] P. Fisher,[(1),(b)] K. Gabathuler,[(2)] H. E. Henrikson,[(1)] D. A. Imel,[(1)] M. Z. Iqbal,[(1),(c)] V. Jörgens,[(3)] L. W. Mitchell,[(3),(d)] B. M. O'Callaghan-Hay,[(1)] J. Thomas,[(1),(e)] M. Treichel,[(3)] J.-C. Vuilleumier,[(3)] and J.-L. Vuilleumier[(3)]

[(1)]*Norman Bridge Laboratory of Physics, California Institute of Technology, Pasadena, California 91125*
[(2)]*Paul Scherrer Institute, 5232 Villigen, Switzerland*
[(3)]*Institut de Physique, A.-L. Breguet 1, 2000 Neuchâtel, Switzerland*
(Received 17 May 1991; revised manuscript received 5 August 1991)

A xenon time projection chamber with an active volume of 207 L has been built to study neutrinoless double β decay in ^{136}Xe. Data were taken in the Gotthard Underground Laboratory, with 5 atm of xenon enriched to 62.5% in ^{136}Xe. From 3380 h of data, no evidence has been found for the 0ν $0^+ \to 0^+$ transition. Half-life limits of $T_{1/2}^{0\nu} > 2.5(4.9) \times 10^{23}$ yr in the mass-mechanism mode and $T_{1/2}^{0\nu} > 1.7(3.2) \times 10^{23}$ yr in the right-handed-current mode, at the 90(68)% C.L., were derived. An upper limit for the Majorana neutrino mass parameter was deduced.

PACS numbers: 23.40.Bw, 14.60.Gh, 27.60.+j

It is well known that neutrinoless double β decay provides a sensitive probe for lepton-number violation and, in particular, Majorana neutrino mass as well as right-handed weak currents [1]. The implications of this so far unobserved nuclear decay have stimulated intense experimental efforts in recent years [2]. Experiments on ^{76}Ge have produced the most stringent limits so far [3]. Owing to the uncertainties in the nuclear-matrix-element calculations [4], it is important that other isotopes are studied as well. In this Letter, we report on a double-β-decay experiment with ^{136}Xe, using a time projection chamber.

The isotope ^{136}Xe is suitable for double-β-decay studies. The 0ν transition energy (2.48 MeV) is large compared to other candidate isotopes, leading to an enhancement of the phase-space factor and decay rate. In addition, xenon is a good proportional counter and drift gas, and thus can act as both source and detector. The natural abundance (8.9%) is appreciable, and xenon enriched in ^{136}Xe can be obtained [5] at a relatively modest price. Experimental searches for double β decay in ^{136}Xe have been carried out by groups from Moscow [6] and Milano [7], using a high-pressure ionization chamber and a multielement proportional counter, respectively.

We have built a time projection chamber (TPC) [8] to study double β decay in ^{136}Xe, in both the 0ν and 2ν channels [9]. The Monte Carlo studies [10], basic design [11], electronics [12], and data-acquisition system [13] of the TPC have been described elsewhere. A schematic diagram of the experimental setup is shown in Fig. 1. The TPC has a cylindrical active volume of 207 L. The operating pressure is 5 bars, with an admixture of 3.9% methane to increase the drift velocity (to 1.36 cmμs^{-1}) and to suppress diffusion of the secondary electrons [14]. Gas purity is maintained at a 0.1-ppm level in electronegative contaminants [9]. Xenon enriched to 62.5% ^{136}Xe is used [5], giving a total of 1.6×10^{25} ^{136}Xe atoms in the active volume. According to Monte Carlo calculations [9,10], the probabilities for a double-β-decay event at 2.5 MeV to be completely contained within the active volume are 25% and 21% for $0\nu\beta\beta$ induced by the mass mechanism and by right-handed weak couplings, respectively. The difference is mainly due to the difference in energy distribution between the two electrons in the two modes [15]. There are 168 readout channels, with 3.5-mm pitch, in each of the X and Y axes. The time evolution of the signals, recorded in 500-ns bins, gives trajectory information in the Z direction. The energy of an event is measured from the integration of the anode signals over the drift time of 51 μs.

To minimize the radioactive background, the chamber has been built with OFHC (oxygen-free high-conductivity) copper with a thickness of 5 cm, while all other components have been selected for their trace radioactivity with a low-background germanium detector. The copper vessel is further shielded by 20–30 cm of lead. The experiment is being conducted at the Gotthard Underground Laboratory, with a 3000-m water-equivalent overburden which attenuates the muon flux by a factor of 10^6.

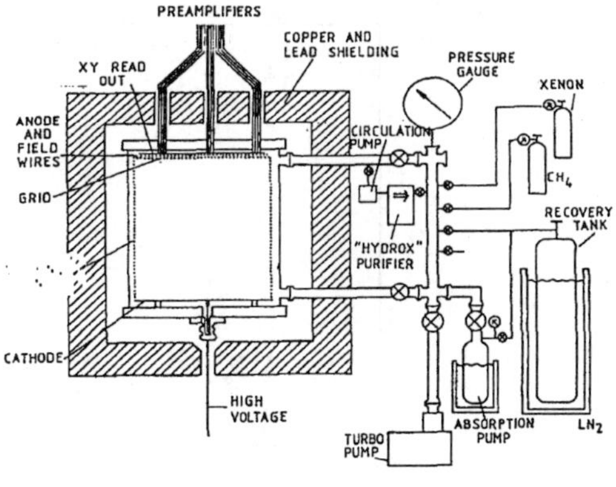

FIG. 1. A schematic diagram of the time projection chamber with the associated setup.

VOLUME 67, NUMBER 10 PHYSICAL REVIEW LETTERS 2 SEPTEMBER 1991

The track reconstruction capability of the TPC provides a powerful means of background rejection. A double β decay is identified as a continuous trajectory with characteristic "end features": high charge depositions (charge "blobs") due to enhanced dE/dx at low energy, and large-angle multiple scattering (that is, staggered trajectories), at both ends. Owing to bremsstrahlung emissions, some of these events have small isolated charge depositions (< 150 keV). The major background is from pair production as well as single-electron events (due to the photoelectric effect, Compton scattering, or β decay) with the emission of energetic secondary ("delta") electrons at the beginnings of their trajectories.

Some typical events recorded by the TPC are shown in Figs. 2(a)–2(c). The XZ and YZ projections as well as the time evolution of the anode signals are displayed. The large black dots indicate signals above a second level threshold, corresponding to charge blobs. The TPC does

(a)
(b)
(c)

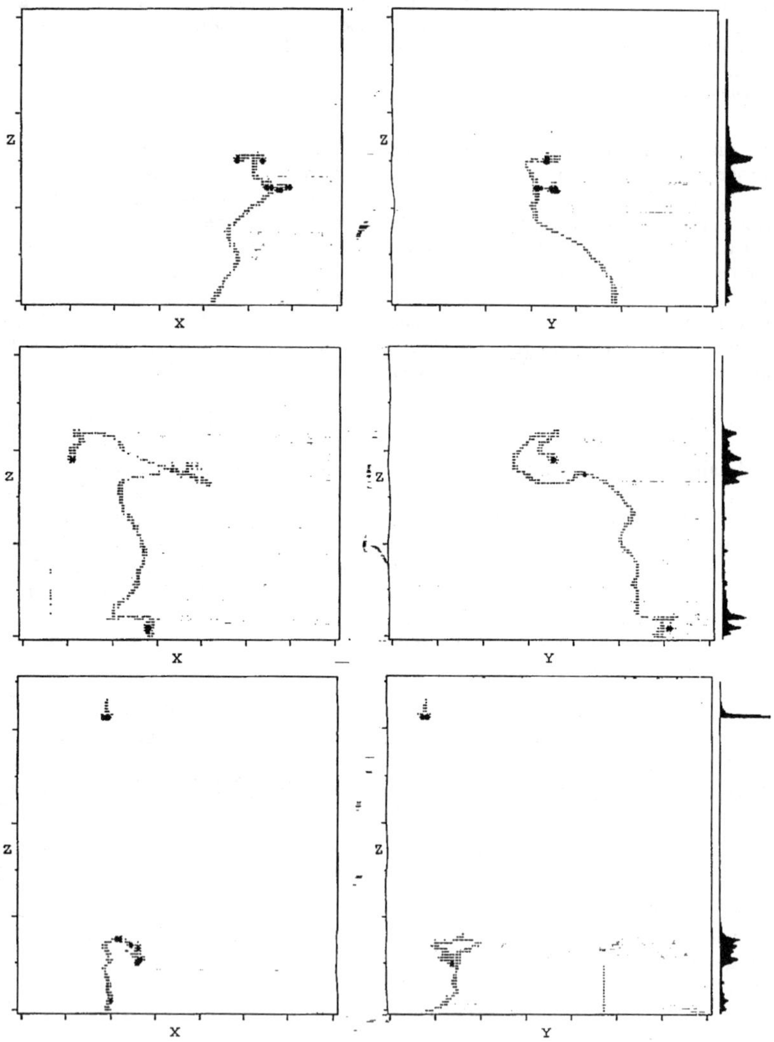

FIG. 2. Typical tracks recorded by the TPC with 5 atm of xenon. (a) Single electron; (b) two-electron candidate event; and (c) β decay followed by emission of an α particle. In (a)–(c) the full range is 60 cm for both X and Y, while the Z calibration is 10.9 cm per unit.

not have a time-zero trigger; the $Z=0$ in the figures corresponds to the first arrival of the signals. The full range for both X and Y is 60 cm. Calibration of the Z axis depends on the drift velocity of the secondary electrons. It is measured from the length of cosmic-ray muons traversing vertically through the detector. Typically, a few such events are recorded per day. The Z calibration is 10.9 cm per unit in the displayed figures, and the variation with time is less than 1%.

A typical single-electron event and a typical candidate for a two-electron event are depicted in Figs. 2(a) and 2(b). They are readily distinguishable by the features their trajectories exhibit at their ends (charge blobs and staggered trajectories). Figure 2(c) shows a β decay followed by the emission of an α particle at the same (X,Y) coordinate 50 μs later. This event is due to the cascade $^{214}Bi \to {}^{214}Po + e^- + \bar{\nu}_e$ ($Q=3.28$ MeV, $T_{1/2}=19.7$ min), followed by $^{214}Po \to {}^{210}Pb + \alpha$ ($Q=7.8$ MeV, $T_{1/2}=164$ μs), and is evidence of trace radon emission in the system. Among the daughter nuclei of the radon isotopes from the ^{238}U and ^{232}Th chains, only this β decay has an end-point energy above 2.5 MeV. However, as demonstrated, this cascade can be singled out by looking for the $\sim$100-μs post-trigger α activity after an initial single-electron event.

The energy resolution and calibration has been studied with various γ sources. A (10–15)% variation in charge multiplication across the effective area of the anode plane is observed. To correct for this effect, the anode plane is subdivided into 45 squares and a gain variation map is made from a measurement with a ^{137}Cs source. The anode signals are then compensated for at each time bin, based on the (X,Y) coordinate of the event at that time. A notable improvement on the energy resolution of 14.8% and 6.6% FWHM at 511 and 1592 keV, respectively, is subsequently achieved. The energy spectrum for a ^{232}Th source, after applying this gain correction, is shown in Fig. 3. The 1592-keV peak is due to double escape (pair production with the total escape of both 511-keV photons) from the incident γ rays of 2614 keV. The overall gain (combined effects of charge multiplication, gas-purity level, and electronics) is constant over the course of data taking to better than 2%. The energy calibration is linear in our range of interest, and is accurate to better than 2%.

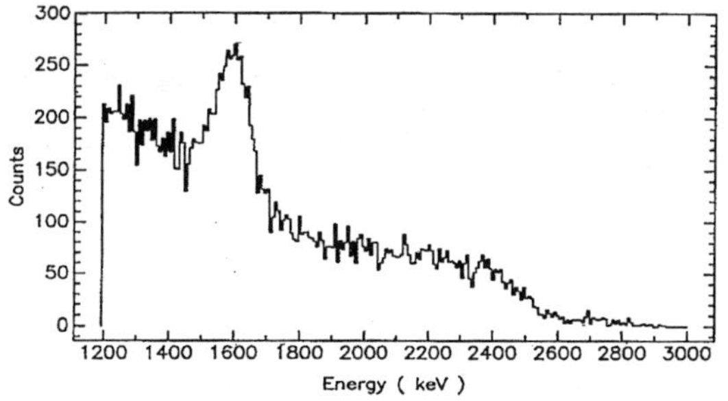

FIG. 3. Energy spectrum of a ^{232}Th source. The prominent line is the double escape peak of the 2614-keV γ rays in ^{208}Pb.

The results of a total of 3380 h of data, with an energy threshold of 1.6 MeV, are presented in this Letter. The two-electron spectrum displayed in Fig. 4 is obtained as follows. An off-line software program identifies and rejects events due to α particles, multiple tracks, cascades, uncontained events, and distinct single electrons, reducing the data size by a factor of 6. Remaining events are then scanned visually. The selection of two-electron events requires charge blobs and staggered trajectories at both ends of a continuous trajectory. Events with isolated charge depositions of more than 150 keV are rejected. For this analysis, those events with sharp vertices in the central portion of the trajectories (which might suggest pair production) are kept. With these analysis procedures, there are typically 36 times as many single-electron as two-electron events above 1.6 MeV. It can thus be deduced that the single-electron rejection efficiency of the TPC is at least 97%. The probabilities of a contained $\beta\beta$ event surviving these cuts have been studied with Monte Carlo simulations and with a smaller data set where all events are scanned visually. They are 81% and 64% for the mass-mechanism and right-handed-current modes, respectively, at the 0ν transition energy. The contributing factors include topological ambiguities (e.g., a track retraces itself), detector effects (inactive or noisy electronics), and the asymmetric energy distribution between the two electrons. The last factor is the major contribution to the difference in the efficiencies between the two mechanisms [15].

As stated above, the measured energy resolution is 6.6% FWHM at 1.6 MeV. We adopt the conservative

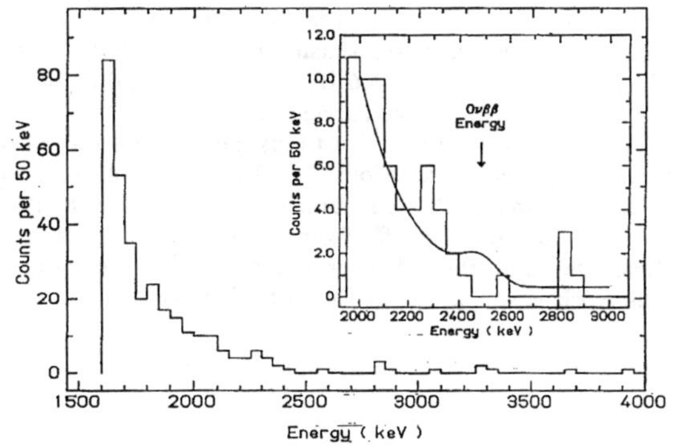

FIG. 4. Energy spectrum of two-electron candidate events, from 3380 h of data. The 90%-C.L. limit curve for a hypothetical 0ν peak at 2481 keV is represented by the solid line in the inset.

approach of taking this to be the resolution at the 0ν transition energy of 2.48 MeV. Based on Poisson statistics, the probability of a peak to be present at the transition energy is evaluated. Assuming an exponential background between 2000 and 2650 keV together with a constant background from 2650 to 3000 keV, we obtain an upper limit of 3.5(1.8) events for a hypothetical 0ν peak at a 90(68)% confidence level. Folding in the overall detector and analysis efficiencies [20% and 13% for the mass-mechanism and right-handed-current (RHC) modes, respectively], we obtain the following 90%-C.L. half-life limits:

$$T_{1/2}^{0\nu}(0^+ \rightarrow 0^+;\langle m_\nu \rangle) > 2.5\times 10^{23}\,\text{yr}\,,$$

$$T_{1/2}^{0\nu}(0^+ \rightarrow 0^+;\text{RHC}) > 1.7\times 10^{23}\,\text{yr}\,,$$

respectively. The corresponding 68% confidence limits are 4.9×10^{23} and 3.2×10^{23} yr, respectively. These numbers represent almost 1-order-of-magnitude improvement over existing limits [6,7]. The 90%-C.L. curve is folded onto the energy spectrum displayed in Fig. 4.

The limit for the Majorana neutrino mass parameter thus deduced depends on which nuclear-matrix-element calculation [4] one uses. Adopting the quasiparticle random-phase approximation by Engel, Vogel, and Zirnbauer [16] (and choosing $\alpha_1' = -375 \pm 15$ MeV fm^3 which reproduces the measured $2\nu\beta\beta$ half-lives in ^{82}Se, ^{100}Mo, and ^{130}Te), the 90%-C.L. upper limit for the mass-mechanism mode (2.5×10^{23} yr) implies

$$\langle m_\nu \rangle < 3.3\text{–}5.0\ \text{eV}\,.$$

This can be compared with $\langle m_\nu \rangle < 2.4$–4.7 eV, derived from the most recent ^{76}Ge results [3] ($T_{1/2}^{0\nu} > 1.2\times 10^{24}$ yr at the 90% C.L., with 3 yr of data) using the same calculation.

Data taking continues on this experiment in the Gotthard Laboratory. With the measured background level of 0.02 count per 100 keV per day (0.01 count keV^{-1} kg^{-1} yr^{-1}) in the 0ν energy range, we expect, with 1 yr of data, sensitivities of 5×10^{23} yr for $T_{1/2}^{0\nu}(0^+ \rightarrow 0^+)$, or 2.3–3.6 eV for $\langle m_\nu \rangle$. Studies of the $2\nu\beta\beta$ channel will also be carried out.

The authors would like to thank Petr Vogel for many contributions and discussions, Val Telegdi for stimulating comments, and J.-P. Bourquin, C. Heche, D. Schenker, and the technical staff from Institut de Physique-Neuchâtel and Paul Scherrer Institute for assistance. This work is supported by the U.S. Department of Energy and the Fonds National Suisse pour la Recherche Scientifique.

(a)Correspondence address: Institut de Physique, A.-L. Breguet 1, 2000 Neuchâtel, Switzerland.

(b)Present address: Department of Physics and Astronomy, Johns Hopkins University, Baltimore, MD 21218.

(c)Present address: Bellcore, 331 Newman Springs Road, NVC 3X-239, Red Bank, NJ 07701.

(d)Present address: Department of Physics, Flinders University, Bedford Park, SA 5042, Australia.

(e)Present address: L-397, Lawrence Livermore National Laboratory, Livermore, CA 94550.

[1] See, for example, F. Boehm and P. Vogel, *Physics of Massive Neutrinos* (Cambridge Univ. Press, Cambridge, 1987).

[2] For a recent review, see M. K. Moe, in Proceedings of the Fourteenth International Conference on Neutrino Physics and Astrophysics, Geneva, 1990, edited by J. Panman and K. Winter [Nucl. Phys. B, Proc. Suppl. (to be published)].

[3] D. O. Caldwell *et al.*, in Proceedings of the Fourteenth European Physical Society Conference on Nuclear Physics, Bratislava, 1990, edited by P. Povinec [J. Phys. G (to be published)]; F. Boehm *et al.*, in *Proceedings of the Twenty-Sixth Recontre de Moriond, Les Arcs, 1991*, edited by O. Fackler and J. Tran Thanh Van (Editions Frontières, Gif-sur-Yvette, 1991).

[4] For a recent review, see T. Tomoda, Rep. Prog. Phys. **54**, 53 (1991).

[5] Supplier: Monsanto Research Corporation (Mound, Oak Ridge).

[6] A. S. Barabash *et al.*, Phys. Lett. B **223**, 273 (1989).

[7] E. Bellotti *et al.*, Phys. Lett. B **221**, 209 (1989).

[8] For an overview of the subject, see *The Time Projection Chamber*, edited by J. A. Macdonald, AIP Conf. Proc. No. 108 (AIP, New York, 1984).

[9] H. T. Wong, Ph.D. thesis, Caltech, 1991 (unpublished).

[10] M. Z. Iqbal, B. M. O'Callaghan, and H. T. Wong, Nucl. Instrum. Methods Phys. Res., Sect. A **253**, 278 (1987).

[11] M. Z. Iqbal *et al.*, Nucl. Instrum. Methods Phys. Res., Sect. A **259**, 459 (1987).

[12] M. Z. Iqbal, B. M. O'Callaghan, and H. T. Wong, Nucl. Instrum. Methods Phys. Res., Sect. A **263**, 387 (1988).

[13] J. Thomas *et al.*, IEEE Trans. Nucl. Sci. **34**, 845 (1987).

[14] M. Z. Iqbal *et al.*, Nucl. Instrum. Methods Phys. Res., Sect. A **243**, 459 (1986).

[15] M. Doi, T. Kotani, and E. Takasugi, Prog. Theor. Phys. Suppl. **83**, 1 (1985).

[16] J. Engel, P. Vogel, and M. R. Zirnbauer, Phys. Rev. C **37**, 731 (1988).

PHYSICAL REVIEW D VOLUME 48, NUMBER 3 1 AUGUST 1993

Search for neutrinoless double-β decay in ^{136}Xe with a time projection chamber

J.-C. Vuilleumier, J. Busto, J. Farine, V. Jörgens, L. W. Mitchell,* M. Treichel, and J.-L. Vuilleumier
Institut de Physique, A.-L. Breguet 1, 2000 Neuchâtel, Switzerland

H. T. Wong,† F. Boehm, P. Fisher,‡ H. E. Henrikson, D. A. Imel,§ M. Z. Iqbal,‖ B. M. O'Callaghan-Hay, and J. Thomas¶
Norman Bridge Laboratory of Physics, California Institute of Technology, Pasadena, California 91125

K. Gabathuler
Paul Scherrer Institute, 5232 Villigen-PSI, Switzerland
(Received 8 March 1993)

A xenon time projection chamber (TPC) with an active volume of 180 liters has been built to study neutrinoless double-β decay in ^{136}Xe. The experiment was performed in the Gotthard Underground Laboratory, with 5 atm of xenon enriched to 62.5% in ^{136}Xe. The experimental details, background considerations, detector performance, and data analysis are discussed. From 6830 h of data, no evidence has been found for the 0ν $0^+ \rightarrow 0^+$ transition. Half-life limits of $T_{1/2}^{0\nu} > 3.4(6.4) \times 10^{23}$ yr in the mass mechanism mode, and $T_{1/2}^{0\nu} > 2.6(4.9) \times 10^{23}$ yr in the right-handed currents mode, at the 90(68)% C.L., were derived, corresponding to an upper limit on the Majorana neutrino mass parameter $\langle m_\nu \rangle$ of about 2.8 eV. Limits on two-neutrino double-β decay of $T_{1/2}^{2\nu} > 2.1 \times 10^{20}$ yr, and on neutrinoless double-β decay with Majoron emission of $T_{1/2}^{0\nu\chi} > 4.9 \times 10^{21}$ yr, both at 90% C.L., were also derived. Accordingly, a limit on the effective Majoron-neutrino coupling parameter of $\langle g_M \rangle < 2.4 \times 10^{-4}$ was deduced.

PACS number(s): 23.40.Bw, 13.10.+q, 14.60.Gh, 27.60.+j

I. INTRODUCTION

As is well known, nuclear double-β ($\beta\beta$) decay is a process in which a nucleus (A,Z) decays spontaneously to a daughter nucleus $(A,Z+2)$ [1]. There are two principal channels for this process: neutrinoless double-β decay ($0\nu\beta\beta$),

$$(A,Z) \rightarrow (A,Z+2) + 2e^- ,$$

and two-neutrino double-β decay ($2\nu\beta\beta$),

$$(A,Z) \rightarrow (A,Z+2) + 2e^- + 2\bar{\nu}_e .$$

Neutrinoless double-β decay provides a sensitive test of the standard model, as it would require violation of lepton number conservation as well as the presence of a massive Majorana neutrino. With the standard left-handed weak couplings, $0\nu\beta\beta$ are restricted to $0^+ \rightarrow 0^+$ transitions between parent and daughter nuclei [the "mass mechanism" (MM) mode]. If, in addition, there exist right-handed weak currents, the transitions thus induced can be $0^+ \rightarrow 0^+, 1^+, 2^+$ [the "right-handed-current" (RHC) mode]. The observation of $0\nu\beta\beta$ would, therefore, provide evidence of new physics beyond the standard model. It would imply (a) the neutrino is a Majorana particle and (b) there is at least one massive neutrino.

Two-neutrino double-β decay, on the other hand, is a second-order weak interaction allowed by the standard model. However, investigations in the $2\nu\beta\beta$ channel are also important, since they can reveal information on the nuclear structure, which in turn will help to interpret the measurements of the $0\nu\beta\beta$ experiments.

In order to derive information on the fundamental parameters in the neutrino sector (masses and mixing matrices) from the measured $0\nu\beta\beta$ decay rates or limits, the phase-space integrals and the nuclear matrix elements must be known. The phase-space factors depend only on the transition energies and can be calculated exactly. The nuclear matrix elements, however, are nontrivial, and their calculations are actively pursued by several groups [2].

Experimentally, $0\nu\beta\beta$ is characterized by a peak at the transition energy for the total-electron-energy spectrum, while $2\nu\beta\beta$ gives a continuous spectrum peaked at about 40% of the transition energy. The theoretical energy spectra are shown in Fig. 1. Experimental searches for double-β decay dated as early as the 1950's, but have stimulated intense interests only in the past decade [3]. There are geochemical measurements for $\beta\beta$ lifetimes in

*Present address: Department of Physics, Flinders University, Bedford Park, SA 5042, Australia.

†Present address: CERN PPE Division, 1211 Geneva 23, Switzerland.

‡Present address: Department of Physics and Astronomy, Johns Hopkins University, Baltimore, MD 21218.

§Present address: Jet Propulsion Laboratory, 4800 Oak Grove Drive, Pasadena, CA 91109.

‖Present address: Bellcore, 331 Newman Springs Road NVC 3X-239, Red Bank, NJ 07701.

¶Present address: L-397, Lawrence Livermore National Laboratory, Livermore, CA 94550.

in Proc. NEUTRINO'96, Helsinki, Finland, June 1996, eds. *K. Enqvist,* 347
K. Huitu and J. Maalampi, W.S., Singapore, (1997) 347 – 351

SEARCH FOR $\beta\beta$ DECAY IN ^{136}Xe GOTTHARD RESULTS

J. FARINE
Physics Institute, Neuchâtel University,
Rue Breguet 1, 2000 Neuchâtel, Switzerland

On behalf of the
Caltech-PSI-Neuchâtel collaboration

The Gotthard Xe TPC has taken data for 6000 hours since a new readout plane was installed in April, 1994. Alpha spectroscopy together with $\alpha\alpha$ and $\beta\alpha$ coincidence techniques lead to estimations of the gas radiopurity at the 10^{-12} g/g level. The differences in count rate reduction after the upgrade indicate that an appreciable fraction of $\beta\beta$ candidates come from the allowed 2ν decay mode. The new 90 % CL limits derived are $T^{2\nu}_{1/2} > 5.5 \times 10^{20}$ yr, $T^{\chi^0}_{1/2} > 1.4 \times 10^{22}$ yr (corresponding to $\langle g_{\nu\chi^0}\rangle < 1.5 \times 10^{-4}$), and $T^{0\nu}_{1/2} > 4.4 \times 10^{23}$ yr at 90 % CL which leads to $m_\nu <$ 1.8–2.8 eV. The near future of the detector is eventually discussed.

1 Introduction

The Gotthard Time Projection Chamber (TPC) has been optimized for the search of 0ν and 2ν double-beta decay in ^{136}Xe, xenon gas being the source and the detector medium. The fiducial volume is 180 liters, filled with a xenon-methane mixture (4 % CH_4 in volume) at 5 bar. With an isotopic enrichment of 62.5 %, the total ^{136}Xe mass is 3.3 kg. The energy resolution is 7 % at 2.48 MeV, the transition energy. The detector is surrounded by shielding layers of copper (5 cm) and lead (30 cm). With a rock thickness of at least 3000 m.w.e, the cosmic muon flux in the Gotthard laboratory is reduced by a factor of 10^6. The detector and its performances are described in more details in ref. [1] and [2] and references therein.

The TPC is now equipped with a newly designed readout plane, made of materials carefully selected for their radiopurity. The wires are crimped in copper needles to avoid dirty soldering. The needles are fixed on copper rings which are mounted on an acrylic frame. Moreover, the amount of material making up the XY plane has been strongly reduced by etching copper strips on a PET foil.

2 Background reduction and new limits

With the new readout, the count rate in single electrons dropped by a factor of 4 above 1800 keV, as shown on Figure 1. Except at higher energies, the

570 in Proc. "WEIN 92", Dubna, USSR, June 1992, ed *Ts. D. Vylov, World Scientific, Singapore* (1993) 570 – 574

^{136}Xe DOUBLE BETA DECAY FROM ITEP

V. ARTEMIEV, E. BRAKCHMAN, M. IVANOVSKY, A. KARELIN, V. KIRICHENKO, V. KNYAZEV, O. KOZODAEVA, V. LUBIMOV, A. MITIN, V. NIKOLAEV, T. TSVETKOVA, O. ZELDOVICH

Institute of Theoretical and Experimental Physics, Moscow, Russia

Abstract

Measurements with a proportional drift detector held 210 g of ^{136}Xe in sensitive volume gave the following results: rejection factor ¿10^7, background level $\sim 0.25\ c/keV \cdot kg \cdot y$ in $0.6 \div 1.6\ MeV$ energy range. There was obtained half-life time $T_{1/2}(2\nu 2\beta) > 9.3 \cdot 10^{19}\ y$ (90% CL). Based on these data detector (13 m^3) with 10 kg ^{136}Xe gives a good hope to obtain $T_{1/2}(0\nu 2\beta) > 10^{24}\ y$ for one year measuring time.

The ^{136}Xe is a suitable candidate for double beta decay studies:

$$^{136}_{54}Xe \rightarrow^{136}_{56} Ba + e^- + e^-(+2\bar{\nu})$$

1) The transition energy 2.48 MeV is far enough from the most iintensive background in the range of $0.3 \div 0.7\ MeV$.

2) Xe can operate effectively as a working gas ion a drift detector. So ^{136}Xe can be simultaneously used as source and detector of the emitted electrons.

The published results on 2β–decay of ^{136}Xe in a track detector are shown in the Table 1.

No evidence is found in these experiments for double beta decay of ^{136}Xe wither in the two neutrino or in the neutrinoless modes. The upper limits are obtained at the level of $> 10^{20}$ and $> 10^{23}$ years correspondingly.

The $2\nu 2\beta$ decay has been directly observed for ^{76}Ge [4], ^{82}Se [5] and ^{100}Mo [6]. No observation of $2\nu 2\beta$ decay in ^{136}Xe is possibly connected with insufficiency of signs for identification of background in mentioned detectors.

Table 1: Comparison of the parameters and results of the ^{136}Xe 2β-decay track experiments

Experiment	*Caltect* [2]	*Milano* [1]	*ITEP* [3]
The detector	TPC	MPC	PDD
Measurement condition	Underground	Underground	See level
Mass ^{136}Xe in kg(mole)	3.6 (26.6)	3 (20.7)	0.210 (1.6)
Measurement time (h)	3380	6210	418
$T_{1/2}(0\nu 2\beta) \cdot 10^{23}\ y$	> 2.5 (90% CL)	> 0.1 (95% CL)	
$T_{1/2}(2\nu 2\beta) \cdot 10^{20}\ y$		> 1.6 (95% CL) > 2.9(84% CL)	> 0.93 (90% CL)

It is obvious that the use of large amount of ^{136}Xe with real detection of the two emitted electrons and the measurement of their energies can provide the best experimental conditions.

Some years ago we proposed a big drift detector in the solinoidal magnet (BDD). To confirm the abilities of the BDD there was constructed a prototype–proportional drift detector (PDD) which is $\frac{1}{4}$ part of BDD. The construction of BDD and PDD, electronics, data acquisition system, the selection event procedure, the Monte–Carlo simulation of 2β decay and background can be found elsewhere [7]. The PDD is situated in magnet and consist of two volumes: one is filled with CH_4, the other held 210 g ^{136}Xe in sensitive volume.

The parameters that were measured for each of two electron events with PDD are following:

1) kinetic energy for each electron in the CH_4–volume by the trajectory in magnetic field;

2) the vertex in a Xe–volume;

3) the sign of a particle by the shape of the trajectory received from strips in the CH_4–volume;

4) energy losses in Xe–volume.

We have finished measurements with ^{1136}Xe at PDD and have got the results.

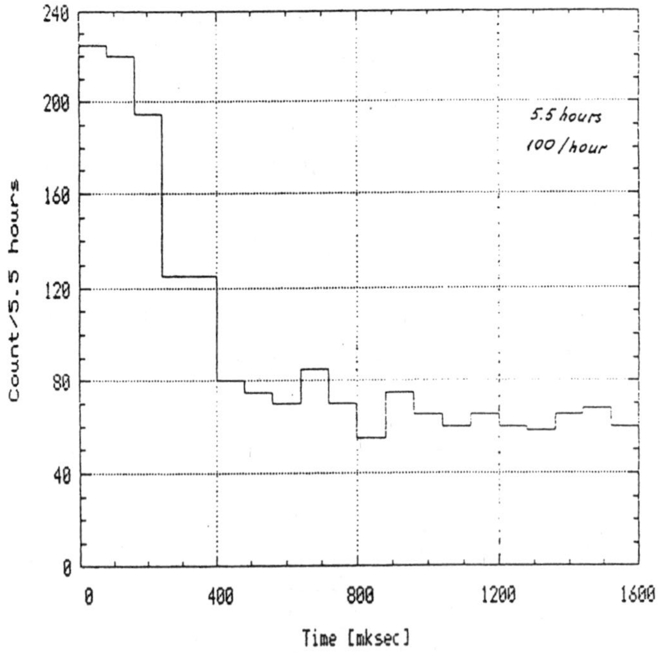

Figure 1: The arrival time distribution of delayed α-particles.

1. Internal background imitating $2\nu 2\beta$ decay was caused by contamination of the Xe-volume by ^{222}Rn from ^{238}U chain. To keep a stable drift regime in the Xe–volume

we applied the gas purification system with absorbents Ni/SiO_2 and molecular sieve. The series of measurement showed that these material increased contamination with the longlived ($T_{1/2} = 3.8\ d$) ^{222}Rn. To estimate the ^{222}Rn contamination level we used the method of the registration of a single β-electron from ^{214}Bi decay and delayed α-particle from ^{214}Po decay. Fig. 1 shows the distribution of the delayed α–particle arrival time in single electron triggers.

A distinct peak with the 160 μsec slope corresponds to the counting rate 100 α/h and the ^{222}Rn concentration 10^{-20} atom/atom. It was found that the level of ^{222}Rn could be reduced when the reactor with Ni/Si_2 was preliminarily pumped out and additional reactor with activated carbon at -70°C was involved in the cycle of cleaning. Analyzing the duration of α–particle signals(Fig. 2) we observed a peak in arrival time distribution which corresponds to the counting rate (4.5 ± 0.9) α/h and the average counting rate over the time interval 5 days is 3 α/y.

It corresponds to reduction of ^{222}Rn concentration by more than 20 times.

2. External γ–radiation background imitates $2\nu 2\beta$ decay mainly by producing Compton electron in the Xe–volume which then undergoes Moller scattering. An external γ–radiation was measured with a NaJ detector (64×64) mm^2 placed inside the magnet without shielding and with a lead shielding. We made a 418-hour exposition with a 10 cm Pb shielding. Ten two–electron events were finaly selected after this run was processed. The two dimentional spectra of the candidate events is shown in Fig. 3.

When the sum energy range was limited to $(0.6 < E1 + E2 < 1.6)$ MeV only 6 events were left. The estimation of background from ^{222}Rn gives 3.8 events in this energy range. So the upper limit for 6 events with average background 3.8 events is $T_{1/2}(2\nu 2\beta) > 9.3 \cdot 10^{19}$ years (90% CL).

The measurements with the prototype show:

1) The background connected with ^{222}Rn contamination decreased to the level $< 10^{-21}$ at/at.

2) The background caused by external γ radiation was reduced by a Pb shielding.

3) The rejection factor is more than 10^7.

4) The average background is 0.25 counts/(kev $\cdot$ kg $\cdot$ y) for energy range $(0.6 \div 1.6)$ MeV. This value is comparable with ^{76}Ge experiments $(0.2 \div 0.6)$ counts/(keV $\cdot$ kg $\cdot$ y) for 0ν energy point 2.0 MeV.

These results encourage us to continue the manufacture of big drift detector (BDD). BDD is intended first of all for the $0\nu 2\beta$ decay mode. The sizes of central volume $(0.2 \times 3.2 \times 3.2)$ mm^3 allow to use the large amount isotopically emriched 2β decay source material for example 10 kg of ^{136}Xe ($4.4 \cdot 10^{25}$ atoms). The Monte–Carlo calculations define that optimum efficiency of registration of $0\nu 2\beta$ mode 70% can be achieved if the concentration of ^{136}Xe is 80% and the magnetic field strength is 1.5 kg. The Monte–Carlo energy resolution in the range of $0\nu 2\beta$ peak is 6% (Fig. 4).

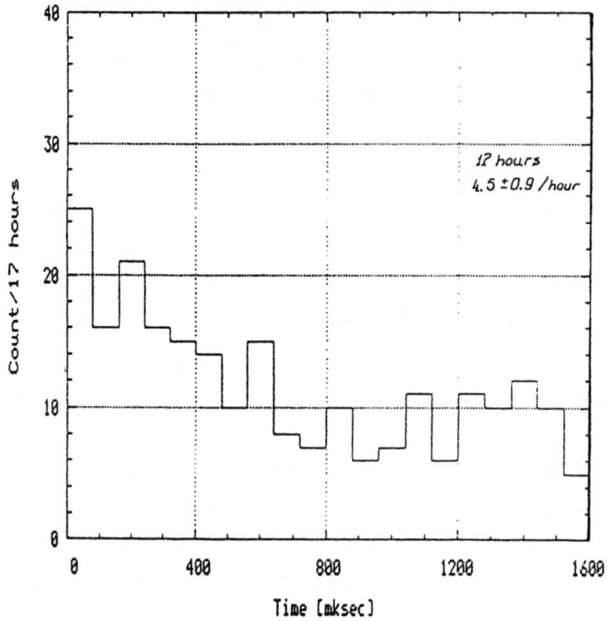

Figure 2: The arrival time distribution of delayed α-particle with using duration information.

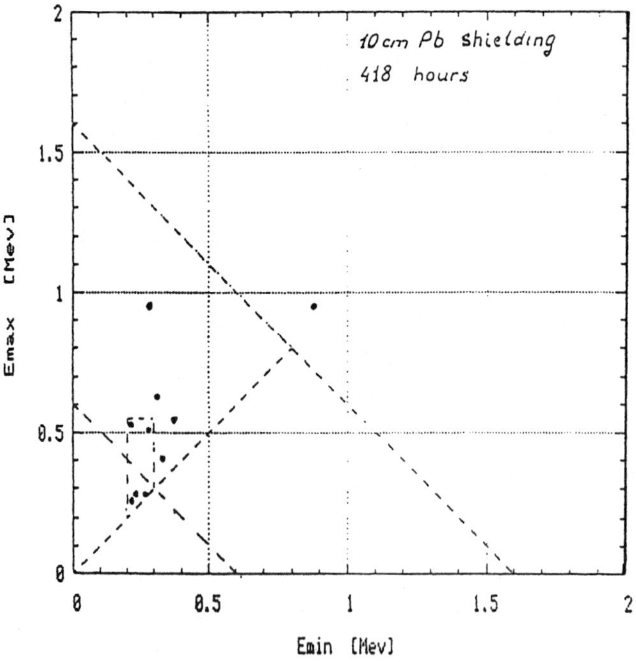

Figure 3: The two-dimensional energy spectrum for the selected two-electron events.

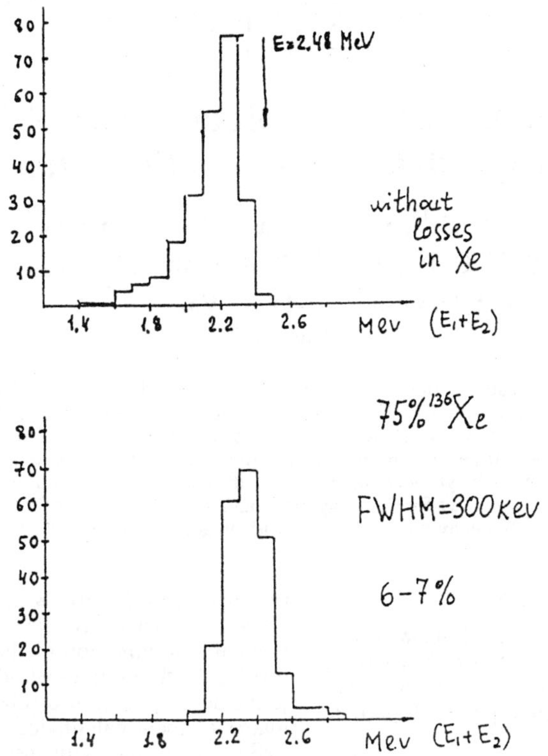

Figure 4: The Monte-Carlo calculation of the $0\nu 2\beta$ decay in the BDD.

So with active mass of ^{136}Xe 5.6 kg and zero background we can reach $T_{1/2}(0\nu 2\beta) > 10^{25}$y for one year measurements.

The estimation of background based on prototype results gives a real possibility to obtain $T_{1/2}(0\nu 2\beta) > 10^{24}$y.

References

1. E. Bellotti et al., *Phys. Lett* **B266** (1991) 193.
2. H. T. Wong et al., *Phys. Rev. Lett.* **67** (1991) 1218.
3. V. Artemiev et al., *Phys. Lett.* **B280** (1992) 159.
4. F. T. Avignone III et al., *Phys. Lett.* **B56** (1991) 559.
5. S. R.Elliot et al., *Phys. Rev. Lett.* **59** (1987) 2020.
6. H. Ejiri et al., *Phys. Lett.* **B258** (1991) 17.
7. V. Artemiev et al., *Nucl. Instr. Methods* **A303** (1991) 309.

Physics of Atomic Nuclei, Vol. 62, No. 12, 1999, pp. 2044–2047. From Yadernaya Fizika, Vol. 62, No. 12, 1999, pp. 2217–2220.

CONFERENCE ON FUNDAMENTAL INTERACTIONS OF ELEMENTARY PARTICLES
Experiment

Investigation of Double-Beta-Decay of ^{100}Mo with the Liquid-Argon Ionization Chamber*

V. D. Ashitkov[1)], A. S. Barabash[1)], S. G. Belogurov[1)], G. Carugno[2)], S. I. Konovalov[1)], F. Massera[3)], I. O. Pilugin[1)], G. Puglierin[2)], R. R. Saakyan[1), 4)], V. N. Stekhanov[1), **], V. I. Umatov[1)], and I. A. Vanishin[1)]

Received April 4, 1999

Abstract—The experiment is located in the Gran Sasso Underground Laboratory (3500 mwe). The full weight of the ^{100}Mo under investigation is 138.7 g. Three measurement runs were done (202, 238, and 313 h). As a result, the limits of half-lives for 0ν and $0\nu\chi^0$ decays of ^{100}Mo were obtained as 3.5×10^{21} and 1.2×10^{20} yr, respectively, at 90% C.L. Some extra events in the energy region 1.6–3 MeV lead to a preliminary result on the $2\nu\beta\beta$ decay of ^{100}Mo, about 8.5×10^{18} yr. We plan to improve the background conditions of the experiment and increase the sensitivity to $(1–2) \times 10^{23}$ yr for 0ν decay mode and to $(1–2) \times 10^{22}$ yr for the $0\nu\chi^0$ decay. In addition, we hope to investigate the 2ν decay mode of ^{100}Mo with an accuracy of about 10%.

1. INTRODUCTION

At present, the neutrinoless double-beta decay ($0\nu\beta\beta$) is one of the best probes for physics beyond the Standard Model of electroweak interactions. Its existence deals with the fundamental aspects of particle physics:

the lepton number nonconservation;

the existence and nature of neutrino mass;

the existence of right-handed currents in the electroweak interaction;

the existence of a Majoron;

the structure of the Higgs sector;

the supersymmetry;

the leptoquarks;

the sterile neutrino existence.

At the moment, only the lower limits of half-lives ($T_{1/2}$) have been obtained experimentally. These limits are used to deduce the upper limits of the Majorana neutrino mass, the right-handed-current-admixture parameter, the Majoron–Majorana neutrino coupling constant, etc. One of the most important uncertainties in the above analysis is the evaluation of the nuclear matrix elements. In connection with the $0\nu\beta\beta$ decay, the detection of double-beta decay with the emission of two neutrinos ($2\nu\beta\beta$), which is an allowed second-order process in the Standard Model, enables the experimental determination of the nuclear matrix elements involved in the double-beta-decay processes. This leads to the development of theoretical schemes for nuclear-matrix-element calculations, both connected with the $2\nu\beta\beta$ decays and with the $0\nu\beta\beta$ decays. In order to search for the double-beta decay with a new technique, a multisectional liquid-argon ionization chamber was suggested [1] (see also [2–4]). For the first step of the experiment, we have selected a nucleus with a sufficiently large 2β-transition energy, ^{100}Mo ($E = 3033$ keV).

2. EXPERIMENTAL SETUP

The experiment is located in the Gran Sasso Underground Laboratory (3500 mwe). The experimental setup consists of a liquid-argon ionization chamber placed in a passive lead shielding, gas system, and manual crane for assembling and disassembling the liquid-Ar chamber. All this equipment was put on a special concrete platform with the dimensions $4 \times 6 \times 0.6$ m^3. To reduce mechanical vibrations, the platform was placed on a rubber layer. A special house was set up around the passive shielding in order to protect the equipment from the influence of high humidity. The electronics and data acquisition system for the liquid-argon chamber were placed in a separate "house" at a distance of 2 m from the platform.

2.1. Liquid-Argon Ionization Chamber

The main construction material of the chamber is titanium. Virtually all insulators are made of teflon. The scheme of the detector is shown in Fig. 1. The chamber

* This article was submitted by the authors in English.

1) Institute of Theoretical and Experimental Physics, Bol'shaya Cheremushkinskaya ul. 25, Moscow, 117259 Russia.

2) Dipartimento di Fisica e INFN, Universita di Padova, via F. Marzolo 8, Padova, 35131 Italy.

3) INFN, Sezione di Bologna, Bologna, 40127 Italy.

4) Laboratori Nazionali del Gran Sasso del'INFN, Assergi (L'Aquila), 67010 Italy.

** e-mail: stekhanov@vxitep.itep.ru

1063-7788/99/6212-2044$15.00 © 1999 МАИК "Наука/Interperiodica"

PHYSICAL REVIEW D VOLUME 51, NUMBER 5 1 MARCH 1995

Two-neutrino double-β decay measurement of ^{100}Mo

D. Dassié, R. Eschbach, F. Hubert, Ph. Hubert, M.C. Isaac, C. Izac, F. Leccia, P. Mennrath, and A. Vareille
Centre d'Études Nucléaires, IN2P3-CNRS et Université de Bordeaux, 33170 Gradignan, France

C. Longuemare and F. Mauger
Laboratoire de Physique Corpusculaire, IN2P3-CNRS et Université de Caen, 14032 Caen, France

F. Danevich, V. Kouts, V.I. Tretyak, Yu. Vassilyev, and Yu. Zdesenko
Institute for Nuclear Research of the Ukrainian Academy of Sciences, Kiev, Ukraine

A.S. Barabash, V.N. Kornoukov, Yu.B. Lepikhin, V.I. Umatov, and I.A. Vanushin
Institute for Theoretical and Experimental Physics, Moscow, Russia

C. Augier, D. Blum, J.E. Campagne, S. Jullian, D. Lalanne, F. Laplanche, F. Natchez, G. Pichenot, and G. Szklarz
Laboratoire de l'Accélérateur Linéaire, IN2P3-CNRS et Université de Paris-Sud, 91405 Orsay, France

R. Arnold, J.L. Guyonnet, T. Lamhamdi, I. Linck, F. Piquemal, and F. Scheibling
Centre de Recherches Nucléaires, IN2P3-CNRS et Université Louis Pasteur, 67037 Strasbourg, France

V. Brudanin, V. Egorov, O. Kochetov, A. Nozdrin, Ts. Vylov, and Sh. Zaparov
Joint Institute for Nuclear Research, Dubna, Russia

H.W. Nicholson and C.S. Sutton
Mount Holyoke College, South Hadley, Massachusetts 01075

(NEMO Collaboration)

(Received 29 June 1994)

From data accumulated over 6140 h with 172 g of enriched molybdenum (1.18 mol yr of ^{100}Mo) with the NEMO 2 detector in the Fréjus Underground Laboratory, a clear $\beta\beta 2\nu$ signal (1433 events) is observed, leading to a half-life, $T_{1/2} = 0.95 \pm 0.04(\text{stat}) \pm 0.09(\text{syst})\ 10^{19}$ yr. The experimental two-electron energy spectrum and the two-electron angular distribution are in agreement with the expected ones. Limits for $\beta\beta 0\nu$ decays to the ground state, excited states (2_1^+ and 0_1^+), and also with Majoron emission are given.

PACS number(s): 23.40.Bw, 14.60.Pq

I. INTRODUCTION

$\beta\beta 0\nu$ decay can occur if the neutrino is a massive Majorana particle, i.e., if it is its own antiparticle so that lepton number conservation is violated and neutrino exchange between two nucleons can take place with the emission of only two electrons. If only left-handed currents are present, the transition is possible because the Majorana mass term can flip the chirality of the emitted left-handed neutrino to a right-handed neutrino which is absorbed. Therefore, the observation of a $\beta\beta 0\nu$ transition would prove the Majorana nature of the neutrino.

In the study of double-β decay processes nuclear physics is important because one cannot have predictions for the fundamental parameters such as the neutrino mass or the coupling to possible right-handed currents in the weak interaction if there are no reliable calculations of the nuclear matrix elements. Many efforts have been devoted to clarify the mechanism by which these processes occur and to perform calculations of the nuclear structure which enters the half-life formulas. As far as the $\beta\beta 2\nu$-allowed process is concerned, a half-life can be measured and nuclear matrix element calculations can be tested. Although there are no clear relations between the matrix elements of the two processes ($\beta\beta 2\nu$ and $\beta\beta 0\nu$), comparisons with experiments in the case of the allowed process may give some confidence in the computations.

During the past four years the NEMO Collaboration has built two detectors NEMO 1 [1] and NEMO 2 [2,3] in a research and development effort devoted to the measurement of the two-electron background in the 3 MeV region ($Q = 3.03$ MeV for the ^{100}Mo $\beta\beta 0\nu$ decay). Toward this end, the NEMO 2 detector was installed in the Fréjus Underground Laboratory [4800 meters of water equivalent (m.w.e.) depth] in August of 1991. The $\beta\beta 2\nu$ signal reported here is a by-product of this research and preliminary results [4] have already been presented with a fraction of the current data sample. The collaboration is now building a new detector called NEMO 3 [5] which will be able to accommodate up to 10 kg of enriched molybdenum and/or other double-β-decay candidates to

NEMO COLLABORATION: LATEST RESULTS AND PERSPECTIVES FOR THE FUTURE

A.S. BARABASH

Institute of Theoretical and Experimental Physics, B.Cheremushkinskaya 25, 117259 Moscow, Russia

and NEMO Collaboration[a]

To investigate double beta decay processes the NEMO collaboration developed an electron tracking detector made of Geiger cells for track reconstruction and plastic scintillators for energy measurements. The prototype, NEMO-2, is described and results on $\beta\beta$ decay of ^{100}Mo, ^{116}Cd, ^{82}Se and ^{96}Zr are given. A brief presentation of the status and design parameters of a new detector NEMO-3, which is approximately 20 times larger, will also be given.

in Proc. NEUTRINO'96, Helsinki, Finland, June 1996, eds. *K. Enqvist, K. Huitu and J. Maalampi, World Scientific, Singapore,* (1997) 374 – 380

1 Introduction

The NEMO collaboration is building the NEMO-3 [1] detector for double beta decay experiments which will be capable of studying $\beta\beta0\nu$ decays of ^{100}Mo and other nuclei with half-lives up to $\sim 10^{25}$ y corresponding to neutrino masses of 0.1 to 0.3 eV. Two prototype detectors, NEMO-1 [2] and NEMO-2 [3], have been constructed as research and development efforts to establish reliable techniques. NEMO-2 is currently operating in the Fréjus Underground Laboratory (4800 m.w.e.). Presented here are the final results for ^{100}Mo and ^{116}Cd, the preliminary results for ^{82}Se and ^{96}Zr, and the present status of the detector NEMO-3.

2 NEMO-2 detector and $\beta\beta$ decay experiments with ^{100}Mo, ^{116}Cd, ^{82}Se and ^{96}Zr

NEMO-2 [3] consists of a 1m^3 tracking volume filled with helium gas and 4% ethyl alcohol.Vertically bisecting the detector is the plane of the source foil (1m $\times$ 1m). The tracking portion of the detector is made of open Geiger cells with octagonal cross sections defined by 100μm nickel wires. On each side of the source there are 10 planes of 32 cells which alternate between vertical and horizontal orientations. The cells provide three-dimensional tracking of

[a] CEN-Bordeaux-Gradignan, France; CFR-Gif/Yvette, France; CRN-Strasbourg, France; Department of Physics-Jyväskylä, Finland; INR-Kiev, Ukraine; ITEP-Moscow, Russia; JINR-Dubna, Russia; LAL-Orsay, France; LPC-Caen, France; MHC-South Hadley, USA; INEL-Idaho Falls, USA.

charged particles by recording the drift time and two plasma propagation times in each cell.

A calorimeter made of scintillators covers two opposing, vertical sides of the tracking volume. Two configurations of the calorimeter have been implemented. The first one consisted of 2 planes of 64 scintillators (12 cm×12 cm×2.25 cm) associated with "standard" photomultiplier tubes (PMTs). This configuration was used in the experiment with ^{100}Mo. The present configuration includes 2 planes of 25 scintillators (19 cm×19 cm×10 cm) with PMTs made of low radioactive glass. The tracking volume and scintillators are surrounded by a lead (5 cm) and iron (20 cm) shield.

The performance and operating parameters are as follows. The threshold for the scintillators is set at 50 keV, the energy resolution (FWHM) is 18% at 1 MeV and the time resolution is 275 ps for a 1 MeV electron (550 ps at 0.2 MeV).

2.1 ^{100}Mo experiment

In this experiment [4] the total running time was 6140 h. The 1 m^2 foil was divided into two parts. The first part was enriched molybdenum (98.4% ^{100}Mo) with a mass of 172 g and ≈40 μm thick, while the natural molybdenum (9.6% ^{100}Mo) part was 163 g and ≈44 μm thick. Backgrounds from internal radioactivity were negligible.

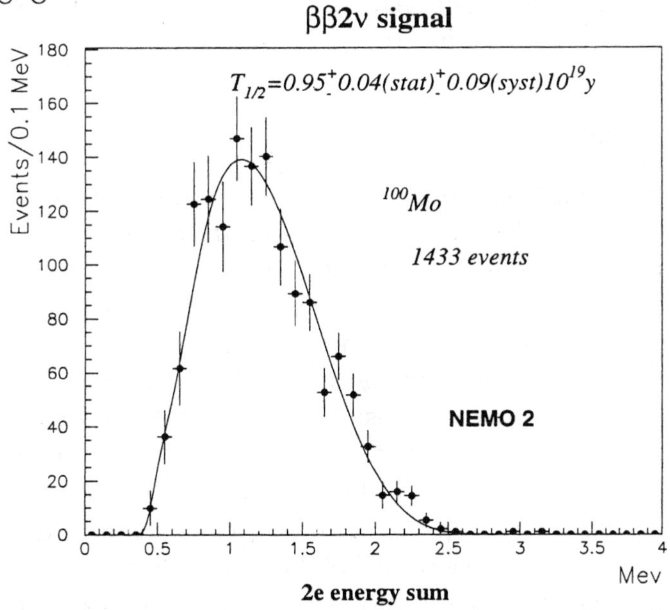

Figure 1: Summed electron energy spectrum of $\beta\beta2\nu$ events in ^{100}Mo.

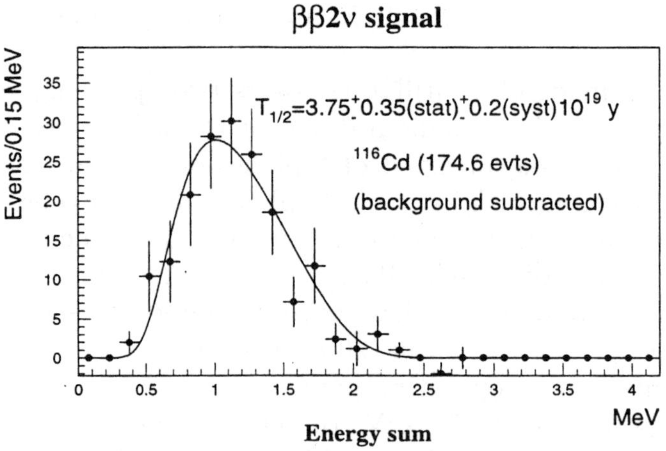

Figure 2: Summed electron energy spectrum of $\beta\beta 2\nu$ events in ^{116}Cd.

Fig. 1 shows the $\beta\beta 2\nu$ spectrum after background subtraction. The solid line is a one parameter fit of the spectrum, leading to the half-life

$$T_{1/2}^{2\nu} = [0.95 \pm 0.04(stat) \pm 0.09(syst)] \cdot 10^{19} y.$$

From the same data, half-life limits on $\beta\beta 0\nu$ and $\beta\beta 0\nu M^0$ decays were obtained at 90% CL: $6.4 \cdot 10^{21}$ y and $0.5 \cdot 10^{21}$ y, respectively.

2.2 ^{116}Cd experiment

An earlier result drawn from 40% of the current data set was already published [5]. Here final results for 6588 h of data taking are presented. The source plane was again divided into two halves, the first one was a 152 g isotopically enriched cadmium foil (93.2% ^{116}Cd) 40 μm thick. The second half was a 143 g foil of natural cadmium (7.58% ^{116}Cd). Radioactive impurities in both foils were measured with HPGe detectors in the Fréjus Underground Laboratory. These measurements accord well with those extracted from NEMO-2 data, and the estimated background from ^{214}Bi, ^{208}Tl, ^{234m}Pa and the neutron flux is only a few 2e events in each foil and can be neglected after subtraction.

In Fig. 2 the $\beta\beta$ energy spectrum in enriched cadmium (174.6 events) is shown after background subtraction (44.4 events). A $\cos\alpha < 0.6$ cut is applied (α is the angle between two electrons). Using the calculated detection efficiency of the $\beta\beta 2\nu$ decay of ^{116}Cd ($\varepsilon = 1.73\%$) one gets the half-life,

$$T_{1/2}^{2\nu} = [3.75 \pm 0.35(stat) \pm 0.21(syst)] \cdot 10^{19} y$$

Half-life limits on $\beta\beta0\nu$ and $\beta\beta0\nu M^0$ decays have been extracted from the data and are $5.0 \cdot 10^{21}$ y and $1.2 \cdot 10^{21}$ y at 90% CL, respectively. The energy windows, number of events, backgrounds, and efficiencies are also given.

2.3 ^{82}Se experiment

The source consists of two nearly symmetric halves. The first half contains 156.6 g of enriched selenium (97.02% is ^{82}Se) and the second part contains 133.7 g of natural selenium in which the ^{82}Se isotopic has an abundance of 8.73%. The sources were produced using a special technique to deposit selenium powder on thin films. The thickness of the foils is $\sim$ 50 mg/cm^2 for enriched and $\sim$ 45 mg/cm^2 for the natural ones.

Radioactive impurities in both foils have again been measured with HPGe detectors in the Fréjus Underground Laboratory before being placed in the NEMO-2 detector. The upper limits on contamination obtained in the enriched selenium for the three isotopes ^{214}Bi, ^{208}Tl and ^{234m}Pa are respectively 4.2, 2.5 and 33 mBq/kg, and in natural selenium these limits are 5, 2 and 16 mBq/kg respectively. Some activity from ^{40}K was found in both samples, specifically 200 mBq/kg in the enriched and 110 mBq/kg in the natural selenium. The NEMO-2 detector was used by itself for purity control of the foils. A so-called "hot" point was discovered in the natural selenium foil, using eγ and 2e channels. This "hot" point ($\sim$ 1.3% of the area) was excluded from the analysis. Using the single electron energy spectra in the range [1.5 - 2.0] MeV, a limit on the difference of contamination in ^{234m}Pa in both foils of 6 mBq/kg is obtained which corresponds to less than two 2e-events.

Presented here are preliminary results after 4028 h of data collection. Fig. 3 shows the energy spectra of 2e-events in enriched and natural selenium foils (respectively 80 and 27 events). A $cos\alpha < 0.6$ cut is applied. Taking into account mass difference of the natural and enriched foils and the contributions from ^{82}Se and ^{40}K in the natural foil an "external" background in the enriched foil of 22 events is calculated. For "internal" backgrounds one can estimate that 7 events are contributed by ^{40}K. Consequently, the $\beta\beta$ signal is 51 events. Using the calculated detector efficiency for the $\beta\beta2\nu$ decay for ^{82}Se ($\varepsilon = 1.7\%$) one gets,

$$T^{2\nu}_{1/2} = [1.2^{+0.3}_{-0.2}(stat) \pm 0.2(syst)] \cdot 10^{20} y.$$

This value is in good agreement with the previous geochemical [6-8] and direct [9] experiment results. Half-life limits on the $\beta\beta0\nu$ and $\beta\beta0\nu M^0$ processes in ^{82}Se were found to be $3.6 \cdot 10^{21}$y and $1.3 \cdot 10^{21}$y at the 90% CL.

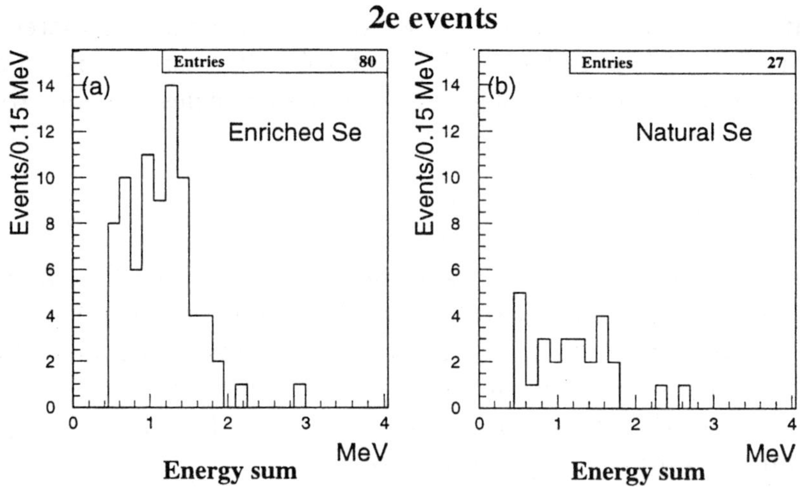

Figure 3: Raw summed electron energy spectra in enriched (a) and in natural (b) selenium.

2.4 ^{96}Zr experiment

The ^{96}Zr double beta decay measurements have been conducted in parallel with the selenium experiment. The source consists of two symmetric halves again and is located in the center of the source plane covering roughly 10% of the available area at the point of greatest geometric acceptance. The mass of the $^{96}ZrO_2$ and $^{nat}ZrO_2$ foils are 20.5 and 18.3 g, respectively. These foils were produced using a technology which binds the zirconium oxide with an organic film. The film thickness with some small variations is 50 mg/cm^2 for enriched and 45 mg/cm^2 for the natural sources. Enrichment of the Zr sample is 57.3% and thus the total mass of ^{96}Zr is 6.8 g.

Radioactive impurities in both foils have been measured with the HPGe detectors in the Fréjus Underground Laboratory before installation in the NEMO-2 detector and then by the NEMO-2 detector itself.

Fig. 4 shows the energy spectra of 2e-events in enriched and natural zirconium, 45 and 8 events after 4028 h. A $cos\alpha < 0.6$ cut is applied. A large part of events in enriched Zr is connected with ^{228}Ac, ^{208}Tl, ^{40}K and others pollutions. Using the informations from HPGe and NEMO-2 measurements, the contributions of these isotopes can be estimated. This work is now in progress. In a "simple" analysis one can use the energy region greater than 1.5 MeV, where 40% of $\beta\beta 2\nu$ events remain but internal and external backgrounds are strongly suppressed. Eight events in enriched and one event in natural Zr are observed. The contribution of the radioactive impurities to the enriched Zr spectrum is estimated to 1 event. If the 6 remaining events in enriched Zr are attributed to $\beta\beta 2\nu$ decay of ^{96}Zr then a very preliminary half-life estimation

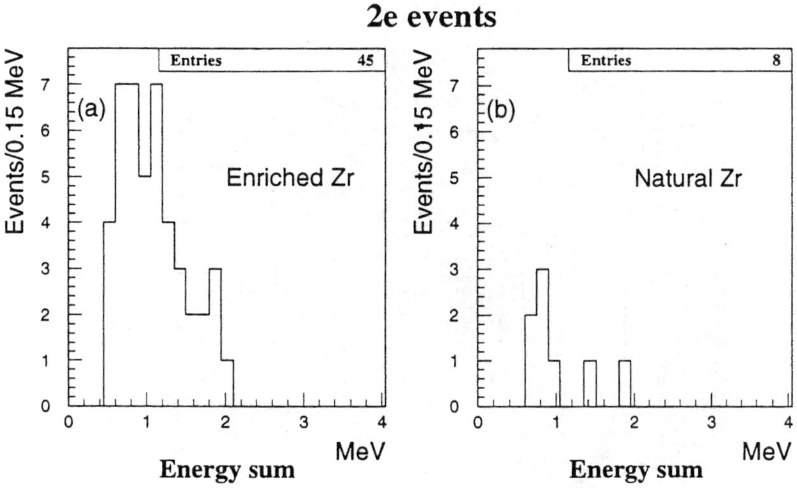

Figure 4: Raw summed electron energy spectra in enriched (a) and in natural (b) zirconium.

is $\sim 3 \cdot 10^{19}$ y. But if these events are due to the background fluctuation then just a conservative limit $T_{1/2}^{2\nu} > 1.9 \cdot 10^{19}$ y (90% CL) can be extracted.

Limits (90% CL) are obtained on $\beta\beta0\nu$ and $\beta\beta0\nu M^0$ decays, $5.5 \cdot 10^{20}$ y and $2.4 \cdot 10^{20}$ y respectively.

The measurements with ^{82}Se and ^{96}Zr will be continued through 1996.

3 NEMO-3 detector

In contrast to the ^{76}Ge experiments, the NEMO experiments[1] use tracking detectors which are not only able to measure the full energy released, but other parameters of the process such as the single electron energy, the angle between electrons, the coordinates of events, etc. The optimal operating parameters of the detector were investigated with the prototype NEMO-2[3,4,5]. Currently the NEMO-3 detector is under construction and will be able to accommodate up to 10 kg of various double beta decay candidates (^{100}Mo, ^{116}Cd, ^{82}Se, ^{130}Te, ^{96}Zr, ^{150}Nd, etc). The lifetime limits after 5 years of measurement will be at the level 10^{25}y for $\beta\beta0\nu$ decay with $< m_\nu >\sim (0.1-0.3)$ eV, and $\sim 10^{23}$ y for $\beta\beta0\nu M^0$ decay ($< g_{ee} >\sim 10^{-5}$), and finally $\sim 10^{22}$ y for $\beta\beta2\nu$ decay.

A general view of the detector's cylindrically symmetric geometry is shown in Fig. 5. The detector consists of a tracking volume filled with helium gas, a thin source foil divides the tracking volume vertically into two concentric cylinders with a calorimeter at the inner and outer walls. The tracking system consists of 6000 Geiger cells 3 m long which are parallel to the detector's vertical axis. Energy and time-of-flight measurements are performed by the

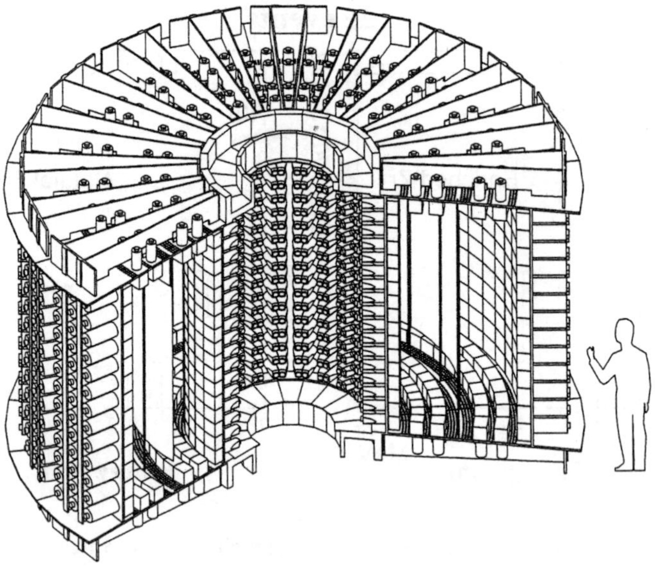

Figure 5: Schematic view of the NEMO3 detector.

plastic scintillators covering the two concentric surfaces discussed above and their associated end caps. The total number of low radioactive photomultipliers will approach 2000.A magnetic field ($\sim$ 30 Gauss) will be used to reject backgrounds connected with pair creation and incoming electrons. A shield consisting of 20 cm of iron and 5 cm of lead will again protect the detector from external radioactivity.

The program with the NEMO-3 detector will commence data collection in 1998.

References

1. NEMO Collaboration, *preprint LAL* 94-29 (1994).
2. D. Dassié *et al.*, *Nucl. Instrum. Methods* A**309**, 465 (1991).
3. R. Arnold *et al.*, *Nucl. Instrum. Methods* A**354**, 338 (1995).
4. D. Dassié *et al.*, *Phys. Rev.* D**51**, 2090 (1995).
5. R. Arnold *et al.*, JETP Lett.**61**, 170 (1995).
6. T. Kirsten in *Nuclear Beta Decay and Neutrinos*, ed. T.Kotani, H.Ejiri and E.Takasugi (World Scientific, Singapore, 1986), p.81.
7. W.J. Lin *et al.*, *Nucl. Phys.* A**481**, 477 (1988).
8. O.K. Manuel, J. Phys. G**17**, 221 (1991).
9. S.R. Elliott *et al.*, *Phys. Rev.* C**46**, 1535 (1992).

RAPID COMMUNICATIONS

PHYSICAL REVIEW C
NUCLEAR PHYSICS

THIRD SERIES, VOLUME 44, NUMBER 3 SEPTEMBER 1991

RAPID COMMUNICATIONS

The Rapid Communications section is intended for the accelerated publication of important new results. Manuscripts submitted to this section are given priority in handling in the editorial office and in production. A Rapid Communication in **Physical Review C** *may be no longer than five printed pages and must be accompanied by an abstract. Page proofs are sent to authors.*

New approach to the detection of neutrinoless double-beta decay

M. K. Moe
Department of Physics, University of California, Irvine, California 92717
(Received 2 April 1991)

A distinctive feature of the double-beta-decay signature that has been neglected in direct counting experiments is the appearance of the daughter atom. The newly created nucleus, usually being stable, is not easily detected. The atomic physics of the daughter, however, may be considerably more accommodating, especially in the case of ionized ^{136}Ba arising from the double-beta decay of ^{136}Xe. The barium ion isolated in the xenon matrix may be detectable by its laser fluorescence. Coincident detection of the ion and the beta particles could well render background nonexistent.

Neutrinoless double-beta decay ($\beta\beta_{0\nu}$) is possible if the distinction between the neutrino and antineutrino is simply a manifestation of two different helicity states of the same particle (the Majorana neutrino), and if a very small mixing of the two states is driven by a nonzero neutrino mass [1]. Massive Majorana neutrinos and a number of other possible consequences of $\beta\beta_{0\nu}$ are at variance with the standard model, and the question of their existence has motivated numerous experimental searches [2] for the $\beta\beta_{0\nu}$ phenomenon.

Several different isotopes have been studied, the best published $\beta\beta_{0\nu}$ result being for ^{76}Ge by the University of California, Santa Barbara-Lawrence Berkeley Laboratory (UCSB-LBL) group [3] who report a lower half-life limit of 2.4×10^{24} yr, and a corresponding upper limit on the effective Majorana mass for the electron neutrino of ~ 1 eV. Yet one would like to probe still smaller mass regions. Grand unified theories favoring a finite neutrino mass give a very wide range for its predicted magnitude. Indeed, in view of proposed nonadiabatic Mikheyev-Smirnov-Wolfenstein (MSW) solutions to the solar neutrino problem [4,5], and various cosmological arguments, it may be that neutrino masses are far too small to ever result in observable neutrinoless double-beta decay. At present the question remains open, and the $\beta\beta_{0\nu}$ searches continue. A reduction of 2 orders of magnitude in detectable mass is not beyond imagination.

The dependence of the minimum detectable effective neutrino mass $\langle m_\nu \rangle_{\min}$ on the source mass and run time for ^{76}Ge $\beta\beta_{0\nu}$ experiments [2(b)] is shown for various combinations of background and isotopic enrichment in Fig. 1. The advantage of *enriched* germanium is clear. Two new experiments [6,7], the larger being ~ 10 kg, are being assembled with germanium enriched to 85% isotope 76, compared to the 7.8% natural abundance used by UCSB-LBL. These experiments should be able to reach a few tenths of an electron volt. With enrichment, energy resolution, and run time then approaching their practical limits, the only remaining parameters available for improved sensitivity in ^{76}Ge are background b (keV kg yr)$^{-1}$, and source mass M (kg). The dependence of $\langle m_\nu \rangle_{\min}$ on these parameters is weak, being proportional to $(b/M)^{1/4}$. An order-of-magnitude improvement in sensitivity to neutrino mass would require a 10000-fold decrease in b/M. To accomplish this decrease entirely with larger sources is economically out of the question, so very substantial improvements in background are necessary to make meaningful strides toward smaller neutrino masses.

There are other $\beta\beta$ isotopes with more favorable matrix elements and phase space [8–10], and corresponding lines as much as an order of magnitude lower in the figure. As a practical matter, affordable, low-background, high-efficiency, large-scale experiments have not been easy to design for these favored isotopes [11].

Among the candidates for double-beta decay, ^{136}Xe may have the greatest potential for very-large-mass exper-

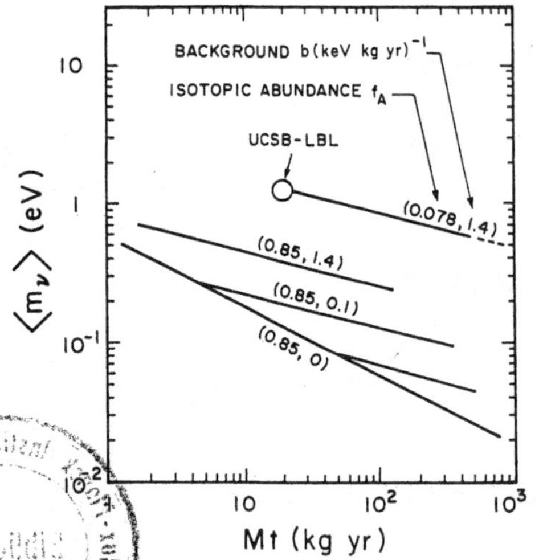

FIG. 1. The minimum detectable neutrino mass vs detector mass and run time for ^{76}Ge experiments with various levels of background and isotopic enrichment. The latest UCSB-LBL result (circle) is shown for comparison. Probing the smallest neutrino masses requires a very large detector mass and very low (preferably zero) background.

iments [12]. The isotope is relatively inexpensive to enrich from its natural abundance of 8.9% to the order of 60% by gas centrifugation; it is a noble gas that can serve simultaneously as source and detector; its $\beta\beta_{0\nu}$ matrix-element, phase-space product is similar to that of ^{76}Ge; and it has a higher transition energy (2479 keV versus 2041 keV for ^{76}Ge). Several experiments have been performed with ^{136}Xe in time projection chambers (TPC's), proportional counters, ionization chambers, and scintillation detectors [13-16]. Some of these experiments have made good progress against background, but all have had to contend with it at some unwanted level.

A feature of double-beta decay previously unexploited in direct counting experiments [17] is the sudden production of two additional protons in the nucleus, which leaves the daughter atom shy two electrons. Decay of ^{136}Xe makes ^{136}Ba^{2+}. As in single-beta decay, often one or more atomic electrons will also be ejected [18]. The daughter atom, therefore, will be born in a double or higher state of ionization. If the xenon detector is normally kept free of ions by a drift field, the appearance of a barium ion together with a 2.5 MeV energy pulse could be sufficiently unique that demanding their coincident detection would eliminate background completely.

The detector proposed is a liquid xenon TPC to be operated in a deep-underground laboratory. The range of ~2 MeV beta particles in the liquid is a few millimeters, so the position of the barium ion would be localized to that order by detection of the beta-particle ionization and a scintillation trigger. The mobility of positive ions in liquid xenon is relatively low, so the barium ion would not move far from its origin in the time required to target it for laser excitation.

Single-isolated Ba$^+$ ions have been successfully detected by their laser fluorescence in the classic experiment of Neuhauser and co-workers [19]. Irradiation of this favorite ion in the blue-green at 493 nm results in red fluorescence at 650 nm. Following xenon decay, the third- and higher-ionization states of barium have sufficient potential to pull electrons from neighboring xenon atoms. Once the ion becomes Ba^{2+} its subsequent behavior is less obvious. The ionization potentials listed in Table I suggest that xenon will not surrender further electrons, and Ba^{2+} will remain stable. In this case the ion's emission and absorption wavelengths would be in the vacuum ultraviolet which, together with the lack of a metastable state, would make detection by laser fluorescence difficult. However, in liquid xenon the gap to the conduction band is slightly below the second-ionization potential for barium [20], and the Ba^{2+} ion is likely to take on an electron to become Ba$^+$. Xenon, being highly polarizable, would also be expected to attach to the ion to ultimately form (BaXe)$^+$.

The spectroscopy of matrix-isolated (BaXe)$^+$ does not appear in the literature. The behavior of this ion is crucial to the proposed detection scheme, and a small experiment is being set up to investigate its spectroscopy in liquid xenon [21].

If a strong fluorescence can be identified, there remain questions of background to be considered. For example, one might ask whether in a large xenon experiment a feeble $\beta\beta_{0\nu}$ spike at the 2479 keV Q value might blend into the high-energy tail of the much stronger $\beta\beta_{2\nu}$ spectrum. The high end of the theoretical $\beta\beta_{2\nu}$ electron sum spectrum for ^{136}Xe is shown in Fig. 2(a) [22]. The spillage of a resolution-smeared version of this spectrum into a $\beta\beta_{0\nu}$ window is shown as a function of full width at half maximum (FWHM) resolution in Fig. 2(b). (The Majoron [1] is assumed to be nonexistent.)

A large liquid xenon TPC should be able to achieve 4% FWHM at 2479 keV without difficulty. Such resolution has already been seen routinely near 1000 keV, and is expected to improve with the square root of the energy [23]. A 4% window centered at Q allows only 2.3×10^{-7} of the $\beta\beta_{2\nu}$ events to spill in. The $\beta\beta_{2\nu}$ spectrum thus limits the $\beta\beta_{0\nu}$ half-life to the $\beta\beta_{2\nu}$ half-life divided by 2.3×10^{-7} (and then multiplied by 0.76 to account for the fraction of the $\beta\beta_{0\nu}$ signal that falls within the FWHM). To take full advantage of this limit one would need enough xenon and run time to produce at least one $\beta\beta_{2\nu}$ count in the $\beta\beta_{0\nu}$ window. At the theoretical ^{136}Xe $\beta\beta_{2\nu}$ half-life of 4.64×10^{21} yr [9], the required source-mass run-time product Mt is ~5000 kg yr. For example, one might run for 5 yr with 1000 kg of ^{136}Xe (a 78 cm fiducial cube of liquid, isotopically enriched to 60%). At ~$16000/kg [24] for enriched ^{136}Xe it is clear that cost is a more seri-

TABLE I. First and second ionization potentials[a] of Ba and Xe (volts).

	I	II
Ba	5.21	10.00
Xe	12.77	21.2

[a]Reference [30].

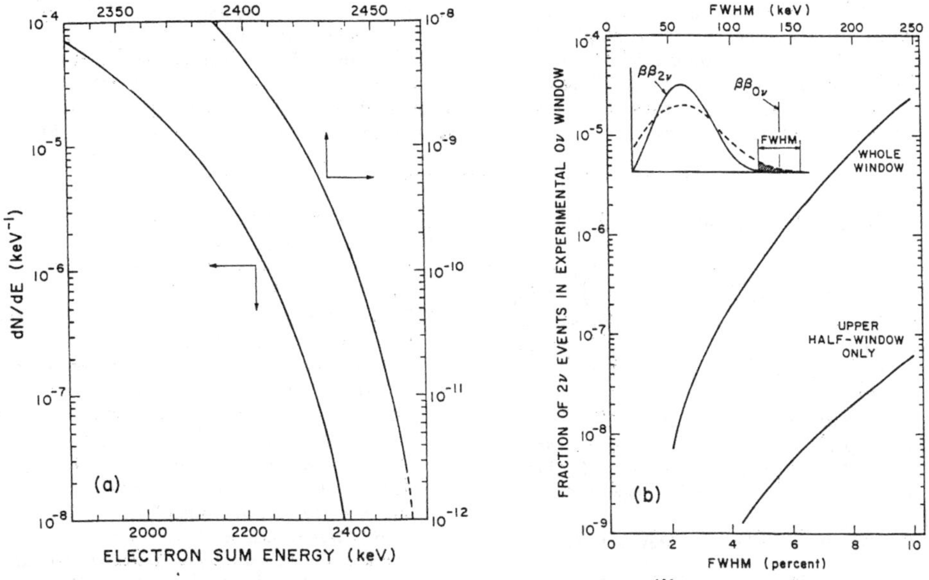

FIG. 2. (a) The high-energy tail of the theoretical $\beta\beta_{2\nu}$ electron sum spectrum for ^{136}Xe, normalized to unity for the whole spectrum from energy = 0 to the Q value at 2479 keV. Calculation courtesy T. Kotani. (b) The fraction of a resolution-broadened $\beta\beta_{2\nu}$ spectrum falling in an experimental FWHM window centered on the expected position of the $\beta\beta_{0\nu}$ spike. For large xenon $\beta\beta_{0\nu}$ experiments, "background" from $\beta\beta_{2\nu}$ must be considered.

ous limitation than interference from $\beta\beta_{2\nu}$.

There are, to be sure, ways other than double-beta decay to create a blob of ionization accompanied by a barium ion. One is Xe(α,n)Ba reactions. For the stable isotopes of xenon the thresholds are between 5.3 and 10.6 MeV—often below the energies of naturally occurring alpha particles of up to 10.5 MeV in the uranium and thorium series. However, the cross sections are strongly suppressed by a Coulomb barrier of 17.5 MeV. There are not enough energetic alpha particles in detector grade xenon to be a problem. Higher-energy alpha particles from spallation by cosmic-ray muons would be vetoed by the muon pulse.

Another source of barium ions accompanied by ionization electrons is the single-beta decay of radioisotopes of cesium. The common fission product, 30 yr ^{137}Cs [which can also arise from ^{136}Xe$(n,\gamma)^{137}$Xe followed by beta decay of ^{137}Xe], has a Q value of 1.2 MeV—too low to be of concern. Cs isotopes with Q values greater than 2.1 MeV are all fission products with mass numbers 136 and above [25]. Of these, three have half-lives greater than 65 sec: ^{136}Cs, ^{138}Cs, and ^{139}Cs. Isotopes with shorter half-lives will not have time to migrate into the fiducial volume unless they are born from fission within the xenon itself, in which case the fission event would veto the beta decay.

Scaling from existing experiments [26] indicates that uranium in the xenon could easily be held to $< 10^{-10}$ g/g. In a laboratory such as the Gran Sasso where the thermal-neutron flux [27] is $\sim 10^{-6}$/cm^2 sec, fission by thermal neutrons on uranium in the xenon is completely negligible. The < 25 spontaneous fissions per year from 10^{-10} g/g of ^{238}U in 1000 kg of xenon would yield < 2 atoms each of ^{138}Cs and ^{139}Cs, and a negligible amount of ^{136}Cs [28]. The resulting number of beta particles in a 4% window at 2479 keV is well below one per year.

The heaviest stable xenon isotope being 136 means there is no path to ^{138}Cs, or ^{139}Cs, via ^{A}Xe$(n,\gamma)^{A+1}$Xe followed by beta decay. Similarly, ^{136}Cs is inaccessible through (n,γ) because there is no stable ^{135}Xe. Although cesium from (α,p), (α,pn), etc., on xenon is strongly suppressed by the Coulomb barrier, spallation alpha particles often do have enough energy to drive these reactions, and the half-lives of the cesium isotopes are too long for a veto by the initiating muon. However, with spallation cross sections for deep-underground muons on the order of 10^{-29} cm^2 per nucleon [29], and a muon flux of ~ 1 (m^2 h)$^{-1}$ at 3600 m water equivalent (e.g., Gran Sasso), high-energy alpha particles are too rare to make significant cesium. Similarly, spallation protons are too few to make troublesome amounts of cesium from (p,n), (p,γ), etc., on xenon, and stopping muons are insufficient to cause significant (μ^+,γ).

Although cesium does not appear to be a background threat, the isotope ^{137}Cs could be purposely introduced to study the efficiency for barium-ion detection. The 2.6 min metastable state in ^{137}Ba would give one a chance to detect the ion and see its presence confirmed by the 0.66 MeV gamma ray or corresponding conversion electron.

Detection of the barium ion in coincidence with the 2.5 MeV energy pulse from the double-beta decay of ^{136}Xe may be a potentially powerful method of eliminating background in the search for the $\beta\beta_{0\nu}$ mode. The utility of the technique depends on the presently unknown spectroscopy of the barium ion in liquid xenon. A test experiment is being assembled. If successful, the method will allow one to relax the requirement for ultrahigh-energy resolution, and concentrate on building a detector of very large mass. A background-free 1000 kg of ^{136}Xe could probe to an effective Majorana mass for the electron neutrino of ~0.01 eV in five years of running.

I am grateful for discussions with V. A. Apkarian, E. Aprile, F. Boehm, P. J. Doe, S. R. Elliott, B. E. Lehmann, W. C. Stwalley, M. A. Vient, K. Winter, R. Zare, and, indirectly, P. E. Toschek, whose comments were relayed to me by K. Winter. This work is supported by U.S. Department of Energy Contract No. DE AT03-76SF00010.

[1] For a recent review, see T. Tomoda, Rep. Prog. Phys. **54**, 53 (1991).

[2] The experiments are reviewed by (a) F. T. Avignone and R. L. Brodzinski, Prog. Part. Nucl. Phys. **21**, 99 (1988); (b) M. K. Moe, in *Proceedings of the 14th International Conference on Neutrino Physics and Astrophysics, CERN, June 1990*, edited by J. Panman and K. Winter [Nucl. Phys. B (Proc. Suppl.) **19**, 158 (1991)].

[3] D. O. Caldwell *et al.*, in *Proceedings of the 12th International Workshop on Weak Interactions and Neutrinos, Ginosar, Sea of Galilee, Israel, 1990*, edited by P. Singer and G. Eilam [Nucl. Phys. B (Proc. Suppl.) **13**, 547 (1990)].

[4] J. N. Bahcall and H. A. Bethe, Phys. Rev. Lett. **65**, 2233 (1990).

[5] S. P. Rosen and J. M. Gelb, Phys. Rev. D **39**, 3190 (1989).

[6] F. T. Avignone *et al.*, in *Weak and Electromagnetic Interactions in Nuclei, Proceedings of International Symposium (WEIN-89), Montréal, 1989*, edited by Pierre Depommier (Éditions Frontières, Gif-sur-Yvette Cedex, France, 1989).

[7] H. V. Klapdor *et al.*, in *Weak and Electromagnetic Interactions in Nuclei, Proceedings of the International Symposium (WEIN-89), Montréal, 1989* (Ref. [6]).

[8] J. Engel, P. Vogel, and M. R. Zirnbauer, Nucl. Phys. A **478**, 459c (1988).

[9] A. Staudt, K. Muto, and H. V. Klapdor-Kleingrothaus, Europhys. Lett. **13**, 31 (1990).

[10] J. Suhonen, S. B. Khadkikar, and Amand Faessler, Department of Physics, University of Jyväskylä Report No. 2, 1991 (unpublished).

[11] One promising approach is the development of cryogenic detectors. Although the technique is not ready for large-scale experiments, it may eventually surpass enriched germanium in $\beta\beta_{0\nu}$ sensitivity. See, for example, E. Fiorini, in *Superconducting and Low-Temperature Particle Detectors*, edited by G. Waysand and G. Chardin (Elsevier, North-Holland, Amsterdam, 1989), pp. 1–26.

[12] T. A. Girard, E. Aprile, and R. C. Fernholz, Phys. Lett. B (to be published).

[13] M. S. Ainutdinov *et al.*, Inst. Expt. Tech. **32**, 1299 (1989).

[14] A. S. Barabash, V. M. Novikov, and B. M. Ovchinivov, Nucl. Instrum. Methods Phys. Res. Sect. A **300**, 77 (1991).

[15] E. Bellotti *et al.*, Phys. Lett. B **221**, 209 (1989).

[16] F. Boehm and M. Z. Iqbal, in *Festival-Festschrift for Val Telegdi*, edited by K. Winter (North-Holland, Amsterdam, 1988), p. 25.

[17] The detection of ^{136}Ba atoms, accumulated over time on an electrode immersed in xenon gas, has been investigated with laser resonance ionization in a radiochemical (*indirect*) $\beta\beta$ experiment sensitive to the sum of the $\beta\beta_{0\nu}$ and $\beta\beta_{2\nu}$ rates. It has been brought to my attention that the *direct* counting approach I describe in this paper has also been discussed independently by Leon Mitchell. The radiochemical work has been published by L. W. Mitchell and N. Winograd, Bull. Am. Phys. Soc. **31**, 1219 (1986); D. M. Hrubowchak *et al.*, in *Proceedings of the Fourth International Symposium on Resonance Ionization Spectroscopy and its Applications, National Bureau of Standards, 1988*, edited by T. B. Lucatorto and J. E. Parks (Institute of Physics Conference Series No. 94, Philadelphia, 1989).

[18] A. H. Snell and F. Pleasonton, Phys. Rev. **107**, 740 (1957).

[19] W. Neuhauser, M. Hohenstatt, P. E. Toschek, and H. Dehmelt, Phys. Rev. A **22**, 1137 (1980).

[20] V. A. Apkarian (private communication).

[21] Single fluorescent molecules in liquid have been detected by E. B. Shera *et al.*, Chem. Phys. Lett. **174**, 553 (1990).

[22] T. Kotani (private communication).

[23] E. Aprile and M. Suzuki, IEEE Trans. Nucl. Sci. **36**, 311 (1989).

[24] H. Rakhorst (private communication).

[25] *Table of Isotopes*, 7th ed., edited by C. M. Lederer and V. S. Shirley (Wiley, New York, 1978).

[26] S. R. Elliott, A. A. Hahn, and M. K. Moe, Nucl. Instrum. Methods Phys. Res. Sect. A **273**, 226 (1988).

[27] A. Rindi, F. Celani, M. Lindozzi, and S. Miozzi, Nucl. Instrum. Methods Phys. Res. Sect. A **272**, 871 (1988).

[28] E. K. Hyde, *Nuclear Properties of Heavy Elements, III Fission Phenomena* (Prentice-Hall, Englewood Cliffs, NJ, 1964).

[29] B. Peters, in *Handbook of Physics*, edited by E. U. Condon and H. Odishaw (McGraw-Hill, New York, 1958), Chap. 9, p. 233.

[30] *Handbook of Chemistry and Physics*, 51st ed., edited by Robert C. Weast (Chemical Rubber Co., Cleveland, 1971).

2.5.3 Scintillation Detectors

ELSEVIER

Nuclear Instruments and Methods in Physics Research A 369 (1996) 164–168

NUCLEAR INSTRUMENTS & METHODS IN PHYSICS RESEARCH Section A

Large volume $CdWO_4$ crystal scintillators

S.Ph. Burachas[b], F.A. Danevich[a], A.Sh. Georgadze[a], H.V. Klapdor-Kleingrothaus[c], V.V. Kobychev[a], B.N. Kropivyansky[a], V.N. Kuts[a], A. Muller[c], V.V. Muzalevsky[a], A.S. Nikolaiko[a], O.A. Ponkratenko[a], V.D. Ryzhikov[b], A.S. Sai[d], I.M. Solsky[d], V.I. Tretyak[a], Yu.G. Zdesenko[a,*]

[a] *Institute for Nuclear Research of the Ukrainian National Academy of Science, 252028 Kiev, Ukraine*
[b] *Institute for Single Crystals of the Ukrainian National Academy of Science, 310001 Kharkov, Ukraine*
[c] *Max-Planck-Institut fur Kernphysik, P.O. Box 103980, D-69029 Heidelberg, Germany*
[d] *Institute for Materials, 290021 Lvov, Ukraine*

Received 28 February 1995; revised form received 20 June 1995

Abstract

A low background scintillation spectrometer based on $CdWO_4$ crystals of large volume (100–200 cm^3) with improved energy resolution has been developed as a result of advances in crystal growth, optimization of light collection and design of a special electronic unit for amplification and shaping of the $CdWO_4$ signals. The energy resolution of crystals with volume 80–150 cm^3 is 10–12% for 662 keV γ-rays. A value of 7–8% at 662 keV, which is comparable to NaI(Tl), has been obtained with small samples (10–20 cm^3). The high purity of the $CdWO_4$ crystals (^{228}Th contamination is less than 8 $\mu Bq/kg$, that of ^{226}Ra less than 13 $\mu Bq/kg$) allows construction of unique spectrometers based on $CdWO_4$ crystals of large volume for environmental control, well logging and measurements of radioactive contamination in various samples with a sensitivity of 10^{-12} to 2×10^{-13} Ci/kg.

1. Introduction

During the last decade high-Z crystal scintillators have been developed stimulated by the practical needs of CT and experimental particle physics. They include bismuth germanate (BGO), $Bi_4Ge_3O_{12}$, gadolinium orthosilicate (GSO), Gd_2SiO_5, cadmium tungstate, $CdWO_4$, lead tungstate, $PbWO_4$, and others, whose distinguished property is a high detection efficiency for X- and γ-rays, as well as excellent operational characteristics (chemical resistance, non-hygroscopicity, etc.). These scintillators can be used for γ-ray registration in well logging applications and in low energy nuclear physics also.

The results of the elaboration of large volume $CdWO_4$ crystals and improvements of their spectrometric characteristics are summarized in this paper. This work has been performed since 1986 at the Institute for Nuclear Researches (Kiev) in cooperation with the Institute for Single Crystals (Kharkov), the Max Planck Institute (Heidelberg) and the Institute for Materials (Lvov).

* Corresponding author. Fax +7 380 44 265 2210, +7 380 44 265 4463.

The luminescence of $CdWO_4$ crystals was first reported in 1948 [1], and their application as scintillators in 1950 [2]. Since that time the scintillation, optical and mechanical properties of these crystals have been widely investigated [3–9]. The main characteristics of $CdWO_4$ are presented in Table 1. This material is non-hygroscopic and

Table 1
Main characteristics of $CdWO_4$ crystal scintillators

Characteristic [units]	Value	Ref.
Effective atomic number Z_{eff}	64	
Density [g/cm^3]	7.9	[2–8]
Melting point [K]	1598	[8]
Cleavage plane	⟨010⟩	
Hardness [Mohs]	4–4.5	
Hygroscopicity	No	
Wavelength of emission maximum[a] [nm]	470/540	[4,8]
Index of refraction	2.3	
Average decay time [μs]	12	[8]
Afterglow (after 3 ms) [%]	0.1	[5]
Light yield [% of NaI(Tl)]	25–40	[7,8]

[a] Two emission components were first observed in Ref. [4] with decay times of 5 and 20 μs and intensity maxima at 470 and 540 nm.

SSDI 0168-9002(95)00675-3

chemically inert. The crystal has to be machined with care because of its cleavability. The important advantages of $CdWO_4$ crystal scintillators, apart from the high detection efficiency for X- and γ-rays mentioned before, are the relatively large light yield (about 40% of NaI(Tl)), radiation resistance and very slight temperature dependence of the scintillation output [3–9].

The main disadvantage of the $CdWO_4$ scintillator is the quite long decay time (12–15 ms), which restricts the value of its maximum counting rate. However, this is not really important in some applications such as, for example, the environmental measurement of radioactive contamination. Moreover, these crystals can be utilized successfully for the study of rare and forbidden processes in nuclear and particle physics. Indeed, the search for the double beta-decay of ^{116}Cd was carried out with the help of $CdWO_4$ scintillators enriched with ^{116}Cd to 83% [10]. In spite of the small volumes of the crystals used (10–20 cm^3), one of the best results on the neutrinoless double beta-decay was achieved with such a crystal [10]. To further advance this research and to improve the sensitivity of low level measurements of radioactivities it is necessary to use crystals of larger volume with better energy resolution. In principle, the high light yield of $CdWO_4$ crystals allows us to obtain an energy resolution that is comparable to that of NaI(Tl) but, in fact, an energy resolution of only 8% (FWHM at an energy of 662 keV) has been reached with a very small crystal (2 cm^3) [7]. For a $CdWO_4$ crystal with a volume of 12 cm^3 the best value was 9.1% (at 662 keV) and it was only 19% for a 50 cm^3 crystal [8]. Such extreme distortion of the energy resolution is a result of the self-absorption of the scintillation light [8].

Summarizing the previous data [2–8] and the results of our tests [9] one can propose the following reasons for the deterioration of the energy resolution of large $CdWO_4$ crystals.

1. Self-absorption and defects of the crystal structure, which form inner scattering centers (optical inhomogeneities). These macrodefects can be caused by deviations from the optimum charge composition, by a significant amount of uncontrollable impurities in the mixture and also by temperature instabilities during crystal growth.

2. The shift between the maximum of the $CdWO_4$ scintillation emission and the spectral sensitivity maximum of standard photomultipliers.

3. Difficulties of light collection from $CdWO_4$ crystals because of the large index of refraction ($n = 2.3$).

4. Problems of output signal registration with standard electronics because of the long duration of the scintillation light pulse.

To overcome these difficulties, research and development was carried out on all mentioned points, which allowed, finally, improvement of the spectrometric characteristics of large volume $CdWO_4$ crystal scintillators.

2. Crystal growth, light collection and electronics

$CdWO_4$ single crystals were grown from oriented seeds by the Czochralski technique from a stoichiometric mixture of high purity CdO and WO_3 oxides. The crystal diameter was controlled during growth by the weight method. To avoid the appearance of micro-pores inside the crystal and, as a consequence, its optical inhomogeneity, the required thermal conditions (radial and axial temperature gradients and their stability) were provided by a specially constructed thermal unit. Heat screens and doubled corundum heat insulation were used. After growth the crystals were cooled slowly and annealing was performed at a constant temperature over about 80 h. The obtained single crystals were sufficiently transparent even for samples of 150 mm length and had a light yellow color.

The scintillation emission intensity of $CdWO_4$ has a maximum at 490 nm, which does not correspond to a maximum of the typical spectral sensitivity of photomultipliers (PMT). To find an appropriate photomultiplier a test of PMT types and individual tubes was carried out. The best result was obtained with Philips XP2412 photomultipliers, but some tubes of the FEY-139 type showed similar characteristics.

It is known that the best energy resolution can be obtained for samples of the correct geometric form (a cylinder is desirable) with a smooth side surface. The crystal surfaces have to be sandpapered, but not polished. The choice of reflecting material and wrapping is also important. For the crystals whose height was less than their diameter we obtained the best energy resolution using a few layers of 5 μm thick Teflon film. For long crystals, wrapping of the side surface with wrappings of different reflecting capability was used. The upper part of the crystals was covered with layers of Teflon film and an aluminized Mylar film to obtain the best reflection. Then, near the PMT photocathode, materials with reduced reflecting capability were used. This approach allows improvement of the uniformity of the light output of the crystal by means of a ≈10% reduction of the total light yield. The resulting improvement of the energy resolution by a factor of 1.2 was achieved due to a non-uniform wrapping for crystals with height greater than 50 mm. The light yield distribution along the axis of the crystals was measured with a collimated beam of γ-quanta from a ^{207}Bi source.

The large decay time is a serious drawback for the practical application of $CdWO_4$ scintillators. The decay constant, with a mean value of 12–15 μs, leads to a required full light collection time of about 30 μs, whereas standard spectrometric electronics imply pulses of a few μs. To solve this problem a special electronic unit was designed and realized as a double-width CAMAC module. This unit contains a leading edge trigger, fast linear amplifier, integrator and output gate. The integrator is opened for 30 μs after the moment when the input signal exceeds the discrimination level. Then, the output gate forms a short output signal with amplitude proportional to the collected charge. The combination of a fast amplifier and an integrator minimizes pile-up effects, providing complete collection of the emitted light during 30 μs. In tests with the ^{207}Bi γ-source no distortion of the energy resolution and no peak shift were found until a counting rate of 3000 counts/s. The electronic unit provides an energy threshold of about 30 keV and a dynamic range of 1:80 for γ-rays detected by $CdWO_4$ scintillators. A built-in pulse shape discrimination circuit allows rejection of the noise pulses and additionally reduces the energy threshold of the spectrometer.

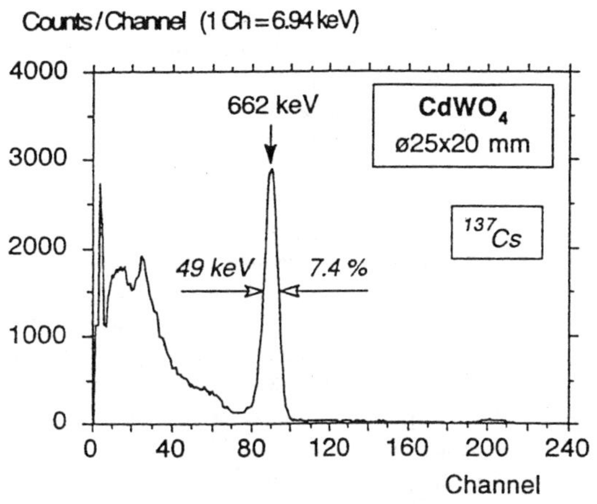

Fig. 1. Gamma-ray spectrum of ^{137}Cs on a "standard" $CdWO_4$ crystal of ∅25 × 20 mm.

3. Energy resolution and background measurements of $CdWO_4$ crystals

Results of energy resolution measurements with different large volume $CdWO_4$ crystals are presented in Table 2. Also shown is the energy resolution of a "standard" $CdWO_4$ crystal of small volume ($V = 10$ cm^3), which is the best published value – 7.5% (FWHM) at an energy of 662 keV (Fig. 1). The calibration spectrum shown in Fig. 2 demonstrates good energy resolution and peak-to-Compton ratio for a crystal with mass 1.1 kg, which exceed considerably the characteristics published previously.

Thus, as a result of the fulfilled R&D on the growth and annealing of high quality crystals, selection of the proper PMT, choice of the complex wrapping and use of special electronics, the energy resolution of large volume $CdWO_4$ crystals has improved to a level that can be compared to the characteristics of NaI(Tl) scintillators.

The intrinsic radioactive impurities of crystal scintillators ultimately determine the sensitivity of the low background installations using these scintillators. Therefore, the background of the developed $CdWO_4$ crystals was measured to test their feasibility for a planned large-scale experiment on the double beta-deçay of ^{116}Cd [10] and also for ultra-low radioactivity detection. The measurements were carried out at the Solotvina Underground

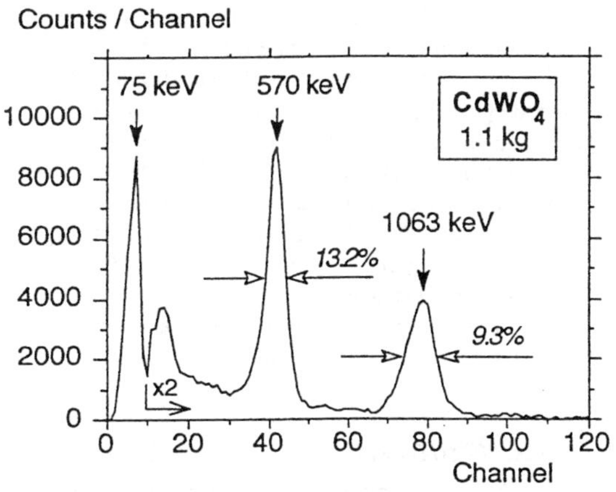

Fig. 2. Calibration γ-ray spectrum of ^{207}Bi measured with $CdWO_4$ scintillators of ∅54 × 65 mm.

Table 2
Energy resolution of $CdWO_4$ scintillators

$CdWO_4$ crystal		Energy resolution [%]		
Dimensions (volume)	Weight [kg]	570 keV	662 keV	1063 keV
∅40 × 30 mm (37.5 cm^3)	0.3	9.7	8.5	6.7
∅50 × 37 mm (73 cm^3)	0.6	11.1	9.8	7.6
∅54 × 65 mm (149 cm^3)	1.1	13.2	11.4	9.3
∅54 × 95 mm (217 cm^3)	1.7	14.9	13.8	11.1
∅25 × 20 mm (10 cm^3)	0.08	8.8	7.5	6.4

Laboratory built in a salt mine at a depth of more than 1000 mwe [11] where the cosmic muon flux is suppressed by a factor of 10^4. Due to the low radioactive contamination of the salt (NaCl), the natural γ background is 30–70 times lower than in chambers built of common materials [11].

In the low background installation a $CdWO_4$ crystal of volume 149 cm^3 was viewed by the PMT through a quartz light-guide 50 cm long. A passive shielding of OFHC copper (5 cm), mercury (7 cm) and lead (15 cm) surrounds the $CdWO_4$ scintillator to reduce the external background. The data acquisition system consists of a microcomputer, magnetic tape recorder and CAMAC crate with electronic modules. The energy calibration was performed with ^{207}Bi.

The background spectrum of $CdWO_4$ (149 cm^3) measured in this installation for 107 h is shown in Fig. 3. There are no peaks and only one peculiarity exists – the sharp increase below an energy of 320 keV which is the spectrum of the fourth-forbidden β-decay of ^{113}Cd (natural abundance 12%) with a half-life of $T_{1/2} = 9.3 \times 10^{15}$ yr and endpoint energy 319 keV [12]. The background rate in the energy region 2–3 MeV is ≈ 5 counts/yr/keV/kg, which is close to that of ultra-low background HP Ge detectors [13]. It is known that the background of the best scintillation setups, based on NaI(Tl) and others, is larger than for semiconductor detectors by one to two orders of magnitude [14]. The very low value of the background rate obtained in our tests is evidence of the high intrinsic purity of the grown crystals. Very small limits of radioactive contamination were calculated from the background spectra: ^{228}Th, 8 μBq/kg; ^{226}Ra, 13 μBq/kg; ^{40}K, 4 mBq/kg; ^{137}Cs, 0.3 mBq/kg; and ^{90}Sr, 3 mBq/kg. Such high purity is an important advantage of the developed $CdWO_4$ crystal scintillators. For example, it was found earlier that BGO crystals contain an activity of 500–600 mBq/kg of ^{207}Bi

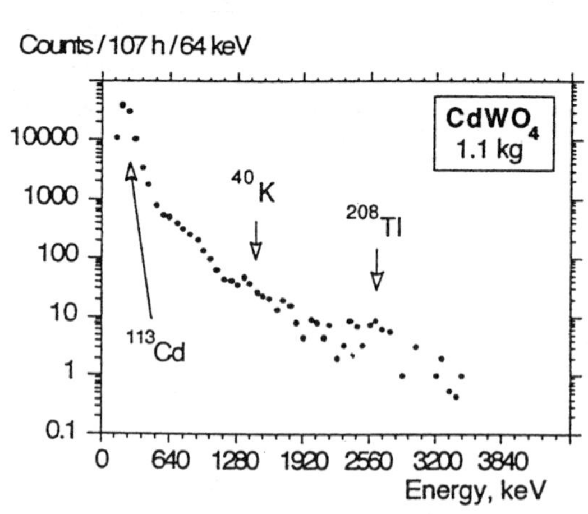

Fig. 3. Background spectrum of a $CdWO_4$ crystal (Ø54 × 65 mm) recorded during 107 h in the low background installation located at the Solotvina Underground Laboratory.

[15] and subsequently cannot be used in low background experiments.

The spectrum in Fig. 3 was also used to estimate the sensitivity of a low background spectrometer with a $CdWO_4$ crystal (149 cm^3) for measuring the radioactive contamination in the environment. The detection efficiency for γ-rays emitted by a 1 kg sample of standard earth was calculated using the CERN code GEANT 3.14 [16]. It was calculated that the sensitivities for the detection of, for example, ^{40}K and ^{208}Tl are 1.2×10^{-12} and 2×10^{-13} Ci/kg, respectively (for a measuring time of 24 h and with a statistical accuracy of 30%). This sensitivity corresponds to the characteristics of the best low background HP Ge spectrometer [13]. In fact, the latter installations provide more complete spectrometric information due to the better energy resolution. But taking into account the simplicity, operational requirements and lower cost, $CdWO_4$ low background spectrometers are preferable for some applications in low level activity measurements.

4. Conclusions

A low background scintillation spectrometer based on large volume $CdWO_4$ crystals (100–200 cm^3) with improved energy resolution has been developed as a result of advances in crystal growth, optimization of light collection and design of a special electronics unit for amplification and shaping of the $CdWO_4$ signals. The energy resolution of crystals with volume 80–150 cm^3 is 10–12% for 662 keV γ-rays. A value of 7–8% at 662 keV has been obtained with small samples (10–20 cm^3), which is comparable to NaI(Tl). The special electronics unit provides constant characteristics of the spectrometer for counting rates up to 3000 counts/s.

The measurements carried out in the Solotvina Underground Laboratory gave the following limits for the internal contamination of the $CdWO_4$ crystals: ^{228}Th, 8 μBq/kg; ^{226}Ra, 13 μBq/kg; ^{40}K, 4 mBq/kg; ^{137}Cs, 0.3 mBq/kg; and ^{90}Sr, 3 mBq/kg. The background rate of $CdWO_4$ crystals of large volume (149 cm^3) is 5 counts/yr/keV/kg in the energy region 2–3 MeV, which is similar to the corresponding feature of the best low background HP Ge detectors. The sensitivity for the detection of ^{40}K and ^{208}Tl contamination in a 1 kg sample of standard earth is 1.2×10^{-12} and 2×10^{-13} Ci/kg, respectively (for a measuring time of 24 h and with a statistical accuracy of 30%). This allows construction of unique spectrometers based on large volume $CdWO_4$ crystals for environmental control, well logging and measurements of radioactive contamination in various samples.

Moreover, considerable progress can be attained with large volume $CdWO_4$ crystals in the planned experiments to study the double beta-decay of ^{116}Cd and other rare decays.

Acknowledgments

The research described in this article was made possible, in part, by grant No. U5400 from the International Scientific Foundation.

References

[1] F.A. Kroger, Some Aspects of the Luminescence of Solids (Elsevier, Amsterdam, 1948) p. 107.
[2] R.H. Gillete, Rev. Sci. Instr. 21 (1950) 294.
[3] G.B. Beard, H.W. Kelley and M.L Malloy, J. Appl. Phys. 33 (1962) 144.
[4] M. Lammers, G. Blasse and D. Robertson, Phys. Status Solidi A 63 (1981) 569.
[5] M.R. Farukhi, IEEE Trans. Nucl. Sci. NS-29(3) (1982) 1237.
[6] L.L. Nagornay and A.S. Cherkasov, Sov. J. Prib. Tekhn. Eksper. 6 (1986) 45.
[7] E. Sakai, IEEE Trans. Nucl. Sci. NS-34(1) (1987) 418.
[8] C.L. Melcher, R.A. Manete and J.S. Schweitzer, IEEE Trans. Nucl. Sci. 36(1) (1989) 1188.
[9] F.A. Danevich, Yu.G. Zdesenko, A.S. Nikolaiko et al., Sov. J. Prib. Tekhn. Eksper. 5 (1989) 80.
[10] F.A. Danevich et al., Phys. Lett. B 344 (1995) 72.
[11] Yu.G. Zdesenko et al., Proc. 2nd Int. Symp. on Underground Physics, Baksan Valley, August 1987, ed. G.V. Domogatsky (Nauka, Moscow, 1988) p. 291.
[12] A. Alessandrello, C. Brofferio, D.V. Camin et al., Nucl. Phys. B 35 (1994) 394.
[13] A Balish et al. (Heidelberg–Moscow Collaboration), Phys. Lett. B (1995) in press.
[14] G. Heusser, Nucl. Instr. and Meth. B 17 (1986) 423.
[15] A.Y. Balish, A.A. Gurov, A.V. Demehin et al., Sov. J. Prib. Tekhn. Eksper. 1 (1993).
[16] GEANT, CERN program library entry W5013, CERN, 1993.

Z. Phys. A 355, 433–437 (1996)

ZEITSCHRIFT FÜR PHYSIK A

Investigation of $\beta^+\beta^+$ and β^+/EC decay of ^{106}Cd

F.A. Danevich[2], A.Sh. Georgadze[2], J. Hellmig[1], M. Hirsch[1], H.V. Klapdor-Kleingrothaus[1], V.V. Kobychev[2], B.N. Kropivyansky[2], V.N. Kuts[2], A. Müller[3], A.S. Nikolaiko[2], F. Petry[1], O.A. Ponkratenko[2], H. Strecker[1], V.I. Tretyak[2], M. Völlinger[1], Yu. Zdesenko[2]

[1] Max-Planck-Institut für Kernphysik, Postfach 103980, D-69029 Heidelberg, Germany
[2] Institute for Nuclear Research, Ukrainian National Academy of Sciences, 252028 Kiev, Ukraine
[3] Istituto Nazionale di Fisica Nucleare, I-67010 Assergi, Italy

Received: 6 February 1996
Communicated by B. Povh

Abstract. A low background scintillation detector with a $CdWO_4$ crystal of 1.046 kg was used to search for $\beta^+\beta^+$ and β^+/EC processes in ^{106}Cd. For the neutrinoless mode the limits $T_{1/2}(0\nu\beta^+\beta^+) \geq 2.2\cdot10^{19}$ y and $T_{1/2}(0\nu\beta^+/EC) \geq 5.5\cdot10^{19}$ y were obtained with 90% C.L. For the possible two neutrino decay limits of $T_{1/2}(2\nu\beta^+\beta^+) \geq 9.2\cdot10^{17}$ y and $T_{1/2}(2\nu\beta^+/EC) \geq 2.6\cdot10^{17}$ y have been determined with 99% C.L.

PACS: 23.90+w; 27.60.+j

1 Introduction

The considerable interest in double beta decay of atomic nuclei is related with the unique role of neutrino physics in modern theories beyond the Standard Model. Double beta decay represents the only way to measure an effective Majorana mass of the electron neutrino and yields the most stringent limits on lepton charge nonconservation, right-handed admixtures in the weak interaction, the neutrino coupling constant with Majorons and other parameters as well.

At present efforts of many experimental groups are aimed mainly to search for $0\nu\beta^-\beta^-$ decay. In the most sensitive experiments lower half-life limits on a level of 10^{21}–10^{24} y were obtained [1] - [8]. In particular, $T_{1/2}(0\nu\beta^-\beta^-) \geq 9.1\cdot10^{24}$ y (90% C.L.) has been reached for ^{76}Ge from which the following limits on an effective Majorana mass of an electron neutrino and parameters of right-handed admixtures in a weak interaction are deduced: $<m_\nu> \leq 0.50$ eV, $<\eta> \leq 6.9\cdot10^{-9}$, $<\lambda> \leq 0.9\cdot10^{-6}$) [1], [2]. The results of investigations of double positron emission ($\beta^+\beta^+$), positron emission/electron capture (β^+/EC) and double electron capture (EC/EC) are significantly more modest. The sensitivity of the best experiments is within 10^{16}–10^{21} years [9] – [15]. At the same time, taking into account the calculation of probabilities of $\beta^+\beta^+$ and β^+/EC decays [16], the search for the neutrinoless modes could help for a refined investigation of the neutrino nature and weak interaction. If the $0\nu\beta^-\beta^-$ decay ever would be observed, the question will arise what mechanism (Majorana mass of the neutrino or right-handed admixtures in weak interaction) gives the main contribution to this process.

There are, in principle, different possibilities to decide this question. One is to measure the angular distribution of the outgoing leptons, in addition to the half life [17] (the angular distribution for the $<\lambda>$-terms differs from the one for the mass mechanism). However, such an experiment would require quite large statistics and moreover can not be done within an experiment using semiconductors, such as ^{76}Ge. Also the observation of $0\nu\beta\beta(0^+\to2^+)$ decays might be helpful. However, in that case very long half-lives are expected. For example (suppose that $0\nu\beta\beta$ decay of ^{76}Ge will be observed with $T_{1/2}\approx10^{25}$ year), taking the coefficients of [18] for the ^{76}Ge $0\nu\beta\beta(0^+\to2^+)$ decay the following half-lives can be estimated: $<\lambda>=9.0\cdot10^{-6}$ ($<\eta>=5.5\cdot10^{-9}$) leads to $T_{1/2}(0^+\to2^+)\simeq9\cdot10^{29}$ years ($T_{1/2}(0^+\to2^+)\simeq5\cdot10^{32}$ years).

On the other hand, according to theoretical calculations, half-lives for $0\nu\beta^+$/EC decay depend strongly on whether the decay is dominated by the mass mechanism or right-handed weak current. For example, the expected half-lives decay of ^{106}Cd for the $0\nu\beta^+$/EC mode (if we insert the values for $<m_\nu>$, $<\lambda>$ and $<\eta>$ quoted above) will be $1.5\cdot10^{27}$ y, $1.2\cdot10^{26}$ y and $2.1\cdot10^{27}$ y, respectively. Therefore, even the non-observation of the $0\nu\beta^+$/EC mode decay could be helpful in obtaining additional information.

There are 34 isotopes in nature which can decay with a decrease of nucleus charge by two units. Besides of $\beta^+\beta^+$ decay, the processes with electron capture can occur also:

$$\beta^+\beta^+:\ (Z, A)\to(Z-2, A)+2e^++(2\nu_e),$$
$$\beta^+/EC:\ e^-+(Z, A)\to(Z-2, A)+e^++(2\nu_e),$$
$$EC/EC:\ 2e^-+(Z, A)\to(Z-2, A)+(2\nu_e).$$

The most interesting nuclei for the experimental search are the ones with a big conversion energy and a high isotopic abundance. Table 1 represents the characteristics of

New results of ^{116}Cd double β decay study with $^{116}CdWO_4$ scintillators

F.A. Danevich, A.Sh. Georgadze, V.V. Kobychev, B.N. Kropivyansky, A.S. Nikolaiko, O.A. Ponkratenko, V.I. Tretyak, S.Yu. Zdesenko, Yu.G. Zdesenko*

Institute for Nuclear Research, MSP 03680 Kiev, Ukraine

P.G. Bizzeti, T.F. Fazzini, P.R. Maurenzig

Dip. di Fisica, Universitá di Firenze and INFN, 50125 Firenze, Italy

(March 1, 2000)

A new phase of ^{116}Cd double β decay experiment is in progress in the Solotvina Underground Laboratory. Four enriched $^{116}CdWO_4$ scintillators with total mass 339 g are used in a set up, whose active shield is made of 15 natural $CdWO_4$ crystals (20.6 kg). The background rate in the energy interval 2.5–3.2 MeV is 0.03 counts/y·kg·keV. The half-life for $2\nu 2\beta$ decay of ^{116}Cd is measured as $T_{1/2}(2\nu) = 2.6 \pm 0.1(\text{stat})^{+0.7}_{-0.4}(\text{syst}) \cdot 10^{19}$ y. The $T_{1/2}$ limits for neutrinoless 2β decay of ^{116}Cd are set as $T_{1/2} \geq 0.7(1.7) \cdot 10^{23}$ y at 90%(68%) C.L. (for transition to ground state of ^{116}Sn), while for decays to the first 2^+_1 and second 0^+_1 excited levels of ^{116}Sn as $T_{1/2} \geq 1.3(2.9) \cdot 10^{22}$ y and $\geq 0.7(1.5) \cdot 10^{22}$ y with 90%(68%) C.L., accordingly. For $0\nu 2\beta$ decay with emission of one or two Majorons, the limits are $T_{1/2}(0\nu\text{M1}) \geq 3.7(5.8) \cdot 10^{21}$ y and $T_{1/2}(0\nu\text{M2}) \geq 5.9(9.4) \cdot 10^{20}$ y at 90%(68%) C.L. Restrictions on the value of the neutrino mass, right-handed admixtures in the weak interaction, and the neutrino-Majoron coupling constant are derived as: $m_\nu \leq 2.6(1.7)$ eV, $\eta \leq 3.9 \cdot 10^{-8}$, $\lambda \leq 3.4 \cdot 10^{-6}$, and $g_M \leq 12(9.5) \cdot 10^{-5}$ at 90%(68%) C.L., respectively.

23.40.–s, 15.80.Mz, 12.60.–i

*Corresponding author: zdesenko@kinr.kiev.ua

2.5.4 Semiconductor Detectors

VOLUME 71, NUMBER 6 PHYSICAL REVIEW LETTERS 9 AUGUST 1993

Experimental Search for Neutrinoless Double-β Decay of ^{100}Mo

M. Alston-Garnjost, B. L. Dougherty,* R. W. Kenney, and R. D. Tripp
Lawrence Berkeley Laboratory, University of California, Berkeley, California 94720

J. M. Krivicich,† H. W. Nicholson, and C. S. Sutton
Mount Holyoke College, South Hadley, Massachusetts 01075

B. D. Dieterle, S. D. Foltz, C. P. Leavitt, and R. A. Reeder
University of New Mexico, Albuquerque, New Mexico 87131

J. D. Baker and A. J. Caffrey
Idaho National Engineering Laboratory, Idaho Falls, Idaho 83415
(Received 25 February 1993)

No evidence for the neutrinoless double-β decay of ^{100}Mo has been found in a search using a segmented Si(Li) detector with source foils enriched to 97% ^{100}Mo. After 0.2664 mol yr of exposure, we report a 1σ lower limit on the half-life for the $J^P = 0^+ \rightarrow 0^+$ transition of 0.44×10^{23} yr.

PACS numbers: 23.40.Bw, 14.60.Gh, 27.60.+j

The existence of the neutrino was proposed over 60 years ago by Pauli, followed shortly thereafter by Fermi's theory of β decay. Yet to this day there are fundamental unanswered questions concerning properties of the neutrino and of weak interactions in general. Neutrinoless double-β decay directly addresses the question of lepton conservation in weak interactions as well as the related question of whether the neutrino has mass and, if it does, whether it is its own antiparticle. As such, it is one of the most sensitive tests of physics beyond the standard model of elementary particles and for this reason has received much attention both theoretically and experimentally during the past decade. Despite considerable experimental progress, there has as yet been no convincing evidence for the existence of this decay mode reported in any element. At the present time, published null experiments in germanium and xenon have obtained neutrinoless double-β decay lifetimes of 1.2×10^{24} yr and 7.8×10^{23}, respectively [1]. As a result, upper limits can be placed on the mass of a Majorana neutrino that are substantially better than those obtained by the more direct β decay end-point method.

The isotope selected here as a neutrinoless double-β decay candidate is ^{100}Mo. The lepton-number violating neutrinoless decay mode, $^{100}\text{Mo}\rightarrow{}^{100}\text{Ru}+2e^-$, is forbidden for massless neutrinos or for Dirac neutrinos (which are distinct from their antiparticles). There have recently been three lower half-life limits reported for the neutrinoless decay of ^{100}Mo, one by an earlier version [2] of this experiment with a limit of 4×10^{21} yr and two others [3,4] of 4.7 and 7×10^{21} yr. ^{100}Mo is selected for its large energy release (which increases the phase space for neutrinoless double-β decay), for its favorable double-β decay matrix element, and for the availability of several moles of the enriched isotope. Assuming comparable matrix elements, the product of phase space and Coulomb correction factor increase the ^{100}Mo neutrinoless double-β decay rate over other commonly used element isotopes for double-β decay searches by factors ranging from 1.3 to 12.3 [5–8]. In addition, the 3.033 MeV energy release shared by the two electrons is energetically above most naturally occurring electron and photon background radiation.

The original apparatus has been described in detail elsewhere [9,10]. Data reported here were obtained from an upgraded detector array which contained 145 Si(Li) detectors, each 7.6 cm in diameter and 0.14 cm thick, arranged in two stacks. One stack with 74 detectors was run without ^{100}Mo source foils for background studies. The second stack, containing 71 detectors, was loaded with 62 ^{100}Mo source foils placed in the gaps between detectors.

Titanium clips were used to provide more positive electrical contacts to the silicon detectors in the upgraded apparatus. Source foils consisted of Mylar bags 6.0 cm in diameter, each containing approximately 1 g ^{100}Mo. The very finely divided pure metallic ^{100}Mo powder had been chemically purified to reduce the Th and U levels to below 1 ppb. Each bag was rolled flat (to within 10%) before being inserted into a gap between detectors. The average thickness of the ^{100}Mo within each bag was 34.4 mg/cm^2, and the thin Mylar walls (4.3 μm) of the bags contributed a negligible amount of additional material. The entire detector array was enclosed in the radioactively clean titanium cryostat of the original apparatus and cooled to 120 K with liquid nitrogen.

Cosmic ray backgrounds are negligible at our underground site in the Consolidated Silver mine in Osburn, Idaho. The site has a rock overburden of 1220 m [3300 m of water equivalent (mwe)]. Shielding of the detectors from natural radioactivity in the surrounding rock consisted of three concentric closed layers of appropriate ma-

0031-9007/93/71(6)/831(4)$06.00

VOLUME 71, NUMBER 6 PHYSICAL REVIEW LETTERS 9 AUGUST 1993

terials to reduce neutrons and γ rays. The neutrons were moderated by a 56 cm thick wax shield and were then captured in a 5.1 to 10.2 cm thick inner layer of 5% borated polyethylene. Any γ rays penetrating the wax and polyethylene were substantially eliminated by absorption in the innermost layer of low activity lead, 25.4 cm thick. The interior of this lead shield was continuously flushed with nitrogen gas from liquid nitrogen boiloff to reduce airborne Rn daughter activity. All the structural parts of the apparatus within the shielding were constructed from materials selected by means of low level counting for low intrinsic radioactivity.

The upgraded detector included electronic latches on each channel to aid in identifying fast β-α decay sequences. The β-α time interval between a ^{212}Bi or ^{214}Bi beta decay and the decay of its Po daughter was measured by a 100 MHz "fast" clock that was started by each event and stopped by any second trigger occurring during the event read-out time of approximately 50 ms. A 1 kHz "slow" clock measured time intervals between events and was used to detect longer-lived decay sequences. Lifetimes measured by these clocks allowed identification of radio contaminants within the cryostat.

The absolute energy calibration of the experiment was checked periodically by placing a sealed ^{228}Th source on the surface of the cryostat near the detectors and identifying the double escape peak at 1.593 MeV in each detector. In addition, before each data run, a digital to analog converter running under computer control was used to inject a series of pulses with precisely known amplitudes into the preamp of each detector to check the electronic gain and linearity of the readout system. Any drifts were recorded in software and corrected on a run by run basis. Throughout the experiment, the energy resolution of a typical detector was between 10 and 20 keV and its energy determination was stable between absolute source calibrations to within the detector's resolution.

The data from this experiment correspond to 3849.5 h live time with 60.63 g of ^{100}Mo in the detector stacks, giving an exposure of 0.2664 mol yr. There were 299 852 triggers above the nominal discriminator threshold of 320 keV. About 90% of the events had energy in only one detector. We ascribe the one-detector events with energies $2.0 < E < 5.5$ MeV to α decays of ^{210}Po ($E_\alpha = -5.3$ MeV), and events below 2 MeV to a combination of ^{210}Po decays and single-β decays from various sources. The ^{210}Po background probably came from ^{210}Pb in solder used to make electrical contacts within the cryostat.

The following cuts, which were optimized by Monte Carlo simulations and studies of the real data, were made to select neutrinoless double-β decay candidates and reject backgrounds: (a) To remove electronic noise, the energy in a detector was ignored if it was less than 110 keV. In addition, for multidetector events any contribution to the event energy sum was ignored if it was less than 1/20 the maximum contribution. (b) To remove α backgrounds, events with energy deposited in only one detector were rejected. (c) To minimize backgrounds from single β decays, which are usually accompanied by γ rays, only events with energy in two or three contiguous detectors were selected. In addition, the energy deposited in the middle detector of a three-detector event had to be greater than 450 keV (the minimum energy deposited by a single electron passing through a detector). (d) Events "tagged" by the fast clock were rejected, since they were usually ^{212}Bi or ^{214}Bi β decays followed by a fast ^{212}Po or ^{214}Po α decay. (e) Events with energy deposited in two detectors, where the energy ratio in the detectors was less than $\frac{1}{3}$, were rejected. This reduced the number of untagged ^{212}Bi events in which the energy of the α from the subsequent fast decay of ^{212}Po was recorded in an ADC but the α arrived before the trigger logic had recovered and so did not stop the fast clock. (f) A possible neutrinoless double-β decay candidate was rejected if there was energy deposited in a detector next to an inoperative detector, since the total energy of the event was then suspect.

After all these cuts there remain 8634 events in the region of the detectors occupied by the ^{100}Mo. This corresponds to about 6% of the total number of events observed in this region of the detector. The energy spectrum for these double-β decay candidate events, for energies above 1.8 MeV, is shown in Fig. 1 (728 events).

To obtain the detection efficiency for observing double-β decay candidates and to estimate the spectral shapes of backgrounds, a Monte Carlo event generating program was written to simulate double-β decays and backgrounds within the geometry of our detector for each

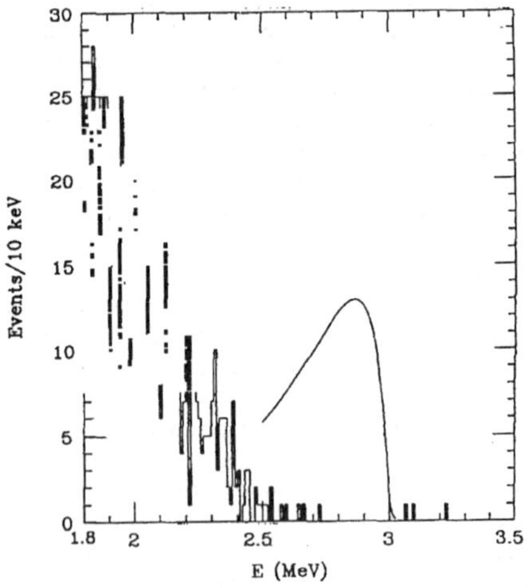

FIG. 1. Energy spectrum of events after all cuts. The corresponding energy spectrum for the Monte Carlo generated $0^+ \rightarrow 0^+$ neutrinoless double-β decays after the same cuts is shown as a solid line. This curve normalized to 500 events for a suitable graphical representation.

data run used in the final analysis [11]. This program was a modification of the CERN program GEANT [12]. The generated events were processed through our standard analysis programs with all the cuts necessary to select double-β decay candidates. The shape of the spectrum for double-β decay candidates in the foils is shown in Fig. 1 and was fitted to an analytical function ($\chi^2/N_{DF} < 1.2$) which was used in fitting procedures to determine the double-β decay half-life described below. The detection efficiency for the $0^+ \rightarrow 0^+$ transition was determined to be $(46.9 \pm 0.4)\%$ for all observed energies, $(36 \pm 0.3)\%$ for energies above 2.5 MeV, and $(20.6 \pm 0.2)\%$ for energies above 2.75 MeV.

To investigate expected β decay backgrounds, energy spectra for the β decays of ^{214}Bi, ^{212}Bi, and ^{208}Tl were generated. The expected energy spectra for α decays were calculated using a simple Monte Carlo. In an attempt to estimate the size and probable sources of backgrounds, the experimental data and simulated data described above were examined carefully. Comparisons were made of event rates in stack 1 and stack 2, and also rates in stack 2 before and after the ^{100}Mo was inserted. An assessment of the ^{238}U and ^{232}Th activity within the cryostat was obtained by studying events tagged by the fast clock during event read out, or by observing rapid multi-α sequences in (a) the β decay of ^{212}Bi in the ^{232}Th chain followed by an α decay of the daughter ^{212}Po, which has a half-life of 0.3 μsec; (b) similarly, the β decay of ^{214}Bi in the ^{238}U chain followed by an α decay of ^{214}Po, which has a half-life of 164 μsec; and (c) the α-α-α rapid decay sequence, $^{220}\mathrm{Rn} \rightarrow {}^{216}\mathrm{Po}(t_{1/2}=56\ \mathrm{s}) \rightarrow {}^{212}\mathrm{Pb}(t_{1/2}=0.15\ \mathrm{s})$ in the ^{232}Th chain. These α sequences were observed as either two or three separate one-detector events in the same or adjacent detectors.

In the α decay sequences, if the contamination is from within the ^{100}Mo, the events have to be seen in the same detector. Because the ^{100}Mo filled bag was thicker than the range of any of the α particles, back-to-back α particles would not give a sequence of events in adjacent detectors. We do, however, see sequences that were in adjacent detectors and can only be explained if most of the ^{232}Th chain α decay sources were not in the ^{100}Mo powder. A possible source of this contamination was the Ti clips (known to contain a minute amount of Th) which made electrical contact with the detectors and were constructed in such a way that they were visible to the detectors above and below them. By comparing the rates of fast clock tagged events for empty Mylar bags and ^{100}Mo filled Mylar bags, we find that some, but not all, of the ^{214}Bi β decays came from contamination in the Mylar bags themselves.

Estimated backgrounds above 2.5 MeV used in the final analysis were obtained from observed tagged ^{212}Bi and ^{214}Bi decays and Monte Carlo generated ^{208}Tl decays. The ^{212}Bi background was estimated to be flat for energies between 2.0 and 3.5 MeV and the ^{208}Tl background above 2.0 MeV was fitted to a straight line with a negative slope. A quadratic was used to fit ^{214}Bi background above 2.0 MeV. All the observed events above 2.5 MeV could be easily explained by various combinations of the estimated backgrounds.

In order to obtain a lifetime limit for $0^+ \rightarrow 0^+$ neutrinoless double-β decay, an extended maximum likelihood method was used to estimate both the number of signal events contained in our data sample and the associated error [13]. The extended likelihood function can be written as the joint probability

$$\mathcal{L}_\varepsilon = \frac{e^{-\mu}\mu^n}{n!}\mathcal{L},$$

where the first factor is the Poisson probability for observing n events in a time and energy interval when μ events are expected, and the second is the conventional likelihood function $\mathcal{L}$ that contains the distribution shape information of the observed data

$$\mathcal{L} = \prod_{i=1}^{n} \{f_1 F_{\mathrm{signal}}(E_i) + f_2 F_{\mathrm{Bi212}}(E_i) + f_3 F_{\mathrm{Bi214}}(E_i) + f_4 F_{\mathrm{Tl208}}(E_i)\}.$$

In the expression for $\mathcal{L}$, the functions F_j are the analytic shapes fitted to the energy spectra of the signal and backgrounds as described above and normalized to unity over the energy interval $2.50 < E < 3.50$ MeV. The functions f_j are the fractions of signal or backgrounds in each observed event and

$$\sum_{j=1}^{4} f_j = 1,$$

where $j=1$ means "signal," etc. In the above expression for $\mathcal{L}$, the functions F_j are evaluated for each event at the appropriate energy E_i and the product is taken over the $n=12$ events observed in the energy interval between 2.50 and 3.50 MeV. $\mathcal{L}_\varepsilon$ can then be reexpressed in terms of the parameters $\phi_j = \mu f_j$ by moving the external factor μ^n into the product. The values of ϕ_j after maximization of $\mathcal{L}_\varepsilon$ are then the numbers of signal and background events in the sample.

MINUIT [14] was used to minimize the negative logarithmic function $-\ln\mathcal{L}_\varepsilon$ expressed in terms of the parameters ϕ_j. For a variety of ϕ_j starting values, including minimizations in which one or two of the background ϕ_j's were held constant, $-\ln\mathcal{L}_\varepsilon$ was always minimized when ϕ_{signal} was pushed to zero with a one-sided error of 0.92 events or less, corresponding to increasing the value of $-\ln\mathcal{L}_\varepsilon$ by $\frac{1}{2}$.

Based on this analysis, a conservative 68% confidence limit on the $0^+ \rightarrow 0^+$ neutrinoless double-β decay can be obtained from the formula

$$t_{1/2} \geq \frac{\ln(2) T\epsilon N_a}{s},$$

where $T\epsilon = 0.2664 \times 0.36 = 0.0959$ mol yr is the product of the live time and detector efficiency, N_a is Avogadro's number, and s is the number of signal events. Since the minimization found zero to be the best value for the signal, we have used the above error value 0.92 for s and have obtained $t_{1/2} > 0.44 \times 10^{23}$ yr. This is a factor of 11 larger than our previous result [2].

An estimate of 6.6 eV for the upper limit on the effective Majorana neutrino mass $\langle m_\nu \rangle$ for the 0.44×10^{23} yr half-life limit obtained in this molybdenum experiment can be calculated using the conservative matrix elements of Engel *et al.* [7]. It should be emphasized, however, that this estimate is both model dependent and sensitive to the approximations used to calculate the matrix elements. Because these matrix elements are not precisely known, it is important to compare mass limits derived from lifetime limits for several candidate isotopes. Published limits on the Majorana neutrino mass derived from germanium and zenon experiments vary from 1.3 to 5.0 eV, the best germanium limit being the most restrictive [1].

We are indebted to the ASARCO Mining Co. and Consolidated Silver Corp. for access to the Silver Summit mine. We thank Philippe Eberhard, Gerry Lynch, Mahiko Suzuki, A. Carl Helmholz, and Michael Moe for valuable discussions and reference information. We thank Alan Smith for radioactive assays and essential advice in our selection of clean materials. We also thank James VanKuiken for a long history of support underground, Michael Long and Chilton Gregory for technical support, Lydia Young for suggesting making clip detector contacts, and Ju Kang, Benjamin Brown, and Kevin Watts for taking shifts in Idaho. This work is supported by U.S. Department of Energy Contracts and Grants No. DE-AC07-76ID0-1570, No. DE-AC03-76SF00098, No. DE-FG04-88ER40395, No. DE-FG05-88ER75444, and No. DE-FG02-90ER40553.

*Present address: Department of Physics, Stanford University, Stanford, CA 94305.

†Present address: 831 Pollard Road, Los Gatos, CA 95030.

[1] M. K. Moe (to be published). The germanium reference (42.) is A. Balysh *et al.*, Phys. Lett. B **283**, 32–36 (1992), and the xenon reference (23a.) is a Neutrino 92 talk by J.-C. Vuilleumier.

[2] M. Alston-Garnjost *et al.*, Phys. Rev. Lett. **63**, 1671 (1989).

[3] H. Ejiri *et al.*, Phys. Lett. B **258**, 17 (1991).

[4] H. Ejiri *et al.*, Nucl. Phys. B (Proc. Suppl.) **28A**, 219 (1992).

[5] D. Bryman *et al.*, Rev. Mod. Phys. **50**, 11 (1978).

[6] S. P. Rosen, in *Proceedings of the 1981 International Conference of Neutrino Physics and Astrophysics, Maui, Hawaii*, edited by R. J. Cence, E. Ma, and A. Roberts (High Energy Physics Group, Department of Physics and Astronomy, University of Hawaii, Honolulu, Hawaii), Vol. 2.

[7] J. Engel *et al.*, Phys. Rev. C **37**, 731 (1988).

[8] D. B. Stout *et al.*, Phys. Rev. Lett. **69**, 1900 (1992).

[9] M. Alston-Garnjost *et al.*, Nucl. Instrum. Methods Phys. Res., Sect. A **271**, 475 (1988).

[10] M. Alston-Garnjost *et al.*, Phys. Rev. Lett. **60**, 1928 (1988).

[11] B. Dougherty *et al.*, BBGEANT, LBL Group A Physics Programming Note No. P-292 (1988).

[12] R. Brun *et al.*, GEANT 3 Users Guide, CERN Data Handling Division Report No. DD/EE/84-1, 1984, revised 1987.

[13] L. Lyons, *Statistics for Nuclear and Particle Physicists* (Cambridge Univ. Press, Cambridge, 1986), p. 91ff.

[14] F. James *et al.*, MINUIT Function Minimization and Error Analysis, Long Write-Up, CERN, Geneva (1989).

REVIEW OF DOUBLE BETA DECAY RESULTS FROM ^{76}Ge

David O. Caldwell

Physics Department, University of California*
Santa Barbara, California 93106, U.S.A.

in Proc. 12th International Conference on "Neutrino Physics and Astrophysics", Sendai, Japan, June 3 - 8, 1986, eds *T. Kitagaki and H. Yuta, World Scientific, Singapore* (1986) 77 - 92

ABSTRACT

Experiments using Ge detectors have obtained lifetime limits for neutrinoless double beta decay in ^{76}Ge which provide the best limits on (1) lepton number conservation, (2) the mass of a light Majorana electron neutrino, (3) the existence of right-handed currents for the conditions in which the "see-saw" mechanism provides neutrino mass, (4) the mass of a heavy Majorana neutrino as a function of its coupling to the electron neutrino, and (5) the coupling of a Majoron to the electron neutrino. Currently six such experiments are in progress and another is in early stages of preparation, representing work in six countries. The status of these experiments, their results, and the interpretation of these results in terms of quantities of interest to particle physics are presented below.

1. INTRODUCTION TO DOUBLE BETA DECAY

Since this is the first contribution on the subject of double beta decay, it may be useful to provide a brief introduction to this extremely sensitive nuclear probe of particle physics. Excellent extensive reviews are available.[1-4] Here we shall concentrate on those aspects of the process which provide an understanding of how double beta decay can yield information on lepton number conservation, light and heavy neutrino mass, right-handed currents, and Majoron-neutrino coupling.

Double beta decay is possible because the pairing energy makes a

*Supported in part by U.S. Department of Energy.

D. Caldwell

nucleus with an even number of neutrons and protons more tightly bound than its odd-odd neighbors. Such a decay for ^{76}Ge to ^{76}Se, by-passing ^{76}As, is shown in Fig. 1. The second-order weak decay, which will be referred to as $\beta\beta_{2\nu}$,

$$^{76}\text{Ge}\rightarrow{}^{76}\text{Se}+2e^{-}+2\bar{\nu}_e, \qquad (1)$$

must occur, although with very small probability.

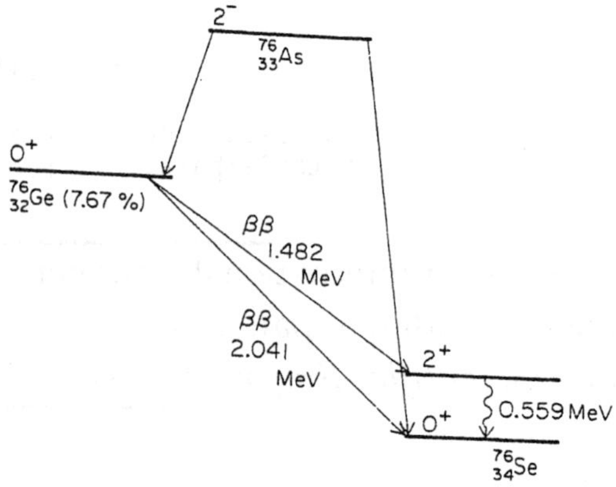

Fig. 1. Double beta decay scheme for ^{76}Ge.

A decay which would be favored by a factor $\sim 10^7$ because of the greater phase space available is

$$^{76}\text{Ge}\rightarrow{}^{76}\text{Se}+2e^{-}. \qquad (2)$$

This process, labeled $\beta\beta_{0\nu}$, clearly violates lepton number and hence may not occur at all. The virtual (anti)neutrino emitted in the first neutron decay must be absorbed as a neutrino by the second neutron to produce an e^-, thus requiring the neutrino to be the same as its anti-particle, a property which defines the neutrino to be of the Majorana type.

Because of parity nonconservation, for a massless neutrino the first neutrino emitted would have helicity opposite to that required by the neutrino being absorbed. The necessary admixture of the opposite helicity can be obtained if the neutrino has a mass, m_ν, and hence the inhibition of this decay is greater the smaller that mass. If right-handed currents (RHC) exist, that mechanism could also provide the required helicity reversal, so the rate of the process increases with increasing admixture of RHC to the normal left-handed currents. A quark-level diagram as an example of RHC-induced $\beta\beta_{0\nu}$ is shown in Fig. 2. Note that with a right-handed, as well as a left-handed W boson involved, both a right-handed and a left-handed electron are emitted. In this example, the back-to-back e_L^- and e_R^- require an angular momentum of one, which is provided by p-wave emission, and the resulting total angular momentum change in the process can then be 0, 1, or 2. Thus the transition from the 0^+ ground state of ^{76}Ge can be to the 0^+ ground state or the 2^+ first excited state of ^{76}Se, as shown in Fig. 1. On the other hand, for double beta decay induced purely by neutrino mass, two left-handed W bosons are involved, producing two left-handed e^-, so their helicities cancel in the back-to-back decay, permitting s-wave emission and giving a total angular momentum of zero. Thus only the $0^+ \to 0^+$ transition is allowed, and this provides a means of distinguishing experimentally between the two mechanisms.

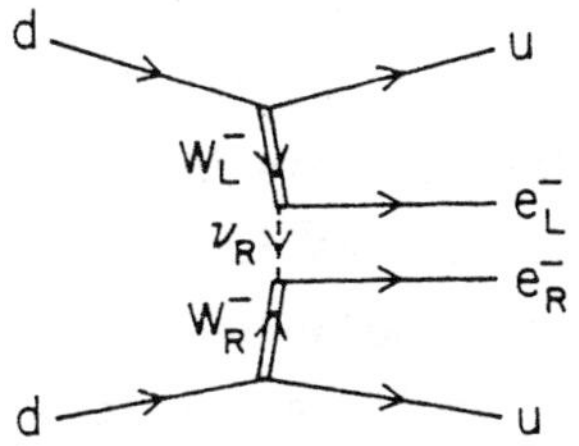

Fig. 2. An example of double beta decay induced by right-handed currents.

A special case of the non-zero neutrino mass decay could exist if there were a very heavy Majorana neutrino, M_ν, which coupled to the electron neutrino (which could then be massless) with some mixing coefficient, U_e^L. The small admixture of M_ν would provide the helicity reversal, and hence a limit on $\beta\beta_{0\nu}$ yields also a limit on a combination of M_ν and U_e^L.

A quite distinct mechanism for inducing $\beta\beta_{0\nu}$ would be possible if a light or massless boson, such as the Majoron,[5] existed. This Goldstone boson, $\chi°$, arising from the breaking of B-L (baryon number minus lepton number) symmetry could induce the decay,

$$^{76}\text{Ge} \to {}^{76}\text{Se} + 2e^- + \chi° , \qquad (3)$$

which is designated $\beta\beta_{0\nu,\chi}$. Experimentally such a three-body spectrum is quite distinguishable from the four-body spectrum of $\beta\beta_{2\nu}$ and of course from the more usual form of $\beta\beta_{0\nu}$, as is shown in Fig. 3.

All of the interesting particle physics results come from the various forms of the $\beta\beta_{0\nu}$ decay, but the values of, or limits,on these quantities are quite dependent on the nuclear matrix elements, which are difficult to calculate. While the $\beta\beta_{0\nu}$ is a short-range, high-energy process in contrast to the long-range, low energy $\beta\beta_{2\nu}$ decay, the latter provides the only check on the nuclear calculations. That the calculations give longer lifetimes than do the measurements for other nuclei has been discussed by Moe and by Pomansky at this Conference. How the

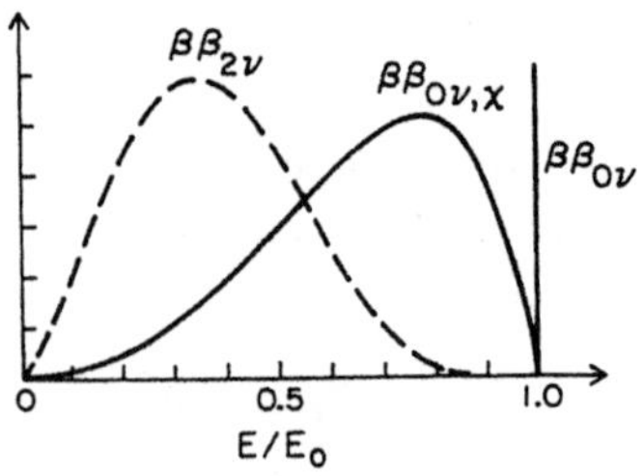

Fig. 3. The summed kinetic energies of the two electrons for $\beta\beta_{2\nu}$, $\beta\beta_{0\nu}$, and $\beta\beta_{0\nu,\chi}$ decay modes (arbitrary scales).

resolution of this issue will affect the interpretation of the $\beta\beta_{0\nu}$ results is totally unclear at this time. However, for a nucleus as light as Ge the lifetime discrepancy is surely no worse than an order of magnitude, and if it were that bad and a similar factor applied to $\beta\beta_{0\nu}$, the quantities of interest would change by a factor of three at most. Unfortunately, $\beta\beta_{2\nu}$ is very difficult to measure in Ge, with the most stringent upper limit of 8×10^{19}y by the UCSB-LBL group[6)] being set by the systematics of the background determination and not by statistics. That limit is still shorter than any calculated value.

2. GENERAL DESCRIPTION OF DOUBLE BETA DECAY EXPERIMENTS

Ge detectors give excellent energy resolution, so that a peak at 2.041 MeV expected for $\beta\beta_{0\nu}$ could be seen with a full width at half maximum of about 3 keV. This helps considerably in reducing background counts, as does the fact that to be a good detector the Ge has to be of unusually high purity. The zone refining and other purification steps eliminate heavier radioactive elements. The sensitivity is quite good, since 7.8% of the Ge is the double-beta-decay candidate nucleus ^{76}Ge, and the nuclear transition is believed to be favorable. An especially big advantage of Ge is that the source is intimately mixed with the detector, and hence the amount of ^{76}Ge can be made as large as can be afforded. With typical electron ranges of the order of one millimeter it is difficult to get large sources when the source and detector are separate.

With these advantages it is not surprising that Ge experiments are being done by many groups around the world. All of them have been struggling with the necessity to reduce backgrounds as much as possible. One component of the background, that due to cosmic rays, can obviously be reduced by going underground, and all experiments have done this. The depths vary greatly, as shown in Fig. 4. After the hadronic component is absorbed, the main residual effect is from muon-produced neutrons

D. Caldwell

which give capture γ rays. Reducing this effect has to be balanced against the problems caused by neutrons from fission in rocks.

The main background problem is ubiquitous radioactivity. Material is selected with as little radioactivity as possible. Specially selected Pb is used as an outer shield, usually associated with some hydrogenous neutron absorber (such as borated polyethylene). Inner shields are often of multiply distilled Hg or of oxygen-free hard copper. Some experiments employ in addition an active shield of NaI. This serves not only to veto activities from the outside, but also to suppress Compton scattered γ's. A γ, usually from material inside the NaI, can Compton scatter in the Ge and leave the requisite 2 MeV, but if it then passes into the NaI that event is eliminated. It is thus essential to have as little material inside the NaI as possible and to have that of as low a Z as can be managed so the Compton γ's are not absorbed before reaching the NaI. The NaI is not as free from radioactivity as the other materials used, and while it is mainly self-vetoing, experiments using NaI tend to have larger and more numerous photopeaks. However, these are at discrete energies, and the bothersome Compton continuum is so well suppressed that these activities do not matter for the $\beta\beta_{0\nu}$ decay. Experiments with NaI try to use the NaI to detect the 559 keV deexcitation γ ray in coincidence with the 1.482 MeV $2e^-$ energy in the Ge from the $0^+ \to 2^+$ transition and hence can set better limits on this $\beta\beta_{0\nu}$ decay.

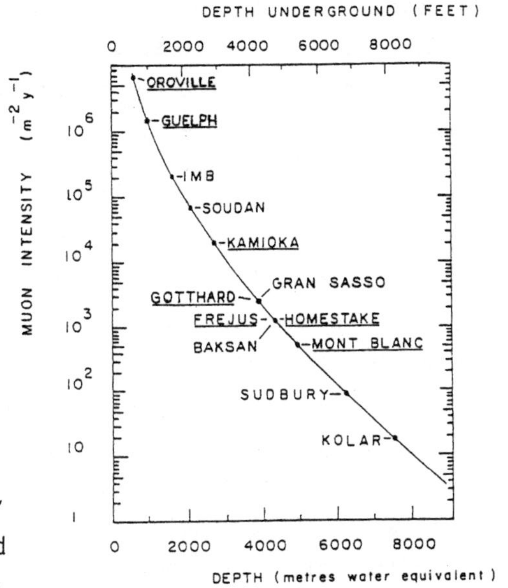

Fig. 4. Depth of underground laboratories, with those underlined having double beta decay experiments.

The background reduction has been so effective that many experiments are now actually limited by activities in the Ge itself, such as ^{68}Ge. Some of these have been induced by cosmic rays when the material was above ground, but others are very likely due to the unfortunate recycling of this valuable substance. Old detectors go back into the melt and contaminate the new product, albeit at a level that other experiments could not detect.

3. STATUS OF SPECIFIC EXPERIMENTS

With this general description as a base, we now proceed to the specific experiments. In each case the lifetime limits obtained for the $0^+ \to 0^+$ (and in several cases the $0^+ \to 2^+$) $\beta\beta_{0\nu}$ transitions will be given, along with background levels and resolutions achieved, counting times, quantity of Ge, and any special features.

The Zaragoza-Bordeaux-Strasbourg experiment[7)] in the Frejus tunnel is shown schematically in Fig. 5. This experiment uses four Ge detectors of about 417 cm^3 active volume inside 19 hexagonal NaI detectors. Five of the latter have not yet been replaced by new scintillators and are quite radioactive, so the background level (40 counts/keV·y·kg) near 2 MeV is still high, but when the NaI is used in coincidence for the $0^+ \to 2^+$ transition the background is only 3 counts/keV·y·kg. Good resolutions have been achieved: 3.1 keV at 2 MeV and 2.6 keV at 1.5 MeV for the Ge and 11% at 0.6 MeV for the NaI. After 2252 hours of counting, lifetime limits of 2.2×10^{22}y ($0^+ \to 0^+$) and 1.8×10^{22}y ($0^+ \to 2^+$) were obtained.

The Cal Tech group used a 90 cm^3 detector of 3 keV resolution inside a plastic scintillator anticoincidence at sea level for 3820 hours and got a lifetime limit of 1.9×10^{22}y.[8)] Collaborating now with SIN and Neuchatel, they moved the detector to the St. Gotthard tunnel where the background was reduced by a factor of 16 to 4 counts/keV·y·kg. In addition to using the detector to test the radioactivity level for components used in the construction of a system of eight Ge detectors of 140 cm^3 each, it was run for 6728 hours to get a $0^+ \to 0^+$ lifetime limit of 6.2×10^{22}y.[9)] The spectrum above 2 MeV for

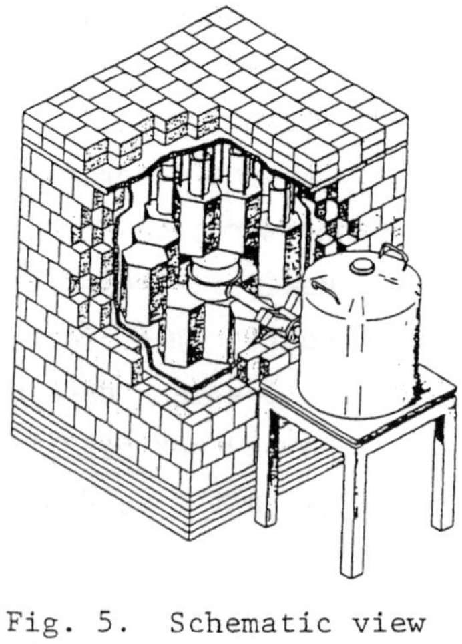

Fig. 5. Schematic view of the Zaragoza-Bordeaux-Strasbourg apparatus.

this run is shown in Fig. 6. The installation of the larger system is now proceeding.

The Osaka experiment[10] was operated first above ground and then moved into the Kamioka mine. A 164 cm^3 detector was used inside a NaI shield, as shown in Fig. 7. Resolutions were 2.7 keV ($0^+ \to 0^+$) and 2.4 keV ($0^+ \to 2^+$). With backgrounds of 6 c/keV·y·kg ($0^+ \to 0^+$) and a coincidence background of 3 c/keV·y·kg ($0^+ \to 2^+$), an 8621-hour run produced limits of 7.4×10^{22}y ($0^+ \to 0^+$) and 6×10^{22}y ($0^+ \to 2^+$). At present the Ge has been replaced by a ^{100}Mo double beta decay experiment using Si detectors, as has been described at this Conference by Ejiri.

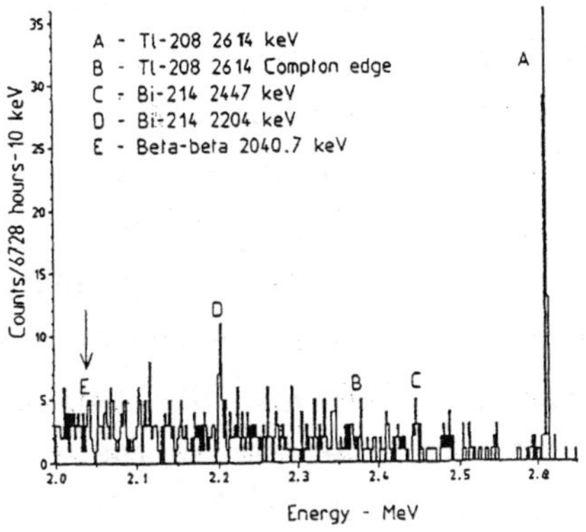

Fig. 6. The spectrum above 2 MeV for the Caltech-SIN-Neuchatel experiment.

The Pacific Northwest Laboratory, University of South Carolina group first did an above-ground experiment[11] which gave a limit of 1.3×10^{22}y in 4054 hours using a 125 cm^3 detector of 3.4 keV resolution inside a NaI anticoincidence and a complex bulk shield. A detector of similar volume and 3.7 keV resolution but much lower intrinsic radioactivity was placed inside Cu and Pb shields in the Homestake mine and a background of 2.4 c/keV·y·kg achieved.[12] By counting for about 8000 hours they obtained a lifetime limit of 1.4×10^{23}y. They are still trying to reduce backgrounds while continuing to count and to prepare a larger (720 cm^3 and eventually twice that volume) detector system.

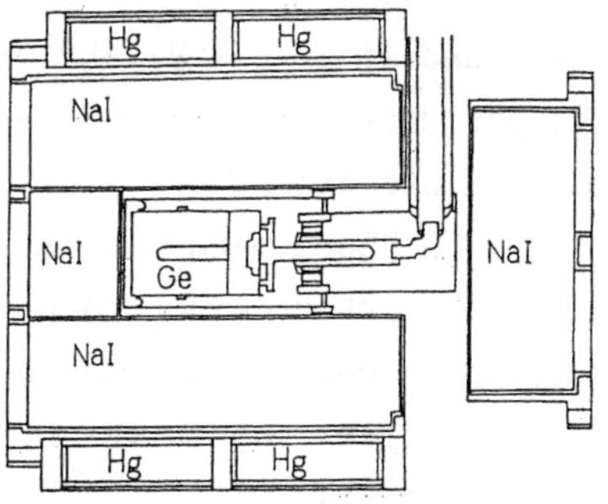

Fig. 7. Cross-sectional view of the Osaka experiment.

The Guelph-Aptec-Queens group used the largest detector then developed of 194 cm^3, which was shielded by Hg, Pb, and a plastic scintillator anticoincidence and placed in a salt mine near Windsor, Ontario. With a background of about 34 c/keV·y·kg at 2 MeV and a reso-

D. Caldwell

lution of 3 keV they obtained in 2363 hours lifetime limits of 3.2×10^{22}y ($0^+\to0^+$) and 1.6×10^{22}y ($0^+\to2^+$).[13] Now with three detectors of similar size and a much improved 2 MeV background of about 2 c/keV·y·kg, a run of 3200 hours has given a $0^+\to0^+$ limit of 1.6×10^{23}y. The data in the vicinity where the $0^+\to0^+$ line (2.041 MeV) should appear is shown in Fig. 8.

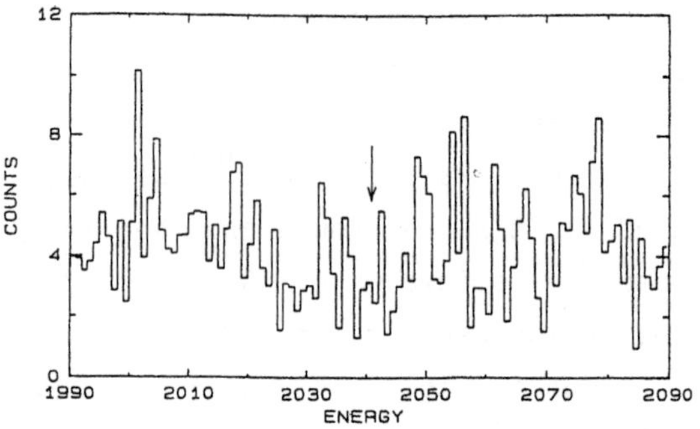

Fig. 8. Data from the Guelph-Aptec-Queens experiment in the region of the 2.041 MeV $\beta\beta_{0\nu}$ line (arrow).

The first Ge experiment[14] was performed by the Milan group. For their more recent results[15] they have used two quite different Ge detectors in the Mt. Blanc tunnel. The first has been operated for 21,000 hours and has a useful volume of 116 cm^3, a resolution of 3 keV, and a background of 24 c/keV·y·kg, and achieved a $0^+\to0^+$ limit of 1.2×10^{23}y. A larger detector (138 cm^3) with about the same resolution but a much improved background (4 c/keV·y·kg) has been operating in the same place for 10,000 hours, and a lifetime of 1.3×10^{23}y has been obtained. If these data are combined, a $0^+\to0^+$ lifetime limit of 1.8×10^{23}y is reached. Since the group is turning to other techniques, the first detector is no longer operating in the Mt. Blanc tunnel, but the second is still collecting data.

The last of these active Ge experiments is that of the University of California Santa Barbara, Lawrence Berkeley Laboratory collaboration. They set out from the beginning to build a multidetector array completely enclosed in a 15-cm-thick assembly of ten NaI counters. However, the cryostats were constructed to house units of two Ge detectors each, and the first experiment was done with just one such unit, which was operated above ground. Using 336 cm^3 of Ge of 4 keV combined resolution with a background of 11 c/keV·y·kg, they obtained a $0^+\to0^+$ limit of 5×10^{22}y in 1618 hours.[16] The experiment was then moved into a powerhouse built into a mountain at the Oroville, California dam and two more detectors were installed for a total of 658 cm^3. The combined resolution

D. Caldwell

was improved to 3.7 keV, and the average background (which dropped by a factor of two in six months) was reduced to 2.2 c/keV·y·kg. In 3550 hours of counting, a $0^+\to 0^+$ limit of 2.5×10^{23}y was obtained.[6] Using 3404 hours of data, a limit of 5×10^{22}y was achieved for the $0^+\to 2^+$ transition. With 2456 hours of data the already mentioned $\beta\beta_{2\nu}$ 90% C.L. limit of 8×10^{19}y was reached, and a 90% C.L. limit of 6×10^{20}y was obtained for the $\beta\beta_{0\nu,\chi}$ decay.

The UCSB-LBL experiment has since been improved by adding two more detectors for a total of about 990 cm³. These were operated for about 5000 hours, and for a small part of that time the total complement of 8 detectors (1.3 liters) was functioning. The resolution has been improved to 3.4 keV and the time-averaged background to 1.7 c/keV·y·kg. For 7.8×10^6 hour·cm³ (about 47,000 detector hours) the $0^+\to 0^+$ limit has become 4.1×10^{23}y and the $0^+\to 2^+$ limit, 1.3×10^{23}y. The spectra in these two energy regions are shown in Fig. 9 and 10. Counting continues, and it is hoped to make some obvious improvements in the detectors to improve backgrounds.

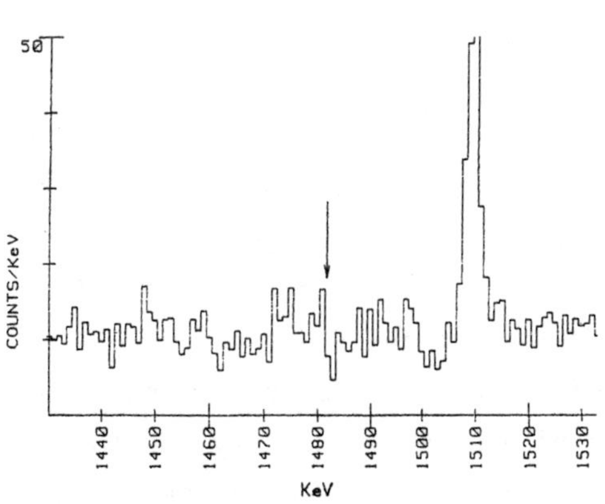

Fig. 9. Ge spectrum with NaI coincidence in the region of the expected $0^+\to 2^+$ peak (arrow) for the UCSB-LBL experiment.

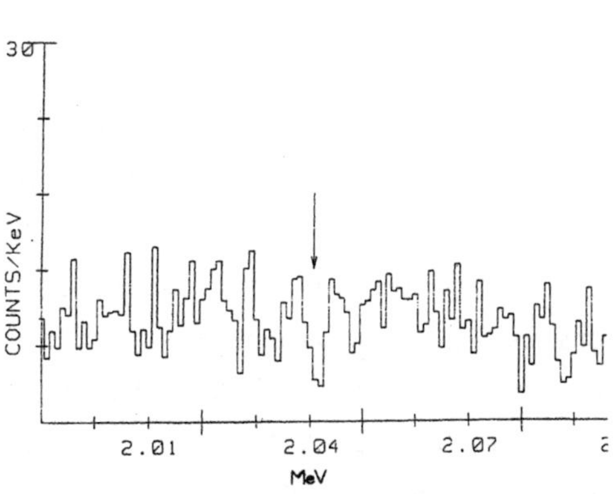

Fig. 10. Ge spectrum with NaI anticoincidence in the region of the expected $0^+\to 0^+$ peak (arrow) for the UCSB-LBL experiment.

A summary of recent results from each of the experiments is given in the table below for the $\beta\beta_{0\nu}$ $0^+\to 0^+$ transition. Also shown is the most recent amount of Ge used and background levels obtained. Note that

D. Caldwell

the lifetime limits are half-lives generally given for 1σ (68% C.L.). Most are obtained by taking the square root of the number of background counts in a broad region in the vicinity of the expected peak. Some are done by a maximum likelihood method which will give a higher limit if there is a dip where a peak is expected. Thus the data in Figs. 9 and 10 would give much higher limits if this procedure were used.

TABLE I: Recent $\beta\beta_{0\nu}$ Results for the $0^+ \to 0^+$ Transition in ^{76}Ge

Group	Background c/keV·y·kg	Ge kg	$T_{1/2}$ Limit 10^{23}y (1σ)
Zaragoza-Bordeaux-Strasbourg	40	2.2	0.2
Caltech-SIN-Neuchatel	4	0.48	0.6
Osaka	6	0.88	0.7
Pacific Northwest-South Carolina	2.4	0.67	1.4
Guelph-Aptec-Queens	∿2	3.0	1.6
Milan	24,4	1.4	1.8
UCSB-LBL	1.7	7.1	4.1

While the limits have improved appreciably in recent years, further improvements will come slowly. Quantities of Ge will not undergo very large increase, backgrounds will improve but not by large factors, and the limit improves only as the square root of the counting time. The one way there could be a significant change in the prospects for Ge experiments is if the Moscow-Leningrad-Heidelberg collaboration succeeds in fully utilizing the large quantity of enriched ^{76}Ge they and they alone are able to obtain. The promised 15 kg of ^{76}Ge would be worth several hundred million dollars in the West. If that could all be used effectively, they would have a factor 20 advantage over the UCSB-LBL experiment. There is normally a large loss of material in making the crystal and a lesser loss in making detectors from the crystal. Also, separated isotopes tend to have radioactive contamination. However, if these problems can be overcome, the next big step for Ge experiments should be taken by this group.

D. Caldwell

4. INTERPRETATION OF THE $\beta\beta_{0\nu}$ RESULTS

The limit of 6×10^{20}y for the $\beta\beta_{0\nu,\chi}$ decay in the UCSB-LBL experiment has been evaluated by Doi, Kotani, and Takasugi[3] to give 90% C.L. limits for the coupling of the Majoron to the electron neutrino. The result, however, depends on the matrix elements used, with values varying from $<1.1\times10^{-3}$ for the Los Alamos[2] matrix elements to $<4.1\times10^{-4}$, using those of Heidelberg.[17] A somewhat lower limit[3] from ^{130}Te depends on additional assumptions, but the present more direct results is more stringent than those from ^{48}Ca[18] and ^{150}Nd.[19]

The variation in these two results indicates only a part of the problem regarding interpreting the lifetime limits in terms of interesting particle physics quantities. Already in the first section of this paper the discrepancy between measured and calculated $\beta\beta_{2\nu}$ lifetimes was noted, and this casts doubt on the $\beta\beta_{0\nu}$ calculations. Conservatively one might scale up the results given below by $\sim$3, but it is quite likely that no such factor is needed.

An especially important caveat applies to deducing a neutrino mass limit from the data. If there is a mixture of neutrinos, these can have opposite CP eigenstates, producing a cancellation which would make the effective neutrino mass for double beta decay, $<m_\nu>$, much less than it might be for, say, single beta decay.[20]

To understand the right-handed-current parameters it is useful to employ a simplified Hamiltonian for $\beta\beta_{0\nu}$ decay:

$$H_W = \frac{-G_F\cos\theta_c}{\sqrt{2}} [j_L^\mu(J_{L\mu}^\dagger+\eta_{LR}J_{R\mu}^\dagger)+j_R^\mu(\eta_{RL}J_{L\mu}^\dagger+\eta_{RR}J_{R\mu}^\dagger)] + \text{h.c.} \qquad (4)$$

Here the $j^\mu_{L(R)}$ are the components of the leptonic current, with L and R standing for left- and right-handedness, while the J's are similar components of the hadronic current, and G_F is the weak coupling constant measured in μ decay, with θ_c the Cabibbo angle. Note that RHC parameters η_{LR}, etc. are defined so that the first subscript refers to the leptonic current and the second to the hadronic current. Another notation used is $\kappa=\eta_{LR}$, $\eta=\eta_{RL}$, and $\lambda=\eta_{RR}$.

For the table below, two more caveats must be stated. First, the

D. Caldwell

$0^+ \to 0^+$ lifetime limit used, 4×10^{23}y, from the UCSB-LBL group is only at the 68% C.L., so that value has a significant uncertainty. Second, the parameters $<m_\nu>$, η_{RL}, and η_{RR} are treated as being independent, with only one at a time taken to be non-zero. In this way the sensitivity of each parameter to the lifetime limit is clearly shown. If there were a positive effect, it is likely that more than one of these would be involved. In particular, Kayser, Petcov, and Rosen[21] have shown that for any gauge theory if RHC exist, then $m_\nu \neq 0$.

TABLE II: Neutrino Mass and RHC Parameter Limits for $T_{1/2}>4\times10^{23}$y

Parameter	Los Alamos[2]	Tübingen[22]	Heidelberg[17]
$<m_\nu>$	2.0	1.5	0.8
$<\eta_{RR}>$	3.3×10^{-6}	2.9×10^{-6}	--
$<\eta_{RL}>$	4.9×10^{-6}	3.5×10^{-8}	

The differing values for the same parameter give an idea of the variations in the matrix elements used, with the exception of $<\eta_{RL}>$. The more recent Tübingen-Jülich calculation includes the effects of using a relativistic p-wave electron wave function and of including nucleon recoil. These impressively small values for the RHC parameters are many orders of magnitude more stringent[3] than those from other experiments for the case of neutrino mass generation by the see-saw mechanism,[23] in which all right-handed Majorana neutrinos are heavier than all left-handed leptons.

The $\beta\beta_{0\nu}$ experiments can also set limits on these heavy Majorana neutrinos, as discussed in the first section. A fourth-generation neutrino could be either a Dirac or a Majorana particle, but extra neutrinos, such as required in the low-energy limit of string theories or in left-right symmetric models, are most frequently Majorana particles. The limits obtained for left-right symmetric theories involve knowledge of the ratio of left-handed to right-handed W boson mass and are given elsewhere.[24] Here the case in which only left-handed W bosons are involved is considered. The Haxton-Stephenson[2] calculation is used to give the straight-line limit seen in Fig. 11, in which the square of the mixing coefficient, U_e^L (coupling of the heavy neutrino to the light or massless electron neutrino), is plotted against the heavy neutrino mass, M_ν. Limits from other experiments are taken from the compilation of Gilman and Rhie.[25] Values below and to the right of the lines are allowed. While the limit from $\beta\beta_{0\nu}$ is much more stringent than those from other experiments, with one small exception, the other limits apply to Dirac neutrinos as well.

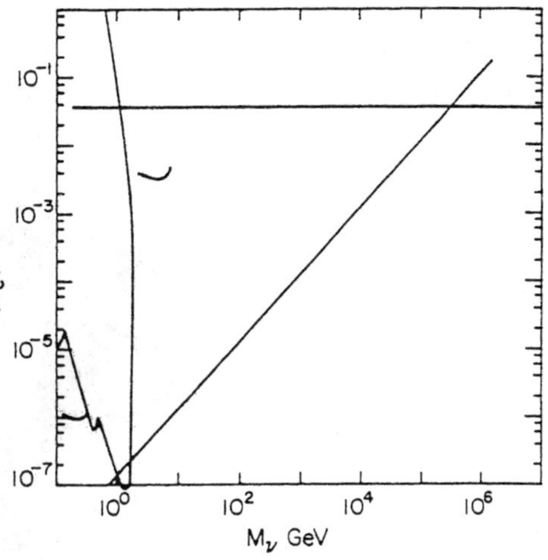

Fig. 11. Limit on the mass of a heavy Majorana neutrino as a function of its probability for mixing with an electron neutrino.

While the limitations imposed by $\beta\beta_{0\nu}$ on light Majorana neutrino mass are relatively well known, those on right-handed currents are much less known. However, the limitation $\beta\beta_{0\nu}$ places on the mass and coupling of heavy Majorana neutrinos seems not to be known even to many experts, despite this not being a new subject.[1] It is necessary to call this to the attention particularly of theorists for two very different reasons. First, the calculations which went into Fig. 11 can be improved considerably. For instance, finite nuclear size effects are very important and have been taken into account quite approximately. Second, the M_ν vs. U_e^L limits can be very useful in restricting models, and even reviews of this subject ignore the $\beta\beta_{0\nu}$ contribution. It may well be that the limitations $\beta\beta_{0\nu}$ imposes on M_ν may turn out to be more important than those it imposes on m_ν.

So far I have covered rather predictable topics, but I wish to close with something less expected. The UCSB-LBL experiment is located outside a small town in the foothills of the Sierra mountains called Oroville, and the people of that town have shown a lot of interest in the experiment. For example, one sees bumper stickers on cars saying, "I brake for neutrinos." The Chamber of Commerce even sponsored a contest for school children to draw their idea of what a neutrino looks like. This was won by an eight-year old, Jim Barnes. For a conference devoted to neutrinos this is an appropriate place to reveal for the very first time a picture of a neutrino:

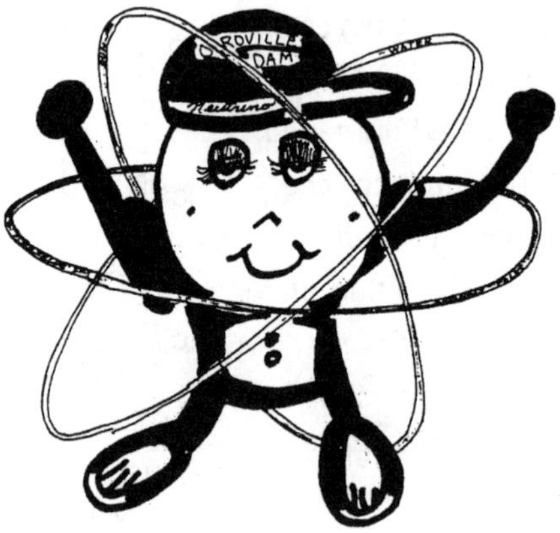

Its mass has not yet been measured!

REFERENCES

1. H. Primakoff and S.P. Rosen, Ann. Rev. Nucl. Part. Sci. 31, 145 (1981).

2. W.C. Haxton and G.J. Stephenson, Jr., Progr. in Part. and Nuc. Phys. 12, 409 (1984).

3. M. Doi, T. Kotani, and E. Takasugi, Prog. of Theo. Phys. Supp. No. 83 (1985).

4. J.D. Vergados, Phys. Rep. 133, No. 1 and 2 (1986).

5. Y. Chikashige, R.N. Mohapatra, and R.D. Peccei, Phys. Lett. 98B, 265 (1981); G.B. Gelmini and M. Roncadelli, Phys. Lett. 99B, 411 (1981).

6. D.O. Caldwell, R.M. Eisberg, D.M. Grumm, D.L. Hale, M.S. Witherell, F.S. Goulding, D.A. Landis, N.W. Madden, D.F. Malone, R.H. Pehl, and A.R. Smith, Phys. Rev. 33D, 2737 (1986).

7. F. Leccia, Ph. Hubert, D. Dassie, P. Mennrath, M.M. Villard, A. Morales, J. Morales, and R. Nunez-Lagos, Nuovo Cimento 78A, 50 (1983); Nuovo Cimento 85A, 19 (1985).

D. Caldwell

8. A. Forster, H. Kwon, J.K. Markey, F. Boehm, and H.E. Henrikson, Phys. Lett. 138B, 301 (1984).

9. P. Fisher, Caltech preprint (1986).

10. H. Ejiri, N. Takahashi, T. Shibata, Y. Nagai, K. Okada, N. Kamikubota, T. Watanabe, T. Irie, Y. Itoh, and T. Nakamura, Nuc. Phys. A448, 271 (1986).

11. F.T. Avignone, III, R.L. Brodzinski, D.P. Brown, J.C. Evans, Jr., W.K. Hensley, J.H. Reeves, and N.A. Wogman, Phys. Rev. Lett. 50, 721 (1983).

12. F.T. Avignone, III, R.L. Brodzinski, D.P. Brown, J.C. Evans, Jr., W.K. Hensley, H.S. Miley, J.H. Reeves, and N.A. Wigman, Phys. Rev. Lett. 54, 2309 (1985).

13. J.J. Simpson, P. Jagam, J.L. Campbell, H.L. Malm, and B.C. Robertson, Phys. Rev. Lett. 53, 141 (1984).

14. E. Fiorini, A. Pullia, G. Bertolini, F. Cappelani, and G. Restelli, Phys. Lett. 25B, 602 (1967) and Nuovo Cimento 13A, 747 (1973).

15. E. Bellotti, O. Cremonesi, E. Fiorini, G. Ligouri, A. Pullia, P. Sverzellati, and L. Zanotti, Phys. Lett. 146B, 450 (1984).

16. D.O. Caldwell, R.M. Eisberg, D.M. Grumm, D.L. Hale, M.S. Witherell, F.S. Goulding, D.A. Landis, N.W. Madden, D.F. Malone, R.H. Pehl, and A.R. Smith, Phys. Rev. Lett. 54, 281 (1985).

17. K. Grotz and H.V. Klapdor, Phys. Lett. 153B, 1 (1985).

18. J.D. Vergados, Phys. Lett. 109B, 96 (1982); erratum, Phys. Lett. 113B, 513 (1982).

19. A.A. Klimenko, A.A. Pomansky, and A.A. Smolnikov, Proc. "Neutrino-84" Conf., Dortmund, W. Germany, 161 (1984).

20. M. Doi, T. Kotani, H. Nishiura, K. Okuda, and E. Takasugi, Phys. Lett. 102B, 323 (1981); L. Wolfenstein, Phys. Lett. 107B, 77 (1981) and Nuc. Phys. B186, 147 (1981); B. Kayser and A.S. Goldhaber, Phys. Rev. D28, 2341 (1983); M. Doi, T. Kotani, H. Nishiura, and E. Takasugi, Prog. Theor. Phys. 69, 602 (1983); B. Kayser, Phys. Rev. D30, 1023 (1984); S.M. Bilenky, N. Nedelcleva, and S.T. Petcov, Nuc. Phys. B247, 61 (1984).

21. B. Kayser, S.T. Petcov, and S.P. Rosen, to be published.

22. T. Tomoda, A. Faessler, K.W. Schmidt, and F. Grümmer, Nuc. Phys. B153, 1 (1985).

23. M. Gell-Mann, P. Raymond, and R. Slansky in "Supergravity", ed. by P. van Niewenhulzen and D. Freedman (North-Holland, Amsterdam, 1979), p. 317; T. Yanagida, Proc. Workshop on Unified Theory and Baryon Number in the Universe, ed. by Sawada and Sugamoto (KEK, 1979).

24. R.N. Mohapatra, University of Maryland preprint, 1986.

25. F.J. Gilman and S.H. Rhie, Phys. Rev. D32, 324 (1985).

VOLUME 59, NUMBER 4 PHYSICAL REVIEW LETTERS 27 JULY 1987

Limits on Neutrinoless $\beta\beta$ Decay Including That with Majoron Emission

D. O. Caldwell, R. M. Eisberg, D. M. Grumm, and M. S. Witherell
Physics Department, University of California, Santa Barbara, Santa Barbara, California 93106

and

F. S. Goulding and A. R. Smith
Lawrence Berkeley Laboratory, Berkeley, California 94720
(Received 5 March 1987)

Neutrinoless double-β decay would require two new phenomena, of which one is lepton-number nonconservation. Among several candidates for the other are nonzero neutrino mass and the emission of a Goldstone boson, such as the majoron. For the former possibility our Ge-detector array now sets a limit for the $0^+ \to 0^+$ transition in ^{76}Ge of $T_{1/2} > 5\times 10^{23}$ yr from the fluctuation in the background or almost a factor of 2 longer by maximum likelihood at the 68% confidence level (C.L.). For the latter process we have a 90%-C.L. limit of 1.4×10^{21} yr, a result in disagreement with a recent possible observation of majoron emission.

PACS numbers: 23.40.Bw, 14.60.Gh, 14.80.Gt

If neutrinoless double-β decay ($\beta\beta_{0\nu}$) were observed,[1-3] it would provide information on physics beyond the standard model in at least two areas: violation of lepton-number conservation plus one or more of a list which includes light neutrino mass, right-handed currents, heavy neutrino mass, supersymmetric particles with R-parity nonconservation,[4] and the existence of a massless Goldstone boson such as the majoron.[5] The last of these would have particularly wide-ranging consequences, since the majoron would result from the spontaneous breaking of baryon- minus lepton-number symmetry, a process also giving mass to light majorana neutrinos. Thus the recent reports[6] of the possible observation of neutrinoless $\beta\beta$ decay induced by majoron emission have aroused widespread interest. We present here results in conflict with those reports, based on an experiment with about an order of magnitude more sensitivity and with lower backgrounds.

The nucleus ^{76}Ge is a candidate for $\beta\beta$ decay and constitutes 7.8% of normal Ge, which can be made into an excellent detector of electron energy. The $\beta\beta$ decay would then be observed as the sum of the energies of the two electrons emitted. The $\simeq$0.1% energy resolution of a Ge detector is particularly useful in the search for the $\beta\beta_{0\nu}$ decay, $^{76}\text{Ge} \to {}^{76}\text{Se}+2e^-$, which would give a spike at the end-point energy of 2.041 MeV. The energy resolution of the Ge is of little use in searching for $\beta\beta_{2\nu}$ decay, $^{76}\text{Ge} \to {}^{76}\text{Se}+2e^- +2\bar{\nu}_e$, which has a four-body decay spectrum peaking at about 0.65 MeV, or the $\beta\beta_{0\nu,B}$ decay, $^{76}\text{Ge} \to {}^{76}\text{Se}+2e^- +B$, giving a three-body spectrum which peaks at about 1.55 MeV. Here B is the massless Goldstone boson which we shall hereafter refer to as a majoron.

Meaningful results for the $\beta\beta_{0\nu,B}$ decay are possible since remarkably low backgrounds have been achieved by placing the experiments underground with good passive shields and making strong efforts of varying success to reduce intrinsic radioactivities. The main distinction in method of background suppression is whether or not an active NaI shield is used. The experiments of Leccia *et al.*,[7] Ejiri *et al.*,[8] and Caldwell *et al.*[9] use NaI, and those of Forster *et al.*,[10] Fisher,[10] Simpson *et al.*,[11] Fiorini and co-workers,[12] and Avignone[13] do not. NaI and associated phototubes are not as free of radioactivity as other materials near the Ge detectors, and despite the self-vetoing, systems with NaI tend to display more and larger full-energy peaks than do those without NaI. However, the peaks themselves do not interfere with observing $\beta\beta$ decay. Rather it is the low-energy tails from those peaks which raise backgrounds, and the NaI provides typically an order of magnitude suppression of the Compton tail. This comes about not only because the Compton scattered photon in most cases exits from the Ge and enters the NaI to veto the event, but also because many of the initial photons are part of a cascade decay, and any of the other time-coincident photons can also veto the event. This suppression is important because the number of counts in a Compton tail can be many times the number of counts in the peak, and this peak/Compton ratio is difficult to model, since it is very dependent on the location of the source relative to the Ge and the presence of any intervening material.

In addition to the Compton tail, there is another source of counts below each peak which should also be taken into account and is easily confused with multiple Compton scattering because of its shape. This effect occurs in a closed-end coaxial cylindrical Ge detector of the type used in $\beta\beta$-decay studies because it has a large exposed open surface at one end of the cylinder which is usually protected by a coating,[14] resulting in a surface which is generally not electrically neutral and thus distorts the internal electric field lines. If any of the γ-ray interactions occurs in the region of distorted field, incomplete charge collection results.[15-17] While only (10–20)% of signals from single interactions would be degraded in this manner, a typical high-energy γ ray undergoes $\simeq 3$ interactions in the detector, so that around half the signals may correspond to energies which are too low. Caldwell *et al.* have made extensive measurements with sources in different locations with respect to the Ge to determine the shapes of the low-energy tails of peaks resulting from both Ge detector charge-collection inefficiencies and Compton scattering.

The Caldwell *et al.* experiment,[9] 200 m underground in the powerhouse of the Oroville, California, Dam, has used between four and eight Ge detectors, averaging about 160 cm^3 (0.9 kg) of fiducial volume each, inside a 15-cm-thick complete NaI shield, which in turn is inside borated polyethylene surrounded by a 20-cm-thick Pb shield. The data presented here for the majoron-induced decay represent a total mass times livetime of 6.7 kg yr. This is by far the largest data sample available for studying this process.

The recording of background events in a Ge detector in which energy is deposited in a NaI crystal or a second Ge detector provides a powerful diagnostic tool. As an example important to the results presented here, we have found in this way that ^{68}Ga provides a significant source of background. The ^{68}Ga activity probably originates from ^{70}Ge$(n,3n)$ ^{68}Ge, produced by cosmic rays when the Ge is above ground or from used Ge detectors going back into the pool of starting material. The ^{68}Ge decays by electron capture (280-day half-life), giving a characteristic Ga x ray. The daughter, ^{68}Ga, decays with a 68-min half-life, emitting a 1.89-MeV β^+. If the positron's annihilation γ's are also absorbed in the Ge detector, a spectrum extending to 2.91 MeV is produced. The ^{68}Ga decay dominates from 0.8 to 1.7 MeV the spectrum which results when the annihilation γ's escape and are registered in the NaI. Confirmation of this important source of background is obtained from (1) the size of the Ga x-ray peak, and (2) the rate at which the background level has decreased since the detectors were placed underground. The quantity of ^{68}Ga will vary from detector to detector, but all detectors used in $\beta\beta$ decay experiments in which sufficiently low-energy measurements have been recorded show the characteristic Ga x-ray peak.

In the energy region of major interest, 1.5 to 3.0 MeV, our data can be accounted for mainly by ^{68}Ga activity and the tails of identified peaks. Since the ^{40}K peak at 1.461 MeV is an order of magnitude larger than any other peak, it is safer to confine the analysis to energies above that peak. Furthermore, the main sensitivity to majoron-induced decay is above 1.5 MeV. However, we can model the ^{40}K tail sufficiently well that the results obtained below do not change appreciably if a wider energy region is included. We have accumulated sufficient counts that 33 γ-ray lines can be identified with known nuclides between 1.5 and 3.0 MeV. Many of these are small and might be ascribed to background fluctuations in data sets with fewer counts, or without Compton suppression. If peaks are not identified and data are averaged over large energy intervals, as is done in looking for majoron-induced decays, then not only are the peak counts subsumed into the average background, but also the far larger number of counts in the tails are not properly identified. In our fit to the data with the peaks subtracted, the tail contributions provide roughly half the background, while the ^{68}Ga activity supplies about $\frac{1}{5}$, although this contribution is both energy and time dependent. The remainder of the background we take as a constant, although replacing it by a term linear or even exponential in energy does not change the results significantly. This $\frac{1}{4}$ to $\frac{1}{3}$ of the data represents our ignorance. With more statistics we would very likely identify other peaks or β spectra.

The resulting fit is shown as a solid curve in Fig. 1, where the data minus peaks are plotted in 50-keV bins. The agreement with the data is good, considering the complexity of the background, the difficulty in modeling the low-energy continuum below each subtracted peak, and the possibility of small peaks that are missed. *Thus no majoron-induced decay is needed to explain the data.* If we fix the contributions from the radioactive backgrounds and allow the number of events in the majoron term to vary, the result is 0 ± 200 events. The error here is purely statistical, however, and does not reflect the uncertainty in the background size and shape. The error in the normalization of the continuous background below peaks is $\pm 10\%$. In addition, different shapes for this

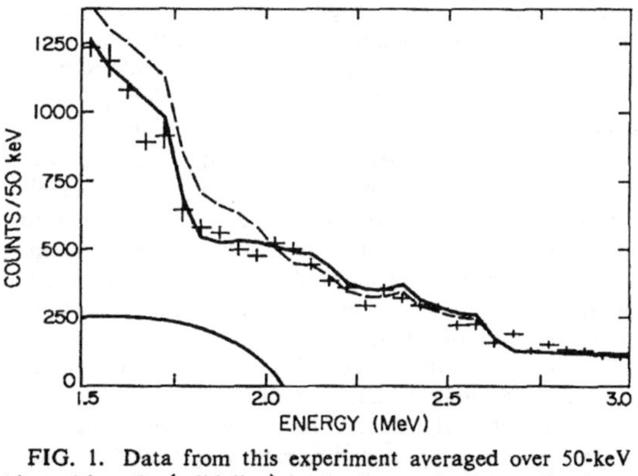

FIG. 1. Data from this experiment averaged over 50-keV bins with a fit (solid line) to the known backgrounds plus a constant term. The dashed curve shows the spectrum including a majoron-induced decay of $T_{1/2} = 6 \times 10^{20}$ yr, with the two background terms fixed at the low end of their allowed range. The other curve shows the expected size and shape of the majoron-induced decay alone.

VOLUME 59, NUMBER 4 PHYSICAL REVIEW LETTERS 27 JULY 1987

background were tried in the fit. The effect of these uncertainties is a systematic error of ± 1200 events in the majoron spectrum. The error in the normalization of the ^{68}Ga contribution is $\pm 20\%$, corresponding to an error in the majoron component of ± 500 events.

The systematic error in the majoron term is ± 1300 events, and the total error, including statistical, is ± 1320 events. The systematic error dominates with this amount of data. This is also reflected in the fact that the errors would need to be increased from the statistical by a factor of about 2 to give a χ^2 per degree of freedom of 1. Using the total error, including systematic, we get a 90%-confidence-level upper limit of 2200 events, which yields $T_{1/2}(\beta\beta_{0\nu,B}) > 1.4\times 10^{21}$ yr. This would correspond[3] to a limit on the coupling of the majoron to the electron neutrino of $< 7\times 10^{-4}$ with use of the matrix elements of Ref. 2 and $< 3\times 10^{-4}$ with use of those of Grotz and Klapdor.[18] These limits on the coupling are probably more stringent than those from ^{48}Ca,[19] ^{82}Se,[20] ^{130}Te,[3] and ^{150}Nd,[21] but uncertainties in nuclear matrix elements make comparisons difficult.

If we try to fit the data by including a contribution of majoron-induced decay at the level of $T_{1/2}(\beta\beta_{0\nu,B}) = 6\times 10^{20}$ yr suggested by the data of Ref. 6, we get the dashed curve of Fig. 1. However, this requires that the ^{68}Ga and tail contributions be as small as the errors allow. Even taking into account the systematic errors in these backgrounds, the probability for the fit with a majoron-induced component corresponding to $T_{1/2} = 6\times 10^{20}$ yr is 10^{-4}.

Our result is clearly in disagreement with that of Ref. 6. It is interesting that one of our eight detectors considered alone shows a spectrum remarkably similar to that of Ref. 6. In both cases there is prominent evidence for an α at 5.3 MeV, which gives a large continuum extending down to the vicinity of 2 MeV. By comparing this spectrum with that for our other detectors with much lower α background, we see that the degraded α's provide a contribution to the background which gradually falls toward lower energies, combining with the normal rising background seen in Fig. 1 to produce an almost flat background down to about 2 MeV. The rise in counts toward lower energy then appears significant. This detector has a considerably larger level of background than any of our others, and data from it were not included in the majoron analysis.

In addition to the misleading flat background at higher energies making the rise at lower energies look more important in the data of Ref. 6, we believe that the authors are also being misled by believing that "the only background in the spectrum above the 1461-keV γ-ray peak is the broad 5.3-MeV α peak and its degraded continuum." While they exclude ^{68}Ga as being the *sole* source of the rise in the spectrum, they certainly have some of this activity. They also may not identify other γ-ray peaks because of insufficient statistics, and in their case with no Compton suppression the tails of those peaks would contribute many more counts than the peaks themselves. One other issue is that the data do not show the steep rise from the end point toward lower energies expected from majoron-induced decay, as shown by the curve in Fig. 1. This is the case for both the data of Ref. 6 (although the statistics in this case obscure the point) and our own, which has an order of magnitude more mass times lifetime in this energy range.

The $\beta\beta_{0\nu}$ decay, which would give a spike at 2.041 MeV, is much easier to observe and to interpret than the spectrum of the $\beta\beta_{0\nu,B}$ decay. Because our limit for this decay has improved considerably since our latest publication,[9] which then gave the most stringent limit from any experiment, we also present this new result here. For a data sample of 8.4 kg yr, the time and detector averaged background in the vicinity of 2.04 MeV is now 1.4 counts/keV·kg·yr. On the basis of the fluctuation allowed in the background around the expected peak, for which our FWHM resolution is 3.36 keV, we get a 68%-confidence-level lifetime limit $T_{1/2}(\beta\beta_{0\nu}) > 5\times 10^{23}$ yr, using the analysis described previously.[9] It is conventional to use instead a maximum-likelihood analysis, but there is a significant dip in the energy region where the peak is expected, so that using the analysis procedure recommended by Aguilar-Benitez *et al.*,[22] we obtain a much larger lifetime limit of 9×10^{23} yr.

These results are to be compared with the best published result from a single experiment[12] of 3.3×10^{23} yr by maximum likelihood. Even 5×10^{23} is a factor of about 2 better than a recently published[13] "world limit" which includes some of our data. For the present result, adding the data from all other experiments changes the limit by less than the uncertainty in assigning a limit from our experiment alone.

The interpretation of this limit in terms of physically interesting quantities will be left to a subsequent publication, except to indicate the sensitivity of the result by its effect on a limit for light majorana neutrino mass. This result gives a *lower* limit on the effective mass of the neutrino for the $\beta\beta_{0\nu}$ process, but were a positive result observed at this lifetime value (and we choose to use 5×10^{23} yr to be conservative), then a neutrino would have to exist with this or a *larger* neutrino mass.[23] To show the effect of using different nuclear matrix elements (calculated in the references given) to interpret the result, we give the neutrino-mass limits to more significant figures than the uncertainty in the lifetime warrants: $\langle m_\nu \rangle <$ 1.8,[2] 1.3,[24] and 0.7 eV.[18]

This result for light neutrino mass emphasizes the power of double-β decay for probing physics beyond the standard model. Unfortunately, so far only minimum lifetime limits have been obtained, and our results do not support a discovery as exciting as neutrinoless double-β decay induced by majoron emission. Fisher *et al.*[25] have reached a similar negative conclusion.

VOLUME 59, NUMBER 4 PHYSICAL REVIEW LETTERS 27 JULY 1987

This work was supported in part by the U.S. Department of Energy, and the results could not have been obtained without the invaluable help of D. L. Hale, D. A. Landis, N. W. Madden, and R. H. Pehl.

[1]H. Primakoff and S. P. Rosen, Annu. Rev. Nucl. Part. Sci. **31**, 145 (1981).

[2]W. C. Haxton and G. J. Stephenson, Jr., Prog. Part. Nucl. Phys. **12**, 409 (1984).

[3]M. Doi, T. Kotani, and E. Takasugi, Suppl. Prog. Theor. Phys. **83**, 1 (1985).

[4]R. N. Mohapatra, University of Maryland Report No. 86-185, 1986 (unpublished).

[5]Y. Chikashige, R. N. Mohapatra, and R. D. Peccei, Phys. Lett. **98B**, 265 (1981); G. B. Gelmini and M. Roncadelli, Phys. Lett. **99B**, 411 (1981); H. M. Georgi, S. L. Glashow, and S. Nussinov, Nucl. Phys. B **193**, 297 (1981).

[6]F. T. Avignone, III, R. L. Brodzinski, H. S. Miley, and J. H. Reeves, unpublished; F. T. Avignone, III, in Proceedings of the 1987 Annual Meeting of the Division of Particles and Fields of the American Physical Society, Salt Lake City, UT, 1987 (to be published), and in Proceedings of the Telemark IV Meeting, Ashland, WI, 1987 (to be published); The New York Times, 14 January 1987, p. 6; M. M. Waldrop, Science **235**, 534 (1987).

[7]F. Leccia, Ph. Hubert, D. Dassie, P. Mennrath, M. M. Villard, A. Morales, J. Morales, and R. Nunez-Lagos, Nuovo Cimento **78A**, 50 (1983), and **85A**, 19 (1985).

[8]H. Ejiri, N. Takahashi, T. Shibata, Y. Nagai, K. Okada, N. Kamikubota, T. Watanabe, T. Irie, Y. Itoh, and T. Nakamura, Nucl. Phys. **A448**, 271 (1986).

[9]D. O. Caldwell, R. M. Eisberg, D. M. Grumm, D. L. Hale, M. S. Witherell, F. S. Goulding, D. A. Landis, N. W. Madden, D. F. Malone, R. H. Pehl, and A. R. Smith, Phys. Rev. Lett. **54**, 281 (1985), and Phys. Rev. D **33**, 2737 (1986).

[10]A. Forster, H. Kwon, J. K. Markey, F. Boehm, and H. E. Henrickson, Phys. Lett. **138B**, 301 (1984); P. Fisher, in *Proceedings of the Twenty-First Rencontre de Moriond on Massive Neutrinos in Particle Physics and Astrophysics, Tignes, Savoie, France,* edited by O. Fackler and J. Trân Than Vãn (Editions Frontières, Gif-sur-Yvette, France, 1986), p. 615.

[11]J. J. Simpson, P. Jagam, J. L. Campbell, H. L. Malm, and B. C. Robertson, Phys. Rev. Lett. **53**, 141 (1984).

[12]E. Fiorini, A. Pullia, G. Bertolini, F. Cappelani, and G. Restelli, Phys. Lett. **25B**, 602 (1967), and Nuovo Cimento **13A**, 747 (1973); E. Bellotti, O. Cremonesi, E. Fiorini, G. Liguori, A. Pullia, P. Sverzellati, and L. Zanotti, Phys. Lett. **146B**, 450 (1984), and Nuovo Cimento **95A**, 1 (1986).

[13]F. T. Avignone, III, R. L. Brodzinski, D. P. Brown, J. C. Evans, Jr., W. K. Hensley, J. H. Reeves, and N. A. Wogman, Phys. Rev. Lett. **50**, 721 (1983), and **54**, 2309 (1985), and Phys. Rev. C **34**, 666 (1986), and Phys. Rev. D **35**, 1713 (1987).

[14]W. L. Hansen, E. E. Haller, and G. S. Hubbard, IEEE Trans. Nucl. Sci. **27**, 247 (1980).

[15]M. G. Strauss, I. S. Sherman, R. W. Bannon, IEEE Trans. Nucl. Sci. **21**, 296 (1974).

[16]J. M. Jaklevic and F. S. Goulding, IEEE Trans. Nucl. Sci. **19**, 384 (1972).

[17]F. S. Goulding, Nucl. Instrum. Methods **142**, 213 (1977).

[18]K. Grotz and H. V. Klapdor, Phys. Lett. **153B**, 1 (1985).

[19]J. D. Vergados, Phys. Lett. **109B**, 96 (1982), and **113B**, 513 (1982).

[20]S. R. Elliott, A. A. Hahn, and M. K. Moe, Phys. Rev. Lett. **56**, 2582 (1986).

[21]A. A. Klimenko, A. A. Pomansky, and A. A. Smolnikov, in *Proceedings of the Eleventh International Conference on Neutrino Physics and Astrophysics, Dortmund, West Germany, 1984,* edited by K. Kleinknecht and E. A. Paschos (World Scientific, Singapore, 1985), p. 161.

[22]M. Aguilar-Benitez *et al.* (Particle Data Group), Phys. Lett. **170B**, 1 (1986); see p. 55.

[23]B. Kayser, in Proceedings of the Twenty-Third International Conference on High Energy Physics, Berkeley, California, 1986, edited by S. Loken (World Scientific, Singapore, to be published).

[24]T. Tomada, A. Faessler, K. W. Schmidt, and F. Grümmer, Nucl. Phys. **A452**, 591 (1986).

[25]P. Fisher, F. Boehm, E. Bovet, J.-P. Egger, K. Gabathuler, H. Henrickson, and J.-L. Vuilleumier, California Institute of Technology Report No. CALT-63-487, 1987 (to be published).

J. Phys. G: Nucl. Part. Phys. 17 (1991) S137–S144. Printed in the UK

Double Beta Decay — Present and Future

David O. Caldwell*

Department of Physics, University of California, Santa Barbara, CA 93106, USA

ABSTRACT: The best lifetime limit (1.2×10^{24} years at the 90% C.L.) for neutrinoless double beta decay, $\beta\beta_{0\nu}$, comes from the UCSB/LBL experiment. This corresponds to a Majorana neutrino mass of about 1–3 eV/c^2, which can now be stated as an upper limit, since the possible cancellations due to mixed neutrinos of opposite CP eigenstates have been severely restricted by laboratory and cosmological constraints. Significant mass limit improvements should become possible with superconducting detectors, but if recent results from the solar neutrino experiments are confirmed, $\beta\beta_{0\nu}$ may not be detectable in the laboratory.

1. INTRODUCTION

Processes which are forced to be second order in the weak interaction have provided some of the most stringent constraints on particle physics ideas which attempt to go beyond the Standard Model of the Quantum Chromodynamic and Electroweak gauge theories. Examples of such sensitive processes are K^0–$\overline{K^0}$ mixing and neutrinoless double beta decay. Limits on the latter process have provided severe constraints on such new physics as lepton-number violation, electron neutrino mass, right-handed currents, heavy Majorana neutrinos which mix with the electron neutrino, Majorons, and masses of supersymmetric particles. More generally, it has served to restrain many theoretical speculations.

Thus neutrinoless double beta decay, $\beta\beta_{0\nu}$, has played an important role, even though the experimental evidence for it has all been negative. After mentioning the status of the UCSB/LBL experiment which has given the leading $\beta\beta_{0\nu}$ lifetime limit for some years, we will discuss the likelihood that the constraints imposed by that limit are even more stringent than has usually been assumed. While the $\beta\beta_{0\nu}$ process can continue to restrain some theoretical ideas, it is now possible that new experimental results on solar neutrinos may remove a lot of the motivation for continuing to search for this second-order weak decay.

2. STATUS OF THE UCSB/LBL EXPERIMENT

The UCSB/LBL double beta decay experiment [1], which is 600 m.w.e. underground in the powerhouse of the Oroville Dam in the foothills of the Sierra Mountains of California, has had up to eight 0.9 kg Ge detectors. These have been used to search for the process $^{76}Ge \rightarrow {}^{76}Se + 2e^-$, which would give a spike from the summed electron energies at 2.041 MeV. The detectors were inside the cavity formed by 10 blocks of 15-cm-thick NaI, providing an active veto with a 30 keV threshold. The NaI is in turn inside a very pure Pb shield of 20-cm thickness. The detectors, mounted in groups of two on single-crystal silicon cold fingers (4-mm thick) to keep the detectors near liquid-nitrogen temperature, were enclosed (as were the cold fingers) in an ion pumped electroformed

* Supported in part by the U.S. Department of Energy.

New Possibilities in a Double Beta Decay Experiment Using Enriched ^{76}Ge Inside of an Active Si(Li) Shielding

L.A. Popeko[1], *A.V. Derbin*[1], *I.A. Kondurov*[1], *V.V. Martynov*[1], *H.V. Klapdor*[2], *and J.Metzinger*[2]

[1]Leningrad Nuclear Physics Institute, Gatchina, Leningrad district, 188350, USSR
[2]Max-Planck-Institut für Kernphysik, D-6900 Heidelberg, F. R. Germany

At present some direct neutrinoless double beta-decay experiments are in progress at a sensitivity level of 10^{23} y. The presently best limit for ^{76}Ge of $T^{0\nu}_{1/2} \geq 3.9 \cdot 10^{23}$ y [1] corresponds to a Majorana neutrino mass $m_\nu \leq 1$ eV [2].

Common feature of present experiments is the application of germanium detectors inside low background passive or active shielding. A decisive quantity for such experiments is the background counting rate of a germanium detector per energy resolution interval at the energy of 2.041 MeV. Measurements in underground laboratories at levels deeper than 4000 m water equivalent (w.e.) eliminate practically the cosmic rays. The residual background of $\sim 10^{-4}$ (keV • h • 100 cm^3)$^{-1}$ is due to the natural radioactivity of the low background shielding.

As possibilities to improve the experimental sensitivity have been discussed:
- increase of the active volume of the germanium detector used;
- use of isotopically enriched ^{76}Ge as detector material;
- use of a passive shielding of semiconductor purity.

The possibilities of active scintillation shielding [3] are practically exhausted.

A new way to improve the sensitivity is to construct an active silicon shielding which consists of a great number of Si(Li) detectors.

Recently a Si(Li) multidetector for ($\tilde{\nu}$,e) scattering experiments was proposed [4]. The detector contains 400 kg of Si(Li) modules with an active volume of a 100 cm^3 each. The modules are packed closely without any support materials and are surrounded by an 80 mm thick mercury shielding in titanium containers. These containers are placed inside a vacuum cryostat at liquid nitrogen temperature. The outside preamplifiers are connected with the Si(Li) modules by HP copper conductors. The main feature of the detector is application of semiconductor purity components (the contaminations of ^{238}U and ^{232}Th are less than 10^{-20} Ci/g).

A background test has been made of a 25 kg detector model consisting of 305 modules. Figure 1 (curves 5 and 6) shows the experimental data at a level of 30 m w.e. The background index is 10^{-4} (keV • h • 100 cm^3)$^{-1}$ [4]. This result is not changed when testing at the level of 1100 m w.e., i.e., the background counting rate is caused by the materials nearest to the detector.

The result obtained was used for the calculation of the background index of a 200 cm^3 HP Ge detector placed in the centre of the 400 kg Si(Li) multidetector

in Proc. WEIN'86, Heidelberg, Germany, July, 1986, *ed. H.V. Klapdor, Springer-Verlag Berlin, Heidelberg,* (1986) 703 – 704

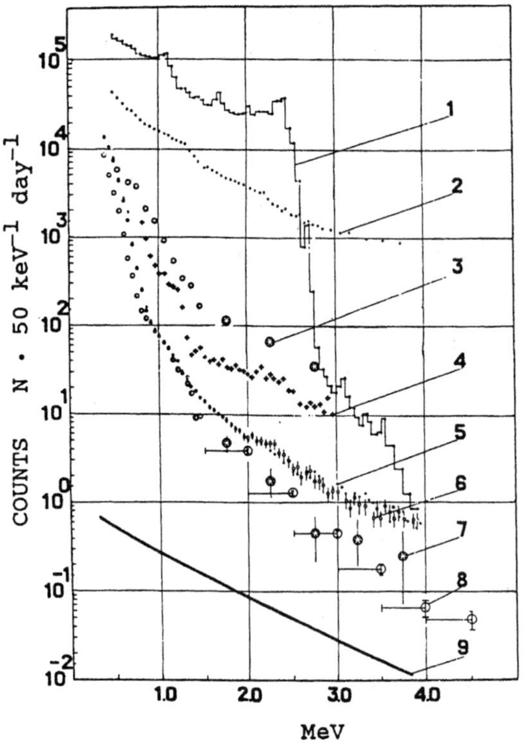

Fig. 1. The results of the background test of the 25 kg Si(Li) active shielding:
1 - the calibration spectrum of ^{24}Na for the whole detector;
2 - the background spectrum of the whole detector at ground level;
3 - the background of the central module;
4 - the background spectrum at a level of 30 m w.e., the preamplifier components are inside the first layer (45 mm) of the passive shielding;
5,6 - the background with the preamplifier components outside the 200 mm passive shielding and after increasing the active shielding;
7 - the same as curve 6 but for the central module;
8 - the results of the Reines's $(\bar{\nu},e)$ scattering experiment;
9 - the calculated recoil electron spectrum of the $(\bar{\nu},e)$ scattering experiment for a neutrino flux of $6 \cdot 10^{12}$ cm^{-2} s^{-1}.

which consists of 2000 modules. The index calculated is 10^{-5} (keV · h · 100 $cm^3)^{-1}$. In case of using 7.6 kg enriched ^{76}Ge as HP detector material inside this active shielding the background counting rate will be 30 counts/y, which corresponds to a sensitivity level of $T_{1/2} \geq 10^{25}$ y for the neutrinoless double beta-decay process.

References

[1] D.O. Caldwell et al., this conference
[2] H.V. Klapdor, Proceed. VI Moriond Workshop on Massive Neutrinos in Astrophysics and Particle Physics, Tignes, France, 25.1-1.2.1986, Editions Frontiéres, Gif-sur-Yvette, p. 597, K. Grotz, H.V. Klapdor, Phys. Lett. 153B (1985) 1 and Nucl. Phys. A, in press (1986)
[3] F.T. Avignone et al., Nucl. Instr. Meth. A239, 207 (1985)
[4] A.V. Derbin, L.A. Popeko, A.V. Cherny and G.A. Shishkina, Pisma v GETF 43, 164 (1986) (in Russian)

Modern Physics Letters A, Vol. 5, No. 17 (1990) 1299 – 1306

NEW RESULTS IN THE ITEP/YePI DOUBLE BETA-DECAY EXPERIMENT WITH ENRICHED GERMANIUM DETECTORS

A. A. VASENKO, I. V. KIRPICHNIKOV, V. A. KUZNETSOV, and A. S. STAROSTIN

Institute for Theoretical and Experimental Physics, Bolshaya Cheremushkinskaya, 25, Moscow 117259, USSR

and

A. G. DJANYAN, V. S. POGOSOV, S. P. SHACHYSISYAN, and A. G. TAMANYAN

Yerevan Physical Institute, Markaryan 2, Yerevan 375036, USSR

Received 23 April 1990

The search for double beta-decay of ^{76}Ge was carried out with a detector fabricated of enriched material (85% abundance of ^{76}Ge compared with 7.8% natural abundance). Measurements have been performed by the ITEP/YePI team in the Avan salt mine, 245 meters underground, situated in Yerevan, Armenia. Evidence for two-neutrino double beta-decay of ^{76}Ge with half-life of $T_{1/2}(2\nu) = (9 \pm 1) \cdot 10^{20}$ y was obtained. New limits for neutrinoless double beta-decay, $T_{1/2}(0\nu) > 1.3 \times 10^{24}$ y, and double beta-decay with majoron emission $T_{1/2}(0\nu, B) > 1 \times 10^{22}$ y were obtained at 68% CL from mean background fluctuations. Limit for 0ν-decay derived by the maximum likelihood method was $T_{1/2} > 2.0 \times 10^{24}$ y.

1. Introduction

There are three different principal modes of double beta decay, the search for which is being intensively performed:

$$(A, Z) \to (A, Z+2) + 2e \qquad 0\nu\text{-decay},$$
$$(A, Z) \to (A, Z+2) + 2e + 2\nu \qquad 2\nu\text{-decay},$$
$$(A, Z) \to (A, Z+2) + 2e + B \qquad 0\nu, B\text{-decay}.$$

The three processes can be distinguished experimentally by the sum energy spectra of emitted electrons.

The ITEP/YePI collaboration performed searches for the double beta decay of ^{76}Ge with a semiconductor Ge-detector fabricated of enriched material (85% of ^{76}Ge as compared to 7.8% natural abundance).[1] The measurements were started in 1987.[2] Germanium detectors seem to be almost ideal for searching for 0ν-decay due to excellent energy resolution and high efficiency. Measurements of the life-time limit for ^{76}Ge have thus far provided the most stringent constraints on the theory.[3–5]

The operation of a Ge-detector of highly enriched material was the next step in the experimental technique for double beta decay investigations. Since the number of the expected events is proportional to the quantity of ^{76}Ge, background to the total volume, such a detector has a much higher signal-to-background ratio and efficiency. The search for 2ν-decay (or 0ν, B-decay) with natural Ge-detectors has no obvious advantage compared to the other methods, due to the broad $2e$ energy distribution. Yet the ITEP/YePI experiment provided the opportunity to compare the spectra obtained with enriched and natural Ge crystals. It resulted in an essentially increased sensitivity, and direct evidence for two-neutrino decay of ^{76}Ge was found. Simultaneously, the upper limit for the life-time for 0ν, $\beta\beta$-decay was improved.

2. The Apparatus

The experimental set up (Fig. 1) included an array of three Ge-detectors inside a NaI(T1) active anticoincidence shield. All the three crystals were mounted in a

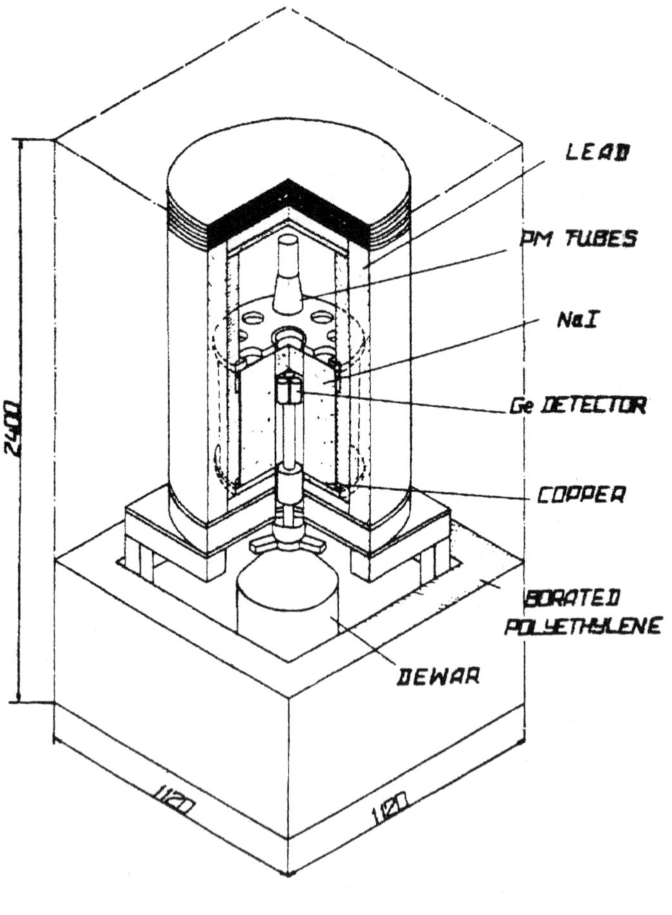

Fig. 1. The experimental setup.

Table 1.

Isotope	Natural crystal	Enriched crystals
^{70}Ge	20.5	< 0.1
^{72}Ge	27.4	< 0.2
^{73}Ge	7.8	< 0.2
^{74}Ge	36.5	15.0
^{76}Ge	7.8	84.6

common cryostat. Parts of the cryostat were made of oxygen-free copper, titanium and teflon. The titanium cryostat cover had a thickness of 0.7 mm. The active shield assembly had the form of a cylinder with a central hole. The dimensions of the NaI are: 450 mm in height, 400 mm in mean diameter with a 135 mm diameter central hole. The set consisted of 8 light-isolated sections in a common copper vessel. The ninth NaI crystal closed the central hole. Nine PM tubes looked through these sections. To minimize the quantity of material between the Ge-detectors and the sodium iodine crystals, 0.1 mm of mylar was used for the central hole walls and the bottom wall of the central NaI crystal.

The cryostat and the active shield were surrounded by 6 cm of borated polyethylene, which in turn was surrounded by a 10 cm lead shield and 8 cm of borated polyethylene. The measurements were performed in the Avan salt mine, 245 meters underground (645 m.w.e.), situated in Yerevan, Armenia. The detector set with two enriched crystals, 106 cm^3 and 109 cm^3, and one of natural germanium, 115 cm^3, has operated since the beginning of 1988.

Isotopic abundances of the germanium samples (%) are given in Table 1. A more detailed description was given elsewhere.[6]

3. Neutrinoless Double Beta-Decay Results

The existence of 0ν-decay would be seen as a narrow peak at the summed energies of the two electrons at 2041 keV. The peaks width was determined by the energy resolution of the apparatus alone.

The summed spectrum of the two enriched detectors (17937 h total live time) in the vicinity of the expected peak is shown in Fig. 2. The energy resolution was measured with the FWHM of the 1764 keV γ-ray (^{214}Bi) peak and for the summed cumulative spectrum; it was 4.0 keV. The life-time limit for the $0^+ \to 0^+$ transition, obtained from the square root of the mean background (2.2 c/keV · kg · y, 2030-2050 keV), was

$$T_{1/2} > 1.3 \times 10^{24}\,\mathrm{y}\ (68\%\ \mathrm{CL})\,,$$

while that derived from the maximum likelihood method was

$$T_{1/2} > 2.0 \times 10^{24}\,\mathrm{y}\,.$$

The setup with two enriched Ge-detectors was the first of a new generation of apparatus for double beta-decay searches. Comparison of the ITEP/YePI apparatus[6] with the other existing devices[3,4] is given in Table 2.

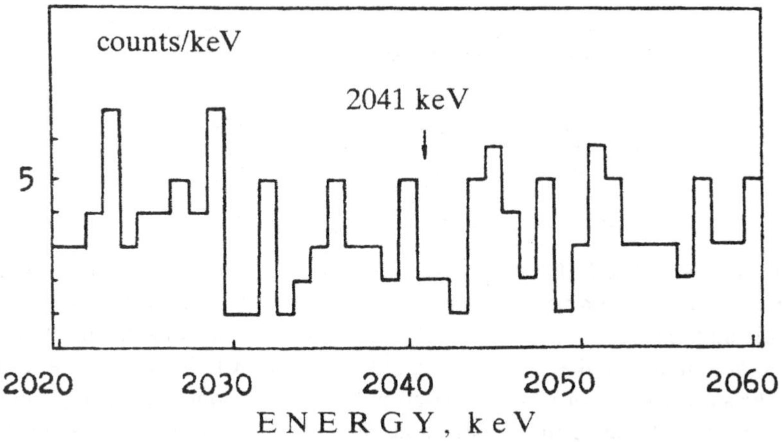

Fig. 2. The summed spectra of the enriched detectors (17937 h total live time) near the expected 2041 keV peak due to 0ν-decay.

Table 2.

	Detect. volume, cm^3	Resol. (FWHM) keV	Mean background (2000-2100 KeV) c/keV·y·kg (natur. Ge)	c/keV·y·kg (^{76}Ge)	Accumul. statist. y·kg (^{76}Ge)	Lifetime limit 10^{24}y, (68% CL)
CT/SIN/NEU	1140	3.3	2.9	37.0	0.306	0.27
UCSB/LBL	1300	3.3	1.4	18.0	0.944	0.7[a]
ITEP/YePI	215	3.7	2.2	2.6	0.866	1.3

[a] Mean background square root.

4. Evidence for Double Neutrino Decay

The expected $2e$ spectrum due to the two-neutrino $\beta\beta$-decay of ^{76}Ge has a broad maximum at 0.6 – 0.7 MeV; the effect is practically zero at 1.5 MeV. On the other hand, the experimental spectra show a sharp increase of background below 0.7 MeV. Therefore the search for the effect was limited to within the energy interval 0.4 – 1.5 MeV.

The procedure included a comparison of the spectra, simultaneously measured with detectors fabricated of enriched and natural Ge. To provide maximum similarity of background conditions for all the detectors, the three crystals were mounted very close together, without any intervening material between them, inside a common cryostat. The crystals have practically the same volumes and geometrical dimensions.

Careful analyses of the two main possible sources of background, such as gamma-radiation of construction materials, and beta-decay due to intrinsic impurities or cosmogenic-induced processes were performed.

It was concluded that in the energy range of interest, 0.7 – 1.5 MeV, the background was dominated by electrons due to Compton scattering of gamma quanta.

The high resolution of the detectors was extremely useful in the analysis of the gamma-ray spectra. The most intense was the 1.46 MeV gamma line of ^{40}K. Gamma-radiations of the members of the U and Th chains, ^{60}Co isotopes, and ^{137}Cs (below 0.7 MeV) were the main source of the remaining part of the background.

To determine the admixtures of U and Th in the crystals, special data collection was made in which the energy range was expanded to 8 MeV. Limits for these contaminations were found by counting 4 – 5 MeV alpha-particles. The limits are $< 5 \times 10^{-14}$ g/g for U^{238} and $< 1.5 \times 10^{-13}$ g/g for ^{232}Th. Possible background due to these contaminations was found to be negligible.

A surplus of counts in the enriched detector spectrum could also be produced by the capture of neutrons throughout the reaction

$$^{76}\mathrm{Ge}\ (n, \gamma)\ ^{77}\mathrm{Ge} \rightarrow {}^{77}\mathrm{As} + e^- ; \quad (T_{1/2} = 11.3\ \mathrm{h}\ ; E_\beta = 2.90\ \mathrm{MeV})\ .$$

The ^{77}Ge production rate was calculated taking into account two mechanisms of neutron creation inside the apparatus. These were cosmic ray muons and (α, n) processes due to the decays of the U and Th chain daughter elements. The estimate yields an upper limit for the Ge production rate of 4×10^{-4} c/h/det, which should be compared with the experimentally observed excess in the collected spectra, 0.1 – 0.2 c/h/det., in the 0.7 – 1.2 MeV energy range.

Because of the significant difference between the ^{76}Ge abundances of the crystals, the spectrum collected with the natural Ge detector could be assumed practically as the background. A possible surplus due to 2ν-decay was deduced by finding the difference

$$N(E) = F(E) - k(E) \cdot f(E)\ ,$$

where $F(E)$ and $f(E)$ are the spectra of enriched and natural Ge detectors, and $k(E)$ is the relative efficiency for the background.

The values of $k(E)$ for all combinations of the detectors were measured using a set of radioactive sources imitating the continua of the background spectra with 10 – 15% accuracy. A set including ^{40}K, ^{226}Ra, Th, ^{60}Co, and ^{137}Cs was used for this purpose.

The procedure has been performed several times throughout the whole cycle of data collection; the final accuracy of the $k(E)$ measurements was determined to be about 2%.

The adopted procedure avoided almost completely the difficulties due to the uncertainties in the detector active volumes. The absolute value of the lifetime (or limit) was dependent on the enriched detector volumes, and an additional uncertainty which did not exceed the uncertainty of the detector volume measurements which were typically 5 – 10%.

The difference, $N(E)$, between the summed enriched detector spectra and that of the control natural Ge crystal is given in Fig. 3a. It included 17259 h total live time of measurements with the enriched detectors (summed) and 8470 h with the control detector. The solid line shows the calculated $2e$ spectrum for $T_{1/2}(2\nu) = 9 \times 10^{20}$ y. The difference between the spectra of identical (enriched) detectors is

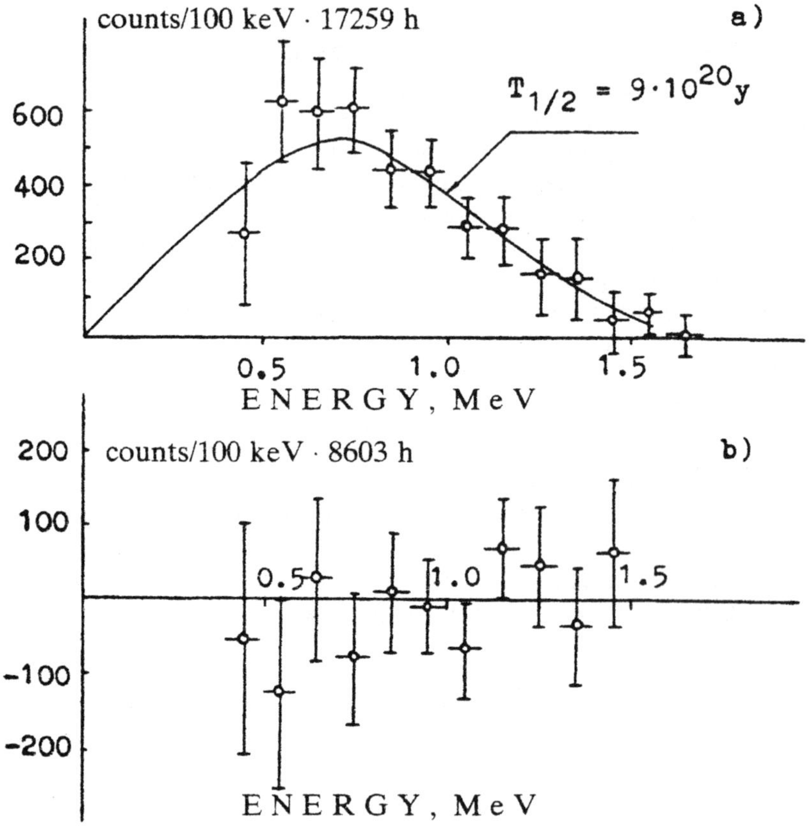

Fig. 3. (a) Difference between the summed spectrum of the enriched detectors (17259 h) and the control detector spectrum (8470 h). (b) Difference between spectra of the identical detector spectra (enriched). The live times are 8603 h and 8656 h, respectively.

given in Fig. 3b (8603 h and 8656 h). Since $N(E)$ should be zero if the backgrounds are the same, the relative efficiency $k(E)$ was determined correctly.

A comparison of the above results and the analysis of the origin of the background gave assurance that the observed effect could not be explained by any known background, but should be attributed to the two-neutrino $\beta\beta$-decay of ^{76}Ge. The best fit to the experimental results was achieved for

$$T_{1/2}(2\nu) = (9 \pm 1) \times 10^{20} \mathrm{y} .$$

5. A Search for Decay with Majoron Emission

In 0ν, B-decay, the summed electron energy is distributed over a broad spectrum which peaks at 1.5 MeV. To search for this process, a method different from that described above was used. Comparison of the spectra was performed in the energy range 1.5 – 2.0 MeV, where only 40% of the expected events would be observed

[redacted]

while the background would be several times smaller as compared with the energies below the intense 1.46 MeV gamma-line of ^{40}K.

Since the radiation component of the background was dominated by γ-rays from U and Th chains, a combination of only two gamma-ray sources, Ra and Th, was used for the relative detector efficiency, $k(E)$, measurements. Two corrections were included in the calculations. Firstly the additional events due to the "tail" from 2ν-decay in the enriched detectors were subtracted. This correction was wholly determined by the measured ^{76}Ge life-time, $T_{1/2} = 9 \times 10^{20}$ y.

The second correction was due to ^{68}Ga positron decay in the natural Ge-crystal. The value of this correction was estimated by examining the background change in the 1.8 – 2.1 MeV energy range over an one and a half year period (the energy interval was chosen as the most sensitive to this decay).

There is a decrease in time of the number of events in the natural Ge-crystals, which can be attributed to ^{68}Ga beta-decay with the life-time of $T_{1/2} = 280$ d. The fraction of this additional background was about 2 – 3% (averaged over the whole cycle of the data collection) in the 1.8 – 2.1 MeV interval, and several times smaller below 1.7 MeV.

The final results were shown in Fig. 4 including 11961 h of summed live time data for the enriched detector spectra and 7501 h of data for the control detector.

Within the accuracy of the data, double beta-decay with majoron emission was not observed. A life-time limit at 68% CL was $T_{1/2} > 1 \times 10^{22}$ y.

6. Conclusion

The data for 0ν-decay and 0ν, B-decay allowed limits to be placed on the effective Majorana neutrino mass $\langle m_\nu \rangle$ and on $\langle g_B \rangle$, the coupling constant of the majoron to the electron neutrino. As the results are strongly dependent on the theoretical nuclear matrix elements, the two sets of the estimates are presented. Matrix elements given by Haxton and Stephenson,[7] and by Engel, Vogel, and Zirnbauer[8] (Table 3) were used for this purpose.

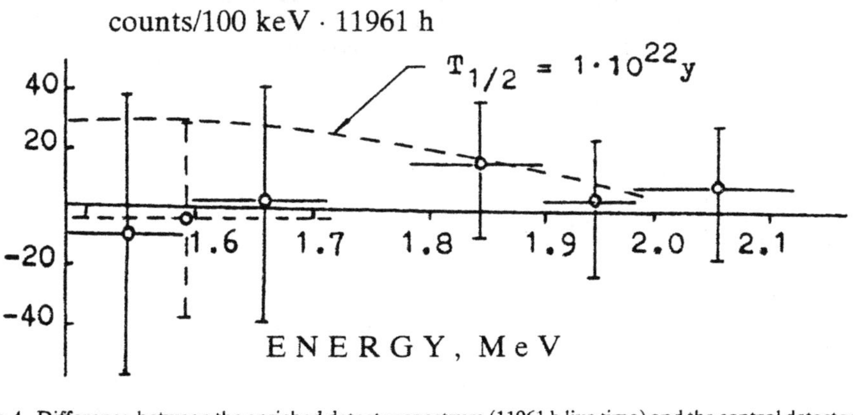

Fig. 4. Difference between the enriched detector spectrum (11961 h live time) and the control detector spectrum (7501 h).

Brief Review

Table 3.

	$\langle m_\nu \rangle$	$g_B \times 10^4$
Haxton, Stephenson	< 1.4	< 2.3
Engel, Vogel, Zirnbauer	< (6 – 8)	< (8 – 10)

The calculations of Doi, Kotani, and Takasugi were also used for our $\langle g_B \rangle$ estimates.[9]

Theoretical predictions for $2\nu\beta\beta$-decay matrix elements are rather indefinite. Calculations in the framework of different theories fall within the ranges of lifetimes 1×10^{20}y – 4×10^{21}y for Ge^{76}.[7–11]

The situation is not understood completely; new ideas and calculations are necessary.

Acknowledgments

The authors are grateful to Professor F. T. Avignone for his advise and critical reading of the manuscript.

References

1. I. V. Kirpichnikov and A. S. Starostin, preprint ITEP-189, Moscow, 1983.
2. A. A. Vasenko, I. V. Kirpichnikov, V. A. Kuznetsov, A. S. Starostin, G. E. Marcosyan, V. M. Oganesyan, V. S. Pogosov, A. G. Tamanyan, and S. P. Shachysisyan, *Proc. of the Second Int. Symp. on Underground Physics*, Baksan Valley, USSR, Aug. 17-19, 1987. Moscow, Nauka, 1988.
3. D. O. Caldwell, *Int. J. Mod. Phys.* **A4** (1989) 1851.
4. F. T. Avignone, III and R. L. Brodzinski, *J. Prog. Part. Nucl. Phys.* **21** (1988) 99.
5. I. V. Kirpichnikov, in *Lepton and Photon Interactions* (World Scientific, 1990).
6. A. A. Vasenko, Yu. N. Vereshagin, I. V. Kirpichnikov, V. A. Kuznetsov, V. N. Prusakov, A. I. Rudnew, A. S. Starostin, and A. V. Tichoniro, *Sov. PTE* (1989) 56.
7. W. C. Haxton and G. I. Stephenson, *Prog. Part. Nucl. Phys.* **12** (1984) 409.
8. J. Engel, P. Vogel, and M. R. Zirnbauer, *Phys. Rev.* **C37** (1988) 731.
9. M. Doi, T. Kotani, and E. Takasugi, *Prog. Theor. Phys. Suppl.* **83** (1985) 1.
10. K. Grotz and H. V. Klapdor, *Phys. Lett.* **B199** (1987) 475.
11. T. Tomoda and A. Fassler, *Phys. Lett.* **B199** (1987) 475.

Invited talk presented at

International Symposium on
GAMMA-RAY LINE ASTROPHYSICS
Dec. 10 - 13, 1990, Paris-Saclay, France

SPECTROSCOPY WITH ENRICHED DETECTORS: DOUBLE BETA DECAY AND PERSPECTIVES IN ASTROPHYSICAL γ-RAY SPECTROSCOPY AND IN DARK MATTER DETECTION.

H.V. Klapdor-Kleingrothaus

Max-Planck-Institut für Kernphysik, Heidelberg, Germany

Abstract: A new era, of second generation experiments, using detectors made from enriched material is starting at present in double beta research. This allows to strongly improve the present limits on the neutrino mass. We review the present status, in particular of the HEIDELBERG-MOSCOW experiment, by which for the first time evidence has been given that the technology of enriched HP Ge detector production can be mastered nowadays. We further explore possible future application of this technology in astrophysics, in particular dark matter search and high-resolution c-ray spectroscopy in future ESA and NASA satellite missions. Use of enriched ^{73}Ge detectors for the former would allow to extend the present LEP limits for spin-interacting WIMPs considerably and use of enriched ^{70}Ge detectors in the satellite experiments would allow to drastically reduce the b-background in these missions.

1. Introduction

The recognition of the close connection between the laws of microphysics (nuclear and particle physics) and macrophysics (astrophysics and cosmology) is one of the most important discoveries of this century (see Fig. 1 and [1-4]).
The barriers between these 'classical' disciplines are becoming in some sense more and more meaningless. Nuclear physics, e.g.,may give - as non-accelerator particle physics - important contributions to beyond standard model physics

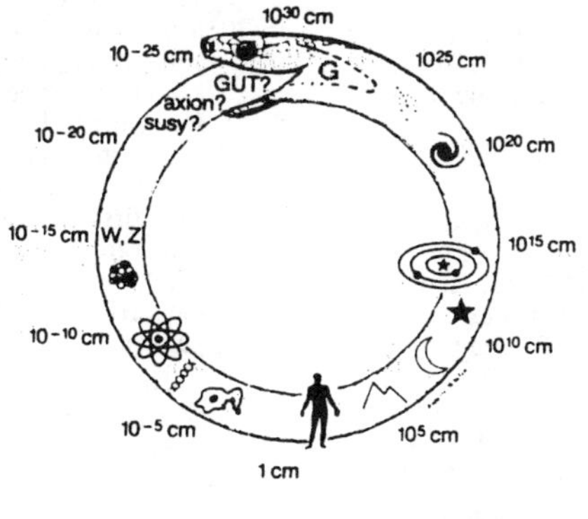

Fig. 1: Illustration of the interconnection between different physical disciplines (after Glashow, see [2]).

(double beta decay, proton decay,...). One of the best examples for this situation is given by the neutrino (Fig.2). The neutrino plays - by its nature (Majorana or Dirac particle) and its mass - a key role for the structure of modern particle physics theories (GUTs, SUSYs, SUGRAs,...), which is at least as important as the role it played at the time of Pauli and Fermi for the understanding of the weak

Nuclear Physics B (Proc. Suppl.) 28A (1992) 207–209
North-Holland

NUCLEAR PHYSICS B
PROCEEDINGS
SUPPLEMENTS

THE HEIDELBERG-MOSCOW DOUBLE BETA DECAY EXPERIMENT WITH ENRICHED ^{76}GE: FIRST RESULTS

H.V. KLAPDOR-KLEINGROTHAUS+
Max-Planck-Institut für Kernphysik
Heidelberg, Germany

Abstract
Status and perspectives of the HEIDELBERG-MOSCOW double beta decay experiment which is undertaken in the GRAN SASSO laboratory, are presented. Of the 16.9 kg of enriched (86%) ^{76}Ge in hands of the cooperation, 6.5 kg have been converted into detectors. A 1 kg enriched detector was installed in July 1990, a 2.9 kg enriched detector, the largest HP Ge detector ever produced, with efficiency of 118% has been installed by september 1991. Present limits for $0\nu\beta\beta$ decay to the g.s. and first excited state of ^{76}Se, respectively, are: $T_{1/2}^{0\nu}$ $(0^+ \rightarrow 0^+) > 1.1$ (2.0) x 10^{24} y, and $T_{1/2}^{0\nu}$ $(0^+ \rightarrow 2^+) > 2.2$ (3.6) x 10^{23} y with 90% (68%) c.l.
These values are the most stringent directly measured $0\nu\beta\beta$ half life limits known up to now. They correspond to an effective neutrino mass limit of $<m_\nu> < 1.5$ (1.1) eV. The background around 2 MeV is β=0.47 counts/kg y keV for the total array, and 0.23 counts/kg y keV for the 2.9 kg detector. Concerning the matrix elements for $\beta\beta$ decay, a major step beyond the frequently used QRPA model has been made by introducing the Operator Expansion Method (OEM)..

1. INTRODUCTION

The neutrino mass m_ν is one of the key quantities for the structure of grand unified theories [1,2]. Predictions for m_ν cover the wide range between 10^{-11} and several eV. Double beta ($\beta\beta$) decay yields at present the sharpest limits on the electron neutrino mass. The future (next 5 years) of $\beta\beta$ experiments will be dominated by use of enriched detectors, ^{76}Ge playing a particularly favourable role here [3,4], and enriched source material, such as ^{136}Xe (see [5]), ^{100}Mo [6],... Such experiments will probe the neutrino mass in the next years down to about 0.1 eV. They are complementary to solar neutrino experiments like GALLEX and SAGE, which are sensitive to a difference of masses squared of neutrinos of different flavour, but do not measure the electron neutrino mass directly.

For deduction of an (effective) neutrino mass (limit) from a measured 0ν decay rate (limit) calculation of nuclear matrix elements is required. Neglecting right handed weak currents

$$\left[T^{0\nu}_{1/2}(0^+_i \rightarrow 0^+_f)\right]^{-1} = (M^{0\nu}_{GT} - M^{0\nu}_F)^2 G_1 \frac{< m_\nu >^2}{m_e}$$

After the major step of recognizing the importance of g.s. correlations for their calculation [7], in recent years the main groups used the QRPA model [8-12]. It was found that though in most cases the uncertainties of the model were tolerable in the case of $0\nu\beta\beta$ decay (see Fig. 3) this was not so for 2ν decay (see Fig. 4 and [11]). For the latter only reliable lower limits of the half-lives were predictable. The problem of QRPA - extreme dependence of the calculated rate on the choice of the renormalization of the particle-particle force, g_{pp}, seems recently to have been solved by applying the socalled Operator Expansion Method (OEM) [13]. In the next section we give the status of our experiment, in the subsequent section some first results of OEM.

2. STATUS OF THE HEIDELBERG-MOSCOW EXPERIMENT

Use of enriched ^{76}Ge (86%) instead of detectors from natural Ge (containing 7.8% of ^{76}Ge) could explore the half-life of $0\nu\beta\beta$ decay up to $\approx 10^{25}$ years, and correspondingly the neutrino mass down to $\approx 10^{-1}$ eV, probing a class of left-right symmetric GUT models with a right-handed Majorana mass term of about 1 TeV, based on $SU(2)_L \otimes SU(2)_R \otimes U(1)$.

The HEIDELBERG-MOSCOW $\beta\beta$ experiment [3,4] makes use of 16.9 kg of ^{76}Ge metal enriched to 86%, corresponding to 14.5 kg of the isotope ^{76}Ge. The full amount of ^{76}Ge has been transferred from Moscow to Heidelberg. Up to now one enriched detector of $\approx$ 1 kg, another of 2.9 kg (the largest ever produced Ge-detector) and a detector of 2.5 kg have been produced. The first detector is running in the GRAN SASSO Underground Laboratory in Italy since end of July 1990, the second since September 1991. Figure 1 shows the HEIDELBERG-MOSCOW $\beta\beta$-Laboratory built generously by the INFN in the GRAN SASSO.

The results of the first 346 days x kg of measuring with the first detector are $T_{1/2}$ $(0^+ \rightarrow 0^+) > 1.1 \times 10^{24}$ (90% confidence limit) or $> 2.0 \times 10^{24}$ years (68%

+) for the HEIDELBERG-MOSCOW collaboration:
M. Beck, J. Bockholt, J. Echternach, G. Heusser, M. Hirsch, H.V. Klapdor-Kleingrothaus*, F. Petry, A. Piepke, U. Schmidt-Rohr, A. Staudt, H. Strecker, K. Zuber (MPI Heidelberg)
A. Balish, S.T. Belyaev*, A. Demehin, A. Gurov, I. Kondratenko, V. Lebedev, (Kurchatov Institute, Moscow)
A. Müller (INFN, Gran Sasso)

*) speakers of the collaboration

Fig. 1
The ββ-laboratory of the HEIDELBERG-MOSCOW experiment in the GRAN SASSO

confidence limit).
The corresponding limits for the neutrino mass are m_ν < 1.5 and < 1.1 eV. The HEIDELBERG-MOSCOW experiment thus yields now the most stringent limit for the $0\nu\beta\beta$ half life of ^{76}Ge and the sharpest limit for the neutrino mass from detector experiments (compare table 1).

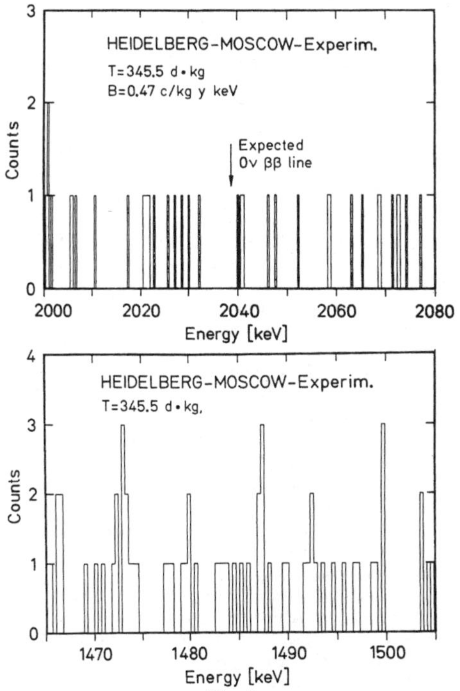

Fig. 2
Details of the spectrum of the HEIDELBERG-MOSCOW collaboration in the region of neutrinoless double beta decay to the ground (upper part) and first excited (lower part) state of the daughter nucleus after a measuring time of 346 days x kg.

The background level which characterizes the quality of the setup is 0.47 events/kg year keV (Fig. 2) in an 80 keV interval around the hypothetical 2038.5 keV $0\nu\beta\beta$-line for the total spectrum, and 0.3 events/kg year keV for the 2.9 kg detector. There is also no indication of a $\beta\beta$ transition to the first excited state of ^{82}Se which should occur at 1479.5 keV. We deduce a half-life limit $T_{1/2}(0^+ \rightarrow 2^+) > 2.2 \times 10^{23}$ (90% c.l.) or $> 3.6 \times 10^{23}$ years (68% c.l.), i.e. far beyond the half-life of 2.5×10^{22} years claimed by [14].
A further 2.5 kg enriched detector is already in operation on surface. In total we hope to have 10 kg of enriched detectors in operation in the Gran Sasso by end of next year. These would correspond to a non-enriched Ge-experiment of at least 1.2 tons. After five years of measurement we hope to come close to the limit of ≈ 0.1 eV for the neutrino mass.

Table 1:

Experiment	a	N¹⁾ [mol]	B²⁾ [c/keV·y·mol]	FWHM³⁾ [keV]	$T_{1/2}$ [10^{24}y] 90% c.l.	$\langle m_\nu \rangle$ [eV] $\langle\eta\rangle \cdot 10^8$ $\langle\lambda\rangle \cdot 10^6$
Caltech-Neuchatel PSI [18]⁴⁾	0.078	6.3	2.3	3.2	0.29	3.7 3.6 5.7
UCSB-LBL [19]⁴⁾	0.078	7.3	1.1	3.3	1.2(0.8)*	1.6 1.6 2.5
Yerevan-ITEP [20]⁴⁾	0.85	13.3	0.19	3.7	1.0	1.7 1.7 2.8
Heidelberg-Moscow⁴⁾	0.86	42.2	0.04	3.2	1.1	1.5 1.4 2.6
Milano [21]⁵⁾	0.64	32.3	1.38	124.	0.02	14.9 11.1 20.3
Caltech-Neuchatel PSI [22]⁵⁾	0.63	26.6	0.003	164.	0.25	3.3 2.4 4.5

1) source strength. Amount of decaying isotope
2) average background at the decay energy in units of the amount of the decaying isotope
3) energy resolution at the decay energy
4) ^{76}Ge experiments
5) ^{136}Xe experiments
$\langle m_\nu \rangle$: effective Majorana neutrino mass
$\langle\eta\rangle$: effective right-right-handed admixture to the weak interaction
$\langle\lambda\rangle$: effective left-right-handed admixture to the weak interaction
*) if the remeasured decay energy is taken into account ref [20] obtains this result. Our analysis of this experiment yielded a similar result.

3. CALCULATION OF ββ MATRIX ELEMENTS, OEM.

The method of OEM (Operator Expansion Method) [13] does not explicitely use the intermediate energy spectrum. This is the essential advantage over QRPA, since in this way the dependence on the pp force (which affects the distribution of β strength in the intermediate nucleus) is drastically reduced.

The method allows (by ignoring many-particle scattering terms) to write the matrix element for 2ν decay in the form

$$M_{GT} \equiv \left\langle 0_F^+ \left| \sum_{i \neq j} \mathcal{M}_{ij} \tau_i^- \tau_j^- \right| 0_I^+ \right\rangle = M_0 + M_1,$$

$$\mathcal{M}_{ij} = \frac{12\big(v_\sigma(r) - v_\tau(r)\big)\Omega_0(ij)}{\Delta^2 - 16\big(v_\sigma(r) - v_\tau(r)\big)^2}$$

$$+\frac{4\left(2v_{\sigma\tau}(r)-v_{\sigma}(r)-v_{\tau}(r)\right)\Omega_1(ij)}{\Delta^2-16\left(2v_{\sigma\tau}(r)-v_{\sigma}(r)-v_{\tau}(r)\right)^2}$$

Thus the matrix element involves the bare nucleon-nucleon interaction without any adjustable parameter. The wave functions of initial and final states we take from QRPA. Fig. 4 shows the result for ^{100}Mo, the results for some other nuclei are given in [13]. Work on $0\nu\beta\beta$ decay matrix elements using OEM is in progress. That the dependence on the pp force is overestimated by QRPA had been shown recently also by comparing

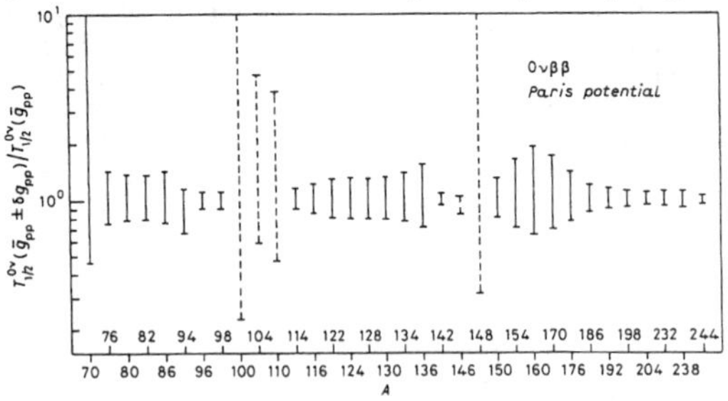

Fig. 3
The uncertainty of QRPA-calculated $0\nu\beta\beta$ rates originating from the limited knowledge of the particle-particle force for the potential double beta emitters (from [12]).

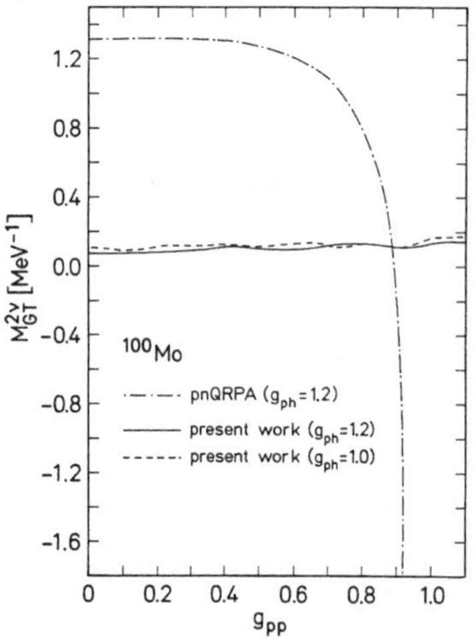

Fig. 4. The matrix element $M_{GT}^{2\nu}$ of $2\nu\beta\beta$ decay of ^{100}Mo as function of g_{pp} in QRPA and in OEM (from [13]).

QRPA and shell model for ^{48}Ca [15]. The essential over-simplification of QRPA is projection of all types of correlations on spin-isospin correlations and changing them simultaneously in one and the same direction.

4. APPLICATION OF $\beta\beta$-TECHNOLOGY TO DARK MATTER SEARCH AND γ-LINE ASTROPHYSICS

The results of the HEIDELBERG-MOSCOW experiment show for the first time that the technology of production of enriched HP Ge detectors can be handled. This has promising consequences for dark matter search and galactic γ-spectroscopy satellite missions (see [4,16,17]). Enriched ^{76}Ge and ^{73}Ge detectors would allow to improve the LEP limits on Dirac- and Majorana WIMPs (weakly interacting massive particles). Use of enriched ^{70}Ge detectors in the NASA project NAE (Nuclear Astrophysics Explorer) or the ESA project INTEGRAL (International Gamma Ray Astrophysics Laboratory) would allow to reduce the β-background dominating in orbit, by an order of magnitude [17]. These satellite missions aim at the study of a variety of astrophysical questions by high-resolution γ-spectroscopy. An agreement to build such detectors from enriched ^{70}Ge and to explore their effectiveness in balloon experiments in 1992 has recently been made between the NASA/ESA groups, and the HEIDELBERG-MOSCOW double beta cooperation. 8.8 kg of ^{70}Ge enriched to 96% have been produced till September 1991.

REFERENCES:

1. P. Langacker, in: "Neutrinos", Springer, Heidelberg, New York, 1988, ed. H.V. Klapdor, p. 71.
2. K. Grotz, H.V. Klapdor, 'The Weak Interaction in Nuclear, Particle and Astrophysics', Adam Hilger, Bristol, New York, 1990.
3. Heidelberg-Moscow-Cooperation: H.V. Klapdor-Kleingrothaus et al., Proc. Internat. Sympos. on Weak and Electrom. Interactions in Nuclei, Montreal, 1989 (WEIN '89).
4. H.V. Klapdor-Kleingrothaus, Proc. 14th Europhys. Conf. on Nucl. Phys. - Rare Nuclear Decays and Fundamental Processes, Bratislava, Oct. 1990. J. Phys. G 17 (1991) S129 and S537.
5. H.T. Wong, J. Phys G. 17 (1991) Suppl.
6. H. Ejiri et al., Phys. Lett. 258B (1991) 17
7. K. Grotz, H.V. Klapdor, Nucl. Phys. A460 (1986) 395.
8. P. Vogel, M.R. Zirnbauer, Phys. Rev. Lett. 57 (1986) 314.
9. O. Civitarese, A. Faessler, T. Tomoda, Phys. Lett 194B (1987) 11,T. Tomoda, A. Faessler, Phys. Lett 199B (1987) 475.
10. K. Muto, H.V. Klapdor, Phys. Lett. 201B (1988) 420
11. K. Muto, E. Bender, H.V. Klapdor, Z. Phys. A334 (1989) 177 and 187.
12. A. Staudt, K. Muto, H.V. Klapdor-Kleingrothaus, Europhys. Lett. 13 (1990) 31.
13. X.R. Wu, A. Staudt, H.V. Klapdor-Kleingrothaus, Phys. Lett. B, in press 1991.
14. J. Busto et al., Nucl. Phys. A513 (1990) 291.
15. K. Muto, E. Bender, H.V. Klapdor-Kleingrothaus, Z.Phys. in press
16. H.V. Klapdor-Kleingrothaus, Proc. Int. Symp. on Gamma-Ray-Line Astrophys., Saclay, Dec. 1990.
17. N. Gehrels, Nucl. Instr. Meth. A292 (1990) 505.
18. F. Boehm et al., Proc. 26 Rencontre de Moriond, Ed. Frontières (1991) p. 85
19. D.O. Caldwell et al., Nucl. Phys. B (Proc. Suppl.) 13 (1990) 547
20. A. A. Vasenko et al., Mod. Phys. Letters A5 (1990) 1299, A.A. Kirpichnikov, priv. comm.
21. E. Bellotti et al., Phys. Lett. 266B (1991) 193
22. H.T. Wong et al., Phys. Rev. Letters 67 (1991) 1218

Nuclear Physics B (Proc. Suppl.) 35 (1994) 351–353
North-Holland

NUCLEAR PHYSICS B
PROCEEDINGS
SUPPLEMENTS

Introductory remarks - workshop session on double beta decay

H.V. Klapdor-Kleingrothaus[1]

Max-Planck-Institut für Kernphysik, Heidelberg, Germany

A brief overview is given on status and perspectives of double beta decay research.

Double beta decay is of major importance for particle and nuclear physics (Table 1). It is, like proton decay, one of the few eminent non-accelerator experiments which may probe grand unification scales far beyond present and future accelerator energies. Among the different methods to investigate the neutrino mass, which plays a key role for the structure of modern particle physics theories (GUTs, SUSYs, SUGRAs, ...) and at the same time is one of the favoured candidates for non-baryonic dark matter in the universe (Fig. 1), it is the most sensitive one (for a Majorana ν-mass). It provides unique information about right-handed weak currents, together with SN87A the sharpest lower limits on the mass of right-handed W_R-bosons, and at the same time it yields limits on the masses of heavy neutrinos of left-right symmetric models. It

Table 1: ββ-decay and particle physics (from [1])

Observable	restrictions
0ν: via ν-exchange: neutrino mass, light neutrino ($\geq 0,1$ eV), heavy neutrino (GeV)	Beyond standard model and $SU(5)$ early universe, matter–antimatter-asymmetry, dark matter, $L-R$–symm. models (e.g. $SO(10)$), seesaw-mechanism
via right–handed weak currents	$V+A$-interaction, $W_R^{\pm}$–masses
via Higgs exchange	
via photino-, gluino-, zino- (gaugino-) exchange	SUSY models: limits for squark and slepton masses beyond accelerator range
$0\nu\chi$: existence of majoron	mechanism of $B-L$ breaking • explicit • spont. breaking of local/global $B-L$ symm.
$0\nu\chi\chi$: majoron model	SUSY models, zino mass
indirect: WIMP-detect.: SUSY particles, heavy neutrinos, cosmions, ...	beyond standard model dark matter

[1]Spokesman of the HEIDELBERG-MOSCOW collaboration

SSDI 0920-5632(94)00486-F

probes SUSY models by 0νββ transitions induced by photino, gluino and zino exchange and yields limits on slepton and squark masses beyond the ranges of accelerators. By looking for Majoron-accompanied 0νββ decay it probes the mechanism of B-L-symmetry breaking. Usually ββ-experiments allow simultaneously to investigate non-baryonic dark matter also by looking for WIMPs, cosmions,...

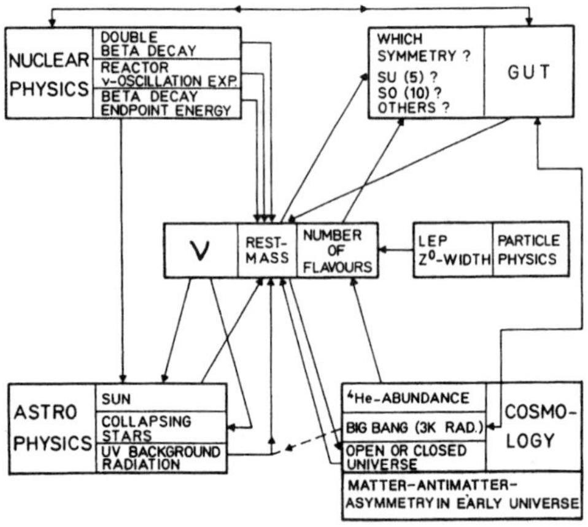

Fig. 1: The neutrino and its role in micro- and macrophysics

The present status of ββ research concerning half-life limits for 0νββ-decay and deduced neutrino mass limits is shown in Fig. 2. In recent time we have seen major progress of the field in various points:

- experiment:
  second generation experiments, using large amounts of enriched ββ-emitter material (HEIDELBERG-MOSCOW experiment (see [2-5] and Maier, these proc.), Gotthard experiment, NEMO experiment, ...), which may probe the neutrino mass down to ~ 0.1 eV in a few years (see [2]).

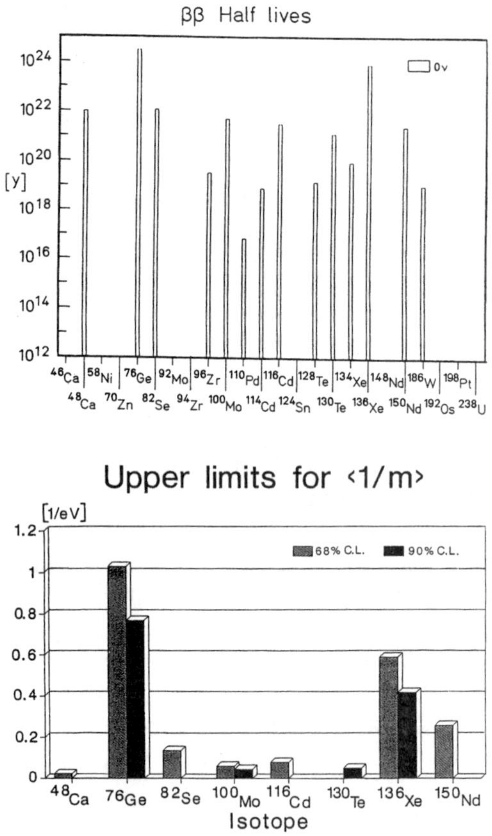

Fig. 2: Measured 0νββ half-life limits (a) and deduced mass limits (b) (from [2])

pulse shape analysis (see Petry, these proc.), background reduction and its quantitative treatment down to a μBq level (see Maier, these proc.) with corresponding positive consequences for dark matter detection (see Beck, these proc.)

- nuclear theory: the problems of QRPA in calculation of 2νββ matrix elements have been probably overcome by the operator expansion method (OEM [5-

7]).

We may have further seen hints of a possible nascent $0\nu\beta\beta$ peak (see [3-5] and Maier, these proc.) which require further attention. If a real peak is developing, one more year of measurement would show it unambiguously. Since a statistically significant $0\nu\beta\beta$-like peak from one $\beta\beta$-emitter alone would not be sufficient as proof of $0\nu\beta\beta$ decay, as a major project a 'control' experiment using $^{116}CdWO_4$ scintillator detectors is under preparation by a HEIDELBERG-KIEV cooperation. Other options are discussed by M. Moe (these proc.).

The perspectives of the most promising $\beta\beta$ experiments are shown in Fig. 3. Only for

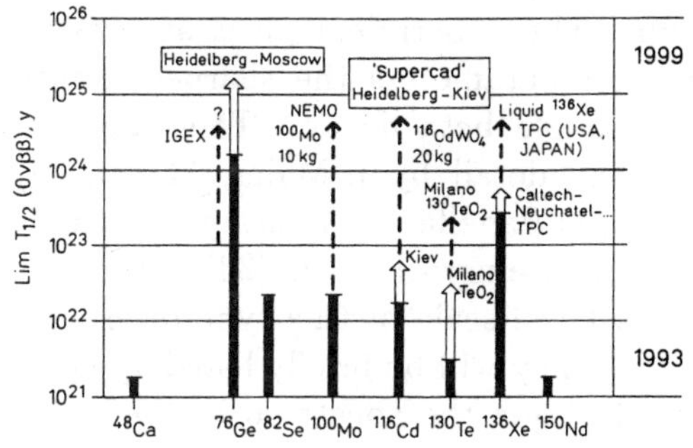

Fig. 3: Present situation (thick solid lines) and perspectives of the most promising $\beta\beta$ experiments (see text).

the isotopes shown $0\nu\beta\beta$ half-life limits > 10^{21} y have been obtained. The thick solid lines correspond to the present, 1993, status, open bars and dashed lines to 'safe' and 'less safe' expectations for 1999, respectively. From Figs. 2,3 it seems that experiments with enriched 'active' detectors (source = detector) will show the largest sensitivity for some further years.

We seem to look forward into an exciting future of our field of research.

Finally it might be mentioned that the new $\beta\beta$-technology has found application in high-resolution balloon and satellite γ-ray astronomy (see Bockholt et al., these proc.).

[1] H.V. Klapdor-Kleingrothaus and A. Staudt, 'Non-Accelerator Particle Physics', (A. Hilger: Bristol, New York) 1994, in print

[2] H. V. Klapdor-Kleingrothaus, Proc. WEIN '92, Dubna, Russia, World Scientific, Singapore 1993, p. 201

[3] H.V. Klapdor-Kleingrothaus, Proc. PANIC, Perugia, Italy, June 28 - July 2, 1993, in press

[4] H.V. Klapdor-Kleingrothaus, at Frontier Topics in Nuclear Physics, Predeal, Romania, August 24 - Sept. 4, 1993 (in press in Plenum Press)

[5] H.V. Klapdor-Kleingrothaus, Proc. Internat. School on Nucl. Physics: Neutrinos in Cosmology, Astro, Particle and Nuclear Physics, Erice, Sept. 1993, (in press in Plenum Press)

[6] X.R. Wu, A. Staudt, H.V. Klapdor-Kleingrothaus, C.R. Ching, T.H. Ho, Phys. Lett 272B (1991) 435

[7] M. Hirsch, X.R. Wu, H.V. Klapdor-Kleingrothaus, C.R. Ching, T.H. Ho, Phys. Reports in press and Z. Phys. A 345 (1993) 163

ELSEVIER Nuclear Physics B (Proc. Suppl.) 48 (1996) 216–222

NUCLEAR PHYSICS B
PROCEEDINGS
SUPPLEMENTS

Double Beta Decay – Physics at Beyond Accelerator Energies

H.V. Klapdor-Kleingrothaus*[a]

[a] Max-Planck-Institut für Kernphysik, P.O.Box 10 39 80, D-69029 Heidelberg, Germany

Double beta decay yields – besides proton decay – one of the most promising possibilities to probe beyond standard model physics at beyond accelerator energies. The possibilities include the neutrino mass, SUSY models, compositeness, leptoquarks, right-handed W bosons and others. We discuss the status and future perspectives of $\beta\beta$ research with enriched ^{76}Ge detectors, in particular of the HEIDELBERG-MOSCOW experiment, including applications some double beta technology can find in the search for dark matter.

1. Introduction

The potential of double beta decay includes investigation of the neutrino mass, of the parameter space of SUSY models, of right-handed W bosons, compositeness, leptoquarks, Majorons, and others. For these topics double beta decay is comfortably competitive to high-energy accelerators [1–5,19].

2. The HEIDELBERG-MOSCOW experiment

2.1. Status

The HEIDELBERG-MOSCOW experiment [17,19] is now exploring the sub-eV range for the mass of the electron neutrino. With five enriched (86% of ^{76}Ge) detectors of a total mass of 11.5 kg taking data in the Gran Sasso underground laboratory the experiment has reached its final setup. The experiment gives at present for most parameters the sharpest limits from double beta decay. They are discussed in detail by B.Maier, M. Hirsch, H. Päs, O.Panella, E. Takasugi and H.V. Klapdor-Kleingrothaus in [33]. These results set the scale of this type of experiments. They will be briefly listed up here. Fig. 1 shows the spectrum in the $0\nu\beta\beta$ region taken in a measuring time of 13.6 kg y [17,19].

Half-life of neutrinoless double beta decay
The deduced half-life limit for $0\nu\beta\beta$ decay is

$$T_{1/2}^{0\nu} > 7.4 \cdot 10^{24} y \;\; (90\% C.L.) \qquad (1)$$

$$> 12.7 \cdot 10^{24} y \;\; (68\% C.L.) \qquad (2)$$

Neutrino mass
-Light neutrinos: The deduced upper limit of an (effective) electron neutrino Majorana mass is, with the matrix ele-

*Spokesman of the HEIDELBERG-MOSCOW cooperation

0920-5632/96/$15.00
PII: S0920-5632(96)00243-5

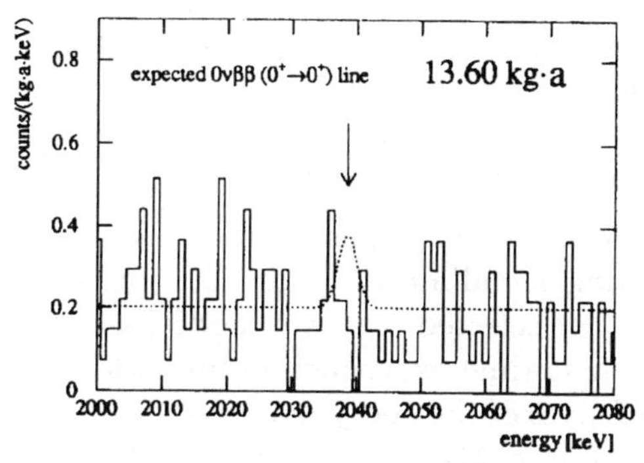

Figure 1. *HEIDELBERG–MOSCOW experiment: Region of interest for $0\nu\beta\beta$ decay after subtraction of the first 200 days of measurement of each detector, leaving 13.60 kg y of measuring time. The dotted curve corresponds to the signal excluded with 90%C.L. It corresponds to* $T_{1/2}^{0\nu} > 7.4 \cdot 10^{24}$ *y.*

ment from [12]

$$\langle m_\nu \rangle < 0.56 eV \;\; (90\% C.L.) \quad (3)$$

$$< 0.43 eV \;\; (68\% C.L.) \quad (4)$$

This is the sharpest limit for a Majorana mass of the electron neutrino so far.

–Superheavy neutrinos: For a superheavy *left*-handed neutrino we deduce ([17]) exploiting the mass dependence of the matrix element a lower limit

$$\langle m_H \rangle \geq 5.1 \cdot 10^7 GeV \quad (5)$$

For a heavy *right*-handed neutrino the relation obtained to the mass of the right–handed W boson is shown in Fig. 2 (see [6]).

Right–handed W boson

For the right–handed W boson we deduce [6] a lower limit of

$$m_{W_R} \geq 1.1 TeV \quad (6)$$

SUSY parameters

New constraints on the parameters of the minimal supersymmetric standard model with explicit R-parity violation are deduced [3,5] from the $0\nu\beta\beta$ half-life limit, which are more stringent than those from other low–energy processes and from the largest high energy accelerators (Fig. 3).

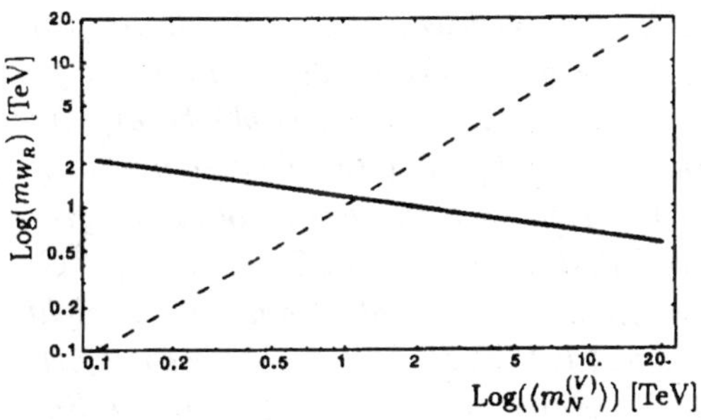

Figure 2. *Area excluded from the HEIDELBERG–MOSCOW experiment (below the curves) in the plane of the right–handed W boson mass versus the mass of a heavy right-handed neutrino. The full line is the constraint from $0\nu\beta\beta$ decay, the dotted line is the requirement of vacuum stability (from [6])*

Compositeness

Evaluation of the $0\nu\beta\beta$ half-life limit for exchange of excited Majorana neutrinos

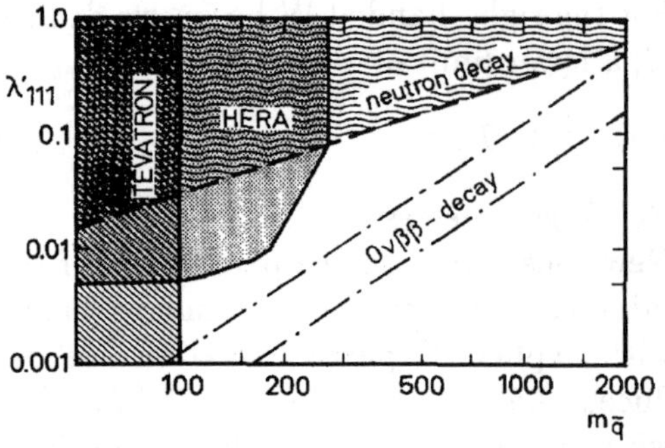

Figure 3. *Comparison of limits on the R–parity violating MSSM parameters from different experiments in the* λ'_{111}*–*$m_{\tilde{q}}$ *plane. The dashed line is the limit from charged current universality according to [27]. The vertical line is the limit from the data of Tevatron [28]. The thick full line is the region which might be explored by HERA [29]. The two dash-dotted lines to the right are the limits obtained from the half-life limit for* $0\nu\beta\beta$ *decay of* ^{76}Ge*, for gluino masses of (from left to right)* $m_{\tilde{g}}$ *=1TeV and 100 GeV, respectively. The regions to the upper left of the lines are forbidden ([3]).*

ν^* yields under some assumptions [34] as lower mass bound of an excited neutrino

$$m_{\nu^*} > 5.9 \cdot 10^4 TeV \tag{7}$$

This is the most stringent bound so far. The bounds deduced on the compositeness scale in different models are roughly of the order of magnitude as those coming from high energy experiments (see Panella and Takasugi [33]).

Half–life of $2\nu\beta\beta$ *decay*

The experiment produced for the first time a high statistics $2\nu\beta\beta$ spectrum ($\sim$ 20000 counts). The deduced half–life is [19]

$$T_{1/2}^{2\nu} = (1.77^{+0.14}_{-0.12}) \cdot 10^{21} y \tag{8}$$

Majoron–accompanied decay

Fitting simultaneously the 2ν spectrum and one selected Majoron mode yields for the first time experimental limits for the half–lives of the decay modes of the newly introduced Majoron models (C. Burgess et al. [14–16], Päs et al. [7]).

The small matrix elements and phase spaces for these modes (see Päs et al. [7]) already determined that these modes by far cannot be seen in experiments of the present sensitivity if we assume typical values for the neutrino–Majoron coupling constants around $\langle g \rangle = 10^{-4}$.

2.2. Perspectives

The HEIDELBERG–MOSCOW experiment will probe the neutrino mass within 5 years down to the order of 0.1 eV (Fig. 5). This limit will be reached taking into account the current background of 0.1 counts/kg y keV in the $0\nu\beta\beta$ region and a further reduction by a factor of $\sim$ 5 by digital pulse shape analysis (DPSA). The new DPSA method which we developed [20] allows for the first time in a very efficient, and reliable way to discriminate between multiple site (MSE) and single site events (SSE) (see [32]). Examples of the second class are the interaction of a beta particle, of the first class multiple Compton scattering events. Fig. 4 shows the result of the first application of this method with one of the enriched detectors in the Gran Sasso for a measuring time of 156 kg d. The energy of the central count in the SSE spectrum is (2038.5 ± 3.6) keV corresponding exactly to the $Q_{\beta\beta}$ value. According to its shape the pulse is a clear single site event and thus a clear double beta candidate. The strong reduction of the background by the new DPSA method, with the potential of reducing the background in the $0\nu\beta\beta$ region to ≤ 0.02 counts/kg y keV, will be essential for the further experiment.

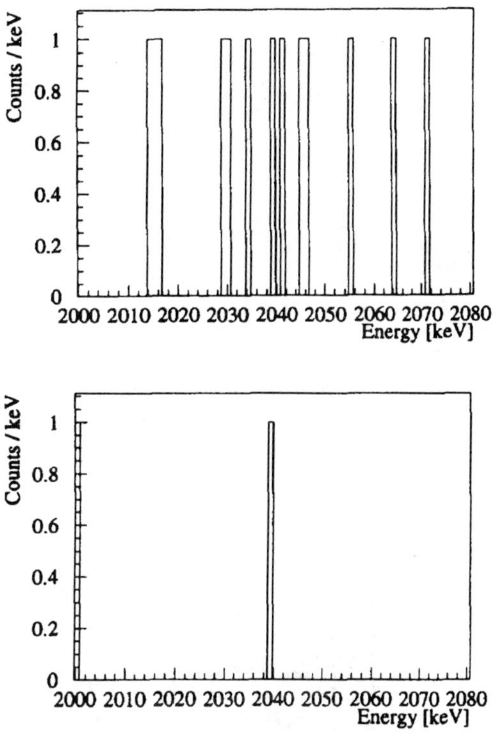

Figure 4. *First test application of digital pulse shape analysis with an enriched ^{76}Ge detector in the Gran Sasso laboratory in a measuring time of 166 kg d. a) MSE spectrum b) SSE spectrum, demonstrating a drastic reduction of the background (from [20])*

3. General Perspectives

Figs. 5a,b show the future perspectives of $\beta\beta$ decay experiments for the next decade. They show the present results and aims of the most promising double beta decay experiments in comparison with the HEIDELBERG-MOSCOW experiment. For a detailed discussion we refer to [19]. As pointed out recently by Raghavan [21], even use of an amount of about 200 kg of enriched ^{136}Xe or 2 tons of natural Xe added to the scintillator of the KAMIOKANDE detector or similar amounts added to BOREXINO would hardly lead to a sensitivity larger than the present ^{76}Ge experiment.
It is obvious that the HEIDELBERG-MOSCOW experiment will give the sharpest limit for the electron neutrino mass till the end of the decade and longer.

4. Dark matter search with enriched Ge detectors

The best laboratory limits on dark matter (WIMPs) are obtained at present by search with Germanium detectors. The HEIDELBERG–MOSCOW experiment allowed for the first time a search for dark matter with isotopically enriched material [22]. The existing cross section limits for WIMP masses above ~ 150 GeV were improved compared to other recent

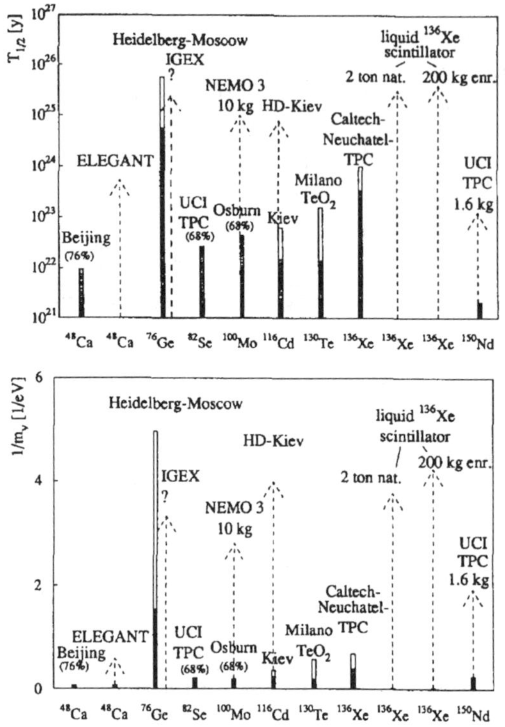

Figure 5. *Present situation, 1995, and expectation for the near future until the year 2000, of the most promising $\beta\beta$–experiments concerning accessible half–life (a) and neutrino mass limits (b). The filled bars correspond to the present status, open bars correspond to "safe" expectations and dashed lines correspond to long–term planned or hypothetical experiments.*

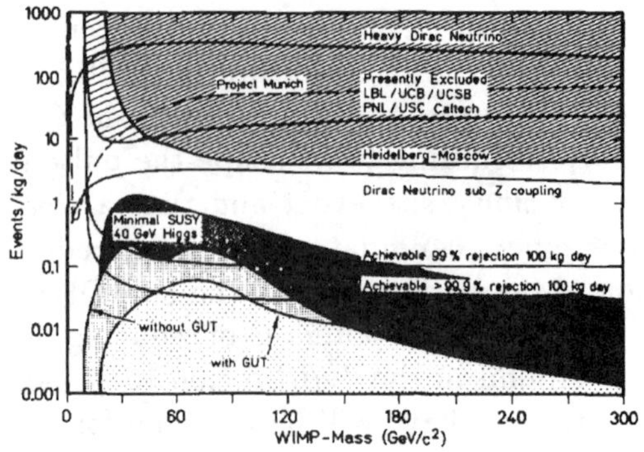

Figure 6. *Current detection bounds for WIMPs with Ge ionisation detectors (shaded regions) and anticipated increase in the sensitivity with Ge cryodetectors The figure shows also expected values for spin–independently and spin–dependently interacting WIMPs in various GUT models (after [25,30,18]).*

work, and Dirac neutrinos could be excluded as the dominant component of the dark halo in the mass range 26 GeV to 4.7 TeV. The measured limit rules out also heavy sneutrinos as dark matter in scenarios of a minimal supersymmetric standard model [23]. The potential of ^{76}Ge dark matter detectors for search for neutralinos in relation to the non–zero spin ^{73}Ge has been carefully investigated recently [24]. For dark matter search the progress obtained with enriched ^{76}Ge is shown in Figs. 6,7. A major step in sensivity improvement is expected on long terms from cryogenic detectors (see Fig. 6 and [25]). Similarly interesting and realizable on shorter time scale could be the new HEIDELBERG project planning to use ionisation Ge detectors in a special new configuration (see Fig. 7 and [31]).

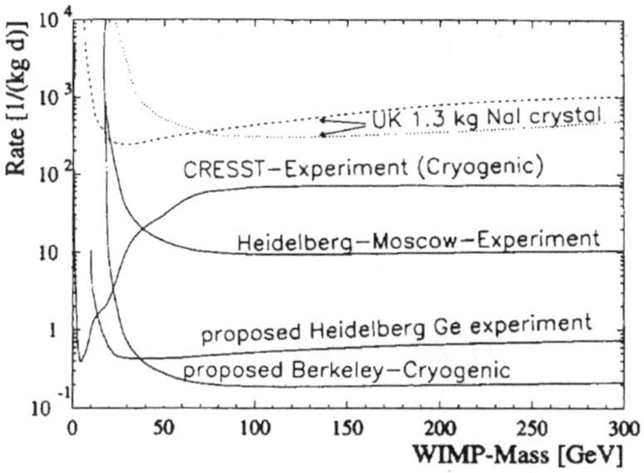

Figure 7. *Current detection bounds for WIMPs from the HEIDELBERG-MOSCOW experiment [22], and the UK experiment [26], the present claimed goal of the Berkeley cryo detector project and the expectation for the new HEIDELBERG project using ionisation Ge detectors in a special configuration [31]*

5. Conclusion

Double beta decay has a broad potential for providing important information on modern particle physics beyond present and future high energy accelerator energies which will be competitive for the next decade and more. This includes SUSY models, compositeness, left–right symmetric models, leptoquarks, and the neutrino mass. For the latter double beta decay now is particularly pushed into a key position by the recent possible indications of beyond standard model physics from the side of solar and atmospheric neutrinos, dark matter COBE results and others. New classes of GUTs basing on degenerate neutrino mass scenarios [8–11] which could explain these observations, can be checked by double beta decay in near future. The HEIDELBERG–MOSCOW experiment among the new $\beta\beta$ experiments as the first now yields results in the sub–eV range.

REFERENCES

1. H. V. Klapdor–Kleingrothaus, A. Staudt, Non–Accelerator Particle Physics, IOP Publ., Bristol, Philadelphia, 1995; and Teilchenphysik ohne Beschleuniger, Teubner Verlag, Stuttgart, 1995
2. W. Buchmüller and G. Ingelman, Proc. Workshop Physics at HERA, Hamburg, Oct. 29–30 (1991); Proc. Beyond the Standard Model III, IV, 1992, World Scientific Singapore
3. M. Hirsch, H.V. Klapdor–Kleingrothaus, S.G. Kovalenko, Phys. Rev. Lett. 75 (1995) 17
4. M. Hirsch, H.V. Klapdor–Kleingrothaus, S. Kovalenko, in preparation
5. M. Hirsch, H.V. Klapdor–Kleingrothaus, S.G. Kovalenko, in Proc. *Int. Workshop on Double Beta Decay and Related Topics*, Trento, 24.4.–5.5.95, World Scientific Singapore, Ed.: H.V. Klapdor–Kleingrothaus and S. Stoica
6. M. Hirsch, H.V. Klapdor–Kleingrothaus, in Proc. *Int. Workshop on Double Beta Decay and Related Topics*, Trento, 24.4.–5.5.95,

World Scientific Singapore, Ed.: H.V. Klapdor–Kleingrothaus and S. Stoica
7. H. Päs *et al.*, in Proc. *Int. Workshop on Double Beta Decay and Related Topics*, Trento, 24.4.–5.5.95, World Scientific Singapore, Ed.: H.V. Klapdor–Kleingrothaus and S. Stoica
8. D.G. Lee, R.N. Mohapatra, Phys. Lett. **B 329** (1994) 463
9. S.T. Petcov, A.Yu. Smirnov, Phys. Lett. **B 322** (1994) 109
10. A. Ioanissyan, J.W.F. Valle, Phys. Lett **B 322** (1994) 93
11. R.N. Mohapatra, S. Nussinov, Phys. Lett. **B 346** (1995) 75
12. A. Staudt, K.Muto, H.V. Klapdor–Kleingrothaus, Europhys. Lett. **13** (1990) 31
13. F. Simkovic, this volume
14. C.P. Burgess, J.M. Cline, Phys. Lett. **B 298** (1993) 141; Phys. Rev. **D 49** (1994) 5925
15. P. Bamert, C.P. Burgess, R.N. Mohapatra, Nucl. Phys. **B 449** (1995) 25
16. C.D. Carone, Phys. Lett. **B 308** (1993) 85
17. HEIDELBERG–MOSCOW collab., Phys. Lett. **B 356** (1995) 450
18. H.V. Klapdor–Kleingrothaus, Progr. Part. Nucl. Phys. **32** (1994) 261
19. H.V. Klapdor–Kleingrothaus, in Proc. *Int. Workshop on Double Beta Decay and Related Topics*, Trento, 24.4.–5.5.95, World Scientific Singapore, Ed.: H.V. Klapdor–Kleingrothaus and S. Stoica
20. F. Petry, H.V. Klapdor–Kleingrothaus, B. Maier, subm . to Nucl. Instr. Meth., 1995
21. R. S. Raghavan, Phys. Rev. Lett. **72** (1994) 1411
22. HEIDELBERG–MOSCOW collab., Phys. Lett. **B 336** (1994) 141
23. T. Falk, A. Olive, M. Srednicki, Phys. Lett. **B339** (1994) 248
24. V. A. Bednyakov, H.V. Klapdor–Kleingrothaus, S.G. Kovalenko, Phys. Lett. **B 329** (1994) 5, Phys. Rev. **D 50** (1995) 7128
25. B. Sadoulet, Nucl. Phys. **B** (Proc. Suppl.) 35 (1994) 117
26. J.J. Quenby *et al.*, Phys. Lett. **B 351** (1995) 70
27. V. Barger, G.F. Guidice, T. Han. Phys.Rev. **D 40** (1989) 2987
28. D.P. Roy, Phys.Lett. **B 283** (1992) 270
29. J. Butterworth, H. Dreiner, Nucl. Phys. **B397** (1993) 3 and H. Dreiner, P. Morawitz, Nucl. Phys. **B428** (1994) 31
30. D.O.Caldwell, Progr. Part. Nucl. Phys. 32 (1994) 109
31. Y. Ramachers, *et al.* (HEIDELBERG-MOSCOW Collab.), Proc. Second Workshop of 'The Dark Side of the Universe', Rome, Nov. 13–14, 1995
32. J. Hellmig *et al.*, this volume
33. H.V. Klapdor–Kleingrothaus, S. Stoica (eds.), Double Beta Decay and Related Topics (World Scientific, Singapore) 1996
34. E. Takasugi, in Proc. *Int. Workshop on Double Beta Decay and Related Topics*, Trento, 24.4.–5.5.95, World Scientific Singapore, Ed.: H.V. Klapdor–Kleingrothaus and S. Stoica

PHYSICAL REVIEW D VOLUME 55, NUMBER 1 1 JANUARY 1997

Heidelberg-Moscow $\beta\beta$ experiment with ^{76}Ge: Full setup with five detectors

M. Günther, J. Hellmig, G. Heusser, M. Hirsch, H. V. Klapdor-Kleingrothaus,* B. Maier, H. Päs, F. Petry, Y. Ramachers, H. Strecker, and M. Völlinger
Max-Planck-Institut für Kernphysik, Heidelberg, Germany

A. Balysh, S. T. Belyaev,* A. Demehin, A. Gurov, I. Kondratenko, D. Kotel'nikov, and V. I. Lebedev
Russian Science Center Kurchatov Institute, 123 182 Moscow, Russia

A. Müller
Istituto Nazionale di Fisica Nucleare, I-67010 Assergi, Italy
(Received 20 December 1995)

The full setup of the Heidelberg-Moscow double β decay experiment is presented. This experiment gives at present the most stringent upper bound, improving the neutrino mass limit into the sub-eV range. Out of 19.2 kg of 86% enriched ^{76}Ge five crystals were grown with a total mass of 11.51 kg. Since February 1995 all five detectors, corresponding to 10.96 kg active mass, are in regular operation in the Gran Sasso underground laboratory, four of them in a common shield. No signal is observed for the neutrinoless double β decay ($0\nu\beta\beta$). The measured data from the first three enriched detectors with a statistical significance of 13.60 kg yr result in a new half-life limit of $T_{1/2}(0^+\rightarrow 0^+)>7.4\times 10^{24}$ yr (90% C.L.). With this limit a Majorana mass of the neutrinos larger than 0.6 eV (90% C.L.) is excluded. From the data taken in the previously operated setup with three enriched detectors in a common shielding and a statistical significance of 10.58 kg yr new results are extracted for the two neutrino double β decay ($2\nu\beta\beta$) of ^{76}Ge. The procedure of a quantitative and model-independent description of the background via a Monte Carlo simulation is outlined in some detail. The combined result is $T^{2\nu}_{1/2}=[1.77^{+0.01}_{-0.01}(\mathrm{stat})^{+0.13}_{-0.11}(\mathrm{sys})]\times 10^{21}$ yr. Further on the results concerning new Majoron models and the impact on SUSY parameters are briefly reviewed. Future improvements on the background with the application of digital pulse shape analysis are discussed and an outlook on the future of $\beta\beta$ research is given. [S0556-2821(96)01523-8]

PACS number(s): 23.40.Bw, 12.60.Jv, 14.60.St

I. INTRODUCTION

The interest in neutrino physics is constantly increasing, because of the fact that physics beyond the standard model of particle physics can be probed. Several indications such as the solar ^{7}Be-neutrino problem, the atmospheric ν_μ deficit, and mixed dark matter models give hints of nonvanishing ν masses as predicted in grand unification theories (GUTs). These indications could be explained in GUT scenarios with degenerate neutrino masses of 0.1–2 eV [1–3].

The mass region assumed in these models can be tested in second-generation $\beta\beta$ experiments like the Heidelberg-Moscow experiment using large amounts of enriched $\beta\beta$-emitter material and thus bring $\beta\beta$ decay into some key position in modern neutrino physics. These experiments investigate the nature of the neutrino (Majorana or Dirac particle) and give at present the most stringent limits on a nonzero Majorana neutrino mass and right-handed weak currents (RHCs). In $\beta\beta$ decay usually four decay modes are discussed [4,5]:

(A) $2\nu\beta\beta \quad {}^{Z}_{A}X\rightarrow {}^{Z+2}_{A}X+2e^-+2\bar{\nu}_e$,

(B) $0\nu\beta\beta \quad {}^{Z}_{A}X\rightarrow {}^{Z+2}_{A}X+2e^-$,

(C) $0\nu\chi\beta\beta \quad {}^{Z}_{A}X\rightarrow {}^{Z+2}_{A}X+2e^-+\chi$,

(D) $0\nu 2\chi\beta\beta \quad {}^{Z}_{A}X\rightarrow {}^{Z+2}_{A}X+2e^-+2\chi$.

Decay mode (A) can be understood as a process of second order Fermi theory, while the observation of the processes (B)–(D) would require physics beyond the standard model. In Fig. 1 the experimental signatures of the different decay

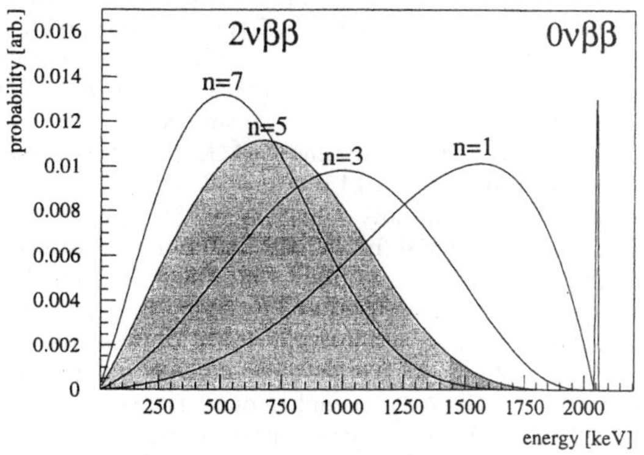

FIG. 1. Spectral shapes of the different investigated double-β decay modes; the continuous spectra are classified by their spectral index n. The spectral index for $2\nu\beta\beta$ decay is $n=5$.

*Spokesmen.

TABLE I. Technical parameters of the five enriched detectors.

Detector number	Total mass [kg]	Active mass [kg]	Enrichment in ^{76}Ge [%]	FWHM[a] at 1332 keV [keV]
enr No. 1	0.980	0.920	85.9±1.3	2.22±0.02
enr No. 2	2.906	2.758	86.6±2.5	2.43±0.03
enr No. 3	2.446	2.324	88.3±2.6	2.71±0.03
enr No. 4	2.400	2.295	86.3±1.3	2.14±0.04
enr No. 5	2.781	2.666	85.6±1.3	2.55±0.05

[a]Full width at half maximum.

modes are shown for the isotope ^{76}Ge. In the case of the neutrinoless double-β decay (B) only two electrons are present in the final state of the decay, resulting in a peak at the Q value as the expected signal for detection. Since in all other decay modes additional particles are emitted, continuous spectral shapes are expected for these decay modes. To classify the different spectral shapes, the spectral index n is used, which corresponds to the power in which the energy is included in the phase space integral. The spectral index of the $2\nu\beta\beta$ decay is $n=5$.

The observation of the neutrinoless double-β $0\nu\beta\beta$ decay (B) would require massive Majorana neutrinos or a contribution of RHC's to the $0\nu\beta\beta$ amplitude. Up to now this decay mode has not been observed. Therefore an upper limit for the effective Majorana neutrino mass can be deduced from a measured half-life limit and calculated matrix elements [6]. This article will focus on the first results of the final setup of the Heidelberg-Moscow experiment with five enriched Ge detectors and the evaluation of decay modes (A) and (B). A first short announcement on these decay modes has been given in [10,11]. Furthermore, the new Majoron-accompanied neutrinoless double-β decays (C) and (D) proposed in [7] will be discussed briefly (for details see [8]).

II. EXPERIMENTAL SETUP AND MEASURED DATA

The three main advantages of the experiment are the excellent energy resolution of germanium detectors, which favors the search for the expected $0\nu\beta\beta$ peak at 2038.56±0.32 keV [12], the large size of the detectors, concentrating the background in the peaks, and the fact that the source is equal to the detector, allowing large source strengths.

Overall there is 19.2 kg of enriched Ge in the Heidelberg-Moscow experiment [4,10,13–16] available with an ^{76}Ge isotopic abundance of 86% compared to 7.8% in natural Ge. Out of the raw material five p-type high-purity Ge semiconductor detectors have been built with a total mass of 11.51 kg. The enrichment we measured for each crystal separately by accelerator mass spectroscopy using residues from the crystal fabrication. All five detectors are now in regular operation in the Gran Sasso underground laboratory, which provides a shielding of 3500 meter of water equivalent (mwe). The sensitivity of this setup corresponds to an experiment with natural Ge of more than 1.2t. The active mass of 10.96 kg is equivalent to a source strength of 125.5 mol ^{76}Ge nuclei, which is at present the largest source strength of all double-β experiments. The main detector parameters are listed in Table I.

In the construction of the cryostat, mainly made of electrolytical Cu, only selected and cleaned low-level materials were used. Underground storage of these materials was applied to minimize the activation due to cosmic rays. The final assembly of the detectors is done in a clean-room environment to avoid any surface contaminations. During the installation of the detectors in the Gran Sasso laboratory, great care is taken to work close to clean-room conditions.

All detectors except detector enr No. 4 are operated in a common Pb shielding of 30 cm, which consists of an inner shielding of 10 cm radiopure LC2-grade Pb followed by 20 cm of Boliden Pb. The whole setup is placed in an air-tight steel box to build up pressure inside with radiopure nitrogen in order to suppress the ^{222}Rn contamination of the air. The steel box is centered inside a 10-cm boron-loaded polyethylene shielding to decrease the neutron flux from outside. Detector enr No. 4 is installed in a separate setup, which has an inner shielding of 27.5 cm electrolytical Cu, 20 cm lead, and boron-loaded polyethylene shielding below the steel box. To check the stability of the experiment, a calibration with a ^{228}Th and a ^{60}Co source is done weakly.

In Table II the background numbers are compared for the different data acquisition periods. The improvement from setup to setup is obvious by the decreasing background numbers. The installation of the boron-polyethylene shielding in the setup with three enriched detectors resulted in a decrease of the overall counting rate by (7.5±0.5)% and in the evaluation interval of the $0\nu\beta\beta$ decay by (22±13)%. The performance of the chosen setup in the experiment is shown in Fig. 2 for the case of the newly installed detector enr No. 4. The effect of going deep underground and building up a suitable shielding results in a background reduction factor of four orders of magnitude in comparison to the low-level laboratory in Heidelberg with 15 mwe. Another factor of 10 is gained with the installation of the N_2-flushing system.

Figure 3 shows the combined sum spectrum of the five enriched detectors in the Heidelberg-Moscow experiment with a statistical significance of 17.70 kg yr. All given statistical significances refer to the active mass of the experiment. Because of the large peak-to-Compton ratio of the large detectors, external γ activities are easily identified, shifting their background from the Compton continuum into the peaks. The background identified immediately by the measured γ lines in the background spectrum consists of (1) primordial activities of the natural decay chains from ^{238}U, ^{232}Th, and ^{40}K, (2) anthropogenic radio nuclides, like ^{137}Cs, and (3) cosmogenic isotopes, produced by activation due to cosmic rays. The activity of these sources in the setup is measured directly and can be located due to the measured and simulated relative peak intensities of these nuclei. Hidden in the continuous background are the contributions of (4) the bremsstrahlungs spectrum of ^{210}Bi (daughter of ^{210}Pb), (5) elastic and inelastic neutron scattering, and (6) direct muon induced events.

The impact of (4) on the background spectrum is determined indirectly with a separate activity measurement, while for (5) the comparison in the measurement in the setup with three enriched detectors in a common shielding with and without neutron shielding was used. Out of the measured coincidence spectra of all detectors, the influence of muons can be estimated. External α and β activities are shielded by

TABLE II. Development of the background numbers in the different data acquisition periods for the enriched detectors.

Detector number	Measuring time [days]	Date start end	Shielding Cu	Pb	boron -polyethylene	Background [counts/keV yr kg] 100–2800 keV	2000–2100 keV
			first low-level setups				
enr No. 1	387.6	8/90–8/91	×			9.74	0.56
		1/92–8/92					
enr No. 2	225.4	9/91–8/92		×		6.09	0.29
			common shielding for three detectors				
enr No. 1	382.8			×		7.81	0.22
enr No. 2	383.8	9/92–1/94		×		4.86	0.22
enr No. 3	382.8			×		6.67	0.21
enr No. 1	263.0			×	×	6.96	0.20
enr No. 2	257.2	2/94–11/94		×	×	4.48	0.14
enr No. 3	263.0			×	×	6.06	0.18
			Full setup four detectors in common shielding, one detector separate (since 2/95 enr No. 5 with digital pulse shape analysis)				
enr No. 1	203.6	12/94–8/95		×	×	7.06	0.14
enr No. 2	203.6	12/94–8/95		×	×	4.20	0.17
enr No. 3	188.9	12/94–8/95		×	×	5.50	0.20
enr No. 5	48.0	12/94–1/95		×	×	7.55	0.23
enr No. 4	147.6	1/95–8/95	×			6.62	0.43

the 0.7-mm inactive zone of the p-type Ge detectors on the outer layer of the crystal. The enormous radiopurity of HP-germanium is proven by the fact that the detectors enr No. 1, enr No. 2, and enr No. 3 show no indication of any α peaks in the measured data. Therefore no significant contribution of the natural decay chains can be located inside the crystals. Nevertheless, it has to be mentioned that detectors enr No. 4 and enr No. 5 seem to be slightly contaminated with ^{210}Pb on the level of a few μBg/kg either inside the crystal or on the bottom surface. This contamination was identified by a measured α peak in the background spectrum at 5.305 MeV of the daughter ^{210}Po and the constant time development of the

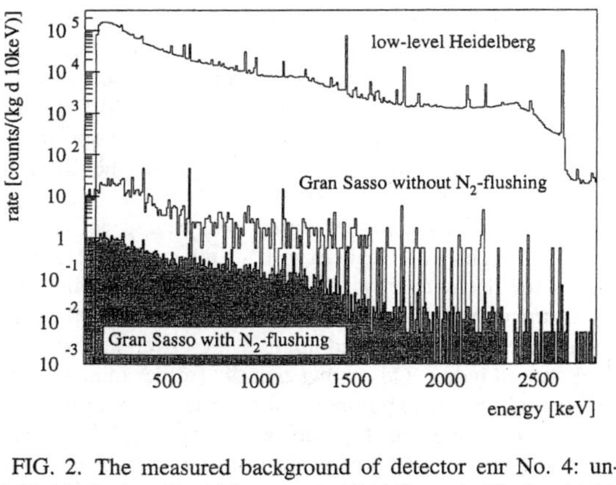

FIG. 2. The measured background of detector enr No. 4: unshielded in the low-level laboratory in Heidelberg, inside the shielding at Gran Sasso without and with N_2 flushing.

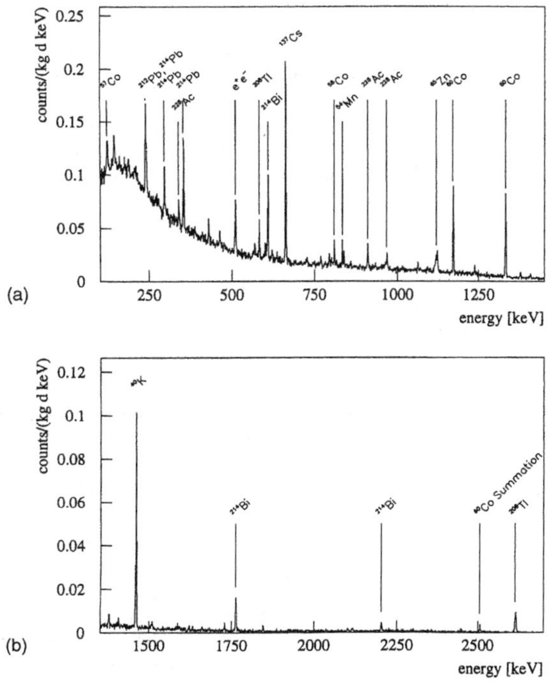

FIG. 3. Combined integral spectrum of the enriched ^{76}Ge detectors with a statistical significance of 17.70 kg yr.

TABLE III. Full data of the experiment and used data after decay of initial activities for the evaluation of the $0\nu\beta\beta$ decay.

Detector	Statistical significance [kg yr]	Statistical significance without first 200 days [kg yr]	Background 2000–2080 keV [counts/keV yr kg]
enr No. 1	3.15	2.18	0.18±0.03
enr No. 2	7.87	7.02	0.20±0.02
enr No. 3	5.40	4.40	0.21±0.02
enr No. 4	0.93		
enr No. 5	0.35		
Σ	17.70	13.60	0.203±0.014

peak counting rate. There is no contribution to the background in the interesting evaluation areas of the experiment due to this activity.

All measured background numbers will improve further on by the decay of short-lived—typical half-lives are ~1 yr—cosmogenic background activities inside the Ge crystal and the cryostat system. This background was produced during the exposure and transportation of the detectors at sea level by the activation through high energy cosmic radiation. The amount of activation was calculated well with the exact history of exposure and agrees within the uncertainties with the measured cosmogenic activities. With this result the influence of not directly identified isotopes can be estimated and accounts dominantly for the declining background.

Since the $2\nu\beta\beta$ spectrum is superimposed on all the background components mentioned above, it is very important to understand the exact composition of the measured background in detail. Further on, the signal-to-background ratio must be of the order of 1:1 for a clear and reliable signal detection. Fulfilling these requirements, the background can be unfolded from the measured spectrum and the result will not be effected by large uncertainties of unknown and unsure background components as in previous experiments [17,18]. For the detection of the $0\nu\beta\beta$ signal, only the achieved integral background in the expected peak area is important, because the clear signal of a line cannot be smeared out by the existing background. Here is the stability of the experiment fundamental in order to maintain the advantage of an excellent resolution of our detectors at the Q value of the $0\nu\beta\beta$ decay in the summed spectrum of all data taken. Extrapolation from the ten strongest γ lines in the background spectrum in Fig. 3 yields an energy resolution at the ground state transition of the $0\nu\beta\beta$ decay at 2038.56 keV of 3.59±0.26 keV in the sum data of all five enriched detectors. This shows the good stability of the experimental parameters during the years of data acquisition when compared to the achieved detector resolutions with calibrated sources in Table I.

III. RESULTS FOR THE $0\nu\beta\beta$ DECAY

The combined spectrum for the evaluation of the $0\nu\beta\beta$ decay with 13.60 kg yr contains all data taken, except the first 200 days of measurement with each detector. These initial data of each detector were removed similar to [17], to avoid the implications of any short-lived radioactive impurities. No further cuts were applied. In Table III the statistical significances of the full data and the data for the evaluation are listed. For the evaluation the symmetric energy interval 2000–2080 keV around the expected $0\nu\beta\beta$ signal is chosen. The effect of the first 200 days of data taking is clearly shown in Fig. 4. The first data of the detectors enr No. 4 and enr No. 5 will be included soon in the final evaluation, when the detectors are taking data beyond the first 200 days in the Gran Sasso laboratory.

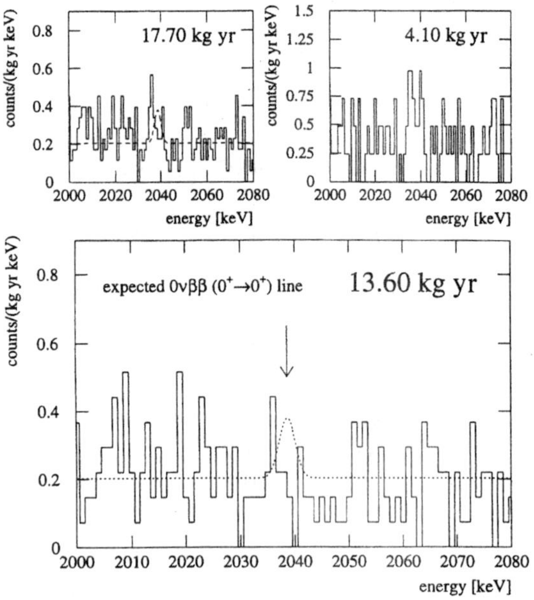

FIG. 4. Region of interest in the combined spectrum for a hypothetical $0\nu\beta\beta$ peak; the inserted curve corresponds to the excluded signal with $T_{1/2}^{0\nu}>3.4\times10^{24}$ yr (90% C.L.) for all 17.70 kg yr data taken and $T_{1/2}^{0\nu}>7.4\times10^{24}$ yr (90% C.L.) for 13.60 kg yr, respectively. The difference spectrum with 4.10 kg yr consists of the first 200 days of measurement of each detector.

The extrapolated resolution at the energy of the hypothetical $0\nu\beta\beta$ peak is 3.45±0.29 keV in the sum spectrum of the three first detectors in the final evaluation with 13.60 kg yr. The 3σ peak interval centered at 2038.56 keV contains 23 events. With the evaluated background left and right of this interval in a 30-keV range, we expect in the peak region 23.86±1.61 events. The achieved background between 2000 and 2080 keV (see Fig. 2) is 0.203±0.014 [counts/keV yr kg]. Additionally the same region of the spectrum is shown for all data taken with 17.70 kg yr and the difference between these data with 4.10 kg yr.

The effect of the higher initial background activities results in a slightly higher background index and a clear structure at the decay energy. This structure is dominating in the spectrum with 4.10 kg yr, where only the first 200 days of each detector were summed up; note the different scale of the background index. This shows that the former indication of a $0\nu\beta\beta$ peak in this experiment on a 2σ level could be identified as short-lived background activities (e.g., ^{56}Co), since the real signal must be constant in time. At present we find no positive evidence for a signal of the $0\nu\beta\beta$ decay ($0^+\rightarrow0^+$-transition) at 2038.56 keV.

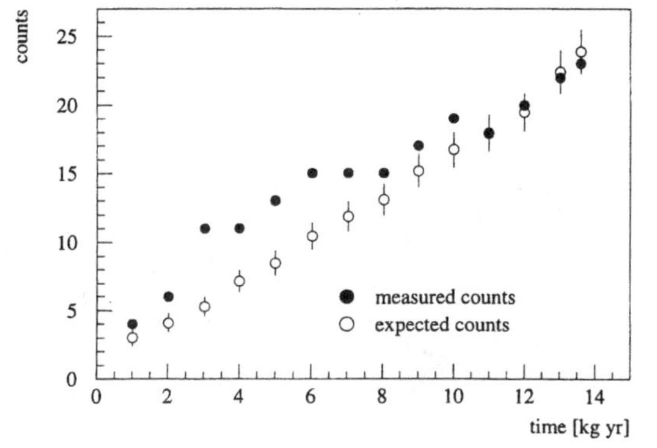

FIG. 5. Number of expected background counts (open circles with error bars) and measured total number of counts (solid circles) in the 3σ energy range of a $0\nu\beta\beta$ ($0^+\rightarrow0^+$) transition of ^{76}Ge as a function of measuring time.

In Fig. 5 the measured and expected number of events in the $0\nu\beta\beta$ region are shown, the latter being the measured background in the energy interval from 2000 to 2080 keV, excluding the 3σ peak region of the hypothetical $0\nu\beta\beta$ line, extrapolated into the 3σ peak region.

Since there is no signal present, we extract a half-life limit for the $0\nu\beta\beta$ decay using the method recommended by the Particle Data Group [19]. With the measured resolution and the measured background we can exclude 8.82 (5.12) events with 90% (68%) C.L., resulting in a half-life limit of

$$T_{1/2}^{0\nu} > 7.4(12.7)\times10^{24} \text{ yr} \quad \text{with } 90\% \ (68\%) \text{ C.L.} \tag{1}$$

To investigate the dependence of the obtained limit on the position in the spectrum, the 3σ peak interval was moved around the measured Q value at 2038.56 keV [12] between 2030 and 2048 keV. The evaluation yielded a variation of the half-life limit between 5.8 and 10.3×10^{24} yr with 90% C.L. and thus demonstrates a rather smooth background level in this energy range. The lower half-life limit of Eq. (5) is converted with the matrix elements [6] into an upper limit for the Majorana mass of the neutrinos, neglecting RHC's:

$$\langle m_\nu\rangle < 0.6(0.5) \text{ eV} \quad \text{with } 90\% \ (68\%) \text{ C.L.} \tag{2}$$

If RHC's are included in the evaluation, we determine the following upper limits with 90% C.L.:

$$\langle m_\nu\rangle < 0.7(0.6) \text{ eV} \quad \text{with } 90\% \ (68\%) \text{ C.L.,} \tag{3}$$

$$\langle \eta\rangle < 6.4(4.9)\times10^{-9} \quad \text{with } 90\% \ (68\%) \text{ C.L,} \tag{4}$$

$$\langle \lambda\rangle < 1.1(0.8)\times10^{-6} \quad \text{with } 90\% \ (68\%) \text{ C.L.} \tag{5}$$

For comparison we give in Table IV a complete list of the neutrino masses obtained when using the matrix elements (given in the convention of [6,20]) of the various theoretical groups.

We see that the results essentially vary within a factor of 2 (see the discussion in [6]), with the exception of the last two cases, which, however, are "special" in the following sense. The calculation of Engel, Vogel, and Zirnbauer [25] is the *only* one, which does *not* use a realistic nucleon-nucleon force, but a schematic force. The calculation of Pantis *et al.* [24] includes p-n pairing, but seems to contain still some serious inconsistencies.

TABLE IV. Neutrino mass limits from neutrinoless double-β decay of ^{76}Ge for different matrix elements.

Ref.	$M_{GT}^{0\nu}-M_F^{0\nu}$	$\langle m_\nu\rangle$	Comment
[6, 20]	4.186	0.56 eV	QRPA with Paris potential
[21]	4.33	0.54 eV	QRPA with Bonn potential
[22]	7.58	0.31 eV	VAMIR code
[23]	5.02	0.47 eV	shell model in weak-coupling limit
[24]	3.04	0.78 eV	QRPA, no pn pairing
[24]	1.34	1.76 eV	QRPA, pn pairing
[25]	1.71	1.38 eV	QRPA schematic force $\alpha' = -390$ MeV, $g_A = 1.25$

By the continuous improvement of the background in the course of the experiment, the evaluated limits of the half-life for the $0\nu\beta\beta$ decay up to now show a linear dependence of time instead of the expected square root behavior. With some years of data in the full setup, the expected square root will take over. Figure 6 shows the time development up to now of the half-life limit and of the resulting Majorana neutrino mass limit with 90% C.L.

Double-β decay yields beyond information on the neutrino mass important restrictions on further parameters of beyond standard model physics, which are competitive or even more sensitive than limits from high-energy accelerators. These include SUSY models, compositeness, leptoquarks, right-handed W bosons, and others. For details we refer to [5,26,9,27,28].

IV. MONTE CARLO SIMULATION

While for the evaluation of the $0\nu\beta\beta$ signal no manipulation of the background has been done, for the evaluation of the $2\nu\beta\beta$ decay and various Majoron-emitting decay modes a detailed knowledge of the composition of the experimental background is required. All measured continuous background sources have to be taken into account in the background model, which leaves, after subtraction from the originally measured spectra, the residual spectra with the spectral shapes searched for. The measured data of the setup with three enriched detectors in a common Pb shielding between

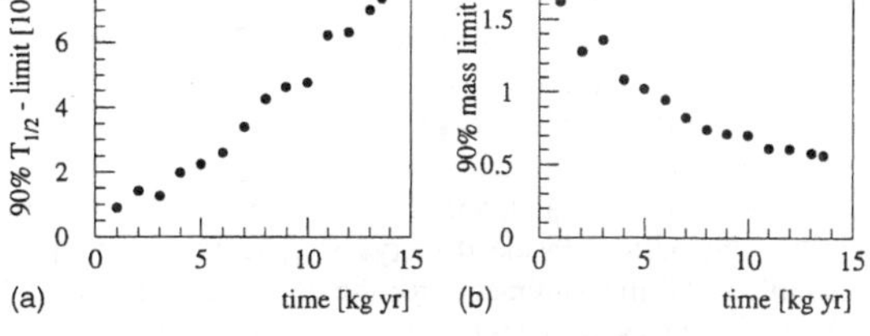

FIG. 6. Half-life limit (left) and resulting Majorana neutrino mass limit (right) with 90% C.L. for the $0\nu\beta\beta$ ($0^+\rightarrow0^+$) transition of ^{76}Ge as a function of measuring time.

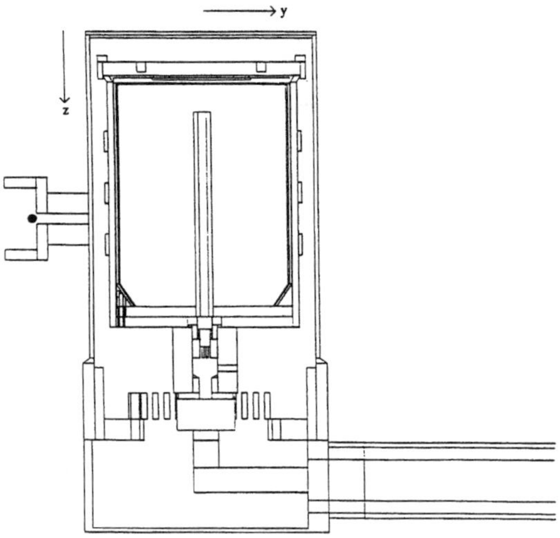

FIG. 7. Implemented geometry of a scanning measurement with a collimated ^{133}Ba source in axial direction at 9 cm below end cap of detector enr No. 2 (the source is symbolized inside the collimator with a black dot).

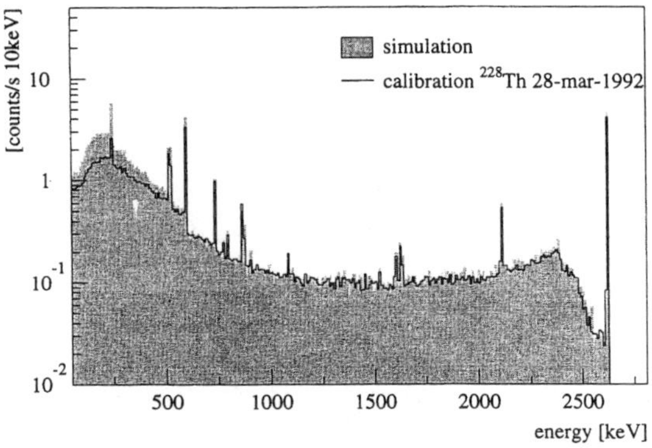

FIG. 8. Comparison of measured and simulated calibration of detector enr No. 2 within the Gran Sasso setup from 28-mar-1992.

9/1992 and 11/1994 has been used to determine the background in detail and to achieve a more sophisticated understanding of the various background sources. The overall statistical significance of the detectors enr No. 1, enr No. 2, and enr No. 3 in this data acquisition period is 10.576 kg yr. To unfold the background a Monte Carlo background model based on the CERN code GEANT3 was developed. All radioactive sources used are available with the complete implemented decay scheme taken from [29] for the simulation. In the case of β activities the correct energy distribution of the particles is included separately, especially for the decay of ^{210}Bi. The background model consists only of measured and clearly identified background activities.

A. Calibration

The implemented geometry of the setups and detectors was tested by comparing the results with measured spectra of calibrated sources in defined locations. The deviation between simulation and measurement represents the error of the simulated detector response, caused by the code itself and small deviations in the implemented geometry in comparison to the real setup.

Each detector was scanned with a collimated ^{133}Ba source in axial (z) and radial (y) directions. The programmed geometry of such a measurement is shown in Fig. 7. The source ^{133}Ba with low-energy γ's was chosen, because the γ beam can be collimated, the measured peak intensities are very sensitive to absorbing materials (e.g., crystalholder or inactive zone of the crystals), and the energy is mainly deposited via the photoeffect inside the crystal, giving a good spatial resolution. All measurements for the detectors enr No. 1 (15 y and 23 z measurements), enr No. 2 (14 y, 19 z), and enr No. 3 (30 y, 42 z) were simulated. For the evaluation we used the four strongest γ lines, which resulted in a combined systematical error of the simulated detector response of 7.1% for enr No. 1, of 5.1% for enr No. 2, and of 7.1% for enr No. 3.

To test the full energy range of natural radioactivity between 100 and 2700 keV, various other sources were simulated. As an example, an original calibration measurement in the Gran Sasso setup with ^{228}Th is compared with the simulation in Fig. 8. The agreement is quite good, although it has to be noticed that in the low-energy region the simulation gives some more events than really measured. This is explained with the uncertainty (within 2 mm) of the position of the source in the setup. In this particular case the source is partly covered by the Pb shielding; therefore, low energies are more likely to be absorbed. Above 500 keV the deviation between measurement and simulation is for the peak counting rates and the integral energy evaluation less than 10%.

Within the energy range for evaluation in this experiment, Monte Carlo simulations can be used to understand the present background in a very efficient way, which is shown with the simulated calibration measurements.

B. Background model

The measured activities in the setup are based on 47 identified γ lines in the spectra of the three enriched detectors and two separate activity measurements of ^{40}K and ^{210}Pb in the LC2-Pb. A uniform distribution of the activities inside a certain volume and material is assumed in the Monte Carlo simulation. In principle, the setup provides five major possible locations: LC2-Pb shielding, detector chamber, copper parts of the cryostats, plastic parts of the cryostats, and the Ge crystals themselves. Other materials or locations in the detectors, e.g., low-level tested steel screws or wires, are negligible, because of their small mass or volume and measured contamination levels.

The interaction and influence between each of the detectors activities with the neighboring detectors is fully included in the background model. The simulated coincidence spectra of the final background model are in good agreement with the measured coincidence spectra. This proves the validity of this approach to build up a uniform background model for three detectors simultaneously.

The simulated background spectra were not just normalized to the measured peak counting rate, because of the in-

TABLE V. Measured activities of all background components included in the uniform background model for three enriched detectors in a common shielding between 9/1992 and 11/1994 (p.d.=per detector located); the errors include the normalization and systematical misplacement errors.

Background component	Localization	Background activity [μBg/kg]		
		enr No. 1	enr No. 2	enr No. 3
^{210}Bi bremsstrahlung	LC2-Pb		360000±30000	
^{238}U decay chain	detector chamber		126±39mBq/m^3	
^{232}Th decay chain	copper p.d.	108.2±12.5	16.3±9.6	105.3±35.7
^{54}Mn	copper p.d.	22.6±5.8	16.5±2.5	22.5±3.3
^{57}Co	copper p.d.	20.5±8.3	30.3±5.0	42.1±6.6
^{58}Co	copper p.d.			56.4±10.8
^{60}Co	copper p.d.	92.5±12.2	72.3±5.2	56.0±5.2
^{54}Mn	Ge crystal	4.5±1.7	3.1±0.9	2.5±0.7
^{57}Co	Ge crystal	2.5±0.8	1.4±0.4	3.7±0.6
^{58}Co	Ge crystal			6.1±2.2
^{65}Zn	Ge crystal		6.7±2.5	25.0±3.6
^{40}K	LC2-Pb		271.3±32.3	
^{40}K	copper enr No. 3			696.1±48.1
^{137}Cs	copper p.d.	220.2±24.7	65.2±19.0	176.4±21.3
^{207}Bi	copper p.d.	20.8±11.3	4.0±2.2	10.2±5.5
^{125}Sb	copper p.d.	50.5±18.0	20.4±7.3	79.1±28.2
^{134}Cs	copper p.d.	9.7±14.6	4.4±6.0	11.9±17.9

teraction in between the detectors. To be consistent with the approach of a uniform background model, a system of linear equations for the simulated and measured peak counting rates was used to determine the correct normalization for the simulation.

The dominant background activity at energies below 500 keV is caused by ^{210}Pb contained in the inner LC2-Pb shielding, which contributes through the bremsstrahlung of its daughter ^{210}Bi (Q value=1.16 MeV). In this region almost half of the measured events are due to this contamination. The absolute activity of 0.36±0.03 Bg/kg was determined separately by low-level α spectroscopy of the decay of the ^{210}Bi daughter ^{210}Po. This intrinsic activity is by far the largest contamination present in the setup. Because of this fact and the location in the large volume of the LC2-Pb shielding, this is the only component which has to be normalized to the measurement with a factor greater than 1, namely, 2.03. Since the resulting bremsstrahlung spectrum is continuous and a greater binwidth of the spectra for the evaluation is selected, the impact of the latter on deduced results is not affected by this. All other identified background activities, whose fraction in the evaluation interval is larger, are normalized with factors at least less than 0.04 to the measured spectra. Therefore the statistical errors of the simulation are negligible in comparison to the measurement.

To include the natural decay chains of ^{238}U (14 γ lines) and ^{232}Th (9 γ lines) in the background model, the peak intensities of the measured γ lines are used. Under the assumption of radioactive equilibrium and a uniform distribution of contamination, the location was determined by comparing the simulated and measured relative peak intensities. In the case of the ^{238}U decay chain, the detector chamber showed the best agreement, while for the ^{232}Th decay chain it is more realistic to locate the activity in the copper cryostat of each detector. The error of a possible misplacement is included in the systematical error of the background model. A placement in the Ge crystal could be ruled out by the absence of any α lines in the high-energy spectra.

During the minimized exposure to the cosmic radiation when the materials were above ground, the copper of the cryostat and the Ge crystals were activated. In the measuring period the cosmogenic activities of ^{54}Mn, ^{57}Co, ^{58}Co, ^{60}Co, and ^{65}Zn are identified through their characteristic γ lines. Since most of the isotopes decay totally or partly by electron capture (EC), it is possible to distinguish between inner activities in the Ge crystal (due to the shifted γ lines with the added energy of the deexcitation x ray) and outer activities in the copper (unshifted γ lines). The measured peak intensities were used to normalize the simulated contributions to the background model.

The primordial ^{40}K content in the LC2-Pb was determined by neutron activation and is in good agreement with the measured data of enr No. 1 and enr No. 2. A second ^{40}K contamination is present in the copper of enr No. 3.

The anthropogenic activities of ^{137}Cs, ^{207}Bi, ^{125}Sb, and ^{134}Cs are surface contamination of the copper parts of the individual detectors. Because of their spectral shape, the deviation of a closer or more distant placement affects the simulated detector response not severely. To cover this uncertainty the error of a possible misplacement is again included in the systematical error.

All measured activities of the background model, given as initial activities at the beginning of the measuring period, and the localization of these contaminations are listed in Table V. To include all background components in a conservative and careful way, the individual detector contaminations are placed per detector system (p.d.=per detector located). Their spectral influence on the measured spectrum is

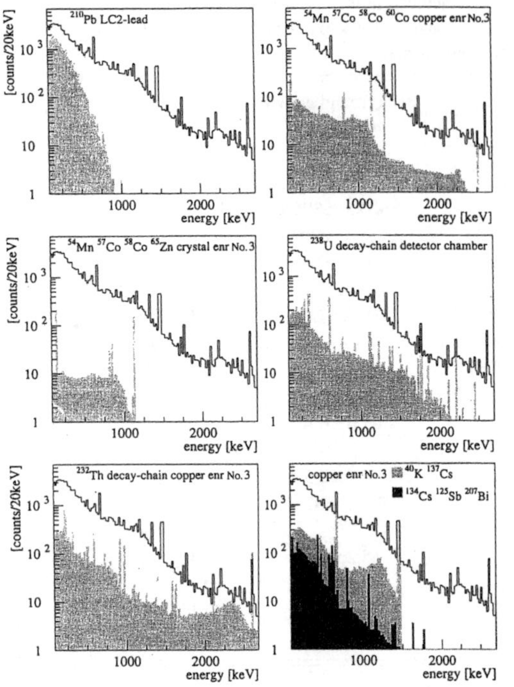

FIG. 9. The simulated background components (shaded areas) as parts of the original measured spectrum of detector enr No. 3 (solid histogram).

shown in the case of enr No. 3 in Fig. 9. The main contribution at energies below 500 keV with about 1/3 of all measured events is due to ^{210}Pb. Other significant parts of the background are caused by the natural decay chains, as well as by ^{40}K and ^{60}Co in the copper.

To avoid statistical fluctuations in the low-energy part, the evaluation interval for the $2\nu\beta\beta$ decay was chosen to be 500–2040 keV, covering 73.9% of the signal. In this way the subtraction of the background model from the original spectrum with larger errors at energies below 500 keV does not affect the results on the $2\nu\beta\beta$ decay, whose fraction on the measured events is rather small in this region. The contribution due to the located activities, e.g., in detector enr No. 3 to the spectra of enr No. 1 and enr No. 2, is shown in Fig. 10. The uniform background model includes all the influences in between the detectors.

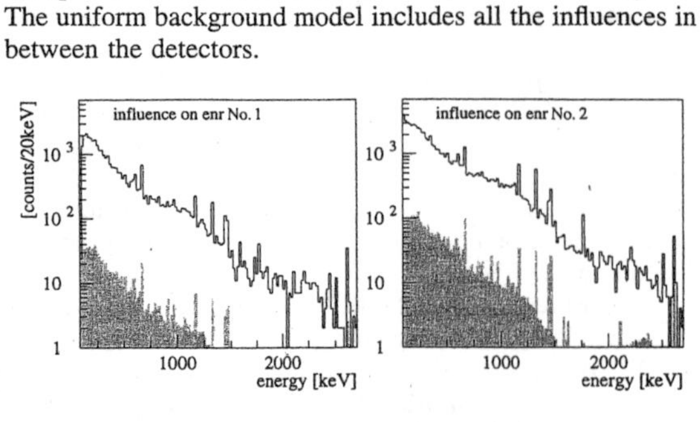

FIG. 10. Influence of activities located in detector enr No. 3 (shaded area) on the spectra of the neighboring detectors (solid histogram).

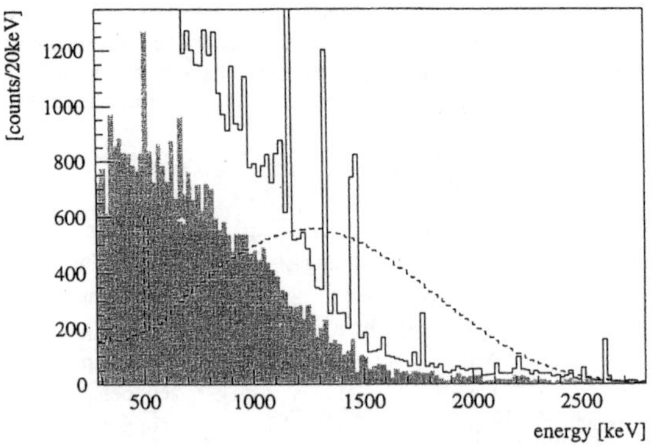

FIG. 11. Summed raw data of the three enriched detectors with a measuring time of 645 d (solid histogram), residual spectrum after subtracting all simulated background components (shaded area), and upper contribution of ^{68}Ge estimated from the 511-keV line (dashed line).

To account for the residual background due to neutrons, muons, and not directly identified background activities, a phenomenological straight line is introduced. In Fig. 11 the summed data of all three detectors are shown together with the result after subtracting all previously discussed background components. In the residual spectrum the shape of the $2\nu\beta\beta$ decay with a maximum at 700 keV is already visible. At energies above 2 MeV there are still some events left, which are due to the additional background sources mentioned above. The expected shape of all these further contributions shows in a first order approximation an increase towards low energies. To model the impact of these sources, a straight line, where the slope and intercept are determined between 2.1 and 2.8 MeV for each detector separately, is used and extended to lower energies. Overall, the contribution of this background line is 2.8% of the measured events and is consistent with the expected spectral shape of these possible background sources.

The excess of 511-keV events is treated separately. In the model only (31–50)% of the annihilation line are reproduced depending on the detector. This gives a strong hint of the influence of muons in the measurement (not explicitly simulated) and goes along with the fact that in the simulated coincidence spectra, integrally, there are slightly less events contained than there are in the measured data. Measured high-energy coincidences due to muons are not present at all in the simulated data and confirm this assumption. Out of this information the muon flux can be estimated to be of the order of the muon flux measured by the MACRO Collaboration with 2.3×10^{-4} m^{-2} s^{-1} [30]. In the near future an active shielding will be placed on top of the setup to measure the effect of the muons directly in coincidence with the detectors inside.

The excess of 511-keV events can be used to derive the possible influence of ^{68}Ge in the measurement, too. ^{68}Ge can be produced by cosmic activation inside the crystals via spallation in the reaction ^{70}Ge$(n,3n)^{68}$Ge. This is already strongly suppressed by the enrichment of our material in ^{76}Ge, corresponding to a deenrichment in ^{70}Ge. The still possible intrinsic contamination of ^{68}Ge could in principle con-

TABLE VI. Fraction of the background components in the evaluation interval 500–2040 keV of the $2\nu\beta\beta$ decay and the entire spectrum together with the total measured counts in these regions in t=645 d, the signal-to-background (S:B) ratio; the simulated model counts, the normalization (Δnorm), and systematical misplacement error (Δsyst).

Background component	Fraction 500–2040 keV [%] enr No. 1	enr No. 2	enr No. 3	100–2800 keV [%] enr No. 1	enr No. 2	enr No. 3
^{210}Bi bremsstrahlung	5.8	6.1	5.2	35.9	35.9	31.3
^{238}U decay chain	9.5	9.1	7.9	8.7	8.1	7.1
^{232}Th decay chain	8.9	2.1	9.6	9.2	2.2	9.8
^{54}Mn copper	0.7	0.8	0.7	0.4	0.5	0.5
^{57}Co copper				0.2	0.4	0.4
^{58}Co copper			0.6			0.5
^{60}Co copper	10.7	12.7	6.8	5.3	6.8	3.6
^{54}Mn Ge crystal	0.4	0.5	0.3	0.2	0.2	0.1
^{57}Co Ge crystal				0.2	0.1	0.3
^{58}Co Ge crystal			0.2			0.1
^{65}Zn Ge crystal		0.5	1.3		0.2	0.6
^{40}K LC2-Pb	7.0	7.6	6.4	3.7	4.3	3.5
^{40}K copper enr No. 3			5.8			3.5
^{137}Cs	6.7	3.5	6.4	6.6	3.2	6.0
^{207}Bi	1.7	0.6	1.0	1.3	0.4	0.7
^{125}Sb	0.5	0.4	1.0	1.4	0.9	2.4
^{134}Cs	0.7	0.5	1.0	0.5	0.4	0.7
511-keV line	0.4	0.6	0.9	0.1	0.2	0.3
straight line	5.3	5.7	5.5	2.5	3.0	2.8
influence on enr No. 1		2.7	1.5		2.7	1.4
influence on enr No. 2	2.0		1.8	1.7		1.5
influence on enr No. 3	2.2	4.2		2.2	3.9	
overall counts	9924	22268	25096	32803	65699	74480
S:B	1:1.66	1:1.37	1:1.78	1:4.0	1:2.8	1:3.4
model counts	6195	12854	16071	26272	48262	57356
Δnorm	345	521	704			
Δsyst	132	525	456			

tribute to the measurement via the decay of its daughter ^{68}Ga with a Q value of 2.921 MeV. The only evident signature of this background source is the characteristic x ray with 10.4 keV. The present settings of the thresholds of the detectors do not allow one to measure this line. Assuming that all excess events are due to ^{68}Ge, meaning the β^+ decay of ^{68}Ga, the resulting contribution is shown in Fig. 11. This estimate is obviously in contradiction with the measured data. Therefore we conclude that no significant contributions to the $2\nu\beta\beta$ background from ^{68}Ge are present and give the following upper bounds, determined for each detector separately: 6 μBg/kg, for enr No. 1, 0.1 μBg/kg for enr No. 2, and 54 μBg/kg for enr No. 3. It has to be mentioned that part of the excess of the 511-keV line might be explained with a ^{106}Ru contamination (especially via the 511-keV γ line from the decay of the daughter nuclei ^{106}Rh) with a half-life of 386 d; making the previously obtained limits conservative.

The fraction in percent of 19 different background components on the total measured spectrum in the evaluation interval and the entire spectrum is given in Table VI. The total number of events measured in 645 d serving as 100% is also listed together with the signal-to-background ratio (S:B): 1:1.7 for enr No. 1, 1:1.4 for enr No. 2, and 1:1.8 for enr No. 3. The S:B ratio is evaluated under the assumption that all events not explained by the background model are due to $2\nu\beta\beta$ decay. These ratios are slightly lower than in [10], because the peaks, with a rather significant fraction of counts, are included in this evaluation, whereas before the S:B ratio was only determined for the continuous parts of the spectrum. Finally, the normalization error and systematical misplacement uncertainty are given for the model per detector. A very sensitive test for the model was performed with the comparison of the simulated and measured coincidence spectra. All lines and the spectral shapes (despite the effect due to muons) are reproduced within the errors and confirm the validity of the uniform background model. The overall model consists of 19 different background components, which are located in 46 distinct locations and produce in total 132 simulated detector-specific spectra.

C. Results for the $2\nu\beta\beta$ decay

The bin width of the spectra was chosen to be 20 keV per channel to avoid statistical fluctuations when subtracting

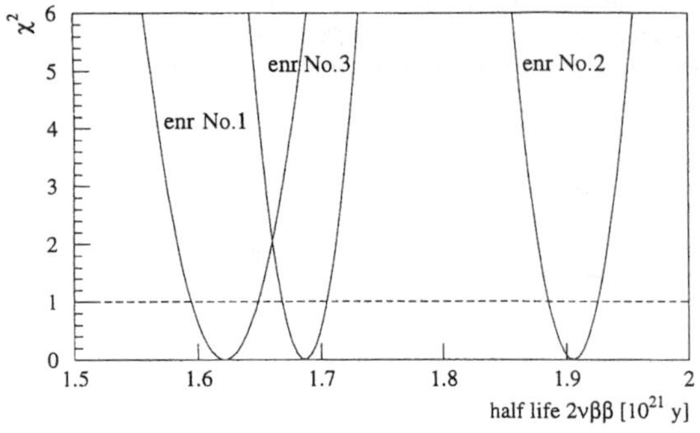

FIG. 12. Resulting χ^2 distribution out of the maximum-likelihood fit for the evaluation of the $2\nu\beta\beta$ decay in the residual spectra after subtraction of the background of the three enriched detectors (the dotted line corresponds to the 68% error).

simulated γ lines from the measured spectra. It is assumed for this evaluation that the resulting difference of the measured and simulated background spectra contains only the $2\nu\beta\beta$ signal.

The result of a maximum-likelihood fit of the data and the theoretically expected $2\nu\beta\beta$ spectrum then results in the following half-lives for the three detectors with 68% C.L.:

$$T_{1/2}^{2\nu}(\text{No. }1)=[1.62^{+0.03}_{-0.03}(\text{stat})^{+0.18}_{-0.14}\times(\text{norm})^{+0.07}_{-0.06}(\text{syst})^{+0.13}_{-0.11}(\text{sim})]\times 10^{21}\ \text{yr}, \quad (6)$$

$$T_{1/2}^{2\nu}(\text{No. }2)=[1.91^{+0.02}_{-0.02}(\text{stat})^{+0.12}_{-0.10}\times(\text{norm})^{+0.12}_{-0.10}(\text{syst})^{+0.12}_{-0.10}(\text{sim})]\times 10^{21}\ \text{yr}, \quad (7)$$

$$T_{1/2}^{2\nu}(\text{No. }3)=[1.69^{+0.02}_{-0.02}(\text{stat})^{+0.17}_{-0.14}\times(\text{norm})^{+0.10}_{-0.08}(\text{syst})^{+0.13}_{-0.10}(\text{sim})]\times 10^{21}\ \text{yr}. \quad (8)$$

The statistical error originates from the parabolic behavior of the logarithmic likelihood ratio, which corresponds to a χ^2 function [32], which is shown in Fig. 12 for the three detectors. With the normalization error due to the errors of the measured area of the γ lines, the systematical misplacement error of background activities, and the error of the simulated detector response, all possible implications to the result are included. The result for the detector enr No. 2 is shown in Fig. 13.

The combined result for the three detectors with the uniform background model with 68% C.L. is

$$T_{1/2}^{2\nu}=[1.77^{+0.01}_{-0.01}(\text{stat})^{+0.13}_{-0.11}(\text{syst})]\times 10^{21}\ \text{yr}. \quad (9)$$

For comparison two former results for the $2\nu\beta\beta$ decay are shown in Fig. 13, too. The obtained result is slightly higher than in [10] with $(1.42^{+0.13}_{-0.13})\times 10^{21}$ yr, because the interaction of the detectors between each other was not taken into account in this earlier result of the Heidelberg-Moscow experiment. The curve with $(0.92^{+0.07}_{-0.04})\times 10^{21}$ yr from [17], which is corrected to $(1.2^{+0.2}_{-0.1})\times 10^{21}$ yr in [33], still deviates significantly from the result presented here, because the background assumptions made in this experiment were only qualitative and incomplete.

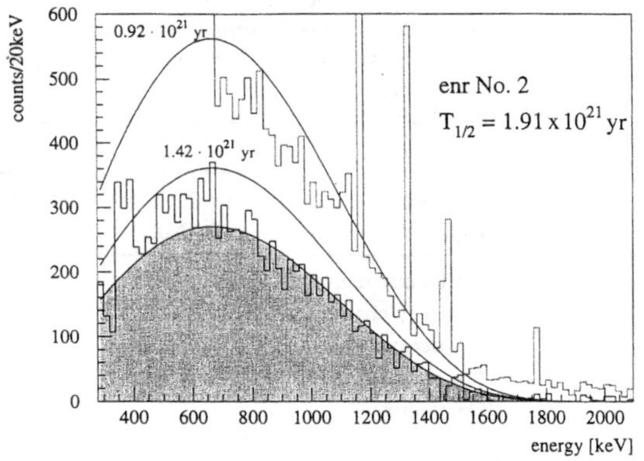

FIG. 13. Result for the evaluation of the $2\nu\beta\beta$ decay of ^{76}Ge in detector enr No. 2: original spectrum (dotted histogram), residual spectrum after subtraction of background model (solid histogram), and fitted $2\nu\beta\beta$ spectrum (shaded area); for comparison, two former $2\nu\beta\beta$ results are shown [10,17].

The presence of isotopes inside crystals like ^{68}Ge and ^{90}Sr with $Q=2.284$ MeV of its daughter ^{90}Y and ^{234m}Pa with $Q=2.29$ MeV, which might in principle influence or even simulate a $2\nu\beta\beta$ signal [10], is excluded by the spectral shape and end point of their continuous simulated spectra [31]. These sources do not have significant γ lines from which they could be identified. In the case of ^{77}Ge and ^{60}Co in the crystal and other cosmogenics, the spectral shape does not fit at all and the calculations of the production via cosmic rays show in no case significant contributions, while all measured cosmogenic activities are reproduced within the errors.

The expected constant integral counting rate in the evaluation interval for the resulting spectra after background subtraction was tested and confirmed for 20 points of time during the measuring period. Based on the uniform background model and the determined activities in Table V for each point of time, the background model was calculated backwards, corresponding to the measured spectra at this time. In order to investigate the $2\nu\beta\beta$ counting rate in between two time points, the previously measured spectra and background model are subtracted from the result of the given time point. The difference between measured data and background model then gives the specific $2\nu\beta\beta$ counting rate in between two time points. The result is shown in Fig. 14 for each detector separately and summarized for all three enriched detectors together.

Inspite of the statistical fluctuations in the time development of the $2\nu\beta\beta$ counting rates in the data of each detector, the expected constant stability for the investigated signal can be seen. Combining the detectors, thus reducing the statistical errors, the effect of the installation of the neutron shielding after 383 d in this measuring period is obvious, too (compare Table II). Overall, we have 5.56 ± 0.24 $2\nu\beta\beta$ events per [d] and [kg] active Ge crystal, corresponding to a half-life of $T_{1/2}=1.74\times 10^{21}$ yr. Taking only the first 14 time points up to the installation of the neutron shielding with 5.84 ± 0.30

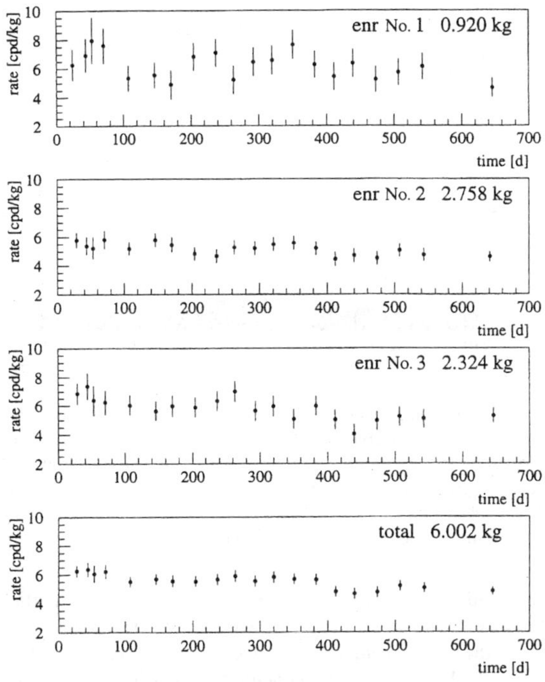

FIG. 14. Time development of the $2\nu\beta\beta$ counting rate in the evaluation interval 500–2040 keV after subtraction of the background model for 20 time intervals for each detector separately, and summarized.

$2\nu\beta\beta$ events counts/d [(cpd)/kg], a half-life of $T_{1/2}=1.65\times10^{21}$ yr is deduced, while the last 260 d on its own gives 4.91 ± 0.20 $2\nu\beta\beta$ events [cpd/kg] and a half-life of $T_{1/2}=1.97\times10^{21}$ yr. The latter result is certainly not correct, since the influence of the straight line was calculated linearly back in time, although it is expected that the slope of this straight line has decreased in the measuring phase with neutron shielding. Therefore the half-life is overestimated.

In order to avoid this complication, the whole evaluation of the $2\nu\beta\beta$ decay was performed again only for the time the neutron shielding was installed. The background activities were calculated back to this time point and normalized to the spectra. A new phenomenological straight line was adapted to this part of data, which showed, as expected, a smaller slope as evaluated in the full set of data. The result of the maximum-likelihood fit yields the following half-lives for the $2\nu\beta\beta$ decay with 68% C.L.: for enr No. 1 $[1.63^{+0.03}_{-0.03}(\text{stat})^{+0.25}_{-0.19}(\text{syst})]\times10^{21}$ yr, for enr No. 2 $[1.96^{+0.02}_{-0.02}(\text{stat})^{+0.23}_{-0.18}(\text{syst})]\times10^{21}$ yr, and for enr No. 3 $[1.78^{+0.02}_{-0.02}(\text{stat})^{+0.27}_{-0.21}(\text{syst})]\times10^{21}$ yr. All these deduced half-lives are slightly larger than the ones evaluated in the total measuring time. Especially in the case for enr No. 3, the half-life is longer, because this detector was installed new in this measuring period, but the initial cosmogenic activities have decayed by the time the neutron shielding was installed. Within the errors these results approve the results deduced before, even giving hints of a slightly longer half-life. Since the results here are still dependent on the overall uniform background model, the previous obtained half-lives are decisive.

TABLE VII. Different majoron modes, their leptonic charge, spectral index n, and the new bounds on the half-life and coupling constants derived in this work (from [8]).

Modus	GB	L	n	$T_{1/2}$>(90% C.L.)	g<(90% C.L.)
$\beta\beta\phi$	no		1	7.91×10^{21}	2.3×10^{-4}
$\beta\beta\phi$	yes		1	7.91×10^{21}	2.3×10^{-4}
$\beta\beta\phi$	no	−2	1	7.91×10^{21}	2.3×10^{-4}
$\beta\beta\phi$	yes	−2	3	5.85×10^{21}	0.18
$\beta\beta\phi$	gauge boson	−2	3	5.85×10^{21}	0.18
$\beta\beta\phi\phi$	yes and no		3	5.85×10^{21}	4.1
$\beta\beta\phi\phi$	no	−1	3	5.85×10^{21}	4.1
$\beta\beta\phi\phi$	yes	−1	7	6.64×10^{21}	3.3

The Heidelberg-Moscow experiment reached with 21 115 $2\nu\beta\beta$ events in the evaluation interval of 500–2040 keV the highest statistics of all all $\beta\beta$ experiments for the direct detection of the $2\nu\beta\beta$ decay mode. With a detailed understanding of all major background components, the evaluation is not dependent on any cuts applied to the measured data, unlike the case of other experiments. Actually, the raw data are used for a full background consideration.

D. Results for the Majoron decay

In the ordinary Majoron model [34], the Majoron is a massless scalar Goldstone boson, which is associated with the spontaneous breaking of B-L symmetry and the existence of Majorana masses of the neutrinos. In this framework only the singlet Majoron is still possible, since doublet and triplet Majorons are ruled out by the CERN electron e^+e^- collider LEP [35] by the measured width of the Z^0 resonance. Nevertheless, "fine-tuning" is required for the singlet Majoron to be consistent with the existing bounds on neutrino masses and to still preserve an observable rate for Majoron-emitting double-β decays.

Therefore new Majoron models have been invented avoiding the unnatural "fine-tuning" [7]. In these models the terminus Majoron stands in a more general sense for a light or massless boson coupling to neutrinos. The new properties of these Majorons are that they can carry units of leptonic charge, that there can be Majorons which are not Goldstone bosons and that decays with the emission of two Majorons can occur. The last case can be fermion mediated or scalar mediated, although only for fermion-mediated decays is a contribution to double-β decay expected [7,8].

In Table VII all considered models are listed with their leptonic charge and spectral shape index n of the sum energy of the emitted electrons (see Fig. 1, $n=1$ corresponds to the singlet Majoron). The spectral index is defined from the phase space of the emitted particles, $G\sim(Q_{\beta\beta}-T)^n$, where $Q_{\beta\beta}$ is the Q value of the decay and T the sum energy of the two electrons.

For the evaluation of the half-life limits, the data of the detector enr No. 2 was used with the corresponding background model previously discussed. This detector was chosen, because of its size and the best background of all detectors. In order to cover all Majoron modes sufficiently and to minimize the systematic errors of the background model an evaluation interval of 300–2040 keV was selected. With a

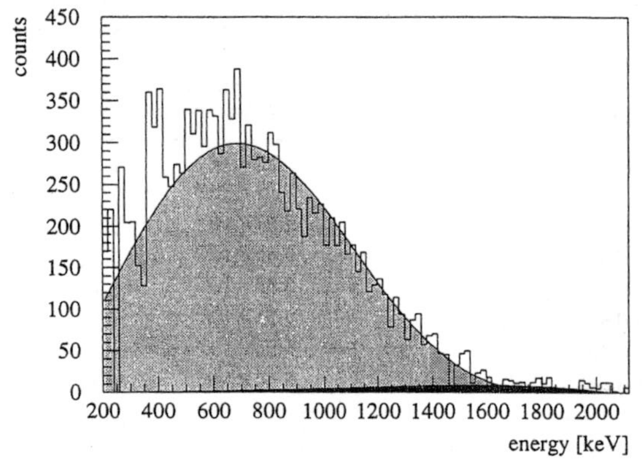

FIG. 15. "Ordinary Majoron" ($n=1$) in the area of fit: 300–2040 keV, yielding a half-life bound of $T_{1/2}>7.91\times10^{21}$ yr with (90% C.L.).

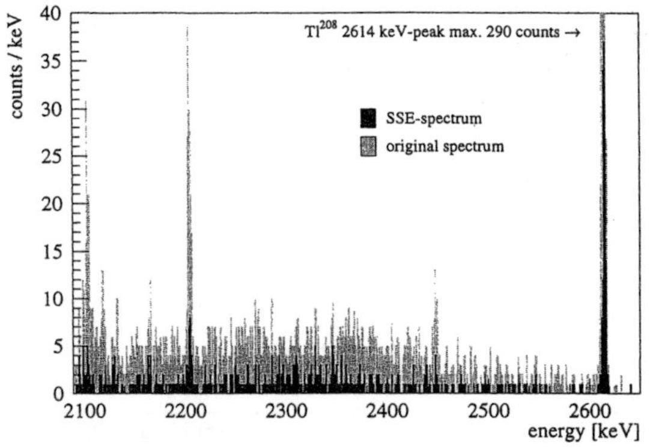

FIG. 16. Comparison of a background spectrum from the detector enr No. 5 measured without any shielding between 2090 and 2650 keV and the derived SSE spectrum. The SSE spectrum contains integral a factor of 4 less counts than the original spectrum. The 2614-keV peak of ^{208}Tl has a peak maximum of 290 counts in the original spectrum (taken from [31]).

simultaneous maximum-likelihood fit of the $2\nu\beta\beta$ decay and one Majoron-emitting decay, the half-life limits given in Table VII were determined.

The result of the data fit for the ordinary Majoron with $n=1$ is, for example, shown in Fig. 15. The experimental spectrum is shown as a histogram, while the light-grey-shaded area is the best fit for the $2\nu\beta\beta$ decay. The dark-shaded area corresponds to the Majoron.

It has to be mentioned that the deviation from zero of the Majoron-emitting modes is, for $n=1$, 0.29σ, for $n=3$, 0.90σ, and for $n=7$, 0.35σ, while the result for the $2\nu\beta\beta$ decay is consistent within the errors with the evaluation mentioned in the previous section. The $2\nu\beta\beta$ half-life extracted in all two-parameter fits varies in the 1σ range of the result obtained in an exclusive $2\nu\beta\beta$ decay evaluation.

With the matrix elements and phase spaces given in [8], the evaluated half-life limits can be converted into upper limits for the various effective Majoron neutrino coupling constants (see Table VII). Note that the surprisingly weak limits obtained for all of the new Majoron models are caused by the small values of the corresponding nuclear matrix elements and phase spaces and are independent of the isotope under consideration [8]. The limits for the new Majoron modes were deduced for the first time.

V. OUTLOOK

A. Digital pulse shape analysis

For the first time a new method of digital pulse shape analysis (DPSA) with the new enriched detector enr No. 5 has been applied [36]. This reduces the background described above further by a considerable factor.

The idea of this electronic approach is to decrease the number of background counts based on the search for characteristic variations, due to different interaction mechanisms of the incident radiation, in the course of the preamplifier output pulse. The DPSA in Ge detectors is capable of distinguishing between single-site events (SSE), which deposit energy locally at one site in the crystal, and multiple site events (MSE's), which are spread out over the entire volume of the detector. Examples of the first class are a single Compton-scattered escape event and the interaction of a β particle because of the small range of electrons in the crystal.

In the experiment the signal of a $0\nu\beta\beta$ event is a SSE, while the background is characterized by MSE's due to multiple Compton scattering in this energy region. The developed DPSA [36] was tested with regular Ge detectors before it was finally installed and checked with the new detectors enr No. 5. In Fig. 16 a background spectrum of enr No. 5, measured in Heidelberg without any further shielding, and the resulting SSE energy spectrum after the application of DPSA are depicted. The integral number of counts in the Compton region between 2260 and 2360 keV in the original spectrum is 1462 counts and 315 counts in the remaining SSE spectrum. This is an improvement of about a factor of 5. The peak at 2614 keV is not removed completely due to the effect that 10% of the MSE's appear like SSE's and the deviation of DPSA from maximum reachable discrimination efficiency. The discrimination efficiency has been determined to be 80% [37]. It is difficult to derive a general quantity which characterizes the improvement obtainable with DPSA, since it is dependent on the composition of the background.

The measurement with DPSA applied to enr No. 5 in the Gran Sasso underground laboratory started in February 1995 in the setup with four enriched detectors. The first data show a discrimination factor for the background similar to that in Heidelberg. From the in total 15 measured counts in the energy range of 2000–2080 keV within 155.9 kg d with detector enr No. 5, only two counts remain after switching on the DPSA, one at precisely the energy of (2038.8 ± 3.6) keV [36]. Further measurements will show whether this event is a real $0\nu\beta\beta$ candidate.

In general, we can expect that an improvement of the half life limit for $0\nu\beta\beta$ decay by at least a factor of 2 can be obtained with the application of DPSA. This would allow, provided in all detectors DPSA would be installed, with the present setup to reach in 1 year a half-life limit for $0\nu\beta\beta$ decay of about 2×10^{25} yr.

B. Future and conclusion

The Heidelberg-Moscow experiment is the now the *only* $\beta\beta$ experiment exploring the sub-eV range for the mass of the neutrinos. With five enriched detectors with a total mass of 11.51 kg operated in the Gran Sasso underground laboratory, the experiment has reached its final setup. The experiment will probe within about 5 years the neutrino mass down to $\sim$0.2 eV.

The best presently existing limits besides the Heidelberg-Moscow experiment, obtained with the isotopes ^{48}Ca [38], ^{82}Se [39], ^{100}Mo [40], ^{116}Cd [41], ^{130}Te [42], ^{136}Xe [43], and ^{150}Nd [44], with half-life limits above 10^{21} yr, are shown in Fig. 17.

Other double-β decay setups presently under construction or partly in operation such as NEMO [45], the Gotthard ^{136}Xe TPC experiment [46], the ^{130}Te cryogenic experiment [42], a new ELEGANT ^{48}Ca experiment using 64 g of ^{48}Ca [47], a hypothetical experiment with an improved UCI TPC [44] assumed to use 1.6 kg of ^{136}Xe, etc., will not reach or exceed the ^{76}Ge limits. As pointed out recently by Raghavan [48], even the use of an amount of about 200 kg of enriched ^{136}Xe or 2 tons of natural Xe added to the scintillator of the Kamiokande detector or similar amounts added to BOREXINO (both primarily devoted to solar neutrino investigation) would hardly lead to a sensitivity larger than the present ^{76}Ge experiment. An interesting future candidate might be a ^{150}Nd bolometer exploiting the relatively large phase space of this nucleus (see [44]). The way outlined by [49] proposing a TPC filled with 1 ton liquid-enriched ^{136}Xe and identification of the daughter by laser fluorescence may not be feasible in a straightforward way.

It is obvious that the Heidelberg-Moscow experiment will give the sharpest limit for the electron neutrino mass for the next decade. For further improvements beyond the region of <0.1 eV, one has to think of *very large* experiments with *much larger* source strength.

Concluding second-generation $\beta\beta$ experiments have, after the introduction of new interpretations of the problems of solar and atmospheric neutrinos and of dark matter, involving degenerate neutrino mass scenarios, reached a key position in modern neutrino physics. They will be able to check these new interpretations in the not too far future. The Heidelberg-Moscow experiment has reached a leading position among these new $\beta\beta$ experiments and as the first of them now yields results in the sub-eV range.

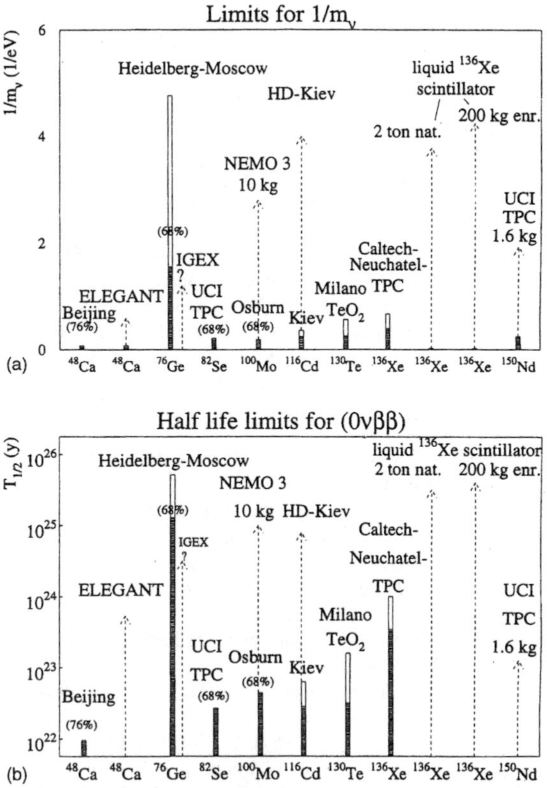

FIG. 17. Present situation, 1995, and expectation for the near future until the year 2000 and beyond, of the most promising $\beta\beta$ experiments concerning accessible half-life (a) and neutrino mass limits (b). The solid bars correspond to the present status, open bars correspond to "safe" expectations for the year 2000, and dashed lines correspond to long-term planned or hypothetical experiments.

ACKNOWLEDGMENTS

This work was supported by the Bundesministerium für Forschung und Technologie der Bundesrepublik Deutschland, the State Committee of Atomic Energy of Russia, and the Istituto Nazionale di Fisica Nucleare of the Italian Republic. The generous support of Professor E. Belotti, Professor N. Cabibbo, Professor L. Maiani, and Professor P. Monacelli is gratefully acknowledged. The authors want to thank Dr. T. Raudorf of EG&G Ortec for good cooperation and Professor V. Prusakov and his co-workers for performing the isotope separation. M.H. was supported by the Deutsche Forschungsgemeinschaft (Grant No. 446 JAP-113/101/0 and KI 253/8-1). B.M. was supported by the Human Capital and Mobility program of the European Community (Grant No. ERBCHBGCT928183).

[1] D. G. Lee and R. N. Mohapatra, Phys. Lett. B **329**, 463 (1994).
[2] S. T. Petcov and A. Smirnov, Phys. Lett. B **322**, 109 (1994).
[3] A. Ioannissyan and J. W. F. Valle, Phys. Lett. B **332**, 93 (1994).
[4] H. V. Klapdor-Kleingrothaus, Prog. Part. Nucl. Phys. **32**, 261 (1994); H. V. Klapdor-Kleingrothaus, in *Proceedings of the IV International Symposium on Weak and Electromagnetic Interaction in Nuclei (WEIN'95)*, Osaka, Japan, 1995, edited by H. Ejiri, T. Kishimoto, and T. Sato (World Scientific, Singapore, in press), p. 174.

[5] H. V. Klapdor-Kleingrothaus and A. Staudt, *Non-Accelerator Particle Physics* (IOP, Bristol, 1995).
[6] A. Staudt, K. Muto, and H. V. Klapdor-Kleingrothaus, Europhys. Lett. **13**, 31 (1990).
[7] C. P. Burgess and J. M Cline, Phys. Rev. D **49**, 5925 (1994); P. Bamert, C. P. Burgess, and R. N. Mohapatra, Nucl. Phys. **B449**, 25 (1995).
[8] M. Hirsch, H. V. Klapdor-Kleingrothaus, S. G. Kovalenko and H. Päs, Phys. Lett. B **372**, 8 (1996); J. Hellmig *et al.*, in *Proceedings of the International Workshop on Double Beta Decay and Related Topics*, edited by H. V. Klapdor-Kleingrothaus and S. Stoica (World Scientific, Singapore, 1996).
[9] M. Hirsch, H. V. Klapdor-Kleingrothaus, and S. G. Kovalenko, Phys. Rev. Lett. **75**, 17 (1995); Phys. Lett. B **352**, 1 (1995); M. Hirsch, H. V. Klapdor-Kleingrothaus, and S. G. Kovalenko, Phys. Rev. D **53**, 1329 (1996).
[10] Heidelberg-Moscow Collaboration, A. Balysh *et al.*, Phys. Lett. B **322**, 176 (1994).
[11] Heidelberg-Moscow Collaboration, A. Balysh *et al.*, Phys. Lett. B **356**, 450 (1995).
[12] J. G. Hykawy *et al.*, Phys. Rev. Lett. **67**, 1708 (1991).
[13] H. V. Klapdor-Kleingrothaus, Internal Report No. MPI-H-1987-V 17 (Proposal), Heidelberg, 1987 (unpublished).
[14] Heidelberg-Moscow Collaboration, A. Balysh *et al.*, Phys. Lett. B **283**, 32 (1992).
[15] Heidelberg-Moscow Collaboration, M Beck *et al.*, Phys. Rev. Lett. **70**, 2853 (1993).
[16] Heidelberg-Moscow Collaboration, M. Beck *et al.*, Phys. Lett. B **336**, 141 (1994).
[17] F. T. Avignone *et al.*, J. Phys. G **17**, 181 (1991).
[18] A. A. Vasenko *et al.*, Mod. Phys. Lett. A **5**, 1299 (1990).
[19] Particle Data Group, J. J. Hernández *et al.*, Phys. Lett. B **239**, 1 (1990).
[20] K. Muto, E. Bender, and H. V. Klapdor, Z. Phys. A **334**, 187 (1989).
[21] T. Tomoda and A. Faessler, Phys. Lett. B **199**, 475 (1987).
[22] T. Tomoda, A. Faessler, K. W. Schmid, and F. Grümmer, Nucl. Phys. **A452**, 591 (1986).
[23] W. C. Haxton and G. J. Stephenson, Prog. Part. Nucl. Phys. **12**, 409 (1984).
[24] G. Pantis, F. Simkovic, J. D. Vergados, and A. Faessler, Phys. Rev. C **53**, 695 (1996).
[25] J. Engel, P. Vogel, and M. R. Zirnbauer, Phys. Rev. C **37**, 731 (1988).
[26] *Proceedings of the International Workshop on Double Beta Decay and Related Topics* [8].
[27] M. Hirsch, H. V. Klapdor-Kleingrothaus, and S. G. Kovalenko, Phys. Lett. B **378**, 17 (1996); **352**, 1 (1995).
[28] M. Hirsch, H. V. Klapdor-Kleingrothaus, and S. G. Kovalenko, Phys. Rev. D **54**, R4207 (1996); Phys. Rev. Lett. **17**, 75 (1995); Phys. Rev. D **53**, 1329 (1996); Phys. Lett. B. **372**, 181 (1996).
[29] *Nuclear Data Sheets* (Academic Press, Duluth, MN).
[30] MACRO Collaboration, H. C. de Marzo *et al.*, Nucl. Instrum. Methods Phys. Res. A **314**, 380 (1992).
[31] B. Maier, Ph.D. thesis, University of Heidelberg, 1995.
[32] S. Baker and R. D. Cousins, Nucl. Instrum. Methods Phys. Rev. A **221**, 437 (1984).
[33] F. T. Avignone *et al.*, Prog. Part. Nucl. Phys. **32**, 223 (1994).
[34] Y. Chikashige, R. N. Mohapatra, and R. D. Peccei, Phys. Rev. Lett. **45**, 1926 (1980); Phys. Lett. **98B**, 265 (1981).
[35] J. Steinberger, Phys. Rep. **203**, 345 (1991).
[36] F. Petry, H. V. Klapdor-Kleingrothaus, and B. Maier (unpublished); F. Petry, Ph.D. thesis, University of Heidelberg, 1995.
[37] J. Hellmig, Ph.D. thesis, University of Heidelberg, 1996; Hellmig *et al.* (unpublished).
[38] Ke You *et al.*, Phys. Lett. B **265**, 53 (1995).
[39] S. R. Elliott *et al.*, Phys. Rev. C **46**, 1535 (1992).
[40] M. Alston-Garnjost *et al.*, Phys. Rev. Lett. **71**, 831 (1993).
[41] F. A. Danevich *et al.*, Phys. Lett. B **344**, 72 (1995).
[42] A. Alessandrello *et al.*, Phys. Lett. B **335**, 519 (1994).
[43] J.-C. Vuilleumier *et al.*, Phys. Rev. D **48**, 1009 (1993).
[44] M. K. Moe *et al.*, Prog. Part. Nucl. Phys. **32**, 247 (1994); M. K. Moe *et al.*, in *Neutrino 94*, Proceedings of the 16th International Conference on Neutrino Physics and Astrophysics, Eilat, Israel, edited by A. Dar *et al.* [Nucl. Phys. B (Proc. Suppl.) **38**, 36 (1995)].
[45] NEMO Collaboration, D. Lalanne *et al.*, in *TAUP 93*, Proceedings of the Third International Workshop on Theoretical and Phenomenological Aspects of Underground Physics, Assergi, Italy, edited by C. Arpesella *et al.* [Nucl. Phys. B (Proc. Suppl.) **35**, 369 (1994)].
[46] V. Jörgens *et al.*, in *TAUP 93*, p. 378.
[47] K. Kume *et al.*, ELEGANT Collaboration, in *Proceedings of the International Workshop on Double Beta Decay and Related Topics* [8].
[48] R. S. Raghavan, Phys. Rev. Lett. **72**, 1411 (1994).
[49] M. K. Moe, Phys. Rev. C **44**, R931 (1991).

Eur. Phys. J. A 6, 463–469 (1999)

THE EUROPEAN
PHYSICAL JOURNAL A

Digital pulseshape analysis by neural networks for the Heidelberg-Moscow-Double-Beta-Decay-Experiment

B. Majorovits, H.V. Klapdor-Kleingrothaus

Max-Planck-Institut für Kernphysik, Postfach 103980, 69029 Heidelberg, Germany

Received: 15 October 1999
Communicated by B. Povh

Abstract. The Heidelberg-Moscow Experiment is presently the most sensitive experiment looking for neutrinoless double-beta decay. Recently the already very low background has been lowered by means of a Digital Pulseshape Analysis using a one parameter cut to distinguish between pointlike events and multiple scattered events. To use all the information contained in a recorded digital pulse, we developed a new technique for event recognition based on neural networks.

1 Introduction

The question of a nonvanishing neutrino mass is still one of the most outstanding open problems in modern physics. Especially after the latest striking hints for neutrino oscillations from the Super-Kamiokande experiment [1] it has become very important to verify these results independently. Neutrinoless double-beta ($0\nu\beta\beta$) decay, which violates Lepton-number and B-L conservation by two units, is one of the most promising tools for the search of a finite neutrino mass and some other physics beyond the standard model [2]. Furthermore it seems to be the only possibility to distinguish between the Majorana- and the Dirac-nature of the neutrino. If $0\nu\beta\beta$-decay is observed neutrinos have to be of Majorana-type and have a finite mass. The atmospheric neutrino problem confirmed by the Super-Kamiokande collaboration [1] has brought degenerate neutrino models back to attention again [3], where all neutrinos have a mass in the order of $\mathcal{O}$(eV). The newest generation of $0\nu\beta\beta$-decay experiments, especially the Heidelberg-Moscow-Experiment [4] started already now to test this mass range.

2 The Heidelberg-Moscow-Experiment

The Heidelberg-Moscow-Experiment is presently the most sensitive experiment looking for $0\nu\beta\beta$-decay [4]. Out of 19.2 kg enriched ^{76}Ge five p-type High-Purity Germanium (HPGe) crystals were grown, which are now operated as p-type detectors in the Gran Sasso Underground Laboratory with an active mass of 10.96 kg in an extremely radiopure surroundings. The experiment has a background rate of 0.2 counts/(kg keV y) in the energy region between 2000 keV and 2080 keV, where the expected signal, a sharp peak at 2038.56 keV (the Q-value of $\beta\beta$-decay) lies. Since 1995 an additional background reduction has been achieved through the use of Digital Pulseshape Analysis (PSA). Due to the fact that the shape of the detected pulse is dependent on the type of interaction a distinction between multiple scattered Compton events and single interaction events is possible. A $0\nu\beta\beta$-decay event would appear as a Single Site Event (SSE), since the mean free path of the two electrons emitted by the decay is smaller than the time resolution of the detector allows to distinguish due to the low drift velocities of the electron-hole pairs. This means that Multiple Site Events (MSE) in the energy region of $0\nu\beta\beta$-decay can be regarded as background. To distinguish between the two interaction types a one-parameter method was developed at that time, based on the fact that the time structure of the pulse shapes in Germanium detectors are mainly dependent on the locations of the various events of a count within the HPGe-crystal. For MSE's one therefore expects a broader pulse in time than for SSE's since the initial locations of the electron-hole pairs are distributed over a larger area of the crystal and the overall detection time therefore increases. With this method a reduction of the background by a factor of three in the area of the expected signal could be reached [5,6].

Nevertheless a large amount of information is neglected with this method since only one parameter serves as the distinguishing criterion. Furthermore the method relies on a statistical correction of the measured SSE pulses since the efficiency of the method is substantially smaller than 100% resulting in a loss of information about the single events. For this reason we developed a new method based

PHYSICAL REVIEW C VOLUME 59, NUMBER 4 APRIL 1999

Neutrinoless double-β decay of ^{76}Ge: First results from the International Germanium Experiment (IGEX) with six isotopically enriched detectors

C. E. Aalseth,[1] F. T. Avignone III,[1] R. L. Brodzinski,[2] J. I. Collar,[1,*] E. Garcia,[3] D. González,[3] F. Hasenbalg,[1,†] W. K. Hensley,[2] I. V. Kirpichnikov,[4] A. A. Klimenko,[5] H. S. Miley,[2] A. Morales,[3] J. Morales,[3] A. Ortiz de Solórzano,[3] S. B. Osetrov,[5] V. S. Pogosov,[6] J. Puimedón,[3] J. H. Reeves,[2] A. Salinas,[3] M. L. Sarsa,[3,‡] A. A. Smolnikov,[5] A. S. Starostin,[4] A. G. Tamanyan,[6] A. A. Vasenko,[4] S. I. Vasiliev,[5] and J. A. Villar[3]

[1]*University of South Carolina, Columbia, South Carolina 29208*

[2]*Pacific Northwest National Laboratory, Richland, Washington 99352*

[3]*University of Zaragoza, 50009 Zaragoza, Spain*

[4]*Institute for Theoretical and Experimental Physics, 117259 Moscow, Russia*

[5]*Institute for Nuclear Research, Baksan Neutrino Observatory, 361609 Neutrino, Russia*

[6]*Yerevan Physical Institute, 375 036 Yerevan, Armenia*

(The IGEX Collaboration)

(Received 25 June 1998)

The International Germanium Experiment (IGEX) has six HPGe detectors, isotopically enriched to 86% in ^{76}Ge, containing approximately 90 active moles of ^{76}Ge. Three detectors of 2 kg each operate in the Canfranc Underground Laboratory (Spain) with pulse-shape analysis electronics. One detector (~0.7 kg active volume) has been operating in the Baksan Low-Background Laboratory for several years, and two additional similar detectors will operate in Baksan. A maximum likelihood analysis of 74.84 active mole years of data yields a lower bound $T^{0\nu}_{1/2} \geqslant 0.8\times 10^{25}$ yr (90% C.L.), corresponding to $\langle m_\nu \rangle < (0.5-1.5)$ eV, depending on the theoretical nuclear matrix elements used to extract the neutrino mass parameter. [S0556-2813(99)01604-0]

PACS number(s): 23.40.−s, 14.60.Pq, 27.50.+e

I. INTRODUCTION

In the standard model of particle physics neutrinos are strictly massless, although there is no theoretical reason for such. On the experimental side, there is no compelling evidence that neutrinos have nonzero masses, although the results of several experiments with solar, atmospheric, and terrestrial neutrinos lead to inconsistencies in the standard theory, unless it is assumed that neutrinos indeed have mass(es).

The anomalous electron to muon ratio observed in underground experiments on atmospheric neutrinos in Soudan and Kamioka, the solar neutrino deficit seen in the Homestake, SAGE, GALLEX, Kamiokande, and Superkamiokande experiments, and the excess electron events seen in the Liquid Scintillator Neutrino Detector (LSND) experiment at Los Alamos could all be explained within the frame of neutrino oscillations. This scenario requires nonzero neutrino masses and mixing between the various neutrino flavors.

Dark matter models of galaxy formation, which have ~75% nonbaryonic cold dark matter, 20% hot dark matter, and 5% dark baryons, fit a large body of data rather well, properly matching observed spectral power at all scales of the universe. A light neutrino could constitute hot dark matter and at the same time solve the neutrino oscillation problem if it has a mass of a few eV.

In the standard model it is assumed that neutrinos and antineutrinos are different particles, but no experimental proof exists thus far. Nuclear double-β-decay experiments address the following questions: (1) are neutrinos self-conjugate and (2) do they have Majorana mass? Various theoretical neutrino mass scenarios reconcile both the matter and vacuum oscillation solutions of the solar neutrino problem and allow the possibility of neutrinoless double-β decay with an effective Majorana neutrino mass in the range of 0.1–1 eV, the target sensitivity of IGEX.

II. DOUBLE-β DECAY

Neutrinoless $\beta\beta$ decay is the only practical way to determine if neutrinos are Majorana particles. In an extension of the black-box theorem of Schechter and Valle [1], Kayser, Petkov, and Rosen showed that in the context of any gauge theory, the observation of 0ν $\beta\beta$ decay would constitute unambiguous evidence that at least one neutrino eigenstate has nonzero mass [2]. In the standard model, weak interactions preserve lepton number exactly, whereas neutrinoless $\beta\beta$ decay would violate it. Additionally, a neutrinoless decay mode could reveal the existence of majorons, scalar Goldstone bosons that result from a spontaneous symmetry breaking mechanism that results in the generation of neutrino mass. These and other features make the search for neutrinoless double-β decay invaluable for exploring non-standard-model physics.

The transition probability of the lepton number violating 0ν $\beta\beta$ decay driven by a Majorana neutrino mass term is expressed as [3]

*Present address: PPE Division, CERN, Geneva 23, Switzerland.
†Present address: Universitaet Bern, CH-3012 Bern, Switzerland.
‡Present address: TUM, Garching, Germany.

2.5.5 Cryogenic Detectors

Physica B 169 (1991) 388–399
North-Holland

Cryogenic thermal detectors in subnuclear physics and astrophysics

E. Fiorini

Dipartimento di Fisica dell' Universita' di Milano e Sezione di Milano dell' INFN, Milano 20133, Italy

The plenary talk was given by the author.

Some subjects are considered which are very relevant to particle physics and astrophysics, like unification of forces and the related problems of neutrino mass, lepton number non-conservation, solar neutrinos and cosmic dark matter. It is shown how low-temperature thermal detectors can be essential in overcoming the experimental difficulties in this field. The "state-of-the-art" in superconducting tunneling junctions and strips, superheated superconducting granules and bolometers and their specific advantages over conventional particle detectors is discussed.

1. Introduction

The growing interconnection between subnuclear physics and astrophysics has been strongly stimulated by the progress and hopes in the unification of forces. About a century after electric and magnetic interactions had been beautifully unified by Maxwell, brilliant theoretical predictions and unambiguous experimental evidence have led to the unification of the weak and electromagnetic forces. The Grand Unified Theories (GUT) are now trying to unify also the strong to the electroweak interaction under the experimentally based assumption that the "strengths" of all these forces become comparable at energies between 10^{23} and 10^{24} eV. Full unification with the gravitational force, like the one predicted in the theory nicknamed SUSY (for Super Symmetry), would require 10^{28} eV. These energies are, and will be, much larger than those reachable with particle accelerators (about a trillion electrovolts at present). Here astrophysics and cosmology come to our help: instead of looking at increasing energies, like physicists are doing since the origin of subnuclear physics, let us go back to the Big Bang. The expansion of the universe began at an incredibly high temperature and energy, which then decreased gradually with time. It took only 10^{-45} and 10^{-35} seconds to "reduce" the temperature to the values corresponding to SUSY and GUT, respectively. Incidentally, it took only one second to reach the temperature corresponding to one MeV, but more than 15 billion years to reach our present situation where the temperature of the cosmic background radiation is as low as 2.7 K! We may investigate what happens at unification energies by searching for "relics" of the Big Bang: light nuclei like ^{4}He, magnetic monopoles, dark matter, cosmogenic neutrinos, etc. Another way is to search for rare events predicted by GUT and SUSY even at "our" energies, like those induced by non-conservation of the baryon or lepton numbers, or to investigate neutrinos from the sun. There is therefore an entire class of phenomena which cannot be studied with the most powerful particle accelerators, but where only the universe can be used as a laboratory. The aim of the present paper is to show the great help that this "physics without accelerators" can receive from the recently developed cryogenic detectors based on the thermal effects induced by single particles. The low rate of the events searched for usually

Low temperature detectors in non-accelerator particle physics

Andrea Giuliani
INFN and Dipartimento di Fisica dell'Università, Via Celoria 16, 20133 Milano, Italy

ABSTRACT: Status of art on phonon mediated particle detection with dieletric absorbers is reported. The results and the techniques which could lead to start soon pilot experiments in dark matter search, double beta decay search and x-ray astrophysics are described.

1. INTRODUCTION

Experimental investigation of a number of crucial questions in elementary particle and astrophysics requires the development of a new class of detectors. Open problems are neutrinoless double beta decay, coherent neutrino scattering, rest mass neutrino determination and search for dark matter candidates. Low temperature detectors in their various versions [1] - bolometers with semiconductor or superconductive thermistors, tunnel junctions, phonon imaging detectors and superconducting granules - can improve our knowledge of the mentioned phenomena: the test detectors realized up to now show infact that these techniques can provide better energy resolution, lower energy threshold, wider material choice than conventional detectors, and exhibit the unique feature to be sensitive to non-ionizing events. In this review I will treat only particle detection mediated by phonons produced in a dieletric absorber, referring to the literature for quasiparticle detectors and superconductive grains [1].

2. PRINCIPLES OF PHONON MEDIATED PARTICLE DETECTION

Several types of low temperature particle detectors were conceived and realized in the last years. However, only those devices which, for their experimental performances, promise a rather prompt employment in a fundamental physics experiment will be considered here.

2.1 Bolometers as perfect calorimeters

Historically, the first proposed bolometers were considered perfect calorimeters [2,3], i.e. devices able to thermalize thoroughly the energy released by the impinging particle: this approach is naïve, but is useful to introduce the terminology about bolometers and to illustrate the excellent potential performances which characterize this new kind of particle detectors.

0954-3899/91/0S0309 + 15 $03.50

Paper presented at the First Int. Symp. on Lepton and Baryon Number Violation, Trento, 1998 302

The CUORE project: a large observatory for Neutrinoless Double Beta Decay and other rare events

Andrea Giuliani [1], *on behalf of the CUORE* [2] *collaboration*

INFN, Sez. di Milano - Via Celoria 16 - I-20133 - Milano, Italy

Abstract. We propose constructing a 750 kg array of 1000 TeO_2 bolometers to search for neutrinoless Double Beta Decay, Cold Dark Matter and Solar Axions. The physical motivations and the technical feasibility of the project are briefly discussed.

1. Introduction

We propose a large, high granularity, close-packed array of bolometric detectors to be operated underground (Gran Sasso INFN Laboratories, Italy), named "CUORE" (which stands for Cryogenic Underground Observatory for Rare Events), aiming first of all at Double Beta Decay (DBD) [1] study of ^{130}Te: other interesting applications are searches for WIMPs [2] and solar axions undergoing coherent Bragg conversion to photons by the Primakoff effect [3]. Due to the high proposed mass (750 kg) and to the innovative technique (bolometry), this experiment represents a revolutionary approach to rare event investigation.

2. ^{130}Te: a promising candidate for Double Beta Decay study

Double Beta Decay [1] is a rare nuclear process consisting of the spontaneous transmutation of an even-even nuclide (A,Z) into a nuclide (A,Z+2),

[1] E-mail: Andrea.Giuliani@mi.infn.it

[2] WEB site: http://hpbbgs.lngs.infn.it/halla/index_CUORE.html

accompanied by the emission of two electrons and, in the standard electroweak frame, of two electron antineutrinos. An occurence of the decay without neutrino emission could be naturally interpreted in terms of new neutrino physics, implying a finite mass and a Majorana nature for the neutrino. Other mechanisms beyond the Standard Model could induce the neutrinoless transition [4].

Since no neutrinoless process has been observed up to now, results are expressed in terms of a limit on the half-life for a given nuclide: from this limit, through nuclear physics calculations, it is possible to extract a value for the relevant weak-interaction-physics parameter, the so-called *effective neutrino mass*, a mixture of neutrino-eigenstate masses usually indicated as $< m_{\nu_e} >$. The neutrinoless DBD rate scales as $< m_{\nu_e} >^2$. A systematic uncertainty of a factor typically 2-3 on $< m_{\nu_e} >$ comes from the different nuclear matrix evaluations.

The present technology, thanks to the experiments which study DBD of ^{76}Ge [5, 6], can push the limit on $< m_{\nu_e} >$ down to $\sim$ 0.1 eV. To improve substantially this goal, a new concept experiment (as in the case of the GENIUS project [4], based on ^{76}Ge) must be designed, by choosing carefully the candidate nucleus and the experimental technique. As we shall show in the following, we think that the study of ^{130}Te using the bolometric technique is one of the most promising approach for new generation DBD experiments.

2.1. ^{130}Te Double Beta Decay probability

The DBD rate is given by the product of three factors: $< m_{\nu_e} >^2$, a phase space term and a nuclear matrix element term. The phase space term, which is exactly evaluable, grows fast with the transition energy E_0 (roughly as E_0^5): therefore, high transition energies are advisable. As a consequence, all the candidates whose transition energy is below 2 MeV are not interesting from an experimental point of view. ^{130}Te has a transition energy of 2.528 MeV, leading to a phase space 6.9 times larger than that of ^{76}Ge.

When taking into account also the nuclear matrix elements, one must consider that some disagreement is present among calculations from different authors [7](see fig. 1). On the average ^{130}Te DBD remains anyway about four/five times faster than ^{76}Ge. One more important point is that for ^{130}Te there is no serious spread among the rates proposed by the various authors (unlike for example ^{100}Mo).

2.2. ^{130}Te availability

In a new generation DBD experiment, at least 10^{27}–10^{28} nuclei must be contained in the detector. At this level, the availability of the nuclide,

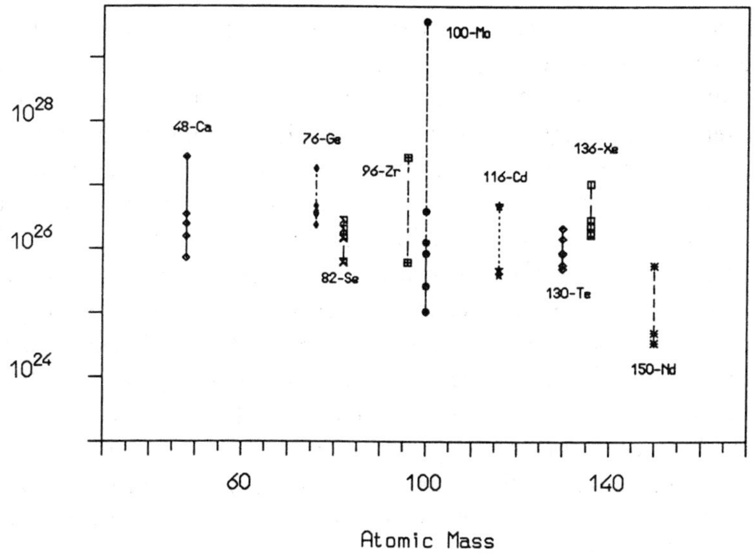

Figure 1. DBD half-life [y] evaluated assuming $< m_{\nu_e} >= 0.1$ eV for various nuclei according to different nuclear matrix calculations [7].

both from the practical and economical point of view, is a critical point. In this respect, ^{130}Te is in a unique position: its natural isotopic abundance is indeed 34%, much higher than for the other interesting nuclei. This implies that a significant experiment can be performed even *without isotopic enrichment*, which is often economically prohibitive in large quantities. Taking into account both isotopic abundance and decay rate, 750 kg of TeO_2 (which contain 202 kg of ^{130}Te) are equivalent to the effectiveness of between 700 and 1000 kg (depending on the nuclear matrix elements) of Ge detectors isotopically enriched to 86 % ^{76}Ge.

2.3. Te compatibility with the bolometric technique

A powerful method to investigate neutrinoless DBD is the so called "source = detector" technique [8], applied to several high sensitivity DBD experiments. In this approach, the isotopes under study are contained in the detector itself: neutrinoless DBD signature would be therefore a peak in the detector energy spectrum in correspondence of the DBD transition energy. Basic requirements for this technique are a high energy resolution and a very low background.

We propose the bolometric technique [9] for ^{130}Te study. A bolometer [10] is a device that measures the energy released to it by a single particle as a temperature rise of the whole detector, measured by means of an extremely sensitive thermometer thermally coupled to the detector absorbing part. In order to have high temperature signals in macroscopic amounts of material for nuclear energy depositions, it is necessary to operate the detector at very low temperatures ($\ll 1$ K), exploiting the strong reduction of specific heats with temperature in ordinary materials. Best materials for this application are dielectric diamagnetic single crystals, whose specific heat scales as T^3 at low temperatures according to the Debye law. We propose 750 g mass TeO_2 single crystals for the CUORE project. A crucial aspect of the device is the thermometric element, which should be in our case a semiconductor doped thermistor, suitable for a multiplication of the channels, mainly because it requires a standard, room temperature, reasonable cost electronics. Reliable and reproducible results are obtained with the Neutron Transmutation Doped (NTD) Ge thermistors developed by E. Haller in Berkeley.

3. Present performances of TeO_2 bolometers

The operation of bolometric detectors with large sensitive volumes requires high sensitivity thermistors, very low temperatures (~ 10 mK), long term stability and single detector reproducibility. The pluriennial work of the Milano group has shown that the bolometric technique is mature in all these aspects and that a large channel multiplication is now viable.

3.1. Single detector mass and resolution

Many tests were performed on 340 g TeO_2 detectors, operated with NTD Ge thermometric elements: presently, the state-of-art of detector construction and signal processing allows to reach in a reproducible way 1.5-2 keV FWHM intrinsic resolution. Only one test was up to now performed on a 750 g mass crystal, as foreseen for the CUORE modules. An energy resolution of 9 keV at 2615 keV was obtained, as can be appreciated in the calibration spectrum reported in fig. 2. The measurement was however performed in unsatisfactory microphonic noise condition, due to naive temporary cryogenic problems. We are therefore confident that this result can be improved in future tests.

3.2. Detector response stability

Cryogenic set-ups exhibit instabilities which can affect bolometer performances: an active mechanism to stabilize detector response is therefore

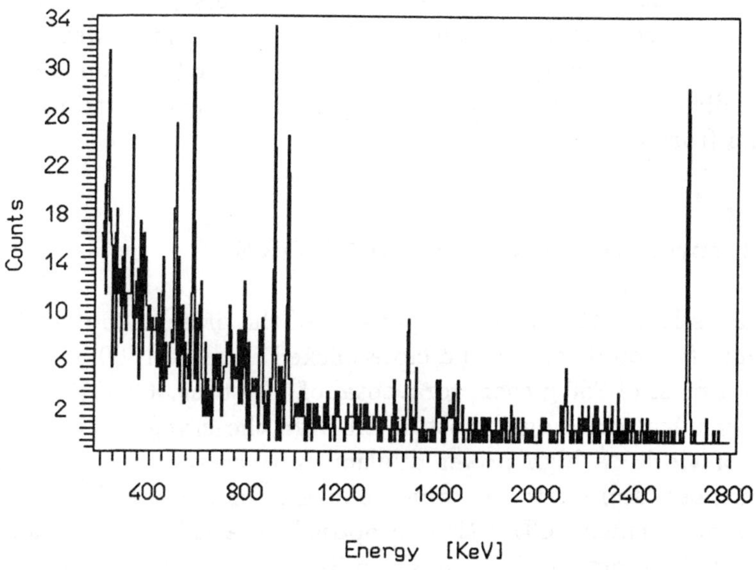

Figure 2. ^{232}Th calibration spectrum obtained with a 750 g TeO_2 detector.

necessary. We use a doped silicon resistance, with a metallic behaviour, as a heater which delivers calibrated energy amounts to the crystal through joule power. The heater pulses are identical to particle pulses: through a correlation between pulse height and detector operation point it is possible to correct the amplitude of all physical pulses in a very effective way, as explained in [11].

3.3. Reproducibility of the array elements

The Milano group is now operating a 20 detector array [12] whose single TeO_2 module has a mass of 340 g. The array performances are very satisfactory and show that the reproducibility problem is almost completely solved, thanks to the construction procedures described in [13]: the base temperature is 8 mK for all the elements within 1 mk spread; to optimize detector resolution, the mixing chamber is slightly heated up to about 10 mK; at the optimum bias point the operating temperatures of the various detectors range from 11 to 14 mK. In these conditions the typical voltage-to-energy conversion is around 250 μV/MeV with a spread of about a factor 2. The detector intrinsic resolution (which corresponds to the real resolution at low energies) is between 1.5 and 2.5 keV FWHM. At high energies (around

2 MeV) , the effective resolution is a factor 2-3 worse due to not yet completely understood systematic effects. This array, in a few week operation (corresponding to 0.35 kg·y), has allowed to set a limit on neutrinoless DBD half-life of 7.7×10^{22} y, corresponding to effective neutrino mass limits ranging from 2.5 to 5.2 eV [12, 7].

4. The future: CUORICINO and CUORE

CUORE constitutes the natural expansion of the present 20 detector array experiment. It should consist of a close-packed array of 1000 TeO_2 bolometers, with a mass of 750 g each, for a total of 0.75 tons. CUORE represents a severe technical challenge, most of all from the cryogenic and bolometric point of views: a high power dilution refrigerator must be developed on the purpose, similar to the ones required for cooling down multi-ton gravitational antennas. CUORE is proposed by a still open collaboration, including at the moment: Milano, Gran Sasso and Firenze, Italy; Berkeley and South Carolina, USA; Leiden, the Netherlands; Neuchatel, Switzerland; Zaragoza, Spain. The total cost of CUORE should be around 8 M$, most of which for the crystals (6 M$), while the rest is mainly due to the refrigerator and the electronics. The CUORE approach is flexible: other materials could be easily studied in addition to or in place of the TeO_2 crystals.

4.1. Preliminary structure of CUORICINO and CUORE

In a very preliminary approach, CUORE should consist of seventeen towers. Each tower consists of a stack of 15 modules. Each module, which is the smallest independent unit, contains 4 crystals. The whole arrangement is such that the inert material is minimized and the crystals are separated each other only by a few millimiters. The detector is included in a 110 cm high, 75 cm diameter cylindrical volume, which will corresponds to the evacuated experimental space of the dilution refirgerator. The operation temperature should be around 10 mK.

A single CUORE tower could be cooled down in the refrigerator presently housing the 20 element array, representing at the same time a test of the CUORE principle and a new very powerful DBD experiment, containing 5.7×10^{25} ^{130}Te nuclei. The CUORE collaboration is proposing this project with the name "CUORICINO", which means "small CUORE".

4.2. Potential of CUORE in Double Beta Decay search

For quoting CUORE sensitivity in DBD search, a reliable background level should be estimated. A Monte Carlo evaluation of the background on the basis of our preliminary structure and of the typical contamination levels of the employed materials (mainly copper and PTFE) is in progress. For the moment, we prefer to discuss the sensitivity *as a function* of the background level. If we had in CUORE the same background as in the 20 detector array (0.5 counts/keV kg day in DBD transition energy region), the effective neutrino mass 5 y sensitivity would be around 0.2 eV. This is however a very pessimistic assumption, since there is still a lot of work to do about radiopurity of cryostat components; furthermore, the CUORE structure, due to a much better volume-to-surface ratio, provides a natural self-shielding. One has to consider also that the detector granularity could help very much in background suppression, by operating the modules in anticoincidence.If we could improve the present background by a factor 100 (it would remain anyway 50 times worse than the background claimed by GENIUS [4]), a 5 y sensitivity down to 6×10^{-2} eV could be reached for effective neutrino mass.

As far as WIMPs and axion searches are concerned, the CUORE potential [14] will depend mainly on the threshold and the low energy background achievable. In the present 20 detector array, a threshold of 5 keV looks possible. Analyses are in progress on these topics.

References

[1] Moe M K and Vogel P 1994 *Ann. Rev. Nucl. Part. Sci.* **44** 247

[2] Mosca L 1996 *Nucl. Phys.* (Proc. Suppl.) **48** 34

[3] Creswick R J *Phys. Lett.* (in press)

[4] Hellmig J and Klapdor- Kleingrothaus H V 1997 *Z. Phys.* **A359** 361-372

[5] Baudis L et al. 1997 *Phys. Lett.* **B407** 219

[6] Aalseth C E et al. 1996 *Proc. XVII Int. Conf. on Neutrino Phys. and Astrophys., Helsinki, Finland* (World Scientific) p 361

[7] Suhonen J and Civitarese O *Phys. Rep.* (in press)

[8] Dell'Antonio G F and Fiorini E 1960 *Suppl. Nuovo Cim.* **17** 132

[9] Fiorini E and Niinikoski T O 1984 *Nucl. Instrum. Methods* **224** 83

[10] Twerenbold D 1996 *Rep. Prog. Phys.* **59** 349

[11] Alessandrello A et al. 1998 *Nucl. Instrum. Methods* **A412** 454-464

[12] Alessandrello A et al. *Phys. Lett.* (in press)

[13] Alessandrello A et al. 1997 *Proc. LTD-7, VII Int. Work. on Low Temperature Detectors, Munich, Germany* (Munich: Max Planck Insitute of Physics) p 249

[14] Alessandrello A et al. 1998 *Proc. PASCOS-98, VI Int. Symp. on Particles, Strings and Cosmology, Boston, MA, USA* (in press)

Journal of Low Temperature Physics, Vol. 93, Nos. 3/4, 1993

Superconducting Granule Detectors

Klaus Pretzl

Laboratory for High Energy Physics
University of Bern, CH-3012 Bern, Switzerland

The present status of the development of superheated superconducting granules (SSG) as possible detectors for neutrinos and dark matter is reviewed. Future perspectives will also be given.

INTRODUCTION

The idea to use SSG as a particle detector goes back to 1967 [1]. The potential use of this device for the detection of dark matter particles, solar neutrinos, neutrino less double beta decays, low energy reactor or accelerator neutrinos as well as its possible application as photon imaging devices motivated many groups to further develop this technique (table 1). A review of previous work on SSG detector development can be found in Ref.[2]. This paper will concentrate on more recent work in this field including some of the results presented at this conference.

SSG DETECTOR PRINCIPLE

A SSG detector consists of millions of tiny superconducting granules with a typical diameter of several micrometers diluted in a dielectric material with volume filling factors up to 30%. The grains serve as target and detector material at the same time. Metastable type I superconductors are used for grain materials. Metastability provides an essential presupposition for a SSG detector, because of the sharp superheating (Hsh) and supercooling (Hsc) phase boundaries (Fig.1) and because of the sudden phase transition.

The detector principle is very simple. The energy loss (ionization or nuclear recoil) of a particle interacting in a granule can sufficiently heat the granule and induce a phase transition (granule flip) from the superconducting to the normal state. Assuming uniform heating of the entire granule (global heating), the temperature change experienced

0022-2291/93/1100-0439$07.00/0

University of Bern, Laboratory for High Energy Physics
preprint BUHE-9906

CRYOGENIC CALORIMETERS IN ASTRO AND PARTICLE PHYSICS

K.PRETZL

Laboratory for High Energy Physics of the University Bern, Sidlerstr. 5 CH 3012 Bern, Switzerland E-mail:Pretzl@lhep.unibe.ch

The development of cryogenic calorimeters was originally motivated by the fact that very low energy thresholds and excellent energy resolutions can be achieved by these devices. Cryogenic devices are widely used in double beta decay experiments, in cosmological dark matter searches, in x-ray detection of galactic and extragalactic objects as well as in cosmic background radiation experiments. An overview of the latest developments is given.

1 Introduction

Cryogenic calorimeters open up the possibility of measuring processes with energy transfers as low as eV with very high accuracy. Their development has recently been motivated by the quest for the dark matter in the universe and for the missing neutrinos from the sun. Other research areas have also benefitted from these developments, such as the double β-decay experiments, neutrino mass experiments, x-ray spectroscopy in astrophysics, single photon counting spectrometry , mass spectrometry of large molecules (DNA sequencing) as well as x-ray microanalysis for industrial applications.

2 First Ideas and Attempts

In 1935 F.Simon [1] suggested measuring the energy deposited by radioactivity with cryogenic calorimeters. Later in 1949 D.H.Andrews, R.D.Fowler and M.C.Williams [2] reported the detection of individual α-particles using a superconducting strip. G.H.Wood and B.L.White [3] detected α-particles with a superconducting tunnel junction (STJ) detector. H.Bernas et al. [4] used superheated superconducting granules (SSG) for beta radiation. A.K.Drukier

2.5.6 Other Large Source Strength Detectors

VOLUME 72, NUMBER 10 PHYSICAL REVIEW LETTERS 7 MARCH 1994

New Approach to the Search for Neutrinoless Double Beta Decay

R. S. Raghavan
AT&T Bell Laboratories, Murray Hill, New Jersey 07974
(Received 9 November 1993)

Sub-eV Majorana neutrino masses $\langle m_\nu \rangle$, can be explored by a new approach to neutrinoless double β decay using ^{136}Xe in a Xe gas-loaded, multiton liquid scintillator installed in a very low background detector such as the Kamiokande facility. With enriched ^{136}Xe, a readily implementable, 10 ton detector experiment can establish an $\langle m_\nu \rangle = 0.45$ eV at 3σ in 1 yr (or exclude an $\langle m_\nu \rangle < 0.23$ eV in 2 yr). A 100 ton detector can extend the limit to $\langle m_\nu \rangle < 0.1$ eV, compared with the present limit of $\langle m_\nu \rangle < 1.3$ eV.

PACS numbers: 29.40.Mc, 14.60.Pq, 23.40.Bw

Neutrinoless double β decay $(0\nu 2\beta)$, represented by the process $A(Z) \rightarrow A(Z+2) + 2\beta^-$, violates lepton conservation and demands two nonstandard neutrino properties: (1) $\nu_e = \bar{\nu}_e$, i.e., a Majorana type neutrino and (2) a nonzero Majorana neutrino mass $\langle m_\nu \rangle$ which introduces an admixture of opposite helicity [1]. Thus, $(0\nu 2\beta)$ decay is one of the few direct pointers of physics beyond the standard model and offers a rare experimental tool for the direct observation of a *small* neutrino mass. For these reasons, $(0\nu 2\beta)$ decay has been intensively searched for in recent years. The best limit achieved on $\langle m_\nu \rangle$ is < 1.3 eV with ~ 7 kg yr of ^{76}Ge in enriched Ge detectors [2]. A breakthrough into the regime of sub-eV masses (suggested indirectly as the likeliest by present limits on neutrino oscillations), requires a large source mass m and very low background b [the sensitivity to $\langle m_\nu \rangle \propto (m/b)^{-1/4}$]. Just these criteria are also basic for recent designs of very low energy (< 1 MeV) solar neutrino detectors which thus offer a possible new framework for high sensitivity $\beta\beta$ studies [3]. As described in this Letter, this idea can indeed be directly realized in practice in the case of ^{136}Xe since large masses of Xe gas can be loaded into a large-scale liquid scintillation detector such as Borexino [4]. Such an experiment is, in fact, less demanding than the detection of < 1 MeV solar neutrinos because of the higher energy of the ^{136}Xe $(0\nu 2\beta)$ line signal at 2.45 MeV, which also opens other facilities such as Kamiokande (K-II) where this approach can be readily implemented by installing a Ze scintillator setup.

A 100 ton scale Xe scintillator can ultimately lead to a limit $\langle m_\nu \rangle < 0.1$ eV with enriched ^{136}Xe and 0.25 eV with natural Xe, thus introducing a new level of sensitivity for the search for $(0\nu 2\beta)$ decay. As a first stage experiment, a small 10 ton scintillator loaded with enriched ^{136}Xe is sufficient to establish a Majorana mass $\langle m_\nu \rangle = 0.45$ eV with 3σ precision in 1 yr (or set a limit $\langle m_\nu \rangle < 0.23$ eV in 2 yr). Sub-eV mass sensitivity in this speedily realizable experiment is particularly attractive at present in view of recent observations of suggestive $(0\nu 2\beta)$ line features consistent with a mass $\langle m_\nu \rangle \sim 1$ eV [5]. The standard lepton-number *conserving* $(2\nu 2\beta)$ decay of ^{136}Xe producing a continuous 2β spectrum with an end point at 2.45 MeV can be $\sim 10^3$ times faster than the $[0\nu 2\beta(\langle m_\nu \rangle = 1$ eV)] decay. In the Xe scintillator approach, these signals will occur at unprecedented rates of hundreds/day at a high signal/noise; note that this rare process was first detected at all (in ^{82}Se) only recently [6] and is yet unobserved in Xe.

The marked advance in sensitivity is made possible by two factors: First, the Xe scintillator approach facilitates, for the first time, observation of a $(0\nu 2\beta)$ source mass of tons (rather than kg). With a solubility as high as 2% of Xe in organic liquid scintillators [7], a Xe source mass of up to 2 ton can be observed in a 100 ton detector, compared to ~ 6 kg (enriched Xe), the maximum employed so far in gaseous Xe detectors [8,9]. In addition, the detection efficiency is $\sim 100\%$, compared to $\sim 20\%$ in gas detectors. Second, the specific background b/m is lowered many orders of magnitude (compared to Ge) because of the high radiopurity of liquid scintillators [10] and the massively shielded environment inherent in a direct-counting facility for low energy solar neutrinos. These two factors more than offset the much poorer energy resolution (compared to Ge) and the relatively low concentration of Xe in the scintillator. A noble gas such as Xe is one of the few types of dopants in a liquid scintillator that guarantees no degradation of chemical/radiopurity and detector response. Indeed, an inert gas is normally loaded into the scintillation liquid to improve and maintain optical performance. Gas loading also offers the key advantage of a source in/source out facility which allows a direct blank measurement without affecting the spectroscopy quality of the detector or the background. A Xe $(0\nu 2\beta)$ experiment thus fits naturally and elegantly with the technique of large-scale liquid scintillation spectroscopy.

Based on theoretical nuclear matrix elements for ^{136}Xe [11], the neutrino mass follows as $\langle m_\nu \rangle \approx [2.7/T_{1/2} \times (0\nu 2\beta)]^{1/2}$ ($T_{1/2}$ = half-life in units of 10^{24} yr). Reference [8] has reported the best limit (90% C.L.) of $T_{1/2} \sim 0.25 \times 10^{24}$ yr for the ^{136}Xe $(0\nu 2\beta)$ decay; thus a limit $\langle m_\nu \rangle < 3.3$ eV. Large quantities of high purity, natural Xe as well as isotropically enriched ^{136}Xe are commercially obtainable at reasonable cost [12]. The main contaminant in these samples as observed in $(0\nu 2\beta)$ experiments so far [8,9] is radioactive 10.8 yr ^{85}Kr which β decays with an end point of 670 keV. The activity is not observed above background after rejecting mass numbers

0031-9007/94/72(11)/1411(4)$06.00

below 124 (by diffusion or ultracentrifugation) [9]. Large quantities of such mass-rejected samples are available [12]. The only other activities in well-sealed, mass-rejected Xe gas samples are small amounts of sea-level cosmogenic ^{3}H and 36-d ^{127}Xe (which emits only a single 390 keV γ ray) [13].

There are several options for implementing a Xe scintillator double β decay experiment, the most attractive being K-II, used so far as a water Cerenkov detector [14] for solar neutrino studies now nearing completion. It can be readily adapted by placing up to a few hundred tons of liquid scintillator in a transparent vessel in the center of the water tank and using the present photon detection and other systems unchanged. A 4 ton liquid scintillation counting test facility (CTF) [15] will operate in 1994 at Gran Sasso. The Borexino facility itself is a possibility following completion of its solar neutrino objectives (a concurrent $\beta\beta$ study is contingent on more sensitive data on ^{85}Kr content of Xe samples). The intensively studied Borexino design [4] is basic to all the options; we can thus apply it to illustrate practical figures of merit in these options.

Borexino consists of 300 ton of liquid scintillator contained in a transparent vessel of radius $R=4.25$ m, at the center of a 17 m diam$\times$17 m water tank. The scintillation light is observed by 1700 photomultipliers (PM) arranged at $R=6.25$ m. A ^{136}Xe ($0\nu2\beta$) event triggers $\sim$500 PM. The total charge in the PM and their relative trigger times yield the energy and spatial location of the events, with Δ(FWHM)$\sim$250 keV and $\Delta r/r(1\sigma)\sim$8 cm at 2.5 MeV. Nuclear events are detected *calorimetrically*; i.e., only the summed energy is observed for prompt cascade nuclear decays, Compton showers, or 2β emission. The good spatial resolution allows a relatively well-defined fiducial volume (FV), thus an (off-line) choice of the ($0\nu2\beta$) source mass. The nominal FV is set at $R=3$ m ($\sim$100 ton of scintillator) which leaves a 1.25 m active buffer shield. For the Xe ($0\nu2\beta$) studies, a hardware trigger at $\sim$500 keV will reject most of the events due to ^{85}Kr.

An important part of the unshieldable background $B(I)$ is set by radiocontaminants in the scintillator. The energy of the ^{136}Xe ($0\nu2\beta$) line, 2.45 MeV, is particularly favorable, freeing interference from lower energy sources such as ^{7}Be, ^{210}Pb, and ^{40}K (of major concern in Borexino). Of the conceivable naturally radioactive contaminants, *only* ^{214}Bi (and, to a small extent, ^{212}Br [16]) are relevant to signals in the 2–3 MeV region. The Bi events are, however, just the type that can be ideally tagged because they are followed at the same spatial location within a few hundred μs by $\sim$8 MeV α's which appear in the scintillator as $\sim$0.8 MeV events. The delayed coincidence tag can remove these events by a factor >500 [17], creating in effect a 2–3 MeV gap in the background spectrum. The efficient taggability of the ^{214}Bi background also minimizes the problem of ubiquitous radon, since the only relevant radon decay product is ^{214}Bi. Long-lived (<10 s) cosmogenic activity via muon capture, spallation, or by reactions due to muon-induced neutrons on C and H in the scintillator, results mostly in 20 min ^{11}C which β^+ decays with a total energy of $(1.02\gamma\gamma+0.96)=1.98$ MeV, well short of the 2–3 MeV gap. Activities induced in the Xe additive ($\sim$10–20/(ton Xe) yr [13]) produce only a relatively small background in the signal window.

The background profile in a Xe-filled Borexino is illustrated in Fig. 1. The Monte Carlo simulation [18] generates events in the full scintillator volume due to all contaminant decays in the U, Th chains and from ^{40}K (assuming as in Borexino, U/Th at 10^{-16} g/g and K at 10^{-12} g/g), reconstructs the scintillation response, and filters out the spectrum of events occurring within a chosen FV. Figure 1 shows the 2–3 MeV gap [19] due to the coincidence tags. The residual Bi events in the gap are only $\sim$10/MeV yr (100 ton) smaller than electron scattering events of ^{8}B solar neutrinos ($\sim$20/MeV yr) and resolution-smeared spillage from the ($2\nu2\beta$) signal continuum ($\sim$20/MeV yr) [20], for a total of $B(I)\sim$50/MeV yr (100 ton). Clearly, the Borexino-grade purity assumed above can be relaxed significantly. The background from external sources, B(E), depends on the FV radius R and arises from materials closest to the FV, viz., the inner vessel and the adjacent shield water. The latest

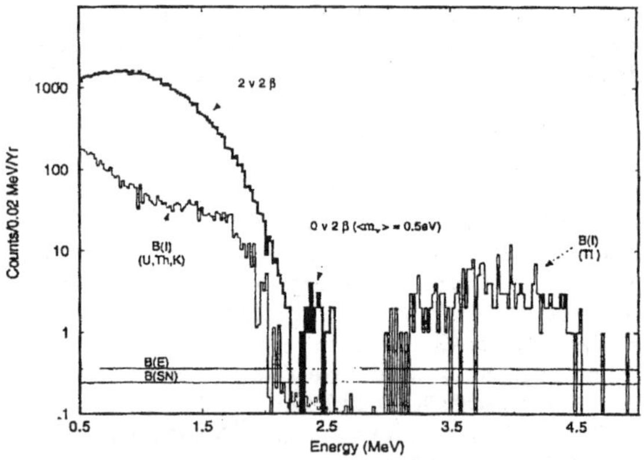

FIG. 1. Monte Carlo simulation of 2β signals and background in the Borexino design option (see line 1 of Table I and text) with 58 ton FV of Xe scintillator (2 wt.% natural Xe) or equally for 90% enriched ^{136}Xe loaded, K-II option of a 6 ton FV scintillator (line 6 of Table I) with 10$\times$ higher impurity concentration. Shown are the spectral profiles of background (thin lines) due to internal sources B(I) and estimated levels from external sources B(E) and solar neutrinos B(SN) and the signals (thick lines) due to the $2\nu2\beta$ continuum and a possible $0\nu2\beta$ peak due to a neutrino mass $\langle m_\nu\rangle=0.5$ eV. The only analysis cuts applied in the B(I) profile are tags of the ^{212,214}Bi delayed β-α coincidences.

designs employ a thin (~1 mm) membrane type vessel [15] and the possibility of an outer membrane vessel enclosing a 0.5 m thick buffer of high-purity (water or organic) liquid. With these improvements, the limiting background, due mostly to fixtures such as the PM and their supports, is estimated at $B(E,R=3\text{ m})\sim 180$/MeV yr [21]. The total specific background is thus $B(R)\Delta=[B(I)+B(E,R)]\Delta=60$/yr (100 ton) for $R=3$ m. With a more optimal choice of $R=2.5$ m (60 ton) [and with $B(I)\propto R^{-3}$, $B(E,R)\propto\exp(-\delta R/21.5\text{ cm})$], we estimate $B\Delta\sim 12$/yr (60 ton). Figure 1 is based on this choice of design parameters.

In a Xe-K-II experiment, the larger water shield at K-II can accommodate the higher rock radioactivity. The muon flux and thus the cosmogenic ^{11}C are 10× higher, still posing no interference. The smaller PM coverage produces a worse energy resolution, $\Delta\sim 350$ keV. The background from the PM is $x\sim 5$ smaller [22]. Overall, we estimate $B(E,R=3\text{ m})\sim 40$/MeV yr. $B(I)$ may be higher because of added spillage of $(2\nu 2\beta)$ events due to the poorer resolution, thus $B(I)\sim 70$/MeV yr. Thus, for a 300 ton Xe-K-II scintillator, $B\Delta(R=3\text{ m})\sim 40$/yr (100 ton). For the optimum $R=2.5$ m as above, $B\Delta\sim 16$/yr (60 ton). A smaller total mass (100–200 ton) can be chosen in K-II, reducing the total Xe inventory without serious loss of sensitivity.

The use of enriched Xe (90%) is of great practical interest since $\langle m_\nu\rangle$ sensitivities possible in the above 100 ton devices can be attained with only ~10 ton of scintillator contained in vessels $R<1.5$ m. Note that Fig. 1 based on a 60 ton (FV) design applies equally to a 6 ton (FV) scintillator with enriched Xe with a scintillator purity of 10^{-15} g/g (10× worse than assumed in the 60 ton simulation). These purities are readily attained in organic scintillators [10]. Indeed, for purities even worse, Fig. 1 shows that the contaminant background is smaller than those from other sources. Because of the smaller R (and thus the larger water shielding), the value of B(E) is set only by residual background due to the vessel material (and the high-purity buffer) rather than the more external fixtures such as PM and their supports. We estimate $B\Delta\sim 18$/yr for a 6 ton FV in a 10 ton vessel, approaching $B\Delta$ values reached above for an equivalent 60 ton FV but a total mass of 300 ton. The low-mass regime with enriched Xe at Kamiokande clearly offers the highest sensitivity in terms of mass to background ratio.

Another possibility for experiments in the "low-mass" regime is the CTF detector. The CTF follows the basic design of Borexino but is dimensioned smaller, in a 11 m diam×11 m water tank [15]. The scintillator mass, contained in a vessel of $R=1$ m, is 4 ton. The water shield thickness is ~1 m smaller and the PM placed closer to the FV than in Borexino. Thus, in the CTF, the background is higher, and it is dominated by B(E) from fixed external sources. The energy resolution is $\Delta\sim 300$ keV. On the basis of simulations of background in the CTF [15] (with a high-purity buffer), $B(R)\sim B(E,R=0.8\text{ m})=270$/MeV yr or $B\Delta\sim 80$/yr (2 ton). The source size is limited to this level since larger masses entail a sharply rising $B(E,R)$.

The double-beta decay signals, $(0\nu 2\beta)$ as well as 2ν, are also simulated in Fig. 1 for Borexino with a 58 ton FV loaded with ~1.2 ton natural Xe or a 6 ton FV at K-II with 90% enriched Xe (and a 10× worse scintillator purity). The 2ν continuum assumes a $T_{1/2}=4\times 10^{21}$ yr [23] while the 0ν peak assumes a $\langle m_\nu\rangle=0.5$ eV. The large 2ν signal and its high signal/background are evident. Two-ν decay to the first excited state (1.31 MeV) of ^{136}Xe (not shown) will also contribute a much weaker continuum extending (by calorimetric summation of the γ ray) from 1.31 MeV to the common end point of 2.45 MeV. The 0ν peak is situated in the middle of the 2–3 MeV gap with residual background B(SN) from solar neutrinos and B(E) from external sources. Clearly, the peak due to a 0.5 eV neutrino can be statistically secured in this 1 yr spectrum.

With the above design options and the likely backgrounds, the attainable neutrino mass sensitivities are summarized in Table I. The design regime of low detector masses 4–14 ton using enriched Xe covers the same sensitivity region as the 100–300 ton detectors with natural Xe. The cost of Xe (last column) is approximately

TABLE I. Majorana neutrino mass sensitivities in Xe scintillator design options.

Option	R' Scint. vessel (m)	M' Scint. mass (ton)	M FV mass (ton)	$B\Delta$ (FV) (/yr)	$\langle m_\nu\rangle$ Limit (90% C.L./2 yr) (eV)	$\langle m_\nu\rangle$ Min. det. (3σ/1 yr) (eV)	Cost of Xe (10^6 \$)
Borexino	4.25	300	58	12	0.21	0.42	3.9 (Nat.)
Xe-K-II	3.5	180	100	74	0.25	0.45	2.3 (Nat.)
Hi-Mass	3	100	58	100	0.36	0.63	1.3 (Nat.)
					0.11	0.2	13 (Enr.)
Xe-K-II	1.5	14	9	24	0.2	0.38	4.5(Enr.)
(Lo-Mass)	1.33	10	6	18	0.23	0.45	3.2(Enr.)
	1	4	2	10	0.35	0.7	1.3(Enr.)
CTF	1	4	2	80	0.58	1.0	1.3(Enr.)

VOLUME 72, NUMBER 10 PHYSICAL REVIEW LETTERS 7 MARCH 1994

the same for the two regimes but the outlay and technical basis for the larger detectors are significantly larger. Thus, a first phase program with the 4–14 ton designs with enriched Xe at Kamiokande offers the fastest route to a breakthrough into the sub-eV Majorana mass regime, with the potential to set a best limit of $\langle m\nu \rangle < 0.2$ eV and establish an $\langle m_\nu \rangle < 0.4$ eV in 1 yr. A second phase in the 100 ton scale using major quantities of enriched Xe can extend the search to $\langle m_\nu \rangle < \sim 0.1$ eV. To our knowledge, this is the only directly implementable program yet conceived for the measurement of such tiny neutrino masses.

I wish to thank T. Kovacs for his help with the simulations, and F. Boehm, D. Caldwell, H. Klapdor-Kleingrothaus, J. Martoff, M. Moe, and A. Suzuki for helpful conversations.

[1] S. P. Rosen *Neutrino 88*, edited by J. Schneps *et al.* (World Scientific, Singapore, 1989), p. 78.

[2] H. Klapdor-Kleingrothaus (Heidelberg-Moscow Collaboration); the limiting mass sensitivity in this experiment is claimed to be 0.2 eV (private communication).

[3] This idea was briefly mentioned in R. S. Raghavan, *Proceedings of the 25th International Conference on High Energy Physics, Singapore*, edited by K. K. Phua and Y. Yamaguchi (World Scientific, Singapore, 1991).

[4] Borexino: Proposal for a Real-Time Detector for Low Energy Solar Neutrinos, edited by G. Bellini, M. Campanella, D. Giugni, and R. S. Raghavan [INFN report, 1991 (unpublished)].

[5] Suggestive $(2\sigma)(0\nu 2\beta)$ line features are seen in ongoing experiments on ^{76}Ge; A. Piepke *et al.*, in Proceedings of the European Physical Society Conference on High Energy Physics, Marseille, July 1993 (to be published).

[6] S. R. Elliott, A. A. Hahn, and M. Moe, Phys. Rev. Lett. **59**, 2020 (1987).

[7] The solubility of Xe in benzene is 2.2 wt.% at 16 °C [*Solubilities of Inorganic and Organic Compounds*, edited by H. Stephen and T. Stephen (Macmillan, New York, 1963), Vol. I, p. 586]. Thus a ∼2% solubility is expected in general for the aromatic liquids used commonly as scintillators.

[8] H. Wong *et al.*, Phys. Rev. Lett. **67**, 1218 (1991).

[9] E. Bellotti *et al.*, Phys. Lett. B **221**, 209 (1989); and in *New and Exotic Phenomena '90*, edited by O. Fackler and J. Tran Thanh Van (Editions Frontières, Dreux, France, 1990), p. 23.

[10] A. de Bari *et al.*, Radiopurity of the Borexino Scintillator, AT&T Bell Labs Technical Memo 11121-921015-37, 1992 (unpublished).

[11] R. Engel, P. Vogel, and V. Zirnbauer, Nucl. Phys. **A478**, 459c (1988).

[12] The price of natural Xe is ∼$600/kg and that of enriched Xe, ∼$15 000/kg. Mass-rejected samples of Xe are available in large quantities; H. Spicer (Isotech) (private communication).

[13] J. Martoff (private communication).

[14] K. Hirata *et al.*, Phys. Rev. D **44**, 2241 (1991).

[15] Borexino Progress Report, INFN report, 1993 (unpublished).

[16] Events from ^{212}Bi, with a decay energy of only 2.25 MeV, can occur in the 2–3 MeV interval since some daughter α's appearing at ∼0.9 MeV with a mean life of ∼440 ns can be summed within the detector time resolution. An estimated fraction (2.5%) untagged is included in the simulation of Fig. 1.

[17] Rejection factors > 500 are attained with a tag time of ∼1.6 ms (∼10 half-lives) and a tag radius of search $=6\times\Delta r(1\sigma)$; the few Hz trigger rate in the entire detector poses no limitations. Even with a base trigger level at 0.5 MeV, simulations show that 0.8 MeV α's lost below this level will be $< 2\times 10^{-3}$.

[18] The code is a local version of CRONOS, S. Bonnetti *et al.*, Nucl. Instrum. Methods Phys. Res., Sect. A **239**, 314 (1993).

[19] The spectrum above the gap is due to ^{208}Tl which appears only above 3.2 MeV because of the calorimetric summing of the Tl decay cascades (these events will also be largely removed by a different delayed coincidence tag; see Ref. [4]).

[20] M. Moe, Phys. Rev. C **44**, R993 (1991).

[21] This background is about 5× smaller than that estimated in Ref. [4] for a thick acrylic shell design (à la SNO) and shield water of 10^{-13} g/g purity.

[22] The 50× higher Th in the K-II PM [which determines $B(E,R)$ at 2–3 MeV], the 50% fewer PM, and the ∼1 m additional water shield are taken into account in this estimate.

[23] A. Stauat *et al.*, Europhys. Lett. **13**, 31 (1990).

超微弱素粒子反応計測施設

― 素粒子と宇宙の謎に迫る ―

１９９５年　１０月

東北大学理学部附属泡箱写真解析施設

Paper presented at the First Int. Symp. on Lepton and Baryon Number Violation, Trento, 1998 189

Solar Neutrinos, Atmospheric Neutrinos and Proton Decays in Super-Kamiokande, and KamLAND Project

Atsuto Suzuki[1]

Research Center for Neutrino Science,
Tohoku University,
Aoba, Sendai, 980-8578, JAPAN

Abstract. Results on solar neutrinos, atmospheric neutrinos and proton decays from about one year operation of Super-Kamiokande are summarized. The solar neutrino deficit and atmospheric neutrino anomaly are reconfirmed with high statistics data. In particular a strong suggestion of neutrino oscillations has been obtained in the zenith-angle distributions of atmospheric neutrinos. No candidate events of proton decays are observed in the present data sample.
Shown also here the project of a 1000 ton liquid scintillator experiment, KamLAND.

1. Introduction

Experimental verification of GUT's is one of the most revolutionary subjects in the present-day particle physics and cosmology. In Kamiokande prospective indication concerning the finite neutrino masses were obtained as solutions to the solar neutrino deficit and atmospheric neutrino anomaly. As for proton decays the simple SU(5) GUT was already rejected, and the stringent constrainrs was also imposed on the minimal SUSY SU(5) GUT.

Under such an experimental situation, 50,000 ton water Cherenkov experiment, Super-Kamiokande started on schedule in April 1, 1996. Much

[1] E-mail: suzukia@awa.tohoku.ac.jp

Spectroscopy of Double-Beta and Inverse-Beta Decays from ^{100}Mo for Neutrinos

H. Ejiri[1], J. Engel[2], R. Hazama[1], P. Krastev[3], N. Kudomi[4], and R.G.H. Robertson[1]
[1]*Nuclear Physics Laboratory and Dept. Physics, University of Washington, Seattle, WA 98195, USA*
[2]*Dept. Physics and Astronomy, University of North Carolina, NC 27599, USA*
[3]*Dept. Physics, University of Wisconsin, WI 53706, USA*
[4]*Research Center for Nuclear Physics, Osaka University, Ibaraki, Osaka 567-0047, Japan*

Spectroscopic studies of two β-rays from ^{100}Mo are shown to be of potential interest for investigating both the Majorana ν mass by neutrinoless double β-decay($0\nu\beta\beta$) and low energy solar ν's by inverse β-decay. With a multi-ton ^{100}Mo detector, coincidence studies of correlated $\beta\beta$ from $0\nu\beta\beta$, together with the large Q value($Q_{\beta\beta}$), permit identification of the ν-mass term with a sensitivity of $\sim$ 0.03 eV. Delayed coincidence studies of the inverse β and the successive β-decay of the short-lived ^{100}Tc, together with the low threshold energy and the large measured GT strengths for the inverse β-decay, make it possible to detect in realtime individual low energy solar ν in the same detector.

PACS number(s): 23.40.-s, 95.55.Vj, 14.60.Pq, 26.65.+t, 27.60.+j

Neutrino mass is a key issue of current neutrino(ν) physics. Recent results with atmospheric [1] [2], solar [1] [3], and accelerator [4] neutrinos strongly suggest ν oscillations due to non-zero ν-mass differences and flavour mixings. Neutrino oscillation measurements, however, do not give the ν masses themselves, essential parameters for both particle physics and astrophysics. The ν mass relevant to the accelerator-ν oscillation is in the eV range [4], while the mass associated with the atmospheric-ν effect is of the order of 0.05 eV [1]. Neutrino mass of astroparticle interest is in the range of 1$\sim$0.01 eV [5]. Thus it is of great interest to study directly ν mass with sensitivity down to $\sim$0.03 eV.

Double beta decay may be the only probe presently able to access such small ν masses. Actually, observation of neutrinoless double beta decay ($0\nu\beta\beta$) would identify a Majorana-type electron ν with a non-zero *effective* mass $< m_\nu >$ [6]- [9] . Calorimetric measurements of total $\beta\beta$-energy spectra have been made on ^{76}Ge , ^{130}Te and other isotopes [6] [10] [11] . They give upper limits on $< m_\nu >$ in the sub-eV to eV region. GENIUS is under consideration to search for ν mass in the 0.01 eV range [10]. The $0\nu\beta\beta$ process is, in fact, sensitive not only to the ν mass ($< m_\nu >$) but also to a right-handed weak current and other terms beyond the Standard Model(SM) [6] [7] [8]. Spectroscopic studies of the energy and angular correlations for two β-rays are useful to identify the terms responsible for $0\nu\beta\beta$. Spectroscopic measurements for two β-rays have been made on ^{82}Se, ^{100}Mo ,^{136}Xe and on others [6] [12]- [15]. They give upper limits of a few eV on $< m_\nu >$. NEMO III is now under construction to study $< m_\nu >$ in the sub-eV region [16].

Solar neutrinos have been studied for more than 30 years [17]. Low-energy solar-ν studies, so far, have been carried out with ^{71}Ga and ^{37}Cl detectors [3]. They are non-realtime and inclusive measurements that do not identify the ν sources in the sun. Realtime spectroscopic studies of low energy solar ν are important for studies of the electron-ν mass and the solar-ν problems [17]- [21].

The present Letter shows that it is possible with ^{100}Mo to carry out both spectroscopic studies of $0\nu\beta\beta$ with a sensitivity of the order of $< m_\nu > \sim 0.03$ eV, and realtime exclusive studies of low energy solar ν. The $\beta\beta$ detector can also be used for studying the solar ν by measuring the inverse β-decay and the following single β-decay. A large-scale ^{100}Mo $\beta\beta$ detector would provide an excellent opportunity also for realtime spectroscopic (exclusive) studies of solar ν.

The unique features of this approach are as follows:

1)The β_1 and β_2 with the large energy sum of $E_1 + E_2$ can be measured in coincidence for the $0\nu\beta\beta$ studies, while the inverse β-decay induced by the solar ν and the successive β-decay are measured in delayed coincidence for the low energy solar-ν studies. The ^{100}Mo is just the isotope that satisfies the conditions for the $\beta\beta - \nu$ and solar-ν studies, as shown in Fig. 1.

2)The large $Q_{\beta\beta}$ gives a large phase-space factor $G^{0\nu}$ to enhance the $0\nu\beta\beta$ rate and a large energy sum of $E_1+E_2 = Q_{\beta\beta}$ to place the $0\nu\beta\beta$ energy signal well above most backgrounds(BG) except ^{208}Tl and ^{214}Bi. The energy and angular correlations for the two β-rays can be used to identify the ν-mass term.

3)The low threshold energy of 0.168 MeV for the solar-ν absorption allows observation of low energy sources such as pp and ^{7}Be. The B(GT) value to the 1^+ ground state of ^{100}Tc is large [22] [23]. Thus ^{100}Mo has large charged-current (CC) capture rates even for low energy solar ν's, as shown in Table 1. The solar-ν sources are identified by measuring the inverse-β energies. Since only the ^{100}Tc ground state can absorb ^{7}Be ν and pp ν, the ratio of ^{7}Be to pp is independent of the ambiguity($\sim$ 15%) of the $B(GT)$ value.Thus ^{100}Mo is very effective for studying the solar-ν problems, as will be detailed in a future paper.

4)The measurement of two β-rays (charged particles) enables one to localize in space the decay-vertex positions for both the $0\nu\beta\beta$ and solar-ν studies. The tightly localized event in space, together with the appropriate time and energy window, are key points for selecting $0\nu\beta\beta$ and solar-ν signals and for reducing correlated and accidental BG.

The $0\nu\beta\beta$ transition rate $R_{0\nu}$ for $< m_\nu >$ is given by

$$R_{0\nu} = G^{0\nu}(M^{0\nu})^2 | < m_\nu > |^2, \qquad (1)$$

where $G^{0\nu}$ is the phase space factor and $M^{0\nu}$ is the matrix element [6] [7] [8], both relatively large for ^{100}Mo.

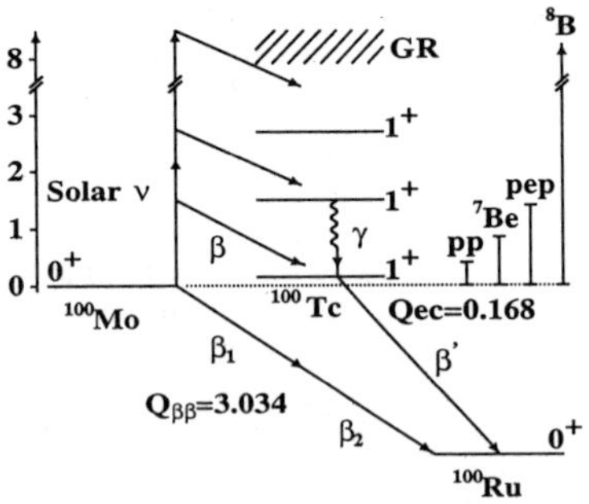

FIG. 1. Level and transition schemes of ^{100}Mo for double beta decays ($\beta_1\beta_2$) and two beta decays ($\beta\beta'$) induced by solar-ν absorption. GR is the Gamow-Teller giant resonance. $Q_{\beta\beta}$ and Q_{ec} are given in units of MeV.

TABLE I. Inverse β strengths of $B(GT)$ and $B(F)$ and solar-ν absorption rates R_ν for ^{100}Mo.

State	Spin parity	E_{ex}(MeV)	$B(GT)^a, B(F)^a$
G.S.	1^+	0	0.33 ± 0.04
1 st	1^+	1.4	0.13 ± 0.02
2 nd	1^+	2.6	0.23 ± 0.03
IAS	0^+	11.2	16.0
GT	1^+	13.3	26.0 ± 4.3
Source	$E_\nu^{(max)}$(MeV)	$E_\beta^{(max)}$(MeV)	R_ν/SNU^b
pp	0.42	0.25	639 ± 85
pep	1.44	1.27	13 ± 2
^{7}Be	0.86	0.69	206 ± 35
^{8}B	14.02	13.85	27(23)^c± 4
^{13}N	1.20	1.03	22± 3
^{15}O	1.74	1.57	32 ± 4

E_{ex}, $E_\nu^{(max)}$, $E_\beta^{(max)}$ are the excitation energy, maximum ν erergy, and maximum β-ray energy, respectively. GT is the GT giant resonance.
a) Normalized to the full Fermi strength for the isobaric analogue state (IAS). GT strength is $(g_A/g_V)^2 \times B(GT)$.
b) Standard-solar-model(SSM) capture rates based on BP98 [17] with errors from those of $B(GT)$. c) Rate for the states below the effective neutron threshold energy.

Simple shell model configurations are involved in the ^{100}Mo decays and accordingly the $2\nu\beta\beta$ rate is measured to be large [13] [14]. Then a large $0\nu\beta\beta$ transition rate of the order of $50 \cdot 10^{-36}$/sec may be expected for the possible ν-mass of $< m_\nu > = \sim 0.03$ eV [8] [9].

The $0\nu\beta\beta$ event is measured by setting the energy window at the $Q_{\beta\beta}$ value in the sum energy of $E_1 + E_2$ and the prompt time window for the $\beta\beta$ coincidence. The $0\nu\beta\beta$ event rate per year is

$$Y = 19 \cdot R \cdot N_0 \cdot A^{-1} \cdot \epsilon, \qquad (2)$$

where R=$R_{0\nu}$ is the $0\nu\beta\beta$ rate in units of 10^{-36}/sec, N_0 is the total ^{100}Mo in tons, A=100 is the mass number, and ϵ=$\epsilon_{0\nu}$ is the efficiency for detecting the $0\nu\beta\beta$ event. The efficiency depends on the energy and time windows and on the energy and angle cuts for two β-rays, which are strongly peaked at 180°. Using a detector with a 7% fullwidth at half maximum(FWHM) energy resolution and 2-ns time resolution, one may get $\epsilon_{0\nu} \sim 0.5$.

The background comes mainly from $2\nu\beta\beta$ events falling in the $0\nu\beta\beta$ energy window. That rate is $Y_{2\nu} = 19 \cdot R_{2\nu} \cdot N_0 \cdot A^{-1} \cdot \epsilon_{2\nu}$, where $R_{2\nu}$ is the $2\nu\beta\beta$ transition rate and $\epsilon_{2\nu}$ is the efficiency. The measured rate [13] is used for $R_{2\nu}$, and $\epsilon_{2\nu}$ is evaluated as $\epsilon_{2\nu} = 1.5 \cdot 10^{-6}$ by a Monte Carlo simulation. Most of BG β-rays due to radioactive impurities such as ^{238}U and ^{232}Th chain isotopes have energies lower than $Q_{\beta\beta}$ of ^{100}Mo. Thus, they do not fall into the energy window. The BG rate is $Y_B = N \cdot 3.15 \times 10^7 \cdot b \cdot \epsilon_B$, where N is the total Mo in tons, ϵ_B is the efficiency, and b is the radioactive impurity in units of Bq/ton. Using natural Mo with 9.6% abundance of ^{100}Mo, $N \sim 100$ tons of Mo is needed for 10 tons of ^{100}Mo. The efficiency is evaluated as $\epsilon_B \sim 2.7 \cdot 10^{-4}$ for the sum of contributions from ^{238}U and ^{232}Th to the $0\nu\beta\beta$ energy and time window. The accidental coincidence BG events can be neglected because of the short time window for the $\beta\beta$ coincidence.

The lower limit (sensitivity) on $< m_\nu >$ to be measured can be obtained by requiring that the number of $0\nu\beta\beta$ events has to exceed the statistical fluctuation of the BG events in y years, i.e. $Y_{0\nu} \cdot y > \sqrt{(Y_{2\nu} + Y_B) \cdot y}$. The obtained limits are shown in Fig. 2. Sensitivities of the order of $< m_\nu > \sim 0.03$ eV are achieved at realistic impurities of the order of $b \sim 10^{-2}$ (Bq/ton) for ^{238}U and ^{232}Th.

The solar-ν signal rate per year is given, as in the case of $0\nu\beta\beta$, by equ.(2), where R=R_ν is the solar-ν capture rate in unit of SNU and ϵ=ϵ_S is the detection efficiency. The efficiency depends on the energy and time windows, which are set for individual ν-sources. Since the inverse β-decay induced by the solar-ν absorption is followed by β-decay with a half life of 16 sec, the time window can be set from $t_1 = 1$ sec to $t_2 = 31$ sec. The starting time of 1 sec is long enough to reject most correlated BG such as the $2\nu\beta\beta$, β-rays followed by conversion electrons, scatterings of single β-rays, etc. The stopping time of 31 sec is short enough to reduce most of the accidental coincidence BG. One may assume $\epsilon_S \sim 0.5$ as a realistic value.

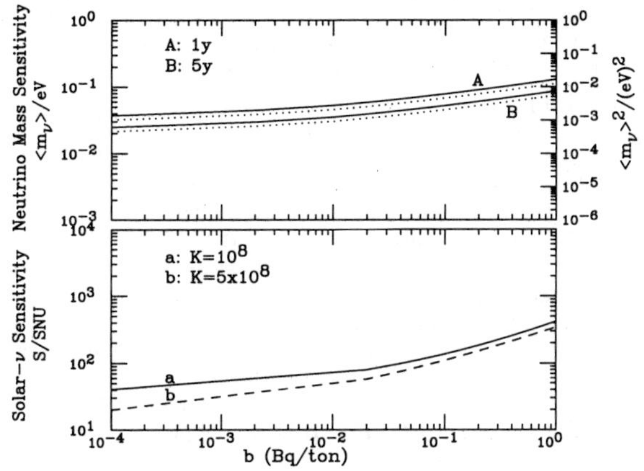

FIG. 2. Sensitivities to Majorana-ν mass and solar ν with 10 tons of ^{100}Mo and 100 tons of Mo as a function of the RI impurity b. Top: Solid and dotted lines for $\beta\beta$ with the nuclear matrix elements in Refs. [8] and [9], respectively, and A and B denote one and five year measurements, respectively. Bottom: Solid (a) and dashed (b) lines are for one year observations of solar ν with K=10^8 and K=$5 \cdot 10^8$, respectively.

For solar ν detection the major backgrounds come from $2\nu\beta\beta$ and radioactive impurities of ^{238}U and ^{232}Th. Here we evaluate BG's for the ^{7}Be-ν energy window. The correlated BG comes mainly from the successive β-decays of ^{214}Pb$\rightarrow$^{214}Bi $\rightarrow$^{214}Po through the short-lived ^{214}Bi with $t_{1/2}$ = 19.9 min. Most of the β-rays in these decays are accompanied by γ-rays. Thus they can be reduced by requiring anti-coincidence with the γ-rays. The correlated BG rate is $Y_{DC} \sim 8 \times 10^8 \cdot N \cdot b \cdot \Delta T$ with ΔT being the time window interval of 10^{-6} year.

The accidental coincidence BG due to ^{238}U and ^{232}Th impurities is $Y_{AC} \sim N^2 \cdot b^2 \cdot 10^{14} \cdot \Delta T/K$, where K is the number of unit cells to be localized in the detector. The position resolution, i.e. the localization, is given by K^{-1}. The two β-rays from $2\nu\beta\beta$ contribute also to the accidental coincidence BG. The rate is $Y_{2\nu} \sim N_0^2 \cdot 1.4 \times 10^{15} \cdot \Delta T/K$. The sensitivity to the solar-ν rate S to be measured for one year is obtained by setting $Y_S = \sqrt{Y_{DC} + Y_{AC} + Y_{2\nu}}$, as shown in Fig. 2. The sensitivity of the level of $S \sim 50$ SNU is obtained with realistic impurities of $b \sim 10^{-2}$ Bq/ton and $K = 10^8$ for 100 tons of natural Mo.

Schematic energy spectra expected for $0\nu\beta\beta$ and solar ν, together with the $2\nu\beta\beta$ BG ones, are shown in Fig. 3. Here scintilator modules with realistic energy resolutions are assumed as an example of the detector for measuring the β-ray energies.

The present spectroscopic method is complementary to the calorimetric methods with a Ge detector [10] and the liquid scintillation detectors under construction [19].

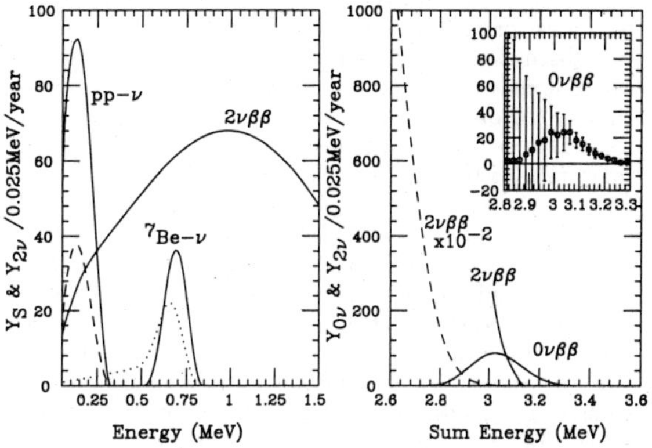

FIG. 3. Schematic energy spectra with 10 tons of ^{100}Mo for one year measurement. Energy resolutions of $\Delta E/E = \sqrt{15/E(keV)}$ are assumed. Right hand side: Sum energy spectra for $0\nu\beta\beta$ and $2\nu\beta\beta$ for $< m_\nu >= 0.1$ eV with the matrix element in Ref. [8]. The insert shows the $0\nu\beta\beta$ spectrum with $< m_\nu >$=0.05 eV after correction for contribution from $2\nu\beta\beta$ with statistical errors. Left hand side: Inverse β spectra for ^{7}Be ν and pp ν on the basis of SSM [17], and possible $2\nu\beta\beta$ background with K=5$\cdot$ 10^8. Dotted and dashed lines are those with 0.05 g/cm^2 Mo foils.

It is of interest to compare the present $\beta-\beta$ studies for low energy solar ν with other proposed studies [18], [20], [21]. The solar-ν capture rates for ^{100}Mo are based on measured values and are larger by more than a factor of 2$\sim$3 than others. The major BG for ^{100}Mo comes from the accidental coincidence $\beta\beta$ decay, while that of ^{115}In comes from that of the single β decay. The accidental rate is proportional to the square of the $\beta\beta$ or β-decay rate ($R_{\beta\beta}$, R_β), the time window ΔT, the number of the source nuclei in the unit cell (N_0/K). The square $(R_{\beta\beta})^2$ for the ^{100}Mo is much smaller than R_β^2 for ^{115}In, while the $\beta - \beta$ time window ΔT for ^{100}Mo is much longer than the $\beta - \gamma$ one for ^{115}In. The localization factor N_0/K for ^{100}Mo with two β-ray(charged particle) detection can be much smaller than that for ^{115}In with $\beta - \gamma$ detection. Then, ^{100}Mo can be better than ^{115}In by a couple of orders of magnitude in the signal to background ratio. The ^{176}Yb uses 50 ns delayed soft ($\sim$0.1 MeV) γ-rays for tagging [21], while the present ^{100}Mo uses 16 s delayed hard($\sim$1.5 MeV) β-rays. Thus they are complementary in technical points of BG reduction, but there is no appreciable $2\nu\beta\beta$ background in ^{176}Yb.

The cosmogenic isotopes to be considered are long lived ^{93}Mo, and short lived ^{99}Nb and ^{100}Nb. Although ^{93}Mo isotopes are not removed chemically, they decay by emitting very low energy X-rays and conversion electrons. Their energies can be lower than the detector threshold. ^{99}Nb and ^{100}Nb are produced by fast neutrons at underground laboratories and decay within tens of seconds by emitting β-rays. They are estimated to give negligible contributions to the present energy and time windows.

The main point of the present Letter is to describe a new idea for ν-mass and solar-ν studies, with quantitative discussions for the basic rates. In case of natural Mo, one may use around 100 tons of Mo with the impuritiy of the order of 10^{-2} Bq/ton for ^{238}U and ^{232}Th or less. This purity level has been achieved for Ni and other materials for the Sudbury Neutrino Observatory [24]. If one would use, as an example, an ensemble of segmented scintillator and Mo-foil modules, Mo foils with thickness of 0.05~0.1 g/cm^2 might be needed to get adequate β-ray yields and energies for low energy solar-ν(see Fig. 3). The foil thickness effect on the $0\nu\beta\beta$ is not serious since the two β rays are emitted with higher energies in opposite directions. The scintillator module would be made so as to get a position resolution of the order of 10 cm^2 on the foil. Since the energy resolution is a key element, one may instead consider bolometric high resolution detectors with appropriate segmentation(localization).

Not only is ^{100}Mo useful for both solar-ν spectroscopy and search for $0\nu\beta\beta$, it also has potential for studying supernova ν. Other promising isotopes include ^{150}Nd , ^{82}Se and ^{136}Xe, if suitable 1^+ states exist in ^{150}Pm, ^{82}Br and ^{136}Cs. The ^{136}Xe may be particularly interesting in view of existing Xe detector technology.

The authors thank Nuclear Physics Laboratory and Institute for Nuclear Theory, University of Washington for supports and discussions, and RH is supported by a Japanese Society for Science Promotion Fellowship.

[1] Super-Kamiokande Coll., Y. Fukuda, et al., Phys. Rev. Lett. **82** (1999) 1810, 2430, 2644. *ibid*, Phys. Rev. Lett. **81** (1998) 1562, **77** (1996) 1683.

[2] IMB Coll., R. Becker-Szendy et al, Nucl. Phys. Proc. Suppl. **38**B, (1995)331; Soudan-2 Coll., W. W. H. Allison et al., Phys. Lett. **B 391** (1997) 491; MACRO Coll., M. Ambrosio et al., Phys. Lett. **B434** (1998) 451.

[3] Gallex Coll., W. Hampel, et al., Phys. Lett. **B388** (1996) 384; SAGE Coll., J. N. Abdurashitov, et al., Phys. Rev. Lett. **77** (1996) 4708; B. T. Cleveland et al., Astrophysics J,**496** (1998) 505; R. Davis, Prog. Part. Nucl. Phys. **32** (1994) 13.

[4] C. Athanassopoulos et al., Phys. Rev. Lett. **75** (1995) 2650, **81** (1998) 1774.

[5] H. Minakata and O. Yasuda, Phys. Rev. **D56** (1997) 1692, V. Barger and K. Whisnant, Phys. Lett. **B456** (1999) 194, S. M. Bilenky et al., UWThPh-1999-41, W. Hu et al., Phys. Rev. Lett **80** (1998) 5255, E. Ma, Phys. Lett, **B456** (1999) 201.

[6] W. C. Haxton and G. J. Stephenson Jr, Prog. Part. Nucl. Phys. **12** (1984) 409, M. Doi et al., Prog. Theor. Phys. **83** (Suppl.)(1985) 1, M. Moe and P. Vogel, Ann. Review Nucl. Science **44** (1994) 247, and refs. therein.

[7] H. Ejiri, Int. J. Modern Phys. E, Vol.**6** . No 1 (1997) 1; H. Ejiri et al., J. Phys. Soc. Japan Lett. **65** (1996) 7.

[8] A. Faessler and F. Simcovic, J. Phys. **G 24** (1998) 2139.

[9] T. Tomoda, Rep. Prog. Phys. **54** (1991) 53.

[10] L. Baudis et al. Phys. Rev. Lett.**83** (1999) 41 ; H.V. Klapdor-Kleingrothaus et al., J.Phys.G **24** (1998) 483.

[11] A. Alessandrello et al., Phys. Lett. **B433** (1998) 156.

[12] S. R. Elliott et al., Phys. Rev. **C46** (1992) 1535.

[13] H. Ejiri et al., Phys. Lett. **B258** (1991) 17; H. Ejiri et al., Nucl. Phys. **A611** (1996) 85;

[14] D. Dassie et al., Phys. Rev. **D51** (1995) 2090; S. R. Elliott et al., J. Phys. G: Nucl. Part. Phys. **17** (1991) 5145; A. De Silva et al., Phys. Rev. **C56** (1997) 2451.

[15] R. Luescher, et al., Phys. Lett.B **434** (1998) 407.

[16] F. Piquemal, NEMO III collaboration, Nucl.Phys.**B** (Proc.Suppl.) **77** (1999) 352.

[17] J. N. Bahcall and M. Pinsonneault, Rev. Mod. Phys. **64** (1992) 885, and **67** (1995) 781; J. N. Bahcall et al., Phys. Lett. **B433** (1998) 1, and refs therein.

[18] R. S. Raghavan, Phys. Rev. Lett. **37** (1976) 259

[19] L. Oberauer,Nucl. Phys. **B** (Proc.Suppl.) **77** (1999) 48; A. Suzuki, *ibid* 171.

[20] W. C. Haxton, Phys. Rev. Lett. **60** (1988) 768; J. Engel, et al., Phys. Rev. Lett. **67** (1991) 426. K. Lande, et al., Nucl. Phys. **B** (Proc.Suppl.) **77** (1999) 13.

[21] R. S. Raghavan, Phys. Rev. Lett. **78** (1997) 3618.

[22] A. Akimune, et al., Phys. Lett. **B394** (1997) 23.

[23] A. Garcia et al., Phys. Rev. **C47** (1993) 2910.

[24] R.G.H.Robertson, Prog. Part. Nucl. Phys. **40** (1998) 113.

2.6. The Future of Double Beta Decay

2.6.1 GENIUS

Paper presented at Beyond the Desert 1997: Accelerator and Non-Accelerator Approaches 485

Double Beta Decay – Physics Beyond the Standard Model now, and in future (GENIUS)

H.V. Klapdor–Kleingrothaus

Max–Planck–Institut für Kernphysik
P.O.Box 10 39 80, D–69029 Heidelberg, Germany

Abstract. Nuclear double beta decay provides an extraordinarily broad potential to search for beyond Standard Model physics, probing already now the TeV scale, on which new physics should manifest itself. These possibilities are reviewed here.
First, the results of present generation experiments are presented. The most sensitive one of them – the Heidelberg–Moscow experiment in the Gran Sasso – probes the electron mass now in the sub eV region and will reach a limit of ~ 0.1 eV in a few years. Basing to a large extent on the theoretical work of the Heidelberg Double Beta Group in the last two years, results are obtained also for SUSY models (R–parity breaking, sneutrino mass), leptoquarks (leptoquark-Higgs coupling), compositeness, right–handed W boson mass and others. These results are comfortably competitive to corresponding results from high–energy accelerators like TEVATRON, HERA, etc.
Second, future perspectives of $\beta\beta$ research are discussed. A new Heidelberg experimental proposal (GENIUS) is presented which would allow to increase the sensitivity for Majorana neutrino masses from the present level of at best 0.1 eV down to 0.01 or even 0.001 eV. Its physical potential would be a breakthrough into the multi-TeV range for many beyond standard models. Its sensitivity for neutrino oscillation parameters would be larger than of all present terrestrial neutrino oscillation experiments and of those planned for the future. It could probe directly the atmospheric neutrino problem and even the large angle solution of the solar neutrino problem. It would further, already in a first

Z. Phys. A 359, 361–372 (1997)

ZEITSCHRIFT
FÜR PHYSIK A

A large scale double beta and dark matter experiment: On the physics potential of GENIUS

H.V. Klapdor-Kleingrothaus, M. Hirsch

Max-Planck-Institut für Kernphysik, P.O. 10 39 80, D-69029, Heidelberg, Germany (e-mail: klapdor@enull.mpi-hd.mpg.de, M.Hirsch@mpi-hd.mpg.de)

Received: 24 September 1997
Communicated by B. Povh

Abstract. The physics potential of GENIUS, a recently proposed double beta decay and dark matter experiment is discussed. The experiment will allow to probe neutrino masses down to $10^{-(2-3)}$ eV. GENIUS will test the structure of the neutrino mass matrix, and therefore implicitly neutrino oscillation parameters comparable or superior in sensitivity to the best proposed dedicated terrestrial neutrino oscillation experiments. If the 10^{-3} eV level is reached, GENIUS will even allow to test the large angle MSW solution of the solar neutrino problem. Even in its first stage GENIUS will confirm or rule out degenerate or inverted neutrino mass scenarios, which have been widely discussed in the literature as a possible solution to current hints on finite neutrino masses and also test the $\nu_e \leftrightarrow \nu_\mu$ hypothesis of the atmospheric neutrino problem.

GENIUS would contribute to the search for R-parity violating SUSY and right-handed W-bosons on a scale similar or superior to LHC. In addition, GENIUS would largely improve the current $0\nu\beta\beta$ decay searches for R-parity conserving SUSY and leptoquarks.

Concerning cold dark matter (CDM) search, the low background anticipated for GENIUS would, for the first time ever, allow to cover the complete MSSM neutralino parameter space, making GENIUS competitive to LHC in SUSY discovery. If GENIUS could find SUSY CDM as a by-product it would confirm that R-parity must be conserved exactly.

GENIUS will thus be a major tool for future non-accelerator particle physics.

PACS: 23.40; 14.60.Pq; 95.35.d

1 Introduction

The question whether neutrinos do have finite rest masses or not is still an open one. Despite more than forty years of active research knowledge on neutrino masses is rather scarce, only upper limits have been firmly established [1].

Neutrinoless double beta ($0\nu\beta\beta$) decay is a sensitive probe for physics beyond the standard model (for recent reviews see [2–4]). The currently most stringent limit on $0\nu\beta\beta$ decay, i.e. the one given by the Heidelberg-Moscow collaboration [3–5],

$$T_{1/2}^{0\nu\beta\beta} \geq 1.1 \times 10^{25}\ \text{y} \quad 90\%\text{C.L.} \tag{1}$$

can be translated into an upper bound on the effective Majorana neutrino mass of $\langle m_\nu \rangle \leq 0.5$ eV. This particular $0\nu\beta\beta$ decay experiment, based on the use of enriched ^{76}Ge, thus for the first time has pushed neutrino mass searches into the sub-eV range [5].[1]

The project GENIUS (GErmanium in liquid NItrogen Underground Setup), which has been proposed and first presented by [4] and for which a more detailed experimental description has been published recently [7], is designed to reach a sensitivity of the order of $T_{1/2} \approx 6 \times 10^{27}$ years for the neutrinoless double beta ($0\nu\beta\beta$) decay of ^{76}Ge in about one year of measurement. It is based on the idea to use about one ton of 'naked' enriched high-purity ^{76}Ge detectors in liquid nitrogen, where the nitrogen acts as shielding.

Do be definite, using the matrix elements of [6] one can estimate a theoretical half-life for $0\nu\beta\beta$ decay of ^{76}Ge in the usual mass mechanism of

$$T_{1/2} = 2.3 \times 10^{24} \left(\frac{\langle m_\nu \rangle}{1\text{eV}}\right)^2 \text{y}. \tag{2}$$

Thus, GENIUS should be able to probe $\langle m_\nu \rangle$ down to about 0.02 eV in about one year. The final sensitivity of GENIUS, after ten years of measurement, is estimated to be $\langle m_\nu \rangle \simeq$ 0.006 eV [7]. An upgraded version of GENIUS, using 10 t of enriched material, would even allow to test neutrino masses down to $\langle m_\nu \rangle \simeq (1-2) \times 10^{-3}$ eV [7]. It is the aim of this paper to study the potential of this proposed experiment for particle physics.

We will show that measuring neutrino masses as small as (few) $\times 10^{-3}$ eV puts double beta decay in a position to test various hypotheses about the neutrino mass spectrum which have been put forward to explain the puzzling existing data from neutrino oscillation experiments. For example, a negative search for $0\nu\beta\beta$ decay by GENIUS would rule out degenerate neutrino mass scenarios unless neutrinos are maximally mixed and have different relative signs in their CP-phases at the same

[1] While the half-life limit quoted above is free from theoretical assumptions, the limit on the neutrino mass is not. Numbers quoted here and in the following are derived using the nuclear matrix element calculation of [6]. The matrix elements have a typical uncertainty of about a factor of 2

time.[2] The impact of GENIUS on neutrino mass models is discussed in Sects. 2-4.

Besides light neutrinos, neutrinoless double beta decay can occur via several other mechanisms, such as for example heavy right-handed neutrino exchange in left-right symmetric models [8, 9], by the exchange of gluinos, sqaurks, etc. in R-parity violating [10, 11] and R-parity conserving supersymmetry [12], as well as via leptoquark exchange [13]. The impact of GENIUS on these models is also discussed and compared to other future experiments in Sect. 5.

Its ultra-low background allows GENIUS to act also as a very efficient dark matter detector [4, 47, 7]. It would allow, for the first time ever, to cover the complete MSSM neutralino parameter space. GENIUS thus should either confirm or rule out the present favorite particle physics candidate for CDM, the neutralino, and test the hypothesis of R-parity conservation to unanticipated levels, as discussed in Sect. 6.

This paper then closes with a short summary and outlook.

2 Neutrino masses and double beta decay

For unmixed neutrinos the limit obtainable by GENIUS would render the electron neutrino definitely uninteresting for cosmology, ruling out ν_e as a sensible hot dark matter candidate. However, $\langle m_\nu \rangle$ is in general not equal to the electron neutrino mass. Once neutrinos have finite masses it is natural to assume that mass eigenstates are no longer weak interaction eigenstates, and mixing among different neutrino generations occurs, analogously to the observed mixing in the quark sector. Allowing for finite mixing among different generations for $\langle m_\nu \rangle$ one has to define:

$$\langle m_\nu \rangle = \sum_j{}' U_{ej}^2 m_j, \tag{3}$$

where the prime indicates that the sum extends over light neutrino mass eigenstates ($m_j \leq 10$ MeV) only.

Taking into account finite mixing, although complicating the analysis considerably, leads to interesting information once the sub-eV range is explored by double beta decay experiments. The consequences of (3) for GENIUS are discussed in detail in the next Sects..

2.1 *General definitions for the neutrino mixing matrix*

One can define a general mixing matrix for neutrinos, which in the Dirac case can be written exactly in the form of the CKM matrix:[3]

[2] Since oscillation experiments measure only differences of squared masses the absolute mass scale can not be fixed by these experiments. GENIUS is the only proposed experiment which could probe *masses* in the meV range up to now. However, recall that $0\nu\beta\beta$ decay measures *Majorana* neutrino masses only

[3] This Sect. assumes three neutrino flavors. Including a fourth, sterile neutrino is straightforward

$$U = \begin{pmatrix} c_{12}c_{13} & s_{12}c_{13} & s_{13}e^{-i\delta} \\ -s_{12}c_{23} - c_{12}s_{23}s_{13}e^{i\delta} & c_{12}c_{23} - s_{12}s_{23}s_{13}e^{i\delta} & s_{23}c_{13} \\ s_{12}s_{23} - c_{12}c_{23}s_{13}e^{i\delta} & -c_{12}s_{23} - s_{12}c_{23}s_{13}e^{i\delta} & c_{23}c_{13} \end{pmatrix}, \tag{4}$$

where $s_{ij} = \sin\theta_{ij}, c_{ij} = \cos\theta_{ij}$ and δ is a CP-violating phase. This matrix can be decomposed into a product of three unitary matrices:

$$U = U_{23} \cdot U_{13} \cdot U_{12}, \tag{5}$$

where

$$U_{23} = \begin{pmatrix} 1 & 0 & 0 \\ 0 & c_{23} & s_{23} \\ 0 & -s_{23} & c_{23} \end{pmatrix}, \tag{6}$$

$$U_{13} = \begin{pmatrix} c_{13} & 0 & s_{13}e^{-i\delta} \\ 0 & 1 & 0 \\ -s_{13}e^{i\delta} & 0 & c_{13} \end{pmatrix}, \tag{7}$$

$$U_{12} = \begin{pmatrix} c_{12} & s_{12} & 0 \\ -s_{12} & c_{12} & 0 \\ 0 & 0 & 1 \end{pmatrix}. \tag{8}$$

In the Majorana case, however, the matrix (4) has to be generalized, since for Majorana neutrinos one can have $n = N(N-1)/2$ CP-violating phases [14]. For Majorana neutrinos, therefore U is given by:

$$\begin{pmatrix} c_{12}c_{13} & s_{12}c_{13}e^{-i\delta_{12}} & s_{13}e^{-i\delta_{13}} \\ -s_{12}c_{23}e^{i\delta_{12}} - c_{12}s_{23}s_{13}e^{i(\delta_{13}+\delta_{23})} & c_{12}c_{23} - s_{12}s_{23}s_{13}e^{i(\delta_{23}+\delta_{13}-\delta_{12})} & s_{23}c_{13}e^{i\delta_{23}} \\ s_{12}s_{23}e^{i(\delta_{23}+\delta_{13})} - c_{12}c_{23}s_{13}e^{i(\delta_{23}+\delta_{13})} & -c_{12}s_{23}e^{i\delta_{23}} - s_{12}c_{23}s_{13}e^{i(\delta_{13}-\delta_{12})} & c_{23}c_{13} \end{pmatrix} \tag{9}$$

Equation (9) is the most general Majorana neutrino mixing matrix for three generations.

Note, however, that the CP-violating phases can be defined only up to arbitrary factors of i, i.e. up to rotations by factors of $\pi/2$.[4] Because of this freedom, one can find different conventions for (9) in the literature. The conventions used in [15] correspond to the replacements $\delta_{12} \rightarrow -2\beta$ and $\delta_{13} \rightarrow -(\beta + 3\gamma - 2\delta)$ in (9). In the following the conventions of [15] are used. The effective neutrino mass, measured in $0\nu\beta\beta$ decay, is then given by,

$$\langle m_\nu \rangle = |c_{12}^2 c_{13}^2 m_1 + s_{12}^2 c_{13}^2 m_2 e^{2i\beta} + s_{13}^2 m_3 e^{i(\beta+3\gamma-2\delta)}|. \tag{10}$$

Thus, in general $\langle m_\nu \rangle$ is a function of a priori seven unknown quantities.

[4] Observables, such as $\langle m_\nu \rangle$ can depend on relative signs, but never on the absolute sign of the mixing matrix

2.2 *Considering two generations only*

It is instructive to analyze the simpler case of two generations first, before doing a full analysis of (10).[5] In the formalism given in the previous Sect. this simply corresponds to setting either s_{13} or s_{12} equal to zero. It should be noted, that both cases are logically equivalent, as far as our analysis of double beta decay is concerned. Only the case $s_{13} = 0$ is therefore discussed. The case $s_{12} = 0$ can be obtained from the formulae given in this Sect. by obvious replacements.

The effective neutrino mass in this scenario is then

$$\langle m_\nu \rangle = |c_{12}^2 m_1 + s_{12}^2 m_2 e^{2i\beta}|. \tag{11}$$

Let us first discuss the CP-conserving case. In this case $e^{2i\beta}$ reduces to $\eta = +1, -1$. Furthermore, for the masses one can discuss two extreme cases: A) $m_1 \ll m_2$ and B) $m_1 \approx m_2$. The case A) is usually denoted as "hierarchical" (a typical realization being for example the simplest version of the see-saw mechanism [16]), while B) is usually called "degenerate" scenario. (Degenerate scenarios have been widely discussed in the literature recently, see for example [17], or the review in [18].)[6]

The most simple case is found assuming $c_{12}^2 m_1 \ll \eta s_{12}^2 m_2$, since then the value of η is insignificant. $\langle m_\nu \rangle$ can be written as

$$\langle m_\nu \rangle = \frac{1}{2} m_2 \left(1 - \sqrt{1 - \sin^2 2\theta}\right), \tag{12}$$

which allows one to express the double beta decay observable in terms of the usual neutrino oscillation parameters,

$$\Delta m_{21}^2 \approx m_2^2 = \frac{4\langle m_\nu \rangle^2}{(1 - \sqrt{1 - \sin^2 2\theta})^2}. \tag{13}$$

Being a bit more general, as assumed in case A), one should keep the corrections of the order (m_1/m_2), such that

$$\langle m_\nu \rangle = m_2 |(\frac{m_1}{m_2}) + \frac{1}{2}\left(1 - \sqrt{1 - \sin^2 2\theta}\right)\left(\pm 1 - (\frac{m_1}{m_2})\right)|. \tag{14}$$

Also,

$$\Delta m_{21}^2 = m_2^2 - m_1^2 = m_2^2 \left(1 - (\frac{m_1}{m_2})^2\right). \tag{15}$$

Rewriting (14)

$$m_2 = \frac{\langle m_\nu \rangle}{|(\frac{m_1}{m_2}) + \frac{1}{2}\left(1 - \sqrt{1 - \sin^2 2\theta}\right)\left(\pm 1 - (\frac{m_1}{m_2})\right)|}, \tag{16}$$

one sees that one can express the $0\nu\beta\beta$ decay observable in terms of oscillation parameters, once some assumption about the ratio m_1/m_2 is made. Typical assumptions would be either the quadratic seesaw ($m_1 : m_2 = m_e^2 : m_\mu^2$) or the linear seesaw relations ($m_1 : m_2 = m_e : m_\mu$). Degenerate models, see below, correspond to $m_1/m_2 \approx 1$ (see footnote 7). Note,

[5] Considering only two generations is an assumption used in most studies of neutrino oscillations, although it is valid only for appearance experiments, but should not be used in disappearance experiments, as for example the current solar neutrino experiments

[6] Again, one could have a third case $m_2 \ll m_1$, which is logically equivalent

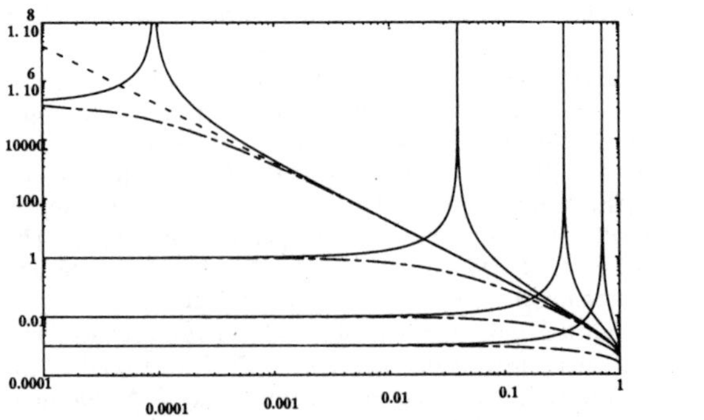

Fig. 1. Oscillation parameters and neutrinoless double beta decay in two generation scenarios, under the assumption of a limit on the effective neutrino mass of $\langle m_\nu \rangle \leq 0.01$ eV. The *dotted curve* corresponds to the "complete hierarchy" case, when only the heavier mass eigenstate contributes to $0\nu\beta\beta$ decay. *Dot-dashed* and *full curves* are for $\eta^{CP} = \pm 1$, respectively, for various assumed values of $R = (m_1/m_2)$. From top to bottom (left to right) for $\eta^{CP} = +1$ ($\eta^{CP} = -1$): $R = (m_e/m_\mu)^2$, 0.01, 0.1 and 0.3. The first case corresponds to the classical quadratic seesaw, large ratios, say $R \geq 0.3$, to the recently often discussed "degenerate" models. *Regions above the lines* would be excluded if $0\nu\beta\beta$ decay is not found

that in case of a positive value for η (16) gives always more stringent limits than using (13), while for negative η the results are very similar to those for positive η except for the small region in the parameter space, where

$$m_1 = m_2 \tan^2 \theta. \tag{17}$$

The sensitivity of $0\nu\beta\beta$ decay on neutrino oscillation parameters using (13) and (16) for various $R := (m_1/m_2)$ for an assumed limit on the effective mass of $\langle m_\nu \rangle \leq 0.01$ eV is shown in Fig. 1. The figure nicely illustrates that the limit where $R \to 0$, corresponding to the "complete hierarchy" models, is something of a worst-case scenario for $0\nu\beta\beta$ decay in terms of sensitivity on oscillation parameters, while for the degenerate neutrino mass scenarios, favored in many recent papers [17, 18], $0\nu\beta\beta$ decay is especially sensitive. Non-observation of $0\nu\beta\beta$ decay at the level of sensitivity of GENIUS with 1 ton would either require a negative CP-phase between the two neutrinos and maximal mixing at the same time or definitely rule out degenerate neutrino mass scenarios with neutrinos as hot dark matter candidates.

Let us now consider case B) $m_1 \approx m_2$. $\langle m_\nu \rangle$ in this case is simply:

$$\langle m_\nu \rangle = m|\cos^2\theta \pm \sin^2\theta|, \tag{18}$$

which depending on the CP-sign gives

$$\langle m_\nu \rangle = m \qquad if \quad \eta = +1, \tag{19}$$

$$\langle m_\nu \rangle = m|\sqrt{1 - \sin^2 2\theta}| \quad if \quad \eta = -1. \tag{20}$$

Thus, in the degenerate, two-generation scenario with positive CP-sign, double beta decay measures directly the average neutrino mass, while for a negative CP-sign we can express $\langle m_\nu \rangle$ again in terms of oscillation parameters.

Let us finally briefly discuss the case of an arbitrary CP-violating phase β. The consequences of an arbitrary β are most

364

Table 1. The 13 different logical possibilities for ratios of neutrino masses. Note, that all four variants have been discussed in the literature

Scenario	Comment	Relation
A)	Hierarchy	$m_i \ll m_j \ll m_k$
B)	Partial degeneracy	$m_i \ll m_j \approx m_k$
C)	Inverted partial degeneracy	$m_i \approx m_j \ll m_k$
D)	Complete degeneracy	$m_i \approx m_j \approx m_k$

easily seen for case B), the degenerate scenario. $\langle m_\nu \rangle$ is then given by

$$\langle m_\nu \rangle = m|\sqrt{1 - \sin^2 \beta \sin^2 2\theta}|. \quad (21)$$

Obviously the lesson to be learned from (21) is twofold. If $0\nu\beta\beta$ decay is not discovered one can always stick to the CP-conserving case, since arbitrary values of β lead always to limits which are more stringent than for the CP-conserving case of $\sin \beta = 1$. And, second, if effects of finite neutrino masses are discovered in both, neutrino oscillation experiments and double beta decay, the $0\nu\beta\beta$ decay data can give a measurement of CP-violation in the neutrino sector. One should keep in mind, however, that this discussion is based on the 2-generation scenario.

2.3 More complex: Three generations

The effective neutrino mass in full generality is given by (10). Again it is easier to consider the CP-conserving case, when the CP-phases can take only discrete values of $(\eta_1, \eta_2, \eta_3) = (+,+,+); (+,+,-); (+,-,+)$ and $(+,-,-)$. For an, albeit not complete analysis including CP-violation, see ref. [15].

Having three, a priori unknown and different masses contributing to $\langle m_\nu \rangle$ there are a number of different logical possibilities. They are summarized in table 1.

The first case of table 1 corresponds to the classical seesaw expectation, if $i = 1$, $j = 2$ and $k = 3$ [16]. However, recently also other cases have been discussed in great detail in the literature. Especially, the completely degenerate models have attracted some attention, because particle physics models for such a scenario can be realized relatively easily [17].[7]

Since practically all models of neutrino masses have been invented to explain currently known hints from experiments, most studies concentrated on the case where neutrino masses are ordered in the same way as those of the charged leptons, i.e. $m_1 < m_2 < m_3$. However, this need not be the case in general. For example, the electron neutrino could be heavier than the muon neutrino etc. Nevertheless this discussion will take the same attitude and only briefly comment on the other cases with inverted mass hierarchy near the end of this section.

Having specified all possibilities, it is seen that the complete degenerate scenario, scenario D), is relatively simple to analyze. For $\langle m_\nu \rangle$ one has

$$\langle m_\nu \rangle = m|c_{12}^2 c_{13}^2 \pm s_{12}^2 c_{13}^2 \pm s_{13}^2|. \quad (22)$$

[7] One could call mass models where mass ratios are not exactly, but approximately equal to one "quasi-degenerate". We will use this terminology in the following, whenever ratios of masses are in the range of $0.01 \leq R_{ij} \leq 100$

Depending on the relative CP-signs double beta decay is sensitive to

$$\langle m_\nu \rangle = m \text{ for } (+,+,+) \quad (23)$$

$$\langle m_\nu \rangle = m|1 - 2s_{13}^2| \text{ for } (+,+,-) \quad (24)$$

$$\langle m_\nu \rangle = m|1 - 2s_{12}^2 c_{13}^2| \text{ for } (+,-,+) \quad (25)$$

$$\langle m_\nu \rangle = m|1 - 2c_{12}^2 c_{13}^2| \text{ for } (+,-,-) \quad (26)$$

Non-observation of $0\nu\beta\beta$ decay defines allowed bands of mixing angle combinations in the plane (s_{12}, s_{13}). In the case where all CP-signs are positive, $\langle m_\nu \rangle$ coincides with m[8].

Figure 2 visualizes current and future constraints on completely degenerate 3-generation scenarios for assumed values of the average neutrino mass m for different choices of the relative CP phases. The horizontal bands correspond to the case of $(+,+,-)$, the bands extending to $\sin^2 \theta_{13} = 0$ to $(+,-,+)$ and the bands extending to $\sin^2 \theta_{13} = 1$ to $(+,-,-)$. Note that an analysis similar to the one in Fig. 2. has been done recently in [15].

However, for a more complete analysis, one should not fix the average neutrino mass at some preferred value. Instead, constraints on neutrino mixing parameters should be calculated as a function of the average neutrino mass. This is shown in Fig. 3., for an assumed limit on $\langle m_\nu \rangle$ of $\langle m_\nu \rangle \leq 0.01$ eV. The figure clearly illustrates that only if large cancellations between the different contributions from the different mass eigenstate occur, $0\nu\beta\beta$ decay will not occur for neutrinos in the interesting mass range of hot dark matter. On the other hand, if all CP-phases would be positive, $0\nu\beta\beta$ decay will either be observed or no neutrino mass can be larger than 0.01 eV.

Next, it is easy to realize that case A) can be also very simple: If the hierarchy is very strong, say as in the quadratic seesaw models, it is sufficient to go back to (13).

In case the degeneracy is not complete or if there is only partial degeneracy, one should keep corrections, scaling out the presumably largest mass,

$$\langle m_\nu \rangle = m_3|c_{12}^2 c_{13}^2 \left(\frac{m_1}{m_3}\right) \pm s_{12}^2 c_{13}^2 \left(\frac{m_2}{m_3}\right) \pm s_{13}^2|$$
$$= m_3|c_{12}^2 c_{13}^2 R_{13} \pm s_{12}^2 c_{13}^2 R_{23} \pm s_{13}^2|. \quad (27)$$

For the various partially degenerate models $\langle m_\nu \rangle$ can then be easily obtained by setting the corresponding R_{ij}'s equal to one and zero. For example, if m_1 is the mass eigenstate which is dominantly an electron neutrino in flavor space and it is assumed that $m_1 \ll m_2 \simeq m_3 \equiv m$, one finds,

$$\langle m_\nu \rangle = m|s_{12}^2 c_{13}^2 \pm s_{13}^2|.$$

All other logically possible cases can be derived analogously.

Let us finally briefly comment on inverted hierarchy models. It is clear that if neutrino masses would not follow the same pattern as the masses of the charged leptons, i.e. the electron neutrino being heavier than the muon and tau neutrinos, $0\nu\beta\beta$ decay would be especially well suited to search for

[8] This case is already on the edge of being excluded, as far as neutrinos as hot dark matter candidates are concerned

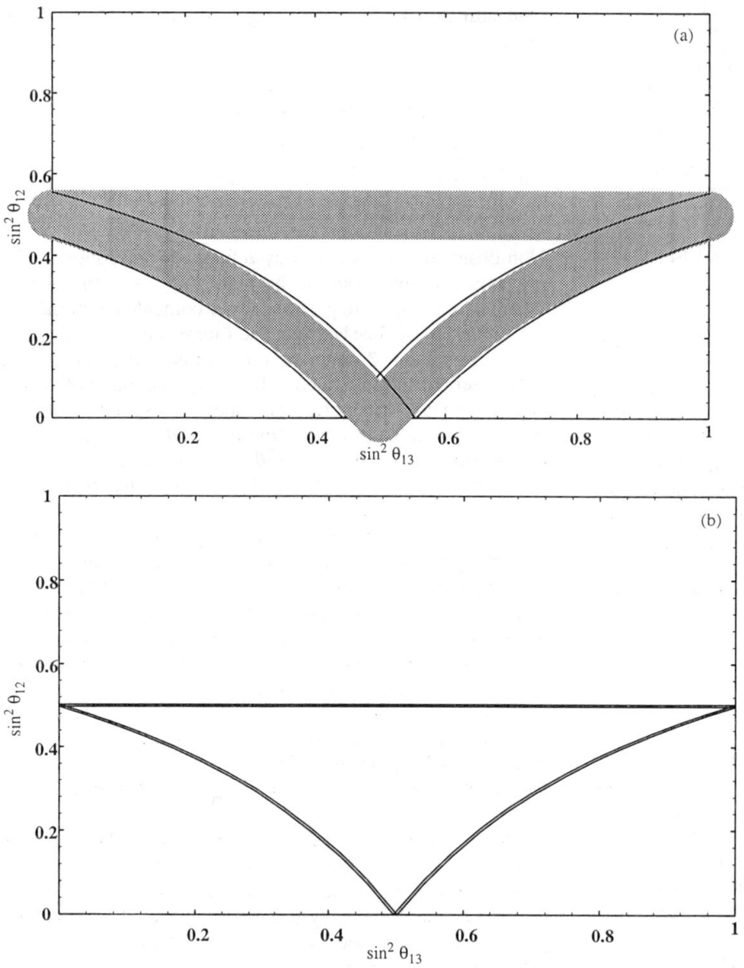

Fig. 2 Allowed bands in the plane ($\sin^2\theta_{12}-\sin^2\theta_{13}$) for a completely degenerate 3-generation scenario for different combinations of the CP-eigenvalues, see text. **a** current limit $\langle m_\nu \rangle \leq 0.5$ eV and assumed average mass of m = 4.5 eV. **b** GENIUS sensitivity of $\langle m_\nu \rangle = 0.01$ eV and average mass of m = 2 eV. Note, that in the case of GENIUS the allowed ranges are even smaller than the thickness of the lines shown

neutrino masses. Under such a - rather strange - assumption non-observation of $0\nu\beta\beta$ in GENIUS simply would rule out any neutrino masses above $\langle m_\nu \rangle$, i.e. inverted neutrino mass models could not provide dark matter candidates.

To summarize the discussion on the three-generation scenarios, it can be stated that only a limited number of possible cases for neutrino mass ratios can exist. Limits on all cases can be derived by an appropriate reformulation of the definition of $\langle m_\nu \rangle$. $0\nu\beta\beta$ decay is especially sensitive to degenerate or quasi-degenerate models. For strongly hierarchical neutrino mass models it is possible to do the analysis by simply going back to the simpler two generation scenarios.

3 Currently existing experimental data on neutrino masses

3.1 Upper limits on neutrino masses

The most model-independent measurements of neutrino masses come from kinematic searches. The particle data group [1] currently quotes the following limits:

$$m_{\nu_e} \leq 15 \text{ eV}, \tag{28}$$

$$m_{\nu_\mu} \leq 170 \text{ keV}, \tag{29}$$

$$m_{\nu_\tau} \leq 23 \text{ MeV}. \tag{30}$$

In addition, there is the constraint on the effective Majorana neutrino mass from the Heidelberg-Moscow double beta decay experiment, which yields [3–5]:

$$\langle m_\nu \rangle \leq 0.5 \text{ eV} \tag{31}$$

3.2 Hints on finite neutrino masses

More interesting than limits may be the existing hints on finite neutrino masses. One should keep in mind, however, that these are only hints: Not a single positive measurement of neutrino masses is currently firmly accepted.

At present there are four independent indications of finite neutrino masses. These are *i*) the solar neutrino problem [19], *ii*) the atmospheric neutrino problem [20], *iii*) the LSND measurement [21, 22] and *iv*) neutrinos as candidates for hot dark matter [23, 24].

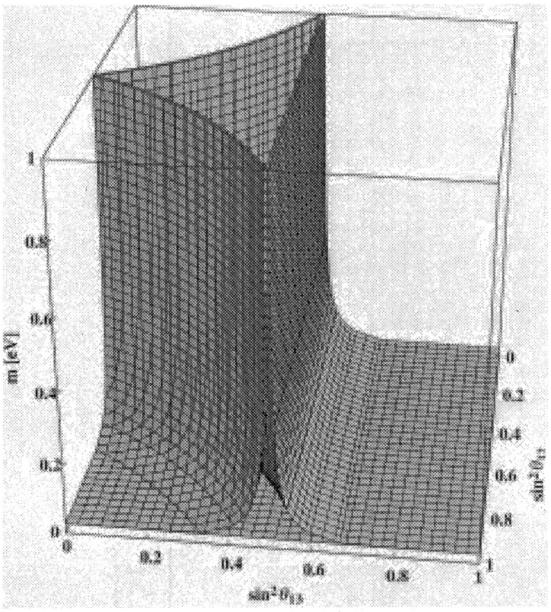

Fig. 3. 3-dimensional exclusion plot for the 3-generation degenerate neutrino scenario. Note that the range of m is from 0 to 1 eV. Excluded are all combinations of neutrino parameters which do not lie in between the three small areas enclosed by two *shaded planes* each. The three ($\times$ 2) different planes correspond to the three cases for different combinations of CP-phases, compare to Fig. 2

The first three are oscillation experiments and, therefore, by themselves give *no information about absolute values of neutrino masses*. It is important to note that only the assumption of neutrinos being the hot dark matter fixes the overall scale of neutrino masses. Recall that for neutrinos being interesting candidates for the hot dark matter, at least one neutrino should have a mass in the (few) eV range.

The solar neutrino problem can be explained by the disappearance of electron neutrinos. Since no appearance experiment exists for solar neutrinos the electron neutrinos can oscillate into any flavor: μ or τ neutrinos, or even new, sterile neutrinos. As is well-known, there are three solutions to the solar neutrino problem in terms of neutrino parameters. There are two MSW solutions;[9] the small-angle solution with $\Delta m^2 \approx (5\times10^{-6}-1\times10^{-5})\,eV^2$ and $\sin^2 2\theta \approx$ (few) 10^{-3}, and the large-angle solution with $\Delta m^2 \approx 10^{-(4-5)}\,eV^2$ and $\sin^2 2\theta \approx$ (few) 10^{-1}. In addition there is the so-called "just-so" solution of vacuum oscillations with parameters $\Delta m^2 \approx 10^{-(10-11)}$ and $\sin^2 2\theta \geq 0.5$. These values are valid assuming a two-flavor oscillation scheme only. In reality the situation is more complex. For a quasi three-generation analysis see, for example, the work of Fogli et al. [25]. Note, however, that also [25] does not deal with a *complete* three-generation analysis. Instead, [25] assumes arbitrarily that the neutrino masses follow either a partially degenerate pattern or are strongly hierarchical.

The atmospheric neutrino problem, on the other hand, could be explained either by the appearance of electron neutrinos or the disappearance of muon neutrinos, i.e. one can either have $\nu_\mu \rightarrow \nu_e$ or $\nu_\mu \rightarrow \nu_{\tau,s}$. In terms of Δm^2 and $\sin^2 2\theta$ both hypotheses require similar values: $\Delta m^2 \approx 0.1$ eV and $\sin^2 2\theta \geq 0.3$.

The LSND [21] measurement directly searches for $\bar{\nu}_\mu \rightarrow \bar{\nu}_e$ oscillations. An observed excess of $\bar{\nu}_e$ events has been interpreted by the LSND collaboration as evidence for oscillations with an oscillation probability of approximately $P_{\bar{\nu}_\mu\rightarrow\bar{\nu}_e} \simeq (0.31 \pm 0.12 \pm 0.05)$ %. The LSND collaboration has also searched for $\nu_\mu \rightarrow \nu_e$ oscillations [22], which gave a value for the oscillation probability consistent with the number given above, although with larger errors, due to the smaller statistics of the sample.

If neutrinos constitute the hot dark matter in the universe, at least one flavor should have a mass of the order of $m_\nu \approx$ (few) eV [23, 24]. Masses below the eV scale would render neutrinos cosmologically uninteresting.

Mainly due to these four indications in the past few years a huge number of neutrinos mass models have been constructed. A recent review can be found in [18].

4 GENIUS sensitivity for neutrino mass and oscillations compared to other experiments

As should be clear from the discussion given above, a comparison between double beta decay and neutrino oscillation experiments always requires that some assumptions about neutrino mass ratios are made. In addition, as has also been mentioned above, in general one should analyze the complete three generation problem in order to get reliable results. However, most oscillation experiments have been analyzed for two generations only, although such an approach is only appropriate for appearance experiments. Nevertheless, for the ease of comparison most of the discussion will stick to the two generation case.

Let's first discuss the case $\sin^2\theta_{12} = 0$. In this case double beta decay can be compared in sensitivity to experiments searching for $\nu_e - \nu_\tau$ oscillations. Figure 4 gives a summary of currently existing limits and future experimental sensitivity in the $\Delta m^2_{13} - \sin^2(2\theta_{e\tau})$ plane. The (background) figure is taken from [26].

The shaded area in Fig. 4 is the currently excluded region by the BUGEY reactor experiment [27]. The thin line is the sensitivity of the currently running CHORUS/NOMAD experiments at CERN. Dotted and dash-dotted lines are for future accelerator experiments, for details see [26]. In addition, the region where neutrinos are relevant for the dark matter is indicated. One can see that information on tau neutrinos at present is rather scarce, mainly limited to large mixing angles.

The two thick (solid and dashed) lines are sensitivities for future double beta decay experiments if one assumes a strong hierarchy between m_{ν_e} and m_{ν_τ}. The dashed line is for GENIUS with one ton while the full line assumes a limit on $\langle m_\nu \rangle$ of $\langle m_\nu \rangle = 0.001$ eV (GENIUS with 10 tons). While already GENIUS 1 ton is competitive to NAUSICAA-CERN it would be necessary to go down to 10^{-3} eV, i.e. GENIUS with

[9] It is very interesting to note, that the mass ranges of the solar neutrino solutions are fixed by the chlorine data alone. Taking only the Kamiokande and gallium data one can explain the solar neutrino problem for practically *any* Δm^2

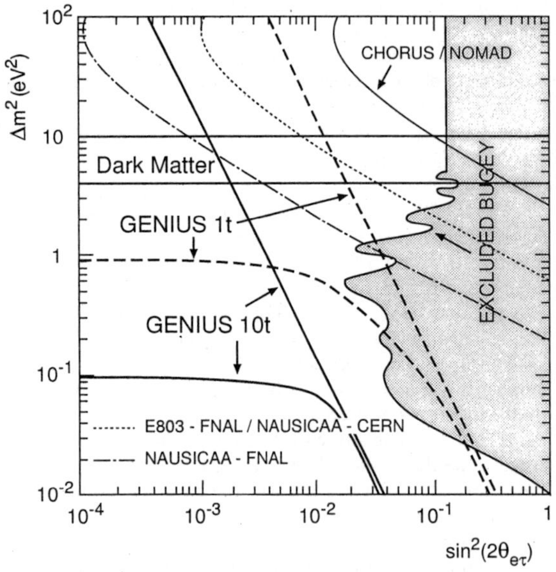

Fig. 4. Current limits and future experimental sensitivity on $\nu_e - \nu_\tau$ oscillations. The *shaded area* is currently excluded from reactor experiments. The *thin line* is the estimated sensitivity of the CHORUS/NOMAD experiments. The *dotted* and *dash-dotted thin lines* are sensitivity limits of proposed accelerator experiments, NAUSICAA and E803-FNAL [26]. The *thick lines* show the sensitivity of GENIUS (*broken line*: 1 t, *full line*: 10 t), for two examples of mass ratios. The *straight lines* assume that the lighter mass eigenstate has zero mass (R = 0), while the *lines* leading to the left assume R = 0.01.

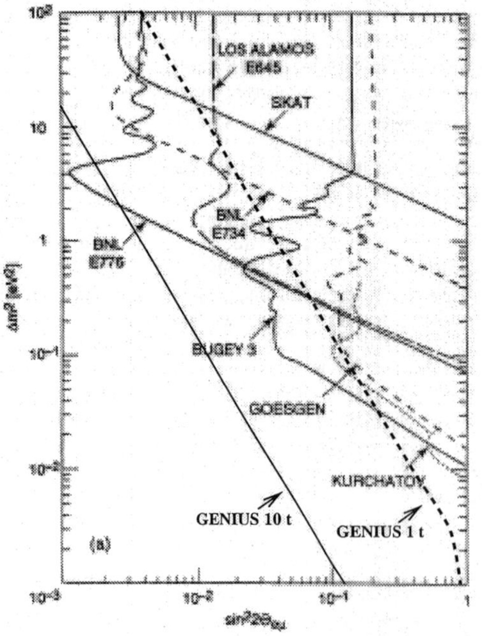

Fig. 5. Current limits on $\nu_e - \nu_\mu$ oscillations. Various existing experimental limits from reactor and accelerator experiments are indicated, as summarized in [28]. In addition, the figure shows the expected sensitivities for GENIUS with 1 ton (*thick broken line*) and GENIUS with 10 tons (*thick, full line*) assuming R = 0 (worst case)

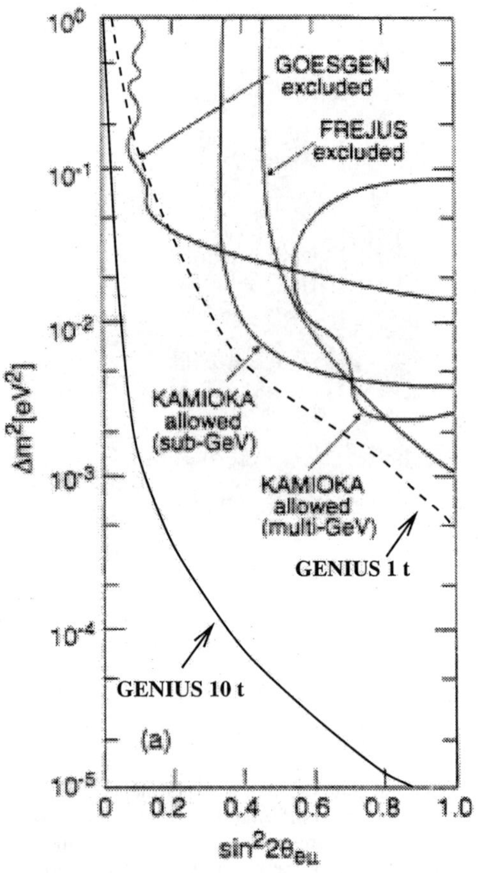

Fig. 6. Oscillation parameters which solve the atmospheric neutrino problem for $\nu_e \leftrightarrow \nu_\mu$ oscillations. In addition the best currently existing reactor constraints are shown. GENIUS would be able to test the atmospheric neutrino problem already with 1 ton

10 tons, to really cover a significant new part of the parameter space - if neutrino masses are strongly hierarchical. For quasi-degenerated models, for example R = 0.01 already, GENIUS with 1 ton would be more sensitive than all currently, planned future accelerator neutrino oscillations experiments.

One can draw similar curves for $0\nu\beta\beta$ decay experiments assuming $\sin^2\theta_{13} = 0$, for $\nu_e - \nu_\mu$ oscillations, see Fig. 5. The (background) figure is taken from [28]. It shows a number of reactor and accelerator data, together with the GENIUS $0\nu\beta\beta$ decay experimental sensitivities for strong $m_{\nu_e} - m_{\nu_\mu}$ hierarchy for $\langle m_\nu \rangle$ = 0.01 (dashed) and $\langle m_\nu \rangle$ = 0.001 (full) lines. While the GENIUS 1 ton sensitivity is sufficient (even for this worst case $m_{\nu_e} \ll m_{\nu_\mu}$) to extend the reach on Δm^2 down to smaller values at large mixing angles, for a large part of the parameter space there exists already a number of constraints. In addition, upcoming oscillation experiments will also be sensitive to smaller Δm^2. A $0\nu\beta\beta$ decay experiment with $\langle m_\nu \rangle \leq 0.001$ (GENIUS, 10 tons), on the other hand, would have a sensitivity better than all existing or planned

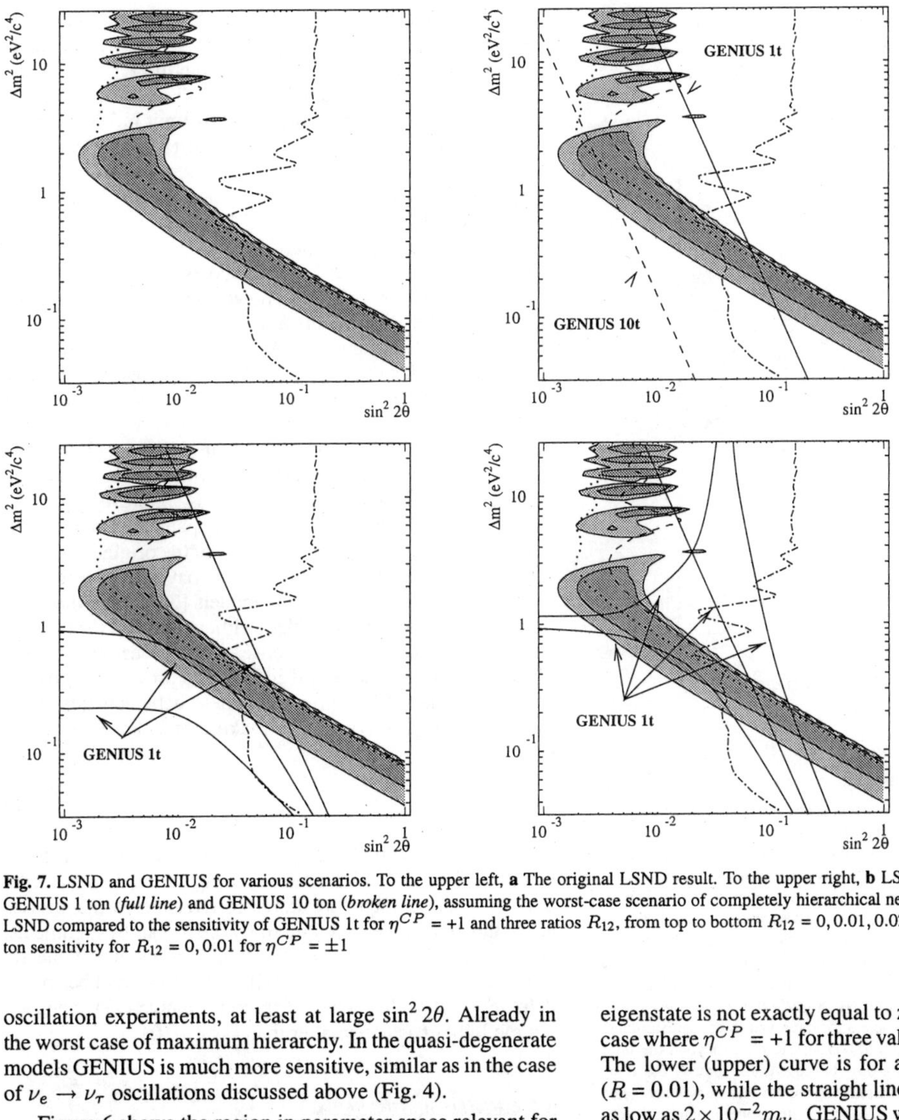

Fig. 7. LSND and GENIUS for various scenarios. To the upper left, **a** The original LSND result. To the upper right, **b** LSND compared to the sensitivity of GENIUS 1 ton (*full line*) and GENIUS 10 ton (*broken line*), assuming the worst-case scenario of completely hierarchical neutrino masses. To the lower left, **c**: LSND compared to the sensitivity of GENIUS 1t for η^{CP} = +1 and three ratios R_{12}, from top to bottom $R_{12} = 0, 0.01, 0.02$. To the lower right, **d**: GENIUS 1 ton sensitivity for $R_{12} = 0, 0.01$ for $\eta^{CP} = \pm 1$

oscillation experiments, at least at large $\sin^2 2\theta$. Already in the worst case of maximum hierarchy. In the quasi-degenerate models GENIUS is much more sensitive, similar as in the case of $\nu_e \to \nu_\tau$ oscillations discussed above (Fig. 4).

Figure 6 shows the region in parameter space relevant for the atmospheric neutrino problem for $\nu_e \leftrightarrow \nu_\mu$ oscillations. The original figure is again taken from [28]. The dashed line is for $\langle m_\nu \rangle \leq 0.01$ (GENIUS 1t) the full line for $\langle m_\nu \rangle \leq 0.001$ eV (GENIUS 10t), again for the worst strong hierarchical neutrino mass scenario. Even in this case GENIUS with 1 ton would already be able to test the $\nu_e \leftrightarrow \nu_\mu$ hypothesis.

Figure 7 compares GENIUS double beta decay to the LSND results. The original figure (upper left) is taken from [21]. The case where neutrino masses are strongly hierarchical, including $0\nu\beta\beta$ decay constraints for $\langle m_\nu \rangle = 0.01$ (full) and $\langle m_\nu \rangle \leq 0.001$ (dashed) line, is shown to the upper right. It would need an experiment sensitive to $\langle m_\nu \rangle = 10^{-3}$ eV (GENIUS 10 tons) in this scenario to probe the most parts of the relevant parameter space.

The lower two figures show the LSND region plus GENIUS, but now also for cases where the lower neutrino mass eigenstate is not exactly equal to zero. To the left is shown the case where $\eta^{CP} = +1$ for three values of the ratio $R = m_1/m_2$. The lower (upper) curve is for an assumed R of $R = 0.02$ ($R = 0.01$), while the straight line is for $R = 0$. Even for m_{ν_e} as low as $2 \times 10^{-2} m_{\nu_\mu}$ GENIUS with 1 ton would be sufficient to find $0\nu\beta\beta$ decay, if the LSND result is to be explained in terms of neutrino oscillations. The figure to the right shows the influence of η^{CP}. The three curves for GENIUS are for $R = 0$ (middle curve), the other two lines are for $R = 0.01$ and $\eta^{CP} = \pm 1$, respectively. Negative η^{CP} allows certain ranges of oscillation parameters to be consistent with non-observation of $0\nu\beta\beta$ decay.

Finally, Fig. 8 shows a summary of currently available information on neutrino oscillation parameters, including possible $0\nu\beta\beta$ decay sensitivities for $\eta^{CP} = +1$: The full lines are for GENIUS 1 ton, while the dashed lines assume $\langle m_\nu \rangle \leq 10^{-3}$ eV (GENIUS 10 tons). The lines are for R = 0, 0.01, 0.1, respectively. According to this result already GENIUS 1 ton tests all degenerate or quasi-degenerate neutrino mass models in any range where neutrinos are interesting for cosmology and also would test the atmospheric neutrino problem

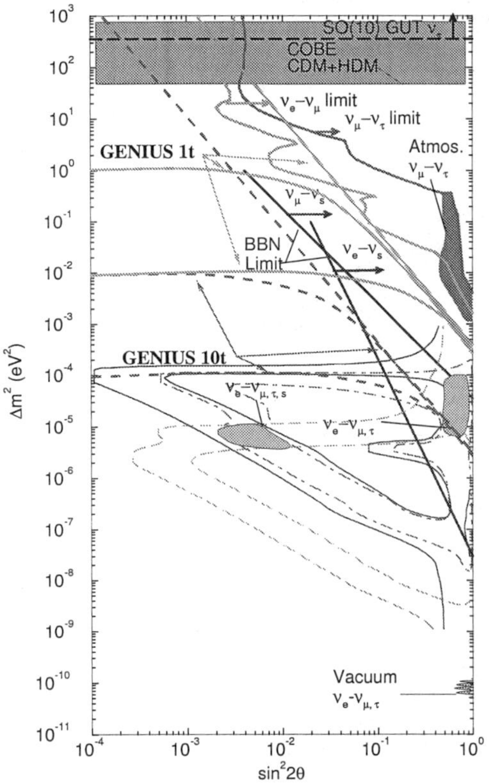

Fig. 8. Summary of currently known constraints on neutrino oscillation parameters. The (background) figure without the $0\nu\beta\beta$ decay constraints can be obtained from http://dept.physics.upenn.edu/ www/neutrino/solar.html. Shown are the vacuum and MSW solutions (for two generations of neutrinos) for the solar neutrino problem, the parameter range which would solve the atmospheric neutrino problem and various reactor and accelerator limits on neutrino oscillations. In addition, the mass range in which neutrinos are good hot dark matter candidates is indicated, as well as limits on neutrino oscillations into sterile states from considerations of big bang nucleosynthesis. Finally the thick lines indicate the sensitivity of GENIUS (*full lines 1 ton, broken lines 10 ton*) to neutrino oscillation parameters for three values of neutrino mass ratios $R = 0, 0.01$ and 0.1 (from top to bottom). Regions beyond the *lines* would be excluded by not observing $0\nu\beta\beta$ decay. While already the 1 ton GENIUS would be sufficient to constrain degenerate and quasi-degenerate neutrino mass models, and also would solve the atmospheric neutrino problem if it is due to $\nu_e \leftrightarrow \nu_\mu$ oscillations, the 10 ton version of GENIUS could cover a significant new part of the parameter space, including the large angle MSW solution to the solar neutrino problem, even in the worst case of $R = 0$

if it is due to $\nu_e \leftrightarrow \nu_\mu$ oscillations. GENIUS in its 10 ton version would directly test the large angle solution of the solar neutrino problem.

5 GENIUS and other physics beyond the SM

5.1 *GENIUS and left-right symmetry*

$0\nu\beta\beta$ decay can be sensitive to the possible existence of right-handed W-bosons if heavy right-handed neutrinos exist [8, 9]. The current limit on m_{W_R} from absence of $0\nu\beta\beta$ decay is about 1.1 TeV (for a heavy right-handed neutrino mass of 1 TeV) [9], better than any existing direct constraint and comparable to the theoretical limit derived from the $K^0 - \bar{K}^0$ mass difference. Note that the latter has large theoretical uncertainties, which makes improvements of this limit rather difficult.

If GENIUS is able to reach down to $\langle m_\nu \rangle \leq 0.01$ eV, it would at the same time be sensitive to right-handed W-boson masses up to $m_{W_R} \geq 8$ TeV (for a heavy right-handed neutrino mass of 1 TeV) or $m_{W_R} \geq 5.3$ TeV (at $\langle m_N \rangle = m_{W_R}$). Such a limit would be comparable to the one expected for the LHC, see for example [29], which quotes a final sensitivity of something like $5-6$ TeV. Note, however that in order to obtain such a limit the experiments at the LHC need to accumulate about $100 fb^{-1}$ of statistics. A 10 ton version of GENIUS could even reach a sensitivity of $m_{W_R} \geq 18$ TeV (for a heavy right-handed neutrino mass of 1 TeV) or $m_{W_R} \geq 10.1$ TeV (at $\langle m_N \rangle = m_{W_R}$).

As is well-known, left-right symmetric models in general do not fix the scale of left-right symmetry breaking, i.e. there is no upper limit to the W_R mass. Interestingly, however, recently Kuchimanchi and Mohapatra and others [30] have studied supersymmetric versions of the original left-right symmetric model and found that these have some very interesting features. Such SUSY-LR models [30] can at the same time solve the strong CP-problem without the need for an axion and have automatic R-parity conservation, a desirable feature of supersymmetric models, if SUSY is to provide a cold dark matter candidate. The "price" for these achievements, however, is that the LR symmetry breaking scale has to be rather low,

$$m_{W_R} \leq \frac{g M_{SUSY}}{f}, \tag{32}$$

where g is the usual $SU(2)$ gauge coupling, f is the Yukawa coupling of the right-handed heavy neutrino and M_{SUSY} is the scale of supersymmetry breaking. Now, if superpartner masses are in the TeV range, as required by the hierarchy arguments, also the right-handed W-bosons should have a mass typically of the order of a few TeV. GENIUS (1 ton) should be able to definitely test such models, if the right-handed neutrinos are not too light (f is not smaller than say $1/10$).

5.2 *GENIUS and $R\!\!\!/_p SUSY$*

As is well-known, $0\nu\beta\beta$ decay at present gives already very stringent limits on R-parity violating supersymmetry [11]. Improving the half-life limit for $0\nu\beta\beta$ decay by more than three orders of magnitude, as expected for GENIUS, would then also improve the limits on λ'_{111} by considerable factors.

This is shown in Fig. 9, where the two full lines to the left are the current TEVATRON limit and the region of sensitivity of HERA. The full line to the right is the expected sensitivity of the LHC (in the limit of large statistics). The three broken lines are (top to bottom) the current constraint and estimated sensitivity of GENIUS 1 ton and GENIUS 10 ton, all for a gluino mass of 1 TeV. If squarks are exceptionally heavy, $m_{\tilde{q}} \geq 1$ TeV, LHC could not compete with GENIUS. However, for typical squark masses below 1 TeV, LHC can finally probe down to smaller couplings than the double beta decay experiment. However, one should keep in mind that

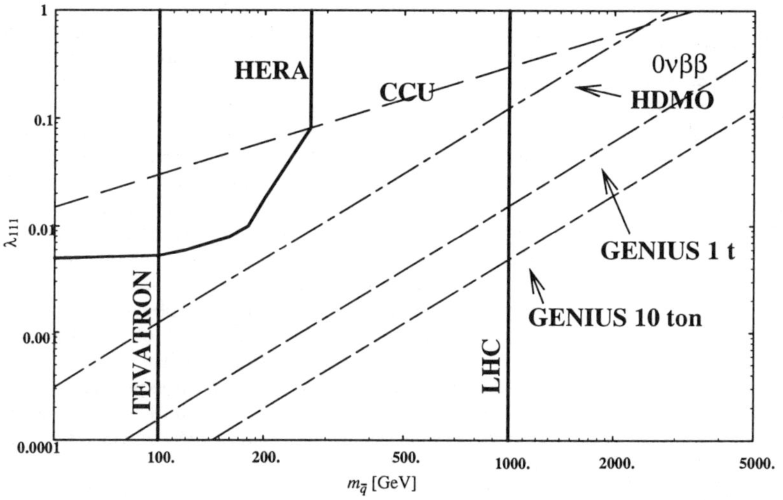

Fig. 9. Comparison of sensitivities of existing and future experiments on R_p SUSY models in the plane $\lambda'_{111} - m_{\tilde{q}}$. Note the doubly logarithmic scale! Shown are the areas currently excluded by the experiments at the TEVATRON, the limit from charged-current universality, denoted by CCU, and the limit from absence of $0\nu\beta\beta$ decay from the Heidelberg-Moscow collaboration ($0\nu\beta\beta$ HDMO). In addition, the estimated sensitivity of HERA and the LHC is compared to the one expected for GENIUS in the 1 ton and the 10 ton version. The figure is essentially an update from [11]

LHC can probe squark masses up to 1 TeV only with several years of data taking. Lower statistics shifts the line for LHC to the left.

5.3 GENIUS and R-parity conserving SUSY

It has recently been realized [12], that R-parity violation is not a necessary ingredient in supersymmetric models for $0\nu\beta\beta$ decay to occur. Instead, extending the MSSM to include Majorana neutrino masses automatically implies that the scalar neutrino has a (B-L)-violating "Majorana" mass, too [12]. Such a "Majorana" sneutrino mass might have interesting consequences for future e^+e^- colliders [12, 32], like for example the NLC.

In such SUSY models with Majorana masses $0\nu\beta\beta$ decay proceeds through the usual mass mechanism, as well as through box diagrams involving loops of supersymmetric particles, for details see [12].

Limits on the (B-L) violating "Majorana" sneutrino mass $\tilde{m}_M$ from the absence of $0\nu\beta\beta$ decay in the Heidelberg-Moscow experiment have been derived [12].

Since limits on $\tilde{m}_M$ scale as $(T_{1/2})^{1/4}$ GENIUS with 1 ton (with 10 tons) would test (B-L) violating "Majorana" sneutrino masses lower by factors of about 7 (20), compared to the present constraints.

5.4 GENIUS and leptoquarks

Leptoquarks have received renewed attention recently, especially due to the unexpected findings at HERA. If leptoquarks exist, in general they should interact with the SM Higgs boson [13]. Such a LQ-Higgs interaction would induce $0\nu\beta\beta$ decay, for LQ-Higgs couplings of the order $\mathcal{O}(1)$ at an unacceptable rate. Absence of $0\nu\beta\beta$ decay thus at present already puts stringent constraints on LQ models.

The scaling of the current limits to the GENIUS (1 ton) sensitivity in such LQ models is simple. Limits on the lepton number violating parameters, defined in [13], improve as $\sqrt{T_{1/2}}$. This means that for LQs in the range of 200 GeV LQ-Higgs couplings down to (few) 10^{-8} could be explored. Putting it the other way round, if leptoquarks interact with the standard model Higgs boson with a coupling of order $\mathcal{O}(1)$ either $0\nu\beta\beta$ decay must be found or LQs must be heavier than (several) 10 TeV(s).

5.5 GENIUS and composite neutrinos

Although currently there are no hints that quarks and leptons are composite particles, one might speculate that, when exploring higher energy ranges one might hit an energy scale Λ_C at which a new level of substructure becomes visible, see for example [31, 33]. If neutrinos are composite particles, there then should exist excited neutrinos which couple to the ordinary leptons [31, 33].

Panella and Srivastava [34] pointed out, that if the excited neutrinos are of Majorana nature, there should be a contribution to $0\nu\beta\beta$ decay. A recent analysis [35, 36] then has shown that the absence of $0\nu\beta\beta$ decay in the Heidelberg-Moscow experiment implied a lower bound on the mass of the excited neutrino, of the order of

$$m_N \geq 3.4 m_W \quad (33)$$

for a coupling of order $\mathcal{O}(1)$ and $\Lambda_C \simeq m_N$. Here, m_W is the W-boson mass.

GENIUS in the 1 ton version would allow to improve this limit up to

$$m_N \geq 0.9\text{TeV} \quad (34)$$

if $0\nu\beta\beta$ decay is not found.

6 GENIUS and Cold Dark Matter

Weakly interacting massive particles (WIMPs) are candidates for the cold dark matter in the universe. The favorite WIMP

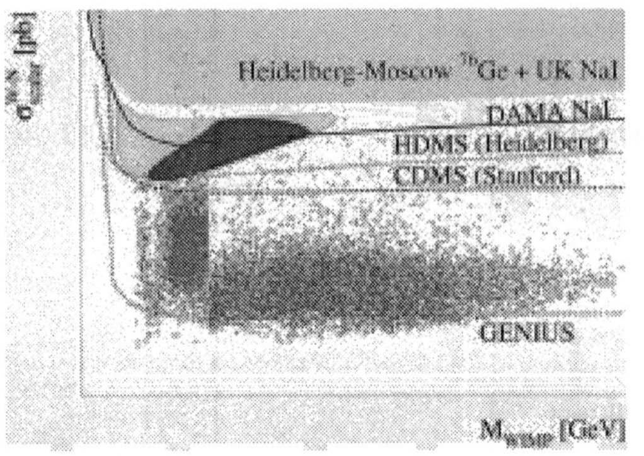

Fig. 10. WIMP–nucleon cross section limits in pb for scalar interactions as function of the WIMP–mass in GeV. *Hatched region*: excluded by the Heidelberg-Moscow collaboration [39] and the UKDMC NaI experiment [40]; *solid line*: DAMA result for NaI, see [41]. Also shown are possible results from upcoming experiments. *Dashed lines* for HDMS [42], CDMS [43] and GENIUS. These experimental limits are compared to expectations (*scatter plot*) for WIMP–neutralino cross sections calculated in the MSSM framework with non universal scalar mass unification [37]. The 90 % allowed region [44] (*light filled area*), which is further restricted by indirect dark matter searches [45] (*dark filled area*), could be easily tested with a 100 kg version of the GENIUS experiment (see [47, 7])

candidate is the lightest supersymmetric particle, presumably the neutralino. The expected detection rates for neutralinos of typically less than one event per day and kg of detector mass [37, 38], however, make direct searches for WIMP scattering experimentally a formidable task.

Figure 10 shows a comparison of existing constraints and future sensitivities of cold dark matter experiments, together with the expectations for neutralino scattering rates [47]. Obviously GENIUS will be the only experiment, which could seriously test the MSSM predictions over the whole SUSY parameter space [7, 47]. In this way, GENIUS could compete even with LHC in the search for SUSY, see for example the discussion in [46].

It is interesting to note, that if WIMP scattering is found by GENIUS it could be used to constrain the amount of R-parity violation within supersymmetric models. The arguments are very simple. Due to the fact that neutralinos are abound in the galaxy even today, neutralino decays via R-parity violating operators would have to be highly suppressed.

The details depend, of course, on the neutralino mass and composition. However, finding the neutralino with GENIUS would imply typical limits on R-parity violating couplings of the order of $10^{-(16-20)}$ for any of the λ_{ijk}, λ'_{ijk} or λ''_{ijk} in the superpotential. A positive result of the CDM search at hand, one could thus finally safely conclude that R-parity is conserved.

7 Summary

This paper gives some first discussion on the physics potential of GENIUS, designed to reach $\langle m_\nu \rangle \leq 0.01$ eV with 1 ton and $\langle m_\nu \rangle \leq 0.001$ eV with 10 tons of enriched ^{76}Ge. Besides the neutrino mass, $0\nu\beta\beta$ decay can be used to explore many models beyond the SM. GENIUS would definitely be a breakthrough into the multi-TeV range for many models currently discussed in the literature.

One of the most interesting features of a $0\nu\beta\beta$ decay experiment in the $(1 - 10)$ meV range is that it might provide interesting constraints on neutrino oscillation parameters. We have discussed several scenarios and analyzed the impact of GENIUS on the corresponding models. A main result is that for all degenerate or quasi-degenerate neutrino mass models a negative result of GENIUS would mean that neutrinos can not be the hot dark matter in the universe. Only if the ν_τ is in the eV range, the ν_e and ν_μ being lighter by at least factors of hundreds and the $\nu_\tau - \nu_e$ mixing angle small at the same time GENIUS with 1 ton would not find double beta decay.

GENIUS would already in its 1 ton version test the $\nu_e - \nu_\mu$ solution of the atmospheric neutrino problem, in its 10 ton version it could do an independent check of the large angle MSW solution to the solar neutrino problem.

It can be concluded that a $0\nu\beta\beta$ decay experiment probing the neutrino mass down to $\langle m_\nu \rangle \leq (0.01 - 0.001)$ would provide very interesting information, most of which could not be obtained by other experiments.

Besides neutrino masses, GENIUS allows to test also other beyond standard model physics. A prominent example are left-right symmetric models. Here, the sensitivity of GENIUS (1 ton) is already comparable to the one of LHC, while a 10 ton version of GENIUS would be clearly superior to LHC in the search for right-handed W-bosons. For R-parity violating supersymmetry, GENIUS would probe regions in parameter space similar, but not as big as those tested by the LHC. In addition, GENIUS would allow to improve the leptoquark and compositeness searches by considerable factors.

GENIUS could also check expectations of supersymmetric models by searching for cold dark matter and compete in this way with the LHC in the search for supersymmetry. Even if SUSY would be first found at LHC, GENIUS (even in its 1 ton version) would be the ultimate test for R-parity conservation in supersymmetry, and by detecting neutralino scattering could solve one of the most puzzling problems in astrophysics.

Thus we conclude, that GENIUS has the ability to provide a major tool for future particle physics.

References

1. Particle Data Group, Phys.Rev. D **54** (1996) 1-720. Updates can be found at: http://www-pdg.lbl.gov/
2. For recent reviews see, for example: Proc. of the Int. Workshop on Double Beta Decay and Related Topics, Trento, Italy, April 24-May 5, 1995, ed. by H.V. Klapdor-Kleingrothaus and S. Stoica, World Scientific, Singapore, 1996
3. H.V. Klapdor-Kleingrothaus, Proc. *NEUTRINO'96*, Helsinki, June 1996, World Scientific, Singapore, 1997, p. 317
4. H.V. Klapdor-Kleingrothaus, Proc. of the Int. Conf. Beyond the Desert, June 8-14, 1997, Castle Ringberg, Germany, eds. H.V. Klapdor-Kleingrothaus and H. Päs, IOP, Bristol, Philadelphia, in press
5. Heidelberg-Moscow collaboration, M. Günther et al., Phys.Rev. **D55** (1997) 54; and L. Baudis et al., Phys. Lett. **B 407**, 219 (1997)
6. K. Muto, E. Bender and H.V. Klapdor, Z. Phys. A 334 (1989) 177, 187, and A. Standt, K. Muto, H.V. Klapdor-Kleingrothaus, Europhys. Lett. **B 13** (1990) 31
7. J. Hellmig and H.V. Klapdor-Kleingrothaus, Z. Phys. A this volume
8. J.C. Pati and A. Salam, Phys. Rev. D **10** (1974) 275; R.N. Mohapatra and J.C. Pati, Phys. Rev. D **11** (1975) 566,2558; G. Senjanovic and R.N. Mohapatra, Phys. Rev. D **12** (1975) 1502; 2R.N. Mohapatra, Phys. Rev. D **34** (1986) 909

9. M. Hirsch, H.V. Klapdor-Kleingrothaus and O. Panella, Phys. Lett. **B374** (1996) 7
10. R. Mohapatra, Phys.Rev. D **34** (1986) 3457
11. M. Hirsch, H.V. Klapdor-Kleingrothaus and S.G. Kovalenko, Phys. Lett. **B372** (1996) 181; Erratum Phys.Lett. **B381** (1996) 488; Phys.Rev. **D53** (1996) 1329; Phys.Rev. Lett. **75** (1995) 17; Phys.Lett. **B 352** (1995) 1
12. M. Hirsch, H.V. Klapdor-Kleingrothaus and S.G. Kovalenko, Phys. Rev. D in press and hep-ph/9707207; M. Hirsch, H.V. Klapdor-Kleingrothaus and S.G. Kovalenko, Phys. Lett. **B 403**(1997) 291, Phys. Lett. **B 398** (1997) 311
13. M. Hirsch, H.V. Klapdor-Kleingrothaus and S.G. Kovalenko, Phys. Rev. **D54** (1996) R4207
14. M. Doi, T. Kotani and E. Takasugi, Progr. Theor. Phys. Suppl. **83**, 1 (1985).
15. H. Minakata and O. Yasuda, hep-ph/9609276 and Phys. Rev. **D56** (1997) 1692
16. M. Gell-Mann, P. Ramond and R. Slansky, in Supergravity, ed. F. van Nieuwenhuizen and D. Freedman, North Holland, Amsterdam, 1979, p. 315; T. Yanagida, Proc. of the Workshop on Unified Theory and the Baryon Number of the Universe, KEK, Japan, 1979; S. Weinberg, Phys. Rev. Lett. **43** (1979) 1566
17. D.O. Caldwell and R.N. Mohapatra, Phys. Rev. **D50** (1994) 3477, *ibid* **D48** (1993) 3259; D.G. Lee and R.N. Mohapatra, Phys. Lett. **B329** (1994) 463; R.N. Mohapatra and S. Nussinov, Phys. Lett. **346** (1995) 75; A. Ioanyssian and J.W.F. Valle, Phys. Lett. **B332** (1994) 93; P. Binetruy, S. Lavignac, S. Petcov, P. Ramond, Nucl.Phys. **B496** (1997) 3 H. Minakata, O. Yasuda, Phys.Rev. **D56** (1997) 1692 A. S. Joshipura, Phys. Rev. **D51** (1995) 1321
18. A.Yu. Smirnov, Proc. of the 28th International Conference on High energy physics, 25 - 31 July 1996, Warsaw, Poland, eds Z. Adjuk and A.K. Wroblewski, World Scientific, Singapore p. 288; and hep-ph/9611465
19. The literature on the solar neutrino problem is practically inexhaustible (more than 150 hits in a hep-ph search!). A near-complete list and much useful information can be obtained from http://dept.physics.upenn.edu/ www/neutrino/solar.html
20. R. Becker-Szendy et al. (IMB-Collaboration), Phys. Rev. **D46** (1992) 3720; K.S. Hirata et al. (Kamiokande Collaboration), Phys. Lett. **B280** (1992) 146 and Y. Fukuda et al., Phys. Lett. **B335** (1994) 237, Y. Suzuki, Proc. Int. School on Neutrinos, Erice, Sicily, Sep. 1997, Plenum Press
21. C. Athanassopoulos et al., LSND collaboration, Phys. Rev. **C54** (1996) 2685, Phys. ReV. Lett. **77** (1996) 3082
22. C. Athanassopoulos et al., LSND collaboration, nucl-ex/9706006
23. G.F. Smoot et al., Astrophys. J. **396** (1992) L1, M. Davis et al., Nature **359** (1992) 393, A.N. Taylor and M. Rowan-Robinson, Nature **359** (1992) 396, E.L. Wright et al., Astrophys. J. **396** (1992) L13
24. J. Primack, Proc. Dark Matter in Astro and Particle Physics, Heidelberg, Germany, 16.-20. Sept. 1996, eds. H.V. Klapdor-Kleingrothaus, Y. Ramachers, World Scientific, Singapure 1997, and J. Primach et al., Phys. Rev. Lett. **74** (1995) 2160
25. G. L. Fogli, E. Lisi abd D. Montanino, Phys.Rev. **D54** (1996) 2048 and hep-ph/9605273
26. M. C. Gonzalez-Garcia, hep-ph/9510419
27. BUGEY Collaboration, B. Achkar et al., Nucl Phys. **B424** (195) 503
28. G. Gelmini and E. Roulet, Rep. Prog. Phys. **58** (1995) 1207
29. T. G. Rizzo, hep-ph/9612440
30. R. Kuchimanchi and R. N. Mohapatra, Phys.Rev.Lett. **75** (1995) 3989; R. N. Mohapatra and A. Rasin, Phys.Rev.Lett. **76** (1996) 3490 and Phys.Rev. **D54** (1996) 5835
31. N. Cabibbo, L. Maiani and Y. Srivastava, Phys. Lett. **B139** (1984) 459
32. M. Hirsch, H.V. Klapdor-Kleingrothaus, St. Kolb and S.G. Kovalenko, to Phys. Rev. **D** in press
33. K. Hagiwara, S. Komamiya and D. Zeppenfeld, Z. Phys. **C29** (1985) 115
34. O. Panella and Y.N. Srivastava, Phys. Rev. **D52** (1995) 5308
35. O. Panella et al., hep-ph/9701251, O. Panella, Proc. of the Int. Conf. Beyond the Desert, June 1997 Castle Ringberg, Germany, eds. H.V. Klapdor-Kleingrothaus and H. Päs, IOP, Bristol
36. E. Takasugi, hep-ph/9706240 and Proc. of the Int. Conf. Beyond the Desert, June 1997. Castle Ringberg, Germany, eds. H.V. Klapdor-Kleingrothaus and H. Päs, IOP, Bristol
37. V. Bednyakov, H.V. Klapdor-Kleingrothaus and S.G. Kovalenko, Phys. Rev. **D50** (1994) 7128; V. Bednyakov, H.V. Klapdor-Kleingrothaus, S.G. Kovalenko and Y. Ramachers, Z. Phys. **A357** (1997) 339; V. Bednyakov, H.V. Klapdor-Kleingrothaus and S.G. Kovalenko, Phys. Rev. **D55** (1997) 503;
38. G. Jungman, M. Kamionkowski and K. Griest, Phys. Rep. **267** (1996) 195
39. M. Beck et al., Heidelberg-Moscow Collaboration, Phys. Lett. B **336** (1994) 141
40. P.F. Smith et al., Phys. Lett. **B379** (1996) 299
41. R. Bernabei et al., Phys. Lett. **B389** (1997) 757
42. L. Baudis et al., HDMS collaboration, Nucl. Inst. and Meth. **A 385** (1997) 265
43. P.D. Barnes et al., Nucl. Inst. and Meth. **A 370** (1996) 233
44. R. Bernabei et al., ROM2F/97/33; P. Belli, TAUP97, Laboratori Nazionali del Gran Sasso, Sep. 7, 1997
45. A. Bottino et al., hep-ph/9709292
46. H. Baer and M. Brhlik, hep-ph/9706509
47. H.V. Klapdor-Kleingrothaus, Y. Ramachers, Proc. Int. Conf. Beyond the Dessert, June 8-14 (1997), Castle Ringberg, Germany, eds. H.V. Klapdor-Kleingrothaus and H. Päs, IOP, Bristol, Philadelphia, in press

J. Phys. G: Nucl. Part. Phys. **24** (1998) 483–516. Printed in the UK PII: S0954-3899(98)88423-5

TOPICAL REVIEW

Future perspectives of double beta decay and dark matter search—GENIUS

H V Klapdor-Kleingrothaus, J Hellmig and M Hirsch
Max-Planck-Institut für Kernphysik, PO 10 39 80, D-69029, Heidelberg, Germany

Received 14 October 1997, in final form 12 November 1997

Abstract. The recent results from the HEIDELBERG–MOSCOW experiment have demonstrated the large potential of double beta decay to search for new physics beyond the standard model. To increase by a major step the present sensitivity for double beta decay and dark matter search, much bigger source strengths and much lower backgrounds are needed than used in experiments under operation at present or under construction. We describe here a project which would operate one tonne of 'naked' enriched germanium-detectors in liquid nitrogen as shielding in an underground set-up (GENIUS). It improves the sensitivity of neutrino masses to 0.01 eV. A 10 tonne version would probe neutrino masses even down to 10^{-3} eV. The first version would allow us to test the atmospheric neutrino problem, the second at least part of the solar neutrino problem. Both versions would allow, in addition, significant contributions to testing several classes of GUT models. These are especially tests of R-parity breaking and conserving supersymmetry models—including sneutrino masses—leptoquark masses and mechanism and right-handed W-boson masses comparable with LHC. The second issue of the experiment is the search for dark matter in the universe. The full MSSM parameter space for the prediction of neutralinos as dark matter particles could be covered already in a first step of the full experiment using only 100 kg of ^{76}Ge or even of natural Ge making the experiment competitive with LHC in the search for supersymmetry.

1. Introduction

Searches for rare events, nuclear double beta decay [1, 2], and nuclear recoils from elastically scattered weak interacting massive particles (WIMPs) [3–6] are performed to discover new particles and test new particle physics theories [7, 8]. This type of experiment is in contest with high-energy accelerator experiments in the investigation of physics at very high energies. Two topics are of greatest interest in high-energy physics and astrophysics. Test of the existence of a so-called supersymmetry (SUSY) is one, if not the, major aim of the large hadron collider (LHC) [9–11], which will dominate the high-energy physics research in the next decade. Second, dark matter, which manifests itself by its gravitational force, has puzzled astrophysicists for a long time [3–6]. A very close connection between both issues in addition to SUSY which could be responsible for the cold dark matter (neutralinos), could be established by non-zero neutrino masses as candidates for hot dark matter, and especially degenerate neutrino mass scenarios [12–14] can explain the recent observations by the COBE satellite for dark matter [15, 16]. Our new germanium in nitrogen underground set-up (GENIUS), first proposed and presented by [1], and described in detail also in [17, 18], is an experiment which is optimized to address both issues. It is a large step beyond the HEIDELBERG–MOSCOW experiment [19], which is the most sensitive existing double beta decay experiment at present and for the next years and which has also given the most

Eur. Phys. J. A **3**, 107–108 (1998)

THE EUROPEAN PHYSICAL JOURNAL A

Short note

Ice shielding in the large scale GENIUS experiment for double beta decay and dark matter search

H.V. Klapdor-Kleingrothaus[1], Yu.G. Zdesenko[2]

[1] Max-Planck-Institut für Kernphysik, D-69029 Heidelberg, Germany
[2] Institute for Nuclear Research, 252650 Kiev, Ukraine

Received: 30 July 1998
Communicated by B. Povh

Abstract. We suggest here the use of ice as shielding material in the large scale GENIUS experiment for the ultimate sensitive double beta decay and dark matter search. The idea is to pack a working volume of several tons of liquid nitrogens, which contains the "naked" Ge detectors, inside an ice shielding. Very thin plastic foil would be used in order to prevent leakage of the liquid nitrogen. Due to the excellent advantages of ice shielding (high purity and low cost, self-supporting ability, thermo-isolation and optical properties, safety) this could be another possible way of realization of the GENIUS project.

PACS. 29.40.Wk Solid-state detectors – 95.35.+d Dark matter (stellar, interstellar, galactic, and cosmological – 95.55.Vj Neutrino, muon, pion, and other elementary particle detectors; cosmic ray detectors

The GENIUS project, which would operate one to ten tons of high purity Ge (enriched in ^{76}Ge and natural) semiconductor detectors, was proposed [1–3] with the aim to increase by a major step the present sensitivity for double beta decay and dark matter search [3–4]. The expected enhanced sensitivity for the neutrino mass is in the range of 10^{-2}–10^{-3} eV, which would allow to check the atmospheric neutrino puzzle and at least part of the solar neutrino problem. Besides that the GENIUS experiment would result in additional significant contributions to testing several classes of GUT models. Some of them are tests of R-parity breaking and conserving supersymmetry models – including sneutrino masses –, leptoquark masses and mechanism and right-handed W-boson masses comparable to LHC. The second issue of the experiment is the search for dark matter in the universe. The full MSSM parameter space for prediction of neutralinos as dark matter particles could be covered even in the first step of the experiment using only about 100 kg of natural Ge detectors making the GENIUS project competitive to LHC in the search for supersymmetry [1–3].

It is well known that the ultimate sensitivity of the super- low background experiments for double beta decay and dark matter search is limited by the available source strengths (mass of the source) from one side and by the detector background from another side. The first origin of background is due to secondary cosmic rays and can be eliminated by the proper deep underground site for the experiment. The second part of the background (most crucial for sensitivity) is determined by the radioactive impurities in the detector itself, in the materials used for detector mounting and shielding, and in the surroundings.

In order to overcome both sensitivity limitations (source strengths and background) the main idea of the GENIUS project is to operate with a large amount of "naked" HP Ge detectors placed directly in liquid nitrogen serving as cooling and shielding medium simultaneously [1–3]. Indeed, this solution could allow to minimize the quantity of materials needed for the mounting of the crystals to a negligible level, and liquid nitrogen could be purified to a very high level. As it was shown using GEANT Monte Carlo simulations [1–3] the required demands to the radioactive contamination in the liquid nitrogen are on the level of 1×10^{-15} g/g for ^{40}K and ^{238}U; 5×10^{-15} g/g for ^{232}Th and 0.05 mBq/m^3 for ^{222}Rn. All these requirements (except for radon) are less stringent than those which have been already achieved in the Counting Test Facility for the Borexino experiment: $(2-5) \times 10^{-16}$ g/g for ^{232}Th and ^{238}U contamination in the liquid scintillators [5].

In accordance with Monte Carlo simulations [1–3] the required dimensions of the liquid nitrogen shield, which could fully suppress the radioactivity from the surroundings (measured, for instance, in the Gran Sasso Underground Laboratory) should be of about 10 m in diameter and 10 m height. These dimensions could be somewhat less (9 m in diameter and 9 m height) for the Solotvina Underground Laboratory located in a salt mine [6], since

2.6.2 The Physics Potential of Future Double Beta Decay for Beyond Standard Model Physics

Paper presented at the First Int. Symp. on Lepton and Baryon Number Violation, Trento, 1998 251

Double Beta and Dark Matter Search - Window to New Physics beyond the Standard Model of Particle Physics

H.V. Klapdor–Kleingrothaus

Max–Planck–Institut für Kernphysik
P.O.Box 10 39 80, D–69029 Heidelberg, Germany

Abstract. Nuclear double beta decay provides an extraordinarily broad potential to search for beyond Standard Model physics, probing already now the TeV scale, on which new physics should manifest itself. These possibilities are reviewed here.
First, the results of present generation experiments are presented. The most sensitive one of them – the Heidelberg–Moscow experiment in the Gran Sasso – probes the electron mass now in the sub eV region and will reach a limit of ~ 0.1 eV in a few years. Basing to a large extent on the theoretical work of the Heidelberg Double Beta Group in the last two years, results are obtained also for SUSY models (R–parity breaking, sneutrino mass), leptoquarks (leptoquark-Higgs coupling), compositeness, right–handed W boson mass, test of special relativity and equivalence principle in the neutrino sector and others. These results are comfortably competitive to corresponding results from high–energy accelerators like TEVATRON, HERA, etc. One of the enriched ^{76}Ge detectors also yields the most stringent limits for cold dark matter (WIMPs) to date by using raw data.
Second, future perspectives of $\beta\beta$ research are discussed. A new Heidelberg experimental proposal (GENIUS) is described which would allow to increase the sensitivity for Majorana neutrino masses from the present level of at best 0.1 eV down to 0.01 or even 0.001 eV. Its physical potential would be a breakthrough into the multi-TeV range for many beyond standard models. Its sensitivity for neutrino oscillation parameters would be larger than of all present terrestrial neutrino oscillation experiments

and of those planned for the future. It could probe directly the atmospheric neutrino problem and the large angle, and for almost degenerate neutrino mass scenarios even the small angle solution of the solar neutrino problem. It would further, already in a first step using only 100 kg of natural Ge detectors, cover almost the full MSSM parameter space for prediction of neutralinos as cold dark matter, making the experiment competitive to LHC in the search for supersymmetry.

1. Introduction – Motivation for the search for double beta decay – and a future perspective: GENIUS

Double beta decay yields – besides proton decay – the most promising possibilities to probe beyond standard model physics beyond accelerator energy scales. Propagator physics has to replace direct observations. That this method is very effective, is obvious from important earlier research work and has been stressed, e.g. by [Rub96], etc.. Examples are the properties of W and Z bosons derived from neutral weak currents and β–decay, and the top mass deduced from LEP electroweak radiative corrections.

The potential of double beta decay includes information on the neutrino and sneutrino mass, SUSY models, compositeness, leptoquarks, right–handed W bosons, Lorentz invariance and the equivalence principle in the neutrino sector, and others (see Table 1). The recent results of the Heidelberg–Moscow experiment, which will be reported here (see also [Kla98a]), have demonstrated that $0\nu\beta\beta$ decay probes already now the TeV scale on which new physics should manifest itself according to present theoretical expectations. To give just one example, inverse double beta decay $e^-e^- \to W^-W^-$ requires an energy of at least 4 TeV for observability, according to present constraints from double beta decay [Bel96, Bel98]. Similar energies are required to study, e.g. leptoquarks [Buc91, H195, Hir96a, Bav95, Leu94, Cho94, Blü94].

To increase by a major step the present sensitivity for double beta decay and dark matter search, we describe here a new project proposed recently [Kla98a, Kla98b] which would operate one ton of 'naked' enriched **GE**rmanium detectors in liquid **NI**trogen as shielding in an **U**nderground **S**etup (GENIUS). It would improve the sensitivity from the present potential of at best ~ 0.1 eV to neutrino masses down to 0.01 eV, a ten ton version even to 0.001 eV. The first version would allow to test a $\nu_e \to \nu_\mu$ explanation of the atmospheric neutrino problem, the second directly the large angle solution of the solar neutrino problem, and, for degenerate neutrinos even the small angle solution. The sensitivity for neutrino oscillation parameters would be larger than for all present accelerator neutrino oscillation experiments, or those planned for the future. GENIUS would further allow one to test the recent hypothesis of a sterile neutrino and the underlying idea of a shadow world (see section 2). Both versions of GENIUS would definitely be a breakthrough into the multi-TeV range for many beyond standard models currently discussed in the literature, and the sensitivity would be comparable or even superior to LHC for various quantities such as right–handed W–bosons, R–parity violation, leptoquark or compositeness searches.

Another issue of GENIUS is the search for Dark Matter in the universe. The full MSSM parameter space for predictions of neutralinos as cold dark matter could be covered already in a first step of the full experiment using only 100 kg of ^{76}Ge or even natural Ge, making the experiment competitive to LHC in the search for supersymmetry.

2. Double beta decay and particle physics

We present a brief introductory outline of the potential of $\beta\beta$ decay for some representative examples, including some comments on the status of the required nuclear matrix elements. The potential of double beta decay for probing neutrino oscillation parameters will be addressed in section 4.2.

Double beta decay can occur in several decay modes (Figs. 1–3)

$$ {}^A_ZX \rightarrow {}^A_{Z+2}X + 2e^- + 2\overline{\nu}_e \tag{1}$$

$$ {}^A_ZX \rightarrow {}^A_{Z+2}X + 2e^- \tag{2}$$

$$ {}^A_ZX \rightarrow {}^A_{Z+2}X + 2e^- + \phi \tag{3}$$

$$ {}^A_ZX \rightarrow {}^A_{Z+2}X + 2e^- + 2\phi \tag{4}$$

the last three of them violating lepton number conservation by $\Delta L = 2$. Fig. 3 shows the corresponding spectra, for the neutrinoless mode (2) a sharp line at $E = Q_{\beta\beta}$, for the two–neutrino mode and the various Majoron–accompanied modes classified by their spectral index, continuous spectra. Important for particle physics are the decay modes (2)–(4).

The neutrinoless mode (2) needs not be necessarily connected with the exchange of a virtual neutrino or sneutrino. *Any* process violating lepton number can in principle lead to a process with the same signature as usual $0\nu\beta\beta$ decay. It may be triggered by exchange of neutralinos, gluinos, squarks, sleptons, leptoquarks,... (see below and [Päs97]). This gives rise to the broad potential of double beta decay for testing or yielding restrictions on quantities of beyond standard model physics (see Table 1), realized

254

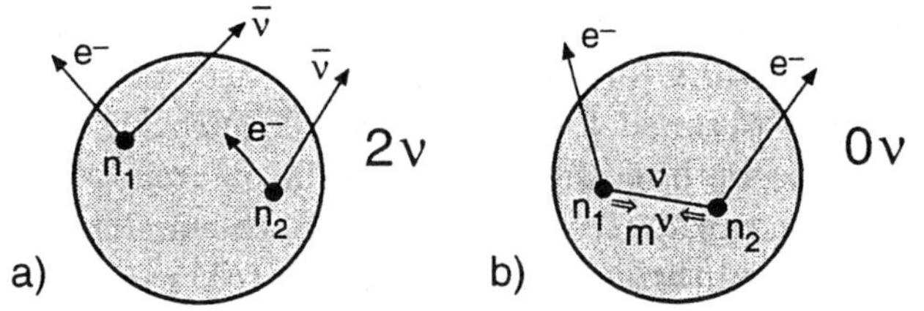

Fig. 1 *Schematic representation of* 2ν *and* 0ν *double beta decay.*

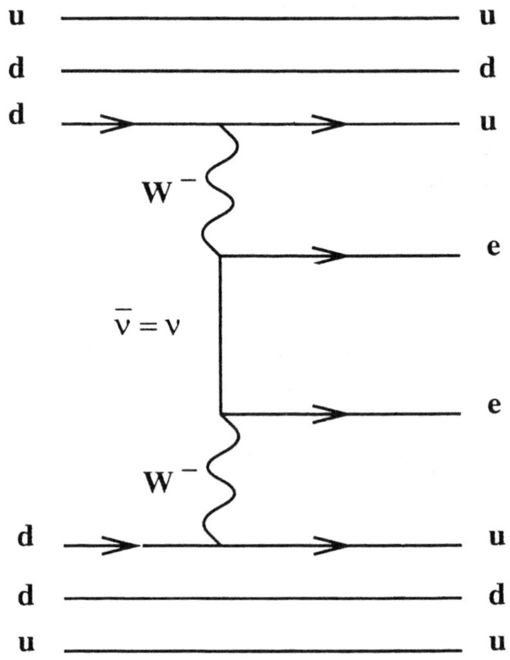

Fig. 2 *Feynman graph for neutrinoless double beta decay triggered by exchange of a left–handed light or heavy neutrino*

Observ.	Restrictions	Topics investigated
0ν:	via ν exchange:	Beyond the standard model and SU(5)
	Neutrino mass	model; early universe, matter–antimatter
	Light Neutrino	asymmetry, Dark matter
	Heavy Neutrino	L–R -symmetric models (e.g. SO(10)), compositeness
	Test of Lorentz invariance and equivalence principle	
	Right handed weak currents	$V+A$ interaction, $W_R^{\pm}$ masses
	via photino, gluino, zino (gaugino) or sneutrino exchange:	SUSY models: Bounds for parameter space beyond the range of accelerators
	R-parity breaking, sneutrino mass	
	via leptoquark exchange	leptoquark masses and models
	leptoquark-Higgs interaction	
$0\nu\chi$:	existence of the Majoron	Mechanism of (B-L) breaking -explicit -spontaneous breaking of the local/global B-L symmetry new Majoron models

Table 1 *$\beta\beta$ decay and particle physics*

and investigated to a large extent by the Heidelberg Double Beta Group in the last two years. There is, however, a generic relation between the amplitude of $0\nu\beta\beta$ decay and the $(B-L)$ violating Majorana mass of the neutrino. It has been recognized about 15 years ago [Sch81] that if any of these two quantities vanishes, the other one vanishes, too, and vice versa, if one of them is non–zero, the other one also differs from zero. This Schechter-Valle-theorem is valid for any gauge model with spontaneously broken symmetry at the weak scale, independent of the mechanism of $0\nu\beta\beta$ decay. A generalisation of this theorem to supersymmetry has been given recently [Hir97, Hir98a]. This Hirsch–Klapdor-Kleingrothaus–Kovalenko–theorem claims for the neutrino Majorana mass, the $B-L$ violating mass of the sneutrino and neutrinoless double beta decay amplitude: If one of them is non–zero, also the others are non–zero and vice versa, independent of the mechanisms of $0\nu\beta\beta$ decay and (s-)neutrino mass generation. This theorem connects double beta research with new processes potentially observable at future colliders like NLC (next linear collider) [Hir97, Hir98].

2.1. Mass of the (electron) neutrino

The neutrino is one of the best examples for the merging of the different disciplines of micro– and macrophysics. The neutrino plays, by its nature (Majorana or Dirac particle), and its mass, a key role for the structure of modern particle physics theories (GUTs, SUSYs, SUGRAs,...) [Kla95, Kla97, Lan88, Gro90, Moh91, Kla97a, Kla97c, Kla98b]. At the

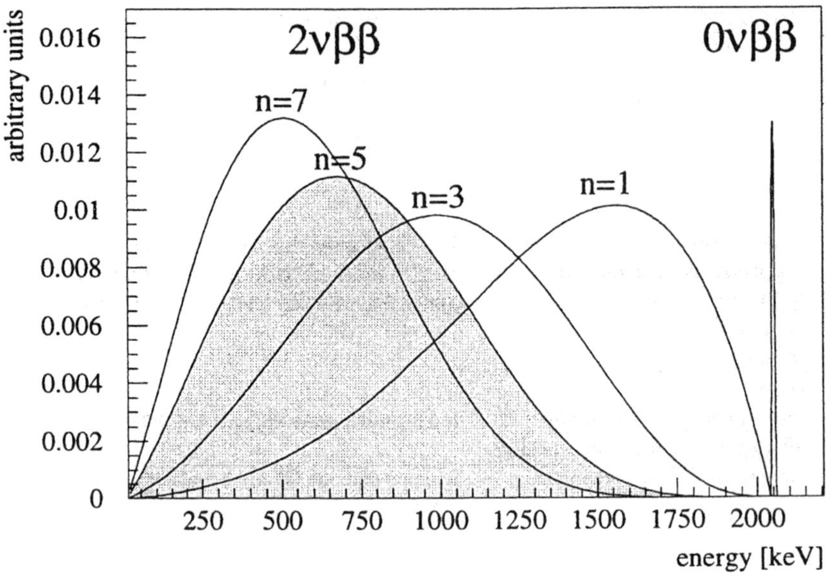

Fig. 3 *Spectral shapes of the different modes of double beta decay, n denotes the spectral index , n=5 for $2\nu\beta\beta$ decay (see text)*

same time it is candidate for non–baryonic hot dark matter in the universe, and the neutrino mass is connected – by the sphaleron effect – to the matter–antimatter asymmetry of the early universe [Kuz90]. Neutrino physics has entered an era of new actuality in connection with several possible indications of physics beyond the standard model (SM) of particle physics: A lack of solar (^7Be) neutrinos, an atmospheric ν_μ deficit and mixed dark matter models could all be explained simulaneously by non–vanishing neutrino masses. Recent GUT models, for example an extended SO(10) scenario with S_4 horizontal symmetry could explain these observations by requiring degenerate neutrino masses of the order of 1 eV [Lee94, Moh94, Pet94, Ioa94, Fri95, Moh95, Pet96, Val96]. For an overview see [Smi96a]. Such degenerate scenarios are the more general solution of the well-known see-saw mechanism, of which the often discussed strongly hierarchical neutrino mass pattern is just a special solution (see [Moh97]). If the atmospheric neutrino data are excluded but LSND [Ath95, Ath96], HDM and solar neutrino constraints are kept, they could be explained by an inverted mass texture [Raf96, Cal95], where $m_{\nu_e} \simeq m_{\nu_\tau} \simeq 2.4eV >> m_{\nu_\mu}$.

This brings double beta decay experiments into some key position, since with some second generation $\beta\beta$ experiments like the HEIDELBERG–MOSCOW experiment using large amounts of enriched $\beta\beta$–emitter material the predictions of or assumptions in such scenarios can now be tested. If the first of the above scenarios of neutrino mass textures is ruled out by tightening the double beta limit on m_{ν_e}, then the only way to understand *all* neutrino results may require an additional sterile neutrino [Cal93, Pel93], coupling only extremely weakly to the Z–boson. Then the solar neutrino puzzle would be explained by the $\nu_e - -\nu_S$ oscillation, and atmospheric neutrino data by $\nu_\mu - \nu_\tau$ oscillations, and the $\nu_{\mu,\tau}$ would constitute the hot dark matter (HDM) of the universe. The request for a light sterile neutrino would naturally lead to the concept of a shadow world [Ber95]. This assumes exact duplication of the Standard Model in both the gauge and the fermion content (the shadow sector), yielding three extra sterile neutrinos ν', the only interaction connecting known and shadow sector being gravitation. Mixing of the ν and ν' will occur by Planck scale effects. Such a scenario could explain all *four* present indications for non-vanishing neutrino mass [Moh97]. The expectation for the effective neutrino mass (see below) to be seen in double beta decay would be $\langle m_{\nu_e} \rangle \simeq 0.002eV$ [Moh97a]. Thus it could be checked by the new Genius project (see section 4.2.2). Interestingly in such a scenario the ν'_μ, ν'_τ having masses of ~ 2 keV could act as warm or cold dark matter in the universe [Moh97].

At present neutrinoless double decay is the most sensitive of the various existing methods to determine the mass of the electron neutrino. It further provides a unique possibility of deciding between a Dirac and a Majorana nature of the neutrino. Neutrinoless double beta decay can be triggered by exchange of a light or heavy left-handed Majorana neutrino (Figs. 1,2). For exchange of a heavy *right*–handed neutrino see section 2.3. The propagators in the first and second case show a different m_ν dependence: Fermion propagator $\sim \frac{m}{q^2 - m^2} \Rightarrow$

$$a) \quad m \ll q \rightarrow \sim m \quad \textit{'light' neutrino} \tag{5}$$

$$b) \quad m \gg q \rightarrow \sim \frac{1}{m} \quad \textit{'heavy' neutrino} \tag{6}$$

The half–life for $0\nu\beta\beta$ decay induced by exchange of a light neutrino is given by [Mut88]

$$[T^{0\nu}_{1/2}(0^+_i \rightarrow 0^+_f)]^{-1} = C_{mm}\frac{\langle m_\nu \rangle^2}{m_e^2} + C_{\eta\eta}\langle \eta \rangle^2 + C_{\lambda\lambda}\langle \lambda \rangle^2 + C_{m\eta}\frac{m_\nu}{m_e}$$
$$+ C_{m\lambda}\langle \lambda \rangle \frac{\langle m_\nu \rangle}{m_e} + C_{\eta\lambda}\langle \eta \rangle \langle \lambda \rangle \tag{7}$$

or, when neglecting the effect of right–handed weak currents, by

$$[T^{0\nu}_{1/2}(0^+_i \rightarrow 0^+_f)]^{-1} = C_{mm}\frac{\langle m_\nu \rangle^2}{m_e^2} = (M^{0\nu}_{GT} - M^{0\nu}_F)^2 G_1 \frac{\langle m_\nu \rangle^2}{m_e^2} \tag{8}$$

where G_1 denotes the phase space integral, $\langle m_\nu \rangle$ denotes an effective neutrino mass

$$\langle m_\nu \rangle = \sum_i m_i U^2_{ei}, \tag{9}$$

respecting the possibility of the electron neutrino to be a mixed state (mass matrix not diagonal in the flavor space)

$$|\nu_e\rangle = \sum_i U_{ei}|\nu_i\rangle \tag{10}$$

The effective mass $\langle m_\nu \rangle$ could be smaller than m_i for all i for appropriate CP phases of the mixing coefficiants U_{ei} [Wol81]. In general not too pathological GUT models yield $m_{\nu_e} = \langle m_{\nu_e} \rangle$ (see [Lan88]).

η,λ describe an admixture of right–handed weak currents, and $M^{0\nu} \equiv M^{0\nu}_{GT} - M^{0\nu}_{F}$ denote nuclear matrix elements.

Nuclear matrix elements:

A detailed discussion of $\beta\beta$ matrix elements for neutrino induced transitions including the substantial (well–understood) differences in the precision with which 2ν and $0\nu\beta\beta$ rates can be calculated, can be found in [Gro90, Mut88, Mut89, Sta90]. After the major step of recognizing the importance of ground state correlations for the calculation of $\beta\beta$ matrix elements [Kla84, Gro86], in recent years the main groups used the QRPA model for calculation of $M^{0\nu}$. The different groups obtained very similar results for $M^{0\nu}$ when using a realistic nucleon–nucleon interaction [Mut89, Sta90, Tom87], consistent with shell model approaches [Mut91, Hax84], where the latter are possible. Some deviation is found only when a non–realistic nucleon–nucleon interaction is used (e.g. δ force, see [Vog86] and also [Vog96]). Also use of a by far too small configuration space like in recent shell model Monte Carlo (SMMC) calculations [Rad95] can hardly lead to reliable results. Recent so–called large scale shell model calculations [Cau96] also fail to fulfill the Ikeda sum rule by about 40-60%, thus predicting too small matrix elements. On the other hand refinements of the QRPA approach by going to higher order QRPA (see [Sto96, Suh96])lead only to minor changes for the $0\nu\beta\beta$ ground state transitions. The most recent QRPA calculations including renormalization [Sim97] do not fulfill the Ikeda sum rule by 30 %. The calculated matrix elements are (correspondingly ?) about 40 % smaller than earlier calculations fulfilling the sum rule properly [Mut89, Sta90]. The consequences of high-lying GT strength (in the GTGR and in the Δ resonance) have already been studied earlier [Gro86].

Since the usual QRPA approach does ignore deformation, some larger uncertainty in these approaches may occur in deformed nuclei. This shows up for example in different results obtained for ^{150}Nd by QRPA and by a pseudo SU(3) model as used by [Hir95d]. Calculation of matrix elements of all double beta emitters have been published by [Gro85, Sta90]. Typical uncertainties of calculated $0\nu\beta\beta$ rates originating from the limited knowledge of the particle–particle force, which is the main source of the uncertainty in those nuclei where this QRPA approach is applicable, are shown in [Sta90]. They are of the order of a factor of 2.

2.2. Supersymmetry

Supersymmetry (SUSY) is considered as prime candidate for a theory beyond the standard model, which could overcome some of the most puzzling questions of today's particle physics (see, e.g. [Hab93, Moh92, Kan97]). Accelerator experiments have hunted for signs of supersymmetric particles so far without success. Lower limits on masses of SUSY particles are at present in the range of 20–100 GeV [PDG96], mainly from experiments at LEP and TEVATRON.

Conservation of R–parity has been imposed ad hoc to the minimal supersymmetric extension of the standard model (MSSM) to ensure baryon number and lepton number conservation. SUSY particles differ then from usual particles not only in their masses but also in R–parity, assigned to be $R_P = 1$ for usual particles and $R_P = -1$ for SUSY particles. This assumption, however, is not guaranteed by supersymmetry or gauge invariance. Generally one can add the following R–parity violating terms to the usual superpotential [Hal84].

$$W_{\not R_P} = \lambda_{ijk} L_i L_j \overline{E}_k + \lambda^{'}_{ijk} L_i Q_j \overline{D}_k + \lambda^{''} \overline{U}_i \overline{D}_j \overline{D}_k, \tag{11}$$

where indices i, j, k denote generations. L,Q denote lepton and quark doublet superfields and $\overline{E}, \overline{U}, \overline{D}$ lepton and up, down quark singlet superfields. Terms proportional to λ, $\lambda^{'}$ violate lepton number, those proportional to $\lambda^{''}$ violate baryon number. From proton decay limits it is clear that both types of terms cannot be present at the same time in the superpotential. On the other hand, once the $\lambda^{''}$ terms being assumed to be zero, λ and $\lambda^{'}$ terms are not limited. $0\nu\beta\beta$ decay can occur within the $\not R_p$ -MSSM through Feynman graphs such as those of Fig. 4. In lowest order there are alltogether six different graphs of this kind. [Hir95, Hir95c, Hir96c]. Attention has, therefore , been focused also on SUSY theories with R–parity violation, in which $0\nu\beta\beta$ decay can proceed by exchange of supersymmetric particles like gluinos, photinos,... Thus $0\nu\beta\beta$ decay can be used to restrict R–parity violating SUSY models [Hir95, Hir96c, Moh91, Hir95c, Moh86]. From these graphs one derives [Hir95] under some assumptions

$$[T^{0\nu}_{1/2}(0^+ \to 0^+)]^{-1} \sim G_{01}\left(\frac{\lambda'^2_{111}}{m^4_{\tilde{q},\tilde{e}} m_{\tilde{g}\chi}} M\right)^2 \tag{12}$$

where G_{01} is a phase space factor, $m_{\tilde{q}\tilde{e}\tilde{g}\chi}$ are the masses of supersymmetric particles involved: squarks, selectrons, gluinos, or neutralinos. λ'_{111} is the

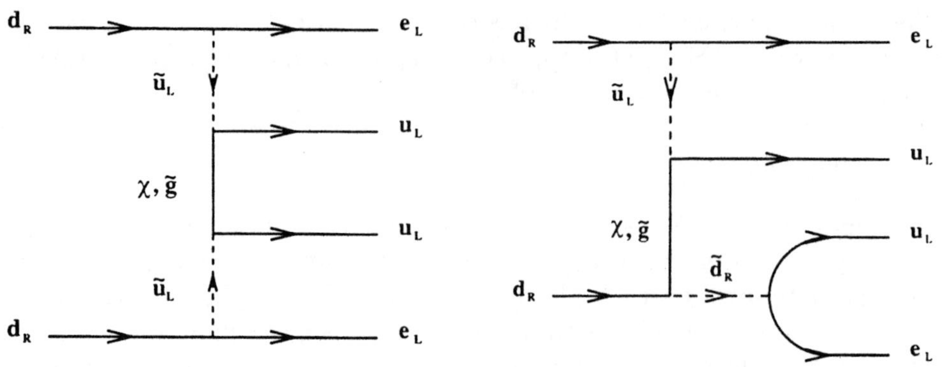

Fig. 4 *Examples of Feynman graphs for* $0\nu\beta\beta$ *decay within R–parity violating supersymmetric models (from [Hir95]).*

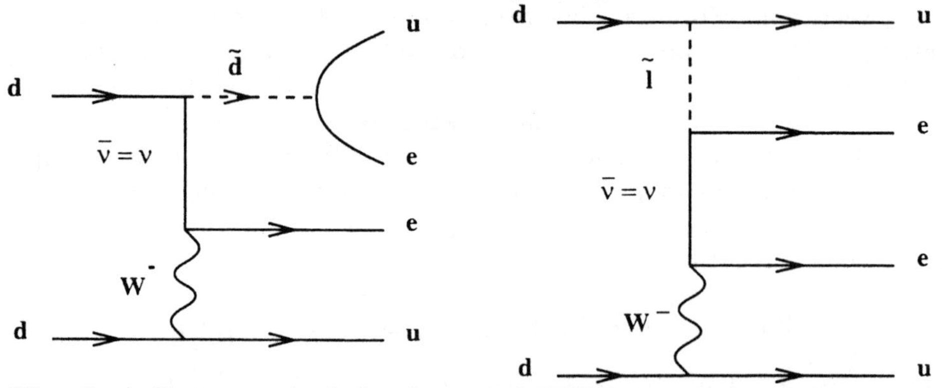

Fig. 5 a) *Feynman graph for the mixed SUSY–neutrino exchange mechanism of* $0\nu\beta\beta$ *decay. R–parity violation occurs through scalar quark exchange.* **b)** *As figure 1, but for scalar lepton exchange (from [Hir96]).*

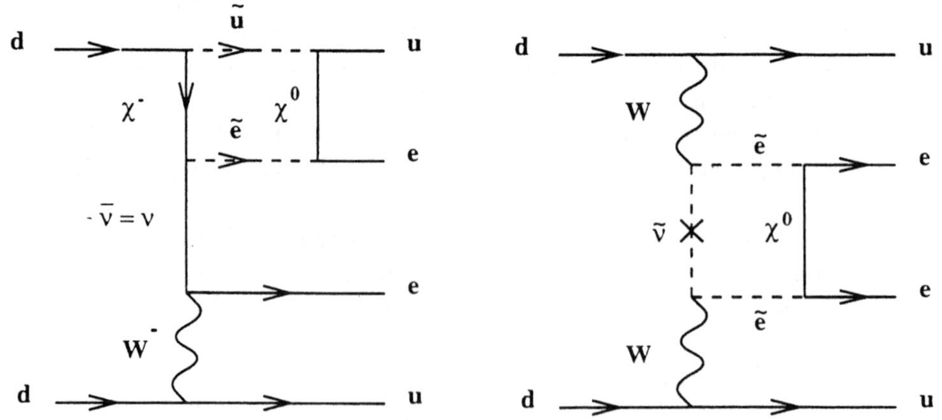

Fig. 6 *Examples of* R_P *conserving SUSY contributions to* $0\nu\beta\beta$ *decay (from [Hir97a]).*

strength of an R–parity breaking interaction (eq. 11), and M is a nuclear matrix element. For the matrix elements and their calculation see [Hir96c].

It is also worthwile to notice that $0\nu\beta\beta$ decay is not only sensitive to λ'_{111}. Taking into account the fact that the SUSY partners of the left and right–handed quark states can mix with each other, one can derive limits on different combinations of λ' [Hir96, Moh96a, Bab95]. Graphs allowing such information are as those shown in Fig. 5. The dominant diagram of this type is the one where the exchanged scalar particles are the $\tilde{b} - \tilde{b}^C$ pair. Under some assumptions (e.g. the MSSM mass parameters to be approximately equal to the "effective" SUSY breaking scale Λ_{SUSY}), one obtains [Hir96]

$$\lambda'_{11i} \cdot \lambda'_{1i1} \leq \epsilon'_i \Big(\frac{\Lambda_{SUSY}}{100GeV}\Big)^3 \tag{13}$$

and

$$\Delta_n \lambda'_{311} \lambda_{n13} \leq \epsilon \Big(\frac{\Lambda_{SUSY}}{100GeV}\Big)^3 \tag{14}$$

With the known values of the SM quark and lepton masses follows $\epsilon'_{1,2,3} = 6.4 \cdot 10^{-5}$, $3.2 \cdot 10^{-6}$ and $1.1 \cdot 10^{-7}$, and $\epsilon = 6.5 \cdot 10^{-8}$. For an overview on our knowledge on λ'_{ijk} from other sources we refer to [Kol97a] and [Bha97].

Also R–parity *conserving* softly broken supersymmetry can give contributions to $0\nu\beta\beta$ decay, via the $B - L$–violating sneutrino mass term, the latter being a generic ingredient of any weak–scale SUSY model with a Majorana neutrino mass [Hir97, Hir98]. These contributions are realized at the level of box diagrams [Hir98] (fig. 6). The $0\nu\beta\beta$ half-life for contributions from sneutrino exchange is found to be [Hir98]

$$[T_{1/2}^{0\nu\beta\beta}]^{-1} = G_{01} \frac{4m_p^2}{G_F^4} \Big| \frac{\eta^{SUSY}}{m_{SUSY}^5} M^{SUSY} \Big|, \tag{15}$$

where the phase factor G_{01} is tabulated in [Doi85], η^{SUSY} is the effective lepton number violating parameter, which contains the $(B - L)$ violating sneutrino mass $\tilde{m}_M$ and M^{SUSY} is the nuclear matrix element [Hir96d].

2.3. Left–Right symmetric theories – Heavy neutrinos and right–handed W Boson

Heavy *right–handed* neutrinos appear quite naturally in left–right symmetric GUT models. They offer in some natural way via the see–saw mechanism explanation for the small neutrino masses compared to other fermions and can explain also naturally parity violation. However the symmetry breaking scale for the right–handed sector is not fixed by the theory and thus the mass of the right–handed W_R boson and the mixing angle between the mass eigenstates W_1, W_2 are free parameters. $0\nu\beta\beta$ decay taking into account contributions from both, left– and right–handed neutrinos have

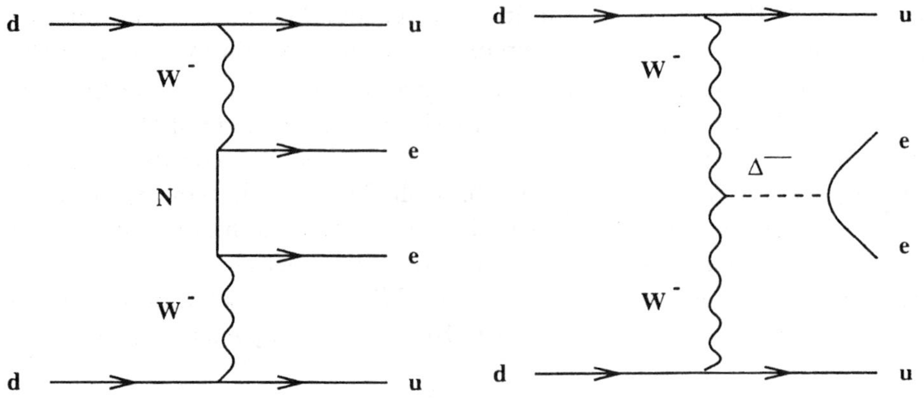

Fig. 7 a) *Heavy neutrino exchange contribution to neutrinoless double beta decay in left right symmetric models, and* **b)** *Feynman graph for the virtual exchange of a doubly–charged Higgs boson, see text (from [Hir96d]).*

been studied theoretically by [Hir96d, Doi93]. The former gives a more general expression for the decay rate than introduced earlier by [Moh86b].

$0\nu\beta\beta$ decay proceeds through the diagram shown in Fig. 7a, where N denotes the heavy right–handed partner of the ordinary neutrino. In order to preserve the unitarity of the cross section in *inverse* $0\nu\beta\beta$ decay, LR models must according to [Riz82] include an additional Higgs triplet. This then gives rise to a second contribution to $0\nu\beta\beta$ decay, shown in Fig. 7b. From the Feynman graphs of Fig. 7 it is obvious that the amplitude will be proportional to [Hir96d]

$$\left(\frac{m_{W_L}}{m_{W_R}}^4\right)\left(\frac{1}{m_N}+\frac{m_N}{m^2_{\Delta_R^{--}}}\right) \tag{16}$$

Eq. (16) and the experimental lower limit of $0\nu\beta\beta$ decay leads to a constraint limit within the 3–dimensional parameter space ($m_{W_R} - m_N - m_{\Delta_R^{--}}$). The most conservative (weakest) limit on m_{W_R} is obtained in the limit, where the mass of the Δ^{--} goes to infinity (see section 3 below). If adding information on the vacuum stability, an absolute lower limit on the mass of the right–handed W–boson can be obtained.

2.4. Compositeness

Although so far there are no experimental signals of a substructure of quarks and leptons, there are speculations that at some higher energy ranges beyond 1 TeV or so there might exist an energy scale Λ_C at which a substructure of quarks and leptons (preons) might become visible [Pan96, Moh92, Sou92](Fig. 8).

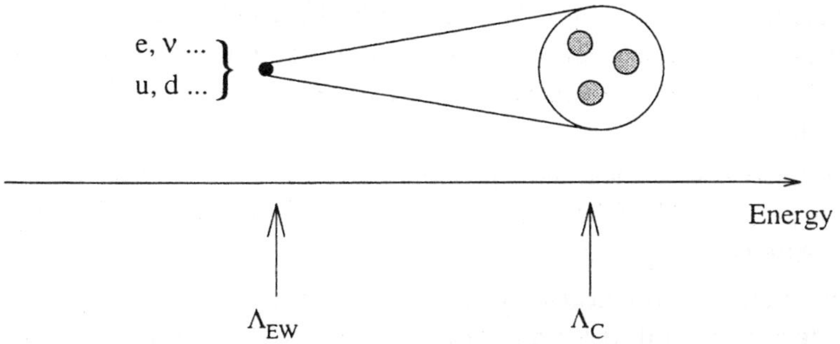

Fig. 8 *The idea of compositeness. At a (still unknown) energy scale* Λ_C *quarks and lepton might show an internal structure*

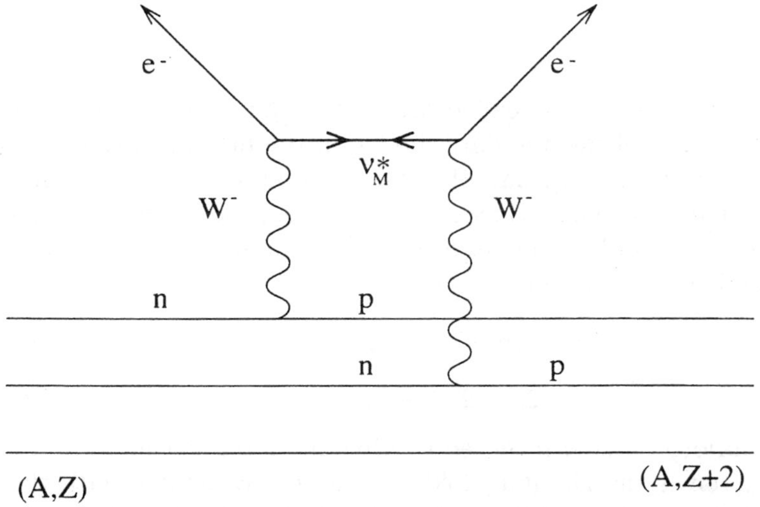

Fig. 9 *Neutrinoless double beta decay* ($\Delta L = +2$ *process) mediated by a composite heavy Majorana neutrino.*

The main consequences of compositeness of quarks and leptons are (1) modifications to the gauge boson propagators and the interaction vertices with fermions, and additional effective four–fermion interactions through constituent exchange (2) highly massive excited states which couple to the ordinary fermions through gauge interactions. This is discussed in detail in [Pan96]. Lower bounds on the compositeness scale have been deduced from accelerator experiments at LEP [ALE93], Fermilab [CDF91], HERA [H195] and from a theoretical analysis of the effect of contact interactions in the leptonic τ decay [Dia93]. They are all in the range of $\Lambda_C \geq 1.6$ TeV.

The *masses* of the excited leptons (l^*) and quarks (q^*) should not be lower than the compositeness scale Λ_C. Already in 1982 it was shown [Ren82] that precise measurements of the anomalous magnetic moment of the electron give bounds on the masses of the excited states and thus the compositeness scale.

Limits on the masses of excited leptons from accelerators are $m_{e^*} > 127$ GeV [Adr92], $m_{e^*,\nu^*} > 91$ GeV [ALE92], $m_{\nu^*} > 180$ GeV [Der95, Rau94] $m_{q^*} > 540$ GeV [Abe94].

A possible low energy manifestation of compositeness could be neutrinoless double beta decay, mediated by a composite heavy Majorana neutrino (Fig. 9), which then should be a Majorana particle.

Recent theoretical work shows (see [Pan96, Tak96, Pan97, Tak97]) that the mass bounds for such an excited neutrino which can be derived from double beta decay are at least of the same order of magnitude as those coming from the direct search of excited states in high energy accelerators (see also section 3).

2.5. Majorons

The existence of new bosons, so–called Majorons, can play a significant role in new physics beyond the standard model, in the history of the early universe, in the evolution of stellar objects, in supernovae astrophysics and the solar neutrino problem [Geo81, Fri88, Kla92]. In many theories of physics beyond the standard model neutrinoless double beta decay can occur with the emission of Majorons

$$2n \rightarrow 2p + 2e^- + \phi \tag{17}$$

$$2n \rightarrow 2p + 2e^- + 2\phi. \tag{18}$$

In the classical Majoron model invented by Gelmini and Roncadelli in '81 [Gel81], the Majoron is the Nambu–Goldstone boson associated with the spontaneous breaking of the $B - L$–symmetry and so generates Majorana masses of neutrinos. This was expected [Geo81] to give a sizeable contribution to double beta decay. It was, however, ruled out, as also the doublet Majoron [Aul82] by LEP [Ste91] since it should contribute the equivalent of two neutrino species to the width of the Z^0. On the other hand, Majoron models in which the Majoron is an electroweak isospin singlet [Chi81, Ber92] are still viable. The drawback of the singlet Majoron is that it requires a severe finetuning in order to preserve existing bounds on neutrino masses and at the same time get an observable rate for Majoron accompanied $0\nu\beta\beta$ decay.

To avoid such an unnatural fine–tuning in recent years several new Majoron models were proposed [Bur93, Bam95, Car93], where the term Majoron denotes in a more general sense light or massless bosons with couplings to neutrinos.

The main novel features of these "New Majorons" are that they can carry leptonic charge, that they need not be Goldstone bosons and that emission of two Majorons can occur. The latter can be scalar–mediated or fermion–mediated. Table 2 shows some features of the different Majoron models according to [Bam95, Car93]. L denotes the leptonic charge, n the spectral index defining the phase space of the emitting particles, M the nuclear matrix elements. For details we refer to [Päs96, Bur96].

The half–lifes are according to [Moh88, Doi85] in some approximation given by

$$[T_{1/2}]^{-1} = |<g_\alpha>|^2 \cdot |M_\alpha|^2 \cdot G_{BB_\alpha} \tag{19}$$

for $\beta\beta\phi$-decays, or

$$[T_{1/2}]^{-1} = |<g_\alpha>|^4 \cdot |M_\alpha|^2 \cdot G_{BB_\alpha} \tag{20}$$

for $\beta\beta\phi\phi$–decays. The index α indicates that effective neutrino–Majoron coupling constants g, matrix elements M and phase spaces G differ for different models.

Nuclear matrix elements:

According to Table 2 there are five different nuclear matrix elements. Of these M_F and M_{GT} are the same which occur in $0\nu\beta\beta$ decay. The other ones and the corresponding phase spaces have been calculated for the first time by [Päs96, Hir96b]. The calculation of the matrix elements show that the new models predict, as consequence of the small matrix elements very large half–lives and that unlikely large coupling constants would be needed to produce observable decay rates (see Table 3).

2.6. Sterile neutrinos

Introduction of sterile neutrinos has been claimed to solve simultaneously the conflict between dark matter neutrinos, LSND and supernova nucleosynthesis [Pel95] and light sterile neutrinos are part of popular neutrino mass textures for understanding the various hints for neutrino oscillations (see section 2.1) and [Moh96, Moh97, Moh97a]. Neutrinoless double beta decay can also investigate several effects of *heavy* sterile neutrinos [Bam95a].

If we assume having a light neutrino with a mass $\ll$ 1 eV, mixing with a much heavier (m $\geq$ 1 GeV) sterile neutrino can yield under certain conditions a detectable signal in current $\beta\beta$ experiments.

In models with two (or more) sterile neutrinos, the sterile neutrinos can mix appreciably even in the limit $m_{\nu_e} \to 0$ and so can be potentially visible in many processes [Pil93]. Neutrinoless double beta decay proceeds in these models through the virtual exchange of the heavier (i.e. GeV scale or higher) neutrinos. Fig. 10 shows the mass ranges leading to a $0\nu\beta\beta$ signal close to observability (shaded areas).

case	modus	Goldstone boson	L	n	Matrix element
IB	$\beta\beta\phi$	no	0	1	$M_F - M_{GT}$
IC	$\beta\beta\phi$	yes	0	1	$M_F - M_{GT}$
ID	$\beta\beta\phi\phi$	no	0	3	$M_{F\omega^2} - M_{GT\omega^2}$
IE	$\beta\beta\phi\phi$	yes	0	3	$M_{F\omega^2} - M_{GT\omega^2}$
IIB	$\beta\beta\phi$	no	-2	1	$M_F - M_{GT}$
IIC	$\beta\beta\phi$	yes	-2	3	M_{CR}
IID	$\beta\beta\phi\phi$	no	-1	3	$M_{F\omega^2} - M_{GT\omega^2}$
IIE	$\beta\beta\phi\phi$	yes	-1	7	$M_{F\omega^2} - M_{GT\omega^2}$
IIF	$\beta\beta\phi$	Gauge boson	-2	3	M_{CR}

Table 2 *Different Majoron models according to [Bam95]. The case IIF corresponds to the model of [Car93].*

model	$T_{1/2}(< g >= 10^{-4})$	$T_{1/2}(< g >= 1)$	$T_{1/2exp}$
IB,IC,IIB	$4 \cdot 10^{22}$	$4 \cdot 10^{14}$	$1.67 \cdot 10^{22}$
ID,IE,IID	10^{38-42}	10^{22-26}	$1.67 \cdot 10^{22}$
IIC,IIF	$2 \cdot 10^{28}$	$2 \cdot 10^{20}$	$1.67 \cdot 10^{22}$
IIE	10^{38-42}	10^{22-26}	$3.37 \cdot 10^{22}$

Table 3 *Comparison of half-lives calculated for different $< g >$-values for the new Majoron models with experimental best fit values, see section 3.1 (from [Hir96b])*

2.7. Leptoquarks

Interest on leptoquarks (LQ) has been renewed during the last few years since ongoing collider experiments have good prospects for searching these particles [Buc87]. LQs are vector or scalar particles carrying both lepton and baryon numbers and, therefore, have a well distinguished experimental signature. Direct searches of LQs in deep inelastic ep-scattering at HERA [H196] placed lower limits on their mass $M_{LQ} \geq 225 - 275$ GeV, depending on the LQ type and couplings.

In addition to the direct searches on LQs, there are many constraints which can be derived from the study of low-energy processes [Dav94]. Effective 4-fermion interactions, induced by virtual LQ exchange at energies much smaller than their masses, can contribute to atomic parity violation, flavour-changing neutral current processes, meson decays, meson-antimeson mixing and some rare processes.

To consider LQ phenomenology in a model-independent fashion one usually follows some general principles in constructing the Lagrangian of the LQ interactions with the standard model fields. In order to obey the stringent constraints from (c1) helicity-suppressed $\pi \to e\nu$ decay, from (c2) FCNC processes and from (c3) proton stability, the following assumptions

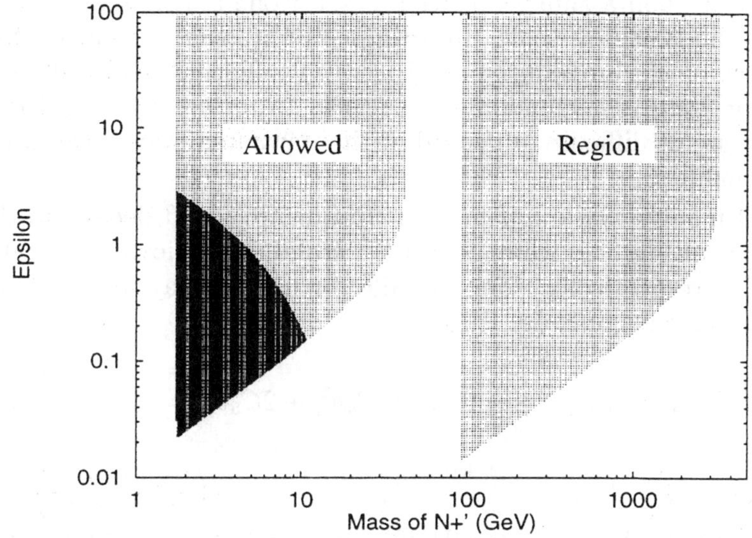

Fig. 10 *Regions of the parameter space ($\epsilon - M_{N'_+}$ plane yielding an observable signal (shaded areas) (from [Bam95a]). Darker area: 'natural' region, lighter shaded: Finetuning needed, to keep m_{ν_e} below 1 eV. $M_{N'_+}$: mass eigenstate, ϵ: strength of lepton number violation in mass matrix*

are commonly adopted: (a1) LQ couplings are chiral, (a2) LQ couplings are generation diagonal, and (a3) there are no diquark couplings.

Recently, however, it has been pointed out [Hir96a] that possible LQ-

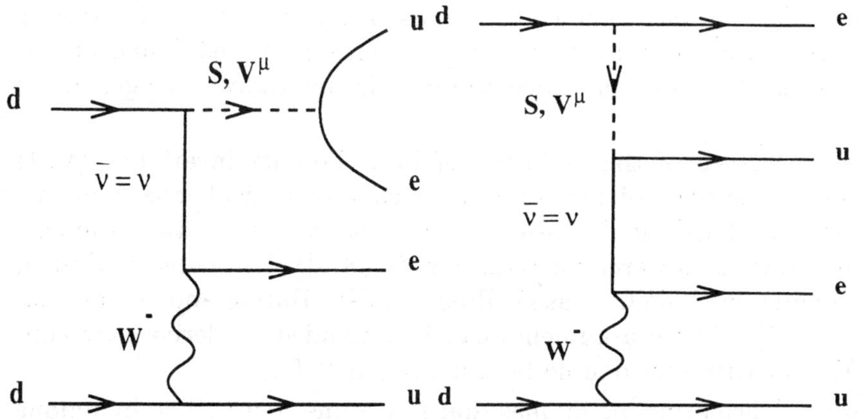

Fig. 11 *Examples of Feynman graphs for $0\nu\beta\beta$ decay within LQ models. S and V^μ stand symbolically for scalar and vector LQs, respectively (from [Hir96a]).*

Higgs interactions spoil assumption (a1): Even if one assumes LQs to be chiral at some high energy scale, LQ-Higgs interactions introduce after electro-weak symmetry breaking mixing between LQ states with different chirality. Since there is no fundamental reason to forbid such LQ-Higgs interactions, it seems difficult to get rid of the unwanted non-chiral interactions in LQ models.

In such LQ models there appear contributions to $0\nu\beta\beta$ decay via the Feynman graphs of Fig. 11. Here, S and V^{μ} stand symbolically for scalar and vector LQs, respectively. The half–life for $0\nu\beta\beta$ decay arising from leptoquark exchange is given by [Hir96a]

$$T^{0\nu}_{1/2} = |M_{GT}|^2 \frac{2}{G_F^2} [\tilde{C}_1 a^2 + C_4 b_R^2 + 2C_5 b_L^2]. \tag{21}$$

with $a = \frac{\epsilon_S}{M_S^2} + \frac{\epsilon_V}{M_V^2}$, $b_{L,R} = \frac{\alpha_S^{(L,R)}}{M_S^2} + \frac{\alpha_V^{(L,R)}}{M_V^2}$, $\tilde{C}_1 = C_1 \left(\frac{\mathcal{M}_1^{(\nu)}/(m_e R)}{M_{GT} - \alpha_2 M_F} \right)^2$.

For the definition of the C_n see [Doi85] and for the calculation of the matrix element $\mathcal{M}_1^{(\nu)}$ see [Hir96a]. This allows to deduce information on leptoquark masses and leptoquark–Higgs couplings (see section 3.2).

2.8. Special Relativity and Equivalence Principle

Special relativity and the equivalence principle can be considered as the most basic foundations of the theory of gravity. Many experiments already have tested these principles to a very high level of accuracy [Hug60] for ordinary matter - generally for quarks and leptons of the first generation. These precision tests of local Lorentz invariance – violation of the equivalence principle should produce a similar effect [Wil92] – probe for any dependence of the (non–gravitational) laws of physics on a laboratory's position, orientation or velocity relative to some preferred frame of reference, such as the frame in which the cosmic microwave background is isotropic.

A typical feature of the violation of local Lorentz invariance (VLI) is that different species of matter have a characteristical maximum attainable speed. This can be tested in various sectors of the standard model through vacuum Cerenkov radiation [Gas89a], photon decay [Col97], neutrino oscillations [Gla97, Gas89, Hal91, Hal96, But93] and $K-$physics [Ham98, Goo61]. These arguments can be extended to derive new constraints from neutrinoless double beta decay [Kla98f].

The equivalence principle implies that spacetime is described by unique operational geometry and hence universality of the gravitational coupling for all species of matter. In the recent years there have been attempts to constrain a possible amount of violation of the equivalence principle (VEP) in the neutrino sector from neutrino oscillation experiments

[Gas89, Hal91, Hal96, But93]. However, these bounds do not apply when the gravitational and the weak eigenstates have small mixing. In a recent paper [Kla98f] a generalized formalism of the neutrino sector has been given to test the VEP and it has been shown that neutrinoless double beta decay also constrains the VEP. VEP implies different neutrino species to suffer from different gravitational potentials while propagating through the nucleus and hence the effect of different eigenvalues doesn't cancel for the same effective momentum. The main result is that neutrinoless double beta decay can constrain the amount of VEP even when the mixing angle is zero, *i.e.*, when only the weak equivalence principle is violated, for which there does not exist any bound at present.

3. Double Beta Decay Experiments: Present Status and Results

3.1. Present Experimental Status

Fig. 12 shows an overview over measured $0\nu\beta\beta$ half–life limits and deduced mass limits. The largest sensitivity for $0\nu\beta\beta$ decay is obtained at present by active source experiments (source=detector), in particular ^{76}Ge [HM95, HM97, Kla94, Kla97] and ^{136}Xe [Ger96]. The main reason is that large source strengths can be used (simultaneously with high energy resolution), in particular when enriched $\beta\beta$ emitter materials are used. Geochemical experiments, though having contributed important information to double beta decay, have no more future in the sense that their inherent background from $2\nu\beta\beta$ decay cannot be eliminated.

Only a few of the present most sensitive experiments may probe the neutrino mass in the next years into the sub–eV region, the Heidelberg–Moscow experiment being the by far most advanced and most sensitive one, see Fig. 12b. No one of them will pass, however, below $\sim 0.1 - 0.2$ eV (see section 4.1) A detailed discussion of the various experimental possibilities can be found in [Kla95, Kla96, Kla96a]. A useful listing of existing data from the various $\beta\beta$ emitters is given in [Tre95].

3.2. Present limits on beyond standard model parameters

The sharpest limits from $0\nu\beta\beta$ decay are presently coming from the Heidelberg–Moscow experiment [Kla87, HM95, Kla94, HM97, Kla97, Kla98a, Kla98b]. They will be given in the following. With five enriched (86% of ^{76}Ge) detectors of a total mass of 11.5 kg taking data in the Gran Sasso underground laboratory, and with a background of at present 0.06 counts/kg year keV, the experiment has reached its final setup and is now exploring the sub–eV range for the mass of the electron neutrino. Fig. 13 shows the spectrum taken in a measuring time of 42 kg y.

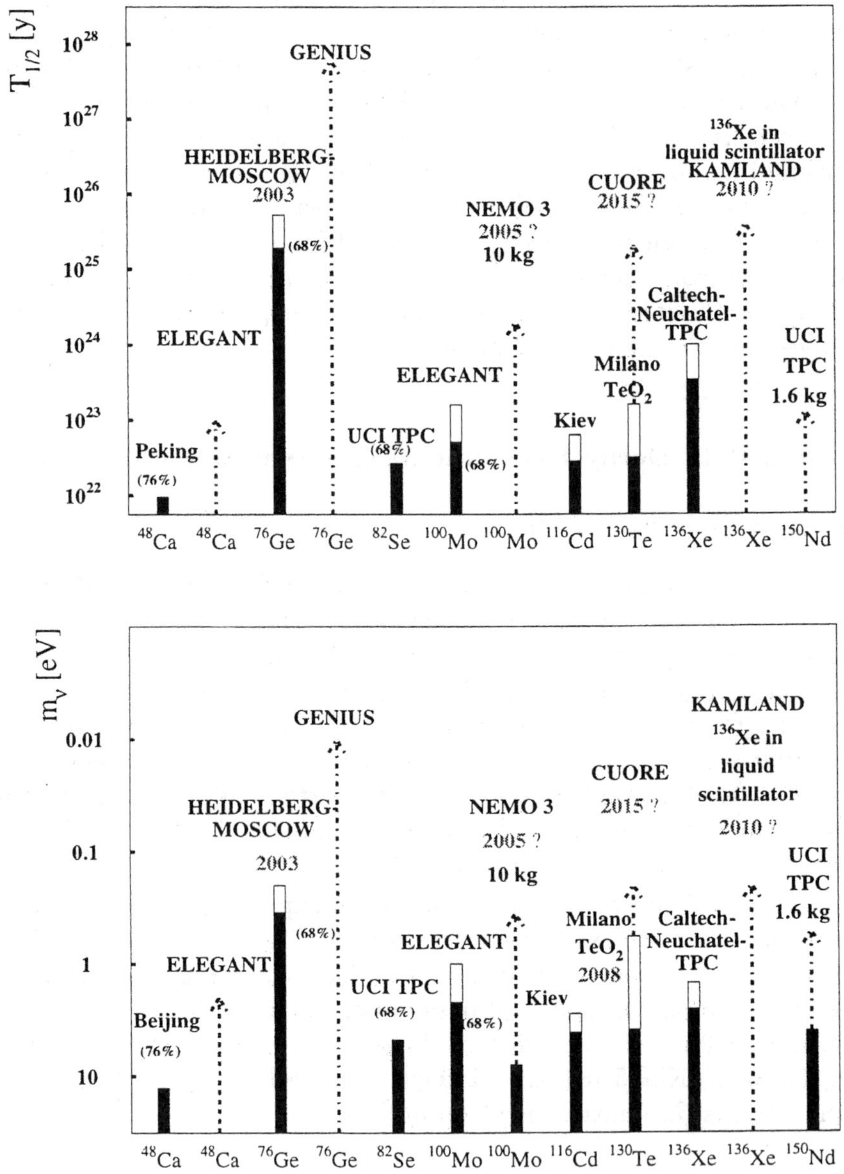

Fig. 12 *Present situation, 1998, and expectation for the near future and beyond, of the most promising $\beta\beta$-experiments concerning accessible half life (a) and neutrino mass limits (b). The filled bars correspond to the present status, open bars to expectations for running experiments, dashed lines to experiments under construction and dash-dotted lines to proposed experiments.*

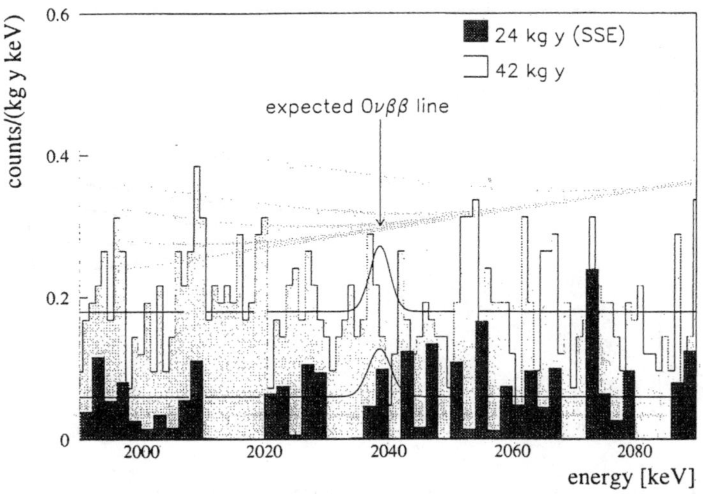

Fig. 13 *Integral spectrum in the region of interest after subtraction of the first 200 days of measurement of each detector, leaving 42 kg y of measuring time. The darkened histogram corresponds to data accumulated meanwhile using a new pulse shape analysis method (SSE) [Hel96] in a measuring time of 24 kg y. The two solid curves correspond to the signal excluded with* 90%*C.L. (by the 42 kg y measurement and the 24 kg y SSE measurement). They correspond to* $T^{0\nu}_{1/2} > 1.3 \cdot 10^{25}$ *y (upper curve) and* $T^{0\nu}_{1/2} > 1.6 \cdot 10^{25}$ *y, respectively.*

Half–life of neutrinoless double beta decay

The deduced half–life limit for $0\nu\beta\beta$ decay is, for the full data,

$$T^{0\nu}_{1/2} > 1.3 \cdot 10^{25} y \quad (90\% C.L.) \tag{22}$$

$$> 2.2 \cdot 10^{25} y \quad (68\% C.L.) \tag{23}$$

and for the 24 kg y of measurement with pulse shape analysis

$$T^{0\nu}_{1/2} > 1.6 \cdot 10^{25} y \quad (90\% C.L.) \tag{24}$$

$$> 2.8 \cdot 10^{25} y \quad (68\% C.L.) \tag{25}$$

Neutrino mass

Light neutrinos: The deduced upper limit of an (effective) electron neutrino Majorana mass is, with the matrix element from [Sta90]

$$\langle m_\nu \rangle < 0.42 eV \quad (90\% C.L.) \tag{26}$$

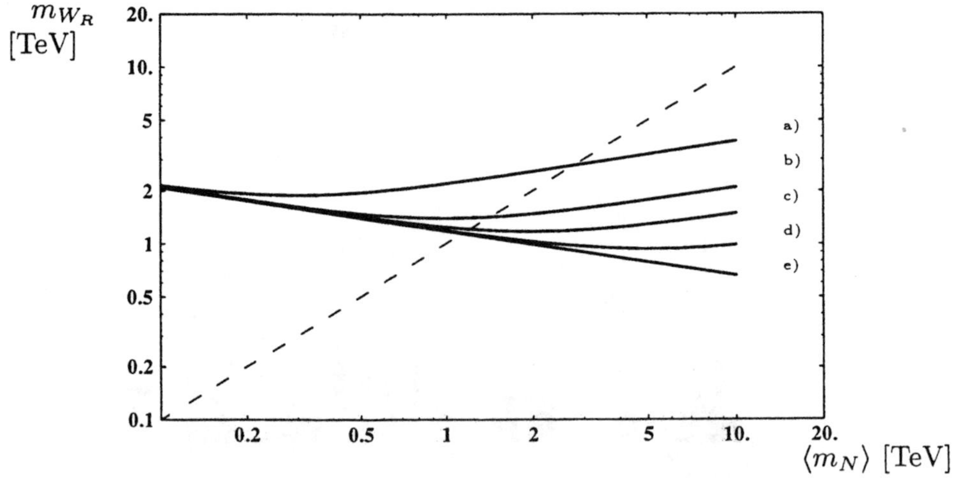

Fig. 14 *Limits on the mass of the right-handed W-boson from neutrinoless double beta decay (full lines) and vacuum stability (dashed line). The five full lines correspond to the following masses of the doubly charged higgs, $m_{\Delta^{--}}$: a) 0.3, b) 1.0, c) 2.0, d) 5.0 and e) ∞ [TeV] (from [Hir96d]).*

$$< 0.33 eV \quad (68\% C.L.) \tag{27}$$

and from the 24 kg y with pulse shape analysis (SSE)

$$\langle m_\nu \rangle < 0.38 eV \quad (90\% C.L.) \tag{28}$$

$$< 0.29 eV \quad (68\% C.L.) \tag{29}$$

This is the sharpest limit for a Majorana mass of the electron neutrino so far.

Superheavy neutrinos: For a superheavy *left*–handed neutrino we deduce [HM95] exploiting the mass dependence of the matrix element (for the latter see [Mut89, Bel96, Bel98]) a lower limit

$$\langle m_H \rangle \geq 100 TeV. \tag{30}$$

For a heavy *right*–handed neutrino the relation obtained to the mass of the right–handed W boson is shown in Fig. 14 (see [Hir96d]).

Right–handed W boson

For the right–handed W boson a lower limit of (Fig. 14)

$$m_{W_R} \geq 1.2TeV \tag{31}$$

is obtained [Hir96d].

SUSY parameters – R–parity breaking and sneutrino mass

The constraints on the parameters of the minimal supersymmetric standard model with explicit R–parity violation deduced [Hir95, Hir96c, Hir96] from the $0\nu\beta\beta$ half–life limit are more stringent than those from other low–energy processes and from the largest high energy accelerators (Fig. 15). The limits are

$$\lambda'_{111} \leq 3.9 \cdot 10^{-4} \Big(\frac{m_{\tilde{q}}}{100GeV}\Big)^2 \Big(\frac{m_{\tilde{g}}}{100GeV}\Big)^{\frac{1}{2}} \tag{32}$$

with $m_{\tilde{q}}$ and $m_{\tilde{g}}$ denoting squark and gluino masses, respectively, and with the assumption $m_{\tilde{d}_R} \simeq m_{\tilde{u}_L}$. This result is important for the discussion of new physics in the connection with the high–Q^2 events seen at HERA. It excludes the possibility of squarks of first generation (of R–parity violating SUSY) being produced in the high–Q^2 events [Cho97, Alt97, Hir97b].

We find further [Hir96]

$$\lambda'_{113}\lambda'_{131} \leq 1.1 \cdot 10^{-7} \tag{33}$$

$$\lambda'_{112}\lambda'_{121} \leq 3.2 \cdot 10^{-6}. \tag{34}$$

For the $(B-L)$ violating sneutrino mass $\tilde{m}_M$ the following limits are obtained [Hir98a]

$$\tilde{m}_M \leq 2\Big(\frac{m_{SUSY}}{100GeV}\Big)^{\frac{3}{2}} GeV, \quad \chi \simeq \tilde{B} \tag{35}$$

$$\tilde{m}_M \leq 11\Big(\frac{m_{SUSY}}{100GeV}\Big)^{\frac{7}{2}} GeV, \quad \chi \simeq \tilde{H} \tag{36}$$

for the limiting cases that the lightest neutralino is a pure Bino $\tilde{B}$, as suggested by the SUSY solution of the dark matter problem [Jun96], or a pure Higgsino. Actual values for $\tilde{m}_M$ for other choices of the neutralino composition should lie in between these two values.

Another way to deduce a limit on the 'Majorana' sneutrino mass $\tilde{m}_M$ is to start from the experimental neutrino mass limit, since the sneutrino contributes to the Majorana neutrino mass m_M^ν at the 1–loop level proportional to $\tilde{m}_M^2$. This yields under some assumptions [Hir98a]

$$\tilde{m}_{M_{(i)}} \leq (60-125)\Big(\frac{m_{\nu(i)}^{exp}}{1eV}\Big)^{1/2} MeV \tag{37}$$

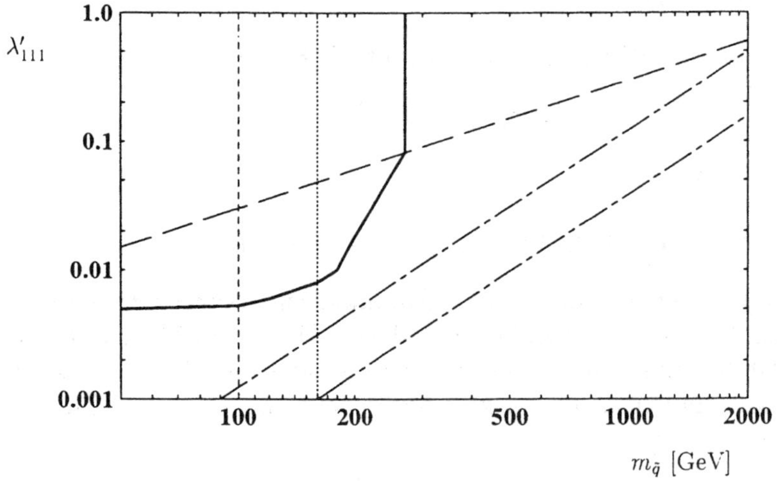

Fig. 15 *Comparison of limits on the R–parity violating MSSM parameters from different experiments in the* λ'_{111}*–*$m_{\tilde{q}}$ *plane. The dashed line is the limit from charged current universality according to [Bar89]. The vertical line is the limit from the data of Tevatron [Roy92]. The thick full line is the region which might be explored by HERA [But93a]. The two dash–dotted lines to the right are the limits obtained from the half–life limit for* $0\nu\beta\beta$ *decay of* ^{76}Ge*, for gluino masses of (from left to right)* $m_{\tilde{g}}$ *=1TeV and 100 GeV, respectively. The regions to the upper left of the lines are forbidden. (from [Hir95])*

Starting from the mass limit determined for the electron neutrino by $0\nu\beta\beta$ decay this leads to

$$\tilde{m}_{M_{(e)}} \leq 22MeV \tag{38}$$

This result is somewhat dependent on neutralino masses and mixings. A non–vanishing 'Majorana' sneutrino mass would result in new processes at future colliders, like sneutrino–antisneutrino oscillations. Reactions at the Next Linear Collider (NLC) like the SUSY analog to inverse neutrinoless double beta decay $e^-e^- \rightarrow \chi^-\chi^-$ (where χ^- denote charginos) or single sneutrino production, e.g. by $e^-\gamma \rightarrow \tilde{\nu}_e\chi^-$ could give information on the Majorana sneutrino mass, also. This is discussed by [Hir97, Hir98a, Hir98]. A conclusion is that future accelerators can give information on second and third generation sneutrino Majorana masses, but for first generation sneutrinos cannot compete with $0\nu\beta\beta$–decay.

Compositeness

Evaluation of the $0\nu\beta\beta$ half–life limit assuming exchange of excited Majorana neutrinos ν^* yields for the mass of the excited neutrino a lower bound of [Pan97, Tak97].

$$m_N \geq 3.4 m_W \tag{39}$$

for a coupling of order O(1) and $\Lambda_c \simeq m_N$. Here, m_W is the W–boson mass.

Leptoquarks

Assuming that either scalar or vector leptoquarks contribute to $0\nu\beta\beta$ decay, the following constraints on the effective LQ parameters (see section 2.7) can be derived [Hir96a]:

$$\epsilon_I \leq 2.8 \times 10^{-9} \left(\frac{M_I}{100\text{GeV}}\right)^2, \tag{40}$$

$$\alpha_I^{(L)} \leq 3.5 \times 10^{-10} \left(\frac{M_I}{100\text{GeV}}\right)^2, \tag{41}$$

$$\alpha_I^{(R)} \leq 7.9 \times 10^{-8} \left(\frac{M_I}{100\text{GeV}}\right)^2. \tag{42}$$

Here, different effective LQ couplings have been introduced. They are defined as:

$$\epsilon_I = 2^{-\eta_I} \left[\lambda_{I_1}^{(L)} \lambda_{\tilde{I}_{1/2}}^{(R)} \left(\theta_{43}^I(Q_I^{(1)}) + \eta_I \sqrt{2} \theta_{41}^I(Q_I^{(2)})\right) - \lambda_{I_0}^{(L)} \lambda_{\tilde{I}_{1/2}}^{(R)} \theta_{13}^I(Q_I^{(1)})\right] \tag{43}$$

$$\alpha_I^{(L)} = \frac{2}{3+\eta_I} \lambda_{I_{1/2}}^{(L)} \lambda_{I_1}^{(L)} \theta_{24}^I(Q_I^{(2)}) \tag{44}$$

$$\alpha_I^{(R)} = \frac{2}{3+\eta_I} \lambda_{I_0}^{(R)} \lambda_{\tilde{I}_{1/2}}^{(R)} \theta_{23}^I(Q_I^{(1)}). \tag{45}$$

$\eta_{S,V} = 1, -1$ for scalar and vector LQs. $\theta_{kn}^I(Q)$ is a mixing parameter defined by

$$\theta_{kn}^I(Q) = \sum_l \mathcal{N}_{kl}^{(I)}(Q) \mathcal{N}_{nl}^{(I)}(Q) \left(\frac{M_I}{M_{I_l}(Q)}\right)^2, \tag{46}$$

where $\mathcal{N}^{(I)}(Q)$ are mixing matrix elements which diagonalize the LQ mass matrices for the scalar $I = S$ and vector $I = V$ LQ fields with electric charges $Q = -1/3, -2/3$, for complete definitions see [Hir96a]. Common mass scales M_S of scalar and M_V of vector LQs are introduced for convenience.

Since the LQ mass matrices appearing in $0\nu\beta\beta$ decay are (4×4) matrices [Hir96a], it is difficult to solve their diagonalization in full generality algebraically. However, if one assumes that only one LQ-Higgs coupling is present at a time, the (mathematical) problem is simplified greatly and one can deduce from, for example, eq. (40) that either the LQ-Higgs coupling must be smaller than $\sim 10^{-(4-5)}$ or there can not be any LQ with e.g. couplings of electromagnetic strength with masses below $\sim 250 GeV$. These bounds from $\beta\beta$ decay are of interest in connection with recently discussed evidence for new physics from HERA [Hew97, Bab97, Kal97, Cho97]. Assuming that actually leptoquarks have been produced at HERA, double beta decay (the Heidelberg–Moscow experiment) would allow to fix the leptoquark–Higgs coupling to a few 10^{-6} [Hir97b]. It may be noted, that after the first consideration of leptoquark–Higgs coupling in [Hir96a] recently Babu et al. [Bab97b] noted that taking into account leptoquark–Higgs coupling reduces the leptoquark mass lower bound deduced by TEVATRON – making it more consistent with the value of 200 GeV required by HERA.

Special Relativity and Equivalence Principle

Violation of Lorentz invariance (VLI): The bound obtained from the Heidelberg–Moscow experiment is

$$\delta v < 4 \times 10^{-16} \quad \text{for} \quad \theta_v = \theta_m = 0 \tag{47}$$

where $\delta v = v_1 - v_2$ is the measure of VLI in the neutrino sector. θ_v and θ_m denote the velocity mixing angle and the weak mixing angle, respectively. In Fig. 16 (from [Kla98f]) the bound implied by double beta decay is presented for the entire range of $sin^2(2\theta_v)$, and compared with bounds obtained from neutrino oscillation experiments (see [Hal96]).

Violation of equivalence principle (VEP): Assuming only violation of the weak equivalence principle, there does not exist any bound on the amount of VEP. It is this region of the parameter space which is most restrictively bounded by neutrinoless double beta decay. In a linearized theory the gravitational part of the Lagrangian to first order in a weak gravitational field $g_{\mu\nu} = \eta_{\mu\nu} + h_{\mu\nu}$ ($h_{\mu\nu} = 2\frac{\phi}{c^2} diag(1,1,1,1)$) can be written as $\mathcal{L} = -\frac{1}{2}(1+g_i)h_{\mu\nu}T^{\mu\nu}$, where $T^{\mu\nu}$ is the stress-energy in the gravitational eigenbasis. In the presence of VEP the g_i may differ. We obtain [Kla98f] the following bound from the Heidelberg–Moscow experiment, for $\theta_v = \theta_m = 0$:

$$\begin{aligned} \phi\delta g &< 4 \times 10^{-16} \text{ (for } \bar{m} < 13\text{eV)} \\ \phi\delta g &< 2 \times 10^{-18} \text{ (for } \bar{m} < 0.08\text{eV).} \end{aligned} \tag{48}$$

Here $\bar{g} = \frac{g_1+g_2}{2}$ can be considered as the standard gravitational coupling, for which the equivalence principle applies. $\delta g = g_1 - g_2$. The bound on the VEP thus, unlike the one for VLI, will depend on the choice for the Newtonian potential ϕ.

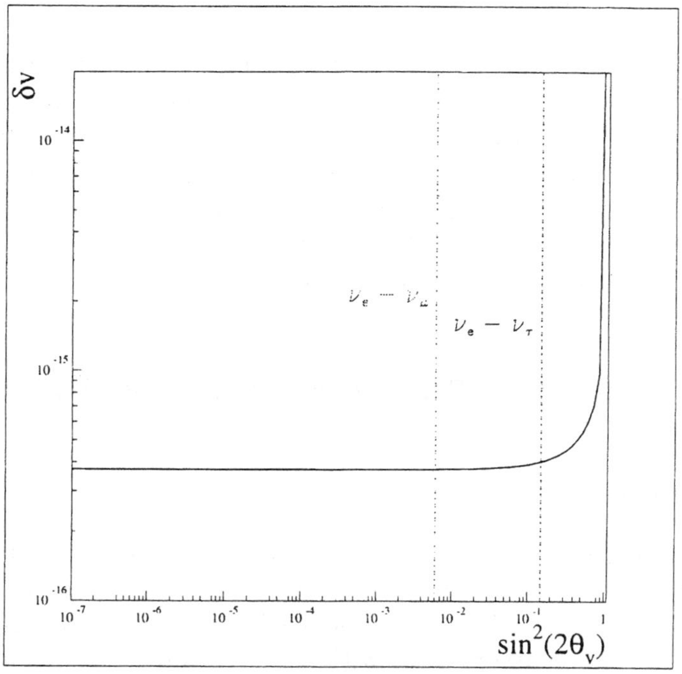

Fig. 16 *Double beta decay bound (solid line) on violation of Lorentz invariance in the neutrino sector, excluding the region to the upper left. Shown is a double logarithmic plot in the* δv*–*$\sin^2(2\theta)$ *parameter space. The bound becomes most stringent for the small mixing region, which has not been constrained from any other experiments. For comparison the bounds obtained from neutrino oscillation experiments (from [Hal96]) in the* $\nu_e - \nu_\tau$ *(dashed lines) and in the* $\nu_e - \nu_\mu$ *(dashed-dotted lines) channel, excluding the region to the right, are shown (from [Kla98f]).*

Half–life of $2\nu\beta\beta$ *decay*

The Heidelberg–Moscow experiment produced for the first time a high statistics $2\nu\beta\beta$ spectrum ($\sim$ 20000 counts, to be compared with the 40 counts on which the first detector observation of $2\nu\beta\beta$ decay by [Ell87] (for the decay of ^{82}Se) had to rely. The deduced half–life is [HM97]

$$T^{2\nu}_{1/2} = (1.77^{+0.01}_{-0.01}(stat.)^{+0.13}_{-0.11}(syst.)) \cdot 10^{21} y \qquad (49)$$

This result brings $\beta\beta$ research for the first time into the region of 'normal' nuclear spectroscopy and allows for the first time statistically reliable investigation of Majoron–accompanied decay modes.

278

Majoron–accompanied decay

From simultaneous fits of the 2ν spectrum and one selected Majoron mode, experimental limits for the half-lives of the decay modes of the newly introduced Majoron models [Bur96] are given for the first time [Päs96, HM96].

The small matrix elements and phase spaces for these modes [Päs96, Hir96b] already determined that these modes by far cannot be seen in experiments of the present sensivity if we assume typical values for the neutrino–Majoron coupling constants around $\langle g \rangle = 10^{-4}$ (see table 3).

4. Double Beta Experiments: Future Perspectives – the GENIUS Project

4.1. The known experiments and proposals

Figs. 12a,b show in addition to the present status the future perspectives of the main existing $\beta\beta$ decay experiments and includes some ideas for the future which have been published. The HEIDELBERG–MOSCOW experiment will probe the neutrino mass within 5 years down to the order of 0.1 eV. The best presently existing limits besides the HEIDELBERG-MOSCOW experiment (filled bars in Fig. 12), have been obtained with the isotopes: ^{48}Ca [You95], ^{82}Se [Ell92], ^{100}Mo [Als89], ^{116}Cd [Dan95], ^{130}Te [Ale94], ^{136}Xe [Vui93] and ^{150}Nd [Moe94]. These and other double beta decay setups presently under construction or partly in operation such as NEMO [NEM94, Bar97], the Gotthard ^{136}Xe TPC experiment [Jör94], the ^{130}Te cryogenic experiment [Ale94], a new ELEGANT ^{48}Ca experiment using 30 g of ^{48}Ca [Kum96], a hypothetical experiment with an improved UCI TPC [Moe94] assumed to use 1.6 kg of ^{136}Xe, etc., will not reach or exceed the ^{76}Ge limits. The goal 0.3 eV aimed at for the year 2004 by the NEMO experiment (see [Piq96, Bar97] and Fig. 12) may even be very optimistic if claims about the effect of proton-neutron pairing on the $0\nu\beta\beta$ nuclear matrix elements by [Pan96b] will turn out to be true, and also if the energy resolution will not be improved considerably (see Fig. 1 in [Tre95]). Therefore, the conclusion given by [Bed97c] concerning the future SUSY potential of NEMO has no serious basis. As pointed out by Raghavan [Rag94], even use of an amount of about 200 kg of enriched ^{136}Xe or 2 tons of natural Xe added to the scintillator of the KAMIOKANDE detector or similar amounts added to BOREXINO (both primarily devoted to solar neutrino investigation) would hardly lead to a sensitivity larger than the present ^{76}Ge experiment. This idea is going to be realized at present by the KAMLAND experiment [Suz97]. An interesting future candidate was for some time a ^{150}Nd bolometer exploiting the relatively large phase space of this nucleus (see [Moe94]). The way outlined by [Moe91] proposing a

TPC filled with 1 ton of liquid enriched ^{136}Xe and identification of the daughter by laser fluorescence seems not be feasible in a straight-forward way. However, another way of using liquid ^{136}Xe may be more promising [Cli97].

It is obvious that, from the experiments and proposals, the HEIDELBERG-MOSCOW experiment will give the sharpest limit for the electron neutrino mass for the next decade. It is also obvious from Fig. 12 that *none* of the present experimental approaches, or plans or even vague ideas has a chance to surpass the border of 0.1 eV for the neutrino mass to lower values (see also [Nor97]). At present there is only one way visible to reach the domain of lower neutrino masses, suggested by [Kla98a] and meanwhile investigated in some detail concerning its experimental realization and and physics potential in [Kla97d, Hel97, Kla98b, Kla98c].

4.2. Genius – A Future Large Scale Double Beta and Dark Matter Experiment

The idea of GENIUS is to use a large amount of 'naked' enriched **GE**rmanium detectors in liquid **NI**trogen as shielding in an **U**nderground **S**etup. Use of 1 (in an extended version 10) tons of enriched ^{76}Ge will increase the source strength largely, removing all material from the vicinity of the detectors and shielding by liquid nitrogen will lead to a drastic background reduction compared to the present level. Using Ge detectors in liquid nitrogen has been discussed already earlier [Heu95]. That Ge detectors can be operated in liquid nitrogen has been demonstrated recently in the Heidelberg low level laboratory [Hel97]. The natural site for GENIUS would be the Gran Sasso underground laboratory. The cost of the project would be a minor fraction of detectors prepared for LHC physics as CMS or ATLAS. We give in the next two subsections some results of Monte Carlo simulations of the setup [Hel97] and some estimates of the physics potential [Kla97d] (see also [Kla98b, Kla98c]).

4.2.1. Realization and Sensitivity of GENIUS A simplified model of GENIUS is shown in Fig. 17 consisting of about 300 enriched ^{76}Ge detectors with a total of one ton mass in the center of a 9 m high liquid nitrogen tank with 9 m diameter. Figs. 18, 19 show the results of Monte Carlo simulations, using the CERN GEANT code, of the background [Hel97], starting from purity levels of the nitrogen being in general an order of magnitude less stringent than those already achieved in the CTF for the BOREXINO experiment. The influence of muons penetrating the Gran Sasso rock on the background can be reduced comfortably through coincidences between the Germanium detectors from the muon induced showers. The count rate in the region of interest for neutrinoless double beta decay is 0.04 counts/ keV · y · ton (Fig. 18). Below 100 keV the background count rate is about

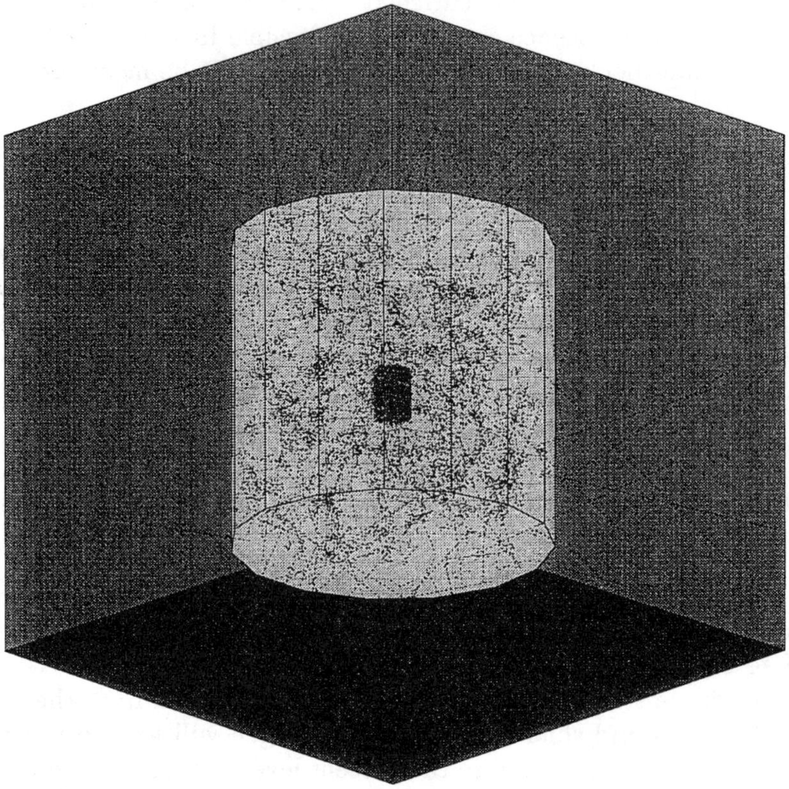

Fig. 17 *Simplified model of the GENIUS experiment: 288 enriched ^{76}Ge detectors with a total of one ton mass in the center of a 9 m high liquid nitrogen tank with 9 m diameter; GEANT Monte Carlo simulation of 1000 2.6 MeV photons randomly distributed in the nitrogen is also shown.*

10 counts/keV · y · ton. Two neutrino double beta decay would dominate the spectrum with $4 \cdot 10^6$ events per year.

Starting from these numbers, a lower half–life limit of

$$T_{1/2}^{0\nu} \geq 5.8 \cdot 10^{27} \qquad (68\% C.L.) \tag{50}$$

can be reached within one year of measurement (following the highly conservative procedure for analysis recommended by [PDG94], which has been used also in the derivation of the results given in section 3.2, but is not used in the analysis of several other $\beta\beta$ experiments). This corresponds – with the matrix elements of [Sta90] – to an upper limit on the neutrino mass of

$$\langle m_\nu \rangle \leq 0.02 eV \qquad (68\% C.L.) \tag{51}$$

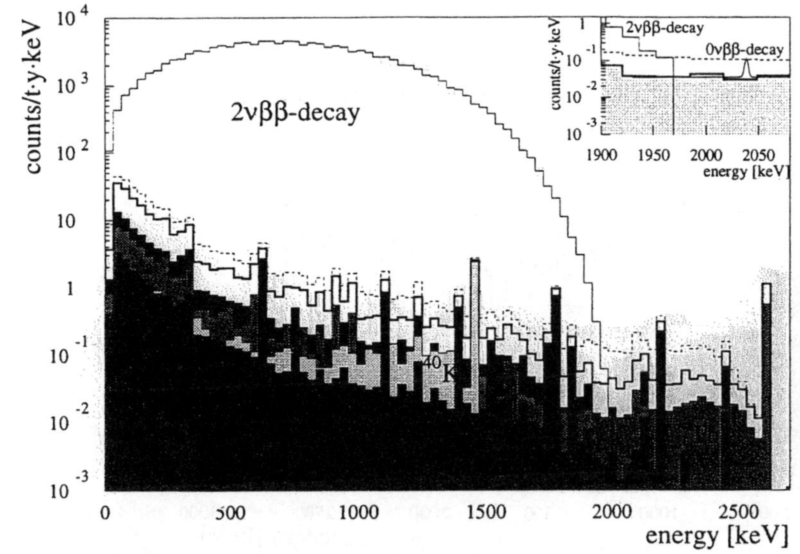

Fig. 18 *Monte Carlo simulation of the background of GENIUS. Simulations of U/Ra, U/Th and ^{40}K (shaded), ^{222}Rn (black histogram) activities in the liquid nitrogen; the sum of the activities is shown with anticoincidence between the 288 detectors (thick line) and without (dashed line); the $2\nu\beta\beta$-decay dominates the spectrum with 4 million events per year (from [Kla98c]).*

Figure 20 shows the obtainable limits on the neutrino mass in the case of zero background. This assumption might be justified since our assumed impurity concentrations are still more conservative than proved already now for example by Borexino. The final sensitivity of the experiment can be defined by the limit, which would be obtained after 10 years of measurement. For the one ton experiment this would be:

$$T_{1/2}^{0\nu} \geq 6.4 \cdot 10^{28}\, y \quad \text{(with 68\% C.L.)} \tag{52}$$

and

$$\langle m_\nu \rangle \leq 0.006 eV \quad \text{(with 68\% C.L.)} \tag{53}$$

The ultimate experiment could test the $0\nu\beta\beta$ half life of ^{76}Ge up to a limit of $5.7{\cdot}10^{29}$y and the neutrino mass down to $2{\cdot}10^{-3}$eV using 10 tons of enriched Germanium.

4.2.2. The Physics Potential of GENIUS
Neutrino mass textures and neutrino oscillations: GENIUS will allow a

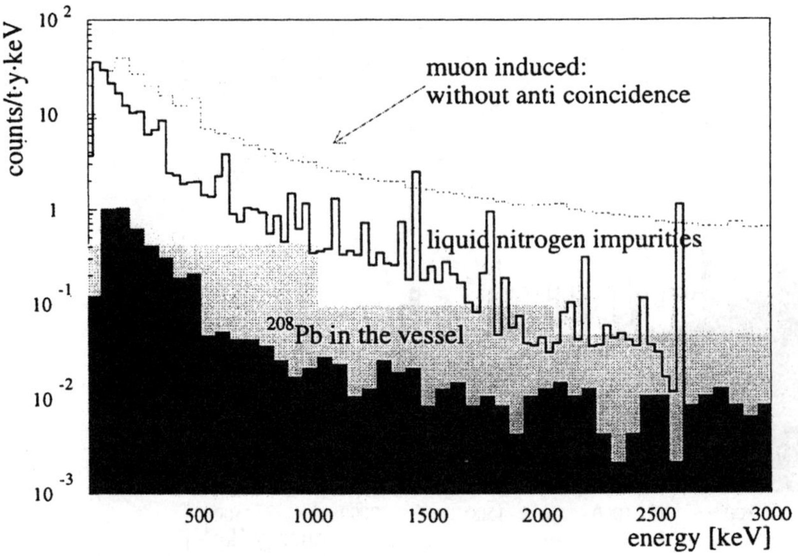

Fig. 19 *Background from outside the nitrogen: 200 GeV muons induced events (dashed line) and single hit events (filled histogram); decay of* ^{208}Tl *in the steel vessel (light shaded histogram) and the background originating from the nitrogen impurities for comparison (thick line) (from [Kla98c])*

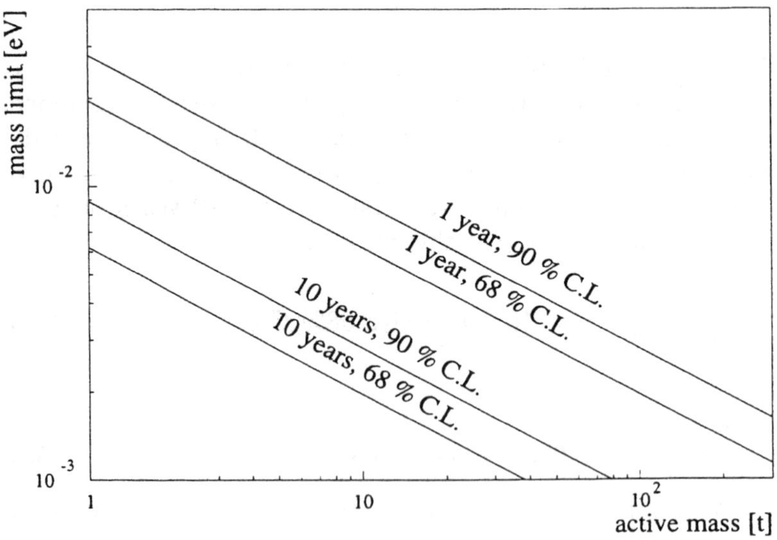

Fig. 20 *Mass limits on Majorana neutrino mass after one and ten years measuring time as function of the active detector mass; zero background is assumed (from [Hel97])*

large step in sensitivity for probing the neutrino mass. It will allow to probe the neutrino mass down to $10^{-(2-3)}$ eV, and thus surpass the existing neutrino mass experiments by a factor of 50-500. GENIUS will test the structure of the neutrino mass matrix and thereby also neutrino oscillation parameters [1] superior in sensitivity to the best proposed dedicated terrestrial neutrino oscillation experiments. Even in the first stage GENIUS will confirm or rule out degenerate or inverted neutrino mass scenarios, discussed in the literature as possible solutions of current hints to finite neutrino masses , and also test the $\nu_e \leftrightarrow \nu_\mu$ hypothesis of the atmospheric neutrino problem. If the 10^{-3} eV level is reached, GENIUS will even allow to test the large angle MSW solution of the solar neutrino problem. It will also allow to test the hypothesis of a shadow world underlying introduction of a sterile neutrino mentioned in section 2.1. The figures 21–25 show some examples of this potential. Fig. 21 compares the potential of GENIUS with the sensitivity of CHORUS/NOMAD and with the proposed future experiments NAUSIKAA–CERN and NAUSIKAA–FNAL, looking for $\nu_e \leftrightarrow \nu_\tau$ oscillations, for different assumptions on m_1/m_2.

Already in the worst case for double beta decay of $m_1/m_2 = 0$ GENIUS 1 ton is more sensitive than the running CERN experiments. For quasi–degenerate models, for example $R = 0.01$ already, GENIUS 1 ton would be more sensitive than all currently planned future accelerator neutrino experiments.

The situation of $\nu_e \leftrightarrow \nu_\mu$ oscillations (assuming $sin^2\theta_{13} = 0$) is shown in Fig. 22. The original figure is taken from [Gel95]. While the GENIUS 1 ton sensitivity is sufficient (even in the worst case of $m_{\nu_e} << m_{\nu_\mu}$) to extend to smaller values of Δm^2 at large mixing angles, GENIUS 10 ton would have a sensitivity better than all existing or planned oscillation experiments, at least at large $sin^2 2\theta$. In the quasi–degenerate models GENIUS would be much more sensitive – similar to the cases shown in Fig. 21. Fig. 23 (background from [Gel95]) compares the double beta worst

[1] The double beta observable, the effective neutrino mass (eq. 10), can be expressed in terms of the usual neutrino oscillation parameters, once an assumption on the ratio of m_1/m_2 is made. E.g., in the simplest two–generation case

$$\langle m_\nu \rangle = |c_{12}^2 m_1 + s_{12}^2 m_2 e^{2i\beta}|, \tag{54}$$

assuming CP conservation, i.e. $e^{2i\beta} = \eta = \pm 1$, and $c_{12}^2 m_1 << \eta s_{12}^2 m_2$,

$$\Delta_{m_{12}}^2 \simeq m_2^2 = \frac{4\langle m_\nu \rangle^2}{1 - \sqrt{1 - sin^2 2\theta}} \tag{55}$$

A little bit more general, keeping corrections of the order (m_1/m_2) one obtains

$$m_2 = \frac{\langle m_\nu \rangle}{|(\frac{m_1}{m_2}) + \frac{1}{2}(1 - \sqrt{1 - sin^2 2\theta})(\pm 1 - (\frac{m_1}{m_2}))|}. \tag{56}$$

For the general case see [Kla97d].

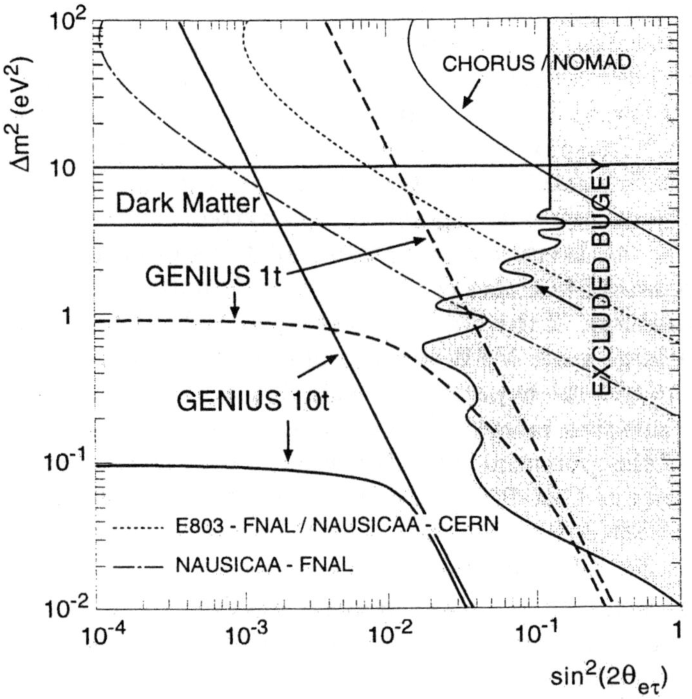

Fig. 21 *Current limits and future experimental sensitivity on $\nu_e - \nu_\tau$ oscillations. The shaded area is currently excluded from reactor experiments. The thin line is the estimated sensitivity of the CHORUS/NOMAD experiments. The dotted and dash-dotted thin lines are sensitivity limits of proposed accelerator experiments, NAUSICAA and E803-FNAL [Gon95]. The thick lines show the sensitivity of GENIUS (broken line: 1 t, full line: 10 t), for two examples of mass ratios. The straight lines are for the strongly hierarchical case (R=0), while the lines bending to the left assume R=0.01. (from [Kla97d])*

case of strong hierarchy ($m_1/m_2 = 0$), to the KAMIOKANDE allowed range for atmospheric neutrino oscillations. GENIUS 1 ton would already be able to test the $\nu_e \leftrightarrow \nu_\mu$ oscillation hypothesis. Fig. 24 shows the potential of GENIUS for checking the LSND indication for neutrino oscillations (original figure from [Ath96]). Under the assumption $m_1/m_2 \geq 0.02$ and $\eta = 1$, GENIUS 1 ton will be sufficient to find $0\nu\beta\beta$ decay if the LSND result is to be explained in terms of $\nu_e \leftrightarrow \nu_\mu$ oscillations. This might be of particular interest also since the upgraded KARMEN will not completely cover [Dre97] the full allowed LSND range. Fig. 25 shows a summary of currently known constraints on neutrino oscillation parameters (original taken from [Hat94]), but including the $0\nu\beta\beta$ decay sensitivities of GENIUS

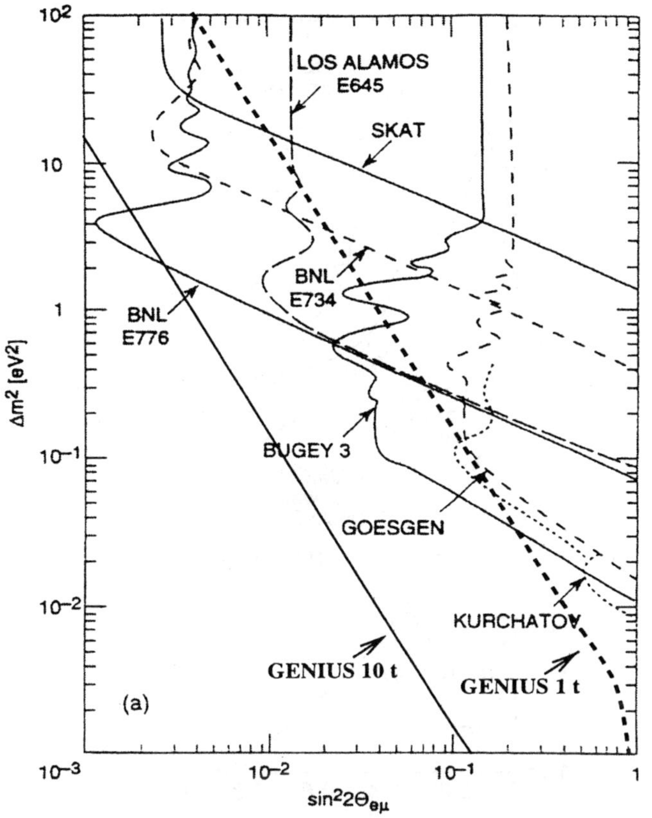

Fig. 22 *Current limits on $\nu_e - \nu_\mu$ oscillations. Various existing experimental limits from reactor and accelerator experiments are indicated, as summarized in ref. [Gel95]. In addition, the figure shows the expected sensitivities for GENIUS with 1 ton (thick broken line) and GENIUS with 10 tons (thick, full line) (from [Kla97d])*

1 ton and GENIUS 10 tons, for different assumptions on m_1/m_2 (and for $\eta^{CP} = +1$). It is seen that already GENIUS 1 ton tests all degenerate or quasi-degenerate ($m_1/m_2 \geq\sim 0.01$) neutrino mass models in any range where neutrinos are interesting for cosmology, and also the atmospheric neutrino problem, if it is due to $\nu_e \leftrightarrow \nu_\mu$ oscillations. GENIUS in its 10 ton version would directly test the large angle solution of the solar neutrino problem.

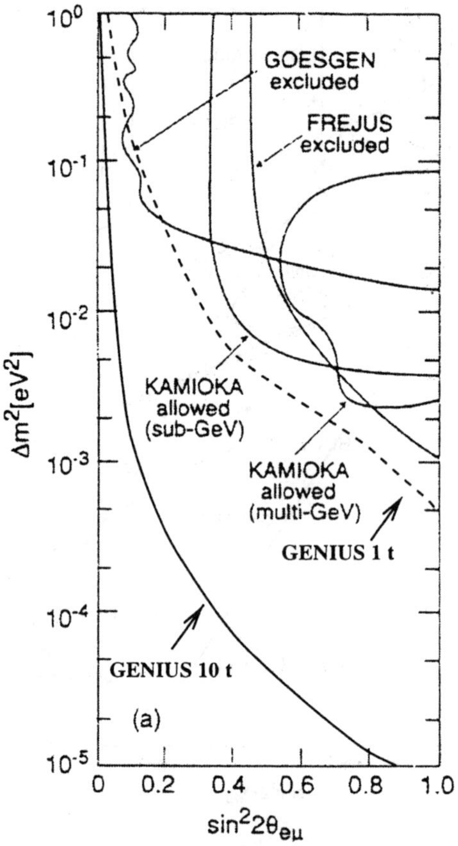

Fig. 23 *Oscillation parameters which solve the atmospheric neutrino problem for $\nu_e \leftrightarrow \nu_\mu$ oscillations. In addition the best currently existing reactor constraints are shown. GENIUS would be able to test the atmospheric neutrino problem already with 1 ton, already in the shown, worst strong hierarchy scenario ($m_1/m_2 = 0$) (from [Kla97d]).*

GENIUS and left–right symmetry: If GENIUS is able to reach down to $\langle m_\nu \rangle \leq 0.01$ eV, it would at the same time be sensitive to right-handed W-boson masses up to $m_{W_R} \geq 8$ TeV (for a heavy right-handed neutrino mass of 1 TeV) or $m_{W_R} \geq 5.3$ TeV (at $\langle m_N \rangle = m_{W_R}$). Such a limit would be comparable to the one expected for LHC, see for example [Riz96], which quotes a final sensitivity of something like 5 – 6 TeV. Note, however that in order to obtain such a limit the experiments at LHC need to accumulate

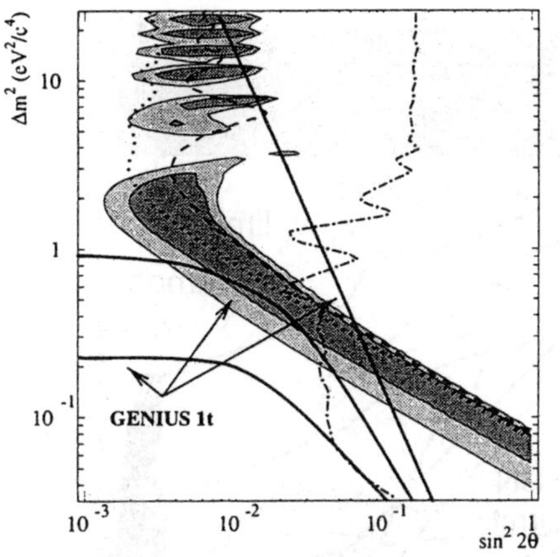

Fig. 24 *LSND compared to the sensitivity of GENIUS 1t for* $\eta^{CP} = +1$ *and three ratios* R_{12}*, from top to bottom* $R_{12} = 0, 0.01, 0.02$ *(from [Kla97d]*

about $100 fb^{-1}$ of statistics. A 10 ton version of GENIUS could even reach a sensitivity of $m_{W_R} \geq 18$ TeV (for a heavy right-handed neutrino mass of 1 TeV) or $m_{W_R} \geq 10.1$ TeV (at $\langle m_N \rangle = m_{W_R}$).

This means that already GENIUS 1 ton could be sufficient to definitely test recent supersymmetric left–right symmetric models having the nice features of solving the strong CP problem without the need for an axion and having automatic R–parity conservation [Kuc95, Moh96].

GENIUS and R_p*-violating SUSY:* The improvement on the R–parity breaking Yukawa coupling λ'_{111} (see section 2.2) is shown in Fig. 26, which updates Fig. 15. The full line to the right is the expected sensitivity of the LHC – in the limit of large statistics. The three dashed–dotted lines denote (from top to bottom) the current constraint from the Heidelberg–Moscow experiment and the sensitivity of GENIUS 1 ton and GENIUS 10 tons, all for the conservative case of a gluino mass of 1 TeV. If squarks would be heavier than 1 TeV, LHC could not compete with GENIUS. However, for typical squark masses below 1 TeV, LHC could probe smaller couplings. However, one should keep in mind, that LHC can probe squark masses up to 1 TeV only with several years of data taking.

GENIUS and R_p*-conserving SUSY:* Since the limits on a 'Majorana–like' sneutrino mass $\tilde{m}_M$ scale with $(T_{1/2})^{1/4}$, GENIUS 1 ton (or 10 tons) would test 'Majorana' sneutrino masses lower by factors of about 7(20), compared with present constraints [Hir97, Hir98a, Hir97b].

288

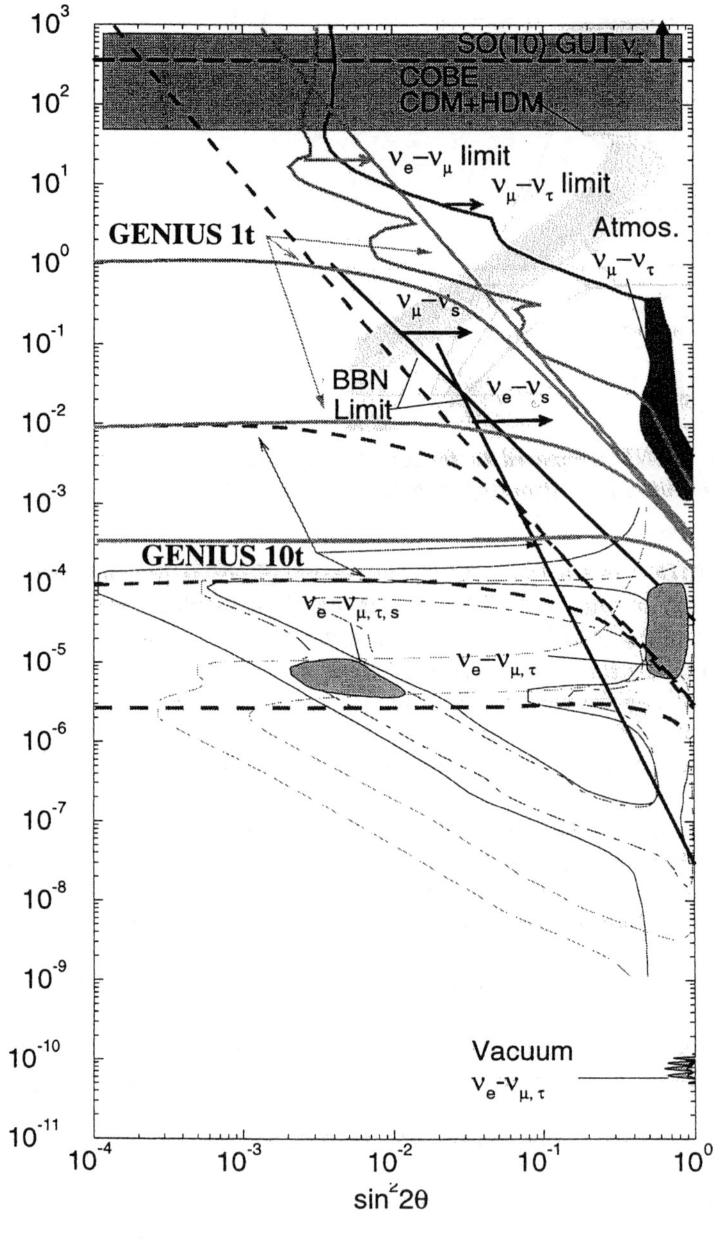
SO(10) GUT ν_τ
COBE
CDM+HDM
$\nu_e-\nu_\mu$ limit
$\nu_\mu-\nu_\tau$ limit
GENIUS 1t
Atmos.
$\nu_\mu-\nu_\tau$
$\nu_\mu-\nu_s$
BBN
Limit
$\nu_e-\nu_s$
GENIUS 10t
$\nu_e-\nu_{\mu,\tau,s}$
$\nu_e-\nu_{\mu,\tau}$
Vacuum
ν_e-$\nu_{\mu,\tau}$
Δm^2 (eV^2)
$\sin^2 2\theta$
10^3
10^2
10^1
10^0
10^{-1}
10^{-2}
10^{-3}
10^{-4}
10^{-5}
10^{-6}
10^{-7}
10^{-8}
10^{-9}
10^{-10}
10^{-11}
10^{-4}
10^{-3}
10^{-2}
10^{-1}
10^0

Fig. 25 *Summary of currently known constraints on neutrino oscillation parameters. The (background) figure without the $0\nu\beta\beta$ decay constraints can be obtained from http://dept.physics.upenn.edu/ www/neutrino/solar.html. Shown are the vacuum and MSW solutions (for two generations of neutrinos) for the solar neutrino problem, the parameter range which would solve the atmospheric neutrino problem and various reactor and accelerator limits on neutrino oscillations. In addition, the mass range in which neutrinos are good hot dark matter candidates is indicated, as well as limits on neutrino oscillations into sterile states from considerations of big bang nucleosynthesis. Finally the thick lines indicate the sensitivity of GENIUS (full lines 1 ton, broken lines 10 ton) to neutrino oscillation parameters for three values of neutrino mass ratios $R = 0, 0.01$ and 0.1 (from top to bottom). For GENIUS 10 ton also the contour line for $R = 0.5$ is shown. The region beyond the lines would be excluded. While already the 1 ton GENIUS would be sufficient to constrain degenerate and quasi-degenerate neutrino mass models, and also would solve the atmospheric neutrino problem if it is due to $\nu_e \leftrightarrow \nu_\mu$ oscillations, the 10 ton version of GENIUS could cover a significant new part of the parameter space, including the large angle MSW solution to the solar neutrino problem, even in the worst case of $R = 0$. For $R \geq 0.5$ it would even probe the small angle MSW solution (see [Kla98g, Kla98e]).*

GENIUS and Leptoquarks: Limits on the lepton–number violating parameters defined in sections 2.7, 3.2 improve as $\sqrt{T_{1/2}}$. This means that for leptoquarks in the range of 200 GeV LQ–Higgs couplings down to (a few) 10^{-8} could be explored. In other words, if leptoquarks interact with the standard model Higgs boson with a coupling of the order $\mathcal{O}(1)$, either $0\nu\beta\beta$ must be found, or LQs must be heavier than (several) 10 TeV.

GENIUS and composite neutrinos GENIUS in the 1(10) ton version would improve the limit on the excited Majorana neutrino mass deduced from the Heidelberg–Moscow experiment (eq. 32) to

$$m_N \geq\sim 1.1(2.3) \quad TeV \tag{57}$$

4.2.3. GENIUS, special relativity and equivalence principle in the neutrino sector

The already now strongest limits given by the Heidelberg–Moscow experiment discussed in section 3.2 would be improved by 1–2 orders of magnitude. It should be stressed again, that while neutrino oscillation bounds constrain the region of large mixing of the weak and gravitational eigenstates, these bounds from double beta decay apply even in the case of no mixing and thus probe a totally unconstrained region in the parameter space.

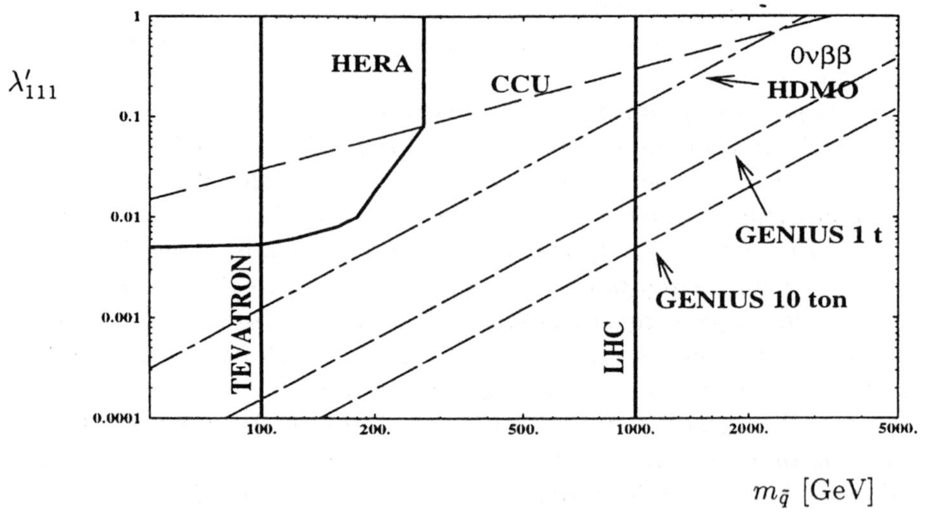

Fig. 26 *Comparison of sensitivities of existing and future experiments on $R\!\!\!/_p SUSY$ models in the plane $\lambda'_{111} - m_{\tilde{q}}$. Note the double logarithmic scale! Shown are the areas currently excluded by the experiments at the TEVATRON, the limit from charged-current universality, denoted by CCU, and the limit from absence of $0\nu\beta\beta$ decay from the Heidelberg-Moscow collaboration ($0\nu\beta\beta$ HDMO). In addition, the estimated sensitivity of HERA and the LHC is compared to the one expected for GENIUS in the 1 ton and the 10 ton version. The figure is essentially an update of Fig. 15.*

4.2.4. GENIUS and dark matter

Neutrinos as hot dark matter

If neutrinos have masses in the range of a few eV, they would be good candidates for the hot dark matter in the universe. Of course, from the dark matter argument itself it does not follow which neutrino has to be in this mass range. Clearly, if a neutrino with sizeable mixing angle to the electron neutrino in this mass range exists, one expects GENIUS to find $0\nu\beta\beta$ decay.

However, if the ν_τ is in the eV range, the ν_e and ν_μ being lighter by at least factors of hundreds and the the $\nu_\tau - \nu_e$ mixing angle small at the same time GENIUS with 1 ton would not find double beta decay. In the case of quasidegenerate models or degenerate models, on the other hand, $0\nu\beta\beta$ decay should be found by GENIUS, unless the CP–phases between the different mass eigenstates take on some special combinations and have a relative minus sign, see the discussion in [Kla98c].

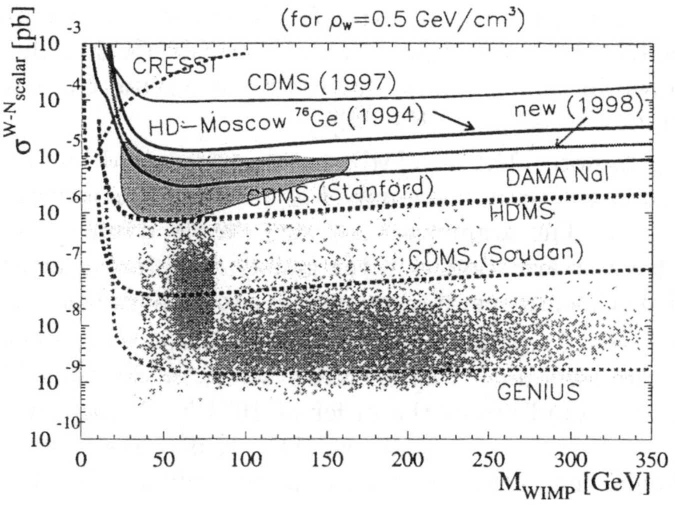

Fig. 27 *WIMP–nucleon cross section limits in pb for scalar interactions as function of the WIMP–mass in GeV. Regions beyond solid lines are excluded by experiment [HM94, HM98, Ber97, Ake97]. Further shown are expected sensitivities of experiments under construction (dashed lines for HDMS [Bau97, Kla97e], CDMS [Ake97], CRESST and for GENIUS). These limits are compared to theoretical expectations (scatter plot) for WIMP–neutralino cross sections calculated in the MSSM framework with non–universal scalar mass unification [Bed97b]. The 90 % allowed region claimed by [Ber97a] (light filled area), which is further restricted by indirect dark matter searches [Bot97] (dark filled area), could already be easily tested with a 100 kg version of the GENIUS experiment.*

Cold Dark Matter

Weakly interacting massive particles (WIMPs) are candidates for the cold dark matter in the universe. The favorite WIMP candidate is the lightest supersymmetric particle, presumably the neutralino. The expected detection rates for neutralinos of typically less than one event per day and kg of detector mass [Bed94, Bed97a, Bed97b, Jun96], however, make direct searches for WIMP scattering experimentally a formidable task.

Fig. 27 shows a comparison of existing constraints and future sensitivities of cold dark matter experiments, together with the theoretical expectations for neutralino scattering rates [Bed97b]. Obviously, GENIUS could easily cover the range of positive evidence for dark matter recently claimed by DAMA [Ber97a, Bot97]. It would also be by far more sensitive than all other dark matter experiments at present under construction or proposed, like the cryogenic experiment CDMS. Furthermore, obviously GENIUS will be the only experiment, which could seriously test the MSSM predictions over the whole SUSY parameter space. In this way, GENIUS could compete even with LHC in the search for SUSY, see for example the discussion in [Bae97]. It is important to note, that GENIUS could reach the sensitivity shown in Fig. 26 with only 100 kg of *natural* Ge detectors in a measuring time of three years [Kla98d].

It is interesting to note, that if WIMP scattering is found by GENIUS it could be used to constrain the amount of R-parity violation within supersymmetric models. The arguments are very simple [Hir97c]. Due to the fact that neutralinos are abound in the galaxy even today, neutralino decays via R-parity violating operators would have to be highly suppressed.

The details depend, of course, on the neutralino mass and composition. However, finding the neutralino with GENIUS would imply typical limits on R-parity violating couplings of the order of $10^{-(16-20)}$ for any of the λ_{ijk}, λ'_{ijk} or λ''_{ijk} in the superpotential (eq. 11). A positive result of the CDM search at hand, one could thus finally safely conclude that R-parity is conserved.

5. Conclusion

Double beta decay has a broad potential for providing important information on modern particle physics beyond present and future high energy accelerator energies which will be competitive for the next decade and more. This includes SUSY models, compositeness, left–right symmetric models, leptoquarks, the neutrino and sneutrino mass and tests of Lorentz invariance and equivalence principle in the neutrino sector. Based to a large extent on the theoretical work of the Heidelberg Double Beta group, results have been deduced from the HEIDELBERG–MOSCOW experiment for these topics and have been presented here. For the neutrino mass double beta decay now is particularly pushed into a key position by the recent possible indications of beyond standard model physics from the side of solar and atmospheric neutrinos, dark matter COBE results and others. New classes of GUTs basing on degenerate neutrino mass scenarios which could explain these observations, can be checked by double beta decay in near future. The HEIDELBERG–MOSCOW experiment has reached a leading position among present $\beta\beta$ experiments and as the first of them now yields results in the sub–eV range. We have described a new idea and proposal of a future double beta experiment (GENIUS) with highly increased sensitivity based on use of 1 ton or more of enriched 'naked' ^{76}Ge detectors in liquid nitrogen. This new experiment would be a breakthrough into the multi-TeV range for many beyond standard models. The sensitivity for the neutrino mass would reach down to 0.01 or even 0.001 eV. The experiment would be competitive to LHC with respect to the mass of a right–handed W boson, in search for R–parity violation and others, and would improve the leptoquark and compositeness searches by considerable factors. It would probe the Majorana electron sneutrino mass more sensitive than NLC (Next Linear Collider). It would yield constraints on neutrino oscillation parameters far beyond all present terestrial $\nu_e - \nu_x$ neutrino oscillation experiments and could test directly the atmospheric neutrino problem and the large and, for degenerate models, even the small angle solution of the solar neutrino problem. GENIUS would cover the full SUSY parameter space for prediction of neutralinos as cold dark matter and compete in this way with LHC in the search for supersymmetry. Even if SUSY would be first observed by LHC, it would still be fascinating to verify the existence and properties of neutralino dark matter, which could be achieved by GENIUS. Concluding GENIUS has the ability to provide a major tool for future particle- and astrophysics.

Finally it may be stressed that the technology of producing and using enriched high purity germanium detectors, which have been produced for the first time for the Heidelberg–Moscow experiment, has found meanwhile applications also in pre-GENIUS dark matter search [HM94, Fal94, Kla97e, Bau97] and in high–resolution γ-ray astrophysics, using balloons and satellites [Kla91, Kla94, Bar93, Bar94, Boc94, Kla97b].

References

[Abe94] F. Abe *et al.* (CDF Collab.), Phys. Rev. Lett. 72 (1994) 3004

[Adr92] O. Adriani *et al.*, Phys. Lett. **B 288** (1992) 404

[Ake97] D.S. Akerib et al. (CDMS collab.), preprint astro-ph/9712343

[ALE92] ALEPH Collab., Phys. Rep. 216 (1992) 343

[ALE93] ALEPH Collab., Z.Phys. **C 59** (1993) 215

[Ale94] A. Alessandrello et al., Phys. Lett **B 335** (1994) 519

[Als89] M. Alston-Garnjost et al., Phys. Rev Lett. **71** (1993) 831

[Alt97] G. Altarelli, J. Ellis, G.F. Guidice, S. Lola, M.L. Mangano, preprint hep-ph/9703276

[Ath95] C. Athanassopoulos *et al.* (LSND–Collaboration), Phys. Rev. Lett. **75** (1995) 2650 and D. Smith, in [Kla96]

[Ath96] C. Athanassopoulos *et al.*, LSND collab., Phys. Rev. **C 54** (1996) 2685, Phys. Rev. Lett. **77** (1996) 3082

[Aul82] C.S. Aulakh and R.N. Mohapatra, Phys. Lett. **B 119** (1982) 136

[Bab95] K.S. Babu, R.N. Mohapatra, Phys. Rev. Lett. **75** (1995) 2276

[Bab97] K.S. Babu et al., preprint hep-ph/9703299 (March 1997)

[Bab97b] K.S. Babu *et al.*, preprint hep-ph/9705414v2 (1997)

[Bae97] H. Baer. M. Bhrlik, hep-ph/9706509

[Bam95a] P. Bamert, C.P. Burgess, R.N. Mohapatra, Nucl. Phys. B 438 (1995) 3

[Bam95] P. Bamert, C.P. Burgess, R.N. Mohapatra, Nucl. Phys. **B 449** (1995) 25

[Bar89] V. Barger, G.F. Guidice, T. Han. Phys.Rev. **D 40** (1989) 2987

[Bar93] S.D. Barthelmy *et al.*, AIP Conf. Proc. 280 (1993) 1166

[Bar94] S. D. Barthelmy et al., Astrophys. J. **427** (1994) 519

[Bar96] P.B. Barnes *et al.*, Nucl. Instr. Meth. **A 370** (1996) 233

[Bar97] A.S. Barabash, Proc. NEUTRINO 96, Helsinki, June 1996, World Scientific, Singapore (1997), p. 374

[Bau97] L. Baudis, J. Hellmig, H.V. Klapdor–Kleingrothaus, A. Müller, F. Petry, Y. Ramachers, H. Strecker, Nucl. Inst. Meth. **A 385** (1997) 265

[Bav95] E. Baver, M. Leurer, Phys. Rev. **D 51** (1995) 260

[Bed94] V. Bednyakov, H.V. Klapdor–Kleingrothaus, S.G. Kovalenko, Phys. Lett. **B 329** (1994) 5 and Phys. Rev. **D 50** (1994) 7128

[Bed97a] V. Bednyakov, H.V. Klapdor–Kleingrothaus, S.G. Kovalenko, Phys. Rev. **D 55** (1997) 503

[Bed97b] V. Bednyakov, H.V. Klapdor–Kleingrothaus, S.G. Kovalenko, Y. Ramachers, Z. Phys. **A 357** (1997) 339

[Bed97c] V. Bednyakov, V.B. Brudanin, S.G. Kovalenko, Ts. D. Vylov, Mod. Phys. Lett **A 12** (1997) 233

[Bel96] G. Belanger, F. Boudjema, D.London, H. Nadeau, Phys. Rev. **D 53** (1996) 6292

[Bel98] G. Belanger, these Proceedings

[Ber92] Z.G. Berezhiani, A. Yu. Smirnov, J.W.F. Valle, Phys.Lett. **B 291** (1992) 99

[Ber95] Z. Berezhiani, R.N. Mohapatra, Phys. Rev. **D 52** (1995) 6607

[Ber97] R. Bernabei *et al.*, Phys. Lett. **B 389** (1997) 757

[Ber97a] R. Bernabei *et al.*, ROM2F/97/33; P. Belli at TAUP, Gran Sasso, Sept. 7 (1997)

[Bha97] G. Bhattacharyya, in [Kla98]

[Blü94] J. Blümlein, E. Boos, A. Pukhov, preprint DESY 94-072 (April 1994)

[Boc94] J. Bockholt, H.V. Klapdor–Kleingrothaus, Nucl. Phys. **B** (Proc. Suppl.) 35 (1994) 403

[Bot97] A. Bottino *et al.*, hep-ph/9709292

[Buc87] W. Buchmüller, R. Rückl and D. Wyler, Phys.Lett. B191 (1987) 442.

[Buc91] W. Buchmüller and G. Ingelman, Proc. Workshop Physics at HERA, Hamburg, Oct. 29–30 (1991)

[Bur93] C.P. Burgess, J.M. Cline, Phys. Lett. **B 298** (1993) 141; Phys. Rev. **D 49** (1994) 5925

[Bur96] C.P.Burgess, in [Kla96]

[But93] M. N. Butler *et al.*, Phys. Rev. **D47**, 2615 (1993); A. Halprin and C. N. Leung, Phys. Rev. Lett. **67**, 1833 (1991); J. Pantaleone, A. Halprin, and C. N. Leung, Phys. Rev. **D47**, R4199 (1993); K. Iida, H. Minakata and O. Yasuda, Mod. Phys. Lett. **A8** (1993) 1037.

[But93a] J. Butterworth, H. Dreiner, Nucl. Phys. **B397** (1993) 3 and H. Dreiner, P. Morawitz, Nucl. Phys. **B428** (1994) 31

[Cal93] D.O. Caldwell, R.N. Mohapatra, Phys. Rev. **D 48** (1993) 3259

[Cal95] D.O. Caldwell, R.N. Mohapatra, Phys. Lett. **354** (1995) 371

[Car93] C.D. Carone, Phys. Lett. **B 308** (1993) 85

[Cau96] E. Caurier et al., Phys. Rev. Lett. **77** (1996) 1954

[CDF91] CDF Collab., Phys. Rev. Lett. **67** (1991) 2148

[Chi81] Y. Chikashige, R.N. Mohapatra and R.D. Peccei, Phys. Lett. **B 98** (1981) 265; Phys. Rev. Lett. **45** (1980) 265

[Cho94] D. Choudhury, hep-ph/9408250 (1994)

[Cho97] D. Choudhury, S. Raychaudhuri, preprint hep-ph/9702392

[Cli97] D. Cline, in: Proceedings of the *International Workshop Dark Matter in Astro- and Particle Physics* (DARK96), Eds. H.V. Klapdor–Kleingrothaus, Y. Ramachers, World Scientific 1997, p. 479

[Col97] S. Coleman and S.L. Glashow, Phys. Lett. **B 405**, 249 (1997).

[Dan95] F. A. Danevich et al., Phys. Lett. **B 344** (1995) 72

[Dav94] S. Davidson, D. Bailey and A. Campbell, Z.Phys. C61 (1994) 613; M. Leurer, Phys.Rev.Lett. 71 (1993) 1324; Phys.Rev. D50 (1994) 536.

[Der95] M. Derrik, *et al.* (ZEUS Collab.), Z. Phys. **C 65** (1995) 627

[Dia93] J.L. Diaz Cruz, O.A. Sampayo, Phys. Lett. **B 306** (1993) 395; Phys. Rev. **D 49** (1994) R2149

[Doi85] M. Doi, T. Kotani, E.Takasugi, Progr. Theor. Phys. Suppl. **83** (1985) 1

[Doi93] M. Doi, T. Kotani, Progr. Theor. Phys. 89 (1993) 139

[Dre97] G. Drexlin, Proc. Internat. School on Neutrino Physics, Erice, Italy, Sept. 1997, to be publ. in Plenum Press

[Ell87] S.R. Elliot, A.A. Hahn, M.K. Moe, Phys. Rev. Lett. **59** (1987) 1649

[Ell92] S. R. Elliott et al., Phys. Rev. **C 46** (1992) 1535

[Fal94] T. Falk, A. Olive, M. Srednicki, Phys. Lett. **B339** (1994) 248

[Fri88] J.Friemann, H.Haber, K.Freese, Phys. Lett. **B 200** (1988) 115; J. Bahcall, S. Petcov, S. Toshev and J.W.F. Valle, Phys.Lett. **B 181** (1986) 369; Z. Berezhiani and M. Vysotsky, Phys. Lett. **B 199** (1988) 281.

[Fri95] H. Fritzsch and Zhi-zhong Xing, preprint hep-ph/9509389, Phys. Lett. **B 372** (1996) 265

[Gas89] M. Gasperini, Phys. Rev. **D38**, 2635 (1988); *ibid.* **D39**, 3606 (1989).

[Gas89a] M. Gasperini, Phys. Rev. Lett. **62**, 1945 (1989).

[Gel81] G.B. Gelmini and M. Roncadelli, Phys. Lett. **B 99** (1981) 411

[Gel95] G. Gelmini, E. Roulet, Rep. Progr. Phys. **58**(1995) 1207

[Geo81] H.M. Georgi, S.L.Glashow and S. Nussinov, Nuc. Phys. **B 193** (1981) 297

[Ger96] G. Gervasio, in [Kla96]

[Gla97] S.L. Glashow, A. Halprin, P.I. Krastev, C.N. Leung, and J. Panteleone, Phys. Rev. **D 56**, 2433 (1997).

[Gon95] M. Gonzalez–Garcia, hep-ph/9510419

[Goo61] M.L. Good, Phys. Rev. **121**, 311 (1961); O. Nachtmann, Acta Physica Austriaca, Supp. VI *Particle Physics* ed. P. Urban, (1969) p. 485; S.H. Aronson, G.J. Bock, H-Y Cheng and E. Fishbach, **48**, 1306 (1982); Phys. Rev. **D28**, 495 (1983); I.R. Kenyon, Phys. Lett. **B237**, 274 (1990); R.J. Hughes, Phys. Rev. **D46**, R2283 (1992);

[Gro85] K. Grotz, H. V. Klapdor, Phys. Lett. **B 157** (1985) 242

[Gro86] K. Grotz, H.V. Klapdor, Nucl. Phys **A 460** (1986) 395

[Gro90] K. Grotz, H.V. Klapdor, 'The Weak Interaction in Nuclear, Particle and Astrophysics' (Adam Hilger: Bristol, Philadelphia) 1990

[H195] H1 Collab., Phys. Lett. **B 353** (1995) 578, Z. Phys. **C 64** (1994) 545

[H196] H1 Collab., S. Aid et al., Phys. Lett. B 369 (1996) 173

[Hab93] H. E. Haber, in Proc. on Recent Advances in the Superworld, Houston, April 14-16 (1993), hep-ph/9308209

[Hal84] L. Hall, M. Suzuki, Nuclear Physics **B 231** (1984) 419

[Hal91] A. Halprin and C.N. Leung, Phys. Rev. Lett. **67**, 1833 (1991); Nucl. Phys. **B28A** (Proc. Supp.), 139 (1992); J. N. Bahcall, P. I. Krastev, and C. N. Leung, Phys. Rev. **D 52**, 1770 (1995); R.B. Mann and U. Sarkar, Phys. Rev. Lett **76** (1996) 865;

[Hal96] A. Halprin, C.N. Leung, J. Pantalone, Phys.Rev. **D 53** (1996) 5365

[Ham98] T. Hambye, R.B. Mann and U. Sarkar, Phys. Lett. **B 421**, 105 (1998); Phys. Rev. **D 58**, 025003 (1998).

[Hat94] N. Hata, P. Langacker, Phys. Rev. **D 50** (1994) 632

[Hax84] W.C. Haxton, G.J. Stephenson, Progr. Part. Nucl. Phys. 12 (1984) 409

[Hel96] J. Hellmig, PhD Thesis, Univ. of Heidelberg, 1996

[Hel97] J. Hellmig, H.V. Klapdor–Kleingrothaus, Z. Phys. **A 359** (1997) 351

[Hew97] J.L. Hewett, T.G. Rizzo, preprint hep-ph/9703337v3 (May1997)

[Heu95] G. Heusser, Ann. Rev. Nucl. Part. Sci. **45** (1995) 543

[Hir95] M. Hirsch, H.V. Klapdor–Kleingrothaus, S.G. Kovalenko, Phys. Rev. Lett. 75 (1995) 17

[Hir95b] M. Hirsch, H.V. Klapdor–Kleingrothaus, S.G. Kovalenko, in [Kla96]

[Hir95c] M. Hirsch, H.V. Klapdor–Kleingrothaus, S. Kovalenko, Phys. Lett. **B 352** (1995) 1

[Hir95d] J. Hirsch, O. Castanos, P.O. Hess, Nucl. Phys. **A 582** (1995) 124

[Hir96] M. Hirsch, H.V. Klapdor–Kleingrothaus, S.G. Kovalenko, Phys. Lett. **B 372** (1996) 181, Erratum: Phys. Lett. **B381** (1996) 488

[Hir96a] M. Hirsch, H.V. Klapdor–Kleingrothaus, S.G. Kovalenko, Phys. Lett. **B 378** (1996) 17 and Phys. Rev. **D 54** (1996) R4207

[Hir96b] M. Hirsch, H. V. Klapdor–Kleingrothaus, S. G. Kovalenko, H. Päs, Phys. Lett. **B 372** (1996) 8

[Hir96c] M. Hirsch, H.V. Klapdor–Kleingrothaus, S. Kovalenko, Phys. Rev. **D 53** (1996) 1329

[Hir96d] M. Hirsch, H.V. Klapdor–Kleingrothaus, in [Kla96]; M. Hirsch, H.V. Klapdor–Kleingrothaus, O. Panella, Phys. Lett. **B 374** (1996) 7

[Hir97] M. Hirsch, H.V. Klapdor–Kleingrothaus, S.G. Kovalenko, Phys. Lett. **B 398** (1997) 311 and **403** (1997) 291

[Hir97b] M. Hirsch, H.V. Klapdor–Kleingrothaus, S. Kovalenko, in [Kla98]

[Hir97c] M. Hirsch, H.V. Klapdor–Kleingrothaus, Proc. Int. Workshop on Dark Matter in Astro– and Particle Physics (DARK96), Heidelberg, Sept. 1996, Eds. H.V. Klapdor–Kleingrothaus and Y. Ramachers (World Scientific, Singapore) 1997, p. 640

[Hir98] M. Hirsch, H.V. Klapdor–Kleingrothaus, St. Kolb, S.G. Kovalenko, Phys. Rev. **D 57** (1998) 2020

[Hir98a] M. Hirsch, H.V. Klapdor–Kleingrothaus, S.G. Kovalenko, Phys. Rev. **D 57** (1998) 1947

[HM94] HEIDELBERG–MOSCOW collab., Phys. Lett. **B 336** (1994) 141

[HM95] HEIDELBERG–MOSCOW collab., Phys. Lett. **B 356** (1995) 450

[HM96] HEIDELBERG–MOSCOW collab., Phys. Rev. **D 54** (1996) 3641,

[HM97] HEIDELBERG–MOSCOW collab., Phys. Rev. **D 55** (1997) 54 and Phys. Lett. **B 407** (1997) 219

[HM98] HEIDELBERG–MOSCOW collab., Phys. Rev. **D** Dec 1998

[Hug60] V.W. Hughes, H.G. Robinson, and V. Beltran-Lopez, Phys. Rev. Lett. **4**, 342 (1960); R.W.P. Drever, Philos. Mag. **6**, 683 (1961); D. Newman, G.W. Ford, A. Rich and E. Sweetman, Phys. Rev. Lett. **40**, 1355 (1978); A. Brillet and J.L. Hall, Phys. Rev. Lett. **42**, 549 (1979); J.D. Prestage, J.J. Bollinger, W.M. Itano, and D.J. Wineland, Phys. Rev. Lett. **54**, 2387 (1985); S.K. Lamoureaux, J.P. Jacobs, B.R. Heckel, R.J. Raab, and E.N. Fortson, Phys. Rev. Lett. **57**, 3125 (1986).

[Ioa94] A. Ioanissyan, J.W.F. Valle, Phys. Lett **B 322** (1994) 93

[Jör94] V. Jörgens et al., Nucl. Phys. (Proc. Suppl.) **B 35** (1994) 378

[Jun96] G. Jungmann, M. Kamionkowski, K. Griest, Phys. Rep. 267 (1996) 195

[Kal97] J. Kalinowski et al., preprint hep-ph/9703288v2 (March 1997)

[Kan97] G. Kane, in [Kla98]

[Kla84] H.V. Klapdor, K. Grotz, Phys.Lett. **B142** (1984) 323

[Kla87] H.V. Klapdor–Kleingrothaus, MPI–H 1987, proposal

[Kla91] H.V. Klapdor–Kleingrothaus, Proc. Int. Symposium on γ-Ray Astrophysics, Paris 1990, AIP Conf. Proc. 232 (1991) 464

[Kla92] H.V. Klapdor–Kleingrothaus, K. Zuber, Phys. Bl. **48** (1992) 1017

[Kla94] H.V. Klapdor–Kleingrothaus, Progr. Part. Nucl. Phys. **32** (1994)261

[Kla95] H. V. Klapdor–Kleingrothaus, A. Staudt, Non–Accelerator Particle Physics, IOP Publ., Bristol, Philadelphia, 1995; and Teilchenphysik ohne Beschleuniger, Teubner Verlag, Stuttgart, 1995

[Kla96] H.V. Klapdor–Kleingrothaus and S. Stoica (Eds.), Proc. *Int. Workshop on Double Beta Decay and Related Topics*, Trento, 24.4.–5.5.95, World Scientific Singapore

[Kla96a] H.V. Klapdor–Kleingrothaus, in Proc. *Int. Workshop on Double Beta Decay and Related Topics*, Trento, 24.4.–5.5.95, World Scientific Singapore 1996, Ed.: H.V. Klapdor–Kleingrothaus and S. Stoica

[Kla97] H.V. Klapdor–Kleingrothaus, Invited talk at NEUTRINO 96, Helsinki, June 1996, World Scientific Singapore 1997, p. 317

[Kla97a] H.V. Klapdor–Kleingrothaus, in Proc. Int. Workshop on Non-Accelerator New Physics (NANP-97), Dubna, Moscow region, Russia, July 7-11, 1997 and in Proc. Int. School on Neutrinos, Erice, Italy, Sept. 1997, Progr. Part. Nucl. Phys. **40** (1998) 265-282

[Kla97b] H.V. Klapdor–Kleingrothaus, M.I. Kudravtsev, V.G. Stolpovski, S.I. Svertilov, V.F. Melnikov, I. Krivosheina, J. Moscow. Phys. Soc. 7 (1997) 41

[Kla97c] H.V. Klapdor–Kleingrothaus and K. Zuber, Particle Astrophysics, IOP Publ., Bristol, Philadelphia 1997; and Teilchenastrophysik, Teubner Verlag, Stuttgart, 1997

[Kla97d] H.V. Klapdor–Kleingrothaus, M. Hirsch, Z. Phys. **A 359** (1997) 361

[Kla97e] H.V. Klapdor–Kleingrothaus, Y. Ramachers, in: Proc. Int. Workshop on Dark Matter in Astro– and Particle Physics (DARK96) Sept. 1996, Eds. H.V. Klapdor–Kleingrothaus and Y. Ramachers, Heidelberg, (World Scientific Singapore) 1997, p. 459

[Kla98] H.V. Klapdor–Kleingrothaus and H. Päs (Eds.), Beyond the Desert – Accelerator- and Non–Accelerator Approaches, IOP, Bristol, 1998

[Kla98a] H.V. Klapdor–Kleingrothaus, in [Kla98]

[Kla98b] H.V. Klapdor–Kleingrothaus, Int. J. Mod. Phys. **A 13** (1998) 3953-3992

[Kla98c] H.V. Klapdor–Kleingrothaus, J. Hellmig, M. Hirsch, J. Phys. **G 24** (1998) 483-516

[Kla98d] H.V. Klapdor–Kleingrothaus, Y. Ramachers, Eur. Phys. J. **A 3** (1998) 85

[Kla98e] H.V. Klapdor–Kleingrothaus, H. Päs; in: Proc. of the 6th Symp. on Particles, Strings and Cosmology (PASCOS'98), Boston/USA, 1998

[Kla98f] H.V. Klapdor–Kleingrothaus, H. Päs, U. Sarkar; submitted to Phys. Rev. Lett. 1998

[Kla98g] H.V. Klapdor–Kleingrothaus, Proc. NEUTRINO'98 (Takayama,Japan,June 1998) World Scientific, Singapore

[Kol97] S. Kolb, M. Hirsch, H.V. Klapdor–Kleingrothaus, S. Kovalenko, in [Kla98]

[Kol97a] S. Kolb, M. Hirsch, H.V. Klapdor–Kleingrothaus, Phys. Rev. **D 56** (1997) 4161

[Kuc95] R. Kuchimanchi, R.N. Mohapatra, Phys. Rev. Lett. **75** (1995) 3939

[Kum96] K. Kume (ELEGANT collaboration), in [Kla96]

[Kuz90] V. Kuzmin, V. Rubakov, M. Shaposhnikov, Phys. Lett. **B 185** (1985) 36; M. Fukugita, T. Yanagida, Phys. Rev. **D 42** (1990) 1285; G. Gelmini, T. Yanagida, Phys. Lett. **B 294** (1992) 53; B. Campbell *et al.*, Phys. Lett. **B 256** (91) 457

[Lan88] P.Langacker, in 'Neutrinos' (Springer: Heidelberg, New York), 1988, ed. H.V. Klapdor, p. 71

[Lee94] D.G. Lee, R.N. Mohapatra, Phys. Lett. **B 329** (1994) 463

[Leu94] M. Leurer, Phys. Rev. **D 49** (1994) 333

[Moe91] M. K. Moe, Phys. Rev. **C 44** (1991) R931

[Moe94] M. K. Moe, Prog. Part. Nucl. Phys. **32** (1994) 247; Nucl. Phys. (Proc. Suppl.) **B 38** (1995) 36

[Moh86] R.N. Mohapatra, Phys. Rev. **D 34** (1986) 3457

[Moh86b] R.N. Mohapatra, Phys. Rev. **D 34** (1986) 909

[Moh88] R.N. Mohapatra and E. Takasugi, Phys. Lett. **B 211** (1988) 192

[Moh91] R.N. Mohapatra, P.B. Pal, 'Massive Neutrinos in Physics and Astrophysics' (World Scientific, Singapore) 1991

[Moh92] R.N. Mohapatra, Unification and Supersymmetry (Springer: Heidelberg, New York) 1986 and 1992

[Moh94] R.N. Mohapatra, Progr. Part. Nucl. Phys. **32** (1994) 187

[Moh95] R.N. Mohapatra, S. Nussinov, Phys. Lett. **B 346** (1995) 75

[Moh96] R.N. Mohapatra, A. Rasin, Phys. Rev. Lett. **76** (1996) 3490 and Phys. Rev. **D54** (1996) 5835

[Moh96a] R.N. Mohapatra, in [Kla96]

[Moh97] R. N. Mohapatra, Proc. *Neutrino 96*, Helsinki, 1996, World Scientific Singapore, 1997, p. 290

[Moh97a] R.N. Mohapatra, Proc. Int. School on Neutrinos, Erice, Italy, Sept. 1997, Progr. Part. Nuc. Phys. **40** (1998)

[Mut88] K. Muto, H.V. Klapdor, in 'Neutrinos (Springer: Heidelberg, New York) 1988, ed. H.V. Klapdor, p. 183

[Mut89] K. Muto, E. Bender, H.V. Klapdor, Z. Phys. **A 334** (1989) 177,187;

[Mut91] K. Muto, E.Bender, H.V. Klapdor–Kleingrothaus, Z. Phys **A** 339 (1991) 435

[NEM94] NEMO Collaboration, Nucl. Phys. (Proc. Suppl.) **B 35** (1994) 369

[Nor97] D. Normile, Science 276 (1997) 1795

[PDG94] Particle Data Group, Phys. Rev. **D 50** (1994)

[PDG96] Particle Data Group, Phys. Rev. **D 54** (1996) 1

[Päs96] H. Päs *et al.*, in [Kla96];

[Päs97] H. Päs, M. Hirsch, H.V. Klapdor–Kleingrothaus, S.G. Kovalenko, in [Kla98]

[Pan96] O.Panella, in [Kla96]

[Pan97] O. Panella, in [Kla98]

[Pan96b] G. Pantis, F. Simkovic, J.D. Vergados, A. Faessler, Phys. Rev. **C 53** (1996) 695

[Pel93] J. Peltoniemi, J. Valle, Nucl. Phys. **B 406** (1993) 409

[Pel95] J.T. Peltoniemi, preprint hep-ph/9506228

[Pet94] S.T. Petcov, A.Yu. Smirnov, Phys. Lett. **B 322** (1994) 109

[Pet96] S. Petcov, in [Kla96]

[Pil93] A. Pilaftsis, Phys. Rev. **D 49** (1993) 2398

[Piq96] F. Piquemal *et al.*, in [Kla96]

[Pri95] J.R. Primack, J. Holtzman, A. Klypin, D.O. Caldwell, Phys. Rev. Lett. 74 (1995) 2160

[Rad95] P.B. Radha *et. al*, preprint nucl-th/9510052 (Oct. 1995)

[Raf96] G. Raffelt, J. Silk, Phys. Lett. **B 366** (1996) 429

[Rag94] R. S. Raghavan, Phys. Rev. Lett. **72** (1994) 1411

[Rau94] F. Raupach (H1 Collab.), Proc. Int. Europhys. Conf. on High En. Phys., Marseille, 1993, Edition Frontieres, Gif–Sur–Yvette, France 1994. Eds J. Carr, M. Perrothet

[Ren82] F.M. Renard, Phys. Lett. **B 116** (1982) 264

[Riz82] T.G. Rizzo, Phys. Lett. **B 116** (1982) 23

[Riz96] T.G. Rizzo, hep/ph/9612440

[Roy92] D.P. Roy, Phys.Lett. **B 283** (1992) 270

[Rub96] C. Rubbia, Proc. TAUP 95, Toledo, Sept. 17–21 (1995), Nucl. Phys. **B** (Proc. Supl.) 48 (1996)172

[Sch81] J. Schechter, J.W.F. Valle, Phys. Rev. **D 25** (1982) 2951

[Sim96] F. Simkovic et al., Proc. TAUP 95, Toledo, 17–21 Sept. 1995; Nucl. Phys. **B** (Proc. Supplement) 48 (1996) 257

[Sim97] F. Simkovic *et al.*, Phys. Lett. **B 393** (1997) 267

[Smi96] P.F. Smith *et al.*, Phys. Lett. **B 379** (1996) 299

[Smi96a] A. Yu. Smirnov, Proc. Int. Conf. on High Energy Physics, Warsaw 1996, hep-ph/9611465v2 (Dec 1996)

[Sou92] I.A. D'Souza. C.S. Kalman, Preons, Models of Leptons , Quarks and Gauge bosons as Composite Objects (World Scientific, Singapore) 1992

[Sta90] A. Staudt, K.Muto, H.V. Klapdor–Kleingrothaus, Europhys. Lett. **13** (1990) 31

[Ste91] J. Steinberger, Phys. Rep. **203** (1991) 345

[Sto96] S. Stoica, in [Kla96]

[Suh96] J. Suhonen, in [Kla96]

[Suz97] A. Suzuki, priv. comm. 1997, and KAMLAND proposal, (in Japanese)

[Tak96] E.Takasugi, in [Kla96]

[Tak97] E. Takasugi, in [Kla98]

[Tom87] T.Tomoda, A. Faessler, Phys. Lett. **B199** (1987) 475

[Tre95] V.I. Tretyak, Yu. Zdesenko, At. Data Nucl. Data Tables **61** (1995) 43

[Val96] J.W.F. Valle, in [Kla96]

[Vog86] P.Vogel, M.R. Zirnbauer, Phys. Rev. Lett. **57** (1986)3148

[Vog96] P.Vogel, in [Kla96]

[Vui93] J.-C. Vuilleumier et al., Phys. Rev. **D 48** (1993) 1009

[Wil92] C. M. Will, *Theory and Experiment in Gravitational Physics*, 2nd edition (Cambridge University Press, Cambridge, 1992).

[Wol81] L. Wolfenstein, Phys. Lett **107 B** (1981) 77

[You95] Ke You et al., Phys. Lett. **B 265** (1995) 53

ELSEVIER

23 September 1999

PHYSICS LETTERS B

Physics Letters B 463 (1999) 77–82

Neutrino mass and magnetic moment in supersymmetry without R-parity in the light of recent data

Gautam Bhattacharyya [a,b], H.V. Klapdor-Kleingrothaus [a], Heinrich Päs [a]

[a] *Max-Planck-Institut für Kernphysik, P.O. Box 103980, D-69029 Heidelberg, Germany*
[b] *Saha Institute of Nuclear Physics, 1/AF Bidhan Nagar, Calcutta 700064, India*

Received 22 July 1999; accepted 11 August 1999
Editor: P.V. Landshoff

Abstract

We consider the generation of neutrino Majorana mass and transition magnetic moment by the lepton-number violating λ and/or λ' couplings in R-parity-violating supersymmetric models. We update (and improve) the existing upper limits on the relevant couplings using the most recent data on neutrino masses and mixings, indicating also the possible improvement by the GENIUS project. We study the implication of this update on the induced neutrino magnetic moment. © 1999 Elsevier Science B.V. All rights reserved.

Supersymmetry without R-parity [1] provides an elegant mechanism for generating neutrino (Majorana) masses and mixings. In these models, there are mainly two sources of neutrino mass generation. In one scenario, products of trilinear λ and/or λ' couplings generate a complete neutrino mass matrix through one-loop self-energy graphs [2,3]. In this framework, the diagrams that generate the masses also generate magnetic moments by insertion of photons to the internal lines. In the other scenario, the bilinear R-parity-violating (RPV) terms induce sneutrino vacuum expectation values (VEV's) allowing neutrinos to mix with the neutralinos. In this mechanism only one physical neutrino becomes massive [4]. In the present analysis, although we concentrate mainly on the trilinear L-violating couplings, we also comment on the possible impact of the bilinear parameters. We use the recent data on atmospheric [5] and solar [6] neutrinos, the measurement of ν_e mass in the Troitsk [7] and Mainz [8] tritium beta-decay experiments, and the measurement of the effective neutrino mass $\langle m_\nu \rangle$ in the Heidelberg–Moscow ^{76}Ge experiment [9], to update (and improve) the existing limits [10] on many different combinations of the trilinear couplings from their contribution to neutrino masses. We also calculate the magnetic moment of the neutrino, which has an intimate connection to its mass (for previous studies of neutrino magnetic moments in RPV models, see [11,12]).

Generally two types of magnetic moment may arise: (i) the Dirac-type magnetic moment that rotates ν_{eL} to ν_{eR}, and (ii) the transition magnetic moment that takes ν_{eL} to $\nu^c_{\mu R}$. If one adds a right-handed singlet neutrino to the standard model (SM), a non-zero Dirac mass of the neutrino implies a non-zero Dirac-type magnetic moment, given by [13]

$$\mu_\nu = \frac{3eG_F m_\nu}{8\pi^2\sqrt{2}} = 3\times 10^{-19}\left(\frac{m_\nu}{1\,\mathrm{eV}}\right)\mu_B, \qquad (1)$$

PII: S0370-2693(99)00947-8

Beyond the Desert 1997
Accelerator and Non-Accelerator Approaches

Proceedings of the First International Conference on Particle Physics Beyond the Standard Model, Castle Ringberg, Germany, 8–14 June 1997

Edited by H V Klapdor-Kleingrothaus and H Päs

vii

Contents

Institute of Physics Publishing
Bristol and Philadelphia

Lepton and Baryon Number Violation in Particle Physics, Astrophysics and Cosmology

Proceedings of the First International Symposium on Lepton and Baryon Number Violation (Lepton–Baryon 98), European Centre for Theoretical Physics, Trento, Italy, 20–25 April 1998

Edited by H V Klapdor-Kleingrothaus and I V Krivosheina

Contents vii

Institute of Physics Publishing
Bristol and Philadelphia

World Scientific
Singapore • New Jersey • London • Hong Kong

Proceedings of the Fifth International WEIN Symposium

Santa Fe, New Mexico, USA June 14–19, 1998

Physics Beyond the Standard Model

Editors

P. Herczeg & C. M. Hoffman
Los Alamos National Laboratory

H. V. Klapdor-Kleingrothaus
Max-Planck-Institut fur Kernphysik, Heidelberg

CONTENTS

BEYOND the DESERT 1999

Accelerator, Non-Accelerator and Space Approaches into the NEXT MILLENIUM

Proceedings of the Second International Conference on Particle Physics Beyond the Standard Model, Castle Ringberg, Germany, 6 – 12 1999

Edited by H V Klapdor-Kleingrothaus and I V Krivosheina

Contents

Institute of Physics Publishing
Bristol and Philadelphia

Appendix A

Ten years of Heidelberg–Moscow Experiment

The **Heidelberg-Moscow** experiment is already *for eight years* now THE MOST SENSITIVE double beta experiment worldwide. It has contributed in an extraordinary way to the research in neutrino physics and more general beyond standard model physics, and the limits for the latter are competing with those from the largest high-energy accelerators. It will keep its outstanding position in non-accelerator particle physics for several further years to come, before it may be succeeded by future large projects. Its unique position may justify that we give some historical keypoints of the development of this experiment and its background on the next pages.

The realization of the experiment goes back to first discussions in 1986 about a cooperation between Max Planck Institute für Kernphysik, Heidelberg, Leningrad Institute of Nuclear Physics, Gatchina, and Kurchatov Institute, Moscow (see [Pop-86*-V] and [Leg86*-A]). The first public announcement of the idea was made - by the author of this book - at the first WEIN conference in **Heidelberg 1986**, held on the occasion of the **600th** anniversary of the *University of Heidelberg.* Later conferences of this series at Montreal, Dubna, Osaka and Santa Fe, *WEIN98*, USA, provided centers of active discussions of among others double beta decay and related particle physics. A proposal of the experiment was presented at the *MPI* Heidelberg in *1987* [Kla87**-A]. In the early time we also tried to include *French* colleagues from *Bordeaux* into a collaboration. The search for suitable sites for the experiment included at that time *Baksan (Kaukasus)*, *Solotvina (Ukraine)*, *Gran Sasso (Italy)* and *Frejus (France).* On the basis of exploratory research of the different parties on the optimal background, the decision was finally made to use the shielding suggested and built by the *Heidelberg* group, and to locate the experiment in the *Gran Sasso Underground Laboratory (LNGS)*, and the collaboration between *MPI Heidelberg* and *Kurchatov Institute, Moscow* was formed [Mem88*-A], [Pro88*-A], [Let88**-A]. The president of the *Istituto Nazionale di Fisica Nucleare (INFN)*, at that time *Prof. N. Cabibbo*, and the director of the *Laboratori Nazionali di Gran Sasso*, *Prof. E. Bellotti*, generously supported the installation of the experiment. With the support of the *LNGS* the experimental building of the experiment was built between Halls A and B, into which the first enriched ^{76}Ge detector **(the first high-purity enriched Ge detector worldwide)** was installed in *Gran Sasso* in *July 1990*. First preparational work had been done since *1989* in a provisional tent in Hall C. Into the same year of *1990* fell the edition of the English version of our book, with K. Grotz, on "The Weak Interaction in Nuclear, Particle and Astrophysics", followed by the Russian edition in *1992* and the Chinese version in *1996*. In *1992*, as a by-product of this experiment we provided together with our Russian colleagues two enriched ^{70}Ge detectors, which were launched in the same year in a joint experiment with NASA in the GRIS and HEXAGONE ballon experiments in Alice Springs Australia, to do research into γ-rays for the center of the galaxy [Kla90**-V], [MPG94*-A], [Bar94]. The full amount of five enriched ^{76}Ge detectors of *in total* 11 kg was finally installed in 1995 and were operated since 1996 with a newly developped pulse shape discrimination method. Since this time the full final experiment is delivering data. *1995* was also the year of the edition of the German and English editions of our book, with A. Staudt, on "Non-Accelerator Particle Physics", followed by a Russian edition in *1997*, the year in which also the German and English editions of our book, with K. Zuber, on "Particle Astrophysics" were published.

While during the eighties and the early nineties the Heidelberg group made important contributions to the investigation of nuclear matrix elements for double beta decay, it contributed in the last five years extensively also to the theoretical exploration of the wide potential of double beta decay to study branches of beyond standard model particle physics other then neutrino physics.

Since *1992/93* the experiment yields the sharpest limits on double beta decay. Since some years it also provides the most sensitive raw data in the search for cold dark matter (*WIMPs*) in the universe [HM98*-III]. The present limits - in the year *2000*, the tenth year since operation of the first detector of this experiment in *Gran Sasso* - on the half-life for neutrinoless double beta decay and the Majorana neutrino mass are 3.1×10^{25}years and 0.27 eV, respectively [HM2000**-III]. The sensitivity thus entered into a range where the results decisively influence neutrino mass scenarios and cosmological parameters presently considered on the basis of the most recent neutrino oscillation experiments such as *Superkamiokande* etc. and Cosmic Microwave Background observations.

H.V. Klapdor-Kleingrothaus
Max-Planck Institut für Kernphysik
Heidelberg, Germany
5 July, 2000

Fig. A.1 View on the castle, Altstadt and the city hall, location of the first WEIN Conference, in 1986 in Heidelberg.

Fig. A.2 Heidelberg at the Neckar, at a summer evening during 'castle illumination'.

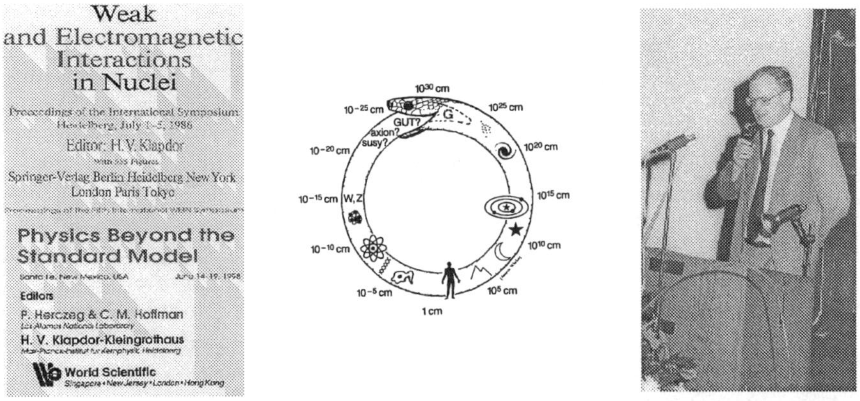

Fig. A.3 Proceedings of WEIN Conferences - WEIN'86 (Heidelberg, 1986) and WEIN'98 (Santa Fe, 1998) (left). The symbol of the first WEIN'86 and second WEIN'89 Conference (middle). Right: Author at WEIN'86.

Fig. A.4 Left: from left to right - the author, Jules Deutsch, Rabindra Mohapatra, Jose W. Valle during WEIN86. Middle: Participants during concert in 'Alte Aula' of Heidelberg University during WEIN86. Right: Session during WEIN86, in the Heidelberg city hall.

Fig. A.5 WEIN'98. Left: The author, having the pleasure of the presence of his parents, at WEIN86. Middle: The first page of the books by the author and by the author with B. Povh, edited in 1988, reflecting further concentration on neutrino physics at Heidelberg. Right: During the WEIN conference, in the great hall of the City Hall.

С С С Р

Институт атомной энергии имени И. В. Курчатова

123182, Москва, площадь им. И. В. КУРЧАТОВА

Prof. Dr. H.V. Klapdor
Max-Planck-Institut
fur Kernphysik
D-6900 Heidelberg 1
FRG

July 8, 1986

Dear Prof. Dr. H.V. Klapdor,

Thank you for your letter in which a collaboration is proposed in the realization of the experiment on investigating the neutrinoless double-beta-decay of germanium-76. Presently, the investigations in this field are headed by Acad. S.T.Belyaev. We plan to carry out the experiment (or the first phase of the work) in one of the deepest salt (NaCl) mines on the territory of the USSR during the period from 1987 to 1989.

Your proposal to prepare HP Ge semiconductor detectors from isotope-enriched germanium seems interesting to us. I suggest a discussion of the collaboration program with the participation of all the interested sides at the I.V.Kurchatov Institute of Atomic Energy in Moscow at some mutually suitable time. I would highly appreciate receiving your proposals on the data of the meeting in the nearest future.

Yours sincerely,

Acad. V.A.Legasov

Fig. A.6 Letter of Academician V.A. Legasov, replying to our proposal of a joint double beta experiment.

1.3 Möglichkeiten des vorgeschlagenen Experiments

An den Anfang dieses Abschnittes sei das Resumé gesetzt, das Prof. Caldwell, der gegenwärtig das weitestgetriebene ^{76}Ge-Experiment betreibt, in seinem Review-Talk [8] auf der Neutrino '86-Konferenz in Sendai/Japan im Juni 1986 gegeben hat (hierzu ist es zweckmäßig, sich Tabelle 2 vor Augen zu halten):

"... While the limits have improved appreciably in recent years, further improvements will come slowly. Quantities of Ge will not undergo very large increase, backgrounds will improve but not by large factors, and the limit improves only as the square root of the counting time. The one way there could be a significant change in the prospects for Ge experiments is if the Moscow-Leningrad-Heidelberg collaboration succeeds in fully utilizing the large quantity of enriched ^{76}Ge they and they alone are able to obtain. The promised 15 kg of ^{76}Ge would be worth several hundred million dollars in the West. If that could all be used effectively, they would have a factor 20 advantage over the UCSB-LBL experiment. ... the next big step for Ge experiments should be taken by this group."

Fig. A.7 Citation out of the review talk presented by D.O. Caldwell at the NEUTRINO'86 conference at Sendai/Japan in June 1986 [Cal86a*-I], given in the Proposal written by the author in 1987 [Kla87**-A], see below.

MAX-PLANCK-INSTITUT FÜR KERNPHYSIK

PROF. DR. H.V. KLAPDOR

D-6900 HEIDELBERG 1

Postfach 10 39 80
Saupfercheckweg 1 (beim Bierhelder Hof)
Telefon Sammel-Nr. (0 62 21) 51 61
Durchwahl (0 62 21) 516 262/259
Telex 4 61 666
Telefax (0 62 21) 51 65 40

Prof. Zichichi
CERN

CH-1211 Genève 23

SWITZERLAND

27th June, 1988

Dear Prof. Zichichi,

As a joint project with our colleagues from Kurchatov Institute in Moscow we are at present preparing a double beta decay experiment using enriched ^{76}Ge as source and detector. A pilot experiment using one enriched detector of 1 kg shall start in a few months from now. The full experiment using about 10 enriched detectors of a total amount of about 10 kg of enriched ^{76}Ge is planned to come into operation in about 1 - 2 years. This would allow to explore the neutrino mass down to a limit of about 0.2 eV.

We would be interested to perform the pilot experiment - and later eventually also the full experiment - in the Gran Sasso laboratory. Would you please inform us about the possibility to obtain corresponding space starting from fall of this year and also to whom we should address an official application.

I look forward to your answer.

Sincerely yours,

H.V. Klapdor

Fig. A.8 Letter to Prof. A. Zichichi exploring the possibility to perform the Heidelberg-Moscow experiment in the Gran Sasso Underground Laboratory, 1988.

Fig. A.9 Seminar on Double Beta Decay at Baksan, Kaukasus, 26 February, 1987.

Fig. A.10 Searching for a site for the project with Leningrad colleagues: the author with Ludvik Popeko (right) and Igor Kondurov at Baksan (Cheget), February 1987. Middle left: Irina Krivosheina from the Baksan Neutrino Scintillation Telescope, which a few days before had observed neutrinos from Supernovae 1987A.

Fig. A.11 For negotiations at Leningrad, November, 1988 (left and middle) and at Moscow, December, 1988.

Fig. A.12 For negotiations in Rome, October 1988 (left), and in Leningrad, November, 1988, (middle), and in Moscow after signing the Protocoll given on the next page (right), 3 December, 1988.

MEMORANDUM

On November 29 and December 2 - 3, 1988 a meeting of the representatives of the IAE and MPI took place at the I.V.Kurchatov Institute of Atomic Energy in connection with signing a protocol about the joint cooperation in search for the double beta decay of Ge-76.

The following problems were discussed at the meeting:
- discussion and working out of the programme and schedule;
- the number and dimensions of crystals of the first batch;
- the status of experimental studies on determination of the background conditions at the IAE and MPI and the prospects for the development of experimental installations.

As a result of the discussion, the sides have agreed upon the following.

(1) During the first stage the IAE prefers to have crystals of the size: ⌀ 50 mm and 70 mm long. The MPI may choose a different size. We have agreed to discuss the details by correspondence. The number and sizes of crystals for the second stage of experiment should be decided upon termination of the first stage.

(2) The experiments on determination of the background conditions for the installations are performed:
- by the Soviet side at the Solotvino underground laboratory of the Institute of Nuclear Physics of the Academy of Sciences of the Ukrainian SSR;
- By the German side in the Grand Sasso tunnel (in Italy).

(3)To ensure the performance of joint operations, the sides have agreed to exchange representatives within the framework of quotas stipulated by the programme of cooperation. The first visit of the IAE representatives to the MPI (FRG) will take place in the first quarter of 1989, while that of the MPI representatives to the USSR, in the second quarter of 1989. During the visit of the Soviet delegation to the FRG, the MPI will organize a meeting with the PGT representatives so that the Soviet delegation could discuss technical problems of the detector design.

(4)The MPI will consider the possibility of shortening the time required for the production of the second batch of crystals. For its part, the IAE is ready to additionally give 2 - 2.5 kg of enriched germanium so as to increase the number of crystals and to decrease the time of their production.

(5)To promote the joint work, the MPI will consider the possibility of giving over to the IAE of a computer and some electronic blocks.

During the meeting the IAE gave over to the MPI 7.49424 kg of 76-isotope-enriched germanium dioxide.

On behalf of
the scientifi supervisor,

Acad. S.T.Belyaev

On behalf of the MPI,
the scientific supervisor,

Prof. H.V.Klapdor

Fig. A.13 Memorandum signed in Moscow, December, 1988.

Südd. Ztg. 19.12.88

Suchaktion im Tunnel

Deutsch-sowjetisches Experiment mit Neutrinos

Anfang dieses Monats wurden siebeneinhalb Kilogramm Germaniumdioxid von Moskau nach Heidelberg gebracht. Es handelt sich um das sehr seltene Isotop Germanium-76. Mit dessen Hilfe wird jetzt ein Experiment begonnen, das für die physikalische Grundlagenforschung wichtig sein könnte. Das Kurchatov-Institut für Atomenergie in Moskau und das Max-Planck-Institut für Kernphysik in Heidelberg wollen die Frage beantworten, ob das Neutrino genannte Elementarteilchen eine Masse besitzt. Leiter der deutschen Forschergruppe ist Hans Volker Klapdor.

Das Neutrino entsteht unter anderem beim radioaktiven, sogenannten Betazerfall von Atomkernen. Dabei bezeichnet man als Betastrahlung das Freiwerden von jeweils einem Elektron bei einem Zerfallsprozeß.

Bei dem russisch-deutschen Experiment soll jedoch der sogenannte doppelte Betazerfall untersucht werden. Beim gewöhnlichen Betazerfall wandelt sich im Atomkern ein Neutron in ein Proton und ein Elektron um. Zusätzlich entsteht ein Neutrino, genauer: ein Antineutrino. Den doppelten Betazerfall gibt es zum Beispiel beim Germanium-76. Ein Neutron im Germaniumatomkern zerfällt und sendet ein Neutrino aus, das als Zwischenprodukt von einem zweiten Neutron eingefangen wird. Bei diesem Vorgang werden zwei Elektronen emitiert; daher der Name „doppelter Betazerfall". Die Zerfallszeit des Germanium-76-Atomkerns wird von der Masse des Zwischenprodukts beeinflußt, ist also davon abhängig, ob das Neutrino im Ruhezustand eine Masse hat. Germanium-76 zerfällt nahezu unendlich langsam und die genaue Zerfallszeit (Halbwertzeit ungefähr fünf mal zehn hoch 23 Jahre) ist äußerst schwer zu ermitteln. Man versucht dies jetzt im Gran-Sasso-Tunnel bei Rom. Die Beobachtung des doppelten Betazerfalls erfordert eine Abschirmung gegen kosmische Strahlungseinflüsse. In der Sowjetunion läuft ein Parallelversuch. Selbst die extrem geringe Masse eines Neutrinos, welche einer Energie von 0,2 Elektronenvolt entspricht, könnte man noch erkennen.

PETER ZAUN

ABGESCHIRMT *von kosmischer Strahlung, im Gran-Sasso-Tunnel bei Rom wollen deutsche und russische Forscher gemeinsam der Frage nachgehen, ob das Neutrino genannte Elementarteilchen im Ruhezustand eine Masse hat; eine Frage, die für die Abschätzung der Massen des Universums entscheidend ist.*

Photo: Klapdor/MPI für Kernphysik/Heidelberg

Fig. A.14 Süddeutsche Zeitung 19.12.1988.

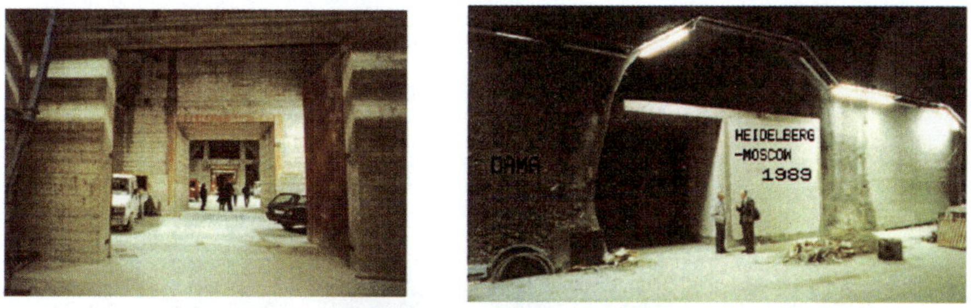

Fig. A.15 In the Gran-Sasso underground laboratory, October, 1988 (left). The building for the HEIDELBERG-MOSCOW experiment being under construction, November 1989. In front Spartak T. Belyaev and Hans V. Klapdor-Kleingrothaus (right).

Fig. A.16 The second part of enriched Germanium (white package on the table) arrived in Heidelberg in March, 1989 (left and right). On the left figure: The two spokesmen of the HEIDELBERG-MOSCOW collaboration, H.V. Klapdor-Kleingrothaus and S.T. Belyaev (right) and A. Staudt, member of the group and coauthor of the book on Non-Accelerator particle Physics. On right figure: with further members of the group - A. Müller (left), V.I. Lebedev, A. Balysh, G. Heusser (from right). Middle figure: The first transport bringing technics, electronics and shielding material to the Gran Sasso Laboratory, starting from MPI Heidelberg in early 1989.

Fig. A.17 The provisional test set-up in a tent in hall C of the Gran Sasso, November 1989. The author with Spartak T. Belyaev.

Fig. A.18 On top: The monument for the first climbers of the Corno Grande (2912m), covering the Gran Sasso Underground laboratory, in the city of Aquila, founded by the German Emperor and King of Sicily (and Jerusalem) Friedrich II in the early 13. century, who also introduced the decimal system in his Empire, i.e. in Europe (foto author, 1990). Down: Castle in L'Aquila 1990.

Unique Detector Involves EG&G, Germany and USSR

Nuclear Europe Worldscan 3–4/1991 51

EG&G Ortec has just produced an ultra-large single crystal of germanium highly enriched in ^{76}Ge. The high concentration of this particular germanium isotope makes possible a new generation of experiments to search for neutrino mass.

While previous measurements have demonstrated that the neutrino, if it does have mass, cannot have enough to be the "missing mass" needed to keep the universe from expanding forever, a non-zero mass for the neutrino will have a major impact on theories of the fundamental structure of matter.

Professor H. V. Klapdor-Kleingrothaus of the Max Planck Institute in Heidelberg has commissioned EG&G Ortec to produce a radiation detector from this unique crystal which will act as both the source of an extremely rare radioactive nuclear decay mode (neutrinoless double-beta decay) and as the means for its detection.

For the theoretical physicist, such a decay mode (which implies lepton number non-conservation) will overthrow the now well accepted Standard Model, which unified two of the four fundamental forces of the universe, and also the simplest of the Grand Unified Theories, such as the SU(5) model, which unified three of these fundamental forces.

The germanium crystal, which weighs 3.5 kg, is virtually priceless, since the production of the polycrystalline germanium, suitably enriched in ^{76}Ge, required a multimillion-dollar, multi-year processing effort in the USSR under the direction of Professor S. T. Belayev of the Kurchatov Institute in Moscow.

Company News

Fig. A.19 Information on the second enriched Ge detector produced for the HEIDELBERG-MOSCOW experiment, at that time the largest high-purity Ge detector worldwide.

Fig. A.20 Left figure: The two spokesmen (right) with part of the Russian team at Gran Sasso, November 1989. Middle: The two spokesmen H.V. Klapdor-Kleingrothaus and S.T. Belyaev at Gran Sasso, November 1989. Right: After finishing construction of the site of the experiment between halls A and B, in Gran Sasso, in July 1990. The author with visiting Prof. Masato Morita in July 1990.

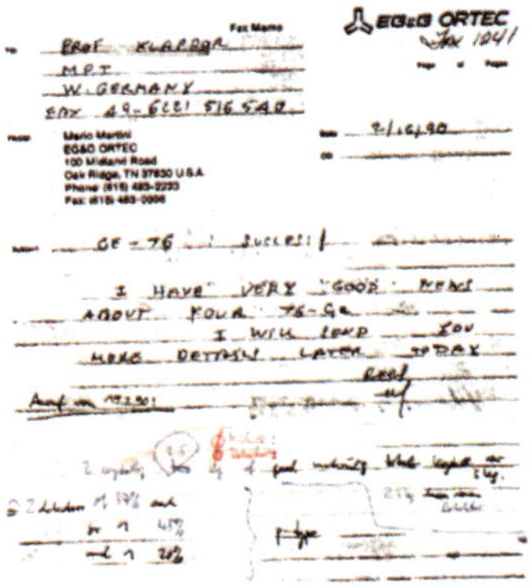

EG&G ORTEC
Fax 1041

PROF. KLAPDOR
M.P.I.
W. GERMANY
FAX 49-6221 516 540

Mario Martini
EG&G ORTEC
100 Midland Road
Oak Ridge, TN 37830 U.S.A.

GE-76 SUCCESS!

I HAVE VERY GOOD NEWS ABOUT YOUR 76-Ge. I WILL SEND YOU MORE DETAILS LATER TODAY.
Regards

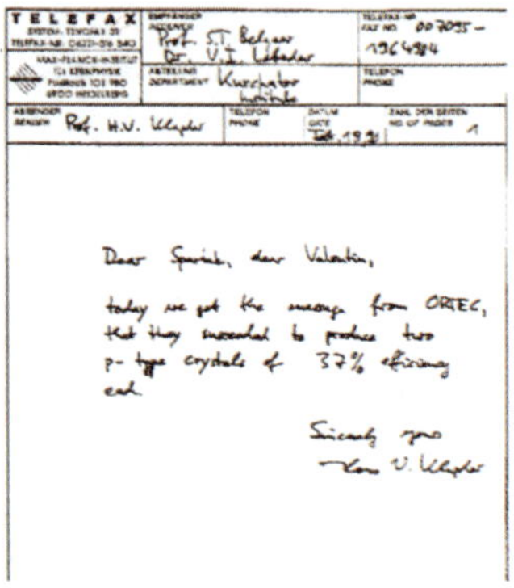

TELEFAX
MAX-PLANCK-INSTITUT FÜR KERNPHYSIK

Prof. S.T. Belyaev
Dr. V.I. Lebedev
Kurchatov Institute
Fax No. 007095-1964884
Prof. H.V. Klapdor

Dear Spartak, dear Valentin,

today we got the message from ORTEC, that they succeeded to produce two p-type crystals of 37% efficiency each.

Sincerely yours
Hans V. Klapdor

Fig. A.21 Left: The letter from ORTEC company informing about the production of the first enriched high-purity ^{76}Ge crystal worldwide (16 February, 1990): "I have very good news about your 76-Ge. I will send your more details later today. Regards, Mario Martini, EG&G ORTEC". Right: The letter from the author, informing the Moscow colleagues. Middle: The first enriched detector in its shielding from electrolytic copper in the Gran Sasso laboratory in July 1990.

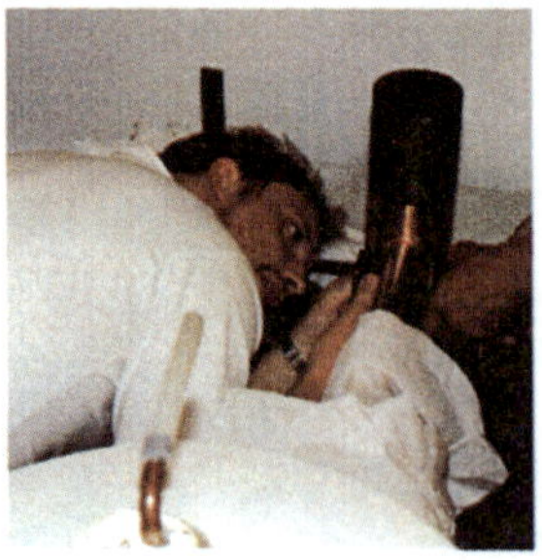

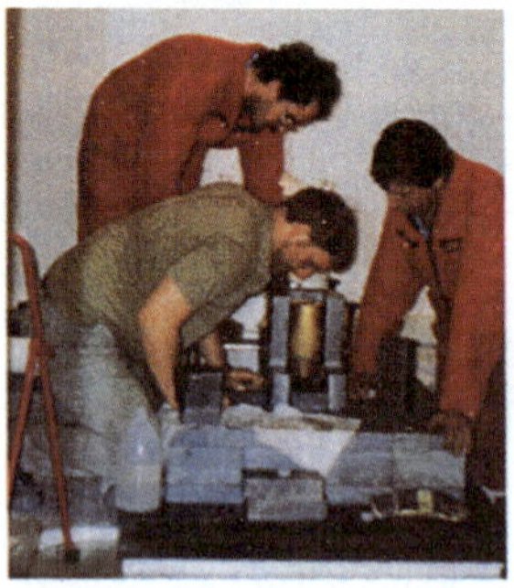

Fig. A.22 During installation of the first ^{76}Ge crystal in Gran Sasso, by the *German team*, July 20 – 27, 1990. Left: Joachim Echternach. Middle: Kai Zuber, J. Echternach and Herbert Strecker. Right: (from left to right) J. Echternach, H. Strecker, Gerd Heusser, Andreas Piepke and K. Zuber.

JUL 25 '90 10:48 NASA/GSFC 301-2869240 PAGE.02

National Aeronautics and
Space Administration

Goddard Space Flight Center
Greenbelt, Maryland
20771

July 24, 1990

Reply to Attn of

Dear Prof. Klapdor:

I very much enjoyed my abbreviated visit to your Institute on July 4. The news on Germanium enrichment was extremely encouraging. Durouchoux and I have been discussing your proposals with our colleagues. Everyone is enthusiastic about test flights with enriched detectors and about using enriched detectors in the INTEGRAL mission. We have agreement from the leaders of the balloon programs to include you on the scientific as well as the technology papers from the balloon flights. Also, the INTEGRAL team has agreed to have MPI-Heidelberg and Russian collaborators on our upcoming (June 1991 - June 1992) Phase A study for the mission. We have arranged to have letters FAXed to you from the balloon and INTEGRAL team leaders. I am FAXing the letter from our balloon collaboration (P.I. - B. Teegarden) with this note.

Best Regards

Neil

Neil Gehrels

Fig. A.23 Letter from NASA, July 1990, on our proposed space collaboration with enriched ^{70}Ge detectors.

Fig. A.24 Launching the **GRIS** ballon experiment, in collaboration of *ESA, NASA, MPI Heidelberg and Kurchatov Institute Moscow*, carrying an enriched ^{70}Ge detector, at Alice Springs, Australia, May 1992. In the foreground of the left photo the author (photos author).

Fig. A.25 Left: The Gran Sasso mountains. Right: Label near the main trasse to Gran-Sasso Laboratory.

Fig. A.26 Left: The first director of LNGS, Prof. Enrico Bellotti (the second from the left), at the first provisional site of the Heidelberg-Moscow experiment in Gran Sasso, in November, 1989 (foto author). Middle: Prof. Nicola Cabibbo, President of INFN, Italy, who supported strongly the HEIDELBERG-MOSCOW experiment in Gran-Sasso, during opening of the Gallex experiment in Gran Sasso, 30. November, 1990. Right: Prof. Piero Monacelli, second director of LNGS, at the Double-Beta Decay and Related Topics Conference, Trento, Italy, 1995, (foto author).

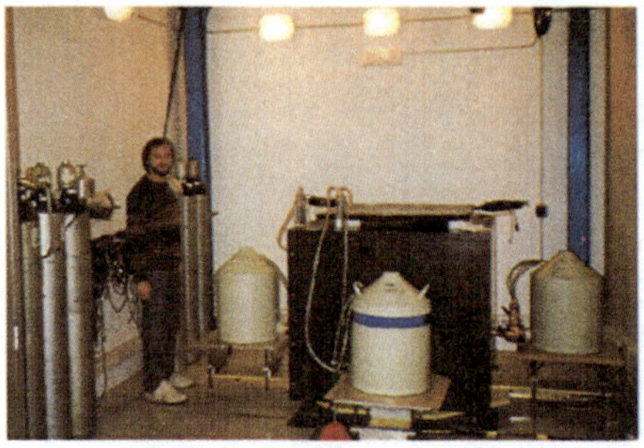

Fig. A.27 Left: A. Müller, at the final setup for four detectors of the Heidelberg-Moscow experiment in the Gran-Sasso, in January 1996 (foto author). Right: During the visit of the President of Italy Carlo A. Ciampi (second from left) at the Gran Sasso Laboratory, September 1999 (passing the hall of the HEIDELBERG-MOSCOW Double Beta and the DAMA Experiments.) The present director of the LNGS, Prof. Alessandro Bettini is the fourth from the left side (white jacket), on right side Prof. H.V. Klapdor-Kleingrothaus (foto I. Krivosheina).

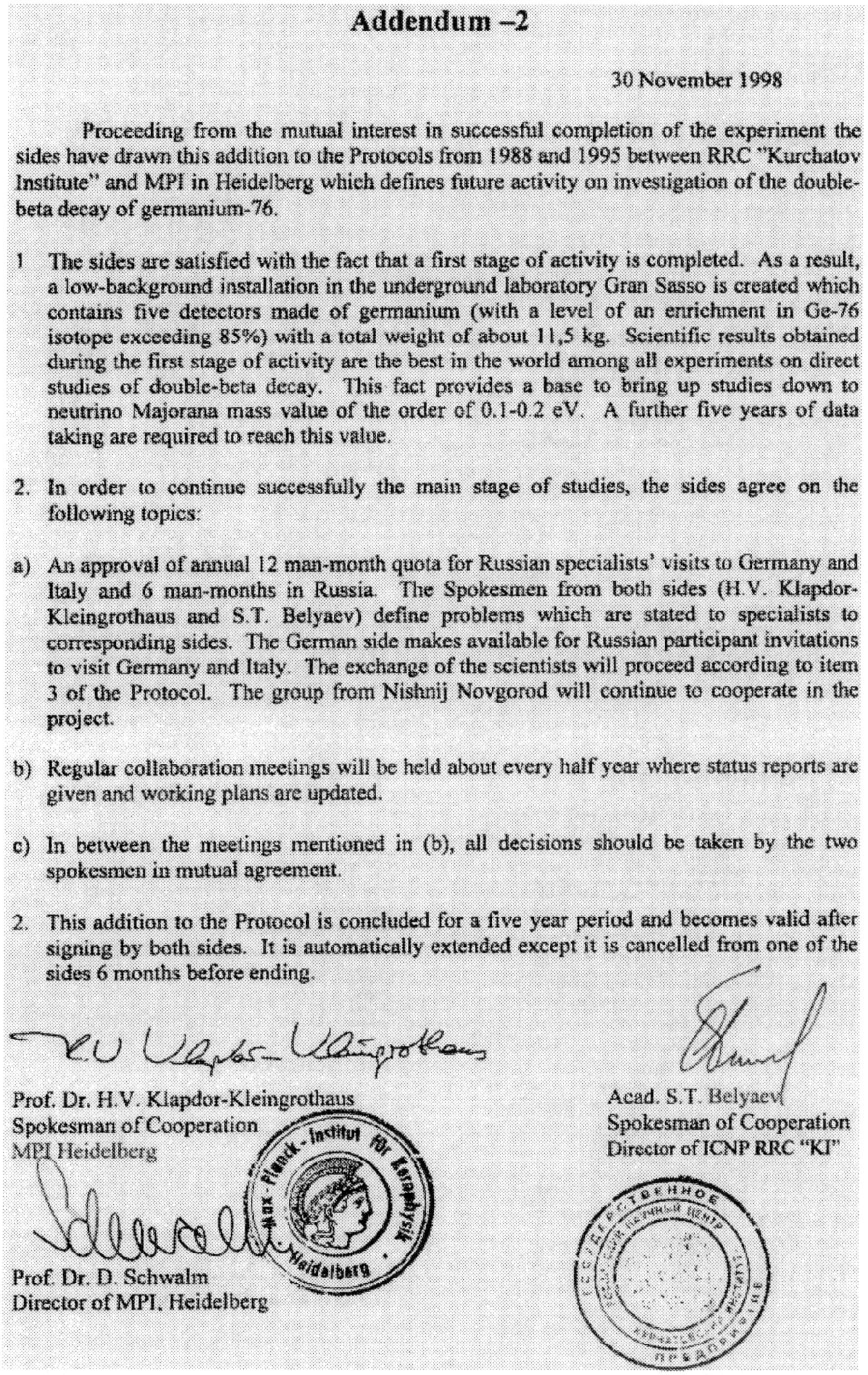

Addendum –2

30 November 1998

Proceeding from the mutual interest in successful completion of the experiment the sides have drawn this addition to the Protocols from 1988 and 1995 between RRC "Kurchatov Institute" and MPI in Heidelberg which defines future activity on investigation of the double-beta decay of germanium-76.

1 The sides are satisfied with the fact that a first stage of activity is completed. As a result, a low-background installation in the underground laboratory Gran Sasso is created which contains five detectors made of germanium (with a level of an enrichment in Ge-76 isotope exceeding 85%) with a total weight of about 11,5 kg. Scientific results obtained during the first stage of activity are the best in the world among all experiments on direct studies of double-beta decay. This fact provides a base to bring up studies down to neutrino Majorana mass value of the order of 0.1-0.2 eV. A further five years of data taking are required to reach this value.

2. In order to continue successfully the main stage of studies, the sides agree on the following topics:

a) An approval of annual 12 man-month quota for Russian specialists' visits to Germany and Italy and 6 man-months in Russia. The Spokesmen from both sides (H.V. Klapdor-Kleingrothaus and S.T. Belyaev) define problems which are stated to specialists to corresponding sides. The German side makes available for Russian participant invitations to visit Germany and Italy. The exchange of the scientists will proceed according to item 3 of the Protocol. The group from Nishnij Novgorod will continue to cooperate in the project.

b) Regular collaboration meetings will be held about every half year where status reports are given and working plans are updated.

c) In between the meetings mentioned in (b), all decisions should be taken by the two spokesmen in mutual agreement.

2. This addition to the Protocol is concluded for a five year period and becomes valid after signing by both sides. It is automatically extended except it is cancelled from one of the sides 6 months before ending.

Prof. Dr. H.V. Klapdor-Kleingrothaus
Spokesman of Cooperation
MPI Heidelberg

Acad. S.T. Belyaev
Spokesman of Cooperation
Director of ICNP RRC "KI"

Prof. Dr. D. Schwalm
Director of MPI, Heidelberg

Fig. A.28 Addendum to Protocol of 1988, between MPI Heidelberg and Kurchatov Institute, Moscow, for continuation of the cooperation on the project "Double Beta Decay", November, 1998.

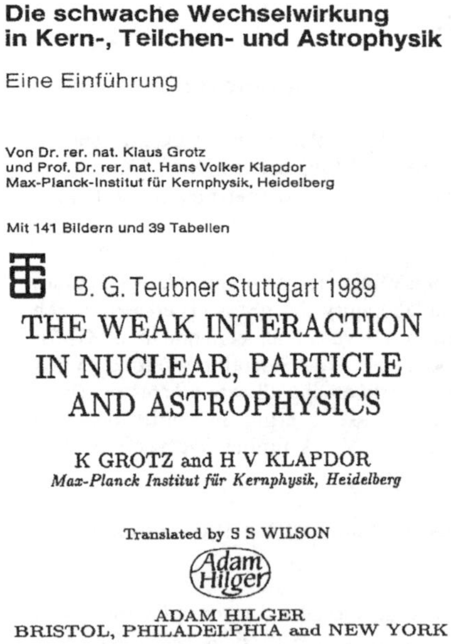
Die schwache Wechselwirkung in Kern-, Teilchen- und Astrophysik

Eine Einführung

Von Dr. rer. nat. Klaus Grotz
und Prof. Dr. rer. nat. Hans Volker Klapdor
Max-Planck-Institut für Kernphysik, Heidelberg

Mit 141 Bildern und 39 Tabellen

B. G. Teubner Stuttgart 1989

THE WEAK INTERACTION IN NUCLEAR, PARTICLE AND ASTROPHYSICS

K GROTZ and H V KLAPDOR
Max-Planck Institut für Kernphysik, Heidelberg

Translated by S S WILSON

ADAM HILGER
BRISTOL, PHILADELPHIA and NEW YORK

Fig. A.29 First book, on 'Weak Interaction..', by H.V. Klapdor-Kleingrothaus and K. Grotz - German and English editions 1989, 1990.

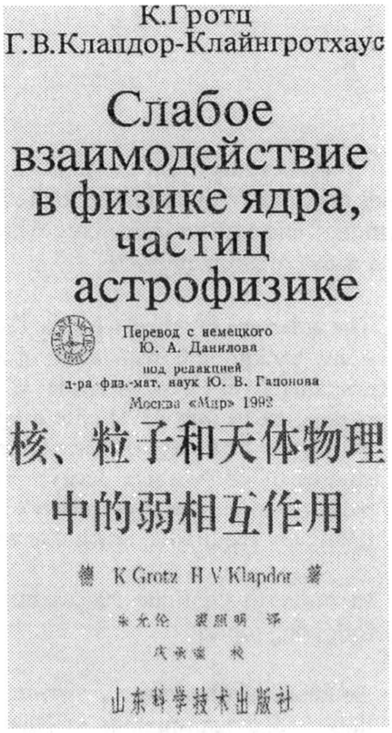
К.Гротц
Г.В.Клапдор-Клайнгротхаус

Слабое взаимодействие в физике ядра, частиц астрофизике

Перевод с немецкого
Ю. А. Данилова
под редакцией
д-ра физ.-мат. наук Ю. В. Гапонова
Москва «Мир» 1992

核、粒子和天体物理中的弱相互作用

K Grotz H V Klapdor

山东科学技术出版社

Fig. A.30 First book, on 'Weak Interaction..', by H.V. Klapdor-Kleingrothaus and K. Grotz. - Russian and Chinese editions, 1992, 1996.

Teilchenphysik ohne Beschleuniger

Von Prof. Dr. rer. nat. Hans Volker Klapdor-Kleingrothaus
Max-Planck-Institut für Kernphysik, Heidelberg
und Dr. rer. nat. Andreas Staudt
Zentrale Forschung Bayer AG, Leverkusen
Mit zahlreichen Abbildungen und Tabellen

B. G. Teubner Stuttgart 1995

Non-accelerator Particle Physics

H V Klapdor-Kleingrothaus

and

A Staudt

Translated by S S Wilson

Institute of Physics Publishing
Bristol and Philadelphia

В. Клапдор-Клайнгротхаус
А. Штаудт

Неускорительная физика элементарных частиц

Перевод с немецкого В. А. БЕДНЯКОВА

МОСКВА
НАУКА • ФИЗМАТЛИТ 1997

Fig. A.31 Second book, on 'Non-accelerator Particle Physics', by H.V. Klapdor-Kleingrothaus and A. Staudt: German, English and Russion editions.

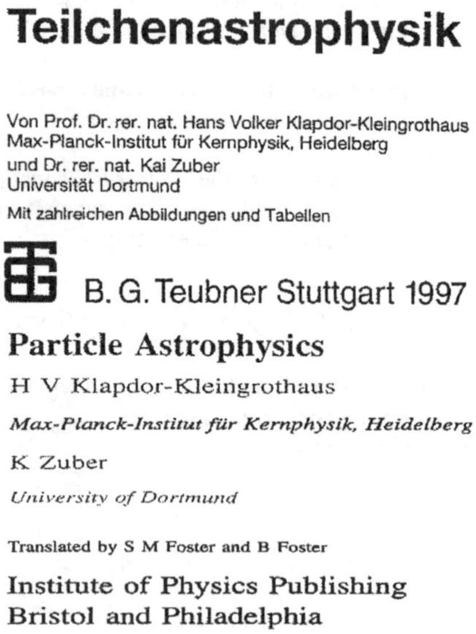
Teilchenastrophysik

Von Prof. Dr. rer. nat. Hans Volker Klapdor-Kleingrothaus
Max-Planck-Institut für Kernphysik, Heidelberg
und Dr. rer. nat. Kai Zuber
Universität Dortmund

Mit zahlreichen Abbildungen und Tabellen

B. G. Teubner Stuttgart 1997

Particle Astrophysics

H V Klapdor-Kleingrothaus

Max-Planck-Institut für Kernphysik, Heidelberg

K Zuber

University of Dortmund

Translated by S M Foster and B Foster

Institute of Physics Publishing
Bristol and Philadelphia

Fig. A.32 Third book, on 'Particle Astrophysics', by H.V. Klapdor-Kleingrothaus and K. Zuber : German and English editions 1997 and paper back 2000.

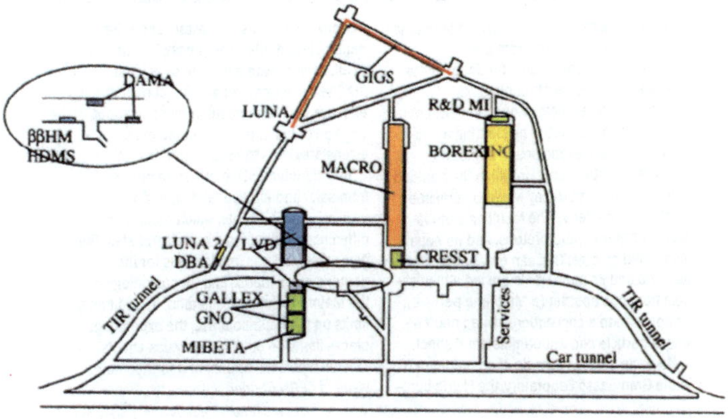

Fig. A.33 Map of the Gran Sasso underground laboratory and approximate location of running experiments. $\beta\beta$ HM and HDMS denote the **HEIDELBERG-MOSCOW** experiment and our HEIDELBERG DARK MATTER SEARCH - **HDMS** Experiment, respectively.

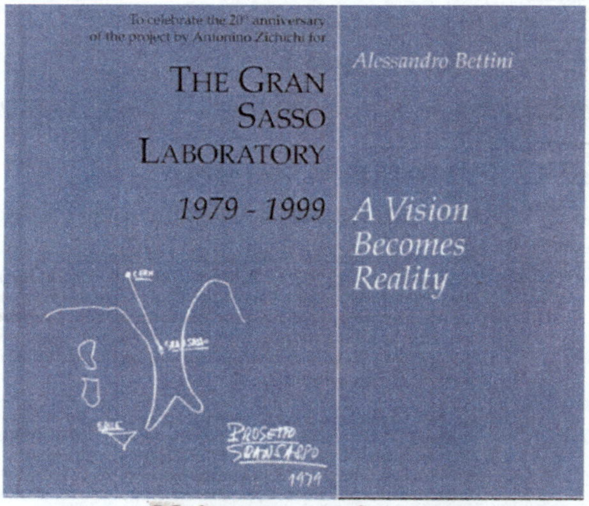

Majorana neutrino mass

. . . As a consequence, the search for the neutrinoless double beta decay is the search for the neutrino Majorana mass. No experiment has ever detected a neutrinoless double beta decay, but limits have been given. The most sensitive limit worldwide is from the Heidelberg Moscow experiment at Gran Sasso, . . .

Fig. A.34 Book edited by Prof. A. Bettini, present director of LNGS, on the Gran Sasso Laboratory in 1999. The sentence is taken from this book, page 27.

NEWS

Majorana mass limit reaches an all-time low

The lowest mass limit, so far, on the Majorana electron neutrino comes from the Heidelberg-Moscow search for double beta decay, after 10 years of running.

In normal beta decay, a nuclear neutron transforms into a proton by emitting a neutrino. A far more exotic possibility is two successive beta decays, in which the neutrino emitted in the first decay would be absorbed by the second decay. The resulting isotope would have two more protons, and no neutrinos would emerge. This can only happen if the neutrino and its antiparticle are indistinguishable from one another (a Majorana particle), as opposed to a conventional Dirac neutrino, whose particle and antiparticle are distinct.

Using an 11.5 kg sample of germanium-76, in the Gran Sasso Laboratory, the Heidelberg-Moscow search has now established that neutrinoless double beta decay, if it happens at all, does so with a lifetime of at least 10^{25} years: a world record. The corresponding effective Majorana neutrino mass (a superposition of the different neutrino mass eigenstates) has to be less than 0.2 eV.

This complements information obtained from solar and atmospheric neutrino oscillation experiments, which determine differences in neutrino mass eigenstates. The new mass limit has implications for the neutrino mass matrix, and for cosmology in the Majorana neutrino scenario. In addition to limits on the neutrino mass, the experiment places limits on other new physics effects.

A full report will appear in a forthcoming issue of *CERN Courier*.

8 CERN Courier *January/February 2000*

Fig. A.35 Note in CERN Courier, January, 2000.

NZZ **Online** NZZ-Archiv

Mittwoch, 19. Januar 2000
Tagesausgabe | Monatsarchiv | Suchen in Tagesausgabe | Suchen im Monatsarchiv

Neue Zürcher Zeitung
Frontpage
Finanzmärkte
Dossiers
English Window
News Ticker
Echo der Zeit
NZZ-Archiv
Service
Anzeigen
Mail/Leserdienst
Hilfe/Index
NZZ-Sites

PDF-Version | Postscript-Version | RTF-Version

NZZ Tagesausgabe

Neue Zürcher Zeitung FORSCHUNG UND TECHNIK Mittwoch, 19.01.2000 Nr.15 65

Warten auf das Unerwartete

Suche nach verbotenem Kernzerfall setzt neuer Physik Grenzen

Die Masse des Neutrinos ist eine Schlüsselgrösse in der Elementarteilchenphysik. Sie liefert Hinweise, wie eine Theorie jenseits des Standardmodells aussehen könnte. Ein Experiment, das nach einem Zerfall von Atomkernen sucht, der gemäss dem Standardmodell verboten ist, liefert die bis heute strikteste Obergrenze für die Neutrinomasse. Von dieser Grenze hängt es ab, ob Neutrinos als Kandidaten für die dunkle Materie in Frage kommen.

Spe. Das Standardmodell der Elementarteilchenphysik, das die heute bekannten Elementarteilchen und die zwischen ihnen wirkenden Kräfte beschreibt, ist über die Jahre hinaus mit immer grösserer Genauigkeit bestätigt worden. Und doch sind vor allem theoretische Physiker aus verschiedenen Gründen mit dieser Theorie unzufrieden. Entwürfe, die über das Standardmodell hinausgehen, gibt es eher zu viele als zu wenige, doch solange handfeste Belege für «neue Physik» fehlen, ist die eine Theorie so gut oder so schlecht wie die andere. Auf Mängel des Standardmodells deutet einzig und allein die Beobachtung hin, dass Neutrinos aller Wahrscheinlichkeit nach eine Masse besitzen. Doch heute weiss man weder, wie gross diese Masse absolut genommen ist, noch kennt man die Natur dieser Teilchen. Bevor man Neutrinos als Leitfaden für die Konstruktion einer verbesserten Theorie der Elementarteilchen heranziehen kann, muss man deshalb mehr über diese Teilchen in Erfahrung bringen.

Ein Experiment, das hier weiterhelfen könnte, ist das vom Max-Planck-Institut für Kernphysik in Heidelberg und vom Kurchatow-Institut in Moskau gemeinsam betriebene Heidelberg-Moskau-Experiment. Seit nunmehr zehn Jahren sucht die von Hans Klapdor-Kleingrothaus geleitete deutsch-russische Gruppe in einem Untergrundlabor in den italienischen Abruzzen nach einem Atomkernzerfall, den es gemäss dem Standardmodell nicht geben sollte. Bis heute ist kein solcher Zerfall beobachtet worden. Trotzdem hat sich das Warten

Fig. A.36 Neue Züricher Zeitung, 19 January 2000.

Die Rätsel um die kleinsten und größten Dimensionen im Kosmos

Gespräch zwischen
Dr. Peter Zaun
und
Prof. Dr. Hans Volker Klapdor

Gesendet vom Norddeutschen Fundfunk (NDR) in der Reihe
'Wissenschaft aktuell' des 3. Programms
am 15.11.1986

Die Rätsel um die kleinsten und größten Dimensionen im Kosmos

Gesendet vom NDR im 3. Progr. in der Reihe
'Wissenschaft aktuell' am 15.11.86

Die Welt des Atomkerns ist ein Labyrinth. Hier ist nicht die Rede vom Reaktorunglück in Chernobyl und der Kernenergie. Es geht um den Kern des Atoms und die Kräfte zwischen den Elementarteilchen. Das ist ein Thema internationaler Forschung, bei dem meist hochspezialisierte Physiker ein ziemlich abgeschlossenes Dasein führen. Ein interessierter Laie kann die Forschungsmeldungen aus der modernen Physik allenfalls nur registrieren jenseits eines wirklichen Verstehens. Dabei hängt die Welt des Kleinsten eng zusammen mit dem Zustand und der Entwicklung des Universums. Für den Atomkern wäre Öffentlichkeitsarbeit nötig, so paradox das angesichts der intensiven Diskussion um die Kernenergie klingt. Im Sommer dieses Jahres fand im Rahmen der 600-Jahr-Feier der Universität Heidelberg ein Kongress über schwache und elektromagnetische Wechselwirkung in Atomkernen statt. Hier soll versucht werden, die wichtigsten Konsequenzen und neue Ergebnisse dieses Spezialisten-Meetings in einem Gespräch vorzustellen. Gesprächspartner von Peter Zaun ist Professor Hans Volker Klapdor vom Max-Planck-Institut für Kernphysik in Heidelberg, Physikpreisträger der Deutschen Physikalischen Gesellschaft des Jahres 1982 und Organisator der Tagung.

ZAUN: Herr Klapdor, das Dilemma mit der Physik des Atomkerns ergibt sich aus der Vielfalt und Komplexität der Phänomene in den kleinsten Bausteinen der Materie. Lassen sich diese Phänomene heute eigentlich überhaupt noch verständlich machen, und läßt sich verdeutlichen, warum schließlich der Erkenntnisstand des Menschen vom Atomkern abhängt. Das klingt vielleicht etwas dramatisch, aber man kann sogar fast sagen, was wir vom Universum wissen, wissen wir vom Atomkern und umgekehrt.

KLAPDOR: Zunächst einmal sollten wir Kernphysik nicht mit Kerntechnologie gleichsetzen. --

Betrachten wir die Kernphysik in einem weiteren Sinne, d.h. unter Einschluß der Elementarteilchen-Physik - die scharfen Trennlinien zwischen beiden Gebieten verschwinden nämlich zunehmend - so hat sich eigentlich - was die Komplexität der Phänomene betrifft - in den letzten 20 Jahren eine gewaltige Vereinfachung ergeben:

Sechs Quarks und sechs Leptonen sind die fundamentalen Bausteine der Materie und haben eine verwirrende Fülle von Hunderten sogenannter Elementarteilchen abgelöst. Zwei Typen dieser Quarks bauen alle bekannten Atomkerne auf.

ZAUN: Die verwirrende Fülle der Elementarteilchen. Das liegt schon eine Zeit zurück. Aber: Was ist mit den W- und Z-Bosonen? Ich erinnere mich, vor 3 Jahren wurden diese neuen Teilchen am CERN entdeckt. Und die Entdeckung wurde gefeiert. Sie suggerierte damals, daß man kurz vor der Lösung des Rätsels der Naturkräfte stand.

KLAPDOR: Der Nachweis der W- und Z-Bosonen war in der Tat ein großer Schritt und zwar in Richtung das Traumes der Physiker von der 'großen Vereinigung', der einheitlichen Beschreibung der Naturkräfte.

W- und Z-Bosonen sind zwei der Teilchen oder besser, der sogenannten Feldquanten, die die Kräfte zwischen den Bausteinen der Materie vermitteln. Ihre Entdeckung ermöglichte die Zusammenfassung von elektromagnetischer und schwacher Kraft zur elektroschwachen Kraft, ähnlich wie man vor 100 Jahren die elektrische und die magnetische Kraft zusammenfaßte.

Die Physiker streben nun die Weiterführung dieser Idee der Vereinigung an, und wollen alle <u>vier</u> aus der Erfahrung bekannten Naturkräfte von der Gravitation bis hin zur Farbkraft zwischen den Quarks zusammenfassen. Die Grund-Idee der aus diesem Bestreben entstandenen 'Großen Vereinigungstheorien' ist, daß es - grob gesagt - eigentlich nur <u>eine</u> Kraft gibt - bei sehr hohen Energien nämlich - wie sie z.B. am Anfang des Universums herrschten! - Zu dieser Zeit waren die verschiedenen heute beobachteten Kräfte wie starke Kernkraft oder elektromagnetische Kraft oder schwache Kraft - die den radioaktiven Zerfall der Kerne z.B. bewirkt - alle gleich. Nur bei den niedrigen Energien und Temperaturen, wie sie heute im Kosmos herschen, kristallisieren sich die vier phänomenologisch bekannten Kräfte heraus.

Diese "Großen Vereinigungstheorien" nun führen zu unmittelbaren Verbindungen zwischen Makro- und Mikro-Kosmos. Die Aufdekkung dieses Zusammenhanges ist sicherlich eine der wichtigsten Entdeckungen unseres Jahrhunderts und eröffnet völlig neue Dimensionen. Sie ist vergleichbar den Entdeckungen eines Kepler oder Kopernicus.

ZAUN: Können Sie einige Beispiele geben für diese Zusammenhänge?

KLAPDOR: Das hervorstechendste Beispiel ist vielleicht die Tatsache, daß es Materie und uns - die ganze belebte Welt - heutzutage überhaupt gibt im Universum, neben Energie in Form von Strahlung. Zunächst mag diese Feststellung erstaunlich scheinen. Beim Urknall - der Geburt unserer Welt - aber ist nach unseren heutigen Vorstellungen zunächst Materie und Antimaterie in gleichen Mengen gebildet worden. Wenn Materie und Antimaterie zusammenkommen, vernichten sich aber bekanntlich beide unter Erzeugung von Strahlung.

Es ist das Verdienst der schwachen Wechselwirkung, daß zu einem sehr sehr frühen Zeitpunkt nach dem Urknall - noch bevor die Grundbausteine der Atomkerne Proton und Neutron entstanden sind - dies anfängliche Gleichgewicht zwischen Materie und Antimaterie sich verschoben hat in Richtung zur Materie. Von da ab gibt es 1 Milliardenstel mehr Materie als Antimaterie -. Nur dieser Rest ist noch vorhanden, der ganze übrige Teil ist kurz nach Ausbildung dieses Ungleichgewichts wieder zerstrahlt.

Ein anderes Beispiel einer engen Verbindung von Mikro- und Makro-Kosmos ist gegeben durch das Neutrino, eines der sechs Leptonen, in seinen verschiedenen Varianten. Darauf sollten wir vielleicht noch zurückkommen.

Ein weiteres Beispiel: Gesetze der Mikrophysik bestimmen etwa die Entwicklung von Neutronensternen, das sind ultradicht gepackte Sterne, die man sich als einen riesigen Atomkern vorstellen kann, und Gesetze der Mikrophysik bestimmen die gesamte Sternentwicklung überhaupt.

4

Die Reihe von Beispielen ließe sich beliebig verlängern. Mikro- und Makrophysik sind also in der Tat auf das engste verwoben und lernen wechselseitig voneinander.

ZAUN: Greifen wir das Neutrinorätsel auf. Es eignet sich vielleicht am besten zur Demonstration der Zusammenhänge zwischen Mikro- und Makrophysik.

KLAPDOR: In der Tat ist das Neutrinoproblem ein Hauptanliegen der modernen Physik. Es hängt z.B. sehr viel davon ab, ob das Neutrino eine Masse hat oder nicht.

-- Neutrinos könnten durch ihre Masse entscheiden, ob unser Universum offen oder geschlossen ist, d.h. ob es bis in alle Ewigkeit expandieren wird oder nach einiger Zeit kollabieren wird zu einem eventuellen neuen Urknall.

-- Die erst im letzten Jahr entdeckte großräumige Struktur des Universums* hängt eng zusammen mit der Frage der sogenannten dunklen Materie im Kosmos. Nicht aus Protonen und Neutronen aufgebaute dunkle Materie macht möglicherweise 90% oder mehr der Masse des Universums aus. Neutrinos sind die besten Kandidaten für diese dunkle Materie.

-- Auf der anderen Seite ist die Masse des Neutrinos eine der Schlüsselfragen für die Struktur der genannten großen Vereinigungstheorien der Kräfte, die im Mikrokosmos wirken.

-- Die großen Vereinigungstheorien wiederum bestimmen die Entwicklung des frühen Universums unmittelbar nach dem Urknall. Sie führen z.B. zwanglos zu der vorhin angesprochenen Asymmetrie.

ZAUN: Das Neutrino ist ein geheimnisvolles Teilchen, das unter anderem beim radioaktiven Betazerfall von Atomkernen entsteht. Das ist sicher. Das eigentliche Rätsel dreht sich um die Frage, hat es eine Ruhemasse oder nicht, d.h. - das ist natürlich nicht physikalisch korrekt gesagt - hat es wenigstens ein Minimal-Gewicht, wenn es ruht, wenn es sich nicht bewegt.

---

*Die jüngst durchgeführte Harvard-Himmelsdurchmusterung ergab riesige Hohlräume mit einem Durchmesser von ~50 Mpc, in deren Wänden die Galaxien angeordnet sind.

5

KLAPDOR: Alle aufwendigen Neutrinoexperimente dienen in der Tat - grob gesagt - schließlich dazu, diese Frage zu beantworten, ob das Neutrino eine Masse hat oder nicht.

ZAUN: Eigentlich macht dieses Neutrino den Physikern doch erstaunliche Schwierigkeiten, wenn man bedenkt, daß es zunächst 25 Jahre gedauert hat, bis es nach seiner theoretischen Vorhersage experimentell direkt nachgewiesen werden konnte und seit diesem Zeitpunkt wieder 30 Jahre vergangen sind, ohne daß die Physiker seine Masse kennen. Viel Masse kann es ja nicht sein, denn wir werden immerhin in der Sekunde von Billionen von Neutrinos durchsiebt, ohne daß wir etwas davon merken.

KLAPDOR: Daß wir nichts von diesen hauptsächlich von der Sonne kommenden Neutrinos merken, liegt daran, daß das Neutrino nur durch die "schwache Kraft" mit uns wechselwirkt. Und entsprechend aufwendig sind deshalb auch die Experimente, mit denen man das Neutrino fassen will. Weltweit gibt es unterschiedliche Konzepte. Es gibt im wesentlichen vier Typen von Experimenten, die sich unterscheiden in der Produktionsweise der Neutrinos, und die sich ergänzende Informationen liefern.

-- Da ist einmal der Betazerfall des Tritiums. Die Form des gemessenen Elektronspektrums hängt von der Masse des zusammen mit dem Elektron ausgesandten Neutrinos ab.

ZAUN: Also es geht um den radioaktiven überschweren Wasserstoff, das Tritium. Das ist ein β-Strahler und zusammen mit den β-Teilchen, das sind Elektronen, werden Neutrinos ausgesendet. Und diese beeinflussen die β-Teilchen. Kann man das so sagen?

KLAPDOR: Ja, in etwa. Die beim Zerfall des Tritiums freigesetzte Energie verteilt sich auf das ausgesandte Elektron und das Neutrino. Und die Energie, die auf das Elektron übertragen wird, hängt auch von der Masse des Neutrinos ab.

-- Die zweite Gruppe von Experimenten benutzt als Neutrinoquelle <u>solare</u> Neutrinos. Sie werden durch thermonukleare Reaktionen im Sonneninneren erzeugt. In diesen Experimenten sucht man nach Veränderungen dieser Sonnenneutrinos auf dem Wege zur Erde. Eines dieser Experimente ist das Gallium-Experiment, das unter Leitung des Max-Planck-Instituts in Heidelberg in einem Tunnel bei Rom das solare Neutrino-Rätsel lösen will. Das 'Rätsel' ist die zu geringe bisher beobachtete Neutrinorate von der Sonne - im Vergleich zur theoretischen Erwartung. Zu diesem Typ von Experimenten gehört ferner das kanadische Projekt eines Solarneutrino-Detektors unter Verwendung von 1000 Tonnen schweren Wassers.

ZAUN: Diese große Menge schweren Wassers war doch gewiß nicht primär für diesen Zweck vorgesehen?

KLAPDOR: Nein, natürlich nicht. Sie ist an sich für die kanadische Schwerwasserreaktorlinie bestimmt.

— Damit kommen wir zur dritten Gruppe, das sind die Reaktor-Experimente. In diesen sucht man wie in den Solarneutrino-Experimenten nach sogenannten Neutrino-Oszillationen. Darunter versteht man die kurzzeitige periodische Verwandlung eines Neutrinotyps in einen anderen - eine solche kann nur auftreten, wenn die Neutrinos eine Masse haben. Man verwendet hier die Brennkammer eines Reaktors als Neutrinoquelle - genauer: als Antineutrinoquelle. Die Neutrinos werden hier durch den Betazerfall neutronenreicher Kerne erzeugt, die durch die Spaltung entstehen. Man beobachtet die Neutrinos dann in verschiedenen Abständen vom Reaktor und sucht nach diesen Neutrinoumwandlungen - beim Reaktor also über terrestrische Entfernungen, während die Solarneutrino-Experimente danach über die Entfernung von einer astronomischen Einheit suchen.

ZAUN: Eine astronomische Einheit, das ist die Entfernung, die dem Abstand Erde-Sonne entspricht?

KLAPDOR: Ja. — Die vierte Gruppe der Neutrino-Experimente betrifft den doppelten Betazerfall bestimmter Kerne. Doppelter Betazerfall ist eine seltene Variante des gewöhnlichen Kern-Betazerfalls. Wieviele Zerfälle pro Zeiteinheit stattfinden, hängt für eine spezielle Zerfallsart des doppelten Betazerfalls direkt von der Masse des Neutrinos ab.

Aus dem Doppel-Betazerfall kommt gegenwärtig die schärfste Grenze für die Neutrinomasse. Zum Beispiel hat man für den Doppelbeta-Kandidaten Germanium 76 eine Lebensdauer von mehr als 10^{23} Jahren bestimmt. Das ist mehr als das Hundert-Milliardenfache des

Alters unserer Universums. Über solche Zeiträume ist das ^{76}Ge also quasi noch stabil! Diese große Lebensdauer entspricht einer winzigen Neutrinomasse, die viel viel kleiner ist als 1 Milliardenstel Gramm.

Trotzdem scheinen beim Doppel-Betazerfall noch experimentelle Fortschritte möglich, die Empfindlichkeit der Reaktorexperimente scheint dagegen bis an die Grenze des heute Möglichen getrieben zu sein.

Ein neues Doppelbeta-Experiment in Zusammenarbeit zwischen dem Max-Planck-Institut in Heidelberg und Laboratorien in Leningrad, Moskau und Bordeaux wird gegenwärtig diskutiert. In ihm soll die Nachweisgrenze um weitere zwei Größenordnungen hochgetrieben werden.

ZAUN: Im Gespräch ist, daß die Russen 8 kg des seltenen Germanium-Isotops 76 zur Verfügung stellen wollen. Wenn es einen freien Markt für dieses kostbare Isotop gäbe, müßte man zirka 200 Millionen Mark dafür bezahlen.

Vielleicht sollte man noch einmal betonen: Das Neutrino ist der beste Kandidat für die fehlende Masse im Universum. Es muß Masse fehlen, weil man sonst nicht den Zusammenhalt der Galaxien untereinander durch Gravitationskräfte erklären kann. Da gilt das Faust'sche Begehren - auf das ganze Universum übertragen - zu sehen, was die Kosmoswelt im Innersten zusammenhält. Und wir stoßen bei solchen Fragen immer wieder auf den Bauplan des Universums, auf seine Struktur. Natürlich geht es dabei auch um die Grundsteinlegung, die Scenarien für die ersten Stunden und Sekunden des Universums. Da spielen wieder die verschiedenen Disziplinen der Kernphysik, Teilchenphysik und Kosmologie zusammen. Und die Urknalltheorie, die Theorie von der Geburt des Universums, ist ja noch nicht abgeschlossen.

KLAPDOR: Ich muß doch einmal sagen, daß gerade diese Verknüpfung der verschiedenen Disziplinen den besonderen Reiz der Arbeit auf diesem Gebiet ausmachen! --

Es ist richtig. - Es gibt noch kein abgeschlossenes Bild des frühen Kosmos. Aber vieles spricht dafür, daß es eine außergewöhnliche Entwicklungsphase im frühen Universum gegeben hat. Wir sagen dazu inflationäre Expansion. Hier hat sich in extrem kurzer Zeit das Universum extrem schnell ausgedehnt. Eine solche Phase inflationärer Expansion würde eine Reihe von Beobachtungen erklären. Dazu gehören die Mikrowellenhintergrundstrahlung - das ist ein Relikt der Urknalls - die Flachheit des Raumes - das heißt, die Geometrie im Weltall von heute ist euklidisch oder wenigstens nahezu euklidisch - d.h. etwas lax gesagt, sie ist im Weltraum so, wie wir sie von der Erde gewohnt sind. Das ist durchaus nicht selbstverständlich - nicht-euklidische Geometrien werden von Mathematikern, wie Sie wissen, seit 150 Jahren studiert - Gauß wäre da als Erster zu nennen und ein "gekrümmter Raum" ist z.B. ein typisches Merkmal der Relativitätstheorie. Aber ich glaube, wir sollten im Rahmen dieses Gespräches nicht näher darauf eingehen.

ZAUN: Wie stellt man sich denn eigentlich diese inflationäre Phase vor. Kann man überhaupt noch von Vorstellung sprechen. Es geht doch dabei um Dinge, die sich in so kleinen Zeitspannen abspielen, die uns genau so unbegreiflich sind wie die riesigen Entfernungen im Universum.

KLAPDOR: In der inflationären Phase sind die Entfernungen zunächst gar nicht riesig. Es geht um die Ausdehnung des Universums sagen wir von Bruchteilen eines Kubikmillimeters zu einem Kubikmeter. Genauer gesagt: Es geht um einen enormen Ausdehnungsschub von einigen zig Zehnerpotenzen.

Das Problem ist die Auslösung der inflationären Phase und deren Beendigung. Man braucht dazu die sogenannte kosmologische Konstante.

ZAUN: Damit hatte doch schon Einstein Probleme?

KLAPDOR: Die kosmologische Konstante ist eine Größe, die die Dynamik des Universums mit beeinflußt. Sie hat in der klassischen Physik keinen Platz, weil damit dem Vakuum eine Energiedichte zugeordnet wird. Aber in der Quantenphysik ist sie eine notwendige Größe.

Eine positive kosmologische Konstante oder positive Energiedichte des Vakuums entspricht einem zusätzlichen Expansionsdruck. Das heißt: Sie könnte, wenn sie groß genug ist, eine inflationäre

Phase auslösen. Die vorhin angesprochenen großen Vereinigungstheorien liefern ziemlich zwanglos eine kosmologische Konstante, die groß genug ist.

Es bleibt allerdings das Problem der Beendigung dieser Phase. Wenn die kosmologische Konstante zur Auslösung einmal sehr groß gewesen sein muß, so ist andererseits aus Beobachtungen sicher, daß sie heutzutage sehr klein ist - die meisten Forscher glauben, daß sie heute Null ist. Es gibt indessen noch keine Theorie, die diesen Übergang befriedigend beschreibt.

ZAUN: Haben wir damit nicht die Verbindung zur dunklen Materie?

KLAPDOR: Das ist genau der Punkt. Eine kosmologische Konstante, die Null ist, verlangt zusammen mit der euklidischen Struktur des Universums, die zwangsläufig aus der inflationären Expansion folgt, zwingend dunkle Materie, die die "handgreifliche" - d.h. aus Protonen und Neutronen aufgebaute - Materie im Weltall an Masse um ein Vielfaches übertrifft.

ZAUN: Kann man ganz allgemein sagen? Die dunkle Materie, dessen bester Kandidat des Neutrino ist, ist untrennbar verknüpft mit der Vorstellung, daß die Geometrie im Universum der der Erde entspricht.

KLAPDOR: Ja, aber nur zusammen mit der Vorstellung, daß die kosmologische Konstante heute exakt Null ist. Das muß aber nicht so sein. Durch die Beobachtung ist nur sichergestellt, daß sie sehr klein ist. Die Annahme, daß sie exakt Null ist, erfolgt eigentlich mehr aus ästhetischen Gründen.

ZAUN: Bedeutet des Ganze nicht, daß die schon angesprochenen "Großen Vereinigungstheorien" zur Erklärung der Kräfte und Struktur im Großen und Kleinen noch nicht abgeschlossen sind?

KLAPDOR: Das ist sicher der Fall und äußert sich eben u.a. darin, daß man die zeitliche Entwicklung der kosmologischen Konstanten theoretisch nicht im Griff hat. Manche Forscher erhoffen sich hier einen Fortschritt von einem Einbezug der Gravitation in die GUTs. Dieser ist bislang nämlich noch nicht gelungen.

ZAUN: Das führt uns zurück zu den Versuchen - man kann wohl sagen - alle Phänomene unter einen Hut zu bringen. Die sogenannten GUTs, die großen Vereinigungstheorien, sind ein Kernthema heute im doppelten Sinne des Wortes. Besonders spannend sind dabei die Versuche zum Zerfall des Protons. Das Proton, das positive schwere Kernteilchen, ist ja schon Schülergenerationen als unveränderbarer Kernbaustein vertraut. Jetzt soll das Proton doch zerfallen können - und von der möglichen Zerfallszeit hängt viel ab, nicht nur für die GUTs. Da spielt sogar Forschungspolitik mit hinein. Die riesigen neuen Beschleuniger zur Erzeugung immer neuer Elementarteilchen werden vielleicht bald überflüssig?

KLAPDOR: Zunächst zum Proton-Zerfall. - Die meisten der GUT-Modelle sagen voraus, daß das Proton - der Kern des Wasserstoffatoms - aus dem das Universum zum größten Teil besteht - instabil ist - wenn auch über unvorstellbar lange Zeiträume. Schon wenn die Lebensdauer des Protons kleiner wäre als 1 Million Mal das Alter des Universums, so würde die durch Zerfall der im menschlichen Körper vorhandenen Protonen verursachte Strahlung bereits lebensbedrohlich. Die gegenwärtige experimentelle Grenze für den Zerfall des Protons liegt bei mehr als 4×10^{32} Jahren. Das ist eine 4 mit 32 Nullen. Dies ist das Ergebnis des deutsch-französischen Experiments, das im Frejus-Tunnel in den französischen Alpen läuft.

Diese beobachtete Grenze für den Protonzerfall schließt eine gewisse Klasse von Modellen der Großen Vereinigung aus.

ZAUN: Das heißt, der Protonzerfall gibt Informationen über solche Vereinigungstheorien. Auch die schon angesprochenen Beschleunigerexperimente haben im wesentlichen das Ziel, Vereinigungstheorien zu prüfen und zu verbessern. Ist da nicht <u>ein</u> Weg zu den Vereinigungstheorien überflüssig? Vielleicht der Weg über die Beschleuniger, die sehr viel kosten? Ich denke an den neuen Ringbeschleuniger bei CERN in Genf, der nach Fertigstellung im Jahre 1988 über 1,2 Milliarden DM gekostet haben wird. Es gibt weitere derartige Projekte.

KLAPDOR: Zunächst ist es für mich gar keine Frage, daß wir die <u>baubaren</u> Beschleuniger brauchen!! Das Problem ist ein anderes.

Wir kommen wahrscheinlich, irgendwann einmal in nicht allzu ferner Zeit, an eine Grenze der Beschleunigertechnologie: Für die 90-ziger Jahre geplante Beschleuniger haben einen Umfang von über 100 km. Selbst ein Beschleunigerring um die ganze Erde aber würde uns - ganz abgesehen von den Kosten - bei weitem noch nicht die gewünschten Energien liefern, um die ganze Energieskala der großen Vereinigungstheorien abzudecken. Neue Experimente unter Verwendung der kosmischen Strahlung könnten im Prinzip vielleicht ein gewisser Ersatz werden. Auf jeden Fall treten die hohen Energien im frühen Universum auf - d.h. wir möchten die bei diesen Energien ablaufenden Prozesse untersuchen und kennen!!

Wir brauchen also andere Wege zu den Vereinigungstheorien. Ein typisches Beispiel, das zu dem anderen Weg gehört, ist die Untersuchung des Proton-Zerfalls, ein weiteres Beispiel ist die Suche nach der Neutrinomasse. Niederenergieexperimente wie diese beiden - und weitere ließen sich aufzählen - werden zwangsläufig - leider in einer indirekteren Form - zur Physik der Vereinigungstheorien in Zukunft mehr als bisher beitragen müssen.

ZAUN: Herr Professor Klapdor, wir sind am Ende des kurzen Ausflugs in die komplizierte Welt der kleinsten Materieteilchen. Darf ich versuchen, das Wichtigste noch einmal ganz kurz zusammenzufassen. Die großen Vereinigungstheorien, die Theorien, die die Phänomene des Universums und der Elementarteilchen erklären sollen, sind noch nicht abgeschlossen. Aber es gibt eine unmittelbare Verbindung von Mikro- und Makrokosmos. Und das heißt: Der einzelne Atomkern und das riesige Universum sind ganz eng miteinander verknüpft.

MPI - 1987 - V 17 September 1987 1

Untersuchung des Doppelbetazerfalls von angereichertem ^{76}Ge zur Bestimmung der Neutrinomasse

Vorschlag eines Experiments

von

Kurchatov Institut, Moskau
Leningrad Institute of Nuclear Physics, Gatchina
Max-Planck-Institut für Kernphysik, Heidelberg

H.V. Klapdor
Max-Planck-Institut für Kernphysik, Heidelberg

Im Folgenden werden das physikalische Ziel, die zu erwartenden Aufwendungen, ein vorläufiger Zeitplan, sowie die Einordnung des Experimentes in den Gesamtrahmen der Forschung des MPI kurz dargestellt.

1. Physikalisches Ziel

1.1. Motivation und Stand der Suche nach einer Neutrinomasse mit kernphysikalischen Methoden. Übersicht.

Es ist bekannt, daß Neutrinos eine Schlüsselrolle für die Lösung des Problems der Vereinigung der starken und elektroschwachen Wechselwirkungen in den Großen Vereinigungstheorien (GUTs, bzw. SUSYs oder SUGRAs) spielen, die andererseits eine der zentralen Fragen der modernen Physik darstellen. Es könnte eine einzigartige Eigenschaft von Neutrinos unter den Elementarteilchen sein, eine Majorana-Masse zu besitzen (Fig. 1). Die Existenz von Neutrinos mit einer Majorana-Masse würde der Brechung der Baryon minus Leptonzahl-Erhaltung entsprechen. Während im minimalen SU5-Modell die Neutrinomasse Null ist, da kein rechtshändiges Neutrino existiert und andererseits SU5-invariante Majorana-Kopplungen mit dem Higgs-Inhalt des minimalen Modells nicht möglich sind, führen erweiterte Modelle (SU5-Modelle mit ...

europhysics news
BULLETIN OF THE EUROPEAN PHYSICAL SOCIETY

Weak and Electromagnetic Interactions in Nuclei

Evolution of nuclear physics into the particle domain

H.V. Klapdor, Heidelberg
(Max-Planck-Institut für Kernphysik)

J.A. Volume 18 Number 2 February 1987

Nuclear physics is presently experiencing a new impetus, addressing itself towards more fundamental physics questions and becoming more and more important as a test of physics beyond the standard model of grand unification. The close connections between the laws of microphysics, astrophysics and cosmology — one of the most significant discoveries of our century — open further exciting perspectives.

Among the most topical fields of activity (see K. Gabathuler, *Europhys. News*, Jan. 1987) are the search for a finite neutrino mass by β-end-point spectroscopy, neutrinoless double beta decay and neutrino oscillations, proton decay, the study of rare and forbidden muon and pion decays and the search for an electric dipole moment of the neutron — all experiments which are complementary to the direct search for new particles and symmetries in high energy experiments — and their relations to grand unification, superstring theories and the evolution of the early Universe. Other topics which not so long ago would have been regarded exclusively as particle physics are the status of electroweak gauge theories, in particular the mixing of quarks by the weak interaction and the origin of the observed CP violation, quarks in nucleons and nuclei, etc. Even traditional fields of nuclear physics carry the same message: the discussions on the positron lines oberved at GSI in collisions of nuclei with high Z; the observation of extremely high spins in super-deformed nuclei; new results on the parity violation in quasi-elastic electron scattering; properties of exotic nuclei and of nuclear matter. But this is now close to astrophysics — the collapse of heavy stars, baryon and lepton number conservation and weak interactions in cosmology.

Weak interactions are of basic importance for the development and structure of the Universe because:
— without weak interactions there would be no heavy elements;
— neutrinos could determine by their mass, whether the Universe is open or closed;
— neutrinos are the best candidates for the non-baryonic dark matter in the Universe, which could represent more than 90% of the mass of the cosmos;
— problems of the weak interaction, such as the mass of the neutrino, are — together with the half-life of the proton — key points for the structure of grand unification theories (GUTs) or supersymmetry theories (SUSYs) which must include the development of the early Universe.

This review of developments in nuclear physics is largely based on discussions at the EPS and IUPAP sponsored International Symposium on Weak and Electromagnetic Interactions in Nuclei, W.E.I.N. 1986, held in Heidelberg, 1-5 July 1986, in conjunction with the 600th anniversary of the University of Heidelberg. About 300 participants worldwide attended the Conference, which was held under the patronage of the Rector of the University, Gisbert zu Putlitz and chaired by the author. Several countries expressed their interest in hosting a similar W.E.I.N. conference in 1989. The Proceedings have been published recently by Springer-Verlag.

Neutrinos

Today, 30 years after the direct observation of the neutrino in the famous Goldhaber experiment, it is still unknown whether the neutrino has a mass or not, and all over the world there are different projects trying to resolve the question.

The European Gallex group (Gallium Experiment) headed by the MPI in Heidelberg plans to detect solar neutrinos in the Gran Sasso tunnel near Rome; R. Mössbauer and his group are looking for neutrino oscillations using the core of a nuclear power reactor as a neutrino source. Both these experiments are looking for differences in the masses of different neutrino types. Solar neutrino experiments with other detectors are expected to yield answers to the recent question of neutrino-matter oscillations.

Two other types of experiment attack the mass question directly. W. Kündig in Zürich studies the beta decay of tritium and has reported an upper limit of 18 eV. A different method is to measure the half-life for neutrinoless double beta decay and here the most active groups are in Osaka, Columbia, Milan, Zaragoza and Santa Barbara.

To deduce the neutrino mass from observations of the decay, additional nuclear structure information is needed, but using the most recent calculations of the nuclear matrix elements done in Heidelberg, the mass of the electron neutrino turns out to be less than 1 eV. This limit corresponds, however, to an effective mass only. If more than one Majorana neutrino couples to the electron, as the interference between the different mass states of the neutrinos in the neutrino exchange amplitude could be destructive, the true mass of the electron neutrino could be substantially larger. If one assumes a second neutrino with a mass not smaller than 100 MeV, a conservative limit for the electron neutrino mass is then 13 eV. Of particular interest are the Osaka group's studies of the $0^+ \rightarrow 2^+$ transitions which in contrast to the $0^+ \rightarrow 0^+$ ground state transitions, can be driven by right-handed currents only, not by a mass term, and further efforts to look for a neutrinoless decay branch with emission of a light boson, the majoron.

To reduce further the limit on the half-life for neutrinoless ββ decay by more than an order of magnitude, a collaboration from Heidelberg, Leningrad, Moscow, and Bordeaux is planning to use enriched ^{76}Ge inside a well-shielded cavity. The Soviet Union will provide the enriched material.

An important constraint for deducing

the neutrino mass is provided by nuclear reaction measurements. In (p,n) charge exchange reactions, β^- strength distributions have a control function not only for the calculation of the double beta matrix elements required for deducing the neutrino mass, but also for the determination of the efficiency of solar neutrino detectors such as the Gallium detector.

Muon Physics

The question of muon-to-electron conversion is as old as the muon itself. When it was realized that the muon does not decay into an electron and a photon, the concept of separately conserved muonic and electronic lepton numbers was introduced as an *ad hoc* hypothesis. Only recently has this question become a very important issue of particle physics. In some extensions of the standard model it leads to the most sensitive mass limit on new particles such as Higgs scalars, leptoquarks and heavy neutrinos.

The signature for coherent μ-e conversion in which the nucleus is left in its ground state is muon capture accompanied by the emission of an electron at an energy $E_e \simeq m_\mu c^2 - B = 104$ MeV, where B is the binding energy of the muonic atom. From the Canadian meson facility Triumf a preliminary upper limit has been reported for the branching ratio, R, of muon-to-electron conversion compared to total muon capture of $R < 4 \times 10^{-12}$ (90% confidence limit).

The experimental results of the European Muon Collaboration (EMC) and other electron, muon and neutrino experiments concerning differences between the momentum distribution of quarks in the nucleus from that in free nucleons (EMC effect) are difficult to explain and new data are awaited from the New Muon Collaboration (NMC) at CERN. According to E. Berger from Argonne, the Drell-Yan process is expected to provide an answer, but there are various models on the market. A Russian-Danish-American collaboration believes that part of the EMC effect could be explained by simple binding effects. An alternative approach which leans on QCD models due to F. Close from the Rutherford Laboratory connects the modification to a change of the confinement scale in nuclei originating from an increase in the size of nucleons, the formation of multiquark clusters or of a large quark bag. This however clashes with the view held by I. Sick of Basel that the size of nucleons should be only a very little larger in nuclei than when free, if at all.

Tests on the Standard Theory, GUTs etc.

In the latest measurements reported on proton life-times, H. Meyer from Wuppertal has given a lower limit for the half-life of the proton decaying into a neutral pion and a positron of 4×10^{32} years. This result rules out the simplest GUT — the minimal SU5 model — which predicts a half-life of less than 3×10^{31} years. Certain modifications to GUT and supersymmetry models, however, could remove this objection.

New measurements have been made of the electric dipole moment of the neutron. The purpose of the experiment is to test CP violation which is observed only in the decay of the neutral kaon. The latest value reported by N. Ramsey from Harvard and V. Lobashev from Moscow is $[-1.8 \pm 2.9] \times 10^{-25}$ cm. (The smallness of the dipole moment can be pictured by enlarging the neutron to the size of the Earth and observing that the radius of the charge distribution at the pole would differ from that at the equator by only 0.01 mm!). Theoretical predictions range from $10^{-19} - 10^{-32}$ cm and the new upper limit leaves roughly four categories of theory still within the experimental bounds: phenomenological milli-weak and super-weak theories, and gauge theories that attribute the T violation to the mixing of six or more quarks, or to the exchange of multiple Higgs bosons.

Extensions of the standard model are seen by R. Peccei (*Europhys. News*, Jan. 1987) to link such apparently disconnected phenomena as small neutrino masses, visible (and invisible) axions, the CP problem and the Universe, and baryon asymmetry . The exciting possibility that the positron peaks observed in collisions of high Z nuclei at GSI originate from the decay of an axion is ruled out by the low probability for rare and forbidden decays of the muon and π meson. This result by R. Engfer and his group at SIN is confirmed by a search for short-lived axions by neutron capture on protons performed with the Grenoble cold neutron beam and also by a search for pseudoscalar axions in isoscalar M 1 transitions in ^{10}B in Amsterdam.

Exotic Nuclei, Nuclear Matter and Astrophysics

New information is being produced on beta decay properties — half-lives, beta-delayed neutron emission and fission rates — which is of great importance for modelling theories that predict beta decay properties of nuclei not accessible to terrestrial laboratories. The latter enter sensitively into our understanding and theoretical "reproduction" of element synthesis in the Universe, and of the age of our Galaxy as deduced from cosmochronometers.

Nuclear electron capture plays a decisive role at several stages in the evolution of massive stars. Electron capture by nuclei of the iron group in the core partly triggers the gravitational collapse of Type II supernova progenitors, once the Chandrasehkar limit has been reached. Also hydrodynamical shock propagation in the post-bound phase and "delayed" explosions depend on the neutrino luminosity behind the shock, which is governed by capture in the homologous core. The weak interaction also controls the life of a neutron star in its primary phase. The collapse scenario of a heavy star could be quite different from the standard collapse scenario — which is a low-entropy collapse — in the case of lepton number violation, mediated *e.g.* by majorons.

Properties of nuclear matter play a significant part also in stellar evolution. A new equation of state for nuclear matter at high density has been used by S. Kahana and collaborators from Brookhaven National Laboratory in hydrodynamical simulations of stellar collapse. These calculations are the first to produce a prompt shock-explosive mechanism for Type II supernovae using realistic evolutionary supernova models.

Over the years, two major classes of mechanism had been proposed: One approach sought an ejection of the stellar mantle and envelope by an explosive shock created in a core bounce after collapse. The second approach would have the exterior regions blown off by the energetic neutrinos somewhat after collapse. Both approaches failed however, to give the desired effect even after correction for the initial omission of important beta-capture channels, which led to a reduction of the core mass.

The new break-through is achieved by theoretically altering the experimentally rather uncertain equation of state for nuclear matter at high density in the direction of softening, and introducing relativistic gravitation into the collapse phase. Both lead to an increase of the energy initially pumped into the shock. This result is important also for understanding the synthesis of heavy elements in the Universe by the r (rapid neutron capture) process in explosive helium-burning.

Weak Interactions and Cosmology

In most current evolutionary models of the Universe only 10% of its mass consists of nucleons. Candidates for the dark matter fall into two categories: hot (neutrino-like) and cold (axion- and photino-like) particles, the first being the more probable solution. The necessity for accommodating non-baryonic dark matter, the recently discovered large-scale structure of the Universe (many large voids with most galaxies distributed on the walls of the voids — and not randomly!) and the early origin of the galaxies together set strict boundary conditions for cosmological models.

Although there is not yet a fully consistent picture of the early Universe, a phase of inflationary expansion in the early cosmos is believed by some (see *e.g.* J. Ellis, *Europhys. News*, Sept. 1985) to offer the most natural explanation for the isotropy of the microwave background radiation, the flatness of space (*i.e.*, its Euclidean metric) and the elimination of "unwanted" particles like monopoles, gravitinos, etc. There remains, however, the problem of the cosmological constant, which has on the one hand to be large to trigger the inflationary expansion, and, on the other, small in order not to contradict present-day cosmological observation. The cosmological constant introduced into the equations of general relativity in 1917 by Einstein to support a stationary Universe and removed by him in 1931, is nowadays a natural consequence of quantum field theory and corresponds to an energy density of the vacuum. A positive vacuum energy (cosmological constant) leads to an expansion pressure.

In principle a large energy density of the vacuum can be produced in the early Universe by the hypothetical Higgs fields which in our understanding are responsible for spontaneous symmetry breaking. A mechanism reducing the corresponding large Λ to a value consistent with the present status of the Universe (*i.e.*, by more than 50 orders of magnitude) is, however, not yet known.

Different views have been expressed over the aesthetical requirement for an exactly vanishing cosmological constant, but there seems to be increasing evidence that the age of the Universe is greater than 15×10^9 years. G. Tammann from Basel estimates $18 \pm 3 \times 10^9$ years from measurements on globular clusters, consistent with the recent Heidelberg figure of about 20×10^9 years from cosmochronometers. This would indicate a non-vanishing cosmological constant for values of the Hubble constant $H_0 \gtrsim 45$ km Mpc^{-1} s^{-1} which by observation is currently constrained to values between around 40 and 100 km Mpc^{-1} s^{-1}.

Record Nuclear Spins

Despite its extensions into particles at one end of the scale and cosmology at the other, nuclear physics has nevertheless its own record book still. J. Sharpey-Schafer and his group from the University of Liverpool have reported the measurement of the highest spin seen in an atomic nucleus, notably 60 $\hbar$ — about 20 units higher than previously. This was discovered in a super-deformed rotational band of ^{152}Dy using a detector system (TESSA III) which allows spectroscopy of a new class of super-deformed prolate nuclei whose major to minor axes have a ratio of 2:1.

'Keine Neutrinos beim Doppel-Betazerfall?'

Prof. Dr. Hans Volker Klapdor

interviewed von

Dr. Peter Zaun

Gesendet vom RIAS Berlin in der Reihe

'Wissenschaft aktuell' des 1. Programms

am 15.7.1987

'Keine Neutrinos beim Doppel-Betazerfall?'

Gesendet vom RIAS Berlin, im 1. Progr. in der Reihe

'Wissenschaft aktuell' am 15.7.1987

Zaun: Die natürliche Radioaktivität gliedert sich in die drei bekannten Zerfallsprozesse. Man spricht von Alpha-, Beta-, und Gamma-Strahlen. Beim Beta-Zerfall wird ein Elektron aus einem Atomkernbaustein herausgeschleudert. Zusätzlich entsteht ein geheimnisvolles Teilchen, welches heute die Physiker in Atem hält: Das Neutrino. Genauer das Antineutrino. Die Reaktionsgleichung für den normalen Betazerfall ist jedem Schüler geläufig. Mittlerweile haben Generationen gelernt, ein Neutron zerfällt in ein Proton, ein Elektron und ein Neutrino. Weniger bekannt ist, daß es auch einen simultanen doppelten Betazerfall gibt. Jetzt erschüttert eine Mitteilung das Gebäude der Elementarteilchenphysik. Es gibt möglicherweise einen doppelten Betazerfall ohne Neutrinos, also Elektronen-Zwillinge ohne Begleiter. Das ist eine sensationelle Entdeckung, wenn sie bestätigt wird, gleichbedeutend mit einem tiefen Einschnitt in der modernen Physik. Die Ergebnisse der Experimente von Arbeitsgruppen aus den USA sind so folgenschwer, daß sie bisher nur in einer Vorpublikation veröffentlicht wurden. Die Autoren sind selber noch skeptisch. Schließlich rüttelt die Entdeckung an einem Naturgesetz. Am Telefon begrüsse ich Herrn Professor Hans Volker Klapdor vom Max-Planck-Institut für Kernphysik in Heidelberg. Herr Professor Klapdor ist Spezialist auf dem Gebiet das atomaren Betazerfalls. Für seine Forschung zum Betazerfall erhielt er 1982 den Physikpreis der Deutschen Physikalischen Gesellschaft. Herr Professor Klapdor, wie stehen Sie zu dieser Meldung aus den USA? Warum könnte diese Entdeckung an einem Naturgesetz rütteln?

Klapdor: Die Entdeckung von neutrinolosem Doppel-Betazerfall wäre eine äußerst aufregende Sache. Sie wäre ein wichtiger Schritt zur Aufdeckung neuer Physik und zwar neuer Physik über das gegenwärtige sogenannte Standard-Modell der elektroschwachen Wechselwirkung hinaus und hinaus ebenso über die einfachsten sogenannten Großen Vereinigungs-Theorien, mit denen man heute versucht, die verschiedenen Naturkräfte einheitlich bzw. vereinheitlicht zu beschreiben. Das Standard-Modell der elektroschwachen Wechselwirkung war sozusagen der erste Schritt hierzu. Mit ihm war es den Physikern Glashow, Weinberg und Salam in den 70iger Jahren gelungen, die sogenannte schwache und elektromagnetische Wechselwirkung zusammenzufassen, und bekanntlich erhielten sie hierfür 1979 den Nobel-Preis. Das Standard-Modell ist bislang in vielen seiner Vorhersagen bestätigt worden, so sind z.B. erst 1983 die von ihm vorhergesagten W- und Z-Bosonen am Europäischen Kernforschungszentrum CERN bei Genf gefunden worden. Die Existenz des jetzt gemeldeten Effekts - ich glaube allerdings, daß hier noch eine gewisse Skepsis angebracht ist, und es gibt bereits auch amerikanische Einwände - würde bedeuten, daß die sogenannte Leptonenzahl nicht erhalten ist, wie die Physiker sagen, und würde zweitens bedeuten, daß das Neutrino ein Masse hat.

Zaun: Die Masse des Neutrinos hängt also zusammen mit dem doppelten Betazerfall. Kann man das so sagen?

Klapdor: Das kann man so sagen.

Zaun: Ich erinnere mich, 1981 wurde in München u.a. gefordert, daß das Neutrino wohl eine Masse hat, daß aber mit dieser Masse verknüpft ist die Existenz eines weiteren geheimnisvollen Teilchens ohne Masse. Das neue Teilchen heißt Majoron nach dem italienischen Physiker Ettore Majorana. Und dieses Teilchen behaupten die Amerikaner jetzt nachgewiesen zu haben. Was bedeutet das für die Teilchenphysik?

Klapdor: Es ist richtig. Die amerikanischen Kollegen behaupten, daß sie einen Doppel-Betazerfall nachgewiesen haben, bei dem ein Majoron erzeugt wird. Dies wäre von Interesse, weil es Aufschluß geben würde über den speziellen Mechanismus, der im Rahmen der sogenannten Eichtheorien den Neutrinos eine Masse gibt. Die Eichtheorien, das ist der Typ von Theorien, der nach heutiger Ansicht allen Naturgesetzen zugrundeliegt.

Zaun: Würde das nicht bedeuten, daß mit der Existenz eines weiteren neuen Teilchens eine Hoffnung zerstört wird? Ich denke daran, daß man gehofft hat, daß die komplexe Vielfalt der Phänomene um den Atomkern sich bildhafter und einfacher darstellen läßt. Gerade in der vergangenen Zeit hat es ja Vereinfachungen gegeben. Die Vereinigung der Naturkräfte schien unmittelbar bevorzustehen. Zumindest bei der Geburt des Universums sollte nur eine einzige Naturkraft wirksam gewesen sein. Die sogenannten Vereinigungs-Theorien sind bisher der scheinbar einzig gangbare Weg, um die Erscheinungen im Mikrokosmos und im riesigen Universum unter ein Dach zu bringen. Wären diese Theorien durch den neutrinolosen Betazerfall gefährdet?

Klapdor: Durch den neutrinolosen Betazerfall, nein, durchaus nicht. Eine nichtverschwindende Neutrinomasse scheidet nur die einfachste dieser Grossen Vereinigungs-Theorien aus. Und diese konnte übrigens auch schon den bestehenden und beobachteten Grad der Stabilität des Protons nicht erklären. Aber damit werden durchaus nicht die in die Grossen Vereinigungs-Theorien ganz allgemein gesetzten Hoffnungen zerstört. Vielleicht sollte man erwähnen, daß eine endliche Neutrinomasse in natürlicher Weise auch zu solchen Modellen im Rahmen der Grossen Vereinigungs-Theorien führt, die auf eine weitere Substruktur der Quarks und Leptonen schließen lassen, also der heute für fundamental gehaltenen Materiebausteine.

Zaun: Also das Gebäude der Physik stürzt nicht ein. Es würde auch diesen sogenannten doppelten neutrinolosen Betazerfall ertragen.

Klapdor: Natürlich. Und die Ergebnisse der amerikanischen Kollegen sind zumindest ein wertvolles Stimulans zu einer verstärkten Suche nach diesem Doppelbetazerfall und nach dem Majoron. Vielleicht können wir hier zu einem weiteren Schritt nach vorn kommen mit dem sehr aufwendigen Experiment, das wir gegenwärtig in Heidelberg zusammen mit der russischen Akademie der Wissenschaften und französischen Kollegen vorbereiten.

Zaun: Also, der doppelte Betazerfall ist demnach keine Domäne der Amerikaner.

Klapdor: Nein. Das gewiß nicht.

Zaun: Vielen Dank nach Heidelberg. Vielen Dank für das Gespräch, Herr Professor Klapdor.

Protokoll

über die Durchführung eines gemeinsamen Experimentes zur Suche nach dem Doppelbetazerfall von ^{76}Ge

zwischen dem

Max-Planck-Institut für Kernphysik

und dem

Kurchatov Institut für Atomenergie

für die Jahre 1988-1992

Das MPI für Kernphysik und das Kurchatov Institut sind, ausgehend vom beiderseitigen Interesse an der Entwicklung der Zusammenarbeit, und im Bewußtsein gemeinsamer wissenschaftlicher Arbeitsrichtungen, auf der Grundlage des Abkommens zwischen dem Bundesminister für Forschung und Technologie der Bundesrepublik Deutschland und dem Staatskomitee für die Nutzung der Atomenergie der Union der Sozialistischen Sowjetrepubliken über wissenschaftlich-technische Zusammenarbeit bei der friedlichen Nutzung der Kernenergie vom 24.04.1987 - in Folgendem übereingekommen:

ПРОТОКОЛ

о проведении совместного эксперимента по поиску двойного бета-распада германия-76 между Институтом атомной энергии им.И.В.Курчатова и Институтом ядерной физики им.М.Планка в 1988-1992г.г.

Институт атомной энергии им.И.В.Курчатова и Институт ядерной физики им.М.Планка,

исходя из взаимной заинтересованности в развитии сотрудничества,

принимая во внимание общность предметов научных исследований,

основываясь на Соглашении между Государственным комитетом по атомной энергии СССР и Федеральным министерством научных исследований и технологии ФРГ от 22.04.1987,

согласились о нижеследующем:

GSI KOORDINIERUNGSSTELLE KERNPHYSIK UND SCHWERIONENFORSCHUN
BEI DER
GESELLSCHAFT FÜR SCHWERIONENFORSCHUNG MBH DARMSTADT

GSI-KKS · 6100 Darmstadt 11 · Postfach 110552

Herrn Professor
Dr. H.V.Klapdor
Max-Planck-Institut für Kernphysik
Postfach 10 39 80

6900 H e i d e l b e r g 1

2. September 1988

Sehr geehrter Herr Professor Dr. Klapdor,

es ist beabsichtigt, Ihren Antrag auf Förderung des Vorhabens "Experiment zur Untersuchung des Doppelbetazerfalls von isotopenangereichertem 76Ge zur Bestimmung der Neutrinomasse" mit insgesamt 1.089.000,-- DM zu fördern. Kassenmäßig teilen sich diese Mittel auf in

1988: 230.000,-- DM
1989: 107.000,-- DM
1990: 752.000,-- DM

Der Bundesminister
für Forschung und Technologie
211-4006-06 HD 555 I

5300 Bonn 2, 18.10.1988
Postfach 200240
Fernruf: 593209
Fernschreiber: 885674
Dienstgebäude:
Bonn-Bad Godesberg
Heinemannstr. 2

[illegible]

Max-Planck-Gesellschaft
zur Förderung der Wissenschaften e.V.
Residenzstr. 1a

8000 München 1

Bearbeitung erfolgt
durch:

Dr. D. Hartwig
GSI Darmstadt
Projektträger für
Mittelenergie-
und Kernphysik
Postfach 110552
6100 Darmstadt 11
Fernruf: 06151/359633

Zuwendungsbescheid

Betreff: Zuwendung aus dem Bundeshaushalt, Einzelplan 30
Kapitel 3003, Titel 89301, Haushaltsjahr 1988
für das Vorhaben: "Experiment zur Untersuchung des Doppelbetazerfalls ..."

Projektleiter: Prof. Dr. H.V. Klapdor

Förderkennzeichen: 06 HD 555 I

Bezug: Antrag des Max-Planck-Instituts für Kernphysik vom 04.03.1988
sowie Ihr Schreiben vom 15.03.1988 - Az.: 42512/0102/59

Der Bundesminister
für Forschung und Technologie

Pressereferat
Postfach 200240
5300 Bonn 2
Telefon
(0228) 593037
(0228) 593042
Telefax
(0228) 593105
Telex 885674

Bonn, den 09. Dezember 1988
111/88

Forschungsergebnisse

Deutsch-sowjetisches Projekt zur Untersuchung der Neutrinomasse gestartet

Deutsche und sowjetische Kernphysiker haben Anfang Dezember 1988 eine Vereinbarung über ein wichtiges gemeinsames Experiment unterzeichnet, mit dessen Hilfe eines der geheimnisvollsten Teilchen der modernen Physik, das Neutrino, untersucht werden soll. Die Masse dieses Teilchens, die mit diesem Experiment bestimmt werden soll, spielt eine zentrale Rolle nicht nur in der Teilchenphysik, sondern auch in der Astrophysik und Kosmologie.

Neutrinos gehören zu den am häufigsten vorkommenden Elementarteilchen im Weltall. Bei den Fusionsprozessen im Innern der Sterne werden so viele Neutrinos erzeugt, daß in jeder Sekunde mehrere Milliarden auf jeden cm^2 der Erdoberfläche treffen. Da sie aber nur äußerst selten mit Materie in Wechselwirkung treten, können sie den Erdkörper fast ausnahmslos durchdringen, ohne Spuren zu hinterlassen.

Neutrinos wurden 1931 von Wolfgang Pauli hypothetisch eingeführt, um Widersprüche beim Beta-Zerfall von Atomkernen lösen zu können. Beim Beta-Zerfall wird ein Neutron des Kerns in ein Protron umgewandelt und sendet dabei ein Elektron und ein Antineutrino aus. Neutrinos (und Antineutrinos) sollen keine Ruhemasse und keine elektrische Ladung besitzen. Erst 1955 konnte ihre Existenz experimentell nachgewiesen werden. Inzwischen sind drei Neutrinoarten bekannt.

- 2 -

Die Frage, ob Neutrinos nicht doch eine winzige Masse besitzen, spielt für Theorien der Elementarteilchen (sogen. Große Vereinigungstheorien) sowie auch für kosmologische Theorien eine große Rolle. Bei ihrer ungeheuer großen Zahl würden die Neutrinos einen erheblichen Teil zur gesamten Masse im Weltall beitragen. Ihre Schwerkraft könnte ausreichen, die Expansion des Universums aufzuhalten und umzukehren. Der Nachweis einer Ruhemasse des Neutrinos hätte auch andere weitreichende Konsequenzen für das physikalische Weltbild. Daher versuchen die Physiker mit verschiedenen Methoden, die Eigenschaften der Neutrinos aufzuklären.

Ein interessantes Konzept beruht auf der Untersuchung des doppelten Beta-Zerfalls, bei dem gleichzeitig zwei Neutronen im Innern eines Atomkerns in Protonen umgewandelt werden. Dies ist ein äußerst seltener Prozeß. Für die Suche nach dem doppelten Beta-Zerfall im Germanium 76-Isotop hat sich eine besonders günstige Konstellation für ein gemeinsames deutsch-sowjetisches Projekt ergeben. Die Sowjetunion verfügt als einziges Land über die erforderliche Menge stark angereicherten Germanium 76. Dieses Material ist wegen des aufwendigen Herstellungsverfahrens sehr wertvoll. Projektpartner auf sowjetischer Seite ist das Kurchatov-Institut für Atomenergie in Moskau.

Auf deutscher Seite bringt das Max-Planck-Institut für Kernphysik in Heidelberg theoretische Erkenntnisse zum doppelten Beta-Zerfall sowie Erfahrungen bei der Entwicklung störungsfreier Detektoren und die erforderliche hochwertige Meßausrüstung ein. Nur durch die Kombination der spezifischen Beiträge aus beiden Ländern kann dieses Projekt aufgegriffen werden.

Anfang Dezember trafen die ersten 7,5 kg des wertvollen Germanium-Isotops in Heidelberg ein. Die Projektarbeit kann jetzt begonnen werden.

Zunächst werden in einer ersten Phase zwei untergrundarme Detektoren aus hochangereichertem Germanium 76 hergestellt. Nach erfolgreichem Test sollen anschließend mehrere dieser Detektoren in ein unterirdisches Laboratorium gebracht werden, damit Störungen durch die kosmische Strahlung vermieden werden. Tief unter der Erde beginnen dann die physikalischen Untersuchungen.

Dieses Vorhaben der reinen Grundlagenforschung wurde auf der Grundlage des Abkommens vom 22.04.1987 zwischen dem Bundesministerium für Forschung und Technologie und dem Staatskomitee für die Nutzung der Atomenergie der UdSSR über wissenschaftlich-technische Zusammenarbeit bei der friedlichen Nutzung, der Kernenergie vereinbart. Die Aufnahme weiterer Partner in das Projekt ist vorgesehen.

Weitere Auskünfte erteilt:

Prof. Dr. H. V. Klapdor
Max-Planck-Institut für Kernphysik
Saupfercheckweg 1

6900 Heidelberg 1.

Phys. Bl. 45 (1989) Nr. 3 85

Aus der Wissenschaft

Deutsch-sowjetisches Projekt zur Untersuchung der Neutrinomasse gestartet

Deutsche und sowjetische Kernphysiker haben Anfang Dezember 1988 ein wichtiges gemeinsames Experiment begonnen, mit dem ein neuer Vorstoß zur Beantwortung der Frage nach der Ruhemasse des Neutrinos unternommen wird. Es soll der Doppel-Betazerfall von ^{76}Ge unter erstmaligem Einsatz mehrerer (8–10) Detektoren aus angereichertem ^{76}Ge untersucht werden. Dieses Vorhaben wurde auf der Grundlage des Abkommens vom April 1987 zwischen dem Bundesministerium für Forschung und Technologie und dem Staatskomitee für die Nutzung der Atomenergie der UdSSR über wissenschaftlich-technische Zusammenarbeit vereinbart. Projektleiter auf deutscher Seite ist Prof. H. V. Klapdor vom Max-Planck-Institut für Kernphysik in Heidelberg, auf sowjetischer Seite Akademiemitglied Prof. S. T. Belyaev vom Kurchatov-Institut in Moskau.

Die Neutrinomasse spielt bekanntlich eine zentrale Rolle nicht nur in der Teilchenphysik, sondern auch in der Astrophysik und Kosmologie [1]. Einerseits ist sie eine der Schlüsselgrößen für die Struktur der großen Vereinigungstheorien (GUTs, SUSYs, SUGRAs). Deren Vorhersagekraft für die Neutrinomasse ist gegenwärtig allerdings außerordentlich gering; die vorhergesagten Werte liegen zwischen 10^{-11} und einigen eV. Da Neutrinos die besten Kandidaten für die nicht-baryonische dunkle Materie im Kosmos sind, könnte sich andererseits an ihrer Masse entscheiden, ob das Universum offen oder geschlossen ist.

Information über die Masse des Elektron-Neutrinos hat man kürzlich insbesondere aus der Präzisionsuntersuchung des Tritium-β-Spektrums und aus der Analyse der Neutrinos aus der Supernova SN1987A erhalten. Daraus wurden obere Grenzen von $m_\nu < 18$ eV bzw. < 25 eV abgeleitet. (Die berühmten Neutrino-Oszillationsexperimente mit Reaktor- bzw. solaren Neutrinos bestimmen dagegen *nicht* die Neutrinomasse selbst, sondern *Differenzen* zwischen den Massen verschiedener Neutrinosorten.)

Die schärfste obere Grenze für eine Majorana-Masse des Neutrinos liefert indessen bereits heute der Doppel-Betazerfall und hier (von den direkten Experimenten) der Zerfall des ^{76}Ge.

Man sucht hierbei nach dem die Leptonenzahl nicht erhaltenden neutrinolosen Zerfall $^{76}\mathrm{Ge} \rightarrow {}^{76}\mathrm{Se} + 2e^-$. Diese Zerfallsart, die bislang noch für *keinen* potentiellen $\beta\beta$-Emitter beobachtet wurde, kann nur auftreten, wenn die Majorana-Masse des Neutrinos von Null verschieden ist, d. h. wenn das Neutrino mit seinem eigenen Antiteilchen identisch ist (Majorana-Teilchen). Das eine der beiden

86 Phys. Bl. 45 (1989) Nr. 3

„gleichzeitig" zerfallenden Neutronen emittiert ein Antineutrino, das als Neutrino vom zweiten Neutron eingefangen werden muß. Das geht nur, wenn die Helizität nicht erhalten ist. Aus der gegenwärtig besten experimentell bestimmten Grenze für die Halbwertszeit des 0ν-Zerfalls von ^{76}Ge von $T_{1/2}(0\nu) > 4{,}1 \times 10^{23}$ Jahren leitet man $m_\nu < 2$ eV ab.

Eine Erhöhung der Empfindlichkeit gerade dieser Experimente ist daher besonders interessant. Bisherige Experimente dieser Art nutzten den Zerfall des ^{76}Ge, das mit 7,8 % im natürlichen Isotopengemisch von Ge enthalten ist. Die verwendeten Ge-Detektoren sind hierbei Quelle der Strahlung und Nachweisinstrument (der erwarteten Elektronen) zugleich. Der effektivste Weg, die Empfindlichkeit des Experiments zu steigern, ist die Erhöhung des Isotopenteils a von ^{76}Ge. Die Empfindlichkeit ist nämlich gegeben durch

$T_{1/2} > (3{,}18 \times 10^{26})\ (a/A)\ \sqrt{Mt/(\Delta E B)}$,

und a ist hier der einzige Parameter, der nicht unter der Wurzel steht. $T_{1/2}[a]$ ist die extrahierte untere Grenze für die Lebensdauer nach der Meßzeit t[a], A das Atomgewicht [g/mol], M die Masse des Ge [kg], ΔE die Energieauflösung im untersuchten Energiebereich (FWHM) [keV], B die Untergrundzählrate bei E = 2,041 MeV (Q-Wert des $\beta\beta$-Zerfalls von ^{76}Ge) [Ereignisse/keV a kg].

Im deutsch-sowjetischen Experiment sollen nun Detektoren aus zu 85 % angereichertem ^{76}Ge eingesetzt werden. Das angereicherte Material, ca. 15 kg $^{76}GeO_2$ mit einem Herstellungswert von etwa 100 Mio. DM, wird von den sowjetischen Partnern gestellt. Die ersten 7,5 kg wurden am 2. Dezember 1988 in Moskau übergeben und nach Heidelberg gebracht, so daß der Bau der ersten Detektoren beginnen kann. Diese sollen Mitte 1989 fertig sein und zunächst in Moskau (bzw. Solotvino/Ukraine) und Heidelberg (bzw. Gran Sasso bei Rom) eingesetzt werden. Diese Vorexperimente sollen darüber Aufschluß geben, welcher der beiden Orte die besseren Untergrundbedingungen für das Hauptexperiment aufweist.

Das auf etwa fünf Jahre veranschlagte Projekt wäre ohne eine deutsch-sowjetische Kooperation nicht möglich gewesen: Die Sowjetunion ist als einziges Land in der Lage, derart große Mengen ^{76}Ge zur Verfügung zu stellen, während am MPI für Kernphysik das nötige Know-how sowohl auf theoretischer als auch auf experimenteller Seite (speziell Low-level-Messungen) vorhanden ist. Mit dem jetzt begonnenen Experiment wird es möglich sein, die Lebensdauer von ^{76}Ge in bezug auf den Doppel-Betazerfall bis zu einem Wert von ca. 10^{25} Jahren zu messen. Das entspricht einer Sondierung der Neutrinomasse bis hinab zu 0,2 eV.

[1] Eine Übersicht über den aktuellen Stand findet man in *H. V. Klapdor* (Hrsg.): Neutrinos. Graduate Texts in Contemporary Physics. Springer, Heidelberg 1988.

Der Bundesminister
für Forschung und Technologie

Pressereferat
Postfach 200240
5300 Bonn 2
Telefon
(0228) 593037
(0228) 593042
Teletex
2283628 = BMFTb
2283893 = BMFTd
Telefax
(0228) 593105

Neue Telefax-Nr.
(02 28)59 36 01

Bonn, 10.10.90
Nr. 69/90

Forschungsergebnisse

Hauptphase eines deutsch-sowjetischen Projektes zur Untersuchung der Neutrinomasse beginnt

Seit Ende Juli dieses Jahres wird mit dem weltweit ersten Detektor aus dem hochreinen Germanium-Isotop (^{76}Ge) im Untergrundlabor des Istituto Nazionale di Fisica Nucleare (INFN) im Gran Sasso in Italien nach der Masse des Neutrino gefahndet. Dieser Detektor wurde im Rahmen eines deutsch-sowjetischen Projektes der Grundlagenforschung hergestellt. Die bisherigen Ergebnisse sind so vielversprechend, daß nunmehr die Hauptphase des Projektes freigegeben wurde, in der Detektoren mit mindestens 10 kg angereichertem Material eingesetzt werden sollen. Der BMFT hat insgesamt 1,1 Mio DM für 4 Jahre zur Verfügung gestellt.

Das Neutrino ist eines der geheimnisvollsten Teilchen der modernen Physik. Die Masse dieses Teilchens spielt eine zentrale Rolle nicht nur in der Teilchenphysik, sondern auch in der Astrophysik und Kosmologie. In dem deutsch-sowjetischen Projekt soll die Masse aus der Untersuchung des doppelten Beta-Zerfalls bestimmt werden, bei dem gleichzeitig zwei Neutronen im Innern eines Atomkerns in Protonen umgewandelt werden. Dies ist ein äußerst seltener Prozeß, von dem eine bestimmte Variante nur dann auftreten kann, wenn die Neutrinomasse nicht verschwindet. Nach dieser Variante, dem sog. neutrinolosen ßß-Zerfall, wird gesucht. Der Einsatz von Detektoren aus angereichertem ^{76}Ge erlaubt eine außerordentliche Erhöhung der Empfindlichkeit bisheriger Experimente.
Für die Suche nach dem doppelten Beta-Zerfall im ^{76}Ge-Isotop hat sich eine besonders günstige Konstellation für ein gemeinsames deutsch-sowjetisches Projekt ergeben. Die Sowjetunion verfügt als

...

- 2 -

einziges Land über die erforderliche Menge stark angereicherten Germanium 76. Dieses Material ist wegen des außerordentlich aufwendigen Herstellungsverfahrens sehr wertvoll.

Projektpartner auf sowjetischer Seite ist das Kurchatov-Institut für Atomenergie in Moskau. Auf deutscher Seite bringt das Max-Planck-Institut für Kernphysik in Heidelberg theoretische Erkenntnisse zum doppelten Beta-Zerfall sowie Erfahrungen beim Bau untergrundarmer Detektoren und die erforderliche hochwertige Meßausrüstung ein. Nur durch die Kombination der spezifischen Beiträge aus beiden Ländern konnte dieses Projekt aufgegriffen werden. Dieses Vorhaben der reinen Grundlagenforschung wurde auf der Basis des Abkommens zwischen dem Staatskomittee für die Nutzung der Atomenergie der UdSSR und dem BMFT aus dem Jahre 1987 vereinbart. Seit Projektbeginn Ende 1988 bis heute wurden 16,8 kg des wertvollen Germanium-Isotops von Moskau nach Heidelberg gebracht. Nach etlichen Fehlschlägen mit anderen Firmen schaffte es die Firma ORTEC (USA) im April 1990, den ersten angereicherten HP (High-Purity) ^{76}Ge-Kristall der Welt mit dem respektablen Gewicht von 2 kg zu fertigen. Im Juli 1990 stand dann der erste Detektor zur Verfügung. Die Experimente werden im großen unterirdischen Laboratorium im Gran Sasso-Massiv durchgeführt. Die Felsmassen des Gran Sasso schirmen störende Strahlung aus dem Weltraum ab. Das vom INFN aufgebaute Labor wurde vom MPI Heidelberg eingerichtet.

Die Hauptphase des Projektes wird bis 1995 dauern. Das Heidelberg-Moskau-Experiment dürfte für die übersehbare Zukunft das empfindlichste Experiment zur Bestimmung der Neutrinomasse über den Doppelbeta-Zerfall werden. Es wird eine Sondierung der Neutrinomasse bis hinab zu ca. 0,1 eV erlauben. Es begegnet einer außerordentlichen internationalen wissenschaftlichen Aufmerksamkeit.

Weitere Auskünfte erteilt
Prof. Dr. H.V. Klapdor-Kleingrothaus
Max-Planck-Institut für Kernphysik
Saupfercheckweg 1
6900 Heidelberg 1

ERFINDER-JOURNAL

Das letzte Rätsel der Physik

Deutsche und russische Wissenschaftler erforschen die Masse des Neutrinos

Mit dem weltweit ersten Detektor aus dem hochreinen Germanium-Isotop (^{76}Ge) wird im Untergrundlabor des Istituto Nazionale di Fisica Nucleare (INFN) im Gran Sasso in Italien nach der Masse des Neutrino gefahndet. Dieser Detektor wurde im Rahmen eines deutsch-sowjetischen Projektes der Grundlagenforschung hergestellt. Die bisherigen Ergebnisse sind so vielversprechend, daß nunmehr die Hauptphase des Projektes freigegeben wurde, in der Detektoren mit mindestens 10 kg angereichertem Material eingesetzt werden sollen.

Das Neutrino ist eines der geheimnisvollsten Teilchen der modernen Physik. Die Masse dieses Teilchens spielt eine zentrale Rolle nicht nur in der Teilchenphysik, sondern auch in der Astrophysik und Kosmologie. In dem deutsch-sowjetischen Projekt soll die Masse aus der Untersuchung des doppelten Beta-Zerfalls bestimmt werden, bei dem gleichzeitig zwei Neutronen im Innern eines Atomkerns in Protonen umgewandelt werden. Dies ist ein äußerst seltener Prozeß, von dem eine bestimmte Variante nur dann auftreten kann, wenn die Neutrinomasse nicht verschwindet. Nach dieser Variante, dem sog. neutrinolosen ßß-Zerfall, wird gesucht. Der Einsatz von Detektoren aus angereichertem ^{76}Ge erlaubt eine außerordentliche Erhöhung der Empfindlichkeit bisheriger Experimente. Für die Suche nach dem doppelten Beta-Zerfall im ^{76}Ge-Isotop hat sich eine besonders günstige Konstellation für ein gemeinsames deutsch-sowjetisches Projekt ergeben. Die Sowjetunion verfügt als einziges Land über die erforderliche Menge stark angereicherten Germanium 76. Dieses Material ist wegen des außerordentlich aufwendigen Herstellungsverfahrens sehr wertvoll.

Projektpartner auf sowjetischer Seite ist das Kurchatov-Institut für Atomenergie in Moskau. Auf deutscher Seite bringt das Max-Planck-Institut für Kernphysik in Heidelberg theroretische Erkenntnisse zum doppelten Beta-Zerfall sowie Erfahrungen beim Bau untergrundarmer Detektoren und die erforderliche hochwertige Meßausrüstung ein. Nur durch die Kombination der spezifischen Beiträge aus beiden Ländern konnte dieses Projekt aufgegriffen werden.

Seit Projektbeginn Ende 1988 bis heute wurden 16,8 kg des wertvollen Germanium-Isotops von Moskau nach Heidelberg gebracht. Nach etlichen Fehlschlägen mit anderen Firmen schaffte es die Firma ORTEC (USA) im April 1990, den ersten angereicherten HP (High-Purity) ^{76}Ge-Kristall der Welt mit dem respektablen Gewicht von 2 kg zu fertigen. Im Juli 1990 stand dann der erste Detektor zur Verfügung. Die Experimente werden im großen unterirdischen Laboratorium im Gran Sasso-Massiv durchgeführt. Die Felsmassen des Gran Sasso schirmen störende Strahlung aus dem Weltraum ab. Das vom INFN aufgebaute Labor wurde vom MPI Heidelberg eingerichtet. □

Neuheiten **6, ISSN 0938–4480** (Dec./Jan. 1990/1991) p.58

Aus der Wissenschaft

H. V. Klapdor-Kleingrothaus

Neuer Vorstoß zur Bestimmung der Neutrinomasse

Deutsch-sowjetisches Doppelbeta-Zerfallsexperiment mit angereichertem ^{76}Ge im Gran Sasso bei Rom angelaufen

Seit Ende Juli 1990 wird mit dem ersten High-Purity (HP)-^{76}Ge-Detektor der Welt, hergestellt im Rahmen eines deutsch-sowjetischen Projekts zur Untersuchung der Neutrinomasse, im Untergrundlabor des Istituto Nazionale di Fisica Nucleare (INFN) im Gran-Sasso-Felsmassiv bei Rom nach der Masse des Neutrinos gesucht. Die Neutrinomasse spielt bekanntlich eine zentrale Rolle nicht nur in der Teilchenphysik, sondern auch in der Astrophysik und Kosmologie (siehe z. B. [1, 2]). Im folgenden gibt H. V. Klapdor-Kleingrothaus vom MPI für Kernphysik in Heidelberg, Initiator und Leiter des Projekts, einen kurzen Überblick zum Stand dieses Experimentes. (Red.)

Das Konzept beruht auf der Untersuchung des Doppelbeta-Zerfalls von ^{76}Ge, bei dem gleichzeitig zwei Neutronen im Innern eines Atomkerns in Protonen umgewandelt werden (vgl. dazu A. Fäßler, Phys. Bl. **45** (1989) Nr. 8, S. 321–325). Dies ist ein äußerst seltener Prozeß, von dem eine bestimmte Variante nur dann auftreten kann, wenn die Neutrinomasse nicht verschwindet. Nach dieser Variante, dem sog. neutrinolosen ββ-Zerfall, wird gesucht. Die besonders einfache Signatur dieses Zerfalls ist eine scharfe Linie im Spektrum der Summenenergie der zwei emittierten Elektronen, die dem Q-Wert des Zerfalls (2041 keV) entspricht.

Bereits gegenwärtig liefern Doppelbeta-Experimente mit Germanium die schärfste Grenze für die (Elektron-)Neutrinomasse überhaupt (ca. 2 eV), schärfer, als aus dem Tritiumzerfall (9 eV) oder dem Neutrinofluß der Supernova 1987A (~25 eV) ableitbar. Dabei ist zu beachten, daß der ββ-Zerfall nur von einer Majorana-Masse des Neutrinos induziert werden kann und gegen eine Dirac-Masse quasi „blind" ist, wogegen der Tritiumzerfall unabhängig von der Natur des Neutrinos (Majorana- oder Dirac-Teilchen) ablaufen würde.

Der natürliche Isotopenanteil des Doppelbeta-Emitters ^{76}Ge in Ge-Detektoren – die Quelle und Nachweisinstrument der Strahlung zugleich sind – ist 7,8 %. Der Einsatz von Detektoren aus angereichertem ^{76}Ge erlaubt es, die Empfindlichkeit dieser Experimente außerordentlich zu erhöhen. Die Empfindlichkeit, ausgedrückt durch die extrahierbare untere Grenze für die Lebensdauer $T_{1/2}$ nach der Meßzeit t (in Jahren), ist nämlich gegeben durch

Prof. Dr. Hans Volker Klapdor-Kleingrothaus, MPI für Kernphysik, Saupfercheckweg 1, W-6900 Heidelberg 1.

Abb. 1: Der weltweit erste HP (High-Purity) ^{76}Ge-Detektor mit einem Anreicherungsgrad von 86% (mit Siliziumkappe) in seiner Abschirmung im Gran-Sasso-Tunnel.

$$T_{1/2}^{(0\nu)} > (3{,}18 \cdot 10^{23})\ (a/A)\ \sqrt{Mt/(\Delta E\ B)},$$

wobei der Anreicherungsgrad a der einzige Parameter ist, der nicht unter der Wurzel steht. Dabei bezeichnen A das Atomgewicht (g/mol), M die Masse des Ge (kg), ΔE die Energieauflösung im untersuchten Energiebereich (FWHM) (keV) und B die Untergrundzählrate bei E = 2,041 MeV (Q-Wert des ββ-Zerfalls von ^{76}Ge) in Ereignissen/Jahr · keV · kg.

Wenn der Untergrund B Null wäre, ließe sich statistisch sogar ein *linearer* Zusammenhang zwischen Meßzeit und abgeleitetem $T_{1/2}^{(0\nu)}$ herleiten, was die Empfindlichkeit nochmals beträchtlich erhöhen würde.

Von Ende 1988 bis heute wurden 16,9 kg von zu 86 % in ^{76}Ge angereichertem Germanium von Moskau nach Heidelberg und weiter zur Reinigung und Kristall-Herstellung in die USA gebracht. Am 9. 4. 1990 traf, nach etlichen mißglückten Versuchen mit anderen Firmen, von der Fa. ORTEC (USA) der erste angereicherte HP (High-Purity) ^{76}Ge-Kristall der Welt in Heidelberg ein – mit dem respektablen Gewicht von 2 kg. Bereits Ende Juli 1990 konnte dann der erste aus dem angereicherten Kristall hergestellte ^{76}Ge-Detektor an den mittlerweile vom italienischen Nationalinstitut für Kernphysik (INFN) mit großem Kostenaufwand eigens für das Heidelberg-Moskau-Experiment gebauten und vom MPI für Kernphysik eingerichteten Low-Level-Experimentierplatz gebracht werden (Abb. 1, 2), womit die erste Phase des Experiments beginnen konnte. Die ca. 15 Tonnen starke Abschirmung des Detektors verwendet in mehrjähriger Arbeit im Low-Level-Labor des MPI getestete Materialien wie ^{210}Pb-armes Blei, Elektrolytkupfer und Silizium in Halbleiterreinheit. Die Abschirmung durch das über dem Labor liegende 1400 m mächtige Felsmassiv entspricht einer Schicht von etwa 3800 m Wasser. In diesem weltweit größten Untergrundlabor läuft übrigens neben Experimenten, die sich mit der Suche nach magnetischen Monopolen (MACRO-Experi-

 0031–9279/91/0303–0206 $ 03.50+.25/0 – © VCH, W-6940 Weinheim, 1991 Phys. Bl. 47 (1991) Nr. 3

Abb. 2: Das im Gran-Sasso-Massiv bei Rom für das Heidelberg-Moskau-Experiment vom italienischen Nationalinstitut für Kernphysik (INFN) errichtete Experimentiergebäude (außen). Im unteren Stockwerk befindet sich der Detektor, im oberen der Computer. Im Vordergrund der Flüssigstickstofftank. Das große Tor im Hintergrund führt zur Halle A, dem Standort von GALLEX.

ment) und dem Nachweis von Neutrinos aus Sternkollapsen (LVD-Experiment) beschäftigen, auch das GALLEX-Experiment, das den Neutrinofluß von der Sonne messen soll.

Das Ergebnis der ersten 108 Tage Meßzeit ergibt einen extrem niedrigen Untergrund von 0,4 ± 0,2 Ereignissen pro Jahr, kg und keV im Bereich der gesuchten Linie aus dem Doppelbeta-Zerfall (Abb. 3), ein Halbwertszeit-Limit von $T_{1/2}^{(0\nu)} > 5{,}66 \cdot 10^{23}$ Jahren und eine obere Grenze für die Neutrinomasse $m_\nu < 2{,}0$ eV (beide mit 90 % Konfidenz-Limit).

Da der Ursprung des Untergrundes, der hauptsächlich aus intrinsischen Kontaminationen der Detektorhalterung und Abschirmung sowie aus Spallationsprodukten der kosmischen Strahlung stammt, weitgehend verstanden ist, erscheint eine weitere Reduzierung möglich. Zur Beurteilung des Untergrunds folgender Vergleich: Für *angereicherte* Detektoren einer Gesamtmasse von 10 kg erwartet man für eine angenommene Halbwertszeit von 10^{24} Jahren 50 Zerfälle pro Jahr.

Da der eingesetzte Detektor mit seinen 985 g Gewicht bezüglich seiner Empfindlichkeit bereits einem Experiment mit 120 kg *natürlichem* Ge entspricht (die größten gegenwärtig laufenden Experimente verwenden weniger als 10 kg natürliches Ge), wird bereits das Experiment mit diesem *einen* angereicherten Detektor in wenigen Monaten die aus jahrelangen Messungen mit natürlichen Germanium-Detektoren abgeleitete weltweit beste Grenze für die Neutrinomasse, gemessen von der UCSB-LBL-Gruppe (D. Caldwell et al.) in Kalifornien, unterschritten haben.

In der zweiten und Hauptphase des Experiments, die bis 1995 dauern dürfte, sollen mindestens 10 kg an angereicherten Detektoren eingesetzt werden. Bereits Ende November 1990 gelang es, einen zweiten angereicherten Kristall von 3,4 kg Gewicht fertigzustellen. Damit hat das Heidelberg-Moskau-Experiment die Chance, für die übersehbare Zukunft das bei weitem empfindlichste Experiment zur Bestimmung der Neutrinomasse über den $\beta\beta$-Zerfall zu werden. Es sollte eine Sondierung der Neutrinomasse bis hinab zu ca. 0,2 eV erlauben (vgl. Abb. 4).

Das Heidelberg-Moskau-Experiment ist komplementär zu den bereits angelaufenen Solarneutrino-Experimenten wie GALLEX und SAGE, die die *Differenz* der Massen von Elektron- und Myon-Neutrino (oder einem anderen Neutrinoflavor) messen, also insbesondere (wenn $m_{\nu e} \ll m_{\nu\mu}$) die Masse des μ-Neutrinos.

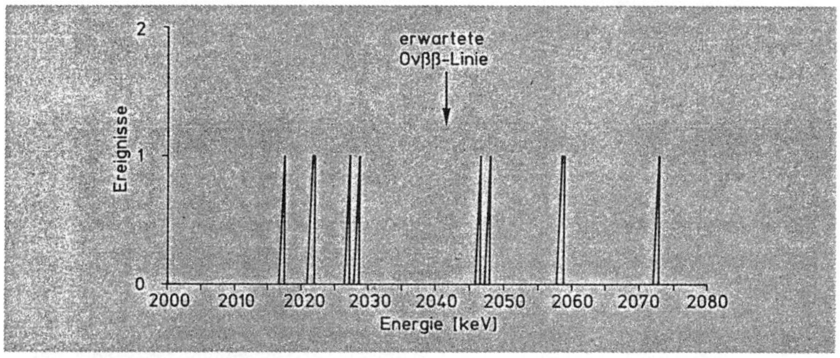

Abb. 3: Das gemessene Spektrum im Bereich der erwarteten 0νββ-Linie bei 2041 keV nach 108 Tagen Meßzeit. Beobachtet wurden zehn Untergrundereignisse. Im 3σ-Bereich um die hypothetische 0νββ-Linie liegt kein Ereignis. Daraus läßt sich ein Limit für die Halbwertszeit von < 5,66 · 10^{23} Jahren und eine obere Grenze für die Neutrinomasse von 2,0 eV ableiten.

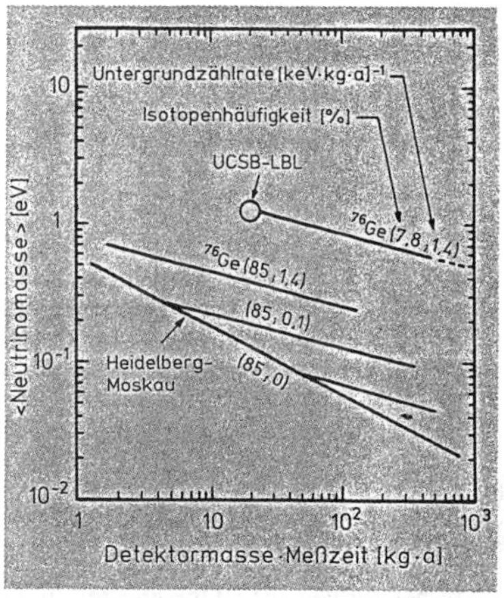

Abb. 4: Die sondierbaren Bereiche der Neutrinomasse mit natürlichem Ge (7,8% ^{76}Ge) und zu 86% an ^{76}Ge angereichertem Germanium als Funktion des Produkts Detektormasse × Meßzeit (kg × Jahr) bei verschiedenen Annahmen über den Untergrund. Eingezeichnet ist auch die gegenwärtige Grenze aus dem UCSB-LBL-Experiment.

Projektpartner sind auf sowjetischer Seite das Kurchatov-Institut für Atomenergie in Moskau, auf deutscher Seite das Max-Planck-Institut für Kernphysik in Heidelberg. Die notwendigen Investitionen werden vom Bundesministerium für Forschung und Technologie getragen.

[1] *H. V. Klapdor* (Hrsg.): Neutrinos. Springer Berlin, Heidelberg 1988.

[2] *K. Grotz, H. V. Klapdor*: Die schwache Wechselwirkung in Kern-, Teilchen- und Astrophysik. Teubner, Stuttgart 1989; englisch bei Adam Hilger, Bristol 1990.

Monatsspektrum

Weitere Eingrenzung der Neutrinomasse über den doppelten Beta-Zerfall

Ein auf fünf Jahre angelegtes deutsch-sowjetisches Experiment, das Ende Juli 1990 im Gran-Sasso-Untergrundlabor bei Rom gestartet worden ist, liefert erste Ergebnisse zur Masse des Elektron-Neutrinos.

Von Hans Volker Klapdor-Kleingrothaus

Seit Wolfgang Pauli vor 60 Jahren das Neutrino postuliert hat, um die Energiebilanz beim Beta-Zerfall eines Neutrons in ein Proton und ein Elektron auszugleichen, ist dieses Teilchen eines der geheimnisvollsten und interessantesten in der Physik geblieben. So weiß man bis heute nicht, ob es eine Masse hat und ob es mit seinem Antiteilchen identisch ist oder nicht (das heißt, ob es sich um ein Majorana- oder ein Dirac-Teilchen handelt). Gerade diese Informationen wären jedoch für theoretische Physiker wie für Kosmologen von größter Bedeutung: Ersteren würden sie helfen, zwischen verschiedenen Modellen zur Vereinheitlichung der vier Grundkräfte der Materie zu entscheiden, und Kosmologen könnten damit die Materiedichte des Universums besser abschätzen.

Der neutrinolose doppelte Beta-Zerfall

In verschiedenen Experimenten ist es bisher nur gelungen, Obergrenzen für die Neutrinomasse (genauer: die Masse des Elektron-Neutrinos; außerdem gibt es noch die weitaus selteneren Myon- und Tau-Neutrinos) zu bestimmen. Das wohl aussagekräftigste Experiment in diesem Zusammenhang ist Ende Juli letzten Jahres im Untergrundlabor des Istituto Nazionale di Fisica Nucleare im Gran Sasso bei Rom als deutsch-sowjetisches Gemeinschaftsunternehmen angelaufen. Dabei soll die Masse des Neutrinos über die Häufigkeit des neutrinolosen doppelten Beta-Zerfalls ermittelt werden; denn dieser Prozeß tritt um so öfter auf, je größer die Neutrinomasse ist, und sollte überhaupt nicht zu beobachten sein, falls das Neutrino, wie von dem derzeitigen Standardmodell der Elementarteilchenphysik angenommen, völlig masselos ist. Ferner setzt er voraus, daß das Neutrino ein Majorana-Teilchen, das heißt mit seinem Antiteilchen identisch ist.

Stark vereinfacht kann man sich den neutrinolosen doppelten Beta-Zerfall so vorstellen, daß das bei einem einfachen Beta-Zerfall ausgesandte Neutrino von einem anderen Neutron im selben Atomkern absorbiert wird und dieses dazu anregt, seinerseits in ein Proton und ein Elektron zu zerfallen. Dem steht allerdings ein grundlegendes Hindernis entgegen: Während das erste Neutron bei seinem Beta-Zerfall nämlich ein Antineutrino mit Rechtsdrall aussendet, kann das zweite Neutron nur durch ein Neutrino mit Linksdrall zum Beta-Zerfall veranlaßt werden.

Daraus ergeben sich zwei fundamentale Bedingungen für das Auftreten des neutrinolosen doppelten Beta-Zerfalls:

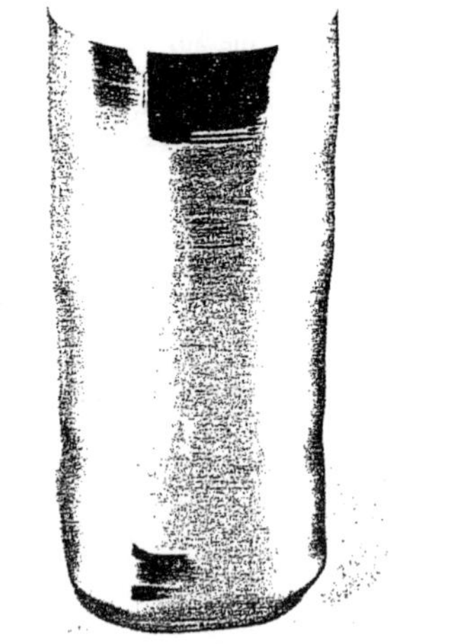

Bild 1: Der weltweit erste hochreine Einkristall aus Germanium, das auf 86 Prozent mit dem Isotop der Masse 76 angereichert ist, wog beachtliche zwei Kilogramm.

Das emittierte Antineutrino darf nicht immer nur, wie die Physiker sagen, rechtshändig sein, und es darf sich nicht von dem entsprechenden normalen Neutrino unterscheiden. Nun hängt die Händigkeit (die Richtung des Dralls relativ zur Flugrichtung) des beim ersten Zerfall ausgesandten Teilchens interessanterweise von dessen Masse ab. Während sich ein masseloses Teilchen nämlich mit Lichtgeschwindigkeit bewegt, bleibt ein mit Masse behaftetes um so weiter hinter dieser Grenzgeschwindigkeit zurück, je schwerer es ist. Ein derart verlangsamtes Teilchen mit Rechtsdrall kann nun aber von einem schnelleren Bezugssystem überholt werden und erscheint dann aus der „Vorderansicht", als hätte es einen Linksdrall.

Diese Erklärung ist nicht wörtlich zu nehmen, sondern nur der Versuch, durch ein anschauliches Bild ansatzweise plausibel zu machen, warum das vom ersten Neutron ausgesandte Antineutrino von dem zweiten um so öfter als linkshändig absorbiert wird – und damit den neutrinolosen doppelten Beta-Zerfall auslöst –, je höher seine Masse ist. Außerdem laufen die beiden Zerfälle in Wahrheit gleichzeitig ab und dürfen strenggenommen nicht einzeln betrachtet werden.

Der Nachweis

Bei welchen Atomen sollte man nach dem ungewöhnlichen doppelten Beta-Zerfall suchen? In Frage kommen alle, bei denen die Energiedifferenz zwischen dem ursprünglichen und dem beim Zerfall gebildeten Kern mindestens der Ruheenergie von zwei Elektronen (1022 Kiloelektronenvolt) entspricht. Außerdem darf bei diesen Atomen nicht der einfache Beta-Zerfall auftreten, da er den doppelten verdecken würde. Das bei dem Experiment im Gran Sasso verwendete Germanium hat den zusätzlichen Vorteil, daß es ein Halbleitermaterial ist und dadurch gleichzeitig auch als Detektor für die emittierten Elektronen dienen kann.

Allerdings ist nur das Germanium-Isotop mit der Masse 76 ein geeigneter Kandidat für den neutrinolosen doppelten Beta-Zerfall, und sein Anteil in natürlichem Germanium beträgt lediglich 7,8 Prozent. Da die Empfindlichkeit des Experiments mit der Menge an zerfallfähigem Material steigt, wurde für das Experiment im Gran Sasso Germanium aus der Sowjetunion verwendet, dessen Gehalt an ^{76}Ge auf 86 Prozent angereichert worden ist. Seit Ende 1988 hat das Kurtschatow-Institut in Moskau 16,9

Bild 2: Der aus dem in Bild 1 gezeigten Einkristall gefertigte Detektor (mit Silicium-Kappe) in seiner Abschirmung aus Elektrolytkupfer und extrem strahlungsarmem Blei im 1400 Meter tief gelegenen Labor unter dem Gran Sasso bei Rom.

Kilogramm davon geliefert. Die amerikanische Firma ORTEC hat daraus Anfang letzten Jahres einen zwei Kilogramm schweren Einkristall gezüchtet (Bild 1). Die Hälfte dieses weltweit ersten hochreinen ^{76}Ge-Kristalls befindet sich, nachdem er zum Detektor umgebaut worden ist, seit Juli letzten Jahres an dem vorgesehenen Experimentierplatz im Gran Sasso unter einem 1400 Meter mächtigen Felsmassiv.

Bei der Suche nach einem so seltenen Ereignis wie dem neutrinolosen doppelten Beta-Zerfall muß jegliche Hintergrundstrahlung möglichst ausgeschaltet werden. Deshalb ist es wichtig, daß das Experiment so tief unter der Erde durchgeführt wird. Zur weiteren Abschirmung ist das Germanium von insgesamt 15 Tonnen an Materialien wie an ^{210}Pb armem Blei, Elektrolytkupfer und Silicium in Halbleiterreinheit umgeben (Bild 2). Damit konnte die Untergrundstrahlung im Bereich der hypothetischen Doppel-Beta-Linie auf 0,6 Ereignisse pro Jahr, Kilogramm Detektor und Kiloelektronenvolt reduziert werden.

Durch den doppelten Beta-Zerfall sollte sich Germanium-76 in Selen-76 umwandeln. Die Energiedifferenz zwischen beiden Atomen beträgt 2041 Kiloelektronenvolt. Den entsprechenden Wert muß die Gesamtenergie der beiden ausgesandten Elektronen haben; im Summenenergiespektrum der Elektronen ist daher ein Peak bei 2041 Kiloelektronenvolt zu erwarten.

Erste Ergebnisse

Ein solcher Peak konnte allerdings auch nach inzwischen 240,22 Experimenttagen nicht beobachtet werden. Unter Berücksichtigung der Empfindlichkeit des Detektors und der Stärke der Hintergrundstrahlung läßt sich daraus berechnen, daß die Halbwertszeit für den neutrinolosen doppelten Beta-Zerfall von ^{76}Ge mindestens $1{,}1 \times 10^{24}$ Jahre (bei 68 Prozent Konfidenzlimit) beträgt.

Diese Untergrenze für die Halbwertszeit kann man – vorausgesetzt, beim Neutrino handelt es sich um ein Majorana-Teilchen – in eine Obergrenze für die Neutrinomasse von 1,46 Elektronenvolt umrechnen. Damit hat das deutsch-sowjetische Experiment im Gran Sasso bereits nach weniger als einem Jahr mit nur einem Bruchteil des Endausbaus den diesbezüglichen Rekordwert nahezu erreicht, den eine Gruppe von der Universität von Kalifornien in Santa Barbara und vom Lawrence-Berkeley-Laboratorium anhand jahrelanger Messungen mit rund acht Kilogramm natürlichen Germaniums erhalten hatte.

Ferner ist diese Untergrenze weitaus niedriger als die aus komplementären Experimenten wie dem Tritium-Zerfall (9 Elektronenvolt) oder dem Neutrinofluß der Supernova 1987A (25 Elektronenvolt) abgeleiteten, wobei letztere allerdings unabhängig davon gelten, ob das Neutrino mit seinem Antiteilchen identisch ist. Überdies schließt das Experiment bereits aus, daß das zur Zeit vieldiskutierte Neutrino mit einer Masse von 17 Kiloelektronenvolt, das möglicherweise beim Zerfall des Tritiums auftritt, ein Majorana-Neutrino sein könnte (siehe „Sein oder Nichtsein: ein Neutrino mit – zu viel – Masse" von Uwe Reichert, Monatsspektrum, Spektrum der Wissenschaft, Juli 1991).

In der Hauptphase des Experiments, die bis 1995 dauern dürfte, sollen mindestens zehn Kilogramm angereicherten Germaniums als Detektoren eingesetzt werden. Bereits Ende November 1990 konnte ein zweiter hochreiner ^{76}Ge-Kristall gezüchtet werden – mit einem Gewicht von 3,4 Kilogramm der größte jemals hergestellte Germanium-Kristall überhaupt. Der daraus gefertigte 2,9 Kilogramm schwere Detektor wurde kürzlich zu dem ersten in den Gran Sasso gebracht. Ein dritter, 2,5 Kilogramm schwerer Detektor ist soeben fertig geworden, und die Herstellung weiterer Kristalle und Detektoren ist im Gange.

Damit hat das deutsch-sowjetische Unternehmen die Chance, für die überschaubare Zukunft das bei weitem empfindlichste Experiment zur Bestimmung der Neutrinomasse über den doppelten Beta-Zerfall zu werden. Es sollte eine Sondierung dieser Masse bis hinab zu etwa 0,2 Elektronenvolt erlauben.

Prof. Dr. Klapdor-Kleingrothaus vom Max-Planck-Institut für Kernphysik in Heidelberg ist Sprecher des Doppel-Beta-Experiments im Gran Sasso.

Science/Business
JOHN AVERY EMISON, SCIENCE WRITER
THE OAK RIDGER, OAK RIDGE, TN. SUNDAY, FEBRUARY 10, 1991 PAGE 1D

Unique crystal 'grown' in OR could play role in basic physics

by John Avery Emison
Oak Ridger staff

A single crystal of an isotope of the element germanium could play a key role in remolding man's understanding of the nature of the universe and the origin of matter.

Moreover, the germanium-76 feed material was refined and the crystal grown right here in Oak Ridge at EG&G ORTEC.

ORTEC is fabricating a large diode from the crystal for delivery to the Max Planck Institute in Germany, according to Mario Martini, manager of detector products at ORTEC.

Martini, who holds an earned doctorate in physics, said Planck will install the diode in a radiation detector in the Gran Sasso Facility in Italy, some 2 kilometers beneath a mountain. It is there that the detector will be used to monitor an extremely rare radioactive nuclear decay mode known as neutrino-less double-beta decay.

A highly shielded setting is needed, Martini explained, to filter out background radiation and detect neutrinos.

When asked to define "neutrino," Martini answered that it is a lepton.

And what is a lepton? "(It) is a particle with a very small mass, or no mass," Martini said.

Determining whether neutrinos have mass will either confirm or overthrow physicists' views of the origin of matter and theories of the physics of the universe, according to Sanford Wagner, a physical chemist who is now ORTEC's advertising manager.

If neutrinos have mass, Wagner said the current understanding of matter formation "is up the flue."

In theoretical physics, ascertaining that neutrinos have mass — as has been theorized by some — will overthrow the now-well-accepted standard model, or TOE (theory of everything), which has unified two of the four fundamental forces of the universe, according to Wagner.

The germanium-76 isotope is thought to be "the one isotope that can have a reaction" with neutrinos, if neutrinos indeed have mass, Wagner said. So part of the crystal's function is to "intercept" neutrinos.

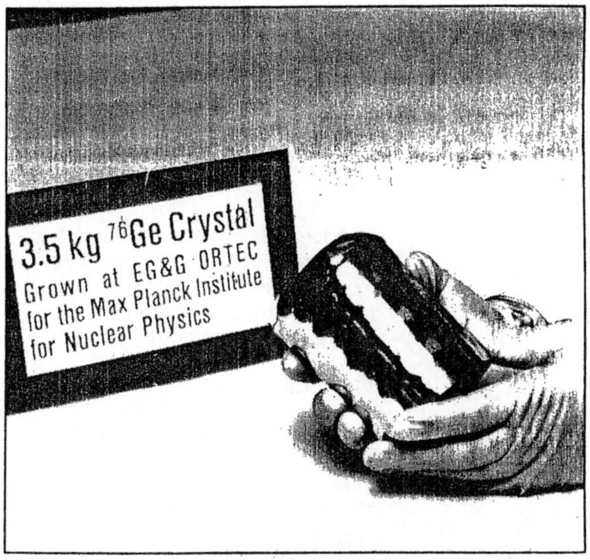

This seven-pound germanium crystal is the most pure substance ever produced by man, ORTEC scientists say. —Photo submitted

But the crystal also has a dual role, Wagner said, in that it will detect the gamma rays that are thought to be given off from the theoretical collision of neutrinos and germanium-76.

Only the purest form of germanium-76 can be used for this sensitive experiment, and Martini tells the highly improbable story about the availability of the purified isotope.

Martini said a German professor, H.V. Klapdor-Kleingrothaus of the Planck Institute, "convinced the Russians to build a huge factory with no military purpose, but to produce germanium-76."

Martini said the feat is even more incredible when one considers that Klapdor-Kleingrothaus also convinced the Russians to turn over the material to him — at no cost to the Planck Institute — when they completed the work.

The Planck Institute is ORTEC's customer for growing the crystal.

Martini said the Russians dedicated a "$50 million facility the size of K-25" for two years to produce enough germanium-76 to eventually grow the crystal.

He said the cost of that effort "boggles the mind. Think of K-25 operating for one year or two; that will give you an idea."

The purity of the material in the crystal is also impressive.

Martini said the crystal has only one part impurity per trillion parts germanium-76. "(It's been) compared to having two grains of salt in a boxcar of sugar," he said.

"It's certainly by far the purest material known to man — not even close — period," Wagner added.

Growing the crystal cost in the range of $100,000 to $200,000, according to Wagner. But he added that the material in the crystal is almost priceless.

Oddly, one of ORTEC's two worldwide competitors for small, more widely used germanium crystal radiation detectors is another Oak Ridge firm, Tennelec, according to Martini. "Mr. Coffey (chairman of Tennelec) has referred to this as germanium valley," Martini said.

The only other competitor anywhere in the world is the Belgian company, Metallurgie Hoboken, according to Martini.

Wagner said germanium-based detectors are used in a variety of radiological sampling devices for environmental studies.

GRAN SASSO
Enriched germanium in action

Two large crystals of carefully enriched germanium, one weighing 1 kilogram and the other 2.9 kilograms, and worth many millions of dollars, are being carefully monitored in the Italian Gran Sasso Laboratory in the continuing search for neutrinoless double beta decay.

In ordinary beta decay, a nuclear neutron decays into a proton, releasing an electron and an antineutrino. The resulting nucleus, lighter but containing an extra proton, is one rung higher in the Periodic Table than its parent.

In double beta decay, the parent and daughter nuclei are two rungs apart, but the transition, involving two interlinked weak interactions, is very difficult. The tell-tale double electron signal was finally seen by an Irvine group in 1987 in the decay of selenium-82 into krypton-82, with a half-life of about 10^{20} years.

This decay is accompanied by two invisible (anti)neutrinos which carry away surplus energy. But another kind of double beta decay might also be possible, where the two neutrinos swallow each other up and are not released.

For this to happen, the normal selection rules governing beta decay and other weak interactions

Signing the agreement for the transfer of the world's largest sample of high purity germanium-76 (86 per cent) from Moscow's Kurchatov Institute for a Heidelberg/Moscow experiment in the Italian underground Gran Sasso Laboratory are (seated) the two collaboration spokesmen – S.T. Belyaev of Moscow (left) and H.V. Klapdor-Kleingrothaus of Heidelberg. Looking on (left to right) are V.I. Lebedev, A. Balish, I. Kondratenko and A. Müller.

would have to be abandoned. The neutrino would also have to behave in a special way – rather than being a conventional 'Dirac' particle with an antiparticle counterpart carrying opposite quantum numbers, it would have to be a 'Majorana' particle, with no distinct antiparticle and with only its spin direction differentiating between electron and positron processes.

However important new physics has come by looking hard for rare processes which violate established weak interaction symmetries – parity in beta decay, CP violation for neutral kaons – while the neutrino, whose very existence was a surprise in itself, has never stopped surprising physicists. So the hunt goes on.

Germanium-76 is a potential double beta decay candidate, and being an electron detector itself, is ideally suited to these studies. The isotope occurs naturally at the level of 7.8 per cent, and germanium detectors, using natural and enriched samples, have established that neutrinoless double beta decay for this isotope, if it happens at all, has a half-life longer than 10^{24} years.

A 1 kilogram detector of high purity (86 per cent) germanium-76 used in the Italian Gran Sasso Laboratory by a Heidelberg/Moscow collaboration has provided limits on neutrinoless double beta decay. With more than 10 kilograms eventually available in the enriched form – the world's largest sample of germanium-76 – the experiment will bring new precision to bear on this hunt.

A Heidelberg/Moscow collaboration now has the world's largest sample of high purity (86 per cent) germanium-76. The 17 kilograms of metal supplied by Moscow's Kurchatov Institute correspond to 14.5 kg of the rare isotope. Further refining and preparation of the final crystals and detectors was carried out in the USA and Western Europe.

An initial 1 kilogram sample, monitored in the Gran Sasso Laboratory over 251 days, shows no characteristic two-electron spike. However several other beta decay processes show up clearly, demonstrating the clean conditions in the underground Laboratory, shielded from cosmic ray background by 1.4 kilometres of rock. The background rate is particularly low in the region where the neutrinoless double beta decay would occur.

A larger (2.9 kilogram) sample was placed in position in September, and all the enriched germanium should be in use by the end of next year.

The Heidelberg/Moscow experiment complements the Gallex and SAGE neutrino experiments now underway, the former in a neighbouring Gran Sasso cavern, the latter in the USSR.

The continued absence of neutrinoless beta decay can be used to deduce limits on the mass of the (electron-type) neutrino. These put the neutrino lighter than 2 electronvolts, a tenfold improvement on mass limits from direct measurements (less than 9.3 electronvolts).

The electron-neutrino mass may well turn out to be zero, but some physicists point out that the limits deduced from neutrinoless double beta decay searches are assumption-dependent. If neutrinoless double beta decay were to be seen, this would certainly imply a non-zero neutrino mass, and the particle would have to be Majorana-type.

Blick in die Forschung

KERNPHYSIK

Suchaktion im Tunnel

Deutsch-russisches Experiment zu Neutrinos und Dunkler Materie

Deutsche und russische Forscher betreiben ein gemeinsames Experiment zur Erforschung von Neutrinos und Dunkler Materie. Von deutscher Seite sind Wissenschaftler des Max-Planck-Instituts für Kernphysik beteiligt. Initiator des Projekts ist Prof. Hans Volker Klapdor-Kleingrothaus vom MPI für Kernphysik in Heidelberg, der in seinem folgenden MPG-Spiegel-Beitrag einen Überblick zum Stand des Experiments gibt.

Im vergangenen Jahr wollte das BMFT von den deutschen Wissenschaftlern erfahren, welche Forschungsaufgaben sie für die wichtigsten halten. In der Kern- und Elementarteilchenphysik, so ergab die »Delphiumfrage«, an der eintausend Fachleute unterschiedlicher Wissensgebiete beteiligt waren, rangiert die Frage, ob Neutrinos eine Masse haben, auf dem ersten Platz. Eine Masse des Neutrinos könnte den Großen Vereinigungstheorien der Teilchenphysik zum Durchbruch verhelfen, an dem die Theoretiker nun schon jahrelang arbeiten. Sie könnte aber auch den Astrophysikern zu ihrer »Dunklen Materie« verhelfen, die benötigt wird, um die Entstehung der Strukturen im frühen Universum in Form von Galaxien und Galaxienhaufen zu erklären, und könnte die Evolution des Universums für die Zukunft bestimmen. Die weltweit besten Auskünfte über eine sogenannte Majorana-Masse des Neutrinos lassen sich aus einem Experiment ableiten, das vor fünf Jahren vom Heidelberger Max-Planck-Institut für Kernphysik und dem Kurchatov-Institut in Moskau ins Leben gerufen wurde und im Gran Sasso Untergrundlabor bei Rom betrieben wird: die Suche nach dem Doppelbetazerfall mit angereichertem Germanium 76. Dieses Experiment ist als erstes seiner Art in den Sub-Elektronenvolt (eV)-Massenbereich des Neutrinos vorgestoßen.

Neutrinos als Sonden »neuer Physik«

Die Neutrinomasse ist eine der Schlüsselgrößen für die Großen Vereinigungstheorien (GUTs = Grand Unified Theories) der Elementarteil-

Blick in die Forschung

Abb. 1: Unterzeichnung der Übereinkunft über den Transfer der weltweit größten Menge von angereichertem Germanium 76 vom Kurchatov-Institut in Moskau nach Heidelberg für das Heidelberg-Moskau Doppelbeta-Experiment. Sitzend die Sprecher der Kollaboration – Prof. S.T. Belyaev aus Moskau (links) und Prof. H.V. Klapdor-Kleingrothaus vom MPI für Kernphysik (rechts). Stehend (von links) Dr. V.I. Lebedev, Dr. A. Balish, Dr. I. Kondratenko und Dr. A. Müller. Fotos: Klapdor-Kleingrothaus

chenphysik und gleichzeitig – als favorisierter Kandidat für nicht-baryonische heiße dunkle Materie im Universum – von außerordentlicher Bedeutung für die Kosmologie. Wie auf der jüngsten NEUTRINO-Konferenz in Eilat/Israel (29.5. bis 4.6.94) besonders deutlich wurde, sind Doppelbeta-Experimente und insbesondere das unsrige in eine Phase ganz besonderer Aktualität eingetreten. Es werden nämlich gegenwärtig mögliche Hinweise auf »neue Physik«, d.h. Physik jenseits des Standardmodells (SM) der Elementarteilchenphysik, diskutiert, die am einfachsten zu verstehen sind, wenn Neutrinos Masse haben.

Diese sind:

- das durch die neuesten Gallex-Ergebnisse erneut in Diskussion geratene Defizit solarer Neutrinos;
- die Anomalie des zu geringen, beobachteten atmosphärischen Myon-Neutrino Flusses (genauer zu großen ν_e/ν_μ-Verhältnisses), die als Effekt von Vakuum-Oszillationen im $\nu_\mu \longleftrightarrow \nu_\tau$-Sektor erklärt werden könnte;
- Modelle dunkler Materie im Universum, die aufgrund der COBE-Daten entwickelt wurden und 70% kalter und 30% heißer dunkler Materie fordern, letztere in Form von Neutrinos mit Massen im Bereich weniger (2–21) eV.

Es wurde Anfang des Jahres von Prof. R. Mohapatra aus Maryland und Prof. J. Valle (Valencia) gezeigt, daß alle diese drei Punkte in einem erweiterten GUT-Modell (SO(10) mit S_4-horizontaler Symmetrie) verständlich werden. Dieses einfachste Szenario für die drei bekannten leichten Neutrinos, welches das solare und atmosphärische Neutrino-Defizit und außerdem die durch COBE nahegelegten Vorstellungen von der Dunklen Materie erklärt, fordert für die drei bekannten Neutrinosorten Massen zwischen ca. 1 und 2 eV – im Gegensatz zu GUT-Modellen, die eine ausgeprägte Massenhierarchie der Neutrinos voraussetzen.

Die Aktualität des laufenden Doppelbeta-Experiments besteht nun darin, daß es die Existenz einer Majorana-Masse von ~1–2 eV in naher Zukunft bestätigen oder verwerfen kann. Die gegenwärtige Grenze, die sich aus der gemessenen Grenze für die Halbwertszeit für den neutrinolosen Doppelbetazerfall von $T^{0\nu}_{1/2} > 5{,}0 \times 10^{24}$ Jahren ergibt, liegt bei 0,7 eV. In den nächsten Jahren ist das Ziel die Sondierung der Neutrinomasse bis hinab zu 0,1 eV (fast zwei Größenordnungen unterhalb der gegenwärtigen Grenzen des besten Tritium-Experiments). Das Experiment kann damit eine zentrale Rolle in der modernen Neutrinophysik spielen.

Es begann mit einem Brief an das Kurchatov-Institut und das Staatskomitee für Atomenergie in Moskau im Jahre 1986. In diesem Brief schlugen wir vor, die dort teilweise bereits vorhandene große Menge des teuren Germanium-Isotops ^{76}Ge für ein gemeinsames Experiment einzusetzen. Es ging dann alles erstaunlich schnell. Im Februar 1987 fanden die ersten persönlichen Verhandlungen in Moskau statt – im Gäste- und Konferenzhaus des Kurchatov-Instituts, noch mit dem obligaten Lenin, einem sowjetischen atomgetriebenen Eisbrecher und anderen Errungenschaften des Instituts an den Wänden. Anfang Dezember 1987 folgte die Unterzeichnung des Vertrages und damit der offizielle Beginn der Kooperation – unter dem Schirm des 1987 in Kraft getretenen Vertrags zwischen dem BMFT und dem Staatskomitee für Atomenergie in Moskau über die wissenschaftlich-technische Zusammenarbeit.

Die ersten 7,5 kg an ^{76}Ge, angereichert zu 86%, wurden per Bahn gleich mit nach Heidelberg genommen. Am Morgen des 6. Dezember 1988 trafen sie in Heidelberg ein, wo auch Reporter der Rhein-Neckar-Zeitung die ungewöhnliche Fracht bereits neugierig erwarteten. Weitere 7,5 kg folgten im März 1989 (siehe Abb. 1), die letzten 2,6 kg schließlich im September 1990.

Doppelbetazerfall und Neutrinomasse

Im Jahre 1935, fünf Jahre nach Postulierung des Neutrinos durch Pauli und nur ein Jahr nach Fermi's Theorie der schwachen Wechselwirkung, untersuchte M. Goeppert-Mayer einen neuen ungewöhnlichen Prozeß, den Doppelbetazerfall. Dieser Prozeß, in dem der zerfallende Atomkern zwei Elektronen und zwei Neutrinos emittiert, wurde schließlich 50 Jahre später nachgewiesen und zwar zunächst über ein geochemisches Experiment von Prof. Till Kirsten und Mitarbeitern, ebenfalls am MPI für Kernphysik, später erstmalig in einem direkten Zählerexperiment von Prof. Mike Moe und seiner Gruppe in Irvine, Kalifornien, im Jahre 1987. Er gibt heute wichtige Aufschlüsse zur Struktur der Atomkerne. Bereits die-

Blick in die Forschung

ser Prozeß ist sehr selten, die Lebensdauer eines Doppelbeta-Emitters beträgt billionenmal das Alter des Universums. Heute suchen wir vornehmlich nach einem anderen, noch selteneren Prozeß, dem sogenannten neutrinolosen Doppelbetazerfall. Für diesen Prozeß ist eine Verletzung der Leptonenzahlerhaltung bzw. von B-L (Baryonenzahl minus Leptonenzahl) Voraussetzung. Er besteht in Umwandlung von zwei Neutronen im Kern unter Austausch eines Neutrinos und unter Abstrahlung zweier Elektronen – plus eventuell eines masselosen weiteren Teilchens, eines sogenannten Goldstone-Bosons, das bei Brechung einer globalen (B-L)-Symmetrie entstehen würde und Majoron genannt wird. Voraussetzungen für die beiden letzteren Prozesse sind gewisse Eigenschaften des Neutrinos: Sie müssen Majorana-Teilchen sein, d. h. Neutrino und Antineutrino müssen identisch sein, und das Neutrino muß eine nichtverschwindende Masse haben. Die meisten Theorien der Großen Vereinigung der Elementarteilchenphysik (GUTs, SUSYs, ...) sagen tatsächlich im Gegensatz zum Standard-Modell vorher, daß das Neutrino ein Majoranateilchen ist und eine Masse hat. Die erwartete Rate für den Zerfallsprozeß ist um so größer, je größer die Masse des Neutrinos ist (proportional dem Quadrat der Neutrinomasse). Würde man diese Prozesse beobachten, was bislang nicht geschehen ist, und hätte das Neutrino also Masse, so hätte dies geradezu umwälzende Konsequenzen für unser physikalisches Weltbild.

Warum gerade angereichertes Germanium 76?

Seit Jahren schon suchen zahlreiche Gruppen in der Welt nach dem neutrinolosen Doppelbetazerfall an verschiedenen Isotopen (es gibt ca. 35 potentielle ββ-Emitter), u. a. auch an Germanium. Bisherige Experimente verwenden allerdings Detektoren aus Germanium in natürlicher Isotopen-Zusammensetzung. Daher ist das ^{76}Ge, das einzige Germanium-Isotop, das Doppelbetazerfall machen könnte, darin nur zu 7,8% enthalten. Die Detektoren sind gleichsam Quelle der (Elektronen-)Strahlung und Nachweisinstrument zugleich. Die Quellstärke ist um so größer, je größer die Anreicherung des ββ-Emitter-Isotops Germanium 76 ist.

Abb. 2: Labor des Heidelberg-Moskau Doppelbeta-Experiments im Gran Sasso Untergrundlabor bei Rom. Im unteren Geschoß befinden sich die Detektoren, im Obergeschoß Computer und Datenauswertung. Vorn ein Flüssigstickstoffbehälter. Das große Tor führt zu Halle A, in der sich u. a. GALLEX befindet.

Weltweit empfindlichstes Experiment

Unser Material und die daraus gebauten hochreinen Detektoren enthalten zu 86% an ^{76}Ge angereichertes Germanium. Die bis Jahresende einsatzfähigen 10 bis 11 kg dieser Detektoren besitzen daher die Empfindlichkeit eines Experiments mit über 1,2 Tonnen an natürlichem Germanium. Dazu muß man wissen, daß die größten Experimente mit natürlichem Germanium in Kalifornien und im Gotthard-Tunnel nur ca. 7,5 kg an Material eingesetzt haben. Unser Experiment ist daher bereits jetzt das weltweit bei weitem empfindlichste und wird dies für die absehbare Zukunft auch bleiben.

Das Germanium 76, das in Form eines weißen Pulvers, GeO_2, zu uns kam, wurde von der Fa. Eagle Pitcher in den USA zu metallischem Germanium reduziert und im Zonenschmelzverfahren gereinigt. Danach wurden daraus Einkristalle und daraus ^{76}Ge-Detektoren hergestellt (von der Fa. Ortec). Der weltweit erste hochreine derart angereicherte Germanium-Kristall (Gewicht 2 kg) traf am 9. April 1990 im MPI ein. Ein weiterer, mit 3,4 kg Gewicht seinerzeit der größte überhaupt jemals hergestellte Ge-Kristall, wurde Ende November 1990 fertig. Gegenwärtig sind im Gran Sasso Untergrundlabor bei Rom drei Detektoren von insgesamt 6 kg aktiver Masse in routinemäßigem Meßeinsatz. Abb. 2 zeigt das Labor des Heidelberg-Moskau-Experiments, das in großzügiger Weise vom INFN (Istituto Nazionale di Fisica Nucleare) gebaut wurde und von unserer Arbeitsgruppe – auch mit Unterstützung des BMFT – ausgerüstet wurde. Es liegt in unmittelbarer Nachbarschaft eines weiteren großen Experiments unter Leitung des MPI für Kernphysik, GALLEX, und anderer Großexperimente wie LVD und dem vorwiegend für die Suche nach magnetischen Monopolen gebauten MACRO. Abb. 3 zeigt den ersten angereicherten Detektor des ββ-Experiments in seiner 15-t-Abschirmung aus extrem strahlungsarmem Blei und Elektrolytkupfer. Ein extrem geringer Untergrund aller Bauteile ist entscheidend für das Gelingen des Experiments, da wir nur eine Ereignisrate von einigen Ereignissen pro Jahr erwarten.

Jahrelange Anstrengungen unserer Gruppe in dieser Richtung im Untergrundlabor des MPI in Heidelberg sowie im Gran Sasso haben es möglich gemacht, das Niveau der Störereignisse auf einen extrem geringen Wert abzusenken. Im Bereich um die erwartete Doppelbeta-Linie registrieren wir eine Untergrundzählrate von 0,2 Ereignissen pro Jahr und kg Detektor. Dies ist besser als in allen anderen großen laufenden Doppelbeta-Experimenten. Wir beobachten bislang keine Signale für einen neutrinolosen Doppelbetazerfall. Wir können aber zeigen, daß die Halbwertszeit für diese Zerfallsart größer sein muß als 5×10^{24} Jahre (also größer als 100-billionenmal das Alter des Universums), und daß die Masse des (Elektron-)Neutrinos kleiner ist als 0.7 Elektronenvolt. Dies sind die gegenwärtig weltbesten Grenzen. Das Elektron, das bislang leichteste massenbehaftete Teilchen, ist rund 500 000 mal so schwer.

In der vollen Ausbaustufe des Experiments mit rund 10 kg an Detektoren aus Germanium 76, die Anfang 1995 erreicht werden dürfte, werden wir eine Empfindlichkeit erreichen, die in etwa fünf Jahren Meßzeit die Neutrinomasse bis hinab zu ca. 0,1 eV abtasten kann. Dies ist weit unter dem vorausgesagten Wert des obengenannten GUT-Modells und erlaubt, auch Voraussagen anderer sogenannter Rechts-Links-symmetrischer GUT-Modelle zu testen.

Abb. 3: Der weltweit erste hochreine angereicherte Germanium 76 Detektor in seiner Abschirmung aus 15 t extrem strahlungsarmen Bleis und Elektrolytkupfers.

Das Heidelberg-Moskau ββ-Experiment und andere Neutrinoexperimente

Neben den Doppelbeta-Experimenten gibt es zahlreiche weitere Aktivitäten, welche die Frage einer Masse des Neutrinos aufklären sollen. Experimente, welche die Neutrinomasse über die Form des Betaspektrums im Tritiumzerfall untersuchen, sind komplementär zu ββ-Experimenten. Im Gegensatz zu den letzteren sind sie empfindlich auch auf eine sogenannte Dirac-Masse der Neutrinos (d.h. im Falle, daß Neutrino und Antineutrino nicht identisch sind). Die Empfindlichkeit unseres ββ-Experiments liegt andererseits fast zwei Größenordnungen über denen der besten Tritiumexperimente, die gegenwärtig eine Grenze für die Neutrinomasse von 7,2 eV angeben.

Auch solare Neutrinoexperimente wie Gallex, Sage, Kamiokande und insbesondere das im Aufbau befindliche Sudbury-Experiment sind komplementär. Sie weisen Neutrinos von der Sonne nach und versuchen, Informationen über die Umwandlung einer der drei existierenden Neutrinosorten (Neutrinoflavors) in eine andere zu liefern (Neutrinooszillationen). Sie könnten so Informationen vorwiegend über die Masse des Myon-Neutrinos oder Tau-Neutrinos liefern – während der ββ-Zerfall die Masse des Elektron-Neutrinos sondiert. Neutrinooszillations-Experimente mit Beschleunigern wie die beiden CERN-Experimente NOMAD und CHORUS wollen in den nächsten Jahren eine aus den solaren Neutrinoexperimenten und einer angenommenen Oszillation zwischen Elektron- und Tau-Neutrino sowie einer angenommenen Massenhierarchie abgeleitete Masse des Tau-Neutrinos von ca. 10 eV prüfen.

Suche nach Dunkler Materie

Das Heidelberg-Moskau-Experiment erlaubt wegen seines extrem niedri-

Blick in die Forschung

Abb. 4: Die Ballon-Kampagne GRIS in Alice Springs (Australien) im Mai 1992 mit einem angereicherten Germanium 70 Detektor an Bord – eine Kooperation von ESA, NASA, Kurchatov-Institut Moskau und MPI für Kernphysik. Nach dem erfolgreichen Flug sollen solche Detektoren auch im für die Jahrtausendwende vorgesehenen ESA-Satelliten-Projekt INTEGRAL für die hochauflösende γ-Astronomie eingesetzt werden.

gen Untergrunds gleichzeitig die empfindlichsten Aussagen zur Existenz kalter dunkler Materie in Form von sogenannten WIMPs (Weakly Interactive Massive Particles). Es wird gegenwärtig angenommen, daß 90% der Masse des Universums aus nichtbaryonischer Materie bestehen sollte. Letztere besteht in einem favorisierten Modell aus einem Gemisch von ~30% heißer dunkler Materie (die relativistisch war zum Zeitpunkt des Ausfrierens aus dem thermodynamischen Gleichgewicht im frühen Universum) und ~70% kalter dunkler Materie (nichtrelativistisch zum Zeitpunkt des Ausfrierens). Eine solche Mischung gibt die beste Erklärung bzw. das beste Verständnis für die Entstehung der Strukturen im Universum auf einem weiten Bereich von Entfernungsskalen sowie der beobachteten Anisotropie der kosmischen Mikrowellen-Hintergrundstrahlung.

Kandidaten für die heiße dunkle Materie sind Neutrinos mit Massen im Bereich von ~2–21 eV. Kandidaten für kalte dunkle Materie sind die WIMPs. Einer der favorisierten Kandidaten ist dabei das Neutralino, das leichteste von supersymmetrischen GUTs vorhergesagte Teilchen. Diese WIMPs, die nur über die schwache Wechselwirkung erfaßbar sind, können im Prinzip durch einen auf die Kerne des Germanium-Detektors übertragenen Rückstoß und dadurch ausgelöste Ionisation nachgewiesen werden.

Das Heidelberg-Moskau-Experiment erlaubt beispielsweise, schwere Dirac-Neutrinos im Massenbereich zwischen 26 GeV und 4,7 TeV als dominante Komponente des dunklen Halos unserer Galaxie auszuschließen. Der erfaßte Massenbereich ist komplementär zu etwa dem im MPI für Physik im Aufbau befindlichen Experiment zur Dunklen Materie mit Kryodetektoren von Susan Cooper und ihrer Arbeitsgruppe, das seine Empfindlichkeit im extrem niederenergetischen WIMP-Bereich hat.

γ-Astrophysik mit Doppelbeta-Technologie

Das Experiment zeigte zum ersten Mal, daß die Technologie der Produktion angereicherter hochreiner Ge-Detektoren handhabbar ist. Diese Technik hat inzwischen eine weitere Anwendung gefunden. In einer Zusammenarbeit der ββ-Gruppe mit NASA und ESA wurden zwei angereicherte ^{70}Ge-Detektoren in zwei Ballon-Experimenten in Alice Springs/Australien im Mai 1992 für hochauflösende γ-Spektroskopie von Strahlung aus dem Zentrum der Galaxis eingesetzt (Abb. 4). Detektoren dieser Art sollen aufgrund der Ergebnisse auch im ESA-Projekt INTEGRAL (International Gamma Ray Astrophysics Laboratory), dem Nachfolgeprojekt des gegenwärtig mit großem Erfolg und unter Beteiligung des MPI für extraterrestrische Physik laufenden Projekts GRO (Gamma Ray Observatory), eingesetzt werden. ■

PARTICLE PHYSICS

New Experiments Step Up Hunt for Neutrino Mass

TOKYO—Some of physicists' fondest hopes ride on the neutrino, a shadowy and seemingly massless particle. If the neutrino does have a trace of mass, swarms of them could account for some of the mass that cosmologists believe is missing from the universe as a whole. A massive neutrino might also point the way to a new theory of elementary particles and forces that would transcend the current Standard Model. Now experiments at underground laboratories around the world are striving to weigh the neutrino—without ever observing it.

The new experiments monitor a radioactive material with sensitive detectors, watching for an excruciatingly rare—perhaps nonexistent—process known as neutrinoless double beta decay. These are far from the only efforts to measure neutrino mass. Other researchers, for example, are trying to catch neutrinos in the act of "oscillating" from one of the three neutrino types—called flavors—to another, a transformation that would be a sure sign of mass. But while the oscillation experiments might be sensitive to neutrino masses as low as thousandths of an electron volt (eV), they can only reveal the difference in mass between two flavors. Neutrinoless double beta decay, in which the energy of electrons flung from a decaying radioactive nucleus is measured, will give an absolute value for the mass of the neutrino. That makes the experiments "an essential part of the program of modern particle physics," says astrophysicist John Bahcall of the Institute for Advanced Study in Princeton, New Jersey.

The catch is that so far, neutrinoless double beta decay has never been observed. Ordinary double beta decay is a rare, although regularly observed, process in which two neutrons in a radioactive nucleus decay into two protons, emitting two electrons, or beta rays, and two antineutrinos (antimatter counterparts of the neutrino). In the elusive neutrinoless form, other protons within the nucleus would absorb the antineutrinos as neutrinos, and only the electrons would escape. By comparing the measured energy of the electrons with the total energy of the process predicted by theory, researchers could calculate the energy—and therefore the mass—of the neutrinos.

Neutrinoless double beta decay can take place only if the neutrino has a nonzero mass. Its failure to appear in experiments so far means the process must be rare—and the neutrino mass minute. Current results already suggest a ceiling of roughly 1 eV—a tiny fraction of the mass of the electron. By monitoring larger quantities of radioactive material with more sensitive detectors while screening out sources of background radiation, the new generation of experiments aims to push the mass sensitivity as low as 0.1 eV. It's a level "people didn't think possible a few years ago," says Rabindra Mohapatra, a theoretical physicist at the University of Maryland, College Park.

PUTTING THE NEUTRINO ON THE SCALE

Name of experiment/ collaboration	Heidelberg-Moscow	ELEGANT V	NEMO-3
Institutions	Max Planck Institute for Nuclear Physics and Russian Science Center Kurchatov Institute	Osaka University Research Center for Nuclear Physics	CNRS; University of Bordeaux; University of Caen; Joint Institute for Nuclear Research, Dubna, Russia; *et al.*
Location of experimental setup	Gran Sasso Underground Laboratory, Italy	Oto Cosmo Observatory, Japan	Frejus Underground Laboratory, France
Emitter material	Germanium-76	Calcium-48	Molybdenum-100
Latest upper limit for neutrino mass	0.48 eV	Beginning operations	Planned opening in 1998
Target level	0.1 eV	0.6 eV	0.1 eV

Earlier this year, scientists from the Max Planck Institute for Nuclear Physics in Heidelberg, Germany, and the Russian Science Center Kurchatov Institute in Moscow reported lowering the upper limit of the possible mass of the electron neutrino to 0.48 eV. Mohapatra notes, however, that uncertainties in the "extremely complicated calculations" mean the results, from Gran Sasso National Laboratory in Italy's Apennines, could be off by a factor of 2.

RESEARCH CENTER FOR NUCLEAR PHYSICS, OSAKA U.

Tunnel vision. Japanese neutrino scientists are setting up shop in this railway tunnel south of Osaka.

Now Osaka University's Research Center for Nuclear Physics in Japan is setting up a double-beta-decay experiment in its new Oto Cosmo Observatory in a never-used railroad tunnel 100 kilometers south of Osaka. Hiro Ejiri, a physicist at Osaka, says that steady winds blowing through the tunnel at the new observatory help reduce its natural radon concentrations, which can result in background events. "We hope to get the sensitivity down to 0.6 eV," he says. That's no better than the Russian-German group has already claimed. But because the two groups are working with different isotopes (see table), Mohapatra and others say the Osaka experiment could provide independent confirmation of the recent Heidelberg-Moscow results.

A second European group, including scientists from France's CNRS, several French universities, and the Joint Institute for Nuclear Research in Dubna, Russia, wants to push the limits even lower. The collaboration, called NEMO, hopes by the end of 1998 to bring a new detector online in the Frejus Underground Laboratory in the Alps along the French-Italian border. Francis Laplanche, a physicist at the University of South Paris and a member of the team, says that the group intends to reduce the upper limit on neutrino mass to 0.1 eV by using a large mass of an isotope of molybdenum, cutting background radiation, and improving detection schemes. The Heidelberg-Moscow group, however, believes it will reach that level first. Hans Klapdor-Kleingrothaus of the Max Planck Institute says the goal is to achieve an upper limit of 0.1 eV within 5 years "just by letting the experiment run."

If any of the experiments actually pin down the mass of the neutrino, physicists would have their first clear clue to a theory beyond the Standard Model. The various Grand Unified Theories make different predictions for neutrino mass, and the results to date, by lowering the ceiling on neutrino mass, are casting doubt on some theories, says Moscow's Alexei Smirnov. "These bounds [on neutrino mass] could forbid some schemes," he says—not a bad payoff for an experiment that never sees its quarry.

–Dennis Normile

International Journal of Modern Physics A, Vol. 13, No. 23 (1998) 3953–3992

STATUS AND PERSPECTIVES OF DOUBLE BETA DECAY — WINDOW TO NEW PHYSICS BEYOND THE STANDARD MODEL OF PARTICLE PHYSICS

H. V. KLAPDOR-KLEINGROTHAUS
Max-Planck-Institut für Kernphysik,
PO Box 10 39 80, D-69029 Heidelberg, Germany

Received 4 February 1998

ELSEVIER

Nuclear Physics B (Proc. Suppl.) 77 (1999) 357–368

NUCLEAR PHYSICS B
PROCEEDINGS
SUPPLEMENTS

Double Beta Decay with Ge-detectors – and the future of Double Beta and Dark Matter Search (GENIUS)

H.V. Klapdor–Kleingrothaus [a]

[a] Max–Planck–Institut für Kernphysik,
P.O. Box 103980, D–69029 Heidelberg, Germany

Nuclear double beta decay provides an extraordinarily broad potential to search for beyond Standard Model physics, probing already now the TeV scale, on which new physics should manifest itself. These possibilities are reviewed here. First, the results of present generation experiments are presented. The most sensitive one of them – the Heidelberg–Moscow experiment in the Gran Sasso, using enriched ^{76}Ge – probes the electron neutrino mass now in the sub eV region and will reach a limit of $\sim$ 0.1 eV in a few years. Basing to a large extent on the theoretical work of the Heidelberg Double Beta Group in the last two years, results are obtained also for SUSY models (R–parity breaking, sneutrino mass), leptoquarks (leptoquark-Higgs coupling), compositeness, right–handed W boson mass and others. These results are comfortably competitive to corresponding results from high–energy accelerators like TEVATRON, HERA, etc. Second, future perspectives of $\beta\beta$ research are discussed. A new Heidelberg experimental proposal (GENIUS) is presented which would allow to increase the sensitivity for Majorana neutrino masses from the present level of at best 0.1 eV down to 0.01 or even 0.001 eV. Its physical potential would be a breakthrough into the multi-TeV range for many beyond standard models. Its sensitivity for neutrino oscillation parameters would be larger than of all present terrestrial neutrino oscillation experiments and of those planned for the future. It would further, already in a first step, cover almost the full MSSM parameter space for prediction of neutralinos as cold dark matter, making the experiment competitive to LHC in the search for supersymmetry.

TEN YEARS OF HEIDELBERG-MOSCOW EXPERIMENT – A FRESH LOOK

HANS V. KLAPDOR-KLEINGROTHAUS† [a]
† *Max-Planck-Institut für Kernphysik*
P.O. Box 103980, D-69029 Heidelberg, Germany

The Heidelberg–Moscow double beta decay project operating now for almost ten years in the Gran Sasso Underground Laboratory in Italy, yields the world–wide most stringent limit on the Majorana neutrino mass. Simultaneously, new limits on other parameters of Beyond Standard Model physics have been obtained on the *TeV* scale, and also the best limits on cold dark matter in the universe using only raw data.

The two fundamental questions to all unified theories of modern particle physics, of neutrino mass and nature, and of the existence of a superworld predicted in supersymmetric theories, probably will be solved underground.

Double beta decay is the only process allowing to solve the question of the nature of the neutrino: Dirac or Majorana particle. Also, it is the common opinion of theorists in the field that the question of the neutrino mass matrix cannot be solved by neutrino oscillation experiments (solar, atmospheric, etc) alone but requires in addition a sufficiently sensitive double beta decay experiment.

Neutrinoless double beta decay is a hypothetical extremely rare radioactive decay mode, in which exchange of a neutrino between two nucleons triggers their decay under emission of two electrons. This process would violate conservation of lepton number L, and, equally important, of baryon number minus lepton number (B–L), which would imply immediately Beyond Standard Model physics.

This mode is only possible if the neutrino has a Majorana mass, which requires that the neutrino is its own antiparticle. – However, by far most Grand Unified Models of particle physics predict Majorana neutrinos. – The decay rate is proportional to an effective neutrino mass $< m >$ squared, which is a superposition of the different neutrino mass eigenstates. Consequently double beta decay yields important information on the parameters of the

[a] Spokesman of the Heidelberg–Moscow Collaboration

in Proc. "Advances in Nuclear Physics", Bucharest, Romania, December 1999, *World Scientific, Singapore* (2000), ed. by *D. N. Poenaru, S. Stoica et al.*

neutrino mass and mixing matrix. The information from double beta decay is particularly important to fix the absolute neutrino mass scale, since neutrino oscillation experiments measure only differences between mass eigenstates.

There are worldwide several double beta experiments running looking for this type of decay for various nuclei, which has, however, not been observed until now. The Heidelberg–Moscow experiment is the by far most sensitive one and is probing for the first time the sub–eV range for the Majorana neutrino mass.

The team from the Max Planck Institute for Nuclear Physics in Heidelberg and the Kurchatov Institute in Moscow constructed a setup of five high-purity Germanium detectors enriched in the isotope Germanium–76 to 86% (natural abundance 7.8%) of total mass of 11.5 *kg*. This results in the largest source strength ever used in a double beta decay experiment. The setup is operated in a heavy shielding by some ten tons of superclean lead and copper, under 1500 *meters* of rock (corresponding to 3500 *m* of water shielding) in the Gran Sasso Underground Laboratory in Italy. The background level reached by the experiment is the lowest worldwide for this kind of experiment – 0.06 *events* per year and kilogramm of detector mass in the energy range of the expected double beta decay signal, which would be a peak in the spectrum produced by the decay electrons, at 2038 keV.

In almost 10 years of measurement, the Heidelberg-Moscow team has obtained a lower limit for the half life of neutrinoless double beta decay of Germanium–76 of several 10^{25} years. This is the world record under all running investigations. With the deduced limit for the Majorana neutrino mass of 0.38 eV, the Heidelberg–Moscow experiment is not only the first one exploring the sub-eV range of the neutrino mass, but also enters in a range of values, which has stringent consequences for neutrino mass models and for cosmological models assuming neutrinos as hot dark matter in the universe (see Fig. 1).

For example, in degenerate neutrino mass models, in which the neutrino mass eigenstates have very close values, the above value excludes models with cold and hot dark matter ($CHDM$ models), and also those including a non–vanishing cosmological constant Λ ($\Lambda CHDM$ models) for the case of the small mixing angle solution of the solar neutrino problem. This means, in this case neutrinos as hot dark matter would have to be of Dirac type. Assuming the bestfits of the large mixing angle solution or the vacuum oscillation solutions, still $CHDM$ models are excluded, while $\Lambda CHDM$ models are still marginally possible (requiring $< m >= 0.15 - 0.30\ eV$). In the case of inverse hierarchical neutrino mass scenarios, with only two neutrinos contributing to hot dark matter, only $LCDM$ models with a small Hubble constant h = 0.5 are

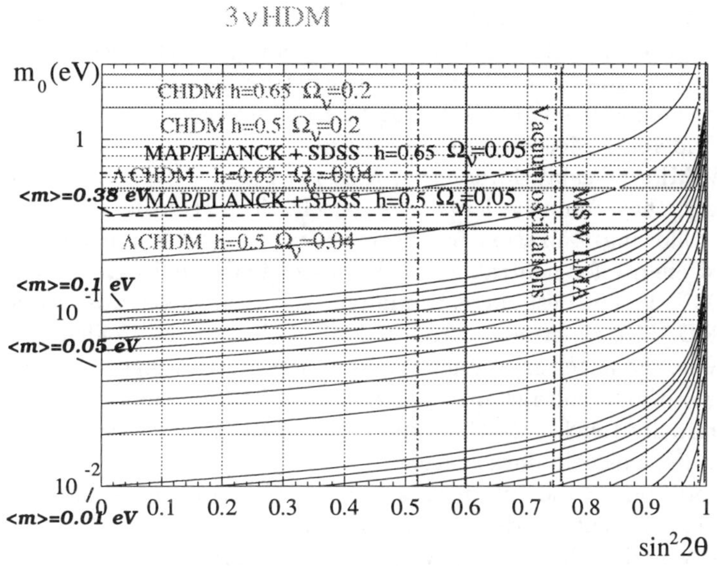

Figure 1. Isomass–lines for the effective neutrino mass $< m >$ measured in double beta decay in the $m_0 - \sin^2(2\theta)$ plane for the case of a degenerate three-neutrino scenario with mass m_o. Also shown are the bestfits for $CHDM$ and $\Lambda CHDM$ for different values of the Hubble constant, the sensitivity of MAP/PLANCK combined with SDSS, and the regions of the MSW LMA and of the vacuum oscillation solutions of the solar neutrino problem. The Heidelberg–Moscow experiment excludes already most of the cosmological models and also, assuming the MSW SMA solution, the whole range of sensitivity of the future satellite experiments MAP and PLANCK for cosmological models involving neutrinos as hot dark matter.

marginally not yet excluded.

Already now, if assuming the small angle solution of the solar neutrino problem to be valid, the experiment would rule out the whole range of sensitivity of the future satellite experiments MAP and PLANCK for cosmological models of the above types involving neutrinos as hot dark matter (see Fig. 1). Fig. 1 also demonstrates that, combined with the neutrino oscillation results obtained in the recent solar and atmospheric neutrino oscillation experiments and with precise determinations of cosmological parameters double beta decay is obviously the only way to obtain precise informations about the neutrino mixing and the absolute mass scale in partially degenerate and degenerate neutrino mass scenarios.

Looking into four–neutrino scenarios, including in addition to the three known neutrino flavours a fourth 'sterile' neutrino, there are as shown recently by Giunti et al., Bilenky et al., and others, only two schemes, that can accomodate the results of *a l l* neutrino oscillation experiments (including the Los Alamos experiment LSND). The first of these two scenarios is ruled out by the Heidelberg-Moscow experiment.

If WIMPs (Weakly Interacting Massive Particles) populate the halo of our galaxy, they could be looked for by elastic scattering off the Ge nuclei and the following ionization by the recoiling nucleus in our detectors. The deposited energy for neutralinos with masses between 10 GeV and 1 TeV is below 100 keV. The best current limits on WIMP nucleon cross sections come from the DAMA experiment, from CDMS and from the Heidelberg–Moscow experiment, the latter experiment yielding the most stringent limits for using raw data without pulse shape analysis. All of these experiments at present just marginally touch only the upper part of the parameter space predicted for neutralinos as cold dark matter.

In addition to the information on the neutrino mass the Heidelberg–Moscow experiment yields information on beyond standard model physics on the TeV scale, where new physics could be expected (Fig. 2).

This is possible since the $\Delta L = 2$ process of neutrinoless double beta decay could occur in *a n y* lepton-number violating theory, so, e.g. by exchange of supersymmetric particles like neutralinos, gluinos, sleptons, etc., and thus the process would yield information on the underlying theories and on properties of the involved particles. It is important to note, and has been proved theoretically already in the early 80's by Schechter and Valle, that independent of the mechanism underlying the double beta decay, its occurrence *a l w a y s* would imply a nonvanishing Majorana neutrino mass.

The half life limit measured in the Heidelberg-Moscow experiment yields *a l o w e r* limit on the the mass of a superheavy left–handed neutrino of $M > 8 \times 10^5\ GeV$, a limit which could be reached only by a far–future 2 TeV Linear Electron–Electron Collider. It yields further an upper limit on the Yukawa coupling Λ'_{111} in the R–parity breaking part of the superpotential of the Minimal Supersymmetric Standard Model (MSSM), which is more stringent than present limits from the TEVATRON and HERA colliders, and which immediately excluded the possibility of squarks of first generation being produced in the recently discussed high–Q squared events at HERA.

The Heidelberg–Moscow result also restricts stringently products of higher generation Yukawa couplings, it yields the sharpest limits for a Majorana–type mass of the sneutrino, the supersymmetric partner of the neutrino, sets limits to compositeness (assuming a substructure of quarks and leptons), which are, as shown recently by Panella et al., more stringent than those from the LEP

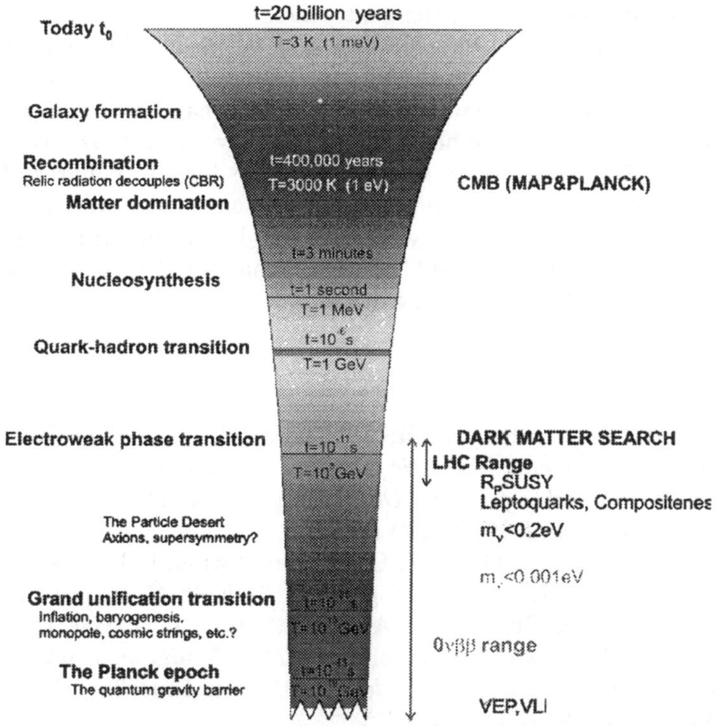

Figure 2. The potential contribution of double β decay to particle physics and cosmology. The energy scales correspond to the right-handed neutrinos in the see-saw mechanism.

II accelerator, yields bounds on violation of special relativity and equivalence principle in the neutrino sector, in the range of small mixing, which cannot be constrained by other experiments, etc.

In summary the Heidelberg-Moscow experiment yields a lot of new information, including the worldwide most stringent information on the Majorana neutrino mass, important for fixing the neutrino mass matrix and cosmological models, including the most stringent limits for cold dark matter (on the basis of raw data), and the sharpest bounds on other beyond standard model physics parameters in the TeV range, where new physics can be expected, and which partly can be covered only by future colliders. For a recent summary of all of these results see [Kla98], [Kla99], [HM99], [HM98], [Kla99a], [Bha99], [Kla2000].

The Heidelberg–Moscow experiment will remain the most sensitive double beta decay experiment also for the next years.

The method of double beta decay could be pushed in another type of setup to a limit of 0.1 eV or ultimately 0.001 eV, which probably would be the ultimate value obtainable in double beta decay experiments. For this purpose the Heidelberg group has proposed the new project GENIUS [Kla98], [Kla99b], [Kla99c], which would have implications for a broad area of physics encompassing particle physics and astrophysics, and which would serve as an important bridge between the physics that will be gleaned from high energy accelerators such as LHC and NLC on the one hand and satellite experiments such as MAP and PLANCK on the other.

References

1. [Bha99] G. Bhattacharyya, H. V. Klapdor-Kleingrothaus and H. Päs *"Neutrino Mass and Magnetic Moment in Supersymmetry without R–Parity in the Light of Recent Data"*, *Phys. Lett.* **B 463** (1999) 77 – 82 and *Preprint hep-ph/* **9907432** (1999)
2. [HM98] HEIDELBERG–MOSCOW Collaboration, L. Baudis, J. Hellmig, G. Heusser, H. V. Klapdor–Kleingrothaus, S. Kolb, B. Majorovits, H. Päs, Y. Ramachers, H. Strecker, V. Alexeev, A. Balysh, A. Bakalyarov, S. T. Belyaev, V. I. Lebedev and S. Zhukov *"New Limits on Dark–Matter Interacting Particles from the Heidelberg–Moscow Experiment"*, *Phys. Rev.* **D 59** (1998) 022001-1 – 022001-5
3. [HM99] HEIDELBERG–MOSCOW Collaboration, L. Baudis, A. Dietz, G. Heusser, H. V. Klapdor–Kleingrothaus, I. V. Krivosheina, S. Kolb, B. Majorovits, V. F. Melnikov, H. Päs, F. Schwamm, H. Strecker, V. Alexeev, A. Balysh, A. Bakalyarov, S. T. Belyaev, V. I. Lebedev and S. Zhukov *"Limits on the Majorana Neutrino Mass in the 0.1 eV Range"*, *Phys. Rev. Lett.* **83** (1999) 41 – 44
4. [Kla98] H. V. Klapdor–Kleingrothaus *"Status and Perspectives of Double Beta Decay – Window to New Physics Beyond the Standard Model of Particle Physics"*, *Intern. Journ of Modern Phys.* **A 13, No. 23** (1998) 3953 – 3992
5. [Kla99] H. V. Klapdor–Kleingrothaus *"Double Beta and Dark Matter Search – Window to New Physics Beyond the Standard Model of Particle Physics"*, in Proc. "Lepton and Baryon Number Violation in Particle Physics, Astrophysics and Cosmology", eds. *H. V. Klapdor–Kleingrothaus and I. V. Krivosheina, International Workshop at ECT,* Trento, Italy, April 20 – April 25, 1998, *World Scientific* (1999) 251 – 301

6. [Kla99a] H. V. Klapdor–Kleingrothaus, H. Päs and U. Sarkar *"Test of Special Relativity and Equivalence Principle from Neutrinoless Double Beta Decay"*, *Eur. Phys. J.* **A 5** (1999) 3 – 6
7. [Kla99b] H. V. Klapdor–Kleingrothaus L. Baudis, G. Heusser, B. Majorovits and H. Päs "GENIUS - a Supersensitive Germanium Detector System for Rare Events", *Proposal* **August 1999 second draft**, *Preprint hep-ph/* **9910205** (1999) and in Proc. "Beyond the Desert: Accelerator, Non–accelerator and Space Approaches into the Next Millenium, BEYOND 2000", International Conference, Castle Ringberg, Tegernsee, Germany, June 6 – 12, 1999, eds. *H. V. Klapdor–Kleingrothaus and I. V. Krivosheina, IOP* (2000) 915 – 1015
8. [Kla99c] H. V. Klapdor–Kleingrothaus *"Double Beta Decay with Ge-detectors – and the Future of Double Beta and Dark Matter Search (GENIUS)"*, in Proc. of International Conference "Neutrino Physics and Astrophysics", NEUTRINO'98, Takayama, Japan, 4 – 9 June, 1998, eds. *Y. Suzuki and Y. Totsuka, Nucl. Phys.* **B 77** *Proc. Suppl.* (1999) 357 – 368
9. [Kla2000] H. V. Klapdor–Kleingrothaus, H. Päs and U. Sarkar *"Effects of New Gravitational Interactions on Neutrinoless Double Beta Decay"*, *Preprint hep–ph/* **0002215** (2000)

Appendix B

The Potential Future - GENIUS

In *1997* - more than ten years after the first discussions of, and *exactly ten years* after the Heidelberg proposal *for* the HEIDELBERG-MOSCOW experiment - the potential successor experiment GENIUS was proposed [Kla97**-A] (see also [Kla99a**-B]) which experimentally opens the range of effective neutrino masses down to ultimate 0.002 eV. This would be simultaneously the ultimate sensitivity, which could be reached in double beta decay experiments. Discussions on a site of the project have been started.

H.V. Klapdor-Kleingrothaus
Max-Planck Institut für Kernphysik
Heidelberg, Germany
5 July, 2000

Fig. B.1 At BEYOND'97 Conference, Castle Ringberg, Germany. From the left to right: Prof. M. Morita (Japan), Prince Luitpold von Bayern, Prof. H.V. Klapdor-Kleingrothaus (Chairman, Beyond the Desert 1997, 1999), Dr. I.V. Krivosheina (Scientific Secretary of BEYOND'99); second row, from left: Prof. G. Börner (München) and Prof. B. Schutz (Potsdam), (9. June, 1997) (foto author).

Beyond the Desert 1997

Accelerator and Non-Accelerator Approaches

Proceedings of the First International Conference on Particle Physics Beyond the Standard Model, Castle Ringberg, Germany, 8–14 June 1997

Edited by H V Klapdor-Kleingrothaus and H Päs

Institute of Physics Publishing
Bristol and Philadelphia

Beyond the Desert 1999

Accelerator, Non-Accelerator and Space Approaches into the NEXT MILLENIUM

Proceedings of the Second International Conference on Particle Physics Beyond the Standard Model, Castle Ringberg, Germany, 6–12 June 1999

Edited by H V Klapdor-Kleingrothaus and I V. Krivosheina

Institute of Physics Publishing
Bristol and Philadelphia

Fig. B.2 Left: Proceedings of BEYOND Conferences in Castle Ringberg, Tegernsee, Germany. At BEYOND'97 for the first time the GENIUS idea was presented, at BEYOND'99 the second draft of the GENIUS proposal has been published. Right: from right to left - Prof. H.V. Klapdor-Kleingrothaus (Chairman, Beyond the Desert 1997, 1999), Prince Luitpold von Bayern, Dr. H. Päs (Scientific Secretary of BEYOND'97), 9. June, 1997) (foto author).

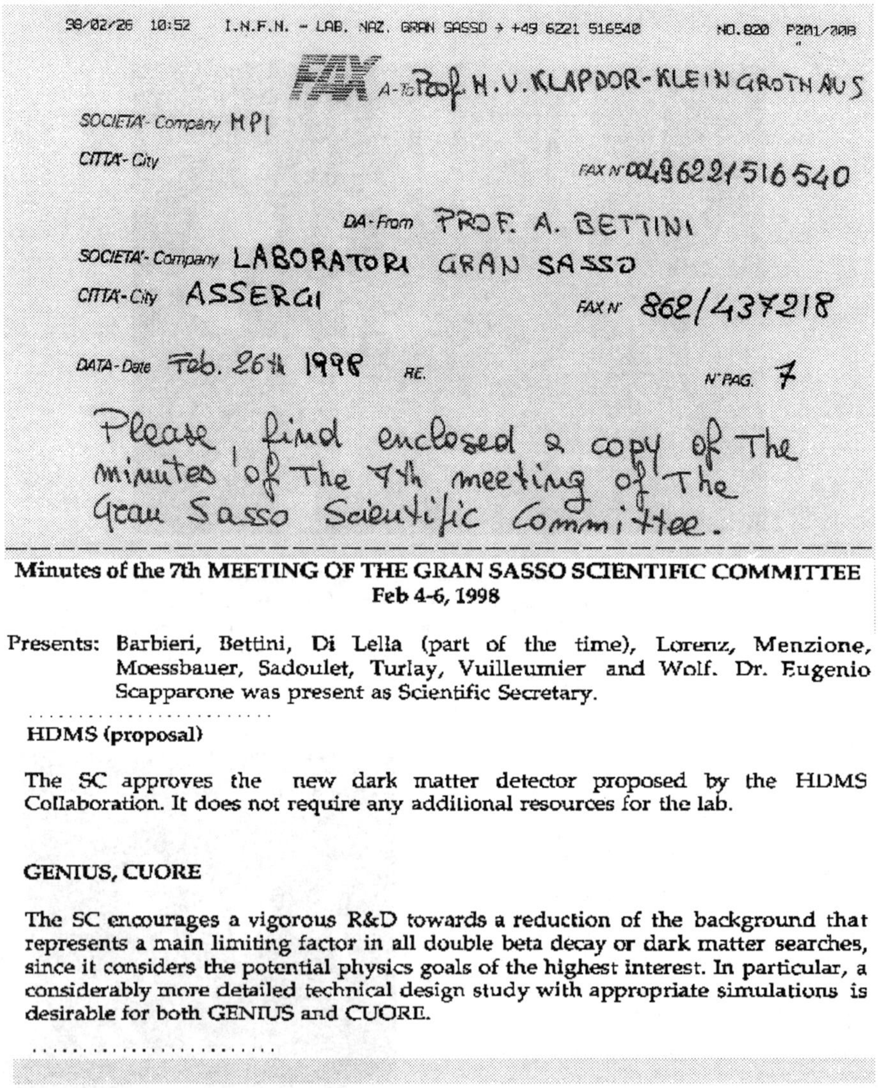

98/02/26 10:52 I.N.F.N. - LAB. NAZ. GRAN SASSO → +49 6221 516540 NO.820 P201/208

FAX A-To Prof. H.V. KLAPDOR-KLEINGROTHAUS

SOCIETA'- Company MPI

CITTA'- City

FAX N° 0049 6221/516540

DA-From PROF. A. BETTINI

SOCIETA'- Company LABORATORI GRAN SASSO

CITTA'- City ASSERGI

FAX N° 862/437218

DATA-Date Feb. 26th 1998 RE.

N° PAG. 7

Please, find enclosed a copy of the minutes of the 7th meeting of the Gran Sasso Scientific Committee.

Minutes of the 7th MEETING OF THE GRAN SASSO SCIENTIFIC COMMITTEE
Feb 4-6, 1998

Presents: Barbieri, Bettini, Di Lella (part of the time), Lorenz, Menzione, Moessbauer, Sadoulet, Turlay, Vuilleumier and Wolf. Dr. Eugenio Scapparone was present as Scientific Secretary.

....

HDMS (proposal)

The SC approves the new dark matter detector proposed by the HDMS Collaboration. It does not require any additional resources for the lab.

GENIUS, CUORE

The SC encourages a vigorous R&D towards a reduction of the background that represents a main limiting factor in all double beta decay or dark matter searches, since it considers the potential physics goals of the highest interest. In particular, a considerably more detailed technical design study with appropriate simulations is desirable for both GENIUS and CUORE.

....

Fig. B.3 Gran-Sasso Scientific Committee, February 1998.

. . .

If neutrinos have masses, the "mass matrix" of the three-state neutrino system can be much more complex than that of (mixed) quark sector. Neutrino mass may be, at least in part, of Majorana type. As a consequence we must include in the experimental plans a next generation of double beta decay experiments to reach lower neutrino mass values.

The MPI Heidelberg group has submitted the GENIUS proposal. Aiming towards a major step forward in sensitivity one must increase the sensitive mass by a large amount and decrease background by a large factor. GENIUS proposes to use "naked" enriched germanium crystals, for a total of 1000 kg, in the middle of a large liquid nitrogen volume. The sensitivity to neutrino mass could go as far as 0.01 eV. In a first phase, with 100 kg enriched or even natural germanium mass, neutralinos can be searched for in all the Minimum Supersymmetric Standard Model parameter space.

303

.... we have clear elements to foresee a bright future for the double beta decay search.

. . .

304

Alessandro Bettini

THE GRAN SASSO LABORATORY *1979 - 1999*

Fig. B.4 Citation from the book by A. Bettini "The Gran Sasso Laboratory 1979 - 1999" pages: 303 – 304.

Physics monitor

Participants at the "Beyond the Desert '97 - Accelerator and Non-accelerator Approaches", conference held in Castle Ringberg, Germany, this summer.

Signals from beyond the desert ?

With physics looking so conventional, when and where will new effects be seen? From 8 -14 June a 'Beyond the Desert 97' conference at Castle Ringberg, Germany, on future aspects and perspectives of particle physics attempted to provide an answer.

The meeting, organized by the Max Planck Institute for Nuclear Physics, Heidelberg, and chaired by H.V. Klapdor-Kleingrothaus set out to analyse the future potential and trends in particle physics in both the accelerator and non-accelerator sectors. It was attended by theorists as well as experimentalists and with both accelerator and non-accelerator fields well represented.

Particle physics is making extreme demands on future accelerators. This route has provided most of the physics discoveries of the past 45 years. On the other hand, non-accelerator physics looking at transient effects (propagator physics) has the advantage of having no energy restrictions. The meeting assessed present physics objectives and current thinking.

Strategies for detection of supersymmetry and new possibilities for CERN's LHC collider and other accelerators were discussed by Daniel Denegri (CERN), Helmut Baer (Tallahassee) and Gordon Kane (Michigan). The potential of future linear electron-positron colliders (e.g. NLC at DESY) was outlined by M. Nojiri (KEK, Japan) and J. Kalinowski (DESY/Warsaw). R. Brinkmann (DESY) presented the already rather advanced project studies for NLC, while D. Cline (Los Angeles) discussed a future muon collider.

Superstring ideas now encapsulated in an "M-theory" seem to be to some extent testable by their "low energy" predictions like extended (R) parity violation, accessible in accelerator and non-accelerator experiments (double beta decay). This was covered by Alain Faraggi (Gainesville), D. Nanopoulos (Texas) and G.Volkov (Moscow). D. Nanopoulos outlined further aspirations of such theories in a special evening lecture 'Superstrings, Quantum Mechanics and Brain Function'.

A possible substructure of quarks and leptons (compositeness) and the search for it at accelerators and in double beta decay was treated by J. Virey (Marseille), E. Takasugi

Limits from double beta decay (the Heidelberg-Moscow experiment) for R-parity violation in supersymmetric models as a function of the squark mass, under two assumptions for the gluino mass (1 TeV and 0.1 TeV). Also given are the limits obtained by the Tevatron and HERA, as well as from neutron decay. Excluded are the areas beyond (or on the left of) the curves.

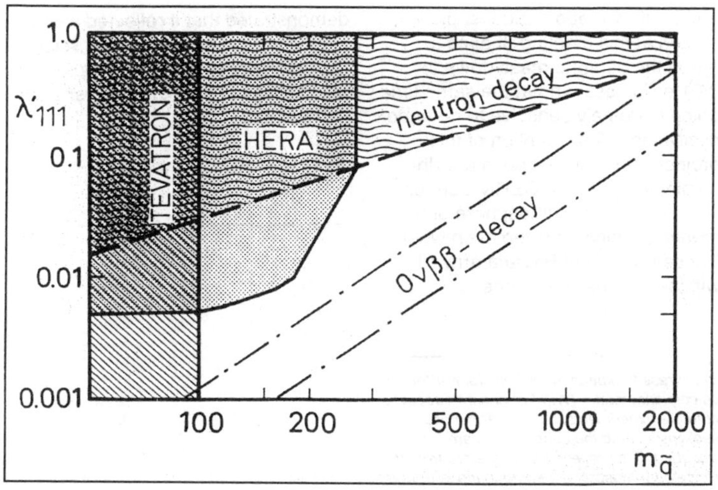

(Osaka) and O. Panella (Perugia), the latter pointing out the large potential of double beta decay in this respect.

The use of neutrinos to test for new theories was stressed by Yu. Smirnov (Trieste/Moscow), Rabi Mohapatra (Maryland) and J. Valle (Valencia). As well as the neutrino mass, M.Hirsch from the Heidelberg double beta group also showed how to obtain information about the supersymmetric neutrino mass. He showed that for first generation sneutrinos, double beta decay is more than competitive with future accelerators - and discussed together with Y. Grossman (SLAC) and S. Kolb (Double Beta, Heidelberg) new phenomenological consequences for future collider experiments at the NLC Next Linear Collider.

In the neutrino sector, the results paralleled those of the recent TAUP Gran Sasso meeting (see page 12).

Status and perspectives of proton decay, neutron-antineutron oscillations and the search for magnetic monopoles were presented by F. Mauri (Pavia) for ICARUS, who announced operation of the first component in Gran Sasso for 1998, Y. Kamyshkov (Oak Ridge), who presented plans for the new neutron-antineutron project in Oak Ridge, and Barry Barish (Caltech), who announced shutdown of MACRO for the year 2001. Bernard Schutz (Potsdam) and F. Fidecaro (Pisa) reported future gravitational wave searches, including experiments in space.

The origin of CP violation was the main topic of a session on fundamental symmetries, and was discussed by Boris Kayser (Arlington), N. Mavromatos (Oxford), and P.Pavlopoulos (CERN). The new analysis by B. Kayser showed that the new experiments aiming at solving the question of the origin of CP violation will have a hard time to really determine all three angles of the unitarity triangle, vital for deciding whether CP violations are inside or beyond Standard Model.

Turning to particle physics and cosmology, E. Kolb (Fermilab), discussed inflation, M. Schmidt (Heidelberg) the electroweak phase transition and Leszek Roszkowski (Lancaster) superparticle mass restrictions from inflation.

Vadim Kuzmin (Moscow) gave a new interpretation of highest-energy cosmic radiation - as decay products of new superheavy dark matter particles. Glennys Farrar presented new results about light supersymmetric gauge particles (gauginos) and their consequences for particle and astrophysics. Rainer Dick (Munich) presented new dark matter candidates from superstring theories. G. Boerner (Garching) presented the cosmological evidence for dark matter and R. Gaitskell (Berkeley) and Yorck Ramachers (Heidelberg) reported on the status of the CDMD and HDMS dark matter searches under construction. These will enter the supersymmetric domain with possible neutralino dark matter, and will go beyond a test phase within the next two years.

One of the highlights of the conference was of course the new HERA high momentum transfer events in positron-proton collisions (April, page 1). Such events could occur in the production of leptoquarks or of squarks - or they could indicate a substructure of quarks and leptons. Here again an interplay between accelerator and non-accelerator experiments opens up.

The recent results of the sensitive Heidelberg-Moscow double beta experiment - reported by the author of this report and speaker of the experiment - contribute directly to the interpretation of the HERA events: The limits for R-parity violation in supersymmetric models deduced from the experiment - considerably more stringent than those obtained from Tevatron and HERA - exclude the formation of first generation squarks.

The double beta experiment also yields a new restriction for the production of leptoquarks. The Heidelberg double beta group shows that the production of leptoquarks with a mass of around 200 GeV at HERA would imply a leptoquark-Higgs coupling in the 10^{-6} region. This would be unexpectedly small and would support the opinion of Harald Fritzsch (Munich - who predicted leptoquarks in the early seventies and named them) that for leptoquarks much larger masses would be expected than are presently accessible at HERA.

More far-reaching connections between non-accelerator experiments and present questions of particle physics could be implied by the planned new Heidelberg Double Beta and Dark Matter project GENIUS (GErmanium in NItrogen Underground Setup) proposed by the author of this report and first presented at this meeting. It aims to probe the neutrino mass down to 0.01 eV or less, and at the same time being extremely sensitive to neutrino oscillations. A description of this project will be published in a forthcoming issue of the CERN Courier.

The stimulating atmosphere of the meeting, which survived even when Prince Luitpold of Bavaria arrived with several barrels of beer, demonstrated that it reflected well the aims and requirements of the scientific community. The unanimous opinion was that the meeting should be repeated.

By H.V. Klapdor-Kleingrothaus

Best present experimental limits for neutralinos as cold dark matter (WIMPs) and perspectives of future experiments. Masses and cross sections beyond the contour lines are excluded. The present best limits are from the Heidelberg-Moscow experiment and later from the DAMA experiment.
The CDMS (Berkeley), CRESST (Munich), and HDMS (Heidelberg) experiments now under construction will probe for the first time the range of supersymmetric expectations for neutralinos. The proposed GENIUS experiment (Heidelberg) will be able to cover the full parameter space of the supersymmetric model predictions for neutralinos.

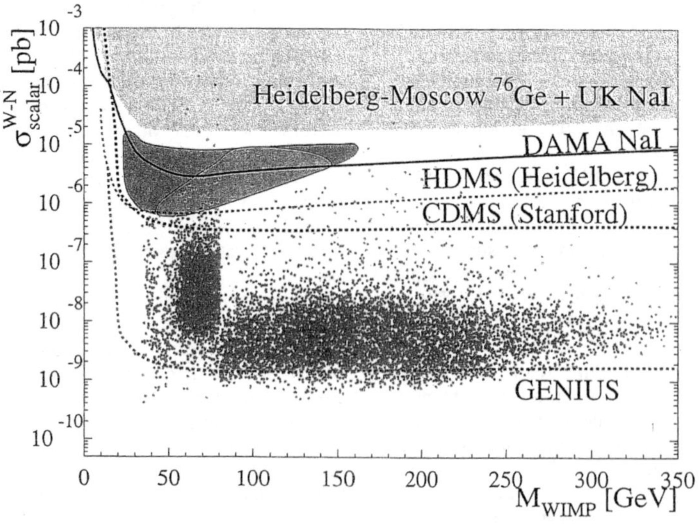

Computing.School@cern.ch http://www.cern.ch/CSC/ The application deadline is 15 May 1998

HEIDELBERG
Debut of GENIUS

With particle physics is making extreme demands on future accelerators, non-accelerator physics looking at transient effects (propagator physics) provides a complementary approach.

Accelerators provide high interaction rates, but are energy limited, while passive experiments have low rates and less energy restrictions.

The recent Castle Ringberg 'Beyond the Desert' meeting (November, page 16) saw the launch of the GENIUS (GErmanium in NItrogen Underground Setup) proposal for a new double beta decay and dark matter search by H.V. Klapdor-Kleingrothaus, head of the Heidelberg specialist double beta decay group.

The idea is to use one ton, or more, of 'naked' enriched germanium-76 detectors in a shielding of liquid nitrogen in a tank of about 10m height and diameter. This should probe the neutrino mass down to 10^{-2} or 10^{-3} eV, and test various variants of supersymmetric models, the atmospheric neutrino problem and some aspects of the solar neutrino problem.

Aside from neutrino physics, it would allow exploration of such ideas as leptoquarks, R-parity breaking and conserving supersymmetry, compositeness, left-right symmetric models etc. in the multi-TeV region.

For cold dark matter, it would cover the whole parameter space of supersymmetric models for neutralinos as dark matter and thus also ranges in which supersymmetry detection in collider experiments has some problems. If CERN's LHC were to observe supersymmetry, it would be fascinating to show the existence of neutralinos as dark matter.

Ranges probed by solar neutrino experiments in comparison to constraints from many other sources, together with predictions of some theoretical models. Also shown is the potential of the GENIUS project using germanium detectors in liquid nitrogen, at various scales. The lines correspond to different assumptions for the neutrino mass hierarchy.

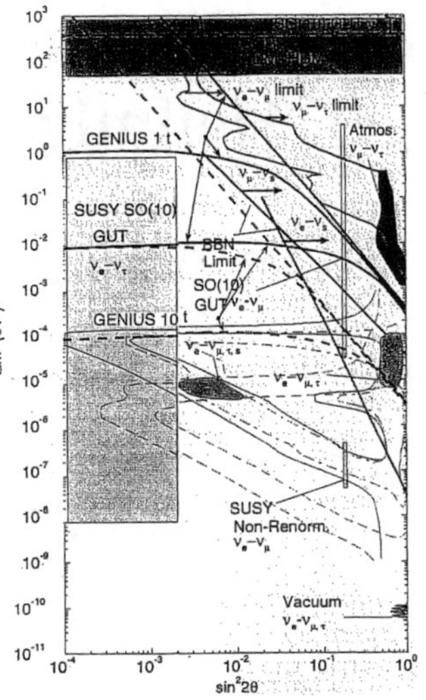

For its GENIUS proposal, the Heidelberg group has demonstrated the feasibility of germanium detectors operating in liquid nitrogen.

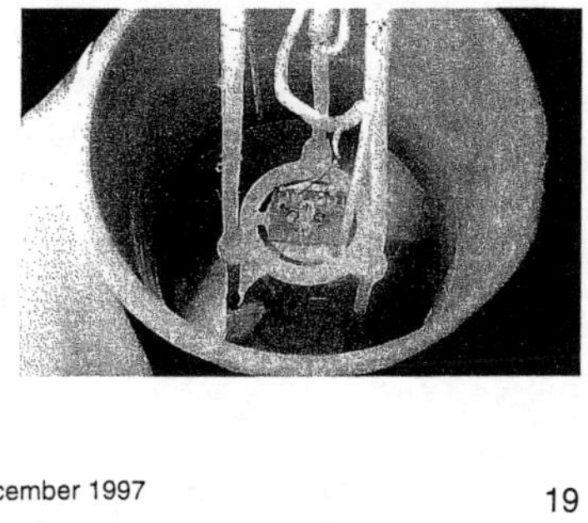

GENIUS – a New Experiment with Large Discovery Potential for Particle and Astrophysics

H.V. Klapdor-Kleingrothaus,
J. Hellmig and M. Hirsch

Max-Planck-Institut für Kernphysik, Germany

Abstract

The recent results from the HEIDELBERG–MOSCOW experiment have demonstrated the large potential of double beta decay to search for new physics beyond the Standard Model. To increase by a major step the present sensitivity for double beta decay and dark matter search much bigger source strengths and a much lower background are needed than used in experiments under operation at present or under construction. We describe here a project which would operate one ton of 'naked' enriched GErmanium-detectors in liquid NItrogen as shielding in an Underground Setup (GENIUS). It improves the sensitivity to neutrino masses to 0.01 eV. A ten ton version would probe neutrino masses even down to 10^{-3} eV. The first version would allow to test the atmospheric neutrino problem, the second at least part of the solar neutrino problem. Both versions would allow in addition significant contributions to testing several classes of GUT models. These are especially tests of R-parity breaking and conserving supersymmetry models - including sneutrino masses -, leptoquark masses and mechanism and right-handed W-boson masses comparable to LHC. The second issue of the experiment is the search for dark matter in the universe. The full MSSM parameter space for prediction of neutralinos as dark matter particles could be covered already in a first step of the full experiment using only 100 kg of ^{76}Ge or even of natural Ge making the experiment competitive to LHC in the search for supersymmetry.

Proposal (first draft)

November 20, 1997

GENIUS

a Supersensitive Germanium Detector System for Rare Events

H.V. Klapdor–Kleingrothaus, L. Baudis, G. Heusser, B. Majorovits, H. Päs

Max–Planck–Institut für Kernphysik, Heidelberg, Germany

The GENIUS Collaboration
Status September 1999

L. Baudis, A. Dietz, G. Heusser, H. V. Klapdor–Kleingrothaus*, St. Kolb, B. Majorovits, H. Päs, F. Schwamm, H. Strecker
Max–Planck–Institut für Kernphysik, Heidelberg, Germany

O.A. Ponkratenko, V.I. Tretyak, Yu.G. Zdesenko
Institute of Nuclear research, Kiev, Ukraine

V. Alexeev, A. Balysh, A. Bakalyarov, S. T. Belyaev, V. I. Lebedev, S. Zhukov
Kurchatov Institute, Moscow, Russia

U. Keyser, A. Paul, S. Röttger, A. Zimbal
Physikalisch-Technische Bundesanstalt, Braunschweig, Germany

A. Yu. Smirnov
Internat. Center for Theoretical Physics, Trieste, Italy

V. Bednyakov
Joint Institute for Nuclear Research, Dubna, Russia

I.V. Krivosheina, V. Melnikov
Institute of Radiophysical Research, Nishnij Novgorod, Russia

P. Nath
Department of Physics, Northeastern University, Boston, USA

R.N. Mohapatra
Department of Physics, University of Maryland, USA

J.W.F Valle
Departamento de Fisica Teorica, University of Valencia, Spain

R. Arnowitt
Department of Physics, Texas A&M University, USA

* Spokesman of the Collaboration

Contents

August 1999

The GENIUS collaboration includes groups from:
MPIK Heidelberg, Germany; Inst. of Nucl. Research, Kiev, Ukraine; Physik. Techn. Bundesanstalt, Braunschweig, Germany; Internat. Center for Theoret. Physics, Trieste, Italy; JINR Dubna, Russia; Inst. of Radiophys. Research, Nishnij Novgorod, Russia; Dep. of Physics, Northeastern Univ. Boston, USA; Dep. of Physics, Univ. of Maryland, USA; Los Alamos Nat. Lab., USA; Dep. de Fisica Teorica, Univ. of Valencia, Spain; Dep. of Physics, Texas A&M Univ., USA

Spokesman of the Collaboration: H. V. Klapdor–Kleingrothaus

MPI-Report MPI-H-V26-1999